Nanotechnology 2008: Microsystems, Photonics, Sensors, Fluidics, Modeling, and Simulation

Technical Proceedings of the 2008 NSTI Nanotechnology Conference and Trade Show

Boston, June 1 – 5, 2008

NSTI Nanotech

The Nanotechnology Conference and Trade Show

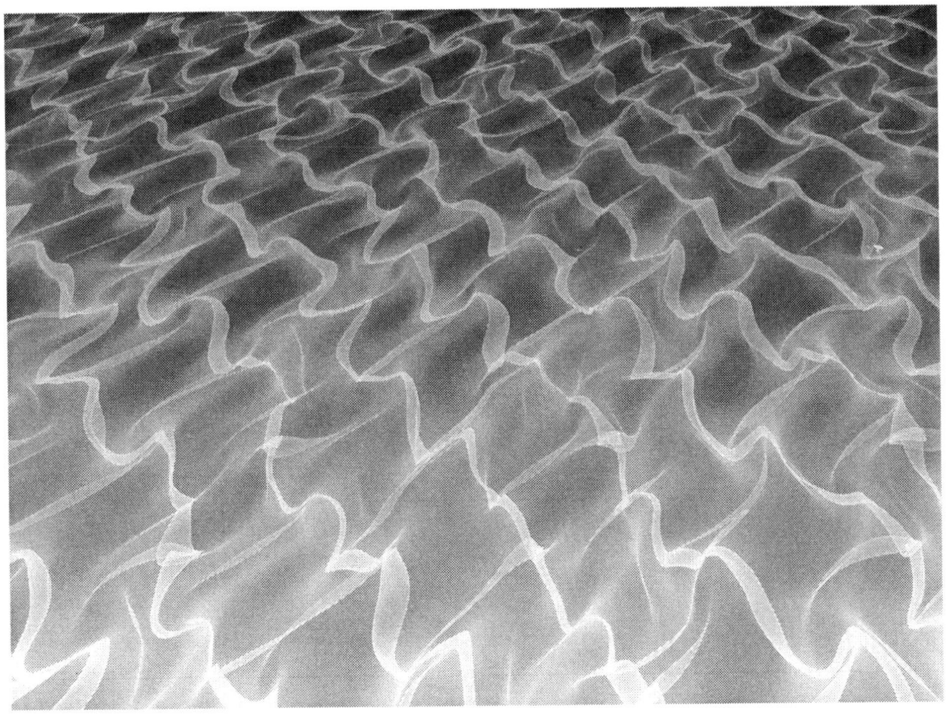

An Interdisciplinary Integrative Forum on Nanotechnology, Biotechnology and Microtechnology

June 1-5, 2008
Hynes Convention Center
Boston, Massachusetts, U.S.A.
www.nsti.org

T0382876

NSTI
Nano Science and Technology Institute

CRC Press
Taylor & Francis Group

Artwork provided by Prof. Eric J. Heller, http://www.ericjhellergallery.com

Caustic I (cover):
Here we focus on the origin of caustics as projections of three- dimensional transparent sheets onto two dimensions. The sheets consist of colored panels; the color combine with each other (by color subtraction for example) in unexpected ways.

Light from a point source is passed through two wavy surfaces, each like the surface of a swimming pool with a few people in it. This focuses the light randomly in a characteristic caustic pattern, here projected by intersection with a plane surface, as on a pool bottom.

Caustics are places where things accumulate; in this case it is light which is accumulating. We often think of focal points as where light gathers after passing through a lens, but more generally for "random" lenses there are much more interesting patterns to examine. In Caustic I, rays of light from a point source have passed through two imaginary, successive layers of water, each with a wavy surface, i.e. a random lens refracting the light. A "sea bottom" ultimately interrupts the refracted rays, which is the plane of the image. The bright patches, or "caustics", are similar to those produced when sunlight shines on a swimming pool, giving the familiar ribbed pattern on the bottom made famous by the painter David Hockney. The swimming pool provides only one refracting surface, but what if there are more? We were led to this question by recent work on the motion of electrons in random landscapes where they are deflected by hills and valleys. The electrons are moving in a thin, two-dimensional layer (see my Transport series), whereas the rays here are moving in three dimensions, and shown interrupted by a two dimensional surface. The electrons are refracted many times, so we wanted to look at light refracted by many wavy surfaces.

Nanotechnology 2008: Microsystems, Photonics, Sensors, Fluidics, Modeling, and Simulation

Technical Proceedings of the 2008 NSTI Nanotechnology Conference and Trade Show

NSTI-Nanotech 2008, Vol. 3

Boston, June 1 – 5, 2008
NSTI **Nanotech**
The Nanotechnology Conference and Trade Show

An Interdisciplinary Integrative Forum on
Nanotechnology, Biotechnology and Microtechnology

June 1-5, 2008
Boston, Massachusetts, U.S.A.
www.nsti.org

NSTI Nanotech 2008 Joint Meeting

The 2008 NSTI Nanotechnology Conference and Trade Show includes:
2008 NSTI Bio Nano Conference and Trade Show, Bio Nano 2008
11th International Conference on Modeling and Simulation of Microsystems, MSM 2008
8th International Conference on Computational Nanoscience and Technology, ICCN 2008
7th Workshop on Compact Modeling, WCM 2008
NSTI Nanotech Ventures 2008
2008 TechConnect Summit
Clean Technology 2008

NSTI Nanotech 2008 Proceedings Editors:

Matthew Laudon
mlaudon@nsti.org

Bart Romanowicz
bfr@nsti.org

Nano Science and Technology Institute
Boston • Geneva • San Francisco

NSTI
Nano Science and Technology Institute

Nano Science and Technology Institute
One Kendall Square, PMB 308
Cambridge, MA 02139
U.S.A.

The papers in this book comprise the proceedings of the 2008 NSTI Nanotechnology Conference and Trade Show, Nanotech 2008, Boston, Massachusetts, June 1-5 2008. They reflect the authors' opinions and, in the interests of timely dissemination, are published as presented and without change. Their inclusion in this publication does not necessarily constitute endorsement by the editors, Nano Science and Technology Institute, or the sponsors. Dedicated to our knowledgeable network of willing and forgiving friends.

ISBN 978-1-4200-8505-1
ISBN 978-1-4200-8507-5 (Volumes 1-3 SET)
ISBN 978-1-4200-8511-2 (Volumes 1-3 CD-ROM)

Additional copies may be ordered from:

CRC Press
Taylor & Francis Group
an informa business
www.taylorandfrancisgroup.com

6000 Broken Sound Parkway, NW
Suite 300, Boca Raton, FL 33487
270 Madison Avenue
New York, NY 10016
2 Park Square, Milton Park
Abingdon, Oxon OX14 4RN, UK

Printed in the United States of America

Nanotech 2008 Proceeding Editors

Sponsors

Accelrys Software, Inc.
Ace Glass, Inc.
Advance Reproductions Corporation
Advanced Materials Technologies Pte Ltd
Agilent Technologies
AIXTRON AG
AJA International, Inc.
ALM
Amuneal Manufacturing Corporation
ANSYS, Inc.
Anton Paar USA
Applied MicroStructures, Inc.
Applied Surface Technologies
Arkalon Chemical Technologies, LLC
Arkema Group
Artech House Publishers
Asemblon, Inc.
ASML
Asylum Research
AZoNano
Banner & Witcoff, Ltd.
Beckman Coulter, Inc.
Bio Nano Consulting
BioForce Nanosciences, Inc.
Bioneer Corporation
Birck Nanotechnology Center - Purdue University Discovery Park
Brookhaven Instruments Corporation
Buchanan Ingersoll & Rooney PC
Buhler AG
BusinessWeek
California NanoSystems Institute
Capovani Brothers, Inc.
Center for Functional Nanomaterials, Brookhaven National Laboratory
Center for Integrated Nanotechnologies, Los Alamos and Sandia National Laboratories
Center for Nanophase Materials Sciences at Oak Ridge National Laboratory
Center for Nanoscale Materials, Argonne National Laboratory
Cheap Tubes, Inc.
Clean Technology and Sustainable Industries Organization (CTSI)
COMSOL, Inc.
Cooley Godward Kronish LLP
Core Technology Group, Inc.
CRESTEC Corporation
CSEM
CVI Melles Griot
Department of Innovation, Industry, Science and Research
Digital Matrix
Draiswerke, Inc.
ETH Zurich
Eulitha AG
Marks & Clerk
Marubeni Techno-Systems Corporation
Massachusetts Technology Transfer Center
Materials Research Society
Merck & Co., Inc.
Microfluidics
Micromeritics Instrument Corporation
Microtrac, Inc.
MINAT 2008, Messe Stuttgart
Minus K Technology, Inc.
Misonix, Inc.
Molecular Foundry at Lawrence Berkeley National Laboratory
Motorola, Inc.
Nano Korea 2008
Nano Science and Technology Institute
nano tech 2009 Japan
Nano Technology Research Association of Korea (NTRA)
NanoAndMore USA, Inc.
NanoDynamics, Inc.
NanoEurope 2008
NanoInk, Inc.
Nanomotion, Inc.
Nanonics Imaging Ltd.
NANOSENSORS™
Nanosystems Initiative Munich (NIM)
Nanotech Northern Europe 2008
nanoTox, Inc.
NanoWorld AG
National Cancer Institute
National Institute of Standards and Technology (NIST)
National Institutes of Health
National Nanomanufacturing Network (NNN)
National Nanotechnology Infrastructure - GATech - Microelectronics Research Center
Natural Nano
Nature Publishing Group
NETZSCH Fine Particle Technology LLC
NIL Technology ApS
Novomer
NTT Advanced Technology Corporation
Olympus Industrial America
OSEC - Business Network Switzerland
Oxford Instruments
Particle Technology Labs, Ltd.
Pennsylvania State University
Photonics Spectra
PI (Physik Instrumente) L.P.
picoDrill SA
Piezo Institute
PVD Products, Inc.
Q-Sense, Inc.
Quantum Analytics
Raith USA, Inc.

European Patent Office (EPO)
Evans Analytical Group
EXAKT Technologies, Inc.
First Nano
Flow Science, Inc.
Food and Drug Administration
Goodwin Procter LLP
Greater Houston Partnership
Greenberg Traurig, LLP
Halcyonics
Headwaters Technology Innovation, LLC
Heidelberg Instruments
Hielscher USA, Inc.
HighNanoAnayltics
Hiscock & Barclay, LLP
Hitachi High Technologies America, Inc.
Hochschule Offenburg University of Applied Sciences
HOCKMEYER Equipment Corporation
HORIBA Jobin Yvon, Inc.
IBU-tec advanced materials GmbH
IDA Ireland
ImageXpert, Inc.
Inspec, Inc.
Institute for Nanoscale and Quantum Scientific and Technological Advanced Research (nanoSTAR)
Institute of Electrical and Electronics Engineers, Inc. (IEEE)
IntelliSense Software Corp.
Invest in Germany
IOP Publishing
Italian Trade Commission
JENOPTIK
Justus-Liebig-University (JLU) in Giessen
KAUST - King Abdullah University of Science and Technology
Keithley Instruments
Kelvin Nanotechnology Ltd.
Kodak
Kotobuki Industries Co., Ltd.
Lake Shore Cryotronics, Inc.
MACRO-M / NanoClay
Research in Germany Land of Ideas
RKS Legal Solutions, LLC
Sandia National Laboratories
Scottish Enterprise
SEMTech Solutions, Inc.
Serendip
Silvix Corporation
SNS Nano Fiber Technology, LLC
SoftMEMS LLC
Sonics & Materials, Inc.
SouthWest NanoTechnologies, Inc.
Specialty Coating Systems
Spectrum Laboratories, Inc.
SPEX SamplePrep, LLC
Springer, Inc.
Sterne, Kessler, Goldstein & Fox P.L.L.C.
Strem Chemicals, Inc.
Sukgyung A-T Co., Ltd.
Surrey NanoSystems
SUSS MicroTec
Swissnanotech Pavilion
Taylor & Francis Group LLC - CRC Press
TechConnect
Technovel Corporation
Tekna Plasma Systems, Inc.
Thinky Corporation
Thomas Swan & Co. Ltd.
UGL Unicco
UK Trade & Investment
UniJet
University Muenster (WWU)
University of Applied Sciences (TFH) Berlin
University of Duisburg-Essen / Center for Nanointegration Duisburg-Essen (CeNIDE)
Veeco Instruments
Wasatch Molecular Incorporated
Weidmann Plastics Technology
Whiteman Osterman & Hanna LLP
Willy A. Bachofen AG
WITec GmbH
World Gold Council

Supporting Organizations

AllConferences.Com
Asia Pacific Nanotechnology Forum (APNF)
AZoNano
Battery Power Products & Technology
Berkeley Nanotechnology Club
BioExecutive International
Biolexis.com
Biophotonics International
BioProcess International
BioTechniques
Biotechnology Industry Organization (BIO)
Business Wire
BusinessWeek
CMP: Circuits Multi-Projets
Drug & Market Development Publications
EurekAlert! / AAAS
Food and Drug Administration (FDA)
Foresight Institute
Fuel Cell Magazine
GlobalSpec
Greenberg Traurig, LLP
IEEE CiSE
IEEE San Francisco Bay Area Nanotechnology Council
ivcon.net
Journal of Experimental Medicine
JSAP - the Japan Society of Applied Physics
KCI Investing
Laser Focus World
Materials Today
MEMS Investor Journal
Microsystem Technologies - Micro- and Nanosystems
mst|news
Nano Today
NanoBiotech News
Nanomedicine: Nanotechnology, Biology and Medicine
NanoNow!
nanoparticles.org
NanoSPRINT
Nanotechnology Law & Business
Nanotechnology Now
nanotechweb
nanotimes
Nanovip.com
Nanowerk
National Cancer Institute (NCI)
National Institutes of Health (NIH)
Nature Publishing Group
Photonics Spectra
PhysOrg.com
R&D Magazine
Red Herring, Inc.
Science Magazine
SciTechDaily Review
Small Times
Solid State Technology
Springer
Springer Micro and Nano Fluidics
The Journal of BioLaw & Business (JB&B)
The Journal of Cell Biology
The Real Nanotech Investor
The Scientist
tinytechjobs
Understanding Nanotechnology
Virtual Press Office

Table of Contents

Sensors & Systems

Lab-on-a-Chip, Micro & Nano Fluidics

MEMS & NEMS

Modeling & Simulation of Microsystems

Computational Nanoscience

Compact Modeling

NSTI Nanotech 2008 Program Committee

TECHNICAL PROGRAM CO-CHAIRS

Matthew Laudon	*Nano Science and Technology Institute, USA*
Bart Romanowicz	*Nano Science and Technology Institute, USA*

TOPICAL AND REGIONAL SCIENTIFIC ADVISORS AND CHAIRS

Nanotechnology	
Clayton Teague	*National Nanotechnology Coordination Office, USA*
Wolfgang Windl	*Ohio State University, USA*
Nick Quirke	*Imperial College, London, UK*
Philippe Renaud	*Swiss Federal Institute of Technology, Switzerland*
Mihail Roco	*National Science Foundation, USA*
Carbon Nanotech	
Wolfgang S. Bacsa	*Université Paul Sabatier, France*
Bennett Goldberg	*Boston University, USA*
Anna K. Swan	*Boston University, USA*
Biotechnology	
Srinivas Iyer	*Los Alamos National Laboratory, USA*
Gabriel A. Silva	*University of California, San Diego, USA*
Micro-Bio Fluidics	
Daniel Attinger	*Columbia University, USA*
Hang Lu	*Georgia Institute of Technology, USA*
Steffen Hardt	*Leibniz Universität Hannover, Germany*
Pharmaceutical	
Kurt Krause	*University of Houston, USA*
Mansoor M. Amiji	*Northeastern University, USA*
Soft Nanotechnology	
Fiona Case	*Case Scientific, USA*
Composites & Interfaces	
Thomas E. Twardowski	*Twardowski Scientific, USA*
Microtechnology	
Narayan R. Aluru	*University of Illinois Urbana-Champaign, USA*
Bernard Courtois	*TIMA-CMP, France*
Anantha Krishnan	*Lawrence Livermore National Laboratory, USA*
Sensors	
Elena Gaura	*Coventry University, UK*
Semiconductors	
David K. Ferry	*Arizona State University, USA*
Andreas Wild	*Freescale Semiconductor, France*
Nano Particles	
Sotiris E. Pratsinis	*Swiss Federal Institute of Technology, Switzerland*
Nano Fabrication	
Warren Y.C. Lai	*Alcatel-Lucent, USA*
Leonidas E. Ocola	*Argonne National Laboratory, USA*
Stanley Pau	*University of Arizona, USA*
Characterization	
Pierre Panine	*European Synchrotron Radiation Facility, France*
Greg Haugstad	*University of Minnesota, USA*
Workshop on Compact Modeling	
Xing Zhou	*Nanyang Technological University, Singapore*

NANOTECH PHYSICAL SCIENCES COMMITTEE

NANOTECH LIFE SCIENCES COMMITTEE

Mansoor M. Amiji	*Northeastern University, USA*
Mostafa Analoui	*Pfizer, USA*
Amos Bairoch	*Swiss Institute of Bioinformatics, Switzerland*
Jeffrey Borenstein	*Draper Laboratory, USA*
Stephen H. Bryant	*National Institute of Health, USA*
Dirk Bussiere	*Chiron Corporation, USA*
Fred Cohen	*University of California, San Francisco, USA*
Tejal Desai	*University of California, San Francisco, USA*
Daniel Davison	*Bristol Myers Squibb, USA*
Tejal Desai	*University of California, San Francisco, USA*
Robert S. Eisenberg	*Rush Medical Center, Chicago, USA*
Michael N. Helmus	*Advance Nanotech, USA*
Andreas Hieke	*GEMIO Technologies, Inc., USA*
Leroy Hood	*Institute for Systems Biology, USA*
Sorin Istrail	*Brown University, USA*
Srinivas Iyer	*Los Alamos National Laboratory, USA*
Brian Korgel	*University of Texas-Austin, USA*
Kurt Krause	*University of Houston, USA*
Daniel Lacks	*Case Western ReserveUniversity, USA*
Jeff Lockwood	*Novartis, USA*
Hang Lu	*Georgia Institute of Technology, USA*
Atul Parikh	*University of California, Davis, USA*
Andrzej Przekwas	*CFD Research Corporation, USA*
Don Reed	*Ecos Corporation, Australia*
George Robillard	*BioMade Corporation, Netherlands*
Jonathan Rosen	*Center for Integration of Medicine & Innovative Technology, USA*
Gabriel A. Silva	*University of California, San Diego, USA*
Srinivas Sridhar	*Northeastern University, USA*
Sarah Tao	*The Charles Stark Draper Laboratory, Inc., USA*
Tom Terwilliger	*Los Alamos National Laboratory, USA*
Vladimir Torchilin	*Northeastern University, USA*
Michael S. Waterman	*University of Southern California, USA*
Thomas J. Webster	*Brown University, USA*
Steven T. Wereley	*Purdue University, USA*

NANOTECH MICROSYSTEMS COMMITTEE

Narayan R. Aluru	*University of Illinois Urbana-Champaign, USA*
Daniel Attinger	*Columbia University, USA*
Xavier J. R. Avula	*Washington University, USA*
Stephen F. Bart	*Bose Corporation, USA*
Bum-Kyoo Choi	*Sogang University, Korea*
Bernard Courtois	*TIMA-CMP, France*
Peter Cousseau	*Honeywell, USA*
Robert W. Dutton	*Stanford University, USA*
Gary K. Fedder	*Carnegie Mellon University, USA*
Edward P. Furlani	*Eastman Kodak Company, USA*
Elena Gaura	*Coventry University, UK*
Steffen Hardt	*Leibniz Universität Hannover, Germany*
Eberhard P. Hofer	*University of Ulm, Germany*
Michael Judy	*Analog Devices, USA*
Yozo Kanda	*Toyo University, Japan*
Jan G. Korvink	*University of Freiburg, Germany*
Mark E. Law	*University of Florida, USA*
Mary-Ann Maher	*SoftMEMS, USA*
Kazunori Matsuda	*Tokushima Bunri University, Japan*
Tamal Mukherjee	*Carnegie Mellon University, USA*
Andrzej Napieralski	*Technical University of Lodz, Poland*

Ruth Pachter	*Air Force Research Laboratory, USA*
Michael G. Pecht	*University of Maryland, USA*
Marcel D. Profirescu	*Technical University of Bucharest, Romania*
Marta Rencz	*Technical University of Budapest, Hungary*
Siegfried Selberherr	*Technical University of Vienna, Austria*
Sudhama Shastri	*ON Semiconductor, USA*
Armin Sulzmann	*Daimler-Chrysler, Germany*
Mathew Varghese	*MEMSIC, Inc., USA*
Dragica Vasilesca	*Arizona State University, USA*
Gerhard Wachutka	*Technical University of München, Germany*
Jacob White	*Massachusetts Institute of Technology, USA*
Thomas Wiegele	*Goodrich, USA*
Andreas Wild	*Freescale Semiconductor, France*
Cy Wilson	*North American Space Agency, USA*
Wenjing Ye	*Georgia Institute of Technology, USA*
Xing Zhou	*Nanyang Technological University, Singapore*

NANO FABRICATION COMMITTEE

Adekunle Adeyeye	*National University of Singapore, Singapore*
Ronald S. Besser	*Stevens Institute of Technology, USA*
Gregory R. Bogart	*Symphony Acoustics, USA*
Chorng-Ping Chang	*Applied Materials, Inc., USA*
Charles Kin P. Cheung	*National Institute of Standards and Technology, USA*
Seth B. Darling	*Argonne National Laboratory, USA*
Guy A. DeRose	*California Institute of Technology, USA*
Zhixiong Guo	*Rutgers University, USA*
Takamaro Kikkawa	*Hiroshima University and National Institute of Advanced Industrial Science and Technology, Japan*
Jungsang Kimg	*Duke University, USA*
Uma Krishnamoorthy	*Sandia National Laboratory, USA*
Andres H. La Rosa	*Portland State University, USA*
Warren Y.C. Lai	*Lucent Technologies, USA*
Sergey D. Lopatin	*Applied Materials, Inc., USA*
Pawitter Mangat	*Motorola, USA*
Omkaram Nalamasu	*Applied Materials, Inc., USA*
Vivian Ng	*National University of Singapore, Singapore*
Leonidas E. Ocola	*Argonne National Laboratory, USA*
Sang Hyun Oh	*University of Minnesota, USA*
Stanley Pau	*University of Arizona, USA*
John A. Rogers	*University of Illinois at Urbana-Champaign, USA*
Nicolaas F. de Rooij	*University of Neuchâtel, Switzerland*
Aaron Stein	*Brookhaven National Laboratory, USA*
Vijay R. Tirumala	*National Institute of Standards and Technology, USA*
Gary Wiederrecht	*Argonne National Laboratory, USA*

SOFT NANOTECHNOLOGY CONFERENCE COMMITTEE

Fiona Case	*Case Scientific, USA*
Greg Haugstad	*University of Minnesota, USA*
Pierre Panine	*European Synchrotron Radiation Facility, France*
Peter Schurtenberger	*University of Fribourg, Switzerland*
Patrick Spicer	*The Procter & Gamble Company, USA*
Donald C. Sundberg	*University of New Hampshire*
Krassimir Velikov	*Unilever Research Vlaardingen, Netherland*

CONFERENCE OPERATIONS MANAGER

Sarah Wenning	*Nano Science and Technology Institute, USA*

NSTI Nanotech 2008 Proceedings Topics

Nanotechnology 2008: Materials, Fabrication, Particles, and Characterization,

NSTI-Nanotech 2008, Vol. 1, ISBN: 978-1-4200-8503-7:

1. Carbon Nano Structures & Applications
2. Nano Materials & Composites
3. Nano Surfaces & Interfaces
4. Nanofabrication & Direct-Write Nanolithography
5. Nanoparticles & Applications
6. Characterization
7. Initiatives, Education & Policy

Nanotechnology 2008: Life Sciences, Medicine, and Bio Materials,

NSTI-Nanotech 2008, Vol. 2, ISBN: 978-1-4200-8504-4:

1. Cancer Diagnostics, Imaging & Treatment
2. Environment, Health & Toxicology
3. Biomarkers, Nano Particles & Materials
4. Drug & Gene Delivery Systems
5. Phage Nanobiotechnology
6. Nano Medicine & Neurology
7. Bio & Chem Sensors
8. Soft Nanotechnology & Polymers

Nanotechnology 2008: Microsystems, Photonics, Sensors, Fluidics, Modeling, and Simulation,

NSTI-Nanotech 2008, Vol. 3, ISBN: 978-1-4200-8505-1:

1. Photonics & Nanowires
2. Sensors & Systems
3. Lab-on-a-Chip, Micro & Nano Fluidics
4. MEMS & NEMS
5. Modeling & Simulation of Microsystems
6. Computational Nanoscience
7. Compact Modeling

Nanotechnology 2008, Vol. 1-3, ISBN: 978-1-4200-8507-5 (hardcopy)

Nnaotechnology 2008, Vol. 1-3 CDROM, ISBN: 978-1-4200-8511-2

Optical Nanotrapping Using Illuminated Metallic Nanostructures: Analysis and Applications

E. P. Furlani, A. Baev and P. N. Prasad

The Institute for Lasers, Photonics and Biophotonics, University at Buffalo SUNY
432 Natural Sciences Complex
Buffalo, NY 14260-3000
efurlani@buffalo.edu,

ABSTRACT

We present a theoretical study of plasmonic-based optical trapping of neutral sub-wavelength particles in proximity to illuminated metallic nanostructures. We compute the dipolar force on the particles using 3D full-wave electromagnetic analysis, and we perform parametric studies of the force as a function of nanostructure geometry, particle size, and field polarization. We discuss advantages and applications of plasmonic nanotrapping.

Keywords: optical nanotrapping, optical nanoparticle manipulation, plasmonic nano-tweezers, plasmonic-enhanced optical manipulation

1 INTRODUCTION

The interest in optical manipulation continues to grow, especially for biological applications where the manipulated objects include viruses, cells and intracellular organelles [1-5]. While micron and sub-micron particles can be manipulated using conventional laser tweezers, the resolution of this approach is diffraction-limited (~250 nm), and the high optical power and focusing of the laser beam can limit the exposure time of a trapped specimen. An alternate trapping method that overcomes these limitations involves the use of plasmonics [6-7]. Specifically, sub-wavelength particles can be manipulated and trapped using the enhanced near-field gradients that exist around illuminated metallic nanostructures. To date, various groups have conducted theoretical studies of plasmonic-based optical trapping involving a metal tip, nanoaperture, and apertureless probe [8-11]. Parallel and selective optical trapping has been demonstrated experimentally using an ordered array of Au disks [12].

The optical force produced by metallic nanostructures can be used to control dielectric or metallic nanoparticles, which can be functionalized to bind with a target biomaterial thereby enabling optical manipulation of nanoscale bioparticles. Plasmonic nano-manipulation holds potential for a number of diverse applications including nanoparticle chemistry, nanorheology, nanoscale bioseparation, ultra-sensitive biosensing, and Lab-on-a-Chip devices.

In this paper we discuss plasmonic-based optical trapping of dielectric nanoparticles in proximity to illuminated metallic nanopillars and nanorings (Fig. 1a). We perform a theoretical study of the dipolar force on particles using 3D full-wave electromagnetic analysis. We consider 2 and 4 nanopillar configurations, and nanoring structure with and without a central plasmonic nanopillar.

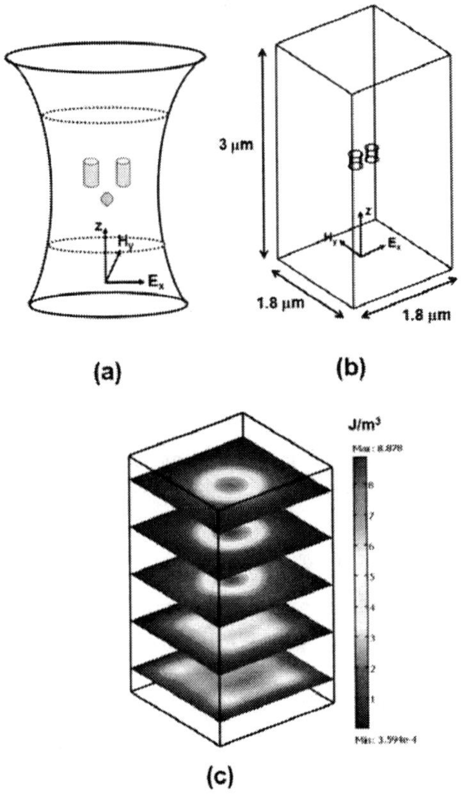

Figure 1: Plasmonic optical manipulation: (a) physical model – a dielectric nanoparticle beneath two metallic nanopillars; (b) the FEA computational domain; (c) time-averaged electric energy density showing focusing at center of computational domain – no pillars present.

We compute the force as a function of the nanopillar configuration, the particle size, and the polarization of the incident field. It should be noted that while the results presented here are theoretical, optical manipulation using an array of tapered nanopillars has been demonstrated in the laboratory wherein the controlled motion of microbubbles in an immersion oil has been achieved with sub-micron precision [13-14].

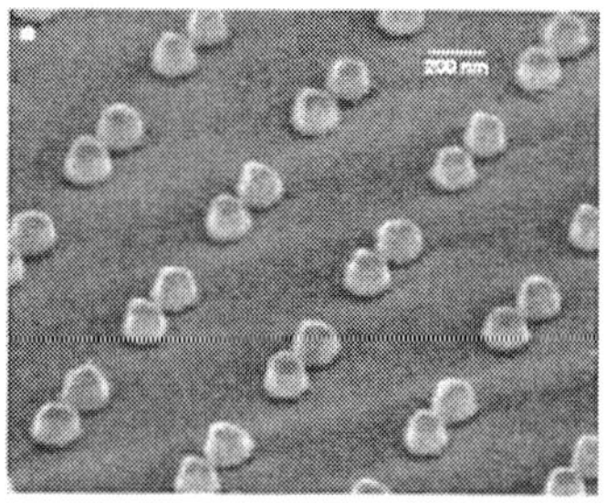

Figure 2: Fabricated array of tapered gold nanopillars (adapted from [14]).

2 ANALYSIS AND RESULTS

We use 3D full-wave time-harmonic finite element analysis (FEA) to study the optical field and dipolar force distribution produced by gold nanopillars and nanorings. We use the COMSOL Multiphysics FEA-based electromagnetic solver for our numerical analysis. The computational domain spans 3 μm in the direction of propagation (z-axis), and 1.8 μm in both the x and y directions (Fig. 1b). We study various nanopillar and nanoring configurations, and for each of these we compute the field and force distribution as a function of geometrical spacing, particle size, and field polarization. In all cases the nanostructures are centered height wise along the z-axis in the computational domain,

We illuminate the nanopillars/nanorings in free-space from below with a uniform plane wave at an optical wavelength λ = 532 nm. The incident field has a prescribed magnitude $|E| = 1\times10^6$ V/m, which corresponds to a CW power of 4 mW. This condition is imposed at the lower boundary (at z = -1.5 μm). We apply scattering (low reflection) boundary conditions at the top of the computational domain (z = 1.5 μm), and at the boundaries perpendicular to the E field (x = y = ± 0.9 μm). The incident field is effectively apertured by the finite x-y cross-section of the computational domain at the lower boundary, and this focuses the field at the center of the computational domain (z = 0) as shown in Fig. 1c. Thus, the field is focused at the center of the nanopillar structure.

We analyze the optical trapping of sub-wavelength particles by computing the time-averaged dipolar force,

$$\langle F_i \rangle = \frac{1}{2}\sum_j \text{Re}\left[\alpha E_{0j}\partial^i\left(E_{0j}\right)^*\right], \qquad (1)$$

where E_{0j} $(j = 1,2,3)$ are the Cartesian components of the optical field, and

$$\alpha = \frac{4\pi\,\alpha_0\varepsilon_0}{\left[1-\alpha_0\left(\dfrac{k^2}{a}-\dfrac{2}{3}ik^3\right)\right]} \qquad (2)$$

is the polarizability of the particle, where $\alpha_0 = R_p^3\dfrac{\varepsilon_r - 1}{\varepsilon_r + 2}$. R_p and ε_r are the radius and relative permittivity of the particle, respectively [15]. The imaginary term in α accounts for the scattering force on a particle, and it is important to note that the sign of this term (i.e. $\pm\frac{2}{3}ik^3$) depends on the convention used in the time-harmonic analysis, i.e. $\exp(i\omega t)$ or $\exp(-i\omega t)$ [16-17]. The COMSOL program uses the former, which is compatible with Eq. (2).

We model the dielectric permittivity of the gold nanostructures using a Drude-like model with the following parameters: the bulk plasma frequency is $1.37\times10^{16}\,\text{sec}^{-1}$, the damping frequency is $4.08\times10^{13}\,\text{sec}^{-1}$, and the high-frequency limit term including contributions from interband transitions is 12.9. We study a variety of nanopillar and nanoring configurations that include a pair of nanopillars along the x-axis (Fig. 1b), a group of four nanopillars centered about the origin (Fig. 6a), and a nanoring with and without an isolated central metallic pillar (Fig. 8a).

2.1 Two Nanopillar System

We first consider a two nanopillar system in which the nanopillars are positioned along the x-axis as shown in Fig. 1a. We fix the dimensions of the nanopillars, radius = 100 nm and height = 200nm. We compute the field and force distribution as a function of pillar-to-pillar spacing, particle size, and field polarization. The time-averaged electric field intensities for TE and TM polarized fields are compared with TE free-space illumination in Fig. 3. These plots show that the nanopillars produce local field gradients that are polarization dependent. They also indicate that particles on the z-axis will be confined in a lateral sense, i.e. F_x and F_y act towards the axis. Plots of the TE and TM time-averaged axial force F_z along the z-axis vs. particle size are shown in Figs. 4a and 4b. The spacing between the pillars is 100 nm. Note that TE polarization (E along x-axis) produces stronger forces than TM polarization (E along y-axis). Particle trapping can occur where F_z changes sign, positive-to-negative, as shown in Fig. 4a. Note that for TM polarization there is no trapping for larger particles (e.g. for $R_p \geq$ 300 nm, Fig. 4b). This is because the scattering force, which is strictly positive, dominates the gradient force for larger particles. The optical force on a 100 nm particle for TE and TM polarization compared to the free-space (scattering) force is shown in Fig. 5.

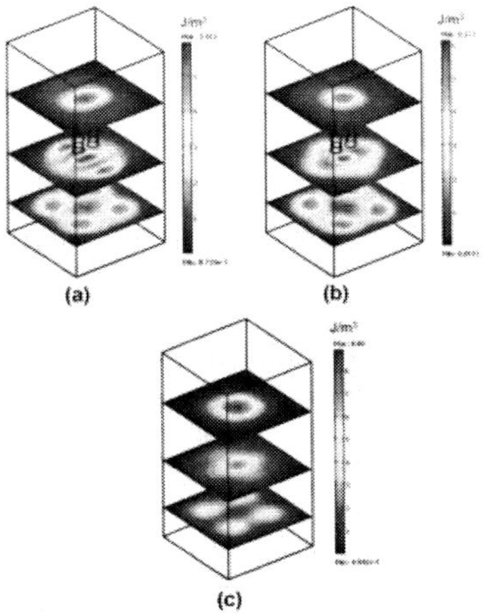

Figure 3: Two nanopillar system - time-averaged electric energy density at various planes: (a) TE analysis; (b) TM analysis; (c) TE analysis - free-space.

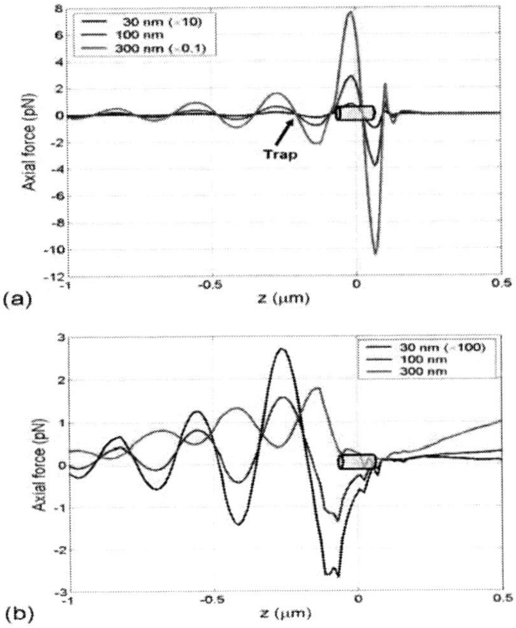

Figure 4: Two nanopillar system - time-averaged dipolar axial force vs. particle radius (a) TE analysis; (b) TM analysis.

2.2 Four Nanopillar System

Next, we study a four nanopillar system in which four gold nanopillars are positioned symmetrically about the axis as shown in Fig. 6a. The nanopillars are the same size as above. The time-averaged gradient force potential ($-|E|^2$) shown in Fig. 6b indicates lateral trapping along the z-axis.

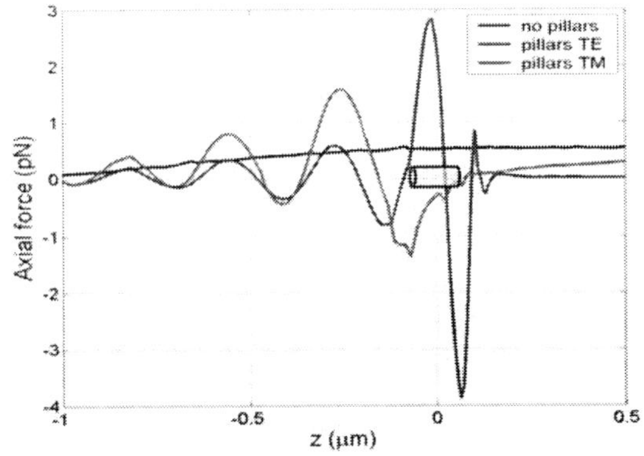

Figure 5: Two nanopillar system - axial force for 100 nm particle.

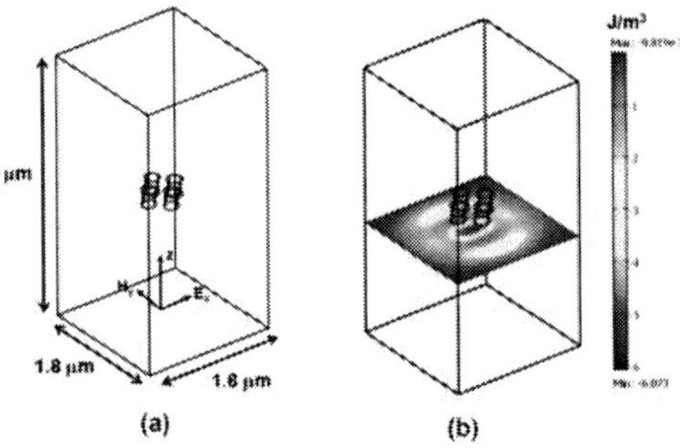

Figure 6: Four nanopillar system: (a) geometry; (b) time-averaged force potential.

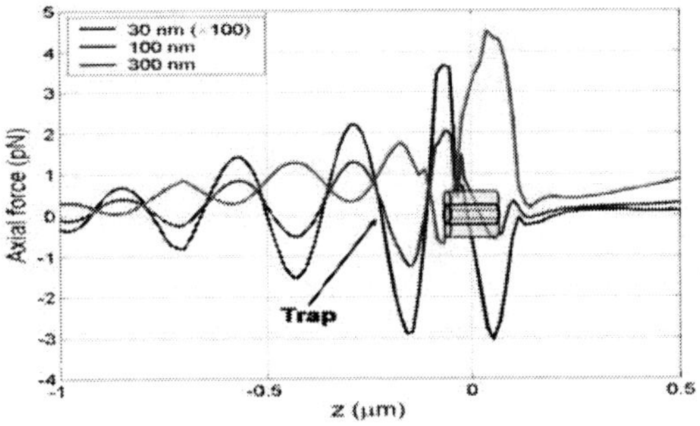

Figure 7: Four nanopillar system: FEA time-averaged dipolar force vs. particle radius.

F_z along the z-axis for different particles sizes is shown in Fig. 7. Note that smaller particles (e.g. R_p = 30, 100 nm) can potentially be trapped, but that larger particles ($R_p \geq$ 300 nm) can not.

2.3 Nanoring System

The last system we study consists of a gold nanoring centered about the origin. The ring has an outer diameter of 600nm and a height of 200 nm. We consider a range of inner diameters, with and without isolated central metallic pillars of varying diameters. Figure 8 shows a nanoring with a 450 nm inner diameter surrounding an isolated central metallic pillar that is 100 nm in diameter. The time-

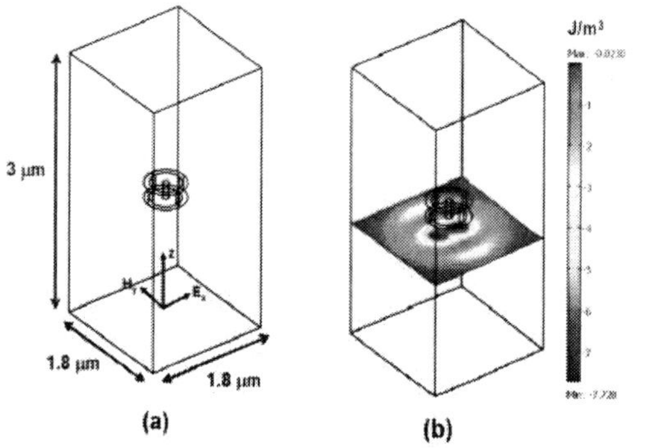

Figure 8: Nanoring analysis: (a) nanoring with a central metallic pillar; (b) dipolar gradient force potential 200 nm below the nanoring.

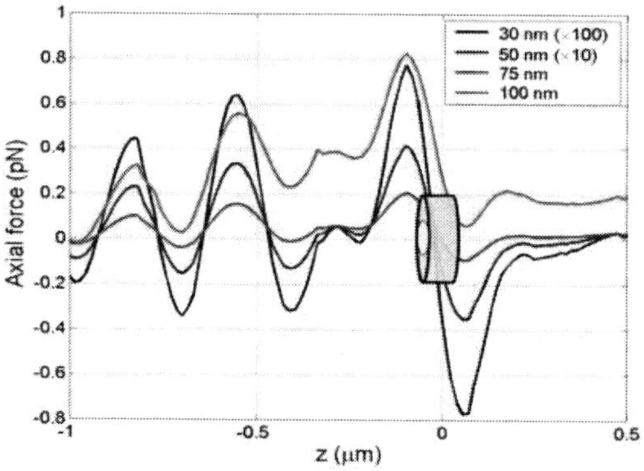

Figure 9: Nanoring system: time-averaged axial dipolar force vs. particle radius.

averaged gradient force potential for this geometry is shown in Fig. 8b. Note that there are two local minima in this plot, which are located along the x-axis on either side of central pillar. Particle trapping occurs at these points. The axial force vs. particle radius for a ring with a 450 nm inner diameter and without the central pillar is plotted in Fig. 9.

In summary, our analysis indicates that all of the nanostructures above provide pNs of trapping force at modest power levels. For the two nanopillar system, the magnitude of the force depends on the spacing of the nanopillars, and the polarization of the incident field. TE polarization and closer spacing produce stronger forces.

3 APPLICATIONS

Plasmonic-based optical manipulation holds potential for a variety of applications, especially in the fields of biophysics and biotechnology. Specific applications include (i), nanoscale size-sensitive bio-separation and filtering (ii) nanoscale biosensing, (iii) virus manipulation, (iv) cellular sorting and patterning, (v) manipulation for cellular microsurgery, (vi) nanobead-based single-molecule DNA sequencing, and (vii) characterization of single biopolymers [10].

4 CONCLUSIONS

Plasmonic-based optical trapping is its infancy and growing rapidly. Research in this area will significantly advance fundamental understanding in fields such as nanophotonics and biophotonics. Plasmonic optical trapping has advantages over conventional laser trapping in that it enables a higher spatial resolution, lower trapping energy, parallel trapping of multiple specimens, and a higher level of system integration, which is important for Lab-on-Chip applications. Novel plasmonic trapping structures and systems can be designed and optimized using commercial electromagnetic software, and this capability will facilitate the development of a new generation of systems for manipulating matter at the nanoscale.

ACKNOWLEDGEMENT

We acknowledge support from the Air Force Office of Scientific Research.

REFERENCES

[1] A. Ashkin, Proc. Natl. Acad. Sci. **94**, 4853, 1997.
[2] C. L. Kuyper and D. T. Chiu, Appl. Spectrosc. **56** (11) 295A, 2002.
[3] A. Ashkin and J. M. Dziedzic, Science **235**, 1517, 1987.
[4] K Svoboda and S. M. Block, Annu. Rev. Biophys. Biomol. Struct. **23,** 247 1994
[5] P. N. Prasad, Introduction to Biophotonics, John Wiley & Sons, NJ, 2003.
[6] H. A. Atwater, Scientific American **296,** 56, 2007.
[7] P. N. Prasad, Nanophotonics, John Wiley & Sons, NJ, 2004.
[8] L. Novotny, R. X. Bian, and X. S. Xie, Phys. Rev. Lett. **79**, 645, 1997.
[9] K. Okamoto and S. Kawata, Phys. Rev. Lett. **83**, 4534, 1999.
[10] P. C. Chaumet, A. Rahmani, and M. Nieto-Vesperinas, Phys. Rev. Lett. **88**, 123601, 2002.
[11] X. Miao and L. Y. Lin, IEEE J. Sel. Top. Quant. Elec. **13**, 1655, 2007.
[12] M. Righini, A. S. Zelenina, C. Girard and R. Quidant, Nature Physics **3**, 477, 2007.
[13] A. R. Sidorov, Y. Zhang, A. N. Grigorenko and M. R. Dickinson, Opt. Comm. **278,** 439, 2007
[14] A. N. Grigorenko, A. K. Geim, H. F. Gleeson, Y. Zhang, A. A. Firsov, I. Y. Khrushchev and J. Petrovic, Nature **438,** 335, 2005.
[15] P. C. Chaumet and M. Nieto-Vesperinas, Opt. Lett. **25**, 1065, 2000.
[16] J. I. Hage and J. M. Greenberg, Astrophys. J. **361**, 251, 1990
[17] Private communication with P. C. Chaumet.

Fabrication and Characterization of Ultra Violet Photosensor based on Single ZnO Nanorod

O. Lupan[*,**], G. Chai[***] and L. Chow[**]

*Department of Physics, University of Central Florida, PO Box 162385, Orlando, USA, lupan@physics.ucf.edu chow@mail.ucf.edu
**Department of Microelectronics and Semiconductor Devices, Technical University of Moldova, Chisinau, MD-2004, Republic of Moldova, lupan@mail.utm.md
***Apollo Technologies, Inc., 205 Waymont Court, 111, Lake Mary, FL 32746, USA, guangyuchai@yahoo.com

ABSTRACT

This article presents the fabrication and characterization of ultra violet (UV) photosensor, which utilizes the semiconducting properties of single ZnO nanorod. We report a study on the ZnO nanorods synthesized by an aqueous solution route in a hydrothermal reactor. Samples were characterized by X-ray diffraction, scanning electron microscopy, optical and electrical studies.

Our cost-effective synthesis method for ZnO nano/microrods permits easy transfer of sample and allows the study of fabrication of novel electronic nanodevices, such as gas and UV nanosensors. Our work is a starting point on the nanofabrication of optoelectronic nanodevices. It is anticipated to have a wide range of applications in space research, warning systems and accurate measurement of UV radiation.

Keywords: ZnO nanorod, nanosensor, UV, photosensor

1 INTRODUCTION

Wide-gap semiconducting metal oxides with nanostructures such as nanorods, nanowires are promising as building nanoblocks in novel nanodevices. These materials have been widely studied in the last few years due to their advantageous features such as good ultra-violet (UV) sensitivity to the ambient conditions [1]. Zinc oxide (ZnO) is an *n*-type metal oxide semiconductor with wide band gap of $Eg = 3.36$ eV at 300 K. Zinc oxide can be used for UV detection owing to its characteristics and radiation hardness, high chemical stability, low cost and flexibility in fabrication [2-4]. These properties enable it to be used also in harsh environment [5].

Recently, ZnO nanorods and nanowires have been extensively investigated for sensors and optoelectronic device applications due to it's compatibility with other microelectronic devices. ZnO nanorods are expected to have good UV response due to their wide band-gap and large surface area to volume ratio, and they might enhance the performance of UV photosensors due to longer photocarrier lifetime and shorter charge carrier transit time.

It has been reported, *n*-type ZnO nanowires arrays/*p*-type Silicon UV photodetector [6], UV sensors based on nanostructured ZnO spheres in network of nanowires [1], multiple ZnO nanowires bridging the gap between the patterned Zn electrodes were studied as UV photodetector [7]. But, fabrication and characterization of a single ZnO nanorod-based photosensor can help in understanding the uniqueness of nanorods for photosensor and enable the design of novel devices. Thus, one-dimensional (1-D) nanorod attracted attention due to their unique properties that strongly depend on their size and morphologies and their possible use as building blocks in near-future UV nanodevices [1,6,7]. Our extensive effort and novel synthesis routes are currently devoted to the controlled synthesis and characterization of transferable ZnO nanoarchitectures. In this report, a new fabrication method of a UV photosensor based on single ZnO nanorod using in situ lift-out technique is presented for the first time. A new type of UV photosensor have been characterized and demonstrated that could detect UV light down to 50 nW cm^{-2} intensity, indicating a higher UV sensitivity than ZnO thin films.

2 EXPERIMENTAL

2.1 Synthesis

The ZnO nanorods in this study were synthesized via aqueous solution deposition technique, which has been previously reported in our work [8, 9]. This technique was found to have advantages of easy scaling and low cost.

All used reagents were of analytical grade and used without further purification. The glass substrates were cleaned according to procedure described in [8].

Zinc sulfate and sodium hydroxide solution were added into 75 ml DI–water under stirring to obtain a transparent solution. Then, the glass substrates and complex solution were transferred inside an aqueous solution in a reactor of 100 ml capacity and sealed. The setup was mounted on a hot plate, and the temperature was increased to 90 °C and kept constant for 15 min and then cooled down naturally to room temperature. Variation of the synthesis conditions such as concentration of precursors and temperature allow certain degree of control on the growth rate and morphology of the obtained nanorods. Finally, the samples were rinsed in deionized water and dried at 150 °C, 5 min.

2.2 Characterization

The crystalline quality and orientations of ZnO nanorods were analyzed by an X-ray diffraction (XRD) using a Rigaku (Japan) 'D/B max' X-ray diffractometer equipped with a monochromatized CuKα radiation source (λ=1.54178 Å). The microstructure of the ZnO nanorods was observed by using scanning electron microscope (Hitachi S800). Transmission electron microscopy micrographs were carried in a FEI Tecnai F30 TEM at an accelerating voltage of 300 kV. The different characterization techniques confirmed that the nanorods are highly crystalline. Current-voltage (*I-V*) characteristics were measured using a semiconductor parameter analyzer with input impedance of 2.00×10^{8} Ω. The UV sensing properties were characterized using a computer-controlled sensing characterization system.

3 NANOFABRICATION OF THE PHOTOSENSOR BY IN-SITU LIFT-OUT TECHNIQUE IN FIB INSTRUMENT

Next the procedure for UV photosensor fabrication is described. A micromanipulator was mounted beside the stage in FIB instrument. For the nanosensor preparation, the glass substrate was used and Al electrodes were deposited as template with external electrodes/connections. The needles used for the lift-out step were electro-polished tungsten wire. The ZnO nanorod has been transferred from initial glass substrate to the Si/SiO_2 substrate in order to avoid charging problems during of the pick-up step.

The next step in our procedure is to scan the surface of the intermediate Si/SiO_2 substrate for suitable ZnO nanorod. Then the W needle was lowered and bringing into the FIB focus and its tip positioned at one end of the ZnO nanorod. In the in-situ lift-out process [9], we found that attachment of single intermediate nanorod on the top of the FIB needle will permit an easy pick-up of the selected nanorod to be further handled. This step makes the fabrication of nanodevice much more efficient.

The needle was moved to one end of the nanorod (Fig. 3a). Then the nanorod was attached to the end of the FIB needle using Pt deposition of 0.5 μm thickness. In this step the nanorod attached to the end of the FIB needle is placed on the desired area with external connections for further sensor fabrication.

The nanorod is cutoff and the needle moves away from the substrate. In the Fig. 3b is shown ZnO nanorod fixed at both ends to the substrate with contacts.

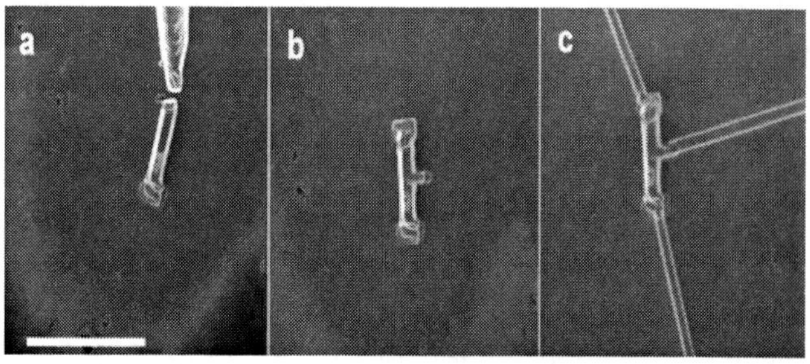

Figure 3: SEM images showing the steps of the in-situ lift-out fabrication procedure in the FIB system. (a) cutted from the W needle and placed the ZnO nanorod on substrate; (b) single nanorod fixed at both ends on substrate; (c) single nanorod welded to three electrode/external connections as final UV nanosensor. The scale bar is 5 μm.

In the last step, the nanorod was fixed to the pre-deposited electrodes/external contacts. Figure 3c show a novel single ZnO nanorod-based sensor fabricated by in-situ lift-out technique in the FIB system. By this technique, different shaped-nanosensors have been fabricated and investigated for their UV sensitivity.

The typical time taken to perform this in-situ lift-out FIB nanofabrication is about 25 min and our success rate is >95%. Also taken in the account that nanorod synthesis was done in 15 min, we substantially improve the fabrication process. This minimizes the time to fabricate nanodevices using FIB and can be extended to other specific nanodevices.

4 RESULTS AND DISCUSSIONS

Figure 1 shows the indexed XRD pattern of the ZnO nanorods in the range of 30-90° shows a predominant sharp peak at 36.2°, which is preferentially oriented (101) plane growth (Fig. 1). It can be seen that all diffraction peaks are caused by crystalline ZnO with the hexagonal wurtzite structure (space group: $P6_3mc(186)$; the lattice constants a = 0.3249 nm, c = 0.5206 nm), which indicates that pure ZnO can be obtained at 90°C for 15 min. The data are in agreement with the JCPDS 036-1451 card for ZnO [10].

The morphology of the ZnO nanorods grown on glass substrate without using any templates was observed by the scanning electron microscopy (SEM). Typical SEM images of the ZnO nanorods are shown in Fig. 2. The SEM images of ZnO nanorods after annealed at 650 °C, 60 sec do not differ from those shown in Fig. 2.

In the inset image, it is clearly seen that ZnO nanorods with an average radius of 150 nm and are uniformly on its length. The medium lengths of ZnO nanorods are about 10 μm.

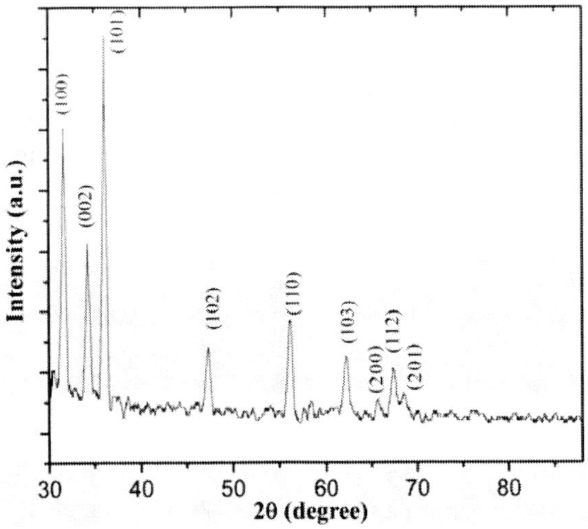

Figure 1: Indexed XRD scan for ZnO nanorods on glass substrate synthesized by the aqueous-solution method showing a wurtzite type structure.

According to our experimental results, the ZnO nanorods obtained by our process can be easily transferred to other substrates and handled by Focused Ion Beam (FIB) system in order to fabricate different nanodevices.

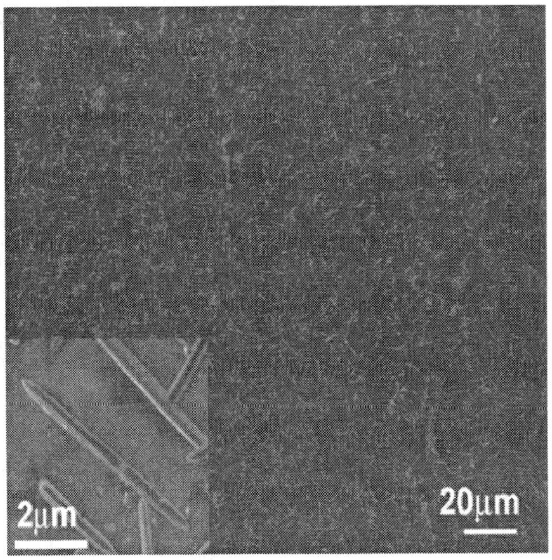

Figure 2: The SEM images of the ZnO nanorods on glass substrates and inset are individual rod image.

The detailed structural characterization of single ZnO nanorod was demonstrated by the high-resolution transmission electron microscopy (HRTEM) image [8].

5 UV SENSING

Next, we measured the *I*-V curves of the ZnO nanorod with connections realized by in-situ lift-out method. Figure 4 outlines the *I-V* characteristics of a three-terminal ZnO nanorod based photosensor in ambient air. The *I-V* measurements were performed by changing the bias voltages from +8 mV to -8 mV and vice versa. The voltage increment and delay time were set to 1 mV and 2 s.

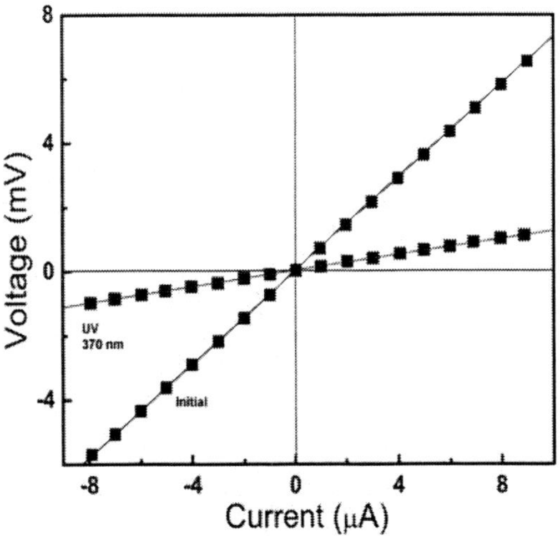

Figure 4: I-V characteristics of the ZnO – single nanorod photosensor without and under UV exposure at 370 nm.

The UV sensitivity was measured using ZnO nanorod device. The fabricated single ZnO nanorod-based sensor was put in a test chamber to detect ultraviolet light. The readings were taken after a UV light was turned on. It was subjected to irradiation with an UV light using a lamp with an incident peak wavelength of 370 nm with conductivity monitoring. The background atmosphere was air. It was found that conductivity change increased linearly with ultraviolet intensity. Due to the fact that the photon energy is higher than the bandgap of ZnO, UV light was absorbed by the ZnO nanorod creating electron-hole pairs, which were further separated by the electric field inside ZnO nanorod contributing to the increase of the conductivity.

When the ZnO nanorod photodetector was illuminated by 370 nm UV light, the conductance increased with a time constant of a few minutes as shown in Fig. 5. When the UV light was turned off, the conductance decreased back within 10% of the initial value (Fig. 5).

Response time constants are on the order of few minutes and after the signal reach the equilibrium value after the UV light was applied. This suggests a reasonable recovery time. The sensor showed relatively fast response and baseline recovery for UV detection.

Several photodetector on single nanorod have been fabricated by in-situ lift-out technique and investigated under identical conditions and was observed similar UV response. Spectral response demonstrates that such photodetector is indeed suitable for detecting UV in the range 300 nm – 400 nm.

The UV response is slow for ZnO nanorod photosensor and can be explained by the adsorption and photodesorption of ambient gas molecules such as O_2 or H_2O [3, 11]. The optical power on the detectors 50 nW so the photoresponsivity at 370 nm is 40 A/W.

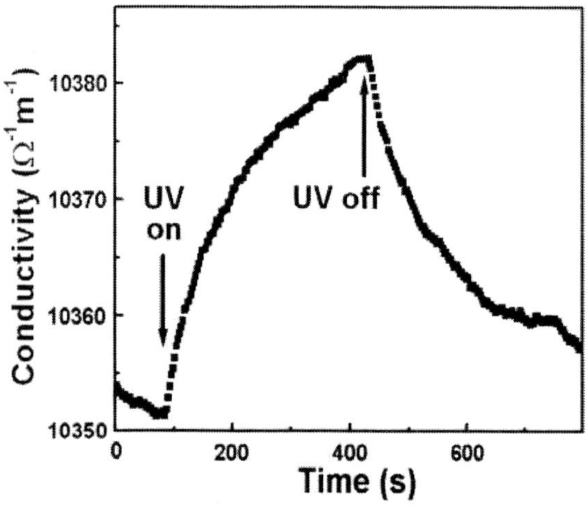

Figure 5: The conductivity response of the single ZnO nanorod-based UV photosensor fabricated by in-situ lift-out technique in the FIB system.

As-grown ZnO nanorod adsorbs oxygen molecules on the surface. Then will take free electrons from the *n*-type ZnO nanorod to form a depletion region. After UV illumination of photosensor the electron-hole pairs will be generated. The holes will recombine with oxygen ions chemisorbed on the surface and eliminate the depletion region [12]. At the same time the electrons produced will contribute to conductivity. Thus the time it takes electrons and holes to recombine will increase. Delay in the recombination of an electron and hole will increase, in this case the number of transit times for electron in the nanorod will be more than one. This phenomenon can lead to the increase in current and results in internal photoconductive gain.

The transit time of the charge carriers though the ZnO nanorod UV sensor can be described by [12]:

$$T_{tr} = \frac{d^2}{\mu_n V_b} \qquad (1)$$

where d - interelectrode spacing, μ_n - mobility of the electron, and V_b is the applied bias voltage. Thus, by using shorter interelectrode spacing d, the photoresponse speed can be improved. Also, by avoiding high temperatures processes during of the fabricating Al electrodes, by avoiding acid medium during the preparation of the sensor we can keep real bias voltage V_b and lower T_{tr} .

In addition, ZnO nanorod has high crystal and optical quality [8], so μ_n will be high. Thus we can shorten the transit time by decreasing recombination of charge carriers.

Thus UV light will hit the nanorod surface and will be generated electron-hole pairs. Electrons must remain free from holes long enough to zip along the nanorod and generate electric current under applied electric field and this will be the detection of light.

6 CONCLUSIONS

In summary, fabrication of single ZnO – nanorod UV photosensor by in-situ lift-out technique in the FIB system is demonstrated. Our technique can fabricate sensors on single nanowire in order to study light detection with single photon sensitivity.

The main advantage of the proposed synthesis is its simplicity and fast growth method. An in-situ lift-out technique has been presented to fabricate single ZnO nanorod –based photosensor. The typical time taken to perform this in-situ lift-out FIB nanofabrication is 25 min. Also taken in the account that nanorod synthesis takes about 15 min, we contribute to overcome some obstacles for nanorods/nanowires sensor production.

This technique has a great potential to be used to fabricate single ZnO nanorod-based photosensor which can help in understanding the uniqueness of nanorods for photosensor and enable the design of novel devices.

In summary, the prototype device provides a simple method for nanorods synthesis and demonstrated possibility of constructing nanoscale photodetectors for nano-optics applications.

Acknowledgments The research described in this publication was made possible in part by Award No. MTFP-1014B Follow-on of the Moldovan Research and Development Association (MRDA) and the U.S. Civilian Research and Development Foundation (CRDF). Dr. L. Chow acknowledges partial financial support from Apollo Technologies Inc and Florida High Tech Corridor Program.

REFERENCES

[1] S. S. Hullavarad, N. V. Hullavarad, P. C. Karulkar, A.Luykx, P.Valdivia, Nanosc.Res.Lett. 2,161,2007.
[2] D.C. Look, Mater. Sci. Eng. B 80, 383, 2001.
[3] J. Law, J.Thong, Appl. Phys. Lett. 88, 133114, 2006.
[4] R. Hauschild, H. Kalt, Appl. Phys. Lett. 89, 123107, 2006.
[5] D. C. Look, D. Reynolds, J. Hemsky, R. L. Jones, J. R. Sizelove, Appl. Phys. Lett. 75, 811, 1999.
[6] L. Luo, Y. F. Zhang, S. S Mao, L. W. Lin, Sensors and Actuators 127, 201, 2006.
[7] J. Law, J.Thong, Appl. Phys. Lett. 88, 133114, 2006.
[8] O. Lupan, L. Chow, G. Chai, B. Roldan, A. Naitabdi, A. Schulte, H. Heinrich, Mater. Sci. Eng. B 145, 57, 2007.
[9] O. Lupan, G. Chai, L. Chow, Microelectronics Journal 38, 1211, 2007.
[10] Joint Committee on Powder Diffraction Standards, Powder Diffraction File No 36-1451.
[11] J. Suehiro, N. Nakagawa, S. Hidaka, M. Ueda, K. Imasaka, M. Higashihata, T. Okada, M. Hara, Nanotechnology 17, 2567, 2006.
[12] W. Yang, R. Vispute, S. Choopun, R. Sharma, T. Venkatesan, H. Shen, Appl. Phys. Lett. 78, 2787, 2001.

Simulation and Fabrication of Large-Area 3D Nanostructures

K. H. A. Bogart,* I. El-kady,* R. K. Grubbs,* K. Rahimian,*
A. M. Sanchez,* A. R. Ellis,* M. Wiwi,* F. B. McCormick,*
D. J.-L. Shir,** and J. A. Rogers**

*Sandia National Laboratories, khbogar@sandia.gov
**University of Illinois, Urbana-Champaign

ABSTRACT

Three-dimensional (3D) nano-structures are vital for emerging technologies such as photonics, sensors, fuel cells, catalyst supports, and data storage. The Proximity-field nanoPatterning[1] method generates complex 3D nanostructures using a single exposure through an elastomeric "phase mask" patterned in x, y, and z, and a single development cycle. We developed a model that predicts the phase mask required to generate a specific desired nanostructure. We have compared this inverse model with experimental 3D structures to test the validity of the simulation. We have transferred the PnP fabrication process to a class-10 commercial cleanroom and scaled-up the processed area to >2000mm^2, tested photopolymer additives designed to reduce resist shrinkage, incorporated atomic layer deposition (ALD) to coat the 3D patterned resist with metals/metal-oxides improve structure robustness, and generated quasi-crystal patterned 3D nanostructures.

Keywords: nanostructure, lithography, quasicrystal, photonic, model

1 INTRODUCTION

Three-dimensional (3D) nano-structures are vital for emerging technologies such as photonics, sensors, fuel cells, catalyst supports, and data storage. Conventional fabrication (repeated cycles of standard photolithography with selective material removal) is costly, time-consuming, and produces limited geometries. Unconventional methods (colloidal self assembly, template-controlled growth, and direct-write or holographic lithography) have uncertain yields, poor defect control, small areas, and/or complicated optical equipment. The Proximity-field nanoPatterning (PnP)[1] method overcomes these limitations by generating complex 3D nanostructures using a simple optic and one lithographic exposure and development cycle. The optic is an elastomeric "rubber phase mask" patterned in x, y, and z with dimensions roughly equal to the exposure wavelength. Exposure through this mask generates a complex 3D light intensity distribution due to diffraction (Abbe theory) and the Talbot effect (self-imaging).[2] The underlying photoresist is thus exposed in certain regions, baked, and developed, producing a 3D network of nanostructures with one lithography cycle. Our goals are to create full models of this process and scale this method to 150mm

2 METHODS AND RESULTS

2.1 FDTD Model and Simulation

We have developed a model using Finite Difference Time Domain (FDTD) methods that predicts the 3D nanostructure resulting from light passing through a phase mask with a given geometry.[3] We have also developed a model to identify the phase mask parameters required to generate a specific desired nanostructure. This "inverse" approach is much more complex than the simplistic modeling of the diffraction pattern produced by passing light through a phase mask. The integrated tool starts with a desired pattern and an initial guess on the PnP mask parameters. Next, the interference pattern is simulated using the mask information and filtered to reveal the expected photoresist burn image, which is then evaluated against the desired pattern. An integrated optimizer makes improvements to the mask parameters and cycles again with a simulation using the new mask parameters. The simulation engine is a high performance, Open MP parallelized FDTD simulator optimized to run on shared memory symmetric multiprocessor (SMP) systems. The product from the simulation is the actual resist burn pattern.

We have compared this model with experimental 3D structures for a hexagonal array of (Figure 1) to test the validity of the forward simulation. The phase mask was patterned in a hexagonal array of posts with diameter (d) = 450 nm, period (p) = 600nm, height (h) 420 nm. SU-8-2 (MicroChem) photoresist was spun twice forming a 4.5 μm

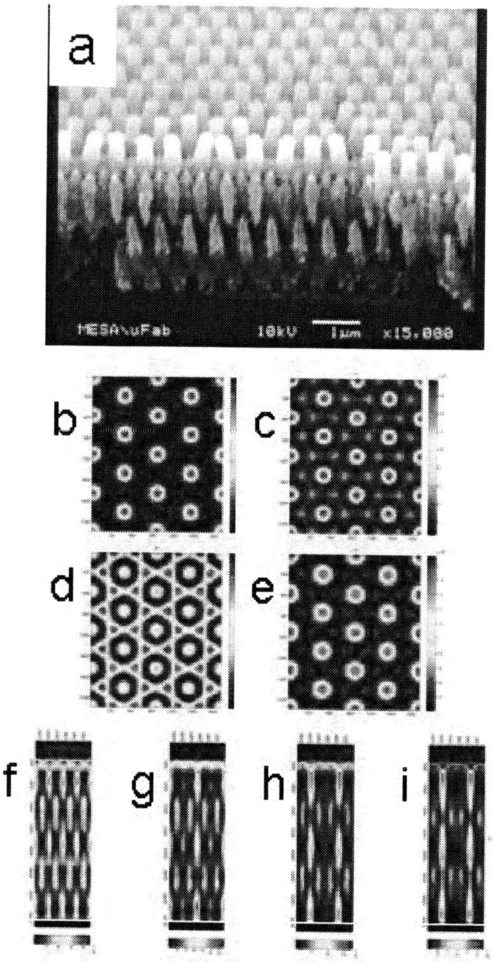

Figure 1. Scanning electron micrograph of an angled cross section of a 3D resist nanostructure made with a hexagonal geometry (1a), corresponding horizontal (x,y) and vertical (x,z) model slices, (1b-e, 1f-i, respectively). The horizontal slices are 160 nm apart and the vertical slices are 80 nm apart.

thick film. Passage of 365 nm light through the phase mask (placed in direct, conformal contact with the resist surface) generated a complex 3D light intensity pattern, which is transferred directly into the photoresist. Normal post-exposure baking, developing, and drying followed [1,2] generating a 3D nanostructure. A scanning electron micrograph of an angled view of the cross section of the 3D resist structure is shown in Figure 1a.

For reference, the surface is defined as the x-y plane and the direction from surface to substrate is z. The 3D structure has an alternating array of resist columns with air gaps, corresponding to the ABAB design of the hexagonal array. The corresponding modeled structure is shown in Figs. 1b-e (horizontal or x,y slices, 160 nm apart) and Figs. 1f-i. (vertical or x,z slices, 80 nm apart). The model used the same phase mask dimensions and exposure wavelength as inputs, and a given threshold value to generate the resist burn. The horizontal model slices show the hexagonal arrangement of resist columns in the 3D resist structure. The vertical model slices show the alternating patterns of resist columns with air holes. The ABAB nature of the resist columns in the resist cross sectional cleave matches the vertical model slice (compare Fig. 1a with Fig. 1f).

2.2 Fabrication and Scale-up

We have successfully transferred the PnP fabrication process to a class-10 commercial clean room and scaled-up the processed area from 490mm^2 to >2000mm^2 using commercial lithography exposure tools. We use a Karl Suss MA-6 contact proximity printer for the exposure. The broad band output of the Hg lamp is narrowed to 364.75±1.25 nm by a multiple thin film narrow bandpass filter and a 350 nm longpass filter. Conventional pre-bake (65°C, 10 min /95°C, 15 min), post-exposure bake (65°C, 20 min) and development in SU-8 developer (MicroChem) processes are used to complete the fabrication. We have produced 3D resist structures for cubic arrays of posts and holes (Fig 2a), hexagonal arrays of posts (Fig 1a) and aperiodic Penrose quasicrystal structures (Fig. 2b). We have also obtained similar comparisons between the aperiodic 3D resist structures and the simulated structures for these geometries.[4]

Our scale-up has recently progressed to exposure of a full 150 mm wafer using the MA6 exposure tool (17600 mm^2). An optical photograph of the wafer is shown in Figure 3a, along with SEM images of 3D nanostructures taken from the center (Fig. 3b) and edge (Fig. 3c) of the 150 mm wafer. No significant difference is observed in the structures across the wafer.

2.3 Chemical Modifications

One of the properties of epoxy-based resists is their shrinkage upon exposure/development. This shrinkage is a function of primarily solvent loss from the resist, compounded by the strong epoxy linkages formed during the cross-linking process. In order to address the issue of resist shrinkage, we have tested photopolymer additives designed to reduce resist shrinkage by replacing a percentage of the resist solvent with reactive solids. We have identified diglycidyl ether, diepoxyoctane, and diglycidylglycidioxyaniline as reactive diluents. We have

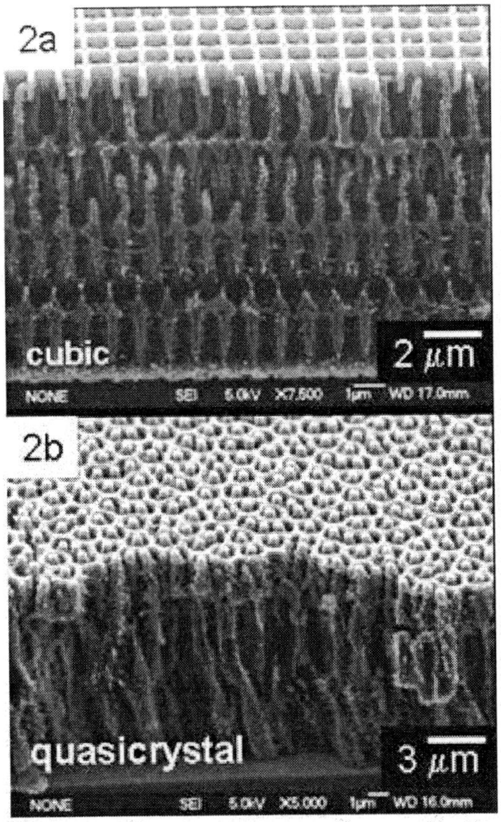

Figure 2. Scanning electron micrographs of 3D resist nanostructures made from a cubic array of holes (2a) and a Penrose quasicrystal array of posts (2b).

successfully fabricated 3D structures with these components at replacement volumes of between 10 and 30 wt%. Measurements of shrinkage reduction are commencing.

PnP 3D nanostructures can also be formed by a 2-photon (2ph) exposure process. In this case, light with a wavelength (λ) double the normal exposure λ (350-400nm) is used, but with much greater energy, ~1TW/cm^2, enabling a 2ph energy absorption that is sufficient to initiate the photoacid generation reactions in the photoresist.[5] Lasers at this power level are often in the near infrared, thus photoresists require a red-shift in sensitization for exposure. We have chemically modified SU8-10 photoresist with Rose Bengal and Uvacure 1600 (photoacid generator) (Fig. 4a) and successfully fabricated 3D nanostructures with a 1-photon exposure at 532 nm (equivalent to doubling of 1064 nm YAG for 2ph mode). Examples of those structures are shown in Figures 4b and 4c.

Epoxy-based photoresists such as SU8, are inherently robust due to the epoxide cross-linking. However, a 3D nanostructure made in SU8 is still an organic resist-based material. We have used atomic layer deposition (ALD) to coat the 3D resist nanostructure to improve the robust nature of 3d structure and also alter the chemical and physical properties of the material.

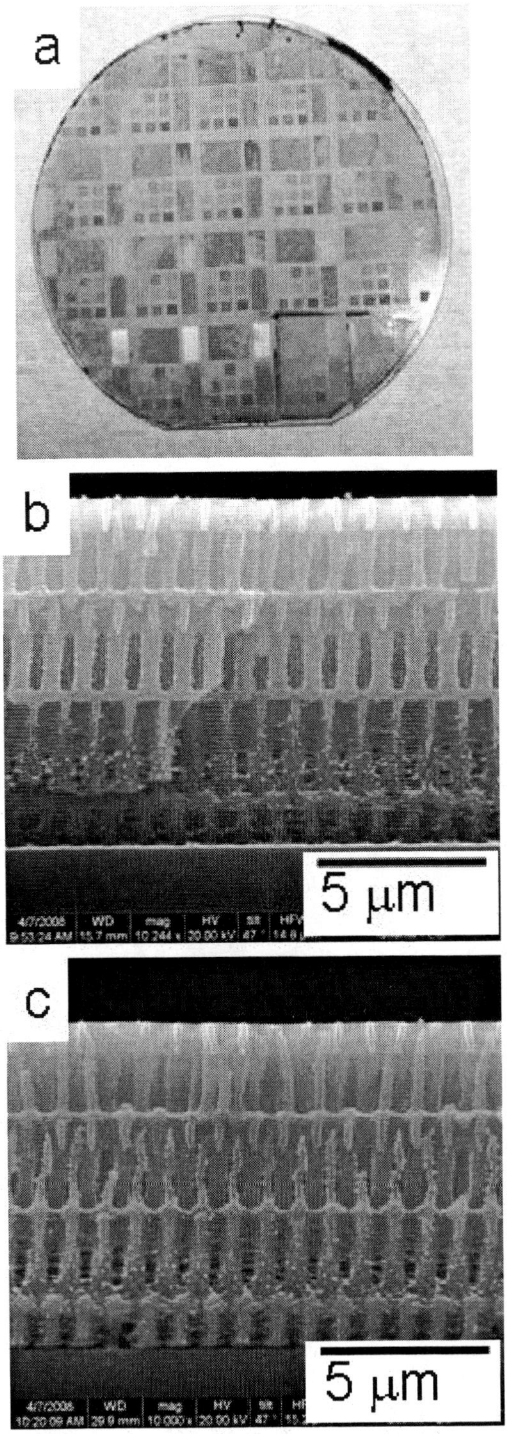

Figure 3. Optical photograph of a 150 mm wafer patterned with PnP 3D nanostructures using a 150 mm phase mask (a), and SEM images of structures from center (b) and edge (c).

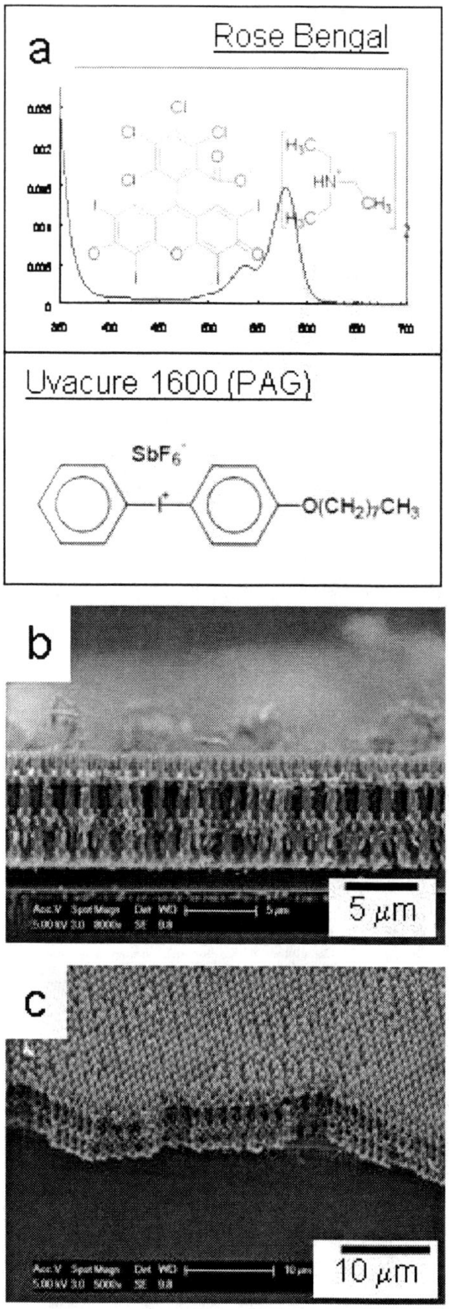

Figure 4. Chemical structure and absorption spectrum for Rose Bengal additive and Uvacure 1600 (3a), SEM images of 3D nanostructures from chemically modified SU8-10 resist exposed at 532 nm in 1-photon mode (3b, 3c).

We have developed a proprietary, graded-temperature deposition approach which does not cause deformation or degradation of the resist structure. Both Al_2O_3 and TiO_2 have been used with this approach, rendering the structure capable of withstanding high ALD deposition temperatures above 250°C. High-temperature ALD materials we have deposited onto these structures include Pt, ZnO, and ZrO_2. Examples of 3D nanostructures made using a 2ph exposure (800 nm) and a square array of posts pattern and coated with Pt using ALD are shown in Figure 5. Deposition of the high-temperature Pt does not degrade the 3D nanostructure. Optical measurements of this structure, with an eye towards photonic crystal properties, are commencing.

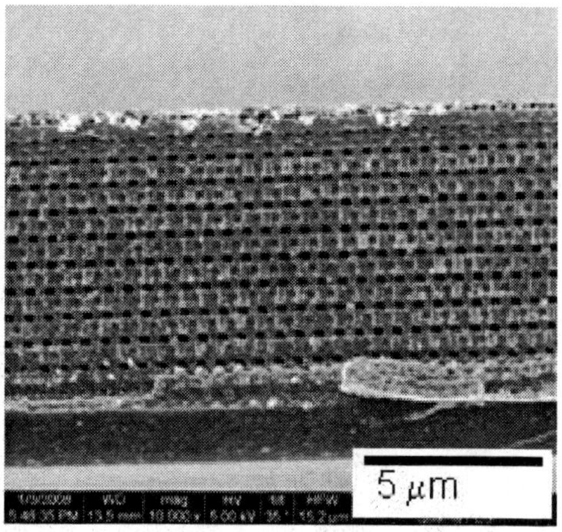

Figure 5. SEM image of PnP 3D resist nanostructure coated with Pt by ALD, showing no degradation of structural features.

3 SUMMARY

The PnP lithography technology, coupled with accurate FDTD modeling, enables predictive simulation and fabrication of 3D nanometer-scale structured materials with specific, desired optical and structural properties.

REFERENCES

1 S. Jeon, J.-U. Park, R. Cirelli et al., Proc. Natl. Acad. Sci. **101**, 12428 (2004); D. J. Shir, S. Jeon, H. Liao et al., J. Phys. Chem. B **111**, 12945 (2007).

2 M. V. Berry and S. Klein, J. Mod. Opt. **43**, 2139 (1996); L. Rayleigh, Philos. Mag. **11**, 196 (1881); H. F. Talbot, Philos. Mag. **9**, 401 (1836).

3 M. F. Su, I. El-kady, M. M. Reda Taha et al., Photon. & Nanostruct. Fund. & Appl. **121**, 1 (2007); M. F. Su, M. M. Reda Taha, C. G. Christodoulou et al., IEEE Photon. Lett. (2008).

4 M. F. Su, D. J.-L. Shir, R. Rammohan et al., manuscript in preparation. (2008).

5 S. Jeon, Y.-S. Nam, D. J.-L. Shir et al., Appl. Phys. Lett. **89**, 253101 (2006).

FROM RADAR to NODAR

C. Falessi*, A. M. Fiorello*, A. Di Carlo **, M. L. Terranova**
* SELEX Sistemi Integrati ,Via Tiburtina 1231, 00131, Rome, ** University Roma2 Tor Vergata
+39 0641502880, cfalessi@selex-si.com , www.selex-si.com

***Abstract*-Photonics and Nanotechnologies are emerging revolutionary technologies that will provide a revolution in the sectors of sensors and radar systems. SELEX Sistemi Integrati intends to face this evolution as a leader by proactively supporting and developing its state of the art. One century after the first RADAR "idea" we registered the NODAR (Nanotechnology Optical Detection And Ranging) trademark. Such NODAR, by including both photonic and nano technologies, aims to implement multifunctional, multirole, multidomain sensors, as well as adaptive, flexible, knowledge-based sensors. NODAR requires proactive studies and developments on: I) nanotechnology vacuum tube amplifiers (TX reverse nano triode); II) wide bandwidth Tx-Rx optical beam forming network by means of optical modulator, combiner, analogue optical receive and programmable true time delay; III) broadband photonic analogue to digital converters; IV) thermal management and interconnection by means of carbon nanotubes; V) nanotechnology infrared and chemical state of art sensors. Finally, the insertion of state of the art innovative technologies requires an integrated "multiscale" approach, by combining materials sciences, photonics, nanotechnologies and production technologies with other based technologies**

***Index Terms*- Nanotechnologies, Photonics, Multiscale.**

I. RADAR

One hundred years ago Christian Hulsmeyer patented in Great Britain the first "idea" containing the "DNA" of the sensor, universally known with the acronym of RADAR (RAdio Detection And Ranging). Then, it took more than 30 years to put the idea into practice. Main achievements in radar performances have been developed during the past century; in particular by Guglielmo Marconi', with the British Home Chain in 1937, afterwards by Prof. Ugo Tiberio with the radar GUFO, and eventually up to the modern multifunctional radars, such as EMPAR or MEADS.

From the '60s, the same Company, named progressively as Selenia, Alenia, AMS, and today SELEX Sistemi Integrati, has developed several radar sensors, characterized by the most innovative technology to operate in a huge spectrum of scenarios.

In this continuous evolution, technologies related to antenna and signal processing have increasingly absorbed all the other ones like bulk transmitter, MF and RF. We envision future digital radar with thousands of heterogeneous nodes, highly adaptive and cooperating in real time, with transmitted power generation elements and digital receivers contained directly in the antenna. Key elements are system and software architecture and communication, software based signal processing. These should provide, in hard real time adaptivity, very demanding algorithms such as adaptive hard real time digital beam forming or knowledge based information fusion among other heterogeneous sensors.

New radar sensors could be "multifunctional" such as the EMPAR (European Multifunctional Phased Array Radar) that is capable to perform, in hard real time adaptivity, search, surveillance and tracking operative modes.

In addition we are developing "multi–domain" sensors in collaboration with Elettronica and SAAB. For instance, in the M-AESA, a MoDs funded multi-domain sensor, the RADAR operative functions, the passive and active electronic warfare and the communications functions are managed and completely fused in the same system. On the contrary, at system level we are evolving towards heterogeneous system of systems.

A system of systems is a "super-system" comprised of elements that are themselves complex and independent systems, which interact to achieve a common goal. The resultant operative function is larger than the sum of the single components functions.

We are developing large integrated systems not only for future ATM but also for command and control in defence and homeland protection applications.

Radars, sensors and systems are continuously evolving: what are the technologies that are needed?

II. EVOLUTION OR INNOVATION?

The trend and roadmap of the sensors, and of the new-generation systems require a continuous technology innovation; however it is required a high effort to maintain the mature technologies at the state of the art. Therefore, we are focusing our attention on the nanotechnologies and photonic innovative technologies because, compared to the evolutionary ones, they have a higher ratio between the obtained results and the required efforts (utility function)

These emerging, innovative and enabling technologies improve weight, size, speed, power consumption, efficiency, and so on. In addition they enable new solutions.

The innovative or revolutionary technologies could open

new frontiers, being "killer technologies" and requiring a "creative destruction" that is the shifting of capital from falling, mature technologies into those technologies at the cutting edge as clearly depicted by Alan Greenspan.

III. NANOSCIENCE AND NANOTECHNOLOGIES

Nanotechnologies represent an emergent domain having a great potential. It is a flurry of activities due to its attainable results, unimaginable performances, and revolutionary applicability. Theirs importance and potentiality are approved by the European Commission with funds of over 3.5 billion euros in the 7th Framework Program.

Nanoscience and nanotechnologies, as stated in Lisbon 2000, "could open a new era, enable, support and drive the 21th century knowledge-based society".

Nanoscience, the new theoretic and descriptive domain, from which derive different nanotechnologies, is rightly defined as a "crucial, horizontal, qualifying science". It allows combining such scientific disciplines that have been wrongly considered separated and different in the past. Nanotechnologies, profiting by interdisciplinary and converging approaches, will contribute to the solution of problems that are typical of the modern society. They will probably give a contribution to medical applications, and to research fields related to food, water, environment, energy production, creation of metamaterials. Nanotechnologies provides optimal performances regarding prognostics, photonics, information technologies (also through organic and inorganic nanodevices), biology, and in the science of cognition.

IV. TOWARDS THE NODAR

SELEX Sistemi Integrati intends to face the evolution of radar, system, and system of systems as a leader by proactively supporting and developing state of the art of enabling and innovative technologies. To do so; a century after the first RADAR "idea", we registerd the NODAR (Nanotechnology Optical Detection And Ranging) trademark. NODAR ® aims to implement multifunctional, multirole, multidomain sensors, as well as adaptive, flexible, knowledge-based sensors, by including both photonic and nano technologies.

The "NODAR", by including both photonic and nano technologies, aims to implement multifunctional, multirole, multidomain sensors, as well as adaptive, flexible, knowledge-based sensors.

We identified the following highly innovative sectors of NODAR:

- Nanotechnology Vacuum Tube Amplifier: TX Reverse Nano Triode by means of million of Carbon NanoTubes. It should able to amplify with high efficiency TeraHz frequencies.
- Wide BandWidth Tx-Rx Optical Beam Forming Network by means of Optical Modulator, Combiner, Analog Optical Receiver, Programmable True Time Delay for Direct Time Beam Forming and Stearing.
- A Broadband Photonic Analog to Digital Converter.
- Nanotechnology Devices for Knowledge Based Signal Processing and for Grid Sensors Signal Fusion. These provide a feasible solution while waiting for the most promising Nano Quantum Processing.
- Carbon NanoTubes for Thermal Management and Interconnection.
- Optical - Nanotechnology Chemical state of art Sensors.
- Integrated "Multiscale" approach, by combining Materials Sciences, Photonics, Nanotechnologies and Production technologies with other technologies based on e.g. information technologies along the nano-micro-meso and macro length.

V. NANO VACUUM TUBE AMPLIFIER

High transmission efficiency is a key element in any Radar, Sensor and System. Actually, the radio frequency high power amplifiers use Gallium Arsenide (GaAs) or Gallium Nitride (GaN) solid state technologies. Unfortunately the recurrent costs are high. Moreover, for very high band applications such as TeraHz, there are gain and bandwidth limits due to the limited electron mobility in the solid state. TeraHz are of growing importance for health, space communication, defence and homeland protection applications,.

The "nano-vacuum tube" is an innovative device for TeraHz regime applications that aim to overcome all the problems related with the scarce miniaturization and the not negligible weight and volume of the standard vacuum electronic devices. As for the standard vacuum tubes, it consists of three main parts: the emitting cathode, the grid and the anode. The innovation is represented by the use of carbon nanotubes as a cold cathode for the emission of the electronic beam and by the introduction of an innovative layout that allows overcoming all the technological problems for the realization of such devices [1].

Carbon Nanotubes can be considered ideal field emitters due to their high aspect ratio, robustness, stability and lack of surface oxides [2]. It allows to obtain the miniaturization of the device and the improvement of the lifetimes when compared with the already existing Spindt type cold cathode devices. However, the critical growth conditions of those materials (high temperature and highly reactive plasma), leads to a difficult integration with standard technological processes. In this context we developed an innovative topology that is devised to limit the effects of the outlined problem. The device foresees the realization of two separate elements, the cathode and the integrated anode-grid structure, that are then packaged using by a vacuum bonding technique. The emitting cathode, made from aligned single wall carbon nanotubes, is realized on a patterned highly conductive silicon substrate that can stand the synthesis conditions (Fig. 1). The grid and the anode are realized on a single substrate separated by a thick silicon oxide layer that works as insulating layer. The grid is covered with an insulating layer in order to reduce the current losses of the device and thus to improve the transparency.

This element is realized using standard lithographic processes. The cathode and the grid-anode plates are bonded together using proper spacers (Fig.2).

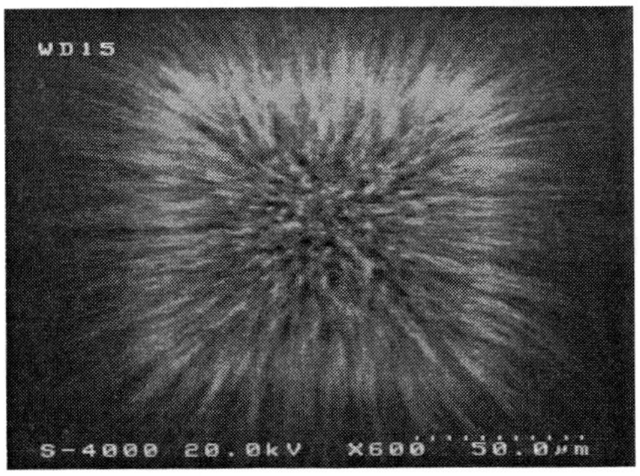

Fig.1: Patterned Cathode

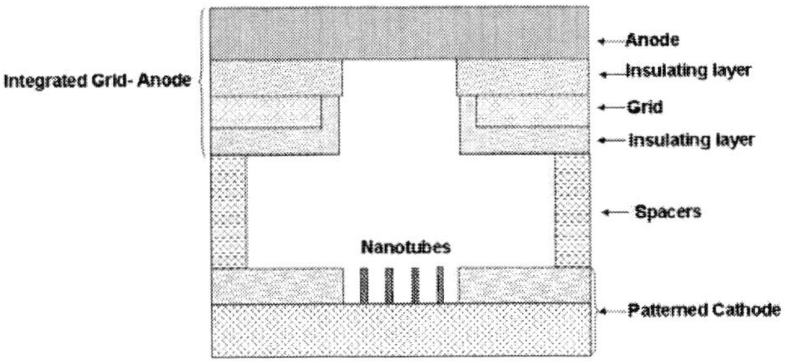

Fig. 2: Vacuum Reverse Triode

This structure has many benefits; as it relaxes many of the technological constraints for realization of conventional CNT base triodes, it allows building micro vacuum diodes with more favourable characteristics, while placing minimum restrictions to the gate structure and the materials used. In this geometry it is straightforward to use different kinds of emitting materials, grown on a variety of substrates, opening the way to new applications for scaled vacuum triodes. Furthermore, it is possible to efficiently tune the working frequency of the device by a proper design of the geometry of the electrodes.

VI. Photonic ADC

There are at least four classes of photonic Analog to Digital Converter (ADC).

In Nyquist-rate photonic ADCs, such as first proposed by Taylor [3] both sampling and quantization are performed optically and the system is analogous to the Nyquist-rate electronic ADC.

In demultiplexing photonic ADCs a high pulse repetition frequency (PRF) train of short optical pulses samples an RF waveform; the pulses are demultiplexed to multiple photodiodes and the current from the photodiodes is quantized electronically.

In oversampling photonic ADCs a high PRF train of short optical pulses and one or more feedback loops are used to yield behavior analogous to electronic oversampling ADCs such as delta sigma modulators.

The time-stretched photonic ADC uses fiber dispersion to lower the frequency of an electronic signal prior to digitization with an electronic ADC.

Other novel ADCs use ultra-stable, short-pulse lasers as part of the sampling process, but the term “photonics-assisted” ADC seems more appropriate for these systems.

Demultiplexing (or time-interleaved) photonic ADCs basically relies on using an optical architecture to:

- Generate a stream of sampling optical pulses
- Modulate the height of the optical pulses by the voltage signal to be sampled through an optical modulator
- Split along multiple (N) parallel channels the samples to be A/D converted
- Perform A/D conversion on each channel with 1/N sampling rate using standard electronic A/DCs
- Recombine the bit stream by digital processing

VII. Carbon NanoTubes for Thermal Management

Increasing attention is being paid to single-walled carbon nanotubes (SWCNT), characterized by very a thermal conductivity that can reach values up to 6000 W m-1 K-1. The good thermal and electrical conductivity make the SWCNTs an excellent candidate material to be used as a heat sink medium to increase thermal dissipation from the chip toward the package and also to build bumps interconnecting the heat sink with the chip (Flip Chip configuration).

In this context SELEX-SI in collaboration with MINASlab is carrying out a research focused on the preparation of SWCNT-based systems for thermal management applications in high power electronic devices.

The performances of the SWCNT as TIM material are measured using the nanotubes in the same configuration foreseen for the working device, i.e. as interface between the microprocessor chip and the heat spreader or between the heat spreader and the heat sink. The activity of material preparation pivots along the following main lines:

i) Setting up of protocols for the preparation of epoxy/SWCNT or polymer/SWCNT nanocomposite layers.

Some investigations of polymer-based nanocomposites have indicated significant increases in the thermal conductivity of the CNT-loaded samples [4]. The controversial values obtained in different experiments can be certainly ascribed to the different capabilities of the testing apparatuses, but the scattering of data can be rationalized by considering, beyond the variations in nanotube characteristics, the homogeneity and reliability of the SWCNT dispersions in the various matrices.

In order to evaluate the efficiency of the nanotubes as TIM in different chemical environments, SWCNT are chemically treated following protocols settled in our laboratory and are incorporated into a variety of polymer or epoxy matrices. Nanocomposites with various amounts of nanotubes and different levels of dispersions are prepared and tested using a test bed for thermal resistance measurements.

Preliminary measurements performed on nanocomposite silicon samples containing SWCNT showed an increase of thermal dissipation up to ~ 30% with respect to the unloaded paste.

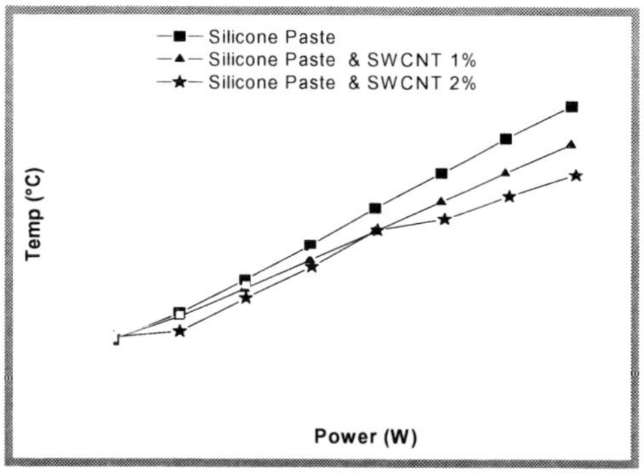

Fig.3: Nano Thermal Conductivity

ii) The growing by CVD techniques of aligned/oriented bundles of SWCNT on areas patterned by lithography and fabrication of bumps for flip-chip interconnections. This research line relies on the expertise reached by MINASlab on the synthesis and manipulation of nanotubes. The employed CVD techniques enable to deposit bundles with specific geometries and arrangements, and to design nanotube-based architectures for assembling network of bumps for flip-chips interconnects.

It is possible to integrate SWCNT bundles with selected orientation on semiconductors and to obtain a direct bonding device-substrate characterized by good thermal and electrical conductivity.

Thermal measurements of the assembled systems can be carried out by means of an innovative electro-optical technique [5] designed for precise measurements of channel temperature in power devices. The main advantages of using this non invasive method are the better spatial resolution and the feasibility to measure the effective temperature of the device due only to the photocurrent.

VIII. Multiscale Approach

New design tools and methodologies are needed to develop more and more complex and multifunctional nanostructured devices and integrate them into systems that are organized according to hierarchical architectures.

The multiscale approach integrates, inside a coherent framework, different layers of scientific and engineering mathematical representations and models, data structures, information and knowledge. The development of complex nanostructured systems like NODAR will benefit from new advanced concurrent and adaptive multiscale methods. Multiscale concurrent and adaptive methods make it possible for the first time to integrate inside a single model the following domains: quantum mechanics, quantum chemistry, multi-particle simulation, molecular simulation, and continuum-based techniques and address, in a unified way, nanoelectronics, nanophotonics, and nanomechanics issues.

A key step for nano electronics and photonics engineering is the development of Integrated Multiscale Multiphysics Science – Engineering Environments. Several civil and military organizations in Italy (NMP), Europe, US and Japan have launched important projects and initiatives along this direction. The strategic objective is to close the gap between classical engineering CAD/CAE systems and specialized atomistic analysis and design environments. New Integrated Frameworks allow for a systematic applications inside nano electronics and photonics design processes of integrated multiscale performance – properties – structures - processing analyses. They represent an important step to overcome classical barriers between engineering and manufacturing design. In this contest, the impact on devices and systems performance of even very small structural and chemical composition variations can be reliably evaluated.

Multiscale sensitivity analyses are an important challenge for nano engineering. New developments are putting the bases for a transition from "multiscale analyses" to a "multiscale design" strategy; this opens the way to a wealth of new nano-based architectural solutions. Multiscale design implies the integration of the classical bottom up approach with the newly developed top down strategy.

IX. Conclusion

Photonics and NanoTechnologies are emerging, innovative, and highly promising; they would enable new solutions and approaches.

Therefore, it is reasonable to predict that the impact of these technologies in the future, particularly as they converge with others inside an Integrated Multiscale Multi-Science – Engineering Environments, will become more and more important as we are entering this "nano-photon century". In order to remain and consolidate our position as a leader, in the field of radar, sensors and systems, we want to take advantage of such technologies through proactive studies and developments in collaboration with many Universities and Research Centres.

The outcome of this complex process will be a new "Quantum Engineering" era.

I. References

[1] F. Brunetti et al "Controlled grow of ordered SWCNT for the…" 4thth IEEE Conference on Nanotechnology, 16-19 Aug. 2004, Pag. 534-536.

[2] J. M. Bonard et al, Solid State Electronics 45 (2001), Pag. 893.

[3] H. F. Taylor, "An optical Analog-to-Digital Converter …". IEEE J Quantum Electronics, Vol 15, No 4, Pag 210-216, Apr. 1979.

[4] C. H Liu et al, AP Letters 2001, 79 Pag. 2252-2254

[5] A. Di Carlo et al, Patent PCT / IB 2005 / 05812

Passive Light Power Control Enabled by Nanotechnology

A.Donval, B.Nemet, R.Shvartzer and M.Oron

KiloLambda Technologies Ltd.,
22a Raoul Wallenberg, P.O.B. 58089, Tel-Aviv 61580, Israel, adonval@kilolambda.com

ABSTRACT

The need to regulate and control light power is relevant not only for sophisticated communication systems but also to everyday optical equipment such as consumer point-and-shoot cameras or even a common car rear-view mirror. We explore the unique capabilities and advantages of nanotechnology in developing next generation non-linear components and devices to control and regulate optical power in a passive way.

We report on passive optical power control devices based on a range of photonic nanostructures. We present the optical fuse, limiter and our next generation solution the Dynamic Sunlight Filter (DSF).

Keywords: nanotechnology, optical power control, limiting, blocking, protection filter

1 INTRODUCTION

The need to regulate and control light power is relevant not only for sophisticated communication systems but also to everyday optical equipment such as consumer point-and-shoot cameras or even a common car rear-view mirror. Regulating optical power levels within various systems, such as cameras, requires today an electronic feedback control or offline data processing, which introduces complex and expensive systems. We explore the unique capabilities and advantages of nanotechnology in developing next generation non-linear components and devices to control and regulate optical power in a passive way.

The design of artificial nanostructured materials for the use in non-linear devices and integrated photonic systems is very challenging as it involves the need to incorporate the nanoparticles, nanomaterials and quantum physics equations. Near-field interactions in artificial nanostructured materials can provide a variety of functionalities useful for optical systems integration. For example, nanoparticles embedded within a dielectric host are known to have a field enhancement effect and therefore lower the threshold of laser induced damage within the material [1,2]. Another example for limiting effect is carbon suspensions and reverse saturable absorber materials [3]. We are taking advantage of the unique capabilities of nanoparticles guest embedded within dielectric host matrices for field enhancement effect in developing next generation of non-linear components and devices to passively control and regulate optical power. Based on our nanotechnology we developed a whole family of Optical Power Control (OPC) components and solutions [4, 5, 6].

We report on our next generation solution, the Dynamic Sunlight Filter (DSF), which is based on our fundamental principles of nanotechnology and nanostructure optics dedicated for sunlight applications. In the normal state, when incident light is below a predefined level the DSF is highly transparent, light just passes through it. As the light level is increased and gets more intense, such as in the case of morning sun, or the headlights of an approaching car facing the rear-view mirror, the DSF transmission decreases accordingly, eventually reaching a darkened state. The darkening effect is selective and is limited only to the intense light areas in the image. This process is reversible and the filter returns to its transparent state once the intensity of light decreases to its normal level.

We present our nanotechnology power control mechanisms as well as a preliminary design of the DSF. We demonstrate power control and regulation in prototype configuration for several device approaches. Finally, we discuss DSF possible applications, our wish list includes, among others, enjoying a cooler room in a sunny summer day by automatically darkening the window to a predefined level, thus passing less heat and resulting in energy saving.

We report on passive optical power control devices based on a range of photonic nanostructures, including mainly nanostructures for spatial field localization to enhance optical nonlinearities. We present the two main optical power control mechanisms: blocking (section 2.1) and limiting (section 2.3), as well as their corresponding nano-scale phenomena. We present device examples of two novel generic optical power control components: fiber optical fuse [7] (section 2.2) and fiber optical limiter [8] (section 2.4). A third one of free-space wideband protection filter is discussed elsewhere [9]. We present also preliminary design for future applications such as optical power regulating of sunlight (DSF) and its possible applications (section 3).

2 OPTICAL POWER CONTROL MECHANISM AND DEVICE EXAMPLES

We developed two main optical power control mechanisms: blocking and limiting. Within the following sections we will discuss the two and their nanostructures based origin.

2.1 Blocking Mechanism Principles

Our blocking mechanism (Figure 1) is enabled by catastrophic breakdown of the material when over power occurs. It is performed by novel nanostructures that are used as threshold trigger at relatively low powers according to the nanoparticles and nanostructure design. The optical blocking mechanism is based on a catastrophic breakdown effect, which occurs at the interfaces between metallic and non-metallic layers in the optical path. These layers are nearly transparent at low input powers. However, the catastrophic breakdown results in significantly enhanced scattering from the layers interface, leading to significant decrease in transmission. This catastrophic breakdown is irreversible in similar way to electrical fuse.

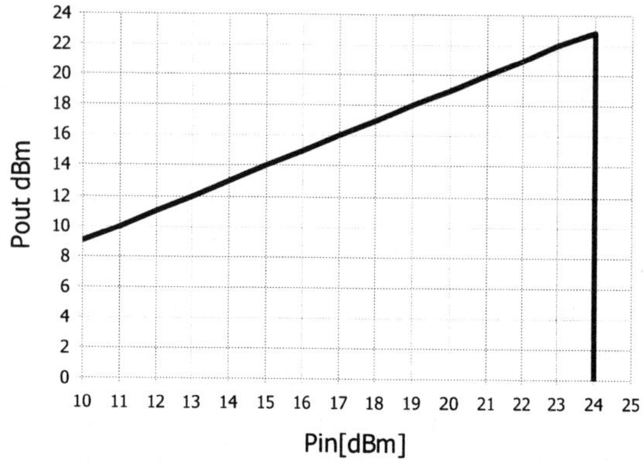

Figure 1: Optical power blocking effect – interruption of optical transmittance by catastrophic breakdown. Based on novel nano-structures that are used as threshold triggered switches

The base materials for this mechanism can be thin layers of only few nanometers size of a certain metal such as gold, in contact with a dielectric layer such as silica, to achieve the required interface. The desired breakdown threshold is then tuned according to the metal thickness, structure and the metal-dielectric interface nature [10]. When embedded guest nanoparticles within a host dielectric material, we can simplify the fabrication technique. We can lower threshold powers down to few tens of mW by taking the advantage of the field enhancement effect of special nanostructures and unique combinations of guest –host pairs.

2.2 Blocking Mechanism for the Optical Fuse

An optical fuse is an inline component that is transparent under low power operation, but becomes permanently opaque when the input power reaches the threshold level. The optical fuse is based on the blocking mechanism as shown in Figure 1. As the fuse action is irreversible, optical fuses are designed to operate at emergency cases, such as networks that are susceptible to undesirable power spikes that arise from amplifiers or external sources that are multiplexed into them.

Figure 2: Optical Fuse In-line device version

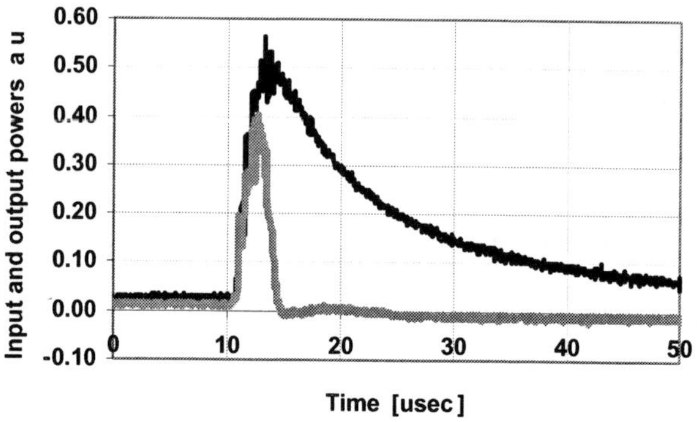

Figure 3: Response time of the optical fuse. The input pulse is depicted in black whereas the output pulse is depicted in gray. As evident, the response time in this case is less than 10 microseconds

In order to measure the fuse response time under different power levels, we developed a custom setup (namely, a programmable optical pulse generator), for creating pulses of different lengths, powers and shapes. Various pulses were input to the optical fuse in order to examine its response. Figure 3 shows an example, where the input is a high power pulse, marked in black. As evident, the output pulse (gray) is blocked after the input pulse exceeds a certain power. Here, the response time is shorter than 10 microseconds.

The response time decreases with the input power. Typically, the response time for powers slightly higher (a few dBs) than the threshold power is few tens of microseconds, whereas for stronger pulses (significantly higher than the threshold power), response times as low as a few nanoseconds were measured.

2.3 Limiting Mechanism Principles

The limiting of optical transmittance (Figure 4) is done mainly by non-linear absorption-induced scattering. The scattering method is based on novel nanostructures and nano-particles inserted in the optical path and are used as the non-linear scattering medium. At low powers, there is only a residual absorption effect (no scattering), which results in relatively small optical transmission loss. However, scattering becomes significant at high input powers, and allows only a fraction of the input power to propagate (see Figure 5).

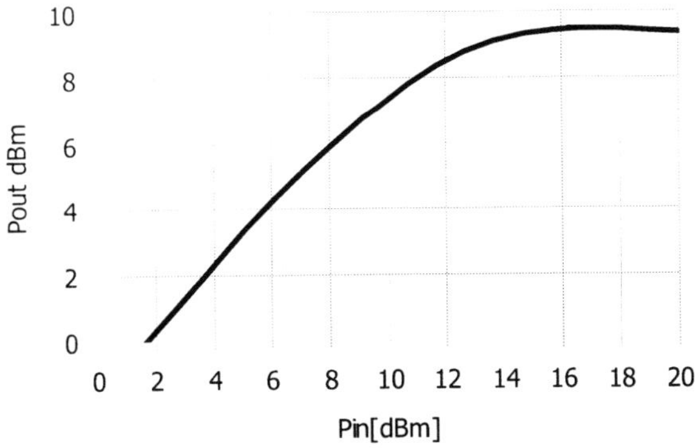

Figure 4: Optical power limiting effect - limiting of optical transmittance mainly by non-linear scattering. Based on novel nano-structures that are used as non-linear scattering medium

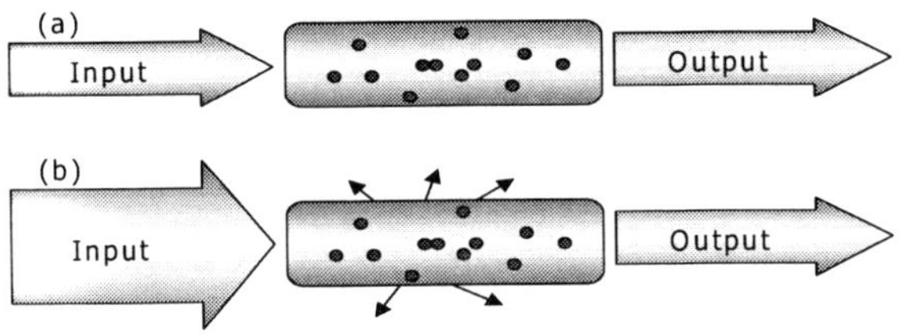

Figure 5: Limiting mechanism: (a) at low input power, only slight attenuation in output power due to absorption. (b) at high input power, large attenuation due to strong scattering induced by the nano-particles.

2.4 Optical Limiter: device example for limiting mechanism

The optical power limiters main function is limiting the output power to a certain level (namely, the limit-power). At low input powers, the limiter is transparent, whereas at input powers higher than the limit-power, the output power is constant. Also, as opposed to the optical fuse, the action of the optical power limiter is reversible; meaning that when the input power drops back, the optical power limiter becomes transparent again. Figure 4 shows an experimental plot of the output power as a function of the input power. Here, the insertion loss at low power is approximately 2dB, whereas the limit power is approximately 7dBm. The maximum CW input power defined is around 14dBm. Note also that the graph describes a few power cycles, confirming the reversibility of the power limiting operation.

Another important property of the optical power limiter is the response time. Fast response is required in order to block the excess power. However, as opposed to the optical fuse where immediate blocking is required, the response of the optical limiter should be slower than the data rate, in order not to affect the transmitted data. Figure 7 presents a measurement of the optical power limiter response time. Here, the input power rises beyond the limit power. First, the output power follows the input power, but stabilizes at the limit power afterwards. The response time is derived from the size of the "hump"..

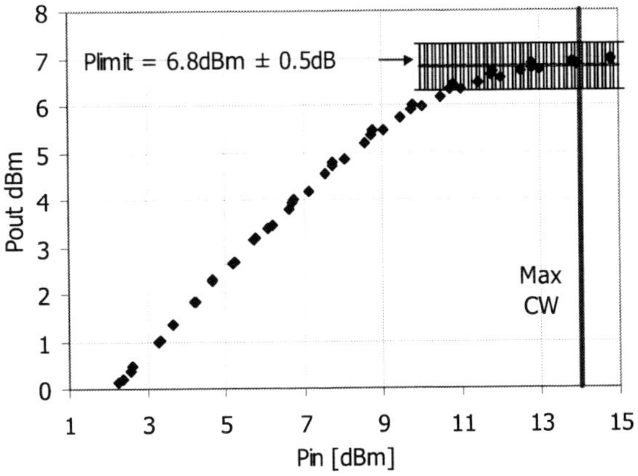

Figure 6: Output power vs. input power cycles as recorded for an of approximately 7dBm optical power limiter

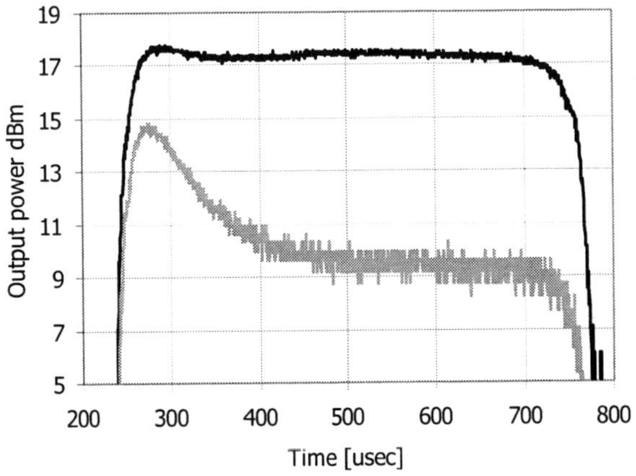

Figure 7: Input pulse (black) and output power (gray) as a function of time. The response time derived from the graph is ~200μsec

2.5 Discussion

We presented several optical power control devices examples: Optical Fuse, Optical Limiter and Wideband Protection Filter. In general, optical power fuses and limiters can regulate and control the optical power in telecommunication networks. They can either replace or complement existing power feedback control loops, as another layer to the electronic layer. The optical power limiter serves either as a protection device or as a power-regulating device. Whereas the optical fuse is designed mostly for protection purposes.

As a power-regulating device, the optical power limiter can serve as a gain- or power-equalizer, or for reducing power fluctuations ("noise eater"). As protection devices, either the optical limiter (at lower power levels) and/or optical fuse (at higher power levels), it can protect detectors or receivers from over-power and even increase the system's dynamic range; The optical fuse can serve also as a laser safety device and even as prevention of catastrophic damage due to effects such as the fiber-fuse phenomenon [11], [12].

Another example of blocking mechanism based device is the Wideband Protection Filter (WPF), which is designed to protect imaging and detection systems that are susceptible to detector saturation or permanent damage caused by powerful light sources or high power lasers in the free space configuration. The WPF is described elsewhere [9].

3 DYNAMIC SUNLIGHT FILTER (DSF)

The need of optical power controlling and regulating implies not only to sophisticate communication systems but also to everyday cameras and even to a common car rear-view mirror. Regulating optical power levels within various systems, such as cameras, requires today an electronic feedback control or after data processing, which introduce complex and expensive systems.

Our next generation of optical power control technology is the Dynamic Sunlight Filter (DSF). DSF technology will enable users to control the amount of light passing the element in a passive way. The DSF element will automatically vary its transparency according to the amount of incident light.

Dynamic Sunlight Filter (DSF) is designed by principles of nanotechnology and nanostructure optics similar to what discussed for our limiting power control mechanism (refer to section 2.3). In the natural state, when incident light is below a predefined level the DSF is highly transparent, so light just pass through (

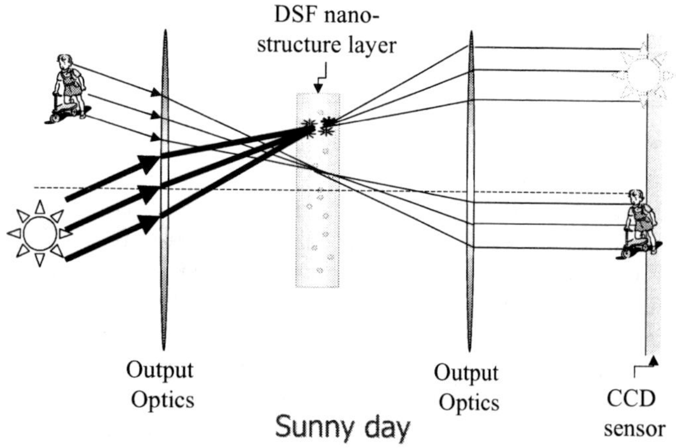

Figure 8). When light increases continuously, such as in the case of sunrise or when a glare from approaching headlights is facing your car rear-mirror, the DSF transmission decreases according to the amount of the incident lights, resulting in a darkened state. The darkening effect is limited only to the over exposed area. The area becomes transparent again, once the amount of light reduces below the required level.

The same effect of automatically transparency decreasing within the glared area is applicable to multiple applications such as cameras, rear-view mirrors, windows, sunglasses and many other. This exciting, cutting-edge technology will allow consumers to benefit a cooling room in a sunny day by an automated window darken itself to the predefined light amount to pass through.

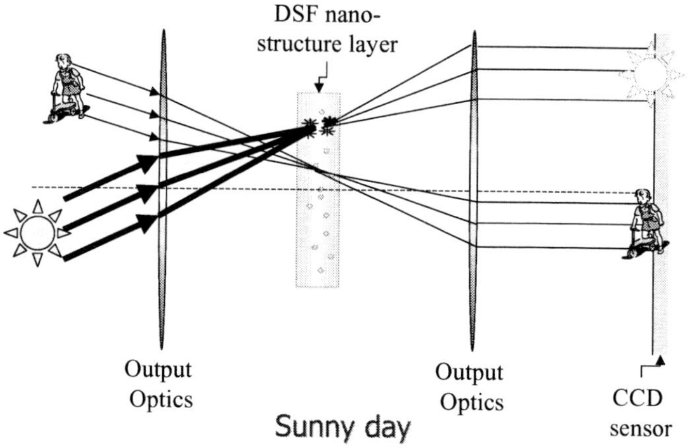

Figure 8: Dynamic Sunlight Filter illustration

REFERENCES

[1] "Establishing Links Between Single Gold Nano-particles Buried Inside SiO2 Thin Film and 351-nm Pulsed-Laser-Damage Morphology" (LLE Review, University of Rochester, Volume 89)

[2] M. Quinten, "Local Fields Close to the Surface of Nanoparticles and Aggregates of Nanoparticles," Appl. Phys. B 73, 245-255 (2001)

[3] D. Vincent, "Optical limiting threshold in carbon suspensions and reverse saturable absorber materials," Applied Optics 40(36), p. 6646, (December 2001)

[4] A. Donval, S. Goldstein, P. McIlroy, R. Oron and A. Patlakh, "Passive components for high power networks," in Optical Components and Devices, Ed.: Simon Fafard, Proc. SPIE 5577 (2004).

[5] R Oron, A. Donval, S. Goldstein, N. Matityahu, M. Oron, A. Patlakh, J. Segal and R. Shvartzer, "Optical Power Control Components in Networks,"

Nat. Fiber Optic Eng. Conf. (NFOEC), paper JWA75 (2005)

[6] A. N. M. Masum Choudhury, Barbara Grzegorzewska, Timothy S. Hanrahan, Tom R. Marrapode, Ariela Donval, Moshe Oron, Ram Oron and Regina Shvartzer, "Dynamic Attenuator – a New Passive Device to Control Optical Power Levels in Networks" Nat. Fiber Optic Eng. Conf. (NFOEC), paper JThA88 (2007)

[7] A. Donval, S. Goldstein, P. McIlroy, R. Oron and A. Patlakh, "Passive components for high power networks," in Optical Components and Devices, Ed.: Simon Fafard, Proc. SPIE 5577 (2004).

[8] R Oron, A.Donval, S.Goldstein, N.Matityahu, M.Oron, A.Patlakh, J.Segal and R. Shvartzer, "Optical Power Control Components in Networks," *Nat. Fiber Optic Eng. Conf.* (NFOEC), paper JWA75 (2005)

[9] A. Donval, B. Nemet, M. Oron, R. Oron, R. Shvartzer, L. Singer, C. Reshef, B. Eberle, H. Bürsing, R. Ebert, "Wideband protection filter: single filter for laser damage preventing at wide wavelength range", in Electro-Optical and Infrared Systems: Technology and Applications, Eds: Huckridge, R. Ebert, Proc. SPIE 6737 (September 2007)

[10] S. Papernov and A.W. Schmid, "Correlations between embedded single gold nanoparticles in SiO2 thin film and nanoscale crater formation induced by pulsed-laser radiation" Journal of Applied Physics Volume 92, Issue 10, pp. 5720-5728 (Nov. 2002)

[11] I. Peterson, "Fibers with flare," Science News **140**(13), 200-201, (Sep. 1991).

[12] R. Oron, "Protecting the optical network," Fiberoptic Product News **18**(9), 20-21 (Sep. 2003).

FIB Generated Antimony Nanowires as Chemical Sensors

Alois Lugstein*, Christoph Schoendorfer*, Youn-Joo Hyun*, Lothar Bischoff**, Philipp M. Nellen***, Victor Callegari***, Peter Pongratz****, and Emmerich Bertagnolli*

* Vienna University of Technology, Institute for Solid State Electronics, 1040 Vienna, Austria
** Research Center Dresden-Rossendorf Inc., POB 510119, D-01314 Dresden, Germany
*** EMPA, CH-8600 Dübendorf, Switzerland
****Vienna University of Technology, Institute for Solid State Physics, 1040 Vienna, Austria

ABSTRACT

Sb nanowires with a homogeneous distribution of diameters of about 25 nm and length up to several microns are synthesized by a FIB induced self-assembling process. We propose a model similar to the vapor-liquid-solid mechanism with Ga acting as catalyst. Thereby FIB processing produces mobile Ga species on the surface which rapidly agglomerate forming catalytic nanoclusters. Sputtered Sb diffuses on the surface and acts as a quasi-vapor phase source. When the solved Sb concentration exceeds saturation, nucleation sites will be formed which initiate the precipitation of the Sb. Individual nanowires transferred onto isolating substrates have been contacted by electron beam lithographically processed Ti/Au pads for electrical characterization. In the contrary to ambient of CO, He and H_2, where no influence is observed, the conductivity of the Sb nanowire is highly sensitive on water and ethanol and a resistivtiy change over 4 orders of magnitude was observed.

Keywords: nanowires, focused ion beam, humidity sensor, self assembling

I. INTRODUCTION

Low-dimensional nanostructures are usually fabricated using either a top down or a bottom up strategy. The former technique is extremely flexible, but suffers from limitations in minimum feature size and uniformity. The latter one, utilizing spontaneous self-ordering effects, is limited by the broad size distribution and the lack of control of the positioning of the self-organized nanostructures. In this context the discovery of the appearance of periodic structures with dimensions in the nanometer regime induced by ion bombardment has attracted growing interest due to the possibility of obtaining a self-organized formation of nanometer structures [1].

Appleton et al. found that heavy-ion implantation leads to the formation of craters with diameters of about 20nm due to morphological instabilities in the amorphous phase of initial crystalline Ge [2]. Wang and Birtcher observed the generation of sponge-like porous structures on Ge [3]. Nitta et al. [4] and Kluth et al. [5] reported the development of an anomalous cellular structure followed by the formation of a network of nanoscale rods on ion irradiated GaSb which is proposed to result from a defect formation mechanism based on movement of the point defects induced by ion implantation.

In this paper we propose a process very similar to the vapor-liquid-solid (VLS) mechanism in which various metals such as Au [6], Fe [7], or Ti [8] catalytically enhance the growth of nanowires. Our study differs from most of the previous reports on VLS grown nanowires in that an intense focused Ga ion beam initiates the growth of nanowires at room temperature without using any additional materials source.

II. EXPERIMENTAL

Thin lamellas of antimony samples with purity > 99.999% were prepared and cleaned by rinsing with acetone and isopropyl alcohol followed by blow-drying with pure nitrogen. The antimony with sufficient initial smoothness was exposed to a 50 keV focused ion beam (FIB) with a diameter of 70 nm and a beam current density of 0.8 A/cm^2. No gases were introduced into the high vacuum chamber and the sample was kept at room temperature. The topographical and compositional evolution of the Sb surfaces irradiated by FIB, is investigated by means of scanning electron microscopy (SEM), high resolution transmission electron micrsocopy (RTEM), selected area diffraction (SAD), and energy dispersive X-ray diffraction (EDX) measurements.

For the purpose of patterning the FIB is scanned over a predefined area in discrete steps with well-defined step size and dwell time, i.e. the time the beam remains on each single spot. Each scan across the selected area deposits an ion fluence which is correlated to the above mentioned parameters. Single pass milling denotes a scanning strategy where the desired ion fluence is deposited within one single scan. For multi pass milling, the beam is scanned several times across the predefined area and the total ion fluence is dependent on the number of scan repetitions of the FIB.

III. RESULTS AND DISCUSSION

Fig. 1a shows a SEM image of the Sb surface after multi pass milling of a (2x2)µm² wide box with a ion fluence of $6.2x10^{18}$ Ga ions/cm². The rim of the several micrometer deep hole is surrounded by a dense network of nanofibers which show very uniform diameters in the range of 25nm. Milling the same box with the same ion fluence in single pass mode leads to the formation of a pattern shown in Fig. 1b. Thereby the FIB scan starts in the upper left of the box and moves along in serpentines with a pixel and line spacing both of 10nm, which guarantees a nearly uniform ion fluence distribution (>99%) [18]. The whole FIB modified area is covered by nanofibers with the exception of the last line scan routed from the lower right to the lower left edge. Nanofibers reach even 2µm beyond the rim of the FIB milled area. Fig. 1c shows the Sb surface after single pass milling viewed under a tilt angle of 75°. The FIB generated nanowire extrusions do not form a plane porous disc as one could assume from the top view SEM image in Fig. 1b.

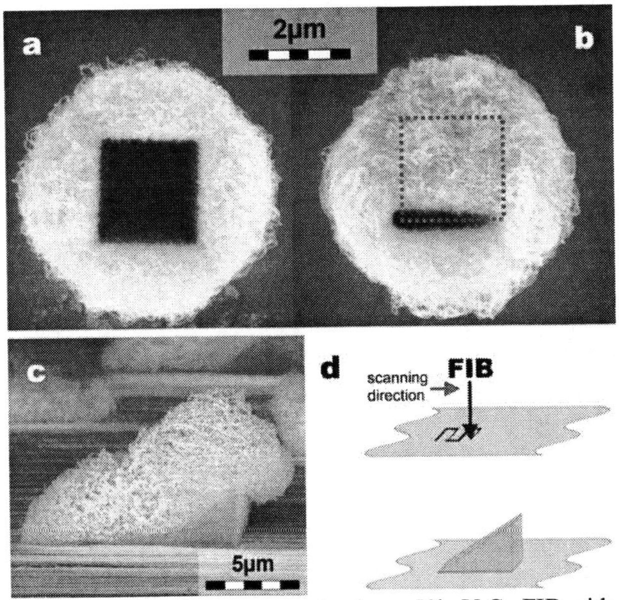

Figure 1: Sb surface processed using a 50keV Ga FIB with an ion current of 200pA. (2x2)µm² milling areas irradiated by an ion fluence of $6.2x10^{18}$ions/cm² in (a) multi pass mode and (b) single pass mode, tilted view SEM image of a (10x10)µm² milling area (c) exposed to an ion fluence of $3.1x10^{18}$ions/cm² processed in single pass mode, schematic sketch (d) visualizing the FIB scanning strategy and the resulting uplifted nanofiber network.

As shown in the schematic of Fig. 1d, the nanofibers appear on a ramp-like base normal to the plane rising along the scan direction of the FIB. The formation of this ramp-like structure is a result of the pixel-by-pixel and accordingly of the line-by-line scanning strategy. Scanning the first line of the predefined milling area leads to nanofiber growth even beyond the ion irradiated region.

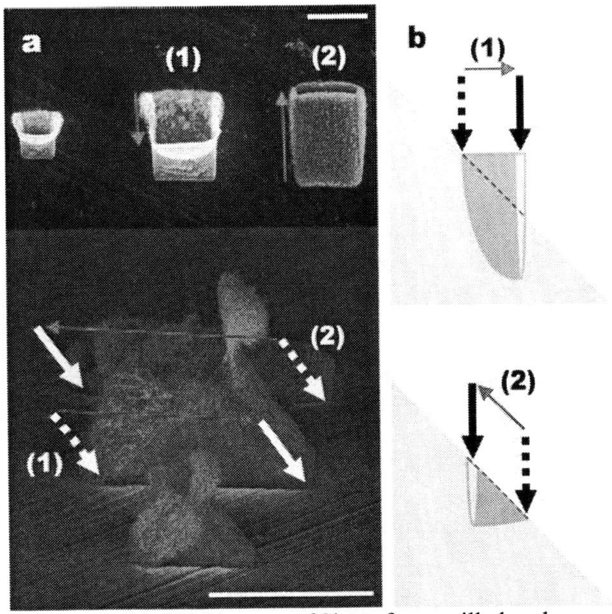

Figure 2: (a) SEM images of Sb surfaces milled under an angle of FIB incidence of 45°. The uppermost top view SEM image and the side view show the resultant pattern achieved under different scanning directions, which are indicated by the red arrows. Track (1) denotes a FIB guidance from higher to lower surface level on the tilted sample, track (2) denotes the opposite direction. The white arrows mark the incident direction of the FIB whereby the dotted arrows denote the start position of the scanning FIB beam. Scale bars, 10µm. (b) Principle sketch to point out the guidance of the FIB and the resulting structure formation.

By the guidance of the FIB through the subsequent lines, nanofibers which were grown on the not yet exposed part in the forefront of the scanning beam are removed by sputtering. Nanofibers in already irradiated zones, i.e. behind the scanning beam, remain unaffected. These nanofibers form a network which is further densified by redeposited Sb. Due to the ongoing FIB scanning this nanofiber network reduces the escape angle for the sputtered Sb and more and more of them are picked up by the network which leads to an upraising of the structures. Accordingly, FIB milling under oblique angles as schematically shown in Fig. 2b should lead to an increase or decrease of the uplifting effect due to the variation of the escape angles for sputtered Sb. The results obtained for FIB processing of a Sb surface tilted by 45° relative to the incident focused ion beam are shown in Fig. 2a. Depending on the scanning direction of the FIB the escape angles of the sputtered substrate material change. The scanning direction from higher to lower levels denoted as track (1) in the schematic of Fig. 2b (escape angle of 135°) results in an upraised structure. Hardly any uplifting can be observed when scanning in the opposite direction denoted as track (2) (escape angle 45°). The smaller (5x5)µm² FIB milled box in Fig. 2a (the left box in the top view and the one in front of the side view image) shows the impact of the box length on this uplifting effect. The height of the uplifting increases with the length of the FIB milled box. For boxes with a length of 15μm or above this height levels out at about 10μm.

Extensive TEM, SAD and EDX of individual nanofibers prove that they are completely amorphous even in the nanometer scale and consist of pure antimony. The TEM image in Fig. 3a shows the exceptional uniform diameters of the nanofibers of about 25nm along their entire length.
The FIB modified samples covered by the nanofibers were annealed in a special furnace setup which allows processing at well-controlled temperature profiles in He atmosphere. Several experiments showed that the temperature ramp is a crucial parameter for the grain size of the resulting re-crystallized structure whereby the diameter and shape of the nanowires remain unaffected by the annealing. The HRTEM image in Fig. 3b displays a Sb nanowire after moderate thermal annealing at 453K for 30min. The diffraction pattern in Fig. 3c shows the most prominent (110) and (120) reflections for Sb with its trigonal crystal structure and the lattice parameter of 0,354nm is consistent with the tabulated value for bulk Sb. Fig. 3d shows the EDX spectrum of the investigated nanowire. The dashed lines mark energies for Ga in the EDX spectrum. As there the peaks for Ga are missing the nanowire seems to consist of pure Sb. The copper signal originates from the sample holder ring.
The observed nanofiber formation, even several microns aside the FIB irradiated surface can not be explained by point defect as proposed by Nitta et al. [9]. We propose a model similar to VLS with Ga acting as catalyst [10]. VLS deals with the fact that a catalytic metal particle on the sample surface - if the ambient temperature is high enough - forms a liquid alloy cluster and serves as the preferential site for adsorption of reactant from the vapor phase. It is supposed that supersaturation is the driving force for nucleation of seeds at the interface between alloy cluster and the substrate surface giving rise to a highly anisotropic growth of nanostructures. Nanowire growth from the base continues as long as the droplet remains in a liquid state and supersaturation is maintained.
At present, we do not understand the origin of the tangling of the nanowires although we note that extensive tangling has been observed previously in Ga based VLS processes [11]. The authors of also stated that Ga droplets could simultaneously catalyze the growth of hundreds of thousands of nanowires.

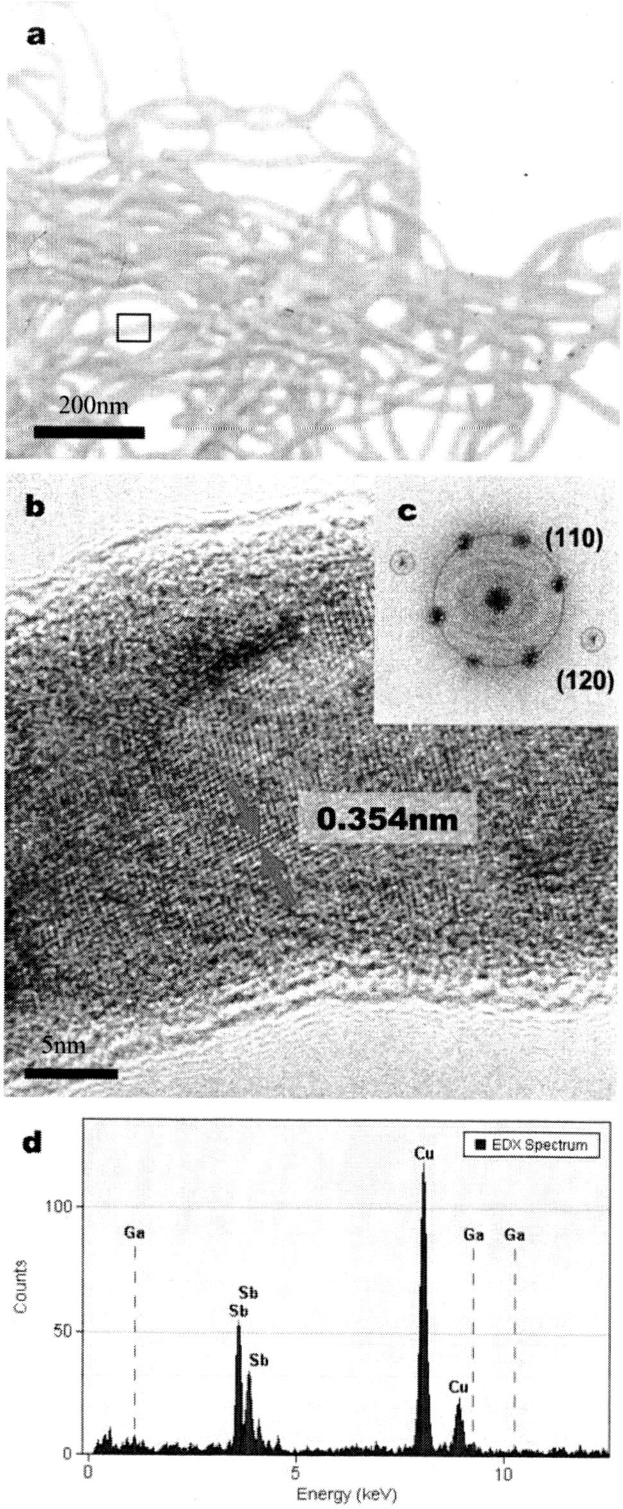

Figure 3: (a) low magnified TEM image. (b) HRTEM micrograph shows the part of a Sb nanofiber (marked by the rectangle in Fig. 3) after annealing at 453K. (c) diffraction pattern ((110) and (120) reflections) of the nanowires after the annealing. (d) EDX spectrum.

NSTI-Nanotech 2008, www.nsti.org, ISBN 978-1-4200-8505-1 Vol. 3

For electrical characterization the FIB generated nanowires are transferred onto isolating substrates. Individual nanowires are contacted with electrodes consisting of 3nm Ti (as an adhesion promoter) and 60nm Au, which is performed by means of electron beam lithography (EBL). During the measurements the ambient humidity showed a great impact on the current-voltage characteristic of a single Sb nanowire, which indicated the sensing potential of this device. Therefore the electrical measurements was performed under well-defined conditions. We investigated the sensing capability of single Sb nanowires for H_2, He, O_2, CO, H_2O and ethanol. To compare the impact of different gases and pressures the gas response or sensitivity of the nanowire sensor is defined as the ratio between Rg (the resistance when the sensor is exposed to a selected amount of gas) and R_0 (the resistance of the device in vacuum).

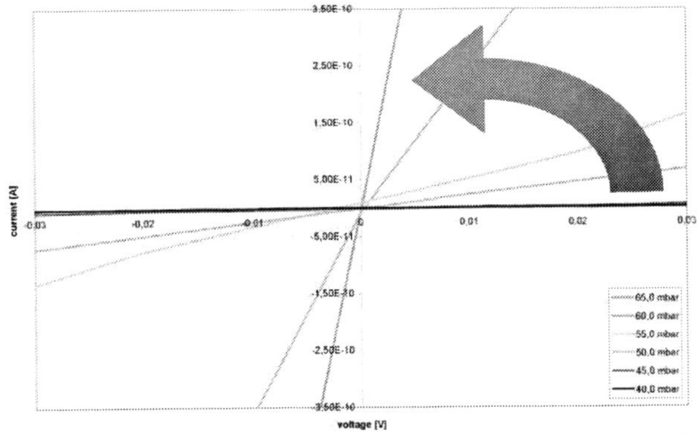

Figure 4: Evolution of the conductance of the Sb nanowire exposed to ethanol vapor.

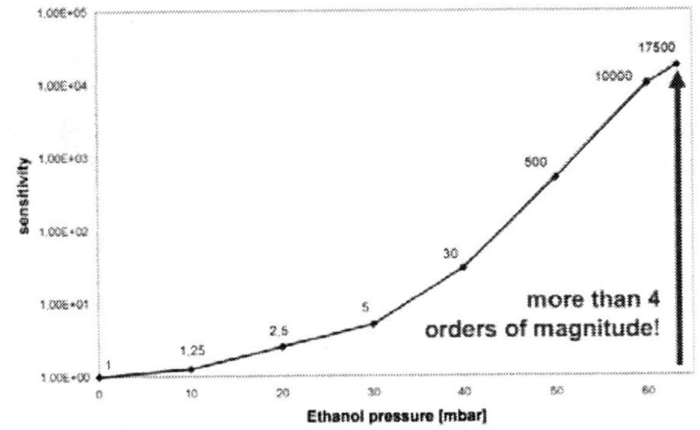

Figure 5: Sensitivity curve of the Sb nanowire sensor. From 0 to 100% saturation of ethanol vapor in the chamber the sensitivity changes by a factor of 17500. All measurements are carried out at room temperature (22°C).

Whereas no response was detected for H_2, He, O_2, and CO, a huge change in the conductance could be observed for ethanol and water. The curves displayed in Fig. 4 shows the current-voltage characteristics of a single Sb nanowire as a function of ethanol pressure. It is impressively shown that the conductance depends on the quantity of the gas to be detected. The graph in Fig. 5 gives an idea of the sensing potential of this Sb nanowire device as the sensitivity increases by more than four orders of magnitude depending on the introduced ethanol amount.

IV. SUMMARY

In summary, the FIB irradiation of Sb with 50keV Ga ions at room temperature leads to the formation of a porous network of pure Sb nanofibers. The as-grown nanofibers are amorphous with remarkable uniform diameters in the range of about 25nm along their entire length. Re-crystallization of the Sb nanofibers could be achieved by moderate thermal annealing at temperatures of about 473K. Depending on the temperature ramp and heating duration finely grained crystallites as well as single crystalline regions along the nanowires can be obtained.
Sensors realized from these Sb nanowires were very sensitive on water vapor and ethanol gas. The ethanol and water sensing are completely reversible processes, which temporarily enhance the electrical conductivity of the Sb nanowire.

V ACKNOWLEDGMENTS

This work is partly funded by the Austrian Science Fund (Project No. 18080-N07) and the Sixth EU Framework Program for Research and Technological Development (FP6) project "Charged Particle Nanotech" (CHARPAN).

REFERENCES

1 Z. X. Jiang, P. F. A. Alkemade, Appl. Phys. Lett. 73, 315 (1998); J. Erlebacher, M. J. Aziz, E. Chason, M. B. Sinclair, J. A. Floro, Phys. Rev. Lett. 82, 2330 (1999); S. Rusponi, G. Constantini, C. Boragno, U. Valbusa, hys. Rev. Lett. 81, 2735 (1998).
2 B. R. Appleton, O. W. Holland, D. B. Poker, J. Narayan, D. Fathy, Nucl. Instr. and Meth. in Phys. Res. B 7-8, 639 (1985).
3 L. M. Wang, R. C. Birtcher, Appl. Phys. Lett. 55, 2494 (1989); L. M. Wang, R. C. Birtcher, Phil. Mag. A 64, 1209 (1991).
4 N. Nitta, M. Taniwaki, Y. Hayashi, T. Yoshiie, J. Appl. Phys. 92, 1799 (2002).
5 S. M. Kluth, J. D. Fitz Gerald, M. C. Ridgway, Appl. Phys. Lett. 86, 131920 (2005).
6 . I. Persson, M. L. Larsson, S. Stenström, B. J. Ohlsson, L. Samuelson, L. R. Wallenberg, Nature 3, 677 (2004).
7 A. M. Morales, C. M. Lieber, Science 279, 208 (1998).
8 T. I. Kamins, R. S. Williams, Y. Cheng, Y. L. Chang, Y. A. Chang, Appl. Phys. Lett. 76, 562 (2000).
9 N. Nitta, M. Taniwaki, Y. Hayashi, T. Yoshiie, J. Appl. Phys. 92, 1799 (2002).
10 M. K. Sunkara, S. Sharma, R. Miranda, G. Lian, E. C. Dickey, Appl. Phys. Lett. 79, 1546 (2001).
11 Z. W. Pan, Z. R. Dai, C. Ma, Z. L. Wang, J. Am. Chem. Soc. 124, 1817 (2002).

Semiconductor Nano-filaments in Fiber

D. S. Deng[1,2], N. Orf[1,2], A. F. Abouraddy[1,2], A. M. Stolyarov[3], Y. Fink[1,2]

[1]Research Laboratory of Electronics, [2]Department of Materials Science and Engineering, Massachusetts Institute of Technology, 77 Massachusetts Avenue, Cambridge, MA, 02139, USA;
[3]School of Engineering and Applied Sciences, Harvard University, Cambridge, MA, 02138, USA

ABSTRACT

One-dimensional structures with nanometre scale have achieved with substantial progresses for device applications in electronics, photonics, photovoltaics and biology. The fabrication for the well-ordered extended semiconductor wires remains highly desirable for the alignment and assembly. Here by relying on optical-fiber drawing technique, we report the semiconductor (Se and As_2Se_3) filament-arrays embedded in our multi-material fiber. These filaments can reach high aspect-ratio on the order of 10^6. This may offer new approach to produce wires with the capability of high-throughput and low-cost.

Keywords: fiber, nano-filaments, semiconductor, high aspect-ratio

1 INTRODUCTION

Optical-fiber thermal drawing from a viscous macroscopic preform is well established in the telecommunications industry and enables the rapid fabrication of kilometre-long silica-glass fibers with precise dimensions [1-3]. With these methods, in the last decade, microstructured fibers incorporating air enclaves have been created, resulting in a large set of novel fiber designs [4]. All these fibers, however, consist of a single material (silica glasses or polymers) with the possible addition of air cavities.

An altogether different class of fibers incorporating multiple materials (*e.g.*, amorphous semiconductor thin films, polymers, and metals) [0] in the same preform can be produced by thermal drawing. These fibers with novel designs and material combinations are enabling unique applications in sensing and flexible electronics [5-12]. However, nothing is known about the extent to which a feature size (such as film thickness) can be reduced in a multimaterial fiber using this fluid processing technique.

2 MULTI-MATERIAL FIBER FABRICATION

Thin films of As_2Se_3 or Se were thermally evaporated onto polyethersulphone (PES) or polysulphone (PSU) films, respectively. The macroscopic preform was fabricated by rolling the glass films and thick polymer claddings onto a PTFE mandrel and consolidating the structure under vacuum for approximately one hour, after which the mandrel was removed.

A conventional optical fiber draw tower consisting of a three-zone furnace to heat the preform to its processing temperature, a feeding mechanism to controllably introduce the preform into the furnace (downfeed speed of $0.003 mm \cdot s^{-1}$), and a capstan to pull the resulting fiber from the preform (set at was $\sim 0.1 m \cdot min^{-1}$) was used to draw tens of meters of fiber from cylindrical preforms measuring 160mm in length and 20mm in diameter. The drawing parameters were fixed to keep a constant draw-down ratio of about 20 between the features sizes of the initial preform and final fiber.

3 STRUCTURE SCALING DOWN

An example of such a fiber is sketched in Fig. 1a, where a solid macroscopic preform contains a micron-scale amorphous semiconductor glass film clad on both sides with a polymer. As in traditional thermal fiber drawing, this preform is heated to the viscous state and controllably stretched into a fiber by applying axial tension. The final fiber cross-section is simply a scaled down version of the initial preform (Fig. 1b, c). The results confirm that the layered structure is preserved

4 EXTENDED FILAMENT

However, as film thickness is reduced, the sub-100 nm layer is breakup in the fiber. We observe the filament arrays embedded in the fiber. Examples of the extracted filaments are given in Fig. 2. Figure 2a shows a 1-mm-long Se filament extracted from a fiber containing 100-nm-thick Se layer that has broken up. The filament is ribbon shaped, with width on the order of 10 μm and thickness on the order of 100 nm. Bundles of As_2Se_3 ribbons extracted from an As_2Se_3/PES fiber with 10-nm thickness are shown in Fig 2b. Individual ribbons have sub-100-nm width and 10-nm thickness. The longest achieved As_2Se_3 ribbons reach the aspect ratio on the order of 10^6.

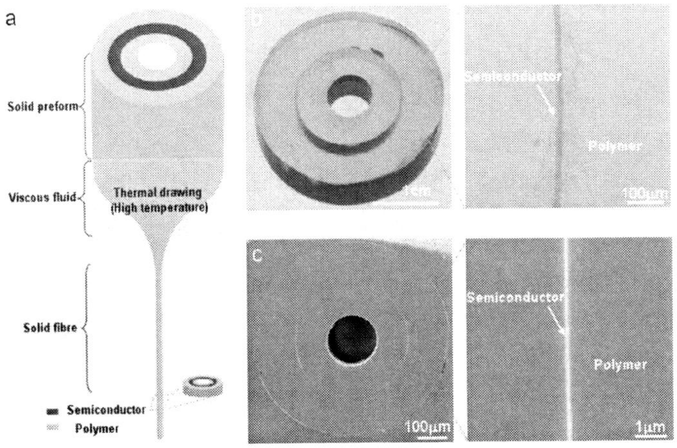

Figure 1 Fiber drawing. **a,** A hollow-core fiber preform is drawn down while maintaining the layer structure through the thermal processing. **b,** An optical microscope image of the layer structure in the preform is shown on the left, and a magnified section of the thin film on the right. **c,** SEM micrograph of the fiber cross section is shown on the left, and a magnified section of the intact semiconductor thin film, right, confirms that the layered structure is preserved.

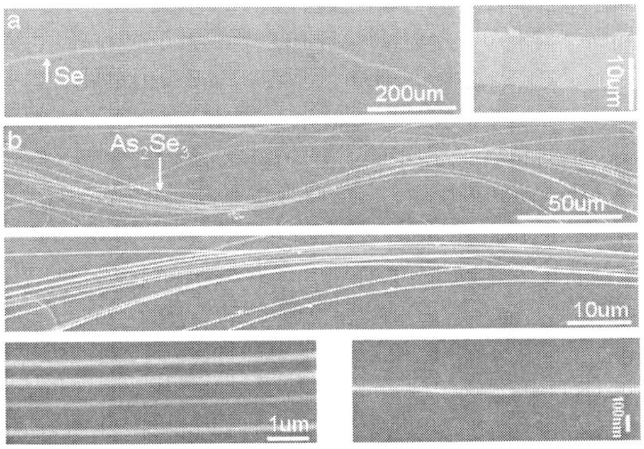

Figure 2 Semiconductor nano-filaments. **a,** SEM micrograph of a Se filament and a magnified section. **b,** (i) A bundle of As_2Se_3 filaments. (ii) Magnified section showing parallel filaments. (iii) A section of a single filament.

5 DISCUSSIONS

The mechanism for these long nano-filaments is not clear so far. What would determine the lower limit of layer scaling down? Why dose break of layer occur? How are these filaments formed? What is the dynamical process during the thermal drawing? All these questions definitely demand more theory efforts.

There are several promising applications for these semiconductor wires. The semiconductors used in this work have attractive optical properties such as a high nonlinearity, high refractive indices, and wide transmission window in the near- and mid-infrared regimes. Additionally, they exhibit intriguing properties such as a high photoconductivity, large piezoelectric and thermoelectric coefficients, and low thermal conductivity.

Exciting directions for future work include crystallizing the amorphous semiconductor nano-ribbons and assembling more diverse materials together in the same fiber.

This work was supported by US DOE, the ARO through the ISN. This work was also supported in part by the MRSEC Program of the National Science Foundation.

REFERENCES

[1] Maurer, R.D. & Schultz, P.C. Fused silica optical waveguide. US patent 3,659,915 (1972).

[2] Keck, D. B., Maurer, R. D. & Schultz, P. C. Ultimate lower limit of attenuation in glass optical waveguides. Appl. Phys. Lett. 22, 307-309 (1973).

[3] Agrawal, G.P. Fiber-optic communication systems 3rd edn (Wiley-Interscience, New York, 2002).

[4] Knight, J. C. Photonic crystal fibers. Nature 424, 847-851 (2003).

[5] Abouraddy, A. F. *et al.* Towards multimaterial multifunctional fibers that see, hear, sense and communicate. Nature Mater. 6, 336-347 (2007).

[6] Bayindir, M. *et al.* Kilometer-long ordered nanophotonic devices by preform-to-fiber fabrication. IEEE J. Sel. Top. Quant. 12, 1202-1213 (2006).

[7] Yeh, P., Yariv, A. & Marom, E. Theory Of Bragg fiber. J. Opt. Soc. Am. 68, 1196-1201 (1978).

[8] Fink, Y. et al. A dielectric omnidirectional reflector. Science 282, 1679-1682 (1998).

[9] Fink, Y. et al. Guiding optical light in air using an all-dielectric structure. J. Lightwave Technol. 17, 2039-2041 (1999).

[10] Hart, S. D. *et al.* External reflection from omnidirectional dielectric mirror fibers. Science 296, 510-513 (2002).

[11] Temelkuran, B., Hart, S. D., Benoit, G., Joannopoulos, J. D. & Fink, Y. Wavelength-scalable hollow optical fibers with large photonic bandgaps for CO_2 laser transmission. Nature 420, 650-653 (2002).

[12] Abouraddy, A. F. *et al.* Large-scale optical-field measurements with geometric fiber constructs. Nature Mater. 5, 532-536 (2006).

Tunnel Field Effect Transistor (TFET) with Strained Silicon Thinfilm Body for Enhanced Drain Current and Pragmatic Threshold Voltage

M. Jagadesh Kumar and Sneh Saurabh

Department of Electrical Engineering, Indian Institute of Technology, New Delhi – 110 016, INDIA.
Email: mamidala@ieee.org Fax: 91-11-2658 1264

ABSTRACT

Quantum tunneling devices are very promising as they have very low leakage current and show good scalability. However, the most serious drawback for tunneling devices hampering their wide-scale CMOS application is their low on-current and high threshold voltage. In this paper, we propose a novel lateral Strained Double-Gate Tunnel Field Effect Transistor (SDGTFET), which not only tackles these problems very well but also shows excellent overall device characteristics. For the first time, using two dimensional simulation, we show that the proposed device improves the on-current by two-order of magnitude without significantly degrading the off-current, lowers the threshold voltage so as to meet the ITRS guideline, improves the average subthreshold swing and also shows good immunity to short channel effects.

***Keywords*: Tunneling, Strained-Silicon, SOI MOSFET, Short-channel Effects, TFET, Two-dimensional simulation.**

1 INTRODUCTION

With the reducing device dimensions, the problem of leakage current becomes more prominent. Novel devices utilizing quantum tunneling phenomenon as the operating principle show good promise. These devices have very low leakage current, have excellent subthreshold swing and exhibit good scalability [1]-[4]. However, the most serious drawback for tunneling devices is their low on-current and high threshold voltage [2]-[4]. This problem needs to be tackled so that these devices could be employed for wide-scale CMOS application. In this paper, we propose a novel lateral Strained Double-Gate Tunnel Field Effect Transistor (SDGTFET), which not only tackles these problems very well but also shows excellent overall device characteristics.

2 DEVICE STRUCTURE AND SIMULATION MODEL

A schematic cross-sectional view of the proposed SDGTFET device is shown in Fig.1. The structure of this device is similar to the conventional Double Gate Tunnel Field Effect Transistor (DGTFET) [3]. However, the silicon body of this device is strained silicon. This device can be fabricated using single-layer strained-silicon-on-insulator (SSOI) technology [5], [6]. The amount of strain in an SSOI is controlled by varying the mole fraction of Ge in the relaxed SiGe buffer layer that is used during its fabrication. The device parameters used in our simulations are as follows: Source doping = $10^{20}/cm^3$, Drain doping = $5x10^{18}/cm^3$, Channel doping = $10^{17}/cm^3$, Gate length = 50 nm, Gate oxide thickness = 3 nm, Silicon body thickness = 10 nm, Gate work function = 4.5 eV, Drain bias, V_{DS} = 1 V. Ge mole fraction in the SiGe buffer layer, x is varied from 0 to 0.5. Using two-dimensional simulation, we show how the modulation of band-structure of silicon (engineered by straining) results in an overall improvement of the device characteristics of an SDGTFET. The simulation model use drift-diffusion model for current transfer. Non-local band-to-band tunneling is turned on in our simulations. A very fine mesh is defined in the simulation structure and especially in the regions where tunneling takes place.

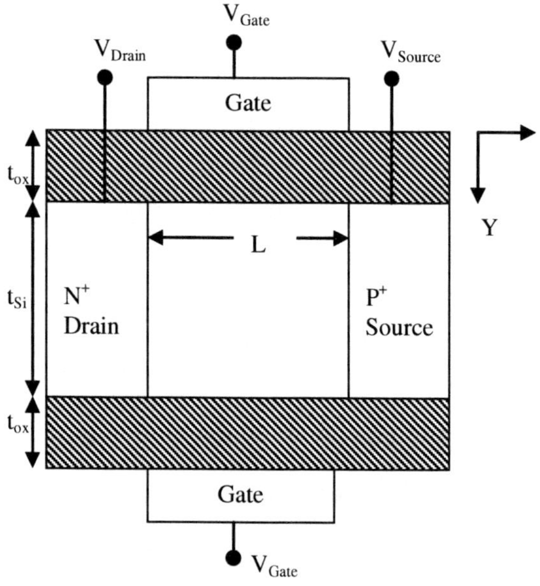

Figure 1 Cross-sectional view of an SGTFET

3 SIMULATION RESULTS

An extensive simulation of this device structure was done so as to study the performance of the device. The transfer characteristics of the SDGTFET at different Ge mole fractions were computed using simulations. The on-current of the device, defined as the drain current at a particular gate voltage, was extracted from these transfer characteristics.

threshold voltage requirement of around 0.3 V. As the Ge mole fraction is increased the threshold voltage of this device decreases. For a mole fraction of 0.5, the threshold voltage can be decreased to around 0.4 V, which is much closer to the abovementioned ITRS guidelines.

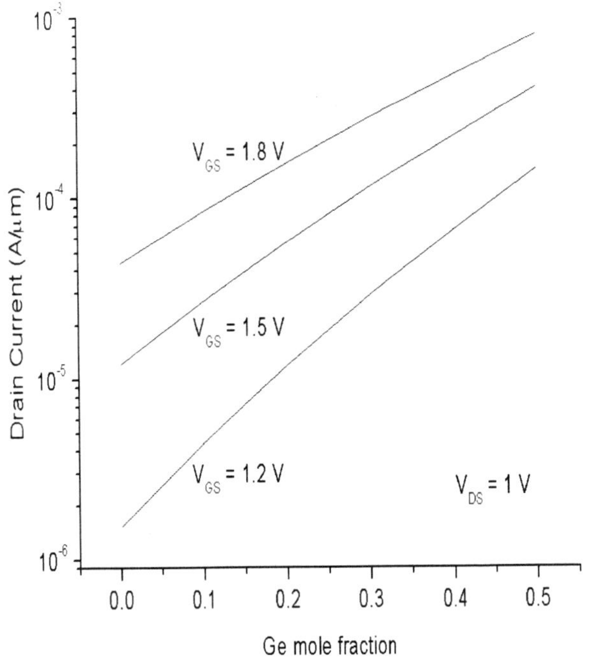

Figure 2 Drain Current of an SDGTFET versus Ge mole fractions at different V_{GS}

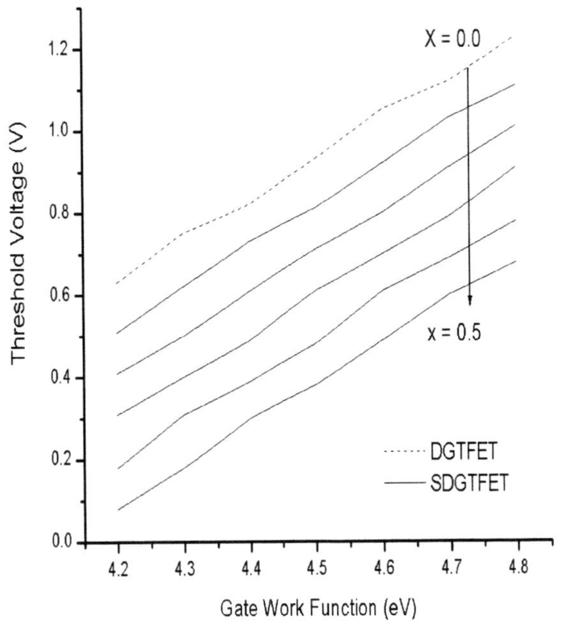

Figure 3 Threshold voltage versus gate workfunction of DGTFET and SDGTFET for different Ge mole fractions

It can be easily seen from Fig. 2 that the on-current of the device improves by two-order of magnitude for a Ge mole fraction of 0.5. The on-current of the unstrained device (Ge mole fraction = 0) is in range of tens of μA/μm which fails miserably to meet the ITRS near-term low power guideline of around 700 μA/μm. However, as the Ge mole fraction is increased, the on-current comes closer to meeting this guideline. In fact, at Ge mole fraction of 0.5, this device is capable of meeting the ITRS guideline. It should be mentioned that though the on-current improves for an SDGTFET, the off-current does not degrade significantly. The ratio of on-current to off-current was extracted from the transfer characteristics. It was found that this ratio improves initially with increasing Ge mole fraction. But at mole fraction greater than 0.2, this ratio starts degrading a little with increasing Ge mole fraction (due to increase in off current). In fact the ratio of on-current to off-current was found to attain a maximum value at a Ge mole-fraction of around 0.2. Also, since the complete device is strained, the improvement in device characteristics can be derived in both NMOS and PMOS type of operations.

The threshold voltage of an SDGTFET was extracted from the simulation results. For an unstrained SDGTFET (Ge mole fraction 0), the threshold voltage is around 0.9 V which is much higher than the ITRS near term low-power

Further reduction in threshold voltage can be brought about by gate work-function engineering. Fig. 3 shows the variation of threshold voltage with the change in gate work-function. However, it should be noted that the reduction in threshold voltage with reduction in gate work-function is accompanied by an increase in off-state leakage. Fig. 4 shows the transfer characteristics of the device at different work function for a Ge mole fraction of 0.2. It can easily be inferred from this figure that there is an increase in off-state leakage at lower work function. Hence a tradeoff needs to be made to get the desired threshold voltage without sacrificing the on-current by off-current ratio.

The subthreshold swing of an SDGTFET was extracted from our simulation results. Since, the subthreshold behavior of tunneling devices is a strong function of gate voltage, classical definition of subthreshold swing cannot be used for these devices. The point subthreshold slope and the average subthreshold slope as defined in [3] were used in our simulation study. It was found that an SDGTFET shows considerable improvement in both point subthreshold slope and average subthreshold slope. In fact, an SDGTFET shows average subthreshold slope better than 60 mV/decade (that of an ideal MOSFET) for a Ge mole fraction of 0.4 or more.

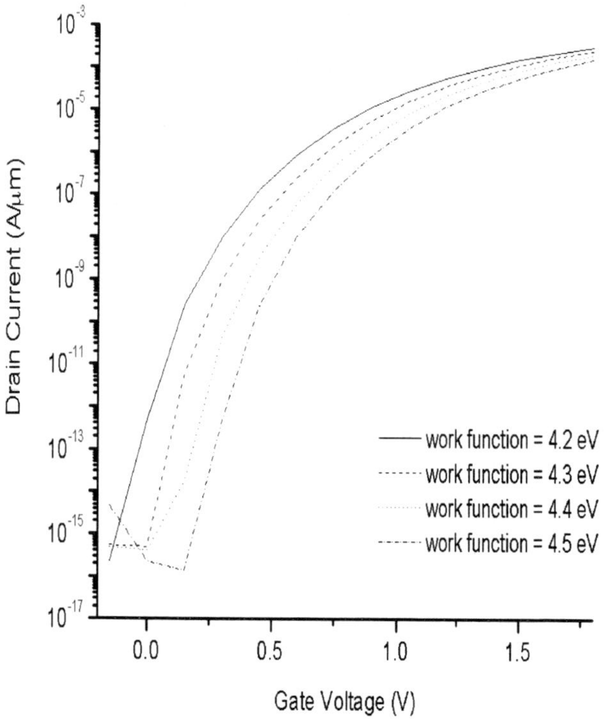

Figure 4 Transfer characteristics of an SDGTFET with x = 0.2 for different gate workfunction

The short channel effects in an SDGTFET were also studied. It was found that an SDGTEFT show negligible threshold voltage roll-off up to 20 nm channel length. The transfer characteristics of an SDGTEFT do not change appreciably with the change in channel length. The primary reason for this is that the tunneling phenomenon is confined to a very narrow region around the source in an SDGTFET. Hence, decreasing the gate length does not have much effect on the transfer characteristics until the drain is too close to the source to impact the tunneling phenomenon. Therefore, it can be concluded that an SDGTEFT is very much immune to short channel effects.

4 CONCLUSIONS

For the first time, using two-dimensional simulation, we have shown that the proposed SDGTFET is a very good candidate for low-leakage CMOS technology. Also, since the device structure is similar to the strained DGFET technology, this can be seamlessly integrated into the existing process flow. Using two-dimensional device simulation we have shown that this device meets the ITRS guidelines in terms of on-current and threshold voltage, has excellent subthreshold slope and negligible short channel effects. Hence this device looks promising to take the center stage of CMOS technology in the times to come.

REFERENCES

[1] P.F. Wang, K. Hilsenbeck, T. Nirschl, M. Oswald, C. Stepper, M. Weis, D. S. Landsiedel, and W. Hansch, "Complimentary tunneling transistor for low power application", *Solid-State Electronics*, vol. 48, pp. 2281-2286, Dec 2004

[2] K. K. Bhuwalka, J. Schulze, and I. Eisele, "A Simulation Approach to Optimize the Electrical Parameters of a Vertical Tunnel FET", *IEEE Trans. on Electron Devices*, vol. 52, pp. 1541-1547, July 2005

[3] K. Boucart and A. M. Ionescu, "Double-Gate Tunnel FET With High-κ Gate Dielectric", *IEEE Trans. on Electron Devices*, vol. 54, pp. 1725-1733, July 2007

[4] W. Y. Choi, B. G. Park, J. D. Lee, and T. J. K. Liu, " Tunneling Field-Effect Transistors (TFETs) with subthreshold swing (SS) less than 60 mV/dec ", *IEEE Electron Device Letters,* vol. 28, pp. 743-745, August 2007

[5] H. Yin, K.D. Hobart, R. L. Peterson, F.J.Kub, S. R. Shieh, T. S. Duffy, and J. C. Sturm, "Fully-depleted Strained-Si on Insulator NMOSFETs without Relaxed SiGe Buffers", in *IEDM Tech. Digest*, pp. 3.2.1-3.2.4, 2003

[6] S.Takagi , T. Mizuno, T. Tezuka, N. Sugiyama, T. Numata, K. Usuda, Y. Moriyama, S. Nakaharai, J. Koga, A. Tanabe, and T. Maeda, "Fabrication and device characteristics of strained-Si-on-insulator (strained SOI) CMOS", Applied Surface Science, vol. 224, pp241-247, March 2004

[7] Semiconductor Industry Association (SIA), International Technology Roadmap For Semiconductors, 2006 Update, http://www.itrs.net

Benchmarking Performance and Physical Limits on Processing Electronic Device and Systems: Solid-State, Molecular and *Natural* Processing Paradigms

Sergey Edward Lyshevski*, Vlad P. Shmerko** and Svetlana N. Yanushkevich**

*Department of Electrical Engineering, Rochester Institute of Technology, Rochester, NY 14623, USA
**Department of Electrical and Computer Engineering, University of Calgary, Calgary, Canada
E-mail: Sergey.Lyshevski@mail.rit.edu Web: www.rit.edu/~seleee

ABSTRACT

This paper examines the physical and technological limits imposed on processing devices which predefine hardware and software solutions. We study various constraints which are imposed. The emerging molecular processing paradigm promises one to ensure benchmarking overall performance in data processing. The soundness of *fluidic* and *solid* solutions is studied. We discuss the use of basic physics, implication of the Heisenberg uncertainty principle, as well as technologies which ultimately may affect various approaches. *Natural* and *engineered* molecular processing solutions are profoundly different. In general, the comparison cannot be made because *natural* processing is not comprehended at the *device* and *system* levels. The information processing, performed by *natural* systems, is a frontier for multidisciplinary research. Therefore, we concentrate on data processing accomplished by *engineered* molecular primitives and platforms. This molecular solution is fundamentally different when compared to solid-state microelectronics due to distinct device physics, system organization, technology, software, etc. The case-studies are reported.

Keywords: electronic device, physical limits, processing

1. INTRODUCTION

For solid-state microelectronics, basic physics and sound fundamentals were largely developed [1, 2]. The established CMOS technology currently progresses to the 45 and 32 nm technology nodes with the expected *effective* field-effect transistor dimensionality ~500×500 nm [3]. However, microelectronics faces many challenges. The major challenges are [1-5]:

1. Fundamental limits at the device, module and system levels are nearly achieved,
2. Stringent technological constraints and affordability rationale emerged and cannot be overcome.

Therefore, sound innovative solutions are sought. *Solid* and *fluidic* molecular processing devices (^{M}devices) and processing platforms (^{M}PPs) were proposed in [6-9]. Molecular processing encompasses novel 3D-topology ^{M}devices, new organization, advanced architecture and *bottom-up* fabrication [9-11].

The *reachable* volumetric dimensionality of *solid* and *fluidic* molecular processing primitives is in the order of 1×1×1 nm. Solid molecular electronic devices (^{ME}device) were studied in [6-9]. For ^{ME}devices, due to synthesis, interface and other constraints, the *achievable equivalent* device cell volumetric dimensionality is expected to be ~10×10×10 nm [9]. These multi-terminal ^{ME}devices can be synthesized and aggregated as $^{\aleph}$hypercells forming functional ^{M}ICs. These ^{M}ICs implement complex combinational circuits, processors, memories, etc. [9-11].

New device physics, innovative organization, novel architecture, enabling capabilities and functionality are the key essential features of molecular and nano electronics. Molecular, atomic and subatomic devices, as compared to semiconductor devices, are distinguished by distinct device physics. ^{M}Devices may:

1. Exhibit exclusive phenomena;
2. Ensure exceptional performance;
3. Provide enabling capabilities;
4. Possess unique functionality.

These ^{M}device may typify, to some extent, processing biomolecular primitives of *natural* PPs (^{N}PPs). However, the device physics is fundamentally distinct. Furthermore, ^{N}PPs perform information processing and are not comprised from ^{M}IC or any *circuit* equivalents. As

- *Natural* biomolecular processing primitives will be coherently examined, assessed and evaluated,
- Information processing of ^{N}PPs will be comprehended,

the benchmarking performance quantitative metrics and qualitative measures of *natural* processing (computing) performance and capabilities may be established.

Due to unsolved problems in information processing and *natural* processing, we focus on ^{M}PPs. The combinational and memory ^{M}ICs can be designed as aggregated $^{\aleph}$hypercells comprised from ^{M}gates and molecular memory cells. From the system-level consideration, ^{M}ICs can be designed within novel organization and enabling architecture which guarantee superior performance and exceptional capabilities.

2. PHYSICAL LIMITS AND ENERGETICS

We study physical limits and research performance estimates. The designer examines the device and system performance and capabilities using distinct performance measures, estimates, indexes and metrics. At the device level, we examine functionality, study characteristics and estimate performance of 3D-topology ^{M}devices. The experimental results indicate that electrochemomechanical transitions in biomolecules and proteins are performed within 1×10^{-6} to 1×10^{-12} sec, and requires $\sim1\times10^{-19}$ to 1×10^{-18} J of energy [9].

To analyze protein energetics, examine the switching energy in microelectronic devices, estimate *solid* ^{ME}devices energetics and perform other studies, distinct concepts are applied. For solid-state microelectronic devices, the logic signal energy is expected to be reduced to $\sim 1\times 10^{-16}$ J [3]. The energy dissipated is

$$E=Pt=IVt=I^2Rt=Q^2R/t,$$

where P is the power dissipation; I and V are the current and voltage along the discharge path; R and Q are the resistance and charge.

The dynamic power dissipation (consumption) in CMOS circuits is analyzed. Using the equivalent power dissipation capacitance C_{pd} (expected to be reduced to $\sim 1\times 10^{-12}$ F [3]) and the transition frequency f, the device power dissipation due to output transitions is $P_T=C_{pd}V^2f$. The energy for one transition can be found using the current as a function of the transition time which is found from the equivalent RC models of solid-state transistors. The simplest estimate for the transition time is $-RC\ln(V_{out}/V_{dd})$. During the transition, the voltage across the load capacitance C_L changes by $\pm V$. The total energy used for a single transition is charge Q times the average voltage change, which is $\frac{1}{2}V$. The total energy per transition is $\frac{1}{2}C_{Load}V^2$. If there are $2f$ transitions per second, one has $P_L=C_LV^2f$. Therefore, the dynamic power dissipation P_D is

$$P_D=P_T+P_L=C_{pd}V^2f+C_LV^2f=(C_{pd}+C_L)V^2f.$$

For ^{M}devices and ^{ME}devices, this analysis cannot be applied. The term k_BT has been used to solve distinct problems. Here, k_B is the Boltzmann constant, $k_B=1.3806\times 10^{-23}$ J/K; T is the absolute temperature. The expression γk_BT ($\gamma>0$) was used to assess the energy, and $k_BT\ln(2)$ was applied with the attempt to assess the lowest energy bound for a binary switching. The applicability of distinct equations must be examined applying sound concepts. Statistical mechanics and entropy analysis coherently utilize k_BT and $k_B\ln w$ within a specific content.

By letting $w=2$, the entropy is $S=k_B\ln 2=9.57\times 10^{-24}$ J/K. Having derived S, one cannot conclude that the minimal energy required to ensure the transition (*switching*) between two *microscopic* states or to erase a bit of information (energy dissipation) is $k_BT\ln 2$, which for T=300K gives $k_BT\ln 2=2.87\times 10^{-21}$ J. In fact, under this reasoning, one assumes the validity of the *averaging* kinetic-molecular Newtonian model and applies the assumptions of distribution statistics, at the same time allowing only two distinct *microscopic* system's states.

The energy estimates are performed utilizing quantum mechanics. To examine the discrete energy levels of an electron in the outermost populated shell, one applies

$$E_n=-\frac{m_eZ_{eff}^2e^4}{32\pi^2\varepsilon_0^2\hbar^2n^2}, \text{ and } E_n=-2.17\times 10^{-18}\frac{Z_{eff}^2}{n^2} \text{ J.}$$

From $Z_{eff}/n\approx 1$, one concludes that ΔE is from $\sim 1\times 10^{-19}$ to 1×10^{-18} J. If one supplies the energy greater than E_n to the electron, the energy excess will appear as kinetic energy of the free electron. The transition energy should be adequate to excite electrons. For different atoms and molecules, as the prospective *solid* ^{ME}devices, the transition (switching) energy is estimated to be $\sim 1\times 10^{-19}$ to 1×10^{-18} J. This estimate is in agreement with biomolecular devices. The energy of a single photon is $E=hc/\lambda$. The maximum absorbance for rhodopsin is ~500 nm. Hence $E=4\times 10^{-19}$ J. This energy is sufficient to ensure transitions and functionality.

Considering an electron as a non-relativistic particle, taking note of $E=\frac{1}{2}mv^2$, we obtain the particle velocity as a function of energy, and $v(E)=\sqrt{\frac{2E}{m}}$.

For $E=1.6\times 10^{-20}$ J, one finds $v=1.9\times 10^5$ m/sec.

Assuming 1 nm path length, the traversal (*transit*) time is $\tau=L/v=5.3\times 10^{-15}$ sec. Hence, ^{ME}devices can operate at a high switching frequency. However, one may not conclude that the device switching frequency to be utilized is $f=1/(2\pi\tau)$ due to device physics features (number of electrons, heating, interference, potential, energy, noise, etc.), system-level functionality, circuit specifications, etc.

Natural, biomolecular, *fluidic* and *solid* devices and systems exhibit distinct performance and capabilities. The advancements are envisioned towards *solid* molecular electronics typifying ^{N}PPs. One can resemble a familiar solid-state microelectronics solution. *Solid* ^{ME}devices and ^{M}ICs may utilize the *soft materials* such as polymers and biomolecules. Figure 1 reports some performance estimates. Here, a neuron is represented as a *natural* information processing/memory module (system) [9].

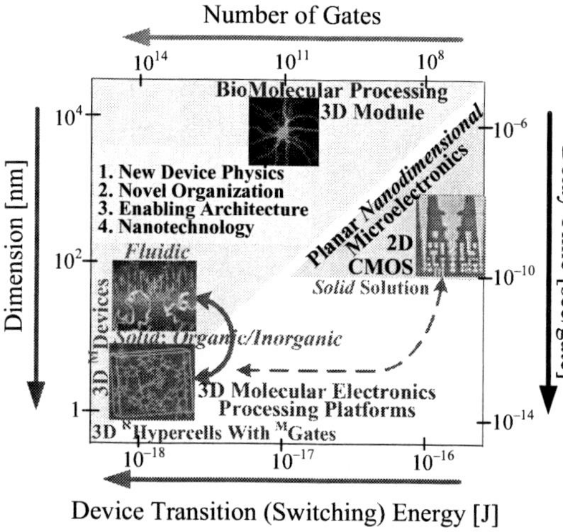

Figure 1. Towards molecular electronics and processing/memory. Revolutionary advancements:
1. From data processing to information processing frontiers;
2. From 2D microelectronics to 3D molecular electronics;
3. From ICs to ^{M}ICs, and to ^{M}PPs typifying ^{N}PPs;
4. From *bulk* to *bottom-up* synthesis and fabrication.

We studied the *microscopic* system energetics applying quantum mechanics. Though these variations can be utilized by *microscopic* systems ensuring overall functionality and soundness of ^{M}devices, the resulting changes and transitions must be observed and characterized. The device, module and system testing, characterization and evaluation are critical, and PPs cannot be designed without these tasks. The Heisenberg uncertainty principle provides the fundamental limits due on the measurements implying constraints on the testability, characterization, etc.

NSTI-Nanotech 2008, www.nsti.org, ISBN 978-1-4200-8505-1 Vol. 3

In particular, the Heisenberg uncertainty principle provides the position-momentum and energy-time limits on the measurements as $\sigma_x\sigma_p \geq \frac{1}{2}\hbar$ and $\sigma_E\sigma_t \geq \frac{1}{2}\hbar$.

Using the energetic estimates, letting σ_E to be 1×10^{-18} J, one obtains $\sigma_t \geq 5\times10^{-15}$ sec. This result complies with the derived *transit* time $\tau=5.3\times10^{-15}$ sec. It should be emphasized, that the Heisenberg uncertainty principle does not define the device functionality and/or physical limits on energy, *transit* time, momentum, velocity, device dimensionality, etc.

3. Design of Processing Platforms: System Organizations and Architectures

We initiate the studies in the design and software developments for envisioned molecular hardware solutions. These problems are frontiers of electrical and computer engineering, computer science and other disciplines, such as physics, chemistry, biology, neuroscience, mathematics, etc. Enormous research efforts are needed, and our intent is to report possible solutions, propose innovative inroads, and report up-to-date major results.

The analysis and design of ^{M}PPs must be three dimensional, and the third dimension Z carries the functional information, that is, $f=f(X,Y,Z)$. There are several known approaches for computing in 3D. The first approach is based on the idea that the 2D computational structures can by layered and assembled by utilizing 2D layers as $f=f_1(X,Y)\times f_2(X,Y)\times\ldots\times f_k(X,Y)$, where functions $f_1, f_2, \ldots, f_k$ correspond to the first-, second-, …, k-th levels; × denotes an assembling operation. This approach is used in VLSI and ULSI. The logic network (with a physical implementation as a planar 2D silicon design) is layered and interconnected (between layers) to achieve the desired functionality. The interconnected layers form the "third dimension". In this approach, the "third dimension" does not carry any functional information about the implemented logic functions. This computational 3D concept is not adequate to the *natural*, *engineered* and other molecular solutions. The second approach is based on the mapping of a logic function into a system with "three coordinates" [12]. This approach requires complex transformations of logic functions with respect to each dimension. The main drawback is that the known techniques of logic network design cannot be used in the representation of logic functions in a 3D space. That is, each logic function requires a 3D expansion [13]. This approach is obscure because it requires a complete revision of the previously developed methods, techniques, tools, etc. The third approach is related to the 3D cellular arrays and their particular case, 3D systolic arrays [10, 11]. These structures are very complicated in design and control. Correspondingly, they have not been fully implemented in practice. However, the linear systolic arrays are found to be tractable and practical [10, 11, 14]. They can be used in the 3D computing structure design.

We are developing a novel computing paradigm which can be applied to arbitrary computing platforms, e.g., conventional and envisioned ^{M}PPs. Furthermore, our concept, supported by a set of efficient methods and tools, is expected to be applicable to assess and study *natural* processing. In the proposed approach, several major problems, which have not been solved by the known methods, are successfully resolved [10, 11]. Hence, we overcome the design complexity problem relaxing the *design complexity limits*. Our design complexity for the homogeneous arrays is $kO(n)$, where n is the number of logic variables. In contrast, the application of conventional methods can lead to the complexity $O(n^3)$. In the proposed design, the homogenous and massive parallelism principles are implemented at the device level without loss of the universality of computing. This is very important for the design and analysis of the processing platforms and their implementation using current and prospective solutions.

To date, we have applied advanced analysis and design methods, some of which were reported in [10, 11]. These concepts, supported by a 3D©NeuralNet toolbox which utilizes computationally-efficient algorithms, are developed to attack super-large-scale problems. That is, design of computing feedforward networks of neurons with 1×10^9 to 1×10^{10} connections is potentially possible by further advancing our toolbox. Various computing architectures were designed. Networks can be embedded into 3D structures using H-trees representation in spatial dimensions [11]. We use the established benchmarks to verify and compare the results with conventional design methods. We replaced logic primitives (AND, OR, NAND, NOR and other gates) by a 3D ^{N}hypercube as illustrated in Figures 2.a and b. A node in the decision tree realizes the Shannon expansion $f=\bar{x}_i f_0 \oplus x_i f_1$, where $f_0=f(x_i=0)$ and $f_0=f(x_i=1)$ for all variables in f. The terminal nodes carry information of computing flow. Figure 2.c illustrates a spatial topology neuron network, consisting of 546 ^{N}hypercubes, which models a neuron with 46 inputs and 32 outputs.

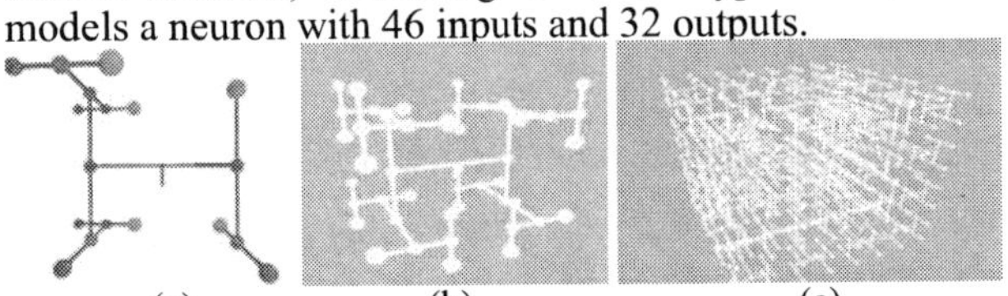

(a) (b) (c)

Figure 2. (a) Optimized spatial topology of a 4-input neuron; (b) Aggregated 5-input neuron; (c) 3D representation of a neuron with 46 inputs and 32 outputs generated by 3D©NeuralNet package

We develop and evaluate a decision diagram-based approach which simplifies the connections or the *path* between two arbitrary neurons in 3D. The alternative to the hypercube-based approach is to describe the logic function between neurons by a set of linear decision diagrams [10, 11]. The linear decision diagrams can be directly mapped into *linear arrays*. A 3D©NeuralNet toolbox can generate computing neural structures based on principles of massive parallelism guarantying *array* processing.

A computing array consists of *operational* and *memory* arrays of primitives as reported in Figure 3. The array specificity is defined by the need for functionality- and efficiency-focused design and optimization at the various design and implementation levels. For example, switches, various logic devices, memory, data transmission and other primitives are characterized by different requirements, specifications, performance, capabilities, etc. Our concept perfectly suits molecular, biomolecular and micro technologies. The proposed solution is expected to be applicable to the *natural* processing. The advantage of the proposed array architecture is a straightforward aggregation of 2D and 3D arrays. This array utilizes specific properties of liner decision diagrams [10, 11]. Table 1 reports the results of our numerical studies. For example, for the sao2 circuit, 3D©NeuralNet generates a 3D 10-input and 4-output neural network with 33 terminal nodes, 100 intermediate (operational) nodes and 133 interconnections.

Table 1. Modeling of 3D feedforward neural networks based on embedding decision diagrams into 3D ^{N}hypercube structures

Benchmarks # I/O		# Terminal nodes	# Intermediate nodes	# Connections
9sym	9/1	99	298	397
clip	9/5	149	448	597
sao2	10/4	33	100	133

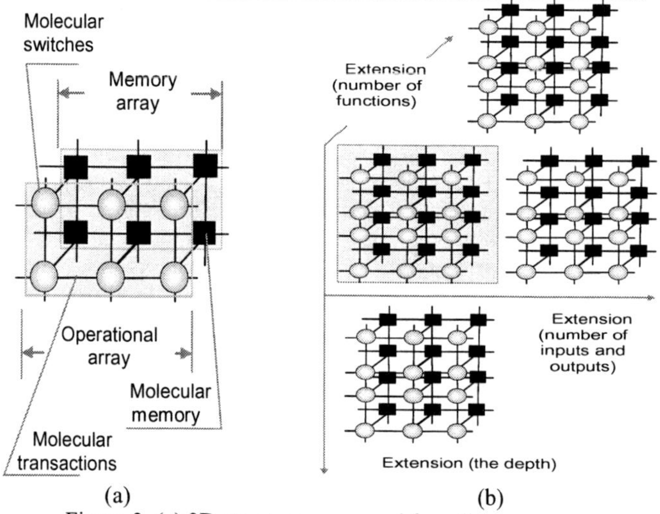

Figure 3. (a) 3D structure composed from linear arrays; (b) Aggregation in 3D space

Our goal is to increase the computational power of 3D neural networks. We represent a neuron by a multiple-valued structure which processes data on radix 3, 4, …,16. A 3D©NeuralNet package can manipulate multiple-valued data structures. A high radix improves performance, capabilities and efficiency of neural networks. We encoded multiple-valued data by binary codes ensuring soundness [10]. The modeling results for the 4-valued neural networks are given in Table 2. In particular, three ICs (c2670 ALU/control, c5315 ALU, and c7552 adder/comparator) are studied. Enhancing information content (quaternary versus binary) does not requires the doubling of computational resources. This result is verified for the 8- and 16-valued data. In our preliminary studies, we did not examine the data reliability which affected by noise, noise/signal ratio and other factors related to data transmission, hardware solutions, functionality, etc.

Table 2. Modeling of 3D neural networks using multiple-valued data structures

TEST	INPUT / OUTPUT	#LEVEL	#CELL	VOLUME
c2670	233/140	40	1374	1375×40×140
c5315	178/123	68	3192	3192×68×123
c7552	207/108	64	4627	4627×64×108

4. Conclusions

For emerging processing devices and platforms, we derived and reported the physical and technological limits. *Natural* and *engineered* molecular processing solutions were examined at the device and system levels. The results contributed to the design and analysis of *engineered* molecular processing devices and platforms. We studied and reported the fundamental differences between data and information processing in *engineered* and *natural* systems, which are the frontiers of a multidisciplinary research [15]. The applications of novel design methods, supported by the developed software tools, were presented to demonstrate feasibility and practicality in synthesis of ^{M}PPs achieving super-large-scale-integration capabilities.

Acknowledgements

The authors sincerely acknowledge a support from the US Department of Energy under the contracts DE-FG02-06 *Three-Dimensional Biomolecular Computing Architectures*.

Disclaimer – Neither the United States Government nor any agency thereof, nor any of their employees, makes any warranty, express or implied, or assumes any legal liability or responsibility for the accuracy, completeness, or usefulness of any information, apparatus, product, or process disclosed, or represents that its use would not infringe privately owned rights. Reference herein to any specific commercial product, process, or service by trade name, trademark, manufacturer, or otherwise does not necessarily constitute or imply its endorsement, recommendation, or favoring by the United States Government or any agency thereof. The views and opinions of authors expressed herein do not necessarily state or reflect those of the United States Government or any agency thereof.

References

1. S. M. Sze and K. K. Ng, *Physics of Semiconductor Devices*, John Wiley and Sons, NJ, 2007.
2. R. S. Muller and T. I. Kamins, *Device Electronics for Integrated Circuits*, John Wiley and Sons, NJ, 2003.
3. *International Technology Roadmap for Semiconductors*, 2005 Edition, Semiconductor Industry Association, Austin, Texas, USA, 2006. http://public.itrs.net/
4. J. E. Brewer, V. V. Zhirnov and J. A. Hutchby, "Memory technology for the post CMOS era", *IEEE Circuits and Devices Magazine*, vol. 21, issue 2, pp. 13-20, 2005.
5. V. V. Zhirnov, R. K. Cavin, J. A. Hutchby and G. I. Bourianoff, "Limits to binary logic scaling – A Gedanken model", *Proc. IEEE*, vol. 91, issue 11, pp. 1934-1939, 2003.
6. J. Chen, T. Lee, J. Su, W. Wang, M. A. Reed, A. M. Rawlett, M. Kozaki, Y. Yao, R. C. Jagessar, S. M. Dirk, D. W. Price, J. M. Tour, D. S. Grubisha and D. W. Bennett, *Molecular Electronic Devices, Handbook Molecular Nanoelectronics*, Eds. M. A. Reed and L. Lee, American Science Publishers, 2003.
7. J. C. Ellenbogen and J. C. Love, "Architectures for molecular electronic computers: Logic structures and an adder designed from molecular electronic diodes," *Proc. IEEE*, vol. 88, no. 3, pp. 386-426, 2000.
8. J. R. Heath and M. A. Ratner, "Molecular electronics," *Physics Today*, no. 1, pp. 43-49, 2003.
9. S. E. Lyshevski, *Molecular Electronics, Circuits, and Processing Platforms*, CRC Press, Boca Raton, FL, 2007.
10. S. N. Yanushkevich, V. P. Shmerko and S. E. Lyshevski, *Logic Design of NanoICs*, CRC Press, Boca Raton, FL, 2005.
11. S. N. Yanushkevich, S. E. Lyshevski, and V. P. Shmerko, *Computer Arithmetics for Nanoelectronics*, CRC Press, Boca Raton, FL, 2008.
12. A. Al-Rabady and M. Perkowski, "Shannon and Davio sets of new lattice structures for logic synthesis in three-dimensional space," *Proc. 5th Int. Workshop on Applications of the Reed-Muller Expansion in Circuit Design*, pp. 165-184, 2001.
13. S. Y. Kung, *VLSI Array Processors*, Prentice Hall, NJ, 1988.
14. S. N. Yanushkevich, D. M. Miller, V. P. Shmerko and R. S. Stankovic, *Decision Diagram Techniques for Micro- and Nano-electronic Design, Handbook,* CRC Press, Boca Raton, FL, 2006.
15. J. von Neumann, *The Computer and the Brain*, Yale University Press, New Haven, CT, 1957.

Three-Dimensional-Topology Processing and Memory Cells

Marina Alexandra Lyshevski* and Sergey Edward Lyshevski**

*Microsystems and Nanotechnologies, Webster, NY 14580-4400, USA
**Department of Electrical Engineering, Rochester Institute of Technology, Rochester, NY 14623, USA
E-mail: E.Lyshevski@rit.edu and Sergey.Lyshevski@mail.rit.edu

ABSTRACT

This paper researches an innovative fundamental concept and proposes a practical solution for envisioned molecular processing platforms (^{M}PPs). At the device level, we report three-dimensional (3D) topology multi-terminal molecular processing primitives (^{M}primitives) which exhibit electrochemomechanical transitions ensuring functionality. These ^{M}primitives can be utilized as logic gates. Aggregated ^{M}primitives comprise neuronal hypercells ($^{\aleph}$hypercells). The networked $^{\aleph}$hypercells form ^{M}PPs within 3D system organization.

Keywords: molecular primitive, neuronal hypercell, polypeptide, processing platform

1. INTRODUCTION

High-performance processing and computing, as forefront challenging areas, have been widely examined in the literature. Electronic, electrical, mechanical, chemical and other logic devices were proposed, studied and utilized in various data processing, computing and memory platforms. Tremendous progress has been achieved by utilizing complementary metal-oxide semiconductor technology to fabricate integrated circuits (ICs) designed using very-large-scale integration methodology. However, emerging fundamental problems and technological limits, associated with solid-state microelectronics and ICs, cannot be overcome, or solutions are unknown [1, 2]. Departing from conventional concepts, we study alternative solutions.

Various *solid* and *fluidic* molecular processing, logic and memory devices (^{M}devices) have been studied [3-7]. Only a few ^{M}devices were synthesized, tested and characterized [3-8]. The major challenges encountered are:

1. Device physics soundness;
2. Synthesis and assembly;
3. Aggregation, integration and compatibility;
4. Testing, evaluation and characterization.

Though some devices promise to exhibit meaningful phenomena and ensure functionality, these devices still pending as long as the aforementioned tasks are performed proving overall device soundness.

The processing (computation) can be accomplished by means of electron transport, conformational changes and other electrochemomechanical transitions. These transitions can be accomplished by molecules or molecular complexes which form ^{M}primitives. The device-level functionality, which is predefined by the device physics, should be supported by synthesis feasibility. We propose to engineer ^{M}primitives using the following innovations:

1. Use the polypeptide backbone as a structural skeleton with side groups which exhibit electro-chemomechanical transitions and interaction ensuring combinational logics (logic functions) and memory storage;
2. Aggregate the side groups of neighboring primitives forming $^{\aleph}$hypercells.

This solution mimics, to some extent, a *natural biomolecular hardware*. However, this does not imply that we mimic or approach *natural* information processing, data processing or memory storage. We study *engineered* ^{M}primitives, $^{\aleph}$hypercells and ^{M}PPs which inherently possess 3D topology and 3D organization features. In general, ^{M}PPs integrate a large class of computing, memory and processing solutions and systems. For example, if the electron transport ensures the device functionality, ^{M}primitive can be classified as a molecular electronics device (^{ME}device), implying that a ^{M}PP can be designed utilizing molecular integrated circuits (^{M}ICs).

2. MOLECULAR PROCESSING

Among the key challenges in devising and the design of ^{M}primitives, the major ones are:

- Device physics, functionality, and processing capabilities;
- Synthesis, e.g., technological feasibility, practicality, yield, etc.;
- Aggregability and integration (assembly, interfacing, compatibility, compliance, interactability, matching, packaging, etc.).

The technological soundness can be achieved by using ^{M}primitives implemented as organic or inorganic molecules (molecular complexes) within the structural backbone formed by polypeptides [8]. It is known that polypeptides are used to form very-complex functional *natural*, organic and hybrid molecular organelles and assemblies such as various proteins, enzymes, hormones and other biological polymers. Though a significant progress has been documented for biomolecules and proteins, there is a need to depart from the attempt to blindly prototype *natural biomolecular hardware* due to immense unsolved fundamental and synthesis problems. For example, one may not be able to utilize *natural* biomolecular and protein functionality, transitions, mechanisms, etc. Our objective is to design *synthetic* ^{M}PPs, including ^{M}ICs, which are entirely distinct as compared to biosystems. The peptide synthesis, custom biosynthesis and wide range of possible structural modifications provide the designer with a needed flexibility and multiplicity maintaining the specificity and soundness. In ^{M}primitives we propose to utilize side groups which must ensure and exhibit:

- Desired device functionality, capabilities and characteristics (for ^{ME}devices, controlled electron transport, switching characteristics, *IV* characteristics, energetics, etc.);
- Assembling, interfacing, networking and interconnecting features.

3. ^{M}PRIMITIVES AND $^{\aleph}$HYPERCELLS

We study envisioned ^{M}PPs which are comprised from networked $^{\aleph}$hypercells formed as aggregated ^{M}primitives. Each ^{M}primitive is engineered utilizing:

- Polypeptide backbone (–N–C–C–)$_n$ which forms structural skeleton;
- Side groups (S_{ijk}).

Consider a 3D-cube-topology $^{\aleph}$hypercell which consists of ^{M}primitives, as depicted in Figure 1. Here, side groups S_{ijk} and the polypeptide backbone (–N–C–C–)$_n$ comprise ^{M}primitives $^{M}P_{ijk}$.

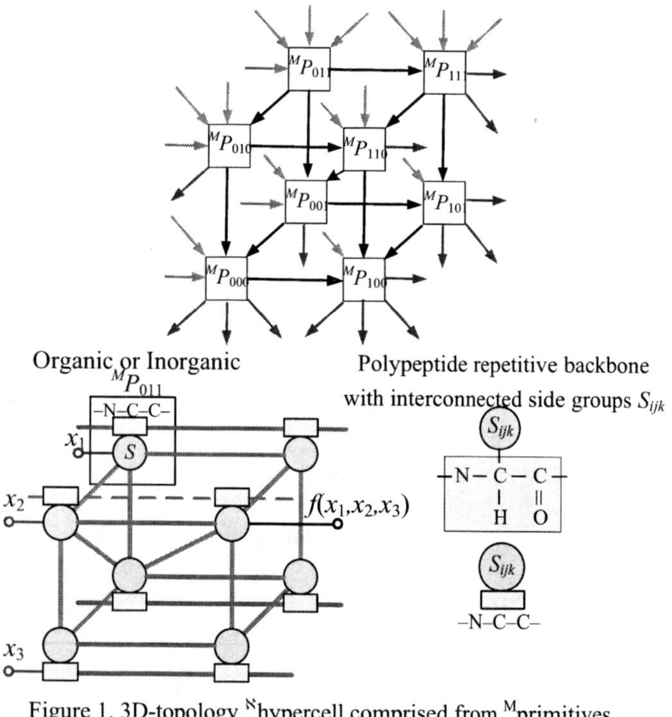

Figure 1. 3D-topology $^{\aleph}$hypercell comprised from ^{M}primitives $^{M}P_{ijk}$ formed by the polypeptide backbone and side groups S_{ijk}

The (–N–C–C–)$_n$ chains provide a structural skeleton (mechanical structure), while side groups S_{ijk} must guarantee the overall functionality within the device physics. Each S_{ijk} may integrate multi-terminal and multi-functional ^{M}devices engineered from organic or inorganic molecules. As shown in Figure 1, the 3D-topology interconnect is accomplished by S_{ijk}. $^{\aleph}$Hypercells can be clustered to form macrocells, thereby forming ^{M}PPs.

Engineered and Natural Processing Platforms – The proposed concept, in general, cannot be manifested to be biomimetic-centered because it is highly unlikely that ^{M}primitives guarantee the functionality and operationability equivalence to biomolecules and *natural* biomolecular aggregates. That is, ^{M}PPs' and *natural* biomolecular platforms' functionality, capabilities, hardware, software and other basic features are entirely distinct. However, we typify the *natural biomolecular hardware*. A significant departure from conventional concepts is achieved providing a viable sound alternative.

4. FUNCTIONALITY OF ^{M}PRIMITIVES

Various quantum, electrostatic, electromagnetic, optical, mechanical and chemical phenomena and effects can be utilized to perform logic operations, implement switching functions and store information. The corresponding side groups must be engineered to guarantee the specified phenomena, effects and transitions. Though electron transport and bond formation/braking are due to electron transitions (transport, exchange, sharing, etc.), they are profoundly different. While the overall device functionality is defined by S_{ijk}, the soundness of spatial topology assembly (geometry and conformation) must be ensured through a coherent backbone–S_{ijk} aggregation.

Various electronic and quantum-effect devices have been utilized focusing the major thrust on the well-defined microelectronic paradigms. Different device-level solutions were examined and extensively investigated. Molecular electronics is within the most promising solutions. This direction affects not only engineering but also life science.

For ^{ME}devices, the expected characteristics and the controlled electron transport are examined by utilizing the methods of quantum mechanics [8]. We investigate various side groups engineered from organic and inorganic molecules which should ensure overall functionality and practicality of multi-terminal ^{ME}devices.

Consider a three-terminal ^{ME}device with the *input*, *control* and *output* terminals, as shown in Figure 2.a. The device physics of this ^{ME}device is based on quantum interactions and controlled electron transport. The applied $V_{\mathrm{control}}(t)$ changes the charge distribution $\rho(t,\mathbf{r})$ and electric field intensity $\mathbf{E}_E(t,\mathbf{r})$ affecting the electron transport. This ^{ME}device operates in the controlled electron-exchangeable environment due to quantum transitions and interactions. The controlled super-fast potential-assisted electron transport may be ensured.

NSTI-Nanotech 2008, www.nsti.org, ISBN 978-1-4200-8505-1 Vol. 3

Consider the electron in the time- and spatial-varying metastable potentials $\Pi(t,\mathbf{r})$. The changes in the Hamiltonian result in:

- Quantum interactions due to variations of the charge distribution $\rho(t,\mathbf{r})$, $\mathbf{E}_E(t,\mathbf{r})$ and $\Pi(t,\mathbf{r})$;
- Changes of tunneling $T(E)$.

The device controllability is ensured by varying $V_{\rm control}(t)$ that affects $\rho(t,\mathbf{r})$, $\mathbf{E}_E(t,\mathbf{r})$ and $\Pi(t,\mathbf{r})$ leading to variations of electron transport. Hence, the device switching, *IV*, *GV* and other characteristics are controlled.

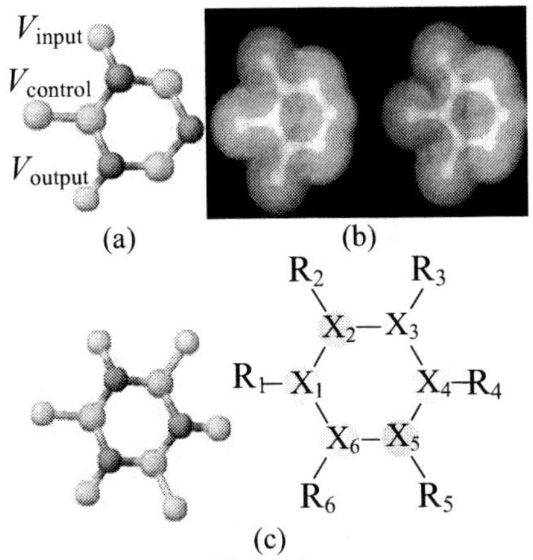

Figure 2.
(a) Side group S_{ijk}: Three-terminal ^{ME}device comprised from a monocyclic molecule with a carbon interconnecting framework; (b) Charge distribution $\rho(\mathbf{r})$; (c) Six-terminal ^{ME}device.

To study the device characteristics, we simplify S_{ijk} to 9 atoms with motionless protons with charges q_i. The radial Coulomb potentials are

$$\Pi_i(r) = -\frac{Z_{eff\,i} q_i^2}{4\pi\varepsilon_0 r}.$$

For carbon, we have $Z_{eff\,C}$=3.14.

Using the spherical coordinate system, the Schrödinger equation

$$-\frac{\hbar^2}{2m}\left[\frac{1}{r^2}\frac{\partial}{\partial r}\left(r^2\frac{\partial\Psi}{\partial r}\right)+\frac{1}{r^2\sin\theta}\frac{\partial}{\partial\theta}\left(\sin\theta\frac{\partial\Psi}{\partial\theta}\right)+\frac{1}{r^2\sin^2\theta}\frac{\partial^2\Psi}{\partial\phi^2}\right]$$
$$+\Pi(r,\theta,\phi)\Psi(r,\theta,\phi)=E\Psi(r,\theta,\phi)$$

should be solved.

One can represent the wave function as

$$\Psi(r,\theta,\phi)=R(r)Y(\theta,\phi)$$

in order to solve the radial and angular equations. We discretize the Schrödinger and Poisson equations to numerically solve these differential equations. The magnitude of the time-varying potential applied to the *control* terminal is bounded due to the thermal stability of the molecule, energetics and other limits. In particular, $|V_{\rm control}|\le V_{\rm control\ max}$, and $|V_{\rm control}|\le 0.25$ V. Figure 2.b documents a three-dimensional charge distribution in the molecule for $V_{\rm control}$=0.1 V and $V_{\rm control}$=0.2 V.

The Schrödinger and Poisson equations are solved using a self-consistent algorithm in order to verify the device physics soundness and examine the baseline performance characteristics. To obtain the current density $\mathbf{j}$ and current in the ^{ME}device, the velocity and momentum of the electrons are obtained using

$$\langle p\rangle = \int_{-\infty}^{\infty}\Psi^*(t,\mathbf{r})\left(-i\hbar\frac{\partial}{\partial\mathbf{r}}\right)\Psi(t,\mathbf{r})d\mathbf{r}\cdot$$

The wave function $\Psi(t,\mathbf{r})$ is numerically derived for distinct values of $V_{\rm control}$. The *IV* characteristics of the studied ^{ME}device for two different control currents (0.1 and 0.2 nA) are reported in Figure 3 [8]. The results documented imply that the proposed ^{ME}device may be effectively used as a multiple-valued primitive in order to design enabling high-radix logics and memories.

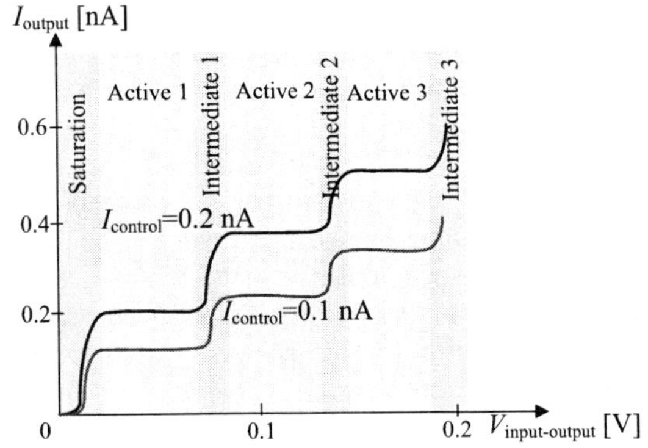

Figure 3. Multiple-valued *IV* characteristics

The traversal time of electron transport is derived using

$$\tau(E)=\int_{\mathbf{r}_0}^{\mathbf{r}_f}\sqrt{\frac{m}{2[\Pi(\mathbf{r})-E]}}d\mathbf{r}\cdot$$

It is found that τ is ~5×10^{-15} sec. Hence, the proposed ^{ME}device ensures super-fast switching.

The reported monocyclic molecule can be used as a six-terminal ^{ME}device as illustrated in Figure 2.c. The use of the device's side groups R_i, shown in Figure 2.c, ensures the variations of the energy barriers, wells potential surfaces $\Pi(t,\mathbf{r})$, interatomic length, etc. The proposed carbon-centered molecular solution, in general,

- Ensures a sound *bottom-up* synthesis at the device, gate and hypercell levels;
- Guarantees assembly and aggregability features to form complex ^{M}ICs;
- Results in the experimentally verifiable and characterizable ^{ME}devices and ^{M}gates.

The studied ^{ME}devices can be utilized in combinational and memory ^{M}ICs. In addition, those devices can be used as routers. In particular, one may achieve a reconfigurable networking, processing and memory. The proposed ^{ME}device can be used as a *switch* or transmission device allowing one to design the neuromorphological reconfigurable ^{M}PPs.

5. Towards Molecular Combinational Logics and Memories

A 2×2 array of SRAM cells is reported in Figure 4. The memories can be implemented using ^{M}NOR and ^{M}NAND gates which can be implemented by the proposed ^{M}primitives and ^{N}hypercells.

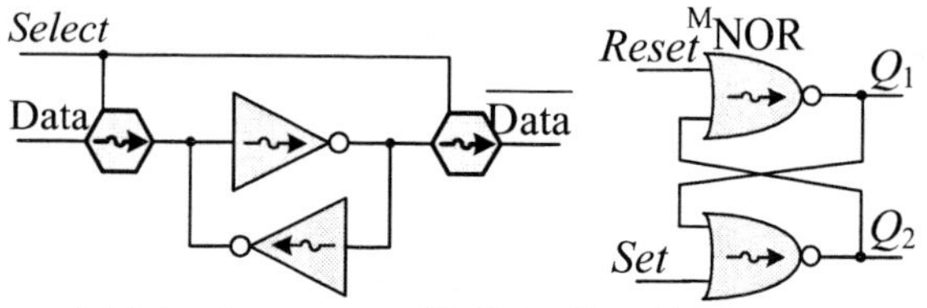

Figure 4. Molecular memory: SRAM cell and 2×2 array of SRAM cells implemented as ^{N}hypercell

Acknowledgements

The second author sincerely acknowledges a support from the US Department of Energy under the contracts DE-FG02-06 *Three-Dimensional Biomolecular Computing Architectures.*

Disclaimer – Neither the United States Government nor any agency thereof, nor any of their employees, makes any warranty, express or implied, or assumes any legal liability or responsibility for the accuracy, completeness, or usefulness of any information, apparatus, product, or process disclosed, or represents that its use would not infringe privately owned rights. Reference herein to any specific commercial product, process, or service by trade name, trademark, manufacturer, or otherwise does not necessarily constitute or imply its endorsement, recommendation, or favoring by the United States Government or any agency thereof. The views and opinions of authors expressed herein do not necessarily state or reflect those of the United States Government or any agency thereof.

6. Conclusions

Fundamental, applied, experimental and technological features of molecular processing were studied. We proposed an innovative and elegant concept. Typifying the *natural* polypeptides and side groups, we utilized the *natural* or *synthetic* polypeptides and side groups ensuring functionality and synthesis features. Examples are reported with the foreseen technology assessments. We proposed an alternative solution to solve a three-fold problem for envisioned ^{M}PPs by: (a) Devising and researching device physics of ^{M}devices; (b) Developing a technology-sound solution; (c) Devising functional and sound ^{M}primitives, ^{M}devices and □hypercells which comprise of ^{M}PPs. We progressed towards 3D ^{M}PPs which promise to ensure neuromorphological reconfigurable data processing. To some extent, we applied the *natural* processing solutions to *engineered* one. Correspondingly, we focused on the transformative science and engineering. The proposed concept promises massive vector processing utilizing robust and reconfigurable neuromorphological organizations, high-radix processing, molecular hardware, etc.

References

1. *International Technology Roadmap for Semiconductors*, 2005 Edition, Semiconductor Industry Association, Austin, Texas, USA, 2006. http://public.itrs.net/
2. J. E. Brewer, V. V. Zhirnov and J. A. Hutchby, "Memory technology for the post CMOS era", *IEEE Circuits and Devices Magazine*, vol. 21, issue 2, pp. 13-20, 2005.
3. J. Chen, T. Lee, J. Su, W. Wang, M. A. Reed, A. M. Rawlett, M. Kozaki, Y. Yao, R. C. Jagessar, S. M. Dirk, D. W. Price, J. M. Tour, D. S. Grubisha and D. W. Bennett, *Molecular Electronic Devices, Handbook Molecular Nanoelectronics*, Eds. M. A. Reed and L. Lee, American Science Publishers, 2003.
4. J. C. Ellenbogen and J. C. Love, "Architectures for molecular electronic computers: Logic structures and an adder designed from molecular electronic diodes," *Proc. IEEE*, vol. 88, no. 3, pp. 386-426, 2000.
5. Handbook on *Nano and Molecular Electronics*, Ed. S. E. Lyshevski, CRC Press, Boca Raton, FL, 2007.
6. J. R. Heath and M. A. Ratner, "Molecular electronics," *Physics Today*, no. 1, pp. 43-49, 2003.
7. J. M. Tour and D. K. James, *Molecular Electronic Computing Architectures, Handbook of Nanoscience, Engineering and Technology*, Eds. W. A. Goddard, D. W. Brenner, S. E. Lyshevski and G. J. Iafrate, pp. 4.1-4.28, CRC Press, Boca Raton, FL, 2003.
8. S. E. Lyshevski, *Molecular Electronics, Circuits, and Processing Platforms*, CRC Press, Boca Raton, FL, 2007.

Quantum Fourier Transform Circuit Simulator

Víctor H. Tellez[1, 3], Antonio Campero[2], Cristina Iuga[2], Gonzalo I. Duchen[3]

[1]Electrical Engineering Departament, [2]Chemical Departament, DCBI, Universidad Autonoma Metropolitana Iztapalapa, Av. San Rafael Atlixco 186, Col. Vicentina, Iztapalapa 09340 D.F. Mexico, email: vict@xanum.uam.mx
[3]SEPI, ESIME Culhuacan, Av. Santa Ana 1000, Col. San Francisco Culhuacan, C.P. 04430, D.F.

ABSTRACT

Quantum Fourier transform is of primary importance in many quantum algorithms. This paper presents the development of a Quantum Fourier Transform Circuit Simulator system that processes classical analog signals and presents the results of the processing data. The data is acquired by an analog to digital classical converter, on a classical computer. The data stored is processed by computer using an algorithm that executes a Quantum Fast Fourier Transform (QFT).

1 INTRODUCTION

In quantum mechanics, quantum information is physical information that is held in the "state" of a quantum system. The most popular unit of quantum information is the qubit, a two-state quantum system. However, unlike classical digital states (which are discrete), a two-state quantum system can actually be in a superposition of the two states at any given time. A quantum bit, or qubit is described by a state vector in a two-level quantum mechanical system which is formally equivalent to a two-dimensional vector space over the complex numbers.

The Quantum Fourier transforms (QFT) plays essential roles in various quantum algorithms such as Shor's algorithms [1, 2, 3] and hidden subgroup problems [2, 4, 5]. Inspired by the exponential speed-up of Shor's polynomial algorithm for factorization [1], many people investigated the problem of efficient realization of QFT in a quantum computer [3, 6, 7, 8, and 9]. Up to now, many improvements have been made. In [6], Moore and Nilsson showed that QFT can be parallelized to linear depth in a quantum network, and upper bound of the circuit depth was obtained by Cleve and Watrous [7] for computing QFT with a fixed error. In reference [8] the actual time-cost for performing QFT in the quantum network was examined. Further, Blais [9] designed an optimized quantum network with respect to time-cost for QFT.

The present work shows a sound processing system based QFT circuit simulator of such sounds. A minimum system based on QFT circuit simulator and acquisition was developed, using a classical computer and the QFT was made on Python compiler. The paper is organized as follows. After the introduction the acquisition system is described, followed by the methodology to processing the sound. Results and conclusions are presented.

2 SYSTEM DESCRIPTION

The analog to digital converter part is based on a 68HC11 microcontroller minimum system; this processor that can perform the required memory, A/D conversion and transfer to Personal Computer. The data received from the Analog to Digital Converter is processing by a personal computer, using the Python compiler, which execute the QFT program and presents the results of the processing.

3 METHODOLOGY

The system requests a subject to emit a sound and it is recorded through the microphone; the program seeks for the suitable tone in order to be recognized as a valid signal. The capture algorithm is executed from the analog to digital converter, and then the processing algorithms are executed from the personal computer.

3.1 The QFT modeling circuit

The processing algorithm, is based on the QFT as an unitary operation on n qubits defined as follows

$$F : |x\rangle \rightarrow \frac{1}{\sqrt{2^n}} \sum_{y=0}^{2^n-1} e^{2\pi i xy/2^n} |y\rangle \tag{1}$$

where xy is a normal ``decimal" multiplication of numbers x and y, which are represented by the quantum registers

$$|y\rangle = |y_{n-1}\rangle \otimes |y_{n-2}\rangle \otimes \cdots \otimes |y_0\rangle \tag{2}$$

where $|x_k\rangle$ and $|y_k\rangle$ are individual qubits.

Comparing this with the notation in the section about Simon Oracle, where $x \bullet y$ meant

$$x \bullet y = x_0 \bullet y_0 + 2x_1 \bullet y_1 + 2x_2 \bullet y_2 + \cdots + 2x_{n-1} \bullet y_{n-1}$$

There we treated **x** and **y** as arrays of bits rather than integer numbers. Of course in computing a single integer number is implemented as an array of bits, but the point is how you interpret this array, and so xy in the Fourier Transform formula is not the same as $x \bullet y$ in the Simon Oracle formula. The former is an integer operation on two scalar numbers and the latter is a binary operation on two binary vectors. The former can be expressed in terms of a binary operation too, but it will not be $x \bullet y$.

Observe that once you know what **F** does to the basis vectors $|x\rangle$, you can figure out what **F** does to any other vector. This other vector can be $\sum_{\infty} f(x)|x\rangle$, which yields the following formula for Quantum Fourier Transform of function f:

$$F\left(\sum_{x=0}^{N-1} f(x)|x\rangle\right) = \sum_{x=0}^{N-1} f(x)F(|x\rangle)$$

$$= \frac{1}{\sqrt{N}} \sum_{x=0}^{N-1} f(x) \sum_{y=0}^{N-1} e^{2\pi ixy/N} |y\rangle \quad (3)$$

$$= \frac{1}{\sqrt{N}} \sum_{y=0}^{N-1} \sum_{x=0}^{N-1} f(x) e^{2\pi ixy/N} |y\rangle \quad (4)$$

From this formula the yth component of **F** is

$$F_y(f) = \frac{1}{\sqrt{N}} \sum_{x=0}^{N-1} f(x) e^{2\pi ixy/N} \quad (5)$$

which is beginning to look quite like a normal Discrete Fourier Transform.

3.2 The QFT Circuit Simulator

The QFT circuit simulator is development on Python compiler, on a personal computer running over Linux operating system. In order to implement the circuit that calculates the QFT:

$$F : |x\rangle \rightarrow \frac{1}{\sqrt{2^n}} \sum_{y=0}^{2^n-1} e^{2\pi ixy/2^n} |y\rangle \quad (6)$$

we shall deploy the trickery of the Fast Fourier Transform. Let us have a look at:

$$e^{2\pi ixy/2^n}$$

This expression is periodic in xy and the period is 2^n. The trick about the Fast Fourier Transform is that it only uses the terms of $e^{2\pi ixy/2^n}$ that correspond to the ``first circle'', i.e., the terms for which $x\ y\ /\ 2^n < 1$. Let us evaluate then $xy/2^n$ while truncating very thing that would go onto the second and third circle:

$$\frac{xy}{2^n} \equiv \frac{1}{2^n}\left(x_0 + x_1 2 + x_2 2^2 + x_3 2^3 + \cdots + x_{n-1} 2^{n-1}\right) \times$$
$$\left(y_0 + y_1 2 + y_2 2^2 + y_3 2^3 + \cdots + y_{n-1} 2^{n-1}\right) = \ldots$$

Here we have decomposed x and y into their binary components, so that each of the x_k and y_k terms is either 0 or 1.

$$= \frac{1}{2^n}\left(y_0\left(x_0 + x_1 2 + x_2 2^2 + \cdots + x_{n-1} 2^{n-1}\right) + y_1 2\left(x_0 + x_1 2 + x_2 2^2 + \cdots + x_{n-1} 2_{n-1}\right) + \cdots\right.$$
$$\left.\cdots + y_1 2^{n-1}\left(x_0 + x_1 2 + x_2 2^2 + \cdots + x_{n-1} 2^{n-1}\right)\right)$$

$$\frac{1}{2^n} = (y_0\left(x_0 + x_1 2 + x_2 2^2 + \cdots + x_{n-1} 2^{n-1}\right) +$$
$$y_1\left(x_0 2 + x_1 2^2 + x_2 2^3 + \cdots + x_{n-2} 2^{n-1}\right) +$$
$$+ y_2\left(x_0 2^2 + x_1 2^3 + x_2 2^4 + \cdots + x_{n-3} 2^{n-1}\right) + \cdots + y_{n-1} x_0 2^{n-1})$$

$$= y_0\left(\frac{x_0}{2^n} + \frac{x_1}{2^{n-1}} + \frac{x_2}{2^{n-2}} + \cdots + \frac{x_{n-1}}{2}\right) + y_1\left(\frac{x_0}{2^{n-1}} + \frac{x_1}{2^{n-2}} + \frac{x_2}{2^{n-3}} + \cdots + \frac{x_{n-2}}{2}\right) +$$

$$+ y_2\left(\frac{x_0}{2^{n-2}} + \frac{x_1}{2^{n-3}} + \frac{x_2}{2^{n-4}} + \cdots + \frac{x_{n-3}}{2}\right) + \cdots + y_{n-1}\frac{x_0}{2} = \ldots$$

There is a special notation, which covers the sums in the brackets:

$$\frac{x_0}{2} \rightarrow (x_0), \quad \frac{x_0}{2^2} + \frac{x_1}{2} \rightarrow (x_0 x_1),$$
$$\frac{x_0}{2^3} + \frac{x_1}{2^2} + \frac{x_2}{2} \rightarrow (x_0 x_1 x_2)\ldots$$

Using this notation:

$$\frac{xy}{2^n} \equiv y_0(x_0 x_1 \ldots x_{n-1}) + y_1(x_0 x_1 \ldots x_{n-2}) +$$
$$y_2(x_0 x_1 \ldots x_{n-3}) + \cdots + y_{n-1}(x_0) \quad (7)$$

So now we can write our Quantum Fourier Transform thusly:

$$F|x\rangle = \frac{1}{\sqrt{2^n}} \sum_{y=0}^{2^n-1} e^{2\pi ixy/2^n} |y\rangle$$

$$= \frac{1}{\sqrt{2^n}} \sum_{y=0}^{2^n-1} e^{2\pi i(y_0(x_0 x_1 \ldots x_{n-1}) + .. + y_{n-1}(x_0))} |y\rangle$$

$$= \frac{1}{\sqrt{2^n}} \sum_{y=0}^{2^{n-1}} e^{2\pi i y_0 (x_0 x_1 \ldots x_{n-1})} |y_0\rangle \otimes e \otimes \cdots \otimes e^{2\pi y_{n-1}(x_0)} |y_{n-1}\rangle$$

Now observe that y_k is either 0 or 1. If it is 0 then the corresponding term is, for example,

$e^{2\pi i 0(x_0 x_1 \cdots x_{n-3})}|0\rangle = |0\rangle$

If it is 1 then the corresponding term is: $e^{2\pi i 1(x_0 x_1 \cdots x_{n-3})}|1\rangle = |1\rangle$

The sum over all possible values of **y** will eventually assign both 0 and 1 to every y_k, therefore the following superposition is equivalent to the above:

$$F|x\rangle = \frac{1}{\sqrt{2}}\left(|0\rangle + e^{2\pi i(x_0 x_1 \ldots x_{n-1})}|1\rangle\right) \otimes \frac{1}{\sqrt{2}}\left(|0\rangle + e^{2\pi i(x_0 x_1 \ldots x2)}|1\rangle\right) \otimes \cdots$$
$$\otimes \frac{1}{\sqrt{2}}\left(|0\rangle + e^{2\pi i(x_0)}|1\rangle\right)$$

And this already points to the way we can implement a QFT circuit. Consider the following circuit:

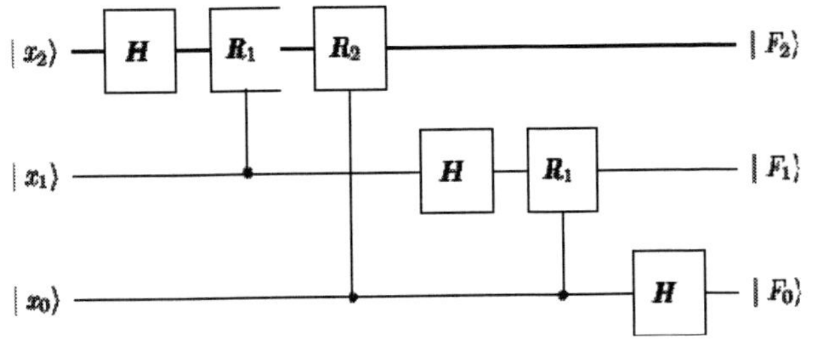

Figure 1 QFT circuit.

Here, as before, **H** is the Hadamard operator and $\mathbf{R}_d$ is a controlled gate defined by:

$R_d = \begin{pmatrix} 1 & 0 \\ 0 & e^{i\pi/2^d} \end{pmatrix}$ Where d is the *distance* between the lines. Let us analyze this circuit step by step: after the first Hadamard gate the top line becomes

$$\frac{1}{\sqrt{2}}\sum_{y=0}^{1}(-1)^{x_2 y}|y\rangle = \frac{1}{\sqrt{2}}\sum_{y=0}^{1} e^{2\pi i x_2 y/2} \quad (8)$$

The second step applies $\mathbf{R}_1$ to the top line under the control of the middle line. Observe that $\mathbf{R}_1$ does nothing to $|0\rangle$ and phase shifts $|1\rangle$. The phase shift factor is $e^{i\pi/2}$ if the control line $|x_1\rangle$ is $|1\rangle$ and there is no phase shifts if $|x_1\rangle = |0\rangle$. We can therefore write that the phase shift inflicted by $\mathbf{R}_1$ on $|1\rangle$ is *always* $e^{i\pi x/2}$, where x is the control signal.

Applying this to states on the top and on the middle line yields

$$R_1|x_1\rangle \otimes \frac{1}{\sqrt{2}}\left(|0\rangle + e^{2\pi i(x_2)}|1\rangle\right) = |x_1\rangle \otimes \frac{1}{\sqrt{2}}\left(|0\rangle + e^{2\pi i(x_2)}e^{i\pi x_1/2}|1\rangle\right) \quad (9)$$
$$= |x_1\rangle \otimes \frac{1}{\sqrt{2}}\left(|0\rangle + e^{2\pi i(x_2/2 + x_1/4)}|1\rangle\right) = |x_1\rangle \otimes \frac{1}{\sqrt{2}}\left(|0\rangle + e^{2\pi i(x_1 x_2)}|1\rangle\right)$$

The third gate applies $\mathbf{R}_2$ to the top line, but this time under the control of the bottom line. This operator, again, will do nothing to $|0\rangle$, but will phase shift $|1\rangle$ by additional $e^{i\pi x_0/4}$ so the state of the whole system now becomes:

$$R_2|x_0\rangle \otimes |x_1\rangle \otimes \frac{1}{\sqrt{2}}\left(|0\rangle + e^{2\pi i(x_1 x_2)}|1\rangle\right) =$$
$$= |x_0\rangle \otimes |x_1\rangle \otimes \frac{1}{\sqrt{2}}\left(|0\rangle + e^{2\pi i(x_1 x_2)}e^{i\pi x_1/4}|1\rangle\right)$$
$$= |x_0\rangle \otimes |x_1\rangle \otimes \frac{1}{\sqrt{2}}\left(|0\rangle + e^{2\pi i(x_0/8 + x_1/4 + x_2/2)}|1\rangle\right) = \quad (10)$$
$$= |x_0\rangle \otimes |x_1\rangle \otimes \frac{1}{\sqrt{2}}\left(|0\rangle + e^{2\pi i(x_0 x_1 x_2)}|1\rangle\right)$$

Reasoning as above we can see immediately that the next two gates applied to $|1\rangle$ will convert it into $\frac{1}{\sqrt{2}}\left(|0\rangle + e^{2\pi i(x_0 x_1)}|1\rangle\right)$. So that now the state of the computer is:

$$|x_0\rangle \otimes \frac{1}{\sqrt{2}}\left(|0\rangle + e^{2\pi i(x_0 x_1)}|1\rangle\right) \quad (11)$$
$$\otimes \frac{1}{\sqrt{2}}\left(|0\rangle + e^{e\pi i(x_0 x_1 x_2)}|1\rangle\right)$$

And finally the single Hadamard transform on the bottom line converts $|x_0\rangle$ to $|0\rangle + e^{2\pi i(x_0)}|1\rangle/\sqrt{2}$, so that in effect the final state of the computer is:

$$\frac{1}{\sqrt{2}}\left(|0\rangle\right) + e^{2\pi i(x_0)}|1\rangle \otimes \frac{1}{\sqrt{2}}\left(|0\rangle + e^{2\pi i(x_0 x_1)}|1\rangle\right) \quad (12)$$
$$\otimes \frac{1}{\sqrt{2}}\left(|0\rangle + e^{2\pi(x_0 x_1 x_2)}|1\rangle\right)$$

But this is a 3-point Quantum Fourier Transform, so the circuit shown above is a QFT circuit.

4 RESULTS

In the table 1 and 2, we can see the results comparing QFT circuit with FFT development in commercial software (MATLAB). These bounds allow simulation for many choices of N and □. However the choices for M and L given in odd integer can usually be improved, and were merely given to show such values can be found. For example, the following table shows, for different N and □ combinations, a triple (g,m, l) of integers, with the choice from line 86 being M = 2g; yet in each case M = 2m and L = 21 is the pair with minimal m satisfying the hypotheses for odd integer Thus choosing M and L carefully may allow lower qubit counts, such as the N = 13, □ = 0.10 case.

Table 1, Values for QFT circuit simulator

□	N=13	N=25	N=51	N=101	N=251	N=501
.001	45,45,28	47,47,28	48,48,29	50,50,29	52,52,30	53,53,30
.01	36,35,21	37,37,22	38,38,23	40,40,23	42,42,23	43,43,24
.05	29,28,17	30,30,17	31,31,18	33,33,18	35,35,19	36,36,19
.10	26,25,15	27,27,15	28,28,16	30,30,16	32,32,17	33,33,17
.20	23,22,13	24,24,13	25,25,14	27,27,14	29,29,15	30,30,15
.30	21,20,12	22,22,12	24,24,12	25,25,13	27,27,13	29,28,14
.40	20,19,11	21,21,11	22,22,12	24,24,12	26,26,13	27,27,13

Table 2, Values for classical FFT

□	N=13	N=25	N=51	N=101	N=251	N=501
.001	43,43,27	46,46,27	48,49,30	54,51,28	51,532,31	52,52,29
.01	35,34,22	36,36,23	39,39,22	41,41,22	41,41,22	42,42,23
.05	30,29,19	31,31,16	32,32,19	34,34,19	34,33,19	38,39,16
.10	25,25,16	26,26,15	27,28,15	31,32,15	31,31,15	32,32,16
.20	23,2113	25,24,14	26,24,15	26,26,14	30,30,16	31,32,16
.30	20,20,11	22,21,10	26,25,13	27,26,14	28,26,12	28,27,16
.40	21,14,16	23,27,12	23,24,14	25,23,13	27,28,14	26,28,12

5 CONCLUSIONS

We processes the same acquiring sound on classical and commercial software using the FFT (Matlab), and the results are presented on table 1 and 2. And from the statistical results, we can see on comparison that there is a similar result using the QFT circuit simulator as the similar FFT. The problems that we resolved is the time for processing using the QFT, is because this algorithm have very cost on memory and resources on hardware, and the time for processing is almost 3 and four hours. So, we recommend using a good processor with enough memory for then.

The age of Quantum Information Processing has arrived, and the solution for different applications is high. Right know we can say that the using for QFT for processing gives us the next advantage:

- Massive parallelism, this for the superposition theory
- Reversible logic
- When the QFT can be done, the processing time will be faster

6 REFERENCES

[1]. *P.W.Shor*, SIAM J.Comput 26, 1484 (1997).

[2]. M.A.Nielesn, I.L.Chuang, *Quantum Computation and Quantum Information* (Cambridge University Press, Cambridge, England, 2000).

[3]. *S.Beauregard*, Circuit for shor's algorithm using 2n+3 qubits, quant-ph/0205095.

[4] *M.Mosca and A.Ekert*, The hidden subgroup problem and eigenvalue estimation on a quantum computer, quantph/9903071.

[5]. *M.Ettinger and P.Hoyer*, On quantum algorithms for non-commutative hidden subgroups, quant-ph/9807029.

[6]. *C.Moore and M.Nilsson*, SIAMJ. Comput 31, 799 (2001).

[7]. *R.Cleve and J.Watrous*, Fast parallel circuits for the quantum fourier transform, quant-ph/0001113.

[8]. *A.Saito, K.Kioi, Y.Akagi*, N.Hashizume, and K.Ohta, Actual computational time-cost of the quantum fourier transform in a quantum computer using spins, quantph/0001113.

[9]. *Ethan Bernstein and Umesh Vazirani*, Quantum complexity theory, SIAM Journal on Computing, 26 (1997), no. 5, 1411–1473.

[10].*Thomas Beth, Markus P¨uschel*, and Martin R¨otteler, Fast quantum Fourier transforms for a class of non-abelian groups, Proc. of Applied Algebra Algebraic Algorithms, and Error Correction Codes (AAECC-13, Springer-Verlag, 1999, volume 1719 in Lecture Notes in Computer Science, pp. 148–159.

[11]*I. L. Chuang and M. A. Nielsen*, Quantum computation and quantum information, Cambridge University Press, Cambridge, 2000.

[12].*R. Cleve, E. Ekert, C. Macchiavello, and M. Mosca*, Quantum algorithms revisited, Proc. Roy. Soc. Lond. A 454 (1998), 339–354.

Effect of swift heavy ion on CdS quantum dots embedded in PVA matrix and their applications.

G Gope[*], D Chakdar, D K Avasthi ,M Paul, S S Nath
Dept. of Physics. National Institute of Technology Silchar, Assam, India.
IUAC, Aruna Asaf Ali Marg, New Delhi- 67,India
*email- goutam_gope@indiatimes.com,Tel:+919864821547

Abstract: We report synthesis of CdS quantum dots by chemical route at room temperature. In this technique CdS specimen are produced by simple chemical reactions where polyvinyl alcohol (PVA), acting as matrix, plays the key role in controlling particle growth during synthesis. These samples have been irradiated by 100MeV C^{+6} swift heavy ion (SHI). After that, the samples have been characterized by High Resolution Transmission Electron Microscopy (HRTEM), Atomic Force Microscope (AFM), UV/VIS absorption spectroscopy and impedance analyzer.

***Topic area :** Nanotechnology*

***Key words:** Quantum dot, Blue shift, Impedance analysis, Nano tuned device, SHI*

Introduction: Recently [1-15], preparation of semiconductor quantum dots and their applications in optoelectronics and electronics are the frontier area of research. Sophisticated methods are available to synthesize quantum dots but due to manifold advantages Chemical route is a very popular method at present [1]. Here in this paper it is an attempt to report synthesis of CdS quantum dots by chemical route and its impedence study.

Experimental: To synthesize CdS quantum dot [13-15], over coated with SiO_2 , 5 grams Polyvinyl Alcohol (PVA) are dissolved into 100 ml double distilled (D/D) water. The mixture is taken in a three necked flask fitted with thermometer pocket and N_2 inlet. The solution is stirred in a magnetic stirrer (stirring rate of 200 rpm at constant temperature of 70^0 C) for 3 hours. Thus, a transparent water solution of PVA has been prepared. The solution is degassed by boiling N_2 for 3 to 4 hours. Similarly, $CdCl_2$ solution is made by dissolving 5 gms of $CdCl_2$ in 100 ml D/D water. Next PVA solution and few drops of HNO_3 is added to the mixture and stirred at the rate of 250 rpm at a constant temperature of 55^0C while 2Wt % aqueous solution of Na_2S is put into it drop wise unless the whole solution turns into yellow color. The solution is kept in dark chamber at room temperature for 12 hours for its stabilization followed by its casting over glass substrate and drying in oven at 50^0 C. This film contains CdS quantum dots[6] embedded in PVA matrix. The prepared sample has been characterized by Transmission Electron Microscopy, (using JEM 1000 C XII), Optical absorption spectroscopy (using Perkin Elmer Lamda 35 1.24),Atomic Force Microscopy(NanoScope. IIIa) and impedance analyzer (Solartron SI 1260).

For irradiation, sample is cut in 1cm X 1cm size and put in irradiation chamber under high vacuum condition (4.6×6^{-10} Torr) for irradiation with 100-MeV C^{+6} ion beam (^{64}C, 1PnA) using four ion doses of 1×10^{11}, 5×10^{11}, 5×10^{12} and 10^{13} ions/cm^2, available from the 15UD tandem pelletron accelerator at IUAC, New Delhi, India

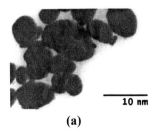

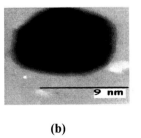

(a) (b)

Fig 1: HRTEM images of CdS quantum dots (a)before and (b) after irradiation.

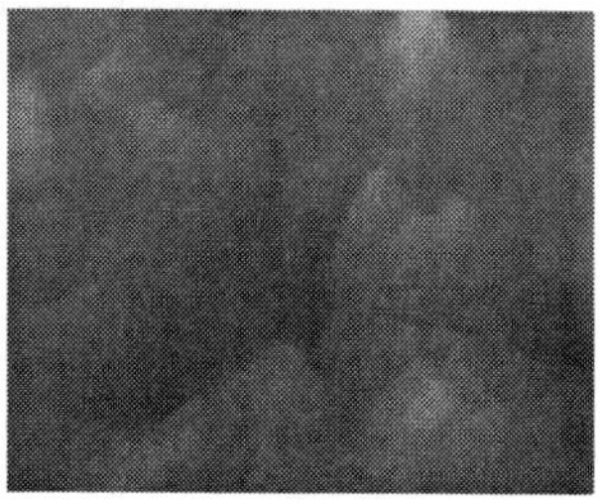

(a) (b)

Fig 2 : AFM images of CdS quantum dots (a)before and (b) after irradiation

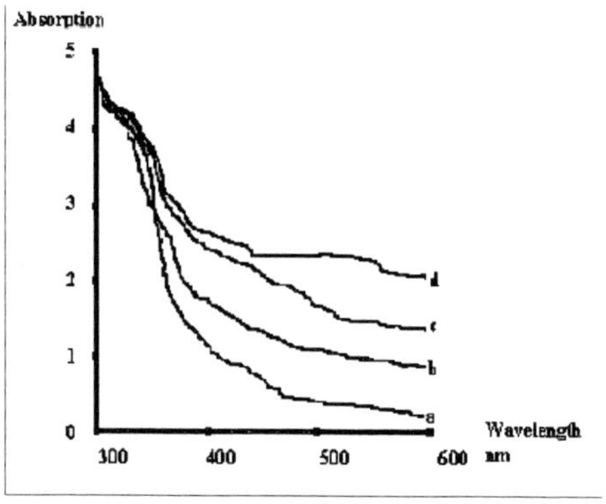

Fig 3: UV/VIS absorption spectra of CdS specimens. a, b, c, and d stand for virgin sample and samples irradiated by 1st, 2nd and 3rd dose respectively

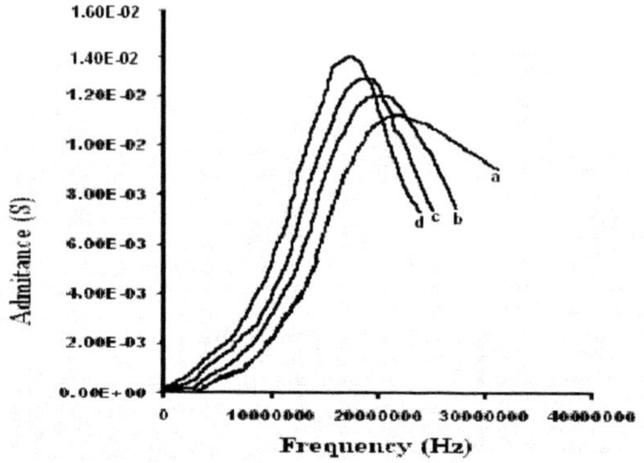

Fig 4: Impedance curve of CdS QD pristine(a)and irradiation (b),(c) and (d) samples

Sample	Absorption edge (nm)	Size (nm)
Virgin	375	7.6
1st dose	380	7.7
2nd dose	379	7.7
3rd dose	380	7.7

Table 1: Absorption spectroscopic data of CdS specimens

Sample	Size	Shape
Virgin	8.8	Spherical
1st dose	8.6	Spherical
2nd dose	8.5	Elliptical
3rd dose	9	Elliptical

Table 2: TEM data of CdS samples

Discussion:

It is already reported[12] that quantum dot impedance is basically due to capacitance which varies directly with particle size. After ion irradiation, particle agglomerates resulting in formation of bigger particle[15] of larger capacitance (fig 4). Due to this phenomena, quantum dot impedance changes resulting in modification in admittance Vs frequency curves of irradiated samples.

Conclusion:

CdS quantum dots prepared by chemical route lies approximately within 9 nm. Impedance analysis infers that CdS quantum dot behaves as nano tuned device. The critical frequencies (equivalent resonant frequency) vary depending on the size, shape and material of the prepared sample. After ion irradiation, particle agglomeration takes place resulting in formation of bigger particle of larger capacitance. Due to this phenomena, quantum dot impedance changes.

Acknowledgement:

Authors thank Prof. A Choudhury (Vice Chancellor, Gauhati University, Guwahati, Assam, India) and Dr. D K Avasthi (Scientist G, IUAC, Delhi, India) for their valuable suggestions and assistances during this work.

References:

1. Mohanta D et al. *Journal of Applied Physics* 2002.92(12).7149p.
2. Biswas S et al. *Synthesis and Reactivity in Inorganic, Metal Organic and Nano-Metal Chemistry* 2006. 36 (1). 33 p.
3. Mohanta D et al. *Bulletin of Mate*rial *Science* 2003.26(3). 289 p.
4. Mohanta D et al. *Indian Journal of Physics* 2001. 75 (A) 53 p.
5. Zhao Jialong, Bardecker A Julie *et al.*; *Nano Letter* 2006. 6(3). 463 p.
6. Barik S et al. *Solid State Communication* 2003127. 463 p.
7. Lee Woo Jun et al. *Japanise Journal of Applied Physics* 2005. 44(10). 7694p.
8. Bhattacharjee B et al. *Thin Solid Film* 2006. 514(1). 132 p.
9. Figliozzi P, Sun L et al.; *Physical Review Letter* 2005. 94. 047401p.
10. Nanda J J et al. *Journal of Appled Physics* 2001. 90(5). 2504p.
11. Tang J et al. *Material Research Society* 2004. 796.
12. Charles P P et al. *Quantum wells, Wires and Dots, Introduction to Nanotechnology* 2003. A John Wiley & Sons. Inc. Hoboken. New Jersey.
13. Nath S S, Chakdar D, Gope G, *Nanotrends-A journal of nanotechnology and its application,* ,Vol 02, Issue 03, 2007.
14. Nath S S, Chakdar D, Gope G, Talukdar A K, Avasthi D K, J. Of Dispersion Science Technology, 30 (2009) 4, (in press).
15. Nath S S, Chakdar D, Gope G,Avasthi D K, Internet J of Nanotechnology,Vol 2, No 1.

Numerical Analysis on Core-Shell Based GeSi Nanowire MOSFETs

Jin He[1,2], Yue Fu[2], Feng Liu[2], Lining Zhang[2], Jie Feng[2], and Chenyue Ma[2], Xing Zhang[1,2]

[1] School of Computer& Information Engineering, Peking University Shenzhen Graduate School, Shenzhen, 518055, P.R.China
[2] School of Electronic Engineering and Computer Science, Peking University, Beijing, 100871, P.R.China
Tel: 86-10-62765916 Fax: 86-10-62751789 Email: frankhe@pku.edu.cn

ABSTRACT

This paper investigates the transport properties of the core-shell based silicon/Germanium nanowire MOSFETs by a numerical method. Coupling Poisson's equation to Schrödinger's equation for electrostatics calculation, and electron structure to current transport equation for channel current computation, the electronic structure, quantized energy levels, relevant wave functions and charge distribution are solved self-consistently by a finite numerical method. Based on these findings, the transistor performances, including the capacitance characteristics and drain current, are further predicted.

Keywords: Non-classical MOSFETs; Device physics; with core-shell GeSi nanowire MOSFET.

1 INTRODUCTION

With the scaling of transistors to the 32nm regime, it has been anticipated that conventional scaling approach will soon hit the limit [1]. To continue the scaling trends for nano-electronic technologies, nanotubes and nanowire are considered as alternatives to conventional metal-oxide-semiconductor field-effect transistors (MOSFETs) [2] due to their unique one-dimensional electronic structure to reduce short-channel-effects (SCE) and scattering of conducting carriers. In particular, the symmetric concentric structure of nanowire created a cylindrical symmetry that is different from other structures being pursued in the industry like FinFET [3] and Trigate MOSFETs [4]. Many popular effects such as carrier confinement, strain and stress in these kinds of transistors would have a concentric dependence rather than the planar coordinate structure.

In particular, heterostructures with Ge/Si core/shell nano-wire [5] has been demonstrated to provide "hole gas" with high mobility at its Ge core that give rises to a high performance p-channel like FETs. The concentric nanowire structure has the potential to provide superior performance compared with other nano-transistor structures. However, the transport mechanism in the concentric nanowire transistors is not clear due to the lack of understanding on the detailed physics that govern their behavior. In addition, there is no design tool for simulation. This prevents the evaluation when these nanowire transistors are used in actual circuit. The study of high-performance nanowire FETs has also been hindered by difficulties in producing uniform nanowire in a top-down approach with controllability and reproducibility. In this work, we propose a detailed numerical study on the device physics that govern the transport mechanism of Silicon/Germanium nano-wire transistors with the core-shell structure and evaluate the transistor performance from the self-consistent numerical solutions between the electrostatics, carrier refinement, and the current transport.

2 NUMERICAL APPROACH

Conduction in the nanowire transistor is governed by 2-D quantum mechanical confinement, which determines that conventional modeling approach that based on Charge-Sheet Approximation cannot be used. Fortunately, the symmetrical properties of concentric nanowire structures simplify the boundary conditions so that the solution of Poisson equation and Schrödinger equation can be expressed in one-dimensional form. As a result of a previous research in our group, a quantum simulator based on the solution of Poisson equation and Schrödinger equation has already been developed [6]. This tool will be extended to model the effects of electron wave function penetrations through the multiple layers in the concentric nanowire structures. The numerical results will serve as a verification tool to the characteristics and performance of the nanowire MOSFETs with the core-shell structure.

In the theoretical formulation, we have two main equations: Schrödinger and Poisson equations, each will be later turned into discrete format for the finite element method. Basic transport equations including current density equation, continuity equation and Einstein relation are also involved. As is discussed above, the simplification resulting from symmetrical properties of this SiGe nanowire MOSFETs with core-shell structure leads to a one-dimensional continuous Schrödinger equation:

$$\left(-\frac{\hbar^2}{2}\frac{d^2}{dx^2}\frac{1}{m(x)}+qU(x)\right)\cdot\Psi_i(x)=E_i\cdot\Psi_i(x), i=1,2,3,......,N \tag{1}$$

where the effective mass $m(x)$ is taken as energy-independent and lposition-dependent.

In order to apply the finite element method to Eq.(1), this equation is turned into a discrete form using central difference method [7]:

$$\frac{-\frac{\hbar^2}{2}\cdot\frac{\Psi(x_{k-1})}{m^-(x_k)}}{\Delta x_{k-1}\cdot\overline{\Delta x_k}}+\left[\frac{\hbar^2}{2}\cdot\frac{\frac{1}{m^+(x_k)}}{\Delta x_k\cdot\overline{\Delta x_k}}+\frac{\hbar^2}{2}\cdot\frac{\frac{1}{m^-(x_k)}}{\Delta x_{k-1}\cdot\overline{\Delta x_k}}+q\cdot U(x_k)\right]\cdot\Psi(x_k)-\frac{\frac{\hbar^2}{2}\cdot\frac{\Psi(x_{k+1})}{m^+(x_k)}}{\Delta x_k\cdot\overline{\Delta x_k}}=E\Psi(x_k) \quad (2)$$

where:

$$m^-(x_k)=\frac{1}{2}[m(x_{k-1})+m(x_k)],$$

$$m^+(x_k)=\frac{1}{2}[m(x_k)+m(x_{k+1})],\quad \overline{\Delta x_k}=\frac{1}{2}[\Delta x_{k-1}+\Delta x_k]$$

Similarly, the initial form of one-dimensional continuous Poisson equation is expressed as:

$$\frac{d}{dx}[\varepsilon(x)\cdot\frac{d}{dx}U(x)]=-\rho(x) \quad (3)$$

where $\rho(x)$ is the charge density.

Change it into discrete form:

$$U(x_{k-1})\cdot\frac{\varepsilon_{k-0.5}}{\Delta x_{k-1}\cdot\overline{\Delta x_k}}-U(x_k)\cdot[\frac{\varepsilon_{k-0.5}}{\Delta x_{k-1}\cdot\overline{\Delta x_k}}+\frac{\varepsilon_{k+0.5}}{\Delta x_k\cdot\overline{\Delta x_k}}]+U(x_{k+1})\cdot\frac{\varepsilon_{k+0.5}}{\Delta x_k\cdot\overline{\Delta x_k}}=-\rho(x_k) \quad (4)$$

where: $\varepsilon_{k-0.5}=\frac{1}{2}[\varepsilon(x_k)+\varepsilon(x_{k-1})]$, $\overline{\Delta x_k}=\frac{1}{2}[\Delta x_{k-1}+\Delta x_k]$

Based on the Fermi-Dirac distribution, the quantized carrier distribution is calculated by:

$$n(x)=\int g(E)\cdot f(E)dE=\sum_{l,k} f_0(\varepsilon_l,k_z)\cdot|\Psi(x)|^2$$

$$=\sum_l g_l\cdot\left(\frac{\sqrt{2m(x)kT}}{2\pi\hbar}\cdot\int_0^{\infty}\frac{\varepsilon^{-\frac{1}{2}}}{1+\exp(\varepsilon-\xi)}d\varepsilon\right)\cdot|\Psi(x)|^2,\ \xi=\frac{E_F-\varepsilon_l}{kT} \quad (5)$$

From Eq.(5), we have

$$\therefore n(x_k)\approx\sum_l g_l\cdot\left(\frac{\sqrt{2m(x_k)kT}}{2\pi\hbar}\cdot\int_0^{15000}\frac{1}{1+\exp(\varepsilon^2-\xi)}d\varepsilon\right)\cdot|\Psi(x_k)|^2 \quad (6)$$

where $\xi=\frac{E_F-\varepsilon_l}{kT}$

Here the upper limit of 15000 is an approximation to substitute for $+\infty$ in the accurate expression of the integral. Since the function: $f(\varepsilon)=\frac{1}{1+\exp(\varepsilon^2-\xi)}$ approaches zero much quicker than the function: $f(\varepsilon)=\frac{1}{\varepsilon^2}$ does when ε approaches $+\infty$, this approximation is ideal, which has also been proved by computational experiments.

Based on the charge conservation concept, the total charge in the space charge layer is written as:

$$\rho(x_k)=q\left[p(x_k)-n(x_k)+N_D^+(x_k)-N_A^-(x_k)\right] \quad (7)$$

By solving Eq.1-4 numerically using a self-consistent method [8] that includes two solvers, one for Poisson solution for the set-up of initial guess of potential distribution, and the other for potential distribution with the knowledge of electron concentration as a function of position after it is derived from the Schrodinger solver, we can get $p(x_k)$, $n(x_k)$, $\psi(x_k)$, $U(x_k)$ and $E(x_k)$ so as to calculate the drain current.

The current can be calculated by applying current continuity equation. Supposing that the electron mobility is a constant, according to Pao-Sah equation [9], the discrete form of current density equation is:

The corresponding discrete form is:

$$I_{ds}=\frac{\mu_n W q_0}{L}\sum_k n(V_k)V_k(A) \quad (8)$$

where $n(V_k)$ is the bulk electron distribution derived from the linear electron carrier distribution calculated in Eq.(6) by dividing it with the unit area applied in the discrete coordinate system. Here μ_n is viewed as a constant in this presented model for simplification. Later we would depend on Eq.(8) to simulates the electrical characteristics of this presented model with core-shell structure.

3 RESULTS AND DISCUSSION

For the studied nanowire structure to be recognized for its anticipated electrical applications, a clear view of energy band structures, wavefunctions and energy levels is indispensable and is thus obtained first. For simplification we assume that there are no interface states at the boundaries of the materials. Shown in Fig.1-2 are the results of electron wave functions computed from a 1-D Si/Ge nanowire MOSFEET model with core-shell structure, with a silicon radius (R_{Si}) of 10nm, a germanium radius (R_{Ge}) of 5nm and a gate oxide thickness (T_{ox}) of 2nm, under different gate voltages. Actually the energy level degenerates into many subbands. For instance, in the case of a gate voltage of 0V, there are 16 and 5 subbands for heavy and light electrons respectively. However, since only the wavefunctions of the first few energy levels are comparatively significant, the rest are not shown in Fig.1-2. As to the case where V_{gs}=0V, only the first 6 of heavy electron wave functions and the first 3 of light electron wave functions plotted, as is shown in Fig.1a. From this series of wavefunctions versus gate voltage profiles, it is clear that the gate voltage (V_{gs}) has a powerful control over the electron distributions: while the effective V_{gs} approximates zero, there are several wavefunctions in the similar orders. When the effective V_{gs} increases, only one or two wavefunctions stand out and all the other wavefunctions are overwhelmed. This agrees with the device physical theory that electrons are attracted and trapped in the quantum well in the inversion layer under a high positive gate voltage, thus forming an electron-dominated channel. The higher the gate voltage, the higher are the electron concentrations in the inversion layer and the nearer the peaks to the semiconductor surface. Since the induced electrons distribute in several sub-energy levels to result in the current transport, such a split distribution on the number of energy bands, that is, a limitation on conduction channels, can benefit electrical performance through a reduction in scattering [10].

Fig.3 illustrates the comparison of electron concentration distributions with and without quantum mechanical effect (QME) and for different R_{Ge} with the same gate voltage and R_{Si}-R_{Ge}; different parameter Rsi-R_{Ge} with the same R_{Ge} and different R_{Ge} changing with the same R_{Si}. It is concluded from these figures that the classical device physics excluding QME predicts an electron distribution with sharp peak very close to the semiconductor surface. In contrast, the quantum based physics picture demonstrates that the electron concentration peaks always have a distance from the semiconductor surface. It is very interesting that the electron distribution shows two peaks either in the classical or the quantum physics picture, which will benefit the channel current transport. This is why the core-shell structure of SiGe nanowire has a series transport advantages over the traditional nano-wire structure. On the other hand, such distribution characteristics also explain why the charge-sheet approximation is no longer suitable for the calculation of drain current in a nanowire MOSFETs, as the quantum mechanical effect is no longer negligible. Moreover, it is also found that the peak of the surface electron concentration also increases with the increase of the silicon radius.

Last but not the least, it seems that the slight variance of R_{Ge} does not affect the electron distribution much, especially that in the quantum well of the channel. This indicates that SiGe nanowire MOS capacitance is not very sensible to the proportion of R_{Si}-R_{Ge} to R_{Ge}, as long as the whole semiconductor layer thickness remains. Also, it reveals the inefficacy of the charge-sheet approximation in nanowire device.

The conductance band profile is plotted in Fig.4 for the gate voltage of 2V. Compared with the case of the zero gate voltage, it is observed from Fig.4 that the quantum well becomes shallower slightly, and its depth decreases significantly from 3.15V to 1.39V. It is also observed that the height of the barrier at the interface between Si and Ge also lowers slightly from 0.5 V to 0.37V. This result coincides with the increase of electron concentration when a gate voltage of 2V is applied to the SiGe nanowire MOSFETs.

Fig.5 plots the capacitance characteristics of a core-shell nanowire MOSFET. It is observed that the core-shell SiGe nanowire MOSFET capacitance presented here has a near-static capacitance characteristic resembling that of a classical bulk MOSFET very much, in that both have a depletion region where the capacitance increase with the increasing V_{gs}, and a strong inversion region where the gate capacitance approaches saturation with the increase in V_{gs}. These characters suggest similar capacitive applications of the presented model to those of a bulk MOSFET in the integrated circuit.

Fig.6 plots the numerical I_{ds}-V_{ds} curves, compared with the classical model without quantum mechanical effects, while Fig.7 depicts the I_{ds}-V_{gs} curves calculated from Eq.(8) for different combinations of R_{Si} and R_{Ge}. These figures show that all the drain current curves, either from the quantum based transport or the classical device physics, share the clear separation of three operation regions: the subthreshold, the linear and the saturation region. However, it seems that the quantum based transport can carry out a larger channel current than the classical one. This accords with the observed high performance of SiGe nano heterostructures. [11-12] Last but not the least, I_{ds} rises with the rise of either (R_{Si}-R_{Ge}) or R_{Ge}. All these findings imply the methods of controlling the drain current by geometry parameters.

4 CONCLUSIONS

In summary, the characteristics and performance of the SiGe nanowire MOSFET with core-shell structure has been studied in this paper by solving the Poisson-Boltzmann equation and the Schrodinger Equation self-consistently, coupled to the channel carrier transport equation. The energy band and the wavefunctions are obtained first, and then analyzed in details, followed by the study of the drain current characteristics and gate capacitance. It is shown how the quantum mechanical effects and different geometry combinations of R_{Si} and R_{Ge} affect the micro electron energy level, carrier distribution and the macro channel current transport. All these provide a useful result for the device engineers how to design and optimize the performance of the SiGe nanowire MOSFET with core-shell structure for the beyond 10nm generation integrated circuit application.

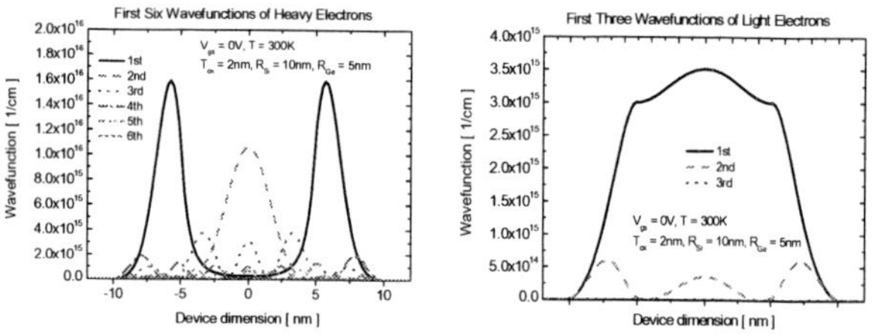

Fig.1 First few wavefunctions of heavy electrons and light electrons of SiGe Nanowire MOSFET while V_{gs}=0V.

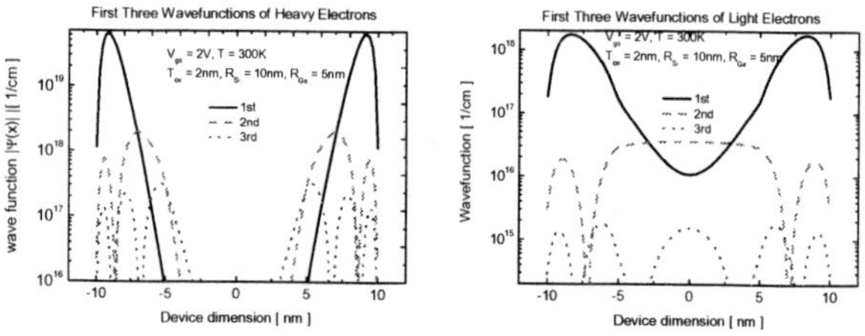

Fig.2 First few wave functions of heavy electrons and light electrons of SiGe Nanowire MOSFETs while V_{gs}=2V.

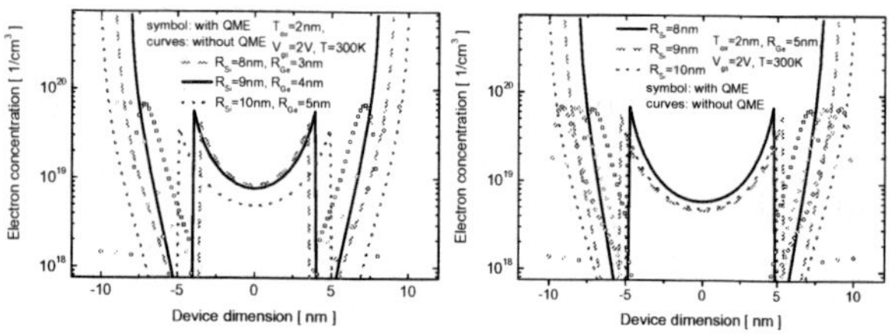

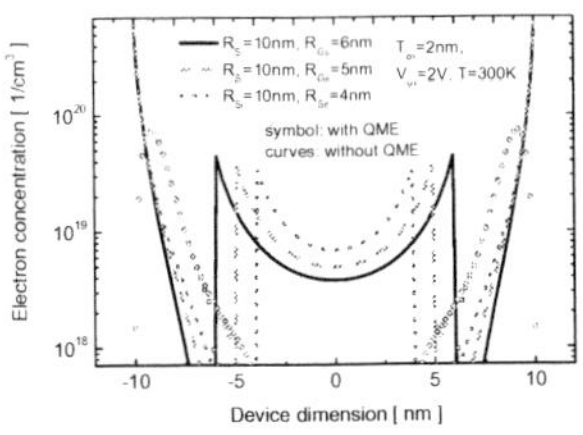

Fig.3 The comparison of electron concentration distribution along the vertical direction with and without QME for different combination of RSi and RGe.

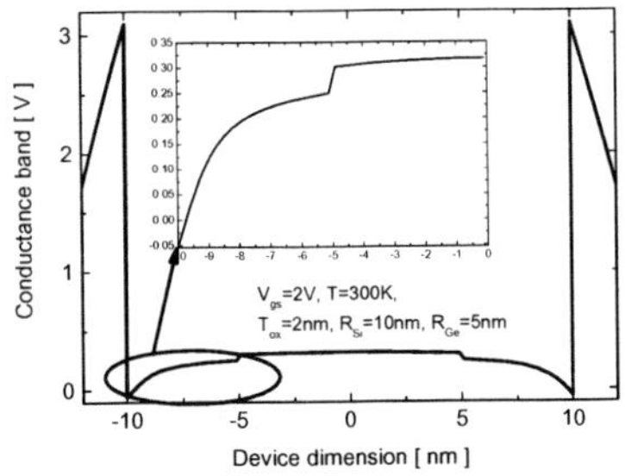

Fig.4 The conductance band profile along the vertical direction of SiGe nanowire MOSFETs with V_{gs}=2V.

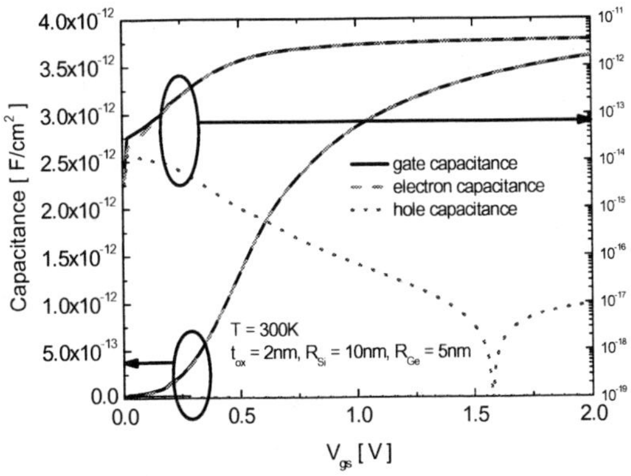

Fig.5 gate capacitance, electron capacitance and hole capacitance versus gate voltage.

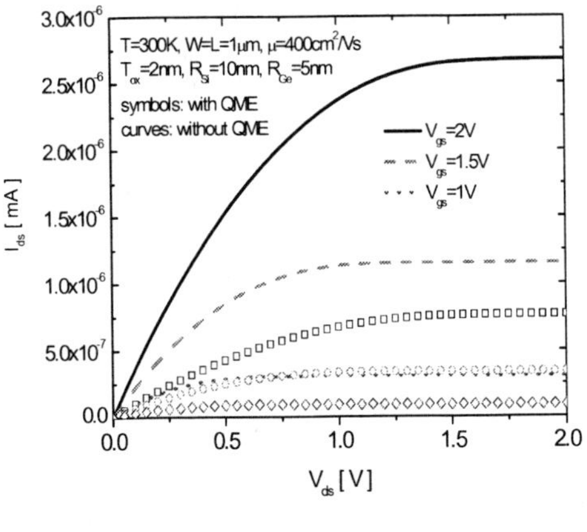

Fig.6 I_{ds}-V_{ds} curves with R_{Si}=10nm, R_{Ge}=5nm.

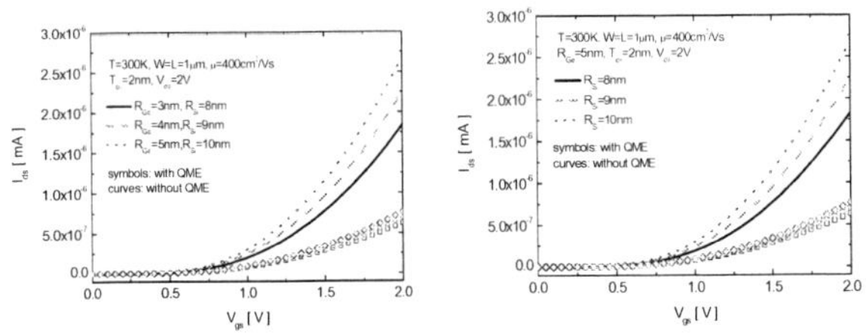

Fig.7 Ids-Vgs curves of different RGe and the same RSi-RGe, of different RSi-RGe and the same RGe.

ACKNOWLEDMENTS

This work is subsided by the special funds for major state basic research project (973) and National Natural Science Foundation of China (NNSFC: 90607017).

REFERENCES

[1] J.-P. Colinge, "Multiple gate SOI MOSFETs", Solid State Electron., vol. 48, no.6, pp. 897-905, Jun. 2004.

[2] Lieber, C.M. "Nanoscale science and technology: Building a big future from small things", MRS Bull. 28, pp. 486-491(2003).

[3] Digh Hisamoto, et al., "FinFET—A Self-Aligned Double-Gate MOSFET Scalable to 20nm", IEEE Trans. Electron Devices, Vol.47, No.12, December 2000.

[4] A. Breed, K.P. Roenker, "Dual-gate (FinFET) and Tri-Gate MOSFETs: Simulation and Design", Semiconductor Device Research Symposium, 2003 international, 10-12, pp.150-151, December, 2003.

[5] Jie Xiang, et al., "Ge/Si nanowire heterostructures as high-performance field-effect transistors", Nature Letters, Vol441, 10.1038/nature04796.

[6] Tsz Yin Man, "One Dimensional Quantum Mechanical Transport in Double-Gate MOSFET", a thesis submitted to the HKUST in Partial Fulfillment of the Requirements for the Degree of Ph.M. in Department of Electrical and Electronic Engineering, May 2003.

[7] Lingquan Wang, et al., "A numerical Schrodinger-Poisson solver for radially symmetric nanowire core-shell structures", Solid State Electronics, vol50, Issue 6, pp.1150-1155, June 2006.

[8] H. C. Pao and C. T. Sah, "Effects of diffusion current on characteristics of metal-oxide (insulator)-semiconductor transistors", Solid State Electron., vol. 9, no. 10, pp. 927-937, 1966.

[9] Jie Xiang, et al., "Ge/Si nanowire heterostructures as high performance field-effect transistors", Nature, vol.441, no.25, pp.489-493, May, 2006.

[10] Lu, W., et al., "One-dimensional hole gas in germanium/silicon nanowire heterostructures". Proc. Natl. Acad. Sci. USA 102, 10046–10051 (2005).

[11]Bryllert T, et al., "Vertical high mobility wrap-gated InAs nanowire transistor". In: DRC proceedings, 2005, p. 157–8.

Kinetic Monte Carlo Study on Transient Enhanced Diffusion *Posterior to* Amorphization Process

Soon-Yeol Park, Young-Kyu Kim, and Taeyoung Won

Department of Electrical Engineering, School of Information Technology Engineering,
Inha University, Incheon, Korea 402-751
twon@hsel.inha.ac.kr

ABSTRACT

We report our theoretical investigation on the suppression of boron diffusion in the silicon substrate *posterior to* PAI (pre-amorphization implant). We numerically investigated the defect-generating behavior of silicon atoms and the subsequent effect on the transient enhanced diffusion of boron as a new species for pre-amorphization implant (PAI). Our kinetic Monte Carlo (KMC) simulation revealed that Si-PAI produces more interstitials than the case of Ge-PAI whilst Ge-PAI makes interstitial move further up to the surface than the Si-PAI case during the annealing process, which results in the suppression of the boron transient enhanced diffusion (TED).

Keywords: transient enhanced diffusion, kinetic monte carlo, pre-amorphization implant.

1 INTRODUCTION

As devices scale down to deca-nanometer regime, the scaling scenario more stringently requires a shallow source/drain junction profile. The junction depth control is one of the key factors which alleviate the short channel effect for the deca-nanometer node technologies, especially for the high performance device. Impurity atoms which have been implanted into the silicon lattice experience various kinds of scattering events and finally stop their penetration when they lose their energy during the scattering process. It is well-known that dopant diffusion during the subsequent annealing process which is called transient enhanced diffusion (TED) deepens the junction depth[1~3]. To prevent these TED effects, pre-amorphization implant (PAI) has been considered as one of the efficient remedies in semiconductor industry. The germanium atom has been favorably employed for the PAI process because it reduces the channeling and TED diffusion of B atoms for the formation of a shallow junction. However, since the size of the germanium atom is larger than that of silicon atom, the lattice structure of the silicon substrate is considered to experience the deformation in a more severe and irregular manner. Furthermore, it has been well-known that the boron atoms experience some kind of impurity scattering. In this work, we theoretically investigated the pros and cons for the silicon PAI as an alternative to the traditional Ge-PAI process because silicon atom is the same species with the silicon substrate and the size of the silicon atom is smaller than that of the germanium, which might be expected to cause less lattice deformation.

2 COMPUTATIONAL DETAILS

2.1 Ion Implantation

In order to investigate the diffusion phenomena of boron after pre-amorphization process, we need an as-implant B profile for Si-PAI as well as Ge-PAI. In this work, we employed the BCA code for the initial as-implant dopant profile, which is based upon the Kinchin-Pease model. Kinchin-Pease model is a computationally efficient damage model based on the modified Kinchin-Pease formula proposed by Norgett et al[4]. In a simplified manner, this model accounts for damage generation, damage accumulation, defect encounters, and amorphization. The basic assumption of the Kinchin-Pease models is the nuclear energy loss which turned into point defects and the number of Frenkel pairs which is created proportionally to the nuclear energy loss. The nuclear energy loss is deposited locally and induces local defects.

2.2 Thermal Annealing Simulation

After implant process, we implement annealing process by using our KMC code[5,6]. In the KMC method, a physical system which consists of many possible events evolves as a series of independent event occurring. Each event has its own event rate. Event rate is calculated from the equation (1). Here, E_b presents the migration energy for the barrier against the jump event of the mobile species or a binding energy for clusters. In addition, ν_0 is the attempt frequency which is simply the vibration frequency of the atoms. Typically, the attempt frequency is the order of 1/100 fs. These parameters can be obtained from ab-initio calculation or experimental data.

$$\nu = \nu_0 \exp\left(\frac{-E_b}{K_B T}\right) \tag{1}$$

Our problem is the consideration about the thermally activated events in a thermal annealing simulation after ion implantation. If the probability for the next event to occur is independent of the previous history, and the same at all times, the transition probability will be a constant which is called Poisson process. To derive the time dependence, we can consider a single event with a uniform transition probability r. Let f be the transition probability density which gives the probability rate at which the transition occurs at time t. The change of $f(t)$ over some short time interval dt is proportional to r, dt and f because f gives the probability density that the physical system still remains at time t:

$$df(t) = -rf(t)dt \qquad (2)$$

The solution of equation (2) can be easily obtained as the following wherein r becomes the initial value of $f(t)$.

$$f(t) = re^{-rt}, f(0) = r \qquad (3)$$

Therefore, the simulation time is updated for($t=t+\ \ t$) according to event rates as follows, because an ensemble of independent Poisson processes will behave as one large Poisson process:

$$\Delta t = -\frac{\ln u}{R}, R = \sum_{i=1}^{N} R_i \qquad (4)$$

Here, u is a random number and R is the total sum of all possible event rates (R_i). We select an event according to the event rates.

3 RESULTS AND DISCUSSION

Fig. 1 is a schematic diagram which illustrates the simulated B as-implant profile wherein the dotted line represents the as-implant profile without PAI, the scattered squares designate the as-implant profile with Ge-PAI, and the scattered circles represent the as-implant profile with Si-PAI. It should be further noted that the filled circles and squares represent the cases for PAI with implant energy of 20 keV while the empty circles and squares represent the cases for PAI with implant energy of 40 keV. Referring to the as-implant PAI-free B profile (dotted line) and other scattered curves (circulars and squares), we can see that the pre-amorphization process either with Ge or with Si helps to realize the shallow junction. If we look into the as-implant profiles for each type of PAI species with different implantation energies, we can see that the PAI process with 40 keV is more favorable than the case with 20 keV in terms of the depth of the as-implanted B profile. In other words, higher energy PAI seems to retard the boron channeling more effectively regardless of the species of PAI atoms. If we make comments on the species of PAI implant, the as-implant profile with Si-PAI is shallower than that with Ge-PAI.

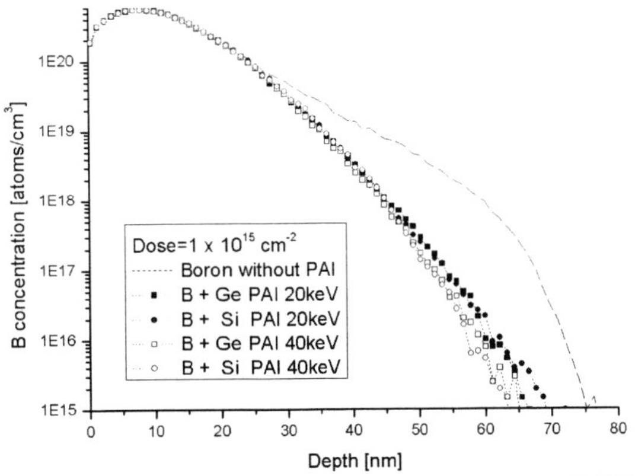

Fig. 1 Boron as-implant profiles. Boron is implanted with energy 2 keV, the dosage of 1 x 10^{15}/cm^2 after Si/Ge PAI is implemented with energy 20, 40keV and the dosage of 1 x 10^{15}/cm^2. All implantation were performed with 7 tilt angle. The dotted line represents the as-implant profile without PAI, the scattered squares designate the as-implant profile with Ge PAI, and the scattered circles represent the as-implant profile with Si-PAI. It should be further noted that the filled circles and squares represent the cases for PAI with implant energy of 20 keV while the empty circles and squares represent the cases for PAI with implant energy of 40 keV.

Fig. 2 and 3 are schematic diagrams which illustrate the simulated B profile for different annealing conditions wherein Fig. 2 corresponds to the thermal annealing for 60 seconds @ 850 ºC while Fig. 3 corresponds to the RTA for one second @ 600 ºC, respectively, for Si-PAI as well as Ge-PAI. PAI implantation was performed with a dosage of 1 x 10^{15} /cm^2 and with energy of 20 keV and 40keV, respectively. Boron implantation is performed with a dosage of 1 x 10^{15} /cm^2 and with energy of 2 keV. The solid line represents the B diffusion profile without PAI while the line with squares correspond to the cases with Ge PAI and the lines with circles represent the cases with Si PAI.

Referring to Fig. 2, we can recognize that there seems to be no significant difference in the diffusion profiles after annealing for 60 seconds. The reasons seem to be due to the fact that the boron implantation energy is relatively low when compared to the annealing time and temperature. This seems partly due to the fact that a low energy implantation induces the statistical inaccuracy and further a long duration annealing at high temperature supply enough energy for boron diffusion. In order to confirm our reasoning, we performed KMC simulation under the different annealing condition, i.e., under the RTA condition.

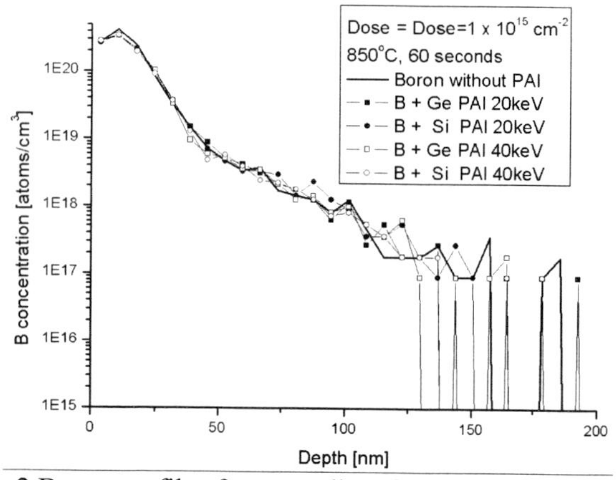

Fig. 2 Boron profile after annealing @850ºC for 60 seconds. The solid line represents the B diffusion profile without PAI while the line with squares correspond to the cases with Ge PAI and the lines with circles represent the cases with Si PAI.

Fig. 3 is a schematic diagram illustrating the diffusion profiles after annealing @600ºC for one second wherein the notations are the same as the ones used in Fig. 2. Referring to Fig.4, we can see that PAI significantly reduces the TED and PAI effect is more pronounced than the case of long time annealing. Moreover, we can see that Si-PAI with energy of 20keV is better than the other PAI cases with respect to the suppression of diffusion profiles.

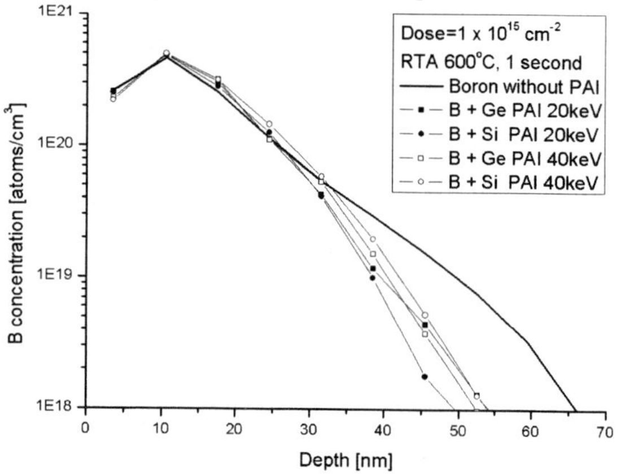

Fig. 3 Boron profile after annealing after annealing @600ºC for one second. The solid line represents the B diffusion profile without PAI while the line with squares correspond to the cases with Ge PAI and the lines with circles represent the cases with Si PAI.

In order to understand the physics behind the suppression of diffusion profiles, we investigated the defect (interstitial and vacancy) distribution with our KMC tool, which is depicted in Fig. 4 and 5. Fig. 4 is a schematic diagram which illustrates the simulated interstitial (I) distribution for Ge-PAI and Si-PAI cases under our KMC simulation. Here, the triangles represent the case with Ge PAI wherein the filled triangle represents the case with 20 keV while the empty triangle represents the case with 40 keV. Furthermore, the diamonds represent the case with Si PAI wherein the filled diamond represents the case with 20 keV while the empty diamond represents the case with 40 keV. Referring to Fig. 4, we can recognize that the interstitial distribution for the Si-PAI produces more amounts of interstitial than the Ge-PAI case near at the surface, which seems to suppress the boron diffusion more effectively than Ge-PAI[7].

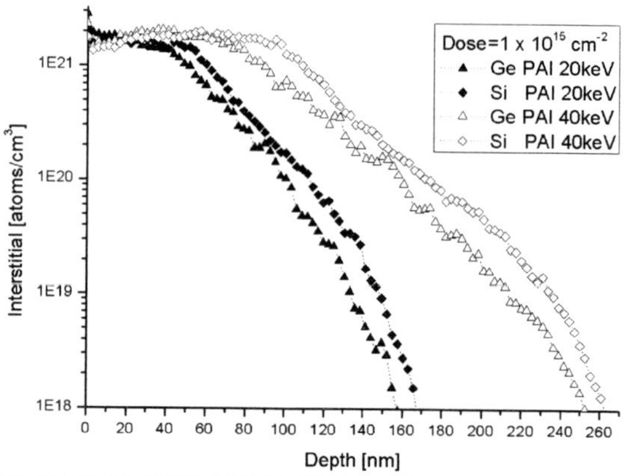

Fig. 4 A plot illustrating the simulated interstitial distributions. The triangles represent the case with Ge PAI wherein the filled triangle represents the case with 20 keV while the empty triangle represents the case with 40 keV. The diamonds represent the case with Si PAI wherein the filled diamond represents the case with 20 keV while the empty diamond represents the case with 40 keV.

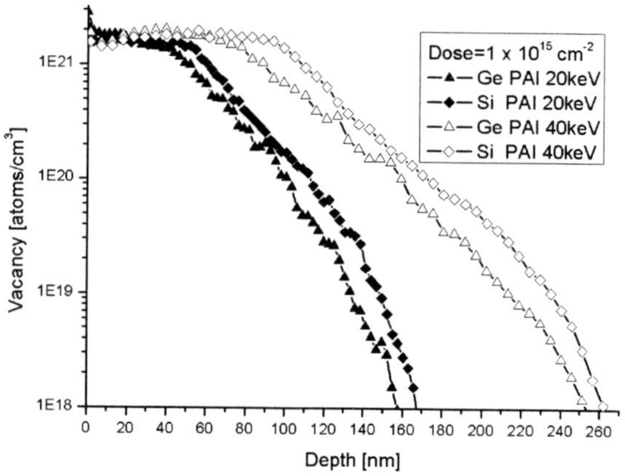

Fig. 5 A plot illustrating the simulated Vacancy distributions. The triangles represent the case with Ge PAI wherein the filled triangle represents the case with 20 keV while the empty triangle represents the case with 40 keV. The diamonds represent the case with Si PAI wherein the filled diamond represents the case with 20 keV while the empty diamond represents the case with 40 keV.

Fig. 5 is a diagram which illustrates the simulated vacancy distribution for Ge-PAI and Si-PAI cases according to our KMC simulation. Referring to Fig. 5, we can see that the vacancy profile is very similar to the interstitial profile.
Further, we simulated the change in the interstitial distribution using Si atom as well as Ge as PAI source with a dosage of 1 x 10^{15}/cm^2, and with energy of 40 keV while the boron implantation energy was 2 keV with a dosage of

$1 \times 10^{15}/cm^2$ during the annealing process as shown in Fig. 6.

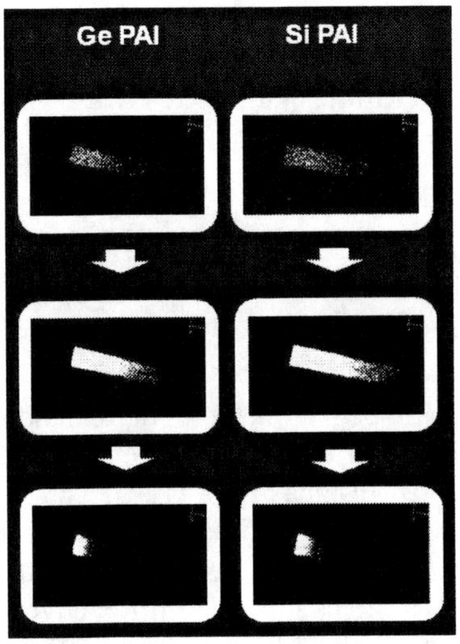

Fig. 6 Interstitial distribution changes in terms of annealing time. Finally, interstitials move toward near the surface in both PAI and both implantation energy.

Fig. 6 is a schematic diagram which illustrates the 3-dimensional interstitial distribution. In this figure, we can see that the interstitials produced by Si -PAI are formed more deeply than those of Ge PAI at the initial stage of the entire procedure while the interstitials move toward the inner depth. Finally, the interstitials are positioned near surface for both samples. The KMC simulation in Fig. 6 reveals that interstitials induced by Ge-PAI are located more near at the surface than the cases of Si-PAI. From this difference in the distribution, we can see that both PAI processes reduce the boron diffusion as the interstitials move up near to the surface.

In order to verify our KMC simulation, we compared our diffusion profiles with experimental data, which is taken from the prior published literature [8]. Referring to Fig. 7, we can see quite a good agreement between the simulated boron profile after PAI as well as the as-implant profile and the SIMS data. The numerical verification for Ge-PAI has already been conducted and reported in our in previous publication [6].

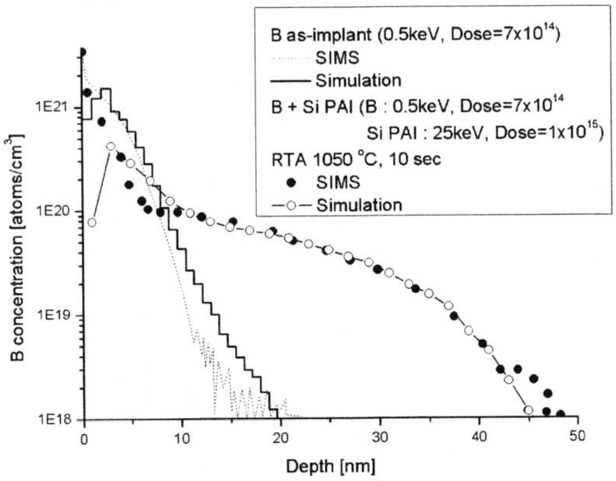

Fig. 7 KMC profile with SIMS data for as-implant B as well as the boron with Si-PAI.

4 CONCLUSION

In this paper, we investigated silicon atom as a new pre-amorphization implant (PAI) sources in addition to Ge atoms. Our KMC simulation revealed that the Si-PAI process produces more amount of interstitial and vacancy, which reduces the boron transient enhanced diffusion (TED). We compared the effects of Si-PAI with those of Ge-PAI under the same annealing condition. From the KMC investigation of the interstitial distribution, we found that Si-PAI produces more interstitials than the case of Ge-PAI while Ge-PAI makes interstitial move further to the surface than the Si-PAI case during the annealing process.

ACKNOWLEDGMENT

This research was supported by the Ministry of Knowledge and Economy, Korea, under the ITRC (Information Technology Research Center) support program supervised by the IITA (Institute of Information Technology Advancement) (IITA-2008-C109008010030).

REFERENCES

[1] P.M. Fahey, P. B. Griffin, and J. D. Plummer, Rev Mod. Phys. 61, 289 (1989).
[2] P. A. Stolk, H-J. Gossmann, D. J. Eanglesham, D. C. Jacobson, C. S. Rafferty, G. H. Gilmer, M. jaraiz, J. M. poate, H. S. Luftman, and T. E. Haynes, J. Appl. Phys. 81, 6031 (1977).
[3] A. Agarwal, H-J. Gossmann, D. J. Eaglesham, S. B. herner, A. T, Fiory, and T. E. Haynes, J. Appl. Phys. 81, 6031 (1997).4. P.M. Fahey, P.B. Griffin, and J. D. Plummer, Rev Mod. Phys. 61, 289 (1998).
[4] M. J. Norgett. M. T. Robinson, and I. M. Torrens, Nucl. Eng. Des., 33, 50, (1975).
[5] Jae-Hyun Yoo, Chi-Ok Hwang, Byeong-Jun Kim, and Taeyoung Won, Simulation of Semiconductor Processes and Devices 2005, pp.71-74, (2005).
[6] Joong-sik Kim, Taeyoung Won, Microelectronic Engineering, 84, 1556, (2007).
[7] A. Ural, P. B. Griffin, and J. D. Plummer, J. Appl. Phys. 85, 6440 (1999).
[8] B. J. Pawlak, T. Janssens, B. Brijs, and W. Vandervorst, E. J. H. Collart, S. B. Felch and N.E.B. Cowern, Appl. Phys. Letters 89, 062110 (2006).

Design and implementation of silicon-based optical nanostructures for integrated photonic circuit applications using Deep Reactive Ion Etching (DRIE) technique

S. Selvarasah*[1], R. Banyal**, B. D. F. Casse**, W. T. Lu**, S. Sridhar** and M. R. Dokmeci*

*Department of Electrical and Computer Engineering,
Northeastern University, Boston, Massachusetts 02115, USA, sselvara@ece.neu.edu
**Department of Physics and Electronic Materials Research Institute,
Northeastern University, Boston, Massachusetts 02115, USA

ABSTRACT

In this paper, we present the fabrication of nano optical elements by means of deep reactive ion etching technique (Bosch process) on a silicon-on-insulator substrate. The nano structures are fabricated in a two step process. The first step consists of direct-writing nanoscale patterns on PMMA polymer by electron beam lithography. These nano patterns are then transferred to the silicon surface by a low temperature and low pressure deep reactive ion etching (DRIE) process using PMMA as a mask. The low temperature and low pressure conditions in the DRIE process minimize scalloping in the nanoscale features. We found that the etch rate is highly dependent on the aspect ratio of the structure. We have used the DRIE method to fabricate a negative-index photonic crystal flat lens and demonstrated the focusing properties of this flat lens using a near-field scanning optical microscope.

Keywords: Photonic Crystals, Deep reactive ion etching, negative refraction, Near Field Scanning Optical Microscope.

1 INTRODUCTION

Photonic crystals (PhCs), a periodic structure of holes made in a dielectric medium with the hole size and lattice spacing comparable to the wavelength scale, are artificially engineered structures capable of manipulating and shaping optical signals in optoelectronic devices [1-4]. Nanoscale photonic crystals is expected to add new functionalities to the next generation of optical integrated circuits by providing novel ways of losslessly routing and modulating light, thereby enhancing the performance of micro-nano-optoelectronic devices. Among semiconductor materials, silicon (Si) has been adopted as the material of choice for PhCs applications [5-9] mainly because of its compatibility with CMOS electronics and its availability. The high-index contrast of core silicon to the cladding silicon dioxide (SiO_2) layer on a silicon-on-insulator (SOI) substrate, allows the confinement of light. The semiconductor industry has developed a plethora of techniques to etch silicon which includes reactive ion etching (RIE) [10,11], inductively coupled plasma (ICP) [12], electron-cyclotron-resonance (ECR) [13], and time multiplexed inductively coupled fluorine plasma, and chemically assisted ion-beam etching (CAIBE) systems [14]. Deep Reactive Ion Etching has been utilized by the MEMS community to etch high aspect ratio structures at the microscale on silicon substrates [15]. Even though DRIE etching has been utilized in realizing various MEMS devices [16], the fabrication of silicon nanophotonic structures and their optical applications has been relatively unexplored. Here we present our recent results on the nanofabrication of negative-index photonic crystal lens on a SOI substrate by DRIE technique, and then demonstrate the negative refractive index fingerprint of a photonic crystal flat lens (i.e. a focusing effect) by near-field scanning optical microscopy (NSOM).

2 DESIGN OF THE FLAT LENS

As illustrated in Figure 1, the flat lens consists of a 5μm wide 500 μm long waveguide, triangular lattice PhC1 (lattice constant a=560nm and hole diameter 2r=480nm) square lattice PhC2 (a=420nm, 2r=0.85a), and a 22 μm wide and 61 μm long air cavity. The widths of the PhC1 and PhC2 are 8 μm and 6 μm, and their lengths are 47 μm, respectively.

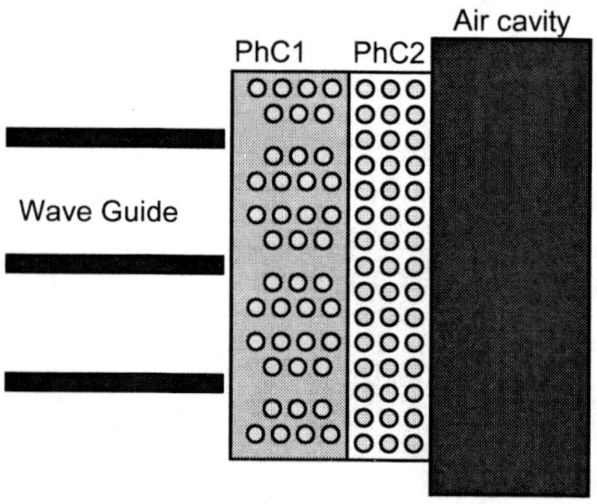

Figure 1: A drawing of the photonic crystal flat lens (PhC2) and adjacent point-source generating waveguides (PhC1). (The dark shaded areas are etched).

---

[1] ECE Department, Northeastern University, 409 Dana Bldg., 360 Huntington Ave., Boston, MA, 02115, USA, Ph: 1-617-373-3518, sselvara@ece.neu.edu

3 FABRICATION OF THE NANOPHOTONIC LENS

The SOI substrates are fabricated using amorphous silicon grown on SiO_2. First, a silicon wafer is cleaned by a standard RCA clean process. Then, a 500nm SiO_2 cladding layer is thermally oxidized on the silicon substrate. An amorphous 600nm silicon guiding layer is then deposited using a plasma enhanced chemical vapor deposition system. A schematic of the high-index contrast waveguide system ($Si/SiO_2/Si$) is illustrated in Figure 2a. A 600nm layer of polymethyl methacrylate (PMMA) layer (A7 950 K) is next spun on the SOI wafer.

The nanostructures are realized with a two-step process: Electron beam lithography (EBL) is used as a primary pattern generator to define patterns on the PMMA resist. The pattern is then transferred to the silicon surface by DRIE with PMMA layer as the mask. The process flow for realizing Si nanophotonic devices is illustrated in Figure 2. The DRIE process for etching nanostructures utilized etch and passivation cycles (SF_6 gas to etch and C_4F_8 to passivate, Bosch process, Surface Technology Systems ASE HRM) with the final device as shown in (Figure 2d).

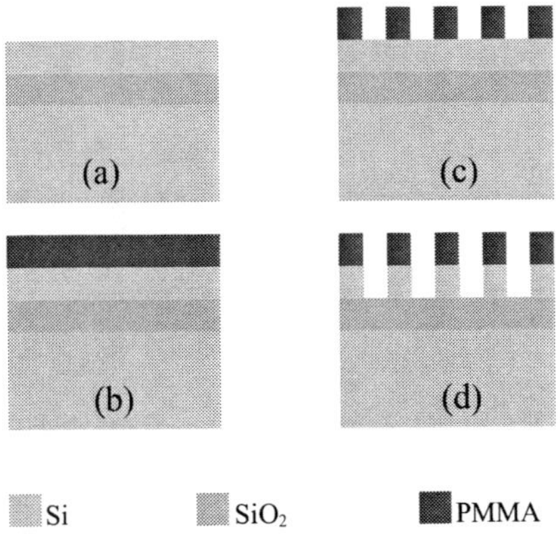

Figure 2: Process flow for the fabrication of a photonic crystal flat lens on SOI.

We have taken into account multiple aspects of nanostructure fabrication (e.g. scalloping, feature dependent etch rates, etc…) and optimized etch parameters to create functional nanophotonic structures. Our study indicates that the etch rate was highly dependent on the aspect ratio of the structures. We also found out that the low temperature and low pressure etching approach significantly reduced the width of scalloping routinely formed during the DRIE process due to the alternating cycles of etch and passivation. Figures 3 and 4 show the scanning electron microscope (SEM) images of arrays of circular patterns and the flat lens.

4 RESULTS

The focusing property of the flat-lens was experimentally investigated with Near Field Scanning Optical Microscope (NSOM). A sketch of the experimental setup is shown in Figure 5.

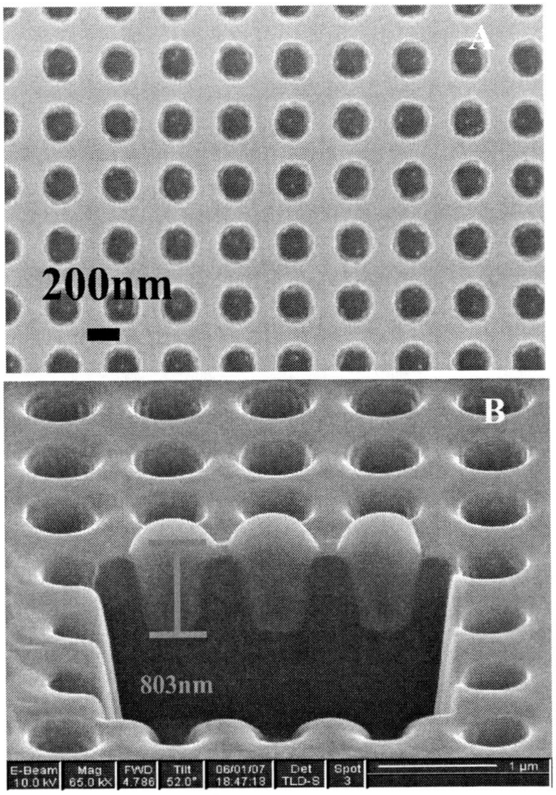

Figure 3: Scanning Electron Microscope (SEM) images of (a) 200 nm circular patterns, (b) cross section of 803nm deep and 500nm wide holes.

The input light from a semiconductor laser (central wavelength of 1550 nm) source carried by a single mode lensed fiber was butt-coupled into the cleaved end of a 5 μm wide planar waveguide fabricated in the backplane of the PhC lens. The single mode lensed fiber mounted on a six-axis micro-positioning stage is capable of focusing light in free space (FWHM < 3.0 μm in air). The other end of the planar waveguide was interconnected to the triangular lattice PhCs1 (as seen in Figure 1) slab. The PhC1 was designed to exhibit photonic band gap at 1550nm with the goal to realize a point source like object. This was accomplished by introducing a series of defects in the form of missing air holes in PhC1 during the fabrication process [17]. Light propagating through the defects in PhC1 creates a point like source in the vicinity of PhC2 (as seen in Figure 1). A high sensitivity infrared camera (Hamamatsu Model C2741) connected to a microscope port aids the initial alignment for optimizing IR light coupling from the optical fiber to the waveguide.

Based on our calculations and of similar devices that are published [18-23], the flat-lens PhC2 is expected to form a point-like image in the backplane. The scattered light from the point source formed near the PhC2-air interface was captured by an IR camera from top. The device and IR measurements are shown in Figure 6a. The focal region was further probed by NSOM technique to obtain a high resolution image.

Figure 4: A) SEM image of a flat lens after the fabrication, B) Magnified image of A.

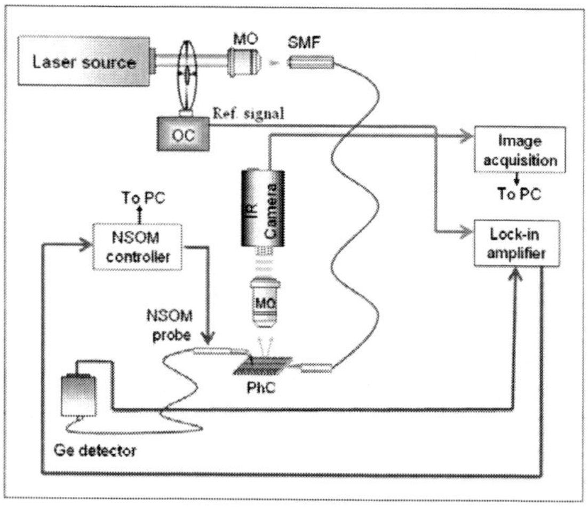

Figure 5: Experimental arrangement for NSOM detection and IR imaging of the PhC lens. MO: Microscope objective; SMF: Single Mode Fiber; OC: Optical Chopper.

The NSOM used a 250 nm aperture diameter, metalized, and tapered fiber probe raster scanned in the air cavity etched next to the PhC2. The output end of the fiber probe was connected to nitrogen cooled Ge detector (North Coast Scientific Corp. Model # EO-817L). Additionally, a typical lock-in amplifier was utilized to optimize the detection scheme. The reconstructed NSOM image is shown in Figure 6b. The focused image appears approximately 1.5 μm away from the trailing edge of the flat lens PhC2. The prominent light focusing features are evident in the overlaid contour plots of the image. The measured full width at half maxima (FWHM) of the focused spot is $\approx 1.8\lambda$. The measured results demonstrate that focusing by a flat lens is possible with the measurements matching our computations [24].

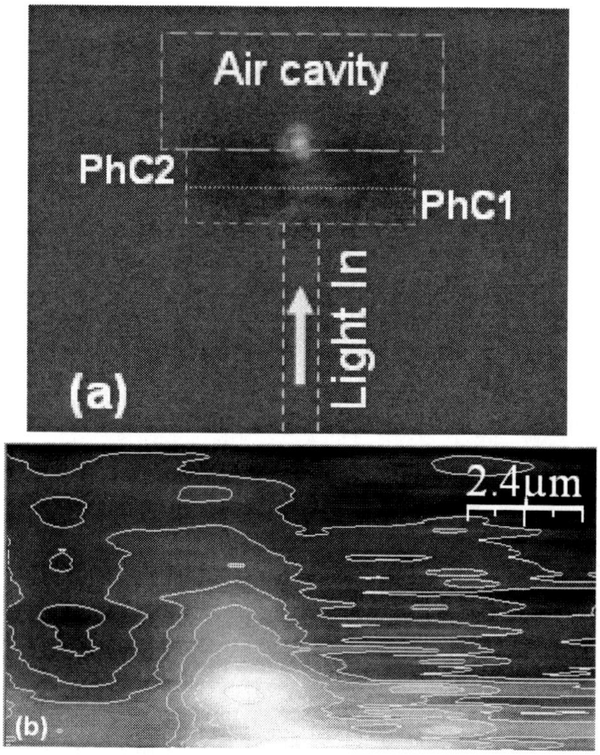

Figure 6: Experimental demonstration of point-source imaging with SOI flat-lens (a) Infrared camera image taken from the top and (b) the corresponding high resolution image of the focal region obtained with a NSOM scan.

5 CONCLUSION

Utilizing a DRIE process, we fabricated a Photonic crystal flat lens on a SOI substrate. Utilizing the flat lens, focusing of light with a wavelength of 1550 nm is demonstrated with the measured results matching our calculations. Demonstrating focusing on silicon enables new opportunities for CMOS based nanophotonic devices.

ACKNOWLEDGEMENT

NSTI-Nanotech 2008, www.nsti.org, ISBN 978-1-4200-8505-1 Vol. 3

The authors would like to thank the Air Force Research Laboratory, Hanscom, MA, for their support, contract # FA8718-06-C-0045.

REFERENCES

[1] S. John, "Strong Localization of Photons in Certain Disordered Dielectric Superlattices", Phys. Rev. Lett., vol. 58, no. 23, pp. 2486-2489, 1987.

[2] E. Yablonovitch, "Inhibited Spontaneous Emission in Solid-State Physics and Electronics", Phys. Rev. Lett., vol. 58, no. 20, pp. 2059-2062, 1987.

[3] J. S. Foresi, P. R. Villenevue, J. Ferrera, E. R. Thoen, G. Steinmeyer, S. Fan, J. D. Joannopoulos, L. C. Kimerling, H. I. Smith, and E. P. Ippen, ''Photonic-band gap mirocavities in optical waveguides,'' Nature (London), vol. 390, pp. 143–145, 1997.

[4] M. Notomi, "Theory of light propagation in strongly modulated photonic crystals: Refraction like behavior in the vicinity of the photonic band gap", Phys. Rev. B, vol. 62, pp. 10696-10705, 2000.

[5] A. Shinya, M. Notomi, I. Yokohama, C. Takahashi, J.-I. Takahashi, and T. Tamamura, "Two-dimensional Si photonic crystals on oxide using SOI substrate", Optical and Quantum Electronics, vol. 34, pp. 113-121, 2002.

[6] W. Bogaerts, D. Taillaert, B. Luyssaert, P. Dumon, J. Van Campenhout, P. Bienstman, D. Van Thourhout, R. Baets, V. Wiaux, and S. Beckx, " Basic structures for photonic integrated circuits in Silicon-on-insulator", Opt. Exp.,vol. 12, issue 8, pp. 1583-1591, 2004.

[7] Y. Wang, Z. Lin, C. Zhang, F. Gao, and F. Zhang, "Integrated SOI Rib Waveguide Using Inductively Coupled Plasma Reactive Ion Etching", IEEE Journal of selected Topics in Quantum Electronics, vol. 11, no. 1, 2005.

[8] R. D. L. Rue, H. Chong, M. Gnan, N. Johnson, I. Ntakis, P. Pottier, M. Sorel, A. M. Zain, H. Zhang, E. Camargo, C. Jin, M. Armenise, and C. Ciminelli, "Photonic crystal and photonic wire nano-photonics based on silicon-on-insulator" New J. Phys., vol. 8 (256). pp. 1-15, 2006.

[9] S. Venkataraman, J. Murakowski, T. N. Adam, J. Kolodzey, and D. W. Prather, "Fabrication of high-fill-factor photonic crystal devices on silicon-on-insulator substrates", J. Microlith., Microfab., Microsyst., vol. 2 no. 4, October, 2003.

[10] J. Arentoft, T. Serndergaard, M. Kristensen, A. Boltasseva, M. Thorhauge, and L. Frandsen, " Low-loss silicon-on-insulator photonic crystal waveguides", Electronics Letters 14th,vol. 38 no. 6,2002.

[11] L. O'Faolain, X. Yuan, D. McIntyre, S. Thoms, H. Chong, R.M. De La Rue, and T.F. Krauss, "Low-loss propagation in photonic crystal waveguides", Electronics Letters 7th, vol. 42, no. 25, 2006.

[12] E. Bennici, S. Ferrero, F. Giorgis, C.F. Pirri, R. Rizzoli, P. Schina, L. Businaro, and E. Di Fabrizio, "a-Si:H based two-dimensional photonic crystals", Physica E 16, pp. 539 – 543, 2003.

[13] M. Tokushima, H. Kosaka, A. Tomita, and H. Yamada, "Lightwave propagation through a 120° sharply bent single line-defect photonic crystal waveguide", Appl. Phys. Lett., vol. 76, no. 8, pp. 952–954, 2000.

[14] M. Lončar, B. G. Lee, L. Diehl, M. Belkin, F. Capasso, "Design and fabrication of photonic crystal quantum cascade lasers for optofluidics", Opt. Exp., vol. 15, no. 8, pp. 4499-4514, 2007.

[15] K.-S. Chen, A. A. Ayón, X. Zhang, and S. M. Spearing, "Effect of Process Parameters on the Surface Morphology and Mechanical Performance of Silicon Structures After Deep Reactive Ion Etching (DRIE)", Journal of Microelectromechanical Systems, vol. 11, no. 3, pp. 264-275, 2002.

[16] C.-H. Choi and C.-J. Kim, The 13th lntemational Confercnce on Solid-State Sensors, Actuators and Mictosystems, pp. 168-171, 2005.

[17] Z. Lu, B. Miao, T.R. Hudson, C. Lin, J. A. Murakowski and D. W. Prather, "Negative refraction imaging in a hybrid photonic-crystal device at near-infrared frequencies", Opt. Exp., vol. 15, no.3, pp. 1286-1291, 2007.

[18] T. Matsumoto, K. S. Eom, and T. Baba, "Focusing of light by negative refraction in a photonic crystal slab superlens on silicon-on-insulator substrate". Optics Letters vol. 31, no. 18, pp. 2786-2788, 2006.

[19] A. Berrier, M. Mulot, M. Swillo, M. Qiu, L. Thylen, A. Talneau, and S. Anand,, "Negative refraction at infrared wavelengths in a two-dimensional photonic crystal", Phys. Rev. Lett., vol. 93, no. 7, pp. 073902, 2004.

[20] P. V. Parimi, W. T. Lu, P. Vodo, and S. Sridhar, "Photonic crystals - Imaging by flat lens using negative refraction", Nature 426, pp. 404-404,2003.

[21] P.V. Parimi, W.T. Lu, P. Vodo, J. Sokoloff, J. S. Derov, and S. Sridhar, "Negative refraction and left-handed electromagnetism in microwave photonic crystals", Phys. Rev. Lett., vol. 92, no. 12, pp. 127401, 2004.

[22] E. Schonbrun, M. Tinker, W. Park, and & J. B. Lee, "Negative refraction in a Si-polymer photonic crystal membrane", IEEE Photonics Technology Letters 17, no.6, pp.1196-1198, 2005.

[23] E. Schonbrun, T. Yamashita, W. Park, and C. J. Summers, "Negative-index imaging by an index-matched photonic crystal slab", Physical Review B 73, pp. 195117, 2006.

[24] W. T. Lu and S. Sridhar, "Flat lens without optical axis: Theory of imaging", Opt. Exp., vol. 13, no. 26, pp. 10673-10680, 2005.

Experiments and Simulations of Infrared Transmission by Transverse Magnetic Mode through Au Gratings on Silicon with Various Air-slot Widths over the Period

Yan-Ru Chen*, and C. H. Kuan

Graduate Institute of Electronics Engineering and Department of electric engineering, National Taiwan University, Taipei, Taiwan, Republic of China

*e-meal:d93943029@ntu.edu.tw

ABSTRACT

Au gratings with various air-slot widths were fabricated by electron beam lithography system (e-beam) and their infrared (2.5~25μm) transmission of the transverse magnetic mode (TM mode) were investigated by the Fourier Transform Infrared Spectroscopy (FTIR). Simulations were used theoretically by the Maxwell equations and the surface impedance boundary condition (SIBC) method. Simulation results explain the relationship between surface plasma and transmission dips.

Keywords: gratings, e-beam, infrared

1 INTRODUCTION

Transmission through one-dimensional metallic gratings has been researched for decades because of their optical characteristics and potential applications in various fields, including beam splitting polarizers and photo-detectors. Advanced nano-technologies that can realize sub-wavelength metal/dielectric structures and the development of calculation methods to analyze their electromagnetic properties have all contributed to the understanding and application of metallic gratings. It is the purpose of this paper to demonstrate experimentally and simulate theoretically the infrared transmission by the transverse magnetic mode (TM mode) through Au gratings with various air-slot widths. Au gratings with various air-slot widths are fabricated on silicon substrate by electron beam lithography system (e-beam), and infrared transmittance (2.5~25μm) by transverse electric mode (TM-mode) are investigated by Fourier Transform Infrared Spectroscopy (FTIR). Moreover, the Maxwell equations and surface impedance boundary condition (SIBC) were used to simulate the transmission through Au gratings with various air-slot widths. Simulation results agree with experiments and support the *surface plasma theory*.

2 EXPERIMENT

At first, the e-beam, which was provided by ELIONIX, was used to pattern various air-slot widths of Au gratings on silicon substrate. In order to measure the whole transmission through gratings, the total area of the Au gratings had to be larger than 2mm×2mm to meet the minimum FTIR spot size. 10nA beam current was used in order to dramatically shorten the exposing time. In our experiment, the area of 3.6mm×3.6mm was exposed after the spin-coated of ZEP520A (e-beam resist). 20nm Au film was evaporated on the sample after develop. Then ZDMAC was used to lift off the un-patterned parts. The fabrication procedures are shown in the sub-plot of Figure 1. In our following experiments, the period of Au gratings was always 4μm. The thickness of Au was always 20nm, which is larger than the skin depth of Au (~10 nm) to prevent the infrared from transmitting. The scanning electron microscopy (SEM), which is provided by ELIONIX, used to observe the plain view of Au gratings. Figure 1 shows the SEM picture of Au gratings with 34% air-slot width on silicon substrate. After the fabrication process, the FTIR system (IFS 66v/S), which was provided by Bruker, was finally used to investigate the TM-mode infrared transmittance of the sample, as shown in the schematic set-up of Figure 2. In this system, infrared is normal incident to the Au gratings. It is noticed that bare silicon substrate is as our reference, so the measured grating transmission was divided by that of bare silicon. In Figure 3 shows the experimental infrared transmission of Au gratings with various air-slot widths, which is already divided by that of bare silicon. The x-axis represents the wavelength (2.5μm~25μm), and the y-axis represents transmission. The percentage of the air-slot width over the period of 4μm, from 25% to 91% (along the dashed lines), is shown by various solid-lines.

3 SIMULATIONS

The Maxwell equations and the surface impedance boundary condition (SIBC) method[3] are both used to simulate the TM-mode infrared transmittance through Au-gratings with various air-slot widths. Figure 2 demonstrates the coordinates and parameters in simulation below. Boundary conditions,

$$E_{\parallel} = i(Z/k_0)\hat{n} \times H_{\parallel}$$

, are applied to relating tangential components of electric and magnetic fields at Au/dielectric interface, where Z is □ $1/n_{metal}$ with n_{metal} being the refraction index of the metal[4], k_0 is the wavenumber of the incident light, which is equal to $2\pi/\lambda$ with λ being wavelength of incident light, and $\hat{n}$ is a dimensionless unit vector, which direction is outward to the Au surface. Because the total area of Au gratings is lager than the FTIR spot size, wave functions outside the gratings can be considered as the plain wave. Figure 2 illustrates the definition of coordinates and the related parameters in later simulation. The tangential electric fields in various regions are expressed as a linear combination of orthogonal modes as follows:

tangential magnetic fields in region 1:

$$\exp(-ik_{y,0,1}(y-h/2))+\sum_{n=-\infty}^{\infty} R_n \exp\{i[k_{x,n,1}x+k_{y,n,1}$$

tangential magnetic fields in region 2:

$$\sum_{m=1}^{\infty} X_m(x)Y_m(y)=\sum_{m=1}^{\infty}[d_m \sin(k_{x,m,2}x)+\cos(k_{x,m,2}x)][a_m$$

with

$$X_m(x)=d_m \sin(k_{x,m,2}x)+\cos(k_{x,m,2}x)$$

and

$$Y_m(y)=a_m \exp[i(k_{y,m,2})y]+b_m \exp[-i(k_{y,m,2})y]$$

tangential magnetic fields in region 3:

$$\sum_{n=-\infty}^{\infty} T_n \exp\{i[k_{x,n,3}x-k_{y,n,3}(y+h/2)]\}$$

where

p is the period of gratings,

h is the Au thickness,

ε_i is the dielectric constant of region i,

$k_{x,n,i}$ is the nth mode x-direction wavenumber in region i (i =1,2, or 3),

$k_{y,n,i}=\sqrt{\varepsilon_i k_0^2-k_{x,n,i}^2}$ is nth mode y-direction wavenumber in region i (i =1, 2, or 3),

a_m/b_m is the magnitude of +y/-y direction wave at y=0 in region 2 (m=1, 2, 3…), and

R_n/T_n is the nth mode reflectance/transmission coefficient (n=0, ± 1, ± 2, ± 3…).

Because the period of Au gratings is p, $k_{x,n,1}$ and $k_{x,n,3}$ are equal to $2n\pi/p$ in region 1 and 3. d_m is mth mode coefficient which will be determined below. ε_1 and ε_2 are the dielectric constant of air, which are equal to 1. ε_3 is the dielectric constant of silicon, which is equal to 11.7. Applying SIBC to the left-hand and right-hand side of Au/air interface results:

$$d_m=(k_0\varepsilon_2 Z/i)/k_{x,m,2}$$ -------- Eq. 1

and

$$\tan(dk_{x,m,2})=2k_{x,m,2}(k_0\varepsilon_2 Z/i)/[k_{x,m,2}^2-(k_0\varepsilon_2 Z/i)^2]$$

-------- Eq. 2

It is known that $k_{x,m,2}$ $(m=2,3...)$ has a solution near $(m-1)\pi/d$ but not $(m-1)\pi/d$, so $k_{x,m,2}$ $(m=2,3...)$ can be found easily by Newton method. An important step is to find $k_{x,1,2}$. $k_{x,1,2}$ is near zero but not zero. We use Taylor series expanded at $k_{x,1,2}$=0 in Eq. 2 and get

$$dk_{x,1,2}=2k_{x,1,2}(k_0\varepsilon_2 Z/i)/[k_{x,1,2}^2-(k_0\varepsilon_2 Z/i)^2]$$

, in which an approximate $k_{x,1,2}$ can be easily found. The found $k_{x,1,2}$ above is used as initial value of the Newton method and we can get more precise $k_{x,1,2}$. Thus we can get $k_{x,m,2}$ (m=1, 2, 3...). Equating the tangential electric and magnetic fields and applying the SIBC conditions at $y=h/2$ and $y=-h/2$ yields the following four equations:

$$1+\sum_{n=-\infty}^{\infty} R_n \exp[i(2n\pi x/p)]=\sum_{m=1}^{\infty} X_m(x)(\varphi_m a_m+\varphi_m^{-1} b_m),\ 0\le x\le d$$

-------- Eq. 3

$$-ik_{y,0,1}+\sum_{n=-\infty}^{\infty} ik_{y,n,1}R_n \exp[i(2n\pi x/p)]=\begin{cases}\sum_{m=1}^{\infty} X_m(x)k_{y,m,2}(\varphi_m a_m-\varphi_m^{-1} b_m),\ 0\le x\le d\\ (k_0\varepsilon_1 Z/i)\{1+\sum_{n=-\infty}^{\infty} R_n \exp[i(2n\pi x/p)]\},\ d\le x\le p\end{cases}$$

-- Eq. 4

$$\sum_{n=-\infty}^{\infty} T_n \exp[i(2n\pi x/p)]=\sum_{m=1}^{\infty} X_m(x)(\varphi_m^{-1} a_m+\varphi_m b_m),\ 0\le x\le d$$

-------- Eq. 5

$$\sum_{n=-\infty}^{\infty} -ik_{y,n,3}T_n \exp[i(2n\pi x/p)] = \begin{cases} \sum_{m=1}^{\infty} iX_m(x)k_{y,m,2}(\varphi_m^{-1}a_m - \varphi_m b_m), & 0 \le x \le d \\ (k_0\varepsilon_1 Z/i)\sum_{n=-\infty}^{\infty} T_n \exp[i(2n\pi x/p)], & d \le x \le p \end{cases}$$

-------- Eq. 6

where $\varphi_m = \exp(ik_{y,m,2}h/2)$.

Thus R_n, T_n, a_m, and b_m can be determined from above equations. Taking the power into consideration, the total power is $|T_0|^2 \times \sqrt{\varepsilon_3}$. Finally the total power is divided by the theoretical transmission of air/silicon interface, $(4\sqrt{\varepsilon_1}\sqrt{\varepsilon_3}) / (\sqrt{\varepsilon_1} + \sqrt{\varepsilon_3})^2$, and final results are shown in Figure 4. From Figure 3 and 4, it is observed that experimental and simulation results are matched. The results obtained above are checked using another method that assumes that the Au/dielectric interface is perfectly conducting. It is found that $|T_0|$ will approach to zero as the mode number is very large. So the SIBC method is more suitable and convergent for calculating infrared transmission through Au gratings with various air-slot widths over the period, compared with the method that assuming Au is a perfect conductor. In the above simulation, the waves in region 1 and 3 are defined as *evanescent waves when* $k_{y,n,i}$ *is an image number* and as *propagation waves when* $k_{y,n,i}$ *is a real number*. Because of their exponential-decayed-in-y-axis properties, they can not be detected by the FTIR photo-detector when their transmission is investigated, as shown in Figure 2. Only the *propagation waves* are considered in Figure 4 because of their propagation properties.

4 DISCUSSION

Simulation results explain the relationship between surface plasma and transmission dips. The major two transmission dips are at the wavelength of 4μm and 13.7μm, respectively at b and e in Figure 3 and 4. At first, the 4μm transmission dip is discussed. In Figure 5, the simulated $|R_0|^2$, $|T_0|^2$, $|R_1|^2$, $|T_1|^2$, $|a_0|^2$, and $|b_0|^2$ versus wavelength of 53% air-slot width over Au gratings in the wavelength 2.5μm to 5.5μm are demonstrated. It is noticed that $|T_0|^2$ and $|T_1|^2$ are divided by the transmission of bare silicon, and $|a_0|^2$ and $|b_0|^2$ are very close and represented by black-solid line. In Figure 5, the a and c (at around 4μm but not 4μm) represents the incident waves transmit through the gratings by the waves with coefficients T_0 and T_1. At a and c, R_1 is approach to zero thus no surface plasma on the air side is induced. At b, $|R_1|^2$ is much lager than a and b, and $|T_0|^2$, $|T_1|^2$, $|a_0|^2$, and $|b_0|^2$ are approach to zero. Surface plasma on the air side is induced at b (4μm). In Figure 6, it shows the schematic of the transition between a, b and c. The arrows represent the waves, and the length of arrows represents the amplitude of the waves. Most incident light can transmit through the gratings at a and c. At b, $|R_1|^2$ is enhanced in a large quantity. The waves with coefficients a_0, b_0, T_0 and T_1 are approach to zero. Surface plasma on the air side is induced at b. The same story happens at transition d, e, and f in Figure 3 and 4. Figure 7 shows the schematic of the transition between d, e, and f. It is shown that the waves with coefficients R_1 are always approach zero. No waves with coefficients R_1 are induced. At e, the waves with coefficients T_1 are induced and the surface plasma on the silicon side is induced. Thus, our results support the surface plasma theory.

5 SUMMARY

To sum up, Au gratings on silicon substrate with various air-slot widths over the period were fabricated experimentally by e-beam and evaporator. The infrared (2.5μm~25μm) transmission through Au gratings was investigated by FTIR system. Furthermore, the Maxwell equations and the surface impedance boundary condition (SIBC) method were used to simulate the experimental results. Simulation results explain the relationship between surface plasma and transmission dips. These experimental and theoretical results can contribute to the understanding of TM-mode infrared transmission properties of metal gratings and the application on the field of infrared photo-detectors, bio-chip sensor, and light-emitting diode polarizers.

REFERENCES

[1] F. J. Garcia-Vidal and L. Martin-Moreno, "Transmission and focusing of light in one-dimensional periodically nanostrucutred metals," Phys. Rev. B 66, 155412(1) -155412(10) (2002)

[2] J. M. Steele, C. E. Moran, A. Lee, C. M. Aguirre, and N. J. Halas, "Metallodielectric gratings with subwavelength slots: Optical properties," PHYSICAL REVIEW B 68, 205103 (2003)

[3] David Crouse and Pavan Keshavareddy, "Polarization independent enhanced optical transmission in one-dimensional gratings and device applications," OPTICS EXPRESS, Vol. 15, No. 4, 1415

[4] M. A. Ordal, L. L. Long, R. J. Bell, S. E. Bell, R. R. Bell, R. W. Alexander, Jr., and C. A. Ward, "Optical properties of the metals Al, Co, Cu, Au, Fe, Pb, Ni, Pd, Pt, Ag, Ti, and W in the infrared and far infrared," 1 April 1983 / Vol. 22, No. 7 / APPLIED OPTICS

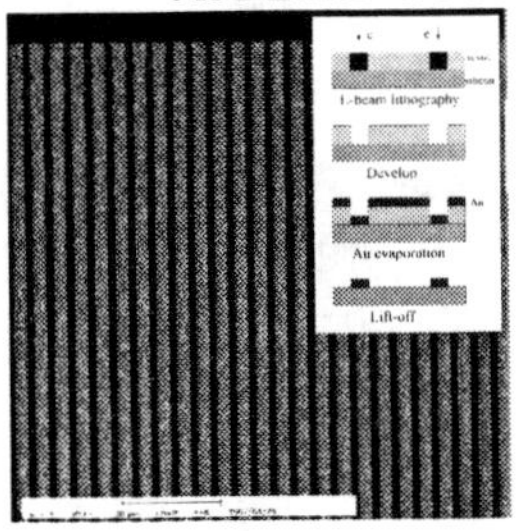

Figure 1: Scanning electron microscopy (SEM) picture of Au gratings with 34% air-slot width on silicon substrate
Sub-plot of Figure 1: Fabrication process of Au gratings on silicon substrate by e-beam and evaporator

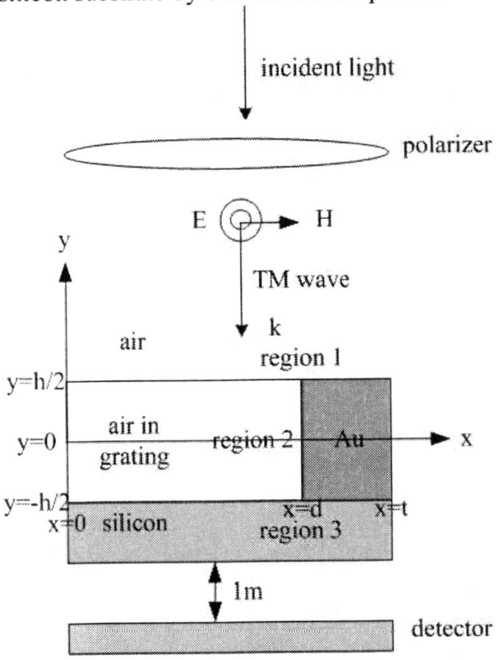

Figure 2: Illustration of infrared transmission measurement and definition of coordinate and parameters in simulation

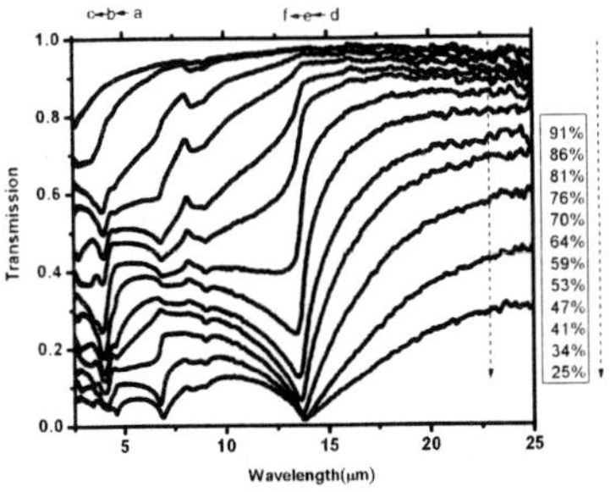

Figure 3: Experimental results of infrared transmission through Au gratings

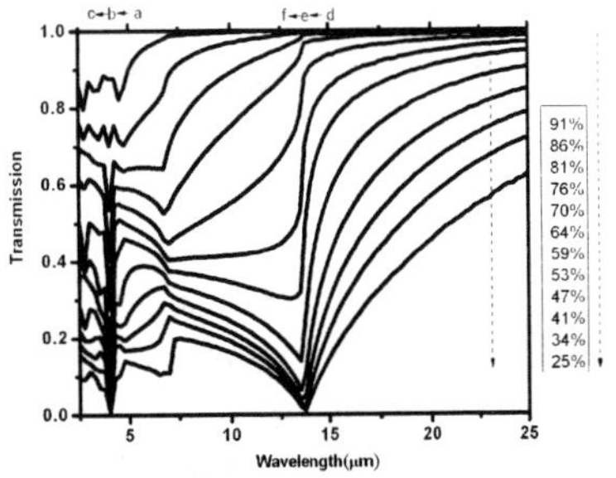

Figure 4: Simulation results of infrared transmission through Au gratings

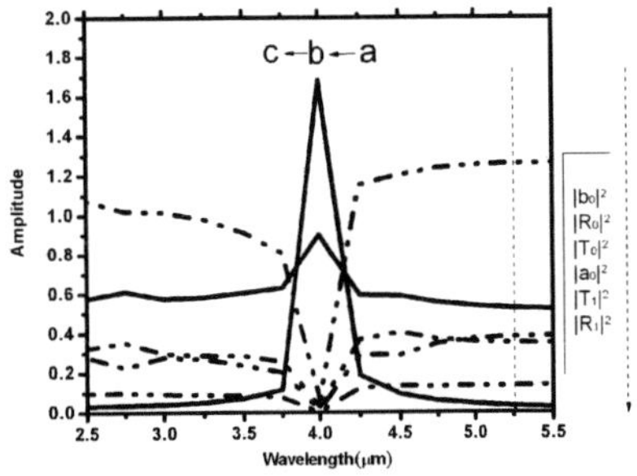

Figure 5: the calculated $|R_0|^2$, $|T_0|^2$, $|R_1|^2$, $|T_1|^2$, $|a_0|^2$, and $|b_0|^2$ versus wavelength of 53% air-slot width

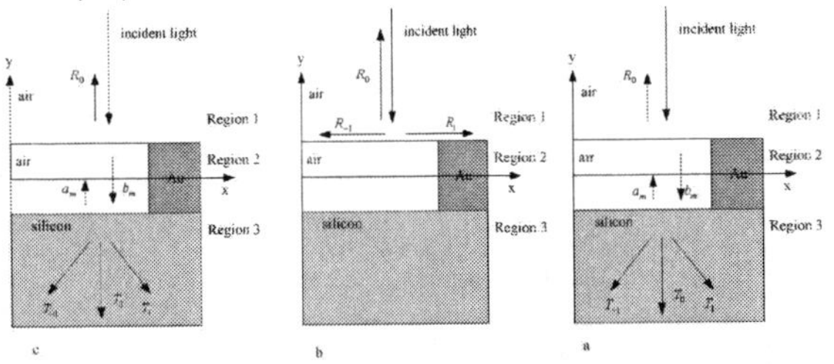

Figure 6: The schematic of the transition between a, b, and c

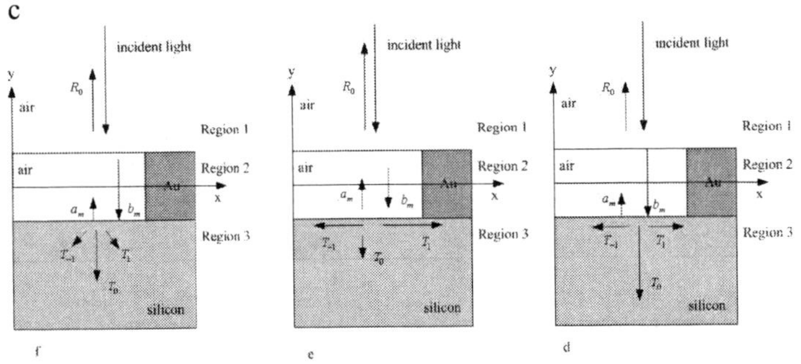

Figure 7: The schematic of the transition between d, e, and f

Optical absorption and luminescence study of ZnS quantum dots

D. Chakdar*, G. Gope, A Talukdar, D K Avasthi, S. S. Nath
Department of Physics, National Institute of Technology Silchar,
Silchar-10, Assam, India
Dept of Chemistry, Gauhati University, Guwahati-14, Assam, India
IUAC, Aruna Asaf Ali Marg, New Delhi- 67,India
*email- ankar53@rediffmail.com
Tel no: +91 9707050545

Abstract: We report here ZnS quantum dots synthesized by chemical method at room temperature. In this technique ZnS quantum dots are produced by simple chemical reactions where Zeolite, acting as matrix, plays the key role in controlling particle growth during synthesis and irradiation. The nanodots exhibit self-assembly dependent luminescence properties such as Zn^{2+} related emission, efficient low voltage electroluminescence (EL). This study demonstrates the technological importance of aggregation based self assembly in semiconductor nanosystems.

Topic area: *Nanotechnology*

Key Words: *Quantum dots, SHI, Blue Shift, Zn vacancy, Electroluminescence*

1. Introduction:

During recent years, fabrication[1-11] of quantum dots by various methods has been an emerging area of research for different optical, electronic, magnetic and spectroscopic applications. Synthesis of CdE, ZnE (E= S, Se, Te etc.) quantum dots by molecular- beam epitaxy technique as well as by simple chemical routes have been found in literature[1-11]. In this present article, we report synthesis of ZnS quantum dots in Zeolite by using chemical method_at room temperature. Zeolite controlls the size and shape of quantum dots during sample fabrication. The samples have been characterized by different techniques to reveal their nano nature. Also, the samples have been tested for their applications in electronics as nano LED by exploring the variation of EL intensity (Brightness) with dc voltage[5].

2. Experimental:

Synthesis of quantum dots by chemical method is a low cost and simple process. To synthesize ZnS quantum dots[6] by chemical route at room temperature, 8 grams of Zeolite are dissolved into 120 ml double distilled water. This mixture is taken in a three necked flask fitted with thermometer pocket and N_2 inlet. The solution is stirred in a magnetic stirrer at a stirring rate of 200 rpm at a constant temperature of 70^0 C for 5 hours. Thus, a water solution of zeolite has been prepared. Similarly, $ZnCl_2$ solution is made by dissolving 7 gms of $ZnCl_2$ in 100 ml double distilled water. The solution is degassed by boiling N_2 for 3 hours. Next, zeolite solution and

NSTI-Nanotech 2008, www.nsti.org, ISBN 978-1-4200-8505-1 Vol. 3

$ZnCl_2$ solution have been mixed and few drops of HNO_3 is added to the mixture followed by moderate stirring while aqueous solution of Na_2S is put into it slowly by means of a dropper unless the whole solution turns white. This solution is kept in dark chamber at room temperature for 14 hours for its stabilization. Now the whole solution is filtered by filter paper and the powder sample of ZnS in Zeolite was prepared.

3. Result and discussion:

High resolution transmission electron microscopy (HRTEM) (using JEM 1000 C XII) shows the surface morphology and particle size of the sample (Fig.1). Optical absorption spectroscopy (using Perkin Elmer Lamda 351.24) also displays a strong blue shift in the absorption edge of quantum dot samples in comparison to that of bulk specimen.

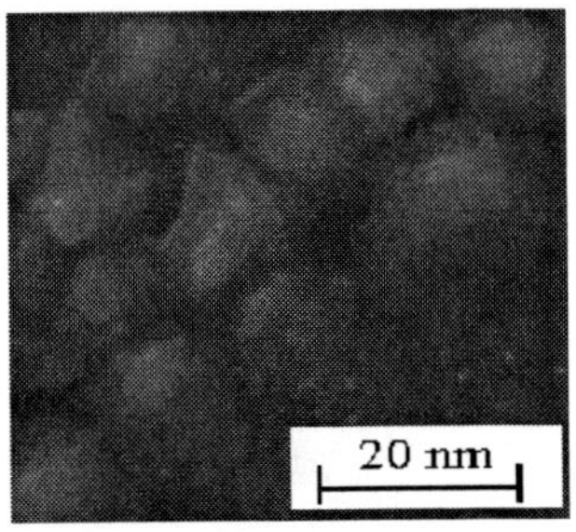

Fig 1. HRTEM Images

By considering strong absorption edge average particle size has been estimated theoretically by using the following hyperbolic band model [2].

$$R = \sqrt{\frac{2\pi^2 h^2 E_{gb}}{m^*(E_{gn}^2 - E_{gb}^2)}}$$

Where, R is the radius of quantum dot, E_{gb} is the bulk band gap, E_{gn} is the quantum dot band gap, h is Planck's constant, m*(for ZnS 3.64×10^{-31} Kg) is the effective mass of the specimen. This model yields the average particle size at around 10 nm. The reason for discrepancies between sizes obtained from HRTEM, and hyperbolic band model[2,6]. To obtain exact size from hyperbolic band model, the particle should be of exactly circularly shape. Form HRTEM images it is observed that the particles are not exactly circular, hence discrepancy occurs.

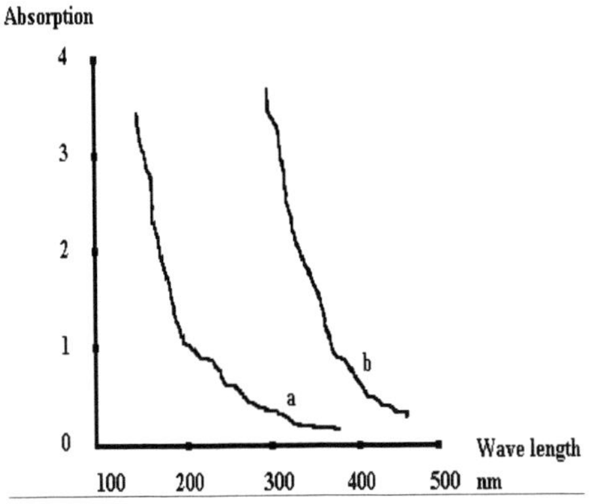

Fig 2. UV/VIS absorption spectre

Photoluminescence study (using HITACHI-F-2005) of the samples show that ZnS quantum dots produce emission [6-10] at 500 nm when excited with the optical signal of 200 nm.

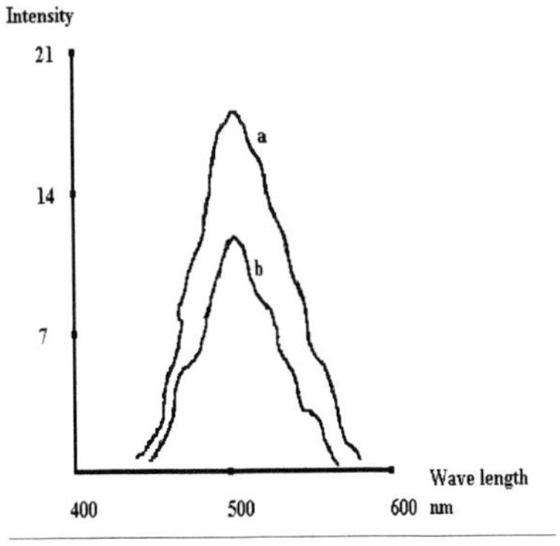

Fig. 3: PL spectra of ZnS: Excitation source200nm (a) for ZnS quantum dot and (b) for ZnS bulk

Fig 4 displays the room temperature EL spectra [8-11]. This experiment reveals that EL intensity is a function of excitation dc voltage at about 4.6 V but the EL emission peak position does not shift significantly and observed at around 500 nm irrespective of the excitation voltage. We believe that the reasons[7] behind emission in both the cases of PL and EL is the same and it is Zn^{+2} vacancies created during fabrication.

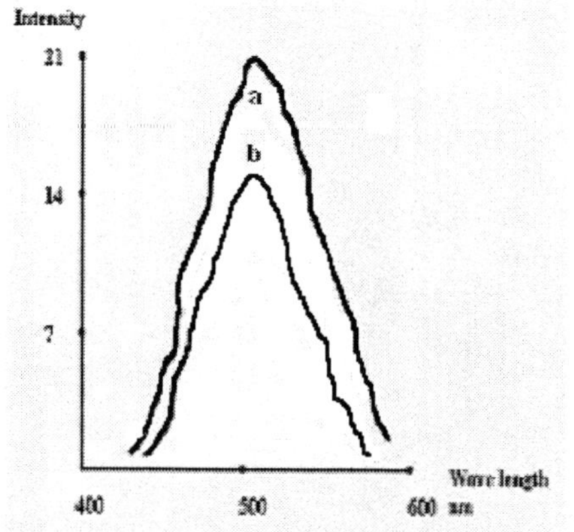

Fig 4: Electroluminescence Curve (a) for ZnS quantum dot and (b) for ZnS bulk

4. Conclusion:

ZnS quantum dots fabricated by chemical method, possesses Zn vacancies. These defects act as the optical sources in the specimen. Thus ZnS quantum dots can act as nano LED in green region .

Acknowledgement:

Authors thank Prof. A Choudhury (Vice Chancellor, Gauhati University, Guwahati, Assam, India) and Dr. H Chander, Dr S Chawla, LMD Group, Division of Electronic Materials, NPL, New Delhi, India for their suggestions and assistance during the work. .

Reference:

[1] Mohanta D, Nath S. S., Mishara N.C and Choudhury A.; *Bull. Mater.* Sci., 2003, Vol. 26, pp 289.

[2]. Nath S S, Chakdar D, Gope G, 2008, , The Internet journal of Nanotechnology, Vol 2, No.1.

[3]. Chen Wei, Wang Zhanguo, Lin Zhaojun and Lin Ianying, , J.Appl. Phys. , 1997,Vol. 82 , pp. 3111,.

[4]. Zhao J J, Zhang, J, *J. Appl. Phys.* 2004,Vol. 96, No. 6, pp 3208.

[5] Zhu Y, Yuan C L, Ong P P , J. Appl. Physics, 2002,Volume 92,pp. 6828.

[6] Nath S S, Chakdar D, Gope G, *Nanotrends-A journal of nanotechnology and its application,* ,Vol 02, Issue 03, 2007.

[7] Biswas S, Kar S, Choudhary S, Synthesis and Reactivity in Inorganic, Metal Organic and Nano-Metal Chemistry, 2006,Volume 36, Issue 1.

[8] Lee Jun woo, Cho K, Kim Hyunsuk, Kim Jim Hyong, Park Byoungjun, Noh Taeyong, Kim Sung Hyun, Kim Sungsig, JJAP, 2005, Vol. 44, No 10, pp. 7694.

[9] Charles P P Jr., Frank O J; *Quantum wells, Wires and Dots*, Introduction to Nanotechnology, A john Wiley & Sons, Inc., Hoboken, New Jersey, ,2003, pp 232.

[10] Xu J., Cui D, IEEE Photonics Technology Lett. , 2005, Vol 17, No 10,.

[11] Nath S S, Chakdar D, Gope G, *Nanotrends-A journal of nanotechnology and its application,* 2007,Vol 02, Issue 03.

Polypyrrole nanoionic composites for solid state electronics

J. H. Zhao[1], R. G. Pillai[2], M. Pilapil[2], M. S. Freund[1,2] and D. J. Thomson[1]

[1]Department of Electrical and Computer Engineering, thomson@ee.umanitoba.ca
[2]Department of Chemistry, michael_freund@umanitoba.ca
University of Manitoba, Winnipeg, MB, CA

ABSTRACT

In this report, a conjugated semiconducting polymer composite was designed to have a mobile charge carrier density that can be controlled through an electric field. The polypyrrole composite in the form of ($PPy^0(DBS^-Li^+)$) was electrochemically grown between two gold electrodes to form a junction. Application of an electric field across the junction should result in the drift of mobile Li^+ cations that are replaced by injected holes to maintain charge balance with immobile (DBS^-) anions. The holes in turn should create regions with higher conductivity and have the overall effect of increasing the current flowing through the junction. The current conduction is bulk limited rather than the electrode limited and space charge limited current (SCLC) is dominant at high potentials. The system displays field and time dependent conductance corresponding to the charge carriers' reconfiguration that has been used to implement a memory in a crossbar configuration that promises better scaling in nanometer-scale electronics.

Keywords: conjugated polymer composite, mobile ions, nanometer crossbar memory.

1 INTRODUCTION

The continuing progress in organic electronics including organic light emitting devices, photovoltaic cells and thin film transistors [1] has recently stimulated considerable research activity directed at developing organic memory electronic devices [2, 3] and a wide variety of memory phenomena and schemes have been extensively exploited. Conductance switching based organic memories, as one of emerging technologies beyond the CMOS era [4, 5], provide supplements or alternatives to the conventional Si-based semiconductor information storage technology with better scaling ability, especially in molecular level electronics [6]. Depending on their specific conduction mechanisms, organic memory approaches can be realized in the forms of flash memory, WORM, DRAM [3, 7, 8].

The conjugated semiconducting polymers are attractive in that their electronic properties (bandgap) can be altered through chemically modifying their molecular structures, i.e., changing their electrical conductivity by 'doping' other anions or cations into the polymer to create p-type or n-type semiconductor in analogy to inorganic counterpart. This characteristic enables a very useful functionality that charge carrier density of the conjugate polymer can be chemically modified through controlling the ions incorporated in the polymer. In particular, when the doped ions become mobile, conductance state can be changed with an electric field driving the mobile ions into and out of the polymer, this process is electrochemically equivalent to the oxidization and reduction of the polymer. This redox process introduces an approach to construct electronic devices [9-12], which often requires the presence of ionic conductors. Based on the mechanism that charge carrier density can be modulated with an electric field by modifying mobile ions distribution, we have conceived a single solid-state polymer system without the presence of electrolyte [13], the conjugated polypyrrole polymer composite containing large immobile anions and highly mobile cations. In this system the drift of the mobile ions controlled by a field can induce a more conducting (oxidized) region, resulting in an electronically asymmetrical junction that can be potentially applied to construct memory devices. Similar electrochemical process has been used to create inorganic solid-state electrolyte memory devices where conduction pathways, 'filamentary conduction', are electrochemically formed and deformed due to the oxidization and the reduction of mobile metallic ions [14]. Here, we report the application of our conjugated polypyrrole composite to submicron/nano level crossbar junctions and demonstrate a dynamic resistive memory in terms of the time and field dependent electrical conduction that can be controlled by modulating field generated charge carriers.

2 PREPARATION OF DEVICE

The polymer based crossbar junctions were created by initially fabricating two layers of perpendicularly crossed gold electrodes between them a sputtered dielectric SiO_2 thin film was sandwiched. A schematic structure and its AFM image of an $Au/SiO_2/Au$ crossbar junction with an electrode separation of about 200 nm and a width of electrodes of 20 μm is shown in Figure 1. The polypyrrole (PPy) thin films were electrochemically grown from the bottom of Au electrodes to the top Au electrodes using an aqueous solution of freshly distilled pyrrole monomer (100 mM) and an electrolyte (100 mM, NaDBS) at a constant potential of +0.65 V vs. Ag/AgCl. The thickness of the polypyrrole films were controlled by passing specific amount of charge (200 mC/cm^2 for 1 μm thick film) [15] during the electrochemical deposition. All electrochemical experiments were performed using a CHI660 electrochemical workstation (CH Instruments).

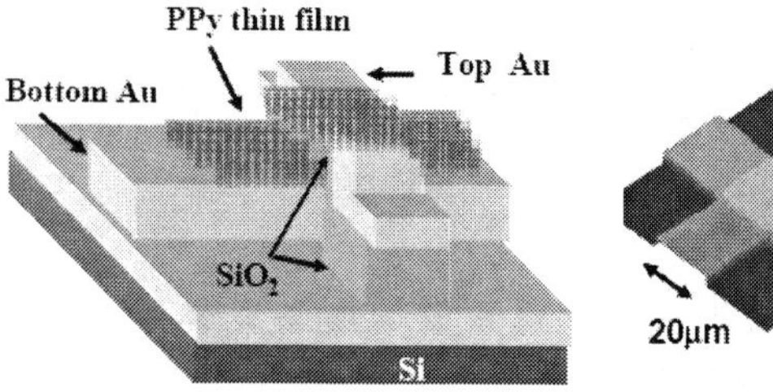

Figure 1. A schematic Au/SO2/Au crossbar junction and its AFM image.

The polymer thin film in the form of $PPy^{+}DBS^{-}$ was firstly formed and it is in an oxidized state, i.e., the P-doped conducting state. This conducting film was then reduced in an electrolyte $LiClO_4$ through incorporating Li cations into the polymer to balance the charge of DBS anions, as a result, this changed the conductivity of the polymer into a less conducting (semiconducting/insulating) state and resulted in the polymer composite in the form of $PPy^{0}(DBS^{-}Li^{+})$ in a charge neutral state. For such PPy composite crossbar junctions their electrical transport properties were characterized under a nitrogen atmosphere with CHI660 electrochemical workstation (CH Instruments) or a Hewlett Packard 4145A semiconductor parameter analyzer.

3 ELECTRICAL CONDUCTANCE

The conjugated polymer based electronic devices are conventionally designed to create diode-like junctions as their counterparts in the Si-based semiconductors, where potential barrier at the interface dominates the electrical conductance, i.e., electrical current passing through the junctions is the electrode limited. However, for most of the PPy polymer system with Au electrodes, the conduction is bulk limited due to the insignificant potential barrier at the interface [16].

For the $PPy^{0}(DBS^{-}Li^{+})$ composite crossbar junctions, the current-voltage behavior was measured and is shown in Figure 2. At low voltage, a linear relationship of the current-voltage indicates an ohmic conduction. With increasing the potential between the electrodes, a nonlinear behavior suggests that the space charge limited current (SCLC) start to become dominant, which is usually proportional to the V^2 and $1/L^3$ in the absence of trapping effect, where V is the applied voltage and L is the distance of the SCLC passing through between the top Au electrodes and the bottom Au electrodes [16]. Further increasing potential drives the electrical conductance into a regime where the field generated charge carriers (FGCC) begin to make a contribution to total current. It is believed that this additional FGCC is produced by the drift of mobile Li+ ions, under the external field and the space charges induced field, left behind the immobile anions (DBS) stabilized region in a high conducting state. The resulting reconfiguration of internal ions species would cause a reduction of the effective conductance path L that the space charge limited current flowing through, thus giving rise to the FGCC current.

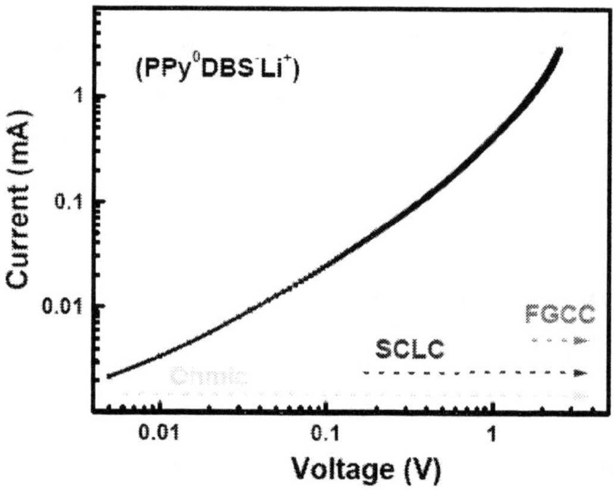

Figure 2. Current-voltage behaviors of the polymer crossbar junction under three voltage ranges.

The FGCC current was evidenced by measuring time dependence of the current flowing through the junctions as shown in Figure 3. At higher applied potentials, the current increased with time before approaching a new equilibrium, the increased current was associated with the reduced effective conductance distance L that was induced from the drift of cations driven by the field.

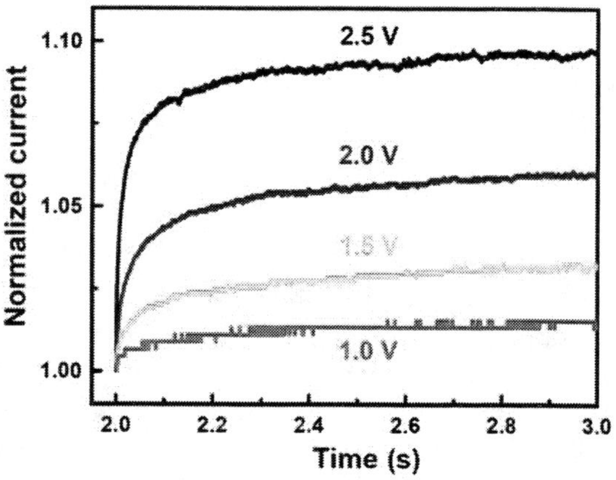

Figure 3. Currents were normalized to their initial values upon applying step voltages, suggesting the occurrence of the field generated charge carriers.

The time dependent conductance related to the field driven ions redistribution were also observed with further measurements as shown in Figure 4, where a delay time (T_0) was introduced between the forward and reverse field as shown in the inset of Figure 4. The effect of the ions redistribution was observed for the time periods of about 1 second. Within a very short field-free time before reversing the applied field, the cations can not return to their original equilibrium positions after the field is removed, the internal configuration of charges and carriers, established under the forward field, does not change significantly. After the forward field was removed for more than 1 second, the ions distribution, established under the forward field, returned to its initial equilibrium state and the current under the reversed field behaves as same as that when the forward field was applied at the first time.

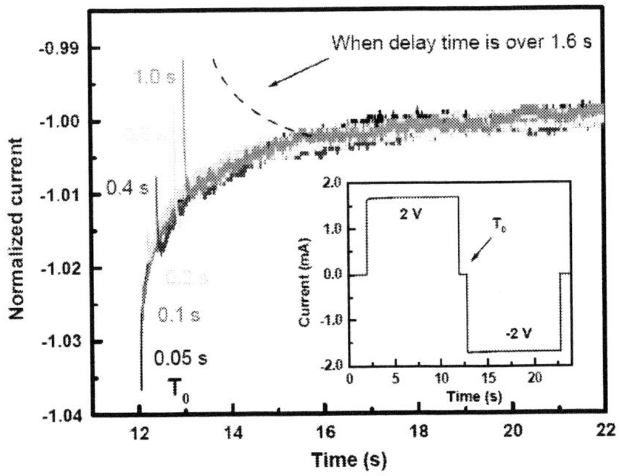

Figure 4. The time dependent currents were normalized to the current at the steady state under different delay times (T_0) introduced between the forward and reversed field, signaling the effect of the field driven ions redistribution.

As experimentally observed and discussed above, the conductance state of the polymer junctions can be determined by the internal configuration of charges and carriers that are pre-established under the field. This can produce some interesting transient current behaviors that can be controlled by reversing the field relative to the previously applied field. These were experimentally observed as shown in the Figure 5 (a). When the magnitude of the reversed field is greater than the previous forward field, the magnitude of current will increase with time as the effective distance L of the junction becomes shorter; when the magnitude of the reversed field is smaller than the previous forward field, the current will decrease with time as the drift of cations increases the effective junction width L. Thus, two transient conducting states can be produced by manipulating the applied field as shown in the Figure 5 (b). The two transient conductance states, represented by the state "1" and the state "0" as shown in the Figure 5 (a), can be employed to construct a dynamic memory cell. In such a cell, state high "1" is set by first applying a voltage across the junction; state low "0" is set by first applying a low/zero voltage to the junction. Reading the two states can be realized by applying a reversed field. When the reading voltage is smaller than that for setting high "1" state, the current will start high then reduce to the steady state at the reading voltage; when the reading voltage applied to the junction is higher than the applied writing voltage (low/zero), the current will start low then increase to the stead state at the reading voltage.

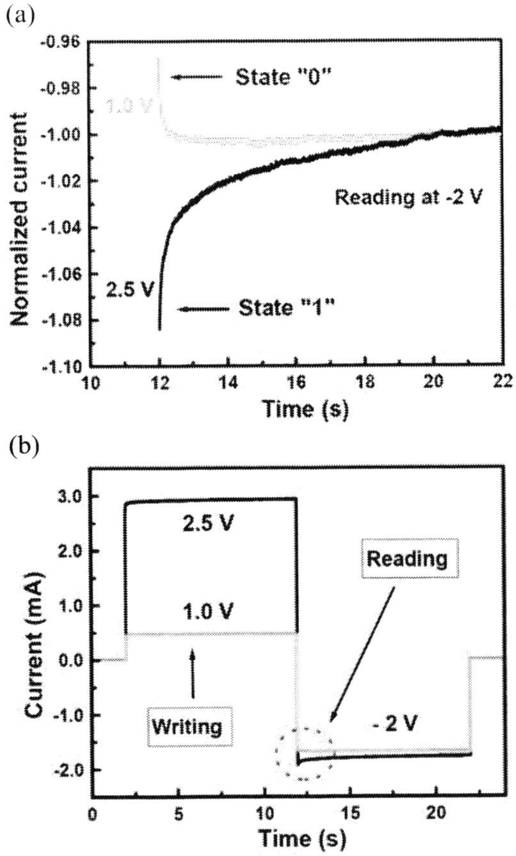

Figure 5. (a) Two opposite transient current states represent two digital states; (b) writing potentials (low (1 V) and high (2.5 V)) switch to reading potential (-2V).

4 A DYNAMIC MEMORY

The writing and reading process of the memory cell was realized by a simple electronic memory circuit as schematically shown in Figure 6 (a). This circuit was constructed to capture the two opposite transient current states that represent two digitized states "0" and "1" written (W) by first applying potential on the junction and then read (R) by reversing potential. Clock1 controls when a read voltage or a write voltage is applied to the junction. In the write state $V_W = 0$ V is applied for writing low state and $V_W = 2.5$V is applied for writing high state. In the read state V = -2 V is applied and the current through the junction is detected and amplified by a transimpedance amplifier. Sample/hold amplifier 1 (S/H1) captures the current signal shortly after the application of the read voltage (0.02s) and sample/hold amplifier 2 (S/H2) captures the signal 0.2 seconds later. The two captured signals are sent to a comparator and the signal of level high "1" or level low "0" is produced in response to the writing high or low. Clock 4 triggers a Flip/Flop to store the information of level high "1" and level low "0". This memory mechanism is interesting in that the stored state is extracted from a self referenced signal and the read function also acts to erase the previous state of the device. Figure 6 (b) shows the waveforms captured with the read-write clock (CLK), the current signal passing through the polymer junction, and the output of the memory cell corresponding to the writing and reading back the bit sequence 11100011100. This circuit is analogous to conventional dynamic random access memory (DRAM) circuits where a reference capacitive line is charged and then differentially compared using an amplifier to a second capacitive line that is either pulled up or pulled down by the charge from a memory cell [17].

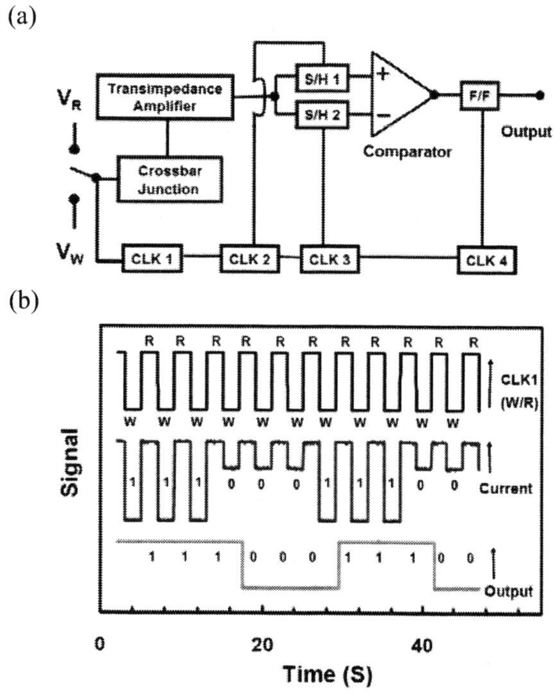

Figure 6. (a) The schematic electrical memory circuit for W/R process; (b) the captured waveforms of W/R clock (CLK), current signal and memory data of 11100011100 bits.

5 SUMMARY

A new approach for polymer based memory elements was explored through designing the conjugated semi-conducting polymer composites based on the ability of electrochemically oxidizing and reducing the polymer by field modulating ions configuration. The demonstrated resistive memory in the crossbar architecture has significant advantages in nanometer-scale electronics since the space charge limited current dominated conductance should result in better scaling with device size [13]. Its fabrication is compatible with standard CMOS circuits and memory cells can be electrochemically fabricated and controlled after all conductor layers have been built without the interface problems introduced by the post process of the contact electrodes.

REFERENCES

[1] S. R. Forrest, Nature, 428, 911, 2004.
[2] J. C. Scott and L. D. Bozano, Adv. Mater. 19, 1452-1463, 2007.
[3] Q.-D. Linga, D.-J. Liawb, E. Y.-H. Teoc, C. Zhuo, D. S.-H. Chang, E.-T. Kanga, K.-G. Neoha, Polymer, 48, 5182-5201, 2007.
[4] K. Galatsis, K. Wang, Y. Botros, Y. Yang, Y-H Xie, J. F. Stoddart, R. B. Kaner, C. Ozkan, J. L. Liu, M. Ozkan, C. Zhou, and K. W. Kim, IEEE Circuit & Devices Magazine, 12-21, 2006.
[5] M. M. Ziegle, Mircea R. Stan, IEEE Trans. Nanotechnology, 2, 217, 2003.
[6] J. E. Green, J. W. Choi, A. Boukai, Y. Bunimovich, E. Johnston-Halperin, E. Delonno, Y. Lou, B. A. Sheiff, K. Xu, Y. S. Shin, H.-R. Tseng, J. F. Stoddart & J. R. Heath, Nature, 445, 414-417, 2007.
[7] J. Ouyang, C.-W Cu, C. R. Szmanda, L. P. Ma and Y. Yang, Nature materials, 3, 918, 2004.
[8] S. MÖller, G. Perlov, W. Jacson, C. Taussig & S. R. Forrest, Nature, 426, 166, 2003.
[9] Z. Gadjourova, Y. G. Andreev, D. P. Tunstall & G. Bruce, Nature, 412, 520-523, 2001.
[10] J. C. deMello, N. Tessler, S. C. Graham, and R. H. Friend, Phy. Rev. B. 57, 12951, 1998.
[11] L. Edman, J. Swensn, D. Moss and A. J. Heeger, Applied Physics Letters, 84, 3744, 2004.
[12] D. A. Bernards, S. Flores-Torres, H. D. Abruña and G. G. Malliaras, Science, 313, 1416, 2006.
[13] R. G. Pillai, J. H. Zhao, M. S. Freund, D. J. Thomson, Adv. Mat. 20, 49-53, 2008.
[14] M. N. Kozici, M. Park, and M. Mitkova, IEEE Trans. Nanotechnology, 4, 331-338, 2005.
[15] E. Smela, J. Micromech. Microng. 9, 1, 1999.
[16] P. W. M. Blom, M. C. J. M. Vissenberg, Materials Science and Engineering, 27, 53-94, 2000.
[17] N. C.-C. Lu, H. H. Chao, IEEE J. Solid-State Circuits, SC-19, 451, 1984.

Acknowledgements

Financial support is gratefully acknowledged from the Canadian Institute for Advanced Research (CIFAR), the Natural Sciences and Engineering Research Council (NSERC) of Canada, the Canada Foundation for Innovation (CFI), the Manitoba Research and Innovation Fund, and the University of Manitoba. Support for contributions from RGP and MSF was provided in part from the Canada Research Chairs Program. JHZ and RGP contributed equally to this work.

Multi-Bit/Cell SONOS Flash Memory with Recessed Channel Structure

Kyoung-Rok Han*, Hyuck-In Kwon** and Jong-Ho Lee*

*School of Electrical Engineering and Computer Science, Kyungpook National University
EEBD11-503, #1370, Sangyuk-Dong, Buk-Gu, Daegu, Korea, jongho@ee.knu.ac.kr
**School of Electrical Engineering, Daegu University, Jilyang, Gyeongsan, Korea, hikwon@daegu.ac.kr

ABSTRACT

A novel device structure formed in recess region was presented and characterized for 4-bit/cell NOR-type nonvolatile memory (NVM) technology. The memory cells were designed to have common control gate and source line in a cell so that memory density is high. The proposed memory cell has a nitride layer formed on the surface of the recessed channel region for a charge storage node. Using channel hot electron (CHE) injection, we observed successfully ΔV_{th} and V_{th} margin of 2.8 V and 2.34 V, respectively, in 2-bit/cell operation mode without programming disturbance by adjacent storage node. The recess depth (x_{rw}) and width (x_{rd}) are 200 nm and 60 nm, respectively. By controlling doping profiles of the localized p-type channel doping in the recessed channel region, we could obtain the V_{th} margin of 1.7 V. Especially the cell size could be shrunk to $1.25F^2$/bit in a 4-bit/cell.

Keywords: multi-bit/cell, SONOS, nonvolatile, flash memory, recessed channel, localized charge trap, hot electron injection

1 INTRODUCTION

Low cost, low power, high density, and high reliability are main issues in Flash memory market. The concept of device scaling has been applied over many technology generations, resulting in consistent improvement in both device density and performance. Localized charge trapping device based on SONOS flash structures has been considered as a very promising candidate for future NVM beyond the floating gate technology, due to its various advantages. Especially, the capability of 2-bit/cell operation from the physically separated storage node is very attractive for ultra-high density memory application. The 2-bit/cell or 4-bit/cell SONOS type NVMs such as NROM™ [1], twin MONOS [2], mirror-bit [3], side wall type (or spacer) [4-5] and double SONOS [6-7] can operate as a multi-bit memory. The spacer-type and split-gate structure made possible physical isolation of bits [2-5], but were not scalable to sub-80 nm due to the difficulties in geometry controllability and performance degradation. To shrink gate length down to sub-100 nm and remove the interference between storage nodes by charge redistribution [8-9], we have proposed compact 2-bit/cell SONOS device with recessed channel structure [10-11]. However, research for higher bit-density than 2-bit/cell has not been performed.

In this paper, we propose an advanced cell concept for 4-bit/cell (or 2-Tr/cell) SONOS memory by adopting recess structure. We investigate the device characteristics of the proposed cell with an L_g of 60 nm using technology computer aided design (TCAD) device simulator [12] and show successful 4-bit/cell operation without having any interference between bits in a cell.

2 DEVICE STRUCTURE

Figure 1 shows schematic of 3-dimensional (3-D) 3×3 array view of proposed flash memory device with recess region. We tried to show key parts (source line, isolated poly gate, recessed channel region, and STI) of the device structure. The recess region is formed on the thin silicon body having wall-type shape. We can achieve high-density 4-bit/cell by sharing one control gate and one common source. The common source line located near the recessed region enables the biasing for bit programming /erasing and operates as a global bit line. The O/N/O gate stack is formed on the surface of the recessed region.

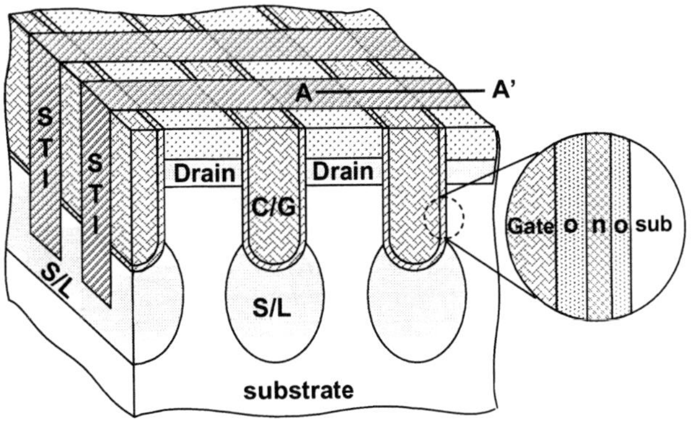

Figure 1: Schematic 3-D 3×3 array view of proposed flash memory device with recessed channel region. Each cell has 2 transistors and 4 storage nodes.

Figure 2 show 3×3 array layout of key layers (a) and schematic cell arrays (b) of our proposed flash memory cells as an example. In figure 2 (b), a cell is consisted of 2 devices which share the gate and the source, and each device in a cell can be operated independently. The bit lines and the common source lines run together vertically, and the word lines run horizontally in this figure. A cell of proposed scheme is controlled by three independent addressing lines like those of conventional 2-bit/cell scheme. However, the memory density can be ~2 times due to the sharing. It is estimated that the cell size is $\sim 5F^2$ or a 1.25 F^2/bit in a cell. Even if we consider peripheral circuit complexity for 4-bit/cell operation, we can implement ~3 times higher bit density compared to conventional 60 nm 1-bit/cell NOR flash technology.

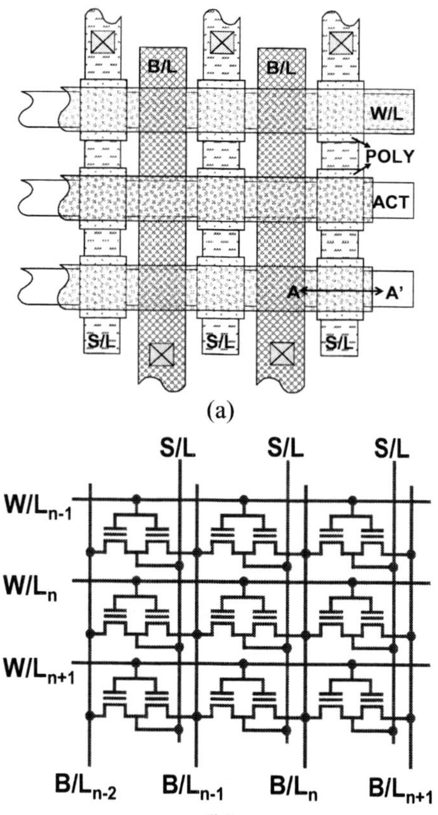

Figure 2: An example of a layout of 3×3 arrays consisted of proposed NOR-type cells (a) and its schematic cell arrays (b). The proposed unit cell has two transistors (2-Tr/cell) which shares one control gate and one source.

Figure 3 shows schematic cross-sectional view of cut along A-A' direction in Fig. 2(a). Our previous flash memory cell has recessed channel structure and floated counter doping near the bottom of the recessed region to achieve excellent 2-bit/cell (1-Tr/cell) operation. By changing the floating counter doping region to the common source region, we can achieve two 2-bit/cell devices without increasing cell area. Therefore, there are 4 storage sites (4-bit/cell) in a cell as shown in figure 3. The width (x_{rw}) and depth (x_{rd}) of the recess region are 60 nm and 200 nm, respectively. The x_{rd} is an important parameter in determining channel length. Long channel length can be possible without increasing x_{rw}, resulting in reasonable bit separation between the source and the drain sides. Thus we can achieve high density and high performance. The ONO layers have corresponding thicknesses of 4 nm (tunnel oxide), 5 nm (nitride), and 6 nm (block oxide), which is equivalent to an oxide thickness of nearly 12 nm. The local channel dopings by boron (p-type, $N_{A,peak}=1\times10^{18}$ cm^{-3}) are localized nearby metallurgical source and drain junctions for V_{th} control, DIBL suppression, and efficient hot carrier generation.

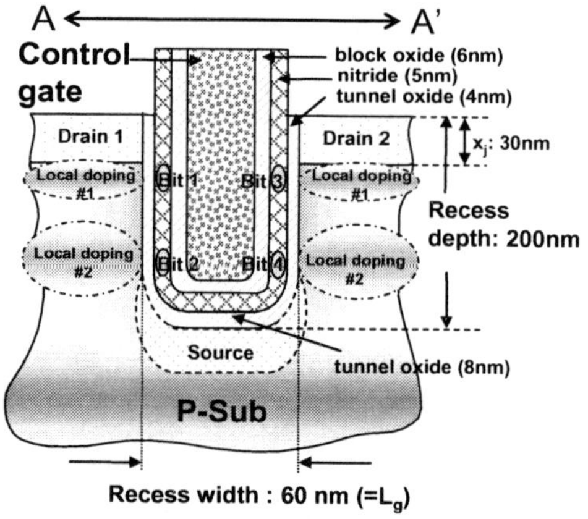

Figure 3: Schematic cross-sectional view of the proposed device structure of cut along A-A' direction. The channel length (L_g) represents the open width (x_{rw}) of the recess region. The x_{rd} and x_{rw} represent recess depth and recess width, respectively.

3 DEVICE SIMULATION AND RESULTS

Figure 4 shows simulated doping profile from the drain to the source as a parameter of ΔR_p (5 and 15 nm) of the source/drain doping profile. Uniform substrate doping is 1×10^{16} cm^{-3} and a peak concentration of a localized channel doping is 1×10^{18} cm^{-3}. ΔR_ps of the localized doping profiles near the drain and source junctions are 20 and 50 nm, respectively, by controlling ion implantation energy.

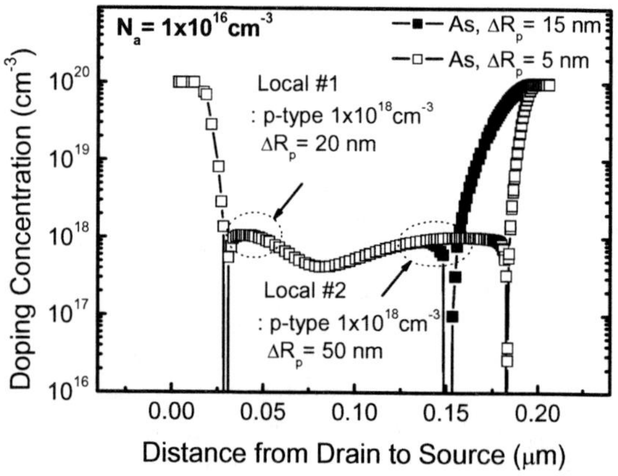

Figure 4: Simulated doping profiles along a vertical channel. Uniform substrate doping concentration is 1×10^{16} cm^{-3}. Two peak channel dopings are located near the drain and the source.

	Bit 1	Bit 2	Bit 3	Bit 4
Schematic	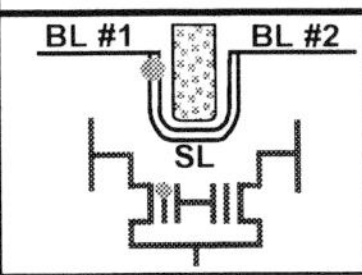	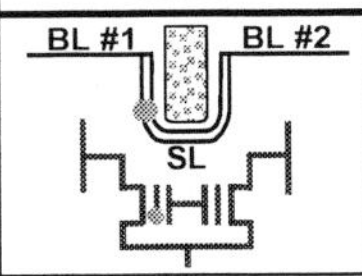	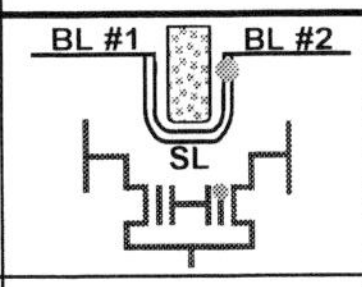	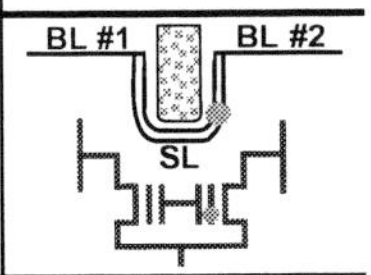
PGM bias (V)	V_G= 5, V_{SL}= 0 V_{BL1}= 3.5 V_{BL2}= F*	V_G= 5, V_{SL}= 3.5 V_{BL1}= 0 V_{BL2}= F	V_G= 5, V_{SL}= 0 V_{BL1}= F V_{BL2}= 3.5	V_G= 5, V_{SL}= 3.5 V_{BL1}= F V_{BL2}= 0
ERS bias (V)	V_G= -6, V_{SL}= 0 V_{BL1}= 3.5 V_{BL2}= F	V_G= -6, V_{SL}= 3.5 V_{BL1}= 0 V_{BL2}= F	V_G= -6, V_{SL}= 0 V_{BL1}= F V_{BL2}= 3.5	V_G= -6, V_{SL}= 3.5 V_{BL1}= F V_{BL2}= 0
Forward/Reverse read (V)	V_G= sw, V_{SL}= 0/1.5 V_{BL1}= 1.5/0 V_{BL2}= F	V_G= sw, V_{SL}= 1.5/0 V_{BL1}= 0/1.5 V_{BL2}= F	V_G= sw, V_{SL}= 0/1.5 V_{BL1}= F V_{BL2}= 1.5/0	V_G= sw, V_{SL}= 1.5/0 V_{BL1}= F V_{BL2}= 0/1.5

Table 1: Summary of a schematic for a bit and bias condition for programming, erasing and forward/reverse read operation. Here 'F' and 'sw' mean floating and sweep, respectively.

It needs to be noted that a sharp peak of the channel doping near the source and drain junctions is very important to increase V_{th} margin for 2-bit/Tr.

The concept of 4-bit/cell operation of proposed device is summarized in Table 1. It shows bias conditions for programming (PGM), erasing (ERS), and forward/reverse read operation for each bit. Here 'F' and 'sw' mean floating and sweep, respectively.

To show internal physics regarding the bit programming, simulated electric field profiles from the drain to source are shown in figure 5 at programming times (t_{PGM}) of 1 ns (before) and 1 μs (after). After programming with V_D=3.5 V and V_G=5 V, V_{th} near the drain increases due to localized charge trapping, which is verified by high peak of the electric field profile near the drain.

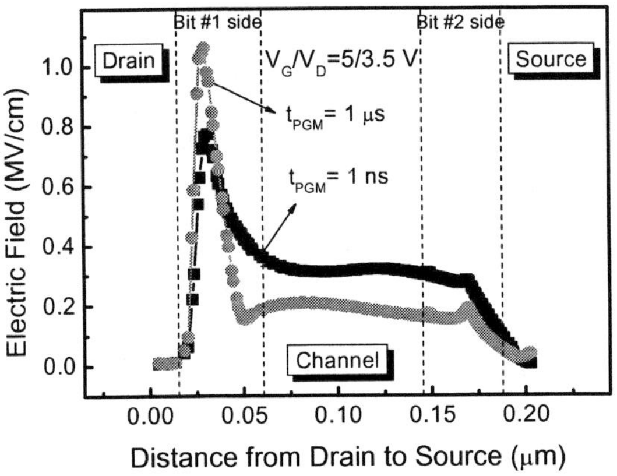

Figure 5: Simulated electric field profiles from the drain to the source before (t_{PGM} of 1 ns) and after (1 μs) programming.

To see interference between two transistors in a cell, we checked programming status of bit #3 when bit #2 is programmed. Figure 6 shows injected charges density on each storage node during bit #2 programming. Here, common gate bias is 5 V, and common source bias is 3.5 V. For the programming of bit #2, the drain2 shown in figure 3 is floated and the drain1 is grounded. Therefore, current flows from the common source to the drain1 and hot carriers are generated for programming bit #2 as a result. In figure 6, we can observe bit charges on only the bit #2. The interference between other storage nodes in a cell is impossible.

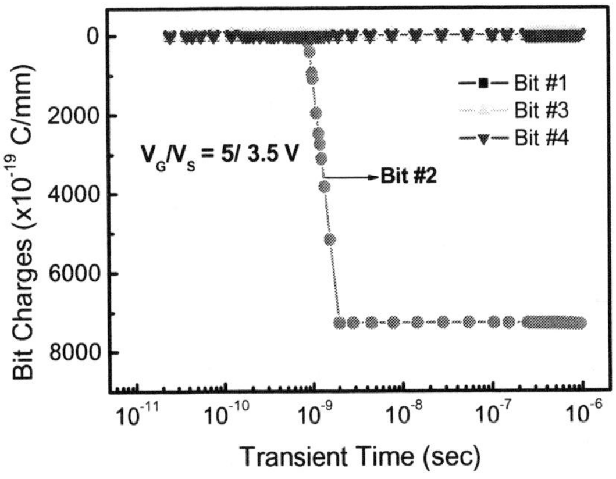

Figure 6: Injected charge density of each storage node for 1μs of bit programming. Applied V_G and V_S are 5 V and 3.5 V, respectively. The drain1 is grounded and the drain2 is floated. There is no bit charge on bit #3 during programming bit #2.

Figure 7 shows log I_D-V_G characteristics of before and after programming by injecting CHE for 1 μs at a fixed V_G of 5 V. In figure (a), V_D is 3.5 V and V_S=0 V. The bias condition in figures (b) and (c) is V_S=3.5 V and V_D=0 V. At V_D=0.1 or V_S=0.1 V after programming, the V_{th} shifts (ΔV_{th}) are about 2.8 V and 1.8 V for a bit #1 (a) and bit #2 (b), respectively. Reverse read and forward read (with V_D=1.5 V or V_S=1.5 V) are carried out immediately after each charge injection. We obtained the V_{th} margins are 2.34 V (a) and 1.4 V (b) which is enough for 2-bit/Tr operation. By making the source doping profile more steep (ΔR_p: 15 nm → 5 nm), the V_{th} margin was improved form 1.4 V (b) to 1.7 V (c).

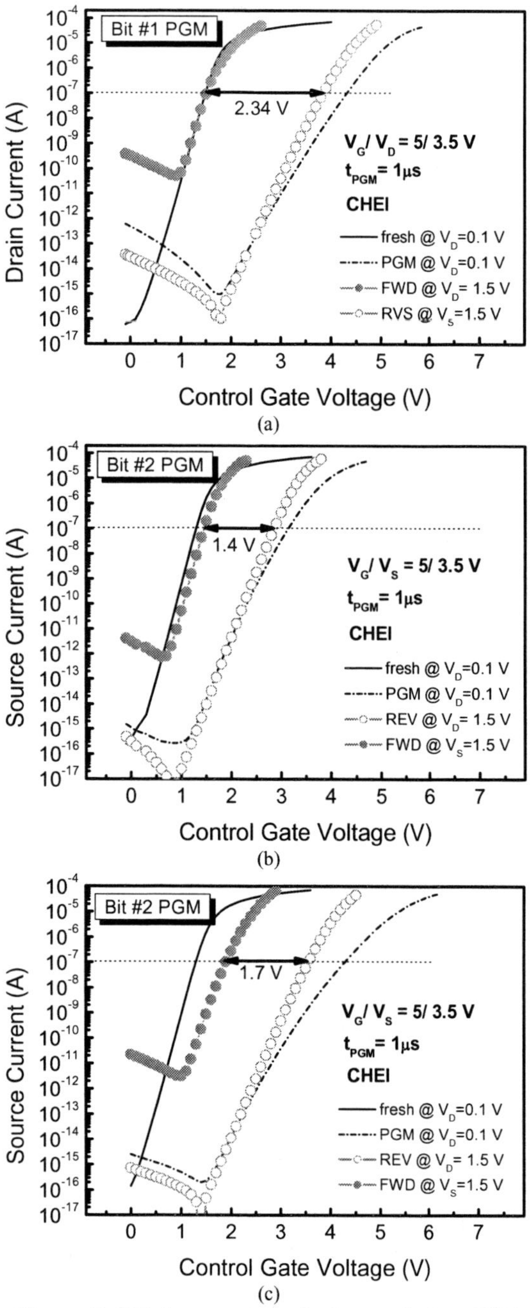

Figure 7: *I-V* characteristics before and after CHE programming of bit #1 (a) and bit #2 (b), (c). ΔR_ps of common source doping profile is 15 nm (b) and 5 nm (c).

The localized channel doping mentioned above also can help the V_{th} increase. However, the read disturbance can be happened from the by doping fluctuation. It needs to be noted that a sharp peak of the channel doping near the source and drain junctions is very important to increase V_{th} margin for 2-bit/Tr.

4 CONCLUSION

We have proposed compact 4-bit/cell SONOS flash memory device that has recessed channel structure. Since two transistors in a recessed region share the gate and the source, integration density could be very high (~3 times compared to conventional 60 nm NOR cell). Enough V_{th} margins (> 1.4 V) for 2-bit/Tr was obtained. We also have shown there is no interference between storage nodes in a cell. Our proposed device structure can be one of most promising candidates for multi-bit operation scaled NVM technology.

ACKNOWLEDGEMENT

This work was supported by "The National research program for the 0.1 Tb Non-volatile Memory Development sponsored by Korea Ministry of Science & Technology" in 2007.

REFERENCES

[1] B. Eitan, P. Pavan, I. Bloom, E. Aloni, A. Frommer, and D. Finzi, *IEEE Electron Device Lett.*, vol. 21, no. 11, pp. 543-545, 2000.

[2] Y. Hayashi, S. Ogura, T. Saito, and T. Ogura, *VLSI Tech.*, p. 122, 2000.

[3] Y. K. Lee, T. -H. Kim, S. H. Lee, J. D. Lee, and B. K. Park, *IEEE Trans. Nanotechnology*, vol. 2, no. 4, pp. 246-252, 2003.

[4] M. Fukuda, T. Nakanishi, and Y. Nara, *IEDM Tech. Dig.*, pp. 909-912, 2003.

[5] B. Y. Choi, B. G. Park, Y. K. Lee, S. K. Sung, T. Y. Kim, E. S. Cho, H. J. Cho, C. W. Oh, S. H. Kim, D. W. Kim, C. H. Lee, and D. Park, *VLSI Tech.*, pp. 118-119, 2005.

[6] C. W. Oh, S. H. Kim, N. Y. Kim, Y. L. Choi, K. H. Lee, B. S. Kim, N. M. Cho, S. B. Kim, D. W. Kim, D. Park, B.-I. Ryu, *VLSI Tech.*, pp. 40-41, 2006.

[7] C. W. Oh, N. Y. Kim, S. H. Kim, Y. L. Choi, S. I. Hong, H. J. Bae, J. B. Kim, K. S. Lee, Y. S. Lee, N. M. Cho, D. -W. Kim, D. Park, B. -I. Ryu, *IEDM Tech. Dig.*, 2006.

[8] H. Sunamura, T. Ikarashi, A. Morioka, S. Kotsuji, M. Oshida, N. Ikarashi, S. Fujieda, and H. Watanabe, *IEDM Tech Dig.*, 2006.

[9] B. -G. Park, Y. K. Lee, B. Y. Choi, D. G. Park, *Proc. of 7th ICSSICT*, pp. 679-684, 2004.

[10] K. -R. Han and J. -H. Lee, *Jpn. J. Appl. Phys.*, vol. 45, pp. L1027-L1029, 2006.

[11] K. -R. Han, Y. M. Kim, K. -H. Park, S. -G. Jung, B. -K. Choi, and J. -H. Lee, *Solid State Device and Materials*, pp. 316-317, 2007.

[12] SILVACO International, ATLAS user manual, Ver. 5.11.3.C.

Logicless Computational Architectures with Nanoscale Crossbar Arrays

B.Moutte*

*George Mason University
4400 University Drive, Fairfax, Va. 22030, bmouttet@gmu.edu

ABSTRACT

Modern computational architectures are reliant on basic logic circuits carrying out Boolean operations such as AND, OR, XOR, etc. Recently efforts have been made to replicate the functions of logic circuits using various nanoelectronic architectures in the form of nanoscale crossbars arrays. However, the interconnections required to be formed so as to create basic computational units such as full adders and more complex arithmetic circuits may be cumbersome to implement in nanoscale crossbars. The present article proposes an alternative computational paradigm formed from a hybrid of nanoscale crossbars and basic analog circuit elements so as to form arithmetic circuitry without the need for replicating basic logic functions.

Keywords: crossbar architectures, molecular electronics, analog computing, nanoelectronics, nanoprocessors

1 INTRODUCTION

Crossbar electronic architectures are currently under experimental investigation by start-up companies including Nantero, which has proposed carbon nanotube ribbons as mechanical switches in a crossbar arrangement [1], as well as more established corporations such as Hewlett-Packard, which has experimented with using rotaxane molecules as molecular switches between crossed nanowires [2]. Initial proposed applications have been in high density memory but, more recently, proposals for implementing basic logic structures and more complex computational structures have been made [3] and techniques are being explored to integrate crossbar architectures with more conventional CMOS circuitry [4]. However, in order to create useful computational architectures the current proposals would require interconnections between multiple sections or tiles of crossbar arrays leading to difficult design and manufacturing requirements. A simpler solution requiring only a single crossbar tile is thus desirable.

Figure 1 illustrates an example of a basic configuration for a nanoscale crossbar consisting of two arrays of orthogonally oriented parallel nanowires layers separated by a rotaxane molecular film such as described in [2] and [4]. Each intersection between the nanowires forms a rectified crosspoint equivalent to a configurable diode which can be set with either a high or low resistance state. The low resistance state may be represented of a binary value 1 while the high resistance state may be represented by a binary value 0 so that each column of the crossbar effectively stores a binary numerical value.

The primary advantage of such a crossbar architecture is that it is easily scalable to nanometer dimensions due to the simplicity of its structure using self-assembly or imprinting lithography without resorting to optical lithography or the use of any masks. The use of opposite doping between the upper and lower wiring plane generates a rectification in the current flow of the crossbar preventing feedback paths and allowing for addressing of an entire row (or column) of crossbar junctions at one time for faster reading or writing of data to the crossbar. However, the benefits of the crossbar design extends beyond the nanoscale and various proposals exist for RRAM (resistive RAM) to achieve two-terminal resistance state non-volatile crossbar memory arrays as a future replacement for SRAM and DRAM by using chalcogenide or conductive polymer materials. Table 1 provides examples of the high and low resistances states and approximate switching times for some of the materials used in the proposed designs as discussed in a variety of sources [1,2,5,6]. It is noted that in the cited sources the resistances are largely dependent on contact area with the material and thus nanoscale wire crossbars expectedly produce higher resistances than microscale wires used in GeSbTe and Cu-TCNQ resistance switches.

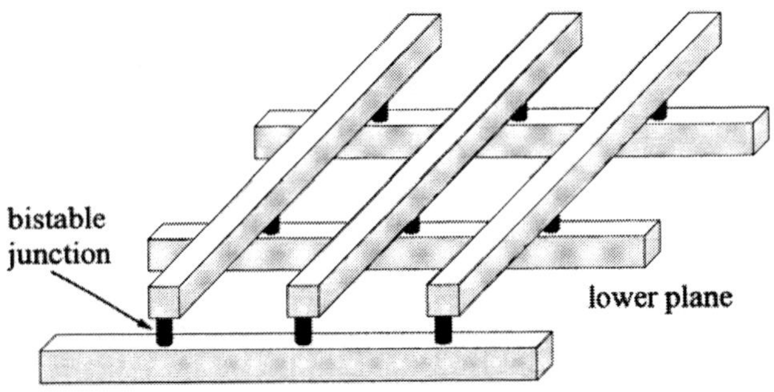

Figure 1: Nanoscale crossbar (from [4])

Resistance Switching Material	Low Resistance (On=1)	High Resistance (Off=0)	Approx. Switching Time
CNT ribbon (from [1])	112kΩ	~10GΩ	~10 ps
Rotaxane (from [2])	10^6-$5\text{x}10^8$Ω	>$4\text{x}10^9$Ω	~1 ns
GeSbTe (from [5])	6-20 kΩ	50-150 kΩ	10-30 ns
Cu-TCNQ (from [6])	~200 Ω	~2 M Ω	200 ns

Table 1: Examples of low and high resistance values and switching time of various resistance switch material.

2 LOGICLESS CROSSBAR ARITHMETIC

A circuit arrangement is illustrated in Fig. 2 for a basic arithmetic processor that used a 16x4 crossbar array connected to operational amplifier and analog-to-digital conversion circuitry. Operational amplifiers are a very basic component to many analog computing systems but have largely been rendered irrelevant in processor design due to the proliferation of digital computing over the past several decades. However, the key benefit of op-amps which make them useful for computation is that when a feedback resistance if provided between the output terminal and the inverting terminal, the generated voltage output V_{out} is a resistance weighted sum of the input voltages, V(j). This is a well known result of the conservation of current at the inverting terminal of the op-amp and is expressed as:

$$\frac{V_{out}}{R_f} = -\sum_j I(j) = -\sum_j \frac{V(j)}{R(j)} \tag{1}$$

where j is a summation index, I(j) represents the input currents, R(j) is the resistance connecting each input voltage V(j) to the inverting terminal, and R_f is the feedback resistance from the output voltage V_{out}.

Connection between a nanoscale crossbar and an op-amp such as shown in Fig. 2 may be achieved by using microscale wires for the rows of the crossbar and nanoscale wires for the columns. The resistances connecting the output of the crossbar rows to the inverting terminal provide a weighting value for each row and proper tuning of these resistances can establish a bit significance for each row. Due to the rectifying junctions within crossbar arrays the current provided by each junction of the crossbar for a particular row j can be expressed in terms of the input voltages V(i) to the columns of the crossbar as:

$$I(j) = \sum_i (V(i) - V_{rect})/(R(j) + r(i,j)) \tag{2}$$

where i is a column summation index, V_{rect} is a threshold voltage required to overcome the rectifying effect, and r(i,j) are the crossbar resistance states.

Provided that the feedback resistance is set to R and the input resistances are set to 2^j R - r, wherein r is the value of the average low resistance state for the resistance switching material, combining equations (1) and (2) produces:

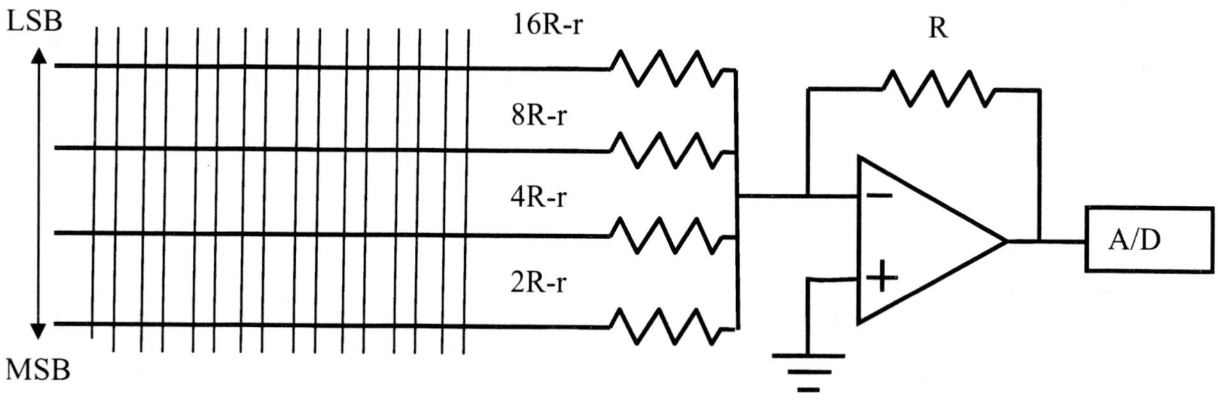

Figure 2: 16x4 crossbar with nanowire crossbar columns and microscale row wire connections to op-amp and A/D circuitry

$$V_{out} = -\sum_{ij} \frac{R}{2^j R - r + r(i,j)} (V(i) - V_{rect}) = -\sum_{ij} T(i,j)(V(i) - V_{rect}) \quad (3)$$

where r(i,j) are the crossbar resistance states that contribute to the input resistances and T(i,j) is used to represent the overall transfer function between the op-amp output voltage and the input voltages of the crossbar.

Assuming an ideal behavior for the crossbar resistance states in which all of the low resistances are uniform in value and equal to r and all of the high resistance states are very large in comparison to R, T(i.j) may either take the value of $1/2^j$ (for r(i,j)=r) or a value close to zero (for r(i,j)>>R). The net effect of the proposed design is that if two bits in a common row are both at a low resistance state the sum increases the bit significance value so that $1/2^j+1/2^j=1/2^{j-1}$,which is equivalent to a carry operation in bit addition. Thus the selection of multiple columns of the crossbar array by a common voltage higher than V_{rect} results in an analog sum in accordance with the stored binary values of the selected columns. The provision of an analog-to-digital converter may then produce the expected digital output representing the binary sum of the stored values. The A/D resolution required for crossbars of different sizes is given by Table 2. Multiplication can similarly be performed by programming each column of the crossbar with a binary value representative of a multiplicand which is shifted between different columns and applying high or low input voltage to the crossbar in accordance with a multiplier.

Crossbar Size	Required A/D Resolution
4x4	>=6 bit
8x4	>=7 bit
16x4	>=8 bit
16x8	>=12 bit
NxM	>= ($\log_2$N+M)

Table 2: A/D conversion for crossbars of different sizes.

3 ADVANTAGES AND LIMITATIONS

The primary limitation of the proposed system for computation may be that analog circuitry typically has microsecond or nanosecond settling times whereas modern logic-based processors possess switching times of the order of picoseconds. However, the current trend of using multi-core processors demonstrates that raw speed is not the only factor when comparing computing power and the potential exists to include dozens or even hundreds of the proposed circuit design on a single chip. In addition, the proposed design offers the advantage of integrating both data storage and data processing in a single circuit. Whereas conventional logic-based processing requires several steps for transferring data between a memory circuit and a processor circuit the proposed circuitry has the potential to perform both functions so that a data retrieval operation may be achieved by selecting one crossbar column while an addition operation may be achieved by selecting two (or more) crossbar columns.

Other limitations include the requirement of a large ratio of high and low resistance states for the resistance variable material used in the crossbar. In order to produce consistent numerical results the total current produced by the high resistance states in any particular row should be significantly less than the current produced by a single low resistance state. This problem can be tempered to some degree by limiting the number of crossbar columns to be substantially less than the ratio between the high and low resistance states and by maintaining a high impedance input for any unselected columns. Fluctuation of current due to spatial or temporal changes between different r(i,j) values may also lead to inconsistent numerical results. However, by providing the weighting resistance R>>r in equation (3) the impact of variations in r may be reduced.

REFERENCES

[1] Rueckes et al., "Carbon Nanotube-Based Nonvolatile Random Access Memory for Molecular Computing," Science, vol. 289, 94-97, 2000.

[2] Chen et al., "Nanoscale molecular-switch crossbar circuits," Nanotechnology, vol. 14, 462-468, 2003.

[3] Snider, "Architecture and Methods for Computing with Reconfigurable Resistor Crossbars," U.S. Patent 7,203,789, 2007.

[4] Ziegler et al., "CMOS/Nano Co-Design for Crossbar-Based Molecular Electronic Systems," IEEE Transactions on Nanotechnology, Vol. 2, No. 4, 217-230, 2003.

[5] Lai, "Current status of the phase change memory and its future," IEDM '03 Technical Digest, 2003.

[6] Potember et al., "Electrical switching and memory phenomena in Cu-TCNQ thin films," Applied Physics Letters, vol. 34, issue 6, 405-407, 1979.

Implementation of Comparison Function Using Quantum-Dot Cellular Automata

Meghanad D. Wagh*, Yichun Sun* and Viswanath Annampedu**

*Department of ECE, Lehigh University, Bethlehem, PA 18015, mdw0, yis205@Lehigh.Edu
** Storage Peripherals Group, LSI Corp., Allentown, PA 18109, Vish.Annampedu@lsi.com

ABSTRACT

A comparison function is important to implement arbitrary large Boolean functions. In this paper we show that a comparison function can be directly and efficiently implemented using QCA. The resultant architecture of an n bit comparator has a delay of $\lceil log_2 n \rceil + 1$ and a complexity of $O(n)$ gates. By duplicating certain majority gates, all crossing wires in the implementation except those at the input level can be eliminated.

Keywords: quantum-dot cellular automata, majority logic, comparison function

1 INTRODUCTION

Comparison is one of the common function used in many important applications including realization of arbitrary Boolean functions. In CMOS technology, one can easily realize an n bit comparison with a carry propagation subtractor having $O(n)$ delay. This speed can be improved to $O(\log n)$ by using a more complex block carry look-ahead subtractor. However, a further dramatic improvement in speed is only possible through use of nanoelectronic technologies such as the Quantum-dot cellular automata (QCA) [1].

QCA can be implemented in many technologies including ferromagnetic and molecular. The molecular QCA is particularly interesting because of its projected density of up to 1×10^{12} devices per cm^2 [2]. A major advantage of QCA over other nanoelectronic architectural styles is that the same cells that are used for making logic gates can be used to build wires carrying logic signals. However, QCA architectures have to rely upon only two basic building blocks, namely a three input majority gate and an inverter. As a result, QCA implementations of only a few logic circuits such as binary adders, multipliers, barrel shifters, serial/parallel converters are currently available [3]–[5].

This paper describes an efficient implementation of the comparison function using QCA. The resultant architecture has optimal delay $O(\log n)$ and a low gate complexity $O(n)$ for an n bit comparison. The architecture can be easily pipelined to improve its throughput.

2 QUANTUM-DOT CELLULAR AUTOMATA

A basic QCA building block can be described as a cell with four quantum dots and two charged particles which occupy the dots. The charged particles can migrate between quantum dots when the barriers between them are lowered by an external *clock*. When the barriers are raised by removing the clock, the particles settle into two possible stable (polarized) positions because of electrostatic forces. These stable states represent logic 0 and 1 as shown in Fig. 1.

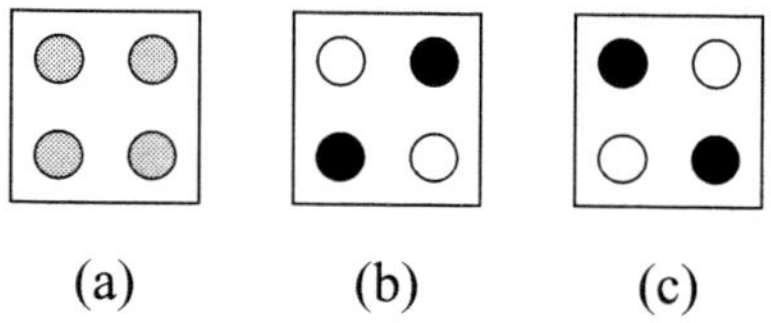

Figure 1: (a) The basic QCA cell (b) a polarized QCA cell representing logic 1 and (c) a polarized QCA cell representing logic 0.

When the clock to a QCA cell is reduced, the polarization state of its surrounding cells determines its own polarization state. This allows conduction of logic values along wires made of QCA cells as well as the design of logic elements such as an inverter and a three input majority gate. The two logic blocks made with QCA cells are illustrated in Figs. 2 and 3.

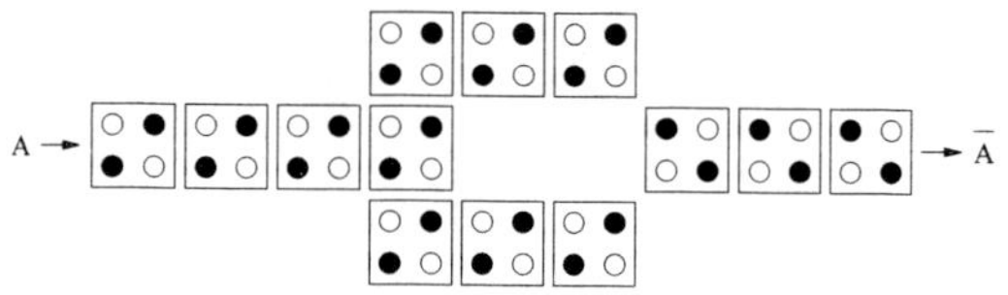

Figure 2: An inverter implementation in QCA.

A three input majority gate outputs a logic 1 when 2 or more of its inputs are 1. By fixing one of the inputs to the majority gate at 1, one can convert the majority gate into a 2-input OR gate. Similarly by fixing one of the inputs to 0, one can turn it into a 2-input AND

NSTI-Nanotech 2008, www.nsti.org, ISBN 978-1-4200-8505-1 Vol. 3

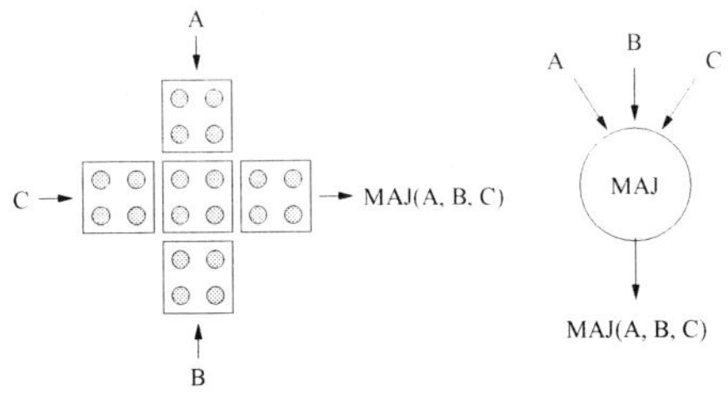

Figure 3: A 3-input majority gate implementation in QCA and its symbol.

gate. It therefore appears that it should be possible to implement any Boolean function in the QCA architecture. Unfortunately use of 2-input ANDs and ORs results in highly inefficient realizations since about 33% of the gate inputs are tied to constant values. Further, the large number of gates makes the circuit layout difficult. It is therefore important to develop designs that allow one to fully utilize the capabilities of the 3-input majority gates.

3 COMPARISON FUNCTION

An n bit comparison function compares two n bit binary strings and outputs a logic one if the value represented by the first string is greater than or equal to that of the second string. The comparison function is important in its own right. But we illustrate here its use in implementing an arbitrary Boolean function. Consider a Boolean function $f(x_{n-1}, x_{n-2}, \ldots, x_0)$ which outputs a logic one only when the n-bit input string $X = x_{n-1}x_{n-2}\cdots x_0$ has a value between X and Y. Let function $C(X)$ denote a comparison function that compares value of string X with X and outputs a 1 only if value of X is greater than or equal to that of X. Since f is 1 only when $X \geq X$ is true but $x \geq Y+1$ is false, it can be expressed as $f = C(X)\overline{C}(Y+1)$. Note that since an inverter as well as a 2-input AND gate can be realized in the QCA technology, so can the function f if one can design the comparison function $C(\cdot)$ in QCA. Similarly if f is 1 anytime the input string X is either between X_1 and Y_1 or between X_2 and Y_2, then that f may be described as $f = C(X_1)\overline{C}(Y_1+1)+C(X_2)\overline{C}(Y_2+1)$ and realized by the QCA. Fig. 4 shows this realization.

It is obvious that any arbitrary Boolean function can be implemented using the strategy described here. Creating such an implementation requires one to determine contiguous groups of 1's in the truth table of the function. The end points of each group are then compared with the input variable string, and the results ANDed. Finally, outputs of all the ANDs are added together to get the function. Clearly, All the comparisons can be done concurrently and all the ANDs evaluate at the same time. Thus the resultant Boolean function realization may have a fairly small depth provided one can find a small depth comparison function implementation.

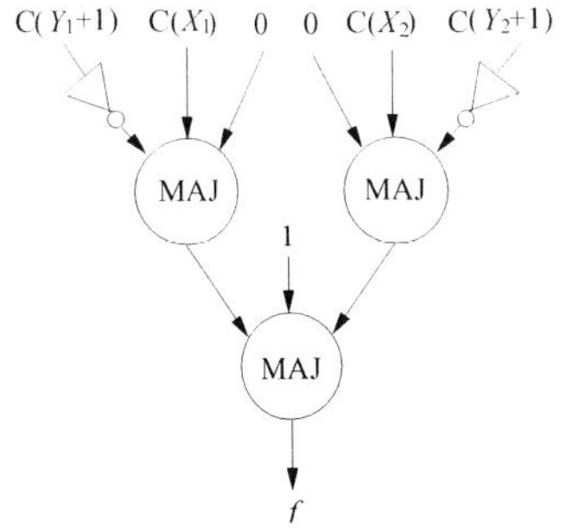

Figure 4: QCA realization of function f which is 1 only when its argument is between X_1 and Y_1 or between X_2 and Y_2. Note that $C(\cdot)$ is a comparison function.

The comparison function $C(B)$ that compares two n bit strings $B = b_{n-1}b_{n-2}\cdots b_0$ and $X = x_{n-1}x_{n-2}\cdots x_0$ can be implemented by subtracting B from X while keeping track of the carry only. Fig. 5 shows a QCA implementation of this strategy.

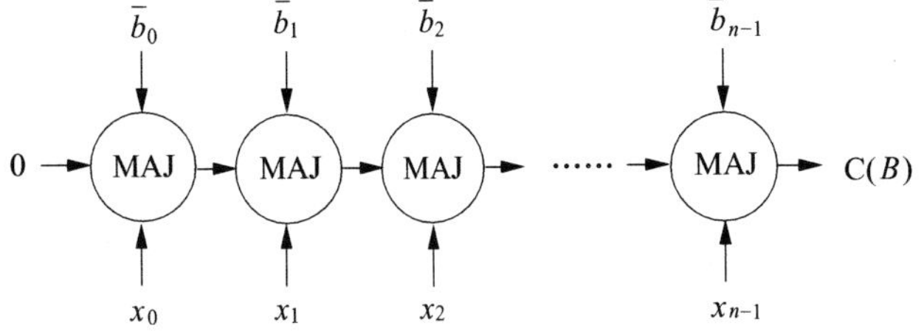

Figure 5: A serial realization of comparison $C(B)$ which outputs a 1 when $x_{n-1}x_{n-2}\cdots x_0 \geq b_{n-1}b_{n-2}\cdots b_0$.

Unfortunately, the architecture in Fig. 5 has an $O(n)$ delay. Minimizing this delay in QCA architectures is important because even in the combinational logic, unlike CMOS, all the gates in QCA implementations need to be clocked. Further, in applications such as the Boolean function implementation, the comparator delay can directly impact the delay of the Boolean function.

In order to achieve the optimal delay, we build the comparison output recursively. Suppose operands X and B are partitioned as $X = [X_1|X_0]$ and $B = [B_1|B_0]$. $X \geq B$ is true if $X_1 > B_1$ or if $(X_1 = B_1)$ and simultaneously $(X_0 \geq B_0)$. However, computing equality of X_1 and B_1 requires XOR gates which are expensive in QCA technology. We therefore define intermediate logical variables $p_i \equiv (X_i > B_i)$ and $q_i \equiv (X_i \geq B_i)$, $i = 0$ or 1. In this new notation, $(X_1 = B_1)$ has the same truth value as $(q_1\overline{p}_1)$. Thus the output of comparison $X \geq B$ can be written as a Boolean expression $p_1 + (q_1\overline{p}_1)q_0 = p_1 + q_1q_0$. Further, using the fact that $p_1 = 1$ implies $q_1 = 1$, one gets $p_1 = p_1q_1$ and $q_1 = p_1 + q_1$. Thus $p_1 + q_1q_0$ can be rewritten as $p_1q_1 + p_1q_0 + q_1q_0$, which is precisely the output of a

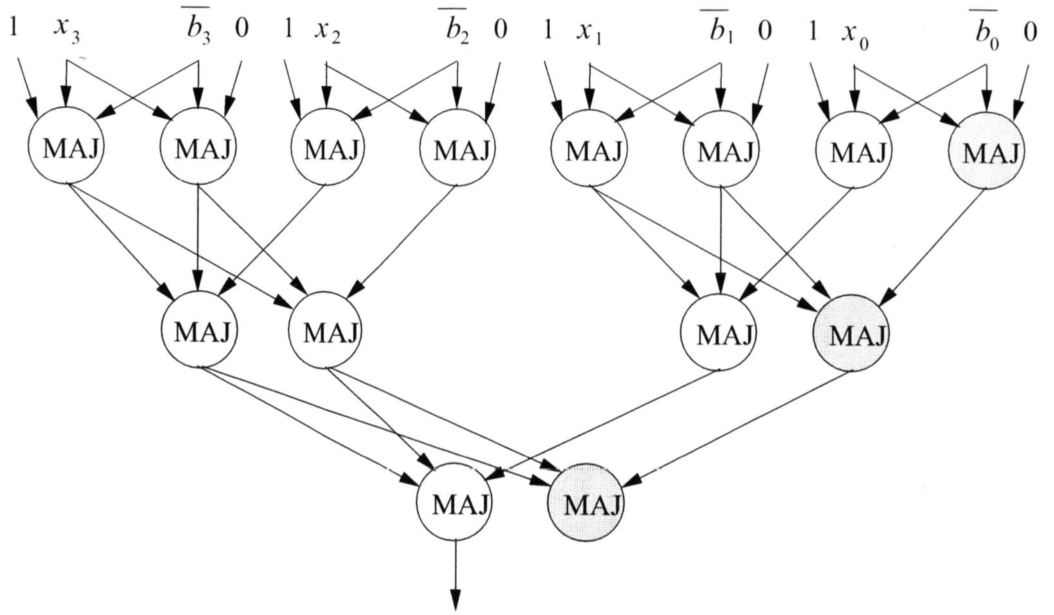

Figure 6: A 4-bit comparator architecture which produces a 1 when $x_3x_2x_1x_0 \geq b_3b_2b_1b_0$. (The majority gates in gray need not be implemented.)

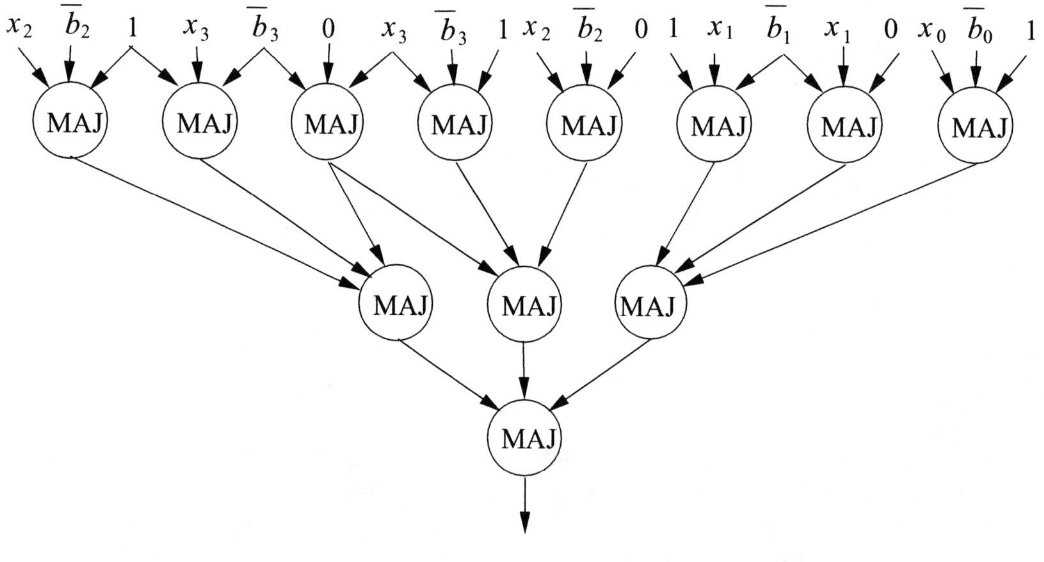

Figure 7: The same comparator as in Fig. 6 after elimination of wire crossings.

NSTI-Nanotech 2008, www.nsti.org, ISBN 978-1-4200-8505-1 Vol. 3

three input majority gate with inputs p_1, q_1 and q_0.

Similarly, the truth value of $X > B$ can also be computed from the intermediate logical variables p_i and q_i. In particular, $X > B$ if either $X_1 > B_1$ or $X_1 \geq B_1$ and simultaneously $X_0 > B_0$. This can be expressed by the Boolean expression $p_1 + q_1 p_0$. Once again using the fact that $p_1 = p_1 q_1$ and $q_1 = p_1 + q_1$, the expression for $X > B$ can be rewritten as $p_1 q_1 + (p_1 + q_1)p_0$. But this says that $X > B$ can be computed with a 3-input majority gate with inputs p_1, p_0 and q_1.

To compute p_i and q_i, the operands X_i and B_i can be partitioned recursively and the same procedure used. This can be continued till each partition is a single bit. Fig. 6 shows an implementation of a 4 bit comparator using only inverters and 3-input majority gates derived by this strategy. In every pair of majority gates in this figure, the gate on the left computes q_i and the one on the right, p_i.

The number of majority gates used in this realization of an n bit comparator is $4n - \lceil log_2 n \rceil - 3$ and the delay of the structure is $\lceil log_2 n \rceil + 1$. The tree-like structure allows for easy separation of clocking zones in QCA architectures. Further, if one of the strings, say B, is constant, then all the $\bar{b}_i$ are known in the design and one can apply 1's and 0's, as appropriate at the inputs that expect $\bar{b}_i$'s.

A two bit comparator architecture layout is illustrated in Fig. 8. It should be noted that for n as small as

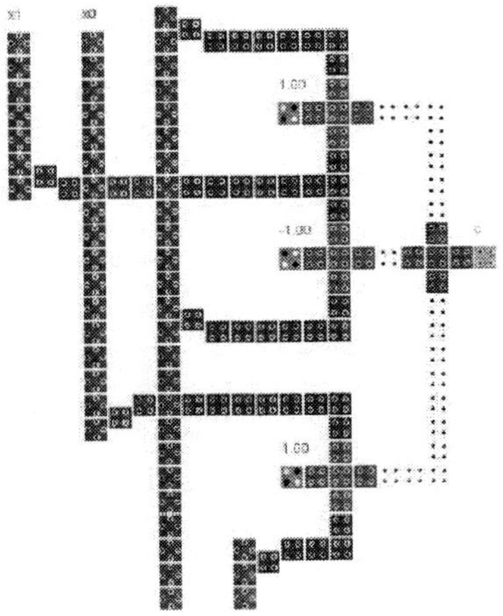

Figure 8: Layout of a 2 bit comparator in QCA technology.

2, the proposed strategy actually will have the same delay but more gates than the sequential strategy of Fig. 5. One may note that the advantage of the new architecture in Fig. 6 can be realized only for larger values of n. It reduces the comparator delay from n to $\lceil log_2 n \rceil + 1$, while increasing the number of majority gates from n to $4n - \lceil log_2 n \rceil - 3$.

However, as can be seen from Fig. 6, the tree architecture has wire crossings at every level of the tree. QCA being a planer architecture, minimizing these wire crossings is important. To achieve this objective, we duplicate certain majority gates as shown in Fig. 7. Note that this conversion does not affect the architecture but it completely eliminates the crossing of wires from all tree levels except the first. The resultant comparators of 2, 4, 8 and 16 bits use only 4, 12, 33 and 89 majority gates respectively.

4 CONCLUSION

This paper provides a new QCA architecture for the comparison function with an $O(\log n)$ delay and $O(n)$ gate complexity for an n bit comparator. Using this comparison realization, one may be able to obtain better (low depth) implementations of some Boolean functions.

REFERENCES

[1] C. S. Lent, P. D. Tougaw, W. Porod, and G. H. Bernstein, "Quantum cellular automata," *Nanotechnology*, vol. 4, pp. 49–57, Jan. 1993.

[2] "The international technology roadmap for semiconductors: Emerging research devices." http://www.itrs.net/, 2005.

[3] K. Walus, G. A. Jullien, and V. S. Dimitrov, "Computer arithmetic structures for quantum cellular automata," in *Proc. 37th Asilomar Conf. Signals, Systems and Computers*, (Pacific Grove, CA), pp. 9–12, Nov. 9–12 2003.

[4] I. Hanninen and J. Takala, "Binary multipliers on quantum-dot cellular automata," *Facta Universitatis Ser.: Elec. Energ.*, vol. 20, pp. 541–560, Dec. 2007.

[5] H. Cho and E. E. Swartzlander, "Adder designs and analyses for quantum-dot cellular automata," *IEEE Trans. Nanotechnology*, vol. 6, pp. 374–383, May 2007.

Modeling and Analysis of a Membrane-Based Randomized-Contact Decoder

Jennifer Long and John E. Savage

Computer Science, Brown University
Providence, RI 02912-1910, jes@cs.brown.edu

ABSTRACT

Decoders are employed to address individual NWs in crossbars (xbars) used for storage and computation. We report on the simulation and analysis of a membrane-based nanowire (NW) decoder proposed by Pribat and Savage [1] that implements the randomized-contact decoder (RCD) introduced by Williams and Kuekes [2] and analyzed by Hogg *et al* [3] and Rachlin and Savage [4]. An RCD has two sets of parallel wires, NWs and mesoscale wires (MWs), that are orthogonal to one another. "Contacts" are made at random at NW/MW junctions. These contacts, which allow a MW to control a NW, assign codewords to NWs, some of which are unusable. In the membrane-based decoder a contact is made by filling a pore with a metallic pin. As the probability of filling pores increases, the probability of creating a MW/NW contact increases monotonically. However, the average number of usable NW addresses achieves a maximum when a small fraction of the pores is filled. The goal is to maximize the average number of usuable NW addresses. We report on the results of our simulations of the membrane-based RCD as well as corroborate the experimental results with analysis of probabilistic models.

Keywords: crossbars, memories, decoders, nanowires

1 Introduction

Nanowire (NW) xbars (see Figure 1) provide a promising basis for nanoscale devices such as electronic memory and logic circuits [5]–[8]. To form an xbar, two groups of parallel nanowires are placed orthogonally with programmable molecules (PMs) between them. The PMs at NW crosspoints are switched between open and a diode by applying a positive or negative electric field. NW xbars can be used as storage units or programmed logic arrays. Storage densities as high as 10^{11} bits/cm^2 have been achieved [9].

To achieve high storage densities a method for controlling NWs with mesoscale wires (MWs) must be used. Given that MWs are an order of magnitude larger than NWs, it does not suffice to attach one MWs to each NW. Three basic techniques have been proposed to control NWs with MWs. The first (FET-based) assumes

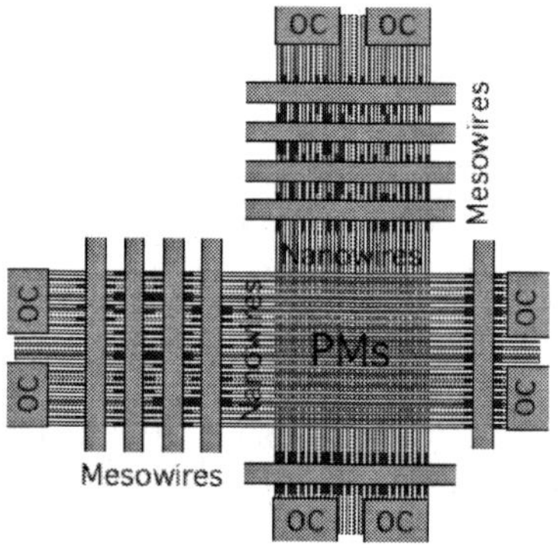

Figure 1: A NW xbar with programmable molecules (PMs) at NW crosspoints. A small number of NWs is connected between ohmic contacts (OCs). A NW is addressed by selecting and OC pair and deactivating all but one NW by applying fields to MWs. Data is stored at a crosspoint by applying a large electric field across it. Data is sensed with a smaller field.

that segments of each NW can be "opened" by electric fields applied by MWs. (They act like field effect transistors.) [5], [6], [10]–[12]. The second (CMOL) assumes that nanoscale pins are grown at crosspoints of a MW xbar that make contact with NWs in a NW xbar by passing through the openings between NWs [13]. The third places two electrodes on either side of a small set of lightly-doped NWs and applies a graduated electric field that depletes the carriers in all but one NW [14].

All three techniques introduce randomness in the connections between NWs and MWs. This is unavoidable when the NW pitch is 15-20 nms or less; here lithography is either too coarse or too expensive. Each technique assigns addresses to NWs that are not predictable in advance. A separate memory is needed to map from contiguous external addresses to internal ones.

Three general methods of realizing FET-based decoders have been proposed. The **encoded nanowire decoder**, assumes that NWs are lightly-doped in some sections along their length and heavily-doped elsewhere and assembled fluidically [10], [15]. Lightly-doped regions are placed in NWs during growth [16]–[18] or exposed by etching NWs that are radially encoded with differentially etcheable shells [12]. The latter method

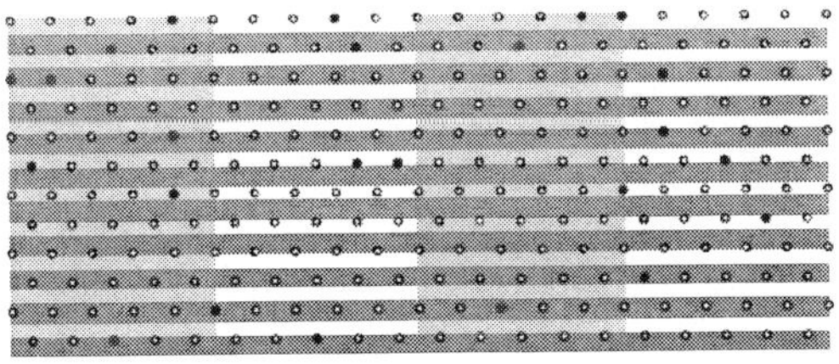

Figure 2: The membrane-based RCD has metal-filled pores placed at random between MWs (light grey) and NWs (dark grey). The red (blue) dots denote filled pores whose centers overlap (don't overlap) a NW.

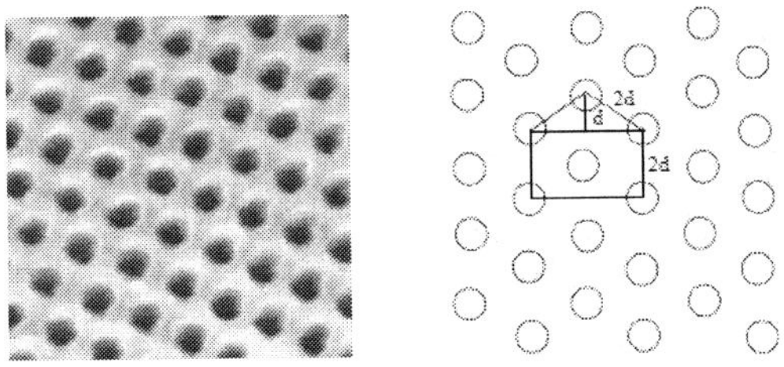

Figure 3: Regular pores in aluminum oxide and a model hexagonal lattice.

avoids misalignment arising in the former. The **mask-based decoder**, assumes that a regular array of uniform NWs [19], [20] is first placed on a chip. High-K dielectric rectangles are deposited between some NWs and MWs [11]. A NW is controllable by a MW if there is rectangle between them. The **randomized-contact decoder (RCD)** (see Figure 2) also assumes uniform NWs. It assumes that a "contact" (equivalent to a FET) is made at random between NWs and MWs.

We report on simulation and analysis of the membrane-based RCD. Membranes can be created by anodizing a thin aluminum film deposited between NWs and MWs to form hexagonal pores. (See Figure 2.) When pores are filled at random with metal or a high-K dielectric, an electric field applied to MWs creates a FET in lightly doped NWs. We investigate the number of usuable NWs resulting from random assembly as a function of the sizes of NWs and MWs and the fraction of filled pores. We believe that RCDs are among the most promising methods of controlling NWs in xbars.

2 The Membrane-Based Decoder

It is assumed that pores form a hexagonal lattice, as shown in Figure 3. The pore lattice is created and rotated to a preset angle relative to the MWs and pores filled at random.

Let λ_{meso}, λ_{nano} be the width and separation of MWs and NWs. (Their pitch is twice their width.) Let $2d_{pore}$ be the center to center pore spacing in the pore lattice (see Figure 3) and let θ be the rotation angle of

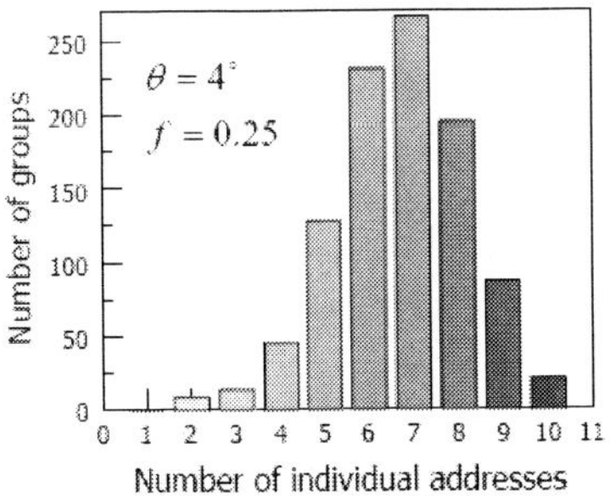

Figure 4: A histogram of N_a, the number of i.a. NWs in 1000 groups of 10 NWs. The rotation angle is 4 degrees and the fraction of pores filled is 0.25.

the pore lattice relative to NWs. Finally, let f be the fraction of filled pores in the lattice and n be the number of runs of the program.

The simulator chooses pores to fill by picking two random integers u and v treated as row and column indices. These integers are mapped to physical positions based on the dimensions of the lattice, its displacement, and the angle between NWs and MWs. Pores are filled until a preset fraction f of the pores are filled.

A pore forms a NW/MW contact if the pore center overlaps both (red dots in Figure 2). A MW **controls** a NW if there is at least one filled pore between them. Let there be N NWs and M MWs. Let ν_i by the ith NW, $1 \leq i \leq N$, and let μ_j be the jth MW, $1 \leq j \leq M$. ν_i has **codeword** $\mathbf{c}_i$ in which the jth component, $c_{i,j}$, is 1 if μ_j controls ν_i and 0 otherwise. If $c_{i,j} = 1$ and an electric field is applied to μ_j, ν_i is turned off.

Codeword $\mathbf{c}_r$ **implies** the codeword $\mathbf{c}_s$ ($\mathbf{c}_r \Rightarrow \mathbf{c}_s$) if for each $1 \leq j \leq M$ $c_{r,j} \Rightarrow c_{s,j}$ where the latter holds if and only if $c_{s,j}$ is 1 whenever $c_{r,j}$ is 1. Otherwise, it is unconstrained. Clearly, $010 \Rightarrow 110$.

If $\mathbf{c}_r \Rightarrow \mathbf{c}_s$, whenever fields are applied to MWs that leave $\mathbf{c}_r$ on, $\mathbf{c}_s$ is also on. Thus, the two NWs cannot be controlled independently. We say that the ith NW with codeword $\mathbf{c}_i$ is **individually addressable (i.a.)** if for no other codeword $\mathbf{c}_s$ does $\mathbf{c}_s \Rightarrow \mathbf{c}_i$. Consider the codewords 010 and 110. The NW with codeword 110 is on when no field is applied to the first two MWs. But in this case, the other NW is also on. To determine if $\mathbf{c}_r \Rightarrow \mathbf{c}_s$, associate with each codeword its set of "1" positions and then test for containment of these sets.

The simulator is given N and M as well as $\rho = \lambda_{meso}/\lambda_{nano}$, $r = \lambda_{nano}/d$ (d is the pore diameter), θ, and f, the fraction of filled pores. It was run 1000 times to produce 1000 sets of N NW codewords. The simulator reports the number of controllable NWs as well as N_a, the number of i.a. NWs. A histogram of N_a is shown in Figure 4 when $\theta = 4°$ and $f = 0.25$. In 992 of 1000 address groups all NWs were controllable. N_a for these controllable groups is 6.71.

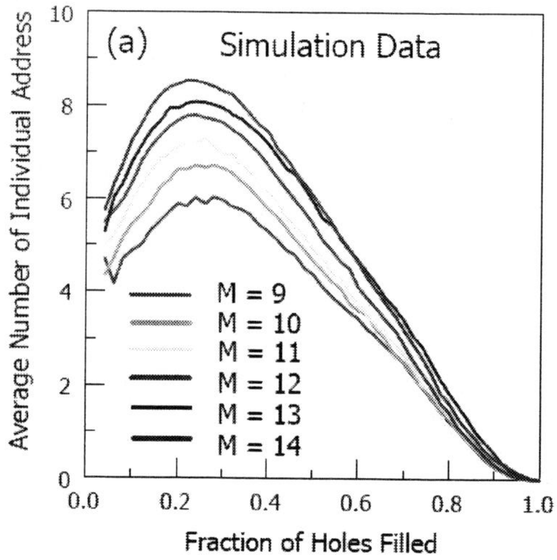

Figure 5: N_a as a function of f, the fraction of pores filled when $N = 10$, $r = 1$, $\theta = 15°$, and $\rho = 1$.

3 Simulation Results and Analysis

The simulator was used first to study the impact of θ and f. θ is important because when $r = 1$ and θ is a multiple of 60°, the number of controllable NWs is zero.

To study the effect of θ, we ran 1,000 simulations with $N = M = \rho = 10$, $r = 1$, and $f = 0.25$ at $\theta = 1°, 61°$ and multiples of 6°. We found that N_c, the average number of controllable NWs, was almost exactly N and N_a was between about 6.8 and 7.0. We conclude that the rotation angle is unimportant in the simulation except if close to a multiple of 60°. Because in practice pores do not form a perfect hexagonal pattern, the rotation angle should not be important.

We studied the effect of the fraction f of filled pores on N_c, and N_a by setting $\theta = 15°$. N_c grows monotonically with f and has a sharp transition from 0 to 1 as f increases from 0.05 to 0.2 or less, depending on the value of M. This phase transition has been seen before [3]. The more pores that are filled, the higher is the probability that a NW can be controlled by a MW.

Figure 5 shows that N_a is increases with M and has a single maximum at about $f = 0.25$. Experiments show the maximum increasing with pore diameter. N_a increases with M because NW codewords are more likely to be i.a. when M is large. The maximum in f reflects the fact that codewords are more likely to be similar (imply one another) when f is small or large.

We now explain the simulation results through probabilistic analysis. Let $a_{i,j}$ denote the number of pores at the ν_i/μ_j junction. Then, $A_i = \sum_{j=1}^{M} a_{i,j}$ is the total number of pores in all junctions between ν_i and MWs. Let $p_{i,j}$ $(q_{i,j} = 1 - p_{i,j})$ be the probability that ν_i is (is not) controlled by μ_j. A NW is controlled if one or more pores between and a MW are filled. It follows that

$$p_{i,j} = 1 - q_{i,j} = 1 - (1-f)^{a_{i,j}}$$

Let P_i $(Q_i = 1 - P_i)$ be the probability of that ν_i is (is not) controlled. It follows $Q_i = \prod_{j=1}^{M} q_{i,j} = (1-f)^{A_i}$ and $P_i = 1 - Q_i = 1 - (1-f)^{A_i}$. A group of NWs is controllable when each NW in the group is controllable. Since pores are filled independently, each NW is independent and the probability P' that all N NWs are controllable satisfies

$$P' = \prod_{i=1}^{M} P_i = \prod_{i=1}^{M} \left[1 - (1-f)^{A_i}\right]$$

Applying the inequality between arithmetic and geometric means, twice and simplifying we have the following.

$$P' \leq \left(1 - (1-f)^{M\bar{a}}\right)^N$$

Here $\bar{a} = (\sum_{i=1}^{N} A_i)/(MN)$ is the the average number of pores per junction.

Computing P' when $f \approx 0.20$, $N = 10$, $M = 10$ and $\theta = 15°$ and $\bar{a} = 3.0$ (obtained from simulations) shows that $P' \leq (1 - (1-0.20)^{30})^{10} = 0.987$ which is slightly larger than the simulation result, 0.975.

Consider next N_a. NW ν_i has codeword $\mathbf{c}_i$. ν_i is i.a. if there is no ν_s such that $\mathbf{c}_s \Rightarrow \mathbf{c}_i$. The probability that that NW ν_i is not i.a. is the probability that for some $s \neq i$, $c_s \Rightarrow c_i$. From the union bound, $P(\nu_i \text{ is not i.a.}) \leq \sum_{s \neq i} P(c_s \Rightarrow c_i)$.

$$P(\nu_i \text{ is i.a.}) \geq 1 - \sum_{s \neq i} P(c_s \Rightarrow c_i)$$

But $c_s \Rightarrow c_i$ if and only if $c_{s,k} \Rightarrow c_{i,k}$ for all $1 \leq k \leq M$ which holds when $c_{s,k} = 1$ and $c_{i,k} = 0$ does not occur. Thus, $P(c_{s,k} \Rightarrow c_{i,k}) = p_{s,k}q_{i,k}$. Recall that $p_{i,j} = 1 - (1-f)^{a_{i,j}}$. $a_{s,k}$ and $a_{i,k}$ may be statistically dependent. However, $p_{s,k}$ and $p_{i,k}$ are statistically independent when $a_{s,k}$ and $a_{i,k}$ are fixed. Thus,

$$P(c_{s,k} \Rightarrow c_{i,k} | a_{s,k} a_{i,k}) = 1 - (1 - (1-f)^{a_{s,k}})(1-f)^{a_{i,k}}$$

To procede we assume that $a_{s,k}$ and $a_{i,k}$ are statistically independent, an assumption tested below. Averaging all choices for $a_{s,k}$ and $a_{i,k}$, we have

$$P(\nu_i \text{ is i.a.}) \geq 1 - (N-1)(1-pq)^M$$

where $q = \sum_{a_{r,k}} (1-f)^{a_{r,k}} P(a_{r,k})$. It follows that $N_a \geq N \cdot [1 - (N-1) \cdot (1-pq)]^M$.

We measured $P(a_{i,j})$ when $N = 10$, $M= 10$, $\rho = 1$, $r = 1$ and $\theta = 15°$. In this case $0 \leq a_{i,j} \leq 6$. We found

x	0	1	2	3	4	5	6
$P(x)$	0.0	0.02	0.38	0.34	0.13	0.07	0.06

The theoretical and simulation results both give maximums in N_a when $f \approx 0.24$. When $f = 0.24$, $q \approx 0.457$ and $p = 1-q \approx 0.543$, and the analytical lower bound to $N_a \approx 4.81$, which is somewhat less than the simulation result of 6.75 but comparable given the estimates that have been made.

4 Conclusions

The membrane-based randomized contact decoder provides a promising implementation of a NW decoder. It obviates the technical difficulty of precisely positioning contacts on the nano-scale. It can also cope with fabrication defects present in the nano-scale xbar. The simulations and analysis presented here advance our understanding of the potential of this technology by showing that the number of individually addressable NWs is largest for a number of MWs between 9 and 14 when the fraction of filled pores $f \approx .25$, a resulted supported by analysis.

Work supported in part by NSF Grant CCF-0403674.

REFERENCES

[1] Didier Pribat and John E. Savage. Membrane-based randomized-contact decoders. Technical report, Computer Science Department, Brown University, February 26 2007.

[2] R. S. Williams and P. J. Kuekes. Demultiplexer for a molecular wire crossbar network, US Patent Number 6,256,767, July 3, 2001.

[3] Tad Hogg, Yong Chen, and Philip J. Kuekes. Assembling nanoscale circuits with randomized connections. *IEEE Trans. Nanotechnology*, 5(2):110–122, 2006.

[4] Eric Rachlin and John E Savage. Nanowire addressing with randomized-contact decoders. In *Procs. ICCAD*, pages 735–742, November, 2006.

[5] P. J. Kuekes, R. S. Williams, and J. R. Heath. Molecular wire crossbar memory, US Patent Number 6,128,214, Oct. 3, 2000.

[6] André DeHon. Array-based architecture for FET-based, nanoscale electronics. *IEEE Transactions on Nanotechnology*, 2(1):23–32, Mar. 2003.

[7] André DeHon, Seth Copen Goldstein, Philip Kuekes, and Patrick Lincoln. Nonphotolithographic nanoscale memory density prospects. *IEEE Transactions on Nanotechnology*, 4(2):215–228, 2005.

[8] André DeHon. Nanowire-based programmable architectures. *J. Emerg. Technol. Comput. Syst.*, 1(2):109–162, 2005.

[9] Jonathan E. Green, Jang Wook Choi, Akram Boukai, Yuri Bunimovich, Ezekiel Johnston-Halperin, Erica DeIonno, Yi Luo, Bonnie A. Sheriff, Ke Xu, Young Shik Shin, Hsian-Rong Tseng, J. Fraser Stoddart, and James R. Heath. A 160-kilobit molecular electronic memory patterned at 1011 bits per square centimetre. *Nature*, 445:414–417, 2006.

[10] André DeHon, Patrick Lincoln, and John E. Savage. Stochastic assembly of sublithographic nanoscale interfaces. *IEEE Transactions on Nanotechnology*, 2(3):165–174, 2003.

[11] Robert Beckman, Ezekiel Johnston-Halperin, Yi Luo, Jonathan E. Green, and James R. Heath. Bridging dimensions: Demultiplexing ultrahigh-density nanowire circuits. *Science*, 310:465–468, 2005.

[12] John E. Savage, Eric Rachlin, André DeHon, Charles M. Lieber, and Yue Wu. Radial addressing of nanowires. *J. Emerg. Technol. Comput. Syst.*, 2(2):129–154, 2006.

[13] K. K. Likharev and D. B. Strukov. Cmol: Devices, circuits, and architectures. In G. Cuniberti *et al.*, editor, *Introduction to Molecular Electronics*, pages 447–477, 2005.

[14] K. Gopalakrishnan, R. S. Shenoy, C. Rettner, R. King, Y. Zhang, B. Kurdi, L. D. Bozano, J. J. Welser, M. B. Rothwell, M. Jurich, M. I. Sanchez, M. Hernandez, P. M. Rice, W. P. Risk, and H. K. Wickramasinghe. The micro to nano addressing block. In *Procs. IEEE Int. Electron Devices Mtng.*, Dec. 2005.

[15] Benjamin Gojman, Eric Rachlin, and John E Savage. Decoding of stochastically assembled nanoarrays. In *Procs 2004 Int. Symp. on VLSI*, Lafayette, LA, Feb. 19-20, 2004.

[16] Mark S. Gudiksen, Lincoln J. Lauhon, Jianfang Wang, David C. Smith, and Charles M. Lieber. Growth of nanowire superlattice structures for nanoscale photonics and electronics. *Nature*, 415:617–620, February 7, 2002.

[17] Yiying Wu, Rong Fan, and Peidong Yang. Block-by-block growth of single-crystal Si/SiGe superlattice nanowires. *Nano Letters*, 2(2):83–86, 2002.

[18] M. T. Björk, B. J. Ohlsson, T. Sass, A. I. Persson, C. Thelander, M. H. Magnusson, K. Deppert, L. R. Wallenberg, and L. Samuelson. One-dimensional steeplechase for electrons realized. *Nano Letters*, 2(2):87–89, 2002.

[19] G. Y. Jung, S. Ganapathiappan, A. A. Ohlberg, L. Olynick, Y. Chen, William M. Tong, and R. Stanley Williams. Fabrication of a 34x34 crossbar structure at 50 nm half-pitch by UV-based nanoimprint lithography. *Nano Letters*, 4(7):1225–1229, 2004.

[20] E. Johnston-Halperin, R. Beckman, Y. Luo, N. Melosh, J. Green, and J.R. Heath. Fabrication of conducting silicon nanowire arrays. *J. Applied Physics Letters*, 96(10):5921–5923, 2004.

Nanostimulation in Brain and Emerging Imaging Capabilities

R. Sharma[1] and A. Sharma[2]
[1]Center of Nanobioscience at University of North Carolina, at Greensboro NC USA
[1]Florida State University, Tallahassee, Florida, USA, r_sharma@uncg.edu
Computation Science lab, Electrical Engineering, MP A & T University, Udaipur, Rajasthan, India, avdhesh_2000@yahoo.com

ABSTRACT

Neural network in brain is a complex system. Its neuroimaging with directionality, composition, ionic dynamicity and nanoscale molecular detection are major nano-stimulations as challenges. Mainly, sodium-potassium pump, glucose dynamics and oxygen status in neurons are major players in the process of neuroactivation and control over brain functions of cognition, thought, speech, movement and sensation including motor-sensory reflex. There is still undefined dogma of close relationship between intracellular sodium pump and glucose uptake with oxygen level in neuron or blood. Possibly different neurotransmitters play active role in this process but still it is dilemma. With advent of nanoscale imaging technology available, possibly this dilemma may be solved by fusion of microimaging techniques like MRI/PET/CT/SPECT using smart imaging contrast materials. Present time common methods include positron emission tomography (PET), functional magnetic resonance imaging (fMRI), multichannel electroencephalography (EEG) or magnetoencephalography (MEG), and near infrared spectroscopic imaging (NIRSI). PET, fMRI and NIRSI can measure localized changes in cerebral blood flow related to neuroactivations, neural computations of any part to evaluate brain behaviour.

Key words: nanostimulation, fMRI,brain,imaging,BOLD

INTRODUCTION

fMRI detects the blood oxygen level–dependent (BOLD) changes in the MRI signal that arise when changes in neuronal activity occur following a change in brain state, such as may be produced, for example, by a stimulus or task. One of the underlying premises of many current uses of functional imaging is that various behaviors and brain functions rely on the recruitment and coordinated interaction of components of "large-scale" brain systems that are spatially distinct, distributed, and yet connected in functional networks. Thus, although the practice of phrenology is dead, identification of the neurobiological substrates associated with various specific functions of the brain is likely to shed light on how the brain determines behavior. In addition, geographic maps identifying the locations of particularly critical areas, such as those involved in producing and understanding language, are of direct importance in clinical assessments and the planning of interventions.

The physical origins of BOLD signals are reasonably well understood, though their precise connections to the underlying metabolic and electrophysiological activity need to be clarified further. It is well established that an increase in neural activity in a region of cortex stimulates an increase in the local blood flow in order to meet the larger demand for oxygen and other substrates. The change in blood flow actually exceeds that which is needed so that, at the capillary level, there is a net increase in the balance of oxygenated arterial blood to deoxygenated venous blood. Essentially, the change in tissue perfusion exceeds the additional metabolic demand, so the concentration of deoxyhemoglobin within tissues decreases. This decrease has a direct effect on the signals used to produce magnetic resonance images. While blood that contains oxyhemoglobin is not very different, in terms of its magnetic susceptibility, from other tissues or water, deoxyhemoglobin is significantly paramagnetic (like the agents used for MRI contrast materials, such as gadolinium), and thus deoxygenated blood differs substantially in its magnetic properties from surrounding tissues [1]. When oxygen is not bound to hemoglobin, the difference between the magnetic field applied by the MRI machine and that experienced close to a molecule of the blood protein is much greater than when the oxygen is bound. On a microscopic scale, replacement of deoxygenated blood by oxygenated blood makes the local magnetic environment more uniform. The longevity of the signals used to produce magnetic resonance images is directly dependent on the uniformity of the magnetic field experienced by water molecules: the less uniform the field, the greater the mixture of different signal frequencies that arise from the sample, and therefore the faster the decay of the overall signal.

The result of having lower levels of deoxyhemoglobin present in blood in a region of brain tissue is therefore that the MRI signal from that region decays less rapidly and so is stronger when it is recorded in a typical magnetic resonance image acquisition. This small signal increase is the BOLD signal recorded in fMRI.. It is typically around 1% or less, though it varies depending on the strength of the applied field; this variability is one reason why higher-field MRI systems are being developed. As can be predicted from the above explanation, the magnitude of the signal depends on the changes in blood flow and volume within tissue, as well as the change in local oxygen tension, so there is no simple relation between the signal change and any single physiological parameter. Thus fMRI does not report absolute changes such as the units of flow obtained with positron emission tomographic (PET) imaging. Furthermore, as neurons become more active, there is a time delay before the necessary vasodilation can occur to increase flow, and for the wash-out of deoxyhemoglobin from the region to occur. Thus the so-called hemodynamic response detected by BOLD imaging is delayed and has a duration of several seconds following a stimulating event [3].

In fMRI, a subject is placed in the magnet of an MRI machine, where various different kinds of stimulus may be administered in a controlled fashion. For example, sounds may be played, visual scenes may be presented, and small motor movements or responses can be recorded. Although conventional imaging methods can be adapted for fMRI [4], most studies are performed using "snapshot" imaging methods, of which echo-planar imaging is the main exemplar [5]. Improvements in gradient coil technology in recent years have permitted the implementation of such ultrafast imaging methods, in which complete cross-sectional images are recorded in substantially less than a second (typically about 50–100 ms). Although these images have poorer resolution and overall quality than the images used for radiological diagnosis, they do not suffer blurring from physiological motion, they permit multi-slice recording of the entire brain in a few seconds, and, in principle, they permit sampling of the hemodynamic response to transient events, as well as multiple recordings of repeated stimuli over a typical experimental time course. The images in a sequence may differ from one another even when there is no change in brain state, because of signal variance and the effects of extraneous "noise" (such as that produced by random voltages within the coils and components used to record MRI signals), so fMRI studies compare sets of images acquired during two or more different conditions using some form of statistical analysis. Multiple recordings of the MRI signal permit a degree of signal averaging that increases the reliability of the results.

Among the many types of study that can be performed, two main experimental paradigms are in common use. In so-called block designs, stimuli are presented in alternating short runs ("blocks") of several seconds' duration, and the MRI signals are then compared for the two types of blocks. For a visual-stimulation task to localize primary visual areas, a subject might view a bright flickering checkerboard for 20 seconds, followed by a dark screen for 20 seconds, with these blocks repeated several times; eight pairs of blocks, for example, would require a total recording that lasts under 6 minutes. During that time, images may be recorded for many different parallel slices (typically 10–20), such that each slice is imaged about every 2 seconds. In this example, 80 images would be acquired for each slice for both conditions (stimulus ON and OFF). Those volume elements (voxels) within the brain that are affected by the stimulus (such as the primary visual cortex) provide a sequence of data points in which the signal alternates in intensity in synchrony with the stimulation because of the BOLD effect [Figure 2]. By detection of which voxels show this alternating pattern, the visual cortex can be identified. The block design is simple to implement and can be used to localize several basic functions including: (i) primary sensory areas, e.g., by presenting auditory stimuli versus silence; (ii) areas involved with simple motor tasks, such as finger tapping; (iii) higher-order visual areas, for example, by presenting faces versus other objects to identify regions that respond preferentially to faces; and (iv) higher-order language areas by presenting speaking versus other complex sounds.

In a block design, voxels not affected by the stimulation should provide a steady set of data points, albeit with some variance due to random and physiological changes not connected with the brain state. Various strategies have been devised for identifying the real, task-related signal changes in the presence of other fluctuations. For example, in comparing the 80 data points in the ON with the 80 data points in the OFF condition for each voxel, a simple Student's t test may suffice for evaluation of the likelihood that any single voxel has responded. Computation of the relevant statistic at each voxel produces a statistical parametric map of each brain slice. In this map, the voxels that on balance appear to have responded (using some statistical decision criterion, such as $P < 0.05$) are portrayed as "activated". For quantitative comparisons between regions, the percentage signal change is often used, but statistical maps are generally used for evaluation of whether changes are reliable. As with any statistical analysis, positive findings should be couched in probabilities, and there is, therefore, some arbitrariness about the threshold criteria used for defining active regions. This is exacerbated by the fact that within a single image there may be over 10,000 voxels, so that a large number of statistical comparisons are necessarily performed simultaneously. Correcting for this issue of multiple comparisons complicates the assessment of reliability and significance, and at present there are no universally accepted strategies for rigorous analysis of data. The presence of variations that are not task related — for example, from patient motion, respiration, and cardiac pulsations — decreases the reliability of fMRI data, and much ongoing research aims to reduce these effects.

The second type of experiment in common use uses transient stimuli that replicate those used in so-called event-related paradigms that have been successfully exploited to record event-related potentials (ERPs) in electrophysiological studies. For testing of many cognitive functions, event-related studies offer a greater variety of more powerful experimental designs. For example, in electrophysiology, a well-known ERP is the P300, which denotes a positive (P) electrical potential that is measurable on the scalp approximately 300 ms after the detection of an "oddball" event, such as a target stimulus that occurs infrequently and at random within an otherwise regular sequence of stimuli (e.g., a change in a sequence of identical tones) [6]. P300s are routinely recorded for diagnostic applications in neurology and psychiatry, and their amplitude and latency are affected by various disorders. If the same set of stimuli is presented with an MRI machine, a transient hemodynamic response is also elicited by each oddball. By averaging of the responses to several "epochs," voxels that show transient fMRI signal changes can be identified. Note that in such an example, although the electrical activity elicited by the oddball stimulus peaks at 300 ms after the event and lasts for substantially less than 1 second, the corresponding BOLD signal may not peak until several seconds later and endures for much longer. Images recorded at multiple time points after the oddball will portray the time course of the blood flow changes produced by the neural activity.

fMRI has found applications in both clinical and more basic neuroscience. Appropriate experiments may now be designed to address specific hypotheses regarding the nature of the distributed systems responsible for various functional responses. For clinical applications, simple mapping of critical sensory and motor functions can be readily performed with subjects lying in the bore of a magnet where they perform simple tasks or experience sensory stimuli in blocks. This is the primary approach for evaluations of the brains of patients prior to neurosurgery (e.g., for the treatment of temporal lobe epilepsy or arterio-venous malformations) or radiation therapy. In several centers, standard protocols have been developed that permit the efficient mapping of auditory, visual, motor, and language areas to inform surgeons of the positions of critical functional areas. fMRI data can readily be integrated with image-guided neurosurgical procedures. The identification of eloquent cortex and the hemispheric dominance of language can be performed using fMRI rather than by invasive procedures such as the Wada test [7].

In neurological applications, fMRI may have a role in studies of patients recovering from stroke, as well as in understanding the extent of occult or asymptomatic losses of cortical functions in degenerative disorders such as Alzheimer disease [8]. In psychiatry, fMRI is being exploited to delineate the neurobiological bases of various cognitive deficits and aberrant behaviors. For these studies, more sophisticated tasks have been developed that require more subtle or specific cognitive responses. Many studies build on the tasks that have been developed over many years in neuropsychological testing of subjects for the identification of different traits or pathologies. Consider the classic Stroop test, in which subjects are required to name the color of various words printed in a colored font. In this test, subjects are instructed explicitly to suppress the (automatic) response of reading the words, and instead to identify only the color of the font. The words displayed may also be the names of the colors of the fonts used, with the word printed sometimes in the congruent color (e.g., "RED" printed in red) and sometimes in an incongruent color (e.g., "RED" printed in blue) [9]. Given the propensity to read the word despite the instruction to name only the color of the font, there is a well-documented attentional conflict when the word and the color are incongruent, leading in behavioral studies to a longer reaction time in arriving at the correct color identification. This is the Stroop effect, and it involves several elements of attention and task monitoring as well as correction of automatic responses. In an event-related Stroop test performed in the magnet, a series of congruent word-color pairs may be presented, and incongruent words can then be interspersed at random. At the appearance of each incongruent word, a set of brain areas begins to activate. The time course of these activations; clearly they depict the recruitment and subsequent recovery of multiple brain regions in one or more well-defined networks [10]. Stroop tests are one of many such psychological tests that may be used to discriminate various subject groups. In fMRI, the patterns of activation revealed by the Stroop test are different, for example, for subjects with depression and disorders such as pathological gambling, and these differences may be quantified. One aim of such tests may be to more accurately identify specific differences and deficits, as well as potentially to monitor the effects of treatments or interventions. An example is the use of fMRI to demarcate subgroups of subjects broadly diagnosed with schizophrenia based on characteristic deficits in specific cognitive domains. For example, in fMRI studies, some subjects demonstrate a failure to activate specific frontal regions during a verbal working memory task [11]. In this test they are presented with sequences of words to rehearse and to remember for several seconds prior to a simple test of word order. Normal subjects show robust activation in prefrontal cortex throughout the rehearsal period, whereas patients may show much-diminished and shorter-lived activation over this interval [Figure 6]. However, in some subjects the pattern of activation appears to normalize following several weeks of appropriate therapy in which the underlying skills required for this task are emphasized [12]. Similar documentation of the effects of therapy or learning have been achieved by several groups in different domains, so that the ability of fMRI to demonstrate the apparent plasticity of the brain is clear. The effects of pharmacologic interventions on cognitive functions may also be revealed via patterns of activation within the brain. For example, acute effects of intravenous administration of ketamine on the circuits activated during P300 stimuli have been quantified. Longer-term changes have also been shown; for example, fMRI studies of language have demonstrated cognitive effects produced by hormone replacement treatments with estrogen [13].

Among the most important applications of fMRI is the study of neurodevelopment and disorders associated with children, where the safety and noninvasive nature of MRI are paramount. For example, fMRI has been used to demonstrate the failure of autistic individuals to recruit the cortical substrate (which includes the fusiform gyrus) used for face processing by normal subjects, providing an anatomic basis for interpreting the lack of affect that such subjects demonstrate when confronted with human faces [14]. In developmental dyslexia, fMRI has been used to identify deficits in specific posterior circuits involving the angular gyrus that account for reading disability [15], and activations within these circuits have been shown to correlate well with reading skill [17] and to respond to educational interventions that focus on specific language and reading skills. fMRI may have an important role in evaluating the benefits of specific learning strategies and other interventions. An important area of further development is the application of advanced data analysis techniques and modeling in order to interpret fMRI results. For mapping of single critical functions (e.g., for neurosurgical planning), a simple but accurate map of activated voxels may be sufficient, but for many other purposes, a knowledge of where activity occurs may not advance our understanding of brain function significantly. However, greater insights into the neural basis of behavior may be obtained by examination of the manner in which regional activities vary with behavioral or other performance or physiological measures, or across tasks, or among subjects within a group, and between groups. To quantify these covariations, more advanced mathematical techniques have been developed, many of which resemble the multivariate statistical methods employed to derive relationships among variables in nonimaging data [17]. By such approaches, the relationship of fMRI activity to specific behavioral measures, and the connectivity between different brain regions, may be derived. Such methods have proven especially helpful in extracting new information on brain systems involved in complex responses from simple maps of activation acquired during different conditions.

fMRI has several limitations. The use of fast imaging reduces the spatial resolution to a few millimeters, somewhat worse than conventional MRI. The BOLD effect is small, and thus the sensitivity is limited, so that fMRI experiments require multiple samplings of brain responses. The temporal resolution is poor and is limited by the nature of the hemodynamic response. Furthermore, the reliability is reduced when there are significant subject motions or physiologically related variations. The origins and influence of various sources of such variance are not yet completely understood. For example, the importance of variations in the blood levels of everyday substances (such as caffeine, nicotine, and glucose), or of hormones (such as estrogen), all of which are likely to affect the BOLD signal, is not well documented. Aging and impaired cerebrovascular supply are also likely to affect the magnitude of the BOLD response. The BOLD effect is an indirect measure of neural activity, and there are couplings at different stages — for example, between electrical activity and metabolic flux, and between neurotransmitter release and energy supply — that are not well understood. Nonetheless, fMRI signals are not simply an epiphenomenon of neural activity, and they clearly are sensitive to subtle changes in the state of the brain, as well as to the effects of learning, of expertise, and of various treatments. fMRI provides an accurate and painless method for mapping of critical functions and likely has a much larger role to play in the management of clinical patients for diverse disorders in the future.

REFERENCES

1. Pauling, L., and Coryell, C.D. Proc. Natl. Acad. Sci. U. S. A. **22**:210-216,1936.
2. Ogawa, S., Lee, T.M., Kay, A.R., and Tank, D.W. Proc. Natl. Acad. Sci. U. S. A. **87**:9868-9872,1990.
3. Robson, M., Dorosz, J.L., and Gore, J.C. Neuroimage. **7**:185-198,1998.
4. Constable, R.T., McCarthy, G., Allison, T., Anderson, A.W., and Gore, J.C. Magn. Reson. Imaging. **11**:451-459,1993.
5. Cohen, M.S., and Weisskoff, R.M. Magn. Reson. Imaging. **9**:1-37,1991.
6. Sutton, S., Braren, M., Zubin, J., and John, E.R.. Science. **150**:1187-1188,1965.
7. Schlosser, M., Aoyagi, N., Fulbright, R.K., Gore, J.C., and McCarthy, G. Hum. Brain Mapp. **6**:1-13,1998.
8. Smith, C.D. et al. Neurology. **53**:1391-1396,1990.
9. Stroop, J.R. J. Exp. Psychol. **18**:643-662,1935.
10. Leung, H.C., Skudlarski, P., Gatenby, J.C., Peterson, B.S., and Gore, J.C. Cereb. Cortex. **10**:552-560,2000.
11. Stevens, A.A., Goldman-Rakic, P.S., Gore, J.C., Fulbright, R.K., and Wexler, B.E. Arch. Gen. Psychiatry. **55**:1097-1103,1998.
12. Wexler, B.E., Anderson, M., Fulbright, R.K., and Gore, J.C. Am. J. Psychiatry. **157**:1694-1697,2000.
13. Shaywitz, S.E. JAMA. **281**:1197-1202,1999.
14. Schultz, R.T. Arch. Gen. Psychiatry. **57**:331-340,2000.
15. Shaywitz, S.E. Proc. Natl. Acad. Sci. U. S. A. **95**:2636-2641,1998.
16. Shaywitz, B.A. Biol. Psychiatry. **52**:101-110,2002.
17. Mencl, W.E. Microsc. Res. Tech. **51**:64-74,2000.
18. Stevens, A.A., Skudlarski, P., Gatenby, J.C., and Gore, J.C. Magn. Reson. Imaging. **18**:495-502,2000.

Low Turn-On Voltage and High Focus Capability for Field Emission Display

Yi-Ting Kuo, Hsiang-Yu Lo and Yiming Li

Department of Communication Engineering, National Chiao Tung University, Hsinchu, Taiwan

ABSTRACT

In this work, the properties of low turn-on voltage and high focused capability for novel field emission display are studied. According to the development of novel structure for surface conduction electron-emitter provided by Tsai et al, the 3D FTDT-PIC method has been used to analyze the properties of this device. We can find the novel structure having a tip around the corner on the left electrode implies that it can produce high electric fields around the emitter apex, and generate high emission current.

Keywords: low turn-on voltage, high focus capability, field emission display, nanogap, FDTD-PIC

1 INTRODUCTION

Nanometer scale gaps (nanogaps) has been wildly used as electrode in molecular electronics [1,2], biosensor [3], and vacuum microelectronics [4]. However, the difficulty from the fabrication limits the development of such technique until the surface conduction electron-emitter (SCE) for the flat panel displays (FPDs) has been provided by Sakai et al [5]. Recently, the field emission efficiency of the SCE device has been studied, and the result shows the structure of the nanogap similar to the type 1 in Fig. 1 will be helpful to generate the high field emission efficiency [6]. After Tsai et al [7] have succeeded in fabricating a new type of SCE device using hydrogen absorption under high pressure treatment. A well-defined gap size and simple process can be given by this method which is accompanied by extensive atomic migration during the hydrogen treatment.

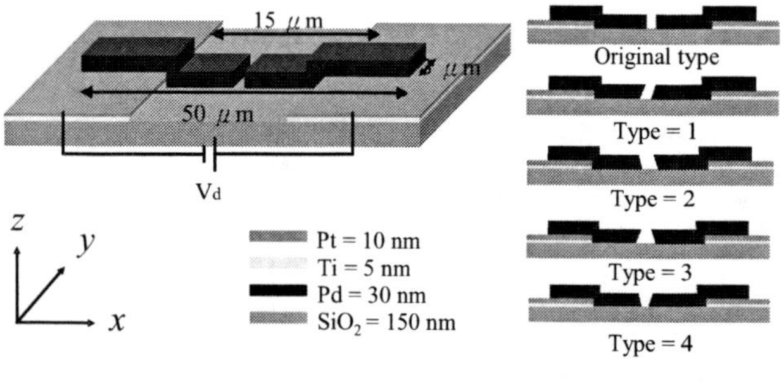

Figure 1: The schematic of the emitter structures and the five types of the nanogap due to the process variation.

To change the angle of the conventional nanogap results in the better emission efficiency using the type 1 in Fig. 1 [6]. The direction of the nanogap in type 1 provides a path for the electron trajectory with lower collision. The similar property is expected for the novel SCE device. In this paper, to explore the electron-emission behaviors in the novel SCE device, the FDTD-PIC simulation technique [6] is employed to solve a set of 3D Maxwell equations coupled with the Lorentz equation. Moreover, the property of the low turn-on voltage and high focus capability in the novel SCE device has been analyzed in this study.

2 SIMULATION TECHNIQUES

Figure 2 represents the procedure of FDTD-PIC. The procedure of PIC starts from a specified initial state, and the simulated electrostatic fields is applied as its evolution in time. Then a time-dependent differential form of Faraday's law, Ampere's law, and the relativistic Lorentz equation are shown as follows:

$$\begin{aligned}
&\frac{\partial \mathbf{B}}{\partial t} = -\nabla \times \mathbf{E}, \\
&\frac{\partial \mathbf{E}}{\partial t} = -\frac{\mathbf{J}}{\varepsilon} + \frac{1}{\mu\varepsilon}\nabla \times \mathbf{B}, \\
&\mathbf{F} = q(\mathbf{E} + \mathbf{v} \times \mathbf{B}), \textbf{ and} \\
&\frac{\partial \mathbf{x}}{\partial t} = \mathbf{v},
\end{aligned} \qquad (1)$$

subject to constraints provided by Gauss's law and the rule of divergence of **B**,

$$\nabla \cdot \mathbf{E} = \frac{\rho}{\varepsilon} \textbf{ and } \nabla \cdot \mathbf{B} = 0. \qquad (2)$$

that **E** and **B** are the electric and magnetic fields, **x** is the position of charge particle, and **J** and ρ are the current density and charge density resulting from charge particles. The full set of Maxwell equations is simultaneously solved to obtain electromagnetic fields. Similarly, the Lorentz force equation is solved to obtain relativistic particle trajectories. In addition, the electromagnetic fields are advanced in time at each time step. The charged particles are moved according to the Lorentz equation using the fields advanced in each time step. The obtained charge density and current density are successively used as sources in the 3D Maxwell equations for advancing the electromagnetic fields. These steps are repeated for each time step until the specified number of time steps is reached. The FDTD procedure can be seen in Fig. 2 and PIC procedure in Fig. 3. In the FE process, the electron emission is modeled by the Fowler-Nordheim (F-N) equation

$$J = \frac{AE^2}{\phi\, t^2} \exp\left(\frac{-Bv(y)\phi^{3/2}}{E} \right). \tag{3}$$

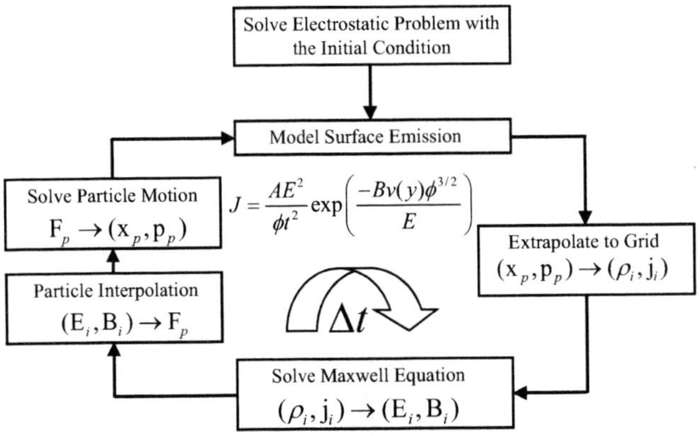

Figure 2: The computational scheme and the corresponding equations for the electron emission simulation.

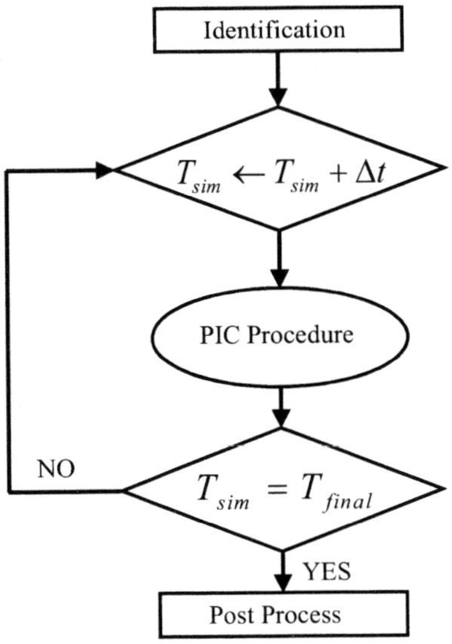

Figure 3: The computational scheme and the corresponding equations for the flowchart of PIC procedure.

3 COMPUTAIONAL RESULTS

The schematic of the novel emitter structures and related SEM images are shown in Fig. 4. The turn-on voltages for different spacing widths of both conventional and novel SCE devices are analyzed using the FDTD-PIC. The results show the novel structure produces the emission current 0.1 mA under lower turn-on voltage than the conventional one, which is represented in Fig. 5. This is because the tip of the novel structure generates the strong electric field intensity, especially for the narrower spacing of the nanogap, and such strong electric field in the vacuum improves the electrons tunneling from the Pd strip into vacuum. Hence for the same electric field intensity, the turn-on voltage for the structure with the tip is lower. Additionally, the turn-on voltage for both cases decreases due to the intensity of electric field becomes larger, which comes from the spacing width for the nanogap decreases. Hence the low turn-on voltage SCE devices can be achieved by reducing the spacing width for nanogaps. Table 1 shows the turn-on voltages for both structures under 30, 60, and 90 nm gap width. When the gap width is large, the turn-on voltage in the novel case is large due to the different heights for the two electrodes. Such electrodes make the electric field be small within the gap. However, as the gap width decreases, the turn-on voltage drops significantly in the novel case because the tip structure increases the electric field. Hence the turn-on voltage of the novel structure is smaller than the conventional one with the gap width smaller than 60 nm.

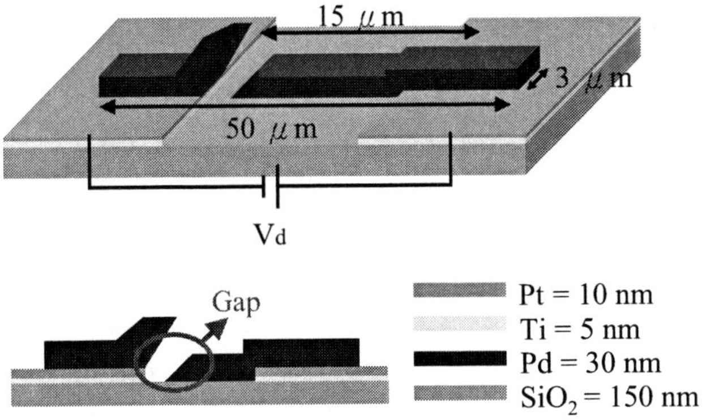

Figure 4: The schematic of the conventional (upper left) and novel emitter (upper right) structures and related SEM images (lower left one is conventional, and lower right one is novel).

Furthermore, the focus capability for both cases is simulated under the applied voltage is 60 V. The both conventional and novel SCE devices with 25 nm nanogaps are provided as examples. Figure 6 and 8 indicate the electron beam on both the x-z and y-z planes for the conventional SCE device. Figure 7 illustrates the electron trajectories near the nanogap on the x-z plane. The electron beam starts to spread on the y-z palne when electrons move far from the nanogap. It is due to the large fraction for the electrons and makes them collided with the driving electrode, such that electrons are scattered back into the vacuum. The simulated current density for the conventional SCE device is presented in Fig. 6. The current density spreads due to the scattered electrons.

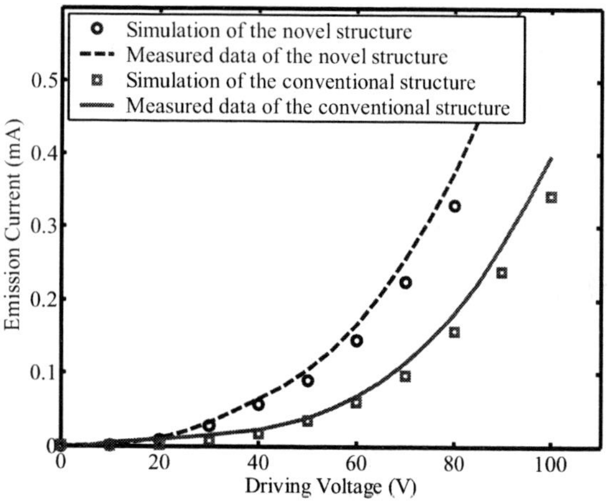

Figure 5: The I-V curves of the SCE devices with the 25 nm nanogap are given by conventional (in red) and novel (in blue) structures.

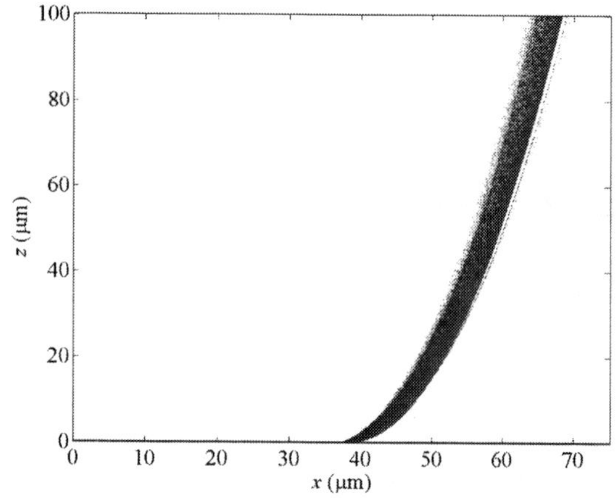

Figure 6: The electron beam on the x-z plane for the conventional SCE device with the 25 nm nanogap under applied voltage is 60 V

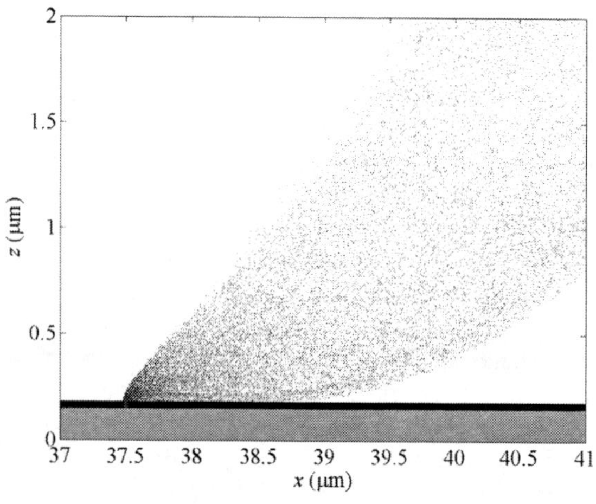

Figure 7: The zoom-in plot of the Fig. 6 near the nanogap.

Method \ Width	30 nm	60 nm	90 nm
Focus Ion Beam	60 V	85 V	160 V
Hydrogen Embrittlement	50 V	80 V	200 V

Table 1: The turn-on voltage for different structures under various gap widths.

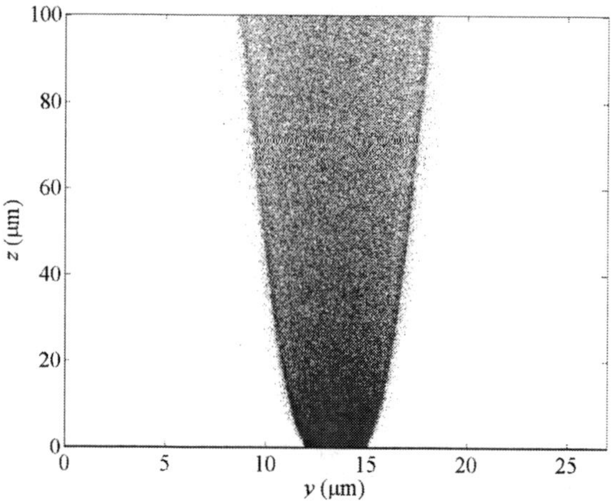

Figure 8: The electron beam on the y-z plane for the conventional SCE device with the 25 nm nanogap under applied voltage is 60 V.

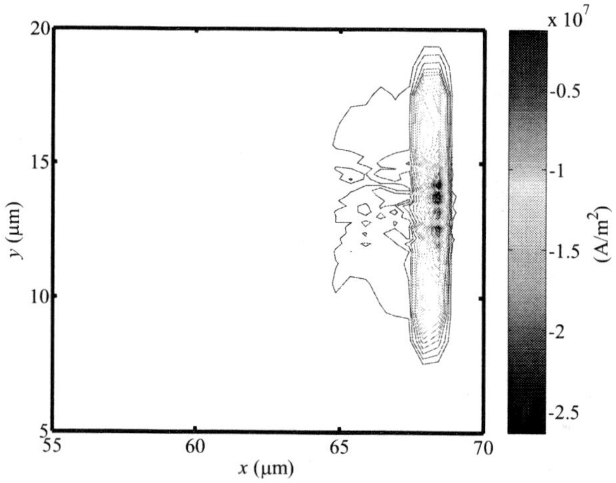

Figure 9: the simulated current density distributions on the anode plates for the conventional SCE structures with the 25 nm nanogap under applied voltage 60 V.

For the novel case, Figure 10 shows the electron trajectories on the x-z plane. The zoom-in plot, as shown in Fig. 11, illustrates the fewer collided particles with the opposite electrode comparing with the conventional case, as shown in Fig. 7. Furthermore, the new structure effectively reduces the fraction between the electrons and the driving electrode, so that the electron beam does not spread, as shown in Fig. 12. As the result, the simulated current density shows the better focus capability for the novel SCE device.

4 CONCLUSIONS

We have applied the 3D FDTD-PIC simulator to analyze the new SCE device provided by Tsai et al [7] and to show the advantages of such new structure. The conventional and novel SCE devices with 25 nm nanogaps are provided to verify the simulation result with the experiment one. The simulation and experiment results agree well to each other. Consequently, the advantages of low turn-on voltage, high focus capability are convinced for the novel SCE device.

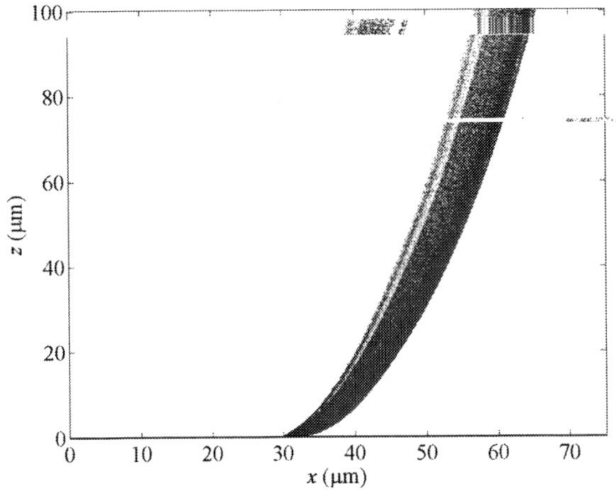

Figure 10: The electron beam on the x-z plane for the novel SCE device with the 25 nm nanogap under applied voltage is 60 V.

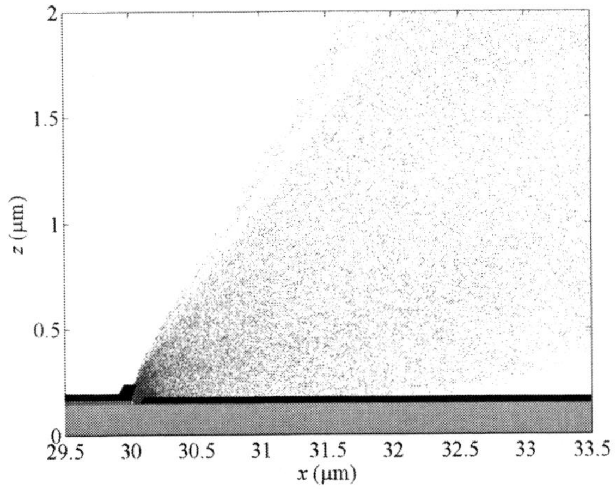

Figure 11: The zoom-in plot of the Fig. 10 near the nanogap.

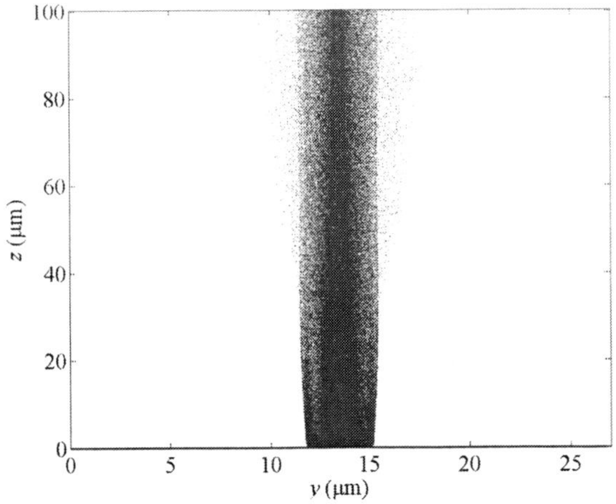

Figure 12: The electron beam on the y-z plane for the novel SCE device with the 25 nm nanogap under applied voltage is 60 V.

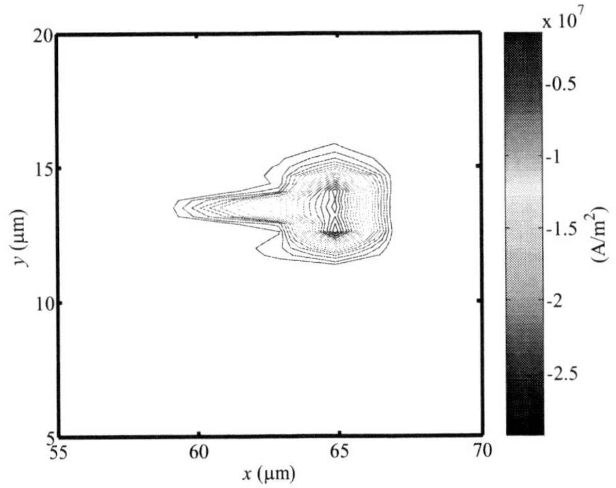

Figure 13: the simulated current density distributions on the anode plates for the novel SCE structures with the 25 nm nanogap under applied voltage 60 V.

REFERENCES

[1] Reed M A, Zhou C, Muller C J, Burgin T P and Tour J M, Science 278 252-4, 1997.

[2] Linag W, Shores M P, Bockrath M, Long J R and Park H, Nature 417 725-9, 2002.

[3] Yi M, Jeong K H and Lee L P, Biosens. Bioelectron. 20 1320-6, 2004.

[4] Lee H I, Park S S, Park D I, Ham S H and Lee J H 1, J. Vac. Sci. Technol. B 16 762-4, 1998.

[5] E. Yamaguchi, K. Sakai, I. Nomura., J. Soc. Inf. Disp. 5, 345, 1997.

[6] H.-Y Lo, Y Li, H.-Y Chao, C.-H Tsai, F.-M Pan, M.-C Chiang, M. Liu, T.-C Kuo, and C.-N Mo, Proc. IEEE Nano, 353, 2007.

[7] C.-H Tsai, F.-M Pan, K.-J Chen, C-Y, Wei, M. Liu, and C-N, Mo, Appl. Phys. Lett. 90 163115, 2007.

OH Radical Oxidation of Ethylene Adsorbed on Si-doped Graphene: A Quantum Chemistry Study

Cristina Iuga[1]

[1]Departamento de Química, Universidad Autónoma Metropolitana, Iztapalapa, México, D.F., México
cristina_iuga@prodigy.net.mx

ABSTRACT

The functionalization of carbon based nanomaterials could have applications in atmospheric chemistry and contaminants removal. In this work, a quantum chemistry and computational kinetics study has been performed on the OH radical addition to ethylene previously adsorbed on a model surface of graphene. Rate constants are calculated and compared with the ethylene + OH gas phase reaction, for which experimental data are available. To the best of our knowledge, this is the first theoretical study of a reaction between a free radical and a molecule adsorbed on a graphene-type nanostructure.

Keywords: ethylene, graphene, coronene, reaction kinetics, Quantum Chemistry, Transition State Theory.

1 INTRODUCTION

Polycyclic aromatic hydrocarbons (PAHs) occur in many natural environments where carbon is present. They are implicated in soot formation, in hydrocarbon combustion, and in the formation of dust grains in the interstellar medium. Coronene ($C_{24}H_{12}$) is the smallest PAH that presents the essential structural elements of graphite. Therefore, theoretical models of graphite, soot, or carbon nanotube walls often employ coronene as models representing a finite section of a carbon surface.

It is well-known that C and Si have completely different bonding characteristics, and that, instead of the characteristic carbon planar structures, silicon prefers 3-dimensional formations. Carbon atoms have been partially replaced by Si atoms in cage materials, and the synthesis of heterofullerenes containing up to 50% silicon atoms[1,2] and of silicon carbon nanotubes [3] has been reported. The presence of Si atoms in extended carbon compounds changes their electronic properties and chemical behavior. In particular, it is expected that extended polycyclic aromatic compounds with silicon defects may influence the atmospheric chemistry of unsaturated compounds.

In the troposphere, alkenes participate in a sequence of reactions which ultimately lead to their breakdown into highly toxic aldehydes, at the same time altering the equilibrium ratio of nitrogen oxides and indirectly producing ozone.[4]

In this work, quantum chemistry methods have been used to study the OH oxidation of ethene adsorbed on silicon doped graphene.

2 COMPUTATIONAL METHODOLOGY

Electronic structure calculations have been performed with the Gaussian03 program.package [5] Geometry optimizations of the stationary points along the coronene-ethylene + OH· reaction path have been performed using the hybrid density functional BHandHLYP [6,7], in conjunction with the 6-311G** basis set. Frequency calculations were carried out for all the studied systems at the same level of theory, and the character of the modeled structure was identified by the number of imaginary frequencies. Zero point energies (ZPE) and thermal corrections to Gibbs free energy at 298.15 K were included in the determination of the relative energies. Reaction Coordinate (IRC) [8] calculations were also performed to confirm that the transition states structures properly connect reactants and products. The rate constants were determined using Transition State Theory, [9,10] as implemented in the Rate 1.1 program .

3 RESULTS AND DISCUSSION

3.1 Graphene Models

A coronene molecule and a Si doped coronene (2Si-coronene) have been used as models to represent graphene sheets. These systems have previously been used [11] o study the adsorption of tiophene on graphemes. These models are expected to retain the fundamental characteristics and chemistry of the systems of interest while, at the same time, rendering high-level calculations feasible. The graphene models are shown in Figure 1. While coronene is planar (a), 2Si-coronene is bent at the center (b).

3.2 Adsorption Complexes

Both adsorption complexes have been fully optimized at the the hybrid functional BHandHLYP/6-311g(d,p) level (Figure 2). The ethylene molecule is allowed to move freely until it reaches the optimum adsorption site. In the complexes, the ethylene molecule is in a plane that is parallel and in the middle of the central ring.

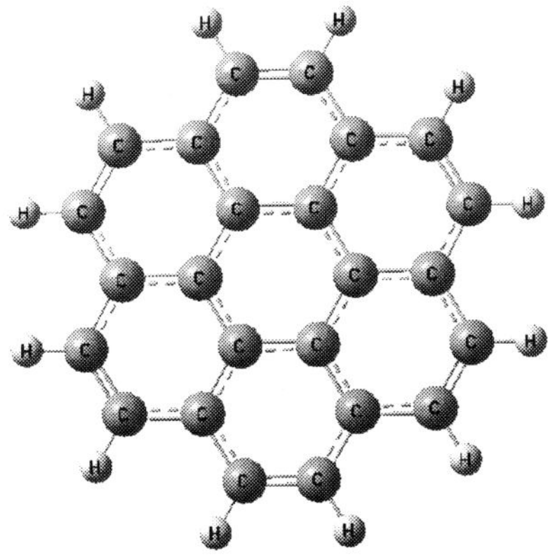

Top view

Side view

a) coronene

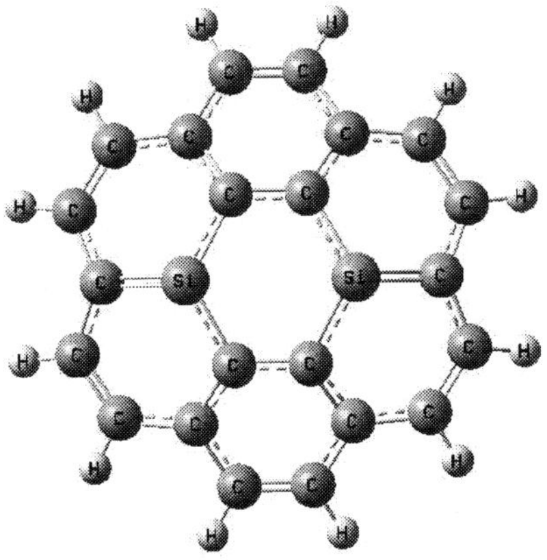

Top view

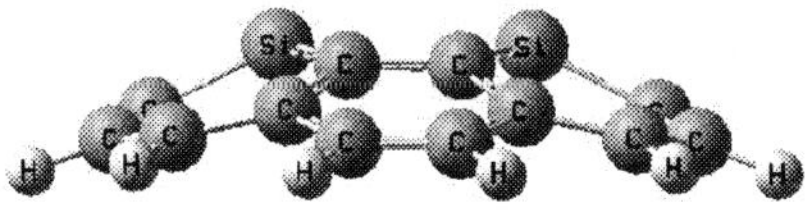

Side view

b) 2Si-coronene

Figure 1. Graphene surface models.

The adsorption energy is defined as the difference between the total electronic energy of the surface-adsorbate complex and the sum of those of the isolated molecule and the model surface, including ZPE corrections:

$$E_{\text{adsorption}} = E_{\text{adsorption complex}} - (E_{\text{molecule}} - E_{\text{surface}}) + \Delta(\text{ZPE}) \qquad (1)$$

The ethylene molecule does not adsorb on coronene at all (a very shallow minimum, of less than 0.1 kcal/mol is obtained when ethene is more than 5 Å away from the surface), while the adsorption energy on 2Si-coronene is about -1.2 kcal/mol. The adsorption complex on 2Si-coronene is shown in Fig. 2, and relevant distances, in Å.

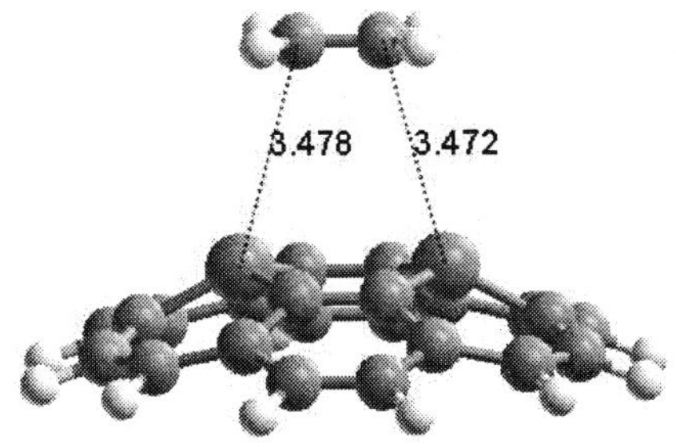

Figure 2. 2Si-coronene adsorption complex.

3.3 Reaction Mechanism

The OH reaction mechanism is assumed to be complex, and similar to the one observed in the gas phase. In the first step, the OH radical hydrogen atom interacts with the ethylene double bond and forms a pre-reactive Van der Waals intermediate, or reactants complex (RC), which is in equilibrium with the separated reactants. In the second step, the formation of a surface-ethylene-OH radical adduct is irreversible. All stationary structures are shown in Figure 3.

$$\text{Step 1:}\quad H_2C{=}CH_2 + OH\bullet \underset{k_{-1}}{\overset{k_1}{\rightleftarrows}} [H_2C{=}CH_2 \text{ ---- } HO]^{\cdot}$$

$$\text{Step 2:}\quad [H_2C{=}CH_2 \text{ ---- } HO]^{\cdot} \xrightarrow{k_2} \text{Adduct}$$

In general, the initial approach of the OH radical to the ethylene molecule is guided mainly by the Coulomb interaction between the positively charged hydrogen atom of the OH radical and the C=O double bond. For the adsorbed ethylene RC structure, shown in Figure 3, one can see that the Van der Waals distances between the surface and the ethylene molecule are slightly relaxed, while an H-π interaction takes place between the ethylene double bond and the radical hydrogen.In the case of the ethylene adsorbed on 2Si-coronene, the transition state is somewhat different to the one observed in the gas phase. It involves the breaking of the ethylene double bond and the simultaneous formation of *two* sigma bonds. One carbon atom attaches to the OH radical (as in the gas phase), while the other forms a covalent bond with coronene. Thus, a very stable radical product is obtained. In the radical adduct, the spin density is distributed over the coronene.

Energy values are calculated relative to the separated reactants, and are given in Table 1, including ZPE corrections. In order to take into account the entropy changes, Gibbs free energies are also included.

The energy profile obtained for the ethylene + OH reaction using the BhandHLYP/6-311G** energies is shown in Figure 3.

3.4 Reaction Kinetics

For the ethylene + OH· reaction, it has been shown that BHandHLYP/6-311G** quantum chemistry calculations followed by conventional Transition State Theory rate constant calculations yield results that agree very well with experimental results in the gas phase. Therefore, one may assume that it is possible to use the same methodology to calculate reliable kinetic data for reactions on graphene surface models, for which experimental data are not available. The effective rate constant is obtained using the following equation:

$$k_{ef} = \sigma \kappa_2 \frac{Q^{TS}}{Q^R} \exp\left[-\frac{E_a^{eff}}{RT}\right] \tag{2}$$

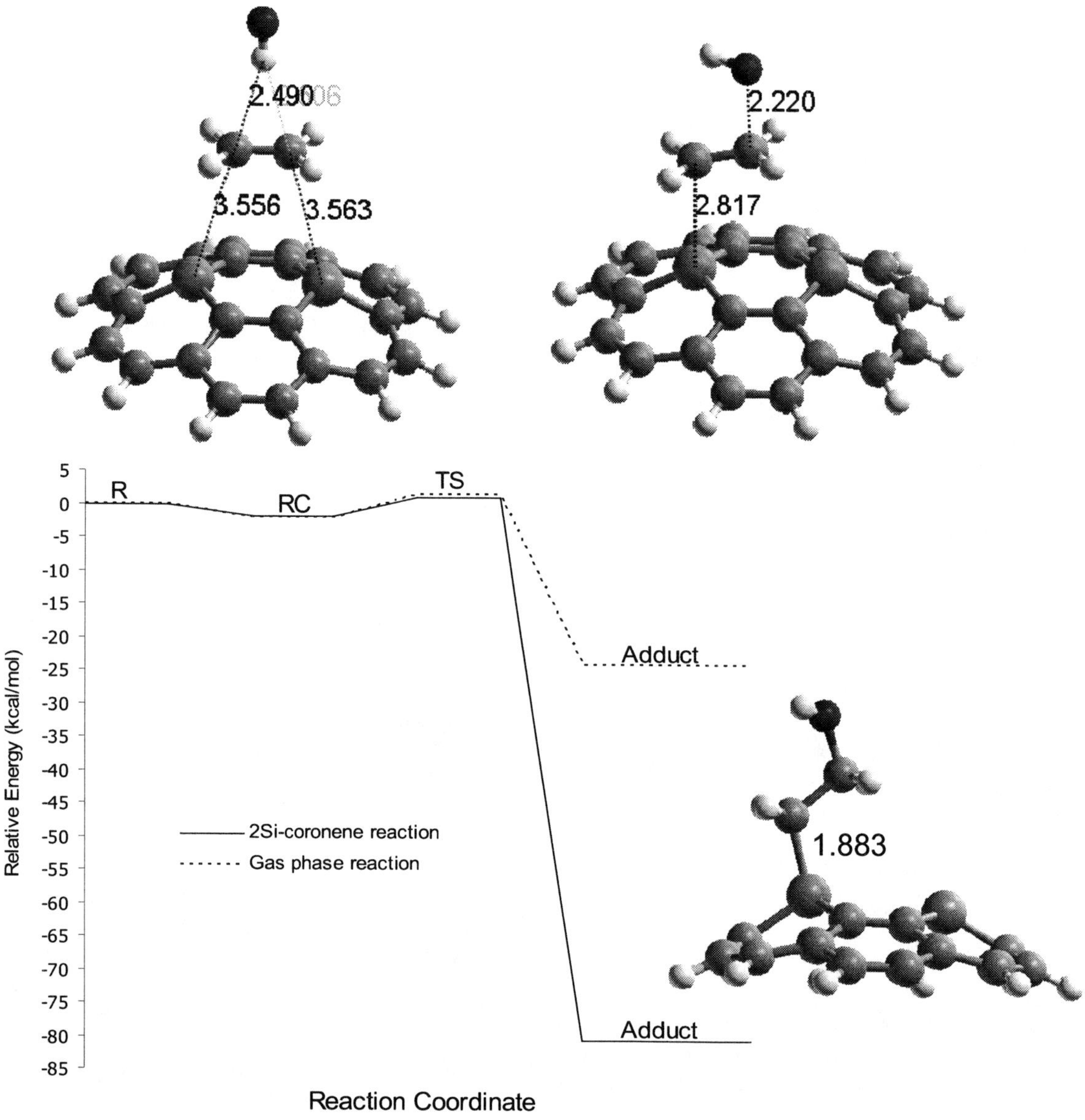

Figure 3. Energy profiles and optimized structures in the reaction ethylene + OH on 2Si-coronene.

Table 1. Relative energies including ZPE in kcal/mol and calculated rate constants (in cm^3/molecule s) at 298 K. In this table, $E_{-1}=E_{RC}-E_R$; $E_2=E_{TS}-E_{RC}$; $E_a^{eff}=E_{TS}-E_R$; and $\Delta E=E_{products}-E_{reactants}$.

	E_{ads}	E_{-1}	E_2	E_a^{eff}	ΔE	ΔG	K_{eq}	k_2	k^{eff}
Gas phase	-	-2.12	3.64	1.52	-24.06	-16.64	1.81×10^{-22}	7.13×10^{8}	1.29×10^{-13}
2Si-Coronene	-1.19	-1.98	2.90	0.92	-80.58	-68.97	5.75×10^{-22}	4.76×10^{7}	2.73×10^{-14}

4 CONCLUSIONS

Physisorption processes of ethylene on coronene and 2Si-coronene have been studied using density functional theory methods. The calculated data show that the presence of silicon atoms favors the interaction of the surface model with ethylene and probably with other similar compounds. Small stabilization energies have been found. The ethylene physisorption on coronene is unlikely to occur, while that on 2Si-coronene leads to a stable addition product. In addition, the rate constant for the oxidation of ethene physisorbed on 2-Si-graphene is found to be an order of magnitude smaller than in the gas phase.

The results reported in this work suggest that silicon defects on extended polycyclic aromatic hydrocarbons, such as graphite, soot, and large-diameter carbon nanotubes, could make them useful in the removal processes of atmospheric pollutants.

Acknowledgment.

The authors are grateful to CONACYT Project No. SEP-2004-C01-46167-Q and to the PIFI 3.3 program for financial support.

5 REFERENCES

[1] M. Pellarin, C. Ray, J. Lerme, J. L. Vialle, M. Broyer, M. Blasé, P. Keghelian, P. Melinon, A. Perez, *Phys. ReV. Lett. 80*, 5365, 1998.

[2] C. Ray, M. Pellarin, J. L. Lerme, J. L. Vialle, M .Broyer, X. Blase, P. Melinon, P. Keghelian, A. Perez, *J. Chem. Phys. 110*, 6927, 1999.

[3] C. Pham-Huu, N. Keller, G. Ehret, M. J. Ledouxi, *J. Catal. 200*, 400, 2001.

[4] B. J. Finlayson-Pitts, N. Pitts, *Atmospheric Chemistry: Fundamentals and Experimental Techniques*; Wiley-Interscience: New York, 1986.

[5] Gaussian 03, Revision D.01, M. J. Frisch, G. W. Trucks, H. B. Schlegel, G. E. Scuseria, M. A. Robb, J. R. Cheeseman, J. A. Montgomery, Jr., T. Vreven, K. N. Kudin, J. C. Burant, J. M. Millam, S. S. Iyengar, J. Tomasi, V. Barone, B. Mennucci, M. Cossi, G. Scalmani, N. Rega, G. A. Petersson, H. Nakatsuji, M. Hada, M. Ehara, K. Toyota, R. Fukuda, J. Hasegawa, M. Ishida, T. Nakajima, Y. Honda, O. Kitao, H. Nakai, M. Klene, X. Li, J. E. Knox, H. P. Hratchian, J. B. Cross, V. Bakken, C. Adamo, J. Jaramillo, R. Gomperts, R. E. Stratmann, O. Yazyev, A. J. Austin, R. Cammi, C. Pomelli, J. W. Ochterski, P. Y. Ayala, K. Morokuma, G. A. Voth, P. Salvador, J. J. Dannenberg, V. G. Zakrzewski, S. Dapprich, A. D. Daniels, M. C. Strain, O. Farkas, D. K. Malick, A. D. Rabuck, K. Raghavachari, J. B. Foresman, J. V. Ortiz, Q. Cui, A. G. Baboul, S. Clifford, J. Cioslowski, B. B. Stefanov, G. Liu, A. Liashenko, P. Piskorz, I. Komaromi, R. L. Martin, D. J. Fox, T. Keith, M. A. Al-Laham, C. Y. Peng, A. Nanayakkara, M. Challacombe, P. M. W. Gill, B. Johnson, W. Chen, M. W. Wong, C. Gonzalez, and J. A. Pople, Gaussian, Inc., Wallingford CT, 2004.

[6] A. D. Becke, *J. Chem. Phys.* 98, 1372, 1993.

[7] K. Raghavachari, J. B. Foresman, J. Cioslowski, J.. V. Ortiz, M. J. Frisch, A. Frisch, A. *GAUSSIAN 98 User's Reference*; Gaussian Inc.: Pittsburgh, PA, 1998.

[8] a) C. Gonzalez, H. B. Schlegel, *J. Chem. Phys.90*, 2154-2161, 1989, b) C. Gonzalez, H. B. Schlegel, *J. Phys. Chem.94*, 5523-5527, 1990, and references therein.

[9] Eyring, *J. Chem. Phys.* 3, 107, 1935.

[10] D. G. Truhlar, W. L. Hase and J. T. Hynes, *J. Phys. Chem.* 87, 2264, 1983.

[11] A.Galano, *J. Phys. Chem. A, 111,* 1677-1682, 2007.

Germanium antimony sulphide nano wires fabricated by chemical vapour deposition and e-beam lithography

C.C. Huang[*], Chao-Yi Tai[**], C.J. Liu[***], R.E. Simpson[*], K. Knight[*], D.W. Hewak[*]

[*]Optoelectronics Research Centre, University of Southampton, Southampton, SO17 1BJ, UK, cch@orc.soton.ac.uk

[**]Department of Optics and Photonics, National Central University, 32001, Jhongli, Taiwan, R.O.C.

[***]Institute of Physics, Academia Sinica, Nankang, 11529, Taipei, Taiwan, R.O.C.

ABSTRACT

Germanium antimony sulphide (Ge-Sb-S) amorphous thin films have been deposited directly onto SiO_2-on-silicon substrates by means of chemical vapour deposition. The Ge-Sb-S films have been characterized by micro-Raman, scanning electron microscopy, energy dispersive X-ray analysis and static tester techniques. Ge-Sb-S nano wires have been patterned and fabricated by e-beam lithography and dry etching techniques. Modeling results show the potential for fast switching of these Ge-Sb-S nano wire structures, making it a possible candidate for the phase-change memory applications.

Keywords: germanium antimony sulphide, nano wires, chemical vapour deposition, e-beam lithography, phase-change memory

1 INTRODUCTION

Chalcogenide materials (which contain sulphur, selenium or tellurium) have been studied since the 1960s. Chalcogenide thin films are very interesting materials because their properties allow a diverse range of applications in the field of optoelectronics [1-3]. However, it was only recently that their special properties generated attention in electronic memory [4]. There is currently a worldwide interest in the development of the next generation computer memory, fuelling research in new materials which can be used to store vast amounts of information.

The preparation of chalcogenide thin films can be performed by techniques which include thermal evaporation [5], sputtering [6], laser ablation [7], sol-gel [8], spin coating [9], and chemical vapour deposition (CVD) [10-13]. Phase-change thin films are predominately grown by thermal evaporation or sputtering techniques, which to date have yielded demonstrator chips for high-density memories but have yet to enter the commercial domain. To do this, and at the same time meet the continued down-scaling of nanoelectronic devices, better control of film deposition and conformal coatings over non-planar structures is necessary. At the same time, reductions in feature size are essential and would allow lower programming currents, leading to improved performance and lower costs [14]. We believe that the CVD technique could provide better yield and device performance, in part from its ability to provide high throughput deposition over irregular surfaces. In addition, CVD technique could also enable the production of thin films with superior quality compared to those obtained by sputtering, especially in terms of purity, conformality, coverage, and stoichiometry control thus allowing the implementation of phase-change films in nanoelectronic devices.

2 APPARATUS AND EXPERIMENTAL METHOD

For the past seven years we have been developing new CVD techniques specially targeted at the deposition of chalcogenide materials, in particular glassy and crystalline thin films and high purity bulk materials. The CVD apparatus we have constructed for Ge-Sb-S amorphous thin film deposition is shown in figure 1. In the process we exploit, Ge-Sb-S amorphous thin films are deposited on SiO_2-on-silicon substrates placed in the quartz tube reactor (25mm O.D. x 500mm long) and heated by an electrical resistance furnace to a temperature of 300°C. The reactive gas, H_2S, and the carrier argon gas for $GeCl_4$ and $SbCl_5$ are delivered through the mass flow controllers (MFC) at a flow rate in the range of 50 ml/min-150 ml/min.

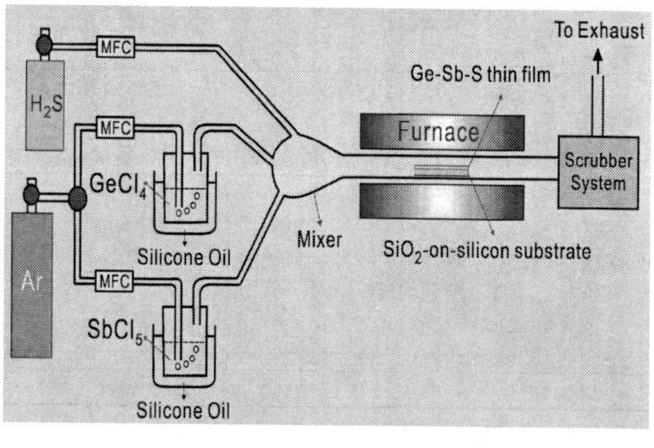

Figure 1: Schematic diagram of CVD system used for Ge-Sb-S thin film deposition.

The Ge-Sb-S thin films were formed at a deposition rate of approximately 10nm/min at a temperature of 300°C. In this way, Ge-Sb-S thin films with 100nm in depth were achieved in approximately 10 minutes. The compositions of Ge-Sb-S thin films were characterized by micro-Raman and energy dispersive X-ray analysis (EDX) techniques whilst a static tester system [15, 16] has been employed to measure the crystallization time. A 658nm laser diode was focused to a diffraction limited spot through a 0.65NA lens on to the samples surface. The change in reflectivity was measure with a second 635nm diode laser. Its optical power at the sample was controlled at 100μW. The percentage increase in reflectivity was collected for pulse times increasing from 5 to 110mW in steps of 5mW whilst the pulse time was increased from 20ns to 500ns. The resultant power, time, reflectivity matrix has been plotted as an intensity map in figure 2. The lighter areas denote an increase in reflection due to crystallization.

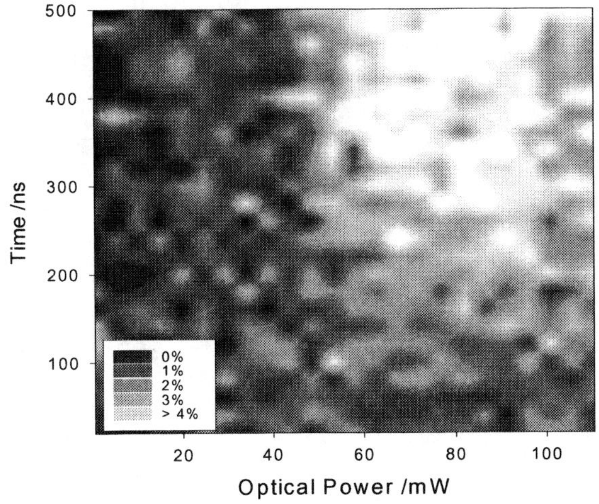

Figure 2: Phase change time as a function of power, as measured directly on a static tester for Ge-Sb-S thin films fabricated by CVD.

3 DEVICE FABRICATION

As part of this work, a prototype adopting the so-called phase-change line memory structure [14] was fabricated by electron beam (e-beam) lithography and reactive ion etching techniques. Prior to the fabrication of the nano wire, electrodes were photolithographically defined in a Ti-W film, which was evaporated by E-gun on the CVD grown Ge-Sb-S thin film. The electro-resist (ZEP520A) was then spin-coated on the sample at 4000 rpm for 35 seconds, obtaining a masking film with thickness of 100nm. After curing of the resist at 180°C for 3 min, the nano-wire was patterned with an e-beam writer (Elionix ELS-7000) operating at 10 keV. With the irradiation dosage keeping at 640 Coulomb/cm^2, the nano-wire pattern was directly created in a dot-by-dot fashion. The electro-resist was then developed and the unwritten regions were used as mask for the subsequent etching process.

Upon etching, SF_6, CH_4, and He were injected into the chamber with gas flow rate controlled at 16 sccm, 4 sccm, and 10 sccm, respectively. The radio frequency (rf) power of 150 W was used in conjunction with a low chamber pressure of 8 mTorr to efficiently generate high density corrosive plasma, resulting in an etch rate of 20nm/min for Ge-Sb-S thin film. Finally, after stripping off the residual resist, a phase-change memory in form of a line cell was constructed. It should be noted that the fabrication process of the present memory cell is completely compatible with current standard Si-CMOS technology.

4 RESULTS AND DISCUSSION

High quality amorphous Ge-Sb-S thin films have been successfully deposited on 2 micron SiO_2-on-silicon substrate at a temperature of 300°C. Scanning electron microscopy (SEM) with energy dispersive X-ray analysis (EDX) technique has been applied to study the morphology and composition of Ge-Sb-S thin films, from which a composition with a Ge:Sb:S molar ratio of 12.2: 27.5: 60.3, respectively, was obtained.

In addition, a micro-Raman was used to characterize the composition and phase structure of Ge-Sb-S thin films deposited by the CVD process. A RENISHAW Ramascope equipped with a CCD camera was used to perform the micro-Raman analysis of Ge-Sb-S thin films. A 633nm He-Ne laser was used to excite the sample and the Raman shift spectrum was measured from 1200cm^{-1} to 100cm^{-1} with a resolution of 1cm^{-1}. The Raman spectrum is shown in figure 3 which also agrees with that in the reference [17]. From the Raman spectrum, the Raman shift peaks of Si, SiO_2, and Ge-Sb-S are found at 520cm^{-1}, 940cm^{-1} and 298cm^{-1}, respectively. In addition the band width of the peak revealed that a stable amorphous phase of the Ge-Sb-S thin film has been demonstrated.

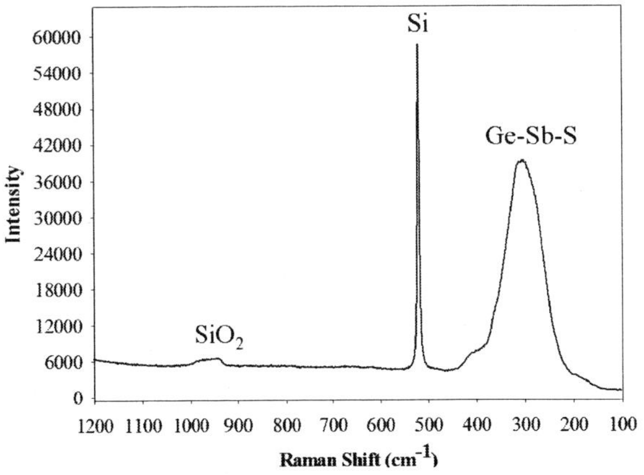

Figure 3: Raman spectrum of Ge-Sb-S thin film deposited on SiO_2-on-silicon substrate at the temperature of 300°C.

The SEM image of the fabricated Ge-Sb-S nano wire phase-change device is shown in figure 4. From the image, a Ge-Sb-S nano wire with the dimensions of 24nm x 100nm x 100nm thickness has been demonstrated.

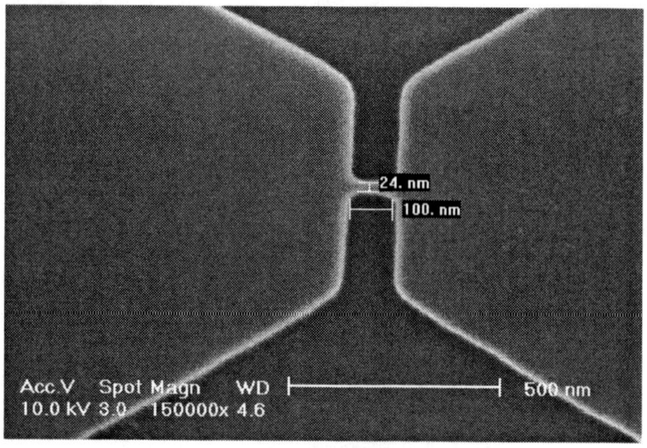

Figure 4: Scanning electron microscope image of Ge-Sb-S nano wire fabricated by CVD, e-beam lithography and dry-etching techniques.

The crystallization time of Ge-Sb-S thin films have been measured with a static tester system. The results in figure 2 show a clear crystallization region for pulse duration greater than 300ns. To invoke the crystallization high optical powers were required. This is partly due to the relatively high crystallization temperature of these materials but also due to the inefficient absorption of the below bandgap laser radiation. The poor optical absorption in the crystalline phase prevented amorphization of written crystalline marks. It is also thought that the small change in optical reflectivity after crystallization is due to the long wavelength of the probe laser.

The thermal properties of a Ge-Sb-S nano wire device have been simulated using a finite-element approach to solve the heat diffusion equation for a structure similar to that shown in figure 4. Figure 5 shows the heating and cooling curves for a device comprising of a Ge-Sb-S cell in its crystalline state. For the results shown in figure 5, the nanowire is contacted by Ti9W1 tapered electrodes with a final width of 300nm reduced from 2.5μm in a length of 1μm. The heating and cooling curves for an input current of 6.3 μA are shown for a cell of increasing length but a constant width of 20nm. The model assumed that the whole arrangement was deposited on a SiO_2 substrate. The electrical conductivity and thermal conductivities were assumed to be $5x10^{-3}$ Ω.m [18] and 0.77 $J.m^{-1}.s^{-1}$[19] respectively.

Due to the lack of published data the specific heat was assumed to be similar to that of Ge-As-S, approximately 1000J.kg-1[20]. To simplify the model, the thermal and electrical properties of the materials were assumed to be independent of temperature. It can be seen that even with a current of 6.3 μA the melting temperature of this material can be achieved efficiently. In order to compare the material, $Ge_2Sb_2Te_5$, has also been simulated in a similar device. The relatively high electrical conductivity of $Ge_2Sb_2Te_5$ had a significant influence on its heating efficiency and a 2.5mA current was necessary in order to achieve its melting temperature, this is in agreement with other results [14] on other Tellurium based films. The results show the importance of scaling on the heat diffusion for such a device. Ge-Sb-S has an order of magnitude lower thermal diffusitivity than that of its tellurium based counterpart [21] and consequently smaller volumes of material are required to achieve the amorphous state. In order to realize comparable quench rates in Ge-Sb-S materials to $Ge_2Sb_2Te_5$, the active phase change area should be less than 50nm by 20nm as can be seen in figure 5.

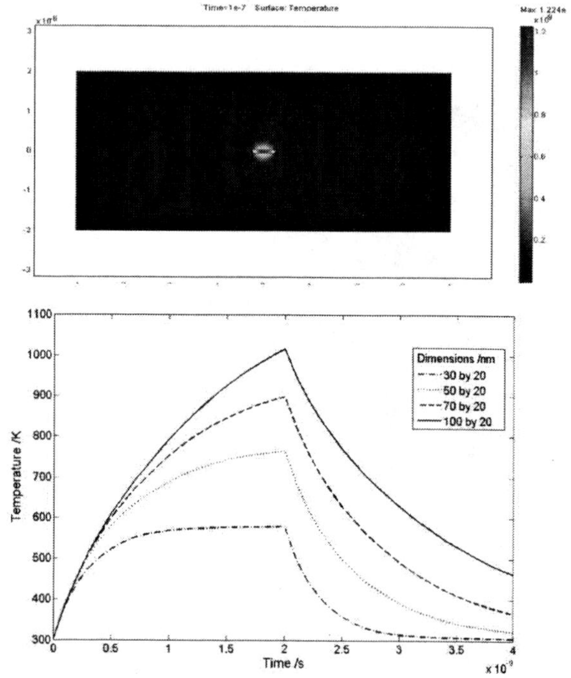

Figure 5: Simulation of temperature at centre of nano wires as a function of time.

5 CONCLUSION

Amorphous Ge-Sb-S thin-films deposited on SiO_2-on-silicon substrate have been successfully fabricated by chemical vapour deposition (CVD) at a temperature of 300°C. The resulting thin films have a composition of $Ge_{12.2}Sb_{27.5}S_{60.3}$, which has been characterized by micro-Raman, SEM and EDX techniques. Single Cell Ge-Sb-S nano-wire phase-change line memory devices have been fabricated on the CVD grown Ge-Sb-S thin film by e-beam lithography followed by reactive ion etching techniques.

The finite element simulations have shown that the high electrical resistivity of these sulphide based films allow a significant increase in the heating efficiency in comparison to $Ge_2Sb_2Te_5$ films.

An attempt to measure the crystallization time of the as deposited films has been made with a static tester. The crystallization time for the Ge-Sb-S film was found to occur after 300ns. This is three times slower than that of $Ge_2Sb_2Te_5$ [22] however the significant increase in heating efficiency and the opportunity to optimize the crystallization rate through changes in the device structure or the active material composition through the use of dopants should be considered. This is the focus of our current research. Finally, the potential for fast switching (<4ns) of Ge-Sb-S nano wire structures makes it a possible candidate for the phase-change memory applications.

6 ACKNOWLEDGEMENTS

The authors would like to acknowledge the technical assistance of Mr. John Tucknott, Mr. Neil Fagan, Mr Mark Lessey and Mr. Trevor Austin and Ilika Technologies Ltd for the use of the static tester. This work was funded by the Engineering Physical Sciences Research Council through our Portfolio Grant EP/C515668/1.

REFERENCES

[1] Ovshinsky S.R., "Reversible electrical switching phenomena in disordered structures", Physical Review Letters, 1968, 21, (20), pp. 1450-1455.

[2] Zakery A., Elliott S.R., "Optical properties and applications of chalcogenide glasses: a review", J. Non-Cryst. Solids, 2003, 330, pp. 1-12.

[3] Kolobov A.V., "Photo-Induced Metastability in Amorphous Semiconductors", (Wiley-VCH, 2003).

[4] Lai S., "Current status of the phase change memory and its future", Electron Devices Meeting, 2003. IEDM '03 Technical Digest. IEEE International, 8-10 December, 2003, pp. 10.1.1-10.1.4.

[5] Marquez E. , Wagner T. , Gonzalez-Leal J.M., Bernal-Oliva A.M., Prieto-Alcon R., Jimenez-Garay R., Ewen P.J.S., "Controlling the optical constants of thermally-evaporated $Ge_{10}Sb_{30}S_{60}$ chalcogenide glass films by photodoping with silver", J. Non-Cryst. Solids, 2000, 274, pp. 62-68.

[6] Ramachandran S., Bishop S.G., "Low loss photoinduced waveguides in rapid thermally annealed films of chalcogenide glasses", Appl. Phys. Lett., 1999, 74, pp. 13-15.

[7] Gill D.S., Eason R.W., Zaldo C., Rutt H.N., Vainos N.A., "Characterization of Ga-La-S chalcogenide glass thin-film optical waveguides, fabricated by pulsed laser deposition", J. Non-Cryst. Solids, 1995, 191, pp. 321-326.

[8] Xu J., Almeida R.M., "Preparation and characterization of germanium sulphide based sol-gel planar waveguides", J. Sol-Gel Science and Technology, 2000, 19, pp. 243-248.

[9] Mairaj A.K., Curry R.J., Hewak D.W., "Inverted deposition and high-velocity spinning to develop buried planar chalcogenide glass waveguides for highly nonlinear integrated optics", Applied Physics Letters, 2005, 86, pp. 94102-94104.

[10] Sleeckx E., Nagels P., Callaerts R., Vanroy M., "Plasma-enhanced CVD of amorphous Ge_xS_{1-x} and Ge_xSe_{1-x} films", J. de Physique IV, 1993, 3, pp. 419-426.

[11] Huang C.C., Hewak D.W.,"High purity germanium sulphide glass for optoelectronic applications synthesized by chemical vapour deposition", Electronics Letters, 2004, 40, pp. 863-865.

[12] Huang C.C, Hewak D.W., Badding J.V., "Deposition and characterization of germanium sulphide glass planar waveguides", Optics Express, 2004, 12, pp. 2501-2505.

[13] Huang C.C., Knight K., Hewak D.W., "Antimony germanium sulphide amorphous thin films fabricated by chemical vapour deposition", Optical Materials, 2007, 29, pp.1344-1347.

[14] Lankhorst M.H.R., Ketelaars B. W. S. M. M., Wolters R. A. M., "Low-cost and nanoscale non-volatile memory concept for future silicon chips", Nature Materials, 2005, 4, pp.347 – 352.

[15] Rubin K. A., Barton R. W., Chen M., Jipson V. B., Rugar D., "Phase transformation kinetics-the role of laser power and pulse width in the phase change cycling of Te alloys" Applied Physics Letters, 1987, 1488.

[16] Smith A.W., "Injection laser writing on chalcogenide films", Applied Optics, 1974, 795.

[17] Takebe H., Hirakawa T., Ichiki T., Morinaga K., "Thermal stability and structure of Ge-Sb-S glasses", J. of the Ceramic Society of Japan, 2003, 111, pp. 572-575.

[18] Pamukchieva V, Levi Z and Savova E: "DC electrical conductivity in $Ge_xSb_{40-x}S_{60}$ glasses and thin films", Semicond. Sci. Technol., 1998, 1309-1312.

[19] Velinov T., Gateshki M., Arsova D., Vateva, E., "Thermal diffusivity of Ge-As-Se(S) glasses", Phys. Rev. B, 1997, 11014-11017.

[20] Baro M.D., Clavaguera N., Surinach S., Barta C., Rysava N., Triska A., " DSC study of some Ge-Sb-S glasses", Journal of Materials Science, 1991, 3680-3684.

[21] Yanez-Limon, J.M., Gonzalez-Hernandez, J., Alvarado-Gil, J.J., Delgadillo, I., Vargas, H., "Thermal and electrical properties of the Ge:Sb:Te system by photoacoustic and Hall measurements", Physical Review B, 1995, 16321-16324.

[22] Yamada N., Ohno E., Nishiuchi K., AkahiraN., Takao M., "Rapid-phase transitions of GeTe-Sb_2Te_3 pseudobinary amorphous thin films for an optical disk memory", Journal of Applied Physics, 1991, 2849.

"Improved Methodology to Nucleate $Zn_xCd_{1-x}Se$ cladded $Zn_yCd_{1-y}Se$ Quantum Dots using PMP-MOCVD for Lasers and Electroluminescent Phosphors"

F. Al-Amoody[1], A. Rodriguez[2], E. Suarez[1], W. Huang[3], and F. Jain[1]
[3]US Military Academy, West Point, NY; [2]Chemistry Department and Institute of Materials Science, [2] Intel Corp., Rio Rancho, NM
[1]Electrical and Computer Engineering Department, University of Connecticut, 371 Fairfield Road, Storrs, CT 06269-2157. FAX: (860)-486-2447; Email: fcj@engr.uconn.edu

Earlier, we have reported growing 3-8 nm CdSe and pseudomorophic $Zn_xCd_{1-x}Se/Zn_yCd_{1-y}Se$ cladded quantum dots (QDs) (x > y) in a novel Photo assisted Microwave Plasma Metalorganic Chemical Vapor Phase Deposition (PMP-MOCVD) reactor [1, Angel *et al.*] Influence of growth parameters including microwave power, ultraviolet intensity, gas phase II/VI [Zn+Cd/Se] molar ratio, temperature of growth, and post-growth processing was investigated. This paper reports improvement in uniformity of dots nucleated by a methodology which preheats the reaction zone prior to the initiation of Microwave Plasma in a metalorganic chemical vapor deposition (MOCVD) The grown dots, shown in Fig. 1 are compared with those reported before [1]. They are also compared with dots prepared by other methods [2, 3, see Fig. 3 and Fig. 4], both using high-resolution transmission electron microscopy (HR-TEM).

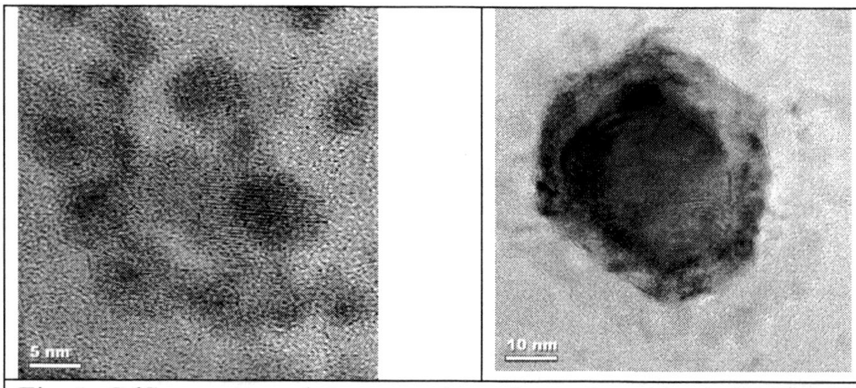

Fig. 1. CdSe quantum dots grown in preheated cavity via PMP-MOCVD.

(a) (b)

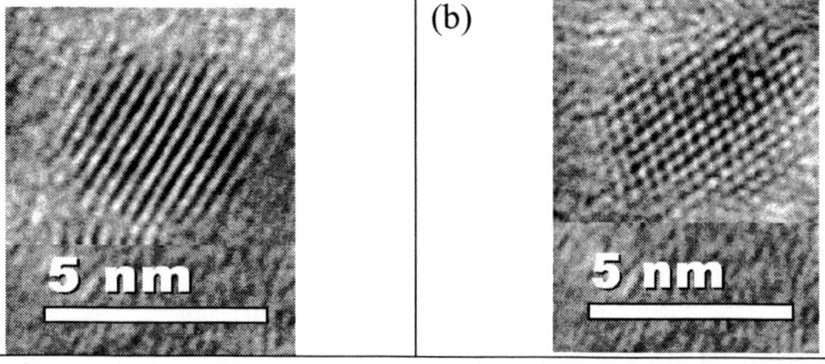

Fig. 2. CdSe quantum dots grown using PMP-MOCVD technique [1].

It has been shown [1] that photoluminescence (PL) peaks and full width at half maximum (FWHM), and X-ray diffraction (XRD) data have been used to calculate dot size. Comparison of X-ray diffraction peaks are shown in Fig. 5 and 6, respectively.

The influence on PL intensity of this new methodology is under investigation. We have also simulated the optical gain of cladded quantum dots including the effect of strain in the cladding for different composition of cladding layer. These results are shown in Figs. 7 and 8. Simulation is based on excitonic model reported by Jain and Huang [JAP, 1999] with some modification.

Symposium: Nanoelectronics and Photonics
Topic Area: Optoelectronics

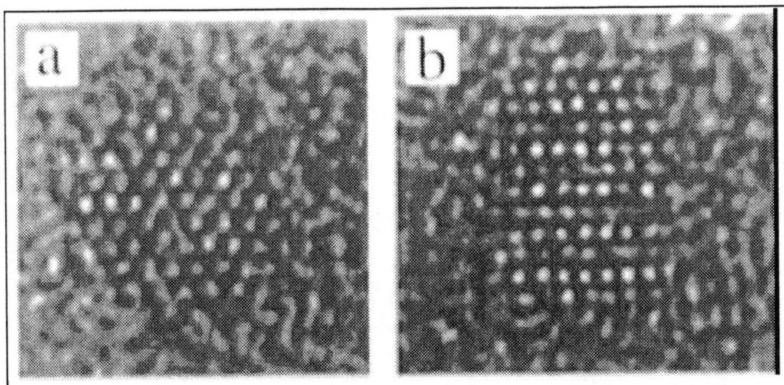

(a) Wurzite c-axis (b) Wurtzite a-axis.
Fig. 3. CdSe quantum dots grown by colloidal method (Peng *et al.*, JACS, 1997, Ref.2].

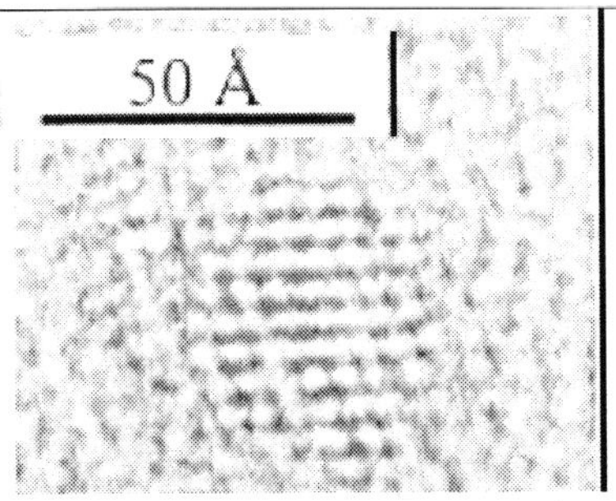

Fig. 4. CdSe quantum dots grow n by colloidal method [Dabbousi *et al.*, JPCB, 1997; ref 3].

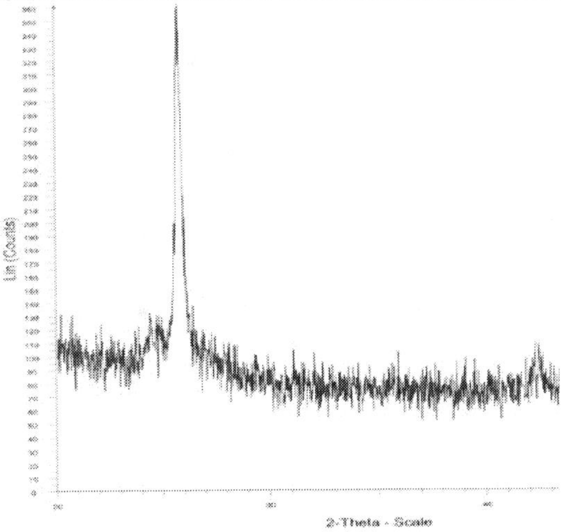

Fig. 5 X-ray diffraction data on dots shown in Fig. 1.

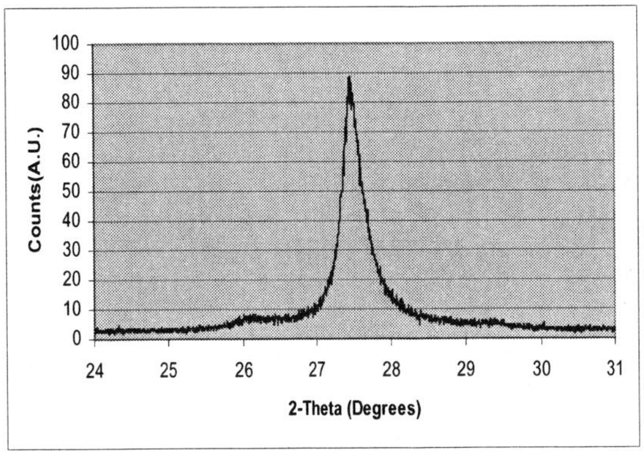

Fig. 6. X-ray peak from QD ensemble shown in Fig. 2 with out preheating [1].

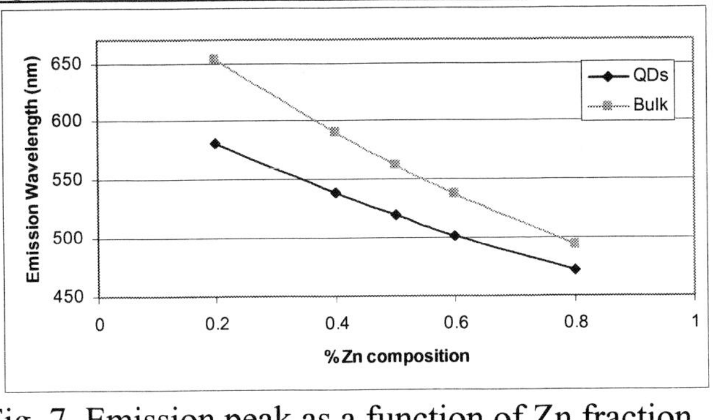

Fig. 7. Emission peak as a function of Zn fraction.

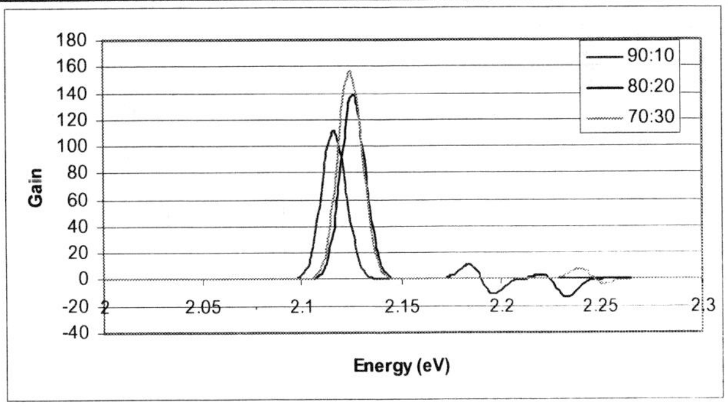

Fig. 8. Optical gain of $Zn_xCd_{1-x}Se$-CdSe QDs as a function of band offset ratio $[\Delta E_c/\Delta E_v]$.

References: [1] A. Rodriguez, R. Li, P. Yarlagadda, F. Papadimitrakopoulos, W. Huang, J. Ayers, and F. Jain, NSTI, Boston Conference, May 8, 2006.
[2] X. Peng, M.C. Schlamp, A.V. Kadavanich and A. P. Alivisatos, J. Am. Chem. Soc., **119**, 7019-7019.(1997).
[3] B.O. Dabbousi, J. Rodriguez-Viejo, F.V. Mikulec, J.R. Heine, H. Mattoussi, R. Ober, K.F. Jensen, M.G. Bawendi. J. Phys. Chem. B, **101**, 9463-9475.(1997)
[4] F. Jain, W Huang, JAP, **85**, 2706-2712 (1999).

Symposium: Nanoelectronics and Photonics
Topic Area: Optoelectronics

Synthesis of Nanostructures in Chlorine-Containing Media

S. Zaginaichenko and A. Dubovoy

Institute for Problems of Materials Science of NAS of Ukraine, Laboratory #67, Kiev-150, P.O.Box 195, 03150 Ukraine, shurzag@materials.kiev.ua

ABSTRACT

This work demonstrates a possibility to produce such materials by the new method for synthesis of carbon nanostructures using arc evaporation of materials in liquid medium. The possibility to produce chlorine-filled nanostructures is illustrated by the example of synthesis in chlorine-containing media. The proposed method may be one of the most efficient methods for synthesis of carbon nanostructures.

Keywords: arc graphite evaporation, dichlorethane, carbon nanostructure, morphology

1 INTRODUCTION

Modification of carbon nanostructures by different chemical elements opens an opportunity for synthesis of materials of a new generation for different applications. Filling carbon nanotubes with one or other element will allow for conferring different mechanical, electrical, magnetic and other physical and chemical properties on the nanotubes.

This work demonstrates a possibility to produce such materials by the new proposed by authors [1-3] method for synthesis of carbon nanostructures using arc evaporation of materials in liquid medium.

The possibility to produce chlorine-filled nanostructures is illustrated by the example of synthesis in chlorine-containing media.

2 EXPERIMENTAL

Synthesis in the liquid phase has been carried out on the setup designed specially for these studies. The setup allows metal and graphite electrodes to be evaporated in the liquid medium at the temperature of medium 4 to 340 K using an electric arc. The arc temperature near a cathode may be as mush as $1.2{\cdot}10^4$ at currents 200 to 300A. The product can be cooled at the rate of 10^{-9} K/c to 4K in the liquid phase.

The electronic control block is simple in operation and gives a possibility to vary and measure voltage and electric current. These changes in their turn allow the action on the conditions of the plasma-chemical process, which proceeds in the reactor, and the profound effect on the morphology and the yield of product.

All the chemical reagents used in synthesis have been subjected to preliminary purification and rectification. Graphite of MPG-7 grade has been used. Preliminary graphite rods have been annealed in vacuum. Metallic rods have been melted repeatedly in an arc furnace in argon medium of spectral purity.

The synthetic products have been investigated by scanning and transmission electron microscopy. The liquid phase has been studied by a spectrophotometer and mass spectrometry.

The variety of properties of different resulting carbon materials is conditioned by the electronic structure of a carbon atom. The electron density redistribution, the formation of electronic clouds of different modifications around the atoms, the hybridization of orbitals (sp^3-, sp^2-, sp- hybridization) are responsible for the existence of different crystalline allotropic phases and their modifications.

The proposed method gives a possibility to produce the wider range of materials by varying conditions of their synthesis. This method allows the change of the chemical composition of electrodes and the medium in which the synthesis is carried out (see Fig. 1). At present time different research groups the world over are engaged in such studies [4-14]. The electrodes may contain or not contain carbon or consist of graphite doped with any element. The liquid phase in its turn may have different chemical compositions that affect significantly the structure and the composition of the forming nano-objects.

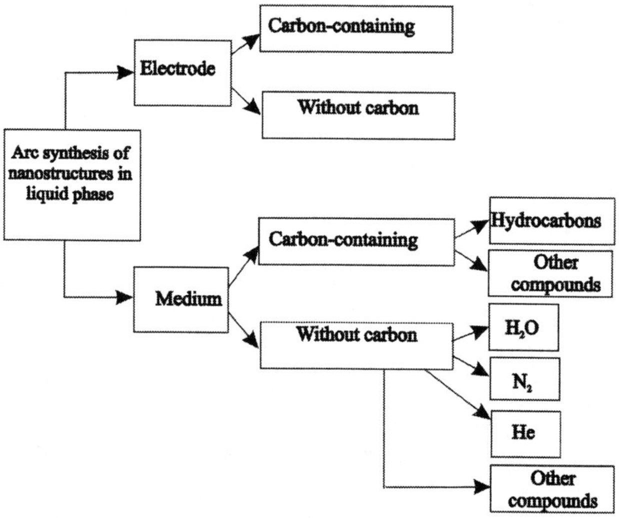

Figure 1: Scheme for possible combinations of medium and electrode materials in synthesis of nanostructures by the arc method in the liquid phase.

In the course of the arc synthesis when carbon atoms derive sufficient amounts of energy, they pass from the graphite surface into the gas phase as separate atoms or groups of atoms [6]. These atoms form in certain technological conditions a new carbon structure determined by the synthesis conditions. As this takes place, the atoms spend the derived energy for constructing this structure. The further existence of this carbon nanostructure, the conservation or the change in the initial morphology and geometrical dimensions of a nucleus are determined by the thermodynamical and technological conditions for the nucleus stay in a particular medium.

During simultaneous evaporation of graphite and different elements they interact or nanostructures are doped with the element.

Nanostructures of different morphologies (see Figs. 2 and 3) and different volume filling are formed by graphite evaporation in chlorine-containing liquids.

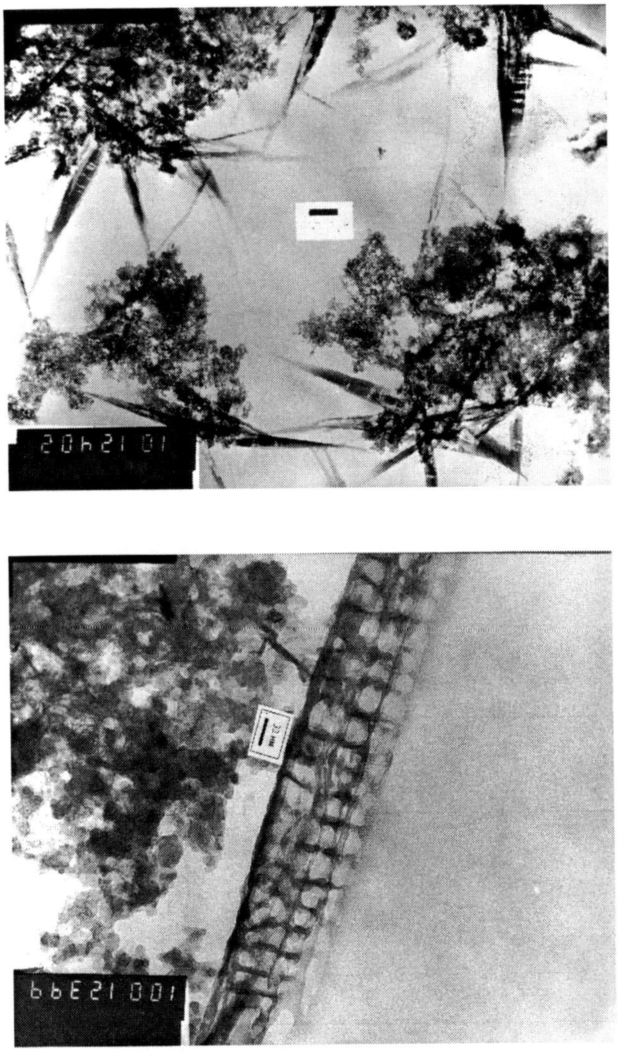

Figere 2: Carbon nanostructures produced in CCl_4.

Encapsulated nanotubes produced in dichloroethane are shown in Fig. 3. Their morphology and volume structure are in complete agreement with the mechanism of the fast formation of a nanotube and its simultaneous encapsulation proposed by Loiseau et al. [6].

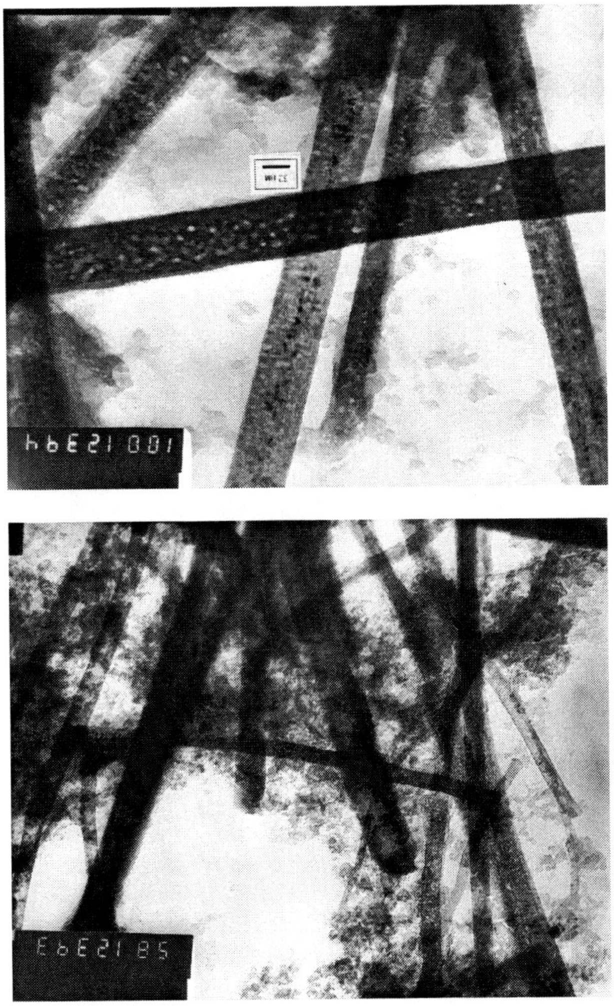

Figure 3: Carbon nanotubesfilled with a chlorine-containing compound (the tubes have been produced in dichlorethane).

3 CONCLUSIONS

Based on the experimental data and the theoretical calculations we have attempted to consider the conditions and the mechanism of the processes proceeding in synthesis of carbon nanostructures.

The possibility to produce chlorine-filled carbon nanostructures by the arc synthesis of nanostructures in the liquid phase has been demonstrated.

All the obtained results are of scientific and practical interest. The produced materials invite further investigations. The proposed method can be one of the most effective methods of fullerenes and nanotubes synthesis.

REFERENCES

[1] A.G. Dubovoy, A.E. Perekos and K.V. Chuistov, "Structure and magnetic properties of small amorphous particles of metallic Fe-15 at.% B alloy", Physics of Metals, V.6, N 5, 1085-1088, 1985.

[2] A.G. Dubovoy, V.P. Zalutskiy and I.Yu. .Ignat'ev, "Structure, magnetic parameters and thermal stability for small amorphous particles and amorphous strips of Fe-15 at.% B", Physics of Metals, V. 8, N 4, 804-807, 1990.

[3] K.V. Chuistov, A.E. Perekos, V.P. Zalutskiy, T.V. Efimova, N.I. Glavatskaya, "The effect of production conditions on the structural state, phase composition and fineness of iron and iron-based powders made by electric-spark erosion", Metal Physics and Advanced Technologies, V. 16, N 8, 865-875, 1997.

[4] K.V. Chuistov and A.E. Perekos, "Structure and properties of small-size metallic particles. 1. Phase-structure state and magnetic characteristics (Review)", Metal Physics and Advanced Technologies, V. 17, N 1, 57-84, 1998.

[5] AYu. Ishlinsky, "Polytechnic Dictionary", Soviet Encyclopedia; Moscow, 611, 1989.

[6] A. Loiseau, N. Demoncy, O. Stephan, C. Colliex and H. Pascard, "Filling carbon nanotubes using an ARC discharge. In: Science and Application of Nanotubes", Kluwer Academic Publishers, New York, 398, 2000.

[7] D.V. Schur, A.G. Dubovoy, E.A. Lysenko, T.N. Golovchenko, S.Yu. Zaginaichenko, AF. Savenko, and et al., "Synthesis of nanotubes in the liquid phase", Extended Abstracts, 8th International Conf. "Hydrogen Materials Science and Chemistry of Carbon Nanomaterials" (ICHMS'2003), IHSE, Sudak (Crimea, Ukraine), 399-402, 2003.

[8] D.V. Schur, A.G. Dubovoy, S.Yu. Zaginaichenko and A.F. Savenko, "Method for synthesis of carbon nanotubes in the liquid phase", Extended Abstracts, An International Conf. on Carbon Providence (Rhode Island. USA), American Carbon Society, 196-198, 2004.

[9] M.V. Antisari, R. Marazzi and R. Krsmanovic, "Synthesis of multiwall carbon nanotubes by electric arc discharge in liquid environments", Carbon, V. 41, N 12, 2393-401, 2003.

[10] L.P. Biro, Z.E. Horvath, L. Szalmas, K. Kertesz, F. Weber, G. Juhasz, G. Radnoczi and J. Gyulai, "Continuous carbon nanotube production in underwater AC electric arc", Chem. Phys. Lett., V. 372, N 3-4, 399-402, 2003.

[11] N. Sano, J. Nakano and T. Kanki, "Synthesis of single-walled carbon nanotubes with nanohorns by arc in liquid nitrogen", Carbon, V. 42, N 3, 686-688, 2004.

[12] J. Qui, Y. Li, Yu. Wang, Z. Zhao, Y. Zhou and Ya. Wang, "Synthesis of carbon-encapsulated nickel nanocrystals by arc-discharge of coal-based carbons in water", Fuel, V. 83, N 4-5, 615-617, 2004.

[13] D. Bera, S.C. Kuiry, M. McCutchen, A Kruize, H Heinrich, M. Meyyappan and S. Seal, "In-situ synthesis of palladium nanoparticles-filled carbon nanotubes using arc discharge in solution", Chem. Phys. Lett., V. 386, N. 4-6, 364-368, 2004.

[14] L.A. Montoro, C.Z. Lobrano Renata and J.M. Rosolen, "Synthesis of single-walled and multi-walled carbon nanotubes by arc-water method", Carbon, V. 43, N 1, 200-203, 2005.

Synthesis of Carbon Nanostructures in Liquid Helium

D. Schur, V. Bogolepov and A. Savenko

Institute for Problems of Materials Science of NAS of Ukraine, Laboratory #67,
Kiev-150, P.O.Box 195, 03150 Ukraine, shurzag@materials.kiev.ua

ABSTRACT

The advantages of arc synthesis of carbon nanostructures in a liquid medium have been investigated. Changes in the chemical compositions of the reagents (electrodes and a medium), the high temperature and pressure of the medium, the high cooling rate and the growth of structures in the reaction zone allow the materials with unique properties to be produced. The production of a purer product makes this efficient method promising for the synthesis of carbon nanostructures for different applications.

Keywords: arc discharge, pyrolysis, graphite, carbon nanotube, liquid helium

1 INTRODUCTION

After the discovery of fullerenes and carbon nanotubes methods of their synthesis has been constantly investigated and improved.

In parallel with the arc method in the gaseous phase and the pyrolytic method of synthesis of carbon nanostructures, since 2000 we have investigated and developed the method of arc synthesis in the liquid phase. This method has a number of advantages over those used at present.

We propose the method of synthesis of carbon nanostructures and composites on their base by arc discharge in the liquid phase. In the eighties we began our work on producing ultradispersed metal powders by the electroerosion method [1-4] and continue it today. Besides carbon nanostructures produced by evaporation of carbon electrodes in the liquid phase, there appears a possibility to produce metal-carbon composites by sublimation of metal in the carbon-containing liquid. In this case the metal nanoparticles form along with carbon nanostructures on their surface.

The main positive features [3, 4] of the method used are as follows:

1. The high temperature in the arc zone, > 4000 K.
2. The high cooling rate of evaporated products, > 10^9 K/s.
3. The high degree of dispersion. The particles range in size from 1 to 100 nm.
4. The high nucleation rate at a low growth rate of a particle.

All these conditions are analogous to those for synthesis of fullerenes and nanotubes by the arc graphite evaporation in the gaseous phase. The proposed method provides a possibility of producing a wide range of materials by varying the conditions for synthesis. This method presents a way of modifying the chemical composition of electrodes and a medium, in which the synthesis is carried out. At present time different research groups the world over are engaged in such studies [9-14]. The liquid phase may be of different chemical compositions that affect the structure and the composition of the produced nanoobjects of under study (see Fig. 1).

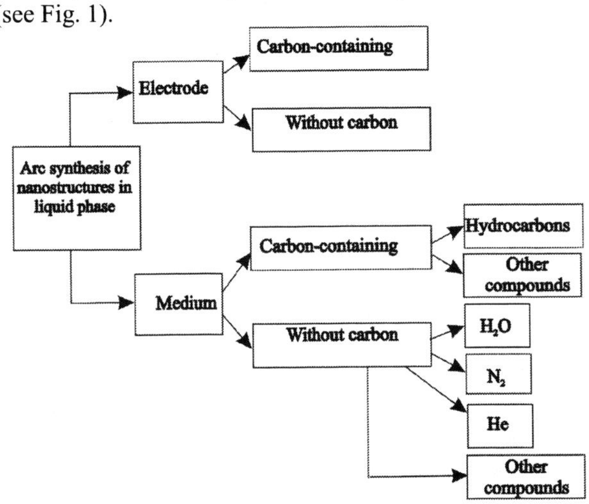

Figure 1: Diagram for possible combinations of medium and electrode materials in synthesis of nanostructures by the arc method in the liquid phase.

Besides the described method, the carbon nanostructures have been produced by the methods of hydrocarbon pyrolysis and arc evaporation of graphite in the gas phase in order to compare physical and chemical peculiarities of these nanostructures formation and morphology.

Notice that earlier the first method was used in the production of different carbon pyrofibers and therefore it is better understood. The second method of arc synthesis attracted a deserved attention of scientists as the method of carbon nanotubes (CNT) synthesis only after Iijima's work was published in 1991. This method requires the explanation for many unintelligible attributes. Vagueness in understanding the mechanism of the nanotubes growth hinders the progress in developing more controllable technologies of synthesis of these materials.

In this work we have investigated physical and chemical peculiarities of the carbon nanostructures synthesis and the effect of the cooling rate (i.e. the residence time of a carbon atom in the reaction zone) on the peculiarities of the product formation and morphology. We have compared the peculiarities of the formation of the nanostructures synthesized by pyrolysis, the arc method in the gas phase and the arc method in the liquid in order to understand the effect of the earliest stages of nucleation on the further process of the nanostructure growth. All the methods are distinct in the time of interaction between reagents.

In the course of the work we have demonstrated the possibility of producing carbon nanostructures in the liquid phase (water, hydrocarbons, dichloroethane, CCl_4, in liquid gases).

2 EXPERIMENTAL

Synthesis of carbon nanostructures by pyrolysis and arc discharge methods in the gas phase has been performed according to the known procedures. Synthesis in the liquid phase has been carried out in the installation specially designed for these studies (see Fig.2). This installation allows the metal and graphite electrodes to be evaporated by an electric arc in the liquid medium in the temperature range from 4 to 340 K. In the vicinity of the cathode the arc temperature may be as much as $1{,}2\cdot10^4$ K at currents of 200-300A (see Fig.3).

Figure 2: The installation for synthesis of nanocarbon structures and Me-carbon composites in the liquid phase.

The electronic control block is simple to operate and provides a possibility of varying and measuring voltage and electric current. These changes in turn make it possible to affect the conditions of the plasmochemical process, which proceeds in the reactor, and the product morphology and yield.

All the chemical reagents used in the synthesis have been subjected to the prior purification and rectification. Type MPG-7 graphite has been used. The graphite rods have been pre-annealed in vacuum. The metallic rods have been remelted repeatedly in an arc furnace in the spectro-pure argon medium.

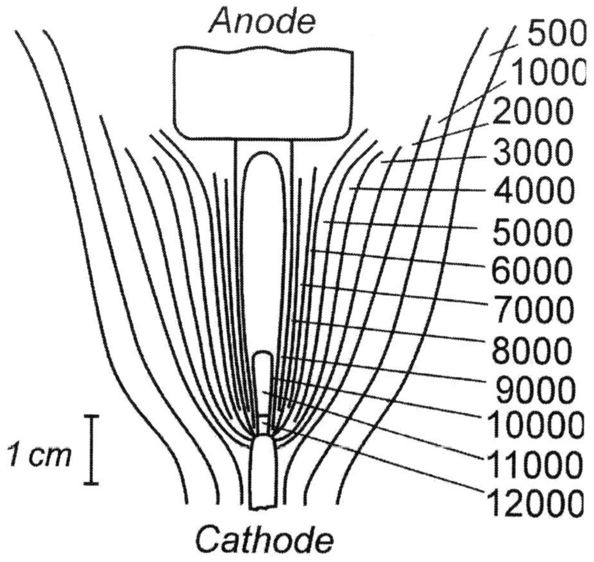

Figure 3: Temperature distribution (in K) in different regions of the electric arc between the carbon electrodes at strength of current equal to 200A [5].

The synthesized products have been washed with hydrocarbons using ultrasound and then investigated using the scanning and transmission electron microscopy. The liquid phase has been studied by the spectrophotometer and mass spectrometry (these results are not discussed in the present work).

3 RESULTS AND DISCUSSION

The product produced by the arc evaporation of graphite in liquid helium has no deposit and without additional purification it contains up to 85-90% of carbon nanotubes. Such results are not always achieved using different methods for nanotubes purification (see Fig. 4).

Different nano-objects can be synthesized under changes of the regime of synthesis and application of catalysts. In the course of synthesis the foamy particles (see Fig. 5) are formed during the evaporation of highly dispersed graphite dust. Their conglomerates range up to 1-5 μm.

4 CONCLUSIONS

From the aforesaid, it may be assumed that the discussed method of carbon nanostructure synthesis is sufficiently productive. It allows the production of nanostructures and nanostructure-based composites of different morphologies and properties for different applications. With this method, many carbon nanostructures that are thought to be the by-products have no time to form or grow because of a high rate of temperature changes in the course of the reaction.

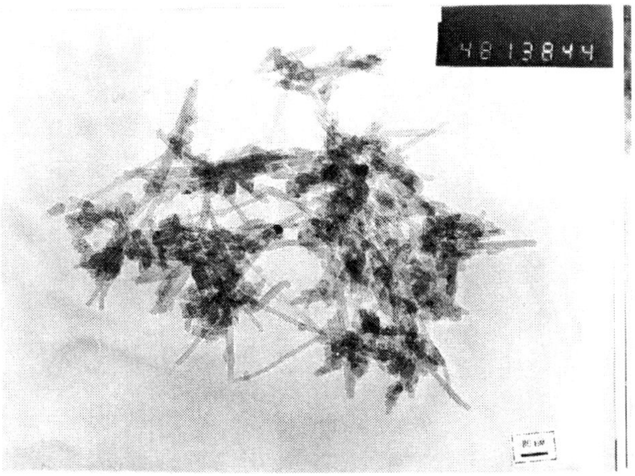

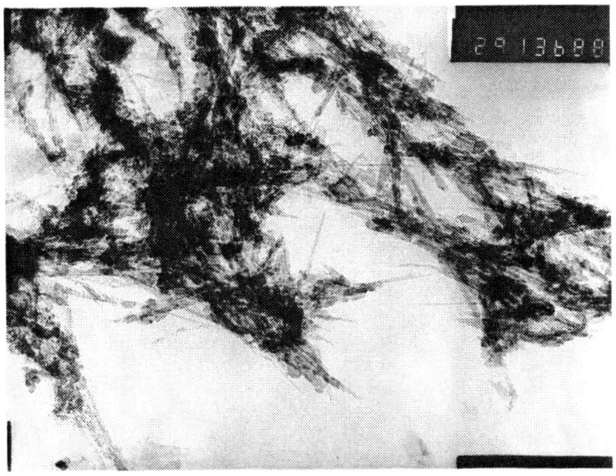

Figure 4: Carbon nanotubes produced in liquid helium (demonstrated without additional preparation: fragmentation, washing, extraction, purification etc.).

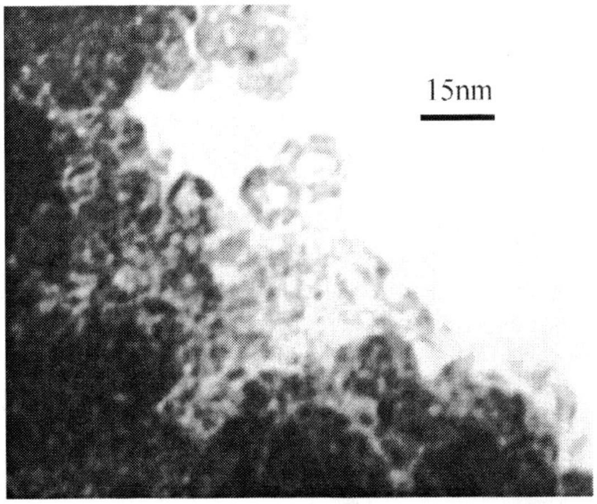

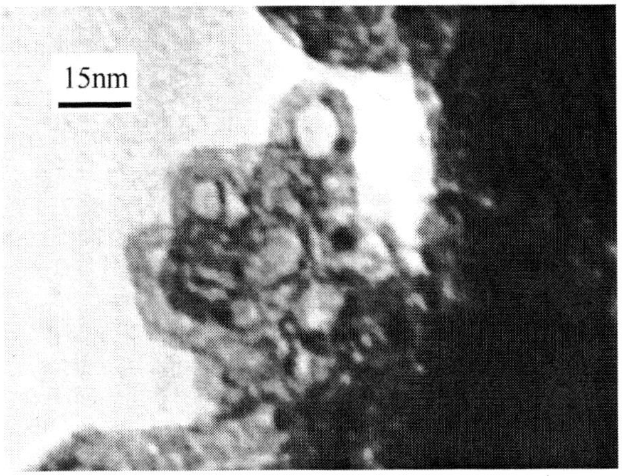

Figure 5: Carbon nanostructures produced in liquid helium.

REFERENCES

[1] A.G. Dubovoy, A.E. Perekos and K.V. Chuistov, "Structure and magnetic properties of small amorphous particles of metallic Fe-15 at.% B alloy", Physics of Metals, V.6, N 5, 1085-1088, 1985.

[2] A.G. Dubovoy, V.P. Zalutskiy and I.Yu. .Ignat'ev, "Structure, magnetic parameters and thermal stability for small amorphous particles and amorphous strips of Fe-15 at.% B", Physics of Metals, V. 8, N 4, 804-807, 1990.

[3] K.V. Chuistov, A.E. Perekos, V.P. Zalutskiy, T.V. Efimova, N.I. Glavatskaya, "The effect of production conditions on the structural state, phase composition and fineness of iron and iron-based powders made by electric-spark erosion", Metal Physics and Advanced Technologies, V. 16, N 8, 865-875, 1997.

[4] K.V. Chuistov and A.E. Perekos, "Structure and properties of small-size metallic particles. 1. Phase-structure state and magnetic characteristics (Review)", Metal Physics and Advanced Technologies, V. 17, N 1, 57-84, 1998.

[5] AYu. Ishlinsky, "Polytechnic Dictionary", Soviet Encyclopedia; Moscow, 611, 1989.

[6] A. Loiseau, N. Demoncy, O. Stephan, C. Colliex and H. Pascard, "Filling carbon nanotubes using an ARC discharge. In: Science and Application of Nanotubes", Kluwer Academic Publishers, New York, 398, 2000.

[7] D.V. Schur, A.G. Dubovoy, E.A. Lysenko, T.N. Golovchenko, S.Yu. Zaginaichenko, AF. Savenko, and et al., "Synthesis of nanotubes in the liquid phase", Extended Abstracts, 8th International Conf. "Hydrogen Materials Science and Chemistry of Carbon Nanomaterials" (ICHMS'2003), IHSE, Sudak (Crimea, Ukraine), 399-402, 2003.

[8] D.V. Schur, A.G. Dubovoy, S.Yu. Zaginaichenko and A.F. Savenko, "Method for synthesis of carbon nanotubes in the liquid phase", Extended Abstracts, An International Conf. on Carbon Providence

(Rhode Island. USA), American Carbon Society, 196-198, 2004.

[9] M.V. Antisari, R. Marazzi and R. Krsmanovic, "Synthesis of multiwall carbon nanotubes by electric arc discharge in liquid environments", Carbon, V. 41, N 12, 2393-401, 2003.

[10] L.P. Biro, Z.E. Horvath, L. Szalmas, K. Kertesz, F. Weber, G. Juhasz, G. Radnoczi and J. Gyulai, "Continuous carbon nanotube production in underwater AC electric arc", Chem. Phys. Lett., V. 372, N 3-4, 399-402, 2003.

[11] N. Sano, J. Nakano and T. Kanki, "Synthesis of single-walled carbon nanotubes with nanohorns by arc in liquid nitrogen", Carbon, V. 42, N 3, 686-688, 2004.

[12] J. Qui, Y. Li, Yu. Wang, Z. Zhao, Y. Zhou and Ya. Wang, "Synthesis of carbon-encapsulated nickel nanocrystals by arc-discharge of coal-based carbons in water", Fuel, V. 83, N 4-5, 615-617, 2004.

[13] D. Bera, S.C. Kuiry, M. McCutchen, A Kruize, H Heinrich, M. Meyyappan and S. Seal, "In-situ synthesis of palladium nanoparticles-filled carbon nanotubes using arc discharge in solution", Chem. Phys. Lett., V. 386, N. 4-6, 364-368, 2004.

[14] L.A. Montoro, C.Z. Lobrano Renata and J.M. Rosolen, "Synthesis of single-walled and multi-walled carbon nanotubes by arc-water method", Carbon, V. 43, N 1, 200-203, 2005.

Preparation and applications of transition metal oxide nanofibres and nanolines

T. Tätte, R. Talviste, M. Paalo, A. Vorobjov, M. Part, V. Kiisk, K. Saal, A. Lõhmus and I. Kink

*Institute of Physics, University of Tartu, and
Estonian Nanotechnology Competence Center,
142 Riia St, 51014 Tartu, Estonia, tanelt@fi.tartu.ee

ABSTRACT

In current paper we demonstrate the applicability of high viscous oligomeric alkoxide concentrates in fabrication of technologically interesting structures like nanometer diameter fibers (down to 200 nm), nanoneedles (tip radii 15-25 nm) and some micrometers wide linear surface structures. The approach is cost-effective and simple as it utilizes low-cost precursor materials (metal alkoxides) combined with low-tech processing (tape casting, fibers pulling, aging and baking).

Keywords: nanofibres, tape-casting, SPM tips, sol-gel, alkoxides

1 INTRODUCTION

During the last decade preparation of transition metal oxide nanofibres (wires), nanoribbons and related structures have attracted considerable attention. To produce materials for potential applications like gas and humidity sensors, optical sensors, filter materials etccheap chemical methods like solution-liquid-solid method, electrospinning, solvothermal method, vapor-liquid-solid template directed synthesis etc have been advised [1,2]. Still, some applications like nanometer diameter optical waveguides need fibres that have higher homogeneity than is achievable by these methods. Conventional taper-drawing technology enables to achieve the desired degree of homogeneity and fibers with diameters down to 20 nm can be made successfully, expressing ultra high smoothness, structural uniformity and perfectly circular cross-section [3]. However, the method has some drawbacks like very high processing temperature (2000K) and low refractive index of silica, which limit the possible technological applications. It is also demonstrated that SnO_2 high aspect ratio crystalline nanoribbons are suitable for optical waveguiding [4]. Furthermore, these methods leave a crucial issue – manipulation of the fibers – practically uncovered.

To overcome these limitations in producing ultra-high uniformity fibres we have developed a new method that utilizes oligomeric metal alkoxides as precursors. In our earlier works we have shown that high viscosity transition metal alkoxide concentrates can be used for making ultra-sharp needles [5] applicable as scanning tunneling and photon imaging microscopy probes [6], and also for making thin and narrow stripes on a substrate [7]. The method is similar to the well-know sol-gel process, where monomeric transition metal alkoxide is hydrolyzed and polymerized, followed by extraction of the solvent and water, and remaining highly viscous polymer concentrate:

Polymerization

$$\underset{\text{metal alkoxide}}{M(OR)_4} + H_2O + \text{solvent} \rightarrow \tag{1}$$

$$\rightarrow (\text{-}\underset{OR}{\overset{OR}{M}}\text{-O-})_{n<10} + ROH + \text{solvent} \xrightarrow{\text{Solvent removal}} \underset{\text{oligomeric concentrate}}{(\text{-}\underset{OR}{\overset{OR}{M}}\text{-O-})_{n<10}}$$

The material can be readily shaped to aforementioned nanometric structures at room temperature and then transformed to solid oxide state as a result of sufficient aging and baking.

In this paper we primarily refer to the possibility to prepare metal oxide nanofibres using highly viscous alkoxide concentrates. In addition we present our results on preparing oxide nanoneedles as possibilitie to prepare very sharp end fibres. Because of low-tech nature and low cost of the method, it may hold valuable impact on relevant technologies in the future.

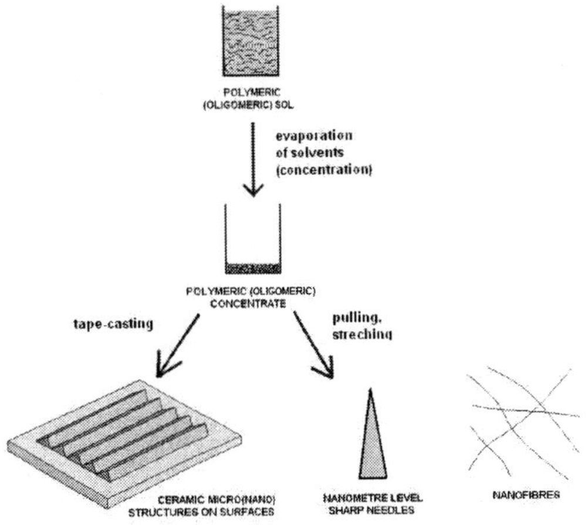

Figure 1. Schematic illustration of the different applications of high viscosity alkoxide concentrates.

2 FIBERS, NEEDELS AND MICROPATTERNS OF TRANSITION METAL OXIDES

Starting with neat liquid metal alkoxides (Ti, Sn, Si, Hf, Zr) the precursor material is made simply by addition of water in an appropriate solvent [5]. As a result, the alkoxide polymerizes (see eq. (1)) to the extent of up to ten monomers. The remaining alcohol and solvent are then extracted from the formed oligomeric mass, and the basic material for fabrication of nanostructures is obtained. The oligomeric concentrate is a highly viscous mass, which can only be stored in dry atmosphere. If introduced to humid air the material continues to polymerize via cross-linking the individual oligomer molecules near to the exposed surface, leading to very quick solidification of the surface. Eventually, the material becomes completely solid, but still containing some organics and water. These can be removed by heating the material to a sufficient temperature, depending on the type of the oxide and desired degree of crystallinity.

Fibers and needles are both made by pulling the concentrate jet in air [5]. Here, the outcome depends on the viscosity of the concentrate, pulling speed and humidity of the surrounding atmosphere [5]. By carefully optimizing the parameters nanometrically sharp needles (Figure 2) and ultra-narrow fibers (Figure 3) can be drawn. As transition metal oxides are typically transparent and can be readily made conductive via addition of appropriate dopants the structures may have many practical applications, e.g. probes for simultaneous STM and photon imaging [6] and others.

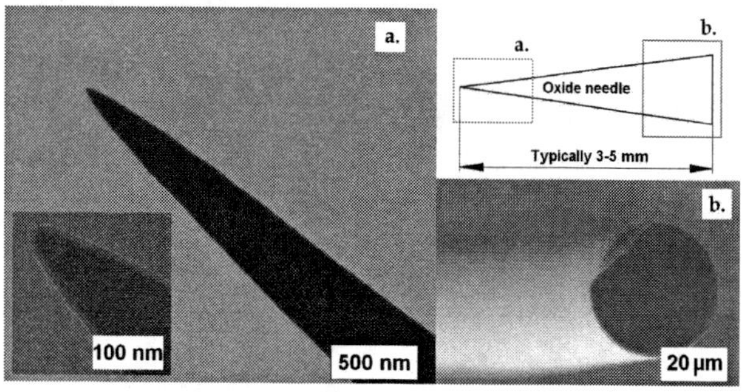

Figure 2. Nanometrically sharp oxide needles (SnO_2).

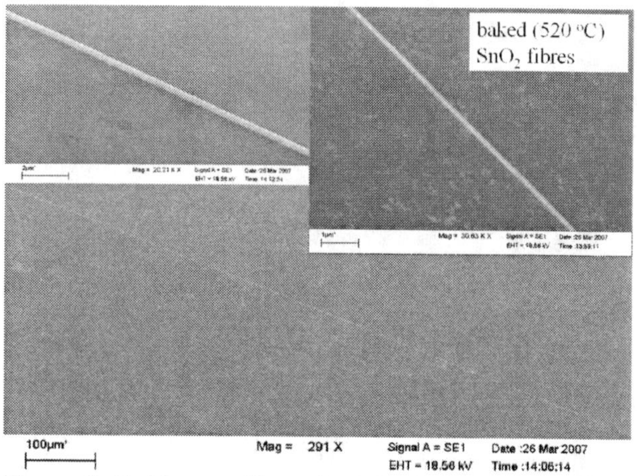

Figure 3. Oxide nanofibers (SnO_2) after 2h baking at 520 °C. The fibers can be drawn up to some cm-s in length.

Nanopatterning is performed by tape casting the precursor to a substrate using an appropriate blade. We prepared the structured blade from a cleaved silicon monocrystal using conventional wet etching technique [7]. The surface typically does not require any special treatment and ordinary glass can readily be applied. Figure 4 shows an example of obtained structures.

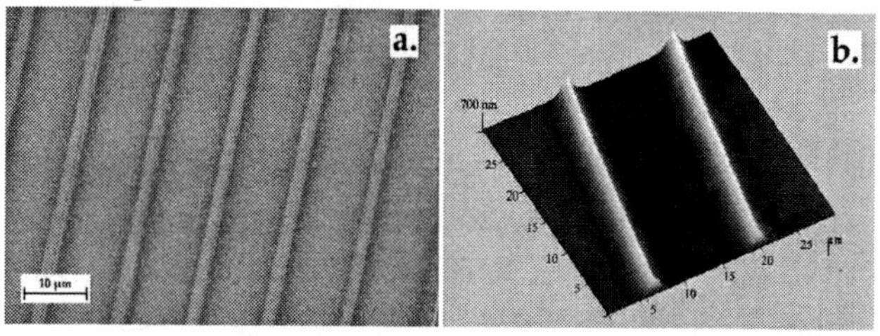

Figure 4. A fragment of a tape casted microstructured film prepared using a blade with triangular cross-section grooves.

The cross-section as can be seen in Figure 4 is determined by the shape of the grooves on the blade, volume loss during aging and baking and flow properties of the precursor. There is no fundamental limit on the lengths of the structures and tens of square centimeters of patterned areas have already been demonstrated in practice. Furthermore, the lines can intentionally be made nonlinear and even applied to uneven substrates.

The described structures can be doped with appropriate dopants in order to modify the useful properties of the material such as conductivity, hardness, fluorescence and others. We have also demonstrated that the structures can be doped by carbon nanotubes in order to prepare transparent CNT/oxide electrodes [8].

3 CONCLUSIONS

Transition metal alkoxide concentrates are promising materials for fabrication of nanometric fibers, needles and micropatterns because of low cost and simplicity of production. The method is focusing more on the lower resolution and flexible production of micro- and nanostructures and is thus complimentary to the conventional lithographic methods. Transition metal oxide nanostructures have useful practical applications in nano-optics, -electronics and -optoelectronics, since optically transparent and conductive materials can be fabricated.

ACKNOWLEDGEMENTS

This work was supported by the Estonian Nanotechnology Competence Centre and Estonian Science Foundation (Grants ETF 7612, 6537, 6163, 6660 and 7102).

REFERENCES

[1] J. G. Lu 2006 Quasi-one-dimensional metal oxide materials – Synthesis, properties and applications. *Material Science and Enginering R* **52** 49-91

[2] J. Xuchuan et al 2004 Ethylene glycol-mediated synthesis of metal oxide nanowires. *Journal of Materials Chemistry* **14**(4) 695-703.

[3] L. Tong and E. Mazur 2008 Glass nanofibres for micro- and nano-scale photonic devices *Journal of Non-Crystalline Solids* **354** 1240-44

[4] M. Law 2004 Nanoribbon waveguides for subwavelenght photonics integration. *Science* **305** 1269-1273

[5] T. Tätte et al 2007 Pinching of alkoxide jets—a route for preparing nanometre level sharp oxide fibres. *Nanotechnology* **18** 125301

[6] V. Jacobsen et al 2005 Electrically conductive and optically transparent Sb-doped SnO_2 STM-probe for local excitation of electroluminescence. *Ultramicroscopy* **104**(1), 39-45

[7] V. Kisand et al 2007 Preparation of structured sol-gel films using tape casting method. *Materials Science & Engineering B* **137**(1-3) 162-165,

[8] M. Paalo et al 2008 Preparation of transparent electrodes based on CNT-s doped metal oxides Technical Proceedings of the 2008 Nanotechnology Conference and Trade Show, Nanotech 2008

Some Properties of Electro-Deposited Fullerene Coatings

O. Ershova, N. Khotynenko, E. Golovko, V. Pishuk, O. Mil'to, L. Kopylova and A. Vlasenko

Institute for Problems of Materials Science of NAS of Ukraine,
Kiev-150, P.O.Box 195, 03150 Ukraine, lab67@materials.kiev.ua

ABSTRACT

This paper represents the results of investigations into thermal oxidation of electrodeposited fullerene-containing anodic coatings, the dependence of oxidation parameters on the composition of working solution. Some electrolysis products have been investigated by IR spectroscopic and X-ray phase methods.

The thermal parameters for oxidation of the coatings produced by electrolysis of TFE solutions with additives and without them are given. The dependence of the chemical composition of the coatings on the type of additive has been established. It has been shown that in heating the coating, deposited electrically from the TFE solution, fullerenes are sublimated and graphitized.

IR transmission spectra of electrolysis products, produced from the TFE solutions with KBr or KOH additives and without them, and also diffractograms of coatings, deposited electrically from the TFE solution without additives and with KOH, are represented. Analysis of IR transmission spectra and diffractograms confirms heterogeneity of the chemical composition of the coatings, deposited electrically from the TFE solution with KBr or KOH additives. This solution is a mixture of the fcc lattice fullerenes and fullerene-containing compounds.

Keywords: fullerene solution, ethanol, toluene, anodic coating, oxidation, electrolysis

1 INTRODUCTION

Earlier we have demonstrated the principal possibility to produce fullerene-containing coatings on metal electrodes by the electrochemical method and the dependence of their structure on the chemical composition of the working solution and experimental conditions [1].

This work represents the results of investigations into thermal oxidation of electro-deposited fullerene-containing anodic coatings, the dependence of oxidation parameters on the composition of the working solution. Some electrolysis products have been investigated by IR spectroscopic and X-ray phase methods. Based on comparison of the obtained results, we have made conclusions about phase and chemical compositions of the substances investigated.

2 EXPERIMENTAL CONDITIONS

The anodic coatings on Ni-electrodes have been investigated. The coatings have been produced by the electrochemical method from the fullerenes solution in toluene (TF). The base electrolyte for this solution was either ethanol (TFE solution), or ethanol with one of the additives of KOH, KBr, $LiClO_4$ type, or KCl (TFE with additives). The difference in potentials of electrodes depended on the composition of the working solution and corresponded 600-1600 V for the TFE solution and 10-80 V for the TFE solution with above additives.

Thermal-gravimetric studies have been performed on the derivatograph Q-1500D in conditions of dynamic heating the coatings in air at 20-1000°C. IR transmission spectra of electrolysis products, which were mechanically separated from the electrodes (powders), have been registered using the two-beam spectrophotometer Specord 75-IR in the range of wave numbers 400 to 4000 cm^{-1}.

3 RESULTS AND DISCUSSION

Oxidation of the coatings starts at $T \geq 200°C$. The temperature range and the temperature of the maximum rate of oxidation depend on the composition of the working solution. In this case the maximum rate of oxidation is observed at $T_{max}=435°C$. Oxidation of the coating produced from the TFE solution with KOH additive is characterized by the appearance the shoulder at $T_1=350°C$ and the sharp peak at $T_{max}=470°C$ in the DTG curve (Fig.1, curve 2).

The observed peculiarities in oxidation of anodic coatings produced from the TFE solutions with additives and without them indicate the difference in their chemical composition. According to the literature data [2, 3], the temperature of the maximum rate of fullerenes oxidation is in the temperature range 420 to 570°C and depends on dispersivity and the degree of fullerenes crystallinity.

Comparison of the experimental thermal-gravimetric parameters with the literature data on fullerenes oxidation allows us to suppose that the anodic coating produced from the TFE solution without additives comprises fullerenes with the oxidation temperature ≈435°C. This is also confirmed by IR spectroscopic and X-ray phase studies of the electrolysis products.

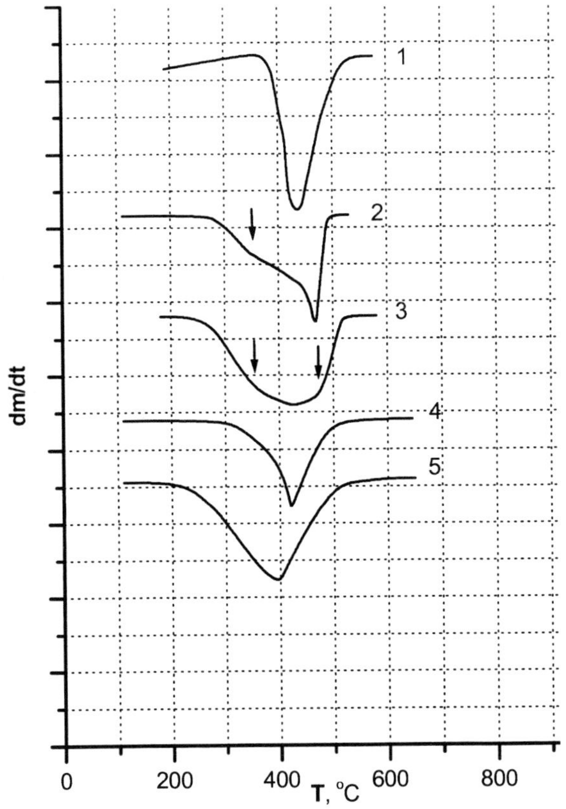

Figure 1: DTG curves for oxidation of anodic coatings on Ni-electrodes. The coatings have been produced by electrolysis of TFE solutions with additives and without them: 1 – without additives; 2 – KOH; 3 – KBr; 4 – $LiClO_4$; 5 – KCl.

Table 1: Parameters for oxidation of anodic products after electrolysis of TFE solutions with additives and without them.

#	Composition of electrolyte	Beginning interaction T,°C	DTG		
			T_{max}, °C	T_1, °C	T_2, °C
1	TFE	200	435		
2	TFE + KOH	280	470	350	
3	TFE + KBr	240	419	350	470
4	TFE + $LiClO_4$	280	419		
5	TFE + KCl	200	395		

4 CONCLUSIONS

1. The thermal parameters for oxidation of the coatings produced by electrolysis of TFE solutions with additives and without them are given. The dependence of the chemical composition of the coatings on the type of additive has been established.
2. It has been shown that in heating the coating that was deposited electrically from the TFE solution fullerenes are sublimated and graphitized.
3. IR transmission spectra of the electrolysis products that were produced from the TFE solutions with KBr or KOH additives and without them, and also diffractograms of the coatings that were deposited electrically from the TFE solution without additives and with KOH are represented. Analysis of IR transmission spectra and diffractograms confirms heterogeneity of the chemical composition of the coatings that were deposited electrically from the TFE solution with KBr or KOH additives. This solution is a mixture of fullerenes of the fcc lattice and fullerene-containing compounds. The coating that was deposited electrically from the TFE solution without additive comprises fullerenes of the fcc lattice and is a mono-phase system.

REFERENCES

[1] D.V. Schur, N.G. Khotynenko, S.Yu. Zaginaichenko, A.F. Savenko, V.M. Adeev, S.A. Firstov, A.P. Pomytkin, L.G. Shcherbakova, A.A. Rogozinskaya, B.P. Tarasov and Yu.M. Shulga, "On the electrodeposition of fullerenes and their compounds from solutions", Nanosystems, Nanomaterials, Nanotechnologies, V. 3, N 1, 215-225, 2005.

[2] E.G. Rakov, "Production techniques of carbon nanotubes", Uspekhi khimii, V. 69, N 1, 41-59, 2000 (in Russian).

[3] E.I. Golovko, O.V. Pishuk, V.A. Bogolepov, N.S. Anikina, G.A. Vojchuk, A.Yu. Vlasenko, S.Yu. Zaginaichenko, E.A. Lysenko and D.V. Schur, "The implementation of derivatographic investigations for nanostructure materials certification", Nanosystems, Nanomaterials, Nanotechnologies, V. 3, N 3, 633-643, 2005.

[4] H. Werner, Th. Schedel-Niedrig, et.al., J. Chem. Soc. Faradey Trans., V. 90, N 3, 403-409, 1994.

[5] M. Wohlers, H. Werner, D. Herein, et.al., "Reaction of C_{60} and C_{70} with molecular oxygen", Synthetic metals, V. 77, N 4, 299-302, 1996.

[6] A.V. Eletsky, B.M. Smirnov, "Fullerenes and carbon structure", Uspekhi fiz. nauk, V. 165, N 9, 977-1009, 1995 (in Russian).

[7] ASTM, X-ray diffraction date cards, 1995.

Optical properties of [001] GaN nanowires- An ab-initio study

B. K. Agrawal, A. Pathak and S. Agrawal

Physics Department, Allahabad University, Allahabad - 211002, India.

E-mail : balkagl@rediffmail.com, balkagr@yahoo.co.in

ABSTRACT

We make a detailed comprehensive study of the structural, electronic and optical properties of the unpassivated and H-passivated GaN nanowires (NWs) having diameters in the range, 3.29 to 18.33 Å grown along [001] direction by employing the first–principles pseudopotential method within density functional theory (DFT) in the local density approximation (LDA). Two types of the NW's having hexagonal and triangular cross-sections have been investigated. All the NW's after relaxation (to achieve minimum energies) show distorted structures where the chains of the Ga- and N- atoms are curved in different directions. The binding energy (BE) increases with the diameter of the NW because of decrease in the relative number of the unsaturated surface bonds. The BE's of the triangle cross-sectional nanowires are somewhat smaller than those of the hexagon cross-sectional nanowires in accordance with the Wulff's rule. As expected, the band gap varies appreciably with the diameter of the NW, at first falls rapidly in the small diameter nanowires and very slowly thereafter in the large diameter nanowires. After atomic relaxation, appreciable distortion occurs in the nanowires where the chains of Ga- and N-atoms are curved in different directions. These distortions are reduced with the diameters of the nanowires. The optical absorption in the GaN NW's is quite strong in the ultra-violet (UV) region but an appreciable absorption is also present in the visible region. The present results indicate the possibility of engineering the properties of nanowires by manipulating their diameter and surface structure. Strong optical absorption around 3.0 eV has been seen for the large diameter uncompensated and H-passivated wire which is in concurrence with the peak observed at 3.25 eV by Kuykendall et al in their photoluminescence emission spectra of the large diameter GaN NW's. The presently predicted GaN nanowires possessing the triangular cross-sections should be observable in the experiments.

Keywords: **structural stability, electronic structure, optical absorption**

INTRODUCTION

The study of the semiconductor nanowires (NWs) recently grown by the different methods [1] will help in the understanding of the fundamental roles of the reduced dimensionality and size. Almost all the physical properties such as mechanical, electrical, optical and magnetic ones etc are affected by the quantum confinement. The NWs have the future interesting applications in the photonics, nano- and molecular electronics, and thermoelectrics [2]. GaN has high melting point, high thermal conductivity and large bulk modulus [3] arising from the strong ionic and covalent bonding. They are useful for generating high-temperature, high-power, and high frequency devices.

CALCULATION AND RESULTS

In our calculations, we chose a soft non–local pseudopotential of Hartwigsen et al [4] within a separable approximation [5] after employing the plane waves [6].

We investigate the GaN NWs in wurtzite (WZ) structure having hexagonal and triangular cross sections having diameters in the range, 3.29 – 15.95 Å grown along the [001] direction. These WZ NWs in the hexagonal and triangular shapes are denoted as WZ-H (m, n) and WZ-T (m, n), respectively where m, n represent the number of atoms in each closely spaced double layer.

The atomic configurations of the hexagon cross-sectional WZ-H(24,24) NW in three dimensions are shown in Figs. 1(a), 1(b) and 1(c), respectively.

For all the studied NWs, the outer most surface Ga atoms relax inward whereas the outermost surface N atoms move outward but by a smaller magnitude. The inner most bulk atoms do not relax appreciably.

Fig. 1(c) reveals that the Ga and N planes are quite distorted in the optimized NW. The atomic chains lying along the [010] direction get curved in different directions. Some of the Ga atomic chains are curved more severely as compared to the N atomic chains. The atoms of the N-plane form a curved sheet like structure having curvature away from the nearest Ga-plane. On the other hand, the atoms of Ga-plane show a quite different configuration. Here, the configuration of atoms of a part of the Ga-plane is like a curved sheet whereas the atoms of the remaining part of Ga-plane also reveal a curved sheet configuration but has curvature in the opposite direction. This behaviour is also seen in the NWs of larger diameters but with reduced curvatures. The behaviour of the optimized triangular cross-sections NWs [Fig. 2(c)] are similar to hexagon cross-sectional NWs.

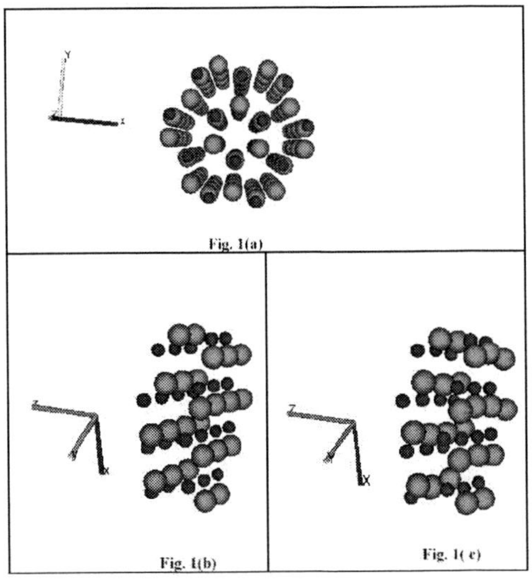

Fig. 1 *Atomic configuration of apart of the hexagon cross-sectionalWZH(24,24)nanowire(NW). Ga (N) atoms are denoted bybigger (smaller) spheres. (a)Optimized (relaxed) structure having the nanowire axis normal to the plainof the paper. (b) and (c) atomic configurations of a unit cellof the hexagon cross-sectionalWZ-H(24,24) NW having [010]direction normal to the plain ofthe paper (b) unoptimised and (c) optimized nanowire structures.*

The variation of the binding energy (BE) with the diameter of the NW is shown in Fig. 3.

The BE increases with the diameter of the NW.

We have also investigated the effect of H-passivation of the atoms lying on the surface facets of the hexagon and triangle cross-sectional nanowires. The NWs passivated by H-atoms are denoted as H-WZ-H(m,n) and H-WZ-T(m,n) for the hexagon and triangle cross-sectional nanowires, respectively where the meanings of the symbols are similar to those taken for unpassivated nanowires.

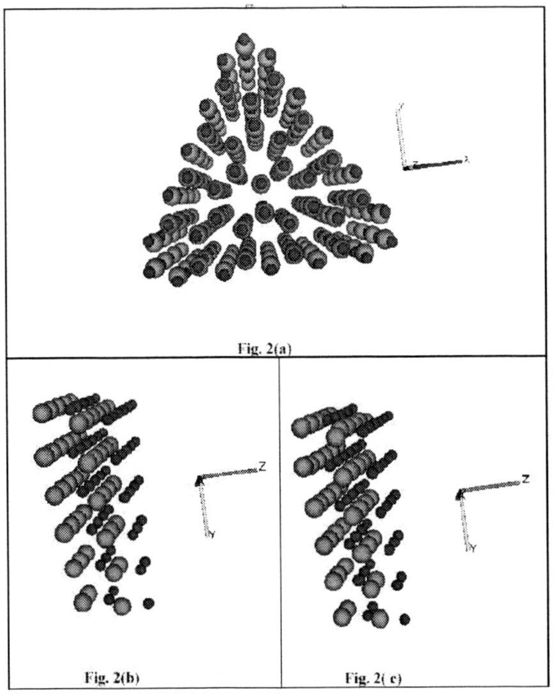

Fig. 2 *Atomic configuration of a part of the triangle cross-sectional WZ-T(50,42)) nanowire(NW). Ga (N) atoms are denoted by bigger (smaller) spheres. (a) Optimized (relaxed) structure having the nanowire axis normal to the plain of the paper. (b) and (c) atomic configurations of a unit cell of the triangle cross-sectional WZ-T(50, 42)) NW having [100]direction normal to the plain of the paper (b) unoptimised and (c) optimized nanowire structures.*

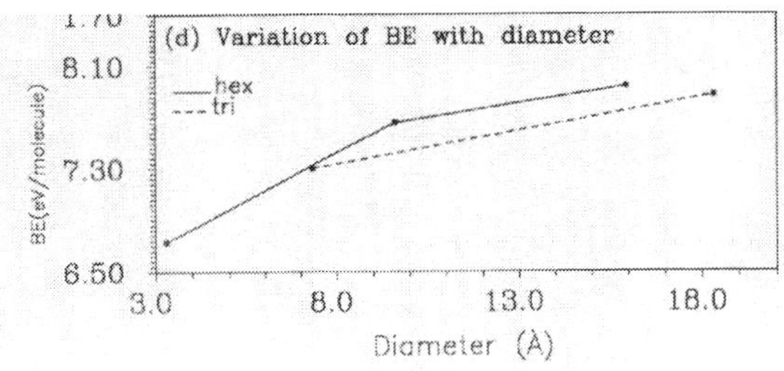

Fig. 3 *Variation of the binding energy with the diameter (D_{NW}) for the hexagonal and triangular nanowires.*

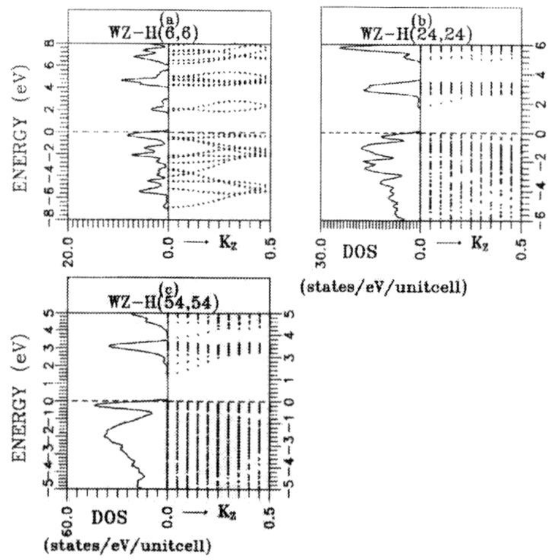

Fig.4 *Electronic structure and density of electronic states for the unpassivated hexagon cross-sectional (a) WZ-H(6,6), (b) WZ-H(24,24) and (c) WZ-H(54,54) GaN nanowires, respectively*

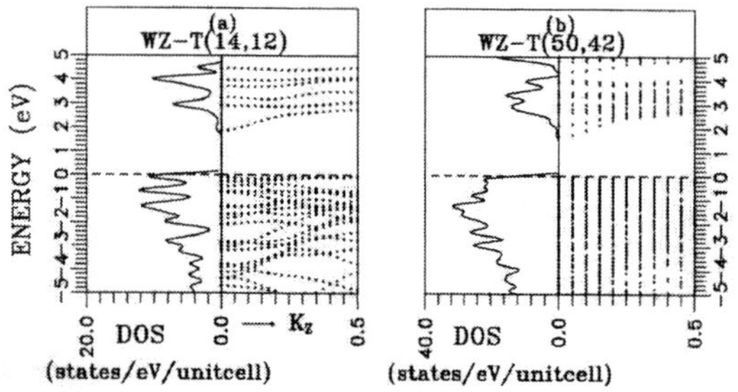

Fig. 5 *Same as for Fig. 4 but for the unpassivated triangle cross-sectionalWZ-T(14,12) and (b) WZ-T(50,42) GaN nanowires, respectively.*

After the passivation of the surface dangling bonds by the H atoms, the distortions of the Ga- and N-planes seen earlier in the unpassivated NW's remain intact but with quite reduced magnitudes especially the surface atoms. Also, the directions of curvatures are changed in the H-passivated NW's. The distortions in the NWs can be attributed to the presence of the strong electrostatic interaction between the cations and anions.

The electronic energy bands alongwith the DOS for various NW's having hexagonal and triangular cross-sections are depicted in Figs. 4 and 5, respectively. In all the electronic structure figures, the origin of energy has been set at the upper most filled valence state. In the unpassivated NW's, there occur surface states in the fundamental electron energy gap.

In all the hexagon and triangle cross-sectional nanowires, the electronic states in the vicinity of the energy gap are comprised of the hybridized s-p-d like orbitals of both the cations and anions but having different relative contributions.

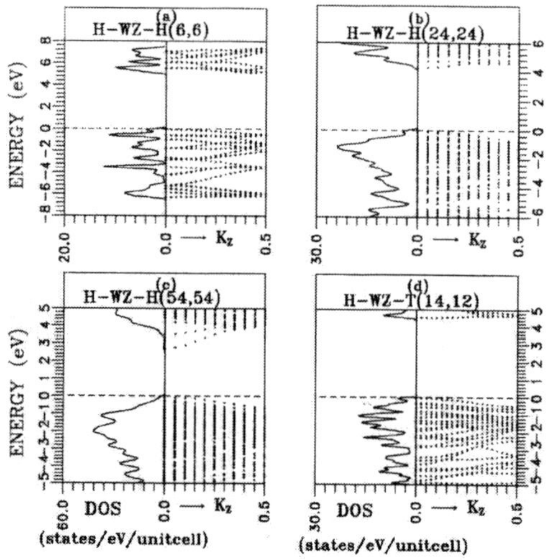

Fig.6 *Same as Fig.4 but for the H-passivated hexagon cross-sectional (a) H-WZ-H(6,6), (b) H-WZ-H(24,24) and (c) H-WZ-H(54,54) and (d) the triangle cross-sectional H-WZ-T(14,12) GaN nanowires.*

The electronic structures and the DOS's for H-passivated hexagon and triangle cross-sections are shown in Fig. 6. For all the H-passivated NWs, the highest occupied doubly degenerate valence state is the hybridized state containing contributions of Ga(p)-, Ga(d)- and N(p)- orbitals alongwith the H(s)-, H(p)- and H(d)-orbitals.

The no-phonon optical spectra for unpassivated hexagon, triangle cross-sectional and H-passivated nanowires are shown in Figs. 7, 8 and 9, respectively.

For unpassivated NWs, one observes an extended absorption commencing in the neighbourhood of 2.0 eV. The commencement of absorption is shifted to the higher energy side with decrease in the diameter of the NW.

For the H-passivated NWs in the absence of the absorption caused by the surface states, the continous optical absorption commences at high energies. For all the NWs, the continous optical absorption starts in the ultra-violet region except the large diameter H-WZ-H(54,54) NW where the absorption is also seen in the violet region.

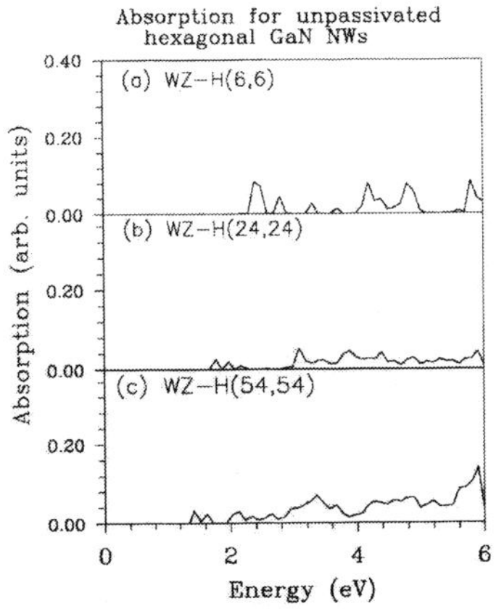

Fig. 7 *No-phonon optical absorption for the unpassivated hexagonal (a) WZ-H(6,6), (b) WZ-H(24,24) and (c) WZ-H(54,54) GaN nanowires.*

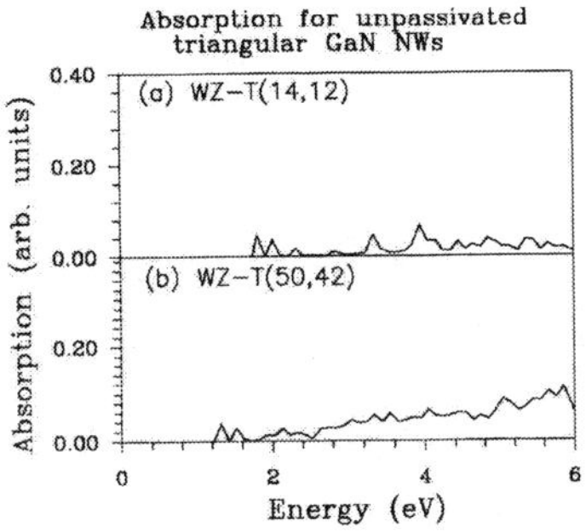

Fig. 8 *Same as Fig. 7 but for the unpassivated triangular (a) WZ-T(14,12) and (b) WZ-T(50,42) GaN nanowires*

CONCLUSIONS

For the H-passivated nanowires, the DOS's at VBM and CBM are small (except the H-WZ-H(6,6) nanowire) because the states contain the s-like contributions of the various atoms.

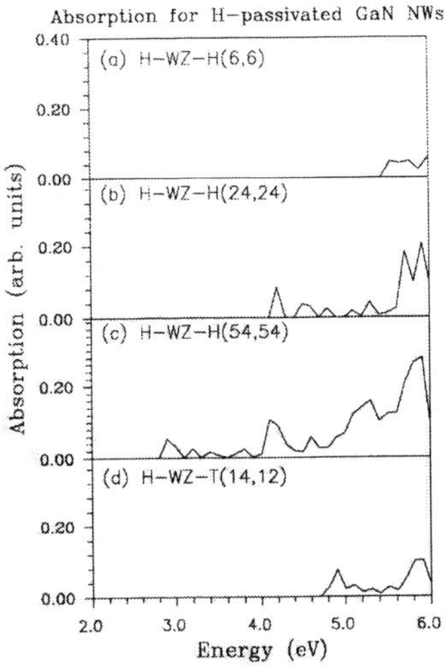

Fig. 9 *Same as Fig. 7 but for the H-passivated hexagonal (a) H-WZ-H(6,6), (b) H-WZ-H(24,24) and (c) H-WZ-H(54,54) and (d) the triangular H-WZ-H(14,12) nanowires. The energy of the optical absorption has been plotted above 2.0 eV.*

The thin hexagonal H-WZ-H(6,6) nanowire shows a indirect band gap of 5.21 eV whereas the thin triangular H-WZ-T(14,12) nanowire has a direct band gap of 4.58 eV.

The no-phonon optical absorption in the GaN NW's is seen to be quite strong in the UV region but appreciable absorption is also present in the visible region.

REFERENCES

[1]. R. S. Wagner and W. C. Elis, Appl. Phys. Lett. **4**, 89 (1964).

[2]. R. F. Seervice, Science **294**, 2442 (2001).

[3]. Properties of Group-III Nitrides, edited by J. H. Edgar, EMIS Data reviews Series (IEE, London, 1994).

[4]. C. Hartwigsen, S. Goedecker and J. Hutter, Phys. Rev. B **58**, 3641 (1998).

[5]. L. Kleinman and D. M. Bylander, Phys. Rev. Lett. **48,** 1425 (1982).

[6]. ABINIT code is a common project of the University Catholique de Louvain, Corning incorporated and other Contributors.

Direct Synthesis of CNT Yarns and Sheets

J. Chaffee, D. Lashmore, D. Lewis, J. Mann and M. Schauer, B. White

Nanocomp Technologies Inc.
162 Pembroke Road
Concord, New Hampshire 03301

ABSTRACT

We have developed an automated CVD process based on gas phase pyrolysis for the synthesis of CNT based yarns and sheets. The materials may be single-walled or dual-walled. Sheet material is now being fabricated in 3 foot by 6 foot panels. Yarns over 1000 m have also been produced at a rate of 150 m/hour per spin system. These products potentially enable a variety of applications which make use of their specific strength, flexibility, and electronic properties. Post processing of the yarns can improve uniformity and a controlled pitch angle of 15 degrees improves properties. Prepregged composites of the textiles with strengths of about 1.8 GPa have been demonstrated. Applications include: very light weight electrical conductors to replace copper wire in some applications, composites, sandwich structures, hybrid batteries, fire resistant coatings, and thermoelectric devices

Keywords: SWCNT, DWCNT, MWCNTS, Yarns, Wires, Sheets, EMI, Strength, Conductivity, Ballistics

1 INTRODUCTION

Carbon nanotubes are generally blended with various matrix materials at volume fractions of about 1 to 5% [1-3] or as minor constituents of battery electrodes [4]. A number of investigators [5, 6] have synthesized yarns mostly pulled from forests grown on silicon substrates. For the most part these have been multi-walled nanotubes. The notable exception was the research done at Cambridge by Professor Alan Windle and colleagues dealing with direct multiwall yarn synthesis also using a CVD process. Windle produces a dual-walled yarn of very light weight (0.04 g/1000 m) [7]. We describe in this paper a means of forming relatively thick yarns directly from SWCNT tube bundles and further a means to post process these nanotube yarns to yield very high strength yarns spun directly from the CVD furnace. The formation of CNT sheets can also be accomplished by drawing tubes from forests as shown by Baughman and colleagues [8]. Alternatively, CNT Bucky papers have been produced: formed from surfactant suspensions of CNTs in water that are subsequently filtered [9, 10].

2 EXPERIMENT

Carbon nanotubes are synthesized using a gas phase pyrolysis process [7, 11-16] whereby a metal organic compound is added to a carbon based fuel in the CVD process. The catalyst particles formed in the furnace are allowed to grow to a certain size then stopped. The carbon decomposition reaction is followed by tube growth throughout the transit time in the reaction chamber. Sometime during the growth process the nanotubes are electrostatically attracted to each other to form bundles [17, 18]. These bundles form a cloud of nanotubes which exit the furnace as a large coherent tube which can be manipulated to form yarns or alternatively to form sheets.

Yarns are formed by causing this cloud of nanotube bundles to impinge on an anchor or substrate from which they can be pulled off by a yarn spinning device. An example of how this works is shown in Figure 1.

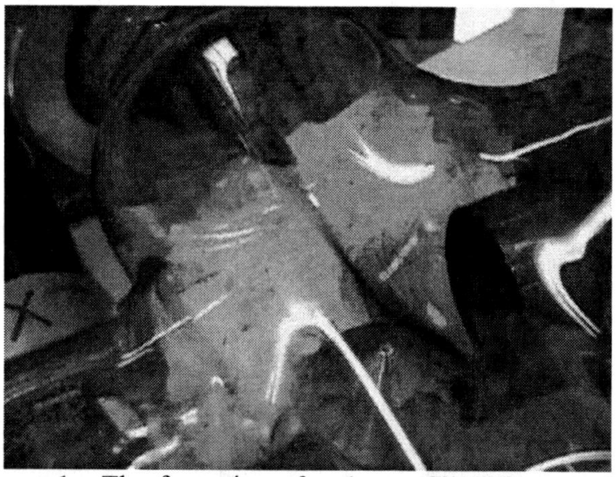

Figure 1. The formation of ~ 1 nm, SWCNT yarns by spinning from a rotating anchor. The CNTs are coming from the furnace on the right and the spinning system is out of the photograph on the top left.

The resultant yarn is then post process into a tightly woven wire whose pitch angle is set to about 10 degrees. A reel of 1 km of this yarn is shown in Figure 2.

Figure 2. A reel of 1 km, 3 tex, SWCNT yarn.

Sheets of either single-walled or dual-walled CNTs, determined by fuel composition and run conditions, are formed in the apparatus of Figure 3.

Figure 3. Apparatus to synthesize DWCNT sheets is shown above. The sheets are formed by nanotubes exiting the reactor not shown, and impinging on the belt located within the large box enclosed within the exhaust hood. The Allen Bradley controller is also not shown.

The output of this box is a CNT textile which is removed from the box by cutting. An example of a typical product is show in Figure 4.

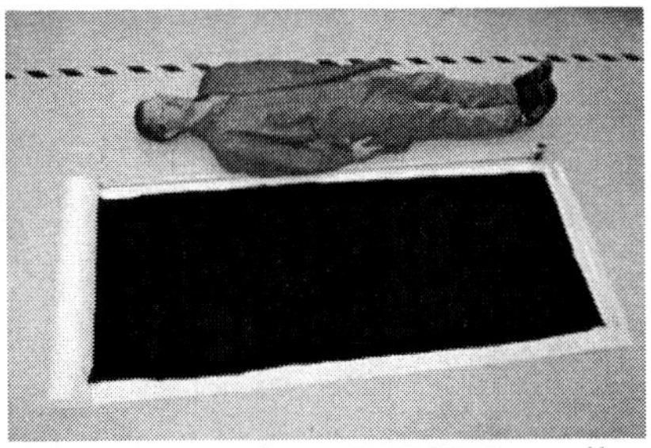

Figure 4. A >6 foot, DWCNT textile compared with a standardized "furnace operator" for scale.

2.1 Growth

The growth of carbon nanotubes is not fully understood. Our concept involves creation of the catalyst in a pre-chamber. In this pre-chamber both the carbon source such as ethanol is blended with an organometallic catalyst such as nickelocene, cobaltocene or ferrocene, and a sulfur containing compound such as thiophene. The function of this pre-chamber is to provide sufficient energy to vaporize the fuel and its constituents, induce the thermal decomposition of the organometallic molecules to create catalyst clusters and allow these catalyst clusters to grow to the desired size. Finally at the end of this pre-chamber, sulfur reacts with the clusters to stop the growth process. No nanotube growth occurs in this portion of the reactor. The catalyst clusters then are injected into the furnace where they catalytically react with the fuel to cause its decomposition. At this point the fuel source may be quite different from the raw ethanol due to decomposition cascade reactions. Growth appears to be initially very fast and to then slow down as the reactants are consumed. Detail models of this growth process will be presented in the future.

2.2 Mechanical Properties

Mechanical properties of both sheets and yarns are determined by tensile testing in a vibration isolated tensile testing instrument. The gauge length is fixed at 10 mm for all samples and the cross head rate is set at 5 mm/minute. Most of our measurements are made without an extensometer so that strain rates reflect system compliance, and even slippage in the grips. No adhesives are used as they seem to diffuse down the specimen.

The tensile strength for a typical yarn is shown in Figure 5.

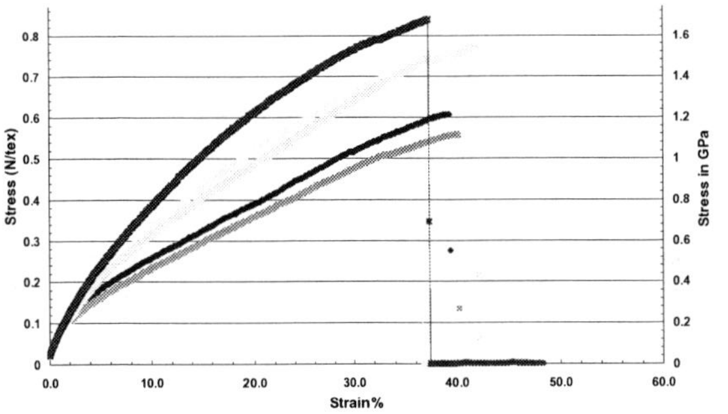

Figure 5. Breaking strength characteristics for SWCNT yarns. Breaking strength is in N/tex shown on the left axis and in GPa on the right axis. Gauge length is 10 mm for all of these samples. These samples have the catalysts retained within the structure and have not been heat treated

2.3 Electronic Properties

Carbon nanotubes and presumably yarns and sheets made of CNTs, conduct electricity quite differently than copper. High frequency effects are also quite different in copper wires electrons are caused to diffuse the skin of the conductor and therefore proximity effects are also very high. Because of the high radial mobility, external magnetic fields, for example those existing in coils, solenoids, electric motors, can affect electron mobility especially at high frequency. Since axial mobility of electrons in CNTs is restricted by the requirement than they have to overcome $1/G^o$ at every junction HF effects are minimal until the frequencies exceed 30 GHz .

Sheets made of CNTs are often compared with Bucky paper. Bucky paper however, is made quite differently, generally by suspending CNTs with the help of a surfactant in a water solution and then filtering out the CNTs (coated with surfactant) to make a thin film resembling paper. A comparison between the best Bucky paper and our dual-walled nanotube sheets is shown in **Table I.**

Table I. Difference between Nanocomp's sheet or ribbon conductors and Bucky Paper.

Property	Bucky Paper	DWCNT Sheet
Tensile strength	74 MPa [19]	>1000 MPa
Modulus	8 GPa	>30 GPa
Resistivity	5×10^{-2} Ω-cm[20]	2.6×10^{-4} Ω-cm (ambient) At T, R decreases
Thermal conductivity	40 (magnetically aligned)	60 Watts/m-°K[21]
Seebeck Coefficient	NA	>70 μV/°K

DC Electrical Characteristics

Electrical characteristics at ambient temperature were measured using the four point contact method at ambient and at elevated temperatures.

Sample temperatures were measured with a calibrated optical technique. The data shown in Figure 6 suggests that even though the CNT conductors under dc conditions can carry more power, they heat up faster under the same conductions as copper. Under some cases the temperature increase can be designed for in the applications especially where other factors such as stress or fatigue are an issue for conductors.

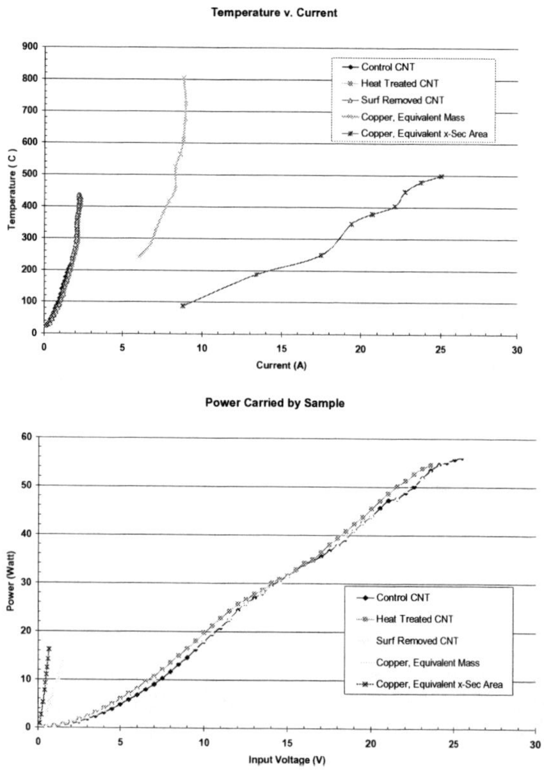

Figure 6. Resistivity data suggesting that even though CNT conductors can carry more power than copper, they need to be protected from oxidation even at temperatures as low as 300 °C. (a) The equivalent mass of copper carries more dc current but overheats to failure. CNT conductors oxidize in air at about 400 °C, in inert environment the CNTs can operate well over 1000 °C. (b) The CNT can carry more power in air before burn up, even at dc conductions!

AC Electrical Characteristics

High frequency behavior of CNTs has been studied by a number of investigators [22-24]. The mean free path of electrons in copper is 0.04 microns whereas the mean free path in CNT ranges from 1 to 60 microns [25]. At high frequency the impedance is controlled by skin effects, eddy currents and proximity effects. Since electron transport in CNT is said to be ballistic, such that the electrons are surface confined, the driving force to cause electron transport to the outer surface of the wire is much higher than for copper, therefore this effect is "pushed" to very high frequencies. Ballistic conduction has been reported for multi-wall nanotubes as well [26]. An example of this remarkable behavior is shown in Figure 7 comparing Litz wire of two different diameters with a CNT wire and a copper wire. The resistance has been normalized for this comparison.

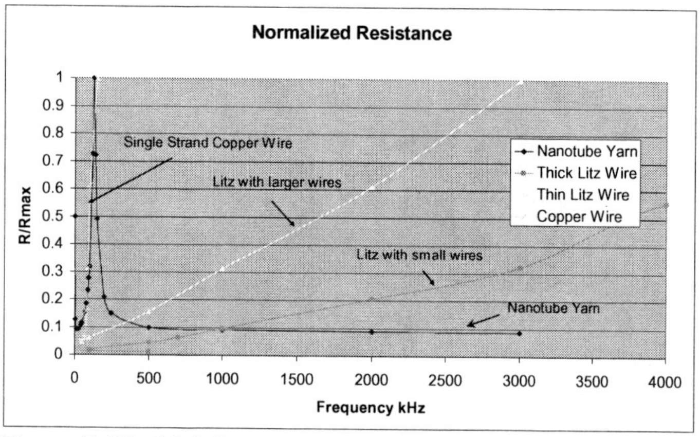

Figure 7. The high frequency response of CNT wire compared with single strand copper and litz wire of two different diameters.

3 SUMMARY

3.1. Carbon nanotube sheets and yarns exhibit dc conductivity worse than copper but at moderate frequency may outperform copper for some applications.

3.2. Strength of CNT yarns is starting to approach that of graphite and if adsorbed species and the catalyst weight is taken into account they already will.

3.3. A continuous process to produce CNT yarn has been developed and production rates exceeding 150 m/hr have been demonstrated.

3.4. A process for producing very large CNT sheets which can be single wall or dual wall has been demonstrated and sheets larger than 3 feet by 6 feet have been produced.

ACKNOWLEDGMENTS

The authors gratefully acknowledge the support of the Office of Naval Research, the Air Force and the Army Natick Soldier Center for partial support of the work presented here.

REFERENCES

1. S. Ghose, K. A. Watson, K. J. Sun, J. M. Criss, E. J. Siochi and J. W. Connell, Compos Sci Technol **66** (13), 1995-2002 (2006).
2. J. Chen, H. Y. Liu, W. A. Weimer, M. D. Halls, D. H. Waldeck and G. C. Walker, J Am Chem Soc **124** (31), 9034-9035 (2002).
3. P. M. Ajayan, L. S. Schadler, C. Giannaris and A. Rubio, Adv Mater **12** (10), 750-+ (2000).
4. Y. H. Lee, K. H. An, S. C. Lim, W. S. Kim, H. J. Jeong, C. H. Doh and S. I. Moon, New Diam Front C Tec **12** (4), 209-228 (2002).
5. M. Zhang, K. R. Atkinson and R. H. Baughman, Science **306** (5700), 1358-1361 (2004).
6. K. L. Jiang, Q. Q. Li and S. S. Fan, Nature **419** (6909), 801-801 (2002).
7. M. Motta, Y. L. Li, I. Kinloch and A. Windle, Nano Lett **5** (8), 1529-1533 (2005).
8. M. Zhang, S. Fang, A. A. Zakhidov, S. B. Lee, A. E. Aliev, C. D. Williams, K. R. Atkinson and R. H. Baughman, Science **309** (5738), 1215-1219 (2005).
9. V. Skakalova, A. B. Kaiser, U. Dettlaff-Weglikowska, K. Hrncarikova and S. Roth, J. Phys. Chem. B **109** (15), 7174-7181 (2005).
10. J. L. Bahr, J. P. Yang, D. V. Kosynkin, M. J. Bronikowski, R. E. Smalley and J. M. Tour, J Am Chem Soc **123** (27), 6536-6542 (2001).
11. M. Endo, ChemTech **18**, 568-576 (1988).
12. A. Oberlin, M. Endo and T. Koyama, J Cryst Growth **32** (3), 335-349 (1976).
13. R. Andrews, D. Jacques, A. M. Rao, F. Derbyshire, D. Qian, X. Fan, E. C. Dickey and J. Chen, Chem Phys Lett **303** (5-6), 467-474 (1999).
14. L. Ci, Y. Li, B. Wei, J. Liang, C. Xu and D. Wu, Carbon **38** (14), 1933-1937 (2000).
15. C. N. R. Rao and A. Govindaraj, Accounts Chem Res **35** (12), 998-1007 (2002).
16. R. Andrews, D. Jacques, D. L. Qian and T. Rantell, Accounts Chem Res **35** (12), 1008-1017 (2002).
17. M. J. O'Connell, S. M. Bachilo, C. B. Huffman, V. C. Moore, M. S. Strano, E. H. Haroz, K. L. Rialon, P. J. Boul, W. H. Noon, C. Kittrell, J. Ma, R. H. Hauge, R. B. Weisman and R. E. Smalley, Science **297** (5581), 593-596 (2002).
18. A. Thess, R. Lee, P. Nikolaev, H. Dai, P. Petit, J. Robert, C. Xu, Y. H. Lee, S. G. Kim, A. G. Rinzler, D. T. Colbert, G. E. Scuseria, D. Tomanek, J. E. Fischer and R. E. Smalley, Science **273** (5274), 483-487 (1996).
19. X. Zhang, T. V. Sreekumar, T. Liu and S. Kumar, J. Phys. Chem. B **108** (42), 16435-16440 (2004).
20. B. Ruzicka, L. Degiorgi, R. Gaal, L. Thien-Nga, R. Bacsa, J. P. Salvetat and L. Forro, Phys Rev B **61** (4), R2468-R2471 (2000).
21. , pp. Measured June 5, 2007 at the University of Texas on a CAMA™ processed Nanocomp textile.
22. Z. Yu and P. J. Burke, Nano Lett **5** (7), 1403-1406 (2005).
23. H. G. Han, Z. Y. Zhu, Z. X. Wang, W. Zhang, L. P. Yu, L. T. Sun, T. T. Wang, F. He and Y. Liao, Phys Lett A **310** (5-6), 457-459 (2003).
24. T. I. Jeon, K. J. Kim, C. Kang, I. H. Maeng, J. H. Son, K. H. An, J. Y. Lee and Y. H. Lee, J Appl Phys **95** (10), 5736-5740 (2004).
25. P. Poncharal, C. Berger, Y. Yi, Z. L. Wang and W. A. de Heer, J Phys Chem B **106** (47), 12104-12118 (2002).
26. C. Berger, Y. Yi, Z. L. Wang and W. A. de Heer, Appl Phys a-Mater **74** (3), 363-365 (2002).

Morphological, Crystalline and Photo-luminescent Property of Zinc Oxide Nanorod Array Controlled by Zinc Oxide Sol-gel Thin Film

Jing-Shun Huang* and Ching-Fuh Lin**

*Institute of Photonics and Optoelectronics, National Taiwan University, Taiwan, Republic of China
No. 1, Sec. 4, Roosevelt Road, Taipei, 10617 Taiwan (R.O.C), f92921047@ntu.edu.tw
** Institute of Photonics and Optoelectronics, Graduate Institute of Electronic Engineering and Department of Electrical Engineering, National Taiwan University, Taiwan, Republic of China,
cflin@cc.ee.ntu.edu.tw

ABSTRACT

Controllable growth of ZnO nanorod array has been systematically studied. Our investigation demonstrates that the annealing treatment of the ZnO sol-gel thin-film have strong influences on the morphology, crystalline and photoluminescence (PL) of the ZnO nanorod arrays grown thereon. As the annealing temperature increases, the size of the ZnO grain increases, and the diameter of thereon ZnO nanorod arrays increases from 60 to 250 nm. Besides, the growth rate of ZnO nanorod is very sensitive to the ZnO grains, and then influences the PL peak at 380 nm. The x-ray diffraction spectra indicate that the thin film annealed at the low temperature of 130 ℃ is amorphous, but the thereon nanorod arrays are high-quality single crystals growing along the c-axis direction with an orientation perpendicular to the substrates. The as-synthesized ZnO nanorod arrays via all solution-based processing enable the fabrication of next-generation nano-devices at low temperature.

Keywords: nanorods, zinc oxide, morphology, sol-gel, hydrothermal, photoluminescence, crystallinity

1 INTRODUCTION

Zinc oxide (ZnO) is a II-VI compound oxide semiconductor with a direct band gap of 3.37 eV and a large exciton binding energy of 60meV at room temperature, exhibiting near-UV light emission, transparent conductivity, and piezoelectricity. It has attracted great interest for promising applications in optoelectronics devices such as room temperature lasers [1], light emitting diodes [2], ultraviolet (UV) detectors [3], field emission displays [4], photonic crystals [5], and solar cells [6, 7], especially in the form of one-dimensional nanowires, nanorods, or nanotubes. In view of this point, controlled growth of ZnO nanostructures in terms of morphology and orientation is of significant importance from the standpoint of both basic fundamental research and the development of novel devices. To date, various synthesis approaches have been demonstrated to fabricate ZnO nanorods, which can be classified into two categories, vapor-phase and solution-phase methods. Vapor-phase processes such as vapor-liquid-solid epitaxy (VLSE), chemical vapor deposition (CVD), and pulse laser deposition (PLD) have some limitations for substrate size and the need for high temperature operation [8, 9]. For these reasons, there is a significant need to develop a low-temperature, large-scale, versatile route to synthesis of ZnO nanorods. Solution-phase methods are appealing due to their low growth temperature, low cost, and potential for scale-up. Recently, growth of ZnO nanorods in aqueous solutions at low temperature was reported by using the hydrothermal process [10]. Hydrothermal process has shown the possibility for applications in light emitting diodes and solar cells with their growth temperature below 100 ℃ and easy scale-up. This aqueous-based technique has also been used successfully to demonstrate the fabrication of large arrays of vertical ZnO nanorods on glass and plastic substrates [11]. This stimulated the study of using ZnO nanorod arrays on plastic substrates for application in flexible electronic devices.

C. Bekeny *et al.* [12] reported that the size of ZnO nanorods varied with the composition of the chemical precursors. M. Guo *et al.* [13] reported that the diameter and length of ZnO nanorod arrays were controlled at different growth temperatures under hydrothermal conditions. X. Wang *et al.* [14] reported that an annealing treatment of the substrate can influence the density of the ZnO nanorod arrays on indium tin oxide (ITO), but there is no study on crystallinity and photoluminescence (PL) of ZnO nanorod arrays. However, systematic research on the effect of quality characteristics of ZnO sol-gel thin films on the growth of ZnO nanorod arrays has rarely been reported.

In this work, we systematically study the feature-controlled ZnO nanorod arrays via hydrothermal method using ZnO sol-gel thin films were used as the seed layers with different pretreatment conditions. Our investigation shows that the vertical alignment, the crystallinity, and the growth rate of ZnO nanorod arrays are strongly dependent on the characteristics of the thin films. Field-emission scanning electron microscopy (FESEM), x-ray diffraction (XRD) pattern, and room temperature PL spectrum were applied to analyze the quality of the ZnO nanorod arrays.

2 EXPERIMENTAL DETAIL

2.1 Preparation of ZnO sol-gel thin films

The ZnO thin films was deposited on silicon substrates by a sol-gel method as previously reported in the literature [15]. A coating solution contained zinc acetate dihydrate [Merck, 99.5% purity] and equivalent molar monoethanolamine (MEA) [Merck, 99.5% purity] dissolved in 2-methoxyethanol (2MOE) [Merck, 99.5% purity]. The concentration of zinc acetate was 0.5 M. The resulting solution was then stirred at 60 ℃ for 2 hours to yield a homogeneous and stable solution, which served as the coating solution after being cooled to room temperature. Then the solution was spin-coated onto p-type silicon (100) substrates at the rate of 1000 rpm for 20 s and then 3000 rpm for 30 s at room temperature. Subsequently, the gel films were preheated for 10 minutes to remove the residual solvent. Then the films were annealed in a furnace at different temperatures ranging from 130 ℃ to 900 ℃ for one hour.

2.2 Growth of ZnO nanorod arrays

After uniformly coating the silicon substrates with ZnO thin films, hydrothermal growth of ZnO nanorod arrays was achieved by suspending these ZnO seed-coated substrates upside-down in a glass beaker filled with aqueous solution of 50 mM zinc nitrate hexahydrate [Sigma Aldrich, 98% purity] and 50 mM hexamethylenetetramine (HMT) [Sigma Aldrich, 99.5% purity]. During the growth, the glass beaker was heated with an oven and maintained at 90 ℃ for 4 hours. At the end of the growth, the substrates were removed from the solution, then rinsed with de-ionized water to remove any residual salt from the surface, and dried under nitrogen gas flow.

2.3 Characterization of ZnO thin films and thereon ZnO nanorod arrays

The general morphologies of the ZnO thin films and thereon ZnO nanorod arrays were examined by FESEM. The crystal phase and crystallinity were analyzed at room temperature by XRD using Cu Kα radiation. The room temperature PL, measured using a Nd:YAG laser at 266nm as the exciting source, was used to characterize the optical properties of the thereon ZnO nanorod arrays.

3 RESULTS AND DISCUSSIONS

3.1 ZnO sol-gel thin films

Figure 1 shows the top view FESEM images for the surface morphologies of ZnO sol-gel thin films with increasing annealing temperatures from 130 to 900℃. At the annealing temperature of 130 ℃, no grain forms and the surface is smooth. At the annealing temperature of 300 ℃, the film contains fine grains and the particle size is about 80nm. Once the annealing temperature increases, the grains become larger and densely packed. It demonstrates that the grain size of the ZnO thin film was changed due to re-distribution of crystalline grain by supplying sufficient thermal energy and the small grain has been joined into great crystalline surface.

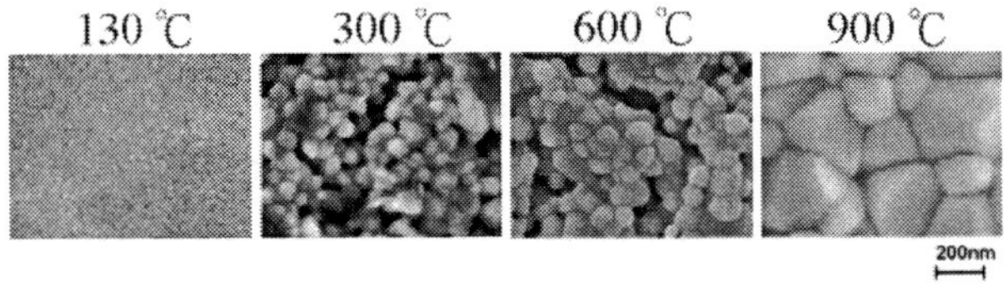

Figure 1: FESEM images of ZnO sol-gel thin films with annealing temperature from 130 to 900 ℃.

Figure 2 gives the XRD patterns of those ZnO thin films annealed at 130, 300, 600 and 900 ℃, respectively. The XRD patterns reveal that the (002) peak intensity varies with annealing temperature. When the ZnO thin films are annealed below 300 ℃, there is no preferred (002) *c*-axis orientation. At the temperature of 600 ℃, (100), (002), and (101) diffraction peaks corresponding to the ZnO wurtzite structure are observed in the XRD pattern, where the preferred (002) *c*-axis orientation dominates. When the annealing temperature increases to 900 ℃, the polycrystalline structure emerges with the (100) and (101) peaks while the intensity of the (002) peak decreases. This result indicates that the crystalline quality of each grain becomes poor [16]. Therefore, it maybe concluded that the preferred *c*-axis orientation initially increases with annealing temperature until it reaches the optimal situation at a certain annealing temperature. Afterwards the *c*-axis orientation intensity decreases gradually.

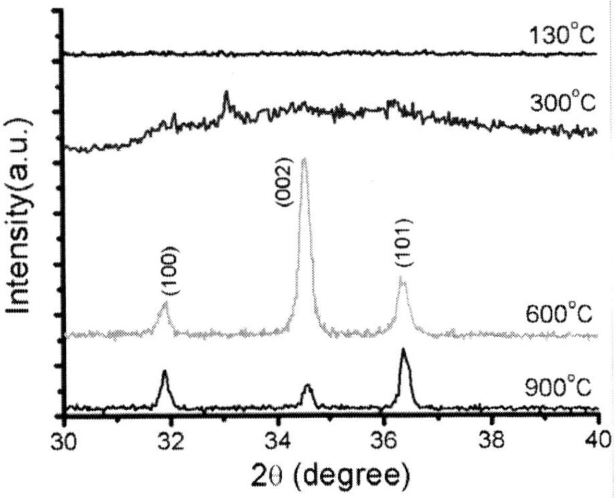

Figure 2: X-ray diffraction spectra of ZnO thin films annealed from 130 to 900 ℃

3.2 ZnO nanorod arrays

The as-grown ZnO nanorod arrays are shown in Figure 3. The obtained ZnO nanorod arrays are typically hexagonal-shaped. As the annealing temperatures of the ZnO thin films increase from 130 to 900 ℃, the diameters of the ZnO nanorod arrays increase from 60 to 260 nm (Figure 4). The reason may be that the high annealing temperature enhances the interaction among the grains and leads the grains to merge together to form bigger ZnO seeds, and thus increases the diameter of the ZnO nanorods thereon. Therefore, the size of the grains is a key factor that affects the nucleation of ZnO nanorod arrays. Furthermore, the ZnO nanorod arrays on the ZnO thin films annealed at 130 ℃ are well-aligned vertically and uniformly, and the well-defined crystallographic planes of the hexagonal single-crystalline nanorods can be clearly identified, providing evidence that the nanorod arrays orientate along the *c*-axis.

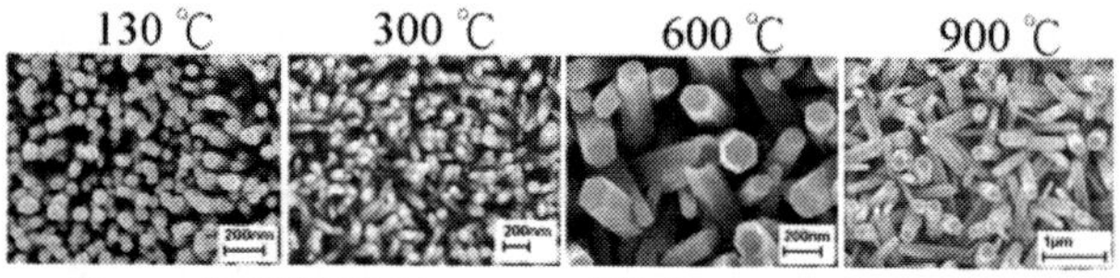

Figure 3: FESEM images of ZnO nanorod arrays grown at 90 ℃ with the thin films annealed from 130 to 900 ℃.

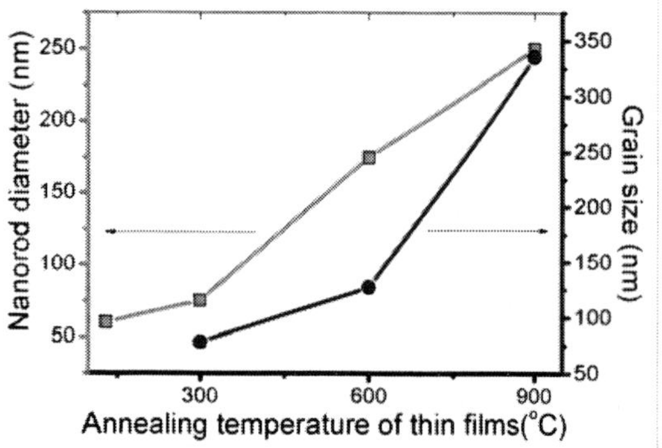

Figure 4: Dependence of the ZnO nanorod diameter (rectangles) and grain size of ZnO thin film (circles) on annealing temperature of ZnO thin films.

Figure 5 shows the XRD patterns of the ZnO nanorod arrays corresponding to those shown in Figures 2. Please note again that the nanorod arrays are grown at the same temperature of 90℃, while the thin films were annealed from 130 to 900 ℃. The peaks in the x-ray diffraction patterns are indexed to the hexagonal phase of ZnO. It is found that no other characteristic peaks corresponding to the impurities of the precursors such as zinc nitrate and zinc hydroxide are observed in the XRD patterns. At the temperature of 130, 300, and 600 ℃, only a very strong (002) diffraction peak and a very weak (101) peak are observed, indicating that the three ZnO samples are all of high *c*-axis orientation. It is noticeable that for the sample annealed at 130 ℃, the XRD pattern shows only the (002) diffraction peak. In addition, the intensity of (002) diffraction peak is strongest, compared to other samples annealed at higher temperatures. This implies its perfect *c*-axis orientation and this result is in accordance with its SEM image. On the other hand, for the sample annealed at 900 ℃, the (002) diffraction peak becomes weak, and at the same time, the (100) and (101) peaks become strong, indicating its tendency toward random orientation. It means that the films annealed at 900 ℃ has worse morphology and hence results in the relatively random orientation of nanorod arrays as observed in the Figure 3.

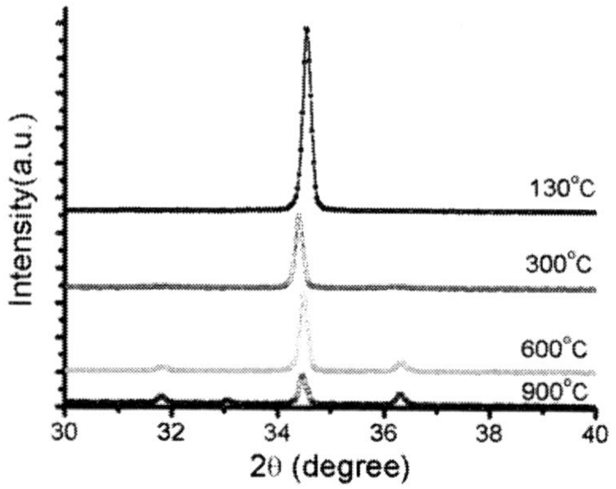

Figure 5: X-ray diffraction spectra ZnO nanorod arrays with ZnO thin films annealed from 130 to 900 ℃.

It is interesting to note that the (002) diffraction peak of the seed layer annealed at 130 ℃ is smaller than the others, while the (002) diffraction peak of thereon ZnO nanorod arrays at 130 ℃ is larger than the others. The previous investigation of the thin films annealed at 130℃ indicates that it is nearly amorphous. However, the growth of ZnO nanorod arrays on amorphous ZnO thin films along the (002) plane is even more notable than that on poly-crystalline thin films. It may be because the poly-crystalline ZnO grains with a certain orientation limit the growth along the (002) plane. In comparison, the amorphous ZnO seed layer does not limit the growth along the (002) plane. This indicates that the ZnO nanorod arrays prepared by the hydrothermal method have preferential orientation along the (002) plane, in particular on the thin films without a certain orientation.

3.3 Annealing effect of ZnO thin films on PL spectra of thereon ZnO nanorod arrays

The room temperature PL characteristics of as-grown ZnO nanorod arrays with different annealing temperatures of thin films are shown in Figure 6. The UV peak increases with annealing temperature. The UV emission of ZnO nanorod arrays corresponding to the near band edge emission is due to the recombination of free excitons through an exciton-exciton collision process. At the temperature of 900 ℃, the defect-related green emission of the nanorod arrays is lower than that of the seed layers. The green emission is also known to be a deep level emission caused by the impurities and structural defects in the crystal such as oxygen vacancies, zinc interstitials, and so on [17].

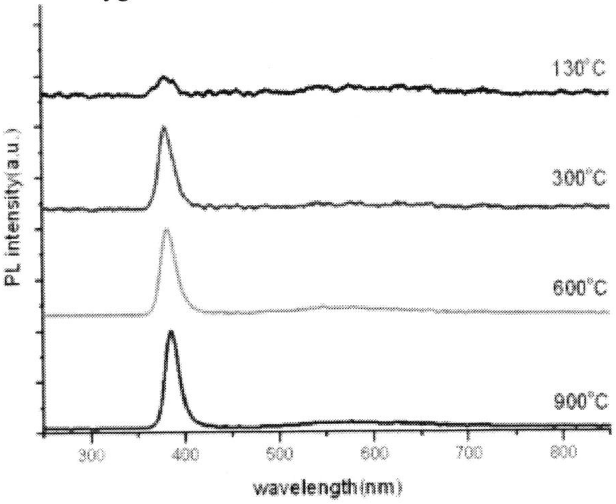

Figure 6: Room temperature PL spectra of ZnO nanorod arrays with ZnO thin films annealed from 130 to 900 ℃.

Moreover, as shown in Figure 7, the sample annealed at high temperature of 900 ℃ has both larger size of ZnO nanorod arrays and stronger PL intensity at 380 nm, while the sample annealed at low temperature of 130 ℃ has both smaller nanorod size and weaker PL intensity. As a result, the growth rate of ZnO nanorods along the c-axis direction is very sensitive to the size of ZnO grains, and then influences the PL peak at 380 nm.

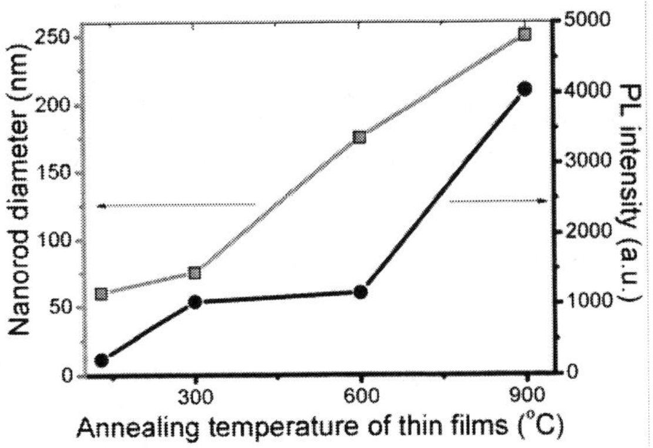

Figure 7: Diameter (rectangles) and PL intensity at 380 nm (circles) of ZnO nanorods as functions of annealing temperature of ZnO thin films.

4 CONCLUSIONS

This work provides a systematic study of controlled growth of ZnO nanorod arrays by using the hydrothermal method. Our investigation demonstrates that the annealing treatment of ZnO sol-gel thin films have strong influences on the diameter and orientation of the ZnO nanorod arrays grown thereon. The annealing temperature of the ZnO thin films can affect the microstructure of the ZnO grains and then the growth of the ZnO nanorod arrays. As the annealing temperature increases from 130 to 900 ℃, the grain size of the thin films increases, and the diameter of thereon ZnO nanorod arrays increases. The thin films influence the nucleation of the ZnO and subsequently affect the diameter and orientation of the thereon nanorod arrays. At the temperature of 130 ℃, the ZnO nanorod arrays align very vertically with growth along the *c*-axis direction. This work provides a route to fabrication of low-cost highly oriented ZnO nanorod arrays at low temperature. These vertical nanorod arrays are highly suitable for use in ordered nanorod-polymer devices, such as solar cells and light emitting diodes.

ACKNOLEDGEMENT

This work was supported by the National Science Council, Taiwan, Republic of China, with Grant NOs. NSC96-2221-E-002-277-MY3 and NSC96-2218-E-002-025.

REFERENCES

[1] M. H. Huang *et al.*, Science 292, 1897 (2001).
[2] S. H. Park, S. H. Kim, and S. W. Han, Nanotechnology 18, 055608 (2007).
[3] C. Y. Lu *et al.*, Applied Physics Letters 89, 153101 (2006).
[4] B. Cao *et al.*, Journal of Physical Chemistry C 111, 2470 (2007).
[5] J. Cui, and U. Gibson, Nanotechnology 18, 155302 (2007).
[6] K. Takanezawa *et al.*, Journal of Physical Chemistry C 111, 7218 (2007).
[7] J. J. Wu *et al.*, Applied Physics Letters 90, 213109 (2007).
[8] M. H. Huang *et al.*, Advanced Materials 13, 113 (2001).
[9] G. Z. Wang *et al.*, Materials Letters 59, 3870 (2005).
[10] L. Vayssieres, Advanced Materials 15, 464 (2003).
[11] J. B. Cui *et al.*, Journal of Applied Physics 97, 044315 (2005).
[12] C. Bekeny *et al.*, Journal of Applied Physics 100, 104317 (2006).
[13] M. Guo *et al.*, Journal of Solid State Chemistry 178, 3210 (2005).
[14] T. Ma *et al.*, Nanotechnology 18, 035605 (2007).
[15] M. Ohyama, H. Kouzuka, and T. Yoko, Thin Solid Films 306, 78 (1997).
[16] X. Q. Wei *et al.*, Materials Chemistry and Physics 101, 285 (2007).
[17] Q. Ahsanulhaq, A. Umar, and Y. B. Hahn, Nanotechnology 18, 115603 (2007).

Synthesis of Uniform ZnO Nanowire Arrays over a Large Area

C-H Wang*, J. Yang**, K. Shum***, T. Salagaj****, Z.F. Ren** and Y. Tu****

*Materials Technology Center, Southern Illinois University at Carbondale
ENGRA-120, Carbondale, IL 62901
** Department of Physics, Boston College, Chestnut Hill, Massachusetts 02467, USA
***Physics Department, Brooklyn College of the City University of New York
2900 Bedford Avenue, Brooklyn, NY 11210
****First Nano, a division of CVD Equipment Cooperation, 1860 Smithtown Ave Ronkonkoma, NY 11779
yi.tu@firstnano.com

ABSTRACT

It has been found that Zinc oxide (ZnO) nanowires have great potential in many applications. Currently, the most commonly used method to grow ZnO nanowire is the vapor transport method. The morphology of the ZnO nanowire is related to the substrate temperature gradient and gas flow dynamic. Previously a uniform ZnO nanowire array could be obtained only on a small area less than 0.5 inch by 0.5 inch due to the temperature gradient between the source and substrate temperatures. This paper reports a novel growth system design that utilized a separated heater to heat up ZnO solid source and uses a three zone furnace to get a uniform temperature across a large area substrate (2 inches by 2 inches). The reacting gas (O2) can be introduced into system in different ways allowing more efficient utilization of the source material. Here we demonstrated the growth of uniform ZnO nanowire array on Si and sapphire substrate and ZnO nanowire on graphite flake over large area. The ZnO source material and substrate can be well controlled in a wide range of temperatures in order to obtain the optimum growth conditions. The reacting gas (O2) can be introduced into system at different locations to improve growth efficiency.

Keywords: ZnO, nanowire, large area, vapor transport method

1 INTRODUCTION

ZnO nanowire is a one-dimensional, single crystalline and self-organized semiconductor [1] [2]. Based on its great properties, ZnO nanowire has been reported as a potential material for many applications, such as nanoscaled electronic, chemical, and photonic devices [3]-[8]. Especially, its wide band gap (3.37 eV) and the large exciton binding energy [9] [10], make ZnO nanowire a promising material for optoelectronic applications [3] [9]-[11]. Currently, vapor transport is the most common method used to grow ZnO nanowire [12] [13]. In this method, ZnO vapor is generated by heating up source material (ZnO powder) at higher temperature zone and transported to lower temperature zone where it reacted with reacting gas and condensed on the substrate [3] [11]-[16]. Banerjee and his coworker's research shows that the morphology of ZnO nanowire primarily related to the process pressure, substrate temperature gradient and gas flow [17] [18]. Previously, the substrate is placed on a furnace zone where temperature drops dramatically, uniform ZnO nanowire array could only be obtained over a small area less than 0.5 inch by 0.5 inch [16] [17]. Obliviously, such a small uniformity area can not meet the requirement for scaling up of ZnO nanowire devices to higher volume production.

In this paper, we have demonstrated the growth of uniform ZnO nanowire arrays on Si and sapphire substrates and ZnO nanowires on graphite flakes over a large area. Photoluminescence measurement over 49 points on the 2inch sapphire substrate also proves the uniformity of ZnO nanowire.

In addition, this method can also be applied to GaN, In_2O_3 and many other nanowire materials as well. It provides an inexpensive and robust process for making nanowire based devices over large area.

2 EXPERIMENT SETUP

ZnO nanowire was synthesized by a vapor transport method on an EasyTube™ 2000ss CVD system (First Nano). The experimental setup is constructed by a solid source heater, a three-zone furnace, gas injector, vacuum pump, and quartz tube. ZnO powder (Alfa Aesar, 99.99%) and graphite powders (Alfa Aesar, 99%) were well mixed as the source material with an atomic ratio of 1:4 and placed at the higher temperature region inside the quartz tube. The source material was placed in a quartz boat placed above the solid source heater which can be heated up to 1100 C. The collecting materials, Si (100) wafer, sapphire wafer or graphite flake (Alfa Aesar, 99.9%) were placed at the lower temperature region about 5 inches away from the source material to collect ZnO nanowire. The carrying gas (Ar) and reacting gas (O_2) were introduced into the quartz tube without or with various lengths (11, 12, and 13 inch) of injectors. All components were placed in the same process tube. Figure 1 shows the schematic of experimental setup. In order to get a uniform temperature across the substrate at the collection area, each zone of the furnace has a respective heater which can be controlled independently to obtain higher and lower temperature regions.

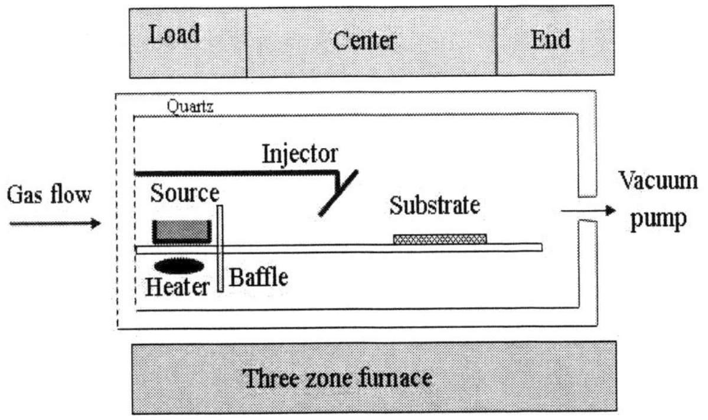

Figure 1: Schematic of experimental setup

During the growth, the load zone and the solid source heater were set at 975 ℃ (actual temperature measured by thermocouple under solid source is around 940 ℃) to vaporize the source material and the center and end zone temperature were set to control the collection area temperature stable between 700 ℃ to 830 ℃. The substrates, a 2 inch Si (100) wafer, or a 2 inch sapphire wafer, or graphite flake on 3 inches by 2 inches quartz boat, was placed at deposition area (about 5 inches away from the source) where temperature was uniform over a distance of 3 inches. Si wafer and sapphire wafer were coated with Au film (1, 2, and 3 nm) as the catalyst. Ccarrier gas (Ar) and reacting gas (O_2) were introduced into the quartz tube by a gas injector that was placed at about 2.5, 3.5, and 4.5 inches away from source with Ar to O_2 flow ratio of 10. Within 30 to 60 minute growth time, the whole tube was maintained at a pressure of 2 or 5 Torr. After the growth phase, the entire substrate was covered by ZnO nanowire and the color turned into gray. The morphology of the as-grown samples was studied by scanning electron microscopy (SEM) and Photoluminescence (PL).

3 RESULTS AND DISSCUSSION

It has been well studied that the morphology of ZnO nanowire is related to the growth temperature. We also obtained these results form our previous experiments, which indicate the temperature uniformity plays an important role in this experiment. Figure 2 shows the system temperature profile was measured by a mobile thermocouple inside the process tube to obtain the actual temperature during growth. By adjusting four individual heaters (solid source, load, center, and end), we successfully controlled the growth temperature uniformity ±5℃ at the lower temperature limit (700 to 760 ℃) and ±1℃ at the higher temperature limit (780 to 820 ℃) over a distance of 3 inches at the collection area.

The source material was vaporized by the below reaction at high temperature, and Zn vapor reacted with O2 to form ZnO vapor and condensed at lower temperature. These reactions are listed below.

$$ZnO_{(s)} + C_{(s)} \xrightarrow{Heated} Zn_{(v)} + CO_{(v)} \qquad (1)$$

$$2Zn_{(v)} + O_{2(v)} \xrightarrow{Condensed} 2ZnO_{(s)} \qquad (2)$$

When the process gases are introduced into the system without an injector, O_2 could react with carbon powder or carbon monoxide gas and form CO_2 at the load zone which reduced the amount of O_2 available to react with Zn vapor and decreased the efficiency of vapor transitive. Moreover, some ZnO vapor will condense at center zone while the temperature starts dropping, and this situation caused a deficiency of ZnO condensing on the stern of substrate. By using a gas injector to introduce the process gases into system directly we avoided the reaction between O_2 and carbon powder and CO.

Although the process pressure range is close to medium vacuum, due to the high gas flow speed, a proper diffusion distance is needed for better uniformity of reactant distribution. However, the injector can not be too far away from substrate otherwise the density of ZnO nanowires will be lower on the rare area than the front area. By adjusting the length of the injector, we have improved the uniformity of ZnO nanowire on the substrate and the efficiency of vapor transport.

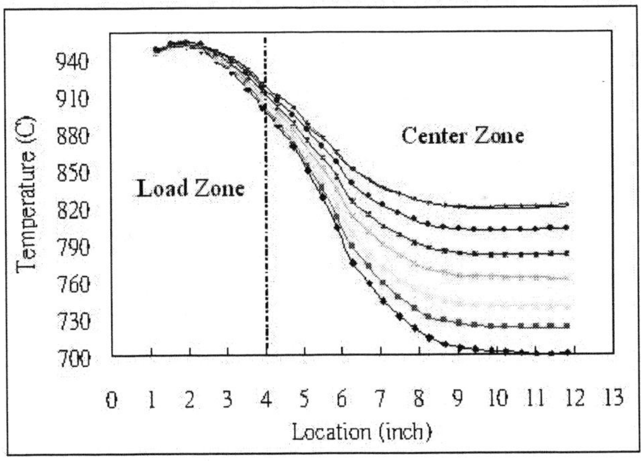

Figure 2: Temperature profile during growth

SEM images have been taken on every 2 mm across the sample, the morphology of ZnO nanowires do not have any sensible change. The cross section SEM images (figure 3 a and b) indicate the ZnO nanowire array grown on the sapphire and Si wafer are vertical aligned to the substrate. The ZnO nanowires dimensions are about 5 μm in length and 50 nm in diameter on all three substrates from 820℃ growth temperature for 45 minutes. However, from the figure 3 c, we can find that using graphite flake substrate to collect material, the density and uniformity of ZnO nanowires is lower than other two samples grown on wafer, and that may come from the non-uniform surface of graphite flake.

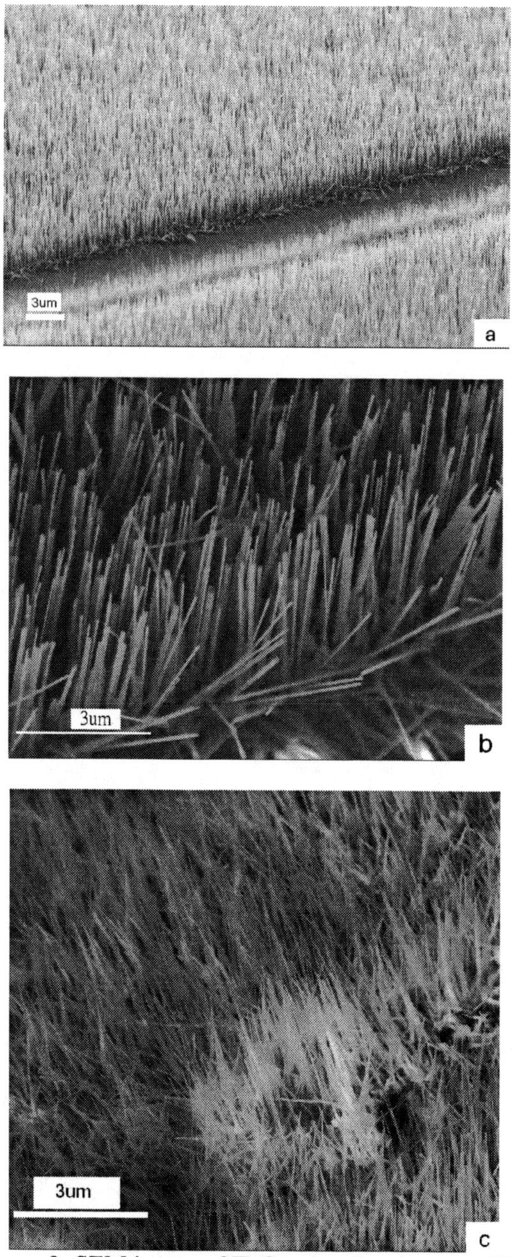

Figure 3: SEM image of ZnO nanowire a) on sapphire wafer, and b) on silicon wafer c) on graphite flake grown at 820℃ for 45 minute

PL was measured by Horiba NanoLog system. Samples at room temperature (RT) were excited by 325 nm light source with 1 nm band pass. The excitation power is about 1 μW with an excitation area of 1 x 5 mm^2, resulting in a very weak excitation power density of 20 μW/cm^2. To shed light on the origin of the broad peak at 515 nm, PL spectra were taken under identical conditions for as-grown samples, O2 - 400 C – 1 h and H2 - 400 C – 1 h annealed samples on Si substrates. As displayed in Figure 4, the PL spectrum from the as-grown sample is almost exactly same as that from the O2 annealed sample. This result indicates the mid-gap emission around 515 nm may not be from the Zn-vacancy related states. However, the excitonic peak from the H2 annealed sample is about 3 times stronger than that from the as-grown sample. The PL peak intensity ratio between the broad peak and the excitonic peak also decreased from 14% to 10% suggesting the H2 annealing effectively reduces interface/surface states.

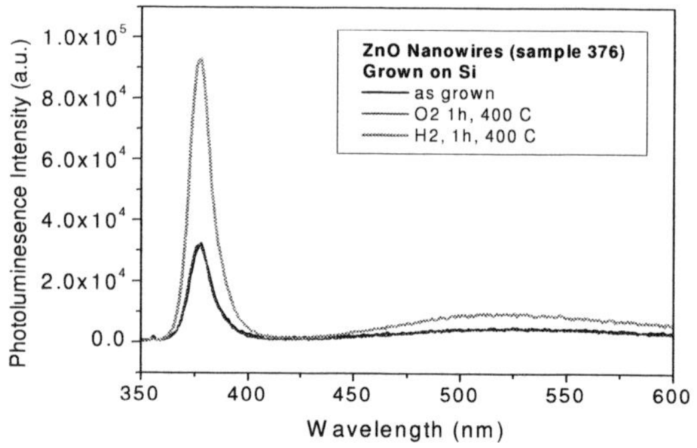

Figure 4: PL spectra at room temperature from the ZnO nanowires grown on Si (100) substrates

The uniformity of ZnO nanowires is studied by scanning PL spectroscopy. Figure 5 shows the excitonic PL intensity of ZnO nanowires grown on a 2" sapphire wafer (0001). The PL intensity is very uniform except a small portion near one of corners. This PL uniformity is expected since the ZnO nanowires growth apparatus described in this paper insures the growth temperature uniformity and optimization reaction gas flow.

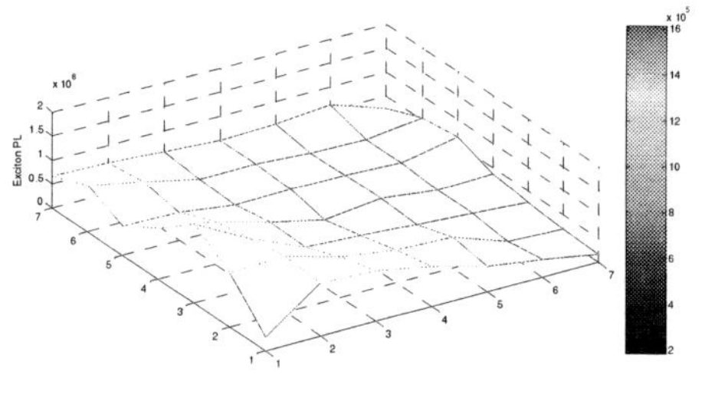

Figure 5: Spatial distribution of PL intensity from excitonic emission of ZnO nanowires grown on a 2" sapphire wafer

4 CONCLUSION

The traditional method can only get a small temperature uniform zone on the substrate, which is the limit in scaling up ZnO nanowire growth. Therefore, how to achieve uniform ZnO nanowires growth over large area until now becomes a major challenge for many applications. In this work, a solid source heater combined with a three zone furnace was used to achieve large temperature uniformity area on substrate. We successfully controlled the growth temperature uniformity ±5℃ at the lower temperature limit (700 to 760 ℃) and ±1℃ at the higher temperature limit (780 to 820 ℃) over a distance of 3 inches at the collection area, which provide uniform environment for ZnO growth. In addition, using a proper length of injector to introduce reaction gases into process tube increased not only the uniformity of ZnO nanowire on the substrate also the efficiency of vapor transport. The SEM images and the PL results show the quality and the good uniformity of ZnO nanowires grown on substrate.

REFERENCES

[1] X. Wang, Q. Li, Z. Liu, J. Zhang, Z. Liu, and R. Wang, Appl. Phys. Lett. 84, 24 (2004)

[2] J. H. Park and J.G park, Appl. Phys. A 80, 43 (2005)

[3] W. Lee, M. C. Jeong, and j. M. Myoung, Nanotech. 15 (2004)

[4] P. Yang, H. Yan, S. Mao, R. Russo, J.Johnson, R. Saykally, N. Morris, J. Pham, R. He, and H. J. Choi, Adv. Funct. Mater.12, 5 (2002)

[5] M. Law, J. Goldberger, and P. Yang, Annual Reviews 34, 83 (2004)

[6] P. C. Chun, Z. Fan, C. J. Chien, D. Stichtenoth, C. Ronning, and J. G. Lu, Appl. Phys. Lett. 89, 133113 (2006)

[7] J. Song, J. Zhou, and Z. L. Wang, Nano Lett. 6, 8 (2006)

[8] Z. L. Wang, and J. Song, Science. 312, 242 (2006)

[9] Y. J. Xing, Z. H. Xi, X. D. Zhang, J. H. Song, R. M. Wang, J. Xu, Z. Q. Xue, and D. P. Yu, Appl. Phys. A. 80, 1527 (2005)

[10] J. Bao, M. A. Zimmler, F. Capasso, X. Wang, and Z. F. Ren, Nano Lett. 6, 8 (2006)

[11] B. D. Yao, Y. F. Chan, and N. Wang, Appl. Phys. Lett. 81, 4 (2002)

[12] X. Wang, J. Song, C. J. Summers, J. H. Ryou, P. Li, R. D. Dupuis, and Z. L. Wang, J. Phys. Chem. B 110, 7720 (2006)

[13] Z. R. Dai, Z. W. Pan, and Z. L. Wang, Adv. Funct. Mater.13, 1 (2003)

[14] M. H. Huang, S. Mao, H. Feick, II, Yan, Y. Wu, H. Kind, E. Weber, R. Russo, and P. Yang, Science 292, 1897 (2001)

[15] Y. Zhang, H. Jia, and D. Yu, J. Phys. D: Appl. Phys. 37, 413 (2004)

[16] Y. X. Chen, M. Lewis, and W. L. Zhou, J. Crystal Growth 282, 85 (2005)

[17] D. Banerjee, J. Y.Lao, D. Z. Wang, J. Y. Huang, D. Steeves, B. Kimball, and Z. F. Ren, Nanotech. 15 (2004)

Low temperature synthesis and characterization of large scale ZnMgO nanowires

Paresh Shimpi, Daniel G. Goberman, Dunliang F. Jian and Pu-Xian Gao*

Department of Chemical, Materials and Biomolecular Engineering
& Institute of Material science, University of Connecticut, Storrs, CT 06269-3136, USA,
* Correspondence e-mail address: puxian.gao@ims.uconn.edu

ABSTRACT

Large scale ZnMgO nanowire array and network with tunable shapes has been successfully synthesized on solid Si substrate and flexible Au coated plastic substrate using low temperature hydrothermal synthesis technique. The diameters of the nanowires have been controlled in the range of 30-600 nm. Field-emission scanning electron microscopy (FESEM) was used to investigate the surface morphologies and orientations of the nanowires. The energy dispersive X-ray spectroscopy (EDS) indicates that the Mg concentration in the ZnMgO nanowires can reach ~2-5 atomic percent. To confirm the alloying process of Mg into ZnO and its distribution during the growth process, Auger electron spectroscopy (AES) was carried out for quantitative analysis of the surface composition profile of typical ZnMgO nanowires.

Keywords: Nanowires, Semiconductor, Synthesis, ZnMgO

1 INTRODUCTION

One dimensional (1D) nanostructures, as a unique low-dimensional nanoscale system, have carried a broad spectrum of extensive research due to their superior properties such as high surface to volume ratio, high crystallinity and quantum confinement effect. For instance, semiconductor oxides such as ZnO have superior chemical and thermal stability as well as specific electrical and optoelectronic properties with a broad range of potential applications[1]. ZnMgO is regarded as an ideal material for ZnO based optoelectronic devices. By alloying with MgO, which has a cubic structure and a broader direct band gap of 7.7 eV, the band gap of ZnO can be remarkably blue-shifted for the realization of light-emitting devices operating in a wider wavelength region[2]. In addition, alloying ZnO with Mg makes ZnMgO a potential candidate for future optoelectronic devices because of its wide band-gap, less lattice mismatch with ZnO as the ionic radius of Mg^{2+} and Zn^{2+} are similar[3,4]. ZnMgO 1D nanostructures[5,6], such as nanowires, nanorods, nanopillars and ZnO/$Zn_{1-x}Mg_xO$ nanoheterostructures[7-10] along an axial or radial direction have been mostly fabricated using vapor phased deposition techniques such as metalorganic vapor phase epitaxy (MOVPE)[5], pulsed laser deposition (PLD)[6], molecular beam epitaxy (MBE)[12], RF magnetron co-sputtering[13], thermal evaporation[14] and vapor transport[15] to obtain high-performance nanometer scale optoelectronic devices. In contrast to vapor phase techniques, there have been significantly fewer reports of ZnMgO NWs grown by wet chemical synthesis approaches due to a significant difficulty to alloy Mg into ZnO lattice at usually much lower processing temperature. However, it is worth noting that wet chemical methods such as hydrothermal synthesis process have several advantages over the other growth processes including catalyst-free growth, low cost, large yield, environmental friendliness and low reaction temperature. Here, we report the synthesis and characterization of large scale ZnMgO NWs on solid Si and flexible polymer substrates by hydrothermal method.

2 EXPERIMENTAL PROCEDURES

Three types of substrates have been used: Si (100) substrates, ZnO nanoparticle-seeded Si (100) substrates and Au-coated flexible plastic substrates. Before Au coating and ZnO seeding, the silicon and plastic substrates were cleaned by deionized water and ethanol 2-3 times and then passed through the plasma cleaning for 40 minutes. For the plastic substrates, a ~20 nm thick Au coating was deposited using a POLARON DC sputtering unit. On the second type of Si substrates, ZnO nanoparticle seeds were made using a sol-gel method. The ZnMgO NWs growth on Si and Au-coated plastic substrates were conducted by keeping the substrates in an autoclave filled with an aqueous solution of $Zn(NO_3)_2.6H_2O$, Hexamethylenetetramine (HMT) and $Mg(NO_3)_2.6H_2O$ in an 1:1:2 ratio at 165 °C for 6 hours. The PH value of the solution before reaction was kept as 7 and after reaction as 9. For the ZnMgO growth on ZnO seeded substrates, a two-step process was used. Firstly, Si substrates were put into solution of $Zn(NO_3)_2.6H_2O$ and HMT in 1:1 ratio at 80 °C for 4 hours, in order to grow a uniform layer of ZnO nanowires on the substrates. Second, the grown ZnO NWs-on Si substrate after the first step was immersed into the solution of $Zn(NO_3)_2.6H_2O$, $Mg(NO_3)_2.6H_2O$ and HMT in an 1:2:1 ratio at 155 °C with PH value 7 in an autoclave for 4 hours. For all the samples, substrates were removed from the solution, rinsed with deionized water and dried in air at 80 °C overnight. The morphology and the orientation of the nanowires were investigated by a JEOL 6335F field emission scanning electron microscope (FESEM) and a BRUKER AXS D5005 (Cu $K\alpha$ radiation, λ=1.540598 Å) X-ray diffractometer (XRD). The chemical composition of ZnMgO NWs on the Si substrates was examined using energy dispersive X-ray spectroscopy (EDXS) in the FESEM and a Phi 595 Scanning Auger spectrometer (AES).

3 RESULTS AND DISCUSSION

3.1 Si substates

Figure 1(a) shows a typical XRD spectrum of the as prepared samples on Si substrates. Three major peaks were resolved to be closely matched to (100), (002) and (101) atomic planes in wurtzite structured ZnO, revealing their good crystallinity. The peak (002) nearly matched to ZnO because of a rather low-concentration of Mg in the NWs, as indicated in the EDS data in figure 2. Figures 1(b) and 1(c) are respectively a low-magnification and high-magnification SEM images showing the grown free-standing ZnMgO micro- and NWs, with a diameter range of ~1 micrometer to ~200-250 nm. The careful observation of the bigger NWs reveals that each individual micro-wire is composed of smaller NWs as the red arrowheads indicated in figure 1(c).

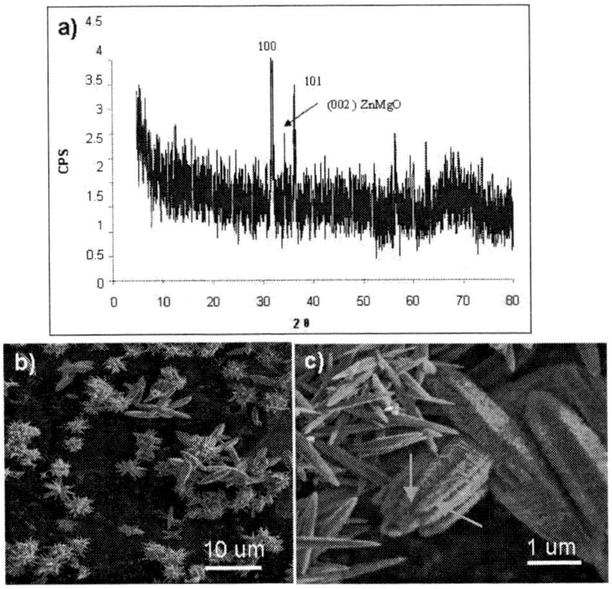

Figure 1: (a) A typical XRD spectrum of free-standing ZnMgO NWs samples. (b) a low-magnification and (c)high-magnification SEM images showing the grown free-standing ZnMgO micro-wires and NWs.

The EDS spectra shown in figure 2(a) and 2(b) are respectively collected from the area with bigger micro wires and smaller nanowires. They both reveal that only C, O, Zn, Si and Mg were present. Both spectra confirmed the presence of Mg in individual micro- and nanowires although the atomic percentage of the Mg is quite low in both cases. In the micro-wires the at. % of Mg was 1.72, which clearly indicates the presence of Mg in NWs. While in the smaller NWs, only 0.41 at. % of Mg was present. The carbon and Si peaks are respectively contributed by a possible contamination during sample preparation and the substrates.

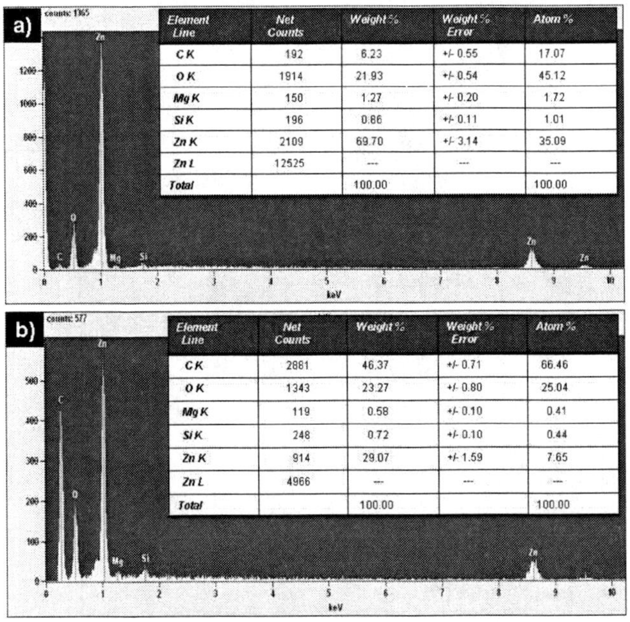

Element Line	Net Counts	Weight %	Weight % Error	Atom %
C K	192	6.23	+/- 0.55	17.07
O K	1914	21.93	+/- 0.54	45.12
Mg K	150	1.27	+/- 0.20	1.72
Si K	196	0.86	+/- 0.11	1.01
Zn K	2109	69.70	+/- 3.14	35.09
Zn L	12525	---	---	---
Total		100.00		100.00

Element Line	Net Counts	Weight %	Weight % Error	Atom %
C K	2881	46.37	+/- 0.71	66.46
O K	1343	23.27	+/- 0.80	25.04
Mg K	119	0.58	+/- 0.10	0.41
Si K	248	0.72	+/- 0.10	0.44
Zn K	914	29.07	+/- 1.59	7.65
Zn L	4966	---	---	---
Total		100.00		100.00

Figure 2: (a) An EDS spectrum of the bigger ZnMgO micro-wires. (b) An EDS spectrum of the smaller ZnMgO nanowires.

3.2 ZnO nanoparticle-seeded Si substrates

Figure 3(a) shows a typical SEM image of the vertically aligned ZnO NWs on the ZnO nanoparticle-seeded Si (100) substrates. The ZnO NWs have a uniform diameter of ~30-40 nm and a length of ~1 μm. The NWs are densely packed and uniform all over the whole silicon substrate. Figure 3(b) shows a typical low-magnification SEM image of the grown NWs after the second step reaction. It is found that the array pattern kept the same, and the length of the NWs remained the same but top portion of each NW changed into cone shape.

The inset in Figure 3(b) reveals the EDXS analysis result from the grown ZnMgO NWs samples corresponding to the dotted square area in the NW arrays. From the shown spectrum, the Mg content was revealed to be ~3.45 at. % (± 0.55 at. % errors). Different area EDXS analysis results have shown that a ~2-5 at. % Mg was generally achieved in the ZnMgO NW arrays after the 2-step hydrothermal processing.

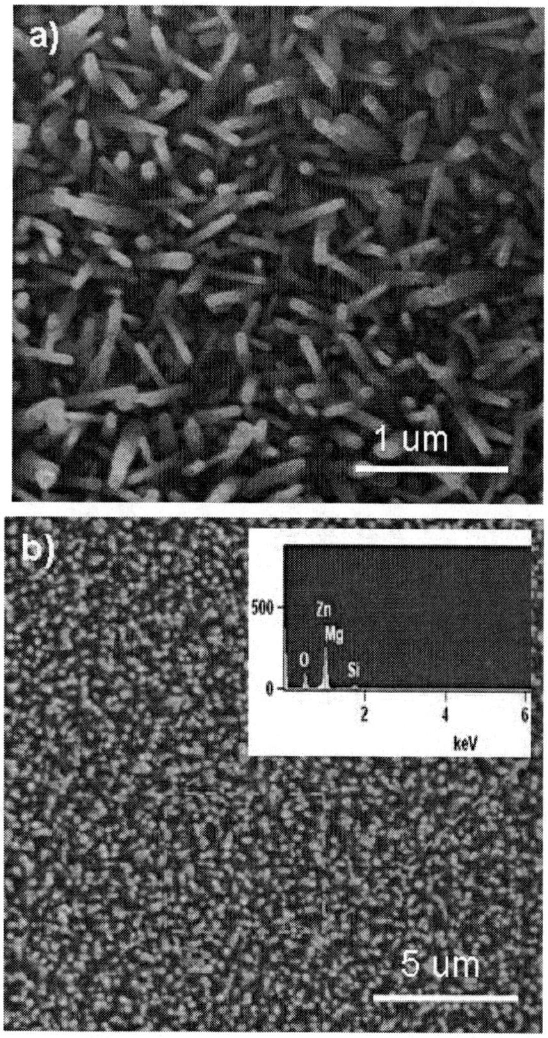

Figure 3: (a) ZnO NW arrays grown on ZnO seed. (b) ZnMgO NW arrays after the second step reaction, inset shows a typical EDS spectrum from dotted square area of NW arrays.

To further confirm the Mg content in the grown NWs, Phi 595 Scanning Auger spectrometer was used for the AES spectra. AES spectra survey obtained from the 2-step grown ZnMgO NW arrays and its chemical composition was determined. Figures 4(a) and 4(b) are AES spectra of the ZMO NWs respectively scanned in a full scan range (40-1100 eV) and the closer view in the Mg peak area. After differential calculus analysis some obvious peaks indicated the presence of Mg, Si, C, O and Zn. The Mg has a ~16.3% atomic concentration, much higher than the one detected in EDS results, which suggested that the Mg alloying in the hydrothermal synthesis here might not be uniform. Further improvement on the processing parameters will be done in the near future.

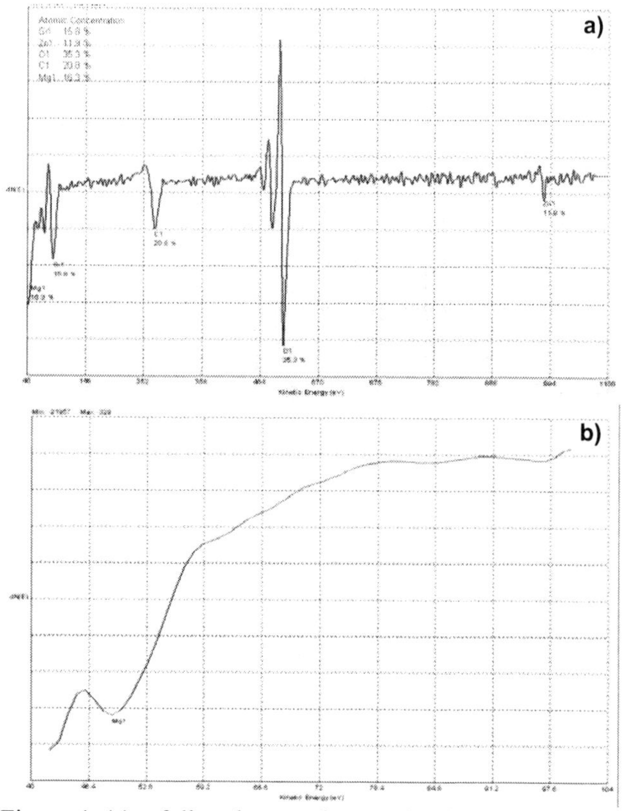

Figure 4: (a) a full scale scanning result of AES spectra. (b) a AES spectrum showing the presence of Mg.

3.3 Au-coated flexible substrates

Figure 5 shows the ZnMgO NWs grown on Au-coated flexible plastic substrates. Sparsely grown ZnMgO NWs have been fabricated normal to the substrates, with a diameter range of 200-600 nm. From the Figures 5(a) and 5(b), most of the aligned NW tips have been interconnected by nanofibres with a diameter range of ~50-100 nm Figure 5(c) shows the clear cone shape of ZMO NW, with top diameter of around 150 to 200 nm and the bottom diameter of around 450 to 500 nm. Figure 5(c) displays the hexagonal structure of the NWs. The EDXS analysis indicated that 1.20 at. % of Mg presented corresponding to the top-square portion of the NWs. The EDXS spectrum also revealed the presence of an excessive carbon amount in this sample, which might suggested that the interconnecting nanofibres might be of carbon-based polymer so as to contribute to the carbon composition.

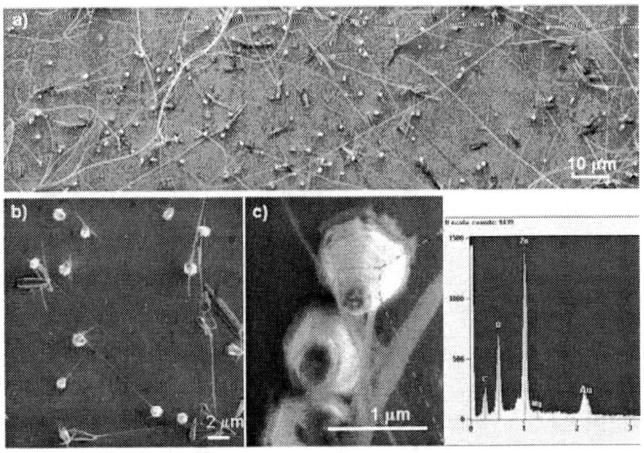

Figure 5: Typical SEM images of (a) ZnMgO NWs interconnected by possibly polymer nanofibres on Au-coated plastic substrates. (b) vertically aligned ZnMgO NWs. (c) closer view of ~1 µm long and cone-shaped ZnMgO NWs with corresponding EDS spectrum for a tip portion.

4 CONCLUSIONS

. In summary, large scale ZnMgO nanowire array and network has been successfully synthesized on solid Si substrate and flexible Au coated plastic substrate using low temperature hydrothermal synthesis technique. The diameters of the nanowires have been controlled in the range of 30-600 nm. Field-emission scanning electron microscopy (FESEM) was used to investigate the surface morphologies and orientations of the nanowires. The energy dispersive X-ray spectroscopy (EDS) indicates that the Mg concentration in the ZnMgO nanowires can reach ~2-5 atomic percent. To confirm the alloying process of Mg into ZnO and its distribution during the growth process, Auger electron spectroscopy (AES) was also carried out for quantitative analysis of the surface composition profile of typical ZnMgO nanowires.

Further detailed transmission electron microscopy (TEM) and SEM structure, morphology and chemical analysis, as well as kinetics and optics studies of the ZnMgO NW array and the interconnected ZnMgO NWs are on-going, which will be discussed elsewhere [16].

ACKNOWLEDGMENT

We greatly appreciate the financial support from the University of Connecticut large faculty research grant and new faculty start-up funds.

REFERENCES

[1] L. Vayssieres, Adv. Mater. 15(5), 464-466, 2003.
[2] A. Ohtaomo, M. Kawasaki and T. Koida, Appl. Phys lett. 72, 2466, 1998.
[3] W. Liu, S. Gu, S. zhu, F. Qin, S. Liu, X. Zhou, L. Hu, R. Zhang, Y. Shi, Y. Zhang, J. crystal growth, 277, 416-421, 2005.
[4] D.K. Hwang, M.c. Jeong, J.M. Myoung, Appl. Surf. Sci. 225, 217-222, 2004.
[5] R. Kling, C. Kirchner, T. Gruber, F. Reuss and A. Waag Nanotehnology, 15 1043, 2004.
[6] M. Lorenz,,E. M. .Kaidashev, Rahm A,T. Nobis, J. Lenzer,G. Wagner, D. Spemann, H. Hochmuth and M. Grundmann , Appl. Phys. Lett., 86, 143113, 2005.
[7] A. Ohtomo and M. Kawasaki Appl. Phys. Lett., 75, 980, 1999.
[8] T. Makino, C.H. China and N. T. Tuan and H. D. Sun, Appl. Phys. Lett, 77, 975, 2000.
[9] W.L. Park, G. Yi and H.M. Jany, Appl. Phys. Lett 79 2022, 2001.
[10] W. Yang, R.D. Vispute and S. choopum, Appl. Phys. Lett, 78, 2787, 2001.
[11] S. Krishnamoorthy, S Iliadis and A Inumpudi, Solid-State Electron, 46, 1633, 2002.
[12] Y.W Heo, M. Kaufman, K. Pruessner , D.P. Norton, F.Ren, M.F. chisholm, P.H. Fleming , solid state electronics 47, 2269-2273, 2003.
[13] J.P. Kar, M.c. Jeong, W.K. , J.M. Myoung, Material science and Eng. B, 147, 74-78, 2008.
[14] G. Wang, Z. Ye, H. He, H. Tang and J. Li., J. Phys D: Appl. Phys., 40, 5287-5290, 2007.
[15] H.-C. Hsu, C.-Y. Wu, H.-M. Cheng, and W.-F. Hsieh., Appl. Phys. Letters, 89, 013101, 2006.
[16] P. Shimpi, and P.-X. Gao, unpublished results, 2008.

Growth α–Fe_2O_3 nanowires with different crystal direction of iron film

Li-Chieh Hsu, Yuan-Yao Li*

Department of Chemical Engineering

National Chung Cheng University, 168 University Rd., Min-Hsiung, Chia-Yi 621, TAIWAN, R.O.C

***Corresponding Author: Yuan-Yao Li, E-mail: chmyyl@ccu.edu.tw**

ABSTRACT

α-Fe_2O_3 nanowires (NWs) were synthesis from different crystal direction of iron film by thermal oxidation in oven. The crystal direction of iron film was successfully control by sputtering 50 nm iron on different direction of MgO substrate. The directions of MgO substrate were (100), (110) and (111). The crystal direction of coated iron film would follow the direction of the MgO. The NWs were fabricated on the directions of (211) and (110) iron film for 10 hours at 350 °C. The α-Fe_2O_3 NWs with the diameter of 10–30 nm and 1μm in length were observed. The α-Fe_2O_3 film was formed on the direction of Fe(100) film after thermal process. The samples were observed and analyzed by field emission scanning electron microscopy (FE–SEM), transmission electron microscopy (TEM) and X–ray diffraction (XRD).

Keywords: α-Fe_2O_3 nanowires

1 INTRODUCTION

α-Fe_2O_3 (hematite) is a semiconductor (Eg=2.1eV) and the most stable iron oxide under ambient environment.[1] The applications of α-Fe_2O_3 nanomaterial were extensively studied for water splitting,[2,3] gas sensor,[4-6] photocatalysts/catalyst,[7-9] solar cell,[10] field emission devices[11,12] and field effect transistors (FET).[13] In addition, the magnetic property of α-Fe_2O_3 crystal also attracts many attentions.[14-17] This is because crystalline α-Fe_2O_3 has the characteristic of antiferromagnet with a Neel temperature (T_N) at 960 K and a Morin temperature (T_M) at 263 K which a magnetic phase will be transited.[18] When the temperature is between TN and TM, the weak ferromagnetic property of α-Fe_2O_3 is observed.[19,20] Many methods have been developed for the fabrication of α-Fe_2O_3 nanowires (NWs) or nanorods including template methods,[14,21] hydrothermal,[18,22,23] and sol-gel mediated reaction,[24] as well as the thermal oxidation of iron.[12,25-27] The thermal oxidation methods normally used a mixture of gas (O2, CO2, N2, SO2 or H2O) or air in the synthesis process and the temperature of oxidation is from 300–700oC within few hours. After the process, α-Fe_2O_3 NWs or nanorods can be grown on the bulk of pure iron.

In general, the growth mechanism of α-Fe_2O_3 NWs was preferred to the tip–growth[25]. The α-Fe_2O_3 NWs were formed on the defect of the Fe film by the diffusion of the oxygen. In previously surveys, most researches were investigated the influence of the grown α-Fe_2O_3 NWs, including the keeping temperature, the gas conditions, the holding time, …et al.. In this study, the NWs were grown on the 50 nm thick iron film at 350 °C for 10 hours. In order to analyze the growth mechanism of the α-Fe_2O_3 NWs, the iron films were sputtering on the three different crystal directions of the MgO substrates. The deposited directions of the Fe films were affected with the crystal directions of the MgO substrates. This is the first time to investigate the relationship between the crystal direction of the Fe film and the growth of the α-Fe_2O_3 NWs. We reported the results in this paper, the α-Fe_2O_3 NWs were grown on the directions of the Fe (211) and Fe (110). As the direction of Fe film was (100), only the α-Fe_2O_3 film was formed after thermal process.

2 EXPREIMENTAL

α-Fe_2O_3 NWs were grown by iron film using the thermal oxidation at 350°C for 10hours. The 50 nm thickness of Iron film was coated on the MgO substrate by direct current (DC) sputtering. The directions of MgO substrate were (100), (110) and (111), respectively. The various MgO substrates were placed into a DC sputter machine, and it was coated while the pressure was achieving 10^{-6} torr. The target was 99.9 % of pure iron. The iron–coated substrates were heated at 350 °C for 10 hours in an oven at the ambient environment. The morphology and the structure of the as–synthesized NWs were characterized by field emission scanning electron microscopy (FE–SEM, HITACH S–4800) and high resolution transmission electron microscopy (HR–TEM, JEOL JSM 3010), respectively. The crystal directions of NWs or films are identified by X–ray diffraction (XRD, RIGAKU Miniflex system using a Cu target, CuKα = 1.5418 Å) spectrometer.

3 RESULTS AND DISCUSSION

Figures 1(a) and 1(b) show the top–view and cross–section of the Fe–coated MgO(110) after thermal oxidation at 350 °C for 10 hours, respectively. As can be seen in

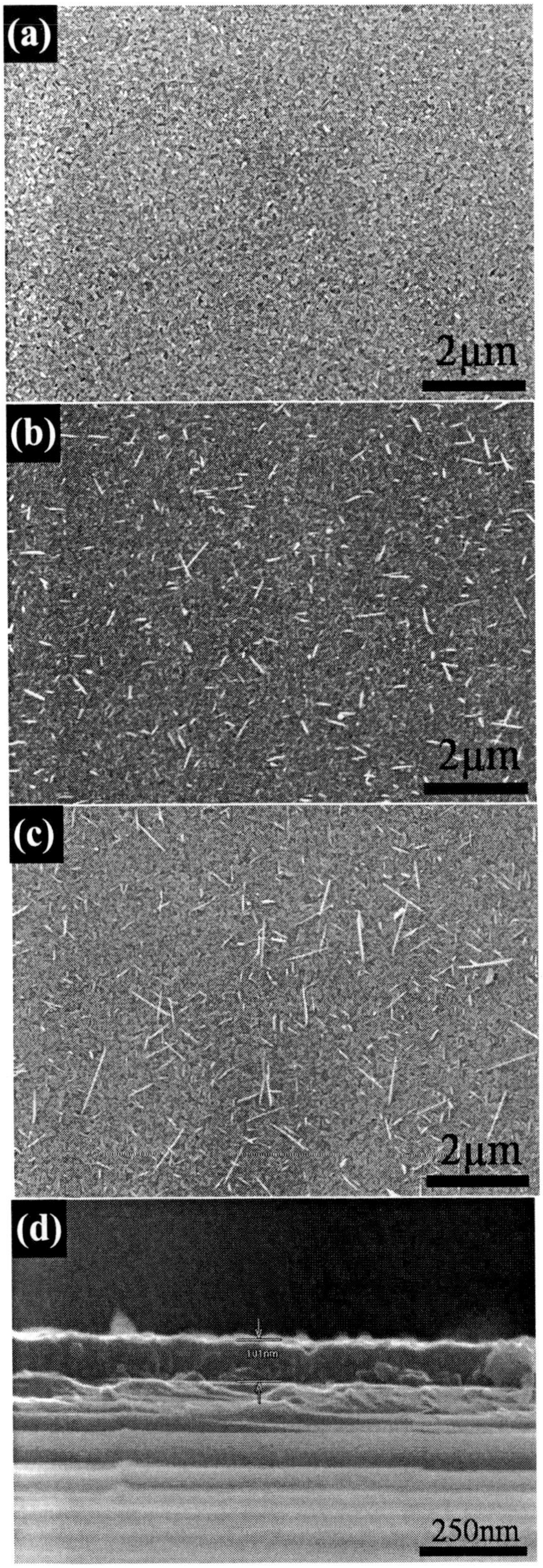

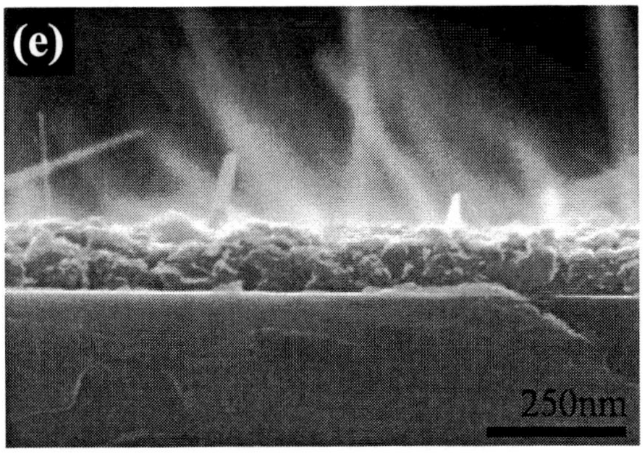

Figure 1: The FE–SEM images of the top view of the Fe - coated (a) MgO(100), (b) MgO(110) and (c) MgO(111) after thermal treatment, the cross - section view of the film Fe - coated (d) MgO(100) and (e) MgO(111) after thermal treatment.

Figure 1(a), there are no wire–like structure on the surface of the film. The morphology of the film was porous structure and very roughly. Figure 1(d) shows the thickness of the film was 100 nm after thermal oxidation. There are no wire–like structures on the surface of the film. Figures 1(b) and 1(c) show the top–view of the Fe coated MgO(110) and MgO(111) substrate. The NWs were grown vertically on the MgO(110) and MgO(111) substrates with the diameters of 10–30 nm and the lengths of 1μm. The porous structures were found on the surface of the films. As can be seen in Figure 1(e), the NWs were vertically growth on the surface of the Fe film. The thickness of film was 90nm.

From Figures 1, we found that, the surface of these films were porous structures. The thicknesses of the films were doubled (~100 nm) compare with the thickness of the original Fe film (50nm). It is well known that the Fe will be oxidized toα-Fe_2O_3 by the diffusion of oxygen. It could explain that the thickness was thicker after thermal process. The film on the MgO(100) was thicker than the films on the MgO(110) and MgO(111), because of the NWs were grown on the film.

Figure 2(a) shows a TEM image of the NWs. The NWs are very straight with a diameter of 10–30 nm and the length of several hundred nm. The high resolution TEM image shown in Figure 2(b) indicates a well crystalline structure of the α-Fe_2O_3 NW with growth direction along [110]. The fringe spacing of 0.251 nm agrees well with the interplanar spacing of (110) of the α-Fe_2O_3 NW. In Figure 2(c), the selected area electron diffraction (SAED) pattern reveals that the crystal structure of the α-Fe_2O_3 NW is the single rhombohedral structure.

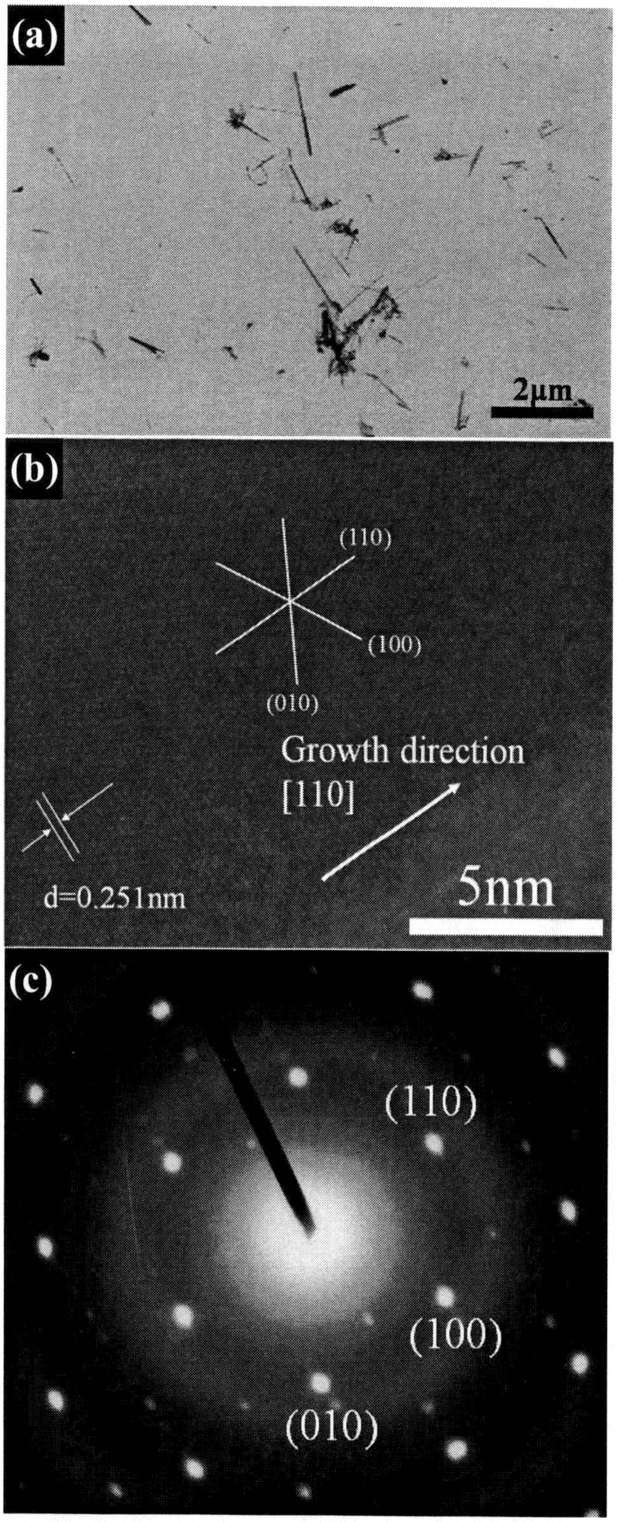

Figure 2: (a) TEM image of the α-Fe_2O_3 NWs, (b) HR–TEM image of the single α-Fe_2O_3NW, (c) SAED pattern of the α-Fe_2O_3 NW.

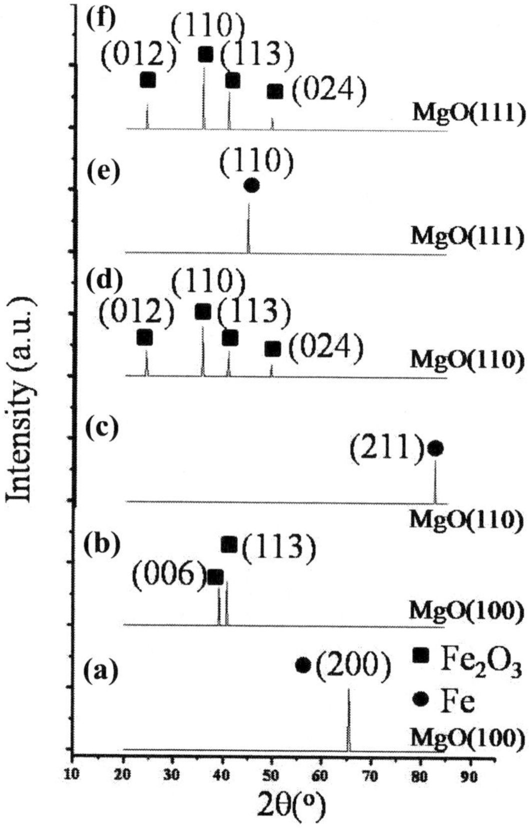

Figure 3: The XRD patterns of Fe–coated MgO(100) (a) before oxidation and (b) after oxidation, MgO(110) (c) before oxidation and (d) after oxidation, and MgO(110) (e) before oxidation and (f) after oxidation.

The XRD patterns of the various Fe–coated MgO substrates were shown in Figure 3. It can be seen that the XRD patterns of the samples are in conformity with rhombohedral α-Fe_2O_3 after thermal treatment. As shown in Figures 3(a) and 3(b), the crystal direction of the Fe film coated on the MgO(100) substrate was (100). The α-Fe_2O_3 peaks of the diffraction planes (006) and (113) were appeared after thermal process. Figures 3(c) and 3(d) shows the Fe(211) was deposited on the MgO(110) substrate. After oxidation, the α-Fe_2O_3 of the (110), (113) and (012) diffraction peak are clear distinguishable. In Figure 3(e), the crystal direction of Fe film coated on the MgO(111) substrate was (110). The α-Fe_2O_3 diffraction peaks of the (110), (113) and (012) were shown in Figure 3(f). From the XRD patterns, the various crystal directions of Fe films were formed by the different directions of the MgO substrates. The three crystal directions of Fe films were oxidized to α-Fe_2O_3. This is the reason why the surfaces of the films we found were porous structures in Figures 1. According to the analysis of the XRD and TEM, the NWs were grown on the Fe film while the crystal direction of α-Fe_2O_3(110) was formed. This is good agreement with the result of the reference.[25] Compared with the Fe(100), the oxygen was easier diffused along the Fe(110) on the surface of the Fe(211) film and Fe(110) film. The films of Fe(211) and Fe(110) were fabricated the α-Fe_2O_3 NWs after the thermal oxidation.

4 CONCLUCTIONS

α-Fe_2O_3 NWs were synthesis from various crystal directions of iron films by thermal oxidation in oven. The crystal direction of iron film was successfully control by sputtering 50 nm iron on different direction of MgO substrate. The directions of MgO substrate were (100), (110) and (111). The crystal direction of coated iron film would follow the direction of the MgO. The NWs were fabricated on the directions of (211) and (110) iron film for 10 hours at 350 °C. The α-Fe_2O_3 NWs with the diameter of 10–30 nm and 1μm in length were observed. The α-Fe_2O_3 film was formed on the direction of (100) iron film after thermal process. Form the results of the XRD and TEM, the NWs were grown on the Fe film while the crystal direction of α-Fe_2O_3(110) was formed. Compared with the Fe(100), the oxygen was easier diffused along the Fe(110) on the surface of the Fe(211) film and Fe(110) film. The controlled growth of the α-Fe_2O_3 NWs promises the more applications in the electrical devices.

5 REFERENCES

[1] R. Dieckmann, Philos. Mag. A, 68, 725-745, 1993.
[2] I. Cesar, A. Kay, J. A. G. Martinez and M. Gratzel, J. Am. Chem. Soc., 128, 4582-4583, 2006.
[3] J. H. Kennedy and M. Anderman, J. Electrochem. Soc., 130, 848-852, 1983.
[4] P. Chauhan, S. Annapoorni and S. K. Trikha, Thin Solid Films, 346, 266-268, 1999.
[5] E. Comini, V. Guidi, C. Frigeri, I. Ricco and G. Sberveglieri, Sens. Actuators, B, 77, 16-21, 2001.
[6] J. S. Han, T. Bredow, D. E. Davey, A. B. Yu and D. E. Mulcahy, Sens. Actuators, B, 75, 18-23, 2001.
[7] B. C. Faust, M. R. Hoffmann and D. W. Bahnemann, J. Phys. Chem., 93, 6371-6381, 1989.
[8] S. N. Frank and A. J. Bard, J. Phys. Chem., 81, 1484-1488, 1977.
[9] T. Ohmori, H. Takahashi, H. Mametsuka and E. Suzuki, Phys. Chem. Chem. Phys., 2, 3519-3522, 2000.
[10] N. Beermann, L. vayssieres, S. E. Lindquist and A. Hagfeldt, J. Electrochem. Soc., 147, 2456-2461, 2000.
[11] T. Yu, Y. W. Zhu, X. J. Xu, K. S. Yeong, Z. X. Shen, P. Chen, C. T. Lim, J. T. L. Thong and C. H. Sow, Small, 2, 80-84, 2006.
[12] Y. W. Zhu, T. Yu, C. H. Sow, Y. J. Liu, A. T. S. Wee, X. J. Xu, C. T. Lim and J. T. L. Thong, Appl. Phys. Lett., 87, 023103, 2005.
[13] Z. Y. Fan, X. G. Wen, S. H. Yang and J. G. Lu, Appl. Phys. Lett., 87, 013113, 2005.
[14] J. J. Wu, Y. L. Lee, H. H. Chiang and D. K. P. Wong, J. Phys. Chem. B, 110, 18108-18111, 2006.
[15] C. Z. Wu, P. Yin, X. Zhu, C. Z. OuYang and Y. Xie, J. Phys. Chem. B, 110, 17806-17812, 2006.
[16] C. H. Kim, H. J. Chun, D. S. Kim, S. Y. Kim and J. Park, Appl. Phys. Lett., 89, 223103, 2006.
[17] L. Liu, H.-Z. Kou, W. Mo, H. Liu and Y. Wang, J. Phys. Chem. B, 110, 15218-15223, 2006.
[18] Y. M. Zhao, Y.-H. Li, R. Z. Ma, M. J. Roe, D. G. McCartney and Y. Q. Zhu, Small, 2, 422-427, 2006.
[19] Y. L. Chueh, M. W. Lai, J. Q. Liang, L. J. Chou and Z. L. Wang, Adv. Funct. Mater., 16, 2243-2251, 2006.
[20] M. F. Hansen, C. B. Koch and S. Morup, Phys. Rev. B, 62, 1124-1135, 2000.
[21] J. Chen, L. N. Xu, W. Y. Li and X. L. Gou, Adv. Mater., 17, 582-586, 2005.
[22] Z. Y. Sun, H. Q. Yuan, Z. M. Liu, B. X. Han and X. R. Zhang, Adv. Mater., 17, 2993-2997, 2005.
[23] L. Vayssieres, C. Sathe, S. M. Butorin, D. K. Shuh, J. Nordgren and J. H. Guo, Adv. Mater., 17, 2320-2323, 2005.
[24] K. Woo, H. J. Lee, J. P. Ahn and Y. S. Park, Adv. Mater., 15, 1761-1764, 2003.
[25] X. Wen, S. Wang, Y. Ding, Z. L. Wang and S. Yang, J. Phys. Chem. B, 109, 215-220, 2005.
[26] Y. Fu, J. Chen and H. Zhang, Chem. Phys. Lett., 350, 491-494, 2001.
[27] L.-C. Hsu, Y.-Y. Li, C. G. Lo, C. W. Huang and G. Chern, Appl. Phys. Lett., 2008.(submitted)

A simple route to synthesize impurity-free high-oriented SiO_x nanowires

Sheng-Cheng Chiu and Yuan-Yao Li*

* Department of Chemical Engineering, National Chung Cheng University, Chia-Yi, Taiwan 621, ROC

ABSTRACT

Impurity-free silicon oxide nanowires were synthesized on a 3 cm x 3 cm alumina plate uniformly by thermal evaporation of a mixture of powder of graphite powders and SiO_x@Si core-shell particles under argon atmosphere with a flow rate of 50 sccm at 1100 °C for 3 hours. The as-synthesized products characterized by field-emission scanning electron microscopy, field-emission transmission electron microscopy and energy-dispersive spectroscopy show that SiO_x nanowires were of diameters ranging from 50-100 nm and the height is up to several millimeters. The growth mechanism of SiO_x nanowires was studied and suggested to be the vapor-solid (VS) mechanism.

Keywords: silicon oxide, nanowires, thermal evaporation, vapor-solid mechanism.

1 INTRODUCTION

Recently, one-dimensional (1-D) silicon oxide (SiO_x) nanostructures (nanotubes, nanorods, nanowires, nanocable and nanobelts) are drawing a significant attention due to its optical properties and exhibit potential applications in photoluminescence (PL), near-field optical microscopy, waveguides and optical devices due to the blue emission and stable properties [1-6]. In the past decade, fabrication of the SiO_x nanowires was carried out by several techniques, including thermal evaporation [7], chemical vapor deposition (CVD) [8, 9], carbon-assisted method [10, 11] etc. In general, there are two fabrication methods: non-catalyst-based and catalyst-based. In the most cases, the formation mechanism of SiO_x nanowires has been due to the vapor-liquid-solid (VLS) growth [12]. The VLS mechanism proposed by Wagner and Ellis in 1964 for silicon whisker growth is the fundamental model of one-dimensional formation.

Among these synthetic routes, both the CVD method and carbon-assisted method are the widely used techniques to synthesize 1-D silicon oxide nanostructures. For instance, Li et al synthesized SiO_x nanowires on Au-coated Si substrate during attempts o fabricate ZnO nanowires [5]. They found that the formation of the SiO_x nanowires depends on a series of experimental conditions, such as substrate coating (Au), the presence of graphite powder, the substrate temperature, the oxygen flow, and the growth time. However, synthesis of SiO_x nanowires still needs utilizing complex experimental step, catalyst such as Au, Fe, Co, Ni, Ge, Pb and other metal alloys etc [6, 13]. Therefore, it is essential to develop a simple synthesis route for the fabrication of impurity-free SiO_x nanowires.

In this study, a novel and simple method for the synthesis of the SiO_x nanowires was developed. The SiO_x nanowires synthesized on the alumina plate by thermal evaporation of the mixture powders (the core-shell SiO_x@Si particles and graphite powders) at 1100 °C for 3 hours. It was successful formed without the participation of catalyst. The growth mechanism of SiO_x nanowires was studied and suggested to be the vapor-solid (VS) mechanism.

2 EXPERIMENTAL

2.1 Preparation of precursor for synthesis of SiO_x nanowires

Si particles (325 mesh, purity 99%, Aldrich) covering with SiO_x thin layers were prepared by the thermal treatment. The powders were heated in oven at 550 °C for 0.5 h. The SiO_x@Si core-shell particles were formed for the synthesis of SiO_x nanowires.

2.2 Synthesis of SiO_x nanowires

0.25 grams of the prepared SiO_x@Si core-shell particles and 0.25 grams of graphite powders were well mixed and loaded on an alumina cup (O.D. 2.8 cm, I.D. 2.4 cm, and height 1.2 cm) covered with alumina plate. The cup was then placed in the center of a quartz tube (O.D. 7.5 cm, I.D. 7.0 cm, and length 120 cm) in a furnace. The experiment started with purging the reactor by Ar (500 sccm) (high-purity argon, 99.99%) for 1 hour. The temperature raised to 1100 °C with a heating rate of about 20 °C/min under a flow of Ar (50 sccm) and maintained for 3 hours. Finally, the furnace cooled down to the room temperature with an Ar atmosphere. A white colored mallow-like material was found on the alumina plate. SiO_x nanowires were therefore formed.

The morphologies of the materials were examined by the field emission scanning electron microscopy (FE-SEM, Hitachi S-4800). The crystalline structure and atomic arrangements of the products were observed by the high resolution transmission electron microscopy (HR-TEM, JEOL JEM-2010) equipped with an energy dispersion spectrometer (EDS). For FE-SEM observation, a white colored mallow-like material scraped from alumina plate was put on the holder and loaded into SEM vacuum chamber. For HR-TEM analysis, as-produced material scraped alumina plate ultrasonically dispersed in ethanol and a drop was placed onto an amorphous carbon-coating copper grid.

3 RESULTS AND DISCUSSION

Figure 1 show the as-produced material synthesized on the alumina plate by thermal evaporation of the mixture powders (the SiO_x@Si core-shell particles and graphite powders) at 1100 °C for 3 hours. The picture is an alumina plate deposited with white mallow-like product. A ruler with millimeter markings underneath the boat was for size reference. It can be seen that the height of the as-produced material is up to several millimeters, shown in Figure 1(a). A top-view picture shown in Figure 1(b) reveals that the diameter of the deposited material is about 24 millimeters the same as the inner diameter of an alumina cup.

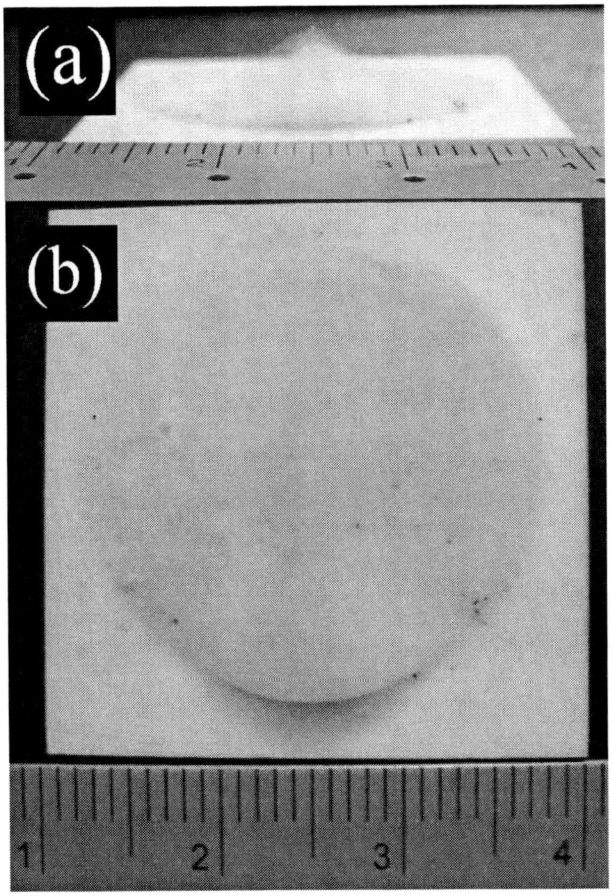

Figure 1. A image of as-produced mallow-like material on the alumina plate (a) side-view, (b) top-view.

Figure 2(a) shows a low magnification SEM image of as-produced one-dimensional nanostructures. It was found that the high purity one-dimensional nanostructures were successful formed without the participation of catalyst. A SEM image of the one-dimensional nanostructures with the diameters of 50-100 nm and the lengths of several hundreds of micrometers was shown in Figure 2(b). As can be seen, there was no impurity observed on/in the one-dimensional nanostructures. The surface of one-dimensional nanostructures revealed smooth morphology. Besides, the end of these straight nanostructures all have a flat tip without catalyst existing.

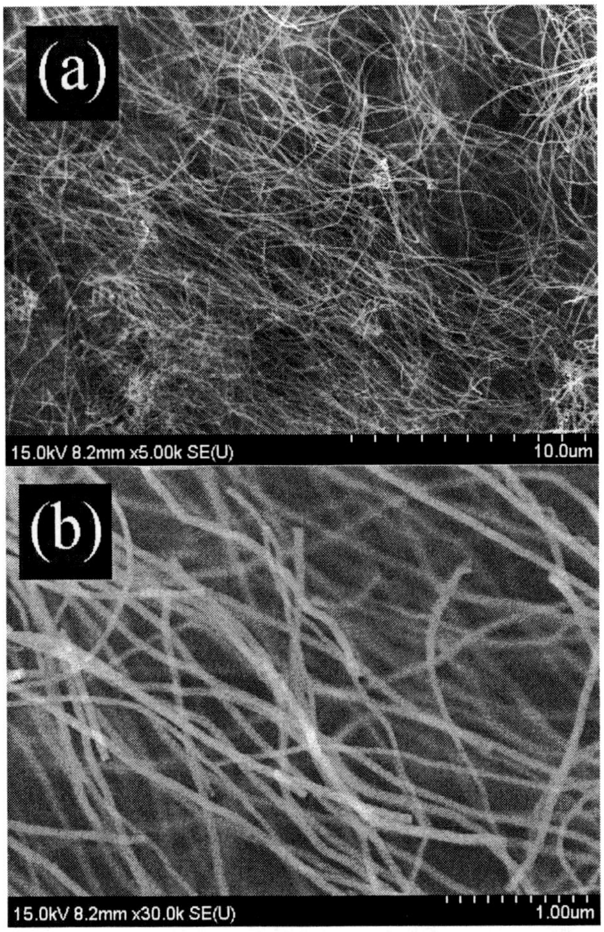

Figure 2. (a) The low-magnification image of as-produced mallow-like one-dimensional nanostructures. (b) FE-SEM image of impurity-free one-dimensional nanostructures.

In the conventional high resolution transmission electron microscopy image of the as-produced material scraped from the alumina plate surface shown in Figure 3(a), a single nanowire is the amorphous structure and has a diameter of about 50 nm. The inset in Figure 3(a) showed the selected area electron diffraction (SAED) pattern. The SAED pattern demonstrates that the nanowire does not reveal any characterization crystalline structure. EDS spectra recorded from a single nanowire demonstrate the chemical composition of Si, O, and Cu (Figure 3b). (Appearance of Cu peak in Figure 3(b) was due to the TEM grid.) A quantitative analysis shown in the inset of Figure 3(b) revealed that the atomic ratio of Si and O is nearly 1:1.1, confirming that the chemical composition of nanowire is SiO_x. Therefore, the as-produced material is defined as SiO_x nanowires. As expected, we don't found catalyst in the SiO_x nanowires in this study.

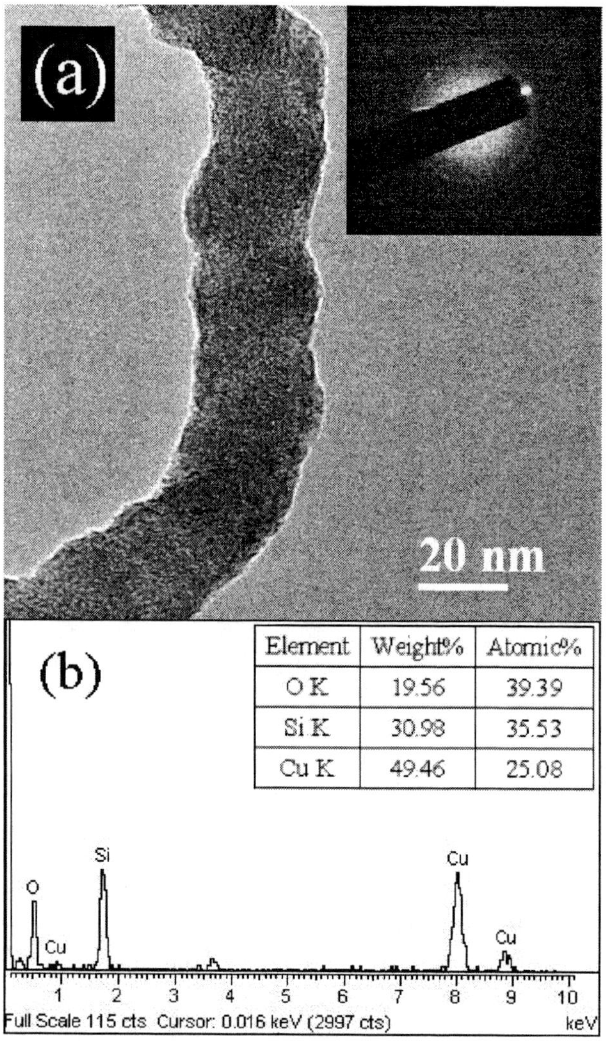

Figure 3. (a) TEM image and SAED pattern of a SiO_x nanowire and. (b) EDS spectrum of the SiO_x nanowire.

Based on the experiment results aforementioned, it is believed that the morphology of as-synthesized SiO_x nanowires via the vapour-solid (VS) mechanism. The possible growth mechanism in this study was proposed. Firstly, it was known that SiO vapor can be generated from a mixture of Si and SiO_2 powders in Ar atmosphere at high temperature. It can be explained as follows [14, 15]:

$$Si_{(s)} + SiO_{2(s)} \rightarrow 2SiO_{(v)} \quad (1)$$

Here s and v refer to the solid state and vapor state, respectively. Simultaneously, the high temperature caused the vaporization of both SiO_x@Si core-shell particles and graphite powders from the mixture powders. The equation known as carbothermal reaction of silica [14, 16, 17], which can be expressed as:

$$SiO_{2(s)} + C_{(s)} \rightarrow SiO_{(v)} + C_{(v)} \quad (2)$$

In second step, the residue O_2 gas shown in Equation (3) was then reacted with the SiO vapor and formed SiO_2 clusters in the gas phase.

$$2SiO_{(v)} + O_{2(g)} \rightarrow 2SiO_{2(s)} \quad (3)$$

Equation (3) is a vapor phase reaction between SiO and O_2 to form SiO_x clusters in gas phase. During the thermal evaporation process, SiO and O_2 deposit on the alumina plate surface and format of SiO_x clusters on the alumina plate, which are considered to be the seed structures for the nanowire growth. According to the research regarding the solid-phase diffusion mechanism reported by A. I. Persson et. al. [18, 19], the seed structures are the solid phase on the alumina plate surface during nanowire growth. By the vaporization onto the alumina plate surface, the SiO vapor thereby incorporated into the SiO_x seed structures and then the nanowires have the grown. Based on the results and the analysis above, it is believed that the as-produced SiO_x nanowires via the vapor-solid mechanism. The SiO_x nanowires with consistent diameter are dictated by the size of the seed particle. Therefore, the growth of SiO_x nanowire in this study is suggested to be a VS mechanism without catalyst.

4 CONCLUSION

In summary, a simple route to synthesize impurity-free high-oriented SiO_x nanowires through thermal evaporation of the mixture powders (the SiO_x@Si core-shell particles and graphite powders) without catalyst method. The height of the as-produced material is up to several millimeters and the diameter of the deposited material is about 24 millimeters. The high purity SiO_x nanowires have the diameters of 50-100 nm and lengths of about several hundreds of micrometers. The growth of SiO_x nanowire in this study is suggested to be a VS mechanism without catalyst.

REFERENCES

[1] Z. W. Pan, Z. R. Dai, L. Xu, S. T. Lee and Z. L. Wang, J. Phys. Chem. B, 105, 2507, 2001.
[2] Y. W. Wang, C. H. Liang, G. W. Meng, X. S. Peng and L. D. Zhang, J. Mater. Chem., 12, 651, 2002.
[3] Z. W. Wang, Z. R. Dai, C. Ma and Z. L. Wang, J. Am. Chem. Soc., 124, 1817, 2002.
[4] R. Q. Zhang, Y. Lifshitz and S. T. Lee, Adv. Mater., 15, 635, 2003.
[5] S. H. Li, X. F. Zhu and Y. P. Zhao, J. Phys. Chem. B, 108, 7032, 2004.
[6] J. Zhang, F. Jiang, Y. Yang and J. Li, J. Cryst. Growth, 307, 76, 2007.
[7] Y. J. Chem, J. B. Li, Y. S. Han, Q. M. Wei and J. H. Dai, Appl. Phys. A, 74, 433, 2002.

[8] J. Q. Hu, Y. Bando, J. H. Zhan, X. L. Yuan, T. Sekiguchi and D. Golberg, Adv. Mater., 17, 971, 2005.
[9] J. Q. Hu, Y. Jiang, X. M. Meng, C. S. Lee and S. T. Lee, Small, 1, 429, 2005.
[10] K. S. Wenger, D. Cornu, F. Chassagneux, T. Epicier and P. Miele, J. Mater. Chem., 13, 3058, 2003.
[11] X. C. Wu, W. H. Song, K. Y. Wang, T. Hu, B. Zhao, Y. P. Sun and J. J. Du, Chem. Phys. Lett., 336, 53, 2001.
[12] R. S. Wagner and W. C. Ellis, Appl. Phys. Lett., 4, 89, 1964.
[13] B. Zheng, Y. Wu, P. Yang and J. Liu, Adv. Mater., 14, 122, 2002.
[14] W. Q. Han, S. S. Fan, Q. Q. Li, W. J. Liang, B. L. Gu and D. P. Yu, Chem. Phys. Lett., 265, 374, 1997.
[15] X. W. Du, X. Zhao, S. L. Jia, Y. W. Lu, J. J. Li and N. Q. Zhao, Materials Science and Engineering B, 136, 72, 2007.
[16] C. H. Dai, X. P. Zhang, J. S. Zhang, Y. J. Yang, L. Cao, and F. Xia, J. Mater. Sci. 32, 2469, 1997.
[17] C. H. Liang, G. W. Meng, L. D. Zhang, Y. C. Wu, and Z. Cui, Chem. Phys. Lett., 329, 323, 2000.
[18] A. I. Persson, M. W. Larsson, S. Stenström, B. J. Ohlsson, L. Samuelson and L. R. Wallenberg, Nature Material, 3, 677, 2004.
[19] J. Wei, K. Z. Li, H. J. Li, Q. G. Fu and L. Zhang, Materials Chemistry and Physics, 95, 140, 2006.

Fabrication of Nanoparticle Embedded Polymer Nanorod by Template Based Electrodeposition

Won-Seok Kang[*], Guang-Hyun Nam[*], Hyo-Sop Kim[*], Kyu-Jin Kim[**], Shin-Won Kang[**] and Jae-Ho Kim*[†]

[*] Department of Molecular Science and Technology, Ajou University, Suwon, 443-749, Republic of Korea, Fax; +82-31-601-8426, Tel; +82-31-219-2950; E-mail: kangfx@ajou.ac.kr

[**] Department of Electronic Engineering, Kyungpook National University, Daegu, 702-701, Republic of Korea, Fax; +82-53-950-7932, Tel; +82-53-950-6829; E-mail: swkang@knu.ac.kr

ABSTRACT

Inorganic semiconductor materials in nano-dimension have drawn intense attention in recent years because of their size-tunable optoelectronic properties arising from the quantum confinement effect. These materials often require to be incorporated into other materials, often polymers, to acquire necessary processability to the various shape and preservation of the quantum confinement effect.

One of the most frequent forms of incorporation is the polymer / semiconductor nanocomposites. This paper describes the templated based synthesis of 5 nm CdSe nanoparticles (NEP) embedded polypyrrole nanorods.

The electrochemical deposition (ECD) to the AAO membrane was used to construct the NEP-rod. In order to preserve quantum confinement effect of CdSe-NEP in the polypyrrole nanorods, distribution and total population of the particles were carefully controlled by concentration of the particle and monomeric pyrrole and electrodepostion parameters including current density and applied potentials.

The transmission electron microscope and near-field scanning fluorescence microscopic measurement showed that NEP-rod are homogeneously dispersed in the polypyrrole nanorod with diameters and length of 200 ~ 320 nm and 15 μm respectively. This structure can be applied in the applications of solar cell, light-emitting diode, and sensors.

Keywords: Nanorod, Electrodeposition, Nanoparticles, Polymer, Nanocomposites

1 INTRODUCTION

Conducting polymers and nanoparticle have attracted much attention because of their physical and chemical properties and technological applications. Embedding of nanoparticle inside conducting polymers has become one of the popular and interesting aspects of nanocomposites because they often exhibit improved chemical and physical properties, and hence can be used in a broader range of applications such as nonvolatile electronic memory,[1] independently addressable remotely triggerable switches and gates,[2] spinnable bioactive coatings,[3] and gas - sensing devices.[4]

An attractive route for preparing nanorod involves electrochemical deposition (ECD) to the AAO membrane.[5] Electrochemical deposition (ECD) synthesis is a very simple process to obtain a nanorod using various materials that have a dielectric pole. Template syntheses, one of the ECD, are widely used for the deposition of nanocylindrical materials including the polymers, metals, semiconductors. Nanomaterials prepared in such templates present an assembly of geometrically and chemically uniform objects.

Thus the studies on the nanorod of polymer composites are increasing. However, to the best of our knowledge, there are no reports on the fabrication of NEP-rod by template based electrodeposition.

In this paper, we synthesized the NEP-rod using the ECD and presented the most suitable condition.

2 EXPERIMENTAL AND RESULT

2.1 Nanorod Manufacturing

The nanorod which was used in this experiment was synthesized by anodic aluminum oxide (AAO) template using electrochemical deposition (ECD) method. AAO was used as Anodisc [13TM] that is sold in Whatman Company by template using electrochemical evaporation. Anodisc [13TM] is usually used as a filter that the pore size is about 230 nm uniformly. The radius of nanorod could not be control if AAO template was used but we can control the length of the nanorod and the nanorod is which the quality of the material fabricated, such as metal, semiconductor or macromolecule and we also can made the nanorod in several material synthesizing. Synthesis process of the NEP-rod is shown in figure 1.

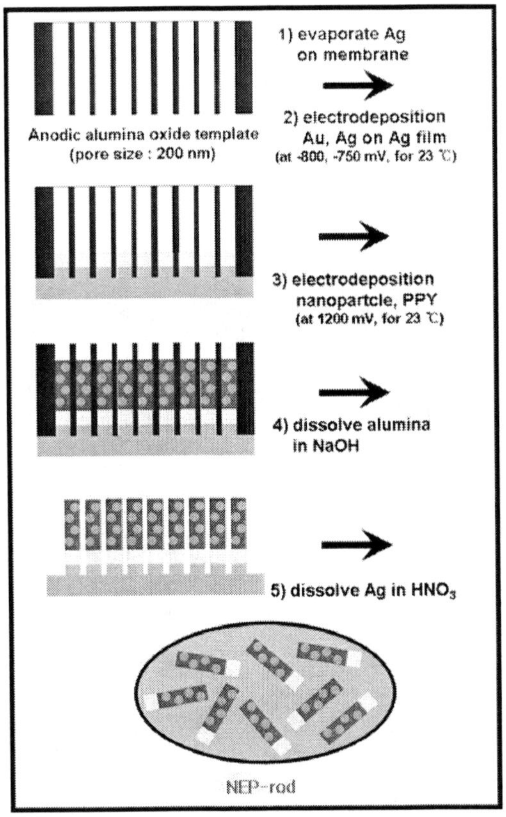

Fig. 1 The experimental schematic for synthesis of CdSe particel embedded polymer nanorod

First of all, an Ag layer (70 or 100 nm thick) was deposited on one side of the AAO template used as working electrode. Pt plate and Ag /AgCl were used as a counter electrode and reference electrode, respectively.

Then, the Ag nanorod was fabricated using Ag plating solution (Technic. Inc., Techni Silver 1025) with the voltage of -800 mV for 5 min. Au nanorod was fabricated using Au plating solution (Technic. Inc., Oretemp 24 Gold Salts) with the voltage of -750 mV for 30 min.

NEP-rod was electro-polymerized usining plating solution (0.1 m pyrrole, 1 mg CdSe nanoparticle and 0.1 m tetra-buthylammonium tetra fluoroborate and in tetra-hydrofuran) with the voltage of 1200 mV for 2 hr. After that, the template was put into 3M HNO_3 solution for 3 hours so as to melt the Ag layer.

Then it dissolve the AAO by NaOH with the concentration of 3 M about 3 hours. Following several centrifugation and rinsing steps, the nanorod fabrication was finally suspended in an aliquot of EtOH.

2.2 Characterization of CdSe particle size

The absorption spectra and photoluminescence spectrum of 32~43 A diameter CdSe nanoparticle samples are shown in Figure 2. CdSe absorptions are shifted dramatically from 485 and 580 nm bulk band gaps. [6]

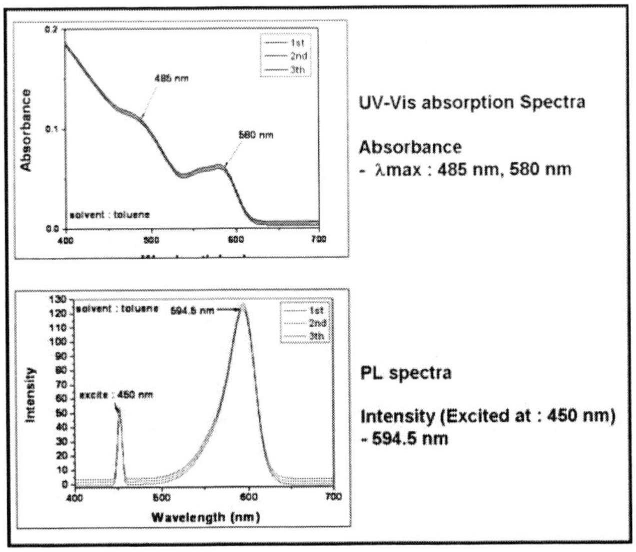

Fig. 2 The theoretical and experimental characterization of the CdSe particle size.

2.3 Analysis of the NEP-rod

The morphologies of the nanorod were analyzed using transmission electron microscopy, field emission scanning electron microscopy (FE-SEM) and optical microscope.

The TEM image of the nanorods was shown in figure 3. These images illustrate how effective the polypyrrole is at filling the CdSe nanoparticle.

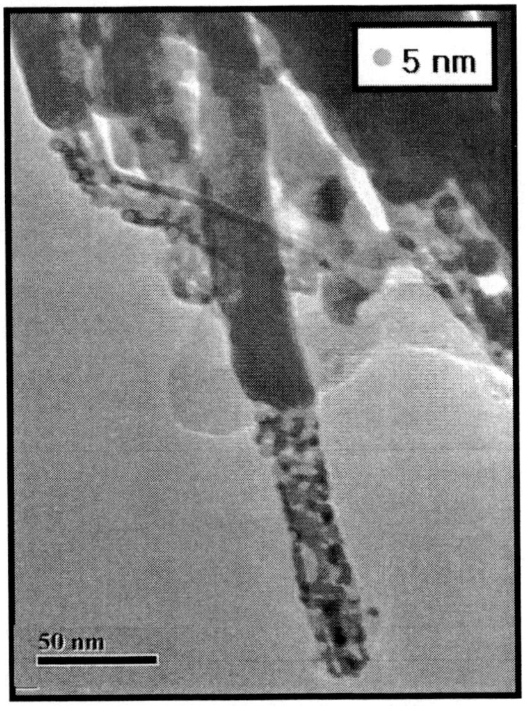

Fig. 3 Transmission electron microscope (TEM) image of NEP-rod

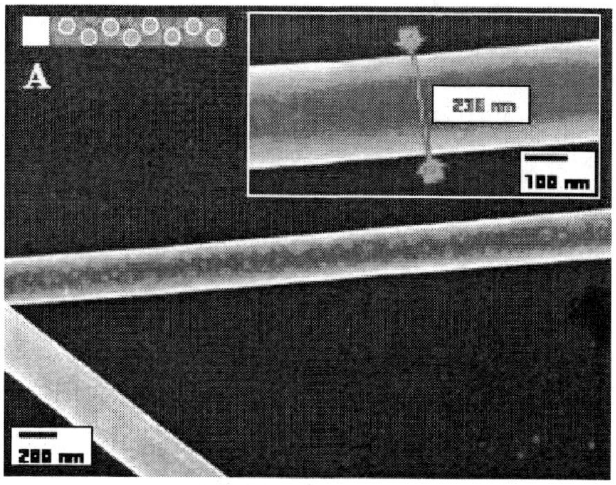

<SEM image>

Fig. 4 (A) Field emission scanning electron microscopy (FE-SEM) image of CdSe particle embedded polymer nanorods.

The manufactured nanorod was 230 nm diameters and 5 μm long. The SEM image of the nanorods was shown in figure 4.

< Fluorescent microscope image>

< Fluorescent microscope image>

Fig. 5 Optical and Fluorescent microscope image of polymer nanorod.

The oprical and fluorescent microscope image of the nanorods was shown in figure 5. Could not confirm a fluorescence in the pure polypyrrole nanorod. Figure 5 show the Polypyrrole nanorod is observed without the CdSe nanoparticle.

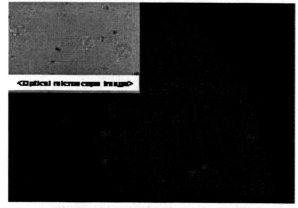

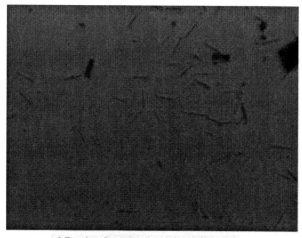

<Optical microscope image>

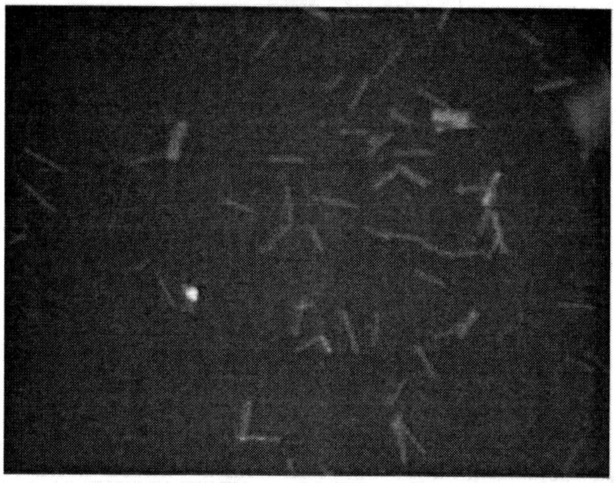

< Fluorescent microscope image>

Fig. 6 Optical and Fluorescent microscope image of CdSe particle embedded polymer nanorods.

Figure 6 show the Polypyrrole nanorod is observed the CdSe nanoparticle. Confirmed the fluorescence of the NEP-rod. In contrast,

3 CONCLUSION

In conclusion, we have described a powerful method for producing nanoparticle/polymer composite nanorod. The concept is versatile and could be extended to diverse composite nanorod with a variety of properties, based on different polymers and nanoparticle. The electropolymerization routes permit the incorporation of additional material within the nanoparticle/polymer composite.

In summary, CdSe/polypyrrole composite nanorod with

embedded structure have been prepared through a electro-polymerization process., and the optical properties.

The transmission electron microscope and near-field scanning fluorescence microscopic measurement showed that NEP-rod are homogeneously dispersed in the polypyrrole nanorod with diameters and length of 230 ~ 250 nm and 5 μm respectively. This structure can be applied in the applications of solar cell, light-emitting diode, and sensors.

4 ACKNOWLEDGNENTS

This research was supported by the Ministry of Commerce, Industry and Energy of the Korean Government and BK21 Program of the Ministry of Education and Human Resources Development for the Molecular Science and Technology in Ajou University, Republic of Korea.

REFERENCES

[1] J. Ouyang, C. Chu, D. Sieves, and Y. Yang, Appl. Phys. Lett. 86, 123507, 2005.

[2] S. R. Sershen, S. L. Westcott, N. J. Halas, and J. L. West, Appl. Phys. Lett. 80, 4609, 2002.

[3] N. Cioffi, L. Torsi, N. Ditaranto, L. Sabbatini, P. G. Zambonin, L. Ghibelli, M. D' 'Alessio, T. Bleve-Zacheo, and E. Traversa, Appl. Phys. Lett. 85, 2417, 2004.

[4] N. Cioffi, I. Farella, L. Torsi, A. Valentini, and A. Tafuri, Sens. Actuators B 84, 49, 2002.

[5] C. R. Martin, Acc. Chem. Res. 28, 61, 1995

[6] C. B. Murray, D. J. Norris, M. G. Bawendi, J. Am. Chem. Soc, 115, 8706, 1993

Magnetic Nanowires based Reciprocal & Non-reciprocal devices for Monolithic Microwave Integrated Circuits (MMIC)

Ryan Marson*, Bijoy K. Kuanr**, Sanjay R. Mishra*, R. E. Camley** & Z. Celinski**

* Physics Department, University of Memphis, Memphis, TN, USA
** Physics Department, University of Colorado, Colorado Springs, CO, USA.

ABSTRACT

The use of nanowires as a tunable stop-band notch-filter (reciprocal device) in a coplanar waveguide (CPW) geometry has been assessed. The stop-band frequency (f_r) is observed to be tunable up to 24 GHz with an applied field (H) of 6 kOe in a parallel configuration and up to 14 GHz with an applied field (H) of 4 kOe in perpendicular configuration. Use of nanowires as microwave resonance-isolators (non-reciprocal devices) are realized and measured. The isolator is designed in the principle of a waveguide based E-plane resonance isolator. The attenuation of the wave in forward and reverse direction results a difference in transmission coefficients that shows a non-reciprocal effect. The isolation is ~ 6 dB/cm at 23 GHz. The bandwidth of the device is relatively large (5-7 GHz) in comparison to ferrite-based devices.

Keywords: Nanowires, NA-FMR, band-stops filter, isolator

1 INTRODUCTION

The dynamic properties of magnetic submicron-size wires are of great importance in the magnetization reversal process and to design high frequency active and passive components. Magnetic microwave devices have been made with dielectric and magnetic substrates [1-4]. Ferrites [1-2] and metallic films [3] are used for their magnetic properties that can be controlled by a magnetic field.

Reciprocal devices such as phase shifters and band-stop filters have been constructed out of both ferrites and metallic films. However, nonreciprocal devices, such as isolators, have only been constructed with ferrites. An isolator is a device allows wave propagation in one direction but has significant attenuation for propagation in the reverse direction. Isolators can be used, for example, in preventing frequency instability of the microwave oscillator by reducing the reflected wave from the load.

Isolators operate in a frequency range that depends on the external field, H, and on the saturation magnetization ($4\pi M_S$). Ferrites are typically characterized by a low saturation magnetization and their operating frequencies are generally below 10 GHz. The frequency range can be increased by using ferromagnetic metals [3] because of their high saturation magnetization (for Fe, $4\pi M_S$=21.5 kG). However, the presence of a metal has drawbacks. For example, the occurrence of eddy currents in the metal results in energy losses that significantly damp electromagnetic waves. Furthermore, the same currents tend to screen out electromagnetic waves, which gives a skin depth on the order of 1 μm at 10 GHz. In this investigation we use a hybrid structure that combines the high $4\pi M_S$ of a ferromagnetic metal (nickel) in the form of nanowires with the extremely high dielectric alumina matrix to develop both reciprocal and nonreciprocal tunable microwave devices.

2 EXPERIMENT & FABRICATION

The Ni nanowires were grown into a hole-pattern of commercial anodized alumina templates (Anodisc 25, Whatman) of 60 μm in thickness and 200 nm in nominal pore size. The templates were used as is. Nickel was electrodeposited into porous alumina templates to create ordered nanowires. To accomplish this, the alumina template was first painted with a gallium-indium eutectic (Aldrich) on one side to allow for electrical conduction. The painted side was placed in contact with a copper plate (32 mm x 51 mm). The copper plate was partially covered with electrical tape to prevent any unwanted deposition. The exposed part of the plate was used for making electrical contact. The copper plate along with the attached template was submerged in a nickel plating solution (Watts Nickel Pure, Technic) which completely covers the top of the membrane. For the anode, a nickel wire (99.98% pure, Goodfellow) of 1.0 mm diameter was used. Electrodeposition was carried out using a potentiostat at 1.5V and a constant current of 100 mA. The deposition time was varied to control the lengths of the nanowires.

After electro-deposition, nitric acid was used to dissolve the conductive layer of GaIn paint. The nanowires were also prepared for TEM measurements by soaking the membrane in a 6M NaOH solution for 30 minutes. The separated nanowires were cleaned several time with deionized water and finally suspended in ethylene glycol solution for the TEM measurements. After TEM imaging, the wires were found to have a typical radius of ~150 nm (Fig.1).

The monolithic microwave devices were fabricated on top of the alumina templates filled with Ni nano-wires. A thick layer of Cu (~ 1.2 μm) was deposited by magnetron sputtering. Photolithography and etching techniques were used to define a coplanar waveguide (CPW) transmission line structure as shown in Figure 2.

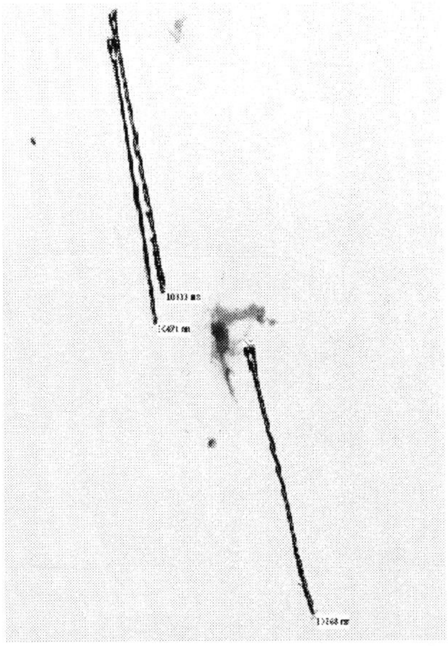

Figure 1: TEM image of nickel nanowire (11 μm length) extracted from the alumina template.

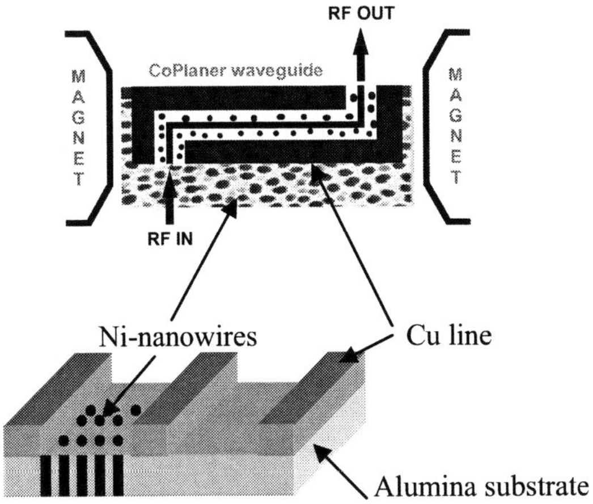

Figure 2: (Top) Design of Filter: Coplanar wave guide transmission line fabricated on top of the alumina template filled with Ni Nanowires. (Bottom) Design of Isolator: nanowires are only on one side of the gap between the signal line and the ground line of the alumina matrix.

The isolator is designed using the general ideas found in a waveguide-based E-plane resonance isolators. The positioning of the nanowires for maximum isolation is observed to be at one side of the gap between the central signal line and the ground line of the CPW. This is achieved by multiple photolithography and etching procedures. The device characterization was done using a vector network analyzer along with a micro-probe station. The frequency was swept from 0.05 to 70 GHz at zero or a fixed external magnetic field (H). Noise, delay due to uncompensated transmission lines connectors, its frequency dependence, and crosstalk which occurred in measurement data, have been taken into account by performing through-open-line (TOL) calibration using NIST Multical® software [9]. The width of the signal lines was 20 μm and the length of the device was 3 mm. The filters were designed for a 50 Ω characteristic impedance [10,11]. The exact resonance frequency (f_r) was obtained from the Lorentzian fits to the experimental transmission (S_{21}) data.

3 RESULTS AND DISCUSSIONS

Figure 3 shows the transmission response of the band-stop notch filter in the CPW geometry for 23 μm length nanowires at two different magnetic fields of 0.25 (red line) and 3.45 (black line) kOe applied perpendicular to the wire-axis. At the ferromagnetic resonance (FMR) frequency there is approximately ~ 6 dB absorption dip and a bandwidth of 6-7 GHz. The larger bandwidth is due to the shape anisotropy of the rods in comparison to the continuous film.

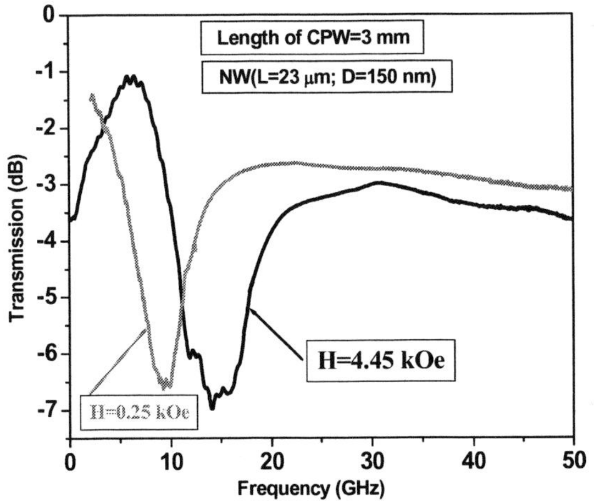

Figure 3: Transmission response measured using a network analyzer system versus frequency at two different magnetic fields applied perpendicular to the nanowires.

The resonance frequency versus magnetic field applied parallel and perpendicular to the nanowire-axis for this filter are shown in Figure 4. These stop-band effects are induced by gyromagnetic resonance phenomenon in the metallic nanowires. The effect occurs even in the absence of an applied dc magnetic field due to the shape anisotropy created by the wire geometry. The zero-field (or lowest field) f_r occurs at 10.5 GHz.

The stop-band frequency can be tuned by the application of magnetic field both in parallel and perpendicular configurations. But the behavior is different in the two cases. For perpendicular case, there are

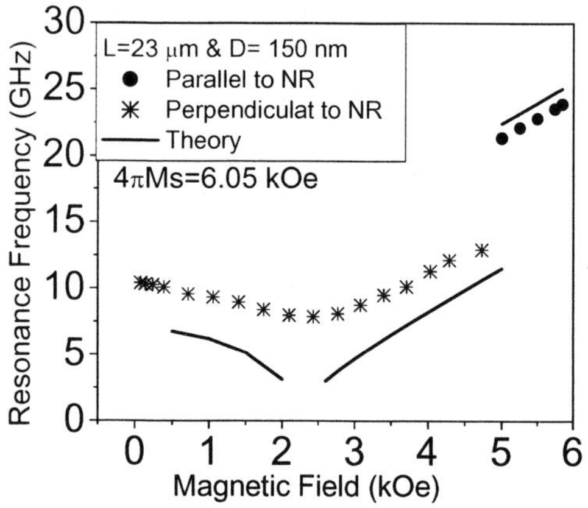

Figure 4: The band-stop notch frequencies versus dc magnetic field, H applied perpendicular (stars) and parallel (dots) to nanowires. The solid lines correspond to the theoretical fit.

two frequency-field regimes. The resonance frequency first decreases with the decrease of applied magnetic field down to a certain value. This point of deviation of frequency-field curve corresponds to the sample leaving its saturation state [6-7]. The kink in the f_r(H) data indicates the effective field (H_{eff}), occurs at ~ 2.4 kOe. This value is very close to the experimentally determined value of H_{eff} for the conventional FMR [12]. For the parallel geometry, the frequency increases with the increase of applied field, without a soft-mode like behavior.

To analyze the frequency-field data, the FMR condition [4] is considered - for an array of infinite single domain nanowires; for field perpendicular and parallel to the wire-axis. The relation for these cases reduced as follows [4, 7];

$$\left(\frac{\omega}{\gamma}\right)_{parallel} = \left(H + H_{eff}\right) \quad (1)$$

$$\left(\frac{\omega}{\gamma}\right)_{\perp} = \sqrt{\left(H_{eff}^2 - H^2\right)} \quad \text{where } H < H_{eff} \quad (2A)$$

$$\left(\frac{\omega}{\gamma}\right)_{\perp} = \sqrt{\left(H - H_{eff}\right)H} \quad \text{where } H > H_{eff} \quad (2B)$$

The eq. (1) corresponds to parallel geometry, when nanowires are parallel to the applied magnetic field. Eq. (2) corresponds to the f_r(H) data when the applied field is perpendicular to the nanowire axis. The eq. (2) has two parts - part (A) corresponds to the applied field value less than the effective field (H_{eff}) and part (B) corresponds to the field value greater than H_{eff}. To fit the f_r(H) data, the effective field (H_{eff}) and gyromagnetic ratio (γ) were used as fitting parameters and the value of $4\pi M_S = 6.05$ kOe (bulk value of Ni) was used. The solid lines in Figure 4 are evaluated from Eqs. (1) and (2). The H_{eff} values obtained from the experimental data increase with decreasing nanowire length (other nanowires not shown here). The H_{eff} values thus obtained are 2.24 and 1.33 kOe for 23 and 50 μm length filters, respectively. The effective gyromagnetic ratio γ'(=γ/2π) are 3.1 and 3.2 GHz/kOe for filters with the 23 μm 50 μm length wires, respectively. We note that the deviations of the experiment from the theory in Fig. 4 show that the nanowires are not in a saturated state over most of the fields examined.

Microwave resonance-isolators (non-reciprocal devices) are realized and measured by a vector network analyzer and a micro-probe system. An isolator is a 2-port device that allows power flow in one direction with little attenuation and a large attenuation in the reverse direction. The scattering matrix of an ideal 2-port isolator is given by:

$$S = \begin{vmatrix} S_{11} & S_{12} \\ S_{21} & S_{22} \end{vmatrix} = e^{j\theta} \begin{vmatrix} 0 & 0 \\ 1 & 0 \end{vmatrix} \quad (3)$$

In an ideal isolator both the ports (1 and 2) are matched, but transmission occurs only in the forward direction and the reverse transmission is blocked by 100%. In our case, the isolator is designed based on the principle of a waveguide based E-plane resonance isolator. Such a result is illustrated in Fig. 5. These isolators operate near the gyromagnetic resonance of the magnetic material.

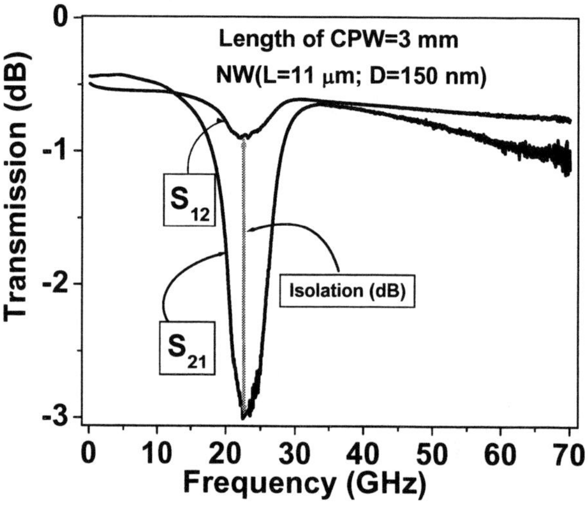

Figure 5: Transmission response measured by Network Analyzer system demonstrating the effect of isolation between port 1 and 2 of the device. The red line shows the level of isolation.

The attenuation of the wave (Fig. 5) propagating from Port 1 to 2 (S_{21}) is large at the gyromagnetic resonance of Ni, while in the return direction from Port 2 to 1 (S_{12}) the attenuation of the signal is very small. The difference between these two coefficients represents the efficiency of the isolator and must be as high as possible. Results on the transmission coefficients show a non-reciprocal effect, which reaches ~ 6.5 dB/cm at 24 GHz (Fig. 5). The bandwidth of the device is relatively large (5-7 GHz) in comparison to typical ferrite-based devices. The device can operate over a wide frequency band (Fig.6) with application of an external magnetic field.

The nanowires are considered to be close to an infinite cylinder, because the diameter is very small in comparison to the length. For this geometry, the demagnetizing factor N_x is in a direction perpendicular to the wire axis and is taken as1/2. The FMR condition for an array of infinite single domain nanowires with field applied parallel to the wire-axis is given by eq. (1). The advantage of this geometry is that the full-height nanowire can be easy to bias by an external permanent magnet. The second advantage of our device is that it has a much broader bandwidth (~ 6 GHz), in comparison to ferrite-based isolators which have a bandwidth of a few hundred MHz. The bandwidth of the device is dictated primarily by the ferromagnetic resonance linewidth of the material, and Ni has a much larger linewidth in comparison to ferrites and garnets. The third advantage of the metallic based isolator over ferrite isolators is that it can be well suited for high power microwave applications. This is because the power-handling capability of a ferromagnetic metal [13] - like Ni - is much higher than YIG or spinel ferrites. The disadvantage of our device is the low value of stop-band rejection (~9 dB/cm). This can be improved by using a microstrip transmission line geometry instead of the coplanar one that is being used here.

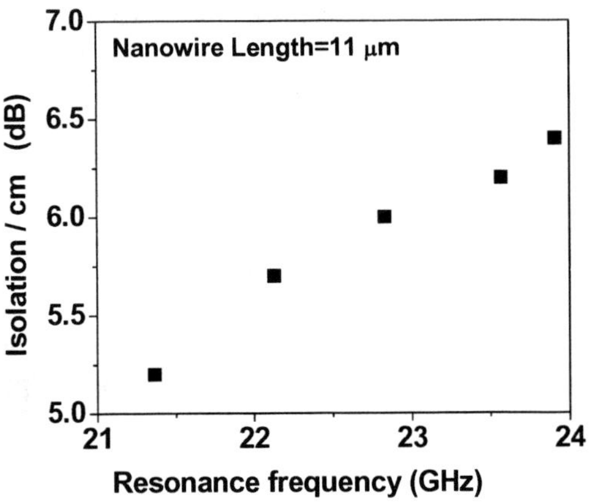

Figure 6: Observed isolation as a function of frequency for the nanowire based coplanar structure. The magnetic field is changed by about 1 kOe over this frequency range.

The performance of broadband isolators can be characterized by the ratio f_{max}/f_{min}, where f_{min} and f_{max} are defined as the edges of the frequency band in which the devices have acceptable operating characteristics. For the most advanced isolators available today this ratio is approximately 3:1. The measured broadband performance of our design is about 2.5:1. This does not represent any fundamental limitation on this ratio, but is due to the unavailability of a larger magnetic field in our laboratory. As the device under present study operates in a quasi-TEM mode, there is no cut-off frequency in the device. A much higher operating frequency (f_{max}) can be achieved by the use of a larger magnetic field. Furthermore the necessary applied fields will be lower than those used in ferrite isolators.

4. CONCLUSION

In summary, we designed, fabricated, and characterized reciprocal and non-reciprocal microwave planar devices using high aspect ratio Ni nanowires, fabricated by electro-deposition inside the pores of an alumina matrix. The field variation of the resonance frequency f_r(H) data provides a measure of the effective anisotropy field (H_{eff}) and the gyromagnetic ratio (γ). The isolation of the device reaches ~ 6.5 dB/cm at 24 GHz with a bandwidth of ~ 6 GHz.

The work at UCCS was supported by DOA Grant No. W911NF-04-1-0247

REFERENCES

[1] D. E. Oates and G. F. Dionne, IEEE Trans Appl. Supercond. 7, 2338 (1997).

[2] I. Huynen, G. Goglio, D. Vanhoenacker, and A. Vander Vorst, IEEE Microwave Guided Waves Lett. **9** 401 (1999).

[3] Bijoy K. Kuanr, R. Camley, and Z. Celinski, Appl. Phys. Lett., **83**, 3969 (2003).

[4] U. Ebels, J. -L. Duvail, P. E. Wigen, L. Piraux, and K. Ounadjela, Phys. Rev. **B 64**, 144421 (2001).

[5]. J-E. Wegrowe, D. Kelly, A. Franck, S. E. Gilbert, and J.-Ph. Ansermet, Phys. Rev. Lett, **82**, 3681 (1999).

[6]. I. D. F. Li, J. B. Wiley, D. Cimpoesu, A. Stancu, and L. Spinu, IEEE Trans. Mag., Vol. **41**, 3361 (2005).

[7]. A. Encinas-Oropesa, M. Demand, L. Piraux, I. Huynen, and U. Ebels, Phys. Rev. **B 63**, 104415 (2001).

[8]. C. A. Ramos, M. Vazquez, K. Niclsch, K. Pirota, J. Rivas, R. B. Wehrspohn, M. Tovar, R. D. Sanchez, and U. Goesele, J. Mag. Mag. Mat. **272-276,** 1652 (2004).

[9]. R. B. Marks, IEEE Trans. Microwave Theo. Tech. **MTT-39**, 1205 (1991).

[10]. Bijoy Kuanr, L. Malkinski, R. Camley, Z. Celinski, J. Appl. Phys., **93**, 8591 (2003)

[11]. B. K. Kuanr, R. Camley, Z. Celinski, Appl. Phys. Lett., **83**, 3969 (2003)

[12] R. Marson, B. K. Kuanr, S. R. Mishra, R. E. Camley, and Z. Celinski, J. Vac. Sci. Technol. B **25,** 2619 (2007).

[13]. B. K Kuanr, Y. Khivinitsev, A. Hutchison, R. E. Camley, and Z. Celinski, IEEE Tran. on Magn., **43**, 2648 (2007).

Phonon-assisted tunneling mechanism of conduction in SnO_2 and ZnO nanowires

P. Ohlckers* and P. Pipinys**

*Vesfold University College, Raveien 197, 3184 Borre, Norway, Per.Ohlckers@hive.no
**D. Gerbutaviciaus 3-18, 2050 Vilnius, Lithuania, pipiniai@takas.lt

ABSTRACT

Non-linear temperature-dependent current-voltages characteristics of SnO_2 nanowires presented by other authors are explained by the model based on phonon-assisted tunneling (PhAT) theory. It is shown that the PhAT model can explain the variation of current with temperature in all measured temperature range from 10 to 200 K over a wide range of electric fields. From the fit of experimental data with the theory the density of states in the interface layer is derived and electron-phonon interaction constant is estimated

Keywords: nanowires, resistivity, tunneling, tin oxide, zinc oxide

1 INTRODUCTION

Recently Ma et al. have presented the results on charge–carriers transport studies of a single SnO_2 [1] and ZnO nanowires [2]. It was found that resistivity of the single SnO_2 nanowire increases slowly with decreasing temperature from room temperature to 50 K, and increases rapidly below 50 K. The temperature dependence of the resistance followed the relation $\ln R \sim T^{-1/2}$. The authors of Ref. [1] asserted that the transport of the electrons in the nanowires is dominated by the Efros-Shklowskii variable-range hopping (ES-VRH) process [3]. However, the characteristic ES's temperature obtained from the slope of relation $\ln\rho \approx$ T of 231 K for SnO_2 nanowire was much higher than in the previously determined for ES-VRH behavior [3, 4]. For the ZnO nanowire case the Coulomb gap in the density of localized states near the Fermi energy was found abnormally big – about 0.8 eV when a few millivolts estimated by other authors [5]

In a recently published paper by Park et al [6] the temperature-dependent *I-V* characteristics in the temperature region of 70-293 K have been explained by Schottky emission but the results at lower temperatures the authors of [6] assigned as Fowler-Nordheim tunneling. Thus, for explanation of temperature dependent *I-V* data and resistivity dependence on temperature in these oxides different mechanisms, such as VRH, Schottky emission and tunneling, are involved.

We would like to show that in Ref. {1, 2, 6] presented data both in high and low temperature region can be explained by the unique phonon-assisted theory has earlier been used to explain similar results obtained for oxides and other dielectrics [7, 8]. Therefore, in this report we present an explanation of current dependence on temperature and electric field measured in ZnO and SnO_2 nanowires [1, 2] and in single ZnO nanorods [6] with the assumption that free charge carriers creation occurs due to the phonon-assisted tunneling from localized states in metal-oxide interface.

2 THEORY AND COMPARISON WITH EXPERIMENT

We assume that a source of charge carriers are the local electronic states in the oxide-electrode interface layer, the electrons from which emerged to the conduction band of the oxide due to electric field induced phonon-assisted tunneling. If electrons released from these centers dominate the current through the crystal then the current *I* will be proportional to the electron released rate *W* and the density of the traps *N*, i.e. $I \propto NW$. On this basis we will compare the current dependencies on field strength measured at different temperatures extracted from Ref. [1] with the calculated tunneling rate *W* dependencies on field strength *E*.

The temperature- and field-dependent tunneling rate *W(E,T)* was computed using the phonon–assisted tunneling theory [9] by the following equation:

$$W_1 = \frac{eE}{(8m^*\varepsilon_T)^{1/2}}[(1+\gamma^2)^{1/2}-\gamma]^{1/2}[1+\gamma^2]^{-1/4}$$
$$\exp\{-\frac{4}{3}\frac{(2m^*)^{1/2}}{eE\hbar}\varepsilon_T^{3/2}\times$$
$$[(1+\gamma^2)^{1/2}-\gamma]^2[(1+\gamma^2)^{1/2}+\frac{1}{2}\gamma]\},\quad \gamma=\frac{(2m^*)^{1/2}\Gamma^2}{8e\hbar E\varepsilon_T^{1/2}}. \qquad (1)$$

Here $\Gamma^2 = 8a(\hbar\omega)^2(2n+1)$ is the width of the center absorption band, with $n=[\exp(\hbar\omega/(k_BT))-1]^{-1}$, where $\hbar\omega$ is the phonon energy, $\hbar$ is Dirac's constant, ε_T is the energetic depth of the center, *e* is the electron charge, and *a* is the electron-phonon interaction constant.

The comparison of *I-V* data measured at low temperatures for SnO_2 nanowire (from Figure 4a in [1]) with theoretical *W(E)* dependencies is shown in Figure 1.

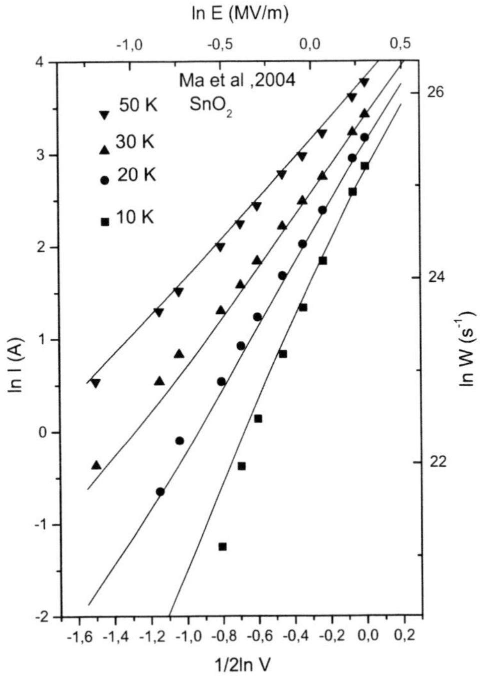

Figure 1: The comparison of *I-V* data for SnO_2 nanowire (from Figure 4a in [1]) with theoretical *W(E)* dependencies calculated using parameters:
a = 1, ε_T = 10 meV, $\hbar\omega$ = 2 meV, m* = 0.65m$_e$

The *W(E)* dependencies were calculated using for the electron effective mass the value of 0.65 m_e. For the phonon energy, the value of 1.6 meV, for the ε_T the value of 14 meV were selected. The electron-phonon coupling constant *a* was chosen so as to get the best fit the experimental data with calculated dependencies, on the assumption that field strength for tunneling is proportional to square root of applied voltage. As is seen from Figure 1, the theoretical *W (E)* dependencies fit well to the experimental data for all range of measured temperatures. It is worthwhile to mention that the theory describes well the measured *dI/dV* dependencies on field strength (see Figure 2).

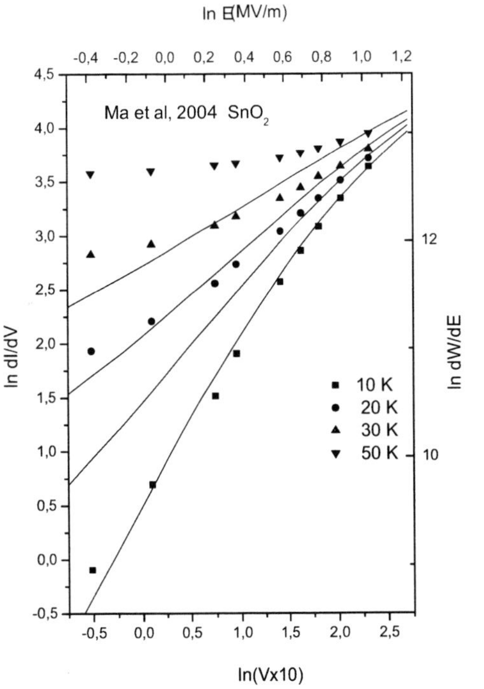

Figure 2: The fit of *dI/dV* dependencies extracted from [1] with *dW/dE* vs *E* calculated for the same parameters as in Figure 1.

In Fig. 3 the results of current dependence on temperature in single ZnO nanorods measured by Park et al [6] fitted to theoretical *W(T)* dependencies are shown. We want to mention that authors of [6] these dependencies above 70 K have explained by Schottky emission, meanwhile the results below that temperature assigned to FN tunneling. As is seen from comparison the phonon-assisted tunneling theory describes well the results over all measured temperature range.

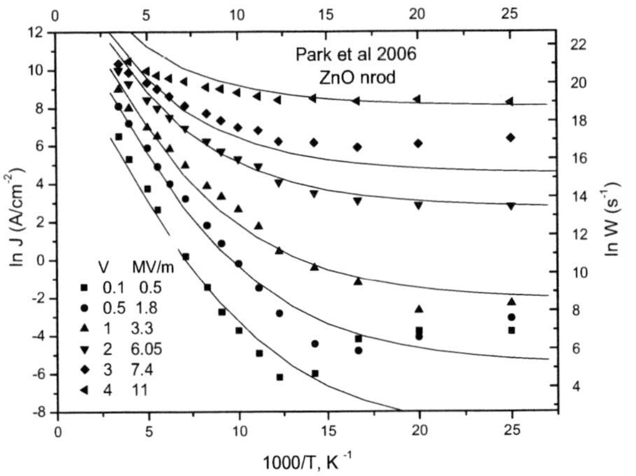

Figure 3: Current vs *1/T* for ZnO nanrods from [6] fitted to *W(T)* vs 1/T calculated using parameters:
a = 1.2; ε_T = 0.12; . $\hbar\omega$ = 16 meV, m* = 0.35m$_e$.

The field strengths for tunneling were found to be in the range from 0.55 MV/m to 11 MV/m. From the relation $I = eSNW$, where S is the area of the electrode, the surface trap density N can be estimated. From the results in Fig. 3 the surface trap density was found to be equal to $N \sim 10^{14}\ cm^{-2}$.

We want to note that on the basis of the phonon-assisted tunneling model is comprehensible not only temperature dependence of current (resistivity), but also temperature independent behavior of current observed in low temperatures region. This assertion is partially confirmed by results presented in Figure 3, and for more clarity in Figure 4 we show the comparison of temperature – dependent conductivity (1/R) from inset in Figure 2 [6] with tunneling rate dependence on T. Below ~ 70 K resistance R changed very little with 1/T meanwhile in the temperature range of 70-293 K lnR linearly increased with *1/T*.

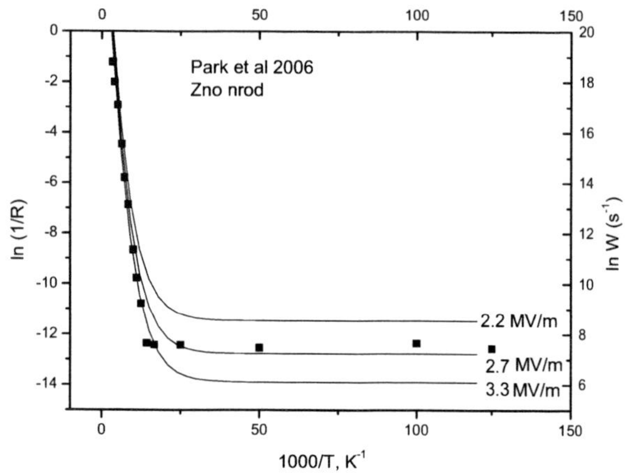

Figure 4: The ln (1/R) vs *1/T* from inset in Fig. 2 [6] fitted to *W(T)* vs *1/T* calculated using the same parameters as in Figure 3.

The fit of these data in coordinates of ln(1/R) versus *1/T* with theoretical ln*W(T)* versus *1/T* dependence shows a very good coincidence of experimental data and theoretical curve. The physical essence of independence of *W(T)* on temperature in the low temperature region is that at low temperatures the phonons are "frozen" and in the regime of low temperatures the phonon-assisted tunneling, as well as F-N ones, becomes temperature-independent process.

3 CONCLUSIONS

In summary, the phonon-assisted tunneling model enables to explain the peculiarities of temperature dependence current in oxide nanostructures, such as strong temperature dependence at high temperatures and less dependence on ones in the region of low temperatures. The variation of current-voltage curves on temperature describes the used equation also well. From the fit of experimental results with theory density of centres, which serve as a source of the free carriers, was found to be equal to $10^{14}\ cm^{-2}$ in ZnO nanorods.

REFERENCES

[1] Y.- J. Ma, F. Zhou, L. Lu and Z. Zhang, Solid State Commun.130, 313, 2004.

[2] Y.- J. Ma, Z. Zhang , F. Zhou, L. Lu, A. Jin and C. Gu, Nanotechnol. 16, 746, 2005.

[3] B. I. Shklovski and A. L. Efros, Electronic Properties of Doped Semiconductors (Berlin, Springer), 1984.

[4] B. Sandow, K. Gloos, R. Rentzsch, A. N. Ionov and W. Schirmacher, Phys. Rev. Lett. 86, 1845, 2001.

[5] V. Yu. Butko, J. F. DiTusa and P. W. Adams, Phys. Rev. Lett. 84, 1543, 2000.

[6] J. Y. Park, H. Oh, J.-J. Kim and S. S. Kim, Nanotechnology 17, 1255, 2006.

[7] P. Pipinys, A. Rimeika, V. Lapeika, Physica Status Solidi (B) 242, 1447, 2005.

[8] P. Pipinys, A. Rimeika, V. Lapeika, J Phys D: Appl. Phys . 37, 828, 2004.

[9] A. Kiveris, Š. Kudžmauskas and P. Pipinys, Phys. Status Solidi A 37, 321, 1976.

Synthesis and Characterization of Ni/Si Nanowires for Electrical Transport

Jae Ho Lee, Michael Carpenter, Eric Eisenbraun, Yongqiang Xue, and Robert Geer

College of Nanoscale Science and Engineering
University at Albany, 255 Fuller Road, Albany, NY 12203
E-mail: RGeer@uamail.albany.edu, Phone/Fax: (518) 956-7003/ 437-8603

ABSTRACT

Self-assembled Si nanowires (SiNWs) have been synthesized and characterized as a template for surface metal silicide formation to investigate confinement of electron transport at the nanowire surface. The SiNWs with diameters ranging from 5 to 180 nm were synthesized via the solid-liquid-solid (SLS) mechanism with a sputtered Au film as catalyst. Post-deposition thermal processing was carried out for silicide formation. Metal-silicide coated wires were dispensed on metal-patterned Si wafers to carry out two-point and four-point electrical conductivity measurements. Electrical contacts were formed via FIB-based Pt deposition. Metal-silicide-coated SiNWs exhibited an improvement in electrical conductivity of several orders of magnitude. Preliminary analyses imply variable thickness of the metal silicide region based, in part, on deposition methodology and thermal process parameters.

Keywords: silicon nanowire (SiNW), solid-liquid-solid (SLS), nickel silicide, electrical conductivity

1 INTRODUCTION

Semiconductor and metallic nanowires have attracted substantial attention for a variety of nanoelectronic applications. In particular, silicon nanowires (SiNWs) may be an attractive alternative to conventionally processed Si transistors if their intrinsic self-assembly can be harnessed to obviate the need for complex lithographic techniques for device fabrication. In addition, SiNWs can potentially function as both the switch (i.e. transistor) and local interconnect (e.g. metal silicide nanowire) to form an inherently integrated nanoelectronic system – potentially on the same self-assembled nanostructure [1-3].

In terms of metal silicide candidates for such nanostructures, nickel silicide (NiSi) possesses several advantages including low resistivity and low formation temperature [4-7]. In fact, for current CMOS technology NiSi has been shown to be a good electrical contact material for gate, source and drain [8-10]. Recent demonstrations of excellent conductance in NiSi nanowires have also highlighted the excellent potential for SiNW-based systems.

Currently, there are several methods available to synthesize silicon nanowires including laser ablation [1, 11], physical vapor deposition [12], thermal evaporation [13], chemical vapor deposition [14-16], solid-liquid-solid (SLS) growth [17, 18], vapor-liquid-solid (VLS) growth [19, 20], and oxide assisted growth [21]. The SLS growth is a relatively straightforward technique to synthesize nanowires without a gas phase precursor such as SiH_4 or $SiCl_4$. Via the SLS process, silicon nanowires can be directly grown on a silicon substrate which acts as the silicon [17].

In the SLS process, a thin metal film is deposited on a single-crystal silicon substrate (e.g. (100) Si wafer) as a catalyst. Heating of the metallized Si results in metal droplet formation. Continuous diffusion of silicon atoms from the substrate to the droplet at elevated temperatures causes saturation of silicon inside the droplet, and subsequent precipitation at the surface of the droplet. In the presence of a negative temperature gradient at the droplet surface (e.g. due to a gas flow) the surface Si precipitate forms a Si growth front resulting in nanowire formation from the catalyst [17].

In this paper, we report investigations of SLS-grown SiNWs as templates for the surface formation of NiSi to investigate confinement of electron transport at the nanowire surface. Silicon nanowires were grown via the SLS approach on Si (100) and (111) substrates using a sputtered Au film as catalyst in an oxygen-filtered Ar ambient. The influence of annealing time on SLS SiNW growth is discussed in term of nanowire diameter. Post-growth Ni deposition is followed by variable-rate thermal processing to investigate the surface morphology of nickel silicide formation.

Following thermal processing, NiSi-coated SiNWs underwent electrical conductivity testing within two-point and four-point probe structures processed within a focused ion-beam scanning electron microscope (FIB-SEM).

2 EXPERIMENTAL DETAILS

The substrates used were p-type (100) and (111) silicon wafers with electrical resistivity in the 1-10 Ω•cm range. Wafers were cleaned with diluted hydrofluoric acid (1%) to remove the native oxide layer and ultrasonicated in acetone to remove organic contamination. The cleaned samples were immediately loaded into a PVD (evacuated to 5×10^{-7} Torr) for sputtering of a 4 nm thick Au catalyst film.

Following Au deposition the silicon samples were placed inside an annealing chamber. The annealing chamber was evacuated to a base pressure of approximately 5 Torr and backfilled with high purity (99.999%) Ar. In order to reduce residual oxygen in the chamber an oxygen filter and a bypass line were configured in the Ar supply.

The total pressure of the system was then raised to atmospheric pressure. The Au/Si sample was annealed at 1000 °C (30 minute ramping time) under Argon (99.999% purity) gas flow at 2,000 sccm. The annealing duration after the temperature ramp was varied from 10 minutes to 120 minutes to investigate the effect of annealing time on the nanowire diameter, length, and overall morphology.

Ni deposition and subsequent thermal processing was carried on as-deposited SLS SiNWs to investigate surface silicide formation. Nickel was deposited on the nanowires via e-beam evaporation (calibrated for an effective blanket film thickness of 150 nm). Nickel deposited SiNW samples were annealed using a rapid thermal annealing system at 550 °C for 5 min and at 600 °C for 5 min with 10 min ramping time to compare resultant SiNW morphologies which were investigated by SEM.

Metal-coated SiNWs and as-grown SiNWs were dispensed on metal-patterned Si wafers to carry out two-point and four-point electrical conductivity measurements. Electrical contacts were formed using Pt deposition within a dual beam FIB-SEM. Structural and compositional properties of these wires were analyzed using SEM, energy dispersive x-ray spectroscopy (EDS), and transmission electron microscopy (TEM).

3 RESULTS

The solid-liquid-solid (SLS) method, as described above, was used to synthesize silicon nanowires. The samples were taken out of the annealing furnace after cooling to room temperature. The effects of annealing duration were investigated. The furnace temperature was ramped up to 1,000 °C for 30 min under 2,000 sccm of Ar flow. After the ramp the furnace temperature was held constant for durations ranging from 10 minutes to 120 minutes. Figure 1 shows SEM micrographs of typical SiNWs synthesized for various anneals (1000 °C). An approximately bimodal diameter distribution was observed. Increased annealing time resulted in larger SiNW diameters. SiNWs diameters ranged from 5-22 nm after 10 minute annealing, 45-52 nm after 60 minute annealing, and 77-180 nm after 120 minute annealing.

Si nanowires exhibiting the smallest diameters of 5 nm observed here have not been reported previously (grown via SLS). Typically, small diameter nanowires were only reported for vapor-liquid-solid (VLS) growth when utilizing nm-thick catalyst layers or distributions of nm-scale catalytic particles. In VLS growth, controlling the diameter of nanowires can be achieved by adjusting the precursor and pressure. However, in conventional SLS growth, diameter control is a challenge since SiNW growth is proceeds rapidly at high temperature. In the work reported here, variation of the anneal duration and Ar gas flow rate serve to immediately quench the growth and help regulate SiNW diameter.

After growth, SiNWs were detached from the initial silicon wafer using ultrasonication in a solvent and distributed on highly oriented pyrolytic graphite (HOPG) substrates to carry out compositional analyses.

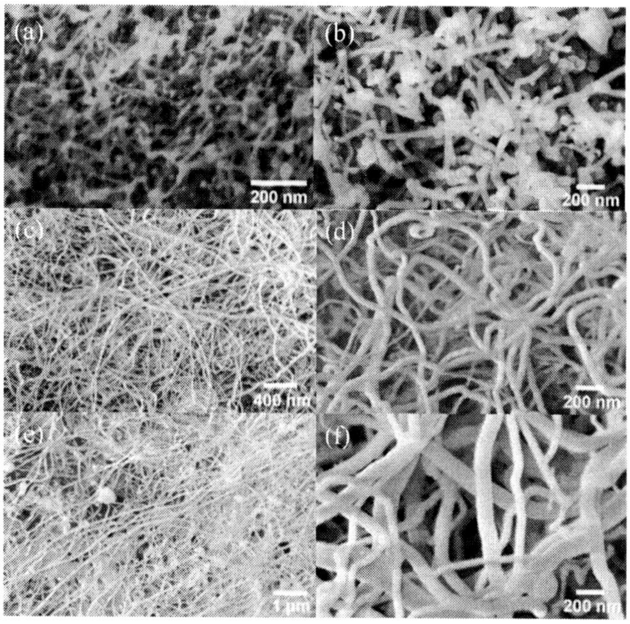

Figure 1: SEM micrograph of nanowires synthesized at 1000 °C for (a) 10 min, (b) 15 min, (c) 45 min, (d) 60 min, (e) 90 min, (f) 120 min annealing times. Deposited Au catalyst layer was 4 nm thick.

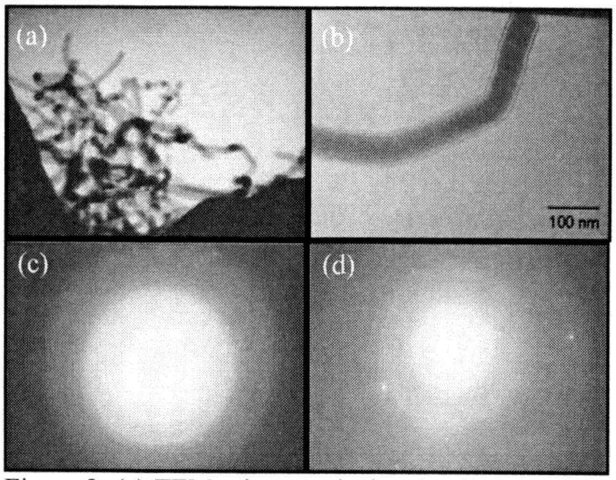

Figure 2: (a) TEM micrograph showing the morphology of a SiNW bundle, (b) Single SiNW, (c), (d) Selected-area electron diffraction (SAED) patterns from SiNWs.

Selected-area electron diffraction (SAED) patterns (Fig. 2(c) and 2(d)) were acquired from individual SiNWs. These implied an amorphous microstructure for the SiNWs evaluated. This implies the possible presence of oxygen in the SiNWs. It should be noted that crystalline nanowires have been typically observed via VLS growth, while SLS methods generally produce amorphous nanowires. This may be due to the fast growth rate and high temperature of the SLS process, which leads to amorphous rather than crystalline growth.

NSTI-Nanotech 2008, www.nsti.org, ISBN 978-1-4200-8505-1 Vol. 3

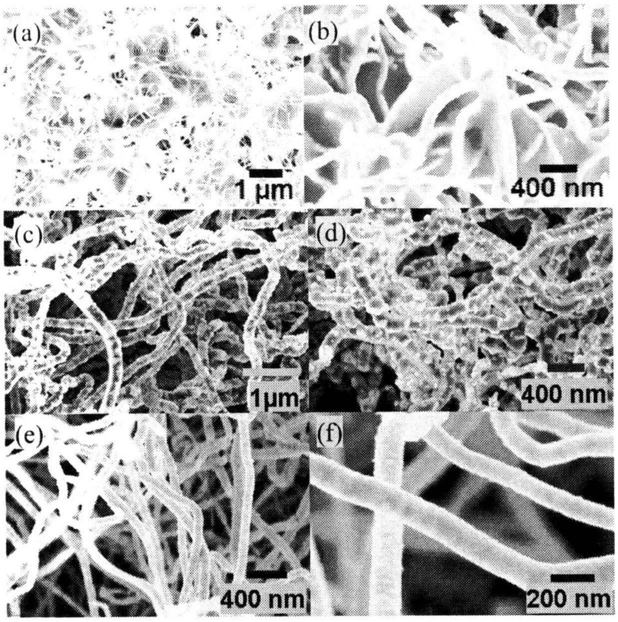

Figure 3: SEM images of Ni-deposited SiNWs: (a), (b) as-deposited; (c), (d) Ni-deposited SiNW after slow annealing at 600 °C for 10 minutes in Ar ambient (c), (d). Note rough surface morphology; (e), (f) Ni-deposited SiNW after rapid thermal annealing at 550 °C for 5 minutes in N_2 ambient. Note smooth morphology.

Nickel was deposited on the nanowires by e-beam evaporation calibrated for an effective blanket film thickness of 150 nm. Post-deposition thermal processing was carried out for nickel silicide formation (550 °C for 5 minutes via RTA and 600 °C for 5 minutes with a 10 minute ramp). Post-anneal nanowire surface morphology was sensitive to anneal temperature and ramp rate. Rapid ramps resulted in an atomically-smooth Ni-SiNW template morphology. Slow annealing resulted in a rough Ni-SiNW surface morphology indicative of nonuniform silicide domain formation, as shown in Fig. 3.

Energy dispersive x-ray spectroscopy (EDS) was performed for post-annealed Ni-deposited SiNWs on an HOPG substrate for compositional analysis. Figure 4 shows EDS spectra of nanowires on the initial silicon substrate, after Ni deposition, after slow annealing, and after RTA (the latter three on HOPG substrates). The results confirm the presence of nickel on individual SiNWs after thermal processing. Note the substantial increase of the La peak (0.85 keV) relative to the Ka peak (7.47 keV) for the annealed Ni-SiNWs. This implies nickel silicide formation [6, 22, 23]. The EDS data in Fig. 4 also confirm the presence of oxygen in the SiNWs.

As-grown silicon nanowires and metal-silicide coated SiNWs were dispensed on SiO_2-coated Si wafers patterned with a metal (Au) electrode pattern to carry out electrical conductivity measurements. Electrical connection between the dispensed SiNWs and the Au electrodes were formed via direct-write FIB-based Pt deposition. Figures 5a and 5b show an SEM micrograph and associated current-voltage characteristic of an as-grown SiNW, respectively. The I-V response of the SiNW is linear and the two-point resistivity of similar SiNWs ranged from 20 Ω·cm to 2×10^5 Ω·cm. The observed electrical resistivity range for as-deposited SiNWs rules out a dominant SiO_2 stoichiometry, although it is not inconsistent with local SiO_x compositions within the nanowire. (No current measurement was possible for open Pt electrodes confirming the insulative properties of the SiO_2 dielectric in the electrical test structure.)

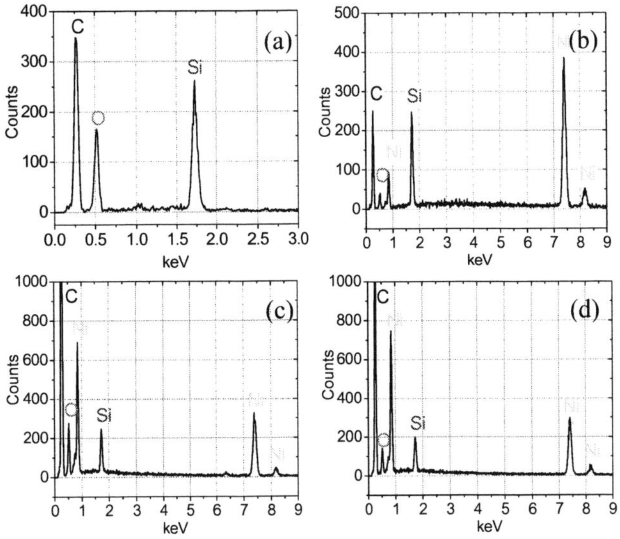

Figure 4: EDS spectra of nanowires: (a) as-grown nanowire on the initial silicon substrate; (b) after Ni deposition (on HOPG); (c) after slow annealing (on HOPG); (d) after rapid annealing (on HOPG).

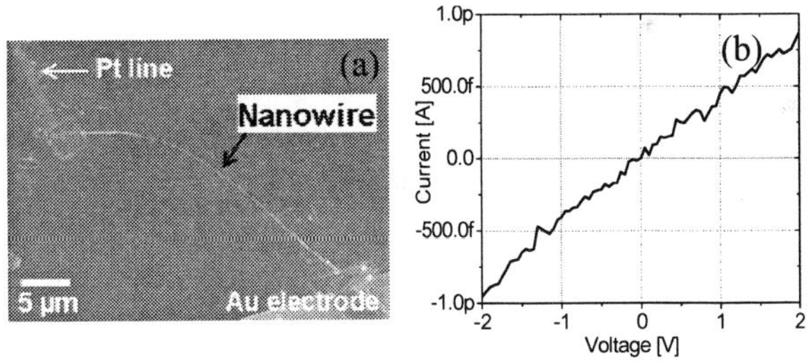

Figure 5: Pt deposition and I-V measurement of dispensed as-grown silicon nanowire. (a) SEM image of the wire connected by Pt lines on Au-patterned oxide substrate, nanowire length was 29 µm and diameter was 168 nm. (b) I-V characteristics.

The resistivity of post-annealed Ni-SiNW wires was measured by two- and four-point electrical conductivity measurements, as shown in Fig. 6. Two- (Fig. 6a) and four- (Fig. 6c) point I-V measurements yield resistance of 2 MΩ and 55 kΩ, respectively, for the 9.2 µm long device with a diameter of 165 nm, which corresponds to a resistivity of 0.47 Ω·cm and 0.013 Ω·cm, respectively.

Fig. 6 (d) shows calculated resistivity from as-grown nanowires and metal-silicide coated wires. It is clear that after post-deposition of nickel on the silicon nanowires the conductivity is increased up to 7 orders.

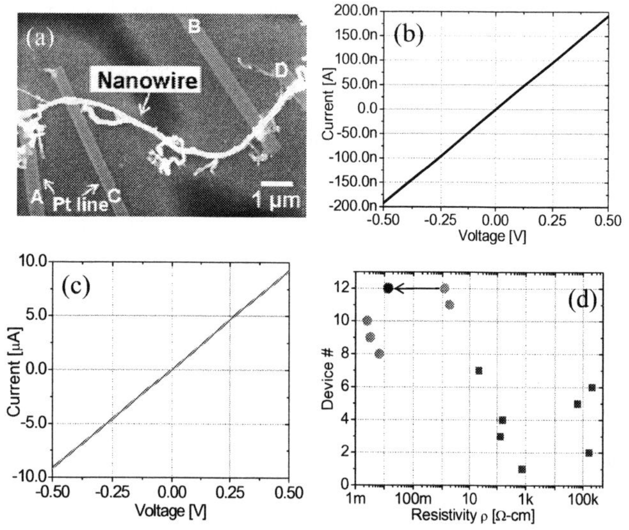

Figure 6: (a) SEM image of metal-silicide coated wire with four terminals, (b) I-V characteristics via two-point measurement, (c) via four-point measurement, (d) calculated resistivity from as-grown nanowires (blue squares) and metal-silicide coated wires (red circles).

4 CONCLUSION

Self-assembled Si nanowires were synthesized and characterized as a template for surface metal silicide formation to investigate confinement of electron transport at the nanowire surface. Silicon nanowires with diameters ranging from 5 to 180 nm were synthesized via solid-liquid-solid (SLS) growth. The diameter of as-grown SiNWs can be controlled, to an extent, through the annealing time.

Post-growth Ni deposition and thermal processing was carried out for nickel silicide formation at the SiNW surface. Post-annealed Ni-SiNW surface morphology was sensitive to anneal temperature and ramp rate. Rapid ramps resulted in an atomically smooth Ni-SiNW surface morphology. Slow annealing resulted in a rough Ni-SiNW surface morphology indicative of nonuniform domain formation. EDS measurements confirm that the nickel remains on the SiNW surface after thermal processing. Also, the substantial increase of the La peak (0.85keV) relative to the Ka peak (7.47 keV) in the EDS spectra for the annealed Ni-SiNWs implies nickel silicide formation.

Metal-silicide coated wires were dispensed on metal-patterned Si wafers to carry out two-point and four-point electrical conductivity measurements. Electrical contacts were formed via FIB-based Pt deposition. Metal-silicide-coated SiNWs exhibited an improvement in electrical conductivity of several orders of magnitude compared with that of as-grown silicon nanowires.

ACKNOWLEDGEMENTS

This work was supported by the Semiconductor Research Corporation Focus Center Research Program, the Nanoelectronics Research Corporation, and the New York State Office of Science, Technology and Academic Research.

REFERENCES

[1] A. M. Morales and C. M. Lieber, Science, 279 (9), 208, 1998.
[2] W. Lu and C. M. Lieber, Nature Mater. 6, 841, 2007.
[3] A. Colli, A. Fasoli, P. Beecher, P. Servati, S. Pisana, Y. Fu, A. J. Flewitt, W. I. Milne, J. Robertson, C. Ducati, S. De Franceschi, S. Hofmann and A. C. Ferrari, J. of Appl. Phys., 102, 034302, 2007.
[4] Y. Wu, J. Xiang, C. Yang, W. Lu and C. M. Lieber, Nature, 430, 61-65, 2004.
[5] X. W. Zhang, S. P. Wong, W. Y. Cheung and F. Zhang, Mat. Res. Soc. Symp. 611, C6.5.1, 2000.
[6] J. Kim and W. A. Anderson, Thin Solid Films, 483, 60, 2005.
[7] G. B. Kim, D. J. Yoo, H. K. Baik, J. M. Myoung, S. M. Lee, S. H. Oh and C. G. Park, J. Vac. Sci. Technol. B 21, 319, 2003.
[8] C. Lavoie, F. M. d'Heurle, C. Detavernier and C. Cabral, Microelectronic. Eng. 70, 144, 2003.
[9] J. A. Kittl, A. Lauwers, O. Chamirian, M. Van Dal, A. Akheyar, M. De Potter, R. Lindsay and K. Maex, Microelectronic. Eng. 70, 158, 2003.
[10] T. Morimoto, T. Ohguro, S. Momose, T. Iinuma, I. Kunishima, K. Suguro, I. Katakabe, H. Nakajima, M. Tsuchiaki, M. Ono, Y. Katsumata, H. Iwai, IEEE Trans. Electron Devices, 42, 915, 1995.
[11] Y. F. Zhang, Y. H. Tang, N. Wang, D. P. Yu, C. S. Lee, I. Bello and S. T. Lee, Appl. Phys. Lett. 72, 1835, 1998.
[12] N.D. Zakharov, P. Werner, G. Gerth, L. Schubert, L. Sokolov and U. Gösele, J. Cryst. Growth, 290, 6, 2006.
[13] D. P. Yu, Z. G. Bai, Y. Ding, Q. L. Hang, H. Z. Zhang, J. J. Wang, Y. H. Zou, W. Qian, G. C. Xiong, H. T. Zhou and S. Q. Feng, Appl. Phys. Lett. 72, 3458, 1998.
[14] J. Westwater, D. P. Gosain, S. Tomiya, S. Usui, and H. Ruda, J. Vac. Sci. Technol. B 15, 554, 1997.
[15] Yi Cui, L. J. Lauhon, M. S. Gudiksen, J. Wang and C. M. Lieber, Appl.Phys. Lett. 78, 2214, 2001.
[16] A. I. Hochbaum, R. Fan, R. He and P. Yang, Nano Lett. 5, 457, 2005.
[17] Maggie Paulose, Oomman K. Varghese and Craig A. Grimes, J. Nanosci. Nanotech. 3, 341, 2003
[18] H. F. Yan, Y. J. Xing, Q. L. Hang, D. P. Yu, Y. P. Wang, J. Xu, Z. H. Xi and S. Q. Feng, Chem. Phys. Lett. 323, 224, 2000.
[19] R. S. Wagner, W. C. Ellis, Appl. Phys. Lett. 4, 89, 1964.
[20] E. I. Givargizov, J. Cryst. Growth, 31, 20, 1975.
[21] R. Q. Zhang, Y. Lifshitz and S. T. Lee, Adv. Mater. 15 (7-8), 635, 2003.
[22] Y. Song and S. Jin, Appl. Phys. Lett. 90, 173122, 2007.
[23] K. S. Lee, Y. H. Mo, K. S. Nahm, H. W. Shim, E. K. Suh, J. R. Kim and J. J. Kim, Chem. Phys. Lett. 384, 215, 2004.

Silicon on Ceramics - A New Concept for Micro-Nano-Integration on Wafer Level

M. Fischer*, H. Bartsch de Torres*, B. Pawlowski**, M. Mach*, R. Gade*, S. Barth**, M. Hoffmann* and J. Müller*

*Institute for Micro- and Nanotechnologies, Technische Universität Ilmenau, Gustav-Kirchhoff-Str. 7, 98693 Ilmenau, Germany, michael.fischer@tu-ilmenau.de
**Hermsdorfer Institut für Technische Keramik e.V., Michael-Faraday-Str. 1, 07629 Hermsdorf, Germany

ABSTRACT

A new integration concept for micro-nano-integration based on a new bonding technique between nano-scaled, modified Black Silicon (BSi) and an adapted, unfired LTCC substrate is presented. The novel technique enables to combine advantages of silicon and ceramic technology, especially electrical and fluidic interconnects from nm- to mm-scale. Current bonding concepts of silicon on ceramics need joining materials like solders, adhesives or glass frits. Alternatively, silicon components can be mounted to a TCE-matched and fired LTCC by anodic bonding, which requires costly surface preparation by polishing. This step is eliminated by the use of the new technique. During a standard lamination process, a self-organized, nano-structured, grass-like silicon surface is joined with the green ceramic body. Pressure assisted sintering allows the co-firing of the composite. Dense contact between the Black Silicon surface and the ceramic, leading to a maximum average bonding strengths of 1775 N/cm², is achieved by optimization of the nano-interface and the lamination procedure. The hermeticity of the interface has been proven showing helium leak rates up to $1.9*10^{-8}$ mbar l/s. Hereby, a gas-tight interface (representing only a 10-nm leakage), which is impermeable for viruses and allows therefore biomedical use, is given. Electrical interconnects between ceramic body and silicon have been realized by means of metallized BSi penetrating into LTCC conductors. Prefabricated silicon-on-ceramic substrates with wiring and vias are compatible to a variety of semiconductor technologies for MEMS manufacturing.

Keywords: silicon on ceramic, black silicon, bonding of silicon to ceramics, wafer level packaging

1 INTRODUCTION

In the recent years, lots of applications based on nano effects and patterns have been investigated. A common difficulty to make use of nano elements in semiconductor devices is the realization of an intelligent and robust connection to the macro world. Tough mechanical, electrical or fluidic coupling of nano elements without affecting its functionality is required. A concept for micro and nano integration based on a fully silicon-ceramic wafer compound material solves this problem. The method is based on a bonding procedure between silicon and a low temperature cofired ceramic (LTCC). A LTCC tape with adapted TCE to silicon (trade name BGK [2]) is joined with a silicon wafer during a standard lamination process [1]. The self-organized, nano-structured silicon surface forms the contact with the green body. A subsequent pressure assisted sintering is utilized to co-fire the composite. This Silicon-On-Ceramic-substrate (SiCer) enables a wide range of design solutions. The functional silicon surface is bonded to prefabricated ceramic tapes with vias, wirings and fluidic channels. After sintering, the ceramic acts as a carrier system with electrical and fluidic connection. To ensure the electronical functionalities of MEMS devices, only a thin silicon layer is necessary. The separation of silicon areas can be easily done by standard silicon etching, while the ceramic works as a natural etching barrier. Consequently, the process enables system packaging on wafer level. Fig. 1 illustrates a scheme of this new concept for micro-nano-integration.

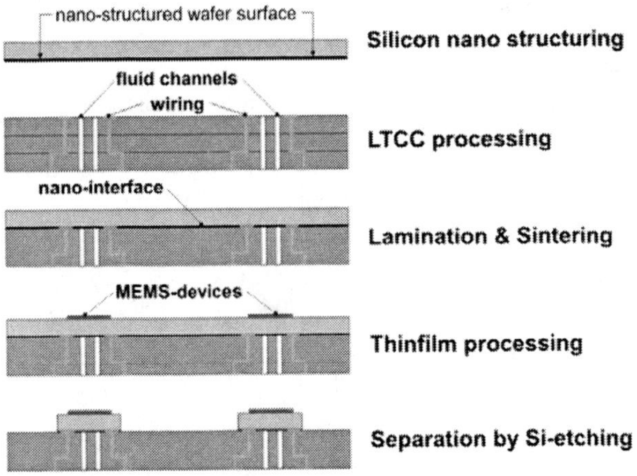

Figure 1: Work flow of the new integration concept

A number of optimization steps in silicon nano structuring and lamination procedures as well as in the adaptation of the BGK powder morphology were accomplished to realize a strong and dense contact between the Black Silicon surface and the ceramic.

2 TECHNOLOGICAL CHALLENGE

During the lamination, the nano patterned grass surface of the silicon wafer penetrates into the ductile and unfired LTCC tape. Because of the highly increased needle surface a form-fit bonding and a material connection is generated during the firing step. Therefore, the lateral shrinking of the ceramic must be avoided by the use of pressure assisted sintering.

2.1 Surface nano structuring of silicon

The nano-textured silicon surface is generated with a self-organized, lithography-free reactive ion etching (RIE) process, which leads to homogeneously distributed needles across a full wafer (Fig. 2). A parallel plate reactor (STS 320) with a cooled wafer electrode and a SF_6 / O_2 – plasma is used.

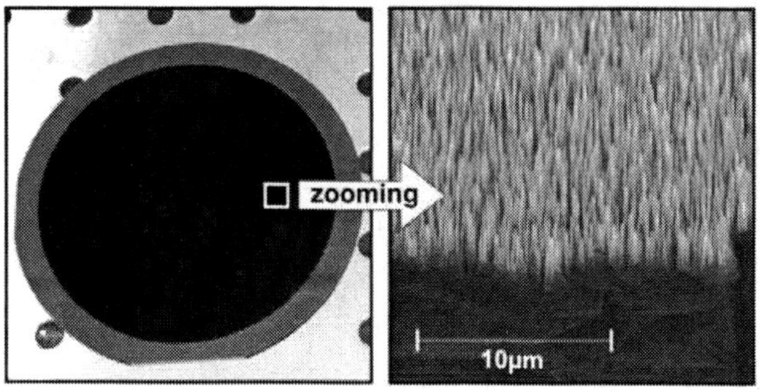

Figure 2: Homogeneously distributed BSi

Depending on the process time, needles up to 2.5 μm length can be achieved and the diameter varies from top to bottom between 5 nm to 400 nm. The needle-like structures have pitch dimensions between 100 nm and 200 nm. For an adequate penetration into the unfired, polymer-bound ceramic tape, the needles are too long and flexible (Fig. 3a). Therefore, an additional plasma treatment with Argon is accomplished for shrinking and thinning out the needles in order to adapt the needle geometry (Fig. 3b) to the powder morphology of the unfired BGK tape.

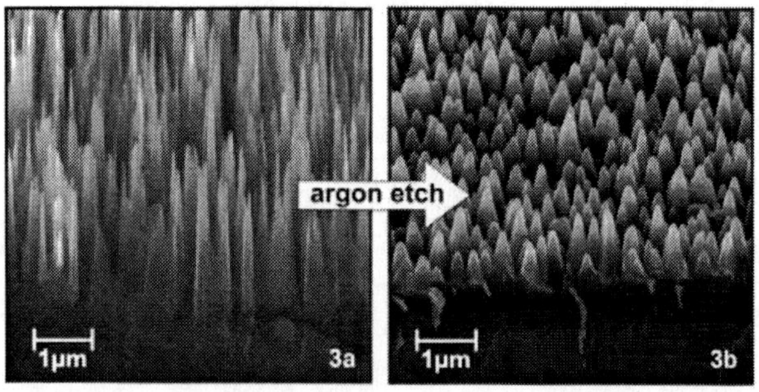

Figure 3: RIE-structured BSi before (a) and after treatment with Ar-plasma (b)

The nano-textured silicon surfaces fabricated as mentioned above can additionally be used as catalyst carrier or surfaces with tunable wetting angles or for room temperature bonding using the Velcro® principle [3]. In this way, the functionality of nano surfaces is implemented in Microsystems.

2.2 Characterization of BGK

BGK is a silicon-compatible LTCC-tape. Due to its powder composition the fired BGK has a silicon adapted thermal expansion of 3,4 ppm/K with a glass transition temperature of 590 °C (Fig. 4). The main constituents of the tape are a boro-silicate-glass, Al_2O_3 and cordierite, necessary for TCE-adaptation, as well as a polymer binder.

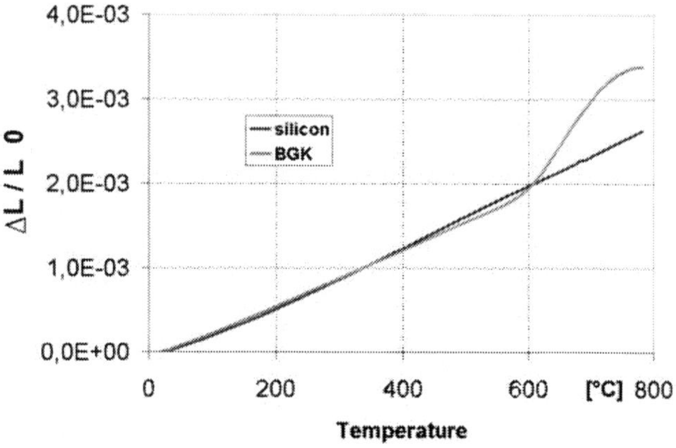

Figure 4: Dilatometer measurements of BGK and Si

BGK has a powder distribution with particle sizes of 50nm up to 2500 nm. However, the standard tape quota of fine particles is low (Fig. 5a) witch can lead to an incomplete filling of the spaces between the needles during lamination and firing, witch results in a non-hermetic interface. In a further development, higher quotas of fine particles were used in the slurry to adapt the BGK-tape (Fig. 5b).

Figure 5: Powder morphology of standard BGK (a) and refined BGK (b)

2.3 Fabrication of the wafer compound

The first step of the bonding procedure is the lamination process, which brings the nano surface into a deep contact with the green BGK - tape. The needles penetrate into the polymer matrix of the green tape and the glass and ceramic particles adhere on the needle surface. Pressure, time, temperature as well as the force rising rate are important lamination parameters and determine the bonding properties of the final compound. Therefore it was necessary to carry out a parameter variation by the use of Design of Experiments to find a lamination optimum. The lamination is carried out in an isostatic press. Pressure-assisted sintering is used to fire the compound. In a furnace press from ATV GmbH, a pressure of 0,5 MPa ensures the tight contact of silicon and ceramic during the temperature profile with a peak temperature of 850°C.

3 APPLICATION ASPECTS

3.1 Mechanical strength of the interface

A modified pull test was performed to determine the bonding strength. The bonded Silicon-On-Ceramic-substrate (SiCer) was cut into 10 mm x 10 mm squares and fixed to aluminum cylinder with epoxy resin. This assembly was pulled until the bond was teared. The analysis of more than 100 samples bonded with the adapted needle geometry of the silicon surface, leaded to an optimal parameter combination of 5.5 MPa, 25 min, 120 °C, 1 kN/min). A maximum average bonding strengths of 1775 N/cm² was determined for these parameter settings. Pressure and lamination time have the strongest influence on the process. An further improvement is achieved by the use of the refined tape. Finer particles fit better into the needle spaces and lead to a tighter interface. The bonding strength increases significantly to values above 5000 N/cm². Figure 6 shows a summary of average bonding strengths realized by common bonding techniques and the SiCer method.

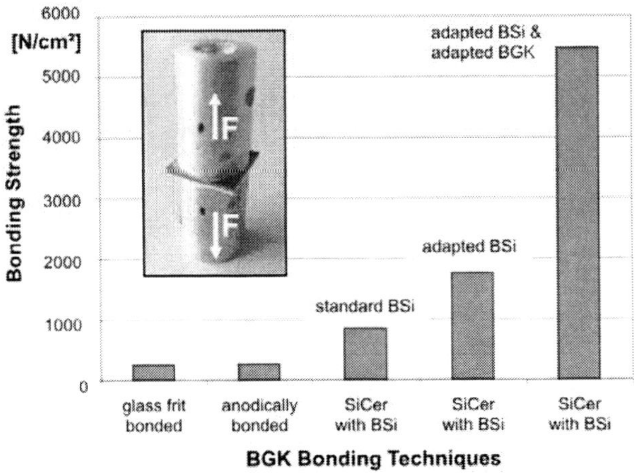

Figure 6: Average bond strength of SiCer compared with other silicon-to-ceramics bonding methods

3.2 Fluidic function

The integration of fluidic components like cavities, membranes or channels into ceramic multilayers requires optimized design and adequate process controlling. Particular the formation of hollows during the pressure assisted sintering is a very sensitive process step, because hollows can be collapsed. Test layouts with different stack sizes were used to investigate the influence of design parameters. For stacks with 7 layers, from which three tapes form cavities, hollows with dimensions up to 3.5 mm can be crack free manufactured without inlets. In order to characterize the tightness of the bond interface and the hollows, test chips with a micro fluidic vacuum chamber and different bonding frames, mounted onto an adapted vacuum flange (KF-series) were fabricated. A leak detection system, commonly used for vacuum devices, was applied to determine a helium leak rate (Fig. 7). First measurements show average leak rates of 2.8*10- 4 mbar l/s. This corresponds to a single leak with 1µm diameter which means the bonding is water-tight and a barrier for bacteria. The minimum leak rate for some samples was 1.9*10-8 mbar l/s. These interfaces were also gastight and impassable for viruses.

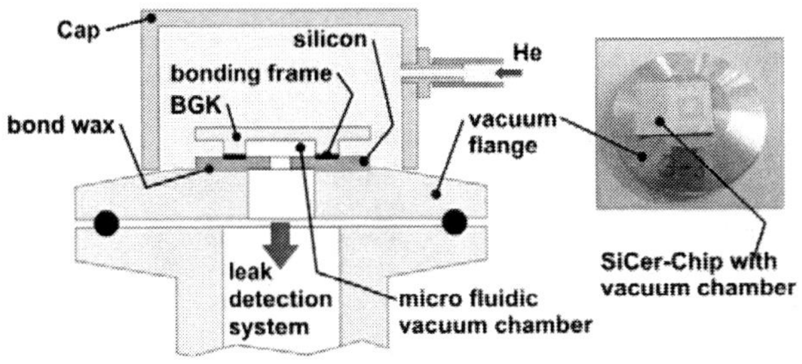

Figure 7: Test structure with leak rate setup

3.3 Electrical function

Electrical connections from the silicon component to LTCC carrier are formed, if metallized Black Silicon penetrates into LTCC metal vias. However, the metallization have to withstand the sintering at 850°C. Furthermore, the used metals should not lead to silicide formation. Figure 8 shows the fabrication. Platinum structures on oxidized BSi-needles, fabricated by a sputtering step and lift-off lithography were laminated to a prefabricated BGK-stack with gold vias (Heraeus TC 7101). After sintering, four-point-measurements were used to determine the sheet resistance of the metallized needle surface in order to calculate the transfer resistance at the interface "via/BSi". An average sheet resistance of 2,2 Ω/□ and a transfer resistance of 3,3 Ω is achieved.

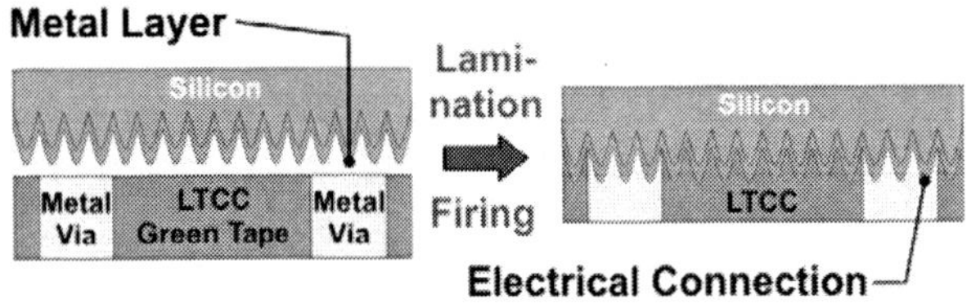

Figure 8: Assembly for electrical interconnects

3.4 System integration

To evaluate the potential of the new integration concept, a fluidic chip cooling system was chosen as a proof of concept (Fig. 9). The system consists of a ceramic carrier, which contains fluidic channels and the wiring for electrical connections. Above the channels a silicon chip with a heating structure and a measuring resistance is aligned. The needle structure forms the bond and laps additionally into the channels. It is expected that the enlarged surface improves the heat exchange.

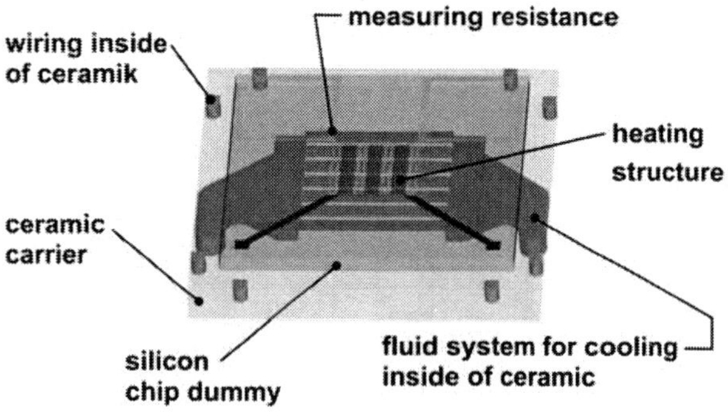

Figure 9: .Fluidic chip cooling system

All process steps have been combined in a first technological run, as depicted in figure 1. The fabrication starts with nano-structuring of the silicon wafer surface. The wafer is 100µm thick and has a diameter of 100mm. Subsequently, the manufactured substrate is covered with a 70 nm thermal oxide to prevent diffusion of metal into silicon at the interface during firing. The LTCC processing starts with cutting and punching of vias and fluid channels into the BGK-tape. After the via filling, conducting paths are screen-printed onto the top BGK-layer using a silver-platinum-paste by Heraeus (TC 7601). The prepared green body is joined with the silicon wafer with lamination parameters mentioned above. After sintering in a furnace press (ATV GmbH) the silicon MEMS processing follows. To build up the heating meander and the measuring resistance a bi-layer metallization of 25 nm chromium and 300 nm gold was patterned by means of lift off lithography. Afterwards the single silicon chips are created during a deep reactive ion etching process in an inductively coupled plasma machine. A patterned thick film resist serves as etching mask and protects the electrical components during the process. After separation of the single ceramic systems using standard wafer dicing, the silicon chips are connected electrically to the ceramic carrier by means of wire bonding. To proof the cooling concept, external pumps must be connected to the ceramic active chip cooling system. Therefore the design was fitted to an existing fluidic connector system. A prototype is depicted in figure 10.

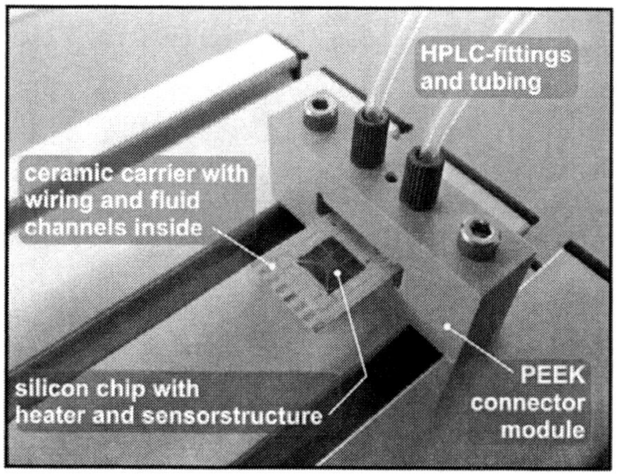

Figure 10: .Fluidic chip cooling system

4 SUMMARY

The capability of a new micro-nano integration concept based on a new silicon-ceramic wafer compound material was proven. A tight and high strength silicon-ceramic-compound was crack free fabricated in a standard wafer format of 4 inch. Electrical and thermal contacts as well as fluidic components can be fabricated from nm to mm scale in one batch using this novel integration concept.

5 ACKNOWLEDGEMENTS

The authors gratefully acknowledge the financial support by Federal Ministry of Education and Research and the Thuringian Ministry of Culture within the framework Mikro-Nano-Integration (project "Nano-SilKe", 16SV3566) and the Initiative "Centre for Innovation Competence", MacroNano®.

REFERENCES

[1] M. Fischer, M. Stubenrauch, M. Hintz, M. Hoffmann, J. Müller, "Bonding of ceramic and silicon – new options and applications", Smart Systems Integration, VDE Verlag, 2007

[2] E. Müller, T. Bartnitzek, F. Bechtold, B. Pawlowski, P. Rothe, R. Ehrt, A. Heymel, E. Weiland, T. Schroeter, S. Schundau, K. Kaschlik, "Development and Processing of an Anodic Bondable LTCC Tape", European Microelectronics and Packaging, Brugge, Belgium, June 2005

[3] M Stubenrauch, M Fischer, C Kremin, S Stoebenau, A Albrecht and O Nagel Author, "Black silicon - new functionalities in microsystems", Journal of Micromech. Microeng. 16 (2006) S.82-S.87

Micro to Nano – Scaling Packaging Technologies for Future Microsystems

T. Braun ([1]), K.-F. Becker ([1]), J. Bauer ([1]), F. Hausel ([2]), B. Pahl ([2]), O. Wittler ([1]), R. Mrossko ([1]), E. Jung ([1]), A. Ostmann ([1]), M. Koch ([1]), V. Bader ([1]), C. Minge ([1]), R. Aschenbrenner ([1]), H. Reichl ([2])

([1]) Fraunhofer Institute for Reliability and Microintegration
Gustav-Meyer-Allee 25, 13355 Berlin, Germany
phone: +49-30/464 03 244 fax: +49-30/464 03 254 e-mail: tanja.braun@izm.fraunhofer.de
([2]) Technical University Berlin, Microperipheric Center

ABSTRACT

As the development of microelectronics is still driving towards further miniaturization new materials, processes and technologies are crucial for the realization of future cost effective microsystems and components. These future systems will not only consist of SMDs and ICs assembled on a substrate, but will potentially integrate also living cells, organelles, nanocrystals, tubules and other tiny things forming a true Heterogeneous System. Futures ICs and passives will also decrease in size, e.g. for RF-ID applications forecast die sizes are smaller than 250 µm, thicknesses less than 50 µm and pitches way below 100 µm, passives, if not directly integrated into the system carrier, will be even smaller. New placement and joining technologies are demanded for reliable and low cost assembly of such applications, as today's packaging technologies only allow the assembly of those small dies and components with a very high effort and for this reason with high cost. With ongoing miniaturization also the protection of the microsystems mostly realized by a polymer needs to be decreased in thickness, yet providing maximum protection. Here, besides mechanical stability, humidity barrier functionality is a key factor for system reliability.

Fraunhofer IZMs approaches towards packaging technologies facing the demands of future nano-based Hetero System Integration are described within this paper, comprising material and process development. Material developments focus on nano-particle enhanced polymers. One example are materials with optimized humidity barrier functionality, where various filler particles are integrated into a microelectronic grade epoxy resin and investigated regarding their barrier properties. Furthermore, the processing of nano-particle filled polymers is illustrated.

Process development comprises touchless handling concepts that are promising for handling miniaturized components, not directly fabricated at the very place where they are needed. Different concepts are under evaluation. Magnetic handling can be regarded as one of the most ripened ones, thanks to the rugged approach explored. Another promising concept is the use of microdroplet manipulation by electrowetting. Results from both concepts show potential for future use. Finally advanced interconnect concepts for low temperature joining by CNT contacts or reactive interconnects are introduced.

In summary an overview on nano-based technologies for heterogeneous system integration is given.

Keywords: nano-particle enhanced materials, touchless handling, electrowetting, heterogeneous integration

1 INTROCUCTION

Microelectronics miniaturization has evolved according to Moore's law since the mid-sixties and since then it was always possible to succeed in meeting its technological challenges. This might be in question for future applications, as indicated schematically by a red brick wall in the SIA roadmap. Fundamentally new approaches will foster further development in microelectronics. High density integration leads to increased interconnect density and thus to a miniaturization of the individual contacts as well. In parallel to this miniaturization of interconnects, the development of an adapted packaging technology is necessary to provide reliable interconnects from the nano- and micro-scale to the meso-world, where microsystems are used. This paper describes exemplarily research and developments in the area of nano-enhanced packaging materials, touchless positioning of smallest component and new advanced interconnection concepts.

2 NANO-ENHANCED PACKAGING MATERIALS

Filler particles are added to polymers in order to modify their properties as e.g. thermo-mechanical, electrical, thermal or diffusion behaviour [1]. Therefore, a wide range of different materials are used as fillers. Main class used for microelectronic encapsulants is SiO_2 with particle sizes in µm range. These particles allow an adaption of encapsulants, adhesives, coatings or base materials for polymer substrates to microelectronic needs mainly in terms of thermo-mechanical and electrical behaviour. Typical encapsulants as e.g. underfiller, glob tops or epoxy molding compounds have a content of 60 to 80 wt.-% SiO_2. Nano-particles may now have the potential to influence the polymer properties significantly only by adding a small amount of fillers.

2.1 Nano-particle enhanced encapsulants with improved moisture resistance

Humidity is always a critical issue in microelectronics where polymers are used as e.g. encapsulants, substrates or adhesives. Water can diffuse into and through the polymer and humidity at bonding interfaces of polymers can lead to a hydrolysis at the interface resulting in delaminations of the polymer. These delaminations can cause two main effects. On the one hand, typically true for encapsulants, mechanical stresses are imposed on interconnects and on the other hand humidity can easily access the interface and can cause corrosion. Both effects have a negative influence on the reliability of the assembly [[2], [3]]. Water in polymers can also lead to a softening and/or to a swelling of the material resulting in degraded thermo-mechanical properties and higher stresses in a package [[4]]. Additionally incorporated water can abruptly evaporate during reflow causing a popcorning of the assembly and therewith an irreversible damage. The introduction of high temperature lead-free soldering processes has even increased the importance of this issue. Therefore, polymers with improved moisture resistance and reduced moisture uptake can be one key factor for miniaturization of low cost plastic packaging solutions with constant reliability compared to standard packages.

One possible particle type for humidity resistance enhancement of polymers in microelectronics application could be Bentonite. Bentonite is an aluminium phyllosilicate clay consisting mostly of montmorillonite. The layered structure of the bentonite particles can be dispersed in a polymer. The separated bentonite discs with diameters in the range of 200 – 300 nm and a thickness around 1 nm can act as diffusion path elongation in a polymer [[5], [6], [7], [8], [9]]. Also water interaction at the particle surface and water uptake of the bentonite itself may influence the diffusion behavior.

Figure 1 shows images from epoxy resins with 2,5 wt.-% content of bentonite I28E from Nanocor (left) and a composition of SiO_2 micro particles and I28E bentonite (right). Both images show well exfoliated and dispersed bentonite particles in a microelectronic grade epoxy resin.

Figure 1: TEM image from exfoliated bentonites 2,5 % (left) and SEM images SiO_2 micro-particles with bentonite

For direct measuring the diffusion of a nano-particle modified encapsulant as close as possible to a microelectronic application a glob top test vehicle with a bare die humidity sensor SHT 01 from Sensirion was designed. The sensor with a moisture depending capacitor structure on top was glued with a standard die attach adhesive on a substrate. The substrate was fully metallized on the edges and on the bottom to avoid moisture ingress through the substrate. A glob top dam material was used to encapsulate the wire bonds to realize a cavity for the nano-filled polymers. This was needed due to the expected high coefficient of thermal expansion and the high variations of viscosities of the materials applied. For the realization of a homogeneous layer above the sensor area a distance bar was used during epoxy curing. Samples were subjected to 85 °C/85 % r.h. atmosphere and signals of the sensor were collected periodically by a multiplexer.

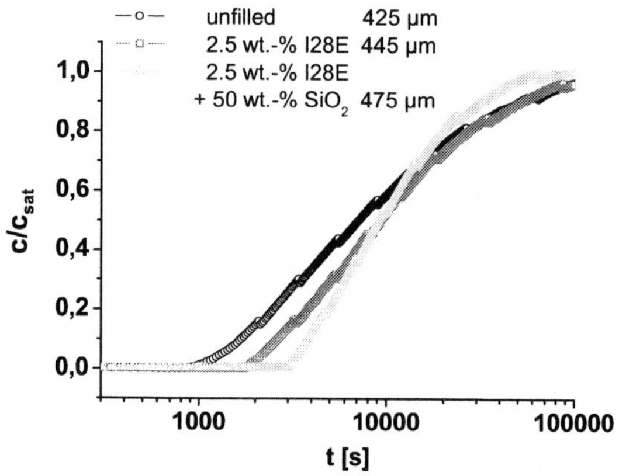

Figure 2: Moisture concentration (C) at sensor surface normalized to moisture saturation concentration (C_{sat}), layer thickness are mentioned for different material combinations

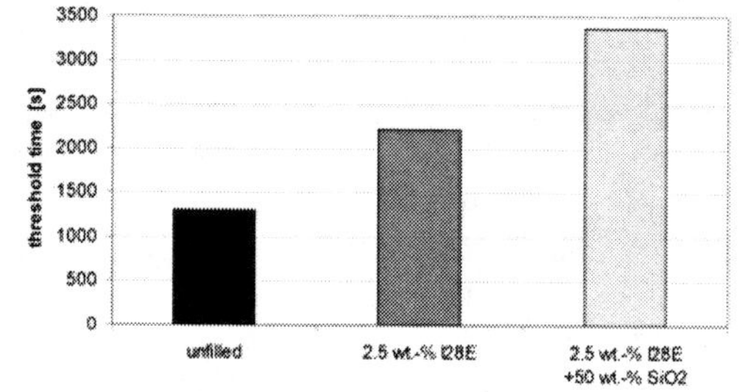

Figure 3: Threshold time when moisture concentration reaches 5 % at sensor surface

First results from diffusion measurements with glob top test vehicle are depicted in Figure 2. The collected moisture concentration data is normalized to the saturation value for all samples. Generally, different behaviour of the material compositions is visible. First the moisture concentration of the unfilled resin increases followed by the bentonite I28E and the combination of I28E with micro-sized SiO_2. In Figure 3 the threshold time values where the concentration has reached 5 % are summarized. Here, a significant improvement of the barrier functionality of the bentonite is visible. The threshold time is around 30 % higher than of the unfilled resin. Due to the high content of micro-sized SiO_2 in combination with the bentonite an additional increase occurs. A barrier enhancement could be therewith verified.

3 TOUCHLESS HANDLING

According to ITRS Roadmap for Assembly and Packaging, where experts identify the demands for system packaging for the next 14 years [10], the following specifications will be true for the next generations of System in Package. SiPs will contain both embedded passives & embedded actives; there will be up to 17 dies stacked or otherwise integrated into a package. From year 2010 the I/O count for high performance applications is estimated to be 3506 and will rise up to 5651 in year 2020, for low cost hand held the I/O number will be a stable 800 from year 2008, for RF packages it will be 200 I/Os. Small BGAs / CSPs in their various geometries will be the dominating package type, with Area Array pitches from 200 µm in 2006 down to 150 µm in 2013 and 100 µm in 2020. The components used will be passives of 600x300 µm² in 2008 down to 200x100 µm² in 2020; dies will be thinned down to 50 µm for substrate level assembly and down to 20 µm for wafer level assembly

Touchless and self-assembly based procedures seem to be the only method for handling miniaturized components, not directly fabricated at the very place, where they are needed. In the next section two promising concepts will be presented allowing touchless handlich of smallest components.

3.1 Magnetic handling

To adapt the principles of magnetic handling to processes useful for microelectronics purposes, research projects have been launched at Fraunhofer institutes IZM and IPK. Here focus is put on the examination of basic technological principles of contact-free positioning of small-sized components by use of magnetic fields. The approach is based on well defined interactions of controlled magnetic fields with magnetically addressable components. The situation of movement of such a magnetically interacting component under exposure of a magnetic field is illustrated in Figure 4.

Differing from other contact-free handling principles, magnetic manipulation not necessarily requires the presence of a coupling medium such as water and hence can be performed in dry state.

The crucial point of this approach is to find a suitable configuration of the magnetic field, which exerts such forces on the component needed for a movement in the wanted position. Besides the design of the magnetic field, e.g. in the form of a "magnetic funnel" (see Figure 4), the magnetic properties of the components are essential for the action of the field. Intense study is in progress to calculate the relationships between field strength and characteristics of the interaction partner such as magnetic permeability and magnetic saturation.

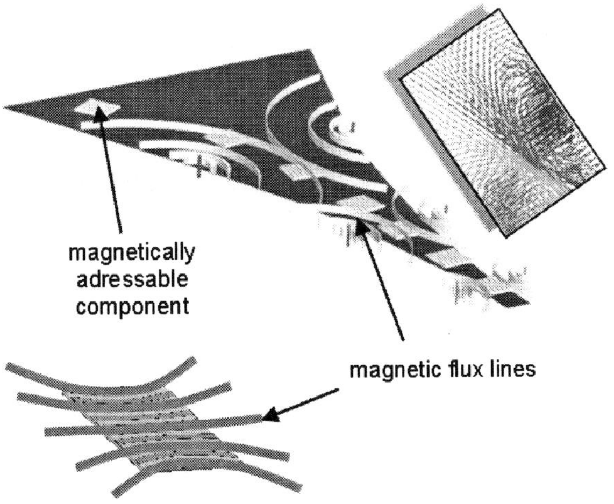

Figure 4: Positioning of a magnetically addressable component by extern magnetic fields

Microelectronic components typically undergo only inadequate interaction with magnetic fields so that magnetically based positioning procedures fail. To overcome this difficulty, magnetically addressable markers must be designed and attached to the components. An effective technical realization of this principle is the application of thin polymer based layers, filled with magnetically interacting micro- or nano-particles, on top of the components. The layers can be applied cost-effectively on wafer level during fabrication of the components by use of e.g. common stencil printing process; results are shown in Figure 5.

Figure 5: Stencil printed magnetically addressable markers obtained by stencil printing on a 4''-wafer; left – overview, right – height profile (maximum 40 µm)

The advantage of this approach is the possibility of generating such patterns of magnetic markers (e.g. lines, curves, dots), that are needed to transform the effect of the magnetic field into the wanted movement of the component. Actual works are focused on the calculation of effective designs of patterns and layer thicknesses of markers and on the development of polymer based magnetically interacting pastes, which can easily be integrated into the process of chip fabrication.

The vision of the complete technological chain of positioning fully disordered components by use of magnetic handling principles could then consist of the sub-processes:

- separation of the clustered components into single parts,
- pre-ordering of the components into tracks,
- rejection of components with wrong z-orientation,
- precise x-y-positioning by use of magnetic fields.

3.2 Microdroplet manipulation by electro-wetting

Based on the "electrowetting on dielectric" effect (EWOD) a contactless handling technology well known from lab-on-chip applications for liquid transport, sorting, mixing and splitting will be used as a basis for microelectronics assembly purposes. Handling shall be feasible for miniaturized components as chiplets, smallest SMDs as well as for nano-scaled building blocs, this is currently investigated by Technical University of Berlin in cooperation with TU Freiburg spin-off IMTEK [11]. Physical principle is a change in droplet contact angle when immersed into an electrical field. By applying a moving e-field to the droplet, it can be guided to a defined spot. Using this effect in combination with conventional circuit board technologies might yield a moderate cost approach to exactly place fluid droplets. The use of component laden droplets might be used for contact-less handling of chiplets or all sorts of nano-components (see Figure 6).

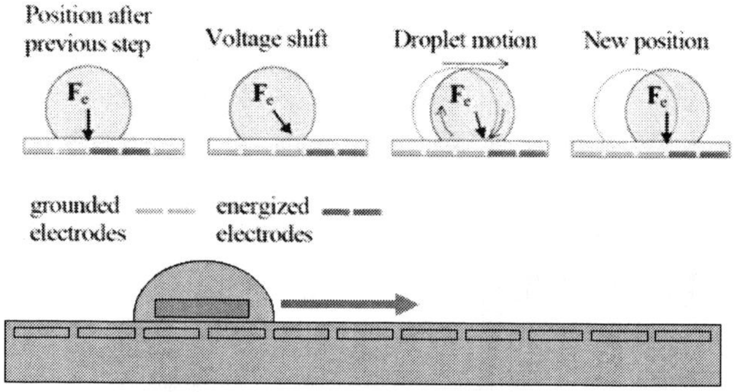

Figure 6: Component transport by electrowetting

The process flow under evaluation starts with positioning of a droplet, containing a component, on a hydrophobic surface of the carrier substrate with rough accuracy takes place (compare Figure 7). Using the mentioned electro wetting effect the droplet will be fast moved until the desired position is reached. The precise placement of the droplet in µm range takes place by means of field gradients and local manipulation of the carrier surface. The assembly is finished with the evaporation of the component containing droplets and the transfer of all components to the final substrate.

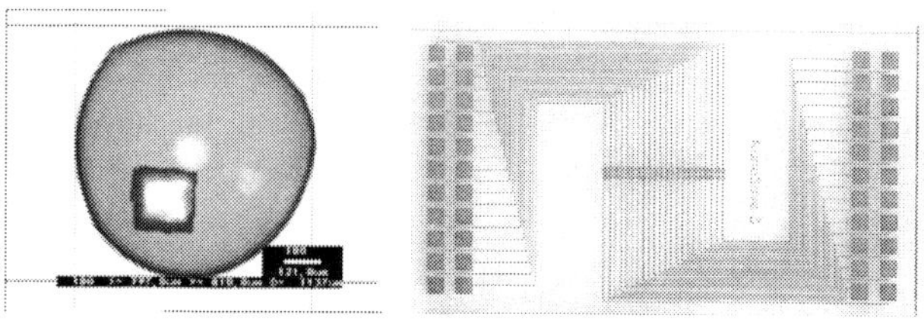

Figure 7: Component in microdroplet (left), electrowetting test vehicle (right)

4 ADVANCED INTERCONNECTS

Interconnection technologies for microelectronics packaging are currently generated at elevated temperatures, either by soldering, adhesive joining or wire bonding. Typical temperatures for adhesive joining are > 120 °C, for soldering > 250 °C and for Au wire bonding > 120 °C. These temperatures are considered too high for the interconnection of temperature sensitive components (e.g. bio chips with degradation temperatures ~ 42 °C) or low cost substrates with glass transition temperatures well below 100 °C. There are some potential solutions for this challenge in the market, e.g. Al wire bonding at RT or adhesive joining using low temperature cure isotropic conductive adhesives with cure temperatures down to 80 °C. However, there are limits to the scalability of this processes, so by further reducing interconnect dimensions a technological limit will be reached. The use of nanoscale functionalized surfaces for interconnect purposes is seen as a promising means to drive interconnection technology towards smallest geometries, in international research a variety of approaches exist (see Figure 8).

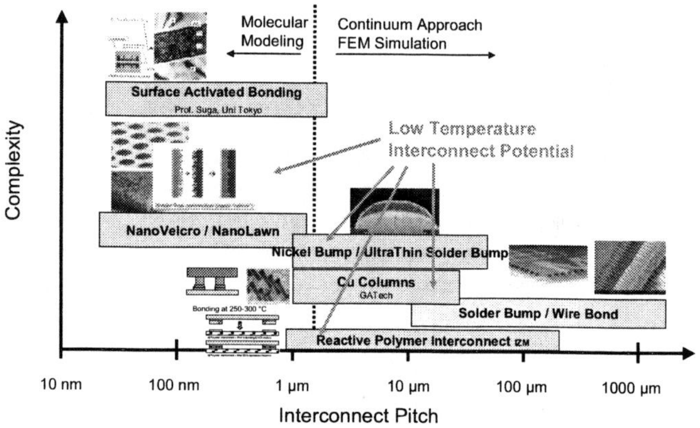

Figure 8: Interconnection Technologies development from micro to nano scale

The use of nanoscale modified surfaces for realization of interconnects in electronics packaging is discussed internationally since several years, but has not yet found their way into industrial applications so far. A mutual approach of Fraunhofer IZM and TU Berlin in Cooperation with the US University of South Florida [UoSF] led to first joining experiments with CNT-covered pads, where weak mechanical and electrical interconnects were created [12].

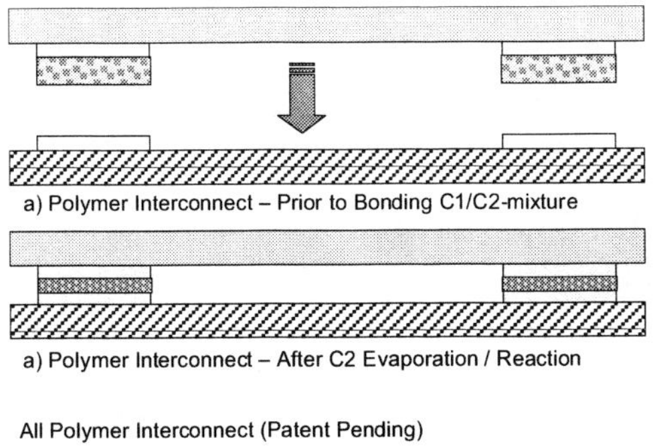

Figure 9: Schematic of all polymer interconnection technology

Additional focus of IZM is put on the realization of low temperature interconnects. IZM is evaluating advanced polymer processing technologies to create polymer based interconnects, that allow an integration of temperature sensitive devices into systems manufactured on wafer level. Technology basis is a polymer bumping technology incorporating conductive and non-conductive structures for interconnects formation at room temperature – the concept is depicted in Figure 9. (Patent Pending) This unique approach allows room temperature joining of electronics components to a silicon base, preventing mechanical stresses and thermal degradation of the structures embedded.

5 CONCLUSION & OUTLOOK

The continuous miniaturization in microelectronics and the development of new functional nanoscaled components, useful for information processing and microelectronics, require the integration of new materials and processes. In this paper the potential of bentonite nano-particle enhanced encapsulants with higher humidity resistance was presented. Touchless and self-assembly based procedures seem to be one method for handling miniaturized components in the future. Two promising processes in this area magneting handling and electrowetting were discussed. Finally, future concepts for low cost and low temperature assembly processes have been introduced. We see this work as building blocks of packaging technology for miniaturized systems developing from a micro to nano scale.

ACKNOWLEDGEMENT

Part of this work has been funded by BMBF (Germany Federal Ministry of Education and Research) via VDI/VDE-IT.

REFERENCES

[1] M. Xanthos (Ed.); Functional Fillers for Plastics; 2005 Wiley-VCH Verlag GmbH & Co. KGaA, Weinheim.

[2] K.-F. Becker, N. Kilic, T. Braun, M. Koch, V. Bader, R. Aschenbrenner, H. Reichl; New Insights in Underfill Flow and Flip Chip Reliability; Proc. Apex 2003; 29.03.-02.04.03, Anaheim, Ca., USA.

[3] T. Braun, K.-F. Becker, J.-P. Sommer, T. Löher, K. Schottenloher, R. Kohl, R. Pufall, V. Bader, M. Koch, R. Aschenbrenner, H. Reichl; High Temperature Potential of Flip Chip Assemblies for Automotive Applications; Proc. of ECTC 2005, Orlando, Fl., USA.

[4] T. Braun, K.-F. Becker, M. Koch, V. Bader, R. Aschenbrenner, H. Reichl; Reliability Potential Of Epoxy Based Encapsulants For Automotive Applications; Proc. of ESREF 2005, Arcachon, F.

[5] O. Becker, G. P. Simon; "Epoxy Layered Silicate Nanocomposites", Adv. Polymer Sci. 2005, 179, 29-82.

[6] A. Sorrentitino, M. Tortora, V. Vittoria; Diffusion Behavior in Polymer-Clay Nanocomposites; Journal of Polymer Science: Part B: Polymer Physics, Vol. 44, 265 – 274, 2006.

[7] H. T. Rana, R. K. Gupta, H. V. S. GangaRao, L. N. Sridhar; Measurement of Moisture Diffusivity through Layered-Silicate Nanocomposites; Materials, Interfaces, and Electrochemical Phenomena; Vol. 51, No. 12; December 2005.

[8] O. Becker, R. J. Varley, G. P. Simon; Thermal stability and water uptake of high performance epoxy layered silicate nanocomposites; European Polymer Journal 40 (2004), 187 – 195.

[9] S. Burla; Barrier Properties of Polymer Nanocomposites during Cyclic Sorption-Desorption and Stress-Coupled Sorption Experiments; PhD College of Engineering and Mineral Resources at West Virginia University; 2006.

[10] International Technology Roadmap for Semiconductors: 2006 Update on Assembly and Packaging

[11] U. Herberth; Fluid manipulation by Means of Electrowetting-On-Dielectrics; Dissertation Albert-Ludwigs-Universität Freiburg im Breisgau, Mai 2006.

[12] K.-F. Becker, T. Löher, B. Pahl, O. Wittler, R. Jordan, J. Bauer, R. Aschenbrenner, H. Reichl; Development of a Scaleable Interconnection Technology for Nano Packaging; Proc. of Nanotech 2005, May 9.-12., Anaheim, Ca., USA.

Contact-less handling of metal sub-micron and nanowires for microelectronic packaging applications

S. Fiedler*, M. Zwanzig*, M. Boettcher**, M. S. Jaeger**, G. R. Fuhr**, and H. Reichl*

*Fraunhofer Institute Reliability and Microintegration (IZM), G.-Meyer-Allee 25, 13355 Berlin, Germany, stefan.fiedler@izm.fraunhofer.de
**Fraunhofer Institute Biomedical Engineering (IBMT), Potsdam, Germany, magnus.jaeger@ibmt.fraunhofer.de

ABSTRACT

Individual sub-micron particles produced by non-lithographic techniques are promising for future microelectronic applications. Being too small to be handled by traditional Pick & Place, they require a contact-less manipulation. We propose a combination of dielectrophoretic and acoustic trapping for the separation and oriented positioning of metal and semiconductor wires on microelectronic substrates. Chiplets, rods, and spheres of different materials can be manipulated in microfluidic channels by dielectrophoresis and ultrasonic standing waves, covering object sizes from the micro- to the nanoscale. Technical principles developed for the biotechnological manipulation of single live cells and other microparticles are adapted to the integration of miniaturized components into a microelectronic periphery. Developing these handling techniques for tiny and delicate individual components in microelectronic packaging ultimately requires the use of carrier liquids that facilitate employing self-assembly strategies.

Keywords: self assembly, dielectrophoresis, acoustic trapping, field cage, particle sorting

1 MOTIVATION

In current microengineering, continuously shrinking sizes of functional elements open up further enhancement of functional density. At the same time, however, they face traditional techniques for microelectronic packaging or MEMS applications. The well-established Pick & Place approaches are limited to a certain size scale. Even with classic SMD, sequential error-free placement of chip components requires sophisticated automata and, hence, rising costs and efforts. In view of the stringent future goals outlined in the ITRS roadmap, the only possible solution lies in the search for alternative and completely new approaches in future component handling, i. e. advanced packaging. Besides miniaturization, the availability of promising materials applicable to information processing diversifies. In order of decreasing organizational level these comprise: neuronal cellular networks; single living cells, like microorganisms, animal cells or valuable plant cells; isolated organelles (e. g. chloroplasts, forisomes [1]); functional biomacromolecular assemblies (e. g. two-dimensional bacteriorhodopsin lipid crystals, i. e. purple membrane [2]), s-layer proteins [3], and different substance-specific channel-forming proteins [4, 5]; molecular motor proteins or gene-engineered variants of the same, and so forth. The inorganic class of materials is made up of different types of micro- and nanotubes, synthetic organic and inorganic particles. Multilayer, i. e. core shell, nanoparticles are thought to become important for future coupling between microelectronics and photonics, exploring evanescent fields generated at resonator-entrapped NPs [6] or even atoms [7]. Not only semiconductor nanowires but also metal nanowires are considered essential elements in future microelectronic packaging [8]. However different these functional materials might be, they all have in common to be not touchable. Therefore, techniques have to be developed that prevent mechanical damage during handling or processing. Another crucial influence on NPs to be avoided is the adsorption of different contaminants, passivating layers, or other unwanted substances from the surrounding atmosphere or carrier liquid. The presence of liquid carrier and shielding media additionally helps to prevent unwanted contamination through electrostatic attraction (ESA), as it is observed in microfabricated silicon microparts.

Where are suitable manipulation techniques to be found? In the biosciences, especially in medical and biotechnological single-cell technologies, a whole range of methods has been developed over the last decades for research and even commercial single-cell applications. Manipulated objects include microorganisms, plant or animal cells [9]. Often, a combination of various physical principles and microfluidics is sufficient to deal with the delicate objects. Such objects are: gene-engineered microorganisms, living single cells, artificially generated cell hybrids, stem cells, progenitor cells, or even somatic stem cells. These cells are either isolated from diluted samples or from complex fluids (e. g. blood or biofermenter broth). These micron-scaled objects are always suspended in a liquid medium which shields against the surroundings and fulfills the function of a nutrient and a buffer. Individual cells are the object of single-cell-based branches in biotechnology, food industry, cellular medicine and comparable fields [10, 11]. Strangely enough, intense related R & D activities remained almost unnoticed by classic microengineering. Since semiconductor technologies have always been open for new materials and inspirations from different disciplines, the observed technological convergence at the sub-micron and nanoscale will doubtlessly facilitate progress in all engaged fields and disciplines.

2 PHYSICAL FORCES AND MANIPULATION PRINCIPLES

What are the general requirements for the massively parallel component processing and assembly necessary in any modern high-throughput screening (HTPS) or future high-throughput assembly (HTPA) scenario? Objects, i. e. components, have to be isolated, sorted, aligned, vectorially oriented, moved, and placed. Technical components (future microelectronic components) have to be joined at their final location for being connected electrically or for being optically addressable. Biotechnical components at the end of the virtual conveyor are placed to give clonal growth (cells, microorganisms) or to be brought into contact with others (cells, macromolecules, or biochemically functionalized microparticles, polymer beads, or NPs). This is a striking technological convergence from a process engineer's point of view.

Like in real macrolife, gradients are the strongest cause for movement in micromanipulation. Since being well-understood, electromagnetic fields are the most important gradient sources in the micro-, sub-micro- and nanoworld. A simplified overview of available methods is summarized in the Table 1.

Force	Principles	Application variants
E-Field	Electrophoresis	Free Solution / Field Flow Fractionation Capillary electrophoresis
	Dielectrophoresis	Negative dielectrophoresis (nDEP) Positive dielectrophoresis (pDEP) Travelling wave DEP Travelling wave pumping
M-Field	Magnetophoresis	Mechanically driven, sliding magnets Electrically driven ferrofluid actors Magnetic tweezers
Optical pressure & Energy absorption	LASER	Laser tweezer / Photonic force microscopy Laser dissection
Acoustic pressure & Hydrodyn. Streaming	Interparticle interaction Acoustic energy	Different types of field flow fractionation (FFF) Acoust. Trapping / Ultrasonic standing waves

Table 1: Physical forces used for contact-free particle handling.

Not all the mentioned principles are helpful in the case of a suspension. It is important to notice that especially magnetic force fields and some variants of dielectrophoresis have to be combined with stringent space restrictions. Usually microfluidic channel networks or capillaries are applied to facilitate directed movement along a given track. The necessary guiding structures can also be provided by "wall-less tubes", i. e. suitably designed electrodes addressed by negative dielectrophoresis. Basic elements of dielectrophoretic guiding and sorting in liquid channel networks have been described earlier: particle enrichment by funnels, particle alignment by longitudinal quadrupole cages, particle trapping and holding inside closed field cages, and particle sorting by temporarily deflecting field barriers [12, 13].

Laser tweezers have also been combined with dielectrophoretic trapping to assemble particles and to measure bonding forces between them [14, 15]. Laser light has been used to engrave particles trapped contact-less in a three-dimensional cage [16] or to fix particle assemblies by initiating polymerization [17]. Some other technological principles important in the discussed context but beyond the table above entered pharmaceutical HTPS, combinatorial chemical analysis [18, 19], cell banking, and cryo-preservation. Prominent examples are:

- Segmented flow techniques based on structured mixtures of immiscible liquids [20];
- Picoliter spotting;
- Microcontact printing for the production of (bio-)chemically patterned substrates;
- Multi(-well) array and panel techniques.

The practically pre-dominant case in future microelectronics packaging will be heterogeneous integration, i. e. the assembly into functional units of discrete tiny components fabricated by different techniques and under different conditions. Beside the above-mentioned bioapplied techniques, self-organization phenomena can be explored. These size scale-independent entropy-driven processes came into the focus of modern microfabrication through artificial structures that occurred seemingly out of nowhere, like patterned films and interfaces [21]. The interest was strengthened by the high resolution power of different commercially available scanning probe microscopy techniques (AFM, AFAM, SEM, KPM) and sophisticated combinations thereof (e. g. crossbeam FIB-REM).

We aim at indicating prospective directions of future work. Our examples may vary in the development stage but are all transferable to broader microtechnical applications in the foreseeable future with a calculable effort. They show a concept for the contact-less manipulation over the important size scales. This concept of converging assembly strategies is developed based on the microtechnology toolbox for the need of both life sciences and future micro- and nanoelectronics.

3 DIELECTROPHORESIS AND ULTRASOUND FOR MANIPULATION AND SORTING OF SUB-MICRON PARTICLES

The dielectrophoretic particle manipulation in liquid carrier media offers wide possibilities for contact-less collection, assembly, sorting, and processing. This was previously shown for living entities, like mammalian cells, algae, pollen grains, yeast, bacteria, and even viruses [22]. By careful tuning of electric conductivity, dielectric constants, electric field (frequency and amplitude), and electrode shape, different tiny objects can be spatially confined, rotated, lifted, and assembled into sophisticated structures [17].

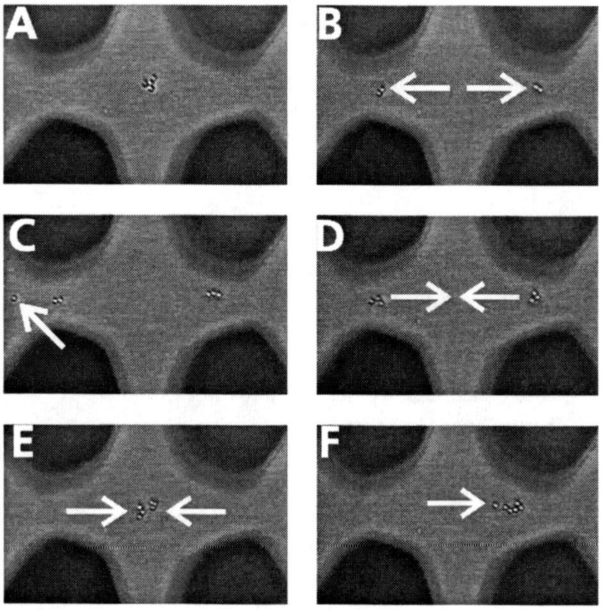

Figure 1: Contact-less assembly of nanoparticle aggregates through dielectrophoretic manipulation. (a) An ensemble of five polymer beads (r = 500 nm) is kept between eight microelectrodes (red) and (b) then separated into two groups by specific electric conditions. (c) Another particle is added from the left. (d) Both groups are joined into one (e) which is then positioned on one side (f).

Both negative dielectrophoresis, i. e. the manipulation at electric field minima, and positive dielectrophoresis have been applied to induce directed particle movement. In a recent research project, we are investigating possibilities to orient and to align metal sub-micron and nanowires (NW). These wires have to be handled with great care, since they are susceptible to mechanical damage which easily causes crystal defects, e. g. the prominent sliding along the <111> crystal plane in fcc metals like Au and Pd or twin formation. In principle, new contact-less assembly techniques must be developed to make the exciting properties of nanowires accessible to future microelectronics.

Nanowires can be produced non-disruptively at the very places where they are to be used. The well-described step edge deposition technique (STD) represents the technologically most conform approach. Since, however, only horizontally oriented edges on the wafer substrate are accessible, generated wires are oriented in plane as well. The carrier substrates encounter high temperatures due to CVD and PECVD deposition conditions. Therefore, we consider individually and externally generated single-crystalline metal wires an alternative for future NW applications. The ballistic electron transport (i. e. conductance) is improved along certain crystal planes and deflected at grain boundaries. Therefore, the electrical conductivity of NWs will be crucially influenced by the electrons' mean free path and, hence, by crystal defects [23]. Future electronic interconnects will require single-crystalline wire interconnects. Such predictable reliability-diminishing effects of electronic devices, like electromigration, might be reduced in the approach we follow.

Acoustic trapping by ultrasonic standing waves are another promising technique for microparticle handling. Different research groups are investigating methods to apply acoustic trapping to biotechnological applications. In essence, standing waves are generated at megahertz frequencies inside fluidic channels. At these frequencies, no cavitation and, hence, no damage to fragile objects occurs. Particles suspended in the coupling medium will be trapped and enriched in nodes and can be moved hydrodynamically in nodal planes or lines. Trapped objects are also accessible to contact-less manipulation through controlling the number, shape, and spatial extension of nodes by the coupling acoustic field and the resonance pattern. The combination of hydrodynamic streaming and dielectrophoresis together with ultrasound has already been shown to be suited for the handling of polymer beads [24]. We aim at extending this approach to the contact-free handling of inorganic particles important for future microelectronic packaging. During the already finished first evaluation phase of a related research project, the basic concept has been approved.

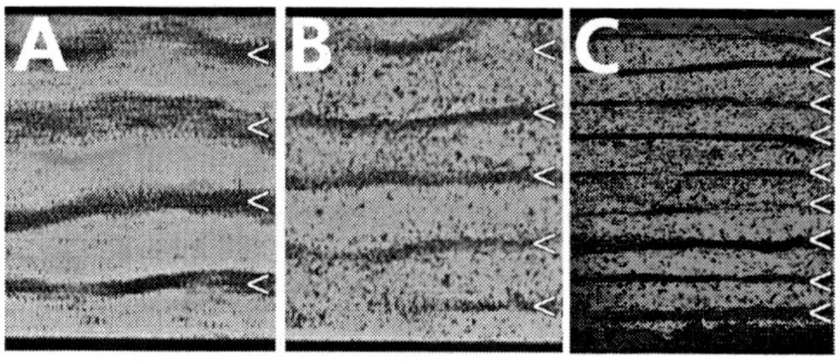

Figure 2: Ultrasonic standing wave trapping of spherical polymer beads (arrowheads) inside a microfluidic channel at increasing frequencies (a to c).

Basic manipulation principles, like aligning, sorting, and assembling are accessible. Their combination must be investigated in more detail in order to finally succeed in the assembly of fully functional nanodevices.

REFERENCES

[1] M. Knoblauch, G. A. Noll, T. Mueller, D. Pruefer, I. Schneider-Huether, D. Scharner, A. J. E. van Bel and W. S. Peters, Nature Materials 2, 600-603, 2003.

[2] N. Hampp, Chem. Rev. 100(5), 1755-1776, 2000.

[3] U. B. Sleytr, M. Sára, D. Pum and B. Schuster, Progress Surf. Sci. 68(7-8), 231-278, 2001.
[4] S. Hutschenreiter, L. Neumann, U. Raedler, L. Schmitt and R. Tampé, ChemBioChem 4(12), 1340-1344, 2003.
[5] E. Tajkhorshid, J. Cohen, A. Aksimentiev, M. Sotomayor and K. Schulten, in: Kubalski and Martinac, "Bacterial Ion Channels and Their Eukaryotic Homologs," ASM Press, Chapter 9, 2005.
[6] M. W. Knight, N. K. Grady, R. Bardhan, F. Hao, P. Nordlander and N. J. Halas, Nano Lett. 7(8), 2346-2350, 2007.
[7] B. Dayan, A. S. Parkins, T. Aoki, E. P. Ostby, K. J. Vahala and H. J. Kimble, Science 319, 1062-1065, 2008.
[8] S. Fiedler, M. Zwanzig, R. Schmidt and W. Scheel, in: Morris and Mallik, "Nanopackaging. Nanotechnologies and Electronics Packaging," Springer, Chptr. 20, 2008.
[9] B. van Duijn, "Signal Transduction - Single Cell Techniques," Springer, 1998.
[10] J. Kehr, Curr. Op. Plant Biol. 6(6), 617-621, 2003.
[11] J. M. Levsky, S. M. Shenoy, R. C. Pezo and R. H. Singer, Science 297, 836-840, 2002.
[12] G. Fuhr, T. Mueller, T. Schnelle, R. Hagedorn, A. Voigt, S. Fiedler, W. M. Arnold, U. Zimmermann, B. Wagner and A. Heuberger, Naturwissenschaften 81, 528-535, 1994.
[13] S. Fiedler, S. G. Shirley, T. Schnelle and G. Fuhr, Anal. Chem. 70(9), 1909-1915, 1998.
[14] C. Reichle, K. Sparbier, T. Mueller, T. Schnelle, P. Walden and G. Fuhr, Electrophoresis 22, 272-282, 2001.
[15] C. Reichle, T. Schnelle, T. Mueller, T. Leya, G. Fuhr, BBA 1459, 218-229, 2000.
[16] G. Fuhr, C. Reichle, T. Mueller, K. Kahlke, K. Schuetze and M. Stuke, Appl. Phys. A 69, 611-616, 1999.
[17] S. Fiedler, T. Schnelle, B. Wagner and G. Fuhr, Microsystem Technologies 2, 1-7, 1995.
[18] E. Reddington, A. Sapienza, B. Gurau, R. Viswanathan, S. Sarangapani, E. S. Smotkin and T. E. Mallouk, Science 280, 1735-1737, 1998.
[19] M. T. Reetz, M. H. Becker, H.-W. Klein and D. Stoeckigt, Angew. Chem. Int. Ed. 38(12), 1758-1761, 1999.
[20] J. M. Koehler, T. Henkel, A. Grodrian, T. Kirner, M. Roth, K. Martin and J. Metze, Chem. Eng. J 101(1-3), 201-216, 2004.
[21] G. M. Whitesides and M. Boncheva, PNAS 99(8), 4769–4774, 2002.
[22] T. Mueller, A. Pfennig, P. Klein, G. Gradl, M. S. Jaeger, T. Schnelle, IEEE Eng. Med. Biol. 22, 51-61, 2003.
[23] M. E. Toimil Molares, V. Buschmann, D. Dobrev, R. Neumann, R. Scholz, I. U. Schuchert and J. Vetter, Adv. Mater. 13(1), 62-65, 2001.
[24] M. Wiklund, C. Guenther, R. Lemor, M. S. Jaeger, G. Fuhr and H. M. Hertz, Lab Chip 6, 1537-1544, 2006.

MEMS Pressure Sensor Array for Aeroacoustic Analysis of the Turbulent Boundary Layer

J. Krause*, R. White*, M. Moeller**, J. Gallman** and R. De Jong***

*Tufts University
200 College Ave, Medford, MA — r.white@tufts.edu
**Spirit AeroSystems, Wichita, KS
***Calvin College, Grand Rapids, MI

ABSTRACT

The design, fabrication, and characterization of a surface micromachined, front-vented, 64 channel (8×8), capacitively sensed pressure sensor array for aeroacoustic testing of the turbulent boundary layer (TBL) is discussed. The array was fabricated using the MEMSCAP polyMUMPs® process, a three layer polysilicon surface micromachining process. An acoustic lumped element circuit model was used to model the system. The results of our computations for the design, including mechanical components, environmental loading, fluid damping, and other acoustic elements are detailed. Theory predicts single element sensitivity of 0.6 mV/Pa at the bandpass output (6 μV/Pa at the preamp output) in the 400-40,000 Hz band. The size of the vent holes has a major impact on low frequency modeled acoustic sensitivity. Laser Doppler Velocimetry shows low frequency electrostatic stiffness and a primary resonant frequency that agree well with the models. Experimental verification of acoustic sensitivity is ongoing.

Keywords: aeroMEMS, microphone array, turbulent boundary layer, polyMUMPs

1 INTRODUCTION

Turbulence has been plaguing transport aircraft designers for over fifty years. A clear and concise definition of turbulence has yet to be made, but Tennekes and Lumley pose seven qualities that characterize turbulence. They present turbulence as being irregular, diffuse, and often associated with large Reynolds numbers. It is a three-dimensional vortical fluctuation following a continuum model and dissipates over time [1]. Several models have been analytically and experimentally obtained to understand the complex nature of turbulence, but as a result of the stochastic nature, a theoretical model is more difficult to obtain. Therefore, using hot wire anemometry, shear stress sensors, and pressure sensors at the microscopic level will help to obtain empirical results describing the phenomena associated with turbulence and more importantly the TBL.

The sources of structural excitation and radiative noise in passenger aircrafts are noise due to the interior environment, the engine, and the fluctuations in wall pressure beneath the TBL. The noise generated by the TBL is considered the most dominant noise source on transport aircrafts [2]. In order to model the structural response of an aircraft, spectral levels at both low and high wavenumbers are needed [3]. The low wavenumber assessment is vital due to the fact that structural resonances take place at low wavenumbers and acoustic noise is generally emitted at low wavenumbers compared to convective turbulent energy [4]. Although low wavenumbers are important for the analysis of acoustic noise generation and structural vibrations, the high convective wavenumbers are where the greatest energy levels are present in the turbulent field, and hence need to be understood. A lack of empirical knowledge as a result of the limits due to conventional instrumentation is one reason for our poor understanding of turbulence [5]. MEMS pressure sensors may alleviate this issue due to their small size and the ability to fabricate multiple microphones in an array. The challenge in MEMS arrays is achieving good matching between elements in the array and across arrays. In addition, due to their small size, the microphones necessarily have low sensitivity.

MEMS pressure sensors have been explored by many researchers over the past 25 years and many review articles can be found on them [6, 7, 8]. Most pressure sensors are developed for auditory applications, biomedical ultrasound arrays, and underwater applications [7]. Few microphones have been developed for aeroacoustic applications, possibly due to the difficulty of surviving the harsh environment. The Interdisciplinary Microsystems Group at the University of Florida Gainesville has done a great deal of work in this area and Martin *et al.* demonstrate a good summary of the previous microphones for aeroacoustic measurement [9].

2 FABRICATION

The fabrication process of the 64 channel capacitive microphone array utilizing the MEMSCAP PolyMUMPs® process along with facilities at Tufts University and Massachusetts Institute of Technology is described. The polyMUMPs process is a foundry process that creates polysilicon structures via surface micromachining with a minimum feature size of 2 μm. The process consists of seven physical layers, including 3 structural, 2 sac-

NSTI-Nanotech 2008, www.nsti.org, ISBN 978-1-4200-8505-1 Vol. 3

rificial and one metal layer. For more information on the PolyMUMPs process, see [10]. *Figure* 1 shows two cross-sections of a single element in the microphone array. The structural diaphragm (top electrode) is made of phosphorus doped polysilicon with a 600 μm diameter and thickness of 3.5 μm. The bottom electrode is 580 μm in diameter, 0.5 μm thick phosphorus doped polysilicon layer that serves as a bottom electrode. The air gap between the diaphragm and the bottom electrode is 2 μm. To prevent stiction during fabrication and while in operation, the bottom of the diaphragm contains 201 dimples, spaced on a 30 μm pitch, with a depth of 1.0 μm. There is also a small corrugation close to the edges of the diaphragm to help relax any tensile residual stresses due to fabrication or deflection.

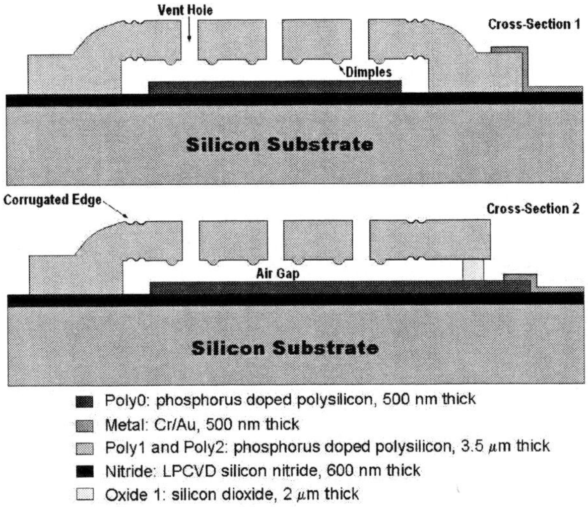

Figure 1: Two cross sections of a single microphone.

There are vent holes through the diaphragm that serve two purposes: (1) introduction of HF etchant to remove the oxide sacrificial layer to create the 2 μm high air gap and (2) act as vent holes to front vent the microphone and allow operation with changing atmospheric pressure. There are 28 holes in the diaphragm with a center-to-center spacing of 100 μm. The vent holes are created using two etches. One etch is a 6 μm diameter hole, while the other is a 4 μm diameter hole. The holes are designed to be concentric and should result in a 4 μm diameter hole through the diaphragm. As manufactured, alignment tolerances result in non-concentric holes, creating irregular shaped holes with a larger area than desired (see *Figure* 2 for an SEM image of this).

The elements are arrayed on a 1 cm × 1 cm chip in an 8×8 pattern. There are 76 bond pads along the edges of the chip for electrical connection. The direction of flow is top to bottom so the flow does not pass across the bond pads. The element center-to-center pitch in the direction of flow is 1.2625 mm (which allows for multiple 8×8 array to be placed end-to-end to determine low wavenumber information through the larger spatial scale), while the pitch across the flow is 1.1125 mm. Packaging uses a pin grid array package to which the MEMS array is wirebonded. Laser cut spacers allow for the MEMS chip to be mounted flush with the package surface.

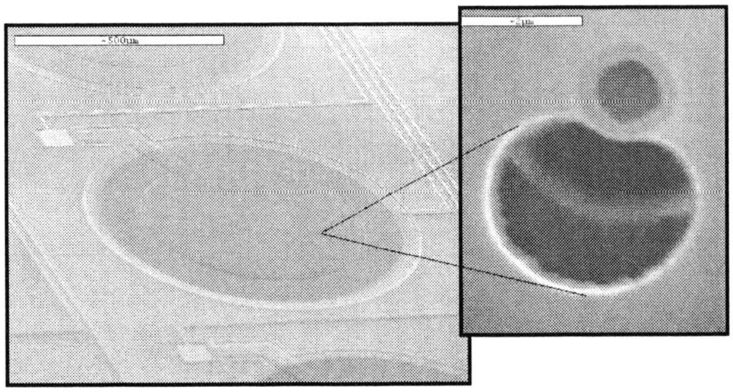

Figure 2: SEM image tilted at a sixty degree angle of an element illustrating the corrugation, wire scheme, vent hole and tunnel concept for static equilibrium of pressure. Diaphragm is 600 μm and vent hole is ≈4 μm.

3 MODELING AND DESIGN

A model for one individual microphone in the array is described. For each element in our design, a MATLAB® script was compiled to examine the response electrostatically as well as to a unit pressure. The parameters of the script were computed following an acoustic lumped element circuit diagram shown in *Figure* 3. The compliance, resistance and mass of the microphone were accounted for in the circuit diagram and then implemented into the MATLAB script. The compliances, resistances and mass loading of the microphones were computed using parameters from [9, 11, 12]. Using Beranek's solutions for environmental loading of the air we compute:

$$R_{A1} = \frac{0.1404\rho c}{a^2} \tag{1}$$

$$R_{A2} = \frac{\rho c}{\pi a^2} \tag{2}$$

$$M_{A1} = \frac{8\rho}{3\pi^2 a} \tag{3}$$

$$C_{A1} = \frac{5.94a^3}{\rho c^2} \tag{4}$$

$$C_{cav} = \frac{V_{gap}}{\rho c^2} \tag{5}$$

where ρ is the density of air, c is the speed of sound, a is the effective radius of the diaphragm (equal to 80% of the actual radius for a circular bending plate), and V_{gap} is the volume of the gap between the diaphragm

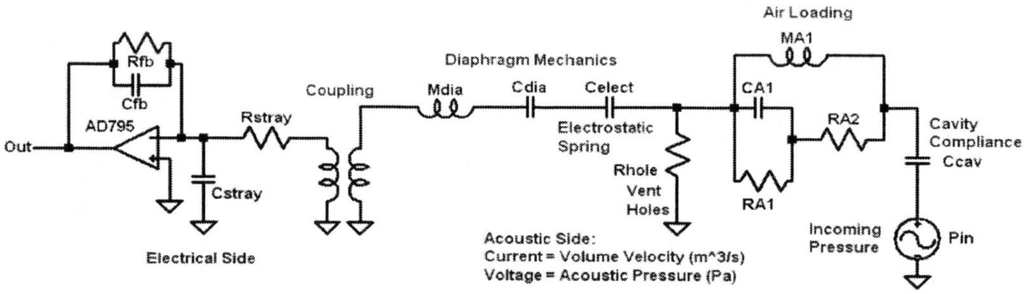

Figure 3: Coupled mechanical-electrical lumped element model.

and bottom electrode. From Martin *et al.* we compute resistance due to the holes in the diaphragm, the compliance of the diaphragm (for a clamped circular bending plate), and the effective mass of the diaphragm (for the first mode of the clamped circular bending plate):

$$R_{through} = \frac{72\mu t_{dia}}{n\pi a_{hole}^4} \tag{6}$$

$$C_{dia} = \frac{\pi a^6(1-\nu^2)}{16Et^3} \tag{7}$$

$$M_{dia} = \frac{9\rho t_{dia}}{5\pi a^2} \tag{8}$$

where μ is the viscosity of air, t_{dia} is the thickness of the diaphragm, n is the number of holes in the diaphragm, a_{hole} is the radius of the holes in the diaphragm, ν is Poisson's ratio, and E is the elastic modulus of the diaphragm. Using Škvor's formula, (S), and calculating a correction factor, (C_f) we can determine the resistance due to the squeeze film damping $(R_{squeeze})$ [9, 13].

$$S = \frac{\pi a_{hole}^2}{C^2} \tag{9}$$

$$C_f = \frac{S}{2} - \frac{S^2}{8 - \frac{1}{4}ln(S) - \frac{3}{8}} \tag{10}$$

$$R_{squeeze} = \frac{12\mu C_f}{n\pi t_{gap}^3} \tag{11}$$

The hole resistance in the circuit model is the series combination of the squeeze film damping, $R_{squeeze}$ and the through-hole damping, $R_{through}$,

$$R_{hole} = R_{squeeze} + R_{through} \tag{12}$$

where C is the center-to-center spacing of holes in the diaphragm. Using the above model for the microphone and using a coupling parameter, N, to relate the pressure to a voltage:

$$N = \frac{V_{bias}\epsilon}{t_{gap}^2} \tag{13}$$

where V_{bias} is the bias voltage applied to the bottom electrode, ϵ is the permittivity of free space, and t_{gap} is the height of the air gap. This coupling parameter gives the acoustic pressure applied to the diaphragm for a given AC voltage on the electrical side, and, equivalently, the current into the electrical side in response to a given volume velocity of the diaphragm.

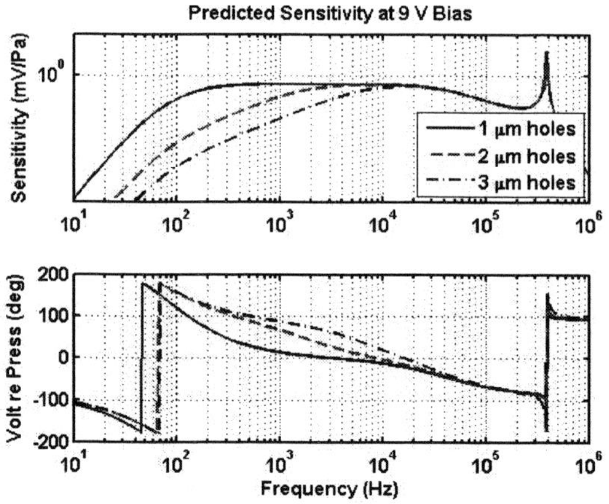

Figure 4: Predicted acoustic sensitivity for a single element with 9 V_{bias}, showing the importance of the vent hole size.

$$P_{electrostatic} = N \cdot V_{ac} \tag{14}$$

$$I = N \cdot U_{dia} \tag{15}$$

The sensitivity (voltage out per Pascal) can be computed as a function of frequency by incorporating the electronics which give the response curve its shape. The model for the receive electronics is a series combination of two single pole passive high pass filters with break frequencies of 60 Hz and 80 Hz, a charge amp with a gain of 100 mV/pC, and a voltage gain stage of 100 with a single pole low pass filter at 40 kHz. The final predicted pressure sensitivity results are shown in *Figure* 4. This is sensitivity at the bandpass output (40 dB above the preamp output in the passband). Varying the size of the vent holes has a major impact on the low frequency response.

4 RESULTS

Laser Doppler velocimetry (LDV) is used to measure the centerpoint vibration of the diaphragm in response to an applied AC voltage plus DC bias. The

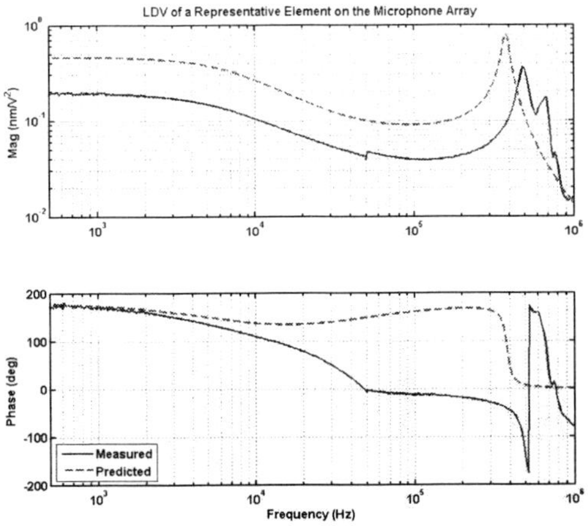

Figure 5: Laser Doppler velocimetry (LDV) measurements as a result of an electrostatic excitation to the microphone.

results of the measurement show a strong, high Q resonance at 480 kHz. The frequency of the resonance is strongly influenced not only by the bending stiffness of the diaphragm, but also by the acoustic stiffness coming from the backing cavity and the environmental acoustic impedance. *Figure* 5 shows a comparison between the measured electrostatic frequency response and the model predictions. The model does a good job of predicting the primary resonance frequency and the shape of the low frequency magnitude curve.

Preliminary tests using prototype electronics (2 channels) in a Faraday cage show acoustic sensitivity that changes with applied bias, as expected. The sensitivity at the bandpass output with 9 V_{bias} is on the order of 1 mV/Pa, similar to the model predictions, but additional testing is required to confirm this.

5 CONCLUSION

A front-vented surface micromachined capacitively sensed microphone for aeroacoustic analysis of the TBL has been fabricated. The microphone utilized the MEMSCAP PolyMUMPs® process as well as facilities at Tufts University's Micro and Nano Fabrication Facility and Massachusetts Institute of Technology's Microsystems Technology Laboratory. Device modeling agrees well with electrostatic testing using laser Doppler velocimetry. Preliminary acoustic testing shows microphone sensitivity, but frequency response measurements of acoustic sensitivity have not yet been obtained. Controlling the geometry of the vent holes is crucial to the low frequency response of the microphones. The high frequency resonance at 480 kHz is strongly influenced by the acoustic compliances coming from the vented air gap and external environmental loading. The next step in the array development is plane wave tube calibration against a reference microphone and array modeling. Packaging for wind tunnel testing is also underway.

References

[1] H Tennekes and J. L. Lumley. *A First Course in Turbulence*. The MIT Press, 1972.

[2] J. F. Wilby and F. L. Gloyna. Vibration measurements of an airplane fuselage structure II. jet noise excitation. *Journal of Sound and Vibration*, 23(4):467–486, August 1972.

[3] W. R. Graham. A comparison of models for the wavenumber-frequency spectrum of turbulent boundary layer pressures. *Journal of Sound and Vibration*, 206(4):541–565, October 1997.

[4] W.L. Abraham and B.M. Keith. Direct measurements of turbulent boundary layer wall pressure wavenumber-frequency spectra. *Journal of Fluids Engineering*, 120:29–39, March 1998.

[5] G. M. Corcos. Resolution of pressure in turbulence. *The Journal of the Acoustical Society of America*, 35(2):192–199, February 1962.

[6] G. M. Sessler. Acoustic sensors. *Sensors and Actuators A: Physical*, 26(1-3):323–330, March 1991.

[7] P. R. Scheeper, A. G. H. van der Donk, W. Olthuis, and P. Bergveld. A review of silicon microphones. *Sensors and Actuators A: Physical*, 44(1):1–11, July 1994.

[8] Lennart Löfdahl and Mohamed Gad-el-Hak. Mems applications in turbulence and flow control. *Progress in Aerospace Sciences*, 35:101–203, 1999.

[9] David T. Martin, Jian Liu, Karthik Kadirvel, Robert M. Fox, Mark Sheplak, and Toshikazu Nishida. A micromachined dual-backplate capacitive microphone for aeroacoustic measurements. *Journal of Microelectromechanical Systems*, 16(6):1289–1302, 2007.

[10] Jim Carter, Allen Cowen, Busbee Hardy, Ramaswamy Mahadevan, Mark Stonefield, and Steve Wilcenski. Polymumps design handbook: a mumps®process. Internet, 2005.

[11] Leo L. Beranek. *Acoustics*. Acoustical Society of America, 1996.

[12] Lawrence E. Kinsler, Austin R. Frey, Alan B. Coppens, and James V. Sanders. *Fundamentals of Acoustics: Fourth Edition*. John Wiley & Sons, 2000.

[13] D. Homentcovschi and R.N. Miles. Viscous damping of perforated planar micromechanical structures. *Sensors and Actuators A: Physical*, 119(2):544–552, April 2005.

Building Micro-Robots: A path to sub-mm^3 autonomous systems

J. Robert Reid, Vladimir Vasilyev, Richard T. Webster

Sensors Directorate, Air Force Research Laboratory
AFRL/RYHA, 80 Scott Dr, Hanscom AFB, MA 01731
James.Reid@hanscom.af.mil

ABSTRACT

A process for realizing sub-cubic millimeter micro-robots is detailed. To date, the process has been used to produce spherical structures ranging from 0.5 mm to 3.0 mm in diameter. A model of a 0.7 mm diameter spherical robot has been developed and an estimated power budget is provided. In addition, a model of the electrostatic actuators that will be used for the robot is covered. This model has been verified through experiments using external actuators.

Keywords: MEMS, Robotics, Self Assembly, Silicon-on-insulator

1 Introduction

While the realization of micro-robotics has been a goal of researchers for several decades [1], it has proven extremely difficult to build systems that combine power collection, power storage, computation, sensing, and actuation in a compact package. Traditional MEMS approaches have tried to do this by first achieving the required functionality and then combining the pieces together to form a robot [2] and systems for the automated assembly of MEMS devices have been demonstrated [3]. In this work, we take the opposite approach and focus on realizing the structure first and then adding the functionality into this structure. We have initiated this work by developing a CMOS compatible process for fabricating sub-cubic millimeter structures. This process has been used to fabricate spherical structures that will eventually integrate electrostatic electrodes and enclose a photovoltaic cell, capacitor, and processor.

2 MICRO-ROBOT ARCHITECTURE

The end goal of this work is to develop a spherical micro-robot with a diameter under 0.7 mm that could serve as the basic unit for programmable matter [4]. However, the fabrication approach being developed is suitable to the realization of a wide range of micro-robotic systems. Fig. 1 provides an artist's rendering of a spherical micro-robot. The robot is formed from an outer shell surrounding an inner core. The outer shell consists primarily of electrodes formed from a thin ($\approx$

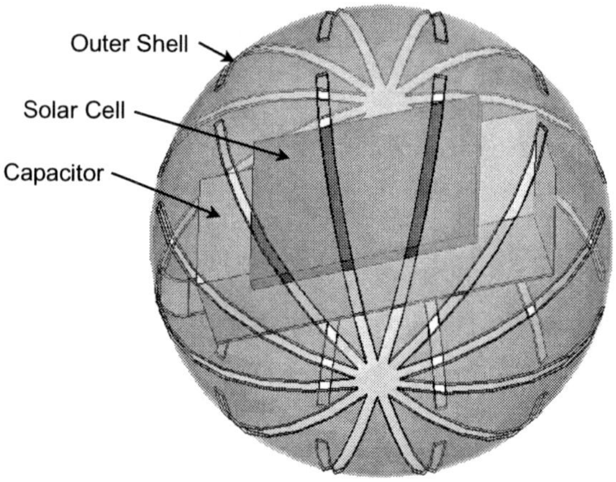

Figure 1: Artist's rendering of a spherical micro-robot. The robot consists of an outer shell containing transparent electrodes for actuation and an inner core containing the electrical circuits, a capacitor, and photovoltaic (solar) cell.

0.25 μm) silicon layer enclosed in silicon dioxide. Also embedded in the oxide, a limited number of transistors ($< 2\%$ of the surface area) will be used to control the bias voltage applied to the electrodes. A short connector composed of oxide enclosed metal wires connects the outer shell to the circuit region of the inner core. This circuit region contains the electrical control and driver circuits for the robot. Finally, a capacitor for energy storage and a photovoltaic cell for energy collection will be located on top of the circuit region.

2.1 Power Collection and Storage

Photovoltaic cells provide a compelling power source for micro-robots since they use a readily available energy source (light), are fabricated using widely available processes, and provide relatively high energy density. In addition to this, high voltage photovoltaic sources can be realized by connecting multiple cells in series. In 2003, Bellew, *et. al*, presented a photovoltaic source with groups of up to 200 individual cells, each sized 400 μm by 400 μm, connected in series to get open circuit output voltages as high as 88.5 V [5]. The maximum power output of 2.01 mW (62.8 μW/mm) with a solar illumination air mass of 1.71 corresponds to an

NSTI-Nanotech 2008, www.nsti.org, ISBN 978-1-4200-8505-1 Vol. 3

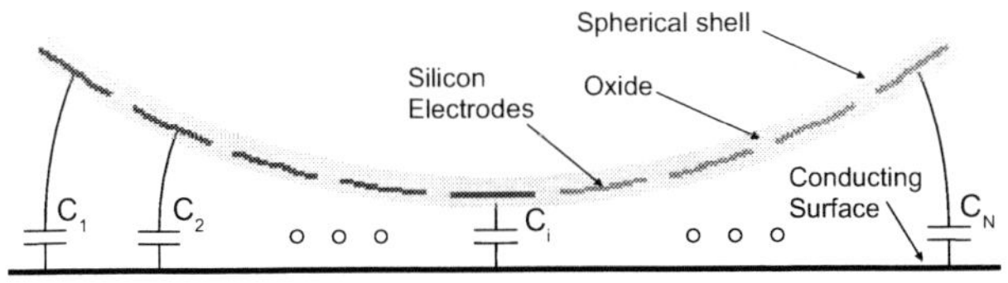

Figure 2: Schematic representation showing the oxide enclosed electrodes. A capacitor is formed between each electrode and the surface that the sphere is located on.

efficiency of 8.3%. This is fairly low compared to efficiencies over 20% commonly achieved with commercial photovoltaic cells, but is reasonable for a research effort. For a spherical micro-robot with a diameter of 0.7 mm, the photovoltaic cell will have a cross section of approximately 0.25 mm^2. Approximately 10% of the energy is lost due to reflection and absorption in the shell. Using an efficiency of 7%, a 0.25 mm^2 photovoltaic cell would provide 13.2 μW. Based on this, the robot is being designed to operate with a power budget below 10 μW.

In order to smooth out power variations and deliver high instantaneous power when needed, a capacitor will be used as a power buffer. Commercial surface mount capacitors are available down to EIA size 0201 with dimensions of 0.6 × 0.3 × 0.2 mm. One such part from Presidio Components provides 0.1 nF of capacitance for voltages up to 125V. Charging this capacitor with the 80 V output of a photovoltaic cell would result in stored power of 0.32 μJ, sufficient to provide nominal operation (10 μW) of the robot for 32 ms.

2.2 Actuation

The spherical micro-robot will be actuated using electrostatic actuators on the exterior shell. Biasing the electrodes relative to each other creates a charge separation that will induce corresponding surface charge on any conducting surface and thus provide an attractive force. Modeling this process is done by evaluating a series of capacitors over a conducting surface as shown in Fig. 2. Without loss of generality, we can arbitrary choose the ground of the robot to have a potential of 0 V. The potential of all of the electrodes is then well defined. The potential of the conducting surface (relative to the robot ground) is calculated as

$$V_{plate} = \frac{\sum_{i=1}^{N_{el}} C_i V_i}{\sum_{i=1}^{N_{el}} C_i} \tag{1}$$

where N_{el} is the number of electrodes and V_{plate} is the voltage induced on the conductive plate when the each electrode is biased with voltage V_i. With all of the voltages defined, the force between each plate and the conducting surface can be calculated by subdividing each electrode into smaller segments and applying the familiar electrostatic force calculation $F_{es} = \epsilon A V_{bias}^2 / g^2$ to each segment. Multiplying the force of each segment by the moment arm from the point of contact and summing all of these values gives the total torque applied to the sphere. For equally spaced electrodes with a fixed pattern of bias voltages, the total torque generated can then be plotted as a function of the number of electrodes. As seen in Fig. 3 (a), for a 0.7 mm diameter sphere, the optimal number of electrodes in 39-40 which corresponds to electrodes covering 9 degrees.

After determining the optimum number of electrodes for a given diameter, it is straight forward to calculate the total torque as a function of voltage. Since it is desired that the robot can move up a vertical surface, it is necessary to calculate the voltage required to overcome the torque that gravity applies to a sphere on a vertical surface. This voltage is plotted as a function of spherical diameter in Fig. 3 (b). The plot shows that limiting the voltage used to 90 V requires that the robot have a diameter under 0.7 mm.

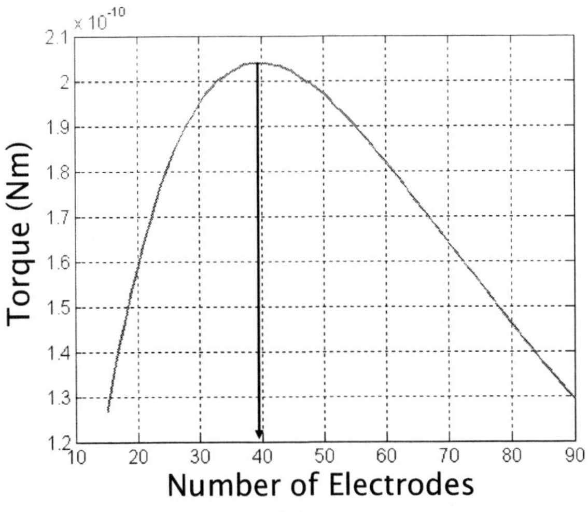

(a)

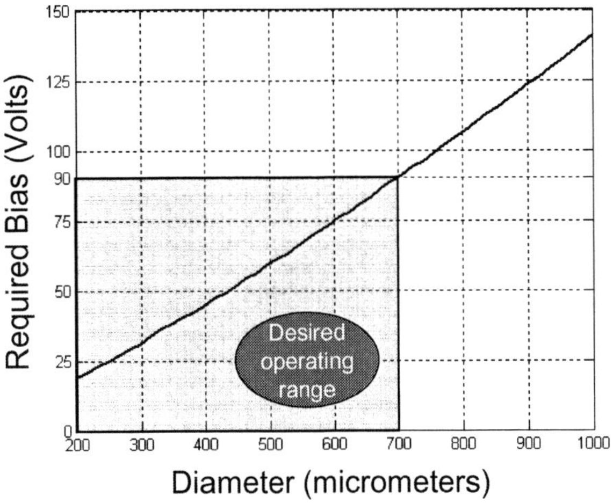

(b)

Figure 3: (a) Torque generated by a 0.7 mm diameter robot. (b) Voltage required to move a sphere vertically up a conductive surface.

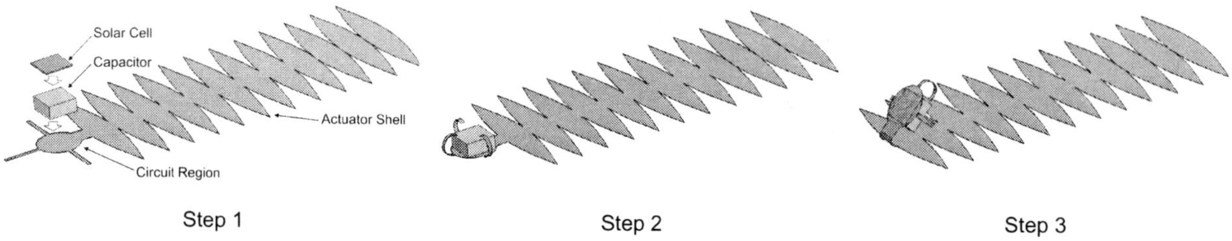

Figure 4: Fabrication of the robot begins with a foundry SOI BiCMOS process. Step 1: The die are post processed to form oxide enclosed circuits of the desired shape and additional devices are flip chip mounted. Step 2 and 3: The die is released by etching away the silicon from under the buried oxide layer causing the released device to roll up into a spherical shell around the bonded devices.

The power required to move the robot can be estimated by the voltage required to charge the drive electrodes. This capacitance is dominated by the inter-electrode capacitance, and therefore can not be known until the design is complete. We estimate that moving one step will require charging a capacitor of $\approx$0.5 pF. To charge to 90 V then requires 2.02 nJ. To move at rate of 5 revolutions per second (1.2 cm/sec) would then require making 200 steps per second, or 0.41 μW. Adding in a safety margin of 10x to account for leakage results in a power requirement of $\approx$4.1 μW.

2.3 Computation

In the first versions of the robot, a small finite state machine will be fabricated in the circuit region. In the long run, an ARM7-TDMI-S can be separately fabricated using a 90 nm CMOS process and bonded onto the circuit region. The ARM7-TDMI-S with 4 kB DRAM and 4 kB flash RAM can be fabricated in 0.16 mm^2. Clocking the core at 20 kHz would require under 2 μW of power. Using more recent 45 nm processes, both the size and power requirements can be reduced.

3 FABRICATION PROCESS

The process for fabricating the robots is the key to realizing highly functional units in a compact size. This process must provide low cost mass production in a highly repeatable manner with tightly packed functionality. Therefore, the process we are developing is based around commercial CMOS processing. These processes are ideal for producing large numbers of repeatable, compact, highly integrated circuits. Adding bonding and micro-machining processes after the CMOS process provides the added functionality required for the cells.

Fabrication begins with the submission of circuit and actuator designs to a commercial foundry with an established silicon on insulator (SOI) bi-polar complementary metal oxide semiconductor (BiCMOS) process. Many SOI processes also include the ability to co-fabricate high voltage circuits that are required for the actuators. In an SOI process, all of the circuits are fabricated on top of a buried oxide (BOX) layer. Further, after the transistors are defined, additional layers of metal and oxide are added on top of the transistor layer. The result is an integrated circuit that is embedded in silicon oxide on both the top and bottom.

Using a process developed in-house at the Air Force Research Laboratory (AFRL), the oxide layers, including the BOX layer, will be patterned and etched down to the silicon handle wafer. This process enables us to create a patterned two dimensional shape on top of a silicon handle wafer as shown in Step 1 of Fig. 4. At this point, individual devices such as a processor, a capacitor, and a photovoltaic cell can be mounted to the circuit region of the device using commercially available flip chip attachment. Next, the silicon layer under the BOX layer is etched out causing the patterned structure to release from the substrate. Since the BOX layer is a highly stressed thermal oxide, and the silicon layers have very low stress, a bending moment is created in the released circuit causing the structure to curl up into a spherical shape. Through the placement of etch access holes, the folding process can be controlled so that initially (Fig. 4: Step 2) small structures are released, then larger structures (Fig. 4: Step 3). The end result is that the structure will roll up into a sphere resulting in the final device illustrated in Fig. 1.

4 RESULTS

While our end goal is the realization of sub-cubic millimeter robots, our efforts to date have focused on developing the fabrication process for creating the spherical shells. Utilizing our in-house release process, a variety of two dimensional patterns have been used to generate spheres such as those shown in Fig. 5. The diameter of the shells as a function of the layers that compose the shell, the layer thicknesses, and the residual stress in the layer is calculated using a simple bending model. Fig. 6 shows the predicted diameter for shells with a 1.1 μm thick BOX layer and varying silicon layer thickness. Also plotted are results for three released shells.

NSTI-Nanotech 2008, www.nsti.org, ISBN 978-1-4200-8505-1 Vol. 3

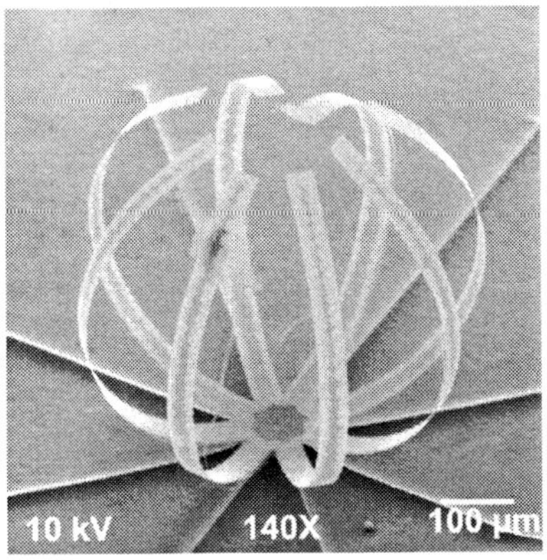

Figure 5: Scanning electron micrograph of a spherical shell formed from glass-silicon-glass layers. Shells with diameters ranging from 0.5-3.0 mm have been produced.

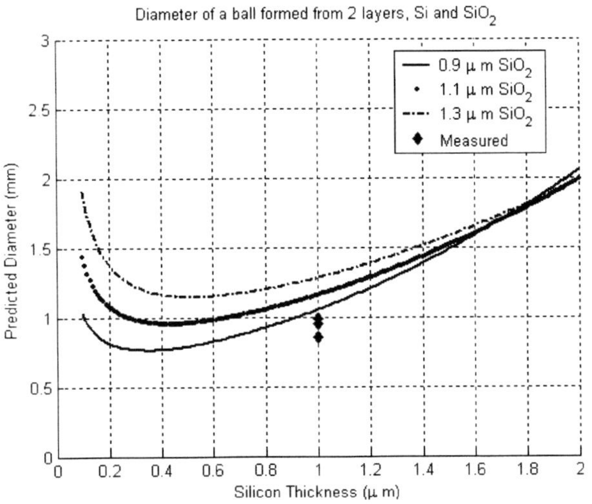

Figure 6: Plot of the predicted and measured sphere diameters.

The thickness of the silicon layer is nominally 1.0 microns, but could range from 0.8-1.05 microns. Generally, the measured values agree with the prediction. As more complex two dimensional patterns such as the layout in Fig. 4 are used, the models will require the use of finite element simulators to accurately predict the diameter.

Using the fabricated shells, we have also demonstrated actuation of the devices using voltages of 80-100 V. The shells are placed on an electrode array. The array has four independent voltages that are repeated. Fig. 7 shows four images of a shell on the external electrode array. As can be seen, the sphere is moving to the biased electrode.

5 CONCLUSIONS

We have developed a process for fabricating spherical shells composed of oxide enclosed silicon. This process is compatible with CMOS circuitry and can therefore be used in the fabrication of very compact micro-robots.

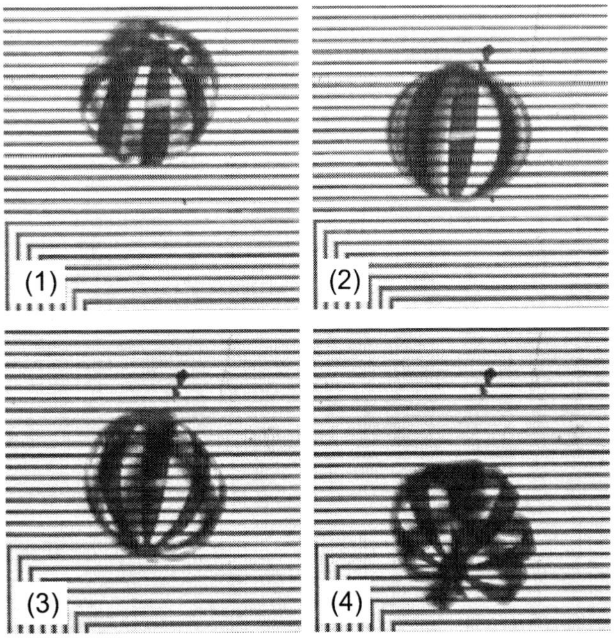

Figure 7: Still frames captured from a movie of a spherical shell rolling under the influence of external electrostatic actuator electrodes.

ACKNOWLEDGMENTS

This work was supported by the Air Force Office of Scientific Research under LRIR 07SN06COR. The views expressed in this article are those of the authors and do not reflect the official policy or position of the United States Air Force, Department of Defense, or the U.S. Government.

REFERENCES

[1] R. Feynman, "There's plenty of room at the bottom," *Journal of Microelectromechanical Systems*, vol. 1, no. 1, pp. 60–66, March 1992.

[2] S. Hollar, A. Flynn, C. Bellow, and K. Pister, "Solar powered 10mg silicon robot," in *IEEE International Conference on Micro Electro Mechanical Systems*, January 2003, pp. 706–711.

[3] J. Reid, V. Bright, and J. Comtois, "Automated assembly of flip-up micromirrors," in *Proceedings of International Conference of Solid-State Sensors and Actuators (Transducers 97)*, vol. 1, June 1997, pp. 347–350.

[4] S. C. Goldstein, J. D. Campbell, and T. C. Mowry, "Invisible computing: Programmable matter," *Computer*, vol. 38, no. 6, pp. 99–101, June 2005.

[5] C. Bellow, S. Hollar, and K. Pister, "An SOI process for fabrication of solar cells, transistors and electrostatic actuators," in *International conference on solid-state sensors and actuators (Transducers 03)*, vol. 2, June 2003, pp. 1075–1078.

Pressure Sensor Data Processing for Vertical Velocity Measurement

Husak,M. - Jakovenko,J. – Stanislav,L.

Department of Microelectronics
Faculty of Electrical Engineering, Czech Technical University in Prague
Technicka 2, CZ – 166 27 Prague 6, CZECH REPUBLIC
husak@feld.cvut.cz

ABSTRACT

The paper describes design and realization of a sensor system for vertical velocity measurement using pressure sensor. The system solution is based on the basic equation for altitude calculation, including the effects of temperature. A linear interpolation function is used to speed up the altitude calculation. Problem of nonlinearity due to the exponential function of air pressure versus altitude is discussed. The transient analysis of the electronic circuit connection was used for vertical velocity simulation. A new method for distortion correction was used, for algorithm simplification and system linearization. New assets are represented by the new electronic circuit blocks. An electronic circuit was designed, taking care of the mathematical functions including compensation functions. The type MPX4115 pressure sensor was used in the system. The system is controlled by an ATMega16 microprocessor.

Keywords: vertical speed, pressure, nonlinearity, altitude, sensor, microprocessor

1 INTRODUCTION

The measurement of vertical velocity is necessary for different use. Typically methods are used in devices that measure the air planes rising or falling. There are several ways how to evaluate the vertical velocity. The solutions are based on principle of evaluating the change of an atmospheric pressure. This is relatively simple method but it has also some drawbacks that can not be omitted. First of all it is a nonlinearity of the exponential dependency of the pressure versus altitude. Also the air temperature plays a significant role in this method.

Compensated methods for measuring the vertical velocity embrace the vertical velocity correction. The correction is calculated from the horizontal velocity change which induces another vertical velocity change. The acceleration is evaluated in the horizontal direction using the information about the dynamic pressure [1]. It means that the kinetic energy (acceleration) is converted to corresponding change of the potential energy (altitude). Vertical velocity thus can be corrected by this value.

No-compensated simple methods are based on measuring the vertical velocity using the information about the change of the static air pressure. In regular time instances the absolute air pressure values are measured. The air pressure difference related to the time interval determines the vertical velocity.

2 PRINCIPLE OF ALTITUDE CALCULATION

The dependence which evaluates the vertical velocity is derived from the exponential form of the barometric equation which relates the air pressure versus the altitude

$$p(z) = p_o e^{-\frac{z}{z_o}}, \qquad z_o = \frac{kT}{gm_o} \tag{1}$$

where p(z) is the air pressure in altitude z, p_o is the sea level altitude air pressure, T (K) is the temperature. Other elements in the equation are constants, m_o (-), g ($m \cdot s^{-2}$), k ($J \cdot K^{-1}$). The exponential element e^{-z} in the equation can be converted to the Taylors polynomial. Omitting the powers from this polynomial bigger then one, it can be found

$$\frac{dz}{dt} = -\frac{z_o}{p_o}\frac{dp(z)}{dt} \tag{2}$$

If the air pressure sensor with the voltage output v(p) is used and if the output voltage is direct proportional to the pressure p(z) it can be obtained simple equation for the vertical velocity – equation 3.

$$\frac{dz}{dt} = -\frac{z_o}{p_o}\frac{dv(p)}{dt} \tag{3}$$

This equation is valid for no compensated method of the vertical velocity measurement. Dependency according the equation 3 can be realized using the differentiator. No-linearity compensation can be done using the circuit with the inverse characteristic (logarithmical circuit). It is essential to solve the temperature compensation as well. The circuit solution is depicted on figure 1. The first stage is the temperature compensated logarithmic circuit which realizes linearization of the input exponential voltage. Following circuits are the amplifier and the differentiator.

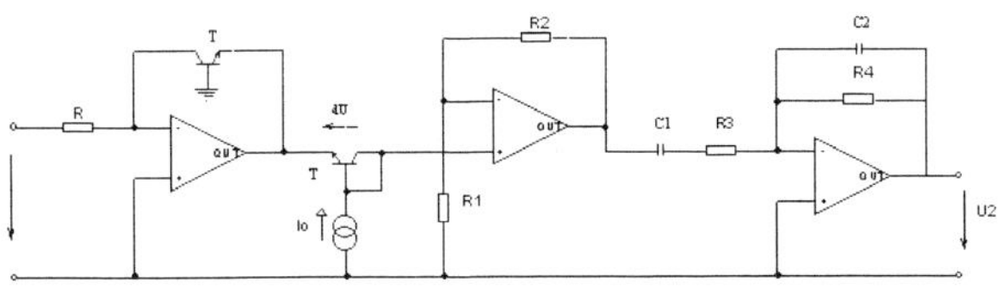

Figure 1: Temperature compensated circuit for the vertical velocity measurement (static measurement).

There can be expressed the output voltage from the simple differentiator with the feedback diode as follow

$$v_2 = -\frac{kT}{q} ln \frac{v_1}{Ri_s} \qquad (4)$$

where R is the input resistor of the differentiator and i_s represents the current given by the diode technological parameters. Transfer characteristic according (4) depends on the temperature. The characteristic have the voltage shift c for different temperatures. The temperature dependency for the current i_s causes small distortion. Output voltage

$$v_2 = f(v_1) + c \qquad (5)$$

Differentiating this equation the constant c disappears. This allows to use the feedback diode (distortion is thus given only by the temperature dependency of the technological factor i_s). The transfer characteristic according the equation (4) is very flat for higher input voltages v_1, so the sensitivity is small. The sensitivity can be increased using the amplifier which multiplies the characteristic by the constant. To obtain more precise calculation of the altitude the equation (1) can be modified to the form (6)

$$z(p) = \frac{T_o}{T_r}(1 - (\frac{p}{p_o})^{\frac{T_r \cdot R}{Mg}}) \qquad (6)$$

where z is the altitude, p is the air pressure, p_o=101.325 kPa is the sea level air pressure according the ISA, Tr=0.0065 $K \cdot m^{-1}$ is the temperature gradient according the ISA, R=8.3 $JK^{-1}mol^{-1}$ is the universal gas constant, M=0.02894 $kg \cdot mol^{-1}$ is the air molar mass, g=9,81 $m \cdot s^{-2}$ is the gravitational constant, T_0=288.15 K is the temperature at the sea level according the ISA. Transient analysis of the circuit from the figure 1 with temperatures 273 K and 333 K indicates that the error can rise up to 20%. That is why it is essential to compensate this no linearity. Software solution and microcontroller were used for the compensation of this no linearity.

3 DESIGN OF SYSTEM

The system consists of several blocks. The analog differentiating network delivers the differentiation peaks at its output. The peaks are read off by the microprocessor. At the same time, the altitude above sea level is measured and its value applied for correction of the measured peak. The design must care for a quick and accurate measurement of the differentiation peaks. The parameters given in [2] were taken as the design basis. The manufacturer gives the output voltage dependence for the MPX4115 pressure sensor as [3]

$$v_{out} = V_{cc}(c_1 p - c_2) \qquad (7)$$

where V_{cc} is the supply voltage and p is the pressure, c_1=0.009 a c_2=0.095. By arrangement of the equation we can obtain the formula for pressure

$$p = \frac{\frac{v_{out}}{V_{cc}} + c_2}{c_1} \qquad (8)$$

By setting the constants into (6) and arrangement we can obtain the formula

$$z(p) = \frac{T_o}{T_r}(1 - (\frac{p}{p_o})^{\frac{T_r \cdot R}{Mg}}) = c_3(1 - (\frac{p}{p_o})^{c_4}) \qquad (9)$$

where the constant c_3=44330.8 and c_4=0.190261. When the ATMega 16 processor is used, a 10-bit converter is available, its resolution can be increased to 12 bits [4].

Derivation of the correction data. The values read for altitude are used for the differentiation peaks correction. Under the assumption of direct proportionality, the differentiating network output (the differentiation peaks value) can be written as

$$v_{peak} = c_5 \frac{dv_{out}}{dt} \qquad (12)$$

where v_{peak} is the output voltage of the differentiating network and c_5 is a constant. The equation (7) applies for the pressure sensor output voltage, and (12) can then be arranged to the form

$$v_{peak} = c_6 \frac{dp}{dt} \qquad (13)$$

where c_6 is a constant. The following formula applies to the vertical velocity

$$\frac{dz}{dt} = \frac{dz}{dp}\frac{dp}{dt} \qquad (14)$$

where the dp/dt term cornensponds to the v_{peak} measured voltage value. To express the dz/dp term, we start with the equation (6), and differentiate it over pressure

$$\frac{dz}{dp} = c_7 p_o^{c_8 p^{-c_8}} \qquad (15)$$

where the values of the calculated constants after setting in are c_7=83.241276 and c_8=0.809739. By setting the converted pressure formula into (15) we get

$$\frac{dz}{dp} = c_7(1 - \frac{z}{c_3})^{-c_9} \qquad (16)$$

By setting the converted equation (6) into (14) we get the formula for the actual dependence of vertical velocity on the differentiation peaks and on the altitude as

$$\frac{dz}{dt} = \frac{v_{peak}}{c_6} c_7(1 - \frac{z}{c_3})^{-c_9} \qquad (17)$$

Marking one part of the formula (16) as a function F(z), we can use this function to generate a table with altitude values and values of the F(z) function as the corresponding corrections

$$F(z) = (1 - \frac{z}{c_3})^{-c_9} \qquad (18)$$

The resulting formula for the real vertical velocity ishas the form

$$\frac{dz}{dt} = c_7 \frac{v_{peak}}{c_6} F(z) \qquad (19)$$

where the F(z) values relating to the particular altitude are stored in the table.

3.1 Hardware of sensor system

An electronic circuit was designed, taking care of the mathematical functions including compensation functions – figure 2. The analog differentiator processes the signal from the pressure sensor. Differentiation peaks appear at its output at nonzero velocities, they are conveyed to the microprocessor through an amplifier. The type MPX4115 pressure sensor was used in the system [3]. The whole system of measurements and calculations is controlled by an ATMega16 microprocessor. The converter is used to read-in the pressure sensor and differentiator data. A display is used to present the measured values. The system is complemented by an audio module connected to the microprocessor. The system communicates with a PC by means of a TTL to RS232 signal level converter.

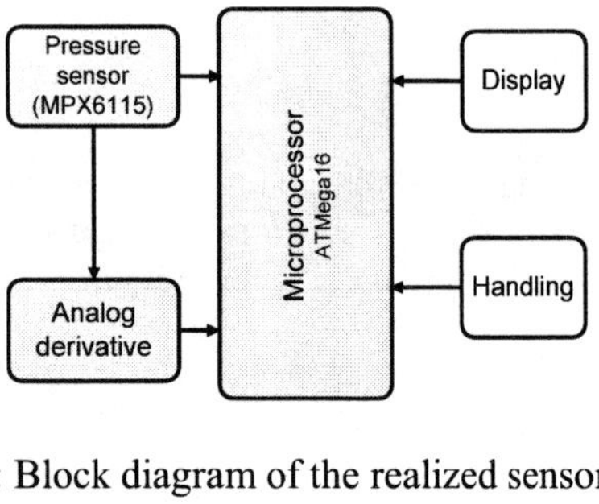

Figure 2: Block diagram of the realized sensor system.

A network with a time constant of 240 ms was used as the analog differentiating circuit. The differentiation peaks at the output are amplified by a non-inverting circuit using a Type-1666 amplifier. In the design it is necessary to pay attention to the parameters directly influencing the differentiating network quality (minimum voltage offset, minimum voltage offset drift, minimum supply current).

A microprocessor Atmel ATMega 16 with a 16 kB program memory was used. Clock signal is controlled by an 8-MHz oscillator, it is also possible to use an external 16 MHz oscillator. The microprocessor contains an 8-channel 10-bit AD converter, used to read data from the differentiating network of the pressure sensor. The microprocessor chip can be programmed in the design sample through the SPI and JTAG interfaces. Correct bypassing must be properly designed of the supply and reference voltages. Correct bypassing has a direct influence on the measurement accuracy.

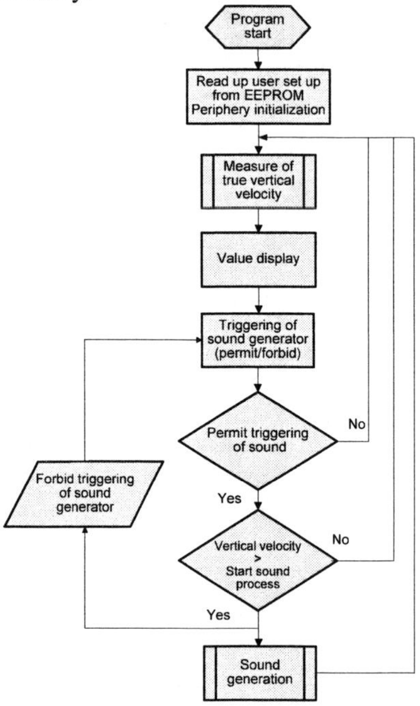

Figure 3: Main system algorithm.

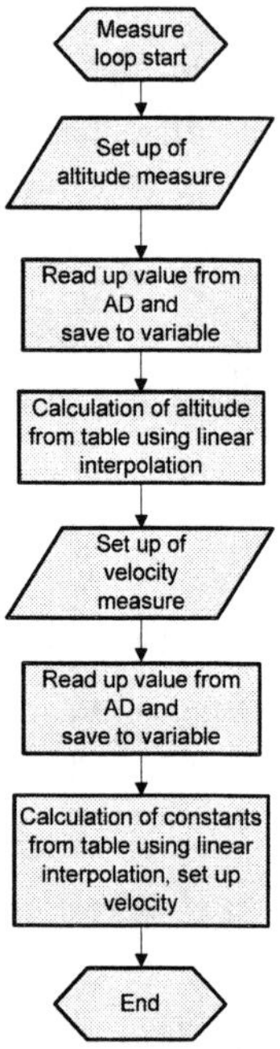

Figure 4: Real vertical velocity algorithm.

3.2 Software of sensor system

The measurements, calculations and other system activities are controlled by the specially designed software using several sub-processors. The main system operation algorithm is shown in figure 3. First, the real vertical velocity value is measured, then it is compared to the threshold rise and descent values. Then the calculations, comparisons and evaluation follow. A second channel is set up in the microprocessor control registers for the AD converter. Using a linear interpolation, the altitude is calculated from the measured value of the converter. The first channel of the AD converter is used for the velocity measurement. Using linear interpolation, the supplementary correction value is calculated from the altitude, and according to the equation that was derived, the real vertical velocity is calculated – figure 4. The drivers for all peripheral circuits were designed in the C language (display, AD converter, control elements etc.).

4 RESULTS OF THE WORK

Certain problems had to be solved due to the integrated AD converter used. To improve the measurement accuracy it is possible, for instance, to use data reading in the low-power microprocessor mode. This process is treated by means of software. A linear interpolation function is used to speed up the altitude calculation from the values read by the AD converter. The calibration of the designed system is performed by means of the SPICE program – figure 5 (simulation of the differentiator response.

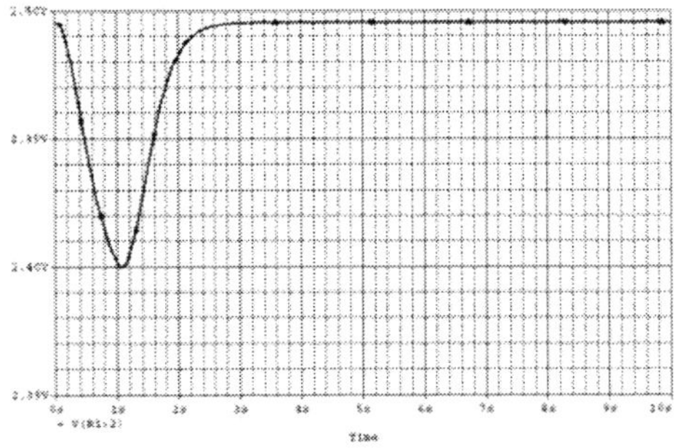

Figure 5: Simulation of the differentiator response corresponding to 1 ms^{-1} velocity.

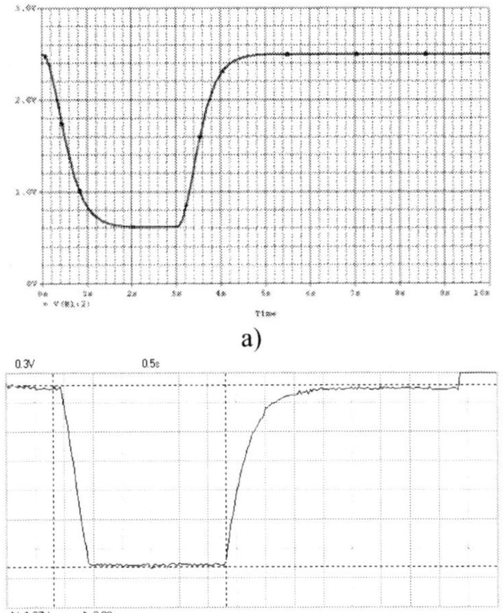

a)

b)

Figure 6: The response form of the calibration section for the velocity 20 m·s^{-1}, a) simulated, b) measured.

The system designed includes distortion compensation, the program takes care of output information linearity. The system is capable to indicate a vertical velocity 0.1 m·s^{-1}. The altitude measurement is auxiliary information for the vertical velocity calculation, the measurement accuracy is 1 m. The system also contains a calibration section. The simulated and measured response form of the calibration section for the velocity 20 m·s^{-1} is shown in figure 6. A trapezoidal signal generator, series-connected with a controlled D.C. source, was used for calibration. A 12-bit DA converter MCP4921 was connected to the ATMega8 processor through the SPI interface. The implemented software makes possible to vary the parameters of the leading edge of the trapezoidal signal generator (corresponding to different velocities) as well as the D.C. level (corresponding to movements in different altitudes).

5 CONCLUSIONS

Problem of nonlinearity due to the exponential function of air pressure versus altitude is discussed in detail. The transient analysis of the electronic circuit connection was used for vertical velocity simulation. An electronic circuit was designed, taking care of the mathematical functions including compensation functions. The type MPX4115 pressure sensor was used in the system. The whole system of measurements and calculations is controlled by an ATMega16 microprocessor. The system communicates with a PC by means of a TTL to RS232 signal level converter. The measurements, calculations and other system activities are controlled by the specially designed software using several sub-processors. Using a linear interpolation, the altitude is calculated from the measured value of the converter. The drivers for all peripheral circuits were designed in the C language (display, AD converter, control elements etc.).

6 ACKNOWLEDGEMENT

This research has been supported by the research program No. MSM6840770015"Research of Methods and Systems for Measurement of Physical Quantities and Measured Data Processing" of the CTU in Prague and partially by the Czech Science Foundation project No. 102/06/1624 “Micro and Nano Sensor Structures and Systems with Embedded Intelligence”

REFERENCES

[1] http://home.att.net/~jdburch/systems.htm
[2] www.sensair.com
[3] Freescale Semiconductor. MPX4115A, MPXA4115A, 2006. www.freescale.com.
[4] www.atmel.com - Application notes: AVR121 (Enhancing ADC resolution by oversampling)

Adaptive Subband Filtering Method for MEMS Accelerometer Noise Reduction

P. Pietrzak*, B. Pękosławski*, M. Makowski* and A. Napieralski*

*Department of Microelectronics and Computer Science, Technical University of Lodz, POLAND
www.dmcs.p.lodz.pl

ABSTRACT

Silicon microaccelerometers can be considered as an alternative to high-priced piezoelectric sensors. Unfortunately, relatively high noise floor of commercially available MEMS (Micro-Electro-Mechanical Systems) sensors limits the possibility of their usage in condition monitoring systems of rotating machines. The solution of this problem is the method of signal filtering described in the paper. It is based on adaptive subband filtering employing Adaptive Line Enhancer. For filter weights adaptation, two novel algorithms have been developed. They are based on the NLMS algorithm. Both of them significantly simplify its software and hardware implementation and accelerate the adaptation process.

The paper also presents the software (Matlab) and hardware (FPGA) implementation of the proposed noise filter. In addition, the results of the performed tests are reported. They confirm high efficiency of the solution.

Keywords: MEMS accelerometer, adaptive subband filtering, NLMS algorithm, vibration measurement, hardware filter

1 INTRODUCTION

Basing on the observation and analysis of a vibration spectrum it is possible to detect broken parts of a machine, determine a type of the failure and predict its future development. In the currently used measurement systems, the information on vibration magnitude is provided by expensive piezoelectric accelerometers. An alternative to these sensors can be silicon microaccelerometers produced using micromachined technologies. These devices are relatively cheap and may be a component of a complex and comprehensive vibration measurement system built as a single IC. Unfortunately, commercially available micromachined accelerometers have some limitations. One of the most significant is a relatively high level of the self-noise observed in the output signal [1]. The noise limits measurement resolution, which makes it impossible to use these sensors for precise diagnostic measurements.

A predominant noise component in the surface technology silicon microaccelerometers with low-noise acceleration detection circuits comes from Brownian motion of the proof mass. A study of the ADXL202 accelerometer produced by Analog Devices showed that amplitude-frequency characteristics of the observed noise signal from the sensor in standstill condition is constant in the frequency range up to about 2 kHz (Figure 1). This was confirmed by the analysis of autocorrelation function of the noise [1].

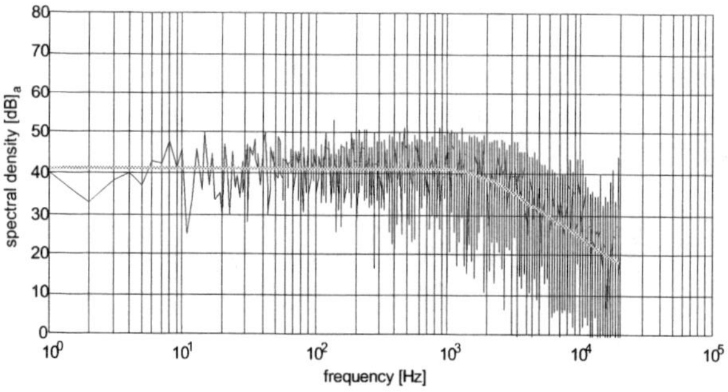

Figure 1: Power density spectrum of the accelerometer noise

Hence, in the given frequency range, the self-noise of the examined accelerometer can be treated as a white noise. The target application engages the use of MEMS accelerometers for turbogenerator vibration measurement. The most valuable sources of information for this type of machines are harmonics of vibration signal. The determination of their level and finding the correspondence with rotational frequency of the respective machine parts enables the identification of a failed part. The above observations indicate the possibility for the implementation of the Adaptive Line Enhancer (ALE) type circuit to reduce noise level.

2 ADAPTIVE LINE ENHANCERS

Adaptive Noise Canceller (ANC) circuits are based on the primary structure of an adaptive filter (Figure 2) [2][3]. The input of ANC adder is fed with the signal $x(n)$ to be filtered, which is a sum of the usable $s(n)$ and interfering $v(n)$ components, where $n \geq 0$ denotes subsequent time instants. It is assumed that the both components are uncorrelated with each other (they are independent).

The necessary condition on which the ANC circuit can work properly is that the $d(n)$ signal, which is applied at the input of the adaptive filter, is correlated with the usable component $s(n)$ and is not correlated with the interfering signal $v(n)$. However, in many real applications there is no possibility to separate the reference signal fulfilling this condition. Then the perfect solution is to apply the ALE filter. Here, the reference signal is the input signal $x(n)$ delayed in time (by d samples). The introduction of the

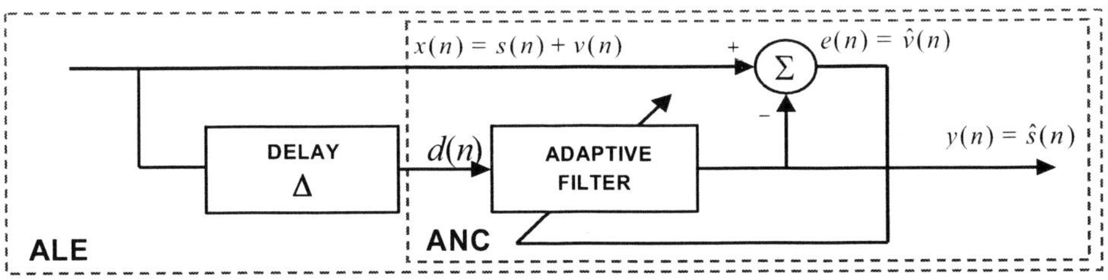

Figure 2: Structure of the Adaptive Noise Canceller (ANC) and Adaptive Line Enhancer (ALE).

delay is aimed to decorrelate the interfering component *v(n)* between the input signal *x(n)* and the reference signal *d(n)*.

The adaptive algorithm for the calculation of respective filter coefficients tends to eliminate the uncorrelated component. As a result, the amplitude-frequency characteristic of the filter adopts the shape of the characteristic of the comb filter, which passes harmonic components present in a signal being processed. The output signal *y(n)* is the estimate of the usable component *s(n)* of the vibration signal.

The analysis of the ALE filters properties shows that their noise reduction efficiency depends not only on the filter parameters but also on the processed signal features. For the ALE filter using the LMS algorithm it was proved [4] that the theoretical SNR improvement is described by the relationship:

$$SNG = \frac{SNR_{out}}{SNR_{in}} = \frac{L}{2} \frac{\sum_{m=0}^{M-1} A_m^2 [1 + (4/L)(\sigma_v^2 / A_m^2)]^{-2}}{\sum_{m=0}^{M-1} A_m^2 \cdot \sum_{m=0}^{M-1} [1 + (4/L)(\sigma_v^2 / A_m^2)]^{-2}} \quad (1)$$

It is thus proportional to the number L of filter coefficients. Additionally, it depends on the amplitude A_m (of power) of the respective harmonics in the processed signal, their number M and the interfering signal power.

As mentioned above, in the target application, the MEMS accelerometers measure turbogenerator vibrations. In the measured signal, for the frequency range of 10 Hz ÷ 6 kHz, one can observe over 100 harmonics. As a consequence, a significant SNR improvement requires the use of filters with large number of coefficients. However, some limitations appear here. The first of them is the necessity for the application of the digital ICs with large computational power to implement the filter. Moreover, in [4], it has been proved that there exists some finite value $L = L_{lim}$, for which the ability of the ALE circuit to reduce noise is the highest. Thus the maximum SNR improvement for the signal containing M harmonics is limited to the value:

$$SNG = \frac{L_{\lim}}{4M} \quad (2)$$

This effect is a result of the adaptive filter transmittance noise, which is caused by an error in the filter coefficients adaptation.

The ALE filter order affects also the frequency resolution, which determines the filter ability to differentiate between the respective harmonics in the sampled signal. In [1], it has been showed that for the signal containing M equidistant harmonics there is a necessity to apply the filter with the number of coefficients satisfying the inequality [2][4]:

$$L > 2M \quad (3)$$

The above considerations indicate that it is possible to obtain high efficiency of the ALE-NLMS filter in terms of the SNR improvement for signals that are sampled with low frequency and whose useable component contains small number of harmonics. This conclusion was a basis for the development of the adaptive subband filtering method for the vibration signal of large rotating machines.

3 ADAPTIVE SUBBAND FILTERING OF NOISE

The principle of operation of the circuit for adaptive signal processing in frequency subbands is illustrated in the below figure.

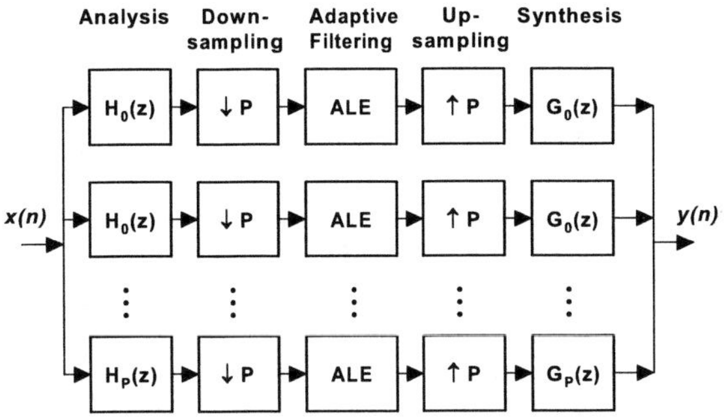

Figure 3: Structure of the system for adaptive signal processing in frequency subbands.

In the applied method, the signal is divided into 16 frequency subbands with the use of the suitably designed [1] bank of bandpass analysis filters. After the filtration, each subband signal occupies 16 times narrower band as compared to the input signal. As a result, all these signals are 16 times up-sampled. The sampling frequency for each signal is reduced by the factor of 16 with the use of reducers (down-sampling). The signals processed in this way are filtered with ALE filters. Upon appropriate processing, the frequency of each subband signal is increased with the use of expanders (up-sampling). The resultant signals are merged together in the bank of bandpass synthesis filters forming output signal of the filter.

The presented method for the signal processing in the frequency subbands enables the use of the ALE filter for the signals containing many harmonics, including a complex signal typical of rotating machine vibrations. The increase in the filter operation efficiency arises from the reduction in the width of the frequency band of the processed signal and from the decrease in its sampling frequency. The band limitation reduces the number of harmonics present in each subband. The decrease in sampling frequency results in higher frequency resolution of the filter.

4 ALGORITHMS OF FILTER WEIGHT ADAPTATION

After the preliminary analyses, it was decided to apply the NLMS (Normalized Least Mean Squares) algorithm [3]. This choice is the trade-off between algorithm and hardware implementation simplicity, on the one hand and weight adaptation precision and speed of convergence to the optimal solution, on the other hand. In the considered algorithm the update of the filter coefficients vector $\boldsymbol{w}$ is performed in accordance with the equation:

$$\mathbf{w}(\mathrm{n}+1) = \mathbf{w}(n) + 2\mu_0 \gamma \mathbf{x}(n) e(n) \tag{4}$$

where: $\mathbf{w}(n) = [w_0, w_1, ..., w_{L-1}]$ – vector of coefficients (weights) for the filter of order L, $\mathbf{x}(n) = [x(n), x(n-1), ..., x(n-L-1)]$ – vector of the subsequent input signal samples, applied to the respective filter weights, μ_0 – adaptation coefficient, γ – normalizing coefficient. Normalizing coefficient value is determined from the relationship [3]:

$$\gamma = \frac{1}{\chi + \|\mathbf{x}(n)\|^2} \tag{5}$$

where: $\|\boldsymbol{x}(n)\|^2$ – squared norm of the vector of L samples of a signal fed into the filter weights, χ – small constant (it is often assumed that $\chi = 0.01$).

In practice, the value of the squared signal norm is calculated basing on the relationship [3]:

$$\|\mathbf{x}(n)\|^2 = \mathbf{x}(n)\mathbf{x}^T(n) \tag{6}$$

Unfortunately, this method is time-consuming and requires the use of many resources when applied in hardware. Considerable hardware resources are also required to perform the division operation that appears in the equation (5).

In order to simplify the selected method for updating the filter coefficients, two modifications of the NLMS algorithm were proposed. The first of them - RP-NLMS takes into account the fact that in theoretical considerations it is often assumed that:

$$\|\mathbf{x}(n)\|^2 \cong L\sigma_X^2 \tag{7}$$

The signal variation σ_X^2 at time instant n can be replaced by its estimate determined recursively [1]:

$$\tilde{\sigma}_X^2(n) = \alpha\tilde{\sigma}_X^2(n-1) + (1-\alpha)x^2(n) \tag{8}$$

Which for the signal with zero mean value is equivalent to:

$$\tilde{P}(n) = \alpha\tilde{P}(n-1) + (1-\alpha)x^2(n) \tag{9}$$

It was suggested that the filtration coefficient α can be determined from the equation which relates its value to the filter order and the scaling factor g accelerating filter response rate for temporary signal amplitude changes [1]:

$$\alpha = \exp(-g/L), \qquad g = 0, 1, 2, ... \tag{10}$$

The use of the presented method considerably simplifies the procedure for finding the value of $\|\boldsymbol{x}(n)\|^2$, since it requires only 4 multiplication operations and one addition (instead of L multiplications and $L-1$ additions required previously).

In the search for further simplification of the algorithm, it was noticed that the decrease in the precision of γ factor value determination does not cause significant deterioration in adaptive filter properties. As a result, it is possible to introduce the modification in the algorithm which consists in quantizing the values of this factor. In order to do it, the subsequent values of γ were determined for certain values of $\|\boldsymbol{x}(n)\|^2$ (in the range from 2^{-7} to 2^{12}) basing on the equation (5) and they were stored in the table. The χ constant could be neglected, thanks to assuming that $\gamma \leq \gamma_{max}$. The choice of the table element used in a given algorithm iteration is made basing on the current value $\tilde{P}(n)$. Here, it is worth to mention that for a given value of μ_0 it is possible to store the whole expression $2\mu_0 / \|\boldsymbol{x}(n)\|^2$ in the table. In the case, when the values of μ_0 and γ are powers of 2, the multiplication by these factors can be replaced in hardware implementation by a bitwise shift operation. This algorithm has been called T-NLMS.

5 APPLICATION OF SUBBAND ADAPTIVE FILTER FOR NOISE REDUCTION IN TURBOGENERATOR VIBRATION SIGNAL

In the first step, the study of proposed method effectiveness was conducted with the use of the modeled signal, which was generated in Matlab (Figure 4a). The performed tests made it possible to verify the correctness of filter operation as well as to determine its properties and behavior in the stationary environment. The stationarity of the test signal allowed finding the maximum amplification error and the time of MSE (Mean Squared Error) minimization. The filtration efficiency was evaluated basing on the value of the SNR improvement factor:

$$SNG = \frac{SNR_{out}}{SNR_{in}} = 10\log\frac{P_{vin}}{P_{vout}}[dB] \quad (11)$$

where: $P_{V\ in}$ – power of noise at the filter input, $P_{V\ out}$ – power of noise at the filter output.

For the exemplar filter configuration (number of frequency subbands: 16, filter order: 128; algorithm: T-NLMS; delay: Δ = 67; adaptation coefficient: μ_0 = 0.125; scaling factor: g = 16; analysis and synthesis filters' attenuation in stopband: 70 dB), the mean ability of noise stochastic component reduction equal to 11 dB was achieved [1]. During tests with the modeled vibration signal, due to its known parameters and stationarity, it was possible to determine the maximum filter amplification error (0.1dB for the exemplar configuration) and time of adaptation to the optimal solution (2080 iterations).

The next step was to verify the method usefulness for the reduction of the level of noise present in a real vibration signal of turbogenerator stator core. For the described filter, the mean SNR improvement of the order of 10 dB was achieved. The effect of filter operation on vibration spectrum is shown in the below figures. The first one additionally presents the circuit usefulness for filtration of complex signals containing periodic components which are not harmonics of the fundamental frequency. The second figure shows the capability of the filter (with large number of coefficients) to detect harmonic components masked by a noise.

(a)

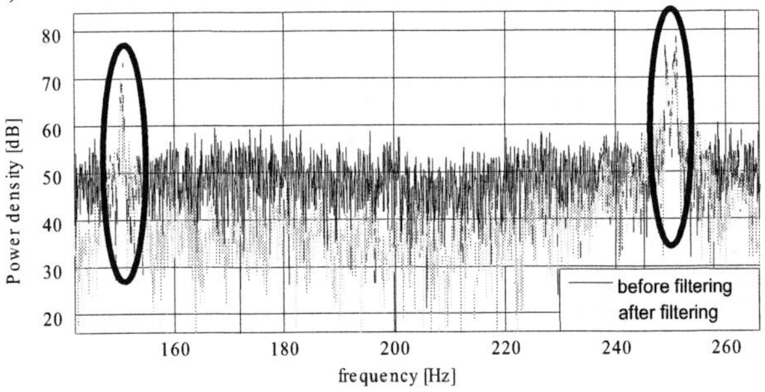

(b)

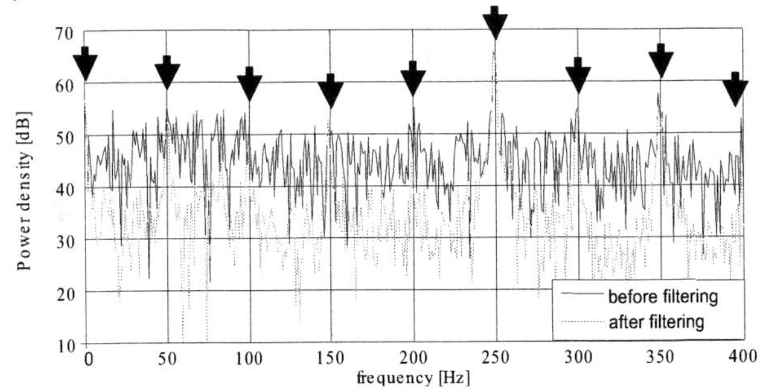

Figure 4: Effect of signal processing using 128-order filter for denoising of components other than 50 Hz harmonics (a) and using 1024-order filter, which enables the detection of low SNR components (b).

The final step of the studies was to implement the denoising filter part, which is responsible for the adaptive noise level reduction in one of the subbands, in FPGA device [1]. The basis for the hardware implementation of this filter was also the T-NLMS algorithm. The filter parameters were the same as for the software implementation. The result of the circuit operation is presented in the Figure 5. It confirms the possibility for hardware implementation of a fully functional adaptive subband filter that reduces the level of noise in periodic vibration signal which is recorded with the use of MEMS accelerometers.

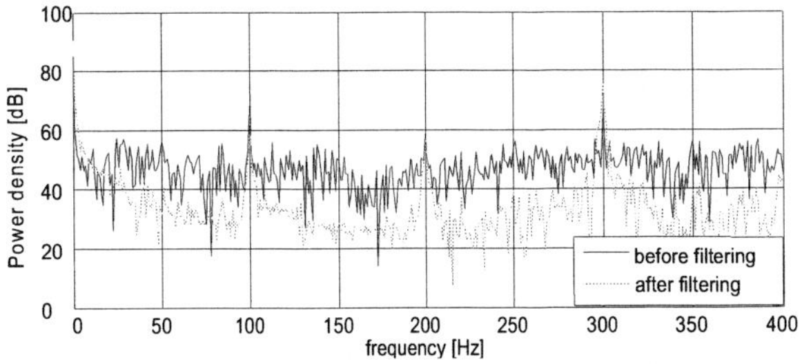

Figure 5: Exemplary result of the hardware-implemented adaptive noise filter operation

6 SUMMARY

The paper describes adaptive method for the reduction of the level of stochastic interference present in the turbogenerator vibration signal that is recorded with the use of micromachined accelerometers. The method applies the Subband Adaptive Line Enhancer circuit. The conducted research indicates the high effectiveness, parametrizability and the possibility for a relatively simple hardware implementation of the method. The proposed method can be employed in rotating machine diagnostic systems that apply harmonic analysis of the vibration signal. The implementation of the method enables the use of MEMS sensors in these systems.

REFERENCES

[1] P.Pietrzak, *Application of Micromachined Accelerometers for Vibration Measurements in Condition Evaluation Systems for Large Rotating Machines*, PhD Dissertation, DMCS TUL, Lodz, Poland, 2006

[2] B.Widrow, J.R.Glover, J.M.McCool, J.Kaunitz, C.S.Williams, R.H.Hearn, J.R.Zeidler, Eugene Dong, R.C.Goodlin, *Adaptive noise cancelling: Principles and applications*, Proc. of the IEEE, v. 63, Issue 12, pp. 1692 – 1716, 1975

[3] Haykin S., *Adaptive filter theory*, Prentice Hall, USA, 1995

[4] R.L.Campbell, N.H.Younan, J.Gu, *Performance analysis of the adaptive line enhancer with multiple sinusoids in noisy environment*, Signal Processing, v.82 n.1, pp. 93-101, January 2002

Microsystems on their Way to Smart Systems

L. Heinze*

* VDI/VDE Innovation + Technik GmbH
Steinplatz 1, 10623 Berlin, Gemany, heinze@vdivde-it.de

ABSTRACT

The aim of this paper is to discuss possible innovation paths from Microsystems to Smart Systems.

Microsystems combine several micro-techniques (Electronics, Mechanics, Optics, Fluidics, etc.) to new and often highly miniaturized products. In the future pure miniaturization will in many cases not be sufficient. As a next step of the technical evolution the implementation of a certain intelligence or smartness into the Microsystems has been considered. This would transform Microsystems into Smart Systems.

Therefore this paper deals with three core questions:
1. What are the requirements of Smart Systems when Microsystems evolve towards Smart Systems?
2. How will the development of Smart Systems influence the future of Microsystems Technologies (MST) and what are important fields of action?
3. What can MST learn from cognitive sciences and how can MST interact with cognitive sciences?

Keywords: Microsystems, Smart Systems Integration, Cognitive Sciences, Convergence of Technologies

1 REQUIREMENTS OF SMART SYSTEMS

Smart Systems in the sense of the European Technology Platform on Smart Systems Integration (EPoSS) are possessing self-diagnosis and sensory and actuatory capabilities. This enables them to describe and to evaluate situations. They can therefore decide and communicate with their environment. Overall these are qualities like self-organization, adaptation, individualization and personalization as well as certain autonomy.

These qualities can only be reached by using new integration technologies. The following techniques and technologies have to be integrated in a miniaturized manner:

- Autarkic energy supply
- RF technologies
- Signal recording and processing
- Preparation of own decisions

The necessary integration technologies consist out of a multitude of single technologies. Regarding the EPoSS Strategic Research Agenda [1] they are marked by:

- Heterogenity e.g. of materials and processes used
- Complexity (in the sense of its reduction!) e.g. of user interfaces and data volumes
- Scale comprehensive demands
- Autonomy e.g. by autarkic energy supply
- Multidisziplinarity e.g by convergence of technologies from physics, chemistry, engineering and cognitive sciences

For the implementation of the qualities mentioned above in future products the use of technologies from nano- and biotechnologies is not enough. This leads to a miniaturization and ubiquitous use of these functions only.

A good example is mobile phones integrating cameras, stereo systems, calendars, maps etc. today (figure 1). This will not make telephones smart in the sense of Smart Systems as noted above! Today the user has to be smart to use a mobile phone.

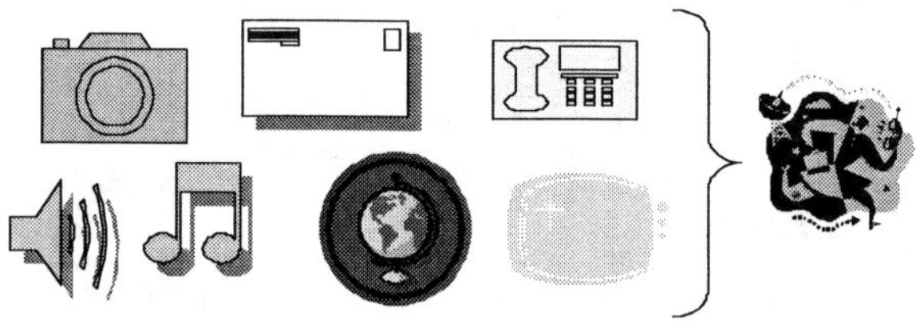

Figure 1: Today's miniaturization needs Smart Users

The same kind of problem exists for at least the majority of today's Microsystems: They are very well miniaturized but need a lot of smartness of the user e.g. when they are applied in the area of process control.

Certain intelligence and smartness may be brought to these systems by using results from cognitive sciences to solve this issue. Otherwise the development of Microsystems Technologies might end in a miniaturized dead-end alley.

On the other hand the implementation of cognitive abilities will lead to complex problems which have to be solved by cooperation of multiple disciplines like micro and nanotechnologies, cognitive sciences and micro-nano inte-

NSTI-Nanotech 2008, www.nsti.org, ISBN 978-1-4200-8505-1 Vol. 3

gration. This interdisciplinary approach might heavily influence the evolution of MST.

2 FUTURE DEVELOPMENT OF MICRO-SYSTEMS TECHNOLOGIES

How will the Smart System's demands influence the future development of MST? MST has integrated a fixed set of micro technologies (electronics, mechanics, optics, fluidics etc.) very successfully for more than 20 years. This has lead to many new and highly innovative miniaturized sensor and actuator systems. New independent engineering disciplines were born from the interdisciplinarity of MST in the past. MST is therefore considered as a motor of Convergence for Engineering Sciences. [2]

The above mentioned requirements of Smart Systems lead to the necessity to create a new knowledge body which overcomes the borders of classical Engineering. This new knowledge body has to incorporate fields like Biology and Cognitive Sciences. In consequence Engineering Science has to undergo not only quantitative rearrangement but also reinvention by using convergence processes. For the achievement of Smart System qualities the integration of several technologies and sciences is necessary. They are summarized by the term Smart Systems Integration (SSI). In fact MST will develop by integration of Biology and Cognitive Sciences towards SSI – and might be transformed into SSI in the end (figure 2).

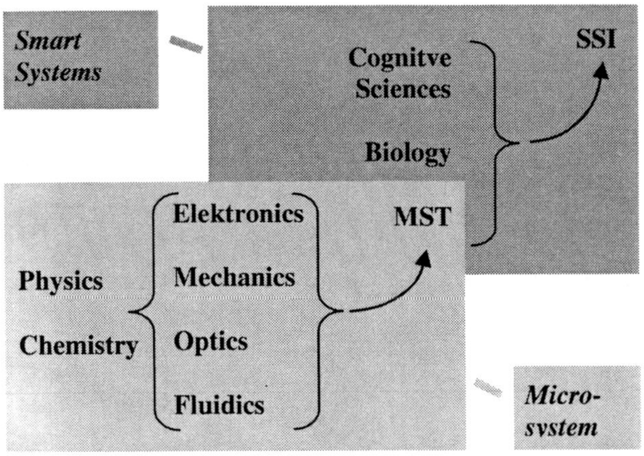

Figure 2: Advancement of MST towards SSI by Convergence of Technologies

From the above mentioned considerations about smart systems six action fields for MST may be derived:

- Integration Technologies
- Autonomy
- Networking of Autonomous Smart Sensor Systems
- Sensory Abilities
- Complexity
- Smartness & Cognition

In the following paragraphs current examples for specific topics are given out of the German Framework program Microsystems. [3]

2.1 Integration Technologies: Micro-Nano-Integration

The integration of nanostructures into micro and nano level as well as the application of effects based on nanotechnology is known as Micro-Nano-Integration. [4]

Activity areas of Micro-Nano-Integration are

- Processes e.g. for contactless assembly (self assembly, self organization) and low-temperature processes (reactive contacts ....).
- Nanomaterials- and stuctures like nano-based materials (e.g. underfiller, glop-top, solder paste, adhisives, ..), Carbon-Nano-Tubes (CNT), nanowires, nanolawn, etc.
- Equipment/ tools for the handling of nano-objects, e.g. cantilevers and grippers.

Results from research done in this activity areas will provide technologies necessary for the highly miniaturized hardware base of Smart Systems.

2.2 Autonomy and Networking of Autonomous Smart Sensor Systems

Autonomous Networked Sensor Systems (ANS) consist of single Microsystems (nodes), which are characterized by integrated sensors, local data (pre)processing, integrated (mostly wireless) communication (broadcasting and receiving) und autarkic energy supply. Often a large and scalable number of sensor systems are involved.

- Sensor networks are restricted by space and energy and self organizing.
- Cooperative data processing delivers precise results and new fields of applications.

Currently the most important problem of real autonomy is the question of energy supply. The energy needed for the receiving and broadcasting of data and information is collected from environment from light, vibration, heat etc. in the ideal case.

In Europe plenty of research is done on wireless sensor networks also with industrial partners. In general the focus lays on miniaturization and optimization towards targeted applications.

ANS will deliver the technologies for the networking of Smart Systems.

2.3 Sensory Abilities

A current topic of sensor technologies is the use of magnetic micro- and nanotechnologies using magnetic materials and effects on the level of micro- and nanostructures. They offer a lot of possibilities for the realization of sensors and actuators and autonomous magnetic Microsystems. Magnetic effects are very useful because of their robustness in harsh environments. In addition they might use much less energy and easy to miniaturize. Well established application areas are:

- Automotive e.g. with magnetic sensors for process control
- Magnetic sensors and contactless data transmission for traffic control
- Automation and process control using magnetic sensors and actuators
- Medicine technology e.g. using magnetic Nanoparticles for Bio-Analytics

Such new sensor systems are necessary to supply smart systems with "eyes" and "ears" to make them smart.

2.4 Smartness & Cognition

For the development of MST towards SSI it will be crucial to use cognitive abilities like thinking, learning and communication (speech). Systems will become smart only by using such abilities. In addition the ability for self organization and at least a certain autonomy are the base for smartness.

By understanding process of thinking and their application of technical system an interesting potential for the implementation of cognitive processes might arise.

Currently a lot of research is done to understand human brain and thinking. In the next step the results have to be transformed to machines where the restrictions in space and energy are not as hard as in miniaturized systems. First steps into this direction become visible in the research on cognitive robotics.

3 MST AND COGNITVE SCIENCES

The author sees in this action field the biggest challenges on the way to SSI because of the efforts which will be necessary to bridge the gap between MST as an engineering science and cognitive sciences as humanities.

A possible path for MST to start implementation of results from cognitive sciences might be the collaboration with the new field of Cognitive Robotics. Examples for Cognitive Robotics might be seen in the work of the German Cluster of Excellence CoTeSys [6] or the European Expert Platform "Feel Europe" [7].

In the shortly created cluster CoTeSys cognitive features for technical systems such as vehicles, robots, and factories are investigated. The aim of the project is to realize cognitive technical systems by equipping them with artificial sensors and actuators and let them act in the physical environment. The systems envisioned shall *"differ from other technical systems in that they perform cognitive control and have cognitive capabilities."* In the European project "Feel Europe" a European expert platform was created for the measurement of human feelings and emotions.

From the above mentioned thoughts the question arises: How might MST interact with Cognitive Robotics? The implementation of cognitive features might be realized either by soft- and/or hardware whereas the software might be seen as the "intelligent" part of the system and the hardware act as the sensory and actuatory part.

It is obvious that Microsystems will be necessary to provide miniaturized sensors to make such systems small enough for common use. An example from automotive industry might be the introduction of lane-keeping assistance systems with miniaturized CMOS cameras where it took over one decade until systems were miniaturized enough for common use. [8]

On the other hand today's Microsystems are small but often lack of smartness. To provide a path to Smart Systems it might be helpful to learn from cognitive robotics. Most common robotic systems have more than enough room for the implementation of software. Therefore it is very good platform to test and implement cognitive abilities into technical systems. The next step will be the stimulation of research for the implementation of cognitive elements into Microsystems.

The target of such research might be seen from figure 3. Until today Microsystems may be seen as structured into a sensory and an actuatory part which is connected by a signal processing unit.

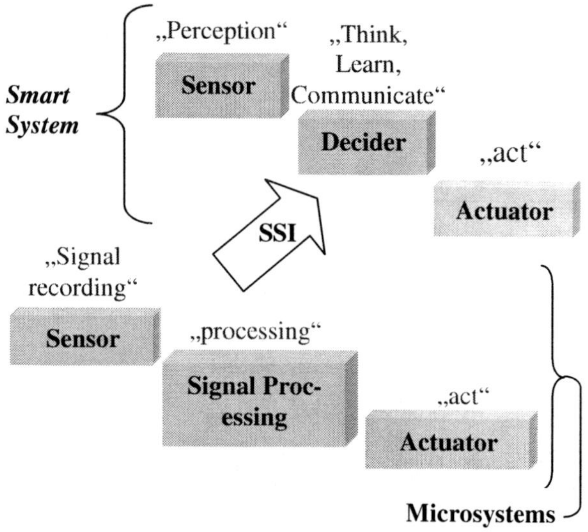

Figure 3: Influence of Cognitive Sciences on the development of Microsystems towards Smart Systems

This signal processing unit is programmed in a more or less fixed manner and therefore relies completely on the smartness of its programmer.

In the future smart system should contain a kind of "decider" unit instead which shall be able to think, learn and communicate. The development of this "decider" will be the central future challenge for the evolution from MST to SSI.

4 SUMMARY

Overall the development of Microsystems towards Smart Systems gives a chance to open a new field of research and innovation based on a bridge between Engineering and Cognition Sciences. This will promote the advancement of engineering sciences towards Smart Systems Engineering. This presentation reflects the possibilities and efforts connected with the convergence of MST and cognitive sciences. It is proposed to start the implementation of cognitive abilities by learning from and supporting cognitive robotics. The central challenge will be the development of a "decider" as cognitive heart of future Smart Systems.

5 ACKNOWLEDGEMENT

This work was supported by the German Ministry of Education and Research in the Framework Program Microsystems 2004 – 2009.

REFERENCES

[1] Strategic Research Agenda of the European Technology Platform on Smart Systems Integration (EPoSS / Version 1.2), February 28th, 2007, http://www.smart-systems-integrtion.org/public/documents/070306_EPoSS_SRA_v1.02.pdf

[2] L. Heinze "Converging Technologies for Smart Systems Integration - The Reinvention of the Engineering Sciences". mstnews 02/2007.

[3] L. Heinze, P. Coskina, H. Strese, B. Wybranski, Whitepaper Smart Systems Integration, April 2008

[4] P. Coskina @ www.mikro-nano.de

[5] H. Strese, L. Heinze @ http://avs.mikro-nano.de/

[6] CoTeSys cluster of excellence Cognition for TEchnical SYStems, http://www.cotesys.org/

[7] http://www.feeleurope.org/

[8] http://www.conti-online.com/ => search for "Lane Keeping Support"

A 20 μm Movable Micro Vehicle

Jangbae Jeon, J.-B. Lee and M.J. Kim

The University of Texas at Dallas
800 West Campbell Road, RL 10, Richardson, Texas 75080, jbjeon@utdallas.edu

ABSTRACT

This paper presents design, fabrication, and characterization of an extremely small (20 x 14 μm) movable silicon 3D micro vehicle. The micro vehicle was fabricated using a combination of focused-ion-beam (FIB) processes for micro/nano parts production, atomic layer deposition (ALD) for friction reduction coating, and a nano manipulator for semi-automated manipulation/assembly of micro/nano parts. A micromachined magnet was used as the frame for the micro vehicle and repeated motion of the vehicle using magnetic force was demonstrated. To the author's knowledge, this is the smallest movable micro vehicle ever been reported.

Keywords: 3D, movable, vehicle, focused-ion-beam, nano assembly

1 INTRODUCTION

Recently, there have been paradigm shifts in many fields of science and engineering to achieve ever smaller multifunctional devices. These kinds of studies and investigations including nanorobots and nanoelectromechanical systems (NEMS) have focused on developing process technologies for movable devices with various motions such as translation, rotation, and oscillation. Most of these studies are still in their primitive and embryonic stages. Even with such relatively young and immature miniaturization endeavor, countless demonstrations have been done to realize so many applications such as electronics, medicine for life science, and instrumentations including sensors. Such examples including logic devices [1], switches [2], and memory devices [3] were fabricated for electronic applications and nanomotor [4], nanotweezer [5], mass sensor [6] and motion sensor [7] were developed in micro/nano scale for instrumentations. Most of the life science applications are still in the stage of conceptual idea demonstrations [8]. It is expected that the commercial markets for nanorobotics and NEMS will drastically grow in the foreseeable future [9].

Miniaturization of robots was one of such types of researches and there have been numerous development activities on realizing small robots for biological [10] and industrial [11] applications. Electromagnetically-driven miniaturized movable vehicles were also investigated [12, 13]. These miniaturized robots and movable vehicles are in millimeter scales and were dependant upon manual assembly processes. It is evident that further miniaturization of robots and movable devices would open unforeseen window of opportunities in various applications. In addition, it would be highly desirable to have automation-friendly assembly processes to reduce the manufacturing time.

This paper presents the design and fabrication of a movable silicon 3D micro vehicle which is 20 μm long, 14 μm wide and has 3 μm diameter wheels and 400 nm diameter axles using semi-automated nano precision assembly process.

2 EXPERIMENT

The sequence for manufacturing of the movable micro vehicle was started with fabrication of various micron/sub-micron parts in silicon wafer using focused-ion-beam (FIB). Semi-automated assembly process with the micron/sub-micron parts was the next step. Since most of the individual parts fabricated in silicon were in micron or smaller scale, scanning electron microscopy (SEM) system equipped with a nano precision manipulation capability was utilized for fabrication and assembly.

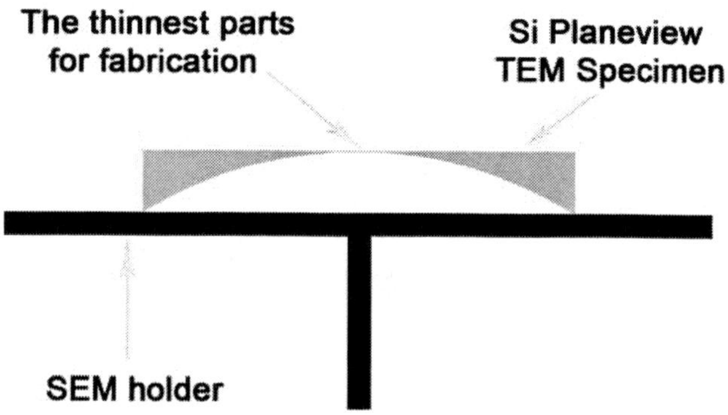

Figure 1. Schematic diagram of a 3 mm diameter silicon disk which was polished and dimpled by typical TEM sample preparation.

2.1 Fabrication of the parts

Various micron and sub-micron scale parts fabrication sequence was started with typical transmission electron microscopy (TEM) sample preparation process. Specifically, mechanical polishing of 3 mm diameter silicon wafer disk was utilized to thin down the silicon disk from approximately 500 μm in thickness to approximately 100 μm. After the mechanical polishing of the silicon disk, the center of the silicon disk was further processed with mechanical dimpling to thin it down to approximately 400 nm. The sample was flipped over and mounted on a scanning electron microscopy sample holder as illustrated in Figure 1. The thin mechanically dimpled part in the middle of the silicon disk is suspended over the SEM sample holder surface for focused ion beam machining.

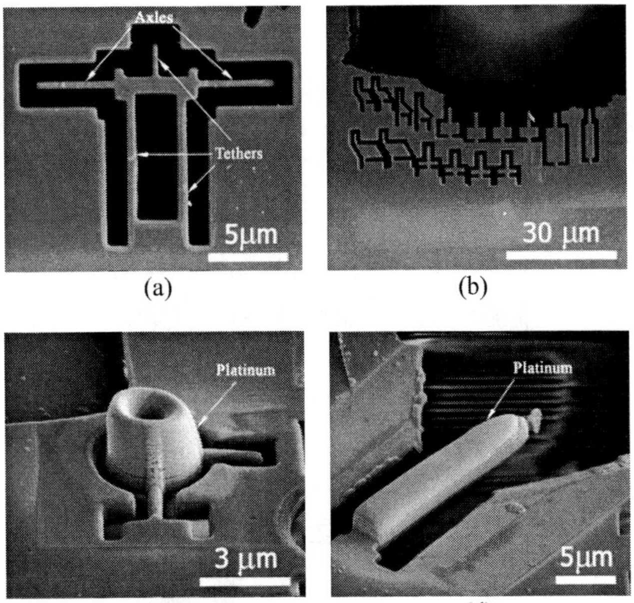

(c) (d)

Figure 2. SEM images of various micro vehicle parts; (a) 400 nm diameter axle; (b) various parts; (c) 3 μm diameter 3D wheel; and (d) frame.

Subsequently, the sample was placed in FEI Nova 200 NanoLab dual beam SEM/FIB system at UTD. Only the thinnest area in the middle of the silicon disk was used for fabrication of the parts. Axles with a part of frame and various body parts for the micro movable vehicle were milled with Ga^+ ion beam with acceleration voltage of 30 kV, probe current of 30 pA, dwell time of 100 ns, and depth of 500 nm (Figure 2-a, b). The minimum feature size of the micro movable vehicle part was 400 nm in axles. 3D wheels and the convex structure of the frame were manufactured by grayscale platinum deposition process by FIB with gas injection system (GIS) (Figure 2-c, d). The wheels have 3 μm outer diameter and 1 μm inner diameter and the frame was 2 μm in thickness. All parts were tethered on the silicon disk before assembly process.

2.2 Semi-automated assembly

The FEI Nova 200 NanoLab dual beam SEM/FIB system is also equipped with a Zyvex F100 nano-manipulation stage, which includes four manipulators with 10 nm positioning resolution. The Zyvex F-100 nano manipulator was extensively used for manipulation and positioning of individual parts during assembly as illustrated in Figure 3. The sample stage in the chamber has 4 degrees-of-freedom (DOF) (X, Y, Z, and Θ) and the actuator on the manipulator also has 4-DOF (X, Y, Z, and Φ), so that the precision robotics system has 5-DOF.

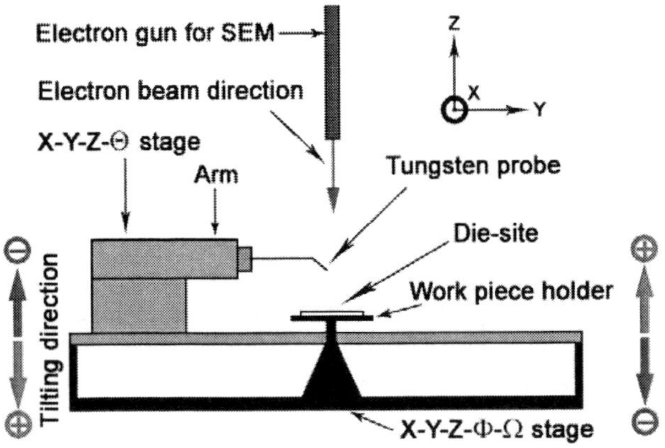

Figure 3. Diagram of the high precision robotics system used for assembly of movable micro vehicle.

In order to realize assembly of the individual micron/sub-micron parts, tungsten probe which is a part of the nano manipulator was manipulated to be positioned on top the edge of the tethered parts and slightly welded with chemical vapor deposition (CVD) of platinum with GIS like glue. After the probe attachment to individual parts, the parts were de-tethered by subsequent ion beam milling. The completely released the parts now attached to the tungsten probe of the nano manipulator were moved, rotated, and assembled in desired location.

The movable micro vehicle assembly was started with placing axles in upright positions (Figure 4-a). For standing the axles parts normal to the wafer surface, two de-tethered

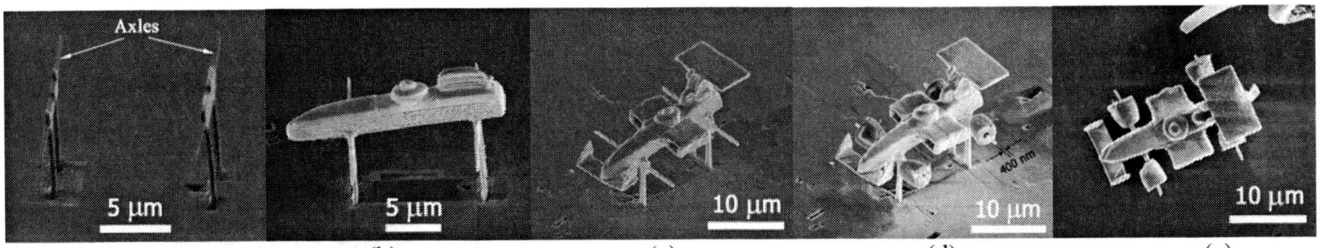

(a) (b) (c) (d) (e)

Figure 4. SEM images showing micro mobile assembly process: (a) axles placed in upright position; (b) frame placed on the axles; (c) body parts welded onto the frame; (d) wheels inserted into the axles; (e) completely de-tethered micro mobile on a silicon wafer axles parts were inserted in the holes which were pre-patterned on a silicon wafer and then secured by platinum welding. Next, the frame was placed on top of the axles and was attached to front and rear axles by platinum welding. Subsequently, platinum deposition was used to fabricate more decorative parts of the frame as shown in Figure 4-b. Onto this frame/axles module, various body parts were welded to create formula-one-like 3D movable micro vehicle. In order to reduce friction between axles and movable wheels, 30 nm thick Al_2O_3 layer was deposited using atomic layer deposition (ALD) [14] on this body of the mciro vehicle and wheels before assembly (Figure 4-c). Figure 5 shows SEM images of the wheels before and after the ALD of the Al_2O_3 layer. The surface of the wheels after ALD deposition of the Al_2O_3 layer was much smoother than the ones before the deposition.

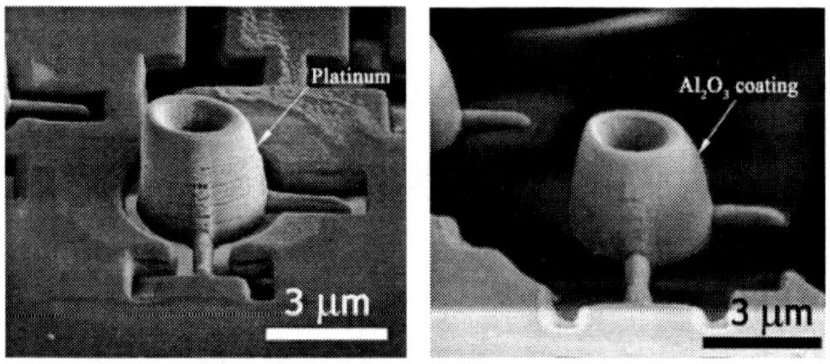

(a) (b)

Figure 5. SEM images of the wheels (a) before and (b) after atomic layer deposition of Al_2O_3 layer.

Finally, the wheels were inserted into the axles (Figure 4-d) and the upright positioned axles were de-tethered to free the micro vehicle (Figure 4-e).

2.3 Driving movable micro vehicle

The completed micro vehicle was placed on a silicon wafer and it was physically pushed by a computer controlled nano manipulator. It was clearly evident that wheels were physically rotated and the micro mobile was reliably and repeatedly driven. A micro mobile without Al_2O_3 coating was tested in order to understand the effect of the friction reduction coating and it was found that the wheels and the axles as well as the wheels and the substrate were stuck and immovable. In order to develop self-driven micro vehicle, commercially available neodymium iron boron (NdFeB) permanent magnet was micromachined to form a magnetic frame for the micro vehicle. A 5 µm thick membrane was prepared by the typical TEM sample preparation process. The frame was ion beam milled and manipulated for assembly. After a piece of the NdFeB magnet was attached to axels for the frame of the micro vehicle, another bigger piece of the magnet was cut and used as the external magnet. As the same piece of magnet was used for the external magnet, the magnetic dipole moment directions of both magnet pieces were known. The magnet-framed micro vehicle was repeatedly driven by external magnetic force using manipulation of the external magnet attached on a nano probe (Figure 6). Since the magnetic domain structure width of the NdFeB permanent magnet is mostly smaller than 5 µm, it was concluded that the magnet-framed micro vehicle which was longer than 10 µm in length was clearly driven by external magnetic field [15].

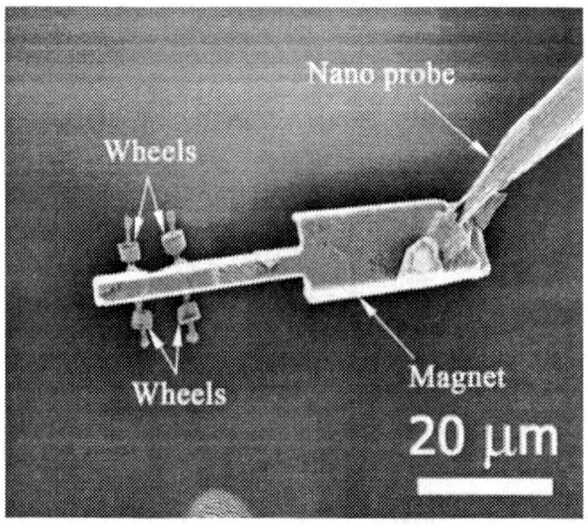

(a)

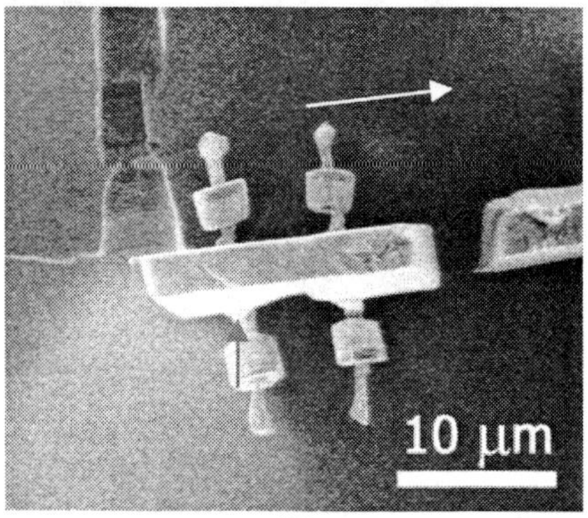

(b)

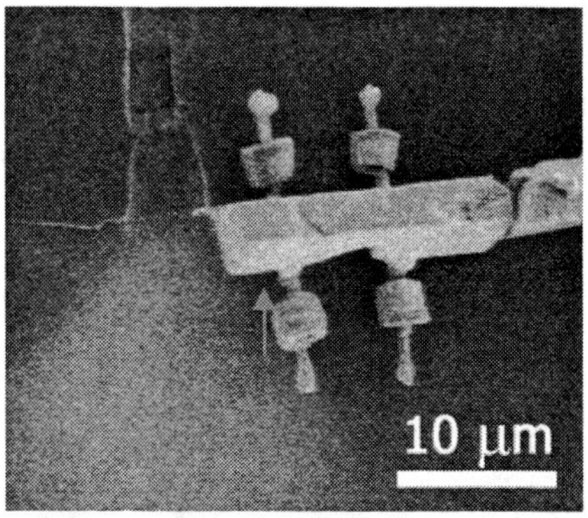

(c)

Figure 6. SEM images of micro vehicle with magnet body that can be driven by external magnetic field: (a) a magnet was attached to a micro vehicle; (b) the magnet was ion beam cut into two pieces; (c) micro mobile with magnet-body can be driven by the external magnetic field.

3 CONCLUSION

We have demonstrated the fabrication sequences and semi-automated assembly processes to realize 3D movable

NSTI-Nanotech 2008, www.nsti.org, ISBN 978-1-4200-8505-1 Vol. 3

micro formula-one-like vehicles with 20 μm in length and 400 nm minimum feature size. Although the fabrication process was a serial process, the time required for the entire manufacturing process for a micro vehicle was less than eight hours. Furthermore, the assembly process has potential to be automated since the nano manipulation system has script-driven capability [16].

We envision that this kind of ultra miniature movable vehicles will certainly open unprecedented opportunities for various nanorobotics and NEMS applications.

REFERENCES

[1] A.P. Graham, G.S. Duesberg, R. Seidel, M. Liebau, E. Unger, F. Kreupl, and W. Honlein, "Towards the integration of carbon nanotubes in microelectronics", Diamond Relat. Mater., 13, 1296-1300, 2004.

[2] Ho Jung Hwang, Jeong Won Kang, "Carbon-nanotube-based nanoelectromechanical switch", Physica E, 27, 163-175, 2005.

[3] Ricky J. Tseng, Jiaxing Huang, Jianyong Ouyang, Richard B. Kaner, and Yang Yang, "Polyaniline nanofiber/gold nanoparticle nanovolatile memory", Nano Letters, 5(6), 1077-1080, 2005.

[4] T.D. Yuzvinsky, A.M. Fennimore and A. Zettl, "Engineering nanomotor components from multi-walled carbon nanotubes via reactive ion etching", Proc. Of AIP Conference, 723, 512-515, 2004.

[5] Philip Kim and Charles M. Lieber, "Nanotube nanotweezers", Science, 286, 2148-2150, 1999.

[6] Y.T. Yang, C. Callegari, X.L. Feng, K.L. Ekinci, and M.L. Roukes, "Zeptogram-scale nanomechanical mass sensing", Nano Letters, 6(4), 583-586, 2006.

[7] G. Abadal, Z.J. Davis, B. Helbo, X. Borrise, R. Ruiz, A. Boisen, F. Campabadal, J. Esteve, E. Figueras, F. Perez-Murano, and N. Barniol, "Electromechanical model of a resonating nano-cantilever-based sensor for high-resolution and high-sensitivity mass detection", Nanotechnology, 12, 100-104, 2001.

[8] Adriano Cavalcanti, Bijan Shirinzadeh, Robert A Freitas Jr., and Tad Hogg, "Nanorobot architecture for medical target identification", Nanotechnology, 19, 015103, 2008.

[9] Margareth Gagliard, "Market research report: nanorobotics and NEMS", BCC Research, NAN042A, 2007.

[10] Paolo Dario, Maria Chiara Carrozza, Benedetto Allotta, and Eugenio Guglielmelli, "Micromechatronics in medicine", IEEE/ASME Trans. Mechatron., 1, 137-148, 1996.

[11] Koichi Suzumori, Toyomi Miyagawa, Masanobu Kimura, and Youkihisa Hasegawa, "Micro inspection robot for 1-in pipes", IEEE/ASME Trans. Mechatron., 4, 286-292, 1999.

[12] Harumi Suzaki, Nobuyuki Ohya, Nobuaki Kawahara, Masao Yokoi, Sigeru Ohyanagi, Takashi Kurahashi, and Tadashi Hattori, "Shell-body fabrication for micromachines", J. Micromech. Microeng., 5, 36-40, 1995.

[13] Akihiko Teshigahara, Masakane Watanabe, Nobuaki Kawahara, Yoshinori Ohtsuka, and Tadashi Hattori, "Performance of a 7mm microfabricated car", J. Microelectromech. Syst., 4, 76-80, 1995.

[14] Corina Nistorica, Jun-Fu Liu, Igor Gory, George D. Skidmore, Fadziso M. Mantiziba, Bruce E. Gnade, and Jiyoung Kim, "Tribological and wear studies of coatings fabricated by atomic layer deposition and by successive ionic layer adsorption and reaction for microelectromechanical devices", J. Vac. Sci. Tech., 23, 836-840, 2005.

[15] Zhen Rong Zhang, Bao Shan Han, Ye Qing He, and Shou Zeng Zhou, "Plate domain structure of sintered NdFeB magnets and dependence on compacting mode", Journal of Material Research, 16(10), 2992-2995, 2001.

[16] Tusher Udeshi and Kenneth Tsui, "Assembly sequence planning for automated micro assembly", The 6th IEEE Intern. Symp. on Assembly and Task Planning, 19-21, 2005.

Development of Active Systems for Military Utilization

J. Zunino III* and H.C. Lim**
*U.S. Army RDE Command, AMSRD-AAE-MEE-M
Bldg 60 Picatinny Arsenal, NJ, james.zunino@us.army.mil
**Physics Department, University Heights,
New Jersey Institute of Technology, Newark, NJ, hcl4186@njit.edu

ABSTRACT

The US Army is transforming into a lighter yet more lethal "objective force", all while fighting wars in the Middle East. Therefore, advanced technologies and materials are being developed and integrated into current and future weapon systems. These weapon systems must be deployable, be 70% lighter and 50% smaller than current armored combat systems, while maintaining equivalent lethality and survivability. To meet these requirements Army scientists and engineers are capitalizing on new technological breakthroughs. Members of US Army ARDEC are developing active materials and sensor systems for use on various military platforms, incorporating unique properties such as self repair, selective removal, corrosion resistance, sensing, ability to modify coatings' physical properties, colorizing, and alerting logistics staff when weapon systems require more extensive repair. The ability to custom design and integrate novel technologies into functionalized systems is the driving force towards the creation and advancement of active systems. Active systems require the development and advancement of numerous technologies across various energy domains (e.g. electrical, mechanical, chemical, optical, biological, etc.). These active systems are being utilized for condition based maintenance, battlefield damage assessment, ammunition assurance & safety, and other military applications.

Keywords: Army, military, sensors, coatings, active systems

1 INTRODUCTION

In the ever-changing world that the U.S. Army's systems must operate in, there is a need for materials and systems that can thrive and survive in an almost infinite variety of environments. Recent conflicts abroad have led the Army to transform into a lighter more lethal force. This transition requires weapon systems to be deployable, be 70% lighter and 50% smaller than current armored combat systems, while maintaining equivalent lethality and survivability [1]. Army scientists and engineers are capitalizing on new technological breakthroughs in nanotechnology, MEMS, etc. to develop materials and active systems to meet these needs.

Along with these new requirements, there is a thrust to transition from scheduled maintenance to condition based maintenance in order to save resources and improve readiness. The need exists for new solutions for structural health monitoring (SHM), armament assurance, and maintenance. The ability to perform prognostic and diagnostic analyses, in real-time, is a key objective for the Department of Defense (DoD). Active coatings, materials, and structures will allow the military to add more advanced capabilities while maintaining weight and lethality requirements.

The ability to custom design and integrate novel technologies into functionalized systems is the driving force towards the creation and advancement of active systems. Members of U.S. Army ARDEC and their partners are developing such active systems for condition based maintenance, battlefield assessment, ammunition assurance & safety, and for other military utilization.

While most materials are designed to operate in predetermined conditions, advances in chemistry, physics, engineering, and other related sciences allow one to create active materials and systems with the ability to react and respond to their surroundings in real-time. These materials, composites, and coatings systems may involve numerous components or layers integrated together combining functionality and capabilities.

These active materials and systems are currently being developed for various military platforms, incorporating unique properties such as self repair, selective removal, corrosion resistance, sensing, ability to modify coatings' physical properties, colorizing, and alerting logistics staff when weapon systems require more extensive repair [2].

The research being performed will directly and indirectly support the warfighter and allow the U.S. Department of Defense to remain in the forefront of active systems technologies.

2 ACTIVE SYSTEMS

Numerous universities and government agencies are currently investigating nanotechnology, active/reactive materials, and active systems. The majority work in specialized areas for a particular need or application. Many believe that active materials and systems should be designed to meet a given goal, or perform a set function, while others feel that active systems should and can be capable of possessing numerous functionalities in one system. U.S. Army ARDEC is employing a multi-disciplined team including experts in physics, chemistry, engineering, and other sciences to develop and integrate revolutionary technologies to meet military needs.

Rather than one "be all end all" system, the approach taken is to develop numerous technologies as solutions for desired requirements. These active systems not only have to meet operational requirements but energy and cost considerations as well. The ability to tailor properties for specific applications is a key feature in the successful implementation of active systems.

Some key areas of research and development at ARDEC include color modifying coatings, flexible electronics, wireless sensor packages, nanotube development, intelligent nano-clays, alternative fuel/power sources, de-painting/self-repair, material modification, and other military capabilities.

2.1 Materials & Coatings Systems

Nanostructured materials yield extraordinary differences in rates and control of chemical reactions, electrical conductivity, magnetic properties, thermal conductivity, and strength. The small feature size allows multiple systems and functions to be incorporated together and embedded into materials such as metals, polymers, paints/films, composites, etc. This gives one the ability to work at the molecular level, atom by atom, to create smart structures with fundamentally new molecular organization and yield advanced materials that will allow for longer service life and lower failure rates. These technologies will allow one to develop customizable material and coating solutions to meet military user requirements.

There is on-going research with nanotubes and their functionalization, development, and production. Single-walled carbon nanotubes (SWCNT) are being implemented into smart coatings and inks to initiate self-healing, active switching, sensing, color modification, and other functionalities. Nanotubes are also being utilized for power/fuel cells development and electroluminescence. Solubility and polymer wrapping of SWCNTs allows these tubes to be functionalized. Such technological advances have allowed for flexible solar cells to be fabricated using nanotube inks (Figure 1).

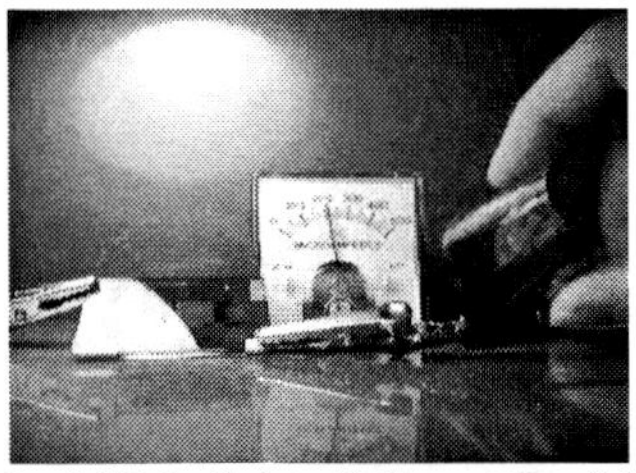

Figure 1. Liquid Ink Solar Cell on Flexible Substrate

There are also research efforts focused on the development of chemistries to enable production of single-walled nanotubes with precise but tunable dimensions (properties). Functionalized nanotubes are also being investigated to increase strength, and other properties in composites and other materials. Increasing the strength to weight ratio of structural materials will allow for better more robust systems to be created.

Besides nanotubes, nanoclays or micronized minerals are being added to current materials and coatings to add additional capabilities. The "intelligent clay" (*i-clay)* can be incorporated into coating/paint systems (Figure 2). These i-clays or smart materials rely on their capabilities to respond to physical, chemical, or mechanical stimuli by developing readable signals. They possess the ability to modify or change their properties and structure, in response to changes in their environment. These changes are often reversible but can be designed to be permanent as well.

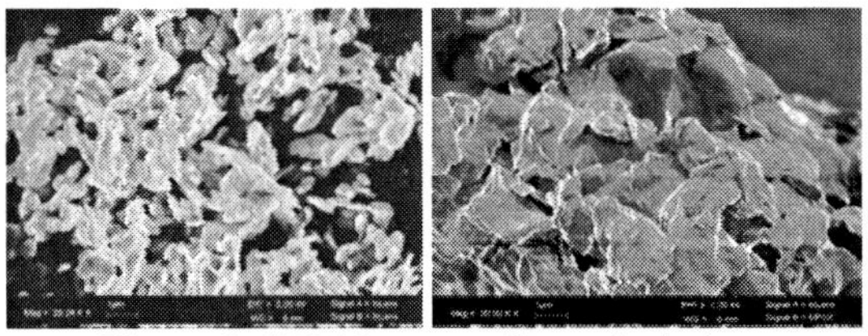

Figure 2. SEM images of sample i-clays

These i-clays can also be incorporated into inks, paints, composites, etc. to add functionalities to current coatings used on Army materiel. Currently the nanoclays (i-clays) are being incorporated in military paint systems to create active coatings. These active coating systems can detect corrosion, humidity, pH, chem/bio agents, etc. via color changes or luminescent properties.

The incorporation of i-clays to act as nanosensors for detecting degradation of coatings/paints as a result of corrosion and/or crack formation for roto-winged aircraft and other Army systems is underway. The i-clay additives are responsive to pH changes (cathodic reaction in oxidative corrosion) or oxidation through color changes as seen in Figure 3.

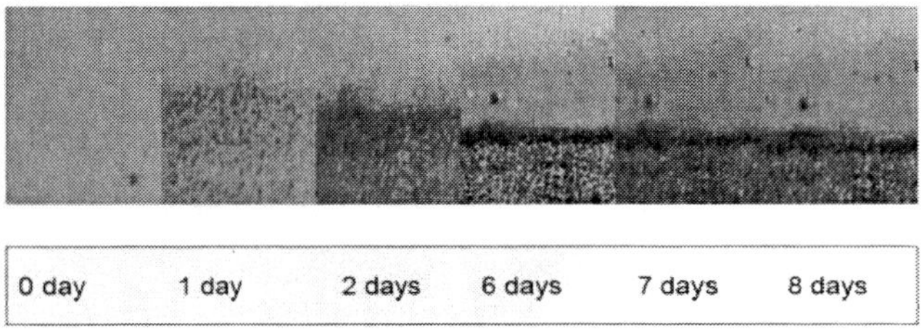

Figure 3. Results of i-clay Smart Coating in Accelerated Corrosion Test

When exposed to potential corrosion environments, the color changes before the actual corrosion and bubbling begins. Beyond the corrosion sensing capabilities, i-clay additives are also being incorporated for barrier properties, de-bonding, and self-repair capabilities. It is hoped that this research will provide solutions to corrosion related problems of military equipment, thus reducing the current multibillion-dollar expense associated with painting/de-painting operations.

Another approach to reduce the DoD's maintenance and painting costs is the development of Teflon-like nanocoatings. The protection of metal surfaces against thermal, chemical, corrosion, and biological injury without major adverse environmental effects remains a challenge. The long-term objective is to develop novel coatings that combine corrosion control, avoidance of chemical and biohazards, and other capabilities to survive harsh military operating conditions.

Metallo-organic composite materials based on a perfluorinated scaffold are of interest to ARDEC. Perfluoroalkyl polymers, such as Teflon, exhibit no C-H bonds and thus are chemically and thermally inert. However, the desired properties of such materials are also obstacles since they do not form strong polymer-metal surfaces bonds and resist the additions of other functionalities. A new, patented class of perfluorinated materials is under development. These materials incorporate metal centers in the central cavity of Phthalocyanine and unlike conventional phthalocyanines all H atoms are replaced by F atoms and -CxFy groups, Figure 4.

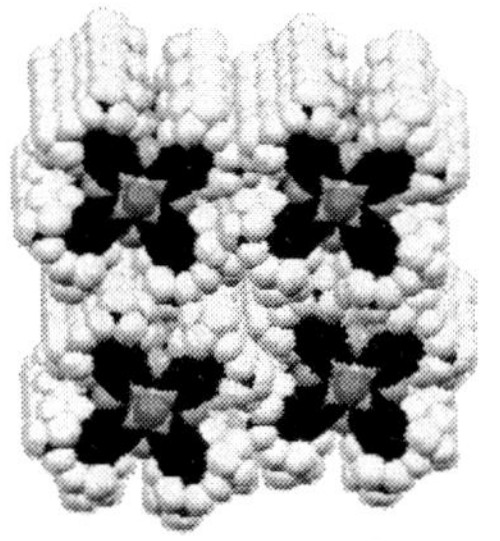

Figure 4. Bundle Assembly of Metallo-Orgnaic Phthalocyanine

These materials are especially of interest since the core is hydrophilic, yet the outer sections are extremely hydrophobic with tested contact angle measurements > 90°. Table 1 illustrates test results of these material systems.

Substrate Coating	Glass Coating effect (°) θ absolute value (°)	Aluminum Coating effect (°) θ absolute value (°)	Steel Coating effect (°) θ absolute value (°)
$F_{64}PcV{=}O$	52 86	41 99	26 138
$F_{64}PcFe$	60 95	50 110	22 132

θ absolute value (°) for H_2O on Teflon 115±4
θ absolute value (°) for H_2O on Polydimethylsiloxane 109

Table 1. Results of Contact Angle Variation Tests

The Teflon-Like properties of metallo-organic polymers developed expressed high thermal, (>300°C), and chemical resistance. Concentrated H_2SO_4, "Piranha solution" (H_2SO_4 /H_2O_2: "O"), Cl_2, and concentrated KOH were used to test chemical resistance [5]. All metal series, except for Mg showed no reaction. Further testing is underway to better understand the properties of these novel materials and their applicability for military utilization.

Another active system being developed involves thermal indicating polymers. Thermal chromic polymers are under development to alert Army logistic staff of dangerous temperature exposures. These polymers are being created and modified to change color when exposed to desired temperature stimuli. An example is a paint band placed on bullets that turn red if the round was exposed to unsafe temperature levels and maybe a safety concern (Figure 5).

Irreversible indication of the exposure of munitions in multiple thermal bands, 145°F-164°F, 165°F-184°F and over 185°F, is possible with thermal polymers. The resulting active coating can be visualy inspected by the human eye to alert if safe temperature ranges were exceeded. More detailed information, including cumulative time of exposure in certain temperature bands, through changes in optical reflectivity can be monitored using a hand-held laser system. The thermal indicating inks and paints can be added directly or added into coating systems to monitor munition items, containers, or any other components where thermal exposure information is desired.

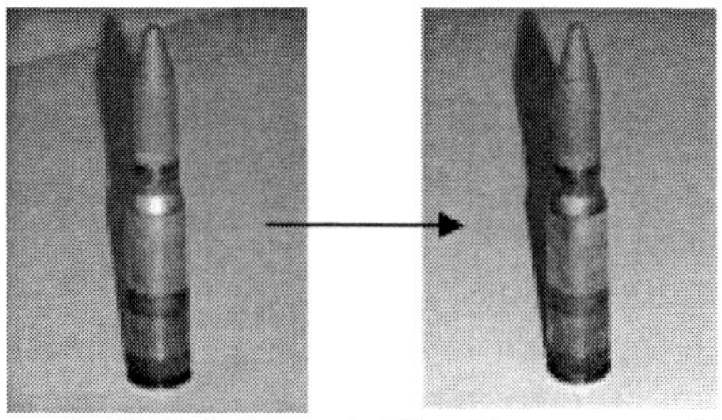

Figure 5. 50 Cal. Round After Exposure at 157°F

These active coatings systems are capable of monitoring elapsed time-temperature/radiation profiles as well as radiation and UV exposure. This is especially of interest for the monitoring of electronic devices and munitions during transportation and storage. Many munitions become unstable when exposed to temperatures beyond their design parameters.

2.2 Active Sensors Systems

Besides active coating systems, active sensor systems are also being developed. Several military programs are developing flexible electronic capabilities for sensing, communication, data collection/storage, and power alternatives. Using novel inks and nano-materials on various substrates has allowed ARDEC to develop several types of active sensors systems. Some of the sensing capabilities include temperature, damage, scratch, flow, pressure, strain, impact, shock, pH, humidity, chem/bio and acoustics. Other sensors' capabilities are under development.

Several different fabrication and manufacturing techniques are used by ARDEC and its partners. Besides techniques common to the development of microelectronics, MEMS, and the like, material printing techniques have been developed (Figure 6). The shift from typical micro-fabrication processes, often requiring cleanrooms or similar environments, to a material printing process greatly reduces the time and cost associated with active sensor development.

A total of 9 individual sensor modules are fabricated via a materials printer as part of the Active Coatings Technologies Program. The sensor modules include a strain sensor that measures the Young's modulus of substrates' bending (Figure 7); humidity sensor that monitors the moisture of the environment; corrosion sensor that detects the salinity of the liquid degrading the structure; fuse sensor for electrical current overload; acoustics sensor for low sound pressure sensing; pressure sensor which measures the actual pressure of a chamber; vibration sensor that can actively monitor the vibration frequency and shocks during transportation; infrared sensor that is capable of detecting near and far IR radiation; and impact sensor that senses the force impulse acting on the vehicle.

Figure 6. Dimatix Printing System.

In order to ensure better film to substrate adhesion condition the flexible substrates used have been thoroughly pre-cleaned with the 3-cycle standard pre-clean procedure with Plasma Enhanced Chemical Vapor Deposition (PECVD) surface roughness modification.

These sensors are fabricated with an aqueous dispersion of the intrinsically conductive piezo-resistive or piezo-electric polymers containing organic solvents and polymeric binders, sintered nano-particles gold, sintered nano-particles silver and nano-particles carbon. The conductive polymer ink is a hole-injection material (HIM) with a conductivity of a minimum surface resistivity depending on reformulation recipe. The sensors also have good photonic stability and good thermal stability of up to ~210°C.

These sensor suites are constructed on the flexible polyimide substrate membranes of 50 micron thicknesses and encapsulated with layers of dielectric and SiNx. The dielectric layers used are flexible polyimide resist ink that is inert to the ambient environment. The sensors with the final dielectric encapsulation layers are annealed at an elevated temperature of 300° C.

The device's sensing range and sensitivity can be modified by varying the sensing element's polymer thickness. Due the nature of the inks, these sensors can be used in harsh environments such as marine (salt water), outdoor (acid rain), rapidly fluctuating relative humidity and thermal shock conditions.

The utility of the sensor suites, for various weapon system applications, with the Army is on-going. Currently, the sensors are planned for integration into the AH-64 Apache Helicopter, and planned transition to unmanned aerial systems as part of RDECOM's ATO-M, "Embedded Sensor Processes for Aviation Composite Structures" [8]. Other variations of the sensors are being transitioned for ammunition surveillance projects, unmanned ground systems, and other Army projects.

3 CONCLUSIONS

Through the advancement of active systems, capabilities can be added to military assets. This will assist the DoD to protect both national and international interests. The overall goal is to develop active systems to be utilized on current military systems and to transition technologies to the field.

The need to protect our current and future military assets is obvious. It is in DoD's best interest to use the latest technologies to advance the protection of these assets. The current and future technological advances made are leading to the development of novel materials and systems that ultimately will allow the military to advance into the twenty-first century and beyond.

Through its R&D efforts, ARDEC is helping to advance the capabilities of the Army by integrating state–of-the-art technology into and on military systems. These technologies will result in new and modernized weapons systems fielded globally that are capable of meeting current and potential challenges.

REFERENCES

[1] *Future Combat System (FCS)*: Article. www.globalsecurity.org.

[2] J. Zunino III, et al. *U.S. Army Development of Active Smart Coatings™ System for Military Vehicles,* NSTI, Nanotech 2005.

[3] W.Feng, J.Zunino, M. Xanthos, S. Patel, M.Young, *Smart Polymeric Coatings-Recent Advances*, Advances in Polymer Technology, Vol. 26, No.1, 1-13, 2007.

[4] Research performed on the *Smart Coatings™ Materiel Program,* U.S. Army Corrosion Office, U.S. Army ARDEC-RDECOM, Picatinny, NJ.

[5] S.Gorun, *Integrated Teflon-like Adhesive Coatings with Long-term Stability and Built-in Functionalities*, Report for U.S. Army ARDEC-RDECOM, Picatinny Arsenal, NJ. 2008.

[6] J. Zunino III, et al. *Development of Active Sensor Capabilities for the U.S. Army's Active Smart Coatings System,* 2005 Tri-Service Corrosion Conference. Department of Defense & NACE.

[7] H.C. Lim, M. Pulickal, S. Liu, G. A. Thomas, J. Zunino, and J. F. Federici, Sensors and Actuators A. "Material corrosion determination via Young's modulus measurement using flexible membrane force sensor," *Sensors and Actuators A.*

[8] Army ManTech Manager, "ATO-M: Embedded Sensor Processes for Aviation Composite Structures," RDECOM Army ManTech / 2007 Brochure, U.S Army RDECOM, ATTN: AMSRD-SS-T, Pg. 12, 2007.

Sensing Weak Magnetic Fields By Living Systems and a Magnetoreception Mechanism For Navigation

Sergey Edward Lyshevski

Department of Electrical Engineering, Rochester Institute of Technology, Rochester, NY 14623-5603, USA
E-mail: Sergey.Lyshevski@mail.rit.edu Web: www.rit.edu/~seleee

ABSTRACT

The fundamentals for sensing weak magnetic fields in biological and *engineered* systems are researched. Our major goals are to study biophysics and apply cornerstone concepts examining existing premises and devising sound alternatives. The reported fundamental findings are applied. The feasibility analysis of proof-of-concept *solid* (silicon) and *hybrid* microdevices are documented.

Keywords: biosystems, magnetic field, sensing

1. INTRODUCTION

Some bacteria, migrating ants, bees, birds, fish, lobsters, salamanders, sea turtles and other living organisms likely exhibit the ability to sense the Earth's magnetic field and utilize the topographical mapping of the geomagnetic field for navigation, homing, foraging, etc. [1-17]. The magnetic properties of the closely-spaced biomineralized magnetite chains (~50 nm in diameter and length magnetites with ~5 nm separation) are utilized by magnetotactic bacteria for propulsion [3]. The iron oxide particles and their complexes are found in various living organisms, some of which are illustrated in Figure 1. These facts led to a hypothesis that intracellular biomineralized iron oxides could interact with the geomagnetic field thereby sensing its direction, variations, intensity and gradient. The cornerstone processes and mechanisms, utilized by living systems to detect the geomagnetic field, have being debated and are under extensive studies [1-17].

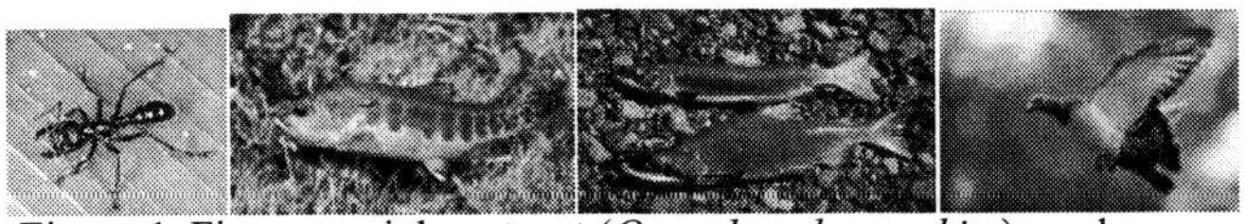

Figure 1. Fire ant, rainbow trout (*Oncorhynchus mykiss*), sockeye salmon (*Oncorhynchus nerka*) and homing pigeon

A great variety of biomineralized iron oxide particles (maghemite γ-Fe_2O_3 and ε-Fe_2O_3, magnetite Fe_3O_4, hematite α-Fe_2O_3 and β-Fe_2O_3, wuestite FeO and other) were found within distinct orientation, patterns, etc. The size, shape, morphology, crystallography, spacing, magnetic moment orientation (single-domain, two-domain, superparamagnetic, etc.), magnetic dipole moment, magnetic and thermal stability, as well as other properties of biomineralized iron oxide particles and clusters vary.

The biomineralized magnetic iron oxides and corresponding receptors could constitute magnetoreceptor cellular assemblies within the peripheral and central nervous systems. Theoretically, these magnetoreceptors can sense the geomagnetic field utilizing the electrochemo-mechanical transitions. In addition, memory storage and retrieval can be accomplished.

2. FEASIBILITY ANALYSIS

The fundamentals of sensing, information retrieval, memory and processing by biosystems remain to be coherently researched performing fundamental and experimental studies. These findings may lead to alternative solutions and re-assessment of basic postulates and premises. The various aspects of magnetoreception (possible mechanisms, phenomena, effects, system organization, etc.) are important due to possible implications to *engineered* systems. We research:

1. Possible biophysics for sensing geomagnetic fields by biosystems, memory storage and memory retrieval;
2. Synthesis and design of *engineered* systems to sense magnetic fields.

Assuming the validity of the magnetic field sensing premise by biosystems, there are many fundamentally distinct phenomena, effects and mechanisms which potentially can be utilized. They range from the quantum mechanics (metastable states, quantization, spin-orbit interaction, etc.) to classical electromagnetics and microfluidics [18]. For example, an electron may have a spin magnetic dipole moment $\mathbf{m}_{spin}$ $\pm 9.27\times10^{-24}$ A-m^2 with the alignment aiding or opposing an external magnetic field. As illustrated in Figure 2, there are small variations of the geomagnetic field which imply very small variations of the *microscopic* system energetics. Though these variations can be utilized by *microscopic* systems ensuring overall functionality and soundness, the resulting changes and transitions may or may not be observed and characterized due to fundamental and technological limits. In particular, the Heisenberg uncertainty principle provides the position-momentum and energy-time limits on the measurements as $\sigma_x\sigma_p \geq \frac{1}{2}\hbar$ and $\sigma_E\sigma_t \geq \frac{1}{2}\hbar$.

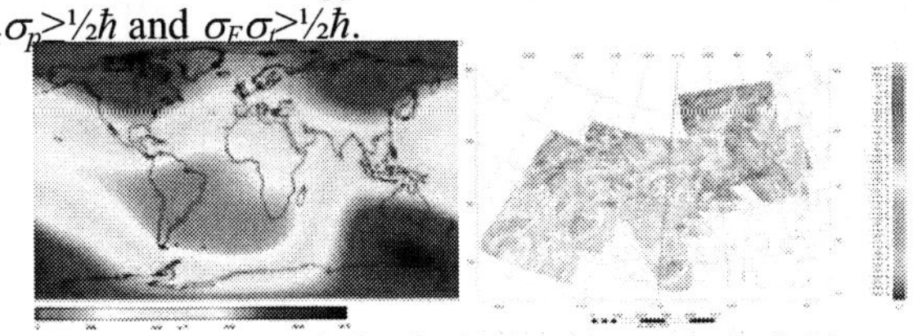

Figure 2. Variation in the Earth's magnetic field

Organic magnets may not exhibit sufficient changes to the mesoscale cellular structures at room temperature. For α-1,3,5,7-tetramethyl-2,6-diazaadamantane-N,N`-doxyl, $[Fe^{III}(C_5(CH_3)_5)_2]^+$[tetracyanoethylene]$^-$, as well as α- and β-$[Fe^{III}(C_5(CH_3)_5)_2]^+$[tetracyanoquinodimethane]$^-$ it was found that the Curie temperature is 1.48K, 4.8K, 2.55K and 3K, the saturation magnetization is 48300, 37600, 34200 and 21600 A/m, while the coersivity is very low.

For *microscopic* systems, the developments may be centered on individual molecules or their assemblies studying specific quantum effects, bond formation/braking, conformation and other phenomena which may be exhibited and utilized to ensure overall functionality at the device and system levels. These premises may be based on unverifiable hypotheses, limiting these developments mainly to theoretical studies. These advancements although having an essential theoretical importance, may not be expected to be materialized as a feasible technology in near future. We focus on the conventional electromagnetics for meso- and macroscopic systems for which well-developed technologies (CMOS, micromachining, synthetic chemistry and other) exist.

To potentially contribute to the biophysics of *natural* systems and apply the results to *engineered* systems, we study the interactive electromagnetic-mechanical phenomena of clustered magnets (magnetic particles) with various molecular (*microscopic*), mesoscopic and macroscipic (*bulk*) receptors' and sensors' assemblies. It is found that weak magnetic field variations result in sufficient changes in the mesascopic system states and quantities. These transitions can be utilized guarantying the overall functionality. The sensing mechanism can be based on the changes of physical quantities (variations of strain, charge, conformation, etc.) caused by the interaction of magnetic clusters, which have the magnetic dipole moment $\mathbf{m}(\mathbf{r})$, with the field $\mathbf{B}$.

3. ELECTROMAGNETICS AND ITS APPLICATION

The magnetic clusters cause electromagnetic interactions. The resulting forces may exert on biomolecular assemblies which can form biological receptors. Single-domain uniform-lattice magnetite (Fe_3O_4) from 30 to 100 nm with the coercivity ~40 mT are found in bacteria [3]. In the pigeon beak, the ~3 nm magnetites are arranged in organized ~1 μm-diameter ferromagnetic or superparamagnetic clusters (assembly of ferrimagnetic or ferromagnetic particles in non-ferromagnetic matrix) within dendrites. In addition, the maghemite clusters occur around the vesicle (diameter ~5 μm) as well as ~10 μm-long bundles of single crystalline uniform square platelets (~1×1×0.1 μm) within the dendrite in the ordered pattern [15]. A ferromagnetic magnetite (the orbital and spin magnetic dipole moments obey $|m_{spin}|>|m_{orb}|$) exhibits a response to an external magnetic field. We consider:

1. Electromagnetic interactions of macroscopic ferromagnetic and superparamagnetic particles/clusters which lead to electromagnetic-mechanically induced transitions in biomolecular assemblies. Single-domain magnetite has been localized in the nervous system of various living organisms, and, correspondingly, may result in the subcellular level of sensory, memory and processing;
2. Microfluidics. The ordered, disordered and controlled dynamic and static behavior of particles (typical size is from ~10 nm to ~10 μm) can be utilized resulting in the possible electromagnetic field-induced viscoelastic, strain-caused and other transitions. The behavior of magnetite clusters can be examined and utilized in *engineered* systems. There are concerns that the arrangement and morphology of the magnetite in the dendrites, receptors and subcellular structures may not comply or be comprehended.

There is contradicting data in magnetic properties of the biomineralized and synthesized iron oxide particles and clusters. For example, the reported coercivity ~10 mT and magnetic dipole moment ~1×10^{-17} A-m^2 for a ~1 μm magnetite cluster were questioned. Maghemite γ-Fe_2O_3 has inherent cation vacancies V in the octahedral positions. From $4Fe_2O_3 \rightarrow 3\{Fe^{3+}O\cdot(Fe_{5/3}{}^{3+}V_{1/3})O_3\}$ one concludes that possible order-disorder at different sites are affected by the synthesis methods resulting in distinct characteristics. Various methods have been reported to synthesize iron oxide particles [19]. For example, ferrous chloride tetrahydrate $FeCl_2\cdot4H_2O$ and ferric chloride hexahydrate $FeCl_3\cdot6H_2O$ can be used. To neutralize the anionic charges on the particles surface, 1N hydrochloric acid HCl is used. The major steps are depicted as the reaction: $FeCl_2+FeCl_3 \rightarrow Fe_3O_4 \rightarrow \gamma\text{-}Fe_2O_3$. In the first step, Fe(II)/Fe(III) with the molar ratio 1:2 are dissolved in water with sonication. The resulting solution is poured into alkali solution. Then, the precipitate is collected using a magnet, and the supernatant is removed from the precipitate by decantation. Deoxygenated water is added to wash the powder, and the solution is decanted after centrifugation. After washing the powder, 0.01M HCl solution is added to the precipitate to neutralize the anionic charges on the particles surface. The resulting magnetite Fe_3O_4 is separated by applying an external magnetic field. The magnetite can be transformed into maghemite crystallites by oxidizing them at ~300^0C by aeration.

Two typical magnetization-applied filed (*M-H*) curves (with and without hysteresis) for ~5 nm maghemite γ-Fe_2O_3 spherical particles, synthesized utilizing distinct procedures, are illustrated in Figure 3. The ferromagnetic (*M-H* curves with hysterisis) and superparamagnetic (no hysterisis) are observed. One recalls that $\mathbf{B}=\mu_0(\mathbf{H}+\mathbf{M})$, $\mathbf{M}=\chi_m\mathbf{H}$, $\mathbf{B}=\mu_0\mu_r\mathbf{H}$, where χ_m is the magnetic susceptibility; μ_r is the relative permeability, $\mu_r=1+\chi_m$. For FeO, the magnetic molar susceptibility $\chi_m V_m$ is 7.2×10^9 cm^3/mol, where V_m is the molar volume. For the organic compounds (C_2H_2, C_6H_6, $C_6H_{12}O_2$, $C_{20}H_{12}$, etc.), the diamagnetic molar susceptibility varies as ~[2.5 20]$\times10^7$ cm^3/mol.

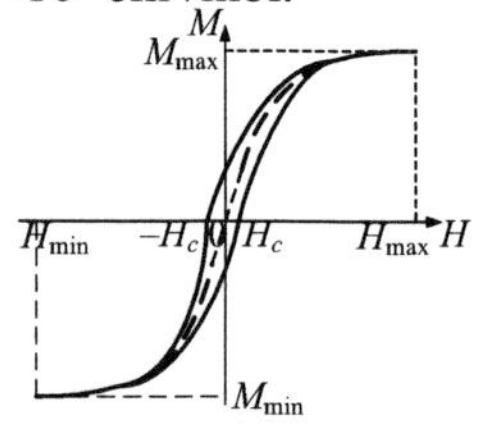

Figure 3. *M-H* curves for ~5 nm maghemite, H_{max} is ~100 A/m

Consider the translational and *torsional-mechanical* motion of magnetic clusters in the magnetic field. The electromagnetic translational and rotational transitions result due to the force and torque developed. The torque $\mathbf{T}$ tends to align $\mathbf{m}$ with $\mathbf{B}$, and $\mathbf{T}=\mathbf{m}\times\mathbf{B}$. For a magnetic rod with the length l and the pole strength Q_m, the magnetic moment is $m=Q_m l$, while the force is $F=Q_m B$. The electromagnetic torque is $T=2F\frac{1}{2}l\sin\alpha=Q_m lB\sin\alpha=mB\sin\alpha$. Thus, $\mathbf{T}=\mathbf{a}_m m\times\mathbf{B}=Q_m l\mathbf{a}_m\times\mathbf{B}$, where $\mathbf{a}_m$ is the unit vector in the magnetic moment direction.

With the average magnetic field of the Earth ~50 μT, which varies ~±0.5 μT, the torque is estimated to be ~1 pN-m. The Newtonian translational and *torsional-mechanical* dynamics are governed by the differential equations $\Sigma\mathbf{F}=m_m\boldsymbol{a}$ and $\Sigma\mathbf{T}=J\boldsymbol{\alpha}$. Here, $\boldsymbol{a}$ and $\boldsymbol{\alpha}$ are the linear and angular accelerations, $\boldsymbol{a}=d\mathbf{v}/dt=d^2\mathbf{r}/dt^2$ and $\boldsymbol{\alpha}=d\boldsymbol{\omega}/dt=d^2\mathbf{r}/dt^2$; m_m and J are the mass and moment of inertia.

Using the pole strength Q_m, the force acting on a magnet is $\mathbf{F}=\mathbf{B}Q_m$. The force between two magnets depends on the shape, magnetization, orientation, etc. The Coulomb law provides the equation for the force. For two magnetic poles we have $\mathbf{F}=\mathbf{a}_r\frac{\mu_0}{4\pi}\frac{Q_{m1}Q_{m2}}{r^2}$, where $\mathbf{a}_r$ is the unit vector along line joining poles; Q_{m1} and Q_{m2} are the pole strengths; r is the distance between poles. The flux density at distance r from a pole with Q_m is $\mathbf{B}=\mathbf{a}_r\frac{\mu_0}{4\pi}\frac{Q_m}{r^2}$.

The magnetization is defined as the *net* magnetic dipole moment per unit volume, e.g., $\mathbf{M}=\mathbf{m}/V=Q_m\mathbf{l}/V$. For a uniformly magnetized cylindrical magnet of length l and cross-sectional area A, we have $M=Q_ml/Al=Q_m/A$. The pole surface density is $\rho_{sm}=Q_m/A=M$. For a cylindrical magnet with length l and radius r_m, the magnetic flux density on the axis is $\mathbf{B}=\frac{1}{2}\mu_0 M\left(\frac{z}{\sqrt{z^2+r_m^2}}-\frac{z-l}{\sqrt{(z-l)^2+r_m^2}}\right)\mathbf{a}_z$. The conformations of the receptors (due to the exhibited electromagnetic force) are studied. The quantitative and qualitative analysis is performed to study the magneto-receptor-centered magnetic field sensing.

4. Information Storage and Retrieval

Assuming the utilization of superparamagnetism, the energy required to change the direction of the particle magnetic moment is comparable to the ambient thermal energy. In ferromagnetic materials, the magnetic moments of neighboring atoms align, resulting in a large internal magnetic field. Superparamagnetism occurs when the temperature of material, composed of ~1 to 10 nm crystallites, below the Curie or Neel temperature (the thermal energy is not sufficient to overcome the coupling forces between neighboring atoms). Hence, the thermal energy is sufficient to change the direction of magnetization of the entire crystallite. Each atom is independently affected by an external magnetic field, and the magnetic moment of the entire crystallite tends to align with the magnetic field. Superparamagnetism establishes a limit on the minimum size of particles, resulting in constraints on the memory (storage) functionality, capabilities and density. The relative motion of the iron oxide particles/clusters with respect to each other can result in the longitudinal or perpendicular data storage (recording) which can be assessed and retrieved, see Figure 4. For the perpendicular storage, the memory density limit is ~1000 bit/μm² which may be sufficient to ensure the geomagnetic field mapping by biosystems. The magnetization of the element should be retained despite thermal fluctuations caused by the superparamagnetic limit. The energy required to reverse the magnetization of a magnetic element is proportional to the size and the magnetic coercivity of the magnet. Biominerilized iron oxides could possess sufficiently large coercivity ensuring thermal stability thereby preventing demagnetization.

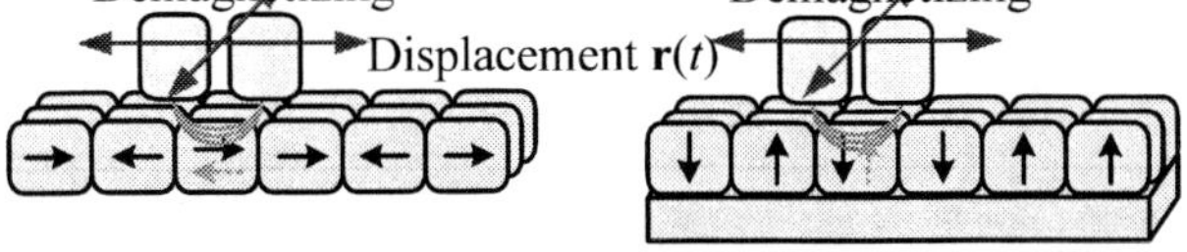

Figure 4. Longitudinal and perpendicular data storage utilizing electromagnetic- or pressure-induced magnetization: Demagnetizing and magnetic storage elements are in the relative motion

5. *Engineered* Magnetic Field Sensors

Researching fundamentals of *engineered* magnetic field sensing devices, we concentrate on basic physics and current technologies to complement theoretical findings. The polymer chemistry and CMOS-centered technology are well-established ensuring high-yield mass-production. Maghemite, magnetite, hematite, wuestite and other oxides were synthesized and characterized. Polymer microcapsules with embedded magnetic particles can be synthesized. The polymer microcapsule's shells are formed as magnetic particles, dispersed in the hydrophobic polymer (for example, NOA prepolymer), are captured into the solid polymer phase at the emulsification step with the subsequent curing and drying. These oxides and microcapsules can be deposited on the movable diaphragm. The magnetic field can be sensed and measured as the membrane deflection or induced *emf*. The micromachined structures, components, proof-of-concept devices and *solid* (silicon) prototypes were designed and fabricated as reported in Figure 5.a [18, 20]. The deflection of the suspended (released) movable structures or diaphragms can be measured by using the variations of capacitance and resistance. For example, the micromachined four polysilicon resistors, which form the Wheatstone bridge, are documented in Figure 5.a [18, 20].

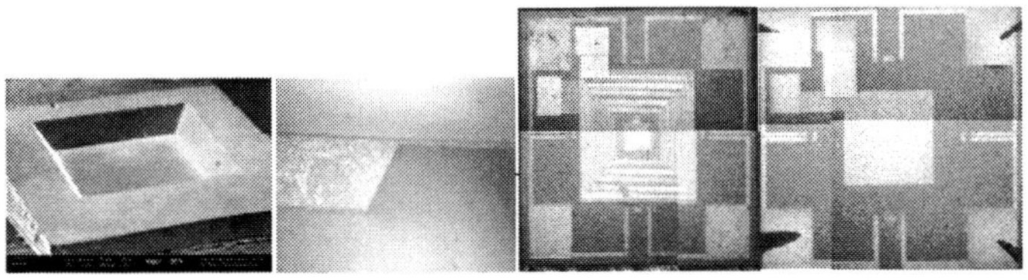

(a)

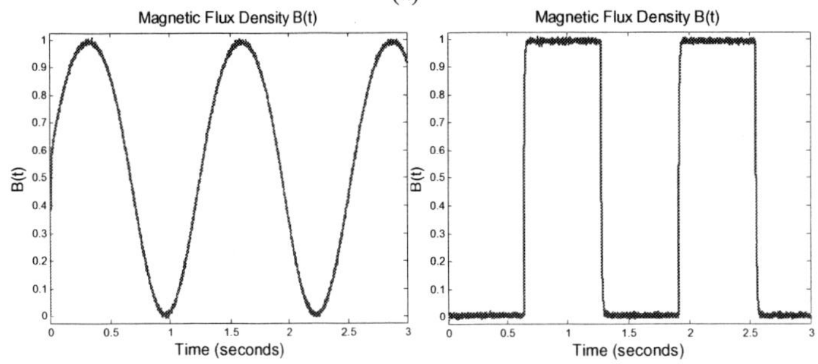

(b)

Figure 5. (a) Etched silicon structure, ~30 μm silicon diaphragm, and micromachined sensors with polysilicon resistors (to measure the force-induced deflection) and Al coils on the silicon diaphragm; (b) Sensing time-varying $B(t)$

The magnetic field is measured by using the force-induced displacement-centered sensing mechanism. Figure 5.b documents the deflection of the suspended diaphragm measuring $B(t)=\frac{1}{2}[\sin(\frac{1}{4}\pi t)-1]$ and $B(t)=\frac{1}{2}[\text{rect}(\frac{1}{4}\pi t)-1]$ mT. The noise $\xi(t)$ can be filtered by ICs. We conclude that it is possible to accurately sense time-varying $\mathbf{B}$. These sensors are key components of navigation systems.

In addition to *solid* concepts, *engineered fluidic* weak magnetic filed sensors are currently under development. We focus on soundness, technological feasibility, practicality and device capabilities. Two possible solutions are briefly reported below.

Various magnetic particles can be coated by polymers and suspended in liquid and solid matrices. For example, the solvent-free surface-functionalized maghemite γ-Fe_2O_3 can be functionalized by a positively-charged organosilane $(CH_3O)_3Si(CH_2)_3N^+(CH_3)(C_{10}H_{21})_2Cl^-$ which forms covalent bonds with the surface hydroxyl groups. A counter anion is $R(OCH_2CH_2)_7O(CH_2)_3SO_3^-$, $R{:}C_{13}$–C_{15} alkyle chain. This ultimately may lead to an *engineered* inorganic apparatus to sense the magnetic field.

Protein complexes, which exhibit magnetic properties, can be synthesized [21-24]. For example, the ferritin protein, which consists of 24 protein subunits, is illustrated in Figure 6. Inside the ferritin shell, iron ions form crystallites with phosphate and hydroxide ions. The resulting complex is similar to the mineral ferrihydrite. Different physiological functionality of ferritin was defined. Ferritin can be used as a marker as well as a precursor. Native ferritin (iron-storage protein) has a spherical shell with an external diameter of 12 nm and an inner core diameter of 8 nm. A *natural* mammalian ferritin protein has an antiferromagnetic 8 nm iron oxyhydroxide core Fe(O)OH with ~4500 Fe^{3+} atoms. This core forms a noninteracting monodispersed superparamagnetic structure. Due to the structural defects and uncompensated surface moments, each iron oxyhydroxide particulate possesses a net magnetic moment due to uncompensated unpaired spins. The ferritin protein and other similar protein cages can be emptied of its contents and mineralized with different complexes [25-27]. With the embedded ferromagnetic maghemite γ-Fe_2O_3, the suspended *synthetic* ferritins can sense the magnetic field utilizing the force generation and displacement mechanisms.

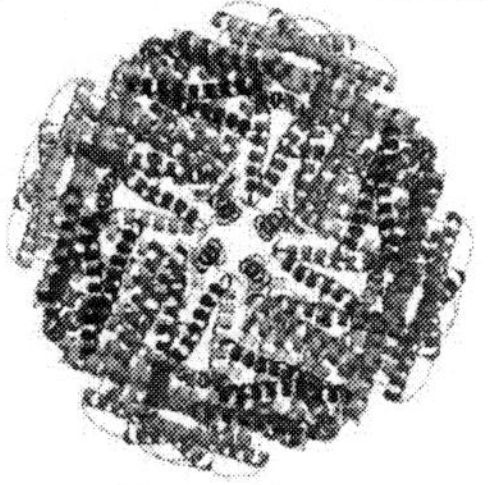

Figure 6. The crystallographic structure of the ferritin

6. Conclusions

Applying fundamentals of science and engineering, we outline and research established and potential paradigms in the sensing of weak magnetic fields. Our overall objective was to advance *engineered* solutions by developing alternative concepts observed in living systems. There is a need to span a wide range of multidisciplinary research activities focused on central issues of biophysics, neuroscience, engineering science and technology. This will lead to synergetic intellectual partnerships to strengthen the fundamentals of theoretical and applied science and engineering. We addressed important issues, studied *natural* systems, proposed possible inroads, and reported alternative solutions. These results promise a profound impact on the ability to create, generate and apply new knowledge contributing to scientific innovations in magnetic field sensing and navigation system designs.

References

1. T. Alerstam, *Bird Migration*, Cambridge University Press, Cambridge, 1990.
2. J. B. Anderson and R. K. Vander Meer, "Magnetic orientation in fire ant *Solenopsis invicta*," *Naturwissenschaften*, vol. 80, pp. 568–570, 1993.
3. R. P. Blakemore, "Magnetotactic bacteria," *Science*, vol. 19, pp. 377-379, 1975.
4. Y. Camlitepe and D. J. Stradling, "Wood ants orient to magnetic fields," *Proc. R. Soc. Lond.* B, vol. 261, pp. 37-41, 1995.
5. J. C. Diaz-Ricci and J. L. Kirschvink, "Magnetic domain state and coercivity predictions for biogenic greigite (Fe_3S_4): A comparison of theory with magnetosome observations," *J. Geophys. Res.*, vol. 97, pp. 17039-17315, 1992.
6. C. E. Diebel, R. Proksch, C. R. Green, P. Neilson and M. M. Walker, "Magnetite defines a vertebrate magnetoreceptor," *Nature*, vol. 406, pp. 299-302, 2000.
7. D. M. S. Esquivel, D. Acosta-Avalos, L. J. El-Jaick, M. P. Linhares, A. D. M. Cunha, M. G. Malheiros and E. Wajnberg, "Evidence of magnetic material in the fire ants *Solenopsis* sp. by electron paramagnetic resonance experiments," *Naturwissenschaften*, vol. 86, pp. 30-32, 1999.
8. G. Fleissner, B. Stahl, P. Thalau, G. Falkenberg and G. Fleissner, "A novel concept of Fe-mineral-based magnetoreception: Histological and physicochemical data from the upper beak of homing pigeons," *Naturwissenschaften*, 2007.
9. J. L. Gould, "The case for magnetic sensitivity in birds and bees (such as it is)," *Am. Sci.*, vol. 68, pp. 256-267, 1980.
10. J. L. Kirschvink, "Magnetite biomineralization and geomagnetic sensitivity in higher animals: and update and recommendations for future study," *Bioelectromagnetics*, vol. 10, pp. 239-259, 1989.
11. J. L. Kirschvink, M. M. Walker, and C. E. Diebel, "Magnetite-based magnetoreception," *Current Opinion in Neurobiology*, vol. 11, pp. 462-467, 2001.
12. S. Mann, N. H. C. Sparks, M. M. Walker and J. L. Kirschvink, "Ultrastructure, morphology and organization of biogenic magnetite from sockeye salmon, *Oncorhynchus nerka*; implications for magnetoreception," *J. Exp. Biol.*, vol. 140, pp. 35-49, 1988.
13. T. P. Quinn, "Evidence for celestial and magnetic compass orientation in lake migrating sockeye salmon fry. *J. Comp. Physiol.* A, vol. 137, pp. 243-248, 1980.
14. H. Schiff and G. Canal, "The magnetic and electric fields induced by superparamagnetic magnetite in honeybees. Magnetoreception: an associative learning?" *Biol. Cybernetics*. Vol. 69, pp. 7-17, 1993.
15. C. Walcott, "Magnetic orientation in homing pigeons," *IEEE Trans. Magnet. Mag.*, vol. 16, pp. 1008-1013, 1980.
16. R. Wiltschko and W. Wiltschko, *Magnetic Orientation in Animals*, Heidelberg: Springer-Verlag, Berlin, 1995.
17. S. Johnsen and K. J. Lohmann, "The physics and neurobiology of magnetoreception," *Nature*, vol. 6, pp. 703-712, 2005.
18. S. E. Lyshevski, *Molecular Electronics, Circuits, and Processing Platforms*, CRC Press, Boca Raton, FL, 2007.
19. R. G. C. Moore, S. D. Evans, T. Shen and C. E. C. Hodson, "Room-temperature single-electron tunnelling in surfactant stabilised iron oxide nanoparticles", *Physics E*, vol. 9, no. 2, pp. 253-261, 2001.
20. I. Puchades, R. Pearson, L. F. Fuller, S. Gottermeier and S. E. Lyshevski, "Design and fabrication of microactuators and sensors for MEMS," *Proc. IEEE Conf. Prospective Technologies and Methods in MEMS Design*, Polyana, Ukraine, pp. 38-44, 2007.
21. F. C. Meldrum, B. R. Heywood and S. Mann, "Magnetoferritin: in vitro synthesis of a novel magnetic protein," *Science*, vol. 257, pp. 522-523, 1992.
22. M. Okuda, K. Iwahori, I. Yamashita and H. Yoshimura, "Fabrication of nickel and chromium nanoparticles using the protein cage of apoferritin," *Bitech. Bioeng.*, vol. 84, pp. 355-358, 2003.
23. M. Okuda, Y. Kobayashi, K. Suzuki, K. Sonoda, T. Kondoh, A. Wagawa, A. Kondo and H. Yoshimura, "Self-organized inorganic nanoparticle arrays on protein lattices," *Nano Lett.*, vol. 5, pp. 991-993, 2005.
24. R. M. Kramer, C. Li, D. C. Carter, M. O. Stone and R. R. Naik, "Engineered protein cages for nanomaterial synthesis," *J. Am. Chem. Soc.*, vol. 126, pp. 13282-13286, 2004.
25. S. Gider, D. D. Awschalom, T. Douglas, K. Wong, S. Mann and G. Cain, "Classical and quantum magnetism in synthetic ferritin proteins," *J.Appl. Phys.*, vol. 79, pp.5324-5328, 1996.
26. T. Douglas and M. Young, "Host-guest encapsulation of materials by assembled virus protein cages," Nature, vol. 393, pp. 152-155, 1998.
27. T. Douglas and V. T. Stark, "Nanophase cobalt oxyhydroxide mineral synthesized within the protein cage of ferritin," *Inorg. Chem.*, vol. 39, 1828-1830, 2000.

Design and Microfabrication of an Electrostatically Actuated Scanning Micromirror with Elevated Electrodes

Mohd Haris*, Hongwei Qu*, Ankur Jain** and Huikai Xie**

*Department of Electrical and Computer Engineering, Oakland University, Rochester, Michigan 48309, USA
**Department of Electrical and Computer Engineering, University of Florida, Gainesville, Florida 32611, USA

ABSTRACT

This paper reports the design and microfabrication of an electrostatically actuated CMOS-MEMS micromirror with elevated electrodes. Two sets of bimorphs are employed to create mismatched vertical comb drives for mirror actuation. Device structural design and fabrication process are detailed and device performance such as scanning angles is simulated using CoventorWare, an integrated MEMS simulator. With a 26 V driving voltage applied to the mismatched comb drives alternately, a rotational angle of over ±12° can be realized. The device chips were fabricated using AMI 0.5 μm CMOS technology through MOSIS. DRIE post-CMOS microfabrication was performed for device release.

Keywords: CMOS-MEMS, electrostatic, micromirror, vertical comb-drive

1 INTRODUCTION

Electrostatic actuation has advantages of easy implementation and low actuation power. Vertical comb drives (VCD) have been studied for sensing and driving of a wide variety of optical and other MEMS devices [1-4]. Micromirrors using VCDs have demonstrated fast scanning speed and low actuation power. Some research groups have reported a 1-Dimensional VCD micromirrors with optical scan angles of 25° at 34 kHz [1] and 40° at 5.8 kHz [2], while other group has reported a 2D VCD micromirror that can scan optical angles greater than 15° in both *x* and *y* directions [3]. Vertical displacement of 7.5 μm [4] and even as high as 55 μm [5] have been achieved through the use of electrostatic VCDs. Angled VCDs have been reported that scanned optical angles greater than ±9° at 18 Vdc [6], and ±3° at 100 Vdc [7]. The driving voltage required by these devices ranges from as low as 10V [5] to as high as 250 V [1].

Existing VCD manufacturing processes are complicated and expensive. Most MEMS VCDs are currently fabricated using non-CMOS compatible processing methods, such as multiple wafer bonding and SOI processing [1-5]. This paper presents a novel design methodology for the fabrication of VCDs using foundry-CMOS compatible processes. This mask-less CMOS-compatible process has the potential to make VCD enabled devices and device arrays less expensive through batch fabrication.

2 DEVICE DESIGN

Figure 1 shows a 3D model of the designed device. Two sets of complementarily-oriented bimorphs elevate the stator electrodes above the corresponding rotor electrodes that are anchored to substrate through a pair of compliant torsional springs [8]. The inset in Fig. 1 shows part of the VCDs formed by the stators and rotors. The uniqueness of this structure also includes the integration of polysilicon heaters in the bimorph beams, as shown in Fig. 2. By passing current in these polysilicon heaters, the bimorph beams, which consist of SiO_2 and aluminum thin films, can be electrothermally driven to tune the elevation height of the stator comb fingers for maximum electrostatic actuation.

To allow rotational motion of the mirror plate, the bimorph sets elevating the stator electrodes are designed stiffer than the torsional mechanical springs connecting the mirror plate to the silicon substrate. A resonant frequency ratio of over 4 was chosen between the bimorph beams and

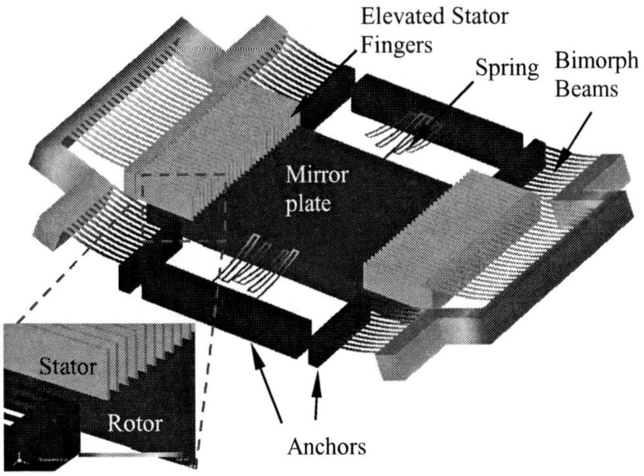

Figure 1. 3D model of the device with inset showing the VCD formed by elevated stator comb fingers and stator comb fingers on mirror plate.

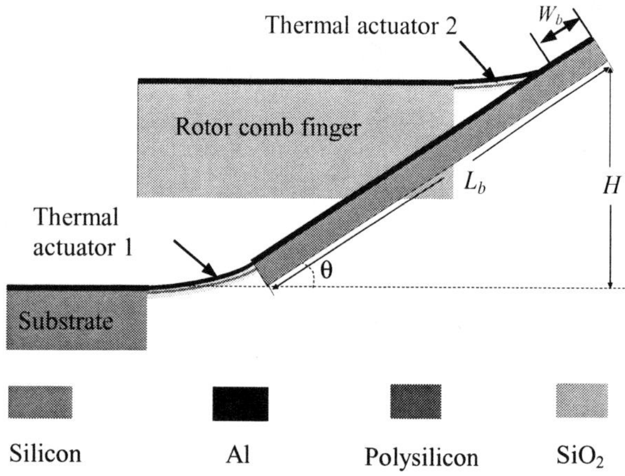

Figure 2. Bimorphs with embedded polysilicon actuators.

torsional springs to ensure stable operation of the device. Major parameters of the device are shown in Table 1.

Table 1: Major parameters of the device.

Parameter	Definition	Value
Mirror Length	l_m	0.4 mm
Mirror Width	w_m	0.4 mm
SCS Thickness	t	60 μm
VCD Finger Gap	g	2 μm
VCD Finger Width	w_f	6 μm
VCD Finger Length	l_f	100 μm
Number of Fingers	N_f	25
Length of bimorph beam	l_b	150 μm
Width of bimorph beam	w	9 μm
Width of connecting beam	w_b	50 μm
Thickness of beam	t	1.8 μm
Number of bimorph beams in LVD	N	24

The two sets of bimorph beams are used as a large-vertical-displacement (LVD) microactuator [9]. They are complementarily oriented in a folded rigid beam such that the curling of the two sets of bimorph beams compensates each other. This unique configuration results in a pure vertical elevation of the attached comb fingers once the comb drives are released from the substrate, as shown in Fig. 2. The elevation, H of the stator comb finger can be determined as

$$H = (L_b - W_b)\sin\theta \quad (1)$$

where, W_b is the width of the rigid beam connecting the two bimorph sets, and θ is the tilt angle of the amplifier beam. The vertical bending and the resultant torsional stiffness of the LVD bimorph beams can be calculated from [6] by

$$k_{z\_LVD} = N\frac{Ewt^3}{4l_b^3} \quad (2)$$

$$k_{\phi\_LVD} = k_{z\_LVD}\left(l_b + \frac{L_b}{2}\right)^2 \quad (3)$$

The required torsional stiffness of the spring connecting the mirror plate to the substrate can be calculated using

$$k_\phi = M\frac{w_m^2}{12}w_R^2 \quad (4)$$

where M is mass of the mirror plate. For the device, the torsional spring constant is calculated to be $k_\phi = 1.2 \times 10^{-8}$ Nm/rad for the desired resonance frequency of 1 kHz. The required length of the standard torsion bar can be determined using Eq. 5 [7]. In order to reduce the length of the spring, folded serpentine spring is used.

$$l_s = \frac{2Gw_s^3 t_s}{3k_\phi}\left[1 - \frac{192}{\pi^5}\frac{w_s}{t_s}tanh\left(\frac{\pi t_s}{2w_s}\right)\right] \quad (5)$$

where G is the shear modulus; t_s and w_s are the thickness and width of the spring, respectively.

When a voltage is applied between the rotor and stator comb fingers, the movable rotor fingers rotate about the torsion axis until the restoring torque generated by the spring balances the electrostatic torque. The electrostatic torque is given by

$$\tau_{VCD} = N_f\frac{\varepsilon_0}{g}\frac{\partial A}{\partial\theta}V^2 \quad (6)$$

The term corresponds to the increase in finger overlap area with changing angular rotation is shown in Fig. 3 and is given by

$$\frac{\partial A}{\partial\theta} = \frac{1}{2}\left[\left(l_f + \frac{l_m}{2}\right)^2 - \left(\frac{1}{2}l_m\right)^2\right] \quad (7)$$

Balancing the electrostatic torque equation with the restoring torque equation of the spring, we get

$$\theta(V) = \frac{N_f\varepsilon_0}{2k_\phi g}V^2\left[l^2 - \left(\frac{l_m}{2}\right)^2\right] \quad (8)$$

The maximum rotation angle of the micromirror can be expressed by [7]

$$\theta_{max} = \frac{t}{l_f + \frac{l_m}{2}} = 12^o \quad (9)$$

The corresponding rotation angle versus the applied

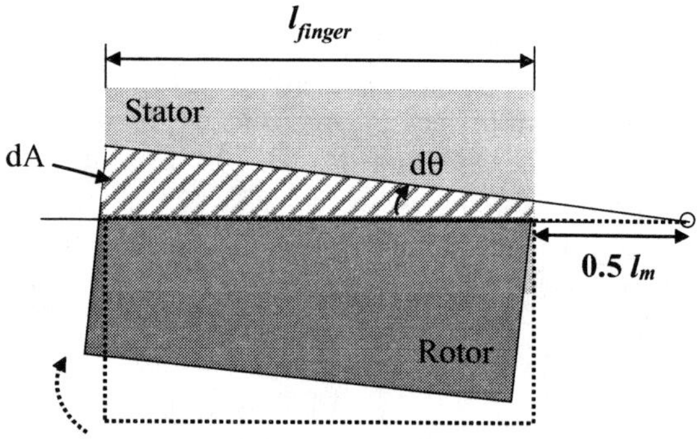

Figure 3. Cross-section of VCD for mirror rotation analysis.

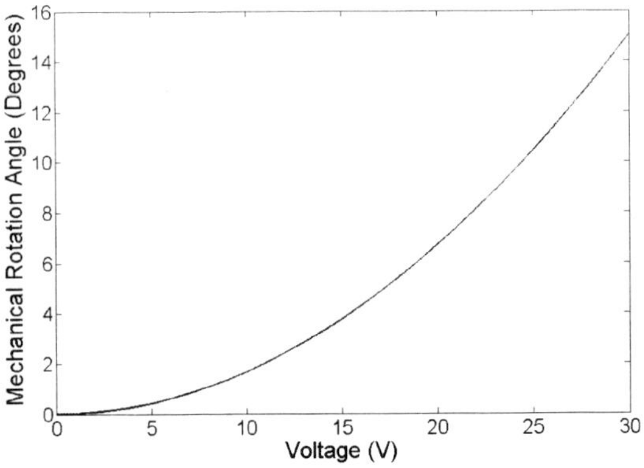

Figure 4. Mirror voltage versus applied dc voltage.

voltage plot as obtained using Eq. (8) is shown in Fig. 4. From Fig. 4, it is predicted that the mirror rotates $\pm 12°$ at 26 V. In order to generate vertical motion, the two sets of electrostatic comb drives (on either end of the mirror plate) should be excited simultaneously. When the same voltage is simultaneously supplied to both VCD sets, the lower rotor fingers displace vertically upward due to the vertical electrostatic force as shown in Fig. 5.

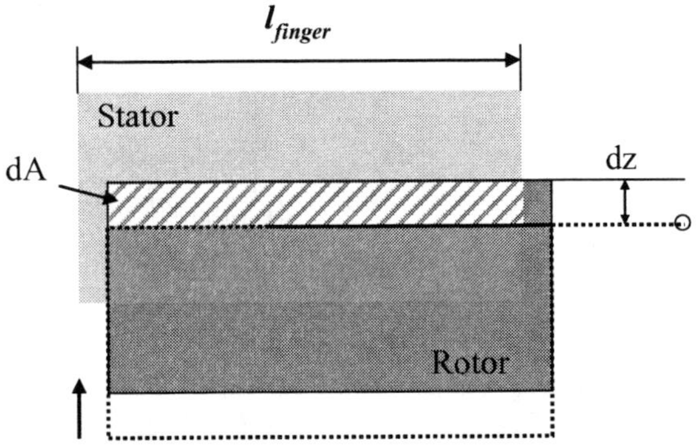

Figure 5. VCD analysis for mirror piston motion

This electrostatic force, generated by one pair of the VCD fingers is given by

$$F = \frac{1}{2} \cdot \frac{dC_z}{dz} \cdot V^2 \quad (10)$$

where ε_0 is the dielectric constant of air between the comb drive, and V is the voltage applied to the comb drive. A factor of $2N_f$ should be applied to this electrostatic force since there are a total of $2N_f$ pairs of comb fingers that contribute towards the vertical actuation.

Fig. 6 shows the relation between the capacitance, and electrostatic force between the rotor and stator when applied a dc voltage from 0.001 V to 32 V.

3 FABRICATION PROCESS

AMI 0.5μm technology with three metal layers was used for the LVD mirror design and fabrication. Because of the

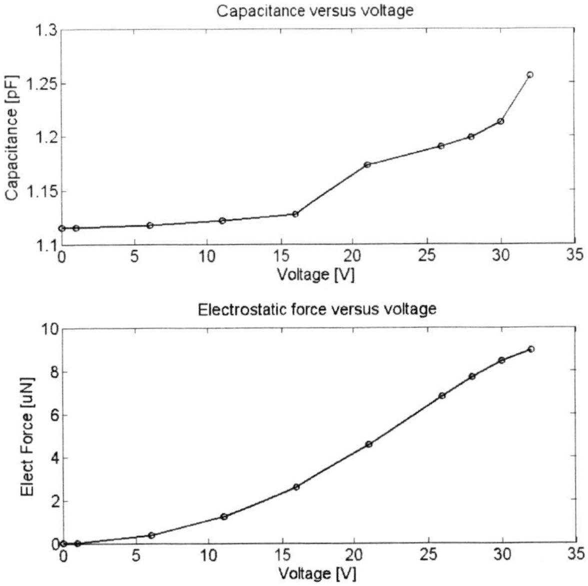

Figure 6. Simulation results of change of the VCD capacitance, electrostatic force versus applied voltage.

difference in the width of LVD bimorph and VCD finger, these two structures cannot be released at the same time. Two release steps were designed and performed to form the bimorphs and VCD fingers separately. When bimorphs are etched and undercut, top metal layer M3 was used as the mask to protect other structures. The device fabrication process flow is illustrated in Fig. 7. It is a maskless post-CMOS micromachining process in which top metal layers were used as etching mask in the dry etch processes [10].

The post-CMOS process starts with the backside etching to define the mirror plate thickness of 50μm. (Fig. 7(a)). Then anisotropic SiO_2 etching is performed to expose the regions for bimorph beams (Fig. 7(b)). A unique wet aluminum (Al) etching is followed to remove the top Al layer M3 (Fig. 7(c)). Next, a deep silicon etch followed by an isotropic silicon etch is performed to undercut the silicon

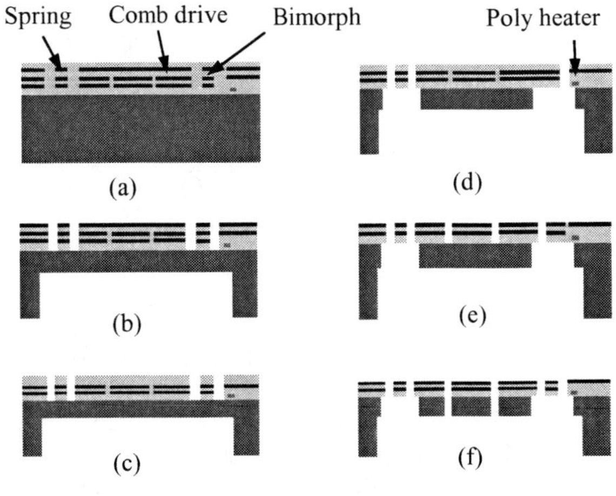

Figure 7. Microfabrication process flow.

underneath the bimorph beams and mirror springs (Fig. 7(d)). This step also electrically isolates the comb fingers on mirror plates and LVD plate from silicon substrate. Next, the second anisotropic SiO_2 etch defines VCD comb fingers (Fig. 7(e)). Finally, a deep silicon etch is performed again to etch through the comb fingers (Fig. 7(f)).

4 FABRICATION RESULTS

Fig. 8 shows the SEM image of the fabricated device. After the release process, the stators of the LVD on both sides were elevated above the substrate with a horizontal surface, validating the effective compensation of the two bimorph sets. Due to the larger tensile stress in the SiO_2 layer in the bimorphs beams, the elevation of VCD was measured as approximately 100 μm from the substrate. The resistance of the functional polysilicon heater was measured as 2.3 kΩ. Because of the fabrication variations caused by the timing-controlled etch, the single crystal silicon under the stator LVD was measured as 40 μm. Due to the large residual stress caused by the foundry CMOS process, and the abrupt release in the plasma etching chamber where high vacuum was present, the mirror plate broke at the connections to torsional springs. The bimorph actuator remained functional. A new microfabrication process is under development to avoid the sudden release of the device in high vacuum chamber that causes the structure damage.

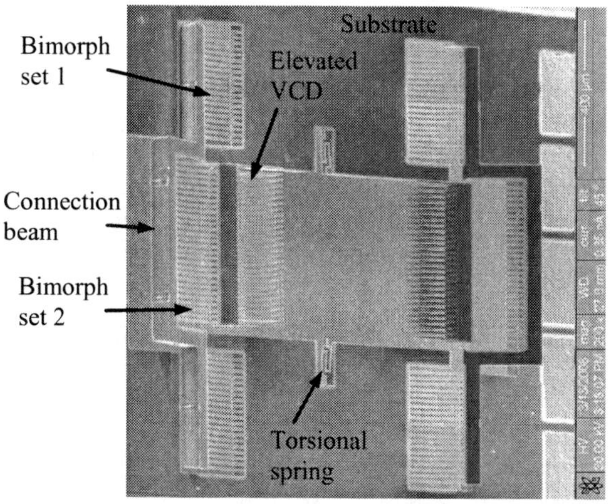

Figure 8. SEM image of the fabricated device.

5 CONCLUSION

An electrostatic micromirror capable of rotational and vertical piston motion has been designed and fabricated. The novel LVD structure elevates the stator fingers 100 μm above the substrate plane. This elevation can be tuned by passing current through the polysilicon actuator embedded in the bimorphs. This mirror can provide bi-directional scanning angles of ±12° at less than 30 V dc.

Maskless post-CMOS process was employed in the device fabrication. Since this mask-less fabrication process is foundry-CMOS compatible, it has the potential to be cost effective over other existing complex processes that are currently used in fabricating VCDs. The bi-directional rotational scanning and vertical piston motion scanning capabilities of this device make it useful in the areas of optical coherence tomography (OCT), adaptive optics and interferometry systems.

ACKNOWLEDGEMENTS

The device microfabrication was performed at the Michigan Nanofabrication Facility (MNF), one of sites in the NSF supported NNIN network. The authors would also like to thank Kai Sun at the Electron Microbeam Analysis Laboratory at the University of Michigan for assistance in SEM imaging of the device.

REFERENCES

1. R. A. Conant, et al., *Technical Digest of the 2000 Solid-State Sensor & Actuator Workshop*, Hilton Head, SC, pp.6-9.
2. U. Krishnamoorthy, et al., *J. of Microelectromechanical Systems*, **12**, pp. 458 – 464 (2003).
3. S. Kwon, et al., *IEEE Journal of Selected Topics in Quantum Electronics*, **10**, pp.498-504 (2004).
4. D. Lee, et al., *Technical Digest of Transducer'03,* pp. 576-579.
5. S. Kwon, etc al., *Technical Digest of the 2002 Solid-State Sensor & Actuator Workshop*, Hilton Head, SC, June 2002, pp.227-230.
6. H. Xie, etc al., *Journal of Microelectromechanical Systems*, **12**, pp. 450 - 457.
7. P. Patterson, etc al., *The Fifteenth IEEE International Conference on Micro Electro Mechanical Systems*, MEMS 2002, pp. 544-547.
8. A. Jain, etc al., *Sensor & Actuators A*, **122**, pp. 9-15 (2005).
9. A. Jain, etc al., *Technical Digest of 2004 Solid State Sensor, Actuator and Microsystems Workshop*, Hilton Head, SC, June 2004, pp. 228-231.
10. H. Xie, L. Erdmann, X. Zhu, K. Gabriel and G.K. Fedder, *J. of Microelectromechanical Systems* **11**, pp. 93-101 (2002).

Deposition of functional PZT Films as Actuators in MEMS Devices by High Rate Sputtering

H.-J. Quenzer*, R. Dudde*, H. Jacobsen**, B. Wagner*, H. Föll**

*Fraunhofer Institute Silicon Technology, D-25524 Itzehoe, quenzer@isit.fraunhofer.de
**University of Kiel, Inst. Materials Science D-24143 Kiel

ABSTRACT

Crack and void free polycrystalline Lead Zirconate Titanate (PZT) thin films in the range of 5 µm to 15 µm have been successfully deposited on silicon substrates using a novel high rate sputtering process. The sputtered PZT layers show a high dielectric constant ε_r between 1000 and 1800 and a distinct ferroelectric hysteresis loop with a remanent polarisation of 17 µC/cm^2 and a coercive field strength of 5.4 kV/mm. A value of $d_{33,f}$ = 80 pm/V for the piezoelectric coefficient has been measured.

Based on this deposition process a membrane actuator consisting of a SOI layer and a sputtered PZT thin film was prepared. The deflection of this membrane actuator depending on the driving voltage was measured with a white light interferometer and compared to the results of **f**inite **e**lement **a**nalysis (FEA). With this approach a transverse piezoelectric coefficient e_{31} = -11 C/m^2 was approximated.

Keywords: MEMS, actuator, PZT, gas flow sputtering

1 INTRODUCTION

Piezoelectric materials are promising candidates for powerful actuators in MEMS devices. The large electromechanical coupling coefficient of Lead Zirconate Titanate ($Pb(Zr_xTi_{1-x})O_3$, PZT) allows the realisation of actuators at high frequencies and low energy consumption. In order to design large force MEMS actuators high quality PZT films with a thickness in the order of 10 µm are demanded.

Despite the fact that the deposition of PZT thin films is a field of intensive research, the fabrication of thicker films is still quite challenging. Common thin film deposition techniques like MOCVD [1], magnetron sputtering [2] and Sol-Gel [3] suffer from low deposition rates and therefore are typically applied for the deposition of layers in or below micron range. Screen printing methods [4] suffer from high firing temperatures above 900°C.

Results are presented for a new high rate sputter process for the deposition of high quality PZT films with a thickness of 5µm to 15µm. As a first demonstrator for a MEMS actuator a PZT film has been produced on top of a silicon membrane produced by a standard KOH wet etch process on a SOI wafer.

2 EXPERIMENTAL

The gas flow sputtering technique is a special physical vapour deposition (PVD) technique, based on a hollow cathode glow discharge and a gas flow driven material transport. In contrast to magnetron sputtering, the target is hollow and arranged perpendicular to the substrate. In this work a circular sputter source with a diameter of 40mm has been used. Therefore the substrate holder was moved linearly to increase the coated area and to improve uniformity of deposited PZT films on the wafer. Experiments have started to exchange the cylindrical sputter source with a set up of parallel sputter plates of 250mm width. With this improved sputter system standard 8" Si-wafers will be processed.

An argon flow through the sputter source transports the eroded metal atoms to the substrate where oxygen is added separately. For deposition of the PZT films pure metallic targets are simultaneously sputtered leading to the desired high sputter rate. The working pressure of 0,5 mbar inside the vacuum chamber for the hollow cathode discharge is maintained by a roots blower unit and a vacuum rotary pump. The gas flow sputter technique is described in more detail elsewhere [5, 6]. Deposition rates of 200-250nm/min have been achieved.

Deposition of ternary or quaternary films by reactive PVD is complicated due to strong differences in the vapour pressure of the elements, their reactivity or by their inmiscibility making target manufacturing impossible. In such cases, gas flow sputtering has strong advantages since targets composed of segments from pure elements can be used. The segments should be arranged along the gas flow direction resulting in a thorough mixing of the sputtered material during gas transport. Changing the size of individual segments, the film stoichiometry can be adjusted. This saves much time compared to experiments using alloy targets. In this work the target was composed of individual metal rings of lead, zirconium, and titanium of appropriate thickness. Oxygen is fed into the space between source and the substrate becoming activated there by the hollow cathode plasma.

During sputtering, the substrates are mounted on a heated wafer holder (chuck), where a bias-voltage is applied. The wafer holder allows substrate temperatures of up to 700 °C during sputtering. In order to obtain the desired crystallographic microstructure of the PZT films, the deposition temperature is of major importance. Experiments showed that a minimum temperature of 550°C is necessary to initiate a crystalline growth of the PZT film, while temperatures of approx. 620°C are required for the deposition of PZT films with good piezoelectric properties. Typical process parameters are summarised in table 1 [7].

Substrate temperature	550–650 °C
Pressure	0.4–0.7 mbar
Argon flow rate	800 sccm
Oxygen flow rate	20 sccm
Source power	600W
Source voltage	400–600V
Bias AC voltage	50–100V at 200 kHz

Table1: typical values of main process parameters

2.1 Material Composition

A crucial aspect in sputtering compound materials consisting of various elements is the stability and reproducibility of the chemical composition of the deposited thin films. Since the piezoelectric properties of PZT films depend very strictly on their exact stoichiometry the repeatability of the composition in the process becomes even more important.

In fact many process parameters influence the stoichiometry like the used power level, target geometry, gas flows and substrate temperature. Therefore process control and optimisation are of high importance.

Material compositions of finished PZT films were determined by electron probe micro analysis (EPMA). While EPMA allows material analysis of PZT films after preparation other techniques have to be used to monitor the material composition during the deposition process. Optical emission spectroscopy (OES) gives an indication of relative material content within the glow discharge during sputtering. Therefore OES was used to monitor the sputtering process and control process uniformity by regulating Ar flow and RF power. The emission spectra were recorded by a grating spectrometer with a sensitive CCD detector (ARC, FC 459180). In several experiments it was verified that the relative emission intensity between the Pb emission line at 406nm and Ti emission lines around 399nm is linearly related to the Pb/Ti ratio in the deposited films. This linearity holds in the vicinity of the operational regime that was adjusted for the sputtering process here. Emission lines from Zr were too weak and noisy to be used as reliable control signal. Therefore OES was used for an optimised process control in PZT deposition leading to improved stability and reproducibility of the sputter process of PZT layers.

3 RESULTS

Thickness of PZT films was determined by SEM (fig. 1). Deposition rates of 200-250 nm/min were observed. These high sputter rates are about 20 - 25 times higher than reported from reactive magnetron sputtering [2] and demonstrate the potential of this sputtering technique. Typical layers had a thickness of 6 µm sputtered in about 30 min. However, even 16 µm thick PZT films were deposited without cracking or delamination within 90 min. Electrical measurements of the relative permittivity ε_r have shown values between 1000 and 1800. A distinct ferroelectric hysteresis loop with a remanent polarisation of 17 µC/cm^2 and a coercive field strength of 5.4 kV/mm has been observed which is comparable to those of screen printed PZT [5]. Measurements of the piezoelectric coefficient $d_{33,f}$ using a double-beam interferometer [8] have shown values of the reverse effect of up to 79.7 pm/V.

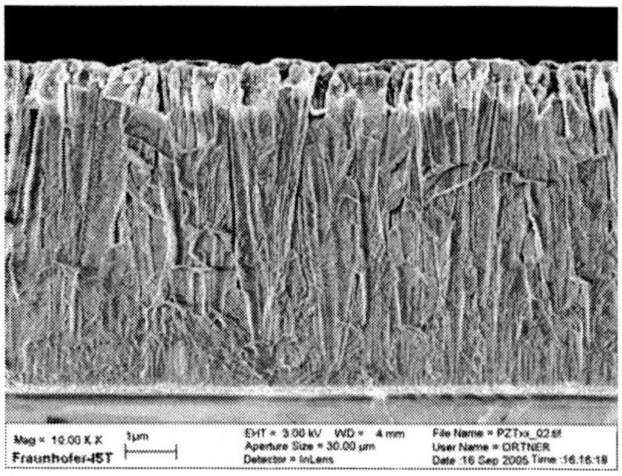

Figure 1: SEM cross section of a 6.6µm thick PZT film

3.1 Membrane actuator

To demonstrate the functionality of the PZT films, a membrane actuator was fabricated with a silicon MEMS process flow. The membrane actuator consists of the active, piezoelectric PZT film with some inter-layers on top of an elastic silicon layer fabricated from a silicon on insulator (SOI) wafer (see fig. 2). The membrane has a quadratic shape with a side length of about 2 mm.

This lateral dimension results in relatively large centre deflections of the membranes thus allowing the use of a commercial white–light interferometer (ATOS Micromap) to measure the bending of the actuated membrane.

The quadratic shape of the membrane is caused by the use of an anisotropic wet etchant (KOH solution, 30%, 80°C) during the membrane etching process. The principle design of the membrane is summarised in figure 2.

After PZT film deposition the PZT material is structured by lithography and wet etching to restrict the actuator material to the position above the membrane structure.

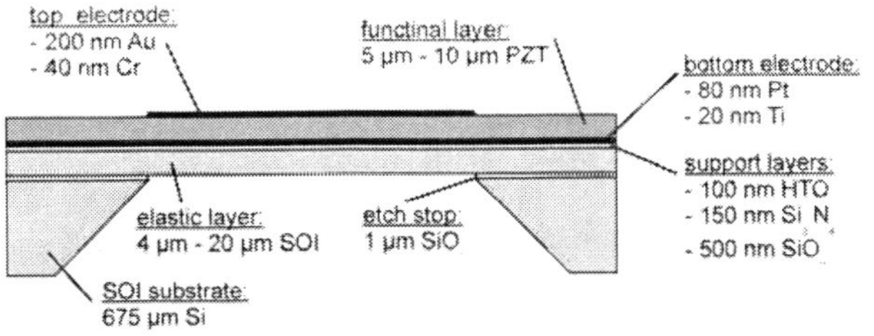

Figure 2: Schematic view of the unimorph membrane actuator.

3.1 Process Flow

150 mm SOI-Wafers with a SOI - thickness of about 20µm and an overall thickness of 675µm are used as substrates. In a first process step a diffusion barrier of 500 nm thermal oxide, 150nm Si_3N_4 and 100nm **H**igh **T**emperature **O**xide (HTO, SiO_2) is deposited in a **l**ow-**p**ressure **c**hemical **v**apour **d**eposition (LPCVD) process. Because of its small lattice mismatch to PZT and its inert behaviour Platinum is used as bottom electrode. To improve the adhesion of Platinum it is recommended [10] to use Titanium as adhesion layer. Both thin films were deposited by e-beam evaporation (Unaxis, BAK). To improve the texture of the Platinum layer the wafers are annealed at 550 °C in a vacuum oven for 30 minutes at a pressure of less than 0.1 Pa. This process leads to a complete (111) texture of the Platinum film.

Afterwards, these wafers are sputter coated with a PZT thin film of approx. 4 -5 µm thickness [5]. For patterning the PZT thin film a wet chemical etching mixture consisting of HCl and HF is used [9].

Tests ended up with an etch rate of the PZT between 1.3µm/min and 1.6µm/min leading to a total process time of approx. 4 - 5 min including over-etching. Due to the HF content and the required etching time the photoresist shows strong delamination effects leading to a poor quality in transferring the pattern into the PZT layer.

Thus in further experiments the photoresist was replaced by **p**lasma **e**nhanced **c**hemical **v**apour **d**eposition (PECVD) silicon nitride hard mask which was previously patterned in a dry etching process. The good adhesion of the Si_3N_4 hard mask on top of the PZT allows an almost perfect isotropic etching of the PZT and leads to a drastically reduction in the under etching of the PZT layer to approx. 5 µm, see figure 3.

Finally the remaining silicon nitride layer is removed by dry etching. Subsequently a top layer is formed by local electroplating of gold with a thickness of 1.8 µm on a plating base consisting of 40 nm Chromium and 200 nm Gold. After removing the plating base the silicon nitride passivation on the rear side of the wafer is patterned by dry etching. This silicon nitride layer acts as hard mask during the deep anisotropic wet etching in KOH solution (fig. 4). Finally the silicon oxide beneath the SOI layer is removed in a HF vapour etch process.

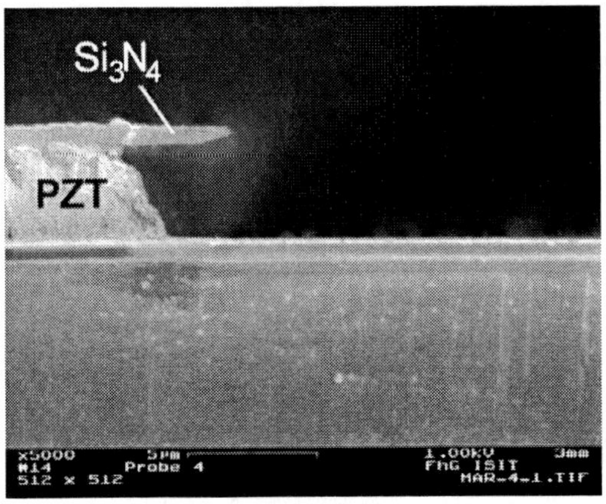

Figure 3: SEM cross section of a 5µm thick PZT layer etched by isotropic wet etching using a Si_3N_4 hard mask.

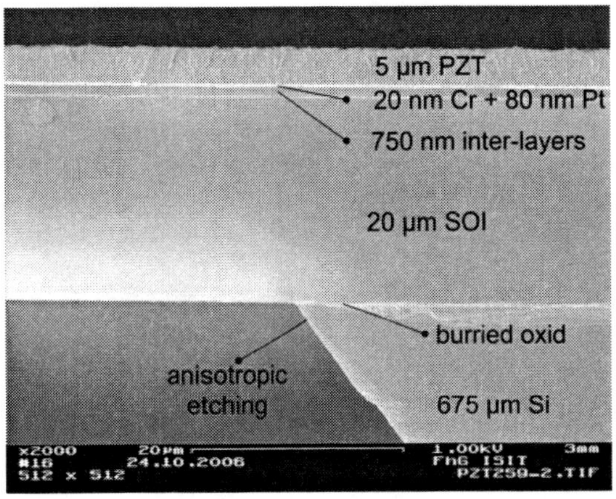

Figure 4: SEM cross section of a membrane: on top of 20 µm silicon (SOI) a stack of 750 nm diffusion barrier layer can been seen. Beneath the PZT layer with a thickness of approx. 5µm the metal layers consisting of 20 nm Titanium und 80 nm Platinum are visible.

3.2 Membrane properties

The poling of the test samples was carried out at 100°C with a voltage of 12.5 kV/mm for 30 min. The curvature and centre deflection of the membranes is measured using a white-light interferometer.

For comparison the deflection vs. the actuation voltage is calculated in a FEA model assuming a transverse piezoelectric coefficient e_{31} of approx. $e_{31} = -11$ C/m^2. This value showed the best fit as compared to the experimentally determined membrane deflection.

Directly after the preparation the membranes of the test samples showed an initial curvature up to 0.5 -1 µm which is very likely stress induced. Obviously the sputter process of the PZT film generates mechanical stress in the PZT layer in the range of approx. 40 MPa. Besides this pre-deflection of the membranes the tests showed minor variations in their responds on the actuation voltage and also a non-linearity in the voltage dependency of the deflections. The variability of the samples prepared is attributed to minor variations in the actual material composition.

4. SUMMARY

Crack and void free polycrystalline PZT thin films in the range of 5 μm to 15 μm have been successfully deposited on silicon substrates using a novel high rate sputtering process. With this sputter process sputter rates of 200-250 nm/min were achieved and complete 6" wafers have been covered with a 12μm thick PZT film within 60 min.

The PZT layers show a high dielectric constant ε_r between 1000 and 1800, a distinct ferroelectric hysteresis with a remanent polarisation of 17 $\mu C/cm^2$ and a coercive field strength of 5.4 kV/mm. Measurements of the piezoelectric coefficient $d_{33,f}$ have shown values of the reverse effect up to 79,7 pm/V.

Based on this new deposition process a simple membrane actuator consisting of a SOI and a sputtered PZT layer was prepared. The deflection of this membrane actuator depending on the driving voltage was measured with a white light interferometer and compared to the results of a FEA model. With this approach a transverse piezoelectric coefficient $e_{31} = -11\ C/m^2$ was approximated.

Still in an early stage of work the Gas Flow Sputtering approach for high rate deposition of PZT thin films has shown its high potential for application in MEMS actuators. Work is ongoing for optimisation and stabilisation of the PZT deposition process and to increase the piezoelectric coefficient $d_{33,f}$ and $d_{31,f}$ respectively. Additionally the process development is extended to process 8" wafers and more methods are developed for PZT film structuring.

REFERENCES

[1] A. C. Jones, et al., "MOCVD of Zirconia and Lead Zirconate Titanate Using a Novel Zirconium Precursor", J. Eur. Cer. Soc. 19, p. 1434, 1999.

[2] P. Muralt, et al., "Fabrication and characterisation of PZT thin-film vibrations for micromotors", Sensors and Actuators A 48, p. 157, 1995.

[3] H. Kueppers, et al., "PZT thin films for piezoelectric microactuator applications", Sensors and Actuators A 97-98, p. 680, 2002.

[4] M. Koch et al., "A novel micropump design with thick-film piezoelectric actuation", Meas. Sci. Technol. 8, p. 49-57, 1997.

[5] H. Jacobsen et al., "Development of a piezoelectric lead titanate thin film process on silicon substrates by high rate gas flow sputtering", Sensors and Actuators A, 133, p. 250, 2007.

[6] K. Ishii, "High-rate low kinetic energy gas-flow-sputtering system", J. Vac. Sci. Technol. A 7, p. 256–258, 1989

[7] H. Jacobsen, K. Prume, H.-J- Quenzer, B. Wagner, K. Ortner, Th. Jung "High-rate sputtering of thick PZT thin films for MEMS", J. Electroceramics, Springer, in press.

[8] K. Prume, P. Muralt, F. Calame, Th. Schmitz-Kempen, S. Tiedke, "Piezoelectric thin films: Evaluation of electrical and electromechanical characteristics for MEMS devices", IEEE Trans. UFFC., 54, p. 8, 2007.

[9] J. Baborowski, "Microfabrication of piezoelectric MEMS, J. Electroceramics", 12, p. 33, (2004).

[10] T. Maeder, L. Sagalowicz, P. Muralt, "Stabilized Platinum Electrodes for Ferroelectric Film Deposition using Ti, Ta and Zr Adhesion Layers", Jpn. J. Appl. Phys., 37, p. 2007, 1998.

UV-Nanoimprint for 9 GHz SAW Filter Fabrication

Nian-Huei Chen, Chen-Liang Liao, Henry J.H. Chen[1], **C-H Lin**[2] and Fon-Shan Huang
Institute of Electronics Engineering, National Tsing Hua University, Hsinchu, Taiwan
[1]Electrical Engineering, National Chi Nan University, Nantou, Taiwan
[2]National Nano Device Laboratories, Hsinchu, Taiwan
E-mail: d9663801@oz.nthu.edu.tw, TEL: +886-3-5715131 ext. 34046, FAX: +886-3-5752120

Abstract

A 8 GHz surface acoustic wave (SAW) filter was fabricated by UV nanoimprint lithography (UV-NIL). The key techniques to produce SAW filter include stamp and interdigital transducer (IDT). For stamp, high aspect ratio HSQ/ITO/glass stamp was first exposed by low dose e-beam writer and the proper post-exposure bake (PEB) temperature. The suitable TMAH concentration, temperature, and etching time were utilized to pattern perfect vertical sidewall of the HSQ stamp. Afterwards, the pattern was transferred on UV-curable resist / $LiNbO_3$ by UV-NIL at room temperature and low pressure. Al/Ti IDTs were then deposited on $LiNbO_3$ for lift-off process. IDTs with feature size 100nm and thickness 20nm can be obtained. Central frequency of SAW filter is as high as 8 GHz. The SEM images depict the fabricated stamps, imprinted features and Al/Ti IDTs. Network analyzer HP8510C was used to examine electrical characterization of SAW filter.

Keywords: surface acoustic wave filter, UV nanoimprint

1. Introduction

SAW devices have been widely implemented for various applications, such as mobile phone and sensor. For emerging wireless communication, radio frequency identification device (RFID) tags used for electronic toll collection system [1] operates at microwave frequency (2.4~5.8GHz) with advantage of fast data transfer rates and long transmission range.

In 2002, Y. Takagaki *et al* [2] described the SAW delay lines with central frequency of 1.488GHz and insertion loss about 35dB fabricated on $LiNbO_3$ by hot embossing nanoimprint. For nanoimprint lithography, stamp plays a key role in imprint procedure. With the concern of stamp deformation and process simplicity, low pressure and room temperature imprinting is available. Ngoc V. Le *et al* [3], in 2005, reported a trilayer (etch barrier/TEOS/PMGI) UV nanoimprint process and performed selective etching process to form the undercut profiles. The Al IDTs were patterned by lift-off process with width 140nm, and thickness 40nm. The main restraint affecting multi-layer resist method is the prices problem. P. Kirsch *et al* [4], in 2006, produced 4.6GHz SAW device on AlN / diamond layered structure by direct e-beam writing lithography and lift-off technique. The IDT made of Al with resolutions down to 500 nm and thickness 100nm can be obtained. However, it is time consumption for IDT fabrication using e-beam lithography. In comparison with e-beam lithography, nanoimprint lithography is a promising technique capable of resolving submicron features.

In this work, we propose a low-cost UV-NIL to fabricate SAW filter on $LiNbO_3$ substrate. First of all, we developed a novel high aspect ratio stamp and used single PR layer to simplify lift-off process. D.P. Mancini *et al* [5] first developed hydrogen silsesquioxane (HSQ) as stamp for UV-NIL. The e-beam dose ranging from 1000 to 2200 $\mu C/cm^2$ were exploited to pattern HSQ film. The minimum feature size of 30nm semidense lines were fabricated. In our investigation, we attempt to use comparatively low e-beam dose on HSQ/ITO/glass substrate. The e-beam exposed HSQ films were post baked before developing in tetramethylammoniumhydroxide (TMAH) solution. TMAH concentration, temperature, and development time are consideration factors to obtain straight sidewall. HSQ pattern were then treated with novel step-like hardbake heating to enhance the hardness [6]. After transferring the pattern on UV-curable resist (PAK-01-200) by UV-NIL and eliminating the residual layer by reactive ion etching (RIE), Al/Ti IDTs lift-off process were realized with high aspect ratio pattern of single layer PAK-01-200 resist. Finally, we utilize optical lithography to define electrode pad on patterned IDT nanowires. High replication fidelity is investigated to make SAW filter characterization reliable with initial design.

2. Experiment

2.1 HSQ/ITO/glass UV stamp fabrication

A 6 inch ITO/glass wafer was first cleaned in KOH solution. The HSQ (FOX-15, Dow Corning) was diluted in methylisobutylketone (MIBK) with the ratio of 2:1 and coated on ITO/glass wafer with thickness about 350nm. Subsequently, HSQ films were soft baked at 120°C for 3 min to serve as e-beam irradiation to transform caged structure into network. The e-beam dose can be therefore reduced [7]. The SAW filter with delay line structure and conventional $\lambda/4$ IDT width design rule was adopted. Various IDT width of 100, 150, and 200nm with a line-width-to-space ratio of 1:1 and the alignment key were designed by L-edit software. A Leica Weprint200 electron beam stepper (beam energy 40KeV, beam size 20nm) with e-beam dose 360$\mu C/cm^2$ was used for exposure. A PEB was carried out at various temperature of 230-280 °C for 2min. Afterwards, the HSQ films were developed in TMAH solution with concentration 25%, TMAH temperature varying from R.T. to 45^{0}C for etch time 10 s. Consequently, a step-like heating cycle to 350 °C was performed to modulate the porosity inside the HSQ stamp for enhancing the hardness [6]. F_{13}-TCS was used as stamp-release layer to lower the surface free energy [8].

2.2 SAW filter fabrication by UV-NIL

The SAW filter fabrication will be described in detail here. The process flow is shown in figure 1. There are three key techniques (1) pattern transferred on PAK-01-200 by UV-NIL, (2) Al/Ti IDTs fabricated by lift-off process, and (3) electrode pad aligned with the IDT for SAW filter fabrication needed to be developed.

For substrate preparation, the piezoelectric wafer considered here is Y-Z lithium niobate ($LiNbO_3$). It was first cleaned in succession of acetone, isopropyl alcohol (IPA), and deionized (DI) water. Then, a low viscosity polymer, PAK-01-200 (Toyo Gosei Co.), was utilized as photocurable resin [9]. Prior to the spinning of PAK-01-200, the $LiNbO_3$ substrate was treated with O_2 plasma to promote the adhesion. After spinning, PR coated $LiNbO_3$ substrate was soft-baked at 80 °C for 2 min. UV-NIL had been done by using Nanonex NX-2000 system. Small imprint pressure of 20psi for 20s and UV light exposure at an intensity of 25mW/cm^2 for 10s were performed to transfer IDTs pattern on PR. The residual resist layer on the bottom of trench was etched by RIE system (Samco PC1000). For fine piezoelectric effect, the interface between IDT and $LiNbO_3$ has to be uncontaminated. Different etching time (40s and 45s) was investigated with O_2 flow rate of 100sccm, Ar flow rate of 3sccm, and RF power 300W. Various thickness of Al/Ti IDTs were then deposited on the RIE etched pattern and lifted off in PG Remover (Microchem) at 65°C bath with ultrasonic vibration. The final electrode pad was defined by optical lithography. Commercial positive tone PR of S1813 (Shipley) was first spin-coated on IDT patterned $LiNbO_3$ wafer. Conventional quartz mask was used to align metal alignment key on patterned IDT substrate by aligner (Karl Suss MJB3) and second lift-off process was exploited to form electrode pad with thickness about 300nm. The fabricated stamps, imprinted features and Al/Ti IDTs after lift-off process were depicted by SEM images. Central frequency, and insertion loss of SAW filter were interrogated by network analyzer HP8510C. The time gating procedure was employed to remove the interferencee with electromagnetic feedthrough and triple transit response [10].

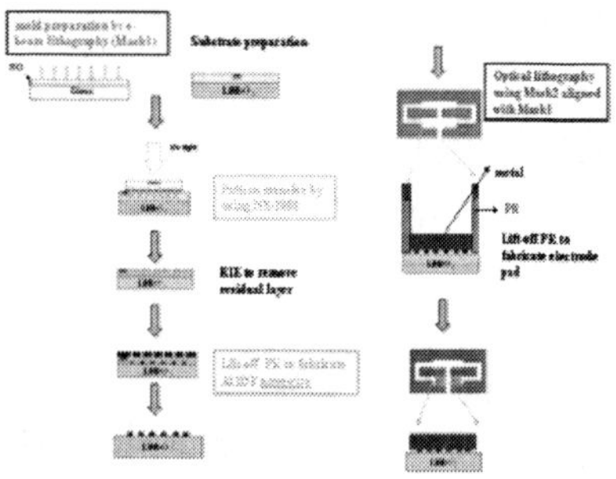

Figure 1. Process flow chart of fabricating SAW filter by UV-NIL and lift-off process

3. Results and Discussions

3.1 HSQ/ITO/glass UV stamp fabrication

The stamps with straight sidewall can achieve vertical replication pattern for applying to following lift-off process. As line width scaling down, the proximity effect is a contributing factor in e-beam lithography. In order to surmount e-beam proximity effect, elevating TMAH temperature with faster vertical etching rate and treating with PEB to prevent from lateral erosion has been presented. Figure 2 shows HSQ stamp with designed IDT width of 150nm which were formed by condition of PEB at 280 °C, and 25% TMAH solution at 45 °C for 10s. From figure 2, HSQ stamp displays a slight overetched profile which appears a height of 312nm but narrower, 135nm in width with an aspect ratio of 2.3. With the same development condition, IDT non-overlaped region with larger spatial period (line width to space ratio of 1:3) was devised for wider width to guard against severe TMAH etching.

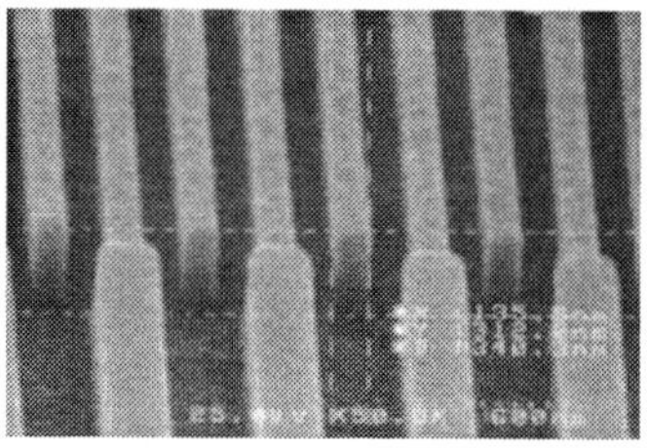

Figure 2. HSQ stamp with designed IDT width of 150nm fabricated by condition of 360 μC/cm^2, PEB at 280 °C, and 25% TMAH solution at 45 °C for 10s.

3.2 SAW filter fabrication

3.2.1 Pattern transferred by UV-NIL on PAK-01-200

During UV-NIL process, with the combination of pressure and UV light, PAK-01-200 refill the stamp features and become solidified. Figure 3 depicts SEM image of transferred pattern of 140nm on PR by the HSQ stamp shown in figure 2. Though, there exists discrepancy between stamp and imprinted pattern. The imprinted pattern exhibits equally width to lithographic design.

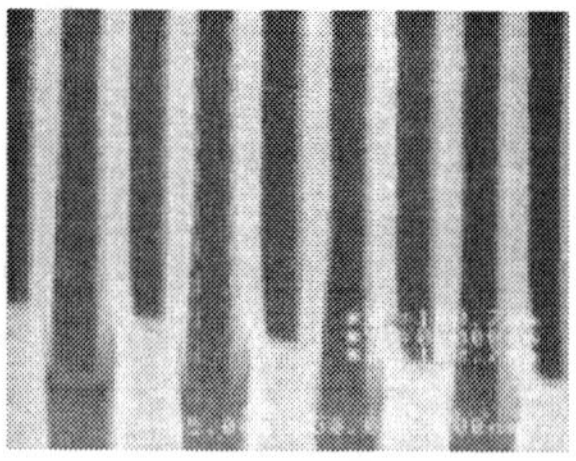

Figure 3. The SEM images of pattern transferring on PAK-01-200 by HSQ stamp shown in figure 3 with pressure of 20psi.

3.2.2 Al/Ti IDTs fabricated by lift-off process

The fabricated HSQ stamp which is 97nm in width and 311nm in height as shown in figure 4(a) was utilized to compress PAK-01-200 with minimum pressure of 15psi. In figure 4(b), the recessed nanotrench with dimension 135nm in width and aspect ratio of 2.55 indicates good replication fidelity. Further, the residual layer of PR left under the compressed region about 30 nm had to be removed by RIE. The SEM pictures in figure 4(c) shows the etching result of patterned PR (figure 4(b)) with RIE process for 40s. It can be observed that residual layer seems not removed totally. The etched structure remained fine profile with trench of 140nm and aspect ratio about 2.4.

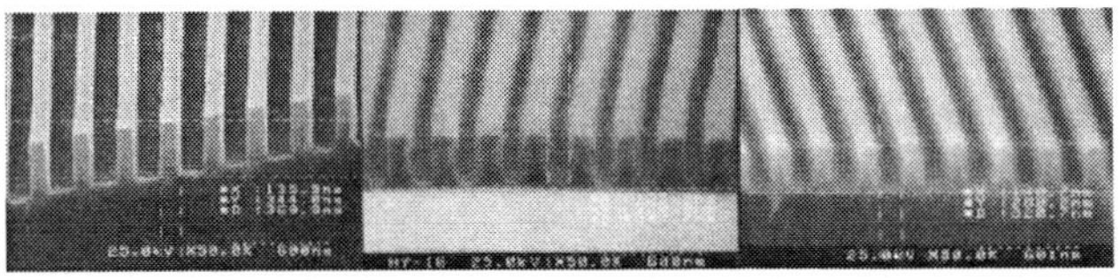

(a) (b) (c)

Figure 4. The SEM images of (a) 135nm HSQ stamp (b) PR transferred pattern before RIE (c) RIE etching for 40s

The pattern transferred PR shown in figure 3 were then used to perform lift-off process. The residual layer of PR left under the compressed region had to be removed by RIE for 40s. SEM images in figure 5 shows the final Al/Ti IDTs with thickness of 20nm and feature size of 136nm with metallization ratio of 0.45. Compared to feature in figure 3, extremely high fidelity and fine geometries can be defined by adequate conditions of RIE and lift-off process.

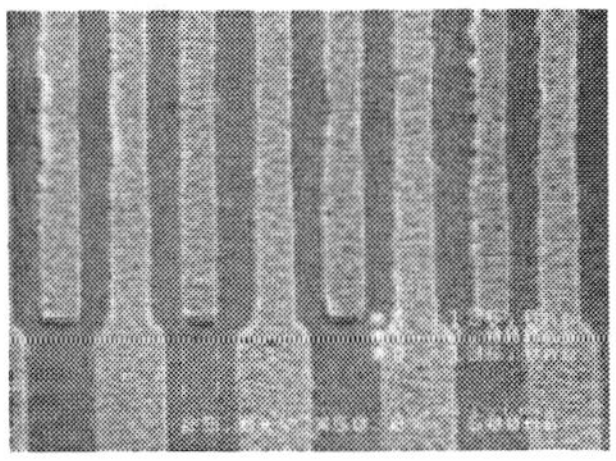

Figure 5. The SEM images of Al/Ti IDTs with thickness of 20nm obtained by RIE and lift-off process using sample in figure 3.

3.2.3 SAW filter fabrication

Optical lithography technique was exploited to define electrode pads for following electrical measurement. The alignment keys formation was accompanied by the fabrication of IDT fingers which were patterned by HSQ stamp (Mask1 shown in figure 1). Conventional quartz mask was then used to align the alignment key on $LiNbO_3$ wafer. The electrical contacts to IDT fingers were patterned by lift-off of a positive tone photoresist (S1813). Figure 6(a) shows the SEM image of a finished SAW filter. All of the ground pads were connected together for common ground circuit. SEM image in figure 6(b) depicts the μm-scaled electrode pad connected with nm-scaled IDT fingers.

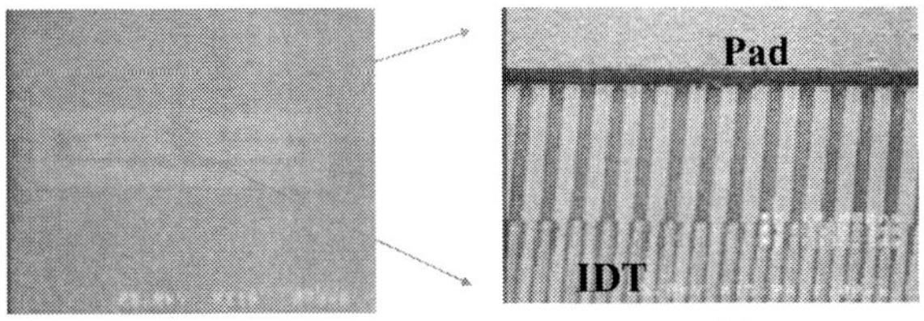

(a) (b)

Figure 6. The SEM images of (a) finished SAW filter profile (b) μm-scaled electrode pad connected with nm-scaled IDT fingers.

3.3 Electrical characterization of SAW filter

Through the piezoelectric phenomenon, the SAW filter employs input IDT to generate surface acoustic wave and transmit along the surface of the elastic underlay toward output IDT which convert acoustic energy to electric energy. With the concern of perfect piezoelectric effect, the interface between IDT and $LiNbO_3$ has to be unpolluted. Therefore, different RIE time for 40s and 45s were performed and frequency responses (S_{21}) of the SAW filters are compared in figure 7(a). For IDT design width of 150nm, SAW filter with 40s and 45s RIE time has central frequency of 5.609, 5.694GHz and insertion loss of 60.9, 23.4dB (figure 7(a)). Compared with ideal central frequency of 5.81GHz, the percentage error of the frequency shift are 3.46% and 2%. From measured frequency response, SAW filter with 40s RIE time represents higher insertion loss. It may be due to an incomplete etching of residual PR which affects the efficiency in energy transformation. A longer RIE time for 45s has demonstrated to improve insertion loss. Figure 7(b) shows the band-pass characteristic (S_{21}) of SAW filter with the ideal central frequency of 8.7GHz (100nm IDT width) treated with 45s RIE time which has central frequency of 7.968GHz, insertion loss of 65.5dB and frequency shift of 8.4%. The poor SAW filter performance is attributed to imperfect interface between IDTs and $LiNbO_3$ caused by incomplete RIE procedure which affects the efficiency in energy transformation. Large frequency shift may due to scratch on IDTs results in electrode open and increases the mass loading. As a result of the mass-loading effect, the loading mass will cause the change of the SAW propagation velocity and the shift of central frequency. With the same RIE time, SAW filter with wider IDT width represents smaller insertion loss. It is caused by RIE procedure that as the aspect ratio of transferred trench PR increasing, the RIE trimming time should increase for the purpose of removing the residual PR entirely.

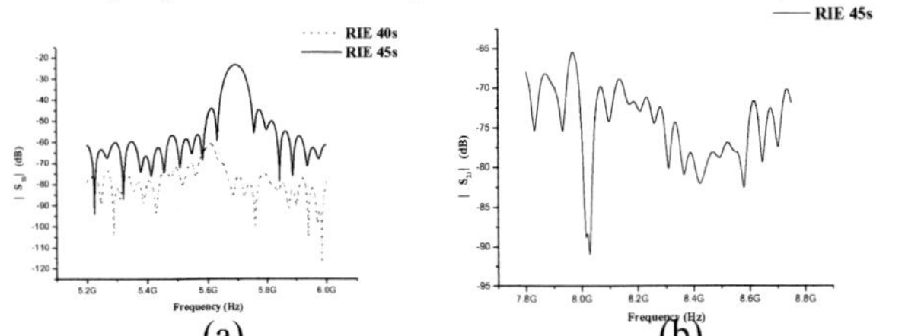

(a) (b)

Figure 7. Frequency response of SAW filter with designed IDT of (a) 150nm (b) 100nm

4. Conclusions

SAW filter with 150, 100nm IDT width for central frequency of 5.694, 7.968GHz and insertion loss about 23.4, 65.5dB are successfully fabricated by UV-NIL. There are 2, 8.4% central frequency mismatch comparing with theoretical numeral. The measured data elucidates excellent agreement with prediction. For SAW filter with 100nm IDTs width, HSQ/ITO/glass stamp with width 93nm and aspect ratio 3.5 was fabricated. After transferring the pattern on UV-curable resist (PAK-01-200) by UV-NIL with low pressure at room temperature, we develop IDTs lift-off process with high aspect ratio pattern of single PR layer and thickness of 20nm Al/Ti IDTs can be obtained. For future study, we will make efforts on optimum design of SAW filter and shrinkage of IDTs width for high frequency communication application.

Acknowledgments

This paper was supported by National Science Council of Taiwan, ROC, under the contract No. NSC 95-2221-E-007-246. The authors also acknowledge GemTech Optoelectronics Corp. for ITO/glass substrate supporting.

References

[1] Wenming Liu, Huansheng Ning, Baofa Wang, Antennas, Propagation and EM Theory 2006, Page(s) 1 – 4 (2006)
[2] Takagaki Y, Wiebicke E, Kostial H, Ploog KH, Nanotechnology 13, 15 (2002)
[3] Ngoc V. Le, Kathleen A. Gehoski, William J. Dauksher, Jeffrey H. Baker, Doug J. Resnickl, and Laura Dues, Proceedings of the SPIE 2005, 5751 (2005)
[4] P. Kirsch, M. B. Assouar, O. Elmazria, V. Mortet and P. Alnot, Applied Physics Letters 88, 223504 (2006)
[5] D. P. Mancini, K.A. Gehoski, E. Ainley, K. J. Nordquist, D.J. Resnick, T. C. Bailey,
S. V. Sreenivasan, J. G. Ekerdt, and C. G. Willson, Journal of Vacuum Science and Technology B 20, 2896 (2002)
[6] N.H. Chen, C.L. Liao, Henry J. H. Chen, C.H. Lin, F.S. Huang, Materials Research Society Symposium 2007, N3.12 (2007)
[7] Ming-Tse Dai, Kai-Yuen Lam, Henry J. H. Chen, and Fon-Shan Huang, Journal of the Electrochemical Society 154, H636 (2007)
[8] Takashi Nishino, Masashi Meguro, Katsuhiko Nakamae, Motonori Matsushita, and Yasukiyo Ueda, Langmuir 15, 4321 (1999)
[9] J. Haisma, M. Verheijen, K. van den Heuvel, and J. van den Berg, Journal of Vacuum Science and Technology B 4, 4124 (1996)
[10] Clement M, Vergara L, Sangrador J, Iborra E, Sanz-Hervas A, Ultrasonics 42, 403 (2004)

Nano-Gap High Quality Factor Thin Film SOI MEM Resonators

D. Grogg, H.C. Tekin, N.D. Badila-Ciressan, D. Tsamados, M. Mazza, A.M. Ionescu

Laboratory of micro/nano-electronic devices
Ecole Polytechnique Fédérale de Lausanne, CH-1015, Switzerland
daniel.grogg@epfl.ch, adrian.ionescu@epfl.ch, phone: +41 21 693 3978

ABSTRACT

Silicon micro-electro-mechanical (MEM) resonators on a 1.25 μm thin SOI substrate are demonstrated to achieve quality factors of 100 000 at 24.6 MHz. Based on a improved fabrication process, 200 nm wide transduction gaps are fabricated, resulting in a strong electrostatic coupling at low bias voltages. Consequently, the motional resistance of the thin resonator is as low as 55 kΩ with a bias voltage of 18 V and values still below 100 kΩ are measured at 14 V.

Keywords: micromachining, resonator, quality factor, microelectromechanical system, electrostatic devices

1 INTRODUCTION

Miniaturization of silicon micro-electro-mechanical (MEM) resonators and co-integration with silicon ICs are main driving forces for the extensive research in this field. Performance characteristics of such miniaturized devices on par with quartz resonators have been demonstrated, mainly on thick substrates [1-3] to reduce the impedance levels and to achieve high quality factors, which in some cases exceed 100 000 [1,3]. Further impedance reduction is obtained with sub-micrometer gaps, using either sacrificial layer etching [2,4] or deep reactive ion etching [5-7] to reach dimensions below what can by obtained by lithographic structuring. Even though such remarkable results have been obtained, most demonstrated integrations for oscillator circuits rely on a two-chip solution [8]. In this work, we build resonators on thin SOI substrates, which will eventually lead to a single chip solution [9]. Such a co-integration will substantially increase the impact of MEM resonators on the RF circuit design. This paper presents the fabrication and the characterization of such 1.25 μm thin resonators with quality factors above 100 000 at 24.6 MHz.

2 DESIGN

The resonator body is made of longitudinally vibrating beams [1] connected in parallel, hence the name parallel beam resonator (PBR). The single beams are 161 μm long, 5 μm wide and separated by 5 μm. The beams are mechanically coupled to avoid independent movements, resulting in one single frequency. The mode shape of the resonator was verified with ANSYS and a synchronous movement of all beams is obtained for the design (Fig. 1). The two resonator designs studied vary by the number of beams connected in parallel and thus their total width. One resonator is 115 μm (12 beams) wide, called PBR 115, and the other has a total width of 195 μm (20 beams), called PBR 195. The parallel beams move in a free-free mode with one nodal point at the center, at which location the anchors are placed on the outermost beams.

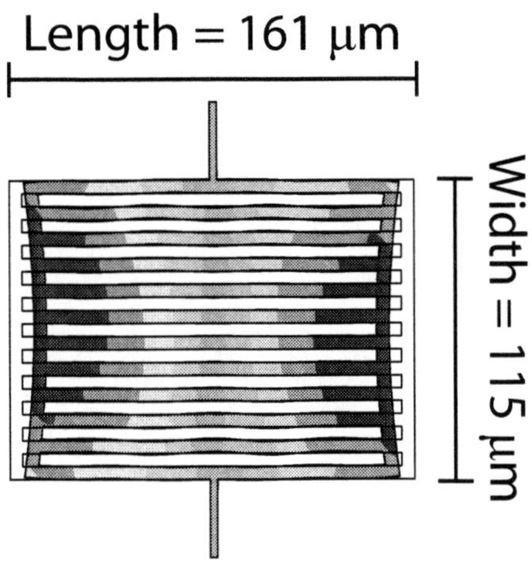

Figure 1: Mode shape of a PBR 115 resonator at its minimum position, simulated with ANSYS. The black line indicates the undeformed shape.

3 FABRICATION

A novel fabrication process of the resonators is described in Fig. 2. The SOI substrate has a 3 μm thin buried oxide and a 1.5 μm thin silicon film with a (100) surface orientation. A hardmask is created on top of the SOI wafer to define the nano-gap according to the method reported [6,7], where the gap is defined by a deposited thin polysilicon layer (Fig. 2a). This process reduces the total thickness of the SOI silicon film from 1.5 μm down to 1.25 μm. The transfer of the hardmask into the SOI device layer is done with a high aspect-ratio SHARP process (Fig. 2b). After releasing in BHF, the structures are dried in a CO_2 supercritical point dryer to avoid sticking and a thermal oxide is grown on the silicon surface (Fig. 2c). A CVD deposited parylene-C layer protects the suspended structures during the following steps, in which openings are etched with an isotropic ICP etch process (Fig. 2d). The top SiO_2 layer and the polysilicon layers are etched on the electrodes and titanium is evaporated onto a two-layer lift-off resist to structure the metal contacts (Fig. 2e). Finally, the resonators are released in an isotropic oxygen plasma (Fig. 2f), which completely removes the parylene layer.

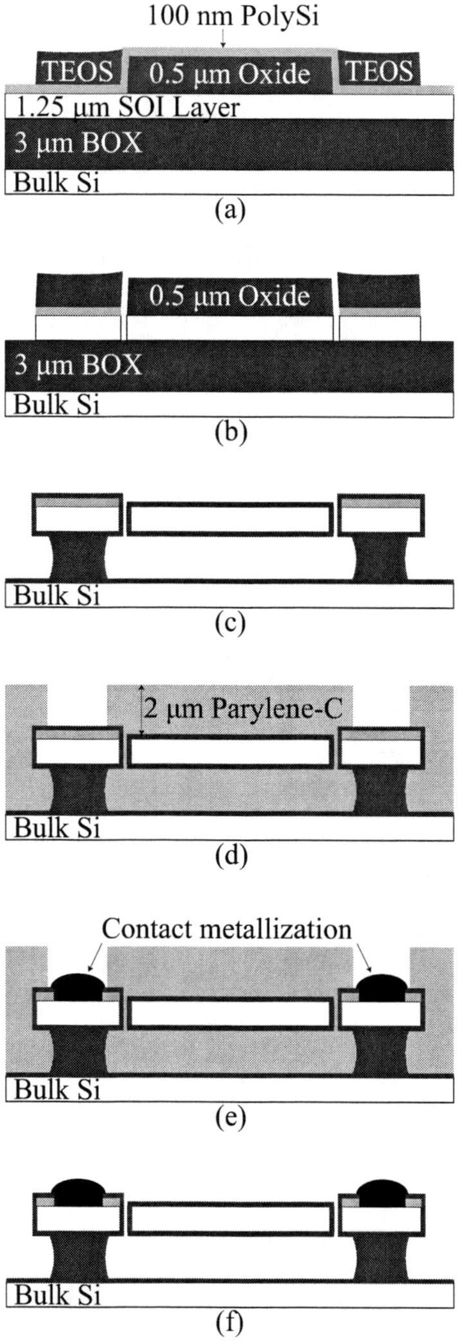

Figure 2: Process flow: (a) A hard mask is prepared, based on a sacrificial layer process (TEOS: Tetra-ethyl-ortho-silicate) (b) followed by a DRIE process. (c) Once released in BHF, CO_2 supercritical point drying avoids sticking and a dry thermal oxide is grown. (d) CVD deposited parylene protects the resonator during contact opening and (e) metallization followed by (f) an O_2-plasma release.

The released resonators are inspected after fabrication with an optical profilometer. These measurements indicate a stress-free completely flat surface for thin oxides grown on the silicon (Fig. 3).

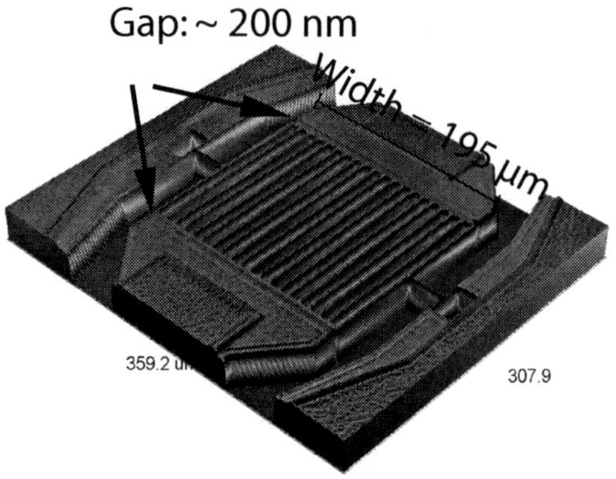

Figure 3: Fabricated parallel beam resonator image taken with an optical profilometer.

4 EXPERIMENTAL RESULTS

The resonators are characterized with a vector network analyzer and a decoupled DC source (Fig. 4). All measurements are taken in a SUSS cryogenic prober chamber, PMC150, under vacuum conditions better then 10^{-5} mbar. This equipment also enables us to test the temperature dependence of the different parameters of interest.

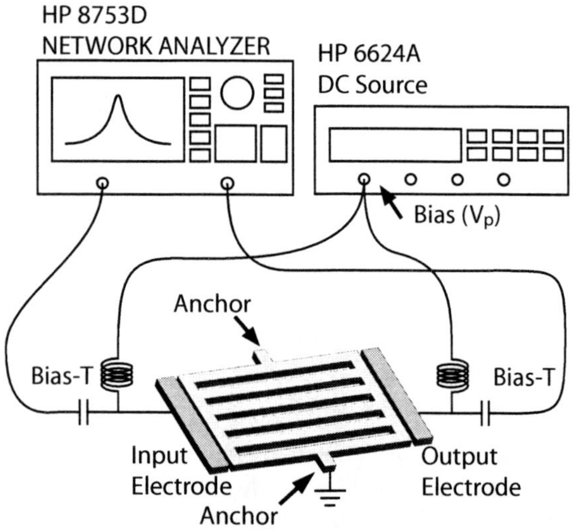

Figure 4: Measurement setup used for the two-port characterization of the parallel beam resonators.

NSTI-Nanotech 2008, www.nsti.org, ISBN 978-1-4200-8505-1 Vol. 3

To avoid vertical pull-in of the resonator onto the substrate, both the bulk silicon and the resonator are grounded. The bias voltage (V_p) is applied through a bias-T on the electrodes to protect the virtual network analyzer.

Fig. 5 shows the scattering parameter S12 for a PBR 195 resonator at a bias of 20 V. The input RF power is -40 dBm. The quality factor is 100 000 with a motional resistance of 55 kΩ at a bias voltage of 18 V. Higher voltages result in a pull-in of the resonator onto the electrode and destruction of the device, reason for the limited tuning range of the motional resistance with the bias voltage. A series of four measurements taken on a second structure of the same type results in a motional resistance of 64 kΩ (Fig. 6) with V_p=14 V, pointing to the high sensitivity of this parameter on gap fabrication tolerances.

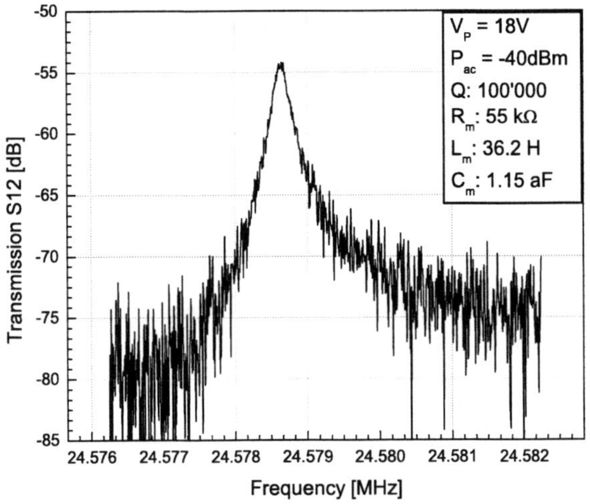

Figure 5: S-parameter measurement of a PBR 195. A quality factor of 100'000 and a motional resistance of 55 kΩ is measured with V_p=14 V.

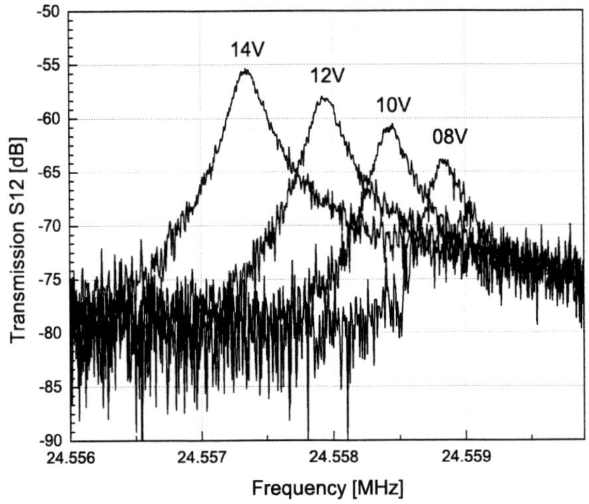

Figure 6: S-parameter of a PBR 195 as a function of bias voltage: the motional resistance is 64 kΩ , with V_p=14 V.

The smaller design PBR 115, with 12 beams, has a very similar transfer characteristic. The measurement in Fig. 7 is taken with a RF power of -30 dBm at bias voltage from 12 V to 20 V. The quality factor is 70 000, lower then for the PBR 195 and the motional resistance is 120 kΩ with V_p=20 V. This higher equivalent impedance level is explained by the smaller electrode surface and the lower Q-factor of the resonator, compared to the PBR 195.

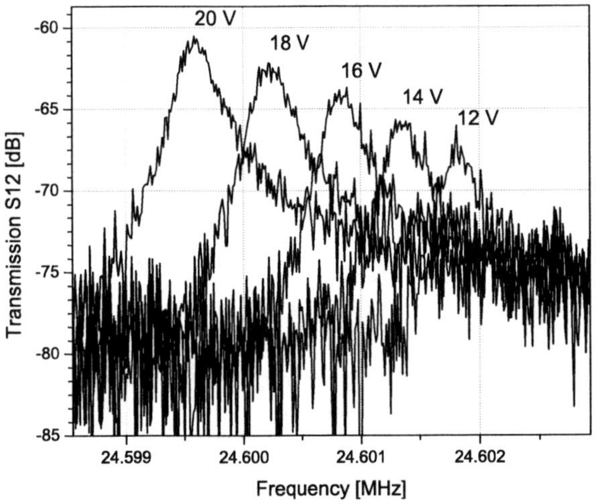

Figure 7: Measured frequency spectra of a PBR 115 with varying bias voltage; R_m=120 kΩ with V_p=20V.

Based on the frequency versus bias voltage characteristics (Fig. 8), an effective gap is extracted through fitting of a resonator model to the measured data. The gap size was found in the range of 190 nm to 210 nm, which corresponds well to the observed motional resistance values. The frequency tuning of the resonator with bias voltage, as given in Fig. 8, is depending on the gap size. For the measured structures the total frequency tuning is >2.5 kHz over a range of 10 V. The difference of the motional resistance for two PBR 195 structures is explained by a difference in effective gap size of 7 nm, due to process variations during the gap etching.

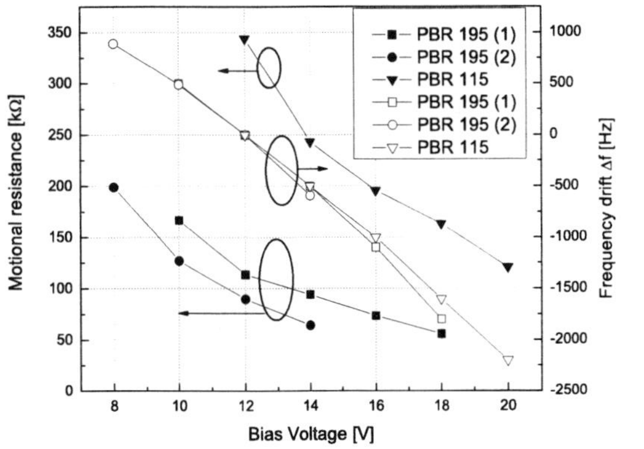

Figure 8: The motional resistance and the frequency tuning versus the bias voltage for three structures. Process variations are strongly influencing the motional resistance of the resonators (PBR 195).

Frequency stability with temperature is an important parameter for any resonator in industrial applications. Fig. 9 gives a comparison of the resonator behavior with respect to temperature for one resonator of each type from 160 K up to 380 K at an interval of 20 K. Both measured resonators behave vary similar, showing a frequency drift which averages at 15.7 ppm/K and 15.3 ppm/K for PBR 115 and PBR 195 respectively, lower then expected from previous results [7].

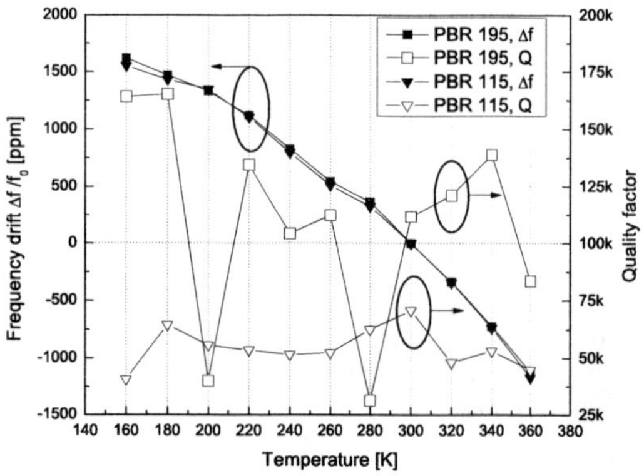

Figure 9: Variation of the resonance frequency and the quality factor with temperature. An unidentified problem caused low Q-factors at 200 K and 280 K for the PBR 195.

The quality factor was extracted for the measurements performing a fit of a Lorentz function to approximately 100 data points around the resonance peak. Even though this method was shown to reduce the influence of noise on the result [10], an unidentified problem occurred on two data points with Q-factors below 40 000 for the PBR 195. Overall, an increasing Q-factor with decreasing temperature is observed, but the precision of the measurement is limited. For the PBR 115 a rather linear trend over the whole temperature range is found, with values for the most part above 50 000.

5 CONCLUSION

A CMOS compatible fabrication process for 1.25 μm thin SOI MEM resonators is presented showing the feasibility to construct resonators with quality factors above 100 000 in thin SOI. The motional resistance is as low as 55 kΩ measured with a low bias voltage of 18 V. Characterization of the resonators in a broad temperature range shows a constantly high quality factor and frequency drift which is dominated by the silicon thermal properties. These results implicate that co-integration of electrostatic MEM resonator and CMOS circuits on thin SOI substrates can be done with extremely high quality factors.

ACKNOWLEDGEMENT

The authors gratefully acknowledge the collaboration with COMELEC SA for CVD deposition of parylene and the staff of the Center of Micro/Nano Technology at EPFL for their advice

REFERENCES

[1] T. Mattila et al., "Micromechanical bulk acoustic wave resonator," in Ultrasonics Symposium, 2002. Proceedings. 2002 IEEE, 2002, pp. 945-948 vol.1.

[2] S. Pourkamali et al.,"Low-Impedance VHF and UHF Capacitive Silicon Bulk Acoustic-Wave Resonators - Part II: Measurement and Characterization," Electron Devices, IEEE Transactions on, vol. 54, pp. 2024-2030, 2007.

[3] V. Kaajakari et al., "Square-extensional mode single-crystal silicon micromechanical RF-resonator," 2003, pp. 951-954 vol.2.

[4] J. R. Clark et al., "High-Q UHF micromechanical radial-contour mode disk resonators," Microelectromechanical Systems, Journal of, vol. 14, pp. 1298-1310, 2005.

[5] R. Abdolvand et al., "Single-mask reduced-gap capacitive micromachined devices," in Micro Electro Mechanical Systems, 2005. MEMS 2005. 18th IEEE International Conference on, 2005, pp. 151-154.

[6] N. D. Badila-Ciressan et al., "Fabrication of silicon-on-insulator MEM resonators with deep sub-micron transduction gaps," Microsystem Technologies, vol. 13, pp. 1489-1493, 2007.

[7] N.D. Badila-Ciressan et al., "Fragmented membrane MEM bulk lateral resonators with nano-gaps on 1.5um SOI," in 37th European Solid-State Devices Research Conference, ESSDERC 2007 Munich, Germany, 2007.

[8] L. Yu-Wei et al., "60-MHz wine-glass micromechanical-disk reference oscillator," 2004, pp. 322-530 Vol.1.

[9] A. Uranga et al., "Fully CMOS integrated low voltage 100 MHz MEMS resonator," Electronics Letters, vol. 41, pp. 1327-1328, 2005.

[10] P. J. Petersan et al., "Measurement of resonant frequency and quality factor of microwave resonators: Comparison of methods," Journal of Applied Physics, vol. 84, pp. 3392-3402, Sep 1998.

Multiband THz detection and imaging devices

L.Popa-Simil*, I.L.Popa-Simil
LAVM LLC, Los Alamos, NM, USA, lps@lavmllc.com*

ABSTRACT

Terahertz imaging finds more and more applications in homeland security, medical, and life sciences applications. While THz sources and detection systems are gaining ground, the ability of electronics conversion systems to keep up with the development of these systems seems to be lagging behind. The THz domain is largely dominated by thermal noise and material characteristics altered by individual and collective modes of molecular vibrations and rotations. Currently, analog-to-digital (ADC) conversion systems are seen as the bottleneck for THz applications. ADC systems available to date have to down-convert the THz frequency into GHz domain and then digitize the signal, adding to the complexity, weight, size, and power requirements of any such system. Using a series of electromagnetic input resonator that selects the band and polarization in series with a plasmon resonator-amplifier the detected electromagnetic signal is applied to a very low current FET. The detection is based on the nonlinear carrier perturbation process that makes the function of a down-converter. A fast ADC is digitizing the detected signal applying it to a set of fast memories. The detection chain repeated for various frequencies and grouped in a single multiband module - eye unit, is further integrated into the imager.

1. INTRODUCTION

The theoretic predictions showed that plasmonic, nano-heterostructures are able to selectively convert Terahertz (IR included) photons into electric detectable signal with the currently available technology of ultra low current or single electron field effect transistors operating at higher temperatures. The range of plasmonic hetero-structures made from an electromagnetic resonator (antenna) and a set of plasmon oscillators with the role of voltage amplifier and/or field propagators, perturbing a lower frequency signal by a nonlinear influence to the gate input of a FET. If the development of such system proves successful, then we have designs that we could apply to data quantification and processing, with the signal generated at the drain of the FET being further applied to a preamplifier stage and to an analog-to-digital converter.

Transitioning from the MHz to the THz domain of the electromagnetic radiation spectrum is just a scaling factor, but a scaling factor that makes a big difference from the materials availability point of view. The THz domain is the region where the manifestation of molecular characteristics (both individual and as a collective) becomes predominant and, while this is commonly perceived as the biggest impediment in the availability of materials for this frequency domain, it may also be the most attractive feature, if used appropriately. Seeing in THz means that one can see and resolve molecules, see their mass and temperature volume distribution, and more. From the airport security point of view it will be possible to see not only the passengers have dynamite in underwear but if they eat healthy food and all the constitutive molecules of the passenger are OK. Due to material characteristics in this domain, currently there are no near-future predictions for availability of materials able to detect large bands as in optics or low GHz domain. However, using current understanding of quantum dot confinement-based phenomena, some very specific resonators with high spectral resolution and sensitivity may be produced.

1.1 State-of-art of existing approaches

The current state of art is gradually attack the THz domain, similar to far IR by gradual approaches:

- Bolo-meter like devices ate the IR end where the photon energy is pretty high of few meV, requiring low or ultra low temperatures to eliminate the electronic noise as the thermal effect to be detected,
- Electronic like devices operating in GHz domain based on advanced junctions and nano-transistors able to operate up to few THz (even made in superconductor structures).
- Ultra-short pulsed laser interferometer devices with optical like imaging devices also operating in low THz domain also known as quantum-cascade lasers and mixers.
- Molecular resonant and tunable devices in MASER class operating mainly in narrow bands.

All the above devices requires large and sensitive cryogenic setups being compatible with static large space applications. The actual literature on this subject is so large with more than 500 patents applications and thousands of publications as I will refrain from giving any reference. None of them qualifies yet for the mobile, real time multispectral imaging and recognition on morphed surfaces or fly eye assemblies operating at normal temperatures and harsh environments.

2. DESIG AND DISCUSSION

The THz detection structure is composed of a chain of functional modules as: input resonator also called antenna, the THz photon induced signal pre-amplification, detection stage, amplifier, analog digital converter, memory buffer and data processing circuit also called imager.

2.1 The receiver input structure

At the core of the THz detection system is the basic structure of a double-gate MOSFET, in which one of the gates is replaced by a metal or semiconductor nanocluster (MNC/SNC), with a role of plasmon resonator. A diagram of the whole structure (including a wavelength-tuned antenna and a quantum dot field transmission chain) is presented below.

The type of material and size for the nanocluster will be selected during the Phase I. Also, while silicon is a good and obvious choice for building the FET given its ubiquity throughout the microelectronics industry, we will not discount alternatives such as gallium arsenide, or gallium nitride, silicon carbide (wide bandgap semiconductors suitable for harsher environments).

In a basic design (without the antenna and the NC quantum

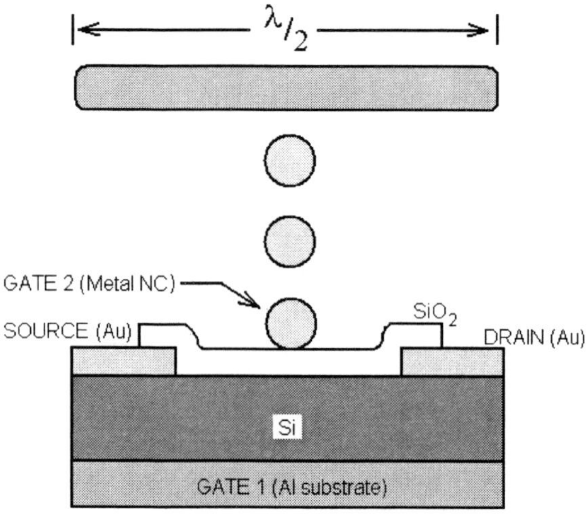

Fig. 1 – The input plasmon structure

dot chain), the NC placed on top of the oxide layer as a second gate to the FET is the primary THz resonator, sensing the electric component of the incoming electromagnetic field. The plasmon resonance of the nano-clusters will perturb the carrier density in the silicon channel of the MOSFET, thus modulating the electric signal passing through the device. However, the nanocluster's plasmon resonance will have only a narrow frequency width situated around the resonant frequency of the NC. The resonant frequency and its width are given by the dimensions and type of material that forms the nano-cluster preamplifier, as well as the refractive index of the surrounding medium (if the nano-cluster is capped by another material that would have to be quasi-transparent to that resonant frequency). We have more elaborate designs to widen the range of frequencies that would excite the NC, therefore increasing the frequency band acceptance of the whole device. Also, the nanocluster will be an omnidirectional signal receiver: if directionality is desired, then the addition of the frequency-tuned antenna and chain of nanoclusters for field propagation becomes a necessity. While Fig.1 show a line antenna, alternatives like a simple dipole or more complex YAGI-like setups could be envisioned. Other types of resonators with polarization or not may be considered, with the condition of being tuned on the nanocluster preamplifier arrays. Single or redundant detection was also be considered.

2.2 The detection electronics

The aim is to produce an elementary cell of a multi-band directive polarized array – which to be further used in creating vision structures like fly eye or by using appropriate optics to get multi-parameter imaging. The directive multi-band vision is an important element in detecting imaged object material's properties.

The structure has the following parts:

- Directive polarized resonator – also called antenna – which may be similar to a yagi structure (see figure 1). The resonator has a reflector, vibrator used to select the polarized photons coming from the right direction. A resonant structure used to capture the photon energy and transform in local oscillation modes, with the capability of frequency selection and resonant amplification.
- A voltage converter-amplifier, made by passive micro-

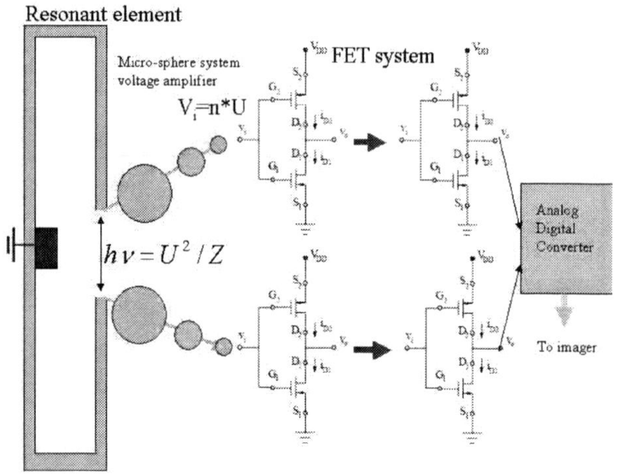

Fig. 2 The detection chain

nano structure elements which takes the resonator voltage and amplifies it driving to a field effect differential active element, where the voltage is transmitted as gate polarization, so that a single photon to become possible to be detected.
- The active amplifier made from a FET structure, running an alternative carrier like frequency in GHz domain, in a differential mode. When the THz voltage is attacking the gate, this is transmitted as a perturbation in the driving frequency, and the differential signal is collected and amplified.
- A signal collection and amplification system that takes the signal from each sensor and transmits it to a data acquisition device that is taking the signals in event mode or in amplitude mode.
- A computer device that is interpreting the acquired data and delivered for further utilization like imaging, detection, material identification

2.3 The plasmon resonator – converter

The single photon energy in the domain of 1mm to 1 μm wavelength, corresponds to wave number range 10-10,000, with energy in the domain of 1.2 meV to 1.2 eV for 1 micron IR. This energy applied for a reasonable time supposing the photon-associated wave has a finite length similar to that measured in optical experiments [1] may drive to a voltage in the picoVolts to nano-Volts range.

Stockman et al. proposes one of the most efficient passive "amplifiers" [2, 3, 4] design based on a chain of spheres. The dimensions of our resonators are in the sub-milimetric domain down to 1 μm and the voltage collector is less than ½ μV. The minimal sequence of "nanolens" [2] drives for a minimal final dimension for the MOS-FET gate in the range of 30 nm up to 10 μm. They proved that the electric potential in the small sphere region / our FET-Gate φ as function of the potential in resonator φ_o and

$$s(\omega) = \frac{1}{1 - \varepsilon(\omega)} \quad (1)$$

being the spectral parameter [5] and,

$$\varphi_0 = -zE_0 \quad (2)$$

$$\varphi(r) = \varphi_0(r) - \int \varphi_0(r') \frac{\partial^2}{\partial r'^2} \times \times \sum_{\alpha} \Phi_\alpha(r) \Phi_\alpha^*(r') \frac{s_\alpha}{s(\omega) - s(\alpha)} d^3 r' \quad (3)$$

Using numerical computations values for voltage gain of about 10^3 or greater were obtained [2].

The combination of low plasmon impedance and shape factor, with high dielectric media impedance makes the efficiency of the plasmon voltage converter very high, while the resonance condition narrows the band-pass, making the whole system to be frequency and polarization selective [6]. The electronic detection system relies on nonlinear carrier perturbation method.

The detection transistors are symmetrically crossed by a high carrier frequency that is perturbed by the gate electric parameter variation making the differential amplifier detect a signal out of electronic noise for each photon up to the high level of signal due to thermal emission. There are various combinations possible to enhance the detection resolution and to extract the useful signal from the thermal noise. The final amplifier stage transfers the signal to a time-of-flow-ultra fast analog digital converter that accesses a special fast multi access memory bench.

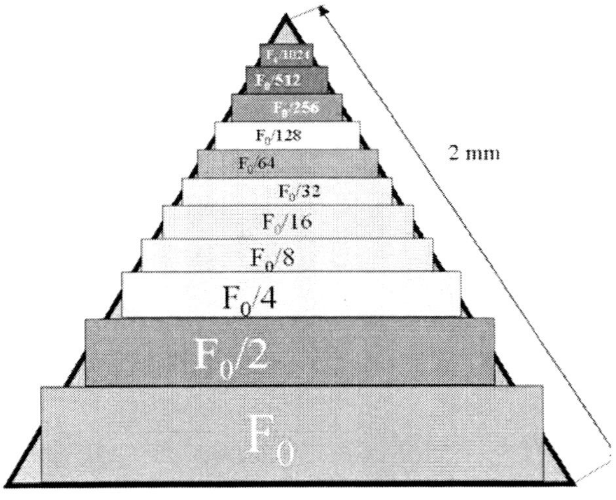

Fig. 3 – Compact multi-band antenna module

2.4 Multiple band detection

This structure may be repeated to detect difference frequency bands, using in correlation the directivity and phase selectivity properties of the resonators with the amplitude signal detection to produce a multi-band detection system, creating a pseudo-chromaticity in the T-ray field. Fig. 3 shows one input resonators grouping alternative, suitable for polarized "antenna". It was also analyzed the possibility of printing the resonator structure on had dielectric materials as diamond, sapphire, silicon carbide to create a hardened structure able to be used in harsh environments or fast moving objects. Any combination of various input resonator is possible as creating detection modulus with triangular, square, hexagonal, etc. cross-sections able to be assembled in various structures.

Because the resonators may be shaped in various aspect ratios there is possible to compact the various waveband centered resonators into a single detection module, called multi-band antenna, as Fir. 3 shows for a Yagi like rectangular structure.

To create an imaging device several procedures have been analyzed as:

- use a optical like lenses to produce a image in a narrow single band on a multi-detector plate
- a fly eye similar configuration made by a multitude of directional detectors
- a computed image generated by a morph detector array similar to the SAR

The T-ray chromatic vision is an in-depth vision selective to molecular parameters having large applications from medical, to scientific and military.

2.5 Data processing and integration

The resonator preamplifier structure has to be followed immediately by the detection module. Due to high frequency the carrier may have in GHz domain, the ADC have to be placed immediately near the module. It is recommended as an integrator circuit with sample-hold to be used, to generate the detection time-frame stability in the input of the ADC. The ADC no dead time is required, and a specific multiple-access buffer memory at high speed is required. When the working frequency is in GHz domain suitable to visualize molecular chemical kinetics the data storage and processing is a challenge to the actual technology. There were analized various data transfer configurations to memory banks due to the fact that the usual acquisition and transfer speed is in the range of several hundreds Giga-bauds. In practice not all molecular processes are evolving in ns level and there are several solutions for the detection electronics:

- to deliver a longer analogical integrator in front the ADC or digital integrator immediately after the ADC with the desired optimal frame rate
- to leave the electronics fast but to process the images by time-line integration.

The first alternative drives to a cheaper solution but more sensitive to S/N ratio, while the second one drives to a more versatile imager but more expensive also. Depending on application the best performance/cost technical solution may be selected. The molecular vision concept comes from the fact that most of the molecular combinations have their vibration mode emission in THz bands. Using a multi-band detection the possibility of performing advanced signature detection by fusion of the various channels with the specific weight is driving to molecular bounds recognition. More the multi-band correction may also determine the average temperatures in the molecule's elemental volume by detecting the temperature driven bands shift.

Special developed software and calibrations are needed to improve the detection accuracy. Knowing the detection vectors and the angular distribution of the input antenna array makes possible the three-dimensional imaging.

2.6 Applications briefing

The THZ detection is providing the following information after appropriate calibration:

- Detection direction and angular distribution of sensitivity
- Polarization,
- Central frequency and bandwidth
- Time of detection, and synchronization
- Amplitude

for each detector chain, inside a multi-band module, and for each module in the form of a 8-12 bit signal for each detection point.

Using this information into a specially programmed array, similarly to neural network several classes of visualizations can be made almost simultaneously:

- Complex 3D-time space information
- Detection of fast transitory evolutions if over the S/N fluctuations or some correlation exists
- Chemical compounds identification and their trend and kinetic rates (reaction, diffusion, etc.)
- Temperature fields as voxel averages per each type of molecule
- Cuasi-static image at the TV broadcasting rate, and more.

With features, and the fact that these rays have various penetration in depth in various materials as clothing up to 0.2 m, walls, soil, etc. the applications are in military domain, security, medicine in metabolism and organs remote imaging, environment, industry, preventive health, chemistry, reverse engineering, space applications and more.

3. CONCLUSIONS

The actual researches in the THz domain showed the possibility of obtaining a multi-band "far-IR" (sub-milimetric EM wave) imager.

The QED calculations play an important role, due to complex interactions between the lattice and photon quanta.

Low temperature electronics – near single-electron-transistor have to be used. The build-up of the multi-band THz sensor opens the way to a wide range of THz imaging applications.

Clustering and grouping the sensors under computer image control offers the possibility of having compact surface detection of the surrounding space in multi-band domain from GHz to THz with a great importance to wide range of application.

The application domain is very large, from military, security to medical and research, in space or terrestrial.

4. ACKNOWLEDGEMENTS

Special thanks to Dr. Muntele Claudiu from Center of Irradiation of the Materials, AAMURI Huntsville, AL for very useful talks and advises.

5. REFERENCES

1. Lange Wolfgang, Getting Single Photons under Control, Measuring the shape of a photon. Atomic, Molecular, and Optical Physics research at Sussex University, 2005. News.
2. Stockmann M.I., L.K., Li X., Bergman D.J.,, An efficient Nanolens: Self-Similar Chain of Metal Nanospheres. SPIE - Metallic Nanostructures and Their Optical Properties II, 2004. 5512(1): p. 87-99.
3. Stockmann M.I., Ultrafast processes in metal-insulator and metal-semiconductor nanocomposites. SPIE - Ultrafast phenomena in Semiconductors VII, 2003. 4992(1): p. 60-74.
4. Lee W., L.D.-B., Kim Y-W., Choi J-W.,, Signal amplification of Surface Plasmon Resonance based on Gold Nano-Particle Antibody Conjugate and its applications to Protein Array. Nano-Science and Technology Conf. -Nanotech, 2005. TP Abstract(1): p. 1-5.
5. Bergman D.J., S.D., Proprieties of macroscopically inhomogeneous media. Solid State Physics, 1992, 46(1): p. 148-270.
6. Popa-Simil L., On the possibility of developing multi-band imaging system for the EM wavelength from 1mm to 2 um. LANL-Comm., 2004. LA-UR-04-4766.

Feasibility Study of Cochlear-like Acoustic Sensor using PMN-PT Single Crystal Cantilever Array

S. Hur[*], S.Q. Lee[**] and W.D. Kim[***]

[*]Korea Institute of Machinery & Materials, Daejeon
305-343, Rep. of Korea, shur@kimm.re.kr
[**] Electronics and Telecommunications Research Institute, Daejeon
305-350, Rep. of Korea, hermann@etri.re.kr
[****]Korea Institute of Machinery & Materials, Daejeon
305-343, Rep. of Korea, wdkim@kimm.re.kr

ABSTRACT

We have fabricated piezoelectric PMN-PT single crystal cantilever array which has the cantilever size of the width of 200 um and the thickness of 10um. The length of cantilever was adjusted with a parameter of resonance frequency. Resonance frequency of PMN-PT cantilevers was measured with laser interferometer and charge sensitivity was measured with charge measuring device. PMN-PT cantilever array was installed on a noise shied case and exposed with sound pressure of specific frequency corresponding to resonance frequency and sensitivity of sound pressure was measured. The experimental results show that the PMN-PT cantilever array exhibits high sensitivity. This implies that the single crystal PMN-PT cantilever array has a potential candidate as cochlear like acoustic sensor.

Keywords: pmn-pt, piezoelectric, cantilever array, resonance, acoustic sensor

1 INTRODUCTION

Researches of artificial sensory system mimicking human and animal senses are increasing with worldwide trend to find the solution from the nature. The mammalian cochlea is an organ that performs the conversion of the incoming mechanical energy into electrical signals in the auditory nerve fibers. Current artificial cochlea has been developed in an effort to restore sensorineural hearing loss of patient. The artificial cochlea, one of the commercialized artificial sensory, consists of microphone for converting sound to electrical signal, signal processor for handling sound signal, inductive coil for transmitting sound signal from outside to inside of body, and electrode array for stimulating nerve cells. Current technologies of the artificial cochlea are difficult to enable the majority of the hearing-impaired people the good benefits due to the expense, inconvenience, frequent recharging requirement due to large power consumption [1, 2]. Piezoelectric materials possess the unique property of being able to generate electric charge when a mechanical force is applied to them, i.e., they are transducers capable of converting mechanical energy into electrical energy. Conceptually speaking, these materials are ideally suited as replacement "organs" for the nonexistent transduction mechanism in patients suffering from profound sensorineural hearing loss [3].

In this paper, we have studied the feasibility of use of PMN-PT($(1-x)Pb(Mg_{1/3}Nb_{2/3})O_3-xPbTiO_3$) single-crystal piezoelectric cantilever array as an alternative of conventional artificial cochlea. The PMN-PT material has been shown to possess piezoelectric coefficient and electro-mechanical coupling responses significantly larger than conventional ceramics [3]. We have fabricated piezoelectric PMN-PT cantilever array which has the cantilever size of the width of 200 um and the thickness of 10um. The length of cantilever was adjusted with a parameter of resonance frequency. Resonance frequency of PMN-PT cantilevers was measured with laser interferometer and charge sensitivity was measured with charge measuring device. PMN-PT cantilever array was exposed with sound pressure of specific frequency corresponding to resonance frequency and sensitivity of sound pressure was measured. The experimental results show that the PMN-PT cantilever array exhibits high sensitivity. This implies that the single crystal PMN-PT cantilever array has a potential candidate as cochlear like acoustic sensor.

2 EXPERIMENTAL PROCEDURE

2.1 Fabrication of PMN-PT cantilever

Figure 1 shows the dimensions of PMN-PT single crystal cantilever and interdigitated electrode. The cantilever is designed with width of 200 μm, thickness of 13 μm and length of 1500 μm. A pair of interdigitated electrodes are designed with comb width of 5 μm and comb gap of 10 μm. The interdigitated electrodes are located at supporting position of PMN-PT cantilever. It is good for getting maximum charge output from interdigitated electrodes or large deflection from free end of cantilever.

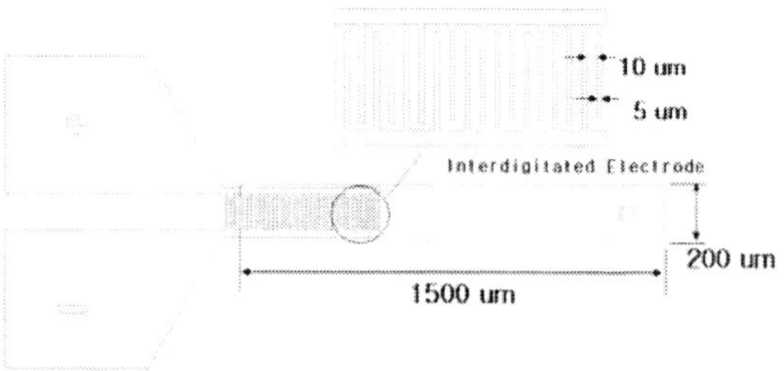

Figure 1: Dimensions of PMN-PT cantilever and interdigitated electrode.

Figure 2 shows the steps required to fabricate a PMN-PT single crystal cantilever type sensor. We use <001>-oriented and poled 20 um-thick PMN-PT single crystal glued on a Si substrate. The sample undergoes a mechanical polishing down to 20~25 μm-thick film. Next, Inductively Coupled Plasma (ICP) etching process is used to thin the film down to 13 μm. For the upper electrode patterning, Au e-beam sputtering is used following the Photo Resistive coating for lift off process. ICP etching process is again applied to define the cantilever shape. As a final step, back side etching with deep RIE etching is used for releasing the cantilever.

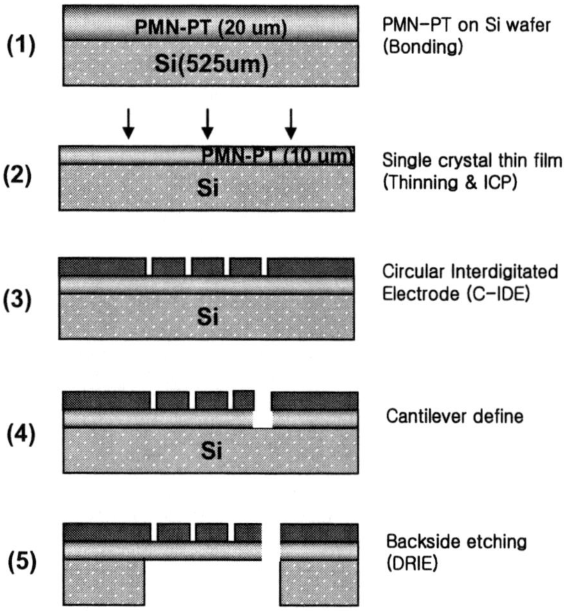

Figure 2: Semiconducting process of PMN-PT single crystal cantilever.

Figure 3 shows the PMN-PT single crystal cantilever fabricated with semiconducting process. Figure 3 (a), (b), (c) and (d) shows length of 900 ㎛, 1000 ㎛, 1300 ㎛ and 1500 ㎛. Wire bonding was performed for each of two cantilevers with same length.

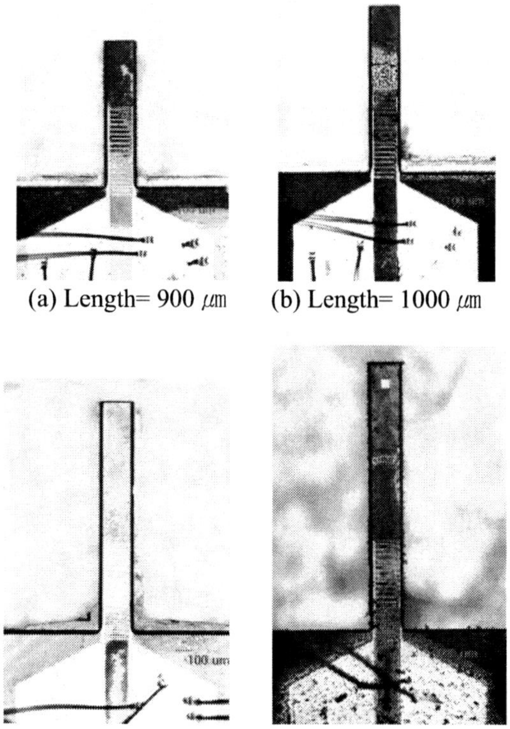

(a) Length= 900 ㎛ (b) Length= 1000 ㎛

(c) Length= 1300 ㎛ (d) Length= 1500 ㎛

Figure 3: The fabricated PMN-PT single crystal cantilever with different length.

2.2 Characterization of PMN-PT cantilever

To measure a resonance frequency of a fabricated PMN-PT single crystal cantilever, the probe was connected with plus and minus electrode of PMN-PT single crystal cantilever. The voltage signal of DC 2.5 V and AC 1.0 $V_{p\text{-}p}$ with a sine wave form between 200 Hz and 4000 Hz was provided on the probe. The electric field was generated between the interdigitated electrodes and the deflection of cantilever was generated with sine wave signal. The experimental setup for measuring the deflection of PMN-PT cantilever was consisted of vibrometer controller(Polytec OFV-5000), Fiber interferometer(Polytec OFV-512) and dynamic signal analyzer(Agilent 35670A). The measured deflection of PMN-PT cantilever was about 13 ㎛ p-p. Also we have measured impedance and phase values using HP 4194A impedance analyzer. Figure 4 shows measuring results of impedance and phase value. The resonance frequencies of the same length of two cantilevers were measured with 4.25 kHz and 4.96 kHz respectively.

Table 1 shows designed and measured values of resonance frequency for PMN-PT single crystal cantilevers using HP 4194A impedance analyzer. It was verified that two cantilevers of the same length has some different resonance frequency. The measured resonance frequency is similar with designed frequency. The density and elastic modulus used for designing PMN-PT single crystal cantilever was 8,200 kg/m^3 and 20.0 GPa.

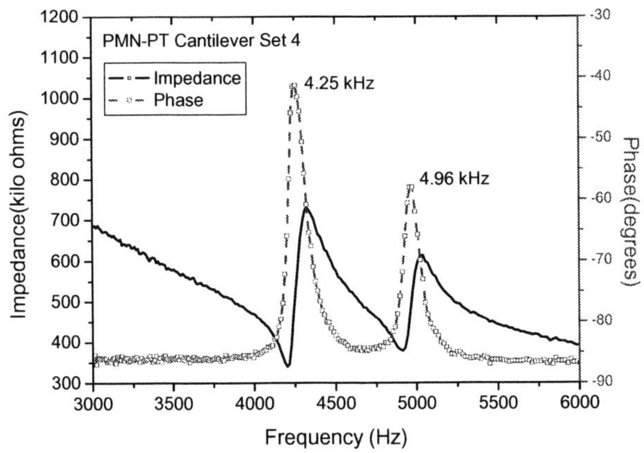

Figure 4: Measurement result of impedance and phase for PMN-PT cantilever set 4 using impedance analyzer.

Cantilever length (μm)	Resonance frequency (measured) (kHz)	Resonance frequency (designed) (kHz)	Remarks
1500	1.74/1.93	1.46	Cant. Set 1
1300	2.41/2.62	1.94	Cant. Set 2
1000	3.26/3.97	3.28	Cant. Set 3
900	4.25/4.96	4.05	Cant. Set 4

Table 1: Measuring results of resonance frequency for PMN-PT single crystal cantilever.

The mechanism of generating a charge in a piezoelectric PMN-PT cantilever is as follows; as PMN-PT cantilever deflects along the interdigitated electrode, causing stress on the cantilever, the variation of stress in the cantilever produces self-generated charges on the interdigitated electrode. The charges are not generated by the static stress, but by the variation of stress. Figure 5 shows an experimental setup of charge measuring system which can measure piezoelectric charges from PMN-PT single crystal cantilever. Charge measuring system was consisted of differential charge amplifier, laser displacement measuring device and precision displacement actuator. A differential charge amplifier has used feedback capacitance and feedback resistance of 5 pF and 22 MΩ. The generated charge signal is amplified by 10 times with instrumental amplifier.

To reduce a measuring noise, the PMN-PT single crystal cantilever was directly installed on the circuit board of charge amplifier. Precise piezoelectric actuator was used to test PMN-PT single crystal cantilever with displacement range of 40 nm to 330 nm using a square wave of 10 Hz. Piezoelectric charge was measured with charge measuring system as the piezo actuator moves up and down while the tip of cantilever just is contacted with piezo actuator. The deflection was measured with laser diode and photo sensitive detector.

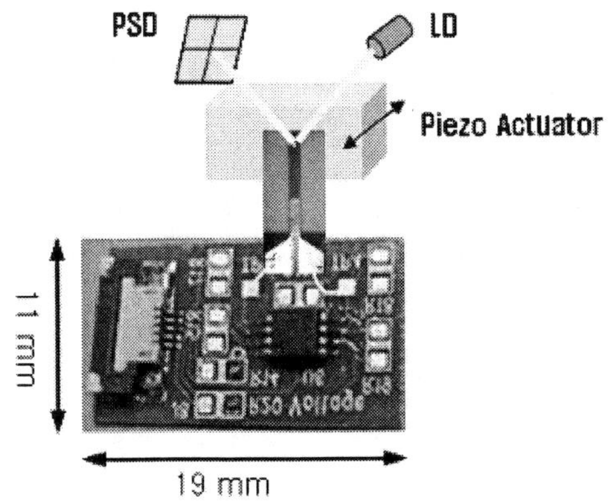

Figure 5: Experimental setup for charge measurement of PMN-PT single crystal cantilever.

3 RESULTS AND DISCUSSIONS

Figure 6 shows a measuring result of charge sensitivity due to a deflection of PMN-PT single crystal cantilever. We measured two samples of PMN-PT single crystal cantilever. The charge signals were linearly proportional to the cantilever deflection. The charge sensitivity of two samples was measured with 2.1 fC/nm and 2.2 fC/nm respectively.

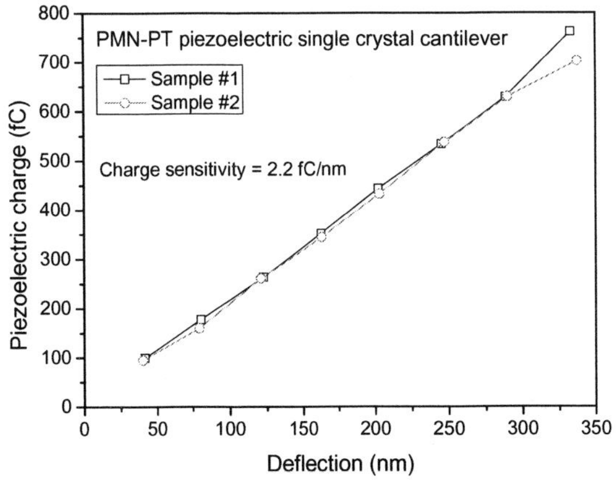

Figure 6: Piezoelectric charge output as a function of PMN-PT cantilever deflection.

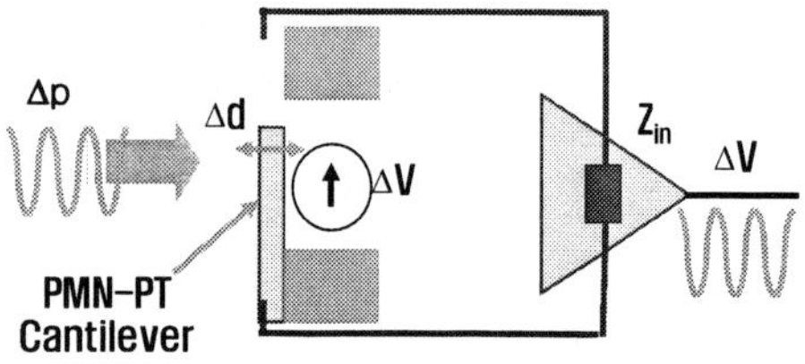

Figure 7: The schematic diagram of acoustic sensitivity measurement for PMN-PT cantilever.

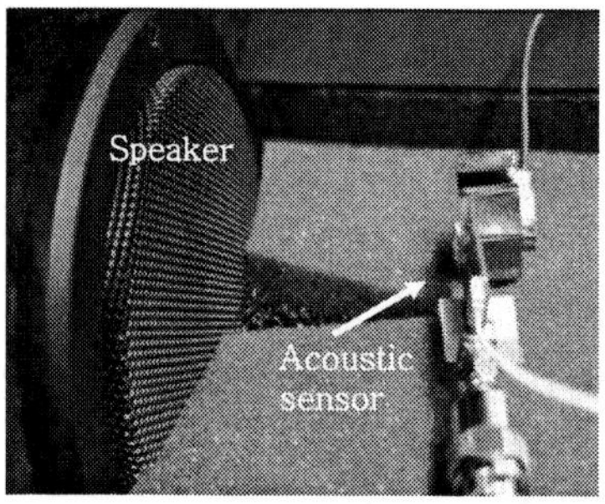

Figure 8: The experimental setup for acoustic sensitivity measurement of PMN-PT cantilever.

Figure 7 shows schematic diagram of acoustic sensitivity measurement for PMN-PT cantilever. When the sound pressure is applied to the cantilever, it is resonated at a specific frequency. It generates the charge signal with the piezoelectric property and is amplified to the output voltage signal. Figure 8 shows the experimental setup for acoustic sensitivity measurement. The PMN-PT cantilever, charge amplifier and reference microphone are installed at 4 cm distance from speaker. This experimental setup was inserted into anechoic test box(type 4232 B&K) to reduce a surrounding noise. Speaker makes the equalized sound pressure of 1 Pa from 1 kHz to 6 kHz with sound signal sweep. The charge signal generated from PMN-PT cantilever was processed with sound processing software.

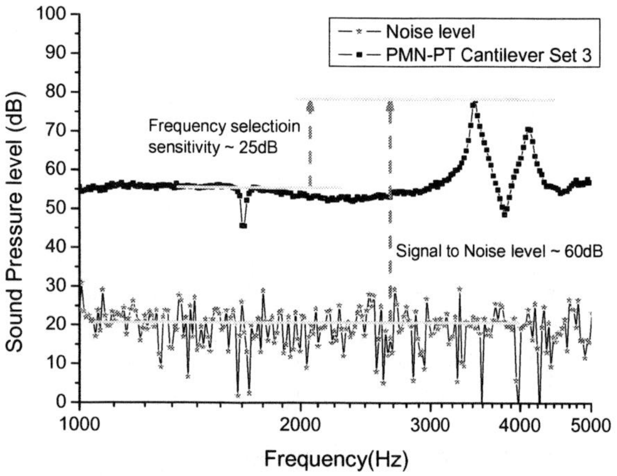

Figure 9: The acoustic sensitivity of cantilever set 3.

Figure 9 shows the acoustic sensitivity and its signal to noise ratio of PMN-PT cantilever set 3. This PMN-PT cantilever can detect the signal with 1000 times of signal to noise ratio and can selectively detect the specific frequency with 18 times higher signal level. Figure 10 shows the experimental results of the acoustic sensitivity for each set of PMN-PT cantilever. The resonance frequencies from 1,741 Hz to 4,960 Hz are obtained from four set of PMN-PT cantilever.

From this study, it was verified that the fabricated PMN-PT cantilever can detect the signal with 1000 times of signal to noise ratio and can selectively detect the specific frequency. In conclusion there is quite a possibility that PMN-PT single crystal cantilever may use as cochlear-like acoustic sensor.

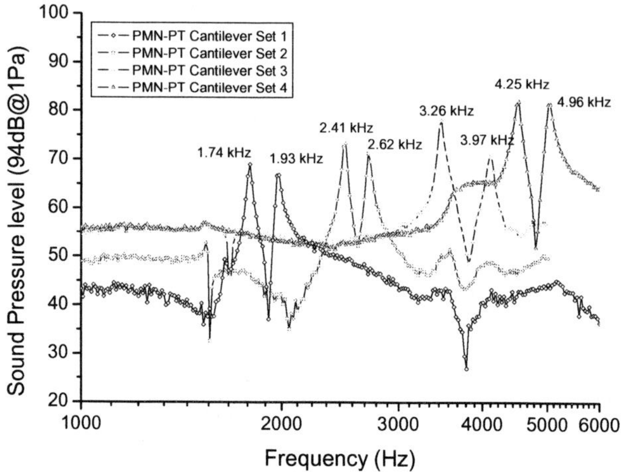

Figure 10: The experimental results of acoustic sensitivity for each PMN-PT cantilever.

REFERENCES

[1] M. Bachman, F.G. Zeng, T. Xu and G.-P. Li, "Micromechanical Resonator Array for an Implantable Bionic Ear," Audiology & Neurotology, 11, 2006.

[2] C. Niezrecki, D. Brei, S. Balakrishnan, and A. Moskalik, "Piezoelectric Actuation: State of the Art ," The Shock and Vibration Digest, Vol. 33, No. 4, 2001.

[3] N. Mukherjee, R.D. Roseman, and J.P. Willging, "The Piezoelectric Cochlear Implant: Concept, Feasibility, Challenges, and Issues," John Wiley & Sons, Inc. 2000.

MEMS based examination platform for neuro-muscular communication in a co-culture system coupled to a multi-electrode-array

U. Fernekorn*, M. Fischer*, M. Klett*, C. Augspurger*, B. Hiebl**, A. Schober*

*MacroNano® Center for Innovation Competence, Dept. for Microfluidics and Biosensors
Technische Universität Ilmenau, Gustav-Kirchhoff-Str. 7, 98693 Ilmenau, Germany
**Forschungszentrum Karlsruhe, Hermann-von-Helmholtz-Platz 1,
76344 Eggenstein- Leopoldshafen, Germany

ABSTRACT

Development of sensor-connected in vitro neuromuscular co-culture systems may provide a useful tool for interpreting mutual signal transfers between the cell species and for future drug screening applications. A glass-based, autoclavable, micro fluidic culturing system consisting of two chambers connected by 50 µm micro capillaries with a diameter of 5 µm was designed. Micro capillaries were generated by two technological approaches. First reactive ion etching was used for structuring the glass capillaries as well as a subsequent bonding to a glass carrier with electrodes by using an UV-activatable adhesive. As second approach a structured SU8-layer was used simultaneously for generating the capillaries and as an intermediate layer for bonding. Chambers and fluidic luer connectors (structures in sub-millimeter scale) were realized by means of a cost effective micro-sandblasting process. The co-culture system was combined with a commercially available multi electrode array. This setup allows stimulation of adherent neuronal cells in one chamber and measurement of action potentials induced in myotubes derived from the myogenic C2C12 cell line in the other. Neuronal processes growing through micro capillaries were detected using immuno-fluorescence staining methods. This hybrid technology represents a new approach for electrophysiological recordings.

Keywords: bio-mems, neuromuscular synapse, mea, co-culture

1 INTRODUCTION

Many aspects about the factors [1] that facilitate the interplay between motoneurons and the muscle fibers innervated have been unravelled by examining tissue specimen or studying different types of co-culture models [2- 4]. Our co-culture model intends to generate motorical end plates derived from cell lines. To examine how these are developed and what type of substances may activate or inhibit this process, electrophysiological recordings of neuromuscular interactions are surveyed on multielectrode arrays. This technique allows to measure neuromuscular activities in spatiotemporal resolution. In the system presented here (Figure1), a commercially available planar multielectrode array (Multi Channel Systems MCS GmbH, Reutlingen, Germany) with 64 electrodes has been combined with a co-culture system that facilitates separated growth of monolayers of neuronal and myogeneic species. Additionally, stimulation of the adherent cells and subsequent recording of the extracellular signals will be examined.

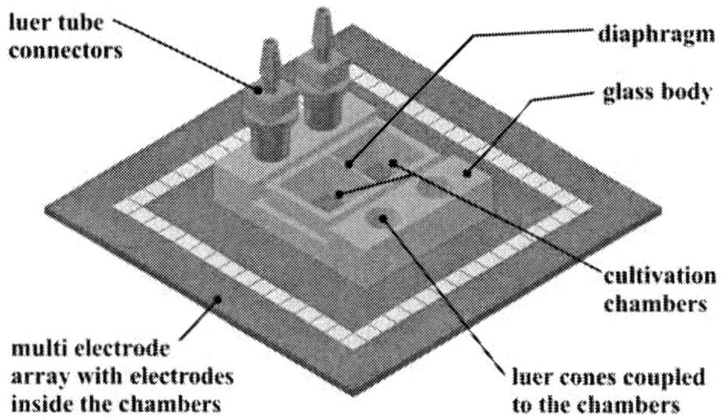

Figure 1: Scheme of the co-culture system

2 TECHNOLOGICAL SETUP

As to obtain a chamber volume of about 300 µl at lateral geometries defined by the MEA, a minimum chamber height of 5 mm is necessary. Glass with an adequate UV-transmission was chosen as construction material to achieve a comfortable illumination for fluorescence microscopy. Therefore a 100 mm wafer-formatted borofloat glass substrate with a thickness of 5 mm was used allowing standard MEMS technologies. Two technological approaches, Reactive Ion Etching (RIE) and SU8-lithography were tested to manufacture the lateral micro capillaries on the bottom of the diaphragm.

2.1 Fabrication of the glass body

To create 5 mm deep structures in glass substrates, micro sand blasting is an effective method with an adequate lateral accuracy and a high ablation ratio compared to plasma etching techniques. Thin steel masks, patterned by lithography and etching are aligned and fixed on both sides of the glass wafer using a bond wax. A compressed air beam containing 30-µm silicon carbide powder is used to scan both sides of the masked glass surface until the chamber structure is broken through. Figure 2 depicts the glass wafer after removing of the steel masks. A grinder fitting the angle of the luer cones was utilized to refinish the connecting holes.

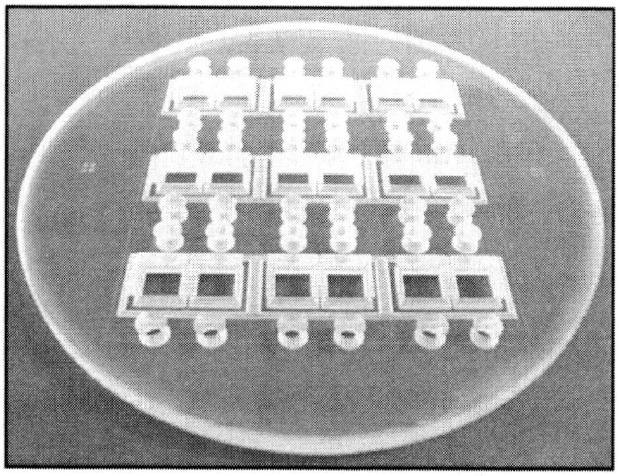

Figure 2: Micro sand blasted glass wafer

2.2 Fabrication of the capillaries using RIE and gluing

As a first approach the micro capillaries which have a dimension of 5 µm x 5 µm x 50 µm were generated by a modified RIE process before micro sand blasting of the glass wafer. A parallel plate reactor (STS 320) with a cooled wafer electrode and a CHF_3 / CF_4 – plasma was used. Thereby an etched chromium layer patterned by a lithography step acts as masking layer. Figure 3 depicts the bottom side of the glass wafer where the micro capillaries are patterned. After the etching process the chromium layer was removed.

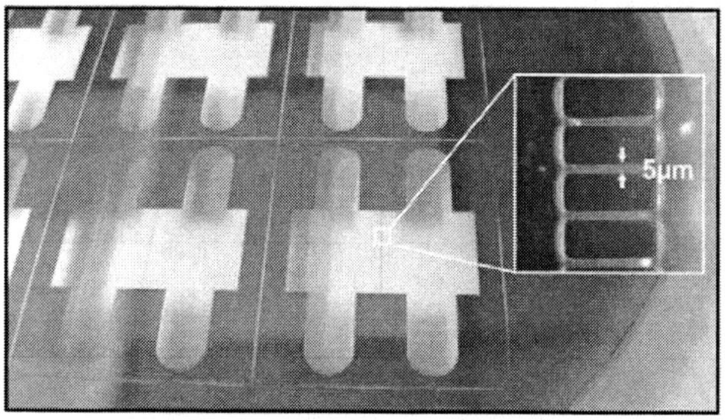

Figure 3: 5-mm glass wafer structured with RIE

Hereafter the processed glass wafer was sand blasted as mentioned above. So as to separate the single glass bodies of the glass substrate standard wafer dicing was used. The bonding procedure of the glass body to the MEA-system was accomplished by gluing. With a microscope the MEA-system and the glass body are aligned and fixed by a clamping mechanism. Using capillary forces the small gap between the two glass components was filled with a liquid UV-cured acrylat adhesive (SurABond HH 051), stable at higher humidity and temperature. Because of the excellent UV-transmission of both glass components the exposure was very simple.

However, after various evaluations, particular after few autoclaving tests and several changes of the medium the glass bodies were de-bonded. Because the overall system is designed for multiple uses it was necessary to verify the second technological approach.

2.3 Fabrication of the capillaries using SU8-lithography and bonding

For simultaneous creation of the 5-µm capillaries as well as the intermediate bond layer SU-8 [5], an epoxy based photoresist was used. This resist from MICRO-CHEM is commonly used for micro electro mechanical systems (MEMS) with high aspect ratios as well as high chemical and thermal stability. The lithography step was directly implemented on the MEA glass carrier. The 49 mm x 49 mm glass substrate with electrodes and a passivation layer was coated with 5-µm thick SU8-2005 layer and patterned by a modified lithography step (Fig. 4a). After an adapted cleaning step, including plasma treatment a single co-culture glass body was aligned and pressed to the MEA glass carrier. Using a heating press with an adequate parallelism a very stable bonding between glass body and MEA system was achieved during the post exposure bake of the resist at 110 °C.

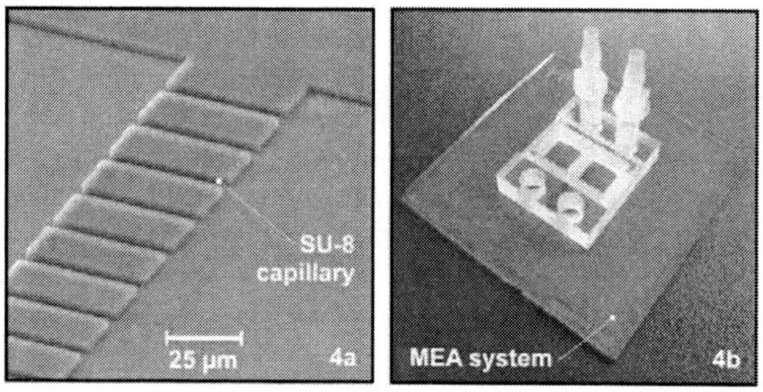

Figure 4: Patterned SU-8 layer before bonding (a), completely bonded co-culture system (b)

3 CELL BASED APPLICATIONS

3.1 Cell Culture

Undifferentiated NG108-15 cells (ATCC) were cultured in Dulbecco's modified Eagle's medium (DMEM, Sigma-Aldrich, St.Louis, MO, USA) supplemented with 10% FCS + Penicillin and Streptomycin + L- Glutamine. To induce the differentiation, the culturing medium was changed to serum free DMEM + Penicillin and Streptomycin + diButyryl- cyclic AMP (dBcAMP, Sigma-Aldrich). C2C12 myogeneic cells (ATCC) were maintained in DMEM + 10% FCS + Penicillin and Streptomycin + L- Glutamine.

Upon changing the culturing medium to DMEM + 2% heat-inactivated horse serum (PAA, Austria), differentiation to myotubes was observed from 5 DIV. MEA dishes were coated with 0.02 % gelatine.

2.5 x 10^4 undifferentiated C2C12 cells were plated to one chamber of the co-culture system and cultured for 9 days. Undifferentiated NG108-15 cells were added to the other chamber. Both cell species were maintained in DMEM + 2% horse serum.

3.2 Transfection

NG108-15 cells were transiently transfected either with the pDsRed-Express-C1 (Clontech, Mountain View, CA, USA) or pZsGreen1-C1 vector (Clontech, Mountain View, CA, USA), respectively. For this process, Lipofectamine 2000 Kit (Invitrogen, Karlsruhe, Germany) was used according to the protocol provided by the supplier. All the expression plasmids contained a G418- resistant gene as described by the supplier. G418 stable cell lines were selected in medium containing 400µg/ml G418 (Geneticin, Invitrogen, Karlsruhe, Germany) and tested by expression of pDsRed or pZsGreen1. Successfully transfected NG108-15 cells were sorted using a Partec CyFlow®space flow cytometer connected to a Partec Particle and Cell Sorter (Partec GmbH, Muenster, Germany).

3.3 Immunohistochemistry

Neuronal processes were stained with antibodies directed against microtubule- associated protein 2a (FITC labelled MAP2a, clone AP20, Abcam, Cambridge, UK), growth- associated protein 43 (GAP43, clone GAP-7B10, Abcam, Cambridge, UK) or neurofilament 200 (NF-H, clone E15, Santa Cruz, CA, USA), respectively. Cells were washed and fixed with ethanol for 5 minutes. Nonspecific binding was blocked using 10 % normal goat serum (PAA, Austria) in PBS. Cells were incubated with primary antibodies at 4°C overnight. After rinsing, cells were exposed to Texas Red labelled donkey anti goat IgG or goat anti-mouse IgG at 4°C, dark and overnight. Cells were examined with an inverted Nikon Eclipse TS100 microscope (Nikon, Japan) equipped with appropriate filters and camera (VDS Vosskühler, CCD1300CB, VDS Vosskühler, Germany). Fluorescent photomicrographs were taken with a magnitude of 40. The images were then processed (Adobe Photoshop, Mountain View, CA, USA) as appropriate for presentation in the figures.

4 RESULTS AND DISCUSION

A commercially available planar multielectrode array (Multi Channel Systems MCS GmbH, Reutlingen, Germany) with 64 TiN electrodes was combined to a glass co-culture system consisting of two cultivation chambers interconnected by micro channels of 5µm width and 50 µm length. These channels serve as a diaphragm. This arrangement enables neuronal processes to grow through the channels. The area size of one chamber comprises 65 mm^2. As the system is autoclavable, it is suitable for repeated measurement approaches. The electrodes are directly located beside the diaphragm (left and right). Via Luer fittings, the co-culture system may be connected to a perfusion system, thus allowing both static and fluidic cultivation conditions. We established a co-culture of the neuroblastoma x glioma cell line NG108-15 [6] and mouse myoblast cell line C2C12 cells [7]. NG108-15 cells have been reported to resemble motoneurons as they are able to synthesize acetylcholine, agrin and neuregulin [8, 9]. These cells are capable of generating functional synapses with myotubes including induction of AChR clusters [8, 10]. Moreover, twitching of myotubes cocultured with NG108-15 neuronal cell has also been described [8]. C2C12 cells provide a well established system to cultivate myotubes and produce acetylcholine receptors. Dye conjugated alpha-bungarotoxin, a snake venom with high binding affinity to nicotinic acetylcholine receptors, was used to visualize attachment of neuronal cells to myotubes (Fig. 5).

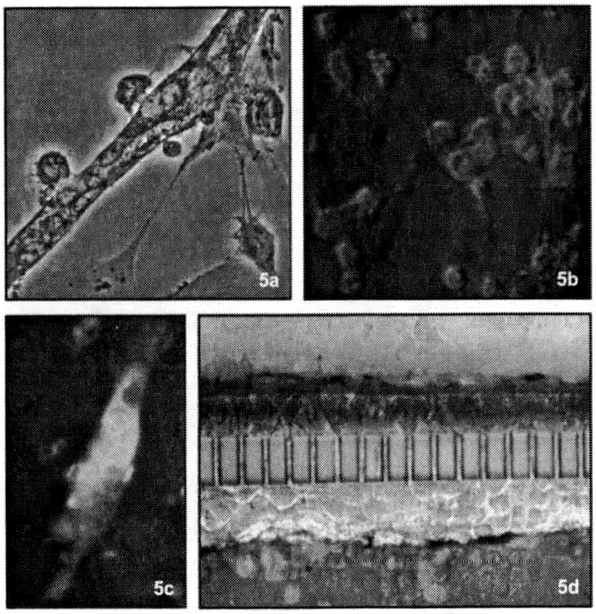

Figure 5: Neurites of NG108-15 cell binding to C2C12 myotubes (a). These cells express neurofilament –H (b) and MAP2a (not shown). Positive staining of FITC conjugated α-bungarotoxin demonstrates the presence of nACh receptors (c) in C2C12 myotubes. Neuronal processes of pDsRed transfected NG108-15 cells growing through microchannels separating the cultivation chambers (d).

Some neurological disorders are correlated with a defective signal transmission between the innervating motoneuron and muscle fibers. Therefore, we are seeking to further understand the electrophysiology of the neuromuscular interface. Our MEA- based co-culture system may provide a new setup for electrophysiological recordings of the cell culture derived end plate potentials.

5 ACKNOWLEDGEMENTS

The authors gratefully acknowledge the financial support by Federal Ministry of Education and Research and the Thuringian Ministry of Culture within the Initiative "Centre for Innovation Competence", MacroNano®. We thank Daniel Hein for SU-8 lithography and Little Things Factory GmbH Ilmenau for support in micro sand blasting.

REFERENCES

[1] T. Meier and B.G. Wallace, BioEssays, 20, 819-829, 1998.
[2] M. Das et al., Neuroscience, 146, 481-488, 2007.
[3] T.F. Kosar et al. Lab on a Chip, 6,632-638,2006.
[4] J.X.S. Jiang et al., J. Biol. Chem., 278;46,45435-45444,2003.
[5] P. Svasek, et al., Sensors and Actuators A 115 (2004) 591–599.
[6] B. Hamprecht, Int. Rev. Cytol., 49,99-170,1977.
[7] H.M. Blau, et al., Science, 230, 758-766, 1985.
[8] N.R. Cashman et al., Dev. Dyn., 194,209-221,1992.
[9] M. Nirenberg et al., Science, 222, 794-799,1983.
[10] N.A. Busis et al., Brain Res., 324, 201-210,1984.

GRAVI™- *Chip*: Automation of Microfluics Affinity Assay Using Magnetic Nanoparticles

Joël Rossier*, Sophie Baranek*, Patrick Morier*, François Vulliet*, Christine Vollet* and Frédéric Reymond*

*DiagnoSwiss S.A., Rte de l'Ile-au-Bois 2, CH-1870 Monthey, Switzerland
j. rossier@diagnoswiss.com

ABSTRACT

This paper presents a novel microfluidic analytical platform called GRAVI™ , with particular emphasis on the benchtop robotised version of this system which is dedicated to fast and automated enzyme-linked immunosorbent assays (ELISA) in low volumes. By dramatically reducing time-to-result (assay time of <10 min, including chip regeneration) and significantly decreasing sample/reagent consumption and cost, this microfluidic biosensor system enables to perform multi-menu immunoassays with simplified robotics and laboratory infrastructure.

Keywords: Microfluidics, microsensors, magnetic nanoparticles, ELISA, electroanalytical methods.

1 INTRODUCTION

For more than twenty years, ELISA has been a reference method in bio-analytics, particularly in *in vitro* Diagnostics (IVD) and in Life Science Research (LSR).

In order to meet the increasing demand for accelerated time-to-results and reduced sample and reagent consumption in ELISA, DiagnoSwiss has developed a microfluidics-based biosensor platform, which can be used as a manually operated stand-alone device or which can be coupled to standard robotic stations. The system, currently introduced under the trade name GRAVI™, comprises a compact instrument (GRAVI™-Cell), electrochemical microchip consumables (GRAVI™-Chip) and a computer software (GRAVI™-Soft) for developing assay protocols, running the analyses and processing the results.

Coupled to a sample preparation robot, the GRAVI™ system fully integrates the following functions:

- Sample preparation (dilution, mixing, etc.)
- Mixture of sample and magnetic bead reagents ("Mix")
- Mix incubation (either in a plate or in the GRAVI™-chip)
- Dispensing of the Mix into the GRAVI™-chip
- Bead capture (by a programmable magnet array integrated in the GRAVI™-Cell) and washing in the microchannels
- Addition of the substrate for enzymatic readout and, finally, in-chip detection
- Robotised removal of the magnet array, chip regeneration and washing.

Either 8 or 16 assays are run in parallel, with one loop protocol taking only 8 minutes. As such, this hands-free system is suitable for a throughput of 60 or 120 tests per hour. Combined with standard robotics, the system allows any combination of volume, order and number of reaction steps (including bead menus). The full menu is programmed by means of the GRAVI™-Soft software controlling both the liquid handler and the GRAVI™-Cell instrument.

2 MATERIAL & METHODS

GRAVI™-Chip: GRAVI™-Chip is a cartridge harboring 8 parallel micro-channels produced in a polyimide flex foil with printed board circuitry [1-2] and comprising 30 uL inlet and outlet reservoirs (see Figure 1).

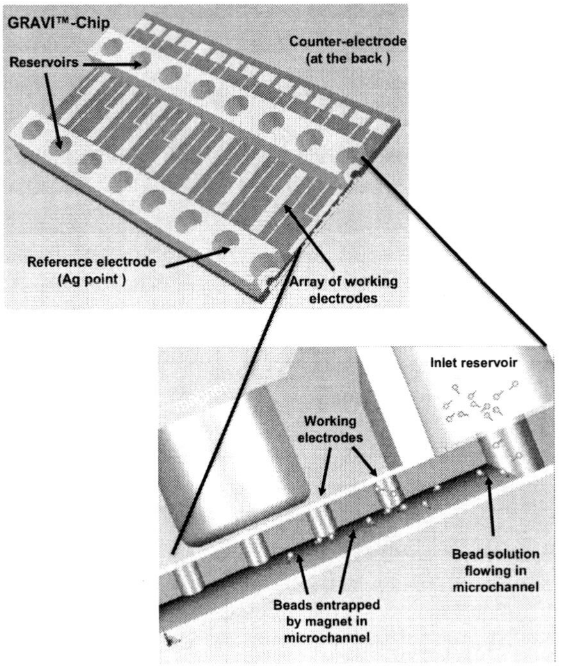

Figure 1: Top: Technical drawing of a GRAVI™-Chip with its network of electrodes and solution reservoirs; Bottom: detailed microchannel cross-section with flowing magnetic beads being captured by a magnet.

An array of microelectrodes positioned along the 1.5 cm micro-channel of 90x250 um^2 section serves for the readout of the enzymatic signal. Multiplexing is feasible by using different enzyme labels, e.g. alkaline phosphatase (ALP) or β-galactosidase.

The GRAVI™-Chip is simply inserted into the USB-powered reader (GRAVI™-Cell) which ensures proper electrical contact and current readout during the enzymatic detection (see Ref. [3] for further details about the detection principles). The GRAVI™-Cell apparatus further comprises a magnet array that can be opened and closed automatically for capture and release of magnetic beads within the microchannels. As shown in Figure 2, the chip is positioned on a tilted plane so that microfluidics is solely driven by capillary force and gravity, thus avoiding the need for any serviceable motors, pumps and tubing.

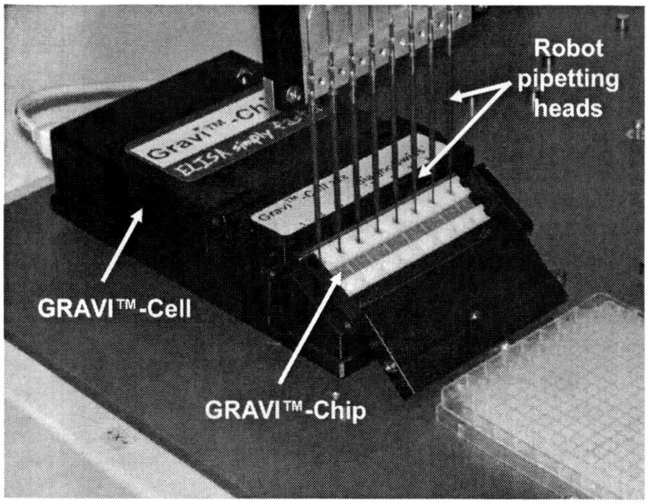

Figure 2: Picture showing a GRAVI™-Cell instrument coupled to a sample preparation robot (Carvo MSP, Tecan, CH)

Correctly placed in the GRAVI™-Cell, the inlet reservoirs of the GRAVI™-Chip is positioned 1 cm above the microchannel outlets, thereby inducing a pressure difference which is sufficient to generate a flow rate of up to 1 $uLmin^{-1}$ [4]. Thus, with an internal channel-volume of 300 nL, this gravity-induced flow allows to replace the total microchannel volume 3 times per minute.

In order to stop the flow (which is notably required for proper signal detection during enzyme-substrate incubation) excess liquid is simply removed from the upper reservoir. As the capillary action is strong at the small dimensions of the microchannels, the capillary forces are larger that the gravity force, thereby preventing emptying of the microchannel and providing efficient valving of the system (which, for instance, ensures stagnation of the solution within the microchannels during the detection).

In order to run bead-based immunoassays, superparamagnetic beads of 1 micron or less are used to carry the affinity capture probes.

Robotic handling: The GRAVI™-Cell is snapped on a plate holder of a sample preparation robot (Cavro MSP9250, Tecan, CH), which is fully controlled by the GRAVI™-Soft computer software. Each fluidic measurement and dispensing step is pre-programmed in a flexible and user-friendly protocol, which can work for hours without need for manual operation.

GRAVI™-Chip for the detection of recombinant human antibody: Aliquots of a mammalian cell culture provided by Selexis (Geneva, CH) containing a theoretical IgG concentration of 400 ug/mL were diluted 100-fold. Calibration samples were then prepared by repeated 2-fold dilutions of the above aliquots so as to obtain IgG standards of 4, 2, 1, 0.5, 0.25, 0.125, 0.063 and 0.031 ug/mL.

Reagents composed of protein A beads (New England Biolabs, Ipswich, MA, USA) at a concentration of 0.015 mg/mL, goat antihuman FAB-ALP conjugate (Sigma, Buchs, CH) at a concentration ratio of 1/1000, washing buffer (Tris pH 9, Tween 0.1%) and p-aminophenyl-phosphate (DiagnoSwiss, CH) as enzyme substrate at a concentration of 10 mM in MAE buffer pH 9 were dispensed in a deep plate. The test consisted in: a) mixing beads, sample and conjugate (5 uL each) in the GRAVI™-Chip inlet; b) permitting flow through the microchannels during 1 min (with the magnet array in place to capture the beads); c) 3 washing of the microchannels three times with 20 uL of washing buffer disposed in the inlet reservoirs; d) adding 20 uL of enzyme substrate solution in the inlet and removing the excess solution after 20 seconds in order to stop the flow. Successive amperometric measurements were then performed in the 8 microchannels simultaneously using the multiplexed potentiostat of the GRAVI™-Cell apparatus, thereby enabling to follow the kinetics of the enzymatic reaction during 30 seconds. Using a fitting procedure integrated in the GRAVI™-Soft, the analyte concentrations were then deduced from the slopes at origin of the obtained current versus time curves.

3 RESULTS AND DISCUSSION

The IgG concentration of mammalian cell culture samples at 8 different concentrations (from 4 ug/mL down to 0.031 ug/mL) have been determined by bead-based immunoassay using the robotised GRAVI™-Cell system.

Eight assays were performed simultaneously within less than 10 minutes (including chip regeneration), and Figure 3 shows the calibration curves obtained for 5 series of 8 assays performed successively in the same chip after regeneration between each series.

As can be seen in Figure 3, the obtained dynamic range covers more than 3 orders of magnitude and the CV is roughly 5 % in the higher concentrations.

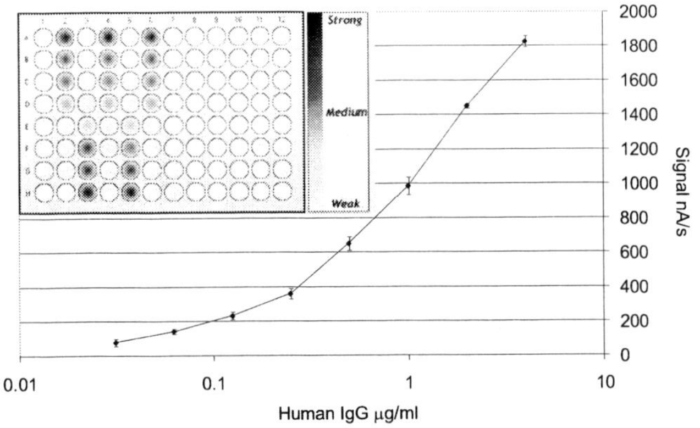

Figure 3: Example of calibration curve for the detection of IgG in samples of mamalian cell cultures using protein A, goat antihuman FAB-ALP conjugate, and p-aminophenyl-phosphate substrate. (Insert: color plate representation for the repeated assays shown in the graph and as dispensed in the microtiterplate).

The GRAVI™-soft also enables to display the results as intensity color plate (see inlet in Figure 3). Each vertical line corresponds to the signals obtained for a row of samples present in the micro-titerplate after the dilution steps processed by the robot as described in section 2. In the present example, a calibration was measured by distributing the samples from high concentration in well A to low concentration in row H of the microtiterplate. The samples are then automatically mixed with the magnetic beads and injected in the microfluidic system for rapid readout. In the second raw, the same calibration is injected, but with the highest concentration in row H and the lowest in row A in order to show that carry over can be reduced to the minimum.

These experiments show that the use of the GRAVI™ platform allows to perform 16 immunoassays in less than 10 minutes. In the 8-channel version, the GRAVI™ system is thus able to perform ~50 tests/h, while the 16-channel option can process about one microtiter-plate per hour.

The concept is amenable to any SBS liquid handler, and different robotic systems can be equipped with the GRAVI™-Cell platform.

4 CONCLUSION

Requiring no maintenance, the pumpless GRAVI™-Chip concept provides an innovative link between rapid and economic microfluidic systems and standard laboratory robotics. Compatible with the conventional 96-well plate format, the GRAVI™ platform can run on standard liquid handling robots used in R&D laboratories. It is a scalable technology which can be declined from portable and bench-top devices to large automated workstations (all with identical assay performances), and it represents a fast, flexible and cost-efficient immunoassay platform.

ACKNOWLEDGEMENT

The authors thank the European Commission for financial support of part of this work (NeuroTAS project of the 6th EU Research program, grant number NMP4-CT-2003-505311).

REFERENCES

[1] J.S. Rossier, F. Reymond and P.E. Michel, Electrophoresis, *23*, 858-867, 2002.

[2] J.S. Rossier, C. Vollet, A. Carnal, G. Lagger, V. Gobry, P. Michel, F. Reymond, and H.H. Girault, Lab Chip, *2*, 145-150, 2002.

[3] P. Morier, C. Vollet, P. Michel, F. Reymond and J.S. Rossier, Electrophoresis, 25, 3761-3768, 2004.

[4] D. Hoegger, P. Morier, C. Vollet, D. Heini, F. Reymond and J.S. Rossier, Anal. Bioanal. Chem., 387 (1), 267-275, 2007.

One-step White Blood Cell Separation from Whole Blood On a Centrifugal Microfluidic Device

Jong-Myeon Park, Byung-Chul Kim, Jeong-Gun Lee & Christopher Ko
Bio & Health Lab, Samsung Advanced Institute of Technology, P.O. Box 111, Suwon, 440-600, Korea Tel : +82-31-280-6948, Fax : +82-31-280-6816, e-mail : jong_myeon@samsung.com

ABSTRACT

In this work, we propose a fully automated WBCs isolation device from biological sample utilizing centrifugal microfluidics on a polymer based CD platform. Using the novel Laser Irradiated Ferrowax Microvalves (LIFM) and liquid density gradient medium on CD platform, the total process of a blood sample loading on the density gradient medium, fractionating to concentrate the WBC, decanting plasma layer, and WBC isolation was finished within 5 minutes with only manual step of adding diluted 100 μL of whole blood. The hematocytometry results showed that the number of WBC was as good as the sample prepared in conventional manual method. Furthermore, As a CD could prepare maximum 24 samples the automatic method is more useful for the high throughput cell isolation as well as enables one to enhance reproducibility, laborious and time-consuming process for the RNA purification.

Keywords: White Blood Cells (WBC), Microvalve, Centrifugal Microfluidics

1 INTRODUCTION

The isolation of intact, high quality RNA is critical to successful gene expression analysis experiments including RT-PCR, RNA mapping, in vitro translation, Q-RT-PCR, RNA labeling, and cDNA library generation. Automation of RNA purification enhances reproducibility and saves time and labor.

For practical reasons, whole blood is often fractionated to concentrate the nucleated cells (i.e., the WBCs) prior to RNA extraction or immunoselection. Removing plasma and red cells eliminates many of the nucleases and inhibitors from blood and reduces the sample volume by at least ten-fold, which allows RNA extraction to be carried out in microfuge tubes (< 2 ml) instead of larger vessels. Also, most protocols for immunoselection of leukocyte subsets are more efficient and cost-effective when carried out on fractionated leukocytes rather than whole blood. Density gradient centrifugation is probably the most common method for fractionating whole blood [1]. Although this commercial available method is very useful in vitro assays, it has many shortcomings because process has many manual steps as well as often time-consuming process. First, blood manually carefully layered onto the medium. Second, isolated WBC is harvested by very carefully pipetting them from the liquid interface after centrifugation. As a result, manual process become reasons of error of the cell separation yield.

We have demonstrated an innovative Laser Irradiated Ferrowax Microvalves (LIFM) that is based on phase transition of ferrowax, paraffin wax embedded with 10 nm sized ferrooxide nanoparticles. Compared to the conventional phase change based microvalves, the control of multiple microvalves was simple by using single laser diode instead of multiple embedded microheaters.15, 36 Furthermore, LIFM is not very sensitive to rotation speed or surface properties. Both Normally Closed (NC)–LIFM and Normally Opened (NO)–LIFM were demonstrated and various fluidic functions such as valving, metering, mixing, and distribution were demonstrated using centrifugal microfluidic pumping. In addition, the response time to open the channel by melting the wax was dramatically reduced from 2 ~ 10 sec. to less than 0.5 sec. because the laser beam effectively heat the nanoparticles embedded in the paraffin wax matrix [2, 3].

Here, we propose a fully automated WBCs extraction device from biological sample utilizing centrifugal microfluidics on a polymer based CD platform. Using the innovative Laser Irradiated Ferrowax Microvalves (LIFM) together with cell separation method using liquid density gradient medium (DGM), we could, for the first time, demonstrate a fully automated WBC separation from whole blood on a CD.

2 MATERIALS AND METHOD

As shown in Fig. 1A, schematic illustration of operation of Laser Irradiated Ferrowax Microvalves (LIFM) on a CD. By adjusting the position of laser beam irradiation, the main channel could be opened for centrifugal pumping.

As shown in Fig. 1B, the polycarbonate (PC) CD is composed of a top layer consisting of various inlet holes and a bottom layer consisting of channels and chambers. The channel width was 1 mm and the depth was 100 μm. The depth of the chamber was 3 mm. The inlet holes and channels were produced by conventional CNC (computer-controlled machine), and the two separate layers were bonded with a double sided adhesive tape. The microfluidic layouts designed by using CAD (Computer-Aided Design) were cut by using a computer controlled vinyl film cutter.

The ferrowax valve is made of a nanocomposite materials composed of 50 % of paraffin wax and 50 % ferrofluids; 10 nm sized ferrooxide nanoparticles dispersed

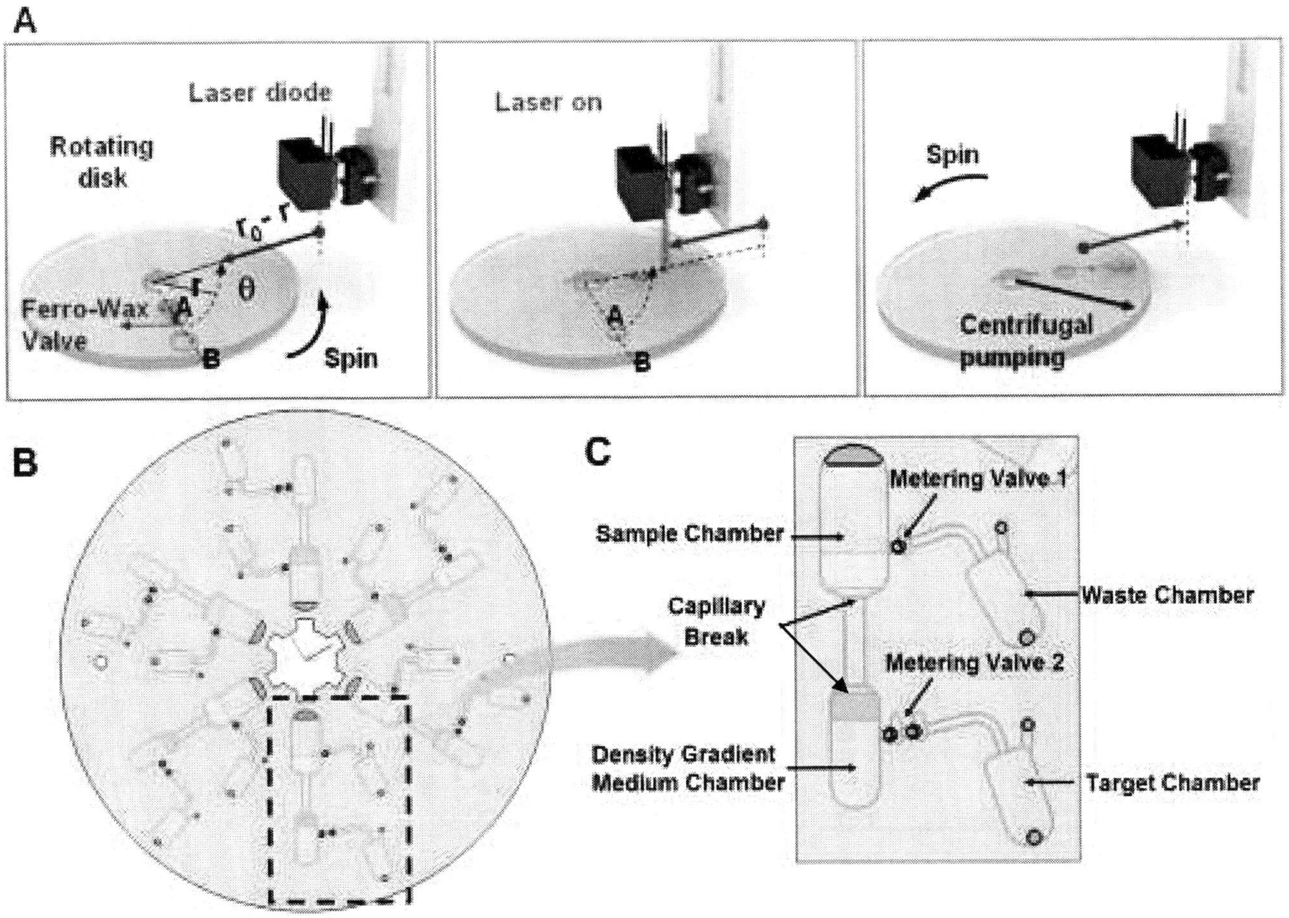

Fig. 1 (A) Schematic diagram of the ferrowax valve operation on a CD. In order to transfer liquid from reservoir A to reservoir B, the CD is rotated with the angle θ to the laser home and the laser diode is moved to the valve position with the distance ($r-r_0$). Then, laser with power of 1.5 W was applied for 1 sec to melt the ferrowax and the disk is spin to pump liquid from reservoir A to reservoir B. **(B)** Schematic diagram of the microfluidic layout on a disc. **(C)** Detailed description and function of the WBC separation microfluidics layout on a disc.

in oil. The detail procedure of the fabrication and characterization of LIFM is previously reported.35

Fractionation of whole blood sample was done by using the liquid density gradient medium (Osmolarity: ± 15 mOsm, LymphoprepTM, Oslo, Norway)

In order to measure number of the isolated WBC, we used hematocytometry.

3 RESULTS AND DISCUSSION

As shown in Fig. 1, a microfluidic layout was designed to fully integrate the WBC separation from whole blood on a CD. Table 1 shows the spin program at each operation step. The images shown in Fig. 2 were obtained during the rotation using the CCD camera and strobe light.

As shown in Fig. 1B ~ 1C. First, 100 μL of whole blood and density gradient medium (DGM) were added to the sample chamber and DGM chamber, respectively, as shown in Fig. 2A. Capillary break region (depth of 50 μm, width of 2 mm) was designed so that sample was not mixed with DGM. In order to separate WBC from the mixture with whole blood and DGM, the CD was spun with an acceleration of 30 Hz s-1 and a maximum speed of 60 Hz s -1 for 300 s (Fig. 2B ~ D). For the purification of the concentrated WBC, at first, metering valve 1 is opened by laser irradiation and waste solution was transferred to waste chamber by centrifugal pumping (Fig. 2E). Then, isolated WBC solution positioned between metering valve 1 and 2 was transferred to target sample chamber with same manner as described above (Fig. 2F). Fig. 2G ~ I show that the WBC separation process was good working. The separated WBC was counted by hematocytometry. The yield of WBC was 28700 +/- 1700 per 100 μL whole blood, which was 54% of the manual preparation (51363 +/- 9513/100 μL whole blood). In case of preparation-to-preparation variation, the automatic isolation showed only 6% of CV. In addition, comparing to the time consuming manual cell

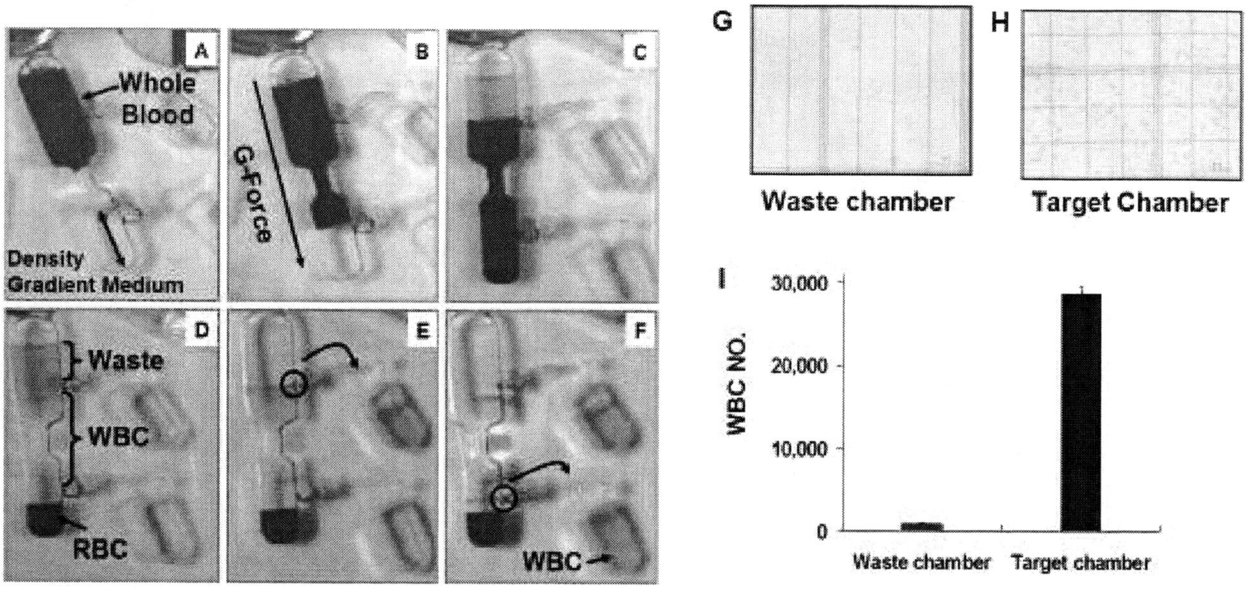

Fig. 2 Photo images captured during the operation of automated WBC separation of the spin and resulting data. **(A ~ F)** Photo images captured during the operation of the spin. **(G ~H)** Microscope view images of the each chamber after completed WBC separation. **(I)** Hematocytometry results from WBC separation by the fully automated method on a CD.

preparation (one hour for three samples), the automatic one took just 5 minutes/CD.

Table 1. A spin program for the microfluidics on a CD

Spin No.	Spin speed (Hz)	Time (sec)	Operation
	-	-	Preload DGM 100 μL to DGM chamber
1	-	-	Input whole blood of 100 μL to sample chamber
2	60	300	Fractionating whole blood
3	30	5	Metering valve 1 opened, transfer plasma decant to waste chamber
4	30	5	Metering valve 2 opened, transfer WBC to WBC chamber

4 CONCLUSION

By combining with the novel LIFM based centrifugal microfluidics platform and a liquid density gradient materials, we could, for the first time, a fully automated WBC isolation platform is developed.

Consequently, automated intact WBC isolation platform enables one to enhance reproducibility and saves time and labour for the RNA purification.

REFERENCES

[1] D.-A. McCarthy, J.-D. Perry, *J. Immunol. Meth.*, 73, (1984) 415-425

[2] J.-M. Park, Y.-K. Cho, B.-S. Lee, J.-G.-Lee and C.Ko, *Lab Chip*, 7, (2007) 557-564

[3] Y.-K. Cho, J.-G Lee, J.-M. Park, B.-S. Lee, Y. Lee and C. Ko, *Lab Chip*, 7, (2007) 565-573.

Detection of Clinical Biomarkers Using a Novel Parabolic Microchip Diagnostic Platform

Duncan Hill, Robert Blue, Lourdes Basabe-Desmonts, Stephen Hearty, Barry McDonnell, Colm McAtamney, Thomas Ruckstuhl*, Richard O'Kennedy and Brian MacCraith

Biomedical Diagnostics Institute, Dublin City University, Dublin 9, Ireland
*Institute for Physical Chemistry, University of Zurich, Winterthurerstrasse 190, 8057 Zurich, Switzerland

ABSTRACT

In this paper we present a novel biosensor platform, designed to enhance the collection efficiency of fluorescence from surface bound assay labels while simultaneously eliminating background signals from the bulk fluid on the chip. The chip consists of nine paraboloid elements moulded into a single piece of polymer, enabling a cheap, mass producible, efficient method of biomarker detection. Herein we demonstrate the detection of the clinically relevant biomarker C-reactive protein (CRP) using a sandwich assay adapted from a traditional ELISA platform. The platform demonstrates significantly enhanced capture of signal from the fluorescent labels leading to higher analytical performance.

1 INTRODUCTION

Detection of biomarkers indicating diseases and conditions such as heart failure [1] has led to the development of increasingly sensitive and accurate assays and biosensors. Typically, clinically relevant biomarkers are isolated and tested in a laboratory-based *in-vitro* setting, but recent years have seen a push towards increased point of care testing, driving the development of cheap, mass producible and reliable test platforms with easy user interfaces [2]. Common transduction methods developed to interface these biomarkers range from optical to electrochemical techniques. Of the optical techniques, the use of fluorescent-based systems are the most common, in which fluorescent labels such as cyanine dyes and quantum dots are attached to antibodies which in turn bind to the analyte of interest.

The goal of this work is to combine the trend towards use of low-cost polymer biochips and integrating this with an optical structure that facilitates the efficient the capture of fluorescence from an assay label. Modelling of an oscillating dipole close to the interface between a lower-index superstrate and higher-index substrate shows that the light does not emit isotropically, but in fact predominantly into the substrate (Fig 1), and of that the majority is emitted into supercritical modes [3]. By capturing only this Supercritical Angle Fluorescence (or SAF) emission, background from the bulk can be eliminated. Additionally the biochip is designed to excite the fluorophores through evanescent wave excitation using Total Internal Reflection Fluorescence (TIRF) to also limit the excitation volume.

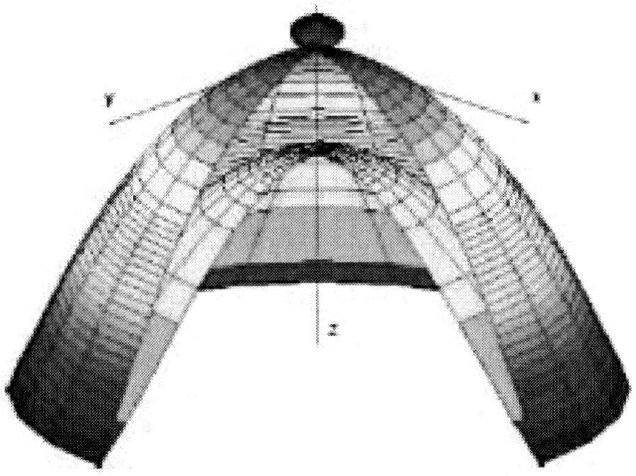

Fig 1: SAF emission from a low index superstrate to high index substrate, with dipole on interface in the x-y plane.

2 MATERIALS AND METHODS

2.1 Optical System

The key element of the biosensor is a platform of paraboloid elements (Fig 2), which is manufactured from a low auto-fluorescence zeonex® polymer, and is designed to collect collimated light from below and reflect it onto a focal point at the upper planar surface from a truncated parabolic element. The incident light is totally internally reflected and so excitation of the fluorophores occurs due to the evanescent wave, which decays exponentially with distance from the surface, providing good surface selection.

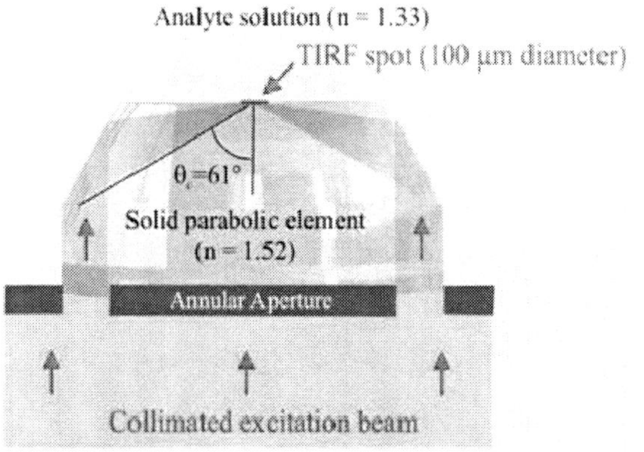

Fig 2: Parabolic element showing TIRF excitation.

The SAF emission that radiates into the substrate is then collected, again by the paraboloid edges (Fig 3), reflecting this into a loss-free collimated beam towards a detector. The magnitude of the detected light is indicative of the amount of fluorophore on the surface and is thus linked to the concentration of the analyte. An annular mask is placed below the paraboloid element to both eliminate excitation light passing vertically through the upper surface and also block non-SAF emission from the bulk liquid.

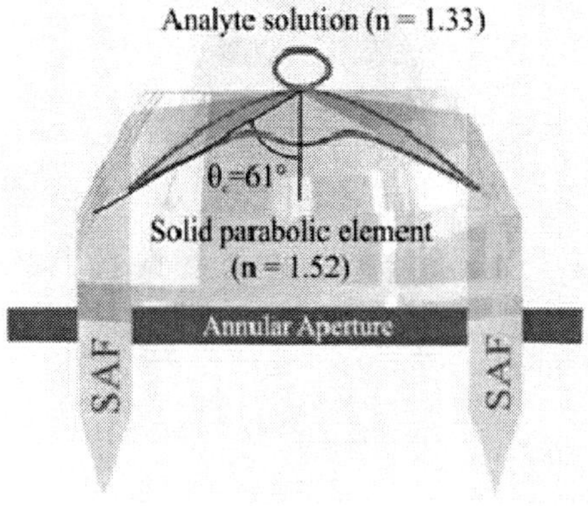

Fig 3: Collection of SAF light by same parabolic element.

The biochip platform consists of a 3x3 array, allowing nine surface reactions to be monitored in real time. The biochips were analysed with a custom designed optical system (Fig 4). A collimated beam of light from a laser diode of wavelength 635nm was reflected via a dichroic mirror onto the biochip, and the emitted fluorescence from the biochip parabolic elements was imaged onto a USB-controlled CCD camera. The system was linked to a computer with custom software developed for image capture and analysis, and a readout could be produced in under 5 seconds.

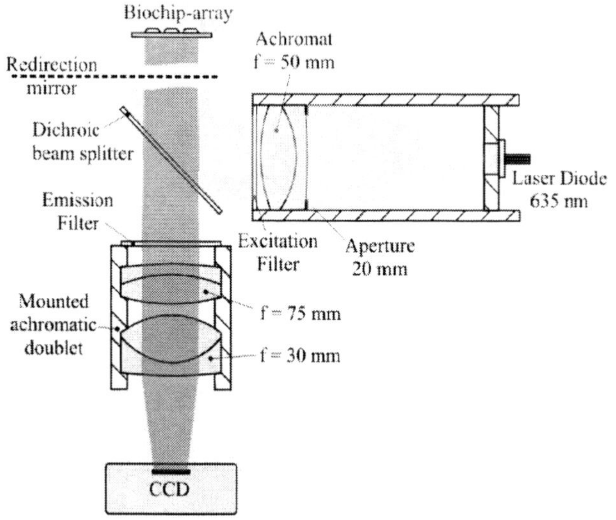

Fig 4: Custom optical setup for illumination and collection

2.2 C-Reactive Protein

C reactive protein is a marker of inflammation and infection produced by the liver and is thought to be strongly linked to the hardening of blood vessels seen in coronary artherogenesis [4] thus allowing it to be used as a prognostic indicator of future cardiac events in both healthy patients and those with a history of acute coronary syndrome [5] As CRP levels are useful in the prediction of thrombotic events such as coagulation of the blood in the heart, arteries, veins or capillaries, they may also offer an alternative to cholesterol measurements for monitoring wellness [6].

2.3 Antibody production and immobilization

The chosen antibody for this assay was an anti-CRP polyclonal antibody, produced through the immunization of a female while New Zealand White rabbit. The polyclonal IgG antibody fraction was purified from the rabbit serum using protein G affinity chromatography. The surface of the chip was functionalised using an amino-dextran coating, as antibodies have amine and carboxyl groups available for attachment. After surface functionalisation, the antibodies were deposited using a non-contact sciFLEXARRAYER™ (Scienion AG, Germany) piezo dispensing system.

2.4 Assay Details

The chips were functionalised and spotted with capture antibody as described in section 2.3. They were then washed in a phosphate buffer solution (PBS) and then blocked in a dilute (5% w/v) solution of milk powder overnight at 4^0C. After another washing step, different concentrations of CRP, again in PBS and additionally different concentrations of CRP in depleted serum were placed on the tops of the chips, covering the paraboloid elements and left to incubate at room temperature for 2 hours, and agitated at 1 hour to remove concentration gradients. After an additional wash step CRP antibody labelled with DyLight™ 647 (Pierce Biotechnology Inc., Rockford, IL) was deposited onto the tops of the paraboloids and allowed to incubate for a further 2 hours at high humidity to stop evaporation of the droplets. The optimum dilution (1/400) of the labelled antibodies was determined empirically. After a final wash step, the chips were then analysed in a GMS 418 array scanner (Genetic Microsystems) to ensure good spot location and quality and then finally the fluorescence measurements were taken in the custom biochip reader.

3 RESULTS AND DISCUSSION

The collimated light from the chip was imaged onto the CCD camera and showed clearly the fluorescence rings (Fig 5). The outer rings appear distorted due to spherical aberrations and additionally are dimmer than the central ring due to the Gaussian beam profile from the laser. The Gaussian profile was accounted for by taking a reading of the illumination intensity through each of the annular rings and normalising to the central ring. The signal was integrated for each ring, giving a total fluorescence reading for that paraboloid, and then the noise background was subtracted from the final value.

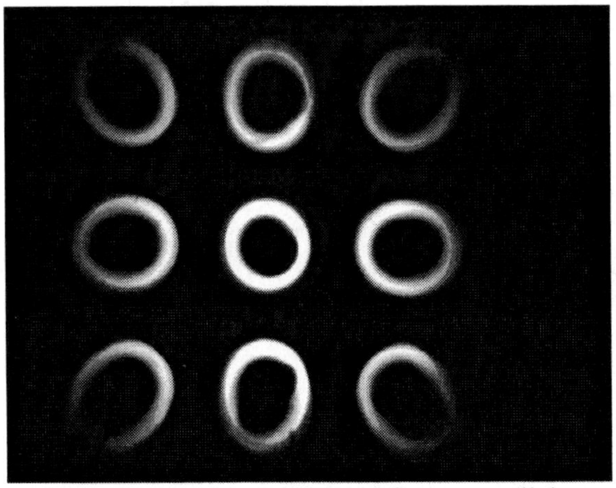

Fig 5: Image of the collimated fluorescent light.

3.1 Surface Coating Optimization

The optimal surface coating was determined by running a limited assay with different surface coating concentrations of capture antibody, ranging from 50µg/mL of antibody to a maximum of 300µg/mL (Fig 6). As the surface coating concentration increases, the slope of the graph increases up to 200µg/mL at which point the surface coating saturates and no further increase is apparent.

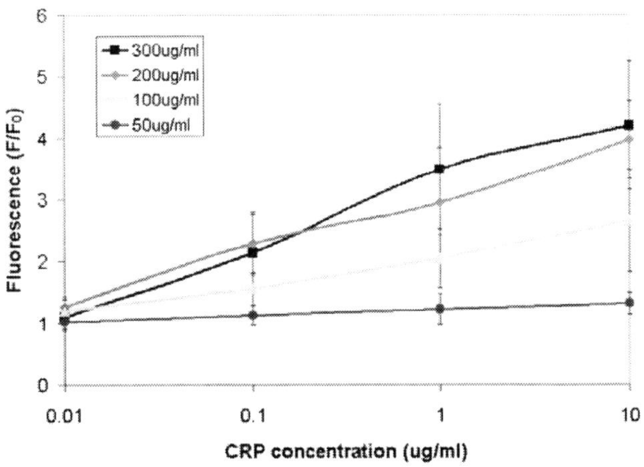

Fig 6: Normalised Fluorescence for multiple surface coating concentrations

This point was chosen as the optimal concentration for maximizing the slope while using the minimal amount of capture antibody. All results are normalized to a control chip which had no CRP added. This surface coating concentration correlated well with an additional experiment in which a fluorescently labeled antibody was deposited directly onto the surface and measured, in this case the fluorescence saturated at around 200µg/mL.

3.2 CRP Assay

The CRP assay was repeated three times at the optimal surface coating concentration of 200µg/mL outlined above, with each chip having one concentration of CRP, and hence nine data points. The CRP range covered the clinically relevant range 0.01 to 10µg/ml. For each assay the result was averaged over the nine parabolas, and then averaged over the three assays giving the final readings shown in figure 7.

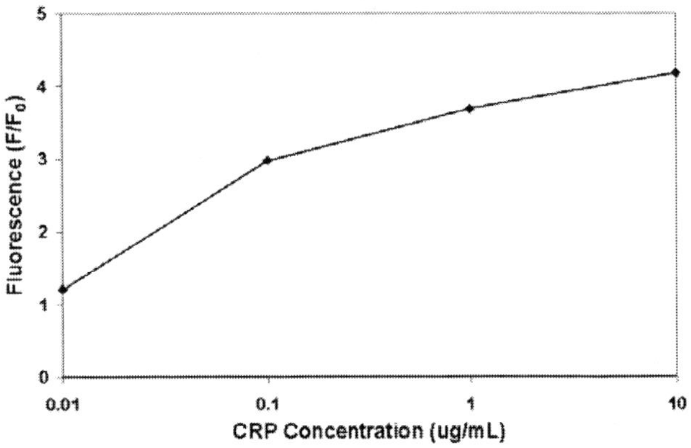

Fig 7: Fluorescence signal versus CRP in buffer assays

The assay was also conducted in duplicate using CRP-depleted human serum (Fig 8) spiked with different concentrations of CRP to verify the sandwich assay performance in a real sample matrix.

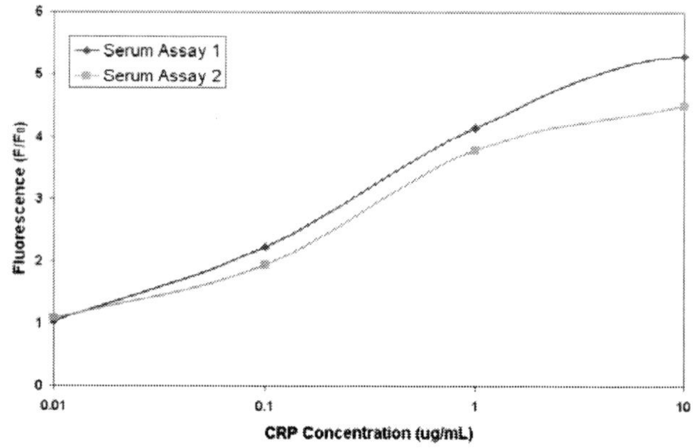

Fig 8: Assays conducted with CRP in blood serum

3.3 Surface Discrimination

In order to demonstrate surface discrimination the central part of four of the annular rings was removed to allow excitation light to pass up through the centre of the parabolic elements. Reaction vessels were then made from perspex and fixed to the top of the chip, with one vessel upon each parabolic element which were then filled with a solution of DyLight 647 dye- labeled antibody and placed on the reader. In this case, there is both direct illumination through the bulk (in addition to the TIRF illumination) and collection of non- SAF light from both the surface and the bulk. Figure 9 shows the resultant fluorescence detected. The four elements to the left have the central mask removed and show additional fluorescence generated from the bulk solution above each parabolic platform. The remaining five parabolas however show only the light from the TIRF excitation and SAF collection, with the optical system blocking and minimizing bulk fluorescence.

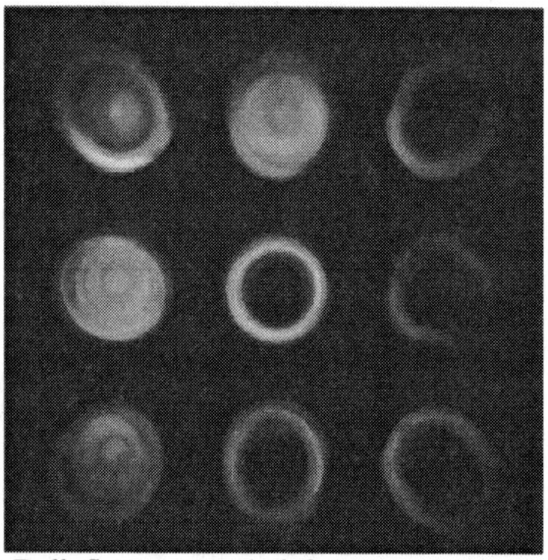

Fig 9: Bulk fluorescence in left 4 "open" annular rings. Bulk fluorescence is eliminated from the remaining 5 platforms due to the parabolic facilitating TIRF excitation.

The design of the system clearly eliminates unwanted signal from the bulk solution and only probes the region of interest on the surface of the parabola.

4 CONCLUSION

A novel biochip platform, which is designed to enhance the collection of fluorescence from biological assays and significantly reduce the signal from non-bound fluorescent labels present in the bulk liquid above the solution was demonstrated. The design of the system is such that it provides a practical method of performing multi-analyte studies in a single measurement, and allows the elimination of background noise. This biosensor represents a significant step towards disposable platforms for the conduction of sensitive and accurate assays for point-of-care applications.

Acknowledgements

This material is based upon works supported by the Science Foundation Ireland under Grant No. 05/CE3/B754".

References

[1] Z. Yang and D. M. Zhou, "Cardiac markers and their point-of-care testing for diagnosis of acute myocardial infarction", Clinical Biochemistry, **39**(8), pp. 771-780 (2006).
[2] K. Schult, A. Katerkamp, D. Trau, F. Grawe, K. Cammann, and M. Meusel, "Disposable Optical Sensor Chip for Medical Diagnostics: New Ways in Bioanalysis", Analytical Chemistry, **71**(23), pp. 5430-5435 (1999).
[3] R. Blue, N. Kent, L. Polerecky, H. McEvoy, D. Gray, and B. D. MacCraith, "Platform for enhanced detection efficiency in luminescence-based sensors", 2005, Electronic Letters, **41** (12), pp. 862-864.
[4] A. Abbate, G. G. Biondi-Zoccai, S. Brugaletta, G. Liuzzo, and L. M. Biasucci, "C-reactive protein and other inflammatory biomarkers as predictors of outcome following acute coronary syndromes", Seminars in Vascular Medicine **3**(4), pp. 375-384 (2003).
[5] M. J. Jarvisalo, A. Harmoinen, M. Hakanen, U. Paakkunainen, J. Viikari, J. Hartiala, T. Lehtimäki, O. Simell, and O. T. Raitakari, "Elevated Serum C-Reactive Protein Levels and Early Arterial Changes in Healthy Children", Arteriosclerosis, Thrombosis, and Vascular Biology, **22**(8), pp. 1323-1328 (2002).
[6] E. T. H. Yeh, "High-sensitivity C-reactive protein as a risk assessment tool for cardiovascular disease", Clinical Cardiology, **28**(9), pp. 408-412 (2005).

One-step Target Protein Detection from Whole Blood in a Lab-on-a-Disc

B. S. Lee, J.-G. Lee, J.-N. Lee, J.-M. Park, Y.-K. Cho, S. Kim and C. Ko

Samsung Advanced Institute of Technology,
Mt. 14-1, Nongseo-dong, Yongin-si, Korea, bio.lee@samsung.com

ABSTRACT

We report a fully integrated, one-step target protein detection utilizing centrifugal microfluidics on a polymer based disc (Lab-on-a-Disc). The design principle is based on microbead-based immunoassay. Since microbeads with high surface-to-volume ratio (50 times higher than a microtiter well) are efficiently mixed in reaction chamber by clockwise and counter-clockwise rotation repeatedly, antibody-antigen reaction is facilitated. So, total reaction time is dramatically reduced from 90 min to 20 min and the final detection value is obtained within 38 min. In addition, the required volume of sample and reagents are reduced 2 ~ 5 times. As model studies, Hepatitis virus B surface antibody(Anti-HBs) detection from whole blood were conducted using Lab-on-a-Disc and compared to commercial ELISA kit. The dynamic range of our method is 6 times wider than the commercial ELISA with the same functional sensitivity (10 mIU/mL).

Keywords: Lab-on-a-Disc, immunoassay, microbead, Anti-HBs

1 INTRODUCTION

There have been significant advances in the lab-on-a-chip development for biomedical applications recently [1-4]. However, the practical applications in clinical diagnostics, such as immunoassay, require processing of complex fluids; e.g. whole blood, conjugate buffer, substrate etc. Due to the complex nature of the sample and many operational steps, most of the sample detection steps still rely on the time-consuming traditional bench top methods. As a result, development of rapid and efficient on-chip specific biomolecule detection, for example protein detection, in "real" sample analysis remains as a major bottleneck for the realization of the true lab-on-a-chip.

Many diagnostic tests and assays use micro-size uniform polystyrene bead, or microspheres, as substrates or supports for immunologically based reaction, such as ELISA [5-16]. One reason is that ELISA and other immunoassay techniques have been miniaturized onto microchip platforms and require very small sample volume. By using the microbead that has large surface-to-volume ratio, detection sensitivity and analysis time have been further improved [9]. In this report, the polystyrene (PS) microbead with the mean diameter of 117 μm is used and its surface is modified by carboxyl functional group (-COOH) to conjugate Hepatitis B surface antigen (HBsAg) covalently on the surface as shown in Fig. 1.

Recently, we reported a novel microvalve system, Laser Irradiated Ferrowax Microvalve (LIFM) [17,18]. Addition of 10 nm-size ferromagnetic particles to wax medium can make very fast and robust microvalves with response time of 12 msec by single laser irradiation (808 nm, 1.5W). Through the LIFMs, fully automated and more complex processing, such as, mixing, target binding, washing, separation, is possible without any external operation.

Here, we report a fully integrated and automated target specific protein detection device utilizing centrifugal microfluidics on a polymer based disc. The design principle

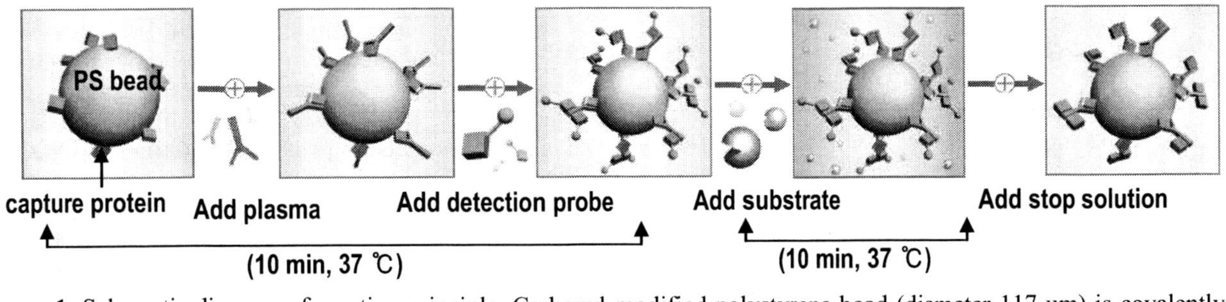

Figure 1. Schematic diagram of reaction principle. Carboxyl modified polystyrene bead (diameter 117 um) is covalently binded by HBsAg. When target sample (whole blood) with Anti-HBs and detection probe are loaded together, the Anti-HBs are captured on the bead, where the detection probes are subsequently binded on the bead. Free Anti-HBs and detection probes are washed out. After addition of substrate and stopping solution, the absorbance is measured at 450 nm and 630 nm as a reference.

is schematically shown in Fig. 1. As a model study, Hepatitis virus B surface antibody (Anti-HBs) detection from whole blood were conducted using Lab-on-a-Disc fully loaded with PS bead, detection probe, washing buffer, substrate, and stopping solution. Total process is finished within 38 min with only one manual input of 120 μl of whole blood.

2 MATERIALS AND METHODS

2.1 Protein Conjugation on Microbead

Our model is to detect Anti-HBs as a target in human whole blood. The first step is to conjugate target specific protein (HBsAg) on the surface of microbeads. As shown in Fig. 2, the microbead (BeadTech, Korea) is carboxylated polystyrene bead with the mean diameter of 117 μm with 1% divinyl benzene crosslinked and its carboxylation is 3.0 mmol/g. The most commonly used bead is sub-micrometer to a few micrometer because of its very high surface-to-volume ratio. However, we used relatively large microbeads for these reasons: (1) one is the confinement in the mixing chamber. Our microchannel has the depth of 100 μm. If the bead diameter is over 100 μm, the bead is easily confined in the mixing chamber. (2) another is easy mixing. Our relatively big bead has heavier than the small bead. Thus, the beads in the mixing chamber are easily mixed by slight vibration. This mixing makes effective binding between the beads and targets. Further more, after washing, the beads are easily settled down and separated from the residue with the low centrifugal force.

Covalent conjugation was employed for the immobilization of HBsAg on the surface of the bead. By means of the covalent conjugation, higher activity and stability are obtained rather than physisorption [13]. HBsAg

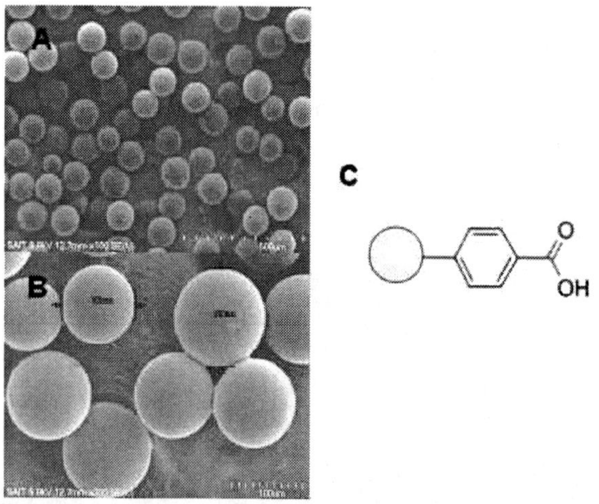

Figure 2. Carboxylated polystyrene beads with the mean diameter of 117 μm measured by HELOS particle size analyzer. The bead is crosslinked by 1% divinyl benzene and its carboxylation is 3.0 mmol/g. (A) SEM image. Scale bar, 500 μm. (B) SEM image. Scale bar, 100 μm. (C) Molecular structure of microbead.

(GreenCross Co., Korea) as a capture protein is recombinant IgG molecule with a molecular weight of 140 kDa.

Covalent conjugation protocols have historically focused on the surface capacity of protein being used, for example, the surface saturation capacity of polystyrene for bovine IgG was calculated at ~2.5 mg/m^2 [14]. The specific surface area to mass ratio for a sphere can be calculated: A/M (m^2/g) = 6/ρD, where ρ is density in g/mL and D is diameter in μm. For polystyrene bead, where ρ = 1.05 g/mL, A/M = 5.7/D. Thus, if D = 117 μm, then A/M = 0.049 m^2/g. Therefore, the maximum binding amount of HBsAg is 0.1218 mg HBsAg/g bead. Usually, the protein concentration should be substantially higher than the calculated value up to 10X. Our optimal concentration is 10 mg HBsAg/g bead (8X).

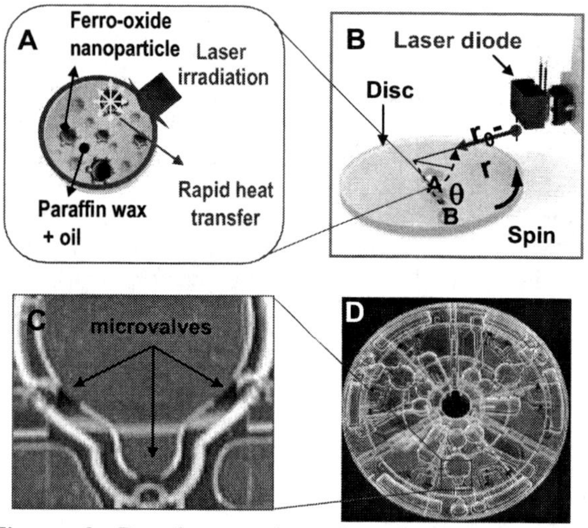

Figure 3. Brand-new microvalves in Lab-on-a-Disc. (A)~(B) Working principle and composition of microvalve. (C) 3 real fabricated microvalve with the dimension of 1 x 0.5 x 0.1 mm. (D) 42 Microvalves are integrated for 3 sample tests in one disc.

2.2 Lab-on-a-Disc Fabrication

One of the main components in our Lab-on-a-Disc is the ferrowax microvalve. The key concept of the microvalve operation mechanism is to use metal nanoparticles as nanoheaters and heat by laser irradiation in order to obtain rapid melting of paraffin wax (Fig. 3A). Once if it is implemented in a Lab-on-a-Disc, it is easily controlled by laser beam (808 nm, 1.5W) in a polar coordinate. For example, in order to transfer 100 μL of solution in reservoir A to reservoir B, as soon as the ferrowax valve is melted by irradiation of laser for 1 sec, the disc is spinning up to 30 Hz (revolution/sec) with an acceleration of 30 Hzsec^{-1}. These basic valve operation steps are repeated many times to control multiple valves on the disc (Fig. 3B). In addition, because the dimension of the microvalve is very small, multiple microvalves

NSTI-Nanotech 2008, www.nsti.org, ISBN 978-1-4200-8505-1 Vol. 3

can be implemented together in one disc. As a model study, our disc with the diameter of 120 mm has 42 microvalves (Fig. 3C and 3D).

3 RESULTS AND DISCUSSION

Microfluidic layout in Fig. 4 is designed to detect a target protein by means of one step ELISA method as shown in Fig. 1. "One step" means binding reaction among capture protein, target protein, and detection protein occurs simultaneously. Thus, capture protein conjugated beads and detection protein can be loaded together in the reaction chamber. The reaction chamber has a design for bead volume and bead separation. Our bead surface area is 3 times more than that in common ELISA kit. The corresponding bead bed volume is 25 μl that is equal to the volume of the lower region of the reaction chamber. When the beads are settled down to remove reagent residue, the beads are packed in the lower region of the reaction chamber and then, the reagent residue is removed easily through the side microchannels.

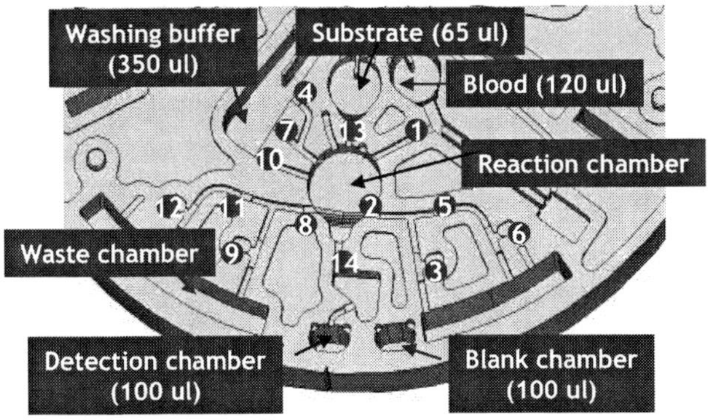

Figure 4. Microfluidic layout and functions in one third of Lab-on-a-Disc. The number in red filled circle shows the position of microvalve. The numbers of 3, 6, 9, 12 are normally opened microvalves and the other numbers are normally close microvalves.

Total process is summarized as follows: After plasma separation in the blood chamber at high rotational speed of 60 Hz, plasma is transferred into the reaction chamber. Then binding reaction proceeds during 10 min (60 min in common ELISA) while continuous mixing occurs inside the reaction chamber by clockwise and counter-clockwise rotation repeatedly.

Mixed residue and washing residue are removed by microchannels connected by the reaction chamber and the two waste chambers. To remove the mixed residue, the normally-closed microvalve (2nd in Fig. 4) is opened by laser irradiation and the residue is transferred into the waste chamber by centrifugal force (40 Hz). After then, the normally-opened valve (3rd in Fig. 4) is closed to enclose the reaction chamber. In the same manner, washing steps (3 times) are done sequentially. During each washing step inside the reaction chamber, efficient mixing is done to remove unbounded proteins from the beads rapidly.

After washing, substrate (TMB) is transferred to the reaction chamber and enzyme (HRP) reaction proceeds for 10 min (30 min in common ELISA). Finally, the final product, whose color is blue, is transferred to the detection chamber. The enzyme reaction is terminated by the stopping solution previously loaded in the detection chamber.

Table 1. Protocol comparision between our method(Immuno LOD) and common ELISA(Genedia Anti-HBs ELISA 3.0, Korea). Required volume of target(plasma) and reagents are 2 ~ 5 times less than the common ELISA. Total reaction time is 20 min (ELISA 90 min) due to highly efficient mixing and large reactive surface area.

Assay Protocol		Immuno LOD	Common ELISA
Surface	Type	**PS bead**	**Well plate**
	Area	3X	1X
Capture protein	Binding	**Covalent**	**Adsorption**
Target (plasma)	Volume (μl)	**50**	**100**
Detection protein	Volume (μl)	25	25
Incubation	Time (min)	**10**	**60**
Washing	Volume (μl)	350	1750
	Times	3X	5X
Substrate	Volume (μl)	65	100
	Time (min)	**10**	**30**
Stopping	Volume (μl)	55	100

Hepatitis virus B surface antibody (Anti-HBs) is one the most frequent test item as a clinical blood analysis after Anti-Hepatitis inoculation, or, before surgery, or at health inspection. So, we chose Anti-HBs as a study model.

Immunoassay protocol for detection of Anti-HBs on Lab-on-a-Disc is optimized for a fully automated analyzer. Table 1 shows the difference between our Lab-on-a-Disc and common ELISA kit. As mentioned earlier, the conjugation method of capture protein (HBsAg) on bead surface in our study is covalent coupling, but adsorption in the common ELISA kit. First of all, reaction time is dramatically reduced from 90 min to 20 min except others' extra operations, such as plasma separation, washing, detection, and valve operation. It is mainly due to highly efficient mixing based on the microbead reaction system. If we apply centrifugal force both in clockwise and counter clockwise direction repeatedly, the microbeads are instantaneously mixed within 1 sec as shown in Fig. 5. Centrifugal force induces large vorticity inside the chamber that makes fluidic rotation. Right at the change of the direction of disc rotation, inertial force still acting on the microbeads dominates. These two forces afford the efficient mixing of microbeads and solution inside the reaction chamber.

other proteins are applicable to the Lab-on-a-Disc if their assay protocols are based on one step ELISA.

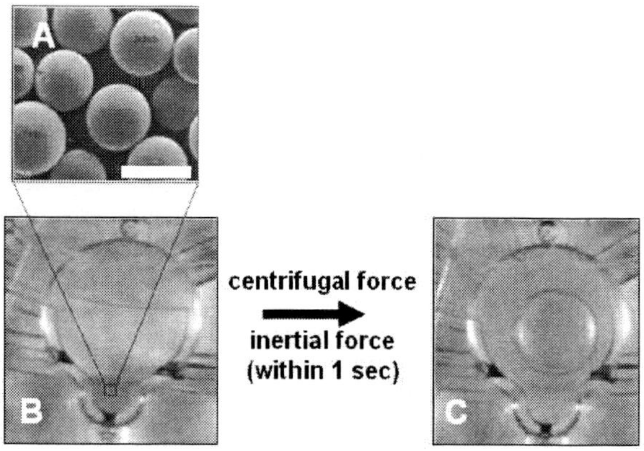

Figure 5. Highly efficient mixing between microbeads and reagents. (A) SEM image of polystyrene microbeads with average of diameter, 117 μm (scale bar 100 μm). (B) Separate state of microbeads. (C) Mixed state of microbeads.

Characterisitics	Immuno LOD	common ELISA
Dynamic Range	0 ~ 1000 mIU/mL	0 ~ 150 mIU/mL
Linearity (R^2)	0.995	0.998
Functional Sensitivity	10 mIU/mL	10 mIU/mL
Precision (CV%)	4.5%	3.3%

Table 2. Performance characteristics for Anti-HBs. Compared to the common ELISA(Genedia Anti-HBs ELISA 3.0, Korea), dynamic range is about 6 times wider with the same functional sensitiviy.

Furthermore, the amount of target sample in the Lab-on-a-Disc is half times less than in the common ELISA kit and other reagents' volume is reduced and optimized for the small volume of the chamber in the disc.

Table 2 shows the performance characteristics for Anti-HBs. Its dynamic range is 0 ~ 1000 mIU/ml with the linearity of R^2 = 0.995, which is 6 times wider than the common ELISA kit. This dynamic range is equivalent to commercial large blood analyzer, such as, 0 ~ 1000 mIU/ml for AxSYM AUSAB (Abbott), 2 ~ 1000 mIU/ml for Elecsys Anti-HBs (Roche), and 1 ~ 1000 mIU/ml for ADVIA Centaur Anti-HBs (Bayer) [19]. The functional sensitivity is 10 mIU/mL (0.69 ng/ml), which is the lower limit of a positive value of Anti-HBs.

4 CONCLUSION

With microbead-based immunoassay on Lab-on-a-Disc, a rapid and sensitive assay of a protein could be realized. By using the fully integrated Lab-on-a-Disc, laborious tasks were eliminated, and assay time was remarkably reduced from 3 hr to 38 min. Additionally, no special skills are needed for the assay.

Even though Anti-HBs detection was demonstrated here, other proteins, such as HBsAg, Anti-HCV etc, can be applied easily to the same Lab-on-a-Disc. It means any

REFERENCES

[1] Haes, A. J.,Terray, A., Collins, G. E., *Anal. Chem.*, 78, 8412 (2006).

[2] Sato, K., Yamanaka, M., Hagino, T., Tokeshi, M., Kimura, H., Kitamori, T., *Lab Chip*, 4, 570 (2004).

[3] Yan, F., Zhou, J., Lin, J., Ju, H., Hu, X., *J. Immuno. Methods*, 305, 120 (2005).

[4] Sato, K., Tokeshi, M., Kimura, H., Kitamori, T., *Anal. Chem.*, 73, 1213 (2001).

[5] Puckett, L. G., Dikici, E., Lai, S., Madou, M., Bachas, L. G., Daunert, S., *Anal. Chem.*, 76, 7263 (2004).

[6] Honda, N., Lindberg, U., Andersson, P., Hoffmann, S., Takei, H., *Clin. Chem.*, 51(10), 1955 (2005).

[7] Molina-Bolivar, J. A., Galisteo-Gonzalez, F., *Polymer Reviews*, 45, 59 (2005).

[8] Bangs, L. B., Meza, M., Microspheres, part 1: Selection, cleaning, and characterization, *IVD Technology Magazine* (1995).

[9] Bangs, L. B., Meza, M., Microspheres, part 2: Ligand attachment and test formulation, *IVD Technology Magazine* (1995).

[10] Kartalov, E. P., Multiplexed microfluidic immunoassays for point-of-care in vitro diagnostics, *IVD Technology Magazine* (2006).

[11] Bangs Laboratories, Technote 205, http://www.bangslabs.com.

[12] Borque, L., Bellod, L., Rus, A., Seco, M., Galisteo-Gonzalez, F., *Clin. Chem.*, 46, 1839 (2000).

[13] Caballero, M., Marquez de Prado, M., Seco. M, Borque, L., Escanero, J. F., *J. Clin. Lab. Analysis*, 13, 301 (1999).

[14] Sanz Izquierdo, M. P., Martin-Molina, A., Ramos, J., Rus, A., Borque, L., Forcada, J., Galisteo-Gonzalez, F., *J. Immuno. Methods*, 287, 159 (2004).

[15] Ramos, J. Martin-Molina, A., Sanz-Izquierdo, M. P., Rus, A., Borque, L., Hidalgo-Alvarez, R., Galisteo-Gonzalez, R., Forcada, J., *Polymer Chemistry*, 41, 2404 (2003).

[16] Siiman, O., Burshteyn, A., Insausti, M. E., *JCIS*, 234, 44 (2001).

[17] Park, J-. M., Cho, Y-.C., Lee, B-. S., Lee, J-. G., Ko, C., *Lab Chip*, 7, 557 (2007).

[18] Cho, Y-.C., Lee, J-.G., Park, G-.M., Lee, B-.S., Lee, Y., Ko, C., *Lab Chip*, 7, 565 (2007).

[19] Yoo, S. J., Oh, H-. J., Shin, B-. M., *Korean J. Lab. Med.*, 26, 431 (2006).

Using Microfluidics for Metabolic Monitoring of Single Embryo Cultures

T. Thorsen*, J. P. Urbanski*, M. T. Johnson**, D. D. Craig*, D. L. Potter*** and D. K. Gardner***

*Department of Mechanical Engineering, Massachusetts Institute of Technology, Cambridge, MA, 02139, USA, thorsen@mit.edu
**Colorado Center for Reproductive Medicine, Englewood, CO, 80110, USA
***Department of Zoology, University of Melbourne, Victoria 3010, Australia

ABSTRACT

In recent years, microfluidics has demonstrated novel methods of culturing and studying cells. Within well-defined volumes comparable to the sample of interest, media, nutrients or toxins may be precisely targeted to individual cells via diffusion or advection, enabled by integrated valves and pumps. Tools, which both enable a repeatable and systematic means of single embryo culture, as well as provide information on metabolic activity of individual embryos, would be of significant utility to the embryologist. To address the need of controlled culture environments and improve embryo health and subsequent positive implantation rates, we have developed programmable microfluidic devices for quantifying metabolite levels in the culture media. Our microfluidic platform provides a high throughput, accurate and repeatable means to non-invasively assess embryo viability, which should be of significant utility to the clinical embryologist.

Keywords: microfluidics, in-vitro fertilization, metabolism

1 INTRODUCTION

It is well established that the developmental potential of preimplantation embryos can be distinguished through evaluation of their metabolism[1-4]. Currently, the most common approach to assess metabolism is to measure changes in concentrations of glucose, pyruvate and lactate in culture media using micropipettes to sample media and perform the assays. This method is technically challenging and labor-intensive. To improve accuracy and throughput of such analyses, we have developed a microfluidic system that can accurately measure concentrations of these three metabolites in nanoliter volumes.

Non-invasive techniques have been pioneered to assess the metabolic activity of pre-implantation embryos during IVF culture[5]. While it is agreed upon that glycolytic rates are correlated with developmental success, metabolic analyses for candidate selection in IVF have not been widely adopted, as existing manual measurement techniques are cumbersome and low throughput. For instance, manual measurements of even ten media samples for a single metabolite using existing microdroplet techniques may take a technician several hours of manned time.

NAD(P)H-based fluorogenic enzymes were used to for automated on-chip assaying of sub-microliter samples of media for select markers correlated with culture viability (glucose, pyruvate and lactate). Microfluidics is particularly suited for manipulation of samples in this volume range and is well poised for automation of the existing protocols. Our metabolite detector platform is scalable to accommodate additional samples and substrates, and most importantly, can be integrated with microfluidic culture systems in the future.

2 EXPERIMENTAL

2.1 Assay Technique

Existing NAD(P)H-linked enzymes assays were used to quantify metabolite concentrations[4, 6]. For every assay it is necessary to combine one part of culture media (typically 1-10nl) with ten parts of a metabolite specific enzyme cocktail. Traditionally, this assay has been performed in microdroplets under mineral oil using custom made glass pipettes. Briefly, droplets containing samples and reagents are dispensed, and allowed to mix by diffusion. Once the reaction has gone to completion (~3 minutes), a fluorescent measurement is manually performed. To increase the throughput of these measurements, an integrated microfluidic device was designed to enable these metering, mixing and measurement operations to be performed automatically.

2.2 Device Architecture

Polydimethylsiloxane (PDMS) devices were fabricated using multi-layer soft lithography which contain channels of micron-scale geometries with integrated pneumatic valves[7] . The chip consisted of a flow layer for manipulating samples, and a complimentary control layer which contained arrangements of valves and pumps. After casting of the two layer PDMS devices, interconnects were punched, devices were cleaned using ethanol and nitrogen, and chips were bonded to glass slides using a plasma treatment. A schematic of the microfluidic architecture is presented in Figure 1. This scalable design accommodates up to ten media samples, as well as up to six metabolite cocktails and wash buffers.

2.3 Device Operation

The assay system was designed to allow 1nl to be sampled from an input channel and subsequently mixed with 10nl enzyme cocktail. Following mixing, NAD(P)H intensity was measured in a detection chamber through an inverted epi-fluorescent microscope equipped with a DAPI filter set and CCD camera.

Standard calibration curves for each metabolite (glucose, pyruvate, and lactate) were automatically generated on-chip in the 0-1mM range. All operations and data acquisition were enabled by an open source JAVA interface[8]. Parallel systems have been developed in both research and clinical labs.

2.4 Sample Preparation

F1 murine embryos were cultured over four days in individual microdroplets at 5% O_2 using sequential G1/G2 media. On days 1 through 4 of culture, 0.5μl aliquots of media were collected from the culture droplet. Samples were stored in a Petri dish under mineral oil to prevent evaporation. When necessary, dishes containing samples were frozen for at -80°C for later analysis.

Calibration standards of media not containing embryos with known amounts of glucose, lactate and pyruvate in the 0 - 1μM range were also prepared. Standards are loaded on chip in parallel with culture samples to relate observed fluorescent intensity to a metabolite concentration.

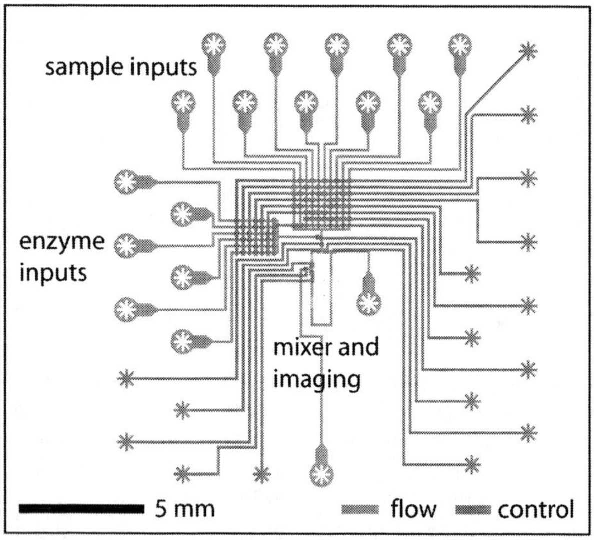

Figure 1: Schematic of the microfluidic metabolite detector. Integrated valves and pumps enable mixtures of samples and enzymes to be automatically generated.

3 RESULTS

3.1 Device Performance

Five-point calibration curves generated for the three metabolites demonstrated linearity from 0.01mM to 1mM (R2>0.999). Figure 2 illustrates an example standard curve obtained in the device using NADH. The microchannels provide a very uniform fluorescent signal and high signal to noise ratio in contrast to the conventional microdroplet assay technique.

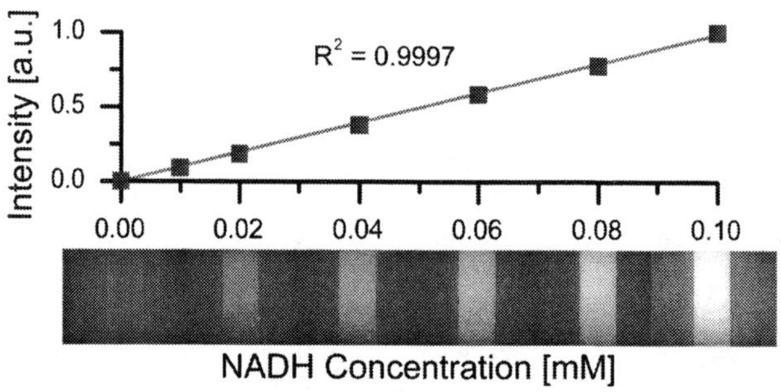

Figure 2: A typical calibration curve overlaid with actual images from the 100um wide microchannels. Similar calibration curves are automatically generated using the on chip mixer for each of the three metabolites of interest.

Each point takes ~1 minute to acquire (including fluid metering and mixing), enabling calibration and parallel sample analysis for three metabolites in approximately ten minutes. In contrast, manual pipetting and measurement generates curves with R2=0.99 at best, has comparable sensitivity and requires at least several fold more operator time.

To contrast the microfluidic approach with the conventional microdroplet technique, several calibration standards with known amounts of glucose, lactate and pyruvate were assayed. Figure 3 compares the results obtained by these two approaches in comparison to the expected concentrations of the metabolite glucose. In general, the microdroplet assays are capable of providing accurate results. However, the microdroplet assays required more than 3h of operator time to obtain results for a single metabolite (a rate of more than five minutes per data point, not including calibration which is required for each enzyme cocktail). In contrast, the microfluidic device performed measurements at a rate of approximately one data point per minute and could process measurements and calibration in parallel. These operations also do not require user intervention.

Figure 4 presents similar measurements performed in parallel using the microfluidic approach with very good repeatability, over ten measurements performed in replicate for each sample and metabolite.

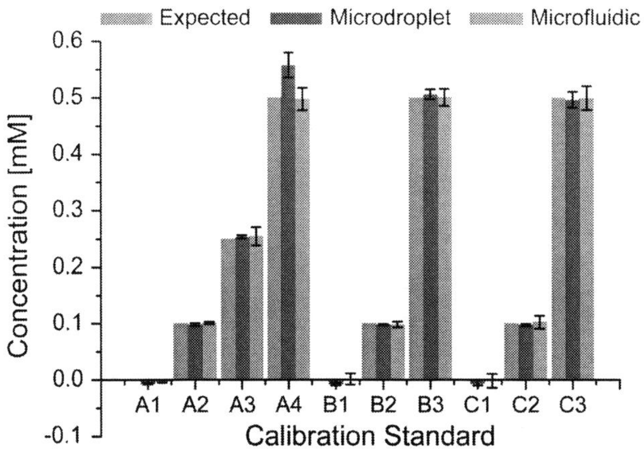

Figure 3: Comparison of glucose measurements obtained using both the conventional microdroplet assay technique and the microfluidic device. Standards A1 through A4 are water based, while standards B and C are based in embryo culture media. Error bars represent one standard deviation on three measurements for the microdroplet approach, and ten measurements on the microfluidic chip.

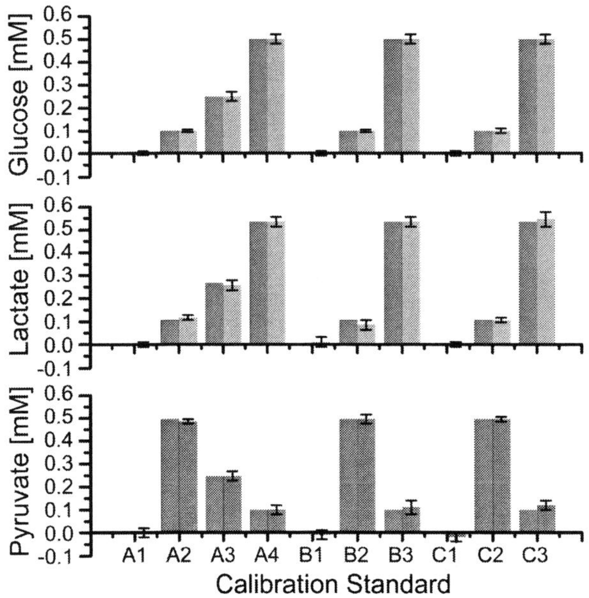

Figure 4: Example results with three metabolites measured in parallel for ten samples demonstrate excellent agreement between expected and measured values. Expected values are indicated in grey and sample names correspond with those in Figure 3. Error bars represent one standard deviation on ten replicate measurements.

3.2 Analysis of Culture Samples

To overcome difficulties in transferring individual 0.5ul media samples from under oil to the microfluidic chip, a dilution approach was employed. Metabolite concentrations of media samples were expected to lie in the 0-1mM range, while calibration studies ensured linearity of measurements of NAD(P)H by-products the 0-10μM range. Thus it was possible to dilute the sample and calibration standards by several orders to increase the working volume.

Media samples were first diluted up to 5μl with water, allowed to mix by diffusion, and then directly collected with a pipette tip. It was possible to inject the sample directly from pipette tips into microbore Tygon tubing, which was then connected to the device input wells. This loading scheme also allowed additional measurements to be performed in replicate sets for each metabolite while avoiding the problems of running out of sample. Calibration studies of different loading schemes enabled robust assays on chip while consuming less than 20nl per measurement.

The error on metabolite assays was also found to be less than 10μM in the 0-1mM range in analysis of multiple murine culture samples. These results suggest that further decreases in analysis run time are possible with reductions in the number of replicate measurements. In conventional assays, measurements are usually performed only in triplicate due to the laborious nature of the work.

4 CONCLUSIONS

A microfluidic device has been developed that greatly improves the accuracy and throughput of metabolite analyses from nanoliter-sized samples. At present, this system can be used as a standalone device for semi-automated determinations of glucose, pyruvate and lactate concentrations in culture media. Work has begun to integrate this device with an on-chip culture system to automate sampling.

The development of a microfluidics-based culture system would undoubtedly revolutionize assisted reproduction. Instead of relying on personnel to monitor the development and transfer embryos to different media, this can be done by using automated, closed chip system. Such high throughput tools will also allow further optimization of embryo culture conditions, and can provide a means to study in depth the correlation between metabolism and embryo viability. We anticipate that an integrated microfluidics approach would be superior to the currently available culture methods. This platform is undergoing continued design and development in research and clinical settings.

REFERENCES

[1] D. K. Gardner and H. J. Leese, "Assessment of Embryo Viability Prior to Transfer by the Noninvasive Measurement of Glucose-Uptake," *Journal of Experimental Zoology*, vol. 242, pp. 103-105, 1987.

[2] D. K. Gardner, T. B. Pool, and M. Lane, "Embryo nutrition and energy metabolism and its relationship to embryo growth, differentiation, and

viability," *Seminars in Reproductive Medicine*, vol. 18, pp. 205-218, 2000.

[3] M. Lane and D. K. Gardner, "Selection of viable mouse blastocysts prior to transfer using a metabolic criterion," *Human Reproduction*, vol. 11, pp. 1975-1978, 1996.

[4] D. K. Gardner and H. J. Leese, "Noninvasive Measurement of Nutrient-Uptake by Single Cultured Preimplantation Mouse Embryos," *Human Reproduction*, vol. 1, pp. 25-27, 1986.

[5] E. A. Mroz and C. Lechene, "Fluorescence Analysis of Picoliter Samples," *Analytical Biochemistry*, vol. 102, pp. 90-96, 1980.

[6] D. Rieger, "Metabolic Pathway Activity," in *A Laboratory Guide to the Mammalian Embryo*, D. K. Gardner, M. Lane, and A. J. Watson, Eds.: Oxford University Press, 2004, pp. 154-164.

[7] M. A. Unger, H. P. Chou, T. Thorsen, A. Scherer, and S. R. Quake, "Monolithic microfabricated valves and pumps by multilayer soft lithography," *Science*, vol. 288, pp. 113-116, 2000.

[8] J. P. Urbanski, W. Thies, C. Rhodes, S. Amarasinghe, and T. Thorsen, "Digital microfluidics using soft lithography," *Lab on a Chip*, vol. 6, pp. 96-104, 2006.

DNA Extraction Chip Using Key-type Planar Electrodes

S. M. Azimi*, W. Balachandran*, J. Ahern**, P. Slijepcevic***, C. Newton***

*School of Engineering & Design, Brunel University, Uxbridge, Middlesex, UB8 3PH, UK, Mohamad.azimi@gmail.com
**Chargelabs, 21 Rhestr Fawr, Ystradgynlais, Wales, SA9 1LD, UK
***School of Health Sciences and Social Care, Brunel University, Uxbridge, Middlesex, UB8 3PH, UK

ABSTRACT

This paper presents a novel method of DNA extraction from whole blood using time varying magnetic field. The novelty of this chip is that both mixing and separation steps are performed in a single chamber in less than a minute with no need for extra microfluidic channels. In order to extract DNA from white blood cells, whole blood is mixed with lysis buffer containing superparamagnetic beads. The mixing chamber is sandwiched between two key-type planar coils. Time varying magnetic field is generated within the mixing chamber to create efficient mixing. This process distributes the magnetic beads both temporally and spatially to achieve the desired mixing effect. Once the white blood cells are lysed, the exposed DNA molecules attach themselves onto the functionalized surface of the magnetic beads. Finally, DNA-attached magnetic beads are attracted to the bottom of the chamber by activating the bottom electrode. DNA molecules are extracted from magnetic beads by washing and re-suspension processes. The extracted DNA output was verified using bench-top PCR and gel electrophoresis.

Keywords: DNA chip, DNA extractor, mixer, magnetic bead

1 INTRODUCTION

Over the past decade, the advent of Micro-Electro-Mechanical Systems (MEMS) has created the potential to fabricate various structures and devices on the order of micrometers. This technology takes advantage of almost the same fabrication techniques, equipment and materials that were developed by semi-conductor industries. The range of MEMS applications is growing significantly and is mainly in the area of micro-sensors and micro-actuators. In recent years, miniaturization and integration of bio-chemical analysis systems to MEMS devices has been of great interest which has led to invention of Micro Total Analysis Systems (μ-TAS) or Lab-on-a-Chip (LOC) systems.

However, whilst there has been a great deal of work in core areas, for example, miniaturizing PCR for expedited amplification of DNA in the microchip format, less effort has been directed towards miniaturizing DNA purification methods. In fact, most of the currently demonstrated microfluidic or microarray devices pursue single functionality and use purified DNA or homogeneous sample as an input sample. On the other hand, practical applications in clinical and environmental analysis require processing of samples as complex and heterogeneous as whole blood or contaminated environmental fluids. Due to the complexity of the sample preparation, most available biochip systems still perform this initial step off-chip using traditional bench-top methods. As a result, rapid developments in back-end detection platforms have shifted the bottleneck, impeding further progress in rapid analysis devices, to front-end sample preparation where the "real" samples are used. A problem with the currently known microfluidic devices is performing efficient chaotic mixing in these platforms, this usually needs existence of moving parts, obstacles, grooves, and twisted or three dimensional serpentine channels. The structures of these components tend to be complex, however, requiring complicated fabrication processes such as multi-layer stacking or multi-step photolithography.

Magnetic beads have been of great interest in both research and diagnostic applications. Functionalized surface of magnetic beads offer a large specific surface for chemical binding and may be advantageously used as a mobile substrate for bioassays and in vivo applications [1]. They may have various sizes ranging from a few nanometres up to tens of micrometres. Due to the presence of magnetite (Fe3O4) or its oxidized form maghemite (c-Fe2O3), magnetic particles are magnetized in the presence of an external magnetic field. Such external field, generated by a permanent magnet or an electromagnet, may be used to manipulate these particles through magnetophoretic forces and therefore, result in migration of particles in liquids. Due to the size and distribution of the small embedded iron-oxide grains, particles lose their magnetic properties when the external magnetic field is removed, exhibiting superparamagnetic characteristics. This additional advantage has been exploited for separation of desired biological entities, e.g. cell, DNA, RNA and protein, out of their native environment for subsequent analysis, where particles are used as a label for actuation.

This paper introduces a novel DNA extractor process using magnetic beads and time varying magnetic field. Extractions of DNA from white blood cells and Lymphoblast GM 607D have been tested using two different schemes.

Whilst dynamic mixing has been used for separation of DNA from whole blood using Y shape micromixer (chip 2), static mixing have been used for separation of DNA from Lymphoblast GM 607D using lysis chamber (chip 1). In both methods, magnetic field has been used for efficient mixing of the lysis buffer and cells. Once the cells are lysed, the exposed DNA molecules attach themselves onto the functionalized surface of the magnetic beads. Finally, DNA-attached magnetic beads are harvested by activating the holding magnet. DNA molecules are extracted from magnetic beads by washing and re-suspension processes. The extracted DNA output was verified using bench-top PCR and gel electrophoresis.

2 MICROFABRICATION

Both chips were built in-house by ChargeLabs, a consultant to the Project.

2.1 Chip 1: Lysis-mixing Chamber

Figure 1 shows a photograph of the assembled chip 1. The device has essentially a sandwich construction, consisting of three basic elements. These are i) an outer substrate layer of PMMA, ii) a central layer of PDMS containing the mixing/lysis chamber, and iii) a second outer PMMA substrate layer, broadly similar to the first.

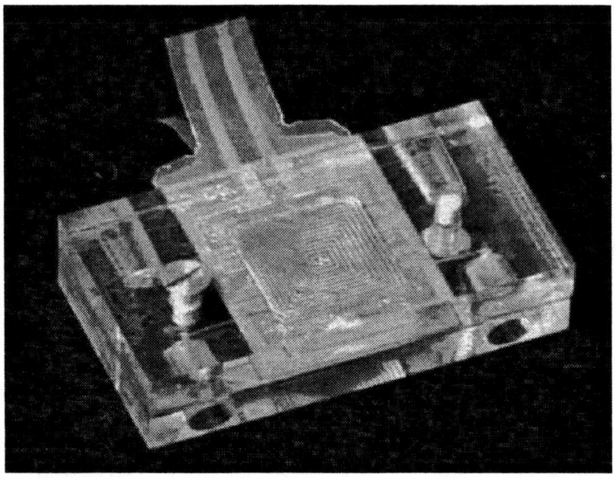

Figure 1: A photograph of assembled fluidic Chip 1.

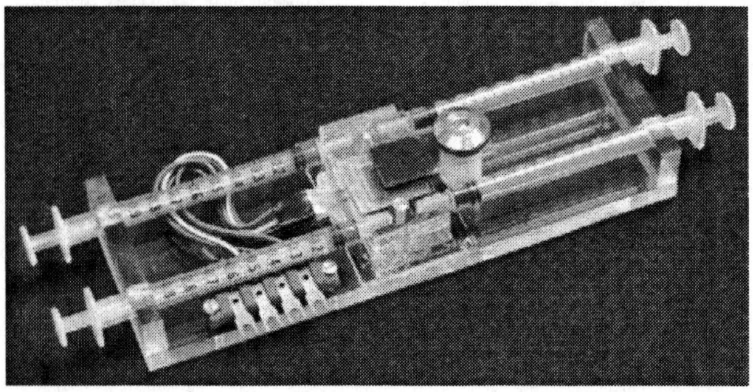

Figure 2: Completed device in docking station.

Both PMMA substrates contain a planar coil array to produce the magnetic fields for the mixing process, centred over the middle lysis chamber. Conventional ferric chloride etching and thermal mask have been applied to etch the coil arrays from Mylar-backed copper foil, a composite consisting of a 30µm layer of copper, held on a 70µm Mylar backing with an adhesive layer some 10µm thick.

Backing these coil arrays are square pieces of high-permeability material to act as magnetic amplifiers. Mu-metal sheet being available, this was employed as having superior permeability characteristics to the more conventional Permalloy. Two pieces approx 13mm square were cut, then flattened and surface lapped to just over 150µm in thickness. They were then finished to 12.50mm square, and their corners radiused to allow seating in a chamber milled with an 800µm cutter.

The substrates were cut from 5.8mm cast sheet of PMMA, machined to 40mm by 22mm overall dimensions, and edge-polished. The centre portions of each were then milled to provide the recess for the planar coil foils and the seating beneath for the mu-metal magnetic amplifiers.

This arrangement completes the lower substrate. The upper substrate additionally carries porting and drillways to allow liquid communication with the vias formed in the central PDMS component. Connexion to these ports is made via standard Luer female tapers formed into the substrate material.

The functional interfaces of both substrates are finally spin-coated with a thin (~6µm) layer of PDMS, resulting, on assembly, in a working chamber entirely faced in this material, whose long-term biocompatibility is well known.

The central PDMS layer consists of a sheet of cast PDMS, 150µm in thickness, with a rectangular cut-out, 12mm by 10mm forming the operational chamber. The shorter sides of this chamber are provided each with two cutout vias, ~250µm in width and some 8mm in length, each terminating in a 1.5mm diameter punching, to connect with the porting on the top substrate. On assembly, the device sections are secured together with two small screws, passing through one substrate and the central layer, and threaded into the second substrate. This allows adequate permanent clamping pressure between the layers of the device, whilst retaining the option of opening the device for cleaning or repair in extreme cases (Figure 1). Finally, the completed microfluidic chip is located in a docking station, which provides an electrical connector for interfacing with the drive electronics, and mechanical support and alignment for up to four standard 0.25" OD syringes (Figure 2).

2.2 Chip 2: Y-shape Micro-mixer

Figure 3 shows a picture of the y-shape micro-mixer. The device has a very simple structure consist of a capillary channel and external electromagnets. Capillary micro-channel was made using warming the glass up to the melting point and subsequent sudden stretching. This resulted in a very uniform capillary channel with an internal diameter of 150 μm and the external diameter of 250 μm.

Electromagnets used in this assembly are external to the capillary micro-channel. The electromagnets were constructed using winding of 60 turns of copper wire of diameter 60 μm onto a Mu-metal core which has been castellated in one end to produce sharp magnetic gradient tips. Castellated tip has been milled down to 200 micron and with the same 250 micron width and length. The teeth of the magnetic core are positioned on both sides wherein they are horizontally offset such that the teeth of one core are positioned between the teeth of the other core. An additional electromagnet has been used down the capillary micro-channel for collecting beads after mixing process. Finally, the completed chip is located on a PMMA substrate with electrical connectors.

3 EXPERIMENTAL SECTION

3.1 Materials

The materials used for the experiments with chip 1 are listed here: Lymphoblast GM 607D 4x106 cell/ml in FCS. (Grown in RPMI 1640 with 2mM glutamine and 15% foetal calf serum). DNA purified from GM 607D lymphoblast cells using Wizard Genomic DNA Purification Kit (Promega). Dynabeads DNA Direct Universal Prod. No 630.06, Kit containing beads in lysis fluid, wash buffer and resuspension buffer. 1.1x mastermix AB-0575/A, Primer F&R D3S3717. 2% agarose gel with 5μl ethydium bromide/100mls, Hyperladder 1 Bioline cat. No. 33025, DNA coated beads.

3.2 Expriment Steps

The chip 1 was washed through with wash buffer and drained. Two Hamilton syringes were loaded to the 10μl marking, one with DNA coated beads the other with FCS and each was attached to an inlet. Two plastic syringes were attached to the outlets, the one diagonal to the serum-containing syringe having its plunger removed. The serum sample was injected into the chip. The outlet syringes were swapped. The bead sample was injected into the chip. The current was switched on for one minute at 3A and 4 Hz to mix the contents of the chamber. Top magnet was switched off, leaving the bottom magnet to attract the beads and hold them. After 2 minutes, 50μl of wash buffer was pushed through an inlet syringe at approximately 1μl/second. This was repeated at the other inlet making a total of 100μl of washings collected in a syringe. The chamber was sucked out gently. The magnet was switched off. The washing procedure was repeated to collect the washed contents of the chamber. The run was repeated with each Hamilton syringe loaded to 20μl. This was repeated with cells in serum and beads in lysis fluid injected into the chamber but the volume injected was 20μl.

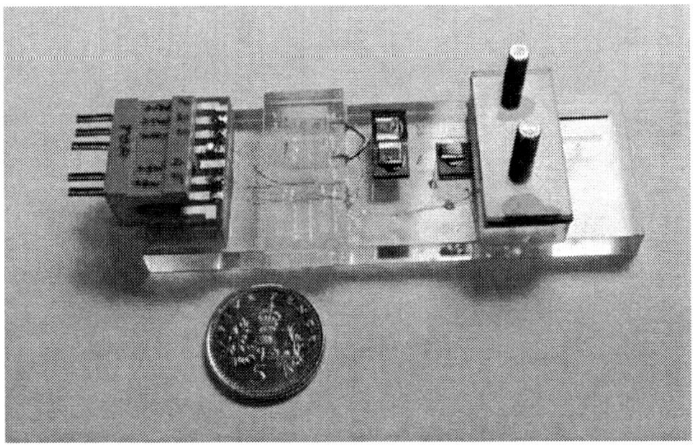

Figure 3: A photograph of chip 2, Y-shape Micro-mixer.

3.3 PCR

A 39 cycles of PCR was performed on the samples with 22μl mastermix, 1μl primers and 2μl samples. The presence of lysis fluid in the samples may act as an inhibitor to PCR therefore; samples with lysis buffer (approximately 100ul) were put on the magnet to remove the beads. The beads were then resuspended in 20μl of resuspension buffer.

4 RESULT AND DISCUSSION

4.1 Extracting DNA from Cells – Chip 1

Figure 4 shows a molecular weight marker of PCR products. Table 1 indicates detail of each column corresponding to Figure 1. Using DNA coated beads, DNA amplification in the harvested bead sample has been shown. This would indicate that the magnet held the beads in the chamber during the wash stage after which they were successfully harvested. This assumes that the washing stage was efficient over the whole area of the chamber. DNA amplification was seen in the wash sample. These beads could have been washed out of the chamber. DNA coated beads would have been washed from the dead space between the end of the inlet syringe and the chamber so DNA amplification was expected here. DNA amplification was present in all the harvested bead samples, both on the beads and free in solution therefore the system was overloaded with DNA and the washing was inefficient. As there are beads in the harvested samples, the magnet seems to have held the beads during washing. Amplified DNA from the wash sample bead may indicate some loss of beads from the chamber during washing.

4.2 Extracting DNA from Blood - Chip 2

Numerical simulation for the structure similar to this chip has clearly shown the chaotic mixing of the magnetic particles [2-3]. A preliminary experiment was carried out on chip 2 to extract DNA from whole blood. The stages are similar to previous section except the whole blood and the lysis buffer containing the superparamagnetic beads were fed into the micro-channel via the two inlets of the Y-shape inlet ports. The switching magnetic field was applied as befor. DNA was amplified from the washed beads from the 20μl whole blood samples run through chip 2 with both heparin and EDTA as blood anticoagulants. DNA was amplified from the washed beads from the heparinised plasma sample obtained from chip 2.

The result obtained is shown in Figure 5 and Table 2. This clearly demonstrates that DNA can be successfully extracted from blood using this chip in a few minutes.

Research is continuing to optimize the chip 1 operating parameters and to quantify the amount of DNA that can be extracted from whole blood in a few minutes.

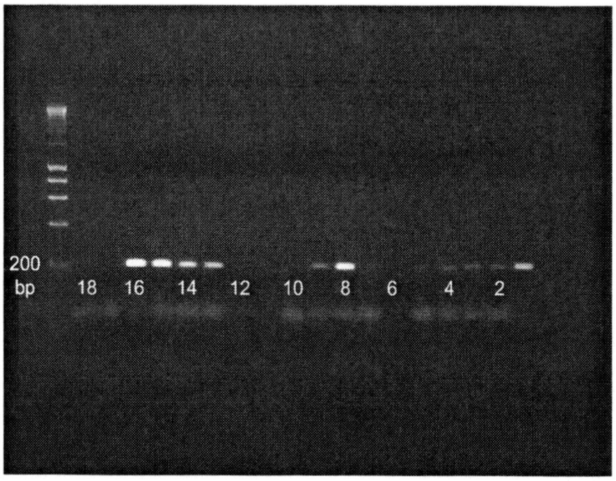

Figure 4: Molecular weight marker of PCR product (chip 1)

Lane	Sample	ProP[1]	PriP[2]
1	DNA/bead wash	++	+/-
2	DNA/bead sample	+	+
3	DNA/bead wash	+	+
4	DNA/bead sample	+	+
5	Cells+beads/ wash/ beads	+/-	+
6	Cells+beads/wash/supernatant	-	-
7	Cells+beads/ sample/ beads	+/-	+
8	Cells+beads/sample/supernatant	+++	+
9	Cells+beads/ wash/ beads	+/+	+
10	Cells+beads/ wash/ beads	+/--	+
11	Cells+beads/wash/supernatant	-	-
12	Cells+beads/wash/supernatant	-	-
13	Cells+beads/ sample/ beads	++	+
14	Cells+beads/ sample/ beads	++	+
15	Cells+beads/sample/supernatant	+++	+
16	Cells+beads/sample/supernatant	+++/+	-
17	dwater	-	+
18	dwater	-	+
[1]ProP: Product Present, [2]PriP: Primers Present			

Table 1: Column information for Figure 4.

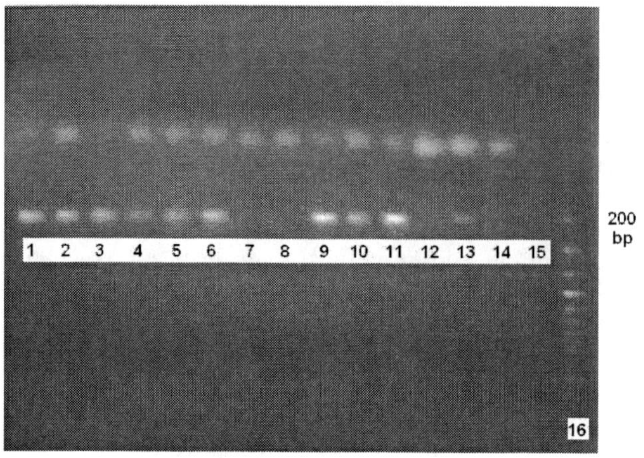

Figure 5: Molecular weight marker of PCR product (chip 2)

Lane	Sample	DNA Amplified
1	EDTA / Blood / Beads	0.5 μL
2	EDTA / Blood / Beads	1.0 μL
3	EDTA / Blood / Beads	2.0 μL
4	EDTA / Blood / Beads	0.5 μL
5	EDTA / Blood / Beads	1.0 μL
6	EDTA / Blood / Beads	2.0 μL
7	EDTA / Serum / Beads	1.0 μL
8	EDTA / Serum / Beads	2.0 μL
9	Heparin / Blood / Beads	0.5 μL
10	Heparin / Blood / Beads	1.0 μL
11	Heparin / Blood / Beads	2.0 μL
12	Heparin / Serum / Beads	1.0 μL
13	Heparin / Serum / Beads	2.0 μL
14	Negative Control	
15	Negative Control + Beads	

Table 2: Column information for Figure 5.

REFERENCES

[1] Martin A. M. Gijs. "Magnetic bead handling on-chip: new opportunities for analytical applications", Microfluid Nanofluid, vol. 1, no. 1, pp. 22-40,2004.

[2] M. Zolgharni, S. M. Azimi, M. R. Bahmanyar, W. Balachandran, "A numerical design study of chaotic mixing of magnetic particles in a microfluidic bio-separator", Microfluid Nanofluid, vol. 3, no. 6, pp. 677-687,2007.

[3] M. Zolgharni, S.M. Azimi, M.R. Bahmanyar and W. Balachandran, "A Microfluidic Mixer for Chaotic Mixing of Magnetic Particles", Nanotech 2007, Vol. 3, pp.336 - 339.

Amperometric glucose sensor for real time extracellular glucose monitoring in microfluidic device

Igor A. Ges, Franz J. Baudenbacher

Department of Biomedical Engineering, Vanderbilt University, 5824 Stevenson Center, VU Station B 351631, Nashville, TN 37235-1631, USA; E-mail: igor.ges@vanderbilt.edu

ABSTRACT

We have combined a microfluidic network and developed a miniature glucose sensor to measure glucose consumption of single cardiomyocytes in a confined extracellular space. The thin film glucose sensor was integrated into a 360 picoliter extracellular space. We used Pt thin film electrodes (20x100 μm^2) as a counter electrode, and a needle type reference electrode (0.45 mm diam) placed in the cell loading port of the microfludic device. Single cardiac cells were trapped in the sensing volume using pressure gradients. Using miniature mechanical valves allowed us eliminate the interference of flow artifacts on the detection of glucose. We have demonstrated the feasibility to measure the glucose consumption of single cardiac myocytes in an enclosed confined extracellular volume using of miniature glucose sensitive electrodes fabricated by spin-coated deposition.

Keywords: glucose sensor, glucose oxidase, spin-coated deposition, microfluidic device, cardiomyocytes.

INTRODUCTION

Microfluidic devices are becoming wide spread in different areas of biology and medicine [1-3]. Microfluidic based lab on a chip devices provide several advantages for cellular analysis systems and cells can be confined in a chemically controlled in-vivo like microenvironment comparable to the cell size. Although, progress has been made incorporating sensors into lab-on-a-chip devices there is a need to develop microsensor arrays for multiple analyte sensing. Glucose and oxygen consumption, acidification and lactate production rates are the most important variables which characterize the metabolic activity of living cells. Measuring multiple metabolites will allow us to monitor the metabolic response of cells chemicals, drugs or toxins in lab on a chip device. The microsensor and microfluidic confinement is necessary to measure of metabolic activity in vicinity of single cells in sub nanoliter volumes reliably with a sufficient signal to noise ratio. To measure glucose consumption rates in small cell culture volumes, we have developed a microfluidic device with active valves, which confines single cardiac myocytes in sub nanoliter volumes on a microfabricated planar thin film glucose sensing electrode. The glucose sensing electrodes were fabricated using thin film platinum microelectrodes spin-coated with glucose oxidase (GOx). Such sensors possess a high sensitivity, long-term stability and can operate in different cell growth media. The development and miniaturization of a stable glucose biosensor are not only of great interest to monitor cell physiology but may lead to an alternative approaches to measure blood glucose levels.

RESULTS AND DISCUSSIONS

A glucose biosensor based on the electrochemical detection of enzymatically generated H_2O_2 was constructed by the spin-coating glucose oxidase and glutaraldehyde cross-linking onto Pt electrodes, followed by an additional coating with a Nafion protection layer to obtain long term stability and sensitivity (Fig.1).

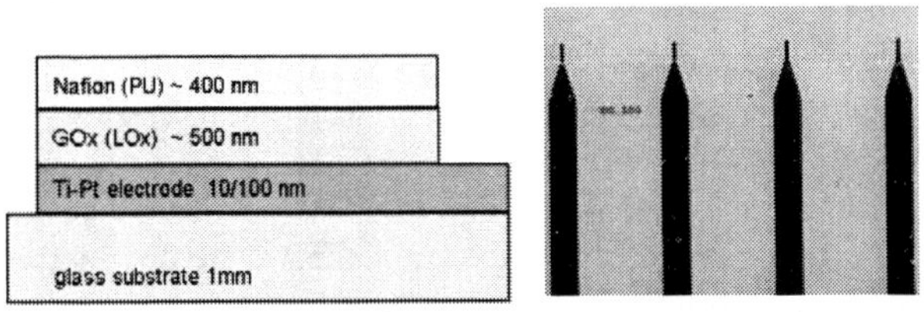

Fig.1. (a) Cross-sectional schematic of the thin-film glucose sensor. (b) Optical image of four glucose sensors on a glass substrate. The working area of sensors was 50x500 and 100x500 μm^2.

Fig.2 shows the layout of the microfluidic device overlaid onto the sensing electrode configuration. The base electrodes for the glucose sensor were fabricated by vacuum evaporation of a Ti adhesion layer and a Pt layer on glass substrates (24x24mm) and standard photolithography. The working areas of electrodes were 50x500 and 100x500 μm^2. The electrode array consists of seven independent electrodes. Usually three of them were used as glucose sensors and were positioned in the cell sensing volume. The others (bare Pt films) electrodes were used as counter electrodes. A needle type electrode (DriRef-450 from WPI Inc., diameter 0.45 mm) was inserted in the cell loading port and used as reference electrode.

The enzyme solution for glucose sensitive electrodes was prepared based on the procedure described by Eklund et al (4). In our protocol, the glucose oxidase

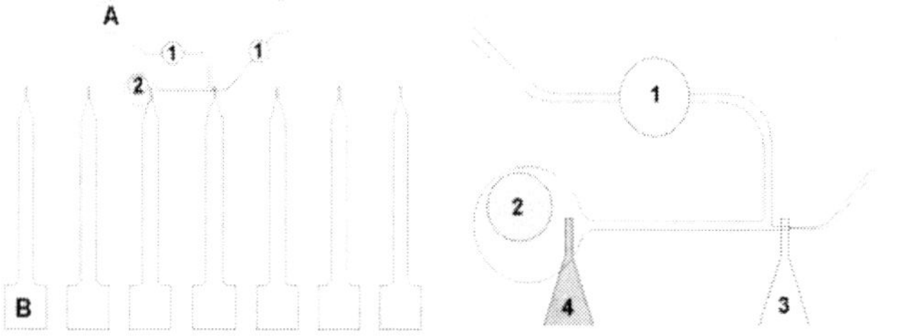

Fig.2. Schematic overview of a NanoPhysiometer for glucose consumption measurements from single CM cells in sub-nanoliter volumes: (A) micro-fluidic channels array; (B) microelectrode array; (1) mechanical valve; (2) Ag/AgCl reference electrode; (3) glucose sensor; (4) Pt counter electrode. The inset (a) shows a detailed view of microelectrodes in the detection volume

(GOx) film solution was prepared by dissolving 2.5 mg GOx and 25 mg of BSA in 250μL of 1mM PBS containing 0.02% v/v Triton X-100. After the BSA and GOx were completely dissolved, 15 μL of 2.5% glutaraldehyde solution was added and thoroughly mixed with a Vortex shaker for 10 min. Each time the fresh enzyme solution was prepared for the deposition of glucose oxidase coated electrodes.

Different types of polymer protective membranes were tested to stabilize the characteristics of the glucose sensors. The requirement for the polymer membranes is the permeability for glucose and oxygen. In our investigation we tested polyurethane (PU) and Nafion membranes because of their good glucose diffusion-limiting behavior and biocompatibility [5]. 3-4 pellets of PU (~ 0.5 g) were dissolve in 5 ml of tetrahydrofuran and stirred for over 2 hours at room temperature. The Nafion films were obtained from 5% Nafion solution.

The electrodes were functionalize by spin coating the enzyme and the protective membrane on freshly cleaned platinum electrodes. The substrates were cleaned by sonicating in ethanol and in acetone and rinsed with running DI water. The enzyme solution (100 μL) was added onto the substrate with a pipette. The substrate was then spun at 1500 rpm for 30s with a conventional spin coater (Laurell Model WS-400A-6TFM/LITE). The resulting film thickness was typically ~0.4 μm. After spin coating the GOx films were dried at room temperature during 1 hour.

The films of polymer membranes were also deposited by spin coating. For the Nafion films the actual conditions were: dispense a 100 μl drop of solution onto the electrodes, spin at a speed = 1000 rpm, acceleration = 300 rpm s^{-1} and time = 30 s. The resulting Nafion film is transparent with a typical film thickness of ~ 0.4 μm. The PU films were deposited at 3000rpm during 30s, which resulted in a 0.7 μm thick film.

For characterizing the glucose sensors in a beaker experiment the surface of the electrodes except working area (100x500 μm, fig.1b) were covered with a thin PDMS film (100 μm) After fabrication the glucose sensors were stored at 4°C before use.

The microfluidic network was fabricated from PDMS (polydimethylsiloxane) by replica molding, using photoresist on a silicon wafer as a master. The master was fabricated by spinning a 20 μm thick layer of photoresist (SU-8 2025, MicroChem Corp, Newton, MA) on a 3"diameter silicon wafer (Nova Electronic Materials, Ltd, Carrolton, TX) and by exposing it to UV light (160 mJ/cm^2) through a metal mask using a contact mask aligner. The photoresist was processed according to the manufacturer's recommendation on the datasheet. An optional 30-minute hard bake at 200°C was performed on a hot plate to increase the durability of the resist. PDMS prepolymer was mixed with the curing agent in a 10:1 ratio by weight and degassed in a vacuum chamber for 20 min. The master was placed in a clean Petri dish, which was filled to a height of approximately 1 cm with PDMS and cured in an oven for 4 hours at 70°C. After curing, the elastomer was mechanically separated from the master and cut into discrete devices. Access holes for the fluidic connections were punched into the PDMS using sharpened blunt-tip 18 gauge needles.

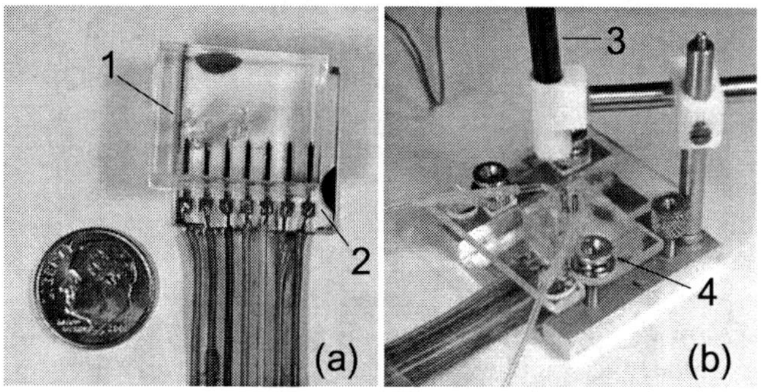

Fig. 3. (a) Alignment of glass substrate with glucose sensors (2) and PDMS microfluidic (1) device (the channel height – 40mm, the cell trap volume – 0.36 nL). (b) Photo of the clamp mechanism (4), which ensures reliable seal between PDMS microfluidic device and glass substrate with thin film glucose sensors; (3) - Ag/AgCl reference electrode diam. 0.45mm.

The PDMS microfluidic device was manually aligned relative to the glucose sensitive electrodes (fig.3a) with a stereo microscope. The PDMS device was sealed to the glass substrate by auto-adhesion (fig.3a), and stabilized with a mechanical clamp (fig. 3b). Glass capillaries were inserted into the access holes to connect the microfluidic channels to syringes/pumps using standard microtubing (0.5mm inner diameter, Cole Parmer). For the trapping of cells in the sensing volume we controlled the syringes connected to the output and control port by hand. A microsyringe pump "Micro 4" (WPI, Sarasota, FL) was used to control the flow of solution during the calibration of the glucose electrodes in the microfluidic devices.

In order to precisely control residual flows after stopping the perfusion in our micro fluidic devices, we developed miniature mechanical screw valves [7]. Our design allowed us to place easily two valves in close proximity to the cell trap (fig.2). The valves were fabricated by drilling a pocket hole above the microfluidic channel into the PDMS. Into the pocket hole we inserted an oversized threaded sleeve and a screw which allowed us to pinch off the microfludic channel beneath the screw.

There are two major reactions involved for electrochemically sensing glucose. The glucose oxidase (GOx) oxidizes the glucose and produces gluconic acid and hydrogen peroxide as described by the following reaction [8-9].

$$\text{Glucose} + O_2 + \text{GOx} \rightarrow \text{gluconic acid} + H_2O_2$$

H_2O_2 is then oxidized electrochemically on the Pt electrode resulting in a current proportional to the glucose concentration.

$$H_2O_2 \rightarrow (0.6\text{V Pt vs Ag/AgCl}) \rightarrow 2H^+ + O_2 + 2e^-$$

The most important advantage of the amperometric detection of H_2O_2 in contrast to a Clark type oxygen electrode detecting the oxygen production is the simplicity of the electrode fabrication, reduced cross-talk and the possibility of miniaturization.

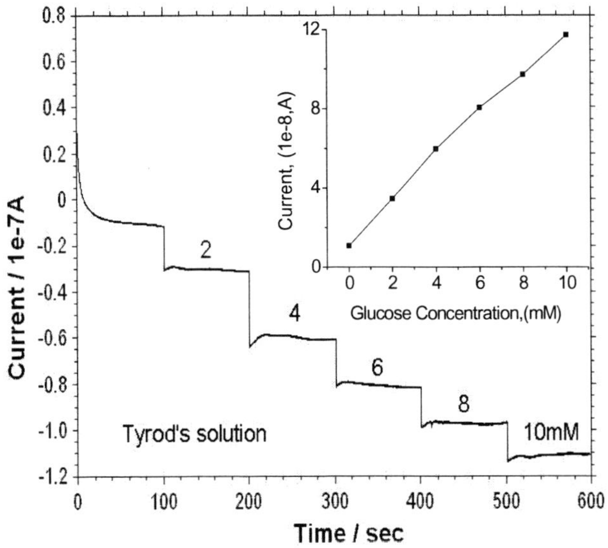

Fig.4. Current-time plots as a function of glucose concentration in Tyrod's (pH7.4) solution. Inset: calibration curve of sensors as a function of glucose concentration

Fig. 4 shows the response of the amperometric glucose sensor obtained by adding glucose to stirred Tyrod's solution. The potential of the working electrode was maintained at 0.6 V with respect to Ag/AgCl reference electrode. The current shows a stable constant stepwise increase in current. The resulting calibration curve for glucose over the physiological concentration range form 0–10mM is presented in the inset of Fig.4. The temporal response of the glucose electrode to a change in glucose concentration is reasonably fast (within 10 s).

The operational stability of our glucose sensors were tested by continuously monitoring the steady-state response in Tyrod's solution at glucose concentrations of 6 and 8 mM at room temperature (Fig.5). With the exclusion of slight low frequency fluctuation which we attribute to temperature fluctuations, the current response didn't change appreciable during a 1 hour period of continuous use, which is sufficient to conduct several experiments with cardiac myocytes. The long-term stability was investigated by successive calibrations. No appreciable loss in glucose sensitivity was observed up to about 3 month.

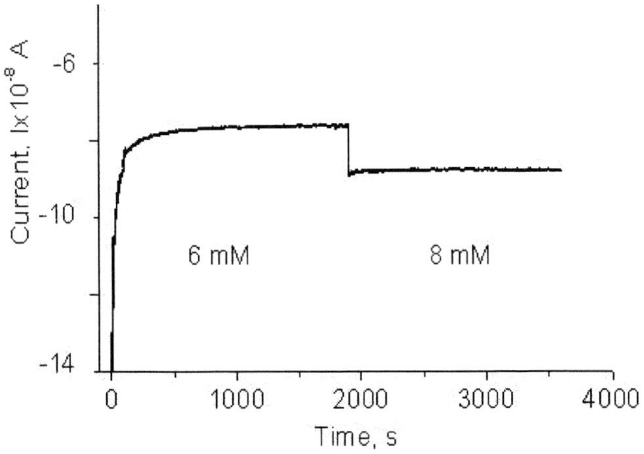

Fig.5. Investigation of long term stability of the glucose sensor in Tyrod's solution (pH7.4) with 6 and 8 mM glucose concentration.

The microfabricated glucose sensors were used in conjunction with a microfluidic device (NanoPhysiometer (NP)) that allows trapping the single cardiac myocytes in a confined sub-nanoliter volume adhered to the microfabricated glucose sensitive thin film electrodes on a glass substrate. The glass slide provides the base layer for the microfluidic network made of polydimethylsiloxane (PDMS) with integrated valves to allow control of the fluidic access to the cell detection volume (Fig.6and 3b).

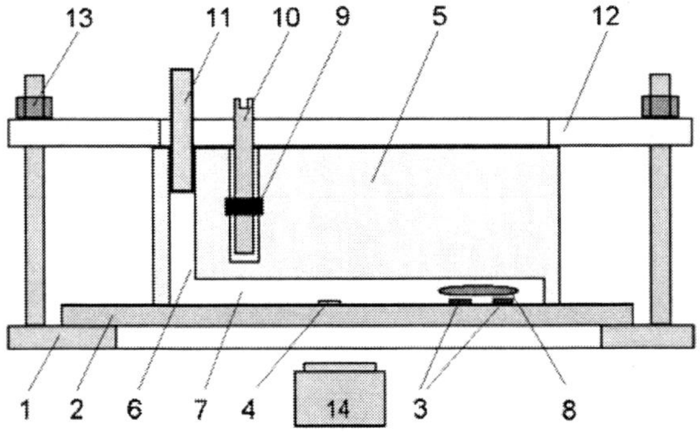

Fig.6. Cross-sectional schematic of the completely assembled NanoPhysiometer. A clamp ensures mechanical stability of the seal between the PDMS microfluidic device and the substrate with the electrode arrays. 1-aluminum

base; 2-glass substrate; 3-glucose sensitive electrode; 4-Pt counter electrode; 5- PDMS microfluidic device; 6-input port; 7-microchannel; 8-single cardiac myocyte cell; 9-thredded sleeve; 10-microvalve screw; 11-Ag/AgCl reference electrode; 12-acylic cover; 13-screw to adjust pressure; 14-objective

The glucose sensitive electrodes were tested in a microfluidic environment to compare their characteristics with the beaker experiments. Fig.7 show that sensitivity and stability of spin coated glucose sensor in the microfluidic channel (100 × 40 μm^2) were practically the same as those obtained from macroscopic experiment in beakers.

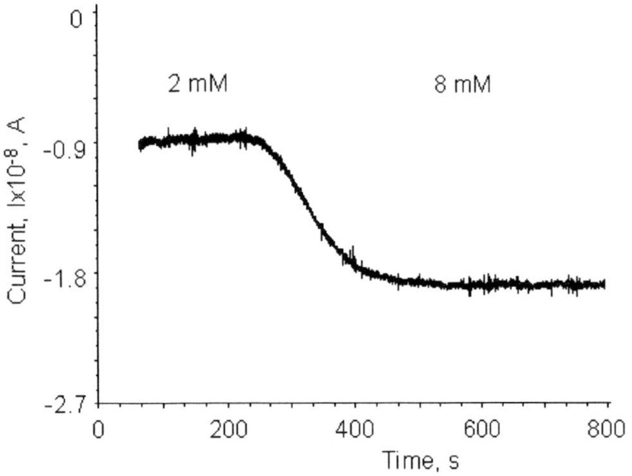

Fig.7 Calibration of glucose sensor in microfluidic channel for two Tyrod's solution with different concentration of glucose (2 and 8 mM). The rate of flow is 2 nL/s.

The design of our microfluidic network allows us to trap single myocytes and to quantify glucose concentration changes in 360 pL volume. Fig.8 shows an example of the measurements of glucose consumption from cardiac myocyte cells in the subnanoliter volume in Tyrode's solution (10 mM glucose, 0.5 mM Ca^{2+}). After initiating the amperometric measurements and closure of the microvalves we observed a rapid change in the current for a period of ~100 s. After this stabilization period we observed a linear decrease of the anodic current over time which corresponded to reduction of the glucose concentration in the sensing volume of the Nanophysiometer. After the removal of the cell, as indicated by the step in the current time plot (Fig 8), the anodic current (glucose concentration) was constant demonstrating that the consumption of glucose by the electrode is negligible compare to the consumption by the myocyte.

In conclusion, we have developed a microfluidic device platform to measure the glucose consumption of single cardiac myo-cytes in confined sub-nanoliter volumes. We plan to use the Nanophysiometer to investigate the metabolism in isolated cardiac cells under condition comparable to ischemia.

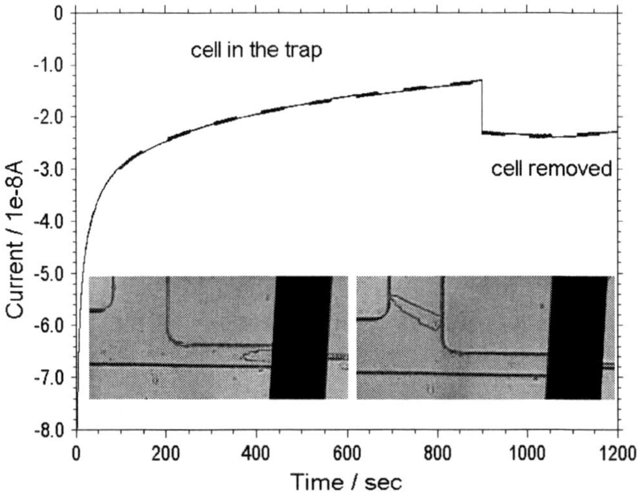

Fig.8. Measurement of the glucose consumption from single cardiac myocytes in the Nanophysiometer. The dark region in the image is the spin coated glucose sensitive electrode (GOD ~ 0.4 μm, Nafion ~ 0.4 μm) with working dimensions of 40x100 μm^2. The inset shows a detailed view of the myocyte in and the removal of the measurement volume during the experiments.

ACKNOWLEDGMENTS

We especially thank Dr. Igor Dzura and the Dr. Knollmann lab for isolating cardiac myocyte cells and helpful discussions. This work has been supported in part by NIH Grant U01AI061223 and the Vanderbilt Institute for Integrative Biosystems Research and Education

REFERENCES

[1] Y. Tanaka et al. Biosensors and Bioelectronics 23 (2007) 449–458
[2] Y. Changqing, et al. Analytica Chimica Acta 560 (2006) 1–23.
[3] X. Cai, et al. Anal. Chem. 2002, 74, 908-914.
[4] S. Eklund, et al. Anal Chem, 76-3,2004p519.
[5] H. Yang, et al. Biosensors & Bioelectronics 17 (2002) 251–259
[6] D.B. Weibel, et al., Anal. Chem.2005, 77, 4726-4732.
[7] I.A. Ges and F.J. Baudenbacher. Biomedical Microdevice, 2008 (in press)
[8] X. Pan, et al. Sensors and Actuators B 102 (2004) 325–330
[9] X. Luo, et al. Electroanalysis 18, 2006, 1131- 1134

Optimizing Multiscale Networks for Transient Transport in Nanoporous Materials

Robert H. Nilson and Stewart K. Griffiths

Physical and Engineering Sciences Center
Sandia National Laboratories, P. O. Box 969, Livermore CA 92007
rhnilso@sandia.gov and skgriff@sandia.gov

ABSTRACT

Hierarchical nanoporous materials afford the opportunity to combine the high surface area and functionality of nanopores with the superior charge/discharge characteristics of wider transport channels. In the present paper we optimize the apertures and spacing of a family of transport channels providing access to a surrounding nanoporous matrix during recharge/discharge cycles of materials intended for storage of gas or electric charge. A diffusive transport model is used to describe alternative processes of viscous gas flow, Knudsen gas flow, and ion diffusion. The coupled transport equations for the nanoporous matrix and transport channels are linearized and solved analytically for a periodic variation in external gas pressure or ion density using a separation-of-variables approach in the complex domain. Channel apertures and spacing are optimized to achieve maximum inflow/outflow from the functional matrix material for a fixed system volume.

Keywords: nanoporous materials, hierarchical, network optimization, nanofluidics, nanoscale transport

1 INTRODUCTION

Nanoporous materials are finding increased use in a range of emerging technologies including hydrogen and methane storage, supercapacitors, desalination membranes, and fuel cell membranes. In these applications, nanoscale pores provide high surface area as well as functionality but may also inhibit device-scale transport needed to rapidly charge and discharge the medium. The opposing goals of nanopore functionality and rapid transport can both be achieved, however, using hierarchical materials having nanopores interconnected with larger-scale transport channels.

Previous work on nanoporous materials has been largely focused on synthesis, material characterization, and measurements of functionality. Modeling of nanopore transport is largely concerned with single-pore processes, while related modeling of network transport has been addressed mainly in other disciplines with the goal of minimizing global flow resistance under steady-flow conditions [1,2,3]. In contrast, we seek to maximize inflow/outflow volumes under transient conditions where flow resistance, capacitance, and desired discharge time are all important.

2 GOVERNING EQUATIONS

Consider a highly idealized nanoporous material comprised of permeable matrix slabs separated by an array of parallel planar channels having uniform aperture and spacing, a_1 and b_1, as illustrated schematically in Fig. 1. The nanopores within the slabs have a characteristic aperture and spacing of a_0 and b_0, bulk porosity of $\phi_0 = a_0 / b_0$, and internal diffusivity of D_0. The nanoporous slabs might alternatively be characterized in terms of their bulk transport properties, such as permeability and porosity, and a characteristic nanopore dimension, a_0, without requiring explicit knowledge of pore morphology.

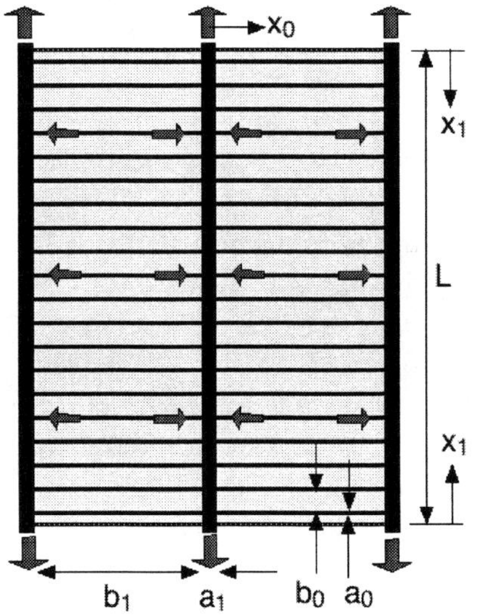

Figure 1: Geometry of multiscale network.

Gas flow or diffusion within the porous matrix slabs can generally be described by a diffusion equation of the form

$$\phi_0 \frac{\partial \rho_0}{\partial t} = \frac{\partial}{\partial x_0}\left(D_0 \frac{\partial \rho_0}{\partial x_0} \right) \qquad (1)$$

where ρ is the density, t is time, x_0 is distance from the slab surface, and D_0 is the diffusion coefficient. In gas transport, ρ is the molar density of the gas and the diffusion rate is limited by collisions among gas molecules and by collisions with pore walls. When the mean free path is small compared to lateral channel dimensions, collisions among gas molecules predominate and $D_0=\kappa P/\mu$ where κ, P, and μ, are the permeability, pressure, and viscosity. However, in nanoscale pores, the mean free path of the gas is often small compared to the pore dimension. In this instance the appropriate diffusivity is given by the Knudsen expression, $D_0=va_0$ in which v is the mean thermal speed. The same model applies in an approximate manner to the porous electrodes of supercapacitors where, ρ is the charge density of the electrolyte within the pores and D_0 is the effective ion diffusivity or mobility.

Transport along the open vertical channels in Fig. 1 is governed by an equation that resembles Eq. (1) but also includes lateral transport between the channels and the adjacent permeable slabs.

$$\phi_1 \frac{\partial \rho_1}{\partial t} = \frac{\partial}{\partial x_1}\left(D_1 \frac{\partial \rho_1}{\partial x_1}\right) + \frac{2D_0}{a_1} \left.\frac{\partial \rho_0}{\partial x_0}\right|_{x_0=0} \qquad (2)$$

Here, x_1 is measured from the bottom inflow/outflow surface and the factor of $2/a_1$ in the last term represents the ratio of surface to cross-sectional area for a rectangular channel. Since the transport channels are typically assumed to be open, the porosity ϕ_1 will be taken as unity. Also, the gas diffusivity is given by $D_1 = P a_1^2/12\mu$ or by $D_1=va_1$ for the viscous and Knudsen regimes, respectively. Thus, wider channels have higher diffusivities. Similarly, the ion transport diffusivity D_1 along wider channels will sometimes exceed D_0 because $a_1>a_0$, reducing the effects of hindered diffusion in nanoscale pores. To facilitate analytical solution we will assume that the diffusion coefficients D_k are constant and we define the diffusivity as $\alpha_k = D_k/\phi_k$ for k=0, 1.

To approximate the discharge/recharge cycle of an energy storage material we seek solutions of the following form appropriate for a sinusoidal variation in the external density having an amplitude and mean of $\Delta\rho$ and ρ_r.

$$\rho^* = \frac{\rho - \rho_r}{\Delta\rho} = e^{i\omega t}\, \rho_0^*(x_0^*)\, \rho_1^*(x_1^*) \qquad (3)$$

Here, $\rho_0^*(x_0^*)$ represents the variation of ρ^* within the matrix slabs, $\rho_1^*(x_1^*)$ is the variation along transport channels, $x_0^* = 2x_0/b_1$, $x_1^* = 2x_1/L$, and $i \equiv \sqrt{-1}$. Substitution of this assumed form into Eq. (1) yields a second order ordinary differential equation for $\rho_0^*(x_0^*)$

$$i\,\lambda_0^2\,\rho_0^* = \frac{d^2\rho_0^*}{dx_0^{*2}} \qquad (4)$$

subject to the boundary condition $\rho_0^* = 1$ at the interface between channels and slabs where $x_0^* = 0$ and the symmetry condition $(\rho_0^*)' = 0$ at the slab midplane where $x_0^* = 1$. The solution is

$$\rho_0^* = \frac{\cosh \lambda_0\sqrt{i}\,(1 - x_0^*)}{\cosh \lambda_0\sqrt{i}} \quad \text{where} \quad \lambda_0 \equiv \frac{b_1}{2}\sqrt{\frac{\omega}{\alpha_0}} = b_1^* \qquad (5)$$

The Fourier modulus λ_0 indicates the ratio of the cross-slab diffusion time, $\tau \sim (b_1/2)^2/\alpha_0$, to the discharge cycle time, $\tau \sim 1/\omega$. Similarly, substitution of Eqs. (3,5) into Eq. (2) yields the following result for the transport channels.

$$\rho_1^* = \frac{\cosh \hat{\lambda}_1\sqrt{i}\,(1 - x_1^*)}{\cosh \hat{\lambda}_1\sqrt{i}} \qquad (6)$$

$$\hat{\lambda}_1 = \lambda_1\left[1 + \frac{\phi_0 b_1^*}{a_1^*}\,\frac{\tanh \lambda_0\sqrt{i}}{\lambda_0\sqrt{i}}\right]^{1/2} \qquad (7)$$

$$\lambda_1 = \frac{L}{2}\sqrt{\frac{\omega}{\alpha_1}} = \frac{L}{2}\sqrt{\frac{\omega}{\alpha_0}\left(\frac{\alpha_0}{\alpha_1}\right)} = L^*\left(\frac{\alpha_0}{\alpha_1}\right) = L^*\left(\frac{a_0^*}{a_1^*}\right)^m \qquad (8)$$

Here, a* and b* and L* have all been normalized by the length scale $\sqrt{4\alpha_0/\omega}$ which also appears in the Fourier moduli, λ_0 and λ_1.

The exponent m appearing in Eq. (8) describes the variation of diffusivity, $\alpha = D/\phi$, with channel aperture, a. As explained earlier, m=2 for a viscous gas flow, m=1 for free molecular gas flow, and m=0 for a simple diffusion process. However, since the surface to volume ratio is 2/a in rectangular channels, each of these exponents will be increased by unity in cases where the capacitance, ϕ, of the channel is associated mainly with the channel surface area, rather than the channel volume. This is often the case in nanoporous gas storage materials where surface adsorption provides the primary capacitance and in supercapacitor devices where electric charge is accumulated in electric double layers having a thickness of one or two molecular diameters. Thus, our example calculations will explore the range $1<m<3$. We exclude m=0, because in this case wider transport channels provide little benefit.

The pore structure is optimized by adjusting the values of a_1^* and b_1^* to obtain the maximum cyclical inflow/outflow from the nanopores subject to a given system volume. Maximizing cyclical flow is equivalent to maximizing the amplitude of the sinusoidal variation in the mean nanopore density, $|\Delta\bar{\rho}_0^*|$. This quantity is readily evaluated by spatial integration of the solution given by Eqs. (3, 5-8) over the nanopore volume.

$$|\Delta\bar{\rho}_0^*| = \mathrm{mod}\left[\frac{\tanh\lambda_0\sqrt{i}}{\lambda_0\sqrt{i}}\frac{\tanh\hat{\lambda}_1\sqrt{i}}{\hat{\lambda}_1\sqrt{i}}\right] \quad (9)$$

Because the smallest scale nanopores generally provide functionality or capacitance unavailable in wider transport channels, we will take credit only for density changes within the nanoporous matrix blocks. Thus, the optimization criterion, Ω, is taken as the product of $|\Delta\bar{\rho}_0^*|$ with the ratio of nanopore volume to total system volume.

$$\Omega = |\Delta\bar{\rho}_0^*|\frac{V_{\text{nanopores}}}{V_{\text{total}}} = |\Delta\bar{\rho}_0^*|\frac{\phi_0}{\phi_0 + a_1^*/b_1^*} \quad (10)$$

Note that increases in a_1^* or decreases in b_1^* help to increase the numerator $|\Delta\bar{\rho}_0^*|$ in Eq. (10) but they also tend to increase the denominator. As a result, Ω has a maximum value for particular choices of a_1^* and b_1^*.

3 RESULTS

Figure 2 illustrates the variation of optimized channel apertures and spacing with normalized system scale. As the scale or frequency becomes larger, the optimal channel apertures and spacing both increase. Although it is relatively obvious that channel apertures would have to increase to maintain necessary transport, increased spacing might not have been anticipated. The wider spacing is apparently needed to avoid the allocation of excessive system volume to transport channels rather than nanopore functionality.

The nearly constant slopes on the logarithmic plot in Fig. 2 indicate power-law variations of the channel dimensions with system scale.

$$b_1^* = B\,L^{*p} \quad \text{and} \quad a_1^* = A\,L^{*q} \quad (11)$$

The powers p and q depend upon m but not on a_0^*. It is quite surprising to see that these power-law relationships remain valid for relatively large values of the Fourier moduli and that they continue to hold even as the channel apertures approach the channel spacing for the case of m=1.

The Fourier moduli, λ_0 and $\hat{\lambda}_1$, are displayed in Fig. 3. As seen in Eq. (9), the quantities $\tanh(\lambda\sqrt{i})/(\lambda\sqrt{i})$ represent the fractional decreases in density amplitude within the pores and along the transport channels. Thus, the observed equality of the optimal moduli for m=2 indicates that the two scales of transport have equal impedance. For other values of m, the two moduli are comparable but not exactly equal. In all cases, the optimal moduli both increase in magnitude as the scale of the system increases, indicating greater impedance and greater reduction in density amplitude.

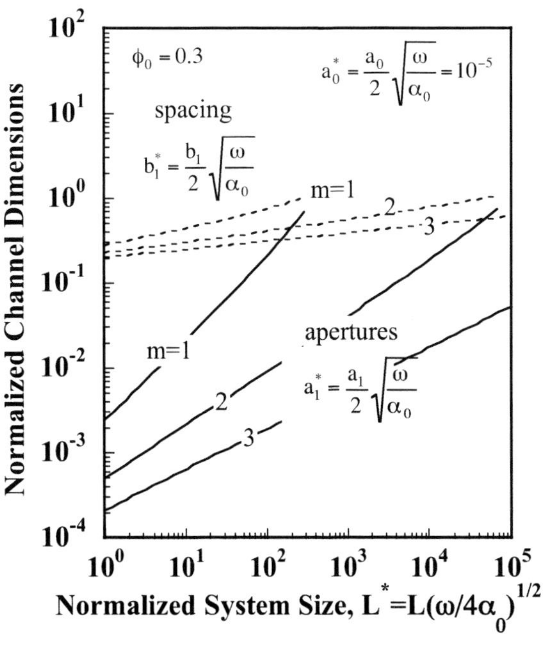

Figure 2: Variation of optimized channel apertures (solid lines) and channel spacing (dotted lines) with system scale. Parameter m describes the variation of transport diffusivity with aperture, $\alpha = D/\phi \sim a^m$.

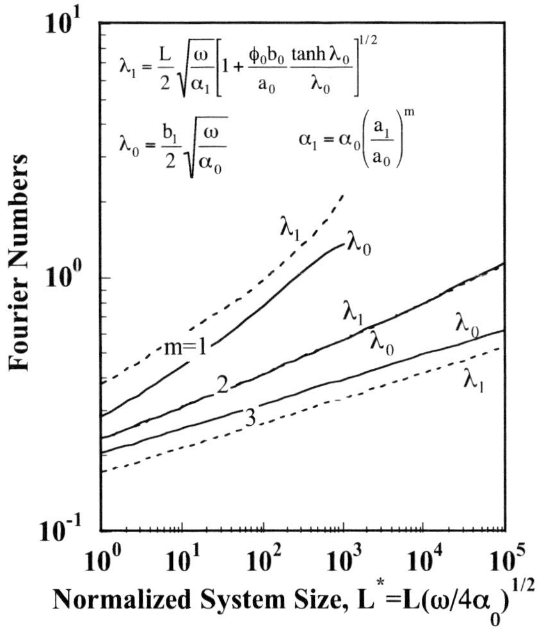

Figure 3: Fourier moduli indicate ratio of transport time to discharge cycle time for nanochannels, λ_0, and transport channels, $\hat{\lambda}_1$. Optimized moduli for the two scales are comparable but not always identical.

Figure 4 depicts the variation of the optimal transport efficiency, Ω, with normalized system scale. A value of $\Omega = 1$ indicates that the density within nanopores precisely tracks the imposed external density variation and that the volume lost to the transport channels is negligible. For m=3, $\Omega > 0.95$ for $L^* < 10^5$, indicating great success in facilitating transport. The results are less impressive for m=1,2 because wider apertures provide less benefit when m is smaller.

Even in the most challenging case with m=1, $\Omega > 0.7$ for $L^* < 10^2$. In this instance, the transport efficiency begins to fall off very sharply for $L^* > 10^3$. This is because the channel apertures are becoming nearly as wide as the channel spacing, as clearly apparent in Fig. 2. For this reason, we did not extend the curves beyond this point.

The uppermost curves in Figure 5 provide a direct indication of the benefit derived by introduction of the transport channels.

$$R \equiv \frac{\Omega_{\text{with channels}}}{\Omega_{\text{without channels}}} \quad \text{where} \tag{12}$$

$$\Omega_{\text{without channels}} = \text{mod}\left[\frac{\tanh L^*\sqrt{i}}{L^*\sqrt{i}}\right] \tag{13}$$

For large values of L^*, $\tanh L^*/L^* \rightarrow 1/L^*$ such that $R \rightarrow L^*$ so long as Ω remains near unity. Even in the worst case of m=1, R=50 for $L^*=100$, indicating a level of discharge performance 50 times better than would have been obtained in the absence of transport channels.

4 SUMMARY

Channel apertures and spacing have been optimized through analytical modeling of diffusion transport in a hierarchical network having transport channels that provide access to nanoscale pores. The model is applicable to viscous gas flow (m=2) and Knudsen gas flow (m=1) as well as ion transport (m=0). These exponents increase by unity when most of the capacitance is associated with channel surfaces. For optimized systems, the impedance of transport channels is similar to that within pores. This observation can be used together with given analytical expressions to compute channel dimensions that maximize transport for prescribed system scales and discharge times.

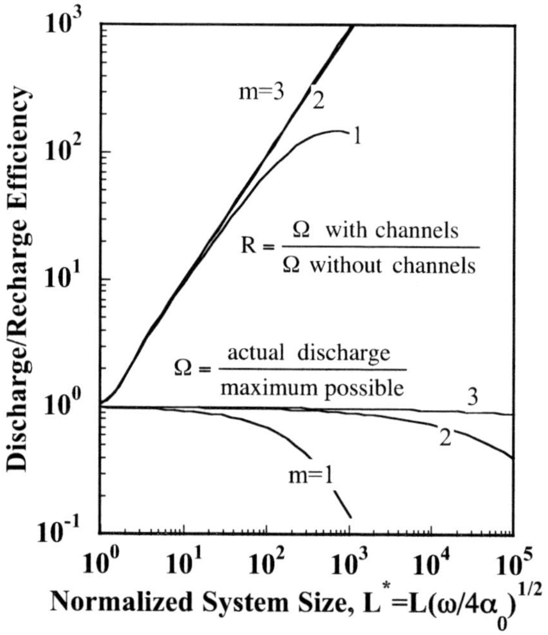

Figure 4: Transport efficiency Ω indicates optimized amount of discharge from nanopores relative to the maximum that might be achieved if nanopore density (or pressure) precisely tracked applied external conditions and if all of the system volume consisted of functional nanoporous material.

REFERENCES

[1] G. B. West, J. H. Brown, and B. J. Enquist, "A general model for the origin of allometric scaling laws in biology," Science 276, 122-126, 1997.

[2] M. Durand, "Architecture of optimal transport networks," Phys. Rev. E **73**, Article No. 016116, 2006.

[3] A. Bejan and S. Lorente, "Constructal theory of generation and configuration in nature and engineering," J. Appl. Phys. **100**, (4), Article No. 041301, 2006.

ACKNOWLEDGMENT

Sandia is a multiprogram laboratory operated by Sandia Corporation, a Lockheed Martin Company, for the United States Department of Energy's National Nuclear Security Administration under contract DE-AC04-94AL85000.

Ultrasound-driven viscous streaming, modelled via momentum injection

J. Packer*, D. Attinger** and Y. Ventikos*

* Fluidics and Biocomplexity Group, Department of Engineering Science, University of Oxford
Oxford, OX1 3PJ, U.K. Tel: +44(0)1865-283452, Email: yiannis.ventikos@eng.ox.ac.uk
** Laboratory for Microscale Transport Phenomena, Department of Mechanical Engineering
Columbia University, New York, NY 10027, USA. Email: da2203@columbia.edu

ABSTRACT

Microfluidic devices can use steady streaming caused by the ultrasonic oscillation of one or many gas bubbles in the liquid to drive small scale flow. Such streaming flows are difficult to evaluate, as analytic solutions are not available for any but the simplest cases, and direct computational fluid dynamics models are unsatisfactory due to the large difference in flow velocity between the steady streaming and the leading order oscillatory motion. We develop a multiscale numerical technique which uses two computational fluid dynamics models to find the streaming flow as a steady problem, and validate this model against experimental results performed by Tho *et. al.* [1].

Keywords: acoustic streaming, CFD, lab-on-a-chip, microfluidic devices, multiscale

1 INTRODUCTION

In addition to the first order oscillating flow generated by a gas bubble in a fluid excited by ultrasound pressure waves, a steady second order flow is generated [2]. This is difficult to model with standard computational fluid dynamics (CFD) techniques, due to the difference in time scales exhibited by the first order oscillating flow (O(kHz) or higher) and the steady second order flow (O(10Hz)) in the configurations considered. Were the second order flow to be determined by standard transient CFD modelling, many thousands of cycles would need to be calculated to determine the nature of the steady flow with reasonable accuracy, as the magnitude of the steady flow is many orders lower than that of the first order oscillating flow.

In this paper, a novel technique for modelling this steady flow is proposed, where the second order flow is modelled directly as a steady state problem, and the forcing for the second order flow is calculated from the modelled first order flow.

2 COMPUTATIONAL FLUID DYNAMICS MODELLING

2.1 First order

The first order model involves a standard CFD approach, where the flow is excited by a moving boundary. In this paper the case of a hemispherical bubble in water is considered, where the bubble wall is displaced sinusoidally, modelling periodic volumetric oscillation. The first order simulation is transient, with three full periods of oscillation modelled. This allows post-transient conditions to be reached, confirmed by comparing the results of the second and third period.

2.2 Second order

The second order CFD model predicts the steady streaming expected for the configuration chosen. The model used is the same as that for the first order, but is a steady model. The moving bubble wall is not modelled, and replaced in the simulation with a static boundary at the mean position. To excite the steady streaming flow, forcing terms are calculated from the first order flow, and added to the fluid volume, as momentum sources, in the region where a viscous sub-layer would exist. The method of calculation of the forcing and the layer thickness is outlined below.

3 CALCULATION OF FORCING

3.1 Theory

The momentum injection is calculated following the analysis of Lighthill [3], finding the forcing from the gradients of the Reynolds stresses. If u, v and w are the flow velocities in the three Cartesian directions and $F_{u,v,w}$ is the force per unit volume in each of these directions and ρ the fluid density, the driving force for the steady streaming is:

$$F_u = -\rho \left(\frac{\partial \overline{uu}}{\partial x} + \frac{\partial \overline{uv}}{\partial y} + \frac{\partial \overline{uw}}{\partial z} \right) \quad (1)$$

$$F_v = -\rho \left(\frac{\partial \overline{vu}}{\partial x} + \frac{\partial \overline{vv}}{\partial y} + \frac{\partial \overline{vw}}{\partial z} \right) \quad (2)$$

$$F_w = -\rho \left(\frac{\partial \overline{wu}}{\partial x} + \frac{\partial \overline{wv}}{\partial y} + \frac{\partial \overline{ww}}{\partial z} \right) \tag{3}$$

Therefore, the mean values over one complete cycle of uu, uv, uw, vv, vw and ww are found for each cell. The differentials of these mean values must then be found in order to find the forcing. Once the differentials are known, as covered in section 3.2.2, the forcing can be calculated for each cell by adding the three appropriate partial differentials.

This forcing only affects the viscous sub-layer adjacent to the boundary [4], [5], as outside this layer the forcing is absorbed into a hydrostatic pressure field [5]. Different authors give slightly different approximations to the thickness of this layer, with Lee & Wang giving the thickness as $\delta = \left(\frac{2\mu}{\rho\omega}\right)^{\frac{1}{2}}$ [5] and Marmottant *et. al.* as $\delta = \left(\frac{\mu}{\rho\omega}\right)^{\frac{1}{2}}$ [4]. Here δ is the thickness, ν the kinematic viscosity, μ the absolute viscosity, ω the excitation frequency and ρ the fluid density. Recalling that both expressions are approximations, the difference is not considered significant. In this paper, Marmottant *et. al.*'s approximation is used.

3.2 Numerical method

3.2.1 Mean values

From the first order model, the flow velocities u, v and w of each cell in the volume around the bubble wall are found at each time-step for one complete period of oscillation, once post-transient conditions have been reached. For each cell, the values of uu, uv, uw, vv, vw and ww are computed and their mean value is estimated for the complete period.

Therefore values of each mean multiple ($\overline{uu}$, $\overline{uv}$, ...) are known at the centre of each cell. These can be treated as scattered data points, but the differentials in the x, y, and z directions are needed.

3.2.2 Numerical differentiation

In order to find the differentials of the mean values at each location, the approach taken is to find the difference in value and difference in position for three surrounding points, and to find the Cartesian partial derivatives from this by solving the set of three equations of the form:

$$\delta V = \delta x \frac{\partial V}{\partial x} + \delta y \frac{\partial V}{\partial y} + \delta z \frac{\partial V}{\partial z} \tag{4}$$

As we know three sets of $(\delta V, \delta x, \delta y, \delta z)$ we can solve at each point for $(\frac{\partial V}{\partial x}, \frac{\partial V}{\partial y}, \frac{\partial V}{\partial z})$. This must be done for each of $V = (\overline{uu}, \overline{uv}, \ldots)$.

We solve this by solving the equation:

$$\begin{pmatrix} \frac{\partial V}{\partial x} \\ \frac{\partial V}{\partial y} \\ \frac{\partial V}{\partial z} \end{pmatrix} = \begin{pmatrix} \delta x_1 & \delta y_1 & \delta y_1 \\ \delta x_2 & \delta y_2 & \delta y_2 \\ \delta x_3 & \delta y_3 & \delta y_3 \end{pmatrix}^{-1} \begin{pmatrix} \delta V_1 \\ \delta V_2 \\ \delta V_3 \end{pmatrix} \tag{5}$$

where the subscripts 1,2 and 3 refer to the values for the three surrounding points. If the three points chosen are nearly collinear or coplanar, this will lead to an ill-conditioned solution. Consequently the solution is found by selecting the three points in close proximity which give a well-conditioned behaviour. The closest 15 points are found and the best combination of three selected. This is found by considering all possible combinations, and finding a parameter which describes the quality of the solution. First the condition number of the matrix

$$\begin{pmatrix} \delta x_1 & \delta y_1 & \delta y_1 \\ \delta x_2 & \delta y_2 & \delta y_2 \\ \delta x_3 & \delta y_3 & \delta y_3 \end{pmatrix}$$

is found for all possible combinations of points. The higher this condition number, the more poorly conditioned the set of equations is. This is then multiplied by the product of the distances to the three points under consideration. The combination with the lowest value of this parameter is chosen, as it is the well conditioned set of points closest to the point at which the differential is required. This technique was shown always to be sufficient for the calculation of the derivatives.

The differentiation method is essentially a forwards difference method, extended to three dimensions and applied to a scattered data field.

3.2.3 Forcing

Once the differentials of the mean values ($\overline{uu}$, $\overline{uv}$, ...) are known for each cell in the viscous sub-layer region, the forcing can be found from equations 1, 2 and 3. The forcing is then used in the second order steady-state CDF model as a momentum injection to force the steady streaming. The forcing for each cell is used within the *CFD-ACE2007 solver (ESI Group)* package, in which the forcing per unit volume for each forced cell is multiplied by the cell volume to find the absolute force, and this force used in the equilibrium equations used by the solver, allowing the streaming flow to be found.

4 VALIDATION

4.1 Comparison with experimental work by Tho *et. al.*

The numerical modelling technique proposed is tested against the experimental results of Tho *et. al.* [1], as their results give both the streaming generated and the bubble motion for different modes of bubble oscillation.

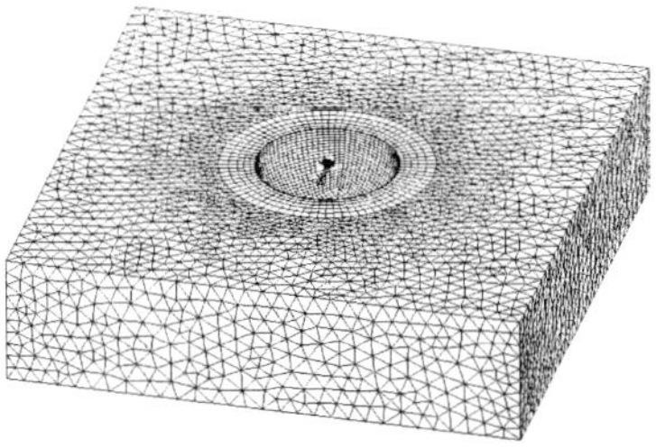

Figure 1: CFD grid

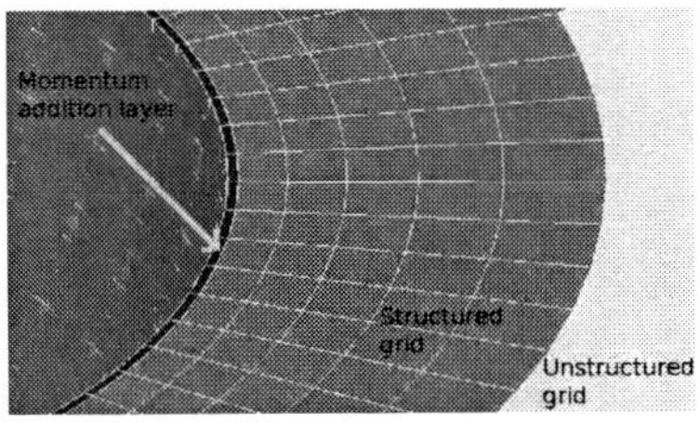

Figure 2: Grid detail

Tho *et. al.*'s experimental conditions correspond to a bubble of mean radius varying between 202 and 274 μm. Several modes of bubble vibration are examined, with case 4 being pure volume oscillation of the bubble. This is the case we chose to present in this paper.

4.1.1 Grid

Tho *et. al.*'s experimental volume is a thin chamber, of height 0.66mm, as described in Figure 1 of Tho *et. al.*'s publication [1]. The hemispherical bubble is on the top wall. Only the region near the bubble is used for our CFD modelling to make the problem more tractable. The grid used is shown in Figure 1. The grid is divided into different volumes so that the required velocities can be output from the first order model, and the forcing applied only to the viscous sub-layer adjacent to the bubble wall in the second order model. These zones are shown in Figure 2.

4.1.2 Boundary conditions

In the first order model, the hemispherical bubble is of radius 270 μm, and the bubble wall is oscillated at 8.658kHz, with a magnitude of 1.41% of the bubble radius, corresponding to case 4 in Tho *et. al.*'s experiments [1]. The boundary condition at the bubble wall is taken as zero slip, since the particles used in the flow visualisation congregate at the interface and allow little slip flow[1]. For comparison a model is also run for zero shear at the bubble wall, which would be expected for perfectly pure fluid, neglecting the viscosity of the bubble gas. The other boundary conditions are the same for the two cases. If the bubble is on the top surface, the top and bottom surfaces have wall boundary conditions (zero tangential and normal flow velocity) and the four edges have fixed pressure boundary conditions, allowing flow between the volume modelled and the large microchamber used experimentally.

4.1.3 Convergence to post-transient conditions

For the first order model, 90 time steps are used per period, and three complete periods modelled. To ensure that the model has reached post-transient conditions, the results of period 2 and 3 are compared and found to be essentially similar, with an average difference of 0.11% between velocities at equivalent time steps within the period.

4.1.4 Data processing

The flow velocities (u,v,w) are found for the volume adjacent to the bubble (the viscous sub-layer volume) and the cells immediately adjacent to this. From the velocity values for the final period (timesteps 181-270), the forcing in the viscous sub-layer is calculated numerically, following the analysis in the section above. All calculations were undertaken with *Octave 2.9*. Due to the grid deformation in the first order model, the position of the cell centres in the layer vary through the period, so their mean positions are used.

4.1.5 Second order model

The same grid and boundary conditions are used for the second order simulation as for the first order transient run, except that the run is steady-state and the bubble wall is maintained in its mean position. The forcing for each cell is added to the viscous sub-layer volume in the simulation, and is the only forcing applied.

4.1.6 Results and discussion

In Tho *et. al.*'s experimental work, the flow velocities are found by a micro-particle image velocimetry (PIV) technique[1]. Tho *et. al.*'s work measures the flow in three planes parallel to the wall on which the bubble is located, which are referred to as the z_1 plane, through the bubble and 75 μm from the wall, the z_2 plane which is 300 μm from the wall and the z_3 plane which is 525 μm from the wall.

Both the velocities predicted in the numerical model, and observed in Tho *et. al.*'s experiment for the volume oscillation case are shown in Table 1. The velocities are seen to be correct to within one order of magnitude, and the accuracy of the predicted flow velocity is better away from the bubble interface. This may be because the PIV method does not pick up the high velocity flow in the small region immediately adjacent to the bubble, as suggested by Tho *et. al.* in section 4.1[1].

Table 1: Comparison of numerical and experimental results. Velocities in mm/s

	Numerical	Tho *et. al.* observed
z_1 plane	1.1	0.3
z_2 plane	0.4	0.3
z_3 plane	0.5	0.3

The pattern of flow predicted by our model, shown in Figures 3 and 4 for the z_1 and z_2 planes is not identical to that observed by Tho *et. al.* for the volume oscillation mode of the bubble, as shown in his Figure 15 [1], but does show interesting similarity with that observed for other modes of vibration, including his case 1 (translating oscillation along a single axis), shown in his Figure 7[1].

A numerical model was also run simulating free-slip conditions at the bubble wall. This gave velocities of an order of magnitude higher than those observed by Tho *et. al.*, suggesting that the assumption of zero slip at the bubble wall due to particle contamination is valid, and a free-slip model invalid for a particle-bearing fluid.

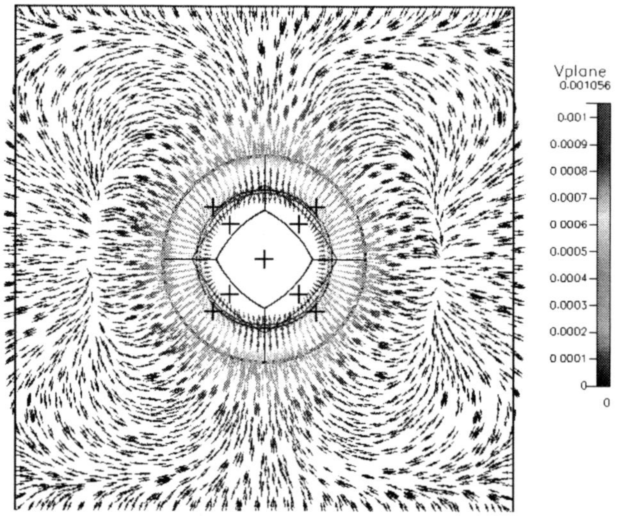

Figure 3: Numerical results, z_1 plane. Velocity in m/s

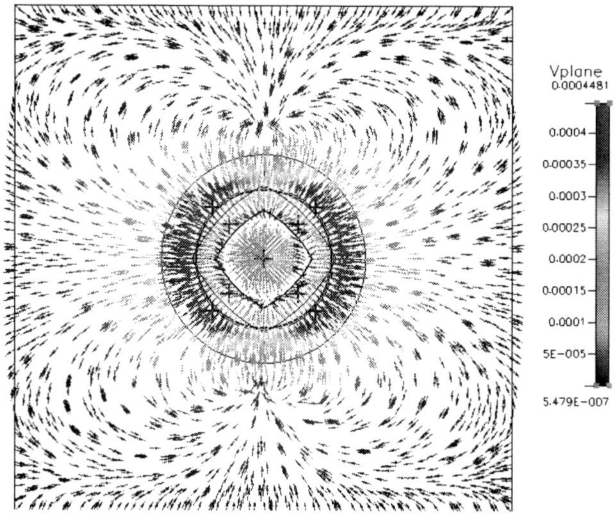

Figure 4: Numerical results, z_2 plane. Velocity in m/s

5 CONCLUSION

The multiscale modelling method described can predict the magnitude of steady streaming flows induced by bubble oscillation to within one order of magnitude. Due to the complex nature of both the pressure field and bubble motion in a real forced oscillation fluid/bubble system, the method cannot capture the detail of the flow pattern that will be generated, but suggests one possible mode of flow. This could be improved by computing a first order model which accurately accounts for both the gas bubble, the fluid and the interaction between them generated by an ultrasonic pressure wave. The numerical calculation of the forcing for the steady flow would remain as set out.

ACKNOWLEDGEMENTS

The authors would like to acknowledge Mr Christopher Fenelly for his contributions to early development work on this methodology, and the ESI Group for allowing the use of *CFD-ACE 2007* in this study.

REFERENCES

[1] Tho, Mannasseh & Ooi, "Cavitation microstreaming patterns in single and multiple bubble systems," J. Fluid Mech., 576, 191-233, 2007.

[2] Xu & Attinger, "Control and ultrasonic actuation of a gas-liquid interface in a microfluidic chip," J. Micromech. Microeng., 17, 609-616, 2007.

[3] Lighthill, "Acoustic Streaming," J. Sound Vib., 63, 391-418, 1978.

[4] Marmottant et. al., "Microfluidics with ultrasound-driven bubbles," J. Fluid Mech., 568, 109-118, 2006.

[5] Lee & Wang, "Outer acoustic streaming," J. Acoust. Soc. Am., 88, 1990.

Fluido-Dynamic and Electromagnetic Characterization of 3D Carbon Dielectrophoresis with Finite Element Analysis

R. Martinez-Duarte*, S. Cito**, Esther Collado-Arredondo***, S. O. Martinez*** and M. Madou*

*Mechanical & Aerospace Engineering Department, University of California, Irvine, 4200 Engineering Gateway, Irvine, CA, 92617, USA, drmartnz@gmail.com
**Universitat Rovira i Virgili, Tarragona, Spain, salvatore.cito@gmail.com
***Tecnológico de Monterrey, Campus Monterrey, Monterrey, Mexico, smart@itesm.mx

ABSTRACT

The following work presents the fluido-dynamic and electromagnetic characterization of a 3D electrodes array to be used in high throughput and high efficiency Carbon Dielectrophoresis (CarbonDEP) applications such as filters, continuous particle enrichment and positioning of particle populations for analysis. CarbonDEP refers to the induction of Dielectrophoresis (DEP) by carbon surfaces. The final goal is, through an initial stage of modeling and analysis, to reduce idea-to-prototype time and cost of CarbonDEP devices to be applied in the health care field. Finite Element Analysis (FEA) is successfully conducted to model velocity and electric fields established by polarized high aspect ratio carbon cylinders, and its planar carbon connecting leads, immersed in a water-based medium. Results demonstrate correlation between a decreasing flow velocity gradient and an increasing electric field gradient toward electrodes' surfaces which is optimal for CarbonDEP applications. Simulation results are experimentally validated in the proposed applications.

Keywords: carbon, dielectrophoresis, c-mems, high flow rate, simulation.

1 INTRODUCTION

Even when the field of bioparticle separation has advanced significantly in recent years with the wide use of techniques such as Fluorescence-Activated Cell Sorting (FACS) or MACS (Magnetically-Actuated Cell Sorting) such techniques require the use of specific, and often expensive, tags to achieve high selectivity. An ideal solution would be a technology which eliminates the need of such tags while maintaining a high throughput separation process. We believe Dielectrophoresis (DEP), the induction of a force, F_{DEP}, on a polar particle immersed in a polar media by a non uniform AC or DC electric field, could be such solution. A huge advantage of DEP is the selection of the targeted particle, by inducing either an attractant or repellent F_{DEP} force to the nearest electrode surface, using only its intrinsic dielectric properties which are solely determined by the particle's individual phenotype such as its membrane morphology. Such advantage eliminates the need of specific tags linked to magnetic beads or fluorophores to discriminate targeted particle types potentially reducing the cost of each assay.

As an enhancement to current DEP devices, we propose the use of Carbon Dielectrophoresis (CarbonDEP). Carbon DEP refers to the use of carbon surfaces to induce DEP. Carbon surfaces offer better electrochemical and biocompatibility properties than other conductive materials [1] as well as lower costs. The use of volumetric (3D) structures allows higher throughput than traditional planar DEP devices [2] towards rates comparable to current separation techniques.

In the following work we present initial models and analysis of a proposed CarbonDEP array to characterize its advantages. The use of Finite Element Analysis is towards implementing a design and fabrication methodology leading to shorter development times of CarbonDEP applications in the health care field. In contrast to previous work where 3D electrodes arrays made out of a perfect conductor and immersed in DI water [3] or only simplified planar geometries in conductive media are simulated [4], we have modeled and analyzed both electric and flow velocity fields independently in an array of polarized carbon 3D electrodes and their 2D connection leads immersed in a water-based medium. Simulated geometry is a replica of physical device. Experimental validation and examples of applications are also described.

2 MATERIALS AND METHODS

Complete experimental device fabrication procedure is detailed elsewhere [2]. Briefly, a 5 X 29 array of carbon electrodes, and their connecting leads, is obtained through the C-MEMS technique [1]. CarbonDEP array features 60 μm high post electrodes with a diameter of 25 μm. The gap between electrodes in the axis parallel to the channel wall is 40 μm while it is 100 μm in the perpendicular one. A 500 μm wide, 65 μm high channel was then fabricated around the electrode array using SU-8 (MicroChem Corp.). Finally, the device was sealed and electrical connections were made. Channel features SU-8 on the four walls.

The geometry implemented for simulations is based on the experimental device. Given the symmetry of the

CarbonDEP array, its length was reduced from 5 X 29 to 5 X 5 to reduce computational power requirements.

2.1 Electromagnetic Analysis

Electromagnetic simulation was conducted using COMSOL v3.3 (COMSOL) running in a Workstation having Solaris 9 (Sun Microsystems) as operating system (OS). Processor used was a Sun Blade 1500 @ 1 GHz. 2 GB of RAM and up to 5 GB of Virtual Memory were at hand.

Analysis is completed by separately meshing connecting leads and carbon posts using Triangular 2D meshes and implementing coupling through identity boundary conditions. Carbon structures were considered to have an equal resistivity at all points of 1.07×10^{-4} W-m [4]. Electrodes are assumed to be in contact with a water medium of conductivity equal to 2 mS/m. Excitation voltage was 4 V_{pp}. Boundary conditions are those of an insulator in all channel walls and in channel regions furthest away from electrode array.

2.2 Fluido-dynamic Analysis

Fluido-dynamic simulation was conducted using Fluent 8.3 (ANSYS, Inc.).

Analysis is done by applying a 3D model and numerically solving the Navier-Stokes equation in a structured grid using an absolute velocity formulation with a relaxation factor of 0.3 for the pressure, density Body forces and momentum variables. Problem has been solved in steady condition as a fully developed flow assuming a Newtonian flow with water as a fluid. Boundary conditions are those of No-slip imposed on the channel walls and carbon posts' surface. A 3D pressure-based laminar approach has been considered with a mass-flow channel inlet of 1.6×10^{-7} kg/s and an outflow boundary condition at the outlet. Such condition assumes a zero normal gradient for all flow variables except pressure. Velocity and pressure have been coupled using the standard SIMPLE model that uses a relationship between velocity and pressure corrections to enforce mass conservation and to obtain the pressure field. For the numerical discretization, a second order upwind scheme has been used.

3 RESULTS

3.1 Electromagnetic

Isometric views of the Electric field distribution at 10 μm from the channel floor (Fig. 1 A) and at 60 μm (Fig 1 B) are shown together with a detailed top view of the 30 μm cross section of the array (Fig. 1 C). As expected, at 10 μm one can discern a fairly strong electric field gradient induced by the connection leads. At 60 μm, top of the channel, the electric field gradient induced by the leads completely vanquishes and the one generated by the carbon posts predominates. Fig. 2 is a X-Z cross section at point A on Fig. 1C clearly showing how the effect of the leads diminishes as distance from the channel floor increases.

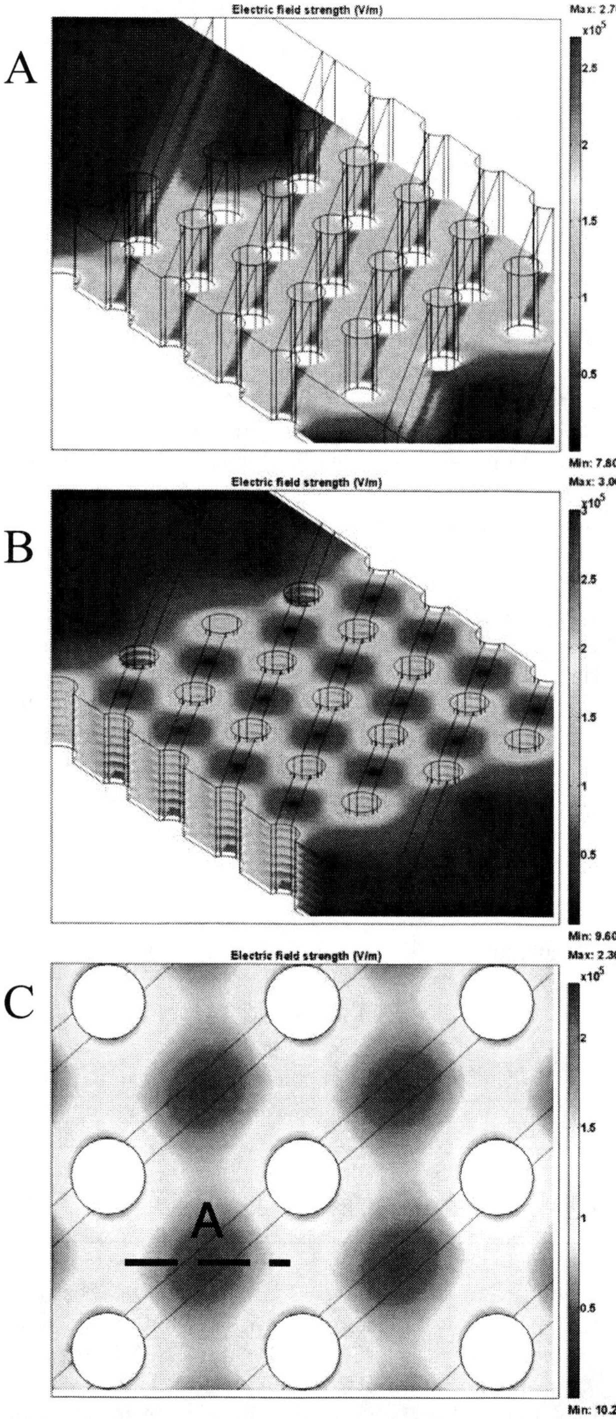

Fig 1. Electric field distribution in a Carbon electrode array contained in a SU-8 channel at 10 μm from channel bottom (A), at 60 μm (B) and top view of 30 μm cross section (C). Excitation Voltage of 4 V_{pp} with medium conductivity of 2mS/m.

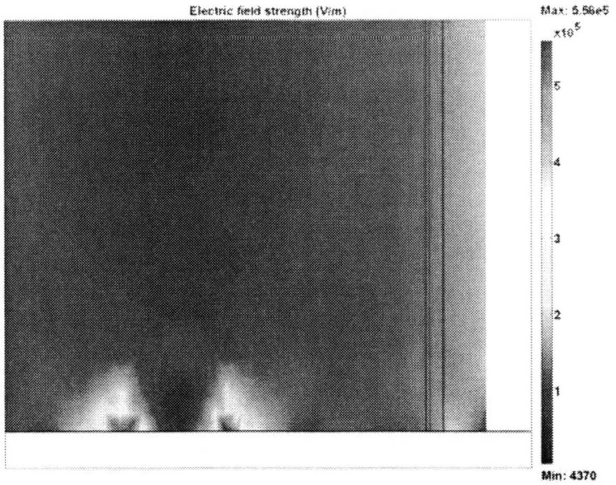

Fig. 2. X – Z cross section on electric field distribution in a Carbon electrode array. Note how the electric field gradient induced by connection leads in channel bottom quickly vanishes at around 20 μm from channel floor.

3.2 Fluido-dynamic

3D Flow velocity field is shown in Fig. 3 A. X-Y, X-Z and Y-Z cross sections are shown in Fig. 3 B, C and D respectively. Analysis clearly shows a 3D parabolic flow profile with highest velocities in between electrodes in the plane perpendicular to channel wall (Fig. B and D).

4 DISCUSSION

Following DEP theory, it is known how particles exhibiting Positive DEP at a given frequency are attracted to high electric field gradient areas while particles exhibiting Negative DEP get repelled to more stable low electric field gradient areas. Based on such principles one designs an electrode array to induce desired electric field distributions optimized for specific applications.

Important is to note how the electric field gradient increases towards the electrode surfaces'. Furthermore, thanks to the different electrode arrangement in the X and Y axis, higher electric field gradients coincide with lower velocity magnitude areas. Such correlation proves to be optimal for DEP trapping since F_{DEP} is at its maximum where velocity magnitude is at its minimum. Such distribution also restricts the trapping of desired particles, and the subsequent particle cluster, to areas that would minimally disturb the path lines and maintain the laminar flow.

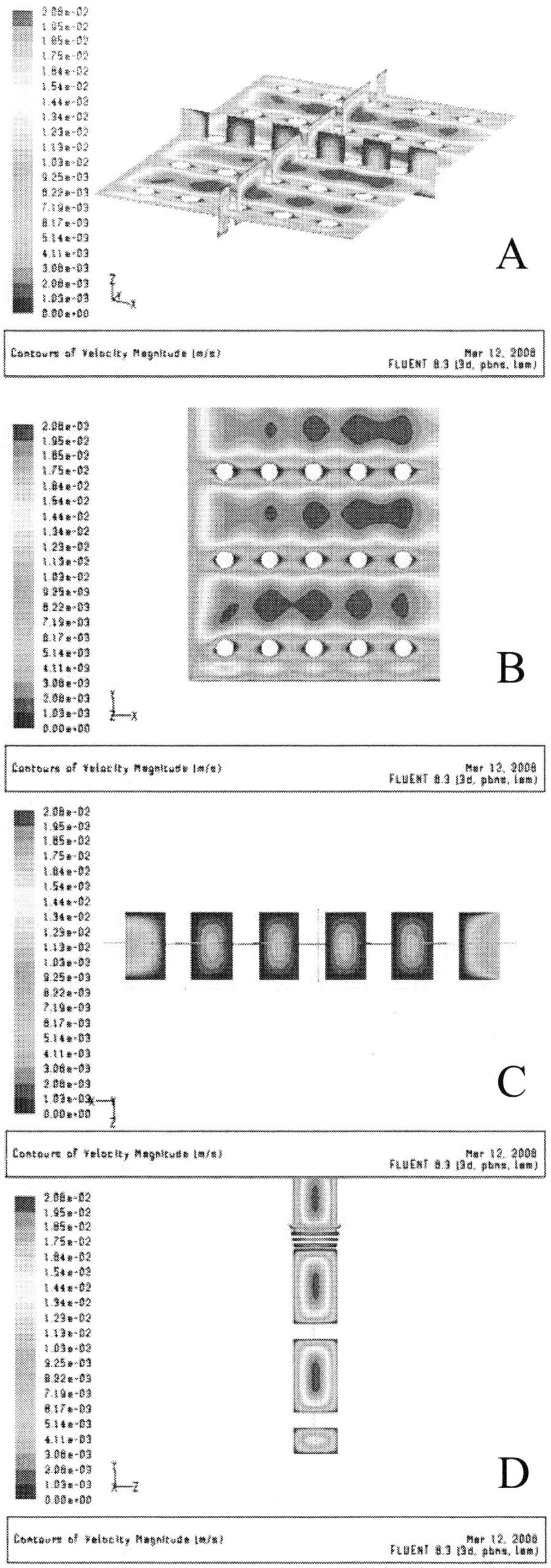

Fig. 3. Simulation results showing the flow velocity field in a Carbon electrode array for a 10 μl/min flow. 3D model (A), X – Y (B), X – Z (C) and Y –Z (D) cross sections.

5 EXPERIMENTAL VALIDATION AND APPLICATIONS

Fig. 4 shows the trapping of yeast against a 10 μl/min flow flowing from the left. It can be clearly seen how yeast is trapped in those areas predicted by the electromagnetic analysis.

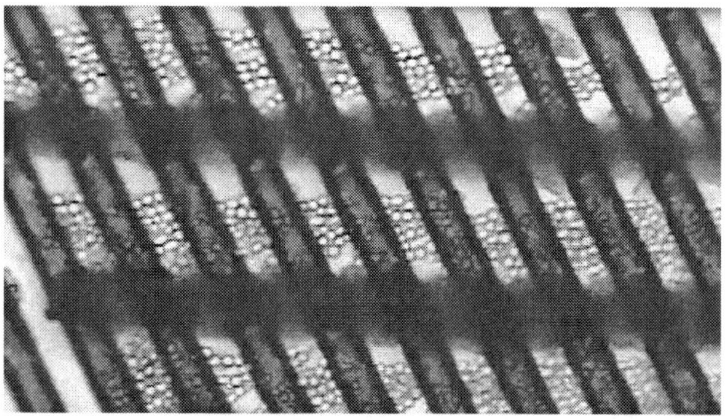

Fig. 4. Yeast trapped in high electric field gradient areas in a CarbonDEP array as predicted by electromagnetic analysis (Fig. 1 A and B). Optical Image focused on channel bottom (compare to Fig. 1A). Dark lines represent tightly trapped yeast which are out of focus (Compare to Fig. 1B).

5.1 Filter

Fig. 5 shows the filtering of viable yeast, by DEP trapping to carbon electrode surface, from non viable yeast at a 10 μl/min flow. Experimental details and Filter efficiencies for different flow rates are described elsewhere [ ]. The correlation of a decreasing flow velocity gradient and an increasing electric field gradient towards electrode surface improves viable yeast trapping.

Fig. 5. Viable yeast filtered from Non viable yeast against a 10 μl/min flow using a polarized CarbonDEP array at 5 MHz and 10 V_{pp}. Medium conductivity of 51 mS/m.

5.2 Continuous Enrichment

Fig. 6 shows the continuous enrichment of viable yeast by Positive Dielectrophoresis Focusing. The principle works when the hydrodynamic force overcomes the DEP trapping force. Since laminar flow is established in the channel and path lines have been shown to be minimally disturbed by our array geometry, viable yeast flushes away contained in those flow lines co-linear with the polarized electrode rows. Such principle allows continuous separation at higher flow rates than those achieved when implementing separation by trapping, such as in a filter, but requires more complicated geometries for enriched population retrieval.

Fig. 6. Continuous enrichment of viable yeast at 20 μl/min with a polarized CarbonDEP array at 1 MHz and 10 V_{pp}. Medium conductivity of 51 mS/m.

6 CONCLUSIONS

We have demonstrated how our 3D CarbonDEP array 1) achieves higher throughput by inducing positive and negative DEP regions in the bulk of the channel and not only in the vicinity of channel walls as in the case of 2D DEP systems, 2) induces an increasing electric field magnitude gradient towards the electrodes' surfaces, 3) establishes a decreasing flow velocity gradient towards the electrodes' surfaces, 4) establishes an optimal correlation between a decreasing flow velocity gradient and increasing electric field gradient toward electrodes' surfaces and 5) do not create physical traps by minimally disturbing the flow path lines. Furthermore we demonstrate such advantages in the detailed applications.

ACKNOWLEDGEMENTS

This work was supported in part by the University of California Institute for Mexico and the United States (UC MEXUS), award number UCM- 40117 .

REFERENCES

[1] C. Wang, et al, Journal of Microelectromechanical Systems, **14**, 2, 348, 2005.

[2] R. Martinez-Duarte, et al, Proceedings of the 11th International Conference in Miniaturized Systems for Health and Life Systems: microTAS 2007, Paris, France, **1**, 828, 2007.

[3] B.Y. Park, M.J. Madou, Electrophoresis, **26,** 3745, 2005.

[4] B.Y. Park, M. Madou, Journal of the Electrochemical Society, **152**, 12, J136, 2005.

Generation of picoliter and nanoliter drops on demand in a microfluidic chip

Jie Xu and Daniel Attinger

Laboratory for Microscale Transport Phenomena, Department of Mechanical Engineering, Columbia University, New York, NY 10027, USA, da2203@columbia.edu

ABSTRACT

In this work, we introduce the novel technique of *in-chip drop on demand*, which consists in dispensing picoliter to nanoliter drops on demand directly in the liquid-filled channels of a polymer microfluidic chip, at frequencies up to 2.5 kHz and with precise volume control. The technique involves a PDMS chip with one or several microliter-size chambers driven by piezoelectric actuators. In this article, the drop formation process is characterized with respect to critical dispense parameters such as the shape and duration of the driving pulse, and the size of both the fluid chamber and the nozzle. Several features of the in-chip drop on demand technique with direct relevance to lab on a chip applications are presented and discussed, such as the precise control of the dispensed volume, the ability to merge drops of different reagents and the ability to move a drop from the shooting area to a contraction.

Keywords: microfluidics, drop on demand, ink-jet

1 INTRODUCTION

The concept of lab on a chip, where a tiny fluid microprocessor performs complex analysis and synthesis tasks relevant to chemistry or biology has drawn great interest since the early 1990's [1, 2]. Several achievements have demonstrated that shrinking a chemical or biological laboratory into a microchip could have significant benefits such as increased sensitivity, fast response time, low reagent and sample consumptions, as reviewed in [2-5]. The ability to dispense and control small liquid volumes in the microchannels is critical for the lab on a chip technology, and several techniques address this issue. For instance, pinching by electrokinetic process [6] or the volumetric change induced by a piezoelectric actuator [7] have been used to inject individual liquid plugs in a miscible liquid. Also the segmented flow technique has been developed: it uses syringe pumps feeding two branches of a T-connection [8-10] or two concentric channels [11, 12] and forms a train of drops inside microchannels. The segmented flow has the advantage over the electrokinetic pinching that the diffusion between the minute amount of liquid dispensed and the carrying fluid is minimal. Both methods are suited for performing complex, multistep analysis or synthesis [13, 14]. However, the ability to dispense and manipulate a single particle rather than a train of particles would be welcome in order to handle expensive reagents or analyze individual biological cells. Indeed, George Whitesides, the Harvard professor who fathered soft microfluidics, mentioned this need in 2007. Cited in Nature Methods [15] he describes the use of bubbles in microfluidic chips, and concludes: "There is a particular bit of the puzzle that needs to be added, which will not be hard to do but it has not been done yet—that is, bubble on demand". While the well-established ink-jet technology dispenses individual drops, on demand, *into the atmosphere* [16], few processes are able to generate particles on demand *into a microfluidic chip*: individual drops can be generated using voltages of 1kV producing a Taylor cone at a water-oil interface with subsequent breakup [17]; also, the back-and-forth motion of an interface can be carefully controlled with a syringe pump to generate a single drop [18]; further, a single bubble has been generated by thermally expanding gas in a chamber until the protruding meniscus is broken into a bubble by a shear flow [19]. While successful at generating a single particle, these techniques rely on complex external actuation and the accuracy of the timing and generated volume was either low [17, 18] or not quantified [19]. In this paper, we present a technique to precisely dispense a single drop on demand in a microfluidic chip with a local piezoelectric actuator.

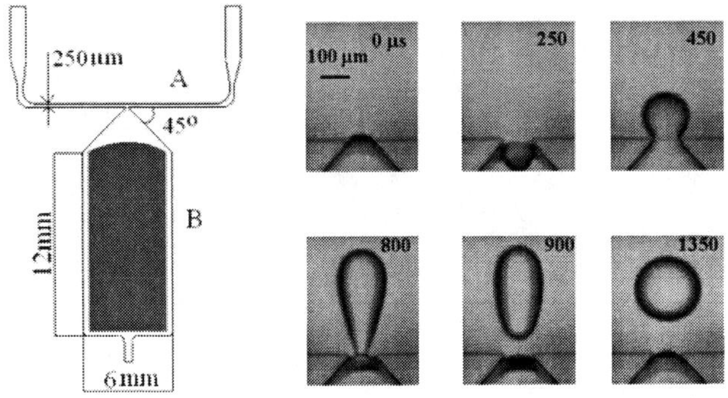

Figure 1: (a) A typical chip geometry consisting of oil-filled channel *A* and water-filled chamber *B*. A piezoelectric bimorph actuator is glued to the chamber *B* allows the release of an aqueous drop on demand in the channel A. (b) Stages depicting of the formation of a 1nL drop from a 50 μm nozzle.

2 DESIGN, FABRICATION AND SETUP

The typical design of a microfluidic chip used in our study is shown in Figure 1a: it involves one or several μL-volume reagent chambers such as *B* connected via a 25-100 μm nozzle to a main channel *A*. The main channel is filled with hexadecane while each chamber B can be filled with water. Since the water-hexadecane system is immiscible, a stable meniscus forms at the nozzle opening. The height of the channels is in the 50-100 μm range. The chip is sealed with a thin membrane. A piezoelectric actuator is placed on top of each chamber, to modify the chamber volume and release an aqueous drop, on demand, in the main channel. This drop generation process is shown in Figure 1b. Drop volumes as small as a few pL and the internal flow associated with the motion of drops in a channel [7] enhance mixing and diffusion.

Microfluidic chips were fabricated using soft lithography [20]. First, a 50-100 μm layer of *SU-8* is cured on top of a wafer with patterns transferred from a transparency mask. The chip is then manufactured from the master using PDMS Sylgard 184 Kit (Dow Corning). The channels are sealed by a thin 180 μm membrane made from spin-coated PDMS. The piezoelectric actuators are commercially available bimorph actuators made of two PZT layers bonded on a thin brass layer, with a total thickness T of 0.51 mm, with lengths and widths slightly smaller than the chamber dimensions as shown in Figure 1a and as given in Table 1. One actuator is then taped on top of each chamber, using a 90 μm layer of double-sided tape.

The experimental setup involves the microfluidic chip described above and a 20MHz function generator (Agilent, 33120A) coupled to a 1MHz 40W amplifier (Krohn-Hite, 7600M), which generates high-voltage driving pulses for the actuators glued on the microfluidic chip. We use Olympus IX-71 microscope and a high-speed camera (Redlake MotionXtra HG-100K) as sensing system.

Symbol	Physical property	Typical Value
γ	Surface tension at the water-hexadecane interface	52.5 mJ/m^2 [21]
d_{31}	Piezoelectric strain coefficient	190×10^{-12} Pa
Y	Piezoelectric elastic modulus	6.2×10^{10} Pa
ρ	PZT density	7750 kg/ m^3
E	Electric field applied across actuator	400 kV/m
L, B, T	Actuator length, width, thickness	12-20, 3-4, 0.51 mm

Table 1: Physical properties and typical values

3 ANALYSIS AND CHARACTERIZATION

3.1 Motion of the Actuator

An important parameter in the actuation design is the eigenfrequency of the actuator, which limits the speed of deformation. The eigenfrequency f_n of a piezoelectric bimorph with L>>w is given in [22] for two types of boundary conditions: anchored at one end as $f_n = (0.16T/L^2)\sqrt{Y_{11}/\rho}$, or anchored at one end and with the other end immobile along the z-direction, as $f_n = (0.48T/L^2)\sqrt{Y_{11}/\rho}$. While none of the boundary conditions is exactly the experimental conditions, the latter was found in better agreement to our measurements.

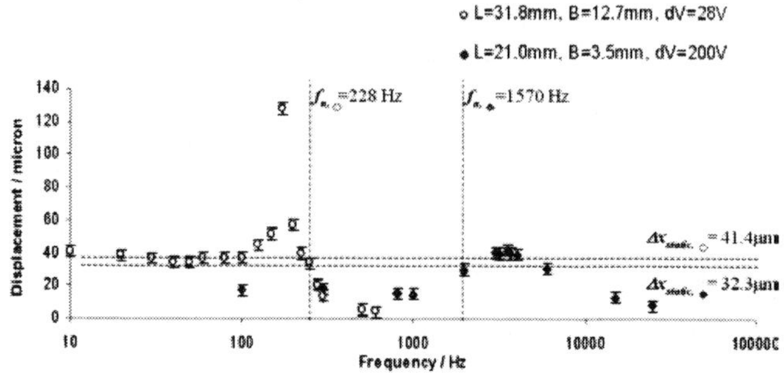

Figure 2: Influence of the excitation frequency on the amplitude of the actuator motion. The empty circles and full lozenges denote two types of boundary conditions. The dashed lines denote the theoretical values for natural frequency and static displacement.

Using an Optem long distance microscope objective and a high-speed camera, we measured the temporal deformation of an actuator driven by a single rectangular pulse of amplitude dV and duration $(2f)^{-1}$. Figure 2 summarizes these measurements, showing the maximum observed displacement as a function of the frequency f of the driving pulse. A first series of measurements, shown by empty circles, is made for a relatively large bimorph clamped at one end, with dimensions given in Figure 2. Theoretical values are also plotted as dashed lines for both the maximum static displacement and the eigenfrequency. The agreement is relatively good in terms of resonance frequency and static (low frequency) displacement. The lower resonance frequency observed experimentally can be explained by the difficulty to experimentally anchor one end of the actuator in a perfect way because we used a C-clamp. A second series of measurements is made with a smaller actuator attached via double-sided tape to a 180 μm thin PDMS layer. The two ends of the PDMS layer are then anchored firmly between two C-clamps, each clamp being about 1.5 mm away from the corresponding end of the piezo. While both the actuator size and configuration are close to the design of the microfluidic chip, the configuration is close but not exactly corresponding to the second type of boundary condition presented above: this might explain why the measured static displacement and resonance frequencies are different, both being larger than the theoretical values. Importantly, the visualization shows that the actuator does not freeze its motion once the driving pulse vanishes, but keeps oscillating at its natural frequency for about 6 periods. This behavior due to the relatively large size and inertia of the actuator, and the very soft, thin PDMS sealing layer contrasts with existing piezoelectric drop on demand dispensers and the relative modelings [23-26], where the chamber walls are much stiffer, typically made of glass [23] or silicon [27].

3.2 Effects of Driving Pulse on the Drop Formation

Characterization experiments reported in Figure 3 describe how the drop volume is influenced by the nozzle size, the pulse shape and the pulse duration. All the data in Figure 3 was obtained with an actuation voltage of +/- 200V. The chamber lengths used for the respective 50 and 100 μm nozzle case were 12mm and 20mm, respectively. The shape of the pulse corresponds to an initial expansion of the chamber for a time t_1 followed by compression for a time t_2. Pulses with $t_1=0$ were also successful at generating a drop: they correspond to simple compression of the chamber. Figure 3 shows that drops with volumes from 25 pL to 4.5 nL can be generated by varying the pulse shape and the nozzle size. For a given nozzle geometry, the drop volume can be controlled by the pulse shape within one order of magnitude. Pulses with durations too different from an optimum duration will not produce any drop. For the 50 μm nozzle we also observe some dual-dispense states where two smaller drops are simultaneously produced, by the *Mickey Mouse instability* process described in section 4 and shown in Figure 4d. Also, in Figure 3, crosses demonstrate how the drop volume can be controlled by changing the ratio between the expansion time and the compression time, while keeping the total actuation time constant.

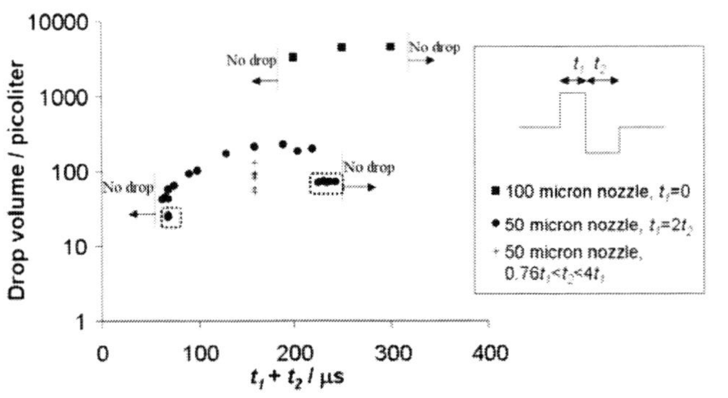

Figure 3: Drop volumes as a function of dispense parameters. The horizontal axis denotes the total pulse length, which involves the chamber expansion followed by the chamber compression, with respective duration t_1 and t_2. The channel height is the same as the nozzle width. The dotted rectangles show *Mickey Mouse* dispenses.

The results in Figure 2 and Figure 3 suggest that the optimum pulse duration to produce drops corresponds to the natural frequency f_n of the actuator. Indeed, the second equation presented in section 3.1 predicts values of f_n of 1.57 kHz, respectively 4.74 kHz for the actuators of the chambers with respectively the 100 μm and 50 μm nozzle. The corresponding single pulse duration for the 100 μm nozzle would be $t_2=1/(2 f_n)$=318 μs, a time close to the 180-300 μs interval effective at producing drops in Figure 3. Similarly, the corresponding total pulse duration for the 50 μm nozzle, in the case of a pulse with $t_1=t_2$, would be $t_1+t_2=1/f_n=211$ *μs*, a time close to the 60-220 μs interval effective at producing drops in Figure 3. The fact that the pulse duration estimated theoretically is at the higher end of the interval of experimentally successful durations might simply indicate that the actual value of f_n is slightly higher than the theoretical value, a fact shown in Figure 2.

We also tried to quantify the maximum dispense rate by repeating the driving pulse with smaller and smaller time interval between pulses. Experiment shows that drops are still generated even if the time interval is reduced to 0s, which corresponds to applying the generation pulse continuously. For example, in a case where a 400 μs pulse, shaped as in Figure 3 with $t_2 = 2t_1$, is applied continuously to the piezoactuator and a train of drops were successfully generated at 2.5 kHz. In addition, we studied the uniformity of the drop volumes. At a dispense rate of 6.2 Hz, 20 drops generated had an average volume of 1023 pL and a standard deviation of 16 pL, which corresponds to less than 2%.

4 FEATURES RELATED TO LAB ON A CHIP

The in-chip drop on demand technique has the potential to perform in-chip reagent mixing, transport and multistep reactions. Three features of the *in-chip drop on demand* technique with direct relevance to lab on a chip applications are presented in Figure 4.

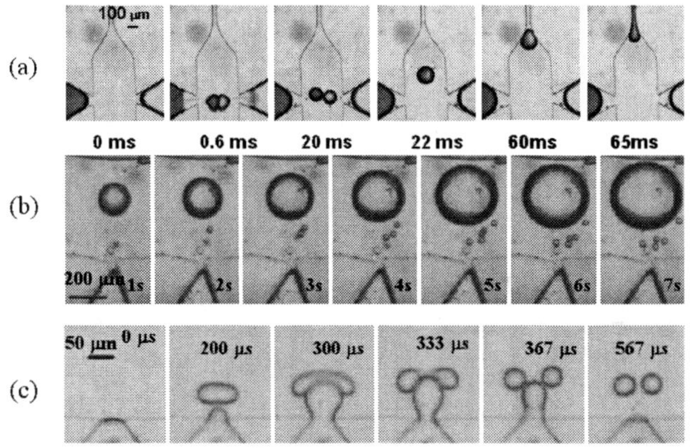

Figure 4: Three features of in-chip drop on demand with relevance to lab on a chip applications: (a) merging and mixing of two different reagents and transport to a contraction (b) digital control of drop volume, and (c) *Mickey Mouse* dispense, where two drops of small volume are generated simultaneously by a single pulse.

Figure 4a shows the ability of mixing different reagents into a single drop. The nozzle on the left generates a drop of ink while the right nozzle generates a pure water drop. Coalescence occurs and then the drop is transported into a contraction which can enhance the mixing inside the drop. This feature enables precise-controlled reactions in drops.

Figure 4b shows the second feature, which is the ability to digitally control the dispensed drop volume, by generating additional drops that coalesce with the original drop. The first frame shows a 500 pL drop whose volume is increased to 3.5 nL by 6 successive increments of 500 pL. This coarse, digital way to control the drop volume can be coupled with the finer, analog volume control by modifying the pulse parameters in order to exactly dispense the desired quantities over a wide range of volumes.

Figure 4c show the third feature, which is the ability to generate a dual-state drops, while applying a single excitation pulse to the actuator. This occurs when an initially generated drop is hit by a strong subsequent excursion of the meniscus. During the process, the meniscus breaks the initial drop into two half drops while briefly assuming the shape of Mickey Mouse, the well-known cartoon character (367 μs). Therefore, we call this type of dispense the *Mickey Mouse* dispense.

5 CONCLUSION AND OUTLOOK

The *in-chip drop on demand* technique presented in this article allows the individual generation of drops of aqueous reagents in a microfluidic chip with a temporal precision of one millisecond, and at rates higher than one kHz. The ability to precisely trigger the drop generation time will allow the coordination of the generation of drops with events occurring in the chip, such as the detection of chemical reaction or temperature changes, or the transit of biological cells and other particles. The drop volume can be controlled from 40 pL to 4.5 nL by varying the pulse shape, the chip geometry, or by merging several drops together.

Acknowledgements:

This work has been supported partially by NSF grants 0449269 and 0701729.

REFERENCES

1. Manz, A., et al., Advances in Chromatography, 1993. **33**: p. 1-66.
2. Lion, N., et al., Electrophoresis, 2003. **24**(21): p. 3533-3562.
3. Verpoorte, E. and N.F. De Rooij, Proceedings of the Ieee, 2003. **91**(6): p. 930-953.
4. Stone, H.A., A.D. Stroock, and A. Ajdari, Annual Review of Fluid Mechanics, 2004. **36**: p. 381-411.
5. Whitesides, G.M., Nature, 2006. **442**(7101): p. 368-373.
6. Bai, X., et al., Lab Chip, 2002(2): p. 45-49.
7. Guenther, A., et al., Lab Chip, 2004. **4**: p. 278-286.
8. Song, H., D.L. Chen, and R.F. Ismagilov, Angewandte Chemie-International Edition, 2006. **45**(44): p. 7336-7356.
9. Gunther, A., et al., Langmuir, 2005. **21**(4): p. 1547-1555.
10. Thorsen, T., et al., Physical Review Letters, 2001. **86**(18): p. 4163-4166.
11. Jensen, M.J., H.A. Stone, and H. Bruus, Physics of Fluids, 2006. **18**(7).
12. Anna, S., N. Bontoux, and H. Stone, App. Phys. Lett., 2003. **82**(3): p. 364.
13. Fair, R., Microfluid Nanofluid, 2007. **3**(245-281).
14. Ismagilov, R.F., et al., Faseb Journal, 2007. **21**(5): p. A42-A42.
15. Blow, N., Nature Methods, 2007. **4**: p. 665-668.
16. Le, H., JOURNAL OF IMAGING SCIENCE AND TECHNOLOGY, 1998. **42**(1): p. 46-92.
17. He, M., J. Kuo, and D. Chiu, App. Phys. Lett., 2005. **87**: p. 031916.
18. He, M., et al., Anal. Chem., 2005. **77**: p. 1539-1544.
19. Prakash, M. and N. Gershenfeld, Science, 2007. **315**: p. 832-835.
20. Xia, Y.N. and G.M. Whitesides, Annual Review of Materials Science, 1998. **28**: p. 153-184.
21. Lee, S., D. Kim, and D. Needham, Langmuir, 2001. **17**: p. 5537–5543.
22. Smits, J. and W. Choi, IEEE Transactions Ultrasonics Ferroelectrics Frequency Control, 1991(38): p. 256-270.
23. Dijksman, J.F., Journal of Fluid Mechanics, 1984. **139**: p. 173-191.
24. Bogy, D.B. and F.E. Talke, ibm journal research and development, 1984. **28**(3): p. 314-321.
25. Hayes, D.J., D.B. Wallace, and M.T. Boldman. *ISHM Symposium 92 Proceedings*. 1992.
26. Wallace, D.B., ASME publication 89-WA/FE-4, 1989.
27. Nguyen, N. and S. Wereley, *Fundamentals of Microfluidics*. 2002: Artech House.

Hollow Atomic Force Microscopy Probes for Nanoscale Dispensing of Liquids

André Meister, Joanna Przybylska, Philippe Niedermann, Christian Santschi, and Harry Heinzelmann

CSEM Swiss Center for Electronics and Microtechnology
2002 Neuchâtel, Switzerland, andre.meister@csem.ch

ABSTRACT

To enable the printing of nanometer sized droplets with volumes in the femto and attoliter range and sub-micron droplet spacing, a nanoscale dispenser (NADIS) based on an atomic force microscopy probe has been developed and microfabricated. The probe consists of a cantilever with a hollow core, which is connected to a reservoir located in the chip. The hollow cantilever acts as a microfluidic channel that connects the reservoir to the dispensing tip located at the free end of the cantilever. The tip possesses an opening at its apex with a typical size of 200 nm, realized by focus ion beam milling. The transfer of liquid from the tip opening to the surface occurs by contacting the probe tip with the sample surface and is driven by capillary pressure alone. To overcome the serial manner of writing by using a single probe, arrays of hollow AFM probes were fabricated. The feasibility to dispense droplets in parallel has been demonstrated.

Keywords: AFM, nanopatterning, microfluidic spotting, microarray

1 INTRODUCTION

The demand for specific tools intended for the deposition of small amount of material on a nanoscale at predefined locations is continuously increasing. Several dispensing tools for liquids, such as ink-jet or pin-spotting heads have been developed. They are currently widely used for various applications, such as for writing microarrays (or biochips) used in proteomics to determine the presence and/or amount of proteins in biological samples. However, such printing tools have some limitation regarding the volume of deposited material (sub-picoliters and above) [1] and the dimension of the spotted dots (typically 10 to 100 μm). Decreasing both the volume of the deposited liquid and the spot size is an important issue regarding an economic use of the dispensed liquid. Furthermore, microarrays with a higher spot-density need less biological analyte to perform an affinity assay.

Nanoscale dispensing (NADIS) is a versatile method developed to deposit small amounts of material on a substrate. The method is based on the atomic force microscope (AFM) technology. In NADIS, the dispensing method uses specifically fabricated probes. Like AFM probes, the NADIS probes are made of a flexible cantilever with a sharp tip. The tip is hollow, and a small aperture is located at the apex of the tip. The volume inside the hollow tip is filled with liquid. Due to capillarity, the liquid doesn't flow through the aperture by gravity alone. Once a NADIS probe in brought in contact with the sample, the liquid will wet the sample surface, and after withdrawal of the probe, a small volume of liquid will remain on the sample (see Figure 1). The volume of the remaining droplets can be as small as a few tens of zeptoliters (1e-21 liter), depending on several parameters such as tip aperture size [2]. Since the NADIS probe is driven by a standard AFM instrument, the droplets can be dispensed with a high spatial accuracy. The NADIS technology allows thus to deposit on a predefined location on-demand single ultrasmall droplets.

Figure 1 Conceptual sketch of nanoscale dispensing (NADIS). A specifically designed AFM probe containing liquid in its apertured tip is brought in contact with the sample surface. During the withdrawal of the probe, a small droplet remains on the sample surface.

In the first generation NADIS probes described above, the liquid was loaded directly on the cantilever. Due to the small amount of loaded liquid, in ambient environment the dispensing was limited to liquid with low volatility, such as glycerol or tetraethylene glycol. If such a first generation NADIS probe is loaded with water, the water evaporates within a few seconds. However, the aim is to use NADIS for a large variety of liquids, especially like water-based solution used as buffer for biological molecules.

In order to overcome this evaporation, a new generation of NADIS probes has been developed and microfabricated. The second generation NADIS probes are made of a hollow cantilever, with large reservoir located in the chip body. This allows first an easier loading due to the larger size of the reservoir, and second permits also to load a larger volume. The hollow core inside the cantilever acts as microfluidic channel connecting the reservoir to the tip. A schematic sketch is shown in Figure 2. The location of the reservoir on the chip rather than on the cantilever enables furthermore the integration of a microfluidic system connected to the chip.

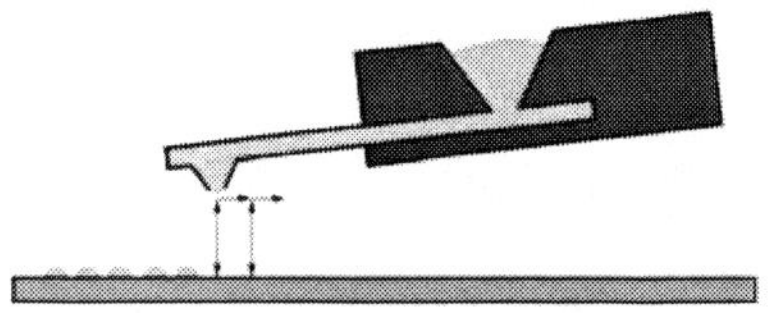

Figure 2: In the second generation of NADIS probe, the reservoir is located in the chip body, and the liquid is driven towards the tip by a microfluidic channel inside the cantilever (sketch not to scale).

The applications of the NADIS probes are various, going from the patterning of a surface with biomolecules or nanoparticles suspended in the dispensed liquid [3], to the local modification of responsive surfaces induced by the presence of the liquid [4].

2 MICROFABRICATION

2.1 Microfabrication process

Whereas the first generation NADIS probes were based on a commercially available cantilever, the second generation NADIS probes are entirely microfabricated using CSEM facilities. The microfabrication relies on the thermal fusion bonding of two pre-processed silicon (Si) wafers. Such a process allows the fabrication of hollow structures without using a sacrificial layer. The process is depicted in Figure 3.

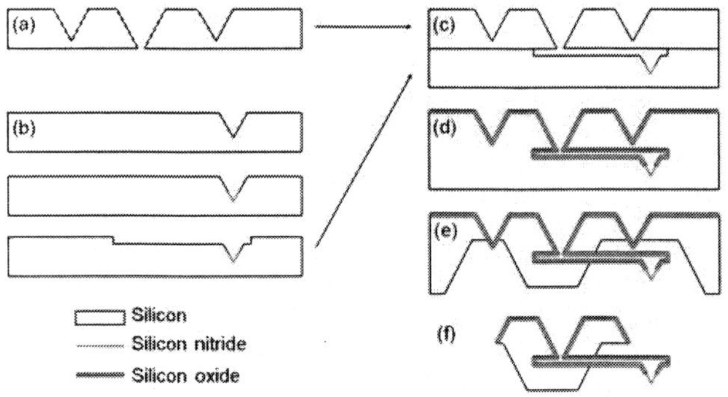

Figure 3: Process flow for the microfabrication of NADIS probes with hollow cantilevers.

(a) A first Si wafer is structured by standard photolithography and anisotropic KOH etching in order to create both, the reservoirs and a rectangular shaped V-groove that defines the future chip. (b) A second silicon wafer is processed to etch the pyramidal shaped pits by wet etching. A 100 nm thin silicon nitride (Si_3N_4) layer is than deposited on the wafer, and structured by dry etching in order to remain only at the bottom of the pits. The microfluidic channels are processed by dry etching. (c) The two pre-structured wafers are cleaned, brought together and aligned. (d) The two wafers are bonded by thermal fusion bonding and a 1 μm silicon dioxide (SiO_2) layer is grown by thermal oxidation on the exposed Si surfaces. (e) The hollow cantilever is released by wet etching. The depth of the etching is adjusted to release partly the peripheral V-groove. To perform the wet etching, the wafer was placed in a single-side etching chuck so that the cantilever side with the reservoirs was not exposed to the KOH. This avoids also an exposition of the microfluidic channels to the etching liquid. (f) The chip is extracted from the finished wafer. The released probes were then coated with a metallic layer (Au + Cr as adhesion layer) on both cantilever side. The back side metallization increases the reflectivity for the AFM laser beam (used to detect the cantilever deflection), whereas the tip side metallization avoids charging effects during the next and final step. Finally, the tip aperture is opened using focused ion beam (FIB) milling.

The hollow cantilever is made of SiO_2, and the tip apex is made of Si_3N_4. The presence of Si_3N_4 at the tip is motivated by the aim to deposit ultrasmall droplets, which requires first a small tip aperture, and thus a thin tip wall (the Si_3N_4 layer is much thinner than the SiO_2 layer), and second a sharp tip apex. A tip made by a SiO_2 layer grown thermally would present a blunt tip apex, which would increase the wetted area during the deposition of the droplets (see Figure 4).

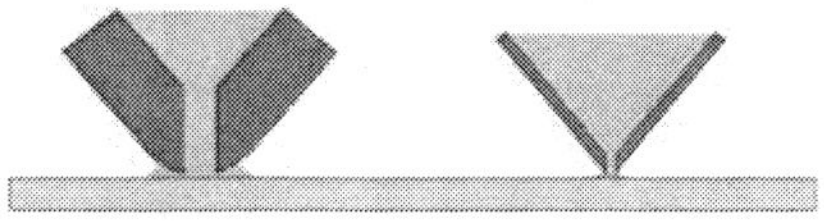

Figure 4: Compared to a thick and blunt SiO_2 tip apex, a thin and sharp Si_3N_4 tip apex is better suited for the deposition of ultrasmall droplets.

2.2 Design

Several types of probes were fabricated, including cantilevers of different lengths, single- and double-beam cantilevers, single cantilevers and one-dimensional cantilever arrays.

Cantilevers with lengths up to 500 μm were designed. The longer hollow cantilevers, despite their area moment of inertia which is much larger than for plane cantilevers, have an estimated spring constant similar to a contact-mode plain cantilever.

A NADIS probe with a single-beam cantilever is connected to one reservoir. But NADIS probes having a U-shaped two-beam cantilever structure, with one beam connected to an inlet reservoir and the other beam to an outlet reservoir, were also designed. Such double-beam cantilever structure would allow to rinse them by flushing a liquid from the inlet reservoir to the outlet reservoir.

And finally, NADIS probe arrays have been designed. Single- and double-beam levers were considered for the NADIS cantilever arrays. All cantilevers in an array can be connected to one (single-beam cantilever) or two (double-beam cantilever) reservoirs, so that all dispense the same liquid. An example of such a design is shown in Figure 5. To allow the delivery of a different liquid from each cantilever within one array, each cantilever has to be connected to specific reservoirs via independent fluidic networks, one for each cantilever. Each single-beam cantilever is connected to its own reservoir, and each double-beam cantilever is connected to two own reservoirs, the inlet reservoir and the outlet reservoir. The number of reservoir integrated on the chip is therefore directly proportional to the number of cantilever, and has thus a direct influence on the chip size (see Figure 6).

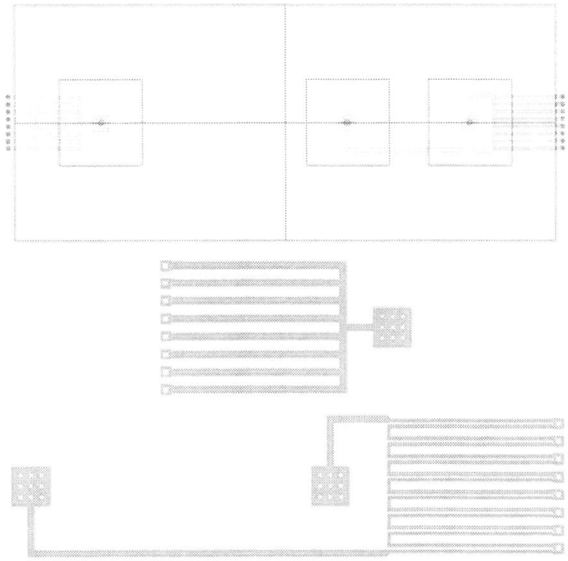

Figure 5: Top: chip design containing two 8x1 arrays of NADIS probes, one with 8 single-beam cantilevers connected to one reservoir (left), and one with 8 double-beam cantilevers connected to an inlet and one outlet reservoir (right). The chip size is 1.6 mm by 3.8 mm. Bottom: detail of both microfluidic networks used in the chip depicted at the top.

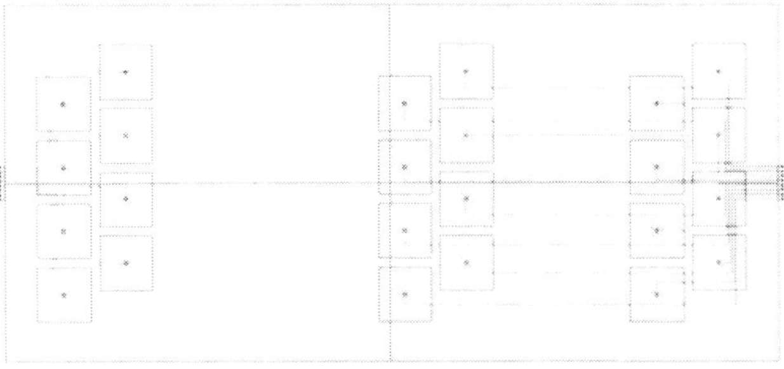

Figure 6: Chip design containing two 8x1 arrays of NADIS probes, one with 8 single-beam cantilevers (left), and one with 8 double-beam cantilevers (right), with a dedicated microfluidic network for each cantilever. The chip size is 3.8 mm by 8.5 mm.

3 RESULTS AND DISCUSSION

3.1 Microfabricated probes

Figure 7 represents scanning electron microscope (SEM) micrographs of the wafer containing the reservoirs and the V-grooves before the thermal fusion bonding. This wafer corresponds to the step (a) in Figure 3. Some examples of microfabricated probe arrays are shown in Figure 8, and Figure 9 depicts tip apertures made by FIB milling.

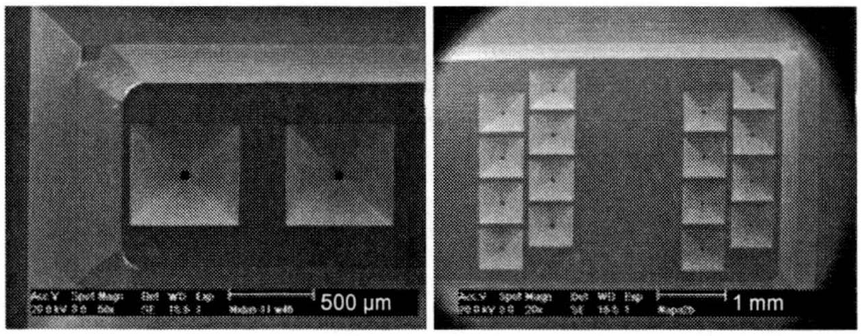

Figure 7: SEM micrographs of the wafer containing the reservoirs.

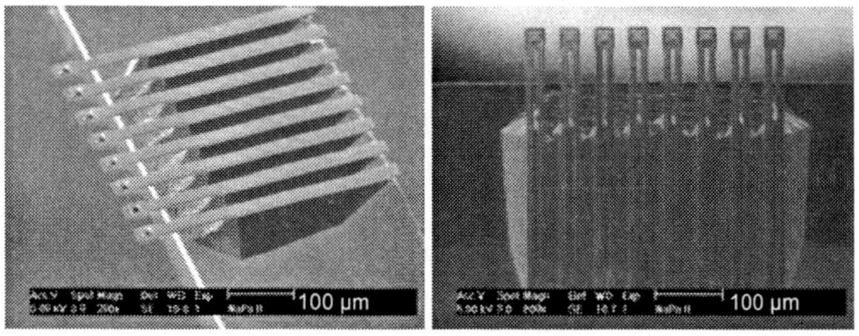

Figure 8: NADIS array with short cantilevers (SEM micrographs).

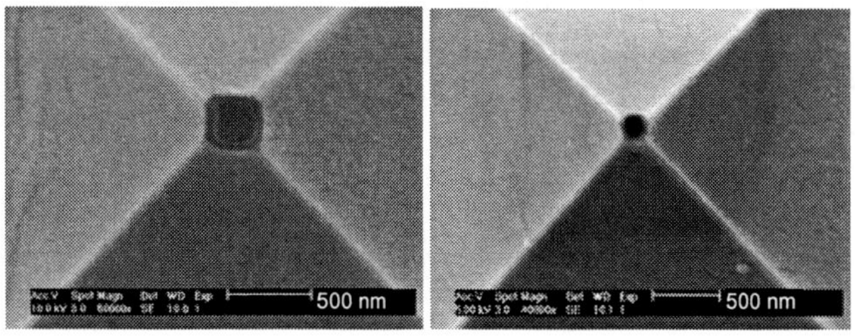

Figure 9: SEM micrographs of two different apertures milled by focused ion beam at the Si_3N_4 tip apex.

During microfabrication, one observation made was that during the drying following the last rinsing, capillary forces bent the long and flexible cantilevers towards the wafer with exceeding forces, resulting in a rupture of the longer cantilever. This phenomenon occurred as well for the single cantilevers as for the cantilever arrays. Currently, investigations are ongoing to avoid this problem. One of the investigated ways was to replace the last wet etching with dry etching, which turned out to be successful for the release (see Figure 10). Unfortunately, the dry etching removed also the thin Si_3N_4 tip. However, a correction to eliminate this problem has been implemented in the new ongoing microfabrication run.

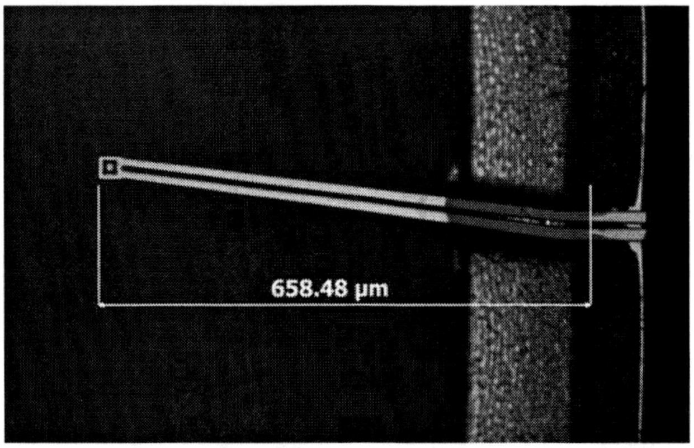

Figure 10: Optical image of a long cantilever released by dry etching. The dimension corresponds to the release length.

3.2 Experimental results

All nanoscale dispensing experiments were performed in ambient environment on standard AFMs (Veeco, Digital Instruments, Dimension 3100 and MultiMode).

The filling of the hollow cantilever was characterized by measuring the cantilever resonance frequency. A filled cantilever, due to the presence of the additional mass of the liquid, has a resonance frequency which is shifted to a lower value compared to the empty cantilever.

The transfer of liquid is characterized by measuring two successive deflection-distance curves. During the first probe-sample contact, a droplet is deposited on the surface. In the deflection-distance curve describing the second contact, a snap-in position far from the surface is recognizable in the extending curve. The snap-in is due to the fact that, when the tip enters in contact with the previously deposited droplet, a capillary force bends the cantilever towards the sample surface. The distance between the snap-in position and the sample surface corresponds approximatively to the height of the droplet deposited during the first contact. The presence of such snap-in positions describes thus a successful transfer of liquid. An example with glycerol is given in Figure 11. However, such characterization could not be demonstrated with water, mainly due to the almost instantaneous evaporation of the deposited droplet.

Examples of depositions using the hollow NADIS cantilevers are depicted in Figure 12. For those experiments, NADIS probes released by dry etching, i.e. without Si_3N_4 tips, were used, resulting in droplets with large dimensions due to the large probe aperture. The difference in droplet dimensions between both pictures is due to a difference in sample surface wettability (hydrophilic on the left hand side, and hydrophobic on the right hand side)

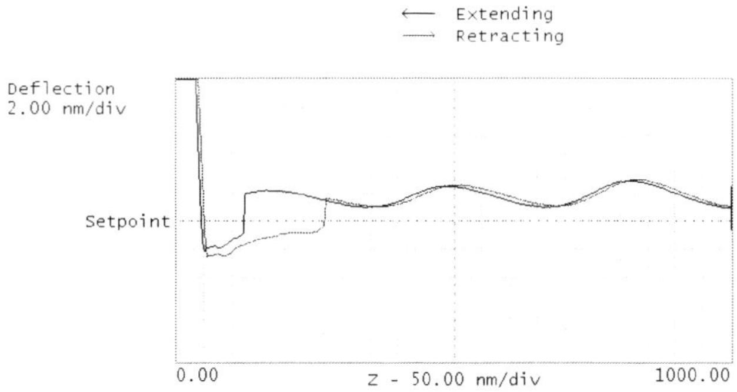

Figure 11: Typical deflection-distance curve that characterizes the transfer of liquid. Note the snap-in position at ~70 nm from the sample surface

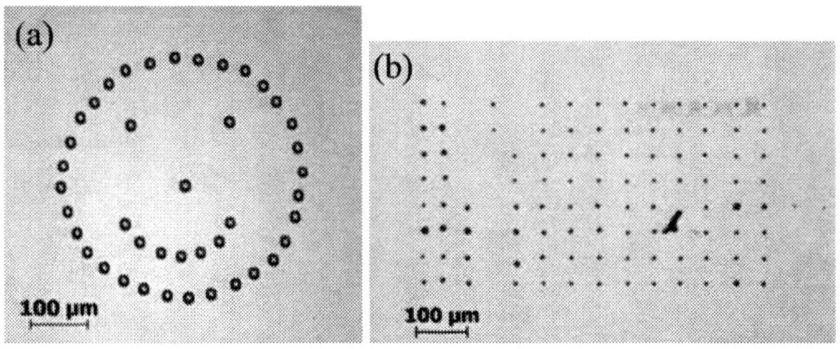

Figure 12: Optical images of an array of dispensed glycerol droplets (a) deposited with a single probe, (b) deposited with a probe array of 8 NADIS cantilever

4 ACKNOWLEDGMENTS

The authors would like to thank the technical staff of COMLAB, the joint microfabrication facility of CSEM and University of Neuchâtel. The partial support of the Swiss Federal Office for Education and Science (OFES) in the framework of the EC-funded project NaPa (Contract no.NMP4-CT-2003-500120) is gratefully acknowledged.

REFERENCES

[1] U. Demirci, G.G. Yaralioglu, E. Hæggström and B.T. Khuri-Yakub, IEEE Transactions on Semicond. Manufac., 18, 709, 2005.

[2] A. Fang, E. Dujardin and T. Ondarçuhu, Nanolett., 6, 2368, 2006.

[3] A. Meister, M. Liley, J. Brugger, R. Pugin and H. Heinzelmann, Appl. Phys. Lett., 85, 6260, 2004.

[4] A. Meister, S. Krishnamoorthy, C. Hinderling, R. Pugin and H. Heinzelmann, Microelectron. Eng., 83, 1509, 2006.

Electrostatic Induced Inkjet Printing System for Micro Patterning and Drop-On-Demand Jetting Characteristics

Jaeyong Choi*, Yong-Jae Kim**, Sang Uk Son*, Youngmin Kim*, Vu Dat Nguyen***, Sukhan Lee*, Doyoung Byun*** and Han Seo Ko**

* School of Information and Communication Engineering, Sungkyunkwan University, Suwon, Kyunggi-do, S. Korea, steinkopf@skku.edu, ssu2003@skku.edu, zmkim@skku.edu, lsh@ece.skku.ac.kr

** School of Mechanical Engineering, Sungkyunkwan University, Suwon, Kyunggi-do, S. Korea, warriorkim@skku.edu, hanseoko@yurim.skku.ac.kr

*** Department of Aerospace and Information Engineering, Konkuk University, Seoul, S. Korea, dat_bk@yahoo.com, dybyun@konkuk.ac.kr

ABSTRACT

Printing technology is a very useful method in the several process of industrial fabrication due to non-contact and fast pattern generation [1]. To make micro pattern, we investigate the electrostatic induced inkjet printing system for micro droplet generation and drop-on-demand jetting. In order to achieve the drop-on-demand micro droplet ejection by the electrostatic induced inkjet printing system, the pulsed DC voltage is supplied from 1.4 to 2.1 kV. In order to find optimal pulse conditions, we tested jetting performance for various bias and pulse voltages for drop-on-demand ejection. For investigated drop-on-demand micro pattern characteristic, conductive inkjet silver ink used. In this result, we have successful drop-on-demand operation and micro patterning. Therefore, our novel electrostatic induced inkjet head printing system will be applied industrial area comparing conventional printing technology.

Keywords: electrostatic induced inkjet printing system, drop-on-demand, pattern, micro droplet, micro dripping mode

1 INTRODUCTION

Printing technology is considered to be a key technology even in the field of electronics, materials processing, and bio applications [2]. Inkjet printing is an interesting patterning technique for electronic devices because it requires no physical mask, less environmental issue, and low fabrication cost, and provides good layer-to-layer registration. It also has the potential to reduce display manufacturing costs and enable roll-to-roll processing for flexible electronics [3]. The conventional inkjet printing systems are based on mechanically or thermally pushing out the liquid in a chamber through a nozzle by actuators, such as thermal bubble and piezoelectric actuators [4]. However, thermal bubble actuator has the heat issues when the array of nozzle make in a large area and piezoelectric actuator is difficult to make droplet smaller than nozzle size [5].

Alternatively, we focused on the EHD theory which may be the type of electrostatic induced actuator based on the direct manipulation of liquid by an electric field that appears to be more promising [6]. It has 10 jetting modes [7] that one of these, micro dipping mode, has a good characteristic to make inkjet printing system. Recently the interference effect is investigated for multi-nozzles [8] and the super-hydrophobic nozzle is fabricated and applied to electro-spray [9].

This paper presents the drop-on-demand operation by the novel electrostatic induced inkjet printing system and patterning lines by conductive nano silver ink, as an alternative to the thermal and piezoelectric print heads described above.

2 EXPERIMENT SYSTEM

Fig. 1 shows the schematic of the electrostatic induced drop-on-demand inkjet printing. That consists of glass capillary tube nozzle which size is 170 μm, electrode, and high pulse voltage power.

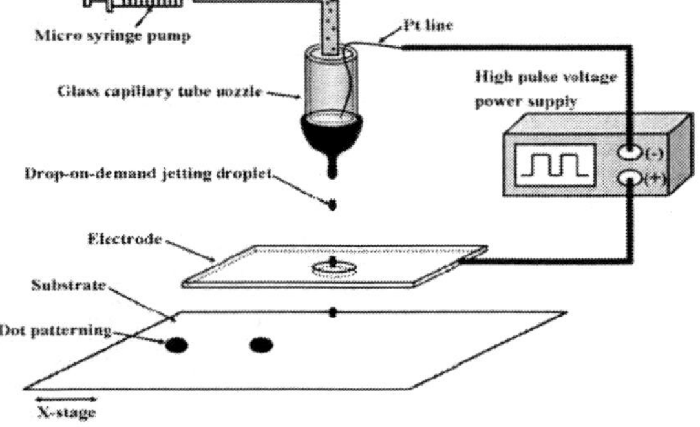

Fig. 1: The Schematic of the electrostatic induced drop-on-demand inkjet printing

The electric voltage signal applied to a ring-shaped upper electrode plate, located against the pole inside the nozzle as the ground, allows a micro-dripping mode droplet ejection to take place. The intense electric field between the electrode and the ground induces the liquid meniscus at the interface to form a micro droplet due to electrostatic force. When the force is stronger than the surface tension on the liquid meniscus, the liquid breaks up into micro droplets which are ejected. A high pulse voltage power supply (maximum voltage of 3.0kV) was used with a relay switch to control electrostatic field. Liquid was supplied by a micro-syringe pump, and an electrode was placed under the nozzle. A linear motor was used to move glass and PET substrate for forming dot and line droplet pattern.

Fig. 2(a) depicts photographs of experiment set-up and the assembly capillary inkjet head that is made of Pt-wire electrode and glass capillary tube packaged on acrylic board. Fig. 2 (b) shows the front and top view of assembly capillary inkjet head. The measurement equipments consist of a high speed camera (IDT XS-4) with a micro-zoom lens and a LED lamp was used to visualize droplet ejection. The high speed camera can image 8000 frames in a second at a 160 x 1280 resolution. Fig. 2(c) shows the fabrication procedure of electrode which extracts the liquid meniscus. The hole is made by sand blaster after deposition of Al layer on the glass wafer.

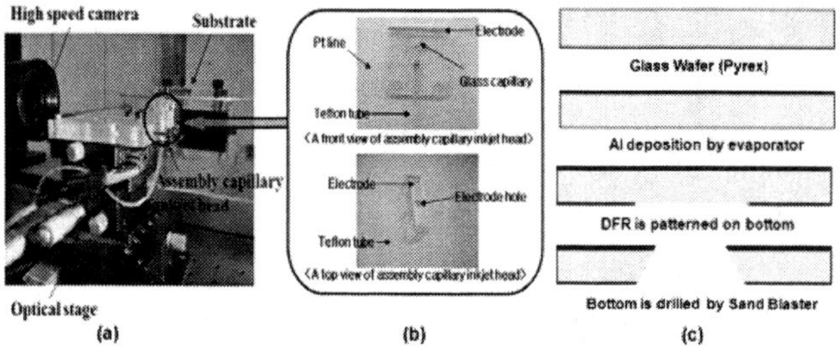

Fig. 2: The photograph of (a) experimental set-up, (b) assembly capillary inkjet head and (c) the electrode of fabrication procedure

3 RESULT

It is important to observe and distinguish dripping mode, which represents jetting drop by drop after swelling on the nozzle tip due to the absence of an electric field and micro dripping mode, where a mono-disperse droplet is formed directly at the meniscus apex, changing the operating voltage. In general, micro dipping mode is appeared under the higher voltage than that for dripping mode.

Fig. 3 shows the images of droplet ejection observed by high-speed camera at various supplied DC voltages of 1.4~2.1 kV. The flow rate of 1 μl/min is supplied, and the gap between the capillary nozzle and the electrode is 2.0 mm. The inner and outer diameters of the glass capillary are 100 μm and 170 μm, respectively.

In the dripping mode at 1.4 kV, the droplet size is bigger than the nozzle size. However when the operating voltage increases above 1.5 kV, the micro dripping mode comes to make tiny droplet.

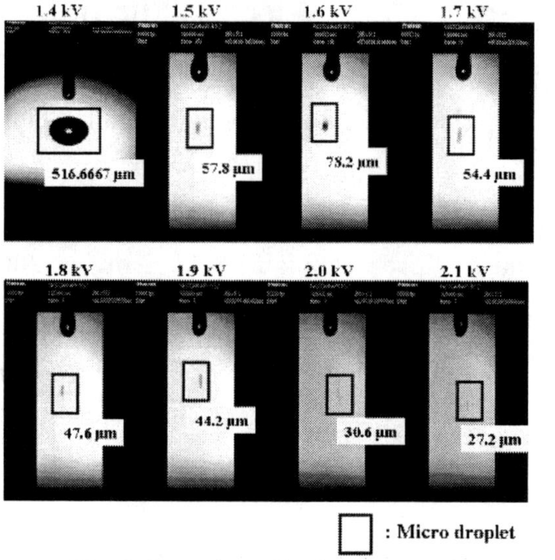

Fig. 3: Snap shot of jetting image using the conductive nano silver ink

We could find that the optimal operating voltage is 2.1kV for the smallest droplet size. Also, droplet ejection is able to take place at frequencies ranged from a few tens Hz to several thousands Hz, giving uniform droplet sizes of about 30 μm. Furthermore to investigate the optimal pulse conditions for the drop-on-demand jetting, we carried out more experiments varying the bias and pulse voltages. By means of the bias voltage of 1.4 kV, forms the shape of semicircular meniscus, and then using the pulse of 0.7 kV, micro droplet is ejected. External pulse is given to the power supply and makes 0.5 kHz and 0.7 kHz square wave high voltages.

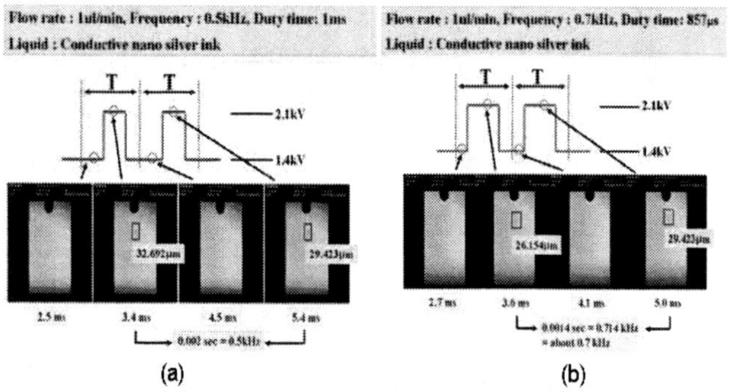

Fig. 4: The jetting images of drop-on-demand result using conductive nano silver ink; (a) 0.5 kHz and (b) 0.7 kHz

Fig. 4 shows the sequential images of drop-on-demand operation using conductive nano silver ink. From these fundamental studies to find the optimal conditions for jetting, we could carry out drop-on-demand operation using conductive nano silver ink and then try to pattern lines on a substrate.

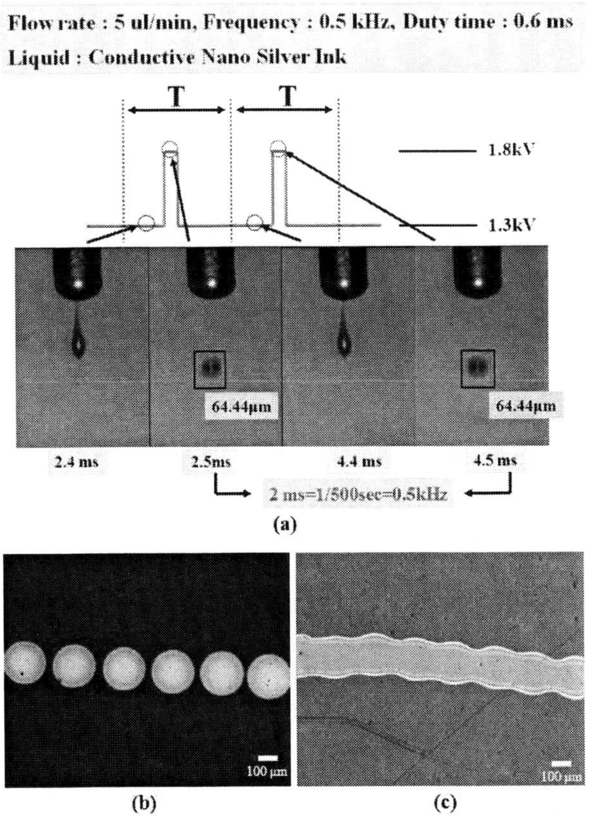

Fig. 5: (a) the jetting images of drop-on-demand (0.5 kHz), (b) the dot pattern of drop-on-demand jetting, (c) the line pattern of drop-on-demand jetting

Fig. 5 (a) shows one example of patterning of nano silver ink by means of drop-on-demand ejection at the jetting frequency of 0.5 kHz. To investigate the effect of flow rate, in figures 5 and 6, we supplied 5 μl/min and 15 μl/min, respectively. As flow rate increases, liquid is supplied fast so we can increase jetting frequency. However the optimal voltage for drop-on-demand operation is 1.8 kV with the bias voltage of 1.3 kV. Fig. 5 (b) and (c) show dots and a line patterned on the PET substrate. The patterning was carried out with drop-on-demand operation at jetting frequency of 0.5 kHz and linear motor speed of 100 mm/sec. The width of the patterned line is 205 μm with 5 μm standard deviations. Even if the diameter of the droplet is around 65 μm, the width of the line is printed as around 3 times larger. It is a good reason that the glass and PET substrate are hydrophilic surface, the droplet is able to be in spread so that size of dot or the width of the line become much larger than size of droplet. Therefore, further research for tuning the condition for tiny droplet and treating the surface of the substrate is needed in order to pattern the tiny lines using our novel electrostatic induced inkjet printing system.

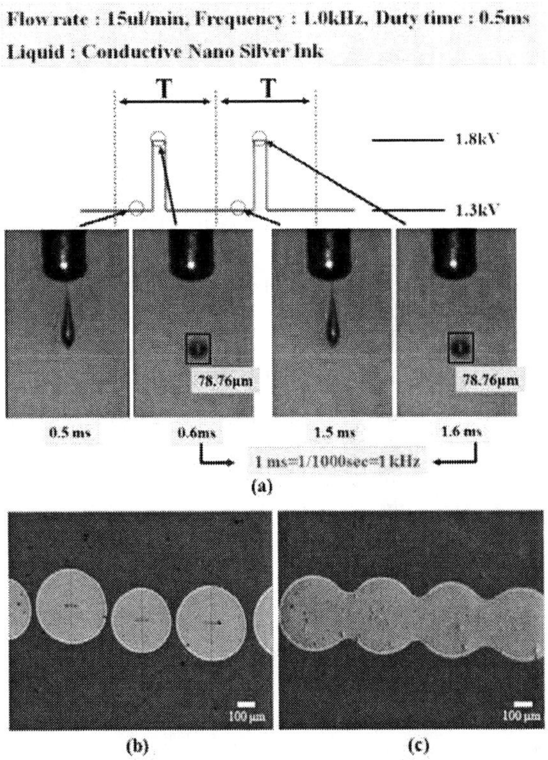

Fig. 6: (a) the jetting images of drop-on-demand (1 kHz), (b) the dot pattern of drop-on-demand jetting, (c) the line pattern of drop-on-demand jetting

As shown in Fig. 6 (a), the condition of drop-on-demand ejection is also 1.8 kV even for 15 μl/min allowing jetting frequency of 1 kHz. Fig. 6 (b) and (c) show dots and a line patterned on the PET substrate. The width of the line is around 350 μm with standard deviations of 22.54 μm. The uniformity looks good in Fig. 5 (b), while the uniformity is not good in Fig 6 (b). It may depend on the flow rate and the following instability. We need further studies associated with the reliability issues and precise control of operating conditions to enhance the uniformity.

4 CONCLUSION

We present the drop-on-demand droplet ejection and micro size pattern using our novel electrostatic induced inkjet printing system. The assembly capillary inkjet head system is composed of capillary tube and electrode, and is packaged by acrylic board. We investigated the optimal conditions of voltage, pulse, and flow rate for drop-on-demand electrostatic droplet ejection of conductive ink. To make micro-dripping mode for tiny droplet ejection, it is needed to apply 1.8 kV to 2.1 kV. Using these conditions, the dots and lines are fabricated by the inkjet system. Our novel electrostatic induced inkjet printing system may possibly be considered as one of alternatives which is able to be applied in industrial printing and to overcome the limitation of conventional print heads such as thermal bubble or piezoelectric inkjet heads which droplet size is similar nozzle size.

5 ACKNOWLEDGMENT

This work was supported by a grant from the National Research Laboratory program, Korea Science and Engineering Foundation Grant (R0A-2007-000-20012-0) and Seoul R&BD program. VDN acknowledges partial support from the Korea Research Foundation (KRF-2007-211-D00019).

REFERENCES

[1] Jin-Won Song, Joondong Kim, Yeo-Hwan Yoon, Byung-Sam Choi, Jae-Ho Kim and Chang-Soo Han, "Inkjet printing of single-walled carbon nanotubes and electrical characterization of the line pattern", Nanotechnology, Vol. 19, 095702, 2008.

[2] Y. S. Choi, D. H. Park, "Electrochemical Gene Detection Using Multielectrode Array DNA Chip", JKPS(Journal of Korean Physics Society), Vol. 44, pp. 1556-1559, 2004.

[3] Jang-Ung Park and John A. Rogers, "High-resolution electrohydrodynamic jet printing", Nature Materials, Vol. 6, pp. 782-789, 2007.

[4] Hue P. Le, "Progress and Trends in Ink-jet Printing Technology", Journal of Imaging Science and Technology, Vol. 42, pp. 49-62, 1998.

[5] Jaeyong Choi, Yong-Jae Kim, Sang Uk Son, Youngmin Kim, Sukhan Lee, Doyoung Byun and Han Seo Ko, "Pattern Characteristic by Electrostatic Drop-On-Demand Ink-jet Printing Using Capillary Inkjet Head System", The proceeding of NSTI-Nanotech 2007, pp. 403-406, 2007.

[6] Sukhan Lee, Doyoung Byun, Daewon Jung, Jaeyong Choi, Yongjae Kim, Ji Hye Yang, Sang Uk Son, Si Bui Quang Tran and Han Seo Ko, " Pole-type ground electrode in nozzle for electrostatic field induced drop-on-demand inkjet head", Sensors and Actuators A, Vol. 141, pp. 506-514, 2008.

[7] A. Jaworek and A. Krupa, "Classification of the modes of EHD spraying", J. Aerosol Sci., Vol. 30, pp. 873-893, 1999.

[8] S. B. Q. Tran, D. Byun, S. Lee, "Experimental and Theoretical Study of Cone-jet for Electrospray Micro-Thruster Considering Interference Effect in Array of Nozzles", J. Aerosol Sci., Vol. 38, No. 9, pp. 924-934, 2007.

[9] D. Byun, Y. Lee, S. B. Q. Tran, V. D. Nguyen, S. Kim, B. Park, S. Lee, N. Inamdar, H. H. Bau, "Electrospray on super-hydrophobic nozzle treated by Ar and oxygen plasma", Applied Physics Letter, Vol. 92, 093507, 2008

A Novel Parallel Flow Control (PFC) System for Syringe-Driven Nanofluidics

H. Liang, Wook Jun Nam and Stephen J. Fonash

Center for Nanotechnology Education and Utilization
The Pennsylvania State University, University Park, PA, USA, hzl114@psu.edu

ABSTRACT

Active nanofluidic flow control is accomplished by utilizing the parallel flow control (PFC) configuration and it allows the primary problems of nanofluidic systems, including interfacing and measuring, to be overcome. PFC uses flow in a syringe-driven micro-channel, which interfaces with the "outside world", to set up the pressure gradient across a nano-channel. Based on the size-scale differences between these nano- and micro-channels arranged in parallel, nano- to micro-channel flow rate ratios of 10^{-4}:1 and smaller are easily attainable thereby allowing the attainment of a broad range of fine nanofluidic flow control. Long residence time and non-uniform flow rate issues are easily avoided by driving a relative large flow rate through the micro-channel. Direct, real-time flow rate measurements in the nano-channel are achieved by having an additional serpentine measurement micro-channel in series with the nano-channel. Nano-channel flow rates as low as ~0.5 pL/s are measured.

Keywords: nanofluidic flow control, nanofluidics, nano-total analysis systems, miniaturized chemical processing

1 INTRODUCTION

The use of nanofluidics for new sensing and chemical reaction approaches has yet to be exploited due to challenges that must be overcome[1]. The basic hindrances to nanofluidics utilization lie in integrating, interfacing, and controlling nanofluidics to realize active flow control and flow characterization. At present, most of the nanofluidic systems are based on a diffusion mode of fluid manipulation[2,3] which can be impracticable for many situations. Pressure-driven flow using a syringe pump is simpler and has a broad range of applicability for producing transport in nano-scale channels since it simply relies on mechanical pressure. However, previously explored nano-scale pressure-driven approaches have either had excessive sample residence times or system complexity issues[4,5] due to the extremely tiny volumes of nanofluidic systems (on the order of femto-liter).

This work uses a novel active nanofluidic flow control configuration to address the primary problems of nanofluidic systems including interfacing and measuring. We term this approach, which is seen in Fig. 1, parallel flow control (PFC). It uses flow in the syringe-driven micro-channel which is in parallel with nano-channel and which interfaces with the "outside world", to set up the pressure gradient across the nano-channel to control the nanofluidic flow. It also uses a serpentine micro-channel in series with the nano-channel to measure the nanofluidic flow.

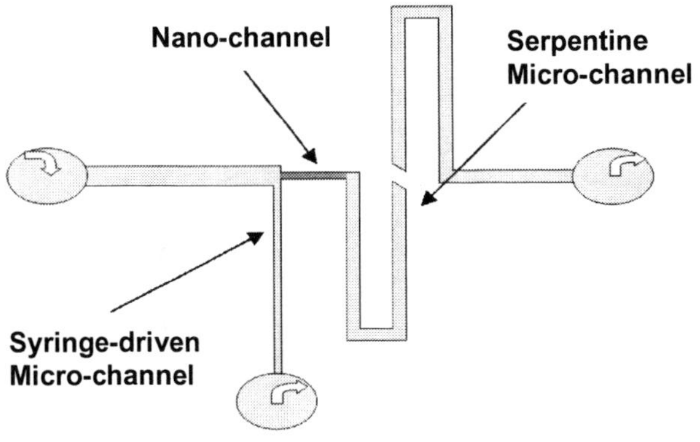

Figure 1. Schematic of the parallel flow control (PFC) approach for nanofluidic flow control. The pressure gradient across the nano-channel is controlled by the syringe-driven micro-channel flow in parallel with nano-channel. The serpentine micro-channel is in series with nano-channel to measure the nanofluidic flow rate by tracking the water-air interface.

2 DEVICE FABRICATION

The devices in this work were fabricated using two-step etching and glass bonding techniques in a class-10/100 cleanroom. The fabrication processes are CMOS compatible allowing potential application in future integrated micro-/nano-total analysis systems. The overall fabrication processes began with dry etching the nano-channel (100nm high, 15 μm wide, and 200 μm long) into a glass wafer. Then both syringe-driven micro-channel (10 μm high, 10 μm wide, and 4350 μm long) and serpentine micro-channel (10 μm high, 20 μm wide, and 8cm long) were patterned in alignment with the nanofluidic channel using wet chemical etching (6:1 buffered oxide etchant). Finally, after drilling the accessing ports through the glass wafer, it was bonded with another blank glass wafer to finish the fabrication. A completed, actual PFC device is seen in Fig. 2 with DI water infused.

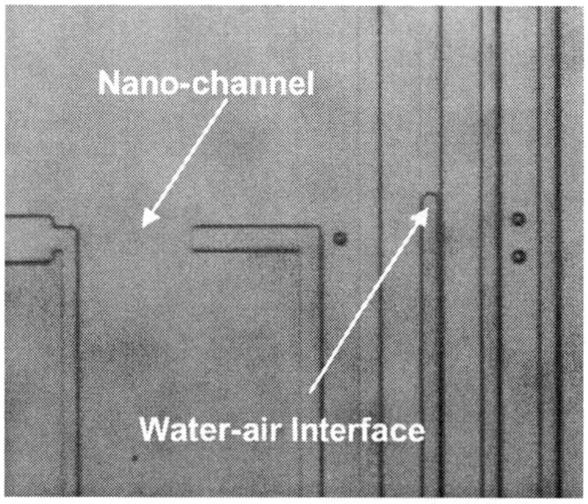

Figure 2. Bonded PFC nanofluidic flow control system with water infused by a syringe pump. The water-air interface is visible in the serpentine micro-channel in series with nano-channel.

3 THEORETICAL ANALYSIS

PFC is able to control the nano-channel flow by using the flow through the syringe-driven micro-channel. Therefore, an assessment of the flow rate ratios between nano- and micro-channel available using PFC can be established by using the simple flow model sketched in Fig. 3.

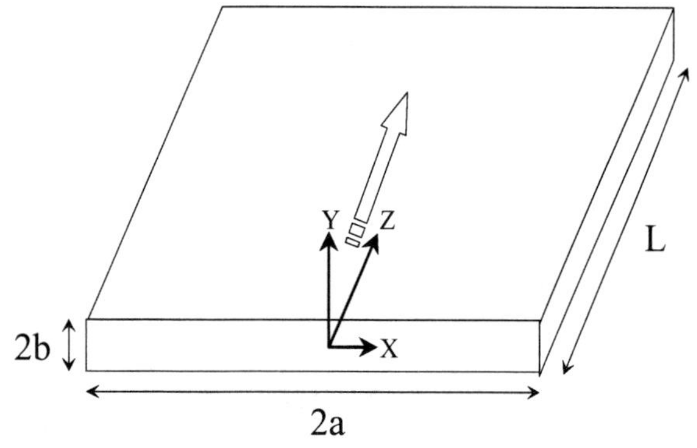

Figure 3. Schematic of fluidic flow model. Here 2a and 2b are channel width and height respectively.

Assuming a single one phase Newtonian fluid in the channel shown and a constant pressure gradient $-dp/dz = \Delta p/L$ along the flow direction, where Δp is the pressure drop along a given channel and L is the channel length, it follows from mass conservation and momentum conservation that the flow rate ratio Q_n/Q_m between nano- and micro-channel can be written as[6]

$$\frac{Q_n}{Q_m} = \frac{b_n^3 \sum_{n=1,3,5,...}^{\infty} \left[a_n - \frac{2b_n}{n\pi} \tanh(n\pi a_n / 2b_n) \right] \frac{L_m}{n^4}}{b_m^3 \sum_{n=1,3,5,...}^{\infty} \left[a_m - \frac{2b_m}{n\pi} \tanh(n\pi a_m / 2b_m) \right] \frac{L_n}{n^4}} \quad (1)$$

where a and b is the channel width and height respectively.

We can see from Eq. (1) that this flow rate ratio can be used to control a very precise (fine) flow through a nano-channel in parallel to a micro-channel. The flow rate ratio of Q_n/Q_m can be of the order of 10^{-4} or smaller for nano- and micro-channel. This is analogous to controlling the electric current through one large resistor by connecting a smaller resistor in parallel (Fig. 4).

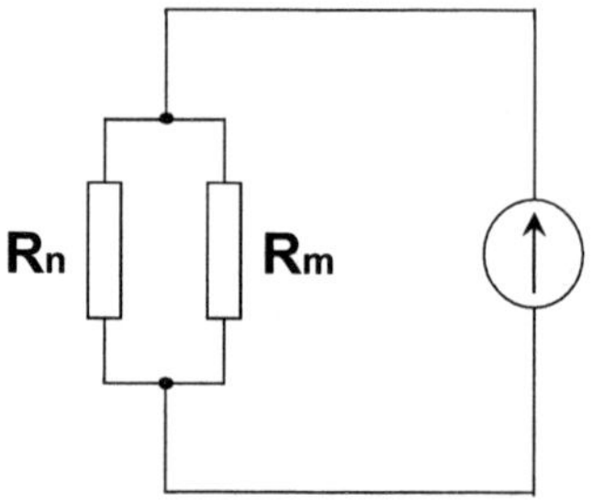

Figure 4. The electric circuit model corresponding to the nanofluidic flow control system. Rn and Rm represent the flow resistance of nano-channel and syringe-driven micro-channel, respectively. The current source represents the flow rate source controlled by a syringe pump. The flow through the nano-channel can be controlled by the flow through the syringe-driven micro-channel with the flow rate ratio decided by Eq. (1).

4 EXPERIMENTAL RESULTS

An assessment of PFC approach was undertaken by comparing the calculated nano-channel flow rate using the flow rate ratio of the nano- and micro-channels (~$7.7\text{x}10^{-5}$:1 from Eq. (1)) with experimental flow rate measurements determined by optically tracking the water-air interface in the serpentine micro-channel (Fig. 2). For this experiment, DI water was infused through the system at different flow rates controlled by a syringe pump. The results are shown in Fig. 5. As can be seen from this figure, either way of assessing the nano-channel flow rate shows extremely fine nanofluidic flow rates and flow control is possible with the PFC approach.

The fact that the experimental nano-channel flow rate is about 22 times slower than the calculated flow rate shown in Fig 5 is due to dimension control during the etching. That is, isotropic wet etching of the syringe-driven micro-channel and of the serpentine micro-channel do not produce the rectilinear cross-section shown in Fig.3 and used in Eq. (1). This wet etching produces an undercut profile and thereby makes the widths of these micro-channels larger than that designed and the channel cross-section far from rectilinear. For example, based on the isotropic etching geometry and etching depth into the glass substrate (10 μ m), the actual serpentine micro-channel cross-section area is estimated to actually be 78.5% larger than what was designed. This larger serpentine micro-channel width dimension resulting from isotropic wet etching means that the actual nano-channel flow is larger than that given by the experimental curve of FIG. 5. Using this increased cross-sectional area value leads to the "adjusted experimental values" curve seen in FIG. 5. This processing-caused increase in the syinge-driven micro-channel cross-section has a corresponding impact on the calculated curve of FIG. 5 because Eq (1) is no longer valid since the rectilinear cross-section of Fig 3 does not apply. By using a computational analysis[7] employing the correct cross-section shape and size to replace Eq. (1), we can re-determine the Q_n/Q_m ratio, which is multiplied times the syringe flow rate to get the calculated nano-channel flow rate. For the case of the device of Fig. 2 this re-calculated Q_n/Q_m decreased to 2.37×10^{-5}. Use of this re-calculated ratio results in the "adjusted calculated values" curve also shown in FIG. 5. As seen, the adjusted experimental flow rate is about 4 times slower than the adjusted theoretical value. We believe this difference is due to some combination of wetting effects, interactions with the walls[4], and evaporation effects in the serpentine micro-channel[8]. As seen from Fig. 5, the smallest nano-channel flow rate is ~0.5 pL/s. While not the lower limit of the PFC flow rates achievable, this value already compares well with the previously reported, smallest flow rate of 30 pL/s[8].

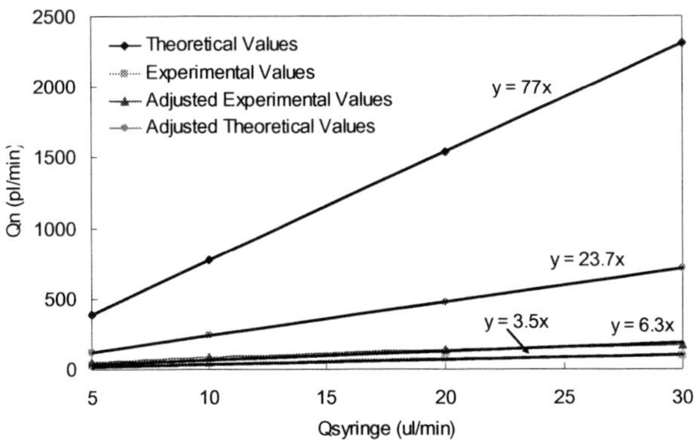

Figure 5. Nano-channel flow rates as a function of syringe, or equivalently, flow-driving micro-channel rates. Shown are the experimental values obtained by tracking flow in the serpentine micro-channel, the theoretical values calculated from Q_n/Q_m and the syringe flow rates, and the same curves adjusted for fabrication-caused dimension changes.

5 SUMMARY

Our novel active nanofluidic flow control configuration allows the primary problems of nanofluidic systems including interfacing and measuring to be successfully solved. We show that such a parallel approach makes it possible to have a broad range of fine flow rate control through the nanofluidic channel even with a system driven by a simple syringe pump. Flow rate differences between theoretical and experimental values were seen and these are shown to be due primarily to fabrication issues. Effects such as electrostatic interactions with the channel walls, wetting, and evaporation effects were shown to be secondary, at least for the DI water solution used.

ACKNOWLEGEMENTS

We thank Nadine B. Smith for allowing us to use the K&S 983 dicing saw in preparing our devices. We also acknowledge use of facilities at the PSU site of NSF NNIN. The project was supported in part by NSF Grant No. DMI-0615579.

REFERENCES

[1] G. Hu and D. Li, Chemical Engineering Science. 62, 3343, 2007.
[2] R. Karnik, K. Castelino, C. Duan and A. Majumdar, Nano Letter. 6, 1735, 2006.
[3] A. Malave, M. Tewes, T. Gronewold and M. Lohndorf, Microelectronic Engineering. 78, 587, 2005.
[4] F. H. J. Van Der Heyden, D. Stein and C. Dekker, Physical Review Letters. 95, 116104, 2005.
[5] E. Tamaki, A. Hibara, H. B. Kim, M. Tokeshi and T. Kitamori, J. Chromatogr. A. 1137, 256, 2006.
[6] J. P. Brody, P. Yager, R. E. Goldstein and R. H. Austin, Biophysical Journal. 71, 3430, 1996.
[7] P. Nath, S. Roy, t. Conlisk, and A. J. Fleischman, Biomedical microdevices. 7, 169, 2005.
[8] K. J. A. Westin, C. H. Choi and K. S. Breuer, Experiments in Fluids. 34, 635, 2003.

Integrated Thermal Modulation and Deflection of Viscous Microjets with Applications to Continuous Inkjet Printing

E. P. Furlani[1], A. G. Lopez[1], K. C. Ng[1], and C. Anagnostopoulos[2]

[1]Eastman Kodak Research Laboratories
1999 Lake Avenue, Rochester, New York 14650-2216
edward.furlani@kodak.com

[2]Department of Mechanical Engineering and Applied Mechanics
University of Rhode Island
Kingston RI 02881

ABSTRACT

We present a novel CMOS/MEMS microfluidic device that enables the controlled production and redirection of streams of picoliter-sized droplets at frequency rates in the hundreds of kilohertz range. Droplet generation and jet manipulation are achieved using voltage-controlled thermal modulation, at modest temperatures and with no moving elements. We discuss the fabrication and operating physics of the device and we compare experimental performance data with 3D CFD simulations. We discuss applications of this device to continuous inkjet printing.

Keywords: micro-drop generator, thermal jet deflection, thermo-capillary instability, continuous inkjet printing, Marangoni instability

1 INTRODUCTION

Microfluidic devices are finding increasing use in a broad range of applications that involve the production and controlled delivery of micro-droplets. The most notable and commercially successful of these is inkjet printing wherein streams of picoliter-sized drops are ejected at high repetition rates onto a media to render an image. Inkjet printing can be broadly divided into two distinct printing methods; drop-on-demand (DOD) printing and continuous inkjet printing (CIJ). In DOD printing, droplets are produced as needed to form the image. In CIJ printing, droplets are produced continuously but only a fraction of these are used to form the image. The unused droplets are deflected and guttered prior to reaching the image, and this unused ink is recycled to the printhead. In this presentation we discuss a novel CMOS/MEMS microfluidic device that enables the controlled production and redirection of streams of picoliter-sized droplets at frequency rates in the hundreds of kilohertz range [1-4]. This device it well suited for CIJ printing as well as numerous other applications.

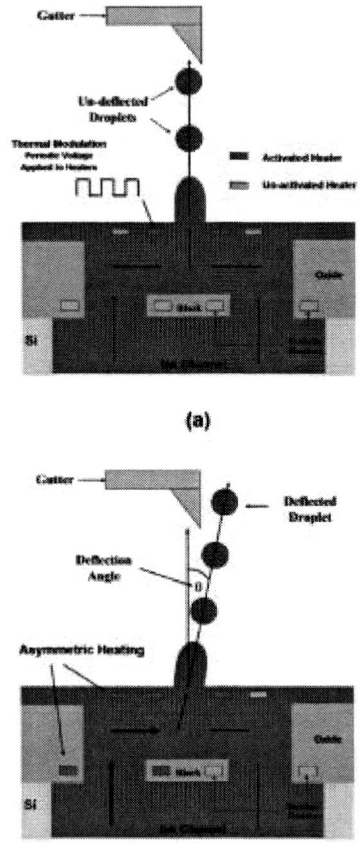

Figure 1: Integrated thermal modulation and deflection of a microjet: (a) symmetric modulation of heaters near the orifice produces a straight stream of droplets, which are recirculated; (b) asymmetric modulation of heaters near the orifice and in the blocking structure produces a deflected stream of droplets.

2 DEVICE PHYSICS

Our microfluidic device consists of a pressurized reservoir that feeds a micro-nozzle manifold with hundreds of active orifices, each of which produces a continuous jet of fluid. An integrated cylindrical blocking structure is suspended beneath each orifice (Figs. 1 and 2a). This structure splits the flow from the reservoir into two opposing flows that merge immediately beneath the orifice to form the jet. Droplet generation and jet deflection are achieved using voltage-controlled thermal modulation of the fluid. This is achieved using individually addressable resistive heater elements that are integrated into the nozzle plate around each orifice, and also into the suspended blocking structure as shown in Figs 1 and 2a. The heaters are configured to enable symmetric or asymmetric heating. Modulated symmetric heating produces a straight stream of droplets whereas asymmetric heating causes the stream to deflect (Fig. 1).

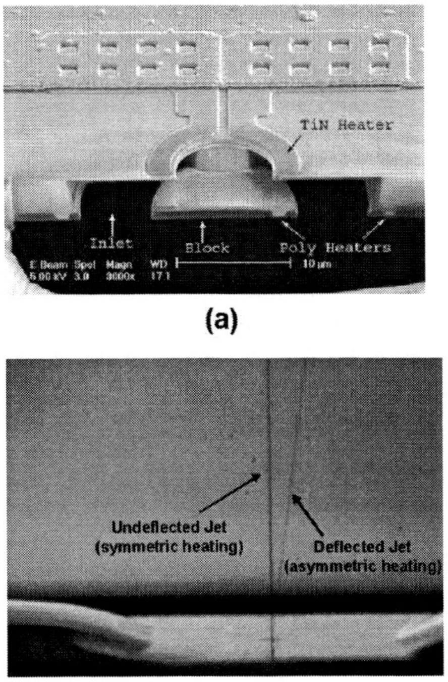

Figure 2: Experimental characterization: (a) fabricated MEMS/CMOS nozzle structure; (b) two adjacent jets (aligned into the page), the front jet is symmetrically modulated producing a straight stream of droplets, the second jet is asymmetrically modulated producing a deflected stream of droplets.

To modulate a jet, a periodic voltage is selectively applied to the embedded heaters, which causes a periodic diffusion of thermal energy from the heaters into the fluid. Thus, the temperature of fluid, and hence the temperature dependent fluid properties, especially viscosity and surface tension, are modulated near the orifice. This thermal modulation causes two distinct effects; droplet generation and jet deflection.

Droplet production is primarily caused by the modulation of surface tension at the orifice [5-8]. It is well known that slender jets of fluid are inherently unstable and breakup into droplets when subjected to the slightest perturbation. Periodic heating near the orifice produces a modulation of surface tension σ, which produces the perturbation. Specifically, to first order, the temperature dependence of σ is given by $\sigma(T)=\sigma_0-\beta(T-T_0)$, where $\sigma(T)$ and σ_0 are the surface tension at temperatures T and T_0, respectively. The pulsed heating near the orifice modulates σ at a wavelength $\lambda = v_0\tau$, where v_0 is the jet velocity and τ is the period of the heat pulse. The down-stream advection of thermal energy gives rise to a spatial variation (gradient) of surface tension along the jet. This produces a shear stress at the free-surface, which is balanced by inertial forces in the fluid, thereby inducing a Marangoni flow towards regions higher surface tension (from warmer regions towards cooler regions). This causes a deformation of the free-surface (slight necking in the warmer regions and ballooning in the cooler regions) that ultimately leads to instability and drop formation [7-8]. The drop volume can be adjusted on demand by varying τ, i.e., $V_{drop}=\pi r_0^2 v_o \tau$, where r_0 is the jet radius. Thus, longer pulses produce larger drops, shorter pulses produce smaller drops, and different sized drops can be produced from each orifice as desired.

Jet deflection is primarily caused by the asymmetric heating of the blocking structure as shown in Fig. 2b. This causes a modulation of viscosity, which results in an increase in the fluid velocity near the activated heater. Thus, there is an imbalance in fluid momentum in the opposing flows that form the jet. This imbalance is carried by the jet through the nozzle resulting in a deflection of the jet away from the heated side as it leaves the orifice. The jet can be deflected from one side of the orifice to the opposite side as needed by applying voltage to the appropriate heaters on either side of the blocking structure. Modulated symmetric heating produces a straight stream of droplets whereas asymmetric heating causes the stream to deflect. The ability to redirect the jet/droplets is useful for applications such as continuous inkjet printing where only a fraction of the generated droplets are used to render an image, and the unused droplets are guttered and recirculated to the reservoir. The integrated CMOS-based thermal modulation and deflection capability of our device represents distinct advantages over conventional continuous inkjet printing systems that rely on piezoelectric driven drop generation and electrostatic deflection of charged droplets.

3 DEVICE FABRICATION

Micro-nozzles with a recessed blocking structure and embedded heaters as shown in Fig. 2a were fabricated using silicon VLSI technology and MEMS fabrication techniques. Arrays of such nozzles were fabricated on standard CMOS wafers containing the logic and drive circuits are in process. Following is the fabrication sequence for lateral flow nozzles on wafers that have not gone through a CMOS process.

(Steps 1-12 would normally be part of the CMOS process)
1) Grow thermal oxide 4000 A.
2) Low-pressure chemical vapor deposition (LPVCD) doped polysilicon 4000A.
3) Define bottom heaters. Dry etch 4000A Poly using chlorine chemistry.
4) Thermal oxidation of polysilicon heater (500 A oxide).
5) Deposit LPCVD TEOS ($Si(OC_2H_5)_4$) 2000A.
6) Deposit LPCVD borophosphosilicate glass (BPSG) 5000A.
7) Define and open contacts to Poly heaters.
8) Sputter TiW/Al (2000A/6000A) for electrical interconnect.
9) Define and dry etch TiW/Al using chlorine chemistry.
10) Deposit 5 μm PECVD oxide.
11) Chemical mechanical polishing (CMP) of oxide (remove approximately 1 μm).
12) Define fluid inlets and partially dry etch oxide, 1.5 μm.
(Steps 13-30 MEMS portion of the process)
13) Define blocks and dry etch remaining oxide, 3.5 μm, to reach the Silicon surface at the inlets.
14) Coat, expose, develop, and bake 10 ∝m polyimide.
□ Polyimide HD8000 by HD Microsystems
15) CMP polyimide using Cabot W-A400 slurry (Al_2O_3).
- Tool by Strausbaugh model 6EC Planarizer.
- Two steps: Removal rate: 5.5 ∝m/minute and 0.5 ∝m/minute.

16) Megasonic bath clean to remove CMP slurry. H_2O: H_2O_2: NH_4OH 40:2:1
17) Deposit bottom passivation layers PECVD TEOS/ Nitride/ TEOS (2500A/ 2000A/ 2500A).
18) Sputter Ti/TiN (50A/600A) as top heater material.
19) Define and dry etch Ti/TiN top heaters using fluorine chemistry.
20) Deposit PECVD TEOS/Nitride (1500A/1500A).
21) Define and open contacts to TiN heaters.
22) Sputter Al (6000A) for electrical interconnect.
23) Define and dry etch Al using chlorine chemistry.
24) Deposit 3000A PECVD Nitride top passivation layer.
25) Define and etch nozzle bore by dry etching Nitride/ Oxide/ Oxide/ Nitride/ Oxide stack (4500A/ 1500A/ 2500A/ 2000A/ 2500A).
26) Define and open contact to bond pads by dry etching dielectric stack.
27) Thin wafer down to 300 μm.
- Tool by Strausbaugh model 7AA-II
- Removal rate: 150 ∝m/minute.

28) Define fluid channels in the backsides of wafers.
29) Etch through wafer with STS tool using Bosch process.
30) Remove sacrificial polyimide using oxygen chemistry.

4 DEVICE CHARACTERIZATION

We characterized fabricated devices to determine the jet deflection as a function of heater power and drive frequency. All measurements were made using water. We first measured static jet deflection as a function of heater power for two different heating modes. In the first mode, an incrementally increasing voltage was applied to the heater on the left side of the orifice, and the steady state jet deflection was measured at each voltage level. In the second mode, the same procedure was applied using the heater on the left side of the blocking structure. In both cases we found that deflection increased with applied power (Fig.3). Furthermore, for a given power, a larger deflection was achieved using the heater in the blocking structure. The reason for this is because this heater creates a more effective imbalance of fluid momentum than the orifice heater.

We also measured the jet deflection as a function of drive frequency. To do so, we applied an alternating voltage to left and right side heaters in both the orifice and

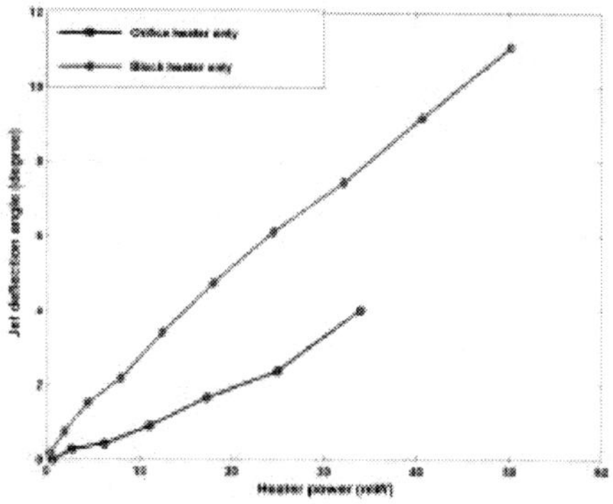

Figure 3: Jet deflection vs. heater power.

the blocking structure. The voltage was synchronized so that the left side heating was 180° out of phase with respect to the right side heating. The periodic alternating heating caused the jet to oscillate back and forth and simultaneously create droplets downstream. The deflection angle vs. drive frequency along with a strobbed picture of the deflected streams of droplets that evolve from the oscillating jet are shown in Fig. 4. The deflection angle decreases with frequency because of the thermal time constant required for an activated heater to cool to ambient temperature. Note

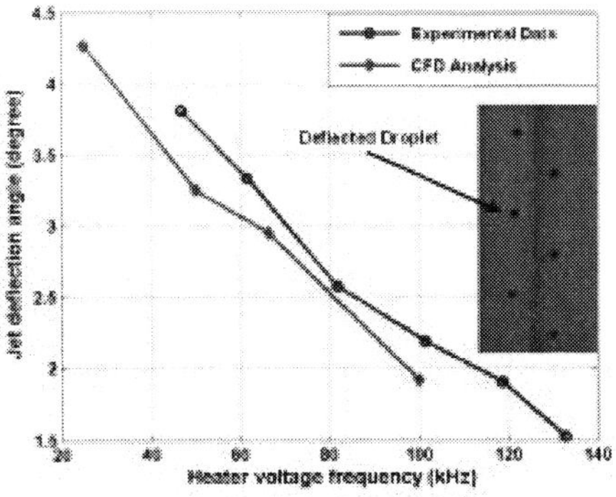

Figure 4: Jet deflection vs. frequency (experimental data and CFD analysis); strobbed image of deflected streams of droplets.

that a working deflection of 1.5 degrees is obtained at a frequency of 132 kHz.

5 DEVICE MODELING

We modeled the performance of the device using 3D CFD. The FLOW-3D software was used for the analysis. The 3D CFD computational domain is shown in Figure 5. The jet deflection angle is predicted by computing the center of mass of the fluid in a control volume near the top of the jet. The predicted deflection vs. frequency data is compared to with corresponding experimental data in Fig. 4. The CFD analysis tended to under predict the measured deflection data. We attribute this to the limited size of the computational domain and the imposition of boundary conditions that did not fully reflect the effects of the neglected physical domain.

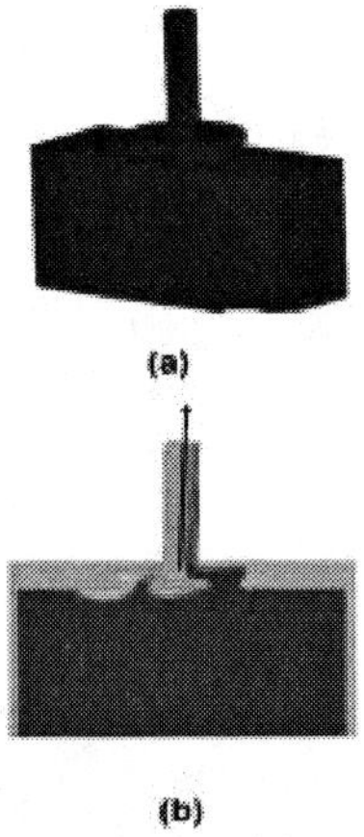

Figure 5: 3D CFD analysis: (a) perspective of internal fluid flow; (b) temperature distribution in the fluid and jet deflection due to asymmetric left-sided heating.

6 CONTINUOUS INKJET PRINTING

A CIJ printhead consists of arrays of orifices wherein each orifice produces a continuous stream of droplets. Only a fraction of the droplets are used to form the image; unused droplets are deflected, guttered and recycled. In a conventional CIJ system, droplet generation is achieved using a piezoelectric transducer, and deflection is achieved using an electrostatic field and charged droplets. The device presented above has distinct advantages in this regard as it enables both droplet production and deflection in an integrated package, thereby eliminating the need for the piezoelectric and electrostatic components altogether. The device has additional advantages in that it enables system miniaturization, higher print resolution, lower production costs, and higher reliability as compared to conventional CIJ systems.

REFERENCES

[1] C. N. Anagnostopoulos, J. M. Chwalek, C. N. Delametter, G. A. Hawkins, D. L. Jeanmaire, J. A. Lebens, A. Lopez and D. P. Trauernicht, "Micro-jet nozzle array for precise droplet and steering having increased droplet deflection," Proc. Transducers 03 Conf. P 368-371, 2003.

[2] J. M. Chwalek, D. P. Trauernicht, C. N. Delametter, R. Sharma, D. L. Jeanmaire, C. N. Anagnostopoulos, G. A. Hawkins, B. Ambravaneswaran, J. C. Panditaratne, and O. A. Basaran, "A new method for deflecting liquid microjets," Phys. of Fluids **14**, 6, 37-40, 2002.

[3] D.P. Trauernicht, C.N. Delametter, J.M. Chwalek, D.L. Jeanmaire, and C.N. Anagnostopoulos, "Performance of Fluids in Silicon-Based Continuous Inkjet Printhead Using Asymmetric Heating," Proc. IS&T NIP17: Int. Conf. on Digital Printing Technologies, pp. 295-298, 2001.

[4] C.N. Delametter, J.M. Chwalek, and D.P. Trauernicht, "Deflection Enhancement for Continuous Ink Jet printers," U.S.Patent 6,497,510, Issued Dec. 24, 2002

[5] E. P. Furlani, "Temporal instability of viscous liquid microjets with spatially varying surface tension," J. Phys. A: Math. and Gen. **38**, 263-276, 2005.

[6] E. P. Furlani, "Thermal Modulation and Instability of Viscous Microjets", proc. NSTI Nanotechnology Conference 2005.

[7] E. P. Furlani, B. G. Price, G. Hawkins, and A. G. Lopez, "Thermally Induced Marangoni Instability of Liquid Microjets with Application to Continuous Inkjet Printing", Proc. NSTI Nanotechnology Conference, 2006.

[8] E. P. Furlani and K. C. Ng, "Numerical Analysis of Nonlinear Deformation and Breakup of Slender Microjets with Application to Continuous Inkjet Printing", Proc. NSTI Nanotechnology Conference, 2007.

Micro bioreactor for muliwell plates with an active micro fluidic system for 3D cultivation

F. Weise*, C. Augspurger*, M. Klett*, S. Giselbrecht**, G. Schlingloff**, K.-F. Weibezahn**, A. Schober*

* Technische Universität Ilmenau, MacroNano® Center for Innovation Competence, Germany
frank.weise@tu-ilmenau.de
** Forschungszentrum Karlsruhe, Institute for Biological Interfaces, Germany

ABSTRACT

Recently it has been shown that 3D cultivation of cells and tissues show great potential for bio/medical research [1]. Within this scope are developments of micro bioreactors with micro fluidic support structures [2]. We have designed a micro system - suitable for parallelization - to cultivate cells three dimensionally. This system is based on the CellChip bioreactor from FZK (Forschungszentrum Karlsruhe). Our system is miniaturized in such a way that it is implementable into a 24 well plate. A single channel system with an integrated micro pump is manufactured as a prototype. The aim of our system is reduction of costs, animal experiments and the acceleration of drug screening. This system uses CellChips to grow cells organ-like, i.e. three dimensionally. In this chip the cells are located in containers. Up to 2 million cells like HepG2 can be cultivated in this system. The system is characterized with respect to oxygen consumption of the cells and cell viability. 24-channel-systems are under construction.

Keywords: three dimensional cell culture, micro pump, micro-fluidics, bioreactor, multi-well-plate screening

1 DESIGN AND FUNCTIONAL PRINCIPLE

Based on the CellChip bioreactor form FZK (Fig.1), we have designed and build a novel miniaturized system to cultivate cells three dimensionally.

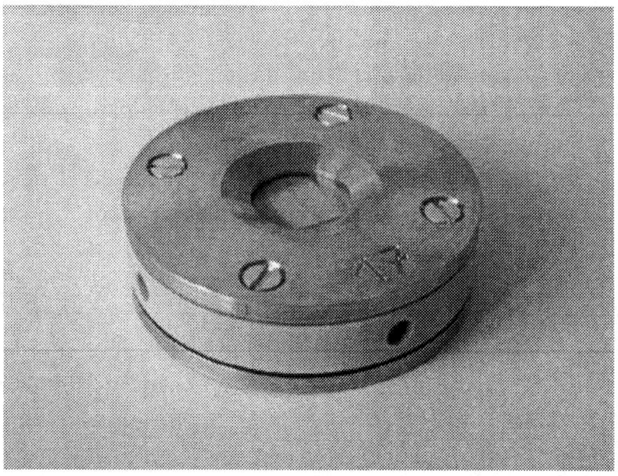

Fig. 1 Bioreactor with CellChip from FZK

A closed loop of culture medium is integrated in our micro bioreactor. Also a micro pump is used to generate a fluid flow in the system. The pump is actuated by compressed air. In Fig. 2 you can see a schematic of our micro reactor.

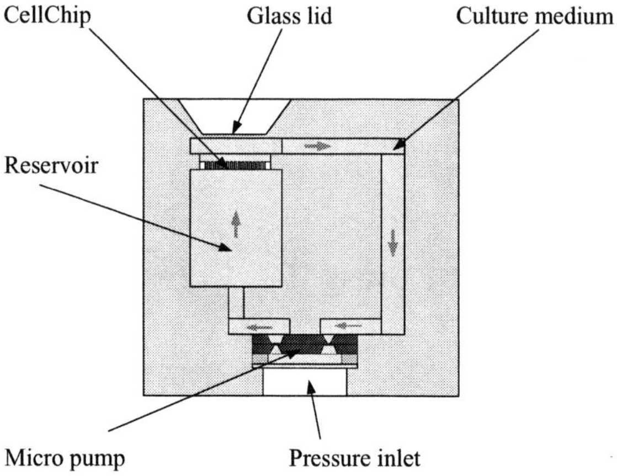

Fig. 2 Schematic of the micro bioreactor

A new smaller CellChip is used to cultivate cells in our system. The dimensions of this CellChip (Fig. 3) are 8 x 8 x 1 mm^3. The chip contains 169 micro containers to carry cells. The size of these containers is 300 x 300 x 300 μm^3. A glass lid above the CellChip allows for microscopical analysis of the cells. If necessary, we can exchange the glass lid with a transparent plastic lid. By this way we can increase the oxygen content in the culture medium, but optical resolution decreases.

Fig. 3 The new smaller CellChip

1.1 The modular body

In order to be able to gain diverse information on fluidic parameters, biological variables etc. we needed a flexible experimental setup. Therefore we choose a modular design for the bioreactor. As material we used polycarbonate, because it is transparent and can easily be machined with a milling machine.

The inner part of the system carries the CellChip. This carrier contains a reservoir of culture medium and fluidic channels. Its size is given by the grid dimensions of a 24 well plate. The carrier is housed by a body, which – in case of the single channel bioreactor – is directly connected to a module either with an internal micro pump or with connectors for an external pump. Fig. 4 shows the micro bioreactor with the integrated micro pump.

Fig. 4 Single channel micro bioreactor with a blue tracer

1.2 The micro pump

Like other bioreactors, also the micro bioreactor can be operated with an external pump for supplying the cells with culture medium. This works for a single channel system but is not practicable for a multi well system. Therefore we have developed a robust internal micro pump which can be integrated into the system. By this means, each channel of a multi well system is controlled by its own micro pump.

The pumping principle is based on a membrane pump. Our pump has two micro machined check valves. They are made of silicon. The membrane of the pump consists of PDMS (Sylgard 184). Without the housing the pump has a size of 5 x 5 x 1 mm^3.

Our micro pump can reach a maximum flow of 1ml per minute. Also, this pump can drive gases and fluids or a mixture of both. This is of advantage, especially if gas bubbles cannot completely be avoided. In addition, the actuation of the pump with compressed air facilitates gas exchange between the culture medium and the compressed air and hence, to some extent, allows oxygenation of the culture medium.

1.3 Multi well plates

The single channel system is the first step towards a multi channel system (multi well plate). A possible design of such a multi channel system is shown in Fig. 5

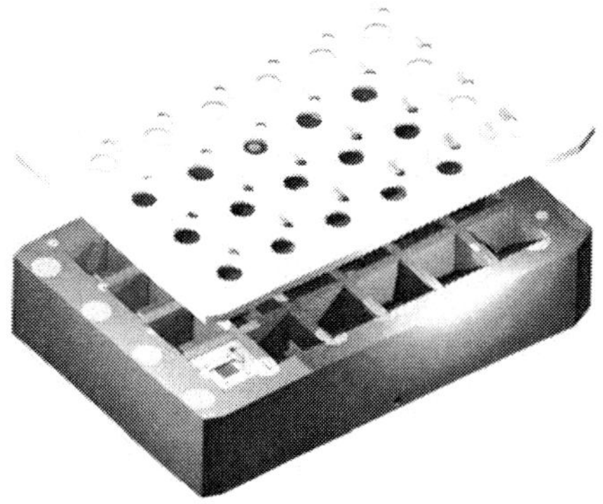

Fig. 5 CAD model of a 24 channel bioreactor system

One challenge of a multi well system lies in the fact that all channels need to have identical characteristics, i.e. every micro pump must have the same flow rate while being addressed by only one shared driving pressure connection. With micro technologies we should be able to address this challenge since it is possible to produce micro valves with small tolerances.

2 METHODS

2.1 Cell Culture

HepG2 cells (ATCC) were cultivated in Minimum Essential Medium (Sigma, Taufkirchen, Germany) with 10% fetal calf serum, 100 U/ ml penicillin, 100 µg/ ml streptomycin, 2 mM L-glutamine and 1 mM sodium pyruvate. CellChips with a micro-structured area of 0.5 x 0.5 cm^2 were used for three-dimensional cultivation of the cells. 1.5 x 10^6 cells were inoculated per single-channel-bioreactor and per CellChip as described elsewhere [4]. Total volume was 1 ml per bioreactor. Bioreactors with and without cells were incubated at 37°C, 5% CO_2.

2.2 Determination of Viability

Viability of cells was compared for three different culture systems: CellChips within actively perfused bioreactors, CellChips within non-perfused bioreactors and non-perfused CellChips placed in a multi well plate (no bioreactor housing). After 48 h (day 2) of cultivation, cells were enzymatically (trypsin) removed from the CellChips and viability of cells was determined by trypan-blue exclusion using a hemocytometer.

In order to get more detailed information on the cellular viabilities on day 2, cells were then plated as monolayers.

After 24 h (day 3) adherence was checked microscopically and viability was estimated as percentage of adherence.

2.3 Determination of oxygen content

For determining the oxygen content of the medium during cell cultivation fiberoptical oxygen minisensors (PreSens, Regensburg, Germany) were used. The sensor was implemented into the main chamber of the bioreactor. Media were incubated at 37°C, 5% CO_2 for 2 h before filling into the bioreactors. Oxygen contents of an actively perfused bioreactor (perfusion via a pneumatically driven micro pump) and a non-perfused (static) bioreactor were compared. As control a non-perfused bioreactor without cells was used. Oxygen content was measured every minute and data were logged using appropriate software (OXY, PreSens).

3 RESULTS AND DISCUSSION

Viability of HepG2 cells cultivated in a perfused bioreactor was higher than that of the other systems (tab.1) on day 2 and also day 3. Adherence of the cells on day 3 was very good for cells removed from the perfused bioreactor and good for cells removed from non-perfused CellChips placed in a 24-well plate. Cells removed from the non-perfused bioreactor adhered poorly.

culture system	day 2	day 3
perfused bioreactor	75	75
non-perfused bioreactor	10	< 10
non-perfused CellChip without bioreactor housing	65	< 50

Table 1: Viability (%) of HepG2 cells after three dimensional cultivation within and after removal from different culture systems (day 2) as well as after replating as monolayers (day 3).

For long-term cultivation of cells within a closed bioreactor system appropriate oxygenation is vital. Therefore the oxygen content within the bioreactors was monitored while cultivating HepG2 cells. A non-perfused bioreactor without cells served as control and demonstrated maximum oxygen content as well as stability of the sensor signals (fig 6). Oxygen content did not change significantly within perfused bioreactors, which demonstrates sufficient oxygenation of the system. The pneumatically driven pump increases oxygen diffusion through the silicone membrane into the medium.

In contrast, oxygen content within a non-perfused system dropped to about 50 % of the initial value (4.8 mg/l) before rising again to 3.4 mg/l. The latter may be explained by a decreased viability of the cells and therefore reduced oxygen consumption which allows an overall increase of oxygen content over time.

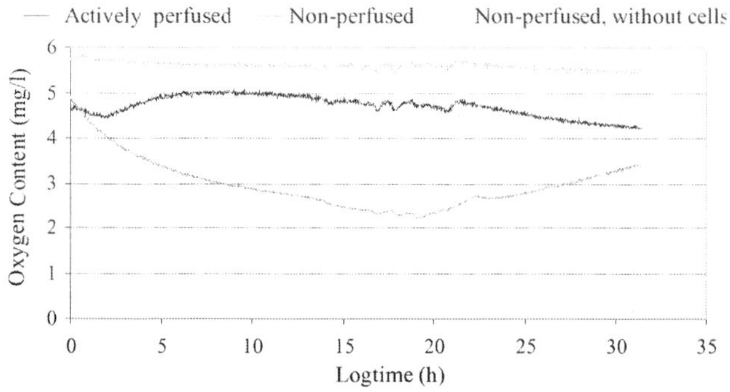

Fig. 6: Oxygen content of medium within an actively perfused bioreactor ($1.5x10^6$ HepG2 cells), a non-perfused bioreactor ($1.5x10^6$ HepG2 cells) and a non-perfused bioreactor without cells.

In conclusion, oxygenation in a non-perfused bioreactor is not sufficient and leads to strongly decreased cell viability. Both, oxygen supply and cell viability, can be improved by active perfusion of the system by a pneumatically driven micro pump as presented in this work.

4 ACKNOWLEDGEMENTS

The authors gratefully acknowledge the financial support provided to this study by Federal Ministry of Education and Research and the Thuringian Ministry of Culture within the Initiative "Centre for Innovation Competence", MacroNano®.

REFERENCES

[1] L. A. Kunz-Schughart, J.P. Freyer: The use of 3-D Cultures for High-Throughput Screening: The multicellular Spheroid Model, Journal of Biomolecular Screening 9(4); 2004

[2] Eschbach E., Chatterjee S.S., Nöldner M, Weibezahn K.F.: Microstructured scaffolds for liver tissue cultures of high cell density: morphological and biochemical characterization of tissue aggregates. Journal of Cellular Biochemistry, 95(2005) S.243-55

Weibezahn, K.-F., Knedlitschek, G., Dertinger, H., Schubert, K., Schaller, T.: International Patent (1993): Cell culture substrate. *International Publication Number* WO 93/07258.

[3] F. Zhao, T. Ma: Perfusion Bioreactor System for Human Mesenchymal Stem Cell Tissue Engineering: Dynamic Cell Seeding and Construct Development, Published online 13 May 2005 in Wiley InterScience (www.interscience.wiley.com). DOI: 10.1002/bit.20532, Biotechnology and Bioengineering, Vol. 91, No. 4, August 20, 2005

Feng Zhao, Pragyansri Pathi, Warren Grayson: Effects of Oxygen Transport on 3-D Human Mesenchymal Stem Cell Metabolic Activity in Perfusion and Static Cultures: Experiments and

Mathematical Model, Biotechnol. Prog. 2005, 21, 1269-1280

[4] C. Augspurger: Untersuchungen zum Differenzierungszustand dreidimensionaler Gewebekulturen im CellChip-basierten Bioreaktor am Beispiel der humanen Hepatokarzinom-Zelllinie HepG2 und primärer Rattenhepatozyten. [thesis] University of Leipzig, Germany (2007).

Discrete Volume Controllable Microdispenser Module

Sungjun Lee and Junghoon Lee

School of Mechanical and Aerospace Engineering, Seoul National University, and Institute of Bioengineering
151-744, Seoul, Korea, jleenano@snu.ac.kr

ABSTRACT

We report a microdispenser that can meter and mix sub-nanoliter liquids in a discrete quantity. Our device is based on the structural stopping valve used in passive dispensers. A vent channel was proposed, designed for the autonomous pausing of a metered liquid segment and the seamless joining of the subsequent segments. While the previous microfluidic devices could meter only a single drop [2-4], lacking in mixing multiple segments, our approach enables the metering/mixing of different liquids in a discrete ratio. This module will find applications in bio-engineering and pharmaceutical research including large-scale cell differentiation experiment and drug discovery platforms.

Keywords: micro fluidics, passive dispenser, structural valve. lab-on-a-chip, microTAS.

1 INTRODUCTION

Like a micropipette in a macro scale experiment, fluid dispensing module is one of the vital elements of bio or chemical experiment to deliver certain volume of samples or to control sample's concentration. However, as the scale of bio-engineering platforms becomes smaller and automated with the use of lab-on-a-chip approach, it becomes challenging to deliver samples with various concentrations and volume.

Continuous active mixing at different ratios was achieved in a micro scale fluidic system, but dispensing of a small volume is inherently impossible [1]. For proper working of active fluidic system, it is necessary to fill up every channel and reservoir. Therefore, dispensing and delivering are inherently impossible. A passive fluidic system could overcome this problem, however, it need a complicated fabrication [2] or can dispense only a pre-fixed volume in a single segment [3]. This raises critical problems when a large number of pipetting is required.

We report a volume controllable microdispenser module that can meter and mix sub-nanoliter liquids in a discrete quantity. Our device is based on the structural stopping valve used in the passive dispensers [3, 4]. By adding an additional structural valve and a vent channel, we could give a controllability of metering volume to the previous passive dispenser module. The additional structural valve prevents liquid leaking through the vent channel. The vent channel prevents air trapping between metered liquid segments. We could meter several liquid segments and merge them. By this way, we could dispense certain volume of liquid with discrete manner.

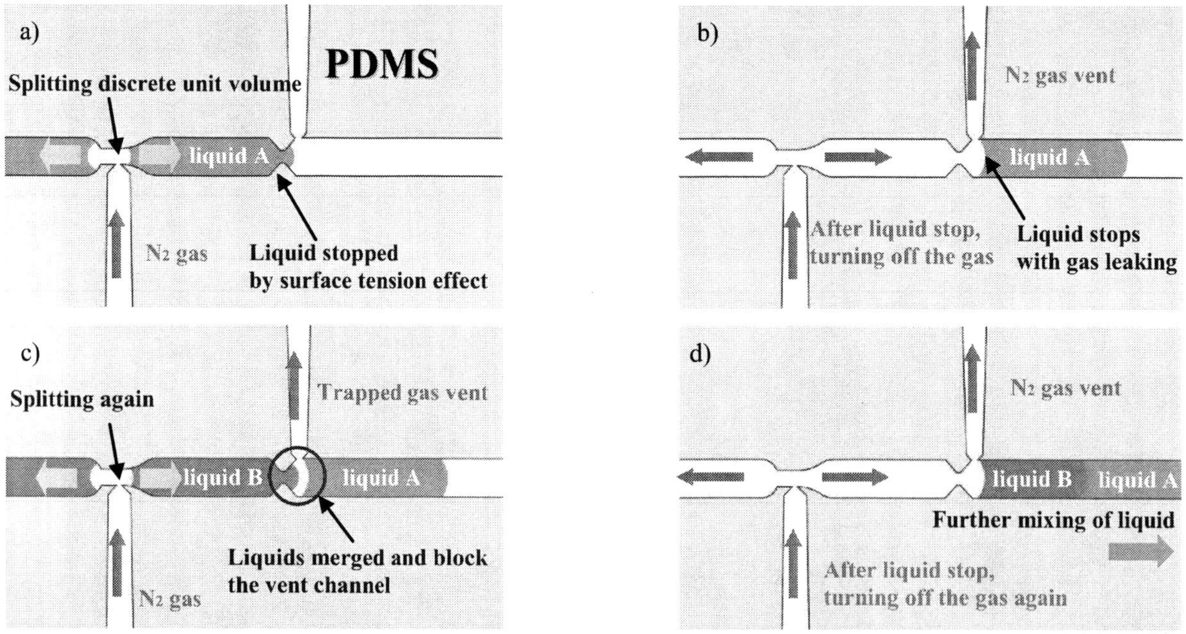

Figure 1: Schematics of device operation. Channels are made of PDMS (hydrophobic)

2 WORKING PRINCIPLE

This work is based on the principle of the structural stopping valve in the passive dispenser. Surface tension is described as

$$\Delta P = 2\sigma_l \cos(\theta_c)\left[\left(\frac{1}{w_1}+\frac{1}{h_1}\right)-\left(\frac{1}{w_2}+\frac{1}{h_2}\right)\right]$$

where, w_1, h_1 and w_2, h_2 are the width/height of the wide micro channel and narrow micro channel respectively, θ_c is the contact angle, σ_l is the surface tension of the liquid/air interface and ΔP is the pressure required to push the liquid into the narrow channel [3, 4]. When the contact angle, θ_c is higher than 90 °, i.e., the surface is hydrophobic, the narrower channel needs a higher pressure to overcome the surface tension barrier. Also, channel geometry influences effective contact angle. Thus, by varying channel geometry and dimension, we can stop flow at a certain point necessary.

We added an additional passive valve and a vent channel essential for pausing the movement of the metered segment, and merging next ones. Figure 1 explains the sequence of the device operation driven by a pressure source. Liquid is metered by graduating reservoir, and the metered liquid is delivered by pneumatic actuation. The vent channel is narrower than the main one with the neck-shape entrance designed to effectively block the liquid flow. Thus, the liquid segment flows only through the main channel while the pressurizing gas escapes thorough the vent channel. The liquid segment stops at the branch point waiting for the next one to join. Through the repetition of the sequence, mixing ratio can be controlled in a discrete manner.

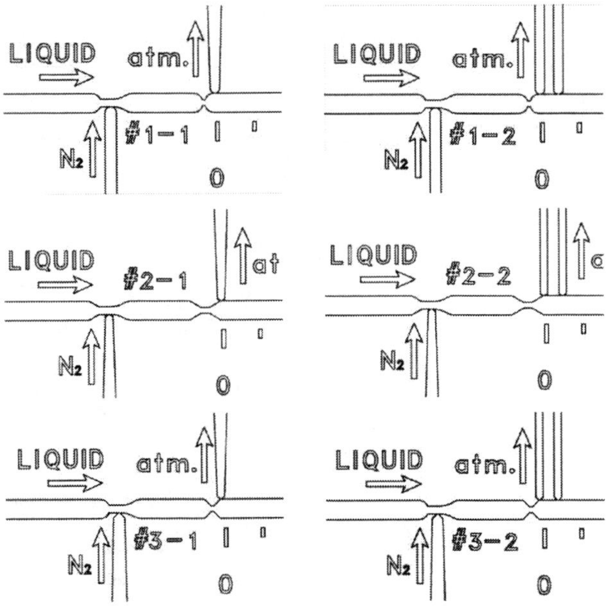

Figure 2: tested design valves shape and dimension was varied. Design number 1-1 showed best performance

3 FABRICATION

Hydrophobic channels were fabricated with PDMS. As PDMS is inherently hydrophobic, we could easily make the dispenser module by the replica molding process without any hydrophobic surface treatment. The mold was made of AZ4620 photoresist (10 μm thick) using photolithography. Then the channel was replicated from the mold and bonded onto a PDMS-coated glass slip by oxygen plasma treatment.

Various designs were tested and optimized to obtain the best working module (Figure 2). The round shape structural valves (1-1 & 1-2) showed better blocking performance than the linear shape valves, but were difficult to fabricate reproducibly. Additional vent channels (1-2, 2-2, & 3-2) did not affect the performance a lot.

The fidelity of fabrication was important. Little difference from the design could cause malfunctions. For example, the fidelity of the lithographic process was critical in controlling channel geometry accurately and preventing a malfunction of devices. Figure 3 explains the failure cases when the device was not fabricated as designed. The neck shape of structural valve easily fabricated wider than designed, and as a result, the wider passive valve failed to block the liquid effectively.

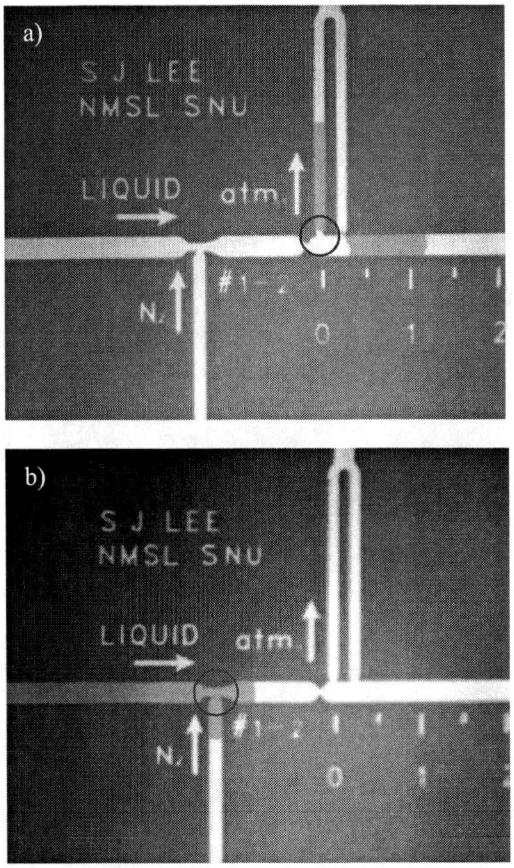

Figure 3: Failure cases: a) liquid blocks vent line; b) liquid blocks N_2 line (Liquids colored using computer graphics)

4 EXPERIMENTAL RESULTS

We successfully demonstrated the volume controllable fluidic dispenser modules. Figure 4 shows the working sequence of the module that operates reliably for many runs of experiments. In Figure 4 (a) it is observed that the liquid segment was stopped by surface-tension effect. This filling sequence was followed by pressurizing the N_2 line with a pressure higher than in the liquid line. The pressure made the segment start moving towards the outlet, but the segment stopped at the branch point because of the vent channel (Figure 3 (b), (c)). When this sequence was repeated the approaching segment joined the stopped one, and proceeded further automatically (Figure 3 (d), (e), (f)). By repeating this series of operations with different liquids, certain volume of mixture can be dispensed at a specific ratio.

Evaporation of liquid was observed during the experiment. The miniscule amount of liquid quickly evaporated because of the porosity of PDMS and high surface to volume ratio on microscale. The loss of liquid by evaporation is serious when the dispenser needs to be accurate. This problem could be overcome by non-porous coating of channel. For instance, parylene or Teflon could be available. Otherwise, this device can be fabricated with totally different materials such as glass since the basic working principle is universal.

5 DISCUSSION AND SUMMARY

It should be emphasized that the key function of the device comes from the operation with the vent channel and the surface wettability. Since the vent channel had a structural valve which is always in an off condition, liquid could not leak through the vent channel. Pressuring gas, on the other hand, could escape through the vent channel without any resistance. Thus the merging of metered segments was allowed without air trapping between them. By repeating this dispensing sequence, we could control the dispensing volume of liquid in a discrete manner without any bubble trapped.

We designed and fabricated a volume controllable micro dispenser module. The operation of the module was demonstrated. Essentially our device was based on the passive dispenser, but key modifications were added. The vent channel with a structural valve enabled a controllable dispensing of liquid volume and removal of bubbles. While the previous dispenser could meter only a pre-fixed volume of liquid, our device can digitally control the dispensing ratio during operation.

This paper also dealt with issues found during the study of the fabricated devices. The fabrication had to be very accurate and repeatable. When the device was fabricated with small error, correct function was not guaranteed. Especially, high quality photolithographic process was crucial for device performance. Liquid evaporation was occurred at a faster rate than expected and degraded the accuracy of dispensing.

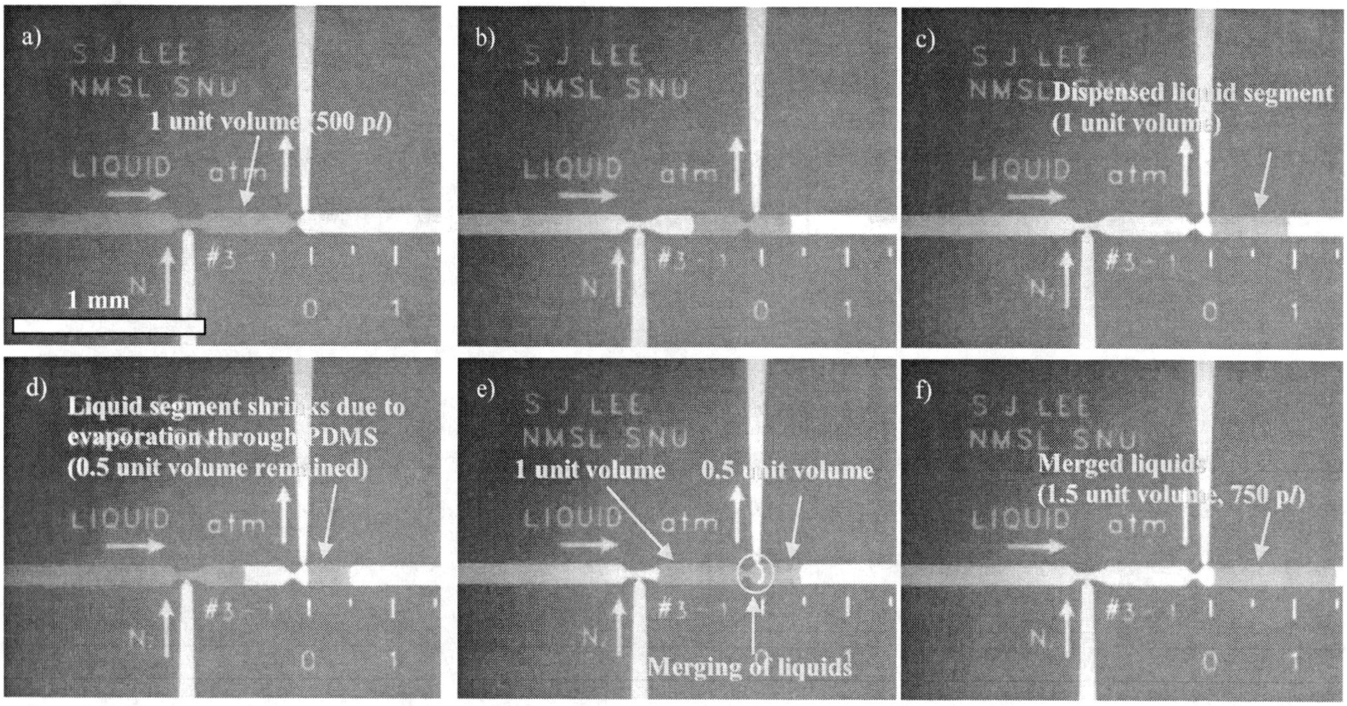

Figure 4: Microphotograph of volume controllable microdispenser: a) reservoir filled; b) splitting liquid column with pneumatic actuation; c) exact stopping right after passing vent channel; d) second reservoir filled, initially dispensed liquid drop does not move; e) merging two liquid drops with pneumatic actuation; f) exact stopping right after passing vent channel

Our dispensing module will find applications in a lab-on-a-chip and micro-TAS systems. Especially, when a large number of pipetting is required, for example, in a large-scale cell differentiation experiment and pharmaceutical research, this module will be useful. Unlike conventional microfluidic mixing modules that offer fixed ratios, our dispenser module can vary the mixing ratio during operation like a micropipette in macro scale experiment.

Acknowledgement

This research was performed for the Intelligent Robotics Development Program, one of the 21st Century Frontier R&D Programs funded by the Ministry of Commerce, Industry and Energy of Korea.

REFERENCES

[1] Choong Kim et al, "The serial dilution chip for cytoxicity and cell differentiation test", Proc. IEEE MEMS 2007, pp. 517-520.

[2] K. Handique et al, Proc. SPIE Conf. Micromachined Devices, pp. 185-195.

[3] A. Puntambekar et al, "Fixed-volume metering microdispenser module", Lab on a Chip 2 (2002), pp. 213-218.

[4] C.H. Ahn et al, Proc. Micro-Total Analysis Systems 2000, pp. 205-208.

Fabrication and Characterization of Thermo-pneumatic Peristaltic Micropumps

Hsin-Hung Liao and Yao-Joe Yang

Department of Mechanical Engineering,
National Taiwan University, Taipei, Taiwan
TEL: +886-2-33662712 Email: yjy@ntu.edu.tw

ABSTRACT

In this paper, the fabrication and characterization of thermopneumatic peristaltic micropumps are presented. Micropumps with three different designs are fabricated using soft lithography techniques. The optimal operating conditions, such as duty ratios, operating frequencies and backpressure, are obtained. The maximum flow rate occurs at a driving frequency of 1.5Hz with a duty ratio of 40%. Under zero backpressure, the maximum flow rates of the 3, 5 and 7-chamber devices are very close, while the devices with larger numbers of pumping chambers exhibit better pumping performance under higher backpressure.

Keywords: peristaltic micropump, thermopneumatic, MEMS, microfluidics, PDMS

1 INTRODUCTION

Microfluidic systems, realized by MEMS technologies, are widely employed for many biomedical applications. The advantages of these microfluidic devices include high throughput, rapid response, small size, low cost, low power consumption and reduction in reagents. In each microfluidic system, micropumps usually serve as the key components to drive and control small volumes of chemical and biological working fluids. For displacement micropumps [1], periodical volume strokes of fluids are generated by using existing actuating mechanisms to push or pull one or more moveable boundaries. The actuating mechanisms include piezoelectric [2], electromagnetic [3], electrostatic [4], pneumatic [5], thermopneumatic [6,7], and so on. One of the most popular types of displacement micropump is the peristaltic-type pump. A peristaltic micropump generates peristaltic motions of the diaphragms, which are arranged in series, for squeezing the fluid in a desired direction.

The actuation mechanism using thermopneumatic principle can generate large deflections of diaphragms with relatively low driving voltages and simple driving sequences. Therefore, a microsystem which employs thermopneumatic mechanism potentially can be realized in a very small package. Furthermore, thermopneumatic actuators can be easily fabricated by soft lithography micromachining process [8]. In this work, we fabricate and characterize three types of peristaltic micropumps based on the thermopneumatic mechanism. We will also measure the pumping performances of the devices, including the relationship between flow rate vs. duty ratio, the relationship between the flow rate vs. driving frequency (under zero backpressure), and the relationship between the flow rate vs. backpressure.

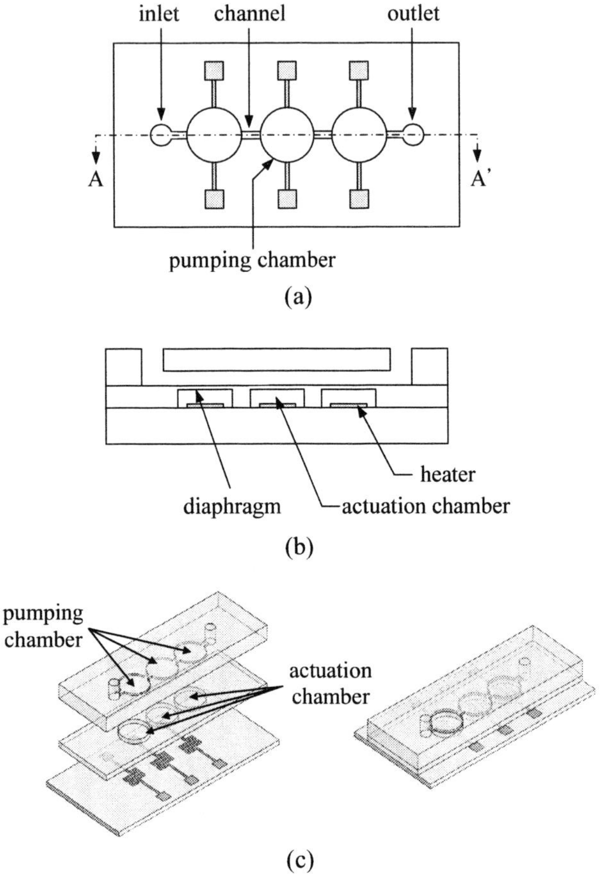

Figure 1. The schematic view of the micropump: (a) top view; (b) cross-sectional view; (c) 3D schematic illustration.

2 DESIGN

Figure 1 shows the schematic of the peristaltic micropump with three pumping chambers (the 3-chamber device). The micropump consists of three different layers: the channel layer, the chamber layer and the heater layer. The channel layer and the chamber layer are fabricated with PDMS elastomer, whereas the heater layer is realized by patterning metal film on a glass substrate. When the voltage is applied to the heater, the temperature of the air inside the *actuation chamber* increases, and consequently the air volume expands and the diaphragm is deformed to squeeze the fluid in the *pumping chamber*. As the temperature of the air inside the actuation chamber reduces due to natural cooling, the air volume decreases and the deformation of the diaphragm decreases. Figure 2 shows the working principles of the actuation sequences studied in this work. By controlling the movement of the diaphragms in sequence, the fluid can be conveyed.

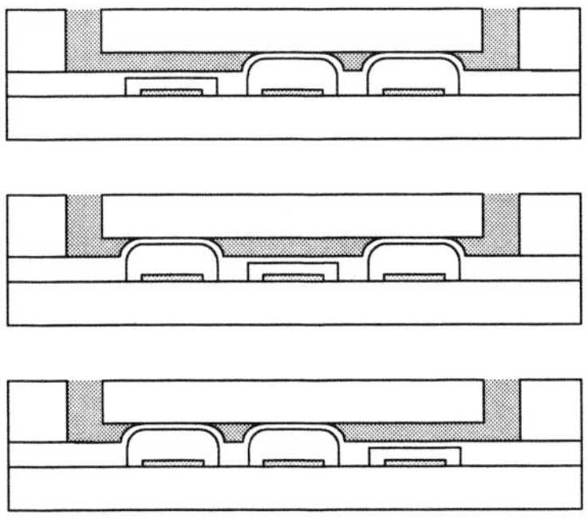

Figure 2. The actuation sequence for the peristaltic micropump.

3 FABRIACATION PROCESS

Figure 3 shows the fabrication process of the micropump. The channel and chamber layers are fabricated with PDMS by soft lithography technology. Figure 3(a)-3(d) show the fabrication process of the channel layer. PDMS prepolymer and curing agent (Sylgard® 184, Dow Corning Corp.) are mixed at 10:1 ratio. After stirred thoroughly and degassed in a vacuum chamber, the prepared PDMS mixture is poured onto a patterned SU-8 master (GM 1070, Gersteltec Sarl) of 50 μm-thick (Figure 3(a)-3(b)). After cured at 90°C for 60 min, the cured PDMS layer is peeled off from the master substrate (Figure 3(c)). The through-holes for the inlet and the outlet are punched using stainless steel pipe (Figure 3(d)). The chamber layer can be fabricated by the similar process, as shown in Figure 3(e)-3(g). The patterned SU-8 master is 180 μm in thickness (Figure 3(e)). Since the thickness of the chamber layer has to be controlled around 230 μm, the prepared PDMS mixture is poured and spun at 300 rpm for 10 sec (Figure 3(f)). After softly cured at 90°C for 30 min, the chamber layer is peeled off from the master (Figure 3(g)).

Figure 3(h)-3(j) show the fabrication process of the heater layer. A lift-off process is employed. Thick positive photoresist (AZ4620, Hoechst) is spin-coated (5,000 rpm, 50sec) and patterned on a glass wafer (Corning® 1737) (Figure 3(h)). A 200Å thick titanium and 3000Å thick gold layer is deposited by electron beam evaporation, then the heater pattern is formed by removing the photoresist (Figure 3(i)-3(j)).

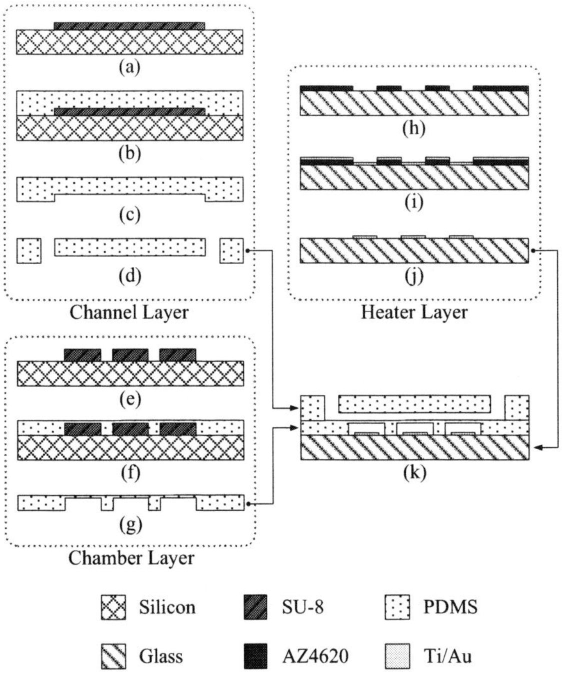

Figure 3. The fabrication process of the micropump.

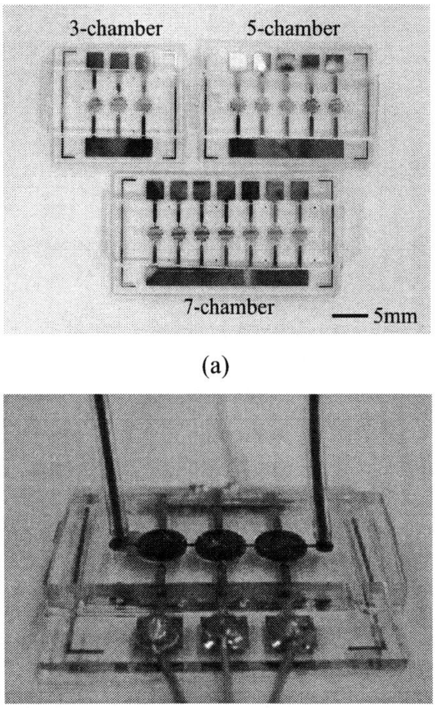

(a)

(b)

Figure 4. The photograph of the fabricated micropump: (a) the 3-chamber, 5-chamber, and 7-chamber micropumps; (b) the micropump after connecting polyethylene tube.

After oxygen plasma treatment (pressure 350 mTorr, RF Power 10.5W, 90 sec.), the channel layer and the chamber layer are bonded and then cured at 80°C for 5 minutes. The heater layer is also bonded to the other side of chamber layer after the similar oxygen plasma treatment (Figure 3(k)). Finally, polyethylene (PE) tubes are connected to the inlet/outlet and electrical wires are soldered on the pads of the heaters. Figure 4(a) shows the picture of the fabricated 3-chamber, 5-chamber and 7-chamber micropumps. Figure 4(b) is a 3-chamber device connected with PE tubes

4 EXPERIMENT AND RESULTS

Figure 5 shows the schematic of the experiment setup for measuring the flow rates of the micropump. The inlet and the outlet of the micropump are connected to reservoirs using polyethylene tubes (PE20, I.D. 0.38 mm, O.D. 1.09 mm, INTRAMEDIC®). The de-ionized water is used as the working fluid for the experiments. By controlling Δh shown in the figure, the backpressure between inlet and outlet can be adjusted. The heating voltage for each heater is controlled by a FET switch, which receives the time sequences generated by a PC-based data acquisition card.

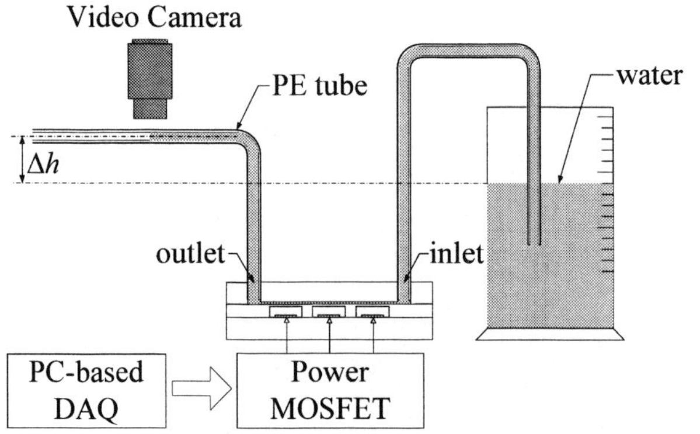

Figure 5. The setup for the measurement of the flow rates of the micropump.

Note that in the following experiments, the applied voltage to the heater inside the thermopneumatic actuation cell is fixed at 5V. The meniscus displacement of the fluid in the tube that is connected to the outlet is recorded by a video camera with a macro lens. From the meniscus displacement, the flow rate is able to be calculated.

Figure 6 shows the timing diagrams of the actuation sequence. In each actuation cycle, the time period T is evenly divided into *three* phase durations (T_{phase}). The duty ratio is defined as:

$$D = \frac{T_{heat}}{T_{phase}} \ (\%) \qquad (1)$$

where T_{phase} is the time duration of each phase, and T_{heat} is the heating time in the duration. Both T_{phase} and T_{heat} are indicated in Figure 6. Note that the driving frequency f is equal to $1/T$.

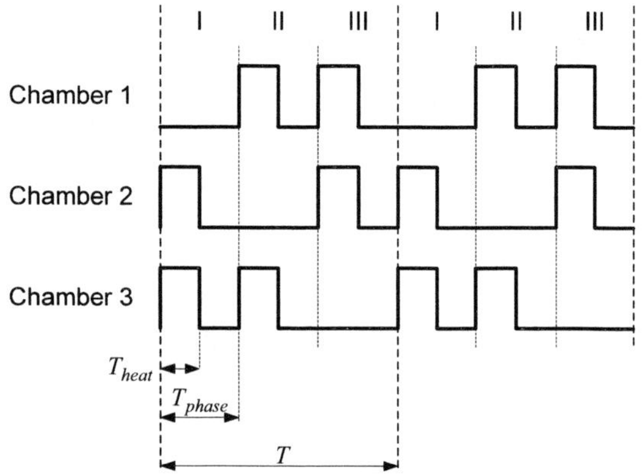

Figure 6. The timing diagram of the actuation sequence.

Figure 7 shows the measured flow rate versus the duty ratio for the 3-chamber device, under the condition of zero backpressure (i.e., $\Delta h = 0$ in Figure 5). As the driving frequency is either 1Hz or 2Hz, the maximum flow rate occurs when the duty ratio is about 40%.

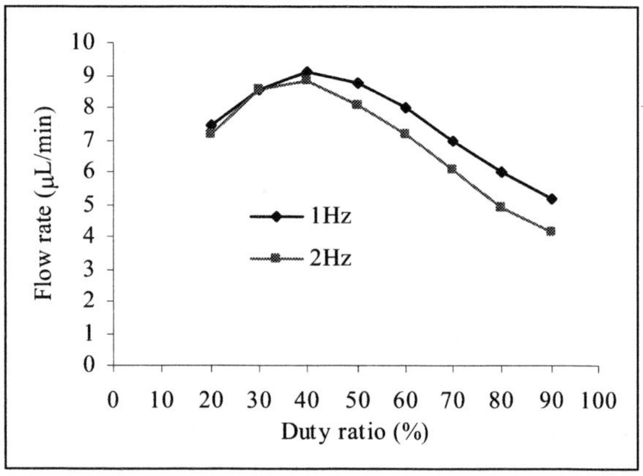

Figure 7. The measured flow rate vs. duty ratio.

Figure 8 shows the measured flow rate versus the frequency for the 3-chamber, 5-chamber and 7-chamber micropumps under zero backpressure. At low frequency (i.e., less than 1.5Hz), the flow rate increases with the frequency. However, the flow rate starts to decrease as frequency is greater than 1.5Hz. It is because the actuation amplitude of the diaphragm decreases with frequency due to insufficient heating. The measured maximum flow rate of these three types of micropumps occurs around 1.5Hz. The corresponding maximum flow rates for these three types of devices are very close. These results indicate that the flow rate of the micropumps is not affected by the numbers of the serial pumping chambers.

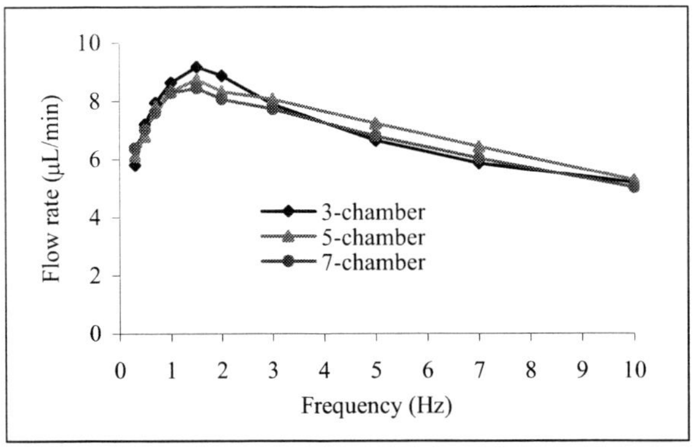

Figure 8. The measured flow rate vs. frequency for three different devices.

Figure 9 shows the measured relationship between the backpressure and the flow rate at 1.5Hz. The flow rate decreases linearly as the backpressure increases. In other words, the pumps with larger numbers of chambers exhibit better pumping flow rates under higher backpressure. The phenomenon can be explained by the numbers of *actuated pumping chambers* for each device during operation. The *actuated pumping chamber* is the pumping chamber which is under heating so that its diaphragm is deformed and the fluid inside the chamber is squeezed out. During operation, the pump with larger number of *actuated pumping chambers* has higher flow impedance, so it can resist higher backpressure. Obviously, the device with larger number of pumping chambers has larger number of *actuated pumping chambers*, so it has better pumping flow rates under backpressure.

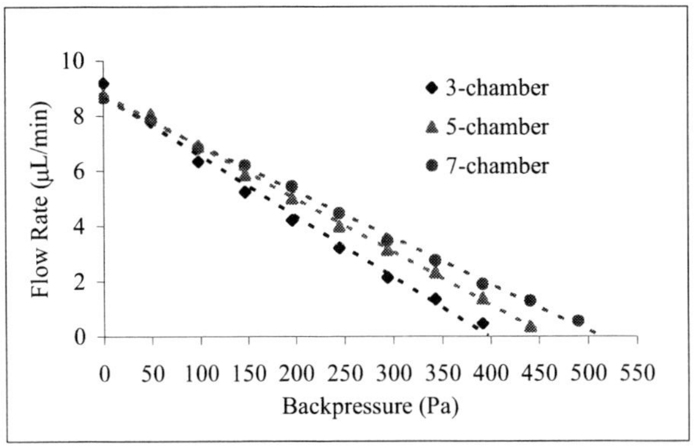

Figure 9. The measured flow rate vs. backpressure for three different devices.

5 CONCLUSIONS

In this work, the design, fabrication and measurement of thermopneumatic peristaltic micropump systems were presented. Micropumps with 3, 5 and 7 serial pumping chambers were fabricated using the soft lithography techniques. The pump systems can be operated with 5V input voltage. The optimal operating conditions, such as operating frequencies, backpressure and duty ratios, were obtained. The maximum flow rate occurs at the driving frequency of 1.5 Hz with a duty ratio of 40%. The pumps with larger numbers of pumping chambers have better pumping performance under the condition of higher backpressure. The advantages of the proposed micropump include simple design, easy fabrication, low operating voltage and biological compatibility.

ACKNOWLEDGEMENT

The authors would like to thanks Prof. Lung-Jieh Yang, Tamkang University (TKU) MEMS group, for providing manufacturing process equipments.

REFERENCES

[1] D. J. Laser and J. G. Santiago, "A review of micropumps," *Journal of Micromechanics and Microengineering*, Vol. 14, 2004, pp. 35-64

[2] L. S. Jang, Y. J. Li, S. J. Lin, Y. C. Hsu, W. S. Yao, M. C. Tsai and C. C. Hou, "A stand-alone peristaltic micropump based on piezoelectric actuation," *Biomedical Microdevices*, Vol. 9, 2007, pp. 185-194

[3] S. Haeberle, N. Schmitt, R. Zengerle and J. Ducrée, "Centrifugo-magnetic pump for gas-to-liquid sampling," *Sensors and Actuators A*, Vol. 135, 2007, pp. 28-33

[4] A. Machauf, Y. Nemirovsky and Uri Dinnar, "A membrane micropump electrostatically actuated across the working fluid," *Journal of Micromechanics and Microengineering*, Vol. 15, 2005, pp. 2309-2316

[5] O. C. Jeong and S. Konishi, "Fabrication and drive test of pneumatic PDMS micro pump," *Sensors and Actuators A*, Vol. 135, 2007, pp. 849-856

[6] W. Inman, K. Domansky, J. Serdy, B. Owens, D. Trumper and L. G. Griffith, "Design, modeling and fabrication of a constant flow pneumatic micropump," *Journal of Micromechanics and Microengineering*, Vol. 17, 2007, pp. 891-899

[7] O. C. Jeong, S. W. Park, S. S. Yang and J. J. Pak, "Fabrication of a peristaltic PDMS micropump," *Sensors and Actuators A*, Vol. 123-124, 2005, pp. 453-458

[8] Y. Xia and G. M. Whiteside, "Soft lithography," *Angewandte Chemie International Edition*, Vol. 37, 1998, pp. 550-575

Discrete Mixing of Nanoliter Drops in Microchannels

Minsoung Rhee[*], Mark A. Burns[*,**]

[*]Department of Chemical Engineering, the University of Michigan
2300 Hayward St. 3074 H.H. Dow Building, Ann Arbor, MI 48109-2136, minsoung@umich.edu
[**]Department of Biomedical Engineering, the University of Michigan
2200 Bonisteel Blvd, Ann Arbor, MI 48109-2099, maburns@umich.edu

ABSTRACT

The drop mixing occurs in three different regimes (diffusion, dispersion, and convection-dominated) depending on various operational parameters collectively expressed by the Péclet number ($Pe = U_d \cdot d/D$) and the drop dimensions. Introducing the modified Péclet number ($Pe^* = Pe \cdot d/L$), we present asymptotic curves to predict the mixing time and the required drop displacement distance for mixing at a given Pe^*. COMSOL simulations and on-chip experiments were performed to verify the theoretical limits. The simulated mixing results for the three different regimes show distinctly different mixing behaviors as predicted in the theoretical modeling. In our experimental work, we used a PDMS microchannel with a unique membrane air bypass valve (MBV) to precisely control the mixing and merging site. Finally, we show that experimental, simulation, and theoretical results all agree and confirm that mixing can occur in fractions of a second to hours. The presented work guides understanding of discrete drop mixing and show that efficient mixing in microfluidic systems can be achieved in a simple mixing scheme.

Keywords: drop mixing, micromixer, lab-on-a-chip

1 INTRODUCTION

In many microfluidic applications, rapid mixing is a major challenge because the time for this diffusive mixing often exceeds processing times for other steps. This slow transport time scale may be a bottleneck for many high-throughput microfluidic tests and has negative implications on microfluidic device fabrication [1-3].

A variety of micromixers have been described in the literature [4] to accelerate mixing. Micromixers can be categorized as passive or active. Passive micromixers rely on geometrical properties of channel shape to induce complicated fluid-particle trajectories and thus greatly enhance mixing. The mixing process in these systems is dominated by diffusion or chaotic advection. Various passive mixers that have been reported include techniques such as hydrodynamic focusing [5], packed-bed mixing [6], multi-lamination [7], injection [8], chaotic advection [9], and droplet mixing [10]. Passive micromixers often require deliberate fabrication of channel geometry, but do not require external actuators. In contrast, active micromixers accommodate one or more of the externally-generated disturbance forces.

Mixing in microfluidic systems occurs not only in a continuous flow but also between two drops in a discrete or batch operation. While most micromixers are used in continuous-flow systems, several studies have been reported on microfluidic mixing in batch systems [3, 11]. As opposed to continuous-flow systems, mixing in batch systems can be enhanced by convection [12]. During the course of drop transportation, the internal circulation streamlines of liquid in a moving discrete drop [13] allows convective mixing as well as molecular diffusion.

2 THEORETICAL MODELING

We assume a pressure driven flow in a slit-type microchannel (w>>d) (Fig. 1a). Because of this aspect ratio, we can further assume that the mass transfer is independent of the channel width direction (z-axis) and thus use a two-dimensional model. The liquid inside a discrete drop moving in a microchannel experiences internal circulation within itself [13] (Fig. 1b). The striations produced by the circulation create diffusion distances that are a fraction of the channel depth thus greatly reducing diffusion times.

When the combined sequential drops move along the channel, the solute can either be transported convectively or diffusively to a region of different concentration. The governing transport equation for the slit-channel model is simplified as follows [12].

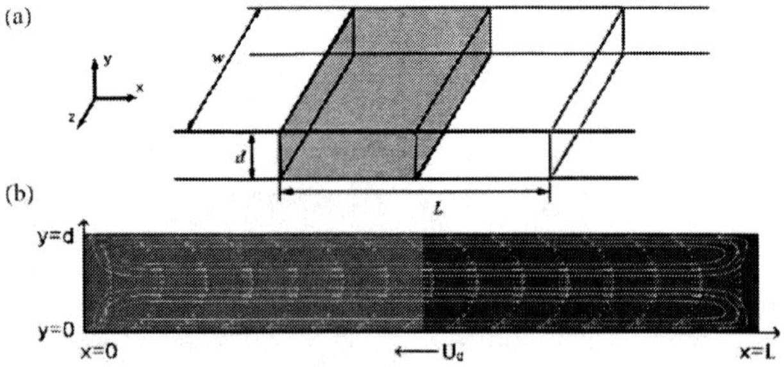

Figure 1. Discrete Drop Mixing Model. (a) An axially-arranged discrete drop placed in a slit-type microchannel (b) Internal circulation streamlines. The arrows represent the velocity vectors in the x-y plane.

$$\frac{\partial c}{\partial t}+U(y)\frac{\partial c}{\partial x}=D\left(\frac{\partial^2 c}{\partial x^2}+\frac{\partial^2 c}{\partial y^2}\right) \quad (1)$$

where $U(y)=0.5U_d\left(1-3\left(\frac{y}{d/2}\right)^2\right)$. (2)

3 RESULTS AND DISCUSSION

3.1 Analytical Solutions

The relative importance of convection to diffusion in mass transport is usually expressed as the Péclet number (Pe), where $Pe=U_d d/D$, which is useful to estimate the mixing time. Although a direct analytical solution to the Eq (1) is not available, the equation can be further simplified to estimate the mixing time when the Péclet number is within certain ranges: (i) when Pe is very small, (ii) when Pe is in an intermediate range, and (iii) when Pe is very large.

Diffusion-dominated Mixing.

When the Péclet number is very small ($Pe \ll 1$), convection is very slow compared to diffusion and the solute is transported entirely by diffusion. We assume that the y-axis dependence can be neglected and Eq (1) can be simplified and gives a cosine series solution as follows.

$$\theta_1=\frac{1}{L/2}\int_0^{L/2}\theta(x)dx=0.5+\sum_{\substack{n=1\\ odd}}^{\infty}\frac{4}{(n\pi)^2}\exp\left(-\frac{t}{\tau_L}n^2\pi^2\right) \quad (3)$$

$$\theta_2=\frac{1}{L/2}\int_{L/2}^{L}\theta(x)dx=0.5-\sum_{\substack{n=1\\ odd}}^{\infty}\frac{4}{(n\pi)^2}\exp\left(-\frac{t}{\tau_L}n^2\pi^2\right). \quad (4)$$

where $\tau_L=L^2/D$ and θ is the average dimensionless concentration of the left(1) or right(2) domain in Figure 1c, which is a function of the axial position (x) and time (t).

We assume that the effective mixing time t_{mix} is the time that elapsed until the drop is mixed 90% ($\theta_1<0.55$ and $\theta_2>0.45$, respectively). The shortest t_{mix} that satisfies the above conditions is given by $0.21\tau_L$. The mixing time can be further non-dimensionalized with the reference diffusion time $\tau_D=d^2/D$ and the aspect ratio, $\varepsilon=L/d$.

$$\tau_{mix}=\frac{t_{mix}}{\tau_D}=0.21\frac{\tau_L}{\tau_D}=0.21\frac{L^2/D}{d^2/D}=0.21\varepsilon^2 \quad (5)$$

In Figure 2a, the diffusion-dominated mixing time is plotted at various drop aspect ratios. Note that the mixing time is neither dependent on the drop velocity nor Pe but is proportional to the square of the drop aspect ratio, L/d. Since the mass transport is caused entirely by diffusion, the mixing time becomes independent of Pe, but as the drop size gets longer, the mixing time increases because the length for diffusion along the drop increases.

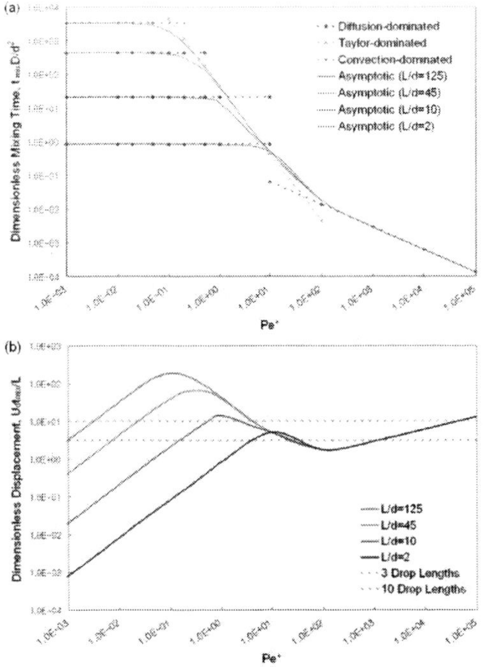

Figure 2. Analytical Quantification of Mixing. (a) Three different mixing regimes and asymptotic curves for mixing time at various drop aspect ratios. (b) Asymptotic curves for drop displacement required for mixing at various aspect ratios.

Dispersion-dominated Mixing.

For $1 \ll Pe < \pi^2\varepsilon$, the mixing can be described by Taylor dispersion. For slit-type channels, the Taylor dispersion coefficient, D_{TD}, is given by the following relationship [14]:

$$D_{TD}=\frac{U_d^2 d^2}{210D}=\frac{1}{210}Pe^2 D. \quad (6)$$

The diffusion coefficient in the right-hand side of Eq (1) can be replaced with D_{TD} without significant loss of accuracy because Pe is much greater than the unity. The replacement of D results in a mixing time of

$$\tau_{mix}=\frac{t_{mix}}{\tau_D}=0.21\frac{\tau_{TD}}{\tau_D}=0.21\frac{L^2/D_{TD}}{d^2/D}=44.1\frac{\varepsilon^2}{Pe^2} \quad (7)$$

When the transport is dominated by Taylor dispersion, the mixing time is proportional to the inverse square of Pe, indicating that the mixing is enhanced as the drop velocity increases or the drop aspect ratio decreases.

To better explain the correlation between mixing and the aspect ratio, a modified Péclet number has been suggested. The main idea behind this modification is that when convective mixing occurs, convection and diffusion usually develop in different directions. Thus, for a system with a high aspect ratio, the Péclet number provides

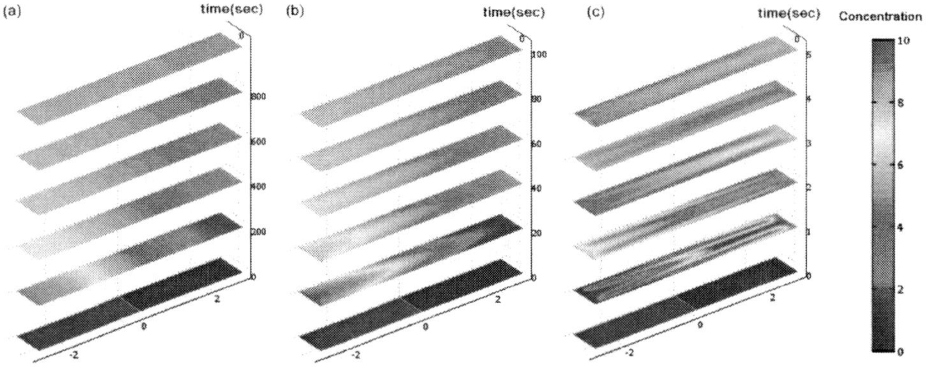

Figure 3. COMSOL 2D Simulation. Concentration variation in time from bottom to top in a straight channel (ε=7) for (a) Pe*=0.4, (b) Pe*=6, and (c) Pe*=132

insufficient information about the importance of convection and diffusion. The modified Péclet number is defined by,

$$Pe^{*}=\frac{diffusive\ time\ scale}{convective\ time\ scale}=\frac{d^2/D}{L/U_d}=\frac{Pe}{\varepsilon}. \quad (8)$$

With Pe*, Eq (15) can be now rewritten by

$$\tau_{mix}=\frac{t_{mix}}{\tau_D}=44.1(Pe^{*})^{-2}. \quad (9)$$

and the dependency of the dispersion-dominated mixing time on Pe* can be seen in Figure 2a. The mixing time is independent of the aspect ratio as a function of Pe*.

Convection-dominated Mixing.

When the drop velocity is sufficiently fast (Pe*>> π^2), multiple layers develop due to the internal circulatory flow. After a certain mixing time t_C, the drop travels N= $U_d t_C/L$ drop lengths and makes n=N/3=$U_d t_C/3L$ internal circulations. The internal circulations cause the solute layers to fold over themselves every half circulation, and thus the number of foldings (N_F) correspond to N_F=2n or N_F=2N/3. Such circulations would create 2.25N_F or 9N_F/4 layers. Thus, the number of layers (N_L) can be calculated as a function of the number of drop lengths that the drop travels using the equation

$$N_L=\frac{9}{4}N_F=\frac{3}{2}N. \quad (10)$$

Mixing occurs when the solutes diffuse between two adjacent layers. When there are N_L layers within the depth d, the interlayer distance, s, is approximated as s=d/2N_L=d/3N. If the interlayering continues until the solute diffuses to complete mixing (90%), the radial diffusion reference timescale 0.21s^2/D must be balanced with t_C (i.e., t_C=0.21s^2/D). The mixing time is then calculated as follows:

$$\tau_{mix}=\frac{t_{mix}}{\tau_D}=\frac{t_C}{\tau_D}=(0.21)^{1/3}\left(\frac{1}{3}\right)^{2/3}\left(\frac{L/U_d}{d^2/D}\right)^{2/3}=0.286(Pe^{*})^{-2/3} \quad (11)$$

Drop Displacement Required for Mixing.

Another problem of great relevance to experimentalists is to find the optimal geometry to give the lowest possible mixing time. A dimensionless displacement for mixing, λ_{mix}, can be defined as the ratio of the displacement versus the drop length,

$$\lambda_{mix}=\frac{U_d t_{mix}}{L}=\tau_{mix}\left(\frac{d^2/D}{L/U_d}\right)=\tau_{mix}(Pe^{*}). \quad (12)$$

Figure 2b plots the estimated displacement required for mixing. When Pe* is very small, the displacement is proportional to Pe* or the drop velocity, because the drop is constantly moving during slow diffusive mixing. In contrast, the required displacement in the Taylor dispersion-dominated mixing regime (π^2/ε<<Pe*<< π^2) decreases with Pe*, since the convective flow greatly improves mixing efficiency. Finally, when the flow is too fast (Pe*>> π^2), the required displacement increases algebraically with Pe*, although the mixing time decreases.

3.2 Drop Mixing Simulations

Computer simulations were performed to confirm the theoretical modeling results. In Figure 1b, the upper and lower walls move in the negative x-direction with the

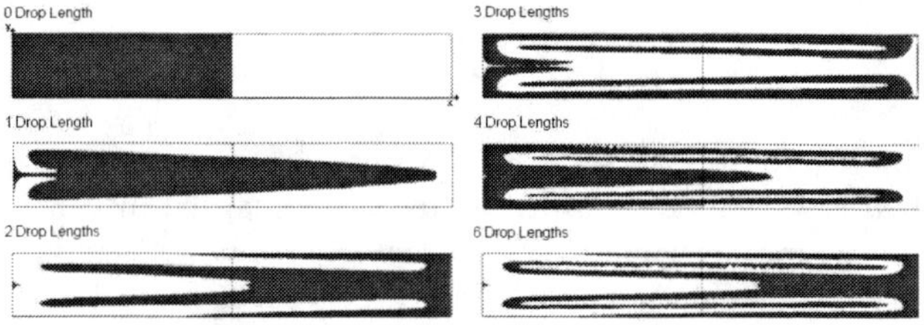

Figure 4. Simulated interlayering progress for ε=7 and Pe*=2000.

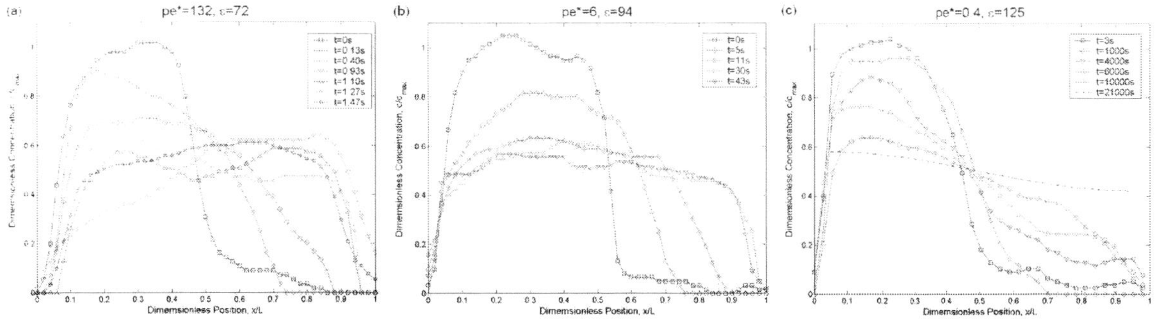

Figure 5. Average concentration profile at different times. (a) Convective Mixing Progress (Pe*=132). (b) Dispersive Mixing Progress (Pe*=6). (c) Diffusive Mixing Progress (Pe*=0.4).

velocity $-U_d$. This motion simulates a drop moving through a straight channel at the velocity of U_d. Combined with the no-slip boundary conditions at the front and back ends of the drop, the internal circulatory flows inside the drop were simulated.

In Figure 3, the mixing progress in a straight channel for (a) Pe*=0.4, (b) Pe*=6, and (c) Pe*=132 were simulated, respectively. Note that the mixing for Pe*=0.4 and Pe*=6 is dominated by diffusion and Taylor-dispersion, respectively. No apparent sign of convective transport by internal circulation is observed. In contrast, interlayering is observed in Figure 3c indicating that convection has a significant role in mixing for Pe*=132. Figure 4 illustrates the simulated interlayering progress for Pe*=2000. The layers remain for an extended time because the convective transport is much faster than the diffusive transport.

3.3 Experimental Verification

In experiments, one can alter the mixing condition (Pe*) by controlling the drop velocity. Figure 5a shows the experimental mixing progress at a high Pe* (~132), starting from the initial state when the two drops fuse and an interface develops, to the state of complete mixing. Note that the merged drop rapidly mixes within a few seconds.

In the intermediate range of Pe*(~6) (Fig. 5b), however,

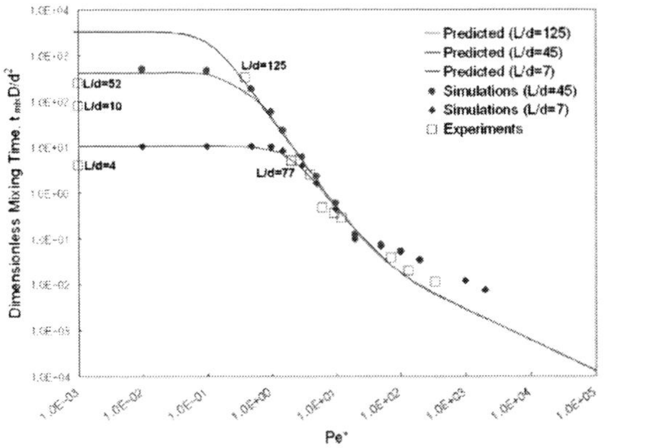

Figure 6. Mixing time comparison among theoretical predictions, simulations, and experimental results.

the mixing occurs more slowly. When Pe* is small (~0.4) the merged drop has been displaced 63 mm for ~10000 seconds back and forth along the channel but the mixing was quite incomplete (Fig. 5c). In Figure 6, we show that our experimental results are in good agreement with the theoretical modeling as well as simulation results.

ACKNOWLEDGMENT

The authors would like to acknowledge the National Institutes of Health (P01-HG001984, R01-AI049541, R01-GM-37006-17) for partial funding of this work.

REFERENCES

[1] Song, H.; Chen, D. L.; Ismagilov, R. F. *Angew. Chem. Int. Ed. Engl.* **2006**, *45*, 7336-7356.
[2] Burns, M. A. *Science*. **2002**, *296*, 1818-1819
[3] Burns, M. A. *et al. Science* **1998**, *282* (5388), 484-487
[4] Nguyen, N.; Wu, Z. *J. Micromech. Microeng.* **2005**, *15*, R1–R16
[5] Knight, J. B.; Vishwanath, A.; Brody, J. P.; Austin, R. H. *Phys. Rev. Lett.,* **80**, 3863–3866.
[6] Lin, Y.; Gerfen, G. J.; Rousseau, D. L.; Yeh, S. R. *Anal. Chem.* **2003**, *75*, 5381–5386.
[7] Schwesinger, N.; Frank, T.; Wurmus, H. *J Micromech. Microeng.* **1996**, *6*, 99–102.
[8] Okkels, F.; Tabling, P. *Phys. Rev. Lett.* **2004**, *92*, 038301.
[9] Bertsch, A.; Heimgartner, S.; Cousseau, P.; Renaud, P. *Lab Chip*. **2001**, *1*, 56–60.
[10] Burns, J. C.; Ramshaw, C. *Lab Chip*, **2001**, *1*, 10–15.
[11] Thorsen, T. ; Maerkl, S. J.; Quake, S. R. *Science,* **2002,** *298*, 580-584.
[12] Handique, K.; Burns, M. A. *J. Micromech. Microeng.* **2001**, *11*, 548-554.
[13] Duda, J. L.; Vrentas, J. S. *J. Fluid Mech.* **1971**, *45*, 247–260.
[14] Taylor, G., *Proc. R. Soc. London, Ser.* A. **1953**, *219*, 186-203.

Modeling Magnetic Transport of Nanoparticles with Application to Magnetofection

E. P. Furlani and K. C. Ng
The Institute for Lasers, Photonics and Biophotonics, University at Buffalo SUNY
432 Natural Sciences Complex
Buffalo, NY 14260-3000
efurlani@buffalo.edu,

ABSTRACT

We present a model for predicting the transport and accumulation of magnetic nanoparticles in magnetophoretic systems. The model involves the solution of a drift-diffusion equation that governs the concentration of nanoparticles in a fluidic chamber. We solve this equation numerically using the finite volume method. We apply the model to the magnetofection process wherein the magnetic force produced by a rare-earth magnet attracts magnetic carrier particles with surface-bound gene vectors toward target cells for transfection. We study particle accumulation as a function of key variables.

Keywords: magnetofection, magnetic particle transport, drift-diffusion analysis, gene transfection, magnetic biotransport, magnetophoresis, gene delivery

1 INTRODUCTION

Magnetic nanoparticles are used in a broad range of bioapplications, primarily as carriers for biomaterials such as cells, proteins, antigens and DNA [1]. The use of biofunctional magnetic particles enables selective immobilization and transport of a biomaterial using an applied magnetic field. Moreover, magnetic biotransport enables accelerated delivery of biomaterial thereby overcoming diffusion-limited accumulation. This enables applications such as magnetofection, which involves the transfection of cells in a culture. In this process magnetic nanoparticles with surface-bound gene vectors are directed towards target cells using an applied magnetic field [2]. In a typical *in vitro* magnetofection system target cells are located at the bottom of a fluidic chamber (well of a culture plate), and a rare-earth magnet beneath the chamber provides a magnetic force that attracts the biofunctional particles towards the cells as shown in Fig. 1. The magnetic force accelerates the nanoparticle transport and enables rapid process times with significantly improved transfection rates.

In this paper we study the transport and accumulation of magnetic nanoparticles in a magnetophoretic system that consists of a fluidic chamber positioned above a cylindrical rare-earth magnet as shown in Fig. 1 We model particle transport using a drift-diffusion equation that governs the particle concentration within the chamber. We solve this equation numerically using the finite volume method (FVM), and we apply boundary conditions that mimic the magnetofection process. We use the model to study particle transport and accumulation. Our analysis indicates that the rate of accumulation at the base of thechamber can be controlled by choosing different sized particles and/or by adjusting the spacing between the magnet and the chamber. The model provides insight into the physics of particle transport and is useful for optimizing the performance of novel magnetofection systems.

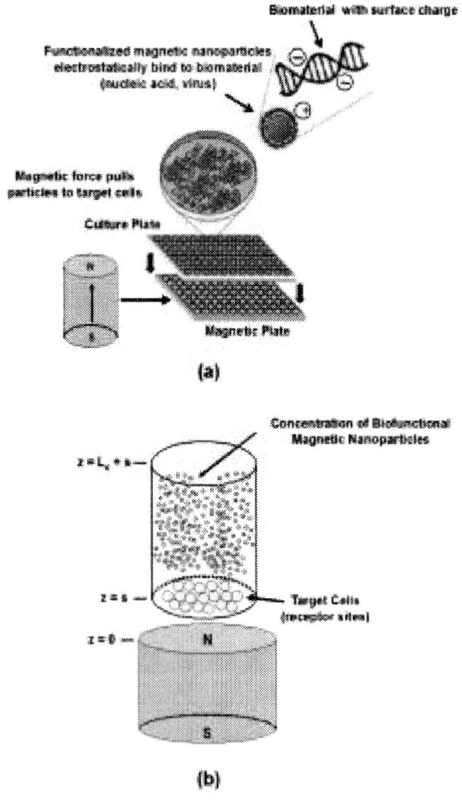

Figure 1: Magnetofection: (a) array of cell cultures positioned above an array of cylindrical rare-earth magnets (adapted from [6]), (b) the magnetic force pulls magnetic nanoparticles with surface bound gene vectors towards the cells.

2 THEORY

We model the transport of the nanoparticles using a drift-diffusion equation for the particle volume concentration $c(t)$ [3-4],

$$\frac{\partial c}{\partial t} = \nabla \bullet (\mathrm{D}\nabla c - \mathbf{U}c) \tag{1}$$

where $D = \mu kT$, $\mu = 1/(6\pi\eta R_{p,hyd})$ is the mobility of a particle with an effective hydrodynamic radius $R_{p,hyd}$, which takes into account surface-bound biomaterial, and η is the fluid viscosity. $\mathbf{U} = \mu\mathbf{F}$ is the drift velocity and $\mathbf{F}$ is the total force on the particle. In our analysis we take into account the fluidic $\mathbf{F}_\mathrm{f}$, magnetic $\mathbf{F}_\mathrm{m}$ and gravitational (buoyancy) $\mathbf{F}_g$ forces. We solve Eq. (1) numerically using the finite volume method (FVM). We use analytical expressions for the force components (drift velocity).

2.1 Magnetic Force

The magnetic force is given by

$$\mathbf{F}_\mathrm{m} = \mu_f \left(\mathbf{m}_{\mathrm{p,eff}} \bullet \nabla \right) \mathbf{H}_\mathrm{a} \tag{2}$$

where μ_f is the permeability of the transport fluid, $\mathbf{m}_{\mathrm{p,eff}}$ is the "effective" dipole moment of the particle, and $\mathbf{H}_\mathrm{a}$ is the applied magnetic field intensity at the center of the particle. A model for $\mathbf{m}_{\mathrm{p,eff}}$ has been developed that takes into account self-demagnetization and magnetic saturation of the particle, i.e. $\mathbf{m}_{\mathrm{p,eff}} = \mathrm{V}_p f(\mathrm{H}_\mathrm{a})\mathbf{H}_\mathrm{a}$, where [4-8]

$$f(\mathrm{H_a}) = \begin{cases} \dfrac{3(\chi_\mathrm{p} - \chi_\mathrm{f})}{(\chi_\mathrm{p} - \chi_\mathrm{f}) + 3} & \mathrm{H_a} < \left(\dfrac{(\chi_\mathrm{p} - \chi_\mathrm{f}) + 3}{3\chi_\mathrm{p}} \right) \mathrm{M_{sp}} \\[2ex] \mathrm{M_{sp}} / \mathrm{H_a} & \mathrm{H_a} \geq \left(\dfrac{(\chi_\mathrm{p} - \chi_\mathrm{f}) + 3}{3\chi_\mathrm{p}} \right) \mathrm{M_{sp}} \end{cases} \tag{3}$$

V_p is the volume of the particle, χ_p and χ_f are the magnetic susceptibilities of the particle and fluid, respectively, and $\mathrm{M_{sp}}$ is the saturation magnetization of the particle.

2.2 Magnetic Field

We obtain an analytical expression for the field distribution by treating the magnet as an equivalent current source (section 3.3 of [9]). A cylindrical magnet that is uniformly magnetized along its axis produces the same field as a sheet of current that circulates around its circumference. We obtain the field distribution above the magnet by decomposing the "equivalent" current sheet into infinitesimal current loop elements and integrating the field contributions from the individual elements. The field solution for a current loop is well-known (p 263 in [10]). If the magnet is magnetized to saturation M_s, and centered about the z-axis with its top surface at $z = 0$ as shown in Fig. 1b, the applied field is given by

$$H_{ar}(r,z) = \frac{M_s}{2\pi} \int_{-L_m}^{0} \Pi_r(r,z)dz \qquad H_{az}(r,z) = \frac{M_s}{2\pi} \int_{-L_m}^{0} \Pi_z(r,z)dz, \tag{4}$$

where

$$\Pi_r(r,z) = \frac{z}{r\left((R_m + r)^2 + z^2\right)^{1/2}} \left[\frac{\left(R_m^2 + r^2 + z^2\right)}{\left((R_m - r)^2 + z^2\right)} E(k) - K(k) \right], \tag{5}$$

and

$$\Pi_z(r,z) = \frac{1}{r\left((R_m + r)^2 + z^2\right)^{1/2}} \left[\frac{\left(R_m^2 - r^2 + z^2\right)}{\left((R_m - r)^2 + z^2\right)} E(k) + K(k) \right], \tag{6}$$

In these expressions $K(k)$ and $E(k)$ are the complete elliptic integrals of the first and second kind, respectively, and $k^2 = \frac{4rR_m}{(R_m + r)^2 + z^2}$ where L_m and R_m are the length and radius of the magnet. We substitute Eqs. (4)-(6) into Eq. (2) and obtain an analytical expression for $\mathbf{F}_\mathrm{m}$, which is used in the numerical solution of Eq. (1).

3. MAGNETOFECTION

We use Eq. (1) to study the magnetofection process. Specifically, we predict the accumulation of magnetite (Fe_3O_4) nanoparticles at the base of a cylindrical fluidic chamber positioned above a rare-earth NdFeB magnet. The chamber has a radius R_c = 2 mm and length L_c = 3 mm, and is positioned 1 mm above the magnet, which has a radius R_m = 2.5 mm. length L_m = 5 mm, and is magnetized to saturation, $M_s = 8\times10^5$ A/m. The dimensions and parameters above are representative of a standard 96 well culture plate system that is used for magnetofection. We assume that the fluid in the chamber is nonmagnetic ($\chi_\mathrm{f} = 0$) with a viscosity and density equal to that of water, $\eta = 0.001$ N·s/m^2, and $\rho_f = 1000$ kg/m^3. Fe_3O_4 nanoparticles have a density $\rho_p = 5000$ kg/m^3 and a saturation magnetization $\mathrm{M}_{sp} = 4.78\times10^5$ A/m. Without loss of generality, we assume that the hydrodynamic radius of a particle is the same as its physical radius $R_{p,hyd} = R_p$.

We solve Eq. (1) subject to an initial condition in which there is a uniform particle concentration throughout the chamber. We apply a zero-flux Neumann boundary condition at the top of the chamber and a Dirichlet condition at the bottom. The latter condition mimics the magnetofection process wherein nanoparticles that reach the bottom of the chamber are removed from the computation as it is assumed that they bind with receptor sites on target cells, and therefore no longer influence particle transport. It is assumed that there are a sufficient number of receptors to accommodate all of the particles in the chamber. We compute particle accumulation by summing the number of particles that reach the base of the chamber during each time step.

In our first study, we fix the particle radius R_p = 50 nm and compute particle accumulation, with and without a magnetic field (Fig. 2a). The analysis indicates that the rate of magnetically induced accumulation is orders of magnitude faster than that of diffusion-limited (no field) accumulation, which is consistent with experimental data shown in Fig. 2b [11]. We have performed similar calculations using a series of magnet to chamber spacings, and have found that the rate of accumulation increases substantially with decreasing separation.

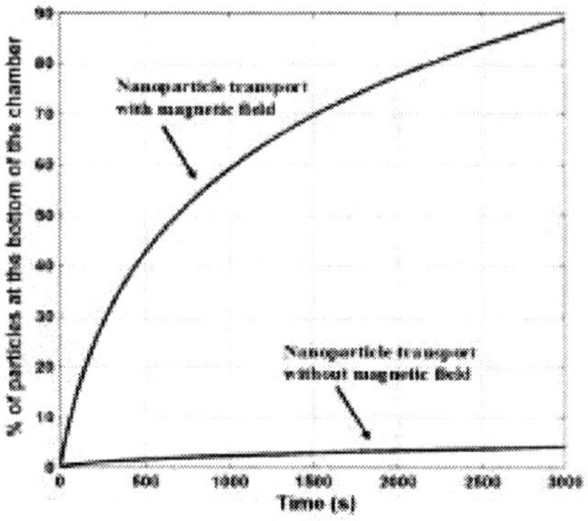

(a)

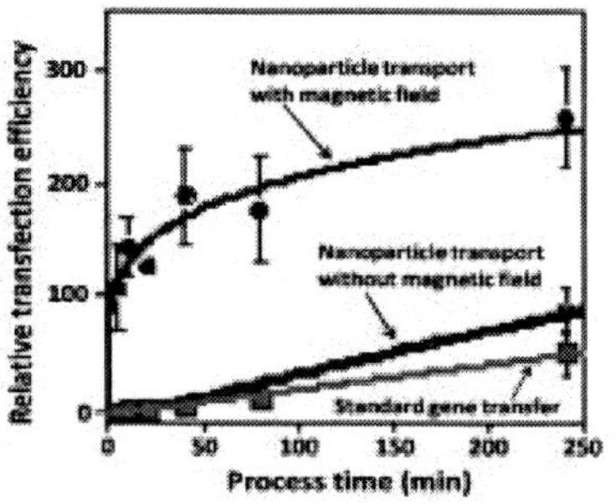

(b)

Figure 2: Nanoparticle accumulation: (a) predicted % of particles at the base of the chamber, (b) measured relative transfection efficiency, with and without an applied field (adapted from [11]).

Next, we compute the distribution of accumulated particles at the base of the chamber. We set R_p = 100 nm. The radial distribution of accumulated particles represented in terms of the particle volume concentration per unit area at the base of the chamber is shown in Fig. 3. Note that a higher concentration of particles occurs near the z-axis (centerline) above the magnet. This is due to the magnetic focusing of the particles during transport. There is also a local maximum in the concentration in an annulus around the centerline, which is due to the off-axis peaks in the axial force F_{mz} near the magnet. In summary, the magnetic particles accumulate towards the center of the chamber away from the edge of the magnet, where they are repelled inward by the magnetic force. This implies that higher transfection efficiencies would be achieved when the fluidic chamber has a smaller radius than the magnet.

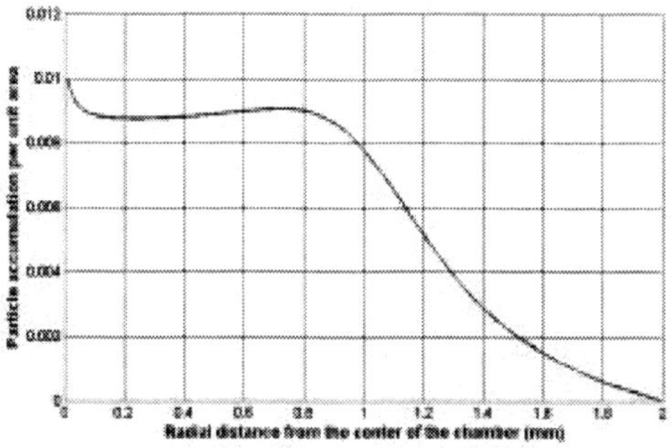

FIG 5

Figure 3: Distribution of particle accumulation per unit area at the base of chamber.

4 PARTICLE SATURATION

The drift-diffusion equation (1) does not account for particle volume saturation, i.e. it does not limit the number of particles that can physically occupy a given volume at a given time. We have developed a numerical method for addressing this issue, and have adapted it to the FVM. In the FVM the physical domain is discretized into a system of computational cells with each cell centered on a computational node. In 1D, the FVM discretization of Eq. (1) at an interior node is

$$c_i^{n+1} = c_i^n + \frac{\delta t}{\delta z}(F_{i+1/2} - F_{i-1/2}) \quad (i = 1,2,\ldots,N_z) \qquad (7)$$

where $\delta z = L_c / N_z$ is the length of a computational cell and L_c is the length of the chamber. c_i^n and c_i^{n+1} are the values of the concentration at the i'th computational node at time steps n and n+1, respectively. $F_{i\pm1/2}$ is a discretized version of the particle flux $D\frac{\partial c}{\partial z} - Uc$ at the edges of the computational cell $z_{i\pm1/2}$. As noted, we need to solve Eq. (7) taking into account particle accumulation/saturation, as each computational cell can only hold a finite number of finite-sized nanoparticles. We apply a two-step algorithm to account for. In the first step, we choose an initial time increment $\delta t_{initial}$ and use this in Eq. (7) to compute a temporary concentration at each computational node, i.e.,

$$c_{itemp}^{n+1} = c_i^n + \frac{\delta t_{initial}}{\delta z}(F_{i+1/2} - F_{i-1/2}) \quad (i = 1,2,\ldots,N) \qquad (8)$$

NSTI-Nanotech 2008, www.nsti.org, ISBN 978-1-4200-8505-1 Vol. 3

Next, we check to see if any of the c_{itemp}^{n+1} values are greater than the saturation concentration, which we denote by c_{sat}. If so, we identify the largest such temporary value and denote its index by $i_{\max}$, and its value by $\max(c_{itemp}^{n+1})$. We use this maximum value to compute a new global time increment δt_{new},

$$\delta t_{new} = \left(\frac{c_{sat} - c_i^n}{\max(c_{itemp}^{n+1}) - c_i^n} \right) \delta t_{initial} \tag{9}$$

In the second step of the algorithm, we recompute the concentration at each computational node using δt_{new} (i.e., substitute δt_{new} for $t_{initial}$ in Eq. (8)). It follows that $c_{i_{\max}}^{n+1} = c_{sat}$, and once a node is saturated we set the particle flux through its boundaries to zero for all subsequent times i.e. $c_i^{n+1} = c_{sat} \Rightarrow F_{i\pm1/2} = 0$. The same algorithm can be used for 2D and 3D drift/diffusion analysis. Figure 4 illustrates a 1D implementations of the algorithm showing computational grids with regions of particle saturation. In this figure, a permanent magnet provides a magnetic force that moves the nanoparticles from left to right, producing regions of saturation on the right-hand-side of the computational domain. All boundaries are walled, i.e. zero flux. A 1D simulation showing particle accumulation is

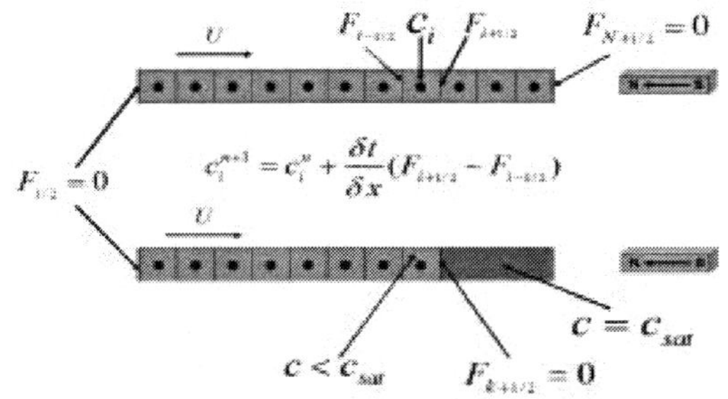

Figure 4: 1D FVM with particle saturation.

shown in Fig. 5. In this analysis the particles are initially distributed in the left half of the computational domain, which has closed (zero-flux) boundaries, and the magnetic force attracts particles from left to right.

5 CONCLUSIONS

Magnetic nanoparticles are increasingly used in bioapplications for the controlled transport of biomaterials. We have presented a model for predicting the transport and accumulation of such particles, and we applied it to the magnetofection process. We have found that magnetically-induced particle transport enables accumulation that is orders of magnitude faster than diffusion-limited accumulation, which is consistent with experimental observations. The model presented here provides a fundamental understanding of magnetophoresis at the nanoscale and enables optimization of novel magnetofection systems.

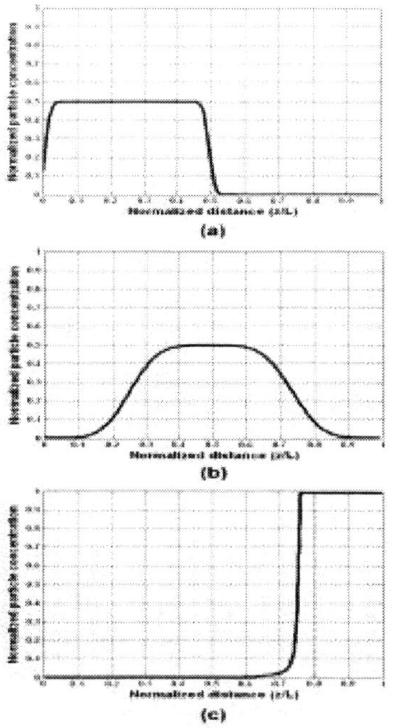

Figure 5: 1D particle concentration distribution analysis : (a) initial distribution, (b) intermediate distribution, (c) final distribution showing volume saturation.

REFERENCES

[1] Q. A. Pankhurst, J. Connolly, S.K. Jones, and J. Dobson, J. Phys. D: Appl. Phys. **36**, R167, 2003.

[2] U. Schillinger, T. Brilla, C. Rudolph, S. Huth, S. Gersting, F. Krötz, J. Hirschberger, C. Bergemann and C. Plank, J. Magn. Magn. Mat. **293**, 501, 200.

[3] D. Fletcher, IEEE Trans. Magn. **27** (4) 3655, 1991.

[4 E. P. Furlani, J. Appl. Phys. **99**, (2) 024912, 2006..

[5] E. P. Furlani and K. C. Ng, Phys. Rev. E **73**, 061919, 2006.

[6] E. P. Furlani and Y. Sahoo, J. Phys. D: Appl. Phys. **39**, 1724, 2006.

[7] E. P. Furlani and Y. Sahoo, J. Phys. D: Appl. Phys. **39**, 1724, 2006.

[8] E. J. Furlani and E. P. Furlani, J. Magn Magn. Mat. **312** (1), 187, 2007.

[9] E. P. Furlani, Permanent Magnet and Electromechanical Devices; Materials, Analysis and Applications, Academic Press, NY, 2001.

[10] J. A. Stratton, Electromagnetic Theory, McGraw-Hill, NY, 1941.

[11] C. Plank, F. Scherer, U. Schillinger, M. Anton and C. Bergemann, European Cells Mat. **3**, (Suppl. 2) 79-80, 2002.

Micron Particle Segregation in Lower Electric Field Regions in Very Dilute Neutrally Buoyant Suspensions

Z. Qiu[*], Y. Shen[**], S. Tada[***] and D. Jacqmin[****]

[*]The City College of the City University of New York, New York, NY 10031, ltscltsc@hotmail.com
[**]Zhengzhou University of Light Industry, Zhengzhou 450002, China, hnshenyan@126.com
[***]National Defense Academy, Kanagawa 239-8686, Japan, stada@nda.ac.jp
[****]NASA Glenn Research Center, Cleveland, OH 44135, fsdavid@tess.grc.nasa.gov

ABSTRACT

When a dilute suspension of neutrally buoyant poly alpha olefin particles and corn oil is exposed to a highly non-uniform AC electric field generated by a spatially periodical electric electrode array, particles are first transported by dielectrophoresis force to the lower electric field region. However, after these particles enter the lower electric field region, they don't form a staple particle strip, instead, are segregated to form island-like structures that it is experimentally testified are suspended somewhere above each ground electrode. Given a suspension of 0.1% (v/v) and an applied voltage of 5kV/100Hz, the DEP induced particle transport time is 10 minutes, the particle segregation time is 30 minutes, and so a driving force for the particle segregation, if any, is estimated as 1% of the DEP force. Two applications are proposed of this finding that are to monitor a surviving rate of cells after electroporation, and sort out dead cells from human being breast cancer cells before perfusion respectively.

Key Words: dielectrophoresis, segregation, electroporation, cell surviving rate, island-like structure

1 INTRODUCTION

Techniques of dielectrophoresis (DEP) based separation and manipulation still have further application potentials in biotechnology. Two previous papers of ours reported DEP-induced transport of particles in dilute suspensions in non-uniform electric fields [1][2]. The single particle model works well because the interactions among particles are so weak that they are negligible. This paper will report an unexpected finding of island-like structure formation of poly alpha olefin particles in corn oil in the lower electric field region.

2 EXPERIMENTAL

2.1 Setup and Suspension

As shown in Figure 1A, the DEP chamber consists of a transparent cover with one side ITO coated and grounded, a Teflon gasket as its four-side wall, and an insulating bottom plate with an electrode array. Its cavity is 150x70x3(mm). Eight pairs of 1.6x1.6mm square brass bars act as the ground and high voltage electrodes that are inlaid into the insulating plate in parallel with a uniform gap of 2 mm. The top surface of each electrode is flattened and polished with fine sand paper and metal polishing agents. The 3mm thick and 5mm wide close gasket of Teflon constitutes its four-side wall. The chamber cover has two layers of a clear

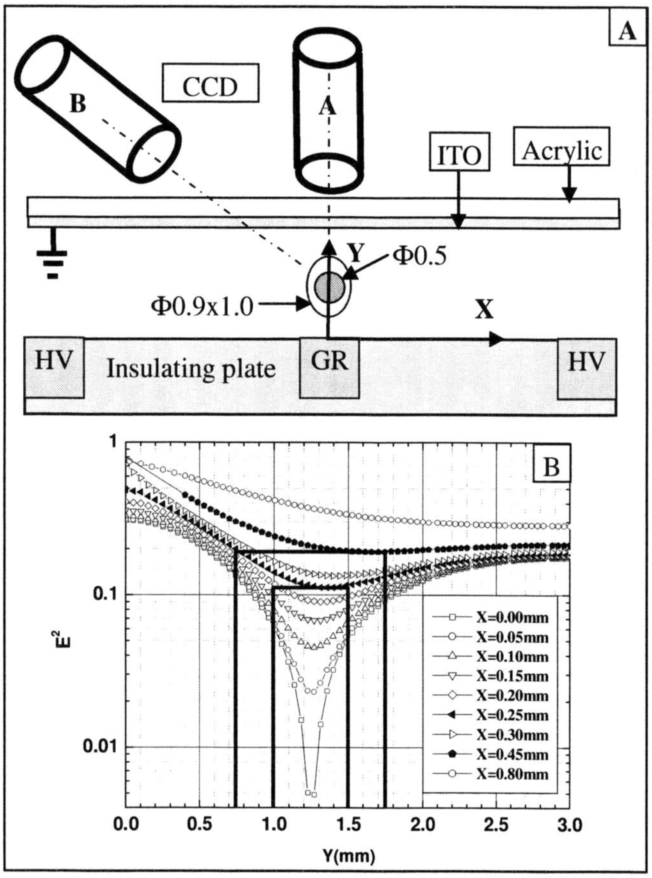

Figure 1 A: The cross section of the DEP chamber with a spatially periodical electrode array. The unit cell area is 3.0x3.6mm. HV and GR denote electrically energized electrodes and ground electrodes respectively. B: The spatial distribution of electric field. The lowest electric field is located at (0, 1.245mm). X-axis and Y-axis are shown in the Figure 1A and Z-axis is along the electrode direction. No drawings are scaled.

acrylic plate and ITO-coated glass. The ITO-coated glass and all the ground electrodes are electrically connected together and grounded. The electrode array is energized by a high voltage AC power supply. Mono-sized poly alpha olefin particles of diameter of 86.7μm are used for experimental studies. The particles have the exactly same density of $0.92 g/cm^3$ as corn oil only at 19°C. The dielectric mismatch factor (β) between particles and corn oil was measured with 4-volt excitation voltage as -0.143 in the frequency rang from 50 to 10000Hz. The viscosity of corn oil (η) is 0.060 Pas at 19°C. A CCD camera was used to record particle distribution with time under an AC electric field from the direction of A or B shown in the Figure 1A.

2.2 Experimental Results

Experiments of particle transport, aggregation and segregation induced by an AC field were carried out with the DEP chamber and the suspension described above. A typical set of experimental photos is shown in Figure 2 for a suspension of 0.1% (v/v) under 5kV/100Hz. The dynamic responses of the particles to the high gradient ac electric field have two stages. First, the particles are transported to the lower electric field region because the dielectric mismatch factor of the suspension is negative, -0.143. Particle dynamics in this stage is reflected in the images of A to D in Figure 2. The single particle model works well for this stage. In the second stage when the particles reach the lower electric field region, they don't form a uniform staple column, instead, are further segregated to form island-like structures, as shown in the images of E to H.

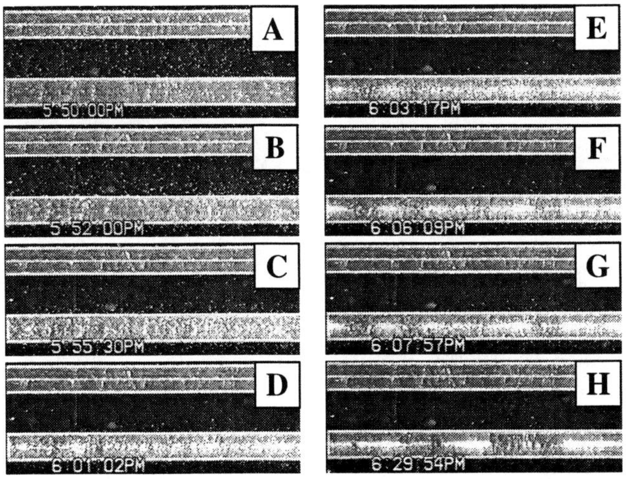

Figure 2: Particle transport, aggrergation and segregation under 5kV/100Hz in the DEP chamber. The suspension consists of poly alpha olefin particles and corn oil at 0.1% (v/v). All the images were taken when the camera was in Position A in Figure 1A. The voltage was applied to the suspension from 5:50:00pm. Time on each image was real time. Images of A to D show the particle transport to the lower field region and images of E to H show the particle aggregation and segregation in the lower field region.

For a better understanding of two dynamics processes and their relations, we numerically simulated the final particle distribution after the suspension was exposed to the electric field for a long enough time [3]. In this simulation, we did not consider the particle aggregation and segregation although we used a continuous model that is very close to the single particle model for a dilute suspension. We first calculated the spatial distribution of particle concentration with time under an externally applied electric field. When the time was approaching to infinite, we got the final distribution of particle concentration. The distribution of mean particle concentration was defined as $<\Phi(X)>=\frac{1}{3}\int_0^3\Phi(X,Y,t)dY$ and $<\Phi(Y)>=\frac{1}{3.6}\int_0^{3.6}\Phi(X,Y,t)dX$. $<\Phi(X)>$ against X and $<\Phi(Y)>$ against Y are shown in Figure 3, data for the suspension of 0.1% initial particle concentration under 5kV/100Hz are expressed by the solid squares. All the particles are finally collected within a circle with a diameter of 0.5mm and its center is at (0, 1.245mm) in Figure 1A. For the suspension of 1% initial concentration under 5kV/100Hz, the distribution3 of mean concentration against X and Y are shown in Figure 3, expressed by the empty circles. The particles are finally collected in an ellipse with long and short axes of 1.00 mm and 0.90mm and its center is at (0, 1.245mm) also, as shown in the Figure 1A.

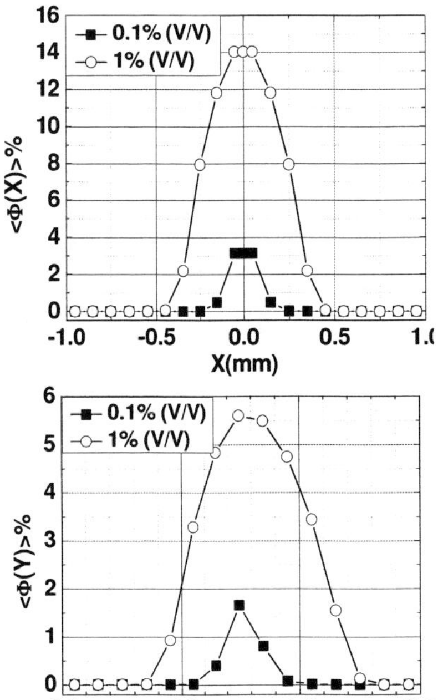

Figure 3: Mean particle concentration $<\Phi(X)>$ versus X and $<\Phi(Y)>$ versus Y calculated based on a continuous model when the exposure time t was long enough [3]. The excitation voltage of 5kV/100Hz and β = - 0.15 were used in the above calculations. The solid squares refer to the 0.1% initial concentration suspension and the empty circles refer to 1%. X- and Y-axes are shown in Figure 1A.

We set the CCD camera in Position B in Figure 1A so as to determine locations of island-like structures relative to the ground electrode. Figure 4 shows that these island-like structures are suspended somewhere above the ground electrode, but don't touch the ground electrode surface. The experiment shows that the island-like structures are in the lower electric field region shown in Figure 1B, which is in qualitative agreement with the calculations.

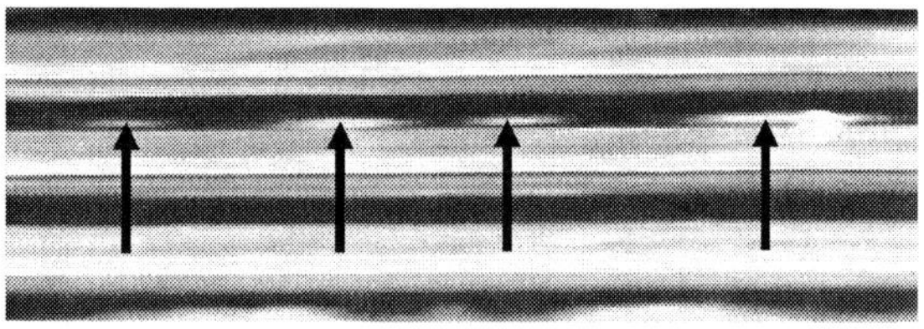

Figure 4: Island-like structures are located somewhere above the ground electrode, as each arrow points. The continuous brighter stripes are the electrode surfaces. This image was taken when the camera was in Position B in Figure 1A. This experiment shows that the island-like structures are in the lowest electric field region.

The particle instability always exists while formation of island-like structures is conditional in the suspension. Effects of initial concentrations of suspensions on the particle segregation are scrutinized with experiments. It was experimentally verified that it depends on the initial concentration whether island-like structures are formed in a suspension or not. The experiments were performed with the suspensions of 0.1% to 1.126% initial concentrations. A critical initial concentration for the island-like structure formation is experimentally determined as 1%. When the initial concentration of suspensions is less than 1%, particles are completely segregated into the island-like structures and there are pure corn oil regions between the island-like structures of particles, as shown in the image H in Figure 2. When an initial concentration is higher than 1%, no island-like structures occur in such a suspension, instead, particles form staple strips in each lower field region. The below experiment confirms existence of the critical initial concentration of 1%. Shown in Figure 5A is a result for the suspension of 1.126% initial concentration at the 611th second after the suspension was exposed to a voltage of 5kV/100Hz. No island-like structures appear and only uniform straight stripes of particles are formed as shown in Figure 5A. The width of particle stripes is 0.618 of the electrode width. Being contrary to what is shown in Figure 5A, Figure 5B shows that particle instability does exist in the same experiment when some amount of particles was removed from one end of the ground electrode with help of the edge effect of the electrode array in the chamber. Clear wavy structures are added on each particle stripe above each ground electrode. The convincing evidence of particle instability is that particle stripes were not uniformly reduced from the central part to the end of the ground electrode; instead these stripes have some sawtooth waveforms on them. The thickest of stripes was measured as 0.561 of the ground electrode width which is less than the limit value of 0.9/1.6=0.5625.

It should be pointed out that all the experiments except for this one in Figure 5 were carried out with a close gasket and that the boundary condition of zero particle flux is valid.

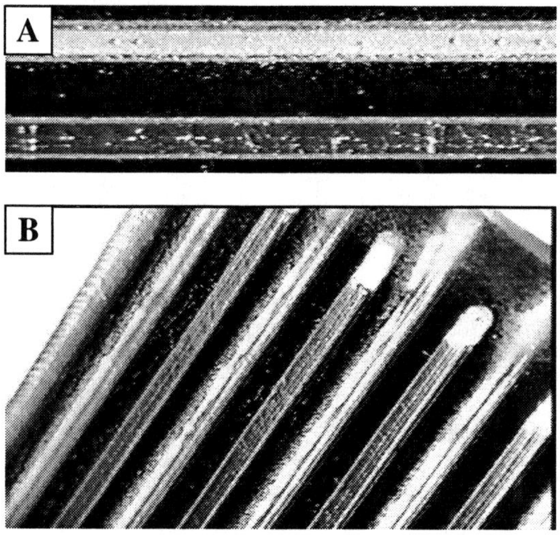

Figure 5 A: Image A shows that uniform stripes of particles were formed in the suspension of the initial concentration of 1.126% (v/v) at the 611th second after 5kV/100Hz was applied to the suspension. The width of the particle stripes is 0.618 of the electrode width. B: Wavy structures are developed on each particle stripe after some amount of particles was removed from one end of the ground electrode with help of the edge effect of the electrode array. The widest of particle stripes is 0.561 of the electrode width. Image B was taken at the 70th minute after 5kV/100Hz was applied to the suspension.

3 MECHANISMS RESPONSIBLE FOR ISLAND-LIKE STRUCTURE FORMATION

What force drives the particle segregation in the lower field region? Why are the island-like structures formed in dilute suspensions of less than 1% initial concentration? How does transverse (that means the DEP force is normal to the electrode direction) DEP force influence island-like structure formation or particle instability in suspensions of different initial concentrations?

Let's look at the distribution of E^2 shown in Figure 1B again. There are deep valleys in the curve families of E^2 versus (X, Y) and their centers are located at (Xmm, 1.245mm). The valleys rapidly become shallow with increase in X and disappears at X = 0.45mm. It is found that the valleys of E^2 area should be an ellipse with long and short axes being 1.00mm and 0.90 mm, respectively and the center is at (0, 1.245mm). It will shed light on the formation of island-like structure or development of particle instability to compare the distribution of electric field with the final particle concentration distributions of the 0.1% and 1% initial particle concentration suspensions under the same excitation voltage of 5kV/100Hz.

For the suspension of the initial concentration of 0.1% under 5kV/100Hz, particles are always distributed within the deep valley no matter before or after the particles are segregated. As shown in Figure 2, although island-like structures are formed, the particles in the island-like structures are still in the deep valley and so these particles only see the very weak transverse DEP force. However, according to the estimation from Figure 3, when the initial particle concentration of a suspension is increased to 1%, the distribution of particles reaches the rim of the valley. This means that, if the initial particle concentration is further increased, particles not only take up the valley but outside it also. As shown in Figure 1B, there exists a very huge difference of the square of electric field intensity (E^2) between inside and outside the valley. Particles outside the valley experience the much stronger transverse DEP force and will be forced to enter the valley as many as possible. Even though there is the particle instability along the electrode direction in the suspension of initial particle concentration of 1% or higher, no island-like structures are formed or even the particle instability cannot be developed in such a suspension because the transverse DEP force in the transit region from inside the valley to outside the valley is strong enough to suppress the particle instability completely. As a result, there appear no island-like structures or development of the particle instability in such a suspension and the particles only form uniform straight stripes, which have been testified by the experiments. As shown in Figure 5A, for the suspension of the initial concentration of 1.126%, the stripe width is 0.618 of the electrode width that is larger than the limit value of 0.9mm/1.6mm = 0.5625, and so the particles take up both inside and outside the valley and no island-like structures or particle instability was developed. However, as shown in Figure 5B, the particle stripes become thinner by removing some particles with help of the edge effects of the electric field, the particle instability or pre-island-like structure appears because the widest of stripes is 0.560 of the electrode width, slightly less than the limit value of 0.563. Also, as shown in Figure 2, the width of particle stripes in the images of E to H is always less than the limit of 0.563 shown in Figure 3 and so particle instability and island-like structure are always developed to be observable.

Furthermore, it is necessary to estimate the relative strength of a driving force responsible for the particle segregation to the transverse DEP force. A possible driving force can be estimated as $F_{DRIVE} = 6\pi\eta a V = 6\pi\eta a L / \tau_{seg}$, where τ_{seg} is the segregation time, a the particle radius and L a characteristic distance for particles to move during the segregation. The transverse DEP force can be written as $< F_{TDEP} >= 2\pi\varepsilon_0\varepsilon_l \beta a^3 < \nabla E^2 >$, where $< F_{TDEP} >$ is to take average over the involved region and ε_l. is the dielectric constant of corn oil. Therefore, the ratio is given as $\frac{< F_{seg} >}{< F_{TDEP} >} = \frac{3\eta L}{\varepsilon_0 \varepsilon_l \beta \tau_{seg} a^2 < \nabla E^2 >} \approx 0.01$. It demonstrates that, compared with the average transverse DEP force, a possible driving force responsible for the particle segregation is so weak! Fortunately, because there is the lowest electric field region where the transverse DEP force has the least effect on particle instability or island-like structure development, we are able to observe such a tantalizing phenomenon due to so weak a driving force!

4 CONCLUSIONS AND FURTHER WORK

The particle instability is recognized in the neutrally buoyant suspensions of poly alpha olefin particles and corn oil, and island-like structures are observed in the suspensions of less than 1% initial concentration when the suspension was exposed to an AC gradient electric field that is generated by a spatially periodical electrode array. It is testified that the island-like structures are located in the lowest electric field region above each ground electrode. In short, experiments demonstrate that there is particle instability in the suspension; when the initial concentration is less than 1%, the particle instability is developed to form island-like structures; when the initial concentration is higher than 1%, the particle instability is suppressed by the strong transverse DEP force and particles form uniform straight stripes above each ground electrode.

So far, we have not had a clear physics picture about how the particle instability in the suspension is initialized. We are going to perform 3-D MD simulations so as to crack the nut. In the mean time, we proposed two applications of this finding. One application is to monitor a surviving rate of cells in a real time domain after electroporation which is one crucial parameter to electroporation techniques like target drug delivery and cancer target therapies. The other application is to sort out dead cells from human being breast cancer cells before perfusion.

The authors gratefully acknowledged that this work was in part supported by NASA under grant No NAG3-2698 and the PSC-CUNY Professional Development Fund. Some experiments were done in the Fluid Mechanics Laboratory of the Levich Institute of the City College of the City University of New York. The numerical simulations of particle concentration distribution were performed in the NASA Glenn Research Center.

REFERENCES

[1] Z.Y. Qiu, N. Markarian, B. Khusid, and A. Acrivos, J. Appl. Phys. 92, 2829, 2002.

[2] A. D. Dussaud, B. Khusid and A. Acrivos, J. Appl. Phys. 88, 5463, 2000.

[3] A. Kumar, Z.Y. Qiu, A. Acrivos, B. Khusid and D. Jacqmin, Phys. Rev. E, 69, 021402, 2004.

Limitations of DNA High-frequency Anchoring and Stretching

H. Dalir*, T. Nisisako**, T. Endo**, Y. Yanagida** and T. Hatsuzawa**

*Department of Mechano-Micro Engineering, Tokyo Institute of Technology,
R2-6, 4259 Nagatsuta-cho, Kanagawa 226-8503, Japan, dalir.h.aa@m.titech.ac.jp
**Precision and Intelligence Laboratory, Tokyo Institute of Technology,
nisisako.t.aa@m.titech.ac.jp, endo.t.aa@m.titech.ac.jp
yanagida.y.aa@m.titech.ac.jp, hatsuzawa.t.aa@m.titech.ac.jp

ABSTRACT

This paper reports measurements that characterize the immobilization of 48 kilobase-pair lambda DNA onto lifted-off microelectrodes by high-voltage and high-frequency dielectrophoresis. Measurements of voltage- and frequency-dependent immobilization of DNA onto microelectrodes by dielectrophoresis show significant reduction in the response as the frequency increases from 200 kHz to 1 MHz or decreases from 200 kHz to 100 kHz and also as the electric field is lower or higher than 0.4 Vp-p/m. We found that the immobilization and elongation of the DNA molecules is restricted by the geometry of the gap, and that by decreasing the electrode gap size, the DNA molecules have less chance for both immobilization and stretching. The produced electrodes with both random microscopic peaks and modified smooth edges are utilized to show the effect of electrode edge roughness. The results imply that more DNA molecules can be immobilized by microelectrodes having rough edges.

Keywords: DNA manipulation, DNA stretching, high frequency measurements, gap geometry, edge roughness

1 INTRODUCTION

Manipulation of single biomolecules such as DNA, RNA, and proteins facilitated significant advances in biology [1-5]. Technological advances in nanotechnology and high resolution visualizing systems facilitate single-molecule manipulation. Through high-resolution fluorescence microscopy, real-time restriction [6] and replication [7] of a single DNA molecule has been observed, and moreover, the transport properties of a single DNA molecule were obtained [8].

DNA molecules can be immobilized by various means including electrostatic, hydrodynamic, or magnetic forces. Washizu used dielectrophoresis and electroosmotic flow to immobilize DNA molecules [9]. Stretching by hydrodynamic force was performed either with free molecular ends or with one end immobilized to a solid surface [10,11].

Besides Atomic Force Microscopy and Scanning Tunneling Microscopy, a variety of techniques has been used including electric, magnetic and optical traps allowing to move and position nano-scale objects and molecules [12-14]. We present here an electric field based technique for the orientation and positioning of DNA molecules on micro-electrode silicon devices. Moreover, we have studied the parameters affecting the dielectrophoretic immobilization of DNA and it's dependencies as a function of the surface, micro-electrode, and bulk solution properties. We have optimized those parameters and been able to achieve more constant DNA immobilization.

The double-stranded DNA molecule is a long macromolecule consisting of two strands of deoxyribonucleotides held together by hydrogen bonding. Under slightly basic conditions, the phosphate groups within the backbone deprotonize and the DNA molecule is negatively charged. This charge leads to the formation of a counterion cloud surrounding the molecule to an extension given by the Debye length. A DNA molecule has essentially no net permanent dipole moment since the two helices forming the double-strand DNA point in opposite directions. However, the counterion cloud can be displaced in the presence of an electric field and it is expected to strongly increase the polarizability of the molecule (ionic polarizability). The suspended molecule can then be treated as a dielectric medium of given volume and shape placed in a continuum solution of different dielectric properties [15]. Electric manipulations of DNA molecules in microfabricated structures based on the induced dipole moment of the molecules have been carried out since the last decade [13]. Such devices allow to work with relatively high applied electric potentials (in the MV/m range) using low voltage sources. Furthermore, due to the small size of the structures, sample cells with small volumes (down to a few pico liters) can be used. So far, a relatively high number of molecules was manipulated within such devices [13, 16 and 17].

2 MATERIALS AND METHODS

2.1 Device Fabrication

Electrodes were fabricated on a thin Silicon wafer (100 μm Thickness). Photoresist (OFPR-800) was spun (see Fig. (1)), patterned by channel mask layer, and developed

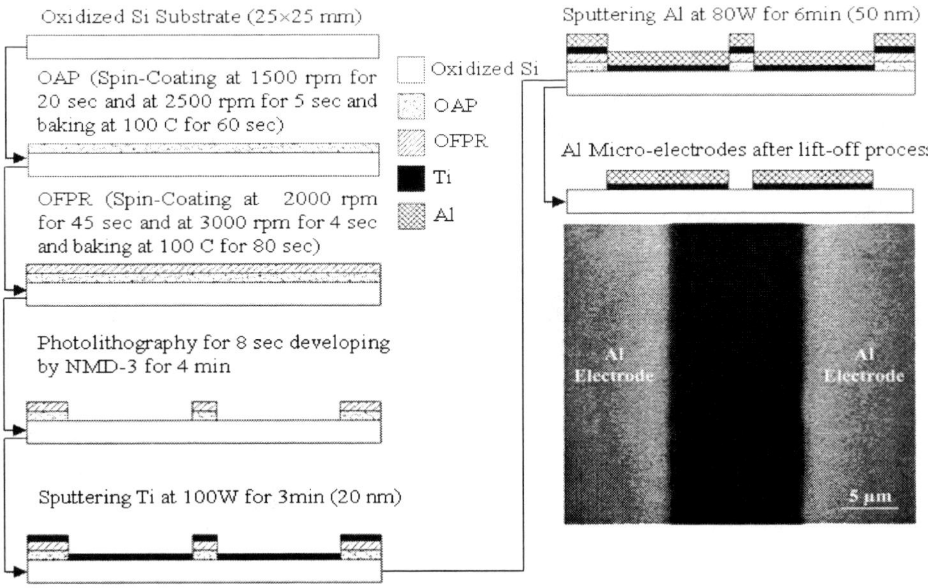

Fig.1 Schematic process of micro-fabricated electro-immobilization device.

in NMD-3 solution. Then, the 20 nm-thick titanium and 50 nm aluminum layers were deposited on silicon surface and were lifted-off by Acetone. The wafer was rinsed in DI water, dried by air, and diced into each pieces. Some wafers were annealed at 210o C for 45 min to improve the edge of electrode.

2.2 DNA Sample Preparation

Plain µ-phage DNA was prepared in this work. The DNA was labeled with fluorescent marker YOYO-1 at a dye-to-base-pair ratio of 1:5. DI water (pH 8, conductivity 2 µS/cm) was used as sample buffer. Labeled DNA molecules were mixed with DI buffer solution to make final concentration of DNA sample at 1 ng/µL.

2.3 Electric Field Application

Alternating current (ac) voltage was generated from a function generator (see Fig. (2)). The frequency was regulated from 1 Hz to 2.3 MHz, and the amplitude ranged from 2 to 25 Vp-p. Waveforms were monitored by an oscilloscope. The gradient of electric field around the electrode was estimated by electric field simulation (See Fig. (3)).

2.4 Fluorescence Microscope

A fluorescence microscope was used to observe the motion of single DNA molecules. A high-magnification X100 was used to get high resolution. The molecular behavior was visualized using a digital CCD camera. A 100-W mercury lamp with FITC filter was used as an excitation source.

3 DNA DIELECTROPHORETIC PRINCIPLES

In the presence of an electric field $\widetilde{E}$, a DNA molecule in solution will bear an induced dipole moment given by $\widetilde{p} = \alpha V\widetilde{E}$ where α is the polarizability of the molecule per unit volume and V is the volume. The polarizability depends on the permittivities of the molecule and of the solution. The counterions cloud is expected to strongly enhance the polarizability of the molecule and the induced dipole moment will thus depend on the solution used and in particular on its ionic strength and pH. However, working with solutions containing a high concentration of ions would facilitate electrochemical reactions at the electrode-solution interface. Such a situation is detrimental to the experiment and can cause irreversible damages to the electrodes due to the relatively high fields applied. An advantage of the technique used here is that it allows working with ac electric fields in the frequency range of a few kHz up to a few MHz. Combining this with relatively low-conductivity buffers (typically below 100 µS/cm) allows to limit the voltage drop at the electrode-solution interface and helps reducing electrochemical reactions.

The manipulations of the molecules are based on the interaction of the induced dipole moment and the applied electric field. The translational motion of the molecules is caused by the application of a non-uniform electric field to the solution which is realized in our case by the smooth edged electrodes. In a non-uniform field, a neutral object of polarizability α and volume V will undergo a force proportional to the square of the applied field $\widetilde{E}$ and given

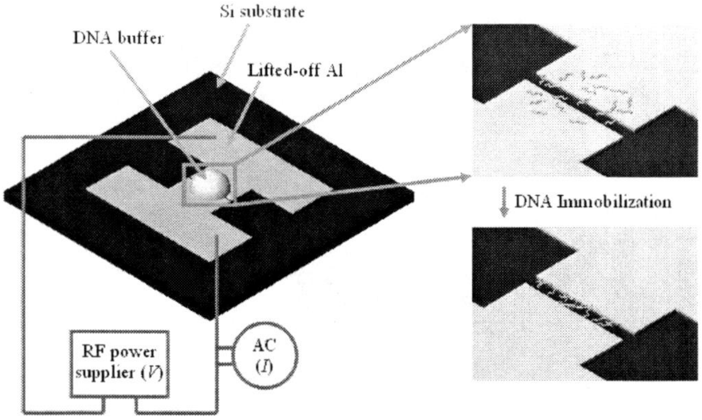

Fig.2 Electric field application. The frequency was regulated from 1 Hz to 2.3 MHz, and the amplitude ranged from 2 to 25 Vp-p

by $\widetilde{F}_d = 0.5\alpha V\widetilde{\nabla}|E|^2$. For a dielectric ellipsoid of volume $r^2 l$ (r: radius, l: length) and permittivity ε_2 immersed in a medium with permittivity ε_1, the force at equilibrium can be estimated by [8]

$$\widetilde{F}_d \approx r^2 l \varepsilon_2 \varepsilon_1 \widetilde{\nabla}|E|^2 \tag{1}$$

as long as $\varepsilon_2 >> \varepsilon_1$ which is expected for DNA in an aqueous buffer. A rough estimate of the electric field value E_{th} necessary to overcome thermal fluctuations can be obtained by integrating Eq. (1) and comparing the result to $k_B T$. Since the DNA molecule is a long, flexible polymer, we will consider here the force exerted on one segment of the molecule. Taking the permittivity of water and using a radius of 1 nm for a λ-DNA molecule and a persistence length of 100 nm, we obtain $E_{th} < 10^7 V/m$ showing the necessity to use relatively high electric fields. In this particular case where the dielectric object is more

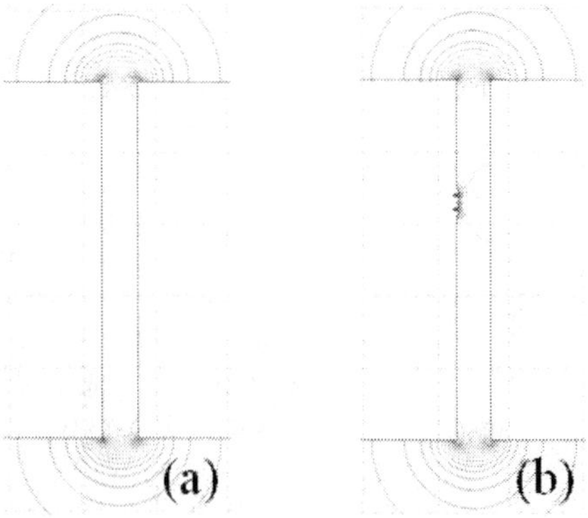

Fig.3 Simulation of the gradient of electric field around the electrodes (a) a smooth electrode edge and (b) a rough electrode edge.

polarizable than the medium, the force will be directed towards the highest intensity point of the electric field (positive dielectrophoresis).

The interaction of the induced dipole moment with the applied electric field will also cause electrically non-symmetrical molecules to sense a torque $\widetilde{T} = \widetilde{p} \times \widetilde{E}$. High aspect-ratio molecules will thus tend to align themselves with their longest axis parallel to the applied field. In the case of a long flexible molecule such as DNA, the orientation effect is expected to take place for every segment of the molecule, resulting in an uncoiling of the molecule to its full length.

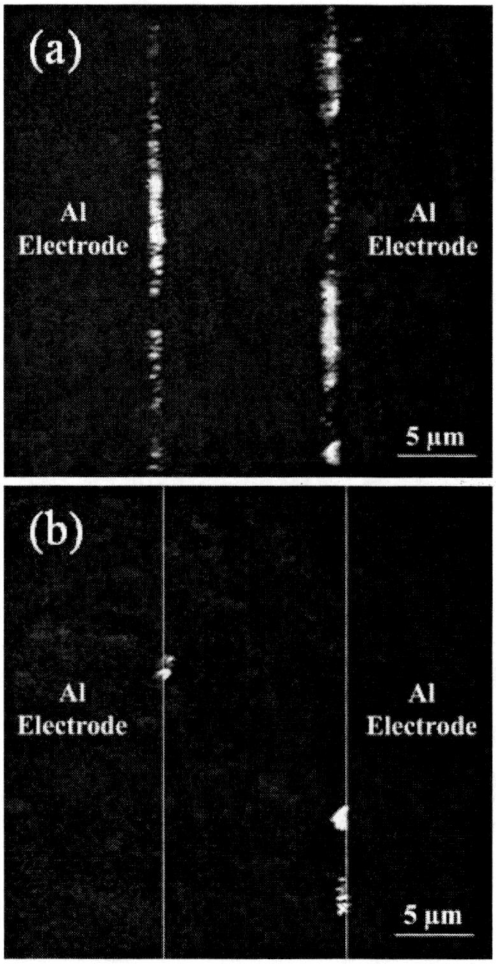

Fig.4 DNA immobilization using (a) a rough electrode edge and (b) a smooth electrode edge. Many molecules were attracted to the random peaks of the rough electrode edge while just a few number of DNAs were immobilized on electrode edges after annealing process.

4 RESULTS AND DISCUSSIONS

The device we constructed for DNA immobilization uses dielectrophoresis to capture the DNA molecules. By using an electric field generated by a straight electrode (high field gradient), we can control the movement as shown in Fig. (4). Generally, DNA molecules are attracted to the higher field gradient region at high frequency (~200 kHz). The directional movement of the molecules is frequency dependent in accordance with their dielectric properties. That is, the transition frequency at which the dielectrophoretic force reverses from positive to negative will depend on the molecule's dielectric properties. Molecules were repelled from high field gradient regions if frequencies higher than 700 kHz were used. DNA immobilization is a strong function of solution conditions such as polymer concentration, conductivity, and pH values. A low-conductivity, pH 8 buffer is optimal for DNA immobilization. Experiments were performed with solutions of pH 4, 8, and 10, but immobilization of DNA could only be obtained in the pH 8 solution, indicating that dielectrophoretic immobilization occurs best near physiological pH values. Low-conductivity solutions are also best for dielectrophoretic immobilization. High-conductivity solutions produce a large amount of Joule heating, and resulting molecular movement depends on the temperature increase in the fluid and not on the dielectric force. Moreover, a high level of bubble generation was observed that disrupted the immobilization process. Electrode edge smoothness plays an important role in attracting DNA molecules to the intended sharp area. Rough electrode edges provide many small points along the electrode's edge that generate high electric field gradients resulting in molecular movement. Fig. (4a) shows how DNA molecules are attracted to the rough edges of the electrode, while Fig. (4b) shows how the number of immobilized DNA molecules is reduced due to annealing process.

In conclusion, we have presented a silicon-based device allowing a satisfactory immobilization and capture of a relatively small number of DNA molecules in free-flow.

We expect that such devices will be useful for the handling of minute amounts of molecules in a variety of experiments involving DNA separation, amplification or detection. The further development of integrated systems such as biosensors or "lab-on-a-chip" devices for biochemical experiments requiring several steps will certainly make use of high intensity electric fields for an efficient handling of the molecules involved.

Moreover, the optimum conditions for robust and controllable DNA immobilization have been investigated. By optimizing the solutions conditions, electric field strength, and shape, and the surface properties, the success of DNA immobilization can approach 90%. Molecular manipulation by dielectrophoresis is a relatively common technique, but there are still unexplored topics in this area. Additional studies are needed to investigate other parameters such as temperature, surface property, and buffer ion dependence on stretching. With the design of more reliable micro-fabricated DNA immobilization and stretching devices, integrated systems may be possible that perform multiple analysis operations on a single DNA molecule extracted from a single cell or other minute samples.

REFERENCES

[1] C. Bustamante, Z. Bryant and S. Smith, Nature, 421, 423, 2003.

[2] M. Wang, Curr. Opin. Biotechnol., 10, 81, 1999.

[3] B. Maier, T. Strick, V. Croquette and D. Bensimon, Single Mol., 2, 145, 2000.

[4] J. Han and H. Craighead, Anal. Chem., 74, 394, 2002.

[5] J. Han, S. Turner and H. Craighead, Phys. Rev. Lett., 86, 1394, 2001.

[6] B. Schafer, H. Gemeinhardt, V. Uhl and K. Greulich, Single Mol., 1, 33, 2000.

[7] B. Maier, D. Bensimon and V. Croquette, Proc. Natl. Acad. Sci. U.S.A., 97, 12002, 2000.

[8] D. Wirtz, Phys. Rev. Lett., 75, 2436, 1995.

[9] M. Washizu, Y. Nikaido, O. Kurosawa and H. Kabata, Electrost., 57, 395, 2003.

[10] J. Allemand, D. Bensimon, L. Jullien, A. Bensimon and V. Croquette, Biophys. J., 73, 2064, 1997.

[11] D. Bensimon, A. Simon, V. Croquette and A. Bensimon, Phys. Rev. Lett., 74, 4754, 1995.

[12] R. Akiyama, T. Matsumoto and T. Kawai, Surface Science, 418, 73, 1998.

[13] M. Washizu and O. Kurosawa, IEEE Trans. Ind. Applicat. 26, 1165, 1990.

[14] S. Smith, L. Finzi and C. Bustamante, Science 258, 1122, 1992.

[15] L. Landau and E. Lifshitz, Electrodynamics of Continuous Media, Pergamon, Oxford, 1963.

[16] S. Suzuki, T. Yamanashi, S. Tazawa, O. Kurosawa and M. Washizu, IEEE Trans. Ind. Applicat. 34, 75, 1998.

[17] C. Asbury and G. van den Engh, Biophys. J. 74 1024, 1998.

A High Throughput Multi-Stage, Multi-Frequency Filter and Separation Device Based on Carbon Dielectrophoresis

R. Martinez-Duarte*, J. Andrade-Roman**, S. O. Martinez** and M. Madou*

*Mechanical & Aerospace Engineering Department, University of California, Irvine, 4200 Engineering Gateway, Irvine, CA, 92617, USA, drmartnz@gmail.com
**Tecnológico de Monterrey, Campus Monterrey, Monterrey, Mexico, smart@itesm.mx

ABSTRACT

We present a multi-stage, multi-frequency carbon dielectrophoresis (CarbonDEP) device for high throughput filtering and separation applications in the health care and biotechnology fields. CarbonDEP refers to the use of carbon surfaces to induce Dielectrophoresis (DEP). Up to X high aspect ratio carbon electrode arrays are sequentially embedded in a microfluidic channel and excited with different optimized signals, provided by a custom engineered multi-channel, programmable signal generator, in their frequency and magnitude to obtain X dielectrophoresis-active filters, or a super filter with X stages. Particle separation is achieved by sequentially releasing and collecting the particles previously trapped at different arrays. We demonstrate the feasibility of a 2-stage system by separating 3 particle populations. We also discuss how such system is best applied when working with specific sample types.

Keywords: carbon, dielectrophoresis, 3D, c-mems, multi-stage, multi-frequency

1 INTRODUCTION

Even when the field of bioparticle separation has advanced significantly in recent years with the wide use of techniques such as Fluorescence-Activated Cell Sorting (FACS) or MACS (Magnetically-Actuated Cell Sorting) such techniques require the use of specific, and often expensive, tags to achieve high selectivity. An ideal solution would be a technology which eliminates the need of such tags while maintaining a high throughput separation process. We believe 3D Carbon Dielectrophoresis (CarbonDEP) could be such solution. A huge advantage of DEP is the selection of the targeted particle using only its intrinsic dielectric properties which are solely determined by the particle's individual phenotype such as its membrane morphology. Such advantage eliminates the need of specific tags linked to magnetic beads or fluorophores to discriminate targeted particle types potentially reducing the cost of each assay.

As an enhancement to current DEP devices, we propose the use of Carbon Dielectrophoresis (CarbonDEP). Carbon DEP refers to the use of carbon surfaces to induce DEP. Carbon surfaces offer better electrochemical and biocompatibility properties than other conductive materials [1] as well as lower costs. The use of volumetric (3D) structures allows higher throughput than traditional planar DEP devices [2] towards rates comparable to current separation techniques.

Even when carbon surfaces can be obtained with a variety of fabrication techniques, we have chosen the C-MEMS technique [1] given its exquisite control of device dimensions achieved by photolithography and its ability to generate very high aspect ratio structures. As the field matures the need for still cheaper fabrication techniques will be evident to allow widely used applications.

The novelty of our work resides in the use of volumetric (3D) CarbonDEP to implement high throughput, re-configurable, multiple independent dielectrophoresis-active filters to be used in filtering and particle separation applications. While filtering is readily implemented, Particle separation is achieved by sequentially releasing and collecting the particles previously trapped at different arrays. Such system might be applied for the rapid separation of several particle populations contained in a mix by only flowing through such mix once. Previous work on carbon in DEP applications by our group includes DEP trapping of latex beads and yeast-cells with planar and volumetric electrodes [2, 3-4]. Higher throughput and efficiency of volumetric carbon electrodes over planar ones when filtering a targeted population have been validated in [2]. On the other hand, a multi-stage DEP system has been reported in [5] where complex fabrication methods to obtain planar aluminum electrodes aided by 30-µm SU-8 structures acting as physical traps have been exploited.

2 MULTI-STAGE, MULTI-FREQUENCY DIELECTROPHORESIS SYSTEM

A general schematic of our system is presented in Fig. 1. Multiple signals, optimized in their magnitude and frequency, are obtained by a multi-channel generator while different electrode array geometries are tailored in electrode size, gap, shape and height to each application.

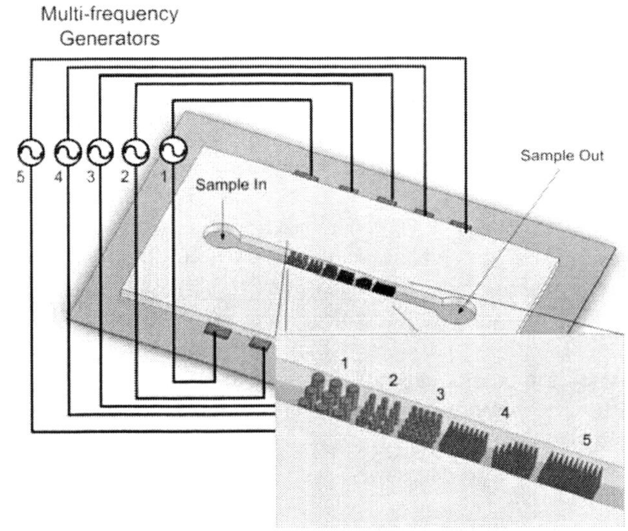

Fig. 1. A schematic drawing of our proposed multi-stage, multi-frequency CarbonDEP system.

2.1 Device

A 2-stage DEP device has been fabricated. Two CarbonDEP arrays, of 20 X 5 electrodes each, and their connecting leads are fabricated by the pyrolysis of two-step photolithographically defined SU-8 (MicroChem Corp.) structures (Fig. 2 A) following standard C-MEMS techniques. Posts are 70 µm high with a 25 µm diameter. Posts separation in the is 100 µm in the Y axis and 45 µm in the X one. In order to protect connecting leads from peeling off while immersed in an aqueous media, a thin layer of SU-8 was coated and patterned around the electrodes (Fig. 2B). Such layer also serves the purpose of providing a planar surface for the adhesion of the micro channel. A micro channel, 600 µm wide and 100 µm high, was cut on double-sided pressure-sensitive adhesive and aligned to a previously drilled polymer cover. Arrangement was then aligned to carbon electrode array and sealed with a mechanical press. Plastic fluidic ports were installed and sealed with epoxy. Device has supported flow rates of up to 8000 µl/min with no leakage. Electrical connections are made with silver conductive paint and common solder resulting in an average contact resistance of 120 Ω. A cross section of completed device, containing 3D electrodes, is shown in Fig. 3C.

2.2 Multi-Frequency Generator

Our engineered multi-channel system is based on a Direct Digital Frequency Synthesizer architecture. A Xilinx Spartan-3E500 FPGA (Field Programmable Gate Array) with 50 MHz crystal has been chosen to synthesize the sinusoidal signals. The output of each channel from the FPGA is then injected to two fast video Digital to Analog Converters (DAC). The first DAC establishes the reference voltage to the second DAC while the second DAC is used to reconstruct the sine wave.

The reconstructed sine wave is worked across the system under a maximum normalized voltage of $1V_p$. This reconstructed sine wave is then filtered through a first order Butterworth high pass filter at 60 Hz for removing offset, and then through a passive RLC fifth order Chevyshev 0.1 dB low pass filter with a cut frequency at 8 MHz for removing aliases. In order to obtain desired power levels, an output amplifier is interfaced between CarbonDEP device and multi-frequency generator.

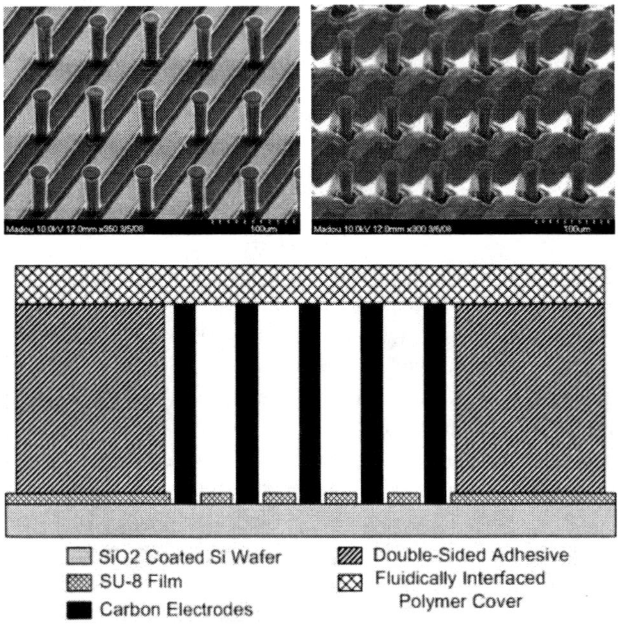

Fig. 2, A) Carbon electrodes obtained through polymer pyrolysis, B) An example of a protective SU-8 layer coated and patterned around carbon electrodes and C) Fabricated device cross section.

3 MATERIALS AND METHODS

3.1 Particles and Media

Yeast culture was obtained by dissolving 200 mg of Yeast (S. cerevisiae, Sigma-Aldrich) in 10 ml sterile YPD medium (MP Biomedicals) and incubated aerobically at 30°C with 150 RPM rotation for 18 hours. Such culture was then diluted into 100 ml of the equivalent media and incubated as before for a further 24 hours. Prior to experiment, Yeast were washed once and re-suspended in DI water. A stock of 8 um Latex particles (Duke Scientific) was prepared by diluting with DI water original stock provided by the supplier.

Experimental sample was composed by 4 ml of re-suspended yeast and 2 ml of latex particles. Measured concentrations by direct counting give 1.84×10^6 yeast cells per ml and 1.76×10^5 latex particles per ml.

Composition is thus 91.3% Yeast, 8.7% Latex. Measured sample conductivity was 1 mS/m.

3.2 Experiment

After fabrication and prior to all experiments, up to 3 ml of a Bovine Serum Albuminum (BSA) solution at 0.5% was flowed into our devices. Prior to beginning each experiment channel was cleared of particles by flushing DI water at 4000 µl/min.

45 µl of experimental sample were flowed in at desired flow rate (10, 20 and 40 µl/min) while electrode arrays were polarized with a 10 V_{pp} sinusoidal signal. First array was polarized at 5 MHz in order to trap viable yeast and repel non viable yeast. Second array was polarized at 500 KHz to trap non viable yeast as well as viable yeast. It is assumed most of viable yeast gets trapped at first array.

4 RESULTS

Fig. 3 shows the separation of yeast particles in the different arrays against a 10 µl/min flow. The array on the left of the electrode gap was polarized at 5 MHz, to trap viable yeast only, while the one in the left was at 500 KHz, to trap both viable and non viable. A clear cut difference in trapped particles can be seen in both figures. Since yeast has been re-suspended in DI water, it is expected most of them have gone non viable giving rise to such an overwhelming trapped population of non viable yeast on the right side.

Fig. 4 shows the trapping of non viable yeast as flow rate increases from 10 µl/min to 20 and 40 µl/min. It can be seen that even when the trapped particle density decreases still a fair amount of non viable yeast remains trapped suggesting the use of higher flow rates, to achieve higher throughputs, to conduct separation.

5 DISCUSSION

Even when the feasibility of a 2-stage CarbonDEP array to separate 3 different particle populations: viable, non viable yeast and latex beads has been qualitatively demonstrated, such separation can also be readily achieved by a single electrode array polarized at different frequencies at different times. The power of the proposed multi-stage, multi-frequency system becomes obvious when working with a sample containing a mix of particles having a DEP spectrum such as that shown in Fig. 5. By closely examining such figure one can conclude is possible to separate up to 4 different particle populations with a 2-stage DEP device such as the one being proposed. By polarizing the first electrode array encountered by the incoming particle mix at a frequency around point A, Particle 1 and 3 will get trapped while Particle 2 and 4 get repelled. If second array is polarized at a frequency close to point B, Particle 1 and 2 will get trapped while those unlikely non-trapped Particle 3 as well as Particle 4 get repelled. At this point we have Particle 3 trapped in the first array, Particle 2 trapped in the second array, Particle 1 trapped in both arrays and Particle 4 have just flowed by and is being retrieved at the channel outlet. If we now change the frequency of the second array to point A, Particle 2 gets repelled and is flushed away for retrieval while Particle 1 and 3 are still being trapped. The next step is to change the polarizing frequency of the whole array to point B to repel Particle 3 which is then retrieved as well. The final step is to turn off all electrodes to release and collect Particle 1. Such "optimal" sample composition might be found when purifying biological cells from ceramic and/or metallic particles, or by choosing the right suspending medium given specific particles.

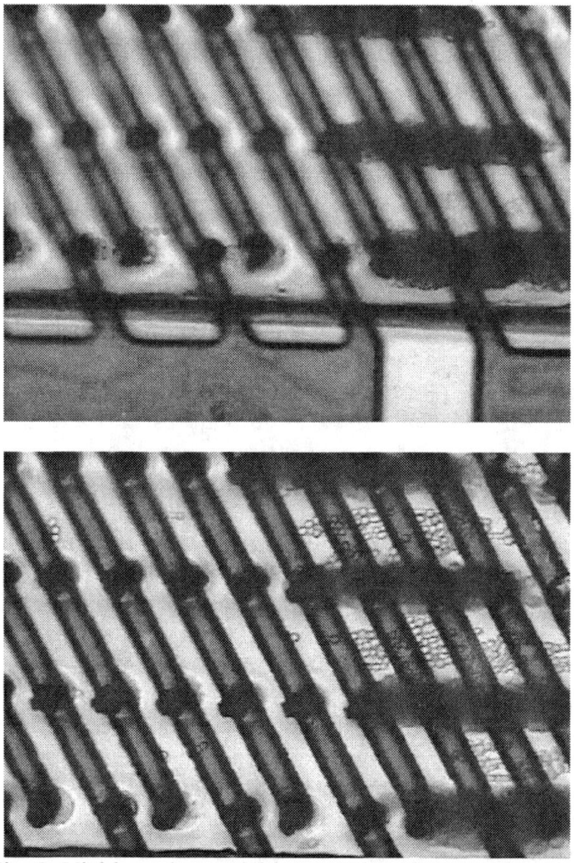

Fig. 3. Viable and non viable yeast trapping by different arrays polarized each at optimized frequencies. Top caption focused on top of electrodes. Bottom caption focused on channel floor. Sample flowing from the left at 10 µl/min. Note how the number of non viable yeast overwhelms that of viable ones.

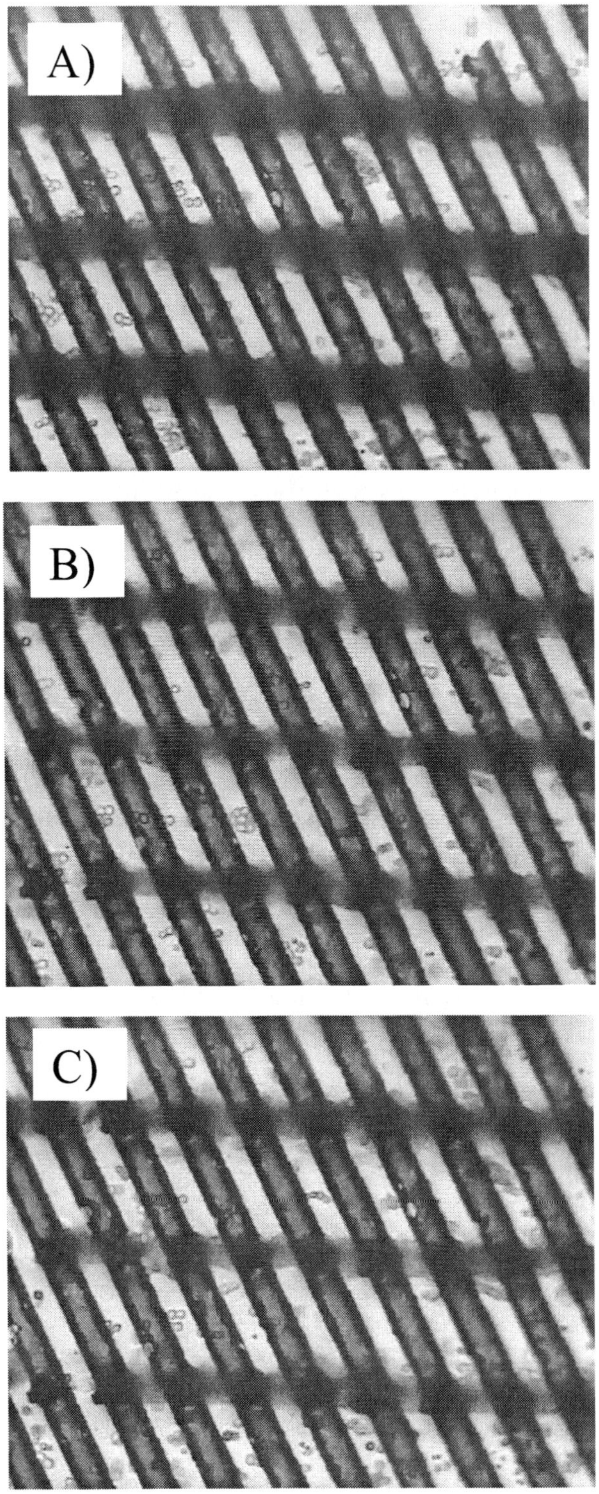

Fig. 4. Non viable yeast filtering against a A) 10 μl/min, B) 20 μl/min and C) 40 μl/min flows. Note how trapped particle density diminishes as flow rate increases but still a fair amount are trapped at 40 μl/min.

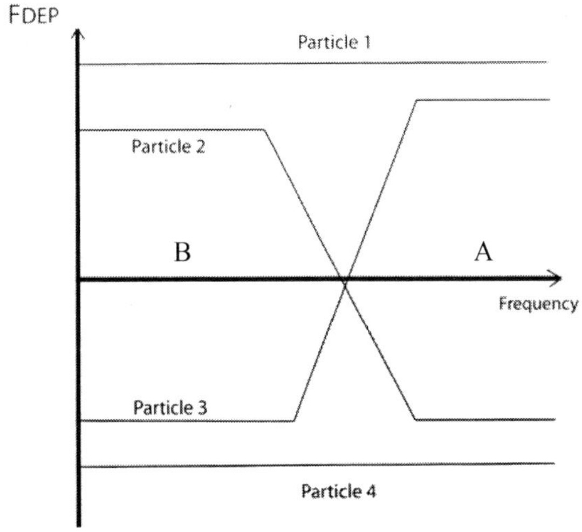

Fig. 5. Our proposed approach would be best applied when separating particles showing an "optimal" DEP as the one shown.

6 CONCLUSIONS

We have fabricated a 2-stage DEP system featuring a 3D CarbonDEP device and a Multi-channel, Multi-frequency signal generator. We have also demonstrated the feasibility of a 2-stage system to separate 3 different particles and how filter throughput does not significantly degrade as flow rate increases. Finally, we discussed how the application of our system is better suited for separating particle mixes showing an "optimal" DEP spectrum. In this case our 2-stage 3D CarbonDEP system would be capable to separate 4 different populations.

ACKNOWLEDGEMENTS

This work was supported in part by the University of California Institute for Mexico and the United States (UC MEXUS), award number UCM-40117. Authors would like to thank Dr. Horacio Kido at UC Irvine and Dr. David Smith at Oncotech (Tustin, CA) for facilitating yeast cultures.

REFERENCES

[1] C. Wang, et al, Journal of Microelectromechanical Systems, **14**, 2, 348, 2005.

[2] R. Martinez-Duarte, et al, Proceedings of the 11th International Conference in Miniaturized Systems for Health and Life Systems: microTAS 2007, Paris, France, **1**, 826, 2007.

[3] B.Y. Park, M.J. Madou, Electrophoresis, **26,** 3745, 2005.

[4] B.Y. Park, M. Madou, Journal of the Electrochemical Society, **152**, 12, J136, 2005.

[5] B. Liu, et al, Proceedings of the 13th International Conference on Solid-Sate Sensors, Actuators and Microsystems: Transducers 2005, Seoul, Korea, **2**, 1733, 2005.

Lateral-Driven Continuous Dielectrophoretic Separation Technology for Blood Cells suspended in a Highly Conductive Medium

Jinsun Seo[*], Jinhee Jung[*], Song-I Han[*], Ki-Ho Han[*], Youngdo Jung[**], and A. Bruno Frazier[**]

[*]Inje University, 607 Obang-dong, Gimhae, GyongNam, South Korea, mems@inje.ac.kr
[**]Georgia Institute of Technology, Atlanta, GA, USA, bruno.frazier@ece.gatech.edu

ABSTRACT

This paper presents lateral-driven continuous dielectrophoretic (LDEP) microseparator for separating red and white blood cells suspended in a highly conductive dilute whole blood. The LDEP microseparators enable the separation of blood cells based on the lateral DEP force generated by a planar interdigitated electrode array placed at an angle to the direction of flow. Experimental results showed that the divergent type of LDEP microseparator can continuously separate out 87.0% of the red blood cells (RBCs) and 92.1% of the white blood cells (WBCs) from diluted whole blood, while the convergent type can separate 93.6% of the RBCs and 76.9% of the WBCs within 5 minutes simply by using a 2-MHz 3-Vp-p AC voltage to create a gradient electric field in suspension medium with a high conductivity of 17 mS/cm.

Keywords: blood cells, cell separation, lateral dielectrophoresis, microfluidics

1 INTRODUCTION

Many studies [1-2] have shown that the dielectrophoretic (DEP) method is one of foremost technologies for separating target cells in a heterogeneous cell mixture. This includes leukocytes [3], erythrocytes [4], and cancer cells [5-6] from blood, cancer cells from hematopoietic CD34+ stem cells [7], neuronal cells [8], and live and dead yeasts [9]. The main advantages of the DEP method are its ease of integration with microfluidic systems [10], the lack of requirement for a tagging material, such as magnetic beads [11] or fluorescent probes [12], and a high selectivity at separating rare cells [13].

Although the DEP method has proven itself as an outstanding technology for the highly specific separation of target cells, conventional DEP technology has limited throughput and requires complicated manipulation of fluids due to the discontinuous separating procedure. In addition, since the discontinuous DEP microseparators operate with a positive DEP force, the target cells become trapped on the electrodes, thereby causing unwanted adhesion between the cells and the electrodes. The trapped cells may not be released easily and can be affected by prolonged exposure to the high-gradient electric field.

The most significant drawback is that conventional DEP microseparators operate with a controlled low-conductive suspension medium compared with the physiological solution. This is necessary to increase the separability of cells for the typical < 200 kHz DEP crossover frequency, the frequency at which the DEP force traverses zero. Even in a low-conductive medium, the DEP affinities between some mammalian cells are not significantly different [14]. New separation methods that combine DEP and other physical or chemical phenomena have been developed to increase cell separability. These include DEP field-flow fractionation (FFF) [15] and DEP-magnetophoretic FFF [16]. Unfortunately, biological analysis must frequently be carried out in a physiological medium, which typically has a conductivity > 10 mS/cm. Therefore, the low conductivity of the suspension medium is one of the critical limitations of conventional DEP microseparators.

This paper presents the design, fabrication, and characterization of lateral-driven continuous DEP (LDEP) microseparators for separating blood cells from whole blood diluted with phosphate-buffered saline (PBS) solution representing the physiological condition. The continuous microseparators herein can separate blood cells based on lateral negative DEP forces acting on blood cells. We derive a simplified theoretical model of a lateral-driven continuous DEP microseparator and compare it with a finite element analysis using commercially available software, ANSYS (ANSYS, Inc., USA). In this study, we propose continuous divergent and convergent type microseparators based on the lateral negative DEP force. We reported on the experiments that we conducted to produce quantitative measurements of the relative percentages of red blood cells and white blood cells separated using both types of LDEP microseparator.

2 THEORY AND DESIGN

A cell, which is passing over the interdigitated electrode array with an angle of θ between the electrode and direction of flow, is driven in the lateral direction, as shown in Fig. 1. The lateral force is determined by the magnitude of the DEP force on the cells; this is determined by the applied frequency and the DEP levitation height. The typical mass densities of human RBCs and WBCs are approximately 1130 and 1050~1080 kg/m^3, respectively [17, 18]. Therefore, the DEP levitation height [19], which is settled by z-directional DEP force and gravitation, of RBCs is typically lower than that of WBCs, and the DEP force acting on RBCs will be stronger than that acting on WBCs.

We designed divergent type (Fig. 2(a)) and convergent type (Fig. 2(b)) lateral-driven continuous DEP micro-separators. In the divergent type of separator, although both the WBCs and RBCs are driven to the edges of the microchannel, as explained in Fig. 1, the lateral DEP force acting on RBCs is stronger than that acting on WBCs, as explained previously. Therefore, the WBCs diffuse toward the center of the microchannel away from the high-density stream of RBCs at both edges of the microchannel. Consequently, the WBCs and RBCs are separated continuously to the central outlet #2, and to the two outermost outlets #1 and #3, respectively, as shown in Fig. 2(a). In the convergent type of microseparator, the WBCs diffuse to the edge of the microchannel away from the high-density stream of RBCs at the center of the microchannel. Consequently, the WBCs and RBCs are separated continuously to the two outermost outlets #1 and #3 and the central outlet #2, respectively, as shown in Fig. 2(b).

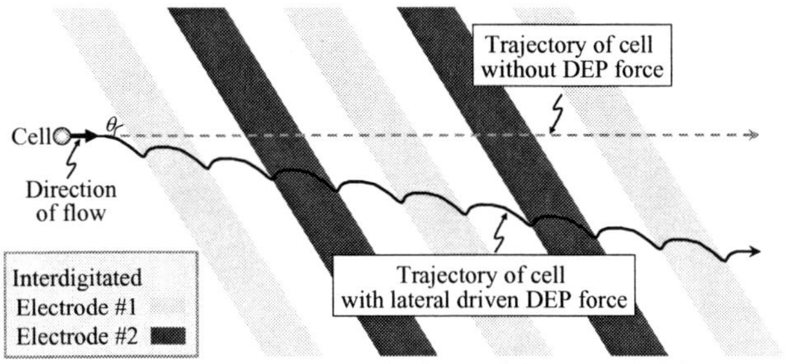

Figure 1: The working principle of the lateral-driven continuous DEP microseparators.

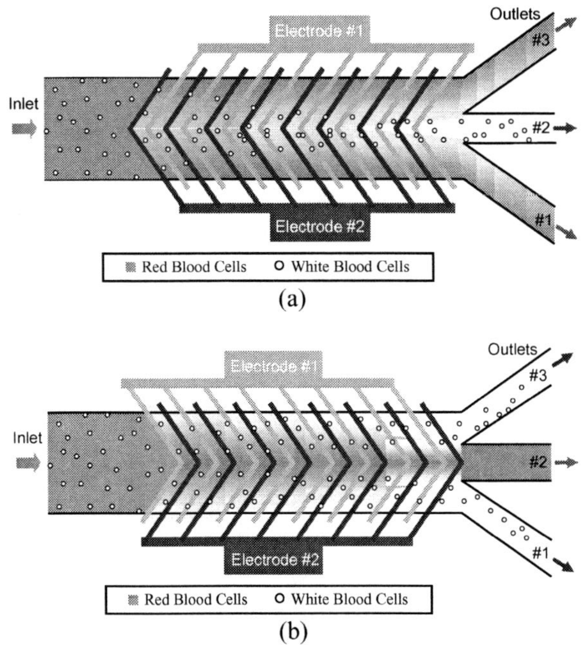

(a)

(b)

Figure 2: Illustrations of (a) divergent and (b) convergent type lateral-driven continuous DEP microseparators with an interdigitated electrode array.

3 MICROFABRICATION PROCESS

The microfabrication process for the lateral-driven continuous DEP microseparator used 0.7-mm-thick Borofloat™ glass slides (Howard Glass, USA), and the polydimethylsiloxane (PDMS) mold as the primary construction materials, along with metal evaporation and glass-to-PDMS bonding, as shown in Fig. 3. The DEP electrodes were made of Cr/Au (200Å/2000Å) evaporated and patterned on the glass substrate. A 100 μm layer of SU-8 2050 photoresist (Microchem, USA) was spun and patterned to create a mold for the microchannel on a glass master. The polymer mold was created using stereolithography (Viper SI2, 3D Systems, USA) and was used for pouring the liquid PDMS. The PDMS mold was completed by assembling the glass master and polymer mold. The liquid phase PDMS, made by mixing the resin and curing agent in a 10:1 ratio (Sylgard 184, Dow Corning, USA), was poured into the PDMS mold and cured for 60min at 80°C in a vacuum oven. The glass substrate with the planar interdigitated electrode array and the PDMS replica were treated with oxygen plasma. These were then aligned and bonded. Finally, to reduce the adherence of the blood cells to the microchannel walls, the microchannel surface was coated with Pluronic-F108 surfactant (BASF, USA) for 24 h. Figs. 4(a) and 4(b) show the fabricated divergent and convergent type lateral-driven continuous DEP microseparators, respectively.

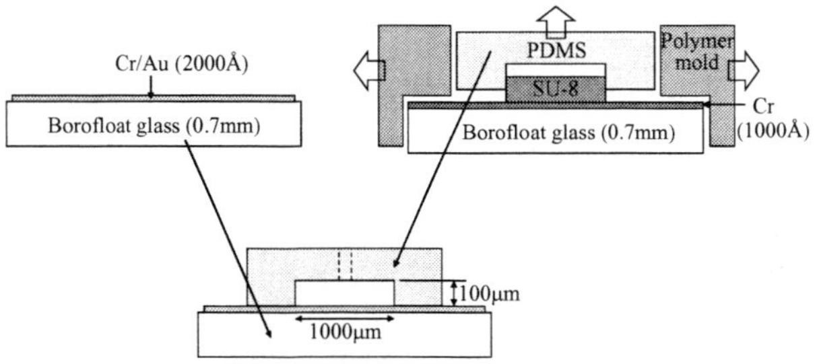

Figure 3: Microfabrication process based on glass and PDMS materials.

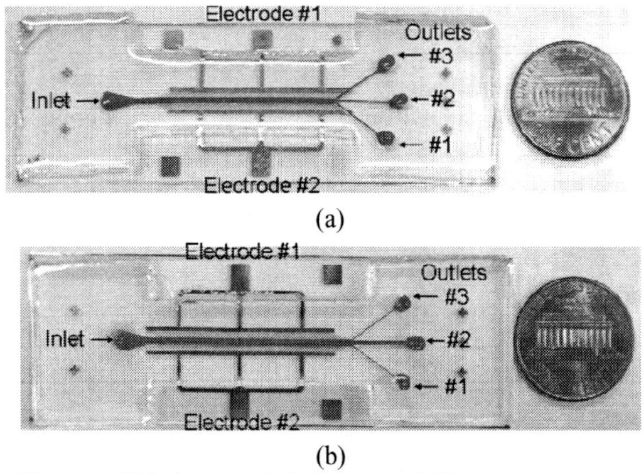

(a)

(b)

Figure 4: Fabricated (a) divergent and (b) convergent type lateral-driven DEP microseparators.

4 EXPERIMENT

4.1 Materials and methods

For the lateral-driven continuous DEP microseparator, we used a 2-MHz sinusoidal voltage of 3 Vp-p from a function generator (AFG3021, Tektronix, USA) to create the DEP force acting on blood cells, and we used a syringe pump (KD100, KD Scientific, USA) to provide controlled flow of the blood sample through the microchannel. A gas-tight glass syringe (81227, Hamilton, USA) minimized the variation in flow velocity. The syringe was connected to the inlet through 0.25-mm inner diameter capillary tubing (Teflon® FER 1/16" tubing, Upchurch Scientific, USA) to push the blood sample in the microchannel. To count the WBCs flowing into each outlet and capture images of blood cells passing through the microchannel, we used a microscope (ME600, Nikon Instruments, USA) with a fluorescence detector (Y-FL, Nikon Instruments).

The blood sample was prepared from anti-coagulated (Heparin-Agarose, H-1027, Sigma Diagnostics, USA) human whole blood diluted in a ratio of 1:5 with PBS solution (Gibco® 10010, Invitrogen, USA). Using a conductivity meter (inoLab® 740, Nova Analytics, Germany), we ensured that its conductivity was 17 mS/cm, the conductivity of physiological medium. A fluorescent probe was added to the WBCs by incubating them at 37°C for 20 min with a cell-permeable nucleic acid fluorescent dye (S-7575, Invitrogen, USA).

4.2 Divergent type LDEP microseparator

Figs. 5(a) and (b) show images of RBCs and WBCs with the fluorescent probe passing through the microchannel of the divergent type DEP microseparator at a volumetric flow rate of 50 µl/h (*i.e.*, 4×10^4 cells/s) with an applied sinusoidal voltage of 2 MHz, 3 Vp-p. The images show that the RBCs were drawn closer to the edges of the microchannel and flowed into the two outermost outlets (#1 and #3), while the WBCs were concentrated at the center of the microchannel and flowed into the central outlet (#2). Fig. 6, which shows the relative separation percentage of RBCs and WBCs at each outlet measured three times using a hemocytometer, shows that the divergent type DEP microseparator separated out 87.0% of the RBCs through the outermost outlets, and 92.1% of the WBCs through the central outlet at a flow rate of 50 µl/h.

Although the divergent type LDEP microseparator were successful for the continuous enrichment of nucleated cells from peripheral blood, the separation efficiencies of 87.0 and 93.6% for RBCs may be relatively low as the LDEP microseparator can stand alone for a number of possible cell-based assays downstream. To increase the separation efficiency, the LDEP microseparator could be improved further through design optimization, including the angle between the electrodes and direction of flow, the electrode width and spacing, and the dimensions of the microchannel.

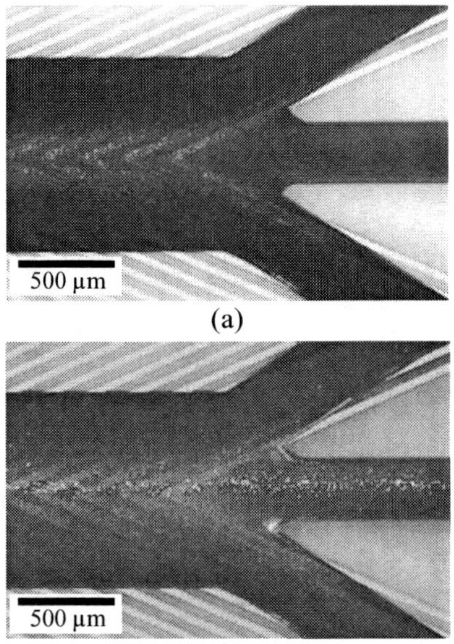

Figure 5: (a) RBCs and (b) WBCs with fluorescent probes passing through the microchannel of the divergent type DEP microseparator at a flow rate of 50 µl/h.

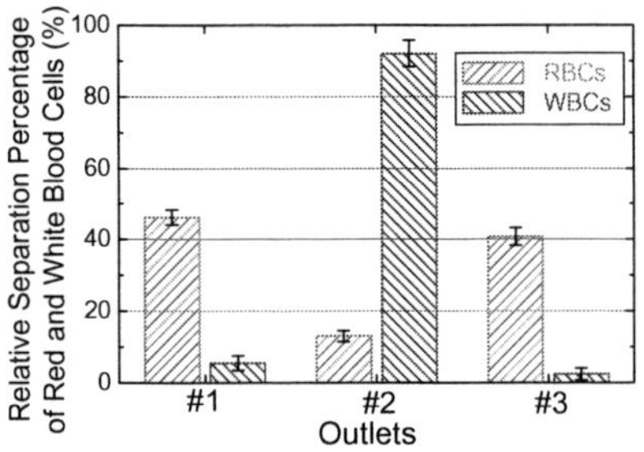

Figure 6: The relative separation percentage of RBCs and WBCs at each outlet of the divergent type microseparator.

4.3 Convergent type LDEP microseparator

Figs. 7(a) and (b) shows images of RBCs and WBCs with the fluorescent probe passing through the microchannel of the convergent type DEP microseparator at a volumetric flow rate of 50 µl/h with an applied sinusoid voltage of 2 MHz, 3 Vp-p. Unlike the divergent type, the images show that the RBCs were concentrated toward the center of the microchannel and flowed into the central outlet (#2), while the WBCs were drawn closer to the edges of the microchannel and flowed into the two outermost outlets (#1 and #3). Fig. 8, which shows the measured relative separation percentage of RBCs and WBCs at each outlet, shows that the convergent type DEP microseparator separated out 93.6% of the RBCs through the central outlet, and 76.9% of the WBCs through the outermost outlets at a flow rate of 50 µl/h.

These experimental results demonstrate that both the divergent and convergent type lateral-driven continuous DEP microseparators can separate RBCs and WBCs within 5 minutes with continuous operation.

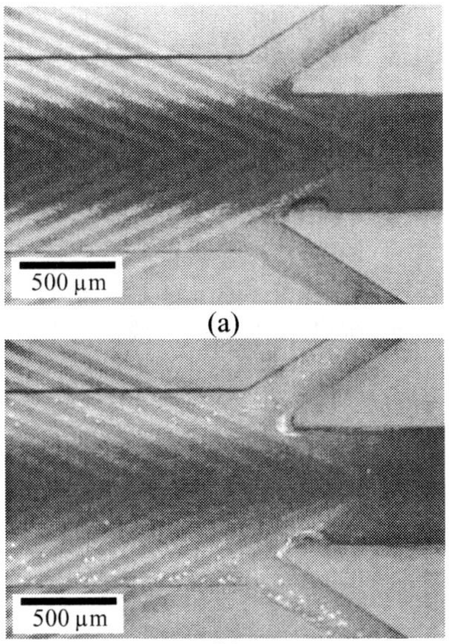

Figure 7: (a) RBCs and (b) WBCs with fluorescent probes passing through the microchannel of the convergent type DEP microseparator at a flow rate of 50 μl/h.

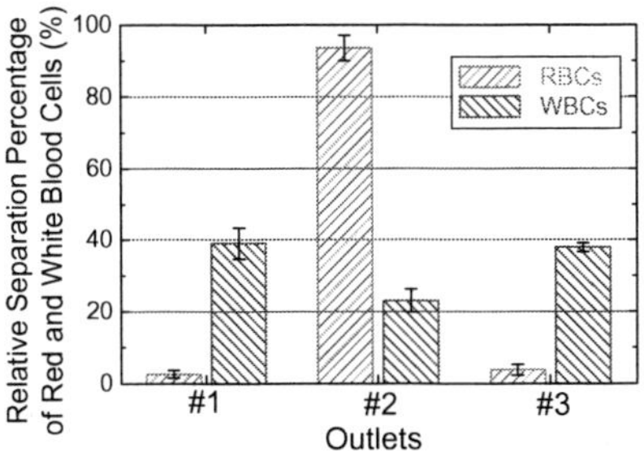

Figure 8: The relative separation percentage of RBCs and WBCs at each outlet of the convergent type microseparator.

5 CONCLUSIONS

We designed and fabricated divergent and convergent types of DEP microseparators based on the lateral-driven DEP force. We demonstrated that they were successful in separating RBCs and WBCs in a suspension medium with a high conductivity of 17 mS/cm. We developed a line charge model to simplify the theoretical analysis of the lateral-driven DEP phenomenon generated by a planar interdigitated electrode array. By comparing this with simulation results, we confirmed that the theoretical model is a simple and useful tool for estimating the lateral DEP force acting on cells in a planar interdigitated electrode array. Experimental results show that our divergent type microseparator can continuously separate out 87.0% of the RBCs and 92.1% of the WBCs from diluted whole blood, while the convergent type can separate 93.6% of the RBCs and 76.9% of the WBCs, within 5 minutes simply by using a 2-MHz 3-Vp-p AC voltage to create a gradient electric field in suspension medium with a high conductivity of 17 mS/cm.

ACKNOWLEDGMENT

This work was supported by the Korea Research Foundation Grant funded by the Korean Government (MOEHRD, Basic Research Promotion Fund) (KRF-2007-313-D00995).

REFERENCES

[1] P. R. C. Gascoyne and J. Vykoukal, Electrophoresis, 23, 1973-1983, 2002.

[2] P. R. C. Gascoyne and J. Vykoukal, Proceedings of the IEEE, 92, 22-42, 2004.

[3] J. Yang, Y. Huang, X.-B. Wang, F. F. Becker and P. R. C. Gascoyne, Biophysical Journal, 78, 2680-2689, 2000.

[4] A. R. Minerick, R. Zhou, P. Takhistov and H.-C. Chang, Electrophoresis, 24, 3703-3717, 2003.

[5] F. F. Becker, X.-B. Wang, Y. Huang, R. Pethig, J. Vykoukal and P. R. C. Gascoyne, Proceedings of the National Academy of Sciences, 92, 860-864, 1995.

[6] Y. Huang, S. Joo, M. Duhon, M. Heller, B. Wallace and X. Xu, Analytical Chemistry, 74, 3362-3371, 2002.

[7] M. Stephens, M. S. Talary, R. Pethig, A. K. Burnett and K. I. Mills, Bone Marrow Transplantation, 18, 777-782, 1996.

[8] S. Prasad, X. Zhang, M. Yang, Y. Ni, V. Parpura, C. S. Ozkan and M. Ozkan, Journal of Neuroscience Methods, 135, 79-88, 2003.

[9] I. Doh and Y.-H. Cho, Sensors and Actuators A, 121, 59-65, 2005.

[10] E. T. Lagally, S.-H. Lee and H. T. Soh, Lab on a Chip, 5, 1053-1058, 2005.

[11] X. C. Hu, Y. Wang, D. R. Shi, T. Y. Loo and L. W. C. Chow, Oncology, 64, 160-165, 2003.

[12] A. Y. Fu, C. Spence, A. Scherer, F. H. Arnold and S. R. Quake, Nature Biotechnology, 17, 1109-1111, 1999.

[13] Y. Huang, J. Yang, X.-B. Wang, F. F. Becker and P. R. C. Gascoyne, Journal of Hematotherapy and Stem Cell Research, 8, 481-490, 1999.

[14] P. R. C. Gascoyne, Y. Huang, R. Pethig, J. Vykoukal and F. F. Becker, Measurement Science & Technology, 3, 439-445, 1992.

[15] J. Rousselet, G. H. Markx and R. Pethig, Colloids and Surfaces A: Physicochemical and Engineering Aspects, 140, 209-216, 1998.

[16] J. Vykoukal, P. R. C. Gascoyne, R. Weinstein, A. Gandini, D. Parks and R. Sawh, In Proceeding of Micro Total Analysis Systems 2002; Y. Baba, S. S., Ed.; Kluwer Academic Publishers: The Netherlands, Vol. 1, 323-325, 2002.

[17] R. C. Leif and J. Vinograd, Proceedings of the National Academy of Sciences, 51, 520-528, 1966.

[18] D. Fisher, G. E. Francis and D. Rickwood, "Cell separation: A practical approach," Oxford University Press: New York, 1998.

[19] Y. Huang, X.-B. Wang, F. F. Becker and P. R. C. Gascoyne, Biophysical Journal, 73, 1118-1129, 1997.

Characterization of DNA Transport through a Semipermeable Membrane with the Effect of Surfactants

S. W. Leung[*], C. Trigo[**], S. Bartolin[***], C.K. Daniels[****], and J.C.K. Lai[*****]

[*]Department of Civil and Environmental Engineering, College of Engineering, and Biomedical Research Institute, Idaho State University, Pocatello, ID 83209, USA, leunsolo@isu.edu

[**]Department of Civil and Environmental Engineering, College of Engineering, Box 8060, Idaho State University, Pocatello, ID 83209, U.S.A, bartcarm@isu.edu

[***]Department of Civil and Environmental Engineering, College of Engineering, Box 8060, Idaho State University, Pocatello, ID 83209, U.S.A, christian_trigo7@hotmail.com

[****]Department of Biomedical and Pharmaceutical Sciences, College of Pharmacy, and Biomedical Research Institute, Idaho State University, Pocatello, ID 83209, USA, cdaniels@otc.isu.edu

[*****]Department of Biomedical and Pharmaceutical Sciences, College of Pharmacy, and Biomedical Research Institute, Idaho State University, Pocatello, ID 83209, USA, lai@otc.isu.edu

ABSTRACT

In current advances of biotechnology, various models of introduction of DNA into cells to alter their gene expression have important biomedical and bioengineering applications. However, the molecular mechanisms underlying how DNA's cross cell and nuclear membranes are poorly understood.

Surfactants are known to influence functions of many proteins in membranes, cells and tissues; however, the natural configurations of DNA's are more stable and are expected to behave differently compared to proteins, especially proteins with smaller size. Information for surfactants on DNA transport is nearly non-existent.

We previously systematically reported how surfactants of different hydrophilicities affected three metabolically important enzymes (namely, glutamate dehydrogenase (GDH), lactate dehydrogenase (LDH), and malate dehydrogenase (MDH)) of various molecular masses and their transport behaviors through a semipermeable membrane at a pH range (6.5-7.4) and concentrations relevant to body functions.

In this study, we employed a similar approach to investigate how membrane pore size, surfactant properties (anionic, cationic, size, non-ionic), pH, would affect the membrane transport of herring DNA. All these factors would modulate DNA transport to certain extent.

Results of this research study would have many implications and applications in bioengineering and cell signaling, further research is on-going and needed.

Keywords: surfactant, DNA, membrane, permeation, transport

1. INTRODUCTION

In the past years, there is an increasing interest in the application of artificial membranes with biomedical and biotechnological purposes. The idea lies in the utilization of these artificial membranes to mimic the function of biological channels, such as those found in living cells. Since most of our body functions are the result of the interactions of different membrane systems, a clear understanding of how important components (proteins, ions, or DNA) for our correct body functions permeate through our cells, and the mechanisms to control the selectivities and permeabilities of those species across these membranes may help to design more efficient and advance membrane systems, such as those in hemodialysis, detoxification of body fluids, and artificial organs. Likewise, the pharmaceutical industry may take advantage of this knowledge in the separation of these species that they rountinely carry out.

Unlike the already published research work with respect to either DNA permeation or use of DNA to modify the permeation of certain species through a semipermeable membrane, which used DNA as driving force for the application of an external potential in order to change the permeation rate of other species; our work focused on acquiring a better understanding of surfactants effect on DNA permeation in our body functioning. Thus, since surfactants are naturally-created in our body, they may by themselves alter the DNA molecular transport in our body, leading to either beneficial or adverse health effects.

We have, systematically and stepwise, studied how 2 cationic surfactants of different molecular weight, a non-ionic, and 2 anionic surfactants of different hydrophobicities might influent the interfacial mass transport of DNA coming from herring sperm through a semipermeable, artificial membrane at three different pHs that are important to human body functions. We consider that the results obtained from this experimental approach would help to design more effective treatment applications involving either DNA alone or with other compounds, such as proteins that are having important influence on DNA.

2. MATERIALS AND METHODS

The DNA for this study of molecular transport came from Herring sperm (Sigma), the DNA contained 6.1 % of sodium. The cationic surfactants used were C-573 (low molecular weight) and C-581 (high molecular weight) (Cytec Industries, Inc.). The anionic surfactants were IB-45 (hydrophilic) and TR-70 (hydrophobic) (Cytec Industries, Inc.). Non-ionic surfactant was Triton-X 100 (Sigma). Experiments setup and surfactant concentrations were similar to studies that we previously reported [1,2].

3. RESULTS

3.1 Effect of pH

Similar to the effect that we observed for protein transport, .the DNA that we used also revealed pH differential by passive permeation; the permeation rate increased with the increase of pH value, as it is shown in Figure 1 below.

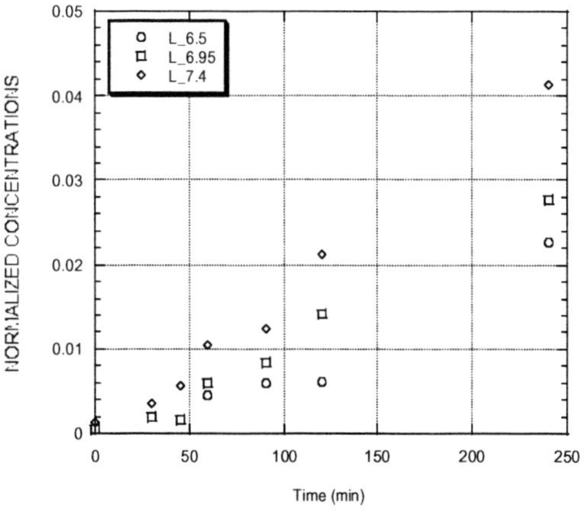

Figure 1: Effect of pH on DNA in the absence of surfactant with 1 micron membrane.

3.2 Effect of Surfactants

As shown in Figure 2, effect of surfactants to DNA permeation appeared to be random initially. This may be due to the fact that only a small amount of DNA was able to pass through the pores when 1 micron of membrane was used in the experiments, the relatively error was high in the initial measurements at time interval of less than 120 minutes. At the time interval of 240 minutes, the differential mass permeabilities were becoming more obvious. The non-ionic surfactant, Triton-X 100, appeared to impede the DNA permeability, as well as the hydrophobic surfactant, TR-70. Both cationic surfactants and the hydrophilic anionic surfactant, IB-45, increased the permeability of DNA, this observation was consistent with what we had reported for the effect of these surfactants to proteins' permeabilities: cationic surfactants behave similarly as hydrophilic anionic surfactant.

3.3 Effect of Membrane Pore Size

As shown in Figure 3, the pore size of the membrane had definitive effect to the interfacial transport of DNA. For the membrane pore size of 0.03 μ, there was no detectable DNA permeated from the high concentration cell to the low concentration cell after 240 minutes. For the pore

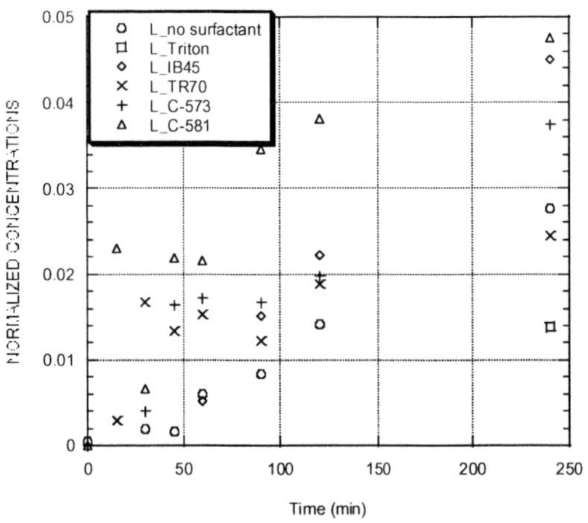

Figure 2: Effect of various surfactants on DNA permeation at pH 6.95 with 1 micron membrane.

size between 0.1 to 1 μ, the overall mass permeated from high to low concentration for the interfacial transport was about the same, although mass permeated for the membrane with the 1 μ-pore was slightly higher in spite of the fact that the pores were 10 times larger than the 0.1 μ membrane. As expected, the 5 μ membrane allowed much higher DNA permeated through the system. One might presume that the molecular radius of the DNA is much smaller than 5 μ and thus the DNA can easily pass through the membrane channels. However, the microfluidic phenomenon and how the DNA interacts with the membrane within the channel, when the pore size is minimal, remain interesting.

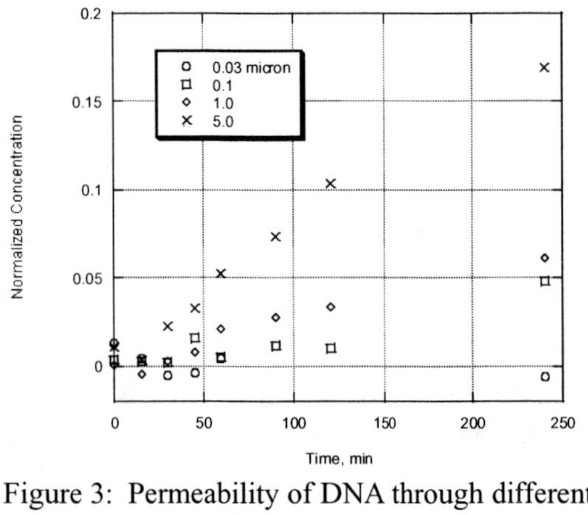

Figure 3: Permeability of DNA through different pore sizes of membrane at pH 6.95 in the absence of surfactant.

4. CONCLUSIONS

Interfacial DNA transport between cells and organs is not well understood, and is very important in tissue engineering and other bioengineering applications. Our exploration toward this interesting, yet challenging, field is just the beginning of many discoveries to be awarded.

5. REFERENCES

[1] S. Leung and J. Lai, "Differential Effects of Anionic Surfactants on Activities of GDH, LDH, and GDH", Biochemical Engineering Journal, 25(1): 79-88, 2005

[2]J.C.K. Lai and S. W. Leung, "Protein and Cell Signaling with Biomaterials: Interfacial Transport". In Encyclopedia of Biometerials and Biomedical Engineering (Wnek GE & Bowlin GL, eds.), pp.1-11, Marcel Dekker. New York, NY, Mar.27, 2006

6. ACKNOWLEDGEMENTS

This study was supported by an USAMRMC Project Grant (Contract #W81XWH-07-2-0078) and NIH Grant #P20 RR016465 from the Idaho INBRE Program of the National Center for Research Resources.

New Application of Nanoparticle based 21 Tesla MRI of Mouse Brain

Structural Segmentation and Volumetrics

Rakesh Sharma,
Department of Biomedical Engineering and National High Magnetic Field laboratory,
Florida State University, Tallahassee, FL 32310, rs05h@fsu.edu

ABSTRACT

The intact mouse brain high resolution MRI is becoming state of art. Two major approaches are used: 1. ultrahigh 21 Tesla magnetic field MRI in visualizing anatomical structural details; 2. the use of nanoparticle based datasets revolve around the creation of a probabilistic brain atlas as a normalization anatomical templet useful in variety of applications. One of the major applications is localization of anatomical landmarks and visualization of brain structures to identify them with validation process. In brain, cortex, hippocampus structures such as dentate gyrus made of granule cells are widely accepted as possible MRI visible structures. These major anatomical sites predict the drug or neuropharmaceutical compounds induced morphological changes. Coregistration of MRI visible neuroanatomical regions with histological identification provides finer details and different MRI contrast mechanisms.

Key words: Mice, brain,MRI,nanoparticle

1.2. Nanoparticle imaging contrast agent Preparation

In general, polymeric nanoencapsulation methods, which combine sonication and nonsolvent temperature induced crystallization steps include (a) providing active agent nanoparticles having an average diameter between about 5 and about 100 nm; (b) treating said active agent nanoparticles (e.g., a superparamagnetic material) with an anionic surfactant to form modified active agent nanoparticles; (c) mixing the modified active agent nanoparticles with a solution of a polymer in a solvent at a first temperature, which is greater than the melting temperature of the polymer and less than the boiling point of the solvent to form a first mixture, said mixing comprising the use of sonication; (d) mixing a non-solvent with the first mixture to form a second mixture, the non-solvent being a non-solvent for the solvent and for the polymer and having a boiling point greater than the melting temperature of the polymer; (e) sonicating the second mixture to form an emulsion; and (f) cooling the emulsion to a second temperature and at a rate effective to precipitate polymeric nanoparticles comprising the polymer with the modified active agent nanoparticles dispersed therein.

Figure 1: (on left) 35 nm sized paramagnetic iron oxide myoglobin particles.(on right) A comparison of nanoparticle with RBC is shown.

1.3 Advantages of ultrahigh resolution MRI microimaging:

Recently, several advantages of ultrahigh magnetic field have been reported to achieve high resolution microimaging [1]. Major advantages are high ultra high contrast, high SNR, multicontrast, high resoluiton DTI etc. However, inhomogeneity also linearly grows at high magnet field strength. Disadvantages of ultrahigh magnet field strength are not reported as safe technique or remains to establish the disadvantages of high magnetic fields.

1.4 Application of ultra high resolution MRI of mice brain

Notable examples are following: 1. Mice brain atlas in localization of site-specific damage: Alcoholic encephalopathy; 2. Transgenic knock-out mice brain and effect of alcohol, colchicine etc.

2. MATERIAL AND METHODS

2.1 Microimaging of brain:

Bruker 21 Tesla magnet and Micro-2.5 microimaging system is available in imaging studies. USPIO iron-oxide nanoparticle injected mice excised brains immersed in a perflourinated solution can generate T2*-weighted datasets with a three dimensional MSME and gradient-echo pulse sequence (TE/TR = 7.5/150 ms) at a 15 μm isotropic resolution in 5 hours and T2 and diffusion tensor weighted MR data sets with a 3 dimensional spin-echo pulse sequence

2.2 Validation of brain structures:

The mouse brains (n = 3) were perfusion-fixed for MRI microscopy. The dentate gyrus of dorsal hippocampus in both left and right sides showed collateral control. The susceptibility matching eliminates both extraneous proton signal and coil loading. (see the Figure 1).

The axial, coronal and sagittal planes in 2D slices reconstruct the 3D brain atlas by volume rendering. The main features at ultrahigh resolution are: distinct gray and white matter tissues and neuron proton density in different structures (shown in Fig 2).

2.3 Mice brain atlas:

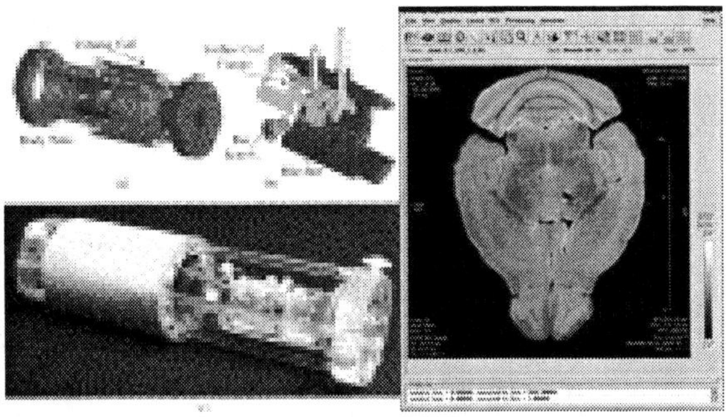

Figure 2: (Top on left) A Rf coil with microimaging; (Bottom on left) The attachment of animal holder for live Animal experiments; (on right)T2* weighted high resolution mice brain image is shown with clear layers of cortex, central semi ovale, hippocampus, frontal, parital and occipital lobes with cerebellum

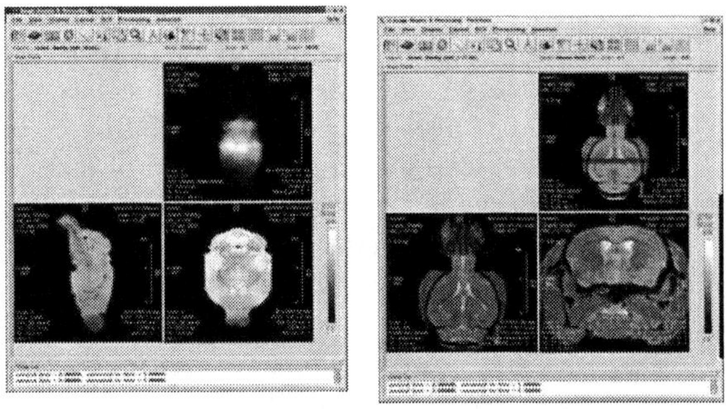

Figure 4: (on left panels) A 3D Gradient echo image set axial, coronal and sagittal planes. (on right panels) 3D-FLASH image set, are shown in 3 plan with clear structures within 15 seconds. Notice the less and clear contrast as benefit over Gradient Echo images

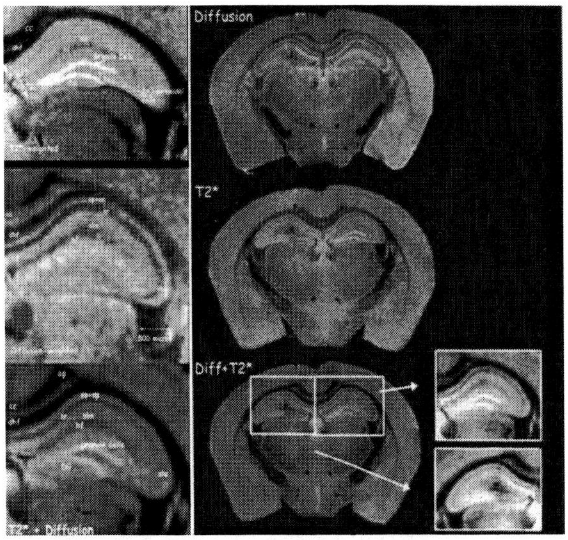

Figure 3: (on left panels) Enlarged section of dentate gyrus by Diffusion (top), T2*(middle) and Diff+T2*(bottom) (on right panels) Whole hippocampus is shown by diffusion-weighted MRI microimages (top) display the stratum oriens/stratum pyramidale (so+sp), stratum radiatum (sr), stratum lacunosum moleculare (slm) and lucidum (slu); T2* weighted images (middle); and Co-registered T2* - and diffusion-weighted MRI microimages (at bottom) provided the better information of hippocampus fissure (hf) and corpus callosum (cc).

***Presentation at NHMFL/COM open house on 12Feb. 2005**

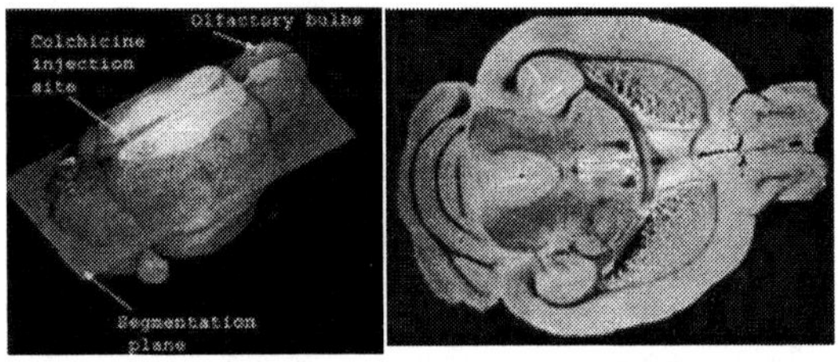

***Figure 5: (a) 3D reconstruction showing the CI site (3D GE);(b) axial slice through the 3D dataset at the segmentation plane of the top image. The elimination of the granule cells in CI hemisphere(red arrow) is evident in contrast to the contralateral control(yellow)**

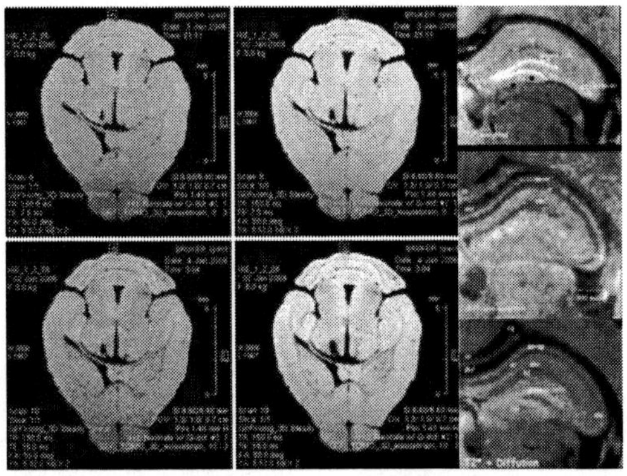

***Figure 6: T2* weighted, diffusion weighted and combined MRI images of a coronal view. The hyperintense line corresponds to granule cells of the dentate gyrus. In the hemisphere shows disruption of hippocampus layers is slightly apparent. Bottom: Comparison of normal hippocampus shows atrophy.**

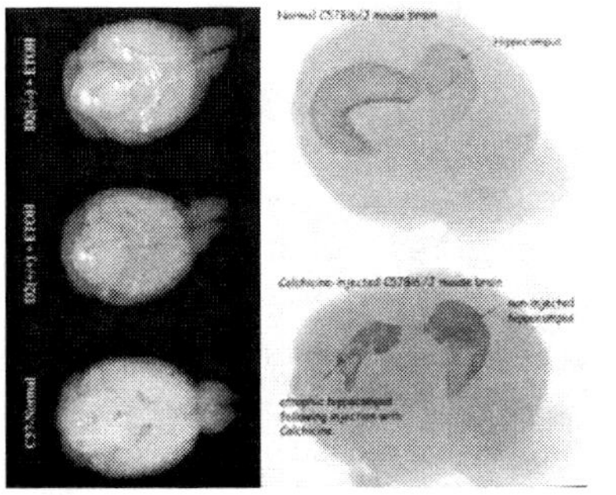

Figure 7: Cerebrospinal fluid (CSF) filled ventricular spaces have been highlighted in color. D2 knockout mouse brain shows large ventricles compared to the D2 (+/+) mouse brain(middle) following alcohol exposure. The control brain displays smaller ventricles than either of treatment groups. It suggests that D2 receptor modulate EtOH induced toxicity.

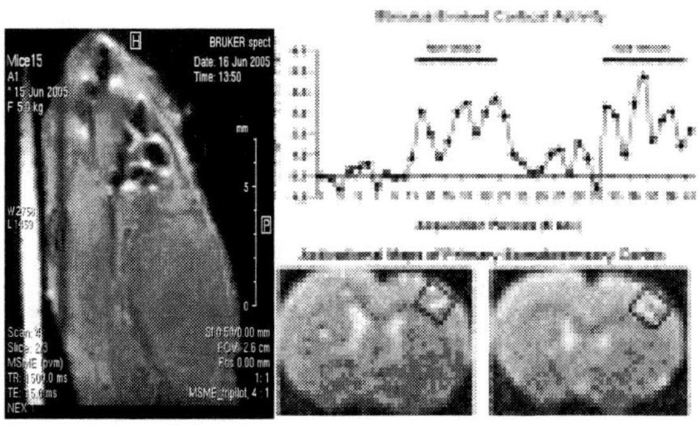

Figure 8: (on left) A nanoparticle filled tube with nanoparticle injected mice brain is shown. Notice the signal loss in air cavities.
(on right) A possibility of functional MRI is shown using live mice using light blinking event and neuroactivation in brain.

4. APPLICATION OF MRI MICROSCOPY:

4.1. D2(-/-) and D2 (+/+) Mutants and Influence of Ethanol

Dopamine (DA) transgenic mice have been genetically engineered and used to study the relevance of specific DA receptors for the acute response to drug abuse as well as predisposition, toxicity and relapse and neurodegenerative diseases. The dopamine appears to play a critical role through all phases of central nervous system ontogeny such as cell proliferation, neural migration growth maturation and synaptogenesis. However, it is not known how DA influences brain maturation and how dysfunction of certain DA receptors might influence overall brain morphology in DA transgenic mice using high resolution MR microscopy as part of longitudinal study.

4.2 D2 knockout mice (D2 (-/-) and overexpressed (D2 +/+) C57BL6/J mice

The animals were exposed to ethanol for 10 months. After exposure (age about 20 months) the mice were euthanized and formalin-perfusion fixed. MR microscopy was performed on the excised brains for gross volume measurement and identification of MRI visible regions.

4.3 Future attempts for Functional MRI Imaging

At low magnetic field, mice fMRI experiments showed promise to evaluate the brain functions and neuroactivation events. As of today, the possibility of 21 Tesla fMRI experiments is unknown.

5. MOLECULAR IMAGING WITH MRI

There is rapidly growing interest in making radiological imaging techniques sensitive to specific molecules or biological processes. The marriage of the wealth of information being generated from molecular genetics, the large number of mouse models of human diseases, and imaging is the engine driving the development of molecular imaging. Nuclear imaging techniques, optical imaging techniques, and MRI have all had aspects of molecular imaging as part of their makeup for a number of years; however, there is renewed vitality as more and more is learned about the pathophysiology of disease and more and more targets for therapeutic intervention become available.

Molecular imaging of the brain offers significant challenges, primarily due to the problem of delivering agents through the blood-brain barrier. Nonetheless, several recent studies demonstrate the potential. For example, amyloid plaque can be imaged in mice with targeted MRI contrast after disrupting the blood-brain barrier.

In transgenic mice engineered to produce a large number of amyloid plaques specific accumulation of the MRI agent enabled detection of plaques as an example. This strategy of coupling MRI contrast agents to peptides or antibodies that recognize specific targets is rapidly growing area in molecular imaging. Creative ways to get large molecules through the blood-brain barrier will be crucial to the general success of this strategy.

Another very promising area for molecular imaging is to monitor cell migration *in vivo*. Here the idea is to label a specific cell population, either *in vitro* or *in vivo*, with MRI contrast agents and then to follow the movement of these cells in the animal. Typically nanometer-sized iron-oxide-based contrast agents are used and some form of endocytosis is used to get these particles into cells. The advantage of these iron oxide particles is that it has been shown that single cells can cause sufficient contrast change to be detected by MRI. There have been recent examples of MRI-based cell tracking to study diseases of the brain.

5. CONCLUSION

MRI microscopy is extremely valuable in in vivo localization and identification of brain anatomical structures in the excised mouse brain. The present study demonstrates the hippocampus structures using T2* and diffusion weighted contrasts at 21 Tesla with details of neuroanatomical features throughout the brain. Within the hippocampus, site-specific, dose-specific, dose-dependent injections of colchicine selectively destroyed the granule cell layer of the dentate gyrus. The lesion volume measurement technique, however, suffers from the artifact of hyperintense cellular layer of hippocampus. The anisotropy of water diffusion in these ex vivo samples may visualize the mossy fiber pathways of the hippocampus. The possibility of fMRI at 21 Tesla is controversial due to mainly uncertain effects of ultrahigh magnetic fields on live or dead brain or any other tissues.

6. ACKOWLEDGEMENTS

National High Field Magnetic Field Lab facility to user investigator is appreciated along with sharing mice brain data.

REFERENCES

[1] Pautler RG, Mouse MRI:Concepts and Applications in Physiology. Physiology 19: 168-175, 2004.

[2] Burton RAB, Plank G, Schneider JE, Grau V, Ahammer H, Keeling S, Lee J. et al. Three-dimensional models of individual cardiac histoanatomy: Tools and Challenges. Ann NY Acad Sci. 1080:301-319, 2006.

[3] Alan P. Koretsky. New Developments in Magnetic Resonance Imaging of the Brain. NeuroRx. 2004 January; 1(1): 155–164.

Study on Fluorescent Detection Glucose Molecule Marked with ZnSe Nano Crystalline in Microchannels

Zhong-chen Bai, Xin Zhang and Shui-jie Qin

Laboratory for Photoelectric Technology and Application,
Guizhou University Guiyang 550025, China
yufengvc@163.com, Ie.xzhang@gzu.edu.cn, sjqin@mail.gzst.gov.cn

ABSTRACT

Shell-core structure of ZnSe nano crystalline cluster is used for marking glucose molecules in this paper. Fluorescent detection is successfully realized in microchannels. Mercaptoacetic acid is used for packing ZnSe nano crystalline cluster, which is dissolved in the water. Carboxyl of mercaptoacetic is used for connecting glucose molecule, and mixture solution of glucose molecule marked with ZnSe nano crystalline cluster is inleted by capillary force in muti-microchannels with diameter of 30 microns. Exciting light with the wavelength range from 265nm to 420nm is used for stimulating glucose molecules marked with ZnSe nano crystalline cluster in multi-microchannels, and fluorescent detection is carried out for glucose molecules. It is concluded that shell-core structure of ZnSe nano crystalline cluster is feasible for marking glucose molecules. The new method will lead to new application in many fields such as biologic analysis, bio-sensor, biologic detection and spaceflight, and so on.

Keywords: nano crystalline, bio-sensor, microchannles, fluorescent detection, nano cluster

1 INTRODUCTION

Nano crystal[1] and nano crystalline cluster have been interested as their potential application in biology and medical analysis. Nano crystalline cluster [2-5] used for fluorescent marking has many merits in contrast with traditional dye fluorescent such as small size in surface, specific chemistry characterization, and the range wide of frequency spectrum of exciting light, continuous distribution, symmetrical distribution of emissive spectrum, wide-narrow of spectral line, adjustable color, and so on. Based on these photochemical stability and spectral characterization, nano crystalline cluster is rapidly developed in biologic analysis and detection. In general, nano crystalline cluster can be dissolved in the water or buffer solution[6]. In the last few decades, nano crystalline cluster can transmit a lot of photons and fluorescent stability, which has been a hot topic in biologic analysis and detection, particularly, nano crystalline cluster of shell-core structure can be widely applied.

Glucose molecules is cardinal element of blood sugar and urine glucose. Diabetes can be discovered and forestalled early by detecting glucose molecules. Glucose molecules exists widely in biologic organism. Detecting glucose molecules in biologic organism has significant in studying on energy transition between two biologic body, and unscramble life origin. Detecting glucose molecules in food can protect health of person from aggrieving of disease, which has a significance.

Usually, glucose molecules is detected by many methods, such as glucose oxidized enzyme methods(GOD), oxidative electrode rate methods, fluorescent methods, and so on. All these methods consume lots of sample with low detecting sensitivity. Especially, detecting experiment of blood glucose and urine glucose in body need extract blood many times, which can causes pain to patients. In scale of microchannels determine consumption of sample, the size of microchannels can affect sensitivity of glucose molecules. In this paper, microchannels as the detecting container content a few sample and satisfy easily detecting demand, withal high sensitivity. Recently, transparent material multi-microchannels with improving of processing microchannels technology was applied more and more in biologic analysis and detection[7]. Microchannles with diameter of 30 microns was applied for detecting container, fluorescent detection glucose molecule marked with ZnSe nanocrystallinein cluster is implemented commendably in multi-microchannels in this paper.

2 EXPERIMENTAL

2.1 Setup for fluorescent detection

Shell-core structure Nano crystalline cluster is indicated in figure 1. Many active molecule groups in surface of shell can connect detecting molecules by adsorption or chemical bonding methods, and that ZnSe nano crystalline cluster photoluminescence characterization is used for exciting light. Moreover bring fluorescent is collected by CCD. Sequentially, fluorescent detection glucose molecule marked with ZnSe nano crystalline cluster is realized successfully in microchannels. Solution of Figure 1 is inhaled in multi-microchannles by capillary force. The range of wavelength from 265nm to 420nm is used for exciting mixture solution of microchannels, and fluorescence is produced at the same time. CCD is used for observing all the course of fluorescent bring.

The setup of fluorescent detection is indicated in Figure 2. Exciting light in the range of wavelength from 265nm to 420nm illuminates glucose molecules which is marked with ZnSe nano crystalline cluster in multi-microchannels in the stage at 45 degree. Fluorescent that is bring by exciting ZnSe nano crystalline cluster overpasses objective lens and two-way color selective mirror, and is filtered by optical filter. The picture is obtained by CCD and computer system.

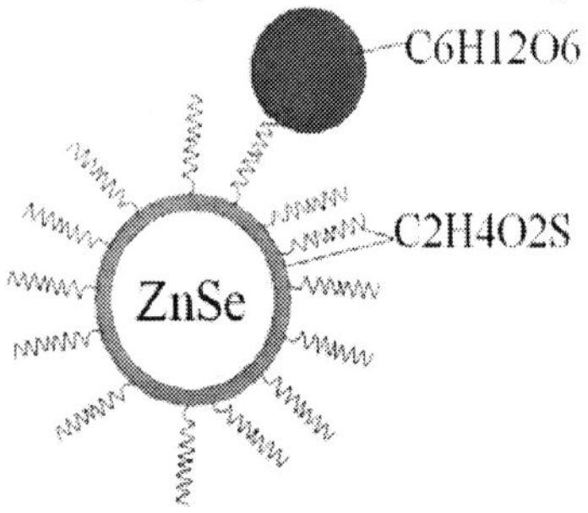

Figure.1 Schematic of detected molecules marked with ZnSe nanocrystalline cluster

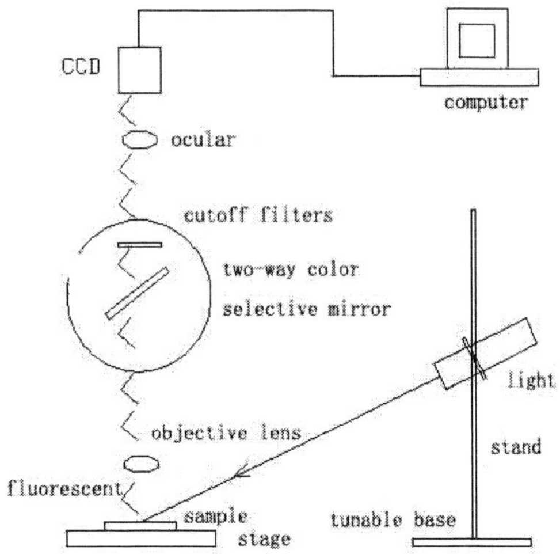

Figure.2 Schematic of the setup for fluorescent

2.2 Preparing experimental sample

2.2.1 Preparation of ZnSe nano crystalline cluster

Se(selenium) and Na_2SO_3(sodium carbonate) are dissolved in water of 90℃ and agitated by magnetic force four hours, which can gain Na_2SeSO_3 solution-1; $Zn(AC)_2$(zinc acetate) is dissolved in water and add to $C_2H_4O_2S$(mercapto-acid).Next, adding NaOH (sodium hydroxide), and adjusting PH=8. Last, adding to solution-1 by N_2 protected and stirring by magnetic force nine hours at 90℃, which can synthesize ZnSe nano crystalline cluster of shell-core structure.

2.2.2 Preparation of auxiliary sample

Glucose crystal is accurately cranked out 10ml of 0.5mol/L solution-2. Solution-2 of 1ml is diluted 10 times by de-ionized water, which is cranked out 0.05mol/L solution-3. Between solution-3 1ml and 1ml of ZnSe packed $C_2H_4O_2S$ solution are mixed. Then standing 45 minutes in test tube, which is made solution-4. Transparent material multi-microchannels with inside diameter of 30 microns and tube wall of 5 microns is provided by xi'an institute of optics and precision mechanics of GAS in this experiment.

2.2.3 Inlet sample

One end of transparent material multi-microchannels is submerged solution-4. Solution-4 can be inleted slowly microchannels by capillary force. When microchannles is filled of solution-4, microchannels is taken out and fixed carefully on the stage, observed under microscope. If microchannels without solution-4 is black in microscope. The reason is that light is refracted between air interface and microchannels wall. However, microchannels with solution-4 is bright in microscope. The reason is refractive index of solution-4 is close to microchannels wall comparing with air.

3 RESULTS AND DISCUSSION

3.1 Fluorescent observed results

Glucose molecule marked with ZnSe nano crystalline is observed in multi-microchannels by CCD. Adjusting the stage of Figure 2, solution of glucose molecules marked with ZnSe nano crystalline cluster is illuminated by exciting light at 45 degree of the horizontal direction and ZnSe nano crystalline cluster radiates light. CCD display technology is used for recording experimental results (Figure 3).

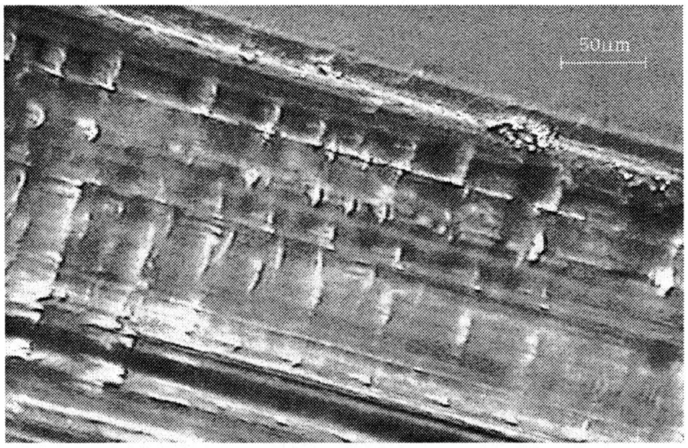

Figure.3(a) Fluorescent molecules excited by the range of 265nm-420nm with 470nm filter brings fluorescence in microchannels

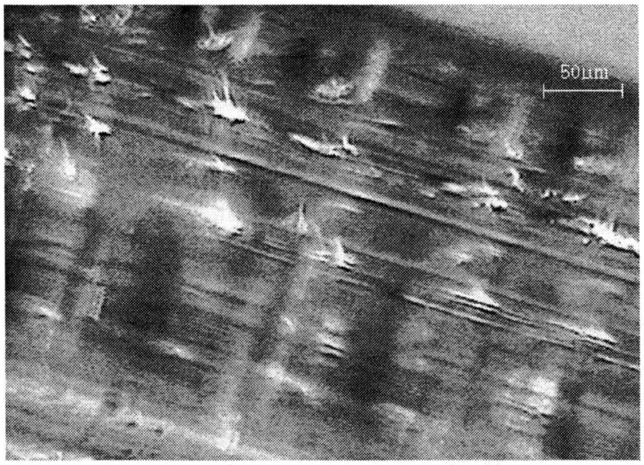

Figure.3(b) Fluorescent molecules excited by the range of 265nm-420nm without filter brings fluorescence in microchannels

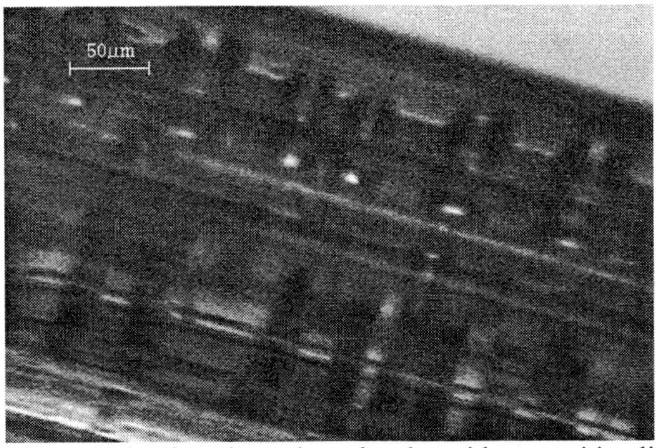

Figure.3(c) Microchannels molecules without exciting light

We can observe remarkably glucose molecules marked with ZnSe nano crystalline cluster from (a), (b), (c) in Figure 3, and distinct effect of fluorescent.

3.2 Experimental results effected by wavelength of exciting light

Absorb curves and fluorescent curves of ZnSe nano crystalline cluster are indicated in Figure 4. we can see from Figure 4(a): when exciting light in the range of wavelength is more than 400nm, ZnSe nano crystalline cluster can not absorb any exciting light, but less-than 400nm, ZnSe nano crystalline cluster can absorb exciting light and absorb continuously; we can see from Figure 4(b): line width of fluorescent of ZnSe nano crystalline cluster is narrow and in the range of wavelength from 300nm to 550nm has fluorescent peaks which is very important to marking molecules.

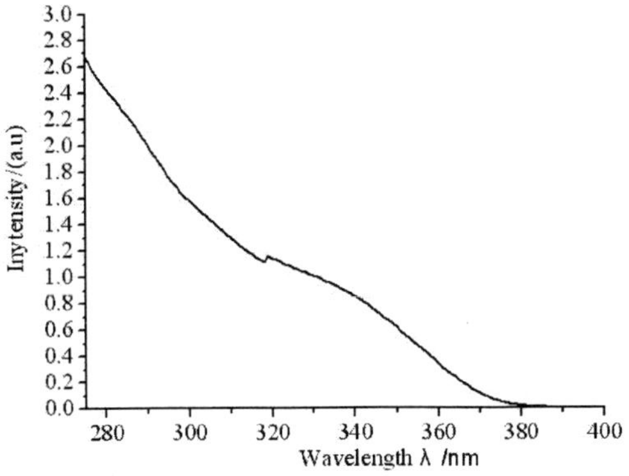

Figure.4(a) PH=8, absorb curves of ZnSe solution

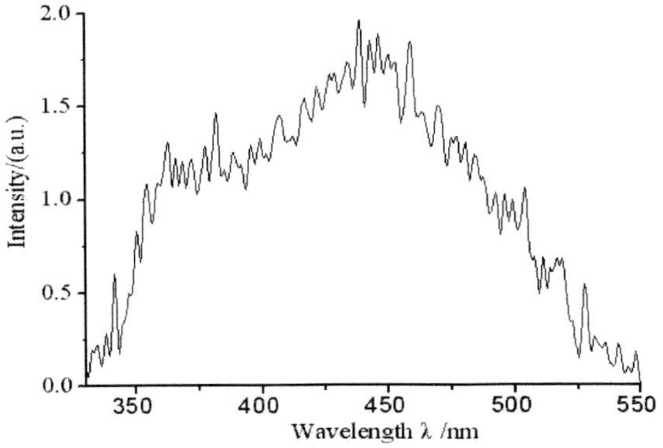

Figure.4(b) PH=8, fluorescent curves of ZnSe solution

Exciting light in the range of wavelength from 265nm to 420nm can be in accord with Figure 4(a). Selecting different filters can come out different rusults in this experiment. When high pass filter of 470nm is used for filtering background light, we can get green light of Figure 3(b); when without any filter, we can get blue light of Figure 3(c), which is received by CCD. The reason is that main peak of ZnSe nano crystalline cluster lies round 430nm.

3.3 Experimental result affected by exciting light intensity

Experimental results is effected seriously by intensity of exciting light in the experiment. Adjusting distance between exciting light and fluorescent molecules can adjust exciting light intensity. The result shows that when distance between exciting light and fluorescent molecules is less-than 30mm, fluorescence is very strong by CCD receiver, but it has fluorescent quenching; when distance between exciting light and fluorescent molecules is more than 110mm, because of lower power of light, a few molecules can not bring fluorescent at once. Exciting light lasts 20 second, weak fluorescent can be brought. Adjusting distance between exciting light and fluorescent molecules from 30mm to 110mm, fluorescent is brought and fluorescent intensity changes from strong to weak, later on from weak to strong. The picture that glucose molecules marked with ZnSe nano crystalline is claer. In addition; the material of microchannels can scatter light which affects results of experiment.

4 CONCLUSION

In this work, shell-core structure of ZnSe nano crystalline cluster was used for marking glucose molecules. Our results suggested that shell-core structure of ZnSe nano crystalline cluster can mark glucose molecules easily. Furthermore fluorescent intensity is stronger, receiving fluorescent color be adjusted easily. We also found that exciting light power and wavelength have an important effect on experimental results. Changing size of ZnSe nano crystalline cluster changes absorb curves and fluorescent curves of ZnSe solution. The water-dispersible ZnSe nano crystalline cluster has potentials in large-scale application of biologic analysis, bio-sensor, biologic detection, voyage and spaceflight[8-10], and so on.

ACKNOWLEDGEMENTS

This work was supported by Laboratory. for Photoelectric Technology and Application of Guizhou university in China and the National Nature Science Foundation of China Programs (Grant No. 50375031)

REFERENCES

[1] Wei-Qi Huang, Li Xu, Ke-Yue Wu, et.al, "Enhancement of photoluminescence emission in low-dimensiona structures formed by irradiation of laser". J.Appl.Phys.,**102**,053517(5),2007.

[2] Xu Rong-hui, WANG Yong-xiang, Xu Wan-bang et. al, "Study on Synthesis and Fluorescent Property of CdS Nanocrystals", Journal of Synthetic Crystals, **35**,1007-1011,2006.

[3] Sung Ju Cho, Dusica Maysinger, Manasi Jain, et al, "Long-Term Exposure to CdTe Quantum Dots Causes Functional Impairments in Live Cells", American Chemical Society, **23**,1974-1980, 2007.

[4] Huang Feng-hua, Peng Yi-ru, "Preparation and Fluorescence Property of ZnS Quantum Dots. Chinese .Journal of Synthetic Chemistry",**12**,529-531,2004.

[5] Narayan Pradhan, David M. Battaglia,, Yongcheng Liu,,et al, "Efficient, Stable, Small, and Water-Soluble Doped ZnSe Nanocrystal Emitters as Non-Cadmium", Biomedical Lbbels.Nano Letters, **7**,312-317,2007.

[6] Sha Xiong, Shihua Huang, Aiwei Tang, et al, "Synthesis and luminescence properties of water-dispersible ZnSe nanocrystals", Materials Letters, **61**,5091-5094, 2007.

[7] S.J Qin, Wen J. Li, Michael Y. Wang, "Fabrication of Subbmicro Channels in Quartz Cubes Using LASER-Induced Splitting", International Journal of Nonliner Sciences and Numerical Simulation ,**3**,763-768,2002.

[8] Shengyuan Yang, M. Taher A. Saif, "MEMS based force sensors for the study of indentation response of single living cells",Sensors and Actuators,**135**,16-22,2007.

[9] O. J. Pallotta, G. T. P. Saccone, R. E. Woolford, "Bio-sensor system discriminating between the biliary and pancreatic ductal systems", Med Biol Eng Comput, **44**,250-255,2006.

[10] Montserrat Calleja, Javier Tamayo, "Low-noise polymeric nanomechanical biosensors", Applied Physics Letters, **88**,11390(3),2006.

'Lab On A Chip' Label Free Protein Sensor Systems Based on Polystyrene Bead and Nanofibrous Solid Supports

V. Kunduru*, J. Grosch*, S. Prasad*, P.K. Patra**, S. Sengupta***

*Portland State University, ECE Dept, PO Box 751, OR USA, sprasad@pdx.edu
**Rice University, TX, USA, prabir@rice.edu
*** University of Massachusetts Dartmouth, MA, USA, ssengupta@umassd.edu

ABSTRACT

A demonstration of the study of polystyrene solid supports on platform based technologies for label - free immunoassay based reactions with improved sensor performance is presented here. Nanomaterials based biosensor was developed by coupling sensing elements with sensitive electronic capture equipment. Thus we constituted a 'Lab on a chip' device. The prototype sensor was evaluated for an inflammatory biomarker protein – C reactive protein (CRP). Polymer composites were used to provide both surface area for protein functionalization and efficient means of electron transduction. Important biosensor features such as the dynamic range for linear response to the analyte of interest and rapid response time are highlighted.

Keywords: *polymer, polypyrrole, protein, sensor, 'lab on a chip'*

1 INTRODUCTION

Current commercially available immuno - assay type detection techniques involve high throughput detection of the analyte using the corresponding enzyme labeled antibody [1]. Most quantitative immuno assays use fluorescence/luminescence detection schemes that display the quantity of analyte present as a function of the intensity of light emitted. Expensive and bulky optical imaging equipment is required for observing the labeled immunological detection results. Most of these technologies face several challenges like assay miniaturization, system robustness, microtiter plate logistics and other fluid handling complications [2-4].

Our research involves the study of polystyrene solid supports on platform based technologies for label - free immunoassay based reactions with improved sensor performance. Appropriate combination of suitable sensing materials with efficient sensing techniques is the focus of this research for the development of protein sensors. We compare the efficiency of polystyrene beads vs electrospun polystyrene nanofibers in terms of analyte detection using different functionalization schemes. A silicon based, photolithographically fabricated microelectrode array [5] (MEA) coated with biofunctionalized polystyrene provides the platform for accessing the electrical signature of the analyte of interest. The presence of nanosensing sites on the polystyrene surface enhances anchorage of the capture antibodies relevant to the experimental antigen of interest owing to their very large specific surface area. A comparative study of the sensitivity of polystyrene bead vs. polystyrene nanofibrous web is conducted to determine the efficiency of the protein sensor with respect to surface area of the sensing material. The surface chemistries of the synthetic polystyrene beads and nanofibers are modified for biomolecule reception by use of various functionalization schemes. In order to enhance the sensitivity we incorporated a layer of conductive polymer that acts as the electron transducer between the protein monolayer and the electronic capture system. We have evaluated the system performance in both the nanostructures using the inflammatory protein CRP as a model system. Important biosensor features like linear dose response to the analyte of interest and rapid response time are highlighted. Issues of repeatability, sensitivity and scalability are addressed in terms of making the prototype biochip commercially viable. The polystyrene based protein prototype sensor chip is encapsulated in a biocompatible flow cell system to constitute a hand held, portable, 'lab on a chip' protein sensor device [6].

1.1 'Lab on a chip'

Well established fabrication techniques that are adapted from the semiconductor industry such as micromachining, injection and replica molding, soft lithography, wet etching and photolithography in an effort to miniaturize fluid handling systems to palm held 'Micrototal Analysis Systems' or 'Lab on a chip' devices [7] which can perform a myriad of tasks associated with a standard laboratory. Microfabricated 'Lab on a chip' microfluidic devices are significantly smaller in size than conventional fluid manipulation systems, rendering them portable and extremely useful in the areas of nanobiotechnology [8, 9], bioanalysis [10] and pharmaceutical [11], medicine and diagnostics. Advances in nanotechnology have impacted research in biotechnology with the development of 'smart devices' capable of molecular manipulation [8]. Microfabricated bioanalytical devices offer highly efficient platforms for genomic, proteomic and metabolic studies. Polymer based microfluidic devices have emerged based on the principles employed in silicon and glass based chips for various applications. Microfluidic devices using biocompatible and cost effective polymer materials for their construction are replacing their more expensive micromachined silicon and glass counterparts as disposable and effective alternatives [7]. In the current research work, we have developed devices that have micro and/or nanostructured active sensing surfaces on to which the biomolecules of interest are adsorbed to yield biomolecule specific electrical perturbations that can be measured in a consistent manner by utilizing the advantages afforded by micro fluidics.

1.2 Analyte of Interest

Measuring the amount of this CRP in the bloodstream can help doctors predict the risk of a heart attack. Globally conducted studies have showed that higher the concentrations of CRP in the blood, greater are the risks of suffering a heart attack. Thus, CRP can be well perceived as a disease biomarker for inflammatory coronary disease. In the current application we are investigating the development of a miniaturized model test system that exploits the electro-activity of CRP to identify low concentrations of CRP in a rapid manner from test samples. The existence of the fundamental coupling between the electronic supports and biological entities in this biosensor system, gives rise to the electrical signatures required for identification of the analyte of interest. Such an electrical contact between biomolecules and electrodes is essential for the functioning of most bio-electronic systems.

2 MATERIALS AND METHODS

2.1 Microfabrication of Base Microelectrode Array (MEA)

The planar microelectrode array consisted of four microelectrodes, each having a narrow edge, which extends to the middle of the surface and spaced 45 degrees from each other. The distended pads are 300μm wide for easy manipulation using the micromanipulators, and the narrow extensions are 50μm each. Two of the four electrodes are selected to generate electric fields around them when a gradient voltage is applied across them. Standard photolithography methods were used to fabricate the microelectrode array [12].

2.2 Polymer Solid Supports

Electrospun polystyrene nanofibers are laid out as an intertwined fiber matrix on the sensing area of the chip. Both functionalized and SO_3H functionalized fibers are utilized to prepare the fiber matrix as the antibody functionalizing surface. The wires are approximately 500nm – 1μm in average diameter and several hundreds microns in length. An electroactive polymer – polypyrrole (doped, 5 wt. % solution in water, *Aldrich*) was used to create a polystyrene – polypyrrole composite to increase the electron transduction to the underlying gold electrodes. The polymer matrices were allowed to incubate in antibody solution for 30 minutes at room temperature before administering it as a mat on the gold sensing area. A baseline signal was recorded and changes in conductance values were noted. Corresponding impedance values were later used for validating the conductance response.

2.3 Polystyrene - Polypyrrole Conductive Matrix

Polypyrrole is used under an applied field to give an additional, and more favorable, conductive pathway. Pyrrole has low energy electrons in the highest occupied molecular orbital (HOMO) attributed to its conjugated aromatic character. The aromaticity of pyrrole is attributed to the planar delocalized pi electrons following the

Hückle rule of 4n+2 pi electrons. Figure 1 describes the ring structure of monomer pyrrole.

A sufficiently powerful electric field can remove an electron from the HOMO resulting in an ionized species. Similarly a conjugated polymer of pyrrole under an electric field will lose HOMO electrons becoming a conducting pathway along its entire length. Lengths of polypyrrole adsorbed onto the CRP/Anti CRP complex may leech electrons from the protein complex resulting in increased conductivity.

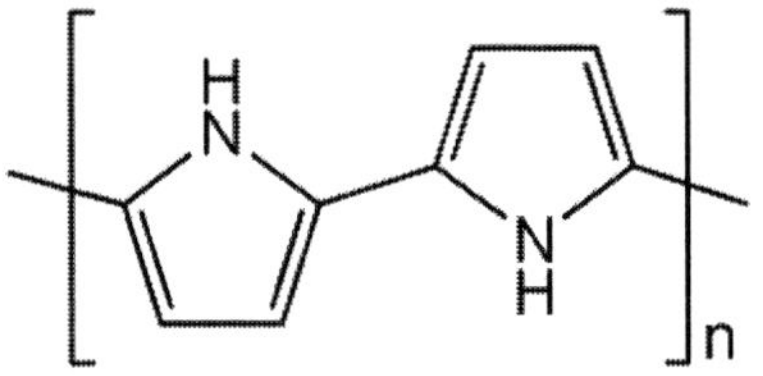

Figure 1: Chemical ring structure of pyrrole monomer *(adapted from www.wikipedia.com)*

2.4 Polystyrene Bead Matrix

Functionalized and unfunctionalized polystyrene bead mats were used spin coated and desiccated on the sensing area of the MEA chip. Plain, unfunctionalized beads were 15μm in diameter and functionalized beads with COOH groups as functional terminations, also of 15μm diameter were used (*PolySciences Inc*) to create a bead bed for functionalization of antibody. Polypyrrole was incorporated into the polystyrene bead system and allowed to coat the beads before the addition of the antigen.

3 DISCUSSION

The polymer matrix was saturated with the protein receptor Anti-CRP and the analyte of interest/antigen is introduced to this bed of antibody. The binding event of Anti - CRP and CRP causes a shift from the baseline signal due to release of electrons [6]. Conductance measurements were executed using a femtoammeter *(Keithley 6430 Subfemtoammeter)* to calculate the extent of electron conductivity through the solution. Impedance spectroscopy was employed to understand the transient behavior of the protein system under non homogenous electric fields. AC oscillating voltage of 0.05V was applied in conjunction with a DC bias voltage of 0.2V together with a frequency sweep from 40 Hz – 100KHz to collect various impedance responses of the layered protein system. Solutions containing protein populations behave as layers capable of developing surface/interfacial charges and increasing electrical double layer gradients. Such potential gradients arising from the interfacial charges give rise to capacitor – like effects, thus contributing to overall system impedance.

4 RESULTS

Impedance spectroscopy studies were conducted first with the control experiment response of the various schemes of polymer layer and that of proteins on a plain chip without underlying polymer matrix as seen in figure 2. Figures 3 and 4 show the impedance responses of functionalized and unfunctionalized polystyrene nanofibers in the presence of polypyrrole. Size matching between the nanoscale spaces in the polystyrene nanofiber mat allows for protein binding. Polypyrrole enhances electron transduction resulting in less significant observable difference between baseline and experimental measurements.

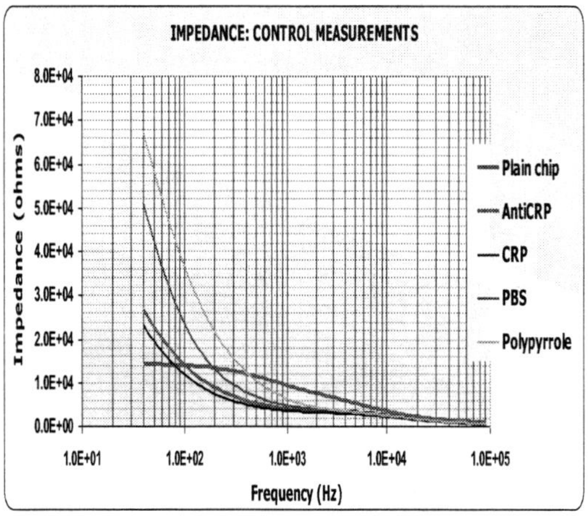

Figure 2: Impedance spectroscopy of control experiments

Similar responses are observed with functionalized and unfunctionalized polystyrene bead mats. Such responses inform the validity and importance of polypyrrole as an efficient electron transporter. However, the absence of clear differences in responses from baseline and experimental steps in the case of functionalized polymer matrices may make the use of polypyrrole meaningless when using as a sensor.

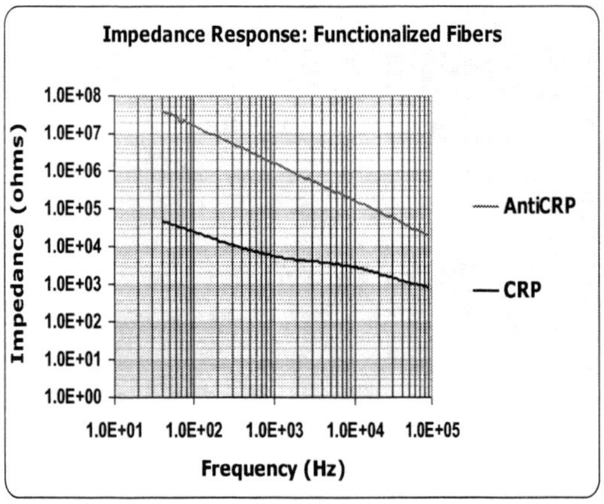

Figure 3: Impedance spectroscopy of Functionalized polystyrene nanofibers with polypyrrole.

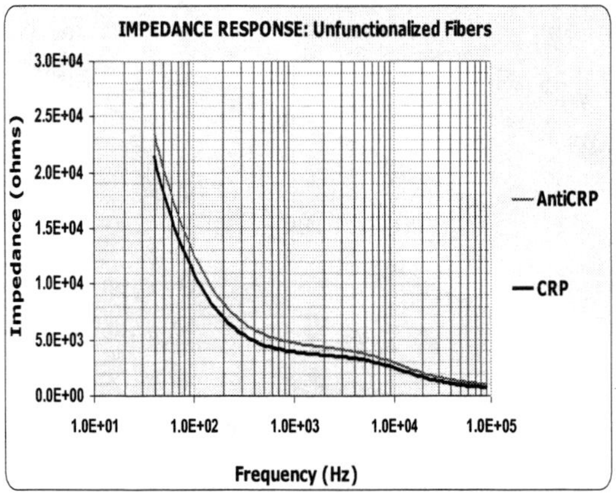

Figure 4: Impedance spectroscopy of Unfunctionalized polystyrene nanofibers with polypyrrole.

5 FUTURE WORK

Further experiments with various types of protein systems with electrical properties different from cardiac disease biomarker CRP are proposed to be studied and analyzed. Such a study would validate the impedance responses obtained from the Anti CRP - CRP immunocomplex functionalization with polystyrene matrices acting as solid supports, thus rendering this method of detection and analyses, universal to all protein immunocomplexes in common.

REFERENCES

[1]. Portsmann T, K.S.T., *Enzyme immunoassay techniques: an overview. 25 years of immunoenzymatic techniques*. Vol.150. 1992, Amsterdam: Elsevier. 5-21.

[2]. *Immunoassay: A Survey of Patents, Patent Applications, and Other Literature 1980 - 1991*, ed. B.S. Judith Sigmond, Marten Terpstra. 1992: Taylor and Francis.

[3]. Craciun, A.M., et al., *Evaluation of a Bead-based Enzyme Immunoassay for the Rapid Detection of Osteocalcin in Human Serum*. 2000. p. 252-257.

[4]. Volkov, S.K., *Immunoassay of alkaloids*. Russian Chemical Reviews, 1993. **62**(8): p. 787-798.

[5]. Vindhya kunduru, S.K.P., Shalini Prasad. *Platform based Detection Technologies from Micro scale to Nanoscale*. in *2006 MRS Spring Meeting*. 2006. San Francisco.

[6]. Vindhya, K. and P. Shalini, *Electrokinetic Formation of 'Microbridges' for Protein Biomarkers as Sensors*. 2007. **12**(5): p. 311-317.

[7]. Chiu, G.S.F.a.D.T., *Disposable microfluidic devices: fabrication, function, and application* BioTechniques, 2005. **38**(3): p. 429-446.

[8]. Lee, S.J. and S.Y. Lee, *Micro total analysis system (μ-TAS) in biotechnology*. Applied Microbiology and Biotechnology, 2004. **64**(3): p. 289-299.

[9]. Ugaz, V. and J. Christensen, *Electrophoresis in Microfluidic Systems*, in *Microfluidic Technologies for Miniaturized Analysis Systems*. 2007. p. 393-438.

[10]. Khandurina, J. and A. Guttman, *Bioanalysis in microfluidic devices*. J Chromatogr A, 2002. **943**(2): p. 159-83.

[11]. Huikko, K., R. Kostiainen, and T. Kotiaho, *Introduction to micro-analytical systems: bioanalytical and pharmaceutical applications*. Eur J Pharm Sci, 2003. **20**(2): p. 149-71.

[12]. Vindhya Kunduru, S.P. *Electrokinetic Alignment of Polymer Microspheres for Biomedical Applications*. in *Materials Research Society (MRS) Spring 2007* 2007. San Francisco, CA.

Nanomonitors: Electrical Immunoassays for Protein Biomarker Profiling

M. G. Bothara*, R. K. Reddy*, T. Barrett**, J. Carruthers*** and S. Prasad*

*Electrical and Computer Engineering Department, Portland State University,
160-11 Fourth Avenue Building, 1900 SW Fourth Avenue, Portland, OR, USA, sprasad@pdx.edu
**Department of Veteran Affairs, Oregon Health Sciences University Portland OR, barretth@ohsu.edu
***Department of Physics, Portland State University, Portland, OR, USA,

ABSTRACT

The objective of this research is to develop a "point-of-care" device for early disease diagnosis through protein biomarker characterization. Here we present label-free, high sensitivity detection of proteins with the use of electrical immunoassays that we call Nanomonitors. The basis of the detection principle lies in the formation of an electrical double layer and its perturbations caused by proteins trapped in a nanoporous alumina membrane over a microelectrode array platform. High sensitivity and rapid detection of two inflammatory biomarkers, C-reactive protein (CRP) and Myeloperoxidase (MPO) in pure and clinical samples through label-free electrical detection were achieved. The performance metrics achieved by this device makes it suitable as a "lab-on-a-chip" device for protein biomarker profiling and hence early disease diagnosis.

Keywords: point-of-care, protein detection, electrical immunoassays, nanopores, ELISA

1 INTRODUCTION

Recent research in the field of proteomics has revealed that proteins can be utilized as biomarkers that facilitate disease diagnostics [1-4]. New trends have shown that a combination of biomarkers can significantly improve the reliability of disease detection [5-7]. Increased number of biomarkers for a disease can also give access to early detection capability. Fast and multiplexed protein detection techniques with high sensitivity and selectivity are imperative to realize the true potential of biomarkers in healthcare. Conventional immunoassay techniques, including enzyme linked immunosorbent assay (ELISA) have not been able to achieve this goal [6-10]. These techniques have several limitations such as the need for use of labels, time of detection in several hours, large volume of reagents, issues with concurrent multiple protein detection due to the associated expense and cross-reactivity [6-10].

Within the last two decades, label free techniques such as surface plasmon resonance, piezoelectric oscillators and electrochemical devices have been developed with higher speed of detection, lower cost and medium throughput [11-13]. The trend over the past few years is towards high throughput systems with device scaling approaching the molecular level to minimize the current limitations of low signal strength, signal variability due to cross reactivity and non-specific binding, which are not completely addressed by the existing label free techniques. Thus, in the last decade, nanotechnology has been applied to biomolecule detection as it provides the advantages of size matching with proteins thereby resulting in higher sensitivity due to increased signal to noise ratio and improved surface area to volume which in turn decreases signal variability. Nanotubes, nanowires and nanospheres are some of the nanomaterial architectures that have been utilized to improve the performance of label free biosensors by targeting the current limitations [13].

This paper describes a device that uses a nanoporous alumina (Al_2O_3) membrane in conjunction with microfabricated gold, Au, electrode platform to form an array of nanoscale well structures to selectively localize proteins onto metallized measurement surfaces in confined volumes. The performance parameters of the device are compared with the traditional assay methods showing that apart from being a label-free technique, it can also provide several improvements such as highly increased speed of detection on the order of minutes as compared to several hours for ELISA, significant reduction in volume of reagents to a few μl, large reduction in cost per assay and the reduction in the size of assay thus making it a candidate for a clinical diagnostic "lab-on-a-chip" device that we call nanomonitor (NM).

2 MATERIALS AND METHODS

2.1 Device Fabrication

The NM comprises of two parts: the microelectrode array base platform that is fabricated using standard photolithography principles and a nanoporous alumina membrane over layer.

The base platform comprises of an array of metallized circular measurement/sensing sites where the binding of the protein molecules occurs in a controlled manner (Figure 1a). Each sensing site is constituted of a working electrode (WE) – and a counter electrode (CE), wherein the surface area ratio of the CE to WE is 225:1 (Figure 1b). Active protein sensing happens at these electrodes (sensing sites), which are in turn connected, to input/output pads for electrical read-out (Figure 1c). Both the electrodes were designed to be circular in shape to attain maximum surface area of interaction and to avoid any possible edge effects. The outer diameter of CE is 150 μm and that of WE is 10 μm, separated by a gap of 5 μm, as shown in Figure 1b.

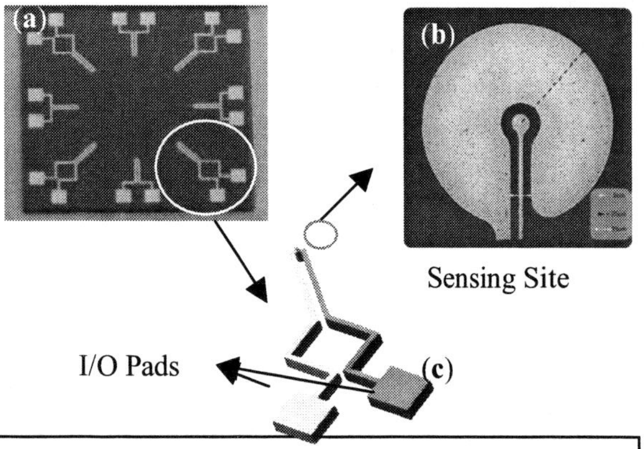

Figure 1: **(a)** Nanomonitor device **(b)** Sensing Site - Working and Counter electrodes layout **(c)** I/O pads connected to the sensing sites (electrodes) through interconnects for electrical read-out

The second part of the NM is the nanoporous Al_2O_3 membrane. A 250 nm thick layer of aluminum is thermally evaporated on the microfabricated platform. This is used as the anode for the electrophoresis reaction that results in a nanoporous Al_2O_3 membrane (Figure 2).

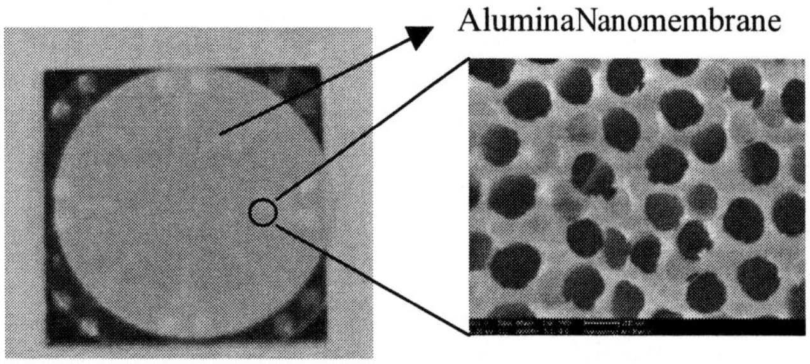

Figure 2: Nanoporous alumina membrane over the microfabricated platform. SEM micrograph showing the nanopores in the membrane.

The pore diameter can be tailored to match the protein size so that each pore forms a well corresponding to a single antibody-antigen conjugate. At 200 nm diameter there are approximately quarter million nanowells on a single sensing site. The platform is fabricated with eight sensing sites and hence can be readily used for multiplexed detection [14].

2.2 Device Operation

Nanomonitor device works on the principle of formation of the electrical double layer. An ionic buffer solution is always present between the electrodes and is used as the basic platform and the medium for protein detection. An electrical double layer occurs whenever an array of charged particles and oriented dipoles are present near the liquid/metal interface. When an electrode is charged, it attracts oppositely charged species and forms a neutral region around the electrode (Figure 3). This neutral layer creates other solvent ions in solution. The inner layer, which is closest to the electrode is called inner Helmholtz plane (iHp) and it contains solvent molecules, specifically adsorbed ions. The next layer is called outer Helmholtz plane (oHp) and the layer after this is called the diffuse layer [15]. When a protein binds at the metal/liquid interface, it perturbs the surface charge distribution at the inner Helmholtz layer as the protein is electrically charged. With more proteins binding at the interface, the associated surface charges also changes significantly. Hence, in the NM device the measurement of the protein biomolecule binding occurs by measuring the surface charge perturbations at the electrical double layer resulting in a measurable electrochemical capacitance change.

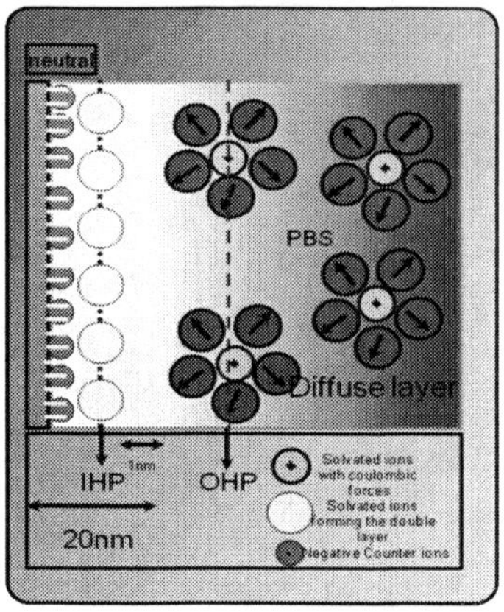

Figure 3: Charge distribution across the liquid/electrode interface forming the double layer

2.3 Methods

Two inflammatory biomarkers, C-reactive protein (CRP) and Myeloperoxidase (MPO), were identified as proteins of study for the development of the device and to demonstrate the detection. It has been suggested that both these biomarkers have pro-inflammatory and pro-coagulant effects, hence making them ideal as study proteins for early disease diagnosis of cardiovascular diseases [16, 17]. Proteins CRP and MPO (antigens) specifically bind with their antibodies or protein receptors, anti-CRP and anti-MPO, respectively. This property of protein binding was incorporated to bring in high selectivity of detection in this device.

The NM was first cleaned in 75% ethanol and de-ionized (DI) water, followed by a DI water rinse and blow drying under dry nitrogen to create a bacteria free device. The nanomembranes/metallic measurement sites were then functionalized by coating a layer of the chemical cross-linker dithiobis (succinimidyl propionate) (DSP), a homo-bifunctional, amine-reactive cross-linker. At this point, the capacitance was measured between the two electrodes to mark the baseline measurements for instrument calibration. Immediately the electrodes were saturated with 3 ul solution of antibodies, as the NHS groups tend to hydrolyze very quickly. There was an increase in impedance due to the increasing capacitive reactance formed between the antibodies and the electrodes. However, impedance saturated at a particular concentration of proteins, which corresponded to what was called the saturation concentration when almost all the nanopores are occupied with antibodies for protein binding (Figure 4).

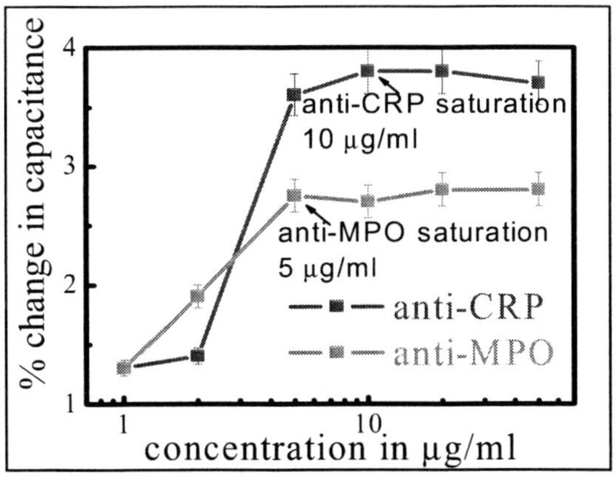

Figure 4: Antibody saturation measurements for proteins CRP and MPO.

After saturating the sensing site with antibodies, a block step with Bovine Serum Albumin (BSA) was performed to block all unoccupied sites and avoid non-specific binding of proteins. A 5% BSA/PBS solution was used to wash the nanowells for few minutes and then left to incubate at room temperature for 15 minutes. After this block step, the protein to be detected was inoculated onto the sensing site (Figure 5).

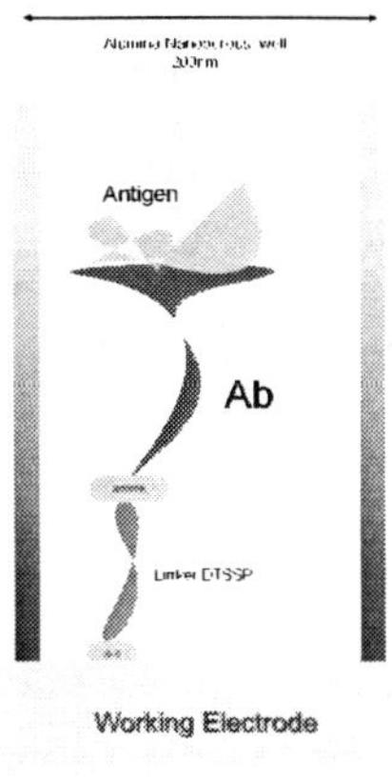

Figure 5: View inside a single nanowell showing the series binding of species.

The protein solutions were prepared by diluting in the pure protein in 1X PBS as well as in human serum (U.S. Biological, MA) to demonstrate the detection from physiological fluids. The capacitance measured after the block step marked the final reference value for quantification of protein characterization. The change in capacitance measured after the exposure to the proteins corresponds to the action of protein binding. Hence the magnitude of change in capacitance can be directly attributed to the protein concentration present. This correlation between the change in capacitance and protein concentration is plotted and used for device calibration for detection. In the case of multiplexed detection, the alternate sites were inoculated with different antibodies and the corresponding change in capacitance induced by the respective protein on each of the sites was recorded for their respective protein characterizations.

For protein detection in clinical samples, known concentrations of proteins were spiked in the serum, Human Serum Albumin (HSA) and detected using the same protocol as mentioned above. HSA is constituted of many competing proteins which also includes CRP, the concentrations of which are known. After recording the concentration of CRP already present in HSA, appropriate spiking with known concentrations of CRP and/or dilution with 1X PBS control buffer is done to achieve a 50% CRP spiked serum by volume. The change in capacitance measured while detecting the CRP in the serum was compared with the reference created by the pure sample detection.

3 RESULTS

This section shows the various responses observed while detecting the proteins CRP and MPO.

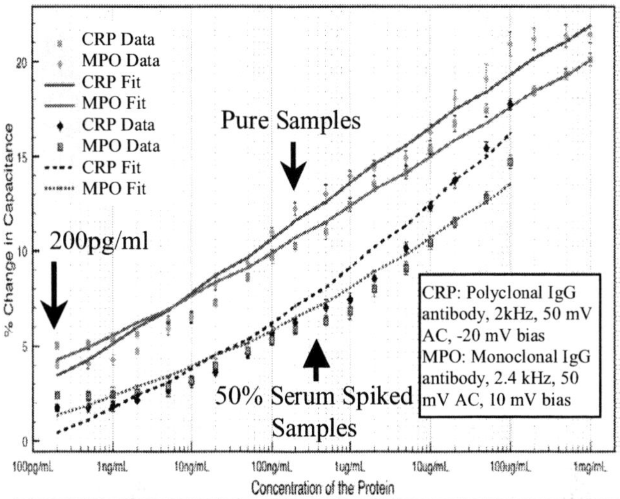

Figure 6: Dose response curves for proteins CRP and MPO when detected in pure and serum samples.

Figure 6 shows the dose response curves for the proteins CRP and MPO detected in pure form and from samples containing 50% serum by volume. Figure 8 shows the response for various concentrations of multiplexed samples. The values in these plots were obtained at 4 kHz, where the change in capacitance was found to be the highest. The detection limit was found to be ~200 pg/ml for CRP and ~500 pg/ml for MPO. The time taken for the detection of proteins was about 120 – 180 seconds. Figure 7 also shows the linear dynamic range of detection extends from ~200 pg/ml to ~100 μg/ml, which is about 6 orders of magnitude.

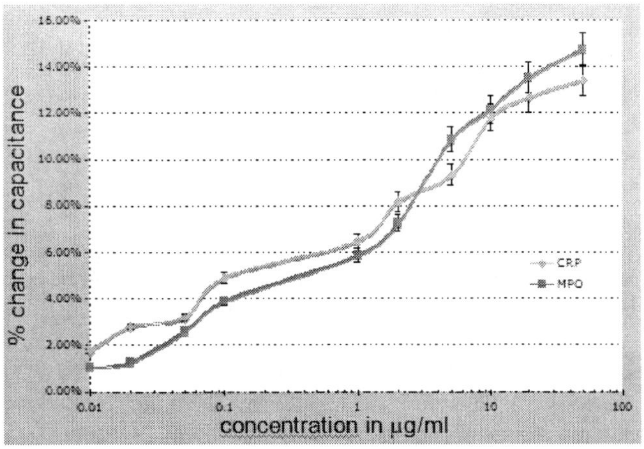

Figure 7: Multiplexed detection dose response curves for proteins CRP and MPO

4 DISCUSSION

A highly sensitive device for protein detection was designed and fabricated using label-free electrical immunoassay techniques. In addition to being highly portable, it is quite inexpensive and very quick in detection when compared to current predominant labeled immunoassay techniques like ELISA. The lower limit of detection achieved by ELISA is better than what is achieved by the NM device. This is attributed to the presence of some non-specific binding or cross-reactivity occurring when competing antigens are present. Competing antigens are those whose structures and chemical properties are very similar in nature and hence they create a noise margin which currently inhibits the limit of detection to 200 pg/ml. The second reason would be because of the background noise present from the substrate, the external circuitry and the environment.

5 CONCLUSIONS AND FUTURE WORK

Multiplexed detection of protein biomarkers from small volumes in a rapid manner is important for cost effective disease diagnosis. The NM device technology has been demonstrated here to detect multiple proteins in pure and serum samples. To address the issues due to non-specific binding, many more nanopores could be introduced over the sensing site for more protein trapping and reducing signal variability. This increases the amount of antibodies present, hence increasing the average signal read out from protein binding. This increases the overall signal-to-noise ratio and hence would help achieve a lower limit of detection.

One of the primary objectives of this research when it started was to make a handheld device. So to move towards that destination, an external electrical circuitry will be built to process the signal read from the device and give an electrical read-out in the form of an electrical display or an LED based output display.

It has the potential for "point-of-care" disease diagnostics, as it is highly portable and inexpensive. There is a great promise for this technology in the areas of laboratory clinical diagnostics, healthcare and in pharmaceutical sectors.

REFERENCES

[1] F. Darain, D. S. Park, J. S. Park and Y. B. Shim, *Biosensors & Bioelectronics,* Vol. 19, pp.1245-52, 2004.

[2] J. Hahm, and C. M. Lieber, *Nano Letters,* Vol. 4, No. 1, pp. 51-4, 2004.

[3] J. M. Nam., C. S. Thaxton, and C. A. Mirkin, *Science*, Vol. 301, No. 5641, pp.1884−86, 2004.

[4] O. Niwa, M. Morita, and H. Takei, *Anal.Chem.,* Vol. 62, pp. 447-56, 1990.

[5] M. D. Abeloff, J. O. Armitage, A. S. Lichter and J. E. Niederbuber, *Clinical Oncology,* Churchill Livingstone, New York, 2000.

[6] S. F. Chou, W. L. Hsu, J. M. Hwang and C. Y. Chen, *Biosens. Bioelectron.*, Vol. 19, pp.999−1005, 2004.

[7] P. Arenkov, et al., *Anal. Biochem.*, vol. 278, pp.123−31, 2000.

[8] Prasad, S., Zhang, X., Ozkan, C.S., Ozkan, M. Neuron-based microarray sensors for environmental sensing. Electrophoresis. 2004. 25(21-22): p.3746-60.

[9] Prasad, S and Quiano, J., Biosens. Bioelectron. 2004. 21(7): p.1219-29.

[10] Cui, Y., Wei, Q.Q., Park, H.K. and Lieber, C.M. Nanowire nanosensors for highly sensitive and selective detection of biological and chemical species. Science. 2001. 293: p.1289−92.

[11] Campagnolo, C. et al., Real-Time, label-free monitoring of tumor antigen and serum antibody interactions. J. Biochem. Biophys. Methods, 2004. 61: p.283−98.

[12] Etzioni, R. et al., The case for early detection. Nat. Rev. Cancer. 2003. 3: p.243−52.

[13] Zheng, G., Patolsky, F., Cui, Y., Wang, W.U., and Lieber, C.M., Nature. 2002. 23(10): p.1294-01

[14] A. P. Li, F. Muller, and U. Gosele, *Electrochem. Solid-State Lett*, 3:131, 2000.

[15] J. R. MacDonald, *Journal of Chemical Physics*, 22:1317-1322, 1954.

[16] Paul M Ridker, *Am Heart J*, 148(1 Suppl):S19-26, 2004.

[17] Svati H Shah and L Kristin Newby, *Cardiol Rev*, 11(4):169-179, 2003.

Ultra-Sensitive Electrochemical Detection of *E. coli* Using Nano-Porous Alumina Membrane

N. N. Mishra*, B. Filanoski*, M. Fellegy*, E. Cameron*, S. K. Rastogi*, W. C. Maki*, and G. Maki*

Center for Advanced Microelectronics and Biomolecular Research,
University of Idaho, Post Falls, 83854, ID, USA, nmishra@cambr.uidaho.edu

ABSTRACT

A combinational method has been developed for ultra-sensitive detection of *E. coli*. The method combined biological, nano-fabrication and electrochemical techniques to achieve highly specific and ultra-sensitive detection of bacterial pathogen. *E. coli* K12 was used as the target bio-agent to demonstrate the detection model. Nano-porous alumina membrane was chosen as filtration substrate to separate *E. coli* from the test samples because of their desirable physical properties. The surface of the membrane was blocked and not to bind to protein molecules. Bacterophage M13 was genetically modified with a histag on its minor coat protein III. This phage was used to infect its specific bacterial host *E. coli* K12 and replicated inside of the host. The nano-porous alumina membrane was used to filter exceeded phages and separate infected *E.coli* from the test sample. The replication time of bacterophage M13 was investigated. After a short period of culture, the replicated bacterophage was released from the host cell. The phage was increased 10^2 -10^5 fold in 2 -4 hrs. These baterophage were captured by anti-histag antibody immobilized on the detection electrode and detected electrochemically. Various surface modification approaches on the detection electrode were tested to improve the efficiency of target recognition and detection sensitivity. This combinational method can be applied to the detection of various pathogenic bacteria.

Key words: nano-porous alumina membrane, *E. coli*, phage, electrochemical, detection

1 INTRODUCTION

Food borne illnesses contribute to the majority of infections caused by pathogenic microorganisms. Toxin producing *E. coli* can cause stomach illness, bloody diarrhea, dehydration or even death in serious cases. It can contaminate ground beef if fecal matter or other contaminated meat is mixed during the meat slaughter or packing operations. Among these pathogenic *E. coli*, *E. coli* O157:H7 is firmly associated with hemorrhagic colitis [1]. As a result of this association, *E. coli* O157: H7 was designated as an enterohemorrhagic *E. coli*. Based on a 1999 estimate, 73,000 infections and 61 deaths occur in the United States each year [2]. In year 2007, several recalls of *E.coli* O157:H7 contaminated ground beef involved three companies and over 20 million pounds products which resulted in huge economic loss. Current tests for the detection of pathogenic *E. coli* depend on the time consuming culture method. PCR amplification is another commonly used method. However, it cannot identify alive or died bacteria and lacks accuracy due to contamination problems. Reliable and accurate molecular biology methods are time consuming and have resulted in delayed recalls of contaminated meat products. Developing rapid and accurate detection technologies will prevent news headlines like, "The Department of Agriculture and Cargill Meat Solutions Corp. today recalled more than 1 million pounds of ground beef products because they may be contaminated with *E. coli* O157:H7 bacteria. The recall comes just weeks after one of the largest meat recalls in history -- more than 20 million pounds. The recall by the Topps Meat Co. sent the New Jersey firm out of business". It is clear that rapid and accurate detection of contaminated food will save lives and prevent economic disasters in agriculture. Detection of food borne pathogens has led to increased research interest. New technologies in the biosensor development are improving the sensor characteristics such as sensitivity, reliability, simplicity and economic viability [3,4,5]. Here we report a combinational method for ultra-sensitive detection of *E. coli* K12. This detection method can be applied to other pathogenic bacteria including *E. coli* o157:H7.

2 METHODS

2.1 Electrochemical Detection

The detection system contains three parts: target recognition, signal amplification and electrochemical detection. Bacterophage was used for both target recognition and signal amplification through its natural features (see electrochemical detection model in Figure 1.).

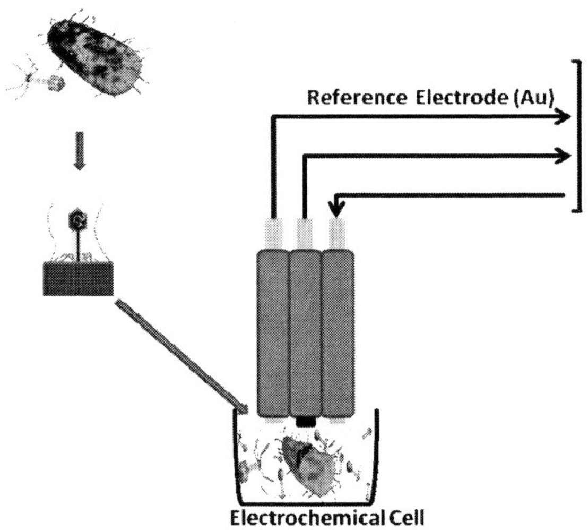

Figure 1. An overview of the detection mechanism Bacterophages infect their host bacteria specifically and are replicated in the host cell. The host bacteria release the replicated phages into culture medium. These bacterophages are captured by their specific antibody on the detection electrode and detected electrochemically.

2.2 Genetic Modification of Phage M13

A peptide phage display cloning vector from New England Biolab was used to insert a 60 base pairs double strand oligonucleotide which codes a poly-histidin peptide at the N-terminal of coat protein gene III. The sequence of the oligonucleotide is: 5'-CATGCCCGGGTACCTTTCTATTCTCACTCTCA TCATCATCATCATCACTCGGCCGAAACATG-3'. Two single strand oligonucleotides corresponding to a partial sequence of the top and bottom strands with 15 base pairs overlap were designed and synthesized. The sequence of the top strand is: 5'-GTA CCT TTC TAT TCT CAC TCT CAT CAT CAT CAT CAT CAC TC -3'. The sequence of the bottom strand is: 5'- GGC CGA GTG ATG ATG ATG ATG ATG AGA GTG AGA ATA GAA AG -3'. Annealing and primer extension were performed to obtain a double strand oligonucleotide with a completed a sequence which was inserted into vector between restriction sites Acc65 I and Eag I. General molecular cloning approaches were used to transform ligation sample into *E. coli* K12. Plagues were isolated from the agar plate and re-suspended in PBS. This phage sample was used to infect host bacteria for its mass production.

2.3 Characterization of Histaged Phages M13

To characterize genetically modified bacterophage M13, nickel-HRP conjugate was used to detect poly-histidin on phage surface. 0.5 ul genetically modified and unmodified M13 phage samples were spotted on nitrocellulose membrane. 0.5 ul of histaged GFP was used as positive control. After blocking, the membrane was incubated with nickel-HRP conjugate. Chemiluminescence detection was used to confirm the presence of histag.

2.4 Phage infection and generation time studies

Over night culture of *E. coli* K12 was diluted in LB broth at 1: 100, 1:10,000, 1:100,000 ratios. Spectrum OD600 reading was taken to estimate bacterial density. Each diluted samples was plated on LB agar plates to determine the number of bacteria in the culture. Bacterophage M13 infection was carried out by incubating bacterial samples with histaged phage M13 at room temperature for 40 minutes. The exceeded phages were removed by using 0.2 nm spin filter. Infected *E. coli* K12 were then recovered and cultured in LB medium. To determine phage replication time course, the samples from 2, 4.5 and 6 hrs cultures were plated on *E. coli* K12 top agar plates and incubate at $37^{o}C$ for over night. Plagues were counted from over night cultures.

2.5 Surface modification of detection electrode

Immobilization of antibody on the detection electrode was tested on gold wires. Sonication was used to clean the gold surface before modifications. Cleaned gold wires were soaked in a PBS solution containing 10 μg/ml mouse IgG or anti-biotin antibody at room temperature for 60 minutes. Goat anti-mouse IgG and alkaline phosphatae (AP) conjugate or biotin labeled AP was used to confirm mouse IgG immobilized on the gold wires. Chemiluminescence detection was recorded by CCD imaging. After confirmation, the same approach of surface modification was used to immobilize a monoclonal anti-histag antibody on the gold detection electrodes.

2.6 Electrochemical detection of bacterophage M13

A pure *E. coli* K12 culture was diluted to100 bacteria/ml. One mili-liter this diluted sample was filtrated through nano-porous membrane with the hole size of 200 nm. Bacterophage M13 infection and replication was carried out as described above. The antibody modified detection electrode incubated with the phage samples replicated from infected *E. coli* K12. Electrochemical signals generated from the interaction of histaged bacterophage M13 and anti-histag antibody were detected.

3 RESULTS AND DISCUSSION

3.1 Genetic Modification of Bacterophage M13

A short double strand oligonucleotide was inserted into a phage display vector M13KE (Fig. 2 left) at the cloning site of between Acc65 I and Eag I. The inserted sequence of this oligonucleotide is : **5'-** <u>GGTACC</u>TTTCTATTCTCACTCT**CATCATCATC ATCATCAC**T<u>CGGCCG</u>-3'. The bold sequence codes a polyhistidin peptide, and the underline sequences are restriction sites Acc65 I and Eag I. This vector was transformed and replicated in *E. coli* K12. Genetically modified phage M13 was confirmed by chemiluminescence test. Fig. 2 (right) showed that histag coat protein III in M13 was detected by Nickel-HRP conjugates. Unmodified phage M13 did not generate a light signal.

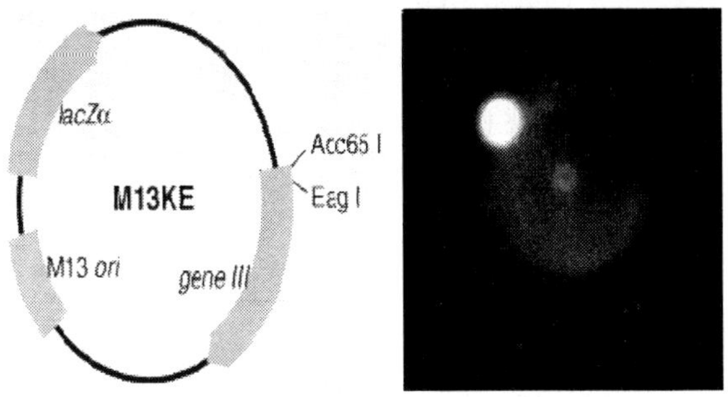

Figure 2. Characterization of genetically modified bacterophage M13 The map of phage displry vector is on the left. Chemiluminescence detection of histaged M13 is showed on the right. Top left is positive control of histaged GFP. Top right is negative control of unmodified M13. Bottom center is modified phage M13.

3.2 Investigation of Bacterophage M13 Generation Time

Results from the study of M13 generation time were show in Figure 3. As less as 7-10 infected *E. coli* K12 bacteria produced 200, 3,600 and 667,000 new phages in 2, 4.5 and 6 hrs respectively. Bacteria concentration was determined on agar LB culture plates. The result showed that in 2, 4.5 and 6 hrs time period, the replication times are 70, 518 and 95,300 fold. This result has been repeated several times.

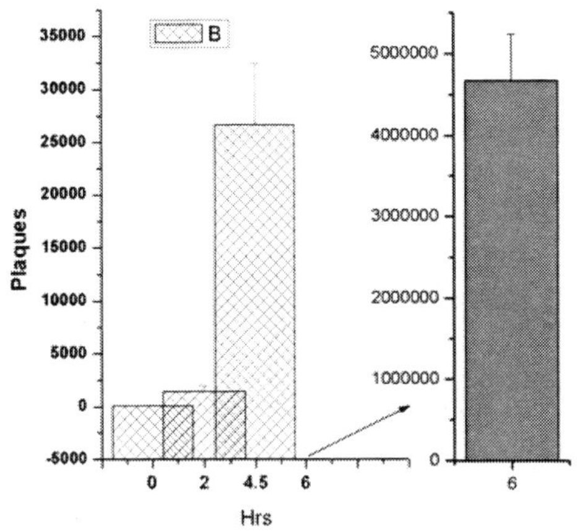

Figure 3. Bacterophage M13 generation time Phage M13 replicated in 7-10 infected *E. coli* K12 cells. The numbers of replicated phages are showed in: 0 hr, 2 hrs, 4.5 hrs and 6 hrs (red).

3.3 Antibody Immobilization

Antibody immobilization on gold wire was confirmed by chemiluminescence assay. Fig.4 showed that mouse IgG was detected by goat anti-mouse alkaline phosphatase (AP) conjugates, and anti-biotin antibody was detected by biotin labeled AP. In both cases, unmodified gold wire did not produce chemiluminescence signal.

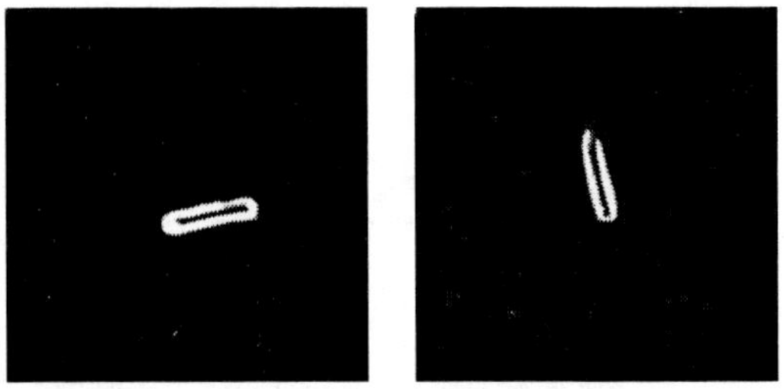

Figure 4 Chemiluminescence detection of antibodies immobilized on gold wire, left: mouse IgG on gold wire, right: anti-biotin antibody on gold wire.

3.4 Detection of *E. coli* K12

Detection of *E. coli* K12 was demonstrated through the processes of bacterophage M13 infection, replication and detection. Alumina nano-porous membrane showed a great advantage over other commercial membranes. Bacteria stuck on the filtration membrane are common problems when using nitrocellulose or PVDF membrane. That reduces biological activity of the cells and affects phage replication in the host. In contrast, alumina materials are not sticky to biological samples. It is easier to re-suspend filtered bacteria into culture medium.

A set of micro-electrodes were fabricated on a glass wafer for electrochemical detection. Fig. 5 is an image of the detection device. The results of electrochemical detection of histaged bacterophage M 13 were showed in Figure 6.

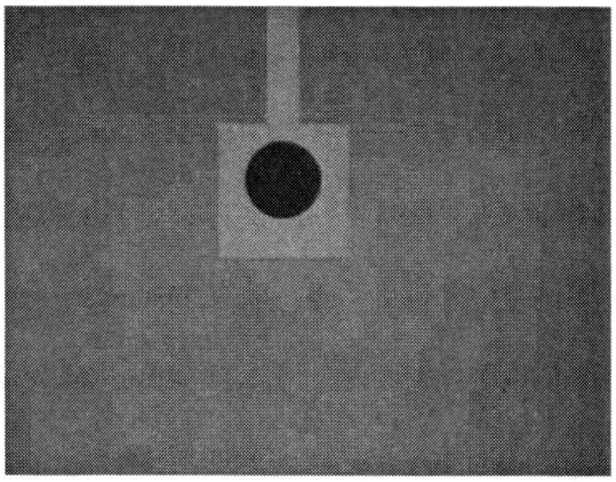

Figure 5. Microscopic images of the 300 micron gold detection (black circle).

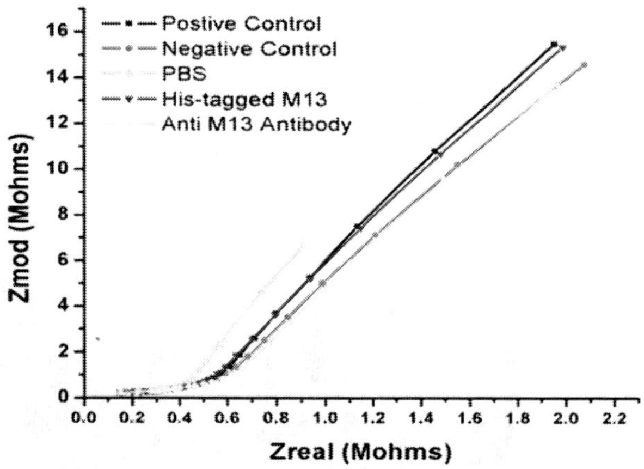

Figure 6. Electrochemical nyquist plot of bacterophage M13 detection on 300 micron antibody coated electrode. The spectra clearly shows the specific detection of M13 phage at AC 100 mV rms.

4. CONCLUSION

The advances in biotechnology, bio-molecular electronics, nano-technology and other sciences have led to the spawning of new and rapid methods of food pathogen detection that have the potential for on-site testing and thereby revolutionizing food product quality control. The advantages of using phage electrochemical detection method are: quicker, sensitive, easy to process and simple devices. This detection model can be applied to the detection of other pathogenic bacterial contamination in food, e g. detection of *E. coli* O157:H7 by its specific bacterophage KH1, and detection *Salmonella* by its specific bacterophage P22.

5. ACKNOWLEDGEMENTS

Authors gratefully acknowledge the assistance in device fabrication of Michael Skvarla from Cornell Nanofabrication Facility at Ithaca, NY. This work was supported in part by This work was supported in part by USDA grant CSREES 3447917058, NASA grant NNX06AB17G and Hatch grant IDA00709-STH.

REFERENCES

[1] Riley LW, Remis RS, Helgerson SD et al. Hemorrhagic colitis associated with a rare Escherichia coli serotype. *New England Journal of Medicine* 308:681-685, 1983

[2] Frenzen PD, Drake, A, et al.Economic cost of illness due to Escherichia coli O157 infection in the United States. *Journal of Food Protection*, 68: 2623-2630, 2005

[3] Patolsky, F., Timko, B. P., Zheng, G. and Lieber, C. M. Nanowire-based nanoelectronic devices in the life sciences *MRS Bull* 32:142-149, 2007

[4] Atora, K., Chand, S., and Malhotra, B. D. Recent developments in bio-molecular electronic techniques for food pathogens Analytica Chimica Acta 568:259-274, 2006

[5] Barbaro, M. et al. Fully electronic DNA hybridization detection by a standard CMOS biochip *Sensors and Actuators B: Chemical* **118**, 41-46 (2006)

----

[1] Nirankar Mishra, CAMBR, University of Idaho
721 Lochsa St,Post Falls, ID 83854 Tel: 208 262 2047
Fax: 208 262 2001 nmishra@cambr.uidaho.edu

Ferromagnetic Resonance Biochip for Diagnosing Pancreatic Cancer

E. Casler*, S. Chae*, V. Kunduru*, M. Bothara*, E. yang*, S. Ghinoea**, P. Dhagat**, S. Prasad*

*Portland State University, Biomedical Microdevices and Nanotechnology Lab
Portland, OR, USA, sprasad@ece.pdx.edu
**Oregon State University, Applied Magnetics Laboratory

ABSTRACT

The objective of the paper is to develop a "lab-on-a-chip" device for early disease (pancreatic cancer) diagnosis by using ferromagnetic resonance (FMR). Magnetic microbeads, which are functionalized for target molecules (antigens), are immobilized by antigen-antibody reactions on the surface of a microwave circuit. These magnetic labels are detected inductively using FMR, which detects a single bead with a sensitivity of 1-10 μV/V. This method has distinctive advantages compared to other conventional immunoassay techniques; it requires a small sample volume, is non-invasive, cost effective, and easy to implement. It also does not alter the native properties of the antigen and antibody complex.

Keywords: biomarker, cancer, microbead, immunoassay

1 INTRODUCTION

Modern medicine's detection of pancreatic cancer comes at a stage when preventative medicine is no longer a viable option. Several methods exist for early stage detection such as tomography and invasive biopsy. These solutions, however, are expensive, dangerous, and/or unreliable [1,2].

A non-invasive method exists by detecting the associated antigen in a blood sample from the patient. Currently several techniques are available to achieve this goal and they can be categorized into two groups: label required and label-free. Label-free methods have several advantages, but are notoriously difficult to implement. There are three label required techniques: radioactive, luminescent, and magnetic [3-10]. They are typically used to detect the target molecules. Luminescent labels, specifically fluorescent or Chemi-luminescence, covalently link to antibodies and function as labels to create a visually detectable signal proportional to the amount of antigen in the sample. This method, however, presents several problems. One of these is the confirmation of the antigen and antibody bonds, where visual detection can be unreliable, bulky, and expensive. The samples used must be large to ensure a detectable signal. Also, this method often alters native properties of the antigens, and puts constraints on the size of the equipment that prevents rapid detection due to transportation of analytes to the test site. Radioactive labels are not suitable for the application due to requirements of the radioactive waste disposal and limited shelf life.

On the other hand, magnetic labels, in the form of micron-sized beads, allow for a cost effective and deterministic alternative toward detecting biomarkers using novel immunoassay principles. The protein sensor device would provide a high-throughput operation, use small sample sizes, be highly transportable, and provide higher process control over interactions, resulting in rapid and accurate detection of diseases. Bead-based detection also has the desirable trait of offering a high surface area to volume ratio for the attachment of the antigens.

Not all magnetic detection methods offer the same benefits as FMR detection. Electromagnetic induction uses a similar "sandwich" assay, measures changes in inductance in its micro-patterned coil, and has the advantage of a rather simple integration with CMOS technology, but suffers from a lack of sensitivity [11, 12]. Similarly, the low sensitivity of devices based on Anisotropic Magnetoresistance (AMR) outweighs the simplicity of its fabrication [13]. AMR, like Tunneling Magnetoresistance (TMR) and Giant Magnetoresistance (GMR) have risen in popularity of late due to their thin-film design, which is conducive to Integrated Circuit (IC) integration. All three technologies detect the presence or absence of beads, similarly, by changes in the resistance of the sensor based on fluctuations in the magnetic field caused by the immobilized beads. While they may integrate well and have sufficient sensitivity, the fabrication of the multiple layers of films is more costly and complicated than the processes already developed for FMR detection. In addition, TMR sensors lose their sensitivity as the functional area is increased [14-18].

Ferromagnetic resonance (FMR) detection is characterized by its high sensitivity and simple, inexpensive fabrication [19]. The functionalized beads are commercially available and become immobilized when they flow over an antibody-activated sensor. Detection with FMR uses frequencies in the gigahertz range and at resonance; the effective permittivity is increased in the magnetic material. Leveraging both of these observations increases the signal strength associated with the detection event. Our aim is to use the principle of ferromagnetic resonance in the detection of magnetically labeled pancreatic cancer biomarkers in order to offer an alternative diagnostic solution that lowers the cost, and increases the sensitivity, specificity, portability, and reliability of the test through simpler sample processing and the low cost of integrated circuit manufacturing technology.

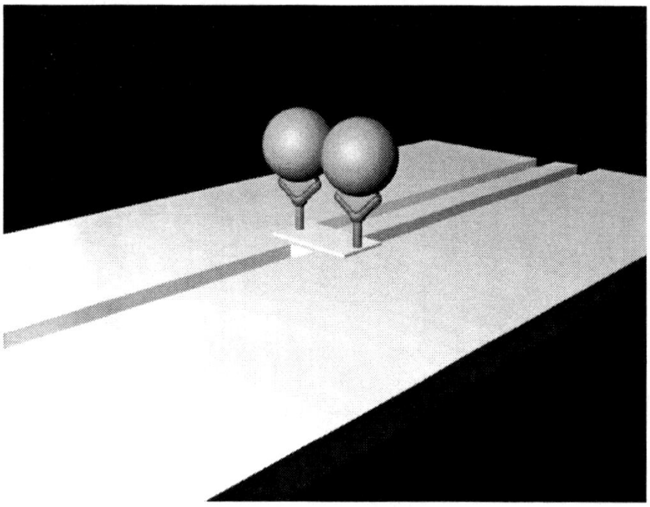

Fig. 1: Representation of beads immobilized on functional area of FMR device.

2 MATERIALS & METHODS

The substrate used for experimentation was a gold sputtered silicon wafer. Streptavidin and biotin solutions were prepared using concentrated solutions from Vector Labs®. Two drops of each added to 0.5 ml of 1X Phosphate Buffered Saline (PBS) in individual eppendorf tubes.

2.1 Characterization of Beads

In order for FMR detection to be possible, there are certain magnetic and compositional properties that must be met, namely that the magnetic particles possess a magnetic susceptibility of at least 1.6 and a ferrite content of at least 27% by weight. We chose the Dynabeads® M-450 Epoxy produced by Invitrogen®, which are uniform, super-paramagnetic polystyrene beads with a diameter of 4.5 μm. The 4.5 μm beads have a surface area of 6.287e-11[m^2] and a volume of 4.666e-17[m^3]. While the magnetic properties of the beads are required for the FMR detection, other properties, such as their diameter were chosen due to product availability.

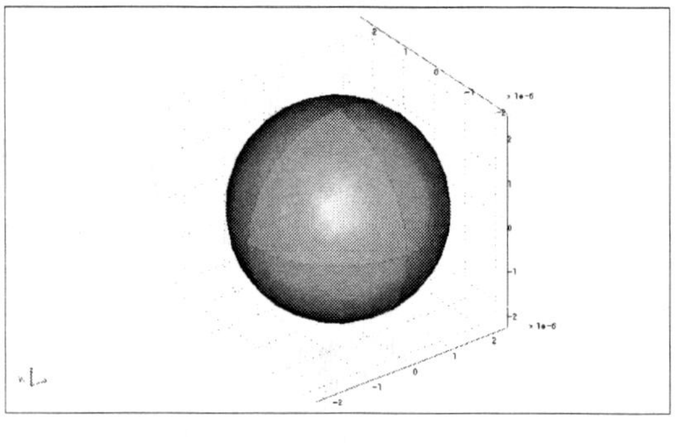

Fig 2: Simulation of 4.5μm diameter bead

2.2 Surface coating and functionalization

The setup for the experiment entailed the independent functionalization of the gold sensing area and the magnetic beads in order for the immobilization of the beads on the surface to occur. The gold substrate was saturated with 50 μl of a streptavidin solution for 30 minutes at room temperature, allowing for physi-chemical adsorption to the surface. Similarly, a 1 μl solution of Dynabeads M-450 in distilled water with a bead density of 4 x 10^8 beads/ml was added to a 50 μl solution of saturated biotin and then agitated for 15 seconds using a vortex in order to re-suspend the particles. The solution was then allowed to incubate at room temperature for 30 minutes, before the beads were introduced to the streptavidin-coated surface. The streptavidin-biotin coupling was left undisturbed for an additional 30 minutes in order to ensure covalent bonding between the proteins before a washing step was conducted with 50 μl of 1X PBS solution. Excess PBS wash was removed using chemical wipes.

For the control experiments the 50 μl of saturated biotin was replaced with 50 μl of 1X PBS solution. The 1X PBS solution was chosen in order to demonstrate that a lack of specific ligand coupling would preclude the immobilization of magnetic particles to the gold surface.

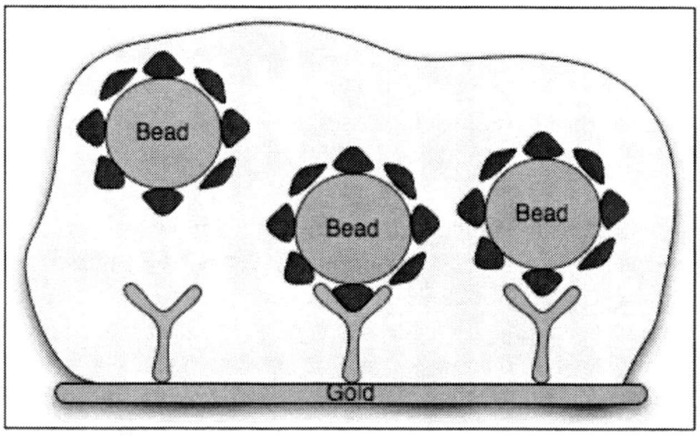

Fig. 3: Symbolic representation of bead immobilization through the coupling of streptavidin, orange, and biotin, blue.

3 RESULTS

The experimental results were quantified using optical detection of the presence of magnetic particles on the surface of the gold substrate after the washing step had concluded. A representative viewing area was chosen and the number of beads present counted and recorded. The results were as expected, the functionalization of the magnetic beads with biotin enhanced their immobilization on the surface, purportedly due to the coupling with the streptavidin saturating the chip. On average, trials with biotin-coated beads resulted in 70 beads remaining in a viewing window at 40x magnification, whereas the control experiments, using naked beads, averaged about 30 beads in the same viewing space. The results are illustrated in Fig. 4(a) and 4(b). A green filter in the microscope was used to improve image clarity.

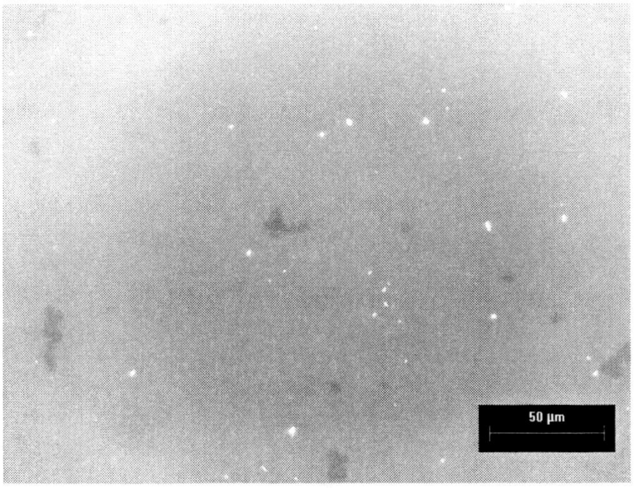

Fig. 4(a): Control Trial: Optical micrograph at 40x magnification of naked beads immobilized on gold surface.

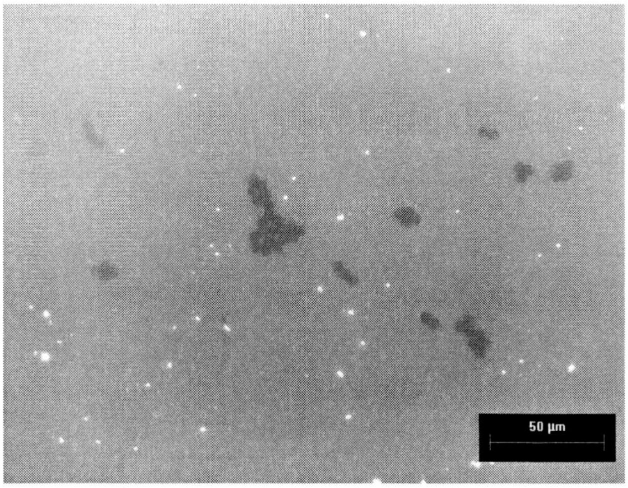

Fig. 4(b): Experimental Trial: Optical micrograph at 40x magnification of biotin-coated beads immobilized on gold surface.

4 DISCUSSION

Our results are demonstrative of the concept that magnetic particles can be immobilized with the current protein chemistries and can presumably be quantified using FMR detection. Beads with alternate surface reactive groups are being explored in order to reduce the incubation period necessary for the covalent coupling of the cancer protein biomarker, versus the current physi-chemical adsorption, to the bead's surface.

Currently, only optical techniques are being employed to quantify the results, which are not as precise, nor quick as the proposed magnetic detection scheme. Trials need to be conducted that incorporate FMR detection on samples that contain beads immobilized by ligands.

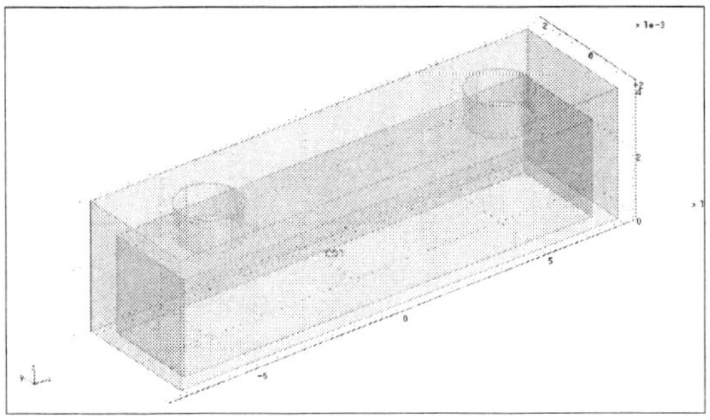

Fig. 5: Model of proposed microfluidic device, simulated dimensions are - exterior: 15x4x4 mm, thickness: 0.5 mm, & channel diameter: 2 mm.

We intend to incorporate a microfluidics device in order to optimize the wash steps and to ensure precise, localized fluid placement onto the functional area of the device. The microfluidic device is a rectangular-parallelepiped-shaped with an inlet and outlet for the deposit and removal of solutions during trials, such as the one shown in Fig. 5. The design of the proposed device is currently being evaluated through the development of a Comsol Multiphysics® model in order to optimize fluid flow and the deposition of magnetic material.

We are applying the incompressible Navier-Stokes equations to simulate Laminar flow of the solutions. The microfluidic device must be small enough to prevent turbulent flow, but big enough to cover the functionalized area (50µm x 100µm), and long enough to apply relatively consistent horizontal force on the functionalized area.

Lastly, the focus of future experimental trials will be to optimize the protein concentration needed to meet the goal of minimizing the potential patient sample volume. This will be accomplished by using a proven, reproducible protocol over a range of antigen concentrations. The expectation is that there will be a minimum concentration necessary to reliably immobilize a sufficient number of beads, for FMR detection, to the antigen-binding site of the biotinylated antibody present on the surface.

REFERENCES

[1] Sager, W.D, zur Nedden, D., Lepuschutz, H., Zalaudek, G., Bodner, E., Fotter, R., Lammer, J., Computer tomographic diagnosis of pancreatitis and pancreatic cancer, Computertomographie, 1(2), 52-58, 1981.

[2] Kosur,i, K., Muscarella, P., Bekaii-Saab, T.S., Updates and controversies in the treatment of pancreatic cancer, Clin Adv Hematol Oncol., 4(1), 47-54, 2006.

[3] G. Shan, W. Huang, S. Gee, B. Bucholz, J. Vogel, and B. Hammock., "Isotope-labeled immunoassays without radiation waste", *Proc. Natl. Acad. Sci. USA.*, vol. 97, no. 6, pp. 2445-2449, 2000.

[4] A. Marco, I. Marucci, M. Verdirame, J. Pérez, M. Sanchez, F. Peláez, A. Chaudhary, and R. Laufer, "Development and validation of a high-throughput radiometric CYP3A4/5 inhibition assay using tritiated testosterone", *Drug Metab Dispos*, vol. 33, no. 3, pp. 349-358, 2005.

[5] R. Ekins and S. Dakubu, "The development of high-sensitivity pulsed light, time-resolved fluoroimmunoassays", *Pure & Appl. Chem.,* vol. 57, no. 3, pp. 473-482, 1985.

[6] S. Nie, D. Chiu, and R. Zare, "Real-Time Detection of Single Molecules in Solution by Confocal Fluorescence Microscopy", *Anal. Chem.*, vol. 67, no. 17, pp. 2849-2857, 1995.

[7] M. Medina and P. Schwille, "Fluorescence correlation spectroscopy for the detection and study of single molecules in Biology", *BioEssays,* vol. 24, no. 8, pp. 758-764, 2002.

[8] F. Mallard, G. Marchand, F. Ginot, and R. Campagnolo, "Opto-electronic DNA chip: high performance chip reading with an all-electric interface", *Biosensors Bioelectron.*, vol. 20, no. 9, pp. 1813-1820, 2005.

[9] K. Shapsford, Y. Shubin, J. Delehanty, J. Golden, C. Taitt, L. Shriver-Lake, and F. Ligler, "Fluorescence-based array biosensors for detection of biohazards", *J. Appl. Microbio.*, vol. 96, no. 1, pp. 47-58, 2004.

[10] Z. Foldes-Papp, U. Demel, and G. Tilz, "Ultrasensitive detection and identification of fluorescent molecules by FCS: Impact for immunobiology", *Proc. Natl. Acad. Sci. USA*, vol. 98, no. 20, pp. 11509-11514, 2001.

[11] S. Baglio, S. Castorina, and N. Savalli, "Integrated Inductive Sensors for the Detection of Magnpp. 372-384, 2005.

[12] M. Lany, G. Boero, and R. Popovic, "Superparamagnetic microbead inductive detector", *Rev. Sci. Instrum.*, vol. 76, no. 8, pp. 084301-084301-4, 2005.

[13] M. Miller, G. Prinz, S. Cheng, and S. Bounnak, "Detection of a micron- sized magnetic sphere using a ring-shaped anisotropic magnetoresistance-based sensor: A model for a magnetoresistance-based biosensor", Appl. Phys. Lett., vol. 81, no. 12, pp. 2211-2213, 2002.

[14] D. Baselt, G. Lee, M. Natesan, S. Metzger, P. Sheehan, and R. Colton, "A biosensor based on magneto resistance technology", Biosensors Bioelectron., vol. 13, no. 7, pp. 731-739, 1998.

[15] G. Li, V. Joshi, R. White, S. Wang, J. Kemp, C. Webb, R. Davis and S. Sun, "Detection of single micron-sized magnetic bead and magnetic nanoparticles using spin valve sensors for biological applications", J. Appl. Phys., vol. 93, no. 10, pp. 7557-7559, 2003.

[16] W. Shen, X. Liu, D. Mazumdar, and G. Xiao, "In situ detection of single micron-sized magnetic beads using magnetic tunnel junction sensors", Appl. Phys. Lett., vol. 86, pp. 253901-253901-3, 2005.

[17] R. Edelstein, C. Tamanaha, P. Sheehan, M. Miller, D. Baselt, L. Whitman, and R. Colton, "The BARC biosensor applied to the detection of biological warfare agents", *Biosensors Bioelectron*, vol. 14, no. 10, pp. 805-813, 2000.

[18] J. Rife, M. Miller, P. Sheehan, C. Tamanaha, M. Tondra, and L. Whitman, "Design and performance of GMR sensors for the detection of magnetic microbeads in biosensors", *Sens. Actuators A,* vol. 107, no. 3, pp. 209-218, 2003.

[19] S Ghionea, P Dhagat, A Jander, "Ferromagnetic Resonance Detection for Magnetic Microbead Sensors", accepted for publication in IEEE sensors

Chemically Modified Poly(Acrylic Acid) Monolayers in Alumina Nanoporous Membranes for Cancer Detection

R. Kishton[1], N. A. Monfared[2], M. Bothara[3], S. Varadarajan[1], S. Prasad[3], and S.V. Atre[2]

[1]University of North Carolina, Wilmington, NC, varadarajans@uncw.edu
[2]Oregon State University, Corvallis, OR, sundar.atre@oregonstate.edu
[3]Portland State University, Portland, OR, sprasad@pdx.edu

ABSTRACT

In this paper we have developed one of the key building blocks of the clinical proteomics test technology. This building block is a nanostructure that enables the confinement of protein biomolecules based on size-matched trapping. The prototype system is a nanoporous alumina membrane that is fabricated using a standard two-step electrochemical anodization process. We have chemically modified these membranes with poly(acrylic acid) to understand the effects of this chemical modification on the protein detection sensitivity. We plan on evaluating this prototype system in identifying cancer markers for early breast cancer detection to experimentally evaluate the effect of nanoscale-confined spaces on protein biomarker detection.

Keywords: nanoporous alumina, nanomembrane sensors, chemical functionalization, electrical measurements

1. INTRODUCTION

Table 1 summarizes the current state of technology in the building blocks of the manufactured systems comprising of nanostructured materials for biosensing [1]. There are three basic building blocks for such systems (i) nanostructures for binding organic species from solution, (ii) device architecture, and (iii) measurement techniques. The focus of this paper is in identifying an appropriate device architecture to enhance protein detection capabilities. With respect to nanomaterial assembly towards the development of functional systems for interrogating organic species a wide range of device architectures have been investigated, from planar configuration and three-dimensional structures to stacked devices. We report a multi-scale device configuration to identify the effects of nanoscale-confined spaces in protein detection based on poly(acrylic acid) derivatized nanoporous membranes.

Parameter		Key Attributes
Techniques	Optical	Measures amplitude, energy, polarization, decay time and/or phase
	Surface Plasmon	Measures resonant oscillation of surface electrons
	Piezoelectric	Oscillation frequency of crystal varies with its mass
	Acoustic	Measures acoustic resonance of quartz
	Electrical	Amperometric or potentiometric devices
Nanostructures	Particles	Metal and metal semiconductor isolable particles coated for preventing agglomeration, Semiconductor quantum dots, colloidal quantum dots, nanodiamonds
	Tubes	Conductance perturbations due to biomolecule binding in carbon nanotubes
	Wires	High surface-to-volume ratio increases interaction with the biomoleules leading to variations in electronic properties
	Pores	Nanoporous membranes with uniform, rigid, open-pore structure enhance biomolecule entrapment
Device Design	Microarrays	Parallel single analyte biosensors for multi analyte sensing arrays
	3D	3D microarrays for applications like gene expression and clinical diagnostics involving high diversity biomolecules
	Stacks	Vertical arrangement of microarrays leads to formation of stacks

Table 1: Examples of Biological Sensor Platforms

2. EXPERIMENTAL

2.1. Alumina Membrane Fabrication

The prototype system comprised of a nanoporous alumina membrane that was fabricated using anodized alumina, since it's electrically insulating, can be very chemically stable and can be easily incorporated into the CMOS fabrication technique. Anodized alumina is biocompatible [2]. Purified aluminum is used as the positive electrode or anode in an acid electrolytic bath comprising of 0.1 M Oxalic acid. Anodization results in the formation of porous alumina. The dimensions of the pores are controlled by varying the temperature of the bath, applied DC voltage and the time of anodization. The porous membrane fabricated for the electrical detection application comprised of pores 200 nm in diameter and 250 nm in depth as shown in Figure 2. The porous membrane is embedded in anisotropically etched silicon/silicon dioxide substrate. These pore arrays behave like a miniaturized micro-titer well plate with picoliter volume and can be thought of as nanowells. Protein binding is expected to occur within these nanowells. In order to improve

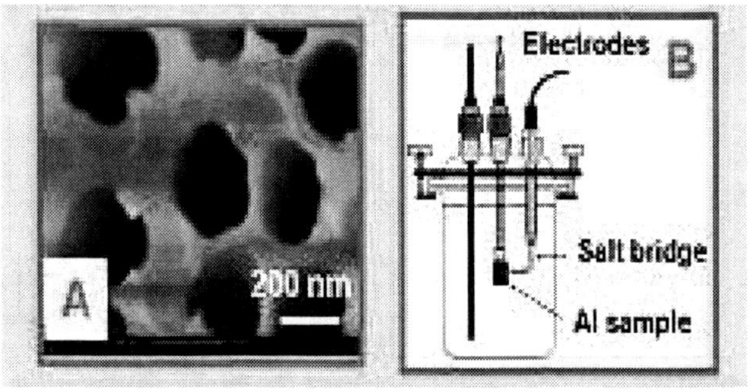

Figure 2: A nanoporous alumina membrane (A) produced by us in a 2-step anodization process (B). Pores of 200 nm diameter were produced.

2.2. Device Fabrication

The fabricated nanotemplates were aligned with the prefabricated silicon substrates with gold micro patterns. These membranes were aligned in such a manner that the gold occludes the nanotemplate in specific regions. The silicon substrates with the gold micro patterns were made using standard photolithography methods. An optical micrograph of the fabricated structure is show in Figure 3. The fabrication process required four steps and used one optical mask. (1) A polished silicon wafer (4 inch diameter and 500μm thick) with a thermally deposited oxide that functioned as the dielectric was used as the base platform. The dielectric provided electrical isolation among sensing sites. (2) Sensing sites were defined using standard mask based photolithography techniques. (3) By thin film deposition techniques a chrome (20 nm) adhesion layer and a gold (150 nm) measurement layer were deposited. The gold layer was laid over the chrome layer. (4) Finally, using wet etching techniques the sensing sites were physically isolated. This formed the base of the nanomonitor. Figure 3 is the representation of the base micro-fabricated platform.

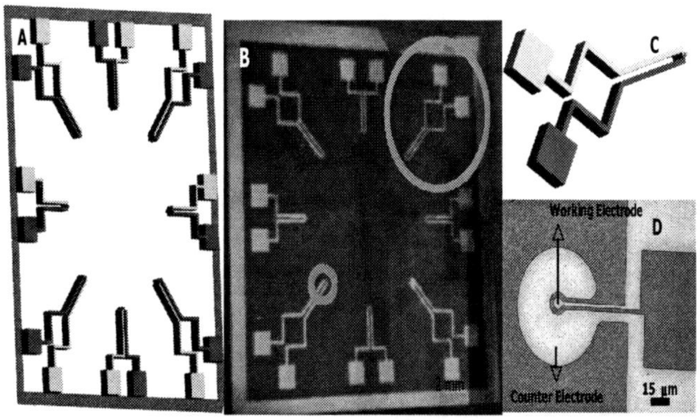

Figure 3: (A) schematic representation of the physical isolation of eight sensing sites (B) Optical micrograph of the micro fabricated platform. (C) Schematic representation of the geometry of the electrode leads from each sensing site. (D) Optical micrograph indicating the working and counter electrode on a single sensing site. The counter electrode is 10 times larger than the working electrode.

2.3. Nanoporous Membrane Functionalizaton

We have previously shown [3] that poly(acrylic acid) (PAA) readily adsorbs from dilute solution (10^{-3} %) on to the intrinsically basic surface of the hydrated native oxide of aluminum to form a robust, uniform monolayer film (0.5-1 nm thick) as represented in Figure 4(A-B). Analysis of the infrared spectrum revealed that both free, undissociated $-CO_2H$ groups and surface-bound $-CO_2^-$ groups were present, in the approximate ratio of 1:2. The free $-CO_2H$ groups could be easily derivatized in quantitative yields as amides and esters, as observed with other approaches to derivatize polymer surfaces with $-CO_2H$ groups to tailor the surface chemistry. The infrared spectra of surfaces of PAA adsorbed on Al evaporated on Si wafers showed the ease of functionalization as stearyl amide and p-nitrobenzyl esters are shown in Figure 4(C). This work indicates a facile approach to create nanowells with controlled chemistry by reacting the free $-CO_2H$ groups with suitable amine-linked molecular recognition sites on the nanoporous alumina membranes. Our objective here was to functionalize the nanowell the estrogen receptor α (ERα) binding ligand, estradiol. ERα is over-expressed in the early stages of breast cancer. We synthesized an estradiol tethered butyl amine using published procedures (Figure 5) [4]. The amine was subsequently covalently linked to the PAA-detivatized membranes (Figure 6).

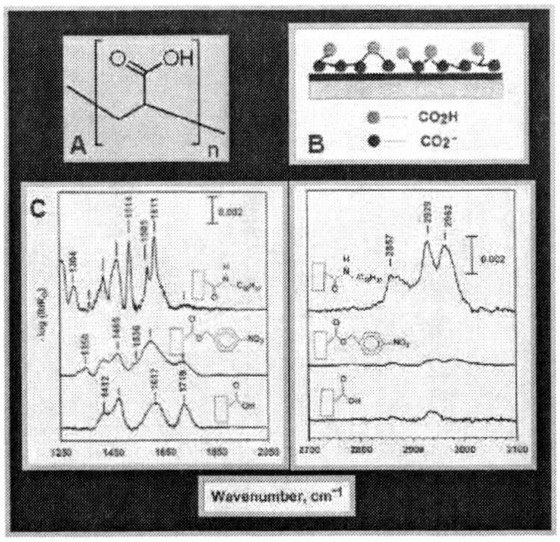

Figure 4: Poly (acrylic acid) (**A**) adsorbs on alumina via carboxylate groups (**B**) leaving free carboxylic acid groups for derivatization into esters and amides as monitored by FTIR (**C**).

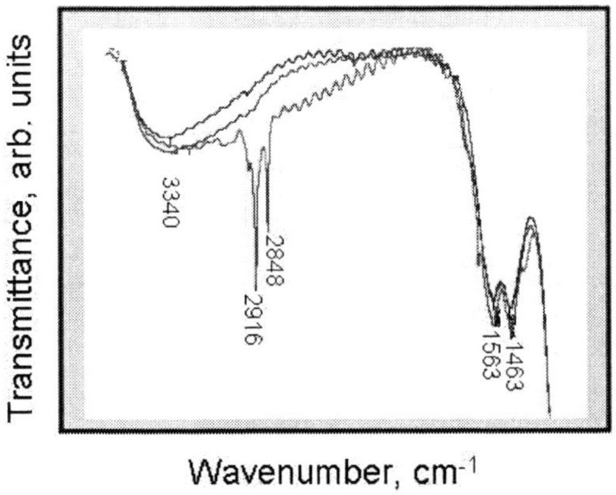

Figure 6: Poly (acrylic acid) derivatized with the estradiol-tethered amine by carbodiimide coupling, as monitored by FTIR. The blue line is the underivatized nanomembrane. The purple line is the poly(acrylic acid)-covered nanomembrane. The red line is the estradiol-linked polymer monolayer on the nanomembrane surface.

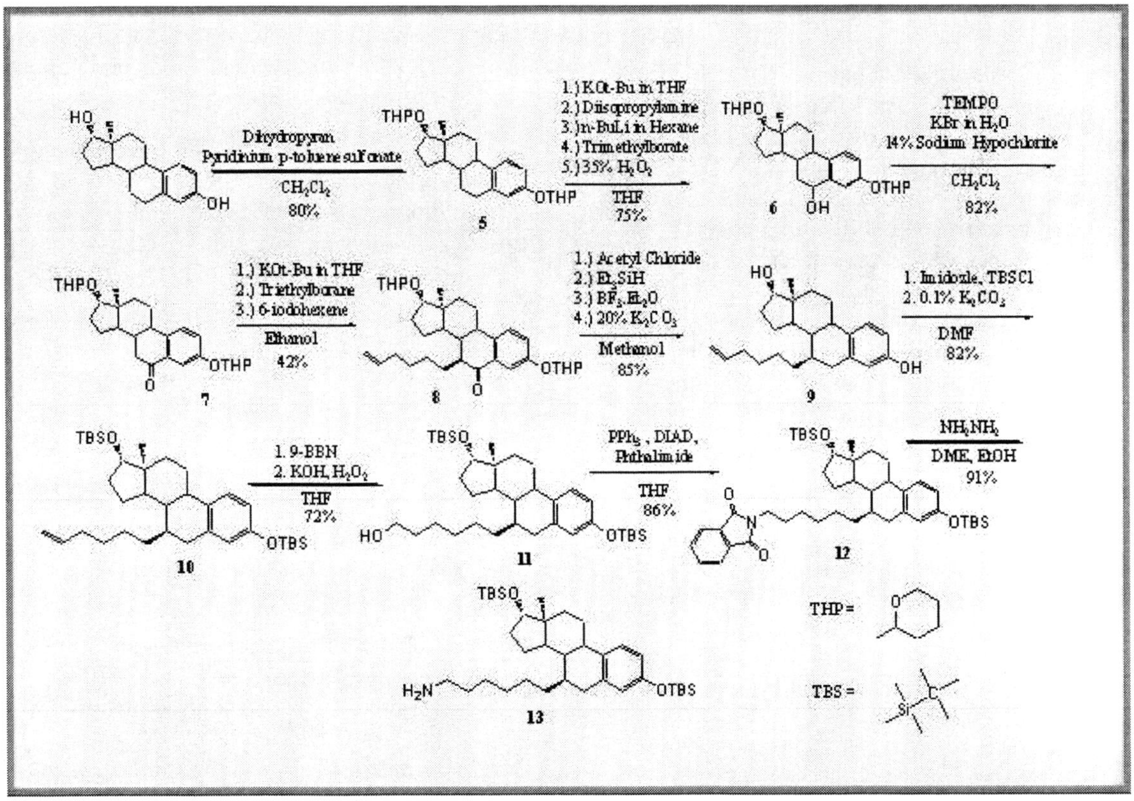

Figure 5: Synthesis sequence for generating estradiol tethered to various substituents. The $-(CH_2)_4-$ tether with amine end group shown in the above sequence has been successfully synthesized by us.

3. RESULTS

The derivatized nanomembranes described in Section 2 will be tested for binding to the cancer biomarker, ERα, by electrical impedance spectroscopy (EIS). The prototype system as shown in Figure 7 works on the principle of double layer capacitive measurement. The other electrochemical measurement techniques namely: trans-conductance and conductance measurements, are limited by the need of redox reactions at the surface for optimal charge transfer. This makes the reactions hard to control and regulate. The non-Faradaic impedance due to the capacitance of the electrical double-layer formed at the electrode surface is sensitive to reactions and is the basis of nanomonitors. This technique is advantageous since it does not require addition of any redox probes. Furthermore, capacitance measurements at different bias voltages and frequencies can reveal much information about dielectric and charge environment at the interface.

The nanomonitors comprised of multiple sensing sites with each sensing site containing approximately quarter million nanowells. The physical dimensions of the nanowell (200nm wide and 250nm in depth) were controlled during the fabrication process such that a single antibody is trapped in an individual nanowell. So, theoretically at antibody saturation, a quarter million antibodies were estimated to be trapped on a single sensing site. In each well the following phenomena are postulated to happen: the antibodies are in the size range of 1-10nm.These antibodies when inoculated flow to the bottom of the well due to capillary forces and they fall within the inner Helmholtz layer of the double layer thereby causing a perturbation and cause a change in the capacitance. The charge associated with the antibody modifies the double layer. When the antigen is added to the sensing site, this further modifies the interface and the formation of the immuno-complex changes the charge distribution causing a change in the capacitance measured. Each nanowell is located on the pre-charged sensing site. Since so many wells are interrogated simultaneously, this would improve the signal to noise ratio. Individual nanowells with trapped biomolecules are electrically equivalent to multiple capacitors connected in parallel. Hence the equivalent capacitance obtained from a single sensing site is the sum of the individual capacitors associated with each nanowell. This results in signal amplification, which is relevant during the detection of lower concentration (less than 10 ng/ml) that in turn improves the limit of detection. In addition as the capacitance is averaged over multiple nanowells this reduces the variability in measurement during the testing of replicates, thus improving the robustness of the nanomonitor technology.

The capacitance was measured from each sensing site. Each site was comprised of a counter and working electrode. The capacitance was measured from the working electrode with respect to the counter electrode. The two electrodes were connected to an Impedance analyzer (HP 4194A) that directly measured the capacitance values.

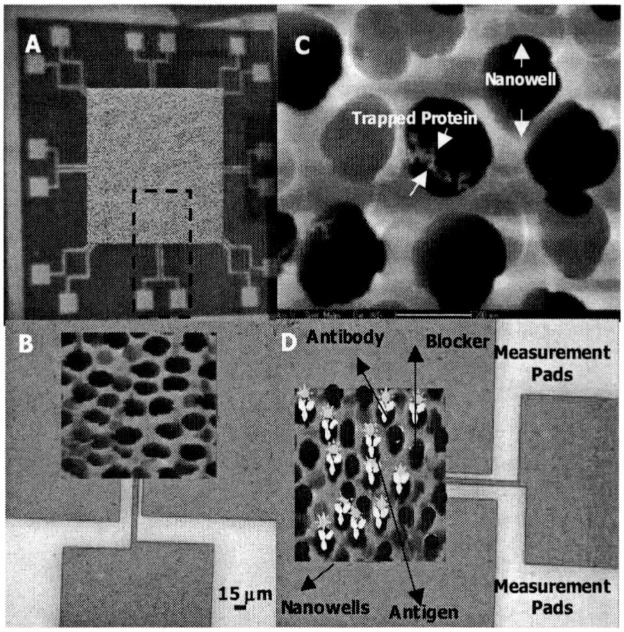

Figure 7: (A) Optical micrograph of the nanomonitor (B) Combination image of a single sensing site: the base is an optical micrograph and the nanoporous membrane is a scanning electron micrograph (SEM). (C) SEM image of the trapped protein within a nanopore (D) Schematic representation of the immuno-complex formation.

In summary, nanoporous membranes were successfully derivatized with poly(acrylic acid) monolayers and estradiol to probe the binding of cancer biomarker, ERα, on a nanomonitor platform.

REFERENCES

[1] J.J. Gooding. Nanoscale Biosensors: Significant Advantages over Larger Devices. *Small*, 2(3):313-315, 2006.

[2] R.K. Reddy. Nanomonitors: Electrical immunoassay for clinical diagnostics implementation of a microfabricated biosensor for the detection of proteins. M.S. Thesis, Portland State University, 2007.

[3] D.L. Allara, S.V. Atre, and A.N. Parikh, "Self-Assembled Monolayers as Polymer Surface Models," *Polymer Surfaces and Interfaces II,* Eds.: W.J. Feast, H.S. Monro, R.W. Richards, John Wiley and Sons Ltd., pp.27-40, 1993.

[4] S.D. Kuduk, F.F. Zheng, L. Sepp-Lorenzino, N. Rosen, and S.J. Danishefsky "Synthesis and Evaluation of Geldanamycin-Estradiol Hybrids" *Bioorg. Med. Chem. Let.* 1999, 9, 1233-1238.

SPRi signal amplification by organothiols-nanopattern

Patrícia Lisboa, Andrea Valsesia, Ilaria Mannelli, Pascal Colpo, François Rossi

European Commission JRC-IHCP ,
Nanotechnology and Molecular Imaging Unit, TP203, Via E. Fermi, 2749, 21027 Ispra (VA,) Italy
patrícia.lisboa@jrc.it

ABSTRACT

A 2D crystalline nanoarray of organothiols has been used for enhancing the detection sensitivity of Surface Plasmon Resonance assays. The nanoarrays have been fabricated by combining colloidal lithography and organothiol self assembly. In particular, immunoreaction produced by nano-spots with carboxylic function in a non-adhesive matrix of Polyethylene oxide (PEO) has been compared to uniform carboxylic surfaces. For the same concentration of analyte, the SPRi signal is much higher by using the 2D crystalline nanoarrayed surfaces.

Keywords: Nanoarray, Surface Plasmon Imaging, Immunosensor

1 INTRODUCTION

The implementation of sensor platforms providing high detection sensitivity is crucial for the design of new analytical devices [1]. Surface Plasmon Resonance (SPR) is a label free technique that is now well-established [2,3]. A constant effort is made for improving the detection sensitivity of plasmonic sensors for instance by using nanoparticles to increase the plasmonic effect in the Localized- SPR (L-SPR) technique [4,5]. Other advantages can be obtained by nanostructuring metallic thin films substrate, because the coupling of the light with Surface Plasmon Polariton SPP modes induces an electric field enhancement [6,7]. Moreover, sensor platforms designed with nanopatterned bio-adhesive/non-adhesive regions have shown increased sensing performance [8-12]. This report presents the study of the contribution of 2D-crystalline organothiol nanoarrays to the SPR imaging (SPRi) sensitivity. Organothiol nanoarray consists in bio-adhesive carboxylic nano-spots distributed in a non-adhesive background of Polyethylene oxide, a type of surface that has already shown a good potential for improved immobilization efficiency of proteins on surfaces [13]. The present study is based on the comparison of the detection of the immunoreaction obtained with nanoarrayed and uniform carboxylic surface by using Human IgG and anti-Human IgG.

2 RESULTS AND DISCUSSION

2.1 Nano-array fabrication and characterization

The gold surface of the SPRi chip was divided in two adjacent areas, one patterned with the nanoarray of two different organothiols (thiolated Polyethylene oxide (PEO) and mercaptohexadecanoic acid (MHD)) and the other uniformly functionalized by MHD.

To produce the nanoarray, three main steps are used. First, the gold coated prism is modified with MHD,. Next, a MHD/Gold nano-pattern is produced by plasma colloidal lithography and gold is finally functionalised with PEO to create a MHD/PEO nano-pattern [13].

The different steps of fabrication were characterized by Cyclic Voltammetry (results not shown). The redox reaction of the couple Fe(CN)63-/ Fe(CN)64- was monitored using the modified surfaces as working electrode with the conditions already explained in our past work [13]. For the bare gold electrode, the voltammogram presents the typical peak shape expected from a flat metallic surface. The functionalisation with MHD creates an insulating layer in the gold electrode reducing dramatically the collected current (no peaks present), thus indicating a good surface coverage. After the polystyrene beads deposition, etching and removal, the voltammogram presents a peak shape with lower definition compared to the bare gold one due to the presence of the MHD nano-areas. Finally after the PEO chemisorption, the surface is again well covered and presents a non-peak shape voltammogram.

Since the RMS roughness of the gold substrate and the RMS roughness of the nano-patterned surface are very close (<1 nm for both the surfaces), the Atomic Force Microscopy (AFM) analysis of the nano-arrayed surface presents a poorly distinguishable pattern from the topography image. Nevertheless, the 2-D Fast Fourier Transform (FFT) of the height function (Figure 1A, inset) clearly shows the presence of a hexagonal array corresponding to the pattern. The nano-pattern is well detected in the phase contrast image (Figure 1B) due to the different chemical properties of the two head group present on the surface. The surface is characterized by a 2D hexagonal lattice of carboxylic spots (9.8% of the total surface area) surrounded by PEO [13].

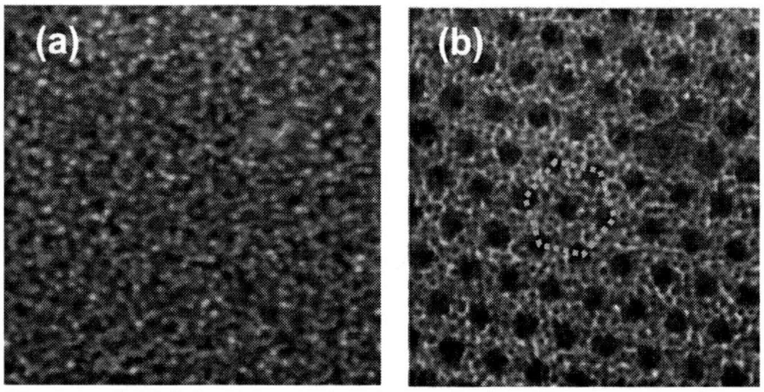

Figure 1-AFM characterization of the Nanoarray: A-topography image (vertical scale 0-5 nm); B- phase image (colour scale: -1° to 5°), the hexagonal arrangement of the COOH spots is highlighted by the dashed red line

2.2 SPRi measurements

The detection principle of the SPRi instrument is based on changes in the reflective index at the interface between gold and dielectric substrate due to the presence of immobilized biomolecules on the gold surface. These changes induce reflectivity variations that are monitored at a fixed angle by a 12 bit CCD camera as a gray level contrast [14].
The momentum matching of the incident light with the Surface Plasmon Polariton (k_{SPP}) at the interface between gold and the dielectric overlayer is ensured by the Kretschmann configuration. The condition for SPP matching with light is expressed by the formula [6,15]:

$$k_{SPP} = \frac{2\pi}{\lambda_0}\left(\frac{\varepsilon_{Au}\varepsilon_d}{\varepsilon_{Au}+\varepsilon_d}\right)^{1/2} = \frac{2\pi}{\lambda_0} n_p \sin(\vartheta) \qquad (1)$$

where λ_0 is the wavelength of the incident light (fixed at 810 nm), ε_{Au} is the dielectric function of the gold, ε_d is the dielectric constant of the dielectric interfacing the gold surface, n_p is the refractive index of the coupling prism and θ is the incident angle of light. Changes in ε_d cause a shift of the SPP momentum, which is detected by the angular detection of reflected light at the surface.
The first step of the SPRi experiment was the immobilization of the recognition element (Human IgG) followed by a blocking step. Reflectivity curves of figure 2 show that the SPRi reflectivity relative change is higher for the nanoarrayed surfaces than for the uniform surface.
Before the recognition of the analyte (anti Human IgG), a control of the interaction specificity was performed using anti-Ovalbumin. For both surfaces, the signal produced by anti OVA adsorption was less than 10% of the recognition signal, indicating a good specificity. After these steps, the recognition of the analyte (anti-Human IgG (Ab specific)) has been performed.

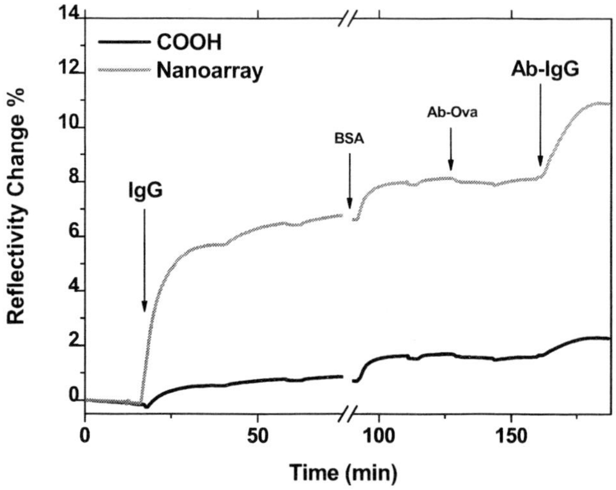

Figure 2-Kinetics of the SPRi immunoreaction experiments for the uniform and nanoarrayed COOH surface.

Since the active surface of the nanoarrayed MHD presents only 9.8% of the total area, one would expect the detection signal to be in the order of 10% of the signal obtained for the uniform MHD surface. On the contrary, the obtained reflectivity changes lead to an amplification of the signal by a factor 4 in the case of the nanopatterned surface. Such a discrepancy cannot not only be explained in terms of a higher efficiency of immobilization of the molecules by the nanopatterned surface, as already observed in other experiments performed with similar surfaces [10,11,16,17]. This abnormal increase of reflectivity is attributed to the 2-D crystalline arrangement of the nanopatterned surface, which leads to the interaction between the SPP modes and the regular spatial modulation of the dielectric constant of the surface above the gold film (2D photonic crystal (2D-PC)). Due to the very close refractive index of the two organothiols constituting the nanoarray (n_{MHD} = 1.45 and n_{PEO} = 1.42), as measured by ellipsometry), modulation has no effect before the exposure of the surface to the protein solution (i.e. the reflectivity curves of the uniformly functionalized COOH surface and the nanoarray are very similar (data not shown). But when IgG molecules adsorb on the nanospots, the surface forms a nanoarrayed substrate characterized by a higher contrast of the refractive index (n_{IgG} = 1.55 and n_{PBS} = 1.33); this nanoarray interacts with the SPP modes at the interface with gold. The interaction between the SPP and PC modes has been observed by many authors on different structured surfaces [7, 18, 19]. The coupling between SPP and PC is favored by the fact that the SPP wavelength is of the same order of magnitude of the lattice constant of the PC (being an hexagonal crystal, the PC has two lattice constant, a_1 = 500 nm and a_2 = 425nm). The SPP wavelength λ_{SPP} can be easily calculated from equation 1 as:

$$\frac{2\pi}{k_{SPP}} = \frac{\lambda_0}{n_p \sin\vartheta} \approx 651 nm \qquad (2)$$

In other words, the matching conditions for the SPP momentum are now [19]:

$$k_{SPP} + i\frac{2\pi}{\vec{a}_1} + j\frac{2\pi}{\vec{a}_2} = \frac{2\pi}{\lambda_0} n_p \sin(\vartheta) \quad (3)$$

where $\vec{a}_1$ and $\vec{a}_2$ are the lattice vectors of the 2D-PC. The contributions of the 2D-PC to the momentum matching conditions are reflected in a change of reflectivity which is enhanced with respect to the one obtained on a non-structured surface. Since the SPRi system detects changes in the reflectivity at a fixed angle (where the derivative of the plasmonic resonance peak is maximum), this leads to an increment of the sensitivity of the SPRi detector.

The formation of the protein nanoarray which is, in the model previously described, responsible for the amplification of the SPRi signal has been confirmed by Atomic Force Microscopy (AFM) analysis. The AFM topographic image of the nanopatterned SPRi prism's surface after IgG-Ab-IgG recognition experiments, in dried conditions, is shown in Figure 3a. The selective immobilization of the proteins in the COOH arrayed nanospots is evidenced by the red circles in Figure 3a (a crystalline line is shown as an illustrative example). Also the profile along the blue line of a single nanospots confirms the formation of protein clusters selectively inside the COOH spots.

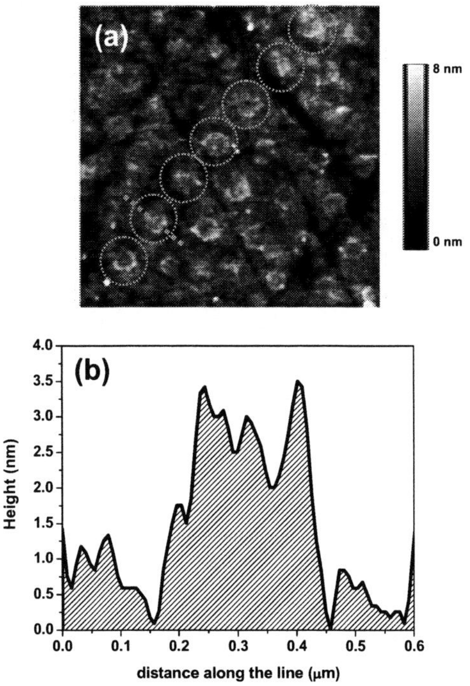

Figure 3- (a) 4 μm * 4 μm AFM scan of the nanopatterned SPRi prism's surface dried after protein absorption experiments. (b) Profile along the blue line in Figure 3a

In order to evaluate the performance of differently functionalized sensor surfaces, dose-response curves were measured with different analyte concentrations (Figure 4) by comparing the nanoarrayed surfaces with uniform COOH functionalized surfaces. In Figure 4 is also reported the Ab-Ova negative control for each concentration explored.

As a first approximation, the sensor response is linear for both the nanoarrayed and the COOH functionalized surfaces. The sensitivity of detection, *S* (defined as the variation of signal produced by an increment in the concentration of the analyte), can be calculated as the derivative of the calibration curve. Calculated values are reported in the inset in Figure 4. The sensitivity of the nanoarray is enhanced by a factor 5 with respect to the COOH one. These results indicate that coupling a commercial SPRi system with nanoarrayed surfaces characterized by bioadhesive motives in a non-bioadhesive matrix amplifies not only the signal by a factor of 4 but improves the sensitivity of the detection of the instrument.

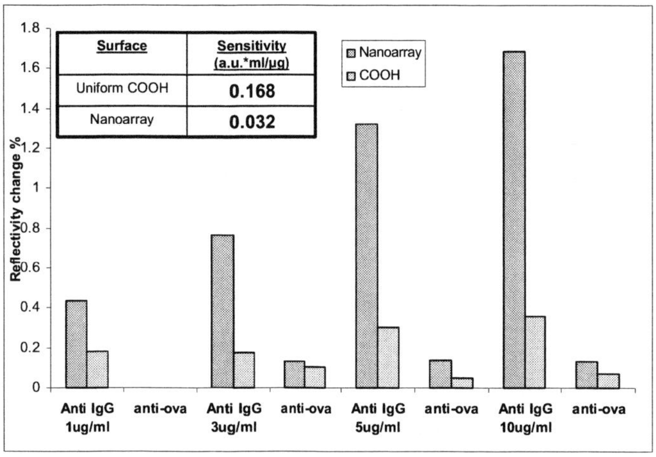

Figure 4-Calibration curves for the nanoarrayed and COOH functionalized surface and calculated sensitivity

3 CONCLUSIONS

This work demonstrates that the modification of the sensor surface with a 2D nano-crystalline pattern improves the SPRi detection. This improvement is due to the 2D crystalline arrangement of the surface that leads to the interaction between the SPP modes and the regular modulation of the dielectric constant of the surface above the gold film modifying the plasmon effect and consequently increasing the reflectivity. These results indicate that SPRi detection performance can be improved by the rational functionalisation of the prism surface with nanopatterns, characterized by spacing of the order of magnitude of the wavelength of the SPP wave. Moreover adhesive – nonadhesive nanopatterns are recognized to be good platforms for the correct immobilization of the biomolecules on biosensing surfaces.

REFERENCES

[1] Turner, A. P. F., Biosensors-Sense and Sensitivity. *Science* **290**, 1315-1317 (2000).

[2] Löfås, S. and Johnsson, B., *J. Chem. Soc., Chem. Commun* **21**, 1526-1528 (1990).

[3] Cooper, M. A., *Drug Discovery Today* **11**, 1061-1067 (2006).

[4] Haes, A. J., Zou, S., Schatz, G. C. and Duyne, R. P. V., *J. Phys. Chem. B* **108**, 109-116 (2004).

[5] Willets, K. A. and Duyne, R. P. V., *Annual Review of Physical Chemistry* **58**, 267-297 (2007).

[6] Barnes, W. L., Dereux, A., and Ebbesen T. W., *Nature* **424**, 824-830 (2003).

[7] Dintinger, J., Klein, S., Bustos, F., Barnes, W. L., and. Ebbesen, T. W, Strong, *Phys. Rev. B* **71**, 035424-1-5 (2005).

[8] Lee, K., Lee, K. B., Kim, E.Y., Mirkin, C. A., and Wolinsky S. M., *Nano-Letters 4,* 1869-1872 (2004).

[9] Rosi, N. L. and C. A. Mirkin, *Chem. Rev.* **105**, 1547-1562 (2005).

[10] Valsesia, A., Colpo, P.,Meziani, T., Lisboa, P., Lejeune, M., Rossi, F., *Langmuir* **22**, 1763- 1767 (2006).

[11] Agheli, H., Malmstroem, J., Larsson, E. M., Textor, M., and Sutherland, D. S**.,** *Nano Letters* **6**, 1165-1171 (2006).

[12] Frederix, F., Bonroy K., Reekmans, G., Laureyn, W., Campitelli, A., Abramov, M. A., Dehaen, W. and Maes, G., *J. Biochem. Biophys. Methods* **58**, 67-74 (2004).

[13] Lisboa, P., Valsesia, A., Colpo, P., Gilliland, D., Ceccone, G., Papadopoulou-Bouraoui A., Rauscher, H., Reniero, F., Guillou, C. and Rossi, F., *Applied Surface Science* **25**, 4796–4804 (2007).

[14] Guedon, P., Livache T., Martin, M., Lesbre, F., Roget,A., Bidan, G. and Levy, Y., *Anal.Chem.***72**, 6003-6009 (2000).

[15] Homola, J. , *Vol. 4* (Springer, Berlin -Heidelberg, 2006).

[16] Valsesia A., Mannelli, I.,. Colpo, P., Bretagnol, F., Ceccone, G. and Rossi, F., *NSTI Nanotechnology Conference and Trade Show - NSTI Nanotech 2007 Technical Proceedings, Vol. 4* ,586 – 589.

[17] Krishnamoorthy, S., Wright, J.P., Worsfold, O., Fujii, T. and Himmelhaus, M., *NSTI Nanotechnology Conference and Trade Show - NSTI Nanotech 2007 Technical Proceedings, Vol 4,* 590-593.

[18] Hibbins, A.P. and Sambles, J. R., *Appl. Phys.Lett.*, **80**, 2410-2412 (2002).

[19] Wedge, S., Giannattasio, A. and Barnes, W.L*Organic Electronics* **8**, 136-147 (2007).

Purification of Nanoparticles by Hollow Fiber Diafiltration

D. Bianchi, D. Serway and W.A. Tamashiro

Spectrum Laboratories, Inc.
18617 Broadwick St. Rancho Dominguez, CA, USA, wtamashiro@spactrumlabs.com

ABSTRACT

Hollow Fiber Diafiltration (Hollow Fiber Tangential Flow Filtration) is an efficient and rapid alternative to traditional methods of nanoparticle purification such as ultracentrifugation, stirred cell filtration, dialysis or chromatography. Hollow Fiber Diafiltration can be used to purify a wide range of nanoparticles including liposomes, colloids, magnetic particles and nanotubes.[1,2] Hollow Fiber Diafiltration is a membrane based method where pore size determines the retention or transmission of solution components. It is a flow process where the sample is gently circulated through a tubular membrane. With controlled replacement of the permeate or (dialysate), pure nanoparticles can be attained. Hollow Fiber Diafiltration can be directly scaled up from R&D volumes to production. By adding more membrane fibers and maintaining the operating parameters, large volumes can be processed in the same time with the same pressure, and flow dynamics as bench-scale volumes.

Keywords: hollow fiber, diafiltration, filtration, purification, tangential flow filtration

MATERIALS AND METHODS

A 100kD polysulfone hollow fiber module (Spectrum Labs MiniKros® Sampler Plus P/N M4AB-260-01P) was selected based upon the unreacted by-product molecular weight (<50kd) and the polymeric nanoparticle (250kD). The tubular geometry of the hollow fiber is beneficial to particle applications due to the phenomenon known as tubular pinch effect where particles migrate to the center of the hollow fiber where flow velocity is the highest.[3] 1200 mL of an 8% polymeric nanoparticle solution was used in a bench scale diafiltration. Discontinuous diafiltration was selected for the application due to the high viscosity, 500 cps. Initial dilution of the product would provide the best performance for permeate flow rate and process pressures and achieve better mixing of the diafiltration buffer with the polymeric solution. A bench scale Tangential Flow Filtration development system with data acquisition software and pressure indicator (Spectrum Labs KrosFlo® Research II P/N SYR2-S21-01N) was used for the process.

RESULTS

1200 mL of the solution was poured into a 4 L TFF process reservoir (Spectrum Labs P/N ACTO-4PP-01N) and diluted to 2400 mL with DI water. The feed pump was set to a feed flow rate of 950 mL/min to begin the process. This feed flow rate was selected to provide an even laminar flow process with a shear rate of $4000 s^{-1}$. Figure 1 illustrates the relationship between flow/fiber and shear rate. Retentate backpressure was applied using a gauge down tubing diameter approach to reach a transmembrane pressure (TMP) of 20 psig with a feed-to-retentate pressure drop of 7 psig.

$$TMP = \frac{P_{feed} + P_{retentate}}{2} - P_{permeate} \qquad (1)$$

Permeate collection began in a graduated cylinder to monitor process volumes and permeate flux. When the collected permeate volume was equal to the initial volume (1200mL), samples of the process were taken for analysis and the product was diluted twofold again to 4%. This dilution volume is referred as one diafiltration volume. This procedure was repeated nine times. The data acquisition program collected pressure and flow rate data results, critical to the direct scaling to larger volumes, throughout the process and provided a process chart in real time (Figures 2 and 3).

Figure 1
Correlation of Flow Per Fiber and Shear
for Three Standard Lumen ID fibers

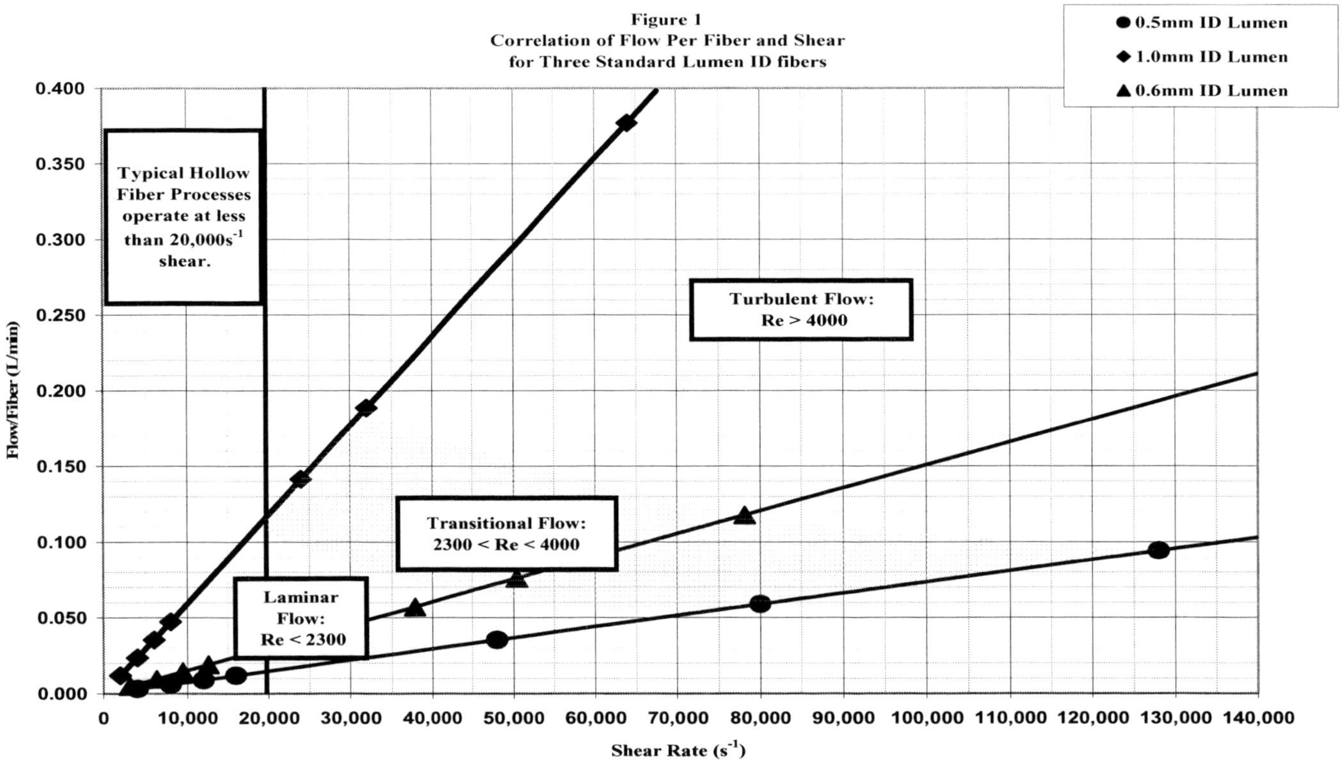

Figure 2
Pressure Profiles vs. Volumetric Throughput
Discontinuous Diafiltraton of Polyacrylic Nanoparticle

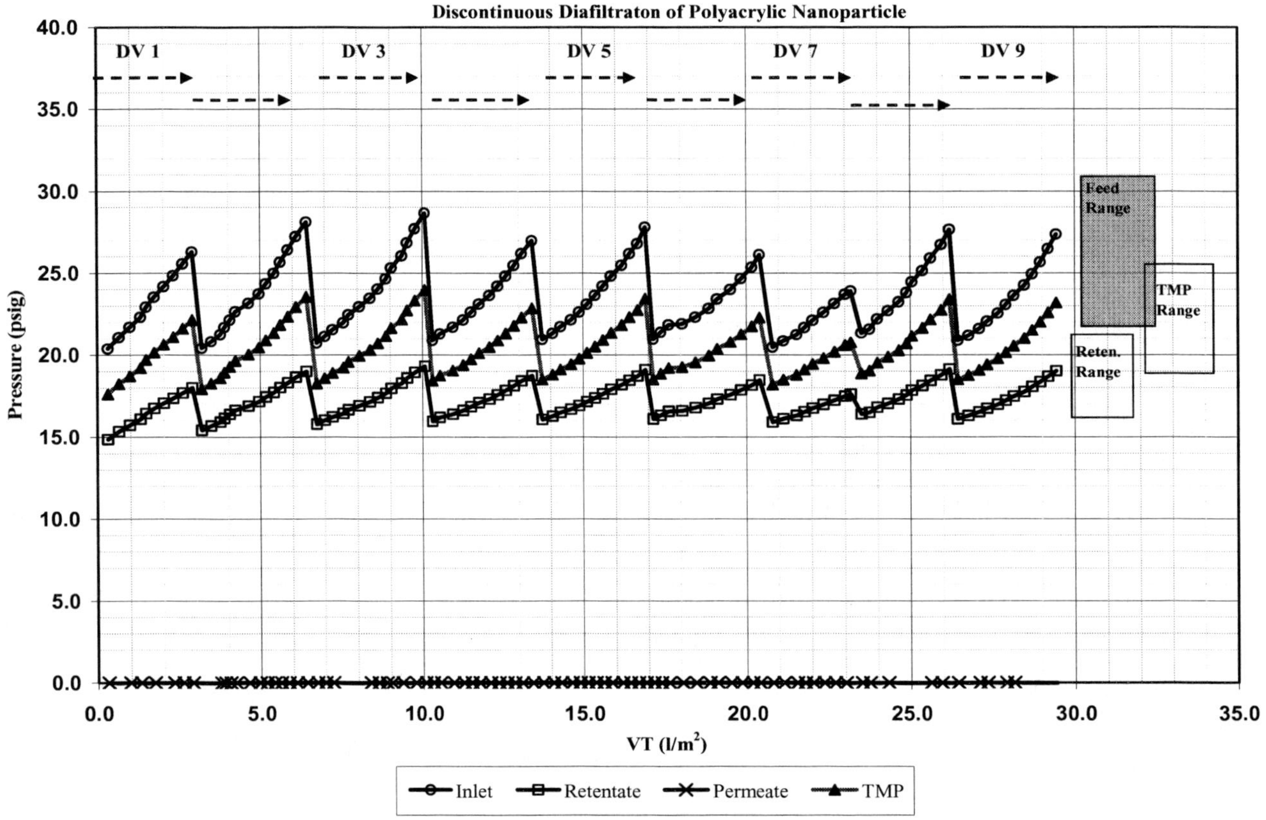

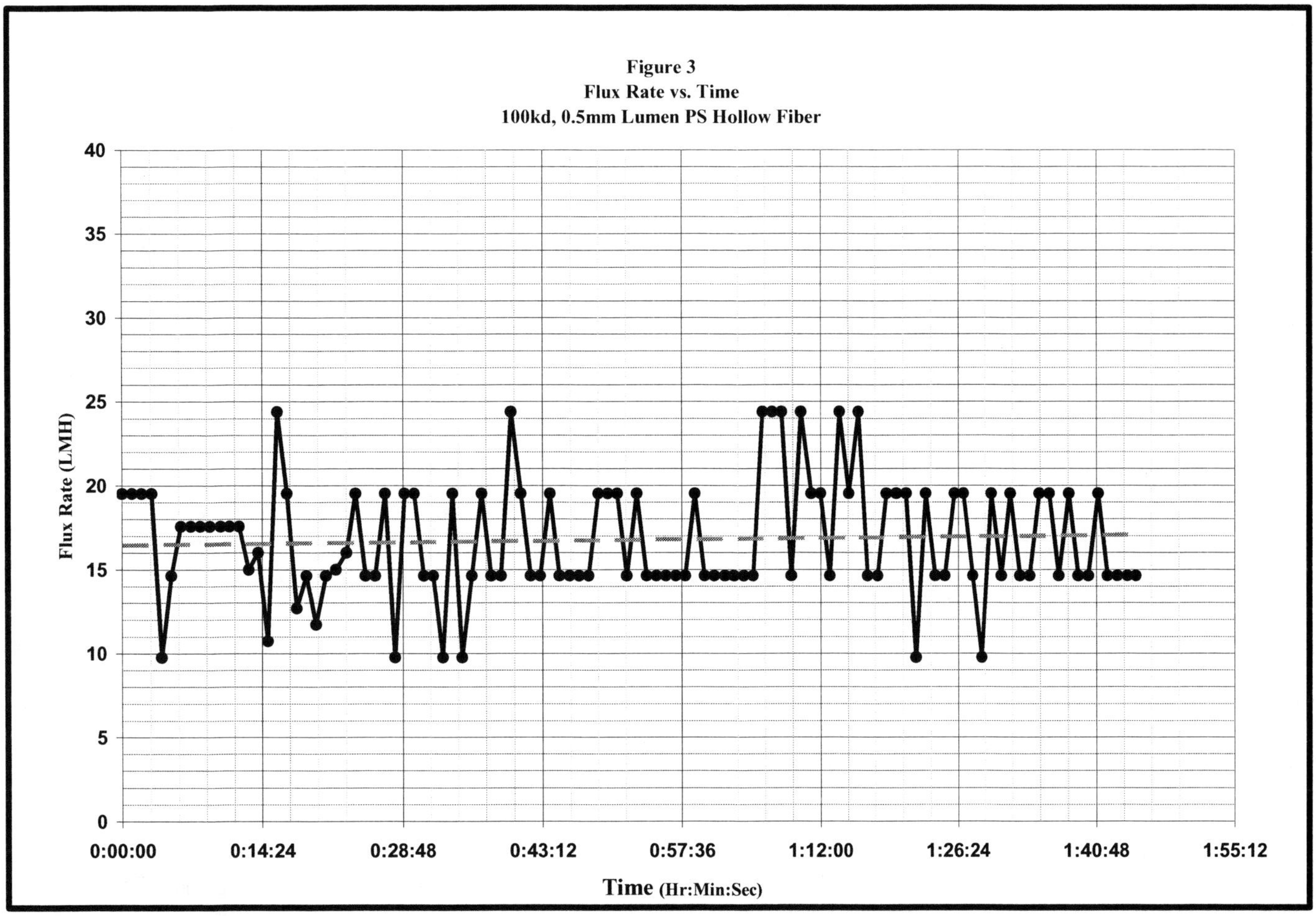
Figure 3
Flux Rate vs. Time
100kd, 0.5mm Lumen PS Hollow Fiber
Flux Rate (LMH)
40
35
30
25
20
15
10
5
0
0:00:00
0:14:24
0:28:48
0:43:12
0:57:36
1:12:00
1:26:24
1:40:48
1:55:12
Time (Hr:Min:Sec)

DISCUSSION

As expected, the process pressures cycled with each dilution, climbing during the concentration from 4% to 8% and then falling back to their initial values with each dilution. This increase in pressure is caused by the increase in viscosity and polymer concentration at the membrane surface during the concentration and is typical for Tangential Flow Filtration processes. The permeate flow rate per unit of membrane area, also referred to as the process flux, ranged from 15-20L/M^2/hr following the concentration of the polymeric nanoparticle. As indicated by the sample analysis, the removal efficiency of the by-product (rejection coefficient) was nearly 50% higher due to the shift in concentration factor based on the viscosity of the process material therefore impacting the permeability of the solute removed as expressed below:

$$Y = 1 - (V_O / V_R)^{1+n(R-1)} \quad (2)$$

Y= % Permeate Solute Removed
V_O / V_R*=Volume Change Ratio*
n=number of stages.
R=Rejection Coefficient

The removal efficiency of diafiltration is dependent upon how easily the small molecular weight by product passes through the membrane. The concentration of the retained material at the membrane surface during the TFF (Tangential Flow Filtration Process) plays a significant role in the transmission of the smaller molecules. Transmission is calculated by measuring the concentration of the byproduct in the permeate stream and dividing it by the concentration of the byproduct in the retentate stream.

$$\%T = \frac{conc_{PERM}}{conc_{FEED}} * 100\% \quad (3)$$

Analysis by HPLC and functional assays showed that the transmission of the by-product in this experiment started at 59% and decreased with each diafiltration volume to 43%. Many diafiltration processes with less viscous starting solutions exhibit transmission of 90 to 100% and require only 4-5 diafiltration volumes to reduce small MW molecules to minimum levels.

The results of several R&D scale tangential flow filtration diafiltration experiments prompted the rapid scale-up of the process to a 12L pilot scale batch volume. To determine the membrane area required for the scale-up, assuming product concentrations were to be identical at both scales, a linear factor of process volume/membrane area was used.

Therefore, for the 12L pilot scale volume, 100kD polysulfone hollow fiber module with 0.57M^2 of membrane (Spectrum Labs MiniKros® P/N K5AB-050-01P) was used. With a count of approximately 1800 fibers, and 20cm fiber length a shear of 4000 sec^{-1} was again utilized at a feed flow rate of 5.2 LPM.

Permeate flux, process time and final product yield of the 12L process reproduced the bench scale measurements. The only anticipated difference was that slightly higher feed-to-retentate ΔP was measured at the pilot scale due to the longer fiber length (20cm vs 12cm at the R&D scale). The following table presents the scalability results.

Area	Shear (s^{-1})	ΔP @ 4% (psid)	DP@8% (psid)	Flux (LMH)
615cm^2 (bench)	4000	4	8	16
0.57M^2 (pilot)	4000	6	11	13

REFERENCES

[1] A.G. Rinzler et.al " Large-scale Purification of single-wall carbon nanotubes, process, product, and characterization" Appl. Phy. A 67,29-37 (1988)

[2] Scott Sweeney et. al. "Rapid Purification and size separation of gold nanoparticles via diafiltration" J. Am.Chem Soc. 128, 3190-3197 (2006).

[3] Mark C. Porter, Handbook of Industrial Membrane Technology, Noyes Publications. pp 186-187. (1990).

Gaseous Flows and Heat Transfer through Micro- and Nano-channels

Hongwei Sun*, Pengtao Wang*, Sai Liu*, Minghao Song* and Mohammad Faghri**

*University of Massachusetts-Lowell, Lowell, MA 01854, Hongwei_Sun@uml.edu
** University of Rhode Island, Kingston, RI 01880

ABSTRACT

The overall object of this paper is a systematic study of gaseous flows and thermal transport in two-dimensional micro- and nano-channels using direct simulation Monte Carlo (DSMC) method. In the flow study, a validation of DSMC code was conducted by simulating a continuum flow in microchannel and the results show that the discrepancy of friction coefficient from theoretical prediction is well below 5%. Then, the effects of compressibility and rarefaction on the flows were investigated through simulating flows with (a) same outlet Knudsen number (Kn) but different pressure drop ratios (=1.3 and 4.5) and (b) low pressure drop ratio (=1.9) but different Kn numbers (=0.043 and 0.083), respectively. For the situation (a), it was found that the high pressure drop flow (pressure ratio: 4.5) show a 15% higher friction coefficient than that of a fully developed flow while the low pressure drop flow (pressure ratio: 1.5) is consistent with incompressible flow prediction. The inspection for the velocity profile development shows that when pressures drop increase along the channel, the center-line velocity become flatten and the velocity gradients near the wall are higher compared with parabolic velocity profile. However, for the situation (b), the rarefactions actually reduce the friction coefficients by 22% (Kn: 0.083) and 36% (Kn: 0.043). An apparent velocity slips along the channel wall exist for both flows. We also studied gaseous flows in microchannels with different surface roughness. The DSMC results show that both relative surface roughness and roughness distribution play very important roles in microchannel flows. High magnitude and densely distributed surface roughness induce higher friction coefficient than that of smooth channels.

In the thermal transport study, we simulated gaseous flows in micro/nano channels under uniform wall temperature (500K) boundary condition. Both temperature distribution and the effects of rarefaction on Nusselt number (Nu) are discussed by comparing with those of fully developed flows.

Keywords: DSMC, thermal transport, gaseous

1 INTRODUCTION

The rapidly emerging Microelectromechanical Systems (MEMS), Nanoelectromechanical Systems (NEMS), and other microscale devices with applications in diverse fields such as molecular biology, space propulsion, particle physics, require a fully understanding of flows in micro- and nano- scales to achieve desired performance. The governing equations based on continuum assumption may become inaccurate to describe their flow behaviors and the simulation technologies based on molecular or atomic interactions like Molecular Dynamics (MD), DSMC etc. become very powerful. The flow chart in Fig.1 clearly illustrates their relationships between these technologies [1].

The Knudsen number (Kn), which is the ratio of the mean free path to the characteristics length of a system, is commonly used to define different gas states. When Kn is in the range of 0.01 to 0.1, the flow is in the slip flow regime where the continuum-based equations with the slip flow boundary condition are valid. For Kn in the range of 0.1 to 3, the flow is in the transition regime where the continuum flow theory is not valid. For Kn >3, the flow is in the free molecular regime where the collisionless Boltzmann equation can be used to predict the flow behavior.

The DSMC method of Bird [2] is a well-developed technology. This technique models thousands or millions of "simulated molecules" activities to obtain a description of gas flows. A detail description is followed. Researchers have been using DSMC to simulate flows in microscale devices and acquire a large amount of valuable results [3-8]. In these simulations, the molecules were reflected either specularly or diffusely from the wall surface depending on the assumption of surface conditions. However, research by Davis [9], on a rarefied gaseous flow in channels with one corrugated wall show lower values of flow rate than if the wall were diffusely reflected. Another study by Usamo et. [10] reported a large reduction of flow conductivity caused by surface roughness in the transient region using DSMC. Our experiment also shows that not just inter-molecular behaviors are important, physical situations of microchannel surface play an important role [11]. Therefore, the overall object of this paper is a systematic study of gaseous flows in two-dimensional micro- and nano-channels in terms of the effects of compressibility, rarefaction, and surface roughness which are usually neglected in conventional flow analysis, using direct simulation Monte Carlo (DSMC) method.

2 METHODOLOGY

DSMC method was developed by Bird in earlier 1960s and is a particle-based simulation method that is based on molecular chaos of kinetic theory. In DSMC, a small number of simulated molecules, which represent a large number of real molecules, are used to reduce the computational requirement. Molecular movement and collisions are uncoupled by using a small time step on the order of a fraction of mean collision time. It consists of four primary processes: 1) particle movement, 2) particle indexing, 3) cross-referencing, and 4) collision simulation and sampling of the flow field. These are described in detail by Sayegh et al. [7].

In the process of particle movement, the distance traveled by a molecule is obtained by the product of the molecular velocity and the time step. The new position of a molecule determines if it passes to an adjacent cell, or reflects from the wall boundary, or collides with another molecule, or exits from the flow field. The molecules in the flow field are then indexed and cross-referenced prior to the modeling of the collisions and the sampling of the flow-field. The so-called No Time Counter (NTC) method is used to determine the number of the molecular pairs (NP) for the collisions as follow:

$$NP = \frac{1}{2} N_{smc} \bar{N}_{smc} R_{r,s} (\sigma_T c_r)_{\max} \frac{dt}{V_{cell}}$$

Where N_{smc} represents the number of simulated molecules in a cell and $R_{r,s}$ is the ratio of real molecules to simulated ones. □□is the total collision cross section and c_r is the relative speed velocity. The interactions of the molecular pairs are described by the Variable Hard Sphere (VHS) model. The □□ and c_r are calculated in this model by:

$$c_r = (u_r^2 + v_r^2 + w_r^2)^{1/2}$$

Where

$$u_r = (2Rf - 1) \cdot c_r$$

$$v_r = (1 - (2Rf - 1))^{1/2} \cos(2\pi Rf) \cdot c_r$$

$$w_r = (1 - (2Rf - 1))^{1/2} \sin(2\pi Rf) \cdot c_r$$

and

$$c_r \sigma_T = \left[\frac{\{ 2kT_{ref} / (m_r c_r^2) \}^{\omega - 1/2}}{\Gamma(5/2 - \omega)} \right]^{1/2}$$

Rf is a random number. k is Boltzmann constant and T_{ref} reference temperature. m_r is the reduced mass of molecule. □ is an empirical constant. The collisions between molecules and the wall are modeled using the diffuse reflection model that assumes the molecules are reflected from the wall surface according to the Maxwell distribution. Finally, the macroscopic variables such as pressure and temperature are calculated by sampling the microscopic variables such as molecular position and velocity.

2.1 Parameters Calculation

The pressure is calculated using an ideal gas equation that is based on the gas number density, n, and the absolute temperature, T, as:

$$P = nkT \tag{1}$$

The DSMC results revealed that the pressure is uniform over the depth of the channel when the flow is fully developed. Therefore, only the centerline pressure is presented in this research.

The friction coefficient is obtained by two different methods. In the first method, it is obtained from DSMC simulation results using the following expression:

$$f_D = \frac{D_H}{\Delta L} \frac{P_2 - P_1}{0.5 * (\rho_2 u_2^2)} \tag{2}$$

where the P_1 and P_2 are pressures at two points, ΔL is the distance between these two points and ρ_2 and u_2 are density and average velocity at point 2. In the second method, the friction coefficient is calculated based on the slip flow theory for incompressible fully developed flow between parallel plates [12]. The Navier-Stokes equation is solved with slip boundary condition to obtain an expression for the friction coefficient. The momentum accommodation coefficient is assumed to be 1.0, which is reasonable for most engineering surfaces:

$$f_L = \frac{96}{(1 + 6Kn)\mathrm{Re}} \tag{3}$$

3 SIMULATION AND DISCUSSION

The modeled microchannel with surface roughness is sketched in Fig. 2. The "surface roughness" is modeled by an array of rectangular blocks. The geometry is described by the "roughness" height (ε), the distance between "roughness" (s), the channel height (H), and the channel length (L). The Kn number is defined as λ/H where λ is the mean free path of gas molecules. Relative surface roughness and roughness distribution are characterized by the ratios of ε to D_h (D_h is hydraulic diameter, =2H) and ε to s, respectively.

In order to check the accuracy of the DSMC method, nitrogen continuum flow through a smooth channel with an aspect ratio (=L/H) of 1.5 was simulated and the friction coefficient results are reported in Fig. 1. The solid line in this figure represents the value of fRe based on a fully developed flow between parallel-plates. In spite of the fluctuation due to DSMC's statistical nature, there is a good agreement between the DSMC results and the continuum flow theory. The maximum average error is less that 5 percent.

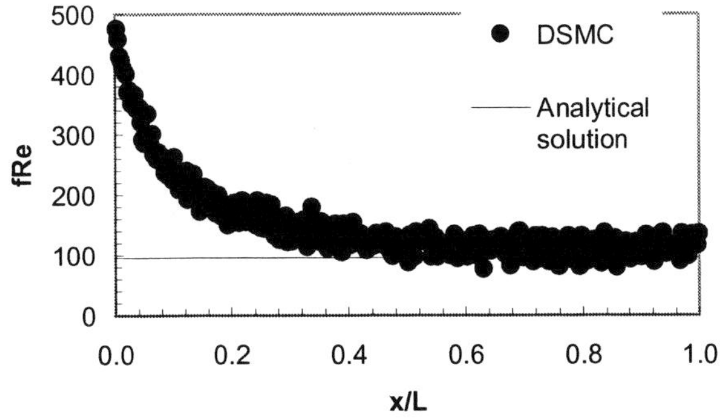

Fig. 1 Comparison of DSMC results with analytical results [12]

3.1 Effects of Surface Roughness on Flows

To highlight the effect of surface roughness, several cases with low Mach number (Ma<0.2) and similar Kn (Kn~0.08) have been performed to minimize compressibility and rarefaction effect. It has been shown the effect of compressibility is negligible for a short microchannel when Ma is lower than 2.0 [12]. The DSMC results for the production of friction factor and Reynolds number (fRe) versus relative surface roughness (ε/D_h) are plotted in Fig. 2. In the mean time, the prediction results from slip flow theory (f_DRe) (Eq. 3) are also presented as a comparison. The value of ε/D_h ranges from 2.4% to 12%. As seen in this figure, the slip flow theory accurately predicts the incompressible nitrogen flow in smooth microchannel and its results show good agreement with DSMC results.

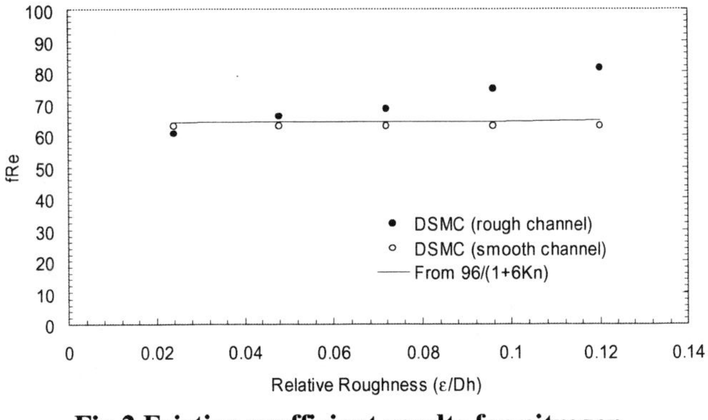

Fig.2 Friction coefficient results for nitrogen flow in smooth and rough channels

However, for nitrogen flows in rough channels under the similar conditions, the friction coefficient increase with surface roughness. The increase in friction coefficient reaches as high as 26.5% while the ε/D_h increasing from 2.4% to 12%. The velocity fields for ε/D_h of 2.4%, 7.2% and 12.5% are also presented in Fig. 5. When the ε/D_h is small (2.4%), the nitrogen flows smoothly through the roughness except for two small recirculation zones behind and in front of the "roughness" block. With the increases in ε/D_h, one large recirculation zone is formed as the replacement of two small ones, as shown in case (b) and (c). It is believed that the increase of 26.5% in fRe is mainly due to the formation of recirculation zone when the ε/D_h increase up to 12%.

3.2 Effects of Rarefaction on Flows

When the characteristic length in the channel is comparable to gas molecular mean free path, which is characterized by an appreciable value of Kn, the interaction between molecules and wall become important. This change can be reflected on the change in fRe. A microchannel with relative surface roughness of ε/D_h =12% and distribution of ε/s= 0.48 was chosen for studying the effect of rarefaction while Kn change from 0.02 to 0.08.

The inlet velocities are adjusted to reduce compressibility effect so that Ma is below 0.02 for all cases. The fRe versus Kn is shown in Fig. 3 which shows that fRe decrease as Kn increase. It means that surface roughness effect become less important for high Kn flows. The main reason may be that the interactions between gas molecules and channel walls are reduced for high Kn flows. At the same time, the momentum transport between molecules is also smaller for rarefied gas than gas under normal condition. The Fig. 4 is the amplified flow fields for cases of Kn=0.02 and Kn=0.08. It is apparent that a large recirculation zone existed between roughness blocks. The recirculation zone causes the additional pressure loss and further increases the value of friction coefficient.

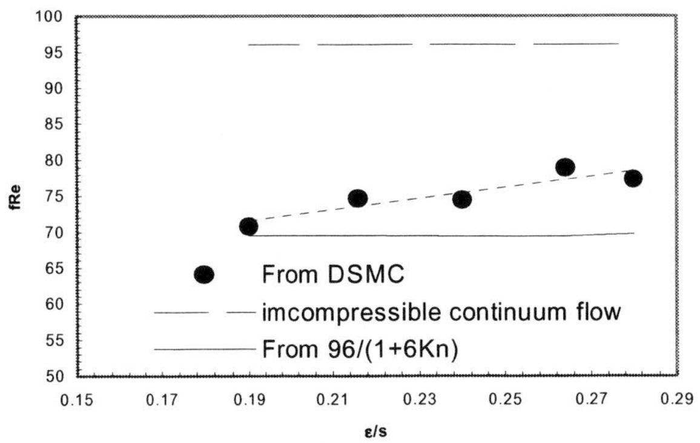

Fig.3 Friction coefficient vs Kn in rough channel

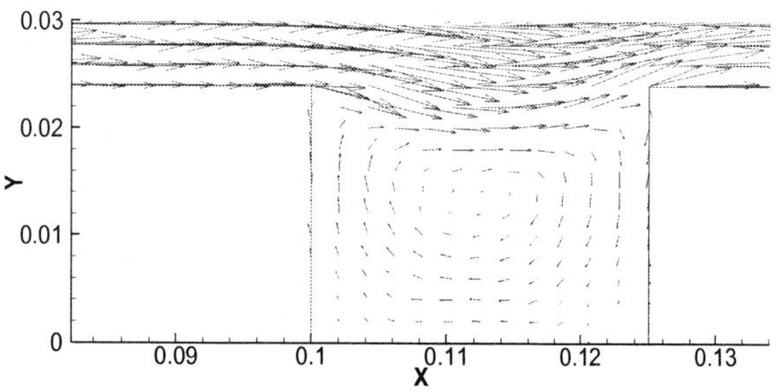

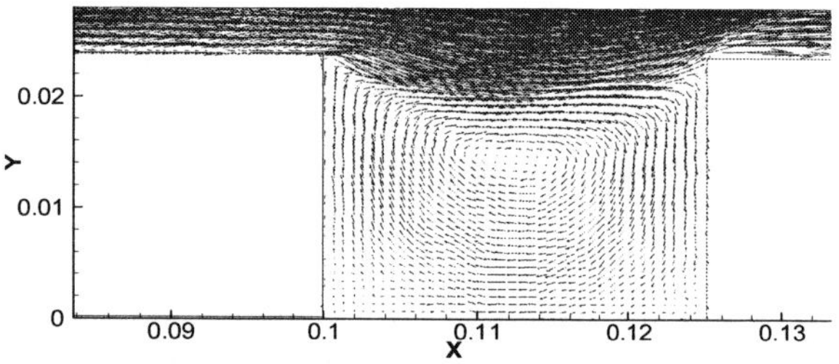

Fig. 4 Flow velocity fields for ε/D_h =12% and ε/s= 0.48 for Kn of 0.02 (up) and 0.08 (down)

3.3 Effects of Rarefaction on Thermal Transport

To study rarefaction effect on thermal transport in micro- and nano- channels, we simulated nitrogen flow through a microchannel with constant wall temperature of 500K. The value of Kn is 0.11, which lies in slip flow region. The variation of local Nusselts number (Nu) and temperature profiles along the microchannel is shown in Figs. 5 and 6.

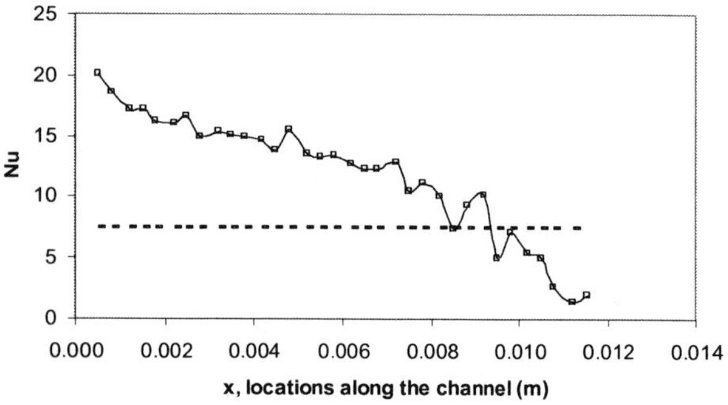

Fig. 5 Local Nusselt number along microchannel

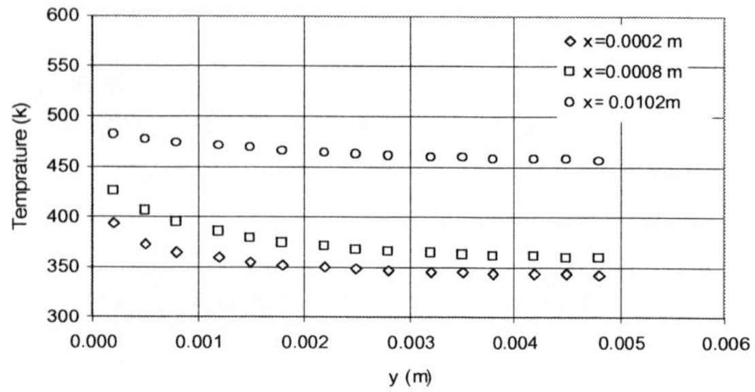

Fig. 6 Temperature profiles at three locations

In the developing flow region, the Nu is high. But in fully developed region, instead of maintaining a value of 7.54 which is predicted by conventional continuum flow theory, the Nu continue decreasing until the gas temperature reach wall temperature, which is apparently the result from rarefaction. More cases are being simulated to quantify the Kn effect.

FERENCES

[1]Gag-el-Hak, "Flow Physics", MEMS handbook, CRC, 2001.

[2]G. Bird, "Molecular Gas Dynamics and the Direct Simulation of Gas Flows, Oxford Science Publications, 1994.

[3] E. S. Piekos and K. S. Breuer, Numerical Modeling of Micro-channel Devices Using the Direct Simulation Monte Carlo Method, *Journal of Fluid Engineering*, Vol. 118, pp. 464-469, 1996.

[4] C. K. Oh, E. S. Oran, R. S. Sincovits, Computations of High-speed, High Knudsen Number Micro-channel Flows, *Journal of Thermophysics and Heat Transfer*, Vol. 11, No. 4, pp. 497-505, October-December 1997.

[5] C. K. Oh, R. S. Sinkovits, B. Z. Cybyk, E. S. Oran, J. P. Boris, Parallelization of Direct Simulation Monte Carlo Method Combined with Monotonic Lagrangian Grid, *AIAA Journal*, pp. 1363-1370, 1996.

[6] C. Mavriplis, J. C. Ahn and R. Goulard, Heat Transfer and Flow Fields in Short Micro-channels Using Direct Simulation Monte Carlo, *Journal of Thermophysics and Heat Transfer*, Vol. 11, No. 4, pp. 489-496, 1997.

[7] R. Sayegh, M. Faghri, Y. Asako and B. Sunden, Direct Simulation Monte Carlo of Gaseous Flow in Micro-channel, *Proc. of the ASME National Heat Transfer Conference*, Albuquerque, NM, August 15-17, 1999.

[8] W. Liou and Y. Fang, Heat Transfer in Microchannel Devices Using DSMC, J. Microelectromech. Systm, 10, pp. 274-279, 2001.

[9] D. H. Davis, L. L. Levenson, and N. Milleron, Effect of "Rougher-than-Rough" Surfaces on Molecular Flow through Short Ducts, *Journal of Applied Physics*, Vol. 35, No.3, pp.529-532, March 1964.

[10] M. Usami, T. Fujimoto, and S. Kato, Mass-flow Reduction of Rarefied Gas by Roughness of a Slit Surface, *Trans. Jpn. Soc. Mech. Eng.*, B, pp.1042-1050, 1988.

[11] Stephen E. Turner, Hongwei Sun, Mohammad Faghri and Otto Gregory, Gas Flow Through Smooth and Rough Micro-channels, *Twelfth International Heat Transfer Conference*, Grenoble, France, August 18-23, 2002.

[12] Hongwei Sun, Mohammad Faghri, Effects of Rarefaction and Compressibility of Gaseous Flow in Micro-channel Using DSMC, *Numerical Heat Transfer*, Part A, 38:153-168, 2000.

Enhanced Fluid Transport Through Carbon Nanopipes

M Whitby*, M Thanou* and N Quirke*

* Chemistry Department, Imperial College London, UK

ABSTRACT

Experimental measurement of fluid flow and diffusion through nanoscale channels is important both for determining how classical theories of fluid dynamics apply at very small length scales and with a view to constructing practical nanofluidic devices. In this study, we observe water flow enhancement of more than 250% in relatively large 271 +/- 31 nm diameter carbon nanopipes with plasma induced surface modification of the carbon walls. Our findings have application in the development of biomedical devices both for sensing and for delivery of therapeutic drugs.

Keywords: nanopipes, nanotubes, carbon, nanofluidics, flow, plasma

1 INTRODUCTION

Understanding the flow of liquids through pipes and channels with dimensions approaching and below 100nm is important both for developing our theoretical models at the nanoscale and for building practical nanofluidic devices [1]. Useful applications are envisaged in process engineering and in nanomedicine. There is now a growing body of evidence based on both theoretical and experimental studies that novel physical phenomena occur when fluids flow through nanoscale pipes, particularly those composed of carbon. Specifically, several recent experimental and theoretical studies [2, 3, 4, 5] have shown that fluids including water can be driven through nanoscale carbon pipes at rates up to five orders of magnitude faster than predicted by the conventional macroscale hydrodynamic equation:

$$Q_0 = \frac{\pi R^4}{8\eta}\frac{\Delta p}{L}$$

This is the Hagen-Poiseuille equation in which Q_0 is the flow rate, R is the pipe radius, η is the viscosity of the fluid, and Δp is the pressure drop between the ends of the pipe with length L. The underlying theory assumes non-slip flow with a parabolic velocity profile across the tube. Molecules of the liquid have zero velocity at the walls and travel fastest at the centre.

The Hagen-Poiseuille equation can be modified by inclusion of a slip coefficient as follows:

$$Q_E = Q_0(1 + \frac{4l_s}{R})$$

Where Q_E is the enhanced flow rate resulting from slip at the fluid/wall interface such that molecules of the liquid do have a positive velocity at the wall. The slip coefficient l_s is the distance outside the tube at which the extrapolated parabolic velocity profile tends to zero. It is also known as the slip length.

Recently we have reported results of pressure driven flow measurements for decane, ethanol and water through well aligned carbon nanopipe arrays with a pore diameter of 43 +/- 3 nm [6]. Observed transport rates were more than four orders of magnitude faster than predicted using classical theory. These findings are in accordance with the similar recent work reported by other groups for smaller carbon nanotubes and indicate that enhanced nanofluidic flow regimes can also occur in larger carbon nanopipes produced via CVD (chemical vapour deposition) in carefully prepared AAO (anodic aluminium oxide) templates.

2 FLUID FLOW MEASUREMENTS

In this study we investigated pressure driven flow of water through the inner pores of carbon nanopipes with relatively large diameter 271 ±31 nm. In channels at this scale the influence of interactions at the fluid/wall interface can begin to dominate bulk flow properties. Our interest was to investigate the effect of chemical surface modification of these carbon pipes to determine whether this might modulate transport rates.

2.1 Materials

The nanopipe arrays used in our study were produced using an established CVD method [7, 8, 9] in commercially available AAO templates (Whatman Anodisc 13mm diameter, nominal pore size 200nm). The template consists of a 60 μm thick film of alumina covered with a well-ordered, high density array of open pores produced by electrochemical etching of the original aluminium metal foil. Amorphous carbon was deposited on the template by flowing ethylene gas (30% in helium) for six hours at a rate of 160 sccm at a temperature of 675°C. SEM (scanning electron microscopy) was used to characterise the size and density of the resulting carbon nanopipe array. Pore diameter was measured to be 271 ±31 nm and pore density $1.6 \pm 0.2 \times 10^9$ per cm^2. Integrity of the membranes was confirmed by inspection in an optical microscope.

2.2 Experimental Methods

To measure pressure at various flow rates through the pipes, individual membranes were mounted in a brass adapter ring and fixed in place with chemically resistant epoxy (Araldite 2014) to form a pressure-tight seal. The assembly was then fitted inside a stainless steel syringe filter (VWR 402078401)) with a coarse grid metal backing support. Double distilled water, degassed under vacuum and previously filtered through 20nm pores, was then driven through the nanopipe array using a syringe pump (Harvard Apparatus, Model 22 2400-01). Nominal flow rates were verified by periodic weighing of the transported fluid. A stainless steel pressure gauge tee (Upchurch Scientific, part number U-433) was located between the syringe pump and the filter holder to allow connection of a pressure sensor (Omega, PX-603-500G5V). Fluid pressure behind the membrane was monitored using a digital panel meter (Omega, DP25B-S-230) calibrated to kPa using a mercury manometer and reading zero at one atmosphere. The stabilized pressure at five different flow rates was measured and standardized to a membrane area of 1 cm^2 before calculating averages and standard deviations. A schematic of the experimental system is shown in Figure 1 below.

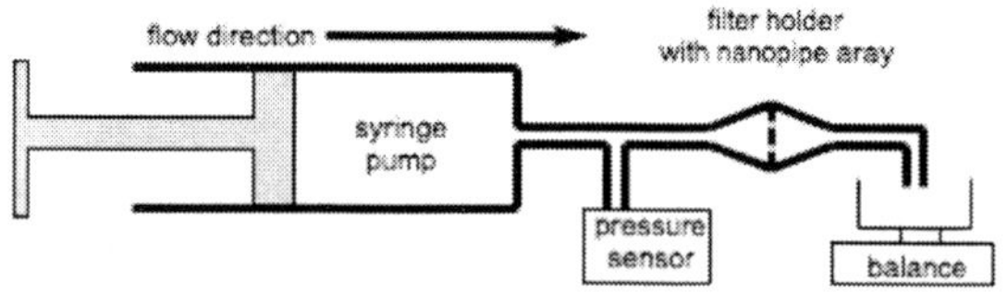

Figure 1. Schematic diagram showing experimental setup for measuring pressure at a series of imposed flow rates.

To assess the effect of chemical surface modification of the carbon nanopipes on water transport rates, intact carbon nanomembranes were exposed to an air plasma for 30 seconds. The instrument used was a March Plasmod running at maximum power (nominally 150W) and at an RF frequency of 13.54 MHz. Wettability is a measure of surface chemistry and the contact angle for a 0.005ml drop of distilled water was compared for the treated and untreated halves of the membrane. The water drop had a high contact angle (~ 110°) on the untreated carbon. On the treated membrane it immediately and completely wetted the surface (contact angle 0°). In other words, brief exposure to air plasma caused the carbon surface to change from somewhat hydrophobic to highly hydrophilic.

To determine the effect of this surface modification on fluid flow, three pairs of treated and untreated membrane fragments were mounted in brass adapters. The pressure driven flow of water was then measured in the manner previously described. Good quality data was obtained for all three treated samples and for two of the controls. Results for the remaining control was discarded due to pressure-induced cracking. This precluded plotting of standard deviations for this series.

2.3 Results

The preliminary data are presented in Figure 3. An average flow enhancement of 250% was seen for the carbon membranes exposed to air plasma. Preliminary XPS data confirms a marked increase in OH and COOH groups in the treated samples compared to the untreated controls.

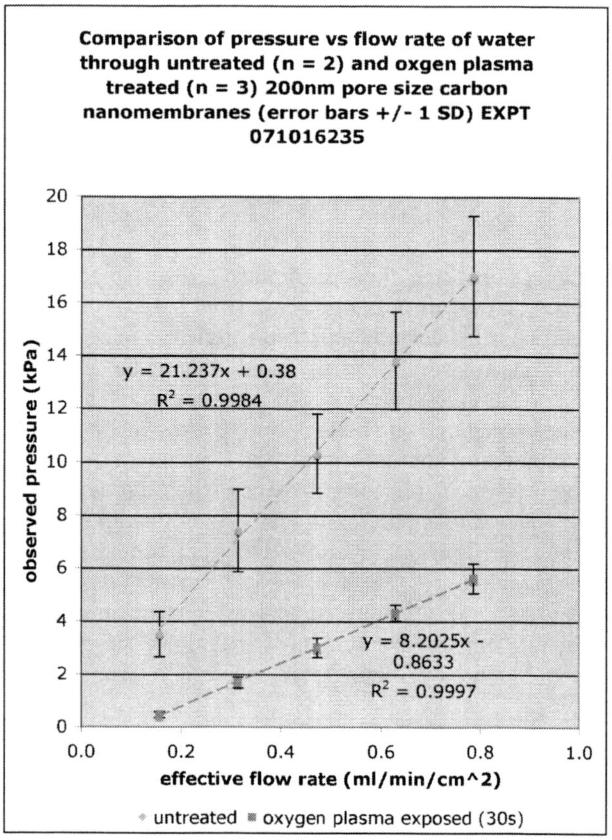

Figure 2. Plot of imposed flow rate against observed pressure for water passing through plasma treated carbon nanopipe arrays compared to untreated controls. The flow rate through the surface modified samples is > 250% faster.

2.4 Discussion

A possible explanation of the observed flow enhancement is that plasma treatment might simply have enlarged the physical size of the pores. Plasma etching is a well know technique in the semiconductor industry and exposure to oxygen plasma can quickly remove organic material including elemental carbon. To test this possibility, the treated membranes used in the experiment were examined using SEM and compared with the untreated samples to see if there was evidence of pore widening. An example micrograph is show below in Figure 3.

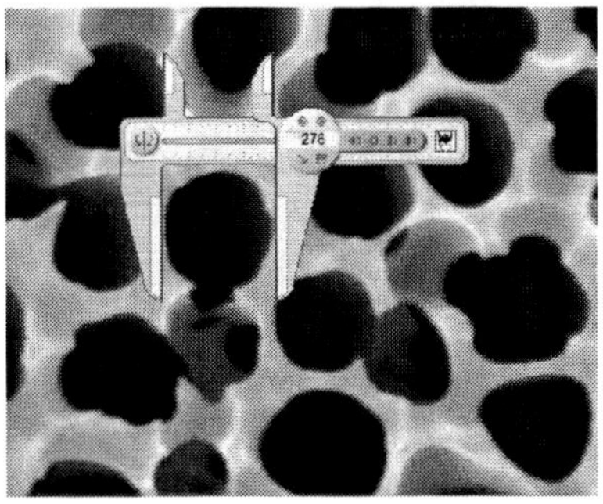

Figure 3. SEM micrograph of plasma treated carbon nanopipe array showing one surface of the membrane with open pores. The caliper scale is in nm.

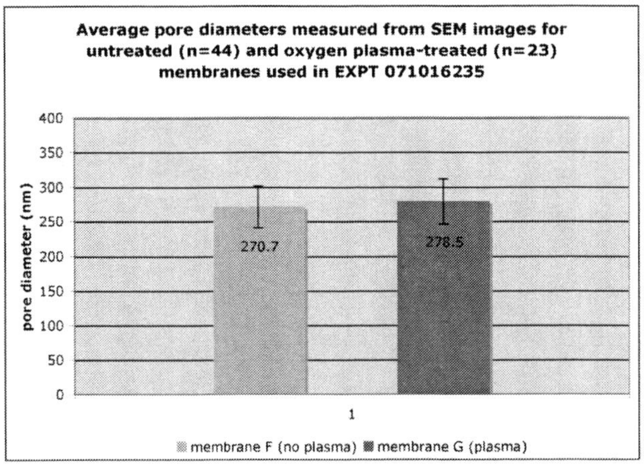

Figure 4. Chart comparing measured diameters of plasma treated and untreated carbon nanopipe pores.

As can be seen from Figure 4 above, there was only a small (< 3%) difference between average pore diameters of the treated and untreated samples. Furthermore this difference was well within one standard deviation of both sets of values and cannot be considered significant. Even if it did reflect a real difference, the resulting increase in flow due to the wider channel would account for less than one tenth of the observed enhancement.

We therefore consider more likely the alternative explanation that increased hydrophilicity of the carbon nanopipes has changed interaction with the water resulting in the observed raised flow rate. Simulation studies and more detailed XPS analysis are currently underway to elucidate the possible mechanisms.

3 CONCLUSIONS

Transport rates of pressure-driven water through 271 ±31 nm carbon nanopipes was measured and flow enhancement of more than 250% was found for carbon surfaces chemically modified by prior exposure to an air plasma. XPS analysis of the treated and untreated surfaces confirms a substantial increase in OH and COOH groups on the plasma treated nanopipes, consistent with an increase in hydrophilicity. This previously unreported method of achieving flow enhancement may be useful in designing future nanofluidic devices.

4 ACKNOWLEDGEMENTS

N Quirke and M Whitby acknowledge support from the EPSRC through grant P09017 "Experimental Nanofluidics". M Whitby acknowledges the support of the EPSRC through a doctoral training award, the Genetic Therapies Centre at Imperial College and RGB Research Ltd for laboratory facilities. XPS analysis kindly provided by M-L Saboungi and R Benoit at Centre de Recherche sur la Matière Divisée, CNRS, Orleans, France.

REFERENCES

[1] Whitby, M.; Quirke, N. Nature Nanotechnology 2007, 2, (2), 87-94.
[2] Majumder, M.; Chopra, N.; Andrews, R.; Hinds, B. J. Nature 2005, 438, (7064), 44-44.
[3]. Holt, J. K.; Park, H. G.; Wang, Y. M.; Stadermann, M.; Artyukhin, A. B.; Grigoropoulos, C. P.; Noy, A.; Bakajin, O. Science 2006, 312, (5776), 1034-1037.
[4] Joseph, S.; Aluru, N. R. *Nano Lett.* **2008**.
[5] Supple, S.; Quirke, N. *Physical Review Letters* **2003,** 90, (21)
[6] Whitby, M; Cagnon, L.; Thanou, M.; Quirke, N. (submitted)
[7] Che, G.; Lakshmi, B. B.; Martin, C. R.; Fisher, E. R.; Ruoff, R. S. *Chemistry of Materials* **1998,** 10, (1), 260-267.
[8] Masuda, H.; Fukuda, K. *Science* **1995,** 268, (5216), 1466-1468.
[9]. Nielsch, K.; Choi, J.; Schwirn, K.; Wehrspohn, R. B.; Gosele, U. *Nano Letters* **2002,** 2, (7), 677-680.

Electrokinetic flows in highly charged micro/nanochannel with Newtonian boundary slip condition

Myung-Suk Chun, Jang Ho Yun and Tae Ha Kim

Complex Fluids Lab., Korea Institute of Science and Technology (KIST)
Hawolgok-dong, Seongbuk-gu, Seoul 136-791, Republic of Korea, mschun@kist.re.kr

ABSTRACT

We have developed the explicit model incorporated together the finite difference scheme for electrokinetic flow in rectangular microchannels encompassing Navier's boundary slip. The externally applied body force originated from between the nonlinear Poisson-Boltzmann field around the channel wall and the flow-induced electric field is employed in the equation of motion. It is evident that the fluid slip counteracts the effect by the electric double layer (EDL) and induces a larger flow rate. Particle streak imaging by fluorescent microscope has been applied to microchannels designed to allow for flow visualization of dilute latex colloids underlying the condition of simple fluid. We recognized the behavior of fluid slip at the hydrophobic surface of polydimethylsiloxane (PDMS) wall, from which the slip length was evaluated for different conditions.

Keywords: electrokinetics, boundary slip, poisson-boltzmann, navier-stokes, nernst-planck, hydrophobic channel

1 INTRODUCTION

The physics of micro/nanofluids has become an area of intense interest both scientifically and technologically. The long-range nature of viscous flows and the small dimension inherent in confined spaces imply that the influence of boundaries is quite significant. Among the boundary effects, we should focus on the hydrodynamic slip at a solid-liquid interface and the electrokinetic phenomena [1-3]. The flow enhancement will benefit during the transport, since friction increases with the surface-to-volume ratio. We first provide the explicit model for rectangular microchannels with solvophobic smooth surfaces. Hydrophobic materials have become attractive for use in MEMS fabrications, and the surface of channel wall frequently has inhomogeneous properties. The slip length was obtained with the particle streak velocimetry using dilute colloids with variations of the shear rate and suspension pH.

2 BASIC CONSIDERATIONS

Many studies have contributed to the slip behavior in narrow channels, in which the Navier's fluid slip occurs in hydrophobic surfaces as depicted in Fig. 1. A slip length β inferred from measurements is the local equivalent distance below the solid surface at which the no-slip boundary condition would be satisfied if the flow field were extended linearly outside.

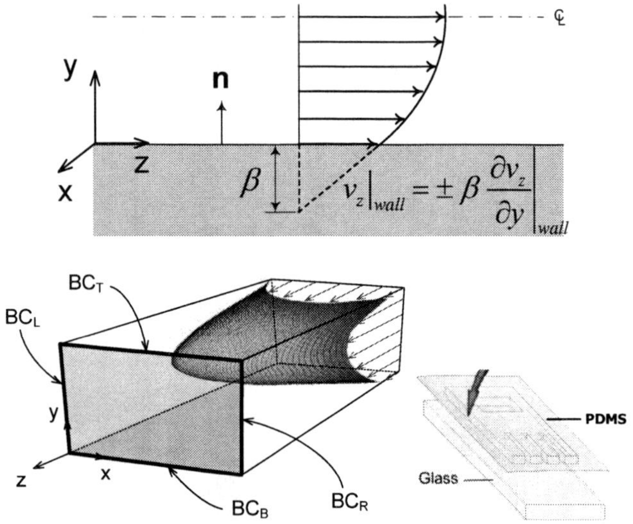

Figure 1: Fluid slip at hydrophobic surface (upper), and channel with different BCs (lower).

2.1 Electrokinetic Flow

For incompressible laminar flow, the velocity of ionic fluid is expressed as $\mathbf{v} = [0,0,v_z(x,y)]$, the pressure $p = p(z)$, and the flow-induced electric field $\mathbf{E} = [0,0,E_z(z)]$. Neglecting gravitational forces, the body force per unit volume ubiquitously caused by the z-directional action of flow-induced electric field E_z on the net charge density ρ_e can be written as $F_z = \rho_e E_z$ [4]. The E_z is defined by the flow-induced streaming potential ϕ as $E_z = -\, d\phi(z)/dz$. With these identities, the Navier-Stokes equation reduces to

$$\eta\left[\frac{\partial^2 v_z}{\partial x^2}+\frac{\partial^2 v_z}{\partial y^2}\right]=\frac{dp}{dz}-\rho_e E_z \, . \tag{1}$$

Based on Eq. (1), the slip boundary condition at the hydrophobic surface is commonly expressed as $v_z\big|_{wall} = \pm\beta\left(\partial v_z/\partial y\right)\big|_{wall}$ in Fig. 1.

When the charged surface is in contact with an electrolyte, the electrostatic charge would influence the distribution of nearby ions. Consequently, an electric field is established and the positions of the individual ions in solution are replaced by the mean concentration of ions. For a rectangular channel, the nonlinear Poisson-Boltzmann (P-B) equation governing the electric potential field is given as [5]

$$\frac{\partial^2 \Psi}{\partial x^2} + \frac{\partial^2 \Psi}{\partial y^2} = \kappa^2 \sinh \Psi \cdot \qquad (2)$$

Here, the dimensionless electric potential Ψ denotes $Ze\psi/kT$ and the inverse EDL thickness κ is defined by $\kappa = \sqrt{2n_b Z_i^2 e^2 / \varepsilon kT}$, where n_b is the electrolyte ionic concentration in the bulk solution at the electroneutral state, Z_i the valence of type *i* ions, e the elementary charge, ε the dielectric constant, and kT the Boltzmann thermal energy. The n_b ($1/m^3$) equals to the product of the Avogadro's number N_A (1/mol) and bulk electrolyte concentration (mM). The Boltzmann distribution of the ionic concentration of type *i* (i.e., $n_i = n_b \exp(-Z_i e\psi/kT)$) provides a local charge density $Z_i e n_i$. We determine the net charge density ρ_e ($\equiv \Sigma_i Z_i e n_i = Ze(n_+ - n_-)$), as follows

$$\rho_e = Zen_b \left[\exp(-\Psi) - \exp(\Psi) \right] = -2Zen_b \sinh \Psi \cdot \qquad (3)$$

Substituting Eq. (3) into Eq. (1) yields

$$\frac{\partial^2 v_z}{\partial x^2} + \frac{\partial^2 v_z}{\partial y^2} = -\frac{1}{\eta}\frac{\Delta p}{L} + \frac{2Zen_b \sinh \Psi}{\eta}\frac{\Delta \phi}{L} \qquad (4)$$

where L is channel length, $\Delta p = p_0 - p_L$, and $\Delta\phi = \phi_0 - \phi_L$.

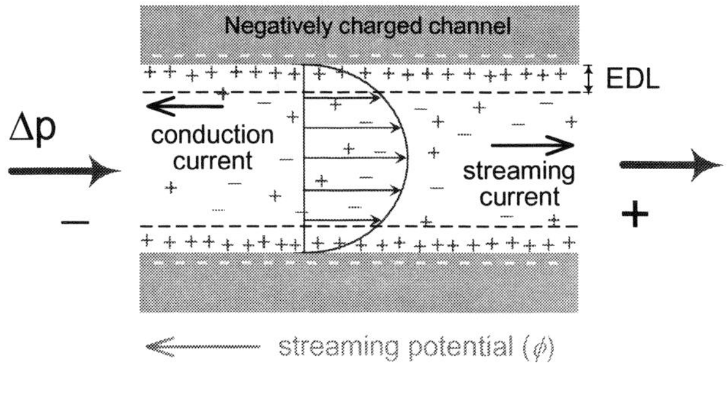

Figure 2: Electrokinetic streaming potential in channels.

In order to analyze the velocity profile, the net current conservation is applied in the microchannel taking into account the Nernst-Planck equation. This equation describes the transport of ions in terms of convection and migration resulting from the pressure difference and electric potential gradient, respectively. Ions in the mobile region of the EDL are transported through the channel, commonly causing the electric convection current (i.e., streaming current) I_S. The accumulation of ions provides the streaming potential difference $\Delta\phi$ (= $E_z L$). This field causes the conduction current I_C to flow back in the opposite direction. In this case, the net current I consists of I_S and I_C, and it should be zero at the steady state, viz. $I \equiv I_S + I_C = 0$ [6].

2.2 Finite Difference Scheme

Basic procedures of the present finite difference method are analogous to the previous work [4]. To obtain the solution of the nonlinear P-B equation with the boundary conditions imposed as $\Psi = \Psi_{s,L}$ at x = 0, $\Psi = \Psi_{s,R}$ at x = W, $\Psi = \Psi_{s,B}$ at y = 0, and $\Psi = \Psi_{s,T}$ at y = H, the five-point central difference method is taken on the left-hand side of Eq. (2). The sinh Ψ on the right-hand side can be linearized as $\sinh \Psi_{l,m}^k + \left(\Psi_{l,m}^{k+1} - \Psi_{l,m}^k\right)\cosh \Psi_{l,m}^k$, where *k* means the iteration index and the grid index *l* and *m* = 1, 2, … , N.

Illustrative computations are performed considering a fully developed flow of the aqueous electrolyte fluid in a 10 μm square microchannel with Δp/L = 1.0 bar/m. The grids of 101×101 meshes were built within the channel and the convergence criterion is given to satisfy the accuracy requirement. The dielectric constant ε is given as a product of the dielectric permittivity of a vacuum and the relative permittivity for aqueous fluid. The fluid viscosity are taken as 1.0×10^{-3} kg/m·sec, at room temperature. For 1:1 type electrolyte, κ^{-1} (nm) is expressed as [*fluid ionic concentration* (M)]$^{-1/2}$/3.278. The EDL thicknesses correspond to 9.7, 96.5, and 965 nm for bulk electrolyte concentrations of 1.0, 10^{-2}, and 10^{-4} mM, where thinning of the EDL means the decrease of electrostatic repulsion.

As shown in Fig. 2, computations were performed with asymmetric variations of both slip length and long-range repulsion, because the hydrodynamic and electric properties depend on the material of the wall. The fluid slip induces a higher flow velocity, while the presence of EDL retards the flow rate. If the slip is absent, a higher apparent viscosity and a higher friction factor would be predicted.

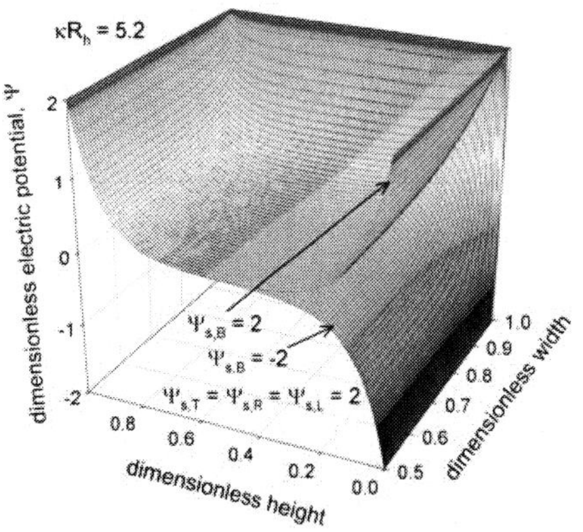

Figure 3: Simulation result of electric potential profile, where hydraulic radius R_h is 5μm.

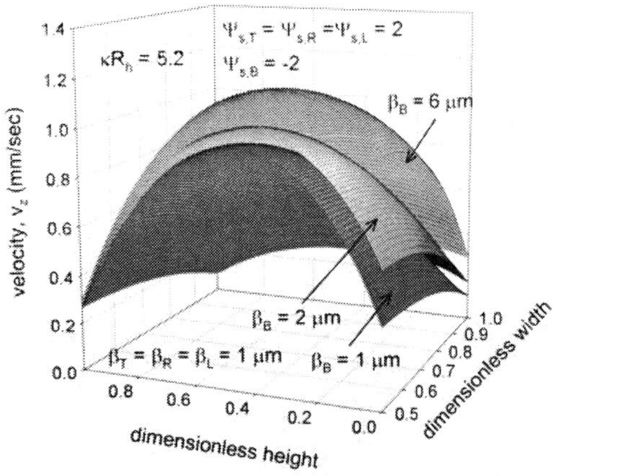

Figure 4: Simulation result of velocity profile, where hydraulic radius R_h is 5μm.

Further, we consider the electrokinetic flow in the serpentine channel that constitutes another source of the Taylor dispersion. The resultant velocity profiles are computed with variations of geometry curvature and electric surface potential.

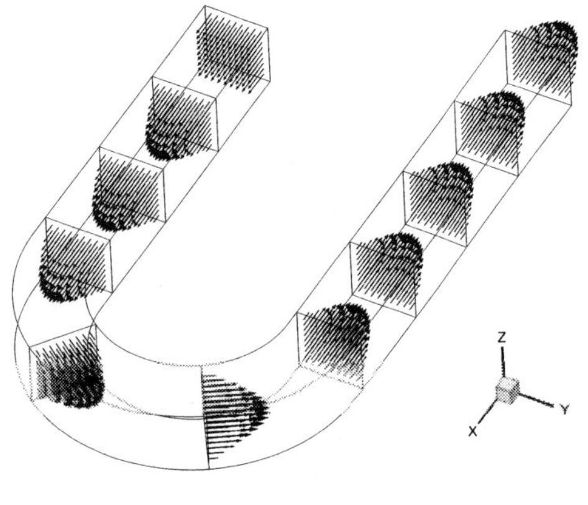

Figure. 5: Flow evolutions in curved rectangular channel.

3 FLOW VISUALIZATION EXPERIMENTS

Experimental observations provide benchmark data for the verification of theoretical study and developing novel micro processes [7]. Microfabrication procedures based on the MEMS micromachining are employed to prepare the microfluidic-chip using molded PDMS and glass cover. The velocity profile of dilute latex colloids was obtained in the channel of PDMS-glass as well as PDMS-PDMS chip shown in Fig. 6, by employing the inverted fluorescent microscope (Nikon, TE-2000) with particle streak velocimetry (PSV) on a parallel uniaxial flow field [8]. The motion of the bulk fluid in particle-based flow velocimetry is inferred from the observed velocity of marker particles. Seeding of the flow field was achieved with fluorescent polystyrene latex (Sigma L-5280, MO) of radius 1.05 μm and density of 1.003. A ratio of particle size to channel width is quite small, and the serpentine channel has also been considered besides the straight one.

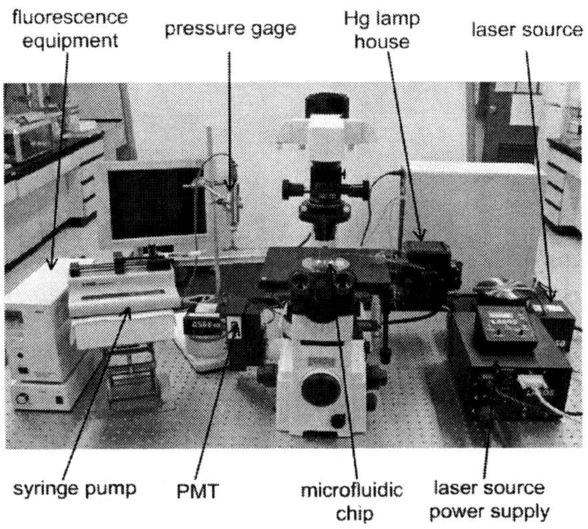

Figure 6: Experimental setup.

4 RESULTS

In Fig. 5, we observe the fluid slip at the hydrophobic surface of PDMS wall, which allows evaluation the value of slip length for different suspension conditions. The PDMS-PDMS channel resulted in a higher flow velocity than PDMS-glass one.

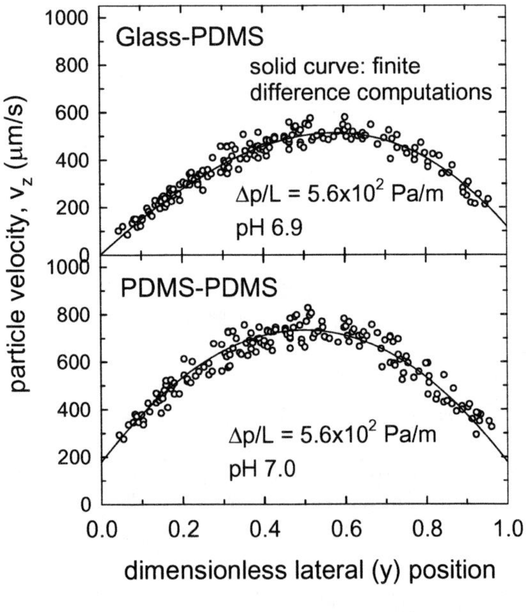

Figure 7: Uniaxial velocity profile in slit-like channel of glass-PDMS and PDMS-PDMS chip at 0.5 mM KCl electrolyte fluid.

The wettability increases by the introduction of surface charge, therefore, the slip length has a trend to decrease as the pH increases causing stronger electrokinetic effects, as shown in Fig. 8. When the fluid slip is absent, a higher friction factor would be predicted in view of the electroviscous effect. In Fig. 9, variations of slip length at PDMS wall are identified as around 1 μm for low shear rate (less than about 50 s^{-1}). We found that, as the nominal shear rate increases, the slip length increases but it appears to be independent of pH.

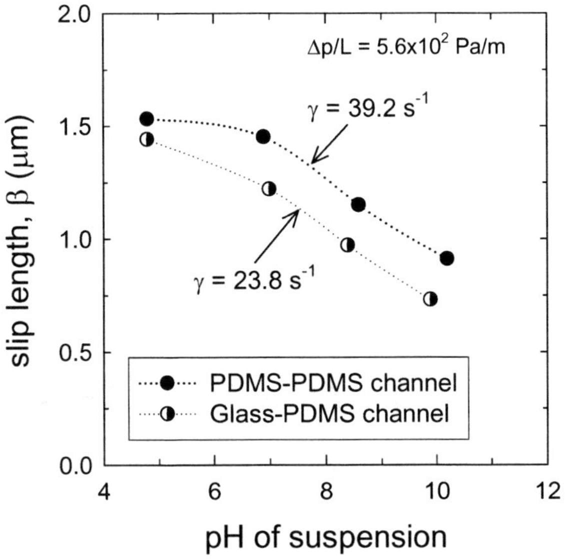

Figure 8: The variation of the slip length for different suspension pH.

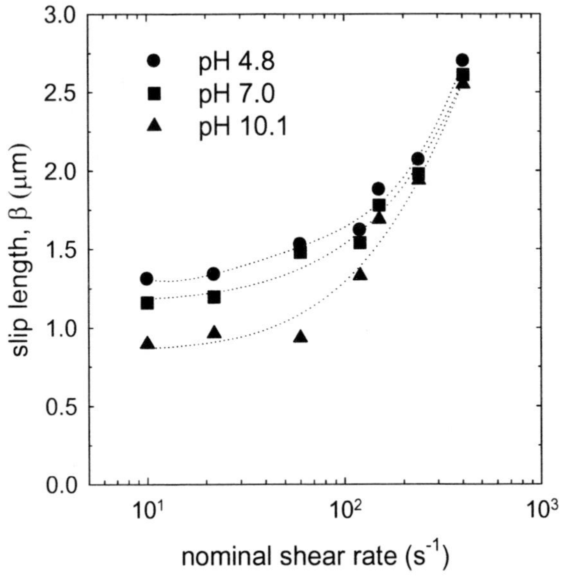

Figure 9: The variation of the slip length for PDMS-PDMS channels for different shear rate and suspension pH.

5 CONCLUSIONS

We developed the numerical scheme for electrokinetic flow with Navier's boundary slip in rectangular channels. The Stokes flow in confined spaces is influenced by the EDL, therefore, the fluid behavior in microchannels deviates from that described by the laminar flow equation in general. Newtonian fluid slip induces a larger flow velocity and then a lower friction factor would be predicted. The validity of the velocity profile determined by flow visualization was justified by comparing with the computational results, where a good agreement was found. Newtonian fluid slip at hydrophobic surfaces has been observed at the length scale of several micrometers, and it is available to obtain the dependency of slip length on the shear rate and suspension pH.

ACKNOWLEDGEMENTS

This work was supported by the KOSEF (R01-2004-000-10944-0) as well as the KIST (μTAS Research: 2E19690).

REFERENCES

[1] D.C. Tretheway and C.D. Meinhart, Phys. Fluids 14, L9, 2002.

[2] E. Lauga and H.A. Stone, J. Fluid Mech. 489, 55, 2003.

[3] P. Joseph, C. Cottin-Bizonne, J.-M. Benoit, C. Ybert, C. Journet, P. Tabeling and L. Bocquet, Phys. Rev. Lett. 97, 156104, 2006.

[4] M.-S. Chun, T.S. Lee and N.W. Choi, J. Micromech. Microeng. 15, 710, 2005.

[5] R.F. Probstein, "Physicochemical Hydrodynamics: An Introduction", Wiley, 185-198, 1994,

[6] C. Werner, H. Körber, R. Zimmermann, S. Dukhin and H.-J. Jacobasch, J. Colloid Interface Sci. 208, 329,1998.

[7] M.-S. Chun, M.S. Shim, N.W. Choi, Lab Chip 6, 302, 2006.

[8] M.H. Oddy, J.G. Santiago and J.C. Mikkelsen, Anal. Chem. 73, 5822, 2001.

Size Effect on Nano-Droplet Spreading on Solid Surface

Nahid Sedighi*, Sohail Murad** and Suresh K. Aggarwal***

*, *** Dept. of Mech. Eng., University of Illinois at Chicago, 842 W. Taylor St., 2039 ERF, Chicago IL, 60607, USA, Tel:312-413-8563, Fax: 312-413-0447, email: *nsedig1@uic.edu, ***ska@uic.edu
** Dept. of Chem. Eng. University of Illinois at Chicago, USA, murad@uic.edu

ABSTRACT

MD simulations have been performed to study the behavior of nano-sized droplets impacting on a solid surface. Three droplet sizes of 6, 9 and 12 nm have been examined. The dynamic contact angle and the droplet maximum spreading rate were analyzed according to the droplet size and the nature of the substrate. The surface properties were varied from wettable to partially wet and non-wettable surfaces. It is found the dynamic contact angle changes significantly as the droplet size changed. While the equilibrium contact angle is less affected by change in the drop size. The wetting radius is increased by increasing the drop diameter for all kind of surfaces. We have observed the wetting radius varys with drop size according to a power law, with the power of nearly 0.5. The dynamic contact angle is affected by size with the same power of 0.5. For partially wet and non-wettable surfaces, there was observed a shift in time that delays the process of wetting with the power of 2/3 with drop size.

Keywords: MD simulation, size effect, impact, nano droplet, spreading

1 INTRODUCTION

Wetting of liquid droplets on solid surfaces is important due to the variety of practical applications ranging from surface coating, and painting to industrial drop depositions, agricultural sprays, biotechnological and many other applications. With the development of micro and nano technology, the contact angle and spreading of micro and nano droplets has attracted much attension recently. The understanding of wetting phenomena is particularly important in the micro and nanofluidic applications. Such as lithography techniques and labs on chip.

Wetting is characterized by the contact angle θ, and the classical Young's equation, is used to describe the equilibrium angle between liquid/vapor and the liquid/solid interfaces at ideal condition. Many of the practical applications require the knowledge of the rate of wetting processes, and the maximum radius of droplet spreading. The spreading of a drop is affected by many parameters. Considering the spreading of a liquid drop on a flat solid surface, when the liquid drop is placed in contact with the solid surface, capillary forces drive the interface spontaneously towards equilibrium. As the drop spreads, the contact angle is changing from its initial value of 180° to its equilibrium angle θ_e. At this stage, a unifying approach describing the dynamics of wetting is still missing. [1]. At equilibrium, while the Yaung's equation is well understood for a macroscopic droplet, this knowledge is not obvious for micro and nano scale droplet [2]. One of the problems that appear in the Young equation is that, the contact angle does not depend on the mass of the liquid droplet. However several authors have shown that this property is not strictly confirmed and in many cases the contact angle depends on the size of the droplet [3]. Many of the literatures on droplet wetting has provided correlations for predicting the spreading behavior and equilibrium state of fluid droplets on solid surfaces. Particular attention has been given to predicting the maximum spread ratio due to the frequent appearance of this parameter in the predictive models. Several works have been done on the dependence of contact angle on size of the droplet [4]. Which results in modifying the Young equation for line tension.

$$\cos\theta = \frac{\gamma_{sv} - \gamma_{sl}}{\gamma_{lv}} - \frac{\tau}{\gamma_{lv} r_B} \tag{1}$$

where γ refers to the interfacial tension and subscript l, s and v refer to liquid, solid and vapor respectively, θ is the equilibrium contact angle and τ is the line tension [5], and $1/r_B$ shows the curvature of the three phase contact line. Equation (1) shows that contact angle is not an intensive quantity and varies with respect to drop size for small r_B or micro-sized droplets, and it approaches toward the classical Young equation for large r_B [5]. Although there are numerous experimental reports for line tension, there is no consistency in magnitude and even the sign of τ. The reported magnitude is from 10^{-11} to 10^{-5} N and both positive and negative values for τ have been reported [6].

These issues made us to reconsider the problem of wettability and its relation to the size of the droplet, to predict the wetting behavior in micro and nano levels. Our aim is to develop this argument and clarify the actual understanding of these phenomena at sub-microscopic level. In this work we study the wetting dynamics of a nano-sized droplet by Molecular dynamic simulation. We have analyzed the dynamic of spreading by measuring the contact angle and wetting radius of the droplet and their variation with time. The focus of the study is to find the effect of size on spreading radius, contact angle, and the time dependent variation in droplet deformation, and their correlation with the original diameter of the droplet.

2 MD SIMULATION

We studied the spreading behavior of a nano-sized liquid argon droplet on a solid surface by MD simulation. The simulation domain is a 3-D box that contains 44048 total argon atoms, which constitute the droplet, solid surface and ambient gas atoms. All the atoms are initially placed at the lattice site of a face-centered cube (FCC) in the simulation system. The density and initial temperature of the system were specified to correspond to the state condition being investigated. The interaction between molecules are expressed by the Lennard-Jones potential

$$\phi(r) = 4\varepsilon[(\sigma / r)^{12} - (\sigma / r)^{6}], \qquad (2)$$

where σ and ε are the characteristics length and energy parameters of Lennard-Jones potential. The potential was truncated at a cut off distance 3σ. The dimensions of the simulation box are approximately, $140 \times 140 \times 280\ nm$. The liquid density and temperature are $0.75\sigma^{-3}$ and 0.72 ε / k, respectively. The fluid particles can move freely in the 3-D system and periodic boundary conditions are employed at the system boundaries. The surface-surface interactions were also modeled with the LJ potential. Each surface atom was attached to the lattice site with a simple harmonic potential, with a spring constant of $K = 200\varepsilon / \sigma^2$, and is allowed to oscillate due to thermal fluctuations around its lattice position. For the result shown, we used $\rho_w = 0.75\sigma^{-3}$. The equations of motion were integrated using Gear's fifth order predictor-corrector algorithm. The time step is $\Delta t = 0.005\tau = 1.078 \times 10^{-2}$ ps , $\tau = (m\sigma^2 / \varepsilon)^{1/2}$ with *m* the mass of argon atom. The reduced temperature is kept at $T^* = 0.72$. The system was stabilized for approximately 500 time steps. After starting the simulation the coordinates of all molecules were subsequently sampled every 500 steps for later analysis. The droplet initially has a spherical shape, and is placed in the middle of the surface with an initial impact velocity of about 4 *m/s*. The droplet is then spread under the influence of the initial impact velocity and the interaction energy of the liquid-surface molecules. The surrounding vapor or ambient gas has an initial density of $\rho_a = 0.0167\rho_f$. Several sets of MD simulations were performed. Three droplets with different sizes of 6, 9 and 12 *nm* have been considered. The wetting and spreading of each droplet is examined on several surfaces with surface properties from high to low surface energies. To provide a different condition of wettability for the surface the interaction energy of surface ε_w is changed. In this way wettable, partially wet, and non-wettable surfaces are generated. where ε_w is the reduced interaction energy of the surface with respect to liquid $\varepsilon^* = \varepsilon_w / \varepsilon_f$ for simplicity it is written as ε_w. At first the simulations focused on the dynamic of droplet spreading and contact angle evolution. An algorithm [7] was developed to track the interface based on the computed density profiles. The liquid-vapor interface is located at the position where the density falls to about 0.25% of the bulk value. Having the interfacial points, the contact angle was determined from the slop of the curve passing through the interfacial points. Latter we observed the flattening and recoiling of the drop in the early stages of the spreading and measuring the maximum spread and time required to reach to the maximum diameter and the other stages of spreading.

3 RESULTS AND DISCUSSION

This research has focused on predicting the magnitude and time of spread parameters such as, point of maximum spread, recoil and equilibrium state of nano-droplets on the solid surface, and how they deform based on the impact parameters of diameter and surface tension. For the results reported in this paper, all variables are reduced with argon parameters ($\varepsilon = 1.67 \times 10^{-21}$ J, $\sigma = 3.405$ Å, and $m = 39.948$ amu). The spreading process of liquid droplet on surfaces can be divided into three different stages [8]. During the first stage (the advancing stage), the liquid drop spreads until it reaches a maximum radius. During this stage, fluid front layer grows ahead of the droplet. The second stage, which observed in some cases, is the stage that contact angle remains constant for a while from the time that spreading stops until recoiling starts. In the final stage (recoiling stage), the drop contact radius decreases toward equilibrium condition, or goes toward zero for bouncing droplets. In this stage the contact or base radius experience some oscillation until the drop reaches to equilibrium or it bounce back from the surface. Figure 1, shows the spreading behavior of a liquid droplet with 6 *nm* diameter in three different surfaces, wettable $\varepsilon = 1$, partially wettable $\varepsilon = 0.15$, and nonwettable $\varepsilon = 0.05$ surfaces. In this figure, the lateral droplet spreading diameter which is normalized with the initial droplet diameter is plotted with respect to time. The initial velocity in all cases is $V_0 = 4$ *m/s*. Figure 2 presents the variation of the equilibrium contact angle for three droplet sizes. It shows the contact angle is increasing with size for wettable surface and decreasing with size for non-wetatable surfaces.

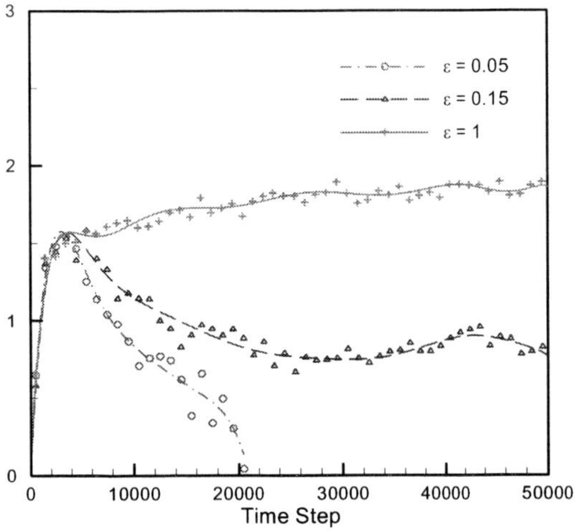

Figure 1: Maximum spread ratio of drop impacting on surfaces with different wetting properties .

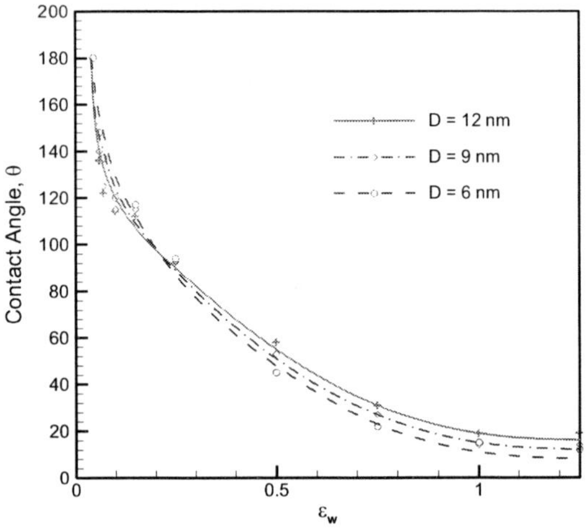

Figure 2: equilibrium contact angle variation with size of drop for different surfaces.

Figure 3a, presents the temporal variation of contact angle for three different droplet sizes spreading on a wettable surface with relative wall interaction energy of $\varepsilon_w = 1$. The contact angle decreases rapidly from 180 toward it's equilibrium value. The magnitude of the contact angle has slightly increased by increasing the droplet size. The time of wetting processes are the same for all drop sizes and it is not observed any time shifting in spreading for watable of surfaces. In Figure 3c, the maximum spreading ratio $[D_m / D_0]_R$ for wettable surface is plotted with time. As drop spreads the wetting diameter increases until it reach to its maximum value. The process in this stage is very fast and it takes only a few thousand time steps to reaches to maximum value. After it reached to the maximum value, the droplet stops from spreading for a short time and it continued spreading with a rate much less than the original rate and eventually reaches to equilibrium after about 35000 time steps. The results show that the magnitude of spreading diameter for the three droplets are different and depends on the size of the drop. This dependence was investigated and it was found there exists a power relation between D_m / D_0 and original diameters of the droplets (D_2 / D_1). In the other words we can write $\frac{(D_m / D_0)_1}{(D_m / D_0)_2} \equiv (D_{0_1} / D_{0_2})^n$. In our analyses, we obtained $n = 1/2$ for the wetting surface. the scaled D_m / D_0 with power ½ is plotted in Figure3d. The increase in the magnitude of the contact angle also showed the same correlation with the droplet size as the maximum spreading correlation. The result is shown in Figure 3b.

The dynamic contact angle for the case of partially wetting surface is demonstrated in the Figure 4a. In this case the relative surface interaction energy is $\varepsilon_w = 0.15$. In this case also the same as previous case contact angle decreases rapidly from 180 degree especially in the first few thousand time steps. By decreasing the contact angle the maximum spreading diameter increases (advancing period), Figure 4c, until it reaches to its maximum value (pick point of the curve). At this time the contact angle has reached to a minimum value and remains at this position for a considerable time. Then, the contact angle increases and drop moves backward toward its original shape. At the same time, the spreading diameter decreases from its pick value and the droplet recoil back (recede). Receding period is relatively fast but slower than the advancing period. After about 40000 time steps the contact angle reaches to equilibrium, and from this point the contact angle remains relatively constant with slightly oscillation around its equilibrium value. While contact angle increases during the

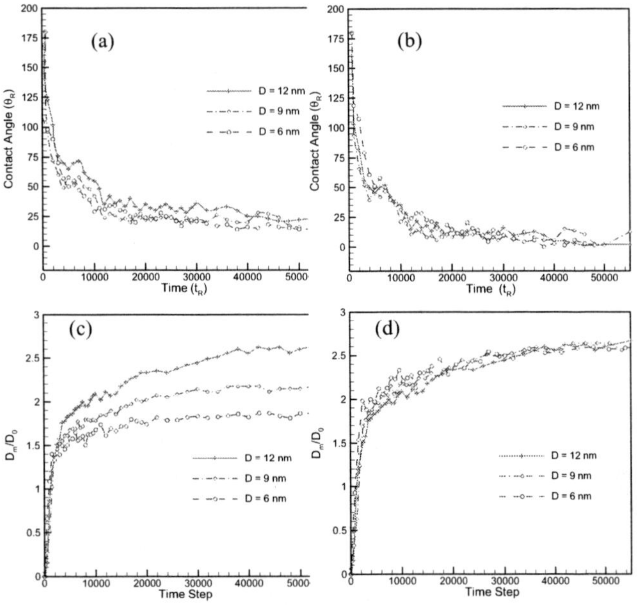

Figure 3: Maximum spread ratio and contact angle of drop impacting on wettable surfaces , and their correlation with drop size, $\varepsilon_w = 1$.

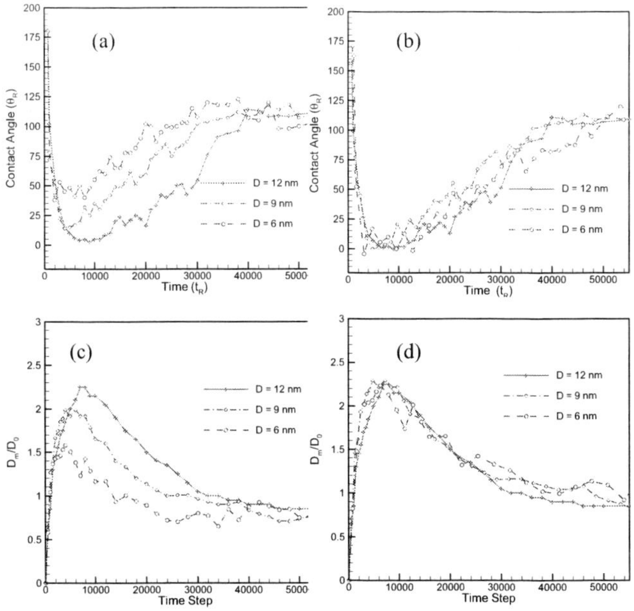

Figure 4: Maximum spread ratio and contact angle of drop impacting on partially wettable surfaces , and their correlation with drop size, $\varepsilon_w = 0.15$.

receding period and approaches the equilibrium, the wetting diameter decreases and approaches to a constant value at equilibrium, Figure 4c. Our observation from figures 4a and 4c indicate that both dynamic contact angle and maximum spreading diameter, depend on the size of the droplet significantly. while the equilibrium contact angle is slightly depends on the size of the droplet. The final wetting diameter is slightly different for all sizes of the droplets. By

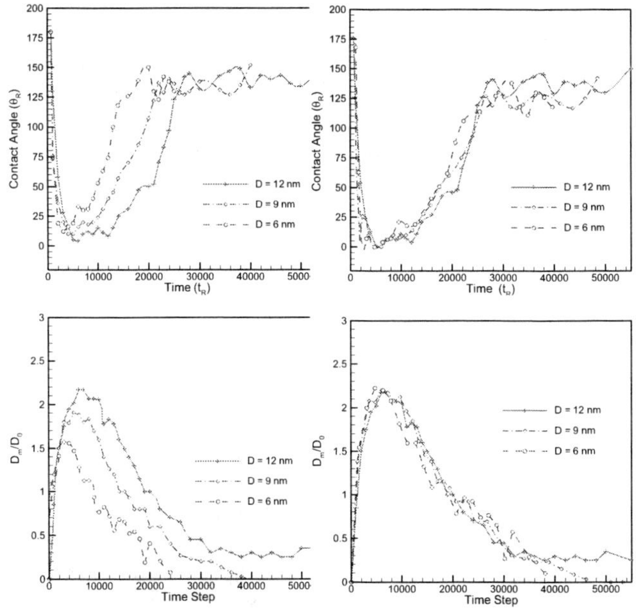

Figure 5: Maximum spread ratio and contact angle of drop impacting on non- wettable surfaces , and their correlation with drop size, $\varepsilon_w = 0.05$.

analyzing the results we could determine the dependence of the contact angle and the wetting diameter to the size of the droplets. For the magnitude of the contact angle it was found $\theta_1 / \theta_2 \equiv (D_{0_1} / D_{0_2})^n$ and $n = 1/2$ in our simulation. The wetting diameter scales with the size as $\dfrac{(D_m / D_0)_1}{(D_m / D_0)_2} \equiv (D_{0_1} / D_{0_2})^n$, with $n = 1/2$ which is the same relation as for the previous case of the wetting surface. We also observe there is a time difference between the events in this case. The time has shifted to the right side of the curve (time is increasing) as the droplet becomes larger. This increase in the time of events is seen in both the dynamic contact angle and the wetting diameter curves. The time shift was analyzed to have a correlation with the droplet size as the $\dfrac{t_1}{t_2} \equiv (D_{0_1} / D_{0_2})^n$, where $n = 2/3$ in this case. The correlation of time and size for wetting diameter also has the same relation as for dynamic contact angle. Figures 4b and 4d shows the scaled values of contact angle and spreading diameter. The results of the simulation for the non-wetting surface where $\varepsilon_w = 0.05$, are plotted in Figures 5. The plots show the same kind of variation as for the partially wetting case for both contact angle and wetting diameter. In this case, the slop of the curves for the non-wetting case are significantly more than the slop of the curves for the partially wetting case for receding period. After receding period, the droplet is oscillating for a while and eventually it bounced back at the end of the simulation. We observed the same kind of correlations for the contact angle and wetting diameter with the original size.

REFERENCES

[1] J. De Coninck, M. J. de Ruijter, M. Voue, Current Opinion in Colloid & Interface Sci, 6, 49-53, 2001

[2] M. Voue, J. De Coninck, Acta Mater, 48, 4405-4417, 2000

[3] P. Letellier, A. Mayaffre, M. Turmine, J. of Colloid and Interface Sci., 314, 604–614, 2007

[4] A. Amirfazli, D. Chatain, A.W. Neumann, Colloids and Surfaces, 142, 183-188, 1998

[5] D. Li, F.Y.H. Lin, A. W. Neumann, J. Colloid Interface Sci. 142, 224, 1991.

[6] S. Vafaei, M.Z. Podowski, Advanced Colloid Interface Science. 113, 133, 2005.

[7] N. Sedighi, S. Murad, S.K. Aggarwal, Atomization and Sprays, (In Press), 2008.

[8] V.M. Starov, S.A. Zhdanov, M.G. Velarde, Langmuir 18 , 9744, 2002

Experimental and Numerical Studies of Droplet and Particle Formation in Electrohydrodynamic Atomization

Jinsong Hua[*] and Liang Kuang Lim[**]

[*] Institute of High Performance Computing, 1 Science Park Road, #01-01 The Capricorn, 117528, Singapore, huajs@ihpc.a-star.edu.sg
[**] Department of Chemical and Biomolecular Engineering, National University of Singapore, 4 Engineering Drive 4, 117576, Singapore liangkuang.lim@nus.edu.sg

ABSTRACT

Electrohydrodynamic Atomization (EHDA) is an effective method to produce micro-droplets/particles in a controllable manner. In this paper, both experimental and numerical approaches are adopted to investigate the mechanism of various EHDA spray modes under different operating conditions. Especially, Computational Fluid Dynamics based numerical simulation is applied to investigate the Taylor cone-jet formation process. The numerical simulations solve the full Navier-Stokes equations for both liquid and gas phases, tracking the liquid / gas interface using a front tracking method, and taking into account the effects of electric stress on the interface. The operating parameters, e.g. the ring electric potential and surface charging density, are varied to study their effects on the Taylor Cone jet formation and the droplet size. The results show that the size of droplets / particles fabricated by the EHDA method can be controlled by adjusting the operating parameters.

Keywords: Electrohydrodynamic Atomization (EHDA), Computational Fluid Dynamics (CFD), two-phase flow, micro-fluidics, droplet and particle.

1 INTRODUCTION

Electrohydrodynamic atomization (EHDA) has been observed and documented for over a century [1] and recently there is a renewed interested in harnessing its ability to produce monodispersed liquid droplets for the fabrication of the monodispersed polymeric particle [2]. An electrohydrodynamic atomization system consists of a nozzle, a high voltage DC power supply attached to the nozzle to raise its electric potential to kilovolt range, and an earthed ground plate placed directly beneath the nozzle to act as the counter electrode in the system. When a liquid solution is pumped through the nozzle and the solution emerges the from the nozzle, electrical stresses on the surface of the solution accelerates the liquid, and a cone like structure, termed the Taylor Cone, is formed at the tip of the nozzle. A fine liquid jet then emerges from the tip of the Taylor Cone at high speed [3]. The jet may further break into micro-droplet due to the stability.

For the fabrication of polymeric particles, a solution containing organic solvent and polymeric solute is sprayed using the electrohydrodynamic spraying system. As the jet travels away from the nozzle, the volatile solvent will evaporates from the solution surface. Depending on the stability of the jet in the electrical field surrounding the nozzle, particles will be formed if the jet breaks into droplets before the solvent has fully evaporated.

In order to control the electrohydrodynamic spraying process properly to obtain the desired droplet and particle, it is essential to understand the interaction mechanism between electrical field and liquid flow, which may result in various patterns in the Taylor Cone formation and liquid jetting. The liquid jet stability in electrical field has been previously studied theoretically [4,5] and experimentally [2,3]. Recently, work has been done to investigate the formation of the Taylor Cone and jet through Computational Fluid Dynamics Simulation [6]. The recent development of Front Tacking / Finite Volume method [7] for multiphase flow provides a robust method for such simulations and analysis.

In this paper, both experimental and numerical approaches are adopted to investigate the formation of various EHDA spray modes under different operating conditions. First, we investigated the different spray modes in the EHDA process under various operating conditions. The spray modes observed in experiments is qualitatively revealed by the numerical simulations. Then, the numerical simulation is applied to investigate the formation process of Taylor cone and jet as well as the droplet size under the same operating conditions. The simulation results indicate that the numerical method proposed here can make reasonable prediction on the electrohydrodynamic spraying process, as well as the jet diameter and droplet size.

2 EXPERIMENTAL METHOD

The present experimental system is shown in Figure 1. A 29-gauge spinal tap needle from Becton Dickinson was flatted at the tip, and connected to the liquid pump to disperse liquid. It also serves as an electrode and is connected to a Glassman high voltage DC power supply (V_n). The needle nozzle has an outer diameter of 340 micron and an inner diameter of 220 micron. The second electrode was in the form of a ring. It was made from copper tube with outside diameter of 2 mm and formed into a ring with a diameter of 40 mm. The center of the ring electrode was placed 10 mm above the tip of the nozzle. The ring electrode was connected to a separated Glassman high voltage DC power supply (V_r). This arrangement enabled the nozzle electric potential and the ring electric potential to be varied independently. An additional 29 gauge spinal tap needle was located 100 mm below the tip of the nozzle. It was connected to ground and used as the counter electrode to the nozzle and the ring electrode. Thus, the electric potential settings to the nozzle electrode and the ring electrode could be adjusted independently to fine tune the electric field in the vicinity of the nozzle tip, which may contribute to the proper control of EHDA process.

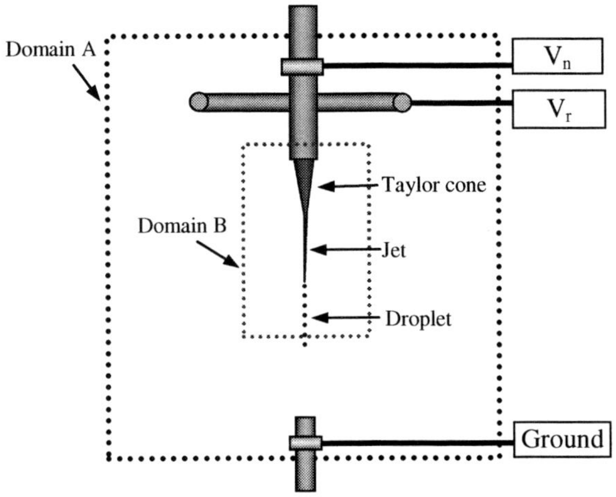

Figure 1: Schematics showing the experimental setup.

Ijsebaert [8] has previously made use of the ring electrode in his study of the production of aerosols, but the ring electric potential was set to a constant and was thought only to have the effect of focusing the spray. Here the ring acted as the means to control the electrical field strength nearby the tip of the nozzle where the Taylor Cone and jet will be formed.

3 NUMERICAL SIMULATION METHOD

3.1 Front Tracking Method

For the EHDA system, the system consist of two phase, a liquid phase which includes the Taylor Cone, jet and droplets, and a gas phase which is the ambient air. The novelty of the Front Tracking Method [7] is that the liquid-gas interface is considered to have a finite thickness of the order of the mesh size, instead of having zero thickness. In this transition zone, the physical properties of fluids change smoothly from the value on one side of the front to the value on the other side. Hence, there is no need to treat the two phases in different solution domains. There exists only one simulation domain, and a "one-fluid" formulation is adopted to describe the two-phase flow problem. The singular discontinuities on the interface such as interface tension and surface charge are distribute to the nearby control volume cell using Peskin interpolation for the solution of the governing equations.

Assuming the both liquid and gas phases to be incompressible, the conservation of mass equation can be simply written as

$$\nabla \cdot \vec{u} = 0, \tag{1}$$

where $\vec{u}$ is the velocity field.

The Navier-Stokes equation would need to be modified to take into account the additional stresses on the interface so that it and can be written as

$$\frac{\partial \rho \vec{u}}{\partial t} + \nabla \cdot \rho \vec{u}\vec{u} = -\nabla p + \nabla\left[\mu\left(\nabla \vec{u} + \nabla \vec{u}^T\right)\right] + \sigma^F \kappa \vec{n}\,\delta\left(\vec{x} - \vec{x}_f\right) + \rho \vec{g} + \vec{\sigma}^E, \tag{2}$$

where, ρ, p, μ, g and t are density, pressure, viscosity and gravitational acceleration and time, respectively. σ^F, κ and $\vec{n}$ denotes the interfacial tension coefficient, curvature and unit outward normal, respectively. $\delta(\vec{x} - \vec{x}_f)$ is the delta function where the value is one on the grid point where the interface lies and zero everywhere else. σ^E stands for the electric stress. Detail description of the numerical methods for solving the above governing equations will not be presented here, but can be found elsewhere [9].

3.2 Calculation of the electric field

The distribution of electric potential (ϕ) within the solution domain can be solved using the following equation as well as the electric field,

$$\nabla \cdot \varepsilon^* \nabla \phi = \rho^c \text{ and } \vec{E} = -\nabla \phi. \tag{3}$$

where ε the fluid electric permittivity and ρ^c is the space change density. ρ^c is non zero due to the distribution of electrical charge on the interface (q), and ε is non constant since it will change as it crosses the moving interface from the liquid phase to gas phase.

The electrical volume stress [10] in equation (2) can be calculated as,

$$\vec{\sigma}^E = -\frac{1}{2}\vec{E} \cdot \vec{E}\nabla \varepsilon + \rho^c \vec{E}. \tag{4}$$

3.3 Simulation strategy and boundary conditions

Accurate calculation of the electrical field strength near the tip of the nozzle is important to the success in modeling the EHDA process. Ideally, the calculation of the electrical field should be conducted in a large solution domain including the nozzle, ring electrode and the ground needle. However, it is not effective to calculate all governing equations for fluid flow and electric field in a domain big enough to encompass all these entities (domain A as shown in Figure 1). This is due to the fact that the electric field in the far field away from the nozzle will not be affected significantly by the presence of liquid jet and the detail flow field in the far field is not important to the EDHA process. Hence, we propose another modeling strategy. We first calculate the electric potential field only in large domain A. The nozzle electric potential and the ring electric potential are both taken into account, and the simulation geometry matches the actual production facilities. Once the electric potential field is calculated, the electric potential value on the boundary of the CFD simulation domain (Domain B as shown in Figure 1) is extracted from the calculated results, and applied to the CFD simulation as the boundary conditions for electric field calculation. The numerical tests have shown that if the electrical charge density on the Taylor Cone and jet in the CFD simulation is kept within a reasonable range, this method has negligible errors.

Figure 2 shows the small domain of an axis-symmetric cylindrical coordinate system for the detail CFD simulation of EHDA process. The length scale is normalized using the nozzle inner diameter. The bottom wall is the symmetry axis. The electric potential boundary conditions on the top, left and right wall are derived from the electric field calculation in the large domain A. The interfacial charge density is assumed to be constant for the model input, and is distributed to the control volume where the liquid gas boundary lies. Since there is no pre-simulation knowledge of the charge density on the liquid-gas interface, the charge density is estimated by trial and error and the comparison of the profiles of the Taylor Cone and jet obtained from the CFD simulations with those observed in the experiments.

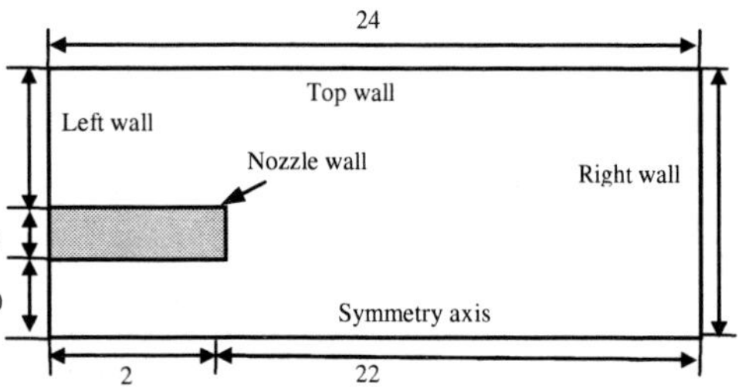

Figure 2: CFD simulation domain.

4 RESULTS AND DISCUSSION

4.1 Experimental observation and simulation on typical EDHA process

A typical case of Electrohydrodynamic Atomization of pure Dichloromethane is chose for this study. The electric potentials on the nozzle and ring electrodes are set to be 8.0 kV and 8.9 kV, respectively. The liquid flow rate is 6 ml/h. The interfacial charge density is determined to be 2.05×10^{-5} C/m^2 by fitting the simulated Taylor Cone and Jet profile to the experimental data. A Single Taylor Cone Single Jet mode was obtained from experiment and simulation result is also shown in Figure 3. The numerical simulation was able to replicate the major qualitative feature of the Taylor Cone and Jet reasonably well. The Taylor Cone is first formed near the tip of the nozzle. From the tip of the Taylor Cone, a jet is formed that has a diameter of around one order of magnitude smaller than the diameter of the nozzle. As the jet travels downward, surface instability sets in and droplets are then formed.

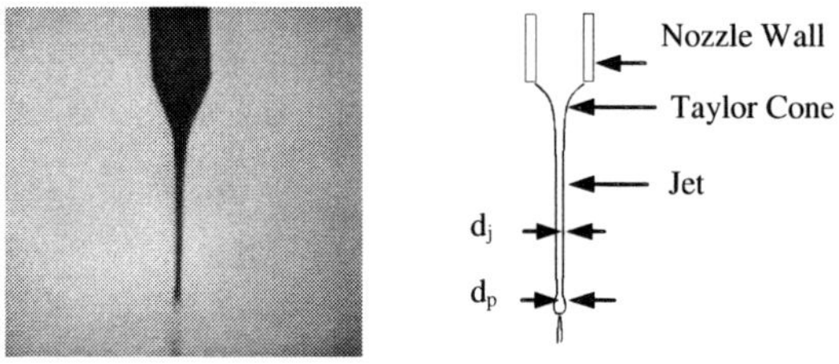

Figure 3: EHDA Taylor Cone and Jet observed in experiment (left) and predicted by simulation (right).

At the end of the jet where droplets are formed, CFD simulation has shown the formation of the unstable wave like structure before the breakup of the jet (Figure 4). This is similar to the formation of droplets in surface tension driven Rayleigh disintegration. Hohman [11] has reported the experimental observation of similar structure on the jet in EHDA. As the neck diameter decreases due to surface tension, the jet will break into droplets of similar size.

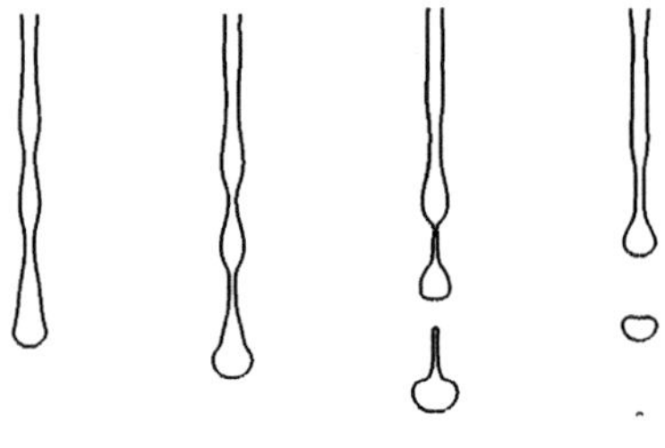

Figure 4: The evolution of unstable wave-like structure at the end of the liquid jet before pinch-off to form droplets. The interval between each frame is about 24 microseconds.

4.2 Effect of the ring electric potential on the Taylor Cone, jet and droplet formation

One of the most interesting observations made in the course of the study for droplet formation is the effect of the electric potential applied to the ring electrode on the EHDA process. In experiments, it was observed that in the Single Taylor Cone Single Jet mode, the cone angle decreases accordingly as the ring electric potential is increased.

Attempts have also been made to replicate such observation using CFD simulation. From the CFD simulation results, it is interesting to note that a constant charge density on the liquid-gas interface would not be able to give reasonable prediction on the changes of the Taylor cone angle. The simulation results would fit the experimental results best only when the interface charge density is reduced as the ring electric potential is reduced. This suggests that the ring electrode, although having no direct contact with the sprayed solution, have an effect on the liquid charging process.

As the ring electric potential changes, a change in the diameter of the jet is also observed in the CFD simulation results. As the ring electric potential is increased, there is an increase in the jet diameter as shown in Figure 5. The droplets size increases when the ring electric potential is increased.

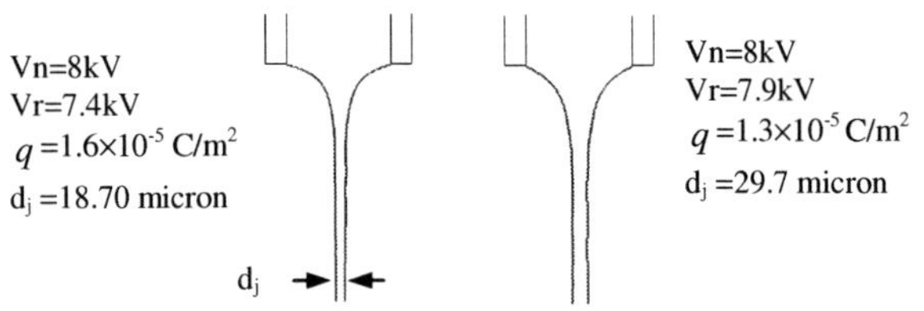

Figure 5: Changes in the jet diameter as the ring electric potential is increased. The liquid used is Dichloromethane and has a flow rate of 6ml/h. d_j represents the diameter of the jet.

5 CONCLUSION

The Electrohydrodynamic Atomization process has been studied through experiments and simulations. The ring electrode, acting as an additional electric potential field source, is found to be effective in controlling both the spray mode, and the droplet size. The Front Tracking has been shown to be a suitable approach for the simulation of the Electrohydrodynamic Atomization process, including Taylor Cone formation, jet formation and the droplet formation. Both the changes in the spray mode and the changes in droplet size can be reasonably predicted.

The simulation results also indicate that the interface charge density plays an important role in the EHDA process. It affects simulation accuracy on predicting the formation of Taylor Cone, jet and droplet. Hence, development of a first principle model to simulate the charging process on interface is highly essential in the future work.

ACKNOWLEDGEMENTS

This work was supported by Science and Engineering Research Council (SERC), Singapore and National University of Singapore under the Grant number R279-000-208-305. In addition, the authors would like to thank Prof. Wang Chi-Hwa and Prof. Kenneth A. Smith for their valuable technical discussion and suggestions on this research.

REFERENCES

[1] B. Vonnegut and R. L. Neubauer, "Production of monodisperse liquid particles by electrical atomization," *J. Colloid Sci.*, **7**, 616-622, 1952.

[2] J. W. Xie, L. K. Lim, Y. Y. Phua, J. S. Hua and C. H. Wang, "Electrohydrodynamic atomization for biodegradable polymeric particle production," *J. Colliod & Interface Sci.*, **302**, 103-112, 2006.

[3] M. Cloupeau and B. Prunet-Foch, "Electro-hydrodynamic Spraying Functioning Modes: A Critical Review," *J. Aerosol Sci.*, **25**, 1021-1036, 1994.

[4] A. J. Mestel, "Electrohydrodynamic stability of a slightly viscous jet," *J. Fluid Mech.*, **274**, 93-113, 1994.

[5] M. M. Hohman, M. Shin, G. C. Rutledge and M. P. Brenner, "Electrospinning and electrically forced jets. I. Stability Theory," *Phys. Fluids*, **12**, 2201-2220, 2001.

[6] O. Lastow and Balachandran, W., "Numerical simulation of electrodydrodynamic (EHD) atomization," *J. Electrostatic*, 64, 850-859, 2006.

[7] G. Tryggvason, B. Bunner, A. Esmaeeli, D. Juric, N. Al-Rawahi, W. Tauber, J. Han, S. Nas and Y.-J. Jan, "A front tracking method for computations of multiphase flow," *J. Compu. Physics*, **169**, 708-759, 2001.

[8] J. C. Ijsebaert, K. B. Geerse, J.C.M. Marijnissen, J. W. J. Lammers and P. Zanen, "Electrohydrodynamic atomization of drug solution for inhalation purposes," *J. Appl. Physiol.*, **91**, 2735-2741, 2001.

[9] J.S. Hua and J. Lou, "Numerical simulation of bubble rising in viscous liquid," *J. Comput. Phys.*, **222**, 769-795, 2007.

[10] D. A. Saville, "Electrohydrodynamics: the Taylor-Melcher leaky dielectric model," *Ann. Rev. Fluid Mech.*, **29**, 26-64, 1997

[11] M. M. Hohman, M. Shin, G. C. Rutledge and M. P. Brenner, "Electrospinning and electrically forced jets. II. Applications," *Physics of Fluid*, **12**, 2221-2236, 2001.

A microfluidic system based on capillary effect for high-throughput screening of yeast cells

M. Mirzaei, A. Queval and D. Juncker

BioMedical Engineering Department, McGill University
3775, rue University, Montréal, QC H3A 2B4
Canada

ABSTRACT

Microfluidic systems have been developed and used successfully for high-throughput cell-based assays. However, these systems require complicated peripherals and a computer to control the flow of liquid which limits their adoption by biologists. Here, we propose a simple-to-use high-throughput microfluidic system based on capillary effects for single cell entrapment, rapid exchange of reagents, and multiplexed assays. The system consists of (i) an autonomous capillary microfluidic system molded in PDMS, (ii) coverslips with patterned arrays of microwells sealed to the PDMS chip, (iii) a custom-built holder that can be mounted onto an inverted confocal microscope.

Keywords: High-throughput screening, microfluidic system, capillary effect, yeast

1 INTRODUCTION

Microscale science and microtechnology enable us to design, manufacture and create microfluidic systems with dimensions ranging from millimeters down to micrometers. Microfluidic systems are being increasingly used for cell biology since they provide small footprints yet offer complex functions and allow changing the fluidic environment of cells for high-throughput cell-based assays, such as high-content screening (HCS). HCS requires a special control of environmental parameters, delivery of multiple reagents, advanced microscopy, and multi-parameter readouts; using high magnification microscopy and consequently is expensive.

Microfluidic systems are promising for use in HCS because they can be used for screening multiple individual cells in a parallel, fast and flexible manner for different types of cells and conditions. For the fabrication of microfluidic devices, polydimethylsiloxane (PDMS) is widely used. It can be easily molded, it is optically transparent, and has low toxicity and high permeability to oxygen and CO_2 all of which make it suitable for cell culture purposes.

Microfluidic systems with orthogonally crossed sets of microchannels offer the opportunity for studying the interaction of large numbers of molecules with cells in a combinatorial manner. Khademhosseini *et al.* presented a soft lithographic method to fabricate multiphenotype cell arrays by capturing cells within an array of reversibly sealed microfluidic channels. In this study they designed a PDMS array of five parallel channels with independent inlets and a common outlet. Hence a combination of 25 different conditions can be tested [1]. The system is promising for HCS however it has some limitation. The channels are more than 200 μm wide, limiting imaging processes and a syringe pump was required to control the flow. The system is not compatible with high magnification imaging while experimenting. In another study Wang *et al.* reported the development of a multilayer elastomeric microfluidic array for the high-throughput cytotoxicity screening of mammalian cells [2]. This platform allowed cell-seeding as "rows" and combinatorial exposure of cells to toxins delivered along "columns". The system is versatile, but it is complicated because it needs to be connected to a multichannel pressure controller unit requiring manifolds, valves and multiple connections.

While each of these studies presents new concepts for improved microfluidic cell culture, the overall challenge of making a practical, cost-effective platform for cellular assays has yet to be solved. Most of these systems are complicated and need specific peripheral equipment and a computer to run.

Here, we present a setup with a microfluidic capillary system (μCS) [3], a coverslip with arrays of microwells and a custom built holder that allows real time high magnification imaging using a microscope. The system can be used for patterning different "rows" of living yeast cells, and exposing them to different "columns" of chemicals. The system facilitates studying the effects of drugs on the immobilized cells in microwells because it allows high magnification imaging while exchanging reagents.

2 PROTOCOLS

μCS were fabricated by standard microfabrication procedures by replicating a master wafer with photopatterned SU-8 into PDMS. The μCS used in this study had microchannels that were 50 μm wide, 10 μm heights, and 500 μm lengths with a 50 μm gap between adjacent microchannels. The microchannels were coated with 1 mg/ml bovine serum albumin (BSA) in phosphate buffered saline (PBS) prior to use.

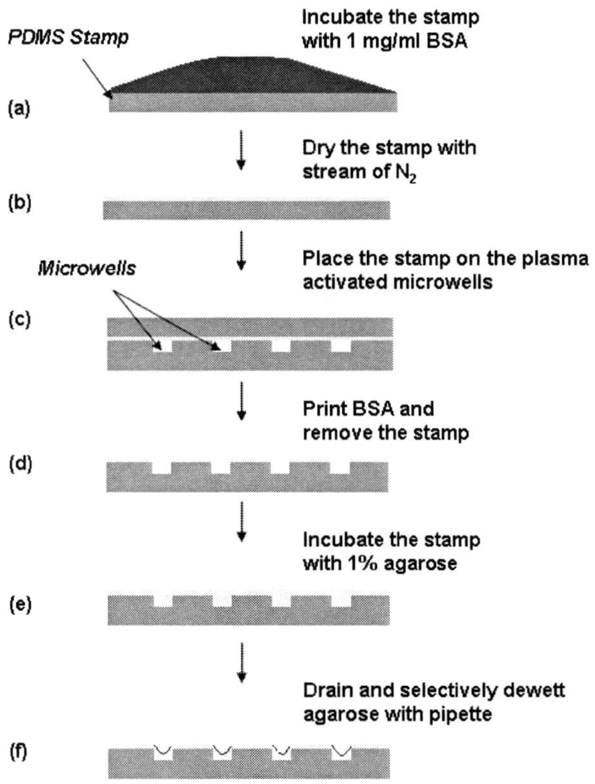

Figure1. Schematic illustration for the preparation of microwell arrays. The surface of the microwells was printed with 1mg/ml BSA to prevent adhesion of the cells on the substrate. Microwells were then selectively incubated with 1% agarose in distilled water, the solution drained followed by gelation of the agarose in the bottom of the microwell. The microwells can be stored in humid condition for several days' prior to use.

The microwells had diameters ranging from 10-50 μm and there is a 100 μm distance between each well. The microwells were formed by pressing a PDMS stamp with posts onto a coverslip coated with PDMS prepolymer which was cured in placed. Air plasma was used for 20-30 s to render microwell arrays and microchannels hydrophilic. Using microcontact printing BSA was selectively printed onto the ridges of the microwell arrays [4]. Microwells were filled with dilute agarose solution using selective dewetting. Gelation occurred after partial evaporation of the water so that the agarose formed only at the bottom of the wells (figure1).

The substrate was sealed to a μCS (with drilled inlet and outlet). Yeast cells were flowed through the microchannels and adhered selectively in the microwells. Remaining liquid was removed by applying a vacuum at the outlet for less than 10 seconds (figure 2).

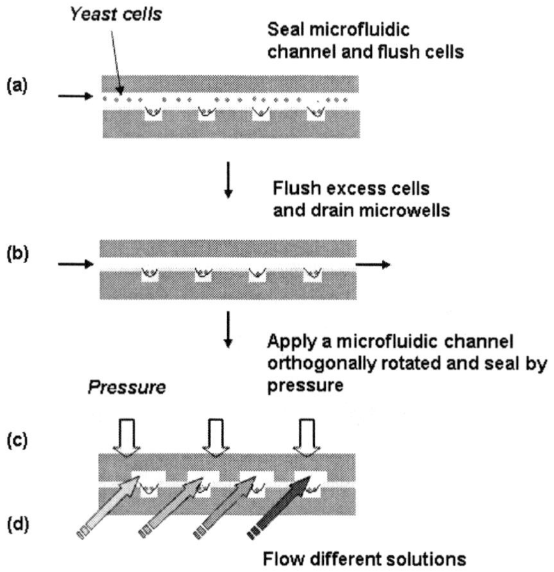

Figure 2. Schematic illustration of the selective cell deposition inside microwells and subsequent exposure to different solutions. (a) μCS was reversibly sealed on the microwells and cells were flowed through independent channels. (b) Cells lying outside of the microwells were flushed out by draining the liquid with a vacuum. Another μCS was placed orthogonally and sealed by applying light pressure. Different chemicals were delivered to different channels and flowed by capillary force.

Next the PDMS with µCS was detached, and a new chip was placed orthogonally over the microwells, so as to expose different cell "rows" to different "columns" of solutions.

3 RESULTS AND DISCUSSION

In this study we designed a microfluidic system consisting of a µCS, a coverslip with arrays of microwells and a custom built holder that allows real time high magnification imaging using a microscope.

According to the process described in figure 1, microwells selectively were filled with agarose (figure 3). Plasma treatment and hydrophilization of the microwells were critical for the selective dewetting and filling of all the microwells with very high yield. We found that BSA was necessary to prevent adsorption of agarose to the top surface during draining of the agarose solutions. If the agarose was left on the top surface, yeast cells would stick to it. BSA coating allowed complete removal of yeast from the channels as shown in figure 4. In addition complete removal of cells from chip facilitates orthogonal placement and sealing of the second µCS. Different solutions can then be flushed over the cells (figure 5).

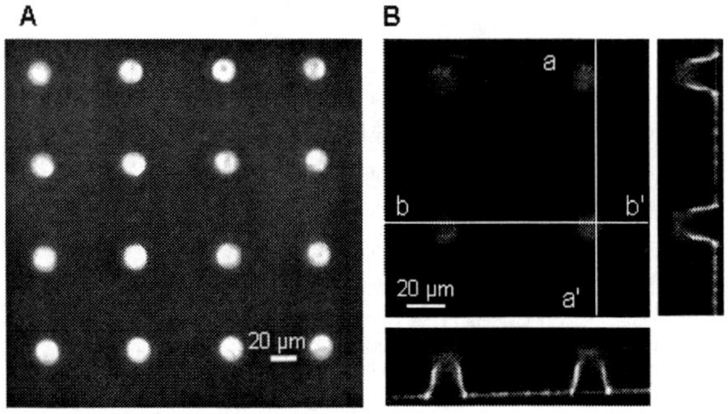

Figure 3. Selective coating of microwells with agarose. (A): microwells filled with 1% agarose solution mixed with trace amounts of fluorscein for visualization of the agarose at the bottom of the microwells. (B): confocal image of the bottom of the microwells. (aa', bb'): cross section of the microwells. Microwells with 17 µm depths which were only filled up to 1/3 with agarose.

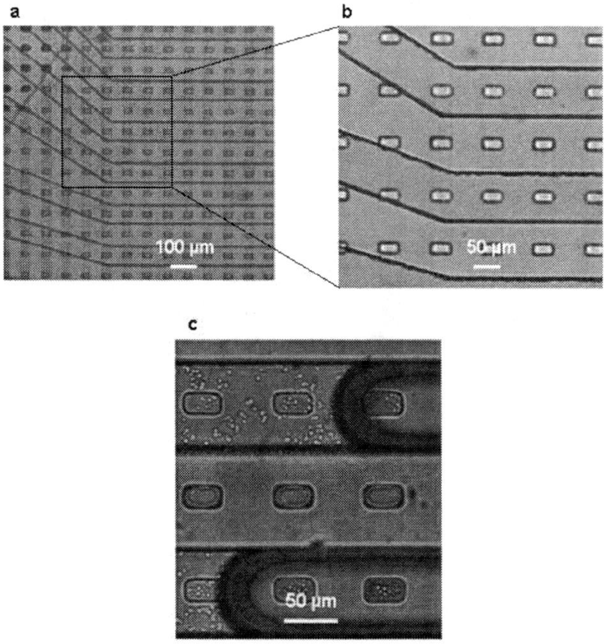

Figure 4. Microchannels sealed to microwell array for cell seeding. (a, b): arrays of microwells sealed to a capillary system. (c): microchannels are filled with cells suspension, after 10 min incubation the liquid was drained. Yeast cells seeded in the channel and were selectively removed by draining of the liquid. The yeast cells adhere to the agarose in the well but not to the BSA on the surface of the substrate.

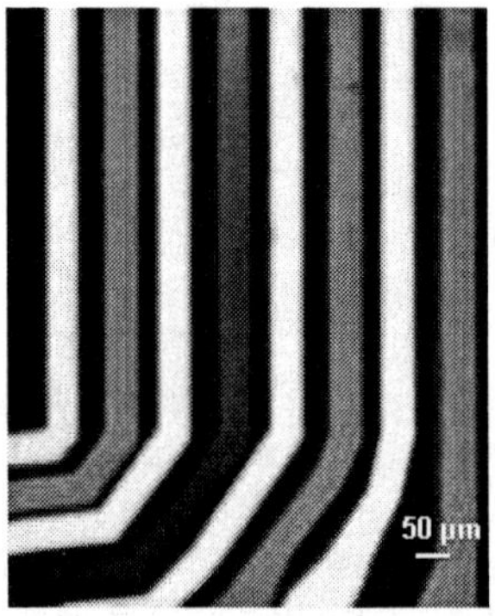

Figure 5. Delivery of different "columns" of the chemicals to different "rows" of the cells. Overview of microchannels filled with Fluorescein and Rhodamine.

4 CONCLUSION

In this study we introduced a simple-to-use high-throughput microfluidic system based on capillary effects for yeast cell entrapment, rapid exchange of reagents, and multiplexed assays. Using this system we were able to expose the cells to different solutions while simultaneously imaging them with a microscope. Our chips feature 16 μCS which by crossing can create 256 combinations if 16 different cell lines and 16 different solutions are used. Because our system is compatible with real time high magnification microscopy, it is well suited for HCS and using yeast cells.

5 ACKNOWLEDGMENT

This work was supported by research grants from NSERC and FQRNT. M. M. acknowledges a Ph.D. Scholarship from the Iranian Ministry of Health and Education.

REFERENCES

1. Khademhosseini A, Yeh J, Eng G, Karp J, Kaji H, Borenstein J, et al. Cell docking inside microwells within reversibly sealed microfluidic channels for fabricating multiphenotype cell arrays. Lab on a chip 2005 Dec;5(12):1380-1386.
2. Wang Z, Kim MC, Marquez M, Thorsen T. High-density microfluidic arrays for cell cytotoxicity analysis. Lab on a chip 2007 Jun;7(6):740-745.
3. Juncker D, Schmid H, Drechsler U, Wolf H, Wolf M, Michel B, et al. Autonomous microfluidic capillary system. Analytical chemistry 2002 Dec 15;74(24):6139-6144.
4. von Philipsborn AC, Lang S, Bernard A, Loeschinger J, David C, Lehnert D, et al. Microcontact printing of axon guidance molecules for generation of graded patterns. Nature protocols 2006;1(3):1322-1328.

Dielectrophoretic characterization and separation of metastatic variants of small cell lung cancer cells

A. Menachery*, S. Chappell**, R. J. Errington***, D. Morris*, P.J. Smith**, M. Wiltshire**, E. Furon**, J. Burt*

* School of Electronic Engineering, Bangor University, Bangor, LL57 1UT, UK
** Department of Pathology, School of Medicine, Cardiff University, Cardiff, CF14 4XN, UK
***Department of Medical Biochemistry and Immunology, School of Medicine, Cardiff University, Cardiff, CF14 4XN.
Email-a.menachery@bangor.ac.uk

ABSTRACT

In this study, the ability of Dielectrophoresis (DEP) to recognize phenotypic variants of small cell lung cancer (SCLC) cells has been tested. The phenotypic variations of the SCLC cells are associated with the cell-surface decorations, which in turn, provide the cells with anti-adhesive properties allowing detachment from the primary tumour and the subsequent metastatic spread of such tumours. Dielectrophoretic separation chambers have been used to sort the phenotypes and so form a cytometric tool for the analysis, identification and monitoring of tumour heterogeneity.

Keywords: Dielectrophoresis, Electrokinetics, Cancer Cells, Cell Membrane Capacitance

1 INTRODUCTION

Dielectrophoresis (DEP) has been shown to be a useful technique for the manipulation of particulate matter and is also being widely used to understand the dielectric properties of cell suspensions. DEP has been commonly used as a cell sorter and one of the main advantages that it provides, is the ability to manipulate hundreds of particles in parallel without the need to fluorescently label them. DEP based flow through systems allow us to discriminate between particles with very subtle differences in their dielectric properties. Several studies have been undertaken to determine a role for DEP-based devices in cancer research [1]. There is accumulating evidence that changes in the physicochemical properties of the cell surface, including charge, could contribute to the efficiency of the metastatic process. More recently DEP has been used to differentiate drug accumulation phenotypes in breast cancer cell lines using cytoplasmic conductivity [2], so opening possibilities in the area of drug discovery where DEP devices may be used to determine the functional impact of drug candidates.

Changes in the expression of polysialic acid (PSA) polyanionic decorated cell surface molecules have been increasingly associated with the dissemination of cells from the primary tumour mass [3]. By increasing expression of PSA and its associated negative charge, the adhesiveness of the cell is reduced, which promotes detachment from the primary tumour and hence metastatic spread. The pleiotropic effects of the neural cell adhesion molecule NCAM, a major target molecule for PSA decoration, appear to reflect the ability of NCAM to regulate membrane-membrane contact required to initiate specific interactions between other molecules. Indeed when NCAM with a low PSA content is expressed, adhesion is increased and contact-dependent events are triggered. Our biological hypothesis is that the phenotypic variations of the SCLC cells are associated with the extent of cell-surface decoration provided by the α-2, 8-linked polysialic acid (PSA) and the neural cell adhesion molecule (NCAM). Here we have used dielectric properties to investigate changes in DEP properties associated with phenotypic variants of small cell lung cancer cells (NCI H69) which we call H69-adherent and H69-suspension variants.

2 THEORY AND SIMULATIONS

Dielectrophoresis can be defined as the movement of polarisable particles in the presence of non uniform electric fields. Taking the simplistic view of a cell as a spherical homogeneous particle, the DEP force on the cell is given by

$$F_{DEP} = 2\pi\varepsilon_m r^3 Re\,(CM)\,\nabla E^2 \qquad (1)$$

Where r is the radius of the cell, ε_m is the absolute permittivity of the medium, E is the electric field acting on a cell and CM is the Clausius-Mossotti factor, which describes the Maxwell-Wagner relaxation that occurs at the interfaces between dissimilar dielectrics such as the particle and medium and is given by

$$CM = \frac{\varepsilon_p^* - \varepsilon_m^*}{\varepsilon_p^* + 2\varepsilon_m^*} \qquad (2)$$

Where ε_p^* and ε_m^* are the effective complex permittivities of the cell and the medium respectively.

The Clausius-Mossotti (*CM*) factor is a measure of the effective polarisability of the cell and is distinctive of a particle being dependent on the cell contents and morphology. *CM* is a frequency dependent variable and by changing the frequency of the electric field it can exhibit shifts from a negative to a positive value. This, in turn, causes the direction of the dielectrophoretic force experienced by a particle to change direction. If the real part of *CM* is positive then particles move towards the regions of high electric field experiencing what is termed positive DEP. In negative DEP, particles move way from high field regions towards regions of low field intensity.

For cells that exhibits a change from negative to positive DEP as frequency of the electric field is increased, the change in direction occurs at a single frequency known as the cross-over frequency (f_{cr}). At f_{cr}, the dielectric properties of the cell equal those of the medium and so the cell appears transparent to the externally applied electric field and experiences zero DEP force. fcr can be approximated to [4]:

$$f_{cr} = \frac{1}{2\pi}\sqrt{\frac{(\sigma_p - \sigma_m)}{(\varepsilon_p - \varepsilon_m)}\frac{(\sigma_p + 2\sigma_m)}{(\varepsilon_p + 2\varepsilon_m)}} \quad (3)$$

Where σ_p, σ_m and ε_p, ε_m are the conductivities and permittivities of the particle and the medium respectively. When DEP is used in combination with fluid flow, additional forces of drag, buoyancy have to be considered. The drag on a spherical particle is given by:

$$F_{Drag} = 6\pi\eta r(u - v) \quad (4)$$

Where η is the viscosity of the medium, u is the fluid velocity and v is the particle velocity. Since cells are denser than their aqueous suspending media, the buoyancy force is given by:

$$F_{By} = \frac{4}{3}\pi r^3(\rho_p - \rho_m)g \quad (5)$$

Where ρ_p and ρ_m are the densities of the particle and medium respectively and g is the gravitational constant.

Considering a two dimensional model with stable forces for computational simplicity, the steady state velocity in the horizontal (x) and vertical (y) directions are given by [5]:

$$v_x = \frac{F_{DEPx}}{6\pi\eta r} + u \ , \quad v_y = \frac{F_{DEPy}}{6\pi\eta r} - \frac{2r^2}{9\eta}(\rho_p - \rho_m)g \quad (6)$$

To obtain a better understanding of the influence of the influence of fluid velocity on the dielectrophoretic capture of cells on microelectrodes, two dimensional simulations of the fluid and particle velocities within the microfabricated dielectrophoresis chambers used in this work were created using Comsol Multiphysics 3.2. Within the devices the fluid flow channel had a height of 150μm with interdigitated electrodes laid out on the bottom surface of the chamber with a width and interelectrode separation each of 20μm. The electrodes were energised with sinusoidal voltages such that the potential difference between adjacent electrodes had a peak value of 6V. Within the simulations *CM* was set to a value of 0.5, allowing particles to experience positive DEP towards the high electric field regions around the edges of the electrodes. Electric field simulations, an example of which is shown in figure 1, predict very localized high field intensities at the electrode edges with a rapid reduction in intensity with increasing distance from the electrode edge such a field pattern gives rise to a strong dielectrophoretic force towards the electrode edges.

The arrows in figure 1 indicate the direction and magnitude of fluid flow along the channel and exhibit a classic laminar-type flow profile. Particle trajectories are indicated by lines starting on the left hand side of the figures and have been calculated using equation 6. The simulations show that for a low fluid velocity of 1μm/s particles up to approximately 75μm above the electrodes experience a net DEP-dominated vertical force towards the electrodes. Simulations at 10μm/s, shown in figure 1, reveal the DEP-dominated vertical force extends only to a height of 60μm with particles above this height typically unaffected by the electric field. The observations from these and similar simulations have been used to optimize the experimental parameters such as fluid velocity, electrode geometry and energising voltage for the selective capture and enrichment of cells.

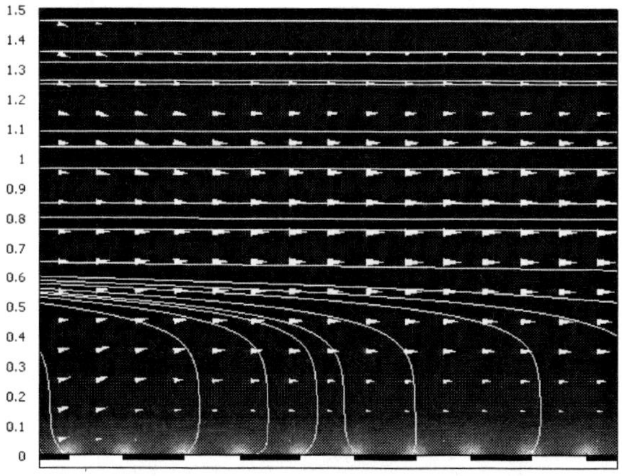

Figure 1: Particle trajectories at a fluid flow rate of 10 microns per second

3 EXPERIMENTAL

3.1 Cell Preparation

Small cell lung cancer (SCLC) cells (NCI-H69) [6] were grown in RPMI-1640 media (GIBCO) with 10% FBS under standard culture conditions at 37°C, 5% CO_2 in a humidified incubator. NCI-H69 cells grow as loose aggregates in suspension culture, but consistently express a sub-population (< 5%) of variants that show substrate adherence. A simple enrichment protocol for substrate adherence was used to generate an adherent phenotypic variant (H69-adherent).

Cells were washed and resuspended three times in conductivity buffer prior to use for DEP. The conductivity buffer (8.6% sucrose, 0.3% dextrose and 0.1% BSA in water) was calibrated at room temperature against a 1413uS/cm standard solution (Hanna instruments, Hungary), to 150μS/cm by adding PBS.

3.2 Devices and Methods

Quadrupole electrodes were used to dielectrophoretically characterise the two phenotypic variants of SCLC cells used in this work. These electrodes, patterned in an 80nm gold and 5nm chromium film on a glass substrate using photolithography, consisted of 4 electrodes with semicircular tips arranged at 90° to each other with an interelectrode separation of 150μm between opposite electrodes..

Flow experiments were undertaken in a dielectrophoretic separation chamber similar to that shown in figure 2, consisting of a glass substrate onto which approx $2cm^2$ array of interdigitated microelectrodes with a 20μm width and spacing was photolithographically fabricated. The channel structure was fabricated from a 150μm thick adhesive backed polymer cut to shape using femtosecond laser micromachining [6]. An upper polycarbonate plate was used to encapsulate the channels. After assembling the substrate, channel forming layer and upper plate the device was passed through a heated roller to ensure good bonding between the layers. Additionally, the upper plate contained 4 drilled holes over which NanoPort (Upchurch Scientific) fluidic interconnect ports were bonded to form microfluidic inlet and outlets to the separation chamber. Fluid flow through the chamber was controlled using a peristaltic pump (Gilson Minipuls 3).

Figure 2: Dielectrophoretic fluidic separation chamber.

3.3 Dielectrophoresis Experiments

SCLC Cells of each phenotype were characterised using the quadrupole electrodes. The electrodes were energised using sinusoidal voltages of 3Vpp with a 180° phase difference between adjacent electrodes. Cells were characterised to identify the crossover frequency for each phenotype by observing the direction of dielectrophoresis for electric field frequencies ranging from 1kHz to 10MHz. The crossover frequency was measured by finding the frequency at which the cells exhibited zero dielectrophoretic motion.

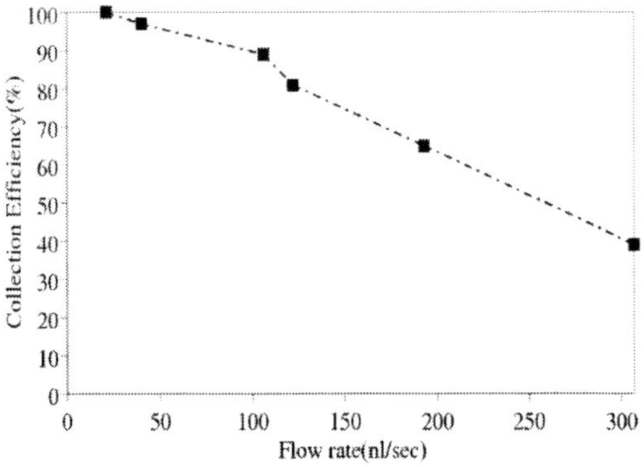

Figure 3: Effect of flow rate on collection efficiency.

Experiments to selectively collect and enrich samples of mixed phenotype were undertaken using the dielectrophoretic separation chambers. Prior to these experiments, the chambers were characterised for collection efficiency at different fluid flow velocities. The flow rate to achieve a desired level of cell collection was determined by fixing the electrode voltage and frequency and adjusting the fluid velocity to achieve good capture of a sample of adherent phenotype cells on the electrodes. Figure 3 shows the influence of fluid flow rate on the collection efficiency of the chambers.

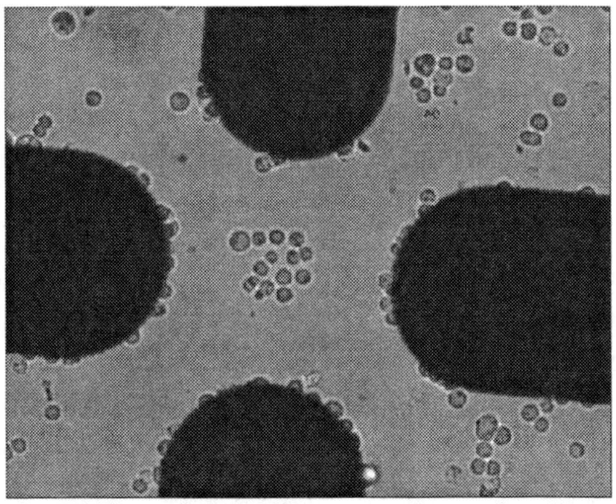

Figure 4: Differential DEP separation of adherent and suspension cells from a mixed population of SCLC cells.

4 RESULTS AND DISCUSSION

Two SCLC cell phenotypes were studied, one an adherent line and the other a line showing predominately suspension characteristics. Using the quadrupole electrodes the distribution of crossover frequencies for each phenotype was measured. For the adherent line observations of >50 cells found the crossover frequency ranged from 30kHz to 40kHz. For the suspension phenotype a larger range of crossover frequencies was observed ranging from 30kHz to 60kHz. Additionally it was observed that a proportion of the suspension phenotype exhibited adherent-like properties and that within a population of suspension cells, choosing an electric field frequency of 40kHz it was possible to selectively separate cells displaying a adherent-like properties. Such separation is shown in figure 4 where a sample of suspension cells is exposed to an electric field of 40kHz. Adherent-like cells are seen dielectrophoretically collected at the edge of the electrodes while suspension-like cells remain uncollected.

Using the separation chamber a mixed population of adherent and suspension cells were fractionated by flowing the cells through the chamber at a flow rate of 50nl/s while applying a 3Vpp 40kHz voltage to the chamber electrodes. At this frequency cells showing adherent characteristics would be expected to collect on the electrodes while suspension-like cells would experience a negative dielectrophoretic force and be carried through the chamber with the fluid flow. Cells exiting the chamber while the voltage was applied to the electrodes were collected. When the entire sample had been passed through the chamber the voltage was removed and additional suspending medium was passed through the chamber. This allowed cells trapped on the electrodes to be collected as a separate fraction. Both fractions were cultured in 96 well plates for 1 day. Post culture observation of the fractions revealed that the dielectrophoretically collected fraction exhibited adherent characteristics while the second fraction exhibited suspension characteristics. This experimental result supports the hypothesis that cells within the suspension population display phenotypic variations ranging from adherent-like to suspension like characteristics.

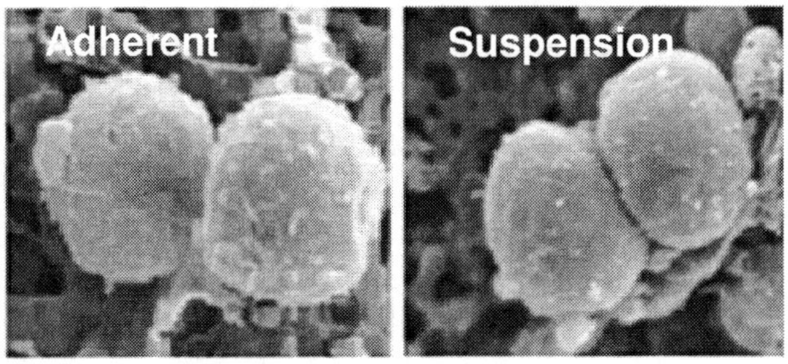

Figure 5: Cell Electron Micrographs shows greater membrane texturing in the adherent phenotype

For viable cells equation 3 can be approximated to [4]

$$f_{cr} = \frac{\sqrt{2}\sigma_m}{2\pi r C_{mem}} \qquad (7)$$

Where r is the cell radius and C_{mem} is the cell membrane capacitance per unit area. Microscope measurements give the radius of both cell types to be 10μm ± 1μm. Using the range of crossover frequencies measured for each cell gives membrane capacitances for the adherent cells of 17-22mF/m^2 while for the suspension cells the membrane capacitance is 11-22mF/m^2. Previous studies [7] have shown that membrane capacitance mainly depends on membrane surface topography. Extra surface features can increase the membrane surface area and hence the membrane capacitance. Scanning electron micrograph studies, similar to those shown in figure 5, of the cell phenotypes used in this work reveal that cells exhibiting adherent characteristics have a greater degree of surface decoration than those exhibiting suspension-like characteristics. This in turn leads to a larger membrane capacitance and subsequently a lower range of crossover frequencies. It is our biological hypothesis that the phenotypic variations of the SCLC cells studied here are associated with the cell surface decoration provided by the 2,8 linked polysialic acid (PSA) and the neural cell adhesion molecule (NCAM).

5 CONCLUSION

This study of H69 phenotypic variants has shown that dielectrophoresis has the ability to selectively sort cell types with subtle phenotypic differences. Electrokinetic studies have shown that within a single phenotype differences can be identified using dielectrophoretic examination leading to the possibility of cytometric devices for analysis, identification and monitoring of tumour heterogeneity.

REFERENCES

[1] P.R.C. Gascoyne, X. Wang, Y. Huang, F.F. Becker, "Dielectrophoretic separation of cancer cells from blood" IEEE Transactions on Industry Applications **33**, 670, 1997

[2] H. M. Coley, F. H. Labeed, H. Thomas, M. P. Hughes "Biophysical characterization of MDR breast cancer cell lines reveals the cytoplasm is critical in determining drug sensitivity" ,Biochimica et Biophysica Acta, **1770**, 601, 2007

[3] E. Ruoslahti and F.G. Giancotti, "Integrins and tumour cell dissemination" Cancer Cells **1** ,119 , 1989

[4] R Pethig and M Talary, "Dielectrophoretic detection of membrane morphology changes in Jurkat T-cells undergoing etoposide-induced apoptosis", Nanobiotechnology IET, **1** 2,2007

[5] D Holmes, N.G. Green, H. Morgan, " Microdevices for dielectrophoretic flow - through cell separation", Engineering in Medicine and Biology Magazine,IEEE **22**, 85 ,2003

[6] J P H Burt, A D Goater, A Menachery, R Pethig and N H Rizvi, "Development of microtitre plates for electrokinetic assays" J. Micromech. Microeng. **17**, 250, 2007

[7] X. Wang , F. F. Becker, P. R.C. Gascoyne, " Membrane dielectric changes indicate induced apoptosis in HL-60 cells more sensitively than surface phosphatidylserine expression or DNA fragmentation", Biochimica et Biophysica Acta **1564**, 412, 2002

External Field Induced Non-Uniform Growth of Micron Composite Materials in Insulating Liquid

S. Tada[*], Y. Shen[**] and Z.Y. Qiu[***]

[*]National Defense Academy, Kanagawa 239-8686, Japan, stada@nda.ac.jp
[**]Zhengzhou University of Light Industry, Zhengzhou 450002, China, hnshenyan@126.com
[***]The City College of City University of New York, NY 10031, USA, ltscltsc@hotmail.com

ABSTRACT

In the present study, motion and aggregation of polarizable particles in a suspension were simulated numerically when the suspension was exposed to a spatially non-uniform AC electric field generated by the electrode array in a rotating flow chamber. Particles are dielectric spheres of the diameter d of 89.6 μm and density ρ_p of 3.75 g/cm^3, and suspending liquid is corn oil. The suspending liquid flow in the chamber is treated as the fully developed laminar flow. The Langevin equation was applied to describe Newtonian dynamics of the suspension under combined electric and flow fields. The simulations reproduced most of our published data about motion and clustering of particles, and also provide more details the experiments can not provide about involved interactions of electrostatics and hydrodynamics, and their effects on motion and clustering of the particles, which is important to designs of composite nano/micron materials with help of external fields.

Keywords: non-uniform growth, Langevin equation, particle clustering, AC electric field

1 INTRODUCTION

Electric manipulation of fluid flow and particle motion are currently becoming major techniques for separation, filtration, orientation and sorting of particles and cells in the fields of nano- and bio-technologies. The development of a new generation of electrohydrodynamic apparatuses is associated with the advent of modern electronics making feasible the wide use of AC field rather than DC field. The use of AC in controlling suspension flow has several definite advantages; 1) application of an AC field employs polarization forces which is independent of particle charges, 2) the evolution of the AC-field driven phenomena, such as particle cluster formation, can be manipulated by varying electric field strength and frequency. In the present study, we focused on the motion and aggregation of polarizable particles suspended in insulating liquid under a spatially non-uniform AC electric field. Numerical simulations based on the single particle model of dielectrophoresis were performed to examine the kinetics of suspension flow exposed to a strong non-uniform AC field considering dielectrophoresis force, dipole-dipole interactions and hydrodynamics interaction.

2 MODEL DESCRIPTION

2.1 Governing Equation

The Langevin equation as governing equation is used to describe Newtonian dynamics of suspension of dielectric particles of diameter d in the insulating liquid exposed to an AC electric field shown in Figure 1;

$$m\frac{d^2\boldsymbol{r}_i}{dt^2} = \boldsymbol{F}_i - 3\pi\eta d\frac{d\boldsymbol{r}_i}{dt} + \boldsymbol{R}_i(t) + \frac{4}{3}\pi(\rho_p - \rho_f)\left(\frac{d}{2}\right)^3 \boldsymbol{g} \quad (1)$$

where m is the particle mass, $\boldsymbol{F}_i$ the external forces including all electric forces on particle i. The second term in the RHS is the Stokes' drag force, $\boldsymbol{R}_i$ is a Brownian force. In the present calculation, Brownian force is negligible compared with strong electric interactions. The last term is the time-dependent gravitational force due to rotation of the chamber. $\boldsymbol{F}_i$ is given by the expression

$$\boldsymbol{F}_i = \sum_{i \neq j} \boldsymbol{f}_{ij} + \boldsymbol{h}_i \quad (2)$$

where $\boldsymbol{f}_{ij}$ is the non-zero time-average Columbic force acting on particle i by particle j, and $\boldsymbol{h}_i$ is the non-zero time-average dielectrophoretic force due to the gradient of the applied AC electric field. The $\boldsymbol{f}_{ij}$ is obtained from

$$\boldsymbol{f}_{ij} = -\frac{\partial \Phi_{ij}}{\partial r}\boldsymbol{e}_r - \frac{1}{r}\frac{\partial \Phi_{ij}}{\partial \theta}\boldsymbol{e}_\theta \quad (3)$$

where, Φ_{ij} is the dipole-dipole interaction between i and j-th particles

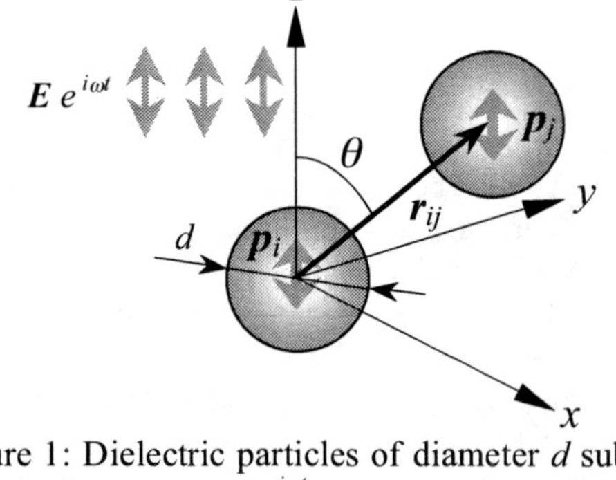

Figure 1: Dielectric particles of diameter d subjected to the AC electric field $Ee^{i\omega t}$.

$$\Phi_{ij} = \frac{1}{4\pi\varepsilon_0\varepsilon_f}\left(\frac{\boldsymbol{p}_i\cdot\boldsymbol{p}_j}{r_{ij}^{\,3}} - \frac{3(\boldsymbol{p}_i\cdot\boldsymbol{r}_{ij})(\boldsymbol{p}_j\cdot\boldsymbol{r}_{ij})}{r_{ij}^{\,5}}\right), \quad \boldsymbol{r}_{ij} = \boldsymbol{r}_i - \boldsymbol{r}_j \tag{4}$$

where the dipole moment of i-th particle, $\boldsymbol{p}_i$, is defined as

$$\boldsymbol{p}_i = 4\pi\varepsilon_0\varepsilon_f\left(\frac{d}{2}\right)^3 \mathrm{Re}(\beta)\boldsymbol{E}_{rms}(\boldsymbol{r}_i) \tag{5}$$

and the $\boldsymbol{h}_i$ is

$$\boldsymbol{h}_i = 2\pi\varepsilon_0\varepsilon_f\left(\frac{d}{2}\right)^3 \mathrm{Re}(\beta)\nabla\boldsymbol{E}_{rms}(\boldsymbol{r}_i)^2 \tag{6}$$

where $\mathrm{Re}(\beta)$ is the real part of the dielectric mismatch factor between particles and corn oil. $\boldsymbol{E}$, ε_0, ε_f are the AC electric field amplitude, vacuum permittivity, the relative dielectric constant of corn oil, respectively.

2.2 Numerical Simulation setup

A schematic illustration of the flow chamber is shown in Fig.2. The parallel plate chamber measures 60(width) × 120(length) × 3.0(height) mm and is driven to rotate around the center axis parallel to the longitudinal direction (x axis; Fig.2 A) at 4 rpm to prevent particles from setting down onto the "bottom" (in Fig.2 A). In the flow channel, linear, flat electrodes (11 cm long, 1.6 mm wide) are arranged in parallel at 2 mm intervals along the flow direction (Fig.2 B, C). The mean velocity of laminar flow of the insulating liquid (density; ρ_f=0.92 g/cm^3) in the channel was 2 mm/s. To save the computation resources, simulations were performed in a unit cell as indicated "calculation domain" (Fig.1). In this unit cell, since the flow of suspension is fully developed, the velocity profile in the flow channel has only a single component, velocity in the x direction U, expressed as the function of the z coordinate only

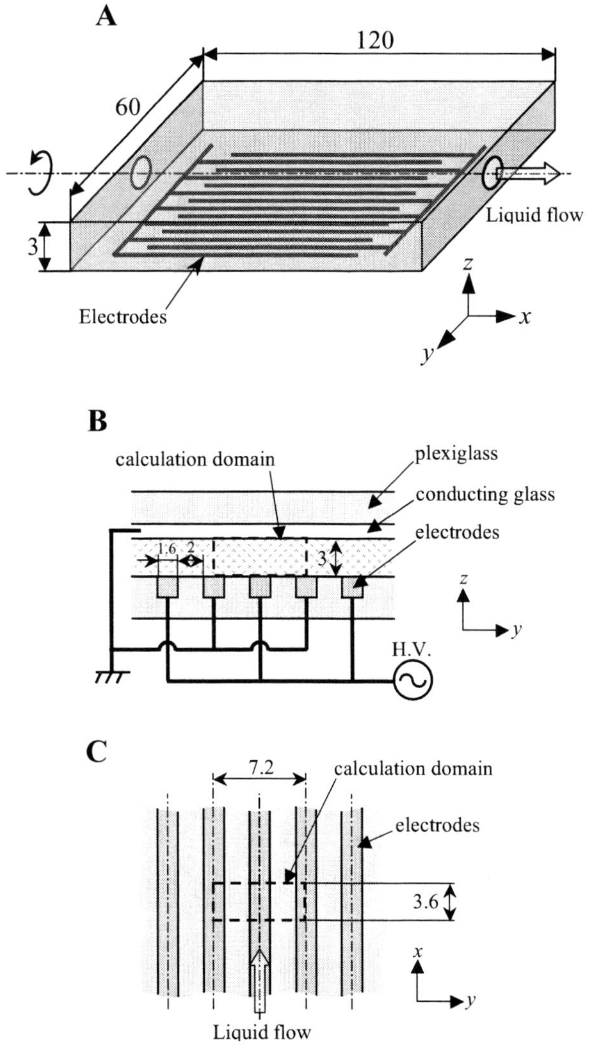

Figure 2: A. The parallel plate channel equipped with an electrode array, B. Cross-sectional view of the channel, C. Top view of the electrode array in the channel.

$$U = 6\frac{Q}{A}z^*(1 - z^*) \tag{7}$$

where Q, A, z^* is the volumetric flow rate of the suspension, the cross-sectional area of the channel, and the z coordinate normalized to the channel height. Therefore, the time derivative of the field-induced displacement of particle (the second term in the RHS of Eq.(1)) is expressed as

$$\frac{d\boldsymbol{r}_i}{dt} = \boldsymbol{u}_i - U\boldsymbol{e}_x \tag{8}$$

where $\boldsymbol{u}_i$ is the velocity of particle i and $\boldsymbol{e}_x$ is the unit vector in the x direction.

For the governing equation, a periodic B.C. was applied over x-z planes on both lateral side of the unit cell perpendicular to the channel top and bottom faces. In the periodic B.C.s, the cubical simulation box is replicated throughout space between walls to form an infinite lattice. At the inlet boundary, a particle leaving the unit cell through the outlet boundary was replaced by another particle that simultaneously enter the unit cell through the inlet cross-section at a position with randomly taken y and z coordinates. Once a particle reached the electrode edges, it was considered to be trapped and its further displacement away from the bottom surface was not allowed. To maintain the volume fraction of particles in the liquid phase, a particle was allowed to enter the flow unit through the

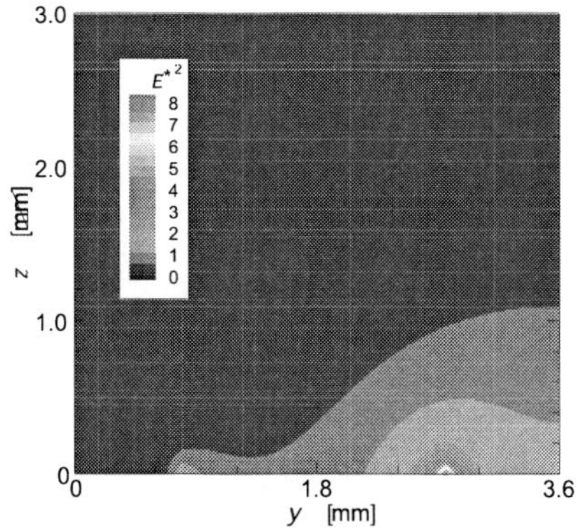

Figure 3: Electric field, $(E^*)^2$, distribution in the y-z plane. The symbol "*" means that values are dimensionless. Note that the actual computational region was $0<y<3.6$ mm, $-0.8<z<3.0$ mm.

inlet boundary at a position with randomly taken y and z coordinates as soon as a particle was trapped at the electrode edges. For the electric field distribution in the flow channel, since there is no variation along the flow direction, the 2D electric field approximation over a *y-z* plane was adopted. At the beginning of the simulation, the Laplace equation of the 2D electric potential distribution in the *y-z* plane of the channel cross section was solved using a finite difference method. The finite difference mesh sizes of the electric field calculation in *y* and *z* direction were $\Delta y = \Delta z < d/2$. As the electric field is symmetric about the center axis (*z* axis) of the calculation domain, half the symmetric area was computed (Fig.3). Forces in Eq.(2) were calculated using the converged electric field distribution. Initially, all simulations was conducted for the fixed number of particles which we assumed were at random in the cubic box. The size of this cubic box was determined by the volume fraction of particles.

In order to perform the simulation, we have developed an integrator, a modification of BBK method [1], for the integration of Eq.(1). The shifted-force version of the usual electrostatic potential was used for the wall-particle and interparticle force calculations to prevent particles from penetrating one another or channel walls. The shifted-force potential and its first derivative go to zero continuously at the cutoff radii of r_c=0.6d and 1.1d, for wall-particle and interparticle repulsive interactions, respectively. For the dipole-dipole interaction energy, the value of cutoff r_c=6.0d which ensures the value of potential reduces less than 1×10^{-8} of its value at the particle surface, was applied. The periodic B.C. was also applied to particles placed by the periodic boundary to eliminate surface effect from the computation

3 RESULTS AND DISCUSSION

Figure 4 shows the snapshots of time course of transient behavior of particles at time t=40, 120 200 280 s. The top panel in Fig.4 shows particles distribution viewed through the channel cross-section, whereas the bottom panel is viewed from the top. The shaded regions indicate the electrodes. The AC electric field applied was V_{rms}=3 kV with frequency 100 Hz, thus the real part of the particle polarization is 0.34[2]. The particles volume fraction of the

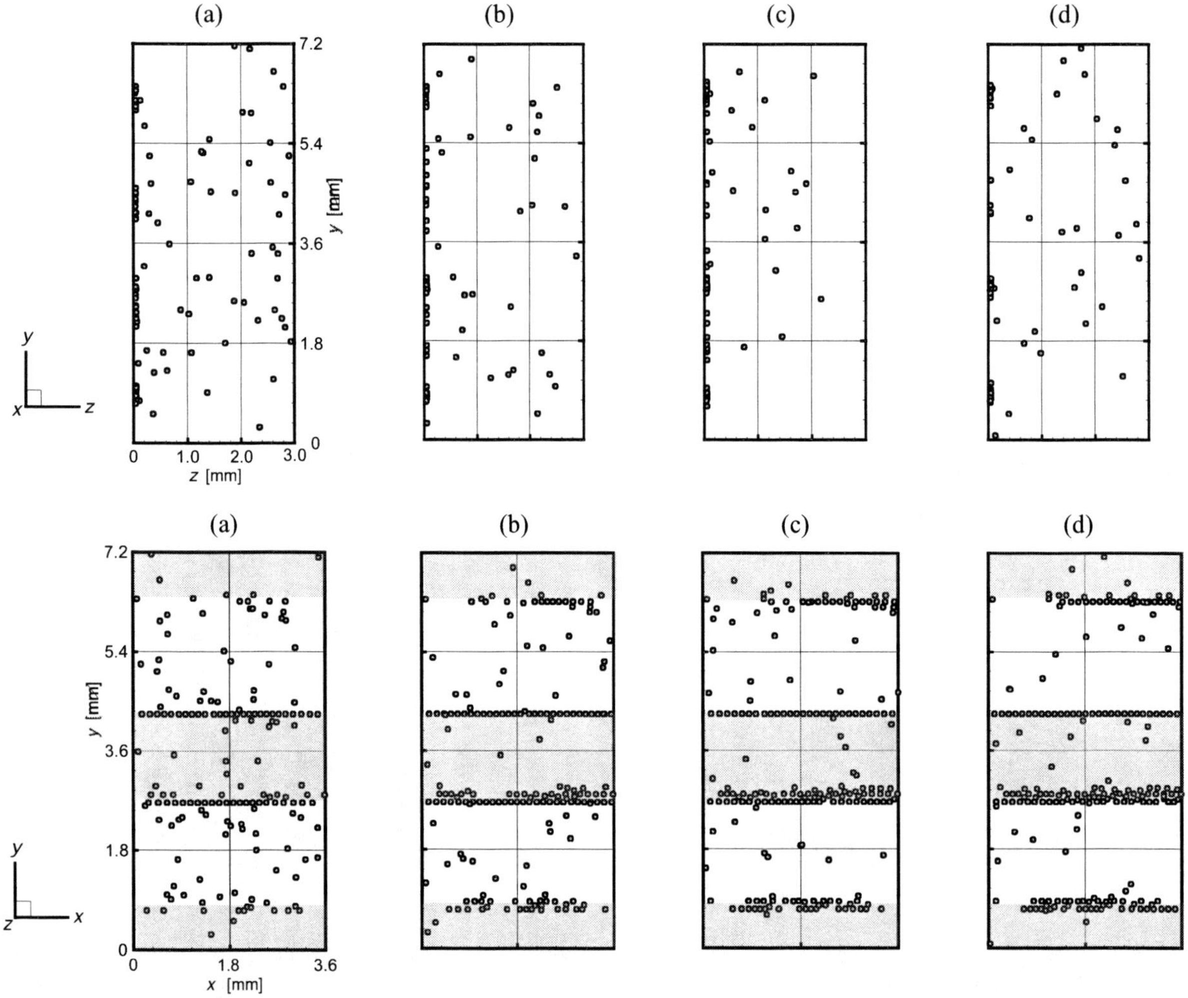

Figure 4: Transient of field-induced particles distribution. Depicted are viewed through the channel cross-section (top panel) and through the channel top (bottom panel) at time t = 40 s (a), 120 s (b), 200 s (c), and 280 s (d).

suspension was 1×10^{-3}. The viscosity of the insulating liquid (corn oil) used was 59.7 cP (296 K). The iteration time interval used in the simulation was 1.0 μs. The 124 CPUs (64-bit Opteron) cluster computer in the Levich Institute (CCNY) was used for the calculations.

In the presence of non-uniform strong AC electric field, on the one hand, particles begin to aggregate along electrodes edges where the gradient of the electric field is the highest because the dielectrophoretic force acting on particles is proportional to ∇E_{rms}^2: on the other hand, these particles do not form a uniform column or layer along the electrodes because the Columbic interaction force among the particles, which is proportional to E_{rms}^2, while inversely proportional to $|\boldsymbol{r}_{ij}|^4$, either pulls neighboring particles together or pushes them away depending on the relative orientation of the particles and the local electric field direction. Consequently, aggregated particles further formed the array of bristles like structures along the longitudinal direction of electrodes.

Results show that particles initially accumulating along the electrode began to cluster with time, and around at time t=200 to 280, formed cluster-like aggregated patterns along electrodes edges. In Fig.4, it is also shown that particles formed a stripe-like pattern along a high voltage electrode edge while on another high voltage electrode edge, particles formed non-uniform aggregated cluster-like structures. In the present simulation, it was found that the pattern of cluster growth depends on the configuration of particles aggregated along electrode edges, that is formed in earlier stage of particle accumulation. Another finding is that particles reached the bottom surface of the channel rolled around the bottom in the downstream-wise direction. Some of them joined the clusters, however, most of them left out the unit cell due to drag force of flow and were hardly seen having built bristles pattern along the lines of electric force perpendicular to the longitudinal direction of electrodes as were observed in the experiment [2] (Fig.5). A longer time-range simulation of particles clustering may be needed to predict the bristles array formation.

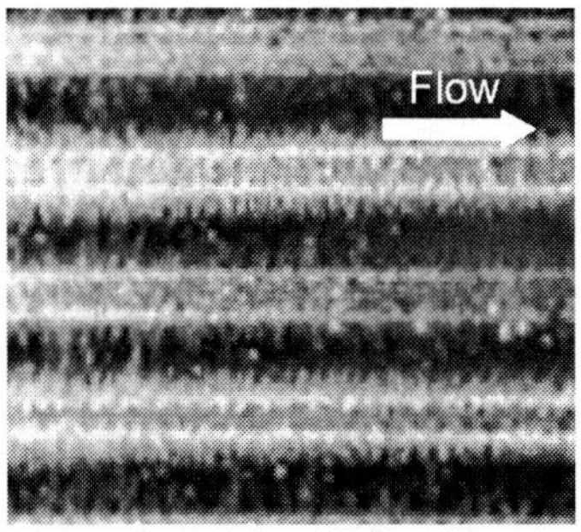

Figure 5: Experimental observations for particle clusters induced by combined dielectrophoretic force and dipole-dipole interactions as well as hydrodynamic interaction in the flowing suspension.

4 CONCLUSIONS AND FUTURE WORK

A suspension of positively polarizable particles subjected to a high-gradient strong AC electric field undergoes a heterogeneous aggregation along the electrode array installed along the flow direction in the rotating flow channel. In the present study, numerical simulations based on the model of Langevin equation of particle kinetics taking into the effects of dipolar interaction between particles and dielectrophoretic force induced by a strong non-uniform AC field were performed to predict the time evolution of particle aggregation and clustering at electrodes. Numerical computation demonstrated that particles began to aggregate in line along the electrode edges, energetically the most favorable site on electrodes, in earlier stage of cluster formation, then new particles coming from a more distant part in the channel joined the vertices of clusters. The results also have similar trends as observed in the experiment, the earlier stage of the bristles array formation along electrode edges. A longer time range simulation is now on going on the parallel computer. Results would provide clearer physics picture on the corresponding phenomena.

The present numerical simulations were performed on the 124 CPUs cluster computer in the Levich Institute of the City College of City University of New York.

REFERENCES

[1] C.L. Brooks, A. Brunger and M. Karplus. Stochastic boundary conditions for molecular dynamics simulations of st2 water. Chem. Phys. Lett., 105:495, 1984.

[2] Z.Y. Qiu, N. Markarian, B. Khusid and A. Acrivos, Positive dielectrophoresis and heterogeneous aggregation in high-gradient ac electric fields. J. Appl. Phys. 92, 2829, 2002.

A Self-vortical Micromixer and its Application on Micro-DMFC

C. Lin* and C. Fu*

*Institute of NanoEngineering and MicroSystems, National Tsing-Hua University
No.101, Kuang-fu Rd., Sec.II, Hsinchu 300, Taiwan, R.O.C., chunyou_lin@yahoo.com.tw

ABSTRACT

Microfluidic micromixer is a very important component of microfluidic biochip for the fluidic mixing procedure of small volume biological sample and reaction reagents. In addition, microfluidic micromixer is possible to be used in a micro-DMFC (direct methanol fuel cell).

In a micro-DMFC, methanol fuel and water management is a critical issue to enhance the fuel cell performance and keep fuel cell operation for an extended period of time. In order to maintain the specific concentration of methanol fuel, some of micro-DMFC designs contain external methanol fuel and water mixing tank for premixing, and then feeding into micro-DMFC. However, methanol fuel and water mixing tank is difficult to be miniaturized and integrated into a micro-DMFC. In this paper, we demonstrate the micro-DMFC, with an internal methanol fuel mixing function, is able to mix methanol fuel and water by the self-vortical micromixer inside itself.

Keywords: micromixer, fuel cell, micro-DMFC

1 INTRODUCTION

Microfluidic micromixer is an important component used in lots of the micro-chip microfluidic systems for biochemistry analysis, drug delivery and sequencing or synthesis of nucleic acids [1]. For example, microfluidic biochip use a micromixer in the fluidic mixing procedure of small volume biological sample and reaction reagents [2]. In addition to micro-chip microfluidic systems, microfluidic micromixer is possible to be used in a micro-DMFC.

Fuel cell is an electrochemical power generator that converts the chemical energy of a chemical reaction to electricity. The electrochemical reaction equations of a DMFC are listed as below:

Anode reaction: $CH_3OH + H_2O \text{-->} CO_2 + 6\,H^+ + 6e^-$
Cathode reaction: $6\,H^+ + 3/2\,O_2 + 6\,e^- \text{-->} 3\,H_2O$
Overall reaction: $CH_3OH + 3/2\,O_2 \text{-->} CO_2 + 2\,H_2O$

Fuel cell consists of an electrolyte membrane (proton exchange membrane, PEM), anode components comprising of anode catalyst layer, gasket, anode plate and flow field plate, as well as cathode components consisting of cathode catalyst layer, gasket and cathode plate. PEM supports hydrogen proton to migrate from anode components to cathode components. Electron generates from anode reaction and migrates from anode plate to cathode plate. Electricity is able to generate by connecting anode plate and cathode plate.

A DMFC has higher power density and relatively high energy-conversion efficiency than a lithium ion battery [3]. Therefore, a DMFC is able to miniaturized to become a micro-DMFC as a favorable candidate for the increasing energy demand of portable electronics, such as cellular phones, laptop computers and PDAs, etc. Nevertheless, in a micro-DMFC, methanol fuel and water management is more difficult than a DMFC.

The methanol fuel with specific concentration as a reductant is fed to the anode components for oxidation reaction on the anode of a fuel cell. The oxygen from air as an oxidant is supplied to the cathode components for reduction reaction on the cathode of a fuel cell. Therefore, the power performance of a micro-DMFC is very sensitive with both the concentration of providing methanol fuel to anode components and the flux of providing oxygen to cathode components.

In order to maintain the specific concentration of methanol fuel, some micro-DMFC designs have external methanol fuel and water mixing tank for premixing, and then feeding into a micro-DMFC. However, methanol fuel and water mixing tank is difficult to be miniaturized. For miniaturization of micro-DMFC, in this study, we propose an internal mixing tank of micro-DMFC and develop the new passive micromixer, the self-vortical micromixer, and integrated this micromixer into a micro-DMFC.

2 EXPERIMENTAL

2.1 Micromixer and Fuel Cell Design

Micromixers could be categorized into passive micromixers and active micromixers. Active micromixers mix microfluids by either actuating movable parts or using external forces, such as pressure disturbance, electrical field, magnetic and acoustic vibration, to achieve mixing effect. Passive micromixers use complicated shapes and structures along microchannel via complicated micromachining processes, and usually require the mixing channels with considerable length and novel design to enhance geometric stirring [4-5].

Figure 1(A) and (B) show the schematic diagram of microfluidic mixing principle of the self-vortical micromixer developed in this study. In the self-vortical micromixer, two fluids separately inject into two inlet reservoir and separately flow into two inlet channel simultaneously, and then the two fluids are able to generate the self-vortical mixing phenomenon in the circular mixing chamber. The mixed fluid mixture of two fluids is able to flow into the mixing channel and achieve complete mixing.

The self-vortical micromixer is made by assembling the upper flow field plate and lower flow field plate. Figure 1(C) presents the upper flow field plate containing two inlet channels both with a width of 500μm and a depth of 150μm, one circular mixing chamber with a diameter of 2000μm and a depth of 150μm, as well as two inlet reservoirs both with a diameter of 5000μm and a depth of 20mm. The lower flow field plate contains one serpentine micro-channel with a width of 500μm and a depth of 150μm, and one outlet reservoir with a diameter of 5000μm and a depth of 20mm.

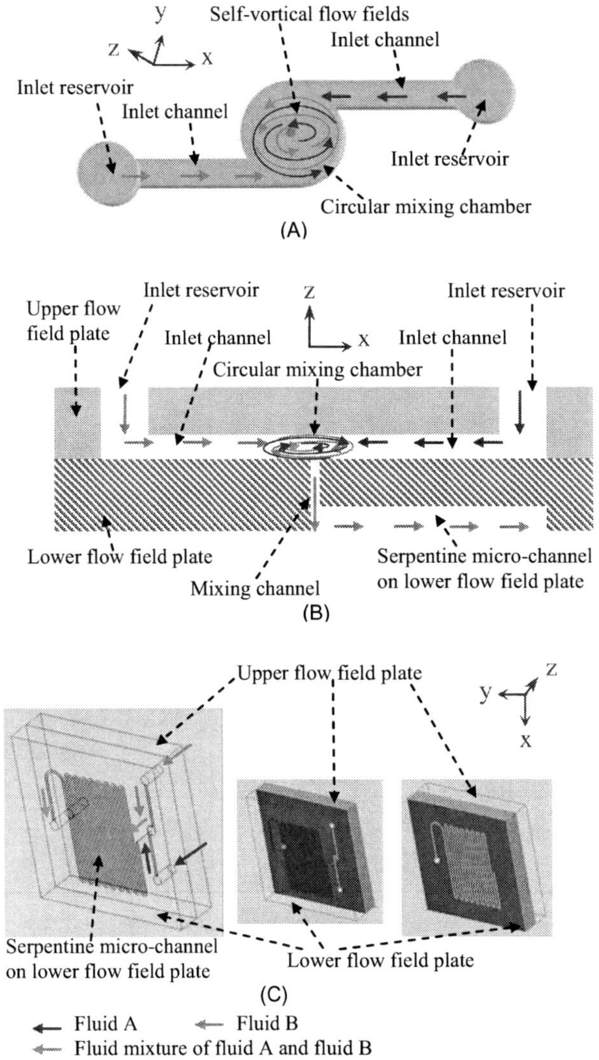

Figure 1: The microfluidic mixing principle and design of the self-vortical micromixer. (A) Top-view of the microfluidic mixing principle of the self-vortical micromixer. (B) Cross-section view of the microfluidic mixing principle of the self-vortical micromixer. (C) The designed self-vortical micromixer integrated into the serpentine micro-channel on lower flow field plate of the micro-DMFC.

Figure 2 shows the assembly picture of the micro-DMFC integrated with the self-vortical micromixer developed in this study. The self-vortical micromixer consists of upper flow field plate and lower flow field plate.

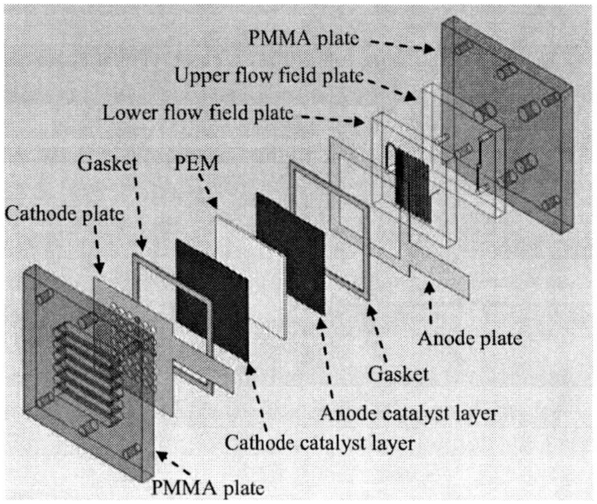

Figure 2: The assembly picture of the micro-DMFC.

2.2 Micromixer CFD Simulation

The computational fluid dynamics (CFD) simulation results can predict the flow velocity profiles and the flow streamlines of microfluidic system [6]. Figure 3 illustrates the CFD simulation result of the micromixer by utilizing the commercial CFD package software program (CFD-ACE+, CFD Research Corporation, CA, USA).

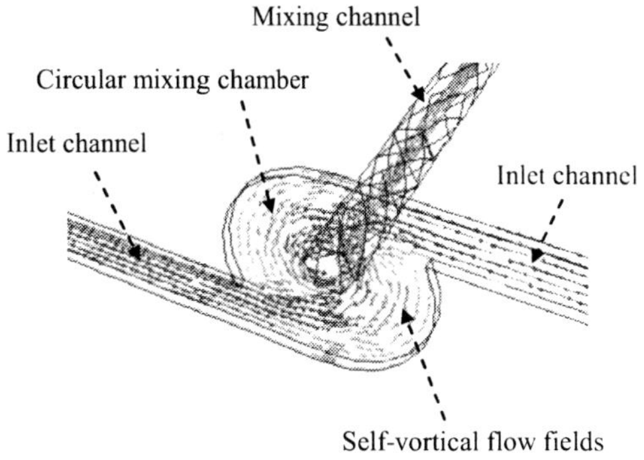

Figure 3: The CFD-ACE+ simulation results of the flow velocity profiles and the flow streamlines in the circular mixing chamber of the self-vortical micromixer. The directions of the vectors represent the directions of flow velocities.

2.3 Micromixer and Fuel Cell Fabrication

The self-vortical micromixer, developed in this study, consists of the upper flow field plate and the lower flow field plate. We used the polydimethylsiloxane (PDMS) material to fabricate both upper flow field plate and lower flow field plate by the PDMS replica molding method [6]. The PDMS made upper flow field plate has the micro feature patterns and involves two inlet channels, one circular mixing chamber and two inlet reservoirs; on the other hand, the PDMS made lower flow field plate has the micro feature patterns and includes one serpentine micro-channel and one outlet reservoir. These micro feature patterns mentioned above on the replication mold were fabricated by thick-film photoresist photolithography process, using a chrome photo mask. Next, PDMS pre-polymer was poured on the replication mold, hardened after baking process, and then we peeled the hardened PDMS from the replication mold to separately made the PDMS made upper flow field plate and the PDMS made lower flow field plate. Finally, we used PDMS oxygen plasma treatment for bonding of the PDMS made upper flow field plate and the PDMS made lower flow field plate.

As shown in Figure 2, we developed the micro-DMFC which was assembled by the following parts: PMMA plate (anode side), upper flow field plate, lower flow field plate, anode plate, gasket (anode side), anode catalyst layer, PEM, cathode catalyst layer, gasket (cathode side), cathode plate and PMMA plate (cathode), Figure 4 shows the photograph of the assembled micro-DMFC.

Figure 4: The photograph of the micro-DMFC.

2.4 Micromixer Testing

As shown in Figure 5, we demonstrate the fluidic mixing performance of the self-vortical micromixer. At the same time, a blue color dye fluid and a yellow color dye fluid were separately injected into two inlet channels by syringe pump and then flowed into the circular mixing chamber to form swirly flow fields and further generated the self-vortical flow fields. The blue color dye fluid and yellow color dye fluid were mixed in the downstream of the mixing channel by the self-vortical flow fields.

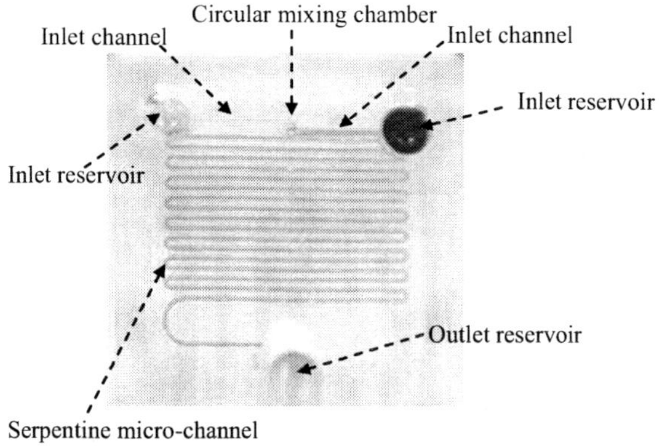

Figure 5: The testing result of the self-vortical micromixer.

As shown in Figure 6, we characterized the mixing efficiency for the self-vortical micromixer by using the definition of the mixing index [7].

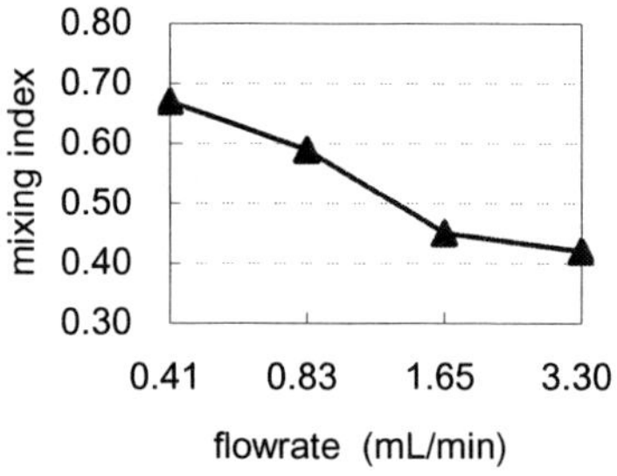

Figure 6: The testing results of mixing index for two different color dye fluids mixed in the circular mixing chamber of the self-vortical micromixer. The lower mixing index means the better mixing efficiency.

2.5 Fuel Cell Testing

Methanol solution with a concentration of 9.0% was prepared to test the performance of the micro-DMFC. Two pumps with same flow rates of 2.4mL/min were used to separately deliver the 9.0% methanol solution and DI water into one of the two inlet reservoirs of the self-vortical micromixer inside the micro-DMFC. After fluidic mixing by the self-vortical micromixer, the mixed fluid mixture of 9.0% methanol solution and DI water had a concentration of 4.5%. The mixed fluid mixture was transported to the serpentine micro-channel (area was 2.45 cm^2) on the flow field plate of the micro-DMFC and flowed into anode components of the micro-DMFC. During the specific concentration of methanol fuel feeding, the cathode components of the micro-DMFC were exposed to ambient air for oxygen supplied by the air natural convection.

An fuel cell load test system was used to measure polarization curve at the polarization condition: scan voltage was 600 mV to 0 mV, voltage decrease step was 10 mV, and load interval time was 1 minute.

3 RESULTS AND DISCUSSION

The micro-DMFC shows an open circuit voltage (OCV) of 700mV. Figure 7 explains the polarization curves of current and voltage characteristic from the polarization test of single cell of micro-DMFC.

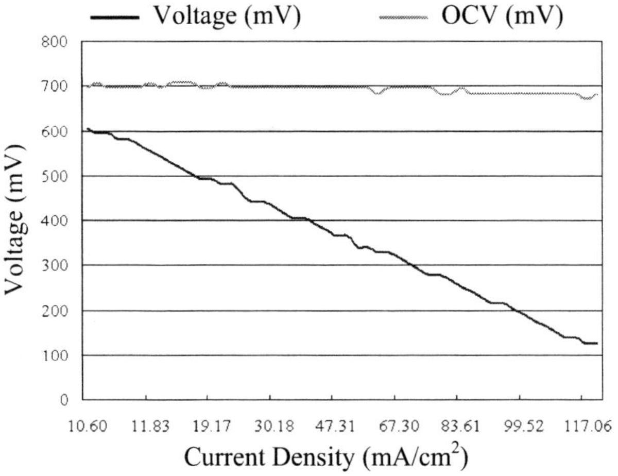

Figure 7: The current and voltage characteristic from the polarization test of a single cell of the micro-DMFC.

The power performance testing results reveal that the micro-DMFC had a maximum power density of 21.27 mW/cm^2. Figure 8 presents the power density curve of power performance of single cell of micro-DMFC.

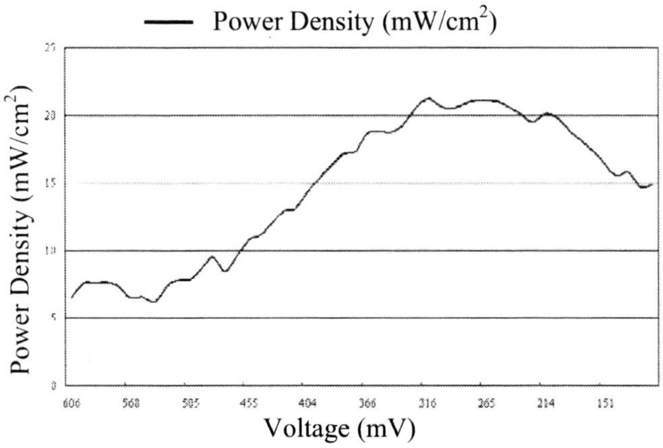

Figure 8: The power performance of single cell of the micro-DMFC.

4 CONCLUSION

In this paper, we demonstrate design and testing results of the self-vortical micromixer integrated with the micro-DMFC. It shows the self-vortical micromixer is able to be operated in a micro-DMFC system.

5 ACKNOWLEDGEMENTS

The authors would like to acknowledgement helps from Micro Base Technology Corp. (www.microbase.com.tw) and providing facilities to test the micro-DMFC.

REFERENCES

[1] Nam-Trung Nguyen and Zhigang Wu, "Micromixers-a review," J. Micromech. Microeng., 15, R1-R16, 2005.

[2] Larry C. Waters, Stephen C. Jacobson, Natalia Kroutchinina, Julia Khandurina, Robert S. Foote and J. Michael Ramsey, "Microchip device for cell lysis, multiplex PCR amplification, and electrophoretic sizing," Anal. Chem. 70, 158-162, 1998.

[3] G.Q. Lu and C.Y. Wang, "Development of micro direct methanol fuel cells for high power applications," Journal of Power Sources, 144, 141-145, 2005.

[4] R.H. Liu, M.A. Stremler, K.V. Sharp, M.G. Olsen, J.G. Santiago, R.J. Adrian, H. Aref, D.J. Beebe, "Passive mixing in a three-dimensional serpentine microchannel," Journal of Microelectromechanical Systems, 9, 190-197, 2000.

[5] A.D. Stroock, S.K.W. Dertinger, A. Ajdari, I. Mezic, H.A. Stone, and G.M. Whitesides, "Chaotic mixer for microchannels," Science, 295, 647-651, 2002.

[6] Sung-Jin Part, Jung Kyung Kim, Junha Park, Seok Chung, Chanil Chung and Jun Keun Chang, "Rapid three-dimensional passive rotation micromixer using the breakup process," J. Micromech. Microeng. 14, 6-14, 2004.

[7] Hsin-Yu Wu and Cheng-Hsien Liu, "A novel electrokinetic micromixer," Sensors and Actuators A, 118, 107-115, 2005.

Pressure driven robust micro pump

F. Weise, C. Augspurger, M. Klett and A. Schober

Technische Universität Ilmenau
MacroNano® Center for Innovation Competence, frank.weise@tu-ilmenau.de

ABSTRACT

For several applications such as small-scale bioreactors there is a lack of suitable commercially available micro pumps. Absolute requirements would be very small flow rates, small size and tolerance of bubbles. Therefore we developed a membrane-type micro pump of modular design. We have developed two different systems: One consists of two check valves and a pumping membrane. The other is a peristaltic pump consisting of a fluidic channel chip, a PDMS-membrane and a cap chip with a spring. Both systems can be integrated into different housings e. g. for several biological applications. The dimensions without housing are only 5 x 5 x 1 mm³ and – depending on the frequency and actuation pressure – we can generate a pump rate up to 1000µl/min. Since the pump can deliver both, gas and liquids, it is absolutely bubble-tolerant. This is of advantage, especially for applications where gas bubbles cannot completely be avoided.

Keywords: micro pump, diaphragm pump, bubble tolerance, PDMS, micro valve

1 DESIGN

1.1 Introduction

The construction of a new micro pump was motivated by the fact that no commercial pump available met the requirements of our application. The pump should be able to address each well of a multi well plate separately. This means ideally every well being equipped with its own pump. Therefore size and connection points of the micro pump are given by the layout of the multi well plate and we designed a pumping chip of 5 x 5 mm² size. Further, for our application the pump must be able to deliver liquids as well as gases at flow rates of up to 400 µl/ min and it should work at back pressures of up to 20 kPa.

Based on these requirements we chose to develop a membrane-type pump. The fundamentals of bubble tolerant micro pumps are described in [1]. Due to the small size of our pump we also have a small diaphragm. In order to make the pump bubble tolerant we need a large stroke of the membrane. Since this cannot be generated by a piezo bimorph – due to physical limits – we have decided to actuate the pump by compressed air. Actuation by compressed air has the additional advantage of promoting diffusion (gas exchange) between compressed air and the medium to be pumped, thereby allowing aeration, oxygenation etc.

1.2 Pressure source

For actuation of our pump we need a source that can generate compressed air or vacuum. Hence we built a device with an external pressure intake. This device can generate pressure square pulses with defined frequencies from 0.1 to 99.9 Hz. Over pressure and vacuum can be adjusted separately by pressure regulators.

1.3 Housing

For using the pump in a labaratory, we need macroscopic connection points. In most cases suitable connectors are Luer connectors. For an easy exchange of the pumping device this is only clamped in the casing. Fig. 1 shows a complete pump with inlet, outlet pipe and the feeding pipe. With another cap this assembly could be used for testing valves. The body could also have some internal fluidic circuits instead of external connections.

Fig. 1 Micro pump with housing

1.4 Membrane pump with check valves

The micro pump is subdivided into a valve chip, a diaphragm, a sealing and a housing. Fig. 2 shows the valve chip with a membrane on top. The material of the membrane and sealing is PDMS (Sylgard 184).

Fig. 2 Valve chip with membrane

1.4.1. Simulation of the valve

In order to find a valve with small pressure drop in flow-through direction and a tiny backflow we simulated the behavior of the valve with CFD-ACE+ (ESI-Software). This is a CFD (Computational Fluid Dynamics) program. Fig. 3 shows a simulated valve in flow-through direction with the inlet being pressurized with 10kPa.

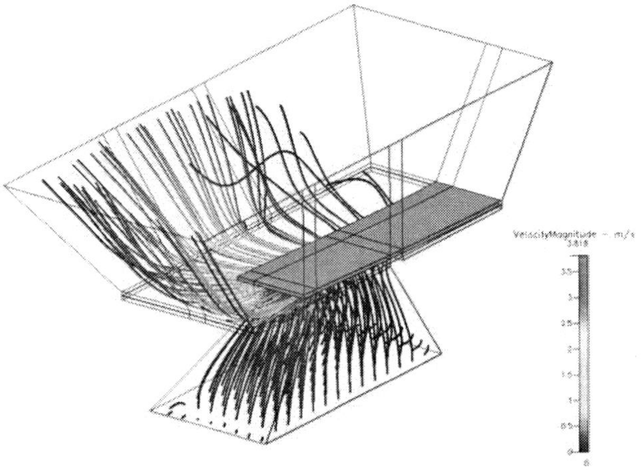

Fig. 3 Micro check valve with streamlines

The simulation showed that thickness of the flap valve should be approx. 10µm and the opening should have a size of 300 x 300 µm². Other valve parameters are defined by the design of the housing (e.g. limitations of size).

1.4.2. Fabrication of the valve chip

The valve chip is produced with micro technologies: The material used is silicon which was masked by microlithography. The wafer was etched in potassium hydroxide and then the valve flaps were etched with a reactive ion etching process. Finally two identical wafers were bonded with a silicon fusion bonding process.

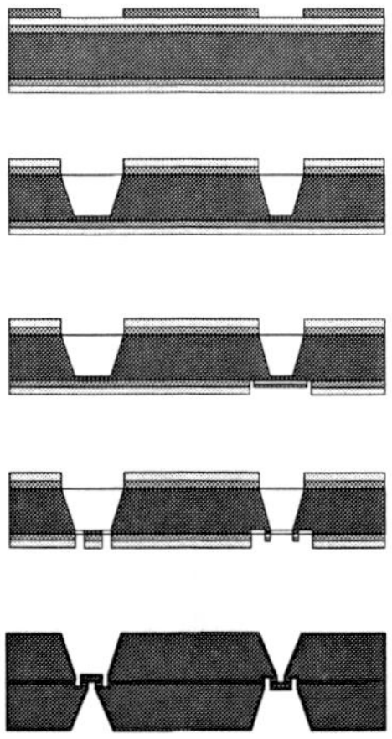

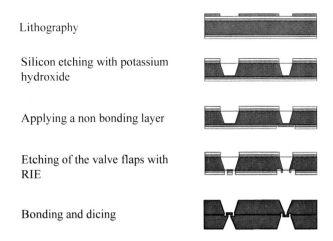

Fig. 4 Fundamental process steps

Although from the technological view it is a problem to create regions of non bonding areas (fig. 4), these were necessary to get a flap valve that can be opened. There are two ways to solve this problem: First, the valve seat can be dimensioned very small (few microns) [2] and second, you can create a layer inhibiting the bonding. We choose the second way, because of a better sealing with a wider valve seat compared to a smaller one. Our experiments show, that the bonding can be inhibited with a thin poly-silicon layer.

1.5 Peristaltic micro pump

Such as the membrane pump also the peristaltic pump is actuated by pneumatics but the working principle is different. In case of the peristaltic pump the function of the valve is integrated into the diaphragm. The deflection membrane opens and closes the inlet and outlet of the pump. An advantage of this pump is that we can control the pumping direction by changing the actuation frequency.

2 RESULTS AND DISCUSSION

2.1 Membrane pump with check valves

First we have characterized the valves. Fig 5 shows a selection of valves. Valve 3 and valve 4 belong to the same pumping chip (inlet and outlet) while valve 1 and valve 2 belong to different pumping chips. Valve 3 and 4 would ideally have identical characteristics. The discrepancies shown in fig. 5 can be explained be tolerances of wafer thickness and etch rates, which result in slightly different characteristics.

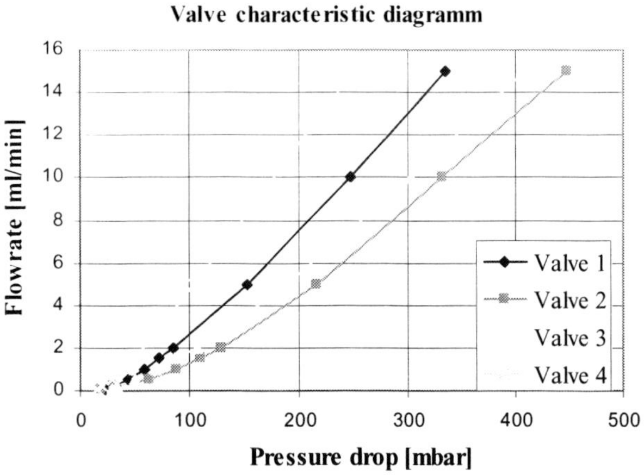

Fig. 5 Valve characteristics

The next test characterizes the flow rate of the pump. In the following diagram (fig. 6) flow rate vs. actuation frequency is shown. The feeding pressure in this case is +20 kPa and -20 kPa respective the atmospheric pressure and water was used as pumping medium. With air as pumping fluid we would get similar characteristics.

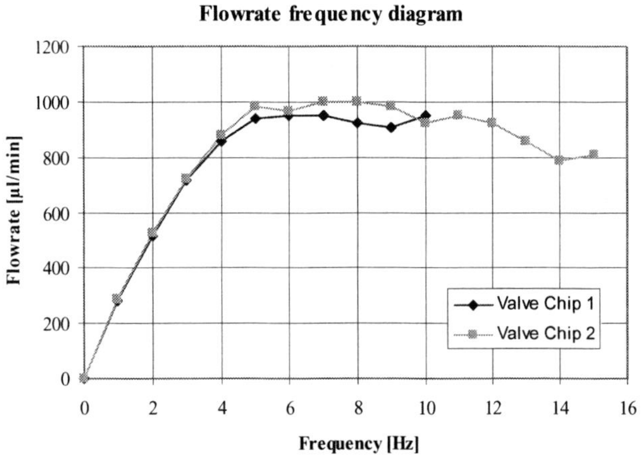

Fig. 6 Flowrate at +20 kPa and -20 kPa actuation pressure

For applications where lower flow rates are needed these can be achieved by decreasing the actuation pressure. Fig. 7 shows the flow rates using different valve chips at an actuation pressure of +15 kPa and -6 kPa.

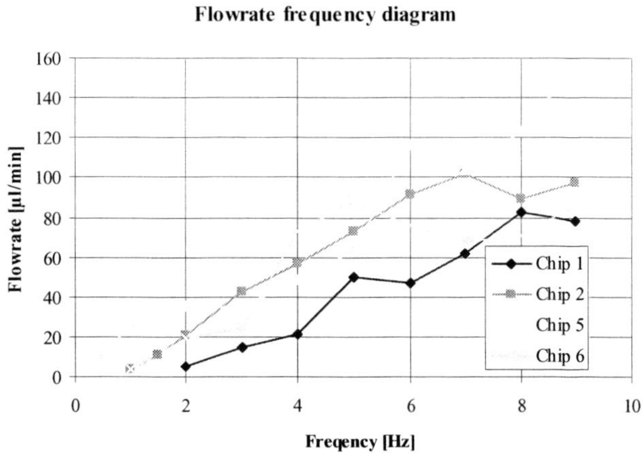

Fig. 7 Flowrate at +15 kPa and -6 kPa actuation pressure

Our goal is to develop a micro pump which can work reliably for several days without any interruption. Although our pump shows very promising characteristics, concerning long-term stability there are still some points to optimize. The valve may get blocked by longer particles such as fuzz. Therefore one way of optimization might be using a filter at the inlet. Another way could be the use of a softer material like silicone as a flap valve.

The future work is directed towards an increase of reliability of the micro pump.

3 ACKNOWLEDGEMENTS

The authors gratefully acknowledge the financial support provided to this study by Federal Ministry of Education and Research and the Thuringian Ministry of Culture within the Initiative “Centre for Innovation Competence”, MacroNano®.

REFERENCES

[1] M. Richter, R. Linnemann, P. Woias: Robust design of gas and liquid micropumps, Sensors and Actuators A68, pp. 480-486, 1998

[2] R.Zengerle, M.Richter: Simulation of microfluid systems, Journal of Micromechanics and Microengineering, 1994

[3] D.C.S. Bien, S.J.N. Mitchell and H.S. Gamble, Fabrication and characterization of a micromachined passive valve, Journal of Micromechanics and Microengineering, 2003

Direct Current Dielectrophoretic Characterization of Erythrocytes: Positive ABO Blood Types

S.S. Keshavamurthy*, K.M. Leonard*, S.C. Burgess**, A.M. Minerick*

*Dave C. Swalm School of Chemical Engineering, Box 9595
Mississippi State University, Mississippi State, MS 39762
**College of Veterinary Medicine, Institute for Digital Biology, Life Sciences Biotechnology Institute and Mississippi Agriculture and Forestry Experiment Station, Mississippi State University, Mississippi State, MS, USA

ABSTRACT

The adaptation of medical diagnostic applications into micrototal analytical systems (μTAS) has the potential to improve the ease, accessibility and rapidity of medical diagnostics. This work adapts direct current dielectrophoresis (DC-DEP) to a medical diagnostic application of sorting blood cells where an insulating obstacle is used to produce a non-uniform electric field. Initial efforts are focused on achieving separation of positive ABO red blood cells. Two dependencies will simultaneously be explored: blood type and blood cell size. Fluorescent polystyrene particles of three different sizes will be tested and compared against the separation and collection of actual blood cells into different sample bins. Further, continuous separation of red blood cells according to blood types and collection into specific bins will be explored. This developed technique is directly applicable for use in a portable device for easy and rapid blood diagnostics.

Keywords: Direct current Dielectrophoresis (DC-DEP), Red blood cells, Antigens, ABO blood type, Microdevice

1 INTRODUCTION

The use of microfluidics in channels and chambers whose cross-sectional dimensions are microns in scale [1] is directly amenable to analysis of biological samples for medical diagnosis. Such systems are commonly called μTAS or micro Total Analytical Systems with wide analytical applications such as biomedical devices, tools for chemistry and biochemistry, and systems for fundamental research [2]. Microdevices have been fabricated out of silica or glass, but thermal plastics such as polymethylmethacrylate (PMMA), polyolefins, polythylene terephthalate, carbonate or elastomers such as poly(dimethylsiloxane) (PDMS) are also widely used [2]. Such microdevices commonly utilize electrokinetics to move analytes. One specialty electrokinetic tool, dielectrophoresis (DEP), utilizes a spatially non-uniform AC electric field to exert forces on polarizable particles or cells. Dielectrophoresis has a number of advantages over linear DC electrophoresis. Electrophoresis works for surface charged particles in DC electric fields whereas DEP enables precise manipulation of polarizable particles with widely varying electrical properties. Further, dielectrophoresis is expected to play a major role in medical microdevices due to operational simplicity, low voltage electric fields, small sample volumes, and would not require skilled medical technicians to operate thus enabling device portability. The electric field strength employed for dielectrophoresis suits biological specimens as advanced by Herbert Pohl in his book "Dielectrophoresis: The behavior of neutral matter in non-uniform electric fields" published in 1978 [3]. In this book two main components were identified which contribute to the nonlinear DEP force: AC electric fields and spatial nonuniform electric field density. In AC dielectrophoresis, frequency dependencies as well as field strength dependencies are utilized to manipulate particles in precise fashions.

Direct current dielectrophoresis (DC-DEP) has been explored recently and only utilizes the spatially non-uniform electric field component, thus facilitating motion of cells. Spatial non-uniformities in the electric field are created using various insulating obstacles strategically positioned in lab-on-a-chip device channels through which a DC field is passed. The benefits associated with using an insulating obstacle were noted by Kang et.al. [4] and can be summarized as below:

1. Insulators are less prone to fouling than electrodes embedded in microdevices,
2. No metal components are involved which reduces the complexity of fabrication of devices,
3. Structure is mechanically robust and chemically inert,
4. Gas evolution due to electrolysis around the metal electrodes is avoided inside the channel.

Initial work in DC-DEP, conducted by Cummings and Singh in 2003, contained an array of insulating posts where two operating regimes were observed, namely streaming and trapping DEP [5]. Insulating posts were adopted by Lapizco-Encinas et.al. (2004) to concentrate and sort live and dead *E.coli* [6]. Efforts to explore other insulating materials for electrodeless DEP included cyclo-olefin polymers by Mela et.al. in 2005 [7], and an oil droplet obstacle [8] in 2006. Thwar et.al. (2007) demonstrated dielectrophoretic potential wells using pairs of insulating oil menisci to shape the DC electric field [9].

The extension of DC-DEP to red blood cell analysis is novel. Previous research by the authors on blood cell behaviors in AC dielectrophoresis [10, 11] showed measurable spatial separation for A^+, B^+, AB^+, and O^+ red blood cells likely due to the blood type antigen expressed on the membrane surface.

Blood type in humans is determined based on the antigens expressed on the red blood cell membrane. Blood typing is a life-essential step prior to blood transfusions. The antigens on the surface of donor blood must match the receiver's blood type or adverse immune responses can cause death in the recipient. After accidents, natural disasters, wars and terror attacks blood transfusions comprise a time sensitive, essential, and important aspect of stabilizing victims to save lives. However, current blood typing technologies require time, a clinical environment, antibody assays for each blood type, and medical technological methodologies, which are not readily portable to emergency sites. This paper analyzes a novel technique of using DC dielectrophoresis as a tool for distinguishing the positive blood types of the ABO system (A^+, B^+, AB^+, O^+). Dielectrophoretic characterizations in lab-on-a-chip devices would be far more portable than current assay techniques.

In DC-DEP, successful separation of cells into bins is dependent on the deflection from an insulating obstacle (Figure 1).

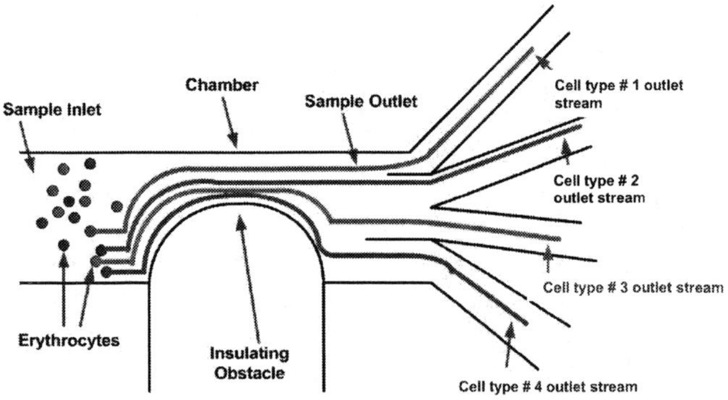

Figure 1: Insulating obstacle in a channel immediately followed by channel bifurcations for sorting.

Y. Kang etal. (2007) [12] showed that the DC-DEP force acting on a cell is proportional to cell size. Therefore, two dependencies will simultaneously be explored: blood type and blood cell size. This is possible because blood type antigens are expressed only on the red blood cells (erythrocytes) while whole blood is comprised of three different cells ranging in size from 3 μm to 12 μm: platelets, erythrocytes, and leukocytes. This technique is directly applicable for portable blood diagnostics.

2. THEORY

In DC dielectrophoresis, an insulating obstacle creates a non-uniform electric field. The resulting dielectrophoretic force acts on a polarizable particle causing it to move within the electric field gradient. This movement is governed by the particle's polarizability relative to the surrounding medium's polarizability [13]. The non-uniform field force can be derived from the net dielectric force:

$$F = (p \cdot \nabla)E \quad (1)$$

Where p, the dipole moment vector can be broken down into particle effective polarizability, α, volume, v, of the particle and the applied electric field, E ($p = \alpha v E$). The Claussius-Mossotti factor, α estimates the effective polarizability term for spherical particles [3, 13] and is a ratio of complex permittivities $\tilde{\varepsilon}$, and is given by $\tilde{\varepsilon} = \varepsilon - i\,\sigma/\omega$, where 'ω' represents frequency, 'ε' the dielectric constant and 'σ' the electrical conductivity of the medium. The imaginary part $\left(i\,\sigma/\omega\right)$ is out of phase with the dielectrophoretic field and does not affect dielectrophoresis as it depends on the real part of the Claussius-Mossotti factor. In case of DC electric field, when there is no frequency component involved, DC-DEP can be estimated as the residual of Claussius-Mossotti factor when frequency approaches zero resulting in a dielectrophoretic force given by,

$$F_{DEP} = \frac{1}{2} v \frac{\sigma_p - \sigma_m}{\sigma_p + 2\sigma_m} \nabla E^2 \quad (2)$$

As a result of this dielectrophoretic force phenomenon, many configurations and operating conditions are possible.

Previous work by the authors suggest that dielectrophoresis of red blood cells depends on blood type [11]. Diluted whole blood was tested at a sinusoidal AC frequency of 1 MHz with an electric field strength of 0.025 V_{pp}/μm. Dielectrophoretic movement of each blood type followed distinct trends with time as quantified with four parameters: total cell count, vertical movement, horizontal movement, and distance. All parameters revealed an attenuated O+ response that was also verified statistically at a 95% confidence interval. This might be due to the lack of functionalized antigens on type O membranes [11].

In this work, a DC field is used to generate an electric field gradient instead of the previous AC field. This dielectrophoretic field is generated inside a microchannel using a DC source and a rectangular obstacle fabricated in the channel. A spatially dense non-uniform field is created near the obstacle as the DC field lines diverge around the obstacle. Since DC-DEP forces the particle away from the high field density regions, it experiences a repulsive force when it moves around the corner of the obstacle, thus facilitating particle motion according to its polarizability.

3. MATERIAL AND METHODS

The parameters changing in this research are: voltage of the DC source, device dimensions, blood type, and time of blood storage. Blood samples will be analyzed in the dielectrophoretic field on day 0, 1, 3 and 5 in storage at 5°C. The important steps in the process are device design, microdevice fabrication by soft lithography, experimentation (including microsample preparation), image analysis and quantification via total cells in each bin.

3.1 Red blood cells

Effective polarizability depends on the permittivity as discussed in the theory section. Charged proteins and cytosol molecules present in the cell membrane impact the ability of a cell to conduct charges and cell membranes impact the ability of charges to penetrate the cell [3, 14]. Previous researchers have shown that cell dielectric properties depend on cell shape and structure [15] thus enabling DEP forces to spatially separate cells [16]. Red blood cell membranes are nonconductive ($\sigma \leq 1$ µS/m) [17] while the interior is conductive ($\sigma = 0.53$ to 0.31 S/m) and varies due to hemoglobin and cytoplasm molecules [18]. RBCs are biconcave in shape, 6 to 8 microns in diameter, and change in response to solvent conditions, pH and temperature [19].

Human blood types are classified based on membrane surface antigens and blood plasma antibodies [20]. Type A blood expresses antigen A, Type B blood expresses antigen B, Type AB blood has both A and B antigens while Type O blood has neither antigens [20]. The A and B antigen differ only in their modifications with the side chain whereas the backbone remains the same. The A antigen terminates in an α 1,3 – linked N-acetylgalactosamine while B terminates in an α 1,3 – linked galactose. The presence or absence of the Rhesus (Rh) factor antigen determines the positive (present) and negative (absent) blood types [21]. The 8 ABO / Rh blood types are A^+, B^+, AB^+, O^+, A^-, B^-, AB^-, and O^-.

3.2 Device Design

Device designs were drawn in AutoCAD (Autodesk, Inc) and printed on transparencies with a resolution of 32,512 dpi (Fine-Line imaging, Inc.). The design consisted of one input (300 µm wide) and 4 output channels (100 µm wide). A rectangular obstacle was positioned in the input channel 250 to 1000 µm prior to the bifurcation to the four outlet channels. The rectangular obstacle was varied from 100 to 150 µm in width and 200 to 250 µm in height. The rectangular obstacle is shown in figure 2 (a) and the complete microdevice design is shown in figure 2 (b).

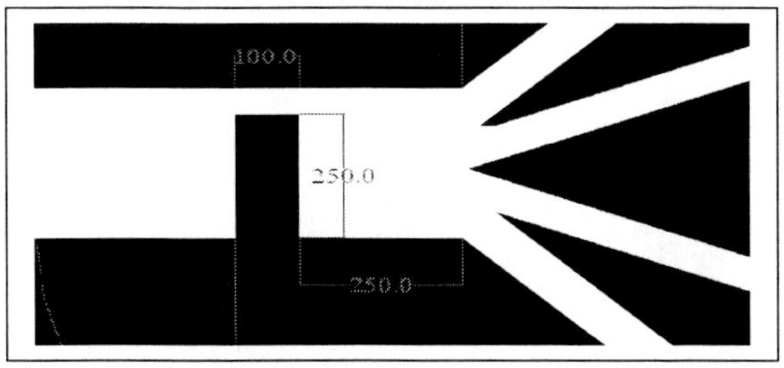

Figure 2 (a): Rectangular obstacle design

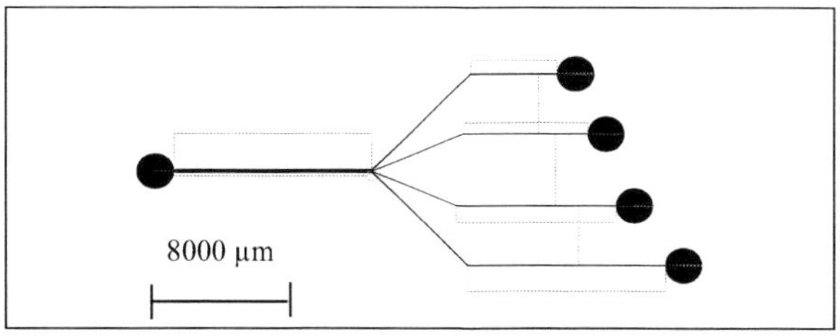

Figure 2 (b): Microdevice with input and output channels

3.3 Fabrication

The device was fabricated as a lab-on-a-chip with a fluid injection port and four outlet ports. This microdevice consisted of a glass slide and lithographically patterned microchannels of poly(dimethylsiloxane) (PDMS).

Standard techniques of soft lithography were used to pattern the microdevice [2]. Some modifications included increasing the exposure time to UV light from 2 to 5 minutes to get the smallest features like the rectangular obstacle on the SU-8 (negative photoresist) spin coated Silicon wafer of 50 µm thick. The SU-8 manufacturer, Microchem, suggests rinsing patterned silicon wafer with isopropyl alcohol (IPA) followed by air/N_2 drying to stop development, but this procedure yielded white residue. Microchem characterizes this residue as undeveloped SU-8 photoresist and recommends further immersion in the developer solution. Li suggests using Dynasolve 185 (Dynaloy LLC) [22] to eliminate residues. Figure 3 shows the patterned silicon wafer.

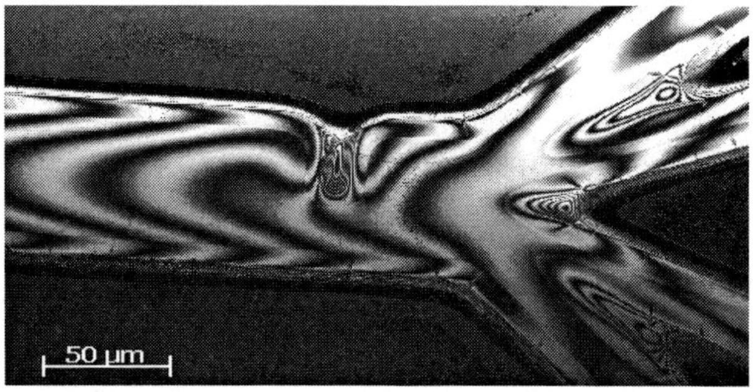

Figure 3: Patterned Silicon wafer with obstacle

PDMS was cast on the patterned silicon wafer to produce the positive relief of Figure 2 design. Sylgard 184 pre-polymer base and curing agent were mixed at a 10:1 ratio, mixed to a milky consistency, and poured onto the patterned silicon wafer and degassed for 2 hours inside a vacuum chamber to remove suspended bubbles. The degassed PDMS assembly was cured in a 65°C oven for 24 hours. The PDMS layer was peeled from the patterned silicon wafer and sealed onto a glass slide. Biopsy punches were used to create holes for inlet and outlet ports which accommodated tubing for pumping to the PDMS microdevice. Irreversible sealing can occur after UVO (UV, ozone) treatment [2]. Covalent bonds between the glass slide and PDMS can withstand pressures of 30 to 50 psi [2].

3.4. Operation

Fluorescent polystyrene particles (Bangs Laboratory, Inc) of three different sizes 2.28 μm (yellow, 540 or 600 nm), 5.49 μm (red, 660 or 690 nm) and 10.35 μm (green, 480 or 520 nm) roughly corresponding to platelets, erythrocytes and leukocytes respectively were selected for initial analysis. These beads will be tested as a size dependence reference and compared against the separation and collection of actual blood cells into different bins in the microdevice shown in figure 2.

Whole blood will be obtained via venipuncture by trained phlebotomists, drawn into vacutainers (Becton Dickinson) containing 1.8 mg K_2 EDTA per mL of blood, and stored at 4°C in a Biosafety Level 2 refrigerator. Whole blood will be diluted 1:60 with 0.14M PBS (0.1 S/m) and introduced via ports to the microdevice mounted on a microscope. A DC power supply will deliver a voltage of 1125 V; corresponding to 0.025 V_{pp}/μm electric field strength. High-resolution video microscopy is used to record bright field video of cell motion and counts within the microdevice. Each experiment runs approximately for 5 minutes followed by image and cell density analysis.

4. RESULTS AND DISCUSSION

Cell movement is recorded between the rectangular obstacle and the bifurcation point. Cell counts are tallied in each exit port at the completion of the run for each sample. Experiments are run over day 0, 1, 3 and 5 of blood storage for each blood type [23]. Total cell counts will be tabulated after the experimental run (4 minutes) to know the percent of cells travelled to each exit port. The port having the maximum cells would likely correspond to a specific blood type. From previous AC dielectrophoresis studies by the authors, type A had the maximum deflection followed by type B, where as type O had an attenuated response in the nonuniform electric field. Type AB had deflections ranging between type O and types A and B. On this basis, authors are predicting the same pattern of deflection to be followed while the blood types separate at the bifurcation point into channels. This deflection pattern is mostly due to the existence of antigens on the membrane surface.

5. SUMMARY

The use of microdevices for blood typing can aid in increasing the speed and efficiency of emergency medical diagnosis. DC -dielectrophoresis shows potential for this purpose. This work provides evidence that dielectrophoretic responses are influenced by the expression of antigens on the surface of cells, which has prospects for future studies of the dielectrophoretic behavior of other biological cells. The current work is limited because it does not account for all the antigens on the blood cell surface, only ABO antigens are involved in this study.

REFERENCES

1. Lin, B., Long, Z., Liu, X., Qin, J., *Biotechnology Journal*, 2006, 11, 1225-1234.
2. McDonald J.C., Duffy D.C., Anderson J.R., Chiu D.T., Wu H., Schueller J.A., Whitesides G.M., *Electrophoresis*, 2000, 21, 27-40.
3. Pohl, H., "*Dielectrophoresis: The behavior of neutral matter in nonuniform electric fields*", New York, NY, Cambridge University Press, 1978.
4. Kang K, Kang Y, Xuan X, Li D., *Electrophoresis*, 2006, 27, 694-702
5. Cummings E.B., Singh A.K., *Analytical chemistry*, 2003, 75, 4724-4731.
6. Lapizco-Encinas B.H., Simmons B.A., Cummings E.B., Fintschenko Y., *Analytical chemistry*, 2004, 76, 1571-1579.
7. Mela P., Van der berg A., Fintschenko Y., Cummings E.B., Kirby B.J., *Electrophoresis*, 2005, 26, 1792-1799.
8. Barbulovic-Nad I., Xuan X., Lee J.S.H., Li D., *Lab on a chip*, 2006, 6, 274-279.
9. Thwar P.K., Linderman J.J., Burns M.A., *Electrophoresis*, 2007, 28, 4572-4581
10. Minerick A.R., Zhou R., Takhistov P., Chang H.C., *Electrophoresis*, 2003, 24 (21), 3703-17
11. Keshavamurthy S.K., Daggolu P.R., Burgess S.C., Minerick A.R., *Electrophoresis*, in review.
12. Kang Y., Li D., Kalams S.A., Eid J.E, *Biomed Microdevices*, 2007
13. Minerick, A. R. "DC Dielectrophoresis in Lab-on-a-Chip Devices." In: Li, Dongqing (ed). *Encyclopedia of Micro- & Nanofluidics*. Springer, Berlin Heidelberg New York (in press) 2008.
14. Pethig, R., *Crit. Rev. Biotech.*, 1996, 16, 331-348.
15. Huang, Y., Wang X.B., Becker F.F., Gasocyne P.R.C., *Biochim. Biophys. Acta*, 1996, 1282, 76-84
16. Pethig R., G.H. Markx, *Trends Biotechnol.*, 1997, 15(10), 426-432
17. Gasocyne P., Mahidol C., Ruchirawat M., Satayavivad J., Watcharasit P., Becker F.F., *Lab on a Chip*, 2002, 2, 70-75
18. Gimsa J., Muller T., Schnelle T., Fuhr G., *Biophys. J.*, 1996, 71, 495-506
19. Daniels G., Bromilow I., *"Essentials guide to Blood Groups"*, Blackwell Publishing, 2007
20. Patenaude S.I., Sato N.O.L., Borisova S.N., Szpacenko A., *Nature structural biology*, 2002, 9, 685 – 690.
21. Sheffield P.W., Tinmouth A., Branch D.R., *Transfusion Medicine reviews*, 2005, 19, 295 – 307.
22. Li Y., Dalton C., Crabtree J.H., Nilsson G., Kaler K.V.I.S., *Lab on a chip*, 2007, 7, 239-248
23. Patenaude S.I., Sato N.O.L., Borisova S.N., Szpacenko A., *Nature structural biology*, 2002, 9, 685 – 690.

Ceramic Microarrays for Aggressive Environments

Sachin Laddha*, Carl Wu*, Sundar V. Atre*, Shiwoo Lee*, Kevin Simmons**, Seong-Jin Park and Randall M. German ***

*Oregon Nanoscience and Microtechnologies Institute,
106 Covell Hall, Oregon State University, OR 97330 sundar.atre@oregonstate.edu
**Pacific Northwest National Laboratory, Richland, WA 99354
***Center for Advanced Vehicular Systems,
Mississippi State University, MS 39372, USA

ABSTRACT

Ceramic injection molding (CIM) is a cost-effective technique for producing small, complex, precision parts in high volumes from nanoparticles. To have a good understanding of the CIM process and to provide the necessary data for simulation studies, detailed characterization of the powder-polymer mixture (feedstock) is essential. In this paper, the characterization of feedstocks consisting of alumina nanopowder (average particle size of 400 nm) with ethylene-propylene/wax (Standard Mix) and polyacetal binder systems (Catamold AO-F, BASF) for micro-ceramic injection molding (μCIM) is reported. It was found that the wax-based binder system had lower viscosity and heat capacity as well as greater pseudo-plasticity compared to the polyacetal binder system. However, the results from Moldflow simulations inferred that the Catamold AO-F filled the microcavities (50μm) more efficiently than the Standard Mix.

Keywords: nanoparticles, ceramic microfabrication, powder injection molding, microchannel arrays

1. INTRODUCTION

The role of powder technologies for the net shape production of complex engineering components from metal and ceramic materials continues to grow [1]. One way of net-shaping such components is the use of ceramic injection molding (CIM) which is advantageous as far as shape complexity, materials utilization, energy efficiency, low-cost production, and mass manufacturing are concerned [2].

A major area of application for CIM is in microfluidic systems [3]. Small particles are required for both geometric and performance attributes. Curiously, numerous shape-forming routes for nanoscale powders have been reported which are shown in Figure 1 [3], yet advantages are poorly realized from these powders. In powder systems, the rule of thumb is that the powder size must be smaller than 5% of the feature dimension to reduce wall effects.

Material homogeneity is a critical issue in μCIM because it results in various molding defects in metal or ceramic mircroparts as shown in Figure 2. In this study, an experimental platform and modeling approaches have been used to carry out the research on ceramic microchannel arrays (MCA's) of μCIM.

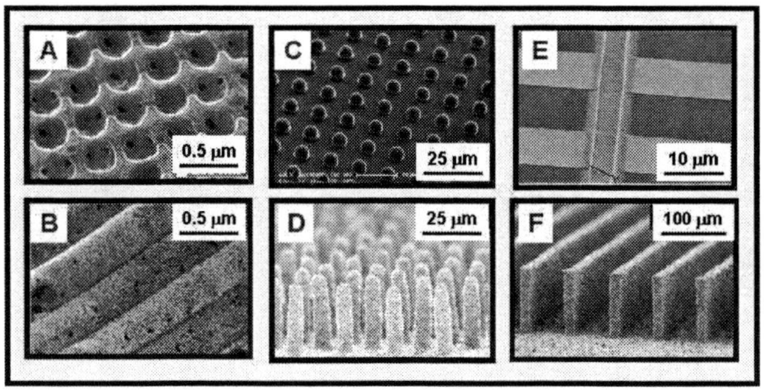

Figure 1: Micromolded nanoparticulate suspensions with 100 nm-20 μm features: (A) SiO_2 photonic bandgap structures, (B) Au nanotube catalysts, (C) SiC tribological microstamps, (D) PZT piezoelectric (E) SnO_2 gas sensors, and (F) Al_2O_3 microfluidic reactor.transducers, (E) SnO_2 gas sensors, and (F) Al_2O_3 microfluidic reactor.

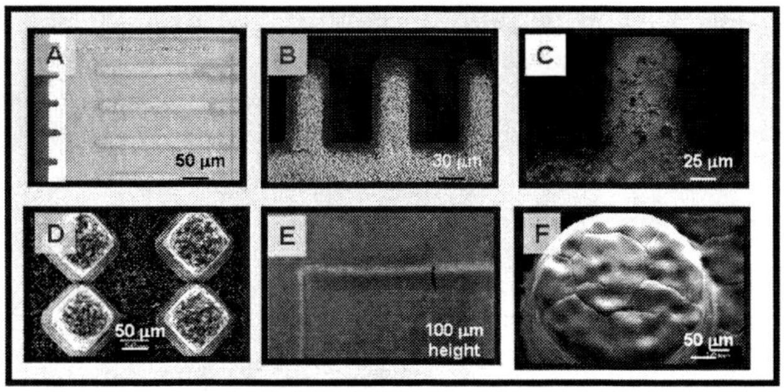

Figure 2: Material inhomogeneity in micromolding nanoparticulate suspensions: [A] incomplete mold fill, [B] ejection crack, [C] porosity distribution, [D] large particle size, [E] bloating during binder removal, and [F] grain growth and cracks during sintering.

The main objective of this research is to investigate the material heterogeneity issue in microcavities through the development of the μCIM process for MCAs. MCAs have been chosen for this study since they have been widely used as the major component and design feature for many microsystems in a large variety of applications, such as microfluidics, micros optics, and micro heat exchangers.

2. EXPERIMENTAL METHODS

2.1 Part Design

Alumina MCA's were designed as the experimental samples for conducting the present study. Figure 3 shows the molded samples of alumina MCA'. The study sample comprised of two structures, the bulk substrate and the microchannel arrays. The bulk size of the small parts with 50μm ribs was 2.5 x 8 x 1.5 mm (width x length x thickness), with an approximate volume of 48 mm^3 and the weight of 0.12g. The small part was also designed with the aspect ratio of 2:1 for the microchannel walls and the large flow path ratio of L/T = 80 (length/thickness).

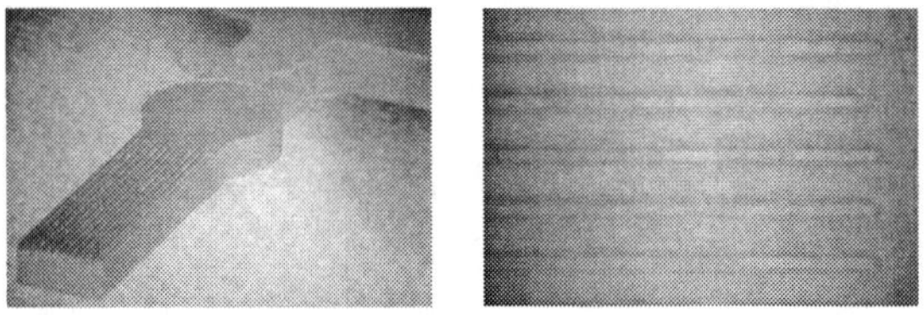

Figure 3: Alumina MCA's fabricated by μCIM in this study

2.2 Materials

Although there is a relatively wide range of materials available for μCIM, it is necessary to focus on powders of small particle size. The starting powder requirements for μCIM are much more stringent than that used for conventional CIM. For example, the particle size should be at least about one order of magnitude smaller than the minimum internal dimension of the micro part. The feedstocks used in this research are the commercially available alumina-polyacetal feedstock (Catamold AO-F, BASF) and an in-house product, Standard Mix, comprising of alumina-propylene/paraffin wax, both containing about 56 vol.% solids loading. Both these feedstock contain alumina powder A16 SG, supplied by Almatis, with an average diameter of 400 nm.

2.3 Processing

Moldflow software was used to simulate the process. The experiment platform used for micro powder injection molding study consisted of a state-of-the-art PIM machine, ALLROUNDER 270C from Arburg. The machine had a clamping force up to 800 kN. A precision micro mold system was designed with multiple cavities for different MCA with the rib sizes of 50μm. The cavities were designed with gas vents and vacuum paths to improve degassing for fast melt fill, which are critical in the micro injection molding process to reduce voids and micro feature short shot. The cavities were fitted with temperature and pressure sensors to monitor the mold filling

After several trial runs, the basic injection molding process for alumina MCAs was set. Among these process parameters, three of them, volume flow rate, holding pressure and mold temperature were chosen as the DOE control factors to explore the process effects on molded part homogeneity issues.

2.4 Characterization

A few material test methods were explored in this research for studying the green ceramic material homogeneity and mold filling behavior. They include rheological behavior (viscosity), thermal properties (specific heat and thermal conductivity) and pressure-volume-temperature (PVT) behavior.

The rheological characteristics of the feedstock were examined on a Gottfert Rheograph 2003 capillary rheometer at different shear rates and temperatures. The testing was carried out in accordance with ASTM D 3835. The temperatures were between the highest melting temperature and the lowest degradation temperature of the binder system. The barrel of inner diameter of 1 mm and die length as 20 mm was used. The pre-heating time was kept as 6 minutes.

Specific heat measurements were carried out on a Perkin Elmer DSC7 equipment in accordance with ASTM E 1269. The testing was done on 11.85 g of sample with the initial temperature of 190 ^{0}C and final temperature of 20 ^{0}C. The cooling rate was kept constant of 20 ^{0}C/minute.

A K-System II Thermal Conductivity System was used to evaluate the thermal conductivity of the feedstocks. The testing was carried out in accordance with ASTM D 5930. The initial temperature was 190 ^{0}C and final temperature was 30 ^{0}C. The probe voltage was kept as 4 V and acquisition time of 45 s.

A Gnomix PVT apparatus was used to find the PVT relationships of the feedstock materials. The test was carried out in accordance with ASTM D 792. The pellets were dried for 4 hours at 70 ^{0}C under vacuum. The measurement type used was isothermal heating scan with a heating rate of approximately 3 ^{0}C/minute.

3. RESULTS AND DISCUSSION

3.1 Rheological Studies

Figure 4 shows the relation of viscosity versus shear rate at a range of temperatures. The viscosity of both the feedstocks decreases with an increase shear rate and temperature. Further, no dilatant behavior was observed. Normally, feedstocks that exhibit shear-thinning flow behavior during molding ease mold filling and minimize jetting. The rheological data was fitted to a modified Cross-WLF equation and used for further analysis in a Moldflow package.

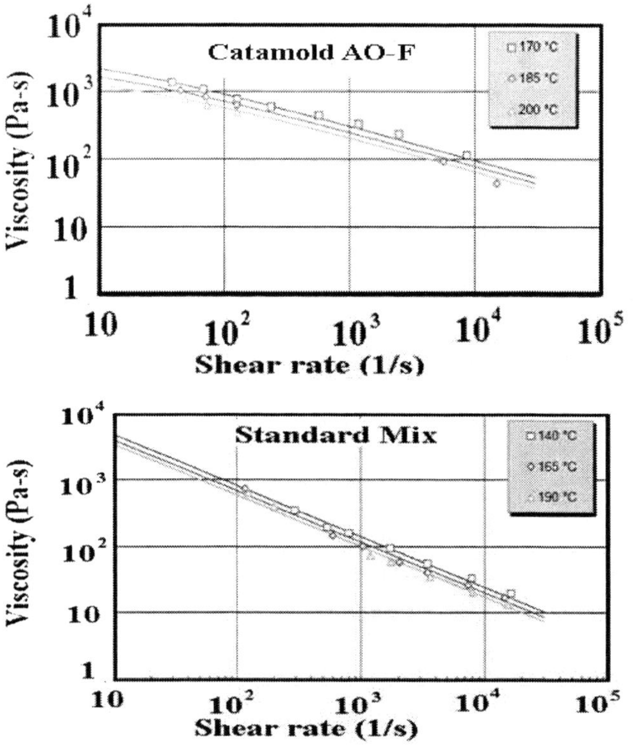

Figure 4: Relationship between viscosity (Pa-s) and shear rate (s^{-1}) for Catamold AO-F and Standard Mix at a melt temperature of 190 ^{0}C.

3.2 Pressure-Volume-Temperature Behavior

The pressure-volume-temperature (PVT) behavior of the Catamold AO- F and Standard Mix (Figure 5), gives the specific volume changes of the melt in cavity as a function of the cavity pressure and temperature. It helps to understand the compression and temperature effects during a typical injection molding cycle. The hold pressure should be chosen after appropriately referring to the PVT-diagram so that the residual cavity pressure is near atmospheric pressure before demolding.

If the hold pressure is too high, the part will still be under pressure when the mold temperature has been reached, which may cause part ejection and relaxation problems. The higher slope in the PVT plot for the Standard Mix (Figure 5) implies a higher tendency for shrinkage in the final part.

3.3 Specific Heat

The Catamold AO-F containing polyacetal has a higher specific heat value compared to the Standard Mix. Figure 6 shows the comparison of the specific heat of the two feedstocks as a function of temperature.

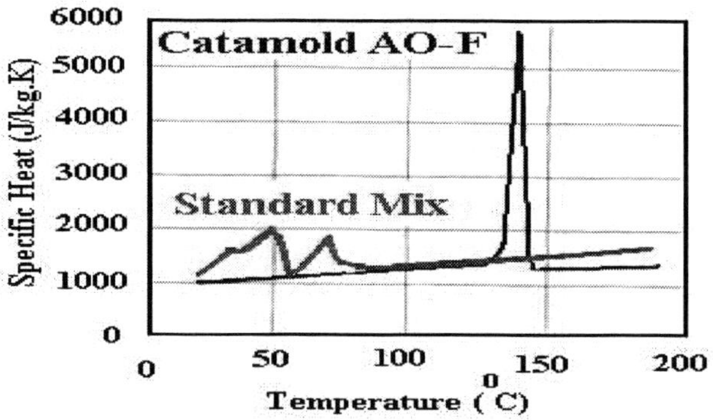

Figure 6: Specific heat (J/kg.K) as a function of temperature for Catamold AO-F and Standard Mix.

3.4 Thermal Conductivity

Figure 7 shows a comparison of the thermal conductivity of the two feedstocks. The Catamold AO-F has a higher thermal conductivity than the Standard Mix. This leads to slower removal of heat from the Catamold AO-F than the Standard Mix during the molding cycle, when the melt and mold temperature are kept constant for both the feedstocks. This can lead to faster and more uniform filling of microchannels for Catamold AO-F compared to the Standard Mix.

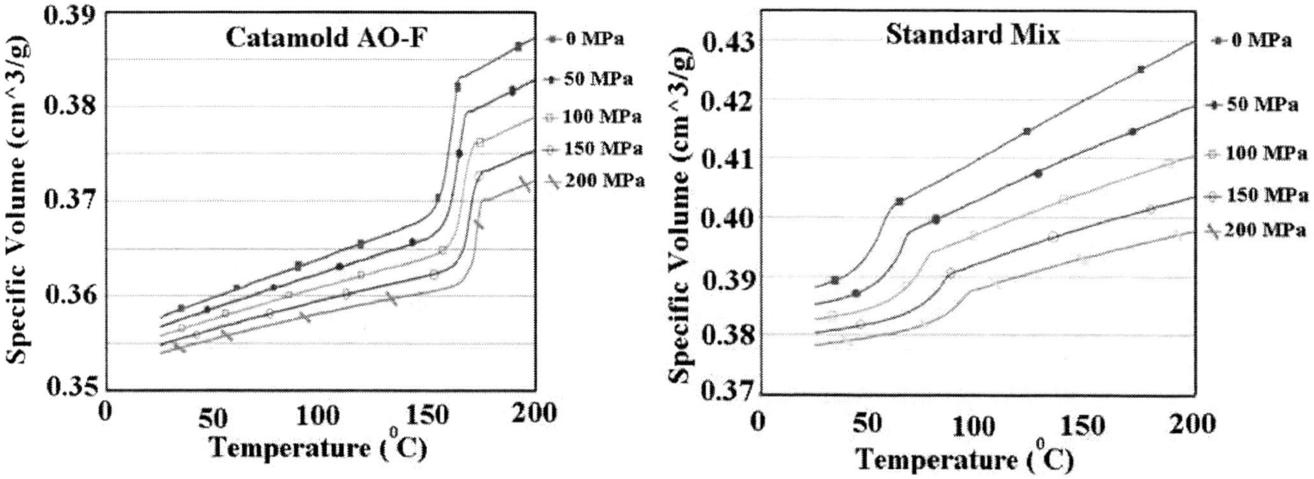

Figure 5: PVT relationships for Catamold AO-F and Standard Mix

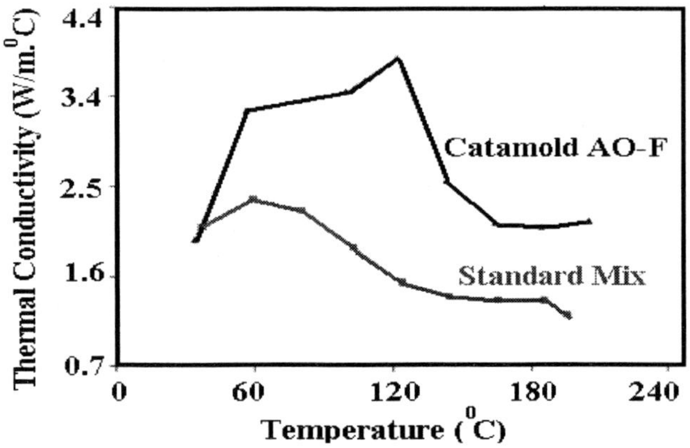

Figure 7: Thermal conductivity of Catamold AO-F and Standard Mix as a function of temperature.

3.5 Progressive filling of microchannels

Figure 8 shows the progressive filling of microchannels using Moldflow simulations for Catamold AO-F at a melt temperature of 190 °C. Section A in Figure 8 shows that the bulk is filled half way before the microchannels began to fill. Section B and C maintains the same momentum as section A. D shows that all the microchannels are filled at the end of the mold-filling cycle which may cause uneven shrinkage in the part.

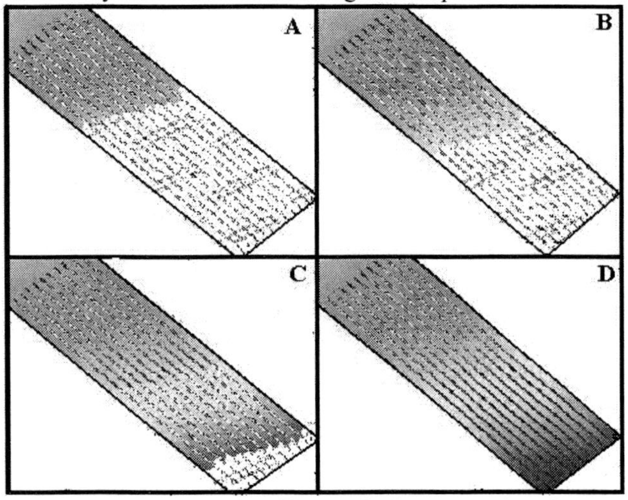

Figure 8: Progressive filling of microchannels using Catamold AO-F at a melt temperature of 190°C from Moldflow simulations.

3.6 Comparison of feedstocks in microchannels

The progressive filling behavior of MCA's discussed in Section 3.5 were quantified and shown in Figure 9. It was found that the amount of feedstock flowing in the microchannels per unit area and per unit time was greater for Catamold AO-F than the Standard Mix. However, the flow in the bulk for both the feedstocks remained nearly the same. This may be due to the difference in thermal and rheological properties of the feedstocks seen in section 3.1 through section 3.4.

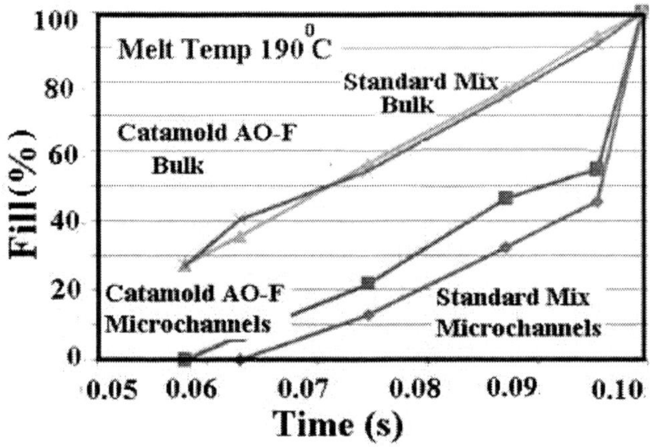

Figure 9: Comparison of Catamold AO-F and Standard Mix in microchannels and bulk at a melt temperature of 190°C

4. CONCLUSIONS

This research emphasizes the effect of polymer on the mold filling behavior of the powder-polymer suspension during the fabrication of MCAs by CIM. It was observed that several material properties determined the mold filling behavior of powder-polymer suspensions. Future work will involve analyzing the influence of mold filling behavior on the occurrence of microscale and macroscale defects.

ACKNOWLEDGEMENT

This material is primarily based on research sponsored by Hewlett-Packard. Additional funding was provided by Air Force Research Laboratory under agreement number FA8650-05-1-5041. The views and conclusions contained herein are those of the authors and should not be interpreted as necessarily representing the official policies or endorsements, either expressed or implied, of Air Force Research Laboratory or the U.S. Government.

REFERENCES

[1] K.F. Ehmann, D. Bourell, M.L. Culpepper, T.J. Hodgson, T.R. Kurfess, M. Madou, K. Rajurkar, and R. E. DeVor "WTEC Panel Report on International Assessment of Research and Development in Micromanufacturing," Based on the findings of a study panel sponsored by NSF, ONR, DOE, and NIST-ATP, October 2005.

[2] R. M. German, *Powder Injection Molding Design & Applications: User's Guide,* IMS Publishers, State College, PA, 2003.

[3] R.A. Saravanan, L. Liew, V.M. Bright, R. Raj, "Integration of Ceramics Research with the Development of a Microsystem," *Journal of the American Ceramic Society*, vol. 86, pp 1217-1219, 2003.

Understanding Conduction Mechanisms in Nano-Structures

V. H. Gehman, Jr., K. J. Long, F. Santiago, K. A. Boulais, A. Rayms-Keller

Electromagnetic and Sensors Systems Department, Naval Surface Warfare Center, Dahlgren Division, Dahlgren, VA, USA., dlgr_nswc_Q23@navy.mil

ABSTRACT

Many emerging technologies require the understanding of charge transport mechanisms through nanostructures. Our previous research has revealed unusual and enhanced conduction properties in pores whose width is significantly less than 1μm over a range of +10V to -10V. The conduction current, as a function of voltage, of a NaCl electrolyte within a 150-nm pore has been measured and characterized using an atomic force microscope (AFM). A double-layer conduction model has been devised to explain the experimental results.

Keywords: atomic force microscopy, nano-conduction, nanopore, double-layer model, electrolyte

1 BACKGROUND

The Naval Surface Warfare Center, Dahlgren Division, has been involved in interface-research issues with liquid dielectrics for three decades [1]. Modeling efforts for water electrolytes have progressed through traditional mechanisms to include charge injection [2] as well as double-layer models [3], [4], [5], and [6]. Recently work has extended into the field of nanotechnology and resulted in measurement of conductance through water-electrolyte-wetted nanostructures [7], [8].

Traditional theoretical investigations into water-electrolyte conductance include classic works [9], [10] and double-layer calculations [11]. More recent investigations into nano-pores include impressive experimental measurement of diffusion and transport [12] (under similar conditions to our work) as well as sophisticated theoretical electrochemistry-based investigations that focus on the effects of double layers and surface potentials on charge transport in nanopores [13]. Still, it would be useful to know if simpler calculations can explain the existing experimental data, especially given the large number of unknown variables in actual nanopore experiments, including actual profile through the material, concentration gradients within the electrolyte, surface charge distribution along the nanopore wall, etc.

2 MEASUREMENT

The nanopores used were samples produced by the Naval Research Laboratory and kindly donated to NSWC Dahlgren Division. The nanopores are well documented [14]. The conduction measurement using the nanopores can be described quite simply but is difficult to do and involves considerable art as well as science. There are some dynamic processes going on simultaneously which will affect results including the rate of solvent evaporation and drift of the Atomic Force Microscope (AFM) tip across the surface of the sample. Fluctuation in the data is significant from experiment to experiment, even on the same sample. Yet, such experimental probes are necessary to characterize the unique behavior found in nanopores.

The procedure consists of:

1. Preparing the electrolyte solution.
2. Cleaning or preparing the nanopore sample.
3. Electrowetting the sample to insure filling the nanopore with the electrolyte solution. (This often requires voltage ~ 100 VDC.)
4. Quickly transporting the sample to the AFM and conducting a contact scan.
5. Simultaneously capturing conductance data during the contact-contour surface scan.
6. Quickly deciding on a representative nanopore. Positioning the tip over the selected pore and conducting a semi-DC sweep of voltage over a range of roughly plus-minus 12 Volts.

3 MODEL

Numerous models have been proposed for conduction in nanostructures. In general, they are adaptations of Debye-Hückel theory [15], [16], [17]. The model chosen to analyze the nanopore conduction behavior seen in this paper is an adaptation of previous pulsed-power research by one of the authors. Instead of relying on an ionic atmospheric to represent "drag" on an ionic conductor moving in a dielectric, polar fluid under the impetus of a semi-DC electric field, the mobility value is assumed to be enhanced directly. Other modifications are certainly possible and reasonable, including concentration alterations, mobility as a function of electric field (the Wien effect), etc. But the first-order correction has been assumed to be a mobility enhancement in the double layer near the walls of the nanopores. Therefore, two regions of conduction exist within the cylindrical nanopore: the bulk region (with standard conduction values) and the double layer region near the edge where enhanced mobility is fit to the data. Please see Figure 1.

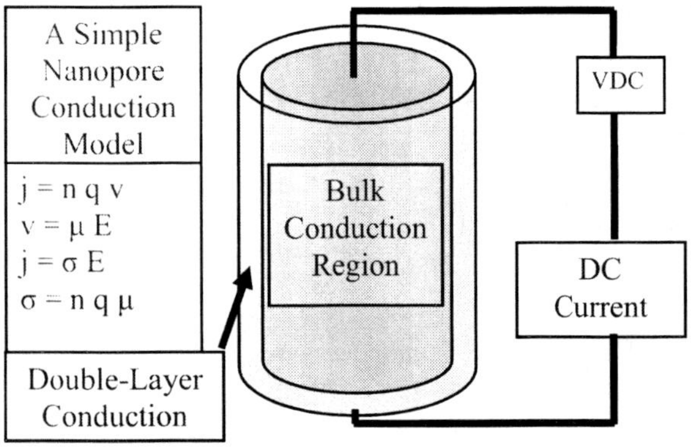

Figure 1: Schematic of a simple conduction model for a nanopore.

The key to the validity of the enhanced-conduction double-layer model is whether 'reasonable' values for the double-layer thickness and enhanced mobility can account for experimental observation. Those parameters are fitted through a trial-and-error process bounded by physical expectations for water electrolytes. Some expected boundaries include:

1. Limit of the enhanced mobility equal to the ratio of the dielectric constant in the bulk to that in the double layer. In practice this means less than a factor of 78 at room temperature, but since the dielectric constant of the double layer is more often calculated to be between 2 and 6, an upper limit of 40 would be better.
2. Limit the thickness of the double layer to some reasonable multiple of the Debye Length in a water electrolyte, which is 3.033 nm at room temperature in 0.01M NaCl.

$$\lambda_D = \sqrt{\frac{\varepsilon k T}{e^2 \sum_{i=1}^{N} Z_i^2 n_i}}, \qquad (1)$$

with λ_D as the Debye Length,

ε = dielectric constant of water, 78 at room temperature of 25C,

k = Boltzmann constant = 1.380658 x 10-23 J/K,

T = Temperature in K, room temperature = 298.15K,

e = electronic charge = 1.60217733 x 10-19 Coulombs,

Z_i = valence of the ith conducting ion in the water electrolyte,

n_i = number density of the ith conducting ion in the water electrolyte, in terms of number of ions per cubic meter.

For this calculation there are assumed to be four conducting ions in the electrolyte: Na+, Cl-, H3O+ and OH-. Although the mobilities of all are within an order-of-magnitude, the number density for the hydronium and hydroxyl ions are five orders-of-magnitude less than the sodium and chloride ions for 0.01M salt. Therefore, the conductance is dominated by sodium and chlorine.

4 ANALYSIS

Figure 2 shows two experimental data curves as well as two theoretical calculations. The experimental data curves do not overlap and give a good indication of the difficulty of the measurement. As stated previously, the AFM tip must be positioned over one of the 150-nm, water-electrolyte-filled nanopores and held there while a slow, semi-DC sweep of voltage and corresponding current measurements are made through the tip. While the tip may be drifting, it is also the case that water is evaporating. Both of these phenomena affect the current measurement. Furthermore, the two curves are not symmetric about 0 Volts, which may mean that there are some other voltage potentials present (e.g., electrochemical mismatch between the platinum-coated silicon tip and the copper grounding plate). It is hardly surprising that the data reflect a degree of uncertainty.

The two theoretical calculations in Figure 2 do not fit the data well at all. The traditional theory represents a calculation using bulk values for 0.01M salt water within a right-circular cylinder 150-nm in diameter and 2-mm long. The traditional theory also has a voltage correction built in to account for the platinum-copper potential difference noted above. However, the asymmetry of the data is still not well explained. The double-layer calculation assumes a double-layer thickness equal to a Debye Length in 0.01M salt water, 3.033 nm, and a mobility enhancement of 10 times. Clearly parameter changes are necessary to fit the data.

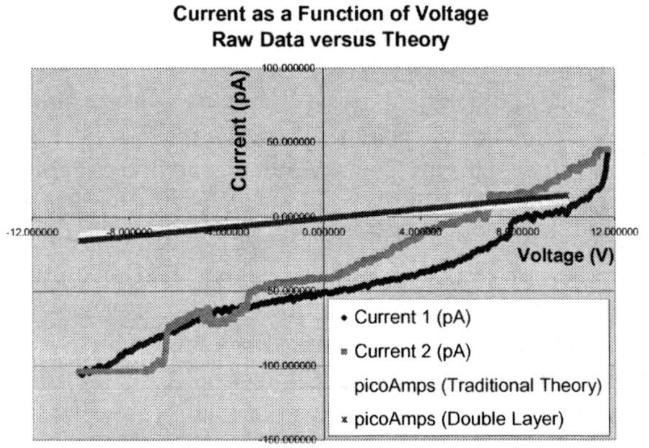

Figure 2: Experimental data plotted versus theoretical calculations of current versus voltage for a nanopore. Double-layer parameters are mobility enhancement is 10 times and Debye Length is 3.033 nm.

Figure 3 displays the same experimental curve and traditional calculation. The new double-layer calculation increases the mobility enhancement to 78 times while keeping the thickness at 3.033 nm. The slope of the line is better but not acceptable yet.

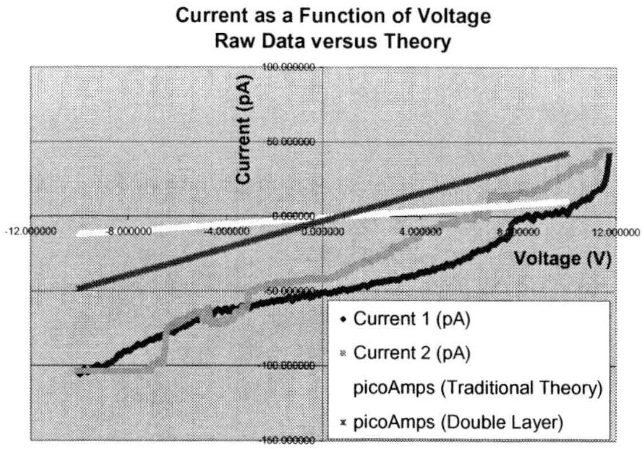

Figure 3: Experimental data plotted versus theoretical calculations of current versus voltage for a nanopore. Double-layer parameters are mobility enhancement is 10 times and Debye Length is 3.033 nm.

Figure 4 maintains most of the parameters as Figure 3, but the double-layer thickness is increased roughly three times to 10 nm. These parameters clearly overestimate the conductance, which is comforting since they represent the outer limits of acceptability for our model of dielectric-dominated conductance effects.

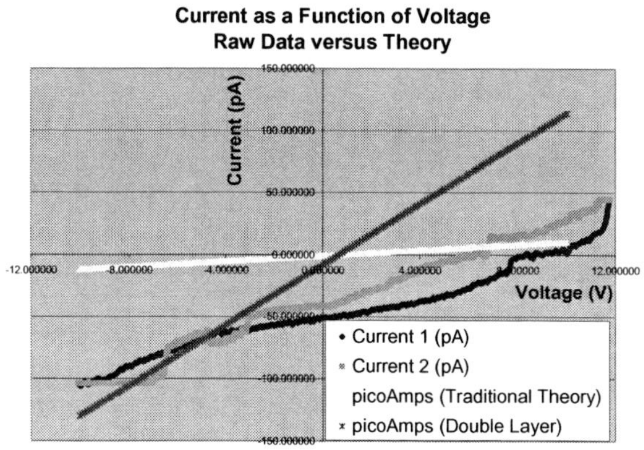

Figure 4: Experimental data plotted versus theoretical calculations of current versus voltage for a nanopore. Double-layer parameters are mobility enhancement is 78 times and Debye Length is 10 nm.

Figure 5 keeps the 10-nm double-layer thickness but reduces the mobility enhancement factor to 40 times. Here the slope is quite acceptable but the voltage offset is not present.

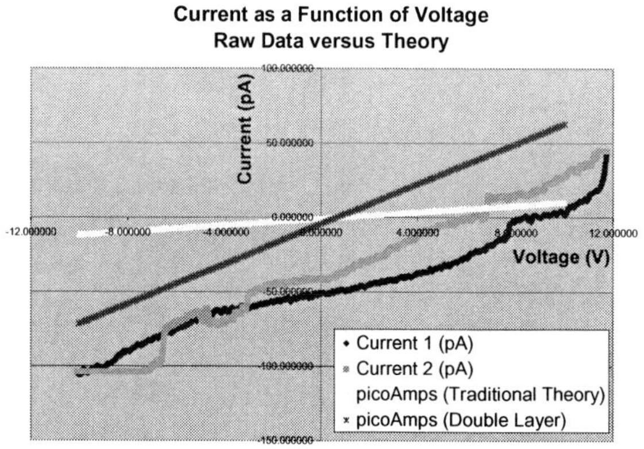

Figure 5: Experimental data plotted versus theoretical calculations of current versus voltage for a nanopore. Double-layer parameters are mobility enhancement is 40 times and Debye Length is 10 nm.

Finally, Figure 6 represents a small tweak of the double-layer thickness from 10 to 6 nm. While a least-squares fit has not been done yet, it is clear that this fit is roughly as good as or perhaps a little worse than Figure 5.

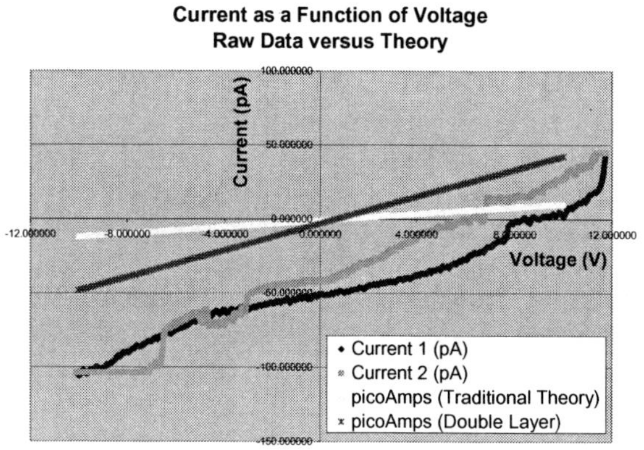

Figure 6: Experimental data plotted versus theoretical calculations of current versus voltage for a nanopore. Double-layer parameters are mobility enhancement is 40 times and Debye Length is 6 nm.

5 CONCLUSIONS

The intent of this paper was to determine if an enhancement of the ionic mobility within a small, reasonable double layer next to the wall could adequately explain experimental data for conductance measurements over 150-nm diameter nanopores filled with 0.01M salt water.

Clearly the data can be explained by this mechanism. However, no other conclusive proof for this mechanism has yet been collected. Other hypotheses by other researchers are clearly still in play. In fact, much remains to be done for this hypothesis including:

1. Accounting for the voltage offset in the experimental data.
2. Allowing the theory to provide for other phenomena observed in past electrochemical and pulsed-power experiments, including:
   a. electric-field-enhanced conductance,
   b. electric-field-enhanced dissociation,
   c. improved calculation for dielectric constant and thickness within the nanopore, especially as a function of surface charge,
   d. measurement of surface charge within the nanopore and any other properties that will affect conductance,
   e. measurement of conductance for smaller nanopores with theoretical analysis, and
   f. determination of the classical-quantum boundary for small nanopores wherein classical water-electrolyte theory fails to explain conductance.

Enhanced conductance within nanopores is still an important phenomenon with many potential uses in technology as well as lessons-to-learn in science. Characterization of conductance is the key to future utilization.

6 ACKNOWLEDGEMENTS

The authors would like to gratefully acknowledge our collaborators at the Naval Research Laboratory, Dr. Charles Merritt and Dr. Jeff Long, and at the Naval Surface Warfare Center, Carderock Division, Dr. Jack Price and Dr. Norris Lindsey. The authors would also like to thank our sponsors; namely the In-house Laboratory Independent Research (ILIR) Program managed by Dr. Jeff Solka of Dahlgren and funded by the Office of Naval Research (ONR) as well as the Discretionary Technical Investment Program (DTIP) managed by June Drake of Dahlgren.

REFERENCES

[1] V. Gehman and D. Fenneman, "Experiments on Water-Capacitor Electrode Conditioning by Ion Bombardment", 1980 Fourteenth Pulsed Power Modulator Symposium, Orlando, Florida, June 3-5, 1980.

[2] M. Zahn, Y. Ohki, D. Fenneman, R. Gripshover and V. Gehman, "Dielectric Properties of Water and Water/Glycol Mixtures for Use in Pulsed Power System Design", Proceedings of the IEEE, September 1986, p 1182-1221.

[3] Gehman, Jr., V.H., Berger, T.L., and Gripshover, R.J., "Theoretical Considerations Of Water-Dielectric Breakdown Initiation For Long Charging Times," Proceedings of the 6th IEEE International Pulsed Power Conference, Arlington, Virginia, June 29-30 and July 1, 1987.

[4] Gehman, Jr., V.H. and Gripshover, R.J., "Water-Based Dielectrics for High-Power Pulse Forming Lines," NATO Advanced Study Institute on The Liquid State and Its Electrical Properties, poster session, Sintra, Portugal, July 5-17, 1987.

[5] Bowen, S.P., Gehman, Jr., V. H., and Gripshover, R.J., "Microscopic Theory Of Dielectric Breakdown In Liquids," Poster Paper at the Gordon Conference on Dielectric Phenomena, Holderness School, Plymouth, NH, July 25-29, 1988.

[6] Gehman, Jr., V. H., Impulse Electrical Breakdown of High-Purity Water, Doctoral Dissertation at Virginia Tech, May 1995.

[7] Boulais K., Santiago F., Long K., Gehman, Jr. V.H., "Conduction Measurements of Electrolyte Filled Nanostructures Using Conducting AFM", 2005 Fall Meeting of the Materials Research Society, Boston, Massachusetts, November 27 to December 1, 2005.

[8] K. J. Long, V. H. Gehman, F. Santiago, K. A. Boulais and A. Rayms-Keller, "Conduction Measurements of Electrolyte-Filled Nanochannel Glass Using Conducting AFM," Submitted to the Seeing at the Nanoscale IV International Conference, University of Pennsylvania, Philadelphia, PA, 17-20 July 2006.

[9] H. Frohlich, Theory of Dielectrics, London, England, Oxford University Press, 1968.

[10] F. Booth, "The Dielectric Constant of Water and the Saturation Effect", Jour. Chem. Phys., Vol. 19, No. 4, 391. April 1951.

[11] J. O'M. Bockris, M.A.V. Devanathan and K. Muller, "On the Structure of Charged Interfaces", Proc. Roy. Soc. (London), A274, 55, 1963.

[12] J.V. Macpherson, C.E. Jones, A.L. Barker and P.R. Unwin, "Electrochemical Imaging of Diffusion through Single Nanoscale Pores", Analytical Chemistry, Vol. 74, No. 8, 1841, April 15, 2002.

[13] S. Pennathur and J.G. Santiago, "Electrokinetic Transport in Nanochannels. 1. Theory", Analytical Chemistry, Vol. 77, No. 21, 6772, November 1, 2005.

[14] R.J. Tonucci, B.L. Justus, A.J. Campillo and C.E. Ford, "Nanochannel Array Glass", Science, Vol. 258, pp.783-785, 30 October 1992.

[15] P. Debye and E. Hückel, Physik. Z., 24, (1923), 311.

[16] L. Onsager, Physik. Z., 27, (1926), 388.

[17] G. Kortüm, Treatise on Electrochemistry, Second, Completely Revised English Edition, pp. 186-191, Elsevier Publishing Company, 1965, LOC 63-19824.

Viscosity measurements of PEO solutions by AFM with long nanowires

Mahdi Hosseini, Mehdi M. Yazdanapanah, Sohel Siddique and Robert W. Cohn
ElectroOptics Research Institute & Nanotechnology Center
University of Louisville, Louisville, Kentucky 40292
rwcohn@uofl.edu

ABSTRACT

Long, constant diameter nanowires on the end of atomic force microscope (AFM) probes are expected to provide clean force distance (F-D) measurements that are easily interpretable. For measurements of the viscosity of low molecular weight liquids this proves to be true. Comparable values are found for glycerol when the nanowire is drawn from the liquid at a constant velocity over microns in scan length and when the nanowire insertion depth changes no more than that due to thermal Brownian vibration fluctuations, as is used for the *Q* damping measurement method — which also matches macroscopic viscometry methods. However, for measurements of the high molecular weight random chain polymer poly(ethylene oxide) (PEO) the differences are dramatic, with the *Q* damping method giving values that are orders of magnitude smaller than the linear drag force measurement. Since the deflection amplitude is sub-nanometer, the unusually low viscosity might be due to displacements that are smaller than the entanglement length of the polymer. Additional preliminary investigations are reported that do show somewhat increased viscosity with 10X increased vibration amplitude. Also time dependence of the measurements appears to be related to adsorption of monolayer coatings of PEO on the needle.

Keywords: AFM, viscosity, drag force, *Q* damping

1 INTRODUCTION

With increasing practical applications of microfluidics, lab-on-a-chip, and micro-reactors, it is becoming increasingly important to develop *in situ* nanoscale sensors of fluid mechanical properties; *e.g.* surface tension, contact angle, evaporation rate, viscosity. Principles and methods from atomic force microscopy, with suitably designed probes, can be adapted to sense such properties in small confined environments or with miniscule quantities of sample. For the measurement of viscosity, a long, constant diameter probe tip is highly desirable because drag force is proportional to insertion depth, which enables values of drag force to be produced that are in the detectable range of the AFM detection circuitry. Additionally, the constant diameter means that the wetting force due to surface tension is constant with insertion depth (due to the constant length of the contact line). With standard tapered AFM probe tips the wetting force (as well as the change in drag force) changes dramatically with insertion depth.

A few groups have reported on the use of (custom formed) constant diameter probe tips to measure surface tension [1]. Mechler *et al.* [2] have utilized the lateral force mode (friction mode) of the AFM to scan carbon-nanotube tipped probes in glycerol/water films. Torsional bending of the cantilever is then related to the drag force. The force-distance (F-D) mode, where the probe is scanned perpendicular to the surface and the force is determined from the bending of cantilever, also provides a direct and general method to measure liquid forces [3].

For the determination of viscosity we use a ten's of microns long, constant diameter metallic nanowire (or "nanoneedle") on the AFM probes to measure drag force (Fig. 1). The needle is custom grown on the cantilever by a room temperature process [4]. Drag force is then measured by drawing the needle with a known velocity and at a known insertion depth from the liquid while recording the F-D response. Viscosity is then determined from the well-known expression for longitudinal drag on a cylindrical object.

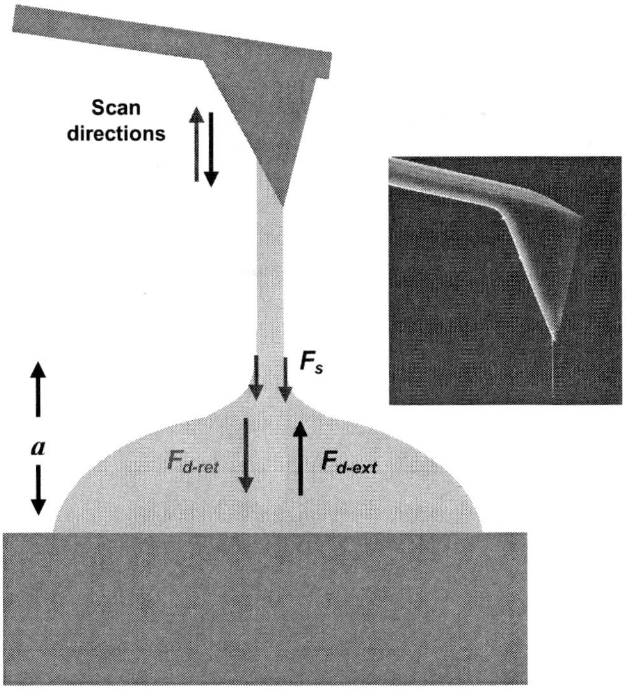

Figure 1. Viscosity measurement setup with nanoneedle-tipped probes. The inset is an SEM image of a needle-tipped probe. Forces considered in the measurements are shown for retraction (red) and extension (blue) scans.

Drag force on a needle, and thereby viscosity, also can be related to the damping it produces on an attached mechanical resonator. The AFM cantilevers generally do have high Q and the AFM detection circuitry is sensitive enough to measure damping changes in the amplitude vibration spectrum of the cantilever due to thermal (Brownian) fluctuations in the cantilever position. These features are part of the standard AFM sensing mode referred to as the Q damping method.

These two measurement methods are found to give the same results for low molecular weight liquids, in correspondence with macroscopic viscometry measurements. However, for highly viscous solutions containing the random chain polymer PEO, we find that the Q damping method under-reports viscosity by orders of magnitude. This paper reports on our initial measurements of both low molecular weight glycerol/water solutions and of aqueous PEO solutions, as well as experimental evaluation of some factors that affect the values of viscosity in the Q damping method. It will be quite interesting and physically informative if the Q damping method turns out to provide detailed information on interactions between polymer chains at molecular length scales that are smaller than the length of the polymer chains.

2 MEASUREMENT PRINCIPLES AND EXPERIMENTAL METHODS

Figure 1 shows the experimental setup, which is identical for both the Q damping method and the constant velocity/linear drag force method.

2.1 Viscosity from constant velocity scans

The AFM is configured to produce F-D scans of retraction and extension as the needle is translated normal to the surface of the liquid. Drag force F_d sensed by the cantilever deflection is oppositely directed from the scan motion, while surface tension induced force F_S is directed downwards (for contact angle less than 90 °, which was the case for all liquids studied.) The total force, F_T applied to the needle during the retraction is

$$F_T = -(F_S + F_{d\text{-}ret}) \tag{1}$$

where $F_{d\text{-}ret}$ is the drag force during the retraction. For long cylindrical needles the drag force in the axial direction of the needle is [5]

$$F_d = \frac{4\pi\eta a v}{\ln(2a/r) - 0.81} \qquad r \ll a \tag{2}$$

where a is the submerged length of the needle, r is the radius of the needle and v is the scan speed. The surface tension force applied to the needle is

$$F_S = 2\pi r \sigma \cos(\theta) \tag{3}$$

where σ is the surface tension and θ is the contact angle of the liquid surface. The F-D curve records the total force applied to the cantilever at any immersion depth of the needle. When the needle breaks free of the liquid, a stepwise change in the force curve equal to F_S occurs [1]. This step in the force curve locates the reference height of the liquid, thereby providing the immersion depth of the needle throughout the scan. The viscosity is determined by repeating the experiment at different scan velocities. The slope of a graph of drag force at a constant insertion depth gives the viscosity of the liquid.

2.2 Viscosity from Q damping

The Q damping mode is a standard feature of the Asylum MFP3D AFM used in this study. Mechanical excitation is normally produced by Brownian thermal fluctuations. In some experiments we introduced an electrically generated white noise to the cantilever holder's piezoelectric shaker, to produce somewhat larger amplitude fluctuations. When a needle is partially inserted into a liquid (Fig. 1), the Q of the resonator is reduced to

$$Q = \frac{\sqrt{Mk_c}}{R} \tag{4}$$

where M is the effective mass of the cantilever, k_c is the cantilever spring constant and R is the drag coefficient. (Note that damping due to the liquid generally reduces the Q substantially, thereby eliminating the consideration of air, and other sources of damping on the cantilever.) The drag coefficient in the axial direction of the needle is calculated by equation (2). From equations (2) and (4), Q can be related to viscosity η as

$$\frac{1}{Q} = \frac{1}{\sqrt{Mk_c}} \frac{4\pi\eta a}{[\ln(2a/r) - 0.81]} \tag{5}$$

The value of M is established by using a reference liquid of known viscosity. Typically we use water which is 1 cP at 20 °C. Then the value of η is the only unknown parameter in equation (5). The needle is especially convenient in that equation (5) can be fit for several values of the insertion depth a, which can be used to average out measurement errors.

Cantilever characteristics [6] and the characteristics of the attached nanoneedles [7a-e], are described in the references.

3 EXPERIMENTAL MEASUREMENTS

3.1 PEO viscosity by constant speed scans

The method from Sec. 2.1 is used together with equations (1) and (2) to evaluate the viscosity of a 4 wt % PEO

(1,000,0000 MW) aqueous solution. The needle [7a] is inserted a few microns into the liquid and retracted from the liquid at a constant velocity while recording the F-D curve is recorded. Figure 2 shows the total force F_T for retraction speeds from 12 to 98 μm/s at a 2 μm insertion depth. The slope of the linear fit corresponds to a viscosity of 2,200 cP. A larger sample of the same liquid sample is measured to be 2,342 cP in a Brookfield LV-DV II+ cone and plate viscometer. In the AFM measurements the laboratory temperature varied between 21 and 24 °C and in the cone-plate viscometer the laboratory temperature varied between 21 and 23 °C.

Additionally, the linear fit gives a wetting force F_S = 62.74 nN at zero scan velocity. The surface tension of 3 wt % PEO and higher has been reported to be 62.6 mN/m. [8] From equation (3), the receding contact angle is calculated to be 48.3 °.

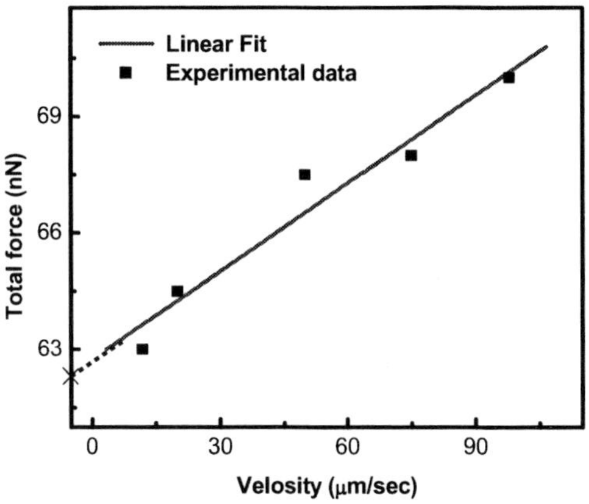

Figure 2. Measurement of total force and it resolution into drag force and wetting forces for 4 wt % PEO.

3.2 Glycerol and PEO by *Q* damping

First a series of glycerol/water solutions were measured from 0 to 100 wt % glycerol. The needle is inserted to several depths *a* into the liquid, and *Q* is recorded for each depth. The data is fit to equation (5) as a function of the unknown value of viscosity. The insertion depths used are 1.5 μm or greater which ensures that the needle is a long slender cylinder, as is assumed by the model drag force in equations (2) and (4). As described above, the published value for water (0 wt % glycerol) is used as a reference value to establish the value of effective mass M=1.3×10^{-9} gm used in equation (5) for the needle-tipped cantilever used [7b]. Fits to three of these solutions are shown in Fig. 3a and the determined viscosity is plotted in Fig. 3b, along with all the measured viscosities. We also measured and graphed the identical solutions for cone-and-plate viscometer measurements (at shear rates from 1350/s for pure water down to 1.5/s for pure glycerol). The plots also show published viscosity values [9]. The viscometer data and AFM data generally track each other, as well as the general trends in the published data at 20 and 30 °C. The published data does make evident the need to regulate the temperature for highly accurate measurements.

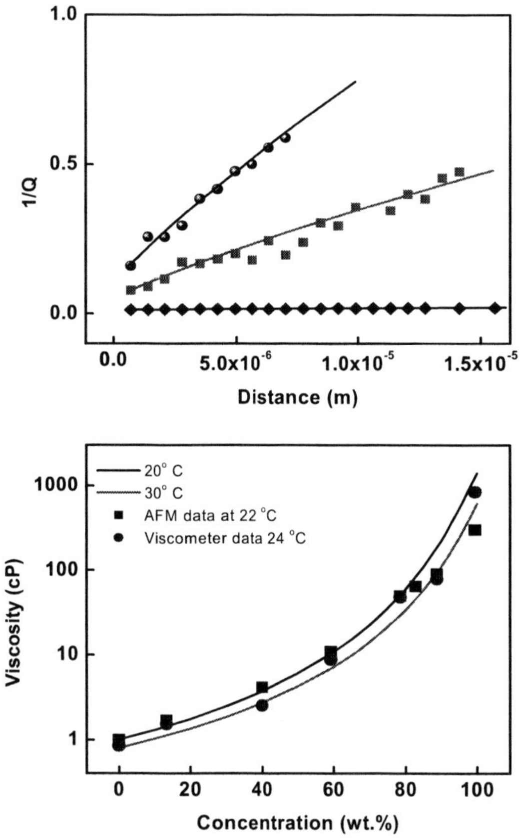

Figure 3. Thermal *Q* damping measurements of solutions of aqueous glycerol. (a) Reciprocal of *Q* as a function of insertion distance. Experimental data and best fit curves to equation (5) for 13.5, 80 and 99.5 wt % glycerol (from lowest to highest curve). (b) Values of viscosity from best fits of equation (5) to data sets of the type in (a). Plot also includes cone and plate viscometry measurements and published values [9]. The relative standard deviation for each measured data point varies from 4 % (for the highest concentration) to 20 % (for pure water).

PEO, especially of high molecular weight, is difficult to completely dissolve and mix in water. At 4 wt % and higher we observe that the solution becomes cloudy. At these concentrations the polymer can coat the needles [7c] to a noticeable degree, which can form a large blob or long string if the tip is not frequently rinsed clean in pure water. Solutions above 4 wt % proved to be too difficult to work with by the AFM method. These problems are not evident in the viscosity measurements using the cone-and-plate viscometer. The resulting AFM and viscometer measurements are compared in Fig. 4. The AFM measurements depart dramatically from the viscometer measurements with increasing concentration. The largest value of viscosity measured by AFM is only ~5 cP for 4 wt % PEO, which is far lower than the viscosities measured for glycerol solutions in Fig. 3. The *Q* damping experiment was repeated using a softer cantilever (k_c<0.1 N/m) in order to produce an increased thermal amplitude fluctuation. For PEO of 3 wt % the measured value for viscosity in Fig. 4 increases, but still is far below that for the cone and plate viscometer. This increase may be due to increase in the amplitude of the cantilever vibration. Because the cantilever is actuated by Brownian thermal fluctuations the cantilever displacement is sub-nanometer, which suggests the possibility that the cantilever does not fully engage the entanglements between the random chains of the PEO.

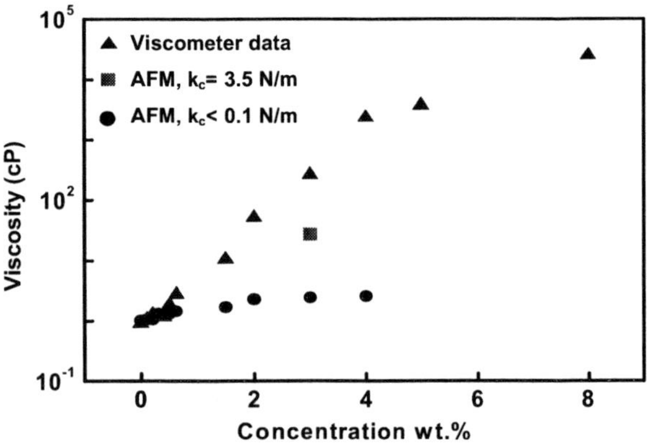

Figure 4. Viscosity measurements of PEO solutions.

In order to explore this we further actuated the cantilever by applying a white noise source to piezoelectric shaker, increasing the cantilever [7d] displacement by a factor of 10X. While this increase in amplitude showed no change in the measured viscosity of glycerin, in 2 wt % PEO it showed an increase from 2 cP to 4 cP. This can be seen in the increased slope of $1/Q$ versus insertion depth of the needle into the liquid in Fig. 5a. The viscosity calculated from a macroscopic drag force equation is still smaller than the macroscopic viscosity by orders of magnitude. With the current AFM setup it is not possible to extend the cantilever oscillation amplitude further.

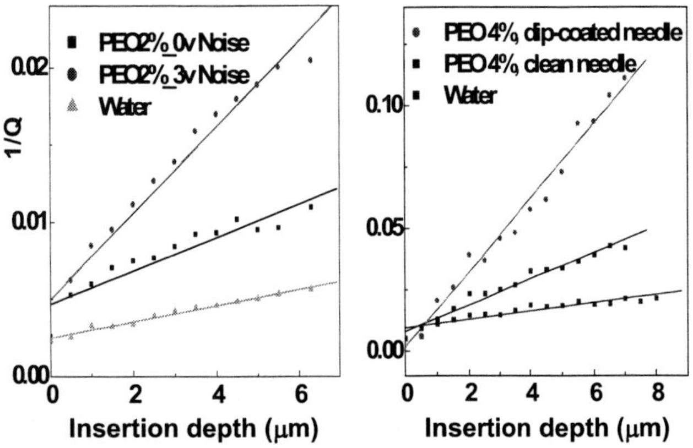

Figure 5. Q damping measurements of PEO solutions for various insertion depths. (a) Q damping by the application of white noise to a piezo shaker (red curve) is greater than with purely thermal actuation (black curve). (b) Q damping increases when the needle adsorbs PEO (red curve) compared to a freshly cleaned needle (black curve).

A second set of experiments with 4 wt % PEO shows that the viscosity changes from 3.1 cP when the needle [7e] is first immersed in the liquid, to 8.9 cP 10 minutes later. Since no coating is observable in SEM images, this suggests that only a few monolayers of PEO are adsorbed to the needle, which nonetheless influences the interaction of the needle with the liquid. Table 1 summarizes the results for various types of measurements on PEO.

Table 1. Viscosity of aqueous PEO by various methods

AFM Method	PEO (wt %)	Viscosity by AFM (cP)	Viscosity by viscometer (cP)
Q (thermal)	2	2	52
Q (electrical)	2	3.8	52
Q (fresh)	4	3.1	2,342
Q (10 min)	4	8.9	2,342
Linear drag	4	2,200	2,342

4 CONCLUSIONS

Probing polymeric solutions with nanoneedles has identified differences between the linear drag force and Q damping methods that raise the intriguing possibility of being able to measure continuous transitions between molecular scale and macroscopic viscosity.

REFERENCES

[1] M. M. Yazdanpanah, M. Hosseini, S. Pabba, S. M. Berry, V. V. Dobrokhotov, A. Safir, R. S. Keynton, R. W. Cohn *SEM Annual Conference, Session 91 Nanocomposite Characterization*, Springfield MA, June 3 - 6, 2007.

[2] A. Mechler, B. Piorek, Lal R. Ratnesh and S. Banerjee, Applied Physics Letters 85, 3881, 2004.

[3]M. Radmacher, M. Fritz, J. P. Cleveland, D. A. Walters, and P. K. Hansma, Langmuir 10, 3809, 1994.

[4] M. M. Yazdanpanah, S. A. Harfenist, A. Safir and R. W. Cohn, Journal of Applied Physics 98, 073510, 2005.

[5] R. G. Cox, Journal of Fluid Mechanics 44, 791, 1970.

[6] Unless otherwise noted the cantilever that supports the needle is a BS-Multi75 (Budget Sensors, Sofia, Bulgaria) with resonance frequency 75 ± 15 kHz, spring constant 3.5 ± 1 N/m, and Q (in air) of between 180 to 220.

[7] The probes are specialized with (a) a 30 μm long × 480 nm diameter needle, (b) a 20 μm long × 362 nm diameter parylene coated needle (including coating thickness), (c) a 57.5 μm long × 388 nm diameter parylene coated needle, (d) a 48.48 μm long × 532 nm diameter needle on a cantilever of 350 ± 15 kHz resonance frequency and 28.5 ± 1 N/m spring constant, (e) 33.57 μm long × 376 nm diameter needle. The cantilever used for (e) has a spring constant of 25 N/m and is from Budget Sensors.

[8] R. Crooks, J. C. Whitez and D. V. Boger, Chemical Engineering Science 56, 5575, 2001.

[9] Lide, D. R. *CRC Handbook of Chemistry and Physics*, CRC Press: Boca Raton, 2002.

A lattice Boltzmann study of the non-Newtonian blood flow in stented aneurysm

Yong Hyun Kim, Sam Farhat, Xiafeng Xu, Joon Sang Lee

Department of Mechanical Engineering
Wayne State University
Detroit, MI, 48202, USA

ABSTRACT

The analysis of a flow pattern in cerebral aneurysms and the effect of strut shapes and stent porosity in 2D and 3D model are presented in this paper. The efficiency of a stent is related to several parameters, including porosity and stent strut shapes. The goal of this paper is to identify numerically how the stent strut shape and the porosity affect the hemodynamics properties of the flow inside an aneurysm. The lattice Boltzmann method (LBM) of a non-Newtonian blood fluid is used. To ease the code development, a scientific programming strategy based on object-oriented concepts is developed. An extrapolation method for wall and stent boundary conditions is used to resolve the characteristics of a highly complex flow. The reduced velocity, vorticity magnitude, and shear rate were observed when the proposed stent shapes and porosities are used. The rectangular strut shape stent is observed to be optimal and decrease the magnitude of the velocity by 89.25% in 2D model and 53.92% in 3D model in the aneurysm sac. Our results show how the porosity and stent strut shapes play a role and help us to understand the characteristics of stent strut design.

Key words: lattice Boltzmann method, extrapolation method, strut shape, porosity,

INTRODUCTION

Studies of aneurysm models have shown complex hemodynamic changes in an aneurysm after the placement of a stent across the aneurysm neck [1-3]. The hemodynamic properties of the blood inside an aneurysm are of great importance in the case of aneurysms by stents. Recently, stents have been used in minimal invasive treatments of aneurysms, as an alternative to by-pass surgery. Due to its limited permeability, the stent modifies the blood flow in the aneurysm. It is thought that the resulting flow stagnation promotes the formation of a stable thrombus in the aneurysm sac leading to its permanent occlusion. Therefore, the stent should modify the blood circulation in the aneurysm but not stop it. In particular, the optimal design of the stent structure is not studied with a function of strut shape of the stent in an aneurysm.

The flow over obstacles of different shapes and sizes has been studied due to its importance to engineering applications. Enhanced surfaces significantly alter the structure of the flow. It was found that shape and size of the rib affected the friction factor significantly [4].

Several experimental and numerical studies have been reported [2, 5 ,6]. The existence of large coherent vortex structures in lateral aneurysm model is emphasized. However, they do not discuss well the reduced flow mechanism by stents.

The main goal of this paper was to propose the effect of stent porosity and the effect of strut shape stent with a non-Newtonian extension to the standard LB model that incorporates correct blood rheology in case of small shear rate. This study was organized as follows. In the 2D model with ghost cell extrapolation method, the effects of stent porosity cases were studied with the steady flow condition. The second was that the effects of various strut shapes of stents were considered and analyzed by the flow reduction/enhancement in an aneurysm sac with the pulsatile flow condition in 2D model. In 3D model with bounce-back method, the flow characteristics were analyzed with or without a stent. In 2D and 3D models, averaged flow characteristics and shear rate with non-Newtonian fluid for the aneurysm were considered.

NUMERICAL METHOD

A. The lattice Boltzmann method

The LBM used in this study is the incompressible lattice Bhatnagar-Gross-Krook (BGK) model (D2Q9) [7] and the incompressible lattice BGK model (D3Q19) [8].

An incompressible viscous fluid is assumed in the present study. For a non-Newtonian fluid, effective viscosity μ is found to vary with local shear rate $\dot{\gamma}$. Carreau Model [9] was used in this study.

B. Boundary Conditions

In this study, two boundary methods for walls were used. The bounce-back method for 3D simulation was easy for implementation and supported the idea that LBM was ideal for simulating fluid flows in complicated geometries. But this method is only the first order in numerical accuracy at boundaries [10] unless the boundary is at the center between lattice nodes.

The extrapolation scheme for 2D simulation is of the second-order accuracy, and it is consistent with the accuracy of the D2Q9 model [11].

Results

A. Validation

Simulation capability to a variety of configurations with respect to the type of flow conditions to obtain a comprehensive comparison with analytical data was conducted.

The first validation test of the 2D extrapolation method of LBM was presented. The aneurysm model in this study was compared with the numerical study in Hirabayasi et al [12]. The velocity magnitude was compared at the aneurysm neck and the difference was less than 8% in 95.51% of porosity case.

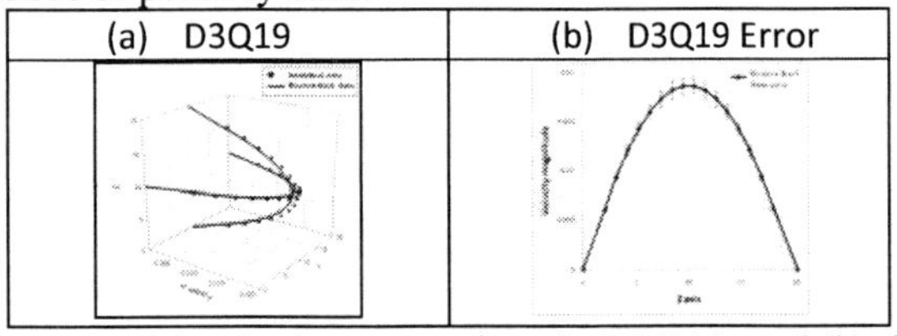

Figure 1. (a) velocity comparison in D3Q19, and (b) the error of velocity in D3Q19 model

The second validation of the fully developed flow in 3D model was shown in Fig. 1(a). The fully-developed channel flow driven by a constant pressure gradient was shown in Fig. 6(b). The profile of the analytical velocity (u_e) [13] is expressed as parabolic. $u_e = -\frac{R^2}{4\rho\nu}\frac{dp}{dx}\left[1-\left(\frac{r}{R}\right)^2\right]$,

where the non-dimensional values of pressure gradient, $dp/dx = 2.026\times10^{-5}$, kinematic viscosity, $\nu = 0.026$, density , $\rho = 1$, and R is the radius of the channel, respectively. The Reynolds number of the flow was 30. The $20\times20\times80$ ($x\times y\times z$) lattices were used. The profiles of the velocity showed that the lattice Boltzmann solutions matched the analytical solutions within a 4.7 percent as shown in Fig. 1(a) and (b).

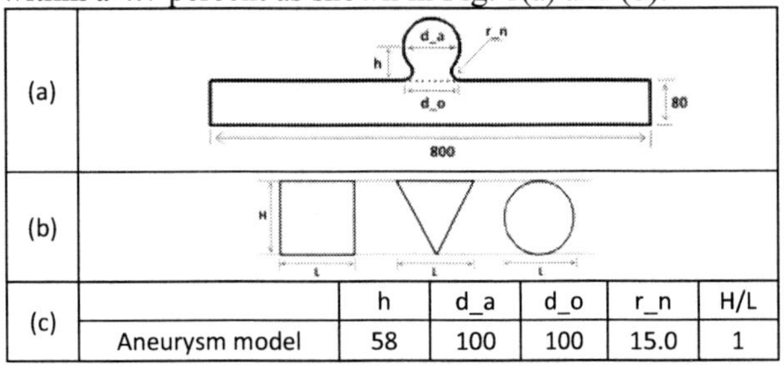

Figure 2. (a) 2D aneurysm model , (b) strut shapes, and (c) aneurysm parameter

B. Flow patterns and vortex

The computational domain for D2Q9 model was shown in Fig. 2. In the cerebral artery, the Reynolds number within the aneurysm is less or about 30 [2] and the velocity magnitude is low (1cm/s). The relaxation time, $\tau = 0.58$, was chosen to produce a constant kinetic viscosity, $\nu = 0.026$, for infinite shear viscosity. The average density of the system was $\rho = 1.0$. The simulation size was 800×188 lattice sites, with a channel width made of 80 sites.

For unsteady flow, at the entry of the artery, the pulsatile inlet velocity as shown in Fig. 3 was used. The parabolic velocity profile at inlet was used. The maximum Reynolds number, used in this study was 20 based on the center velocity, 0.0065m/s, at t=0.3s (peak time). One million time steps were used in one pulsatile period and the time step, Δt, was $10^{-6}s$.

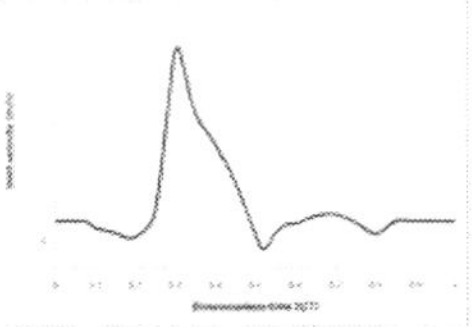

Figure 3, Inlet pulsatile profile

The geometric dimensions of struts were shown in Fig. 2(b). The ratio (H/L) of stents is 1, where H is the height of strut and L is the width of strut. The no-slip boundary condition on the stent surface was used as for the wall.

For this study, we consider 4 cases of unsteady condition in each strut shape and 4 cases of steady state conditions. These cases included three different strut shapes in Fig. 2(b) and four different porosity cases for each strut shape as shown in Table 1.

Table 1, Porosity cases

Case	Porosity (%)	Pore size
Case 1	95.51	35
Case 2	89.53	15
Case 3	83.59	11
Case 4	77.78	9

The Hemodynamic effects of stent implantation on a saccular aneurysm, its flow reduction to the aneurysm sac and critical non-Newtonian flow properties with respect to the strut shape and the stent porosity were studied.

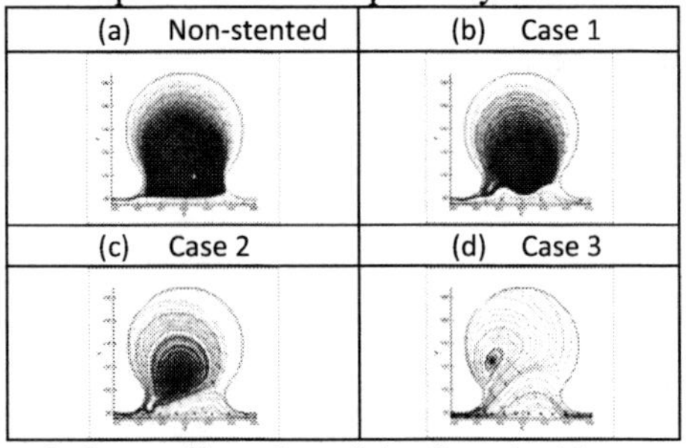

Figure 4, Streamline plot of the flow inside aneurysm with different porosity cases

To understand the effect of stents on the flow characteristics, mean velocity reduction, is defined as $\overline{v}_r = (\overline{v}^{ns} - \overline{v}^{st})/\overline{v}^{ns}\times100$,where $\overline{v}^{ns}$ and $\overline{v}^{st}$ are the averaged non-stented velocity and the averaged stented velocity in the aneurysm sac, respectively [12]. For more analysis, the mean velocity reduction can be replaced to the mean vorticity reduction, $\overline{\omega}_r$ and mean shear rate reduction rate, $\overline{\dot{\gamma}}_r$.

Another effect of the stent is introduced in this study and the so-called mean viscosity increase rate defined as $\overline{\mu}_c = (\overline{\mu}^{st} - \overline{\mu}^{ns})/\overline{\mu}^{st}\times100$, where $\overline{\mu}^{ns}$ and $\overline{\mu}^{st}$ are the averaged non-stented viscosity and the averaged stented viscosity in the aneurysm sac.

The Effect of Porosity

Fig. 4 showed the variation of the flow patterns in the stented aneurysm of a circular strut shape with the steady flow condition. In Fig. 4(a), the vortex in the non-stented aneurysm was driven directly by the flow in the parent vessel and was rather important. On the other hand, in Fig. 4(b), (c) and (d), the vortex was reduced with the stent.
In Table2, it was observed that small velocity magnitudes in the low porosity case compared with all other cases. The mean velocity was decreased by up to 86.8% in the case 3.
Fig. 4 and Table 2 suggested that this reduction was related to the stent porosity. The numerical results of Hirabayashi et al [12], using different stent porosity in various aneurysm sizes, showed that the low stent porosity was better for velocity reduction compared to the high porosity case.

Table 2, velocity reduction at the aneurysm neck

Strut shape	Porosity	$\bar{V}_r$	Umax
Circular shape	C1	46.3	0.016591
	C2	82.0	0.005897
	C3	86.8	0.004007
Without stent		0	0.025305

Stent porosity model can be occluded easily because of the large number of stent struts and the small velocity at the aneurysm neck. The numerical simulation [12] with LBM showed that the stent porosity alone was not sufficient to characterize uniquely the flow reduction in the aneurysm.

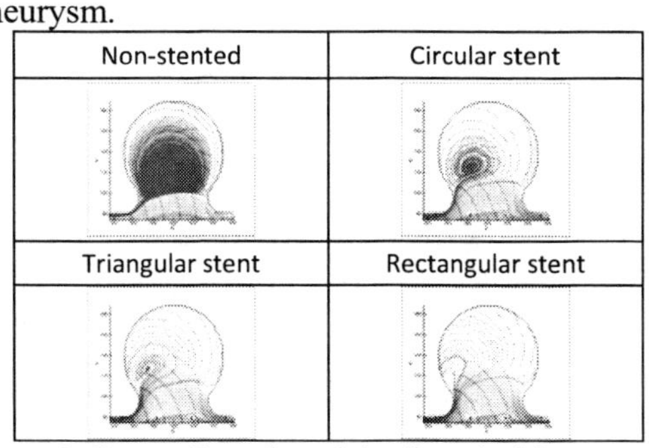

Figure 5, Streamline plot of the flow inside aneurysm with different strut shape stent

The Effect of Stent strut shapes

Different flow patterns in an aneurysm sac with different stent strut shapes were shown in Figs. 5 and 6. Figs. 5 and 6 showed that the velocity was significantly reduced and affected not only in the low porosity cases but also in various strut shape stents. The vortex was significantly reduced in the aneurysm sac in all strut shape stents Fig. 5 showed that the vortex was more reduced in the rectangular strut shape stents than in the other shape stents.
The u-velocity magnitudes were shown in Fig. 6. Since the u-velocity magnitude was higher than the v-velocity magnitude, the u-velocity magnitudes were analyzed in this study. It showed that the maximum velocity of the rectangular strut shape stent was reduced up to 25.9% compared with circular shape and 89% compared with non-stented model as shown in Fig. 6.
The mean flow characteristics reduction or enhancement rate at t=0.3s was shown in Fig. 7. It was observed that the rectangular strut shape stents was very effective to reduce flow into an aneurysm sac more than other strut shapes. The mean velocity reduction rate of the rectangular strut shape stent was up to 92% in case 4(Fig. 7(a)). By the reduced velocity, it was observed that the highest viscosity enhancement rate in the rectangular strut shape stents occurs (Fig. 7(b)). The width of rectangular strut shape stent was observed to be more effective on velocity reduction inside the aneurysm sac than the other stent strut shapes, Because the longer boundary layer in the rectangular strut shape stent was created along the parent artery and keeps the flow from moving into the aneurysm sac.

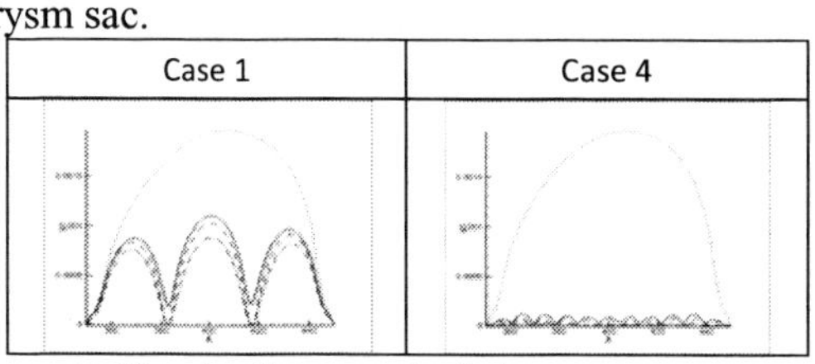

Figure 6, U-velocity magnitude in an aneurysm neck.
(Rigid - circular, Dash - triangular, and Double dot - rectangular shape)

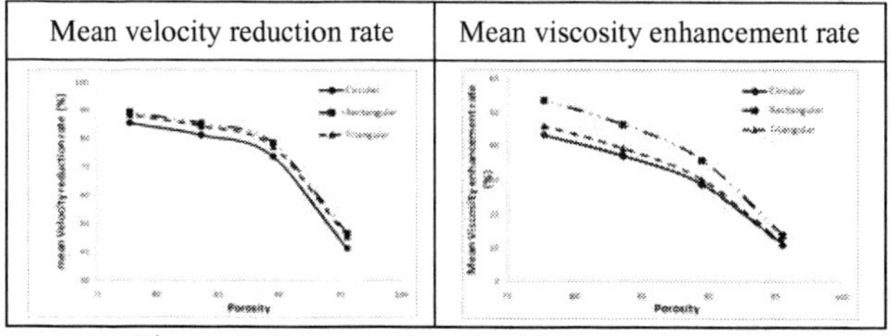

Figure 7, Mean values inside an aneruysm

3D aneurysm model

In this simulation, the 3D aneurysm model with a bounce-back boundary method was considered as shown in Fig. 8. In order to obtain the appropriate Reynolds number, $\mathrm{Re}=75$, for the parent vessel flow. The relaxation time ($\tau=0.58$) was chosen to produce a constant kinetic viscosity ($\nu=0.026$) for infinite shear viscosity. The average density of the system was $\rho=1.0$. The non-Newtonian fluid was used.
The constant pressure gradient $dp/dx \cong 4.17e-6$ is used rather than a pulsatile flow in order to simplify the study of flow reduction by stents. For the inlet and outlet, we used periodic boundaries.
In this study, the 2d aneurysm simulation showed that the rectangular strut shape stent was more effective than other shape stents. Therefore, in the 3D aneurysm model (84.9% porosity), the flow characteristics were compared with rectangular strut shape stent with non-stented model.
Velocity was compared with a non-stented aneurysm model (Fig. 9). Velocity was measured on A-A' (along with the parent vessel) and B-B' sections (across the parent vessel) in an aneurysm neck (Fig. 9). The velocity in an aneurysm neck was observed that stented model had

lower velocity, up to 85% reduction on A-A' section and up to 76.20% on B-B' section. The velocity iso-surface explained that flow into an aneurysm sac was prevented by the stent in Fig. 9.

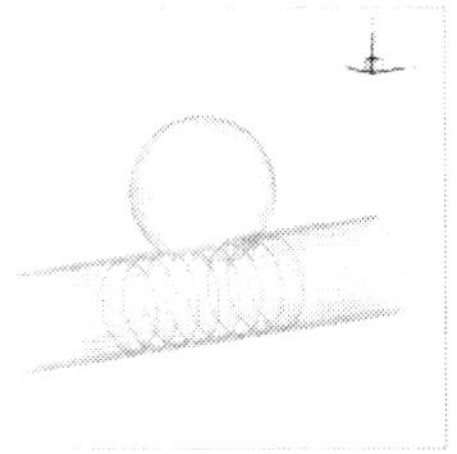

Aneurysm diameter	60
Parent vessel diameter	40
Porosity	84.98 %
Stent strut diameter	2
H/L	1
Re	75
dp/dx	4.17e-6
$200 \times 60 \times 96$ lattice	

Figure 8, 3D aneurysm geometry

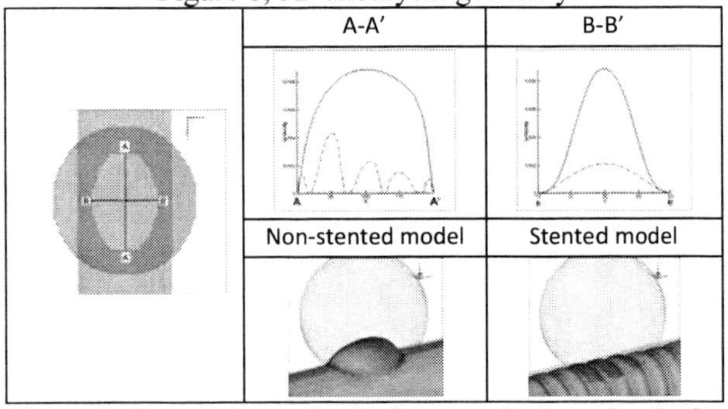

Figure 9, Comparison of velocity magnitude and velocity iso-surface

Table 3, Measured parameters in the 3D simulation

	$\overline{V}_r$	$\overline{\omega}_r$	$\overline{\mu}_r$
Non-stented	2.485e-4	5.198e-5	4.749e-3
Rectangular stent	1.145e-4	1.951e-5	5.522e-3

In Table 3, this velocity reduction affected the other flow characteristics, e.g. vorticity, and visocisty, in the aneurysm sac. It was observed that shear rate was reduced on a stented aneurysm as well as vorticity.

Fig. 10 (a) showed the variation of the flow patterns in non-stented and stented aneurysms. The vortex was reduced with a stent. The secondary flow was also reduced in stented aneurysm. A low secondary flow region had a high viscosity and was related to the low vorticity region [14]. The viscosity in the aneurysm sac was increased with reduced flow by the stent as shown in Fig. 10 (b).

Discussion

A fluid solver based on a lattice Boltzmann method (LBM) is developed and used in 2D and 3D numerical simulations. The flow in aneurysms with different porosity and different shape stents is demonstrated to examine the flow inside the aneurysm. Three shapes of stents and four different porosity cases were investigated in a 2D aneurysm model and the rectangular strut shape was used in a 3D aneurysm model.

The reduced flow and smaller mean velocity magnitude in the aneurysm sac were observed when the low porosity case used. With lower porosity, the velocity was reduced by up to 86.8% with a circular strut shape stent. However, the stent porosity is not enough to describe the effect of the stent alone.

The strut shape, as well as porosity, was to describe the effect of the stent on the velocity reduction and shear rate reduction. So they must be taken into account to fully quantify the role of the stent. The rectangular stent in the unsteady flow condition was observed to be optimal and decreased the magnitude of velocity in an aneurysm neck. The reduction or enhancement rate of flow characteristics of a rectangular strut shape was effective to reduce the flow in an aneurysm, compared with other strut shape and non-stented model.

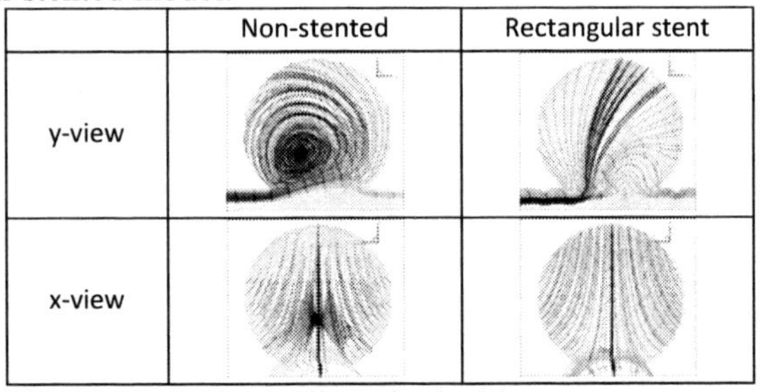

Figure 10, Streamline plot inside an aneruysm

By the small porosity and rectangular strut shape, it is possible to eliminate the strong vortex completely and to create a high viscosity region. An efficient stent should be designed so as to decrease the direct influence of the main flow. For a given porosity, the rectangular shape strut would seem preferable.

The results shown here already indicate that the current fluid solver based on the LBM is a very promising method for blood flow analysis, particularly when complicated geometries were used. More realistic artery geometry and blood models will be used in the future study.

References

1. G Geremia, M Haklic, L Brennecke, *AJNR* 15, 1223-31, 1994
2. BB Lieber, AP Stancampiano, AK Wakhloo, *Ann. Biomed. E ng.* 25, 460-469, 1997
3. F Turjman, TF Massoud, C Ji, G Guglielmi, F Vinuela, J Rob ert, *AJNR* 15, 1087-1090, 1994
4. JC Han, LR Glicksman, WM Rohsenow, *Int. J. Heat Mass Tr ansf.* 21, 1143-1155, 1978
5. M Aenis, AP Stancampiano, AK Wakhloo, BB Lieber, ASME J Biomech Eng, 119, 206-212, 1997
6. SCM Yu, JB Zhao, Med Eng Phys, 21, 133-141. 1999
7. S Chen, GD Doolean, Ann Rev Fluid Mech, 30, 329-6, 1998
8. M Yoshino, Y Hotta, T Hirozane, E Endo, J non-Newtonian F luids mech., 147, 69-78, 2007
9. RB Bird, RC Armstrong, o Hassager, Dynamics of Polymeric Liquids, Vol 1, Fluids Dynamics, 2nd Ed., John Wiley & Sons, New York, 1987
10. L.S. Luo, *Phys. Rev. E.* 62, 4982-4996, 2000
11. S Chen, D Martinez, R Mei, *Phys Fluids,* 8, 2527-36, 1996
12. M Hirabayashi, M Ohta, DA Rufunacht, B Chopard, Future Generation Computer Systems, 20, 925-934, 2004
13. Bruce R. Munson, Donald F. Young, Theodore H. Okiishi, *F undamentals of Fluid Mechanics, 5th Edition* (2006)
14. YH Kim, JS Lee, Multiphase non-Newtonian effects on puls atile hemodynamics in a coronary artery, Int J Num Meth Flui d, 2008, In press.

Multiscale Design of a Laser Actuated Micro Bubble Array Acoustic-Fluidic Microdevice for Bioanalytical and Drug Delivery Applications

Z.J. Chen, A.J. Przekwas
CFD Research Corporation, Huntsville, AL 35805, USA
zjc@cfdrc.com, ajp@cfdrc.com

ABSTRACT

Next generation fluidic micro-nano devices will require precise control of force fields within a device and their interaction with a biological sample. Current force field designs use mechanical structures such as cantilevers, optical fields (e.g. laser tweezers), or electrokinetic fields (e.g. dielectrophoretic devices). All of them suffer from design complexity, small forces, long actuation times, and others. We propose a novel opto acoustic device for fast generation of 3D spatially distributed large force fields within the micro/nano fluidic device. The basic concept is to dynamically produce a field of oscillating micro/nano bubbles which will generate desired 3D pressure and/or flow fields within the device. This concept can be used for bioanalytical applications in vitro exploration of novel drug formulations, novel drug delivery modalities, e.g. to cells and tissues, and for micro/nano surgical applications. Wang et. al. showed experimentally that a micro bubble induced by laser pulse can generate a net streaming flow (Wang, 2004). Their results provide a basic understanding how the blood clot could be broken, emulsified and removed. It is obvious that the bubble formation, growth and collapse are the keys to understand the bubble dynamics and its interaction with the medium, such as blood clot, biological cells, or bio device.

Our overall concept as well as the device has been designed based on multiscale simulation of bubble dynamics, laser beam physics, fluid flow, cavitations, and acoustics. The paper presents the concept and preliminary simulation results for selected micro and nano biodevices.

Mathematical modeling of bubble dynamics has been studied theoretically (Feng, 1997) as well as specifically for biomedical applications. A micro bio device using a single bubble generated by laser pulse has been developed before (Hebert, 2001; Esch, 2001). The major innovation of our work is to explore space/time controlled bubble array for the generation of desired force fields.

For laser induced gas bubble in water, the bubble growth and collapse are function of initial bubble size. The initial bubble radius as well as its formation/collapse dynamics is function of laser pulse energy. Therefore a question that one will face is how to determine the laser pulse energy to achieve desired initial bubble size and subsequent bubble dynamics. This paper will present two essential modeling steps: 1) a novel analytical analysis on the relation between initial gas bubble radius and laser pulse energy and 2) CFD simulation results of bubble behaviors in water.

1 GAS BUBBLE RADIUS

Figure 1 shows the configuration of a spherical gas bubble with radius R in liquid water.

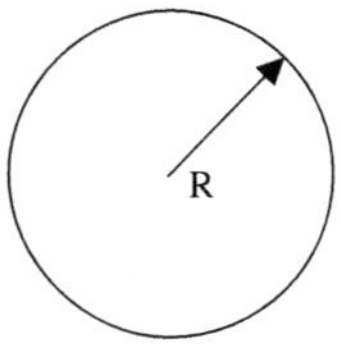

Figure 1. Gas bubble and surrounding liquid water.

The bubble radius is described by Keller-Mikes bubble model (Brennen, 1995):

$$\left(1-\frac{1}{c}\frac{dR}{dt}\right)\cdot R\frac{d^2R}{dt^2}+\frac{3}{2}\left(\frac{dR}{dt}\right)^2\left(1-\frac{1}{3c}\frac{dR}{dt}\right)= \left(1+\frac{1}{c}\frac{dR}{dt}\right)\frac{P_v}{\rho_l}+\frac{1}{c}\frac{R}{\rho_l}\frac{dP_v}{dt} \quad (1.1)$$

The vapor pressure P_v is determined by

$$P_v = P_i\left(R_i/R\right)^{3\cdot\kappa} - P_0 - \frac{2\sigma}{R} - \frac{4\cdot\mu}{R}\frac{dR}{dt} \quad (1.2)$$

The pressure P_i inside bubble is initial gas pressure and determined by ideal gas law. In equations (1.1) to (1.2), c is speed of sound in water, t is time, ρ and μ are density and viscosity of water respectively, and σ is surface tension at bubble/water interface, and κ is the polytropic index of gas whose value is 1.33 in our analysis.

2 INITIAL BUBBLE RADIUS

The initial gas bubble radius can be determined by balancing the laser pulse energy and liquid water internal thermal energy:

$$\int\{\rho_l c_p (T_B - T_i) + \rho_l L\} dV = \int E_{aV}(r,z) dV \quad (2.1)$$

Where, T_B is the water boiling temperature, T_i is the water initial temperature, L is the latent heat of water, E_{aV} is water volumetric absorbed laser energy which is defined by Beer's law for Gaussian spatial distribution:

$$E_{aV}(r,z) = \alpha \frac{2E_p}{\pi w^2} \exp\left(-\frac{2r^2}{w^2}\right) \exp(-\alpha \cdot z) \quad (2.2)$$

Where, r and z are radial coordinate and longitudinal distance from the fiber end respectively, E_p is the peak laser energy, α is water absorbed coefficient and w = w(z) is the beam radius at r = 0:

$$w^2(z) = w_o^2 \left[1 + \left(\frac{M^2 \cdot z}{z_R}\right)^2\right] \quad (2.3)$$

Where, w_o is due to losses at the cladding of the fiber, we assume loss factor F_{cc} to be 5%, so we have

$$w_o = f_r \sqrt{\frac{-2}{\log(F_{cc})}} \quad (2.4)$$

f_r is equal to the $1/e^2$ intensity radius of the Gaussian pulse. And

$$z_r = \frac{\pi w_o^2 n}{\lambda_o} \quad (2.5)$$

n is the refractive index of water whose value is 1.33, λ_o is the laser wave length, its value is 532 nm. M^2 is the laser mode parameter which varies from 1 (single mode) to 16 (multimode).

Equation (2.1) states the energy balance. It assumes that the laser pulse energy heats up the liquid water so that the certain amount of liquid water immediately becomes gas bubble at same volume. The gas bubble V_i generated by laser is then calculated as

$$V_i = \frac{\int E_{aV}(r,z) dV}{\rho_l c_p (T_B - T_i) + \rho_l L} \quad (2.6)$$

The solving process is iterative to obtain the volume V_i. After substituting equation (2.2) into equation (2.6) and assuming the bubble to be spherical shape, we have the final equation for bubble radius:

$$F(r) = R_i^3 - \frac{6\alpha E_p}{\pi \cdot w_o^2 \{\rho_l c_p (T_B - T_i) + \rho_l L\}} \times \int_0^{R_i} \exp\left(-\frac{2r^2}{w_o^2}\right) \cdot r^2 dr \quad (2.7)$$

We assume R_i value at the beginning, and integrate equation (2.7) numerically to see if there is a value R so that the function F(r) reaching to zero, then this R is the initial bubble radius. Newton-Raphson iteration method can be used to determine the radius.

Figure 2 shows the results for equation (2.7). For simplicity, we set specific heat of water to be a constant, 4000 J/(kg K), density of water is 998 kg/m^3, latent heat of water is 2260 kJ/kg (Holman, 2001). We can see that the radius increases when laser pulse energy increases.

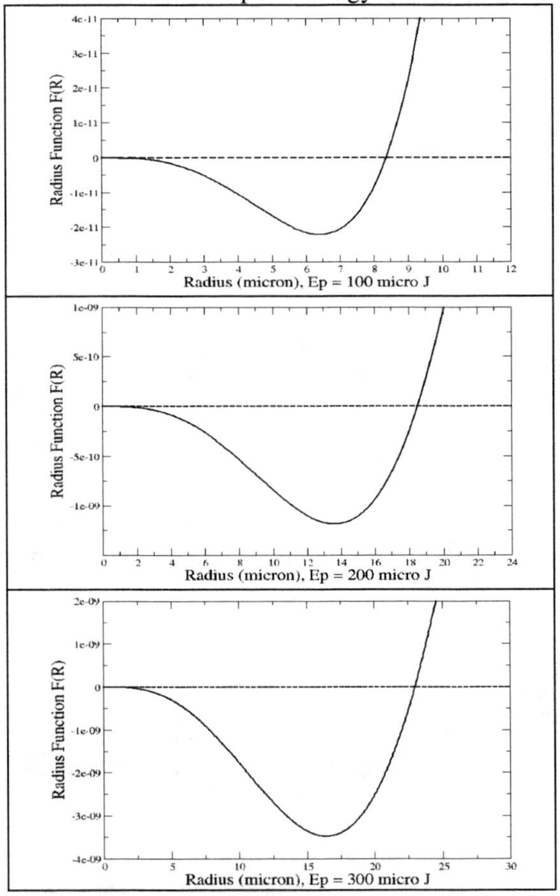

Figure 2. Initial bubble radius at different laser pulse energy E_p.

3 SIMULATION OF BUBBLE BEHAVIOR

Once the laser pulse energy specified, the initial bubble size can be obtained by equation(2.7) and the bubble radius at any time can be obtained by equation(1.1). This is an ODE equation and can be integrated by standard ODE method. The result is shown in figure 3.

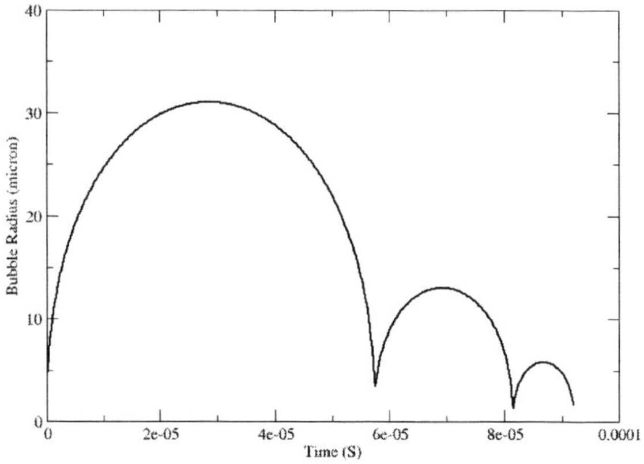

Figure 3. Bubble's radius at different times. The collapse of bubble is shown here.

We have applied the above technique into full CFD simulation on bubble behaviors. By arranging bubble array in the device, such as shown in figure 5, a desired pressure and/or flow fields within the device can be generated. Figure 4 shows pressure field in water induced by single bubble.

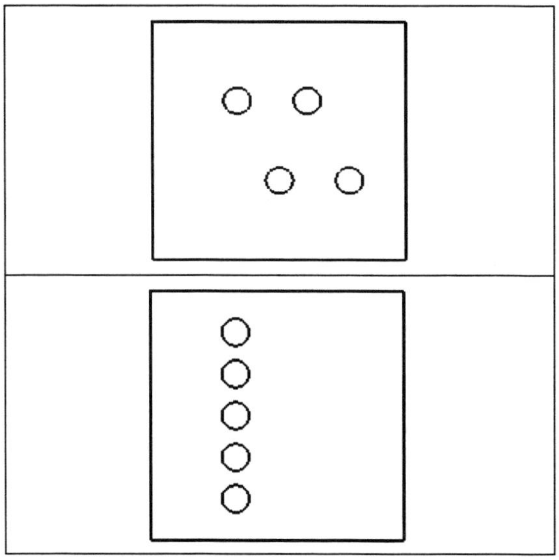

Figure 5. Bubble arraies for desired flow field.

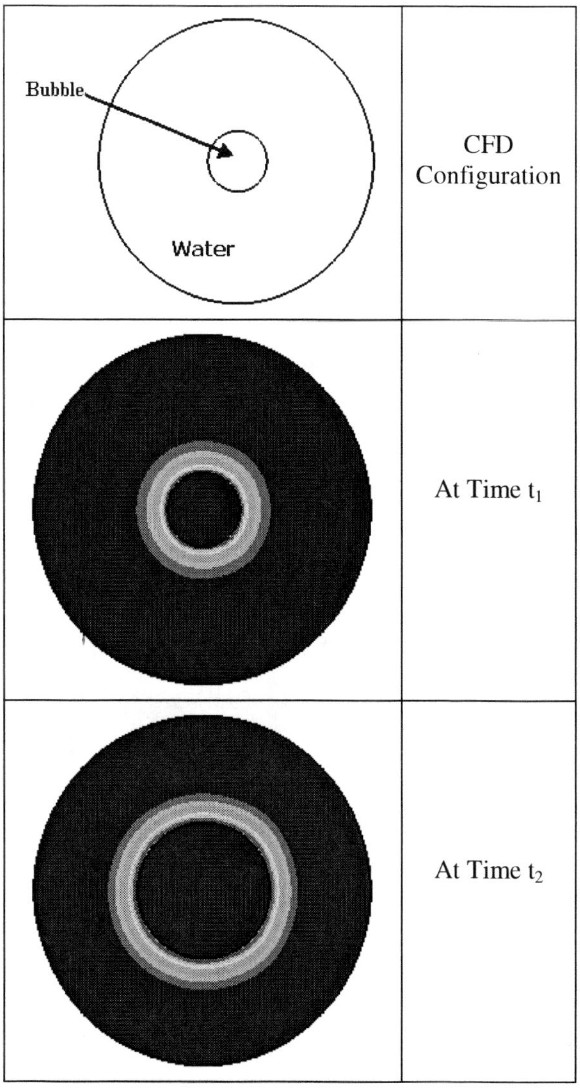

Figure 4. Pressure field in water induced by single bubble's oscillation.

4 CONCLUSION

We have proposed a concept to cerate a desired pressure and/or flow field in micro/nano device by using laser induced gas bubble technique. In the application, the initial gas bubble size is critical for a given laser impulse energy. We have analytically obtained the initial gas bubble size based on the laser energy released in water. We also performed CFD simulation of bubble's behaviors including bubble collapse. By arranging a array of bubbles induced by laser pulse, a certain flow field patter can be achieved. Such device can be applied in many bio areas, such as in vitro exploration of novel drug formulations, drug delivery modalities, e.g. to cells and tissues, and for micro/nano surgical applications.

REFERENCE

Brennen, C.E., Bubble Dynamics, Oxford Press, New York, Oxford, 1995.

Esch, V.C., Tran, Q.Q., Anderson, R.R., Hebert, S.J., Levine, M.A. and Sucgang, E.U., Flow Apparatus for the Disruption of Occlusions, US Patent Specification 6, 139, 543, 2001.

Feng, Z.C., Leal, L.G., Nonlinear Bubble Dynamics, annual Review of Fluid Mechanics, Vol. 29,: 201-243, 1997.

Hebert, S.J., Levine, M.A., Sucgang, E.U., Tran, Q.Q. and Esch, V.C., Flexible Flow Apparatus and Method for the Disruption of Occlusions, US Patent Specification 6, 210, 400, 2001.

Holman, J.P., Heat Transfer, McGraw-Hill College, 2001.

Wang, G.R., Santiago, J.G., Mungal, M.G., Young, B. and Papademetriou, S., A laser Induced Cavitation Pump, J. Micromesh. Microeng., Vol. 14, 1037-1046, 2004.

A Higher-Order Approach to Fluid-Particle Coupling in Microscale Polymer Flows

B. Kallemov†, G. H. Miller† and D. Trebotich‡
† Dept. Applied Science, University of California, Davis, CA 95616, USA
‡ Center for Applied Scientific Computing, Lawrence Livermore National Laboratory, P.O. Box 808, L-560, Livermore, CA 94551, USA
{bkkallemov,grgmiller}@ucdavis.edu, treb@llnl.gov

ABSTRACT

To simulate polymer flows in microscale environments we have developed a numerical method that couples stochastic particle dynamics with an efficient incompressible Navier-Stokes solver. Here, we examine the convergence properties of the stochastic particle solver alone, and demonstrate that it has second order convergence in both weak and strong senses, for the examples presented.

Keywords: stochastic particle dynamics, RATTLE, particle-fluid coupling

1 INTRODUCTION

The dynamics of a continuum fluid with discrete embedded polymers is important for certain microfluidic applications, (e.g., so-called lab-on-a-chip PCR reactors) and for modeling viscoelastic phenomena in the dilute limit. Toward this end we proposed a fluid-particle coupling strategy [7] that uses Brownian dynamics to approximate molecular-level fluid–polymer interactions. In subsequent work (e.g., [3]) the time-stability of the scheme was improved, and constraints such as the non-crossing constraint for polymer-polymer interaction were considered. In this short paper we address the accuracy of our scheme, which has not been previously reported. We work here in the framework of a freely-jointed chain (no polymer-polymer interactions), we consider the fluid velocity field to be prescribed, and we do not consider any rigid domain boundaries. In the context of rigid constraint dynamics (*vs.* soft penalty method constraints) these omitted interactions will diminish the order of the local discretization error.

Recently, [8] proposed a weak second-order stochastic particle dynamics approach that is broadly similar to ours as described in [7], [3]. Our approach differs from theirs in our handling of the fluid–particle coupling, and our use of a Duhamel type discretization that recovers certain limiting behavior, thereby permitting longer stable time steps. In this paper we show that our approach is not only weak second-order accurate, but also second-order strong.

We model a polymer as a collection of coupled point masses, each subject to the Langevin equation of motion

$$m_\alpha \ddot{\mathbf{x}}_\alpha = m_\alpha \gamma(\mathbf{u} - \dot{\mathbf{x}}_\alpha) + \mathbf{F}(\mathbf{x}_\alpha) + \sigma \boldsymbol{\xi}_\alpha(t). \tag{1}$$

Here $\mathbf{x} = \mathbf{x}(t)$ is the position of αth particle with mass m_α, $\mathbf{u}$ is fluid velocity, $\mathbf{F}(x)$ is the interparticle force, $\gamma > 0$ the friction coefficient and $\boldsymbol{\xi}(t)$ is a white noise representing stochastic thermal bombardment by the solvent. The constant σ is given by $\sqrt{2m_\alpha \gamma k_B T}$ with k_B being Boltzmann's constant and T the temperature. The stable numerical integration of (1) can require a very small time step, especially in a highly viscous fluid where the relaxation time $1/\gamma$ can be vanishingly small.

In this work we will use Kramers' freely-jointed polymer model, which represents a polymer as point masses governed by (1) with the interparticle force $\mathbf{F}$ chosen to enforce the constraint of fixed interparticle spacing. The general idea is to add into equations of motion constraint forces that can be expressed as

$$\mathbf{G}_\alpha = -\sum_\beta \lambda_{\alpha\beta}(t) \nabla_\alpha \theta_{\alpha\beta} \tag{2}$$

$$\theta_{\alpha\beta} = \|\mathbf{x}_\alpha - \mathbf{x}_\beta\|^2 - a^2 = 0 \tag{3}$$

where particles of index β are neighbors of particle α, and $\lambda_{\alpha\beta}$ are Lagrange multipliers chosen to satisfy the constraints, and a =const. is the spacing between adjacent particles. This is usually performed by applying the SHAKE algorithm [5] or its velocity version RATTLE [1]. The fluid velocity $\mathbf{u}$ can be determined from a form of the incompressible Navier-Stokes equations with a particle coupling term [7]. For the purpose of developing the particle solver, we will take $\mathbf{u}$ as prescribed.

2 NUMERICAL METHOD

A numerical method for the integration of (1) was given without proof in [7]. Here, the derivation of those equations is given. We begin by expressing the second-order SDE as a system of first-order equations:

$$\begin{aligned} d\mathbf{x}(t) &= e^{-\gamma t}\mathbf{z}(t)dt \\ d\mathbf{z}(t) &= \gamma e^{\gamma t}\mathbf{u}(t, \mathbf{x}(t))dt + \frac{\sigma}{m} e^{\gamma t} d\mathbf{W}, \end{aligned} \tag{4}$$

and

$$\mathbf{x}(t) = \mathbf{x}(0) + \int_0^t e^{-\gamma s}\mathbf{z}(s)ds \tag{5}$$

$$\mathbf{z}(t) = \mathbf{z}(0) + \int_0^t \gamma e^{\gamma s}\mathbf{u}(s, \mathbf{x}(s))ds + \int_0^t \frac{\sigma}{m} e^{\gamma s} d\mathbf{W}_s$$

where $\mathbf{z} = \mathbf{v}e^{\gamma t}$, $\mathbf{W}(t)$ is a standard Weiner process, and $d\mathbf{W} = \boldsymbol{\xi} dt$.

We then expand our equations of motion in an Itô-Taylor series, using the Itô calculus for stochastic ODEs [2]:

$$\begin{aligned} &Y = U(t, X(t)) && (6)\\ &dX(t) = f(t)dt + g(t)dW \\ &dY(t) = \left[\frac{\partial U}{\partial t} + \frac{\partial U}{\partial X} f(t) + \frac{1}{2}\frac{\partial^2 U}{\partial X^2} g^2(t)\right] dt + \frac{\partial U}{\partial X} g dW. \end{aligned}$$

Application of this stochastic chain rule requires care to account for all dependence on stochastic variables. In real systems, the fluid $\mathbf{u}$ is driven by a nonlinear stochastic coupling. Additionally, every fluid element undergoes thermal fluctuation, whether expressed explicitly as a Brownian force or not. However, the average magnitude of such fluctuations in a given volume scales as the inverse of the number of atoms in that volume. At the scales of length with which we are concerned, the continuum fluid motion $\mathbf{u}$ is smooth. Thus, in our analysis, the stochastic dependence of $\mathbf{u}$ is through the particle position $\mathbf{x}$ only: $\mathbf{u} = \mathbf{u}(t, \mathbf{x}(t))$.

With this assumption, application of the Itô formula to the $\mathbf{W}$-dependent integrands of (5) gives

$$\begin{aligned} e^{-\gamma s}\mathbf{z}(s) &= \mathbf{z}(0) + \int_0^s \Big[-\gamma e^{-\gamma s_1}\mathbf{z}(s_1) + \qquad (7)\\ &\quad \gamma \mathbf{u}(s_1, \mathbf{x}(s_1))\Big] ds_1 + \frac{\sigma}{m}\int_0^s d\mathbf{W}_{s_1} \\ \gamma e^{\gamma s}\mathbf{u}(s, \mathbf{x}(s)) &= \gamma \mathbf{u}(0, \mathbf{x}(0)) + \int_0^s \Big[\gamma^2 e^{\gamma s_1}\mathbf{u}(s_1, \mathbf{x}(s_1)) \\ &\quad + \gamma e^{\gamma s_1}\frac{D\mathbf{u}(s_1, \mathbf{x}(s_1))}{Ds_1}\Big] ds_1. \end{aligned}$$

Substituting expansions (7) into (5) gives

$$\begin{aligned} \mathbf{x}(t) &= \mathbf{x}(0) + t\mathbf{z}(0) + \int_0^t \int_0^s \Big[-\gamma e^{-\gamma s_1}\mathbf{z}(s_1) + \\ &\quad \gamma \mathbf{u}(s_1, \mathbf{x}(s_1))\Big] ds_1 ds + \frac{\sigma}{m}\int_0^t \int_0^s d\mathbf{W}_{s_1} ds \qquad (8)\\ \mathbf{z}(t) &= \mathbf{z}(0) + \gamma t \mathbf{u}(0, \mathbf{x}(0)) + \int_0^t \int_0^s \Big[\gamma^2 e^{\gamma s_1}\mathbf{u}(s_1, \mathbf{x}(s_1)) + \\ &\quad \gamma e^{\gamma s_1}\frac{D\mathbf{u}(s_1, \mathbf{x}(s_1))}{Ds_1}\Big] ds_1 ds + \frac{\sigma}{m}\int_0^t e^{\gamma s} d\mathbf{W}_s \end{aligned}$$

where $\frac{D}{Dt} = \frac{\partial}{\partial t} + (\mathbf{v}(t)\cdot\nabla) = \frac{\partial}{\partial t} + e^{-\gamma t}(\mathbf{z}(t)\cdot\nabla)$ is the material derivative. Applying the Itô formula (6) again, now to the integrands of (8), gives, after simplification,

$$\begin{aligned} \mathbf{x}(t) &= \mathbf{x}(0) + t\mathbf{z}(0) + \frac{\gamma t^2}{2}[\mathbf{u}(0, \mathbf{x}(0)) - \mathbf{z}(0)] + \\ &\quad \gamma \int_0^t \int_0^s \int_0^{s_1} \Big[\frac{D\mathbf{u}(s_2, \mathbf{x}(s_2))}{Ds_2} + \gamma e^{-\gamma s_2}\mathbf{z}(s_2) - \\ &\quad \gamma \mathbf{u}(s_2, \mathbf{x}(s_2))\Big] ds_2 ds_1 ds + \frac{\sigma}{m}\int_0^t \int_0^s d\mathbf{W}_{s_1} ds \\ &\quad -\gamma \frac{\sigma}{m}\int_0^t \int_0^s \int_0^{s_1} d\mathbf{W}_{s_2} ds_1 ds_2 \\ \mathbf{z}(t) &= \mathbf{z}(0) + \gamma t \mathbf{u}(0, \mathbf{x}(0)) + \frac{t^2}{2}[\gamma^2 \mathbf{u}(0, \mathbf{x}(0)) + \qquad (9)\\ &\quad \gamma \frac{D\mathbf{u}(0, \mathbf{x}(0))}{Dt}] + \int_0^t \int_0^s \int_0^{s_1} \Big[\gamma^3 e^{\gamma s_2}\mathbf{u}(s_2, \mathbf{x}(s_2)) + \\ &\quad 2\gamma^2 e^{\gamma s_2}\frac{D\mathbf{u}(s_2, \mathbf{x}(s_2))}{Ds_2} + \gamma e^{\gamma s_2}\frac{D^2\mathbf{u}(s_2, \mathbf{x}(s_2))}{Ds_2^2}\Big] ds_2 ds_1 ds \\ &\quad + \frac{\sigma\gamma}{m}\int_0^t \int_0^s \int_0^{s_1} (\nabla \mathbf{u}(s_2, \mathbf{x}(s_2)))\cdot d\mathbf{W}_{s_2} ds_1 ds \\ &\quad + \frac{\sigma}{m}\int_0^t e^{\gamma s} d\mathbf{W}_s. \end{aligned}$$

In the notation of [2], repeated application of the Itô chain rule to the $\mathbf{x}$ equation will give rise to multiple integrals of the form

$$\begin{aligned} I_{(1,0)} &= \int_0^t ds_0 \int_0^{s_0} dW_{s_1} \qquad (10)\\ I_{(1,0,0)} &= \int_0^t ds_0 \int_0^{s_0} ds_1 \int_0^{s_1} dW_{s_2} \\ &\cdots \\ I_{\underbrace{(1,0,\ldots,0)}_{n \text{ terms}}} &= \int_0^t ds_0 \int_0^{s_0} ds_1 \ldots \int_0^{s_{n-2}} dW_{s_{n-1}}. \end{aligned}$$

It can be shown [2, proposition 5.2.3] that

$$\begin{aligned} I_{(1,0)} &= \int_0^t (t-s) dW_{s_0} \qquad (11)\\ I_{(1,0,0)} &= \frac{1}{2}\int_0^t (t-s)^2 dW_{s_0} \\ &\cdots \\ I_{\underbrace{(1,0,\ldots,0)}_{n \text{ terms}}} &= \frac{1}{(n-1)!}\int_0^t (t-s)^{n-1} dW_{s_0}. \end{aligned}$$

It follows that repeated application of the Itô chain rule to the $\mathbf{x}$ equation will converge to a single stochastic integral

$$\frac{\sigma}{\gamma m}\int_0^t \left[1 - e^{-\gamma(t-s)}\right] dW_s. \qquad (12)$$

The Itô-Taylor series expansion therefore gives the effective stochastic position and velocity terms

$$\begin{aligned} \mathbf{R}_x &= \frac{1}{\gamma}\int_0^t \left[1 - e^{-\gamma(t-s)}\right] d\mathbf{W}_s \qquad (13)\\ \mathbf{R}_v &= \int_0^t e^{-\gamma(t-s)} d\mathbf{W}_s \qquad (14) \end{aligned}$$

with zero mean and variances:

$$\begin{aligned} E(\mathbf{R}_x \otimes \mathbf{R}_x) &= \mathbf{I}\int_0^t \frac{[1-e^{-\gamma(t-s)}]^2}{\gamma^2}ds \\ &= \mathbf{I}\frac{2\gamma t - e^{-2\gamma t} + 4e^{-\gamma t} - 3}{2\gamma^3} \\ E(\mathbf{R}_v \otimes \mathbf{R}_v) &= \mathbf{I}\frac{1-e^{-2\gamma t}}{2\gamma} \quad (15) \\ E(\mathbf{R}_x \otimes \mathbf{R}_v) &= \mathbf{I}\frac{(e^{-\gamma t}-1)^2}{2\gamma^2}. \end{aligned}$$

Numerically, these stochastic terms are constructed by assuming $\mathbf{R}_v = \sqrt{E(R_v^2)}\mathbf{U}_1$, where $\mathbf{U}_1$ is a vector of uniform standard deviates. Then, $\mathbf{R}_x$ is given by $\mathbf{R}_x = a\mathbf{U}_1 + b\mathbf{U}_2$, where $\mathbf{U}_2$ is an independent vector of uniform deviates, and constants a and b are

$$\begin{aligned} a &= \frac{1}{\gamma}\tanh\left(\frac{\gamma t}{2}\right)\sqrt{\frac{1-e^{-2\gamma t}}{2\gamma}} \\ b &= \frac{1}{\gamma}\sqrt{t - \frac{2}{\gamma}\tanh\left(\frac{\gamma t}{2}\right)} \quad (16) \end{aligned}$$

in order that $\mathbf{R}_v$ and $\mathbf{R}_x$ obey (15).

Taking into account all of the above, and truncating high order terms, we can write our integral equations of motion as

$$\begin{aligned} \mathbf{x}(t+\tau) &= \mathbf{x}(t) + [\mathbf{v}(t) - \mathbf{u}(t,x(t))]\frac{1-e^{-\gamma\tau}}{\gamma} + \\ & \quad \mathbf{u}(t,\mathbf{x}(t))\tau + \frac{\sigma}{m}\mathbf{R}_x \quad (17) \\ \mathbf{v}(t+\tau) &= \mathbf{v}(t)e^{-\gamma\tau} + \mathbf{u}(t,\mathbf{x}(t))(1-e^{-\gamma\tau}) + \frac{\sigma}{m}\mathbf{R}_v \end{aligned}$$

where τ is a time step of approximation. These discrete integral equations correspond exactly to the analytical solution under the assumption of no stochastic force, and constant uniform $\mathbf{u}$. The recovery of this exact limit through the Duhamel form is the principal advantage of our method, enabling $\tau\gamma \gg 1$. In applications with varying $\mathbf{u}$, we use a predictor-corrector formalism ([7]) to time-center the evaluation of $\mathbf{u}$ on particle paths.

By the theorem of [4], the omission of stochastic terms $\mathcal{O}(\tau^{5/2})$ and deterministic terms $\mathcal{O}(\tau^3)$ in our velocity equation gives a theoretical order of accuracy of 2 strong and 2 weak.

Adding constraint forces into our integrator (17) leads to the following

$$\begin{aligned} \mathbf{x}(t+\tau) &= \mathbf{x}(t) + [\mathbf{v}(t) - \mathbf{u}(t,x(t))]\frac{1-e^{-\gamma\tau}}{\gamma} + \\ & \quad \mathbf{u}(t,\mathbf{x}(t))\tau + \frac{\sigma}{m}\mathbf{R}_x + \frac{1}{m}\mathbf{G}(\mathbf{x}(t)) \quad (18) \\ \mathbf{v}(t+\tau) &= \mathbf{v}(t)e^{-\gamma\tau} + \mathbf{u}(t,\mathbf{x}(t))(1-e^{-\gamma\tau}) + \frac{\sigma}{m}\mathbf{R}_v \\ & \quad + \frac{1}{m}\frac{1}{\tau}\left[\mathbf{G}^*(\mathbf{x}(t+\tau)) + \mathbf{G}(\mathbf{x}(t))\right] \quad (19) \end{aligned}$$

Table 1: Weak approximation error and rate of convergence.

$2\tau/\tau$	error	rate
256/128	6.25E-13	2.54
128/64	3.63E-12	2.01
64/32	1.46E-11	1.85
32/16	5.24E-11	2.03
16/8	2.14E-10	1.98
8/4	8.47E-10	

where $\mathbf{G}(\mathbf{x}(t))$ and $\mathbf{G}^*(\mathbf{x}(t+\tau))$ are given by (2). Here, Lagrange multipliers $\lambda_{\alpha\beta}$ in (18) are chosen so that

$$\theta_{\alpha\beta}(\mathbf{x}(t)) = \|\mathbf{x}_\alpha - \mathbf{x}_\beta\|^2 - a^2 = 0 \quad (20)$$

and $\lambda^*_{\alpha\beta}$ in (19) are chosen so that

$$\dot{\theta}_{\alpha\beta}(\mathbf{x}(t+\tau)) = (\dot{\mathbf{x}}_\alpha - \dot{\mathbf{x}}_\beta)\cdot(\mathbf{x}_\alpha - \mathbf{x}_\beta) = 0. \quad (21)$$

The Lagrange multipliers are determined by the `RATTLE` algorithm [1]. See also [8].

3 RESULTS AND CONCLUSIONS

The calculation uses run parameters $\gamma = 10^{10}$/s, $m_\alpha = 2.0\times10^{-18}$kg, $a = 7.0\mu$m, which correspond with lambda-phage DNA subdivided into Kuhn length segments. The velocity field is $u_i = 10^{-3}\cos(10^3 x_i)$.

Because the exact solution is not accessible we define the error as the difference of successive solutions

$$\mathrm{E}(\|r_\alpha^{(2\tau/\tau)}\|) = \mathrm{E}(\|\mathbf{x}_\alpha^{(2\tau)} - \mathbf{x}_\alpha^{(\tau)}\|), \quad (22)$$

and rate of convergence as

$$k = \log_2\left(\frac{\mathrm{E}(\|\mathbf{r}_\alpha^{(4\tau/2\tau)}\|)}{\mathrm{E}(\|\mathbf{r}_\alpha^{(2\tau/\tau)}\|)}\right), \quad (23)$$

where superscript (τ) denotes the time step used.

The results of our numerical computations, presented in Figs. 1&2 and Tables 1&2, suggest that the rate of convergence is indeed second order weak and second order strong for the examples used. (The strong order assessment does not include constraints.) The weak error is measured by measuring the error after averaging over paths; the strong error is the path-wise average error. Calculation is made for a 6-bead polymer, for time $T = 10^{-2}$. We measure the error in the final coordinate of particle $\alpha = 2$ in the chain. Averaging is performed over 10^4 independent paths. E_1, E_2 and E_∞ in Fig. 2 and Table 2 refer to different norms [6].

Work to assess the strong order of convergence with constraint forces is in progress.

Table 2: Strong approximation error E_1, E_2, E_∞ and rate of convergence, without constrains.

$2\tau/\tau$	error E_1	rate E_1	error E_2	rate E_2	error E_∞	rate E_∞
256/128	8.58E-13	1.96	8.73E-13	1.94	1.50E-12	1.61
128/64	3.34E-12	1.99	3.35E-12	1.98	4.56E-12	1.80
64/32	1.32E-11	2.00	1.33E-11	2.00	1.59E-11	1.85
32/16	5.29E-11	2.00	5.29E-11	2.00	5.73E-11	1.96
16/8	2.11E-10	2.00	2.11E-10	2.00	2.23E-10	1.96
8/4	8.46E-10		8.46E-10		8.68E-10	

Figure 1: Weak error.

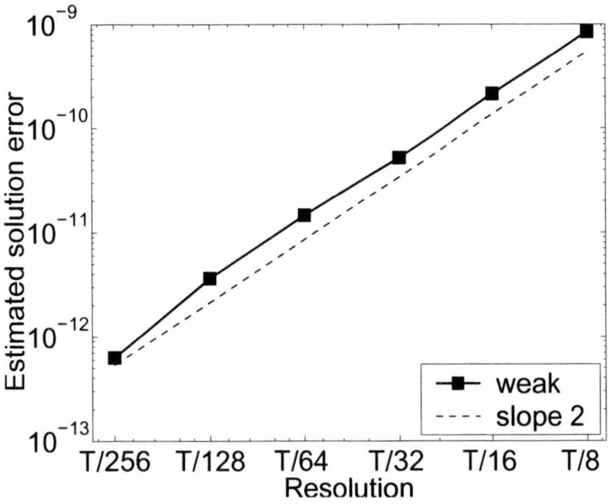

Figure 2: Strong error (no constraints).

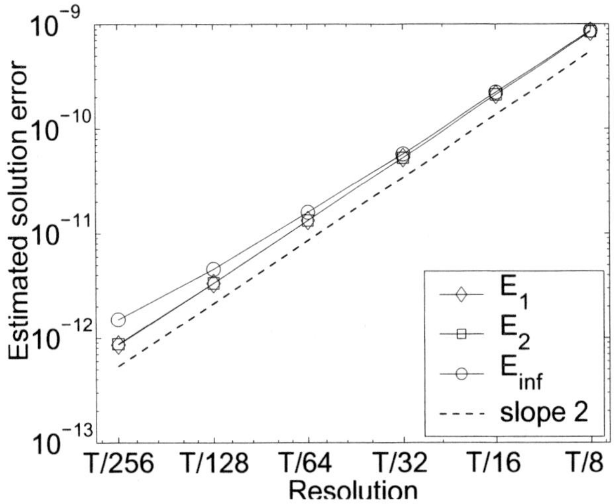

ACKNOWLEDGMENT

B. Kallemov was supported by the Bolashak scholarship of the President of the Republic of Kazakhstan. G. H. Miller was supported by LLNL IUT subcontracts number B550201 and number B553964, and by DOE MICS contract number DE-FG02-03ER25579. Work at LLNL was performed under the auspices of the U.S. Department of Energy by the University of California, Lawrence Livermore National Laboratory under contract No. W-7405-Eng-48.

REFERENCES

[1] H. C. Andersen. `RATTLE`: a "velocity" version of the `SHAKE` algorithm for molecular dynamics calculations. *J. Comp. Phys.*, 52:24–34, 1983.

[2] P. E. Kloeden and E. Platen. *Numerical Solution of Stochastic Differential Equations.* Springer, New York, 1999.

[3] G. H. Miller and D. Trebotich. Toward a mesoscale model for the dynamics of polymer solutions. *J. Comp. Theor. Nanosci.*, 4:1–5, 2007.

[4] G. N. Mil'shtein. A theorem on the order of convergence of mean-square approximations of solutions of system of stochastic differential equations. *Theory Prob. Appl.*, pages 738–741, 1988.

[5] J.-P. Ryckaert, G. Ciccotti, and H. J. C. Berendsen. Numerical integration of the Cartesian equations of motion of a system with constraints: Molecular dynamics of n-alkanes. *J. Comp. Phys.*, 23:327–341, 1977.

[6] J. Stoer and R. Bullirsch. *Introduction to Numerical Analysis.* Springer, New York, 2nd edition, 1993.

[7] D. Trebotich, G. H. Miller, P. Colella, D. T. Graves, D. F. Martin, and P. O. Schwartz. A tightly couple particle-fluid model for DNA-laden flows in complex microscale geometries. *Computational Fluid and Solid Mechanics 2005*, pages 1018–1022, 2005.

[8] E. Vanden-Eijnden and G. Ciccotti. Second-order integrators for Langevin equations with holonomic constraints. *Chem. Phys. Lett.*, 429:310–316, 2006.

Computational Microfluidics for Miniaturized Bio-Diagnostics Devices using the Multiphysics code "TransAT"

C. Narayanan, D. Lakehal*

*ASCOMP GmbH, Zurich, Switzerland, lakehal@ascomp.ch

ABSTRACT

In this paper we have presented new developments achieved on the computational microfluidics front using the dedicated code TransAT. In particular, we have shown how it behaves for the prediction of the Marangoni effects within the Interface Tracking Concept, which can be used in controlling the dynamics of micro-droplets in bio-chips. We have also presented first results of a sub-grid scale ultra-thin film model, capable to mimic the wettability of inner surfaces allowing for a realistic representation of what might be expected in arterial microbubble delivery in gas embolotherapy.

Keywords: Microfluidics, Thin fim, Marangoni, effects

1 INTRODUCTION

Microfluidic devices are now used for such diverse applications as DNA microarrays, drug screening, sensors, and in clinical and forensic analysis. Typical microfluidics flows feature free-surface motion evolving (sometimes) in porous media or as falling films, spreading and dewetting of (complex) liquids on solid or liquid substrates, chemical reaction of binary mixtures, micro-bubbles and beads control and manipulation, phase change or transition. The control of such micro-flow systems is central to future technological advances in emerging technologies, like biological reactors, microreactors, biochannel arrays, and labs-on-chip. It is expected that robust, accurate and fast response computational microfluidics solutions will play a key role in the development in this new business segment. In practical microfluidics applications the flow involves phenomena acting at different time and length scales. At each level of the scale cascade, the physics of the flow is amenable to numerical prediction by scale-specific strategies.

In this paper will present our recent computation results obtained with our CFD code TransAT, in which interfacial flows are treated using the Level Set method. As bio-chips may comprise various components, a new fully automatized version has been developed for microfluidics applications, using IST (Immersed Surfaces Technology) to map complex geometries into a rectangular Cartesian grid. Since IST forces the grid to remain Cartesian and equidistant, high-order schemes (up to 3rd order for flux convection and 3rd order WENO schemes for free surface flows) can maintain their high degree of accuracy. Further, to better resolve boundary-layer regions, near wall flow areas are treated by another new feature, namely the BMR (Block-based Mesh Refinement), in which sub-scale refined blocks are placed around each component. The combination IST/BMR can save up to 70% grid cells in 3D. In this paper we discuss simulation examples treated with this approach (see details in [1]). We will particularly focus on the role played by the Marangoni effects in controlling the dynamics of micro-droplets in bio-chips, and the way ultra-thin film can be predicted using a sub-grid scale model.

2 PREDICITING INTERFACIAL MICROFLUIDCS FLOWS

Interfacial flows refer to multi-phase flow problems that involve two or more immiscible fluids separated by sharp interfaces which evolve in time. Typically, when the fluid on one side of the interface is a gas that exerts shear (tangential) stress upon the interface, the latter is referred to as a free surface. Interface tracking methods (ITM) are schemes capable to locate the interface, not by following the interface in a Lagrangian sense (e.g., by following marker points on the interface), but by capturing the interface by keeping track, in an Eulerian sense (the grid is fixed), of the evolution of an appropriate field such as a level-set function or a volume-fraction field. Examples and classifications are provided in [2]. Application of ITM's to microfluidics flows requires particular attention to the way surface forces and triple-line dynamics are handled.

2.1 Mathematical Formulation

ITMs are based on solving a single-fluid set of conservation equations with variable material properties and surface forces. The coupled fluid and heat transfer equations in incompressible flow conditions take the form

$$\nabla . \mathbf{u} = 0 \quad (1)$$

$$\partial_t(\rho \mathbf{u}) + \nabla .\left(\rho \mathbf{u}\,\mathbf{u} + pI\right) = \nabla .\mu\left(\nabla \mathbf{u}\right) + F_s + F_g + F_w \quad (2)$$

$$\partial_t(\rho CpT) + \nabla .\left(\rho\, CpT\, \mathbf{u}\right) = \nabla .\lambda\left(\nabla T\right) + Q''' \quad (3)$$

where **u** is the velocity vector, ρ is the density, p is the pressure, **I** is the identity matrix, and μ is the dynamic viscosity. The source terms in (2) represent body forces (F_g), surface tension (F_s) defined by

$$F_s = \gamma\, \kappa\, \mathbf{n}\, \delta(\phi) + (\nabla_s \gamma)\, \delta(\phi); \quad \kappa = -\nabla\phi / |\nabla\phi| \tag{4}$$

where **n** stands for the normal vector to the interface, κ for the surface curvature, γ for the surface tension coefficient of the fluid, δ for a smoothed delta function centered at the interface, and ϕ for the phase indicator function (e.g. level sets, or volume of fluid). The second term in the above expression refers to the Marangoni forces, expressing heat-induced variation of the surface tension coefficient. The contact–line wall contribution (F_w) is defined next by (8). In (3), T is the temperature, Cp is the heat capacity, λ is the heat conductivity, and Q''' is the volumetric heat source. In the Level Set method the interface between two immiscible fluids ϕ represents a continuous distance-to-the-interface function that is set to zero on the interface, is positive on one side and negative on the other. The location of the physical interface is thus associated with the zero level. Material properties, body and surface forces in (2) are locally dependent on ϕ, whose equation reads:

$$\partial_t(\phi) + \mathbf{u}.\nabla\phi = \dot{m} / \rho|\nabla\phi|, \tag{5}$$

In (5) above $\dot{m}$ stands for the heat/mass transfer rate, which can be either directly determined using the energy jump across the interface, or modeled using heat transfer correlations [2]. If the mass transfer rate is forced to be active only at the triple line then the source term in (2) is applied only in the cell containing the triple line. Note that, if the mass transfer rate per unit length is known, it is converted to a rate per unit area based on the contact line length and the area of the cell. Material properties (density, the viscosity, heat capacity, thermal conductivity) are updated locally based on ϕ and distributed across the interface using a smooth Heaviside function H(ϕ):

$$\rho, \mu, Cp, \kappa = \rho, \mu, Cp, \kappa|_L\ H + \rho, \mu, Cp, \kappa|_G\ (1-H) \tag{6}$$

The level set function ceases to be the signed distance from the interface after advecting (5). To restore its normal distribution near the interface, a re-distancing equation is solved during a time period τ:

$$\partial_\tau(d) = \mathrm{sgn}(d_0)\left[1 - |\nabla d|\right]; \quad d_0(x,t) = \phi^n(x,t) \tag{7}$$

where **sgn(x)** is the Signum function. The expression above is solved after each advection step of (4), using the 3rd or 5th order WENO schemes. Details about alleviating the mass conservation issue in this approach can be found in [2, 3].

The numerical treatment of wetting dynamics is based on the physical forces associated with triple lines. A triple line force is included in the momentum equation, which could then provide a physically adequate description of wetting dynamics, eliminating the need for any particular boundary condition specifying the contact angle. The triple line force is based on interfacial free energy; accordingly, it contains only two parameters: the interfacial tension between the fluids and the equilibrium contact angle θ_{eq}:

$$F_w = \gamma\left(\cos(\theta_{eq}) - \cos(\theta_{dy})\right) l_t \vec{b}, \tag{8}$$

where θ_{dy} is the instantaneous dynamic contact angle, l_t is the length of the triple line in the cell, and $\vec{b}$ is the unit vector normal to the triple line and parallel to the wall surface. The triple line force is obtained by considerations similar to the derivation of Young's Law and can be referred to as the unbalanced Young force [4]. In the case of the static contact angle treatment, the wall value of the level set is calculated such that the contact angle is always equal to the equilibrium contact angle specified.

2.2 TransAT© Microfluidics Flow solver

The Microfluidics code TransAT© [5] is a multi-physics, finite-volume code based on solving multi-fluid Navier-Stokes equations on structured multi-block meshes. MPI parallel based algorithm is used in connection with multi-blocking. Grid arrangement is collocated and can thus handle more easily curvilinear skewed grids. The solver is pressure based, corrected using the Karki-Patankar technique for weak compressible flows. The Navier-Stokes and level set equations are solved using the 3rd order Runge-Kutta explicit scheme for time integration. The convective fluxes are discretized with TVD-bounded high-order schemes [6]. The diffusive fluxes are differenced using a 2ndorder central scheme. Multiphase flows are tackled using Level Sets [3] and VOF for both laminar and turbulent flows. TransAT deals with phase change, surface tension and triple-line dynamics, Marangoni effects, and micro-film sub-grid scale modeling for lubrication.

3 MICROFLUIDCS FLOW EXAMPLES

3.1 Flows Driven by Marangoni Effects

The problem has been borrowed from [7]. Computations were performed in 2D, and the physical model applied is described in Sec. 2.1. The domain size is Lx = 5.0mm and Ly = 1.5mm. A semi-circular droplet with a diameter of d_0 = 1.5mm is initially placed on the lower wall. The initial temperature is set to 20°C and the fluid is at rest. The temperature driving force is represented by a deficit set between west and east walls of 50°K. The contact angle is 90°. The variation of γ is set linearly, according to

$$\gamma = \gamma_{Tw}\left(T_{crit} - T\right) / \left(T_{crit} - T_{weber}\right) \tag{9}$$

where γ_{Tw} is the surface tension coefficient value at T_{weber} known a-priori, and T_{crit} is the critical temperature. The fluid properties used are listed in Table 1.

Properties	Water	Silicon
γ_{Tw} N/m	0.0728	0.0210
T_{weber} °K	293.0	293.0
T_{crit} °K	327.83	305.83

Table 1: Fluid properties used for our simulations.

Qualitative results are shown in Figs. 1 and 2 for silicon oil and water. The drop is seen to react to the imposed temperature gradient rather vigorously. For water the droplet travels in the negative direction (according to our nomenclature), while for silicon it travels in the positive direction; in both cases the drop is displaced at least one half radius from its initial position. The receding (water) and advancing (silicon) phenomena observed here are very interesting, and corroborate with the results of [7]. This is explained by the properties of the fluids. The velocities of the drops obtained are also in line with the data of [7]: for Silicone oil: V_d = 0.025 m/s, and for water: V_d = -0.25 m/s.

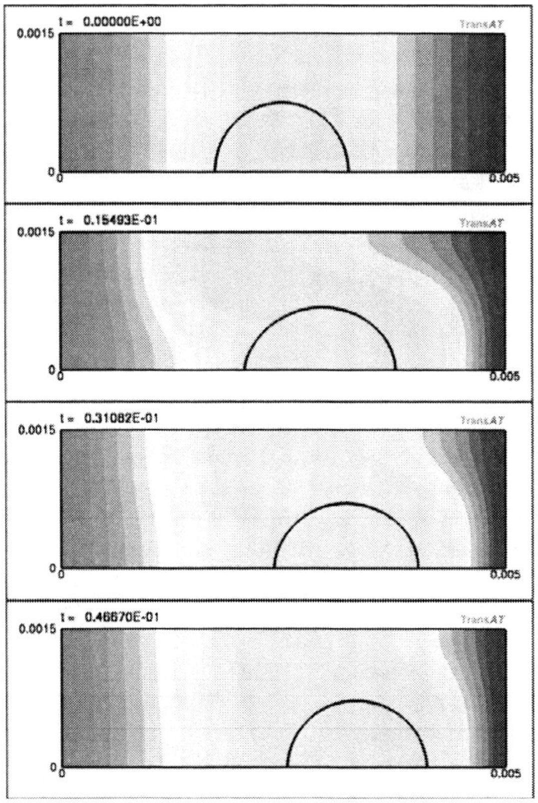

Figure 1: Marangoni driven drop (Silicon).

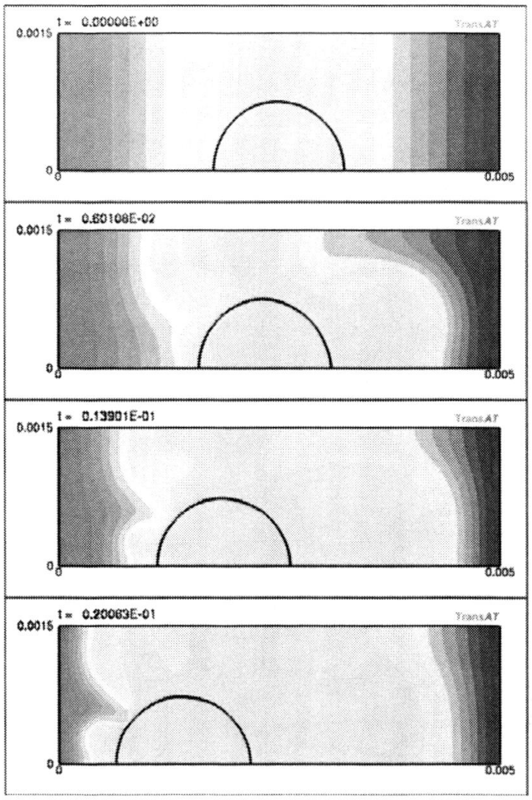

Figure 2: Marangoni driven drop (Water).

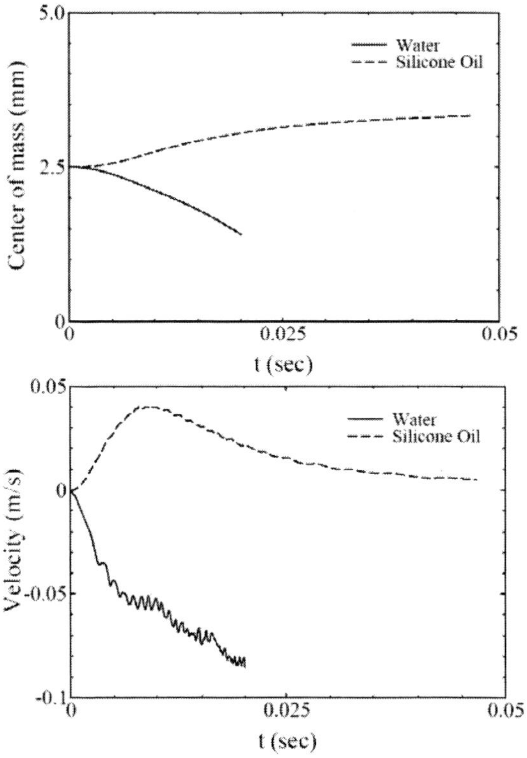

Figure 3: Center of mass displacement (a) and velocity (b).

The drop center-of-mass displacement and velocity are compared in Fig. 3 for both fluids. The plot shows how the center of mass moves recedes (from the initial center of mass 2.5mm) for water and advances for silicon, up to 2.95mm. In [7] the authors report that water moves to the higher temperature side only for 90° contact angle, which has not been checked in our simulation campaign. The drop velocities compared in the 2nd panel of Fig. 3 show clearly the difference between the motion of the two fluids; while the speed rises rapidly for the initial 10ms then decays for silicon, it grow from the start (in the - direction) for water, without marking an inflexion as for silicon oil.

3.2 Thin-film Confined Flows

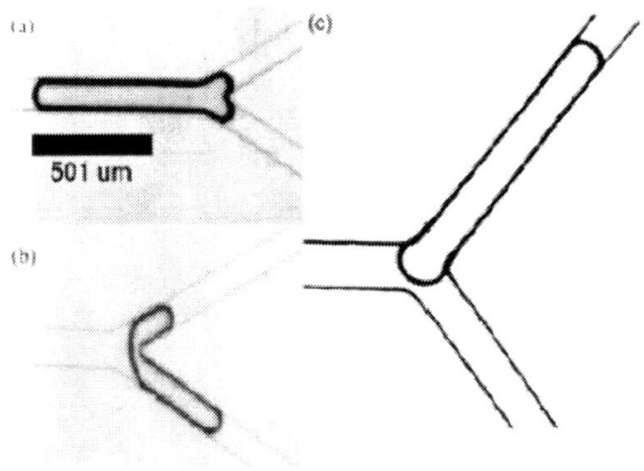

Figure 4: (a) Lodging state, (b) lodging state, (c) bubble lodged in one of the branches [8].

Cardiovascular gas bubbles in arteriole bifurcations have been experimentally addressed to understand the dynamics of their lodging mechanism [8]. The research is motivated by novel gas embolotherapy techniques for the potential treatment of cancer by tumor infarction. To solve such problems, one should resort to detailed multiphysics, interfacial flow simulation. The presence of the junction singular-point controls the bubble dynamics and rupture. In the junction a thin film is formed preventing bubble breakup. This can be predicted only with high-fidelity simulations using with a sub-grid scale model for thin film treatment, and contact angle treatment for wetting.

Present CMFD technology without multiscale treatment will fail to deliver accurate and physical sound results. The bubble will separate producing a non-physical solution (i.e. numerical dry-out) whereas wetting should sustain) as the bubble enters in the junction. Multiscale simulation of thin film flow is necessary here, requiring SGS thin-film models and wetting capabilities. The ultra-thin film SGS model implemented in TransAT for this class of flow is based on a modified model of the Taylor [9] thin-film theory, which originally proposes to model the height of a thin film as a function of the fluid Capillary number $Ca = \mu U / \gamma$; i.e. $\delta / r = 1.337 Ca^{2/3}$, where r stand for the pipe radius, and U for the speed of the bubbles inside the tube.

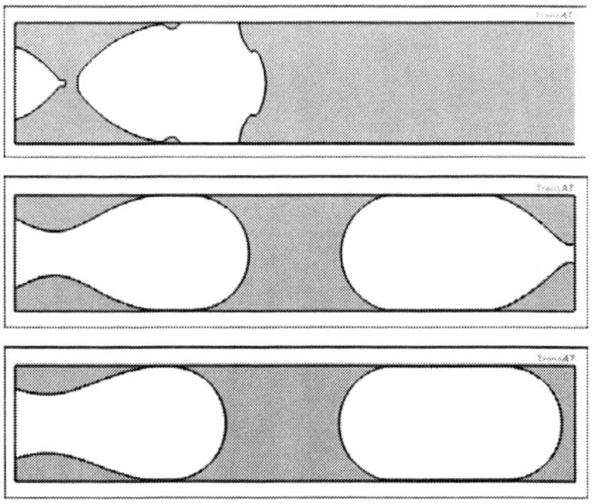

Figure 5: (a) Numerical dry out, (b) Imposed water film thickness of 5micron, (c) 10micron.

The results depicted in Fig. 5 clearly show that without such a SGS thin-film model, numerical dry-out will occur (panel a). Our SGS model is able to take into account the effect of lubrication by imposing a liquid film thickness, thus allowing for a smooth transport of the bubbles at low Capillary number (1mm diameter channel containing air and water, for Capillary number: Ca = 1.5463E-03). Note that the inflow liquid and gas velocities are 0.111 and 0.066 m/s, respectively

REFERENCES

[1] M. Icardi, D. Caviezel and D. Lakehal, The IST for Reliable and Fast Setup of Microfluidics Simulation Problems, Paper 1355, Proc. NSTI, Boston, 2008.

[2] D. Lakehal, M. Meier M. and M. Fulgosi, Interface tracking for the prediction of interfacial dynamics and heat/mass transfer in multiphase flows, *Int. J. Heat & Fluid Flow*, 23, 242, 2002.

[3] M. Sussman, S. Smereka and S. Osher, A Level set Approach for computing incompressible two-phase flow. *J Comp Phys*, 114, 146, 1994.

[4] P.G. de Gennes, Wetting: statics and dynamics, *Reviews of Modern Physics*, 57(3), 827, 1985.

[5] www.ascomp.ch/transat.

[6] B.P. Leonard, A stable and accurate convective modeling procedure based quadratic interpolation, *Comput. Meth. Appl. Eng.*, 19, 59, 1979.

[7] A. Murata and S. Mochizuki, Motion of Droplets Induced by the Marangoni Force on a Wall with a Temperature Gradient, *Heat Transfer--Asian Research*, 33, 81, 2004

[8] A. J. Calderon, Y. S. Heo, D. Huh, N. Futai, S. Takayama and J. L. Bull. Microfluidic model of bubble lodging in microvessel bifurcations. App. Phys. Letters, 89, 244103, 2006.

[9] G. I. Taylor, Deposition of a viscous fluid on the wall of a tube. *J. Fluid Mech. Dig. Archive*, 10, 161, 2006.

Single-phase Multi-component Simulation of RBCs Biconcave Shape Using 2-D Lattice Boltzmann Method

H. Farhat, J.S. Lee

Multi-scale Fluid Dynamics laboratory. Department of Mechanical Engineering.
Wayne State University, Detroit, MI 48202.

ABSTRACT

The dependence of the rheological properties of blood, on RBCs shape, aggregation and deformability has been investigated using hybrid systems which couple fluid with solid models [1-4]. We are presenting a simplistic approach for simulating blood as a multi-component fluid, in which RBCs are modeled as droplets of acquired biconcave shape. We use LBM, owing to its excellent numerical stability as a simulation tool. The model enables good control of the droplet shape by imposing variable surface tension on the interface between the two fluids. The use of variable surface tension is justified by Norris hypotheses, which was supported by recent experiments. This hypotheses states that the shape of the erythrocytes is due to a difference in surface tension caused by the unbalanced distribution of lipid molecules on the surface [5].

Keywords: blood rheology, rbc deformability, rbc shapes, lbm.

1 INTRODUCTION

Biconcave shape and deformability have major influence on the red blood cell delivery functions. Erythrocyte's shape provides higher surface to volume ratio. This is believed to have crucial role in the gas exchange process, which takes place in the body, and especially in the lungs and the periphery [5]. Deformability allows red blood cells to circulate inside microvessels having dimension about half of their size.

Shape and deformability are the subjects of many studies. Of the currently available explanations to these phenomena, two distinct hypotheses are considered.

1.1 Minimal Energy Hypotheses

Classic Canham-Helfrich hypothesis, suggests that the biconcave equilibrium shape of erythrocytes, results from the requirement of minimization of the total bending energy [6].

More recent studies show that the cytoskeleton undergoes constant remodeling in its topological connectivity, in order to achieve zero in plane shear elastic energy [7].

The main goal is to simulate the spectrin network remodeling process and fluidization, and to highlight the network influence on the shape and deformability of the erythrocytes. Spectrin network is modeled as spherical beads connected by unbreakable springs. In some cases thermal and biochemical factors are considered, but hydrodynamic interactions are ignored [7].

The inherent problem of these models, is that the biconcave shape is hardly achieved by carefully selecting the stress-free reference state otherwise cup shape is the dominant [6].

Other models use hybrid systems, which couple the solid model with fluid model. The biconcave shape is achieved by minimizing the total elastic energy of the solid model [4], or by using analytical formula [2].

Generally these models focused on the deformability of the cells during blood motion but failed to recognize the role of the quiescent fluid on the erythrocyte's shape.

1.2 Surface Tension Hypotheses

Surface tension hypothesis suggests that the shape of the erythrocyte is caused by the balance between the spreading tendencies of the lipid molecules of the cell, counteracted by the surface tension forces which cause a fluid object to become spherical in shape, when immersed into another fluid [5].

Experiments using radio autography, shows that labeled cholesterol incorporated into the cell membrane, was concentrated in the periphery of the biconcave disk, while little were found in the flat central area of the erythrocyte [5]. These experiments indicate that the erythrocyte's lipid bilayer is concentrated at the peripheries, which makes these areas less wet-table with corresponding higher surface tension interfacial forces [5].

This theory is further supported by the fact that some substances like bile, free fatty acids, lysolecithin and saponinc; transform the erythrocyte into a sphere, when interacting with it. This process is totally reversible. Washing the spherical RBC with plasma turns it back into its original biconcave shape [5].

Two important facts are worth mentioning about this theory. The first is about the minimal stretching ability of the cytoskeleton as contrasted with its deformability. Since the red blood cell surface area is almost maintained during deformation, it requires less energy from the flow to pass through microvessels. No energy is wasted to stretch the membrane. The second circulation energy saving comes from the fact that the shape of the red blood cell results from the physical interaction between the plasma and the membrane and not from the flow circulation [5].

Another support of this hypothesis comes from fact that red blood cell lipids form immiscible fluid under pressure of

21 dyn/cm for the inner leaflet, and 29 dyn/cm for the outer leaflet [8].

In this paper we are exploring the possibility of using surface tension hypotheses, to answer some basic questions, like the origin of the biconcave shape, and blood cell deformability, but the longer term aim is to try to map the topology of the lipid bilayer during deformation in order to understand the dynamics of this process.

2 NUMERICAL METHOD

Multi-component lattice Boltzmann simulation is a convenient tool for modeling blood under surface tension hypotheses. We use D2Q9, two-components, isothermal, single relaxation, LBGK model. We use guidance from [9], and follow the footsteps of I. Halliday et al [10-12], who did an extensive work to improve the models by Gunstensen et al [13], D'Ortona et al [14], and Listchuk et al [15].

We identify the red and blue fluids momentum distribution functions as $R_i(r,t)$, and $B_i(r,t)$, where r and t are the nodal position and time respectively. The nodal density of the two fluids is defined individually.

$$R(r,t) = \sum R_i(r,t) \tag{1}$$

$$B(r,t) = \sum B_i(r,t) \tag{2}$$

The total momentum distribution function is the sum of the two functions.

$$f_i(r,t) = R_i(r,t) + B(r,t) \tag{3}$$

The main streaming and collision function is the single particle model and is expressed as follows.

$$f_i(r + c_{i,}t + 1) = f_i(r,t) - \omega[f_i(r,t) - f_i^0(\rho, \rho u)] + \emptyset_i(r) \tag{4}$$

$\emptyset_i(r)$ is a forcing term and it is used to include a controlled body force in order to have an interfacial pressure step in the fluid [12]. Another term could be included to enclose a force in the horizontal direction, which causes fluid circulation.

The main idea in two-component LBM is to modify the collision rules, in order to obtain surface tension between the two fluids. This is achieved by applying two-step collision rules. The first step is done by the addition of perturbation to the particle distribution near the interface to create the correct surface-tension dynamics. The second step is the segregation of the two fluids by imposing zero diffusivity of one color into the other [12].

To define the interface between the two fluids a phase field is described as follows.

$$\rho^N(r,t) = \frac{R(r,t) - B(r,t)}{R(r,t) + B(r,t)} \tag{5}$$

In this model we follow Lischuk's interface method [15] where N stands for normal. A surface tension force F(r) is defined.

$$F(r) = -\frac{1}{2}\alpha K \nabla \rho^N \tag{6}$$

Notice that $\nabla \rho^N = 0$, for constant phase field. This means that this force is only applicable on the interface. α is a surface tension parameter. K is the curvature in the phase field, and is obtained from the surface gradients by finite difference method.

F(r) is used to correct the velocity following Guo's methodology [9] as shown in the following equation.

$$u^* = \frac{1}{\rho}(f_i c_i + (1 - f)F(r)), \quad f = \frac{1}{2} \tag{7}$$

The first collision is then applied with the corrected velocities, followed by perturbing the interface using Gusntensen et al [12] perturbation term.

$$c_i^b = \alpha |\nabla \rho^N| \cos 2(\theta_i - \theta_f) \tag{8}$$

The binary collision operator c_i^b depletes mass along lattice links parallel to the interface and adds mass to lattice links perpendicular to it. θ_i is the angle of lattice direction i, θ_f is the angle of the local gradient.

The second collision step is segregation of the two components. The outcome is a solution to a maximization problem with two constraints of which one is.

$$B_i(r,t) = f_i(r,t) - R(r,t) \tag{10}$$

Segregation could be executed numerically as used by Gunstensen et al [12], or by using a set of formulae as described by the model of D'ortona et al [9].

After segregation the two components propagate individually. In our model we use numerical segregation.

3 SIMULATION RESULTS AND DISCUSSION

3.1 Biconcave Shape

The goal in this simulation was to show the surface tension hypotheses in action. We chose for demonstration in Fig.1, 60x60 lattices to model quiescent flow with periodic boundary condition in both directions, and without forcing term to cause circulation. We ran two simulations, first with one droplet corresponding 10 % hematocrit, and second with two droplets, which corresponds to 20% hematocrit. Since the number of lattices determines the magnitude of the fluid body force, and therefore the surface tension as explained in part 2, we expect to use two different set of surface tension parameters and geometry for the two experiments.

The code optimized the variables in order to achieve the required shape and dimensions by running a loop with preset limits for these variables. The program stopped when the ratio stated in table 1 were within a preset range.

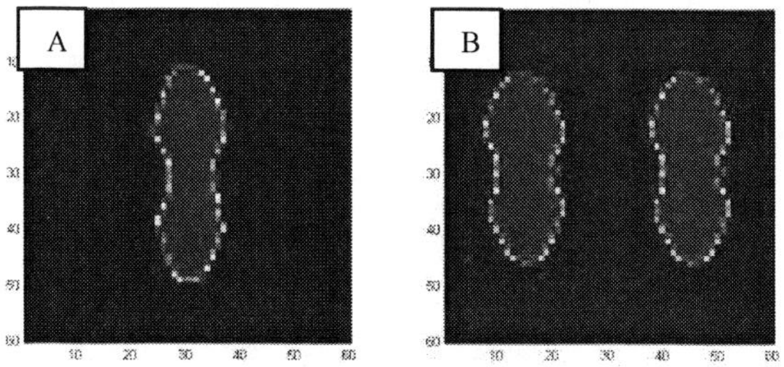

Figure 1: (A) Phase field contour for one cell, (B) for two cells. This figure shows the deformation of the droplets from square to biconcave shape under the influence of interfacial tension force. The only force applied on the two fluids is the controlled body force responsible for creating surface tension on the interface. There is no fluid flow in this simulation beside marginal micro-currents around the interface.

Condition	R1	R2	R3	R4
A	2.8	1.470	0.51	1.44
B	2.25	1.474	0.57	1.285
RBC	3.4	2.5	0.7	2.4

Table 1: R1is the ratio of the RBC length to width. R2 is the ratio of the RBC maximum width to the minimum width. R3 is the ratio of the initial length to final length, and R4 is the initial width to final width. In the case of real RBC the spheroid dimension is 6 μm as stated in [5].

The surface area and the mass of the RBC were conserved during this process. Using the right set of parameters, the droplet would take less than 10000 time steps to acquire its final shape. We used sigma scan pro for measurements. The results could improve if we increase the number of lattices, and widen the scope of the optimization parameters, but this will increase the computational cost.

3.2 Fahraeus-Lindqvist Effect

With a length of 26 lattices for the RBC, 100x60 lattices were used to demonstrate the physical basis of the Fahraeus-Lindqvist effect Fig.2. This effect predicts that the apparent viscosity of blood in long narrow vessels of diameter ranging from 7 μm up to 200 μm, tend to decrease with decreasing diameter. This is due to the presence of a cell- free layer, referred to as plasma-skimming layer near the wall.

Periodic conditions were applied on the vertical boundaries to create a resemblance of the condition of a long tube, and bounce back on the horizontal boundaries, to impose zero velocity on the wall as described in the theory [17].

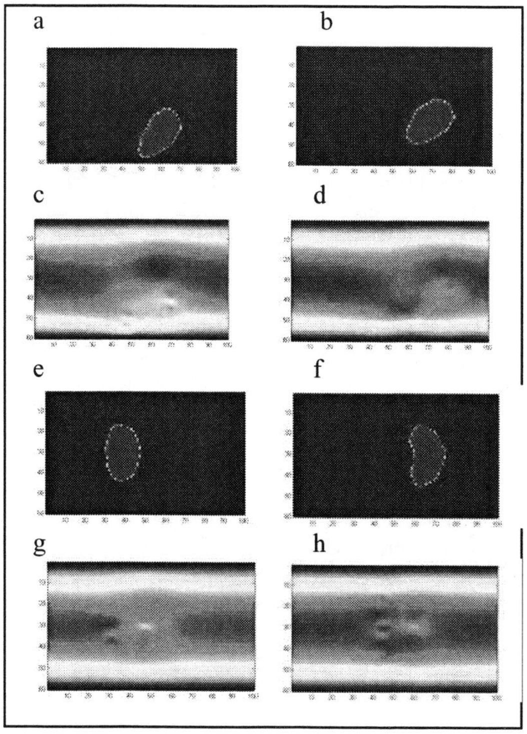

Figure 2: (a, b, e, f) Phase field contours for time steps 5, 8000, 22000, 300000 respectively. (c, d, g, h) Their respective horizontal velocity contour profiles. Looking at the velocity profile and considering subsonic flow conditions, one should consider higher pressure below the droplet, and lower pressure above it. This difference in pressure is responsible for the lift, which drives the droplet towards the center. When the droplet gets to the center, the pressure difference ceases to exist and the lift force vanishes.

This simulation clearly demonstrated in accordance with the theory [17], the tendency of the cell to rotate due to the viscous effects of the wall, and to migrate toward the center because of the Bernoulli force, as evident from the velocity profile.

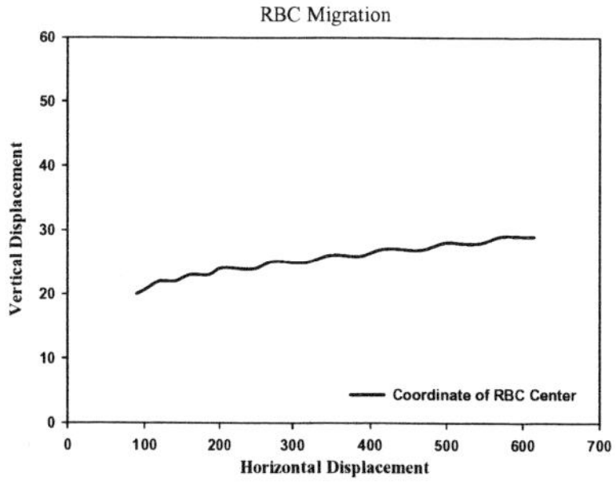

Figure 3: The graph shows the migration of the RBC towards the vertical center of the field. It takes about 30000 time steps for the cell to stabilize and move in a straight line at the center.

3.3 Fahraeus Effect

With a length of 26 lattices for the RBC, 100x60 lattices were used to demonstrate the Fahraeus effect Fig.3. This effect describes the fact that for small tubes, the hematocrit of the tube is smaller than the discharge hematocrit. This is because the RBC mean velocity is higher than the mean blood velocity.

We chose this frame because it corresponds to 20μm tube diameter simulation, which results were shown in [15]. Periodic conditions are applied on the vertical and bounce back on the horizontal boundaries. The mean velocity of the plasma was calculated by averaging all nodal velocities, after separating the RBC velocities from the data. The mean plasma velocity was found to be 0.0372 lu/ts, as compared to 0.0516 lu/ts for the RBC. The ratio of these velocities is 0.72, which is in agreement with the in vitro data recorded by Secomb (2003), and used for validating the model in [16]. We ran another experiment with 100x30, which corresponds to a tube with about 10 μm diameters. We found the plasma mean velocity to be 0.0175 lu/ts, and RBC mean velocity of 0.023lu/ts, giving a ratio of 0.75, which we could not confirm with experimental data.

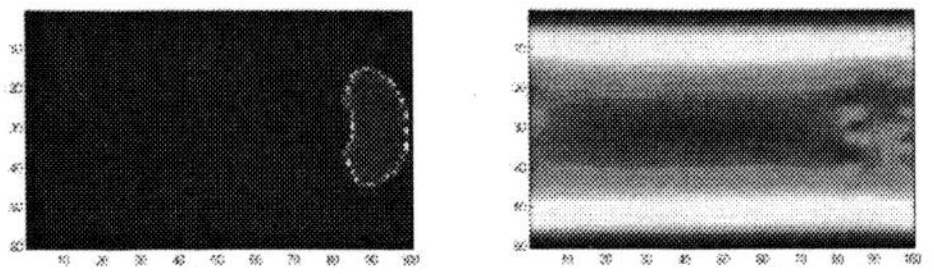

Figure 4: Phase field contour and horizontal velocity profile for 100x60 lattice frames. When the droplet moves in the centre of the flow a Poiseuille condition prevails. This means that the nodal velocities at the center of the flow and especially these of the droplet, will be higher than those of other regions.

3.4 RBC Deformation

100x24 lattices were selected randomly with forcing term of 2.75 E-5 corresponding Re 1.15 to qualitatively demonstrate the capability of the model to simulate variety of shapes resulting from the proximity of the RBC to the wall Fig.4. The objective was to mimic real time shapes obtained from pictures in a rat mesentery [18].

Comparing the simulation results with the real time pictures proves that treating RBC as droplet is more accurate than modeling it as linked springs. No flexible solid model could under real conditions deform in a way, similar to what we were seeing in the pictures. Based on our observation from this simulation we hypothesize that RBC deformation was not caused only by the rigidity of the wall, but rather mainly by the near wall fluid viscous effect. The droplet tends to take a bullet shape in short time after deforming in the vessel. This is due to the fact that the total area and mass of the droplet were conserved, and due to the nature of the flow velocity profile. The forehead of the droplet acquired higher momentum, which led to the formation of bullet shape, thereby spacing the droplet from the walls. This was associated with a reduction in the walls viscous effects, and a loss of the parachute tails. The droplet flow in the microvessels was enhanced. Careful observation of the real pictures shows a gap between the wall and the RBCs. It is worth mentioning that this hypothesis poses a real challenge to be experimentally proven.

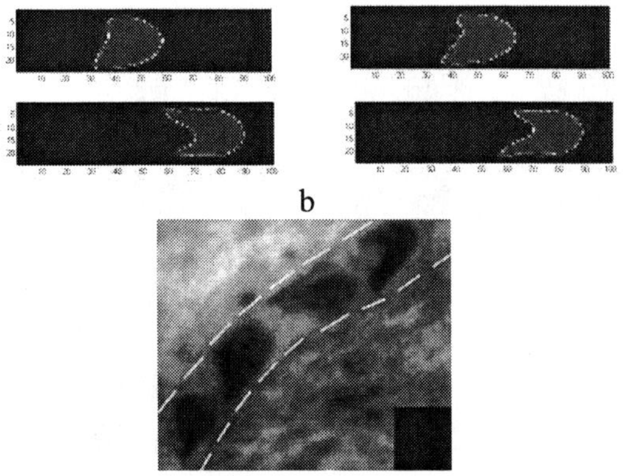

Figure 4: (a) Simulation results. (b) Real time picture of RBC in a rat mesentery. The resemblance of the simulation is quite good, because we picked up snap shots from three different simulations, for which the difference was only with the initial droplet position. One was central; the other was offset toward the bottom of the wall, while the last was offset toward the upper side of the wall.

We enclose a summary of the flow conditions of our simulations in Tab. 2.

Case No	Flow Re	Horizontal BC	Vertical BC
1	0	Periodic	Periodic
2	0.3	Bounce back	Periodic
3	0.15	Bounce back	Periodic
4	1.15	Bounce back	Periodic

Table 2: Flow conditions description for the four mentioned simulations

4 CONCLUSION

We are presenting a two dimensional two-component LBM model, which uses the surface tension hypotheses to explain the reason why biconcave discoid is the resting shape of RBC. The control parameters used to obtain the static shape of the erythrocyte are the same parameters, which we manipulate to control the shape of the red blood cell while streaming through microvessels. It is the author's intention to track the dynamic change in the lipid bilayer density distribution through correlation with these control parameters as future endeavor.

Knowing that two dimensional models have inherent problem of being quantitatively incorrect, we consider this work as more qualitative. This is a step towards moving to simulate same conditions using three-dimensional models.

The model was used as an attempt to validate the surface tension hypothesis, the physics behind the Fahraeus-Lindqvist effect, and the Fahraeus effect. The simulation is capable of reproducing quiescent erythrocyte biconcave shape and track the deformation of the red blood cell in motion.

The data could be used to calculate blood viscosity in microvessels using variety of models, for example the sigma effect model [16].

REFERENCES

[1] M, M.D., I. Halliday, C.M. Care, and Lyuba Alboul, *Modeling the flow of dense suspensions of defornable particles in three dimensions* PHYSICAL REVIEW E, 2007. **75**: p. 066707-(1-17).

[2] Yaling Liu, W.K.L., *Rheology of red blood cell aggregation by computer simulation.* Journal of Computational Physics, 2006. **220**: p. 139-154.

[3] Krzysztof Boryszko. Witold Dzwinel, a.D.A.Y., *Dynamical Clustering of red blood cells in sapillary vessels.* J Mol Model 2003. **9**(16): p. 16-33.

[4] Ken-ichi TSUBOTA, S.W., and Takami YAMAGUCHI, *Simulation Study on Effects of Hematocrit on Blood Flow Properties Using Particle Method.* Journal of Biomechanical Science and Engineering, 2006. **1**(1): p. 159-170.

[5] BRAASCH, D., *Red Cell Deformability and Capillary Blood Flow.* Physiological Reviews, 1971. **51**(4): p. 679-701.

[6] Ju li, G.l., Ming Dao, and Subra Suresh, *Spectrin-Level Modeling of the Cytoskeleton and Optical Tweezers Stectching of the Erythrocyte* Biophysical Journal 2005. **88**: p. 3707-3719.

[7] Ju li, G.l., Ming Dao, and Subra Suresh, *Cytoskeletal dynamics of human erythrocyte.* PNAS, 2007. **104**(12): p. 4937-4942.

[8] S. L. Keller, W.H.P.I., W.H. Huestis, and H.M. McConnell, *Red Blood Cell Lipids From Immiscible Liquids.* PHYSICAL REVIEW LETTERS, 1998. **81**(22): p. 5019-5022.

[9] Michael C. Sukop, D.T.T., Jr., ed. *Lattice Boltzmann Modeling*. 2007, Springer. 172.

[10] I. Halliday, A.P.H., and C.M. Care, *Lattice Boltzmann algorithm for continuum multicomponent flow.* PHYSICAL REVIEW E, 2007. **76**: p. 026708-(1-13).

[11] M, M.D., I. Halliday, C.M. Care, *Multi-component lattice Boltzmann equation for mesoscale blood flow.* JOURNAL OF PHYSICS A: MATHEMATICAN AND GENERAL, 2003. **36**: p. 8517-8534.

[12] M, M.D., I. Halliday, C.M. Care, *A multi-component Lattice Boltzmann scheme: Towards the mesoscale simulation of blood flow.* Medical Engineering & Physics, 2005. **28 (2006)**: p. 13-18.

[13] H.Rothman, A.K.G.a.D., *Lattice Blotzmann model of immiscible fluids.* PHYSICAL REVIEW A, 1991. **43**(8): p. 4320-4327.

[14] D.Salin, U.D.O.a., *Two-color nonlinear Boltzmann cellular automata: Surface tension and wetting.* PhYSICAL REVIEW E, 1995. **51**(4): p. 3718-3728.

[15] S. V. Lishchuk, C.M.C., and I. Halliday, *Lattice Boltzmann algorithm for surface tension with greatly reduced microcurrents.* PHYSICAL REVIEW E, 2003. **67**: p. 03671-(1-5)

[16] Sun, C. and L. L.Munn, *Particulate Nature of Blood Determines Macroscopic Rheology:A2-D Lattice Boltzmann Analysis.* Biophysical Journal, 2005. **88**: p. 1635-1645.

[17] E.Rittgers, K.B.C.A.P.Y.S., ed. *Biofluid Mechanics. The Human Circulation*. First ed., ed. T. Francis. 2007, A CRC PRESS BOOK: Abingdon, UK. 419.

[18] Ekaterina I Galanzha, V.V.T., Vladimir P Zharov, *Advances in small animal mesentery models for in vivo flow cytometry, dynamic microscopy, and drug screening.* World Journal of Gastroenterology, 2007. **13**(2): p. 192-218.

Numerical Modelling and Analysis of the Burning Transient in a Solid-Propellant Micro-Thruster

J.A. Moríñigo* and J. Hermida-Quesada**

National Institute for Aerospace Technology
* Dept. of Space Programmes, morignigoja@inta.es
** Dept. Aerodynamics and Propulsion, hermidaj@inta.es
Ctra. Ajalvir km.4, Torrejón de Ardoz 28850, Madrid, Spain

ABSTRACT

This work investigates an architecture of arrayed micro-thrusters of submillimeter size for spacecraft propulsion. Each single thruster comprises a setup of micro-channels arranged around the solid-propellant reservoir, hence initial heating of both sides of its wall by the hot gas takes place in the transient. Simulation of the unsteady response of the gas and wafer during the grain burning has been conducted under a multiphysics-based approach that fully couples the parts involved and focuses on the early phase of the transient. Preliminary results suggest the beneficial role played by the thermal management provided by the micro-channels, that mitigates the heat loss across the wall. Thus, miniaturization of the propellant reservoir within the submillimeter range is favoured due to the sustainment of higher temperatures ahead of the burning front, by the inner wall. Simulated firings show encouraging performance.

Keywords: combustion in microscale, MEMS, micro-thruster, multiphysics modelling.

1 INTRODUCTION

Two fundamental aspects are inherent to gas flows in propulsion-MEMS: their very low Reynolds numbers, hence turbulence is inhibited; and the high heat loss to the surrounding substrate, because of the large area to volume ratio. It is well known that the use of solid-propellant is advantageous as it provides no moving parts and simplicity to micro-thruster design. In addition, the sizing of arrays of μ-thrusters allows to meet the requirements of very small spacecrafts [1]. Nevertheless, the efficient conversion of the propellant chemical energy into thermal energy in the microscale remains a challenge for this class of micro-systems. To this respect, Rossi *et al.* [2,3] have analysed a variety of reservoir configurations filled with energetic materials and prope-llants. They have identified that thermal loss is responsible of the flame extinction at reservoir sizes of about a millimeter, where combustion becomes unstable or not sustained anymore. Kondo *et al.* [4] reported for a submillimeter rocket tested in vacuum, that the reservoir depressurization at the ignition phase leads to a low rate of success in the combustion starting and thrust generation.

Besides an adequate choice of propellant compositions, the thermal management of the energy released by the combustion influences to a great extent the performance, thus design optimization compatible with microfabrication is of paramount importance. For brevity, this paper presents preliminary results of the thermal management provided by a passage of μ-channels, here referred to as the *heating channels*, intended to thermally isolate the combustion zone and to enhance the heating of the inner walls. The micro-thruster arquitecture investigated is sketched in Fig. 1.

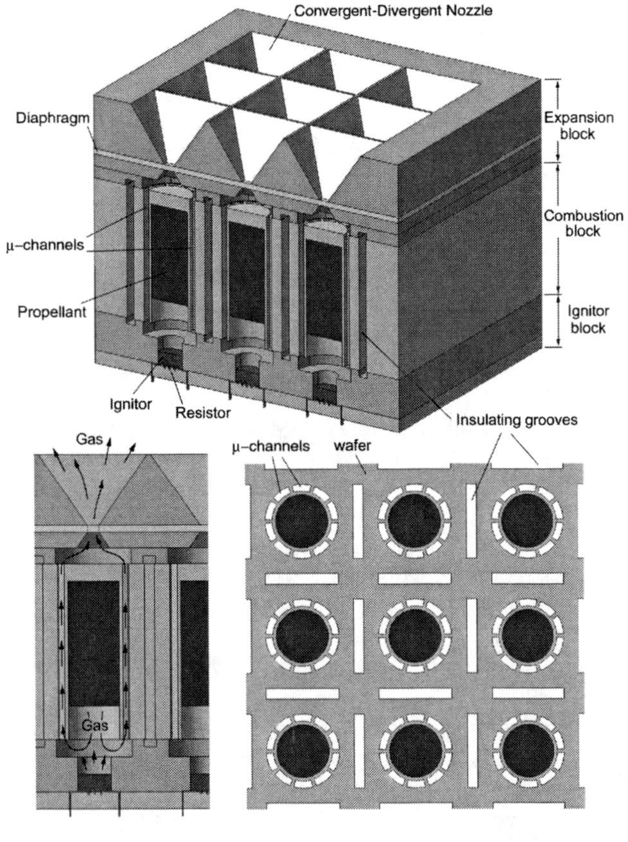

Figure 1: Schematics of a 3x3 array of solid-propellant micro-thruster with *heating channels*. Upper: general view with its first row sectioned; lower: meridian view with flow path indicated and cross-section of combustion block.

2 APPROACH

The computational domain used for the performance assessment is depicted in Fig. 2 for two locations of the propellant front. The passages sizing, of 30μm height, is conservative and compatible with the expected unburned particles sizes reported by other authors. The numerical simulations have been conducted under a multiphysics approach with thermal-fluid coupling between the gas and solids (wafer and propellant) involved. A continuum description of the gas flow based on the Navier-Stokes (NS) equations for laminar, compressible, heat conducting ideal gas is adopted at the microscale. The set of governing equations is solved in axisymmetric form with the FLUENT code, discretized according to a 2nd-order upwind algorithm and supplemented with a 2nd-order slip-flow model [5,6] for the velocity slip and temperature jump boundary conditions (BCs) at the nozzle wall to deal with the moderate rarefaction effects experienced by the gas. NS equations are solved in the gas zone in conjunction with the time-dependent heat diffusion equation in the solids. It is noted that heat conduction from the gas phase to the solid propellant (driven by the temperature normal gradient), is considered as the sole mechanism for ignition. Hence, radiative heat transfer is not included into the modelling, although it may enhance the decomposition rate of the propellant. The thermal balance at the burning surface (here simplified as non-moving and flat) to determine the surface temperature and instant of ignition ($T_w=T_{ign}$), states

$$-k_p \nabla T \cdot \vec{n} = \rho_p r_p q_p + \phi_f + \phi_{ign} \tag{1}$$

being k_p the thermal conductivity of the propellant, q_p the heat released at the surface, ϕ_f the heat flux from the flame and ϕ_{ign} the heat flux from the igniter gas. Other issue of relevance included into the model is a diaphragm rupture condition, applied at the throat section to model the dynamics of the starting pressurization into the fluid cavity caused by the igniter firing prior to the diaphragm bursting. Solid-fluid coupling, heat flux BCs at the gas-propellant and gas-igniter interfaces, slip-flow BCs and diaphragm rupture have been coded in FLUENT as UDF routines. One basic criterion for propellant selection derives from the constrains to its use in actual small spacecrafts. In this work, GAP (Glycidyle Azide Polymer) -based propellant is assumed, suited for microsystems due to its homogeneity and ease of injection properties. The simulations mimic the propellant decomposition provoked by a pyrotechnic paste of ZPP (Zirconium Potassium Perchlorate: ρ=3570kg/m^3, T_f=5018ºK, burn rate (in mm/s): $r=ap^n$, with a=1.85·10^{-4} mm/s/Pa^{-n}, n=0.79, p in Pascal), as depicted in Fig. 1 and set at the lefthand boundary in Fig. 2, to ignite the GAP by providing a high temperature flux. This mechanism has been tested [2] and seems effective to achieve the opening of the nozzle to the near-vacuum ambient. The igniter-BC is modelled with a time dependent massflux law ρr at the adiabatic flame temperature. Albeit the ZPP and propellant products have different composition, it is assumed that thermodynamics and transport properties are identical and the composition is frozen (see Table 1).

GAP-based propellant:	
Density, ρ_p	1528 kg/m^3
Flame temperature, T_f	1991ºK
Auto-ignition temperature, T_{ign}	573ºK
Burn rate (in mm/s), $r_p=ap^n$	a=1.1mm/s/bar^{-n}, n=0.58
Products molecular weight, M	19.7 g/mol
Products specific heat, $C_{p,gas}$	2034 J/kg-K
Thermal conductivity, k_{gas}	0.16 W/m-K
Laminar viscosity, μ_{gas}	6.2 ·10^{-5} kg/m-K
Specific heats ratio, γ	1.26

Table 1: Propellant and combustion gas data

The high heat enthalpy of the igniter released during a very short time interval makes the gas velocity to become supersonic into the μ-channels and to generate shock waves that propagate downstream up to the diaphragm (Fig. 3). The small characteristic time inherent to this phenomenon requires an integration time step of about Δt~10^{-8}s to solve the pressure buildup at the ignition phase. Variation of the integration time-step is carried out in the explicit dual-time stepping method used.

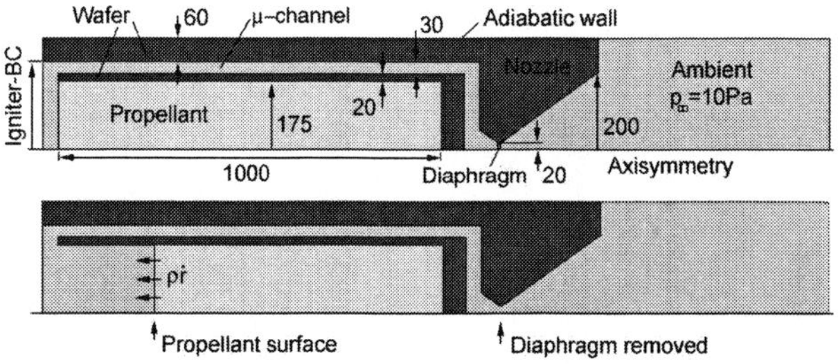

Figure 2: Computational domain of the axisymmetrical simulations (gas, wafer and propellant indicated). Upper: configuration before GAP ignition; lower: configuration at an intermediate instant of burning. Dimensions in μm.

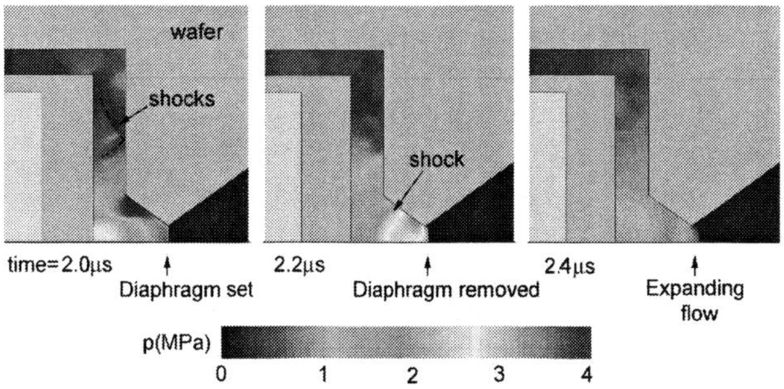

Figure 3: Time sequence of pressure snapshots focusing the diaphragm bursting due to the igniter flux at the early μs'.

3 RESULTS

The pressure peak visible in Fig. 4 is conservative in the sense that the model does not take into account the fluid region filled with air until the diaphragm rupture. Then, since viscosity and thermal conductivity of the combustion gas is notably higher than that of air, pressure peaks will tend to decrease due to the excess of viscous dissipation. After the ZPP flux extinction, steady pressure into the reservoir is rapidly settled down. Interestingly, this pressure level exhibits a significant dependence on the igniter flux lasting (t_{ign}), as it is seen in Fig.4. Regarding this behaviour, it should be noted that the burning surface area to the throat area ratio (A_p/A_{th}) drives the operating pressure p_c according to the steady approximation

$$p_c^{1-n} = \frac{a\rho_p}{\Gamma}\frac{A_p}{A_{th}}\sqrt{R_g T_t}\,, \quad \Gamma = \sqrt{\gamma}\left(\frac{2}{\gamma+1}\right)^{\frac{\gamma+1}{2(\gamma-1)}} \qquad (2)$$

where T_t denotes the averaged stagnation temperature at the nozzle inlet and determines the maximum massflow rate passing through the throat for prescribed geometry and ballistics. In fact, the steady p_c plotted in Fig.4 corresponds to the situation $T_t \ll T_f$, caused by the heat loss into the wafer and grain at large times of the transient. This thermally induced response is not addressed here in detail, requiring further investigation.

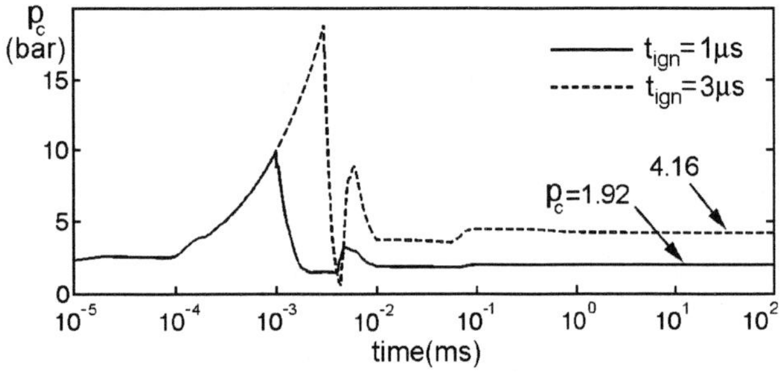

Figure 4: History of averaged pressure into the propellant reservoir for a ZPP heat flux lasting for 1 and 3μs.

Once the grain ignites, the reservoir wall temperature rises due to the heat flux at both sides and its head-end. This increase is illustrated in Fig. 5 for the startup time interval of 3ms, in which the propellant surface recedes a distance about twice the wall thickness (20μm) according to the burning rate law (1.6mm/s at 2bar). For t_{ign}=3μs, the hot gas penetrates further into the μ-channels because of its higher pressure. It is pointed out that the assumption of non-moving surface is only valid for short times after the reference instant corresponding to the surface location.

The thermal response of the μ-thruster shown in Fig. 6 corresponds to 90ms after the ignition and has been computed with the second propellant configuration depicted in Fig. 2. The temperature map shows that the hot gas at the μ-channel inlet enhances the heating of the portion of the reservoir wall ahead of the propellant, though a significant temperature decay occurs downstream the burning front near the wall. Nevertheless, the local temperature in this region remains over the propellant auto-ignition limit, thus it favours the sustainment of the chemical reactions. The progressive cooling of the gas inside the μ-channel is apparent, as well as the drop in stagnation temperature due to the heat transfer, which scales with the difference of the core temperature to the wall temperature, and leads to the bulk temperature T_t~450K upstream the nozzle inlet. Besides, the stagnation pressure drop along the μ-channel is small, under 9%. It is obvious that high p_c is beneficial as it enhances the heating rate at the early phase of the transient and because implies higher thrust delivered. However, the large surface area-to-volume ratio entails a penalty on the specific impulse derived from a lower T_t, which is likely unavoidable. An observation should be pointed out: even with T_t~300K, (cold flow), the smaller molecular mass of the expanding gas would imply a net gain of specific impulse compared with other gases (for instance, more than 27% in the case of N_2).

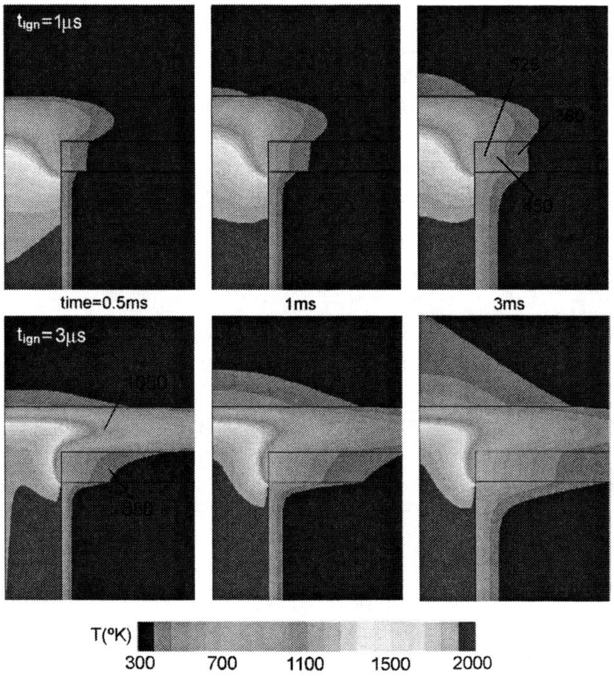

Figure 5: Sequence of temperature maps by the reservoir wall at 0.5, 1 and 3ms after the propellant ignition, for t_{ign}=1μs (upper row) and 3μs (lower row).

The determination of the gas film coefficient h_g in the μ-channels out of the computed flowfield at the beginning and intermediate time instants of the transient, yields values ranging from 1600 to 2800W/m^2-K (with Reynolds number *Re*=90 and 180, respectively and referred to the channel hydraulic diameter). A quantification of its effect on the thermal exchange is shown in Fig.7 for the propellant reservoir with and without *heating channels*. The averaged

T_w on the inner wall is plotted versus time for gas flow conditions at the μ-channel inlet taken from the NS simulations. The comparison of both configurations at the above mentioned h_g, is presented for two wafers made of silicon and glass. With μ-channels, the attained temperature rapidly raises beyond that without μ-channels. In particular, with a wafer of glass, the *heating channels* lead to a heating rate of about 50 to 80K per millisecond at the startup, which represents more than a twofold increase and stresses the improvement provided, at least at the early phase of the transient. Moreover, regarding the reservoir pressure, the gasdynamics equations show that the thermal exchange rate increases roughly linearly with it, so a moderate to high pressure in the reservoir is desirable.

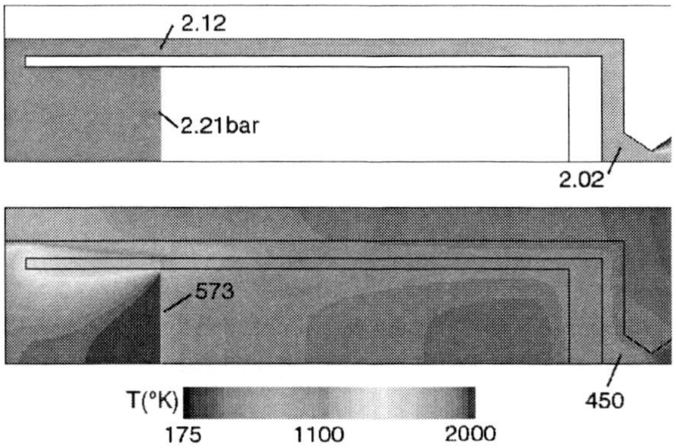

Figure 6: Stagnation pressure (upper) and temperature (lower) map after 90ms of burning transient.

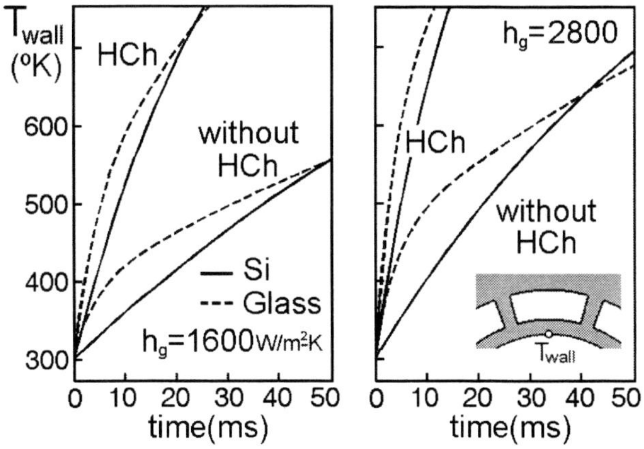

Figure 7: History of temperature on the inner wall of the propellant reservoir, with and without *heating channels* (HCh), made of silicon and glass.

4 CONCLUSIONS

The burning transient of a solid-propellant micro-thruster of submillimeter size with an arrangement of micro-channels around the propellant reservoir, has been simulated following a multiphysics-based approach. The modelling comprises high-speed flow, thermal coupling at the wafer-gas interface and burning surface, slip-flow at the nozzle wall, propellant ignition and diaphragm rupture at the beginning of the transient. The focus of the present investigation is not so much on the ignition times, but on the assessment of the unsteady thermal response of the micro-thruster setup here analysed. Preliminary simulations with GAP-based propellant and frozen composition have shown that the thermal management provided by the micro-channels during the transient reduces the heat loss and leads to the sustainment of higher temperatures in the reservoir. This aspect suggests the broadening of the flame extinction limits, thereby providing longer combustion times and the performance enhancement within the submillimeter range. The addition of a moving, non-flat recessing surface into a 3D model constitutes a next step and a challenge for more accurate predictions.

ACKNOWLEDGEMENT

This work was supported by the Spanish Ministry of Defence, as part of the micropropulsion activities in the Small Satellites Programme. The authors thank Emilio Lago-Llano of INTA for his assistance with some graphical material of this article.

REFERENCES

[1] Youngner D.W., Lu S.T., Choueiri E., Neidert J.B., Black III, R.E., Graham K.J., Fahey D., Lucus R. and Zhu X., "MEMS Mega-pixel Micro-thruster Arrays for Small Satellite Stationkeeping", in Proc. 14th Annual/USU Conference on Small Satellites, North Logan, USA, SSC00-X-2, 2000.

[2] Rossi C., Do Conto T., Estève D. and Larangôt B., "Design, Fabrication and Modelling of MEMS-based Microthruster for Space Application", Smart. Mater. Struct., 10:1156-1162, 2001.

[3] Rossi C., Briand D., Dumonteuil M., Camps Th., Quyên Pham Ph. and de Rooij N.F., "Matrix of 10x10 Addressed Solid Propellant Microthrusters: Review of the Technologies", Sensors & Actuators A, 126:241-252, 2006.

[4] Kondo K., Tanaka S., Habu H., Tokudome S., Hori K., Saito H., Itoh A., Watanabe M. and Esashi M., "Vacuum Test of a Micro-solid Propellant Rocket Array Thruster", IEICE Electronics Express, 1(8): 222-227, 2004.

[5] Zhang K.L., Chou S.K. and Ang S.S., "Performance Prediction of a Novel Solid-Propellant Micro-thruster", J. Prop. Power, 22(1):56-63, 2006.

[6] Moríñigo J.A, Hermida Quesada J. and Caballero Requena F., "Slip-Model Performance for Under-expanded Micro-scale Rocket Nozzle Flows", J. Thermal Science, 16(3):223-230, 2007.

Numerical Analysis of Nanofluidic Sample Preconcentration in Hydrodynamic Flow

Y. Wang*, K. Pant*, ZJ Chen*, W. Diffey**, P. Ashley**, and S. Sundaram*

*CFD Research Corporation, Huntsville, AL, U.S.A., yxw@cfdrc.com
**U.S.Army RDECOM, Redstone Arsenal, AL, USA

ABSTRACT

The phenomenon of sample enrichment due to the presence of an electric field barrier at the micro-nano-channel interface can be harnessed to obtain sample preconcentration that enhances sensitivity and limit-of-detection in sensor instruments. This paper presents a high-fidelity numerical analysis of the electrokinetics at the micro-nano-interface under pressure driven flows. Parametric analysis is performed to capture the critical effect of pressure head and background electrolyte (BGE) ion concentration on the electrokinetic and species transport of the preconcentrator. Our studies demonstrate that ion-polarization and electric field barrier can be established at micro-nano-channel interfaces and substantial sample enrichment ($>10^4$–fold) can be readily achieved using hydrodynamic flow. The results can be used for practical design and operational protocol development of novel nanofluidics-based sample preconcentrators.

Keywords: Nanofluidic, preconcentration, multi-physics

1 INTRODUCTION

Sample preconcentration is recognized as one of the most critical steps in high-performance, integrated lab-on-chip systems. Among various methods currently in use, nanochannel ion-polarization enabled by externally applied electric fields [1,2] or pressure gradients [3] hold great promise due to its capability for obtaining ultra-high enrichment ratios (up to 10^7–fold [1]). Several researchers have reported on experimental studies of nanochannel preconcentration without rigorous analysis to develop a fundamental understanding of the phenomenon. This paper presents the first, high-fidelity numerical study of nanofluidics-based sample preconcentration in hydrodynamic flows. Our analysis shows that both pressure and background electrolyte (BGE) concentration markedly impact the electrokinetics at the micro-nano-channel interface by varying the equilibrium of the species transport therein. Substantial analyte enrichment ($>10^4$–fold) can be achieved using nanofluidic preconcentrators. The modeling framework presented in this paper and our findings can be utilized to guide design and protocol development of sample preparation and biodetection technologies.

2 METHODS

2.1 Nanofluidic Preconcentrator

Figure 1 illustrates the geometry and principle of the nanofluidic preconcentration in hydrodynamic flow. A negatively charged nanochannel bridges two microchannels (or micro-reservoirs). A pressure head (Δp relative to the outlet) is applied at the inlet, transporting background electrolyte (BGE) ions and sample analytes towards the nanochannel. Due to the overlapping electric double layer (EDL) and unipolar nature of the nanochannel, a substantial electric field barrier is induced at the micro-nano interface, which precludes the negatively charged sample analytes from entering the nanochannel. As BGE carrying the analyte continuously flows through the channel, the analyte is enriched at the interface. To achieve adequate analyte enrichment in a reasonable processing time, strong electric field barrier and fast hydrodynamic flow are desired.

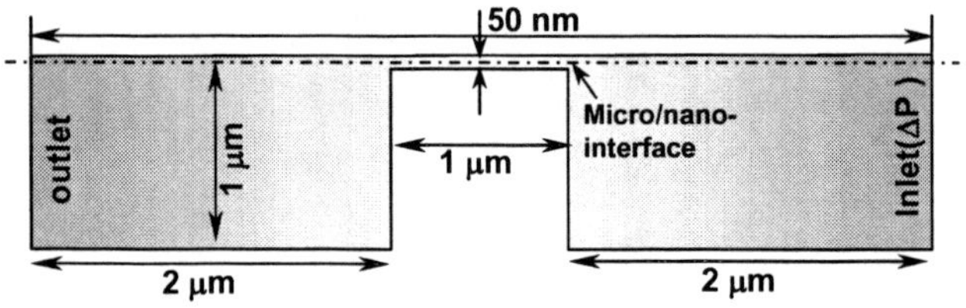

Figure 1. Schematic of nanofluidic preconcentrator.

2.2 Computational Models

Simulations are carried out using CFDRC-developed high-fidelity, multi-physics simulation environment, CFD-ACE+, in which three key modules – fluid flow, electric, and chemistry – were invoked to describe the fluid flow, electric field, and species distribution, respectively. A mathematical description of the models is presented next.

2.2.1 Electrostatic Models

The electric potential and field is governed by Poisson equation

$$\varepsilon_0\varepsilon_r\nabla^2\phi + F\sum_i z_i c_i = 0 \tag{1}$$

where ε_0 and ε_r are, respectively, the electrical permittivity in the vacuum and the relative permittivity; F is the Faraday constant; z_i is the valence and c_i is the molar concentration of the ith ionic species.

2.2.2 Species Transport Models

The species transport in the micro- and nano-channels is described by [4]

$$\frac{\partial c_i}{\partial t} + \nabla \cdot J_i = 0 \tag{2}$$

Here the flux J_i of the ith species is given by Poisson-Nernst-Planck (PNP) equation

$$J_i = -D_i \nabla c_i + u c_i + \omega_i z_i E c_i \quad (3)$$

where the terms on the R.H.S. respectively denote the species flux contributions from molecular diffusion, convection, and electromigration; ω_i is the electrophoretic mobility; D_i is the diffusivity; u is the velocity vector and $E = -\nabla \phi$ is the electric field. Note that Eq. (3) applies to both BGE ions and analytes. The flow of electrical current is a result of the individual flux of BGE ions in the solution, which is given by

$$I_i = F z_i J_i = F\left(-z_i D_i \nabla c_i + u z_i c_i + \omega_i z_i^2 c_i E\right) \text{ and}$$
$$I_{tot} = \sum_i I_i = F \sum_i z_i J_i \quad (4)$$

The three terms on the R.H.S of the first equation signify the current contributions from diffusion, convection, and electromigration of the ith species, respectively. It should be mentioned that while there is no overall current (I_{tot}) in the hydrodynamic case, the current component of the ith (I_i) BGE ion can be non-zero.

2.2.3 Fluidic Models

Viscous, incompressible fluid flow in the micro- and nano-channels is described by the conservation of mass and momentum equations [4]:

$$\frac{\partial u}{\partial t} + (u \cdot \nabla) u = \mu \nabla^2 u - \nabla p / \rho + f_e \quad (5)$$
$$\nabla \cdot u = 0$$

where u, ρ, μ, and p are the fluid velocity, density, dynamic viscosity, and pressure respectively; f_e is the electrostatic body force due to the electrostatic charges (Columbic force) and is expressed as

$$f_e = EF \sum_i z_i c_i \quad (6)$$

2.3 Simulation Methods

As the preconcentrator is normally used to address the analyte at the trace-level concentrations (pico–nano molar), we will assume that the analyte is so dilute that their presence does not alter the electric field, flow, and BGE ionic equilibrium. Following the regular perturbation analysis of Bharadwaj and Santiago [5], our simulation is conducted in two steps. In the first step, Eqs. (1), (2), and (5) are solved in a coupled manner to resolve the flow (u), electric field (E), and BGE ion concentrations (c). In the second step, Eq. (2) is solved for the analyte with u and E from the first step. This approach is most suitable for dilute analyte concentrations, which is typical for a variety of proteomic, genomic, and chemical compound analysis.

Boundary conditions are also supplied for closure of the equations. Electrically, zero potential and zero surface charge density (i.e., zero electric field) are applied at the outlet and the inlet, respectively. A fixed surface charge density (–0.002 C/m^2 [4]) is specified at the nanochannel walls while charges on the microchannels are neglected. For BGE flow, a differential pressure is applied between the inlet and outlet. No-slip boundary conditions are invoked at all channel walls. For BGE ion transport, a constant bulk ion concentration is provided at both the inlet and outlet. For the analyte, a constant concentration is set only at the microchannel inlet. Table 1 lists all the relevant parameters used in the simulation, where KCl is used as the BGE. Given the simulation parameters, Eq. (4) can be simplified as

$$I_{tot} = F\left(-D\nabla\left(c_K - c_{Cl}\right) + u\left(c_K - c_{Cl}\right) + \omega\left(c_K + c_{Cl}\right)E\right) \quad (7)$$

For sake of brevity, a single diffusivity D and mobility ω are used for both K and Cl in Eq. (7). A fully structured computational domain consisting of ~17,000 cells was generated for the entire preconcentrator. Grid checks were performed to ensure mesh-independent results. Specifically, 61×51 grid points in a power law distribution were used to resolve the longitudinal and transverse dimensions of the nanochannel.

Table 1. Simulation parameters for numerical analysis

Case No.	1	2	3	4	5
Electrolyte Conc. (M)	10^{-4}	10^{-4}	10^{-4}	$2X10^{-5}$	10^{-3}
Pressure ΔP (atm)	0.1	0.2	0.4	0.2	0.2

3 RESULTS AND DISCUSSION

In the simulation analysis, case 2 is used as the baseline. The effects of pressure head and BGE ion concentrations are, respectively, captured by cases 1, 2, and 3, and cases 2, 4 and 5. Simulational results are used to interpret the field variation of the BGE and to evaluate sample preconcentrator performance.

3.1 Background Electrolyte

Figure 2 shows the contour plots of K^+ and Cl^- ion concentrations for the baseline case. Note that non-uniform ion concentration develops in the electric double layer (EDL) in the nanochannel. The concentrations of both ions at the two interfaces are different (i.e., ion-polarization [4]).

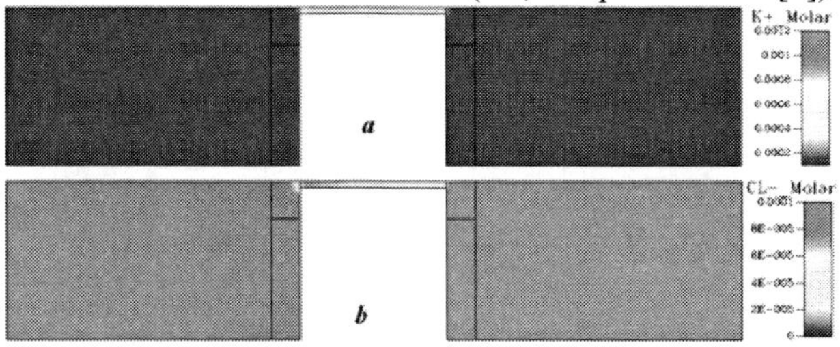

Figure 2. Contour plot of BGE ion concentrations.

Figure 3a shows the transverse profile of K^+ and Cl^- ion concentrations and indicates that a strongly overlapped EDL is established in the nanochannel. The EDL effect is more pronounced for low BGE concentrations and the channel is essentially filled with a unipolar solution of K^+, viz., the solution is dominantly occupied by K^+ ions (note the logarithmic *y*-axis). Figure 3b displays the longitudinal profile of K^+ and Cl^- ion concentrations. In the microchannel, both remain almost constant to satisfy the electroneutrality, i.e., $\sum z_i c_i = 0$. In the nanochannel, the concentration of counter-ion (K^+) is appreciably higher than the co-ion (Cl^-) to neutralize the negative charges on the channel wall. As a result, steep gradients of the BGE ion concentrations form at the micro-nano interface.

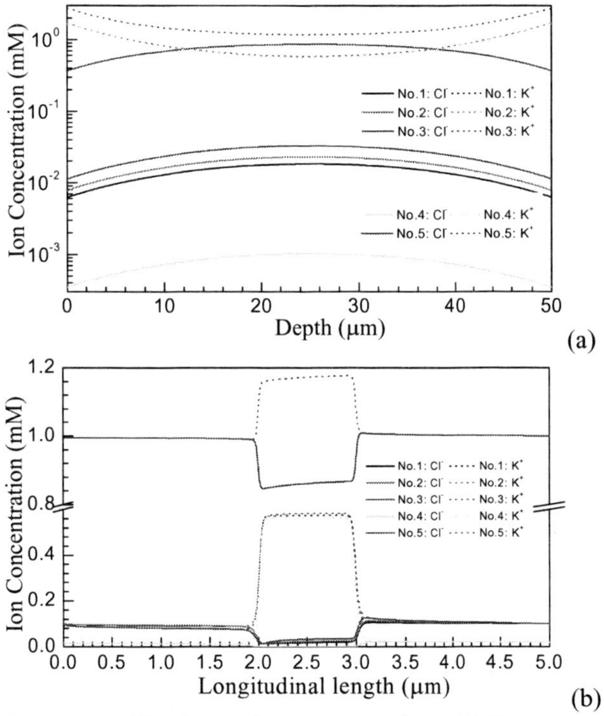

Figure 3. Profile of BGE ion concentrations. (a) Transverse (b) Longitudinal.

Figure 4 shows the pressure distribution along the channel centerline (dash-dot line in Figure 1). The pressure drop in the microchannel is negligible and is nearly linear in the nanochannel. Given the same pressure head (case No. 2, 4, and 5), the pressure drop in the nanochannel is larger for high BGE concentrations (case No. 5). This is attributed to the smaller EDL thickness and smaller concentration difference between K^+ and Cl^- ions at the micro-nano interface, both reducing the electrostatic body force that impedes flow.

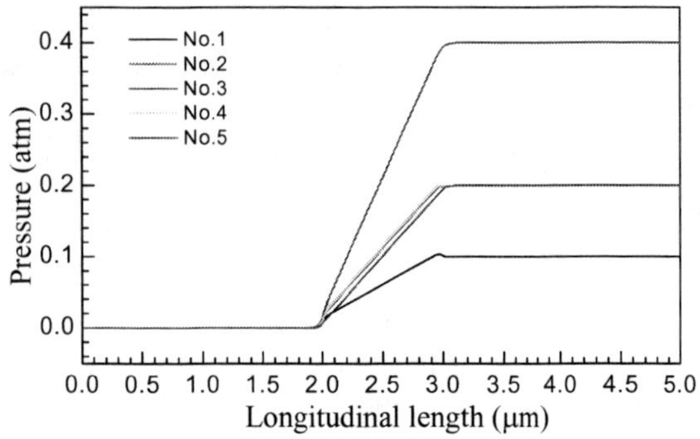

Figure 4. Longitudinal Pressure distribution

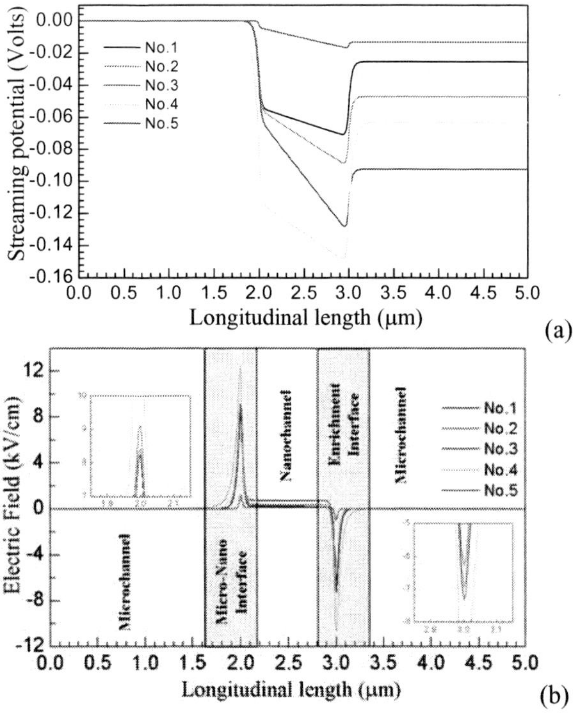

Figure 5. Longitudinal profile of (a) electric streaming potential and (b) electric field.

Figure 5a depicts the longitudinal dependence of the streaming potential and the electric field. The streaming potential is an electric potential generated by the directional movement of the non-electroneutral electrolyte under a pressure gradient through a channel [6]. A larger pressure head and lower electrolytic concentration yield a larger potential because of the increased flux of the non-electroneutral charges. A point of interest is that while the potential difference between the inlet and the outlet is small (0–0.1 Volts), the potential variation along the centerline is non-monotonic and *an abrupt drop/rise occurs at the interfaces, leading to strong electric fields for preconcentration.* This is distinctly different from previous observations on streaming potential (or field) in a single microchannel or nanochannel. The induced electric field (3rd term in Eq. (7)) counteracts the current contributions from molecular diffusion and convection (first two terms in Eq (7)). Figure 5b reveals that the entire nano-preconcentrator can be divided into five sub-domains: two microchannel (electroneutral) segments, nanochannel (non-electroneutral), and two micro-nano interfaces. In the microchannels, electroneutrality produces zero electric field (note I_{tot}=0 in Eq. (7)). In the nanochannel, the gradient of the BGE concentration is almost constant along the longitudinal direction, leading to negligible current contribution from molecular diffusion. Therefore, a constant electric field is induced to counteract the convection current. At the micro-nano-interface, the scenario is more complicated and the currents from diffusion, convection, and migration are all comparable. At the interface to the right (termed the enrichment interface), the diffusive current flux points to the right (see Figure 3b) and needs to be balanced by combined convection and electromigration (to meet I_{tot}=0). Therefore, the electric field is directed to the left (i.e., negative) and its magnitude decreases with an increase in pressure head (viz. convection current becomes stronger), although the overall inlet-outlet potential difference increases. It is the electric field that serves as a barrier that repels negatively charged analytes that approach the interface and enables analyte enrichment. In contrast, at the interface to the left, ion diffusion aligns with convection (both pointing to the left) to oppose the electromigration-induced current (3rd term in Eq. (7)). Hence, the electric field points to the right (i.e., positive). For the same reason, the magnitude of the electric field at the left interface is higher than that at the right interface (see Figure 5b).

3.2 Sample Preconcentration

Figure 6 shows the transient evolution of enrichment ratio (ER) of four sample analytes (A-D) as well as the snapshots of analyte-B preconcentration at the enrichment interface on the right in case No. 3. ER is defined as the ratio of the highest analyte concentration in the computational domain to that at the inlet. The analytes have diffusivity D_{A-D} = {1,2,4,1}×10^{-11} m²/s and mobility ω_{A-D} = {2,2,2,1}×10^{-11} m²/(sV). Our results suggest that ER depends strongly on analyte properties (in particular, mobility). A large electrokinetic mobility results in a stronger electromigration force leading to a higher ER. Depending on analyte properties, ER>10^4X can be successfully achieved.

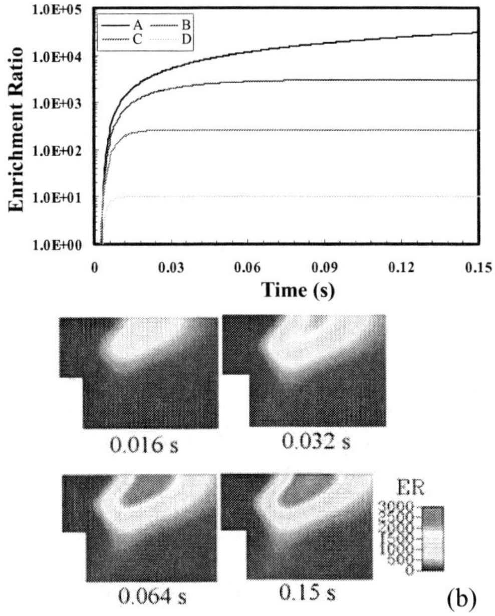

Figure 6. Transient evolution of Enrichment ratio of four sample analytes and contour plot for analyte-B in case No. 3.

Our simulation results (not shown) also indicate that at smaller pressure heads, ER for an analyte reaches a much higher value but at a slower pace (longer enrichment times). This is due to the fact that lower pressure head (or slower flow velocity) reinforces electric field barrier at the enrichment interface while attenuating the convection force that pushes the analyte through the nanochannel.

4 CONCLUSION

In this paper, we have investigated the electrokinetics of a nanofluidic hydrodynamic sample preconcentrator. Multi-physics simulations, which solve the electric field, fluid flow, and species transport in a coupled manner, were carried out. Our analysis clearly demonstrates that a substantial electric field barrier is established at the micro-nano-interface and the equilibrium of electrolytic ion transport is significantly impacted by the pressure head (or pumping flow) and background electrolyte concentration, which can be employed for the development of effective sample preconcentration technologies. Depending on the current flux characteristics, the entire preconcentrator can be divided into five sub-domains. In the microchannel domain, current flux due to diffusion, convection, and electromigration are negligible. In the nanochannel, the longitudinal diffusion is negligible and, the convection-induced current flux is balanced by the electromigration. In the micro-nano interface, all current contributions are important and the diffusion-induced current carries a considerable weight in determining the direction and magnitude of the electric field therein. An interesting point of note is that the field strength in the barrier decreases with increasing pressure head (or pumping flow) and electrolytic concentration (i.e., thinner EDL). Therefore, a tradeoff exists between the enrichment ratio and operating time, which needs to be addressed in practical design and protocol development. Our analysis has also revealed that the enrichment ratio also relies heavily on analyte properties (e.g., mobility and diffusivity of the analytes) and substantial (>10^4–fold) analyte enrichment can be readily achieved.

ACKNOWLEDGEMENT

This research is sponsored by the DARPA and US Army Aviation & Missile Command under grant number W31P4Q-07-C-0035.

REFERENCES

[1] Y.C. Wang *et. al*, *Anal.Chem.*, 77, 4293-4299, 2005.
[2] S.M. Kim *et. al*, *Anal.Chem.*, 78, 4779-4785, 2006.
[3] A. Plecis et al., Proc. MicroTAS'2005, 2, 1038-1041, 2005.
[4] H. Daiguji *et. al*, *Electrochem. Commu*, 8, 1796-1800, 2006
[5] R. Bharadwaj and JG. Santiago. *J. Fluid Mech.* 2005, 543, 57-92.
[6] RF Probstein. An Introduction to Physicochemical Hydrodynamics. John Wiley & Sons, 2nd Ed. 1995.

MICRO-TIP ASSEMBLED METAL CANTILEVERS WITH BI-DIRECTIONAL CONTROLLABILITY

H. Kwon[1,2], M. Nakada[1,2], Y. Hirabayashi[3], A. Higo[4], M. Ataka[2], H. Fujita[2], H. Toshiyoshi[1,2]

[1]KAST, [2]IIS, the Univ. of Tokyo, [3]KITC, [4]RCAST, the Univ. of Tokyo
KAST, East 310, KSP, 3-2-1 Sakado, Takatu-ku, Kawasaki, Japan,
Tel:81-3-5452-6277; Fax: 81-3-5452-6250; E-mail: honamii@hanmail.net

ABSTRACT

Layered-metal micro cantilever with a micro tip was fabricated and characterized for high-density data storage device. The cantilevers were fabricated by the sequential depositions of chromium, gold, and chromium layers by sputtering with their stress and thickness controlled. The reversed sputtering technique was utilized for fine patterning of the cantilevers and hence for good uniformity of cantilever's elevation height after sacrificial release. A micro protrusion was formed at the end position of the cantilever by the newly developed contact shadow mask process. The fabricated cantilevers were driven in the upward direction by the electro thermal Joule heat, and also in the downward direction by the electrostatic force between cantilever and the silicon substrate. A series of small dots as recorded in a thin film photoresist by using the indentation of the fabricated protrusion, which showed that the cantilevers would be applied to high-density data storage devices.

Keywords: metal cantilever, bi-directional controllability, shadow mask, micro tip, peel off, data storage

1 INTRODUCTION

The recording density of hard disk drive is limited by the thermal fluctuation of the magnetic boundary and by the superparamagnetic effect [1]. In the mean time, MEMS (Micro Electro Mechanical Systems) based data storage systems utilizing the atomic force microscopy (AFM) have been studied as one of the next-generation storage technologies [1-5]. The researchers including IBM utilizing MEMS based data storage devices have reported that the data density can be upwards of terabit-per-square-inch range [1-5].

In MEMS based data storage, a micro tip on a cantilever is used to write and read data bits on the recording media. The tips have been usually prepared by the chemical etching of silicon [1, 2] that required careful etch-time control to assure the sharpness of the tip. Tips could also be made by depositing a low stress silicon nitride film on an inverted pyramidal pit that had been formed by silicon crystalline etching technique [4]; this technology required subsequent wafer-bonding and layer-transferring processes to release the tip and cantilever. The cantilevers with the micro tip were driven by utilizing one or combination mechanisms of heat transportation, piezo-resistivity, and light reflection [1-5]. Height control of cantilever tips is needed for the mechanical contract with the recording medium for both writing and reading, and the initial elevation height should be uniformly over the cantilever array to equalize the read/write conditions over the surface. In this paper, we proposed a simple fabrication method of micro tips by the peel off process on a layered-metal cantilever. The cantilever's actuation characteristic was made to be uniform in the array thanks to the newly employed dry etch patterning of the cantilevers.

2 FABRICATION

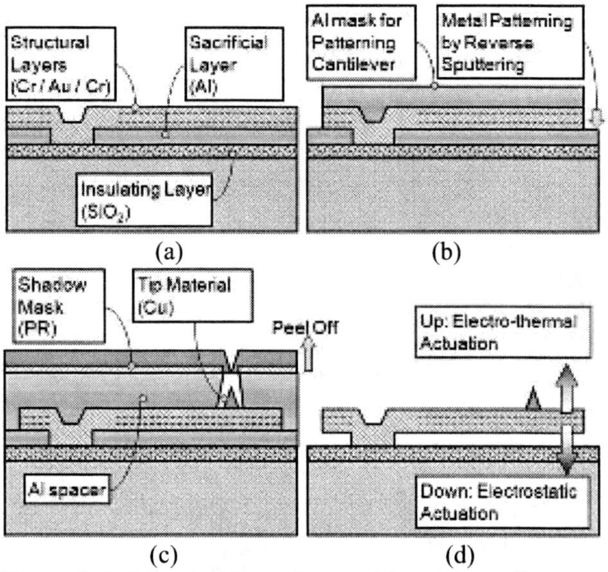

Figure 1: Fabrication steps for metal layered cantilever and micro tip. (a) Insulating layer (SiO_2) and sacrificial layer (Al) are deposited and patterned, which is followed by deposition of structural layers (Cr-Au-Cr). (b) Layered cantilevers were patterned by reverse sputtering with Al mask. (c) Al is deposited on the layers and etched with PR mask, which is used as a shadow mask for continual evaporation of Cu. (d) Cantilever is released by Al removal.

We started the deposition process of glass sputtering of 6000 A (Angstrom) on a silicon wafer for electrical isolation. On the glass layer, an aluminum layer (sacrificial layer) of 6000 A was deposited by vacuum evaporation and patterned in aluminum etchant (Figure 1a). Three metal layers of 1st chromium / gold / 2nd chromium layers were deposited without breaking the vacuum at the thickness combination of 400 A / 5000 A / 1200 A and at the sputtering pressure of 2 / 2 / 20 mTorr, respectively. The argon gas flow rate was 20 sccm to control the internal stresses. On the deposited metal layers, the second aluminum layer was deposited at 6000 A by the thermal evaporation, and was patterned in aluminum etchant. The metal layers were annealed in the oven at 280 degree C for 10 minutes, and then cooled down to room temperature. The cantilever patterns of the second aluminum were transferred to the chromium / gold / chromium layers by the reversed sputtering process at RF 400W for 75 minutes (Figure 1b). The third aluminum layer of 5 microns in thickness was newly deposited on the layers to be used as a tip-spacer (Figure 1c) [6]. The third chromium layer was deposited at 500 A, and patterned for the formation of the contact shadow mask. The through-hole was formed by the wet etching in the aluminum etchant warmed at 60 deg C for 30 minutes. The sharp tip was made by copper deposition (5 um thick or more); the aperture of the contact shadow mask closed itself during metal deposition [6, 7]. The metal deposited on the tip spacer was removed by the peel-off technique utilizing a sticky tape, followed by the removal of the chromium shadow mask layer (Figure 1d and Figure 2).

Figure 2: Peel off process with a sticky tape; Sticky tape clings to the copper evaporated wafer, and peels copper off from the wafer.

The sacrificial aluminum was selectively removed to release of the cantilevers with NMD3™, which is a base liquid of 2.38% TMAH (Tetra Methyl Ammonium Hydroxide). By this fabrication steps, we obtained an array of micro tip equipped micro cantilevers as shown in Figure 3. The height of micro tips was about 2-3 um, and the radius of the apex was about 100 nm at minimum. The bi-directional actuator with a micro tip could be used to develop the AFM-based data storage.

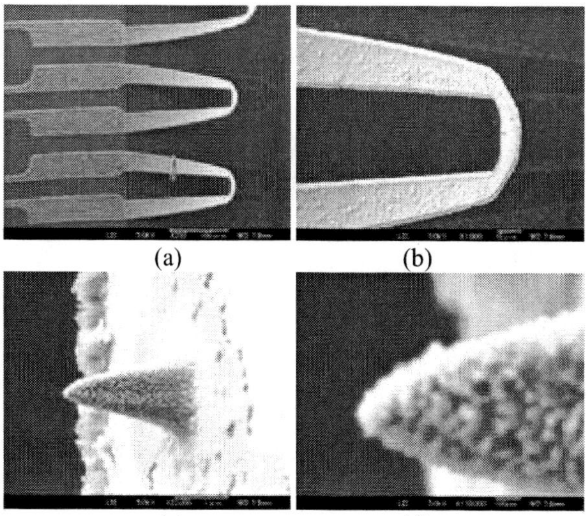
(a) (b)
(c) (d)

Figure 3: Scanning electron microscopic images of (a) the cantilever array, and enlarged views of (b) a cantilever, (c) micro tip at the end of cantilever, and (d) the end part of a micro tip.

3 DRIVING CHARACTERISTICS

We designed the cantilever that deflected upward by the electro-thermal Joule heat and downward by the electrostatic force; this could be done by changing the driving circuits as shown in **Figure 4** [8]. Electro-thermal driving was utilized to write data pits by indentation on the recording media with the micro tips. At the same time, electrostatic driving would be used to equalize the elevation heights of the arrayed tips and to read data bit by the resonance vibration.

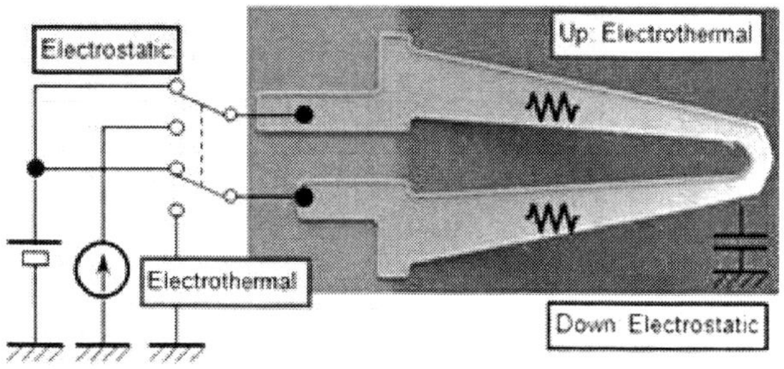

Figure 4: Operating principle of upward and downward operation by electro-thermal and electro-static drive, respectively.

By the prescribed process, we obtained a layered-metal cantilever, in which a chromium layer was positioned on the gold layer for electro-thermal driving. Because the thermal expansion coefficient of chromium (9.5ppm/C) is smaller than that of gold (14ppm/C), the cantilever curled upwards by the Joule heat.

We measured electro-thermal and electrostatic driving with functions of applied power (0.85V x 39mA) and voltage (13.5V) as shown in **Figure 5**. Due to the limitation of optical measurement system, we observed the motion at a spot near the anchor of the metal cantilever, and it was 130 nm and 220 nm by electro-thermal and electrostatic driving, respectively. The actual tip motion was approximately 40 times leveraged, which was thought to be large enough to compensate the tip height variation, by using the FEM analysis.

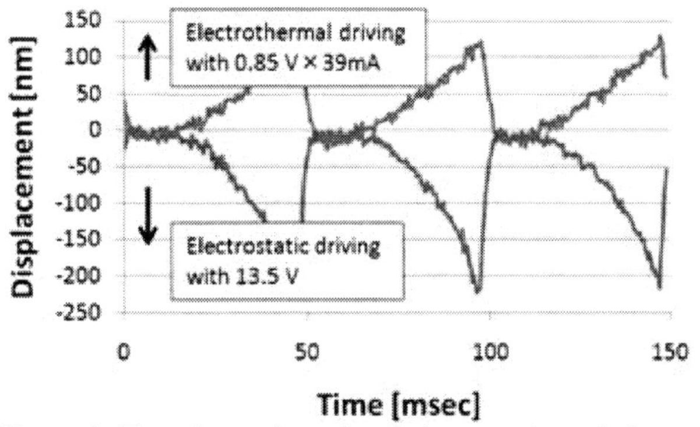

Figure 5: Experimental result of electrostatic and electro-thermal driving of the fabricated Cr/Au/Cr cantilever. The elevation of cantilever was measured at the distance of about 50 um (about one fifth of the length of cantilever) from the anchor because of the measurement limit (numerical aperture) of the laser interferometer.

4 DATA WRITING

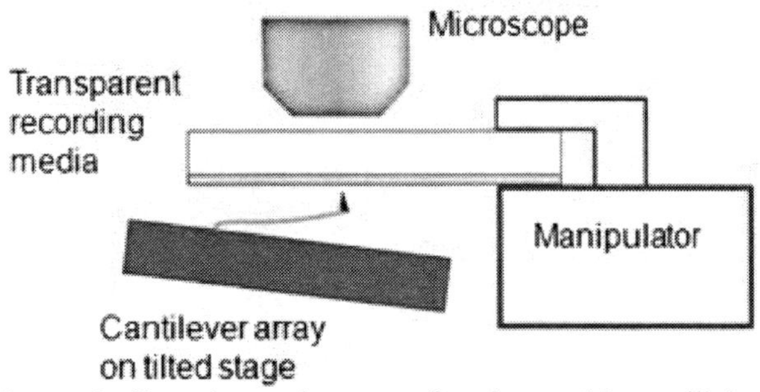

Figure 6: Experimental set up for data writing utilizing fabricated micro cantilever array (not in scale).

Figure 6 shows the experimental set up for writing data pits by utilizing the fabricated cantilever array. On a slide glass, a thin layer of photoresist (S 1805) was spin-coated at the 5000 rpm and baked at 110 degree C. The arrayed cantilevers were heated up to 120 deg C and were brought into contacted with the photoresist on the glass substrate. We obtained data pits separated by the cantilevers pitch (140 um) as shown in **Figure 7 (a)**. **Figure 7 (b)** shows data pits engraved at a 20 um period by the repetitive operation. The size of data bit was about 2 um in diameter, which would be minimized by the control of heating temperature, indenting time and depth.

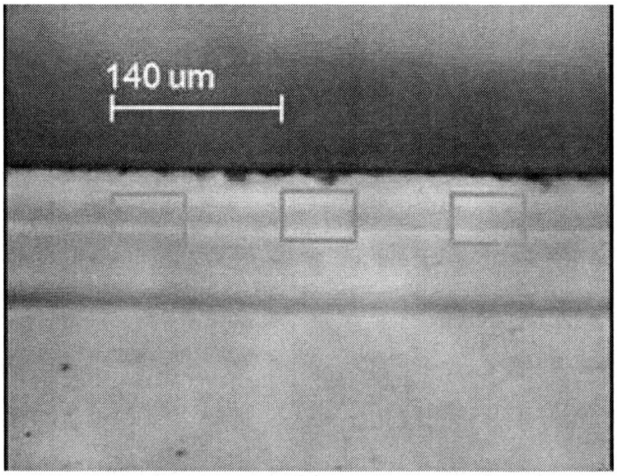

(a)

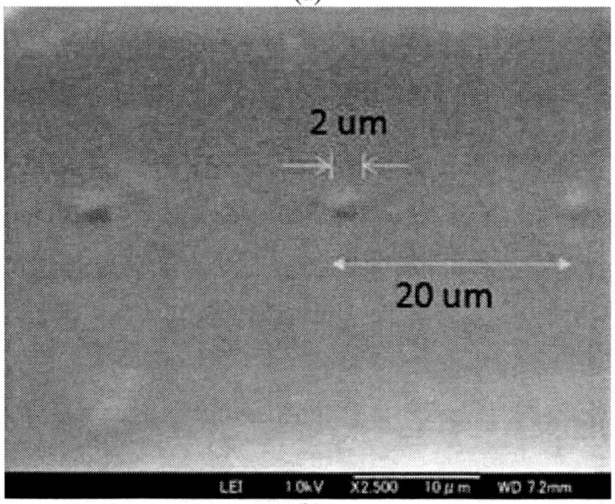

(b)

Figure 7: (a) Microscopic images of the written dot data with the fabricated cantilever array, and (b) the scanning electron microscopic image which magnified right hand side square

Apart from reference 6, we utilized three aluminum layers for the sacrificial layer, the dry etch mask, and for the tip spacer, because aluminum could be etched fast in the etchant, which did not etch the cantilever metals. We once tried photoresist for the tip spacer material but it was burned out during the tip material deposition. In addition to this, the photoresist residue was not completely removed in the following processes. Because of the chemical resistance to the etching solutions, we used copper for the tip and aluminum for the sacrificial layer. The characteristics and processes for the tested metal material are compared in **Table 1**.

Table 1: Characteristics of the utilized materials.

	SiO_2	Al	Cr	Au	Cu
	Insulation	Sacrificial / Mask	Cantilever	Cantilever	Tip
Thick (~um) deposition	△	O	X	△	O
Deposition	Sputter	Sputter or Evaporation	Sputter	Sputter	Evaporation
Endurance to Al etchant	O	X	O	O	X
Endurance to PR developer	O	X	O	O	O
Endurance to O_2 ashing	O	O	X	O	O
Patterning	(BHF)	Al etchant	Reverse sputter	Reverse sputter	Peeling off

5 CONCLUSION

For the AFM-based data storage application, we proposed electro thermal driving to write data bits and electrostatic driving to equalize the heights of the arrayed tips and / or to read data bit by the resonance vibration of the metal layered cantilever. A micro tip was fabricated by contact shadow mask process on the bi-directional actuator with chromium-gold- chromium layers. The driving ranges of cantilever were analyzed to cover the height variation of the arrayed cantilever in chip.

ACKNOWLEDGMENTS

This work was supported in part by a research grant from the Murata Science Research Foundation in the year 2006-2007. The photo masks used in this work were developed using the electron-beam facility at the VLSI Design and Education Center (VDEC) of the University of Tokyo.

REFERENCES

[1] S. Kedler, MEMS based storage systems, presentation material form Coventor™.

[2] W. P. King, T. W. Kenny, K. E. Goodson, G. L. W. Cross, M. Despont, U. T. Durig, H. Rothuizen, G. Binnig, and P. Vettiger, Design of atomic force microscope cantilevers for combined thermomechanical writing and thermal reading in array operation, Journal of microelectrodemechanical systems , Vol. 11, No. 6, pp. 765- 774, Dec. 2002.

[3] H. Nam, Y. Kim, C. Lee, W. Jin, S. Jang, I. Cho, and J. Bu, Integrated nitride cantilever array with Si heaters and piezoelectric detectors for nano-data-storage application, proceedings of IEEE MEMS 2005, pp. 247-250.

[4] A. Chand, M. B. Viani, T. E. Schaffer, and P. K. Hansma, Microfabricated small metal cantilevers with silicon tip for atomic force microscopy, JMEMS, Vol. 9, No. 1, pp. 112-116, Mar. 2000.

[5] C. Hsieh, C. Tsai, W. Lin, C. Liang, and Y. Lee, Bond-and-transfer scanning probe array for high density data storage, Transactions on Magnetics, Vol. 41, No. 2, pp. 989 – 991, Feb. 2005.

[6] H. Kwon, et al., Fabrication of micro-tips by lift off process with contact shadow masking, proceedings of IEEE NEMS 2007, pp. 488-492.

[7] C.A. Spindt, I. Brodie, L. Humphrey, and E. R. Westerberg, Physical prperties of thin-film field emission cathodes with molybdenum cones, Journal of Applied Physics, Vol. 47, no. 12, pp. 5248 – 5263, Dec. 1976.

[8] H. Kwon, M. Nakada, Y. Hirabayashi, A. Higo, M. Ataka, H. Fujita, H. Toshiyoshi, Bi-directionally Driven Metal Cantilevers Developed for Optical Actuation, MOEMS 2007, pp. 49-50.

[9] H. Kwon, M. Nakada, Y. Hirabayashi, A. Higo, M. Ataka, H. Fujita, H. Toshiyoshi, "Micro-tip assembled metal cantilevers with bi-directional controllability," submitted, Nanotech 2008.

Novel synchronous linear and rotatory micro motors based on polymer magnets with organic and inorganic insulation layers

M. Feldmann, A. Waldschik, S. Büttgenbach

Institute for Microtechnology, Technical University of Braunschweig,
Alte Salzdahlumerstr. 203, 38124 Braunschweig, Germany
Tel (+49) 531-391-9768, Fax (+49) 531-391-9751, E-mail: m.feldmann@tu-bs.de

ABSTRACT

In this work, we show the development of several synchronous motors with rotatory, 1D or 2D movements. The synchronous micro motors are brushless DC motors or stepper motors with electrical controlled commutation consisting of a stator and a rotor. The rotor is mounted onto the stator adjusted by an integrated guidance. Inside the stator different coil systems are realized, like double layer sector coils or special nested coils. The coil systems can be controlled by three or six phases depending on the operational modus. Furthermore, inorganic insulation layers were used, which reduced the system thickness. By this means four layers of electrical conductors can be realized especially for the 2D devices. The smallest diameter of the rotatory motor is 1 mm and could be successfully driven.

Keywords: Synchronous micro motors, 1D- and 2D linear actuators, micro coils, UV depth lithography, polymer magnets

1 INTRODUCTION

Due to the development of new technologies, more and more complex MEMS applications can be realized. Especially electromagnetic micro actuators [1, 2] have reached a growing interest in micro technology in addition to commercial applications during the last years. Their basic construction exists of electric conductors and coil systems as well as of soft-magnetic and/or hard-magnetic materials that were fabricated in additive technology via UV-depth lithography and electroplating. For UV-depth lithography photo resists like Epon SU-8, AZ9260, Intervia-3D-N and CAR44 were applied and optimized. Layer thicknesses up to 1 mm and aspect ratios over 60 were achieved (see Fig. 1). Special micro composites were developed. This allowed the fabrication of micro magnets with arbitrary shape and properties, ensuring a complete compatibility to existing process chains. With these potential technologies several synchronous motors with rotatory, 1D or 2D movements were developed. In addition inorganic insulation layers like silicon nitride or silicon oxide were used. The advantage of these inorganic insulation layers in comparison to organic insulation layers like Epon SU-8 is the reduced thickness, whereby four layers of electrical conductors can be realized especially for the 2D devices. Furthermore, for different fields of application inorganic insulation layers are more chemical or thermal resistant.

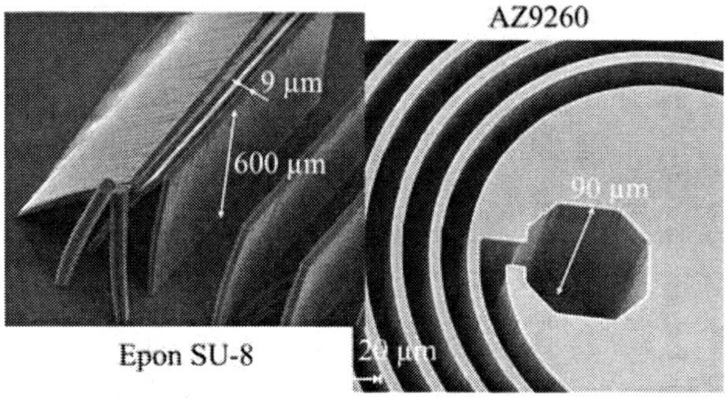

Figure 1: SEM-pictures of Epon SU-8 and AZ9260 photo resist structures.

2 CONCEPT AND DESIGN

The synchronous micro motors are brushless DC motors or stepper motors with electrically controlled commutation. In Fig. 2 the basic setup for an electric motor with disc-shaped rotor consisting of a stator and a rotor is shown. The rotor is made of a SU-8 form, which contains alternate magnets. These magnets were realized by polymer magnets [3, 4, 5] or commercial magnets. Both magnet types have an axial magnetization. The stator consists of double layer coils, which where arranged as sector coils or nested coils. The coils have 6-30 windings per phase depending on the motor size and number of poles. The arrangement of the coils and magnets allows the driving by three or six phases. For the adjustment of the rotor and the stator a centrical arranged circular guidance is integrated.

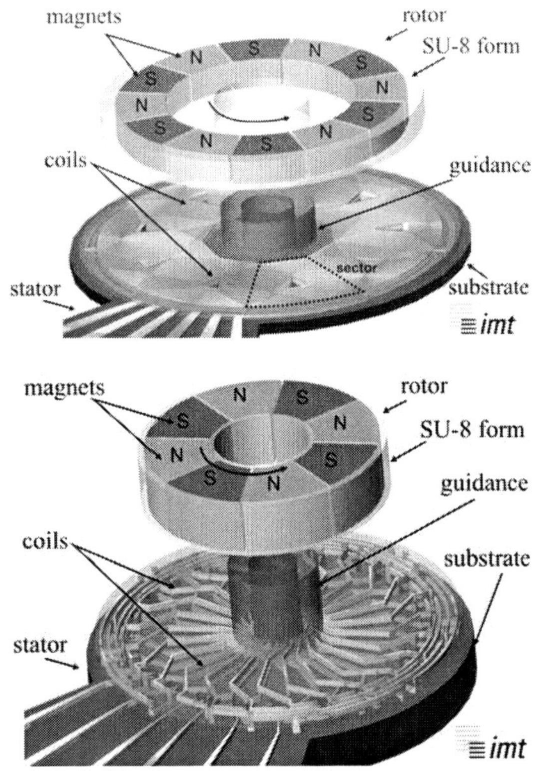

Figure 2: Concepts of rotatory synchronous motors with sector coils (top) and nested coils (bottom).

3 FABRICATION

The principle process chain for fabrication of stator and rotor consists of an iteration loop of single process steps for in layers built-up for complex 3D microstructures. The fabrication process includes UV-depth lithography using AZ9260 of electroforming and Epon SU-8 for insulation, planarization and embedding. Inorganic materials like silicon oxide and nitride where used alternatively to SU-8 for insulation of the different coil layers.

3.1 Stator

The process sequence for fabrication of the stator starts with the lower conductors of the double layer coil. A mould of AZ9260 is patterned and filled with copper by electroplating. After stripping the AZ9260 mould, a SU-8 layer is spun onto these structures as insulation layer. This layer provides openings for through connections to the upper coil layer. Both connections and upper conductors are likewise structured by copper eletrcoplating. A following second SU-8 layer serves on the one hand as insulation between upper conductors and traveler magnets and on the other hand it serves as bearing layer for the traveler. In the last step the circular or linear guidance is made by pattering a 200 µm thick SU-8 layer. The fabricated stators are shown in Fig. 3.

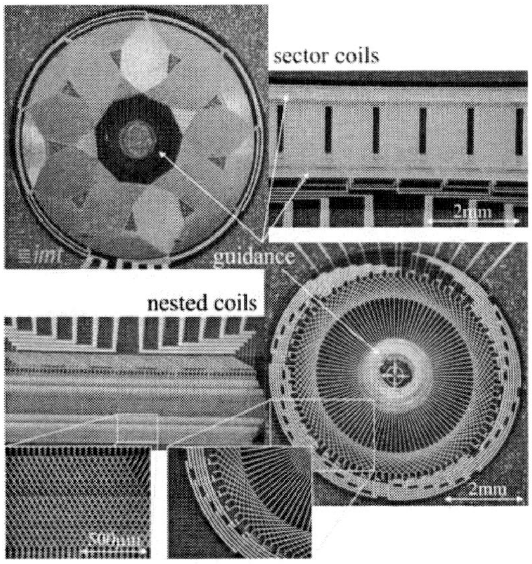

Figure 3: Photograph of different fabricated stators.

For the application of inorganic insulation layers, instead of organic SU-8, investigations with silicone oxide and silicone nitride were carried out. Both films were deposited by a PECVD process with thicknesses of 600-1000 nm for oxide and 100-200 nm for nitride respectively. These layers were masked by patterned AZ9260. Both wet and dry etch were tested. In the wet etching process large undercut could be observed. Pattering silicone nitride could be improved by using dry etching in a barrel etcher because of no undercut has occurred (see Fig. 4).

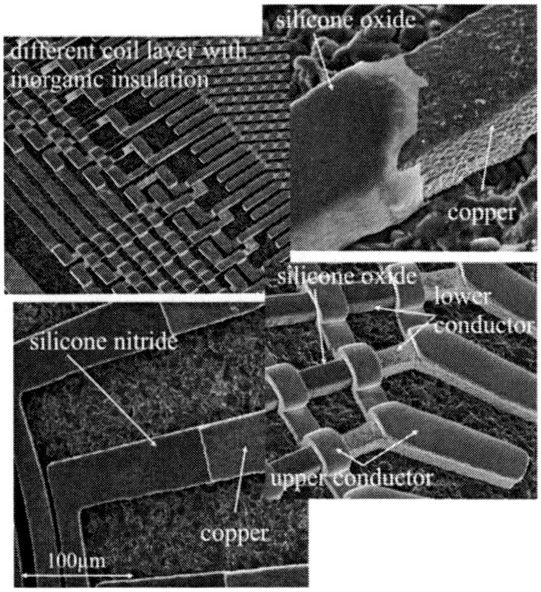

Figure 4: SEM-pictures of lower conductors covered with inorganic insulation layers.

3.2 Rotor

The rotor is realized by SU-8 creating a form for polymer magnets or commercial magnets with high precision adjustment (see Fig. 5). The commercial magnets were mounted inside of this form and hold by a fit. The polymer magnets are filled in these forms and magnetized by special developed magnetization equipment. It is mounted onto the stator adjusted by the integrated guidance.

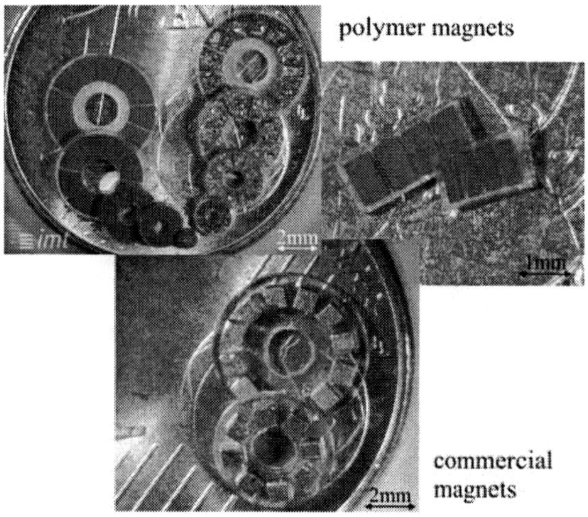

Figure 5: Photographs of different rotors and travelers with integrated polymer magnets (top) and mounted commercial magnets (bottom).

Both rotor types were fabricated to compare their influence of motor properties and performance respectively. The structured polymer magnets show comparatively less magnetic properties. However, the advantage is the flexible formation so that higher area fillings could be achieved. Furthermore, with polymer magnets smallest structures down to several tens micrometers are producible. By that, rotors with a smallest diameter of 1 mm could be realized. The following table shows the realized rotors.

Diameter [mm]	2p	Weight [mg]	Magnets
5,5	12	45	Commercial
5,5	12	24	Polymer magnet
4,5	8	30	Commercial
4,5	8	17	Polymer magnet
3,0	12	6	Polymer magnet
2,0	8	2	Polymer magnet
1,0	8	<1	Polymer magnet

Table 1: Overview of realized rotor configurations.

For the fabrication of the rotor a sacrificial copper layer is electroplated onto the substrate followed by a thin patterned SU-8 layer. After that a 400 μm high SU-8 layer is structured to provide the filling form. The polymer magnet is filled in this form and baked. After baking a polishing process follows to level the compound structure and to remove waste residuals. By etching the sacrificial layer the rotors are detached from substrate. The applied polymer magnets consist of 80wt% barium-strontium ferrite or 90wt% neodymium-iron-boron.

4 MAGNETIZATION

A special magnetization equipment was designed for apply the alternate magnetization into the polymer magnet sectors in axial direction (see Fig. 6). It consists of a ferromagnetic core with a yoke in which special magnetization adapters can be placed. At these adapters milled sectors respective stripes are located for magnetization rotors and linear travelers. A flat coil with 400 windings wound around the core serves for generating magnetic flux. For the magnetization process a rotor is inserted and oriented to the adapter (see Fig. 7) by means of a multi axis positioning stage. By the first current feed every second segment is magnetized. After that the rotor is is rotated at an angle, which corresponds to the pole pitch. By a second current feed the other segments are magnetized in the opposite direction.

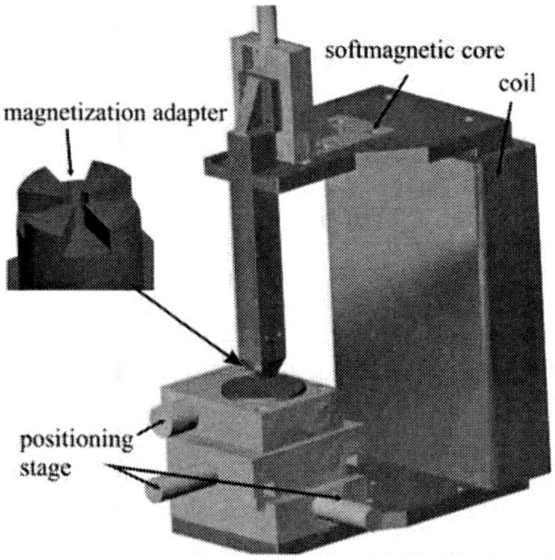

Figure 6: Magnetization equipment.

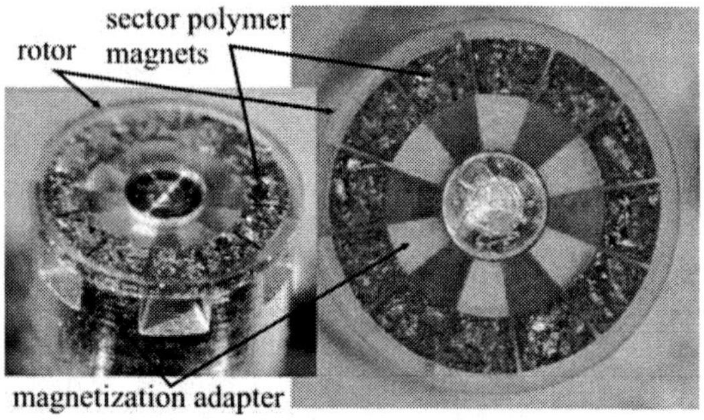

Figure 7: Magnetization adapter with polymer magnet rotor.

5 RESULTS

By the use of the described process sequences stators and rotors or travelers were succesfully fabricated. By means of the guidance structures the components could be mounted and first tests were carried out. The smallest diameter of the rotatory motor is 1 mm and driving currents between 50-300 mA are necessary depending on magnet type. This results in a torque up to 20 μNm and higher depending on the used magnets and size of the motor. In first test the motor could be successfully driven over a long period with a rotating speed over 7000 rpm (see Fig. 8). Generally, the motors with nested coil indicate a smoother movements due to their finer pitch.

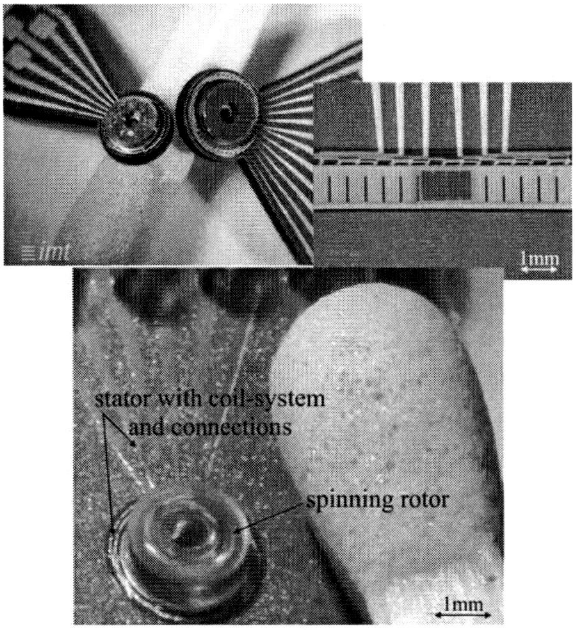

Figure 8: Photograph of different mounted synchronous motors with comparision to a match.

6 CONCLUSION

The purpose of this work was to develop linear and rotatory synchronous motors. Both motor types could be realized succesfully by means of UV-depth lithography, electroplating and by using polymer magnets. The integration of polymer magnets made it possible to miniaturize synchronous motors down to 1 mm rotor diameter. All devices are currently under further investigation to optimize and characterize the operating behavior. In addition, a special packaging is being developed. Also first concepts for integrated magnetic sensors were developed for closed loop control, which allows micro-/nano stepping and rotating. Following investigations also aims to exact measurements of magnetic flux densities generated by the polymer magnet segments.

REFERENCES

[1] Yang, C.; Zhao, X.; Ding, G.; Zhang, C.; Cai, B.: An Axial Flux Electromagnetic Micromotor. Journal of Micromechanics and Microengineering 11 (2001), pp. 113-117

[2] Kleen, S.; Ehrfeld, W.; Michel, F.; Nienhaus, M.; Stölting, H.-D.: Penny-Motor: A Family of Novel Ultraflat Electromagnetic Micromotors. Actuator2000, Bremen, pp. A6.4, 2000

[3] M. Feldmann, S. Büttgenbach: Novel Fabrication Processes for Polymer Magnets and their Utilization to Design Versatile Electro Magnetic Micro Actuators, Eurosensors2005, pp. WC6f.

[4] M. Feldmann, S. Büttgenbach: Novel Versatile Electro Magnetic Micro Actuators with Integrated Polymer Magnets: Concept, Fabrication and Test, Actuator2006, HVG, ISBN 3-933339-08-1, pp. 709-712.

[5] M. Feldmann, S. Büttgenbach: Novel monolithical micro plunger coil actuator using polymer magnets and double layer micro coils, XVIII Eurosensors, Rome, 2004, pp. 34-35.

A Novel High-sensitivity Resonant Viscometer Realised Through the Exploitation of Nonlinear Dynamic Behaviour

W.H.Waugh, B.J. Gallacher and J.S. Burdess

Newcastle University, Newcastle upon Tyne, NE1 7RU, United Kingdom, w.h.waugh@newcastle.ac.uk

ABSTRACT

Existing MEMS viscometers typically measure vibrational characteristics such as resonant frequency, bandwidth and quality factor. In order to significantly improve sensitivity to changes in viscosity the proposed sensor will exploit nonlinear dynamic behaviour and instead measure the frequency separation between singular jump points in the frequency response function. By using a one-mode approximation when excited near resonance, the dynamics of a clamped-clamped slender beam in fluid is that of a standard Duffing oscillator. With harmonic forcing of sufficient magnitude, a bistable region, bounded by amplitude jump points, occurs. The width of this bistable region is dependent on the damping ratio of the system, which is shown to be a function of the dynamic viscosity. It is shown that the sensitivity of the proposed nonlinear viscometer is at least an order of magnitude better than that of conventional devices which measure bandwidth.

Keywords: MEMS, nonlinear, Duffing, viscometer

1 INTRODUCTION

Measurements of viscosity and density allow for the monitoring of fluid quality and processes involving a fluid environment. There are various fields in which such measurements may be required, including oil exploration and production, environmental monitoring, process control, medicine, and the automotive industry.

A variety of density and viscosity sensors have been developed based on the exploitation of resonance. Examples of devices operating in the linear regime include vibrating wire viscometers, oscillating plates and quartz-crystal microbalances. The frequency response of a resonator in a surrounding fluid depends strongly on the properties of said fluid. The resonance frequency and damping (and hence quality factor Q) are influenced by the viscosity and density, and hence can be used to sense these properties.

However, there remain challenges in terms of sensor accuracy, flexibility, and use outside of the laboratory environment [1]. The behaviour of a resonator immersed in a viscous fluid needs careful consideration, as there is no analytical solution to this problem [2].

Nonlinear effects have been observed with resonator devices [3]. Nonlinearities can arise in many different ways and take different forms such as material, geometric, inertia, and friction nonlinearities. In this paper we focus on the geometric nonlinearities which occur due to midplane stretching of clamped-clamped beam resonators of rectangular cross-section. The frequency response of such devices shows the existence of jump phenomenon. The jump points are singularities which can be measured to a high degree of accuracy. By utilising the existence of these jump points, and measuring the frequency difference between them, it may be possible to design novel sensors capable of measuring viscosity and density with increased sensitivity. Previous related work includes investigations into mass detection using nonlinear oscillations [4].

2 BEAM VIBRATION

In order to approximate the behaviour of a vibrating clamped-clamped beam, slender beam assumptions are made, and rotational and longitudinal inertia neglected. The beam is subject to midplane stretching, which introduces a nonlinear term into the beam equation of motion. This nonlinearity has been shown to be the dominant nonlinearity for such beams [5].

With a forcing term $\overline{p}(x,t)$, the equation of motion in vacuum can be shown to be

$$EI\frac{d^4\overline{v}}{dx^4} + \rho A\frac{d^2\overline{v}}{dt^2} + \overline{p}(x,t) - \frac{EA}{2L}\left[\int_0^L \left(\frac{d\overline{v}}{dx}\right)^2 dx\right]\frac{d^2\overline{v}}{dx^2} = 0 \qquad (1)$$

where x is an axial direction, $\overline{v}$ represents displacement on the centre line, L is beam length (along x axis), A is the cross-sectional area, E and ρ are the beam material Young's modulus and density respectively, and I is the second moment of area for the beam.

3 FLUID DAMPING

The behaviour of the beam is now considered for the case of total immersion in an incompressible, homogeneous Newtonian fluid at constant temperature and pressure. Gases may be considered incompressible at Mach numbers of less than 0.3 [6].

As the beam oscillates so an oscillating fluid force is applied to the beam by the surrounding fluid. This force has two components, one acting in phase with the acceleration of the beam, the other in phase with the velocity [7]. These components are known as the added mass and the viscous drag, respectively.

3.1 Added mass

Acceleration of the beam causes the surrounding fluid to accelerate also. The inertia of the entrained fluid is known as the added mass, and it acts with the same sign, frequency and phase as the beam mass. Added mass acts to decrease the natural frequency of the beam.

The added mass M_A for a rectangular beam under free vibration in a fluid has been specified using potential flow theory. It is given by $M_A = \frac{\alpha \rho_f \pi b^2}{4}$ where ρ_f is the fluid density, b the beam width, d the beam depth, and the coefficient α depends upon the width to depth ratio [7]. For a harmonically oscillated beam, the added mass becomes proportional to $1/\omega^2$ [7], where ω is the angular frequency. A full expression for added mass is then

$$M_A = \frac{E\beta^4 \alpha \rho_f \pi b^3 d^3}{48L^4\left(\rho_s bd + \frac{\alpha \rho_f \pi b^2}{4}\right)\omega^2} \tag{2}$$

3.2 Viscous drag

The Keulegan-Carpenter number Kc is a measure for viscous effects. For slightly viscous fluids at $Kc < 1$, the drag coefficient for a smooth circular cylinder is given by [8]:

$$C_D = Kc^{-1}\left[\begin{array}{l}\frac{3}{2}\pi^3(\pi\beta_D)^{-\frac{1}{2}} + \frac{3}{2}\pi^2\beta_D^{-1} \\ -\frac{3}{8}\pi^3(\pi\beta_D)^{-\frac{3}{2}}\end{array}\right] \tag{3}$$

where $Kc = \frac{\mathrm{Re}}{\beta_D}$, $\mathrm{Re} = \frac{\rho_f D}{\mu}\frac{d\bar{v}}{dt}$ is the Reynolds number, $\beta_D = \frac{\rho_f D^2 \omega}{2\pi\mu}$ and D is the characteristic length (here, the diameter of the cylinder.)

Sarpkaya [9] explains that the above formula only strictly holds for $Kc << 1$, $\mathrm{Re}.Kc << 1$, $\beta_D >> 1$. Outside of this range, the drag coefficient must be determined experimentally.

The viscous drag force per unit length is then given by

$$F_D = \frac{1}{2}\rho_f\left(\frac{d\bar{v}}{dt}\right)^2 DC_D$$

If we let $\alpha_D = \frac{\mathrm{Re}\, C_D}{2}$ then we find $F_D = \alpha_D \mu \frac{d\bar{v}}{dt}$.
The damping ratio is then a function of the dynamic viscosity.

A rectangular beam will have a larger drag coefficient than a similarly sized circular cylinder [10].Consequently, the beam drag is approximated by multiplying the cylinder drag by an appropriate amount (again, to be determined experimentally.) In place of α_D then we use Φ, so that the damping coefficient becomes $\Phi\mu$. For characteristic length, use the width of the beam, b.

4 MODIFIED EQUATION OF MOTION

Inserting the drag force terms into equation (1) gives

$$EI\frac{d^4\bar{v}}{dx^4} + (\rho_s bd + M_A)\frac{d^2\bar{v}}{dt^2} + \bar{p}(x,t)$$

$$+\Phi\mu\frac{d\bar{v}}{dt} - \frac{EA}{2L}\left[\int_0^L\left(\frac{d\bar{v}}{dx}\right)^2 dx\right]\frac{d^2\bar{v}}{dx^2} = 0 \tag{4}$$

A solution to equation (4) can be sought by using the modal expansion $\bar{v} = \sum T(t)X(x)$. The resonator will be operated at the fundamental mode of vibration, so it is appropriate to consider only this fundamental mode shape i.e. $\bar{v} = T(t)X(x)$. Using this mode shape, the equation of motion becomes

$$\begin{array}{l}(\rho_s bd + M_A)L\gamma\ddot{T} + \Phi\mu L\gamma\dot{T} + \frac{EI\beta^4\gamma}{L^3}T \\ -\frac{EA\psi_1\psi_2}{2L^3}T^3 = Q(\xi,t)\end{array} \tag{5}$$

where $k = \left(\frac{1}{X}\frac{d^4X}{dx^4}\right)^{1/4}$, $\xi = x/L$, $\beta = kL$,

$Q(\xi,t) = -L\int_0^1 Xp(\xi,t)d\xi$, $\gamma = \int_0^1 X^2 d\xi$, $\varphi_1 = \int_0^1\left(\frac{dX}{d\xi}\right)^2 d\xi$

and $\varphi_2 = \int_0^1 X\frac{d^2X}{d\xi^2}d\xi$.

By making suitable substitutions, equation (5) can be represented as

$$\frac{d^2T}{d\tau^2} + 2\zeta\frac{dT}{d\tau} + T + \lambda T^3 = K\left(e^{ir\tau} + e^{-ir\tau}\right) \tag{6}$$

This equation is in the form of the damped Duffing equation. Rao [11] shows that equations of this type have solutions which exhibit two distinct jump points. These singularities are separated by a frequency difference δF.

5 AVERAGING METHOD

Let a solution to equation (6) be expressed $T(\tau)=\sum_{p=1}^{\infty} A_p e^{ip\tau} + \sum_{p=1}^{\infty} A_p * e^{-ip\tau}$

Harmonic terms are then collected and equated. Letting $A_r = a_r e^{i\phi_r}$ and then taking the real and imaginary parts of the equation, squaring and adding, yields the equation

$$\left[a\left(1-r^2\right)+3\lambda a^3\right]^2+\left(2\zeta ra\right)^2=K^2 \tag{7}$$

The following predictions arise from this equation.

6 MODEL PREDICTIONS

6.1 Effect of added mass

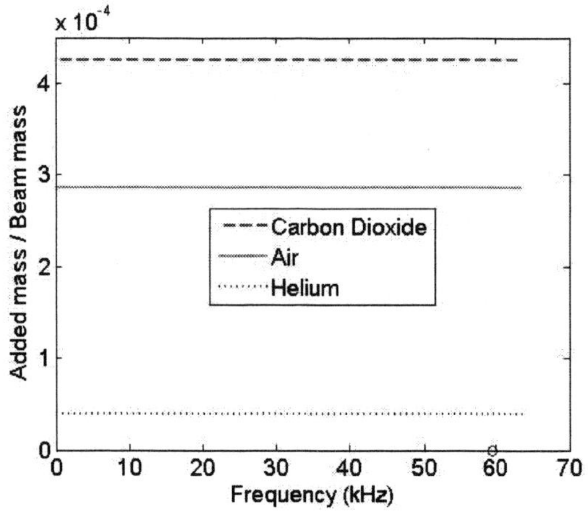

Figure 1: Ratio of added mass to beam mass, for various fluids in vicinity of the natural undamped frequency of the beam.

Figure 1 shows the ratio of the added mass to the beam mass for immersion in a range of gases. It is apparent that added mass is insignificant when fluid density is far lower than the beam density.

6.2 Fluctuation in jump point separation

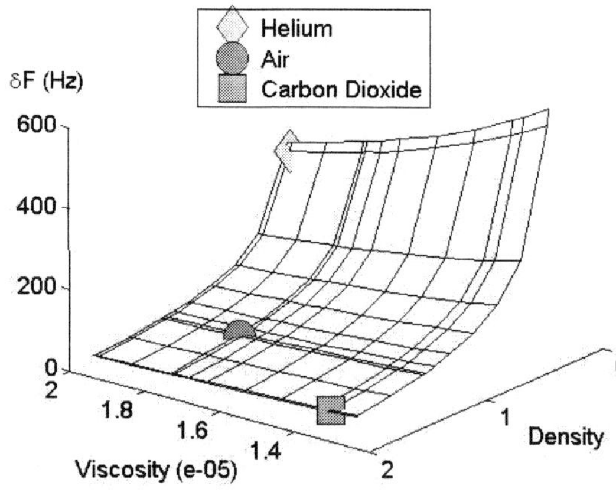

Figure 2: Jump point separation versus density and viscosity, for a copper beam of dimensions L=1400μm, b=60μm, d=30μm.

Figure 2 shows that an increase in viscosity or density has the effect of reducing the jump point separation δF. There exist contours of equal δF across the plotted surface, suggesting that for identification of fluid properties, a second measurement with a differing resonator is required. The intersection of the contours from the two resonators would then indicate the relevant values.

6.3 Sensitivity to changes in fluid properties

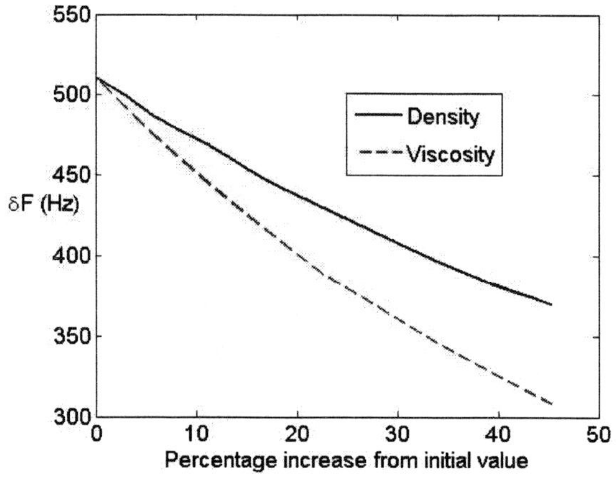

Figure 3: Sensitivity to changes in gas density and viscosity. Initial values of density 0.1625 kgm^{-3}, viscosity 1.37 $kgm^{-1}s^{-1}$

Using the data as for figure 2, it was possible to investigate the relative effect of changing the density and the viscosity by equal amounts. Figure 3 shows the effect on δF of increasing these values from an initial point. Over the range of data under consideration, it is observed that δF is more sensitive to changes in viscosity than density.

6.4 Comparative studies

Many vibrational viscometers use measurements of the peak frequency, bandwidth and/or the quality factor in order to determine the viscous damping. The bandwidth $\Delta\omega$ for a linear oscillator is given by $\Delta\omega \approx 2\zeta\omega_n$, and can therefore be used to measure the damping ratio, and hence the viscous damping.

In the nonlinear regime, it is found that the jump point separation is of comparable sensitivity to changes in viscosity as is the peak frequency. Bandwidth however differs. In figure 4, the jump point separation for a copper beam is compared to the bandwidth of the same beam operated in the linear regime at one-tenth of the forcing.

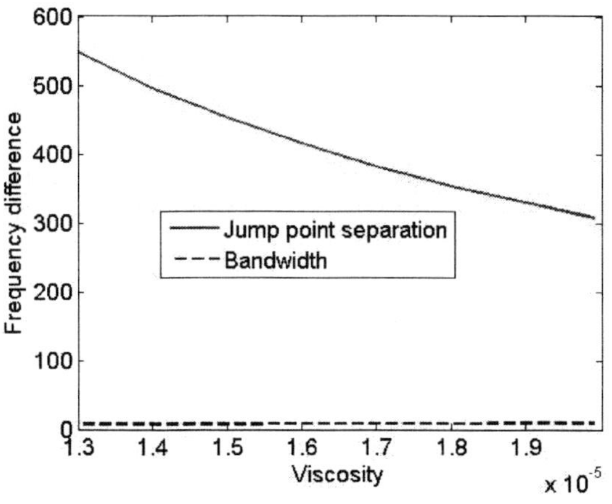

Figure 4: Jump point separation compared to bandwidth for a copper beam resonator of dimensions 1400x60x30μm. Density held constant as that for helium.

Over the range of viscosity considered, the jump point separation is seen to be greater than the bandwidth, plus more sensitive to changes in viscosity. Hence the beam resonator can offer increased sensitivity when operated in the nonlinear regime.

It should also be noted that measurement of the jump points may be more accurate than bandwidth measurements due to their singular nature.

7 CONCLUSIONS

The equation of motion for a clamped-clamped beam in a fluid has been shown to be of the form of the damped Duffing equation. Jump points are seen to occur, and the observation made that these jump points move closer together as the damping is increased. This phenomenon might be utilised in developing a novel sensor which can measure fluid viscosity and/or density.

At low damping, nonlinear beam resonators are shown to be potentially more sensitive to changes in fluid properties than existing resonating devices. Furthermore there may be advantages, in terms of accuracy of measurement, in identifying the two singular points as opposed to bandwidth, peak frequency or peak amplitude.

Further work is required to develop the model to operate in more viscous fluids where nonlinear damping effects may become significant.

REFERENCES

[1] K.A. Ronaldson, A.D. Fitt, A.R.H. Goodwin, W.A. Wakeham, "Transversely oscillating MEMS viscometer: The "Spider"", International Journal of Thermophysics, 27(4), 1677-1695, 2006.

[2] J.E. Sader, "Frequency response of cantilever beams immersed in viscous fluids with application to the atomic force microscope", Journal of Applied Physics, 84(1), 64-76, 1998.

[3] F. Braghin, F. Resta, E. Leo, G. Spinola, "Nonlinear dynamics of vibrating MEMS," Sensors and Actuators, A(134), 98-108, 2007.

[4] E. Buks, B. Yurke, "Mass detection with a nonlinear nanomechanical resonator," Physical Review, 74(4 part 2), 046619, 2006.

[5] M.R.M. Crespo da Silva, "Non-linear-flexural-torsional-extensional dynamics of beams II. Response analysis," International Journal of Solids and Structures, 24(12), 1235-1242, 1988.

[6] R.D. Blevins, "Flow-induced vibration. 2nd ed," Van Nostrand Rheinhold, 7-8, 1990.

[7] R.D. Blevins, "Formulas for natural frequency and mode shape," Van Nostrand Rheinhold, 386-392, 1979.

[8] R.D. Blevins, "Flow-induced vibration. 2nd ed," 209-210.

[9] T. Sarpkaya, " Force on a circular cylinder in viscous oscillatory flow at low Keulegan-Carpenter numbers," Journal of Fluid Mechanics, 165, 61-71, 1985.

[10] C. Zhang, G. Xu, Q. Jiang, "Analysis of the air-damping effect on a micromachined beam resonator," Mathematics and Mechanics of Solids, 8(3), 315-325, 2003.

[11] S.S. Rao, " Mechanical vibrations. 4th ed.," Pearson Prentice Hall, 912-914, 2004.

Ultra-thin Gold Membrane Transducer

Yunho Kim, Misun Cha, Hyungchul Kim, Sungjun Lee, Jaeha Shin, and Junghoon Lee

School of Mechanical and Aerospace Engineering, Institute of Bioengineering
Seoul National University, Seoul, Korea
E-mail: jleenano@snu.ac.kr

ABSTRACT

This work deals with ultra-thin metal-based nano-mechanical transducers that have potentials for extraordinary sensitivity. This nano membrane transducer (NMT) consists of a flat thin film of gold with 30-60 nm thickness freely suspended over large square openings with the side dimensions of 50-500 μm. This work involves the fabrication process and the testing method to obtain pressure sensitivity.

Keywords: NMT, ultra-thin gold film, pressure sensitivity

1 INTRODUCTION

After microcantilever showed potentials as a new approach in sensor technology, many applications have been demonstrated [1-3]. Recently, membrane transducers as chemo-mechanical sensing elements have been introduced, offering advantages over the cantilever approaches [4-6]. Instead of the optical detection method, highly sensitive capacitive detection was readily realized because of the dry cavity under the membrane. The membrane structure not fully immersed in liquid was inherently robust against the flow disturbance compared to the cantilever in liquid.

Membrane sensitivity is affected by the material and the dimension of the membrane. To improve the sensitivity, polymer materials with small elasticity, such as parylene and PDMS, were employed. However, the reliability of the polymer was questioned due to swelling in liquid. It was also difficult to handle the polymer during the fabrication. Also it was hard to immobilize molecules on the polymer surface. To functionalize the surface, other ladder immobilization layer was necessary, possibly leading to reduce the sensitivity.

In this paper, we introduce the use of a pure metal to achieve the high sensitivity and reliability. It is, however, critical to sufficiently reduce the thickness for desirable sensitivity. Therefore our study is focused on dimensional effect of the membrane with fabrication methods. As the membrane thickness is decreased, the sensitivity is highly increased. To achieve the sensitivity level of the polymer membrane with extremely low elasticity, there needs to be a significant reduction of the thickness.

Here we present the use of extremely thin gold that will improve the reliability and immobilization chemistry while good sensitivities are maintained.

2 EXPERIMENT

2.1 Fabrication process

To use the gold membrane, a fabrication process was developed as shown in Figure 1. The gold nano-membrane transducer consists of the upper and the bottom parts.

The gold membrane with the thickness of 30-60 nm was fabricated by anisotropic wet etching process (a). Low stress silicon nitride (Si_3N_4) was deposited on a <100> silicon wafer using the low pressure chemical vapor deposition (LPCVD). The gold layer was deposited on the nitride layer by sputtering (ULTECH, SPS Series). Then, the topside gold layer and backside nitride layer were patterned by photolithography process. By dry etching process, the backside nitride layer was etched (Oxford, RIE 80 plus). For anisotropic wet etching, the wafer was dipped in KOH solution. During this step, single side etching was critical to successfully suspend the extremely thin membrane. After the KOH etching the nitride layer under gold layer was etched by the same method explained above. Then, the gold layer freely suspended as a thin membrane. This membrane was used as a molecular reaction surface on one side, but also used as the upper electrode for capacitive measurement.

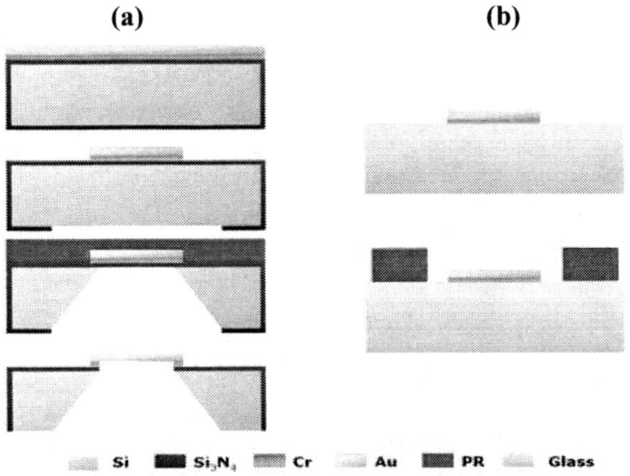

Figure 1. Fabrication process flow of gold TMT
(a) Upper part, (b) lower part

A Lower electrode part was fabricated by metal film patterning on the glass substrate (b). The gold layer was deposited on the glass substrate, and patterned. After patterning the gold layer, photoresist was patterned by photolithography process. The photoresist maintains a gap (1.5 μm) between the upper and the bottom part, creating a capacitor.

2.2 Gold membrane

Figure 2 shows the gold membrane inspected with a light microscope and a scanning electron microscope (SEM, Hitachi S-4800). The gold membranes are 150 x 150 μm^2 and 200 × 200 μm^2 in size. Small size membrane under 100 × 100 μm^2 was easier to fabricate than larger ones. As the size became larger, the yield of the process decreased. For example, in the case of 100 × 100 μm^2 membrane, the yield was over 85 %. However, in the case of 500 × 500 μm^2, the yield was under 40 %.

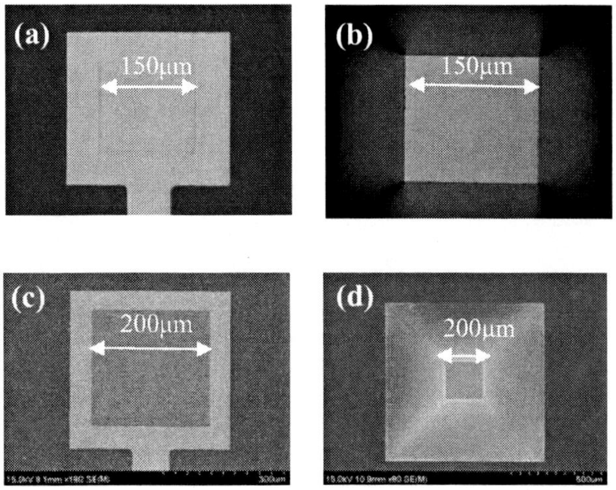

Figure 2. Microscope & Scanning Electron Microscope (SEM) images of gold membranes. (a),(c) Top-side view (b),(d) Backside view

The upper gold membranes in the top view, (a) and (c), show two overlapped squares. The inner squares represent the freely suspended gold membranes and the outer squares are gold layers on the nitride and silicon substrate. With these observations, it is certain that the ultra-thin membrane was created. The backside view, (b) and (d), clearly show the gold membrane suspended, clean and flat.

2.3 Measurement method

The upper gold membrane part and bottom electrode part were bonded together with the photoresist spacer that maintained the gap (1.5 μm). To accurately measure the capacitance between the upper and bottom electrodes, both parts were aligned each other. When the gold membrane deformed, the distance of two electrodes changed. The distance change resulted in the capacitance change which was measured using GLK instruments Model 3000.

We used a compact setup that can exert a hydraulic pressure to the sample. Figure 3 shows the schematic diagram of the setup. The capacitance change was caused by the deformation of gold membrane due to a small hydraulic pressure. Hydraulic pressure was easy to control, and used as a better choice compared with air pressure system which has a pressure fluctuation large enough to break the fine gold membrane.

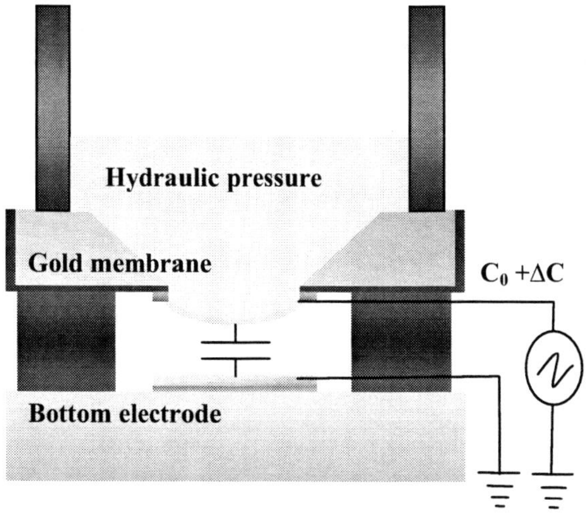

Figure 3. Schematic diagram of hydraulic pressure test to measure capacitance change

3 MODELING AND ANALYSIS

If we assume the square gold membrane as a circular plate with edge fixed conditions, the central deflection can be calculated. Figure 4 is a model for capacitance change of the gold membrane deflection. Upper gold membrane part deforms by pressure. The changed in distance between the two metal plates causes the capacitance variation.

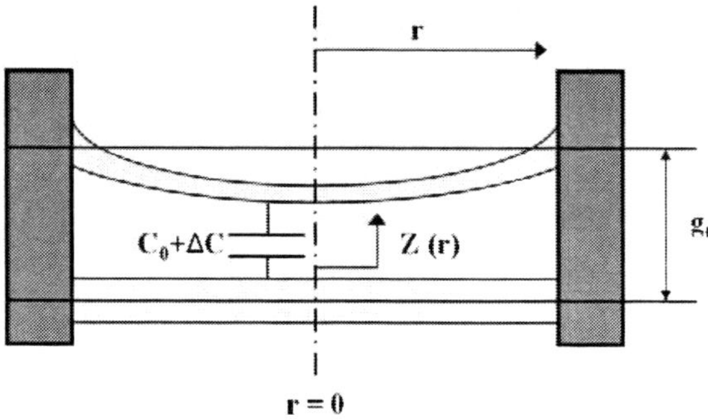

Figure 4. Membrane modeling for capacitance change

r: distance from center, R: radius of membrane
Z(r): gap from the center Z(0): central deflection
g_o: initial gap distance between two electrodes
C_0: initial capacitance between two electrodes
ΔC: capacitance change
ε_o: permittivity of free space

In order to meet the clamped boundary condition, the membrane deflection can be expressed as in (1) with Z(0) the central deflection, and r the distance from the center.

$$Z(r) \cong Z(0)\left[1-\left(\frac{r}{R}\right)^2\right] \qquad (1)$$

The capacitance change due to deflection can be calculated as in (2) using the circular plate deflection.

$$\begin{aligned}\Delta C &= \int dc = \int_A \frac{\varepsilon_0}{g_0 + Z(r)} dA \\ &= \int_0^{2\pi}\int_0^R \frac{\varepsilon_0}{g_0 + Z(0)\left[1-\left(\frac{r}{R}\right)^2\right]} r dr d\theta \\ &= \frac{\pi\varepsilon_0 R^2}{\sqrt{g_0 Z(0)}} \tan^{-1}\sqrt{\frac{Z(0)}{g_0}}\end{aligned} \qquad (2)$$

4 RESULTS

The capacitance changes were measured due to the deformation of gold membrane caused by the hydraulic pressure. In that hydraulic test, we added or subtracted a droplet of water by 50 μl at each step. The hydraulic pressure can be calculated from the water height and membrane area. This amount is equivalent to the water head change of 1.7 mm cross sectional area of 91.56 mm^2. The membrane are varied between 100 × 100 μm^2 to 300 × 300 μm^2. Then, the hydraulic pressure of 50 μl water is estimated to 16.7 Pa.

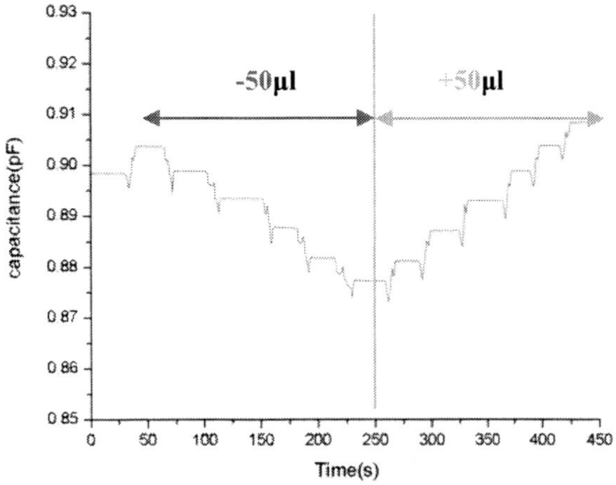

Figure 5. Capacitance changes during the hydraulic pressure test: The volume of water either decreases or increases by 50 μl at each step.

Figure 5 shows the result of test. At every step from 50 s to 250 s, the amount of 50 μl water was subtracted. In this period, total amount of subtracted water was 250 μl and the total capacitance change was 260 fF. In the same manner, the same amount of 50 μl water was added after 250 s. Also, the amount of water at the initial (at 50 s) and the final (at 400 s) states were the same. This result shows the uniform and discrete changes with respect to the varying pressure.

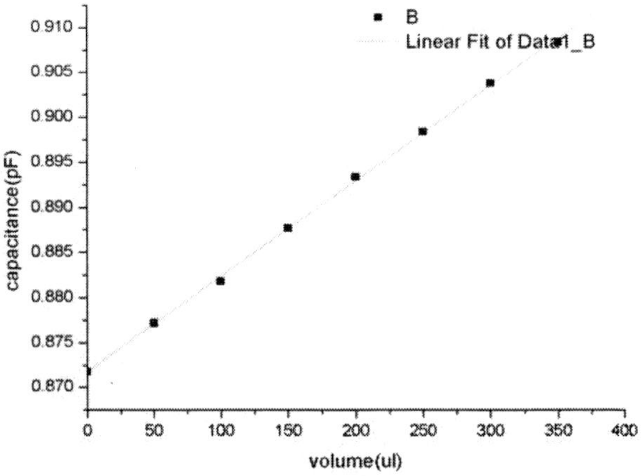

Figure 6. Capacitance according to hydraulic pressure: The capacitance change is 5 fF at each step

The average value of capacitance per unit amount of droplet, 50 μl water was ~5 fF. In Figure 6, the capacitance changed between 0.872 and 0.907 pF when the amount of water varied between 0 to 350 μl. The standard deviation is very small and the capacitance change is linear to hydraulic pressure. The pressure sensitivity calculated using (1) and (2) is 0.3 fF/Pa.

5 DISCUSSION

The noise signal with current setup is ~ 0.5 fF and accordingly the detectable limit of pressure is ~ 1 Pa which is comparable to the state of the art performance (25 mPa) [8]. However, we believe that the noise limit can be substantially reduced with low noise circuit, and our device can easily challenge the best limit of detection. More importantly, the gold nano membrane offers the structure particularly suitable for molecular monitoring through chemo-mechanical sensing mode. Gold surface has broad range of chemistry useful for immobilizing chemicals. Furthermore gold provides stable chemical properties and structural integrity.

6 SUMMARY

Freely suspend gold membrane was fabricated and tested to evaluate mechanical properties. The capacitance change was measured and used to estimate pressure sensitivity. Even though the pressure sensor is not our ultimate goal, this result is promising for the further development in this direction as well as the use for the chemo-mechanical sensing. As a transducer, the ultra-thin gold membrane has many advantages. To achieve the sensitivity demonstrated by the previous polymer devices, the thickness of the gold membrane should be less than 50 nm.

ACKNOWLEGEMENT

This work was supported by the Korea Science and Engineering Foundation (KOSEF) grant funded by the Korea government (MOST) (R0A–2007–000 –10051-0).

REFERENCES

[1] Michael Sepaniak, Panos Datskos, Nickolay Lavrik, Christopher Tipple, "Microcantilever Transducers: A New Approach in Sensor Technology," *Analytical chemistry*, 569A, November 1, 2002

[2] Vincent Tabard-Cossa, et al, "A differential microcantilever-based system for measuring surface stress changes induced by electrochemical reactions," *Sensors and Actuators B*, vol.107, pp.233–241, 2005

[3] Si-Hyung "Shawn" Lim, et al, "Nano-chemo-mechanical sensor array platform for high-throughput chemical analysis," *Sensors and Actuators B*: vol.119, pp.466-474, 2006

[4] R. Rodriguez, J. Chung, K. Lee, J. Lee, "Bio/chemical Sensing by Thin Membrane Trasducers," ASME International Mechanical Engineering Congress and Exposition, Anaheim, California, 2004

[5] Srinath Satyanarayana, Daniel T. McCormick, and Arun Majumdar, "Parylene micro membrane capacitive sensor array for chemical and biological sensing," *Sensors and Actuators B: Chemical*, vol. 115, pp.494-502, 2005

[6] Misun Cha, Ilchaeck Kim, Junbo Choi, Junghoon Lee, "Single-nucleotide polymorphism detection with thin membrane transducer," the proceeding of *Micro TAS*, Tokyo, 2006

[7] Stephen P. Timoshenko, S. Woinowsky-Kreger, "Theory of plate and shells-second edition," McGRAW-HILL International editions, 1959

[8] 6100 Ultra high accuracy digital pressure sensor, http://www.sensorsone.co.uk/products/0/79/6100-Ultra-High-Accuracy-Digital-Pressure-Sensor.html, SensorsONE, 2008

Micromachined Force Sensors for Characterization of Chemical Mechanical Polishing

Douglas Gauthier*, Andrew Mueller*, Robert White*, Vincent Manno*, Chris Rogers*, Sriram Anjur**, and Mansour Moinpour***

* Mechanical Engineering, Tufts University, Medford, MA, USA
** Cabot Microelectronics, Aurora, IL, USA
*** Intel Corporation, Santa Clara, CA, USA

ABSTRACT

Micromachined structures with diameters ranging from 50 - 100 μm have been applied to the measurement of the microscale shearing forces present at the wafer-pad interface during chemical mechanical polishing (CMP). The structures are 80 μm high poly-dimethyl-siloxane posts with bending stiffnesses ranging from 1.6 to 14 μN/μm. The structures were polished using a stiff, ungrooved pad and 3 wt% fumed silica slurry at relative velocities of approximately 0.5 m/s and downforces of approximately 1 psi. Observed lateral forces on the structures were on the order of 5-500 μN in magnitude, and highly variable in time.

Keywords: shear force, sensor, polishing, chemical mechanical planarization, CMP

1 INTRODUCTION

Chemical Mechanical Planarization (CMP) is a critical process for semiconductor manufacturing. As feature sizes continue to shrink, planarity continues to be an important consideration for successful lithography. The CMP process is widely used and, for certain systems, has been characterized experimentally in terms of many of the polishing parameters. Still, a comprehensive model involving the multitude of process variables and their effects on material removal rates, planarity, and defectivity remains elusive [1], [2].

The development and validation of some aspects of this "total CMP model" is hindered by lack of knowledge of *in situ* shear forces present at the micro-scale [3]. Experimental data of shear forces from single asperities would provide a comparison point for further research at this scale. In a more direct sense, knowledge of the local shear forces may be important in designing fragile structures, such as low-k dielectrics, to withstand polish.

In this paper,we detail the development of micromachined shear stress sensors intended for characterizing these *in situ* local contact forces during CMP. Other researchers have investigated the average global shear force *in situ* [4]–[6]. In addition, some groups have investigated micro- or nano-scale forces *ex situ* [7], [8]. However, to the best of our knowledge, the work described in this paper represents the first attempt to measure microscale polishing forces *in situ*.

2 DESIGN AND FABRICATION

The sensor structure is shown in Figure 1. The structures are 80 μm tall poly-dimethyl-siloxane (PDMS) cylindrical posts. The post diameters vary from 50 μm to 100 μm. Each post is recessed in a well which leaves a 50 μm wide empty region around the post. The structure is immersed into the polishing slurry and polished. As microscale features on the polishing pad come into contact with the post top, the post deflects. This deflection is observed through the back of the transparent structure using a high speed microscopy setup. Nearly 98% of the wafer surface is planar PDMS; it is only occasionally broken by the annular well region around sensor posts. This allows the majority of the normal force applied by the polishing pad to be carried by the bulk PDMS, thus not compressing or buckling the sensor posts.

While other researchers using micro post shear stress arrays have determined deflections optically, viewing the sensors from the top (*e.g.* [9]), this is impossible in our application due to the presence of the polishing pad. Thus, we require a transparent sensor and substrate. Our structure also avoids the need for electrical interconnects, which would be difficult to maintain and protect in a polishing environment.

PDMS is chosen for the structure due to the very low elastic modulus, on the order of 750 kPa [10]. This allows deflections of 5-50 μm to be achieved with lateral forces in the range of 4-400 μN (for different diameter posts). The disadvantage of using PDMS is that it is dissimilar from the oxides and metals that are usually polished by the semiconductor industry. We emphasize that the results in this paper are for polishing of PDMS surfaces, and care must be taken when extrapolating these results to other polishing systems.

All microfabrication was conducted in the Tufts Micro and Nanofabrication Facility (TMNF). Sensors were fabricated through a modified two-layer PDMS micromolding process similar to that described in [11]. In this process a master mold consisting of two layers, one silicon dioxide and one SU-8 photoresist, is used. The sil-

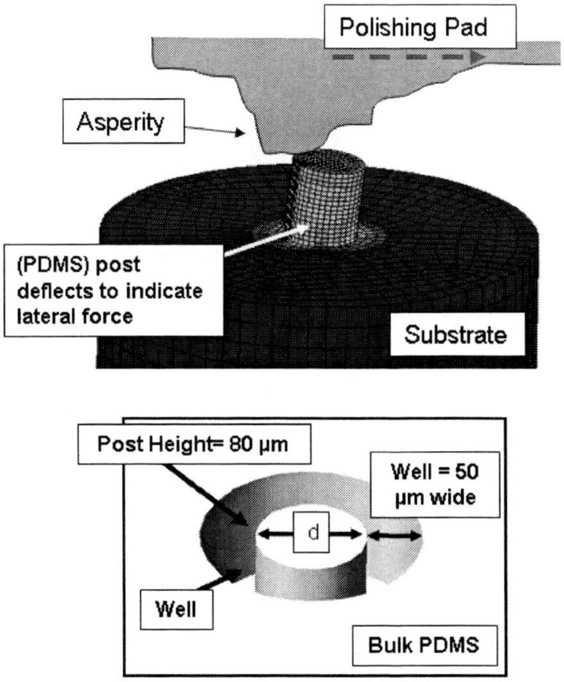

Figure 1: Diagram showing the concept of the recessed micro-post lateral force sensor.

icon dioxide thin film is first lithographically patterned and etched using buffered HF to place fine features (writing and positioning marks) onto the mold. SU-8-100 (MicroChem Corp.) is then spun on at 80 μm thickness to define the main features of the structure. The master mold surfaces are silanized to aid in releasing the PDMS. Dow Corning Sylgard 184 base and curing agent is then mixed in a 10:1 ratio to produce the liquid PDMS which is degassed and poured over the mold. This is cured on a hotplate at 60 ^{o}C for 4 hours. A leveling table is used to ensure that the mold is level, creating a uniform PDMS thickness. After curing, the PDMS structure is pealed off of the mold. A microscope picture is shown in Figure 2. The compliant PDMS structure is bonded to a 0.5 mm thick Pyrex glass wafer (Corning type 7740 Pyrex glass) by exposing both surfaces to a 200 mT, 25 W oxygen plasma for 30s, placing the PDMS and glass in contact, and heating on a hotplate at 60 ^{o}C for 15 minutes. This Pyrex wafer gives a transparent rigid backplate to the structure. A 15 nm thick Chromium film is finally sputtered onto the front side of the PDMS to aid in image contrast, and the entire structure is bonded to a stiff aluminum backing plate. The aluminum plate has windows machined in it for viewing through the backside, and includes mounting hardware for connecting to the polisher shaft.

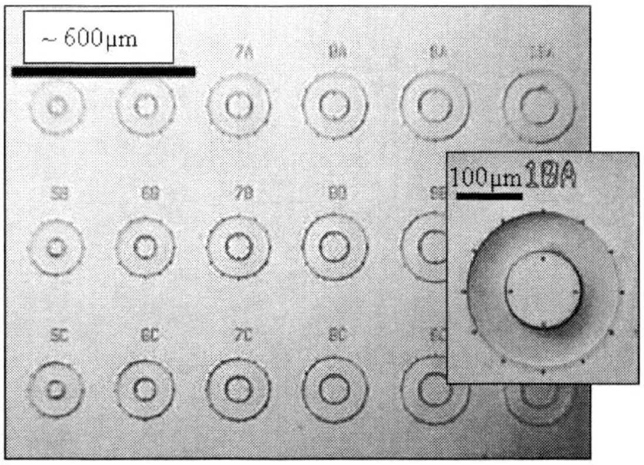

Figure 2: Microscope image of part of the post array, as well as a higher magnification image of a single post in a well.

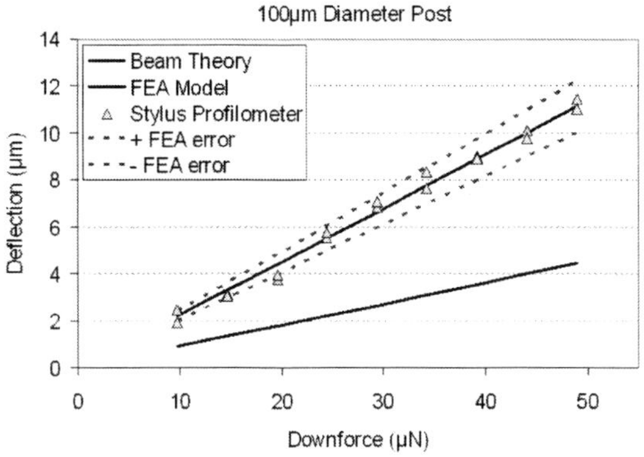

Figure 3: Typical calibration data before O_2 plasma treatment or polishing compard to computational models.

3 CALIBRATION

Post stiffness calibration is carried out using a microscale mechanical testing technique, called MAT-Test, developed by Hopcroft *et al* [12]. The technique utilizes a contact surface profilometer to obtain force deflection curves for small structures. To determine the stiffness of the PDMS sensing posts, calibration posts (not recessed in wells) were fabricated using the same process as the experimental structures. The PDMS was cut and oriented horizontally, so that the stylus tip can travel from the base to the tip of the post. In this fashion, force vs. deflection curves are obtained for various downforces. A Veeco Dektak 6M Stylus Profilometer was used to supply downforces between 10-150 μN.

An example calibration is shown in Figure 3. The result is compared to the expected post stiffness computed for the 3D structure, including the compliance of the base, using linear elastic finite elements with a modulus

Table 1: Post stiffness.

Post Diameter (μm)	Before Treatment (μN/μm)	During Polish (μN/μm)
50	0.8	1.6
60	1.5	3.0
70	2.4	4.8
80	3.6	7.2
90	5.2	10.
100	7.0	14

of 750 kPa and a Poisson ratio of 0.5 [10]. Agreement is very good for this diameter post and for every other diameter post measured. The response is linear across the measured range.

Additional testing (not shown) indicated that the modulus of the PDMS is influenced by O_2 plasma treatment, aging, and polishing. The modulus increased by a factor of 3-5 after O_2 plasma treatment (used in bonding). The modulus reduced by about 10% with age (over the course of 3 days). The stiffness of one 80 μm diameter post was measured after molding (before O_2 plasma treatment or metallization) and then again after polishing, and was found to increase by a factor of 2.

The stiffness of the posts as predicted by finite element analysis and confirmed by experimental calibration prior to O_2 plasma treatment or polishing are given in the first column of Table 1. Due to the observed increase in stiffness, the stiffness used for computing lateral forces from observed deflections *in situ* is twice the FEA computed value, as given in column 2 of the table. It is emphasized that this has only been directly observed on a single 80 μm diameter post, but is assumed to hold approximately true for the other size structures.

4 POLISHING FORCES

Polishing studies were conducted in a tabletop polisher (Struers RotoPol 31) using an ungrooved IC1000 pad (Rodell, Newark, Del.). A fumed silica slurry diluted to 3% by weight particle loading (Cabot Microelectronics, Aurora, IL) was used. We emphasize again that the surface of the wafer and the sensing structures are manufactured out of the low modulus polymer polydimethylsiloxane (PDMS); this is likely to have a significant impact on the polishing forces, as compared to the polishing of stiffer materials. In addition, we emphasize that the wafer is not rotating during these experiments due to limitations with the optical setup. The polishing pad is rotating, but the wafer is not. We are not conditioning the pad.

An optical system consisting of a Phantom v7.0 high speed camera with a 12 bit SR-CMOS sensor coupled to a 15 X relay lens and a 10 X microscope objective is used to determine post deflection during CMP. Light is

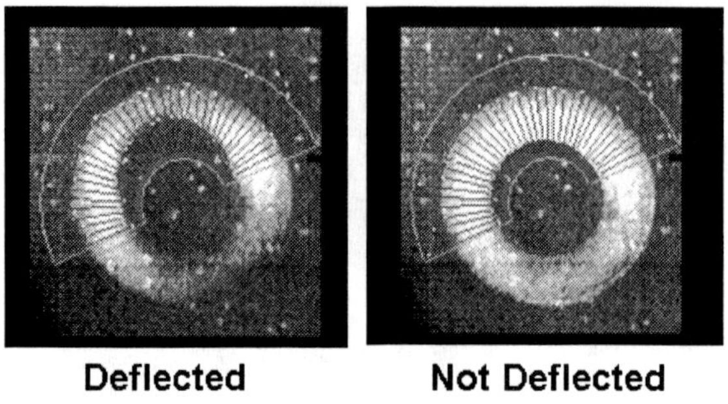

Figure 4: Two example images taken during polishing showing a deflected and not deflected post. Image processing icons are also shown.

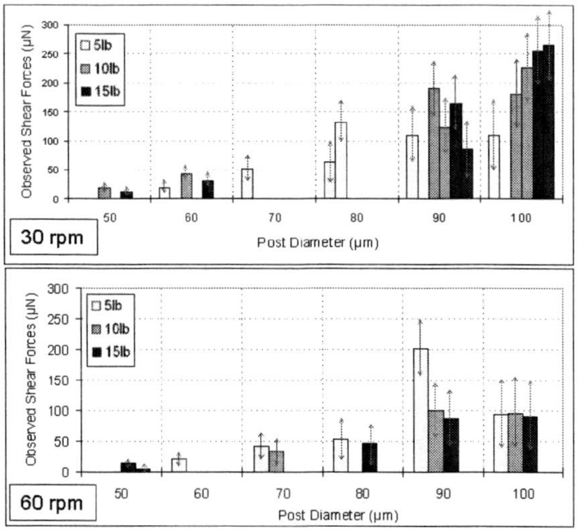

Figure 5: Summary of the *maximum* forces observed during polishing.

provided to the sensor through the microscope objective using a fiber optic light guide and 90 degree soda lime plate beam splitter. The system is mounted on a micropositioning stage for focusing and positioning. For all experiments described here, the camera resolution and speed are set at 512 x 384 pixels and 10,000 frames per second, respectively. At 10,000 frames per second, for a relative pad speed of 0.5 m/s, asperities can be captured at 50μm intervals. Pixel size is 2μm. Image processing is performed on the recorded movies of post deflection to extract the relative motion of the post top. Edge detection finds points along the edge of the well and along the leading edge of the post top. Two circles are fit to these edges. The relative motion of the center of the circles gives the deflection of the post top in two dimensions. Two example images are shown in Figure 4

Measurements were conducted with downforces of 5 lbs, 10 lbs, and 15 lbs. This corresponds to an area average normal load of 0.4 psi, 0.8 psi, and 1.2 psi over

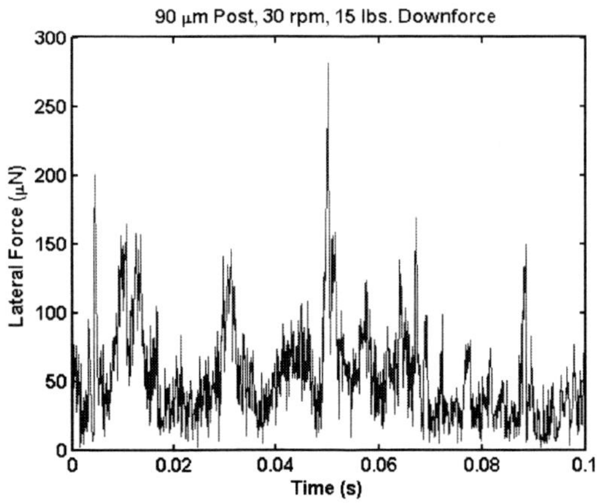

Figure 6: Trace of lateral force magnitude vs. time for a 90 μm diameter post being polished at 30 rpm (0.3 m/s) with 15 lbs downforce (1.2 psi).

the 4" wafer. Two different rotation speeds, 30 rpm and 60 rpm, were used for the polishing platen. The center of the wafer is approximately 10 cm from the center of rotation of the platen, hence 30 rpm corresponds to 0.3 m/s relative velocity, and 60 rpm corresponds to 0.6 m/s relative velocity. Figure 5 shows the magnitude of the maximum observed lateral forces for these various conditions on different sized structures. Increasing the size of the structure leads to an increase in lateral force. Increasing the rotation rate leads to a decrease in lateral force. An increase in downforce does not have a dramatic effect on the observed maximum lateral forces.

A particular time trace of the force on an individual post is shown in Figure 6. The forces are relatively low for much of the time, but periodically spike to higher magnitudes (above 100 μN) for brief periods of time (on the order of 1-10 ms). Similar behavior was observed in most of the test results.

5 CONCLUSIONS

During polishing of PDMS using CMP, the observed lateral forces on 50 - 100 μm diameter structures were on the order of 5-500 μN, with considerable variation of the force level in time. Larger lateral forces were observed for larger diameter structures. Increasing the speed of the polish descreased the lateral forces. Increasing the downforce increased the lateral forces at slow speeds for the largest structures, but had little impact at higher speeds. Force levels were highly variable in time, often maintaining relatively low levels on the order of 50 μN, but periodically spiking to higher magnitudes in excess of 200 μ N. This suggests that the majority of the pad and wafer are not in contact, but that large forces occur in limited regions of contact. Hence one cannot extract the local force magnitude simply from a knowledge of the global coefficient of friction.

REFERENCES

[1] C. Evans, E. Paul, D. Dornfeld, D. Lucca, G. Byrne, M. Tricard, F. Klocke, O. Dambon, and B. Mullany, "Material Removal Mechanisms in Lapping and Polishing," *CIRP Annals-Manufacturing Technology*, vol. 52, no. 2, pp. 611–633, 2003.

[2] E. Paul, "Application of a CMP model to tungsten CMP," *Journal of The Electrochemical Society*, vol. 148, p. G359, 2001.

[3] L. Cook, "Chemical processes in glass polishing," *Journal of Non-Crystalline Solids*, vol. 120, no. 1, pp. 152–171, 1990.

[4] J. Sorooshian, D. Hetherington, and A. Philipossian, "Effect of Process Temperature on Coefficient of Friction during CMP," *Electrochemical and Solid-State Letters*, vol. 7, p. G222, 2004.

[5] J. Levert, F. Mess, R. Salant, S. Danyluk, and A. Baker, "Mechanisms of chemical-mechanical polishing of sio_2 dielectric on integrated circuits," *Tribology Transactions*, vol. 41, no. 4, pp. 593–599, 1998.

[6] J. Lu, C. Rogers, V. Manno, A. Philipossian, S. Anjur, and M. Moinpour, "Measurements of slurry film thickness and wafer drag during CMP," *Journal of The Electrochemical Society*, vol. 151, p. G241, 2004.

[7] G. Basim, I. Vakarelski, and B. Moudgil, "Role of interaction forces in controlling the stability and polishing performance of CMP slurries," *Journal of Colloid And Interface Science*, vol. 263, no. 2, pp. 506–515, 2003.

[8] A. Feiler, I. Larson, P. Jenkins, and P. Attard, "A quantitative study of interaction forces and friction in aqueous colloidal systems," *Langmuir*, vol. 16, no. 26, pp. 10 269–10 277, 2000.

[9] O. du Roure, A. Saez, A. Buguin, R. Austin, P. Chavrier, P. Silberzan, and B. Ladoux, "Force mapping in epithelial cell migration," *Proceedings of the National Academy of Sciences*, vol. 102, no. 7, p. 2390, 2005.

[10] D. Armani, C. Liu, and N. Aluru, "Re-configurable fluid circuits by PDMS elastomer micromachining," *Micro Electro Mechanical Systems, 1999. MEMS'99. Twelfth IEEE International Conference on*, pp. 222–227, 1999.

[11] S. Sia and G. Whitesides, "Microfluidic devices fabricated in Poly (dimethylsiloxane) for biological studies," *Electrophoresis*, vol. 24, no. 21, pp. 3563–3576, 2003.

[12] M. Hopcroft, T. Kramer, G. Kim, K. Takashima, Y. Higo, D. Moore, and J. Brugger, "Micromechanical testing of SU-8 cantilevers," *Proc. JSME Adv. Technol. Exp. Mech*, pp. 735–742, 2003.

Modeling of Microcantilever based Nuclear Microbatteries

B.G. Sheeparamatti [*], J.S. Kadadevarmath [**], and R.B. Sheeparamatti [***]

[*]Basaveshwar Engineering College, Bagalkot, India, sheepar@yahoo.com
[**]Karnataka University Dharwad, India, jagadish_sk59@rediffmail.com
[***]Basaveshwar Engineering College, Bagalkot, India, rajeshwari_sheepar@yahoo.co.in

ABSTRACT

Microcantilevers are extremely versatile and are used as sensors, actuators and in many other microsystems. In this work, simulink model of a microcantilever based actuator, which can be driven by nuclear radiation, is developed. This actuator can be used as micro-battery or micro-generator.

Microcantilever tip portion is exposed to nuclear radiation from lower side and becomes charged because of emission of electrons from the radioactive element. Thus an electrostatic attraction is created between base and the cantilever tip, which gradually bends at the tip and discharges the electrons. Now, electrostatic attraction disappears for a moment and then, the process repeats and thus the cantilever sets into oscillations. The piezoelectric plate on the microcantilever produces electric pulses, and can be used to generate electricity. The mechanical equivalent of the microcantilever is developed. In the electrical model, the force-voltage (or mass-inductance) analogy is considered. The transfer function obtained is used to derive the system performance. The model developed is subjected to different input conditions and the output is observed. The results of the analytical model are compared with the model developed in ANSYS/Multiphysics.

Such microcantilevers offer an attractive power source for MEMS based actuators, which require high power density or long life.

Keywords: microcantilevers, microbatteries, mathematical modeling, microgenerators.

1 INTRODUCTION

Microcantilevers are extremely versatile and are being used as sensors, actuators and in many other microsystems. In this work, mathematical/simulink model of a microcantilever based actuator, which can be driven by nuclear radiation, is developed. This actuator can be used as micro-battery or micro-generator. These microbatteries may not replace chemical batteries but definitely they make a considerable change in power supply for small electronic gadgets.

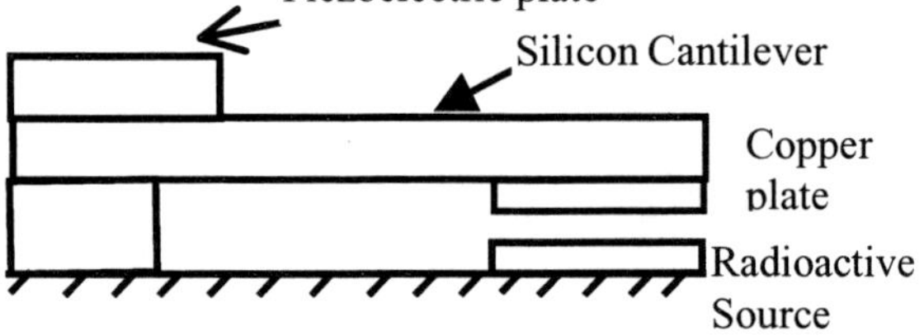

Figure 1: Microbattery

The main idea behind the microbattery is, microcantilever tip portion is exposed to nuclear radiation from lower side as shown in Fig 1[1]. Thereby its lower surface becomes negatively charged because of emission of electrons from the radioactive element underneath. Thus there is an electrostatic attraction between base and the cantilever tip. Once, sufficient number of electrons are collected, gradually the cantilever bends and discharges the electrons either by physical contact / tunneling / gas breakdown. Now electrostatic attraction disappears for a moment and then, the process repeats and thus the cantilever sets into oscillations. This recurring mechanical deformation of the piezoelectric plate kept on the microcantilever produces a series of electric pulses. These pulses can be rectified and smoothed to provide electricity.

2 THEORY

Considering the oscillating microcantilever as a system of single degree of freedom formed by a body of mass m and spring of stiffness k, and a dashpot, equation of vibration is developed. Similarly the mechanical equivalent of the microcantilever is also developed as shown in Fig. 2. In the electrical model, the force-voltage (or mass-inductance) analogy is considered as shown in Fig. 3. Laplace transform of the derived equations are taken for solving and easy analysis of such MEMS systems. The Laplace transform can be used to directly write the transfer function of the electrical equivalent. The transfer function obtained is used to derive the system performance. The model developed is subjected to different input conditions and the output is observed. The results of the analytical model are compared with the model developed in ANSYS/Multiphysics, as the ANSYS/Multiphysics is the finite element (FE) software, which is suited for performing the myriad of physics simulations required for MEMS.

Such microcantilevers offer an attractive power source for MEMS based actuators, which require high power density or long life.

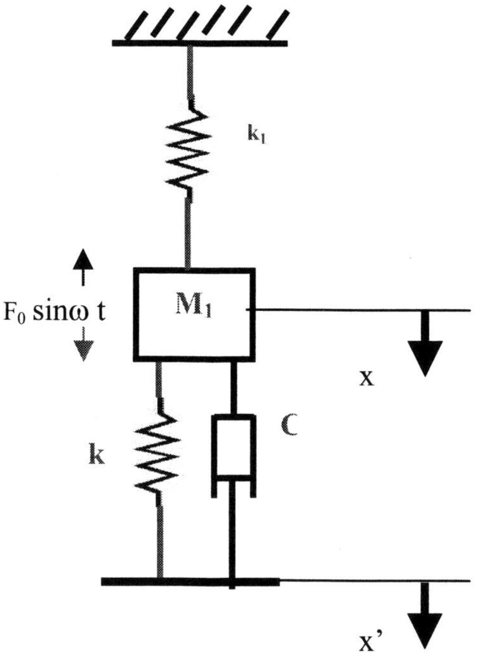

Figure 2 : Equivalent mechanical system

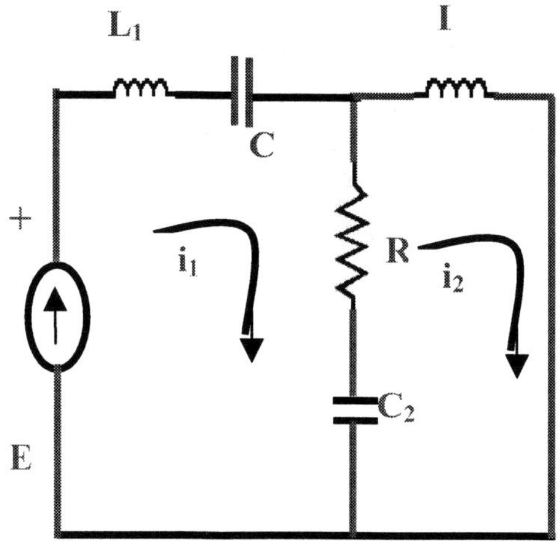

Figure 3 : Equivalent electrical circuit for a damped dynamic vibrating microcantilever with radioactive *stimulus*

3 SIMULINK MODEL

Assuming the silicon cantilever has negligible inertia, and the electrostatic force because of radiating particles leads to continuous oscillations. The capacitance between cantilever and base is given by equation 1, refer Figure 4.

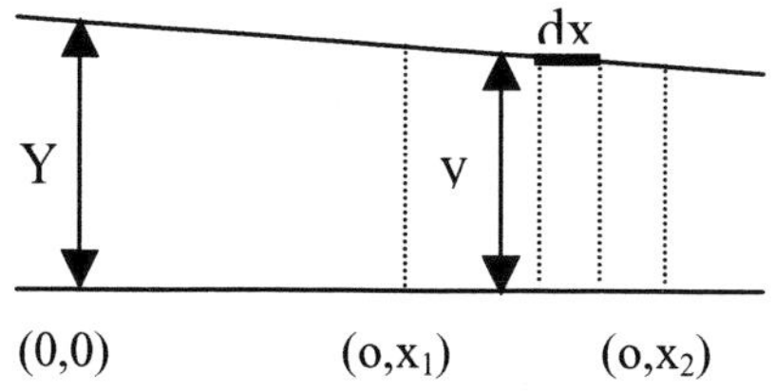

Figure 4: Cantilever Model

$$c = \frac{\omega\varepsilon_0}{a}\left[\log\frac{(y/a - x_1)}{(y/a - x_2)}\right] \qquad (1)$$

The spring force of the cantilever exactly balances the electrostatic attraction force on the cantilever, which is shown in equation 2.

$$k(Y - y_0) = QE \qquad (2)$$

where k is the spring constant, Y is the initial distance, and y_0 is the changed distance.

Assuming a uniform electric field, the capacitor can be modeled as a parallel plate capacitor C and the change on it is given by equation 3, below.

$$Q = CV = \frac{\varepsilon_0 AV}{y} \qquad (3)$$

The MATLAB/Simulink tool enables one to model, simulate and analyze such dynamic systems. In this case a graphical model of the microbattery is created using Simulink model editor. The model depicts the time dependent mathematical relationships among the system inputs, states and outputs. Then using Simulink, the behavior of the system over a specified time can be simulated. The simulink model of the microbattery developed is shown in the Fig. 5. The result of the simulink model is shown in the Fig. 6. The microcantilever is also developed in ANSYS/ Multiphysics software and the output in the form of resonance frequency is shown in the Fig. 7.

4 RESULTS AND CONCLUSIONS

The simulink output showing distance (gap) versus time for an initial distance of 32 microns with a period up to 400 seconds[2]. Figure 6 shows the simulink output of the model. Similarly the ANSYS was used to create the

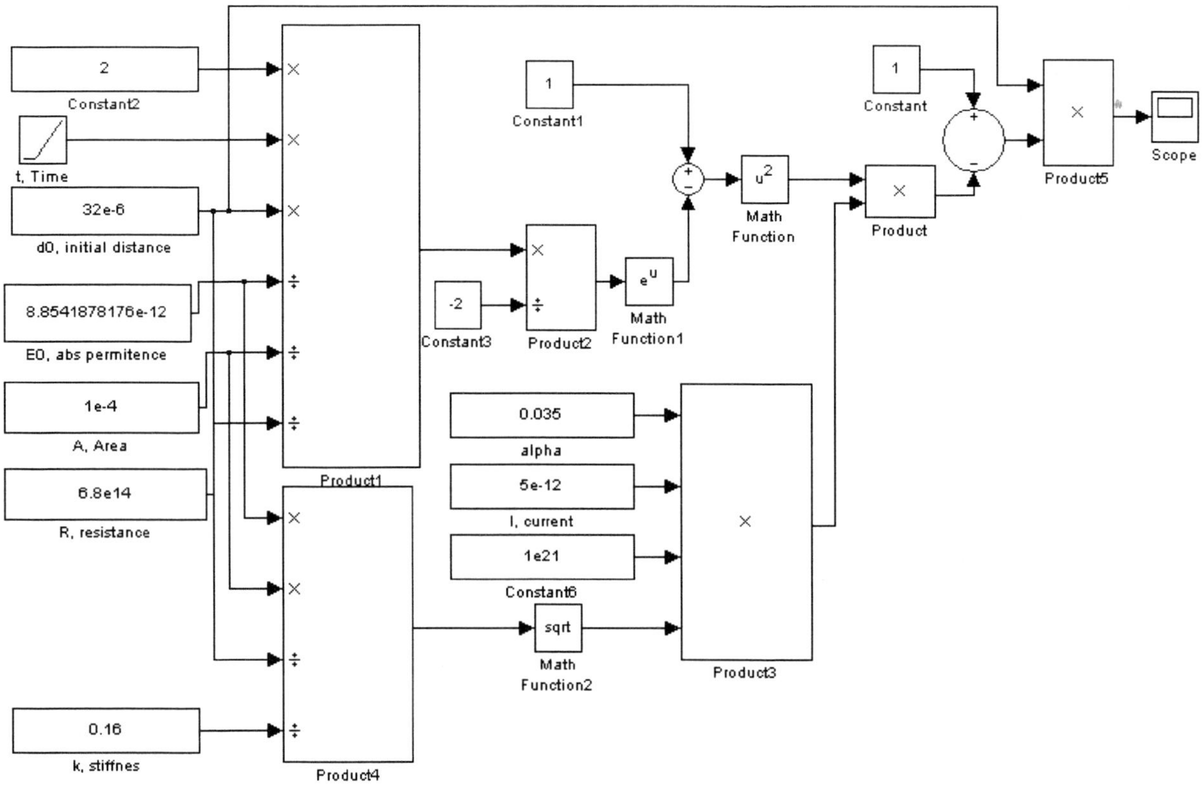

Figure 5 : Simulink model of the Microbattery

cantilever to find the critical oscillating frequencies. The result is shown in Figure 7. Harmonic analysis is carried out with excitation force of F sin ωt where F is in few fractions of micronewtons. In the frequency analysis, the ranges of frequencies considered are, from zero to 1 MHz. It is observed that at about 70 KHz and 430 KHz, maximum amplitudes (resonance) are obtained for a silicon cantilever of 100 micron long, 20 micron wide and 0.5 micron thick. This indicates that it is possible to simulate the behavior of silicon cantilever of the nuclear microbattery referred in this work.

The nuclear powered oscillators may become effective power sources for future personal electronic gadgets. The electrical equivalents, simulink models and ANSYS models developed in this work will help in understanding the microcantilever based nuclear microbatteries. These results help the people working on fabrication of such radiation based microcantilevers. The theoretical results and their analysis developed in this work will be much useful to the people working on design, testing and fabrication of microcantilever based actuators.

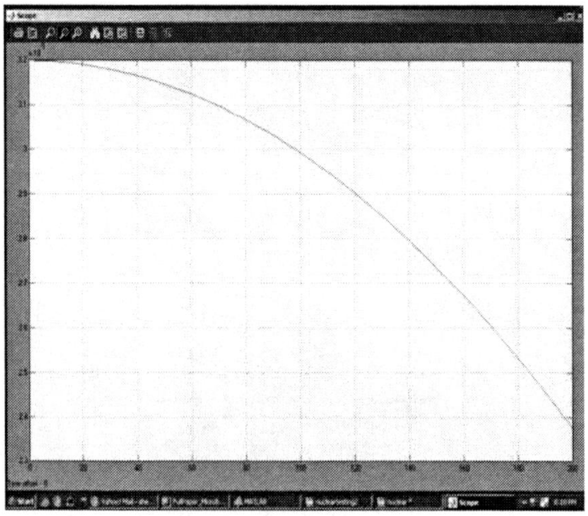

Figure 6 Simulink Output of the Microbattery

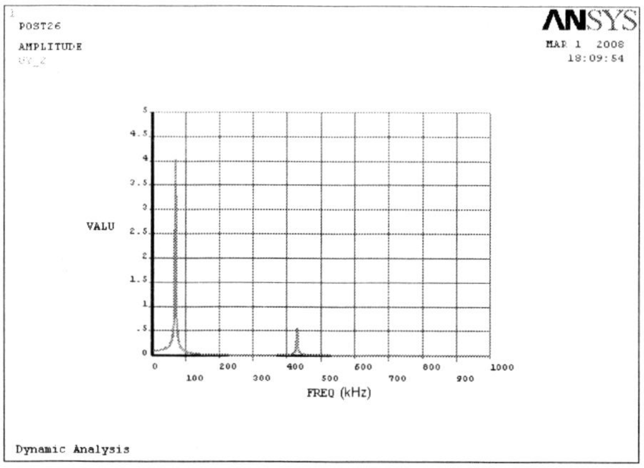

Figure 7 ANSYS/Multiphysics Output

REFERENCES

[1] Amit Lal, James Blanchard, "The Daintiest Dynamos", IEEE Spectrum, 36-41,September 2004

[2] Hui Li, Amit Lal, James Blanchard, Douglas Henderson, "Self-reciprocating radioisotope-powered cantilever", Journal of Applied Physics, Vol.92, Number 2, 1122-1127, 15 July 2002

Shape Memory Alloy and Elastomer Composite MEMS Actuators

P.D. Fallon, A.P. Gerratt, B.P. Kierstead, R.D. White

Department of Mechanical Engineering, Tufts University, Medford, MA, USA

ABSTRACT

A process for fabrication of shape memory alloy MEMS actuators on a elastomeric polymer substrate is described. These actuators are designed for use on innovative soft body robots. Patterned shape memory alloy (nitinol) is sputter deposited on a polyimide mesh structure. The mesh substrate acts as a bias spring to de-twin the shape memory alloy in the martensite phase. Significant portions of the structure have been fabricated, but the nitinol anneal has not been successful, nor has the shape memory effect been demonstrated. Refinement of the fabrication process is ongoing.

Keywords: shape memory alloy, nitinol, actuator, elastomer, polyimide

1 INTRODUCTION

The majority of robots currently built are constructed of rigid members, limiting the capacity of these robots to move through oddly shaped and confined spaces. A new focus in robotics has been the design of soft body robots. Advantages over rigid body robots of comparable size and shape including the capacity to expand and contract upon demand to facilitate movement through and over complex environments environment. Soft bodied robots composed of a soft silicone elastomer material are under developement at Tufts [1]. These soft bodied robots are modeled after the *Manduca sexta* tobacco hornworm caterpillar, which can grow on the order of 10,000 fold during its lifetime [2]. It is of significant interest to develop small scale models of these existing soft body robots. To provide movement to such robots, one option is actuation derived from resistive heating of patterned nitinol, a nickel and titanium shape memory alloy. Previous research has shown the use of nitinol actuators for MEMS devices [3]–[5]. Such actuators develop high forces and strains, and can be fabricated in into flexible sheets, but tend to require relatively high power. The paper describes fabrication of nitinol MEMS actuators.

2 SHAPE MEMORY ALLOY

Shape memory alloys exhibit the phenomenon of shape recovery during phase transitions. Various types of shape memory alloys have been discovered, though for the purpose of this research only nitinol (nickel-titanium) was used. When in a cooled state nitinol displays a martensite crystal structure. When nitinol is heated above its transition temperature it changes to an austenite crystal structure. Nitinol can be trained in one position by raising the temperature of the alloy above its transition temperature while restraining the sample in the desired position. The sample is then cooled while still being restrained and the nitinol shifts to a twinned martensite structure [6]. Once this cooling step is complete the nitinol will revert back to its trained position upon subsequent heating, provided the nitinol has undergone 6-8% strain to de-twin the martensite [7].

For the purposes of this research, a transition temperature greater than air temperature is required. Small shifts in the composition of the alloy can lead to dramatic changes in the transition temperature [8]. As such, an alloy composition of 50% nickel and 50% titanium by atomic weight percentage is desired where the transition temperature is near 100° C. The transition temperature of a specific alloy can be determined through differential scanning calorimetry.

The actuating devices were microfabricated using sputter deposition. Separate nickel and titanium targets were used to co-sputter the nitinol. When determining the processing conditions, the power supplied to the DC magnetron guns must be given special consideration. Initial settings were determined from a literature search [9], [10]. Inductively coupled plasma (ICP-OES) - optical emission spectroscopy (The M&P Lab, Schenectady, NY) was then utilized to analyze the percent composition of a number of samples. Figure 1 shows the results of the elemental analysis. For the desired 50% nickel and titanium by atomic percentage, powers of 70 W and 200 W must be applied to the nickel and titanium targets, respectively.

When nickel and titanium are co-sputtered, the resulting nitinol film is in an amorphous state [3]. For the shape memory effect to be possible, the film must be in a crystalline state. The transition from the amorphous to crystalline state is achieved through an anneal of the film at 480° C for 30 minutes [5]. The precise annealing temperature and time depend on the conditions present during the co-sputtering of the film. The anneal must

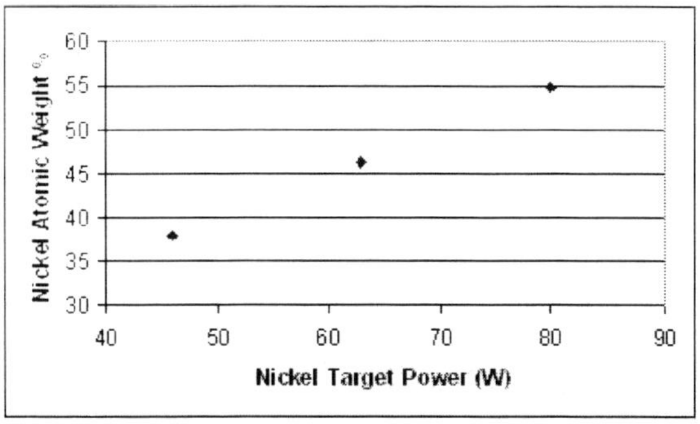

Figure 1: Composition as a function of DC magnetron gun power (Ti gun power = 200 W).

be conducted in an inert atmosphere as the nitinol will oxidize during the anneal if it is completed in air. The transition of the crystal structure can be examined with x-ray diffraction.

3 DESIGN

In order for the nitinol to exhibit its shape memory effect, it must first be strained to de-twin the martensite crystal structure. It is also desirable to laminate or affix the shape memory alloy onto the wall of the soft body robot. The polymer must also be able to survive the high temperature anneal of the nitinol. To accomplish these three things, an elastomeric polymer substrate, HD Microsystems 4100 series photodefinable polyimide, was used. This polyimide was chosen because it does not decompose until a temperature of 600° C is reached, so it can withstand the high nitinol anneal temperature. While polyimide is not the ideal elastomer, it is one of the most chemically and thermally resistant polymers available, making it effective for use as a substrate for nitinol.

A series of devices were designed to examine the effect of various changes in the nitinol wires and the polyimide substrate. A mesh structure was developed for the polyimide layers to make them more compliant than a full sheet. The nitinol meanders across the polyimide mesh structure in an attempt to model a three dimensional spring in two dimensions. The devices range in overall size. The largest are 22 mm long and 10 mm wide layers and the smallest are 7.75 mm long and 3.5 mm wide layers. The nitinol patterned on the polyimide is 1 μm thick and varies in width from a maximum of 0.8 mm to a minimum of 0.1 mm. Representative microscope images are shown in Figures 2 and 3.

4 FABRICATION

Figure 5 outlines the process. A 0.5 mm thick, 100 mm diameter, < 100 > oriented silicon wafer with a 1 μm thermal oxide layer is the starting substrate. To

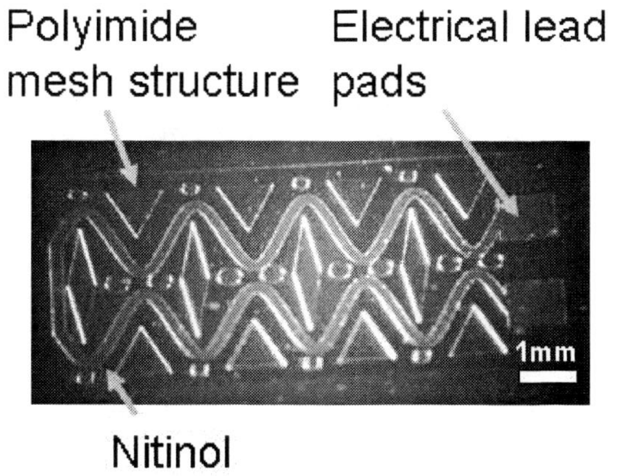

Figure 2: Microscope image of a particular device design prior to the release from the sacrificial layer.

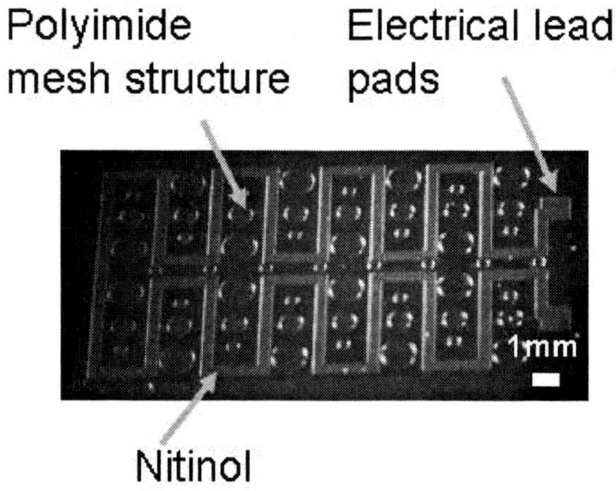

Figure 3: Microscope image of a particular device design prior to the release from the sacrificial layer.

enhance adhesion between the oxide and the polyimide a 100 nm layer of Titanium was sputter deposited (Nanomaster NSC-3000 DC Magnetron Sputter Tool) onto the wafer. The titanium was sputtered at a process pressure of 5 mTorr and a magnetron gun power level of 150 watts. A base pressure of at least $5x10^{-5}$ torr must be met for a quality film. Base pressures above this level resulted in stresses significant enough to cause macro scale cracking in the deposited films, as seen in figure 4.

On top of the adhesion layer a 6 μm layer of polyimide was patterned. This layer of polyimide was spun at 3800 RPM, patterned with photolithography, and cured for 30 minutes at 200° C and 60 minutes at 375° C.

Photoresist (SPR 220-3.0 series) was spun at 2000 RPM and patterned with photolithography (MF CD-

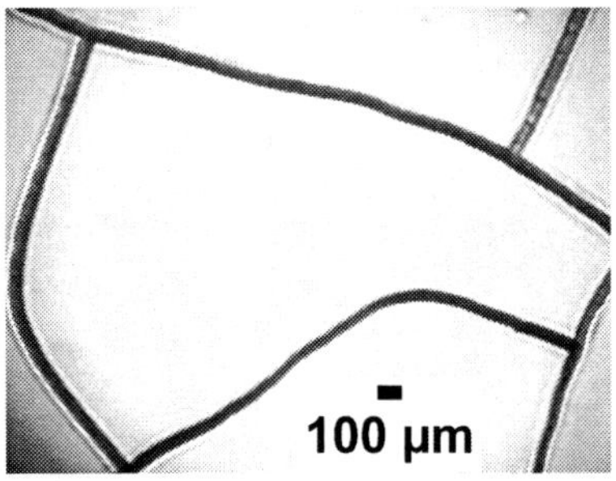

Figure 4: Macro-scale cracking resulting from deposition at a high base pressure.

26 developer). 1 μm of nitinol was sputter deposited. Nickel and titanium targets were co-sputtered at 70 and 200 watts respectively. An acetone and isopropanol liftoff of the nitinol was then performed.

A 6 μm upper layer of polyimide was spun, patterned, and cured with the same conditions as the lower layer. At this point another thick layer of photoresist is spun and patterned with photolithography. The pattern allows for deposition and liftoff of 50 nm chromium adhesion layer and a 300 nm layer of gold for connection of electrical leads.

The devices were then annealed in a nitrogen ambient oven at 480° C. A hydroflouric acid etch of the thermal oxide layer and Ti adhesion layer was performed to release the devices.

5 CONCLUSIONS

The process for the fabrication of shape memory alloy MEMS actuators has been described. The fabrication process has been successfully performed, though shape memory effect has not yet been demonstrated. The devices would be greatly enhanced with a more thermally resistant polymer substrate. Most of the issues with the fabrication process are related to the polyimide mesh not surviving the anneal required to set the crystal structure of the nitinol.

Future work will include a refinement of the fabrication process to achieve a successful demonstration of the shape memory effect. The focus of this effort will be the development of an affective annealing process. Characterization of the devices will also commence once the shape memory effect is demonstrated.

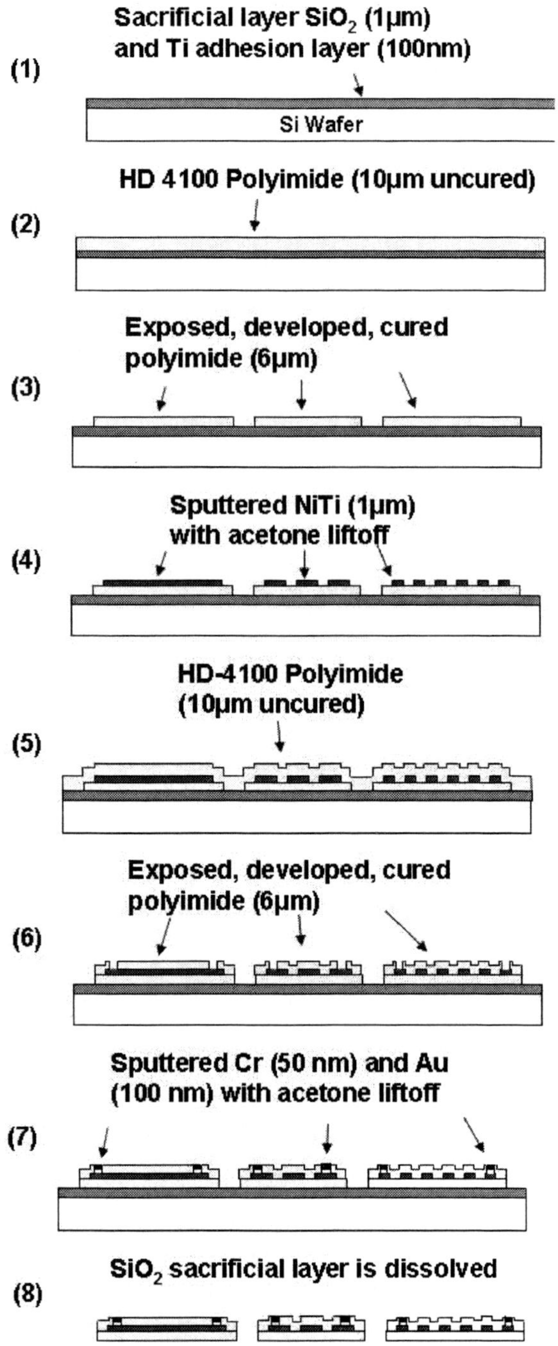

Figure 5: Microfabrication process for producing SMA-elastomer actuator patches.

REFERENCES

[1] Trimmer, B.; Takesian, A.; Sweet, B.; Rogers, C.; Hake, D. & Rogers, D., Caterpillar locomotion: A new model for soft-bodied climbing and burrowing robots, *Proc. 7th Int. Symp. on Technology and the Mine Problem, Monterey, CA: Mine Warfare Association*, 2006.

[2] Mezoff, S.; Papastathis, N.; Takesian, A. & Trimmer, B., The biomechanical and neural control of hydrostatic limb movements in Manduca sexta, *Journal of Experimental Biology*, 207, 3043-3053, 2004.

[3] Fu, Y.; Du, H.; Huang, W.; Zhang, S. & Hu, M., TiNi-based thin films in MEMS applications: a review, *Sensors & Actuators, A. Physical*, 112, 395-408, 2004.

[4] Fu, Y.; Huang, W.; Du, H.; Huang, X.; Tan, J. & Gao, X., Characterization of TiNi shape-memory alloy thin films for MEMS applications, *Surface Coatings and Technology*, 145, 107112, 2001.

[5] Fu, Y.; Dua, H.; Huanga, W.; Zhang, S. & Hua, M., TiNi-based thin films in MEMS applications: a review, *Sensors and Actuators, A.* 112, 395408, 2004.

[6] Fuentes, J. M. G.; Gumpel, P. & Strittmatter, J., Phase Change Behavior of Nitinol Shape Memory Alloys: Influence of Heat and Thermomechanical Treatments, *Advanced Engineering Materials*, 4(7), 437-451, 2002.

[7] Apaev, B. A. & Voronenko, B. I., The Memory Effect in Alloys, *Metal Science and Heat Treatment*, 15(1), 24-29, 1973.

[8] Brandes, E. A., ed., The dependence of the transformation temperature MS on composition of Ti-Ni alloys, after K. N. Melton., *Smithells Metals Reference Book*, 6th edition, Butterworths, London, UK (1983): 1532-36.

[9] Chen-Luen, S. et al., A Robust Co-Sputtering Fabrication Procedure for TiNi Shape Memory Alloys for MEMS., *IEEE Journal of Microelectricalmechanical Systems* 10(1), 69-79, 2001.

[10] Sanjabi, S.; Sadrnezhaad, S. K.; Yates, K. A. & Barber, Z. H., Growth and characterization of Ti_xNi_{1-x} shape memory alloy thin films using simultaneous sputter deposition from separate elemental targets, *Thin Solid Films*, 491, 190-196, 2005.

[11] Brinson, L.C., One-Dimensional Behavior of Shape Memory Alloys: Thermomechanical Derivation with Non-Constant Material Functions and Redefined Martensite Internal Variable, *Journal of Intelligent Material Systems and Structures*, 4, 229-242, 1993.

Development of A 4x4 Hybrid Optical Switch

Bo-Ting Liao and Yao-Joe Yang*

Department of Mechanical Engineering, National Taiwan University
No. 1 Roosevelt Rd., Sec. 4, Taipei, Taiwan, ROC
*E-mail : yjy@ntu.edu.tw

ABSTRACT

In this paper, we present a practical approach to realize a highly accurate but low-cost hybrid optical switch. This hybrid optical switch is composed of a MEMS-based silicon micro-mirror-array structure and a mini-actuator array. The silicon micro-mirror-array structure, which includes vertical mirrors, cantilevers, and trenches for passing light beams, are fabricated by using simple KOH etching process. The wet anisotropic silicon etching technique that is employed in this work greatly reduces the complexity of the fabrication process and thus gives higher fabrication yield. The mini-actuator array consists of commercially-available electromagnetic bi-stable mini-relays and L-shape arms. When the L-shape arm does not contact the cantilever, the mirror that is under zero external force can precisely reflect the light beam. When the L-shape arm pushes up the mirror, the light beam can pass under the mirror. The main advantages of this proposed design include high precision, easy for fiber alignment, high fabrication yield as well as low cost.

Keywords: optical MEMS, optical switches, wet anisotropic etching, bi-stability, self-alignment.

1 INTRODUCTION

With the rapid growth of optical communication networks, the demand of a reliable high-capacity switching system greatly increases. Large-scale optical cross-connect switches are the key components for building complex switching systems with large-port capacities. Recently, MEMS technology has emerged to be the promising solution for developing optical cross-connect switches [1-3]. Optical switches often require the vertical smooth mirrors to reflect optical signals. Silicon mirrors fabricated by using MEMS technology are widely employed for the optical applications. Figure 1 shows the schematic of a typical MEMS optical cross-connect switch. Deep reactive ion etching (DRIE) and wet anisotropic etching are the most popular micromachining techniques to fabricate the silicon mirror. DRIE technique is usually utilized to create high-aspect-ratio structures and is not restricted by the crystal orientation of the silicon wafer. However, DRIE technique is much more expensive than wet anisotropic etching.

Vertical smooth mirrors that are formed by $\{111\}$ planes can be created on a (110) silicon wafer [4]. However, the etched shapes are strongly restricted by the crystal orientation of the silicon wafer. For example, vertical (111) mirrors and V-grooves cannot be simultaneously created on a (110) silicon wafer. Helin *et al.* [5] reported a self-aligned micromachining process that can simultaneously create vertical (100) mirrors and V-grooves on a (100) silicon wafer.

In this work, silicon micro-mirror-array structures for 4x4 hybrid optical switches are designed and fabricated by employing the concept of the self-aligned micromachining process. With our proposed approach, it is possible to simplify the fiber alignment procedure of the packaging process as well as realize an optical switch that possesses the advantages of high accuracy, low cost and high fabrication yield.

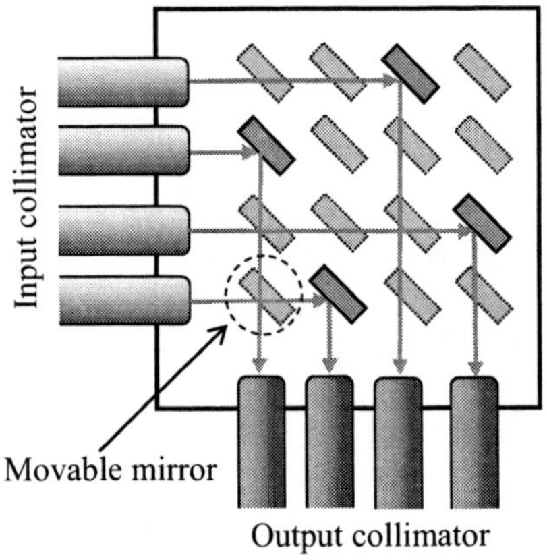

Figure 1 The typical design of the 4x4 optical switch.

2 DESIGN AND FABRICATION

Figure 2 depicts the design of the mask openings for fabricating the micro-mirror-array structures, including vertical mirrors, cantilevers, and light paths. The orientation of the silicon wafer is $<100>$ direction. As shown in Figure 2(b), because the angle between $<100>$ and $<110>$ crystallographic directions is 45°, the mask patterns for each vertical mirror are designed to be aligned with the $<100>$ direction and mask patterns for the light paths are designed to be aligned with the < 110 > direction. The mirror surfaces are designed to be on {100} planes, so the lateral etching rate toward mirror surface is equal to the vertical etching rate toward the wafer surface. Therefore, the surfaces of the etched vertical mirrors are formed on {100} planes and thus can be self-aligned with the optical paths that are along < 110 > direction.

This etching process is also simulated by using the process simulator Etch3D®. With the mask patterns shown in Figure 2(b) and 2(c), the vertical mirror, the cantilever and the light paths can be simultaneously created on a (100) silicon wafer shown in Figure 2(d). The (100) surfaces of the fabricated mirrors are self-aligned to the light paths and thus potentially can reduce the complexity of the alignment procedure.

The vertical mirrors can be individually actuated by the mini-actuator array. The mini-actuator composes of the L-shape arm and the commercially-available electromagnetic bi-stable mini-relays. Each mini-actuator can retain the mirror at either one of the two stable positions without consuming any electrical power, as shown in Figure 3. At the first stable position, the L-shape arm does not contact the cantilever and thus the optical signal can be reflected by the mirror, as shown in Figure 3(a). At the second stable position, the cantilever is pushed up by the L-shape arm and therefore the optical signal can pass under the mirror, as shown in Figure 3(b). It has to be emphasized that the mirror reflects the light beam only when it is under zero external force, so that the stress-free single-crystal silicon mirror structure can precisely reflect the light beam.

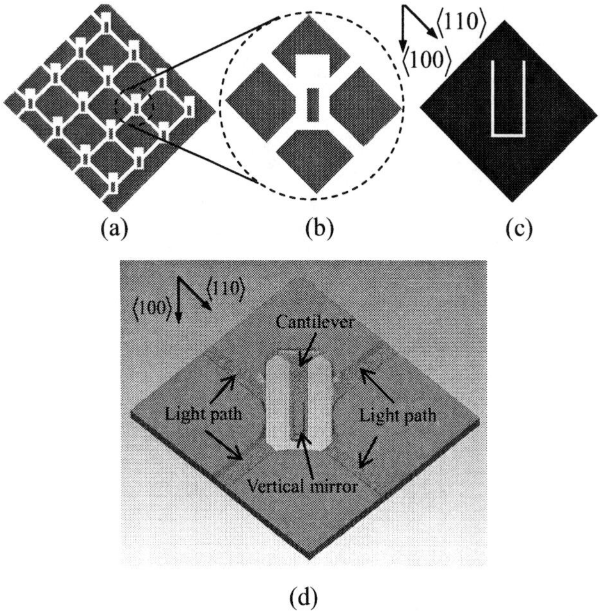

Figure 2. (a) Mask openings for the micro-mirror-array structures. (b) Enlarged view of the mask opening for one mirror. (c) Backside mask openings for the cantilever. (d) Simulated results of the etching process.

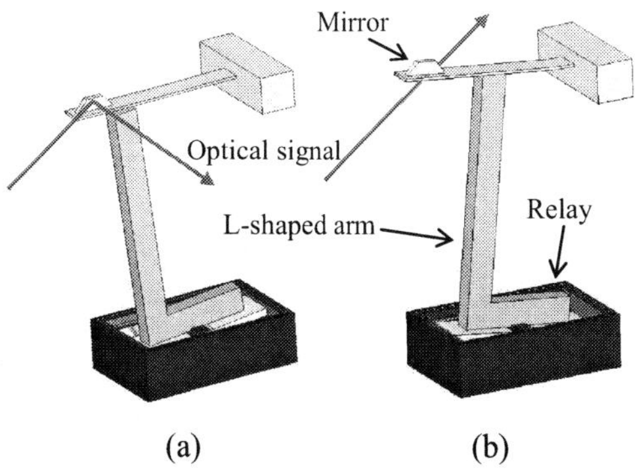

Figure 3. The operational principle of the mirror (a) The first stable position at which the L-shape arm does not contact the cantilever. (b) The second stable position at which the cantilever is pushed by the L-shape arm.

The fabrication process based on KOH etching is illustrated in Figure 4. The starting material is a (100) silicon wafer which is deposited with a silicon nitride layer on the both sides. The silicon nitride layers are used as the etching mask for the device structures. In order to achieve precise alignment with the crystal orientation of the (100) silicon wafer [6], a series of small circles are defined on the backside of the wafer and then etched by KOH etchant for a sufficient time to form pyramidal cavities. These pyramidal cavities will be used as the alignment marks for the subsequent mask patterns of the cantilevers. After the mask openings of the cantilevers are defined on the backside of the wafer, the mask patterns of the mirrors and the optical paths are defined on the front-side of the wafer using a double-side mask aligner. The wafer that is patterned on both sides is then immersed in KOH etchant. Note that the patterns exposed to KOH etching are carefully designed by considering the lateral undercutting width during the etching process. Accordingly, the desired dimensions of the micro-mirror-array structures can be created after the etching process. It has to be emphasized that our proposed process does not require back-side protection of the wafer. As a result, the complexity of the fabrication process can be reduced and the process yield can be improved.

In addition, previous study [7] indicates that high KOH concentration and high etching temperature can lead to the formation of the vertical sidewall while low KOH concentration and low etching temperature will result in the inclined sidewalls on a (100) silicon wafer. On the other hand, KOH solution added with isopropyl alcohol (IPA) can reduce the roughness of the etched surface. Therefore, the concentration of the KOH solution used in this work is 50wt% and isopropyl alcohol is added in the solution. Also, the etching temperature is 75℃.

The fabricated 4x4 micro-mirror-array structure is shown in Figure 5(a). The SEM picture of a fabricated cantilever with a vertical mirror is shown in Figure 5(b). Note that the sidewalls of the cantilever are not vertical due to the fact that the higher etching rate occurs on convex corners. After the device structures are created, the residual nitride layers are removed by using boiling phosphoric acid. Finally, a thin gold layer of about 3000Å is deposited on the device structures for improving the optical reflectivity of the mirrors. The typical dimensions of the fabricated mirrors and cantilevers are $1.5 \times 0.45 \times 0.1 mm^3$ and $6 \times 1.5 \times 0.06 mm^3$, respectively.

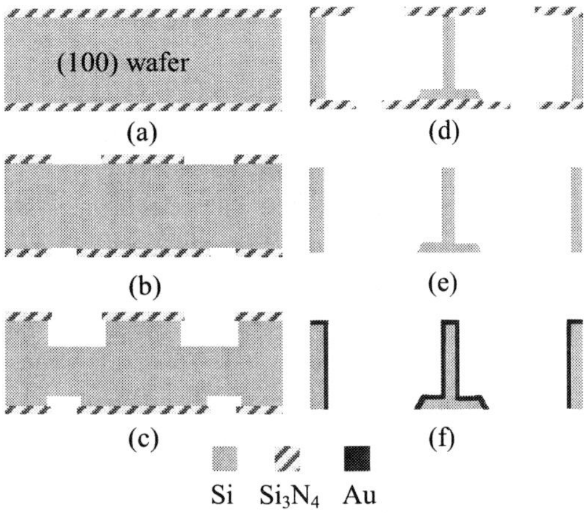

Figure 4. The fabrication process of the micro-mirror-array structures (a) The (100) silicon wafer is deposited with silicon nitride layers by LPCVD. (b) Patterns transferred. (c) The etched trenches are formed at the beginning of the KOH etching process. (d) The device structures are created after the KOH etching process. (e) The residual silicon nitride layers are removed by using boiling phosphoric acid. (f) The device structures are deposited with gold.

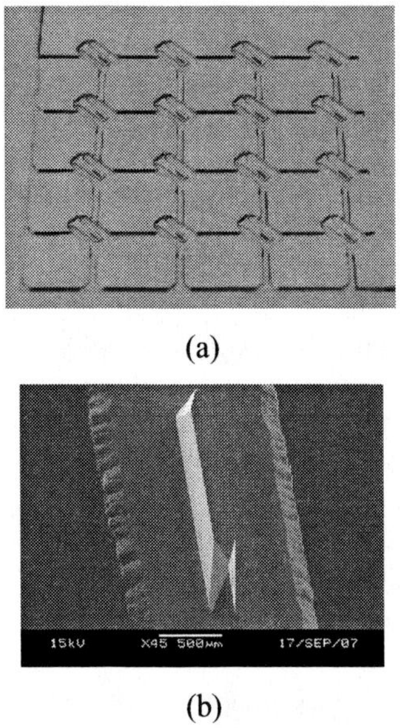

Figure 5. (a) The picture of the fabricated 4x4 micro-mirror-array structure. (b) The SEM picture of the vertical mirror and the cantilever.

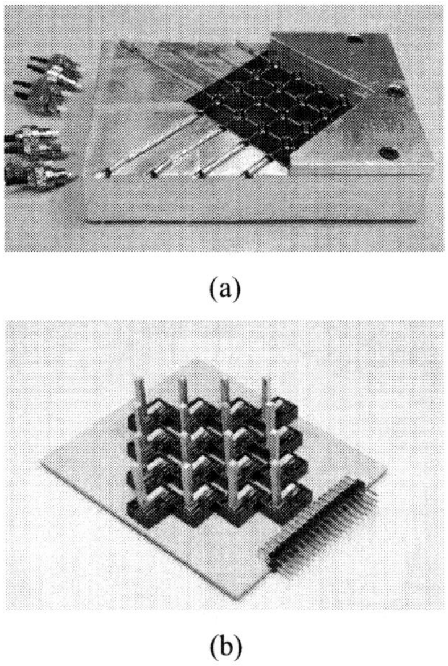

Figure 6. (a)The prototype of the packaged 4x4 hybrid optical switch (b)The relay-based actuator array.

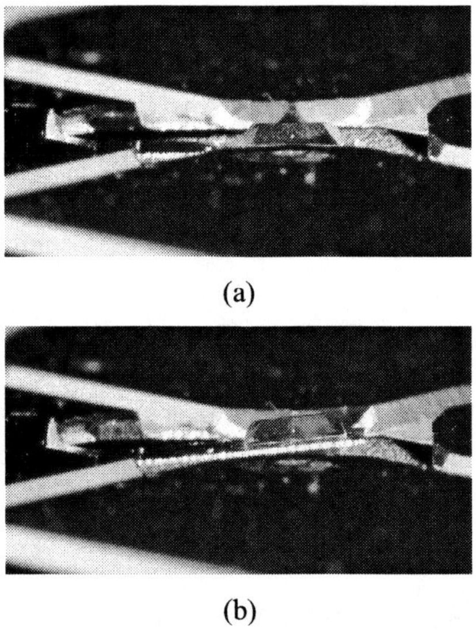

Figure 7. (a)The cantilever is not contacted by the L-shape arm and the mirror is at the first stable position. (b)The cantilever is pushed by the L-shape arm and the mirror is at the second stable position.

3 EXPERIMENTAL RESULTS

The prototype of a packaged 4x4 hybrid optical switch is shown in Figure 6(a). The silicon micro-mirror-array structure is fixed on the aluminum-made housing. The relay-based actuator array, as shown in Figure 6(b), is mounted underneath the micro-mirror-array structure. Due to the well-defined position of the vertical mirrors, the difficulty and complexity of the packaging process can be reduced. The input and output collimators pigtailed with FC/PC connectors are also assembled in the housing.

The commercially-available electromagnetic bi-stable mini-relays (TQ2-L2-5V, Panasonic) are used as the actuators in the array. The relays can be switched between the two stable states with a 5V input voltage. Figure 7(a) shows the mirror is at the first stable position, in which the L- shape arm does not contact the cantilever so that the light beam can be precisely reflected by the stress-free single-crystal silicon mirror. When the relay is in the second stable position, the mirror is pushed up by the L-shape arm, as shown in Figure 7(b). The measured displacement of the cantilever can be up to about 630μm, which is large enough for the light beam to fully pass under the mirror.

The optical performance of the device is measured at the wavelength of 1550nm. The alignment of the collimator is carried out by using six-axis positioners. An infrared card is also used to detect the position of the light spot to determine the optimal position of the collimator. The output optical signal is measured by the power meter to estimate the insertion loss. Once the collimators are aligned to achieve the lowest insertion loss, the collimators are glued on the housing by using UV-glue. The measured insertion losses are about -5dB. The excess losses may be caused by the roughness of the mirrors. Thus, the further study on reducing mirror roughness is now in progress.

4 CONCLUSIONS

The development of a 4x4 hybrid optical switch is presented in this work. This hybrid switch, which consists of a MEMS-based silicon micro-mirror-array structure and a mini-actuator array, possesses the advantages of high precision, high fabrication yield and low cost. The silicon micro-mirror-array structures are realized by using a simple KOH anisotropic etching technique. The vertical mirrors, cantilevers, and optical paths can be fabricated under high etching temperature as well as high etchant concentration. The mirrors are actuated by relay-based actuators and can retain at two stable positions. When the cantilevers are not pushed by the actuators, the mirrors that are under stress-free condition are able to precisely reflect the light beams. When the cantilevers are pushed by the actuators, the mirrors are moved to the second stable position which allows light beams passing through. The maximum measured displacement of the cantilever is about 630μm. The preliminary measurement results of the insertion loss is about -5dB.

ACKNOWLEDGEMENT

This work is sponsored by the National Science Council (contract number: NSC 96-2212-E-002-011).

REFERENCES

[1] D. A. Horsley, W. O. Davis, K. J. Hogan, M. R. Hart, E. C. Ying, M. Chaparala, B. Behin, M. J. Daneman, and M. H. Kiang, "Optical and mechanical performance of a novel magnetically actuated MEMS-based optical switch," *Journal of Microelectromechanical Systems.*, vol. 14, pp.274-284, 2005.

[2] Z. L. Huang, and J. Shen, "Latching micromagnetic optical switch", *Journal of Microelectromechanical Systems.*, vol. 15, pp.16-23, 2006.

[3] J. W. Liu, J. Z. Yu, S. W. Chen, Z. Y. Li, "Integrated folding 4x4 optical matrix switch with total internal reflection mirrors on SOI by anisotropic chemical etching", *IEEE Photonics Technology Letters*, vol. 17, pp.1187-1189, 2005.

[4] H. T. Hsieh, C. W. Chiu, T. Tsao, F. K. Jiang, and G. D. Su, "Low-actuation-voltage MEMS for 2-D optical switches", *Journal of Lightwave Technology*, vol. 24, pp. 4372-4379, 2006.

[5] P. Helin, M. Mita, T. Bourouina, G. Reyne, and H. Fujita, "Self-aligned micro-machining process for large-scale, free-space optical cross-connects", *Journal of Lightwave Technology*, vol. 18, pp. 1785-1791, 2000.

[6] G. Ensell, "Alignment of mask patterns to crystal orientation", *Sensors and Actuators A*, vol. 53, pp. 345-348, 1996.

[7] O. Powell, H. B. Harrison, "Anisotropic etching of {100} and {110} planes in (100) silicon", *Journal of Micromechanics and Microengineering*, vol. 11, pp. 217-220, 2001.

Stacked Coupled-Disk MEMS Resonators for RF Applications

Knut H. Nygaard[1], Christopher Grinde[2], Tor A. Fjeldly[3]

[1]Dept. of Engineering, Oslo University College, Oslo, Norway
[2]Dept. of Engineering, Vestfold University College, Horten, Norway
[3]Dept. of Electronics and Telecommunication, Norwegian University of Science and Technology, UniK – University graduate Center, Kjeller Norway

ABSTRACT

We present a novel, stacked coupled-disk resonator where the silicon discs vibrate in radial contour modes, actuated by electrodes wrapped around the disk periphery. The self-aligned, central stem mediates the coupling and also anchors the stack to the substrate. Operating near-identical discs in the fundamental mode, their acoustic nodes are located at the center, and the coupling strength is determined by the stem diameter. Hence, the stem radius determines the resulting splitting of the resonance. Here, we consider double-disk devices with a radius of 36 μm operating near 150 MHz. The fabrication process for the present resonator is discussed.

Keywords: MEMS resonator, coupled resonators, RF.

1 INTRODUCTION

RF MEMS resonators are of considerable interest for applications in numerous electronic systems, including filters, oscillators, mixers, channel selectors, etc., in areas such as signal processing and transmission [1]. Properties such as high Q, low power consumption, and prospects of on-chip integration pave the way for new, high-performance system architectures. It is, for example, possible to use a MEMS switchable channel bank in a transceiver, making significant power saving possible in addition to further size reduction. Further reduction in power consumption may be achieved by making use of a switchable MEMS resonator oscillator in combination with a micromechanical mixer-filter such as an RF channel selector. This approach may be refined by using a single MEMS resonator oscillator in combination with a micromechanical mixer-filter as an IF channel selector.

In the transmitter section of a transceiver, it is also possible to use the switchable MEMS resonator oscillator in combination with a MEMS RF channel selector before the antenna. MEMS filter banks may also be used as a part of a signal processing scheme by dividing up a given bandwidth into a number of subbands, where each is subject to subsampling. This reduces the bandwidth requirement before an optional AD conversion. Such an implementation saves both space and power.

For RF operation, miniaturized polysilicon disk resonators can be fabricated with resonance frequencies up the GHz range [1]. In particular, circular disk resonators operating in radial contour modes have been shown to combine RF operation with Q factors in the thousands, greatly outperforming comparable transistor implementations. The resonator is actuated by electrodes wrapped around the disk periphery. The self-aligned, central stem anchors the disk to the substrate. The high Q comes from the near-negligible damping from air, thermo-elastic effect, and surface effects. The dominant loss mechanism is instead losses through the stem and into the substrate [2,3]. But even this loss is quite limited since the stem sits at the central acoustic node of the disk. Even this loss may be further reduced by replacing the polysilicon disk with one made from polydiamond, which enhances the acoustic mismatch between the disk and the stem.

2 COUPLED DISK RESONATOR

Here, we propose a novel, stacked coupled-disk resonator where two identical silicon discs are attached to the same stem, as indicated schematically in Fig. 1. The stem mediates the mechanical coupling between the disks, resulting in a splitting of the resonances into a symmetric and an anti-symmetric mode. Because of the acoustic node at the center of the disks, the coupling strength, and therefore, also the mode splitting, will depend on the stem diameter. Hence, a slender stem will bring the modes closer together to give a band-pass filter characteristic, and a wider stem gives well-separated modes where the resonator may be operated as an oscillator at either of the two frequencies.

2.1 Equivalent circuit

The coupled resonator may be described in terms a simplified electric lumped-element equivalent circuit corresponding to that of a 3rd order band-pass filter, see Fig. 2. Here, the coupling is modeled by a shunt capacitor, while the two (nearly-identical) single resonators are shown as series LCR sub-circuits in the top branches of the equivalent circuit. Their elements L_R, C_R, and R_R are derived straightforwardly from the mechanical properties

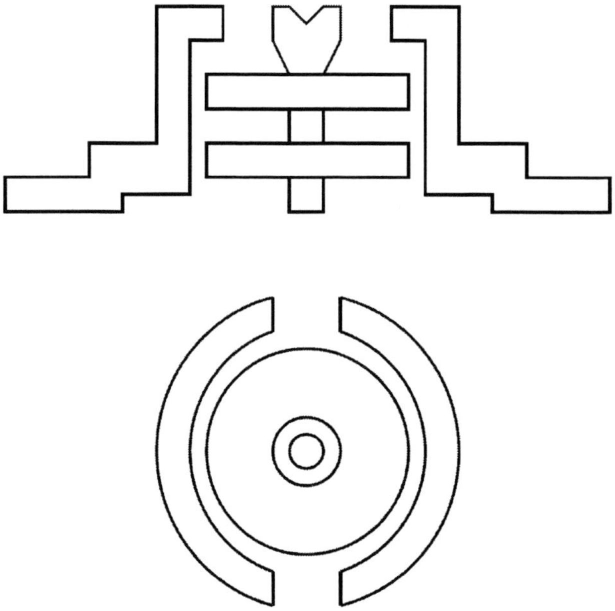

Figure 1: Schematic vertical (top) and horizontal (bottom) cross-sectional views of the proposed double disk resonator.

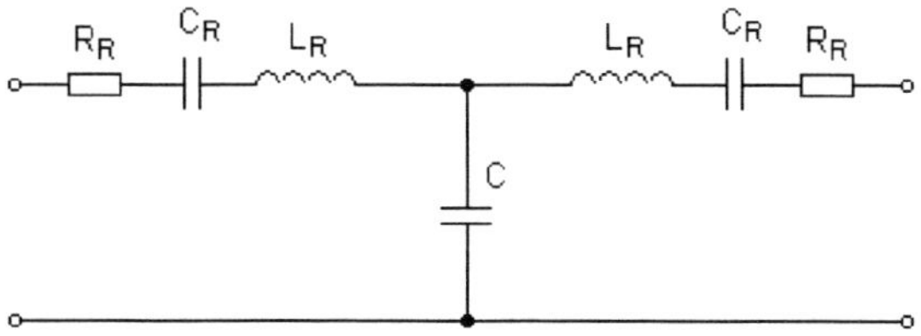

Figure 2: Equivalent circuit of the coupled double disk resonator consisting of the two single-resonator LRC equivalents on top and the coupling network in the middle.

of the resonators [1]. The series motional resistance R_R reflects the loss via the stem to the substrate, and is proportional to $1/Q$ [1]. The single-resonator parameter values are given by

$$L_R = \frac{m_{re}}{\eta_1\eta_2}; \quad C_R = \frac{\eta_1\eta_2}{k_{re}}; \quad R_R = \frac{c_{re}}{\eta_1\eta_2} = \frac{\sqrt{m_{re}k_{re}}}{Q\eta_1\eta_2} \qquad (4)$$

where m_{re}, k_{re} are the equivalent disk mass at the disk perimeter, the equivalent stiffness, and the equivalent electromechanical coupling between the electrodes and the disk, respectively. η_1 and η_2 are the electromechanical coupling factors related to the two ports.

The loss mechanism can be analyzed mechanically by considering separately the disks, the stem sections, and the substrate, and subsequently by connecting them in a self-consistent manner [3]. The vertical vibration of the disk centers propagate throughout the double disk assembly. A part of the associated vibrational energy acts to couple the two disks though the upper stem, giving rise to the coupling capacitor C in equivalent circuit of Fig. 2. Some of energy is transferred to a anchor point where it causes an oscillating deformation of the substrate. We note that the substrate deforms easily because of the relatively low stiffness of the 2 µm SiO_2 layer on top of the silicon substrate (see below). The deformation in the lower stem is small compared to this and can be neglected to lowest order.

The ratio between the reactance of the shunt capacitor C and that of the single-resonator reactance determines the pass-band ripple in the filter. A weak coupling (small C) gives a small mode separation and little ripple. Optimal pass-band filter characteristics are ideally obtained when the total resistance $R_R + R_G$ on the generator (input) side and $R_R + R_L$ on the load (output) side satisfy the requirements for standard filter types such as, for example, Butterworth or Chebyshev filters. However, owing to mechanical constraints, such as the large values of R_R in practical realizations, it may be difficult to achieve ideal filter characteristics, except when the separation between the two vibrational modes is quite small.

Here, we specifically consider a double-disk device with disk diameter of 36 µm, disk thickness of 1.8 µm, stem diameter of 2 µm, and disk-to-electrode gap of 87 nm. The single resonators operate in a fundamental mode near 150 MHz, corresponding well to the LC-product obtained from (4).

Comparisons between simulations based on the presents equivalent circuit, adjusted for a small asymmetry in resonator values, and physical simulations using COMSOL Multiphysics, show that the value of C is about $2x10^{-15}$ F. This gives a mode separation of 41 kHz, as indicated in Fig. 3. From the same comparison we find a Q of about 8000.

In practice, a still lower value of Q (and a higher value of R_R can be expected. This tends to flatten the response between the peaks, to give better pass-band characteristics. A example of a simulation for this case based on the equivalent circuit is shown in Fig. 4.

3 FABRICATION

The process to fabricate a stacked double-disk structure is an extension of that given in [1]. The cross section with the various layers of the device prior to release etch is shown in Fig.3. Starting with a 100 mm Okmetic, <100> n-type 1-5 Ohm cm wafers, the first step is to make a ground plane using $PoCl_3$ doping. This layer is separated from the active components by a thermal oxide about 2µm thick and a layer of 350 nm silicon nitride. Holes through the nitride and oxide for connecting to the ground plane are etched using a combination of wet and dry etches. A 350 nm layer of polysilicon for inter-connects, is deposited, $PoCl_3$ doped, and patterned.

To form the disk structures and the spacers between, three alternating layers of 0.8 µm PVD oxide and two layers of polysilicon are deposited. The polysilicon layers

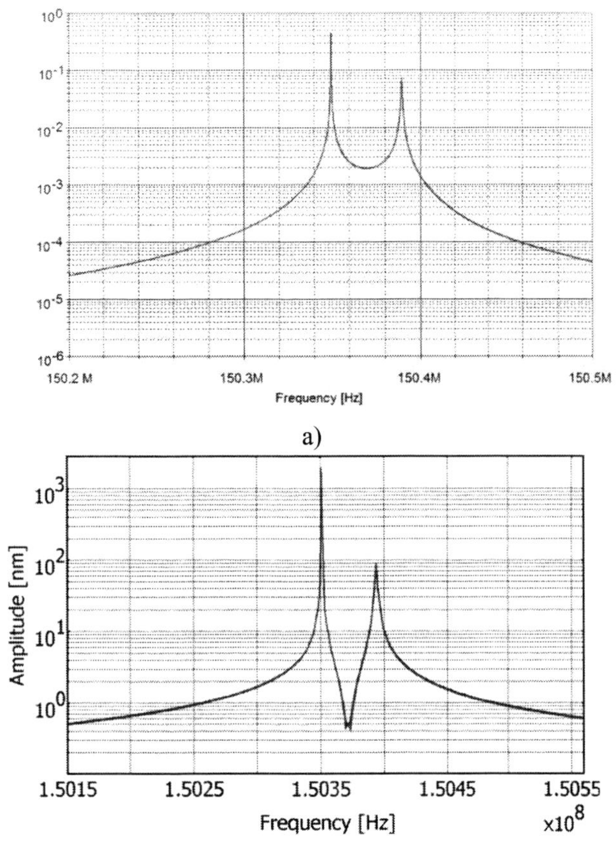

Figure 3: Modeled (a) and simulated (COMSOL) (b) transfer characteristics of the coupled double-disk resonator with disk diameters of 36 μm and stem diameter of 2 μm.

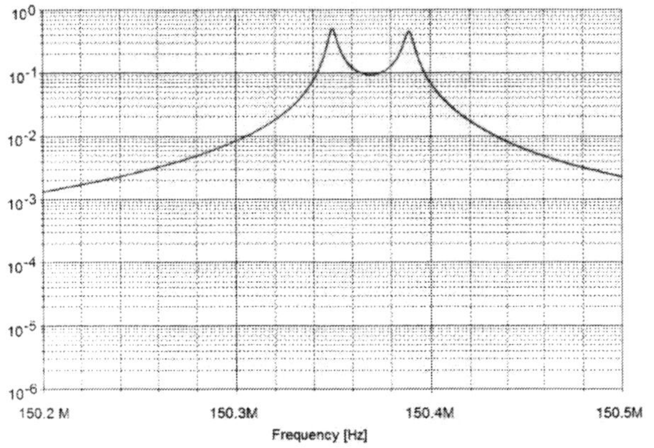

Figure 4: Modeled transfer characteristics of the coupled double-disk resonator with a reduced value of Q.

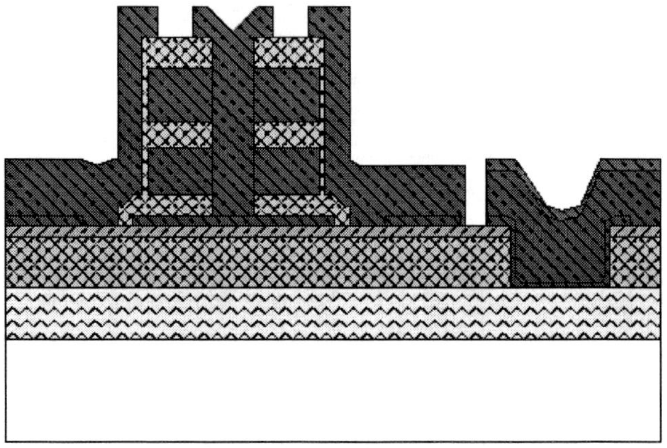

Figure 4: Idealized cross section if the device prior to etching the sacrificial oxide.

are deposited in two 0.75 μm layers with an intermediate $PoCl_3$ diffusion doping step. A sequence of DRIE steps terminating towards the bottom layer of sacrificial oxide is then employed to form the circular disks and the centre hole for the stem.

To form the 100 nm electrode gap, a thermal oxide is grown. In order to remove the oxide in the centre hole, this oxide is patterned and etched with a isotropic dry etch using a negative resist. This is necessary because it would be difficult to expose resist at the bottom of the small diameter centre hole.

A 0.8 μm layer of polysilicon is then deposited in two steps with an intermediate $PoCl_3$ doping to form the electrodes and stem before a thin layer of aluminum for the bonding pads is sputtered and patterned. Using DRIE the last polysilicon layer is etched. Finally, all exposed oxide is etched using HF-vapor phase etching. Processing of devices is in progress at the SINTEF, MinaLAB.

ACKOWLEDGEMENTS

This work was supported by the Norwegian Research Council under contract No. 159559/130 (SMIDA).

REFERENCES

[1] J. Wang, Z. Ren, C.T.-C. Nguyen, IEEE Trans. Ultrason., Ferroelect. and Freq. Contr., vol 51, pp. 1607-1628, 2004.

[2] D. S. Bindel and S. Govindjee, Int. J. Numer. Meth. Engng, vol. 64, pp. 789-818, 2005. 18th IEEE Int. Conf. Micro Electro Mechanical Syst., 2005, pp. 133-136.

[3] Z. Hao and F. Ayazi, Sensors and Actuators A, Vol 134, pp. 582-593, 2007.

Evidence of the existence of complete phononic band gaps in phononic crystal plates

Saeed Mohammadi*, Ali Asghar Eftekhar**, Abdelkrim Khelif**, William D. Hunt*, and Ali Adibi*

*Georgia Institute of Technology, Atlanta, GA, USA, saeedm@gatech.edu, eftekhar@ece.gatech.edu, bill.hunt@ece.gatech.edu, adibi@ece.gatech.edu
**Institut FEMTO-ST, Besancon Cedex, France, abdelkrim.khelif@femto-st.fr

Abstract: **We show, by measuring the transmission through a phononic crystal (PC) plate (slab), the evidence of the existence of large phononic band gaps (PBGs) in two-dimensional PCs made by a hexagonal (honeycomb) array of holes etched through a free standing plate of silicon (Si). A CMOS compatible fabrication process is used on a Si on insulator (SOI) substrate to realize the devices. More than 30dB attenuation is observed for eight periods of the hexagonal lattice a very high frequency (VHF) region that matches very well with the theoretical predicted complete PBG using plane wave expansion (PWE) method. This result opens a new direction in the implementation of high frequency practical PC structures with a possible superior performance over the conventional micromechanical devices used in a variety of applications especially in wireless communication devices, and sensing systems.**

Index Terms — **acoustic devices, filtering, phononic crystals, micromachining, phonons**

1 INTRODUCTION

Phononic crystals (PnCs) [1], [2] are special types of inhomogeneous materials with periodical variations in their elastic properties. PCs to phonons are as photonic crystals are to photons and as semiconductors are to electrons. One of the most interesting phenomena that can be obtained in the PC structure is the existence of frequency ranges in which elastic waves are prohibited from propagation. The existence of these frequency ranges, called phononic band gaps (PBGs), is very important as it can be used to realize fundamental functionalities like mirroring, guiding, entrapment, and filtering for acoustic/elastic waves by creating defects in the PC structure [3]-[5], as in the case of photonic crystals. Possibility of implementation of these functionalities in the PC structures can lead to integrated acoustic devices with superior performance over the conventional electromechanical devices used in wireless communication and sensing systems.

PCs can be categorized as one dimensional (1D), two dimensional (2D), and three dimensional (3D) based on the number of dimensions in which periodicity applies. While 1D PCs provide limited capability to control the elastic waves, 3D PCs are normally found to be difficult or impractical to fabricate. 2D PC structures, however, can provide moderate ability to control the elastic energy while being relatively easier to fabricate. Most of the studies on 2D PCs have been dedicated to the structures that are infinite, or very large in the third dimension, but such structures are not practical to be used in off-the-shelf integrated devices due to their large sizes, and are mostly designed to work at low frequencies (<5MHz).

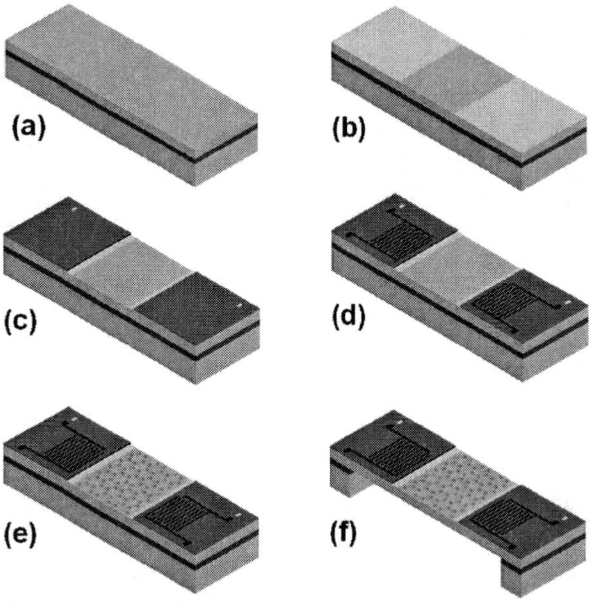

Figure 1. Fabrication steps for Si PC plate characterization, (a) original SOI substrate, (b) lower metal is patterned (c) ZnO layer is deposited and patterned (d) top layer of metal is patterned (e) holes are etched through the device layer (f) lower substrate and the insulator layers are etched by the use of a backside etching technique.

Recently, there has been a growing interest in 2D PC structures that are semi-infinite in the third dimension, and the periodic lattice is used to modify the propagation of surface acoustic waves (SAWs) localized near the free surface of the structure [6]-[8]. Such a structure is interesting as it can be fabricated by the available micromachining technology; however, SAW PC devices are prone to suffer from loss in the form of coupling to the bulk waves in the semi-infinite region [10], [11]. To solve this problem, 2D PC plate (slab) structure was recently introduced [12], in which the third dimension of the structure is limited by two parallel free-standing surfaces, and it was proved that large PBGs are possible to obtain in such structure, though very different from the PBGs of the infinite 2D PC. The structure proposed in Reference [12], however is a solid/solid PC structure, and suffers from fabrication difficulties as high frequency structures with low loss is of interest. In Reference [13] it was shown that it is possible to achieve very large complete PBGs in hexagonal (honeycomb) lattice of perforated holes in a silicon (Si) plate. In this paper we show that such devices are realizable by the use of a CMOS compatible process, and can operate in very high frequencies (VHF), and beyond, and verify the existence of large PBGs previously predicted in Reference [13]. We use interdigital transducers (IDTs) to scan a large frequency range to measure the transmission of elastic waves through the PC structure. A large sharp drop in the transmission through the PC in the range of the predicted PBG is detected. Excellent agreement between theory and experiment is observed.

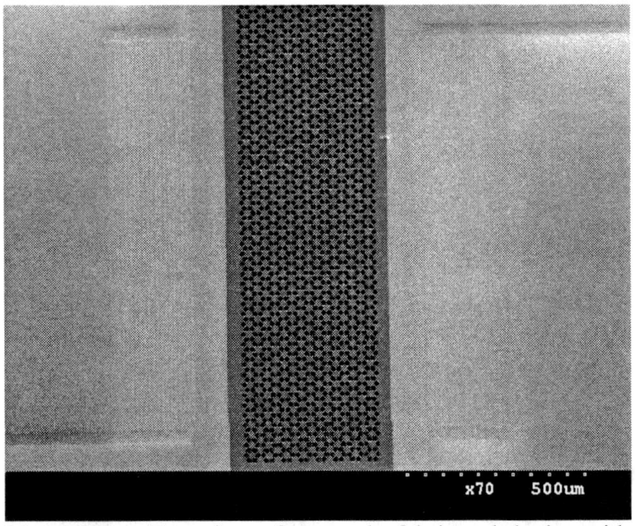

Figure 2. Top view of a sample fabricated device with PC region in the middle, and the transducer electrodes on each side.

2 DEVICE STRUCTURE AND THE FABRICATION PROCESS

The micromachining fabrication steps for devices used in this PnC characterization setup are shown in Figures 1(a)-(f). The process starts with a silicon on insulator (SOI) substrate as shown in Figure 1(a) with the Si layer thickness being 15μm, and a thin layer of metal (~100nm) is deposited and patterned on the Si layer followed by the radio frequency (RF) sputtering and patterning of a piezoelectric zinc oxide (ZnO) layer. Then a second layer of metal is patterned to form the excitation and detection transducers. The fabrication continues by etching the PC holes through the Si layer using a deep plasma etching technique, and finally the lowest substrate and the insulator layers are etched away to form the PC structure with the appropriate acoustic wave transmitters and receivers on the sides.

Scanning electron microscope (SEM) images of the top view, and cross sectional view of a typical device are shown in Figures 2, and 3 respectively. The PC is composed of eight periods of the hexagonal structure with its closest holes 15μm apart (a = 15μm), and the diameter of the holes is approximately 12.8μm as shown in Figure 3(a).

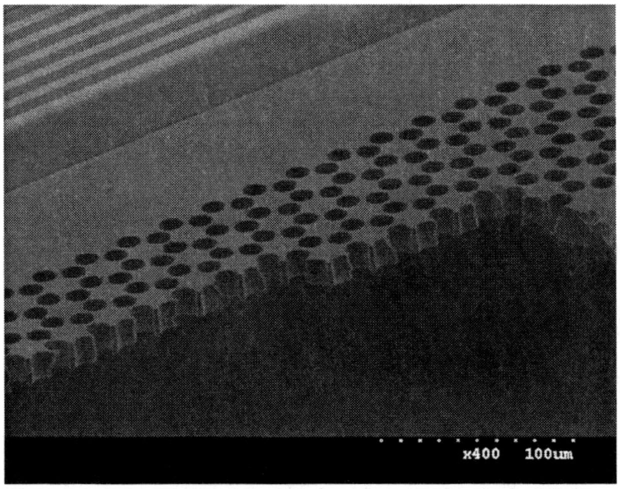

Figure 3. Cross sectional view of the fabricated structure.

Based on these geometrical values, and using the plane wave expansion (PWE) technique discussed in Reference [13] the band structure of the hexagonal Si PC is calculated and shown in Figure 4(c), while Figure 4(b) shows the irreducible part of the Brillouin zone. It can be inferred from Figure 4(c) that the full PBG predicted extends in the frequency range of 115MHz to 151MHz.

3 EXPERIMENTAL RESULTS

On the other hand, several transducers with different geometries are designed and fabricated to scan a wide frequency range around the expected PBG region. A different set of devices with the same frequency coverage, but without the PC holes etched were fabricated to act as a reference. A network analyzer is used to excite the structure by applying a high frequency electrical signal between the first and the second layer of metals producing a strong electric field across the ZnO layer. As the ZnO layer is sputtered by setting appropriate parameters to give highly oriented crystalline nano grains to show proper piezoelectric properties, the elastic energy is induced in the structure, and couples into a wide range of modes in the structure. Based on the spacing distances between the upper layer metal electrode fingers, the frequency of the excited modes would change. A wide range of electrode finger spacing distances is used to make sure various plate modes with frequencies within the PBG are excited. The signals transmitted through the PC structure are collected at each scanned frequency and normalized to the signal transmitted through an intact plate to get the frequency response of the PC structure. The normalized transmission response is shown in Figure 5. As it can be seen in Figure 5, more than 30dB drop in the transmission with a very sharp transition region is apparent in the 119MHz – 150MHz frequency range which is in excellent agreement with the results predicted by theory.

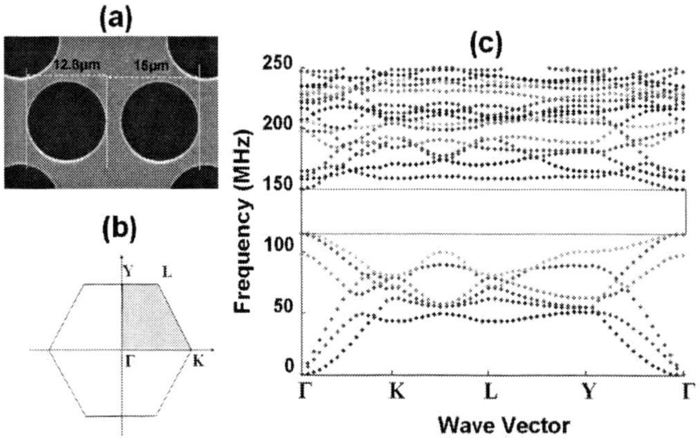

Figure 4. (a) Close SEM image of the PC structure indicating the lattice parameters, a = 15μm, and diameter of 12.8μm (r = 6.4μm). (b) Irreducible part of the Brillouin zone for hexagonal lattice in Si (c) Band structure of the hexagonal PC plate with $a = d =$ 15μm, and r = 12.8μm.

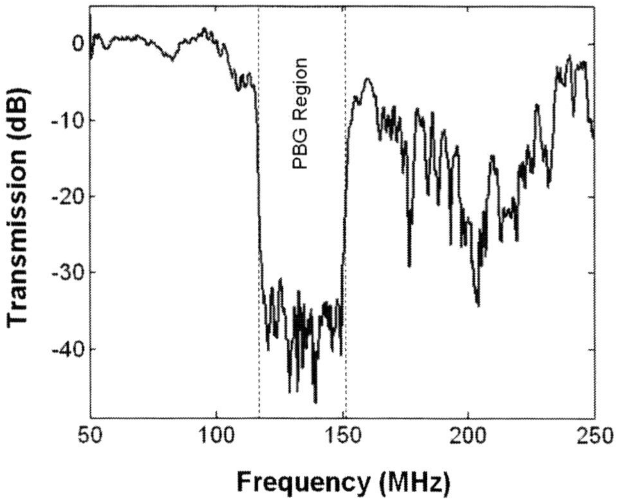

Figure 5. Average transmission through the PC structure as a function of frequency.

IV. CONCLUSION

We verified, by measuring the transmission through a PC plate made by etching eight layers of a hexagonal lattice of air filled holes through a free standing plate of Si, the existence of large PBGs in PC plates made of perforations through a solid plate. A CMOS compatible fabrication process is developed and used on a SOI substrate to realize the PC structure, and the appropriate transmission and receiving transducers. More than 30dB sharp attenuation is obtained within a frequency region that matches very well with the complete PBG region predicted by the PWE method. This result verifies the validity of the theoretical predictions of the existence of large complete PBGs in single crystalline PC plates. This result opens a new direction in the implementation of high-frequency, and high quality PC structures with a variety of applications especially in wireless communication devices and sensing systems.

Acknowledgement

This work was supported by National Science Foundation under Contract No. ECS-0524255 (L.Lunardi), and Office of Naval Research under Contract No. 21066WK (M. Specter). The authors wish to acknowledge Prof. Reza Abdolvand for his valuable inputs in the fabrication.

REFERENCES

[1] M. M. Sigalas, and E. N. Economou, “Elastic and acoustic wave band-structure,” J. sound and vibr., no. 158 (2), pp. 377-382, 1992.

[2] M. S. Kushwaha, P. Halevi, L. Dobrzynski, and B. Djafari-Rouhani, ”acoustic band-structure of periodic elastic composites,”phys. Rev. lett., no. 71 (13), pp. 2022-2025, September 1993.

[3] R. Martinez-Sala, J. Sancho, J. V. Sanchez, V. Gomez, J. Llinares, F. Meseguer, “Sound attenuation by sculpture,” Nature, 378, (6554), pp. 241-241, 1995.

[4] M. Sigalas, “Elastic wave band gaps and defect states in two-dimensional composites,” J. Acoust. Soc. Am., 101, (3), pp. 1256-1261, 1997.

[5] M. Torres, F. R. M. Montero de Espinoza, D. Garcia-Pablos, “Sonic band gaps in finite elastic media: surface states and localization phenomena in linear and point defects,” Phys. Rev. Lett., 82, (15), pp. 3054-3057, 1999.

[6] Y. Tanaka, S. Tamura, “Surface acoustic waves in two-dimensional periodic elastic structures,” Phys. Rev. B 58, 7958-7965, 1998.

[7] T. T. Wu, Z. G. Huang, S. Lin, ”Surface and bulk acoustic waves in two-dimensional phononic crystal consisting of materials with general anisotropy,” Phys. Rev. B 69, 094301, 2004.

[8] S. Mohammadi, A. Khelif, R. Westafer, E. Massey, W. D. Hunt, and A. Adibi, “Full band-gap silicon phononic crystals for surface acoustic waves,” Proceedings of the IMECE 2006, Chicago IL, Oct 2006.

[9] S. Benchabane, A. Khelif, J.-Y. Rauch, L. Robert, and V. Laude," Evidence for complete surface wave band gap in a piezoelectric phononic crystal," Phys Rev. E 73(6), 065601, 2006.

[10] J. H. Sun, T. T. Wu, "Propagation of surface acoustic waves through sharply bent two-dimensional phononic crystal waveguides using a finite-difference time-domain method," Phys. Rev. B 74, p. 174305 2006.

[11] Y. Tanaka, T. Yano, S. I. Tamura, "Surface guided waves in two-dimensional phononic crystals," Wave Motion 44, 501, June 2007.

[12] A. Khelif, B. Aoubiza, S. Mohammadi, A. Adibi, V. Laude, "Complete band gaps in two-dimensional phononic crystal slabs," Phys. Rev. E 74, 046610-1-5 2006.

[13] S. Mohammadi, A. A. Eftekhar, A. Khelif, H. Moubchir, R. Westafer, W. D. Hunt, A. Adibi, "Complete phononic bandgaps and bandgap maps in two-dimensional silicon phononic crystal plates," Electron. Lett. 43, 898, August 2007.

Detection of Plant Cell Compartments and Changes in Cell Dielectric due to Arsenic Absorption via Traveling Wave Dielectrophoresis

S. Bunthawin*, P. Wanichapichart** and A. Tuantranon***

* Membrane Science and Technology Research Center, Department of Physics, Prince of Songkla University, Thailand, sorawuth.b@psu.ac.th

**NANOTEC Center of Excellence at Prince of Songkla University, Thailand, pikul.v@psu.ac.th

***Nano-Electronics and MEMs Laboratory, National Electronics and Computer Technology Center, Thailand, adisorn.tuantranont@nectec.or.th

ABSTRACT

Changes in cell dielectric properties caused by arsenic absorption were detected by means of cell velocity and a frequency causing the cell being repelled from an octa-pair interdigitated electrode. This study found that the velocity spectrum was affected by solution conductivity of which the cells were suspending during the experimentation. An abrupt change in the velocity pattern explained non homogeneous phase, cell wall and the plasmalemma, only if the solution conductivity was small. There was a slightly shift of velocity spectrum towards a lower frequency value with respect to the pretreated arsenic levels. Utilizing our previous Laplace and RC models, curve-fittings with the experimental data revealed that the membrane conductivity was increased with the arsenic levels. Although, arsenic up to 100 ppm prevented cell growth but the velocity spectrum remained similarly to the living cell.

Keywords: cell dielectric property, traveling wave dielectrophoresis, velocity, critical frequency, arsenic

1 INTRODUCTION

Traveling wave dielectrophoresis has shown potential applications in medical diagnostics, drug delivery and cell therapeutics in terms of selectivity, isolation, concentration, purification and separation of bio-particles mixtures [1, 2]. Previous studies were reported using a planar linear interdigitated electrode of one array [1, 3, 4] or two parallel arrays [2, 5, 6], and the driven electric field was generated by sinusoidal quadrature-phase voltages. A phase sequence addressing to the electrodes had been described in details elsewhere [1, 3, 4, 5, 6]. For two parallel arrays interdigitated electrode (TPI), the Clasius-Mossotti factor (CMF) composed of real [Re(CMF)] and imaginary [Im(CMF)] function (see Fig. 1) is governed by complex conductivity and permittivity of the cell related with cell medium. These functions are frequency dependent and hence might affect the cell by either collecting it at the TPI or pushing it -

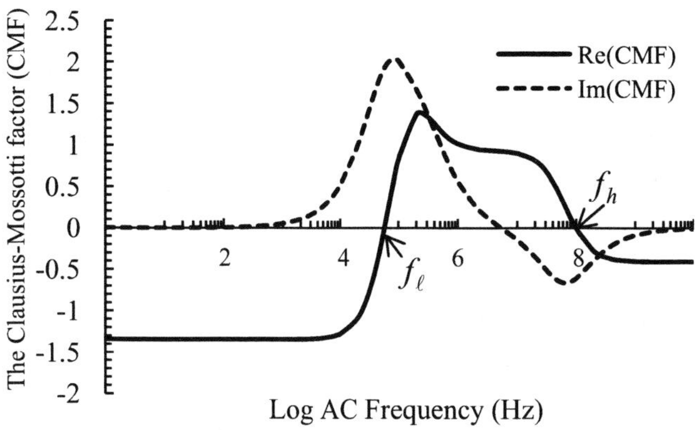

Figure 1: Theoretical plots for real [Re(CMF)] and imaginary [Im(CMF)] part of the CMF.

through electrode central channel, respectively. The negative value of the real function results in cell being repelled from the electrode, representing by a negative velocity, and vice versa. This allows cells with different properties could be separated by means of an appropriate frequency range. However, when dielectric values were predicted, it was time consuming and the model used was questionable, particularly when an ellipsoidal cell was tested. Therefore, our laboratory has investigated for a simpler method, suitable for cell technologists.

Mathematical model has further extended using Laplace method to explain cell velocity and RC methods to express two critical frequencies (Sakshin et. al., to be published). With the RC model, increasing the conductivity of cell suspending medium causes the two frequencies to converge and, finally, join at a critical conductivity, of which the cytoplasmic conductivity is revealed. This work observes cell velocity over a spectrum of field frequency and detects the two cross over values (f_ℓ, f_h). Phytoplankton with several arsenic pretreatment was used as a test model.

2 EXPERIMANTAL AND METHODS

Tetraselmis sp., a spheroid (10.0 ± 0.7 μm × 8.0 ± 0.5 μm), was obtained from the National Institute for Coastal Aqua-culture (NICA), Songkhla, Thailand. The cells were centrifuged and suspended twice in 0.5 M sorbitol using 7,000 rpm for 2 min. Conductivity of the solution (σ_s) was adjusted using 0.1 M KCℓ solution, and the conductivity was measured (Tetracon 325, LF318). An $NaAsO_2$ (Sigma-Aldrich, 99%) arsenic stock solution in distilled water was diluted to a desired concentration from 1-150 ppm for cell pretreatment. After 24 hr, cells were centrifuged and washed as described above. Cell density was 9.2×10^6 and 4.0×10^5 cell/mℓ during experimentation to avoid cell-to-cell interactions. A cell sample was dropped in the middle of a TPI electrode (see Fig. 2a). A commercial available glass slides of 80 × 30 × 1 mm dimension (Marienfeld, Germany) was photo masked and designed using a layout software (AutoCAD). Each gold bar was 200 μm long, 100 μm wide (d_1), and 0.2 μm thick (t). The separation of the adjacent bars on the same array (d_2) is 100 μm and that for the central channel (d_3) is 300 μm. The electrode was energized with four sinusoidal signals (a quadrature phase) of 1.4, 2.8 and 7.1 V (rms), in phase sequence as described by Wang *et al.* (1995). A function generator (Standford Research Systems, Model DS345, and California) was connected to a phase shift unit (PSU), sending 4 signals to an inter-junction unit (IJU) to control the phase sequence (Fig. 2b). During cells undergo dielectrophoresis, velocities ($\vec{v}_{DEP}$) were recorded using a CCD camera (Sony SLV-Japan) which was connected to a microcomputer (Acer, Aspire 4310). The Winfast PVR™ program was employed to store and display the recorded files. A digital stop-watch function of the Winfast program was used to determine cell velocities. Each experiment took place within 5 min., and the f_ℓ, f_h was simultaneously recorded. Electric field derived from an applied signal was numerical calculated through the Quick Field™ program version 5.5.

3 RESULTS AND DISCUSSION

From the Winfast program, all experiments described below perform cell translational velocity towards the TPI electrode in accord with the frequency used under selected σ_s. In all cases, only f_ℓ was detected, due to the limitation of our function generator and it was plotted against the solution conductivity. A series of experimental results was grouped as following.

3.1 Characteristics of Control Cells

When 9.2×10^6 cell/mℓ *Tetraselmis* sp. was subjected

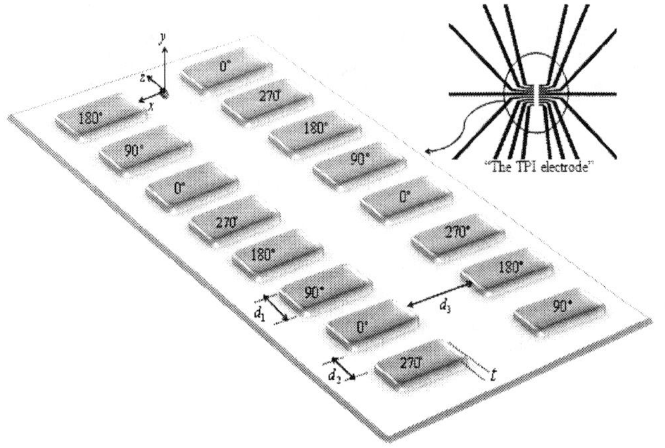

(a)

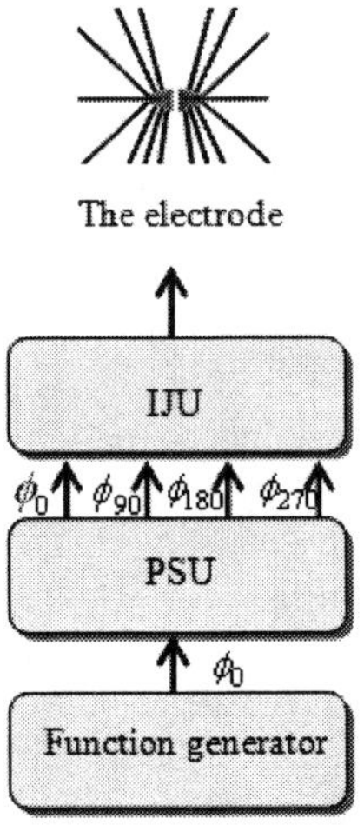

(b)

Figure 2: A configuration of the octa-pair interdigitated electrode and the quadrature phase sequence. (a) Three dimensional view of the electrode on a glass slide (not to scale) (b) Diagram of electrical set up.

to 28 kV.m^{-1} field strength (1.4 V signal), they underwent the positive velocity at frequencies between 50 kHz to 30 MHz. This was observed when varying the medium conductivity (σ_s) from 0.01 to 0.10 S.m^{-1}. Fig. 3a shows that an increasing in the σ_s reduced the velocity magnitude of the spectra and the observed f_ℓ was shifted to a higher value. As is seen, only when 0.01 S.m^{-1} is used it appears a sharp velocity peak at 9.4 μm.s^{-1}. This could be due to the field experiences a non homogeneous phase at about 200 kHz. In this case, it could be the outer shell of the cell (i.e. cell wall and the plasmalemma), which is normally much smaller in its conductivity compared to the cell interior. Evidences showing the finding of non homogeneous particle by electric field were previously reported [7, 8].When the σ_s is increased, the peak disappears and the spectra shows a bell shape with a plateau value of 5.2, 4.4, and 3.5 μm.s^{-1}, respectively. The frequency dependence velocity extends from 500 kHz to 30 MHz or more, since close to zero velocity at the higher critical frequency f_h is not found in all cases. The lower critical frequency f_ℓ is shifted to a higher value at 200, 300, and 500 kHz with respect to the increased σ_s. Fig. 3b shows that the velocity is proportional to the electric field while the f_ℓ is 300 kHz, independent of the increased field. This is also true for the higher σ_s used. Increasing the field strengths further increases the peak velocity to 29 μm.s^{-1}. This study found that minimum field should not be less than 28 kV.m^{-1} otherwise the velocity would be uncertain and rather difficult to measure. Fig. 3c shows that changing the cell density from 4.0×10^5 cell/mℓ to 9.2×10^6 cell/mℓ have no effect on the velocity spectrum. This confirms that electric field strength during experimentation was not affected by the presence of neighboring cells.

3.2 Arsenic Pretreatment

Velocity spectra of arsenic pretreated cells in Fig. 4a shows that greater arsenic content in the cells shifts f_ℓ towards a lower value while the peak velocity is not affected. Those lines are drawn to fit the experimental data, using the appropriate values of dielectric parameters, to verify our previous Laplace model. It appears that the frequency shifting to the lower value is caused by the arsenic and possibly causes pore blockage and, hence, increasing its electric permittivity. This result is in consistency with the f_ℓ *vs.* σ_s plots, using our RC model in Fig. 4b, since the same parameters were utilized fro curve fittings. Theoretically, it shows the same single f_h line, regardless of arsenic content. Noted that at 0.25 S.m^{-1}, the cell velocity was 4 μm.s^{-1} and further reduced towards a negligible value when 0.35 S.m^{-1} solution was used. This implied that critical conductivity of *Tetraselmis* sp. is 0.35 S.m^{-1}, indicating the conductivity of cytoplasm. Our previous results [9] showed that arsenic pretreatment also increased the membrane conductance similar to what found in dead cells, and cell proliferation ceased when the arsenic reached 100 ppm level (see Fig. 5). An experiment using dead cells showed that the f_ℓ diminished from the velocity spectrum. It is, therefore, ascertained that these arsenic pretreated cells are still alive, and the cessation of cell growth could be due to some difficulties in taking up nutrients from its environment. Normally, dead cells possess large membrane conductance, while the increased membrane conductance in this case is caused by an arsenic accumulation in the membrane.

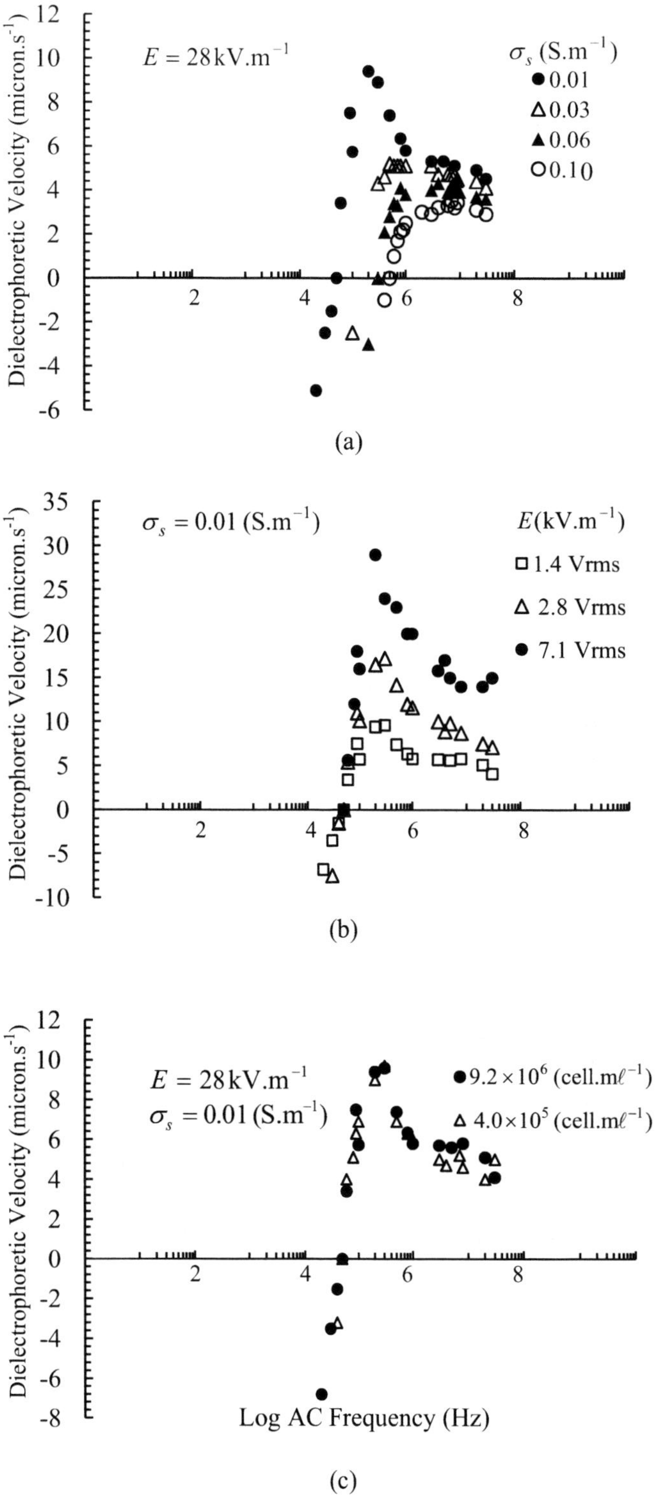

Figure 3: Velocity spectrum (a) effect of the conductivities of the suspending medium (b) effect of the electric field strengths and (c) effect of cell densities.

4 CONCLUSIONS

The home-made TPI electrode could detect the non - homoginuity of a test cell, i.e. cell wall and the plasmalemma of 10 nm thick, by showing a disruption of velocity spectrum only if 0.01 $S.m^{-1}$ sorbitol solution was used. The minimum field of 28 $kV.m^{-1}$ should be utilized for *Tetraselmic* cells so that the velocity was possible for the measurements. After arsenic absorption, the velocity spectrum remains in the similar manner as that of the control, except the membrane increases its conductivity due to arsenic blockage, followed by a cessation of growth if the arsenic level was as high as 100 ppm.

ACKNOWLEDGEMENT

This work is supported by Prince of Songkla University and the National Nanotechnology Center (NANOTEC), NSTDA, Ministry of Science and Technology, Thailand, through its program of Center of excellence network.

REFERENCES

[1] R. Pethig, M. S. Talary, and R. S. Lee, IEEE Engineering in medicine and biology magazine, November/December, 43-50, 2003.

[2] M. S. Talarly, J. P. H. Burt, J. A. Tame, and R. Pethig, J.Phys. D: Appl. Phys., 29, 2198-2203, 1996.

[3] M. P. Hughes, Nanotechnology, 11, 124-132, 2000.

[4] T.B. Jones, IEEE Engineering in medicine and biology magazine, November/December, 33-42, 2003.

[5] X. B. Wang, M. P. Hughes, Y. Huang, F. F. Becker and P. R. C Gascoyne, Biochimica et Biophysica Acta, 1243, 185-194, 1995.

[6] L.M. Fu, G.B. Lee, Y.H. Lin and R.J. Yang, J.IEEE/ ASME Trans. Mechatronics, 9(2), 377-383, 2004.

[7] [a]P. Wanichapichart, S. Bunthawin, A. Kaewpaiboon, and K. Kanchanopoom, Science Asia, 28, 113-119, 2002.

[8] [b]P. Wanichapichart, K. Maswiwat, and K. Kanchanapoom, J. Sci. Technol., 24, 799-806, 2002.

[9] P. Wanichapichart, T. Wongluksanapan , and L. Khooburat, Proceeding of the 2nd IEEE International Conference on Nano/Micro Engineering and Molecular Systems 2007, pp. 1115-1120.

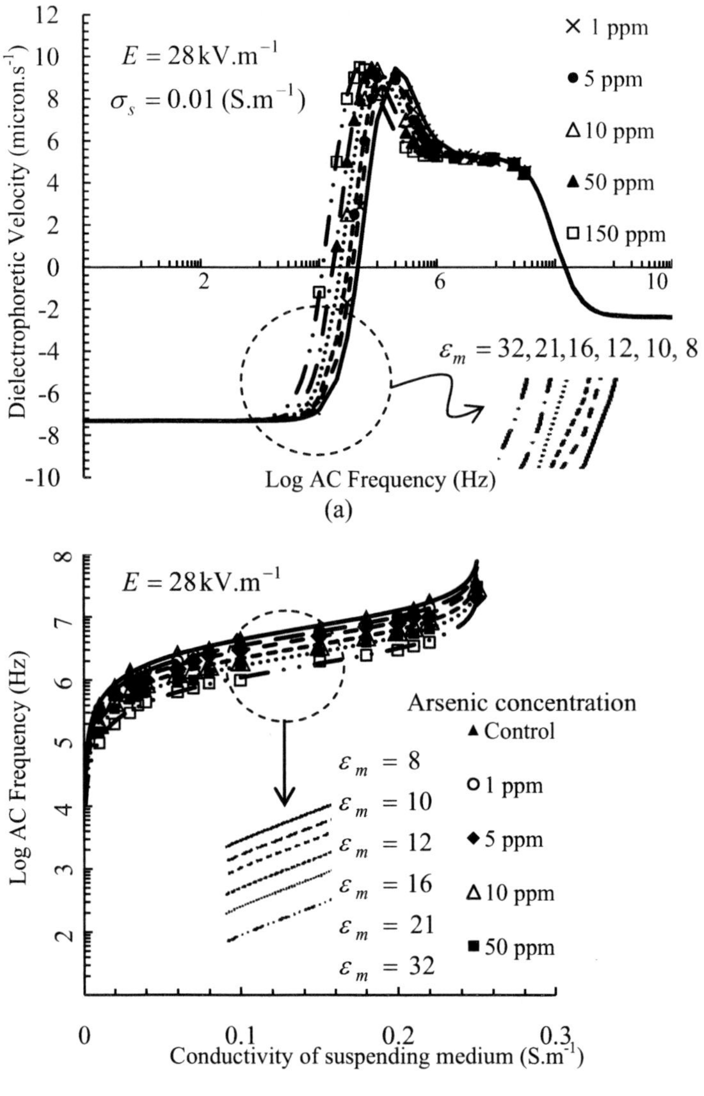

Figure 4: A comparison between control and arsenic pretreated cells (a) cell velocity (b) the lower critical frequency.

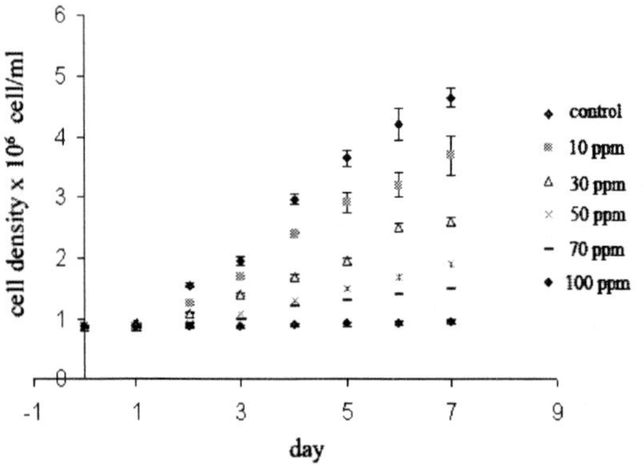

Figure 5: Test for viability by means of cell growth in accord with the arsenic levels after 24 hrs arsenic pretreatment.

Fabrication of nanoscale nozzle for electrostatic field induced inkjet head and test of drop-on-demand operation

Vu Dat Nguyen[*], Doyoung Byun*, Si Bui Quang Tran*, Michael Schrlau**, Haim H. Bau**, Jaeyong Choi*** and Sukhan Lee***

[*]Department of Aerospace and Information Engineering, Institute of Intelligent Vehicle and System Technology, Konkuk University, Republic of Korea
[**]Department of Mechanical Engineering and Applied Mechanics, University of Pennylvania, USA
[***] School of Information and Communication Engineering, Sungkyunkwan University, Republic of Korea,

ABSTRACT

This paper presents the fabrication of microscale to nanoscale nozzle and patterning by means of electrostatic field induced drop-on-demand inkjet printing system. Typically using the nanoscale nozzle we could eject nanoscale droplets and show feasibility to form patterns ranging from micro scale to nano scale on large area substrates at high speed.

Keywords: electrostatic field induced inkjet, micro to nanoscale pattern, drop-on-demand

1 INTRODUCTION

Printing technology is considered to be a key technology even in the fields of electronics [1-3], materials processing, and biotechnology [4]. However, limited edge resolution, thickness homogeneity, and overlay problems still restrict the conventional printing approaches to applications. In general, these limitations have precluded printing patterns smaller than 20 ㎛, and it seems that, despite the great pressure on low-cost manufacturing, these applications have not been implemented yet. Although nanoscale high accuracy is achievable with soft lithography [5], alignment issues and low printing speed remain a great challenge. There is a genuine need for nano-to-macro integration technology for printing. This paper presents the fabrication of nanoscale nozzle and patterning by means of electrostatic field induced drop-on-demand inkjet printing system [6,7], to overcome the limitation and problem of conventional thermal and piezoelectric print heads described above.

When the meniscus of liquid is subject to strong electric field, electric charge is induced on the meniscus and electrostatic force elongates the liquid to form a droplet, as shown in Fig. 1. The electrostatic field induced inkjet printhead can generate very tiny droplet smaller than nozzle size because the droplet is detached from the tip of the meniscus. Therefore, a sub-microscale nozzle is expected to generate nanoscale droplets and write any dot, line, or geometries by controlling the ejection of the droplet.

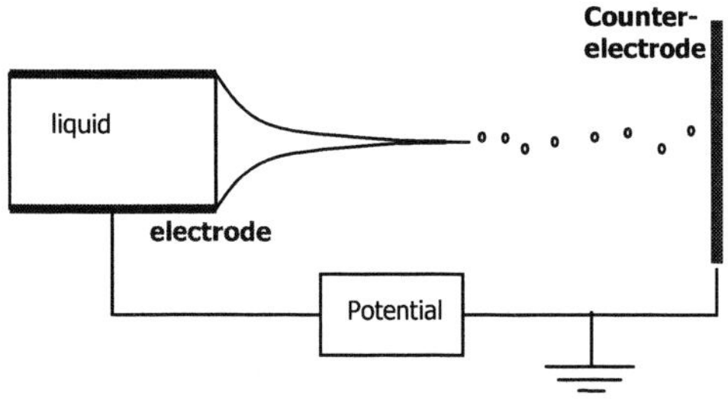

Figure 1 : Principle of electrostatic printing.

Current state-of-the-art electrostatic printing can reach to the resolution of hundreds nanometer [8]. To obtain such a high resolution the nozzle need to be placed as close to the substrate as the order of 100 μm. However, a very thin substrate is needed to get the support from a conductive backed layer. The substrate must be smooth to ensure a constant distance to the nozzle. Once the substrate is an insulator, the drop generation is influenced by suspended charged patterns. Therefore, high resolution electrostatic printing [8] requires either conductive substrates or a composition of a very thin insulator going with a conductive layer.

In this paper, microscale to nanoscale nozzles are fabricated and carry out patterning by means of electrostatic field induced drop-on-demand inkjet printing system. This system has ring-shaped electrode, which is able to focus the electric field at the centre and allow to fly the droplets through the hole and to deposit the following substrate

2 SIMULATION

A micro droplet is formed from the liquid meniscus due the induced electric field. The electric field is obtained by solving the following governing equations.

$$\nabla \cdot \varepsilon \nabla \varphi = -\rho \tag{1}$$

$$\bar{E} = -\nabla \varphi \tag{2}$$

where $\bar{E}$ denotes electric field vector, φ electric potential, ε the permittivity, and ρ charge density. These equations are solved by finite element solver, Comsol Multiphysics (Comsol Inc.) for an axi-symmetric 2 dimensional geometry.

In the direction of ejecting the micro droplet, the liquid meniscus is affected by two forces: an electric field force and a surface tension force [9]. We assumed that the shape of the meniscus is a sphere and the surface tension force f_1 and the electric force f_2 can be described as

$$f_1 = 2\pi r_c \sigma \quad (3)$$

$$f_2 = \int_0^{2\pi} \frac{1}{2} E \cos\theta \rho 2\pi r_d \sin\theta r_d d\theta \quad (4)$$

where E denotes electric field, σ is the fluid surface tension coefficient, ε_0 the permittivity of free space, r_c and r_d denote the radius of the nozzle and the droplet, respectively, and q denotes the net charge of the droplet.

E and ρ can be expressed by

$$E = \frac{q}{4\pi\varepsilon_0 r_d^2} \quad (5)$$

$$\rho = \frac{q}{4\pi r_d^2} \quad (6)$$

If the electric field force is balanced by surface tension force, a critical electric field is obtained [9].

$$E = 2\sqrt{\frac{\sigma r_c}{\varepsilon_0 r_d^2}} \quad (7)$$

Equation (7) indicates that when the electric field on the liquid surface is larger than the critical electric field, the liquid meniscus on the tip of the nozzle will become unstable and the tiny droplet may be generated by the instability.

From this rough calculation, a threshold value of operating voltage for jetting can be estimated. If the droplet radius is 300 μm and 3 μm, the critical electric fields from Equation (7) is 7.39×10^6 V/m and 7.39×10^7 V/m, respectively. Table 1 and 2 show the values of the maximum electric fields, obtained from FEM simulations, on the liquid meniscus according to the various voltages applied to the electrode for cases of 300 μm and 3 μm radii. By comparing the maximum electric fields in Table 1 and 2 with the critical electric fields, the threshold voltage can be estimated as around 2.5 kV and around 250 V.

Table 1. Values of maximum electric field (r_d=300μm, d=2 mm)

Voltage (V)	2500	2800	3000
Max. E (V/m)	7.19×10^6	8.05×10^6	8.63×10^6

Table 2. Values of maximum electric field (r_d=3μm, d=20 μm)

Voltage (V)	100	200	300
Max. E (V/m)	2.88×10^7	5.75×10^7	8.63×10^7

3 EXPERIMENTAL METHOD

3.1 Fabrication of microscale and nanoscale nozzles

On the pursuit of high resolution drop-on-demand patterning, we have fabricated quartz nozzles with inner diameters ranging from several micrometers down to sub-micrometer level. In order to prepare the nanoscale nozzle, a quartz capillary (ID. 700 μm) is pulled out to be extended and cut into micron or nano scale tip. Figure 2 shows a quartz tip with ID of 700 nm. Then, conductive material should be designed to be deposited on the surface of the capillary to enhance electrostatic force around the nozzle tip.

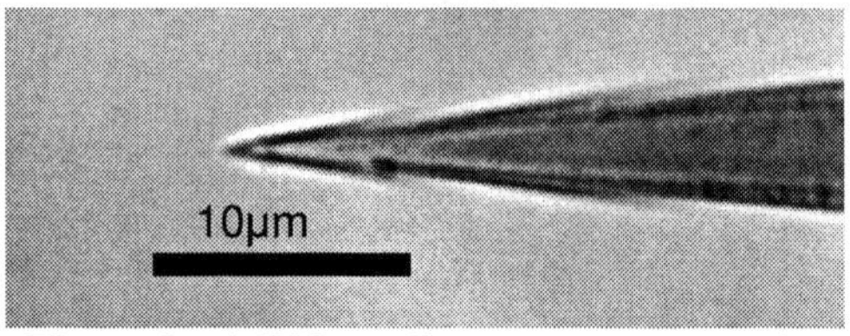

Figure 2 : A 700nm diameter quartz nozzle.

We tried to fabricate conductive nozzles by using three approaches. First, using e-beam evaporator (Thermionics), Au (150nm) layer is deposited on the outer surface of the nozzle, where NiCr (10nm) is coated to improve bonding strength between quartz and gold materials. Second approach is to fabricate a conductive layer in the quartz nozzle using silver nano-particle ink (TEC-IJ-010 from InkTec®). It is filled in the capillary nozzle and then is cured in the vacuum chamber under a temperature of 150 ^{0}C during 1 hour. Silver nano-particles are able to be sintered forming conductive wall inside the nozzle. Third, we used electroless plating [10]. For silver deposition, we prepared 0.2M silver nitrate ($AgNO_3$) in water, and 1.9M glucose ($C_6H_{12}O_6$) in a solution of 30 vol% methanol (CH_4O) and 70 vol% water as a reduction agent. Figure 3 shows the photographs of the coated nozzles.

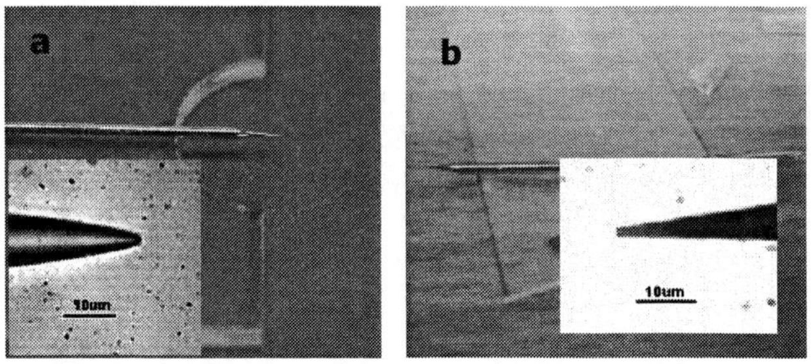

Figure 3 : Picture of tips coated with metal layer a) Au deposited outside b) Silver deposited inside the tips.

3.2 Experiment setup

To conduct a high resolution patterning experiment, a 2-dimensional x-y moving table is used. A conductive substrate is fixed on the moving table and the jetting system is located above the table. Metal ink (TEC-IJ-010 from InkTec®) is supplied through nozzle with flow rate ranging from 0.1 μL/min to 1 μL/min by a syringe pump. Negative potential that control drop-on-demand is provided by a function generator and high voltage amplifier, as shown in figure 4. The distance between the nozzle and the substrate is from 500 μm to 1000 μm. Small distance is favored in terms of accuracy and operating-voltage-reduction. Nevertheless, reducing the distance increases the chance of electrical breakdown. A clearance ranges from 500 μm to 1000 μm is recognized as the best working distance in our experiment. Reaching of charge droplets to the substrate can be detected with a current measurer (Keithley 6485 picoammeter).

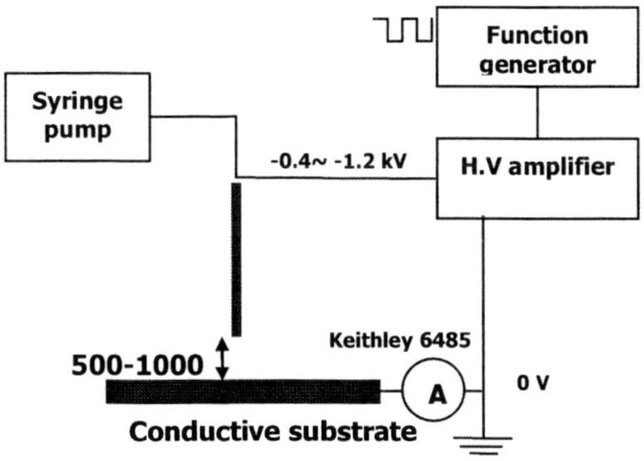

Figure 4 : Experiment setup for high precision patterning.

4 RESULTS AND DISCUSSIONS

4.1 Effects of deposition of metal layer

In this section, we examined the effects of gold exterior coated nozzle and silver interior deposited nozzle. For silver nozzle coated inside, we encounter difficulties from peeling off the silver materials near tip inside due to the electric stress. However, we have some successful jetting droplets and could decrease the operating voltage comparing with Au-coated tips outside. Figure 5 shows photographs of operating jets with the tips externally coated by Au (figure 8a) and internally coated by silver (figure 8b). While those two tips (ID 1 μm) are supplied with the same flow pressure and same distance between electrodes (1mm), the onset voltages are 800V for the tip coated with gold and 720V for the tip coated with silver. When the electrode is deposited outside of the tip, as shown in figure 5 (b), the liquid may be overflown and relatively large jet forms. Therefore, jetting characteristic depends much on the hydrophobic coating of the tip.

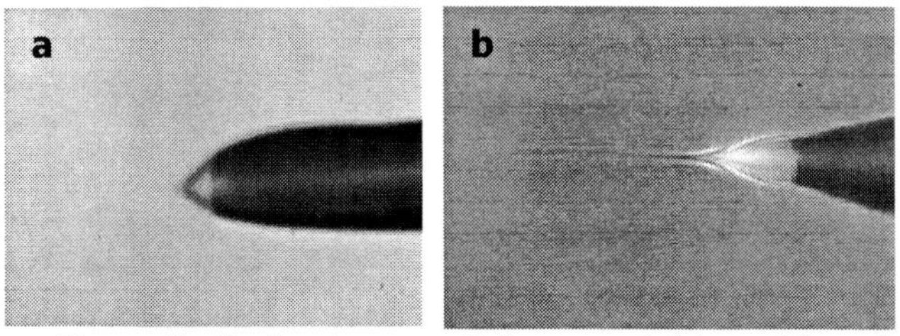

Figure 5 : Jetting images of tips coated (a) outside and (b) inside.

4.2 High resolution patterning experiments

For all patterning, we use nano silver ink (TEC-IJ-010, InkTec®) that contains nano silver particles. Figure 6 shows patterns of dots through a tip of 15 μm in diameter. The diameter of the droplets is around 20 μm, which is patterned on glass substrate. When we consider the hydrophilic surface of the substrate, the size of the droplet is estimated to be less than 10 μm.

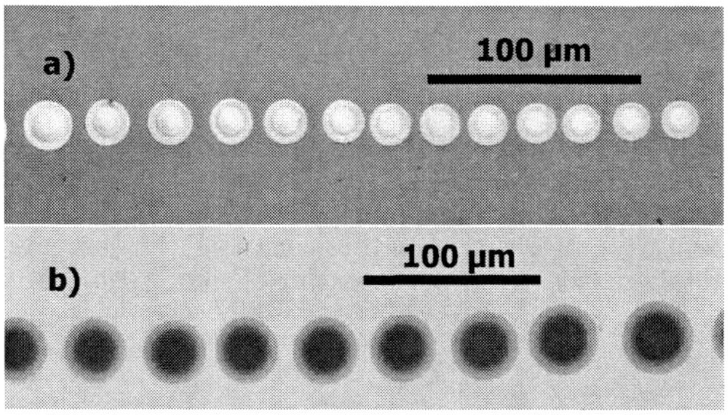

Figure 6 : Pattern dots through tip with 15 μm diameter, a) 10 Hz, and b) 5 Hz.

As the nozzle size scales down, the device is expected to generate fine droplets and offer higher resolution printing. However it is difficult to determine an optimal micro-dripping mode due to the lack of optical observation. Backing up with promising result that is up to 20 μm of resolution, we see the feasibility of scaling down the resolution to nano level as what have been obtained with the nozzle's size.

As the size of meniscus is getting smaller, the surface tension force that keeps the liquid in shape is also higher. When the meniscus breaks down to generate droplets under high electric field, the droplets are charged up with abundant amount of electric that surpasses the Rayleigh limit and are forced to divide into finer drops. Therefore we need further intensive study for investigating the optimal condition for micro-drippping mode. Figure 7 shows a line patterned by spray mode obtained from nozzle of 1 μm diameter. The drops are too fine to be spotted with optical range of our current device.

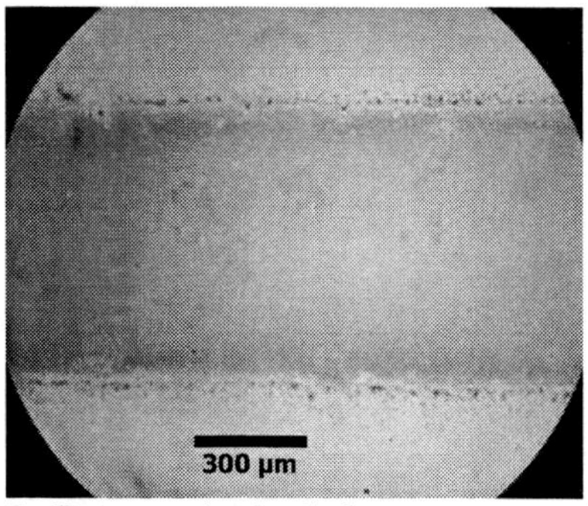

Figure 7. Pattern obtained from spray mode through nozzle of 1 μm in diameter.

In fact, jetting droplets and patterns may be unstable due to very small nozzle, liquid supply of small amount of liquid, alignment of nozzle and electrode. Furthermore, once the liquid is disconnected from the conductive edge of the tip, the edge can be stripped off under high electrostatic stress and lead to working failure. Even if a quartz tip is deposited with Au, high electric potential is applied to the deposit layer and breaks the bonding of metal layer and quartz tip. Therefore we need more researches for developing more stable coating technology and jetting devices.

5 CONCLUSION

In this paper, we show the feasibility of scaling down the nozzle for the electrostatic field induced inkjet head. Currently, we fabricated nozzle tips coated by metal inside or outside and examined high resolution printing. To utilize these tiny nozzles for nanoscale patterning, further examination need to be conducted to obtain the exact optimal working condition.

6 ACKNOWLEDGMENT

This work was supported by a grant from the National Research Laboratory program, Korea Science and Engineering Foundation Grant (R0A-2007-000-20012-0) and Seoul R&BD program. VDN acknowledges partial support from the Korea Research Foundation (KRF-2007-211-D00019).

REFERENCES

[1] B.-J. de Gans, P. C. Duineveld, U. S. Schubert, Adv. Mater. 2004, 16, 203.

[2] Arias, A. C.; Ready, S. E.; Lujan, R. A.; Wong, W. S.; Paul, K.; Chabinyc, M.; Salleo, A.; Street, R. A, Applied Physics Letters 2004, 85 (15), 3304.

[3] Y.Young Noh, Ni Zhao, M. Caironi, H. Sirringhaus, Nature Nanotechnology 2007, 2, 784 – 789.

[4] Maxim E. Kuil, J. P. Abrahams and J. C. M. Marijnissen, Biotech. Journal 2006, 1(9), 969-975.

[5] John, A. Rogers, Ralph G. Nuzzo, Materials Today, February, 50-56, 2005

[6] Sukhan Lee, Doyoung Byun, Daewon Jung, Jae Yong Choi, Yongjae Kim, Ji Hye Yang, Sang Uk Son, Bui Quang Tran Si, Han Seo Ko, Sensors and Actuators, A, Vol. 141, pp. 506-514, 2008.

[7] J.Y. Choi, Y.J. Kim, S.U. Son, Y.M. Kim, S.H. Lee, D.Y. Byun, H.S. Ko, Proceedings of NSTI-Nanotech 2007, pp.403-406

[8] J.U. Park, M. Hardy, S. J. Kang, K. Barton, K. Adair, D. Mukhopadhyay, C.Y. Lee, M. S. Strano, A.G. Alleyne, J.G. Georgiadis, P.M. Ferreira, and J.A. Rogers. High-resolution electrohydrodynamic jet printing, Nature materials, Vol. 6, pp. 782 – 789, 2007.

[9] J. Xiong, Z. Zhou, X. Ye, X. Wang, and Y. Feng, Research on electric field and electric breakdown problems of a micro-colloid thruster, Sensors and Actuators A, Vol. 89, pp. 159–165, 2003.

[10] N. Takeyasu, T. Tanaka, and S. Kawata, Metal Deposition Deep into Microstructure by Electroless Plating, Japanese Journal of Applied Physics, Vol. 44, No. 35, pp. L 1134– L 1137, 2005.

Non-Traditional Dicing of MEMS Devices

S. Sullivan*, T. Yoshikawa**

*DISCO Corp, Tokyo, Japan, scott_s@disco.co.jp
**DISCO HI-TEC America, Santa Clara, CA, USA, yoshikawa@discousa.com

ABSTRACT

Many of the processes that are used in the manufacture of MEMS are taken directly from IC fabrication. Singulating the wafer into individual MEMS is such a process. Most MEMS are singulated using diamond blade dicing barrowed from the IC process. MEMS are many times more sensitive to contamination, vibration, thermal, and electrical shock than ICs. This sensitivity requires modifications to the dicing process. These modifications are often improvements instead of solutions. SD laser dicing is a non traditional singulation process. A laser is focused below the surface of the substrate creating a modified layer. Optical and physical characteristics of the substrate dictate the type of laser used for the SD process. This paper will compare blade dicing to the SD laser process.

Keywords: MEMS, dicing, particles, stealth dicing

1 STEALTH DICING ENABLAING VOLUME PRODUCTION OF MEMS

Stealth Dicing (SD) is being accepted in the volume production of MEMS. As MEMS increasingly becoming a part of everyday life, manufacturing of the MEMS goes from low volume production, in wafer starts/month, to medium volume production.

Through quality improvements and lower cost of ownership SD aids in MEMS moving from R&D and low volume into full volume production. When production ramps up, yields must improve to allow for automated manufacturing. Along with improving yields is the push for lower costs.

According to the well respected Yole Developpement group the predicted CAGR of MEMS devices is 13%. That would give a $10.8 billion market in 2011. [1]

By being a dry and non-contact process SD eliminates or reduces the traditional yield issues of damage and particles associated with traditional blade dicing.

2 THE SD PROCESS

Laser light with a wavelength allowing for a balance between transmission and absorption is focused below the surface of the silicon wafer. At the point of focus the energy converts the single crystal silicon to a microcrystalline structure. This modification of the silicon creates a vertical separation between the two MEMS devices. The modified layer is termed the SD layer. Silicon thickness and a number of other factors such as edge quality determine if one or multiple SD layers will be created.

2.1 Structure of the SD Layer

Figure 1 below shows a single SD layer on a 50µm thick silicon sample. The SD layer is some distance below the surface. 15 micrometers of silicon has been modified vertically.

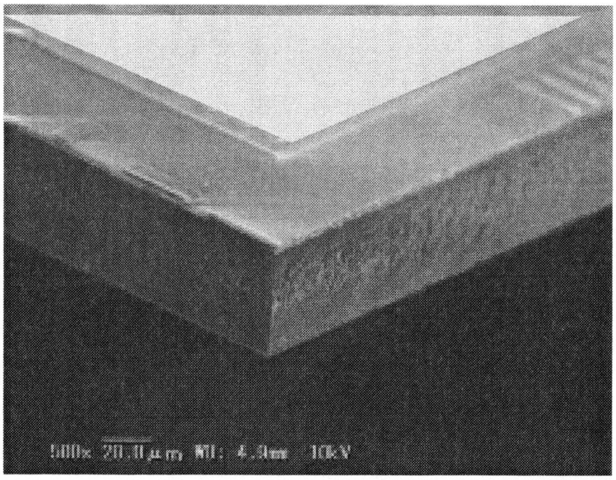

Figure 1: Edge view of a single SD layer.

Figure 2 below shows a thicker silicon sample. At 300μm thick a number of SD layers are necessary. In this case 7 are used.

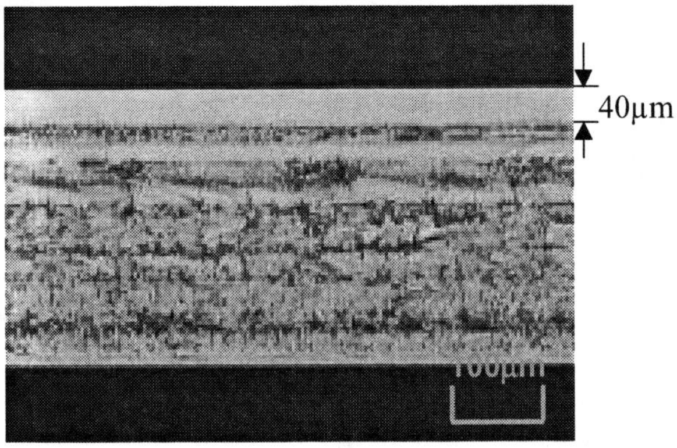

Figure 2: Edge view showing multiple SD layers.

In order to create a clean separation between the MEMS die, vertical height of the SD layer is many times larger than the horizontal component. Figure 3 shows the horizontal component or width of the SD layer.

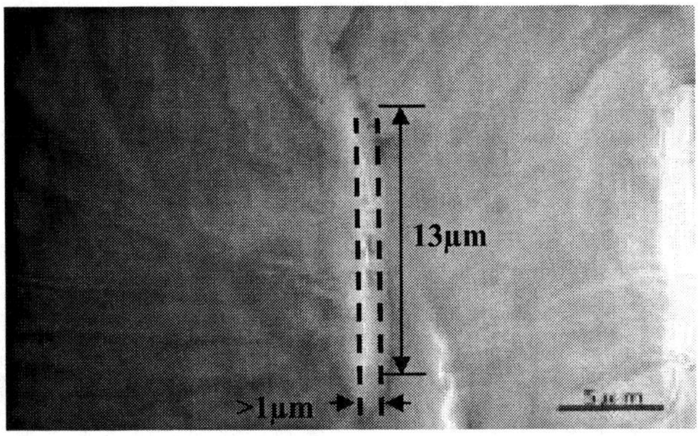

Figure 3: Cross section showing the width of the SD layer.

3 CONVENTIONAL BLADE DICING

Blade dicing is currently the most common way to separate a semiconductor wafer into devices. As with other processes that have been adapted from IC fabrication for the manufacture of MEMS fabrication some changes are necessary.

The typical IC wafer has a passivation layer protecting it from the environment. Additionally ICs are for the most part unaffected by moderate vibration, heating, and electrostatic discharge (ESD). This is not the case for most MEMS. For many MEMS being effected by the environment is their function, they need to have the active mechanical components exposed to the atmosphere. Cantilevers, gears, hinges, bridges, and membranes of the MEMS devices are often extremely fragile and have extremely fine movements. For example a resonator can have a peak-to-peak displacement of 10nm. During dicing, contamination, vibration, heating, and ESD can reduce the performance or cause the device to catastrophically fail.

3.1 Blade Dicing Contamination

Contamination can affect a device in two ways. A fine film of sub micrometer particles can disrupt the function or reduce the reliability of a device. Larger particles can hinder or impede movable parts rendering devices useless.

As the dicing blade cuts through the wafer particles are created. Most of these particles are suspended in water and then carried away by the water being showered on the wafer. For IC's the relatively smooth surface allows the particles to be carried away and in some cases surfactants are added to aid the sheeting action. The particles that remain on the wafer are easily washed away in a spin rinser dryer. Structures of the MEMS often trap the particles and surfactants that would normally rinse off of the wafer. In order to dice with the traditional blade the design of the MEMS may be limited.

To allow the use of standard dicing saws prior to dicing the surface of the MEMS wafer can be covered with a temporary protective film or permanent cap. Another similar alternative is to coat the surface of the MEMS wafer with an oxide or polymer layer that covers the devices and holds the structures in place. After dicing, the film or coating is removed, releasing the MEMS.

3.2 Physical Damage

Due to the fragile nature of MEMS, limiting the amount of stress applied to the wafer is a primary consideration. This is in opposition to eliminating contamination with standard processing. Increasing water flow and pressure is a common way to reduce contamination. With MEMS there is a high likelihood of damage from the water.

Figure 4: Water spray during blade dicing.

Figure 4 shows the amount and force of the water that is applied to IC wafers. Water is supplied to the blade throughout the dicing process using a high-pressure nozzle as shown. Water serves both to remove the heat of dicing and to slough off the material being cut.

As with protecting the MEMS from particles applying a temporary film or capping the wafer will protect the devices from the damaging water. Removing the film from individual die is usually a time consuming and costly manual process. Cap wafers are usually silicon or glass. The cap is an integral part of the device and remains after dicing. Both methods protect the devices from contamination and physical damage done by water spray but not from vibration.

Vibration during dicing and cleaning can damage the fragile and pressure sensitive components of the MEMS. During dicing the blade rotates brining abrasive particles in contact with the wafer. Rotating at roughly 30,000 rpm the blade creates a vibration. Added vibration to this is a spray of water that is directed at the cut area and the wash cycle.

To dampen the vibration force applied to the MEMS devices the wafer is submerged in water during dicing. This method lowers the impact of the water on the surface of the wafer.

3.3 ESD Damage

A sudden discharge of electrostatic energy can damage the electronic components in MEMS devices. Inside of the dicing saw there are a number of stages where charges are built up. Wafer transfer between stations lead to several locations where an uncontrolled discharge can occur. For MEMS dicing it is essential that the dicing saw be equipped with ionizers. Typically the ionizers are fixed and the wafer moves or passes by the ionizer. SEMI E78-1102 is a standard for electrostatic compatibility. Much of the standard can be applied to MEMS processing.

DI (De-Ionized) water is often used when dicing ICs and MEMS. Ions are removed from the water in order to insure a device free of ionic contamination. Some MEMS and most bonding pads are susceptible to contamination by a number of ionic elements. The electrical isolation of the wafer due to the dicing tape, high resistivity of the water and the high rotation speed of the blade combine to cause charge build up. By re-ionizing the water with CO_2 the resistivity of the water is reduced and the bicarbonate has no negative effects on the MEMS.

4 SD QUALITY

The SD process is non-contact and does not involve water. For these two reasons the issues with blade dicing are not of concern. ESD damage from wafer transfer is still a concern and is addressed in the same manner, with ionizers. This is not to say that the SD process does not have any quality issues. When manufacturing a MEMS device using the SD process the key quality issues to be aware of are particles, impulse vibration, die form factor, and double die.

4.1 Particles

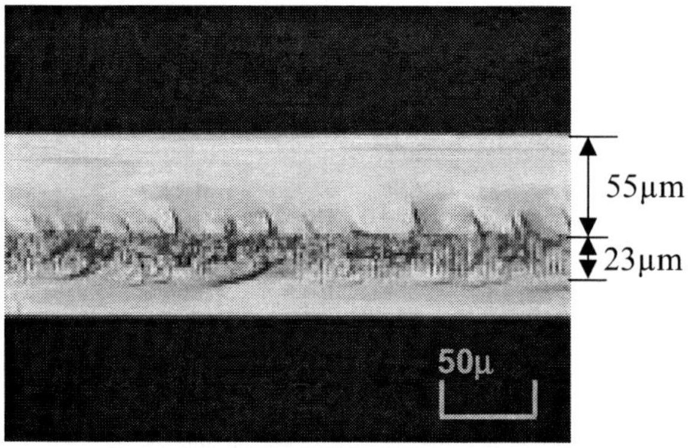

Figure 5: Scalloping.

Figure 5 shows the scalloping on the edge of a die. This occurs when the majority of the silicon in the cut plane is not modified by an SD layer. Quite often a wafer can be diced using a single or a limited number of SD layers. The fewer the number of SD layers the higher the throughput.

Where there is an SD layer the silicon has been mostly if not completely separated. Areas that are not modified are still whole. Where the scalloping appears on the edge of a die is where the silicon between the die was connected. Expanding the tape holding the wafers will break this connecting silicon separating the die. When the silicon breaks often silicon particles are generated. In many cases some of these particles will make it to the surface of the die. For MEMS devices that are sensitive to particles such as silicon microphones it is necessary to increase the number of SD layers in order to reduce the risk of particle contamination.

4.2 Impulse Vibration

As discussed above when die are separated there is an amount of silicon connecting the die. When this connection is severed there is an amount of impulse vibration. The fewer the SD layers the more silicon connects die together. For a device that has had a number of SD layers the amount of silicon connecting it to the neighboring die is minimal leading to the least amount of impulse vibration. As the number of SD layers is reduced the amount of impulse vibration during separation increases.

Only a very few MEMS designs are fragile enough to be affected by impulse vibration. Because there are no standards to measure impulse vibration process specifications do not contain criteria. As of now there has not been any quality issues found due to impulse vibration. With devices becoming more and more sensitive this may change.

4.3 Die Form Factor

Figure 3 above showed how narrow the separation between die can be. This narrow separation allows for a reduction in the dicing street width and an increase in the wafer real estate used for devices. Figure 6 shows die that are expanded following the SD process. There is no silicon lost between the die.

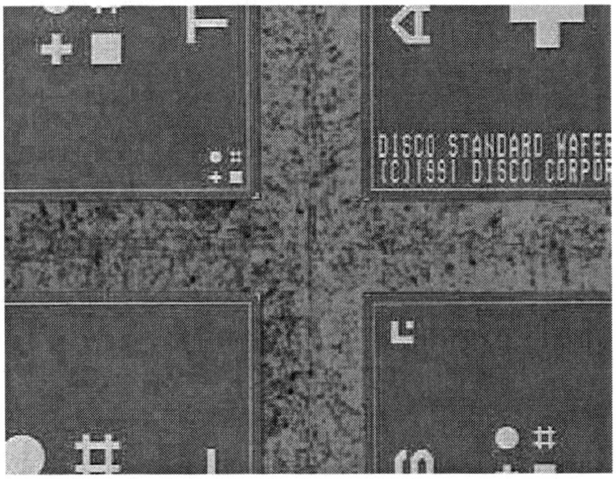

Figure 6: Zero kerf width.

To allow for blade dicing, if the dicing street was 120μm wide, after moving to the SD process the dicing street might be reduced to 20μm wide. Due to the size of many MEMS reducing the street width would dramatically increase the number of die/wafer.

The issue is that because there is no material removed from the dicing process the die have grown. Many processes cannot accept a 100μm variation in die size. When moving to the SD process the wafers have to be designed for the reduced street width.

4.4 Double Die

The force separating the die is simply the force generated by the silicon. Many times there are materials in the dicing street. To separate the materials in the dicing street the separating silicon has to have enough energy to continue the separation through the materials in the dicing street. If there is not enough energy the die do not separate and double die exist.

By removing all materials from the dicing street the problem does not exist. Because this is not always possible the SD process is often modified to create more singulating force. Many times this means adding SD layers at the expense of throughput.

5 CONCLUSION

Up until recently traditional dicing methods were used on MEMS which often limited the design and output of many MEMS devices. In the first years of life a great number of MEMS designs had to be abandoned because they could not be tested or diced. New processes, such as Stealth Dicing is helping bring MEMS into higher volume production and may allow designers even more freedom to create our nano-future.

REFERENCES

[1] Yole Développement, "Global MEMS/Microsystems - Markets and Opportunities, 1, 2007.

FAST METHODS FOR PARTICLE DYNAMICS IN DIELECTROPHORETIC BIOCHIPS

Indranil Chowdhury*, Xiren Wang* and Vikram Jandhyala*

ACELAB, Electrical Engineering, University of Washington, USA,
burunc@u.washington.edu, xrwang@u.washington.edu, vj@u.washington.edu

ABSTRACT

This paper introduces a Schur-complement based boundary element method (BEM) for predicting the motion of arbitrarily shaped three-dimensional particles under combined external and fluidic force fields. The BEM approach presented here relies entirely on modeling the surface of the computational domain, significantly reducing the number of unknowns when compared to volume-based methods. In addition, the Schur complement based scheme leads to a huge reduction in solution time during time-stepping in the microfluidic domain. Parallelized oct-tree based O(N) multilevel iterative solvers are used to accelerate the setup and solution costs.

Keywords: BEM, cell-handling devices, microfluidics

1 INTRODUCTION

Many lab-on-chip (LoC) devices use dielectrophoretic (DEP) manipulation of polarized species inside microfluidic channels [1-3]. Understanding the fluidic and electromagnetic forces in these devices require rigorous treatment of the underlying physics. BEM based system matrix is dense in nature due to the highly coupled interaction between the wall and the particles, especially when the particle size is comparable to that of the channels (Fig. 1). Conventionally, the numerical treatment of such systems is achieved via brute-force computation of the whole fluidic domain during each time-step of the iteration. Hence, during computation of the motion of rigid or deformable particles a large number of time steps are required, where each time step consists of a computationally expensive solution of a dense matrix system. Previous work on Lab-on-chip modeling has been mainly based on finite-element and volume based methods [3-6,9]. The problem with these methods lie in the fact that they need to remesh the whole channel for each time step, while in BEMs only the surface is meshed, which significantly reduces the number of unknowns [6, 17, 18]. Here a scheme based on Schur-complement is presented to accelerate the time-stepping algorithm by partially decoupling wall-particle interactions. Particle motion can be predicted for arbitrarily shaped three-dimensional particles under combined external and fluidic force fields. Parallelized oct-tree based O(N) multilevel iterative solvers are used to accelerate the setup and solution costs [12-16]. In the past BEM techniques have been used to study low Re flows [8, 17], however, fast algorithms for dynamic systems remain a topic of active research [10]. Besides the fluidic fields, DEP fields are produced by on-chip electrodes. A coupled circuit-EM formulation is used for accurate prediction of DEP field distribution that allows circuit control of resulting electromagnetic fields [11] (fig 5). Simulations of particle trajectory in pressure-driven dielectrophoretic LoCs are presented. Evidence of applications of the current methodology to a large class of flow devices [7] for particle transport is presented.

2 INTEGRAL EQUATIONS

The integral representation for incompressible Stokes flow are given by the following expressions [17,18]:

$$u_i(\mathbf{x}_0) = \frac{-1}{4\pi\mu}\int_D G_{ij}(\mathbf{x},\mathbf{x}_0)\, f_j(\mathbf{x})dS(\mathbf{x})$$

$$+\frac{1}{4\pi}\int_D^{PV} u_j(\mathbf{x})\, T_{ijk}(\mathbf{x},\mathbf{x}_0)\, n_k(\mathbf{x})\, dS(\mathbf{x})$$

No-slip and pressure boundary conditions are applied on the surface of the channel, while force and torque balance equations are setup for rigid particles,

$$\oint \sigma \bullet dS = F_{ext}$$

$$\oint r \times (\sigma \bullet dS) = T_{ext}\ ; \quad u_{particle} = U + \Omega \times r$$

U and Ω are the translation and angular velocities. The external forces and torques can be DEP fields for example. The surface of the domain is discretized using triangular patches and subsequently a collocation method is used for solving the unknown traction and velocity fields. The particle velocities can be solved at each time step and the resulting trajectory can be encountered for. However solving the whole dense matrix system for each time step becomes prohibitively expensive and an algorithm to reduce this cost is described below.

The first step in this algorithm is the isolation of the particle under study and is achieved by identifying the patches belonging to the mathematical surface bounding the particle $\mathbf{S}_p$. This surface isolates the problem into two parts – a subset of the problem which is constant over time and another part which is changing over time. All the force and velocity unknowns belonging to the channel constitutes the first part (**D**) which consists of the single and double layer interactions of the channel walls and faces, whereas all the force unknowns belonging to the particle are represented by **A** which consists of the single layer interactions among the particle patches. Notice that this decomposition does not change integral equations of the equivalent problem with a change in the shape or location of the object bounded by $\mathbf{S}_p$. A small change in the shape of $\mathbf{S}_p$ may not require additional patches, however if the shape and size of the particle changes drastically more patches have to be added to the particle surface to maintain a desired level of accuracy. As a result of decomposing the actual problem can be thought of as four different parts – the channel interacting with itself (**D**), the particle interacting with itself (**A**), the channel to particle interaction (**B**) and the particle to channel interaction (**C**). **B** contains both single and double layer interactions while **C** consists of single layer interactions only. The resulting system can be described by the following systems of equations:

$$\mathbf{D}y + \mathbf{C}x = a_1$$

$$\mathbf{B}y + \mathbf{A}x = a_2$$

Since the channel walls are static and unchanging in time the matrix **D** is constructed and stored in the form of its inverse - only once during the first time step. This is followed by updating the Schur's complement of the unchanging part (**D**) which is given by $\mathbf{S} = \mathbf{A} - \mathbf{B}\mathbf{D}^{-1}\mathbf{C}$, at each time step. The unknowns are then computed by back-substitution using a standard Schur complement method. Therefore for each step of the iteration the dimension of the net system is greatly reduced which saves considerable computational cost. It is important to note that the proposed technique bypasses the cost of explicit modeling of the mutual coupling between the channel walls and the particles.

Here the surface traction forces in both the channel and the particle problems are implemented and a simple case where the particle is treated as a rigid body and its shape is constant in time. However the methodology presented here is generally valid for deformable particles whose shape and size change over time. The surface of the channel supports both velocity and force unknowns while the particle surface supports only force unknowns. In order to take into account the translation and rotational velocities of the particle the net external force and torque have to be imposed on its surface. On the channel walls the three components of the velocities are set to zero in order to enforce no-slip conditions, whereas the traction forces on the channel faces are provided by the imposed pressure gradient. Note that the velocities on the channel faces are unidirectional. An appropriate preconditioner can be applied during use of a fast solver when required. The traction forces on particle surface are unknowns and the double layer contribution due to a closed particle is zero outside its boundary. Combining all the interactions, along with the force and torque balance equations, the overall system is written in a matrix form:

$$\begin{bmatrix} \mathbf{D} & \mathbf{C} \\ \mathbf{B} & \mathbf{A} \\ \sum \mathbf{F} & \\ \sum \mathbf{T} & \end{bmatrix} \begin{bmatrix} f_c \\ u_c \\ f_p \end{bmatrix} = \begin{bmatrix} p_1 \\ p_2 \\ F_{ext} \\ T_{ext} \end{bmatrix}$$

where f_c are unknown traction forces on the channel surface, u_c are the velocity unknowns at the faces and f_p are the traction forces on particle surface. The known vector (RHS) represents the single layer contributions due to the imposed pressure at each face and excites the fluid flow. The other parts are sparse since they represent the summation of the surface forces on the particle and don't involve the channel patches. Typically for large complicated channels the number of unknowns on the channel surface greatly exceeds the particle surface. This implies the tacit assumption that the number of patches on particle << number of patches on the channel for the Schur complement based scheme proposed here to be an efficient solver. Fig. 2 shows the speedup due to the proposed scheme. It can be seen that for small number of particles the speedup factor is 5-6 times for each time step.

2.1 Algorithmic Complexity

If N_1, N_2, N_3 are the number of unknowns on the surfaces S_c, S_f and S_p, the cost of solving during the stages of the iterations where the appropriate blocks are pre-computed is given by C_1, where

$$C_1 = N_1^3 + N_1 N_2 (N_1 + N) + N_1^2 + N_1 N_2 + N_2 N_3$$

If changing the excitation on the neighboring geometry is not required the cost can be further reduced to C_1' given by

$$C_1' = N_1^3 + N_1 N_2 (N_1 + N) + N_1^2 + N_1 N_2$$

The cost reduction is therefore given by a factor of N_3^3 / N_i^3.

The expensive step in the iterations however is the computation of the Schur complement,i.e., during computing $\mathbf{D}^{-1}\mathbf{C}$. But as long as the number of particle patches is small, large speedups during time stepping can be achieved. Simulation examples are depicted in figs 1-5.

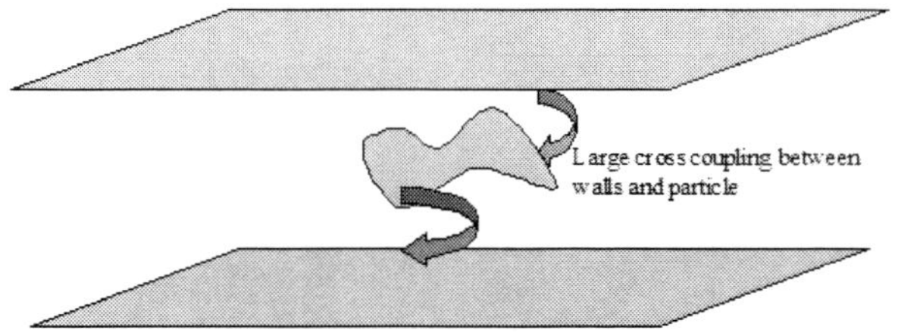

Fig 1: Cross-coupling due to interaction between channel wall and particle

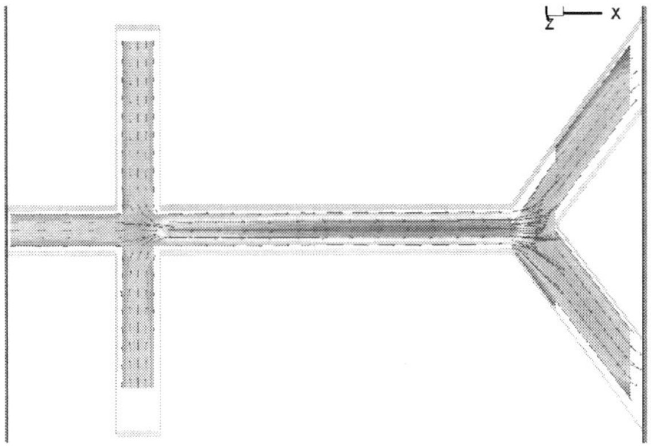

Fig 2: Flow field distribution in a complicated channel topology

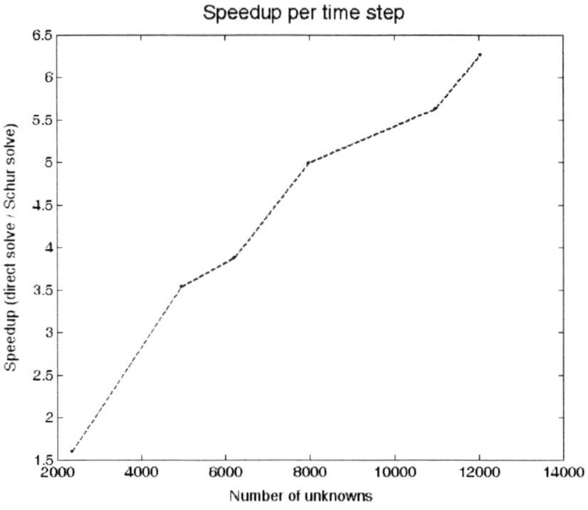

Fig 3: Speedup of solution time for solver

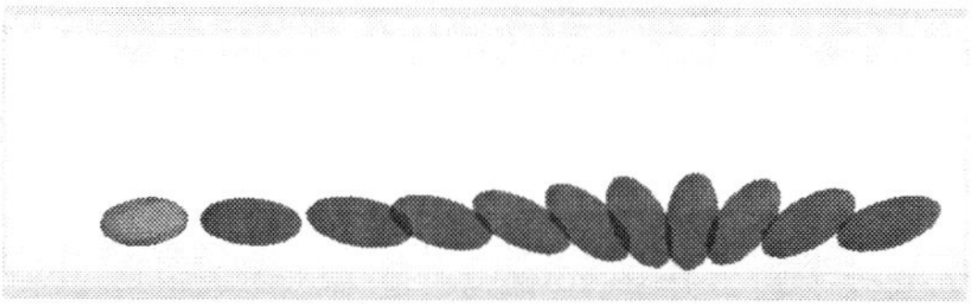

Fig 4: Simulation of platelet motion in a pressure driven micro-channel

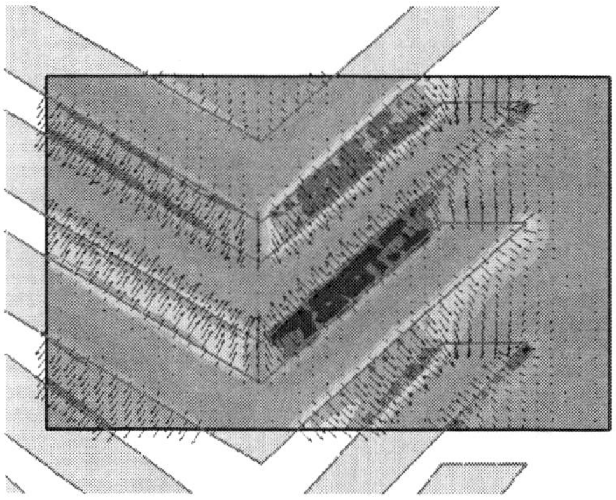

Fig 5 : DEP field distribution due to a V-shaped electrode array

REFERENCES

[1] Weigl BH, Bardell RL, Cabrera CR, *Lab-on-a-chip for drug development,* Adv Drug Deliv Rev. 2003, 3, pp. 349-77.

[2] Gascoyne P. R., Vykoukal, *DEP based Sample Handling in General Purpose Programmable Diagnostic Instruments*, Proc IEEE, vol 92, no. 1(2004).

[3] Haibo, L., Z. Yanan, A. Demir, and R. Bashir, *Characterization and Modeling of Microfluidic Dielectrophoresis Filter for Biological Species*, Journal of MEMS, Vol. 14 No. 1 (2005).

[4] Wang, X, Wang X-B, Becker F. and Gascoyne P. R. C., *A theoretical method of electrical field analysis for dielectrophoretic electode arrays using Green's theorem*, J. Phys. D: Appl. Phys., 29; pp. 1649-1660 (1996).

[5] Clague, D.S., Wheeler EK, *Dielectrophoretic manipulation of macromolecules: The electric field*, Physical Review E, 64 (2): Art. No. 026605 Part 2, AUG 2001.

[6] David Ericksson, *Towards numerical prototyping of labs-on-chip: modeling for integrated microfluidic devices*, Microfluid Nanofluid (2005), 1, pp. 301–318.
[7] Barry Lutz, *Microeddies as Microfluidic Elements:Reactors and Cell Traps*, PhD thesis, University of Washington, 2003.
[8] C. Pozrikidis, *Numerical Simulation of Cell Motion in Tube Flow*, Annals of Biomedical Engineering, 33, No. 2, February 2005, pp. 165–178.
[9] R. Chein and S. H. Tsai, *Microfluidic Flow Switching Using Volume of Fluid Model*, Biomedical Microdevices 6: I, pp. 81-90, 2004.
[10] Dong Liu, Martin Maxey, and George Karniadakis, *A Fast Method for Particulate Micrflows*, Journal of MEMS, 11, no. 6, 2002, pp. 691-702.
[11] Y. Wang, D. Gope, V. Jandhyala and C.J. Shi, *Generalized KVL-KCL Formulation for Coupled Electromagnetic-Circuit Simulation with Surface Integral Equations*, IEEE Transactions on. Microwave Theory Tech., vol. 52, no. 7, pp. 1673-1682, July 2004.
[12] X. Wang, FastStokes: *A fast 3D fluid simulation program for MEMS, PhD thesis*, MIT (2002)
[13] Senturia S. D, Aluru N., White J., *Simulation the behavior of MEMS devices: computational methods and needs*, Comput. Sci. and Engg., IEEE, 4 (1997).
[14] Gope D., Chakroborty S., Jandhyala V., *A Fast Parasitic Extractor Based on Low Rank Multilevel Matrix Compression for Conductor and Dielectric Modeling in Microelectronics and MEMS,*DAC 2004, 794-799.
[15] Nabors K. and White J., *FastCap: A Multipole-Accelerated 3-D Capacitance Extraction Program*, IEEE Transactions on Computer-Aided Design, vol. 10, November 1991, p1447-1459.
[16] I. Chowdhury and V. Jandhyala, *Single multilevel expansions and operators for potentials of the form* $r^{-\lambda}$, SIAM J. Sci. Comput., vol. 26(2005), pp. 930-943.
[17] Youngren G. K., Arcivos A., *Stokes flow past a particle of arbitrary shape: a numerical method of solution*, J. Fluid Mech., 69 (1975).
[18] C. Pozrikidis, *BI and singularity methods for linearized viscous flow,* Cambridge, 1992.

High-stability numerical algorithm for the simulation of deformable electrostatic MEMS devices

X. Rottenberg[1,2], B. Nauwelaers[2] W. De Raedt[1] and D. Elata[3]

[1] IMEC v.z.w., Division MCP, Kapeldreef 75, B3001 Leuven, Belgium
xrottenb@imec.be

[2] K.U.Leuven, ESAT, Kasteelpark Arenberg 10, B3001 Leuven, Belgium

[3] Technion, Faculty of Mechanical Engineering, Haifa 32000, Israel

ABSTRACT

This paper presents a novel high-stability electrically-driven algorithm for the simulation of the electro-mechanical actuation of electrostatic MEMS devices. The stability of this algorithm improves on that of voltage- and charge-drive algorithms. Key in our algorithm are the use of a local charge density as driver for an adapted relaxation algorithm and the adequate selection of the bias node in the mesh. The high stability of this algorithm allows probing the electromechanical equilibrium locus way beyond the V- and Q-drive pull-in instabilities. The new algorithm allows investigating the effect of dielectric charging in deformable electrostatic MEMS devices and especially the narrowing of their equilibrium locus due to dielectric charging non-uniformities. We implement this algorithm in 2D for clamped-clamped beams of rectangular cross-section and take into account, among other things, distributed dielectric thickness, permittivity, rest air gap, actuation electrode and linearly distributed dielectric roughness.

Keywords: dielectric charging, electrostatic actuation, numerical algorithm, high stability

1 INTRODUCTION

Electrostatic actuators are commonly used MEMS building blocks for various applications. In the case of RF-MEMS devices, these actuators are used for example to realize resonators, tunable capacitors and capacitive switches. The general actuation characteristics of typical devices, e.g. parallel-plate or comb-drive actuators, are well known and described in analytical form, e.g. stability or unstability of the voltage-drive actuation of parallel-plate actuators. Designers have however to turn to numerical techniques to assess the actuation characteristics of complex non-standard devices or of standard devices perturbed by parasitic phenomena, e.g. dielectric charging.

In a previous work [2], Rottenberg et al. extended the analytical model for the electrostatic actuation of MEMS devices in the presence of parasitic uniform surface charging of the dielectric [3] in order to account for non-uniform distributions of charges in the bulk of the dielectric and of air gaps at rest in the device. They showed in particular that the spatial covariance of charge and rest air gap distributions breaks the symmetry of the actuation characteristics and that the spatial variance of the charge distribution can produce a narrowing and finally a closure of the pull-out (-in) window that can result in the stiction (self-actuation) of the device. Strikingly, these phenomena can even occur for a total zero trapped charge.

The analytical model developed in [2] assumes rigid plates. Under this assumption, only the mean, variance and covariance of the charge and rest air gap distributions have an impact on the actuation of the device. In actual devices however, the mobile electrodes deform during the actuation. This introduces a non-uniform distribution of the air gaps in the structure, non-present in rest position, and modifies the impact of the trapped charges on the actuation characteristics. This modification was illustrated in [5] for the clamped-clamped beam shown in Figure 1. However, the poor stability of both voltage- and charge-drive numerical algorithms implemented limited the demonstration of the impact of conjunct distributed charging and progressive deformation.

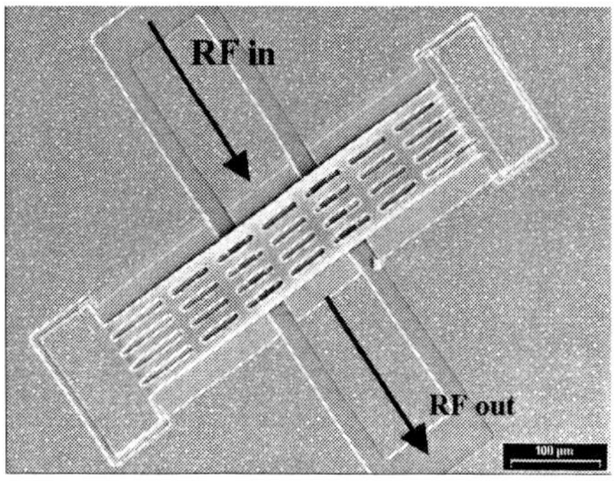

Figure 1 SEM of a typical RF-MEMS shunt capacitive switch in thin-film technology

This paper presents a novel high-stability algorithm for the simulation of the electromechanical actuation of electrostatic MEMS devices. Its stability improves on that of voltage- and charge-drive algorithms. It further offers a purely electrical simulation alternative to previous work [1] based on displacement control of the electromechanical response to drive the simulation. Key in our algorithm are the use of a local charge density as driver for an adapted relaxation algorithm and the adequate selection of the bias node in the mesh. The high stability of this algorithm allows probing the electromechanical equilibrium locus way beyond the voltage-drive and charge-drive pull-in instabilities. The new algorithm facilitates investigation of the effect of dielectric charging in deformable electrostatic MEMS devices and especially the narrowing of their equilibrium locus due to dielectric charging non-uniformities [2]. We implement this algorithm in 2D for clamped-clamped beams of rectangular cross-section and take into account, among other things, distributed dielectric thickness, permittivity, rest air gap, actuation electrode and linearly distributed dielectric roughness.

2 ALGORITHM DESCRIPTION

Figure 1 shows a typical RF-MEMS capacitive shunt switch, consisting of a 1/500/100µm thick/long/wide Al bridge under 50MPa tensile stress. The bridge is suspended 2µm above a 100µm wide actuation electrode coated with a 200nm thick high-k (ε_r=25) dielectric. It is modelled as the perfect clamped-clamped beam of Figure 2 and discretized along its length as depicted in Figure 3 and detailed in [4].

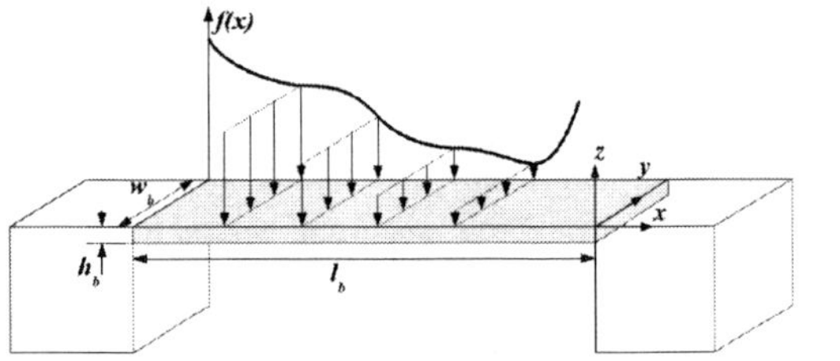

Figure 2 Shunt capacitive switch modelled as an ideal clamped-clamped uniform beam with distributed electrostatic load.

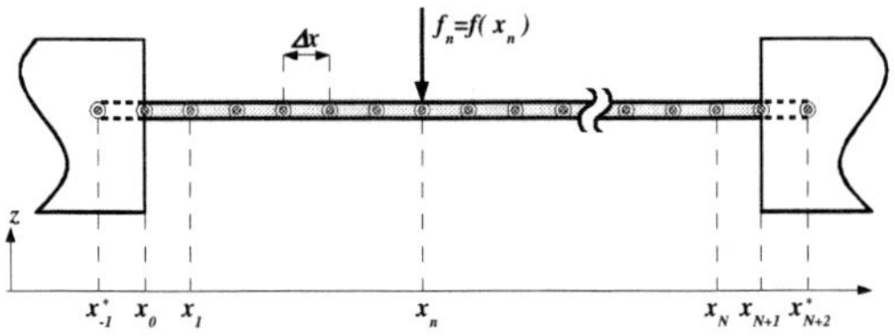

Figure 3 Discretization scheme for the simulation of an ideal clamped-clamped uniform beam.

For such a deformable device, both V- and Q-drive are unstable. Indeed, even under fixed Q_{bias}, charges can reorganize on the bridge and flow in an avalanche towards the valley defined by the deformed device. A more stable actuation scheme is thus needed in order to probe the actuation-profile way through $V_{PI}^{+/-}$ and $Q_{PI}^{+/-}$, respective limits of the stable V- and Q-drives.

A simple numerical way to prevent the destabilizing Q-drive charge redistribution is to fix the local charge on a selected node n, i.e. $q(x_n)=q_n$, instead of fixing the total charge on the device, i.e. Q_{bias}. While the V-drive and Q-drive algorithms imposed a global electrical variable, the novel q_l-drive algorithm imposes a local electrical variable. It is as a result more binding and more stable. Remark that this algorithm is in a way more subtle than that proposed in [1] as it does not impose a local geometrical effort, i.e. the displacement of a node, but a local electomechanical effort. Imposing the bias q_n corresponds indeed to imposing the force applied to the node n.

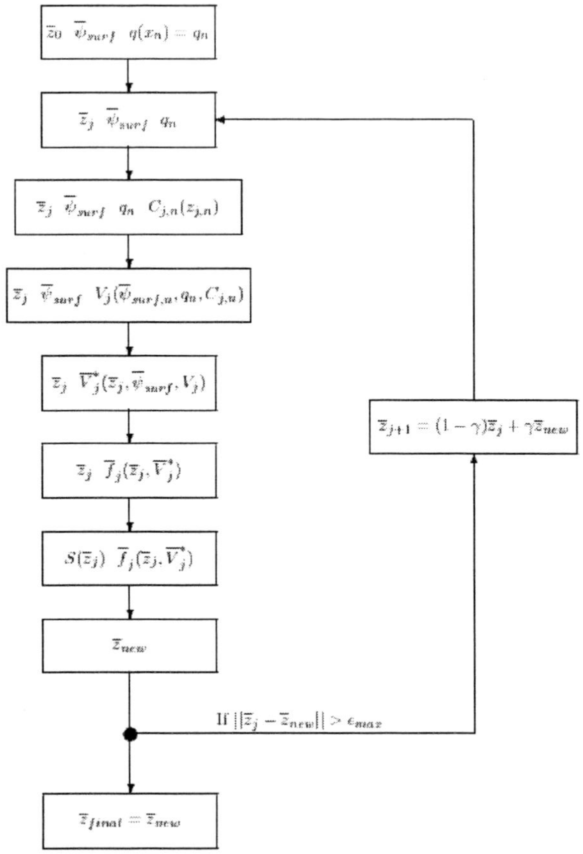

Figure 4 Numerical algorithm for the local-charge actuation of deformable devices with distributed surface charge.

The novel iterative relaxation scheme is detailed in Figure 4. It is a slightly modified version of the V-drive algorithm. From the deformation at a given iteration step, i.e. $\overline{z_j}$, the capacitance $C_{j,n}$ associated with the bias-node n is computed. From this capacitance, the algorithm computes the voltage V_j needed to accommodate the forced charge located on the node n, i.e. $q(x_n)=q_n$, in association with the equivalent surface charge distribution located under the same node, i.e. $\psi_{surf,n}$. With this voltage and the other parameters, the iteration is carried on following a modified V-drive scheme based on [2] and [4]. The distributed effective voltage V^* and resulting electrostatic force are first computed. The displacement $\overline{z_{new}}$ balancing this distributed force is then evaluated. Should the algorithm have converged, the simulation stops. Otherwise, a step is performed from $\overline{z_j}$ in the direction of $\overline{z_{new}}$ and a

NSTI-Nanotech 2008, www.nsti.org, ISBN 978-1-4200-8505-1 Vol. 3

novel iteration starts. At each iteration, the voltage on the beam is thus recomputed to maintain the local charge on the bias node.

A simple physical insight brings a solution to the numerical stabilization of the simulation. Remark nevertheless that imposing the charge on a portion of a metal armature is highly non-physical. The proposed stabilization technique remains thus a numerical trick in opposition to the Q-drive that can be physically implemented with its limitation [5][6].

3 SIMULATION RESULTS

Figure 5 shows the result of the simulation of the device from Figure 1 using the local-charge algorithm in absence of trapped charge in the dielectric.

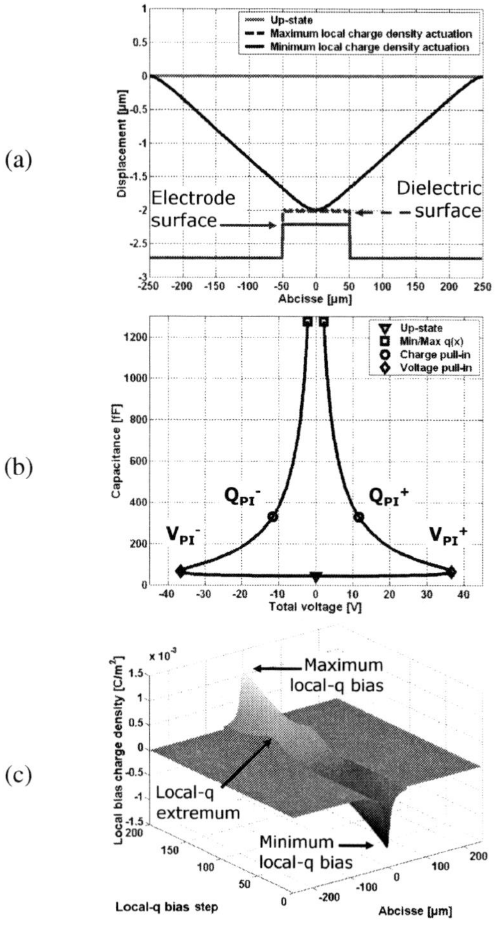

Figure 5 Local-charge drive simulation of a clamped-clamped beam in case of ideal uncharged dielectric layer; (a) beam deformations at maximum opening and at extreme stable positions for the local-q drive algorithm, (b) C-V profile stable for the local-q actuation, (c) charge density on the bridge during the local-q actuation.

Figure 5(a) shows the deflections at rest and at contact with the dielectric layer, demonstrating the stability of our algorithm. Figure 5(b) depicts the symmetric C-V profile up to contact (diamond and round markers show respectively voltage and charge pull-in's). The charge map Figure 5(c) shows that the centre node is a proper bias node as its local-q monotonically increases. In contrast, the local-q above the electrode-edges exhibits two extrema, similar to the Q-drive instabilities. In this case, working with off-centre bias nodes may partially stabilize the simulation. A general strategy for bias node selection consists in choosing the node with the highest rate of change of its local charge density.

In presence of centered linearly distributed trapped charges in the dielectric, the V-drive presented in Figure 6(a) partially depicts the expected narrowed and asymmetric C-V due to the covariance of deflection and charge distributions [2]. At maximal opening, the beam is not flat. The maximal deflection is noticeably different for $V_{PI}^+>0$ and $V_{PI}^-<0$.

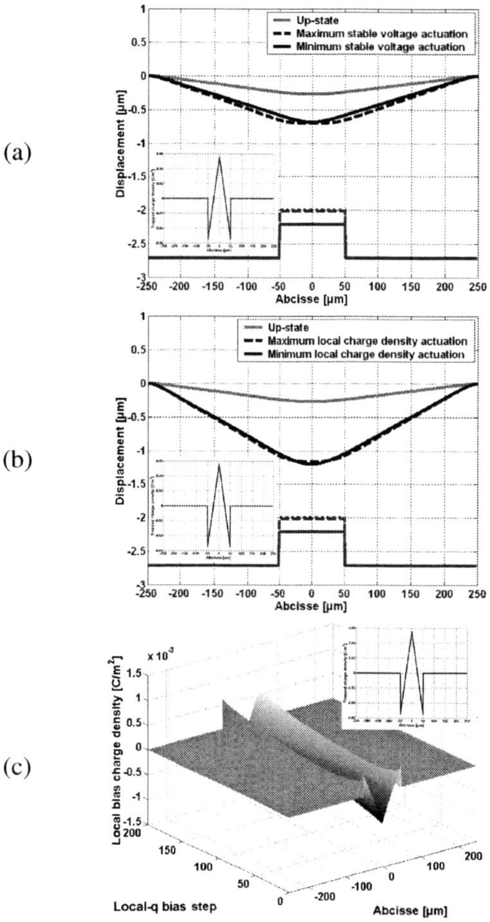

Figure 6 Simulation of a clamped-clamped beam in case of centred linear symmetric equivalent surface charge distribution trapped in the dielectric layer with zero mean; beam deformations at maximum opening and at extreme stable positions for (a) the voltage drive algorithm and (b) the local-q drive algorithm, (c) charge density on the bridge during the local-q actuation.

In comparison, the local-q-drive, Figure 6(b), tracks the equilibrium much further. Positive and negative branches of the locus converge. So do the beam shapes. The charge map Figure 6(c) strongly differs from Figure 5(c). The local-q of each node exhibits an extremum. The bias node had therefore to evolve during the actuation from the centre to the edges respectively at low and high local-q.

Figure 7 presents a comparable but more realistic case of a centered, net positive, cosine charge distribution in the dielectric. The portion of equilibrium locus made accessible by the local-q simulation shows that, as expected, the C-V profile is clearly shifted upwards in voltage, narrowed and asymmetric. Despite this voltage shift, the charge map remains comparable to that in Figure 6(c).

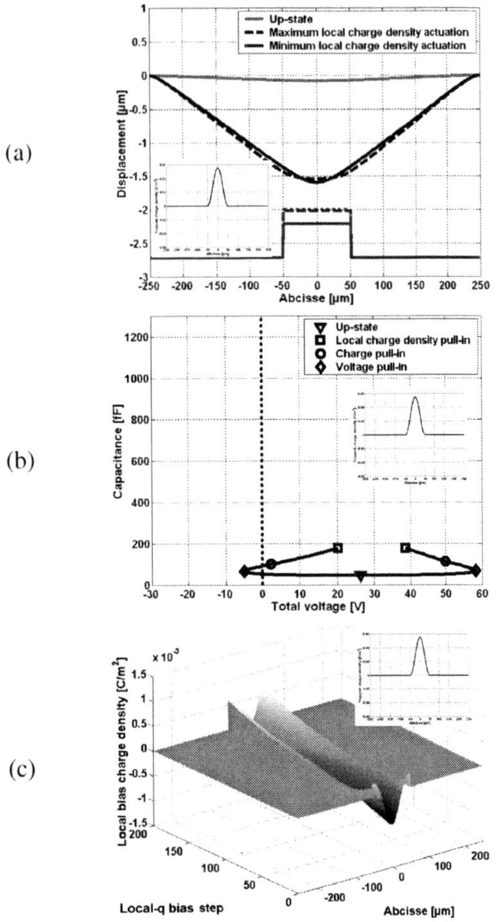

Figure 7 Local-charge drive simulation of a clamped-clamped beam in case of centred positive cosine equivalent surface charge distribution trapped in the dielectric layer; (a) beam deformations at maximum opening and at extreme stable positions for the local-q drive algorithm, (b) C-V profile stable for the local-q actuation, (c) charge density on the bridge during the local-q actuation.

4 CONCLUSIONS

We presented a novel high-stability algorithm for the simulation of the electromechanical actuation of electrostatic MEMS devices. Its stability improves on that of voltage- and charge-drive algorithms. It further offers a purely electrical simulation alternative to previous works based on displacement control of the electromechanical response to drive the simulation. Key in our algorithm are the use of a local charge density as driver for an adapted relaxation algorithm and the adequate selection of the bias node in the mesh. The high stability of this algorithm allows probing the electromechanical equilibrium locus way beyond the voltage-drive and charge-drive pull-in instabilities. The new algorithm facilitates investigation of the effect of dielectric charging in deformable electrostatic MEMS devices and especially the narrowing of their equilibrium locus due to dielectric charging non-uniformities. We implement this algorithm in 2D for clamped-clamped beams of rectangular cross-section and take into account, among other things, distributed dielectric thickness, permittivity, rest air gap, actuation electrode and linearly distributed dielectric roughness.

5 ACKNOWLEDGEMENTS

The authors want to acknowledge financial support from the Flemish government through the IWT SBO-mmWAVE project (Visualization of concealed objects using mmwave systems) under contract nr. 040114.

REFERENCES

[1] D. Elata, O. Boschobza-Degani and Y. Nemirovsky, "An efficient numerical algorithm for extracting pull-in Hyper-Surfaces of electrostatic actuators with multiple uncoupled electrodes", Proc. of MSM 2002, pp. 206-209.

[2] X. Rottenberg, I. De Wolf, B.K.J.C Nauwelaers, W. De Raedt and H.A.C. Tilmans, "Analytical model of the DC actuation of electrostatic MEMS devices with distributed dielectric charging and nonplanar electrodes", JMEMS, vol. 16(5), pp. 1243-1253, 2007.

[3] J. Wibbeler, G. Pfeifer and M. Hietschold, "Parasitic charging of dielectric surfaces in capacitive microelectromechanical systems (MEMS)", Sensors and Actuators A: Physical, pp. 74-80, November 1998.

[4] T. Tinttunen, T. Veijola, H. Nieminem, V. Ermolov and T. Ryhänen,"Static equivalent circuit model for a capacitive MEMS RF switch", Proc. of MSM 2002, pp. 166-169.

[5] X. Rottenberg, B. Nauwelaers, E. M. Yeatman, I. De Wolf, W. De Raedt, H. Tilmans, "Model for the voltage and charge actuations of deformable clamped-clamped beams in presence of dielectric charging", Proceedings of the 16th MicroMechanics Workshop - MME, Chalmers, Göteborg, Sweden, pp.172-175, Sep. 2005.

[6] D. Elata and V. Leus, "Switching time, impact velocity and release response of voltage and charge driven electrostatic switches" Proc. of the Int. Conf. on MEMS, Nano and Smart Systems (ICMENS), pp. 331-334, 2005.

Physically-Based High-Level System Model of a MEMS-Gyroscope for the Efficient Design of Control Algorithms

Ruslan Khalilyulin, Gabriele Schrag, Gerhard Wachutka

Institute for Physics of Electrotechnology, Munich University of Technology, Arcisstrasse 21, 80290 Munich, Germany
Phone: +49-89-28923128, Fax: +49-89-289-23134, ruslan@tep.ei.tum.de

ABSTRACT

We present a high-level model of a dual gimbaled mass gyroscope, which provides an accurate physical description of the impact of external and internal disturbances on the output signal, but which also allows for the efficient analysis and optimization of the full sensor system including the electronic circuitry for drive, control, and signal conditioning. The dynamics of the gyroscope is described by a reduced-order high-level model. Internal disturbances (manufacturing tolerances, e.g.) as well as external impact factors (pressure-dependent viscous damping, shock, vibrations of the housing, etc.) are included in the model equations by introducing physically-based, parameterized functions extracted from detailed FEM or mixed-level simulations. Exemplary simulations investigating the impact of package vibrations on the sensor output signal prove the efficiency of our model and the practicality of our approach.

1 INTRODUCTION

Robust sensor systems designed for applications in harsh environments (e.g., automobiles) have to be equipped with additional "smart functionalities" to compensate environ mental impacts (e.g., ambient pressure variations or package vibrations) or internal imperfections due to manufacturing tolerances.

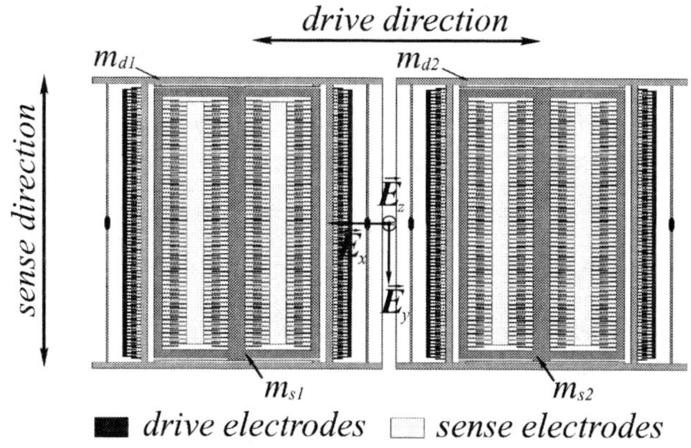

Fig. 1: Structure of the dual gimbaled mass gyroscope.

In order to ensure a proper and reliable sensor operation, adequate concepts for sensor control and signal conditioning must be developed, which are based on a profound understanding of the effects disturbing the sensor output signal. To this end, we derived a high-level model of the dual gimbaled mass gyroscope depicted in Fig. 1, which provides an accurate physical description of the impact of external and internal disturbances on the output signal, but, at the same time, allows for the efficient analysis and optimization of the full sensor system including the electronic circuitry for drive, control, and signal conditioning and, thus, is perfectly suited for the co-design of the transducer elements and the control electronics

2 HIGH-LEVEL GYROSCOPE MODEL

2.1 Modeling Approach

As it would be much too time- and memory-consuming to perform a complete continuous-field FEM analysis of the gyroscope including all possible mechanical configura tions, we focus on the most relevant modes of motion and decompose the device structure into four rigidly moving

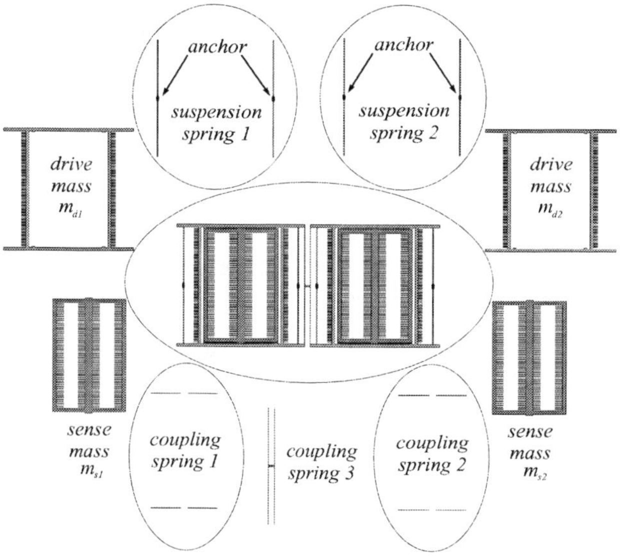

Fig. 2: Decomposition of the gyroscope into four rigidly moving bodies connected by mass-less springs.

mechanical substructures, which are hinged and inter linked by springs (Fig. 2). The dynamics of this simplified structure is described by the Langrangian equations of motion, leading to a reduced-order model of the complete gyroscope with 24 degrees of freedom [1]:

$$\boldsymbol{M} \cdot \underline{\ddot{u}} + (\boldsymbol{C} + \boldsymbol{C}_\Omega) \cdot \underline{\dot{u}} + (\boldsymbol{K} + \boldsymbol{K}_\Omega) \cdot \underline{u} = = -\boldsymbol{M} \cdot \underline{a}_o + \underline{F}_{el} \tag{1}$$

Here $\boldsymbol{K}$ and $\boldsymbol{M}$ denote the stiffness and the mass matrices, $\boldsymbol{C}$ the damping matrix, which accounts for viscous friction due to the surrounding air, $\underline{F}_{el}$ the electrostatic forces actuating the drive masses, and $\underline{a}_0$ the external linear acceleration acting on the substrate frame. The measurand, the angular velocity $\vec{\Omega}$, enters the system through the terms $\boldsymbol{C}_\Omega$ and $\boldsymbol{K}_\Omega$, which comprise the contributions of the Coriolis and the centrifugal force to the stiffness and the damping matrix, respectively.

2.2 Mechanical Model Parameters

The mechanical model parameters like the mass matrix $\boldsymbol{M}$ and the stiffness matrix $\boldsymbol{K}$ were extracted from detailed 3D FEM simulations by calculating the respective inertial forces and spring constants for each rotational and translational degree of freedom. The three fundamental resonance frequencies and the associated mode shapes of the gyroscope as obtained from the harmonic solutions of the homogeneous equation system (1) (i.e. $\underline{a}_o = 0$, $\underline{F}_{el} = 0$) conform well with those calculated by modal FEM analysis (see Fig. 3 and Tab. 1).

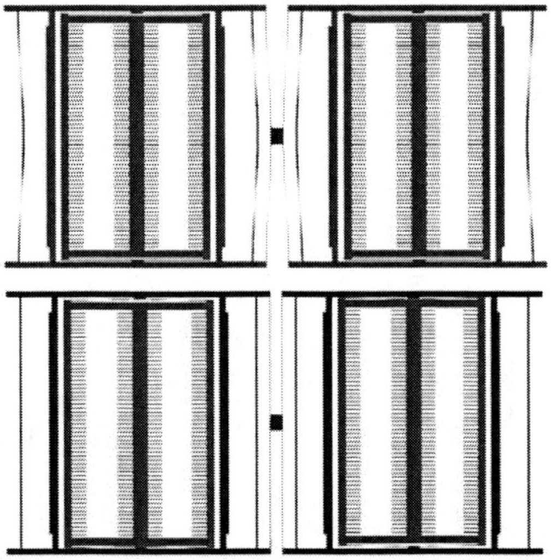

Fig. 3: Results of FEM modal analysis of the gyroscope structure.
Top: anti-phase drive mode at 10.654 kHz
Bottom: sense mode at 10.527 kHz

Resonance frequencies of the gyroscope	***FEM***	***harmonic solution***
in-phase drive mode	*10.301 kHz*	*10.459 kHz*
sense mode	*10.527 kHz*	*10.677 kHz*
anti-phase drive mode	*10.654 kHz*	*10.723 kHz*

Table 1: Three fundamental mechanical resonance frequencies of the gyroscope.

It shows that manufacturing fluctuations such as, e.g., etch bias tolerances have a considerable impact on the mass and stiffness parameters.

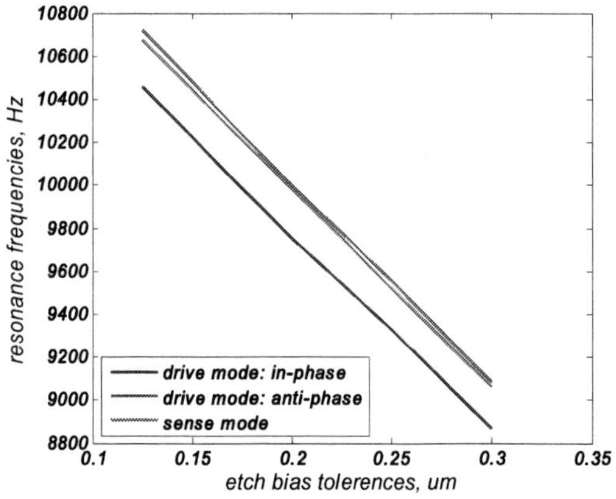

Fig. 4: Impact of manufacturing tolerances on the resonance frequencies of the drive and the sense masses.

The dependencies from these parameter variations have been analyzed (see Fig. 4) and extracted by detailed FEM simulations and, subsequently, fed into the high-level model as parametrized functions.

2.3 Viscous Damping Model

Viscous damping due to the surrounding air is, by its nature, a distributed effect, which cannot be properly described by analytical compact models and which is also difficult to extract from 3D coupled-domain FEM simulations, especially for geometrically complex microstructures. Therefore, we included this effect by introducing modal quality factors $\hat{C}_i$ for all relevant modes of motion:

$$\hat{C}_i = \frac{\sqrt{\hat{K}_i \cdot \hat{M}_i}}{\hat{Q}_i} \tag{2}$$

where $\hat{K}_i$, $\hat{M}_i$ and $\hat{Q}_i$ denote the stiffness, mass and quality factors of the *i*-th eigenmode, respectively. The $\hat{C}_i$ are then converted into the spatial damping matrix $\boldsymbol{C}$ by applying modal transformation techniques.

Since it is computationally quite expensive and for the required accuracy of the system model also not necessary to determine the quality factors for a large number of eigenmodes, we focussed on the two most relevant modes of motion, i.e. the sense mode, which is mainly affected by squeeze film damping, and the drive mode, whose dynamics is dominated by slide film damping.

In detail, the modal quality factor of the drive mode can be determined by:

$$\frac{1}{Q_{drive}} = \frac{1}{Q_{sl\_fr}} + \frac{1}{Q_{comb}} \tag{3}$$

where Q_{sl_fr} represents the quality factor of the drive and sense frames and Q_{comb} the quality factor of the comb drives. For the calculation of both quality factors we apply the following expression for slide film damping given in [2]:

$$Q_i = \frac{\hat{M}_i \cdot \omega_i \cdot h}{\eta_{eff} \cdot A_{eff}} \tag{4}$$

Here ω_i stands for the angular eigenfrequency of the i-th eigenmode (drive frequency), h for the distance between the moving frame and the substrate or, alternatively, the distance between the comb fingers, A_{eff} for the relevant effective area, and η_{eff} is the effective viscosity accounting for gas rarefaction effects in the low pressure regime and/or small structural dimensions [2]:

$$\eta_{eff} = \frac{\eta_T}{1 + 2 \cdot K_n + 0.2 \cdot K_n^{0.788} e^{-Kn/10}} \tag{5}$$

(with η_T = viscosity of air under normal pressure conditions at room temperature and K_n = Knudsen number).

The modal quality factor of the sense mode is dominated by squeeze film damping and was determined by applying the mixed-level modeling approach proposed in [3]. This approach takes advantage from the small Reynolds numbers and the large aspect ratios typically encountered in MEMS structures and reduces the degree of complexity by replacing the non-linear and highly complicated Navier-Stokes equation by the well-known Reynolds' equation [4]:

$$\nabla\left(\frac{\rho h^3}{12\eta_{eff}} \nabla p(x,y)\right) = \frac{\partial}{\partial t}(\rho h) \tag{6}$$

The pressure distribution *p(x,y)* underneath the moving plates is calculated by discretizing this equation to form a fluidic Kirchhoffian network and solving it by the use of a standard circuit simulator. Edge effects and perforations in the structure are taken into account by introducing physically-based compact models at the respective locations. The pressure-dependent quality factors of the sense mode, as they have been extracted from these mixed-level simulations and incorporated in the high-level model, are shown in Fig. 5.

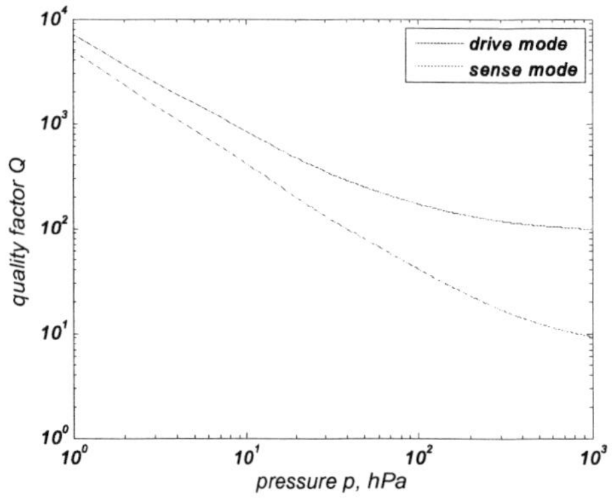

Fig. 5: Pressure-dependent quality factors of the drive and the sense mode of the gyroscope.

3 IMPACT OF ENVIRONMENTAL DISTURBANCES ON THE SENSOR SYSTEM

By their nature, inertial sensors are sensitive to any kind of mechanical forces originating from their environment. Thus, their operation is always affected by disturbing effects arising under real-world operating conditions like shock, cross-talk between sensing and non-sensing axes, and vibrations of the housing. These impact factors enter our sensor model in a physically transparent manner through the parameters $\boldsymbol{C_\Omega}$ and $\boldsymbol{K_\Omega}$ and the right-hand side of the model equations, respectively; hence, it provides the proper basis for studying their influence on the sensor system on the whole and, therefore, it is well suited for the design of adaptive control algorithms, which minimize or even eliminate the environmental disturbances.

As an example, we demonstrate the efficiency of our high-level model by analyzing the impact of housing vibrations on the transient sensor response. To this end, we first consider the package alone and extract its eigenmodes and resonance frequencies from FEM calculations [1]. Tab. 2 shows the results for two exemplary standard package types (PSOIC and QFP).

		PSOIC	*QFP*
in-plane	*x-direction*	*14.654 kHz*	*30.379 kHz*
in-plane	*y-direction*	*14.835 kHz*	*30.387 kHz*
out-of-plane	*z-direction*	*25.666 kHz*	*43.656 kHz*

Table 2: Vibration modes of the PSOIC and QFP packages

The interface circuitry of the gyroscope system has been designed as a digital self-oscillation loop based on Δ-Σ-modulation (see Fig. 6). Among others, it ensures the proper operation of the gyroscope by forcing the drive masses to operate in anti-phase mode.

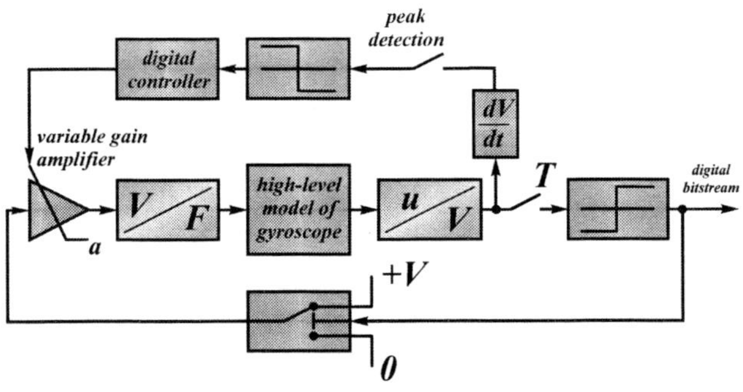

Fig. 6: Block diagram of the drive control loop.

For the compensation of disturbances a so-called "Automatic Gain Control (AGC)" algorithm has been implemented. The basic idea of this approach is to introduce an externally controlled term proportional to $\dot{\underline{u}}$ in the friction force in order to control the oscillation velocity of the drive masses and to adjust it to a constant value [5].

The package vibrations enter the model of the gyroscope system through the term $\underline{a}_0$ in eq. (1). Since the strongest impact on the sensor output signal is to be expected from those vibration modes, which have the same vibration direction as the sense masses, i.e. the y-direction, only these quantities are fed into the gyroscope model, assuming that the vibrations of the package are transferred to the gyroscope without damping losses.

An important result of a full system analysis is that the device loses the ability to compensate the disturbing package vibrations if, due to manufacturing tolerances, slight deviations from a perfectly symmetric sensor layout occur. This becomes evident from the disturbances emerging in the simulated differential output signal of the two sense masses depicted in Figs. 7a and 7b. By the aid of an adaptive control loop these disturbances will be eliminated.

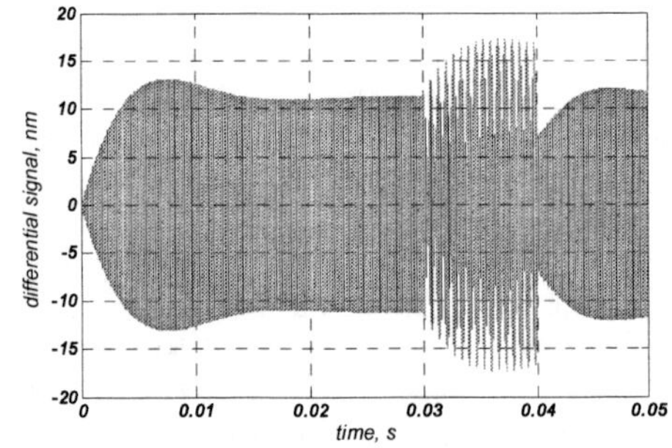

Fig. 7a: Differential output signal of the two sense masses. Impact of the first vibrational eigenmode of the PSOIC housing.

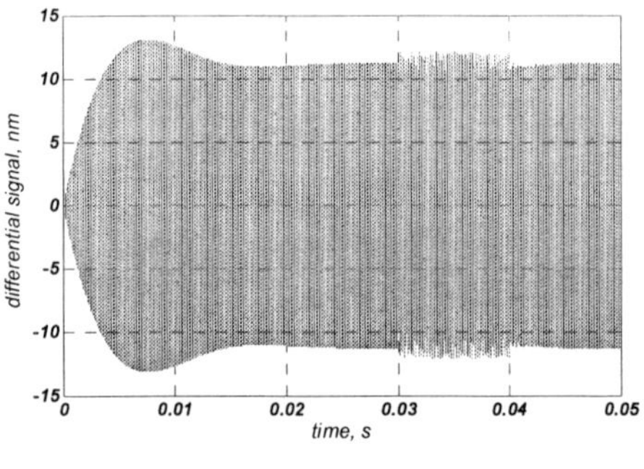

Fig. 7b: Differential output signal of the two sense masses. Impact of the first vibrational eigenmode of the QFP housing.

CONCLUSIONS

We presented a high-level model of a dual gimbaled mass gyroscope that combines computational efficiency with physical transparency on system level, providing a realistic description of the dynamic behavior under real-world operating conditions. Exemplary investigations showing the impact of housing vibrations on the sensor output signal demonstrate the practicality of our approach. It thus provides the proper basis for on-going research activities focusing on the design of control algorithms, which minimize or even eliminate the impact of the most prominent environmental or internal, manufacturing-induced disturbances on the sensor system and, thus, helps to significantly enhance the quality of signal conditioning.

REFERENCES

[1] R. Khalilyulin, F. Salwetter, G. Wachutka: Coupled Package-Device Modeling of an Angular Rate Gyroscope. Book of Abstracts of 18th Micromechanics Europe Workshop (MME 2007), Guimaraes, Portugal, Sept. 16-18, 2007, pp. 297-300.

[2] T. Veijola: Compact Damping Models for Lateral Structures Including Gas Rarefaction Effects. Proceedings of MSM 2000, San Diego, CA, March 27-29, 2000, pp. 162-165.

[3] G. Schrag, G. Wachutka: Physically-Based Modeling of Squeeze Film Damping by Mixed Level System Simulation. Sensors and Actuators A: vol. A97-98 (2002), pp. 193-200.

[4] B.J. Hamrock, Fundamentals of Fluid Film Lubrication, McGraw-Hill, 1994.

[5] R. Khalilyulin, G. Wachutka: Adaptive Control for Reducing the Impact of Damping Effects on the Output Signal of Microgyroscopes. Proc. of the 20th European Conference on Solid-State Transducers (EUROSENSORS 2006), Göteborg, Sweden, Sept. 17-20, 2006, pp.100-101.

Towards an efficient multidisciplinary system-level framework for designing and modeling complex engineered microsystems

J. V. Clark[1,2], Yi Zeng[1], Pankaj Jha[1]

[1]School of Electrical and Computer Engineering
[2]School of Mechanical Engineering
Purdue University, USA, E-mail: jvclark@purdue.edu

ABSTRACT

We present advances in designing and modeling complex microsystems at the network/system-level. For design, we develop a graphical user interface that allows users to quickly configure complex systems in 3D using a computer mouse or pen. And we couple it with a powerful Matlab-based netlist language for design flexibility. For modeling, we apply recent advances in analytical system dynamics and differential-algebraic equations into a framework that facilitates the systematic modeling of multidisciplinary systems with holonomic and non-holonomic constraints. For a test case, we efficiently simulate a microsystem comprising gears, hinges, slider, electronic components, comb drives, and electromechanical flexures. In comparison, it is difficult to model, configure, and simulate such a microsystem using conventional tools.

Keywords: GUI, netlist, DAE, PSugar, Sugar, MEMS, design, modeling, simulation, constraints, multidisciplinary.

1. INTRODUCTION

Distributed-element tools using finite element analysis (FEA) are often used for characterizing new geometries and interactions [1]. However, a significant number of microsystem designers reuse parameterized sets of well-characterized elements. This area of design and modeling led to the development of network/system-level tools for microsystems. Examples include Sugar [2], Nodas [3], Architect [4], and Synple [5]. What such tools have in common are: (1) the ability to abstract the configuration of a system in the form of a netlist, where each line of text describes an element's connectivity to other elements, its orientation, and it modeling parameters; and (2) a significant reduction in the computation time (when compared to FEA) by using computationally-efficient models based on reduced order modeling, matrix structural analysis, and or modified nodal analysis [2]-[5]. Such benefits facilitate the exploration of parameterized design spaces of systems with a multitude of components.

Recently, there has been interest in accommodating the user's need to efficiently design, model, and simulate complex engineered microsystems. Users demand friendlier graphical user interfaces, more flexible netlist capabilities, and less complicated modeling frameworks [2]. In effort toward fulfilling these requirements, in this paper we expand upon our previous effort called Sugar to develop an interactive GUI-netlist and a systematic modeling framework. The present effort is called PSugar. Its architecture is illustrated in Figure 1.

In Section 2, we present our powerful netlist features. In Section 3, we present our novel 3D interactive GUI technology. In Section 4, we present our systematic modeling framework. In Section 5, we discuss differential-algebraic equation solvers. And in Section 6, we provide a simulation example based on a microsystem developed by Sandia National Lab.

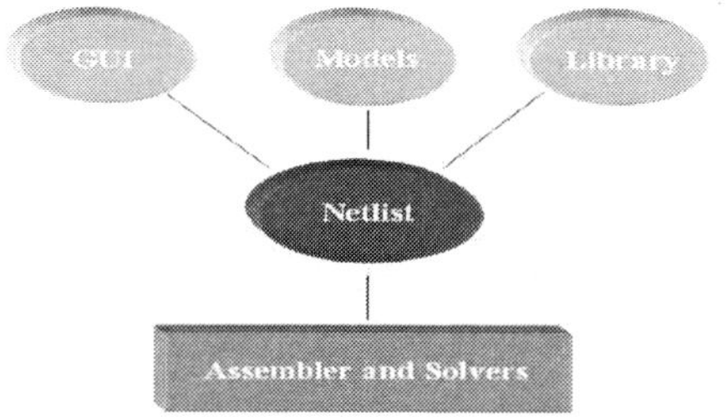

Figure 1: Architecture of PSugar. The ellipses symbolize parameters and constitutive relationships of a prescribed system; the rectangle symbolizes solution algorithms, which assemble and solve the mathematical representation of the system.

2. NETLIST

In this section we present PSugar's netlist capabilities by comparing it to Sugar's netlist capabilities.

What we improve on is the limited capabilities of Sugar's netlist language. For instance, Sugar's netlist syntax has the form

$$\mathit{model}_1 \;\; \mathit{layer}_1 \;\; [\mathit{nodes}_1] \;\; [\mathit{parameters}_1]$$
$$\mathit{model}_N \;\; \mathit{layer}_N \;\; [\mathit{nodes}_N] \;\; [\mathit{parameters}_N]$$

where *model* is the name of the element model, e.g., mechanical flexure, resistor, comb drive, anchor, etc.; *layer* is the material that the element is composed of; *nodes* is a list of element nodes which are the places that the element can be connected to the nodes of other elements; and *parameters* is the list of modeling parameters, e.g. orientation, Young's modulus, residual stress, temperature, etc. Sugar's netlist language allows for subnets and importing parameter values from the Matlab workspace into the netlist. However, some problems with this format are as follows. The format requires the user to continuously switch between two incompatible syntaxes – one for Sugar, and another for Matlab and Sugar functions. The netlist format allows simple arithmetic expressions such as defining constants, allowing nodes to be represented by simple array variables, and nested loops. However, it does not allow sophisticated computations such as those dealing with matrices or those needing access to external functions from within the netlist. This limits design flexibility.

PSugar's netlist format overcomes these issues by being 100% Matlab compatible. Many students and professionals are familiar with Matlab. PSugar's netlist is an m-file with arbitrary format, where the only requirement is that the output must be a Matlab cell structure of the form

$$[\,model_1\ \{nodes_1\}\ \{parameters_1\};\quad model_N\ \{nodes_N\}\ \{parameters_N\}\,];$$

where *model* is the name of the element model; *nodes* is a list of node names; and *parameters* is the list of modeling parameters such as,

{'L' 100e-6; 'force' comb(V,W,H, L,N); 'C' cross(A,B) };

where the designer has unfettered access to all Matlab functions, workspace quantities, and third party plug-ins.

3. GUI

In this section, we present a few of our efficient, interactive GUI features.

Besides using a netlist, another way to configure a system in PSugar is by using its graphic configuration window (GCW). With a computer mouse or pen, drawplanes and elements can be quickly and easily positioned, repositioned, and default values can be edited in the GCW. The GCW is coupled to an interactive netlist window (INW). That is, each element that is graphically configured in the GCW has its corresponding netlist text automatically generated in the INW; and vice versa, a line of netlist text entered into the INW immediately updates the system configuration display in the GCW.

Drawplanes are uniquely determined with 3 coordinates. These 3 coordinates are identified in PSugar by projecting the 2D cursor position upon the 3D xy-, zy-, or xz-planes of the coordinate axis; or other object. For example, Figures 2a-2c illustrate the 3-button-click sequence to configure a drawplane in 3D. Figure 2d shows an element placed on that drawplane using 2 button clicks. By dragging a node about its associated plane, both elements and drawplanes can be readily repositioned. The GCW also has snap-to-grid and snap-to-node options.

The elements or subsystems that are configured onto the GCW are selected through an element menu window (EMW). The set of elements from PSugar's library that are listed in these EMWs are user-definable. For example, a user may choose resistor, capacitor, inductor, opamp, integrator, diode, rectifier, and transistor as button choices for one EMW; and comb drive, beam, anchor, folded flexure, hinge, point mass, and carbon nanotube as button choices for another EMW. E.g., see Figure 3.

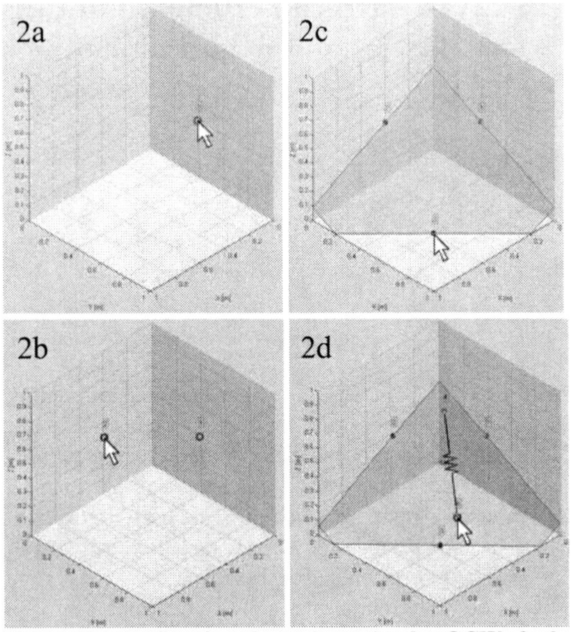

Figure 2: Configuring drawplanes in the GCW. 2a-2c shows a 3-button-click sequence to configure a drawplane in 3D. 2d shows the placement of a resistor element on the drawplane, using 2 button clicks. The 2a-2d sequence took a user a couple of seconds.

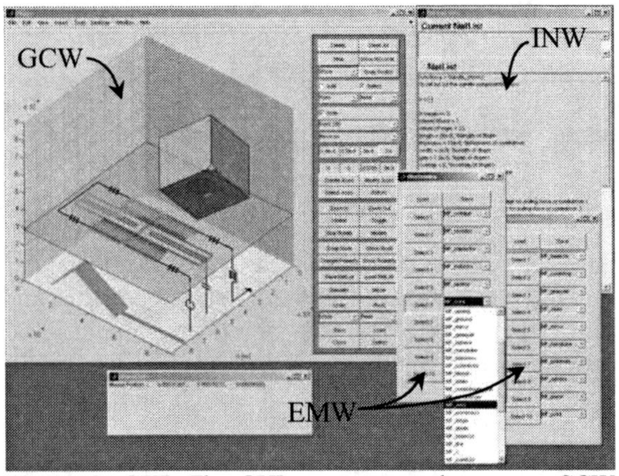

Figure 3: GUI. The GUI components shown are: GCW (graphic configuration window), INW (interactive netlist window), and a couple of EMWs (element menu windows). Elements shown include a hinged mirror with slider, folded flexure, and gear-pair in a local frame.

4. MODELING

In this section we discuss our systematic modeling method, which reduces the modeler's effort and readily accommodates systems subject to algebraic constraints.

In our previous work with Sugar, we represented microsystems as a set of second order differential equations (ODEs) of the form,

$$M\ddot{q} + B\dot{q} + Kq - F_{ext} = 0 \qquad (1)$$

where M, B, K are the system mass, damping, and stiffness matrices, and F_{ext} is the vector of externally applied efforts such as electrostatic, gravitational, noninertial, voltage, stress, etc. [2]. However, ODE solvers are typically not equipped to solve a system of differential-algebraic equations (DAEs) [6], where the system mass matrix may be singular or zero. There are many complex microsystems that are amenable to DAEs, such as those comprising elements without inertia; elements with displacement-, flow-, dynamic variable-, or effort-constraints; or elements with inequality constraints.

The DAE form we use in PSugar is [7],

$$\frac{d}{dt}\nabla_f T^* + \nabla_f D + \nabla_q V - F_{ext} = \nabla_q T^* - \Phi_q^T \mu - \Psi_f^T \kappa \quad \text{(2a)}$$

$$0 = \Phi(q,t) \quad \text{vector of displacement constraints} \quad \text{(2b)}$$

$$0 = \Psi(f,q,t) \quad \text{vector of flow constraints} \quad \text{(2c)}$$

$$0 = \Gamma(e^\gamma, s, f, q, t) \quad \text{vector of effort constraints} \quad \text{(2d)}$$

$$0 = \dot{s} - \Lambda(e^\gamma, \dot{s}, s, f, q, t) \quad \text{vector of dynamic variables} \quad \text{(2e)}$$

where $T^*(f,q,t)$, $D(f,q,t)$, and $V(q,t)$ are kinetic co-energy, content, and potential; Φ, Ψ, Λ, and Γ are displacement-, flow-, dynamic variable-, and effort-constraints; $\mu(t)$ and $\kappa(t)$ are Lagrange multipliers; vector $q(t)$ represents generalized displacements such as the change in translation, rotation, charge, volume, entropy, etc. Comparing (2a) to (1), each term on the left-hand side of (2a) has the same meaning as each term in (1), respectively; and vector $f = \dot{q}$ is flow. However, (2a) allows kinetic co-energy to be a function of q and allows the dynamics to be constrained. Since vectors μ and κ are additional unknowns, relations (2b) and (2c) are required additional equations. (2d) is included to allow for effort e^γ constraints. And (2e) is included to allow for relations that are not represented within the energy functions; e.g., time derivatives of flow $\dot{s} = d\dot{q}/dt$ or time integrations of displacement $s = \int q\, dt$, etc.

Systematic modeling in PSugar is as follows. Each element has a representative parameterized model function containing its energy functions, constraints, and efforts. E.g. a linear 2-node, 12-DOF flexure model function returns the symbolic scalar $V_i = \frac{1}{2} q^{1\times12} K^{12\times12} q^{12\times1}$. The assembler sums all energy functions, e.g. $V = \sum_{i=1}^{N} V_i$, then substitutes the functions into (2a)-(2e) for symbolic differentiation. Hence, in PSugar the modeler's effort is reduced to simply providing energy functions, constraints, and efforts. The common practice of rigorously manipulating a model into a particular form is eliminated.

5. Simulation

In general, the solution of a DAE involves solving a nonlinear algebraic equation at each time step.

For instance, applying linear approximations to (2), such as $\dot{q}_{n+1} \approx (q_{n+1} - q_n)/h_{n+1}$, Euler's method yields

$$\begin{pmatrix} \dfrac{q_{n+1} - q_n}{h_{n+1}} - f_{n+1} \\ M_{n+1} \dfrac{f_{n+1} - f_n}{h_{n+1}} + \Phi_q^T\big|_{n+1} \kappa_{n+1} + \Psi_q^T\big|_{n+1} \mu_{n+1} - \Upsilon_{n+1} \\ \Phi_{n+1} \\ \Psi_{n+1} \\ \Gamma_{n+1} \\ \dfrac{s_{n+1} - s_n}{h_{n+1}} - \Lambda_{n+1} \end{pmatrix} = 0 \quad (3)$$

where h is the step size in time, $M = \nabla_f^2 T^*$, $\Upsilon = Q - (\nabla_f T^*)_q f - (\nabla_f T^*)_t + \nabla_q T^* - \nabla_q V - \nabla_f D$, and the Jacobian $\partial F/\partial y_{n+1}$ can be evaluated using finite differences. There are several pubic domain solvers available for DAEs [6]. The choice of solver usually depends on the differential index of the DAE; that is, the minimum number of times that some or all of the equations would need to be differentiated in time to determine its underlying ODE. However, using an ODE solver to solve the resulting underlying ODE is not preferred, because the solution trajectory often drifts from the solution manifold that is defined by the explicit constraints in the original DAE. Methods such as BDF (Backward differentiation formula) and IRK (implicit Runge-Kutta) improve Euler's method by using higher-order approximations for $\dot{q}_{n+1}$, $\dot{f}_{n+1}$, and $\dot{s}_{n+1}$, and using variable step sizes.

Since Matlab's ode15s and ode23t solvers are only for index-1 DAE systems, we do not use them to directly solve (2), because its index can be as high as 3. Currently, we use any one of a collection of public domain high-index DAE solvers. Future work includes developing a DAE solver that exploits our particular DAE structure.

6. Example

To exemplify our methodology, we simulate a microsystem similar to one that was developed by Sandia National Labs (Figure 5). We chose this particular system as an example because it is difficult model and simulate using conventional tools [8]. Multidisciplinary elements of the microsystem comprise resistors, capacitors, voltage sources, mechanical flexures, hinges, sliders, gears, and electrostatic comb drives. Its representation in PSugar is shown in Figure 6. We configure the system using both the GUI and netlist (e.g. the repetitive comb fingers). The system is represented by 4601 degrees of freedom. Using a Pentium-4, 3GHz processor with 1GB RAM, our BDF DAE solver averages 1.4 seconds per step in Matlab.

Identical ramp voltages applied to the two orthogonal sets of comb drives rotate the smallest gear counter-

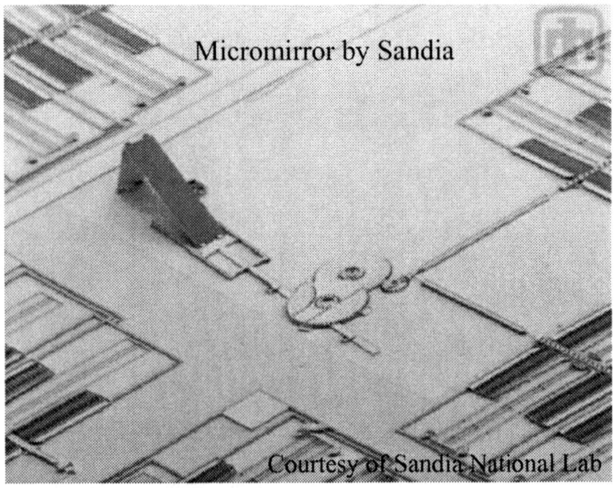

Figure 5: A complex microsystem.

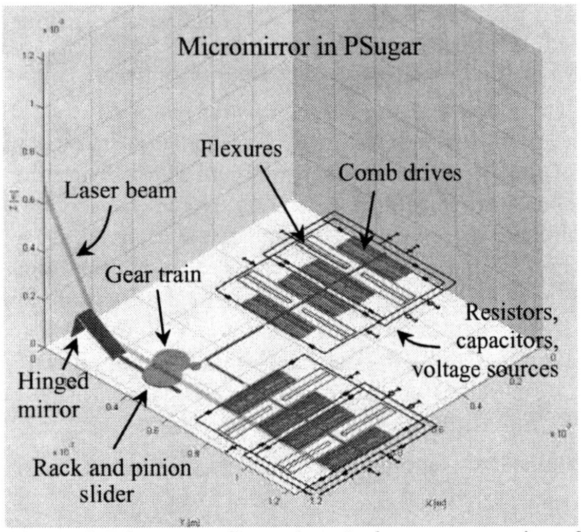

Figure 6: PSugar configuration. A close representation of the Sandia device from Figure 5 configured in PSugar. The red beam in the figure is a laser reflecting off the mirror.

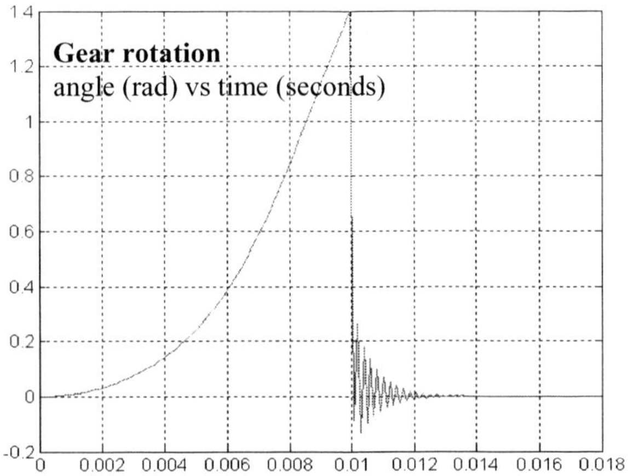

Figure 7: Transient analysis. Simulation of the configuration shown in Figure 6. A pair of linear voltage ramps applied across the two sets of orthogonal comb drives rotates the smallest gear a quarter turn. After the ramps end, the system settles back to its initial state.

clockwise $\sim \pi/2$ radians. As the voltage is removed, the system settles back to its initial state. See Figure 7. Slack between hinges and gear teeth are not modeled. The true geometry and material properties of actual microsystem shown in Figure 5 were not made available at the time of this writing. The actual friction in the hinges, gears, and slider are not known.

CONCLUSION

We presented a couple of design and modeling advancements toward an efficient system-level framework for complex engineered microsystems. Regarding design, we presented our interactive GUI that allows users to quickly and easily configure complex configurations; and we presented our powerful netlist language, which embraces the full flexibility and functionality of the Matlab language. Regarding modeling, we discussed the systematic method we use for representing complex, multidisciplinary components by their energy functions, constraints, and efforts. All code development was done in Matlab.

Some of our future work will involve verifying the performance of complex systems (qualitative results were presented here), developing a solver that fully exploits our DAE structure, and enhancing our library with components, subsystems, and complete systems.

ACKNOWLEDGEMENT

We thank R. A. Layton of Rose-Hulman Institute of Technology and B. Fabian of the University of Washington for insightful discussions on system dynamics.

REFERENCES

[1] S.D. Senturia, "CAD Challenges for Microsensors, Microactuators and Microsystems," *Proceedings of the IEEE*, Vol. 86, No. 8, August 1998, pp. 1611-1626.

[2] J. V. Clark and K. S. J. Pister, "Modeling, Simulation, and Verification of a Advanced Micromirror Using Sugar", *Journal of Microelectromechanical Systems*, Vol. 16, No. 6, 2007, pp. 1524-1536.

[3] G. K. Fedder and Q. Jing, "NODAS 1.3: Nodal Design of Actuators and Sensors", *Proc. IEEE/VIUF Int. Workshop on Behavioral Modeling and Simulation*, Orlando, FL, Oct 1998.

[4] Coventorware, 951 Mariners Island BLVD, Suite 205, San Mateo CA 94404. http://www.coventor.com

[5] IntelliSense Corporation, 600 W. Cummings Park Suite 2000, Woburn MA 01801. http://www.intellisense.com

[6] K. E. Brenan, S. L. Cambell, L. R. Petold, *Numerical Solution of Initial-Value Problems in Differential-Algebraic Equations*, SIAM, 1996.

[7] R. A. Layton, *Principles of Analytical System Dynamics*, Springer, 1998.

[8] D. Sandison, "Keynote Address: Moving MEMS from Novelty to Necessity – a National Security Perspective", *TEXMEMS VII*, El Paso TX, 2005. http://www.uacj.mx/Texmems/Keynotes.htm

Bulk-Titanium Waveguide – a New Building Block for Microwave Planar Circuits

X. T. Huang, S. Todd, C. Ding and N. C. MacDonald

Mechanical Engineering Department
University of California at Santa Barbara
Santa Barbara, CA 93106

ABSTRACT

We developed a new building block for constructing planar microwave circuitry: bulk titanium waveguide. The waveguide is formed by deep trench etching, dielectric gap filling, and planarization. The high aspect ratio of the resulting structure provides superior field confinement, low crosstalk, therefore a small footprint. To enable bulk titanium as a substrate compatible with microfabrication, we developed a suite of techniques that overcome unique manufacturing issues associated with titanium. The propagation loss of the waveguides was measured to be ~0.68dB/mm at 40GHz.

Keywords: titanium, MEMS, waveguide, RFMEMS

1 INTRODUCTION

The majority of microwave planar circuits are based on transmission lines of two basic topologies: Microstrip [1-2] and Coplanar Waveguide (CPW) [3-4]. Microstrip offers low loss, while suffering from large circuit size, high dispersion, poor design flexibility, and the necessity of large via holes. CPW improves upon microstrip by bringing the ground to the same level as the signal, therefore providing low dispersion, smaller circuit size, high design flexibility and no via holes. However, the fields in CPW concentrate on the thin edges and are therefore less confined. This translates into higher conductor loss, cross-talk and poor power handling capability.

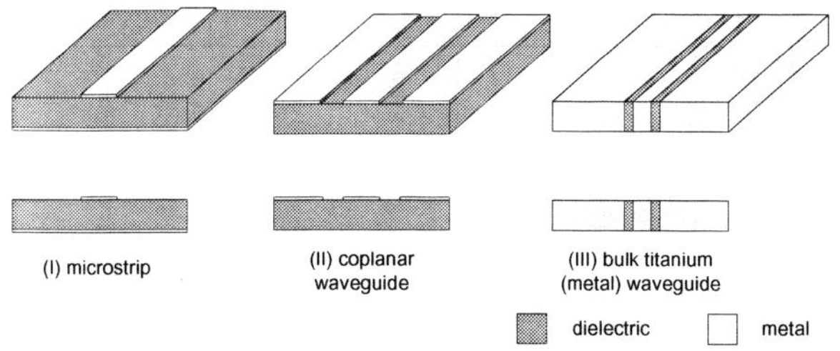

Figure 1. Bulk titanium waveguide compared to conventional planar circuits

We present a new topology for planar microwave transmission line: bulk titanium waveguide. As shown in Figure 1, the waveguide is formed by coplanar titanium regions separated by low-loss dielectric that is planarized to the same level as the titanium. Due to the high aspect ratio of the dielectric cross section, the electric field is confined primarily within the dielectric, therefore reducing losses due to radiation and parasitic coupling. This geometry distributes the surface current across the entire height of the structure, reducing conductor losses while enhancing power handling capabilities. The high aspect ratio in-plane electrical isolation also serves as through-wafer interconnects with a high packing density. When integrated with a bulk titanium package, this waveguide provides a robust, compact, packaging solution for microwave subsystems as well as other microcomponents such as MEMS. We have successfully designed, fabricated and characterized packaged waveguides which measured, at 40GHz, ~0.68dB/mm insertion loss and the impact due to package < 0.1dB. Additionally, we developed a quasi-static model based on conformal mapping that accurately describes the characteristics of the waveguide.

Titanium [5-7] has one of the highest strength to weight ratio among metals. It is also one of the few materials with an endurance limit, a desirable property as a robust packaging material. It is naturally resistant to corrosive environments and is widely used as implants due to its bio-compatibility. The discovery of anisotropic plasma etching of bulk titanium [8-9] has opened new doors to numerous MEMS applications [10-12]. However, the current manufacturing processes for titanium sheets are not tailored for microfabrication. We developed a suite of technologies that overcome issues such as residual stress, thickness variation, embedded defects and surface roughness. With a typical 1"x1" starting substrate with roughness ~100 nm, a thickness variation of ~10 μm, and a bow of >~50μm, our processes achieved a thickness variation of <2μm, a bow of ~7μm, an average roughness as low as 2.8 nm, and μm-scale features with >200μm etching depths.

2 ANALYTICAL MODEL

While numerical simulation typically gives more accurate answers, analytical models provide more physical intuition and conceptual understanding.

The field distribution in the bulk titanium waveguide is illustrated in Figure 2. The waveguide height is h; center conductor width w; gap width s. Device packaging is completed by attaching a metal cavity of height h_p. The electromagnetic signal is confined primarily within the high aspect ratio gaps filled with low loss dielectric material with dielectric constant ε_d.

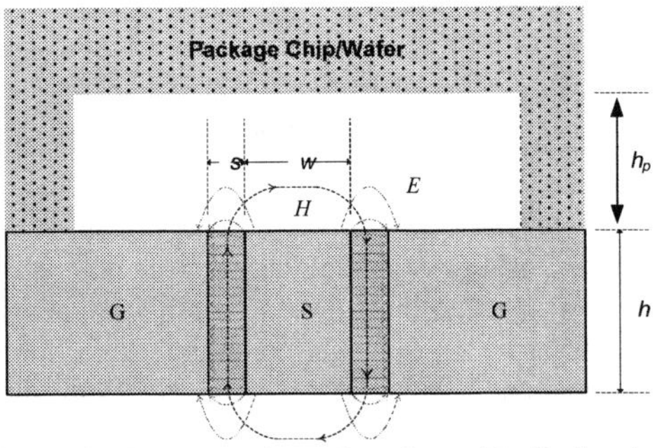

Figure 2. Cross section of packaged bulk titanium waveguide.

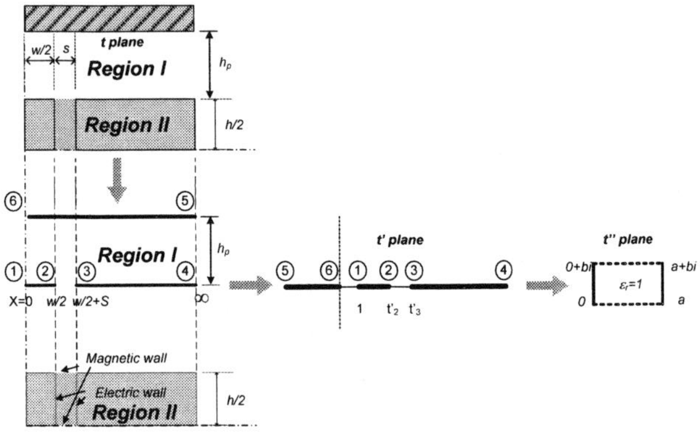

Figure 3. Conformal maps for quasi-static analysis

While parallel plate assumption offers a qualitative description, much more accurate modeling can be obtained by using conformal mapping.

Using a series of Schwartz-Christoffel transformations [13], the different regions of the device are transformed into mathematically solvable geometries. The reverse transformation gives the approximate solution.

At microwave frequencies and for the case in which the loss is primarily due to the conductors, the propagation loss can be simplified:

$$\alpha \approx \frac{\mu_0\mu_d \tan\delta_d \omega\varepsilon_d\varepsilon_0 + \frac{2}{h}\sqrt{\frac{\omega\mu_c\mu_0}{2\sigma_c}}\varepsilon_0\left[\frac{2K(k)}{K'(k)} + \frac{K(k_1)}{K'(k_1)} + \frac{\varepsilon_d h}{s}\right]}{2\sqrt{\frac{\mu_0\mu_d\varepsilon_0 s}{h}\left[\frac{2K(k)}{K'(k)} + \frac{K(k_1)}{K'(k_1)} + \frac{\varepsilon_d h}{s}\right]}} \quad (1)$$

where: $k_1 = \tanh(\pi w / 2h_p) / \tanh[\pi(w+2s)/2h_p]$, $k = w/(w+2s)$, and $K(k)$ is the complete elliptic integral of the first type, $K'(k) \equiv K(k' = \sqrt{1-k^2})$.

In case loss is primarily due to the finite conductivity of the conductors, the additional loss due to the presence of the metal package can be approximated by:

$$\Delta\alpha_{pac} \approx \sqrt{\frac{\omega\mu_c\varepsilon_0}{2\mu_d\sigma_c hs}}\left(\sqrt{\frac{2K(k)}{K'(k)} + \frac{K(k_1)}{K'(k_1)} + \frac{\varepsilon_d h}{s}} - \sqrt{\frac{2K(k)}{K'(k)} + \frac{\varepsilon_d h}{s}}\right) \quad (2)$$

The advantage of the high aspect ratio waveguide is demonstrated in the reduced coupling between waveguide and a metal package reflected in the small change in propagation loss: the change in loss, Δα~0.04dB at 40 GHz using Equation (2).

3 EXPERIMENT

3.1 Titanium planarization

Although titanium has a long history of industrial, medical and military applications [5-7], it had not been used as a MEMS substrate until recently. Titanium sheets as obtained, especially thinner (<200 μm) and smaller (non-wafer scale) samples, may suffer from planarity issues, such as rolling hills, due to the manufacturing processes. The sample surface may contain defects which result in etching difficulties. Samples undergo plastic deformation around the edges during cutting and improper handling. We develop a suite of processes which correct these problems, thus enabling the use of bulk titanium as a MEMS material.

At elevated temperatures, titanium can be annealed to release the stress [7]. We used a stress-anneal straightening process to correct the global deformations in thinner samples.

Figure 4. Lapping and polishing development

We developed a lapping and polishing process for removing the defects embedded near the titanium surface. Shown in Figure 4 is the Engis Hyprez polisher we used and the associated tools for mounting samples.

3.2 Waveguide fabrication

The process starts with a 200μm titanium substrate. As shown in Figure 4, first, deep trenches are etched into the substrate using a CL_2/Ar chemistry in an ICP etcher [9,12]. High density PECVD SiO_2 etch mask is used to ensure uniformity. Once etching is finished and the remaining mask oxide removed, a ~ 1μm Au layer is sputtered. This layer covers the sidewall of the trenches and enhances the waveguide conductivity. The trenches are then filled with BenzoCycloButene (BCB) and cured. The polymer fills the trenches and forms the waveguide dielectric. The first lapping and polishing process is done to remove the excess BCB and expose the trenches. Next, Au bonding pads are deposited and patterned to ready to receive the packaging chip.

Similar to the device chip, the fabrication of the packaging chip is also done by deep etching and Au sputtering. Once both chips are ready, they are bonded together by using Au thermal compression bond at 300°C for 30 minutes. The bonding creates a sealed metal cavity that can be used for future device packaging as well as providing additional shielding against undesirable electromagnetic waves. Last, the chip/wafer assembly is polished to remove excess substrate titanium and the waveguide structure is completed.

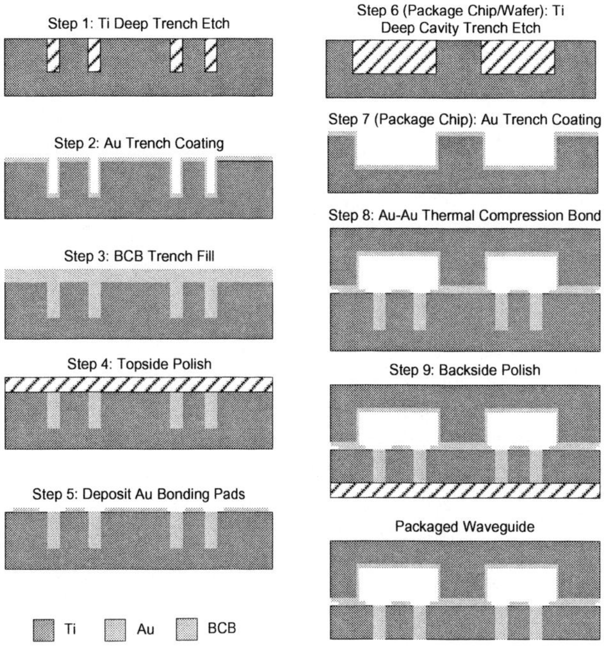

Figure 4. Process sequence for bulk titanium waveguide

Figure 5a shows the structure after trench etching and conformal Au sidewall coating; Figure 5b demonstrates the completed device. The regions isolated by the BCB-filled trenches produced charging in the SEM.

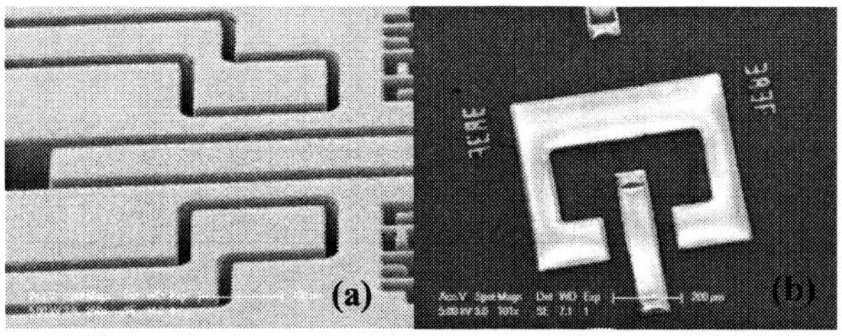

Figure 5. a) waveguide structure after sputtered Au sidewall enhancement; b) completed waveguide

A close-up view is shown in Figure 6. BCB demonstrates good structural integrity after the final planarization step.

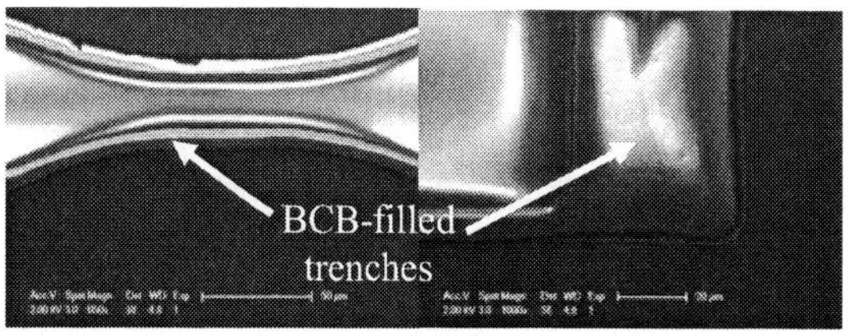

Figure 6. Close-up view of BCB-filled isolation trenches

3.3 Measurement setup

For improved accuracy, a set of TRL calibration devices are fabricated on the same chip. Line lengths ranging from 100μm to 2mm are used to calibrate the entire measurement range of 18GHz to 40GHz, as well as to characterize the propagation loss.

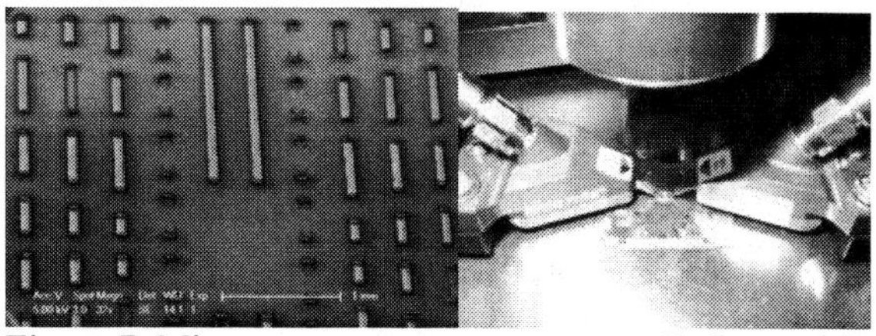

Figure 7. Microwave measurement setup. Left: on-chip TRL calibration devices; Right: packaged device under test

4 RESULTS AND DISCUSSION

4.1 Titanium planarization

Although with a Young's modulus of 108.5 GPa [7], the metallic nature of the titanium makes thinner (<200μm) samples subject to plastic deformation due to cutting and improper handling. Residual stress may also exist resulting from the manufacturing process. The stress-anneal straightening process proves effective in removing such deformations. The average gross deformation of 5 samples, measured by optical interferometry, is reduced from ~52.8μm to ~6.9μm.

Other issues with large impact to subsequent fabrication steps and RF performance include thickness variation and surface defects. The thickness variation of the 200μm thick samples is measured to be ~9.51μm and reduced to ~1.58μm after planarization. Roughness, an important indicator of the surface, is reduced from ~50-100nm to as low as ~2.8nm.

4.2 Loss characterization

The propagation loss of the bulk titanum waveguide is measured and compared to theory in Figure 8. Waveguide lengths range from 100μm to 600μm long with 100μm increments. At 40 GHz, the loss for the 600μm long waveguide is measured to be ~22.3% higher than theory. This can be attributed to the process imperfection, measurement error and the approximations used in the derivation.

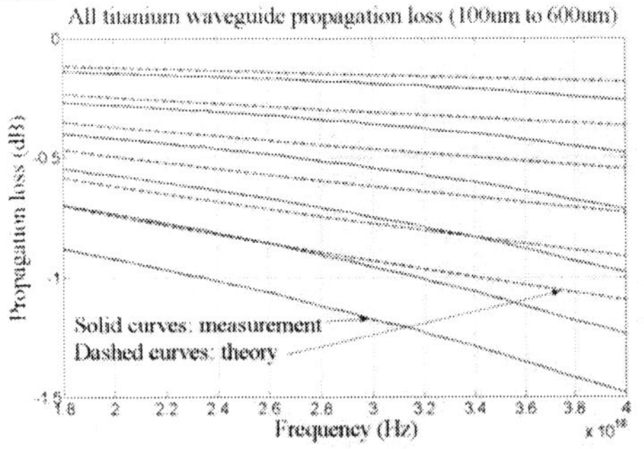

Figure 8. Propagation loss of bulk titanium waveguides. From top to bottom: 100μm to 600μm with 100μm increments. Solid: theory; Dashed: measurement

The loss is significantly reduced by Au coating on the sidewall. As shown in Figure 9, the loss at 40GHz is limited to ~0.68dB/mm. Ripples in the curves may be due to uneven Au thickness on the sidewall.

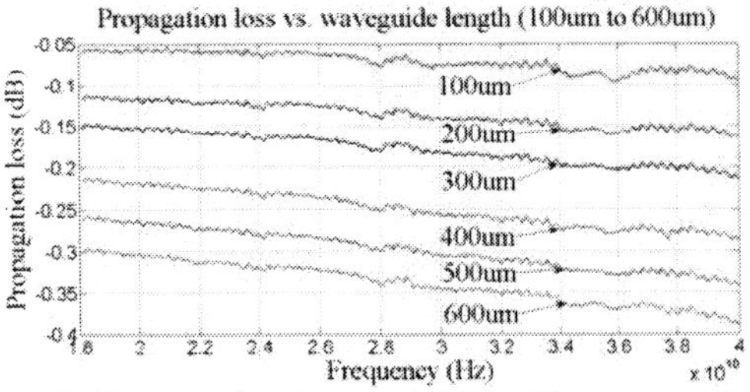

Figure 9. Propagation loss at 40GHz. From top to bottom: 100μm to 600μm with 100μm increments

The additional loss at 40GHz due to packaging is measured to be ~0.06dB, consistent with model prediction.

5 CONCLUSION AND FUTURE WORK

Bulk titanium waveguide has distinct advantages over traditional topologies of microwave planar circuits. The superior field confinement helps reducing undesirable coupling, as demonstrated in the low measured impact due to packaging. Overall circuit area can be condensed due to reduced parasitic coupling. The integrated, high-density, through-wafer interconnects further reduce circuit size while improving performance due to shorter signal paths. For packaging applications, this translates into a compact, titanium packaging well suited for corrosive environments such as sea water. In addition, the bio-compatibility of titanium lends itself to potential implantable biomedical applications.

We have successfully demonstrated the low-loss and low parasitic coupling of bulk titanium waveguide. In order to further explore the opportunities of potential applications based on the bulk titanium RF platform, it is necessary to investigate the fundamental properties of the waveguide in implementations beyond its original ideal waveguide form. The most notable implementation in this case involves discontinuities. Fully understanding these discontinuities is essential for device applications beyond waveguides.

REFERENCES

1. H. A. Wheeler, *Transmission-line properties of parallel wide strips by a conformal-mapping approximation.* IEEE Tran. Microwave Theory Tech., 1964. MTT-12: p. 280-289.
2. H. A. Wheeler, *Transmission-line properties of parallel strips separated by a dielectric sheet.* IEEE Tran. Microwave Theory Tech., 1965. MTT-13: p. 172-185.
3. C. P. Wen, *Coplanar Waveguide, a Surface Strip Transmission Line Suitable for Nonreciprocal Gyromagnetic Device Applications.* Microwave Symposium Digest, G-MTT International, 1969. 69(1): p. 110-115.
4. C. P. Wen, *Coplanar-Waveguide Directional Couplers.* Microwave Theory and Techniques, IEEE Transactions on, 1970. 18(6): p. 318-322.
5. *Titanium.* 2005: Encyclopedia Britannica Concise.
6. G. Leutjering, J. C. Williams, *Titanium.* 2003: Springer.
7. M. J. Donachie, Jr., *TITANIUM: A Technical Guide.* 1988, Metals Park, OH: ASM International.
8. M. F. Aimi, M. P. Rao, N. C. MacDonald, A. S. Zuruzi, D. P. Bothman, *High-aspect-ratio bulk micromachining of titanium.* Nature Materials, 2004. **3**: p. 103-105.
9. E. R. Parker, M.F.A., B. J. Thibeault, M. P. Rao, N. C.MacDonald, *High-Aspect-Ratio ICP Etching of Bulk Titanium for MEMS Applications*, in *Electrochemical Society*. 2004: Honolulu, Hawaii.
10. M. F. Aimi, *Bulk Titanium Micro-electro-mechanical Systems.* 2005, University of California: Santa Barbara.
11. C. Ding, X. Huang., G. Gregori, E. R. Parker, M. P. Rao, D. R. Clarke, N. C. MacDonald. *Development of Bulk Titanium Based MEMS RF Switch for Harsh Environment Applications,.* in *2005 ASME International Mechanical Engineering Congress and Exposition (IMECE 2005)*. 2005. Orlando, Florida.
12. E. R. Parker, B.J.T., M. F. Aimi, M. P. Rao, N. C. MacDonald, *Inductively Coupled Plasma Etching of Bulk Titanium for MEMS Applications.* Journal of the Electrochemical Society, 2005. 152(10): p. 675-683.
13. T. A. Driscoll, L. N. Trefethen, *Schwarz-Christoffel Mapping*. 2002: Cambridge University Press.

The static behavior of RF MEMS capacitive switches in contact

H.M.R. Suy*, R.W. Herfst*, P.G. Steeneken*, J. Stulemeijer** and J.A. Bielen**

* NXP Semiconductors Research, High Tech Campus 37
5656 AE Eindhoven, The Netherlands, hilco.suy@nxp.com
** NXP Semiconductors, Gerstweg 2
6534 AE Nijmegen, The Netherlands

ABSTRACT

A method is presented in which surface topography characterization is combined with the electrical measurement of the contact mechanics under electrostatic loading. Contact characteristics such as the surface separation versus the applied pressure, and the applied pressure versus the contact area, are derived. Based on these results, a contact model is validated. In combination with a compact model of a capacitive switch, this contact model is used to predict the contact behavior of different switch designs.

Keywords: MEMS switches, RF MEMS, contact, contact model, surface roughness

1 INTRODUCTION

Mechanical contact plays a crucial role in MEMS switches, influencing, for example, contact resistance in Ohmic contact switches [1], [2] and reliability (stiction and wear) in general [3], [4]. In this study, we focus on the contact between a membrane and a dielectric in a capacitive switch, and its influence on the capacitance density and slope of the capacitance versus voltage (C-V) curve above pull-in. Depending on the layout of the device, these two parameters affect important switch properties, such as the on-off capacitance ratio, the pull-out voltage, and the capacitance drift in case of dielectric charging. The comprehension of the contact mechanism, and the predictive modeling of such behavior, is thus essential for successful future integration of these devices in, for instance, adaptive antenna matching modules and reconfigurable RF circuits. The goal is to combine surface topography measurements with standard C-V measurements to obtain contact characteristics that can be used to: i) validate contact models more accurately, ii) assist in the contact model parameter extraction, and iii) predict the behavior of other switch designs.

2 METHODS AND RESULTS

Every surface has some degree of roughness. One method to characterize surface topography is Atomic Force Microscopy (AFM), in which a cantilever with a sharp tip is used to scan a surface. Here, the dynamic tapping mode is employed to avoid cantilever stiction. As mentioned in the introduction, the surface topography characterization results (discussed in detail in subsection 2.1) are combined with measurements of the contact mechanics under electrostatic loading (subsection 2.2). These contact mechanics measurements are carried out via C-V measurements on a customized RF setup at a probing frequency of 890 MHz [5]. The combined results of the AFM and C-V measurements are discussed in subsection 2.3 and compared to a contact model.

2.1 Surface Topography Characterization

A Scanning Electron Microscopy (SEM) image of the capacitive switch under study is shown in figure 1. The

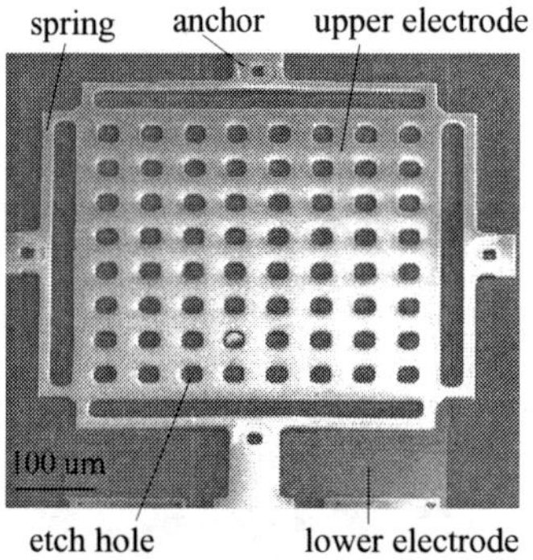

Figure 1: SEM image of the MEMS capacitive switch under study.

switch consists of a perforated upper electrode (membrane) that is suspended by springs over a lower electrode, covered with a dielectric. For various devices, the membrane is removed and flipped so that the bottom is exposed, after which the surface height is measured with AFM on different parts of the membrane. From the AFM data (figure 2), relevant statistical parameters such as the standard deviation, skewness and kurtosis (measure of asymmetry and number of outliers in a statistical distribution, respectively) are calculated, as well as a 2D autocorrelation function. The results (figures 3 and 4) show that the surface height distribution is close to Gaussian, and that no spatial periodic components are present. No dependency on wafer location, switch design, place on the sample, and number of switching cycles, is observed.

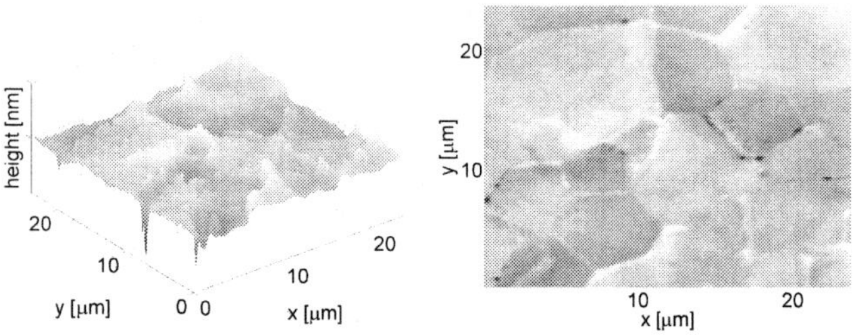

Figure 2: AFM measurement of the bottom of a switch membrane. A planar view is presented on the right, showing the granular structure of the membrane.

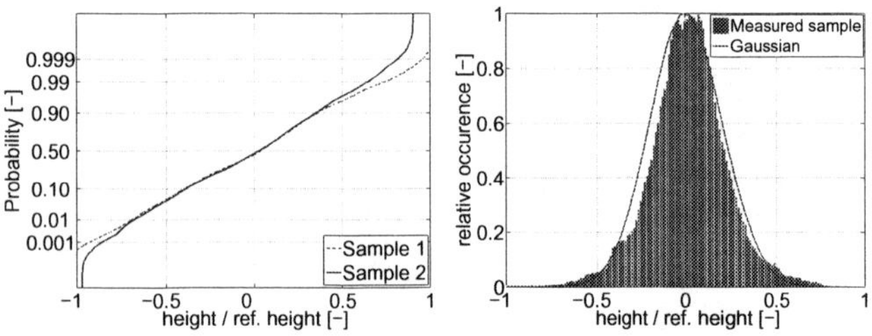

Figure 3: Normal probability plot (left) and histogram (right) of the surface height, normalized to an arbitrary reference height. In the histogram, a comparison is made with an ideal Gaussian distribution, which would produce a straight line in the normal probability plot.

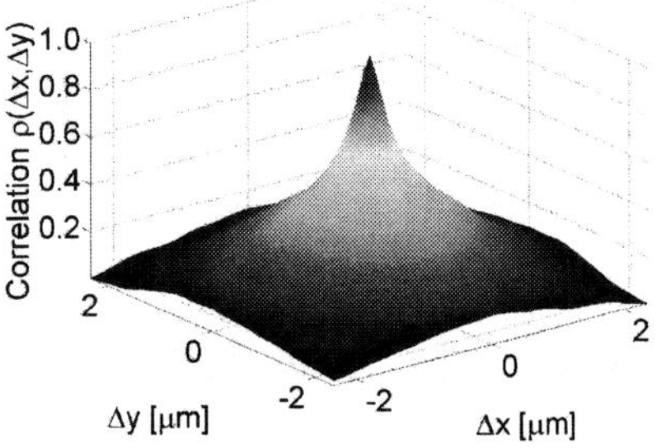

Figure 4: 2D autocorrelation function of the surface height of a sample.

After removal of the membrane, the other contacting surface, the dielectric, is also measured and found to be relatively flat compared to the bottom of the membrane. Because the dielectric material is also much harder than the membrane material, the contact between membrane and dielectric can be regarded as the contact between the membrane and an infinitely hard smooth surface.

2.2 Contact Mechanics Measurement

For seven switches differing in membrane size and spring layout, the C-V curve is measured. The number of switching cycles does not influence the C-V curve, ruling out that plastic deformation occurs. De-embedding is employed to reduce the influence of parasitics. From the de-embedded C-V curve above pull-in, both the applied pressure p_{appl} and the separation g_{eq} between the midplane of the rough membrane and the flat dielectric are estimated. This procedure will be explained using figure 5.

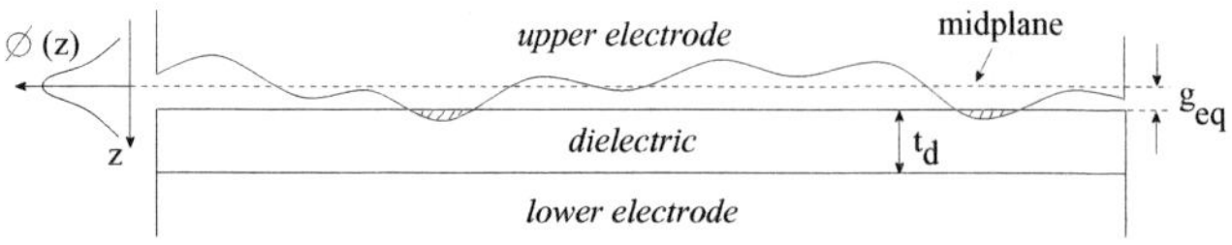

Figure 5: Schematic representation of a rough surface in contact with a flat dielectric.

A rough surface with height distribution $\phi(z)$ is brought into contact with a flat dielectric of thickness t_d, at a midplane distance of g_{eq}. We have seen above that $\phi(z)$ is Gaussian with standard deviation σ:

$$\phi(z) = \frac{1}{\sigma\sqrt{2\pi}} \exp(-z^2/2\sigma^2) . \tag{1}$$

The material in contact (shaded, protruding area) is assumed not to influence the distribution $\phi(z)$ [6], [7]. Because the lateral spatial component of the rough surface (correlation length in figure 4) is much larger than the standard deviation σ, fringing field contributions to the capacitance can be neglected. Also, fringing contributions arising from the edges and finite thickness of the membrane are insignificant in closed-state [8], and the closed-state capacitance equals:

$$C(g_{eq}) = \epsilon_0 A \left[\int_{-\infty}^{g_{eq}} \frac{\phi(z)}{g_{eq} - z + t_d/\epsilon_d} dz + \int_{g_{eq}}^{\infty} \frac{\phi(z)}{t_d/\epsilon_d} dz \right], \tag{2}$$

with ϵ_0 the dielectric constant of vacuum (or air), ϵ_d the relative permittivity of the dielectric, and A the nominal membrane area. Expression (2) can be evaluated numerically and compared to the parallel plate approximation of two flat surfaces separated by a gap g_{eq} and dielectric t_d:

$$C_{pp}(g_{eq}) = \frac{\epsilon_0 A}{g_{eq} + t_d/\epsilon_d} . \tag{3}$$

The results in figure 6 show that for values of σ around the measured one (σ_{meas}), the capacitance can be approximated by the parallel plate capacitance (solid lines). In practice, elastic contact will remain in the upper regions of separation ($2 < g_{eq}/\sigma < 3$), resulting in a contact area that is only a small fraction ($< 1\%$) of the nominal area, but still sufficient to carry the total load [9], [10]. For this region of interest of g_{eq}/σ, the error between (3) and (2) remains below 5%. This error decreases with t_d/ϵ_d, which here approximately equals 3σ.

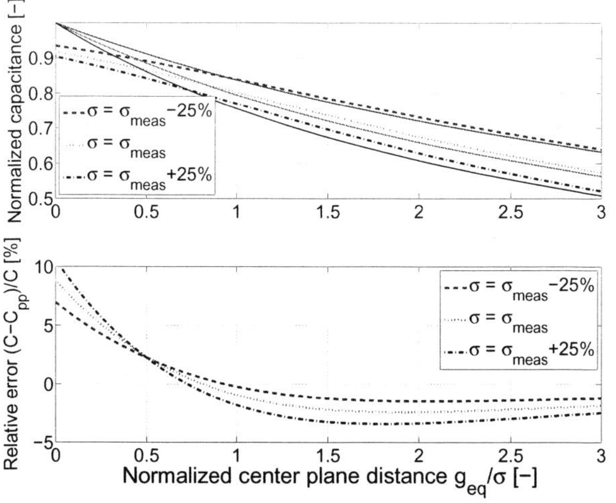

Figure 6: Capacitance of a rough surface C eq. (2) (dashed lines) compared to the parallel plate capacitance C_{pp} eq. (3) (solid lines) for various values of σ.

The surface separation g_{eq} can thus be easily extracted from (3), instead of the more complex (2). Furthermore, the difference in slope dC/dg_{eq} between (2) and (3) for $2 < g_{eq}/\sigma < 3$ is negligible, which means that the electrostatic force can be approximated using (3):

$$F_{el} = \frac{\epsilon_0 A}{2(g_{eq} + t_d/\epsilon_d)^2} V^2 , \tag{4}$$

with V the applied voltage. After rewriting, the electrostatic pressure equals:

$$p_{el} = \frac{F_{el}}{A} = \frac{C_{pp}^2 V^2}{2\epsilon_0 A^2} . \tag{5}$$

In the voltage region above pull-in, the elastic (spring) forces counteracting the electrostatic force are relatively small and partial release is relatively insignificant, so that the applied contact pressure $p_{appl} \approx p_{el}$.

2.3 Contact Model and Combined Results

In literature, many contact models are described, of which the Greenwood model [9] is widely regarded as the classical work in this field. In this model, the rough surface is regarded as a Gaussian distribution of spherical summits with constant radius. Various adaptations to the Greenwood contact model have been made, for example by Whitehouse and Archard [11] to include varying summit radii. Some other contact models use an exponential force expression [12] or fractal theory [13]. Elaborate overviews of contact models are presented in [10], [14], and [15].

The main drawback of many contact models is the number of required model parameters and/or the unclear parameter extraction method. Therefore, the Mikić model [16] is chosen based on its theoretical background, the availability of closed-form expressions, the low number of parameters and their ease of extraction. The contact model is based on the bearing area method [6], [7], in which an estimate of the contact area A_g is obtained:

$$A_g(g_{eq}) = A \int_{g_{eq}}^{\infty} \phi(z) dz = \frac{A}{2} \mathrm{erfc}\left(\frac{h}{\sqrt{\pi}}\right) . \tag{6}$$

Here, $h = g_{eq}/\sigma$ is the dimensionless separation. The contact force F_c is then given by:

$$F_c = H_E A_g = \frac{E\theta}{\sqrt{2}} A_g , \tag{7}$$

with elastic hardness H_E, Young's modulus E and mean absolute slope of the surface roughness θ. The determination of θ is often prone to measurement noise. Therefore, as an alternative, the indentation hardness H is often used as an approximation of H_E. Here, the same method is applied, so that the contact model only has two parameters: the standard deviation σ (measured with AFM) and the indentation hardness H (measured in this research with a Berkovich tip).

The measured separation versus applied pressure characteristic (compliance curve) is shown in figure 7. All seven devices follow the same trend, independent of membrane size and spring layout, confirming the assumption that elastic forces and partial release can be ignored in this analysis. Furthermore, the separation decreases approximately linearly with the logarithm of the applied pressure. Although already observed for macroscopic rough surfaces in tribology [17], [18], this is the first time that such behavior is shown to hold for MEMS switches. The Mikić model prediction of the compliance curve is also plotted in figure 7, showing good agreement.

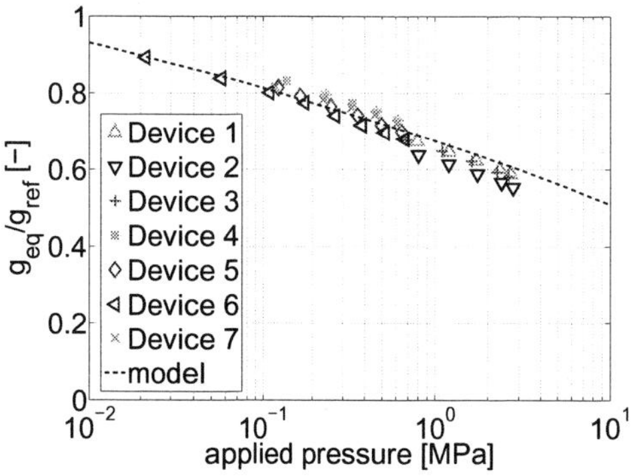

Figure 7: Compliance curve of the separation g_{eq} (normalized to an arbitrary reference gap g_{ref}) versus the logarithm of the applied pressure.

Because AFM measurements are relatively time consuming, four out of seven devices on which C-V measurements are carried out and that cover a wide range of applied pressures, are selected. On these four devices,

AFM measurements are performed. The bearing area is taken from the AFM data by measuring the geometrical intersection area of the rough surface (e.g. figure 2) and a horizontal plane at a distance g_{eq} with image processing techniques. In combination with the compliance curve, the applied pressure versus contact area characteristic is then obtained (figure 8). The Mikić model output (6), (7) is also shown. Despite some spread in

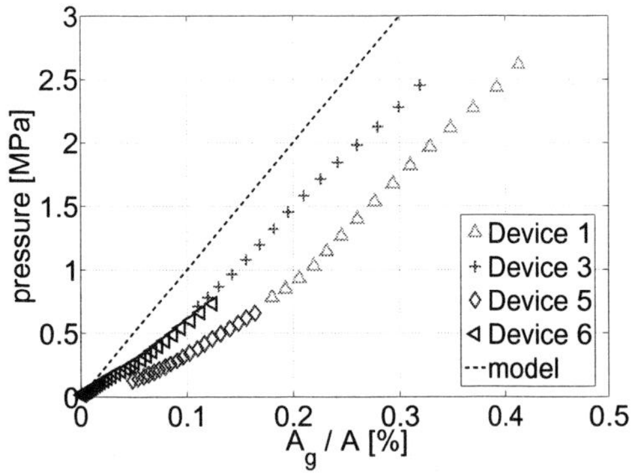

Figure 8: Applied pressure as a function of the bearing area ratio A_g/A.

the measurement results, it can be seen that the applied pressure is approximately linearly dependent on the bearing area ratio, in agreement with Amonton's law of friction [7], [17]. Other sources indicate that a slight deviation from linearity can occur [18], [19]. Here, for the first time, this type of characteristic is evaluated on MEMS structures. The discrepancy between model and measurements in figure 8 can be caused by the samples not being as perfectly Gaussian as the model assumes, and the inaccuracy in the determination of the contact hardness. However, this characteristic can be used to obtain a more accurate estimate of the elastic hardness H_E, which is, as mentioned, difficult to measure.

Finally, the contact model is validated by implementing it in a compact model of the switch [8] and predicting the C-V curve above pull-in for five devices that have designs differing from the seven original ones (figure 9). Both the C-V curve slope and the capacitance density are very well predicted, without the need for fitting.

3 CONCLUSIONS

A method to measure and analyze the static behavior of RF MEMS capacitive switches in contact is demonstated that combines AFM measurements with standard C-V measurements. Important characteristics that have not been characterized in the field of MEMS before, and that are valuable in the evaluation of contact models, have been analyzed. The presented characteristics are in accordance with the behavior as identified in the field of tribology. The Mikić contact model applied here shows

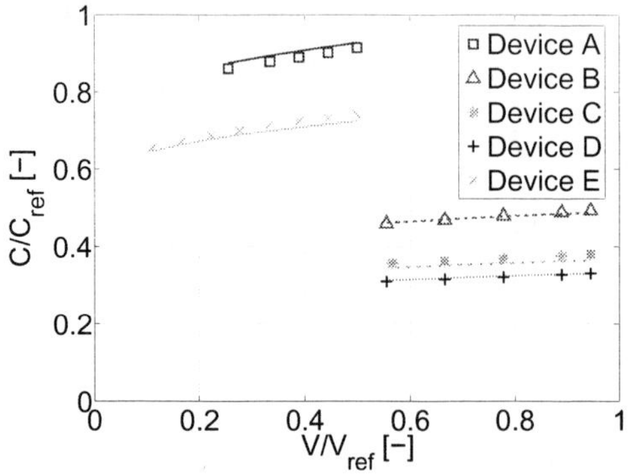

Figure 9: Measured (symbols) and modeled (lines) C-V curves above pull-in for five validation devices, normalized to an arbitrary reference capacitance C_{ref} and voltage V_{ref}.

good agreement with the measurements and is implemented in a compact model of the switch.

ACKNOWLEDGEMENTS

This work was partially supported by the Point-One project MEMSland.

REFERENCES

[1] O. Rezvanian *et al.*, J. Micromech. Microeng. **17**, pp.2006–2015 (2007).
[2] D. Hyman *et al.*, IEEE Trans. on CPT **22**, pp.357–364 (1999).
[3] D.L. Liu *et al.*, Appl. Phys. Lett. **91**, pp.043107-1-3 (2007).
[4] N. Tayebi *et al.*, J. Appl. Phys. **98**, pp.073528-1-13 (2005).
[5] R.W. Herfst *et al.*, IEEE IRPS 2007, pp.417–421.
[6] E.J. Abbott *et al.*, Mechanical Engineering **55**, pp.569–572 (1933).
[7] K.L. Johnson, *Contact Mechanics* (1985).
[8] H.M.R. Suy *et al.*, Nanotech MSM 2007 **3**, pp.65–68.
[9] J.A. Greenwood *et al.*, Proc. Roy. Soc. London **295**, pp.300–319 (1966).
[10] T.R. Thomas (ed.), *Rough Surfaces* (1982).
[11] D.J. Whitehouse *et al.*, Proc. Roy. Soc. London **316**, pp.97–121 (1970).
[12] J.A. Bielen *et al.*, IEEE EuroSime 2007, pp.59–64.
[13] A. Majumdar *et al.*, Journal of Tribology **113**, pp.1–11 (1991).
[14] A. Hariri *et al.*, J. Micromech. Microeng. **16**, pp.1195–1206 (2006).
[15] G. Zavarise *et al.*, Wear **257**, pp.229–245 (2004).
[16] B.B. Mikić *et al.*, Int. J. Heat Mass Transfer **17**, pp.205–214 (1974).
[17] N. Kikuchi *et al.*, *Contact Problems in Elasticity* (1988).
[18] K.L. Woo *et al.*, Wear **58**, pp.331–340 (1980).
[19] I.V. Kragelsky *et al.*, *Friction and Wear* (1965).

Modeling and Design of Electrostatic Voltage Sensors Based on Micro Machined Torsional Actuators

Jan Dittmer[*,**], Antje Dittmer[***], Rolf Judaschke[**] and Stephanus Büttgenbach[*]

[*] Institute for Microtechnology, Technische Universität Braunschweig, GERMANY, j.dittmer@tu-bs.de

[**] Physikalisch-Technische Bundesanstalt, PTB, Braunschweig, GERMANY

[***] Institute for Aerospace, Technische Universität Braunschweig, GERMANY

ABSTRACT

The optimization of design parameters and methods for electrostatic voltage sensors based on torsional actuators is presented. The analytical software model used for this optimization process is discussed and compared to measurement results. Voltage excitation leads to an electrostatic attractive force, inducing a deflection of the actuator which in turn is detected using capacitance measurement. The conventional measurement principle for RMS voltages is based on power dissipation by ohmic resistance. We look at devices with parallel plate capacitances in a rotational design, which allows direct, substitution and compensation measurement of RF voltages. An analytical and a numerical model for this type of sensor are presented, focusing on absolute and relative sensitivity and precision for metrology purposes. The main design parameters are extracted from this model and variation of each parameter is discussed. The influence of fabrication tolerances on the dimensions and subsequently on the measuring results is investigated and an upper limit is given. Using this model, sensor designs are developed which fulfill the requirements for electrostatic voltage sensing for different sensitivity and range requirements. The final designs are checked against fabrication process limitations and an estimate of the maximum total error is given. The results of this work will be verified with devices designed with this theoretical understanding and built in a micromechanical silicon process.

Keywords: RF-MEMS, design, modeling, electrostatic, torsional actuator, RMS-Voltage

1 INTRODUCTION

In this paper we present the theoretical foundation of electrostatic voltage sensors based on torsional actuators. The conventional method for traceable high-frequency voltage metrology is based on power dissipation measurement of ohmic resistances allowing RMS voltage conversion by the square power law. Employing the principle of electrostatic force is an alternative method to solve this problem.

Actuator based voltage sensors rely on the principle of electrostatic force. Two electrodes with opposite charges attract each other. The charge is linked to the voltage by the capacitance of the structure. By elastically suspending one of the electrodes, an equilibrium position between the repelling spring force of the actuator and attracting electrostatic force exists. Due to the quadratically decreasing force as a function of distance, it is necessary to utilize micromachining for the fabrication of this sensor. Comb-like structures are often used in sensors and actuators for this task, allowing surface micromachining, but showing unwelcome parasitic effects. For metrology applications it is preferable to have a well calculable structure with few fringing effects. Therefore, a parallel plate capacitor is chosen for actuation and sensing. Building a torsional actuator with multiple electrodes on each side makes it suitable for voltage compensation and substitution measurements. Voltage excitation on one side leads to a deflection of the actuator which in turn is detected using capacitance measurement, but may also be detected by optical detection.

In this publication, we present at first an overview of the geometrical dimensions and present an analytical steady state solution for this sensor. Subsequently, we build a model for time-domain simulations using Matlab/Simulink with special emphasis on the damping behavior for different damping models. Using these models, the influence of the physical dimension tolerances of the sensor on different measurement principles is discussed. Finally measurement results of real sensors are presented and compared to the model. Similar research has been presented in [1]. In comparison, this publication presents a more general case, where the electrodes do not necessarily have to occupy the full plate area.

The results of this research are based on practical sensors, already presented in [2-4]. Other sensors based on the principle of electrostatic charge have been presented in [5] and [6] for other frequencies and voltage ranges.

2 THEORY

The principle sensor geometry is shown in Figure 1. The sensor consists of a plate with a length of $2L$ and width of B, which is suspended at the center axis with two beams of length l and width w. The plate and beam are made of one material, silicon in this case, with a thickness t. They form a common electric potential. The plate is mounted at an initial distance h_0 above the opposing electrodes. It is assumed throughout the analysis that the plate only moves in torsional direction and exhibits no translational movements. This assumption has been confirmed by the fact that the spring constant orthogonal to the plate is at least two orders of magnitude higher than the rotational spring constant. This has also been verified using finite element simulation. As mentioned in the previous section, the electrodes do not cover the full plate area but only reach from $\alpha_1 L$ to $\alpha_2 L$, with $\alpha_1 < \alpha_2$, $\alpha_1 >= 0$ and $\alpha_2 <= 1$, while still covering the full width of the plate B. The default dimensions for all calculations in this paper are summarized in Table 1.

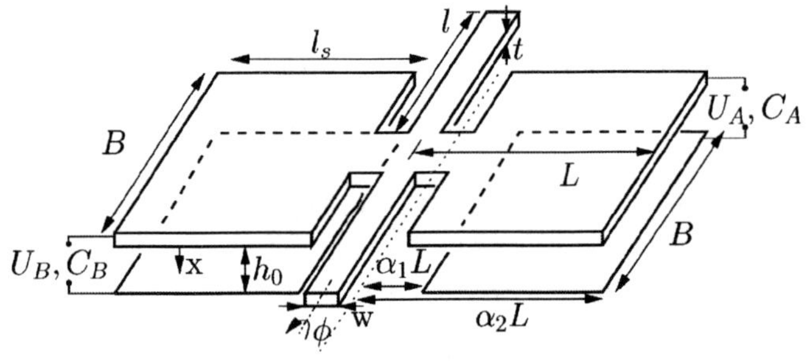

Figure 1: Geometrical model of the torsional actuator showing the important dimensions of the actuator and of the opposing electrodes.

Dimension	Symbol	Value	Unit
Plate length	L	2000	µm
Plate width	B	1000	µm
Gap	h_0	5	µm
Plate thickness	t	20	µm
Length of torsion beam	l	1000	µm
Width of torsion beam	w	20	µm
Start of electrode	$\alpha_1 \cdot L$	1000	µm
End of electrode	$\alpha_2 \cdot L$	2000	µm
Young's modulus, Silicon	E	160	GPa
Poisson's ratio, Silicon	υ	0.2	
Density, Silicon	ρ	2700	kg/m^3
Air viscosity, room temp	µ	$1.79 \cdot 10^{-5}$	Pa s

Table 1: Default dimensions used for simulations and calculations if not noted otherwise.

The following analysis is for the special case of a design with one electrode on each side of the beam, but can also easily be extended for multiple electrodes on each side of the beam. The capacitance and voltage on the side of the plate in positive direction of the angle ϕ are marked with a subscript A (C_A, U_A), the other side with a subscript B (C_B, U_B). In the following sections, we present a fully analytic steady-state model of the sensor, a numerical time-domain simulation and present first results.

The principle sensor design allows three different modes of voltage measurement. Direct measurement takes the deflection angle to calculate the applied voltage. Substitution measurement relies on a known (DC) voltage for which the deflection angle is measured. If the unknown voltage leads to the same deflection, it is known that both voltages are equal. The obvious advantage being that geometrical variances do not influence the result as they stay the same between two measurement cycles. The same advantage applies for compensation measurements, though it requires a third electrode or optical instruments for the detection of the deflection angle.

2.1 Analytical steady-state model

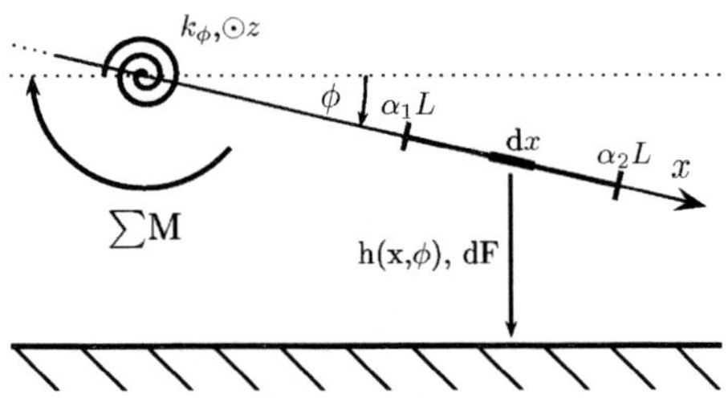

Figure 2: Geometrical model for the derivation of the analytical steady-state model.

For the derivation of the analytical model, we consider a simple 2D model of the geometry as shown in Figure 2. The local distance h along the beam, twisted with an angle ϕ can be calculated with

$$h(\phi, x) = h_0 - x \sin \phi . \quad (1)$$

Assuming a parallel field distribution between the electrodes, which is obviously true because $L >> h_0$, the

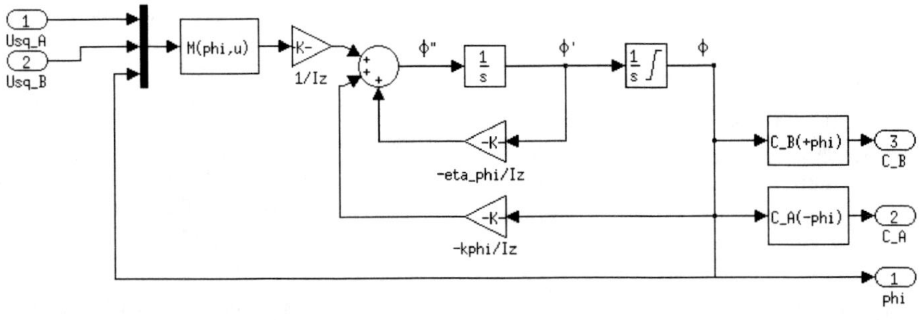

Figure 3: Matlab/Simulink time-domain model of the torsional actuator. Voltage squared inputs on the left and the capacitance and angle output on the right

force dF acting on a short section of the plate, depending on the voltage U, angle ϕ and position along the beam x, can then be derived from the parallel plate capacitance equation as

$$dF(x,\phi,U) = \frac{\varepsilon \cdot B \cdot U^2}{2 \cdot h(\phi,x)^2}. \qquad (2)$$

Integrating the force along the loaded part of the beam from $\alpha_1 L$ to $\alpha_2 L$ gives an expression for the moment $M_U(U,\phi)$ around the rotating axis

$$M_U(U,\phi) = \int_{\alpha_1 L}^{\alpha_2 L} x dF(x,\phi,U) dx. \qquad (3)$$

This can be analytically solved, employing the substitution $\psi = \sin(\phi)\, h_0 / L$ to

$$M_U(U,\psi) = \frac{BL^2 \varepsilon U^2}{2h_0 \psi^2}\left[\frac{1}{1-\psi\alpha_2} - \frac{1}{1-\psi\alpha_1} - \ln\frac{1-\psi\alpha_1}{1-\psi\alpha_2}\right] \qquad (4).$$

The restoring moment $M_k(\phi)$ from the torsional beam is expressed by

$$M_k(\phi) = \phi \cdot k_\phi, \qquad (5)$$

with k_ϕ:calculated from

$$k_\phi = \frac{E \cdot \beta \cdot t \cdot w^3}{l(1+\nu)}. \qquad (6)$$

E, υ are the Young's modulus and Poissons's ratio of the material respectively, and β is a numerical constant depending on the ratio t/w.

Figure 4 shows the generated moment for three distinct voltages over the turning angle and the corresponding restoring spring moment. It can be seen, that only for 1 V and 2 V intersections exist with the restoring moment graph. For 4 V there are no intersections and thus the actuator gets pulled-in up to the mechanical limit. Also, for 2 V clearly two stable positions exist. From this fact is may be concluded that the sensor has a hysteretic behavior.

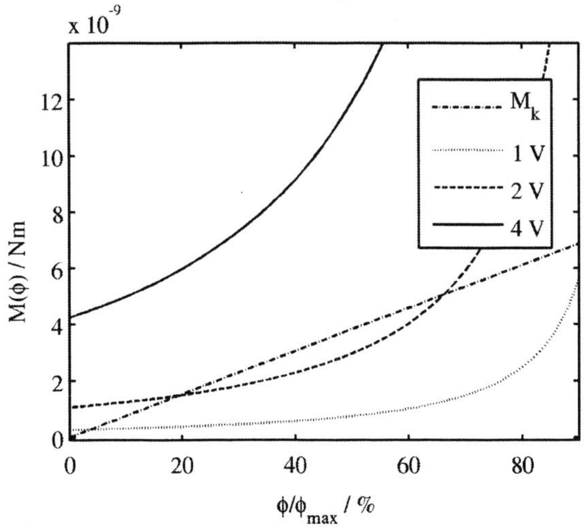

Figure 4: Acting and restoring moments on the actuator for different voltages 1 V, 2 V and 4 V

Solving $M_k = M_U$ for a function $\phi(U)$ is not analytically possible in the general case. Hence a numeric zero crossing detection is used. This gives the solution of the stable position of the actuator depending only on geometrical quantities and the applied voltage U. The solution for varying beam width, with capacitive feedback, can be found in Figure 5. For the highest sensitivity of the system, a low beam width and thus a low spring constant is required. For a high voltage range, on the other hand, a stiffer system is preferable. Depending on the application, different sensor designs have to be used. For compensation measurements, a high sensitivity is necessary, while for direct and substitution measurements the pull-in voltage has to be higher than the maximum voltage to be measured.

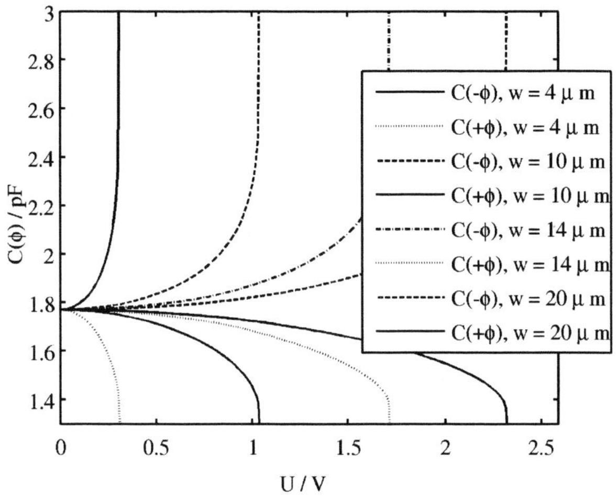

Figure 5: Equilibrium positions of the actuator with capacitive feedback, with the sensing capacitance on the same side (upper part of the figure) and opposite side (lower part of the figure) for varying beam width w.

2.2 Numerical time-domain model

For a time-domain analysis, the analytical model from section 2.1 is not adequate. It is thus extended to incorporate damping and inertia. Equation 7 gives the differential equation with the damping coefficient η and the moment of inertia I_z for the torsional actuator

$$I_z \cdot \ddot{\phi} + \eta \cdot \dot{\phi} + k_\phi \cdot \phi = M_U(U(t),\phi) \qquad (7)$$

The damping coefficient can be analytically derived [7] as can be I_z using a model of a rigid plate rotating around its center. The inertia of the torsional beam can be disregarded, as it has a much lower mass and volume than the plate. Using this model, a Matlab/Simulink simulation has been implemented (Figure 3). Keeping the geometrical dimensions from Table 1, the step response of the system is recorded (Figure 6). For varying damping coefficients it can be seen that a voltage step can result in the excitation of an oscillation or a slow movement to the final equilibrium position with practically no overshoot. It has to be noted that if the overshoot is too big, it is possible that the second equilibrium position (cf. sect. 2.1) may be attained. The damping coefficient can be varied over a big range by having holes of different sizes and geometries in the plate. The theoretical value of the damping coefficient for the presented geometry is $5 \cdot 10^{-7}$ Nms [7]. For electrical voltage metrology purposes, this is fully adequate as an excitation of an oscillation is undesired and the second equilibrium position has to be avoided for valid results.

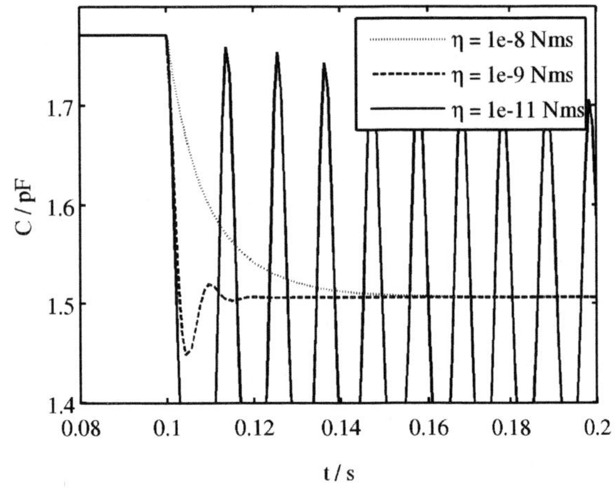

Figure 6: Time-domain step response of the system for different damping coefficients

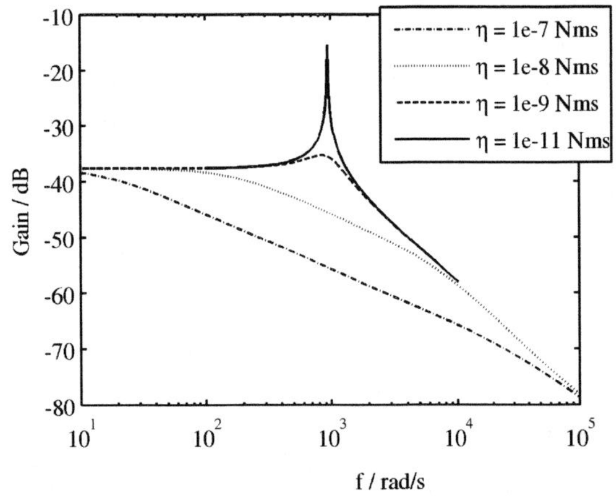

Figure 7: Bode plot for different damping coefficients η, showing a resonant behavior at $f = 10^3$ rad/s

The Matlab/Simulink model also allows an easy determination of the behavior of the model in the frequency domain (Figure 7). After linearizing the model at its operating point, the Bode plot shows that low damping coefficients lead to resonant behavior at the mechanical resonant frequency of the device. This can ultimatively lead to the destruction of the device. Thus, as mentioned before, a high damping coefficient is desirable for the sensor.

3 CONCLUSION

We presented a theoretical overview of electrostatic voltage measurement based on a torsional actuator. After an introduction to the basic setup of the measurement, an analytical steady-state model is presented. This model is afterwards extended for numerical time-domain simulations in Matlab/Simulink allowing the determination of resonant and transient behavior. The model will allow for designing specialized sensors for metrology applications for distinct voltage ranges and measurement applications. Future work will aim at the validation of the simulation results using real sensors. Furthermore, the electrical behavior of the device in the scope of high frequency voltages will have to be modeled as close as possible to investigate the effects with regard to metrology applications.

REFERENCES

[1] R. Sattler et al, „Modeling of an electrostatic torsional actuator: demonstrated with an RF MEMS switch", Sensors and Actuators A 97-98, pp. 337-346, 2002

[2] S. Beißner et al., "Micromechanical Device for the Measurement of the RMS Value of High-Frequency Voltages", Proc. 2nd IEEE Int. Conf. on Sensors, pp. 631-635, Toronto, 2003

[3] J. Dittmer et al., "A Miniaturized RMS Voltage Sensor Based on a Torsional Actuator in Bulk Silicon Technology", Micro- and Nano Engineering, pp. 769-770, Kopenhagen, 2007

[4] J. Dittmer et al, "Aufbau und Charakterisierung eines mikro-elektromechanischen Torsionssensors für die Hochfrequenzspannungsmessung", Mikrosystemtechnik Kongress, pp. 775-758, Dresden, 2007

[5] M. Bartek et al., "Bulk-micromachined electrostatic RMS-to-DC converter: Design and fabrication", Proc. MME, Uppsala, Sweden, 2000

[6] L. J. Fernandez et al., "A capacitive RF power sensor based on MEMS technology", J. Micromech. Microeng., vol. 16, pp. 1099-1107, 2006.

[7] F. Pan et al., "Squeeze film damping effect on the dynamic response of a MEMS torsion mirror", J. Micromech. Microeng, Vol. 8, 1998, pp. 200-208

Anomalous Thermomechanical Softening-Hardening Transitions in Micro-oscillators

T. Sahai*, R. Bhiladvala** and A. Zehnder***

* Department of Theoretical and Applied Mechanics, Cornell University
317 Kimball Hall, Ithaca 14853, NY, USA, ts269@cornell.edu
** Materials Research Institute and Department of Electrical Engineering,
The Pennsylvania State University, rbb16@psu.edu
*** Theoretical and Applied Mechanics, Cornell University, NY, USA, atz2@cornell.edu

ABSTRACT

We report a transition in the response of a doubly-clamped micromechanical oscillator with linear inertial forcing. The experimental response of the oscillator changes, counter-intuitively, from softening to hardening on increasing the power of an incident laser beam. A novel response structure, that simultaneously displays characteristics of both softening and hardening resonances, is observed at intermediate laser powers. Using a dynamic nonlinear thermomechanical model, we show that the transition is explained by opposing responses of linear and cubic stiffness terms to temperature.

1 INTRODUCTION

The micromechanical oscillator used in our study is a doubly-clamped silicon nitride paddle-beam with a square plate at its center (Fig. 1 (a)). A film stack, consisting of a $0.2\,\mu$m thick low-stress silicon nitride layer over a $1.5\,\mu$m thick layer of annealed silicon oxide, is grown on silicon wafers. The stack is patterned and etched to remove the supporting sacrificial oxide layer, leaving suspended, paddle-beam structures on the silicon wafer chip.

A schematic of the experimental setup is shown in Fig. 1 (b). The chip is bonded to a piezoelectric shaker obtained from a commercial buzzer (Radio Shack 273-073) and operated in a vacuum chamber at a pressure of 4×10^{-7} Torr. The oscillators are driven inertially by the piezo using a signal from the tracking generator of a spectrum analyzer (Agilent E4402B). This driving signal is set to sweep over a selected frequency range. Oscillator motion is detected by interferometric modulation of a laser beam (He-Ne 633 nm), focused on the paddle using a microscope objective [2]. Laser intensity modulation is detected and amplified by an AC-coupled photodetector (New Focus 1601) and converted by the spectrum analyzer to show oscillation amplitude variation with frequency. The laser power can be continuously varied and its value recorded by a laser power detector (Graseby Optronics 52575R).

During the experiment, described in Fig. 1 (b), it is observed, while holding the driving piezo amplitude constant, that increasing the power of the sensing laser beam changes the response curve of the first bending

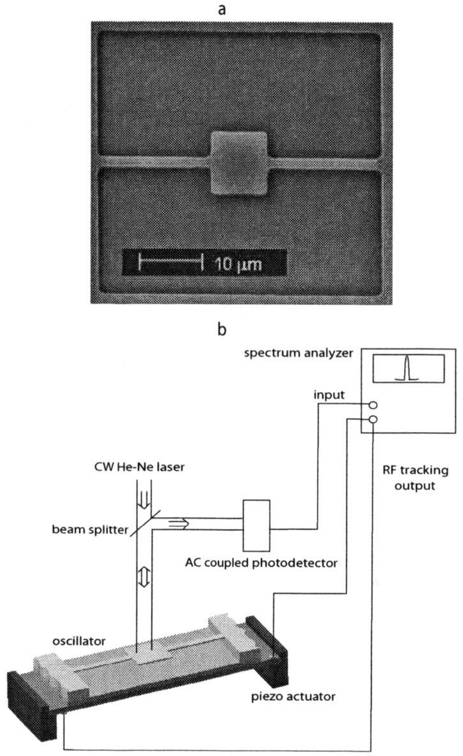

Figure 1: (a) SEM image showing suspended silicon nitride oscillator, $0.2\,\mu$m thick, with a $10 \times 10\,\mu\text{m}^2$ paddle and $18\,\mu$m long beams. (b) The oscillator is inertially driven over a selected frequency range from the spectrum analyzer's tracking generator, using a piezoelectric actuator. Interferometric intensity variation due to the oscillator motion, sensed using the AC coupled photodetector, is converted to spectral information by the analyzer.

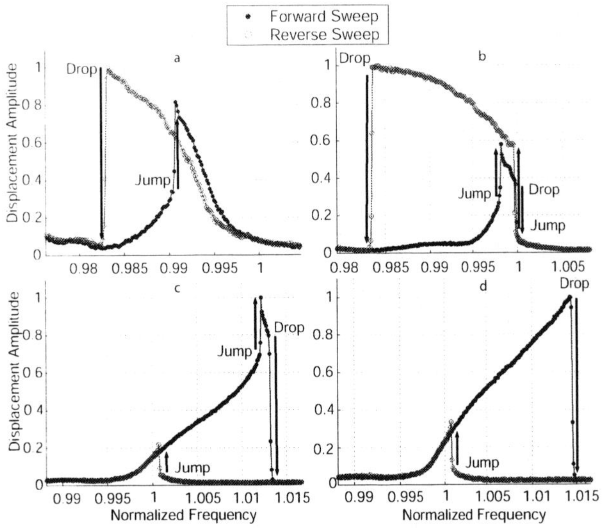

Figure 2: Experimentally obtained (a) softening curve at lowest laser power, is characterized by a shorter peak and a jump for the forward sweep direction (closed circles) and a taller peak and drop for the reverse sweep (open circles);(b) and (c) show transitional curves at increased laser powers;(d) hardening curve seen at highest laser power. Curve in (b) has the softening behavior in (a) with an added drop jump feature of (d). The laser power for curves a:b:c:d is 5 : 6 : 8 : 9 units, with one unit = 45 μW. All plots are normalized to 0.825 MHz. The amplitudes are normalized by the height of the taller peak in each plot. The piezo forcing amplitude is held constant.

mode of oscillation from softening to hardening (Fig. 2); at intermediate laser power a transition is observed, with both softening and hardening characteristics seen in the same response curve. In earlier work, electrostatically driven oscillators have been tuned, using nonlinear forcing through selection of the DC bias voltage, to exhibit either hardening or softening behavior [5], [7] but not both. It is useful to understand the system reported here not only for its striking transitional response which has technological applications, but also for the prediction and control of nonlinearities in optically sensed motion of purely mechanical micro-oscillators with linear forcing.

2 MODELING

The standard Duffing equation [10], a model for oscillation of structures with constant-coefficient linear and cubic stiffness terms, cannot account for the transitions observed in our experiments. Here we show, using a thermomechanical model of the interaction of the sensing laser with the oscillator stiffness and pre-tension, that the transition arises from the opposing response of the linear and cubic stiffness terms to temperature changes.

Large transverse deflections in a doubly-clamped beam cause it to stretch. The resulting tension increases the stiffness against further deformation, hence the beam acts as a nonlinear, hardening spring. In this model, the effect of the pretension and laser heating have also been considered. The estimated residual tensile stress of 50 MPa is found to increase the overall stiffness of the beam, while reducing the relative magnitude of deformation-induced nonlinear stiffening. As the beam heats up due to laser light absorption, this tensile stress is reduced, increasing the nonlinearity of the load-deflection response of the beam (Fig. 3).

The vibration of the paddle-beam structure is modeled by a modified Duffing equation with temperature dependent stiffness terms,

$$m\ddot{x} + c\dot{x} + k_1(T)x + k_3(T)x^3 = F_s \sin(\omega t)\,, \qquad (1)$$

where x, m, c, T are the deflection, mass, damping, and temperature respectively, and $k_1(T)$ and $k_3(T)$ are the linear and cubic stiffnesses of the spring mass system. F_s is the piezo forcing amplitude, ω is the piezo forcing frequency and t is time. Eq. (1) can be normalized as:

$$\ddot{z} + \frac{\dot{z}}{Q} + f(T)z + g(T)\beta_0 z^3 = M \sin(\omega t)\,, \qquad (2)$$

by dividing the entire equation by m, λ (wavelength of the incident laser) and rescaling time by $\omega_0 = \sqrt{k_1(0)/m}$. Note, in the above equation, $\beta_0 = k_3(0)\lambda^2/k_1(0)$, $M = F_s/m\omega_0^2\lambda$, $Q = m\omega_0/c$, $z = x/\lambda$, $f(T) = k_1(T)/k_1(0)$ and $g(T) = k_3(T)/k_3(0)$.

A lumped thermal mass using Newton's law of cooling models the thermal aspects,

$$\dot{T} = -BT + AP_{\text{absorbed}}\,, \qquad (3)$$

where T is the temperature above ambient, of the micromechanical structure, A is the inverse of the lumped thermal capacity and BT is the rate of cooling due to conduction. As in Eq. (2), time in Eq. (3) is rescaled by ω_0. The mass, damping, forcing amplitude and forcing frequency are assumed to be independent of T. The effects of radiation, photon pressure and Casimir forces are found to be negligible and thus not included in the model.

We need to estimate the different parameters that are part of the system of equations. To determine A and B a thermal finite element model is constructed. To simulate laser heating, a heat flux is applied at the center of the paddle. The resulting thermal response of the system yields $A = 0.299\,K/\mu W$ and $B = 0.004$. Similarly, β_0 is estimated to be 0.875 by using large deflection finite element analysis. The quality factor, Q, is estimated to be $\approx$ 1400 from experimental results. Forcing, $M \approx 5 \times 10^{-5}$ is chosen such that the system is

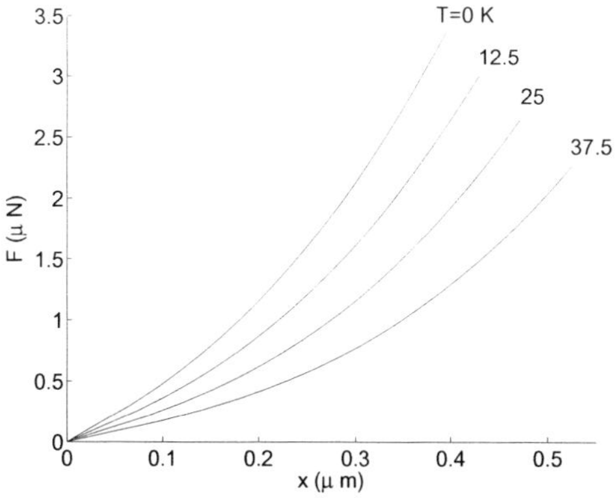

Figure 3: Spring force, F vs. static deflection, x, at different temperatures (above ambient) for a 50 MPa pretensioned beam, loaded at its center. $F = k_1(T)x + k_3(T)x^3$. A least squares fit on the linear and cubic stiffnesses at different temperatures gives functional approximations to $f(T)$ and $g(T)$, where $f(T) = k_1(T)/k_1(0)$ and $g(T) = k_3(T)/k_3(0)$.

pushed into the nonlinear regime. The material properties used for thin film silicon nitride are Young's Modulus $E = 300\,GPa$, Poisson's ratio $\nu = 0.28$, density $\rho = 2900\,kg/m^3$, coefficient of thermal expansion $\alpha = 1.3 \times 10^{-6}\,K^{-1}$, thermal conductivity $k = 3\,W/(mK)$, specific heat capacity $C_p = 700\,J/(kgK)$ and thermal diffusivity $a = k/(\rho C_p) = 1.5 \times 10^{-6}\,m^2/s$ [6], [4], [8]. A value of 50 MPa for beam pretension is calculated by matching the theoretical values of torsional and translation frequencies to the experimentally observed values.

The beam-substrate system forms a Fabry-Perot interferometer [3], [11]. P_{absorbed} is approximated by $P\left[\eta + \gamma \sin^2(2\pi(z - z_0))\right]$, where P is the incident laser power. Using $\eta = 0.133, \gamma = 0.084$ and $z_0 = 0.05$, this approximation is found to agree well with theoretical calculations [11]. So Eq. (3) now becomes:

$$\dot{T} = -BT + AP\left[\eta + \gamma \sin^2(2\pi(z - z_0))\right] . \quad (4)$$

Eqs. (2) and (4) thus serve as the model equations for the first bending mode of the micromechanical oscillator. To determine $f(T)$ and $g(T)$, the nonlinear deflection of pretensioned beams is calculated, as shown in Fig. 3 for different temperature [1]. We find that $f(T) = 1 + cT$ and $g(T) = 1 + b_1T + b_2T^2 + b_3T^3$ with $c = -0.01$, $b_1 = 1.37 \times 10^{-2}$, $b_2 = 2 \times 10^{-6}$ and $b_3 = 2 \times 10^{-5}$, gives a good approximation for the changes in stiffness with temperature.

3 RESULTS

The negative sign of c gives rise to a system with a linear stiffness that decreases with increasing temperature. The nonlinear stiffness, however, increases with increasing temperature. It is the competition between

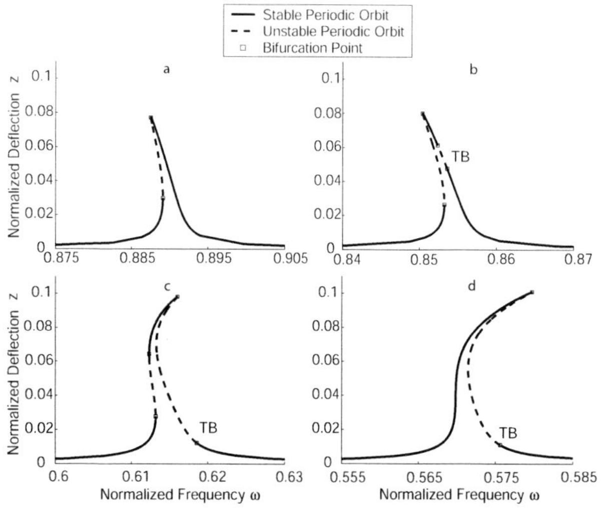

Figure 4: Maximum deflection vs. frequency at various laser powers (model predictions). Solid lines denote stable periodic motions, dotted lines denote unstable periodic motions and open squares denote bifurcation points. (a) Softening at low laser power (b) Transition from softening (c) Transition to hardening (d) Hardening at high laser power. The laser power for curves a:b:c:d is 4: 5 : 12 : 13 units, with one unit = 5 μW.These resonance curves would give jumps and drops similar to Fig. 2. TB denotes a torus bifurcation. The periodic motion at this point loses stability and quasiperiodic motion arises.

these two terms that gives rise to the observed transitions. Using continuation algorithms [9], we compute the resonance curves numerically. At low laser powers the system is softening Fig. 4(a), whereas at high laser powers it is hardening Fig. 4(d). At intermediate laser powers (Figs. 4(b) 4(c)), the resonance curves display behavior similar to the experimental observations. At points labeled TB in Fig. 4 the periodic motion loses stability at a torus bifurcation [10] and a new quasiperiodic motion arises (not depicted in Fig. 4). Quasiperiodic motions are characterized by two incommensurate periodic components. Here, one component is at the driving frequency and the other is close to the natural frequency of the linear oscillator. The change in concavity of the resonance curve can be explained by observing that the softening in the structure arises due to a linear thermal term, the hardening response, however, arises due to a term that is nonlinear in nature.

4 CONCLUSIONS

The model has captured all of the features observed in the experiments. Quantitative matching will require more accurate experimental values for material properties and for post release curvature and residual stress in the fabricated structure. The predicted range of laser power for transition is lower than the experimentally observed values. Experimental values are higher than the actual laser power incident on the oscillator as they exclude losses in some components of the optical path. As seen in Fig. 4, the predicted resonant peak shifts to lower frequencies as the laser power is increased, due to the detuning caused by the linear stiffness $1 + cT$ term. Such large detuning is not observed experimentally, thus the parameter estimation method may be overestimating c.

The softening-hardening transition reported in this work gives a good example of how models built from first principles can be used to gain greater insight into micromechanical systems. These transitions are found to be a consequence of the interaction of the thermal stresses with the stiffness of the device. The heat from the laser changes the pretension in the beam, which changes the linear and cubic stiffnesses, but in different directions. Increasing the temperature of the oscillator decreases the linear stiffness while increasing the cubic stiffness. The competition between these two effects gives rise to the observed transitions. Using the model, oscillators could be designed to operate in the transition region. As seen in Fig. 2(b), the oscillator could be used as a mechanical realization of a bandwidth switch with sharp low and high frequency cutoffs at the jump and drop points.

REFERENCES

[1] R Frisch-Ray, Flexible Bars, Butterworth and Co., 1962.

[2] D. Carr et. al., Journal of Vacuum Science and Technology, B 16,3281-3285, 1998.

[3] K. Aubin et. al., Journal of Micromechanical Systems, 13, 1018-1026, 2004.

[4] C. H. Mastrangelo et. al., Sensors and Actuators, 23, 856-860, 1990.

[5] M. L. Younis and A. H. Nayfeh, Nonlinear Dynamics, 31,91-117, 2003.

[6] B. L. Zink and F. Hellman, Solid State Communications, 129, 199-204, 2004.

[7] W. Zhang et. al., Applied Physics Letters, 82, 130-132, 2003.

[8] M. Gad-El-Hak, The MEMS Handbook, CRC Press, 2002.

[9] E. J. Doedel et. al., International Journal of Bifurcation and Chaos, 1,493-520, 1991.

[10] J. Guckenheimer and P. Holmes, Nonlinear Oscillations, Dynamical Systems and Bifurcations of Vector Fields, Springer, 1996.

[11] E. Hecht, Optics, Addison-Wesley, 1987.

Test ASIC for Real Time Estimation of Chip Temperature

M. Szermer, Z. Kulesza, M. Janicki, A. Napieralski
Department of Microelectronics and Computer Science, DMCS
Technical University of Lodz, TUL
Al. Politechniki 11, 90-924 Lodz, POLAND
E-mail: szermer@dmcs.pl

ABSTRACT

The main goal of this paper is to present the design and operation of an ASIC, which will be used in the research devoted to the development of a real time temperature monitoring system. This circuit will serve two major purposes: gathering information necessary for the conception of sensor placement strategies and the elaboration of robust algorithms for the real time estimation of the circuit temperature as well as the practical verification of the developed solutions.

The first part of this paper will be devoted to the detailed description of the entire ASIC design whereas the second part will present the operation of the circuit based on the computer simulations. The main aspects of proper layout preparation are also included.

Keywords: temperature monitoring, thermal modeling, ASIC

1 INTRODUCTION

The continuous increase of the operating frequency and the miniaturization of electronic circuits augmented the density of dissipated power, which led to the serious thermal problems, even in apparently low power applications. Thus, in order to protect electronic circuits from an overheating and also a destruction, thermal issues should be considered both in the circuit design phase as well as during its operation.

Nowadays, temperature differences in electronic circuits, especially those equipped with efficient cooling systems, can exceed several tens of degrees Kelvin. On the other hand, temperature very rarely can be measured directly where the heat is generated and the placement of a single temperature sensor at some location in the layout will not guarantee that the maximal temperature allowed is not exceeded elsewhere. Thus, there arises a need to devise more sophisticated systems for real time temperature monitoring for overheat protection purposes, which would be based on remote sensor measurements. Such a system should be fast and reliable one but should not occupy too much surface.

2 CIRCUIT DESIGN

2.1 Chip Overview

The entire circuit was designed in the Department of Microelectronics and Computer Science (DMCS) at the Technical University of Lodz. DMCS is one of the very few ASIC design centers in Poland, but at the same time it is the leading one and it has a long-term experience in the thermal and electro-thermal simulation of electronic devices and systems.

The entire circuit was designed in the CADENCE environment and the AMS (AustriaMicroSystems) 0.35 um HV technology, which offers a possibility of combining the standard 3.3 V CMOS logic circuits with the 50 V power MOS transistors. The ASIC consists of the following two main parts:

- analog – a heat source and temperature sensor matrices,
- digital – a control unit, data processing units and a static RAM.

These parts are described in next sections of the paper.

2.2 Analog part

The analog part of the chip is responsible for the heat generation and temperature measurements. It consists of such elements as temperature sensors, heating transistors, comparators, current mirrors and overheating sensor. The heat generation is realized with NMOS High Voltage (HV) transistors. Each heating transistor is the NMOS one with the V_{DS}=50 V and maximal current I_{DS}=21 mA. There is a possibility to adjust the current of each power transistor (circled in Figure 2) with the aid of the three-step current mirror. As temperature sensors, diodes made by proper connection of bipolar transistors are used. Three diodes connected in series are one of the temperature sensors. The output voltage from the temperature sensor is given to the input of a comparator. The comparator with the counter creates a simple ADC. The analog signal from the temperature sensor is converted to the digital one. Next step is processing obtained data with usage digital blocks and finally send it to a PC computer.

In the test chip, 25 temperature sensors are placed near the heat transistors. The idea of placing these sensors is shown in Figure 1. Such a sensors location allows for monitoring the silicon temperature that is dissipated by power transistors as near the heat source as possible.

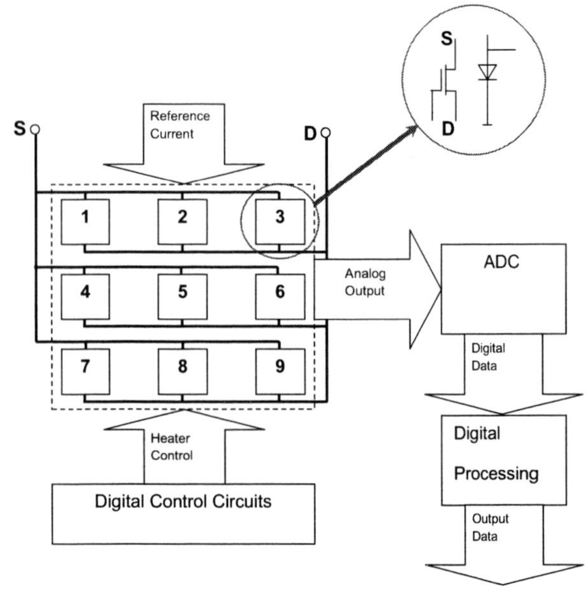

Figure 1. ASIC Block Diagram

Moreover the analog output of each temperature sensor is led out outside the chip to verify the proper processing in the digital part of a chip.

2.3 Digital part

The digital part of the circuit was described as the VHDL code, which was synthesized then to schematics using the tools built into the Cadence environment. Then, the layout was obtained in a fully automatic way with the standard place and route tools. The digital part contains circuits that realize the following functions:

- control of the heat source power dissipation,
- A/D conversion of temperature sensor signals,
- acquisition of measurement data,
- communication with the environment.

The memory of a given size and aspect ratio was synthesized by the manufacturer and provided as a kind of a 'black box'. The main purpose of this memory is to store the measured data, coefficients of digital filters and the temperature estimation results in the real time operation mode of the circuit.

The circuit was designed to work in two modes: the measurement mode and the estimation mode. In the first one, power is dissipated in the circuit and all the signals from the temperature sensors are sent to some external circuits for storing and further processing. In the second mode, the measured data converted to the digital signal will be processed on-chip by the digital filters to produce real time circuit temperature estimates.

The first mode will be used for the investigations aimed at the optimization of circuit thermal model and topology as well as the development of algorithms for real time temperature estimation. Then, all the proposed solutions will be implemented in the chip and thoroughly verified in practice using the second operating mode.

In Figure 4, the block diagram of the digital control unit is presented. It contains following parts:

- main calculation module with memory and control unit subcircuits,
- serial input/output from the main module,
- digital part of the ADCs,
- heat transistors control unit.

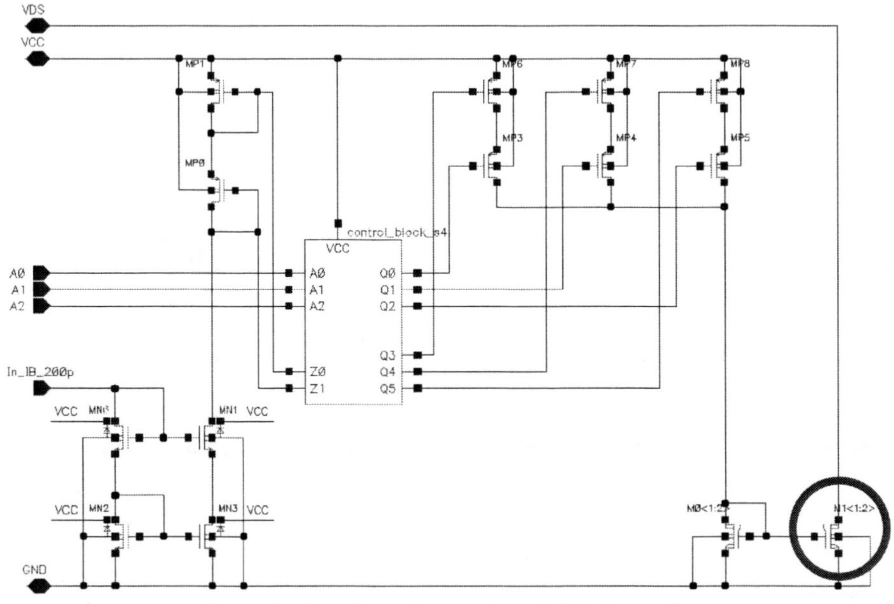

Figure 2. One Heat-Generation Cell Schematic

3 SIMULATION RESULTS

The analog part of the chip, i.e. the heat sources and temperature sensors, was designed from a scratch as a full-custom circuit. The 9 heat sources, HV NMOS transistors, were arranged in a 3x3 matrix. The sources can be switched on and off independently at prescribed power dissipation levels as illustrated in Figure 3 (current levels with multiplication of V_{DS}=50 V). The local chip temperature can be measured with diodes, which are placed in the middle of each heating transistor and close the edges of the chip in 16 other locations marked with circles in Figure 5. The 1 mA forward current flows through each diode, what makes possible to obtain the measurement sensitivity of -1.5mV/K. The voltage signals of each diode can be simultaneously read on the ASIC pins for sensor calibration and method testing purposes.

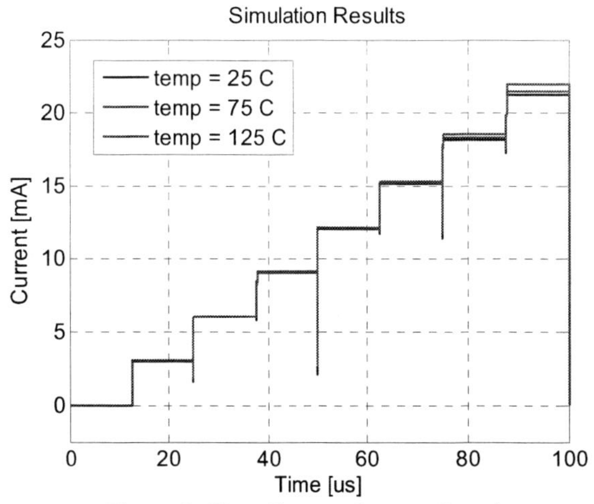

Figure 3. Heat Source Current Levels

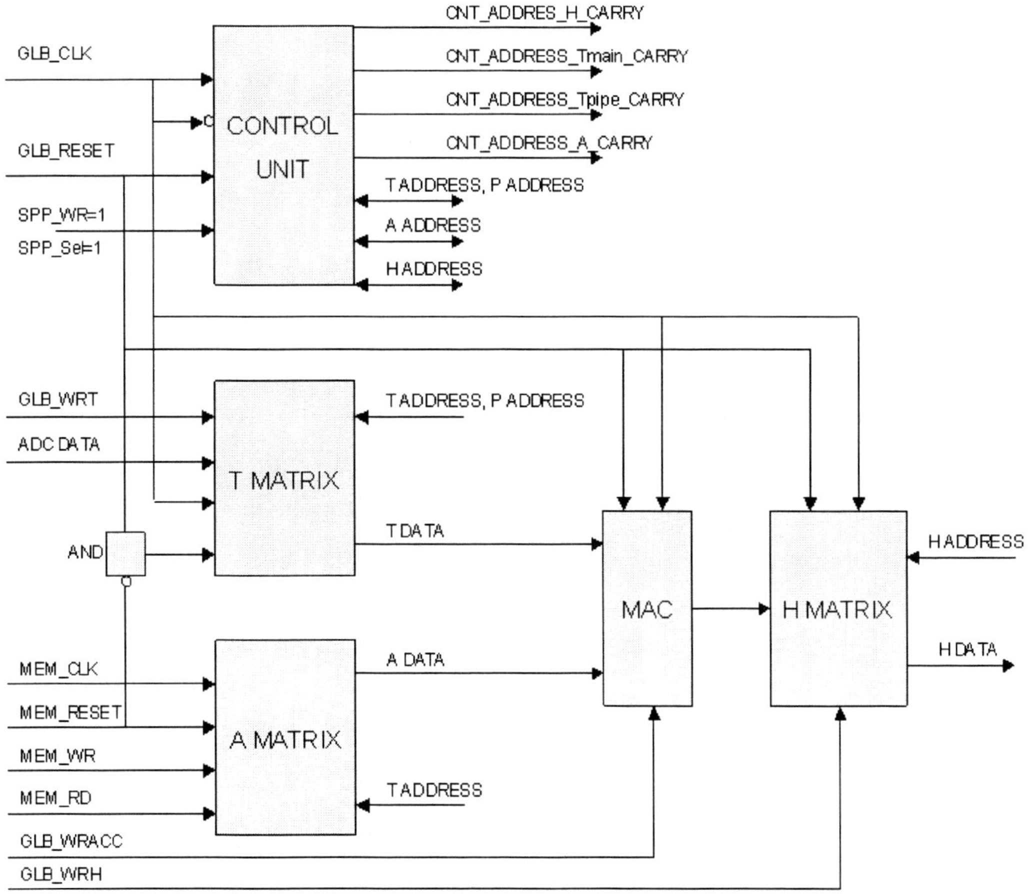

Figure 4. Block Diagram of the Digital Control Unit

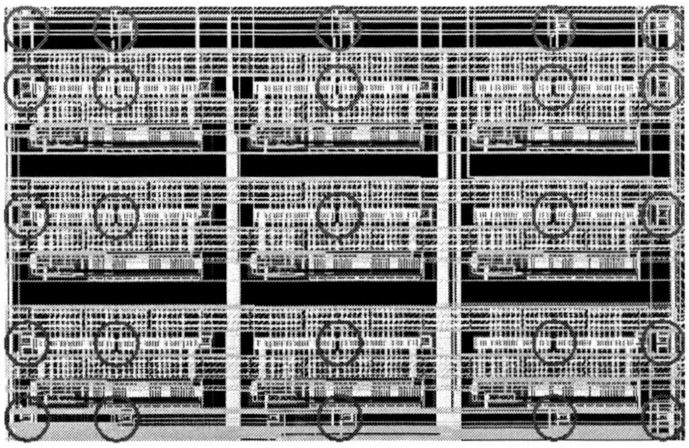

Figure 5. Analog Part Layout
with Marked Temperature Sensors

Due to long time of designing layout and waiting for the manufactured chip, the full verification of the simulations against the measurements is not presented in this paper but will be certainly presented during the conference.

4 LAYOUT

The full test chip layout is presented in Figure 6. The ¾ of the total core area is occupied by the analog part (heat transistors and temperature sensors). To minimize the influence of the analog part on the digital one and vice versa, each block is saved by guard ring. They reduce the disturbances profoundly. The total chip area is about 20 mm^2.

The size of the analog core part, i.e. without pads and digital blocks (simplified floor plan is shown in Figure 5), is 2 mm × 3 mm.

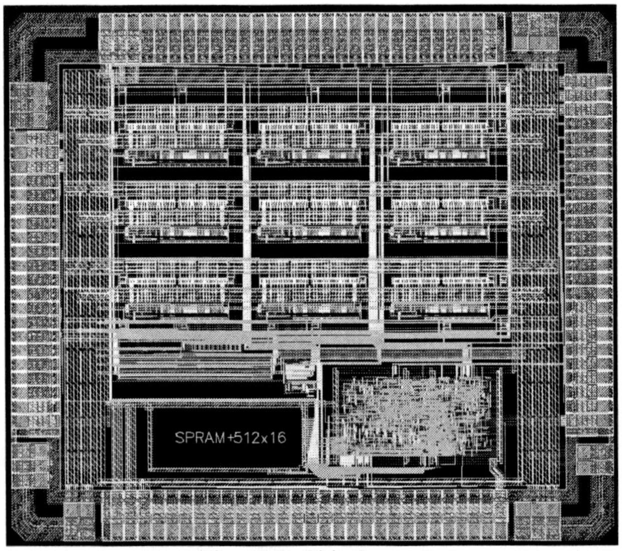

Figure 6. Chip Layout

The entire design of the circuit is currently being finished and was sent for manufacturing. The packaged chips should be delivered till the end of April. This should allow the full evaluation of the design.

5 CONCLUSIONS

In this paper the test ASIC for the temperature monitoring was presented. The power transistors located in a 3×3 matrix generate proper heat, that is measured with usage of diodes. There are two kinds of processing data from sensors: direct output from the chip or send them to the digital processing unit that calculates all necessary information. This approach allows the authors to choose proper temperature sensors location inside the chips that needs temperature-monitoring system.

6 ACKNOWLEDGEMENTS

This research was supported by the grant of the Polish Ministry of Science and Higher Education No. N515 008 31/0331.

REFERENCES

[1] Janicki M., Zubert M., Napieralski A.: "Application of inverse heat conduction methods in temperature monitoring of integrated circuits", Sensors & Actuators: Part A. Physical, Vol. 71, 1998, pp. 51-57

[2] Wójciak W, Napieralski A., Zubert M., Janicki M.: „Thermal Monitoring in Integrated Power Electronics - New Concept", Proceedings of 7th European Conference on Power Electronics and Applications - EPE'97, 8-10 September 1997, Trondheim, Norway, pp. 3.269-3.271

[3] Rencz, M.; Farkas, G.; Szekely, V.; Poppe, A.; Courtois, B.; Boundary condition independent dynamic compact models of packages and heat sinks from thermal transient measurements, 5th Conference on Electronics Packaging Technology EPTC, 10-12 December 2003, pp. 479-484

[4] Rencz, M., Poppe A., Kollar E., Ress S., Szekely V.: Increasing the Accuracy of Structure Function Based Thermal Material Parameter Measurements. IEEE Transactions on Components and Packaging Technology, Vol. 28, No. 1, 2005, pp. 51-57

[5] M. Janicki, G. De Mey, A. Napieralski, Transient Thermal Analysis of Multilayered Structures Using Green's Functions, Microelectronics Reliability, Vol. 42, July 2002, pp. 1059-1064

[6] M. Janicki, A. Napieralski, Analytical Transient Solution of Heat Equation with Variable Heat Transfer Coefficient, Proceedings of 8th International Workshop on Thermal Investigations of ICs and Systems THERMINIC'2002, 1-4 October 2002, Madrid, Spain, pp. 235-240

Simulation of Field-Plate Effects on Lag and Current Collapse in GaN-based FETs

K. Itagaki, A. Nakajima and K. Horio

Faculty of Systems Engineering, Shibaura Institute of Technology
307 Fukasaku, Minuma-ku, Saitama 337-8570, Japan, horio@sic.shibaura-it.ac.jp

ABSTRACT

2-D transient simulations of GaN MESFETs and AlGaN/GaN HEMTs are performed in which a deep donor and a deep acceptor are considered in a buffer layer. It is studied how the existence of field plate affects buffer-related lag phenomena and current collapse. It is shown that in both FETs, the drain lag and current collapse could be reduced by introducing a field plate, because electron trapping in the buffer layer is weakened by it. The dependence on insulator-thickness under the field plate is also studied, suggesting that there is an optimum thickness of insulator to minimize the current collapse.

Keywords: GaN, FET, drain lag, current collapse, trap

1 INTRODUCTION

Even if the gate voltage or the drain voltage is changed abruptly, slow current transients are often observed in GaN-based FETs [1]. This is called gate lag or drain lag, and is problematic for circuit applications. The slow transients mean that dc *I-V* curves and RF *I-V* curves become quite different, resulting in lower RF power available than that expected from the dc operation [2]. This is called power slump or current collapse in the GaN device field. These are serious problems, and many experimental works are reported [1-5], but few theoretical works are made [5-7]. The lags and current collapse can be reduced by surface passivation [3] and by using a field plate [4]. These are considered due to the decrease in surface-state effects. In this work, we have made two-dimensional transient simulations of field-plate GaN MESFETs and AlGaN/GaN HEMTs with a semi-insulating buffer layer in which deep levels are considered, and found that buffer-related lag phenomena and current collapse could also be reduced for the structures with a field plate.

2 PHYSICAL MODEL

Figure 1 shows (a) GaN MESFET and (b) AlGaN/GaN HEMT structures simulated in this study. The gate length L_G and the field-plate length L_{FP} are typically set to 0.3 μm and 1μm, respectively. Polarization charges of 10^{13} cm^{-2} are set at the heterojunction interface, and the surface polarization charges are assumed to be compensated by surface-state charges, as in [8]. As a model for a semi-insulating buffer layer, we use three-level compensation model which includes a shallow donor, a deep donor and a

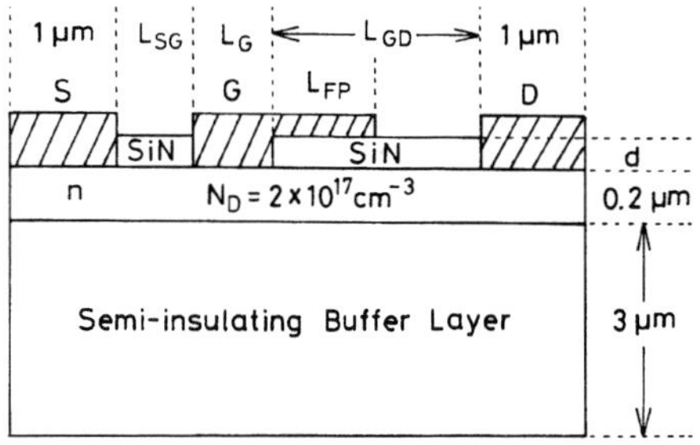

(a)

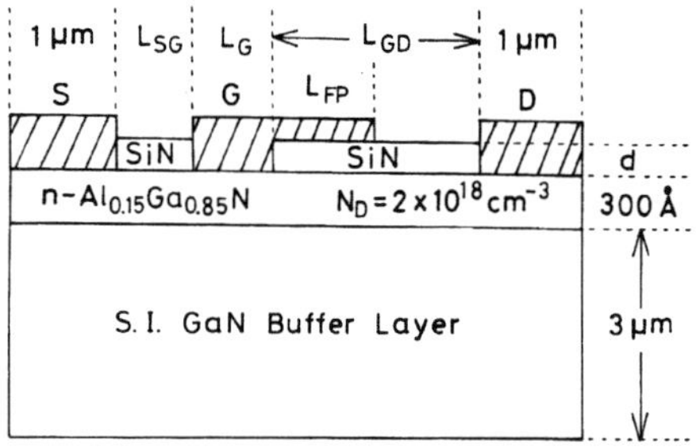

(b)

Figure 1: Device structures simulated in this study. (a) GaN MESFET, (b) AlGaN/GaN HEMT.

deep acceptor [6,7]. Some representative experiments show that two levels (E_C – 1.75 eV, E_C – 2.85 eV) are associated with current collapse in GaN-based FETs [1], and hence we use energy levels of E_C – 2.85 eV (E_V + 0.6 eV) for the deep acceptor and of E_C – 1.75 eV for the deep donor. Other experiments show shallower energy levels for deep donors in GaN [9], and hence we vary the deep donor's energy level E_{DD} as a parameter. Mainly, we present the data for $E_C - E_{DD}$ = 1.0 eV and 0.5 eV here. The deep-donor density N_{DD} and the deep-acceptor density N_{DA} are typically set to 2x10^{17} cm^{-3} and 10^{17} cm^{-3}, respectively. The shallow-donor density N_{Di} is set to 10^{15} cm^{-3}. When $N_{DD} > N_{DA}$, the deep donors donate electrons to the deep acceptors, and hence the ionized deep-donor density N_{DD}^+ becomes nearly equal to N_{DA} under equilibrium, and it acts as an electron trap.

Basic equations to be solved are Poisson's equation including ionized deep-level terms, continuity equations for electrons and holes which include carrier loss rates via the deep levels and rate equations for the deep levels. These are expressed as follows.

1) Poisson's equation

$$\nabla^2\psi = -\frac{q}{\varepsilon}(p - n + N_{\rm D} + N_{\rm Di} + N_{\rm DD}^{+} - N_{\rm DA}^{-}) \qquad (1)$$

2) Continuity equations for electrons and holes

$$\frac{\partial n}{\partial t} = \frac{1}{q}\nabla \bullet J_n - (R_{n,\rm DD} + R_{n,\rm DA}) \qquad (2)$$

$$\frac{\partial p}{\partial t} = -\frac{1}{q}\nabla \bullet J_p - (R_{p,\rm DD} + R_{p,\rm DA}) \qquad (3)$$

where

$$R_{n,\rm DD} = C_{n,\rm DD} N_{\rm DD}^{+} n - e_{n,\rm DD}(N_{\rm DD} - N_{\rm DD}^{+}) \qquad (4)$$

$$R_{n,\rm DA} = C_{n,\rm DA}(N_{\rm DA} - N_{\rm DA}^{-})n - e_{n,\rm DA} N_{\rm DA}^{-} \qquad (5)$$

$$R_{p,\rm DD} = C_{p,\rm DD}(N_{\rm DD} - N_{\rm DD}^{+})p - e_{p,\rm DD} N_{\rm DD}^{+} \qquad (6)$$

$$R_{p,\rm DA} = C_{p,\rm DA} N_{\rm DA}^{-} p - e_{p,\rm DA}(N_{\rm DA} - N_{\rm DA}^{-}) \qquad (7)$$

3) Rate equations for the deep levels

$$\frac{\partial}{\partial t}(N_{\rm DD} - N_{\rm DD}^{+}) = R_{n,\rm DD} - R_{p,\rm DD} \qquad (8)$$

$$\frac{\partial}{\partial t} N_{\rm DA}^{-} = R_{n,\rm DA} - R_{p,\rm DA} \qquad (9)$$

where $N_{\rm DD}^{+}$ and $N_{\rm DA}^{-}$ represent ionized densities of deep donors and deep acceptors, respectively. C_n and C_p are the electron and hole capture coefficients of the deep levels, respectively, e_n and e_p are the electron and hole emission rates of the deep levels, respectively, and the subscript (DD, DA) represents the corresponding deep level.

The above basic equations are put into discrete forms and are solved numerically. We have calculated the drain-current responses when the drain voltage $V_{\rm D}$ and/or the gate voltage $V_{\rm G}$ are changed abruptly.

3 DRAIN LAG

Figure 2 shows calculated drain-current responses of GaN MESFETs when $V_{\rm D}$ is lowered abruptly from 20 V to $V_{\rm Dfin}$, where $V_{\rm G}$ is kept constant at 0 V. Fig.2(a) is for a structure without a field plate ($L_{\rm FP}$ = 0) and Fig.2(b) is for a case with a field plate ($L_{\rm FP}$ = 1μm). Here the thickness of SiN passivation layer d is 0.1 μm. In both cases, the drain currents remain at low values for some periods and begin to increase slowly, showing drain-lag behavior. It is understood that the drain currents begin to increase when the deep donors in the buffer layer begin to emit electrons [6]. It is seen that the change of drain current is smaller for the case with a field plate, indicating that the drain lag is smaller for the field-plate structure. We will discuss below why this reduction in drain lag arises.

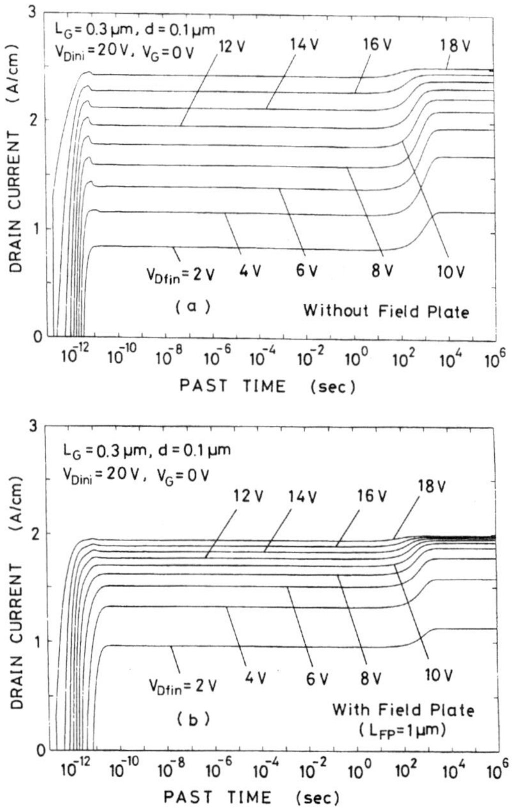

Figure 2: Calculated drain-current responses of GaN MESFETs when $V_{\rm D}$ is changed abruptly from 20 V to $V_{\rm Dfin}$, while $V_{\rm G}$ is kept constant at 0 V. $N_{\rm DD}$ = 2x10^{17} cm^{-3}, $N_{\rm DA}$ = 10^{17} cm^{-3} and $E_{\rm C} - E_{\rm DD}$ = 1.0 eV. (a) Without field plate, (b) with field plate.

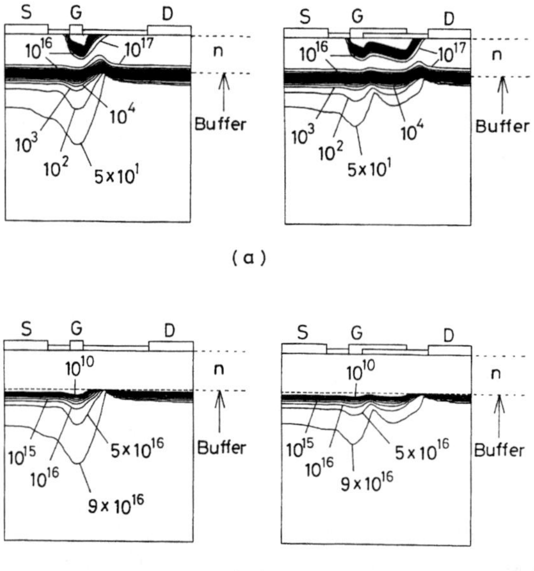

Figure 3: (a) Electron density profiles and (b) ionized deep-donor density $N_{\rm DD}^{+}$ profiles at $V_{\rm G}$ = 0 V and $V_{\rm D}$ = 20 V. d = 0.1 μm. $N_{\rm DD}$ = 2x10^{17} cm^{-3}, $N_{\rm DA}$ = 10^{17} cm^{-3} and $E_{\rm C} - E_{\rm DD}$ = 1.0 eV. The left is for the case without field plate, and the right is for the field-plate structure ($L_{\rm FP}$ = 1μm).

Figure 3 shows (a) electron density profiles and (b) ionized deep-donor density N_{DD}^{+} profiles at V_G = 0 V and V_D = 20 V. The left is for the structure without a field plate, and the right is for the field-plate structure. In Fig.3(a), it is seen that for the structure without a field plate, electrons are injected deeper into the buffer layer under the gate, particularly under the drain edge of the gate region. These electrons are captured by the deep donors, and hence N_{DD}^{+} decreases there as seen in Fig.3(b). As mentioned before, when V_D is lowered abruptly, the drain current remains at a low value for some periods and begins to increase slowly as the deep donors begin to emit electrons (and N_{DD}^{+} increases), showing drain lag. In the case of field-plate structure, as seen in Fig.3(a), electrons are injected into the buffer layer under the drain edge of field plate as well as under the gate. But the injection depth is not so deep as compared to the case without a field plate. This is because the electric field at the drain edge of the gate becomes weaker by introducing a field plate. Hence, the change of N_{DD}^{+} by capturing electrons is smaller for the field-plate structure as seen in Fig.3(b). Therefore, the drain lag becomes smaller for the structure with a field plate.

4 CURRENT COLLAPSE

Next, we have calculated a case when V_G is also changed from an off point. V_G is changed from threshold voltage V_{th} to 0 V, and V_D is changed from 20 V to V_{Don} (on-state drain voltage). The characteristics become similar to those in Fig.2, although some transients arise when only V_G is changed (gate lag). From these turn-on characteristics, we obtain a quasi-pulsed *I-V* curve.

In Figs.4 and 5, we plot by (x) the drain current at $t = 10^{-8}$ s after V_G is switched on for GaN MESFET and AlGaN/GaN HEMT, respectively. Figs.4(a) and 5(a) are for the structures without a field plate and Figs.4(b) and 5(b) are for the field-plate structures (L_{FP} = 1 μm). These curves are regarded as quasi-pulsed *I-V* curves with pulse width of 10^{-8} s. They stay rather lower than the steady-state *I-V* curves (solid lines), indicating current collapse behavior. Note that the gate lag is larger for AlGaN/GaN HEMTs [7]. In Figs.4 and 5, we also plot another pulsed *I-V* curve (Δ), which is obtained from figures like Fig.2 (where only V_D is changed), indicating drain-lag behavior. From Figs.4 and 5, we can definitely say that the lag phenomena (drain lag, gate lag) and current collapse become smaller for the structure with a field plate.

5 INSULATOR THICKNESS DEPENDENCE

We have next studied dependence of lag phenomena and current collapse on the SiN thickness *d*. Figures 6 and 7 show drain-current reduction rate $\Delta I_D/I_D$ (ΔI_D : current reduction, I_D : steady-state current) due to current collapse, drain lag or gate lag, with *d* as a parameter. Figure 6 is for GaN MESFETs, and Figure 7 is for AlGaN/GaN HEMTs.

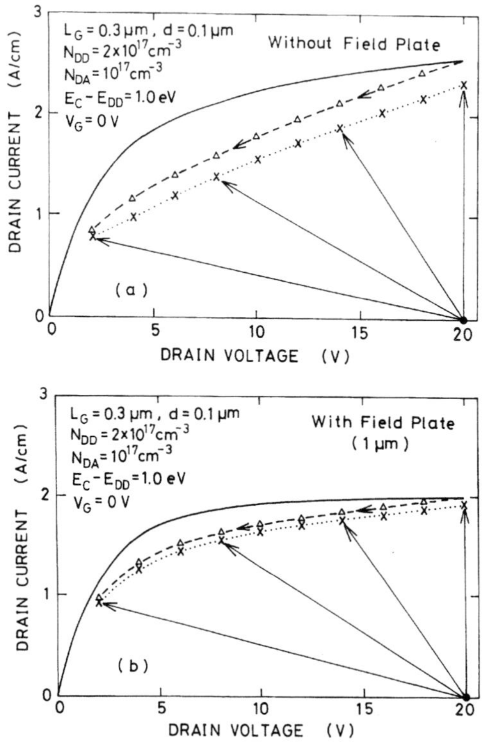

Figure 4: Steady-state *I-V* curves (V_G = 0 V; solid lines) and quasi-pulsed *I-V* curves (Δ, x) of GaN MESFETs. (a) without field plate, (b) with 1 μm-length field plate. (Δ): Only V_D is changed from 20V ($t = 10^{-8}$ s), (x): V_D is lowered from 20 V and V_G is changed from V_{th} to 0 V ($t = 10^{-8}$ s).

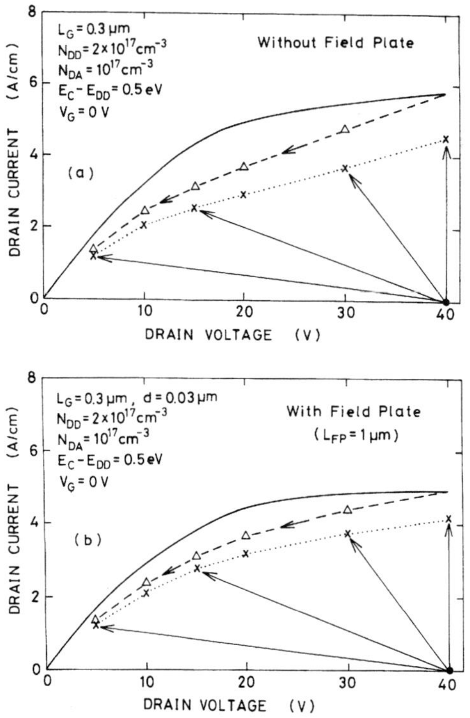

Figure 5: Steady-state *I-V* curves (V_G = 0 V; solid lines) and quasi-pulsed *I-V* curves (Δ, x) of AlGaN/GaN HEMTs. (a) without field plate, (b) with 1 μm-length field plate. The symbols have similar meanings as in Fig.4.

Here d = 0 (L_{FP} = 1 μm) corresponds to a case of L_G = 1.3μm without a field plate. When d is thick, the current collapse and lag phenomena are relatively large because the field plate does not almost affect the characteristics. As d becomes thinner, the current collapse and lag phenomena become smaller, although the rates of current collapse and drain lag increase for very thin d. When d = 0 (L_G = 1.3 μm), that is, without a field plate, the current collapse becomes rather noticeable. (In a previous work [10], we showed that the current collapse was not so dependent on the gate length L_G in the structure without a field plate.) From Figs.6 and 7, we can say that there is an optimum thickness of SiN to minimize the buffer-related current collapse and drain lag in GaN MESFETs and HEMTs.

6 CONCLUSION

Two-dimensional transient simulations of the field-plate GaN MESFETs and AlGaN/GaN HEMTs with a semi-insulating buffer layer have been performed in which a deep donor and a deep acceptor are considered in the buffer layer. Quasi-pulsed *I-V* curves have been derived from the transient characteristics. It has been shown that the drain lag is reduced by introducing a field plate because the trapping effects become smaller. It has also been shown that the gate lag and current collapse are also reduced in the field-plate structure. It is suggested that there is an optimum thickness of SiN passivation layer to minimize the buffer-related drain lag and current collapse in GaN MESFETs and AlGaN/GaN HEMTs.

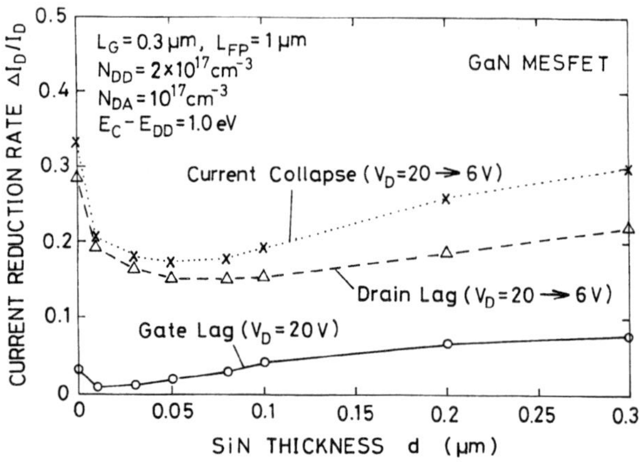

Figure 6: Current reduction rate $\Delta I_D/I_D$ due to current collapse, drain lag or gate lag for GaN MESFETs, with SiN thickness d as a parameter.

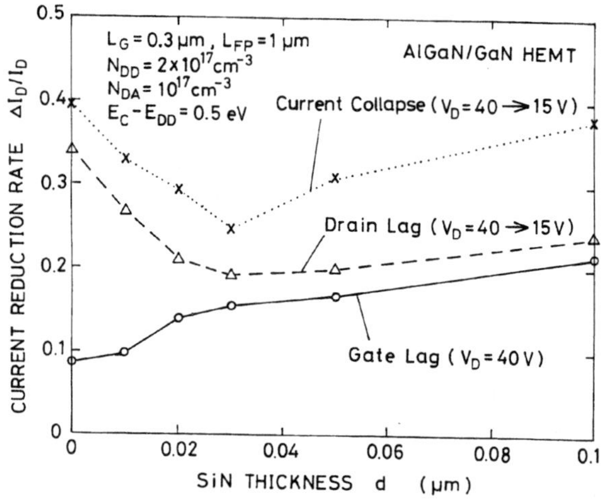

Figure 7: Current reduction rate $\Delta I_D/I_D$ due to current collapse, drain lag or gate lag for AlGaN/GaN HEMTs, with SiN thickness d as a parameter.

REFERENCES

[1] S. C. Binari, P. B. Klein and T. E. Kazior, "Trapping effects in GaN and SiC Microwave FETs", Proc. IEEE, vol.90, pp.1048-1058, 2002.

[2] U. K. Mishra, P. P. Parikh and Y.-F. Wu, "AlGaN/GaN HEMTs — An overview of device operation and applications", Proc. IEEE, vol.90, pp.1022-1031, 2002.

[3] G. Koley, V. Tilak, L. F. Eastman and M. G. Spencer, "Slow transients observed in AlGaN/GaN HFETs: Effects of SiN*x* passivation and UV illumination", IEEE Trans. Electron Devices, vol.50, pp.886-893, Apr. 2003.

[4] A. Koudymov, V. Adivarahan, J, Yang, G. Simon and M. A. Khan, "Mechanism of current collapse removal in field-plated nitride HFETs", IEEE Electron Device Lett., vol.26, pp.704-706, 2005.

[5] J. Tirado, J. L. Sanchez-Rojas and J. I. Izpura, "Trapping Effects in the transient response of AlGaN/GaN HEMT devices", IEEE Trans. Electron Devices, vol.54, pp.410-417, 2007.

[6] K. Horio, K. Yonemoto, H. Takayanagi and H. Nakano, "Physics-based simulation of buffer-trapping effects on slow current transients and current collapse in GaN field effect transistors" J. Appl. Phys., vol.98, no.12, pp.124502 1-7, 2005.

[7] K. Horio and A. Nakajima, "Physical mechanism of buffer-related current transients and current slump in AlGaN/GaN high electron mobility transistors", Jpn. J. Appl. Phys., vol.47, 2008.

[8] S. Karmalkar and U. K. Mishra, "Enhancement of breakdown voltage in AlGaN/GaN high electron mobility transistors using a field plate", IEEE Trans. Electron Devices, vol.48, pp.1515-1521, Aug. 2001.

[9] H. Morkoc, Nitride Semiconductors and Devices, Springer-Verlag, 1999.

[10] K. Itagaki, N. Kobayashi and K. Horio, "Analysis of buffer-related lag phenomena and current collapse in GaN FETs", phys. stat. soli. (c) vol.4, pp.2666-2669, 2007.

Self-Consistent Thermal Electron-Phonon Simulator for SOI Devices

S. M. Goodnick*, K. Raleva** and D. Vasileska*

*Arizona State University, Tempe, AZ 85287-5706, USA
**FEIT-UKIM, Skopje, Republic of Macedonia

ABSTRACT

To investigate the role of self-heating effects on the electrical characteristics of nano-scale devices, we implemented a two-dimensional Monte Carlo device simulator that includes the self-consistent solution of the energy balance equations for both acoustic and optical phonons. The acoustic and optical phonon temperatures are fed back into the electron transport solver through temperature dependent scattering tables. The electro-thermal device simulator was used in the study of different generations of nano-scale fully-depleted (FD) Silicon On Insulator (SOI) devices that are either already in production or will be fabricated in the next 5-10 years. We find less degradation due to self-heating in very short channel device structures due to increasing role of non-stationary velocity overshoot effects which are less sensitive to the local temperature.

Keywords: heating effects, nanoscale devices, phonons, BTE, Monte Carlo method

1 INTRODUCTION

Heat conduction in dielectric materials and most semiconductors is dominated by lattice vibrational waves. The basic energy quantum of lattice vibration is called a phonon, analogous to a photon which is the basic energy quantum of an electromagnetic wave. Similar to photons, phonons can be treated as both waves and particles. Size effects appear if the structure characteristic length is comparable to or smaller than the phonon characteristic lengths. Two kinds of size effects can exist: the classical size effect, when phonons can be treated as particles, and the wave effect, when the wave phase information of phonons becomes important. Phonon-boundary scattering is responsible for a large reduction in the thermal conductivity of a thin silicon layer where the thickness of the film is comparable to or smaller than the phonon mean free path, Λ.

The lateral thermal conductivity of the thin silicon layer decreases as the thickness of the film is reduced. Deviation of the thermal conductivity from the bulk value takes a sharp dive as the thickness of the film is reduced beyond 300 nm, which is the order of magnitude for the phonon mean free path in silicon at room temperature. For example, the thermal conductivity of the 20 nm thick silicon layer is nearly an order of magnitude smaller than the bulk value.

It should also be mentioned that the Fourier heat conduction equation cannot explain the thickness dependency of thermal conductivity in silicon. The impact of phonon-boundary scattering on the thermal conductivity of a thin silicon layer can be predicted using the Boltzmann Transport Equation (BTE) for phonons.

On the other hand, it is well known that heat generation and the associated thermal management in very large scale integrated (VLSI) circuits (or nanoscale devices) is one of the major barriers to further increase clock speeds and decrease of feature size. The modern semiconductor industry benefits greatly from device scaling for the purpose of improving the device performance and reducing the manufacturing cost [1]. It is predicted that, when the device dimension scales down for a factor of F, the power consumption density usually increases by a factor of F^2 or F^3 for the case of constant voltage scaling. Also, for conventional complementary metal–oxide–semiconductor (CMOS) devices, by scaling them down to nanometer dimensions, it was predicted that the characteristic *phonon hot spot region* [2] near the drain would not scale proportionally. The size of the phonon hot spot is on the order of the magnitude of the high electric field region near the drain in a 180 nm device. What happens in nanoscale devices is very difficult to predict. As it will be seen from the presented simulation results, in nano-scale transistors, self-heating has a less detrimental effect when compared to larger structures due to the fact that velocity overshoot dominates carrier transport, and there is less exchange of energy between the electron system and the phonon bath [3, 4]. While this result does not mean that one does not have to be concerned with heating in nanoscale devices; it only means that we mainly have to focus on efficient ways of removing the heat from the device active region with, for example, state of the art Peltier coolers. Note that the predictions regarding heating in devices becomes more complicated with the fact that for channel lengths below 35 nm [5], various non-classical transistor structures will likely take over due to their delivery of higher performance with lower leakage current than traditional scaled CMOS approaches. New transistors, particularly ultra-thin-body (UTB) and double-gate MOSFETs, offer paths to further scaling, perhaps to the end of the 2006 International Technology Roadmap for Semiconductors (ITRS). In UTB-SOI, power consumption is drastically reduced along with leakage current and the devices show promise for high-performance CMOS, microprocessors and system-on-a-chip designs. In UTB-SOI structures [6], control of short-channel effects (SCE) and threshold voltage (V_t) adjustment can be realized with little or no channel doping. However, an issue that has shown to be important for the SOI devices is lattice heating. Heating effects arise in SOI devices because the device is thermally isolated from the substrate by the buried oxide layer. No simulator can properly predict the electrical characteristics of nano-scale devices if it does not treat electron transport correctly; in particular non-stationary velocity overshoot effect. In particular, simulators that rely on energy balance models for the electronic transport typically overestimate or underestimate velocity overshoot due to the improper choice of the energy relaxation times taken from bulk calculations.

In the present work, we solve the Boltzmann transport equation for electrons using the Ensemble Monte Carlo (EMC) method coupled moment expansion equations for the phonons, both acoustic and optical. The coupling of electrons and non-equilibrium phonons has been studied for many years, and was included, e.g. in EMC simulations to study photoexcited carrier relaxation in quantum wells [7,8], and more recently by Alam and Lundstrom [9] to simulate laser diodes. However, these models are essentially momentum space models, and do not address spatially varying systems such as short channel transistors. Recently, there have been studies on describing thermal effects in devices that couple the Monte Carlo/Poisson approach to electro-thermal modeling [10,11] in SOI and Nitride devices. In that work, a somewhat simplistic model for non-equilibrium phonons is taken which does not distinguish the acoustic and the optical phonons as separate subsystems.

Because SOI devices consist of two distinct regions, the silicon device layer and the buried oxide layer (in which the phonons have significantly smaller mean-free paths), the phonon BTE is solved in the silicon layer to accurately model heat transport, but the simpler heat diffusion equation is used in the amorphous BOX because the characteristic length-scale of conduction is much smaller than the film thickness. The two distinct computational regions are coupled through interface conditions that accounts for differences in material properties. For the coupling of the silicon and oxide solution domains, it is necessary to calculate the flux of energy through the interface between the two materials at each point along the interface for every time step.

The boundary conditions used have been chosen based on those typically used in commercial simulators. The Silvaco ATLAS simulation package [12] (THERMAL3D module) states that the only thermal contact should be the substrate. We have performed simulations on our structure to verify this assertion with and without the silicon substrate present, and concluded that due to the large thermal conductivity of bulk Si, a 300K boundary condition on the bottom contact maps well into 300K boundary condition on the bottom of the BOX. In other words, the presence of the bottom silicon substrate does not affect either the electrical or the thermal characteristics of the structure being considered. Also, according to prescriptions given in the Silvaco ATLAS package, the source and drain should be left floating and the only electrode where one should specify isothermal boundary conditions is the gate. In fact, in the paper itself we change the thickness of the gate metal and the boundary condition at the end of the gate has not much influence on the current degradation. In these simulations, except at the boundary, we treat the metal gate as a material characterized with its own thermal conductivity. Since current nano-scale devices use metal gates to avoid poly-silicon depletion, such an assumption isothermal also is seemingly justified. However, to study the efficacy of the gate as a heat sink, we simulate the effect on the current of several different temperatures for the gate.

2 INFLUENCE OF THE BOUNDARY CONDITION ON THE GATE ELECTRODE ON THE ON-CURRENT

To properly solve the phonon balance equations, the device should be attached to a heat sink somewhere along the boundary or finite heat conduction through the surface should be allowed for. In our code, a heat sink is modeled by a simple Dirichlet boundary condition (i.e. constant temperature). We use the gate electrode contact and the bottom of the BOX as heat sinks. Table 1 gives the percentage of the current decrease due to the heating effects with the variation of gate electrode temperature. The calculated results show that the current degradation is more prominent for higher gate temperatures. When the temperature of the bottom of the BOX was set to the same values as given in Table 1, the current degradation was around 1%, so in all other simulations the temperature of the bottom of the BOX was set to 300K.

Table 1 Current variation with gate temperature for 25 nm fully-depleted SOI device structure with SiO2 as gate oxide.

Type of simulation	*Gate Temperature*	***Current Decrease***
thermal	300K	5.1%
thermal	400K	9.18%
thermal	600K	17.12%

Figure 1 show the velocity profile along the channel for the same bias conditions (Vgs=Vds=1.1 V) and different gate temperatures, where, as can be seen, the velocity in the channel decreases with the increase of the gate temperature, but the carriers in the channel are still in the velocity overshoot regime. As seen from the temperature maps of acoustic phonons in Figure 2, the lattice temperature in the source, the channel and the drain region is increasing with the increase of the gate electrode temperature, which means that the increased lattice temperature has larger impact on the decrease of the carrier velocity in the channel.

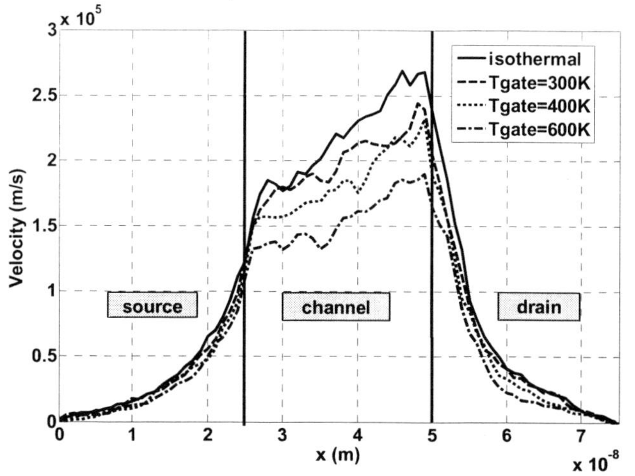

Fig. 1 Velocity along the channel for Vgs=1.1 V and Vds=1.1 V for different gate temperatures. Notice that the electrons are in the velocity overshoot regime.

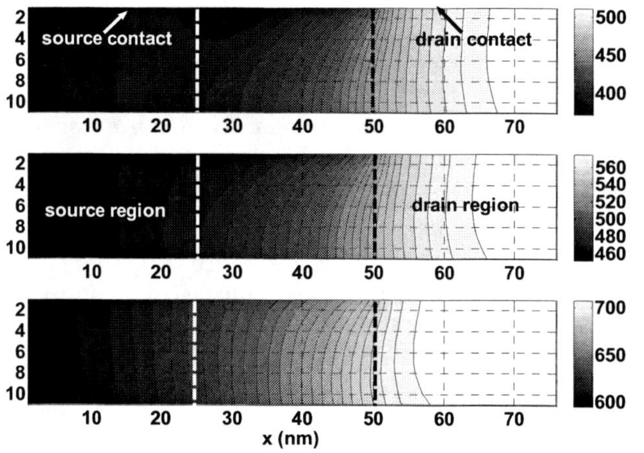

Fig. 2 Lattice temperature profiles in the silicon layer in 25 nm gate-length fully-depleted SOI MOSFET for Vgs=1.1 V and Vds=1.1 V and different gate electrode temperatures (300K, 400K and 600K from up to down). SiO2 is used as gate oxide.

3 THERMAL DEGRADATION WITH SCALING OF DEVICE GEOMETRY

In addition to the previously noted observation regarding the influence of the velocity overshoot, we modeled larger fully-depleted SOI device structures and we also investigated the influence of the temperature boundary condition on the gate electrode on the current degradation due to heating effects. The calculated results show that the current degradation is more prominent for larger devices and for higher gate temperatures. For 80 nm and larger devices, simulated carriers are not in the velocity overshoot regime in the larger portion of the channel (especially near the source end of the channel). Snapshots of the lattice temperature profiles in the silicon layer for these devices when the gate temperature is set to 300K, are given in Figure 3 and 4. From these snapshots one can observe that: a) the temperature in the channel is increasing with the increase of the channel length, b) the maximum lattice temperature region (hot spot) is in the drain and it shifts towards the channel for larger devices. This behavior is more drastic for higher gate temperatures.

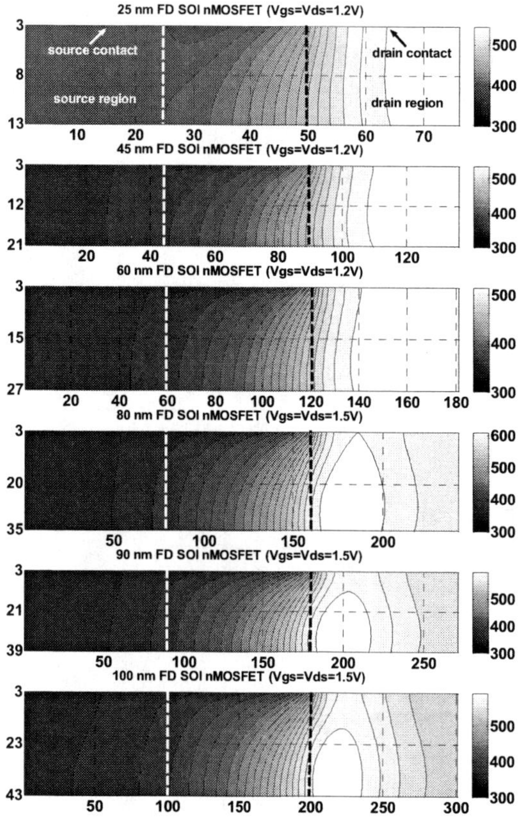

Fig. 3 Lattice temperature profiles in the silicon layer for FD SOI MOSFETs with gate temperature set to 300K. (25 nm –top, 100 nm-bottom).

4 CONCLUSIONS

A self-consistently coupled thermal/Ensemble Monte Carlo device simulator has been developed and applied to the study of fully-depleted SOI devices. We show that the pronounced velocity overshoot present in the nanometer scale device structure considered in the present study minimizes the degradation of the device characteristics due to lattice heating. This observation was also justified with SILVACO Atlas simulations that demonstrated that for larger energy relaxation times, that correspond to the case of more pronounced velocity overshoot, current degradation in the on-state due to thermal effects is on the order of 10%, not to 30% as found in larger device structures in which velocity overshoot does not play significant role.

We also investigate the influence of the gate temperature on the amount of current degradation due to heating effects. Namely, we used the gate contact as a heat sink to solve properly the phonon balance equations. As seen from the temperature maps of acoustic phonons presented in the paper, the lattice temperature in the source, channel and drain region is increasing with the increase of the gate temperature, which means that the increased lattice temperature has larger impact on the decrease of the carrier velocity in the channel. When examining heating in different device technologies, we observed a bottleneck between the lattice and the optical phonon temperature in the channel which is more pronounced for shorter devices, due to the fact that the energy transfer between optical and acoustic phonons is relatively slow compared to the electron-optical phonon processes and the fact that the electrons are in the velocity overshoot (and since the channel is very short, they spent little time in the channel).

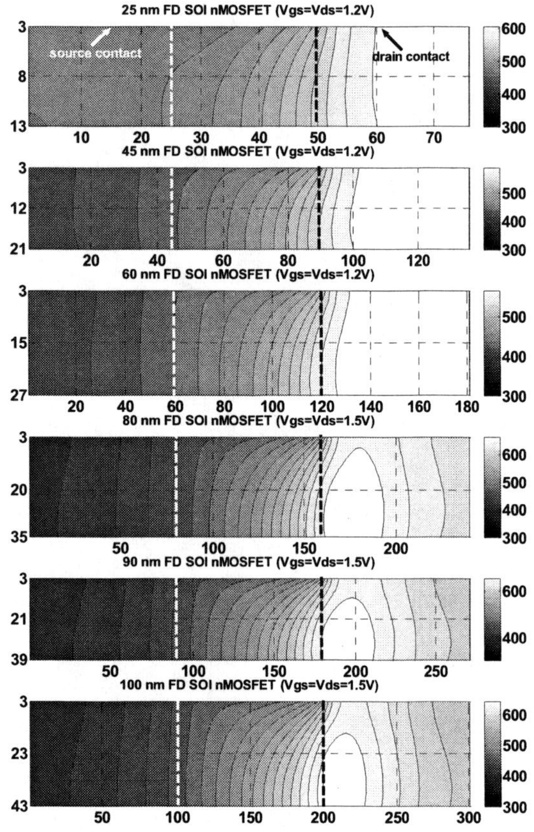

Fig. 4 Optical phonon temperature profiles in the silicon layer for FD SOI MOSFETs with gate temperature set to 300K. (25 nm –top, 100 nm-bottom).

To better understand the phonon temperature bottleneck, different cross-sections of the lattice and the optical phonon temperature profiles in the channel direction were investigated. Briefly, we find that the bottleneck is decreasing from Si/SiO_2 interface to Si/BOX interface. For shorter devices, it exists in the whole channel region, which is not a case for longer devices (thicker Si-layer and longer channel length). From the results we have presented here and those we have published earlier one can conclude that the higher the temperature in the channel and/or the longer the electrons are in the channel, the larger the degradation of the device electrical characteristics is due to the heating effects.

REFERENCES

[1] R. Chau, B. Doyle, M. Doczy, S. Datta, S. Hareland, B. Jin, J. Kavalieros, M. Metz, "Silicon nano-transistors and breaking the 10 nm physical gate length barrier", *Device Research Conference*, pp. 123-126 (2003).

[2] E. Pop, K. Banerjee, P. Sverdrup, R. Dutton and K. Goodson, "Localized Heating Effects and Scaling of Sub-0.18 Micron CMOS Devices", *IEDM Techn. Dig.*, 679 (2001).

[3] K. Raleva, D. Vasileska and S.M. Goodnick, accepted for publication in *Journal of Computational Electronics*, 2008.

[4] Katerina Raleva, Dragica Vasileska, and Stephen M. Goodnick, "The Role of the Temperature Boundary Conditions on the Gate Electrode on the Heat Distribution in 25 nm FD-SOI MOSFETs with SiO2 and Gate-stack (High-K Dielectric) as the Gate Oxide", *accepted for presentation at the ISDRS, Washington DC December 12-14,* 2007.

[5] IA Technology Roadmap for Semiconductors, 2003 (http://public.itrs.net/).

[6] T. Numata and S.-I. Takagi, "Device design for sub-threshold slope and threshold voltage control in sub-100 nm fully-depleted SOI MOSFETs", *IEEE Trans. Electron Devices*, Vol. 51, pp. 2161-2167 (2004).

[7] P. Lugli and S.M. Goodnick, "Non-Equilibrium LO Phonon Effects in GaAs/AlGaAs Quantum Wells", *Phys. Rev. Lett.* Vol. 59, pp. 716-719 (1987).

[8] S. M. Goodnick and P. Lugli, "Hot Carrier Relaxation in Quasi-2D Systems," in Hot Carriers in Semiconductor Microstructures: Physics and Applications, (J. Shah, Ed.), Academic Press Inc., pp. 191-234, 1992.

[9] M. A. Alam and M. S. Lundstrom, "Effects of Carrier Heating on Laser Dynamics – A Monte Carlo Study", IEEE J. Quantum El. Vol. 33, pp.2209-2220 (1997).

[10] T.Sadi, R.W.Kelsall and N.J.Pilgrim, "Electrothermal Monte Carlo Simulation of Submicrometer Si/SiGe MODFETs", IEEE Trans. on Electron Devices, Vol.54, No.2, February 2007.

[11] T. Sadi, R.W.Kelsall and N.J.Pilgrim, "Electrothermal Monte Carlo simulation of submicron wurtzite GaN/AlGaN HEMTs", J.Comput.Electron. (2007) 6:35-39.

[12] Silvaco Inc.

Structure Generation for the Numerical Simulation of Nano-Scaled MOSFETs

C. Kernstock*, M. Karner**, O. Baumgartner**, A. Gehring***, and H. Kosina**
* Global TCAD Solutions, Rudolf Sallinger Platz 1, 1030 Wien, Austria
c.kernstock@globalTCADsolutions.info
** Institute for Microelectronics, TU Vienna,
Gußhausstraße 27–29/E360, 1040 Wien, Austria
*** AMD Saxony, Wilschdorfer Landstraße 101, D-01109 Dresden, Germany

ABSTRACT

An accurate and predictive numerical simulation of MOS transistors in the deca-nanometer channel length regime relies on the precise mapping of the physical device to a simulation model. A quick and accurate method which allows to extract the relevant transistor parameters, based on data which are typically available within a process flow, is presented.

Keywords: MOSFETs, TCAD, numerical device simulation

1 INTRODUCTION

To make TCAD analysis applicable to state-of-the-art MOSFETs in the deca-nanometer channel length regime [1], numerical simulation using advanced transport models [2] is required. An accurate and predictive result relies on the precise mapping of the physical device onto a virtual structure. Besides the use of full process simulation, which requires extensive calibration of the underlying models, this work presents an alternative approach for the quick estimation of the relevant device parameters.

2 STRUCTURE GENERATION

We developed a method to setup simulations with the data available during the production with the aim to assist the device engineer. The topological structure of the devices is directly extracted from TEM images, which are available for each major process step. In contrast to simple template device structures, the full transistor geometry is covered. A device editor allows to identify the device regions and to create the segments as shown in Fig. 1. The doping profiles can be specified using analytical distribution functions or numerically, provided from previously obtained calibrated process simulation or from SIMS measurements (c.f. Fig. 2).

3 GRID GENERATION

The grids for the numerical solution of the PDE system are generated automatically using a basic triangulation of the regions and a quad tree refinement strategy [3]. A segment is divided into four rectangular sub regions. If the quality criterion is not met, a sub region is recursively refined until sufficent accuracy is achieved. This is demonstrated in Fig. 3.

For MOSFETs, two criteria apply: The channel refinement follows the distance to the interface in order to properly resolve the inversion channel. The junction refinement depends on the gradient of the dopant concentrations. This ensures accurate simulation results and good convergence [4]. Simulation grids with the applied refinements are shown in Fig. 4.

4 SIMULATION MODELS

A pMOS transistor is considered. For holes, the energy transport model has been applied:

$$\mathbf{J}_p = -\mu_p \mathrm{k_B} \left(\nabla (pT_p) - \frac{\mathrm{q}}{\mathrm{k_B}} \mathbf{E} p \right) ,$$

$$\mathbf{S}_p = -\frac{\tau_S}{\tau_m} \left(\frac{5\mathrm{k_B}^2}{2\mathrm{q}} \mu_p p T_p \nabla T_p - \frac{5\mathrm{k_B}^2}{2\mathrm{q}} T_n \mathbf{J}_p \right) ,$$

$$\nabla \cdot \mathbf{J}_p = -\mathrm{q} R ,$$

$$\nabla \cdot \mathbf{S}_p = \mathbf{E} \cdot \mathbf{J}_p - \frac{3}{2} \mathrm{k_B} p \frac{\mathrm{T}_p - T_L}{\tau_{\mathcal{E}}} - G_{\mathcal{E}p} .$$

Here, $\mathbf{S}$ denotes the energy flux density, T_p the local hole temperature, $\tau_{\mathcal{E}}$, τ_S, and τ_m the energy, energy flux, and momentum relaxation time, respectively, and $G_{\mathcal{E}p}$ the net energy generation rate [2].

The electrons are treated within a quasi-fermi level (QFL) approximation. For the simulation of the partially depleted SOI MOSFET, a combined QFL scheme was applied [6]. It typically converges within a few iterations for the biased device.

The gate leakage current is calculated within a post processing step by evaluation of the Tsu-Esaki formula [7].

5 RESULTS AND CONCLUSION

The described methodology has been used to investigate a state-of-the-art pMOS transistor fabricated on SOI substrate [5]. Based on a TEM image the material properties are assigned, the dopant profile is defined, and a numerical analysis of the device has been carried out. This allows to inspect distributed quantities like

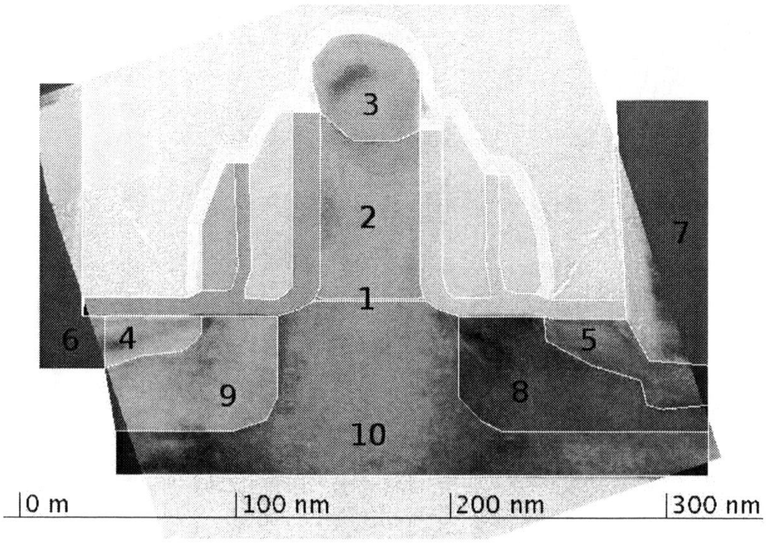

Nr	Name	Material
1	Oxide	SiO_2
2	Gate	Poly-Si
3	GateContact	Silicide
4	SourceContact	Silicide
5	DrainContact	Silicide
6	SourceContact2	Metal
7	DrainContact2	Metal
8	Drain	SiGe
9	Source	SiGe
10	Bulk	Si

Figure 1: The topological structure of the device is directly created from a TEM-image using the segment editor. The silicide regions (4 and 5) have been modeled as conductors. The SiGe regions have a Ge content of 20 %. [5]

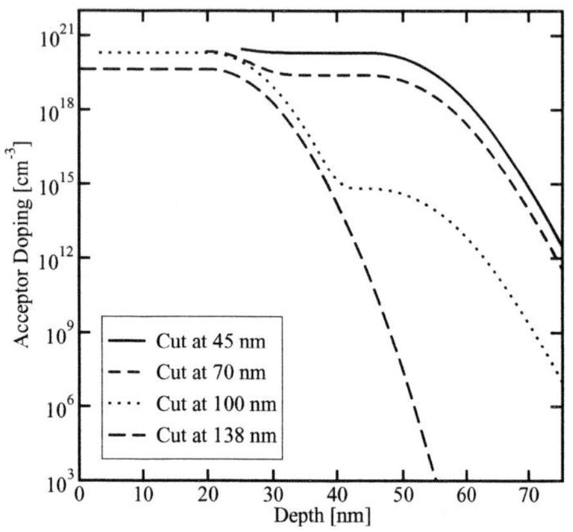

Figure 2: Doping profile displayed on vertical cuts. Depth denotes the vertical distance to the oxide-plane.

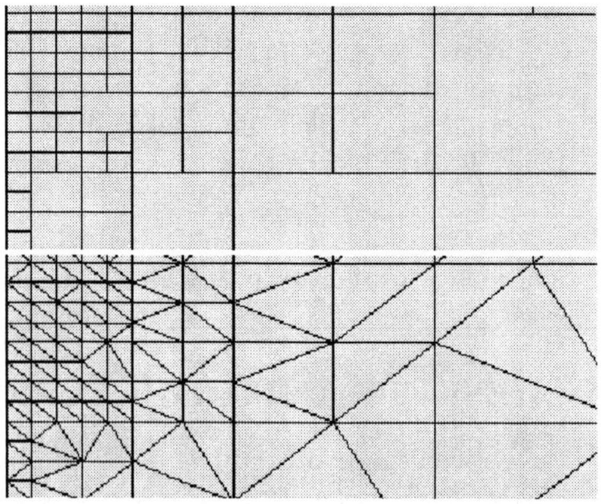

Figure 3: The upper part of the figure shows the recursive quad tree refinement. The corresponding triangularization is shown in the lower part.

carrier concentration as shown in Fig. 5. Proper simulation grids are needed to obtain accurate results with a reasonable number of grid points. This is demonstrated in Fig. 6.

The electric field and the current density of the turned on transistor are shown in Fig. 7 and Fig. 8. The gate capacitance was extracted from a small signal analysis of the device structure (c.f. Fig. 9). It shows the importance of including the spacer and the contact segments. The calculated gate leakage characteristic is given in Fig. 10. The transfer and the output characteristics of the device are shown in Fig. 11 and Fig. 12, respectively.

In conclusion, a quick and accurate TCAD tool which allows to extract the relevant device parameters, based on data which are typically available within a process flow, has been presented.

REFERENCES

[1] International Technology Roadmap for Semiconductors, http://www.itrs.net.

[2] T. Grasser et al., in Proc. Intl. Conf. on Simulation of Semiconductor Processes and Devices, pp. 1–8, 2004.

[3] G. Garreton, Ph.D. thesis, ETH Zurich, 1998.

[4] V. Axelrad, Computer-Aided Design of Integrated Circuits and Systems, IEEE Transactions on Electron Devices, 17, 149 (1998).

[5] M. Horstmann et al., in Proc. Intl. Electron Devices Meeting, pp. 233–236, 2005.

[6] MINIMOS-NT 2.1 Users Guide, Institut fur Mikroelektronik, Technische Universitat Wien, Austria, 2004.

[7] Gehring et al. IEEE Trans.Electr.Dev.Mat.Rel., Vol. 4, No. 3, pp. 306–319, 2004

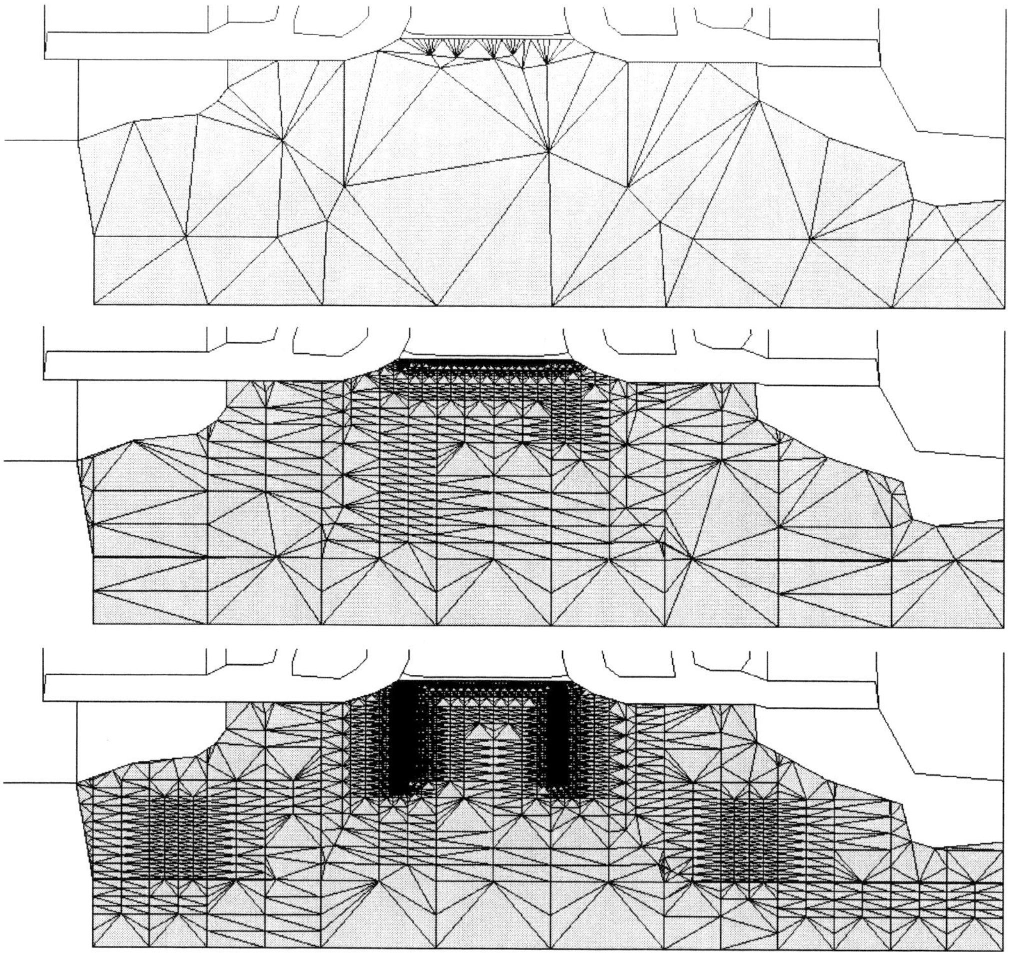

Figure 4: The upper figure shows a minimal triangulation of the geometry. The middle figure shows the channel refinement, and the lower figure the channel and pn-junction refinement.

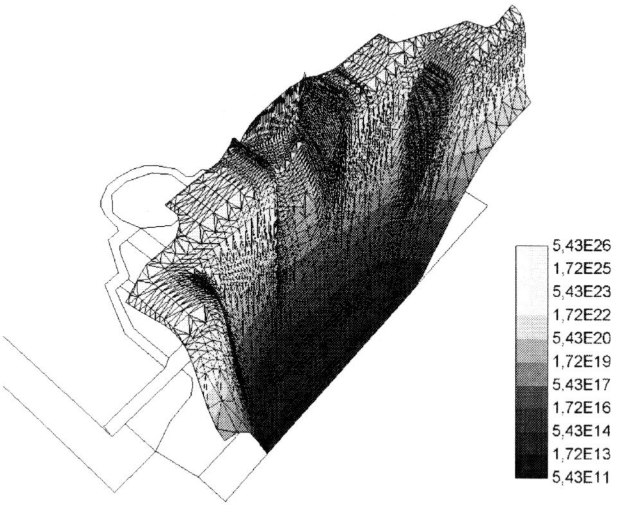

Figure 5: The spatial distribution of the hole concentration in the pMOS transistor ($V_G = -1.5\ V, V_{DS} = -1.5\ V$).

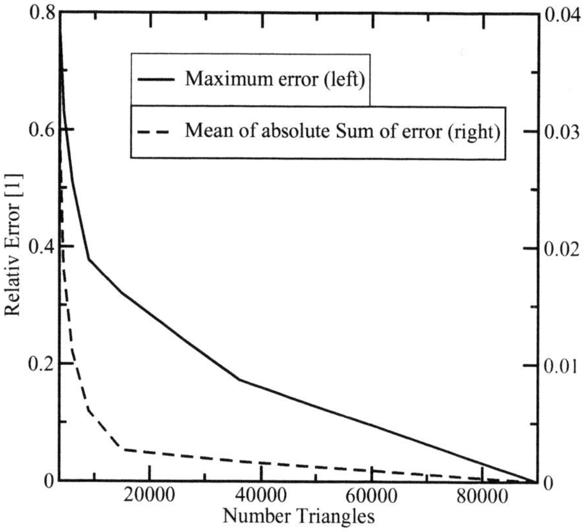

Figure 6: The left and the right axes show the maximum and the mean value of the error in carrier concentration as a function of the number triangles used for the discretization.

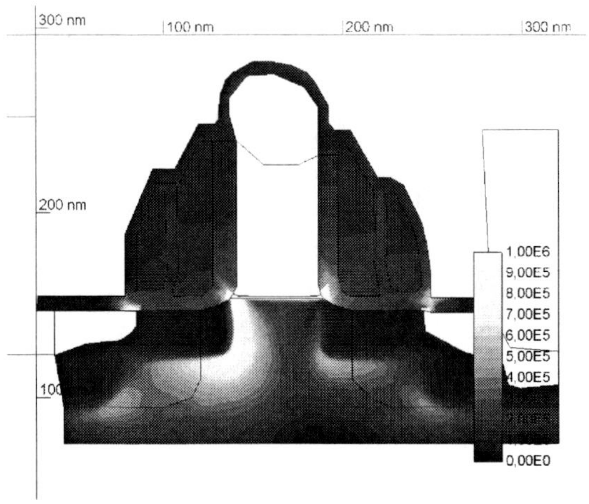

Figure 7: The electric field of the turned on transistor shown at a linear scale.

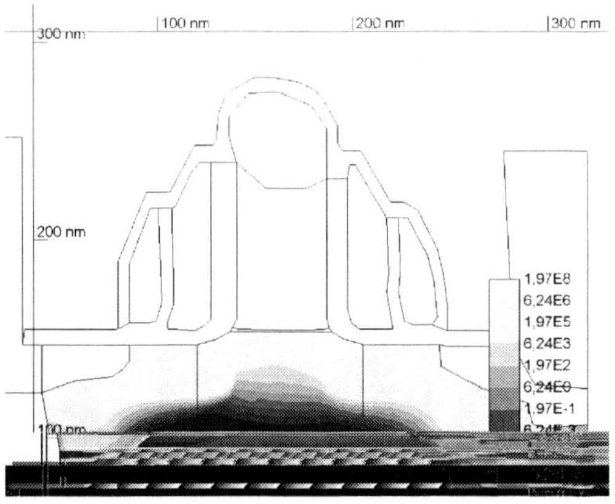

Figure 8: The absolute value of the current density of the turned on transistor shown at logarithmic scale.

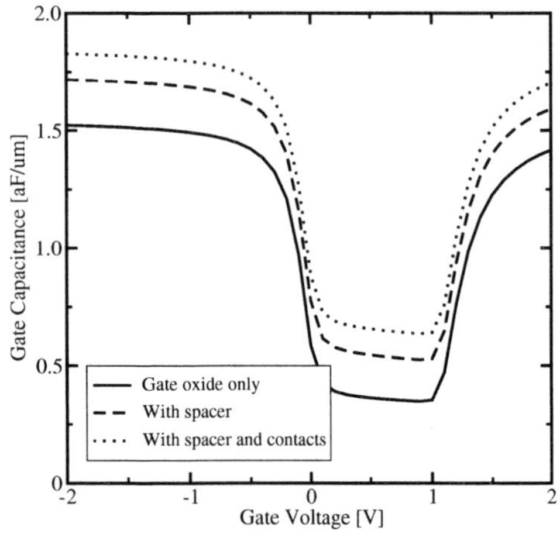

Figure 9: The gate capacitance/voltage characteristic of the structure. Including contacts and spacer to the simulation domain gives rise to an additional coupling.

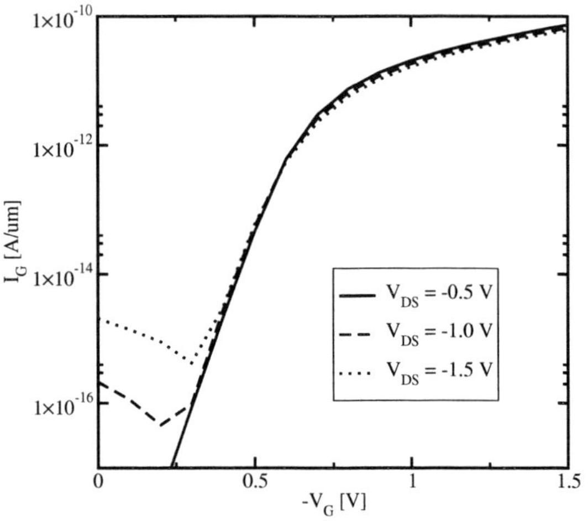

Figure 10: The gate leakage current as a function of the gate bias shown for different drain biases.

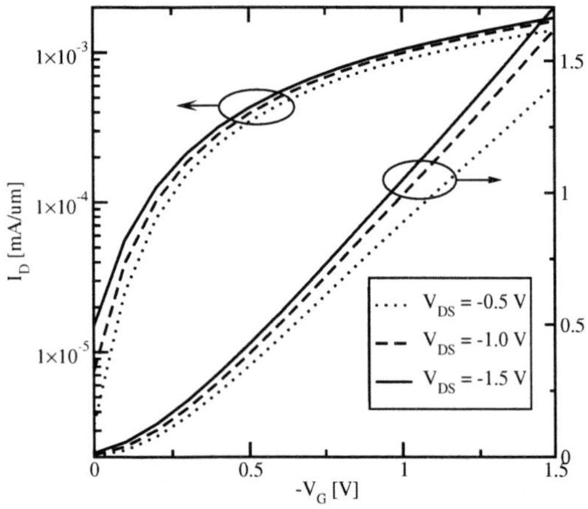

Figure 11: The transfer characteristic of the device on a linear and semi-logarithmic scale.

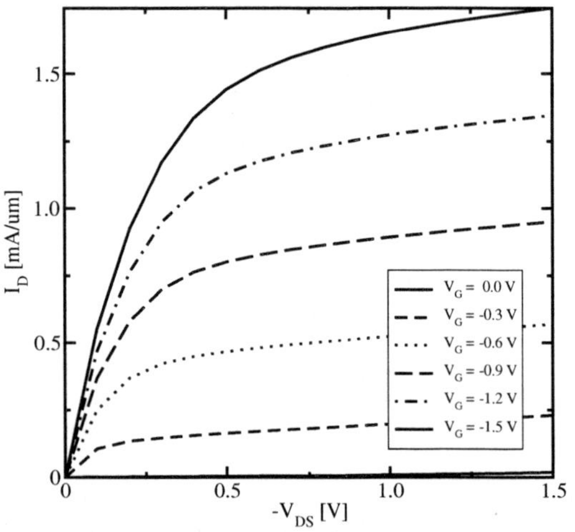

Figure 12: The simulated output characteristics of the pMOS transistor using an energy transport model.

The Immersed Surfaces Technology for Reliable and Fast Setup of Microfluidics Simulation Problems

M. Icardi[1], D. Caviezel, D. Lakehal*

*ASCOMP GmbH, Zurich, Switzerland, lakehal@ascomp.ch

ABSTRACT

In many microfluidic applications, the complexity of the system is such that it requires tremendous efforts to setup the CFD problem, ranging from dimensioning of the computational domain, to grid setup. In setting up a grid, the user must complain to various constraints: adequate near-wall resolution, reduce the aspect ratio and the rate of grid stretching, and control of cell skewness (in the framework of BFC structured grids). These issues can cause a large amount of numerical diffusion errors, which in turn requires reducing convection scheme accuracy to 1stt-order to avoid divergence of the solution. We present here a new technique based on immersed surfaces that alleviates most of the evoked problems with traditional grid generation, permitting reliable and fast setup of microfluidics problems.

Keywords: Microfluidics, CFD, IST, BMR, Interfacial flow

1 INTRODUCTION

As bio-chips may comprise various components (multi channels and components, complex configurations), we have developed a new fully automatized version for microfluidics applications in bio-devices, using the IST (Immersed surfaces Technique) technique to map complex components/geometries into a simple rectangular Cartesian grid. In such a way, the drawbacks of traditional girding are alleviated: aspect ratio, cell stretching and skewness.

IST forces the grid to remain Cartesian and equidistant in all directions, thus high-order schemes (we use up to 3rd order for flux convection and 3rd order WENO schemes for free surface flows) can maintain their high degree of accuracy. Further, to better resolve boundary-layer regions, near wall flow areas are treated by another new feature, namely the BMR (Block-based Mesh Refinement), in which sub-scale refined blocks are placed around each structure or obstructions. The connectivity between blocks can be achieved in parallel (using MPI) up to 8-to-1 cell mapping. The combination IST/BMR can save up to 70% grid cells in 3D. In this paper we report examples of microfluidics bio-devices treated with this approach, without dealing in detail with the flow results. The flow physics simulated by the code TransAT of similar problems is detailed in our companion paper (this volume [1]).

2 INTERFACE TRACKING FOR MICROFLUIDCS FLOW PROBLEMS

Interfacial flows refer to multi-phase flow problems that involve two or more immiscible fluids separated by sharp interfaces which evolve in time. Typically, when the fluid on one side of the interface is a gas that exerts shear (tangential) stress upon the interface, the latter is referred to as a free surface. Interface tracking methods (ITM) are schemes capable to locate the interface, not by following the interface in a Lagrangian sense (e.g., by following marker points on the interface), but by capturing the interface by keeping track, in an Eulerian sense (the grid is fixed), of the evolution of an appropriate field such as a level-set function or a volume-fraction field. Examples and classifications are provided in [2]. Application of ITM's to microfluidics flows requires further attention to the way surface forces are handled.

2.1 TransAT© Microfluidics Flow solver

The Microfluidics code TransAT© [3] of ASCOMP is a multi-physics, finite-volume code based on solving multi-fluid Navier-Stokes equations on structured multi-block meshes. MPI parallel based algorithm is used in connection with multi-blocking. Grid arrangement is collocated and can thus handle more easily curvilinear skewed grids. The solver is pressure based, corrected using the Karki-Patankar technique for weak compressible flows. The Navier-Stokes and level set equations are solved using the 3rd order Runge-Kutta explicit scheme for time integration. The convective fluxes are discretized with TVD-bounded high-order schemes [4]. The diffusive fluxes are differenced using a 2ndorder central scheme.

Multiphase flows are tackled using Level Sets and VOF for both laminar and turbulent flows. The solver incorporates phase-change capabilities, surface tension and triple-line dynamics models, Marangoni effects, and a micro-film sub-grid scale model for lubrication. 3D flows are treated using the IST to map the components into a simple rectangular Cartesian grid. BMR helps refine grids around the flow areas of interest, up to 1-to-8 level connection between the blocks, as explained below. Details of the modeling of interfacial flows can be found in [1].

[1] M. Icardi is a MS student from the Polytecnico di Torino (Group of Prof. Canutto), Italy, hosted by ASCOMP.

2.2 Predictive Performance of ITM's

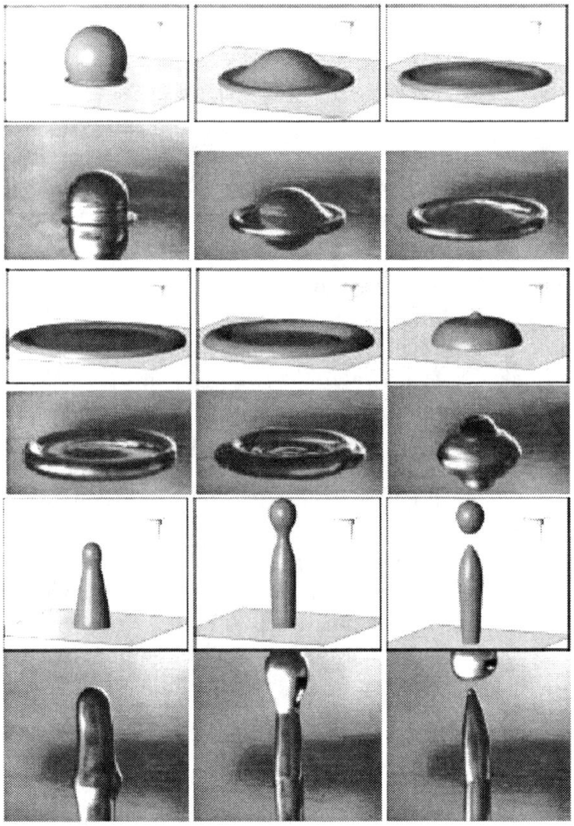

Figure 1: Partial bouncing of microdroplets on a dry surface obtained for CA = 100, We= 52 and Re = 3245 [5]. CFD results (surface level sets in blue) vs. experiments (grey).

Figure 1 above illustrates the predictive performance of the ITM's methods (Level Sets [6]) implemented in TransAT for treating contact-angle driven flows. The bouncing or splashing of liquid droplets on dry surfaces is mainly dependent on the contact angle (CA), Weber and Reynolds number (We and Re), and surface roughness (in this example the surface is smooth). The comparison between the data and TransAT simulations is perfect, including for 3D splashing and bouncing; see detailed comparison in [5].

3 IMMERSED SURFACES TECHNIQUES WITH BLOCK MESH REFINEMENT

The Immersed Surfaces Technology (IST) has been developed at ASCOMP GmbH, although other similar approaches have appeared in parallel. The underpinning idea is inspired from ITM's for two-phase flows (VOF and Level Sets), where free surfaces are described by a convection equation advecting the phase color function.

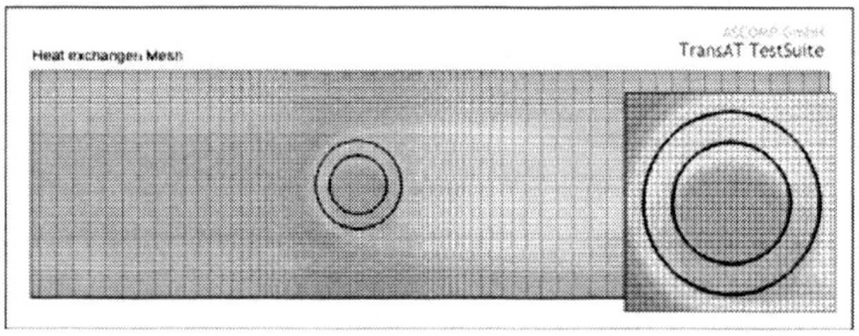

Figure 2: Representation of the tube, $\phi_s = 0$, within IST.

In the IST solid surfaces are described as the second 'phase', with its own thermo-mechanical properties. This new technique differs substantially from the Immersed Boundaries methods [7], in that the jump condition at the solid surface is implicitly accounted for, not via direct momentum forcing on the Eulerian grids. It has the major advantage to solve conjugate heat transfer problems, in that conduction inside the body is directly linked to external fluid convection.

The air-coolant flow past a circular tube half-filled with hot liquid shown in Figure 2 is an illustrative example. The solid is first immersed into a Cartesian mesh. The solid is defined by its external boundaries using the solid level set function, ϕ_s. Like in fluid-fluid flows, ϕ_s function is defined as a distance to the surface; is zero at the surface, negative in the fluid and positive in the solid. The treatment of viscous wall shear is handled as in all CFD codes; wall cells are identified as those in which $0 < \phi_s < 1$. The figure shows the grid and the heat transfer resulting from the coupled convection-conduction; the coolant air from the left removes heat from the tube walls, first convected within the liquid in the tube, then conducted through the tube walls.

4 MICROFLUIDCS GRID SETUP TESTS

4.1 Fluid Handling in a Micro-reactor

Micro-reactor chips consist of an array of on-chip fluid handling modules, and are used as "liquid circuit boards" that can be configured for a variety of biochemical and cell-based assays including: drug formulation, on-chip chemical synthesis and screening, reagent mixing, etc.

The test case presented below is an idealized set-up (Fig. 3), consisting of two inflow sections from which a liquid train (e.g. reagent) is injected, which later on breaks up into micro-droplets. These should coalesce in the center and exit from below. Such systems may enable rapid screening of a wide range of cells and molecules under a variety of assay conditions in a very cost effective way, compared for example to manual manipulation using for instance robotic microliter plate systems. Fast response of CFD here can be invaluable for large data-basis treatment.

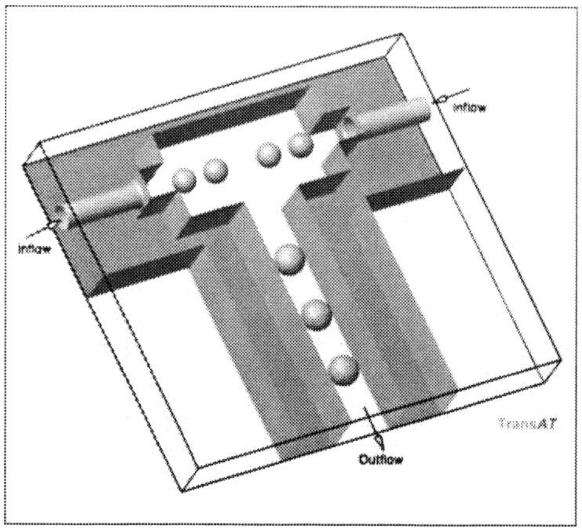

Figure 3: An idealized configuration of a micro-reactor.

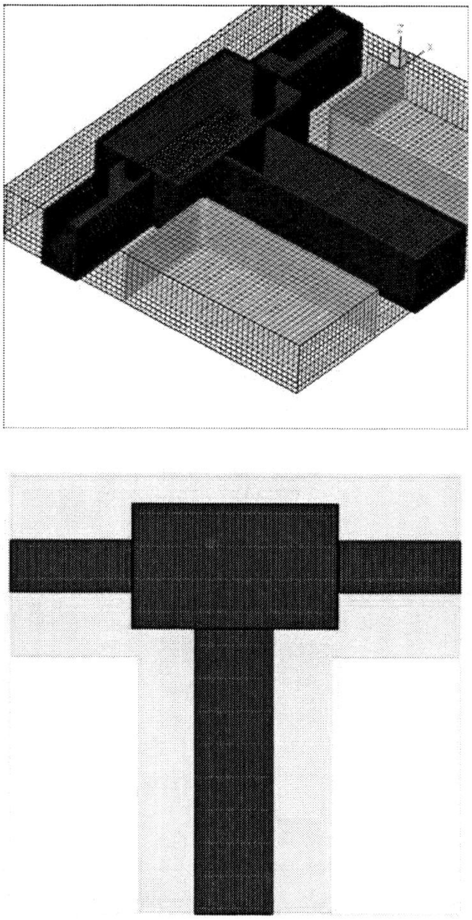

Figure 4: Step 1, cover the domain with a 1st grid. Step 2: create refined blocks then remove the coarse grid.

We start by importing the CAD file representing the systems from a library. The file is then loaded and will be automatically entirely covered by a 3D Cartesian mesh (Fig. 4). The microfluidics code TransAT recognizes all solid surfaces as the zero level of the solid level set function, as well as the number of blocks contained in the CAD file. A coarse grid covers first the entire domain (Fig. 4), and then refined blocks are added automatically in the tubes and channels. Finally the coarse grid is removed.

4.2 Cardiovascular Gas Bubbles in Arterial Y-Junction Bifurcation

Cardiovascular gas bubbles in arteriole bifurcations (Fig. 5) have been experimentally addressed to understand the dynamics of their lodging mechanism [8]. The research is motivated by novel gas embolotherapy techniques for the potential treatment of cancer by tumor infarction; the findings may be useful in developing strategies for microbubble delivery in gas embolotherapy. A prototype has been set as a challenging example for the IST/BMR technique of TransAT. The physics of the flow itself is very complicated, requiring in particular the inclusion of sub-grid ultra thin film in the model (built within the ITM strategy) to avoid numerical breaking of the film as it approaches bifurcation corner. The model developed for the purpose is explained in the companion paper, together with selected results.

The same procedure alluded to previously is repeated here to mesh the Y-junction bifurcation shown in Fig. 5 below. The case represents the situation where a liquid slug may either travels within one branch of the junction, or breaks at the junction corner. Again we start by importing the CAD file of the systems from a library (Fig. 6). The file is then loaded and will be automatically entirely covered by a 3D Cartesian mesh (Fig. 7a). The microfluidics code TransAT recognizes all solid surfaces as the zero level of the solid level set function, as well as the number of blocks contained in the CAD file. BMR refined blocks are then automatically generated around the three branches. Finally the coarse grid can be removed (Fig. 7b). The entire procedure lasts less than 30 minutes; the fluid dynamics simulation time depends on the available computing power.

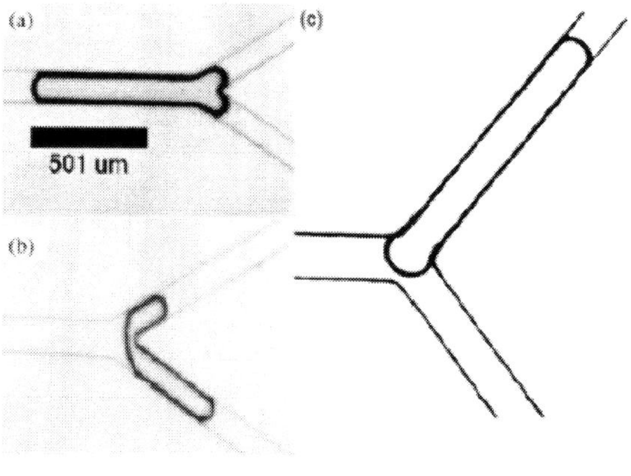

Figure 5: (a) Lodging state, (b) Lodging state, (c) Bubble lodged in one of the branches [8].

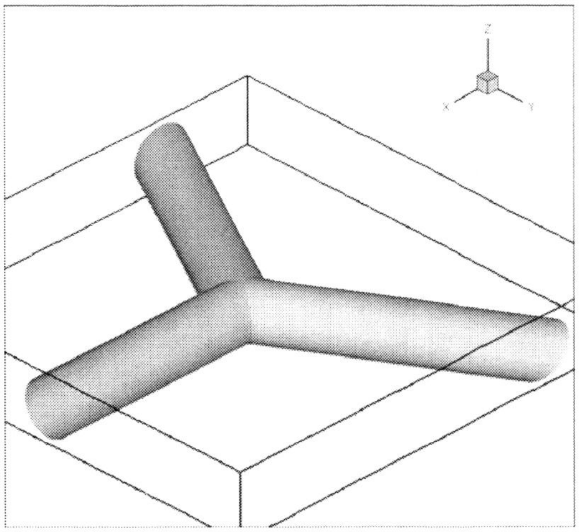

Figure 6:An idealized configuration of Fig. 5.

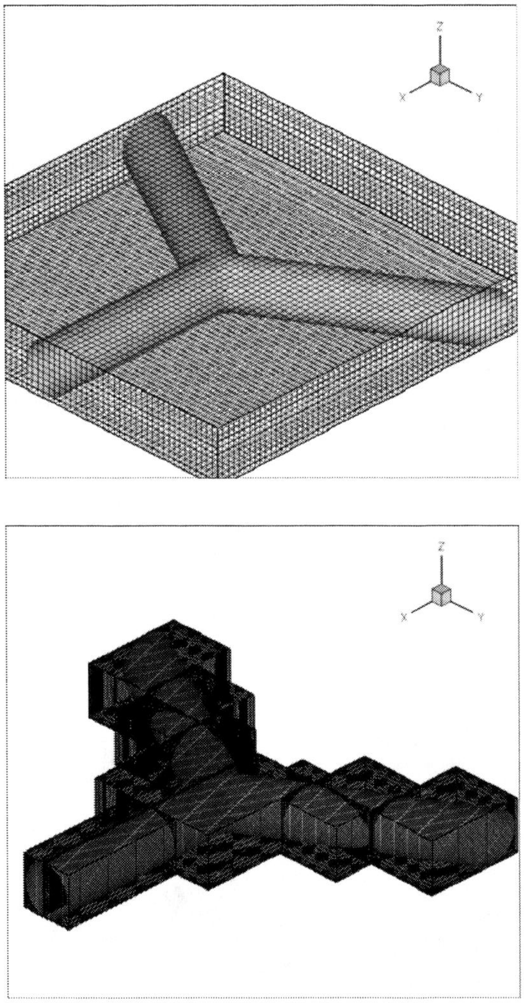

Figure 7: Step 1, cover the domain with a coarse grid. Step 2: Create refined BMR blocks around the braches and remove the coarse grid.

4.3 Liquid Filling of a Lab-on-Chip

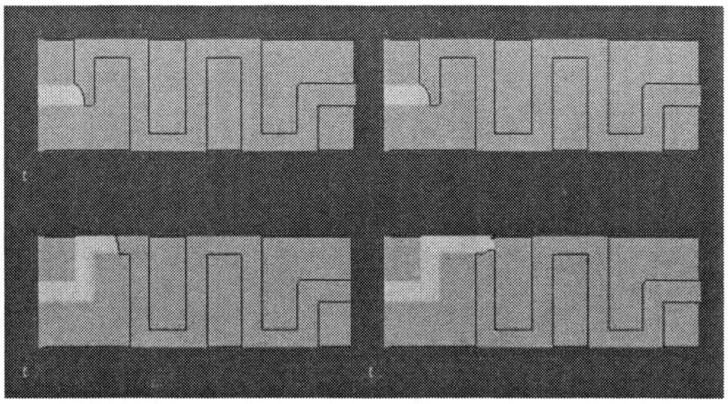

Figure 8: 2D filling of a lab-on-chip by a liquid (CA=90)

The example shown in Fig. 8 above corresponds to the microfluidic filling in lab-on-chip channels (in 2D). The setup of the grid is as explained in the two previous examples; here we present snapshots of the simulated flow (using the Level Set approach). The objective here is to analyse the possibility of small air bubbles entrapment in the cavities, depending on the surface wettability (contact angle). This phenomenon can only be simulated with appropriate wetting modelling of the triple line dynamics, as explained previously; here the contact angle is set statistically to 90 deg.

REFERENCES

[1] C. Narayanan and D. Lakehal, Computational Microfluidics for Miniaturized Bio-Diagnostics Devices using the Multiphysics code "TransAT", Paper 1022, Proc. NSTI, Boston, 2008.
[2] D. Lakehal, M. Meier M. and M. Fulgosi, Interface tracking for the prediction of interfacial dynamics and heat/mass transfer in multiphase flows, Int. J. Heat & Fluid Flow, 23, 242, 2002.
[3] www.ascomp.ch/transat.
[4] B.P. Leonard, A stable and accurate convective modeling procedure based on quadratic interpolation, *Comput. Meth. Appl. Eng.*, 19, 59, 1979.
[5] D. Caviezel, C. Narayanan and D. Lakehal, Adherence and bouncing of liquid droplets impacting on dry surfaces, (available on line DOI 10.1007/s10404-007-0248-2), 2008, Microfluidics & Nanofluidics Journal, Springer.
[6] M. Sussman, S. Smereka and S. Osher, A Level set Approach for computing incompressible two-phase flow. J Comp Phys, 114, 146, 1994.
[7] R. Mittal and G. Iaccarino, Immersed Boundary Methods, *Ann. Rev. Fluid Mech*, 37, 239, 2005.
[8] A. J. Calderon, Y. S. Heo, D. Huh, N. Futai, S. Takayama and J. L. Bull. Microfluidic model of bubble lodging in microvessel bifurcations. App. Phys. Letters, 89, 244103, 2006.

Process Sensitivity Analysis of a 0.25-μm NMOS Transistor Using Two-Dimensional Simulations

Abdol A. Keshavarz*, Laurent F. Dion*, Pierre C. Leonardi**
* Reliability and Process Control Department, STMicroelectronics, Phoenix, AZ
** Device Engineering Department, STMicroelectronics, Phoenix, AZ
Phone: (602) 485-2925, Fax: (602) 485-2925, Email: abdol.keshavarz@st.com

ABSTRACT

The purpose of this work is to present an accurate simulation of the effects of major process changes on some of the most important device parameters of a 0.25-um NMOS transistor. The approach predicts the effects of several critical process steps including poly CD (channel length), threshold-voltage implant, NLDD implant, N+ Source/Drain (S/D) implant, and the final RTA thermal budget on the 0.25-μm NMOS device parameters. Results can be used for the fine-tuning of the device, and its parameter variations as a result of process changes. The modeling approach can reduce expensive and time consuming experiments for device improvements. The simulation model proves to predict accurate results and is currently used for more investigation of the 0.25- μm NMOS Transistor in STMicroelectronics site in Phoenix, AZ.

Keywords: NMOS 2D-simulation, 0.25-μm NMOS, Sensitivity Analysis, Process Effects on 0.25-μm NMOS

1. Key Process Features

CMOS devices are built on N-epi, use shallow trench isolation (STI), and the process utilizes twin-tubs with retrograde wells. Double poly with salicidation on poly and junctions (N+ and P+) are used and the oxide thickness is 50 Angstroms. NMOS source and drain structures receive only Arsenic implants (no Phosphorus). This purpose is to make the device more robust, especially with respect to thermal budget changes. The anti-punch-through implant dopant is Indium, a large-atom dopant that is stable and with minimum diffusion, a critical factor for the short channel device. Only the NMOS is addressed in this work. This device is part of the 0.25-μm technology in STMicroelectronics that utilizes both MOS and Bipolar devices.

2. Methodology

0.25-mm transistors were developed and published in late 80's and early 90's, [1]-[3], with technology features and characterization results. Subsequently, the application of this transistor in the newly developed products showed up [4], [5]. Parallel to the technology development, simulation tools also had to improve to include the new necessary features. These included an overall improvement of the models covering the heavy-ion implant models (In), and accurate simulation of the effects of Arsenic and Indium implants on boron diffusion. This illustrates only some of the challenging problems to be solved before the accurate simulation of the 0.25-μm device was possible.

In this work, two-dimensional process and device simulation tools (ISE from Synopsys [6]) that have advanced Models to address all of the above were used for this work. Latest process simulation models (implant/ diffusion) available in the simulation tool were utilized for this purpose [7]. For Indium profile, Monte-Carlo method was used for highest level of accuracy. For the device simulation part, the accurate mobility models that are crucial for the accuracy of the results (mobility dependence on doping density, saturation velocity, and normal electric field) were used [8]. For the calibration of the results, only gate poly work function was adjusted. Very good agreement between the experimental and the simulated results was seen on the linear, sub-threshold, and saturation regions of the device.

Subsequently, a selected number of major device parameters were extracted from the simulation results. These included the threshold voltage (Vth), saturation current (IdSat), linear current (IdLin), sub-threshold slope (Subth), maximum transconductance (Gmax), device resistance in the linear region (Rlin), and output conductance in saturation region (Gds). Accurate process recipes used for the manufacturing of the 0.25-μm NMOS were used for the simulations.

3. Results and Discussions

Figure 1 shows the main part of the two-dimensional simulated structure as produced by the process simulator. Figures 2 and 3 show the comparison of the simulation and the experimental results (Data) for the major NMOS characteristics. Selected wafer was close to the targets based on all electrical parameters measured in line. Experimental data for the linear region of the device showed some variations on different measured sites, so two measured curves (lowest and highest measured currents) were provided for this region. The simulated sub-threshold region showed very good match to the data, too. Same level of agreement was seen on the saturation region. No optimizations were done to improve the curves in this region. The good agreement between the measured data and the simulation seen on these figures strongly supports and validates the simulation approach.

Table 1 is the summary of the main results. Device parameters and their simulated baseline values are included. The selected process steps, their baseline values and their shifts used for sensitivity analysis are shown too. The simulated change for each device parameter in response to the specified process shift is included.

Overall results indicate that the device is stable with respect to process changes, which is a desirable result. This is in fact expected from the all-Arsenic S/D and its low diffusion coefficients. But changes in poly CD (gate length) are considerable and calculated effects are presented. Effects of threshold-voltage implant (dose and energy) are included too. RTA time variation shows some effects, but it is not significant.

Results for the effects of channel length on device parameters are shown in Figures 4 and 5. These include simulated linear region characteristics, maximum transconductance (Gm), the saturation region characteristics, and the output conductance of the device for different channel lengths.

4. Conclusions

The two-dimensional simulation model is in very good agreement with experimental data. It is a valuable tool for the analysis and improvement of the 0.25-μm NMOS performance. Results indicate that the device is robust and stable with respect to a number of process changes. Poly CD change is the most dominant variable affecting the device parameters. This is an ongoing project and the model is actively being used.

References

[1] W. Chang, B. Davari; M. Wordeman; Y. Taur, C. Hsu, M. Rodriguez, "A high-performance 0.25-μm CMOS technology. I. Design and characterization", IEEE Trans. on Electron Devices, Vol. 39, Apr 1992.

[2] B. Davari, W. Chang, K. Petrillo, C. Wong, D. Moy, Y. Taur, M. Wordeman, J. Sun, C. Hsu, M. Polcari, "A high-performance 0.25-μm CMOS technology. II. Technology", IEEE Trans. on Electron Devices, Vol. 39, Apr 1992.

[3] B. Davari; W. Chang; M. Wordeman; C. Oh; Y. Taur; K. Petrillo; D. Moy; J. Bucchignano; H. Ng; M.. Rosenfield; F. Hohn; M. Rodriguez, "A high performance 0.25 μu m CMOS technology", IEDM Conference Digest, 1988, pages 56-59.

[4] S. Wuu, D. Yaung; C. Tseng; H. Chien; C. Wang, H. Yean-Kuen ; B. Chang, "High performance 0.25-um CMOS color imager technology with non-silicide source/drain pixel", Electron Devices Meeting, 2000.

[5] E. Zervakis, N. Haralabidis, "A fast 0.25um CMOS current-mode front-end stage for solid state detector interfaces", 9th International Conf. on Electronics, Circuits and Systems, 2002. Vol. 1, Page(s): 243- 246.

[6] Sentaurus Work Bench y-2006.06 from Synopsys, Inc.

[7] ISE TCAD Release 10.0, Vol. 2a Process Simulation, Part 8 Dios, Chapter 9 (Implantation), and 10 (Diffusion).

[8] ISE TCAD Release 10.0, Vol. 4a Device Simulation, Chapter 8, Mobility Models.

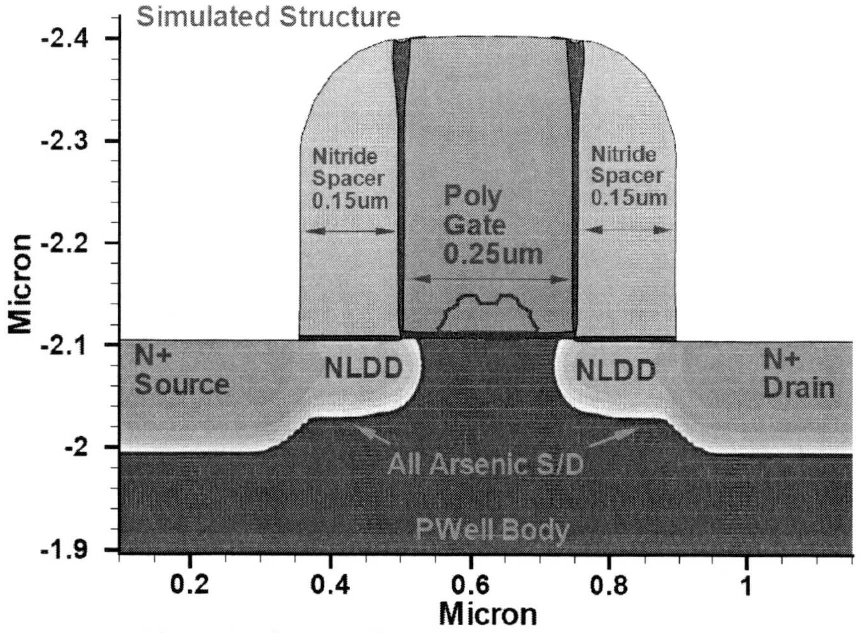

Figure 1. The two-dimensional simulated 0.25μm NMOS.

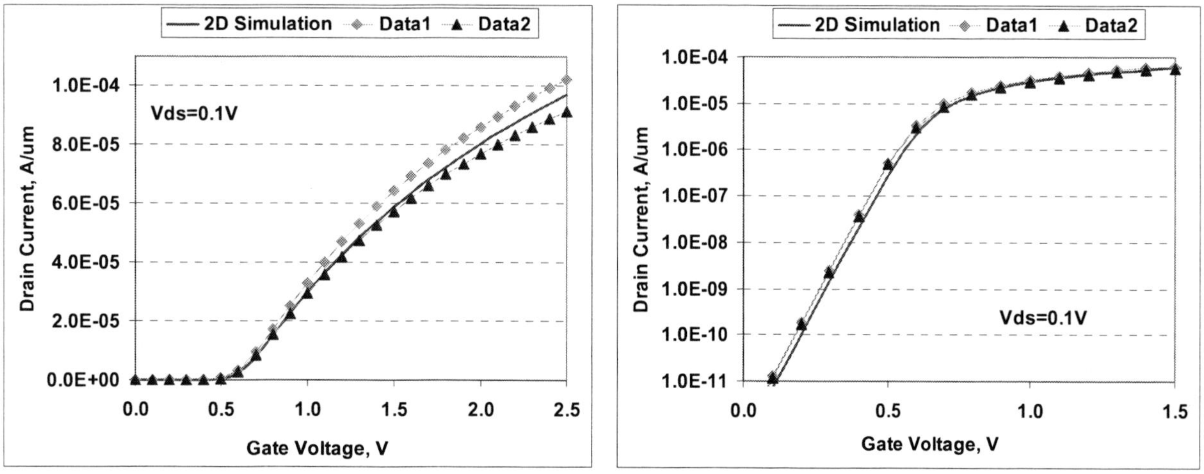

Figure 2. Simulated vs. measured results for the linear and the subthreshold regions.

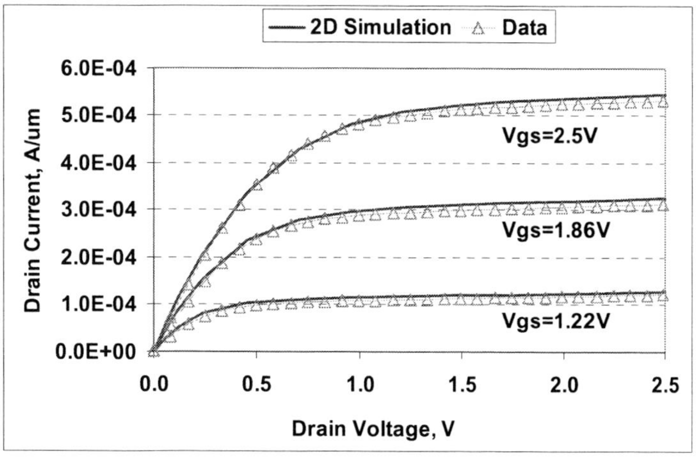

Figure 3. Simulated vs. measured results for the saturation region.

	Device Parameter Name →		VTh, mV	Rlin, Ohm.mm	IDLin, A/mm	Gmax, A/V	IDSat, A/mm	Gds, A/V	Subth, mV/dec
	Parameter Base Value →		550	1029.9	9.7E-05	7.8E-05	5.5E-04	1.7E-02	86.5
			↓ Device Parameter Shifts ↓						
Steps Selected for Analysis	Nominal Value	Process Shift ↓	mV	Ohm	A/um	A/V	A/um	A/V	mV/dec
Channel Lengh	0.25 um	0.05 um	73.86	226.3	-1.7E-05	-2.0E-05	-1.4E-04	-2.9E-03	2.74
Vth Implant Dose	7.0E12 cm-2	0.3E12	10.11	-12.0	1.1E-06	-8.0E-07	-7.0E-06	-2.5E-04	0.15
Vth Implant Energy	20 kev	5 kev	-44.95	-56.3	5.6E-06	4.5E-06	3.3E-05	8.5E-04	-1.52
NLDD As Implant Dose	3.0E14 cm-2	0.5E14	0.81	-29.4	2.9E-06	1.5E-06	7.7E-06	4.8E-04	-0.01
NLDD As Implant Energy	50 kev	5 kev	-1.40	1.6	-1.5E-07	6.1E-07	3.8E-06	8.4E-05	0.01
NPLUS As Implant Dose	3.80E+15	0.3E15	0.05	1.1	-1.0E-07	2.5E-08	1.5E-07	5.3E-06	0.00
NPLUS As Implant Energy	60 kev	5 kev	-0.85	-0.1	1.2E-08	-1.7E-07	-7.7E-09	2.3E-06	-1.87
S/D RTA Anneal	1048 C/20 sec	5 sec	2.93	1.2	-1.1E-07	4.4E-08	6.3E-07	3.4E-05	-1.04

Table 1. Simulated device baseline parameter values and their shifts in response to process changes.

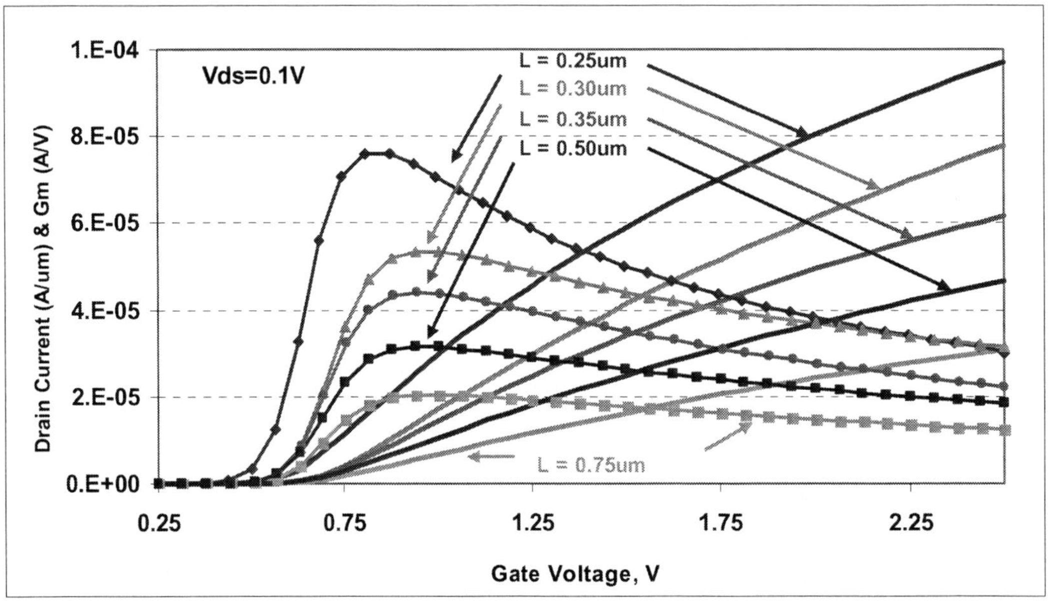

Figure 4. Simulated results for the device linear region for different channel lengths.

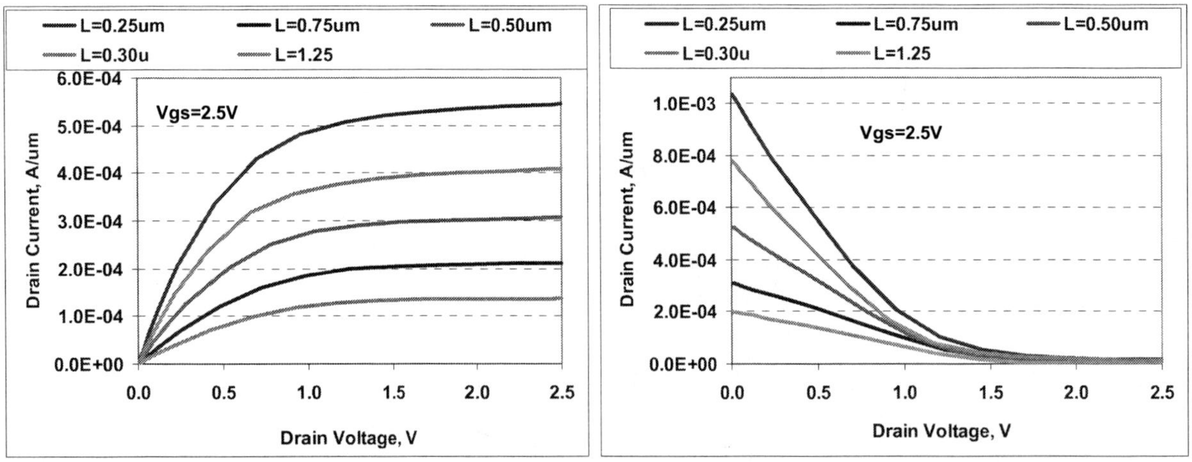

Figure 5. Simulated results for the device saturation region for different channel lengths.

The IMPRINT software: quantitative prediction of process parameters for successful nanoimprint lithography

N. Kehagias[*], V. Reboud[*], C. M. Sotomayor Torres[*,**], V. Sirotkin[***], A. Svintsov[***] and S. Zaitsev[***]

[*]Tyndall National Institute, University College Cork, Lee Maltings, Cork, Ireland
[**]Catalan Institute of Nanotechnology, Campus de Bellaterra, Edifici CM7, ES 08193 - Bellaterra, Spain and Catalan Institute for Research and Advanced Studies ICREA, 08010 Barcelona, Spain
[***]Institute of Microelectronics Technology, RAS, Chernogolovka, Moscow district, 142432 Russia

ABSTRACT

The IMPRINT software is applied for simultaneous calculation of the resist viscous flow in thermal nanoimprint lithography (NIL) and the stamp/substrate deformation. From the presented comparison of calculated and experimental results, it can be concluded that the simulation allows predicting the residual layer thickness with accuracy better than 10%. The obtained results demonstrate the potential of the IMPRINT software as an efficient tool for choosing NIL process parameters and the optimization of the NIL stamp geometry.

Keywords: nanoimprint lithography, stamp and substrate deformation, computer simulation

1 INTRODUCTION

A homogeneous residual layer thickness in thermal nanoimprint can be achieved by optimizing the NIL stamp geometry (the distribution of cavities and protrusions, the stamp cavities depth and the stamp thickness) as well as by choosing NIL process parameters (the initial resist thickness, the imprint temperature and the duration of the imprint) (see [1] for example). This optimization produces the greatest benefit if its implementation is performed before expensive stamp manufacturing starts. To do this requires an effective tool for the simulation of NIL process.

Modeling of the resist viscous flow at the detail level (for a single cavity) is widely covered in the literature (see [2] and references therein). However, approaches for quantitative analysis of resist spreading in large areas are not described.

In this paper, the IMPRINT software for simulation of NIL at the structure-scale level is presented. The software is based on the mathematical model and the coarse-grain numerical algorithm from [3-5]. The IMPRINT software has been specially designed for use on standard Personal Computers, by using the GDS data of the stamp design ("Graphic Data System" is a database file format for integrated circuit layout data exchange). The software demonstrates a high computational performance. Typical simulation times for a test structure with 2×2 mm^2 area (see Section 3) are less than 20 min on an AMD Athlon 64, 2400 MHz processor.

2 MATHEMATICAL MODEL

In [4] the mathematical model for the simultaneous calculation of the resist viscous flow in NIL and the stamp/substrate deformation has been introduced. The model specifies the 2D temporal distributions of the pressure and the normal displacement of the stamp/substrate surface. The model has the following input parameters: the distribution of the stamp relief height $h(x,y)$; the initial resist thickness d_0; the stamp velocity V_{st}; the total force F acting on the stamp; the duration of the imprinting process T; the dynamic viscosity of the resist η.

In the model, at every point in time t, the pressure distribution $P(x,y,t)$ is calculated from the following problem:

$$\begin{aligned}
&\nabla\left\{[D(x,y,t)+h(x,y)]^3\,\nabla P(x,y,t)\right\}=12\eta\frac{\partial D(x,y,t)}{\partial t},\\
&(x,y)\in\Omega_f,\quad t\in(0,T],\\
&P(x,y)=0,\quad (x,y)\in\bar{\Omega}/\Omega_f,\\
&D(x,y,t)=d_0-\int_0^t V_{st}(\varsigma)d\varsigma+\delta_{st}(x,y,t)+\delta_{sb}(x,y,t),
\end{aligned}\tag{1}$$

where Ω is the considered domain of the stamp; Ω_f is the part of Ω, in which all cavities are filled with the resist, $D(x,y,t)$ is the temporal distribution of the residual layer thickness. In (1) the zero value of the pressure on the boundary of the considered domain Ω and in the unfilled cavities corresponds to the imprinting process performed in vacuum.

Note that equation (1) is derived from 3D Navier-Stokes equations with the understanding that the resist has very high viscosity and its motion is largely directed along the substrate surface.

For the calculation of the elastic normal displacement δ_{st} and δ_{sb}, the stamp and the substrate are represented as semi-infinite regions (an elastic medium bounded by a plane). In this situation, the displacements are described by the following expression:

$$\delta(x,y,t)=\frac{1-\sigma^2}{\pi E}\iint_{\Omega}\frac{P(x',y',t)\,dx'dy'}{\sqrt{(x-x')^2+(y-y')^2}},\quad (x,y)\in\Omega,\tag{2}$$

where σ is Poisson's ratio and E is modulus of elasticity [6].

It must be emphasized that a comparison of simulated and experimental results presented in Section 3 demonstrates the validity of the proposed deformation model. However, if required the displacement δ_{st} and δ_{sb} can be specified using alternative models of the stamp and substrate deformation.

By the numerical approximation of the above-described mathematical model (1)-(2), a special finite difference method is applied. The method provides a high precision of simulation results by using a reasonably coarse grid [5].

3 EXPERIMENTS

Below it is given results for imprint processes using a structure from the experimental test stamp of the NaPa project [7]. The test structure measures 2×2 mm^2 and contains arrays of circular protrusions (see Fig. 1). The average value of fill factor (i.e. relation between cavities area and the total area) for the structure is equal to 1/3. However, the local value of fill factor varies significantly from one array to another. Therefore, imprinted samples of the test structure demonstrate distinct inhomogeneities of the residual layer thickness related to the non-uniform deformation of stamp and substrate as well as to incompletely filled cavities.

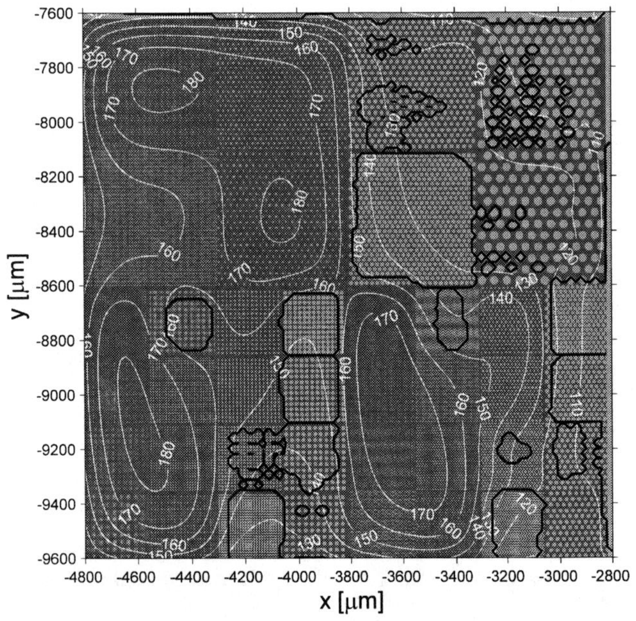

Figure 1: Simulated distribution of the residual layer thickness D for the test structure with cavities of depth h_c=200 nm. The initial resist thickness is 200 nm. White contour lines are numbered in nanometers. Black lines bound areas with incompletely filled cavities. The duration of the imprinting process is 180 s. As background, the relief of the test structure is used (cavities and protrusions are painted light and dark grey, respectively).

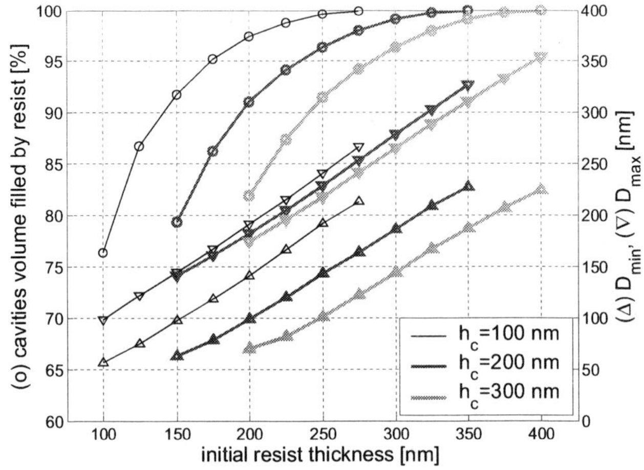

Figure 2: The simulated influence of the initial resist thickness on minimal (D_{min}) and maximal (D_{max}) values of the residual thickness as well as on cavities fillability for the test structure with different cavities depth h_c.

In experiments, silicon stamps and silicon substrate are applied. The experiments were performed for different values of the stamp cavities depth h_c: 100 nm, 200 nm and 300 nm. The stamps were imprinted into resist mr-I 8000 (Micro Resist Technology GmbH). The imprint temperature was 200°C.

By the coarse-grain simulation, a 128x128 pixel grid is applied. For the calculation of the stamp and substrate deformation, elastic properties of single-crystalline silicon are used: modulus of elasticity – 10^{11} Pa, Poisson's ratio – 0.2. The stamp velocity is supposed to be V_{st}=1 nm/s (see in [5] about loading regimes which are selected by modeling). In the simulation, the resist dynamic viscosity is taken to be 3×10^3 Pa·s. This value gave the best fit of calculated residual thickness distribution to the experimental one [8].

Fig. 1 shows an example of simulated results for the test structure. In the figure the non-uniform distribution of the residual layer thickness and numerous areas with incompletely filled cavities are observed. The effect of the initial resist thickness d_0 on the homogeneity of the residual layer thickness and cavities fillability is presented graphically in Fig. 2. By the simulation, the duration of the imprinting process is chosen as $T=(d_0-20\text{ nm})/V_{st}$.

In Figures 3 and 4 the measured and calculated results are compared. The simulation confirms the high residual layer thickness variation ranging (the spacing between minimal and maximal values of the residual thickness is more than 45 nm for h_c=100 nm, 75 nm for h_c=200 nm, and 105 nm for h_c=300 nm) with a precision of 10%.

ACKNOWLEDGEMENT

The partial support of the EC-funded project NaPa (Contract no. NMP4-CT-2003-500120) is gratefully acknowledged. The content of this work is the sole responsibility of the authors.

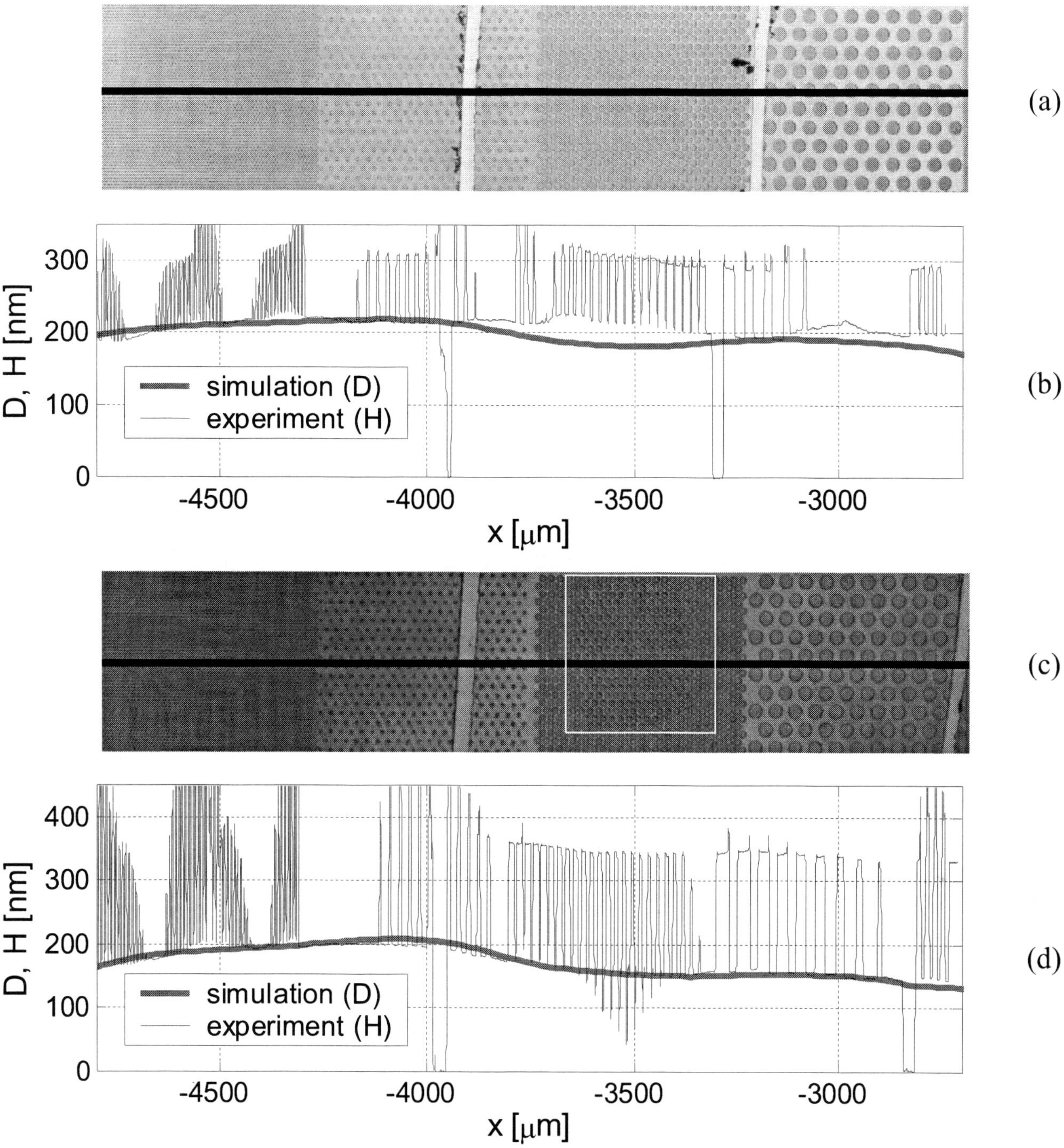

Figure 3: **(a)** and **(c)** The optical microscopy images of the test structure imprinted into resist at two different values of the stamp cavities depth: 100 nm and 200 nm, respectively. The initial resist thickness is 230 nm. The imprint temperature is 200°C. The duration of the imprinting process is 200 s. Black horizontal lines indicate zones of profilometer measurements of resist thickness. Vertical scratches observed in the images have been applied for the determination of the residual layer thickness. White box marks an area with incompletely filled cavities. **(b)** and **(d)** Comparison of measured distributions of resist thickness H and simulated distributions of residual layer thickness D.

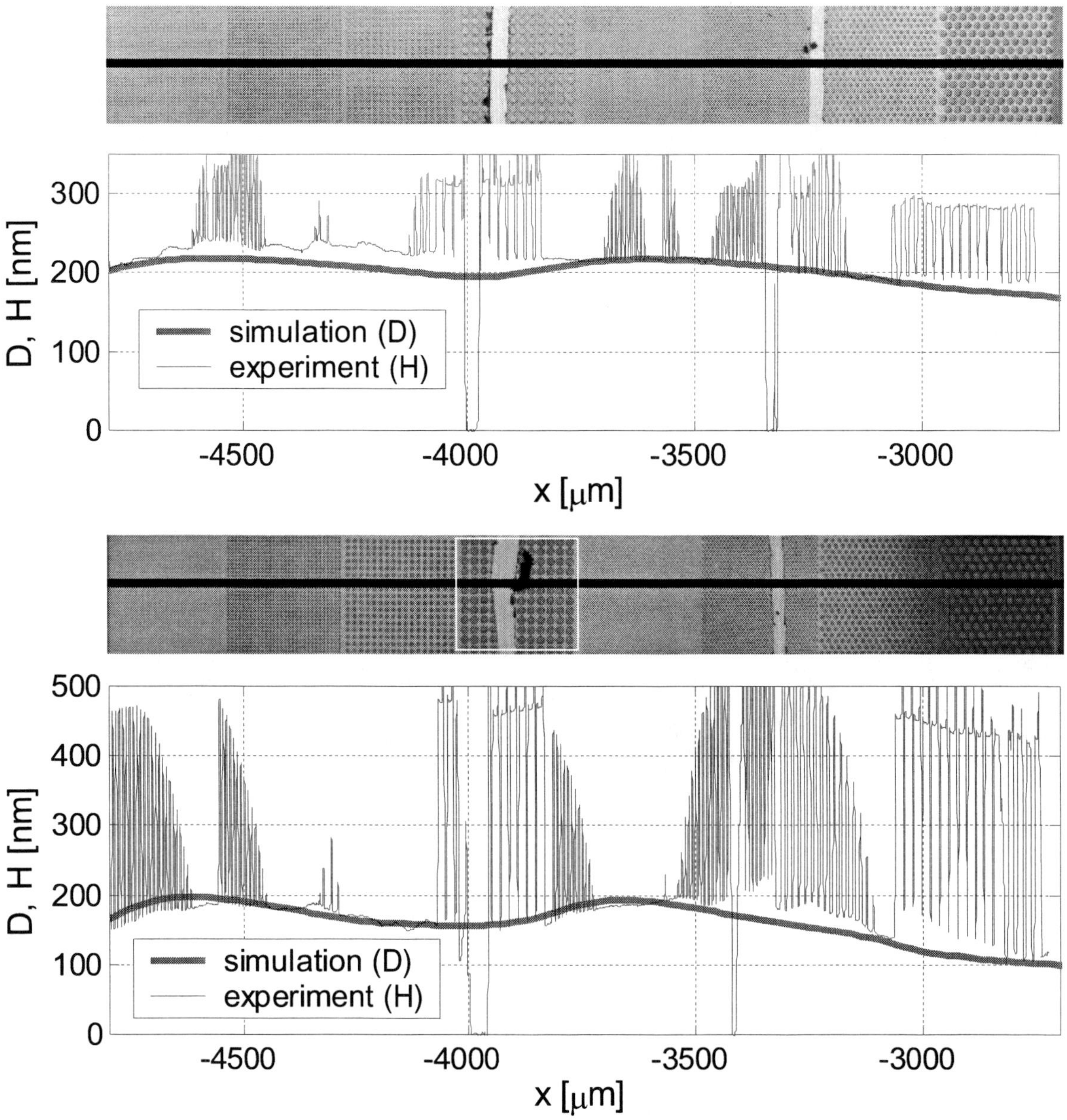

Figure 4: **(a)** and **(c)** The optical microscopy images of the test structure imprinted into resist at two different values of the stamp cavities depth: 100 nm and 300 nm, respectively. Other process parameters are listed in Fig. 3. Black horizontal lines indicate zones of profilometer measurements of resist thickness. White box marks an area with incompletely filled cavities. **(b)** and **(d)** Comparison of measured distributions of resist thickness H and simulated distributions of residual layer thickness D.

REFERENCES

[1] H. Schift and L.J. Heyderman, in: C.M. Sotomayor Torres (Ed.), "Alternative Lithography" (Series: "Nanostructure Science and Technology"), Kluwer Academic Dordrecht/Plenum, 47-76, 2003.

[2] H.D. Rowland, A.C. Sun, P.R. Schunk, W.P. King, J. Micromech. Microeng. 15, 2414, 2005.

[3] V. Sirotkin, A. Svintsov, S. Zaitsev and H. Schift, Microelectron. Eng. 83, 880, 2006.

[4] V. Sirotkin, A. Svintsov, H. Schift and S. Zaitsev, Microelectron. Eng. 84, 868, 2007.

[5] V. Sirotkin, A. Svintsov, S. Zaitsev and H. Schift, J. Vac.Sci. Technol. B 25, 2379, 2007.

[6] L.D. Landau and E.M. Lifshitz, "Theory of elasticity", Pergamon Press, 108, 1986.

[7] NaPa project, Integrated Project, EU Sixth Frame Program (http://www.phantomsnet.net/NAPA/index.php).

[8] N. Kehagias, V. Reboud, C.M. Sotomayor Torres, V. Sirotkin, A. Svintsov and S. Zaitsev, Microelectron. Eng. doi:10.1016/j.mee.2007.12.041

Microfluidic Simulations of Micropump with Multiple Vibrating Membranes

K. Koombua[*], R. M. Pidaparti[**], P. W. Longest[***], and G. M. Atkinson[****]

[*]Department of Mechanical Engineering, Virginia Commonwealth University, Richmond, VA 23284, USA, koombuak@vcu.edu
[**]Department of Mechanical Engineering, Virginia Commonwealth University, Richmond, VA 23284, USA, rmpidaparti@vcu.edu
[***]Department of Mechanical Engineering, Virginia Commonwealth University, Richmond, VA 23284, USA, pwlongest@vcu.edu
[****]Department of Electrical and Computer Engineering, Virginia Commonwealth University, Richmond, VA 23284, USA, gmatkins@vcu.edu

ABSTRACT

A novel design of a micropump with multiple vibrating membranes has been investigated in this study. The micropump consists of three nozzle/diffuser elements with vibrating membranes, which are used to create pressure difference in the pump chamber. The dynamic mesh algorithm in the computational fluid dynamics solver, FLUENT, was employed to study transient responses of fluid velocity and flow rate during the operating cycle of the micropump. The design simulation results showed that the movement of wall membranes combined with the rectification behavior of three nozzle/diffuser elements can minimize back flow and improve net flow in one direction. The maximum flow rate from the micropump increased when the membrane displacement and membrane frequency increased. Based on the performance characteristics from the simulations, the designed micropump is suitable to fabricate for practical applications.

Keywords: Valveless micropump; Nozzle/diffuser elements; vibrating membrane; Computational fluid dynamics

1 INTRODUCTION

A micropump is a primary component of many microfluidic systems and can be used to control the movement of small fluid volumes. Applications of the micropumps include implantable drug delivery systems [1], insulin injectors [2], artificial prostheses [3], liquid cooling systems [4], fuel cells [5], as well as macromolecule and cell analysis [6].

The main disadvantage of the current valveless micropump is overpressure at the outlet can cause a back flow to occur throughout the operating cycle [7]. This reverse flow decreases the average flow rate from the micropump and leads to an increase in energy consumption. In this study, a micropump consisting of three nozzle/diffuser elements with multiple membranes was designed to improve net flow in one direction. The fluidic characteristics of the micropump were analyzed implementing the CFD simulations.

2 MICROPUMP DESIGN

In contrast to the existing valveless micropumps, we propose a new technique to improve unidirectional fluid flow using three nozzle/diffuser elements with multiple vibrating membranes. The proposed design of the micropump is shown in Figure 1. The micropump consists of four components: microfluidic chip, pump chamber, actuator unit, and top cover. The pump chamber of the micropump has an hourglass shape and consists of membranes 1, 2, and 3 and chambers 1, 2, and 3. The dimensions of the pump chamber are approximately 2 mm x 1 mm x 50 μm (see Figure 2). The actuator unit can be made of a piezoelectric material. The movement of this piezoelectric material can be accurately controlled by an applied voltage. Change in length and curvature of the piezoelectric material results in the movement of membranes 1, 2, and 3 of the pump chamber and lead to decrease or increase in a volume of the pump chamber. Movement of these membranes can potentially minimize back flow and improve net flow in one direction.

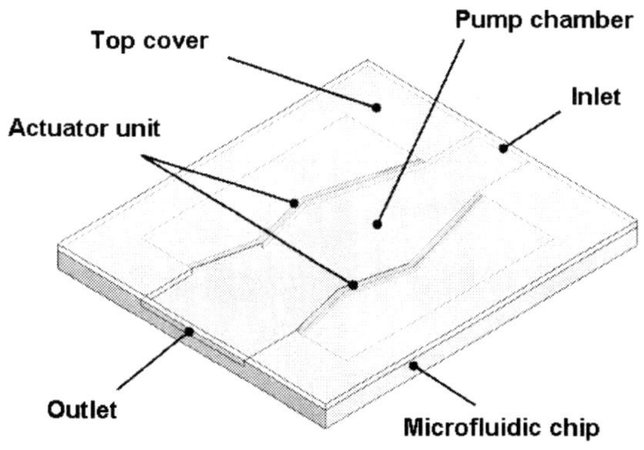

Figure 1: A design of the micropump with multiple vibrating membranes.

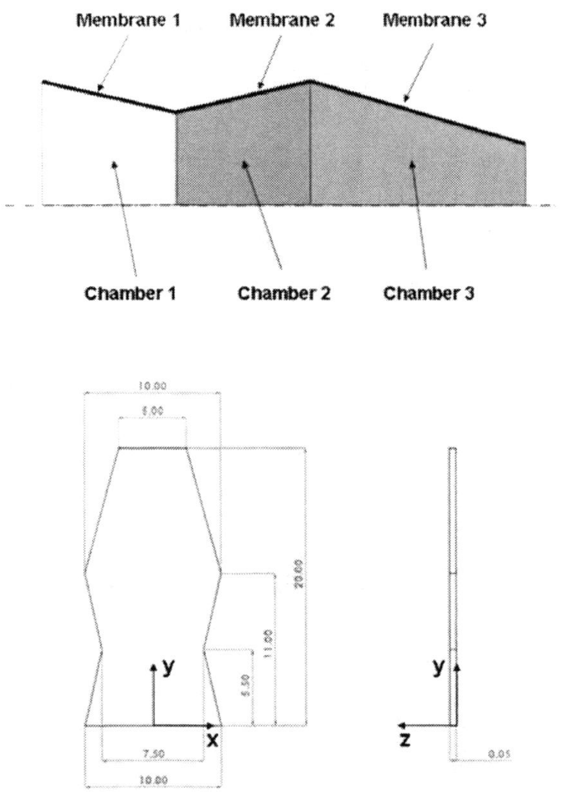

Figure 2: A schematic diagram of the pump chamber (top). Due to symmetry, only a half portion of the pump chamber is shown. Detail geometry of the pump chamber (bottom). All units are in mm.

The unique performance characteristics of the proposed micropump result from a specific sequence of membrane motion. For pumping mode, Fluid moves from chambers 2 and 3 to chamber 1 and to the outlet when membranes 2 and 3 move down and membrane 1 move up. These movements generate low pressure in chamber 1 and high pressure in chambers 2 and 3 (see Figure 3). In supply mode, fluid is drawn from the inlet into chambers 2 and 3 due to low pressure when membranes 2 and 3 move up. At the same time, membrane 1 moves down and generate high pressure in chamber 1. This high pressure minimizes a reverse flow from the outlet into the chamber 1 (see Figure 3).

3 MICROPUMP MODELING

The commercial computational fluid dynamics software FLUENT, was used to simulate a transient response of fluid flow within the micropump. This software uses a finite volume method to discretise the Navier-Stokes equations that describe the exchange of mass, momentum, and energy through the boundary of a control volume, which is fixed in space. The Navier-Stokes equations in the integral formulation that describes the fluid dynamics were shown below [8].

$$\frac{\partial}{\partial t}\int_{\Omega} \vec{W} d\Omega + \oint_{\partial\Omega} \left(\vec{F}_c - \vec{F}_v\right) dS = \int_{\Omega} \vec{Q} d\Omega \qquad (1)$$

where $\vec{W}$ is a vector of conservative variables, $\vec{F}_c$ is a vector of convective fluxes, $\vec{F}_v$ is a vector of viscous fluxes, and $\vec{Q}$ is the source terms.

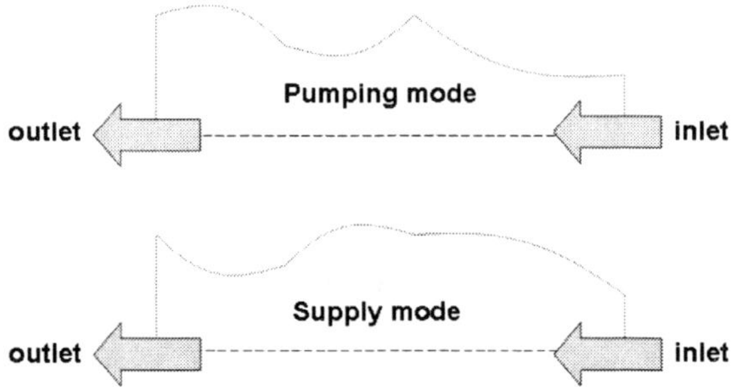

Figure 3: The micropump operating in pumping and supply modes. Arrows show the direction of fluid flow.

The working fluid of the micropump was assumed to be water with density of 998.2 kg/m^3 and viscosity of 959e-06 N.s/m^2. Since the thickness of the micropump was much less than other dimensions of the micropump, the flow can assumed to be constant in the thickness direction. The 2-D model of the half micropump, due to symmetry, was created in pre-processor software, GAMBIT. Triangular elements were used to represent this 2-D model in order to facilitate use of the dynamic mesh algorithm

A user-defined function was written in C complier to control the movement of the membrane. While the fluid-membrane interaction can affect the change in micropump volume, this effect is neglected in this study. The movement of the membrane was modeled using the first mode of vibrating string with two fixed edge. This movement is given by the following expression.

$$u(x,t) = A\sin\frac{\pi x}{L}\sin 2\pi f t \qquad (2)$$

where u is a membrane displacement (m), A is a maximum membrane displacement (m), x is a location along membrane length (m), L is membrane length (m), f is membrane frequency (Hz), and t is time (s).

The characteristic of micropump was evaluated by three input parameters; the maximum membrane displacement A, membrane frequency f, and pressure head ΔP. Pressure head $\Delta P = P_{out}-P_{in}$, where P_{in} and P_{out} are pressure levels at the inlet and outlet of the micropump. Transient simulations of the micropump were carried out at a maximum membrane displacement A from 0.25 to 1.50 mm, membrane frequency f from 1 to 500 Hz, and pressure head ΔP from 0 to 5 kPa. In all cases, the simulations were performed until the flow rate at the outlet of the micropump reached a steady state solution. The solution for each time step was considered to be converged when residuals of mass and all velocity components were less than 10^{-6}.

4 MICROPUMP CHARACTERISTICS

4.1 Velocity Field and Flow Rate

The fluid velocity for the pumping and supply mode of the micropump operating with 1-mm membrane amplitude, 1-Hz membrane frequency and zero back pressure is shown in Figure 4. In contrast to the existing nozzle/diffuser micropump [4, 9], there is no back flow at the inlet during the pumping mode and there is no back flow at the outlet during the supply mode. The transient response of flow rate at the outlet of the micropump is shown in Figure 5. The maximum flow rate at the outlet of the micropump increased and reached a steady state after eight seconds (eight operating cycles). The maximum flow rate was about 36.93 μl/min. It is interesting to note that the flow rate of the micropump in this study was different from that of the nozzle/diffuser micropump, which had a retrograde flow rate throughout the operating cycle [4, 9].

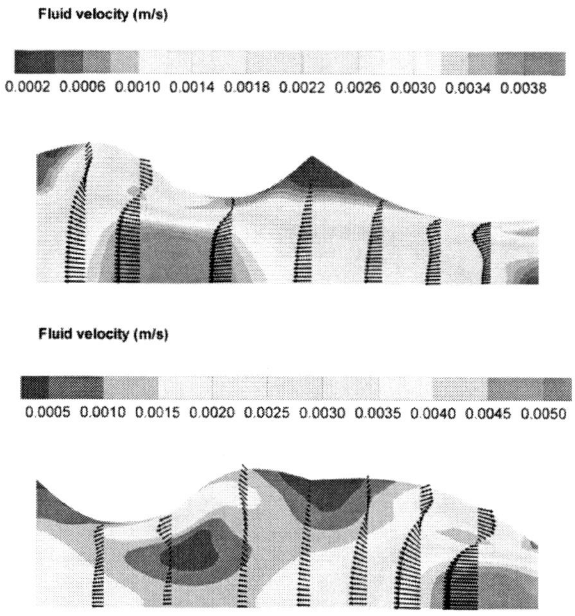

Figure 4: Velocity field within the micropump during the pumping (top) and supply (bottom) modes.

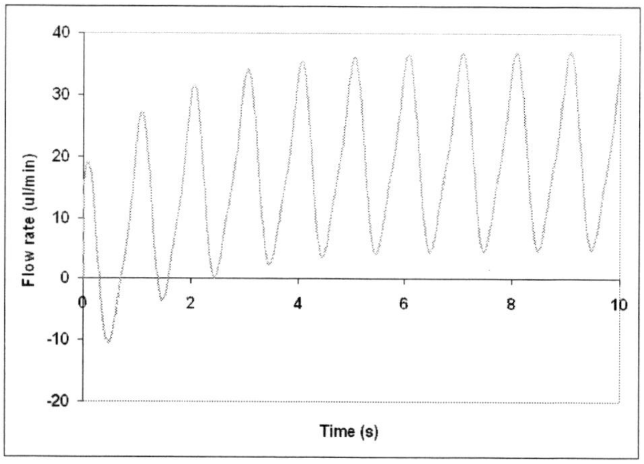

Figure 5: Flow rate at the outlet of the micropump with 1-mm membrane amplitude, 1-Hz membrane frequency and zero back pressure

4.2 Effects of Membrane Displacement

The transient responses of flow rate at the outlet of the micropump with zero pressure head and 1-Hz membrane frequency for four different maximum membrane displacements are plotted in Figure 6. The flow rate for all membrane displacements at the outlet of the micropump reached a steady state after eight seconds. As can be seen from this figure, this micropump should operate with maximum membrane displacement at least 0.50 mm to prevent a negative flow rate at the outlet. The maximum flow rates were 6.27, 15.56, 36.93, and 58.16 μl/min for 0.25-, 0.50-, 1.0-, and 1.5-mm maximum membrane displacement, respectively. The effect of the membrane amplitude on the proposed micropump was similar to that on nozzle/diffuser and peristaltic micropumps [10].

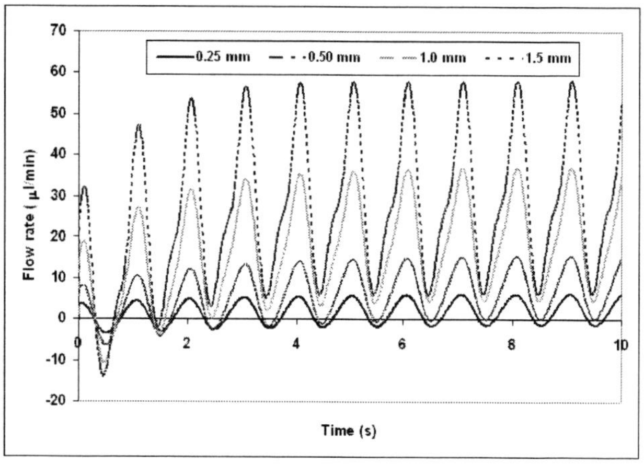

Figure 6: Flow rate at the outlet of the micropump with zero pressure head and 1-Hz membrane frequency for four different maximum membrane displacements.

4.3 Effects of Membrane Frequency

The flow rates at the outlet of the micropump with zero pressure head and 1.0-mm maximum membrane displacement for four different membrane frequencies are plotted in Figure 7. The flow rates for all membrane frequency had the same trend. The flow rate at the outlet of the micropump was a pulse flow and the negative flow rate only appeared in the first two operating cycle. The flow rate at the outlet of the micropump reached a steady state after eight operating cycles. There is no negative flow rate at a steady state. The maximum flow rate at the outlet increased when the membrane frequency increased. The maximum flow rates were 36.93, 435.33, 4731.06, and 24297.73 μl/min for 1-, 10-, 100-, and 500-Hz membrane frequency, respectively. The effect of the membrane frequency on the proposed micropump was similar to that on nozzle/diffuser and peristaltic micropumps [9-12].

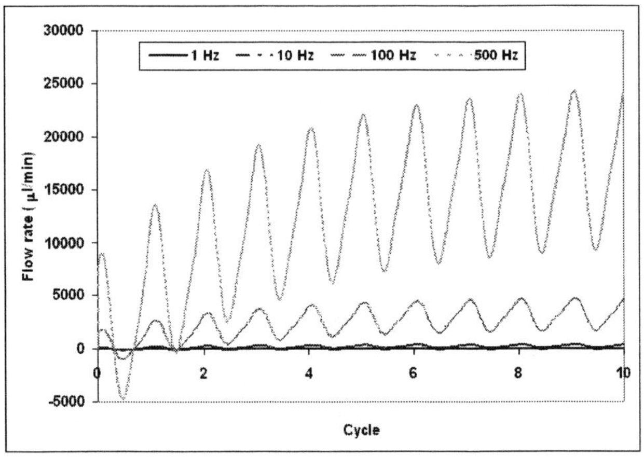

Figure 7: Flow rate at the outlet of the micropump with zero pressure head and 1.0-mm maximum membrane displacement for four different membrane frequencies.

5 CONCLUSION

Performance characteristics of a micropump with multiple vibrating membranes that are used to achieve a unidirectional flow were investigated in this study. This micropump uses the movement of multiple membranes and three nozzle/diffuser elements to improve fluid flow in one direction. The performance characteristics of the micropump were analyzed using a computational-based finite volume method. The fluid velocities and flow rates within the micropump were obtained through CFD simulations. The computational results showed that the maximum membrane displacement and membrane frequency were factors affecting the maximum flow rate of the novel micropump. The maximum flow rate increased when the maximum membrane displacement and frequency increased. Using the proposed design, retrograde flow was eliminated for most conditions that were considered. Based on these design simulations, a prototype of the micropump made of PDMS material will be studied in future.

ACKNOWLEDGEMENTS

The authors thank the US National Science Foundation for sponsoring the research reported in this study through a grant ECCS-0725496.

REFERENCES

[1] Cao, L., Mantell, S., and Polla, D., Sensors and Actuators A, 94, 117-125, 2001

[2] Jang, L., and Kan, W., Biomedical Microdevices, 9, 619-626, 2007

[3] Doll, A. F., Wischke, M., Geipel, A., Goldschmidtboeing, F., Ruthmann, O., Hopt, U. T., Schrag, H., and Woias, P., Sensors and Actuators A, 139, 203-209, 2007

[4] Singhal, V., and Garimella, S. V., IEEE Transactions on Advanced Packaging, 28, 216-230, 2005

[5] Zhang, T., and Wang, Q., Journal of Power Sources, 140, 72-80, 2005

[6] Beebe, D. J., Mensing, G. A., and Walker, G. M., Annual Reviews of Biomedical Engineering, 4, 261-286, 2002

[7] Woias, P., Sensors and Actuators B, 105, 28-38, 2005

[8] Blazek, J., "Computational Fluid Dynamics: Principles and Applications," Elsevier Ltd., 2005

[9] Yao, Q., Xu, D., Pan, L. S., Teo, A. L. M., Ho, W. M., Lee, V. S. P., and Shabbir, M., Engineering Application of Computational Fluid Mechanics, 1, 181-188, 2007

[10] Nguyen, N., and Huang, X., Sensors and Actuators A, 88, 104-111, 2001

[11] Goldschmidtboeing, F., Doll, A., Heinrichs, M., Woias, P., Schrag, H. J., and Hopt, U. T., Journal of Micromechanics and Microengineering, 15, 673-683, 2005

[12] Huang, C., Huang, S., and Lee, G., Journal of Micromechanics and Microengineering, 16, 2265-2272, 2006

Numerical Modeling of Microdrop Motion in a Digital Microfluidic Multiplexer

A. Ahmadi*, H. Najjaran*, J. F. Holzman*, M. Hoorfar*

*School of Engineering, University of British Columbia Okanagan, Kelowna, British Columbia, Canada, V1V 1V7

ABSTRACT

This paper introduces a computationally efficient numerical modeling approach for microdroplet motion within a new Digital Microfluidic Multiplexer structure. This structure offers an enhanced level of controllability and flexibility in drop actuation compared to the existing digital microfluidic systems that rely on addressable (square-matrix) cell structures. The real-time control of microdroplets under the resulting electrical, magnetic and hydrodynamic effects requires a reliable and computationally efficient model of the micron-scale kinematic forces. The proposed numerical model analyzes the time-variant average microdrop velocity using a computational-fluid-dynamic (CFD) code that solves the Navier-Stokes equation for velocity and pressure distributions inside the microdroplet. The estimated average velocity is modified according to the contact-angle hysteresis and an external force (e.g., induced by a magnetic field) using an iterative method.

Keywords: Digital Microfluidic Multiplexer, CFD, Microdroplet, Modeling, Hysteresis

1 INTRODUCTION

The development of micron-scale devices has evolved through a technological trend in the recent decades. This trend started in the microelectronics industry and quickly spread to many new fields, including fluid mechanics and integrated digital microfluidics. With digital microfluidic systems, discrete droplets can be manipulated as an alternative approach to overcome the challenges of continuous flow systems. In contrast to continuous flow, droplet-based systems are compatible with wall-less structures, and droplet manipulation can be carried out on the surface of a planar substrate in a reconfigurable and scalable manner. They can be used as programmable "microfluidic processors" in bio-related areas such as parallel DNA analysis, real-time biomolecular detection and automated drug discovery [1].

Modeling of droplet motion is very important in designing and controlling digital microfluidic systems, and most recent efforts have focused on predicting the bulk velocity of droplets on the chips. While the average velocity of these droplets is important, it has been shown that the internal motion of the droplet can also directly affect the droplet behavior [2]. In most previous studies, one dimensional flow has been assumed for the internal flow, and the vertical component of velocity inside the droplet has been ignored. Beni and Tenan [3], for example, used the Poiseuille flow assumption for flow inside the droplet [3], while Nichols et al. [4] included the effects of an external force on the droplet motion. In the latter, the system was analyzed analytically with the Navier-Stokes equation by neglecting the vertical component of the velocity inside the droplet. In a recent study by Walker and Shapiro [5], a two dimensional three-phase contact line motion is modeled assuming one dimensional Hele-Shaw flow inside the drop.

When the height of the channel is much smaller than the radius of the investigated droplet, an assumption of one dimensional flow inside the drop may lead to a good approximation for the bulk velocity of the drop. However, when the height of the channel is comparable to the radius of the drop, the vertical component of the velocity inside the droplet plays an important role in drop dynamics as the fluid flow becomes two dimensional in nature. Two-dimensional flow strongly affects the bulk pressure distribution, i.e., the vertical component of the pressure gradient at the interfaces, and the ultimate bulk microdrop velocity. Unlike the previous models of drop motion based upon the Poiseuille flow pressure distribution assumption [2, 3, 4], the present work models a micron-scale integrated microfluidic structure with the two dimensional Navier-Stokes equation. External body forces and hysteresis effects are included in the numerical model, and the results are applied to the analysis of the proposed Digital Microfluidic Multiplexer structure. It has been shown that the resulting two-dimensional fluid dynamics dominates the operation of digital microfluidic devices in the investigated micron-sized dimensions, and cannot be ignored.

2 METHODOLOGY

2.1 Description of the Device

A schematic of a Digital Microfluidic Multiplexer device is shown in Figure 1. Localized electrical current are induced in the Digital Microfluidic Multiplexer structure through the application of voltage differences between perpendicular x- and z- addressable electrodes. Regions with identical x- and z-electrode voltages (identically-biased) have no net current whereas regions with differently-biased electrodes will have net current flow from the positive electrode to the negative electrode. Such an arrangement avoids the complexities of integrated address lines in square-matrix cell structures and multidimensional via-lines when large numbers of electrodes with particularly small dimensions are desired. Ultimately, an external body force is applied using an external electric field, and the net motion of the microdrop is dictated by balancing this magnetic force and the internal capillary forces.

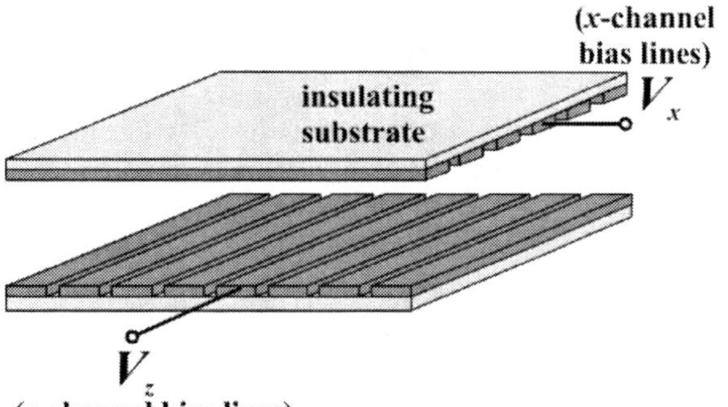

Figure 1: The Digital Microfluidic Multiplexer. The conductive x-channel and z-channel bias lines are used to control the motion of the microfluidic drop within the central fluid control layer [4].

2.2 Governing Equations

With the micron-sized dimensions of the Digital Microfluidic Multiplexer in mind, the main focus of the present work will be on the effects of internal fluid motion inside droplets. The deformation of droplet boundary will not be studied here, and a cylindrical shape will be assumed for the droplet (Figure 2). In this way, the three-dimensional motion of the system can be approximated by a two-dimensional flow within a cross-sectional meridian plane. This assumption is valid for large body forces in the x direction where the flow in the z direction can be neglected (Figure 2).

The analysis begins through the dimensionless Navier-Stokes equations:

$$\frac{\partial P^*}{\partial t^*}+\frac{\partial u^*}{\partial x^*}+\frac{\partial v^*}{\partial y^*}=0 \tag{1}$$

$$\frac{\partial u^*}{\partial t^*}+\frac{\partial u^{*2}}{\partial x^*}+\frac{\partial u^* v^*}{\partial y^*}=-\frac{\partial P^*}{\partial x^*}+\frac{1}{\mathrm{Re}}\left(\frac{\partial^2 u^*}{\partial x^{*2}}+\frac{\partial^2 u^*}{\partial y^{*2}}\right)+F_x^* \tag{2}$$

$$\frac{\partial v^*}{\partial t^*}+\frac{\partial u^* v^*}{\partial x^*}+\frac{\partial v^{*2}}{\partial y^*}=-\frac{\partial P^*}{\partial y^*}+\frac{1}{\mathrm{Re}}\left(\frac{\partial^2 v^*}{\partial x^{*2}}+\frac{\partial^2 v^*}{\partial y^{*2}}\right)+F_y^* \tag{3}$$

where asterisk parameters are the dimensionless quantities ($t^*=tU_\infty/L$, $P^*=P/(\rho U_\infty^2)$, $u^*=u/U_\infty$, $v^*=v/U_\infty$, $x^*=x/L$, and $y^*=y/L$). Here, U_∞ is the reference velocity equal to the bulk droplet velocity, and L is the reference length in the problem (which is in our case equal to the height of the channel, H). The Reynolds number is defined as $\mathrm{Re}=U_\infty L/\nu$, where ν is the kinematic viscosity of the liquid. The term F^* is the dimensionless body force in each direction as defined by $F^*=F_b L/(V\rho U_\infty^2)$, where F_b represents the body force in the problem and V is the volume of the droplet.

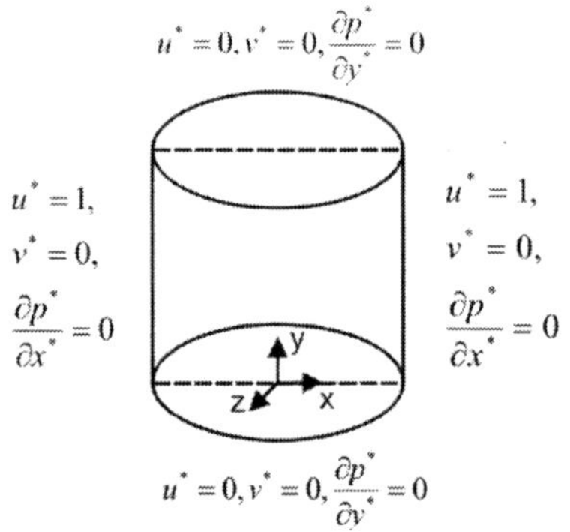

Figure 2: Boundary conditions in the meridian plane

2.3 Boundary Conditions

Boundary conditions for the two-dimensional domain are particularly important. The fluid motion inside the drop is considered with respect to a moving coordinate frame, which is attached to the droplet. In this coordinate frame, the left and right walls will be stationary (since they move with the droplet), and the top and bottom walls will have the bulk velocity of the drop. The hysteresis effect and the resulting pressure distribution in such a system are particularly important (see the following sections). Although considering a moving coordinate system may lead to a correct velocity distribution, the actual pressure distribution cannot be accurately calculated as the Euler equation, relating the pressure field to velocity field, cannot be used in viscous flows. The actual boundary conditions for the meridian plane are shown in Figure 2, and the explanation of Neumann boundary conditions for a given pressure distribution is explained in [6].

2.4 Hysteresis Effect

It is noted that all the cells around a droplet have the same voltage so that the driving force (e.g., induced by an applied magnetic field) is a body force and not of the electrowetting type. Such a body force will cause contact angle hysteresis (i.e. a difference between advancing and receding contact angles) along the three phase contact lines. It is well understood that the interfacial properties is related to the pressure difference across the liquid-vapor interface based on the Laplace equation of the capillarity [3]. Such a condition gives

$$P_A - P_R = \frac{2}{h}\bar{\gamma}, \tag{4}$$

where $\bar{\gamma}$ represents the total friction on the drop. A linear dependence of $\bar{\gamma}$ on $\bar{u}$ is assumed here [3] giving

$$\bar{\gamma} = \alpha + \beta\bar{u}, \tag{5}$$

where α and β represent the static and dynamic friction coefficients, respectively. The value of β is 0.1 Pa·s [7]), and a sufficiently small $\alpha \approx 0$ value is assumed for the static friction where a study of different values of α can be found in [4].

Equation (4) relates the pressure difference of the advancing and receding edges of the droplet to the aforementioned hysteresis effect. This hysteresis effect is included in the CFD code in such a way that the pressure distribution is corrected at each step. The numerical routine starts with an initial average velocity which is based on the analytical solution of the Navier-Stokes equation (by ignoring the vertical component of the velocity), and this initial average velocity is used for calculating our dimensionless parameters. An iterative approach is then used to solve for the flow dynamics inside the droplet, and the resulting velocity distribution yields the ultimate droplet average velocity. This algorithm is shown in Figure 3. The trend of convergence is also shown in Figure 4. The average velocity oscillates between two values; the upper one is the value obtained based on zero vertical velocity and the lower one is for ignoring hysteresis effect.

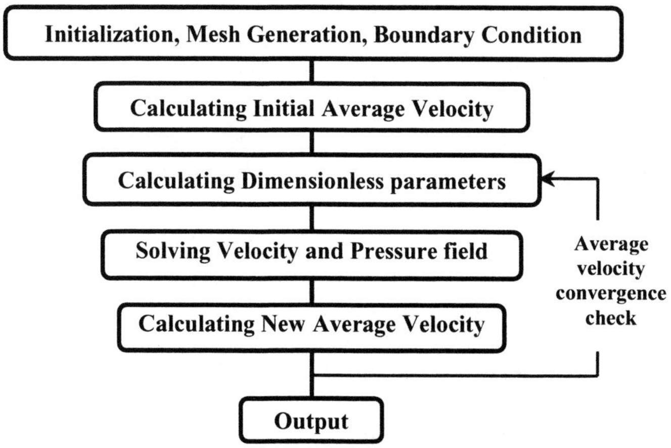

Figure 3: Flowchart of the numerical scheme

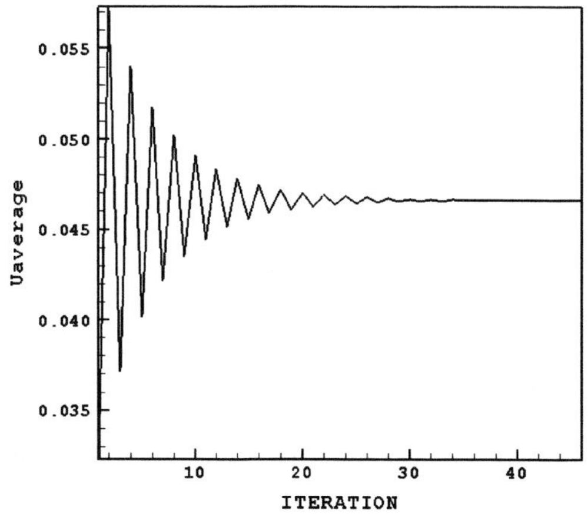

Figure 4: Trend of convergence

3 RESULTS

The no slip condition applied along the walls is apparent in the velocity field shown in Figure 4. It is observed that the stream lines align themselves to satisfy this boundary condition. The displacement of stream lines from their parallel state depends on the kinematic properties of the flow.

The bulk velocity of the droplet can be subtracted from the horizontal component of the absolute velocity to provide insight into the fluid motion inside the droplet (see Figure 5). For this purpose, a moving coordinate frame is considered that travels with the drop. It is apparent that the viscous effects lead to the formation of vortices inside the droplet. The position and strength of these vortices strongly depends on the Reynolds number as well as the applied voltage. It should be noted that most solutions for predicting the flow behavior inside droplets are based on assumptions that ignore the vertical velocity component. The vortices of Figure 5 and the vertical component shown explicitly in Figure 6 demonstrate that such a condition does not hold in these small-scale regimes. The vertical component of the velocity should not be ignored, as it becomes significantly pronounced near the drop corners.

In an effort to investigate the mass conservation inside the drop, the horizontal dimensionless velocity profile for different cross sections inside the droplet is shown in Figure 8. It is observed that the difference between the maximum velocity and average velocity increases drastically along the drop. At the liquid-vapor interface, this difference is zero and at the centre line it reaches its maximum. As it is shown in Figure 9, it is evident that the vertical component of pressure gradient cannot be ignored. As the height of the channel increases, the change in pressure across the channel becomes increasingly important since dimensionless pressure on the centerline increases along the droplet. This is due to the effect of the external body force which compensates the pressure loss due to the viscous shear stress.

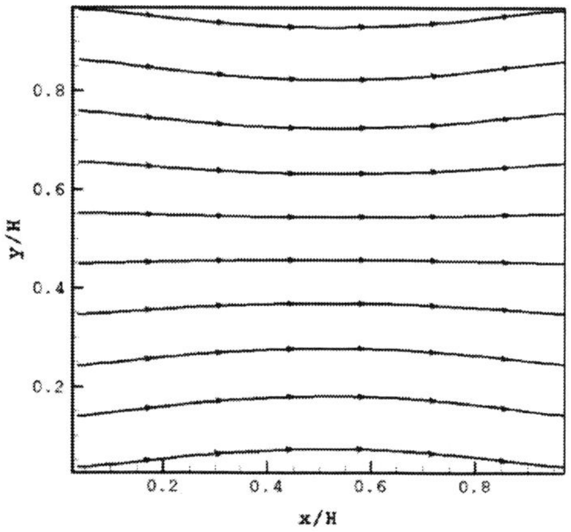

Figure 5: Absolute velocity field as a function of the normalized y/H and x/H

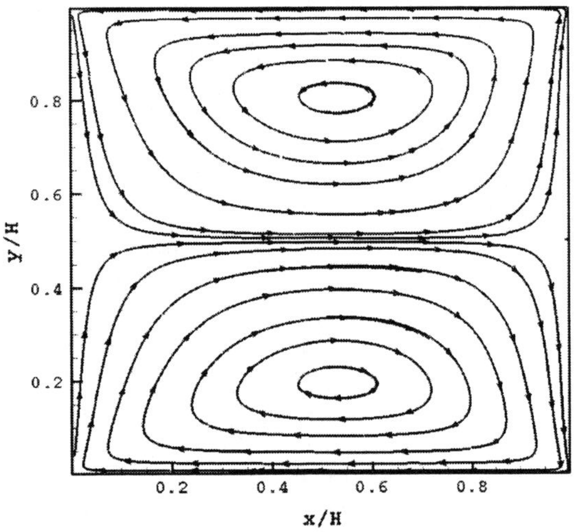

Figure 6: Internal motion of fluid inside the droplet

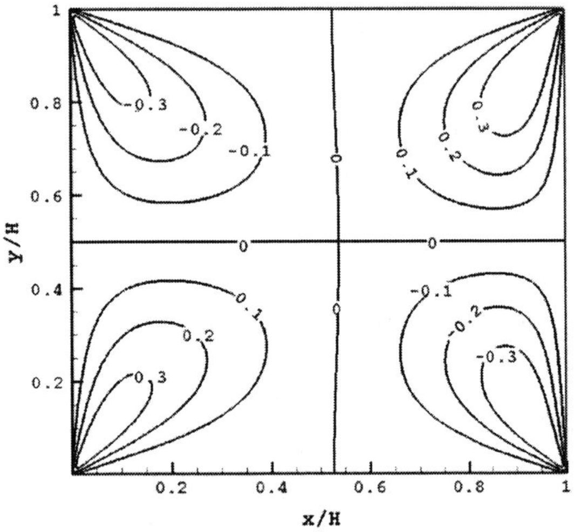

Figure 7: Horizontal dimensionless absolute velocity profile for different cross sections

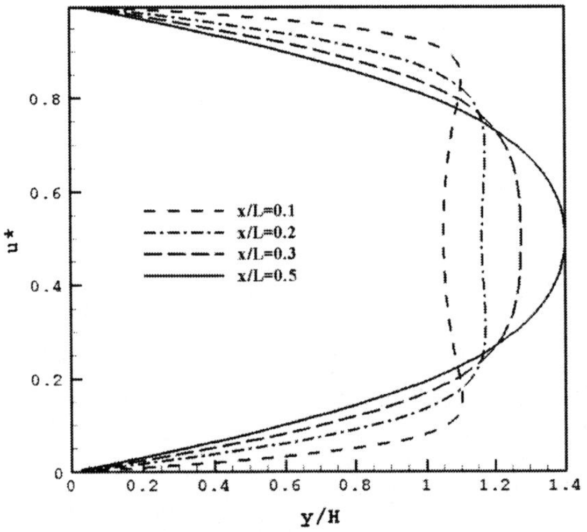

Figure 8: Dimensionless pressure distribution at the advancing cross section of the droplet

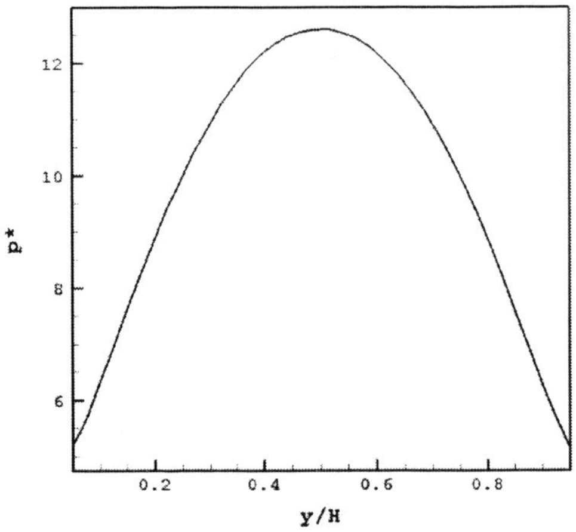

Figure 9: Vertical component of the absolute dimensionless velocity

4 SUMMARY

The effects of fluid motion inside the droplet have been studied. It is shown that vertical component of the velocity inside the droplet cannot be ignored. Moreover, it is shown that, in spite of conventional assumptions considered for solving motion of fluid inside the capillary tube, the vertical component of pressure gradient cannot be ignored. This new pressure distribution directly affects the estimated bulk velocity of the drop. The effects of body forces on drop motion have been studied, and it has been found that the body force increases the pressure along the channel.

REFERENCES

[1] K. Chakrabarty, F. Su, "Digital Microfluidic Biochips, Synthesis, Testing and Reconfiguration Techniques," CRC press, Taylor & Francis Group, 2007.

[2] J. Berthier, "Microdrops and Digital Microfluidics," William Andrew, 2008.

[3] G. Beni, M. A. Tenan, "Dynamics of Electrowetting Displays," J. Applied Physics, Vol. 52, pp. 6011-6015, 1981.

[4] J. Nichols, A. Ahmadi, M. Hoorfar, H. Najjaran, J. F. Holzman, "Micro-Drop Actuation Using Multiplexer Structures," Submitted to ICNMM 2008.

[5] S. W. Walker, B. Shapiro, "Modeling the Fluid Dynamics of Electrowetting on Dielectric (EWOD)," J. Microelectromechanical Systems, Vol. 15, No. 4, 2006.

[6] H. Lomax, T. H. Pulliam and D. W. Zingg, "Fundamentals of Computational Fluid Dynamics", Springer-Verlag, Berlin, 2001.

[7] A. M. Gaudin, A. F. Witt, "Hysteresis of Contact Angles in the System Mercury-Benzene-Water", in: Contact angle: wettability and adhesion, edited by R. F. Gould, Vol. 43, pp. 202–210, 1964.

Finite Element Analysis of a MEMS-Based High G Inertial Shock Sensor

Y.P. Wang*, R.Q. Hsu*, C.W. Wu**

*Department of Mechanical Engineering, National Chiao Tung University,
1001 Ta-Hsueh Road, 300 Hsinchu, Taiwan, anitawu.wlh@msa.hinet.net
**Department of Mechanical and Mechatronic Engineering, National Taiwan Ocean University
2, Pei-Ning Road, Keelung, Taiwan.

ABSTRACT

Conventional mechanical inertial shock sensors typically use mechanisms such as cantilever beams or axial springs as triggering devices. Reaction time for these conventional shock sensors are either far too slow or, in many cases, fail to function completely for high G (>300G) applications.

In this study, a Micro-Electro-Mechanical (MEMS)-based high G inertial shock sensor with a measurement range of 3,000–21,000 G is presented. The triggering mechanism is a combination of cantilever and spring structure. The design of the mechanism underwent a series of analyses. Simulation results indicated that a MEMS-based high G inertial shock sensor has a faster reaction time than conventional G inertial shock sensors that use a cantilever beam or spring mechanism.

Furthermore, the MEMS-based high G inertial shock sensor is sufficiently robust to survive the impact encountered in high G application where most conventional G inertial shock sensors fail.

Keywords: MEMS, high G, inertial shock sensor, spring, proof mass.

1 INTRODUCTION

Inertial sensors have been extensively utilized in science and industry. For high G (>300G) applications, reaction times for conventional mechanical type shock sensors are not fast enough. In some cases the shock sensor structures disintegrate (>5000G). Designing a shock sensor that has a faster reaction time than conventional sensors and a mechanism that is sufficiently robust to survive the impact when a vehicle collides with a hard target is the major goal of this study. Thus, a MEMS high-G inertial shock sensor that has two advantages is presented. Silicon was first chosen as the structure material, as its Young's modulus [1] approaching 190 Gpa, which is close to that of steel (210 Gpa). Moreover, silicon has virtually no mechanical hysteresis, and, thus, is an ideal material for sensors and actuators.

Second, the MEMS process favors production of miniature mechanisms that are always demanding in the application. Trimmer [2] proposed a unique model that demonstrated reducing the scale of a structure, will decrease the time required for displacing a fixed point. Thus, the reaction time of a small inertial shock sensor can be decreased.

2 THEORETICAL ANALYSIS

Fig. 1 presents the proposed micro shock sensor. This sensor uses a Mass-Damper-Spring Dynamic (MDS) System to trigger the mechanism.

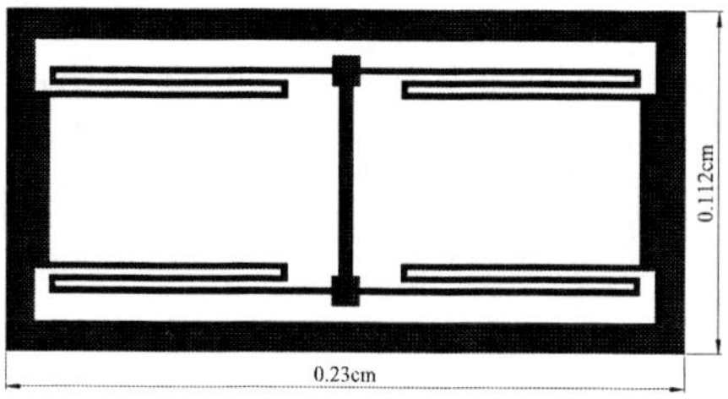

Fig. 1 The micro shock sensor proposed in this study

Fig. 2 is a schematic of the system. The dynamic equation of motion of proof mass can be expressed by one-dimensional lumped-system model given by [3]

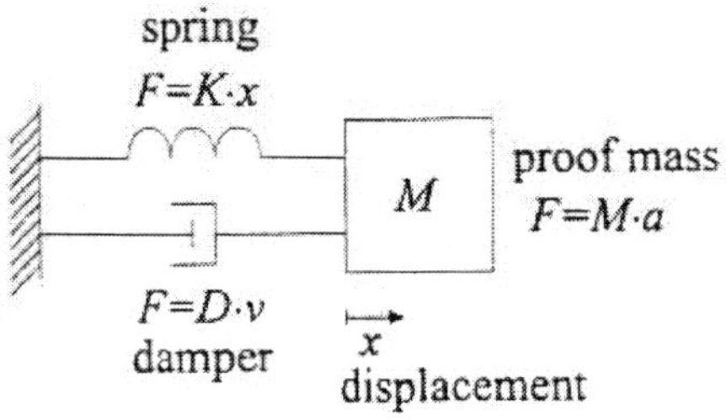

Fig. 2 Mass-Spring-Damping System

$$M\frac{d^2x}{dt^2}+D\frac{dx}{dt}+Kx=F_{ext}=Ma \qquad (1)$$

where F_{ext} is the external force acting on the frame, D is the damping factor, K is the effective spring constant of the elements and M is the proof mass attached to a fixed frame by one or more spring elements. Using the Laplace transformation, the following second-order function for acceleration of the mass is

$$H(s) = \frac{X(s)}{A(s)} = \frac{1}{s^2 + \frac{D}{M}s + \frac{K}{M}} = \frac{1}{s^2 + \frac{\omega_r}{Q}s + \omega_r^2} \quad (2)$$

where $\omega_r = \sqrt{\frac{K}{M}}$ is the resonance frequency and $Ms = \frac{M}{K}$ is the mechanical sensitivity of the system. Thus, system mechanical sensitivity varies with the spring constant and proof mass. Reducing the spring constant or increasing proof mass, increases mechanical sensitivity and shortens reaction time.

3 FINITE-ELEMENT SIMULATION

The spring is divided into four sections and anchored on two sides of the sensor frame structure. The proof mass is located at middle zone of the sensor and linked with the four spring sections. To evaluate system reaction time, 10 different arrangements of spring and proof mass were tested. All proof masses have the same thickness; consequently, the ratio of these masses is equal the ratio of proof mass surface areas. We assume that the proof mass scale in type 4 is 1.0. Fig. 3 presents the different sensor designs considered in this study.

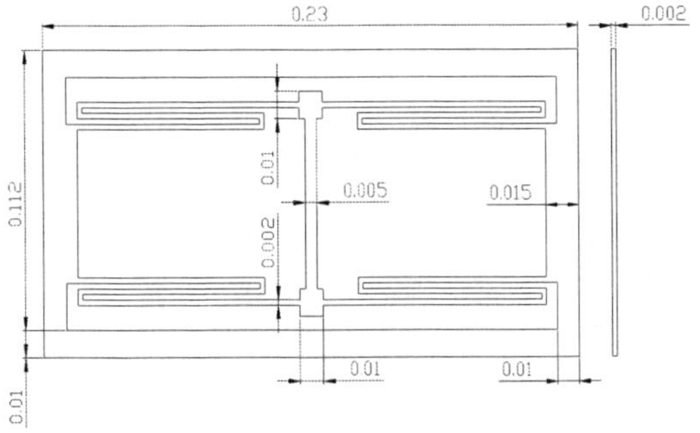

(a) Type 1 (cm)

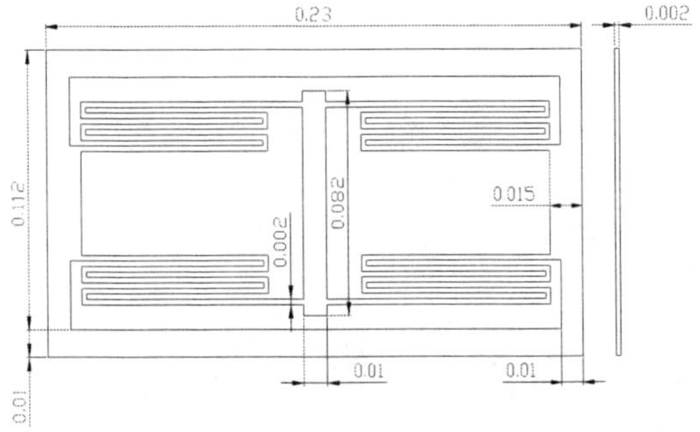

(b) Type 4 (cm)

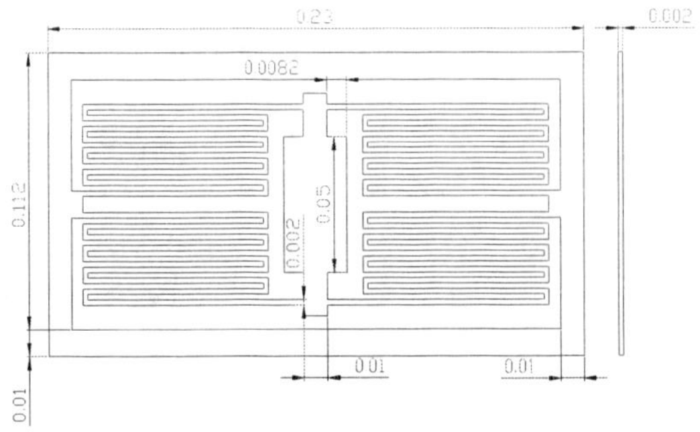

(c) Type 10 (cm)

Fig. 3 Diagrams and dimensions of 4 typical sensors

Table 1 presents their proof masses and coil numbers. Finite element analyses for displacement of the proof mass when the sensor encountered an impact were performed with ANSYS version 8.0 and LS-DYNA [4]. The mesh element adopted for modeling the proposed sensors is type SOLID 164 that is used for 3-D modeling of solid structures and defined by eight nodes with six degrees of freedom at each node, namely, translations, velocities, and accelerations in nodal directions x, y, and z. Only one-half of the sensors were utilized for simulation because of the symmetricity of sensor's form. Fig. 4 shows typical finite element meshes for a spring and proof mass. Table 2 lists the number of nodes and elements in each sensor.

Sensor type	Proof mass scale	Coil number
1	0.62	4
2	1.0	4
3	0.62	8
4	1.0	8
5	0.62	12
6	1.0	12
7	2.24	12
8	0.62	16
9	1.0	16
10	2.24	16

Table 1 The proof mass scale & coil number of the sensor

The spring and proof mass are assumed to be made of silicon, with a modulus of elasticity of 190 GPa, Poisson's ratio of 0.23 and density of 2.3g/cm3. Other assumptions are as follows: (a) the enclosure frame of the sensor is a rigid body; (b) the dimensions of sensor components are sufficiently large for principles of continuum mechanics that are applicable for analysis [5]; and, (c) the air damping effect can be ignored as the shock sensor is packaged in a vacuum environment.

Because of the complication of the spring shape, a direct calculation of the spring constant K is almost impossible. Instead, we use ANSYS to simulate the proof mass displacement under various load. Fig. 5 is an example of the simulation result of the proof mass displacement under load. Fig. 6 is a depiction of the displacement / applied force for type 1-10 sensors. Apparently, $K_1=K_2>K_3=K_4>K_5=K_6=K_7>K_8=K_9=K_{10}$.

In simulation, a series of half-sine waves were applied to sensors. Seven different G values, ranging from 3,000–21,000G, are considered in the simulation in accordance with (Mil-Std-810F) [6].

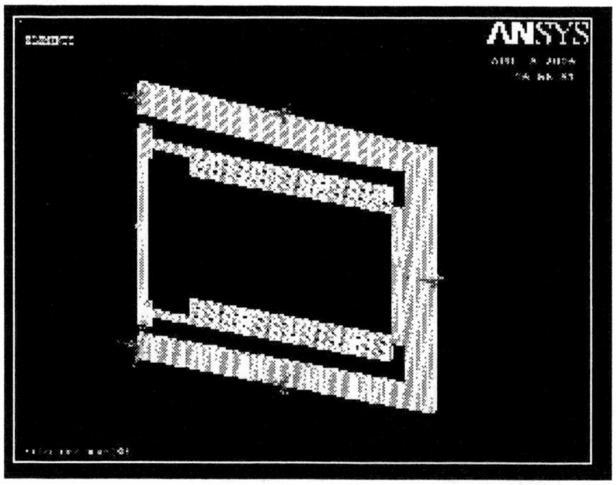

Fig. 4 The finite element meshes of the type 1 sensor

Sensor type	Node number	Element number
1	17100	9518
2	17676	9908
3	19833	10684
4	20373	11050
5	22524	11822
6	23085	12202
7	24159	12900
8	25245	12980
9	25833	13378
10	27003	14140

Table 2 The number of the nodes and elements

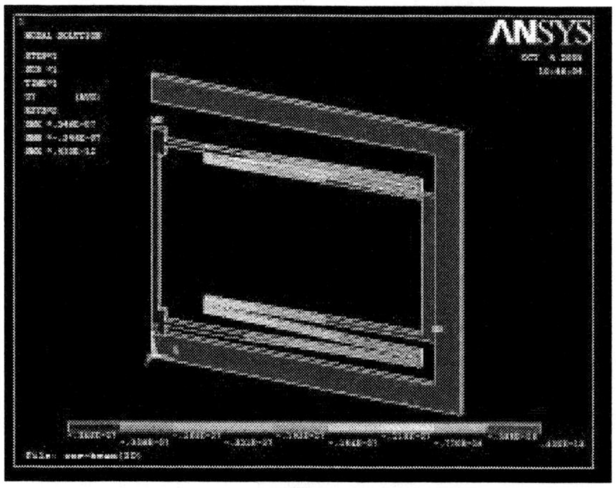

Fig. 5 Proof mass displacement of the type 1 sensor

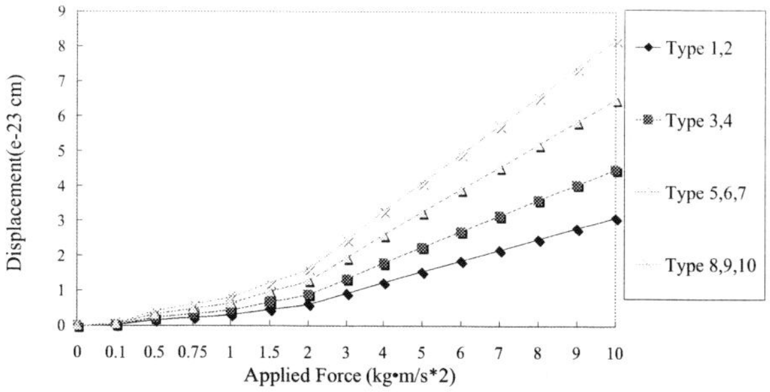

Fig. 6 Displacement vs. applied forces for each sensor

4 RESULTS AND DISCUSSION

In the dynamic simulations of time-domain analysis, a shock wave (G–T curve) is loaded onto an impact sensor, and the responses of the impact sensor are identified by observing the displacement of the proof mass in the impact direction. When the proof mass contacts the top frame (the displacement is 5.0E-03 cm), the impact sensor triggers. When the proof mass does not reach the top frame, the sensor does not trigger.

Simulation results demonstrated that the spring constant was reduced or proof mass increased, the G value required for the sensor to trigger and response time decreased (Table 3).

Two principal categories of reaction times were identified. First, when a proof mass increases from 0.62 to 2.24, and the spring constant remains unchanged, the reaction time is decreased (Fig. 7) and the minimum triggering G value decreases for sensors (Fig. 8). Second, reducing the spring constant, and retaining the proof mass, the reaction time decreased (Fig. 9) and the trigger G value decreased for sensors (Fig. 10).

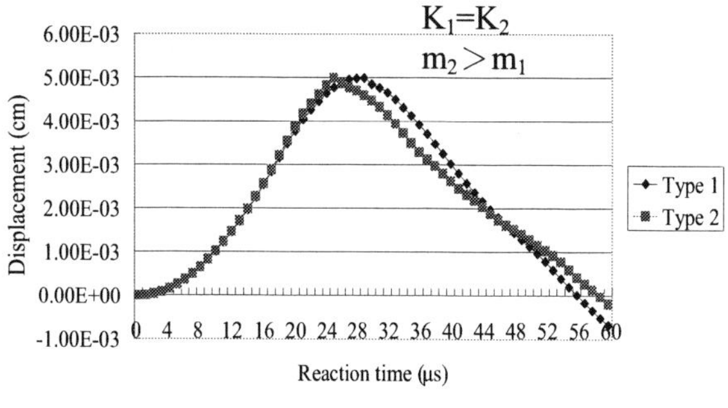

Fig. 7 Reaction time of the sensors at 21000G

Sensor type \ G value	21000	20000	10000	8000	5000	4000	3000
1 (m1 0.62 , K_1)	28.9	×	×	×	×	×	×
2 (m2 1.0 , K_2)	24.9	25.9	×	×	×	×	×
3 (m3 0.62 , K_3)	21.9	23.9	34.9	×	×	×	×
4 (m4 1.0 , K_4)	21.9	22.9	33.9	40.9	×	×	×
5 (m5 0.62 , K_5)	21.9	22.9	31.9	35.9	52.9	×	×
6 (m6 1.0 , K_6)	21.9	22.9	31.9	35.9	48.9	×	×
7 (m7 2.24 , K_7)	21.9	22.9	31.9	35.9	47.9	56.9	×
8 (m8 0.62 , K_8)	21.9	22.9	31.9	35.9	46.9	52.9	×
9 (m9 1.0 , K_9)	21.9	22.9	31.9	35.9	45.9	51.9	×
10 (m10 2.24, K_{10})	21.9	22.9	31.9	35.9	44.9	50.9	65.9

Table 3 The response time (μ s) of the micro-sensors

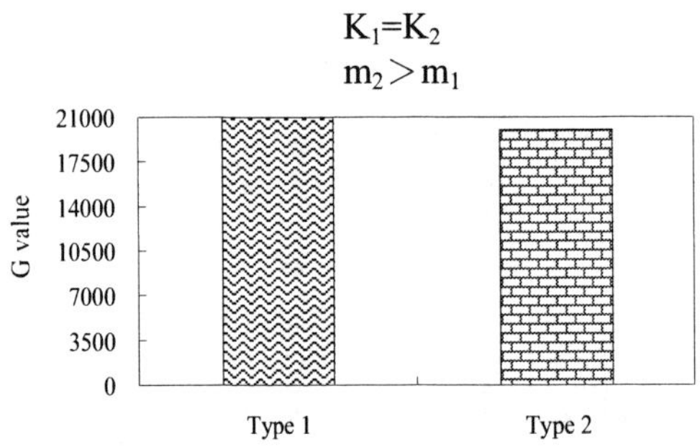

Fig. 8 Minimum G values for the sensors to be triggered

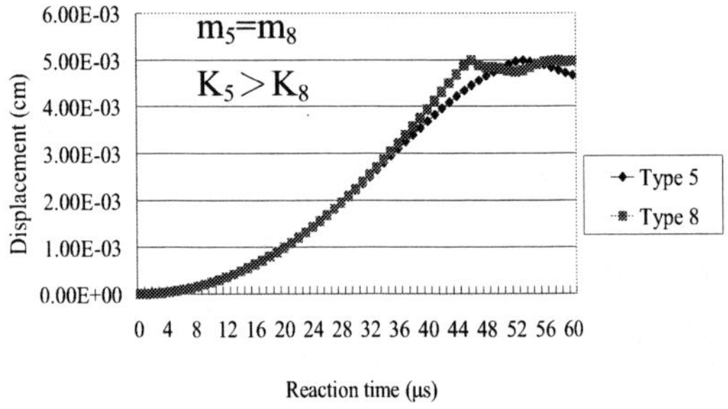

Fig. 9 Reaction time of the sensors at 5000G

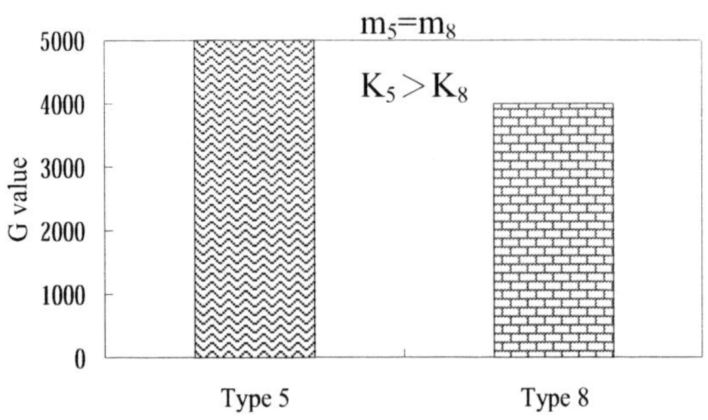

Fig. 10 Minimum G values for the sensors to be triggered

5 CONCLUSION

This proposed shock sensor is intended for use at 3,000–21,000G. Ten different designs were analyzed. Simulation results demonstrated that these MEMS high G inertial shock sensors have faster reaction times than conventional G inertial shock sensors. The shock sensors were sufficiently robust to survive the impact of at least 21,000G, four times higher than that of conventional inertial shock sensors.

REFERENCES

[1]. Donald R. Ask eland, The science and engineering of materials, 1st edn, Taipei, Kai Fa, 1985, ch. 6, pp. 126-127.

[2]. Trimmer, W.S.N, Microrobots and Micromechanical Systems, Sensors and Actuators vol.19 no.3, pp. 267-287, 1989.

[3]. M. Elwenspoek, R. Wiegerink, Mechanical Microsensors, Germany, Springer, 2001.

[4]. Training Manual, Explicit Dynamics with ANSYS/LS-DYNA, 1st edn, U.S.A, SAS IP, 2001.

[5]. Tai-Ran Hsu, MEMS & Microsystems Design and Manufacture, international edition 2002, Singapore, McGraw-Hill, pp. 157-159.

[6]. Military Standard, Mechanical Shock Test, MIL-STD-883E Method 2002.4, US Dept. of Defense, 2004.

Design for Manufacturing integrated with EDA Tools

U. Triltsch, S. Büttgenbach
Technical University of Braunschweig, Institute for Microtechnology (IMT),
Alte Salzdahlumer Str. 203, 38124 Braunschweig, Germany,
u.triltsch@tu-bs.de, s.buettgenbach@tu-bs.de

ABSTRACT

In this paper a process planning and optimization tool, which can be linked to commercially available EDA tools, is presented. A data model was developed, which can be used for product and process design, simultaneously. The main advantage of the system is that arbitrary technologies can be used to build process flows, as a knowledge based validation system ensures the validity of the produced process plan before manufacturing. Being able to combine new materials and technologies with established process flows makes a specific optimization of the overall process possible. A multi-criteria selection method was adapted for the use within microtechnological process chains. This method helps a designer to choose the right technologies for a given problem and is described in detail.

Keywords: design for manufacturing, technology rating, knowledge database, multi-material MEMS, T-CAD

1 INTRODUCTION

The increasing variety of available fabrication technologies and materials for microtechnological devices make the design process more and more demanding. Engineers are no longer bound to only functional aspects of the design but also have to meet cost targets and optimize the process flow itself. A data model was developed, which can be used for product and process design, simultaneously. This data model is accessed by the designer through a process editor [1]. This editor can also be used in fabrication to document the process results and herby guaranty that all relevant data is preserved for later use in quality management (Figure 1).

The main advantage of the system is that arbitrary technologies can be used to build process flows, as a knowledge based validation system ensures the validity of the produced process plan before manufacturing.

Whenever standard MEMS libraries, which are based on fixed foundry processes, are insufficient for a new design, layout and process design can be performed simultaneously (Figure 2). A layout editor is connected to a building block database [2]. Single building blocks, which contain a layout and a process model are used to design multi material micro systems. The single processes are merged using an interactive algorithm [3].

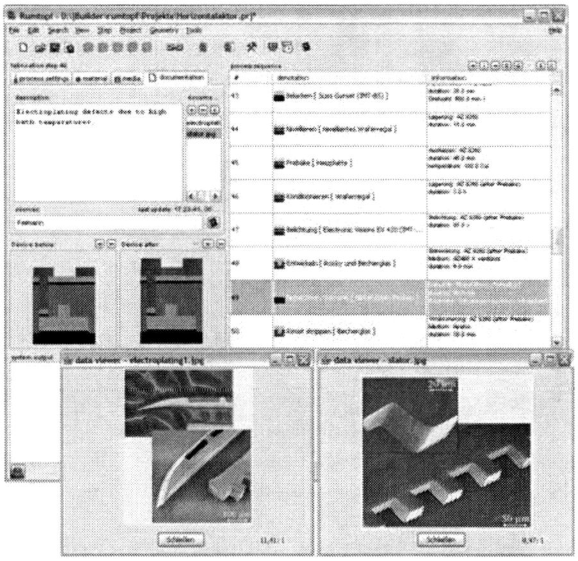

Figure 1: The process flow is edited and monitored in a single tool.

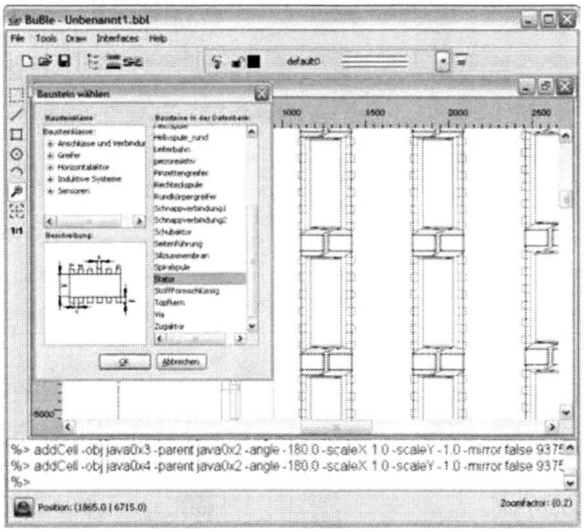

Figure 2: Layout and process are designed simultaneously using a building block editor

Once a satisfying process flow and a design rule compliant layout has been developed, the information can be transferred to SoftMEMS, MEMSPro or Coventor Designer at the push of a button (Fig 3). This enables the designer of microsystems, which are not based on foundry processes, to use the full functionality of specialized MEMS simulation tools for a detailed analysis of the behaviour of the proposed system.

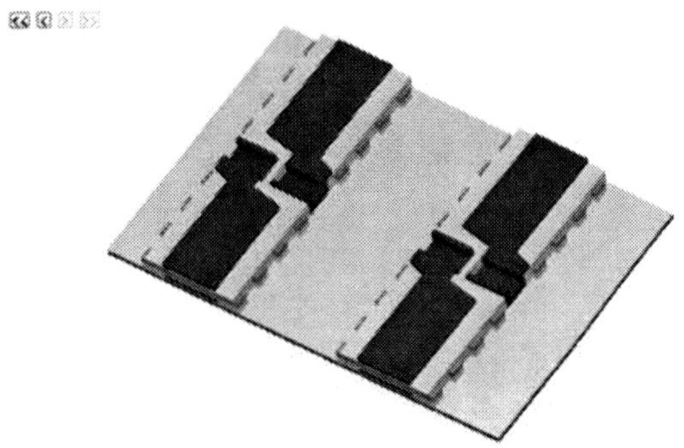

Figure 3: At the push of a button the data is transferred to Softmems MemsPro 3-D Modeller.

Another important feature of the presented system is the connection to specialized process simulators for the wet-chemical or dry-etching of silicon or UV-lithography. Whenever it is crucial to get a more detailed view of the process result such specialized modules are used.

Being able to combine new materials and technologies with established process flows makes a specific optimization of the overall process possible. A multi-criteria selection method was adapted for the use within microtechnological process chains. This method helps a designer to choose the right technologies for a given problem. For example might such factors as the lot size or overall budget influence the choice of technologies and the system assists a designer make the right decision.

The next section of the paper will focus on the description of the rating method.

2 TECHNOLOGY RATING

There are several methods for the rating of single technologies known from literature. All methods can be divided into methods on the operational or the strategic level of technology rating. The strategic level is used for predicting technological impact and market placement for innovative technologies. Such methods are portfolio technologies, s-curves or life cycle models [4].

However, all the strategic methods do not take detailed technological boundary conditions into account. For selecting technologies in existing production scenarios a more detailed view on technologies is needed. Typically methods on the operational level deliver the means to perform a detailed rating for a given problem. One suitable method for the use in microsystem technology is a fuzzy based, analytical hierarchy process as it was introduced by Eversheim [5] for process chains in classical mechanical production environments. He used linguistic fuzzy sets to describe the compliance of technologies with a predefined set of criteria. The use of fuzzy sets makes the mathematical analysis of diffuse information possible. The mapping between a linguistic term and a fuzzy number makes the input for the user easy and comprehensive. Figure 4 shows the use of such linguistic variables for the states exactly 5, approximately 5, between 2 and 8 and approximately between 2 and 8, respectively.

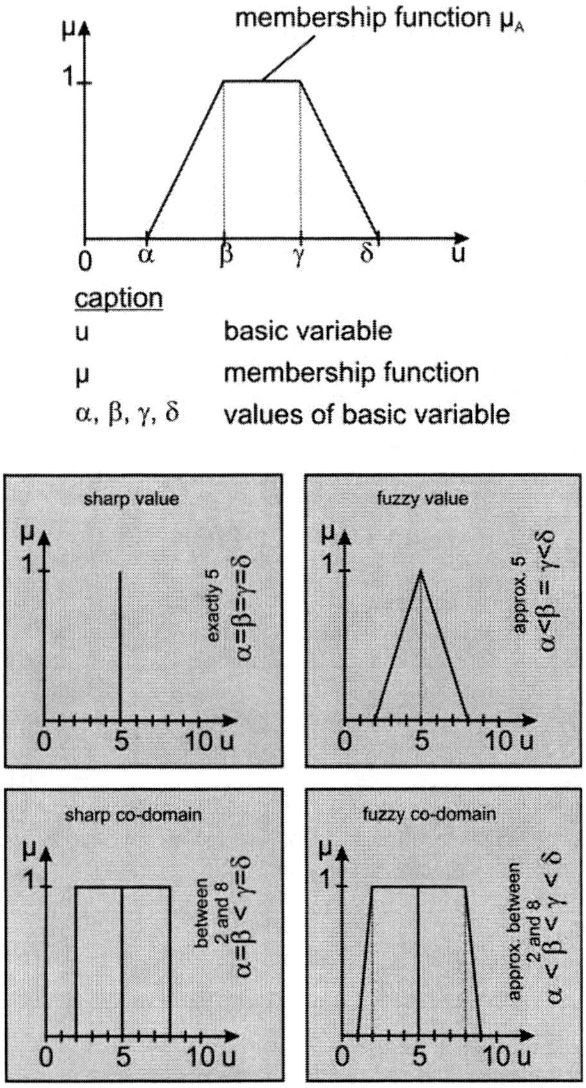

Figure 4: Fuzzy sets can be used for the input of diffuse information and later can be mathematically analyzed to calculate a score for a certain technology.

Eversheim defined a set of criteria, which form the portfolio. All single criteria can be assigned to one of the main parameters 'technological attractiveness' and 'economic benefit'. The criteria for these main parameters will be discussed in the next subsections.

2.1 Economic benefit

This main parameter can be subdivided into three parameters which are influenced by certain criteria, each. These parameters are: economic potential, qualitative benefit and costs of implementation. In microtechnology one can assume that not all process steps are performed within the company, but some are purchased from third party providers. For example, it is common practice that electronic components of micromechanical sensors are fabricated using foundry processes and only the structuring of the mechanical components as well as the packaging is accomplished in the own company. This results in an adaptation of the criteria structure. The 'economic potential' is largely dependent on costs of logistics and lot sizes. Investments are not necessarily a determining factor for the 'costs of implementation', as processes can be purchased from third party suppliers. On the other hand the training effort plays an important role, because employees have to be trained on design rules and the development tools of the suppliers. Looking at the parameter 'qualitative benefit' the transfer of know how is a crucial criteria as the company has to exchange designs with third party suppliers.

2.2 Technological attractiveness

Due to the technological restrictions in microsystem technology the ability to use a certain process is largely dependent on material properties and geometrical boundary conditions. To avoid the elimination of too many processes in early design stages it proved to be inadequate to constrict processes to a certain material. The ferromagnetic core of a fluxgate sensor could consist of several different foil materials, which can be mechanically structured. On the other hand electroplated or sputter deposited permalloy could be used. For the function of the sensor only the material properties are crucial not the material itself. Which material and process combination should be used is depended on all other criteria of the rating. This means that for the parameter 'technological feasibility' the restriction to geometrical data is sufficient to constrict the number of possible technologies. The 'technological potential' can be described by the same criteria as suggested by Eversheim. Special focus has to be given to the parameter 'environmental compatibility', as many of the microtechnological processes make use of poisonous and environmentally hazardous substances.

2.3 Criteria structure

The criteria structure for microtechnological processes is depicted in figure 5. Criteria, which come to the fore in this context, are highlighted in bold letters. This approach uses the technology database (ProcessDB) of the design environment to access technologies which already contain criteria regarding the 'technological attractiveness'. Other criteria can easily be integrated into the database by extending the parameters, which already describe the technologies. Unlike the technological parameters the company specific benefits cannot be extracted from a central technology database. The benefit will vary for a single technology depending on the company. Therefore, a criteria database should be implemented, which only provides links to the technology data.

The next section will describe how the described criteria structure is used to evaluate technologies and place them into the portfolio.

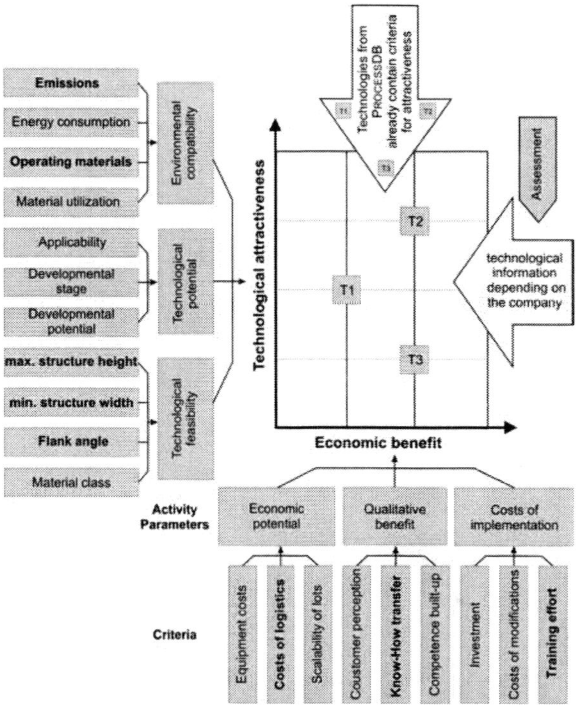

Figure 5: Criteria structure for the rating of microtechnological processes.

2.4 Merging of criteria and rating

All criteria, which describe an activity parameter are weighted and evaluated by the user. For this purpose the user inputs the weight and compliance by linguistic fuzzy sets. The weight of criterion j is given by $w_j = (\alpha_j, \beta_j, \gamma_j, \delta_j)$ and the compliance of an alternative i is specified by the fuzzy number $r_{ij} = (\alpha_{ij}, \beta_{ij}, \gamma_{ij}, \delta_{ij})$. The value of an activity parameter A of an alternative solution i can be calculated by a fuzzy-multi-attributive method according to equation 1:

$$A_i = \frac{\sum_{j=1}^{n} w_j r_{ij}}{\sum_{j=1}^{n} w_j} \quad (1)$$

The resulting activity parameters are then mapped to a rule based rating system as it was introduced by Traeger [6] for fuzzy control systems. This system determines the influence of each single activity parameter on the benefit and the attractiveness of the technology under investigation. Figure 6 shows the rule based rating method for the attractiveness of a technology as an example.

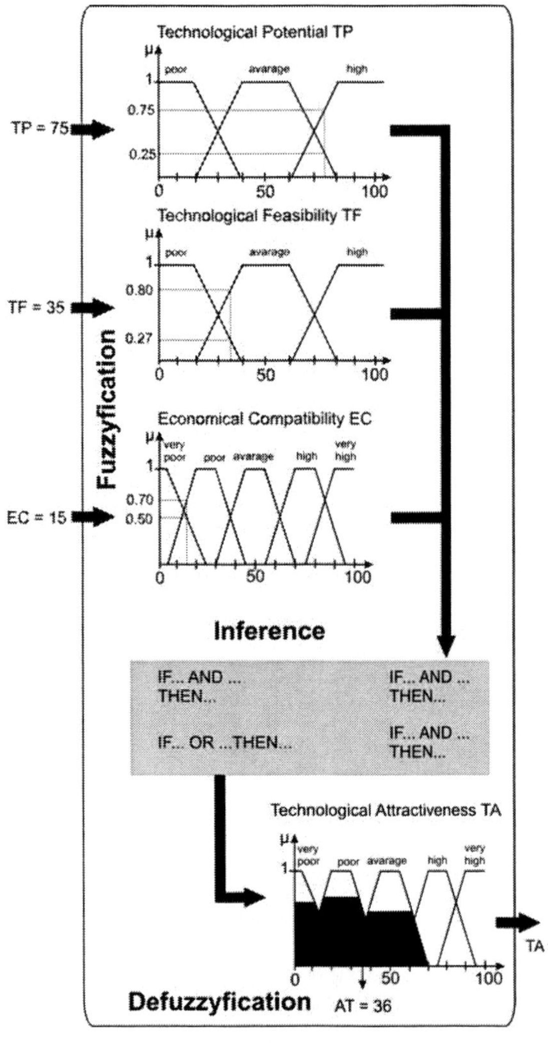

Figure 6: Rule-based rating of individual activity parameters [6].

All rules, which are used by the inference mechanism, are configured using certain fuzzy sets. As the rules are configured only once, the fuzzy numbers that result from equation 1 cannot be used directly. They must be defuzzyficated and then mapped to the predefined sets. The main advantage of this system is that complex rules can be stated without using mathematical expressions. This again is achieved by the use of linguistic fuzzy sets and simple if-then statements. After using the inference mechanism a defuzzification method, e.g. the center of gravity method, leads to a sharp score of an alternative. The sharp values for attractiveness and benefit can then be used to place a technology into the portfolio.

3 SUMMARY AND PERSPECTIVE

A software environment for integrated process and layout design was presented. Due to the capabilities of the system to validate process flows an optimization of process chains by substituting single processes is possible. A method for the rating of the most suitable process for a given design at certain economical boundary conditions was developed. This system uses a fuzzy-based analytical hierarchy process. A special criteria structure for the use in microtechnology was developed and presented.

To make use of the presented method a user interface has to be designed and a data structure for storing company specific criteria will be implemented.

4 ACKNOWLEDGEMENT

The Deutsche Forschungsgemeinschaft (DFG) has financially supported this work within a collaborative Research Center (Sonderforschungsbereich 516) titled, 'Design and Fabrication of Active Microsystems'.

REFERENCES

[1] U. Hansen, U. Triltsch, S. Büttgenbach, C. Germer, H.-J. Franke: *Analysis and Verification of Processing Sequences*, Proc. Nanotech 2003, Vol. 2, San Francisco, 2003, pp. 484-487

[2] D. Straube, U. Triltsch, H.J. Franke, S. Büttgenbach: *Modular software system for computer aided design of microsystems*, Microsystem Technolgies, Vol.12, 2006, April, pp. 650-654

[3] U. Triltsch, S. Büttgenbach, D. Straube, H.-J. Franke: *T-CAD Environment for Multi-Material MEMS Design,* Proc. of Nanotech 2005, Vol. 3, Anaheim, USA, May 2005, pp. 541-544

[4] R.C. Dorf: *The Technology Management Handbook*, Berlin, Springer, 1999

[5] W. Eversheim, G. Schuh: *Integrated Product and Process Design*, Springer, Berlin, 2005

[6] D.H. Traeger: *Introduction to Fuzzy Logic*, Teubner, Stuttgart, 1994

Modeling Voids in Silicon

M. Hasanuzzaman*, Y. M. Haddara* and A. P. Knights**

*Department of Electrical and Computer Engineering
McMaster University, Hamilton, ON, Canada L8S 4K1
**Department of Engineering Physics
McMaster University, Hamilton, ON, Canada L8S 4L7

ABSTRACT

We propose a model for void evolution in Silicon (Si). Si samples were implanted with high energy Si ions. We considered the release of vacancies from the voids to be the rate limiting step for the dissolution of the voids and we account for faster dissolution of smaller voids. The model predicted well the depth distribution of the voids and the evolution of voids under varying anneal temperatures with varying anneal times. We were able to fit the experimental data using a consistent set of fitting parameters and excellent fits were obtained in all the cases.

Keywords: void, modeling, silicon

1 INTRODUCTION

Voids in silicon have important scientific and industrial applications. They are used for gettering transition metals [1], detecting point defect injection [2], controlling carrier life time [3] or producing silicon-on-insulator materials [4]. High energy (MeV) ion implantation can create spatial separation between vacancies and interstitials making a vacancy rich region near the surface and interstitial rich region at the bulk [5]-[7]. The vacancies created during the ion implantation process can aggregate upon annealing to create small vacancy clusters and voids [8]-[9]. Experimental evidences of increased antimony diffusion and shrinkage of interstitial type extended defects upon annealing the high energy implanted samples are reported in the literature [10]-[11] indicating formation of voids in the vacancy rich region. In this paper we focus on the results reported by Venezia *et al.* [5] where voids were created in Si by MeV Si implants. We propose a moment-based model for void growth and dissolution. We have adapted the model from the work on interstitial capture in {311} clusters in [12] and implemented it using commercial process simulator FLOOPS-ISE [13]. We compared the simulated concentration of vacancies in the voids with the previously published results of void formation in Si [5] for annealing in the range 750-1000°C for different anneal times. We begin by discussing model we propose, then describe our simulation approach followed by the simulation results, and finally making our concluding remarks.

2 MODEL

The model we propose here is adapted from the model for {311} defect evolution proposed by Law and Jones [12]. Our proposed model is capable of solving the concentration of vacancies in cavities, C_{void}, the density of voids, D_{void}, and the concentration of vacancies in the small vacancy clusters, C_{sv}. We consider that the capture and release of vacancies in the voids vary in proportion to the number of voids. The release of the vacancy from the voids is the rate limiting step for the void dissolution process. We further consider that the smaller voids dissolve faster than the larger voids. Considering these facts, the equations those govern the void formation mechanism are given by,

$$\frac{dC_{void}}{dt} = k_1 D_{void}\left(C_V - C_{void,eq}\right) - C_{void} D_{rate} \qquad (1)$$

$$\frac{dD_{void}}{dt} = -k_1 D_{void} C_{void,eq} \frac{D_{void}}{C_{void}} - D_{void} D_{rate} \qquad (2)$$

$$\frac{dC_{sv}}{dt} = k_2 C_{sv}\left(C_V - C_{sv,eq}\right) \qquad (3)$$

In these equations, k_1 and k_2 are the reaction rates for the corresponding equations, $C_{void,eq}$ and $C_{sv,eq}$ respectively are the concentrations of vacancies in voids and small vacancy clusters under equilibrium condition, and D_{rate} is the dissolution rate. In Eq. 2 the term D_{void}/C_{void} is inverse of the size of the voids and thus accounts for the smaller voids dissolving faster than larger voids.

3 EXPERIMENTAL DATA AND SIMULATION RESULTS

In this work we considered the experimental results reported by Venezia *et al.* [5] to study the void evolution in Si. We considered the experimental results where the Si samples were ion implanted with 2 MeV Si ions at a dose of 1×10^{16} cm^{-2} and subsequently annealed in the temperature range 750-1000°C for 10 sec-10 min. After annealing, the samples were implanted with Au and annealed at 750°C. Rutherford Back Scattering (RBS) technique was used to measure the Au concentration distribution in the samples. In the presence of vacancies, interstitial Au atoms get trapped into vacancies, one vacancy is required to trap one Au atom. Thus it is reasonable to expect that the concentration of Au in trapped vacancy sites give the estimation of the concentration of vacancies itself. Also Venezia *et al.* reported that the drive-in anneal of Au in the samples did not affect the void distribution in the samples. Kalayanaramana *et al.* [14] reported the ratio of Au to the vacancies to be close to unity for similar experimental conditions used in Venezia *et al.* study. Thus it is reasonable to accept that the Au concentration reported by Venezia *et al.* [5] resembles the concentration of vacancies in the voids. We compared our simulation results with these experimental profiles.

We implemented the model in the commercially available process simulator FLOOPS-ISE [13]. We generated the net excess interstitial and vacancy distributions at the samples for the implanted conditions using the Monte Carlo ion implantation simulator TRIM [15] and used these profiles to initialize the vacancy and interstitial profiles during the simulations. Figure 1 shows the net initial distributions of vacancy and interstitials used in the simulations.

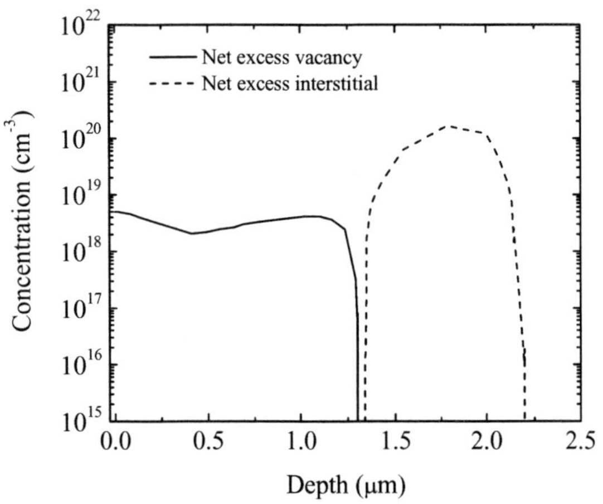

Figure 1: Net initial vacancy and interstitial profiles calculated from TRIM for 2 MeV Si ion implanted with a dose 1×10^{16} cm^{-2} in Si.

FLOOPS-ISE provides well established values for the equilibrium vacancy concentration in Si and we used these values for $C_{void,eq}$. We found that the simulation results of C_{void} are relatively insensitive to the values of k_2 and $C_{sv,eq}$. The remaining two parameters k_1 and D_{rate} were used as the fitting parameters in our model. These fitting parameters were held constant for each temperature.

Figures 2-4 show the simulated C_{void} compared with the experimental profiles. In all the cases we were able to get good matches between the simulated and experimental profile of C_{void}.

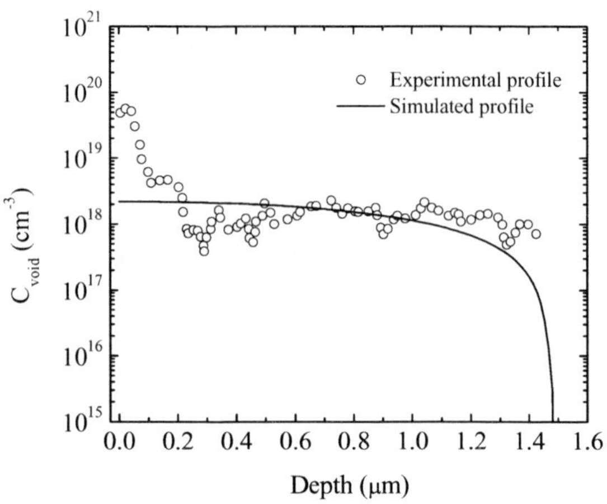

Figure 2: Concentration of vacancies in voids after 10 min anneal at 750°C. Experimental data from [5].

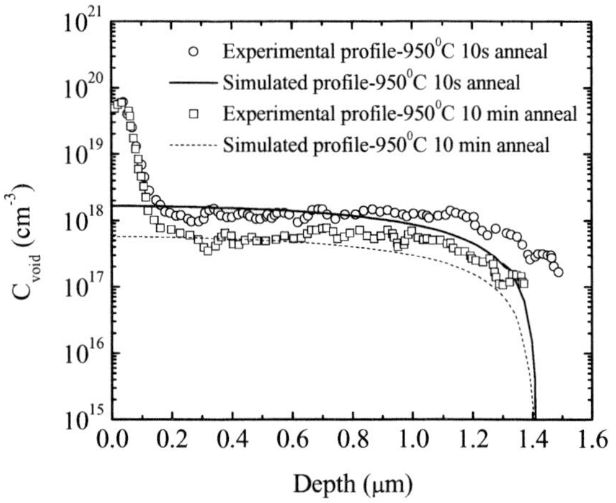

Figure 3: Concentration of vacancies in voids after 10 sec and 10 min anneal at 950°C. Experimental data from [5].

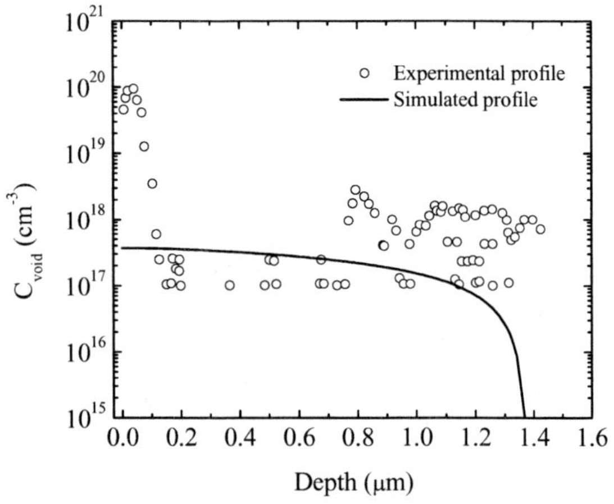

Figure 4: Concentration of vacancies in voids after 10 min anneal at 1000°C. Experimental data from [5].

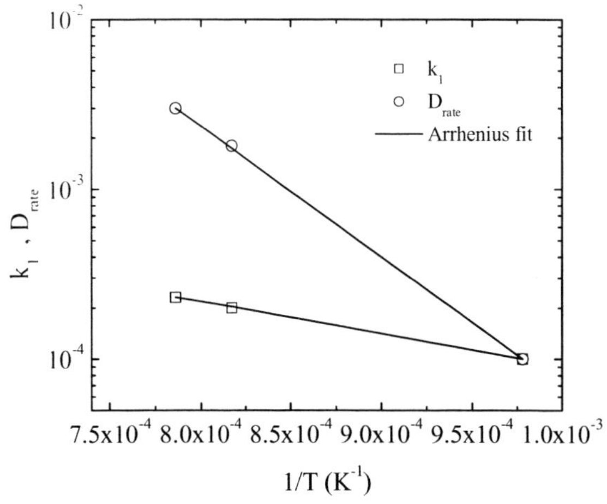

Figure 5: Arrhenius plots for the model parameters.

Figure 5 shows the fitting parameters used during the simulations, these follow Arrhenius relation given by,

$$k_1 = 6.94\times10^{-3}\exp\left(-\frac{0.374eV}{kT}\right) \tag{4}$$

and

$$D_{rate} = 3.747\times10^{3}\exp\left(-\frac{1.537eV}{kT}\right) \tag{5}$$

4 CONCLUSIONS

The model we propose here is able to predict the concentration vacancies in the voids grown in high energy implanted in Si samples. The model is capable of predicting the depth distribution, temperature and time evolution of vacancies in the voids. It is expected that the model will work as a base for modeling more complex void growth and evolution situations in coming days.

REFERENCES

[1] D. M. Follstaedt, S. M. Myers, G. A. Petersen and J. W. Medernach, J. Electron. Mat. 25, 151, 1996.

[2] D. A. Abdulmalik, P. G. Coleman, H. Z. Su, Y. M. Haddara and A. P. Knights, J Mater Sci : Mater Electron 18, 753, 2007.

[3] G. A. Petersen, S. M. Myers and D. M. Follstaedt, Nucl. Inst. Met. Phys. Res. B 127-128, 301, 1997.

[4] V. Raineri and S. U. Campisano, Appl. Phys. Lett. 66, 3654, 1995.

[5] V. C. Venezia, D. J. Eaglesham, T. E. Haynes, A. Agarwal, D. C. Jacobson, H. -J. Gossmann and F. H. Baumann, Appl. Phys. Letts 73, 2980, 1998.

[6] M. Tamura, T. Ando and K. Ohya, Nucl. Inst. Meth. Phys. Res. B 59-60, 572, 1991.

[7] O. Kononchuk, R. A. Brown, S. Koveshnikov, K. Beaman, F. Gonzales and G. A. Rozgonyi, Solid State Phenom. 57-58, 69, 1997.

[8] M. Prasad and T. Sinno, Appl. Phys. Lett. 80, 1951, 2002.

[9] A. Bongiorno, L. Colombo and T. D. De La Rubia, Europhys. Lett. 43, 695, 1998.

[10] D. J. Eaglesham, T. E. Haynes, H. -J. Gossmann, D. C. Jacobson, P. A. Stolk and J. M. Paote, Appl. Phys. Lett. 70, 3281, 1997.

[11] V. C. Venezia, T. E. Haynes, A. Agarwal, H. –J. Gossmann and D. J. Eaglesham, Mater. Res. Soc. Symp. Proc. 469, 303, 1997.

[12] M. E. Law and K. S. Jones, Tech. Dig. Int. Electron Devices Meet., p 511-514, 2000.

[13]FLOOPS-ISE Manual, Release 9.5, Integrated Systems Engineering.

[14] R. Kalayanaramana, T. E. Haynes, V. C. Venezia, D. C. Jacobson, H. -J. Gossmann and C. S. Rafferty, Appl. Phys. Lett. 76, 3379, 1995.

[15] http://www.srim.org , access date October 1, 2007.

Modeling Germanium-Silicon Interdiffusion in Silicon Germanium/Silicon Superlattice Structures

M. Hasanuzzaman*, Y. M. Haddara* and A. P. Knights**

*Department of Electrical and Computer Engineering
McMaster University, Hamilton, ON, Canada L8S 4K1
**Department of Engineering Physics
McMaster University, Hamilton, ON, Canada L8S 4L7

ABSTRACT

We present a model for the interdiffusion of silicon (Si) and germanium (Ge) in SiGe/Si superlattice structures. We considered both the vacancy exchange mechanism and the interstitial diffusion to fit the experimental data available in the temperature range 850-1125°C with Ge fraction up to 27% in SiGe. We considered the effect of Ge fraction on different material properties and we explicitly account for the lattice site conservation during the simulations. We were able to obtain excellent fits for all the cases. The same sets of fitting parameters can successfully describe the interdiffusion mechanism for both inert and oxidizing anneals for samples annealed below ~1075°C, where we find vacancy exchange mechanism dominates the interdiffusion process. For higher temperatures, Si interstitials dominate the interdiffusion mechanism.

Keywords: interdiffusion, SiGe, modeling, superlattice

1 INTRODUCTION

The incorporation of Silicon Germanium (SiGe) in device fabrication in recent years has facilitate increasing the device speed along with the continued device scaling and at the same time using the well matured low cost Si CMOS processing steps. SiGe/Si superlattice (SL) structures are used in applications such as photodiodes, waveguides and photodetectors [1]-[2]. Due to the high temperature fabrication process steps involved, the interdiffusion of Si and Ge in the structures is inevitable, which may lead to detrimental device performance. Thus getting an insight of the interdiffusion phenomena in SiGe/Si SL structures, accurate physically based atomistic models are needed. Few studies [3]-[4] have so far been reported to study the interdiffusion behaviour in $Si_{1-x}Ge_x$/Si SL structures. All these studies report effective diffusivity values for the interdiffusion process. In this work we have used the mathematical model for vacancy exchange mechanism and interstitial diffusion to describe the Si and Ge interdiffusion in $Si_{1-x}Ge_x$/Si SL structures annealed in the temperature range 850-1125°C in inert ambient and where experimental data are available in oxidizing ambient. While in the literature, experimental data are available with very high values of Ge fraction in $Si_{1-x}Ge_x$, we limit our study for the $Si_{1-x}Ge_x$/Si SL structures containing Ge fraction up to 27%. While the vacancy exchange mechanism and interstitial diffusion mechanism have long been thought to play an important role in self and inter-diffusion in semiconducting materials, it is very recently where we quantitatively showed that vacancy exchange mechanism alone is sufficient to describe the interdiffusion phenomena in $Si/Si_{1-x}Ge_x$/Si single quantum well structures at low anneal temperatures and interstitial diffusion mechanism can explain the interdiffusion phenomena at high anneal temperatures [5]. However, till to date, the vacancy exchange mechanism and the interstitial diffusion mechanism are not used quantitatively to describe the interdiffusion mechanism in $Si_{1-x}Ge_x$/Si SL structures.

2 MODELING INTERDIFFUSION FOR LOW TEMPERATURE ANNEALING

Vacancy exchange mechanism suggests that an atom and its adjacent vacant site interchange their position with each other for diffusion to occur. This concept was first proposed by Kirkendall *et al.* [6] in 1939 while describing the interdiffusion phenomena in metallic alloys. In 1942, Huntington and Seitz [7] proposed that the self-diffusion in metals occurs by vacancy exchange mechanism. Later, Smigelskas and Kirkendall [8] and Darken [9] reported further study on the vacancy exchange mechanism and provided a detailed mathematical model describing the phenomena. Though the standard process simulators [10]-[13] are capable to simulate diffusion phenomena, these are applicable for describing the diffusion of dilute dopants only. However, the interdiffusion of Si and Ge in $Si_{1-x}Ge_x$/Si SL structures involve the diffusion of the host atoms; the dilute dopant approximation no longer holds and the lattice site conservation constraint must be explicitly taken into account.

The equations those govern the interdiffusion of Si and Ge by vacancy exchange mechanism in $Si_{1-x}Ge_x$/Si SL structures are given by [5], [14],

$$\frac{\partial C_{Si}}{\partial t} = \frac{\partial}{\partial x}\left[D_{Si}^{\text{int}} \frac{C_V}{C_V^{*,\text{int}}} \frac{\partial C_{Si}}{\partial x} - D_{Si}^{\text{int}} C_{Si} \frac{\partial}{\partial x} \frac{C_V}{C_V^{*,\text{int}}} \right] \quad (1)$$

$$\frac{\partial C_{Ge}}{\partial t}=\frac{\partial}{\partial x}\left[D_{Ge}^{\text{int}}\frac{C_V}{C_V^{*,\text{int}}}\frac{\partial C_{Ge}}{\partial x}-D_{Ge}^{\text{int}}C_{Ge}\frac{\partial}{\partial x}\frac{C_V}{C_V^{*,\text{int}}}\right] \tag{2}$$

$$\frac{\partial C_V}{\partial t}=-\frac{\partial C_{Si}}{\partial t}-\frac{\partial C_{Ge}}{\partial t} \tag{3}$$

In these equations, C_{Si} , C_{Ge}, and C_V respectively are the concentrations of Si, Ge and vacancy. D_{Si}^{int} and D_{Ge}^{int} respectively are the Si and Ge self-diffusivities under inert intrinsic conditions. These parameters in general depend on the concentration of Ge and thus change as a function of time and position in the structure. Equations (1) and (2) respectively take into account the diffusion of Si and Ge under their own concentration gradient and the lattice motion to compensate for the difference in diffusivities. Equation (3) arises from the conservation of lattice site constraint. We also take into account the recombination between vacancies and interstitials by adding a bulk recombination term in (3). Considering that Ge has a catalytic effect on the equilibrium concentration of point defects [15]-[16], we implement these by [11],

$$C_X^{*,\text{int}}(SiGe)=C_X^{*,\text{int}}(Si)\exp\left(\frac{\Delta V_X}{kT}\right) \tag{4}$$

$$\Delta V_X=\Delta V_X^0 .a_{SiGe}\cdot\frac{C_{Ge}}{5\times10^{22}} \tag{5}$$

where X refers to the point defect being considered, and for this case it is vacancy, V. $C_X^{*,\text{int}}(Si)$ is the equilibrium concentration in pure Si and we used the well-established default value given in [11]. The value of ΔV_X^0 is 25.6 for vacancy [11] and a_{SiGe} is calculated using Vegard's rule [17].

In this work we considered the experimental results reported by Griglione [3] and Holländer *et al.* [4]. Both these studies report SL structures grown on either Si or $Si_{1-y}Ge_y$ buffer layer. The studies reported the effective diffusivity values for Ge interdiffusion without regard to the atomistic mechanisms involved. Griglione used FLOOPS-ISE [12] process simulator to extract the effective diffusivity values of the Ge interdiffusion in $Si_{0.85}Ge_{0.15}$/Si SL structures from the experimental SIMS profiles annealed at different temperatures for different anneal times both in inert and oxidizing ambients. On the other hand Holländer *et al.* reported effective diffusivity values for Ge interdiffusion extracted from experimental RBS profiles where the samples were annealed in inert ambient only.

In our work, we compare our simulated results with the reported experimental results. Where the experimental results are not available, we used the effective diffusivity values reported to extract the experimental profiles since the effective diffusivity values were reported to be extracted with reasonable fits with the experimental profiles. We then compare our simulation result with these extracted experimental profiles.

We applied the model given in Eqs. 1-5 using the commercial process simulator FLOOPS-ISE [11]. While simulating the results reported by Griglione, we used the experimentally reported as-grown profile for Ge in the SL structures as the initial Ge profile. However for simulating the results reported by Holländer *et al.* we used a box type initial profile for Ge in the SL structures as the actual as-grown Ge profile was not available. The initial vacancy profile was set equal to the intrinsic equilibrium vacancy concentration at the structure. To calculate the initial profiles of Si we impose the conservation of lattice site constraint,

$$C_{Si}+C_{Ge}+C_V=5\times10^{22} \tag{6}$$

In the simulations, the variations of D_{Si}^{int} and D_{Ge}^{int} as a function of Ge fraction in structures were taken into account. During the simulations we took the values of D_{Si}^{int} in pure Si from [18] and the values of D_{Si}^{int} for 80% Ge in $Si_{1-x}Ge_x$ from [19]. The values of D_{Ge}^{int} in Si for samples annealed below 1000°C were reported by Sharma [20] and the values of D_{Ge}^{int} in Si for higher temperatures and for Ge fractions of 10, 20 and 30% in $Si_{1-x}Ge_x$ were reported by Zangenberg *et al.* [21]. We allowed the values of D_{Ge}^{int} to vary and consider this as a fitting parameter during the simulations. We comment on the values used in the simulations later.

First, we applied the model for simulating anneals in inert ambient. Once we get good matches between the experimental and simulated Ge interdiffusion profiles, keeping the fitting parameters fixed for each anneal temperatures, the model was successful in simulating the Ge interdiffusion profiles for oxidizing ambient anneals without any need of additional fitting parameters.

Figures 1, 2 and 3 show typical fits for Ge interdiffusion in $Si_{0.85}Ge_{0.15}$/Si SL structures grown on $Si_{1-y}Ge_y$ buffer. Figure 4 shows the actual value of D_{Ge}^{int} used in the simulations compared to the published values. For temperatures up to 1075°C, the values are close to the previously published values and for all the cases we were able to get reasonable fits of Ge interdiffusion profiles to the experimental profiles. However, for temperatures 1100 and 1125°C, the values of D_{Ge}^{int} close to the previously published values were not able to give good matches between the experimental and simulated Ge interdiffusion profiles, which indicates that the formulation presented so far is not sufficient to fully describe the diffusion mechanism at these temperatures and this necessitates to consider the contribution of interstitials in diffusion process.

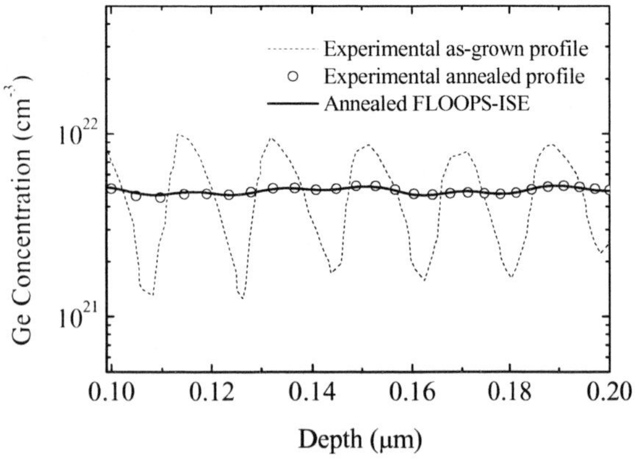

Figure 1: Ge interdiffusion in $Si_{0.85}Ge_{0.15}$/Si SL grown on relaxed $Si_{0.85}Ge_{0.15}$ buffer after 3 min anneal in inert ambient at 950°C. Experimental data from [3].

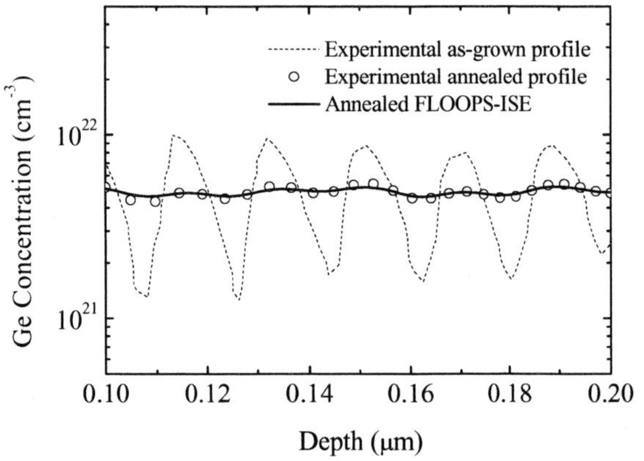

Figure 2: Ge interdiffusion in $Si_{0.85}Ge_{0.15}$/Si SL grown on relaxed $Si_{0.85}Ge_{0.15}$ buffer after 3 min anneal in oxidizing ambient at 950°C. Experimental data from [3].

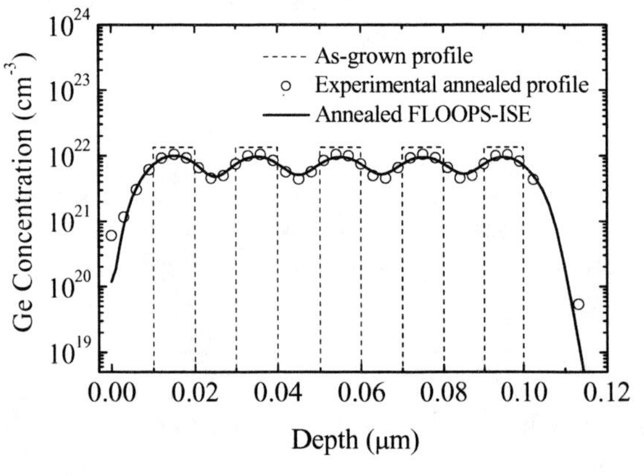

Figure 3: Ge interdiffusion in $Si_{0.73}Ge_{0.27}$/Si SL grown on Si substrate after 100 sec anneal in inert ambient at 1025° C. Experimental data from [4].

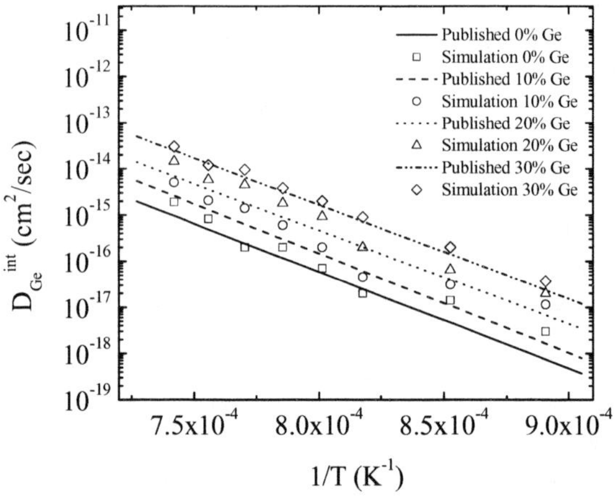

Figure 4: D_{Ge}^{int} values used in the simulations compared with values from literature [20]-[21].

3 MODELING INTERDIFFUSION FOR HIGH TEMPERATURE ANNEALING

To take into account the contribution of interstitials in the interdiffusion mechanism, we have to consider the kick-out mechanism [22] and Frank-Turnbull mechanism [23]. The complete system of equations considering vacancy exchange mechanism, kick-out mechanism and Frank-Turnbull mechanism are given by [5],

$$\frac{\partial C_{Si}}{\partial t} = \frac{\partial}{\partial x}\left[D_{Si}^{\text{int}} \frac{C_V}{C_V^{*,\text{int}}} \frac{\partial C_{Si}}{\partial x} - D_{Si}^{\text{int}} C_{Si} \frac{\partial}{\partial x} \frac{C_V}{C_V^{*,\text{int}}} \right] + k_{f,KO} C_{ISi} C_{Ge} - k_{r,KO} C_{IGe} C_{Si} + k_{f,SiR}\left(C_{ISi} C_V - C_{ISi}^* C_V^* \right) \quad (7)$$

$$\frac{\partial C_{Ge}}{\partial t} = \frac{\partial}{\partial x}\left[D_{Ge}^{\text{int}} \frac{C_V}{C_V^{*,\text{int}}} \frac{\partial C_{Ge}}{\partial x} - D_{Ge}^{\text{int}} C_{Ge} \frac{\partial}{\partial x} \frac{C_V}{C_V^{*,\text{int}}} \right] - k_{f,KO} C_{ISi} C_{Ge} + k_{r,KO} C_{IGe} C_{Si} + k_{f,GeR}\left(C_{IGe} C_V - C_{IGe}^* C_V^* \right) \quad (8)$$

$$\frac{\partial C_{ISi}}{\partial t} = \frac{\partial}{\partial x}\left[D_{ISi} C_{ISi}^* \frac{\partial}{\partial x} \frac{C_{ISi}}{C_{ISi}^*} \right] - k_{f,KO} C_{ISi} C_{Ge} + k_{r,KO} C_{Si} C_{IGe} - k_{f,SiR}\left(C_{ISi} C_V - C_{ISi}^* C_V^* \right) \quad (9)$$

$$\frac{\partial C_{IGe}}{\partial t} = \frac{\partial}{\partial x}\left[D_{IGe} C_{IGe}^* \frac{\partial}{\partial x} \frac{C_{IGe}}{C_{IGe}^*} \right] + k_{f,KO} C_{ISi} C_{Ge} - k_{r,KO} C_{Si} C_{IGe} - k_{f,GeR}\left(C_{IGe} C_V - C_{IGe}^* C_V^* \right) \quad (10)$$

There is another equation to be considered that arises from the conservation of lattice site constraints and is given by Eq. 3. The reaction constants of the above equations are given by,

$$k_{f,KO} = 4\pi . r_{Si} . D_{ISi} \quad (11)$$

$$k_{r,KO} = 4\pi . r_{Ge} . D_{IGe} \quad (12)$$

NSTI-Nanotech 2008, www.nsti.org, ISBN 978-1-4200-8505-1 Vol. 3

$$k_{f,SiR} = 4\pi . r_{IVSi} . (D_{ISi} + D_V) \quad (13)$$

$$k_{f,GeR} = 4\pi . r_{IVGe} . (D_{IGe} + D_V) \quad (14)$$

where the r's are capture radii and we take this value equal to 0.5 nm following [24], D_{ISi}, D_{IGe} and D_V respectively are the diffusivities of the interstitial Si, interstitial Ge and vacancy. We assumed that each of these reactions is diffusion limited.

During the simulations, we used the published values [18]-[21] of D_{Si}^{int} and D_{Ge}^{int}. For D_{ISi}, we used the values given in FLOOPS-ISE [11], used the same value for D_{IGe} considering the similar atomic nature of Si and Ge. We used Eqs. 4-5 to calculate the values of C_V^* and C_{ISi}^* with ΔV_X^0=11.8 for interstitials [11]. We found the simulations to be insensitive to the variations of C_{IGe}^* and thus we set arbitrarily $C_{IGe}^* = 0.01 C_{ISi}^*$. With these comments, there is no other free parameter available.

Figures 5 shows typical fit for Ge interdiffusion profile at 1100°C inert ambient anneal. Excellent fits were obtained for all the cases at 1100 and 1125°C inert ambient anneals.

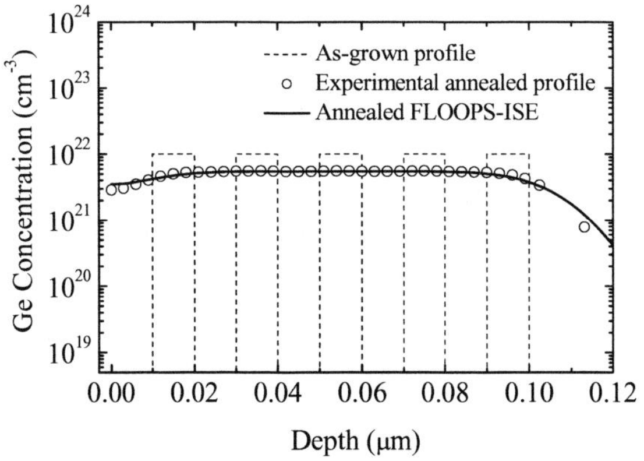

Figure 5: Ge interdiffusion in $Si_{0.80}Ge_{0.20}$/Si SL grown on Si substrate after 1 min anneal in inert ambient at 1100° C. Experimental data from [4].

4 CONCLUSIONS

We have successfully modeled Si-Ge interdiffusion in $Si_{1-x}Ge_x$/Si SL structures grown on relaxed $Si_{1-y}Ge_y$ buffer and Si substrate. During the simulations we considered the well established values of the intrinsic diffusivities of different species and the equilibrium point defect concentrations, including their dependences on the Ge fraction in the structures. We were able to successfully model experimental results on Si-Ge interdiffusion for a wide range of published data in the temperature range 850-1125°C with Ge fraction in $Si_{1-x}Ge_x$/Si SL structures up to 27%. For temperatures up to 1075°C, interdiffusion of Si and Ge mainly occurs by vacancy exchange mechanism. On the other hand for temperature at and above 1100°C, the interdiffusion of Si and Ge is controlled by diffusion of Si interstitials.

REFERENCES

[1] C. Engel, P. Baumgartner, M. Holzmann, J. F. Nützel and G. Abstreiter, Thin Solid Films 294, 347, 1997.

[2] O. Qasaimeh, P. Bhattacharya and E. T. Croke, IEEE Photonics Tech. Lett. 10, 807, 1998.

[3] M. D. Griglione, PhD Thesis, University of Florida, 1999.

[4] B. Holländer, R. Butz and S. Mantl, Phys. Rev. B 46, 4975, 1992.

[5] M. Hasanuzzaman and Y. M. Haddara, J Mater Sci: Mater Electron, Published online, 2007, doi 10.1007/s10854-007-9391-5.

[6] E. Kirkendall, L. Thomassen and C. Upthegrove, Trans. AIME 133, 186, 1939.

[7] H. B. Huntington and F. Seitz, Phys. Rev. 61, 315, 1942.

[8] A. D. Smigelskas and E. O. Kirkendall, Trans. AIME 171, 130, 1947.

[9] L. S. Darken, Trans. AIME 175, 184, 1948.

[10]SUPREM-IV.GS Manual, Stanford University, 1993.

[11]FLOOPS-ISE Manual, Release 9.5, Integrated Systems Engineering.

[12]FLOOPS-ISE Manual, University of Florida, Version 2002.

[13] ATHENA User's Manual, SILVACO International, December 2002.

[14] C. -Y. Tai, PhD Thesis, Stanford University, 1997.

[15] A. Pakfar, Mat. Sci. Engg B 89, 225, 2002.

[16] M. J. Aziz, Appl. Phys. Lett. 70, 2810, 1997.

[17] A. R. Denton, N. W. Ashcroft, Phys. Rev. A 43, 3161, 1991.

[18] R. J. Borg and G. J. Dienes, "An Introduction to Solid State Diffusion," Academic Press, 1998.

[19] P. Laitinen, A. Strohm, J. Huikari, A. Nieminen, T. Voss, C. Grodon, I. Riihimäki, M. Kummer, J. Äystö, P. Dendooven, J. Räisänen, W. Frank and the ISODLE Collaboration, Phys. Rev. Lett. 89, 085902, 2002.

[20] B. L. Sharma, Def. Diff. Forum 70-71, 1, 1990.

[21] N. R. Zangenberg, J. Lundsgaard, J. Fage-Pedersen and A. N. Larsen, Phys. Rev. Lett. 87, 125901, 2001.

[22] U. Gösele, W. Frank and A. Seegar, Appl. Phys. 23, 361, 1980.

[23] F. C. Frank and D. Turnbull, Phys. Rev. 104, 617, 1956.

[24] J. L. Ngau, P. B. Griffin and J. D. Plummer, J. Appl. Phys. 90, 1768, 2001.

Effect of Strain on the Oxidation Rate of Silicon Germanium Alloys

Mohamed A. Rabie[*], Steven Gou[**], Yaser M. Haddara[***], Andrew P. Knights[****], Jacek Wojcik[*****], and Peter Mascher[******]

[*]Electrical and Computer Engineering, McMaster University, Hamilton, ON, Canada, rabiema@mcmaster.ca
Department of Physics and Astronomy, University of British Columbia, Vancouver, BC, sgou@phas.ubc.ca
[***]Electrical and Computer Engineering, McMaster University, Hamilton, ON, Canada, yaser@mcmaster.ca
[****]Engineering Physics, McMaster University, Hamilton, ON, Canada, aknight@mcmaster.ca
[*****]Engineering Physics, McMaster University, Hamilton, ON, Canada, wojcikj@mcmaster.ca
[******]Engineering Physics, McMaster University, Hamilton, ON, Canada, mascher@mcmaster.ca

ABSTRACT

We report on a series of experiments in which a strained $Si_{0.95}Ge_{0.05}$ layer 600Å thick was oxidized along with relaxed SiGe layers and Si samples. In this work, we observed that the oxidation rate of the strained SiGe layer is always much higher than the relaxed model expectations. We also observed that the rate is higher than that of relaxed SiGe layers with higher Ge concentration oxidized at similar temperatures. We attributed this increase in the rate of oxidation to the effect of strain in the SiGe layer. To confirm our hypothesis, we oxidized strained SiGe layers together with relaxed SiGe layers of higher concentration. We observed that the strained SiGe layer compared with relaxed sample has a thicker oxide despite the lower Ge concentration. We conclude that the strain in the SiGe layer causes a further increase in the oxidation rate.

Keywords: SiGe, oxidation, modeling, strain.

1 INTRODUCTION

Thermal oxidation of SiGe is a necessary process step for many applications in which SiGe is used. The oxidation process of SiGe results in a low quality oxide due to an undersirable Ge-rich layer below the oxide as well as high fixed charge density and trap density at the oxide/substrate interface [1 and references therein]. To improve the quality of the oxide, the physics of the oxidation process has to be well understood. One factor that has not been thoroughly considered before is the effect of the strain on the oxidation process. To understand the effect of strain on the oxidation process we conducted a number of oxidation experiments on strained SiGe samples as well as relaxed samples.

It is well established in the literature (see [1-2] and references therein) that the oxidation rate of SiGe increases with the increase in temperature and the increase in the concentration of Ge. Our previously published model [1] was able to quantify this increase to a high degree of accuracy. The experiments we carried showed that the strained SiGe layers oxidize at a higher rate than relaxed SiGe layers at similar temperatures and Ge concentration. The experimental results were also compared to the relaxed oxidation model expectations [1]. We were able to reach a conclusion that strain enhances the oxidation rate of SiGe. The faster oxidation rate can be attributed to a weaker Si-Ge bond in the case of strained SiGe. This new conclusion can help us to understand the mechanism involved in enhancing the oxidation rate of SiGe over that of Si. Three mechanisms have been previously proposed [1]. One of them is the weaker Si-Ge bond compared to Si-Si bond. Our conclusion confirms that the weaker Si-Ge bond is one of the causes of the enhancement in the oxidation rate of SiGe compared to that of pure Si. Understanding the effect of the bonding factor on the reaction rate will help in achieving more accurate physical and mathematical modeling for the SiGe oxidation.

In the following section we describe the experiments that were conducted as well as the experimental challenges we faced in achieving accurate results. The experimental results are shown and discussed in the last section of this paper.

2 EXPERIMENTAL PROCEDURES

Initially, a strained $Si_{0.95}Ge_{0.05}$ was oxidized at 820°C in a tube furnace. The SiGe layer was 600Å thick on top of a Si substrate. The wet oxygen mixture was prepared by bubbling oxygen through boiling deionized water at 1 SCFH flow rate [3].

When the enhancement in oxidation rate of strained SiGe was observed a number of experiments were conducted to make sure that the effect of furnace calibration on the oxidation rate is minimal. That set of experiments resulted in a number of precautions that were taken with the later experiments. The temperature of the DI water was adjusted just below the boiling temperature to ensure that the wet mixture will result in wet oxygen rather than a mixture close to dry O_2. We ensured that the furnace is dry before every oxidation experiments. Water droplets from the furnace resulted in islands in the oxide. The samples orientation in the furnace was always the same to have the same exposure for all the oxidized samples.

After this set of calibration experiments, more oxidation experiments were conducted. In the new set of oxidation experiments Si was always oxidized with SiGe. Since the oxidation rates of pure Si are well established in the literature. Si served as an indicator for the actual furnace temperature. The initial experiment was repeated and the results came close to the first run. A number of other experiments were conducted for various temperatures and for different periods of time. For the experiment done at 780°C, a relaxed CZ grown $Si_{0.905}Ge_{0.095}$ was oxidized with the strained $Si_{0.95}Ge_{0.05}$ to recognize the effect of strain with no doubt.

3 RESULTS AND DISCUSSION

Figure 1 shows the experimental measurements for the oxide thickness of the strained $Si_{0.95}Ge_{0.05}$ compared to the predictions of the relaxed oxidation model [1] at 820°C. It is clear that the rate of oxidation is much higher than the expectations by the model. The results were then compared to the oxidation rate of $Si_{0.86}Ge_{0.14}$ reported in the literature [2] at a close temperature: 800°C. The relaxed model predicts the results reported in the literature to a high degree of accuracy (not shown in the figure). Yet, our experiments show a higher oxidation rate than that predicted by the model. Furthermore, our experimental results show a higher oxidation rate than that reported by LeGoues et al. [2] despite a higher Ge fraction in LeGoues et al.'s experiment (14% compared to 5%). It is well known that the oxidation rate of SiGe increases considerably with the increase in the Ge content [1 and references therein]. It is also well known that the rate increases with the increase in temperature [1 and references therein]. Our previously reported model [1] was able to quantify these effects to a high degree of accuracy. Despite these facts, the oxidation rate for the 5% SiGe at 820°C was higher than that for the 14% SiGe. This contradicts with the model expectations. The higher oxidation rate was attributed to the effect of strain on the oxidation rate.

To justify this conclusion; further experiments were conducted. The results of those experiments are shown in table 1. It is clear from the table that the experimental results for the oxide thickness on the strained $Si_{0.95}Ge_{0.05}$ are much higher than the expectations of the relaxed model. In the table the experimental measurement for the oxide thickness of the relaxed $Si_{0.905}Ge_{0.095}$ is also shown. This sample has been oxidized simultaneously under the same furnace conditions with the strained $Si_{0.95}Ge_{0.05}$. Despite a higher Ge fraction, the oxide thickness was larger for the relaxed sample. This is evidence that the strain enhances the oxidation of SiGe alloys.

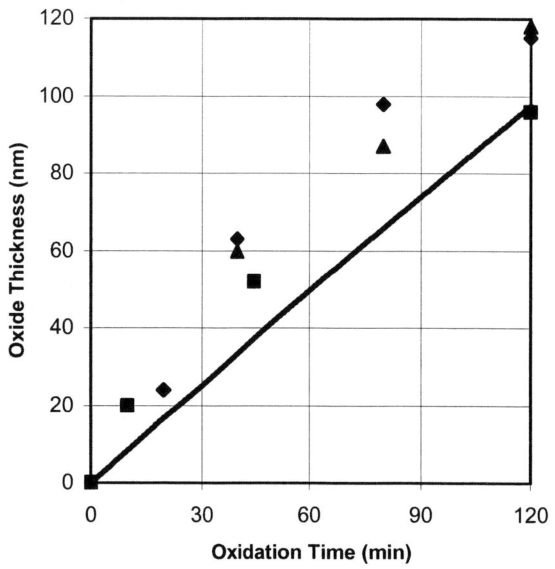

Figure 1: Oxidation rate of strained $Si_{0.95}Ge_{0.05}$ layer 600Å thick at 820ºC (♦: RBS; ▲: ellipsometry) compared to oxidation rate of a relaxed $Si_{0.86}Ge_{0.14}$ layer at 800ºC (■: ellipsometry) [2]. The experimental data is also compared to the relaxed oxidation model [1] expectations for 5% Ge at 820ºC (line).

Three mechanisms were proposed previously to explain the enhancement in the oxidation of SiGe alloys over pure Si [1]. The first mechanism is the change of the path of the oxidation reaction. GeO is formed first and then Ge is replaced by Si forming SiO_2. That path of the reaction is faster than the direct oxidation of Si. The second mechanism is the suppression of the injection of interstitials. Interstitials are injected during the oxidation process of pure Si which results in slowing down the oxidation process. The Ge atom is larger in size than the Si atom. It was experimentally shown that the injection of interstitials is suppressed during the oxidation of SiGe [4]. This will enhance the oxidation rate if the injection of interstitials was a rate limiting step in the case of oxidation of pure Si [1].

The third mechanism which contributes to the enhancement in the oxidation rate of SiGe alloys is the weaker Si-Ge bond compared to Si-Si bond. This means that the Si-Ge bond breaks easier than the Si-Si causing an enhancement in the oxidation rate. Our results support this hypothesis. Strained bonds are weaker than relaxed bonds. We observed an enhancement in the oxidation rate of strained SiGe over relaxed SiGe. This can be attributed to the weaker Si-Ge bond in case of strained SiGe. Similarly, we can conclude that the oxidation of relaxed SiGe is enhanced over pure Si due to the weaker Si-Ge compared to the Si-Si bond.

Temperature (°C)	Time (min)	Experimental Thickness (Å)	Relaxed Model Expectations A (Å)	Closest Relaxed Experimental Measurements (Å)	Relaxed Model Expectations B (Å)
715	30	50	40	–	–
780	50	550	222	342[a]	322
805	40	464	253	520[b]	550
845	120	2262	1435	–	–

Table 1: Results of oxidation of strained $Si_{0.95}Ge_{0.05}$ layer 600Å thick at different times and temperatures compared to the relaxed oxidation model [1] expectations (A) at the corresponding times and temperatures. The results are also compared to the closest relaxed experimental measurements available. The model expectations (B) for those experimental results are also reported.

[a] This result comes from our own experimental measurements for a relaxed CZ grown $Si_{0.905}Ge_{0.095}$ oxidized for the same time and at the same temperature with the strained $Si_{0.95}Ge_{0.05}$ layer indicated in that row.

[b] Oxide thickness for the relaxed $Si_{0.86}Ge_{0.14}$ oxidized for 45 minutes at 800 °C as reported by [2].

REFERENCES

[1] Mohamed A. Rabie, Yaser M. Haddara and Jacques Carette, J. Appl. Phys. 98, 074904, 2005.

[2] F. K. LeGoues, R. Rosenberg, T. Nguyen, F. Himpsel, and B. S. Meyerson, J. Appl. Phys. 65, 369, 1990.

[3] H. Z. Su, "Void formation and vacancy injection in Silicon and Silicon Germanium," Masters Thesis, McMaster University, 2006.

[4] D. A. Abdulmalik, P. G. Coleman, H. Z. Su, Y. M. Haddara and A. P. Knights, Journal of Materials Science: Materials in Electronics 18, 753, 2007.

A Physics-Based Empirical Model for Ge Self Diffusion in Silicon Germanium Alloys

Mohamed A. Rabie* and Yaser M. Haddara**

*Electrical and Computer Engineering, McMaster University,
Hamilton, ON, Canada, rabiema@mcmaster.ca
**Electrical and Computer Engineering, McMaster University,
Hamilton, ON, Canada, yaser@mcmaster.ca

ABSTRACT

We propose a physics-based model for the Ge self diffusivity in SiGe alloys and empirically fit the model to previously reported experimental results [1-3]. The dominant mechanism for Ge self diffusion in SiGe is the vacancy exchange mechanism since the published data on Ge diffusivity are in most cases for experiments done at temperatures lower than 1050 ºC. We added two new terms to modify the regular diffusivity equation to relate the diffusivity to the change in Ge concentration. The first term is a consequence of the change in the point defects disorder entropy [4] as a result of adding more Ge atoms to the SiGe system. The second term is to account for the change in formation and migration enthalpies of the point defects due to the change in Ge content in the alloy.

Keywords: SiGe, self-diffusion, modeling.

1 INTRODUCTION

SiGe alloys are gaining more attention in the industry due to their increased carrier mobility over Si and their compatibility with the cheap silicon process. However, there are some crucial differences between the SiGe and the Si process. One of those is the interdiffusion of Si and Ge in heterostructures and the fact that the self-diffusivity of Si and Ge will vary with the Ge content. In this paper we focus on the Ge self-diffusivity. Currently, there is no general equation to describe the self diffusivity of Ge in SiGe. Such an expression is required for the modeling of several process phenomena in SiGe, including SiGe interdiffusion [5] and SiGe oxidation [6].

The model presented here is able to predict the Ge self diffusivity in SiGe alloys for all various Ge concentrations and for all temperatures. The regular Arrhenius expression for self diffusivity is modified by adding a term to reflect the change in point defects disorder entropy as a result of the increase of Ge concentration in the SiGe alloy. The activation energy is also a function of Ge fraction to account for the change in the formation enthalpy of vacancies and migration enthalpy of the Ge by a vacancy mechanism.

2 THE MODEL

It has been established in the literature [4, 7] that the two main mechanisms for self diffusion in silicon and in germanium and consequently in silicon germanium are the interstitial-based and vacancy-based mechanisms. Therefore, the self diffusivity of Si or Ge in SiGe can be formulated in the same manner the self diffusivity of Si [4] is formulated:

$$D_{Xself} = D_{XI} + D_{XV}$$
$$= f_{XI} d_{XI} \frac{C^*_{XI}}{C_S} + f_{XV} d_{XV} \frac{C^*_{XV}}{C_S} \qquad (1)$$

where X stands for the diffusing species either Si or Ge, D_{XI} and D_{XV} are the diffusivities of X due to interstitial and vacancy based mechanisms respectively, f_{XI} and f_{XV} are correlation factors of diffusion for each mechanism, C_S is the number of lattice sites in the crystal, C^*_{XI} and C^*_{XV} are the equilibrium concentrations of interstitials and vacancies in the SiGe alloy respectively, d_{XI} and d_{XV} are the diffusivities of X species' interstitials and vacancies in the SiGe alloy. The equilibrium concentration of a point defect species can be expressed as [4]:

$$\frac{C_Y}{C_S} = \theta_Y \exp\left(\frac{S_Y^f}{k}\right) \exp\left(\frac{-H_Y^f}{kT}\right) \qquad (2)$$

where H^f_Y is the enthalpy of formation of the point defect Y, S^f_Y is the disorder entropy not associated with configuration and is usually attributed to lattice vibrations, and θ_Y is the number of degrees of internal freedom of the defect on a lattice site. The diffusivities of the point defects in the SiGe alloy can be expressed in terms of an Arrhenius expression.

Self-diffusion in pure Ge is mediated only by vacancies [8]. On the other hand, Ge diffuses in Si by a combination of vacancy and interstitial diffusion mechanisms [9]. However, it has been shown that Ge diffusion is dominated by a vacancy mechanism at low temperatures (below 1050°C) [9]. At high temperatures (above 1050°C), Ge diffusion in Si is dominated by interstitial assisted mechanism. This causes a break in the Arrhenius behavior of Ge diffusion around the 1050°C temperature. It has to be noted that 2 [1, 2] of the 3 studies used for empirical fitting in our work were done at temperature lower than 1050°C. The only study [3] which swapped a wider temperature range (894-1263°C) did not observe a break in the Arrhenius behavior of the Ge diffusion in SiGe alloys. This can be partially attributed to the fact that Ge diffuses by vacancy mediated mechanisms in SiGe with x>=0.5 [10]. It can also be attributed to the limited number of data points reported at high temperatures. The end result is that for the purpose of this study, Ge diffusion for temperatures lower than 1050°C is only considered. Consequently, the diffusion process is proposed to be dominated by vacancy mechanism for all Ge concentrations below 1050°C.

Laitinen et al. [1] claim that the dominant mechanism for the diffusion of Ge in SiGe with Ge content under 25% is the interstitialcy mechanism while the dominant mechanism above this concentration is the vacancy mechanism. The fact that Ge diffusion is dominated by vacancies in both Si and Ge at low temperatures leaves very little room for the validity of the argument used by Laitinen et al.

The above arguments will lead us to reduce the equation (1) to:

$$D_{Xself} = D_{XV} = f_{XV} d_{XV} \frac{C^{*}_{XV}}{C_S} \tag{3}$$

Equation (3) is used instead of equation (1) in this paper to express self diffusivity in SiGe. This is mainly due to the experimental conditions of the reports in the literature. However, given the role of interstitials in the self diffusivity of Si and in the Ge diffusivity in Si, equation (3) is not valid for high temperatures (>1050°C for Ge and >900°C in Si-rich SiGe alloys). A more general expression based on equation (1) is required to explain the complete behavior of the self diffusivity of Ge and Si in SiGe alloys at a wider temperature range.

Using equations (2) and (3), a general expression for the self diffusion of Ge in SiGe can be obtained:

$$D^{SD}_{Ge} = D'_{Ge0} \exp\left(\frac{S^f_V(x)}{k}\right) \exp\left(-\frac{E'_{Ge}(x)}{kT}\right) \tag{4}$$

where D'_{Ge0} is the pre-exponential factor encompassing all the terms that are independent of temperature and Ge concentration in equations (2) and (3), x is the Ge fraction in the SiGe alloy, $S^f_V(x)$ is the disorder entropy of formation of vacancies which is a function of the Ge concentration and $E'_{Ge}(x)$ is the activation energy for the Ge self-diffusivity which is given by the sum of the enthalpy of formation and enthalpy of migration of the vacancies.

The vacancies in Si and in Ge can be viewed as two miscible gases which form a homogeneous mixture when mixed. The entropy of the mixture will be a linear function in the entropies of both gases with the constants proportional to their mole fraction [11]:

$$S^f_V(x) = S^f_{GeV} x + S^f_{SiV}(1-x) \tag{5}$$

where S^f_{GeV} and S^f_{SiV} are the disorder entropies of formation of vacancies in Ge and Si respectively. Similarly, the activation energy of mixing can be obtained from the thermodynamic theory of mixing two homogenous gases [11]:

$$E'_{Ge}(x) = E_{Ge} x + E_{Si}(1-x) \tag{6}$$

where E_{Ge} and E_{Si} are the activation energies for self diffusion in Ge and for Ge in Si respectively. Using equations (5) and (6), equation (4) can be rewritten as follows:

$$D^{SD}_{Ge} = D_{Ge0} \exp(\Delta S_{Ge} x) \exp\left(-\frac{E_{Ge} + \Delta E_{Ge} x}{kT}\right) \tag{7}$$

where $D_{Ge0}=D'_{Ge0}\exp(S^f_{SiV})$, $\Delta S_{Ge}=[S^f_{GeV}-S^f_{SiV}]/k$ and $\Delta E_{Ge}=[E_{Ge}-E_{Si}]$.

3 MODEL PARAMETERS

To obtain the actual values of D_{Ge}, ΔS_{Ge}, E_{Ge} and ΔE_{Ge}, the experimental results reported in the literature for Ge self diffusivity in SiGe alloys [1-3] were all plotted in one graph shown in figure 1. Ge self diffusivity at 43% Ge content as reported by Strohm et al. did not fit the trend shown in the graph for Ge self diffusivity and therefore was not included. Curve fitting was then done using a single equation (equation (7)) for all the data points. The result of the curve fitting is shown in figure 1. The best fit was obtained by using the following values for the model parameters:

$$D^{SD}_{Ge} = 310 \exp(-2.4x) \exp\left(-\frac{4.68\text{-}1.58x}{kT}\right) \tag{8}$$

Using equation (8), we can obtain expressions for the self diffusivity in pure Ge as well as the Ge diffusivity in pure Si. Those expressions can then be compared to the diffusivity equations reported in the literature. Ge diffusivity in pure Si can be easily obtained from equation (8) by setting x to zero:

$$D_{Ge} = 310 \exp\left(-\frac{4.68}{kT}\right) \tag{9}$$

Similarly, Ge self diffusivity in pure Ge can be obtained from equation (8) by setting x to 1:

$$D^{SD}_{Ge} = 28.12 \exp\left(-\frac{3.1}{kT}\right) \tag{10}$$

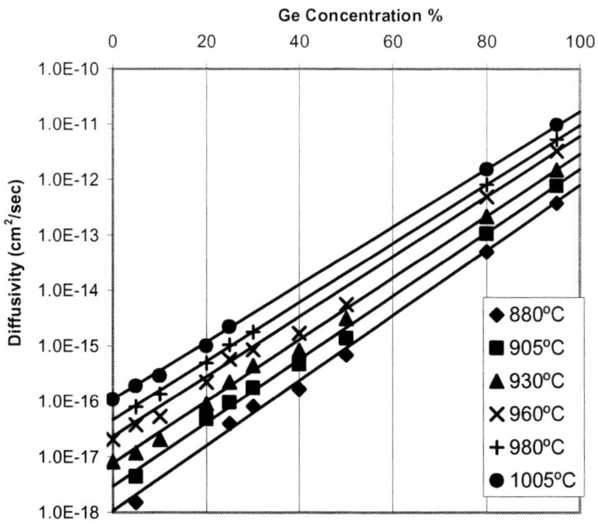

Figure 1: Ge diffusivity as predicted by our model (lines) compared to the experimental measurements done by Zangenberg et al. [2], Laitinen et al. for 80% Ge [1] and extrapolation from Strohm et al. measurements [3] for 5%, 25%, and 95% Ge content.

The prefactor and the activation energy for Ge diffusivity in Si are close to those reported by Zangenberg et al. [2] (D_0 is the same 310cm^2/sec and E_{Ge} was reported by Zangenberg et al. to be 4.65eV compared to our reported value 4.68eV). It was indicated by Zangenberg et al. that the reported value lies within the range of previously reported values. This serves as the first check for our model.

Table 1 shows a comparison between our expectation for the Arrhenius expression for self diffusivity in pure Ge and the reported values in the literature for the same quantity. It is clear from the table that our model predicts a close value for the self diffusivity in pure Ge to those reported in the literature. This serves as another check for the validity for our model.

D_0 (cm^2/sec)	E_{Ge} (eV)	Temperature Range (K)
7.8 (a)	2.97	1039-1201
32 (b)	3.1	1023-1156
44 (c)	3.14	1004-1189
10.8 (c)	3	1004-1189
24.8 (d)	3.14	822-1193
28.12 (e)	3.1	<1323

Table 1: Reported self diffusivities in Ge by (a) Letaw et al. [12], (b) Valenta and Ramasastry [13], (c) Widmer and GuntherMohr [14], (d) Vogel et al. [7] and (e) this work. After [7].

REFERENCES

[1] P. Laitinen, A. Strohm, J. Huikari, A. Nieminen, T. Voss, C. Grodon, I. Riihimäki, M. Kummer, J. Äystö, P. Dendooven, J. Räisänen, and W. Frank., Phys. Rev. Let. 89, 085902, 2002.

[2] N. R. Zangenberg, J. Lundsgaard Hansen, J. Fage-Pedersen, and A. Nylandsted Larsen, Phys. Rev. Let. 87, 125901, 2001.

[3] A. Strohm, T. Voss, W. Frank, J. Räisänen and M. Dietrich, Physica B 308-310, 542, 2001.

[4] P. M. Fahey, P. B. Griffin, and J. D. Plummer, Reviews of Modern Physics 61, 289, 1989.

[5] Mohammad Hasanuzzaman and Yaser M. Haddara, J. Mater. Sci.: Mater. Electron (Accepted)

[6] Mohamed A. Rabie, Yaser M. Haddara and Jacques Carette, J. Appl. Phys. 98, 074904, 2005.

[7] G Vogel, G Hettich and H Mehrer., J. Phys. C: Solid State Phys. 16, 6197, 1983.

[8] M. Werner, H. Mehrer and H. D. Hochheimer, Phys. Rev. B 32, 2930, 1985.

[9] P. Fahey, S. S. Iyer, and G. J. Scilla, Appl. Phys. Lett. 54, 843, 1989.

[10] P. Venezuela, G. M. Dalpain, Antonio J. R. da Silva, and A. Fazzio, Phys. Rev. B 65, 193306, 2002.

[11] Yunus A. Cengel and Michael A. Boles, "Thermodynamics: An Engineering Approach," McGraw-Hill, 648-653, 2002.

[12] Letaw H Jr, Portnoy W and Slifkin L, Phys. Rev. 102, 636, 1956

[13] Valenta M W and Ramasastry C, Phys. Rev. 106, 73, 1957.

[14] Widmer and Gunther-Mohr G R, Helv. Phys. Acta 34, 635, 1961.

Assessment of L-DUMGAC MOSFET for High Performance RF Applications with Intrinsic Delay and Stability as Design Tools

R.Chaujar*, R.Kaur*, M.Saxena**, M.Gupta* and R.S.Gupta*

* Semiconductor Devices Research Laboratory, Department of Electronic Science, University of Delhi, South Campus, New Delhi, India, rishuchaujar@rediffmail.com and ravneetsawhney13@rediffmail.com, rsgu@bol.net.in
**Department of Electronics, Deen Dayal Upadhyaya College, University of Delhi, Karampura, New Delhi, India, saxena_manoj77@yahoo.co.in

ABSTRACT

The paper assesses RF performance of Laterally amalgamated DUal Material GAte Concave MOSFET (L-DUMGAC) MOSFET using ATLAS device simulator. The L-DUMGAC MOSFET design integrates the advantages of dual material gate architecture with the concave MOSFET. In this work, TCAD assessment is done by performing AC simulations at very high frequencies, and evaluating the RF figure of merits in terms of maximum available power gain (Gma), stern stability factor (K) and intrinsic delay. Results reveal that L-DUMGAC architecture exhibits a significant improvement in Gma & K; and an appreciable reduction in intrinsic delay, in comparison to its conventional counterpart: SIngle Material Gate Concave (SIMGAC) MOSFET; hence, strengthening the idea of using L-DUMGAC for switching applications, thereby giving a new opening for high frequency wireless communications.

Keywords: ATLAS, L-DUMGAC, intrinsic delay, stability, and RF.

1 INTRODUCTION

Silicon device technology has become an attractive low cost solution for many high frequency personal communication products. With scaling of the device dimensions into the sub-50-nm regime, the transistors have achieved cutoff frequencies in the range of several GHz, making CMOS technology suitable for wireless communications and other RF applications [1-2]. However, undesirable short-channel effects, the mobility degradation and increased parasitic capacitances drastically reduce the device transconductance, voltage gains and noise performance making the scaled technologies unsuitable for analog/RF applications [3].

Concave MOSFETs [4-5] are known to alleviate the short channel effects and the hot carrier effects appreciably due to the presence of a groove separating the source and the drain regions; and consequentially their respective depletion regions. Further integration of concave MOSFETs with dual material gate architecture [6-7] enhances the driving current capability and suppresses the short channel effects. The resultant integrated device structure is referred to as L-DUMGAC MOSFET [8-9].

This work focuses on the TCAD assessment of L-DUMGAC MOSFET for high performance RF applications in terms of RF figure of merits such as stern stability factor, maximum available power gain and intrinsic delay. Further, the impact of various technological variations such as gate length and negative junction depth (NJD) is explored on these device metrics for improved performance. All simulations have been performed using ATLAS [10] device simulation software. The models activated in simulation comprise the inversion layer Lombardi CVT mobility model along with Shockley–Read–Hall (SRH) and Auger recombination model for minority carrier recombination.

2 SIMULATION RESULTS AND DISCUSSION

Maximum available power gain (Gma) and stability are two of the most important considerations for use in Low Noise Amplifier (LNA) and RF amplifier design.

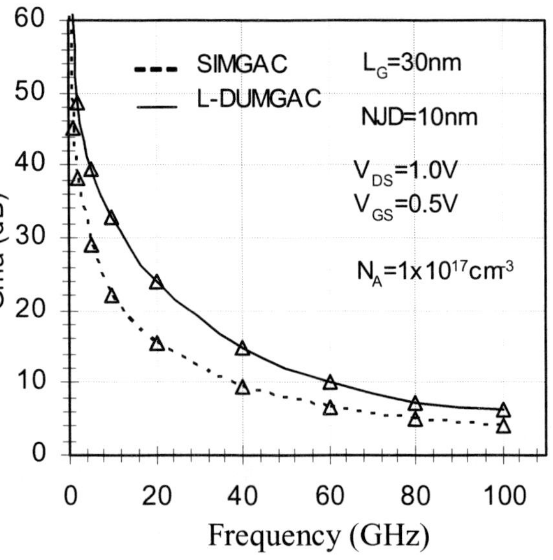

Figure 1: Maximum Available Power Gain, Gma, for SIMGAC & L-DUMGAC MOSFET designs.

Fig.1 reflects the performance enhancement of L-DUMGAC in terms of Gma which indicates the maximum theoretical power gain that can be expected from the device. Improvement in Gma is experienced with L-DUMGAC due to lower parasitics and improved gate control.

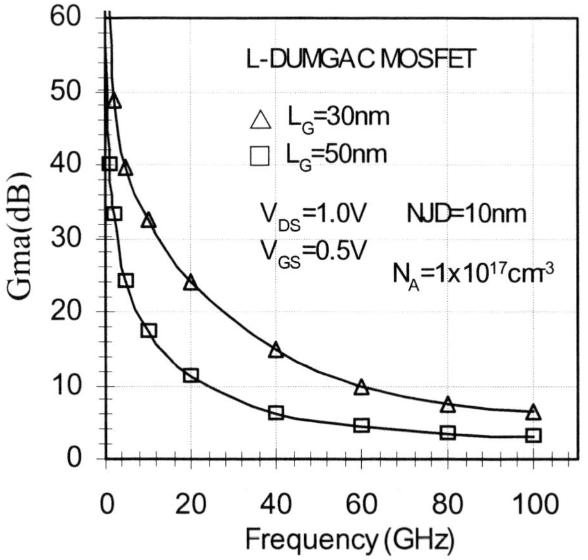

Figure 2: Maximum Available Power Gain, Gma, for L-DUMGAC MOSFET design giving L_G variation.

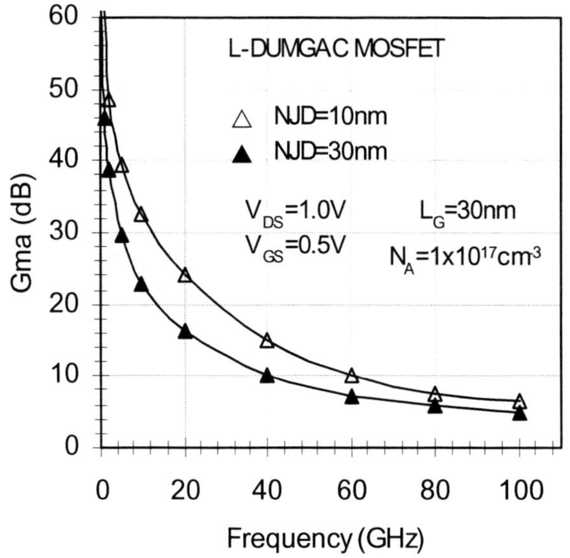

Figure 3: Maximum Available Power Gain, Gma, for L-DUMGAC MOSFET design giving NJD variation.

As the gate length is reduced, the performance metrics further enhance, owing to improved drive current as is depicted from Fig.2. Further, from Fig.3, it is reflected that reduction in NJD also leads to the power gain improvement as a result of higher gate control over the channel. Fig.4 explains the stability of L-DUMGAC and conventional SIMGAC MOSFETs.

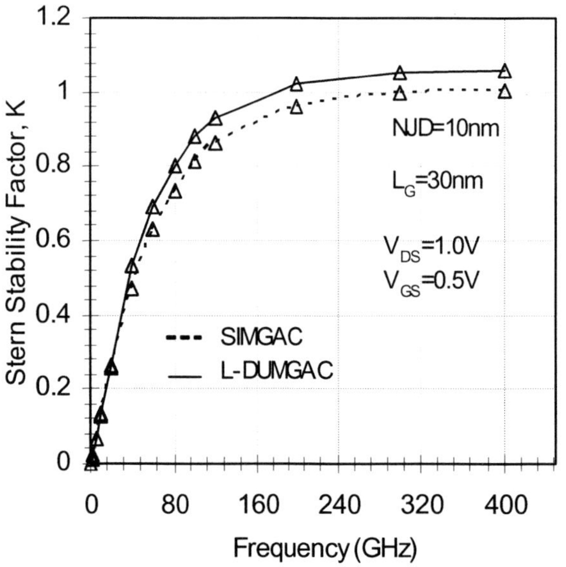

Figure 4: Stern Stability Factor, K, for SIMGAC & L-DUMGAC MOSFET designs.

The Stern Stability Factor (K) predicts the absolute stability of a transistor and is usually less than 1 (i.e. unstable) at LF and higher than 1 (i.e. stable) at HF. As is clear from the figure, K is appreciably higher (or greater than 1), for L-DUMGAC in comparison to SIMGAC where K is just approaching 1; reflecting an oscillatory nature of SIMGAC MOSFET. If K is less than 1, the circuit is potentially unstable and a simultaneous input and output impedance match cannot be obtained for maximum power transfer.

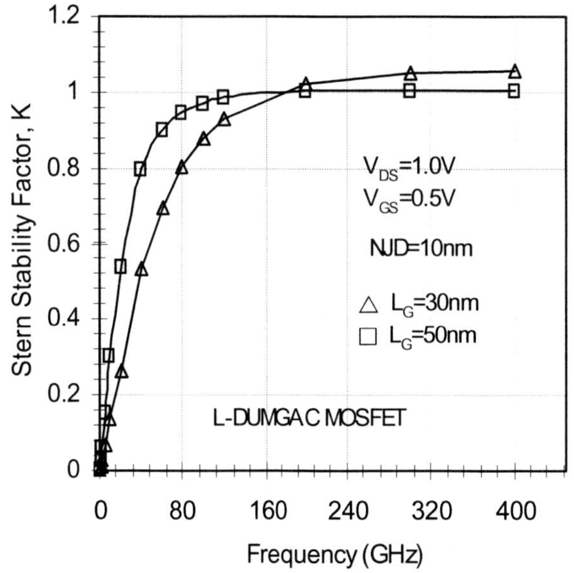

Figure 5: Stern Stability Factor, K, for L-DUMGAC MOSFET design giving L_G variation.

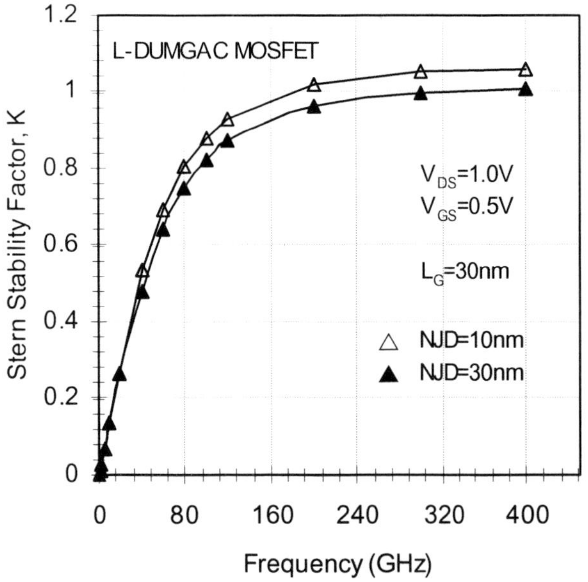

Figure 6: Stern Stability Factor, K, for L-DUMGAC MOSFET design giving NJD variation.

However, if K is greater than 1, the circuit is stable, thereby maximizing the power transfer. Device miniaturization, however, leads to instability, as is predicted from Fig.5, owing to reduced gate control as a result of overlapping of the source and drain fields in the scaled devices. Further, Fig.6 reflects that NJD reduction perks up the K-factor. This is due to increased gate controllability of the device over the channel which hence, results in the maximum power transfer from source to load. Fig.7 compares the SIMGAC and L-DUMGAC devices using the intrinsic delay metric. Intrinsic delay of the L-DUMGAC structure is significantly lower than the SIMGAC MOSFET i.e. incorporation of the DMG architecture leads to an appreciable reduction of intrinsic delay by 63.8%. Further, increase in the gate length intensifies the intrinsic delay resulting in device switching

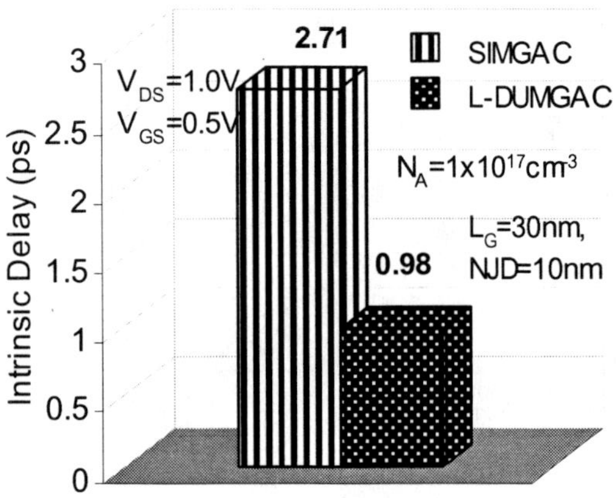

Figure 7: Intrinsic Delay Metric Evaluation for SIMGAC & L-DUMGAC MOSFET designs.

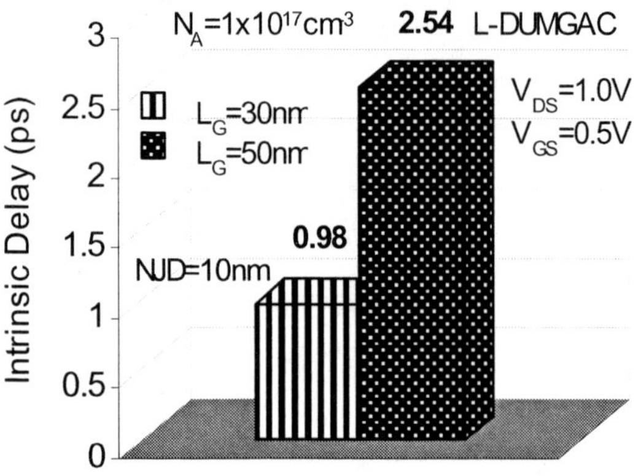

Figure 8: Intrinsic Delay Metric Evaluation for L-DUMGAC MOSFET design giving L_G variation.

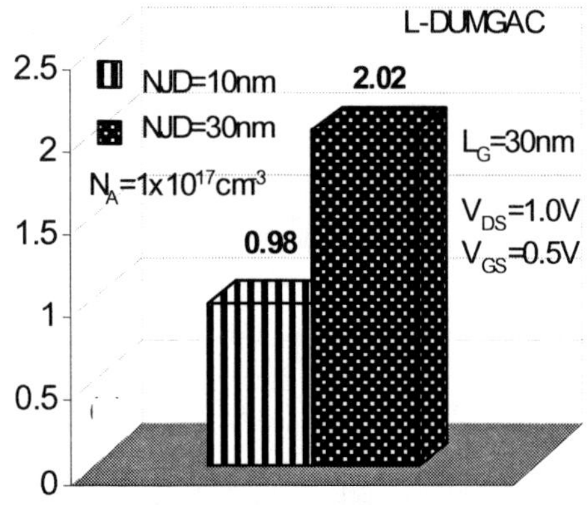

Figure 9: Intrinsic Delay Metric Evaluation for L-DUMGAC MOSFET design giving NJD variation.

deterioration as is clear from Fig.8. Moreover, with decrease in NJD from 30nm to 10nm, the intrinsic delay improves by 51.4% for L_G=30nm as demonstrated in Fig.9. This is mainly due to the decrease in potential barriers at the corners leading to enhanced carrier transport. Thus, L-DUMGAC leads to a decrease in the parasitic behavior resulting in high speed performance and, hence, proves to be a promising candidate in terms of switching speed.

3 CONCLUSION

In this paper, we have systematically investigated the effect of dual material gate architecture on the concave MOSFET for improved transistor RF performance. It has been observed that L-DUMGAC device exhibits significantly improved power gain with an appreciably low intrinsic delay; hence, proving its effectiveness in RF/wireless applications. Moreover, L-DUMGAC MOSFET is found to be more stable than the SIMGAC MOSFET; reflecting improved gate controllability and maximum power transfer from source to load. These results therefore show the potential of L-DUMGAC technology for low noise, high-performance System-On-Chip and RF applications.

4 ACKNOWLEDGMENTS

Authors are grateful to Defense Research and Development Organization (DRDO), Ministry of Defense, Government of India, and one of the authors (Rishu Chaujar) is grateful to University Grant Commission (UGC) for providing the necessary financial assistance to carry out this research work.

REFERENCES

[1] M. Saito, M. Ono, R. Fujimoto, H. Tanimoto, T. I. N. Yoshitomi, H. S. O. T. Momose, and H. Iwai, "0.15 μm RF CMOS technology compatible with logic CMOS for low-voltage operation," IEEE Trans. Electron Devices, 45, 737–742, 1998.

[2] V. Kilchytska, A. Neve, L. Vancaillie, D. Levacq, S. Adriaensen, H. Van Meer, K. DeMeyer, C. Raynaud, M. Dehan, J. P. Raskin, and D. Flandre, "Influence of device engineering on the analog and RF performances of SOI MOSFETs," IEEE Trans. Electron Devices, 50, 577–588, 2003.

[3] D.Buss, "Device issues in the integration of analog/RF functions in deep sub-micron digital CMOS," IEDM Tech. Dig., 423–426, 1999.

[4] P.H.Bricout and E.Dubois, "Short-Channel Effect Immunity and Current Capability of Sub-0.1-Micron MOSFET's Using a Recessed Channel", IEEE Trans. Electron Devices, 43, 1251-1255,1996.

[5] H.Ren and Y.Hao, "The influence of geometric structure on the hot-carrier-effect immunity for deep-sub-micron grooved gate PMOSFET", Solid-State Electronics, 46, 665-673, 2002.

[6] W.Long, H.Ou, J.M.Kuo and K.K.Chin, "Dual-Material Gate (DMG) Field Effect Transistor", IEEE Trans. Electron Devices, 46, 865-70, 1999.

[7] X.Zhou, "Exploring the novel characteristics of hetero-material gate field-effect transistors (HMGFET's) with gate-material engineering," IEEE Trans. Electron Devices, 47, 113–120, 2000.

[8] R.Chaujar, R.Kaur, M.Saxena, M.Gupta and R. S. Gupta, "On-State and Switching Performance Investigation of Sub-50nm L-DUMGAC MOSFET Design for High-Speed Logic Applications", International Semiconductor Device Research Symposium (ISDRS-2007), Maryland, USA, 1892-1893, 2007.

[9] R.Chaujar, R.Kaur, M.Saxena, M.Gupta and R. S. Gupta, "Laterally amalgamated DUal Material GAte Concave (L-DUMGAC) MOSFET For ULSI", Microelectronic Engineering, 85, 566-576, 2008.

[10] ATLAS: Device simulation software, 2002.

A Continuous yet Explicit Carrier-Based Core Model for the Long Channel Undoped Surrounding-Gate MOSFETs

Lining Zhang[1, 2], Jin He[1, 2], Jian Zhang[2], and Jie Feng[2]

[1] School of Computer& Information Engineering, Peking University Shenzhen Graduate School, Shenzhen, 518055, China
[2] Institute of Microelectronics, EECS, Peking University, Beijing, 100871, China
Tel: 86-0755-26032104 Fax: 86-0755-26035377 Email: frankhe@pku.edu.cn

ABSTRACT

An explicit carrier-based core model for the long channel undoped surrounding-gate MOSFETs is presented in the paper. An analytic approximation solution to the carrier concentration is developed from a simplified Taylor expansion of the exact solution of Poisson's equation of the surrounding-gate MOSFETs, instead of resorting to the Newton-Raphson numerical iterative. The analytic approximation not only gives accurate dependences of the carrier concentration on the geometry structures and bias, compared with the Newton-Raphson numerical method, but also is used to develop an explicit current-voltage model of the surrounding-gate MOSFETs combined with Pao-Pah current formulation. The presented explicit model is found to be computationally more efficient than the previous numerical Newton-Raphson iterative while more accurate than the previously published explicit model.

Keywords: device physics, compact model, surrounding-gate MOSFET, carrier-based model.

1 INTRODUCTION

The SRG-MOSFET based circuit design is contingent on the precision of the explicit transistor model involved in circuit simulation. In addition to the precise description of the SRG device characteristics, computation efficiency remains an important constraint for the SRG compact model to efficient circuit simulation. In the recent years, there were some reports investigating the compact models of the non-classical SRG MOSFET device characteristics [1-4]. D.Jimenez and his co-workers developed the β -based model and the result matched well with the 3-D simulation [2]. Following a quite different method, Jin He's group developed a carrier-based non-charge-sheet analytic model for SRG MOSFETs directly from both Poisson equation solution and Pao-Sah current formulation [3-4]. These models provide a fundamental yet solid basement for engineers and circuit designers to understand the SRG device physics and characteristics. However, most developed SRG MOSFET models rely on numerical iteration or table-lookup to solve the fundamental nonlinear implicit equations between the input voltage and the state variables such as β, potentials and carrier [2-4]. We noted that B. Iñíguez et al. presented an explicit charge-based compact model for SRG-MOSFET devices based on the threshold voltage concept and smooth functions in [5]. However, it is observed from the accuracy test that this explicit model is not good enough for high precise requirement of the SRG-MOSFET core model because it gives the prediction with the relative drain current error up to 32% error in the moderate inversion region which is very important for non-classical MOSFET low voltage and low power circuit design.

A SRG MOSFET compact model for circuit simulations requires accurate yet computation efficient core framework. Thus, the aim of SRG-MOSFET compact model formulation is to conflate an accurate description of device characteristics with high computational efficiency. In this brief, an explicit carrier-based SRG MOSFET current-voltage model has been presented from an accurate yet analytic approximation to the carrier concentration solution. Compared with the previous explicit model [5].

2 EXPLICIT MODEL DERIVATION

The undoped SRG MOSFET structure and coordinate system used in the analysis are shown in Fig.1. Following a carrier-based approach [2-4], a complete Poisson equation solution in terms of the silicon center mobile carrier concentration as a function of the gate voltage, quasi Fermi potential, and the geometry structure is derived as shown in [4].

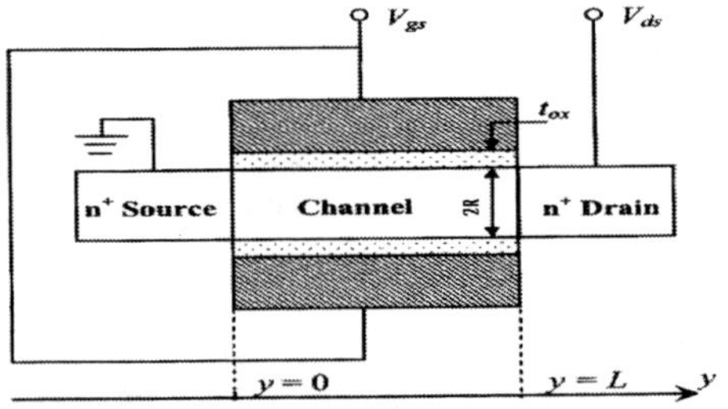

Fig.1 Schematic diagram of a SRG MOSFET

$$V_G-\Delta\phi_i-V_{ch}=$$

$$\frac{kT}{q}\ln\left(\frac{n_0}{n_i}\right)-\frac{2kT}{q}\ln\left[1-\frac{R^2}{8L_i^2}\frac{n_0}{n_i}\right]+\frac{R^2\varepsilon_{si}kT\ln\left[1+\frac{t_{ox}}{R}\right]}{2qL_i^2\varepsilon_{ox}}\frac{\frac{n_0}{n_i}}{\left[1-\frac{R^2}{8L_i^2}\frac{n_0}{n_i}\right]} \quad (1)$$

where $L_i = \sqrt{\frac{\varepsilon_{si} kT}{q^2 n_i}}$ is the Debye length of the intrinsic silicon materials, n_i is the intrinsic silicon concentration with the unit of cm^{-3}, ϕ_s is the electrostatics in $Volt$ at the silicon surface. kT/q and V_{ch} are the thermal voltage and the quasi-Fermi-potential in $Volt$, respectively. n_0 is the induced electron concentration at the silicon center with the unit cm^{-3}. t_{ox} and R are the gate oxide thickness and the SRG MOSFET silicon radius in cm, respectively.

It is evident that the mobile carrier concentration on the silicon film center can be obtained from Eq. (1) for the given gate voltage, quasi Fermi potential and structure parameters such as the oxide layer thickness and the silicon film thickness. However, Eq. (1) is a nonlinear implicit equation, which needs a very accurate solution for complete current-voltage and capacitance model development. Traditionally, this equation was solved by a Newton-Raphson iterative routine or table-lookup method in terms of some intermediate variables such as the carrier concentration and β [1-4]. The numerical iterative and table-lookup methods, however, are not preferred for a compact model due to the need of extensive computation time and memory to store intermediate data. Thus, an accurate analytic approximate solution for the carrier concentration is desired for the explicit current-voltage model development for the SRG MOSFETs.

Here, we follow a recently developed method to derive an analytic approximation solution to the carrier concentration based on a simplified Taylor expansion formulation [6]. Firstly, we transformation Eq. (1) into a normalized formulation for simplicity

$$f(i) = g - \ln\left[i + i^2\right] - \lambda i = 0 \tag{2}$$

where

$$s = \frac{R^2}{8L_i^2}, \lambda = \frac{4\varepsilon_{si} \ln\left[1 + \frac{t_{ox}}{R}\right]}{\varepsilon_{ox}}, i = \frac{R^2}{8L_i^2}\frac{n_0}{n_i}\left[1 - \frac{R^2}{8L_i^2}\frac{n_0}{n_i}\right]^{-1} \tag{3}$$

$$g = \frac{q\left(V_G - \Delta\phi - V_{ch}\right)}{kT} + \ln s \tag{4}$$

In order to obtain an analytic approximation of (2), we use the perturbation method to get the correct functions for an initial guess that is accurate enough. As shown in [6, 7], the correction function can be derived out from the modified Taylor expansion of a function

$$\delta = -\frac{f(i)}{f' - \frac{f''(i) f(i)}{2f'(i)}} \tag{5}$$

(5) will be used to build an analytical approximation solution. The exact first and second derivatives are obtained from (2)

$$f'(i) = \frac{1}{i} + \frac{1}{1+i} + \lambda \tag{6}$$

$$f''(i) = -\frac{1}{i^2} - \frac{1}{(1+i)^2} \tag{7}$$

And then substituting them into (5) gives

$$\delta = -\frac{f(i)}{f' + \frac{f(i)}{2i}\frac{(1+i)^2 + i^2}{(1+i)^2 + \lambda i(1+i)^2 + i(1+i)}} \tag{8}$$

Here, we need a continuous and accurate initial guess before the use of (8), which can be obtained from (2). Since i has non maximum boundary and the term $\ln[1+i]$ has a little contribution in (2) either for the sub-threshold or the strong inversion region. Thus, we can review (2) as a W-Lambert function by neglecting the term $\ln[1+i]$ and it has a simple approximation for the principal branch

$$i_b = \frac{\ln(1 + \lambda e^g)}{\lambda} \tag{9}$$

In physics, i_b presents an asymptotic approximation developed. If δ is a relative small refinement and we define

$$f_0 = g - \ln\left[i_b + i_b^2\right] - \lambda i_b \tag{10}$$

Then δ is obtained from (8). We define

$$i_0 = i_b + \delta \tag{11}$$

As a result, we have

$$i_0 = i_b - \frac{f_0}{f' + \frac{f_0}{2i_b}\frac{(1+i_b)^2 + i_b^2}{(1+i_b)^2 + \lambda i_b(1+i_b)^2 + i_b(1+i_b)}} \tag{12}$$

Further, we define

$$f_1 = g - \ln\left[i_0 + i_0^2\right] - \lambda i_0 \tag{13}$$

$$i = i_0 + \alpha \tag{14}$$

Finally, substituting (13), the first and second derivatives as a function of i_0 into (8) to get the symbol expression of α, we obtain the accurate analytic approximation of the normalized carrier concentration

$$i = i_0 - \frac{f_1}{f'(i_0) + \dfrac{f_1}{2i_0} \dfrac{(1+i_0)^2 + i_0^2}{(1+i_0)^2 + \lambda i_0(1+i_0)^2 + i_0(1+i_0)}} \quad (15)$$

This analytic approximation requires computation of three logarithms, e.g. (9), (10) and (13) and one exponential (9), however, with no iteration. As long as i is obtained, the electron concentration can be calculated based on (3).

Once the carrier concentration solution is obtained, the Poisson equation solution (1) is coupled to the Pao-Sah current formulation can result in the explicit current-voltage and capacitance-voltage model. For a given V_g, n_0 is solved from (1) as a function of V_{ch}. Following Pao-Sah current formulation [7], integrating $I_{ds}dy$ from the source to the drain and expressing V_{ch}/dy as $(dV_{ch}/dn_0)(dn_0/dy)$, the drain current is written as

$$I_{DS} = \mu\frac{W}{L}\int_0^{V_{DS}} Q_i(V_{ch})dV = \mu\frac{W}{L}\int_{n_{0s}}^{n_{0d}} Q_i(n_0)\frac{dV_{ch}}{dn_0}dn_0 \quad (16)$$

where n_{0s} and n_{0d} are solutions of (1) corresponding to $V_{ch}=0$ and $V_{ch}=V_{ds}$, respectively. Note that the dV_{ch}/dn_0 can also be expressed as a function of n_0 by differentiating (1). Substituting these factors into (16), integrating can be performed analytically to yield:

$$I_{ds} = \mu\frac{2\pi\varepsilon_{si}}{L}\left(\frac{2kT}{q}\right)^2 F[n_0]\Big|_{n_{0d}}^{n_{0s}} \quad (17)$$

where

$$F[n_0] = \lambda\left(\frac{R^2 n_0}{8L_i^2 n_i}\left[1-\frac{R^2 n_0}{8L_i^2 n_i}\right]^{-1}\right)^2 /2 + 2\left(\frac{R^2 n_0}{8L_i^2 n_i}\left[1-\frac{R^2 n_0}{8L_i^2 n_i}\right]^{-1}\right) - \ln\left[1+\frac{R^2 n_0}{8L_i^2 n_i}\left[1-\frac{R^2 n_0}{8L_i^2 n_i}\right]^{-1}\right] \quad (18)$$

3 RESULTS AND DISCUSSION

The accuracy of the electron concentration calculation is one key factor for the SRG MOSFET performance prediction and circuit simulation. Comparison of the new analytic approximation (15) with the iterative numerical results of (2) for the carrier concentration calculation is shown in Fig.2 for the different silicon film radius. We note that an excellent accuracy is achieved for the whole operation region for different silicon radius, e.g. $R = 5nm$ or $R = 30nm$.

In order to further demonstrate the accuracy of the developed analytic approximation, the relative error of the electron concentration prediction with the different methods is also shown in Fig.3 for the different silicon film thickness. It is easily found that the relative error introduced by the new analytic approximation (15) is under

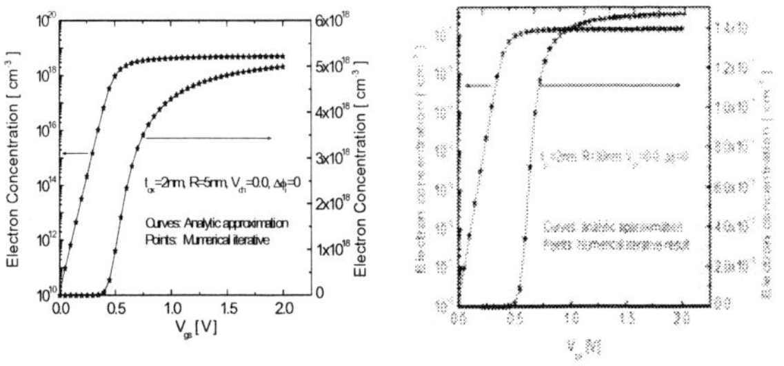

(a) (b)

Fig.2 Comparison of the electron concentration versus the gate voltage obtained from the analytic approximation (Solid curves) and the fully numerical Newton-Raphson method (points) in undoped cylindrical surrounding-gate MOSFETs with the midgap gates for silicon radius R=5nm(a) and R=30nm(b).

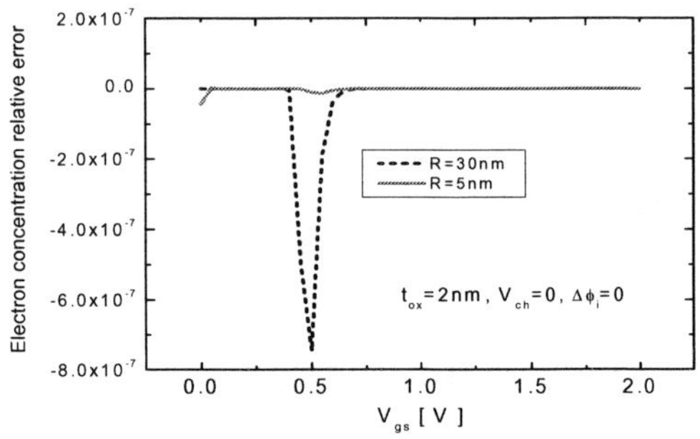

Fig.3 Relative error of the electron concentration calculation based on the analytic approximation and the Newton-Raphson iterative method for the different silicon radius in undoped cylindrical surrounding-gate MOSFETs with the midgap gates.

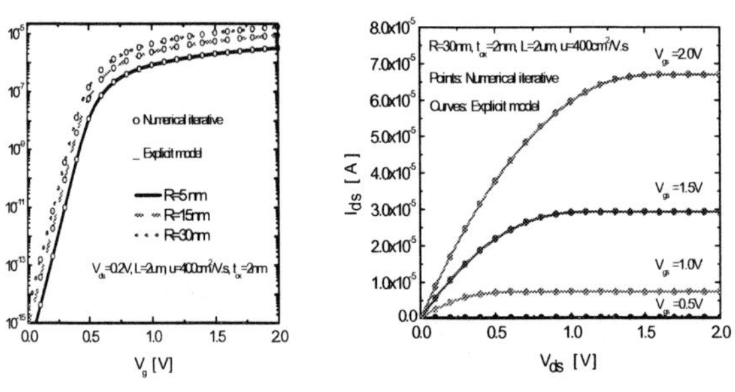

(a) (b)

Fig.4 (a) Ids-Vgs curves and (b) Ids-Vds curve calculated from the explicit model (solid curves), compared with the numerical iterative results (points).

10^{-6} order. Such a precise is satisfactory enough for the compact modeling development of the SRG MOSFETs.

Fig. 4(a) shows the good comparison of the drain current versus gate voltage between the explicit model and the iterative method for three silicon body radius, e.g. $R = 5nm$, $15nm$ and $30nm$. It is easily found that the explicit prediction shows in agreement with the iterative results. Again, it is found that the sub-threshold current is almost proportional to the square of the silicon body diameter because of the "volume inversion" effect. To optimize the device performance, the silicon film body radius should be reduced as much as possible, e.g., a nanowire body to be used, to suppress the off current although it is difficult in fabrication process. Fig. 4(b) is I_{ds}-V_{ds} curves calculated from the explicit model (solid curves), compared with the numerical iterative results (dots). Both match well in both the linear and the saturation region.

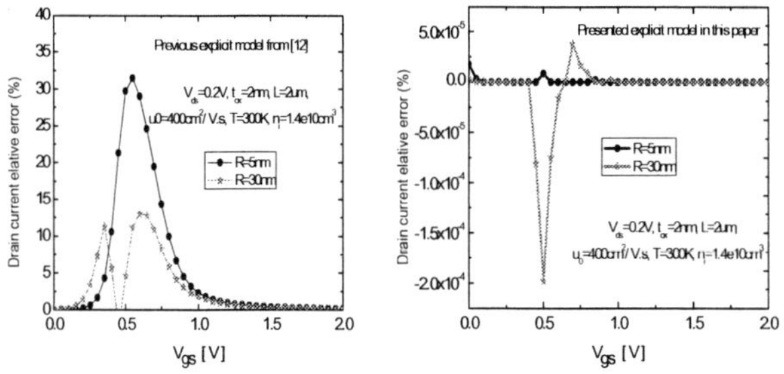

(a) (b)

Fig.5 Predicted drain current relative error from B. Iñíguez et al's explicit model in reference [5] (a) and from the presented explicit model in this paper (b), compared with the iterative method.

We noted that B. Iñíguez et al. presented an explicit charge-based compact model for SRG-MOSFET devices based on the threshold voltage concept and smooth functions in [5]. One interesting issue is to compare the accuracy and computation efficiency of two different explicit SRG models. Such a relative drain current error comparison is shown in Fig.6 for two different silicon body radius. It is observed from Fig.5 that B. Iñíguez et al's explicit SRG-MOSFET model gives the drain current prediction with the different relative error, e.g. the maximum relative error is 13% for $R = 30nm$ while up to about 32% for $R = 5nm$.

4 CONCLUSIONS

An explicit carrier-based model for the undoped surrounding-gate MOSFETs has been developed in this brief by an accurate yet analytic carrier concentration approximation from Poisson equation solution of the SRG MOSFETs. The accuracy of the electron concentration calculation has also been verified compared with the result of the iterative method. It is shown that the predicted current-voltage curves are in complete agreement with the fully numerical iterative results without any fitting parameter. Compared with the previous explicit model and numerical iterative method, the presented explicit model requires no numerical iteration but more accurate and computation efficient, thus it is more suitable to implement SRG MSOFET core model into the circuit simulators for circuit design and application.

ACKNOWLEDGEMENT

This work is support by the National Natural Science Foundation of China (NSFC: 90607017), the Competitive Earmarked Grant HKUST6289/04E from the Research Grant Council of Hong Kong SAR and the International Joint Research Program (NEDO Grant) from Japan under the Project Code NEDOO5/06.EG01.

REFERENCES

[1] S.L.Jang and S.S.Liu, "An analytical surrounding gate MOSFET model." Solid State Electronics, vol.42, No.5, pp.721-726, 1998.

[2] D. Jiménez, B. Iñíguez, J. Suñé, L. F. Marsal, J. Pallarès, J. Roig and D. Flores, " Continuous analytic I-V model for surrounding-gate MOSFETs," IEEE EDL-25, no. 8, pp. 571-573, 2004.

[3] Jin He, Mansun Chan, "physics based analytical solution to undoped cylindrical surrounding-gate (SRG) MOSFETs." 15th IEEE Int. Conf. on Devices, Circuits and Systems, Nov.3-5, pp. 26-28, 2004.

[4] Jin He, Xing Zhang, Ganggang Zhang, Mansun Chan, and Yangyuan Wang, "A carrier-based DCIV model for long channel undoped cylindrical surrounding-gate MOSFETs", *Solid-State Electronics*, Vol. 50, Issue 3, pp. 416-421, March 2006.

[5] B. Iñíguez, D. Jiménez, J.Riog, H.A.Hamid, L. F. Marsal, and J. Pallarès, "Explicit continuous model for long channel undoped surrounding-gate MOSFETs." IEEE Trans on Electron Devices, TED-52, No.8, pp. 1868-1873, 2005.

[6] Jin He, Min Fang, Bo Li, Yu Cao, "A new analytic approximation to general diode equation." *Solid-State Electronics*, vol. 50, 1371–1374, *2006.*

[7] H. C. Pao and C. T. Sah, "Effects of diffusion current on characteristics of metal-oxide (insulator)-semiconductor transistors." *Solid-State Electronics*, vol. 9, pp. 927-937, 1966.

Diode Parameter Extraction by a Linear Cofactor Difference Operation Method

Chenyue Ma[1,2], Bo Li[1], Yu Chen[1] , Lining Zhang[2], Jin He[1,2], and Xing Zhang[1,2]

[1] School of Computer & Information Engineering, Peking University Shenzhen Graduate School, Shenzhen, 518055, China

[2] Institute of Microelectronics, EECS, Peking University, Beijing 100871, P. R. China

Tel: 86-10-62765916 Fax: 86-10-62751789 Email: frankhe@pku.edu.cn

ABSTRACT

The direct extraction of the key static parameters of a general diode by the new method named Linear Cofactor Difference Operator (LCDO) method has been carried out in this paper. From the developed LCDO method, the extreme spectral characteristic of the diode voltage versus current curves has been revealed, and its extreme positions are related to the diode characteristic parameters directly. Two different diodes are applied and the related characteristic parameters such as the reverse saturation current, the series resistance and non-ideality factor have been extracted directly and the results have also been discussed.

Keywords: LCDO, diode, parameter extraction, ideality factor, series resistance.

1 INTRODUCTION

The static parameters of the diode such as series resistance, ideality factor and reverse saturation current play an important role in determination of the terminal current in the simulation and modeling of devices. Hence, a number of methods have been proposed to extract the device parameters of diodes [1-4]. But these methods often rely on the complex algorithm and various approximations.

In this paper, a novel parameter extraction method, named Linear Cofactor Difference Operator (LCDO) has been applied to extract the reverse saturation current, ideality factor and series resistance of the practice diodes. From this method, the unique extreme spectral characteristics of the current-voltage of the diode have been revealed, and then the related diode parameters could be determined from these extreme spectral characteristics. Its mathematical simplicity and physical concept clearness are the main advantages of this method over the traditional approaches, thus this method will be a useful tool in the analysis of the diode characteristics and the extraction of the diode static parameters, as shown in the following discussion.

2 LCDO METHOD APPLICATION

As shown in [5], If a function *f(x)* is strictly monotonic, non-linear, continuous and differentiable over region (x_0, x_1), there definitely exists a point x_p, $x_0 < x_p < x_1$, so that

$$G(x_p) = \frac{\partial G}{\partial x}\Big|_{x=x_p} = 0 \tag{1}$$

Where

$$G(x) = \Delta LCDO(x) = b + K_p x - f(x) \tag{2}$$

is the linear cofactor difference of the measured *f(x)* and $\Delta LCDO(x)$ is the linear cofactor difference operator, *b* and K_p are the LCDO intersection and linear factor, respectively.

The constants b and K_p can be determined via equations

$$G(x_1) = \mathrm{b} + K_p x_1 - f(x_1) = 0 \tag{3}$$

$$G(x_0) = \mathrm{b} + K_p x_0 - f(x_0) = 0 \tag{4}$$

The detail description of this LCDO method principle has been found in [11-13]. Here we apply the application of this method to the extraction of the static parameter of a general diode.

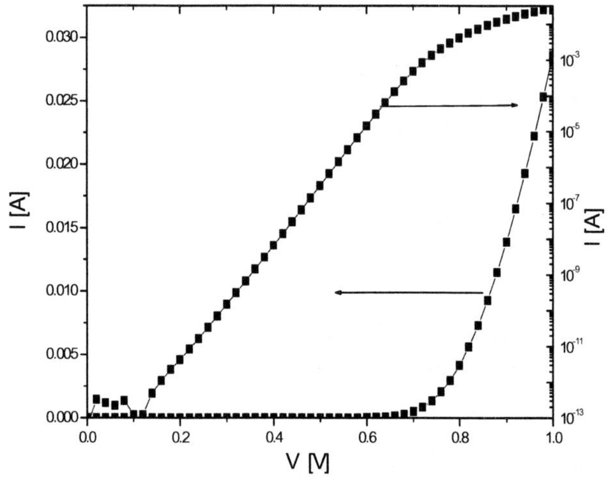

Fig 1 Measured I-V Characteristics of D1.

A measured diode current-voltage characteristics are shown in Fig 1. According to the semiconductor device physics, the current of a diode is frequently modeled by the following single exponential equation [1-5]:

$$I_d = I_s\left[\exp\left(\frac{V - R_s I_d}{nV_t}\right) - 1\right] \quad (5)$$

Where I_s is the reverse saturation current, R_s is the series resistance, $V_t=K_BT/q$ is the thermal voltage, and n is the diode ideality factor.

In Equation (1), the term (–1) is negligible for the forward bias condition; therefore, the Equation (5) can be rewritten as

$$V = nV_t \ln\left(\frac{I_d}{I_s}\right) + I_d R_s \quad (6)$$

The LCDO method is applied onto Equation (6) with b=0 and then the following equation is obtained

$$\Delta lcdoV(I_d) = K_p I_d - nV_t \ln\left(\frac{I_d}{I_s}\right) - I_d R_s \quad (7)$$

Where K_p is a linear cofactor difference factor.

Therefore, we define

$$\frac{\partial \Delta lcdoV(I_d)}{\partial I_d}\Big|_{I_d = I_{dP}} = 0 \quad (8)$$

The extreme spectral of the linear cofactor difference diode voltage versus the current can be obtained. Substituting Equation (8) into Equation (7) at the extreme position point I_{dP}, we can obtain

$$R_s = K_p - \frac{nV_t}{I_{dP}} \quad (9)$$

For a given diode current versus voltage curve, the diode ideality factor can be obtained from Equation (9) by using two different linear cofactor difference operator factors, K_{p1} and K_{p2}.

$$n = \frac{I_{dP1} I_{dP2}(K_{p1} - K_{p2})}{V_t(I_{dP2} - I_{dP1})} \quad (10)$$

Where I_{dp1} and I_{dp2} are current values in the linear cofactor difference diode voltage extreme point positions corresponding to the two different factors K_{p1} and K_{p2} respectively. Consequently, R_s can be determined from Equation (9) and then I_s from Equation (5).

3 RESULTS AND DISCUSSION

In order to apply this method to the practical diode, the measurement of the current-voltage (I-V) characteristics of two different diodes (D1 and D2) has been carried out first. The resultant data are taken at room temperature, using a semiconductor parameter analyzer HP-4156 with a step voltage of 0.02V. Fig 2 shows the linear cofactor difference voltage characteristics under the conditions of the difference linear cofactor difference factor K_p.

$K_{p1} = 43.6$, $K_{p2} = 55.76$, $K_{p3} = 76.5$

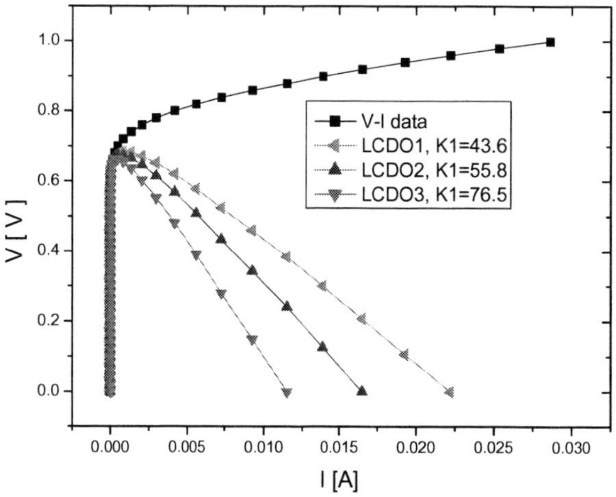

Fig 2 Diode voltage linear cofactor difference versus current for different values of K_p.

These extreme points and corresponding linear cofactor difference diode current are shown in Table 1.

Table 1 LCDO extreme points and linear cofactor difference diode voltage of D1 with different LCDO factors.

Linear cofactor difference operator factor K_p	Extreme point and corresponding LCDO diode current	
	Extreme point I_{dP} (*A*)	LCDO voltage (*V*)
43.6	8.36×10^{-4}	0.684
55.8	8.36×10^{-4}	0.673
76.5	34.8×10^{-4}	0.663

Based on the obtained extreme spectral peak magnitude and positions, the related diode static parameters can be easily extracted from the present formula.

From the extreme points and linear cofactor difference values at these extreme points, we obtain the consistent ideality factor n=1.2 for D1 from Equation (10) under difference combination of K_p. Then we can obtain the series resistance $R_s = 6.28\Omega$ from Equation (9) with any known K_p, n and I_d, even if the LCDO peak values and peak positions are different for different K_p values. Adjusting Equation (5) to the experimental values, we get $I_s = 9.4\times10^{-14} A$

Fig 3 shows the measured and extracted I-V characteristics of D1, which represents that the calculated line matches the measured one. The deviation in the small current region between both is due to the fact that the practical I-V characteristic of this diode does not follow the single exponential Equation (5) in the small current region.

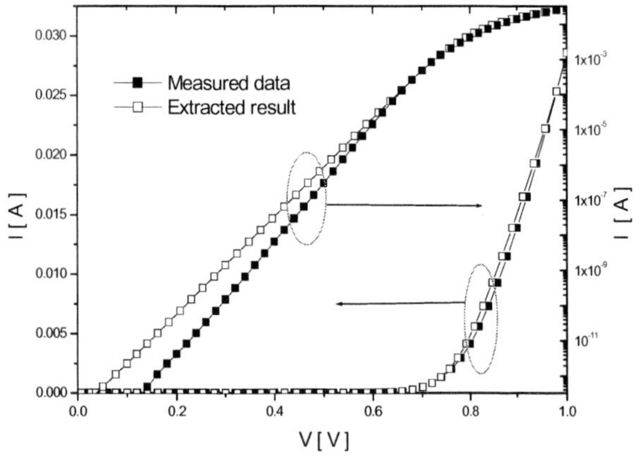

Fig 3: Comparison between measured and extracted I-V Characteristics of D1.

To demonstrate the efficiency of LCDO method, another practical diode (D2) is used for parameter extraction with this method. The measured I-V characteristics is shown in Fig 4.

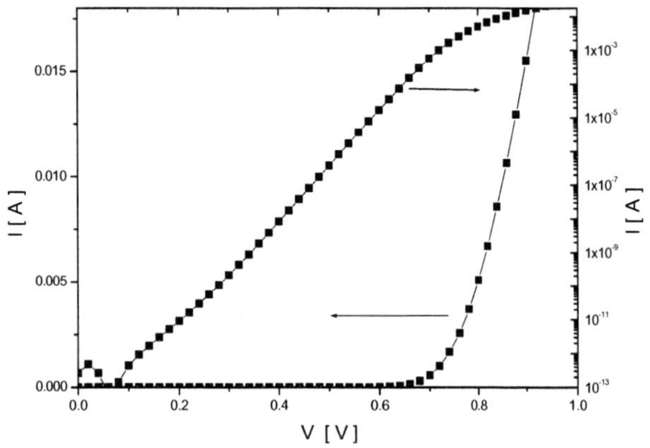

Fig 4: Measured I-V Characteristics of D2.

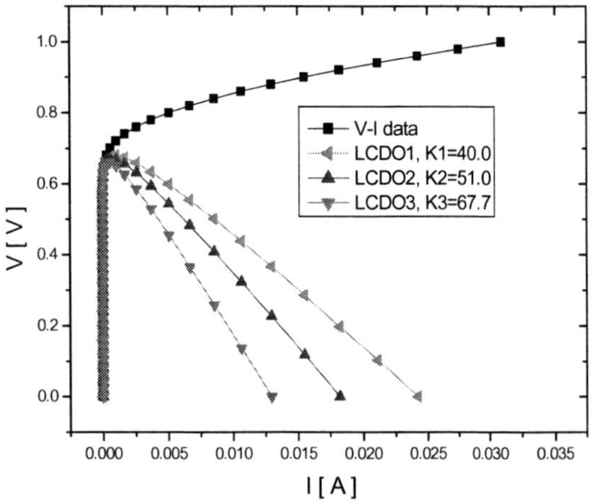

Fig 5: Diode voltage linear cofactor difference versus current for different values of K_p.

Fig 5 shows the diode linear cofactor difference voltage versus current result with the measured voltage-current characteristics for D2 with the linear cofactor difference factors:

$$K_{p1} = 40, K_{p2} = 51, K_{p3} = 67.7.$$

Following the same steps as used in the parameter extraction of D1, we obtain

$$n = 1, R_s = 5.4\Omega, I_s = 1.176 \times 10^{-15} A$$

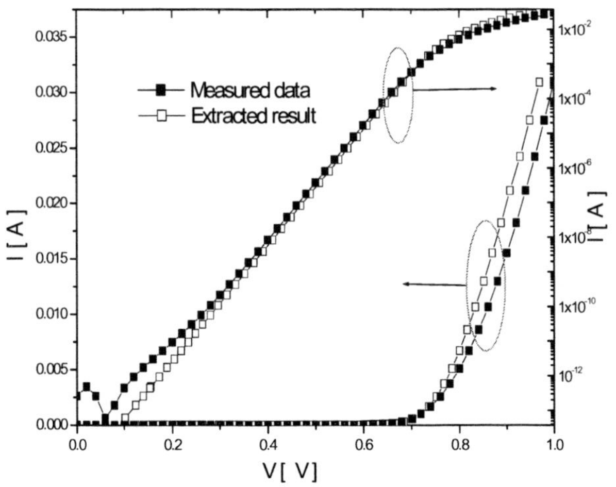

Fig 6: The comparison between the measured and extracted I-V characteristics of D2.

As Fig 6 shows, the extracted result matches the measured one in the most regions. However, there are some deviations either in the small current region or in the large current region. This is due to that the practical diode data does not really follow a simple equation as shown in Equation (5), it involves more complex current mechanisms in the small region such as R-G current and in the large current region such as high-level inject. In these cases, the LCDO method may be applied in separate regions to get better result. Such a treatment, however, would complicate the parameter extraction. So we can conclusion that the LCDO method for extracting the static parameters of diode is useful and effective in most diode operation region.

4 CONCLUSIONS

The extraction of the static parameters of two general diodes has been carried out in this paper by the linear cofactor difference operator method. The principle of this method is to apply the special extreme spectral characteristics of the diode voltage versus the current to obtain the related physical parameters such as the series resistance, ideality factor and reverse saturation current. The extraction practice for two different diodes has been carried out, showing the effective of the method presented.

ACKNOWLEDMENTS

This work is subsided by the special funds for major state basic research project (973) and National Natural Science Foundation of China (NNSFC: 90607017).

REFERENCES

[1] Aubry V, Meyer F. Schottky diode with high series resistance: limitations of forward I-V methods. J.Appl.Phys 1994;76, pp.7973-84.

[2] Norde H. A modified forward I-V plot for Schottkydiode with a high series resistance. J.Appl.Phys 1979; 50: 5052-3.

[3] Bennet RJ. Interpretation of forward bias behavior of Schottky barrier. IEEE Trans Electron Devices 1987; ED-34, pp.935-7.

[4] Werner JH. Schottky barriers and pn junction I-V plots-small signal evaluation. Appl.Phys A 1988; 47, pp. 291-300.

[5] Tor A. Fjeldly, Byung-jong Moon, and Michael Shur, Approximate analytical solution of generalized diode equation, IEEE TRANSACTIONS ON ELECTRON DEVICES, VOL. 38, NO. 8, AUGUST 1991,pp. 1976-1978.

[6] J.C. RanuaÂ rez, F.J. GarcõÂa SaÂ nchez, A. Ortiz-Conde, Procedure for determining diode parameters at very low forward voltage, Solid-State Electronics 43 (1999), pp. 2129-2133.

[7] J.C. RanuaÂ rez, A. Ortiz-Conde, F.J. GarcõÂa SaÂ nchez, A new method to extract diode parameters under the presence of parasitic series and shunt resistance, Microelectronics Reliability 40 (2000), pp. 355-358.

[8] Adelmo Ortiz-Conde, Francisco J. Garc_õa S_anchez, Juan Muci, Exact analytical solutions of the forward non-ideal diode equation with series and shunt parasitic resistances, Solid-State Electronics 44 (2000). pp .1861-1864.

[9] Lee JI, Brini J, Dimitriadis CA, Simple parameter extraction method for non-ideal Schottky barrierdiodes, Electron Lett 1998;34, pp.1268-1269.

[10] DARLING, R.B, Current-voltage characteristics of Schottky barrier diodes with dynamic interfacial defect state occupancy, IEEE Trans. Electron Devices, 1996, ED-43, (7), pp.1153-1160.

[11] Jin He, Wei Bian, Yadong Tao, Feng Liu, Jinhua Hu, Yan Song, Xing Zhang, Wen Wu, and Mansun Chan. Linear Cofactor Difference Extrema of MOSFET's Drain–Current and Application to Parameter Extraction. IEEE Transactions On Electron Devices, vol. 54, No. 4, April, 2007. pp. 874-878.

[12] HE Jin, ZHANG Xing, HUANG Ru ,WANG Yang-yuan, Study on Extraction of Stress Induced Interface Traps in MOSFETs by Linear Cofactor Differernce Subthreshold Voltage Peak Technique, ACTA ELECTRONICA SINICA, Vo.30, No.8, Aug. 2002. pp.1108-1110.

[13] Jin He, Xing Zhang, Ru Huang, and Yangyuan Wang, Linear Cofactor Difference Method of MOSFET Subthreshold Characteristics for Extracting Interface Traps Induced by Gate Oxide Stress Test, IEEE TRANSACTIONS ON ELECTRON DEVICES, VOL. 49, NO. 2, FEBRUARY 2002, pp.331-334.

A Complete Analytic Surface Potential-Based Core Model for Undoped Cylindrical Surrounding-Gate MOSFETs

Jin He[1,2], Jian Zhang[2], Lining Zhang[2], Chenyue Ma[2], and Mansun Chan[3]

[1] School of Computer & Information Engineering, Peking University Shenzhen Graduate School, Shenzhen 518055, China

[2] Institute of Microelectronics, EECS, Peking University, Beijing 100871, P. R. China

[3] Department of ECE, Hong Kong University of Science & Technology, Clearwater Bay, Kowloon, Hong Kong

Tel: 86-0755-26032104 Fax: 86-0755-26035377 Email: frankhe@pku.edu.cn

ABSTRACT

An analytic surface potential-based non-charge-sheet core model for cylindrical undoped surrounding-gate (SRG) MOSFETs is presented in this brief. Starting from the exact surface potential solution of the Poisson's equation in the cylindric surrounding-gate (SRG) MOSFETs, a single set of the analytic drain current expression in terms of the surface potential evaluated at the source and drain ends is obtasined from the Pao-Sah's dual integral without the charge-sheet approximation. It is shown that the derived drain current model is valid for all operation regions, allowing the SRG-MOSFET characteristics to be adequately described from the linear to saturation and from the sub-threshold to strong inversion region without fitting-parameters. Moreover, the model prediction is also be verified by the 3-D numerical simulation.

Keywords: non-classical MOS transistor, surrounding-gate MOSFETs, device physics, surface potential model, non-charge-sheet approximation.

1 INTRODUCTION

Extensive studies on surrounding-gate (SRG) MOSFET modelling have been performed in recent years and the related device physics have been well described by many different models [1-6]. In the channel potential-based SRG-MOSFET models, the closed-form current models are presented in terms of the intermediate variables or the potentials of the surface and centric point at the source and drain ends [3-4]. In the charge-based SRG-MOSFET models, the charge expression is developed for the SRG-MOSFETs based on a smooth function and interpolation [5]. In addition, a carrier-based approach is found to be useful in developing generic compact model for SRG-.

MOSFETs [6]. On the other hand, we also noted that the considerable attention has been focused on developing surface potential-based models in recent compact model formulations [7-8]. At present, there is a general consensus that the surface potential approach not only includes as much device physics as possible but also retains high accuracy and model continuity.

Under such a background, an analytical surface potential-based non-charge-sheet core model for obtaining the $I_{ds}(V_{gs},V_{ds})$ characteristics of the SRG-MOSFETs is proposed in this brief, based on the closed-form solution of Poisson's equation and Pao-Sah's dual integral fot the drain current formulation. We demonstrate in this paper that an analogous formulation as the proposed by J.R.Brews et al. for the single-gate (SG) Bulk MOSFET [9] can be carried out for the SRG-MOSFETs. The model has three distinctive features: (i) A single set of the surface potential voltage equation is obtained from the exact Poisson equation solution in the undoped SRG-MOSFET structure, analogous to that of the bulk MOSFETs, for which the complete surface potential equation is the beginning to develop a continuous model; (ii) the drain current, obtained from the Pao-Sah's dual integral, is described by one continuous function in terms of the surface potentials at the source and drain ends, tracing properly the transition between different SRG-MOSFET operation regions without resorting to non-physical fitting-parameters; (iii) the charge-sheet approximation, typically used in bulk MOSFET models to simplify the Pao-Sah's double integral for the current [9-10], is not invoked, properly capturing SRG-MOSFET's volume inversion effect. In order to complete the model, short-channel effects, quantum effects, low and high field transport, and more, will be added in near future.

2 ANALYTIC MODEL DEVELOPMENT

An ideal long channel SRG MOSFET without second order effects, such as poly-silicon depletion, quantum confinement, short channel effects, drain induced barrier lowering and velocity saturation is considered in this paper for simplicity. It will be possible to incorporate these second-order effects into a complete SRG MOSFET model in the near future after the core model is developed. The coordinate system used in this work is shown in Fig. 1 with r representing the radial distance from the center of the channel and $r = R$ giving the oxide silicon interface. It also assumes that the quasi-Fermi level is constant in the radial direction, so that the current flows only along the channel (y direction). The energy levels are referenced to the electron quasi-Fermi level of the source end since there is no body contact in the undoped SRG MOSFETs.

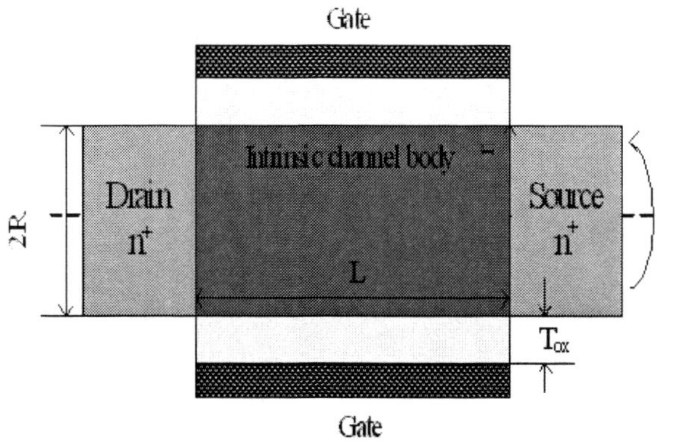

Fig.1 Diagram and coordinate of a undoped surrounding-gate MOSFET device.

. Following the gradual-channel-approximation (GCA), Poisson's equation takes the one-dimensional (1-D) form:

$$\frac{d^2\phi}{dr^2}+\frac{1}{r}\frac{d\phi}{dr}=\frac{kT}{qL_i^2}e^{\frac{q(\phi-V)}{kT}} \tag{1}$$

Where all symbols have common physics meanings as shown in [3-6]

n_i is the silicon intrinsic carrier concentration ($1.14\times10^{10}cm^{-3}$ at room temperature).

n_0 is the induced carrier concentration at the reference coordinate point (at the center of the silicon channel in this study).

$\Delta\phi_i$ is the work function difference between the gate and the channel silicon body.

V is the quasi-Fermi-potential with $V_{ch}=0$ at the source end and $V_{ch}=V_{ds}$ at the drain.

L_i^{-2} =$q^2n_i/kT\varepsilon_{si}$ is the square of the intrinsic silicon Debye length.

Equation (1) satisfies the following boundary conditions at the surface and centric point for the cooridate choice with the origin point at the silicon film middle:

$$\left.\frac{d\phi}{dr}\right|_{r=0}=0,\quad \phi|_{r=R}=\phi_s,\quad \phi|_{r=0}=\phi_0 \tag{2}$$

Then equation (1) can be analytically solved yielding [4-6]:

$$\phi_s=\phi_0-\frac{2kT}{q}\ln\left[1-\frac{R^2}{8L_i^2}\exp\left[\frac{q(\phi_0-V)}{kT}\right]\right] \tag{3}$$

and

$$\left.\frac{d\phi}{dr}\right|_{r=R}=\frac{RkT}{2qL_i^2}\frac{\exp\left[\frac{q(\phi_0-V)}{kT}\right]}{\left[1-\frac{R^2}{8L_i^2}\exp\left[\frac{q(\phi_0-V)}{kT}\right]\right]} \tag{4}$$

From Gauss's law, the following relation exsits between the charge, potential and the gate volgage:

$$C_{ox}(V_{gs}-\Delta\varphi-\phi_s)=Q=\varepsilon_{si}\left.\frac{d\phi}{dr}\right|_{r=R} \tag{5}$$

Where $C_{ox}=\frac{\varepsilon_{ox}}{R\ln(1+t_{ox}/R)}$ and $\Delta\varphi$ is the work-function difference.

Substituting the resuklts from (3) and (4) into (5) leads to

$$C_{ox}\left(V_{gs}-\Delta\varphi-\phi_s\right)=\frac{\sqrt{2}\varepsilon_{si}kT}{qL_i}\exp\left[\frac{q(\phi_s-V)}{2kT}\right]\sqrt{1-\exp\left[-\frac{q(\phi_s-\phi_0)}{2kT}\right]} \tag{6}$$

Moreover, substituting (3) into (6) leads to

$$C_{ox}\left(V_{gs}-\Delta\varphi-\phi_s\right)=\frac{R\varepsilon_{si}kT}{2qL_i^2}\exp\left[\frac{q\left(\phi_s+\phi_0-2V\right)}{2kT}\right] \tag{7}$$

From (7), we obtain the centric potential expression:

$$\phi_0=2V-\phi_s+\frac{2kT}{q}\ln\left[\frac{2L_i^2C_{ox}q\left(V_{gs}-\Delta\varphi-\phi_s\right)}{R\varepsilon_{si}kT}\right] \tag{8}$$

Then (8) is submitted to (3), so we have

$$\frac{q\left(V_{gs}-\Delta\varphi-\phi_s\right)e^{-\frac{q(\phi_s-V)}{kT}}}{kT}\left[\frac{1}{R}+\frac{qC_{ox}\left(V_{gs}-\Delta\varphi-\phi_s\right)}{4\varepsilon_{si}kT}\right]=\frac{\varepsilon_{si}}{2L_i^2C_{ox}} \tag{9}$$

(9) is a fully rigorous surface potential-voltage equation of the SRG MOSFETs [4], which can be solved by the Newton–Raphson (NR) method to get the accurate surface potential value. In order to test the analytic surface potential model, we have compared the result of (9) prediction for long-channel SRG-MOSFETs with the numerical simulation from DESSIS-ISE®. We have assumed a channel length (L) of 1 μm, silicon oxide thickness (t_{ox}) of 2 nm, and a mid-gap gate structure. A constant effective mobility of 400 cm^2/V-s has been used for calculations both in the model and in the simulation..

To apply (9) to current and charge modelling, ϕ_s needs to be evaluated at the source (y=0) and drain (y=L) ends with $V=0$ and $V=V_{ds}$, respectively. The results are separately labelled as $\phi_s=\phi_{SS}$ and $\phi_s=\phi_{SL}$. Fig.2 shows the surface potential versus gate voltage curves calculated from (9) for the source and drain ends, compared with the 3-D simulation. The solution given by (9) is continuously and smoothly valid for all regions of the SRG-MOSFET operation. It is found that the results from (9) shows in an agreement with the 3-D simulation in all operation regions for both the source and drain potentials. Fig.3 plots the comparison of the inversion charge density between the model prediction and the 3-D simulation for different silicon body radii. It is observed from Fig.3 that the agreement between the model and simulation is very good. In addition, the sub-threshold charge increases with the increase of the silicon radius. A unique "volume inversion effect", is also predicted from the presented model, coinciding with non-classical MOSFET device physics.

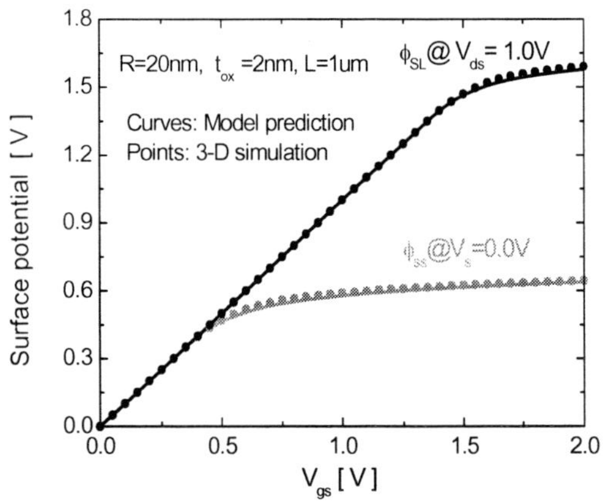

Fig.2. Comparison of source and drain end surface potentials obtained from (9) (solid lines) compared with the 3-D numerical result (points).

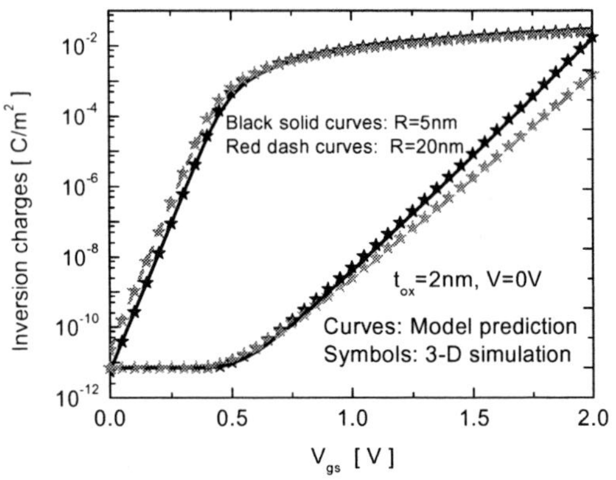

Fig.3. Inversion charge characteristics obtained from (5) (solid and dash lines) based on calculated surface potential, compared with the 3-D numerical simulation (points).or a given V_{gs}, ϕ_s can be solved from (9) as a function of V.

Note that V varies from the source to the drain. The functional dependence of $V(y)$ and ϕ_s (y) is determined by the current continuity equation. From the Pao-Sah's dual integral [10], integrating $I_{ds}dy$ from the source to the drain and expressing dV/dy as $(dV/d\phi_s)(d\phi_s/dy)$, the non-charge-sheet drain current of the SRG-MOSFETs is written as

$$I_{ds} = \mu\frac{2\pi R}{L}\int_0^{V_{ds}} Q(V)dV = \mu\frac{2\pi R}{L}\int_{\phi_{SS}}^{\phi_{SL}} Q(\phi_s)\frac{dV}{d\phi_s}d\phi_s \qquad (10)$$

Where ϕ_{SS} , ϕ_{SL} are the solutions to (9) corresponding to V=0 and V=V_{ds} respectively. The parameter μ is the effective mobility. By using (9) and replacing it into $Q = C_{ox}(V_{gs} - \Delta\varphi - \phi_s)$, the total mobile charge per unit gate area expressed in terms of ϕ_s yields $Q(\phi_s)$. Note that $dV/d\phi_s$ can also be expressed as a function of ϕ_s by differentiating (9). Substituting these factors in (10), we have

$$I_{ds} = \frac{2\pi R\mu C_{ox}}{L}\int_{\phi_{SS}}^{\phi_{SL}}\left[(V_{gs} - \Delta\varphi - \phi_s) + \frac{2kT}{q} - \frac{kT}{q}\left(1 + \frac{C_{ox}qR(V_{gs} - \Delta\varphi - \phi_s)}{4\varepsilon_{si}kT}\right)^{-1}\right]d\phi_s \qquad (11)$$

The integration of (11) is performed analytically to yield:

$$I_{ds} = \frac{2\pi R\mu C_{ox}}{L}\left[\left(V_{gs} - \Delta\varphi)\phi_s - \frac{\phi_s^2}{2}\right) + \frac{2kT\phi_s}{q} + \left(\frac{kT}{q}\right)^2\frac{4\varepsilon_{si}}{RC_{ox}}\ln\left(1 + \frac{C_{ox}qR(V_{gs} - \Delta\varphi - \phi_s)}{4\varepsilon_{si}kT}\right)\right]\Bigg|_{\phi_{SS}}^{\phi_{SL}} \qquad (12)$$

3 RESULTS AND DISCUSSION

From (12), the SRG-MOSFET drain current can be easily computed. In the following, the SRG-MOSFET operation regions are derived from this continuous surface potential-based analytical model:

(i) Linear region above threshold: In this region the drift current component dominates the device performance. Hence, we observe that the total drain current can be approximated by first two terms only above the threshold as shown in (13).

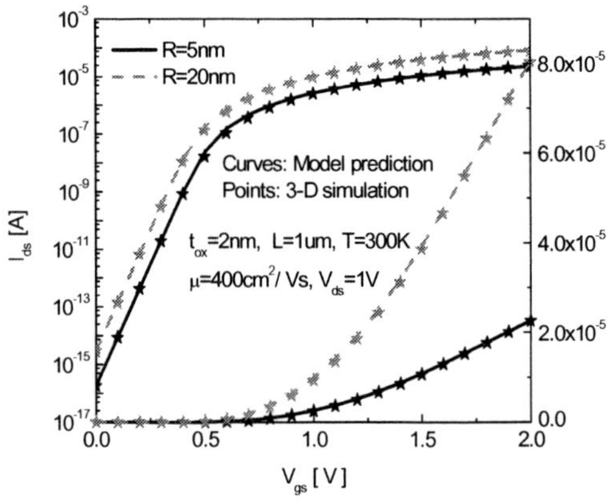

Fig. 4 Transfer characteristics obtained from the surface potential-based model for two silicon film radius (solid and dashed lines), compared with numerical simulations from DESSIS-ISE® (points).

$$I_{ds} \approx \frac{2\pi R\mu C_{ox}}{L}\left[(V_{gs} - \Delta\varphi)(\phi_{SL} - \phi_{SS}) - \frac{1}{2}(\phi_{SL}^2 - \phi_{SS}^2)\right] \qquad (13)$$

This current expression is just the drift component of the traditional surface potential-based bulk MOSFET models, thus dominates in the strong inversion region;

(ii) Subthreshold region: Below threshold the SRG MOSFET current picture has a little difference from the bulk MOSFET model. Here, the first two components are negligible in this region. As a result, the total drain current is described by

$$I_{ds} \approx \frac{2\pi R\mu C_{ox}}{L}\left[\frac{2kT}{q}(\phi_{SL} - \phi_{SS}) + \left(\frac{kT}{q}\right)^2\frac{4\varepsilon_{si}}{RC_{ox}}\ln\left[\frac{4\varepsilon_{si}kT + C_{ox}qR(V_{gs} - \Delta\varphi - \phi_{SL})}{4\varepsilon_{si}kT + C_{ox}qR(V_{gs} - \Delta\varphi - \phi_{SS})}\right]\right] \qquad (14)$$

This drain expression can be simplified into:

$$I_{ds} = \mu \frac{\pi R^2}{L} n_i kT e^{\frac{q(V_{GS}-\Delta\varphi)}{kT}} \left(1 - e^{\frac{-qV_{DS}}{kT}}\right) \quad (15)$$

The sub-threshold current in (15) is proportional to the cross-section area of the SRG-MOSFET and independent of t_{ox}. This is a characteristic of the volume inversion effect that cannot be captured by the standard charge-sheet based models;

(iii) Saturation region: This regime occurs when the contribution of the drain end is little to the drain current. Hence, the drain current is expressed as

$$I_{ds} = \mu \frac{\pi R}{L} (Q_S + \frac{4C_{ox}kT}{q})(\phi_{SL} - \phi_{SS}) \quad (16)$$

The saturation current mainly depends on the source inversion charge density as expected for a bulk MOSFET.

In order to verify the presented drain current model, the comparison of drain current curves between the model prediction and the 3-D simulation is also completed as done for the surface potential and inversion charge. Fig.4 shows the SRG-MOSFET transfer curves and Fig.5 plots the SRG-MOSFET output curves, calculated from the surface potential-based model and the 3-D numerical simulation. Again, good agreement is observed without using any fitting parameter in both figures. Especially, the volume inversion effect of SRG-MOSFET demonstrated in Fig.4 is well described by the presented model, matching the 3-D numerical simulation.

4 CONCLUSIONS

In summary, we have presented an analytical surface potential-based non-charge-sheet core model current-voltage model suitable for compact modeling of undoped (lightly doped) SRG-MOSFETs. All the operation regions and the transitions are correctly described by preserving the physics. In particular, the volume inversion effect, that cannot be captured by using the traditional charge-sheet approximation, is well accounted of in this model.

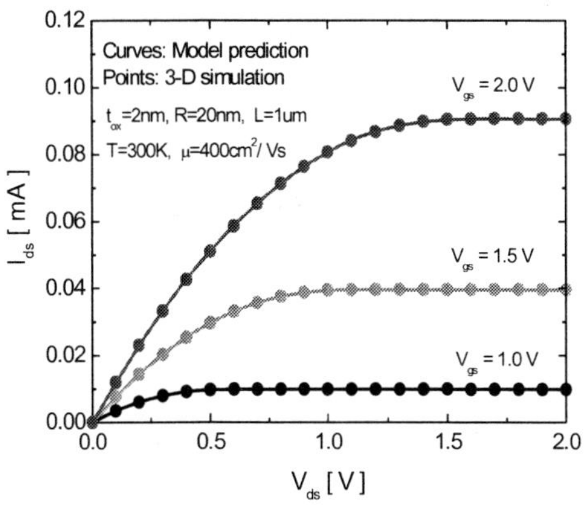

Fig.5 Output characteristics obtained from the surface potential model (solid lines) compared with numerical simulations from DESSIS-ISE® (points).

Acknowledgements

This work is supported by the Scientific Research Foundation for the Returned Oversea Chinese Scholars, State Education Ministry (SRF for ROCS, SEM), and the National Natural Science Foundation of China (90607017). This work is also partially support by the International Joint Research Program (NEDO Grant) from Japan under the Project Code NEDOO5/06.EG01.

References

[1] J. -P. Colinge, "Multiple-gate SOI MOSFETs," Solid-State Electron. vol. 48, no. 6, pp. 897-905, June 2004.

[2] B. Iniguez TA.Fjeldly, A. Lazaro, F.Danneville, and M.J.Deen, "Compact-modelling solutions for nanoscale double-gate and gate-all-around MOSFETs," IEEE Trans. on Electron Devices, TED-53, No.9, pp.2128-2142, Sept. 2006

[3] D. Jiménez, B. Iñiguez, J. Suñé, L. F. Marsal, J. Pallarès, J. Roig, and D. Flores, "Continuous analytic current-voltage model for surrounding gate MOSFETs," IEEE Electron Device Lett. vol. 25, No. 8, pp.571–573, Aug. 2004

[4] Bian Wei, Jin He, Yadong Tao, Ming Fang, Jie Feng, " An analytic potential-based model for undoepd nanoscale surrounding-gate MOSFETs," IEEE Trans. on Electron Devices, Vol. 54, No.9, Sept., 2007

[5] B. Iñiguez, D. Jimenez, J. Roig, H. A. Hamid, L. F.Marsal, and J. Pallares. "Explicit continuous model for long-channel undoped surrounding-gate MOSFETs," IEEE on Trans Electron Devices, TED-52, No.8, pp.1868-1873, Aug. 2005

[6] Jin He, Mansun Chan, Xing Zhang, and Yangyuan Wang, "A carrier-based analytic model for the undoped (lightly doped) cylindrical surrounding-gate MOSFETs," Solid-State Electronics, vol.50, No.4, pp. 416-421, 2006

[7] SRC, "Request for White Papers: Semiconductor Device Compact Modeling for Circuit Design (April, 23, 2003)," http://www.src.org

[8] CMC website of Next Generation MOSFET Model Standard Phase-III Evaluation Results. http://www.eigroup.org/CMC

[9] J.R.Brews, "A charge sheet model of the MOSFET". Solid-State Electronics, vol.21, pp.345-352, 1978

[10] H. C. Pao and C. T. Sah, "Effects of diffusion current on characteristics of metal-oxide (insulator)-semiconductor transistors". Solid-State Electronics, vol. 9, pp. 927-937, 1966

A Novel Dual Gate Strained-Silicon Channel Trench Power MOSFET For Improved Performance

Raghvendra Sahai Saxena and M. Jagadesh Kumar

Department of Electrical Engineering,
Indian Institute of Technology, New Delhi – 110 016, INDIA.
Email: mamidala@ieee.org Fax: 91-11-2658

ABSTRACT

In this paper, we propose a new dual gated trench power MOSFET with strained Si channel using the $Si_{0.8}Ge_{0.2}$ base and graded strained Si accumulation region using compositionally graded $Si_{1-x}Ge_x$ buffer (x varying from 0 to 0.2 from the drift region to the base region) in drift region. We show that the introduction of strain in channel and accumulation region results in about 10% improvement in its drive current, about 20% reduction in the on-resistance and about 72% improvement in peak transconductance with only about 12% reduction in the breakdown voltage when compared with equivalent conventional device. Furthermore, the structure contains two separated gates out of which one controls the inversion charge in the channel region and the other one controls the accumulation charge in the accumulation region. The separate control of the charge in different regions further improves the device performance. We show that by applying a suitably high fixed positive voltage at the accumulation gate improves the device performance parameters up to about 10%.

Keywords: Strained Si, $Si_{1-x}Ge_x$, Trench Gate, Dual gate, Power MOSFET, Compositionally Graded $Si_{1-x}Ge_x$ buffer

1 INTRODUCTION

A trench gate MOSFET is used in many medium to low voltage power applications [1-5]. In various applications the main requirements of a power MOSFET are the low on-state resistance, high drive current, low switching delays and high transconductance with a considerably high immunity for the inductive switching damages. However, while designing a power MOSFET, meeting all these requirements is not possible in conventional structures and the device can not be used efficiently in many applications if designed for one specific application. To overcome these difficulties, we propose a new dual gate trench power MOSFET that provides many advantages over convetional power MOSFETs like (a) high drive current as it has high carrier mobility due to the strained Si in the channel region [6, 7], (b) better inductive switching immunity as it provides the confinement of carriers near the trench side-walls in accumulation layer [5] and (c) higher peak transconductance with additional capability of in-circuit parameter tuning for efficient use in different applications.

2 DEVICE STRUCTURE AND SIMULATION

Fig. 1 shows the schematic cross-sectional view of the proposed DGSCT device. It contains two trenches in such a way that shallower trench is inside the deep trench and touches its boundaries. This structure allows the formation of two gates with strained Si channel as depicted in Fig. 1. The deeper gate (called accumulation gate here after and denoted by G_{Acc}) controls the accumulation charge and the shallower gate (called inversion gate and denoted by G_{Inv}) controls the inversion charge of the strained channel. It uses P-type $Si_{0.8}Ge_{0.2}$ in the body and a compositionally graded N-type $Si_{1-x}Ge_x$ buffer layer as a part of the drift region.

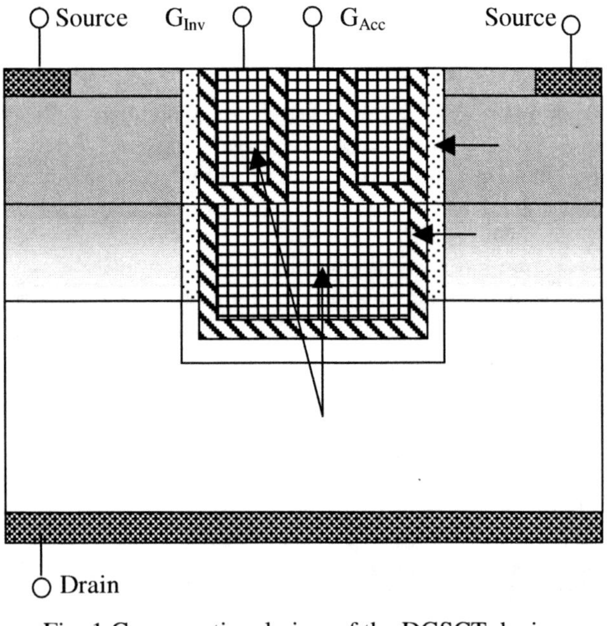

Fig. 1 Cross-sectional view of the DGSCT device

The compositionally graded SiGe buffer has been realized by a 10 layer stack with varying Ge composition from 0 at the drift region to 20% at the body region. Similarly, we have realized the graded strained Si layer adjacent to the graded SiGe buffer layer. The use of graded strain results in the smoothing of energy band discontinuity faced by the carriers while moving from channel to the accumulation region.

Device Parameter	Value
Source/Drain doping	$1x10^{19}$ atoms/cm^3
Source thicknes	100 nm
Body doping	$5x10^{17}$ atoms/cm^3
Body thickness	500 nm
Ge mole fraction in the SiGe buffer layer, x	0-0.2
SiGe buffer layer doping	$1x10^{16}$ atoms/cm^3
SiGe buffer thickness	500 nm
Si Drift region doping	$1x10^{16}$ atoms/cm^3
Si Drift region thickness	2.5 μm
Gate oxide thickness	50 nm
Strained Si thickness	20 nm

Table 1: Device parameters used in simulations

Using 2D numerical simulations performed with ATLAS device simulator [8], we show a comparison of the proposed DGSCT device with the equivalent conventional device that has similar doping concentration and geometry but only single trench with no SiGe region. Various parameters used in simulations are listed in table 1.

3 RESULTS AND DISCUSSION

The DGSCT device has independent control of its two gates. Therefore, it may be operated in two modes, namely, (a) standard mode where we externally short both the gates and (b) extended mode where the different voltages may be applied to get maximum benefit of the separate accumulation gate control. The structure results in various advantages in standard mode and these advantages may further be improved in extended mode, as discussed in the following subsection.

3.1 Standard mode

The use of strained Si causes energy band discontinuity in the body region along the transverse direction to the current flow. This results in the carrier confinement in the channel region. Similarly, the carriers are also confined in the gradually strained accumulation region and this confinement reduces as we go towards the drain side. The carrier confinement near the trench sidewalls results in better gate control of the channel charge and therefore, we get many advantages in DGSCT device when operated in standard mode. We get lower threshold voltage as compared to the conventional device as shown in log I_{DS} vs V_{GS} plot in Fig. 2. It indicates a shift in threshold voltage from 2.1 V in conventional device to 1.5 V in the proposed DGSCT device. The better gate control also results in high transconductance as shown in Fig 3. The peak transconductance of DGSCT device in linear mode of operation (at V_{DS} = 0.1 V) is about 70% larger as compared to the conventional device. The higher mobility of the strained Si also improves the drive current and hence the on state resistance of the device, as depicted in the output characteristics shown in Fig. 4. It amounts to around 20% reduction in on state resistance.

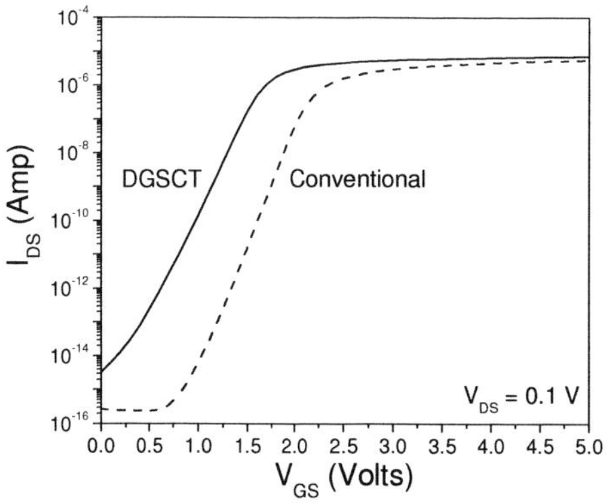

Fig. 2: I_{DS}-V_{GS} Characteristics of DGSCT and Conventional Devices

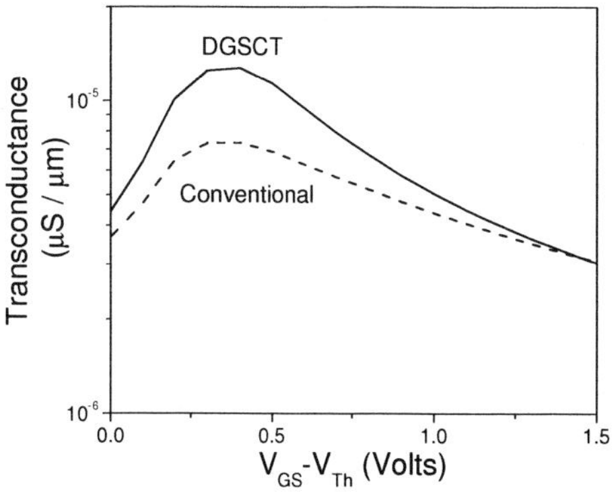

Fig. 3: Transconductance as a function of gate over drive voltage at V_{DS} = 0.1 V

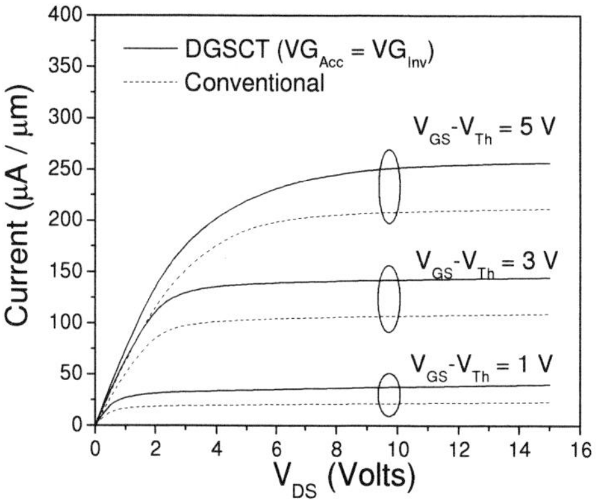

Fig. 4: Output characteristics of DGSCT and conventional devices at different gate overdrive voltages

These improvements we get at the cost of reduced breakdown voltage that reduces by 12% in DGSCT device as compared with the conventional device due to use of SiGe body region, as shown in Fig. 5.

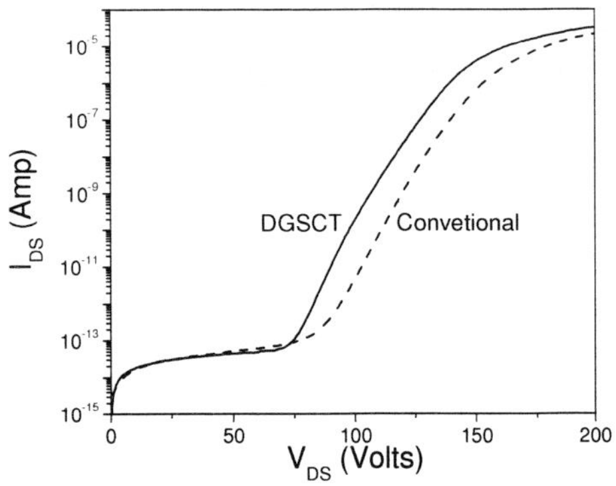

Fig. 5: Log I_{DS} vs V_{DS} characteristics with gate connected to source showing breakdown voltage comparison of DGSCT and conventional devices

3.2 Extended mode

The device parameters may further be improved by keeping accumulation gate at some fixed positive potential.

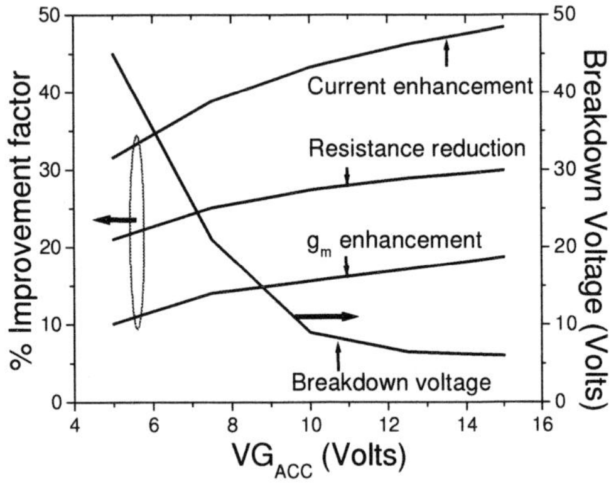

Fig. 6: The effect of VG_{Acc} on various parameters

The effect of accumulation gate voltage on various performance parameters are shown in Fig. 6 for bias conditions of VG_{Inv} = 5.0 V and V_{DS} = 0.1 V on the left Y-axis. We found that the device performance improves by increasing the fixed potential of accumulation gate. This occurs because of the ease in carrier transport from channel region to the accumulation region. However, the accumulation of carriers in the drift region due to high positive G_{Acc} voltage causes the depletion region at the accumulation-body junction to reach up to the source at the V_{DS} much lower than the breakdown voltage obtained in standard mode, resulting in a lower effective breakdown voltage in extended mode. The effect of increasing the accumulation gate voltage on the breakdown voltage of the device is illustrated in Fig. 6 on right Y-axis.

Thus we see that a DGSCT device may be tuned for better performance parameters for a given application if breakdown voltage requirement permits so. For example, for low voltage (~5.0 V) light load applications, one can set the G_{Acc} voltage at 10.0 V to get excellent performance, and for moderate breakdown voltage (20.0 V – 30.0 V) applications the G_{Acc} bias voltage may be reduced accordingly. The same device may be used in standard mode for high breakdown voltage requirement with inductive load switching capability.

4 CONCLUSIONS

For the first time, using two-dimensional simulation, we have demonstrated the strained Si channel in trench power MOSFET produced by using $Si_{1-x}Ge_x$ body improves the device performance parameters such as current drivability, peak transconductance and on-state device resistance. Furthermore, the sectioning of the gate into two parts in order to have separate control of accumulation charge gives the flexibility of applying different voltages at the two gates of thus obtained DGSCT device. We have shown that by using this feature, we can achieve a further improvement in the performance of the device at the cost of reduced breakdown voltage, by suitably adjusting the accumulation gate voltage.

REFERENCES

[1] B. J. Baliga, "Modern Power Devices", *John Wiley and Sons, New York,* 1987.

[2] K. Shenai, *IEEE Trans. Electron Devices,* vol. 39, No. 6, 1435-1443, Jun, 1992.

[3] C. Hu, M. H. Chi and V. M. Patel, *IEEE Trans. Electron Devices,* vol. 31, No.12, pp. 1693-1700, Dec, 1984.

[4] M. H. Juang, W. C. Chueh and S. L. Jang, *Semicond. Sci. Technol.*, vol 21, No. 6, pp. 799–802, May, 2006.

[5] Narazaki, J. Maruyama, T. Kayumi, H. Hamachi, J. Moritani and S. Hine, in *Proc. of ISPS-2OOO,* May 21-25, Toulouse, France, pp. 377-380, 2000.

[6] Y. K. Cho, T. M. Roh and J. Kim, *ETRI Journal,* vol. 28, No. 2, pp. 253-256, Apr, 2006.

[7] P. L. Yajuan, S. Mengsi and Y. X. Li, in *Proc. of ISPS-2OOO,* May 21-25. Toulouse, France, pp. 109-112, 2000.

[8] *Atlas User's Manual: Device Simulation Software*, Silvaco Int., Santa Clara, CA.

Pre-Distortion Assessment of Workfunction Engineered Multilayer Dielectric Design of DMG ISE SON MOSFET

Ravneet Kaur[1], Rishu Chaujar[1], Manoj Saxena[2] and R. S. Gupta[1]
[1]Semiconductor Devices Research Laboratory, Department of Electronic Science, University of Delhi, South Campus, New Delhi-110021, India.
[2]Department of Electronics, Deen Dayal Upadhyaya College, University of Delhi, New Delhi, India

In order to overcome device miniaturization roadblock in sub-100 nm regime, innovative architectural enhancements involving the use of an improved SOI like architecture called Silicon On Nothing (SON) capable of SCE and DIBL suppression [1] onto the existing ISE devices [2], has been considered for SCEs suppression. Further to overcome the electron transport inefficiency [3], Dual Material Gate (DMG) architecture has also been integrated thereby presenting ultimate device architecture of DMG ISE SON for ULSI era. In RF applications minimization of third order intermodulation (IMD3) is vital as it generates harmonics close to the desired signal and cannot be cancelled by push-pull configuration as for second order distortion. In this work, linearity performance of 50nm DMG ISE SON MOSFET as shown in Fig. 1, has been investigated and compared with other Single & DMG taking into consideration the non-equilibrium transport effect implemented via EBT-model activated through ATLAS-2D device simulation software. The work discusses the linearity Figure of Merits -VIP_2, VIP_3 and IMD3 and thereby, provides optimal bias point selection. In order to explore the linearity performance, transconductance g_m as well as the higher order derivatives g_m' & g_m'' as shown in Fig. 2 is studied. g_m is low in the subthreshold region and then rises to a peak between 0.9 and 1.0V, for all devices under consideration. The value of g_m' & g_m'' are found to be lowest around V_{GS} of 0.9-1.0V & 0.6-0.7V respectively for all devices under consideration. The nonlinearity exhibited by these higher-order derivatives of I_{DS}-V_{GS} characteristics determines a lower limit on the distortion and therefore, the amplitude of g_m' & g_m'' should be minimum and from Fig. 2, lowest value where a dip in g_m' &g_m'', are seen for DMG ISE SON. However, gate bias of 0.6V is preferred as at this bias point, g_m'' crucial for improved linearity, attained a minima. Fig. 3 & 4 gives the variation of VIP_2 ($VIP_2 = 4\frac{g_m}{g_m'}$) and VIP_3 ($VIP_3 = \sqrt{24\frac{g_m}{g_m''}}$) with V_{GS} for all the structures. It is seen that VIP_2 peak is highest for DMG ISE among all devices. However, with thinning of channel film (T_{film}), VIP_2 value improves significantly for DMG ISE SON. For the same electrical gate length as SMG ISE, DMG ISE SON exhibits a considerable enhancement in VIP_2 but is lower than DMG ISE and DMG SON. The VIP_3 peak, signifying the second order interaction effect reflecting the cancellation of third-order nonlinearity coefficient by device internal feedback around second-order nonlinearity as shown in Fig. 4, in DMG ISE SON is appreciably higher than the other MOSFETs. The lower gate bias (V_{GS}=0.6V) is desirable as at this biasing condition considerably higher value of g_m is obtained as compared to the values obtained at higher gate bias (V_{GS}=1.4-1.6V). It further implies that a lower gate drive is required to preserve linearity. Thus, DMG ISE SON is more linear than other counterparts.

[1] Monfray et al., *IEDM Tech. Digest*, pp. 645, 2001.
[2] R. Kaur et al., *IEEE Trans. on Electron Devices*, Vol. 54, No.2, pp. 365, February 2007.
[3] W. Long and K. K. Chin, *IEDM Tech. Digest*, pp. 549, 1997.

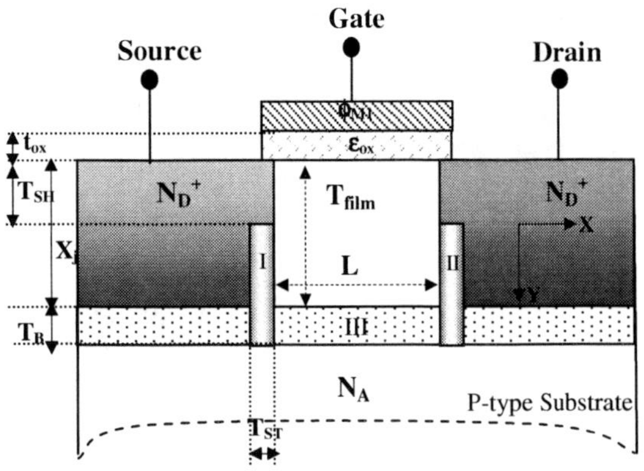

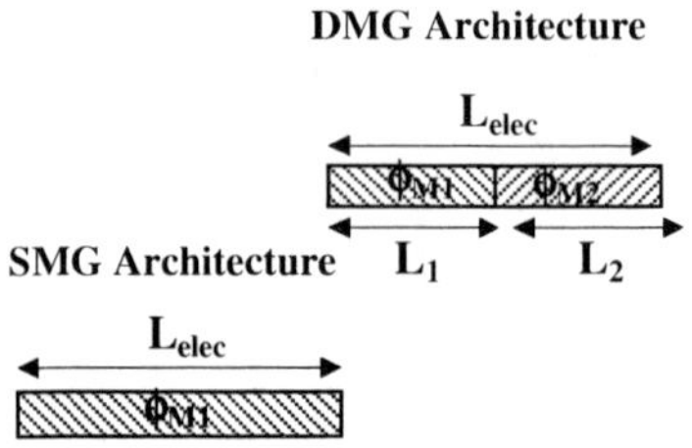

Devices	Material		
	I	II	III
ISE	SiO_2	SiO_2	Si
SON	Si	Si	Air
ISE SON	SiO_2	SiO_2	Air
Bulk	Si	Si	Si

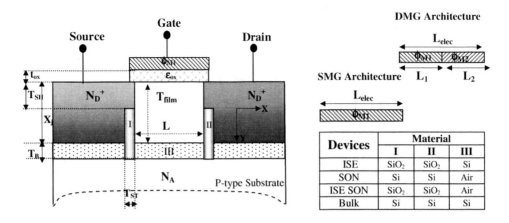

Fig. 1 Schematic cross section of various MOSFET structure under consideration. L_1 = 50 nm, T_{ST} = 10 nm, T_{SH} = 10nm, T_{VH} =75 nm, T_{film} = 30nm, T_B=10nm, X_j = 30 nm, t_{ox} = 3nm, $N_A = 1\times10^{17}$ cm^{-3}. Work function - $q\phi_{M1}$ = 4.77eV & $q\phi_{M2}$ = 4.71 eV.

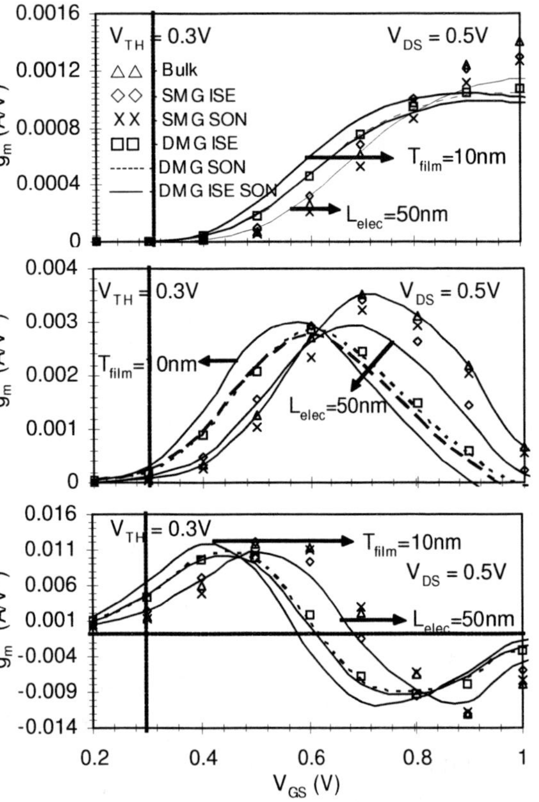

Fig. 2 Simulated (g_m) ,(g_m') & (g_m'') Vs V_{GS} characteristics for various MOSFETs. V_{DS}=0.5V, L=50nm, W=1μm, ε_{ox1}=3.9, T_H=10nm, work function ($q\phi_{M1}$) = 4.77 V for SMG and for DMG, ($q\phi_{M1}$) = 4.77 eV and ($q\phi_{M2}$) = 4.10 eV.

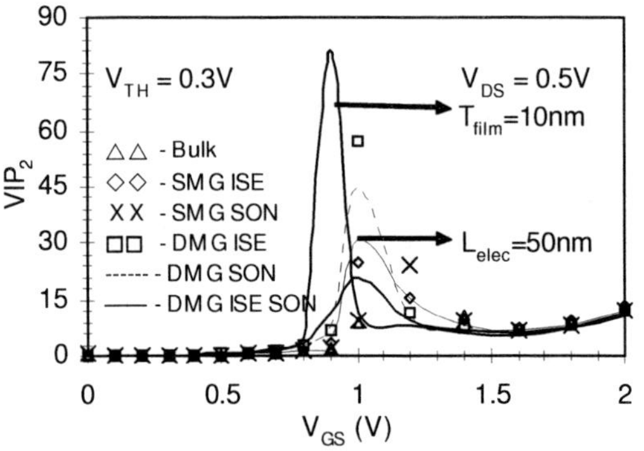

Fig. 3 VIP_2 variation with V_{GS} for various DMG and SMG MOSFETs. L=50nm, T_H=10nm, T_S=20nm, ε_{ox1}=3.9, $q\Phi_{M1}$=4.77eV, $q\Phi_{M2}$=4.10 eV and V_{DS}=0.5V.

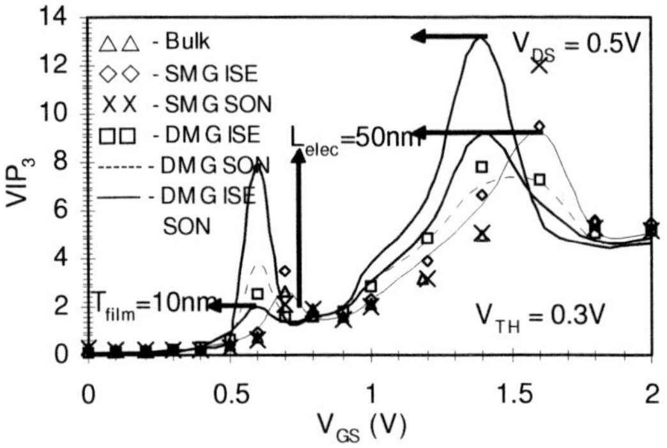

Fig. 4 VIP_3 variation with V_{GS} for various DMG and SMG MOSFETs. L=50nm, T_H=10nm, T_S=20nm, ε_{ox1}=3.9, $q\Phi_{M1}$=4.77eV, $q\Phi_{M2}$=4.10 eV and V_{DS}=0.5V.

Linearity Performance Enhancement of DMG AlGaN/GaN High Electron Mobility Transistor

Sona P. Kumar*, Anju Agrawal**, Rishu Chaujar*, Mridula Gupta* and R.S.Gupta*

* Semiconductor Devices Research Laboratory, Department of Electronic Science, University of Delhi, South Campus, New Delhi, India, sonapkumar@rediffmail.com and rsgu@bol.net.in
** Department of Electronics, Acharya Narendra Dev College, Kalkaji, New Delhi 110019, India, dyuti97@yahoo.com

ABSTRACT

In the work presented, Dual Material Gate (DMG) AlGaN/GaN HEMT is studied for its superior linearity performance as compared to the conventional Single Metal Gate (SMG) HEMT using ATLAS device simulator. The device transconductance, drain conductance, VIP3 and IMD3 have been used for the performance assessment of the DMG HEMT and the results indicate an appreciable reduction in IMD3 and significant increase in VIP3 as compared with the SMG counterpart for its application in low noise amplifiers and 3-G mobile communication.

Keywords: ATLAS, DMG, HEMT, Linearity, Transconductance, RF , VIP3.

1 INTRODUCTION

HEMTs are promising devices for high speed integrated circuits and RF wireless communications [1-3]. Recent experiments indicate that owing to the material properties of GaN, AlGaN/GaN HEMTs demonstrate higher unity-gain cutoff frequency putting the material system in a good position for RF/microwave transistors [4]. During the last few years, outstanding high speed performance has been achieved through better design and gate length reduction. However, reduction of gate length down to deep sub-micron leads to undesirable short channel effects (SCE) such as drain induced barrier lowering (DIBL), hot electron effect and poor carrier transport efficiency. In 1999, Long [5-7] proposed a new structure (DMGFET) wherein the two materials with different workfunctions are amalgamated together to form a gate. The use of the two gate metals leads to a step function in the channel potential and a peak in the electric field distribution at the interface of the two gate metals. This results in reduced SCEs like DIBL, increased transconductance and on current and reduced drain conductance .

For wireless communication applications and RF circuit design, linearity is one of the most important issues. This is because the devices used in the system may produce non-linear distortion and thus degrade the S/N ratio of the system. For short gate HEMTs, the transconductance and output drain conductance become important sources of non-linearity. Thus, the device structure needs to be tailored to improve the RF performance of the devices used in the system. Since, in DMG HEMT [8] peak transconductance occurs at much lower V_{gs} and it exhibits lower drain conductance in the saturation region, hence, it is worth exploring its linearity performance.

This present work sheds light on the improved linearity performance of DMG $Al_{0.2}Ga_{0.8}N$/GaN HEMT in comparison with the corresponding SMG HEMT in terms of IMD3, which represents the third order intermodulation distortion power and mainly determines the harmonic distortion. Other parameters used as linearity metrics are *gm2, gm3, gd2, gd3* and *VIP3*, the extrapolated input voltage at which the first and the third harmonics are equal. All simulations have been performed using ATLAS [9] device simulation software. The models used in simulation include CONMOB and FLDMOB for mobility

2 SIMULATION RESULTS AND DISCUSSION

In the figures shown the solid symbol denote the conventional HEMT and hollow symbols denote DMG AlGaN/GaN HEMT (Fig.1).

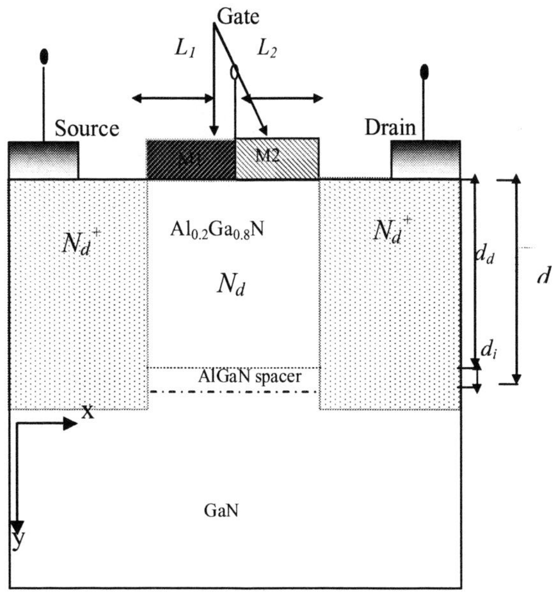

Fig.1 Schematic diagram of Al0.2Ga0.8N/GaN HEMT with $Nd^+=1x10^{26}m^{-3}$, L=L1+L2=120nm, d=30nm, W=20μm

At higher frequencies, g_{m3} and g_{d3} are the dominant nonlinear sources. Harmonic distortion is present due to the nonlinearity of these higher order components and hence

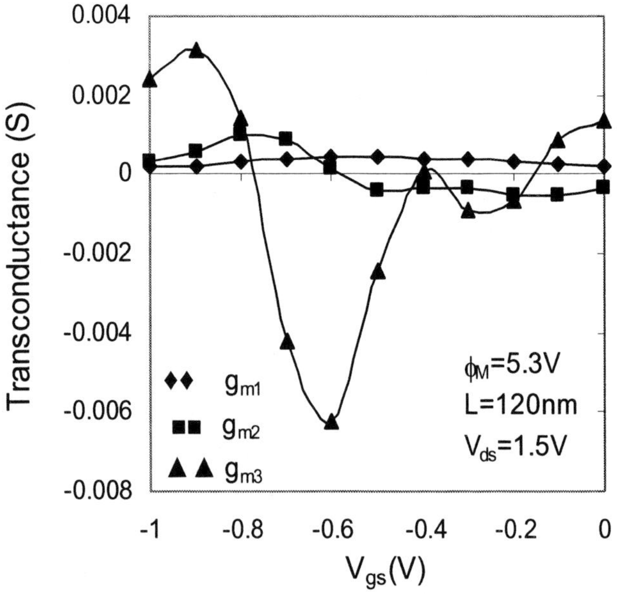

Fig.2 shows the simulated g_{m1},g_{m2} and g_{m3} vs V_{gs} characteristics for conventional HEMT.

their amplitude should be minimized. Fig. 2 and 3 give the variation of g_{m1}, g_{m2} and g_{m3} with gate bias for SMG HEMT and DMG HEMT respectively and clearly depict that peak

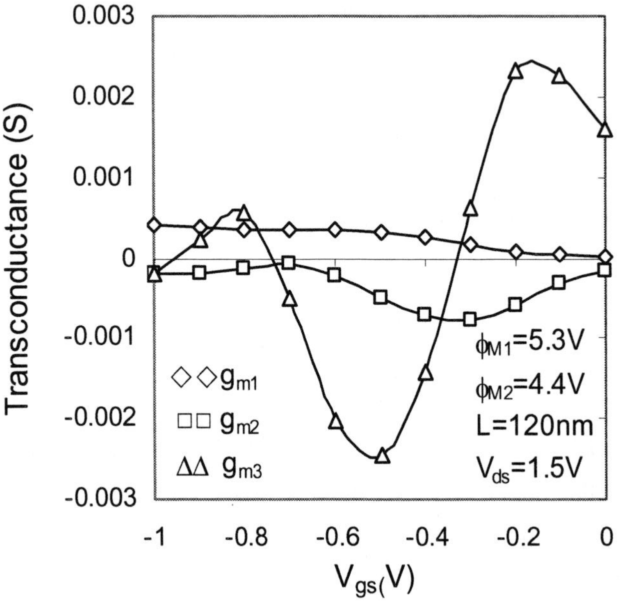

Fig3. shows the simulated g_{m1},g_{m2} and g_{m3} vs V_{gs} characteristics for DMG HEMT.

in g_{m3} occurs at a lower V_{gs} and is less for DMG HEMT . Fig. 4 and 5 give the plot of g_{d1}, g_{d2} and g_{d3} with gate bias for SMG and DMG respectively. Although g_{d3} for DMG HEMT is marginally higher than SMG HEMT, the zero

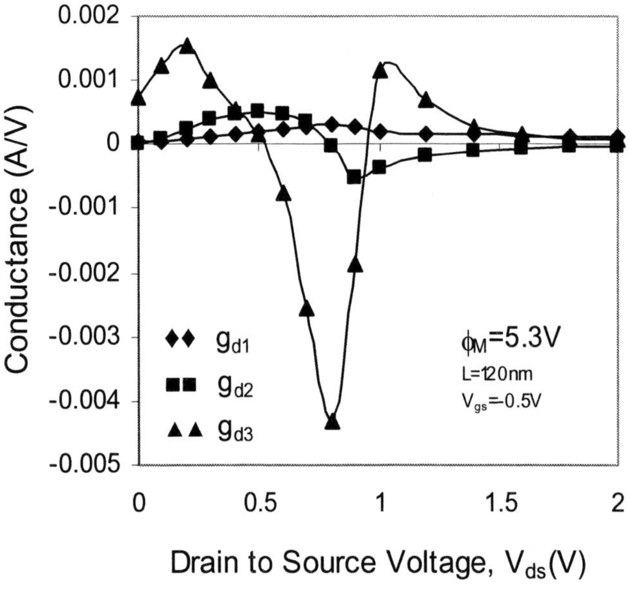

Fig.4. shows the simulated g_{d1},g_{d2} and g_{d3} vs V_{gs} characteristics for conventional HEMTs.

crossover occurs at a much lower V_{ds} implying that a lower V_{ds} is needed to minimize this harmonic distortion.

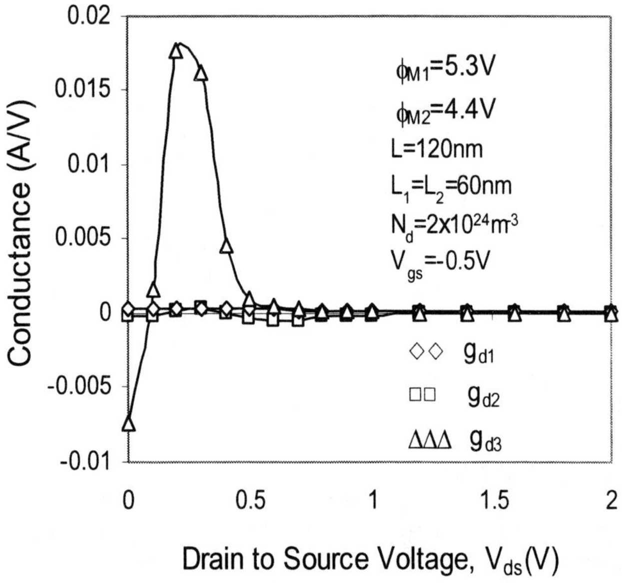

Fig.5 shows the simulated g_{d1},g_{d2} and g_{d3} vs V_{gs} characteristics for DMG HEMTs.

Figure 6 shows the *VIP3* variation with the applied gate bias. For linearity, *VIP3* is used as a first order parameter and a large *VIP3* is required for high linearity. Figure shows

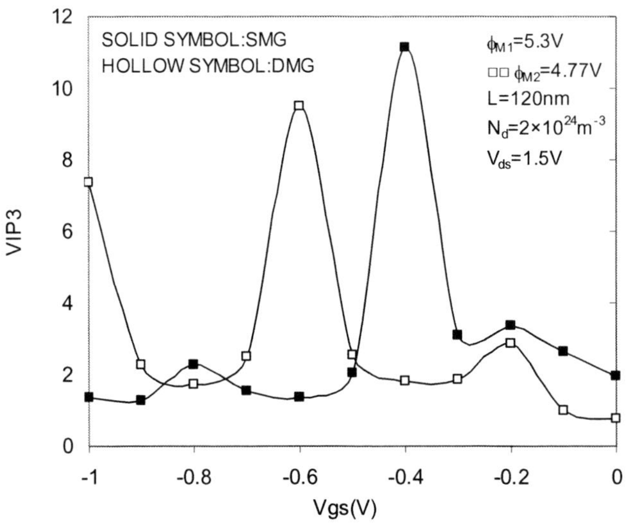

Fig.6 shows the simulated *VIP3* variation with applied gate bias for DMG and conventional HEMTs.

that in DMG HEMTs the singularity in *VIP3* occurs at a much lower V_{gs}. The shift of peak towards higher V_{gs} (as in SMG HEMT) is not desirable as it implies that a higher V_{gs} is needed to preserve the linearity. Fig. 7 shows the plot of third order intermodulation distortion power (*IMD3)* with the gate bias. This linearity parameter should be minimized

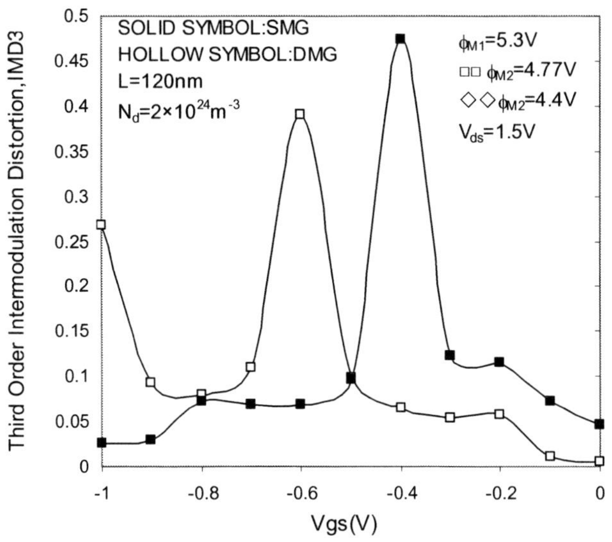

Fig.7 shows the simulated *IMD3* vs V_{gs} characteristics for DMG and conventional HEMTs.

to reduce distortion. IMD3 originates from the nonlinearity exhibited by the transistor's I_{ds}-V_{gs} characteristics and leads to corrupting signals in the wireless systems Figure clearly shows that DMG HEMT exhibits a much lower *IMD3*due to reduction in hot carrier effect. Thus, by applying DMG architecture on the conventional HEMT, we can minimize the linearity degradation and improve the RF performance of these devices.

3 CONCLUSION

In this paper, DMG AlGaN/GaN HEMT has been analysed through ATLAS device simulation. On comparison with the SMG counterpart, we conclude that DMG HEMT is more linear as it exhibits a much higher VIP3 and reduced IMD3 due to better screening which in turn leads to reduced hot carrier effect and better gate control. As a result, DMG HEMT is more suitable for communication systems that require high linearity operation.

4 ACKNOWLEDGMENTS

Authors are grateful to Defense Research and Development Organization (DRDO), Ministry of Defense, Government of India, and one of the authors (Rishu Chaujar) is grateful to University Grant Commission (UGC) for providing the necessary financial assistance to carry out this research work.

REFERENCES

[1] S.P.Kumar, A.Agrawal, S. Kabra, M. Gupta and R. S. Gupta, "An analysis for AlGaN/GaN Modulation Doped Field Effect Transistor using accurate velocity-field dependence for high power microwave frequency applications," *Microelectronics Journal*, 37, 1339-1346, 2006.

[2] S.P.Kumar, A.Agrawal, R.Chaujar, S.Kabra, M. Gupta and R. S. Gupta "Threshold Voltage Model for Small Geometry AlGaN/GaN HEMTs Based on Analytical Solution of 3-D Poisson's Equation," *Microelectronics Journal*, 38, 1013-1020, 2007.

[3] E.R.Srinidhi,A.Jarndal and G.Kompa, "A new Method for Identification and Minimization of Distortion Sources in gaN HEMT Devices Based on Volterra Series Analysis", *IEEE Trans. Electron Device Letters,* 28, 343-345, 2007.

[4] C.H.Oxley, "Gallium nitride: the promise of RF power and low microwave noise performance in S and I band", Solid State Electronics, 48, 1197-1203, 2004.

[5]] W.Long, H.Ou, J.M.Kuo and K.K.Chin, "Dual-Material Gate (DMG) Field Effect Transistor", *IEEE Trans. Electron Devices*, Vol.46, No.5, May, 1999, pp. 865-870.

[6] W.Long, H.Ou, J.M.Kuo and K.K.Chin, "Dual-Material Gate (DMG) Field Effect Transistor", IEEE Trans. Electron Devices, 46, 865-70, 1999.
[7] X.Zhou, "Exploring the novel characteristics of hetero-material gate field-effect transistors (HMGFET's) with gate-material engineering," IEEE Trans. Electron Devices, 47, 113–120, 2000.
[8] S.P.Kumar, A.Agrawal, R.Chaujar, M.Gupta and R.S.Gupta,"Nanoscale HEMT with GME Design for High Performance Analog Applications" , Mini Colloquia on "Compact Modeling of Advance MOSFET structures and mixed mode applications", 33, 2008.
[9] ATLAS: Device simulation software, 2002.

Formal Verification of a MEMS Based Adaptive Cruise Control System

Jairam S* Kusum Lata** Subir K Roy* Navakanta Bhat†

* SDTC, TI India, (sjairam,subir)@ti.com
** CEDT, IISc India, lkusum@cedt.iisc.ernet.in
†ECE IISc India, navakant@ece.iisc.ernet.in

ABSTRACT

A formal verification approach is presented for MEMS based adaptive cruise control (ACC) system. The system consists of a MEMS based gyroscope for measuring speed. The ACC system and the MEMS component are first modeled as a hybrid system, and then validated using a discrete time domain dynamic simulation approach in the Simulink/Stateflow (SS) framework from Mathworks. For its validation using a formal approach, CheckMate [1] a public domain formal verification tool for hybrid systems is used. In this paper we outline our experiences and highlight several issues faced in using CheckMate to carry out a formal analysis. The key contributions of the paper include 1) Formulation of realistic properties to enable formal analysis 2) Techniques to model an open hybrid system in CheckMate (it accepts only closed hybrid systems for formal analysis). 4) Transformations in SS models of the ACC and MEMS gyroscope needed to conform to the CheckMate model. 5) Description of changes necessary in the CheckMate methodology to enable formal analysis. 6) Optimization of the ACC system parameters using formal runs in CheckMate to identify fail-safe regions of operation. 7) Selection of MEMS gyroscope topologies based on optimized ACC system parameters.

Keywords: Hybrid Systems, MEMS Gyroscope, Simulation, Formal and Semi-Formal Verification

1 Introduction

With the design of hybrid systems becoming increasingly complex their validation to ensure fail safe (or safety-critical) behavior is becoming a challenging task. Automated methods based on formal analysis are the only route by which safety criticality can be guaranteed in such systems [2]. This is specially true of embedded hybrid system controllers targeting the automotive domain. Hybrid systems are characterized by continuous time differential equations which work concurrently with discrete time digital systems. Modeling of such hybrid system involves modeling both the discrete behavior, as well as, the continuous time or dynamic behavior. Most approaches to modeling hybrid systems is based on extending finite automata used for modeling discrete behavior to include simple continuous behavior. Validation for safety critical behavior implicitly involves determining the reachable set of states of the hybrid system on this model, and ensuring that it never reaches a state space representing unsafe operation.

In this paper paper we propose a formal verification approach for validation of MEMS based hybrid systems. The methodology is demonstrated on an adaptive cruise control (ACC) system consisting of a MEMS based gyroscope for measuring speed. The organization of the paper is as follows. In Section 2 the ACC system and the MEMS component are first modeled as a hybrid system in the SS framework. The safety critical behavior of the ACC system is validated in this framework through time domain simulation. In Section 3, we introduce the formal analysis approach based on CheckMate [1]. The safety critical behavior of ACC is captured by a set of properties which are then formally verified in CheckMate.CheckMate imposes the following restriction for performing formal analysis. It assumes a hybrid system to be closed. Our ACC system model is open. While the SS framework easily allows modeling of an open hybrid system, it needs some effort to model this in CheckMate. The MEMS gyroscope model in SS uses several continuous time domain dynamic components which do not belong to the set of dynamic components allowed by CheckMate. To enable formal analysis of the MEMS based gyroscope ACC system in CheckMate, we circumvent this problem through the use of a Look Up Table (LUT) to macromodel the gyroscope. In Section 4 we describe the generation and integration of this LUT in CheckMate, as well as, show how an open system can be modeled in it. In Section 5, we present our results, discuss several modeling issues faced in the formal validation process, extensions to the proposed work related to optimization of the ACC system parameters and selection of MEMS gyroscope topologies for these parameters.

2 ACC System Description, Modeling & Simulation Setup

Figure 1 shows the state transition diagram of the MEMS based ACC system. The system behavior consists of four states, *viz*, '*HALT*', '*ACCELARATE*',

'$CRUISE$' and '$RETARD$'. The variables x_p (Proximity of the tracking vehicle to the leading vehicle) and v (velocity of the tracking vehicle) govern the assignments to different states and the transitions between these states. Our ACC system model can easily be seen to be open with respect to the velocity of the leading vehicle. In the ACC system, the control actions depends on the behavior of the leading vehicle resulting from changes in its velocity V_L [3]. As can also be noted in the ACC system model, though x_p is a derived system variable, it is nevertheless treated as an independent system variable, as it too causes control actions to be initiated. Thus, x_p is a system level input to the controller, while the velocity of the tracking vehicle is an intra-system input, sensed by the MEMS based gyroscope whose output is an input to the ACC system. The differential equations corresponding to the $ACCELARATE$ and $RETARD$ states are $\dot{v} = A$ and $\dot{v} = R$, respectively.

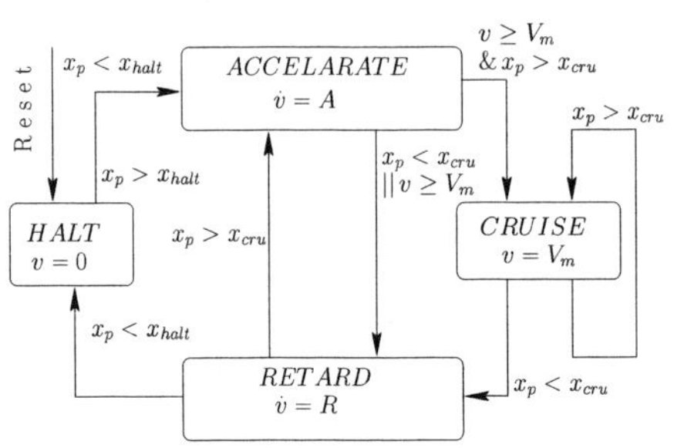

Figure. 1 State Transistion Graph for ACC

We give below, a brief summary of the MEMS gyroscope model. More details can found in [4] and in [7]. Figure 2 illustrates a first order implementation of a gyroscope. The output is modeled as a capacitance, which is seen to be a function of input motion [5]. Thus, the analytical model gives us the value of the capacitance as a function of vehicular angular velocity. The model transfer function $H(s)$ can be computed by abstracting out the system parameters and retaining only the input and output variables. This results in the following expressions:

$$h_1(s) = \mathcal{K}\frac{A\omega^2}{s^2+\omega^2}\frac{s}{(Ms^2+B_ds+K_d)} \quad (1)$$

$$h_2(s) = \frac{1}{(Ms^2+B_ss+K_s)} \quad (2)$$

where $\mathcal{K}$ is a constant, M is the proof-mass, $B_{s|d}$ and $K_{s|d}$ are the respective damping and stiffness co-efficients for drive and sense modes respectively.The output response for an input angular rate in time and frequency domain can be written down as [7]:

$$O(s) = [\Omega(s) \star h_1(s)]h_2(s) \quad (3)$$

$$O(t) = [\Omega(t)\, h_1(t)] \star h_2(t) \quad (4)$$

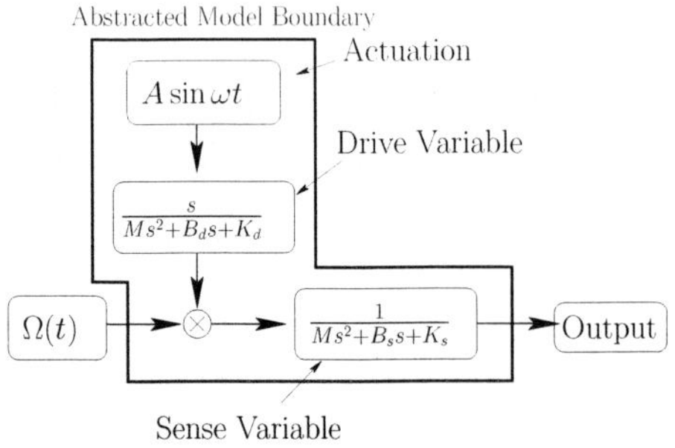

Figure. 2 Mathematical Model of a Gyroscope

The above mathematical model of the MEMS gyroscope and the complete ACC system is implemented in the SS framework as shown below in Figure 3. This setup is used to validate the specifications by real time analysis. The safety critical properties (described in Section 7) are initially analysed using time domain simulation in this platform.

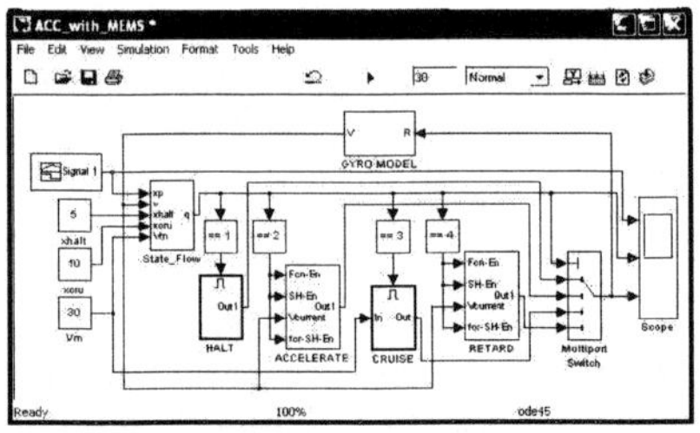

Figure.3 SS Model for ACC

3 Formal Analysis in CheckMate

We give a brief overview of the proposed formal verification approach carried out in CheckMate. For more details on CheckMate we refer to [6]. We then explain the implementation of the ACC system in CheckMate, with our proposed modifications.

For the formal analysis, CheckMate identifies the continuous set of reachable state space for a given set of different dynamics associated with a hybrid system. This is achieved by constructing a flow-pipe using different trajectories over time of the hybrid system originating from a well defined minimal set of initial states chosen from a given initial continuous state space set. This flow pipe represents the set of reachable points in the vector space of state variables of the hybrid system. It then approximates this flow-pipe with overlapping linear polyhedrons [6]. Formal analysis of the ACC system with respect to each property is carried out by using the approximate flow pipe region and the region defined by a property being validated. The formal analysis in CheckMate is performed in two stages; *viz. Explore* and *Verify*. The *Explore* phase performs time domain simulation in SS to store the different trajectories needed to construct a flow pipe. The construction of the flow pipe and its approximation along with property verfication computational geometry based algorithms is carried out in the *Verify* phase.

4 ACC Model in CheckMate system

CheckMate accepts a restrictive hybrid automata model, known as polyhedral invariant hybrid automata (PIHA) [6]. This requires transformation of the general SS model into a restrictive Simulink/Stateflow model by using a subset of its models/blocks allowed by CheckMate to create the equivalent PIHA model for formal analysis. The MEMS gyroscope model in SS uses several continuous time domain dynamic components which do not belong to the set of dynamic components allowed by CheckMate.

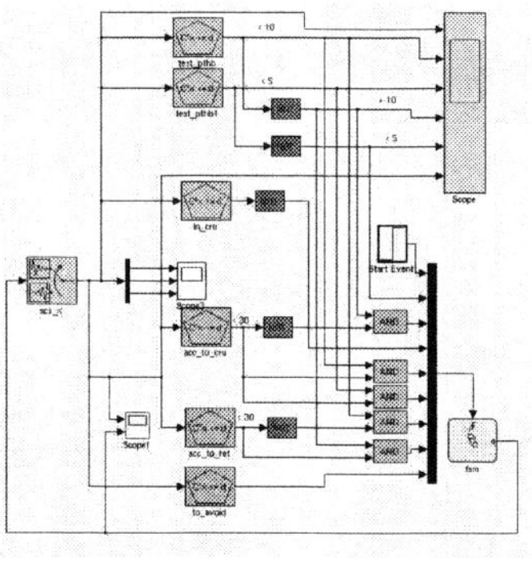

Figure. 4 ACC CheckMate Model

To enable formal analysis of the MEMS based gyroscope ACC system in CheckMate, we circumvent this problem through the use of a Look Up Table (LUT) to macromodel the gyroscope. The data points for the LUT is obtained for a range of velocity values by carrying out dynamic simulation on an exact macro-model of the MEMS gyroscope in the SS framework (Figure 3). To integrate the LUT macro-model of the MEMS gyroscope in CheckMate it is necessary to make changes in its implementation code in Matlab. The LUT is accessed through function calls in CheckMate to get the desired outputs of the gyroscope to obtain the trajectories needed in the formal analysis of the ACC system.

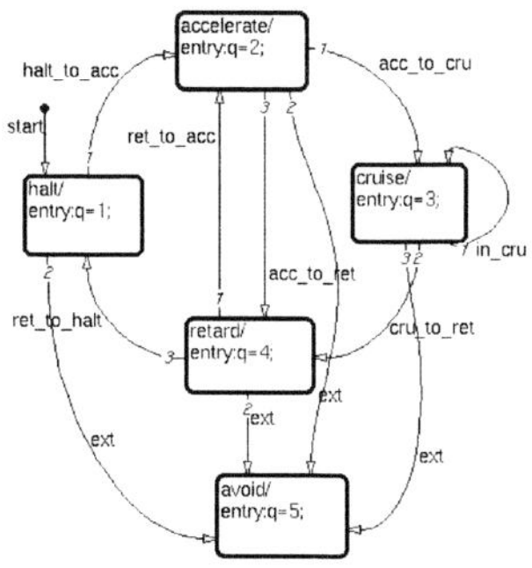

Figure. 5 ACC CheckMate State Transistion Graph

Figures 4 and 5 represent the CheckMate model for the ACC system and its associated state transition graph. The LUT based gyroscope macro-model was included between the switched dynamic block and the PTHB modules in Figure 4. One restriction imposed by CheckMate is that for formal analysis it assumes a hybrid system to be closed. Our ACC system model can easily be seen to be open with respect to the velocity of the leading vehicle V_L, as the control actions in the hybrid automata of the ACC system (Figure 1), depends on the behavior of the leading vehicle resulting from changes in its velocity V_L. While the SS framework easily allows modeling of an open hybrid system, it needs some effort to model this in CheckMate. We model such a scenario by addition of a redudant equation in terms of V_L, V_T and proximity (x_p) in which we render V_L as a parameter (V_T and x_p as the closed system state variables). The equation used is, $\int(V_L - V_T)dt = x_p$.

5 Results, Modeling Issues and Extensions

Table 1 lists the properties that were verified formally in CheckMate. Some of these properties are related to checking for safety critical conditions, such as checking of cruise velocity limit and tracking vehicle velocity limits within halt range.

Table 1: System Property Summary

Functional Specification	Property Validation
$R = 0 : x_p > x_{cru}$	Pass
$A = 0 ; x_p < x_{halt}$	Pass
$State \neq CRUISE \ : x_p < x_{xcruise}$	Pass
$x_p > 0 \, \forall \, t$	Pass

The sensing of velocity introduces error in the control process. This can be comprehended either as an error between the state velocity and the engine sensed velocity, or as a delay in the measured velocity. We use an error approximation to capture this effect of the gyroscope. This also provides another motivation for the design of the ACC system to be more realistic. The reset feature in CheckMate invoked in the sink state of a transition edge, at the end of a state transition, can easily hide and abstract out real life behavior which maybe needed in the optimization route based on the formal analysis. We show below, how we can avoid using this feature and still be able to obtain authentic PIHA models by transforming the hybrid automata. We validate such transformations to the ACC system model by both, simulation and formal analysis.

From the formal analysis results obtained from the proposed model a property was found to be failing. This was mainly because of a step change in velocity from a non-zero value to zero due to a transition from *RETARD* to *HALT* state. As a result of this analysis, the Stateflow graph shown in Figure 5, was refined by addition of an extra retardation state (Figure 6). A lower bound on the retardation rate of the dynamics in this intermediate state was computed. This ensures that the vehicle actually retards to zero velocity before transitioning to the *HALT* state, while at the same time maintains the proximity constraint associated with the *HALT* state. The CheckMate model of the ACC system shown in Figure 6, was constructed for the hybrid automata model shown in Figure 1. The simulation results for this property is shown in Figure 7. The formal property was able to capture a faulty state that captures a crash phenomenon. The refined hybrid automata as shown in Figure 6 was analysed both, formally and with simulation. The property passed in both forms of analysis.

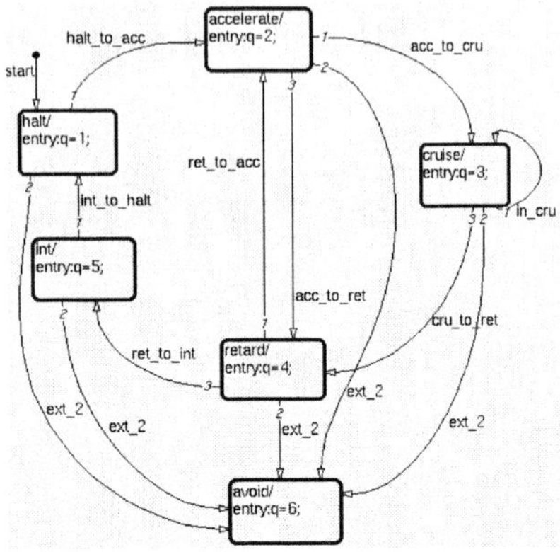

Figure. 6 Refined FSM for ACC CheckMate STG

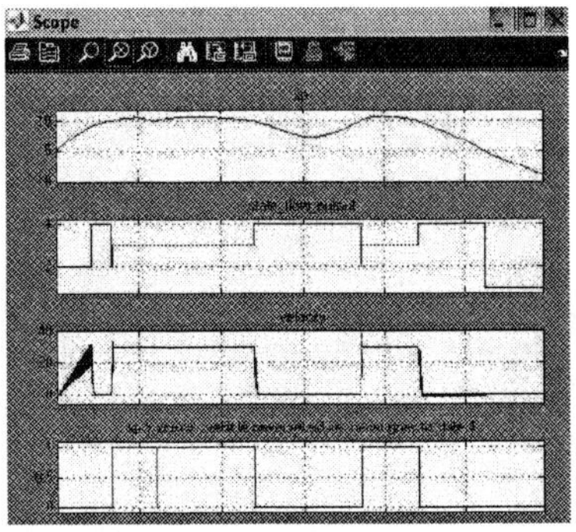

Figure 7 SS Output of ACC for Different Properties

We next, show an extension of our approach in which we optimize the ACC system parameters. We do this by using formal runs in CheckMate to identify fail-safe regions of operation for a given continuous set of initial states. Based on the formal analysis, the lower bound on the retard rate could be computed. The vehicle should halt before the proximity reduces below x_{halt}. We introduce a parameter η, in a new intermediate state which is a function of x_{halt}. The vehicle retards through this state such that at $x = x_{halt}$, velocity is zero. This translates the velocity constraint for the intermediate state as, $V_{final}^2 = V_{initial}^2 - 2R\eta$. Now the upper bound on $V_{initial}$ is V_{cru} and V_{final} should tend to zero at the end of the state. Hence retardation rate can be constrained as $R \geq V_{cru}^2/2\eta$. Hence if a suitable constraint for η is selected (eg. $\eta = 2x_{halt}$) a lower bound on the retardation rate for the new state can be computed. As a result the other parameters of the system (*viz.* accelaration, retardation rates and cruise limits) could be re-computed.

These same parameters can be directly translated to frequency requirements of the MEMS gyroscope that needs to measure this velocity. In [7] it is shown that system parameter variations can result in different topology selection of gyroscopes. This helps in selecting appropriate MEMS gyroscope topologies for the optimized ACC system parameters as the choice of system parameters can be directly mapped to compiler based approaches for topology synthesis of gyroscopes [7]. Given these parameters we translated this into a topology selection problem, through the framework presented in [7].

6 CONCLUSIONS

In this paper we proposed a formal verification approach for MEMS based embedded systems. The approach was demonstrated on a MEMS based adaptive cruise control system. The MEMS component is a gyroscope used for speed measurement. CheckMate, a public domain formal verification tool, was modified and deployed to include MEMS based components for formal analysis. The approach was also used to make topology selections of gyroscopes based on ACC system requirements.

REFERENCES

[1] Checkmate Carnegie Mellon University Website. http://www.ece.cmu.edu/ webk/checkmate/

[2] ”Claire J. Tomlin et al., Computational Techniques for the Verification of Hybrid Systems, Proc. IEEE, 91,7, pp 986-1001, 2003.

[3] Jairam S, Subir K Roy and Navakanta Bhat, A Transformation Based Method for Formal Analysis of Hybrid Systems, Proc. VDAT, Aug 2007, Kolkata India.

[4] Gary Fedder et. al, Integrated Microelectromechanical Gyroscopes, Journal of Aerospace Engg, 6,2, pp 65-75, 2002.

[5] Mohit S et. al, Design, modelling and simulation of vibratory micromachined gyroscopes, Journal of Physics Conference Series, Apr 34, pp 757-763, 2006.

[6] Chutinan A and Krogh B. H,, Computational techniques for hybrid system verification, IEEE Tran. on Automatic Control, 48, 1, pp 64-75, 64-75, 2003

[7] Jairam S and Navakanta Bhat,GyroCompiler: A Soft-IP Model Synthesis and Analysis Framework for Design of MEMS based Gyroscopes, International Conf. on VLSI Design, pp 589-594, Jan 2008.

High-Frequency Characteristic Optimization of Heterojunction Bipolar Transistors

Yiming Li*, Ying-Chieh Chen, and Chih Hong Hwang

Department of Communication Engineering, National Chiao-Tung University, Hsinchu 300, Taiwan,
*E-mail: ymli@faculty.nctu.edu.tw

ABSTRACT

With the downward scaling of IC critical dimension, the speed of SiGe heterojunction bipolar transistors has been increased dramatically. The speed of HBTs is dominated by the base transit time, which may be strongly influenced by the doping profile in the base region and the Ge concentration of base region. Therefore, the determination of the doping profile and Ge concentration of base region is crucial for design of SiGe HBTs in advanced communication circuits. In this study, the design of HBTs is transformed to a convex optimization problem, and solved efficiently by geometric programming approach. The result shows that a 23% Ge fraction may maximize the current gain and a 12.5% Ge can maximize the cut-off frequency, where 254 GHz is achieved. The accuracy of the optimization technique was confirmed by TCAD simulation. This study successfully transforms the device characteristic and manufacturing limitation into a geometric programming model and provides an insight into design of SiGe HBTs.

Keywords: Bipolar Transistors, Impurity Doping, SiGe, HBT, High Frequency, Optimization.

1 INTRODUCTION

SiGe heterojunction bipolar transistors (HBTs) have undergone substantial development for nearly two decades. The speed of SiGe HBTs has increased dramatically, with the consequence of relentless vertical and lateral scaling. The HBTs' speed is dominated by the transit time of base region, which is strongly influenced by the doping profile and Ge concentration in the base region [1-6].The determination of the doping profile and Ge concentration o is crucial for design of SiGe HBTs in advanced communication circuits.

Diverse approaches have been proposed to optimize the base transient time [3-6]. An analytical optimum base doping profile by using variation calculus considering the dependence of diffusion coefficient on base doping concentration is firstly derived [3]. The analytical approach has been extended to consider the dependence of intrinsic carrier concentration on base doping concentration [4]. An iterative approach is proposed by Winterton to obtain the optimum base doping profile [5], where the dependence of mobility and bandgap narrowing on the base doping concentration is further considered by Kumar [6]. However, the solution can not be guaranteed to be the global optimal.

A geometric program (GP) is a type of mathematical optimization problem characterized by objective and constraint functions with special form. Recently, numbers of practical problems, particularly in semiconductor and electrical circuit design, have been found to be equivalent (or can be well transformed) to GP's form [8-13]. For the SiGe HBTs, it has been reported that the triangular Ge profiles are best suited to achieve the minimum base transit time and trapezoidal Ge profiles are best suited to get high current gain in SiGe HBTs [7]. The geometric programming approach has been utilized to simultaneously optimize the Ge-dose and base doping profile in SiGe HBTs [1, 2]. However, the co-optimization of cutoff frequency and current gain in SiGe HBTs is lacked.

In this study, the GP approach is used to obtain the optimal Ge-dose and doping profile to get high cutoff frequency or the high current gain in SiGe HBTs. The design of HBTs is first expressed as a special form of optimization problem, called geometric programming. The background doping profile is adjustable to improve the cutoff frequency and current gain. The result shows that a 23% Ge fraction may maximize the current gain, where a factor, current gain divided by emitter Gummel number, of 1100 is attained. To maximize the cut-off frequency of HBTs, a Ge-dose concentration of 12.5% is used, where the cut-off frequency can achieve 254 GHz. The accuracy of the optimization technique was confirmed by TCAD simulation. This study successfully transforms the device characteristic and manufacturing limitation into a geometric programming model and provides an insight into design of SiGe HBTs.

The paper is organized as follows. In Sec. 2, the design of HBTs and manufacturing limitation are transformed to a geometric programming model. In Sec. 3, the cut-off frequency and current gain are optimized. Finally we draw conclusions.

2 THEORY AND METHODOLOGY

Mathematically, a doping profile tuning problem for the frequency property of SiGe HBTs can be formulated as an optimization problem:

$$\begin{aligned} &\text{Minimize} && \tau_B \\ &\text{s.t.} && N_{\min} \le N_A(x) \le N_{\max}, 0 \le x \le W_B, \\ & && Ge_{AVG} \le 0.23 \end{aligned} \qquad (1)$$

where the base doping profile denoted $N_A(x)$ is a spatial-dependent positive function over the interval $0 \le x \le W_B$. The base doping profile is lower than the doping level of emitter-base junction $N_{\max}$ and higher than background doping $N_{\min}$. W_B is the base width of the transistor. Ge_{AVG} is the average value of Ge fraction. Due to the manufactory limitation, the average value of Ge fraction should be less than 0.23. For SiGe HBTs, the base transit time τ_B as shown in Eq. (1), is given by [6]:

$$\tau_B = \int_0^{W_B} \frac{n_{i,SiGe}^2(x)}{N_A(x)} \left(\int_x^{W_B} \frac{N_A(y)}{n_{i,SiGe}^2(y) D_{n,SiGe}(y)} dy \right) dx, \tag{2}$$

where $n_{i,SiGe}(x)$ is the intrinsic carrier concentration in SiGe, and $D_{n,SiGe}(y)$ is the carrier diffusion coefficient of SiGe. The $D_{n,SiGe}(y)$ can be rewritten as

$$D_{n,SiGe}(y) = (1 + k_{SiGe} Ge_{AVG}) D_n(y), \tag{3}$$

where k_{SiGe} is a constant, and $D_n(y)$ is the carrier diffusion coefficient of Si. In the present work, peak base doping $N_{\max}$ of $1\text{x}10^{19}$ cm^{-3} at emitter edge of base and a minimum base doping $N_{\min}$ of $5\text{x}10^{16}$ cm^{-3} at collector edge has been chosen to include the heavy doping induced band gap narrowing effect in the entire base region. A neutral base width of 100 nm is chosen. We can change Eq. (1)-(3) to a function of $N_A(x)$, Ge_{AVG} and formulate it to GP's form. For a SiGe HBT; the cutoff frequency f_t of a HBT is given by [2]:

$$\frac{1}{2\pi f_t} = \tau_F + \frac{C_{J,BE} + C_{J,BC}}{g_n} + R_C C_{J,BC}, \tag{4}$$

where τ_F is the forward transit time , $C_{J,BE}$ is the base–emitter junction or depletion layer capacitance, $C_{J,BC}$ is the base–collector junction or depletion layer capacitance , g_m is the transconductance, R_C is the collector resistance. For this model, the base transit time is often the major part to determine the value of τ_F and govern the f_t. We can also change (4) as a function of $N_A(x)$. Without loss of generality, we may assume the doping profile to be the form

$$N_A(x) = bx^m, 0 \le x \le 0.05W_B \tag{6}$$

Here we assume m=0 for liner doping within 5 nm base width near the emitter-base junction. To figure out an ideal shape of the optimal doping profile that maximizes the f_t, we consider the optimal problem:

$$\begin{aligned} \text{Minimize} \quad & \tau_B + AN_A(x)^{1/2} G_B (1 + k_{SiGe} Geavg)^{-1} + B \\ \text{s.t.} \quad & N_{\min} \le N_A(x) \le N_{\max}, i = 0,1,\dots,M-1, \\ & N_A(x), x = 0 \le x \le 0.05W_B \\ & Ge_{AVG} \le 0.23 \end{aligned} \tag{7}$$

where A and B are constants, G_B is the base Gummel number, which is also a function of $N_A(x)$. The SiGe HBT with various Ge-dose concentration, 2%, 8%, 12%, are then explored in Fig. 1(a). The device with higher Ge-dose concentration shows a higher f_t. Moreover, the obtained optimzied doping profile changes with different Ge-dose concentration. The dependence of f_t as a function of Ge-dose concentration is plotted in Fig. 1(a). The device with maximum f_t is with 12% Ge dose concentration, as studied in Fig.1(b).

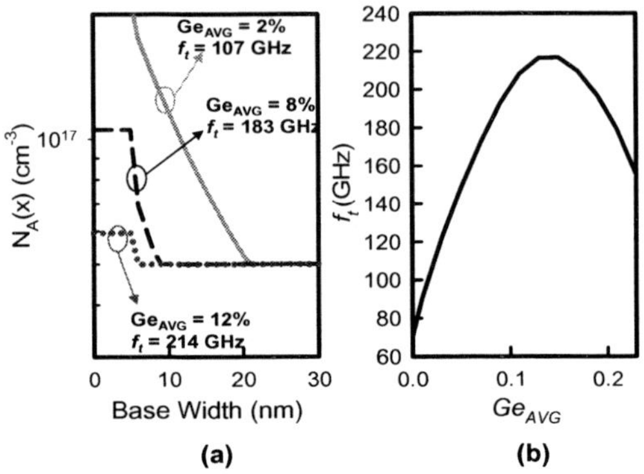

Fig.1(a). Doping profile and the corresponding f_t with 2%, 8%, and 12% Ge-dose. (b) f_t with various Ge-dose.

Besides the optimization of f_t, the current gain, β, is also significantly influenced by the base doping profile. The current gain is defended by the ratio of collector and can be expressed as ratio of Gummel numbers:

$$\beta = \frac{G_{E,Sige}}{G_{B,SiGe}} \tag{8}$$

where $G_{E,SiGe}$ is the emitter Gummel number, and $G_{B,SiGe}$ is the base Gummel number. Since the emitter Gummel number depends mostly on the emitter doping profile, and thus can be treated as a positive constant. For the base Gummel number, the dependence of Gummel number depends on the base doping profile:

$$G_B = \int_0^{W_B} \frac{N_A(x) n_{i0}^2}{D_n(x) n_i^2(x)} dx \tag{9}$$

where n_{i0} is the intrinsic carrier concentration in a undoped Si . $n_i(x)$ is the intrinsic carrier concentration in SiGe. The relationship Eq.(7) and Eq. (8) are then transformed as the current gain constraint and added in the GP model. The background doping of the doping profile is also a factor in optimization of base doping profile. Figure 3 shows the impact o f background doping profile on f_t. As the background doping, N_{min}, is decreased from 5×10^{16} cm^{-3} to 3×10^{16} cm^{-3}, the obtained optimal f_t could be increased from 71 GHz to 85 GHz.. To ensure the accuracy of the optimized doping profile, the doping profile is implemented in the TCAD tool, as shown in Fig. 3, where the solid line shows the optimized doping profile and the dashed line shows the doping profile in TCAD. The f_t in TCAD simulation approaches 70 GHz, which is very similar to the f_t in the GP model, 71 GHz. The result confirms the reliability of the GP model.

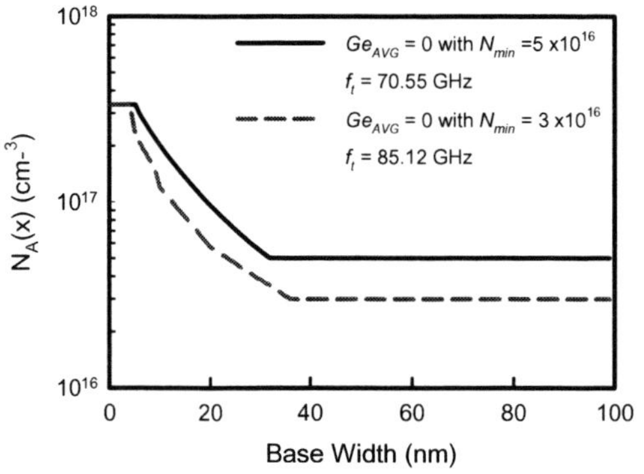

Fig. 2. Doping profile of decreasing background doping to 3×10^{16} cm^{-3} for 0% Ge content.

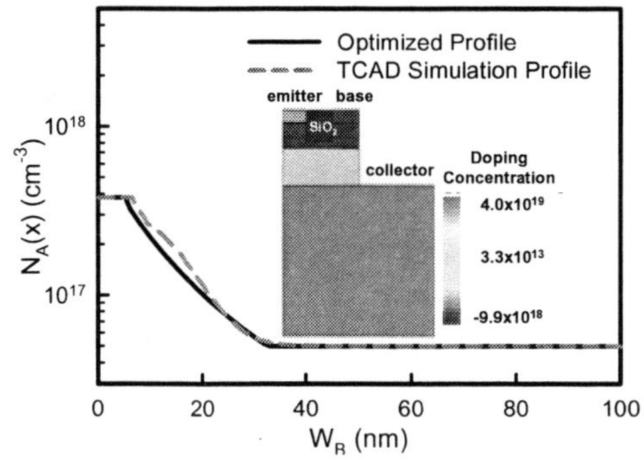

Fig. 3. Two-dimensianal device structure of a SiGe HBTs and the doping profile obtained from GP model and TCAD simulation.

3 RESULTS AND DISSCUSSION

The dependence of f_t and gain on Ge-dose and base doping profile are discussed above. In this section, Due to the strong influence of the shape and content of Ge on the base transit time [4], the f_t and gain of SiGe HBT are co-optimized with subject to the constraint mentioned above.

Figure 4 shows the f_t as a function of background doping and Ge-dose. As expected, the maximum f_t can be obtained at 12% Ge-dose with low background doping. Figure 5 shows the f_t as a function of the current gain. Since the f_t is related to the gain and bandwidth, the obtained f_t will be smaller with higher current gain constraint. As expected, device with higher Ge-dose could provide higher gain and thus release the design constraint. The tuning point, in which the current gain constraint starts to significantly reduce the f_t, is crucial in obtaining the maximum current gain with sufficient f_t. Therefore, by careful selection of the maximum current gain constraint, we could find the optimal current gain constraint, $\beta / G_{E,SiGe} \times 10^{11}$, with sufficient f_t, as shown in Figure 6, where the lower background doping and higher Ge-dose may provide the largest current gain. In Figure 4, it's found that 12.5% Ge-dose and 2×10^{16} cm^{-3} background doping can maximize the f_t. The higheset f_t can reach 254 GHz. On the other hand, for obtaining the maximum current gain in Fig. 6, the Ge-dose concentration is about 23% and the background doping is about 2×10^{16} cm^{-3}, where maximize current gain constraint $\beta / G_{E,SiGe} \times 10^{11}$ = 1100.

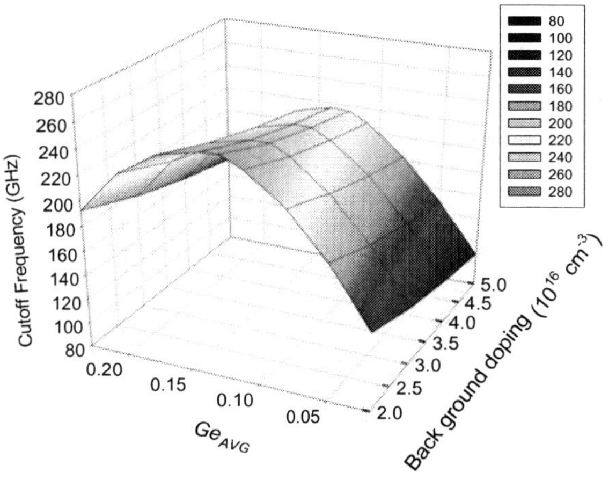

Fig. 4. The f_t as a function of Ge-dose and background doping concentrations.

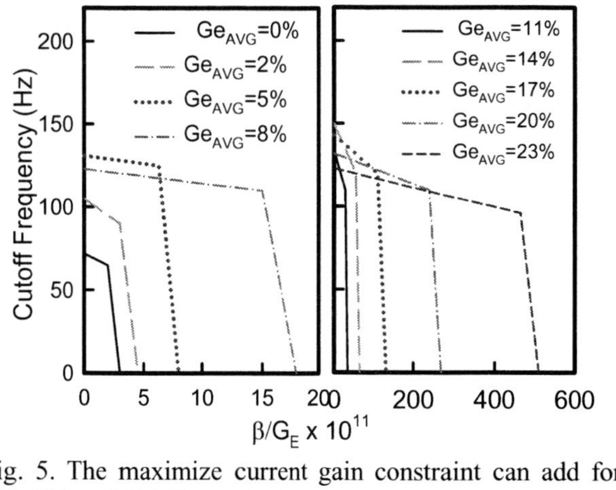

Fig. 5. The maximize current gain constraint can add for 0%, 5%, and 8% Ge content.

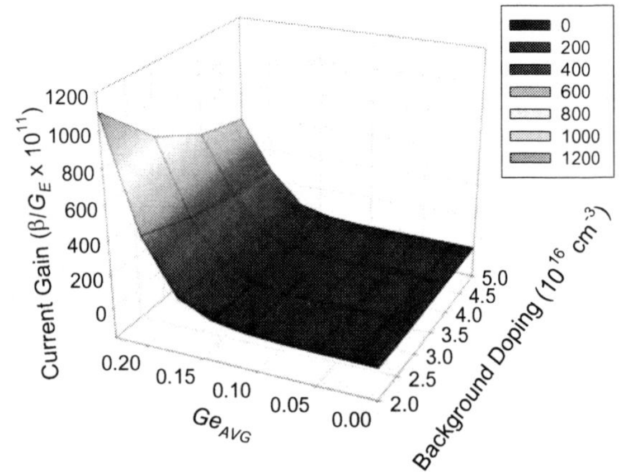

Fig. 6. The maximum current gain constraint can add for every Ge content and background doping to maintain sufficient f_t.

The obtained optimal doping profile and Ge-dose concentration are plotted in Fig. 7. Result shows that for the SiGe HBTs, the triangular Ge profiles are best suited to achieve the minimum base transit time and trapezoidal Ge profiles are best suited to get high current gain in SiGe HBTs, which matches the result in [7].

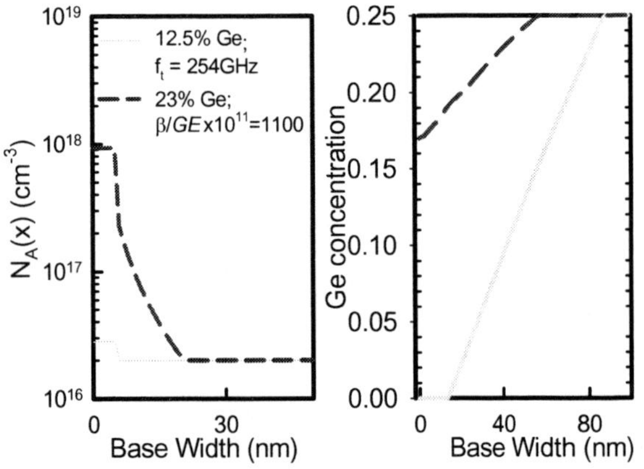

Fig. 7. Optimal Si and Ge doping profile for f_t maximize and maximize current gain constraint.

4 CONCLUSION

In this study, the cutoff frequency and the current gain of SiGe HBT are optimized by geometric programming approach. The design of HBTs is transformed to a convex problem, and solved efficiently. The result shows that a 23% Ge fraction may maximize the current gain and a 12.5% Ge can maximize the cut-off frequency, where 254 GHz cut-off frequency is achieved. The accuracy of the optimization technique was confirmed by TCAD simulation. This study successfully transforms the device characteristic and manufacturing limitation into a geometric programming model and provides an insight into design of SiGe HBTs. We are currently applying the GP approach for multifinger HBTs' optimization.

5 ACKNOWLEDGEMENT

This work was supported by Taiwan National Science Council (NSC) under Contract NSC-96-2221- E-009-210, NSC-95-2221-E-009-336 and Contract NSC- 95-2752-E-009-003-PAE, by MoE ATU Program, Taiwan under a 2006-2007 grant, and by the Taiwan Semiconductor Manufacturing Company, Hsinchu, Taiwan under a 2006-2008 grant.

REFERENCES

[1] Gagan Khanduri, "Simultaneous Optimization of Doping Profile and Ge-Dose in Base in SiGe HBTs," in IEEE Southeast Conf., 579, 2007.

[2] S. Joshi, S. Boyd, and R. Dutton, "Optimal doping profiles via geometric programming," IEEE Trans. Electron Device, 52, 2660-2675, 2005.

[3] A. H. Marshak, "Optimum doping distribution for minimum base transit time," IEEE Trans. Electron Devices, ED-14, 190, 1967.

[4] S. Szeto and R. Reif, "Reduction of f by nonuniform base bandgap narrowing," IEEE Electron Device Lett., 10, 341, 1989.

[5] S. S. Winterton, S. Searles, C. J. Peters, N. G. Tarr, and D. L. Pulfrey, "Distribution of base dopant for transit time minimization in a bipolar transistor," IEEE Trans. Electron Devices, 43, 170, 1996.

[6] M. J. Kumar and V. S. Patri, "On the iterative schemes to obtain optimum base profiles for transit time minimization in a bipolar transistor," IEEE Trans. Electron Devices, at press.

[7] Kyunghae Kim, Jinhee Heo, Sunghoon Kim, D. Mangalaraj, and Junsin Y, "Optimum Ge Profile for the High Cut-off Frequency and the DC Current Gain of an SiGe HBT for MMIC," IEEE Trans. Electron Device, 40, 584, 2002.

[8] K. Kortanek, X. Xu, and Y. Ye, "An infeasible interior-point algorithm for solving primal and dual geometric progams," Math. Program., 76, 155, 1996.

[9] M. Avriel, R. Dembo, and U. Passy, "Solution of generalized geometric programs," Int. J. Numer. Methods Eng., 9, 149, 1975.

[10] C. S. Beightler and D. T. Phillips, Applied Geometric Programming. New York: Wiley, 1976.

[11] R. J. Duffin, "Linearizing geometric programs," SIAM Rev., 12, 211, 1970.

[12] R. J. Duffin, E. L. Peterson, and C. Zener, Geometric Programming—Theory and Applications: Wiley, 1967

[13] J. Ecker, "Geometric programming: Methods, computations and applications," SIAM Rev., 22, 338, 1980.

From MEMS to NEMS: Modelling and characterization of the non linear dynamics of resonators, a way to enhance the dynamic range

N.Kacem[1], S.Hentz[2], H.Fontaine[1], V.Nguyen[1], P.Robert[1], B.Legrand[3] and L.Buchaillot[3]

[1]Microsystems Components Laboratory, CEA/LETI - MINATEC, Grenoble, France
[2]Device simulation and characterization Laboratory, CEA/LETI - MINATEC, Grenoble, France
[3]IEMN, CNRS UMR 8520, Villeneuve d'Ascq, France

ABSTRACT

The resonant sensing technique is highly sensitive, has the potential for large dynamic range, good linearity, low noise and potentially low power. The detection principle is based on frequency change that is induced by rigidity changes in the resonator.
In order to compensate the loss of performances when scaling sensors down to NEMS, it proves convenient to find physical conditions in order to maximise the signal variations and to push the limits of the linear behavior. To do so, a comprehensive model including the main sources of nonlinearities is needed. In the present paper, a process, characterization methods and above all an original analytical model are presented.

Keywords: MEMS, NEMS, Resonator, non linear dynamics, dynamic range, characterization, analytical model

1 Introduction

The permanent quest for cost cuts has led to the use of potential "In-IC" compatible thin SOI-based technologies, which imposes drastic size reduction of the sensors. Combined with the need for in-plane actuation for fabrication and design simplicity, this implies a large reduction in detectability. Moreover nonlinearities [1] occur sooner for small structures which reduces their dynamic range.
On the way from MEMS to NEMS, a "small" MEMS resonant accelerometer [2] has been fabricated. The sensor structure has been designed in order to validate process and characterization choices.
This paper deals with the design of the sensitive element: the resonator. Many studies have presented models on the dynamic behavior of MEMS resonators. Some of them are purely analytical [3, 4] but they include coarse assumptions concerning nonlinearities. Other models [5, 6] are more complicated and use numerical simulations which makes them less interesting for MEMS designers. In the present paper, a compact and analytical model including the main sources of nonlinearities is presented and validated thanks to the characterization of the accelerometer sensing element, an electrostatically driven clamped-clamped beam.

2 Model

A clamped-clamped microbeam is considered (Figure 1) subject to a viscous damping with coefficient $\tilde{c}$ per unit length and actuated by an electric load $v(t) = Vdc + Vac\cos(\tilde{\Omega}\tilde{t})$, where Vdc is the DC polarization voltage, Vac is the amplitude of the applied AC voltage, and $\tilde{\Omega}$ is the excitation frequency.

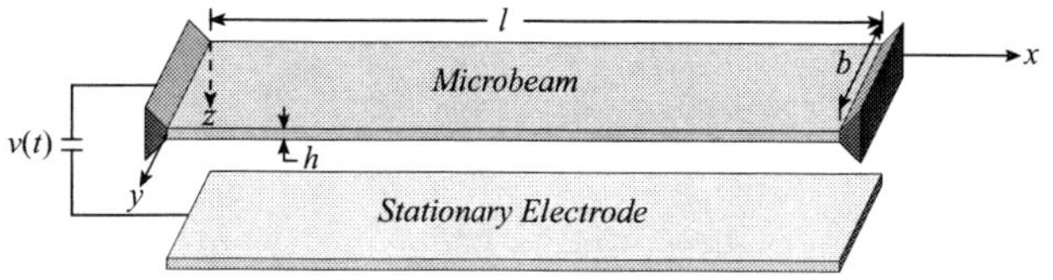

Figure 1: Schema of an electrically actuated microbeam

2.1 Equation of motion

The equation of motion that governs the transverse deflection $w(x,t)$ is written as:

$$EI\frac{\partial^4\tilde{w}(\tilde{x},\tilde{t})}{\partial\tilde{x}^4} + \rho bh\frac{\partial^2\tilde{w}(\tilde{x},\tilde{t})}{\partial\tilde{t}^2} + \tilde{c}\frac{\partial\tilde{w}(\tilde{x},\tilde{t})}{\partial\tilde{t}} =$$

$$\left[\tilde{N} + \frac{Ebh}{2l}\int_0^l\left[\frac{\partial\tilde{w}(\tilde{x},\tilde{t})}{\partial\tilde{x}}\right]^2 d\tilde{x}\right]\frac{\partial^2\tilde{w}(\tilde{x},\tilde{t})}{\partial\tilde{x}^2} \quad (1)$$

$$+\frac{1}{2}\varepsilon_0\frac{bC_n\left[Vdc + Vac\cos(\tilde{\Omega}\tilde{t})\right]^2}{(g-\tilde{w}(\tilde{x},\tilde{t}))^2}$$

where $\tilde{x}$ is the position along the microbeam length, E and I are the Young's modulus and moment of inertia of the cross section. $\tilde{N}$ is the applied tensile axial force due to the residual stress on the silicon or the effect of the measurand, $\tilde{t}$ is time, ρ is the material density, h is the microbeam thickness, g is the capacitor gap width, and ε_0 is the dielectric constant of the gap medium. The last term in equation (1) represents an approximation of the electric force assuming a complete overlap of the area of the microbeam and the stationary electrode including the edge effects by the coefficient C_n [7]. The boundary conditions are:

$$\tilde{w}(0,\tilde{t}) = \tilde{w}(l,\tilde{t}) = \frac{\partial \tilde{w}}{\partial \tilde{x}}(0,\tilde{t}) = \frac{\partial \tilde{w}}{\partial \tilde{x}}(l,\tilde{t}) = 0 \qquad (2)$$

2.2 Normalization

For convenience and equations simplicity , we introduce the nondimensional variables:

$$w = \frac{\tilde{w}}{g}, \; x = \frac{\tilde{x}}{l}, \; t = \frac{\tilde{t}}{\tau} \qquad (3)$$

Where $\tau = \frac{2l^2}{h}\sqrt{\frac{3\rho}{E}}$. Substituting equation (3) into equations (1) and (2), we obtain :

$$\frac{\partial^4 w}{\partial x^4} + \frac{\partial^2 w}{\partial t^2} + c\frac{\partial w}{\partial t} - \alpha_2 \frac{[Vdc + Vac\cos(\Omega t)]^2}{(1-w))^2}$$

$$= \left[N + \alpha_1 \int_0^1 \left[\frac{\partial w}{\partial x}\right]^2 dx\right] \frac{\partial^2 w}{\partial x^2} \qquad (4)$$

$$w(0,t) = w(1,t) = \frac{\partial w}{\partial x}(0,t) = \frac{\partial w}{\partial x}(1,t) = 0$$

The parameters appearing in equations (4) are :

$$c = \frac{\tilde{c}l^4}{EI\tau}, \; N = \frac{\tilde{N}l^2}{EI}, \; \alpha_1 = 6\left[\frac{g}{h}\right]^2$$

$$\alpha_2 = 6\frac{\varepsilon_0 l^4}{Eh^3 g^3}, \; \Omega = \tilde{\Omega}\tau \qquad (5)$$

2.3 Resolution

A reduced-order model is generated by modal decomposition transforming equations (4) into a finite-degree-of-freedom system consisting of ordinary differential equations in time. We use the undamped linear mode shapes of the straight microbeam as basis functions in the Galerkin procedure. To this end, we express the deflection as :

$$w(x,t) = \sum_{k=1}^{n} a_k(t)\phi_k(x) \qquad (6)$$

Where $a_k(t)$ is the kth generalized coordinate and $\phi_k(x)$ is the kth linear undamped mode shape of the straight microbeam, normalized such that $\int_0^1 \phi_k\phi_j = 0$ for $k \neq p$ and governed by:

$$\frac{d^4\phi_k(x)}{dx^4} = \lambda_k^2 \phi_k(x) \qquad (7)$$

$$\phi_k(0) = \phi_k(1) = \phi_k'(0) = \phi_k'(1) \qquad (8)$$

Here, λ_k is the kth natural frequency of the microbeam. We multiply equations (4) by $\phi_k(x)(1-w)^2$, substitute equations (6) into the resulting equation, use equations (7) to eliminate $\frac{d^4\phi_k(x)}{dx^4}$, integrate the outcome from $x = 0$ to 1, and obtain :

$$\ddot{a}_k + c_k\dot{a}_k + \lambda_k^2 a_k$$

$$-2\sum_{j=1}^{n}\{\lambda_j^2 {a_j}^2 + c_j a_j \dot{a}_j + a_j \ddot{a}_j\} \int_0^1 \phi_k {\phi_j}^2 dx$$

$$+\sum_{j=1}^{n}\{\lambda_j^2 {a_j}^3 + c_j {a_j}^2 \dot{a}_j + {a_j}^2 \ddot{a}_j\} \int_0^1 \phi_k {\phi_j}^3 dx$$

$$-\sum_{j=1}^{n}\left\{N a_j + \alpha_1 {a_j}^3 \int_0^1 [\phi_j']^2 dx\right\} \int_0^1 \phi_k \phi_j'' dx$$

$$+2\sum_{j=1}^{n}\left\{\alpha_1 {a_j}^4 \int_0^1 [\phi_j']^2 dx + N {a_j}^2\right\} \int_0^1 \phi_k \phi_j \phi_j'' dx$$

$$-\sum_{j=1}^{n}\left\{\alpha_1 {a_j}^5 \int_0^1 [\phi_j']^2 dx + N {a_j}^3\right\} \int_0^1 \phi_k {\phi_j}^2 \phi_j'' dx$$

$$= \alpha_2 [Vdc + Vac\cos(\Omega t)]^2 \int_0^1 \phi_k dx \qquad (9)$$

Noting that the first mode should be the dominant mode of the system and the other modes are neglected, so it suffices to consider the case $n = 1$. Equation (9) becomes :

$$\ddot{a}_1 + (500.564 + 12.3N)a_1 + (1330.9 + 38.3N){a_1}^2$$

$$+ (927 + 28N + 151\alpha_1){a_1}^3 + 471\alpha_1 {a_1}^4 + 347\alpha_1 {a_1}^5$$

$$+c_1\dot{a}_1 + 2.66c_1 a_1 \dot{a}_1 + 1.85 c_1 {a_1}^2 \dot{a}_1 + 2.66 a_1 \ddot{a}_1$$

$$+1.85{a_1}^2\ddot{a}_1 = -\frac{8}{3\pi}\alpha_2 [Vdc + Vac\cos(\Omega t)]^2 \qquad (10)$$

To analyse the equation of motion (10), it proves convenient to invoke perturbation techniques which work well with the assumptions of "small" excitation and damping, typically valid in MEMS resonators. To facilitate the perturbation approach, in this case the method of averaging [8], a standard constrained coordinate transformation is introduced, as given by:

$$\begin{cases} a_1 = A(t)\cos[\Omega t + \beta(t)] \\ \dot{a}_1 = -A(t)\Omega\sin[\Omega t + \beta(t)] \\ \ddot{a}_1 = -A(t)\Omega^2\cos[\Omega t + \beta(t)] \end{cases} \qquad (11)$$

In addition, since near-resonant behavior is the principal operating regime of the proposed system, a detuning parameter, σ is introduced, as given by:

$$\Omega = \omega_n + \varepsilon\sigma \tag{12}$$

Separating the resulting equations and averaging them over the period $\frac{2\pi}{\Omega}$ in the t-domain results in the system's averaged equations, in terms of amplitude and phase, which are given by:

$$\begin{aligned} \dot{A} &= -\tfrac{1}{2}\epsilon\xi_0 A - \tfrac{1}{8}\epsilon\xi_2 A^3 - \tfrac{1}{2}\epsilon\tfrac{\kappa}{\omega_n}\sin\beta + O(\varepsilon^2) \\ A\dot{\beta} &= A\sigma\epsilon - \tfrac{3}{8}\epsilon\tfrac{\chi_3}{\omega_n}A^3 - \tfrac{5}{16}\epsilon\tfrac{\chi_5}{\omega_n}A^5 + \tfrac{7}{10}\epsilon\omega_n A^3 \\ &\quad + \tfrac{1}{2}\epsilon\tfrac{\kappa}{\omega_n}\cos\beta + O(\varepsilon^2) \end{aligned} \tag{13}$$

Where $\omega_n = \sqrt{500.564 + 12.3N}$, $\xi_0 = c_1$, $\xi_2 = 1.85c_1$, $\chi_3 = 927 + 28N + 151\alpha_1$, $\chi_5 = 347\alpha_1$ and $\kappa = \frac{16}{3\pi}\alpha_2 VacVdc$.

The steady-state motions occur when $\dot{A} = \dot{\beta} = 0$, which corresponds to the singular points of equation (13). Thus, the frequency-response equation can be writen in its implicit form as:

$$\begin{aligned} &\left(\frac{3}{4\omega_n}\chi_3 A^2 + \frac{5}{8\omega_n}\chi_5 A^4 - \frac{7\omega_n}{5}A^2 - 2\sigma\right)^2 \\ &= \left(\frac{\kappa}{A\omega_n}\right)^2 - \left(\xi_0 + \frac{1}{4}\xi_2 A^2\right)^2 \end{aligned} \tag{14}$$

The normalized displacement W_{max} with respect to the gap in the middle of the beam and the drive frequency Ω can be expressed in function of the phase β. Thus, the frequency response curve can be plotted parametrically as shown in figure 2.

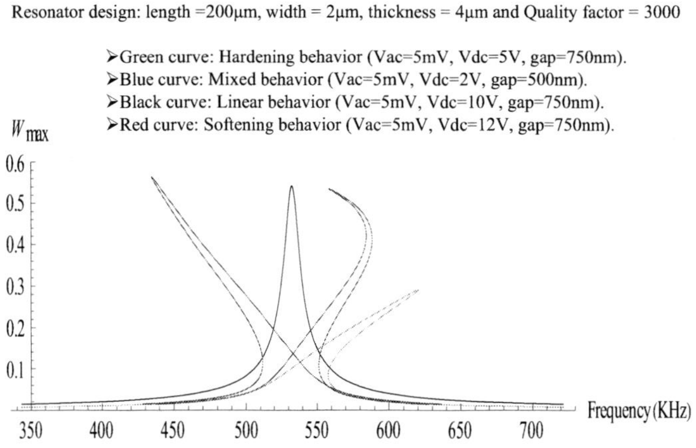

Figure 2: Predicted forced frequency responses

This analytical model enables the capture of all the non linear phenomena in the resonator dynamics and describes the competition between the hardening and the softening behaviors. Specific parameters combination permits the compensation of the nonlinearities in order to enhance the dynamic range of the resonator, i.e. its detectability by obtaining a linear behavior (black curve).

2.4 Motional current

Since the system driving electrode/air/resonator constitutes a capacitor (CMEMS), any mechanical motion of the resonator directly translates into an electrical signal. The total capacitive current generated by this system can be expressed as:

$$\begin{aligned} I_{cap} &= \frac{d}{dt}CV = -\omega_n Vac\sin\left[\omega_n t\right]\int_0^1 \frac{bC_n\epsilon_0}{1 - a_1(t)\phi_1(x)}dx \\ &+ \left(Vdc + Vac\cos\left[\omega_n t\right]\right)\int_0^1 \frac{bC_n\epsilon_0\phi_1(x)a_1'(t)}{\left(1 - a_1(t)\phi_1(x)\right)^2}dx \end{aligned} \tag{15}$$

The total capacitive current was calculated analytically using the results of the reduced order model in displacement and a Taylor series expansion of the capacitance.

3 Manufacturing

The fabrication starts with 200mm SOI wafers ($4\mu m$ Si, $1\mu m$ $SiO2$). The use of DUV lithography combined with deep RIE process has allowed 500nm wide gaps and lines. Some low stiffness beams have been designed, so FH-vapor technique had to be improved to enable the release and protection against in-plane sticking. Main steps are shown in figure 3.

Schema	Description - Equipment
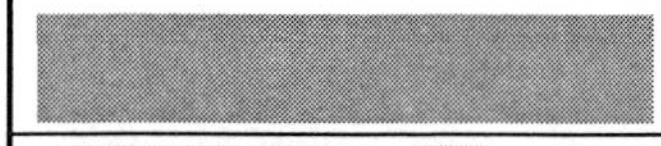	Substrate : SOI wafer - SOI 4μm P 1-10 mohm.cm - Oxide 1000nm - Bulk 725 μm P 7-12 ohm.cm
	Photolithography ASM300 (MEMS Level) Dry etching 4μm of Si (DRIE).
	HF vapor etching of TEOS HF vapor etching of SiO2 (Time control – Lateral release = 5μm)

Figure 3: Simplified process flow

4 Experimental characterisation

For a start, only the resonant beams were released. The fabricated resonators are electrostatically actuated in-plane. Considering the low capacitance variations and the high motional resistance combined with the important parasitic capacitances, tracking the resonance peak purely electrically is rather awkward. An original SEM set-up was developed, coupled with a real-time insitu electrical measurement, using an external low noise lock-in amplifier (Fig.4).

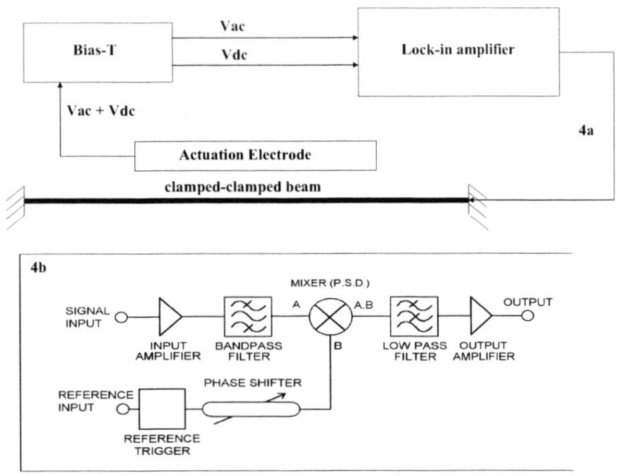

Figure 4: (4a): Connection layout for the electrical characterization. (4b): Block diagram of a lock-in amplifier.

This set-up allows the simultaneous visualization of the resonance by SEM imaging (Fig.5) and the motional current frequency response.

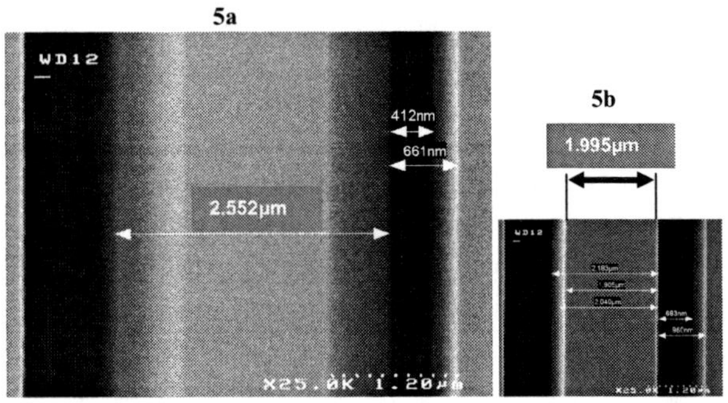

Figure 5: (5a): SEM image of the resonator resonance. (5b): SEM image of the resonator at rest. Dimensions: $200\mu m \times 2\mu m \times 4\mu m$. The gap: around $750\,nm$.

Figure 6 shows one of the first current peaks obtained, slightly non-linear. The residual stress ($\frac{\tilde{N}}{bh} = 38MPa$), the quality factor ($Q = 35000$) and the edge effects coefficient ($C_n = 1.8$) have been measured experimentally. In addition to these 3 experimental parameters, all what we need to compare the experimental and the model results are the geometric and the electric parameters of the resonator (dimensions: $200\mu m \times 2\mu m \times 4\mu m$, $gap = 750nm$, $Vac = 10mV$ and $Vdc = 5V$). First comparisons are in good agreement with model results.

5 Conclusions

As detailed throughout this work, in-plane MEMS resonators was fabricated and electrically characterized thanks to an original SEM-setup. An analytical model including all the sources of nonlinearities was done in order to predict the resonator behaviors. This analytical model based on a reduced order model obtained with Galerkin procedure and solved via the averaging method, has the advantages to be simple and easy to be implemented for MEMS designers. Moreover, using this model, the dynamic range improvement of MEMS resonators is realizable by compensating the hardening and the softening behaviors. In a future work, the whole sensor will be released and characterized under ac acceleration, so its overall dynamic behavior will be studied. Experimental and model results will be compared, which will complete the model validation.

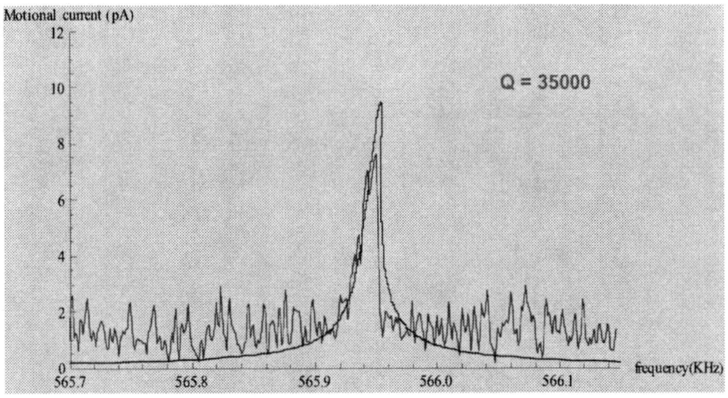

Figure 6: Measured (red) and predicted (blue) motional current frequency responses

REFERENCES

[1] T. Roessig, Ph.D. dissertation, Department of Mechanical Engineering, University of California, Berkeley, 1998.

[2] A. A. Seshia, M. Palaniapan, T. A. Roessig, R. T. Howe, R. W. Gooch, T. R. Schimert, and S. Montague, J. Microelectromech. Syst. 11(6), 784-793 (2002).

[3] H.A.C. Tilmans R. Legtenberg, Sensors and Actuators A 45, pp. 67-84, 1994.

[4] Gui, C., Legrenberg, R., Tilmans, H.A., Fluitman, J.H.J., Elwenspoek, M.:, J. Microelectromech. Syst. 7, 122-127 (1998).

[5] F Najar,S Choura, E M Abdel Rahman, S El-Borgi and A Nayfeh,J. Micromech. Microeng 16 (2006) 2449-2457.

[6] Nayfeh, Ali, Younis Mohammad, Abdel-Rahman Eihab, J. Nonlinear Dynamics 48: 153-163, 2007.

[7] H. Nishiyama and M. Nakamura, IEEE Trans. Comp., Hybrids, Manuf: Technol., vol. 13, pp. 417-423, June 1990.

[8] Nayfeh, A.H.: Introduction to Perturbation Techniques. Wiley, New York (1981).

Modeling and simulation of a monolithic self-actuated microsystem for fluid sampling and drug delivery

Peiyu Zhang, G.A. Jullien
ATIPS Laboratory, ECE, University of Calgary
1-403-2105431, memszhang@yahoo.com

ABSTRACT

A novel MEMS-based microsystem, including a microneedle array with a self-actuated structure for fluid sampling and drug delivery, is modeled, designed and simulated. The self-actuating mechanism and the microneedle array can be fabricated on a monolithic chip with a very reliable fabrication procedure. The microsystem is composed of a microneedle array at the center of the microsystem and an actuating mechanisms of symmetrically arranged Z-shaped PZT unimorph benders. The characteristics of the microsystem are simulated. The simulations show that a large displacement can be accomplished with a relative low actuating voltage. This monolithic microsystem opens up a wide application area for the commercialization of microsystems for fluid sampling and drug delivery.

Keywords: microneedle array, actuator, BioMEMS

1 INTRODUCTION

Hypodermic needles have played an important role in biological fluid extraction and drug delivery systems since they don not have the orally administered drug disadvantages of degradation in the gastrointestinal tract and/or elimination by the liver [1]. But, the disadvantages are also significant such as tissue trauma, insertion pain, difficulty in providing sustained drug release, and the need for expertise to perform an injection. In addition, bolus injections caused by hypodermic needles can lead to high concentrations of drugs being injected into the body and blood stream with the potential of toxic side effects. The need for new drug delivery routes and microbiological sampling has been recognized for some time, especially for new biotechnology drugs that cannot be administered using conventional approaches [2].

Z-shape benders are perfectly suited to actuation of microneedle arrays because they can enlarge and advance the distance of the array. In this paper we discuss the simulation of such an actuator and also include the modal resonance of the system in these simulations. The actuation issue is a required feature and also a considerable challenge for microneedle commercialization. Very few publications on microneedles with actuating mechanisms appear in peer reviewed publications, let alone both microneedle and actuation structures on a monolithic chip.

Microneedles based on MEMS technologies, which are a byproduct of the advances in microelectronics and integrated circuit technologies, can overcome the disadvantages above and provide pathways for drug delivery and fluid extraction across the skin, and offer further potential functionality.

The earliest microneedles were fabricated using solid silicon as microprobes for neural electrical activity recording [3]. Following that work, a variety of micromachined needle designs were reported [4]. Many different approaches have been employed, including surface micromachining [5], bulk micromachining [6], and LIGA techniques, and many materials have been used including silicon dioxide, silicon nitride, metal, plasma, plastic and others.

Actuation of microneedles is an important issue in real applications, but there are few publications in the literature relating to actuation. Microactuators represent major components for MEMS-based microsystems and can be driven by various forces suitable in the micro domain. The common actuation principles are based on: electrostatic [7], piezoelectric [8], electro-thermal [9][10], magnetic [11][12], electrochemical [13] effects at the device level. When choosing a driving mechanism, system level requirements for performance should be considered such as displacement, force, size, power, and energy consumption.

Magnetic actuators can produce both high force and large displacement, but it is more commonly used in macro actuators owing to scaling consideration and difficulties with integration into batch-fabricated micro assemblies. Most microacutuators exploit the other driving principles such as electrostatic forces.

Electrostatic actuation has been widely employed in various microstructures [14][15] where there are no requirements for special materials except for silicon and its oxides, and these are well established micromachining techniques that can be integrated with a variety of silicon-based sensors and circuits to form complete microsystems. However, electrostatic actuators are susceptible to particulates and moisture, requiring high voltages, and very narrow gaps for large forces.

Polysilicon electrothermal microactuators have also shown considerable promise in MEMS applications because of their large displacement and force output capabilities combined with their ability to be driven at CMOS compatible voltages and currents. However, electrothermal actuators consume considerably more power than comparable electrostatic or piezoelectric actuation strategies.

Electrochemical actuation, which is based on the electrolysis of an aqueous electrolyte solution, is a relatively new principle. The reversible chemical reactions lead to gas evolution and gas pressure can be used to change the deflection of a membrane. Integration and package represent the major challenges for this method.

Piezoelectric actuation is another option for microactuators. The conversion of electrical energy into mechanical motion by piezoelectric thin films is a promising technique for microactuation applications. The multimorph piezoelectric actuator has several advantages including low energy losses, fast response, high induced forces, and easy geometrical adaptation in comparison with the electrostatic, electromagnetic, and electrothermic conversion mechanisms, respectively.

Silicon is an attractive choice for microactuators due to well established micromachining techniques available and the ability to be integrated with the variety of silicon-based sensors and circuits to form complete microsystems. Since silicon lacks magnetic, piezoelectric and other such properties, that are often exploited in mechanical actuators, electrostatic and electrothermal actuation are the primary options available.

In this paper, a microneedle array with a self-actuating structure on a monolithic chip is presented. The proposed microneedle array is fabricated by employing a bi-mask technique to facilitate very sharp tips with keen edges, a cylindrical body and side ports. The microneedle array uses a piezoelectric mechanism for actuation, and we present simulation results for the microneedle array.

2 DESIGN AND MODELING

The needle structure mainly defines the needle's properties. Due to the small dimensions of the microneedles, the fluid flow is quite small; thus, the high needle density of a microneedle array is needed. An out of plane microneedle array design can yield high needle density to to provide high fluid flow rate. In addition, the properties of the fabrication material should be also taken into account. Briefly: silicon dioxide is fragile; the strength of metal is good, but thin films of metal formed by sputtering or depositing are soft. The mechanics of microneedle insertion are also critically important for practical applications. Sharper needle tips can be expected to require less force for insertion, but the reduced penetration force comes at the expense of reduced strength near the tip. Only microneedles with the correct geometry and physical properties are able to penetrate skin without breaking or bending during insertion.

The geometries of the needle tip have a great effect on the forces required for insertion and fracture. Insertion force can be shown to be independent of wall thickness; thin-walled hollow needles and solid needles with the same outer tip radii require the similar insertion force [16]. Fracture force increases strongly with increasing wall thickness and increases weakly with increasing wall angle, but is independent of tip radius [16]. Therefore needles with small tip radius and large wall thickness are considered the best choice. Also, the blockage (clogging) problem upon insertion must be taken into consideration when designing microneedle structures.

We present a new monolithic self-actuated MEMS-based microneedle array. The basic structure of the microsystem is shown in Fig. 1

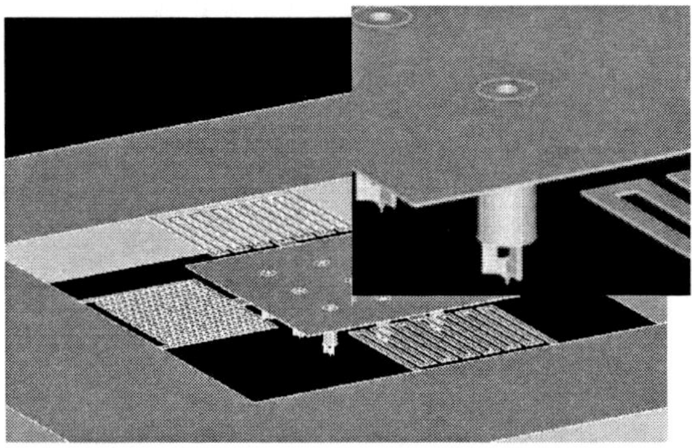

Figure 1 Structure of the microsystem

In order to realize a reliable side port design, and to be able to place the tip on the top of a cylindrical needle base, rather than directly on the wafer surface, the fabrication process employs a bi-mask technique [17][18].

3 FBRICATION

The process flow is shown in Fig. 2.

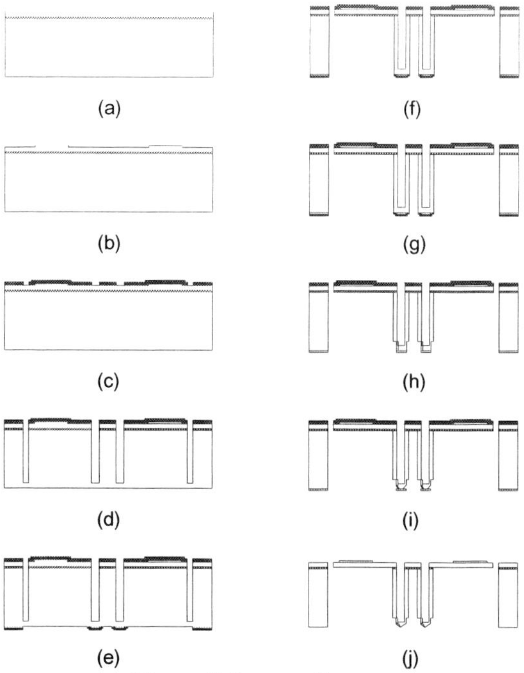

Figure 2 Process flow

In this work, a SOI silicon wafer is used. The PZT film is prepared by sol-gel method and the metal films are deposited as electrodes (Fig. 2a). After the PZT benders and electrodes are formed by etching, a thin film of SiO2 is patterned and the SOI wafer is etched to form the silicon benders (Fig. 2b). Following that, a thin film of SiO2 is deposited and etched as the mask to form the microneedle channels (Fig. 2c). The channels are anisotropically etched by anisotropic ICP, as shown in Fig. 2d. The next step is patterning the bi-mask (Fig. 2e), which is aligned to the center of the hole on the front side of the wafer. An anisotropic ICP step follows, forming the body of the needle (Fig. 2f). Then, the wafer is given another thermal growth of oxide (Fig. 2g), which forms a thin film of SiO2 on the surface of both needle bodies and channels and is used to protect both the needle bodies and channels for the subsequent steps, thereby ensuring the needle thickness. After the upper mask from the bi-mask is removed, the wafer is isotropically etched using ICP. Following that, a second anisotropic ICP is carried out. During this step, the side ports are formed (Fig. 2h). In order to guarantee a sharp needle tip, a third isotropic ICP is employed (Fig. 2i). Finally, the SiO2 is etched (Fig. 2j), and the processing is complete. Fig. 3 shows an SEM micrograph of one of the fabricated microneedle arrays.

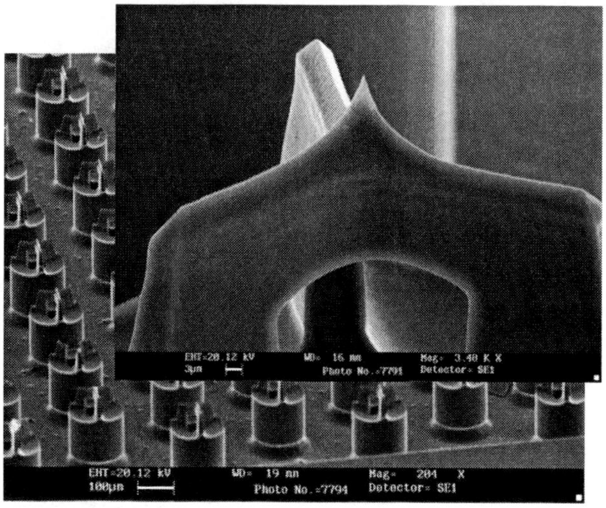

Figure 3 microneedle array

4 SIMULATION AND DISCUSSION

A series of FEM simulations were performed in order to characterize the microsystem in order to prove the adequacy of the actuating mechanisms and the efficiency of the system during actuation. The needle array plate can be actuated up to 700 m and more. Von Mises stress distribution and deformation of microsystem with five-levels are shown in fig.4.

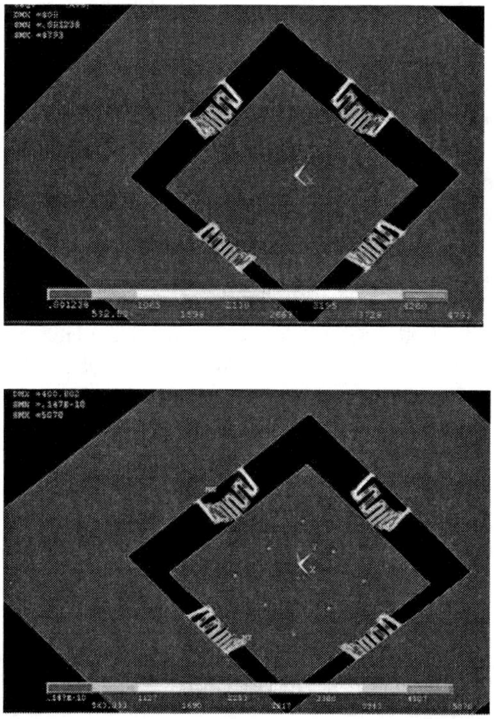

Fig 4 Von mises stress distribution and deformation of microsystem with five-level

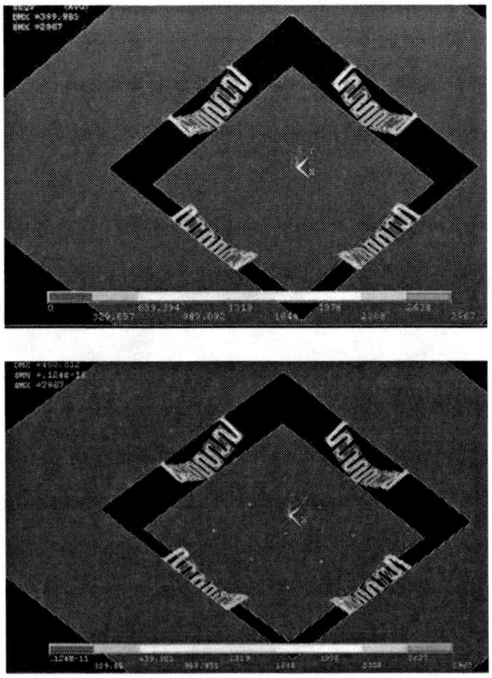

Fig 5 Von Mises stress distribution and deformation of a microsystem with seven-level

The stress distribution during actuation shows that the highest stress is concentrated where the levels cross. Low magnitude stress is found in the area of the needle array plate except at the linkage parts (Fig. 4). The deformation force of self weight is less than 1 nm and can be neglected. At the first mode resonance of the system, the microneedle array moves parallel up and down. The length of levels has a large effect on the stress.

Fig.5 shows the shape and stress distribution of the system, with seven-levels, during operation. The simulation clearly predicts that the needle array plate can maintain its initial shape during actuation and can advance the microneedle array with sufficient length (Fig. 5). The first five modal resonance frequencies of system with five-level are 2792, 4088, 4088, 6337, 10394 separately.

5 CONCLUSION

A self-actuating microsystem for fluid sampling and drug delivery is presented and simulated. The actuating mechanisms and microneedle array are on a monolithic chip to avoid having to use a bonding process. The large displacement can be accomplished through employing Z-shaped PZT unimorph benders. The properties of a target microsystem have also been simulated. This monolithic microsystem appears to provide a secure path towards commercialization.

REFERENCES

[1] D. V. McAllister, M. G. Allen, M. R. Prausnitz, in Annu Rev Biomed Eng. vol. 2, pp. 289-313, 2000.

[2] R. Langer, Nature, vol. 392(6679 Suppl), pp. 5-10, 1998

[3] K. Najafi, K. Wise, T. Michizuki, IEEE Trans. Electron. Devices, vol. ED-32, no. 7, pp.1206-1211, 1985.

[4] P. Zhang, G.A. Jullien, Proceedings. ICMENS, July 2003 P247 - 250

[5] L. Lin, A. P. Pisano and R. S. Muller, Transducers'93, Yokohama, Japan, pp. 237-240, 1993

[6] J. Chen and K. D. Wise, Solid State Sensor and Actuator Workshop, Hilton Head, SC., 1994, pp.256-259

[7] H. Toshiyosh, W. Piyawattanametha, C.T. Chan, and M.C. Wu, JMEMS, vol. 10, no. 2, pp. 205–214, 2001.

[8] J. Baborowski, J. Electroceram. 12: 33–51, 2004

[9] T. Moulton, G.K. Ananthasuresh, Sensors and Actuators A 90: 38-48, 2001

[10] N. D. Mankame and G. K. Ananthasuresh, J. Micromech. Microeng. 11: 452–462, 2001

[11] D. de Bhailís, C. Murray, M. Duffy, J. Alderman, G. Kelly and S. C. Ó Mathúna, Sensors and Actuators A: Physical, 81(1-3): 285-289, 2000

[12] M. Kohl, D. Brugger, M. Ohtsuka and T. Takagi, Sensors and Actuators A: Physical, 114(2-3): 445-450, 2004

[13] Cristina R. Neagu, Johannes G. E. Gardeniers, Miko Elwenspoek, and John J. Kelly, JMEMS, 5(1):2 – 9, 1996

[14] A.A. Yasseen, J.N. Mitchell, J.F. Klemic, D.A. Smith, and M. Mehregany, IEEE Journal of Selected Topics in Quantum Electronics, 5(1): 26–32, 1999.

[15] H. Toshiyosh, W. Piyawattanametha, C.T. Chan, and M.C. Wu, JMEMS, 10(2): 205–214, 2001.

[16] S. P. Davis, B. J. Landis, Z. H. Adams, M. G. Allen, M. R. Prausnitz, J. Biomech., 37(8): 1155-63, 2004.

[17] P. Zhang, G. A. Jullien, 6th International Conference on Modeling and Simulation of Microsystems, San Francisco, 2003, pp. 510-513

[18] P. Zhang, G. A. Jullien, The 2005 International conference on MEMS, NANO, and Smart Systems, Banff, Canada, 24-27 July 2005, pp. 392-395

Characterization and Modeling of Capacitive Micromachined Ultrasound Transducers for Diagnostic Ultrasound

C.B. Doody[1], J.S. Wadhwa[2], D.F. Lemmerhirt[2], and R.D White[1]

[1]Tufts University, Medford, MA
[2]Sonetics Ultrasound, Inc., Ann Arbor, MI

ABSTRACT

This paper describes the characterization and modeling of capacitive micromachined ultrasonic transducers (cMUTs). Combining the use of finite element analysis (FEA) and lumped element modeling, computational models of the transducers were produced. Frequency response plots were generated for both transducers in air and water environments. Transient step response and frequency sweep tests were performed on single array elements using laser Doppler velocimetry. These measurements are compared to model predictions. The computational results for coupled and uncoupled arrays are compared, and show a significant increase in the array bandwidth due to coupling.

Keywords: cMUT, MEMS, FEA, LDV, ultrasound

1 INTRODUCTION

Diagnostic medical ultrasound requires arrays of ultrasound transducers for both transmit and receive operations. Most existing commercial technology uses piezoelectric crystals or piezocomposites. Capacitive micromachined ultrasound transducers (cMUTs) are a competing MEMS technology with some attractive features; particularly the possibility of integrating signal processing, signal routing, and power electronics on chip with the transducers, and also the possibility of increased bandwidth.

The design of cMUTs has been studied since the early 1990's [1]. Researchers have described a variety of cMUT designs and models [1-2]. Both lumped element modeling and finite element analysis (FEA) have been employed [3-4]. Measurements of device response often include transmit and receive frequency response measurements in a water tank, and also device input electrical impedance [1-2]. Laser interferometry has also been used in at least one case to characterize cMUT dynamics [5].

This paper presents a hybrid finite element/lumped element modeling scheme for cMUT arrays, and compares the predictions to laser Doppler velocimetry measurements. The cMUTs being tested were fabricated using layers common to standard commercial CMOS processes. For this project, two designs of micromachined ultrasonic array elements were considered.

2 MODELING

2.1 Lumped Element Modeling

In order to create a computational model of the transducers, a lumped element acoustic model was used. A coupled electrical-mechanical acoustic model of a single cMUT element can be seen in Figure 1 in transmit mode. In the model, V_{ac}, the driving RF voltage, is applied to the element's diaphragm. The output of the model is the diaphragm's volume velocity, U_{dia}. In "receive" mode an external acoustic pressure is applied to the diaphragm resulting in a volume velocity which is converted to a current by the ideal transformer, feeding from there into the receive electronics.

The lumped element acoustic model incorporates environmental loading, diaphragm mass, diaphragm stiffness, the negative electrostatic spring, and backing cavity compliance. Many of these elements were calculated analytically using known acoustic parameters [6-7].

Environmental mass loading comes from the four components for a rigid baffled piston radiating into an infinite half space for $(\omega a/c)<2$ [7],

$$R_{A1} = 0.1404\rho c / a^2 \tag{1}$$

$$R_{A2} = \rho c / \left(\pi a^2\right) \tag{2}$$

$$M_{A1} = 8\rho / \left(3\pi a^2\right) \tag{3}$$

$$C_{A1} = \left(5.94a^3\right) / \left(\rho c^2\right) \tag{4}$$

where a is the effective diaphragm radius, equal to 80% of the physical radius for a bending circular plate. ρ and c represent the speed of sound and density for the environment.

The electrostatic spring, C_{elect} is the only nonlinear element in the model, but it can be treated as a short circuit as long as the bias voltage on the transducer is not approaching the pull-in voltage. For cMUTs with a low vacuum backing cavity, the compliance of the backing cavity may not contribute much stiffness, but may be easily included,

$$C_{cav} = V_{cav} / \left(\rho_{cav} c^2\right) \tag{5}$$

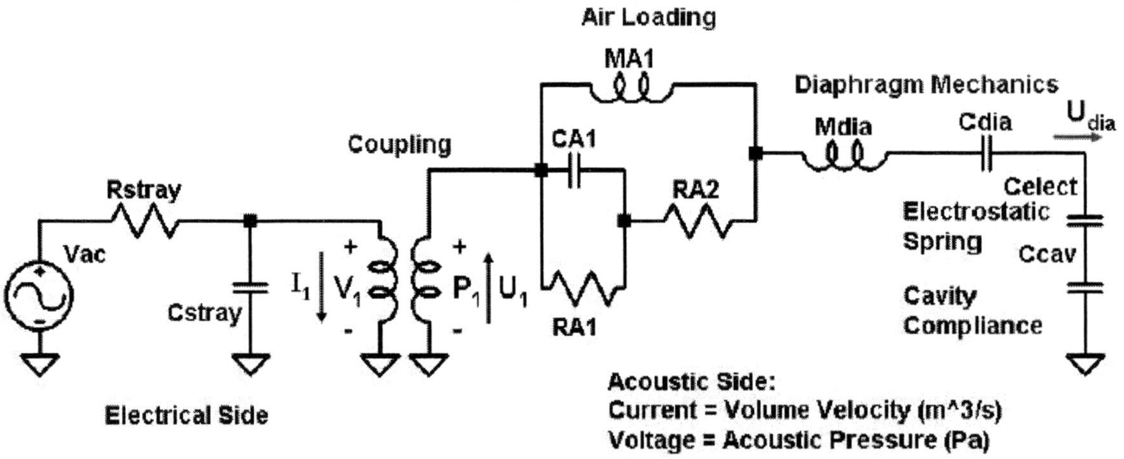

Figure 1: Mechanical-electrical lumped element model of the transducer.

where V_{cav} is the volume of the cavity, c is the speed of sound (not strongly affected by pressure), and ρ_{cav} is the density of the air in the cavity, which may be approximated by

$$\rho_{cav} = \rho_0 \frac{P_{cav}}{P_0} \tag{6}$$

where the density in the cavity is the density of air at atmospheric pressure, ρ_0, multiplied by the ratio of the cavity pressure to atmospheric pressure.

2.2 Finite Element Analysis

Due to the complex cross-sectional geometry of the transducer designs, finite element analysis was used to determine the diaphragm stiffness, effective diaphragm mass, and the electrostatic coupling for the transducer.

Using COMSOL Multiphysics ®, the geometry of each transducer was modeled as an axisymmetric cross-section. A basic layout of the transducer's cross section can be seen in Figure 2. In each case, the transducer is structured as an axisymmetric cross-section, comprised of a bulk silicon base, several thin film dielectric layers, and a passivation layer. Within the diaphragm region rests a top and bottom conductor, as well as a vacuum-filled cavity resting in between the two. The two conductive layers form the variable parallel plate capacitor for electrical-mechanical coupling. The air gap acts as a small mechanical spring.

The diaphragm acoustic compliance, C_{dia}, was calculated using a linear elastic axisymmetric static analysis. The acoustic compliance is the surface integral of the displacement in response to a unit applied pressure on the face of the transducer. An eigenfrequency analysis was then performed on the same model to determine the diaphragm effective mass, M_{dia} according to

$$\frac{1}{\sqrt{M_{dia} \cdot C_{dia}}} = f_1 \cdot 2\pi \tag{7}$$

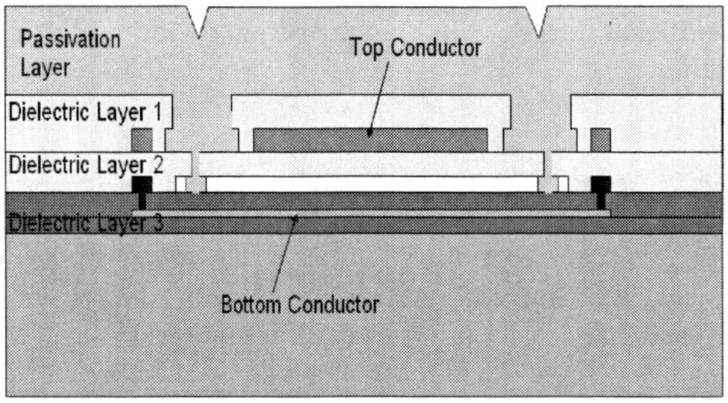

Figure 2: Axisymmetric cross-section of the transducer

where f_1 is the first eigenfrequency in cycles per second. The FEA computation was conducted for a number of different passivation layers, including PECVD nitride, Oxynitride and Parylene-C

Coupling can be computed by considering the parallel plate capacitor formed between the aluminum and the doped polysilicon with the two intervening dielectrics and the vacuum gap,

$$N = \alpha \left(\frac{d_{nom}}{\varepsilon_1} + \frac{d_2}{\varepsilon_2} + \frac{d_3}{\varepsilon_3} \right)^{-2} \cdot \left(\frac{1}{\varepsilon_1} \right) \cdot V_{bias} \tag{8}$$

where d_{nom} is the nominal height of the vacuum gap, ε_1 is the permittivity of free space, d_n and ε_n are the height and permittivity of the other intervening dielectric layers, and V_{bias} is the applied DC bias. Note that N has units of *Pa/V*, or, equivalently, *Amp/(m/s)* (in SI units). It is a bidirectional coupling constant for the ideal transformer.

Incomplete electrode coverage was handled using a nondimensional parameter α, $0 < \alpha < 1$. α is computed using

FEA as the ratio of the volume displacement obtained with a uniform unit pressure applied only on the top electrode to that obtained by applying a unit pressure over the entire face of the transducer.

With these parameters in hand, the volume velocity in response to a given DC bias plus RF drive voltage can be computed. This volume velocity can be translated into a membrane centerpoint displacement by

$$u = \frac{u_{ctr} U_{dia}}{j\omega} \tag{9}$$

where u_{ctr} is the ratio of the centerpoint displacement to the volume displacement taken from the static finite element computation. The pressure at a distance r from the element can also be estimated by treating the element as a baffled simple source, assuming we are in the farfield and there are no reflections,

$$P(r,t) = U_{dia} \frac{\rho f}{r} e^{j\omega\left(t - r/c\right)} \tag{10}$$

3 SINGLE ELEMENT RESULTS

Using the computational model, frequency response plots were generated for a single element of the array in both air and water environments. The models were also tested with different passivation layer materials and thicknesses. A sample frequency response plot can be seen in Figure 3. In air, the primary resonance for this device is predicted to be 6.1 MHz, with a very narrow fractional bandwidth of 0.2%. In the underwater environment, the model predicts a 3.4 MHz center frequency, with a fractional bandwidth of 27% for the same device. As a comparison, experimental results for transmit operation in a water tank indicate an approximate center frequency for this device of 3.3 MHz with 50% fractional bandwidth.

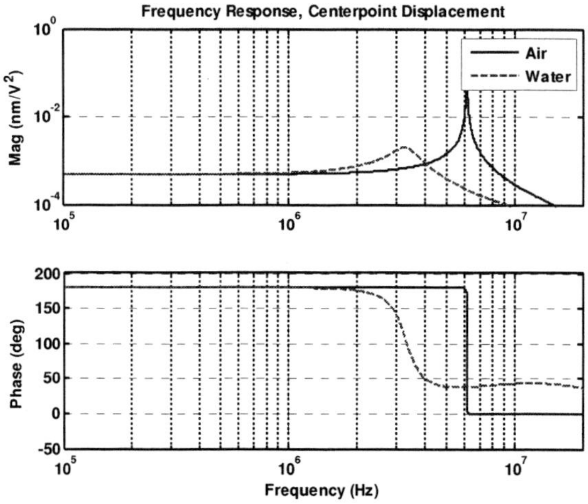

Figure 3: Modeled transmit frequency response for a single element in air and water environments.

Using LDV, a transient step response was measured for a single element in air. The result is shown in Figure 4. This is for a 0 to 10 V step. The step response shows the very high Q of the system when operating in air, and a resonant frequency of 5 MHz, similar to the high-Q 6 MHz resonance predicted by the model. The Q of the system decreases dramatically when submerged, both in computation and in water tank experiments. Model results illustrating this appear in Figure 3.

A frequency sweep test was also performed on a single element transducer, in air, with a 20 V_{PP} RF input. Due to the nonlinearity of the electrostatic drive, the transducer responds at twice the drive frequency. Test results show a peak frequency of 5.0 MHz, with an approximate 0.003 nm/V^2 low-frequency gain.

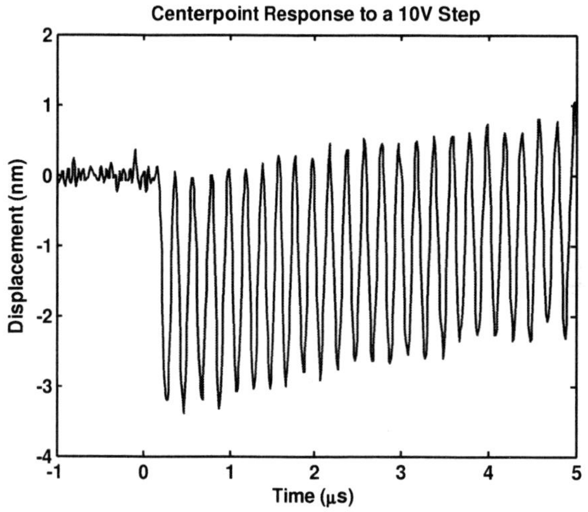

Figure 4: A 10 transient step response of a single transducer element (LDV measurement).

Centerpoint displacement calculated from the computational model was compared to the frequency sweep data obtained from the from the single cMUT element transducer. The computational model predicted a peak frequency of approximately 6 MHz, similar to the 5 MHz obtained from the transducer chip. The experimental results show somewhat larger displacements at low frequencies than the model predicts. A frequency plot comparison can be seen in Figure 5. The magnitude for the model is the amplitude of the centerpoint displacement normalized to the product of the applied DC bias and the amplitude of the RF drive voltage. For the measurement, the displacement is normalized by the RF peak voltage squared over 4, which is equivalent.

4 ARRAY COMPUTATIONS

Array computations have been carried out for a 55 element columnar array. In the coupled computation, each element is forced not only by the electrostatic force but also by the pressures generated by the motion of all other elements in the array. This leads to a matrix computation, with a fully populated transfer function matrix including the phase lag and geometric spreading of the baffled monopole pressure field for each individual element. The array computation shows a considerable increase in bandwidth for the array over the bandwidth of an individual element. Figure 6 compares the predicted pressure for the 55 element columnar array transmitting into water at a distance 7.5 mm from the center of the array for a 40 V_p pure AC drive. The two curves in the plot represent the result when each element in the array transmits in isolation (labeled "uncoupled"), and when the fully coupled solution is computed.

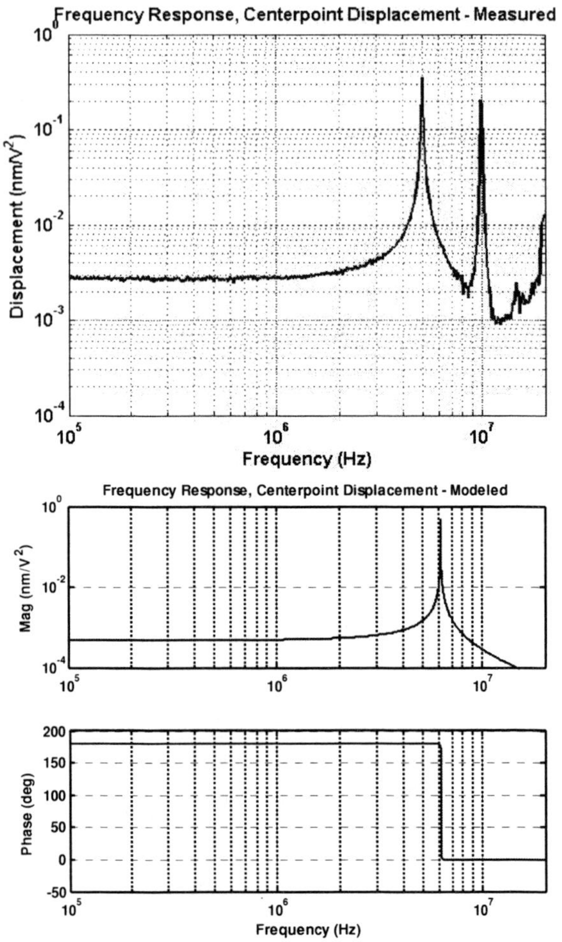

Figure 5: Frequency response comparison between computational work and experimental data.

5 CONCLUSIONS

A method of combining FEA and lumped element modeling for cMUT elements has been described. The modeling method is computationally efficient, and leads to good predictions of the resonant frequency and bandwidth of individual elements in both air and water environments. Array computations have been briefly described. The bandwidth predicted by a fully coupled computation is much wider than the uncoupled result.

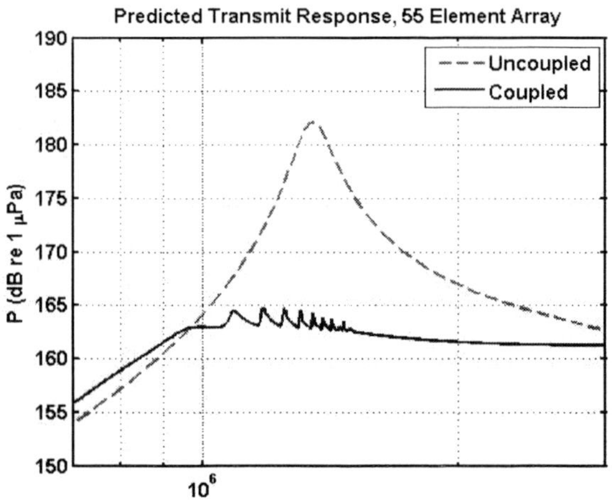

Figure 6: Comparison between coupled and uncoupled computations.

REFERENCES

[1] Ladabaum, I., Jin, X., Soh, H., Atalar, A., and Khuri-Yakub, B. *Surface Micromachined Capacitive Ultrasonic Transducers.* IEEE Transactions on Ultrasonics, Ferroelectrics, and Frequency Control, 1998, **45**(3): p. 678-690.

[2] Jin, X., Ladabaum, I., Degertekin, F., Calmes, S., and Khuri-Yakub, B. *Fabrication and Characterization of Surface Micromachined Capacitive Ultrasonic Immersion Transducers.* IEEE Journal of Microelectromechanical Systems, 1999, **8**(1): p. 100-114.

[3] Lohfink, A., Eccardt, P.-C., Benecke, W., and Meixner, H. *Derivation of a 1D CMUT Model from FEM Results for Linear and Nonlinear Equivalent Circuit Simulation.* IEEE Ultrasonics Symposium, 2003: p. 465-468 .

[4] Yaralioglu, G., Badi, M., Ergun, A., and Khuri-Yakub, B. *Improved Equivalent Circuit and Finite Element Method Modeling of Capacitive Micromachined Ultrasonic Transducers.* IEEE Ultrasonics Symposium, 2003: p. 469-472.

[5] Hansen, S., Turo, A., Degertekin, F., and Khuri-Yakub, B. *Characterization of Capacitive Micromachined Ultrasonic Transducers in Air Using Optical Measurements.* IEEE Ultrasonics Symposium, 2000: p. 947-950.

[6] Kinsler, L., Frey, A., Coppens, A., and Sanders, J. *Fundamentals of Acoustics: Fourth Edition.* John Wiley & Sons, Inc. 2000.

[7] Beranek, Leo. *Acoustics.* American Institute of Physics. 1986.

Modeling and Simulation of Novel Structure for Sub-millimeter Solid-state Accelerometer with Piezoresistive Sensing Elements

Ranjith Amarasinghe. , D.V.Dzung and S.Sugiyama

Ritsumeikan University, 1-1-1, Noji Higashi, Kusatsu, Shiga, 525-8577 JAPAN, ranama@fc.ritsumei.ac.jp

ABSTRACT

This paper presents the modeling and simulation of new structure for solid-state three degrees of freedom (3-DOF) micro accelerometer utilizing piezoresistive effect in single crystal Si. The proposed sensor can detect three components of linear acceleration simultaneously. The sensing structure consists of combined cross-beam and surrounding beams and seismic mass. Therefore, this novel proposed sensor is showing good performance than other miniaturized sensor structures reported thus far.

Keywords: accelerometer, piezoresistor

1 INTRODUCTION

MEMS (Micro Electromechanical Systems) based accelerometer is one of the most important types of the mechanical silicon sensors, since there have been large demands for accelerometers in automotive applications, where they are used for crash detection, and for vehicle stability systems. In addition, due to small size and light weight, they are also used in biomedical and robotics applications for active motion monitoring, and in consumer for stabilization of pictures in camera, head-mounted displays. A substantial study on micromachined accelerometers has been reported so far, with the working principles are mainly based on the piezoresistive effect, capacitance, tunneling effect, resonant, and so on [1]. Among those, the piezoresistive accelerometers have many advantages such as the simplicity of the structure and batch-fabrication process, as well as the readout circuitry, since the resistive bridge (Wheatstone bridge) has low output-impedance. Most popularly reported sensing structure for 3-DOF micro accelerometer are cross-beam type, which consists of a seismic mass suspended on a four small sensing beams in a cross shape. In this paper, a novel structure, which consists of a seismic mass suspended on double surrounding beams combined with four cross beams are used as the acceleration sensing structure.

Reduction of chip size and increase of sensitivity are the important targets of silicon-based sensors, since it increases the total die/wafer, and thus the productivity, and accordingly, decreases the total cost. Small, light weight accelerometers are necessary for many potable devices, such as camcorders, navigation systems, robot motion monitoring systems, and so on. It is sometime a challenge to maintain sensitivity of an accelerometer as chip size is decreased.

When the feature size of the sensor gets smaller, some technical issues become more serious and need to be carefully considered, such as the noise problem and the damping control.

In this paper, the design, modeling and simulation of new structure which has an overall chip size less than 700μm× 700μm for solid-state three degrees of freedom (3-DOF) micro accelerometers utilizing piezoresistive effect in single crystal Si are presented.

2 DESIGN OF THE ACCELEROMETER

The three dimensional (3D) model of the proposed novel structure for 3-DOF acceleration sensor is shown in Figure 1.

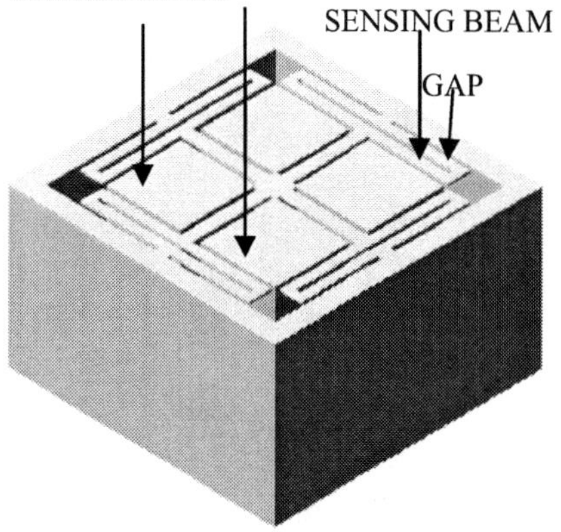

Figure 1: 3D model of the 3-DOF accelerometer.

The sensing beams are made of Si, with a seismic mass is suspended at the middles of the four surrounding beams combined with four cross beams. Si piezoresistors are formed by diffusing boron ions at the suitable places on the surface of the Si sensing beams. When an external acceleration is applied to the sensor, the seismic block will be displaced due to the inertial force. This movement of the seismic mass makes the beams deformed; as a result, the resistance of Si piezoresistors will be changed. The change of resistance will be converted to an output voltage change by a Wheatstone bridge.

2.1 Structural Analysis

In order to decide the optimal positions of the piezoresistors in the sensing beams, it is necessary to perform structural analysis. The structural analysis of the sensing chip consists of two steps. The first step deals with qualitative analysis by classical elasticity theory. The dimensions of the sensing chip were tentatively specified based on the predefined ranges of accelerations acting, e.g. ± 50g (g is the gravitational acceleration), the uniformed sensitivity condition, i.e. sensitivities to three components of acceleration are similar, and the necessary beam-width for interconnection. The Overall size of the sensing chip, includes the sensing area and the frame, which accommodates bonding pads, is 0.7×0.7×0.4 mm3, (Length ×Width ×Thickness).

Simulation using finite element method (FEM) has been performed to verify mechanical behavior of the structure as well as to optimize the design. The commercial package software ANSYS 9.0 has been used for the analysis. Figure 2 shows the FEM model of the accelerometer. The boundary condition with non-displacement of beam ends is applied.

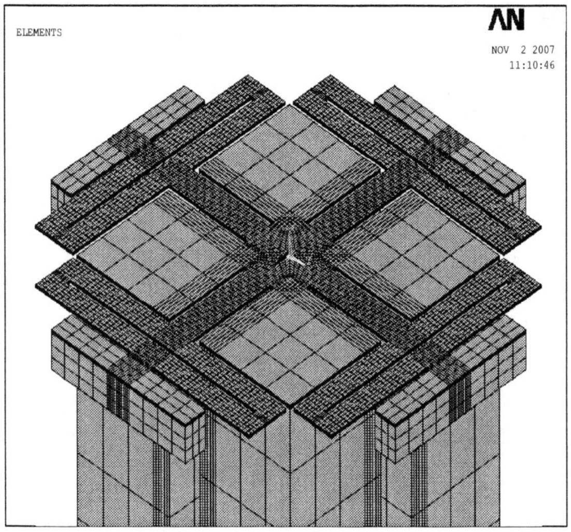

Figure 2: 3D FEM model of the accelerometer

Figures 3 to 10 show the distributions of stress components on the top surface of the beams due to application of accelerations. The FEM result shows that on the top surface of beam the stress components other than the longitudinal one can be neglected.

Modal analysis has been performed for the first three modes. Natural frequency of the first mode (vertical vibration) is f_z = 1.7 KHz, and of the second and third modes (i.e. rotational vibration around X or Y-axis) are f_x = f_y = 1.2 KHz.

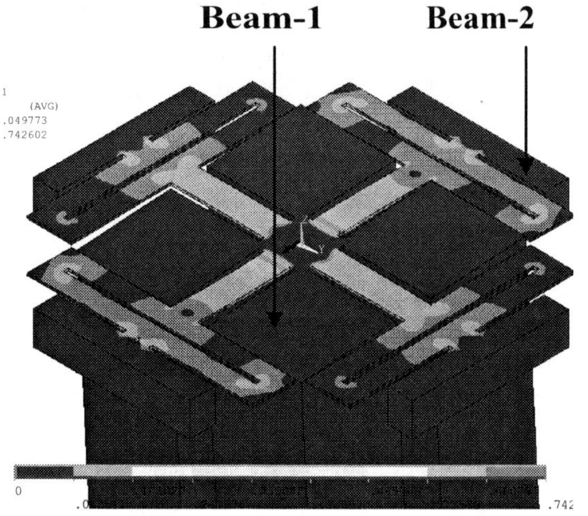

Figure 3. Graphical representation of the stress distribution on surface of X-oriented beam structure due to the application of acceleration Ax, by FEM analysis (ANSYS).

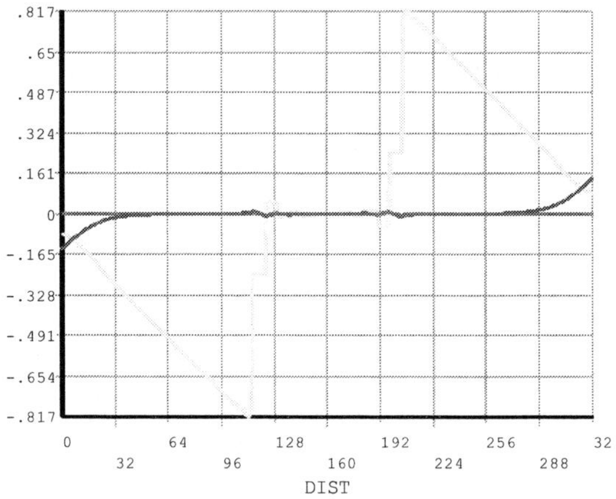

Figure 4. Stress distributions along central axis of beam-1 due to application of vertical acceleration Ax.

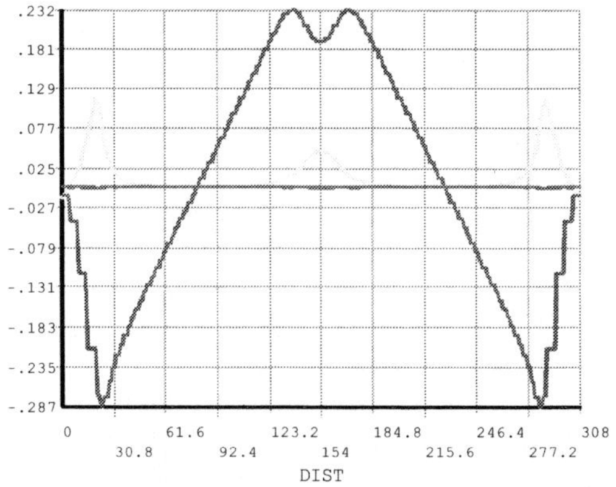

Figure 5. Stress distributions along a sensing inner side of beam-2 due to application of acceleration Ax

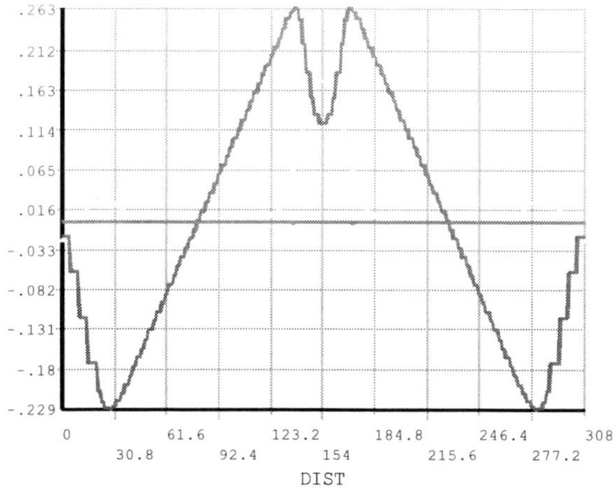

Figure 6. Stress distributions along a sensing outer side of beam-2 due to application of acceleration Ax

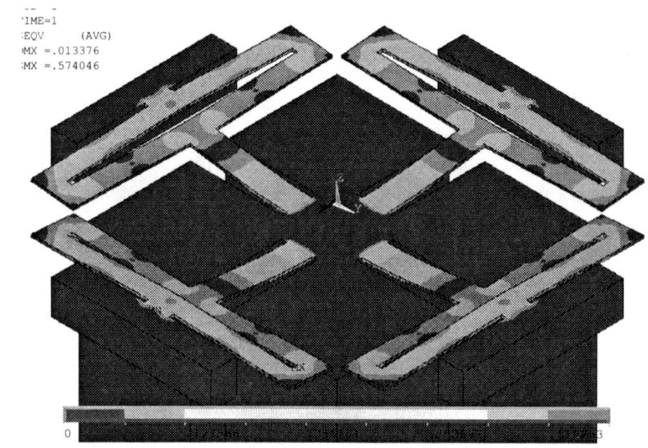

Figure 7. Graphical representation of the stress distribution on surface of beam structure due to the application of acceleration Az, by FEM analysis (ANSYS).

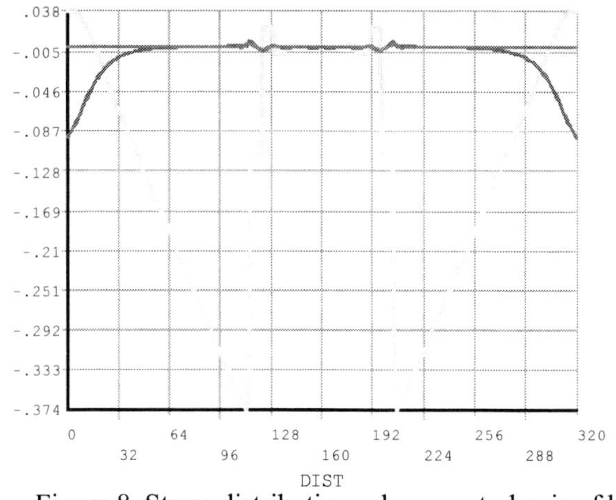

Figure 8. Stress distributions along central axis of beam-1 due to application of vertical acceleration Az.

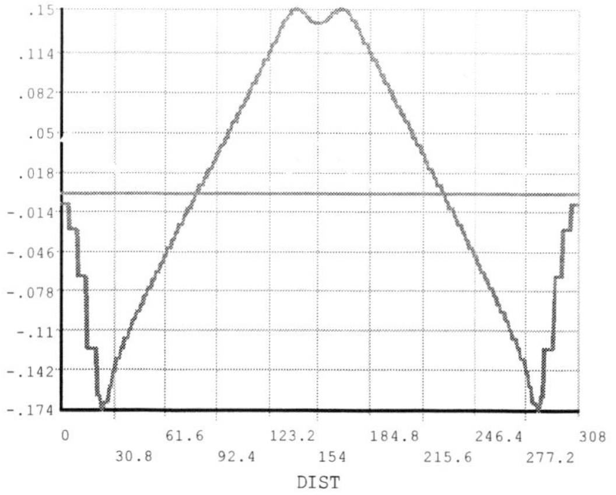

Figure 9. Stress distributions along a sensing inner side of beam-2 due to application of acceleration Az.

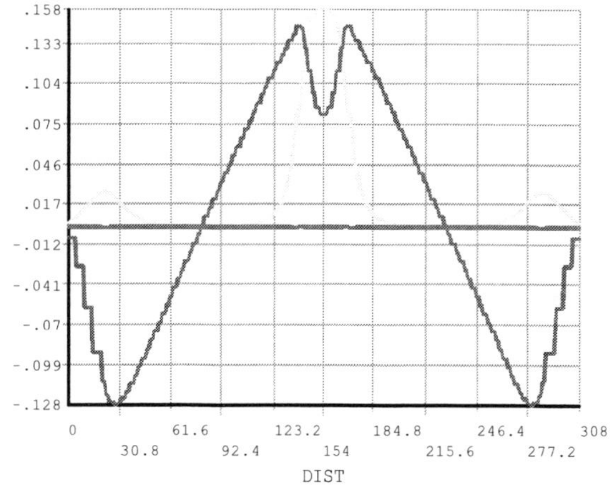

Figure 10. Stress distributions along a sensing outer side of beam-2 due to application of acceleration Ax

3 PIEZORESISTIVE EFFECT

The piezoresistive effect of conventional single-crystalline piezoresistors can be expressed as below. For a three-dimensional anisotropic crystal, the electric filed vector (ε) is related to the current vector (*i*) by a three-by-three resistivity tensor. Experimentally, the nine coefficients are always found to reduce to six and the tensor is symmetrical.
In matrix form this relation can be written as follows];

$$\begin{bmatrix} \varepsilon_1 \\ \varepsilon_2 \\ \varepsilon_3 \end{bmatrix} = \begin{bmatrix} \rho_1 & \rho_6 & \rho_5 \\ \rho_6 & \rho_2 & \rho_4 \\ \rho_5 & \rho_4 & \rho_3 \end{bmatrix} \cdot \begin{bmatrix} i_1 \\ i_2 \\ i_3 \end{bmatrix} \tag{1}$$

The piezoresistance effect can now be described by relating each of the six fractional resistivity changes, $\Delta\rho_i/\rho_i$ (i=1 to 6) to each of the six stress components. Mathematically this yields a matrix of 36 coefficients. By definition, the elements of this matrix are called piezoresistance coefficients, π_{ij}, (i,j=1 to 6), expressed in Pa^{-1}. In this research, two-terminal piezoresistors are formed by masked-diffusion method, and lie on very thin surface layer. Therefore, only two piezoresistance coefficients, i.e. π_{11} and π_{12} are important [2]. π_{11} corresponding to the case the stress parallels with the direction of electric field and current density, thus it is called the longitudinal piezoresistance coefficient, denoted by π_l. Similarly, π_{12} relating to the case the applied stress is perpendicular to the electrical field and current density, hence it is called transverse piezoresistance coefficient, π_t. These two coefficients can be expressed through 3 fundamental piezoresistance coefficients π_{11}, π_{12}, π_{44}, and directional cosines by:

$$\pi_l = \pi_{11} + 2(\pi_{44} + \pi_{12} - \pi_{11})(l_1^2 m_1^2 + l_1^2 n_1^2 + m_1^2 n_1^2) \tag{2}$$

$$\pi_t = \pi_{12} - (\pi_{44} + \pi_{12} - \pi_{11})(l_1^2 l_1^2 + m_1^2 m_2^2 + n_1^2 n_2^2) \tag{3}$$

where l_i, m_i, n_i, ($i = 1, 2, 3$) are directional cosines.
In directions <110> and $<1\bar{1}0>$, these coefficients can be expressed as:

$$1/2(\pi_{11} + \pi_{12} + \pi_{44}) \tag{4}$$

$$1/2(\pi_{11} + \pi_{12} - \pi_{44}) \tag{5}$$

(π_{11}=6.6x10^{-11}Pa^{-1}, π_{12}=6.6x10^{-11}Pa^{-1}, π_{44}=6.6x10^{-11}Pa^{-1})

Based on the theory and equation above, the resistance change can be calculated as a function of the beam stress. The mechanical stresses are constant over the resistors; the total resistance changed is given by:

$$\frac{\Delta R}{R} = \sigma_l \pi_l + \sigma_t \pi_t \tag{6}$$

It is noted that the above equation are only valid for uniform stress fields or if the resistor dimensions are small compared to the beam size.

4 PIEZORESISTORS ARRANGEMENT

Based on the stress distribution in the crossbeam derived from FEM analysis, piezoresistors were placed to eliminate the cross-axis sensitivities, and to maximize the sensitivities to various components of linear acceleration and angular acceleration [2]. Twelve p-type conventional piezoresistors, are diffused along the central-longitudinal axes on the upper surface of n-type silicon crossbeam. The in-plane principal axes of the piezoresistors are aligned with the crystal directions <110> and $<1\bar{1}0>$ of silicon (001) plane. All conventional piezoresistors are designed to be identical as shown in Figure 11.

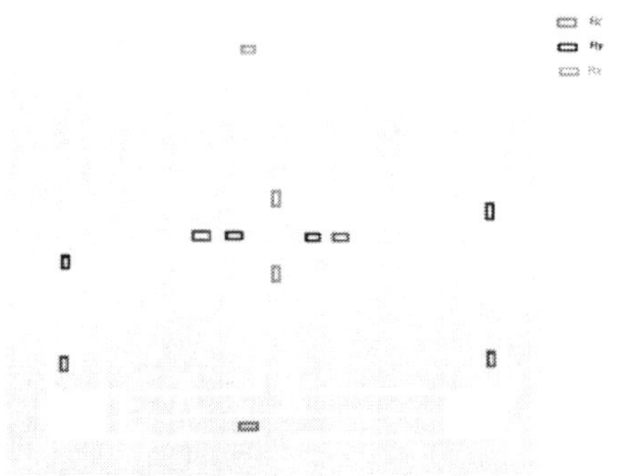

Figure 11. The arrangement of twenty piezoresistors on beam structure of the accelerometer.

	Rz1	Rz2	Rz3	Rz4	Ry1	Ry2	Ry3	Ry4	Rx1	Rx2	Rx3	Rx4
Az	+	−	+	−	−	−	−	−	−	−	−	−
Ay	−	0	+	0	−	+	−	+	0	+	+	0
Ax	−	−	+	+	0	+	+	0	−	+	−	+

Table 1 summarizes the increase (+), decrease (-), or invariable (0) in resistance of piezoresistors due to application of accelerations Ax, Ay, and Az. To detect the accelerations, piezoresistors are connected to form Wheatstone full bridges. The Wheatstone bridges to measure the accelerations Ax and Ay are similar and denoted by Ax-bridge and Ay-bridge, respectively. Similarly, for vertical acceleration Az, we have Az-bridge. The change in resistance of piezoresistors is converted to output voltage by these bridges.

4.1 Sensitivity Estimation

The sensitivity of the accelerometer can be defined as the ratio between the output voltage and the applied acceleration. With the notices that the piezoresistors of one bridge are designed to be identical, and that the transverse piezoresistive effect is very small in comparison with the longitudinal effect, the sensitivities to each components of acceleration *Ax*, *Ay*, *Az* can be calculated, $S_{Ax} = S_{Ay} =$ 0.42mV/g, and $S_{Az} = 0.48$ mV/g.

Acknowledgements

A part of this work was supported by The Monbusho's Grant-in Aid for JSPS Fellows relating to the Japan Society for the Promotion of Science (JSPS) JSPS Postdoctoral Fellowship for Foreign Researchers.

REFERENCES

[1]]R.Amarsinghe, D.V.Dao, T.Toriyama, S.Sugiyama, "Design and Fabrication of Miniaturized Six-Degree of Freedom Piezoresistive Accelerometer" MEMS2005 Conference, pp. 351-354, 2005.

[2] D. V. Dao, T. Toriyam, S. Sugiyama, "Noise and Frequency Analyses of a Miniaturized 3-DOF Accelerometer Utilizing Silicon Nanowire Piezoresistors" Vienna, Austria, IEEE 2004.

Simulation of Constant-Charge Biasing Integrated Circuit for High Reliability Capacitive RF MEMS Switch

K-H. Choi*, J-B. Lee* and C. Goldsmith**

* Erik Jonsson School of Engineering and Computer Science, RL10,
The University of Texas at Dallas, Richardson, Texas 75083, USA, kxc046000@utdallas.edu
** MEMtronics Corporation, Plano, TX, USA, cgoldsmith@memtronics.com

ABSTRACT

This paper presents simulation results of a constant-charge biasing integrated circuit for capacitive RF MEMS switches for improvement in reliability. 1-D nonlinear dynamic electromechanical model of the switch was used to extract the switch model parameters for circuit simulations. The constant-charge biasing integrated circuit is designed to operate to charge, float, and discharge the switch membrane as a capacitive load. Austria Microsystems 0.35μm 50V high voltage CMOS technology was used for the circuit simulation. The simulation results showed that within a mere 300 ns period enough charge to pull down the switch membrane is provided. Due to the constant-charge biasing, the voltage across the dielectric layer was found to be 530 mV after the membrane snapped down to the dielectric layer. The constant-charge biasing scheme results in electric field across the dielectric remains below 27 kV/cm regardless of the membrane position which is substantially lower than the maximum electric field of 2.38 MV/cm in the case of constant-voltage biasing. The circuit simulation results imply that such a constant-charge biasing would substantially improve reliability of the switches.

Keywords: RF MEMS, switch, simulation, constant-charge, integrated circuit, reliability

1 INTRODUCTION

Recently, there have been various studies on the RF (radio frequency) MEMS (Microelectromechanical systems) switches as one of the preferred switching elements for the RF devices [1- 3]. It is commonly known that RF MEMS switches provide lower insertion loss, higher isolation, better linearity, low power consumption, and a smaller size than the conventional electronic switching counterparts (such as *pin* diodes and FETs) [4]. Even though RF MEMS switches showed excellent overall device performances, device reliability and commercial feasibility issues were major stumbling blocks for the RF MEMS switches' penetration into real world applications.

For the capacitive RF MEMS switches, it has been reported that the lifetime of the switches improves on the order of a decade for every 5~7V decrease in applied voltage when a conventional constant-voltage (CV) biasing scheme is used [5]. In order to increase the reliability of capacitive the RF MEMS switches, we previously suggested the constant-charge (CC) biasing method [6].

This study is an extension of our previous work towards the integration of the constant-charge biasing circuit in monolithic integrated circuits (IC) using conventional MOS (metal-oxide-semiconductor) foundry service.

2 CONSTANT-CHARGE BIASING SCHEME

The constant-voltage method that is widely used in capacitive switches can cause high electric fields across the dielectric layer once the membrane is snapped down, so that dielectric charging can easily occur in the layer. The CC basing method can effectively reduce the electric field across the dielectric layer and therefore significantly reduce dielectric charging. Our previous work demonstrated that the electric fields are constant when the CC method is applied as the biasing scheme [6].

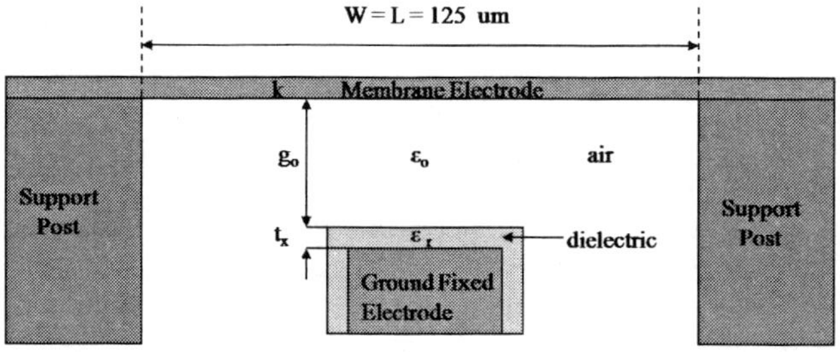

Figure 1. A schematic diagram of a cross-sectional view of a doubly-clamped switch that shows parameters used in the simulation.

Table 1. Model parameters of the capacitive switch

Spring Constant of the Membrane k	4
Initial Gap g_o	3um
Thickness of the Dielectric Layer t_x	0.21um
Absolute Permittivity of Air ε_o	$8.854\text{x}10^{-12}$ F/m
Relative Permittivity of the Dielectric Layer ε_r	6.7
Area of Membrane A = WL	$(125\text{x}10^{-6})^2$

2.1 Pull-In Charge and Voltage

The pull-in charge (Q_{PI}) which is the minimum charge required to pull down the switch membrane is given by [6]:

$$Q_{PI} = \sqrt{2\varepsilon_0 A k g_0}\ , \qquad (1)$$

where A is the area of the electrode, k is the spring constant of the switch membrane and g_0 is the initial gap between the switch membrane and the dielectric layer. The corresponding voltage of the membrane with the constant-charge Q_{PI} is given by:

$$V_Q = \frac{Q_{PI}}{C_0} = \sqrt{\frac{2kg_0^3}{\varepsilon_0 A}}\ . \qquad (2)$$

With the parameters presented in Figure 1 and Table 1, the pull-in charge of is calculated as 2.28 pC and the corresponding V_Q is calculated as 39.5V. To pull the membrane down, the CC biasing method requires a factor of 2.6 times higher voltage than CV biasing method [6].

2.2 An Implementation Method

The constant-charge biasing can be realized by the following sequences: *charging–floating-discharging*. First, the circuit provides high voltage pulse to charge the switch capacitor. Once the membrane switch capacitor is charged enough, the electrostatic force between the membrane and the bottom electrode pulls the membrane down. When the membrane starts to move down, the switch is disconnected from the circuit output node, so that the switch becomes floated. The total charge on the membrane remains constant during the membrane's actuation. As the gap between the membrane and the dielectric layer decreases, the capacitance increases and the voltage decreases since the charge in the switch remains constant. When the membrane reaches to the dielectric layer, the voltage across the membrane and the bottom electrode becomes the lowest. After certain time, the charge stored in the switch membrane can be discharged by the circuit, then the switch moves back to its unbiased initial position due to the mechanical restoring force of the membrane.

3 DYNAMIC MODEL

In order to find the capacitance model of the switch, dynamic analysis using 1-D nonlinear equation of motion with damping factor of air and contact forces of the dielectric surface was carried out. The electrostatic force on the membrane is given by:

$$F_{el} = \frac{1}{2} C'_{sw}(t) \left[\frac{Q_0}{C_{sw}(t)}\right]^2 , \qquad (3)$$

where $C_{sw}(t)$ is the capacitance of the switch and Q_0 is the applied constant-charge.

The equation of motion for switch membrane is given by:

$$m\frac{d^2 z}{dt^2} + b\frac{dz}{dt} + kz = F_{el} + F_c\ , \qquad (4)$$

where m is the mass of the membrane, z is the amount of the displacement of the membrane in the out of the plane direction, b is the variable damping coefficient, and F_c is the contact force of the dielectric surface. The contact force of the surface of the dielectric layer consists of an attractive Van der Waals force and a repulsive nuclear contact force.

The damping coefficient as a function of displacement is derived from Reynolds gas-film equation for rectangular parallel-plate [7] which is given by:

$$b \cong \sqrt{2}\mu_{air} l \left(\frac{w}{g_0 - z}\right)^3 , \qquad (7)$$

where μ_{air} is the viscosity of air ($\approx 1.8 \times 10^{-5}$ kg/m^3), l and w are the length and width of the switch membrane.

Mathematica® was used to solve the nonlinear 2nd order differential equation. Figure 2 shows the gap of the membrane as a function of time for the constant-charge biasing case. It takes approximately 16 μs for the membrane to be snapped down onto the dielectric layer.

As the gap between the membrane and the dielectric layer decreases, capacitance of switch increases. The capacitance as a function of time is shown in Figure 3 for the constant-charge biasing. Initial capacitance of a given switch model is 45 fF. The capacitance becomes greatly increased to be 4.63 pF when the membrane touches the dielectric surface. This time varying capacitance model was used as a load capacitance model for the CC biasing circuit simulation.

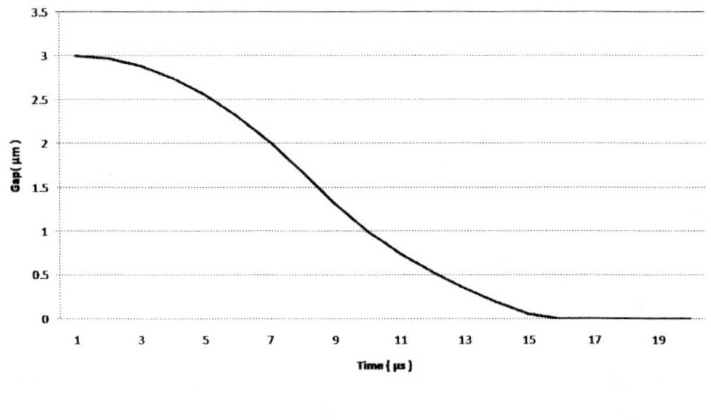

Figure 2. The gap between the membrane and the dielectric layer as a function of time for the constant-charge biasing

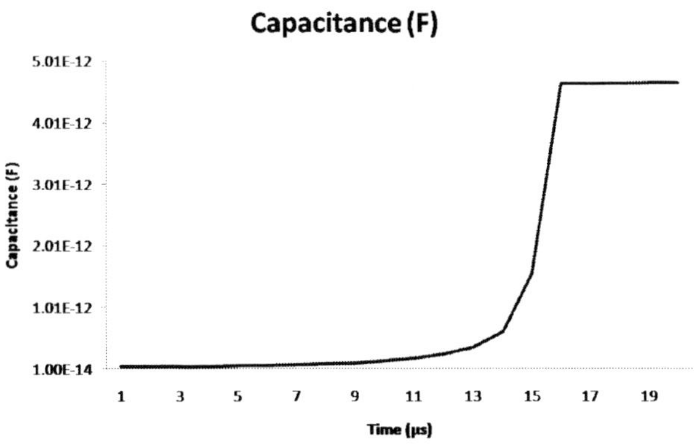

Figure 3. The capacitance of the switch as a function of time.

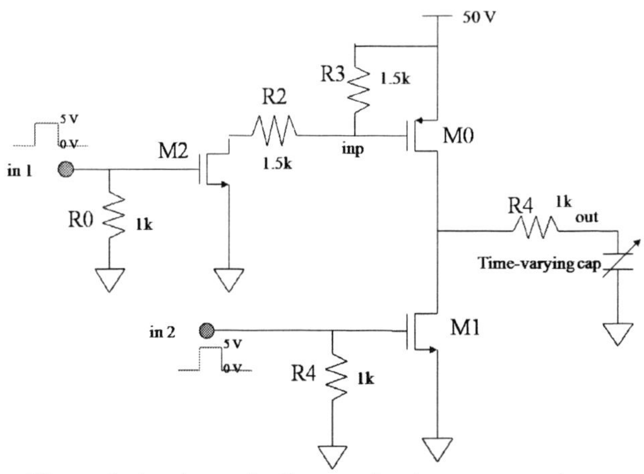

Figure 4. A schematic diagram for the constant-charge biasing circuit.

5 CIRCUIT SIMULATION

The CC-biasing circuit consists of one high voltage PMOS, two high voltage NMOSs, five resistors, and a variable load capacitor which is the capacitive switch itself. Figure 4 shows the schematic diagram of the proposed CC-biasing circuit. M0 is a PMOS transistor which can charge the load switch capacitor with 50 V for the membrane actuation. M1 is an NMOS transistor that can discharge the load to push the membrane back to its original unbiased state. M2 is an NMOS transistor which receives 5 V input signal and generates turn-on pulses for the M0 high voltage PMOS transistor. Table 2 shows the size of components and simulation condition.

Table 2. Components' size and simulation condition

Components	Size (W/L)	Condition
M0 (NMOS)	10μm / 2μm	Process : Typical-Typical (TT) Temperature : 25°C Power Supply : 50V Input Signal Voltage: 5V Simulation Tool: HSPICE
M1 (PMOS)	20μm / 2μm	
M2 (NMOS)	10μm / 2μm	
R0	1 kΩ	
R1	1 kΩ	
R2	1.5 kΩ	
R3	15 kΩ	
R4	1 kΩ	
C0	Variable (45 fF ~ 4.65 pF)	

The designed circuit was simulated with the HSPICE model of transistors provided by Austria Microsystems under typical condition at room temperature of 25°C. The power supply (Vdd) was given by 50 V and the input signal voltage level was set at 5 V to incorporate typical digital control signals. Figure 5 shows the timing diagram of the input/output signals.

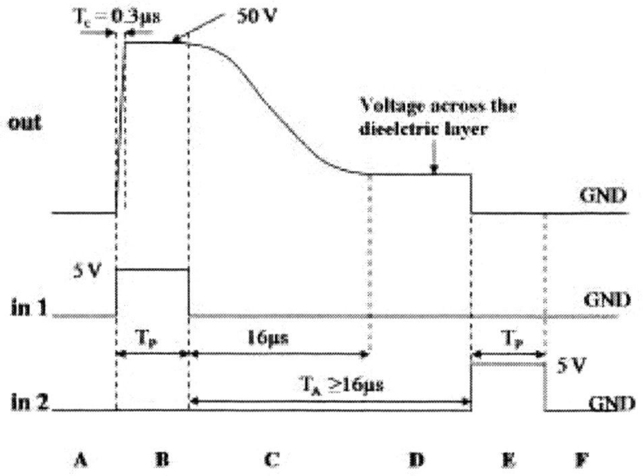

Figure 5. A timing diagram of the input signal and the output signal.

Table 3. M0, M1, output node, and membrane states in the charge-float-discharge cycle.

Phase	M0	M1	Output State	Membrane State
A	OFF	OFF	Floating	No Bending
B	ON	OFF	VDD (Charging)	Start to bend
C	OFF	OFF	Floating	Bending
D	OFF	OFF	Floating	Stick to dielectric layer
E	OFF	ON	GND (Discharging)	Start to recover
F	OFF	OFF	Floating	Recovering

Input signals, ***in1*** and ***in2,*** are the charging and the discharging pluses respectively. The time interval T_A between ***in1*** and ***in2*** should be greater than the switching time of the membrane which is 16 μs for this specific model. The circuit was simulated with various values of pulse width T_P, such as 200 ns, 300 ns, 500 ns, 1 us, and 2 us. It was found that the pulse width smaller than 300 ns was not enough to fully charge the membrane for the pull-in actuation. Any pulse width greater than 300 ns was providing enough charge to the membrane switch for pull-in actuation.

Figure 6 shows that the output load capacitor can be charged in 300 ns and the output voltage is 50 V when the capacitor is charged. After the membrane is snapped down to the dielectric layer, the voltage across the dielectric layer becomes mere 0.53 V. The novel feature of the circuit is that once the actuation of the membrane is started, the output port of the circuit will be floated. As a result, charge stored in the capacitive switch is nearly constant and the voltage automatically decreases as the membrane moves down. Electric field across the dielectric layer was calculated from the simulated voltage waveform throughout the actuation period for the 50V CMOS IC circuit. As shown in Figure 7, electric field across the dielectric is below 27 kV/cm. This is substantially lower than that of the CV biasing which can be found to be 2.38 MV/cm. Consequently, the dielectric charging can be reduced significantly and the lifetime of the switch can be increased.

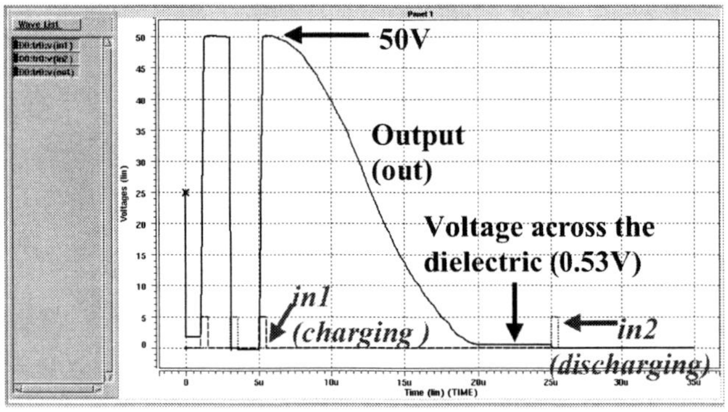

Figure 6. Voltage waveforms of simulation results with the pulse width T_P of 500ns.

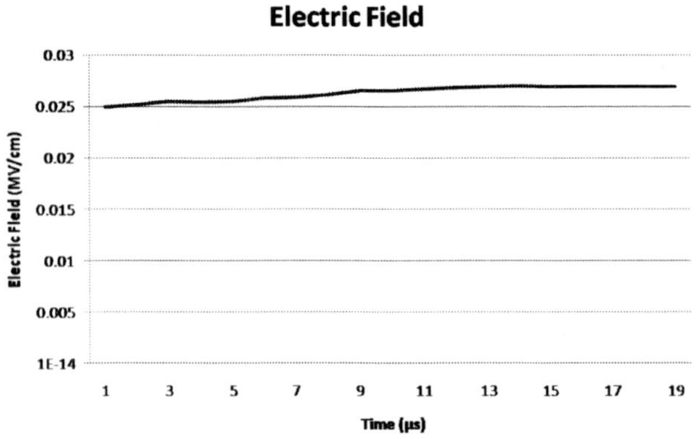

Figure 7. Electric field across the dielectric layer throughout the charging-floating-discharging period.

Figure 8 shows the waveforms of the calculated charges on the load switch capacitor. As a result of the CC biasing, the charge remains nearly constant throughout the charge-float cycle. Initial amount of charge stored in the switch capacitor was found to be 2.31 pC and it was increased gradually to 2.49 pC as the membrane moves down. Charge difference during the membrane actuation is only 0.18 pC. The amount of difference may come from the shared charge in the parasitic capacitors.

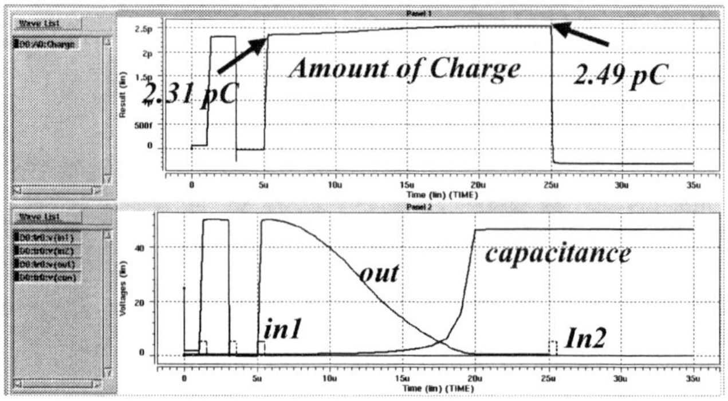

Figure 8. Waveforms of the calculated charges in the switch capacitor. (To be modified)

6 CONCLUSION

We have designed and simulated a constant-charge biasing circuit implementation using Austria Microsystems' 0.35 µm 50 V high voltage CMOS Technology. The proposed circuit can be monolithically integrated with the capacitive RF MEMS switch in high voltage CMOS process. The circuit simulation results in very low hold voltage and near constant charge throughout the actuation period, which implies that the lifetime of the switch can be substantially increased compared to a conventional CV biasing scheme.

REFERENCES

[1] A. Malczewski et al., "X-band RF MEMS phase shifters for phased array applications," IEEE Microwave and Guided Wave Letter, v. 9, n. 12, pp. 517-519, 1999.

[2] B. Pillans, et al., "Ka-band RF MEMS phase shifters," IEEE Microwave and Guided Wave Letter, v. 9, n. 12, pp. 520-522, 1999.

[3] J.J. Maciel, et al "MEMS electronically steerable antennas for fire control radars" Proceedings, IEEE 2007 Radar Conference, p 677-682, 2007.

[4] Y.B. Gianchandani, et al, "COMPREHENSIVE MICROSYSTEMS" ELSEVIER, v.3., Chap. 3.10, pp323-332, 2008.

[5] C. Goldsmith, *et al.* "Lifetime characterization of capacitive RF MEMS switches," 2001 IEEE Int. Microwave Sym., v. 1, pp. 227-230, 2001.

[6] J-B Lee, *et al* "Numerical Simulations of Novel Constant-Charge Biasing Method for Capacitive RF MEMS Switch," Nanotech, v. 2, chap. 8, pp. 396-399, 2003.

[7] J. Muldavin, "Design and analysis of series and shunt MEMS switches," Ph.D. dissertation, U. Michigan, 2001.

Vibration-Actuated Bistable Micromechanism for Microassembly

D.-A. Wang[*] and H.-T. Pham[**]

Institute of Precision Engineering, National Chung Hsing University
250 Kuo Kuang Rd, Taichung, Taiwan, ROC, [*]daw@nchu.edu.tw
[**]cslamvien@yahoo.com

ABSTRACT

This paper proposes a novel method to switch an on-substrate bistable micromechanism. An external vibration is exploited to switch the micromechanism between its bistable positions. There is no need to build any actuators on substrate with this method. The vibration-actuated bistable micromechanism (VABM) is vibrated by shaking the entire substrate with a piezo actuator. The vibration provides a simple means of switching the VABM. Finite element analyses are utilized to obtain the nonlinear spring stiffness of the VABM and an analytical model is derived in order to analyze its dynamic behavior. Prototypes of the VABM are fabricated using elecroforming. A scenario of the VABM for on-substrate fine positioning of micro components is presented.

Keywords: bistable, micromechanism, vibration

1 INTRODUCTION

Bistable micromechanisms (BMs) are gaining attention in MEMS applications such as memory cells [1], switches [2], relays [3], and valves [4]. One advantage of BMs is that no power is required to keep the mechanism in either of its bistable positions [5]. Most researches utilize on-substrate thermomechanical actuators [6,7] for switching between bistable positions, while others use comb-drive actuators [5,8]. In this investigation, an external vibration is exploited to switch BM between its bistable positions. There is no need to build any on-substrate actuator with this method.

This paper describes a design of a BM actuated by external vibration. The design is based on a BM reported by Wilcox and Howell [6], where their BM is actuated by a thermomechanical actuator. Dissimilar to their actuation method, here, the BM is actuated by shaking the entire substrate with a piezo actuator, requiring no need for built-in driving mechanisms such as thermomechanical or comb-drive actuators. This vibration of the BM via an external vibration source provides a simple means of switching BMs. An analytical model of the vibration-actuated bistable micromechanism (VABM) is derived in order to analyze its dynamic behavior. Prototypes of the VABM are fabricated using an electroforming process. A scenario of the VABM for on-substrate fine positioning of micro components is presented.

2 DESIGN

2.1 Operational principle

A schematic of a VABM investigated here is shown in Fig. 1(a). It consists of flexible folded beams and a shuttle. Fig. 1(b) shows a quarter model of the VABM. A Cartesian coordinate system is also shown in the figure. As the shuttle is moving linearly, besides bending, short and long segments of the folded beam are under tension and compression, respectively. The combined bending, compression and tension in folded beams results in bistable behavior of the device.

The concept of the actuation by external vibration is illustrated in Fig. 2. First, the substrate is shaken at frequency f_1 (see Fig. 2(a)) and this vibration causes the BM to move to its second stable position (see Fig. 2(b)). Next, the substrate is shaken at frequency f_2 (see Fig. 2(c)). This external vibration causes the BM to move to its first stable position.

2.2 Modeling

An equation of motion of a lumped parameter model of the BM shown in Fig. 1(a) is derived to analyze its dynamic behavior. The equation of motion for a simple mass-damper-spring system with excitation $F_{ext}(t)$ from external vibration has the form

$$m\ddot{x} + c\dot{x} + kx = F_{ext}(t) \tag{1}$$

where m, c, and k are the mass, viscous damping coefficient, and spring stiffness of the BM. Assuming Couette air flow between the substrate and the BM, c can be expressed as

$$c = \mu A / d \tag{2}$$

where μ is the viscosity of the fluid in the environment. A and d are the planar area of the BM and the gap between the substrate and the BM, respectively. The nonlinear spring stiffness, including terms up to ninth order, is written as

$$k = \sum_{i=0}^{9} k_i x_i \tag{3}$$

This ninth order stiffness function allows for the detailed modeling of the highly nonlinear force-displacement relation of the BM.

When the BM is actuated by shaking the entire substrate with an external vibration $x = Z\cos(\omega t)$ in the x direction, the inertial force $F_{ext}(t)$ exerted to the BM is

$$F_{ext}(t) = m\omega^2 Z\cos(\omega t) \quad (4)$$

where Z and ω are the amplitude and frequency of the external vibration, respectively.

2.3 Analyses

In order to obtain the nonlinear spring stiffness for a VABM, finite element analyses are carried out. Due to symmetry, only a quarter model is considered. Fig. 1(b) shows a schematic of a quarter model with $L_1 = 300$ μm, $L_0 = 900$ μm, $L_s = 200$ μm, $t_1 = 5$ μm, $t_0 = 40$ μm, $t_s = 9$ μm, and $\theta = 2.3^\circ$. The thickness t of the device is 5 μm. In the analyses, the material of the device is assumed to be linear elastic and isotropic. The Young's modulus E is taken as 207 GPa, and the Poisson's ratio ν is taken as 0.31. The commercial finite element program ABAQUS is employed to perform the computations.

A force-displacement curve and a potential energy curve of the VABM are shown in Fig. 3(a) and 3(b), respectively. We selected a ninth order function, Eq. (3), to fit the simulation data in Fig. 3(a). The values of the coefficients of Eq. (3) are listed in Table 1. The two local minimums of the energy curve correspond to the two stable positions of the VABM.

In order to simulate the dynamic behaviors of the VABMs, Park method [11] is used to solve the governing nonlinear differential equations, Eqs. (1)-(4). Fig. 4 shows time responses of the VABM vibrated at different frequencies and amplitudes. As shown in Fig. 4(a), when the external vibration with $\omega / 2\pi = 5.1$ kHz and $Z = 1$ μm is applied to the substrate, the VABM is switched from stable position 1 to stable position 2. When the substrate is vibrated at $\omega / 2\pi = 3.96$ kHz and $Z = 2$ μm, the VABM is found to switch from stable position 2 to stable position 1.

3 FABRICATION AND TESTING

VABMs have been fabricated by a simple electroforming process on glass substrates. Fig. 5 shows the fabrication steps, where only two masks are used. First, a 2 μm-thick titanium seed layer is sputter-coated on the whole glass substrate. Next, a 5 μm-thick photoresist (AZ4620) is coated and patterned to prepare a mold for electrodeposition of a copper sacrificial layer. Following that, a 5 μm-thick photoresist (AZ4620) is coated and patterned on top of the copper sacrificial layer. Into this mold, a 5 μm-thick nickel layer is electrodeposited using a low-stress nickel sulfamate bath with the chemical compositions listed in Table 2. Finally, the photoresist and copper sacrificial layers are removed to release the nickel microstructures.

Fig. 6 shows an OM photo of a fabricated device. The fabricated devices will be tested using the experimental apparatus shown in Fig. 7. A piezo actuator (PPA20M, Cedrat Technologies) is used to shake the substrate. Further experimental work is underway.

4 A VABM FOR MICROASSEMBLY

A scenario of the VABM for on-substrate fine positioning of micro components is presented. The device shown in Fig. 8(a) includes a clamp and two actuating BMs. It is fabricated on a glass substrate. Fig. 8 illustrates a four-step operation of the device. First, the substrate is shaken at the frequency f_1, and this vibration resonates only BM 1 (see Fig. 8(a)), causing BM 1 to move towards its second stable position (see Fig. 8(b)). This motion causes the clamp to push a micro component against an anchored fixture, achieving precise positioning (see Fig. 8(b)). Next, the substrate is shaken at the frequency f_2. This external vibration resonates only BM 2 (see Fig. 8(c)), causing BM 2 to move towards its second stable position. This motion causes the clamp to return to its original position (see Fig. 8(d)), releasing the positioned component.

The force required to move the clamp between its stable positions, R_c, is designed to be smaller than the forces for the BM 1 and BM 2, R_1 and R_2, respectively (see Fig. 9). The subscripts f and b in Fig. 9 represent the forward (from stable position 1 to stable position 2) and backward (from stable position 2 to stable position 1) motion, respectively. For assembly of the micro component, R_{1f} must be greater that the sum of R_{2b} and R_{cf}. For release of the micro component, R_{2f} must be greater that the sum of R_{1b} and R_{cb}. It is also designed that the clamp has a different natural frequency from those of BM 1 and BM 2. Since the dimensions for BM 1 and BM 2 are the same except their shuttle masses, their force-displacement curves are the same as shown in Fig. 9. However, with different shuttle masses, BM 1 and BM 2 have different natural frequencies. Finite element analyses are carried out to obtain the nonlinear spring constants of the BMs. Based on numerical simulations, frequencies and amplitudes for BM 1 and BM 2 are found to move the clamp between its stable positions. Further experimental work is underway.

5 CONCLUSIONS

External vibration is proposed to switch a BM between its stable positions. The frequencies and amplitudes for the dynamic switching are found by solving the nonlinear equation of motion of the BM. The selective vibration provides a simple means of switching the BM. For demonstration of the effectiveness of the BM for microassembly, prototypes of BMs, a clamp, and micro components are fabricated on glass substrates using elecroforming. A scenario of the BM for on-substrate fine positioning of micro components is presented.

REFERENCES

[1] B. Hälg, "On a nonvolatile memory cell based on micro-electro-mechanics," in Proc. IEEE MEMS 1990 Conference, pp. 172-176.

[2] M. Freudenreich, U.M. Mescheder, G. Somogyi, "Design considerations and realization of a novel micromechanical bi-stable switch," in Transducers 2003 Workshop, pp. 1096-1099.

[3] T. Gomm, L.L. Howell and R.H. Selfridge, "In-plane linear displacement bistable microrelay," J. Micromech. Microeng. 12, 257-264, 2002.

[4] B.Wagner, H.J. Quenzer, S. Hoerschelmann, T. Lisec and M. Juerss, "Bistable microvalve with pneumatically coupled membranes," in Proc. IEEE MEMS 1996 Conference, pp. 384-388.

[5] J. Casals-Terre and J. Shkel, "Snap-action bistable micromechanism actuated by nonlinear resonance," in IEEE Sensors 2005, pp. 893-896.

[6] D.L. Wilcox and L.L. Howell, "Fully compliant tensural bistable micromechanisms (FTBM)," J. Microelectromech. Syst., 14, 1223-1235, 2005.

[7] J. Qiu, J.H. Lang and A.H. Slocum, "A curved-beam bistable mechanism," J. Microelectromech. Syst., 14, 1099-1109, 2005.

[8] K.B. Lee, A.P. Pisano and L. Lin, "Nonlinear behaviors of a comb drive actuator under electrically induced tensile and compressive stresses," J. Micromech. Microeng., 17, 557-566, 2007.

[9] W.Y. Tseng and J Dugundji, "Nonlinear vibrations of a buckled beam under harmonic excitation," J. Applied Mechanics, 38, 467-476, 1971.

[10] N. Yamaki and A. Mori, "Non-linear vibrations of a clamped beam with initial deflection and initial axial displacement. I - Theory," Journal of Sound and Vibration, 71, 333-346. 1980.

[11] A.F. D'Souza and V.K. Garg, "Advanced Dynamics: Modeling and Analysis," Prentice-Hall, 1984.

k_0	-9.945×10^{-15}
k_1	6.286×10^{-12}
k_2	-1.678×10^{-9}
k_3	2.445×10^{-7}
k_4	-2.083×10^{-5}
k_5	0.001016
k_6	−0.02369
k_7	−0.001379
k_8	7.784
k_9	0.3754

Table 1: Values of the coefficients of the nonlinear spring stiffness function.

Chemical/Plating Parameter	Amount/Value
Nickel sulfamate ($Ni(NH_2SO_3)_2 \cdot 4H_2O$)	450 g/L
Boric acid (H_3BO_3)	35 g/L
Nickel chloride ($NiCl_2 \cdot 6H_2O$)	4 g /L
Stress reducer	17 ml/L
Leveling agent	17 ml/L
Wetting agent	2 ml/L
Bath temperature	45 °C
Plating current type	dc current
pH of the solution	4
Plating current density	0.113 A/dm^2
Deposition rate	0.071 μm/min
Anode-cathode spacing	100 mm
Anode type	titanium

Table 2: Chemical composition and operation conditions for the low-stress nickel electroplating solution.

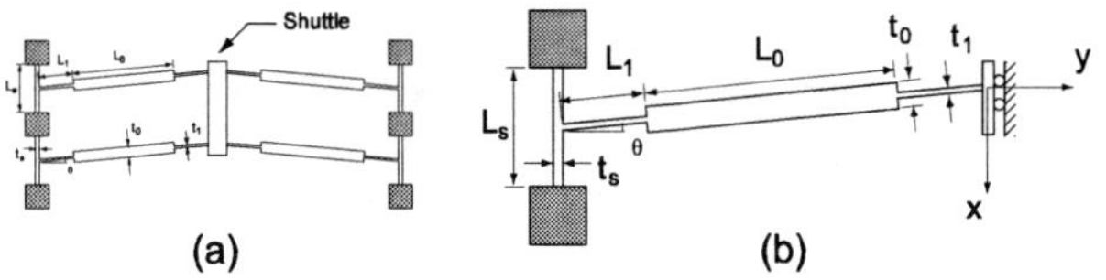

Figure 1: Schematics of a VABM. (a) A full model. (b) A quarter model.

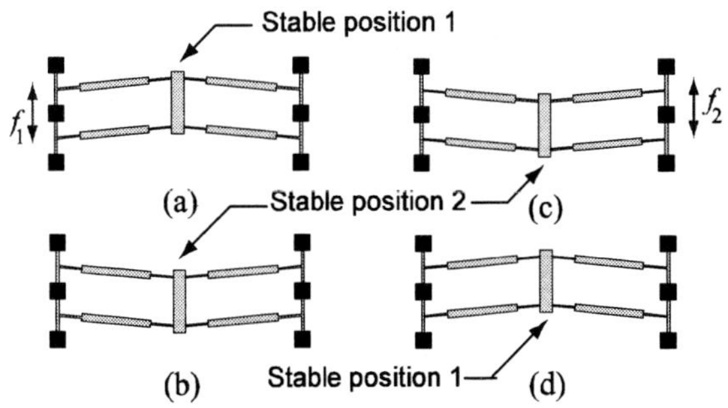

Figure 2: Actuation by external vibration.

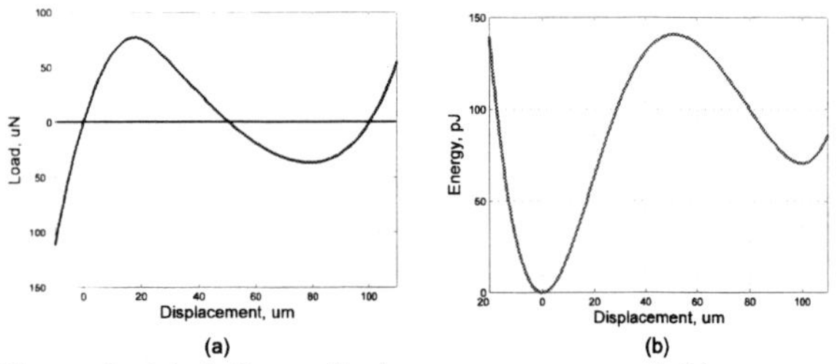

Figure 3: (a) A force-displacement curve and (b) an energy curve of a VABM.

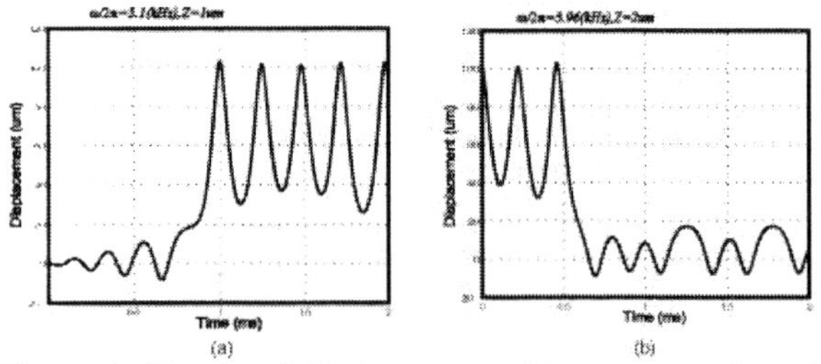

Figure 4: Simulated displacement with respect to time: (a) switching from stable position 1 to stable position 2 and (b) switching from stable position 2 to stable position 1.

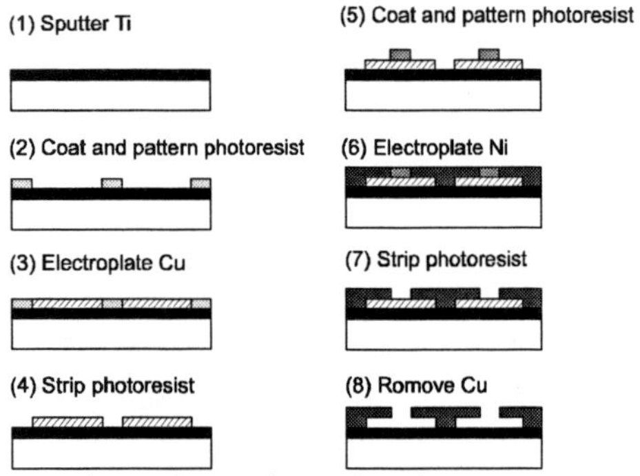

Figure 5: An electroforming process.

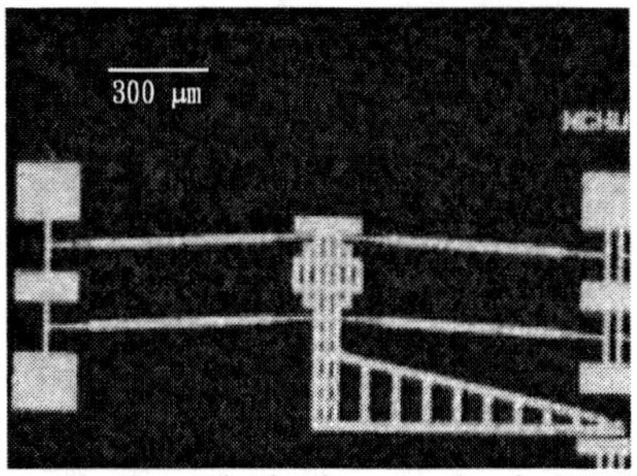

Figure 6: An optical micrograph of a VABM.

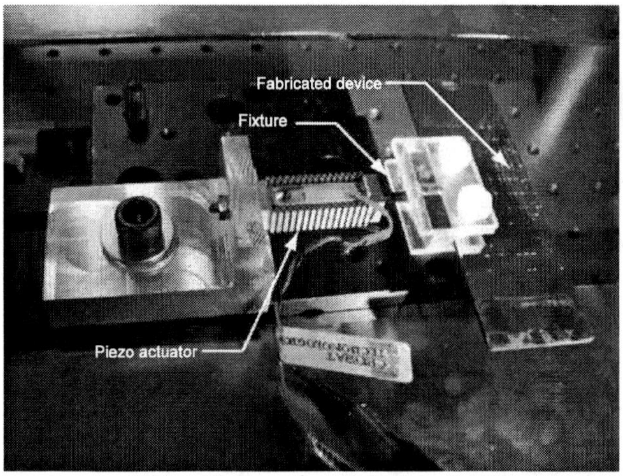

Figure 7: Experimental apparatus.

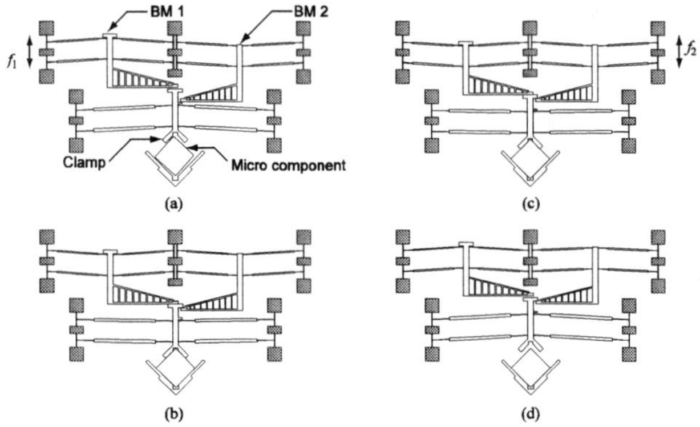

Figure 8: A scenario of a VABM for on-substrate fine positioning of micro components.

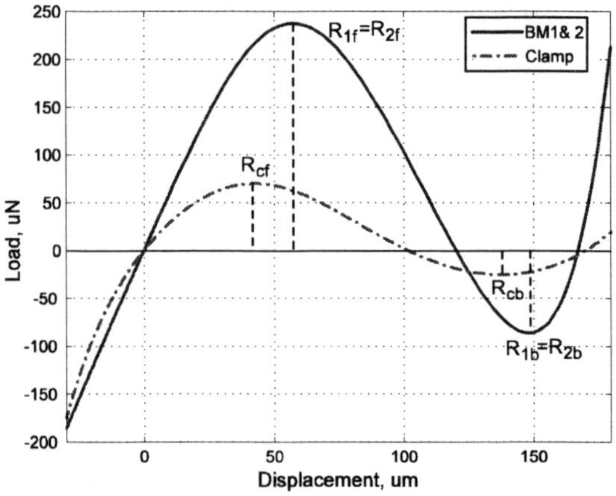

Figure 9: Force-displacement curves of BM 1, BM 2 and the clamp.

A Semiclassical Study of Carrier Transport combined with Atomistic Calculation of Subbands in Carbon Nanoribbon Transistors

D. Rondoni*, J. Hoekstra*, M. Lenzi** and R. Grassi**

* Delft Technical University, Electronics Research Laboratory
Faculty of Electrical Engineering, Mathematics and Computer Science
mekelweg 4, 2628CD Delft, The Netherlands, D.Rondoni@tudelft.nl
** ARCES and DEIS, University of Bologna, Viale Risorgimento 2, 40136, Bologna, Italy

ABSTRACT

In this paper we present a deterministic approach to the study of one-dimensional carrier transport in CNR devices, both in the ballistic regime and when phonon scattering is taken into account. The energy subbands and the potential energy profiles along the longitudinal coordinate are obtained through the atomistic slab-by-slab solution of the transverse Schrödinger equation. The longitudinal transport is modeled through a deterministic solution of the Boltzmann transport equation. Finally, these equations are solved self-consistently with the Poisson equation in order to obtain the electrostatic potential, the charge density and the electrical currents at different biases. Results are presented for different channel widths and transistor geometries and the impact of phonon scattering is discussed.

Keywords: carbon nanoribbons, carrier transport, Boltzmann equation, phonon scattering.

1 INTRODUCTION

Carbon Nanotubes (CNTs) have recieved a very large attention from the nanoelectronic research community during the last years. Infact CNT-FETs exhibit very promising performances, with very high mobilities and carrier velocities [1]. However, despite the big amount of research in the last few decades, many serious technology issues are still to be solved (control on the chirality of the tubes, mixing of planar deposition and cylindrical geometry, circuit patterning, amongst the others). Therefore, at present, it appears that the feasibility of large scale integration of CNT devices will require the development of revolutionary techniques in process technology.

Carbon Nanoribbons (CNRs) exhibit properties similar to CNTs [2], [3], together with a planar geometry that appears in principle more feasible for integration with SiC substrates through high performance litography techniques. Therefore, in the last few years CNRs are being addressed as a valid alternative to CNTs [4]–[6] and simulation studies of CNR-FETs are becoming very important in order to evaluate the performances of such nanodevices.

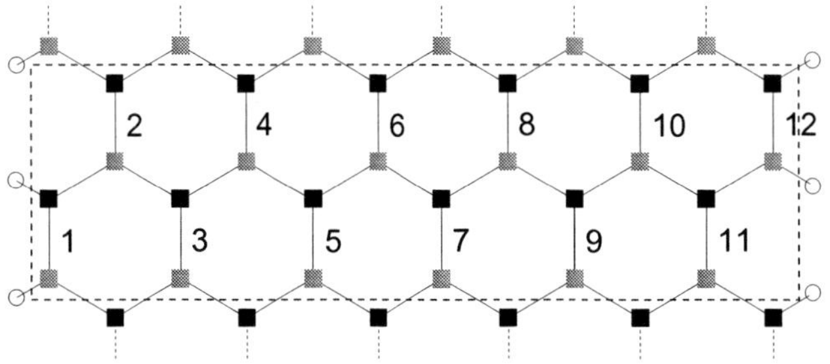

Figure 1: Example of a basic cell for a (12,0) armchair nanoribbon (note that the chiral vector formalism is 90 degrees rotated for CNRs compared to CNTs). The dimers are numbered from left to right.

In this paper we simulate and evaluate the performance of nanometric planar FETs whose channel is made of a narrow (few nanometers wide) armchair semiconducting single-layer nanoribbon. The CNR is treated as a 2D-channel and 1D transport is obtained through the confinement in the transverse direction. The nanoribbon is considered as a serie of basic cells, called slabs (see Fig. 1).

We separate the 2D problem in the transverse and longitudinal directions and solve it self-consistently with the 3D Poisson's equation. In the transverse direction (quantum confinement), a Tight-Binding (TB) Hamiltonian is used to find the energy subbands in each slab of the CNR. The 1D problem of transport along the longitudinal direction is then solved with a deterministic solution of the Boltzmann Transport Equation (BTE), both in the ballistic regime and when scattering is taken into account, based on a simplified dispersion relation that includes first order non-parabolic effects.

2 CONFINEMENT: TRANSVERSE SOLUTION

In this section we describe the TB model used for the transverse solution (lateral confinement). In principle, for every slab l of the ribbon, the energy dispersion relations (subbands) $\varepsilon(k)$, with k the longitudinal wavevector, should be calculated starting from the full TB Hamiltonian and solving the transverse Schrödinger equation with the potential energy repeated periodically throughout the device.

In our recent work [7], we show that the dispersion relations obtained with the non-parabolic expression

$$(\varepsilon_b - \frac{E_g^b}{2})(\frac{1}{2} + \frac{\varepsilon_b}{E_g^b}) = \frac{\hbar^2 k^2}{2m_b^*} \quad (1)$$

where b is the subband index, $E_g^b = E_c^b - E_v^b$ is the energy gap and m_b^* the effective mass, match with excellent agreement, over an extended range of energies, the dispersion relations obtained with a full TB approach. The results described in [7] show that we can perform our calculations over a simplified version of the transverse problem. For a given width, the full subband structure is calculated only once for a slab with flat zero potential, from which the effective masses of the lowest desired subbands are extracted. The eigenvalue TB calculation is then repeated in every slab of the device only for $k = 0$, in order to obtain E_c^b and E_v^b, as well as the eigenfunction χ^b, as a function of the longitudinal coordinate z. Subsequently, the potential profile along z, obtained from the energy minima of the calculated bands, is used, together with the effective masses and non-parabolicity factors, as the input of the BTE solver.

3 TRANSPORT: DETERMINISTIC SOLUTION OF THE BTE

The transport problem along the longitudinal direction is treated with a semi-classical approach. In this work we simulate carrier transport only in the conduction band, therefore we consider only electrons as carriers. In the next future we intend to include holes and extend the model to the valence band.

The electronic current along the device is described by a system of BTEs, having as unknowns the electron distribution probabilities $f_b(z,k)$, where b is the subband index. The steady-state BTE of the b-th subband is written in the form

$$u_b \frac{\partial f_b}{\partial z} - \frac{1}{\hbar}\frac{\mathrm{d}E_b}{\mathrm{d}z}\frac{\partial f_b}{\partial k} = C_{in}^b - C_{out}^b \quad (2)$$

where $u_b = 1/\hbar(\mathrm{d}\varepsilon_b/\mathrm{d}k)$ is the group velocity, with $\varepsilon_b(k)$ the energy-dispersion relation. C_{in}^b and C_{out}^b are the in- and out-scattering integrals, respectively, relative to the state k of the b-th subband, and are integrals of the scattering probabilities, which are written in accordance to Fermi's Golden Rule [8], [9]. Both acoustic (elastic) and optical scattering is considered, intrasubband and intersubband.

For acoustic scattering, in the high temperature and isotropic approximation [9], the scattering probability from state k in band b to k' in band b' is

$$S_{ac}^{bb'}(k,k') = \frac{2\pi}{\hbar}\frac{D_{ac}^2 k_B T I^{bb'}}{v_s^2 \rho}\delta\left[\varepsilon_{b'}(k') - \varepsilon_b(k)\right] \quad (3)$$

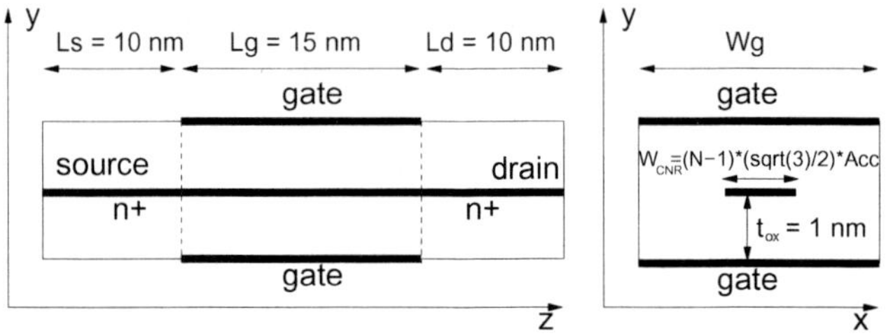

Figure 2: Longitudinal (left) and transverse (right) cross-section of the CNR-FETs analyzed in this work. N is the number of dimers in the basic cell (see Fig. 1) and $A_{cc} = 1.42$ Å is the length of the covalent bonding between carbon atoms in graphene.

with D_{ac} the acoustic deformation potential, v_s the sound velocity in graphene, ρ the graphene superficial mass density, $\delta[x]$ the Dirac delta-function, and $I^{bb'}$ the form factor. For optical phonons, the scattering probability from state k in band b to k' in band b' is

$$S_{op}^{bb'}(k,k') = \frac{\pi D_{op}^2 I^{bb'}}{\rho\omega_{op}}[N_{op}(\hbar\omega_{op}) + \frac{1}{2} \mp \frac{1}{2}]$$

$$\delta\left[\varepsilon_{b'}(k') - \varepsilon_b(k) \mp \hbar\omega_{op}\right] \quad (4)$$

with D_{op} the optical deformation potential, ω_{op} the phonon angular frequency and $N_{op}(\hbar\omega_{op})$ the phonon number given by Bose statistics. The upper and lower signs correspond, respectively, to phonon absorption and emission processes.

The values of the physical constants for the scattering processes are taken from recent literature. Since research on scattering in CNRs is at its very beginning, reliable values are still not avaliable, and therefore some of them have been extrapolated from analogous values of CNTs. In particular, for acoustic phonons the reader may refer to [10], while for optical phonons we focus on longitudinal optical phonons (see [2], [5], [10]) and the value of the energy quantum associated with the optical scattering is $E_{op} = \hbar\omega_{op} = 160\ meV$ (taken from CNTs values [11], [12]). Finally, the form factors $I^{bb'}$ in equations (3) and (4) are defined, in agreement with [8], as

$$I^{bb'} = \int_0^w \left|\chi^b(x)\right|^2 \left|\chi^{b'}(x)\right|^2 dx \simeq \begin{cases} 3/2w & \text{if } b = b' \\ 1/w & \text{if } b \neq b' \end{cases} \quad (5)$$

where w is the ribbon width .

The distribution functions and the charge densities obtained from the BTE are fed into the Poisson's equation to calculate the new electrostatic potential and the whole procedure is iterated until the self-consistent global solution is reached.

4 RESULTS

Simulations have been carried out for CNR-FETs with the structure shown in Fig. 2. Two types of arm-

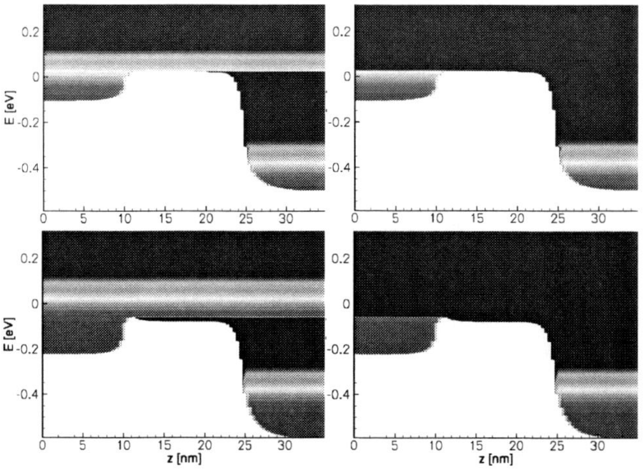

Figure 3: 2D contour plots (in log-scale and (z, E) space) of carrier distribution functions in ballistic transport conditions for the most populated subband of a 12-dimers (top) and 19-dimers (bottom) armchair nanoribbon, positive (left) and negative (right) flows, with $V_{DS} = 0.4V$ and $V_{GS} = 0.4V$. Values of the distribution function go from 10^{-6} (deep blue) to 1 (red).

chair nanoribbons have been considered, one with $N = 12$ (about 1.35 nm wide) and the other with $N = 19$ (about 2.2 nm wide), N the number of dimers in the basic cell. The length of both nanoribbons is 83 slabs, equal to about 35 nm of which 15 nm of undoped channel and about 10 nm source and drain extensions. It is important to point out that the subbands calculated with the TB approach in [7] allow us to restrict ourselves to the two most populated subbands because the third conduction subband of the nanoribbons considered here results very far from the bottom of the conduction band and doesn't contribute to the conduction. Therefore only the two most populated subbands are simulated in the BTE and considered in the transport problem.

The distribution functions in the ballistic case for the most populated subband of the two types of CNR-FETs are depicted in Fig. 3. In the case of flow in the positive direction, at the drain side it can be clearly noticed a population of carriers in equilibrium with the drain and a population of hot carriers coming from the source. In the case of negative flow the carriers coming from the drain cannot overcome the potential barrier and hot carriers at the source are not present. The net flow, calculated amongst all the subbands considered, will give the total current of the device.

In Fig. 4 the effect of scattering is highlighted in the case of the 19-dimers nanoribbon, for positive flow and different subbands. In the case of acoustic scattering only, the carrier density coming from the source decreases towards the drain because the carriers are backscattered either in the same subband or in the other subband. However the energy level of the carriers stays

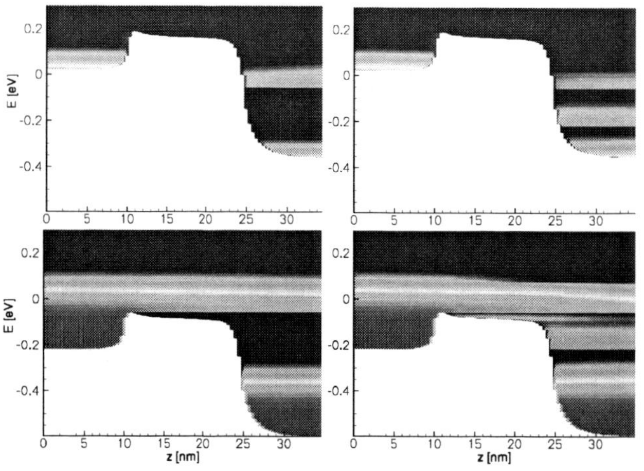

Figure 4: 2D contour plots (in log-scale and (z, E) space) of carrier distribution functions of the positive flow for both subbands (most populated subband in the lower graphs) of a 19-dimers armchair nanoribbon when scattering processes are included. Left graphs include only acoustic phonons, right ones also opticals. $V_{DS} = 0.4V$ and $V_{GS} = 0.4V$. Values of the distribution function go from 10^{-6} (deep blue) to 1 (red).

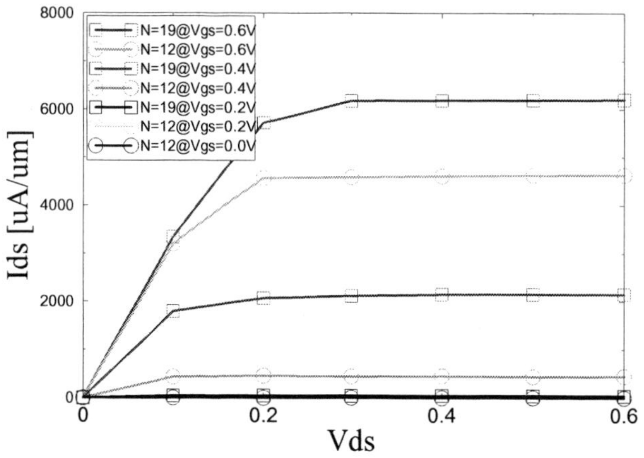

Figure 5: $V_{DS} - I_{DS}$ characteristics of 19-dimers (squares) and 12-dimers (circles) CNR-FETs in ballistic conditions. V_{GS} is varied from $0.0V$ to $0.6V$ in steps of $0.2V$.

the same. Viceversa, if we consider in our calculations also optical scattering, we can notice the effect of energy spreading due to the optical phonon absorption/emission process.

The $V_{DS} - I_{DS}$ output characteristics for ballistic CNR-FETs are depicted in Fig. 5. Both 12-dimers and 19-dimers FETs are reported for a comparison. Results (for similar widths and bias conditions) are in good agreement with calculations performed in other papers [3], [4].

Fig. 6 shows the output characteristics of the 19-dimers FET when V_{GS} is varied and different phonons

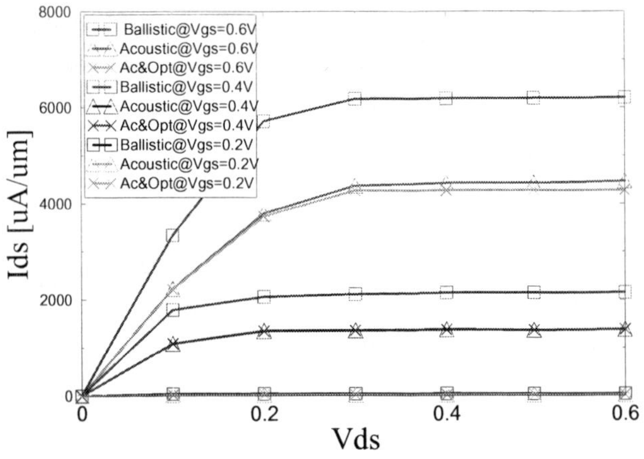

Figure 6: $V_{DS} - I_{DS}$ characteristics of the 19-dimers CNR-FET in ballistic conditions (squares), with acoustic phonons (triangles) and with both acoustic and optical phonons (crosses). V_{GS} is varied from $0.2V$ to $0.6V$ in steps of $0.2V$.

are considered. It is evident that acoustic scattering is the main cause of current degradation and only when the channel is sufficiently open ($V_{GS} = 0.6V$) optical scattering plays an appreciable role. This is due to the fact that only when there are avaliable states *inside* the channel optical scattering is able to backscatter the carriers towards the source, while carriers which are scattered by optical emission *after* the channel are sooner or later collected by the drain (see Fig. 4).

5 CONCLUSIONS

A computational method for the solution of carrier transport in CNR-FETs has been presented. The problem is solved through a deterministic solution of BTEs coupled with an atomistic TB transverse solution of the Scrödinger equation, self-consistently solved with Poisson's equation. The approach exhibits short simulation times compared to Monte-Carlo, where the computational times are quite long due to the statistical noise. In fact, in the case of carrier transport in CNRs the one-dimensionality of the problem makes direct solutions of Boltzmann Equations more convenient, particularly when the number of subbands to consider is limited.

Future work will have to include the transport of positive carriers (solving the BTEs for the valence band) and the edge-roughness scattering, that, especially in very narrow ribbons, plays a strong (if not major [3]) role. Moreover, a study of the device mobility will be useful to further evaluate the reliability and the accuracy of the model.

ACKNOWLEDGEMENTS

This research project has been supported by a Marie Curie Early Stage Research Training Fellowship of the European Community's Sixth Framework Programme under contract number 504195-EDITH.

REFERENCES

[1] S. Hasan, M. A. Alam and M. S. Lundstrom, "Simulation of Carbon Nanotube FETs Including Hot–Phonon and Self–Heating Effects", IEEE Trans. on El. Dev. 54, No. 9 (2007);

[2] D. Gunlycke, H. M. Lawler and C. T. White, "Room temperature ballistic transport in narrow graphene strips", Phys. Rew. B 75, 085418 (2006);

[3] G. Fiori and G. Iannaccone, "Simulation of Graphene Nanoribbon Field-Effect Transistors", IEEE Electron Device Letters 28, No. 8 (2007);

[4] G. Liang, N. Neophytou, D. E. Nikonov and M. S. Lundstrom, "Performance Projections for Ballistic Graphene Nanoribbon Field Effect Transistors", IEEE Trans. on El. Dev. 54, No. 4 (2007);

[5] B. Obradovic, R. Kotlyar, F. Heinz, P. Matagne, T. Rakshit, M. D. Giles, M. A. Stettler and D. E. Nikonov, "Analysis of graphene nanoribbon as a channel material for field effect transistors", Appl. Phys. Lett. 88, 142102 (2006).

[6] M. C. Lemme, Tim J. Echtermeyer, Matthias Baus and Heinrich Kurz, "A Graphene Field-Effect Device", IEEE Electron Device Letters 28, No. 4 (2007);

[7] R. Grassi, S. Poli, E. Gnani, A. Gnudi, S. Reggiani and G. Baccarani, "Tight-binding and effective mass modeling of armchair carbon nanoribbon FETs", ULIS 2008 Conference Proceedings;

[8] D. K. Ferry and S. M. Goodnick, *Transport in Nanostructures*, Cambridge Univ. Press, UK (2001);

[9] M. Lundstrom, "Fundamentals of carrier transport", Cambridge University Press, Cambridge (2000);

[10] D. Finkenstadt, G. Pennington and M. J. Mehl, "From graphene to graphite, a general tight binding approach for nanoribbon carrier transport", Phys. Rev. B 76, 121405(R) (2007);

[11] J. Guo and M. Lundstrom, "Role of phonon scattering in carbon nanotube field-effect transistors", Appl. Phys. Lett. 86, 193103 (2005);

[12] S. O. Koswatta, S. Hasan, M. S. Lundstrom, M. P. Anantram and D. E. Nikonov, "Nonequilibrium Green's Function Treatment of Phonon Scattering in Carbon-Nanotube Transistors", IEEE Trans. on El. Dev. 54, No. 9 (2007);

Comprehensive Examination of Threshold Voltage Fluctuations in Nanoscale Planar MOSFET and Bulk FinFET Devices

Chih-Hong Hwang, Hui-Wen Cheng, Ta-Ching Yeh, Tien-Yeh Li, Hsuan-Ming Huang, and Yiming Li*

Department of Communication Engineering, National Chiao Tung University
1001 Ta-Hsueh Rd, Hsinchu City, Hsinchu 300, Taiwan
*ymli@faculty.nctu.edu.tw

ABSTRACT

Intrinsic fluctuations on device characteristics, such as the threshold voltage (V_{th}) fluctuation is crucial in determining the behavior of nanoscale semiconductor devices. In this paper, the dependency of process-variation and random-dopant-induced Vth fluctuation on the gate oxide thickness scaling in 16 nm metal-oxide-semiconductor field effect transistors (MOSFETs) is investigated. Fluctuations of the threshold voltage for the studied planar MOSFETs with equivalent oxide thicknesses (EOT) from 1.2 nm to 0.2 nm (e.g., SiO_2 for the 1.2 and 0.8 nm EOTs, Al_2O_3 for the 0.4 nm EOT and HfO_2 for the 0.2 nm EOT) are then for the first time compared with the results of 16nm bulk fin-typed filed effect transistors (FinFETs), which is one of the promising candidates for next generation semiconductor devices. An experimentally validated simulation is conducted to investigate the fluctuation property. Result of this study confirms the suppression of Vth fluctuations with the gate oxide thickness scaling (using high-κ dielectric). It's found that the immunity of the planar MOSFET against fluctuation suffers from nature of structural limitations. Bulk FinFETs alleviate the challenges of device's scaling and have potential in the nanoelectronics application.

Keywords: Threshold voltage fluctuation; random dopant; process-variation; gate-length deviation; line-edge roughness; modeling and simulation

1 INTRODUCTION

As the dimension of complementary metal-oxide-semiconductor (CMOS) devices shrunk into sub-90 nm scale, threshold voltage (Vth) fluctuations resulting from such as the short channel effect and random dopant are pronounced [1-12]. The random-dopant-induced fluctuation is mainly from the random nature of ion implantation. The gate-length deviation and line-edge roughness could be attributed to the short channel effect. Fluctuation is getting worse due to serious short channel effect when the dimension of device is further scaled. Consequently, it affects the design window, yield, noise margin, stability, and reliability of ultra large-scale integration circuits. The tolerance of fluctuation has to be controlled even strictly with increases in the number of transistors as technology advances. The use of thin gate oxide is one of effective ways to suppress the process-variation- and random-dopant-induced Vth fluctuation [6]. Our recent work has demonstrated that the V_{th} fluctuation of a 16 nm planar metal-oxide-semiconductor field effect transistor (MOSFET) device could be suppressed by 20%, as the gate oxide scales is scaled down from 1.2 nm to 0.8 nm [6]. However, ongoing scaling of gate oxide thickness may raise problems of process controllability, leakage current, and reliability. The use of a high-κ dielectric is a key to enhance the performance of such devices [13-14]. For devices with vertical channel structures, such as fin-typed FETs (FinFETs), immunity against fluctuation is also fascinating because they possess better channel controllability [7,15-16]. Study of the effectiveness of fluctuation suppression and the mechanism against fluctuations according to these two approaches will be an interesting and benefit the nanodevice technology.

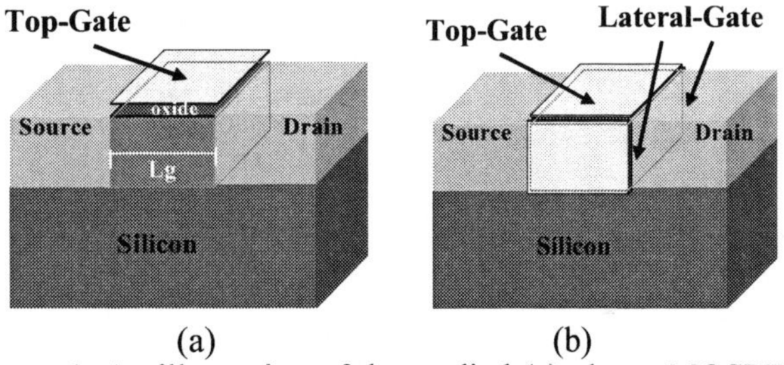

Figure 1: An illustration of the studied (a) planar MOSFET and (b) bulk FinFET.

In this paper, the dependency of process-variation- and random-dopant-induced threshold voltage fluctuation on the gate oxide thickness scaling in 16 nm nano-MOSFETs is examined, as shown in Fig. 1(a). Together with statistically generated process-variation-induced gate lengths and the large-scale doping profiles, the V_{th} fluctuation for each studied devices is computed by solving a set of three-dimensional (3D) quantum correction transport equations [17-18]. We notice that the device's threshold voltage and the mobility of the explored planar device (the case of 1.2 nm EOT) are calibrated with the measured data [6]. Fluctuations of the studied planar MOSFETs with equivalent oxide thicknesses (EOT) ranging from 1.2 nm to 0.2 nm, where Al_2O_3 is for the 0.4 nm EOT and HfO_2 is for the 0.2 nm EOT are then compared with the results for 16 nm bulk FinFETs [7,15-16]. The 16-nm-gate planar MOSFET with the 0.2 nm EOT is demonstrated to offer similar immunity against fluctuation as the 16 nm-gate bulk FinFET device with the 1.2 nm EOT, as shown in Fig. 1(b). The immunity of the planar MOSFET against fluctuation is limited by structural limitations while the multiple-gate FET, such as FinFETs, can overcome challenges with device scaling and is favorable in the era of nanoelectronics.

This article is organized as follows. In Sec. 2, we describe the analyzing technique. We then present and discuss the results. Mechanism of process-variation- and random-dopant-induced fluctuation are shown and discussed. Finally, we draw conclusions.

2 SIMULATION METHODOLOGY AND RESULTS DISCUSSION

The threshold voltage fluctuation is assumed to be contributed from the random-dopant and short channel effect. Effects of gate-length deviation (Lg), and the line-edge roughness (LER) are resulted from process-variation and belong to the short channel effects The random-dopant and short channel effect are independent sources of fluctuation, and the standard deviation of the total threshold voltage, $V_{th,total}$, is expressed by the following relation

$$\sigma^2_{Vth,total} = \sigma^2_{Vth,RD} + \sigma^2_{Vth,Lg/LER}, \quad (1)$$

where $\sigma_{Vth,RD}$ is the random-dopant-induced fluctuation, $\sigma_{Vth,Lg/LER}$ is fluctuations caused by the gate-length deviation and line-edge roughness. In this study, the EOT of planar MOSFET ranges from 1.2 nm to 0.2 nm and the EOT of bulk FinFET is fixed at 1.2 nm. The used dielectric materials are summarized in Tab. 1, where SiO_2 is used for a gate oxide thickness of 1.2 nm and 0.8 nm, Al_2O_3 is for a gate oxide thickness of 0.4 nm, and HfO_2 is for a gate oxide thickness of 0.2 nm. The nominal channel doping concentration of the devices herein is 1.48×10^{18} cm^{-3}. The devices have a 16 nm gate and a workfunction of 4.4 eV.

Material	Dielectric Constant	This Work
SiO_2	3.9	EOT = 1.2 nm / 0.8 nm
Al_2O_3	8-11.5	EOT = 0.4 nm
HfO_2	25-30	EOT = 0.2 nm

Table 1: The used dielectric materials in this study, SiO_2 is used for the cases of the 1.2 and 0.8 nm EOTs, Al_2O_3 is for the case of the 0.4 nm EOT and HfO_2 is for the case of the 0.2 nm EOT

To elucidate the effect of random fluctuations of the number and location of discrete dopants in the device channel, 758 doping islands are initially generated in an 80 nm^3 cube, in which the equivalent doping concentration is 1.48×10^{18} cm^{-3}, as shown in Fig. 2(a). The 80 nm^3 cube is then partitioned into 125 sub-cubes of volume 16 nm^3. The number of dopants may vary from zero to 14, and the average number is six, as shown in Figs. 2(b), 2(c) and 2(d), respectively. These 125 sub-cubes are then equivalently mapped into the channel region of the device for discrete dopant simulation, as shown in Figs. 1 (a) and 1(b). All statistically generated discrete dopants are incorporated into the large-scale 3D device simulation using the parallel computing system [19-20]. Characteristic of each device is obtained by solving a set of Poisson equation, electron-hole current continuity equations, and density-gradient equation [17-18]. This approach enables us to calculate the fluctuations of electrical characteristics that induced by the randomness of the number and position of dopants in the channel region to be investigated. This statistically sound full-scale 3D "atomistic" device simulation technique considers the computational cost and accuracy simultaneously.

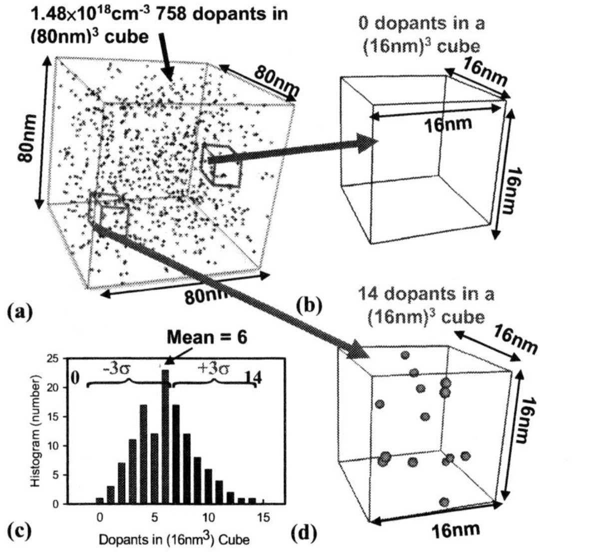

Figure 2: (a) Discrete dopants randomly distributed in (80nm)^3 cube with an average concentration of 1.48×10^{18} cm^{-3} and then partitioned into 125 sub-cubes of (16 nm)^3, where the numbers of dopant in sub-cubes may vary from zero to 14, as shown in (b), (c), and (d). These sub-cubes are then equivalently mapped into channel region of studied devices, Fig. 1, for dopant position/number-sensitive simulation. It means that for each explored device there are 125 cases of the 3D device simulation have to be performed.

Furthermore, we apply the statistical approach to evaluate the effect of process-variation-induced V_{th} fluctuation, $\sigma_{Vth,Lg/LER}$ [21]. The magnitude of the gate-length deviation and the line-edge roughness are extracted from the projections of the ITRS 2005 for different technology nodes. A look-up table of the threshold voltage versus gate length is established, as shown in Fig. 3. It enables us to evaluate the threshold voltage with respect to the deviation of gate-length deviation and line-edge roughness, which following the roadmap of ITRS that $3\sigma_{Lg}$ = 0.9 nm and $3\sigma_{LER}$ = 1.2 nm for the 22nm node and $3\sigma_{Lg}$ = 0.7 nm and $3\sigma_{LER}$ = 0.8 nm for the 16nm technology node, as the inset table of Fig. 3. Thus, we can calculate the standard deviation of threshold voltage resulting from the deviation of gate length and the roughness of line edge. The accuracy of the simulation is verified by comparing the simulated fluctuation results and the measured data of experimentally fabricated 20 nm devices [6]. The threshold voltages of the studied devices are adjusted to 140 mV. The threshold voltage is calculated according to a current criterion of $10^{-7}(W/L)$ (A), where the W and L are the width and length of device, respectively.

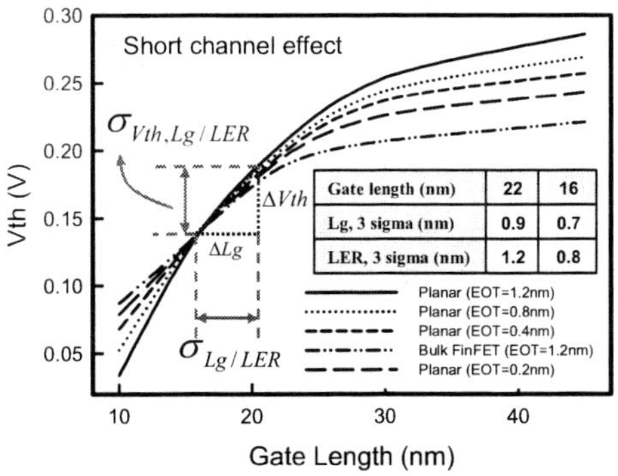

Figure 3: The threshold voltage roll-off of the studied devices, where the variation follows the projection of ITRS 2005 roadmap. These results are used to estimate the V_{th} fluctuation resulting from the gate-length deviation and the line-edge roughness.

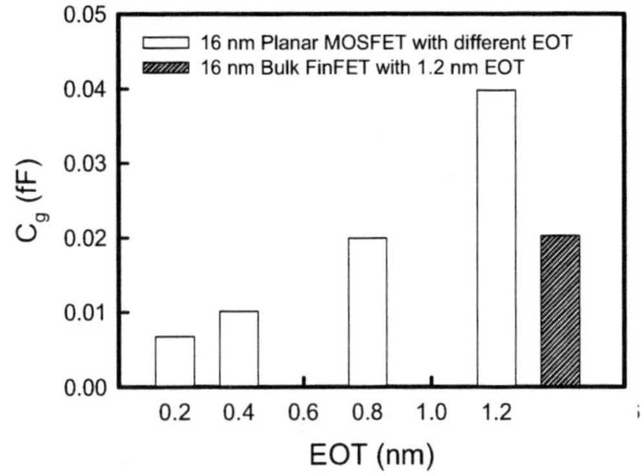

Figure 4: The gate capacitance (Cg) as a function of the EOT, where the solid line shows the planar MOSFETs with various EOT and the square symbol indicates the bulk FinFET device with 1.2 nm EOT.

Figure 4 plots the gate capacitance (C_g) as a function of the EOT, where white bars are the planar MOSFETs with various EOT and the shaded bar indicates the bulk FinFET device with 1.2 nm EOT. The planar MOSFET with 0.4 nm EOT, where Al_2O_3 is used for gate dielectric, exhibits a similar gate capacitance with the bulk FinFET device with 1.2 nm EOT. Since the value of gate capacitance is one of the indexes for the channel controllability of device, the bulk FinFET device with 1.2 nm EOT is expected to have similar immunity against process-variation induced fluctuation with the planar MOSFET with 0.4 nm EOT. This assumption is then verified in Fig. 5, which presents the process-variation-induced V_{th} fluctuation, $\sigma_{Vth,Lg/LER}$, of the planar MOSFETs and bulk FinFETs. The use of thin gate oxide and high high-κ dielectric material is effective in suppression of process-variation-induced V_{th} fluctuation. As expected the process-variation-induced V_{th} fluctuation of the planar MOSFET with 0.4 nm EOT is very similar that of the bulk FinFET device with 1.2 nm EOT. From the viewpoint of process-variation-induced fluctuation, the bulk FinFET device with 1.2 nm EOT exhibits a similar immunity against process-variation-induced fluctuation with the planar MOSFET with 0.4 nm EOT. However, will the trend still valid in the random-dopant-induced fluctuation?

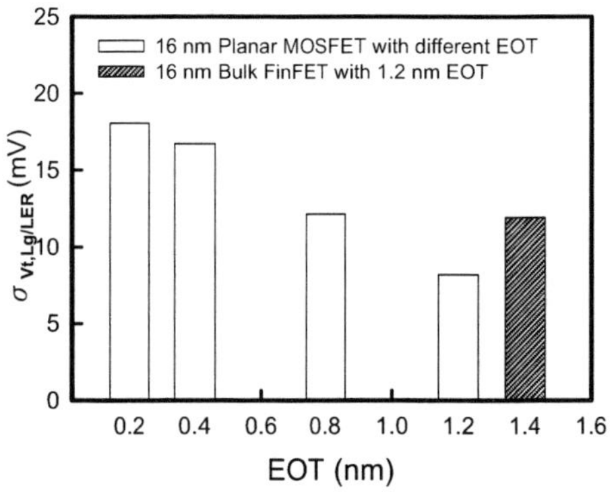

Figure 5: The process-variation-induced threshold voltage fluctuation for the studied devices.

Figure 6 shows the random-dopant-induced V_{th} fluctuation, $\sigma_{Vth,RD}$, of the studied devices. The random-dopant-induced Vth fluctuation decreases significantly as the EOT is scaled down. However, even thought the planar MOSFET with 0.4 nm EOT has a similar gate capacitance with bulk-FinFET with 1.2 nm EOT, the immunity against random-dopant-induced fluctuation of these two devices is rather different. The bulk FinFET device shows a better immunity against fluctuation than expected. It exhibits a similar V_{th} fluctuation as the planar MOSFET device with 0.2 nm EOT, where HfO_2 is used for gate dielectric. The process-variation- and random-dopant-induced V_{th} fluctuations are summarized in Table 2. In this study, the device with best immunity against process-variation- and random-dopant-induced V_{th} fluctuations are the planar MOSFETs with 0.2 nm EOT and the bulk FinFETs with 1.2 nm EOT. Considering the total V_{th} fluctuation according to the relation of equation (1), the V_{th} fluctuation is dominated by random-dopant effect and the bulk FinFETs with 1.2 nm EOT shows a similar immunity against fluctuation with the planar MOSFETs with 0.2 nm EOT. It's found that the immunity of the planar MOSFET against fluctuation suffers from nature of structural limitations and the bulk FinFET device can alleviates the challenges of device's scaling and have potential in the nanoelectronics application.

3 CONCLUSIONS

The threshold voltage fluctuations caused by the random dopant effect, the gate-length deviation, and the line-edge roughness have been calculated and compared for the nanoscale planar MOSFETs and bulk FinFETs. Fluctuations of the studied planar MOSFETs with EOT from 1.2 nm to 0.2 nm were compared with the results for 16 nm bulk FinFETs. Result of this study has confirmed the suppression of fluctuations with the gate oxide thickness scaling. The immunity of the planar MOSFET against random-dopant-induced fluctuation suffers from nature of structural limitations. The bulk FinFETs with 1.2 nm EOT shows a similar immunity against fluctuation with the planar MOSFETs with 0.2 nm EOT. Multiple-gate FETs, such as the examined bulk FinFET may alleviate the challenges of device's scaling and could be a potential candidate in the era of nanoelectronics. We notice that for the bulk FinFETs besides the gate-length deviation, and line edge roughness, the effects of Si thickness variation, Si sidewall roughness, local field enhancement at the top or bottom Si fin are the important sources of fluctuation, which are considered in our future work.

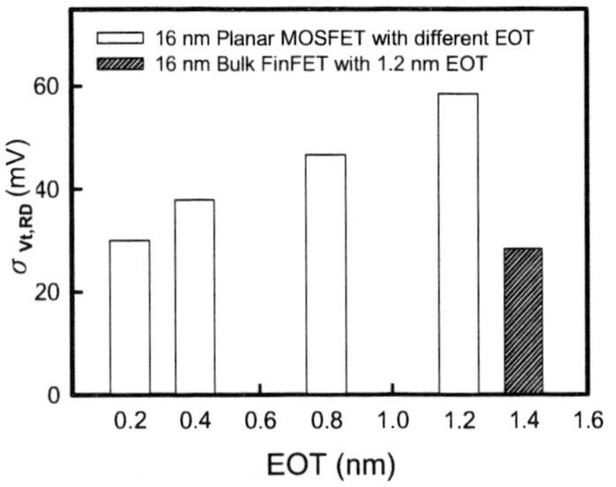

Figure 6: The random-dopant-induced threshold voltage fluctuation for the studied devices.

	Planar MOSFET				Bulk FinFET
EOT	1.2	0.8	0.4	0.2	1.2
Lg/LER	18.1	16.7	12.2	8.18	12.6
RD	58.5	46.6	37.9	29.9	28.3
Total	61.2	49.5	39.8	31.0	30.7

Table 2: Components of the threshold voltage fluctuation (mV) of the explored planar MOSFETs and bulk FinFETs.

ACKNOWLEDGEMENTS

This work was supported by Taiwan National Science Council (NSC) under Contract NSC-96-2221-E-009-210 and Contract NSC-96-2752-E-009-003-PAE, and by the Taiwan Semiconductor Manufacturing Company, Hsinchu, Taiwan under a 2006-2008 grant.

REFERENCES

[1] H.-S. Wong, Y. Taur, and D. J. Frank, "Discrete Random Dopant Distribution Effects in Nanometer-Scale MOSFETs," Microelectronics Reliability, 38, 1447, 1999.

[2] X.-H. Tang, V.K. De, and J.D. Meindl, "Intrinsic MOSFET Parameter Fluctuations Due to Random Dopant Placement," IEEE Trans. VLSI Systems, 5, 369, 1997.

[3] P.A. Stolk, F.P. Widdershoven, and D.B.M. Klaassen, "Modeling statistical dopant fluctuations in MOS transistors," IEEE Trans. Electron Device, 45, 1960, 1998.

[4] Y. Li, and C.-H Hwang, "Discrete-dopant-induced characteristic fluctuations in 16 nm multiple-gate silicon-on-insulator devices", J. Appl. Phys. 102, 084509, 2007.

[5] Y. Li, and S.-M. Yu, "A Coupled-Simulation-and-Optimization Approach to Nanodevice Fabrication With Minimization of Electrical Characteristics Fluctuation," IEEE Trans. Semi. Manufacturing, 20, 432, 2007.

[6] F.-L. Yang, J.-R. Hwang, H.-M. Chen, J.-J. Shen, S.-M. Yu, Y. Li, and Denny D. Tang, "Discrete Dopant Fluctuated 20nm/15nm-Gate Planar CMOS," VLSI Technol. Tech. Symp. Dig., 208, 2007.

[7] Y. Li, and C.-H Hwang, "Electrical characteristic fluctuations in 16nm bulk-FinFET devices", Microelectronics Engineering, 84, 2093, 2007.

[8] F.-L. Yang, J.-R. Hwang, and Y. Li, "Electrical Characteristic Fluctuations in Sub-45nm CMOS Devices," Proc. IEEE Custom Integrated Circuits Conf., 691, 2006.

[9] G. Tsutsui, M. Saitoh, T. Nagumo and T. Hiramoto, "Impact of SOI thickness fluctuation on threshold voltage variation in ultra-thin body SOI MOSFETs," IEEE Trans. Nanotechnology, 4, 363, 2005.

[10] Y. Li and S.-M. Yu, "Comparison of Random-Dopant-Induced Threshold Voltage Fluctuation in Nanoscale Single-, Double-, and Surrounding-Gate Field-Effect Transistors," Jpn. J. Appl. Phys., 45, 6860, 2006.

[11] Y. Li, and S.-M. Yu, "A study of threshold voltage fluctuations of nanoscale double gate metal-oxide-semiconductor field effect transistors using quantum correction simulation," J. Comp. Elect. 5, 125, 2006.

[12] S. Xiong and J. Bokor, "A simulation study of gate line edge roughness effects on doping profiles of short-channel MOSFET devices," IEEE Trans. Elec. Dev., 51, 228, 2004.

[13] Y.T. Hou, T. Low, Bin Xu, M.-F. Li, G. Samudra and D.L. Kwong, "Impact of metal gate work function on nano CMOS device performance," Proc. of Int. Conf. on Solid-State and Integrated Circuits Technology, 1, 57, 2004.

[14] B.-H. Lee, J. Oh, H.-H. Tseng, R. Jammy, and H. Huff, "Gate stack technology for nanoscale devices," Materials Today, 32, 2006.

[15] Y. Li, H.-M. Chou, and J.-W. Lee, "Investigation of electrical characteristics on surrounding-gate and omega-shaped-gate nanowire FinFETs," IEEE Trans. Nanotech., 4, 510, 2005.

[16] Y. Li, and C.-H Hwang, "Effect of Fin Angle on Electrical Characteristics of Nanoscale Round-Top-Gate Bulk FinFETs", IEEE Trans. Electron Device, 54, 3426, 2007.

[17] S. Odanaka, "Multidimensional discretization of the stationary quantum drift-diffusion model for ultrasmall MOSFET structures," IEEE Trans. Computer-Aided Design Integrated Circuit and Sys, 23, 837, 2004.

[18] G. Roy, A. R. Brown, A. Asenov, and S. Roy, "Quantum Aspects of Resolving Discrete Charges in 'Atomistic' Device Simulations," J. Comp. Elect., 2, 323, 2003.

[19] Y. Li, H.-M. Lu, T.-W. Tang, and Simon M. Sze, "A Novel Parallel Adaptive Monte Carlo Method for Nonlinear Poisson Equation in Semiconductor Devices," Math. Comp. Simulation, 62, 413, 2003.

[20] Y. Li, S. M. Sze, and T.S. Chao, "A Practical Implementation of Parallel Dynamic Load Balancing for Adaptive Computing in VLSI Device Simulation," Eng. with Comp., 18, 124, 2002.

[21] Y. Li and Y.-S. Chou, "A Novel Statistical Methodology for Sub-100 nm MOSFET Fabrication Optimization and Sensitivity Analysis," Proc. of Int. Conf. on Solid State Devices and Materials, 622, 2005.

Molecular Dynamics Study on Thermodynamical Properties of Bulk Silicon Under Strain

H. Zhao* and N. R. Aluru**

* University of Illinois at Urbana-Champaign, Champaign, IL, USA, hzhao4@uiuc.edu
** University of Illinois at Urbana-Champaign, Champaign, IL, USA, aluru@uiuc.edu

ABSTRACT

With classical molecular dynamics (MD) simulation, thermodynamical properties such as Helmholtz free energy and internal energy are calculated when the silicon crystal is subjected to a compression/tension and a shear deformation. In order to account for the quantum corrections under strain in the classical MD simulations, we propose an approach where the quantum corrections to the internal energy and the Helmholtz free energy are obtained by the corresponding energy deviation between the classical and quantum harmonic oscillators from the quasi-harmonic approximations. We calculate the variation of thermodynamical properties of bulk silicon with temperature and strain and compare them with results obtained by using the quasi-harmonic approximations in the reciprocal space.

Keywords: silicon, thermodynamical properties, strain effects, Molecular Dynamics.

1 Introduction

With the increasing number of applications of nanoelectromechanical systems (NEMS), silicon is one of the most popularly used materials both as a substrate and as a key device component. To understand the role of silicon as a nanoelectromechanical material, it is important to accurately predict the thermodynamical and mechanical properties of silicon at various length scales, especially when the material is under strain and at high temperature. Although a variety of physical models and simulation techniques [1]–[3] have been developed to understand the material properties of silicon, molecular dynamics technique with empirical potential is one of the robust techniques to understand material behavior because of its simplicity and universality. It is often used to understand fundamental issues governing material behavior and response.

In this paper, thermodynamical properties of silicon crystal subjected to a compression/tension and a shear deformation are calculated using classical MD open source code LAMMPS [4] with the Tersoff interatomic potential [5]. Since MD simulations obey the rules of classical statistical mechanics, quantum corrections are necessary for MD simulations results in order to compare with experimental data and quantum-mechanical calculations. We investigate the extensively used temperature rescaling method [6] and realize that this technique can not accurately predict the quantum effects in the Helmholtz free energy at low temperature under general strain conditions. We suggest an approach where the quantum corrections are obtained by investigating the corresponding energy differences between the classical and quantum harmonic oscillators. The Helmholtz free energy is then computed by the ensemble method [7].

The rest of the paper is organized as follows: in section 2, we describe MD simulations setup, free energy calculations, and quantum correction calculations.In section 3, we present the thermodynamical properties of bulk silicon for both non-strain and strain cases. The variation of pressure with temperature and strain is also shown. Finally, conclusions are presented in section 4.

2 Methodology

2.1 MD Setup

With MD simulator LAMMPS, a silicon cubic structure of 216 atoms ($3 \times 3 \times 3$ unit cells) with periodic boundary conditions in three directions is used for both non-strain and strain cases. Tersoff interatomic potential is employed to determine the interactions between silicon atoms. Nose-Hoover thermostat is adopted for both NPT and NVT ensembles with a time step of 1.0fs. The center of mass of the system is fixed during the simulation to neglect any translational movements. For each specified temperature, we first perform NPT ensemble simulations to determine the zero pressure lattice constant. With the corresponding lattice constant, NVT ensemble simulations are performed to compute the internal energy and Helmholtz free energy. Each simulation runs for $0.5 \sim 1.0$ns to obtain an equilibrium state and an additional $4.0 \sim 6.0$ns for time averaging.

In order to study the thermodynamical properties of silicon under strain, we perform MD simulations with two types of strain: compression/tension and shear. The strain is assigned to the system through the deformation gradient $\mathbf{F}$ by modifying the initial configuration of the system with the relation $\hat{\mathbf{R}}_\mathbf{n} = \mathbf{F}\mathbf{R}_\mathbf{n}$, where $\mathbf{R}_\mathbf{n}$ is the coordinate of atom n under non-strain conditions, and $\hat{\mathbf{R}}_\mathbf{n}$ is the corresponding atom coordinate with the specified strain. Note that the deformed configuration is no longer a cubic structure, but a parallelepiped structure. LAMMPS employs the Parrinello-Rahman method [8] to maintain both the volume and the shape of the parallelepiped simulation box within the NVT ensemble simulations.

2.2 Free Energy Calculations

Thermodynamical properties of a system such as free energy, entropy, chemical potential, etc. can not be computed directly from the phase-space trajectory which is the basic output of most MD simulations, because they are formally related to the accessible phase-space volume. One alternative is to transform these properties, for example, free energy, into a function which can be evaluated using the phase-space trajectory.

At zero pressure conditions, the Helmholtz free energy A is equal to the Gibbs free energy G while the enthalpy H is equal to the internal energy E. Since the Gibbs free energy can be obtained by directly integrating the thermodynamic relation

$$\frac{d}{dT}\left(\frac{G}{T}\right) = -\frac{H}{T^2}, \tag{1}$$

the Helmholtz free energy can be easily computed. But this method is no longer valid under finite pressure condition, e.g. when the system is under strain.

The usual procedure to compute the Helmholtz free energy of a homogeneous crystalline solid is called the ensemble method proposed by Frenkel and Ladd[7]. The basic idea is to construct a reversible path from a state of the known free energy (Einstein crystal with the analytical expression of the Helmholtz free energy) to the system of interest by modifying the potential energy as

$$U_\lambda = (1-\lambda)U + \lambda U_E, \tag{2}$$

where U is the Tersoff potential energy of the silicon structure, U_E is the potential energy of the Einstein crystal, which is a collection of identical independent harmonic oscillators with the same silicon structure. Thus as the parameter λ switches slowly from 0 to 1, the Helmholtz free energy of the system of interest $A = A_{\lambda=0}$ can be calculated by the thermodynamic integration

$$A = A_{\lambda=0} = A_{\lambda=1} + \int_0^1 d\lambda \langle U - U_E\rangle_\lambda, \tag{3}$$

where $A_{\lambda=1}$ is the analytical free energy of the reference Einstein crystal [9], $\hbar$ is the Planck's constant, and $\langle\cdots\rangle_\lambda$ is the canonical ensemble average.

In this paper, we use Eq. (3) to calculate the Helmholtz free energy of bulk silicon for both non-strain and strain cases.

Table 1: Formulation of 1-D quantum harmonic oscillator and classical harmonic oscillator

Quantum	$A_Q(\omega) = \frac{\hbar\omega}{2} + k_BT\ln\left(1 - e^{-\hbar\omega/k_BT}\right)$
System	$E_Q(\omega) = \frac{\hbar\omega}{2} + \hbar\omega\frac{1}{e^{\hbar\omega/k_BT}-1}$
Classical	$A_C(\omega) = k_BT\ln\left(\frac{\hbar\omega}{k_BT}\right)$
System	$E_C(\omega) = k_BT$

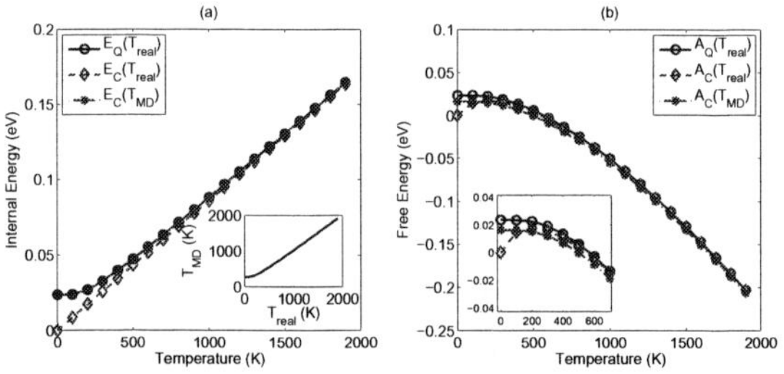

Figure 1: (a): Internal energy of a 1-D Einstein oscillator obtained with the quantum model $E_Q(T_{real})$, the classical model $E_C(T_{real})$ and the classical model with temperature rescaling $E_C(T_{MD})$. (inset) Temperature rescaling relation. (b): Helmholtz free energy of a 1-D Einstein oscillator obtained with the quantum model $A_Q(T_{real})$, the classical model $A_C(T_{real})$ and the classical model with temperature rescaling $A_C(T_{MD})$. (inset) Enlarged view of free energy at low temperatures.

2.3 Quantum Corrections for the Classical MD Simulations

Classical MD simulations obey the rules of classical statistical mechanics. Considering a simple 1-D Einstein oscillator as an example, Table 1 lists the expressions for both the internal energy and Helmholtz free energy when the oscillator is treated as a quantum or as a classical model, where k_B is Boltzmann's constant, T is temperature, ω is oscillator frequency. At $T = 0$K both the Helmholtz free energy A_C and the internal energy E_C of the classical model are zero. However, the corresponding quantum energies A_Q and E_Q are not zero because of the quantum effects at zero temperature. As a result, quantum corrections are required in order to compare the thermodynamical properties predicted by classical MD simulations with quantum simulations, especially for low temperatures.

The basic idea of the temperature rescaling method [6] is to find a one to one mapping of temperature T between the classical system and the quantum system based on the internal energy. Considering a 1-D Einstein oscillator as an example, with the scaling relation between T_{MD} and T_{real} as shown the inset of the left figure in

Fig. 1, the internal energy of the classical model with temperature rescaling matches with the quantum result. However, for Helmholtz free energy, a small deviation between the quantum result and the classical result with temperature rescaling is observed at low temperatures. It is because that the one-to-one mapping of temperature T does not exist for the Helmholtz free energy between the two systems. Even though the above discussion is based on a 1-D Einstein oscillator, we can expect similar behavior when the temperature rescaling method is used to account for quantum corrections in MD simulations of silicon structures based on the Tersoff interatomic potential.

At zero pressure condition, internal energy for the quantum system can be accurately obtained with temperature rescaling method. The Helmholtz free energy can then be calculated by Eq. (1). However, Eq. (1) is not valid when the system is subjected to a strain condition. As pointed out above, the Helmholtz free energy can not be accurately predicted directly with temperature rescaling technique. In this paper, we suggest to extract the quantum correction terms for both the internal energy and the Helmholtz free energy from the energy deviation between the classical model and the quantum model of the harmonic oscillators as

$$E_{qc} = \sum_{n=1}^{3N} \left\{ \frac{\hbar\omega_n}{2} + \hbar\omega_n \frac{1}{e^{\hbar\omega_n/k_BT} - 1} k_BT \right\}, \quad (4)$$

$$A_{qc} = \sum_{n=1}^{3N} \left\{ \frac{\hbar\omega_n}{2} + k_BT \ln\left(1 - e^{-\hbar\omega_n/k_BT}\right) \right.$$
$$\left. k_BT \ln\left(\frac{\hbar\omega_n}{k_BT}\right) \right\}, \quad (5)$$

where for a N atom system, all the $3N$ required normal modes ω_n are obtained from the quasi-harmonic approximation by diagonalizing the $3N \times 3N$ matrix [3]

$$\Phi_{\alpha i\beta j} = \frac{\partial^2 U}{\partial \mathbf{r}_{\alpha i} \partial \mathbf{r}_{\beta j}}, \quad (6)$$

where U is the Tersoff potential energy for the whole system, $\mathbf{r}$ is current atom position, α and β denote atom numbers, and i and j denote Cartesian components.

In this paper, we use the temperature rescaling method in classical MD simulations to obtain the lattice constants. After the lattice constants are determined, we run the classical MD simulations for each specified temperature. The Helmholtz free energy is computed by adding the quantum correction obtained from Eq. (5) to the energy obtained from Eq. (3), which does not include quantum effects. The internal energy is obtained from the time average of the total energy of the system with the quantum correction given by Eq. (4).

Table 2: Thermodynamical properties of bulk silicon

temperature (K)		Internal Energy (eV/atom)	Free Energy (eV/atom)
0	This work	-4.564	-4.563
	QHMK[10]	-4.563	-4.563
	Ref[11]	-4.562	-4.552
300	This work	-4.530	-4.580
	QHMK	-4.531	-4.584
	Ref	-4.535	-4.590
1500	This work	-4.228	-5.102
	QHMK	-4.232	-5.083
	Ref	-4.221	-5.112

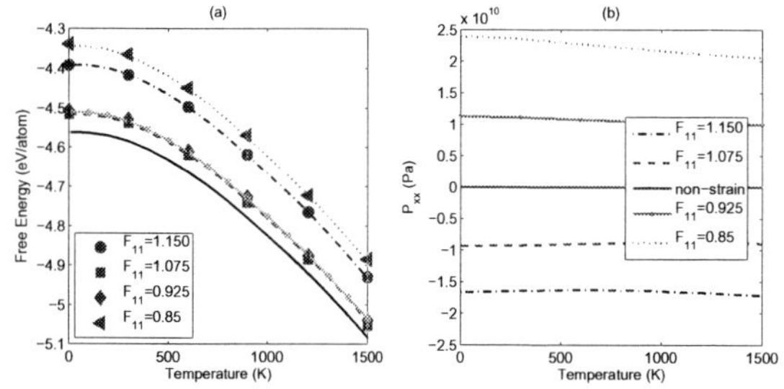

Figure 2: (a): Variances of the Helmholtz free energy with temperature under different compression/tension conditions. (b): Variations of normal pressure with temperature under different compression/tension conditions. The thick solid line is the Helmholtz free energy of bulk silicon under non-strain condition. All the lines are from QHMK calculations. All the symbols are from MD calculations in this work. The error-bar is within the symbol size.

3 Results and Discussion

After the lattice constants are obtained, the internal energy and Helmholtz free energy of bulk silicon are calculated using techniques described in the previous section. Table 2 summarizes the results obtained from MD simulations at 0K, 300K, and 1500K. Two sets of comparison data are also presented in the table. Accounting for quantum corrections via Eqs. (4) and (5), we observe that at low temperatures, the internal energy and the free energy match with the QHMK results quite well. At high temperature, the anharmonicity becomes important. Since QHMK method can not capture any anharmonicity, but MD can, we see a small deviation between the MD results with QHMK results at $T = 1500$K.

To study the strain effect on the thermodynamical and mechanical properties of silicon, we consider two types of strain: compression/tension and shear, by varying the deformation gradient component F_{11} from 0.85

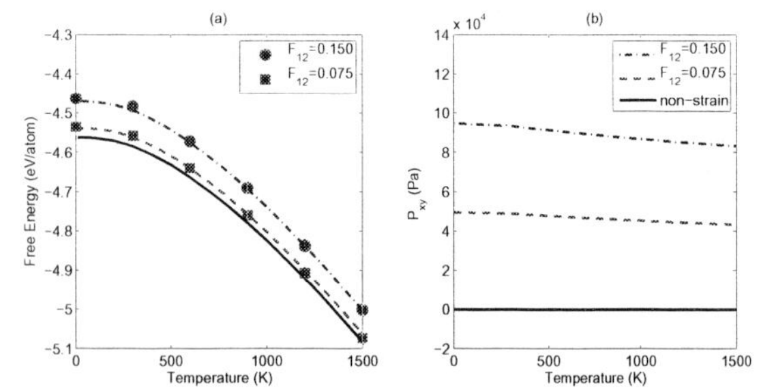

Figure 3: (a): Comparison of the Helmholtz free energy with temperature for different shear magnitude. (b): Variation of pressure p_{xy} with temperature under different shear magnitude. The thick solid line is the Helmholtz free energy of bulk silicon without deformation. All the lines are from QHMK calculations. All the symbols are from MD calculation in this work.The error-bar is within the symbol size.

(compression) to 1.15 (tension) and F_{12} from 0.0 (no deformation) to 0.075 and 0.15 (shear along the x-direction), respectively. Fig. 2(a) and Fig. 3(a) show the variation of Helmholtz free energy with temperature when the silicon crystal is under a compression/tension strain and a shear strain, respectively. The corresponding results from QHMK model are also shown for comparison. We note that the MD results match well with the QHMK results. Fig. 2(b) shows the variation of pressure with temperature under compression/tension in the x-direction. For the temperature range considered, the absolute value of pressure p_{xx} appears to be smaller under tension than under compression, implying that silicon material is softer under tension than under compression. When the material is under constant compression, the value of p_{xx} decreases mildly with temperature indicating that the stiffness of silicon decreases at high temperature under compression. Under tension, we observe that the stiffness of the material does not change very much at high temperature. Fig. 3(b) shows the variation of pressure p_{xy} with temperature when the crystal is under shear deformation. We note that pressure p_{xy} decreases with the increase in temperature, implying that shear modulus of the material decreases with the increase in temperature. These observations are consistent with the results presented in Ref [12].

4 Conclusions

In this paper, we presented results from MD simulations for Tersoff silicon at different temperature and under two types of strain – compression/tension and shear. The Helmholtz free energy is calculated by employing the ensemble method. We investigated the widely-used temperature rescaling method to account for quantum corrections in classical MD simulations and we observed that it may not be accurate for Helmholtz energy calculations at low temperatures. We propose a method where the quantum corrections of both the internal energy and the Helmholtz free energy are obtained from the deviation of the corresponding energy between the classical and quantum harmonic oscillators with the required normal modes obtained from the quasi-harmonic approximation of the Tersoff potential. We show that the computed Helmholtz free energies match with the published QHMK results. Finally, using MD we also computed the variation of pressure with temperature and strain and discussed how this can effect the mechanical properties.

5 Acknowledgments

We gratefully acknowledge support by the National Science Foundation under Grants 0519920, 0601479, 0634750.

REFERENCES

[1] S. Wei, C. Li, and M. Y. Chou, Physical Review B **50** (19), 14587, 1994.

[2] R. LeSar, R. Najafabadi, and D. J. Srolovitz, Physical Review Letters, **63** (6), 624 - 627, 1989.

[3] H. Zhao, Z. Tang, G. Li, and N. R. Aluru, Journal of Applied Physics, **99** (6), 064314, 2006.

[4] S. J. Plimpton, Journal of Computational Physics, **117** (1), 1 - 19, 1995.

[5] J. Tersoff, Physical Review B, **38** (14), 9902 - 9905, 1988.

[6] C. Z. Wang, C. T. Chan, and K. M. Ho, Physical Review B, **42** (17), 11276-11283, 1990.

[7] D. Frenkel, and A. J. C. Ladd, Journal of Chemical Physics, **81** (7), 3188 - 3193, 1984.

[8] M. Parrinello, and A. Rahman, Physical Review Letters, **45** (14) 1196-1199, 1980.

[9] J. F. Lutsko, D. Wolf, and S. Yip, Journal of Chemical Physics, **88** (10), 6525 - 6528, 1988.

[10] L. J. Porter, S. Yip, M. Yamaguchi, H. Kaburaki, and M. Tang, Journal of Applied Physics **81** (1), 96 - 105, 1997.

[11] I. Barin, and O. Knake, "Thermochemical Properties of Inorganic Substances", Springer, Berlin, 1973.

[12] Z. Tang, H. Zhao, G. Li, and N. R. Aluru, Physical Review B, **74** (6), 064110, 2006.

Understanding the interaction of ammonia with carbon nanotubes

Carolina Oliva*, Paul Strodel*, Gerhard Goldbeck-Wood* and Amitesh Maiti**

*Accelrys Ltd, 334 Cambridge Science Park, CB4 0WN, Cambridge, United Kingdom
**Lawrence Livermore National Laboratory, Livermore, CA 94551, United States

ABSTRACT

An important emerging application of Nanotechnology is in the area of gas sensors that can detect harmful small molecules at a level of parts-per-billion or even lower. Nanosensors based on carbon nanotubes can detect different chemical substances (e.g. small molecules like NO_2, NH_3, CO_2, CH_4, and CO, various nerve agents, and a whole host of biomolecules) thanks to the change in electrical conductivity [1] that the nanotube experiences when those molecules adsorb on its surface. In this communication we will focus on the understanding of the interaction of ammonia with carbon nanotubes at the molecular scale. Due to the fact that ammonia physisorbs on carbon nanotubes, adsorption energies that agree with experimental results can only be obtained with a proper description of van der Waals interactions, something that cannot be expected from current Density Functional Theory (DFT) methods. In this study we have used a hybrid QM/MM (Quantum Mechanics/Molecular Mechanics) method that combines DFT to describe the active site with a forcefield to describe the surroundings, to properly model the whole system.

Keywords: QM/MM, NH_3, adsorption energy, carbon nanotubes, bundle

1 INTRODUCTION

Chemical sensors based on carbon nanotubes show very high sensitivity towards the detection of molecular gases such as NH_3 and NO_2. The unique structural features of carbon nanotubes, with a high surface area exposed, confers them their high sensitivity. Nano-sensor devices based on carbon nanotubes work by measuring changes in the nanotubes conductance after gas exposure, and find application in environmental, industrial and medical monitoring studies. To understand the mode of action of carbon nanotubes in those chemical sensors, several experimental and theoretical studies have been performed in the last few years. Experimental studies with only small amounts of ammonia (1%) showed a dramatic decrease in the nanotubes' conductance [1] as a result of charge being transferred from NH_3 to the nanotubes [2]. Photoemission experiments confirmed that ammonia molecules act as charge donors and that the presence of contaminants due to purification processes do not affect the sensitivity towards ammonia [3]. Temperature-programmed desorption (TPD) experiments have also been performed, demonstrating that ammonia can enter the grooves between carbon nanotubes and interact with several of them [4]. The few theoretical studies published in the literature correspond to DFT calculations of NH_3 adsorption [5-8] and, in all cases, ammonia was found to be weakly adsorbed on the carbon nanotube surface. More recently, the role of structural defects in the performance of nanotube-based gas sensors was also studied using DFT [9].

2 METHODS

The QMERA module [10] implemented in Materials Studio® was used to perform the hybrid QM/MM calculations. With this module combined quantum mechanical (QM) and molecular mechanics (MM) calculations using the ChemShell [11] environment can be performed. QMERA combines the accuracy of quantum mechanics with the speed of a force field calculation and, as a consequence, accurate calculations can be carried out on very large systems using little computer time. QMERA uses DMol[3] [12] for the description of the QM region and the GULP [13] force field engine for the MM region. For capping the QM region hydrogen link atoms were used in all cases and to calculate the total energy of the system a subtractive expression was used (see Figure 1). This type of embedding scheme is called subtractive mechanical and neglects the polarization of the QM region due to the MM region, which is not expected to be important in the system under consideration. In this approach, the van der Waals interactions are handled by the force field method.

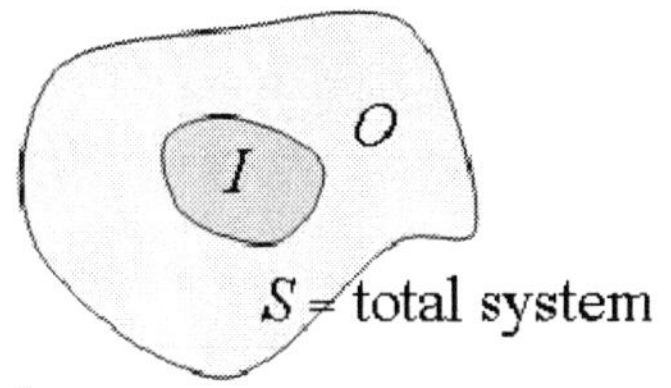

$$E_{tot}^{subtractive} = E_{QM}(I) + E_{MM}(S) - E_{MM}(I)$$

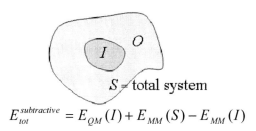

Figure 1. Mechanical subtractive QM/MM scheme used in the simulations.

For describing the QM region, the Perdew-Burke-Ernzerhof exchange-correlation functional (PBE) [14] and a DND basis set were used in all cases. For the MM region the DREIDING [15] force field was employed, which has been used in previous studies of carbon nanotubes giving good results [16]. The charges of the QM and MM regions were assigned using a charge equalization method maintaining charge neutrality in both regions. For modelling the nanotubes, we used (8,0) carbon nanotubes of 30 Å of length and H atoms to terminate both nanotube ends. Depending on the absence or presence of a defect on the nanotube and on the size of that defect, different QM regions were used in the simulations, containing between 28 and 46 atoms from a total of 244 atoms.

To study the adsorption of NH_3 on bundles of carbon nanotubes, the Adsorption Locator [17] module implemented in Materials Studio® was used in this study. The Adsorption Locator module identifies possible adsorption configurations by carrying out Monte Carlo searches of the configurational space of the substrate-adsorbate system as the temperature is slowly decreased (Simulated Annealing). This process is repeated several times to identify further local energy minima. Different force fields are available in Adsorption Locator; COMPASS [18] has already been used in previous studies of ammonia interacting with carbon nanotubes giving good results [9] and, as a consequence, was the one used in this study. To obtain good statistics the calculations were performed at the ultra-fine level, which means that 10 Simulated Annealing cycles decreasing the temperature from 10000 to 100 K were performed (a total of 100000 steps per cycle).

3 RESULTS

3.1 Role of defects

The first step in our study was to model the physisorption of NH_3 on a defect free nanotube. We performed QM/MM calculations to obtain the adsorption energy and the preferred adsorption configuration. In those calculations the QM region included seven phenyl rings on the nanotube plus the ammonia molecule (28 atoms in total). The calculated adsorption energies were 2.9, 3.0 and 4.5 kcal mol^{-1} depending on the orientation of the ammonia hydrogen atoms with respect to the surface of the nanotube (Figure 2). The most stable configuration corresponds to the situation where one hydrogen atom points towards the nanotube surface, which is in agreement with previous theoretical results performed with different *ab initio* methods and models for the nanotube structure [7]. The distance between the hydrogen atoms and the nanotube surface was between 2.7 and 3.0 Å for those configurations.

After studying the adsorption of ammonia on defect-free nanotubes, we performed QM/MM calculations to understand the role of structural defects on NH_3 adsorption. Common defects that have been detected experimentally correspond to pentagon-heptagon pairs (Stone-Wales defects) and monovacancies [19]. For the Stone-Wales (SW) defect the selected QM region contains 46 atoms including the ammonia molecule and for the vacancy defect 35 atoms (see Figure 3). Our QM/MM calculations gave adsorption energies of 2.1 and 10.1 kcal mol^{-1} for the SW and the vacancy defect respectively. It can also be seen in Figure 3 that the ammonia molecule is physisorbed on the SW defect, while the same molecule is chemisorbed on the vacancy defect. The C-N distance for the NH_3 molecule on the vacancy defect is of 1.55 Å, clearly indicating the formation of a covalent bond, while for the SW defect the NH_3 molecule is at around 3 Å from the nanotube surface. Another indication of this different behavior is the amount of charge transferred from the NH_3 molecule to the nanotube in those two different defects. We calculated the Mulliken atomic charges for the QM region and we saw that for a SW-defective nanotube the charge being transferred is around 0.02 e per ammonia molecule, while for the vacancy defect is of 0.47 e per molecule. The calculated amount of charge transfer corresponding to NH_3 physisorbed on the nanotube (defect free or SW defect) is in agreement with previous experimental measurements [2].

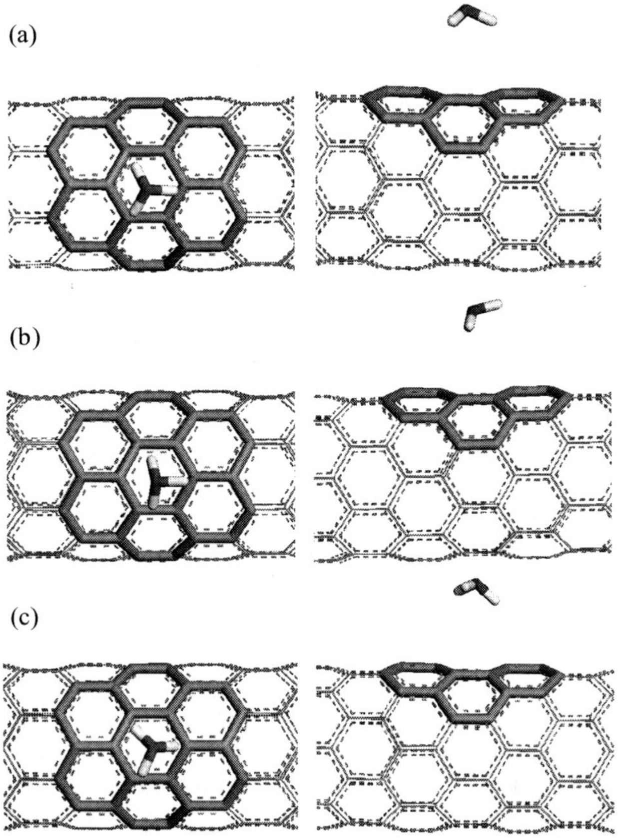

Figure 2. Adsorption configurations of NH_3 on a defect free nanotube. Adsorption energies are (a) 2.9, (b) 3.0 and (c) 4.5 kcal mol^{-1}. The QM region is indicated in stick representation.

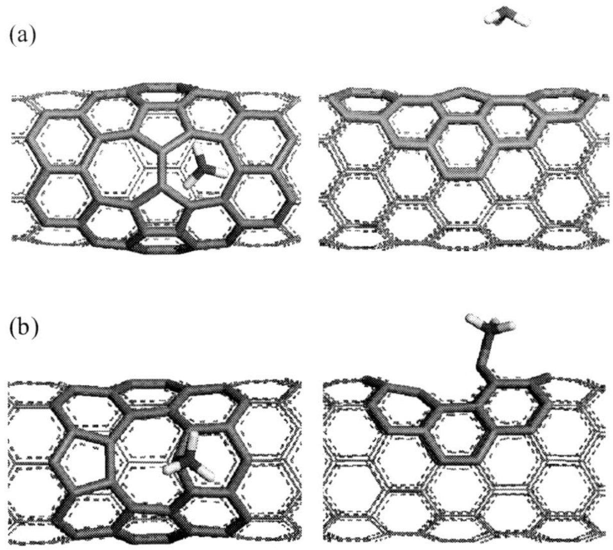

Figure 3. Adsorption of ammonia on a carbon nanotube with a Stone-Wales defect (a) and with a vacancy defect (b). The QM region is indicated in stick representation.

3.2 Bundle effects

Gas adsorption on bundles of carbon nanotubes can take place at different sites: a) internal or endohedral sites, b) interstitial channels, c) external grooves sites, and d) external surfaces [20]. Depending on the adsorption site, the gas molecule will be able to interact with a different number of nanotubes. In this study, we used the Adsorption Locator module in Materials Studio to identify the preferred adsorption site of ammonia on carbon nanotube bundles. To model the bundle we used three carbon nanotubes in a triangular honeycomb configuration. Although 3.35 Å corresponds to the optimum intertube distance found in previous theoretical studies of nanotube bundles [21-22], real bundles contain nanotubes of different diameters which translate in packing defects with the corresponding formation of large intertube channels [23]. Due to that, we used an intertube distance of 5 Å in our simulations. The structure of the nanotubes was not optimized during the calculations, only the ammonia structure was relaxed. The results from the Adsorption Locator calculation are shown in Figure 4.

From the Adsorption Locator results it is clear that the ammonia molecule prefers to be adsorbed in the interstitial channel where it can interact with three carbon nanotubes. The adsorption energy obtained for those different sites are: 4.2 kcal mol^{-1} for the interstitial channel, 2.8 kcal mol^{-1} for external grooves and 1.5 kcal mol^{-1} for external surfaces. Those results are in agreement with previous experimental results which concluded that the ammonia molecule is small enough to access the interstitial channels [4]. From the energy of the two extreme situations we can also estimate the gain in adsorption energy due to bundle effects to be around 2-3 kcal mol^{-1}.

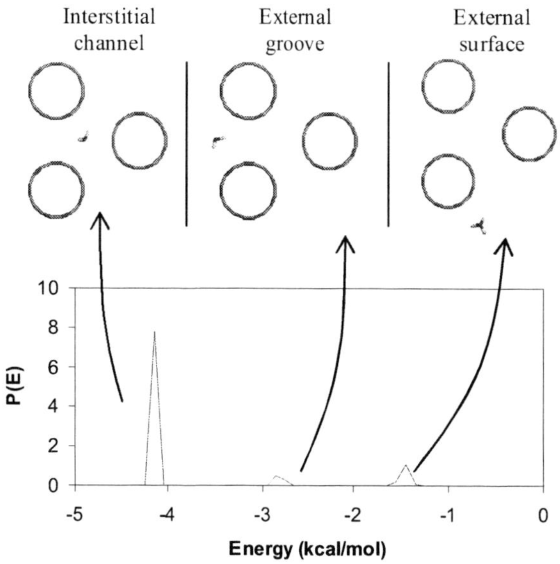

Figure 4. Energy distribution obtained with Adsorption Locator for NH_3 on a bundle composed of 3 nanotubes with an intertube distance of 5 Å. Representative bundle-NH_3 configurations are also shown.

4 SUMMARY

Two different modeling techniques have been applied to understand, at a quantum and molecular level, the interaction of ammonia with carbon nanotubes. A hybrid QM/MM method has been used to study the adsorption of NH_3 on defect free nanotubes, on nanotubes with Stone-Wales defects and on nanotubes with monovacancy defects; and a Monte Carlo sampling method has been used to study the effect of bundles on ammonia adsorption. The calculations show that ammonia is physisorbed on the defect free nanotube and on the SW defect, while it chemisorbs on a vacancy defect. Adsorption energies and the amount of charge transfer have been calculated with our QM/MM method and agree with previous *ab initio* results and experimental observations. The preferred adsorption site of ammonia on carbon nanotube bundles has been identified with our Monte Carlo technique and corresponds to the interstitial channel, again in agreement with previous experimental observations. The stabilization energy due to the adsorption of ammonia on the interstitial channel has been estimated to be around 2-3 kcal mol^{-1}.

REFERENCES

[1] J. Kong, N. R. Franklin, C. Zhou, M. G. Chapline, S. Peng, K. Cho, H. Dai, "Nanotube Molecular

Wires as Chemical Sensors", Science 287, 622, 2000.
[2] K. Bradley, J.-C. P. Gabriel, M. Briman, A. Star, G. Grüner, "Charge Transfer from Ammonia Physisorbed on Nanotubes", Phys. Rev. Lett. 91 218301-1, 2003.
[3] A. Goldoni, R. Larciprete, L. Petaccia, S. Lizzit, "Single-Wall Carbon Nanotube Interaction with Gases: Sample Contaminants and Environmental Monitoring", J. Am. Chem. Soc. 125, 11329, 2003.
[4] M. D. Ellison, M. J. Crotty, D. Koh, R. L. Spray, K. E. Tate, "Adsorption of NH_3 and NO_2 on Single-Walled Carbon Nanotubes", J. Phys. Chem. B. 108, 7938, 2004.
[5] H. Chang, J. D. Lee, S. M. Lee, Y. H. Lee, "Adsorption of NH_3 and NO_2 molecules on carbon nanotubes", Appl. Phys. Lett. 79, 3863, 2001.
[6] J. Zhao, A. Buldum, J. Han, J. P. Lu, "Gas molecule adsorption in carbon nanotubes and nanotube bundles", Nanotechnology 13, 195, 2002.
[7] C. W. Bauschlicher, A. Ricca, "Binding of NH_3 to graphite and to a (9,0) carbon nanotube", Phys. Rev. B 70, 115409-1, 2004.
[8] J. Lu, S. Nagase, Y. Maeda, T. Wakahara, T. Nakahodo, T. Akasaka, D. Yu, Z. Gao, R. Han, H. Ye, "Adsorption configuration of NH3 on single-wall carbon nanotubes", Chem. Phys. Lett. 405, 90, 2005.
[9] J. Andzelm, N. Govind, A. Maiti, "Nanotube-based gas sensors – Role of structural defects", Chem. Phys. Lett. 421, 58, 2006.
[10] For more information visit: http://accelrys.com/products/materials-studio/modules/qmera.html
[11] P. Sherwood, A. H. de Vries, M. F. Guest, G. Schreckenbach, C. R. A. Catlow, S. A. French, A. A. Sokol, S. T. Bromley, W. Thiel, A. J. Turner, S. Billeter, F. Terstegen, S. Thiel, J. Kendrick, S. C. Rogers, J. Casci, M. Watson, F. King, E. Karlsen, M. Sjøvoll, A. Fahmi, A. Schäfer, Ch. Lennartz, "QUASI: A general purpose implementation of the QM/MM approach and its application to problems in catalysis", J. Mol. Struct. (Theochem.) 632, 1, 2003.
[12] B. Delley, "An All-Electron Numerical Method for Solving the Local Density Functional for Polyatomic Molecules", J. Chem. Phys. 92, 508, 1990.
[13] J. D. Gale, "GULP: Capabilities and prospects", Z. Kristallogr. 220, 552, 2005.
[14] J. P. Perdew, K. Burke, M. Ernzerhof, "Generalized Gradient Approximation Made Simple", Phys. Rev. Lett. 77, 3865, 1996.
[15] S.L. Mayo, B.D. Olafson, W.A. III. Goddard, "DREIDING: A generic force field", J. Phys. Chem. 94, 8897, 1990.
[16] A. Maiti, J. Wescott, P. Kung, "Nanotube–polymer composites: insights from Flory–Huggins theory and mesoscale simulations", Molecular Simulation 31, 143, 2005.
[17] For more information visit: http://accelrys.com/products/materials-studio/modules/adsorption-locator.html
[18] H. Sun, "COMPASS: An ab Initio Force Field Optimized for Condensed-Phase Applications, Overview with Details on Alkane and Benzene Compounds", J. Phys. Chem. B. 102, 7338, 1998.
[19] A. Hashimoto, K. Suenaga, A. Gloter, K. Urita, S. Lijima, "Direct evidence for atomic defects in graphene layers", Nature 430, 870, 2005.
[20] G. Stan, M. J. Bojan, S. Curtarolo, S. M. Gatica, M. W. Cole, "Uptake of gases in bundles of carbon nanotubes", Phys. Rev. B 62, 2173, 2000.
[21] J.-C. Charlier, X. Gonze, J.-P. Michenaud, "First-Principles Study of Carbon Nanotube Solid-State Packings", Europhys. Lett. 29, 43, 1995.
[22] S. Zhang, S. Zhao, M. Xia, E. Zhang, X. Zuo, T. Xu, "Optimum diameter of single-walled carbon nanotubes in carbon nanotube ropes", Phys. Rev. B 70, 035403-1, 2004.
[23] W. Shi, K. Johnson, "Gas Adsorption on Heterogeneous Single-Walled Carbon Nanotube Bundles", Phys. Rev. Lett. 91, 015504-1, 2003.

Computational Modeling of Ligands for Water Purification Nanocoatings

Kumar Vanka,* Yao Houndonougbo,* Nathan R. Lien,** James M. Harris,***
Lisa M. Farmen,*** Darren W. Johnson,** Brian B. Laird,* and Ward H. Thompson*,†

*Department of Chemistry, University of Kansas, Lawrence, KS 66045
**Department of Chemistry, University of Oregon, Eugene, OR 97403
and the Oregon Nanoscience and Microtechnologies Institute (ONAMI)
***Crystal Clear Technologies, Inc., 12423 NE Whitaker Way, Portland, OR 97230
†wthompson@ku.edu

ABSTRACT

Electronic structure calculations and molecular dynamics simulations are being used to investigate ligand-based nanocoatings for use in drinking water purification applications. The aim is to use computational chemistry to aid in the development of design principles for water purification systems based on adsorption of inorganic contaminates by inexpensive mesoporous substrates functionalized with low-cost ligands. These systems can provide simple, low-cost water purifiers that meet the needs of underdeveloped countries as well as more specialized applications for developed economies.

Keywords: nanocoatings, multilayers, water purification, metal binding

1 INTRODUCTION

Clean drinking water is not readily available globally, especially in underdeveloped nations. By 2025, the worlds population will reach eight billion with water needs tripling over current levels as more regions move to industrial activity. Women and children are impacted the most from the global water crisis in that the burden of providing water for the family is typically the womans; children bioaccumulate toxins at a rate ten times that of an adult and are the most sensitive to neurotoxins such as lead and mercury. The overriding issue to providing clean drinking water is cost, for an individual and also for a government. For individuals earning less than \$5 per day their water needs must be met for a fraction of that. The technical challenge is a water purifier, which can process water of uncertain origin to at least World Health Organization drinking quality levels for less than \$0.0004 per Liter operating cost.

Crystal Clear Technologies (CCT) has demonstrated ligand-assisted filter media to remove toxic metals, such as arsenic, lead, mercury and others, from water; these metals are serious concerns in water even today, in developed countries and even more so in the developing world. Central to NMX^{TM} technology is the use of ligand-based nano-coatings that attach to a substrate filter media and sequester water-borne metallic contaminates. The unique ligand capability enables CCT to enhance the adsorption capacity of all conventional media including extremely low cost adsorbent media such as boehmite, converting it into a highly effective adsorbent filter media with at least eight times its conventional capacity. The goal is to supply one person's water needs for one year for one dollar.

The water purification systems studied use a metal-oxide substrate (*e.g.*, boehmite or titania with nano-size pores) combined with ligands that bind to both the substrate and metal contaminants, forming nanoscale multilayers. This approach allows for greater flexibility and selectivity in addition to significantly increasing adsorption capacity. The key to improving existing systems of this type and expanding to other toxic metals is the identification of ligand candidates and the understanding of ligand-metal binding energies and structures.

We are addressing this issue with a combination of density functional theory and molecular modeling calculations. These approaches provide ligand-contaminate binding energies and the corresponding geometries. More importantly, by allowing rapid examination of different ligands, these computations aid in the development of design principles for the multilayer nanocoatings of interest. The focus of the present work is on arsenic and selenium as the contaminates and thiol- and carboxylic acid-functionalized ligands. Currently there is limited understanding of the binding and structures of such systems.[1]

2 METHODS

A two-fold approach has been used to investigate the properties of metal-ligand complexes important in water purification nanocoatings. Both electronic structure calculations and molecular modeling simulations have been performed.

Density functional theory (DFT) calculations have been used to determine the lowest energy structure(s) of metal-ligand complexes. The DFT computations reported here have been performed with the Gaussian 03[3] series of programs using the nonlocal hybrid Becke's three-parameter exchange functional (B3LYP).[4] The geometry optimizations have been carried out with the Berny analytical gradient method.[5] The 6-31+G* basis set was used for all the calculations unless otherwise noted. Solvent effects were not considered during the geometry optimization process. The energy difference in solution was corrected from the gas-phase energy by accounting for the solvation energy with a single-point calculation. The solvation energies were obtained using the polarized continuum model (PCM)[6] with water ($\epsilon = 78.39$) as the solvent.

Electronic structure calculations are impractical for modeling a multilayer. Thus, molecular mechanics simulations are also being used; these must be validated by comparison with DFT and X-ray crystallography results. Molecular mechanics conformation search for the arsenic-ligand complexes was performed using the Monte Carlo minimization (MCM) method of Li and Scheraga.[7] At each step, a Monte Carlo atom move is taken followed by an energy minimization. The obtained trial structure is then subjected to the Metropolis acceptance criterion. The MCM method was implemented in the MCCCS Towhee program[8] that was used for all the modeling.

The molecules were modeled using the molecular mechanics universal force field (UFF).[9] The non-bonded interactions are represented by a Lennard-Jones (LJ) 12-6 potential

$$U(r_{ij}) = 4\epsilon_{ij}\left[\left(\frac{\sigma_{ij}}{r_{ij}}\right)^{12} - \left(\frac{\sigma_{ij}}{r_{ij}}\right)^{6}\right] \quad (1)$$

where r_{ij}, ϵ_{ij}, and σ are the separation, LJ well depth, and LJ size, respectively, for interacting atoms i and j. The intramolecular interactions were modeled in the following way: a harmonic potential is used to control bond stretching

$$U_b = \frac{k_{ij}}{2}(r - rij)^2 \quad (2)$$

where k_{ij} is the force constant. The angle bending is represented with a cosine Fourier expansion

$$U_\theta = k_\theta \sum_{n=0}^{m} C_n \cos n\theta \quad (3)$$

where θ is the bending angle, C_n is the expansion coefficient, and k_θ is the force constant. The torsional interactions are similarly described

$$U_\phi = k_\phi \sum_{n=0}^{m} C_n \cos n\phi. \quad (4)$$

The improper torsion is controlled by a cosine Fourier expansion

$$U_\omega = k_\omega(C_0 + C_1 \cos\omega + C_2 \cos 2\omega) \quad (5)$$

where ω is the improper torsion angle, and k_ω is the force constant, and c_i is the expansion coefficient.

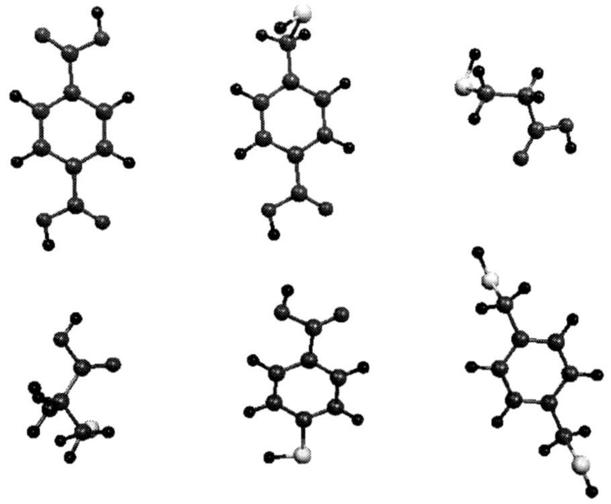

Figure 1: Several of the ligands considered in this work. (C=brown, H=black, O=red, S=yellow.)

3 RESULTS

A requisite property of potential ligands is an ability to bind strongly to a contaminate. This has been investigated for several ligands, some of which are shown in Fig. 1, for both arsenic and selenium using DFT calculations. For example, considering As(III) in aqueous solution as present primarily in the As(OH)$_3$ form, the ligand exchange reactions,

$$As(OH)_3 + n\,LH \;\rightarrow\; AsL_n(OH)_{3-n} + n\,H_2O \quad \Delta E_n, \quad (6)$$

were examined for $n = 1 - 3$. In addition, the structure of the ligand coordination to arsenic was examined through DFT geometry optimizations, as this can have important consequences for the properties of a multilayer formed from the $AsL_n(OH)_{3-n}$ species.

The DFT calculations find that, for every ligand considered, the successive ligand exchange energies are more favorable. That is, AsL_3 is the thermodynamically preferred species. In general, the thiolate functional group binds more strongly than carboxylates. An important consideration in understanding the As-ligand binding is the formation of more complex structures, such as As_2L_n "dimers." It is not yet clear whether formation of a dimer is helpful or harmful in the formation of stable multilayer nanocoatings for sequestering arsenic. However, the structure of the ligand can affect whether dimer formation is favorable or not, *i.e.*, whether AsL_3 or As_2L_n is the most stable species, and what the dimer structure is, *e.g.*, As_2L_2 or As_2L_3. The DFT calculations have shed light on what ligand properties influence these issues; these include flexibility in the functional group linkage to the phenyl ring and steric bulk on the ring. The DFT results are in agreement with the observed stable structures where data is available. The optimized structures are in very good agreement with the corresponding X-ray crystallography data where comparisons are possible.[2] This indicates that DFT can be predictive in describing the geometries of these arsenic-ligand complexes; additional tests of these predictions are underway.

Analogous calculations are being carried out for selenium, which poses additional challenges due to its speciation. For example, Se(VI) is present primarily as selenate, SeO_4^{2-}, a pH > 5. Thus, an analogous ligand exchange reaction to Rxn. 6 would be

$$SeO_4^{2-} + n\,LH \rightarrow [SeO_{4-n}L_n]^{-2+n} + n\,OH^-. \quad (7)$$

Note that, in contrast to the As(III) case, the reactants and products are ions such that solvation free energies for the different species involved can determine the favorability of binding. A similar issue arises for Se(IV) which has a more complicated speciation pattern. Preliminary DFT calculations indicate that some species undergo ligand exchange reactions much more readily than others. In addition, Se_2L_n "dimer" complexes are found to be thermodynamically stable for some species, in analogy to the arsenic chemistry discussed above. Experimental tests of the predicted binding energies and complex geometries are underway including examinations of the speciation dependence and dimerization tendencies predicted by the DFT calculations. Understanding how these binding patterns relate to multilayer nanocoating formation and properties will involve molecular dynamics calculations closely coupled with experimental studies.

Preliminary molecular modeling calculations are encouraging. For example, the structures obtained for some As_2L_3 dimers are in good agreement with DFT-optimized geometries, while others display minor differences. We are exploring whether the latter cases can be improved by simple modifications of the force field.

4 Summary

The computations presented here are generating new insight into the ligand-As and ligand-Se complexes by obtaining successive binding energies for up to three ligands, the characteristics of the lowest energy structures, the energies to form As_2L_n and Se_2L_n complexes, solvation effects, and the dependence of binding upon speciation. These are all important considerations in the design of multilayer nanocoatings for water purification. The calculated structures are in excellent agreement with crystal structures where available.[2] Experimental tests of the predicted binding patterns are currently underway.

The DFT and molecular modeling calculations are informing and will be complemented by larger-scale molecular dynamics simulations that permit the structure of the multilayers to be examined. Specifically, solvent effects can be included more accurately and quantitative information about the competition between energetic and entropic factors important in determining multilayer stability obtained.

REFERENCES

[1] T.G. Carter, E.R. Healey, M.A. Pitt, and D.W. Johnson, *Inorg. Chem.* **44**, 9634 (2005).

[2] W.J. Vickaryous, R. Herges, and D.W. Johnson, *Angew. Chem. Int. Ed.* **43**, 5831 (2004).

[3] Gaussian 03, Revision C.02, M.J. Frisch *et al.*, Gaussian, Inc., Wallingford CT, 2004.

[4] A.D. Becke, *J. Chem. Phys.* **98**, 5648 (1993).

[5] H.B. Schlegel, *J. Comp. Chem.* **3**, 214 (1982).

[6] M. Cossi, V. Barone, R. Cammi, and J. Tomasi, *Chem. Phys. Lett.* **255**, 327 (1996).

[7] Z. Li and H. Scheraga, *Proc. Natl. Acad. Sci. USA* **84**, 6611 (1987).

[8] M.G. Martin, B. Chen, C.D. Wick, J.J. Potoff, J.M. Stubbs, and J.I. Siepmann, *http://towhee.sourceforge.net.*

[9] A. Rappe, C. Casewit, K. Colwell, W.G. Goddard III, and W.M. Skiff, *J. Am. Chem.* **144**, 10024 (1992).

Numerical Simulation of Polymer Phase Separation on a Patterned Substrate with Nano Features

Yingrui Shang*, David Kazmer**, Ming Wei, Joey Mead, and Carol Barry
Department of Plastics Engineering, University of Massachusetts at Lowell
*yingrui_shang@student.uml.edu
**david_kazmer@uml.edu

ABSTRACT

Phase separation of an asymmetric immiscible binary polymer system in an elastic field with the existence of a patterned substrate was numerically studied in 2D and 3D. An unconditionally stable method for time marching the Cahn-Hilliard equation was employed in the numerical simulation. Compared to the conventional interface tracing mechanism, this diffusion controlled system is characterized by a thick interface with a composition gradient. The evolution mechanisms were studied. The evolution of the characteristic length, $R(t)$, of the phase separation morphology patterns was measured with the Fast Fourier Transform method. The results indicated $R(t)^{1/3}$ increases linearly with time. The influence of the material composition, the attraction factor on the template, and the gradient energy coefficient between the two polymers on the result patterns were also observed in this study. Qualitative and quantitative correspondence can be observed between the numerical results and the experiment results.

Keywords: spinodal decomposition, numerical simulation, patterned substrate, discrete cosine transform

1 FUNDAMENTALS

The thermodynamic behavior of a spinodal decomposition of a blend can be described by Cahn-Hilliard Equation [1], the free energy, F, with the consideration of a surface energy on the substrate in our study, for a binary system can be written as:

$$F(C) = \int_V \{f + f_e + \kappa(\nabla C)^2\} dV + \int_S f_S(C,\mathbf{r}) dS \qquad 1$$

where C is the mole fraction of one polymer component, f is the local free-energy density of homogeneous material, f_e is the elastic energy density, and κ is the gradient energy coefficient. Thus the term $\kappa(\nabla C)^2$ is the additional free energy density if the material is in a composition gradient.

The free energy variation on the heterogeneously functionalized substrate is simulated by a free energy term, f_s, which is a function of the composition and the coordinates, $\mathbf{r}$, as well. The surface free energy is added to the total free energy on the surface of the substrate.

The evolution of the composition can be written as the function of local composition:

$$\frac{dC}{dt} = \nabla^2 \cdot \left[M\left(\frac{\partial f}{\partial C} + \frac{\partial f_e}{\partial C} - \kappa\nabla^2 C \right) \right] \qquad 2$$

The Flory-Huggins type of free energy [2] is used to model the bulk free energy density

$$f = \frac{RT}{v_{site}}\left(\frac{C_A}{m_A}\ln C_A + \frac{C_B}{m_B}\ln C_B + \chi_{AB} C_A C_B\right) \qquad 3$$

where C_i is the volume fraction of component i, m_i is the degree of polymerization of component i, T is the temperature in K, R is the ideal gas constant, χ_{AB} is the Flory-Huggins interaction parameters between two components, which is dependent on temperature, and v_{site} is the molar volume of the reference site in the Flory-Huggins lattice model.

The elasticity is assumed isotropic in the domain. According to the Vegard's law [3], the stress-free strain is isotropic and depends linearly on the composition:

$$e_{ij}^0 = \eta(C - C_0)\delta_{ij} \qquad 4$$

where e_{ij}^0 is the stress-free strain, c_0 is the average composition of the domain, δ_{ij} is the Kronecker delta function, and η is the compositional expansion coefficient which is expected to be independent of the composition and the composition gradient [4].

According to the linear elasticity, the stress, σ_{ij} is linear with the change of the strain by Hook's law:

$$\sigma_{ij} = c_{ijkl}\left(e_{kl} - e_{kl}^0\right) \qquad 5$$

where c_{ijkl} represents the isothermal elastic tensor, which is independent of position and composition. The elastic energy then can be expressed as follows, with no external anisotropic elastic applied,

$$f_e = \frac{1}{2} c_{ijkl}\left(e_{ij} - e_{ij}^0\right)\left(e_{kl} - e_{kl}^0\right) \qquad 6$$

The total strain can be evaluated by the local displacement, $\mathbf{u}$ [5].

$$e_{ij} = \frac{1}{2}\left(\frac{\partial u_i}{\partial r_j} + \frac{\partial u_j}{\partial r_i} \right) \qquad 7$$

The displacement of the reference lattice is then solved by the elastic equilibrium. Given the fast relaxation time compared to the rate of morphology evolution, it can be assumed that the system is in elastic equilibrium [6]:

$$\frac{\partial \sigma_{ij}}{\partial r_j} = 0 \quad 8$$

2 NUMERICAL METHODS

The Cahn-Hilliard equation is known for its difficulty to solve due to its non-linearity and the bihamonic term. The cosine transform method is applied to the spatial discretization:

$$\frac{d\hat{C}}{dt} = M\lambda\left(\left\{\frac{\partial f}{\partial C}+\frac{\partial f_e}{\partial C}\right\} - \kappa\lambda\hat{C}\right) \quad 9$$

where $\hat{C}$ and $\left\{\frac{\partial f}{\partial C}+\frac{\partial f_S}{\partial C}\right\}$ represent the cosine transform of the respective terms in Equation 4. λ is the approximation of discrete Laplacian operator in the transform space [7]:

$$\lambda(\mathbf{k}) = \frac{2\sum_i \cos(2\pi k_i) - \sum_i 2}{(\Delta x)^2} \quad 10$$

The vector **k** denotes the discretisized spatial element position in all dimensions. Numerically, $k_i=n_i/N_i$, where n_i is the element in the ith dimension and N_i represents the number of elements in the ith dimension. Δx is the spatial step in the numerical modeling.

By this means, the partial differential equation is transformed into an ordinary differential equation in the discrete cosine space. A semi-implicit method is used to trade off the stability, computing time and accuracy [8, 9]. The linear fourth-order operators can be treated implicitly and the nonlinear terms can be treated explicitly. The resulting first-order semi-implicit scheme is:

$$(1+M\Delta t\kappa\lambda)\hat{C}^{n+1} = \hat{C}^n - M\Delta t\lambda\left\{\frac{\Delta f(C^n)}{\Delta C}+\frac{\Delta f_e(C^n)}{\Delta C}\right\} \quad 11$$

A second-order Adams-Bashforth method [10] was also used for the explicit treatment of the nonlinear term after the first time step,

$$(3+2M\Delta t\kappa\lambda^2)\hat{C}^{n+1} = 4\hat{C}^n - \hat{C}^{n-1} + 2M\Delta t\lambda\left(2\left\{\frac{\partial f(C^n)}{\partial C}+\frac{\partial f_e(C^n)}{\partial C}\right\} - \left\{\frac{\partial f(C^{n-1})}{\partial C}+\frac{\partial f_e(C^{n-1})}{\partial C}\right\}\right) \quad 12$$

The parameters are selected according to the corresponding experiment conditions and the literature. The temperature is 363K (90^{0}C). The degree of polymerization of polymer A and B are 447 and 915. The interaction parameter, χ_{AB}, in the Flory-Huggins type of free energy is evaluated as 0.117 with an empirical function by Kressler et. al. [11]. The gradient energy coefficient, κ, is selected as 6.9e-11J/m [12] for a generic polymer material. The diffusivity, D=1e-20m^2/s, is chosen as a typical value for polymers. The mobility of the system then can be evaluated.

In the elastic model, the reference composition is chosen according to the initial condition. The compositional expansion coefficient is chosen as η=0.02 [12]

The surface energy is simulated as linear to the local composition [13].

$$f_S(C,\mathbf{r}) = s_0(\mathbf{r}) + s_1(\mathbf{r})(C - C_{ref}) \quad 13$$

where $s_0(\mathbf{r})$ and $s_1(\mathbf{r})$ are surface attraction parameters that represent the functionalization of the substrate, and C_{ref} is the reference composition. For non-dimensionalization, C_{ref} and $s_0(\mathbf{r})$ are chosen to be 0. The functionalized substrate energy factor $s_1(\mathbf{r})$ is alternating across the x direction of the substrate in an effort to develop a corresponding alternating strip patterns in the self-assembling polymer near the substrate.

3 RESULTS AND DISCUSSION

The characteristic length is then measured with respect to time. In the numerical modeling, the domain growth is studied by evaluating the pair-correlation functions g_i, where the subscript i and j represents the directions in the domain [14].

$$g_i(d,t) = \frac{1}{N_j}\sum_{k_j=1}^{N_j} g_{i,k_j}(d,t) \quad 14$$

where $g_{i,ki}$ is defined as

$$g_{i,k_j}(d,t) = \langle \psi(\mathbf{k}=(k_i,k_j),t)\,\psi(\mathbf{k}=(k_i,k_j+d),t)\rangle \quad 15$$

$\psi(\mathbf{k})$is the order parameter which equals the composition difference of the two polymers in the element **k**, and d is a positive variable and varies from 1 to k_i. The angle brackets denote the average value of the expression inside the brackets over all the lattice points. The characteristic length in the ith dimension, $R_i(t)$, is the first zero value of $g_i(d, t)$.

The interface of the polymer and the patterned substrate is the focus of this study. For improved model fidelity, the characteristic length is measured by a 2D model of 128×128 elements instead of a 3D model. The phase separation takes place on a neutral substrate and the effect of the elastic field is included. It has been previously established that $R(t)$ increases proportionally to $t^{1/3}$ [14-16]. The value of $R(t)$ is obtained by evaluation of $g_i(d, t)$. $R(t)$ is determined with first order interpolation between the two points when the sign of the pair-correlation function changes, as can be seen in Figure 1.

In Figure 1 the $R(t)$ value without the elastic energy can be fitted to a straight line with the x-axis set to be $t^{1/3}$. The slope of the fitting line is 0.545. The value of $R(t)$ in the system with the consideration of isotropic energy evolves more slower than that in the elasticity free system, as can be observed in Figure 1. In an isotropic elastic system, the $R(t)$ value can also be fitted with a straight line with the respect of $t^{1/3}$, and the slope of the fitting line is 0.370, smaller than that for the elasticity free situation. This result is due to the elastic free energy term added to the local free energy, which slows the minimization of the free energy.

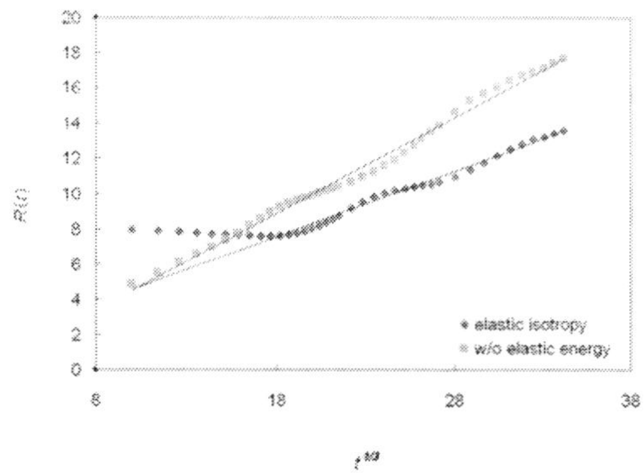

Figure 1 The relationship of the characteristic length, $R(t)$, and the time, t.

To optimize the final pattern according to the functionalized substrate, the characteristic length should be compatible with the pattern scale. The similarity of the final pattern to the pattern on the substrate is measured with respect to time. The compatibility of the final pattern and the patterns designed on the substrate is measured by a factor, C_S

$$C_S = \frac{1}{2}\langle|\psi(\mathbf{k}) - S_k|\rangle \qquad 16$$

where

$$S_k = \begin{cases} \dfrac{s_1(k)}{|s_1(k)|}, & s_1(k) \neq 0 \\ 0, & s_1(k) = 0 \end{cases} \qquad 17$$

$s_1(\mathbf{k})$ is the parameter in the surface energy, which denotes the strength of the surface functionalization. $s_1(\mathbf{k})$ can be varies across the substrate surface. S_k is the qualitative representation of the substrate attraction. Obviously,

$$0 \leq C_S \leq 1$$

The greater the value of C_S, the more compatible the resulting polymer morphology is to the functionalized substrate. The phase separation is simulated numerically in a 128×128 elements 2D model on a heterogeneously functionalized substrate. Figure 2 shows the value $R(t)$ across the direction perpendicular to the strips patterns on the substrate. It can be seen that $R(t)$ in the x direction increases quickly in the early stage and slows beyond a critical value, which is related to the dimension of the pattern strips. The system with isotropic elastic energy has smaller evolution rates but an earlier critical time than the phase separation without elastic energy. It can be seen that the characteristic length can be fitted into straight lines before and after the critical time, both in the situations with and without elastic energy.

In Figure 3 the compatibility factor, C_S is also plotted with $t^{1/3}$. There is also a critical time during the evolution of the compatibility factors. It can be observed that the critical time for C_S is empirically identical with that of $R(t)$ in Figure 2. The compatibility of the phase separation pattern with the substrate increases quickly before the critical time and become stable afterward. After the critical time, $R(t)$ and C_S increase in a much slower pace. In practice, this implies that the optimized annealing time should be close to the critical time, since a longer annealing time does not help to significantly refine the final pattern.

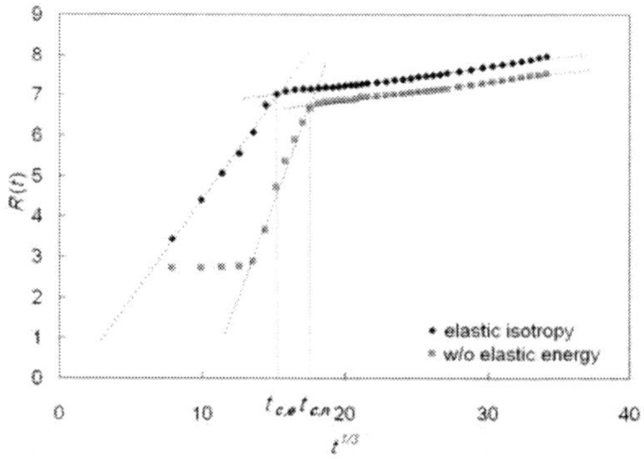

Figure 2 The Characteristic length in x direction (the direction perpendicular to the strips in the pattern)

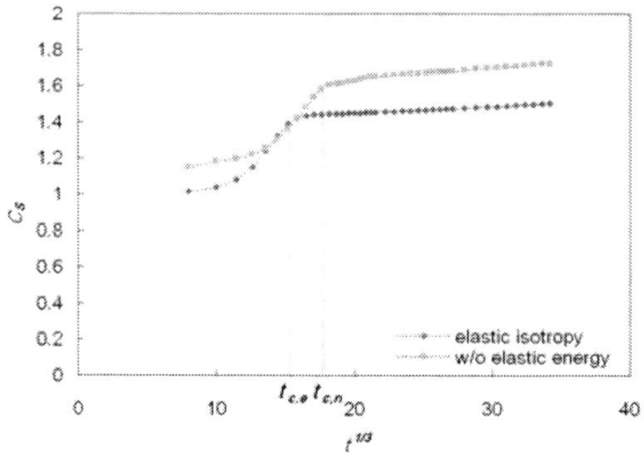

Figure 3 The compatibility of the result pattern to the patterned substrate

The lateral composition profile of phase decomposition with a patterned substrate is observed in a 3D model. A checker board structure is observed in the early stage of the phase separation as reported in other papers [17, 18]. Since the attraction factor is alternating on the substrate, the polymer near the substrate is attracted and attached to the respective area on the substrate, and the neighboring part in the domain to the depth direction is concentrated with the other type of polymer. This effect starts from the interface with the substrate and decays through the thickness direction to the domain. As the spinodal decomposition develops in the bulk domain, the random phase separation overcomes the checker board effect.

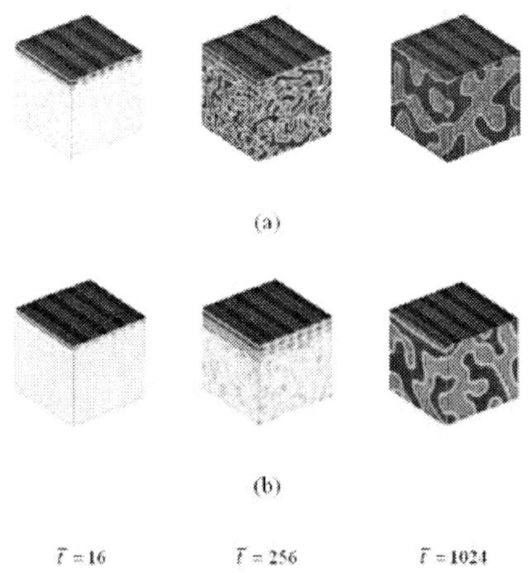

Figure 4 Effect of the heterogeneously functionalized pattern on the phase decomposition. (a) without elastic energy; (b) with isotropic elastic energy.

4 CONCLUSIONS

2D and 3D numerical simulations are developed for the phase separation of a binary immiscible polymer blend. The effect of isotropic elastic energy is investigated. It is observed that in the spinodal decomposition process, the characteristic length, $R(t)$, is proportional to $t^{1/3}$. The involvement of the isotropic elastic energy results in a smaller slope in the $R(t)\sim t^{1/3}$ diagram. The introduction of a patterned substrate with regular strips will induce a critical time in both the $R(t)$ in the direction perpendicular to the strips, and the compatibility of the result pattern and the substrate pattern, C_S. The strips confine the increase of $R(t)$ in the direction perpendicular to the strips. The slope of $R(t)\sim t^{1/3}$ diagram decreases in the second stage.

The lateral composition profile is investigated in a 3D model. A checker board structure is shown in the early stage of the decomposition and decays as the intrinsic value of $R(t)$ in the bulk domain increases.

REFERENCES

[1]. Cahn, J.W. and J.E. Hilliard, *Free Energy of a Nonuniform System. I. Interficial Free Energy.* The Journal of Chemical Physics, 1958. **28**(2): p. 258-267.

[2]. Flory, P.J., *Principles of Polymer Chemistry.* 1953, Ithaca, NY: Cornell University Press.

[3]. L. Vegard, Z., *The constitution of mixed crystals and the space occupied by atoms.* Phys, 1921. **5**(17).

[4]. Denton, A.R. and N.W. Ashcroft, *Vegard's law.* Physical Review A, 1991. **43**(6): p. 3161-3164.

[5]. AG, K., *Theory of structrual transformations in solids.* 1983, New York: Wiley.

[6]. Seol, D.J., et al., *Computer simulation of spinodal spinodal decomposition in constrained films.* Acta Maerialia, 2003. **51**(17): p. 5173-5185.

[7]. Copetti, M.L.M. and C.M. Elliot, *Kinetics of Phase Decomposition Process:Numerical Solutions to Cahn-Hilliard Equation.* Material Science and Technology, 1990. **6**: p. 279-296.

[8]. Eyre, D.J., *An Unconditionally Stable One-step Scheme for Gradient Systems.* Unpublished article, 1998.

[9]. Eyre, D.J., *Unconditionally Gradient Stable Time Marching the Cahn-Hilliard Equation.* Unpublished article, 1998.

[10]. Mauri, R., R. Shinnar, and G. Triantafyllow, *Spinodal decomposition in binary mixtures.* Physical Review E, 1996. **53**(3): p. 2613-2623.

[11]. Kressler, J., et al., *Temperature Dependence of the Interaction Parameter between Polystyrene and Poly(methyl methacrylate).* Macromolecules, 1994. **27**: p. 2448-2453.

[12]. Wise, S.M. and W.C. Johnson, *Numerical Simulations of Pattern-directed Phase Decomposition in a Stressed, Binary Thin Film.* Journal of Applied Physics, 2003. **94**(2): p. 889-898.

[13]. Jones, R.A.L., *Effect of Long-Range Forces on Surface Enrichment in Polymer Blends.* Physics Review E, 1993. **47**(2): p. 1437-1440.

[14]. Brown, G. and A. Chakrabarti, *Surface-directed Spinodal Decomposition in a Two-Dimensional Model.* Physical Review A, 1992. **46**(8): p. 4829-4835.

[15]. Wheeler, A.A., W.J. Boettinger, and G.B. McFadden, *Phase-field Model for Isothermal Phase Transitions in Binary Alloys.* Physics Review A, 1992. **45**(10): p. 7424-7439.

[16]. Jones, R.A.L., L.J. Norton, and E.J. Kramer, *Surface-directed Spinodal Decomposition.* Physical Review Letters, 1991. **66**(10): p. 1326-1329.

[17]. A.Karim, et al., *Phase Separation of Ultrathin Polymer-blend Films on Patterned Substrates.* Physical Review E, 1998. **57**(6): p. 6273-6276.

[18]. Kielhorn, L. and M. Muthukumar, *Phase Separation of Polymer Blend Film near Patterned Surfaces.* Journal of Chemical Physics, 1999. **111**(5): p. 2259-2269.

Towards Rational *de novo* Design of Peptides for Inorganic Interfaces

M.J. Biggs[†] and M. Mijajlovic

Institute for Materials and Processes, University of Edinburgh
King's Buildings, Mayfield Road, Edinburgh, UK.

ABSTRACT

Protein adsorption at inorganic surfaces is highly relevant to nano- and bionanotechnology. Just two examples of great significance are the use of specific peptide sequences to control cell deposition on tissue scaffolds, and the tethering of bio-photosynthetic reaction centers on electrodes to harvest light for power generation and hydrogen production. Understanding of the behavior of proteins in such situations and the design of surface-binding proteins for technological applications is currently limited by the largely empirical approaches used. We are, therefore, developing the application of molecular modeling to the elucidation of the behavior of peptides at fluid/solid interfaces – in this talk, we will provide details of these models and their application to the study of peptides at fluid/solid interfaces and their *de novo* rational design.

Keywords: biosensors, nanoelectronics, nanophotonics, surface binding peptides, tissue engineering.

1 INTRODUCTION

Protein adsorption at inorganic interfaces occurs across science, engineering, medicine and nature [1]. Protein adsorption is, for example, the first step in the body's response to implants such as artificial heart valves [2] that ultimately may lead to complications and even life-threatening reactions such as emboli. Improvements in our understanding of this response are now being exploited to develop new implant technologies [2]. Similar approaches are also underpinning the next generation of tissue scaffolds to improve spatial control over cell adhesion, which is essential for growing all but the simplest tissue [3]. Protein adsorption and migration on solid surfaces are central to bioseparations [4], fouling in the processes industries and beyond [5], and biosensors and bioarrays [6-8]. Examples of protein adsorption important in the natural world include antifreeze proteins (AFPs) that allow some species to survive at sub-zero temperatures by inhibiting ice crystal growth [9], and proteins that aid biomineralization, a process responsible for the formation of all natural inorganic materials such as bones, teeth and egg shells [10]. Such examples from nature are now inspiring new 'biomimetic' technologies. By borrowing ideas from AFPs, a number of groups have in recent years developed peptides that control crystal growth [11], for example, whilst some are developing the use of peptides to self-assemble nanoscale entities to form complex multiscale structures [11].

Experimental challenges mean fundamental understanding of the behaviour of proteins at inorganic interfaces is still in its infancy [11]. For example, the role and actions of various proteins during biomineralization is still much debated [12] – this in part arises from our current inability to determine the three-dimensional conformation of biomolecules at solid interfaces and the associated interactions. This lack of fundamental understanding means the design of peptides that recognise specific inorganic surfaces for technological uses is also very much empirical (*e.g. ref.* [13]), albeit sometimes supported by sequence analysis methodologies [14]. Whilst experiment must undoubtedly play a central role in improving fundamental understanding and in any design of peptide/surface systems, molecular simulation can also play a major role.

In this talk, we will outline how we are applying molecular simulation to the elucidation of the behavior of peptides at fluid/solid interfaces and the rational *de novo* design of peptides that recognize such interfaces with specificity.

2 METHODS

The methodology being developed at Edinburgh for the rational *de novo* design of peptides that preferentially bind at a target solid-fluid interface is shown in Fig. 1. Briefly, a physicochemical-based *in silico* evolutionary process is used to identify sets of peptide sequences that may potentially bind to the target interface. Once identified, these sequences are synthesized and subject to experimental characterization to determine the validity of the predictions. If the predictions are poor, the experimental data is used to improve the models underpinning the *in silico* evolutionary process. If, on the other hand, the predictions prove reasonable, data obtained from sequence analysis is fed back to inform further searches if this is warranted. Data from *in vivo* or *in vitro* combinatorial searches can also be used as input to the process.

An essential component of the *in silico* evolutionary process is the prediction of the binding energy of candidate sequences using physicochemical models of the binding process. As a typical *in silico* evolutionary simulation requires many 1000s of such predictions, an *ab initio* structure prediction approach [16] is used to determine the global free energy minimum of candidate sequences in the bulk phase and at the target interface. The semi-explicit

† Fax: +44-131-650-6551. E-mail: m.biggs@ed.ac.uk.

Langevin dipole approach [17] is used to model solvents [18], which we have shown to be $O(10^2)$ times faster than explicit solvent methods yet is still able to resolve phenomena such as hydrogen bond bridging and solvent structuring between the peptide and solid surface.

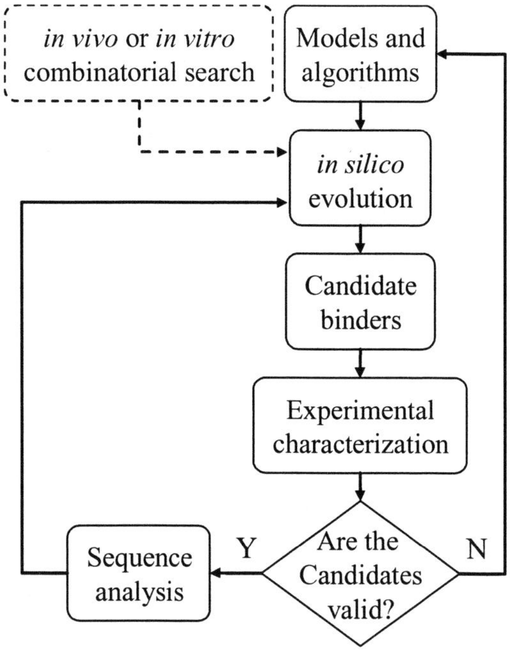

Figure 1. The rational *de novo* design methodology being developed.

3 RESULTS AND DISCUSSION

We have applied the *ab initio* structure and free energy prediction method to a variety of peptides in the gas and liquid phases, and at fluid/solid interfaces [19, 20]. A recent example is the pentapeptide met-enkephalin at the water/graphite interface [20]. The intrapeptide and peptide-graphite interactions were modeled using the Amber94 PE model [21] with the graphite carbon atoms being assumed to be equivalent to the aromatic carbon atoms of this PE model. The peptide-graphite and water-graphite interactions were modeled using the LD-Amber method [22], with the graphite carbon atoms once again being modeled as sp^2 hybridized C atoms; the evaluated water-graphite interfacial energy obtained using this approximation was found to be inline with experimental data.

Fig. 2 shows the lowest energy structure of zwitterionic met-enkephalin in water. The extended structure seen here is in good agreement with other simulation studies [23] and is inline with experimental results which suggest this molecule can take a range of extended structures in dilute solutions [24-28]. The experimental studies also suggest met-enkephalin in water is flexible. Although the EA cannot offer direct evidence for such flexibility or otherwise, the hydrogen-bonding extant in the structure predicted here suggests it is likely to be flexible – there is a single hydrogen bond (HB) between the CO-group of the first glycine and NH-group of the phenylalanine residue, shown as a dashed red line in Fig. 2, whilst the only groups that can support HB bridging are the CO-groups of the phenylalanine residue and C-term, and the NH-groups of the N-term and first glycine residue.

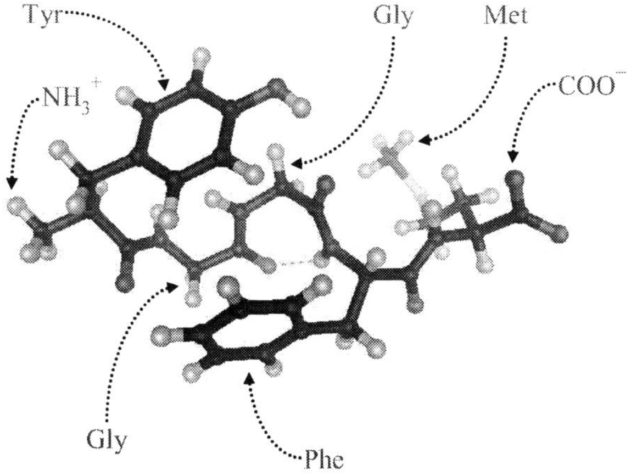

Figure 2. Predicted conformation of zwitterionic form of met-enkephalin in water.

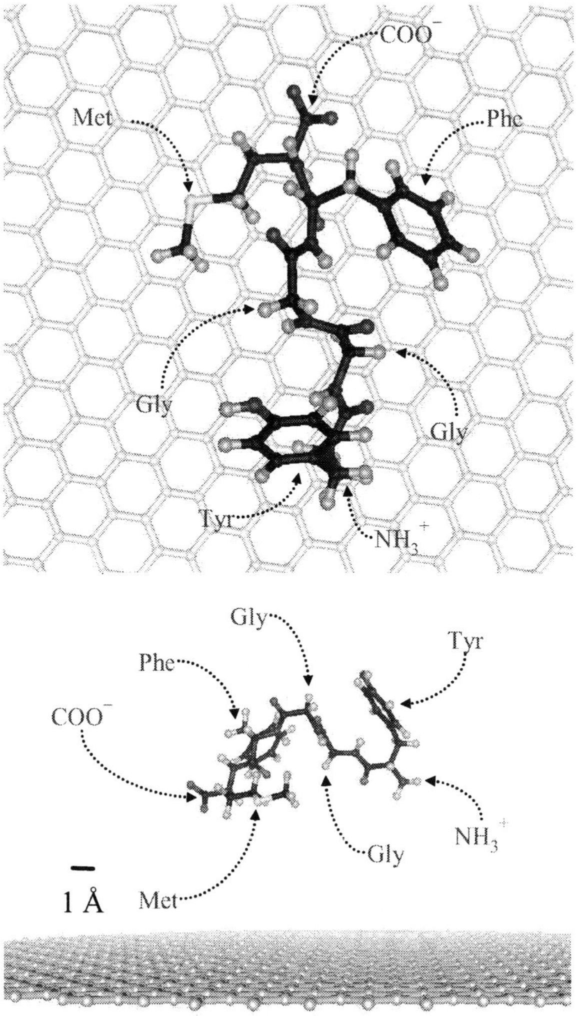

Figure 3. Predicted structure of zwitterionic form of met-enkephalin at water/graphite interface.

Fig. 3 shows the predicted structure of the zwitterionic form of met-enkephalin at the water/graphite interface. This figure shows that the sole HB of the solvated structure in Fig. 2 is absent, leading to an even more extended structure. Although this conformation is preferred to the bulk phase structure by slightly less than 20 kcal/mol, the peptide is located approximately 9 Å from the graphite surface, which corresponds to three layers of water between the surface and the peptide. As suggested by this degree of separation and confirmed in Fig. 4, the adsorbed state is water-mediated rather than *via* dispersive interaction between the peptide and graphite.

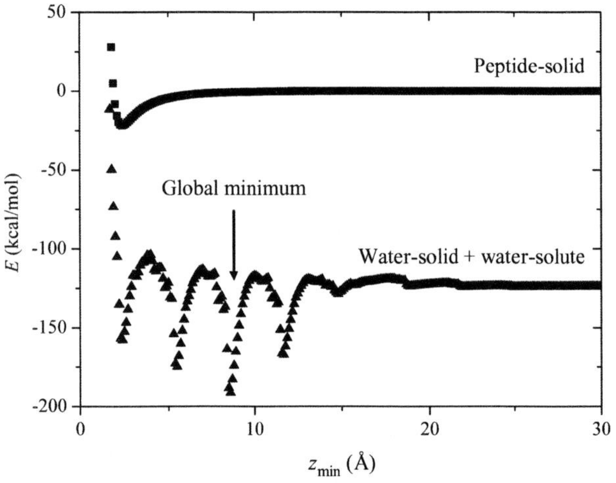

Figure 4. The variation of the energy associated with the peptide-graphite and water-graphite + water-peptide interactions with the shortest distance between the graphite and the peptide in its adsorbed conformation.

Upon removal of the water, Fig. 5 shows that met-enkephalin flattens out and becomes more closely associated with the graphite surface via π-bonding between the graphite and the benzene rings of the phenylalanine and tyrosine residues. The structure is also less extended due to the strong interaction between the oppositely charged N- and C-terms that are no longer shielded by water.

4 CONCLUSIONS

We have developed a means of rapidly predicting the structure and binding energy of peptides at fluid/solid interfaces. We are now extending the approach to various surfaces (e.g. metals where electron polarization is likely to be important) and looking to apply it to identify solid surface binding peptides of high specificity that will facilitate, amongst other things, the self-assembly of nanoelectronic devices such as carbon nanotube based field effect transistors and nanoelectronic systems based on such devices.

ACKNOWLEDGMENTS

MM thanks the University of Edinburgh and UUK for PhD scholarships. MJB thanks the Royal Academy of Engineering and the Leverhulme Trust for partial support of this research.

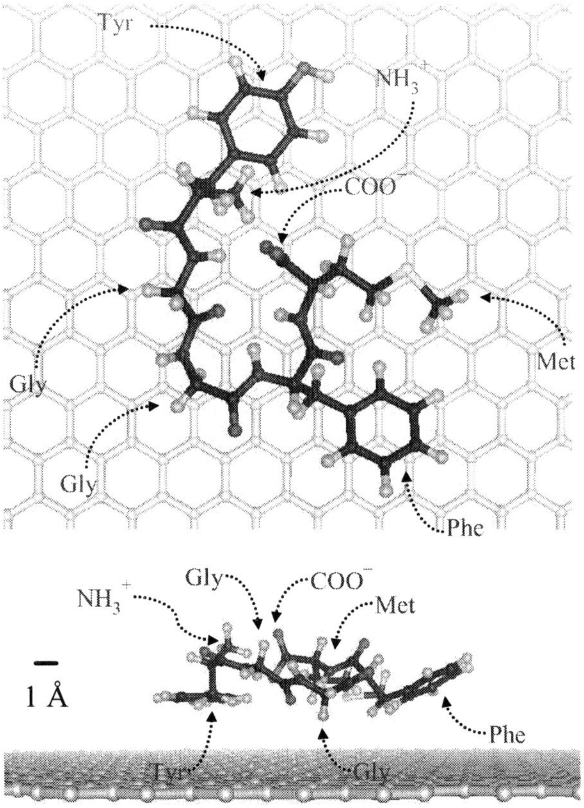

Figure 5. Predicted structure of zwitterionic form of met-enkephalin at water/graphite interface.

REFERENCES

[1] B. Kasemo, Surf. Sci. 500, 656, 2002.

[2] B.D. Ratner and S.J. Bryant, Annu. Rev. Biomed. Eng. 6, 41, 2004.

[3] H. Shin, S. Jo, and A.G. Mikos, Biomaterials 24, 4353, 2003.

[4] T.M. Przybycien, N.S. Pujar and L.M. Steele, Curr. Op. Biotech. 15, 469, 2004.

[5] H.-C. Flemming, Appl. Microbiol. Biotechnol. 59, 629, 2002.

[6] J. Castillo, S. Gáspár, S. Leth, M. Niculescu, A. Mortari, I. Bontidean, V. Soukharev, S.A. Dorneanu, A.D. Ryabov and E. Csöregi, Sensors & Actuators B 102, 179, 2004.

[7] C. Hultschig, J. Kreutzberger, H. Seitz, Z. Konthur, K. Büssow and H. Lehrach, Curr. Op. Chem. Biology 10, 4, 2006.

[8] K.E. Sapsford, Y.S. Shubin, J.B. Delehanty, J.P. Golden, C.R. Taitt, L.C. Shriver-Lake and F.S. Ligler, J. Appl. Microbiol. 96, 47, 2004.

[9] Y.-C Liou, A. Tocilj, P.L. Davies and Z. Jia, Nature 406, 322, 2000.

[10] S. Weiner and L. Addadi, Mater. Chem. 7, 689, 1997.

[11] M. Sarikaya, C. Tamerler, A.K.Y. Jen, K. Schulten and F. Baneyx, Nature Mat. 2, 577, 2003.

[12] S. V. Patwardhan, G. Patwardhan and C.C. Perry, J. Mat. Chem. 17, 2875, 2007.
[13] P. Schaffner and M.M. Dard, Cell. Mol. Life Sci. 60, 119, 2003.
[14] C. Tamerler and M. Sarikaya, Acta Biomat. 3, 289, 2007.
[15] D.J. Osguthorpe, Curr. Op. Struct. Biol. 10, 146, 2000.
[16] D.P. Djurdjevic and M.J. Biggs, *J. Comp. Chem.* 27, 1177, 2006.
[17] J. Florián and A. Warshel, J. Phys. Chem. B 101, 5583, 1997.
[18] M. Mijajlovic and M.J. Biggs, J. Phys. Chem. B 111, 7591, 2007.
[19] D.P. Djurdjevic, "*Ab initio* Protein Fold Prediction Using Evolutionary Algorithms", PhD Thesis, University of Edinburgh, United Kingdom, 2006.
[20] M. Mijajlovic, "*Ab initio* prediction of the conformation of solvated and adsorbed proteins", PhD Thesis, University of Edinburgh, United Kingdom, 2008.
[21] W.D. Cornell, P. Cieplak, C.I. Bayly, I.R. Gould, K.M. Merz, D.M. Ferguson, D.C. Spellmeyer, T. Fox, J.W. Caldwell and P.A. Kollman, J. Am. Chem. Soc., 117, 5179, 1995.
[22] M. Mijajlovic and M.J. Biggs, J. Phys. Chem. B, 111, 7591, 2007.
[23] M. Kinoshita, Y. Okamoto and F. Hirata, J. Am. Chem. Soc., 120, 1855, 1998.
[24] B.P. Roques, C. Garbay-Jaureguiberry, R. Oberlin, M. Anteunis and A.K. Lala, Nature, 262, 778, 1976.
[25] C.R. Jones, V. Garsky and W.A. Gibbons, Biochem. Biophys. Res. Commun., 76, 619, 1977.
[26] M.A. Khaled, M.M. Long, W.D. Thompson, R.J. Bradley, G.B. Brown and D.W. Urry, Biochem. Biophys. Res. Commun., 76, 224, 1977.
[27] M.A. Spirtes, R.W. Schwartz, W.L. Mattice and D.H. Coy, Biochem. Biophys. Res. Commun., 81, 602, 1978.
[28] W.H. Graham, E.S.I. Carter and R.P. Hicks, Biopolymers, 32, 1755, 1992.

Molecular dynamic simulations of interactions between LDH and DNA bio-nanocomplex

P. Tran*, S. Smith*, Y. Wong‡, G. Q. Lu‡
* Centre for Computational Molecular Science And ARCCFN
University of Queensland, Brisbane, 4072, Australia
Email: s.smith@uq.edu.au
‡ARC Centre for Functional Nanomaterials (ARCCFN)
University of Queensland, Brisbane, 4072, Australia
Email: maxlu@cheque.uq.edu.au

***Abstract*—We are using molecular dynamics computer simulation to understand the structure of layered double hydroxides and intercalated DNA strands. The conformational changes of layered double hydroxides depends greatly on the nature of counterions, for example the interlayer spacing can increase from 8.1Å to 19.1Å. We also compare these deformations to previous experimental structures. We observed important modifications of the crystal structure on a and b cell parameters.**

Layered double hydroxides; molecular dynamics calculations; inorganic intercalates; drug delivery,DNA s trands

Layered double hydroxides (LDH) are anionic clay materials[1]. Their importance in areas such as catalysis, medicine and oil-field-technology has greatly increased in the last two decades. Indeed, a variety of organic structures can be intercalated into layered double hydroxides (LDH) structures[2], making LDH nanoparticles highly promising vehicles for drug delivery and gene therapy applications. The nature of the interactions between DNA strands (or other target molecules) and the LDH nanoparticles is, however, poorly understood.

To understand and to be able to predict the properties of LDHs and the interlayer interactions of different intercalates demands an understanding of the structure of the metallic hydroxide layers; the intercalation process and the subsequent reorganization / relaxation. It is difficult experimentally to gain insight into the detailed LDH organization because of structural disorder in the crystalline structures and because of the small size of the particles. The interlayer arrangement of the water and anionic molecules is not fully understood and the structure of hydroxide layers is still under debate. Accurate determination of water molecule positions and geometries is problematic because adsorbed water molecules are not distinguishable from interlayer water in thermal analysis.

The main objective of our work is to develop MD simulations which would help us to define the nature of interactions between the DNA-containing nanoparticles and cell membranes, leading ultimately to a joint computational and experimental attack on questions as to the mechanism [3] of transfection of these particles through the cell membrane. Even though MD simulation techniques, at this stage, do not describe completely the succession of these complex bio-mechanisms, they do provide accurately and, in a very short amount of time, snapshots of model-interactions that contribute insights into interactions and mechanism. These hybrid models combine existing forcefield parameters obtained from mineral and biomolecule databases. The "realistic" MD model will not only support experimental kinetic studies of the transfection process but will also be a key utility for the elaboration of potential new hybrid systems.

Computational techniques have been extremely valuable tools in understanding the structures[4] of bio-organic and polymeric systems. The crystal structure of LDH nanoparticles has not been characterized[5] yet, the closest solved structure for this family of LDH being the original crystalline brucite, $MgOH_2$, which consists of M^{2+} metallic cations coordinated octahedrally by hydroxyl groups. Isomorphic substitution of M^{2+} metal ions by a metallic cation of higher positive charges, M^{3+}, will induce permanent positive charges in the crystal layers. The charge density of the hydroxide layers is hence directly linked to the M^{2+}/M^{3+} ratio of the cationic metallic matrix. Overall charge neutrality is maintained by the intercalation of interlayer anions which are very often labile. DNA is typically negatively charged, which makes it a possible intercalating species. Our present work seeks to characterized the detailed structure and dynamics of a bio-nanohybrid complex structure of the LDH with DNA and RNA strands. Mixed forcefield parameters (Amber and Dreiding) are being used for this nanohybrid complex in order to study the interactions of DNA with the LDH nanoparticles in water. Predicting guest orientation in Layered Double Hydroxide intercalates is very important and will help us to determine and explain the shape of the resulting bio-nanohybrid complex.

I. METHODOLOGY

The structures and energies of the systems considered in this work were calculated using molecular dynamics simulation. This method requires parameters for intra-atomic potentials and for all inter-atomic interactions between LDH and DNA.

A. *Construction of the models*

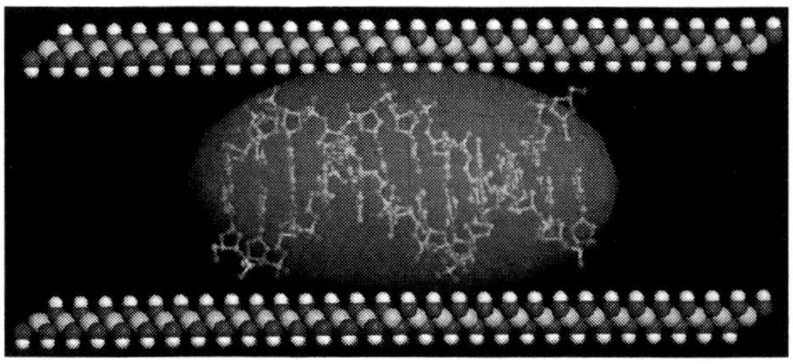

Figure 1: LDH-DNA nanocomplex model

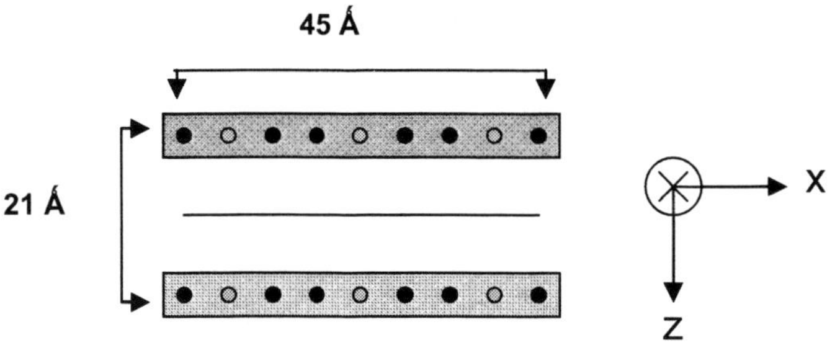

Figure 2. Schematic representation of LDH interlayer along the Z and X axis (Black: Mg^{2+}; Blue: Al^{3+})

The crystal model structure of LDH is constructed using the brucite crystal structure[1]. A $P3_{MI}$ super lattice system was constructed with the brucite cell parameters: a=b=3.148 Å, c=4.779 Å, α=β=90 and γ=120. To obtain the original LDH system, one substitutes a fraction of Mg^{2+} ions with Al^{3+} ions and then allows negatively charged bio-system to intercalate between the gallery layers in order to balance the charge. The model systems typically contains about 450 atoms for the LDH structures and the appropriate DNA anion is then introduced between Mg-Al, also Cl^- counterions are added into LDH layers such that overall charge neutrality is maintained. For each case, the bio-intercalate molecules were placed in the galleries with their longest axis oriented approximatively parallel to the normal interlayer in a bilayer-like arrangement[6]. The geometries of the DNA strain at the start of the simulation were obtained by molecular mechanics energy minimization using AMBER forcefield. The total nonbonded potential interaction energy of the simulated bio-nanocomplex consisted of long-range Coulombic interactions between partial atomic charges and van der Waals interactions, computed using the Ewald summation technique. A direct cutoff radius of 8 Å was used to treat short-range, repulsive van der Waals interactions.

B . *Forcefield and MD conditions*

The intercalated molecules can be anionic structures such as Cl^-, NO_3^- or DNA strands as counterions. The simulations of the modified clay model were carried out using molecular dynamics software AMBER 7. This pioneering calibration study compares favorably with a previous experimental study from Xu *et al.*[7] of the LDH containing NO_3^- and serves as a guide for forcefield parameters for the bio-nanohybrid complex structure. The MD models are inserted into cubic boxes for periodic boundary conditions. We are using parameters described in Table 1 for LDH and AMBER forcefield parameters for the DNA. We only take into account the long range electrostatic interaction for the interatomic interactions.

TABLE 1. PARAMETERS USED TO DESCRIBE INTERATOMIC INTERACTIONS BETWEEN LAYERED DOUBLE HYDROXIDES AND THE GUEST INTERCALATES[8]

	$D_{O,i}$ (kcal/mol)	$R_{O,I}$ (Å)	q_i(e)
Mg	9.03E-07	5.91	1.36
Al	1.33E-06	4.79	1.5752
O	0.1554	3.55	-1.0069
H	-	-	0.425

$$U=\sum_{i,j}\left\{\frac{q_i q_j}{4\pi\varepsilon_0 r_{ij}}+D_{O,ij}\left[\left(\frac{R_{O,ij}}{r_{ij}}\right)^{12}+2\left(\frac{R_{O,ij}}{r_{ij}}\right)^{6}\right]\right\}$$

$$R_{O,ij}=\frac{(R_{O,i}+R_{O,j})}{2};D_{O,ij}=\sqrt{D_{O,i}D_{O,j}}$$

Equation (1)

A. *ANALYSIS and PRELIMINARY RESULTS*

The first trend we have observed from previous results containing respectively Cl^- and NO_3^- counterions was an increase of cell parameters, a and b. The second important observation concerning NO_3^- is that it seems to prefer a planar conformation rather than a "tilted" conformation as it was suggested. NO_3^- undergoes successive rotations without really being "tilted". Early results on DNA strands suggest a huge deformation of the LDH layer with cationic metals being displaced. This is believed to be induced by high-energy states associated with the ordered starting conformation of the Brucite structure coupled with random insertion of DNA within the layer galleries. More refinement of the starting structures is needed in order to approach closer to a minimum energy state.

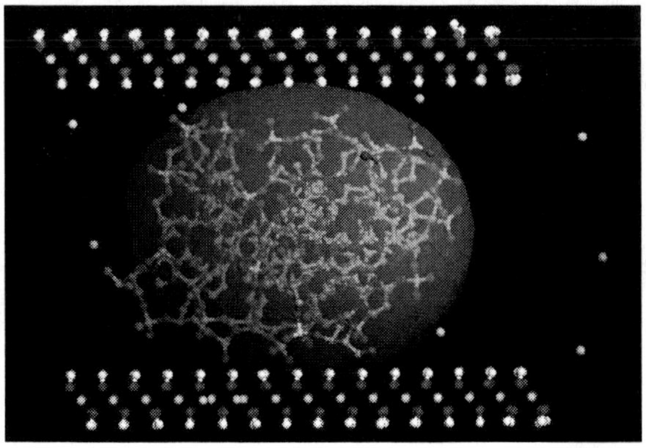

[1] Allmann, R.; Jepsen, H. P., "Structure of hydrotalcite. "*Neues Jahrbuch fuer Mineralogie, Monatshefte* **1969**, (12), 544-51.

[2] Allmann, R., "Crystal structure of pyroaurite. "*Acta Crystallographica, Section B: Structural Crystallography and Crystal Chemistry* **1968,** 24, (Pt. 7), 972-7.

[3] Kwak, S.-Y.; Jeong, Y.-J.; Park, J.-S.; Choy, J.-H., "Bio-LDH nanohybrid for gene therapy. "*Solid State Ionics* **2002,** 151, (1-4), 229-234.

[4] Boek, E. S.; Coveney, P. V.; Skipper, N. T., "Monte Carlo Molecular Modeling Studies of Hydrated Li-, Na-, and K-Smectites: Understanding the Role of Potassium as a Clay Swelling Inhibitor. "*Journal of the American Chemical Society* **1995,** 117, (50), 12608-17.

[5] Choy, J.-H.; Park, J.-S.; Kwak, S.-Y.; Jeong, Y.-J.; Han, Y.-S., "Layered double hydroxide as gene reservoir. "*Molecular Crystals and Liquid Crystals Science and Technology, Section A: Molecular Crystals and Liquid Crystals* **2000,** 341, 425-429.

[6] Newman, S. P.; Di Cristina, T.; Coveney, P. V.; Jones, W., "Molecular Dynamics Simulation of Cationic and Anionic Clays Containing Amino Acids. "*Langmuir* **2002,** 18, (7), 2933-2939.

[7] Xu, Z. P.; Zeng, H. C., "Abrupt Structural Transformation in Hydrotalcite-like Compounds Mg1-xAlx(OH)2(NO3)x.nH2O as a Continuous Function of Nitrate Anions. "*Journal of Physical Chemistry B* **2001,** 105, (9), 1743-1749.

[8] Fogg, A. M.; Rohl, A. L.; Parkinson, G. M.; O'Hare, D., "Predicting Guest Orientations in Layered Double Hydroxide Intercalates. "*Chemistry of Materials* **1999,** 11, (5), 1194-1200.

Quantitative Analysis of HBV Capsid Protein Geometry Based Upon Computational Nanotechnology

Harish Chandra Soundararajan* and Muthukumaran Sivanandham**

* Department of Electronics and Communication Engineering, Email: harish.chandra.s@gmail.com
** Department of Biotechnology, Email: msiva@svce.ac.in
Sri Venkateswara College of Engineering Pennalur, Sriperumbudur - 602105 :: Tamil Nadu, India.

ABSTRACT

Computational biology is an interdisciplinary field that applies the techniques of computer science, functional mathematics and statistics to address problems inspired by biology. Numerous studies have shown that structure of proteins contribute towards their functionality in a direct or indirect way. The structure of viral capsid proteins are critical for hosting and shielding the genetic material and are also crucial for viral entry. In this work, nano-environmental energetics of mutated Hepatitis B capsid protein (HBc) dimer was studied for determining the stability of the protein. Results provide structural information regarding the important residues contributing to Hepatitis B virion (HBV) synthesis. This work illustrates the salience of computational nanotechnology and paves the way for future pharmaceutical applications aimed at destabilization of the capsid-surface protein interactions.

Keywords: Nano-environmental Energetics, Point Mutation, Bioinformatics, Hepatitis B virus, Structure function relationship

1 INTRODUCTION

A variety of computational tools exists for molecular and structural analysis that can perform amino acid mutations, calculate H-bonds and distances between atoms quite easily [2]. Many bioinformatics tools are available to analyze structures in molecular level, to study the different interactions in nano-scale and determine the energy in this nano-environment. The different interactions that exist at this level are charge-charge interaction, hydrophilic-hydrophobic interaction, steric effects and bond stability [5]. Structure function relationship of proteins has been studied and structure has been shown to be closely related to the functionality of the protein [6]. In virions, the capsid proteins are nano-scale structural proteins that protect the nucleic acids of the virus. Hepatitis B virus (HBV) is one of the smallest enveloped animal viruses with a virion diameter of 42nm [1]. The major building block of the HBV is the core protein (HBc) [7]. The core protein contains 183 amino acids. Previous work has been done to record the dimensions of the HBc experimentally [3] [4]. Two bundles, each of length 4.2 nm, run from the inside of the shell to the extremity of the spike. The two bundles are joined at the tip of the spike by a loop. They are the anti-parallel alpha helical hairpins. Each dimer is made of four such helical hairpins which have formed the dimensions of the spike [7]. The total length of the alpha helical spike region is 12.2 nm which corresponds to 81 amino acids. The dimensions of the five alpha helical domains are taken as the five parameters of the HBc protein dimers. Parameter A contains amino acid 27 to 35 and of length ~1.5 nm, parameter B contains amino acid 48 to 78 and of length ~4.2 nm, parameter C contains amino acid 82 to 94 and of length ~4.2 nm, parameters D + E contains amino acid 94 to 110 and amino acid 112 to 128 of total length ~2.3 nm. The tips of the dimer were in contact with the envelope via protrusions that emerged from the inner surface of the envelope [9].

2 METHODS AND METHODOLOGY

2.1 Choice of Model Parameters

All measurements and nano-environmental energetics measurements were carried out in The Swiss-PdbViewer [11], a bioinformatics tool used for computational analysis. The dimension of the alpha helical regions of an individual dimer that were obtained using bioinformatics method are used for comparison between the point mutated HBc protein dimer and the HBc protein dimer present on the wild-type hepatitis B virion. Parameters obtained bioinformatically are given in table1. The HBc protein consists of four chains namely A, B, C and D. Chain C is taken for the comparative study.

2.2 Quantitative Nanoscale Energetics

Initially the five parameters defined above were measured and recorded for the wild-type HBc protein dimer for chain C. Then the point mutations, as given in table 2 were induced in chain C, in their appropriate position. The mutated dimer in this stage is in its most stable form. After this, hydrogen bonds are computed for the dimers. Next molecular surface is computed. Electrostatic potential is then computed. Coulomb's method is used for the computation of the electrostatic potential. Force field energy is then computed and the result contains the total energy with all the energy contributing factors. The summation of all these energy values, provide the total energy of the selected dimers. Nano-environmental energy minimization is then carried out for the entire structure containing the four dimers. Nano-environmental energy minimization is applied to this problem so as to obtain the changed structure after point mutation, because this energy minimization can repair distorted geometries by moving atoms to release internal constraints in nanoscale structures. Moreover, this energy minimization is usually performed by gradient optimization i.e atoms are moved so as to reduce the net forces on them. The minimized structure has small forces on each atom and therefore serves as an excellent starting point for molecular dynamics simulations. After this the final nanoscale structure is obtained and the five nanoscale parameters are measured and compared with the parameters obtained for wild-type hepatitis B virion.

2.3 Domain selection and Point Mutation

The amino acids selected to perform point mutation on the capsid dimer on chain C are listed in table 2. As it can be clearly seen from table 2, the amino acids selected to perform the point mutations are Arginine, Glutamine and Asparagine. This is because, from the force field energy calculation, these amino acids were found to contribute maximum towards the stability of the virion. Moreover, it is always ensured that the replacing amino acid causes maximum structural change for the HBc protein dimer, which is decided based on ΔG_a [8] value. A negative value of ΔG_a is selected because it is destabilizes the HBc protein.

3 RESULTS

Comparing the value obtained in table 1 for chain C with the parameters obtained after point mutation in table 3, the following observations were made. The maximum change in the five parameters combined together (A+B+C+D+E) is obtained for the point mutation Q99H (12.4113nm) followed by R82K (12.3244 nm) and then N75W (12.3137 nm). The difference in value of the five parameters combined together for chain C between mutated and wild-type HBc dimer, varies between 0.01544 nm to 0.02675nm. These five parameters represent the key geometric factor determining the dimension of the dimeric spike present on the nucleocapsid of the HBV. Even though this difference is in the order of 0.01 to 0.03 nm it has significant effect on the structural and functional effects of the HBV virion. In wild type HBV virion, it has

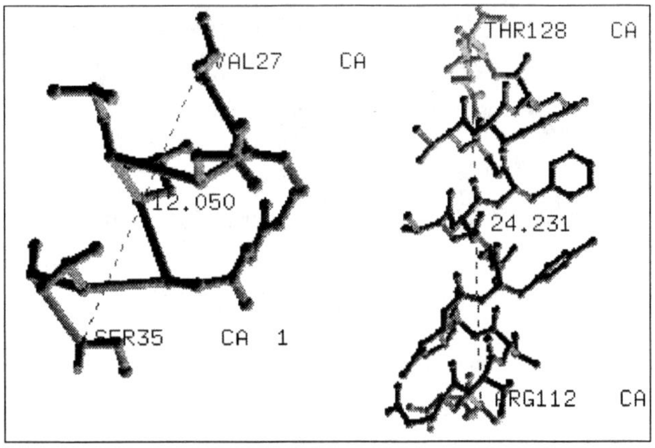

Figure 1: Screenshot providing support to data presented in table 2, parameter A (left) and parameter E (right).

been found that, the tip of the spikes contact the envelope via protrusions, emerging from the inside surface of the HBV envelope. This clearly indicated that contact is formed between the core dimeric spikes and HBs protein [9]. Image reconstruction of the various gapped and compact HBV has identified the tip of the spike as the major interface between HBc protein and the HBs protein. It has been found out that structural changes at the tip of the spike may alter the charge distribution in such a way that the affinity of the nucleocapsid to the envelop enables control of envelopment [9]. In the point mutation analysis in this paper, the dimensions of the spike vary in accordance with the various point mutations made, which in turn ensures that the position of the tip of the spike changes and hence it will affect the functionality of the hepatitis B virion. It has been found by experimental methods that point mutation in ASP78 abolished nucleocapsid formation completely [10]. Moreover, it has been shown that ASP78 is a potential site of contact between the HBc protein dimeric spike and HBs protein [9]. In our bioinformatics study, point mutation is performed in ASP78 with Tryptophan to see the change in the five parameters (table 5). Tryptophan was selected to perform the mutation since it had negative value of ΔG_a (-2.15 kcal/mol) and hence destabilizing. Any other amino acid having negative value of ΔG_a could have been selected but Tryptophan was found to produce the maximum structural change.

Table 1. Parameters of Chain C in wild-type hepatitis B virion obtained bioinformatically.

S.No	Parameter	Amino Acid	Chain	Length
1	A	27 - 35	C	1.2050nm
2	B	48 - 78	C	4.4745nm
3	C	82 - 94	C	1.7443nm
4	D	94 - 110	C	2.2969nm
5	E	112 - 128	C	2.4231nm

Table 2. Amino acid selected to perform point mutation in Chain C.

S.No	Chain	Residue ID	Amino Acid present in wild-type HBV dimer	Amino acid with which point mutation is performed
1	C	39	Arginine	Valine
2	C	57	Glutamine	Glycine
3	C	75	Asparagine	Tryptophan
4	C	82	Arginine	Lysine
5	C	99	Glutamine	Histidine
6	C	133	Arginine	Cysteine

Table 3. Results of nanoscale energetics measurements based upon point mutation analysis on chain C.

S. N 0	Chain	Residue ID	Amino Acid present in wild-type HBV dimer	Amino acid with which point mutation is performed	Accessibility of solvent (in %)	Torsion angle (φ,ψ)	Parameter A (in nm)	Para-meter B (in nm)	Parameter C (in nm)	Parameter D (in nm)	Parameter E (in nm)	Predicted ΔΔG value (kcal/mol)
1	C	39	Arginine	Valine	33.19	-34.1°, -74.9°	1.2348	4.5223	1.6963	2.3604	2.4844	-2.44
2	C	57	Glutamine	Glycine	6.72	-45.2°, -49.8°	1.2514	4.5213	1.6962	2.3607	2.4844	-3.35
3	C	75	Asparagine	Tryptophan	49.86	-153.0°, -2.4°	1.2515	4.5218	1.6953	2.3607	2.4844	-2.57
4	C	82	Arginine	Lysine	27.51	-54.9°, -37.4°	1.2464	4.5476	1.6888	2.3585	2.4811	-0.62
5	C	99	Glutamine	Histidine	12.88	-62.1°, -45.7°	1.2516	4.5217	1.6930	2.4607	2.4843	-1.62
6	C	133	Arginine	Cysteine	24.45	-145.4°, 159.8°	1.2514	4.5218	1.6962	2.3601	2.4833	-1.97

The change in the dimensions of the dimer maybe one of the potential reasons that point mutation at ASP78 abolishes the nucleocapsid formation and envelopment. On comparing the parameters obtained in table 5 with those of wild-type HBc protein dimer. It can be seen that the overall dimension of the spike increases by .1722 nm. It indicates that change in overall dimension of .1722 nm can lead to abolishment of nucleocapsid envelopment. But in our point mutation studies, dimensional changes up to .25 nm has been achieved and recorded (table 6). As it can be observed from table 5, the differences in parameters are always below .065 nm.

Whereas it is observed from table 6 that almost all mutations produce equivalent and in many cases greater change than the change produced by point mutation at D78R. In table 6, Δ(parameter A,B,C,D,E) represent the change in dimension of that parameter (in Å) between the

Table 4. Results obtained for point mutation D78W on chain C.

S.No	Parameter	Amino Acid	Chain	Length
1	A	27 - 35	C	1.2516 nm
2	B	48 - 78	C	4.5300 nm
3	C	82 - 94	C	1.6877 nm
4	D	94 - 110	C	2.3605 nm
5	E	112 - 128	C	2.4843 nm

mutated HBc capsid protein dimer and the dimer present in wild-type HBV virion. It has been shown that structural way that the affinity of the nucleocapsid to the envelop enables control of envelopment and formation [9]. Moreover, point mutation in ASP78 abolished nucleocapsid envelopment and formation [10]. From the above two facts it is clear that differences as small as 0.065 nm influences the suggested morphology and possibly the functionality of HBV in terms of nucleocapsid envelopment and formation. It is understood from table 6 that the point mutations taken in this paper produce structural changes for the HBc protein dimer by changing the position of the tip. The dimension of the spike determines the point of contact of the dimeric spike with the HBs protein. Hence the change in the dimension of the spike also varies the point of contact of the dimeric spike with HBs protein. It can be seen that the dimension of the spike contributes towards the functional attributes of the HBV.

4 CONCLUSION

In this computational approach for determining the structural basis for stability of the capsid protein, nanoscale structural changes have been observed in accordance with the point mutations made. It has been shown with HBV as an example. Nano-environmental energetics helped us in determining the dimensional changes of the alpha helical

Table 5. Comparison of parameters between wild-type HBV and point mutated (D78W) HBV on chain C with difference in parameters.

S.No	Parameter	Wild-type HBV dimer	D78R	Difference in parameter
1	A	1.2050 nm	1.2516 nm	0.0466 nm
2	B	4.4745 nm	4.5300 nm	0.0555 nm
3	C	1.7443 nm	1.6877 nm	0.0566 nm
4	D	2.2969 nm	2.3605 nm	0.0636 nm
5	E	2.4231 nm	2.4843 nm	0.0631 nm

region after point mutation. This study has further shown that experimental HBc protein point mutations change the morphological and possibly the functional property of HBV. It is important to observe these experimental point mutations in the HBV obtained from patients. Based on these observations, therapeutic molecules can be designed to treat HBV infection.

This study further strengthens the fact that structure of the protein is possibly related to its functionality. Nano-scale dimensionality changes of the alpha helical region have been observed with the help of nano-environmental energetics after point mutation. Computational nanotechnology has been successfully used to determine the key geometric domains contributing towards the structural stability of capsid protein with HBV as an example. Laboratory experiments have to be done to confirm this study, which may provide information on the utility of the changes in the diagnostic and therapeutic applications.

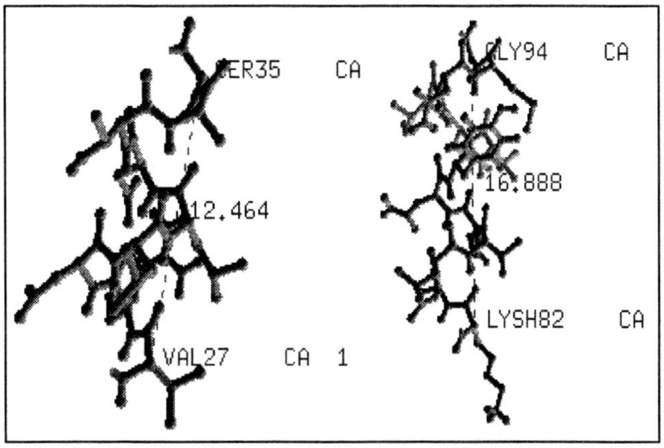

Figure 2: Screenshot providing data for point mutation R82K on chain C. Parameter A (left) and parameter C.

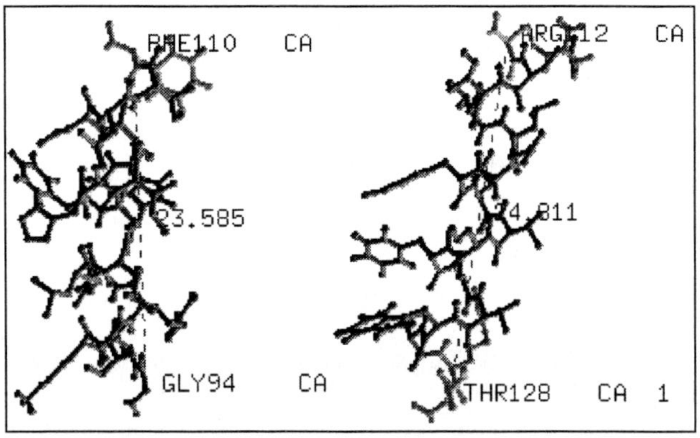

Figure 3: Screenshot providing data for point mutation R82K on chain C. Parameter D (left) and parameter E.

Table 6. Change in parameters for point mutations done on chain C.

Point Mutation	Chain	ΔA (in nm)	ΔB (in nm)	ΔC (in nm)	ΔD (in nm)	ΔE (in nm)
R39V	C	.0298	.0478	.0480	.0635	.0613
Q57G	C	.0464	.0468	.0481	.0638	.0613
N75W	C	.0465	.0473	.0490	.0638	.0613
R82K	C	.0414	.0731	.0555	.0616	.0580
Q99H	C	.0466	.0472	.0513	.0638	.0612
R133C	C	.0464	.0473	.0481	.0632	.0602

REFERENCES

[1] AJ (1996). Hepatitis Viruses. In: Baron's Medical Microbiology (Baron S et al, eds.), 4th ed., Univ of Texas Medical Branch. ISBN 0-9631172-1-1.

[2] Algorithms on Strings, Trees, and Sequences: Computer Science and Computational Biology, D Gusfield 1997

[3] Böttcher B, Wynne SA, Crowther RA.: Determination of the fold of the core protein of hepatitis B virus by electron cryomicroscopy. Nature. 1997 Mar 6;386(6620):88-91.

[4] Wynne SA, Crowther RA, Leslie AG.: The crystal structure of the human hepatitis B virus capsid. Mol Cell. 1999 Jun;3(6):771-80.

[5] Nanotechnology: convergence with modern biology and medicine, Curr Opin Biotechnol. 2003 Jun;14(3):337-46.

[6] Zheng J, Schödel F, Peterson DL.: The structure of hepadnaviral core antigens. Identification of free thiols and determination of the disulfide bonding pattern. J Biol Chem. 1992 May 5;267(13):9422-9.

[7] Zlotnick A, Cheng N, Conway JF, Booy FP, Steven AC, Stahl SJ, Wingfield PT.: Dimorphism of hepatitis B virus capsids is strongly influenced by the C-terminus of the capsid protein. Biochemistry. 1996 Jun 11;35(23):7412-21.

[8] Parthiban V, Gromiha MM, Hoppe C, Schomburg D. Structural analysis and prediction of protein mutant stability using distance and torsion potentials: role of secondary structure and solvent accessibility. Proteins. 2007 Jan 1;66(1):41-52.

[9] Seitz S, Urban S, Antoni C, Böttcher B.: Cryo-electron microscopy of hepatitis B virions reveals variability in envelope capsid interactions. EMBO J. 2007 Sep 19;26(18):4160-7. Epub 2007 Aug 30.

[10] Ponsel D, Bruss V.: Mapping of amino acid side chains on the surface of hepatitis B virus capsids required for envelopment and virion formation. J Virol. 2003 Jan;77(1):416-22.

[11] Guex N, Peitsch MC.: SWISS-MODEL and the Swiss-PdbViewer: an environment for comparative protein modeling. Electrophoresis. 1997 Dec;18(15):2714-23.

Assembly of Nanoscale Scaffolds from Peptide Nucleic Acids (PNA)

Timothy Husk, Jr[1, 2, 4] and Donald E. Bergstrom[1, 2, 3, 4, 5]

[1]*Department of Medicinal Chemistry and Molecular Pharmacology, Purdue University, West Lafayette, Indiana 47907*
[2]*The Walther Cancer Institute, Indianapolis, Indiana 46208*
[3]*Birck Nanotechnology Center, Purdue University, West Lafayette, Indiana 47907*
[4]*Bindley Bioscience Center, Purdue University, West Lafayette, Indiana 47907*
[5]*Corresponding Author, email: bergstrom@purdue.edu Ph: (765) 494-6275*

ABSTRACT

Current models of G-quadruplex (multiple stacking G-tetrads) formation based on DNA are large structures containing at least four consecutive guanine bases. PNAs containing one to four guanines were synthesized using standard Fmoc peptide coupling conditions. Monitoring the assembly of short PNA quadruplexes with variable cation concentrations as well as with various cations shows an independence of PNA G-quadruplexes on cation concentration with short sequences. The strategy for synthesis allows modification of both ends of a PNA sequence. This in theory would allow one to synthesize multivalent nanoparticles containing up to eight additional substituents. A quadruplex containing four strands with four guanine residues in each strand occupies a volume of only about 8 nm^3. Consequently, depending on the modifications, this approach should allow directed-assembly of nanomaterials with volumes in the 10 to 100 nm range.

Keywords: peptide nucleic acid, quadruplexes, scaffold, self-assembly

1 DNA AS A SCAFFOLD

Deoxyribonucleic acid or DNA has been demonstrated to be a useful scaffold for self-assembling nanomaterials [1-5]. Using complementary and non-complementary sequences of single stranded DNA (ssDNA) self-assembly through hybridization yields an unlimited variety of higher order structures. The ability to direct assembly in three dimensions through the design of linear sequences is especially appealing. DNA can also be utilized as a scaffold for the direction of organic reactions through appropriate arrangement of reactants linked to DNA strands. DNA facilitated reactions include conjugate addition, reductive amination, amine acylation, oxazolidine formation, nitro aldol, nitro Michael, Wittig, 1,3-nitrone cycloaddition, Huisgen cycloaddition, and Heck coupling [3].

2 PNA AND QUADRUPLEXES

Peptide nucleic acids, or PNA, which retains the nucleic acid bases adenine, cytosine, guanine, and thymine on a pseudo-peptide backbone, can provide a unique scaffold for the construction of biologically functional nanomaterials. PNA hybridizes to natural nucleic acids by Watson-Crick base pairing and can be used as a carrier for complementary DNA or RNA. PNA has been shown to have a lower tolerance for mismatches as well as increased stability as measured by melting temperature [2, 6, 7]. The PNA backbone, comprised of 2-aminoethylglycine unit repeats, can be modified in much the same fashion as a normal peptide backbone using solid phase synthesis methods as well as standard coupling conditions for Fmoc peptide synthesis. Synthetic modification allows for the direct incorporation of peptides and molecular probes with relative ease. In addition, PNA sequences can be designed to self-assemble into nanoparticles. Compact structures based on G-tetrad formation are currently under study in our lab. Current models of G-quadruplex (multiple stacking G-tetrads) formation based on DNA are large structures containing at least four consecutive guanine bases. Because complementary PNA strands typically self-associate with higher affinity than DNA strands, we initially focused on the tetrad assembly properties of shorter

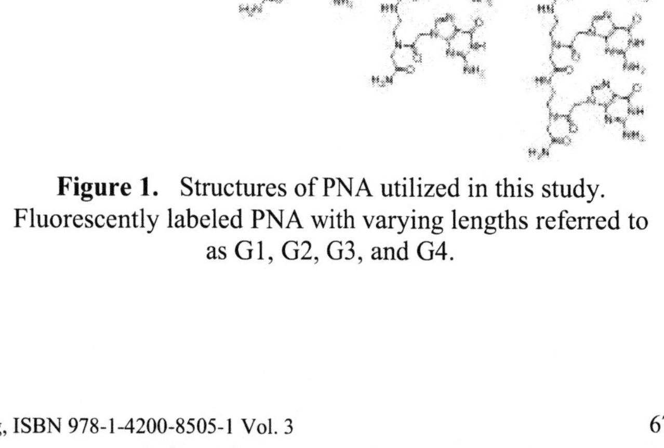

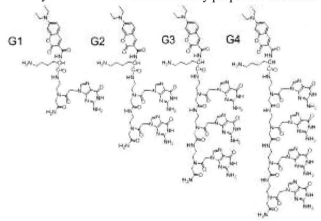

Figure 1. Structures of PNA utilized in this study. Fluorescently labeled PNA with varying lengths referred to as G1, G2, G3, and G4.

sequences. PNAs G1-G4 (Figure 1) were synthesized by the strategy mentioned above. The association and stability of the PNA G-quadruplex has several implications in the nanotechnology field. The strategy for synthesis allows modification of both ends of a PNA sequence. This in theory would allow one to synthesize multivalent nanoparticles containing up to eight additional substituents. A quadruplex containing four strands with four guanine residues in each strand occupies a volume of only about 8 nm^3. Consequently, depending on the modifications, this approach should allow directed-assembly of nanomaterials with volumes in the 10 to 100 nm range. The effects of modifications of the PNA on assembly properties are currently under study.

2.1 Monitoring Quadruplex PNA

The melting or disassociation of a G-quadruplex can be monitored by following the hypochromic effect that occurs with an increase in temperature at 295 nm [8, 9]. While this effect has been reported and studied using DNA, the same effect is also seen with PNA, since the alignment of the transition dipoles within the plane of the bases remains unchanged, even though the backbone is different. With fluorophore modified PNA we have observed this effect with sequences containing from one to four bases (PNA G1-G4, Figure 1). The thermal melting profile for these tetrads is shown in Figure 2. In turn, the assembly of G-quadruplexes can be monitored by a hyperchromic effect that occurs when the quadruplex is assembled. Monitoring the assembly of short PNA quadruplexes with variable cation concentrations as well as various cations shows an independence of PNA G-quadruplexes on cation concentration with short sequences. When comparing to the established relationship between DNA and cation concentration, DNA exhibits a strong dependence and PNA does not exhibit this dependence [10].

3 STABILITY OF PNA QUADRUPLEXES

For analysis of the melting of the synthesized PNA sequences, the sequences were diluted into 10 mM sodium cacodylate buffer pH 7.2 containing 100 mM sodium chloride just before starting the melting experiments. This particular buffer was chosen because it does not absorb in the UV region in question and it does not exhibit a shift in pKa with changes in temperature. The UV absorbance was recorded at a wavelength of 295 nm as the temperature was increased from 25°C to 90°C and subsequently cooled back to 25°C. After holding the sample at 25°C for 100 minutes, the sample was again heated to 90°C. All of the heating and cooling ramps were conducted at a rate of 0.2°C/min. At this wavelength, G-quadruplexes exhibit a hypochromic effect as the temperature is increased to and above the melting temperature of the structure. This unique characteristic of the spectrum can be used to identify and monitor the melting or unfolding of the quadruplex. Various concentrations of PNA strands were chosen for analysis. These solutions were made just before starting the melting experiments.

3.1 Thermal Stability

Initially all of the sequences tested showed the ability to form quadruplexes as evident from the melting curves showing the characteristic hypochromic effect at 295 nm (Figure 2). The initial slope of the melting curves indicates that the addition of guanine bases adds to the low temperature stability of the quadruplex, but the melting temperature is, interestingly enough, unaffected. This would indicate that the mechanism of dissociation is unaffected by the length of the G-tract. This finding is contrary to what has been shown with DNA G-quadruplexes [10, 11]. The properties of PNA_4 quadruplexes are not yet known. It is known that PNA_4 quadruplexes can form and be analyzed via electro-spray ionization mass spectrometry with a guanine tract containing only three guanine residues [12]. In this report, it should be noted that the PNA quadruplex studied displayed a $T_{1/2}$ of 24°C. It should also be noted that this structure is much different in that the lysine residue is placed at the carboxy-terminus as well as the presence of the free acid. The structures that have been synthesized and studied in our laboratory thus far have shown a remarkable increase in stability over the PNA that has been studied previously. DNA G-quadruplexes are stabilized by monovalent cations and destabilized by small cations, such as lithium ions. Because it has been reported that PNA does not follow this trend, at least with long sequences, this concept was then tested with the short PNA sequences.

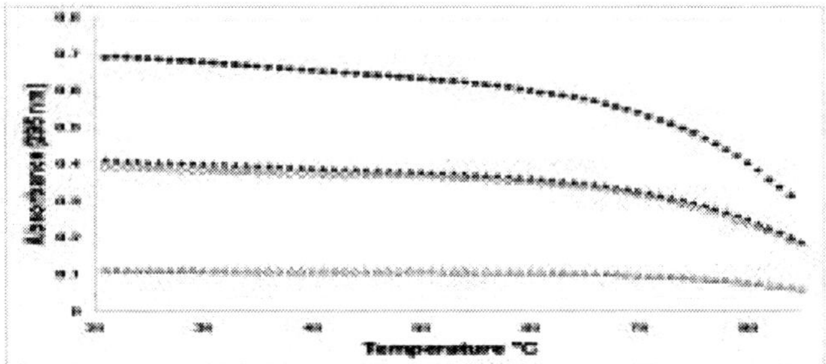

Figure 2. Melting curves for PNA of varying lengths at 100 μM. G1 (filled circles, •), G2 (filled triangles, ▲), G3 (open circles, ○), and G4 (open triangles, Δ). It is evident from the hypochromic effect seen at higher temperatures that a quadruplex is formed in all cases.

4 ION EFFECTS

The stability of the resulting PNA quadruplexes has been tested under various conditions. It has been reported that for PNA quadruplexes containing at least three guanine bases, cation concentration does not add to the stability of the complex [12]. Structures of assembly with less than three guanines have not previously been studied. Also it should be noted that the thermodynamic properties of PNA quadruplexes of such short sequences has not been determined, as is the same with the kinetic parameters of PNA quadruplex assembly.

4.1 Ion Concentration And Assembly

Kinetic data for assembly of quadruplexes at 22°C was recorded. PNA samples that contained no quadruplex component were used for this analysis. The PNA of varying lengths (G2 and G3) was diluted into 10 mM cacodylate buffer pH 7.2. Chloride salts containing either sodium or potassium were added to the sample so that the final concentration was one of the following: 50, 100, 200, 300, 400, or 500 mM. Upon addition of the sample to the salt buffer, the absorbance at 295 nm was recorded. Absorbance readings were taken for 16 hours, and following the conclusion of the time recordings, plots were constructed of the absorbance versus time. These resulting plots showed that no sharp increases in absorbance were recorded. This indicates that the cation concentration did not accelerate the assembly of the quadruplexes at 22°C on this time scale. A slight increase in absorbance was noted in the PNA G3 sample, but this increase was not significant enough to indicate dependence on cation concentration.

4.2 Temperature And Assembly

Because lower temperatures have been shown to increase the rate of quadruplex assembly in DNA [9, 10], the kinetic experiments were repeated at 3.5°C. This decrease in temperature should aid in the assembly of the PNA quadruplexes if the same effect with DNA is carried over to the PNA constructs. Again the PNA samples were added to 10 mM sodium cacodylate buffer containing 100, 300, or 500 mM sodium chloride. The absorbance readings were recorded every five minutes for 16 hours. Once all of the readings were recorded, it was noted that again there were no increases in absorbance for any of the concentrations used with any of the PNA samples (Figure 3), which indicates that the decrease in temperature again did not accelerate the assembly of the PNA quadruplexes on this time scale.

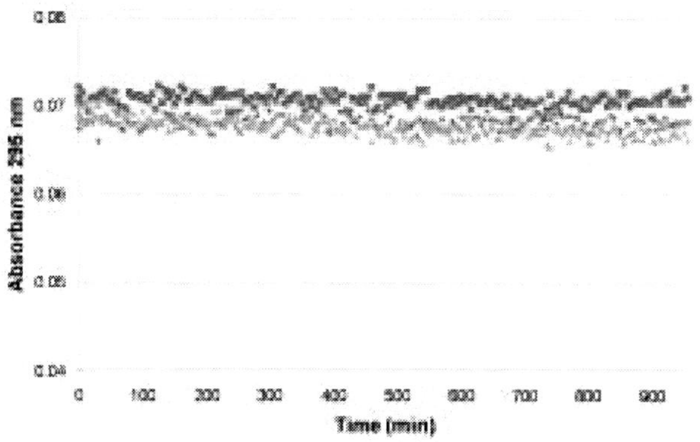

Figure 3. Representative plot (G2 shown) of the assembly of PNA quadruplexes from a 100 μM solution with various sodium ion concentrations (100 mM NaCl Blue Diamond, 300 mM NaCl Red Square, and 500 mM NaCl Green Triangle). It is apparent that there is no assembly occurring at 3.5°C on this time scale.

4.3 Ion Size And Assembly

Because ion size is important for stabilizing quadruplex DNA, the size of the cations present in solution was probed. Again PNA samples were diluted into 10 mM sodium cacodylate buffer pH 7.2 containing 100 mM lithium chloride (destabilizes quadruplex DNA), 100 mM sodium chloride (stabilizes quadruplex DNA), or 100 mM potassium chloride (highly stabilizes quadruplex DNA). The absorbance was again recorded every five minutes for 16 hours at 295 nm. The resulting plots that were obtained indicated that PNAs G1 and G2 did not assemble at 3.5 °C regardless of the cation size. These plots showed no increase or decrease in absorbance over the course of the experiment; however, PNA G3, and to a much larger degree PNA G4, showed a decrease in absorbance over the time of the experiment. The slight decrease was most evident in PNA G4 samples that contained potassium (Figure 4). Interpretation of this result will require additional studies.

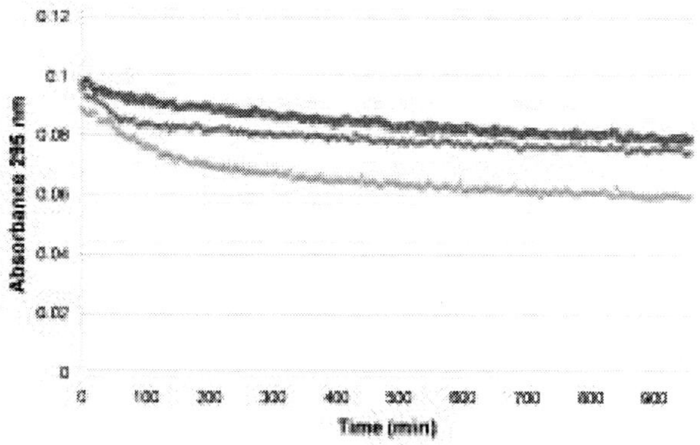

Figure 4. Representative plot (G4 shown) of the assembly of PNA quadruplexes from a 100 μM solution with various cation concentrations (100 mM LiCl Blue Diamond, 100 mM NaCl Red Square, and 100 mM KCl Green Triangle). It is apparent that there is no assembly occurring at 3.5°C on this time scale.

4.4 Ion Concentration And Stability

Because the concentration of the cation did not accelerate the assembly of the PNA quadruplexes at temperatures down to 3.5 °C and time periods of the order of 24 hours, additional experiments were designed to test the effect of salt concentration on the melting curve of the PNA quadruplexes. For this set of experiments, the PNA samples were stored for one week at -20°C in the Tm buffer (10 mM sodium cacodylate buffer pH 7.2) containing 100, 300, or 500 mM sodium chloride to ensure the formation of the quadruplexes took place in the cation concentrations desired. The samples were then thawed at room temperature for 10 minutes before the start of the experiment. The UV absorbance was recorded at a wavelength of 295 nm as the temperature was increased from 25°C to 90°C and subsequently cooled back to 25°C. After holding the sample at 25°C for 100 minutes, the sample was again heated to 90°C. All of the heating and cooling ramps were conducted at a rate of 0.2°C/min. Upon conclusion of this set of experiments, the absorbance readings were plotted as a function of temperature. It was evident from these plots that the cation concentration had no effect on the temperature at which the apparent melting of the PNA quadruplexes occurred, and had no effect on PNA G1 or PNA G2 at all. However, in the temperature range 25 to 40 °C PNA G3 showed an increase in the absorbance at 295 nm suggesting a cation dependent alteration in the quadruplex PNA structure (Figure 5). The increased absorbance remained constant at this higher level until the apparent melting temperature was reached.

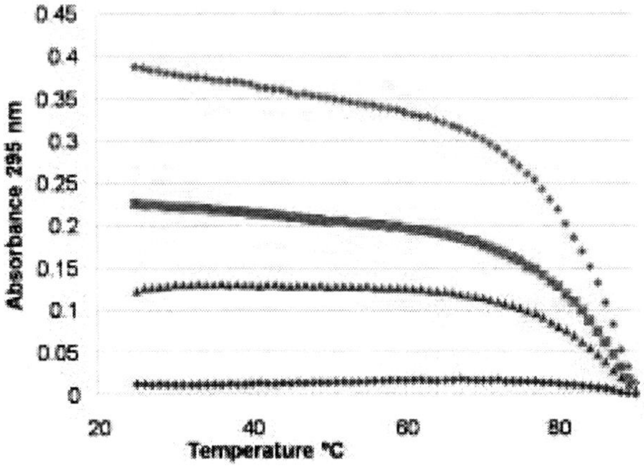

Figure 5. Representative plot of the dissociation or melting of PNA quadruplexes from a 100 µM solution formed in 300 mM sodium ion solutions (PNA G1 Green Diamond, PNA G2 Red Square, PNA G3 Red Triangle, and PNA G4 Purple Diamond).

5 CONCLUSIONS

This is the first report of short sequences of PNA, less than three guanine residues, forming quadruplex structures. The structures that are formed are four-stranded, and they are extremely stable once formed. The stability is demonstrated by the apparent melting temperatures which are greater than 80°C. The assembly of the quadruplexes on a reasonable time scale requires that the samples be frozen. Modification of both ends of the PNA will allow attachment of up to eight individual ligands, which will allow exploitation of the multivalence effect on binding target proteins. It may be possible to decrease the size of the scaffold necessary to produce this arrangement to a single guanine residue given the right conditions, resulting in the minimum possible scaffolding.

REFERENCES

1. Sakurai, K., T.M. Snyder, and D.R. Liu, *DNA-templated functional group transformations enable sequence-programmed synthesis using small-molecule reagents.* Journal of the American Chemical Society, 2005. **127**(6): p. 1660-1661.
2. Ng, P.S. and D.E. Bergstrom, *Alternative nucleic acid analogues for programmable assembly: Hybridization of LNA to PNA.* Nano Letters, 2005. **5**(1): p. 107-111.
3. Li, X.Y. and D.R. Liu, *DNA-Templated organic synthesis: Nature's strategy for controlling chemical reactivity applied to synthetic molecules.* Angewandte Chemie-International Edition, 2004. **43**(37): p. 4848-4870.
4. Cao, R., et al., *Synthesis and characterization of thermoreversible biopolymer microgels based on hydrogen bonded nucleobase pairing.* Journal of the American Chemical Society, 2003. **125**(34): p. 10250-10256.
5. Gartner, Z.J., M.W. Kanan, and D.R. Liu, *Multistep small-molecule synthesis programmed by DNA templates.* Journal of the American Chemical Society, 2002. **124**(35): p. 10304-10306.
6. Egholm, M., et al., *Peptide Nucleic-Acids (Pna) - Oligonucleotide Analogs with an Achiral Peptide Backbone.* Journal of the American Chemical Society, 1992. **114**(5): p. 1895-1897.
7. Nielsen, P.E., et al., *Sequence-Selective Recognition of DNA by Strand Displacement with a Thymine-Substituted Polyamide.* Science, 1991. **254**(5037): p. 1497-1500.
8. Guzman, M.R., et al., *Characterization of parallel and antiparallel G-tetraplex structures by vibrational spectroscopy.* Spectrochimica Acta Part a-Molecular and Biomolecular Spectroscopy, 2006. **64**(2): p. 495-503.
9. Mergny, J.L., A.T. Phan, and L. Lacroix, *Following G-quartet formation by UV-spectroscopy.* Febs Letters, 1998. **435**(1): p. 74-78.
10. Mergny, J., et al., *Kinetics of tetramolecular quadruplexes.* NUCLEIC ACIDS RESEARCH, 2005. **33**(1): p. 81-94.
11. Datta, B., C. Schmitt, and B.A. Armitage, *Formation of a PNA(2)-DNA(2) hybrid quadruplex.* Journal of the American Chemical Society, 2003. **125**(14): p. 4111-4118.
12. Krishnan-Ghosh, Y., E. Stephens, and S. Balasubramanian, *A PNA(4) quadruplex.* Journal of the American Chemical Society, 2004. **126**(19): p. 5944-5945.

Direct Numerical Simulation of Carbon Nanofibre Composites

M. Yamanoi, João M. Maia

IPC – Institute for Polymers and Composites, University of Minho, Portugal
Campus de Azurém 4800-058 Guimarães, Portugal
yamanoi@dep.uminho.pt; jmaia@dep.uminho.pt

ABSTRACT

Carbon nanofibers (CNFs) and carbon nanotubes (CNTs) are expected to yield very good mechanical properties and high thermal and electrical conductivities when embedded into polymeric matrices, especially if one can control fiber orientation and dispersion. Since it is difficult to monitor the behavior of CNF and CNT composites under flow fields, it is hoped that by realistic modeling a better insight into their structure and the consequent relations with macroscopic properties can be gained. We have thus developed a software based on the particle simulation method (PSM) considering van der Waals (VDW) and hydrodynamic (HD) interactions to analyze fiber orientation and dispersion kinetics and rheological properties under simple shear flow. The main results are the following: 1) flocculation in the flow direction is obtained when fibers are rigid; 2) flocculation in the vorticity direction is obtained when fibers are semi-flexible; 3) apparently large flocculation cannot be developed when fibers are flexible. The results of 1) and 2) match well with published experimental results.

Keywords: fiber simulation, coarse grained modeling, van der Waals interaction, hydrodynamic interaction, flocculation

1 INTRODUCTION

It is well known that carbon nanotube (CNT) and carbon nanofiber (CNF) have good mechanical, thermal, and electrical properties and are good candidates for the development of advanced composites with very good mechanical, electrical and thermal properties.

The present work details a direct fiber simulation software based on the particle simulation method (PSM) of Yamamoto et al. [1], that can treat flexible as well as rigid fibers extended in order to be applied to fiber dispersed systems by considering hydrodynamic (HD) interactions, which, to the authors best knowledge has not be developed. The software considers both effects in order to study flocculation of nano-scaled fiber dispersed systems under flow fields.

2 MODELING

This section explains the modeling of flexible fibers – intra-fiber interaction, inter-fiber interaction, HD interaction and dynamics.

2.1 Intra-fiber Interactions

Each individual fiber is modeled by joining single segments and considering stretch, bending and torsion, each of which can be expressed by the following equations [1]:

Stretching force: $$F^s = -k_s(r - r_0) \quad (1)$$

Bending torque: $$T^b = -k_b(\theta - \theta_0) \quad (2)$$

Torsion torque: $$T^t = -k_t(\theta_t - \theta_{t0}) \quad (3)$$

where k_s is the stretching force constant, k_b the bending torque constant, k_t the torsion torque constant, r the segment center-to-center distance, θ the bending angle and θ_t the torsion angle. At equilibrium r_0 is the diameter of fiber and $\theta_0 = \theta_{to} = 0$.

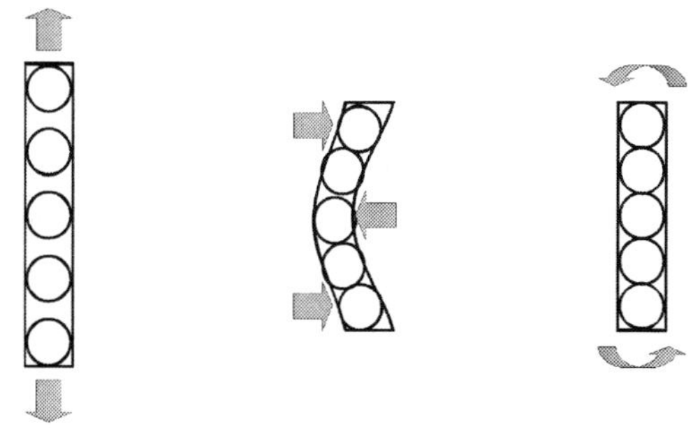

(a) stretch (b) bending (c) torsion

Figure 1: Intra-fiber Interaction

2.2 Van der Waals Interactions

Particle size is of the nano-order and thus it is necessary to consider VDW interactions in the modeling. The VDW potential, V, is expressed by the following equation:

$$V = -\frac{A}{6}\left[\frac{2}{s^2-4} + \frac{2}{s^2} + \ln\frac{s^2-4}{s^2}\right], \quad (4)$$

where A is the Hamaker constant of the system, $s = r/a$, r is the particle center to center distance and a is the radius of the particle. "A" depends on both medium and particles, and can be calculated the following equation,

$$A = \left(\sqrt{A_{11}} - \sqrt{A_{22}}\right)^2, \tag{5}$$

where A_{11} and A_{22} are the Hamaker constants for the particle and the medium, respectively. The VDW forces, $\boldsymbol{F}^{VDW}$, can be calculated by differentiating equation (4) in order to s.

2.3 Hydrodynamic Interactions

HD interaction has been studied before in colloidal dispersed systems. For calculating HD interactions, a $6N \times 6N$ sized system of equations, where N is the number of segments [2], needs to be solved for each time step. This method implies long calculation times, with PSM classifying the HD interaction into one between segments in the same fiber and one between different fibers. However, this treatment is not accurate for condensed systems. The authors [2] then introduced a different method, which does not need to solve system of equations as shown below:

$$\begin{pmatrix} \boldsymbol{F}^h \\ \boldsymbol{T}^h \end{pmatrix} = -\boldsymbol{R}\begin{pmatrix} \boldsymbol{v} - \boldsymbol{U} \\ \boldsymbol{\omega} - \boldsymbol{\Omega} \end{pmatrix}, \tag{6}$$

where $\boldsymbol{F}^h$ and $\boldsymbol{T}^h$ are the hydrodynamic force and torque, respectively, $\boldsymbol{v}$ and $\boldsymbol{\omega}$ are the velocity and the angular velocity of segments and $\boldsymbol{U}$ and $\boldsymbol{\Omega}$ are those of matrix. $\boldsymbol{R}$ is the resistance matrix. In the expression, it is not necessary to calculate the full system of equations, thereby reducing calculation time.

2.4 Collision Interaction

When VDW interaction is important collision effects should be considered and this paper follows the method PSM applied in [3]:

$$\boldsymbol{F}^c = -D_0 \exp\left[G_0\left(1 - \frac{|\boldsymbol{r}_{ij}|}{2a}\right)\right]\boldsymbol{n}_{ij}, \tag{7}$$

where a is the radius of the fiber, D_0 and G_0 are constants, and $\boldsymbol{n}_{ij} = (\boldsymbol{r}_j - \boldsymbol{r}_i)/|\boldsymbol{r}_j - \boldsymbol{r}_i|$.

2.5 Fiber Dynamics

The model considers translational and rotational behaviors of segments and a non-slip condition between segments; thus, the following equations must be solved:

$$m\frac{dv_i}{dt} = \sum_j F_{ij}^s + \sum_j f_{ij} + \sum_j F_j^{VDW} + F_i^h + \sum_j F_j^c \tag{8}$$

$$\frac{2}{5}ma^2\frac{d\boldsymbol{\omega}_i}{dt} = \sum_j \boldsymbol{T}_{ij}^b + \sum_j \boldsymbol{T}_{ij}^t + \sum_j\left(\boldsymbol{f}_{ij} \times a\boldsymbol{n}_{ij}\right) + \boldsymbol{T}_i^h \tag{9}$$

$$\boldsymbol{v}_i + a\boldsymbol{\omega}_i \times \boldsymbol{n}_{ij} = \boldsymbol{v}_j + a\boldsymbol{\omega}_j \times \boldsymbol{n}_{ji} \tag{10}$$

where $\boldsymbol{f}_{ij}$ is the tangential friction force which exerts at the constant point between paired segments in the fiber.

2.6 Fiber Orientation

The orientation of a singe fiber is characterized by the angles θ and ϕ and, the unit vector $\boldsymbol{p}$ directing along the fiber axis, as depicted in Figure 2 [4]. The components of $\boldsymbol{p}$ are given by:

$$\begin{aligned} p_1 &= \sin\theta\cos\phi \\ p_2 &= \sin\theta\sin\phi \\ p_3 &= \cos\theta \end{aligned} \tag{11}$$

The 1-axis is the flow direction, the 2-axis is perpendicular to the 1-axis in the shear plane and the 3-axis is perpendicular to the shear plane. Let ψ be a probability density function describing fiber orientation, being defined in such a way that the probability of a fiber lying within the range $\boldsymbol{p}$ and $(\boldsymbol{p} + d\boldsymbol{p})$ is equal to $\psi(\boldsymbol{p})d\boldsymbol{p}$. In this work, a second order fiber orientation tensor, is used:

$$a_{ij} = \int p_i p_j \psi(\boldsymbol{p})d\boldsymbol{p} = \langle p_i p_j \rangle, \tag{12}$$

where the bracket denotes the orientation averaging.

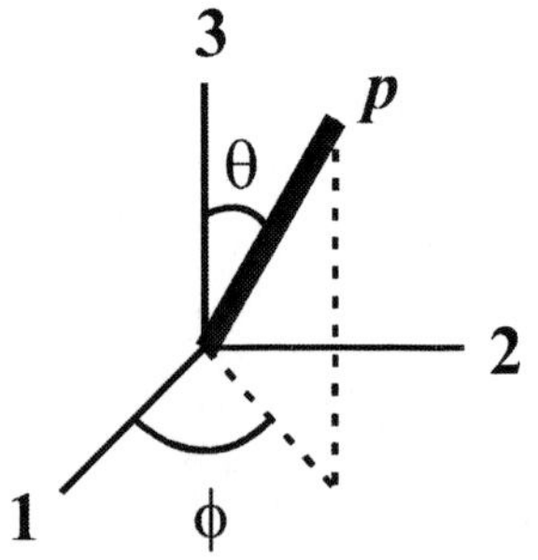

Figure 2: Definition of fiber angles

3 CALCULATION CONDITIONS

This work focuses on the flocculation structures that form under shear flows for different fiber flexibilities. Young moduli, E, of 6, 60, and 600 G Pa were selected for flexible, semi-flexible, and rigid fibers, respectively. Table 1 shows details of the model nano-fiber composites.

Volume fraction, ϕ	0.01
Radius of fiber, a	25 nm
Aspect ratio	20
Number of fiber	50
Shear rate, $\dot{\gamma}$	10
Viscosity of matrix, η_0	1000 $Pa \cdot s$
Young modulus, E	6, 60, 600GPa

Table 1: Materials parameters used for simulation.

4 RESULTS AND DISCUSSION

The randomly oriented, perfectly dispersed, no contact between fibers state shown in the figure 3 was used as an initial condition for the simulations. Figure 4 shows the snapshots of fiber structures of (a) E= 6 GPa, (b) E= 60 GPa, and (c) E= 600 GPa under simple shear flows. Figure 4a shows small round flocculation, consisting of a few fibers. There is not only flocculation but also isolated S-shaped fibers that tumble periodically along their axis. Figure 4b shows cylindrical flocculation from many fibers, with the bundles essentially aligned in the vorticity direction. Figure 4c shows flocculation parallel to the flow direction, with percolation conditions occurring. The percolation volume fraction threshold for rigid fibers of aspect ratio 20 is 0.0415 [6]. Although the present simulation is for a volume fraction 0.01, which is less than the threshold, it is conceivable that VDW interactions effectively decrease it, thus allowing percolation to occur. The behaviors shown in Figures 4b and 4c have also been observed experimentally [5].

Figure 5 shows the orientation of fibers under simple shear flows. At strain $\gamma = 0$, the components of orientation a_{xx}, a_{xx}, a_{xx} are 0.33, from which it is obvious that the initial state is an isotropically oriented one. After the application of the flow field, the maximum oriented state is achieved for every case at a strain γ around 20. After that, differences are observed, with both Figure 5a and 5b showing that a_{xx} decreases and a_{yy} and a_{zz} are gradually increasing (at strain 200 a_{yy} and a_{zz} are almost equal).

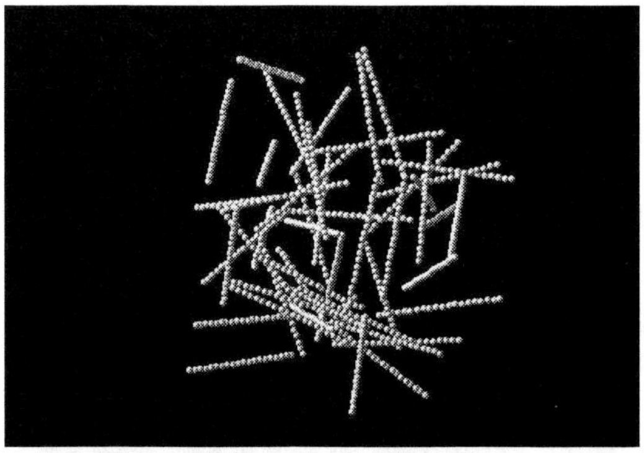

Figure 3: Initial condition of nano-fibers

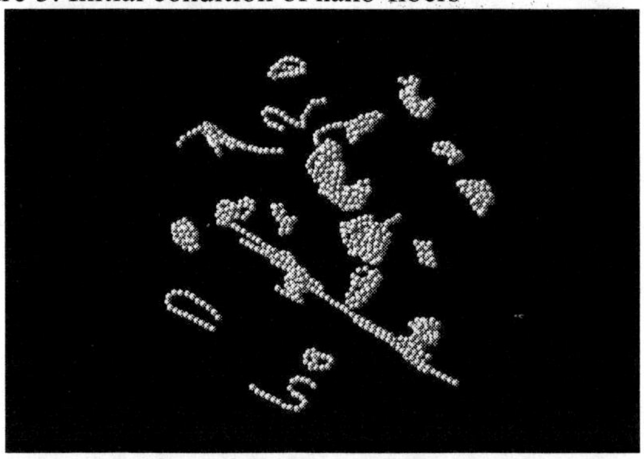

(a) E= 6 GPa

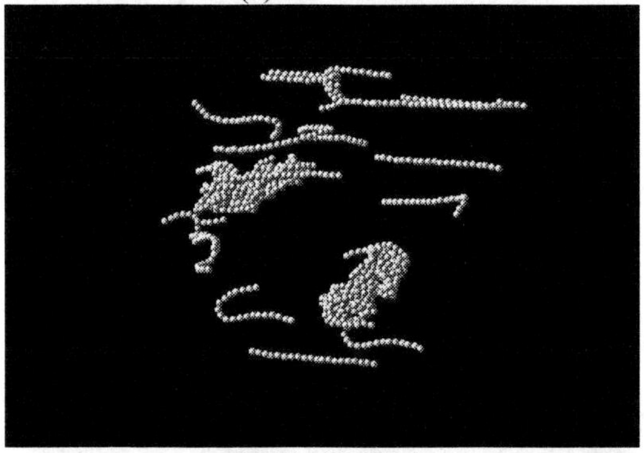

(b) E= 60 GPa

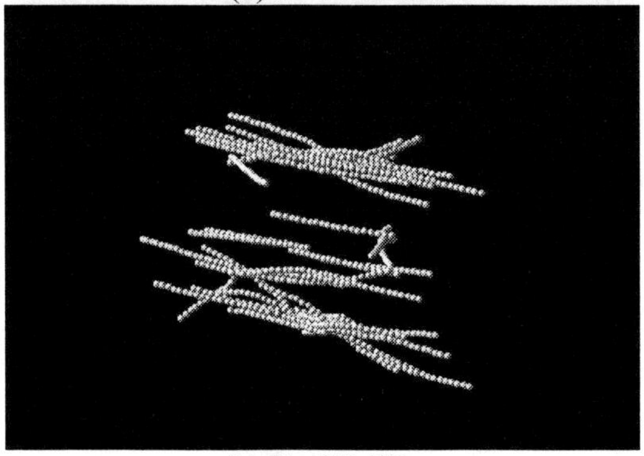

(c) E= 600 GPa

Figure 4: Aggregated structures of nano-fibers at γ=200 under simple shear flows.

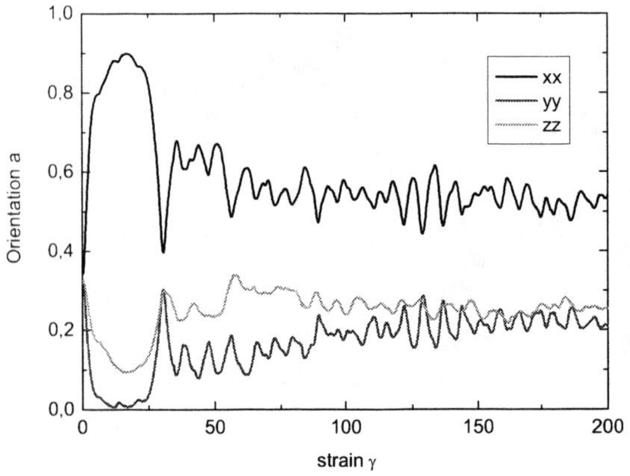

(a) E= 6 GPa

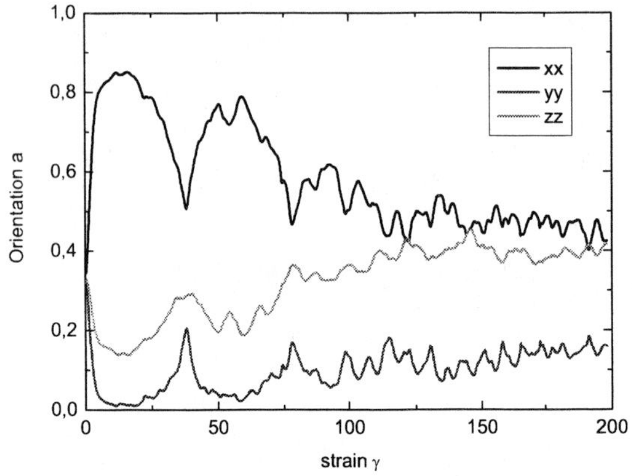

(b) E= 60 GPa

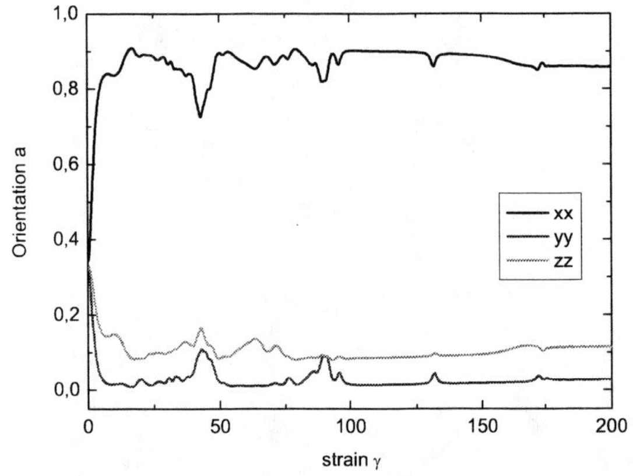

(c) E= 600 GPa

Figure 5: Development of orientation of nano-fibers under simple shear flows.

On the other hand, a_{xx} and a_{zz} are almost same in Figure 5b. In the case of Figure 5c, at small strains orientation is wavy, but after strain of approximately 100 all the components of orientation are constant, meaning that steady state has been reached. Depending on the stiffness of the fibers, development of aggregated structures and orientation are different.

CONCLUSIONS

The main conclusions of the current work are the following: 1) flocculation in the flow direction is obtained when fibers are rigid; 2) flocculation in the vorticity direction is obtained when fibers are semi-flexible; 3) large flocculation cannot be developed when fibers are flexible. The results of 1) and 2) match very well with published experimental results.

Also, in general the results show that the model is useful for analyzing nanoscale fiber dispersed systems and that fiber stiffness is an important factor in determining the flocculation structures that are formed.

REFERENCES

[1] S. Yamamoto and T. Matsuoka, J. Chem. Phys. 98, 644, 1993.
[2] L. Durlofsky, J. F. Brady and G. Bossis, J. Fluid Mech. 180, 21, 1987.
[3] S. Yamamoto and T. Matsuoka, J. Chem. Phys. 102, 2254, 1995.
[4] S. G. Advani and C. L. Tucker III, J. Rheol. 31, 751-784, 1987.
[5] A. W. K. Ma, M. R. Mackley and A. A. Rahatekar, Rheol. Acta. 46, 979, 2007.
[6] E. J. Garboczi, K. A. Snyder, J. F. Douglas and M. F. Thorpe, Phys. Rev. E 52, 819, 1995.

DNA Mechanical Properties: Formulation and Comparison with Experiments

H. Dalir*, T. Nisisako**, Y. Yanagida** and T. Hatsuzawa**

*Department of Mechano-Micro Engineering, Tokyo Institute of Technology,
R2-6, 4259 Nagatsuta-cho, Kanagawa 226-8503, Japan, dalir.h.aa@m.titech.ac.jp
**Precision and Intelligence Laboratory, Tokyo Institute of Technology,
nisisako.t.aa@m.titech.ac.jp, yanagida.y.aa@m.titech.ac.jp
Hatsuzawa.t.aa@m.titech.ac.jp

ABSTRACT

A general Newtonian elastic model for both single- and double-stranded DNA is proposed, and an explicit force-extension formula is introduced to characterize their deformations. The effective elastic properties of the DNA backbone are numerically extracted from the ss-DNA experiments. The mechanical properties of long ds-DNA molecules is then studied based on this model, where the base-stacking interactions and the hydrogen bond forces are also considered. Results are obtained by the explicit force-extension formula and good agreement with the single molecule experiments is achieved.

Keywords: double-helical DNA structure; mechanical properties; single-molecule manipulation; base-stacking interactions; hydrogen bond forces.

1 INTRODUCTION

A thorough investigation of the deformation and elasticity of DNA will enable us to gain a better understanding of many important biological processes concerned with life and growth [1]. A detailed study of DNA elasticity has now become possible due to recent experimental developments, including, e.g., optical tweezer methods, atomic force microscopy, and fluorescence microscopy [2-9]. These techniques make it possible to manipulate single polymeric molecules directly, and to record their elastic responses with a high precision.

Smith *et al.* [5] examined the elastic properties of single- and double-stranded DNA (ss-DNA/ds-DNA) by stretching the immersed lambda DNA (48 kbps) in the aqueous buffer using the dual-beam optical tweezers system. They understood a sharp structural transition of tension under roughly 65 pN of the freely rotating ds-DNA. It was also revealed that the S-form DNA occurs at the yielding point of DNA backbone while the freely rotating ds-DNA is stretched [9]. In another study, an analytical wormlike rod chain (WLRC) model for predicting DNA mechanical response under low stretching force was proposed [10] and its accuracy was then improved in a study by Sarkar *et al.* [11] by considering free energies of the five DNA states experimentally. ZZO model [1], which considers the bending energy and the stacking energy of ds-DNA can successfully describe S-form DNA under high level stretching, but is not able to represent the structural transition from the B-form DNA to the P-form DNA due to its limitation of geometric assumptions [12]. Additionally, these theoretical models mentioned above do not adequately consider the sequence effects on the ds-DNA mechanics.

In the present work, we have tried to obtain a comprehensive and quantitative sequence dependant understanding of DNA elastic mechanical properties considering its double-stranded helical nature, base-stacking interactions originated from the weak van der Waals attractions between the polar groups of DNA adjacent nucleotide base pairs and the hydrogen bond force between complementary bases.

2 THEORY

A general elastic model for both single-stranded and double-stranded biopolymers is proposed (Figs. 1 and 3(a)), and a structural angle α (see Fig. 1) is introduced to characterize their deformations. The geometry of DNA backbone is initially assumed based on the B-DNA helix function. For a single stranded lambda DNA at any point

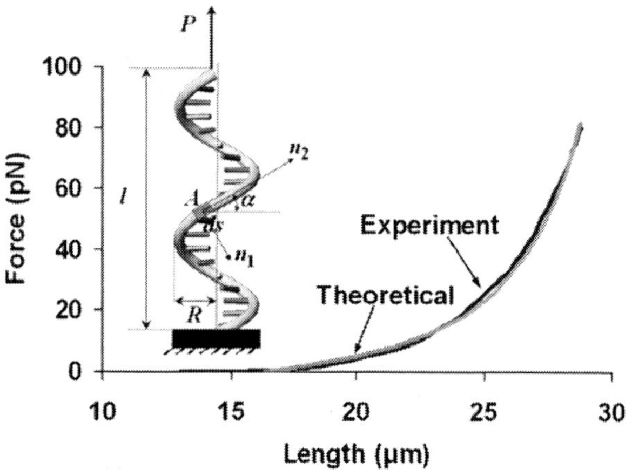

Figure 1. Experimental [5] and theoretical results of single-stranded lambda-DNA in blue and red lines, respectively.

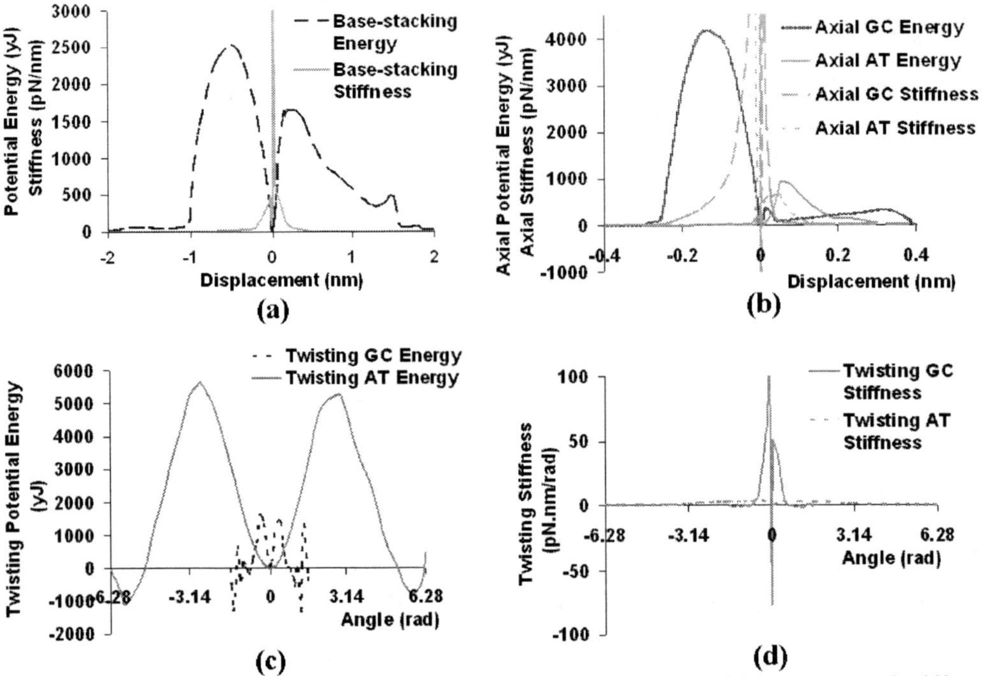

Figure 2. The stiffness of the virtual elements. (a) is the base-stacking energy and stiffness of the virtual springs between the complementary base pairs of the ds-DNA molecule. (b-d) are the AT and CG hydrogen bond axial and torsional energy and stiffness plots.

A, the force P produces a bending moment $M_b = PR\sin\alpha$ and a torque $M_t = PR\cos\alpha$ about the axis n_1 and n_2 respectively. Based on the theory of elasticity, the total energy of the ss-DNA backbone can be expressed in the following form

$$U_{ss} = \frac{1}{2}\int_0^S \left(\frac{M_b^{\ 2}}{EI} + \frac{M_t^{\ 2}}{GI_p} \right) ds , \tag{1}$$

where EI and GI_p are the elastic bending and torsional rigidity of the DNA backbone, respectively. Since the base-stacking and hydrogen bond energy of ss-DNA are negligible, the force-extension ($P - Ex_{ss}$) relation of a ss-DNA then is

$$P = \frac{4\pi^2 N^2 Ex_{ss}}{S^3 \cos^2\alpha} \left(\frac{\sin^2\alpha}{EI} + \frac{\cos^2\alpha}{GI_p} \right)^{-1} , \tag{2}$$

where N is the total number of base pairs and S is the contour length of DNA backbone. The experimental results of ss-DNA in phosphate buffer [5] were used as a benchmark to determine the unknown effective parameters of the elastic backbone. The experimental and theoretical results are shown as the blue and red lines in Fig. 1, respectively. To match the theoretical and experimental data, the effective elastic bending and torsional rigidities were set to $8521\ \text{pN.nm}^2$ and $26\ \text{pN.nm}^2$, respectively. These effective parameters are much larger than the expected ones which is mainly due to the effects of the buffer solution and the ions on the ss-DNA molecules stretching. Moreover, we assume the same effective parameters for ds-DNA molecules under the same experimental conditions which is due to their similar mechanical behaviors [5,8,9,11].

Before constructing the ds-DNA mechanical model, we should discuss another kind of important interactions, namely, the base-stacking interaction between adjacent nucleotide base pairs [13,14] and hydrogen bond interaction between complementary bases [15]. Base-stacking interactions originate from the weak van der Waals attraction between the polar groups in adjacent nucleotide base pairs. Such interactions are short ranged, and their total effect is usually described by a potential energy of the Lennard-Jones form (6-12 potential [13]) which contributes significantly to the stability of the double helix. In a continuum theory of elasticity, the total base-stacking potential energy can be converted into the form of the integration [1]:

$$U_{bs} = \int_0^S \frac{\varepsilon}{r_0} \left[\left(\frac{\sin\alpha_0}{\sin\alpha} \right)^{12} - 2\left(\frac{\sin\alpha_0}{\sin\alpha} \right)^{6} \right] ds , \tag{3}$$

where $r_0 = S/N$ is the backbone arclength between

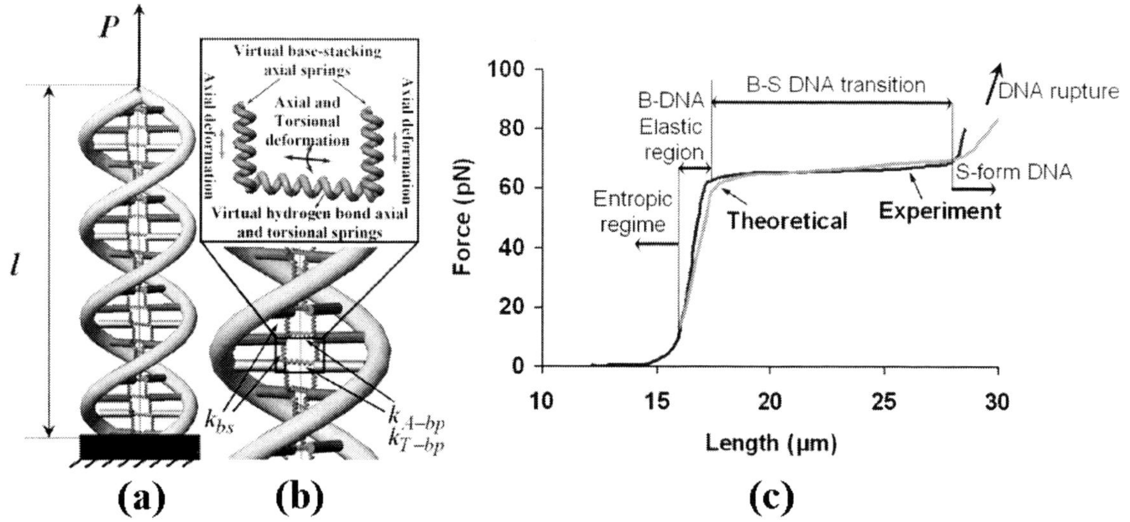

Figure 3. (a) Schematic representation of the double-stranded DNA geometry and loading conditions. (b) The location of virtual springs in the ds-DNA mechanical model. Springs parallel and perpendicular to the DNA axis represent the equivalent base-stacking and hydrogen bond interaction forces between the complementary bases along the ds-DNA double helix, respectively. (c) Good agreement was achieved between the theoretical (red line) and the experimental (blue line) results [5] for the applied forces larger than ~10 pN.

adjacent bases; α_0 is related to the equilibrium distance between a DNA dimer ($r_0 \sin\alpha_0 \approx 3.4\,\mathring{A}$); α can take the values between 0° and 90° and ε is the base stacking intensity which in the average sense can be considered as a constant, with $\varepsilon \approx 14.0\, k_B T$ averaged over quantum mechanically calculated results on all the different DNA dimers [1]. Via the Crotti-Engesser theorem [16], the base stacking stiffness ($k_{bs} = \partial^2 U_{bs}/\partial l^2$) of the Lennard-Jones potential energy could be presented as Fig. 2(a).

The hydrogen bond force is the interaction between complementary bases. Moreover, the GC base pair has three hydrogen bonds and AT has two. In the ds-DNA mechanical modeling, the three/two hydrogen bonds in GC/AT are replaced by only one virtual spring, with the axial and torsional stiffness as a function of the distance (R_i) and the angle (θ_{iHj}) between the donor and the acceptor. The single hydrogen bond energy could be expressed as [15]

$$E(R_i,\theta) = \sum_{R_{ij}} AD_0\left[5(R_0/R_i)^{12} - 6(R_0/R_i)^{10}\right]\cos^4\theta_{iHj}, \quad (4)$$

where D_0 represents the hydrogen bond energies intensity. Through the Crotti-Engesser theorem [16], the axial and torsional stiffness of AT and CG could be presented as Fig. 2 (b-d).

Considering the effects of base-stacking interactions and hydrogen bond forces, the total energy of a ds-DNA molecule under the action of an external force (Fig. 3(a)) is expressed as

$$U_{ds} = U_{backbone} + U_{bs} + U_{Ahb} + U_{Thb}, \quad (5)$$

where $U_{backbone} = \frac{1}{2}\int_0^S \left(M_b^2/EI + M_t^2/GI_p\right)ds$ is the ds-DNA backbones energy with the same ss-DNA effective elastic parameters under similar experimental conditions; $U_{bs} = \int_0^{\frac{Ex_{ds}}{N-1}} 2k_{bs}\xi(N-1)\,d\xi$ is the total base-stacking potential energy along all the adjacent nucleotide base pairs, and $U_{Ahb} = \frac{1}{\pi^2 N}\times \int_0^{Ex_{ds}} \left[\left((N_{GC}/N)k_{A\text{-}GC} + (1 - N_{GC}/N)k_{A\text{-}AT}\right)\chi\right]d\chi$ and $U_{Thb} = \int_0^{Ex_{ds}} \left[\left(N_{GC}k_{T-GC} + (1 - N_{GC})k_{T-AT}\right)\times \left(4(S^2 - \chi^2)^{-0.5}\sin^{-1}(\chi/S)\right)\right]d\chi$ are the axial and torsional parts of the total hydrogen bond energies caused by the variations in distance and angle between the donor and acceptor respectively. Because the GC and AT base pairs have three and two hydrogen bonds respectively, the ds-DNA sequence information has been embedded into the ds-DNA mechanical model based on Eq. (5). We obtain the force-extension ($P - Ex_{ds}$) relation of a ds-DNA chain under uniaxial extension by introducing the work done by the force, P, acting on the chain along its central axis explicitly into the total energy term of Eq. (5),

$$P = \frac{Ex_{ds}}{2R^2S}\left[1+\left[1-\frac{8R^2S}{Ex_{ds}^2}\left(\frac{\sin^2\alpha}{EI}+\frac{\cos^2\alpha}{GI_p}\right)\left(U_{bs}+U_{Ahb}+U_{Thb}\right)\right]^{0.5}\right]\left(\frac{\sin^2\alpha}{EI}+\frac{\cos^2\alpha}{GI_p}\right)^{-1} \quad (6)$$

3 DISCUSSIONS AND CONCLUSIONS

In the simulation, a prescribed displacement is applied to the free end of the freely rotating ds-DNA. The mechanical ds-DNA model is then solved based on Eq. (6) (see Fig. 3(c)). A good agreement was achieved between the numerical and the experimental results [5] considering that the information of ds-DNA at low applied forces (under ~10 pN) is ignored because the mechanical characteristics at large applied forces (10~80 pN) are focused. When the applied forces are lower than ~10 pN (entropic regime), the molecule behaves as an ideal (WLC) polymer of persistence length $\xi \approx 50\,\text{nm}$ [2]. Its elastic behavior is due purely to a reduction of its entropy upon stretching. From 10 pN and up to about 80 pN, DNA stretches elastically as does any material (see Fig. 3(c)), i.e., following Hooke's law: $F \approx \Delta l$ (where $\Delta l = l - l_0$ is the increase in the extension l of the molecule of contour length l_0). However, at about 65 pN a surprising transition occurs, where DNA stretches to about 1.8 times of its crystallographic length. This transition is highly cooperative, i.e., a small change in force results in a large change in extension. To address the possible structural modification in the results of DNA molecule stretching, three stages are considered here.

When the external applying force is in a range of 10~65 pN (first stage), the energy required for twisting the complementary base pairs is higher than the backbones bending and torsional energies and part of the work done by the external force acting on the chain is accumulated in the base-stacking and hydrogen bonds virtual springs. As the applied force is more increased (second stage), the virtual base-stacking springs release most of their absorbed energy because the distance between adjacent base pairs exceeds the limitation. This will enforce the base pairs and the backbones to store more energy which causes the torque of the ds-DNA local structure to overcome the backbone torsional rigidity and B-S conformational transition occurs. The exact structure of the new ds-DNA which is indeed ~80% longer than B-DNA depends on which extremities of the DNA are being pulled ($3'-3'$ or $5'-5'$). If both $3'$ extremities are being pulled, the double helix unwinds upon stretching, but in the case of pulling both $5'$ ends, the helical structure is preserved and characterized by a strong base pair inclination, a narrower minor groove and a diameter roughly 30% less than that of B-DNA [17]. In both cases, if the loading further increases (third stage), the backbones are forced to absorb more energy by stretching as a S-form DNA double helix in a linear manner (Hooke's law). Rupture of the molecule (by unpairing of the bases) has been predicted to occur as the extension is more than twice that of B-DNA [3,18,19].

In conclusion, we have constructed an elastic model for both single- and double-stranded biopolymers such as DNA molecules through experimental single-molecule manipulation. The key progress was that the bending and torsional deformations of the DNA backbones, the base-stacking interactions and the hydrogen bond force between complementary base pairs were quantitatively considered in this model. A general sequence-dependant Newtonian elastic model for both single- and double-stranded biopolymers was proposed, and an explicit force-extension formula was introduced to characterize their force-extension relationships in torsionally relaxed DNA chains. Good agreement was achieved between the theoretical results and experimental data. Based on this robust model, further study may be warranted on the mechanical response of ss- and ds-DNA molecules.

REFERENCES

[1] H. Zhou *et al.*, Phys. Rev. E 62, 1045 ,2000.
[2] S. B. Smith *et al.*, Science 258, 1122 ,1992.
[3] D. Bensimon *et al.*, Phys. Rev. Lett. 74, 4754 ,1995.
[4] P. Cluzel *et al.*, Science 271, 792 ,1996.
[5] S. B. Smith *et al.*, Science 271, 795 ,1996.
[6] T. R. Strick *et al.*, Science 271, 1835 ,1996.
[7] T. R. Strick *et al.*, Proc. Natl. Acad. Sci. USA 95, 10579 ,1998.
[8] J. F. Allemand *et al.*, Proc. Natl. Acad. Sci. USA 95, 14152 ,1998.
[9] J. F. Leger *et al.*, Phys. Rev. Lett. 83, 1066 ,1999.
[10] C. J. Benham, Phys. Rev. A 39, 2582 ,1989.
[11] A. Sarkar *et al.*, Phys. Rev. E 63, 51903 ,2001.
[12] K. N. Chiang *et al.*, Appl. Phys. Lett. 88, ,2006.
[13] W. Saenger, Principles of Nucleic Acid Structure ,Springer-Verlag, New York, 1984.
[14] J. D. Watson *et al.*, Molecular Biology of the Gene ,Benjamin/Cummings, Menlo Park, CA, 1987, 4th ed.
[15] G. A. Jeffery *et al.*, Hydrogen Bonding in Biological Structures ,Springer, Berlin, 1994, 1st ed.
[16] A. C. Ansel *et al.*, Strength and Applied Elasticity, Prentice Hall, Upper Saddle River, NJ, 2003, 4th ed.
[17] A. Lebrun *et al.*, Nucleic Acids Res. 24, 2260 ,1996.
[18] M. Wilkins *et al.*, Nature 167, 759 ,1951.
[19] A. Bensimon *et al.*, Science 265, 2096 ,1994.

Computations on $Li_x@C_{60}$

Zdeněk Slanina,*,** Filip Uhlík*** and Shigeru Nagase*

*Department of Theoretical and Computational Molecular Science, Institute for Molecular Science
Myodaiji, Okazaki 444-8585, Aichi, Japan
**Institute of Chemistry, Academia Sinica, Nankang, Taipei 11529, Taiwan-ROC
***School of Science, Charles University, 128 43 Prague 2, Czech Republic

ABSTRACT

$Li@C_{60}$ and $Li@C_{70}$ can be now produced by the low-energy bombardment method in bulk amounts and thus, their computations at higher levels of theory are also of interest. In the report, the computations are carried out on $Li@C_{60}$, $Li_2@C_{60}$ and $Li_3@C_{60}$ with the B3LYP density-functional treatment in the standard 3-21G and 6-31G* basis sets. In all three species Li atoms exhibit non-central locations relatively close to the cage. The computed energetics suggests that $Li_x@C_{60}$ species could be produced for several small x values if the Li pressure is enhanced sufficiently. This type of metallofullerenes also belongs among potential candidate agents for nanoscience applications including molecular electronics.

Keywords: Endohedral fullerenes; carbon-based nanotechnology; molecular electronics; molecular modeling; molecular electronic structure.

1 INTRODUCTION

There has been a renewed interest [1-21] in systems containing alkali metals and fullerenes, in particular $Li@C_{60}$ and $Li@C_{70}$ produced by low energy ion implantation [11,13,14] in bulk amounts. The vibrational spectra were obtained [13,14] for $Li@C_{60}$ and $Li@C_{70}$. $Li_2@C_{60}$ was also evidenced in observations [11]. This experimental progress makes computations of the species even more interesting. In the report, the computations are carried out on $Li@C_{60}$, $Li_2@C_{60}$ and $Li_3@C_{60}$ using the density-functional-theory (DFT) approach.

2 COMPUTATIONS

The geometry optimizations were carried out with Becke's three parameter functional [21] with the nonlocal Lee-Yang-Parr correlation functional [22] (B3LYP) in the standard 3-21G basis set (B3LYP/3-21G). The geometry optimizations were carried out with the analytically constructed energy gradient as implemented in the Gaussian program package [23].

In the optimized B3LYP/3-21G geometries, the harmonic vibrational analysis was carried out with the analytical force-constant matrix. In the same optimized geometries, higher-level single-point energy calculations were also performed, using the standard 6-31G* basis set, i.e., the B3LYP/6-31G* level (or, more precisely, B3LYP/6-31G*//B3LYP/3-21G). As $Li@C_{60}$ and $Li_3@C_{60}$ are radicals, their computations were carried out using the unrestricted B3LYP treatment for open shell systems (UB3LYP). The ultrafine integration grid was used for the DFT numerical integrations throughout.

Recently, Zhao and Truhlar [24-29] performed a series of test DFT calculations with a conclusion [29] that the MPWB1K functional (the modified Perdew and Wang exchange functional MPW [30] and Becke's meta correlation functional [31] optimized against a kinetics database) is the best combination for evaluations of nonbonded interactions with a relative averaged mean unsigned error of only 11% . The MPWB1K functional is also applied in this report.

3 RESULTS AND DISCUSSION

The UB3LYP approach is preferred here over the restricted open-shel ones (ROB3LYP) as the latter frequently exhibits a slow SCF convergency or even divergency. Although the unrestricted Hartree-Fock (UHF) approach can be faster, it can also be influenced by the so called spin contamination [32] and indeed, this factor was an issue in our previous [16] UHF SCF calculations as the UHF/3-21G spin contamination turned out to be higher then recommended threshold [32]. The spin contamination is measured by the expectation value for the $< S^2 >$ term that should be equal to $S(S+1)$ (i.e., singlet: $S = 0 \rightarrow < S^2 >= 0$, doublet: $S = 1/2 \rightarrow < S^2 >= 0.75$, triplet: $S = 1 \rightarrow < S^2 >= 2$, and so on). As long as the deviations from the theoretical value are smaller than 10 %, the unrestricted results are considered applicable [32]. This requirement is well satisfied in the UB3LYP/3-21G calculations as the $< S^2 >$ expectation value amounts even before annihilation to 0.7552 and 0.7546 for $Li@C_{60}$ and $Li_3@C_{60}$, respectively, i.e. less than 1% higher than the exact value.

Table 1 surveys the computed structural parameters (see Fig. 1) and formal charges in both functionals, B3LYP and MPWB1K, though the results are similar. In all three cases the Li atoms in the optimized structures are shifted from the cage center towards its wall. In particular, in the $Li@C_{60}$ species the shortest computed Li-C distance is 2.26 Å while in a central location (which is a saddle point) the shortest Li-C distance at the UB3LYP/3-21G level is 3.49 Å. Obviously, there must be substantial symmetry reductions owing to the off-centric locations. The symmetry distortion can be seen in the values of the computed rotational constants A, B, C (the rotational constants are proportional to the reciprocal values of the moments of inertia). In the empty fullerene C_{60}, all the three terms are exactly the same owing to the icosahedral symmetry of the cage, namely 0.08331 GHz at the B3LYP/3-21G level. How-

** *The corresponding author e-mail: zdenek@ims.ac.jp*

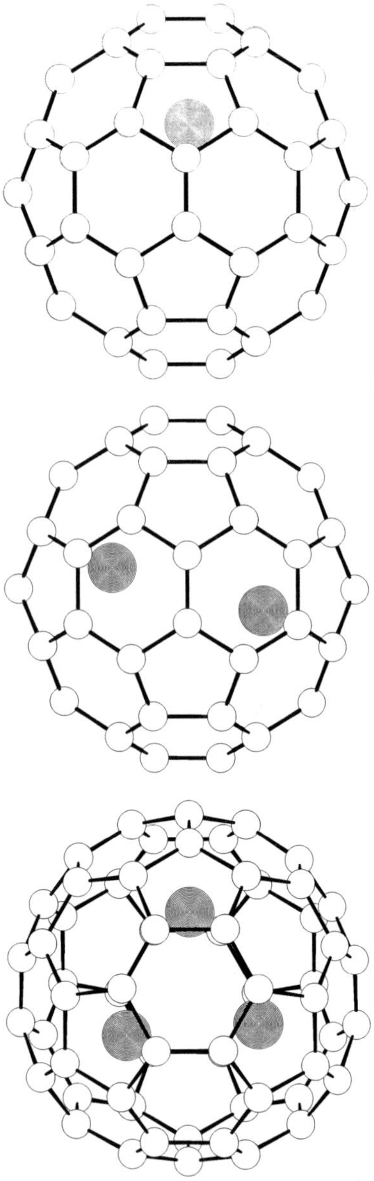

Fig. 1. B3LYP/3-21G optimized structures of $Li_x@C_{60}$ (the Li atoms are darkened).

ever, for $Li@C_{60}$ the UB3LYP/3-21G rotational constants differ to some extent: 0.08305, 0.08296 and 0.08273 GHz. As for the energetics of the centric and off-centric location, the saddle point is placed by some 9.9 kcal/mol higher at the UB3LYP/3-21G level. However, the energy separation is further increased in the UB3LYP/6-31G*//UB3LYP/3-21G treatment, namely to 15.0 kcal/mol.

In the $Li_2@C_{60}$ case, the shortest Li-C distance is even bit shorter, 2.14 Å. On the other hand, the Li-Li separation is computed as 3.29 Å, i.e., substantially longer than the observed value in the free Li_2 molecule (2.67 Å, cf. refs. [33-35]). In the third species, $Li_3@C_{60}$, the shortest computed Li-C contact is even further reduced to 2.05 Å. The Li-Li distances in the encapsulated Li_3 cluster are not equal - they are computed as 2.70, 2.76 and 2.84 Å. Incidentally, while the observed Li-Li

Table 1. B3LYP/3-21G and MPWB1K/3-21G computed shortest Li distances and charges on Li in $Li_x@C_{60}$

Species	Li-C [a] (Å)	Li-Li [b] (Å)	Charge[c]
B3LYP			
$Li@C_{60}$	2.26		1.16
$Li_2@C_{60}$	2.14	3.29	1.10
$Li_3@C_{60}$	2.05	2.70	0.86
MPWB1K			
$Li@C_{60}$	2.27		1.22
$Li_2@C_{60}$	2.10	3.22	1.16
$Li_3@C_{60}$	2.10	2.58	0.90

[a]The shortest Li-C distance. [b]The shortest Li-Li distance. [c]The formal Mulliken charge on Li (the largest value).

distance for free Li_2 is [33-35] 2.67 Å, the B3LYP/3-21G computed value is 2.725 Å (it changes to 2.723 Å at the B3LYP/6-31G* level). Similarly, also the observed values for the free Li_3 cluster are available [36,37], actually for two triangular forms - opened (2.73, 2.73, 3.21 Å) and closed (3.05, 3.05, 2.58 Å). The UB3LYP/3-21G computed distances in the free Li_3 opened cluster are 2.78, 2.78, and 3.30 Å. Hence, there is a good theory-experiment agreement.

Table 1 also presents the largest value of the formal Mulliken charge found on the Li atoms. The term is somewhat decreasing in the $Li@C_{60}$, $Li_2@C_{60}$, and $Li_3@C_{60}$ series with the values of 1.16, 1.10, and 0.86, respectively. Still, the total charge transferred to the cage is increasing in the series: 1.16, 2.21, and 2.46. Thus, the Coulomb attraction between Li and the cage should also increase along the series. This factor can compensate somehow the Li-Li repulsion term.

The vibrational analysis enables to test if a true local energy minimum was found. All the computed frequencies for the structures in Fig. 1 are indeed real and none imaginary (though we could also locate some saddle points not discussed here). Table 2 just presents the lowest three B3LYP/3-21G computed vibrational frequencies ω_i. The related vibrational modes are mostly represented by motions of the Li atoms. Obviously, owing to symmetry reductions upon encapsulation, the symmetry selection rules do not operate any more in the way they simplify the C_{60} vibrational spectra [38]. Hence, the vibrational spectra of $Li_x@C_{60}$ must be considerably more complex than for the icosahedral (empty) C_{60} cage with just four bands in its IR spectrum [38]. This increased spectral complexity has indeed been observed [13,14]. Incidentally, the observed harmonic frequency [33-35] for free Li_2 is 351 cm^{-1} while the computed B3LYP/3-21G term is 349 cm^{-1} (and the B3LYP/6-31G* value 342 cm^{-1}). For the endohedrals, a larger-basis frequency calculations are not yet feasible as, at the B3LYP/3-21G level, the frequency job is about 25-times longer than one geometry-optimization cycle (when computations are carried out with one processor).

Table 2. B3LYP/3-21G and MPWB1K/3-21G computed three lowest[a] vibrational frequencies ω_i in $Li_x@C_{60}$

Species	ω_1 (cm^{-1})	ω_2 (cm^{-1})	ω_3 (cm^{-1})
B3LYP			
$Li@C_{60}$	40.1	105.0	186.5
$Li_2@C_{60}$	96.2	136.4	167.8
$Li_3@C_{60}$	24.7	113.6	155.8
MPWB1K			
$Li@C_{60}$	69.2	138.7	206.0
$Li_2@C_{60}$	75.5	115.5	132.5
$Li_3@C_{60}$	63.0	119.3	171.2

[a]The vibrational motions mostly involving Li atoms.

There is a general stability problem related to fullerenes and metallofullerenes - either the absolute stability of the species or the relative stabilities of clusters with different stoichiometries. One can consider an overall stoichiometry of a metallofullerene formation:

$$\mathrm{xY(g)} + \mathrm{C_n(g)} = \mathrm{Y_x@C_n(g)}. \quad (1)$$

The encapsulation process is thermodynamically characterized by the standard changes of, for example, enthalpy $\Delta H^o_{Y_x@C_n}$ or the Gibbs energy $\Delta G^o_{Y_x@C_n}$. In a first approximation, we can just consider the encapsulation potential-energy changes $\Delta E_{Y_x@C_n}$. Table 3 presents their values for $Li_x@C_{60}$. Their absolute values increase with the increasing number of the encapsulated Li atoms. In order to have some directly comparable relative terms, it is convenient to consider the reduced $\frac{\Delta E_{Y_x@C_n}}{x}$ terms related to one Li atom. Although the absolute values of the reduced term decrease with increasing Li content, the decrease is not particularly fast (so that, a further increase of the encapsulated Li atoms could still be possible). The MPWB1K terms in Table 3 are reported for the standard 3-21G basis set. If the MPWB1K calculations are extended to the 6-31G* basis, the values come closer to the B3LYP functional, e.g., the MPWB1K/6-31G*//MPWB1K/3-21G $\frac{\Delta E_{Y_x@C_n}}{x}$ terms for $Li@C_{60}$ and $Li_2@C_{60}$ read -34.9 and -31.4 kcal/mol, respectively. The computational findings help to rationalize why also the $Li_2@C_{60}$ species could be observed [11]. Although the basis set superposition error is not estimated for the presented values (an application of the Boys-Bernardi counterpoise method would be rather questionable in this situation), the correction terms could be to some extent additive. Interestingly enough, the stabilization of metallofullerenes is mostly electrostatic as documented [39,40] using the topological concept of 'atoms in molecules' (AIM) [41,42] which shows that the metal-cage interactions form ionic (and not covalent) bonds.

Various endohedral cage compounds have been suggested [43,44] as possible candidate species for molecular memories (in addition to carbon cages, non-carbon polyhedrons could also be considered [45] for the purpose).

Table 3. Computed encapsulation potential-energy changes $\Delta E_{Y_x@C_n}$ (kcal/mol) for $Li_x@C_{60}$

Species	$\Delta E_{Y_x@C_n}$	$\frac{\Delta E_{Y_x@C_n}}{x}$ [a]
B3LYP[b]		
$Li@C_{60}$	-28.4	-28.4
$Li_2@C_{60}$	-51.1	-25.6
$Li_3@C_{60}$	-71.0	-23.7
MPWB1K[c]		
$Li@C_{60}$	-54.1	-54.1
$Li_2@C_{60}$	-97.5	-48.7
$Li_3@C_{60}$	-148.3	-49.4

[a]The relative term related to one Li atom. [b]Computed at the B3LYP/6-31G*//B3LYP/3-21G level. [c]Computed at the MPWB1K/3-21G level.

While one approach is built [43] on endohedral species with two possible location sites of the encapsulated atom, another studied concept [44] of quantum computing aims at a usage of spin states of $N@C_{60}$. In fact, the $Li@C_{60}$ system can be considered as yet another potential agent for possible future applications in molecular electronics. At present, however, a still deeper knowledge of various molecular aspects of the endohedral compounds is needed before their tailoring to nanotechnology applications is possible. In addition to the molecular parameters, the production yields and their modulation is also of a great importance. Their production yields could be judged from the computed encapsulation equilibrium constants. The encapsulation equilibrium constants themselves result from both the standard enthalpy and entropy changes connected with the encapsulation process. However, there is still another important factor - the metal pressure. The scheme could be simplified by a presumption that the metal pressure is close to the saturated pressure though, under some experimental arrangements, under-saturated or super-saturated metal vapors are also possible. In any case, the pressure factor [46,47] cannot be excluded from the consideration as it represents one of the essential parameters of metallofullerene formations. Hence, a kind of combination of microscopic and macroscopic system characterizations will be necessary.

ACKNOWLEDGMENTS

The reported research has been supported by a Grant-in-aid for NAREGI Nanoscience Project, for Scientific Research on Priority Area (A), and for the Next Generation Super Computing Project, Nanoscience Program, MEXT, Japan, and by the Czech National Research Program 'Information Society' (Czech Acad. Sci. 1ET401110505).

REFERENCES

[1] R. C. Haddon, A. F. Hebard, M. J. Rosseinsky, D. W. Murphy, S. J. Duclos, K. B. Lyons, B. Miller, J. M. Rosamilia, R. M. Fleming, A. R. Kortan, S. H.

Glarum, A. V. Makhija, A. J. Muller, R. H. Eick, S. M. Zahurak, R. Tycko, G. Dabbagh and F. A. Thiel, Nature, 350, 320, 1991.
[2] B. I. Dunlap, J. L. Ballester and P. P. Schmidt, J. Phys. Chem., 96, 9781, 1992.
[3] C. G. Joslin, J. Yang, C. G. Gray, S. Goldman and J. D. Poll, Chem. Phys. Lett., 208, 86, 1993.
[4] T. Kaplan, M. Rasolt, M. Karimi and M. Mostoller, J. Phys. Chem., 97, 6124, 1993.
[5] Z. M. Wan, J. F. Christian, Y. Basir and S. L. Anderson, J. Chem. Phys., 99, 5858, 1993.
[6] C. G. Joslin, C. G. Gray, S. Goldman, J. Yang and J. D. Poll, Chem. Phys. Lett., 215, 144, 1993.
[7] Z. Slanina and L. Adamowicz, J. Mol. Struct. (Theochem), 281, 33, 1993.
[8] S. A. Varganov, P. V. Avramov and S. G. Ovchinnikov, Phys. Solid Stat., 42, 388, 2000.
[9] Z. Slanina and S.-L. Lee, Chin. J. Phys., 34, 633, 1996.
[10] A. Bol, M. J. Stott and J. A. Alonso, Physica B, 240, 154, 1997.
[11] C. Kusch, N. Krawez, R. Tellgmann, B. Winter and E. E. B. Campbell, Appl. Phys. A, 66, 293, 1998.
[12] S. A. Varganov, P. V. Avramov and S. G. Ovchinnikov, Phys. Sol. Stat., 42, 388, 2000.
[13] A. Gromov, A. Lassesson, M. Jönsson, D. I. Ostrovskii and E. E. B. Campbell, In Fullerenes, Vol. 12: The Exciting World of Nanocages and Nanotubes, PV 2002-12, Eds. P. Kamat, D. Guldi, K. Kadish, The Electrochemical Society, Pennington, 2002, p. 621.
[14] A. Gromov, N. Krawez, A. Lassesson, D. I. Ostrovskii and E. E. B. Campbell, Curr. App. Phys., 2, 51, 2002.
[15] Z. Slanina, F. Uhlík, S.-L. Lee and L. Adamowicz, J. Low Temp. Phys., 131, 1259, 2003.
[16] Z. Slanina, F. Uhlík and T. J. Chow, In Fullerenes, Vol. 13: Fullerenes and Nanotubes: The Building Blocks of Next Generation Nanodevices, PV 2003-15, Eds. D. M. Guldi, P. V. Kamat, F. D'Souza, The Electrochemical Society, Pennington, 2003, p. 569.
[17] E. E. B. Campbell, Fullerene Collision Reactions, Kluwer Academic Publishers, Dordrecht, 2003.
[18] V. N. Popok, I. I. Azarko, A. V. Gromov, M. Jonsson, A. Lassesson and E. E. B. Campbell, Sol. Stat. Commun., 133, 499, 2005.
[19] A. Lassesson, K. Hansen, M. Jonsson, A. Gromov, E. E. B. Campbell, M. Boyle, D. Pop, C. P. Schulz, I. V. Hertel, A. Taninaka and H. Shinohara, Eur. Phys. J. D, 34, 205, 2005.
[20] Z. Slanina, F. Uhlík, S.-L. Lee, L. Adamowicz and S. Nagase, J. Comput. Meth. Sci. Engn., 6, 243, 2006.
[21] A. D. Becke, J. Chem. Phys., 98, 5648, 1993.
[22] C. Lee, W. Yang and R. G. Parr, Phys. Rev. B, 37, 785, 1988.
[23] M. J. Frisch, G. W. Trucks, H. B. Schlegel, G. E. Scuseria, M. A. Robb, J. R. Cheeseman, J. A. Montgomery, Jr., T. Vreven, K. N. Kudin, J. C. Burant, J. M. Millam, S. S. Iyengar, J. Tomasi, V. Barone, B. Mennucci, M. Cossi, G. Scalmani, N. Rega, G. A. Petersson, H. Nakatsuji, M. Hada, M. Ehara, K. Toyota, R. Fukuda, J. Hasegawa, M. Ishida, T. Nakajima, Y. Honda, O. Kitao, H. Nakai, M. Klene, X. Li, J. E. Knox, H. P. Hratchian, J. B. Cross, C. Adamo, J. Jaramillo, R. Gomperts, R. E. Stratmann, O. Yazyev, A. J. Austin, R. Cammi, C. Pomelli, J. W. Ochterski, P. Y. Ayala, K. Morokuma, G. A. Voth, P. Salvador, J. J. Dannenberg, V. G. Zakrzewski, S. Dapprich, A. D. Daniels, M. C. Strain, O. Farkas, D. K. Malick, A. D. Rabuck, K. Raghavachari, J. B. Foresman, J. V. Ortiz, Q. Cui, A. G. Baboul, S. Clifford, J. Cioslowski, B. B. Stefanov, G. Liu, A. Liashenko, P. Piskorz, I. Komaromi, R. L. Martin, D. J. Fox, T. Keith, M. A. Al-Laham, C. Y. Peng, A. Nanayakkara, M. Challacombe, P. M. W. Gill, B. Johnson, W. Chen, M. W. Wong, C. Gonzalez and J. A. Pople, Gaussian 03, Revision C.01, Gaussian, Inc., Wallingford, CT, 2004.
[24] Y. Zhao and D. G. Truhlar, J. Phys. Chem. A, 108, 6908, 2004.
[25] Y. Zhao, B. J. Lynch and D. G. Truhlar, ChemPhysChem, 7, 43, 2005.
[26] Y. Zhao and D. G. Truhlar, J. Phys. Chem. A, 109, 4209, 2005.
[27] Y. Zhao and D. G. Truhlar, Phys. Chem. Chem. Phys., 7, 2701, 2005.
[28] Y. Zhao and D. G. Truhlar, J. Phys. Chem. A, 109, 5656, 2005.
[29] Y. Zhao and D. G. Truhlar, J. Chem. Theory Comput., 1, 415, 2005.
[30] J. P. Perdew and Y. Wang, Phys. Rev. B, 45, 13244, 1992.
[31] A. D. Becke, J. Chem. Phys. 104, 1040, 1996.
[32] W. J. Hehre, L. Radom, P. v. R. Schleyer and J. A. Pople, Ab Initio Molecular Orbital Theory, J. Wiley Inc., New York, 1986.
[33] R. A. Logan, R. E. Cote and P. Kusch, Phys. Rev., 86, 280, 1952.
[34] R. A. Brooks, C. H. Anderson and N. F. Ramsey, Phys. Rev. Lett., 10, 441, 1963.
[35] K. P. Huber and G. Herzberg, Molecular Spectra and Molecular Structure, IV. Constants of Diatomic Molecules, Van Nostrand Reinhold Company, New York, 1979.
[36] J. Blanc, M. Broyer, J. Chevaleyre, P. Dugourd, H. Kuhling, P. Labastie, M. Ulbricht, J. P. Wolf and L. Wöste, Z. Phys. D, 19, 7, 1991.
[37] R. Kawai, J. F. Tombrello and J. H. Weare, Phys. Rev. A, 49, 4236, 1994.
[38] Z. Slanina, J. M. Rudziński, M. Togasi and E. Ōsawa, J. Mol. Struct. (Theochem), 202, 169, 1989.
[39] S. Nagase, K. Kobayashi and T. Akasaka, J. Mol. Struct. (Theochem), 398/399, 221, 1997.
[40] K. Kobayashi and S. Nagase Chem. Phys. Lett., 302, 312, 1999.
[41] R. F. W. Bader, Chem. Rev., 91, 893, 1991.
[42] R. F. W. Bader, J. Phys. Chem. A, 102, 7314, 1998.
[43] J. K. Gimzewski, In The Chemical Physics of Fullerenes 10 (and 5) Years Later, Ed. W. Andreoni, Kluwer, Dordrecht, 1996, p. 117.
[44] W. Harneit, M. Waiblinger, C. Meyer, K. Lips and A. Weidinger, In Recent Advances in the Chemistry and Physics of Fullerenes and Related Materials, Vol. 11 - Fullerenes for the New Millennium, PV 2001-11, Eds. K. M. Kadish, P. V. Kamat, D. Guldi, The Electrochemical Society, Pennington, 2001, p. 358.
[45] J. Zhao, L. Ma, D. Tian and R. Xie, J. Comput. Theor. Nanosci., 5, 7, 2008.
[46] Z. Slanina, X. Zhao, N. Kurita, H. Gotoh, F. Uhlík, J. M. Rudziński, K. H. Lee and L. Adamowicz, J. Mol. Graphics Mod., 19, 216, 2001.
[47] Z. Slanina, F. Uhlík and S. Nagase, Chem. Phys. Lett., 440, 259, 2007.

Molecular Simulation of the Nanoscale Water Confined between an Atomic Force Microscope Tip and a Surface

H. J. Choi*, J. Y. Kim*, S. D. Hong*, M. Y. Ha*, and J. Jang**

*Department of Mechanical Engineering, Pusan National University, Busan 609-735, Korea
**Department of Nanomaterials Engineering, Pusan National University, Miryang 627-706, Korea,
jkjang@pusan.ac.kr

ABSTRACT

Under ambient humidity, water condenses as a meniscus between an atomic force microscope (AFM) tip and a surface, giving rise to a strong capillary force on the tip. We performed a molecular dynamics simulation of the nano-confined water meniscus. We studied the effects of the tip and surface hydrophilicity on the meniscus structure. By changing the tip-surface distance, we have simulated the formation, shrinkage, and breakage of the water meniscus. The nanometer meniscus substantially fluctuates in its periphery due to its instability, in agreement with our previous lattice gas simulation. We obtained the density profile and contact angle of the meniscus. By using these structural parameters, we calculated the capillary force between the tip and surface. Our calculation reproduced the typical behavior of the experimental force-distance curve.

Keywords: water meniscus, capillary force, AFM, molecular dynamics simulation

1 INTRODUCTION

Water naturally condenses between an atomic force microscope (AFM) tip and a surface under ambient conditions. This nanoscale water meniscus gives rise to a significant adhesion force on the order of nanonewtons which must be supplied to retract an AFM tip contacting a surface [1] (so called pull-off force). This meniscus also serves as a nano-channel for molecules to flow from the tip to a substrate in dip-pen nanolithography [2]. Considering the widespread applications of the AFM tip in surface science and nanolithography, it is important to understand the meniscus and the resulting capillary force at the molecular level.

Previously, we have performed Monte Carlo simulations based on a lattice gas model [3,4]. We have explored the roles of surface roughness and hydrophilcity on the capillary force. This coarse grained model, although successful in capturing the qualitative features of the meniscus, cannot provide the full molecular details of the water meniscus. Herein, in order to gain molecular insights on the nanoscale meniscus, we have performed all-atom type molecular dynamics (MD) simulation. We study the molecular features of the water meniscus by examining the density profile of water and its variation as the tip retracts from a surface. We also calculate the capillary force by combining the thermodynamic theory and the structural parameters of the meniscus obtained from the molecular density profile.

2 SIMULATION DETAILS

We use the TIP3P model [5] for water-water intermolecular interaction potential. In this model, three atom-centered point charges (-0.834 on O and +0.417 on H) interact with each other through the Coulomb potential. The oxygen-oxygen interaction has an additional Lennard-Jones (LJ) potential, $U(r) = 4\varepsilon\left[(\sigma/r)^{12} - (\sigma/r)^{6}\right]$ [6]. The LJ parameters ε and σ of oxygen are 0.1521 kcal/mol and 0.35365 nm, respectively. Water molecule is taken to be rigid: the oxygen-hydrogen bond length is 0.09572 nm and the angle between two O-H bonds is 104.52°. The particle-mesh Ewald method with a grid spacing of 0.1 nm [7] is used to calculate the long-range electrostatic interactions between partial charges.

We assume both the AFM tip and surface are made of LJ atoms. The tip geometry is taken to be a hemispherical shell of atoms. To construct the tip, we initially set up a face-centered cubic (FCC) crystal with a lattice spacing equal to 0.321 nm (similar to the LJ σ parameter of carbon). Then we choose lattice points closest to the hemispherical surface with a radius of 13 nm. The resulting AFM tip is made up of 8212 atoms. The surface structure is similarly generated by picking up the top two layers of FCC with the same lattice spacing as the tip. The surface is composed of 4976 LJ atoms. Total of 1906 water molecules are initially positioned regularly (as a cubic lattice) between the AFM tip and surface. The simulation box has a size of 7.5 nm by 7.5 nm by 8.0 nm.

We apply the periodic boundary conditions in the directions parallel to the surface. We use the velocity Verlet algorithm for the propagation of MD trajectories [6]. We use the SHAKE [6] algorithm to keep water molecules rigid. Temperature was held constant (300K) by applying Langevin dynamics [8] method with a damping coefficient of 5 ps^{-1}. The time step is taken to be 1 fs and the typical length of MD trajectory is 0.5 ns. We have studied the formation, shrinkage, and breakage of the water meniscus by varying the distance between the AFM tip and surface from 1.4 nm to 4.2 nm with an increment of 0.1 nm. In addition, we have systematically varied the water-tip and water-surface interaction energies. That is, we have considered two LJ epsilon values for the tip atom, ε_T=0.1 and 2.5 kcal/mol. The case of ε_T =0.1 kcal/mol can be thought to correspond to a hydrophobic tip because it is smaller than the LJ epsilon parameter (0.1521 kcal/mol) of water-water interaction. The tip with ε_T=2.5 kcal/mol can be called hydrophilic. The LJ epsilon parameter of the surface, ε_S, is varied as 0.1, 0.5, 1.0, 2.0, and 2.5 kcal/mol. As for the tip, the surface with ε_S=0.1 kcal/mol can be considered as a hydrophobic surface. All the other tips correspond to hydrophilic surfaces. The LJ interactions between a tip atom and water and between a surface atom and water are calculated by using Lorentz-Bertholet mixing rule [6].

3 RESULTS

Figure 1 shows the simulated water menisci for two different tip-surface distances. In this hydrophilic tip-hydrophilic surface case (both ε_S and ε_T are 2.0 kcal/mol), the meniscus has a concave shape at all the tip-surface distances. The meniscus neck reduces in width as the tip-surface distance increases from 1.4 nm to 3.7 nm. Also, notice the concave curvature of the water meniscus decrease in magnitude as we increase the distance between the AFM tip and the surface.

By collecting MD snapshots like in figure 1 and averaging them, we can calculate the density profile defined as the average number of water molecules at a given position. The density profile ρ is assumed to be cylindrically symmetric so that it is a function of the horizontal distance from the tip center, r, and the height from the surface, h. The initial 0.4 ns of MD trajectory is thrown out for equilibration, and MD snapshots after 0.4 ns are collected for the calculation of the density profile. The density profile is normalized in the sense that its highest value is scaled to be one. The bin sizes of the density profile in the horizontal and vertical directions, dr and dh, are 0.25 nm and 0.15 nm, respectively. Such a density profile is drawn for two tip-surface distances in figure 2. The contour line with a density of half is drawn as a dashed line. Then, assuming the meniscus periphery is a circle, we determined the radius r_1 and the center position of the circle by using the least-squares fitting of the concave part of the half-density contour line.

Figure 2 illustrates how the density profile changes as we vary the tip-surface distance from 1.6 nm to 3.7 nm. Since both the tip and the surface are hydrophilic ($\varepsilon_T = \varepsilon_S$=2.0 kcal/mol), the water meniscus has a concave shape. As the tip (drawn as a curved solid line) retracts from the surface (height of zero), the meniscus neck narrows almost down to 1 nm in half width. Also notice that, for a given tip-surface distance, the density profile changes rather smoothly from a liquid value (=1) to a vapor value (=0). In other words, the meniscus periphery is not infinitely sharp as is often assumed in thermodynamic theories. Instead, the periphery is rather fuzzy due to the inherent thermal fluctuations at the nanoscale. Roughly, the periphery has a thickness of about half nanometer.

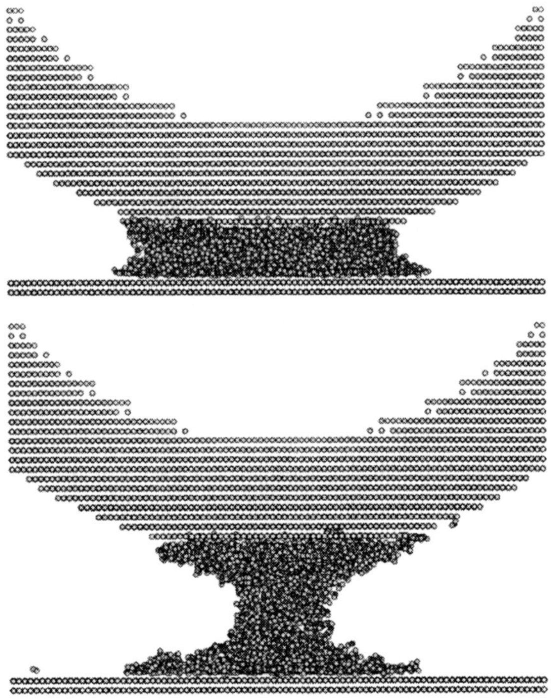

Figure 1: Simulated water menisci for two different tip-surface distances, 1.4 nm (top) and 3.7 nm (bottom). Both the tip and surface are hydrophilic.

Using the meniscus periphery obtained from the density profile, we can estimate the half-width of the meniscus neck r_2. Figure 3 shows how r_2 changes with an increase in the tip-surface distance. As expected, the half width in the neck decreases as the tip-surface distance increases up to the distance of 2.3 nm. As the distance becomes greater than 2.5 nm however, the meniscus half-width fluctuates, even though the overall trend is that the width decreases with raising the tip-surface distance. We think, due to the instability of the water meniscus at the distance over 2.5 nm, the water meniscus width varies rather irregularly.

With the structural parameters of meniscus (r_1 and r_2) obtained from the density profile, we can estimate the humidity of our simulation from the Kelvin equation,

$$RT \ln(p/p_0) = \gamma V\left(\frac{1}{r_1}+\frac{1}{r_2}\right), \tag{1}$$

where γ and V are the surface tension (=0.076 N/m) and the molar volume (=1.8×10^{-5} m^3/mol) of water. R is the universal gas constant, T, the absolute temperature, p_0 the normal vapor pressure of the liquid, and p the pressure. The sign of r_1 is positive for a concave meniscus but negative for a convex one.

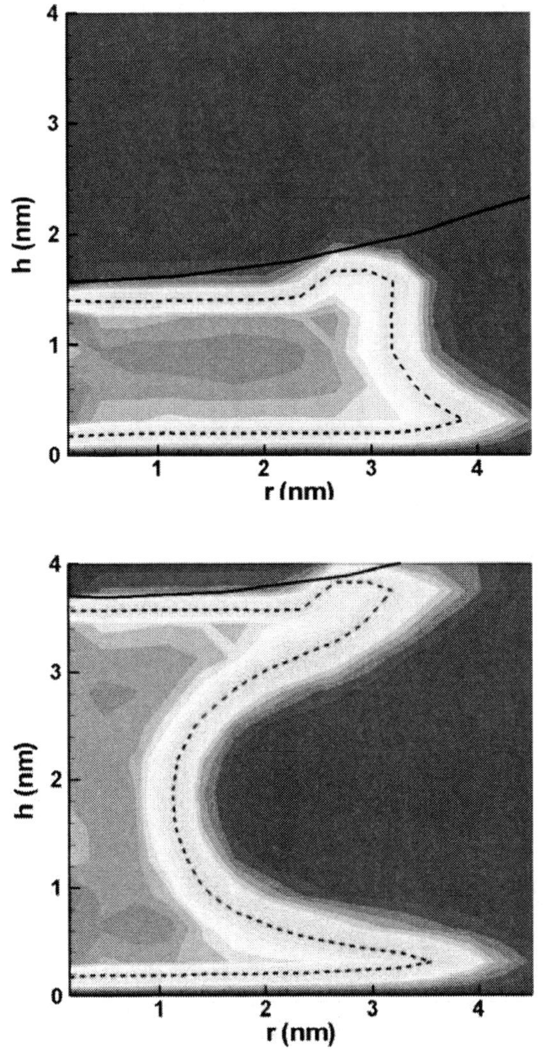

Figure 2: Change in the density profile of the water meniscus as the tip retracts from the surface. For a hydrophilic tip and a hydrophilic surface, the tip-surface distance is varied from 1.6 nm (top) to 3.7 nm (bottom).

The humidity calculated from equation (1) varies by changing the tip-surface distance, and it is found to vary from 60 to 70 percent. The capillary force F for a given tip-surface distance is given by

$$F = \gamma\left(\frac{1}{r_1} + \frac{1}{r_2}\right) \times \pi r_2^2. \tag{2}$$

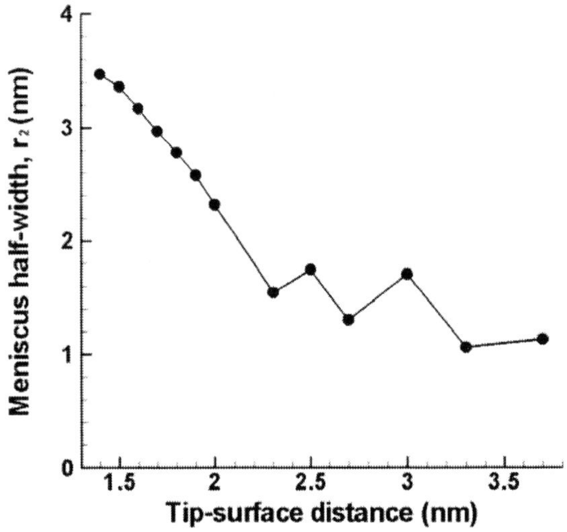

Figure 3: The half-width at the meniscus neck, r_2, vs. the tip-surface distance. Both the tip and surface are hydrophilic (ε_T=2.0 kcal/mol, ε_S=2.0 kcal/mol).

Figure 4 shows the capillary force calculated from equation (2) vs. the tip-surface distance. The figure captures the essential features of a typical force–distance curve obtained in AFM experiment. The capillary force is most attractive at the shortest distance of our simulation, 1.4 nm, and it becomes less attractive as the tip-surface distance increases. At the distance of 3.3 nm, the force becomes zero due to the breakage of the meniscus.

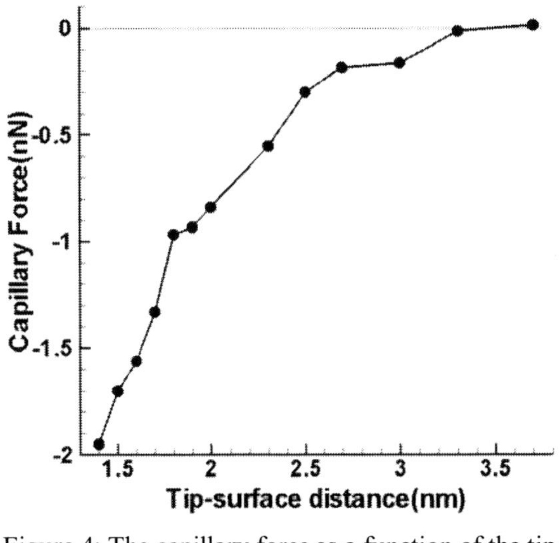

Figure 4: The capillary force as a function of the tip-surface distance. The tip and surface are both hydrophilic ($\varepsilon_T = \varepsilon_S$=2.0 kcal/mol).

Figure 5 illustrates the effects of the surface hydrophilicity on the capillary force. The tip-surface distance is fixed to 1.5 nm. As the surface becomes more hydrophilic with an increase in ε_S, the capillary force becomes more attractive. In the case of a hydrophobic tip (circles), the capillary force change its sign from positive (repulsive) to negative (attractive) at around ε_S =1.8 kcal/mol. As ε_S increases, an initially convex meniscus becomes a concave one, and the capillary force changes from repulsive to attractive. In the case of hydrophilic tip (triangles), the change in sign of the capillary force occurs at a lower value of ε_S (near 0.8 kcal/mol) than in the hydrophobic tip. The figure also shows, at a very low ε_S (=0.1 kcal/mol), the hydrophobicity of surface dominates and the capillary force does not depend much on whether the tip is hydrophobic or hydrophilic. When the surface is strongly hydrophilic (ε_S=2.5 kcal/mol), the capillary force is again dominated by the surface, not quite dependent on the tip hydrophilcity. The force difference between the hydrophilic and hydrophobic tips becomes maximal near ε_S=1 kcal/mol.

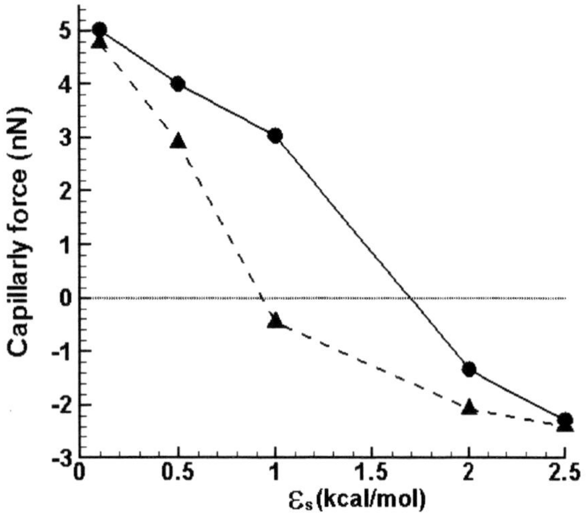

Figure 5: Capillary force as a function of the LJ epsilon parameter of the surface ε_S. Lines represent linear interpolations of data.

4 CONCLUSIONS

We have performed a molecular dynamics simulation of the nanoscale water meniscus formed between an AFM tip and a surface. We examined how the meniscus structure varies as the distance between the tip and surface changes. We obtained the density profile of water for various tip-surface distances and for the tip and surface hydrophilicities. The meniscus periphery obtained from the molecular density profile is not infinitely sharp but instead is fuzzy with a thickness of about half nanometer. Overall, the meniscus width decreases with an increase in the tip-surface distance. Interestingly, beginning at some tip-surface distance (2.5 nm), the meniscus becomes unstable and its width significantly fluctuates as the tip retracts farther from the surface. Overall, the meniscus width decreases as the tip retracts from the surface, eventually becoming zero at a very long tip-surface distance.

From the structural parameters of the meniscus, we were able to calculate the capillary force by using the Kelvin equation. Such a combined approach gives a force distance curve in reasonable agreement with typical AFM experiment. If the surface is strongly hydrophilic or strongly hydrophobic, the capillary force does not depend much on whether the tip is hydrophilic or hydrophobic. The difference in the capillary force becomes maximal for a moderately hydrophilic surface.

We believe our work will serve as a starting point for further investigation of the nanoscale meniscus and force at the molecular level. For example, a combination of the current MD with the density functional theory [9] should give a more accurate estimate of the capillary force.

REFERENCES

[1] J. N. Israelachvili, "Intermolecular and Surface Science", 2nd edition, Academic Press, 1991

[2] C. A. Mirkin, ACS Nano 1, 79, 2007.

[3] J. Jang, J. Sung, and G. C. Schatz, J. Phys. C. 111, 4648, 2007.

[4] J. Jang, M. Yang, and G. C. Schatz, J. Chem. Phys. 126, 174705, 2007.

[5] W. L. Jorgensen, J. Chandrasekhar, J. D. Madura, R. W. Impey, and M. L. Klein; J. Chem. Phys. 79, 926, 1983.

[6] M. P. Allen and D. J. Tildesley, Computer Simulation of Liquids, Clarendon Press, 1987.

[7] T. Darden, D. York, and L. Pederson, J. Chem. Phys. 98 , 10089, 1993.

[8] R. Kubo, M. Toda, and N. Hashitsume, "Statistical Physics II: Nonequilibrium Statistical Mechanics" 2nd Ed., Springer, 1991.

[9] P. B. Paramonov and S. F. Lyuksyutov, J. Chem. Phys. 123, 084705, 2005.

Monte Carlo simulations of 1keV to 100keV electron trajectories from vacuum through solids into air and resulting current density and energy profiles.

Andreas Hieke

GEMIO Technologies, Inc.

3000 Sand Hill Road, Building 1, Suite 170, c/o Quantum Insight, Menlo Park, CA 94025

ah@gemiotech.com, ahi@ieee.org

ABSTRACT

Design and optimization of systems which perform electron beam irradiation under atmospheric conditions requires both the modeling of pure electron optics inside electron beam guns as well as the treatment of electrons passing through solid films and the adjacent gas.

It is shown that in order to simulate trajectories of electrons from an emitting metal cathode, through high vacuum (with effectively no electron-gas collisions), through a solid object (typically a thin foil acting as wall), into air and onto a target a combination of different modeling approaches for the various domains is required.

1. INTRODCTION

Increased awareness to use environmentally friendly methods for surface sterilization, i.e. without the use of chemicals, has rejuvenated interest in methods based on electron beam irradiation. Secondly, electron beam irradiation enables a very high degree of control of the sterilization process and can be employed in cases where other (e.g. heat based) methods are not applicable. This is particularly important in applications such as preparation of spacecraft for life detection missions [1], [2].

While sterilization of bulk material, such as food, with high energy electrons (several MeV) has been used for many decades, surface treatment with medium energy electrons in the presence of gas is only recently gaining attention.

2. METHODS

Design and optimization of systems which perform electron irradiation under atmospheric conditions requires both the modeling of pure electron optics inside electron beam guns as well as the treatment of electrons passing through solid films and the adjacent gas. Thus far, Monte Carlo electron beam simulations have been described either as typical electron optics simulations (i.e. collision free in vacuum), or exclusively in solids [3]-[6] or gases (plasmas) [7]-[9] with emphasis on the electron matter interaction.

By combining different modeling approaches for the various domains a more complex simulation system can be created which permits to model the entire path of electrons from an emitting metal cathode, through high vacuum (with effectively no electron-gas collisions), through a solid object (typically a thin foil acting as wall), into air and onto a target.

3. RESULTS

The discussed method builds partially on [5], although comparison with other sources (NIST) revealed that some of the material data in [5] appear to be erroneous. Fig. 1 shows the energy loss per distance traveled for electrons in three different solids as function electron energy. Effects due to crystallographic structure are neglected. Typical values are on the order dE_{kin}/dr = 1 eV/nm ...100 eV/nm for usually encountered electron beam energies (i.e. acceleration potentials).

Very recently, highly accurate solutions for differential and total elastic scattering cross sections of electrons with $E_{kin} \geq$ 1keV on various neutral atoms have been published which are based on Dirac partial-wave calculation using the ELSEPA code [10] which have also been published in comprehensive form by ICRU in [11]. These data are of critical importance and are being incorporated in the presented simulation system.

Fig. 2 shows as an example of such result as provided in [11] the total cross-section σ for elastic scattering of electrons by aluminum.

Of further importance it the mean free path between elastic scattering events and typical values will range from λ=1nm...100nm dependent on the material and electron energy (Fig. 2). For 100keV electrons in Titanium this value is about 50nm and the previously mentioned energy loss per distance traveled is on the order of 1.3eV/nm.

Fig 4 shows an example of a simulation of a generic electron beam gun producing an electron beam with E_{kin} =20keV. The yellow base represents foil through which electron beam will penetrate.

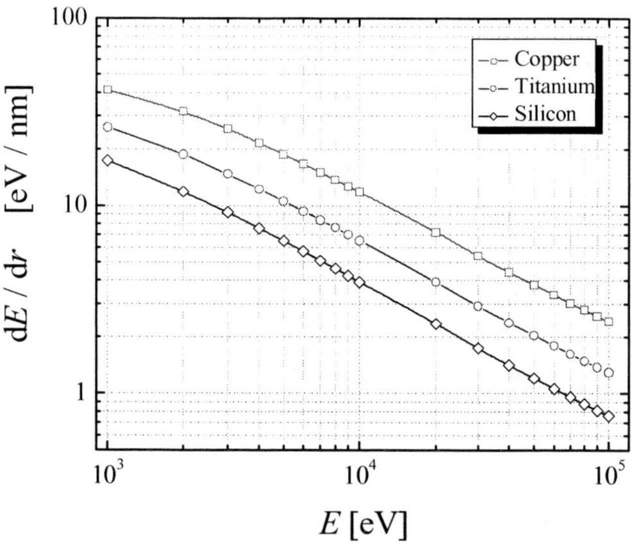

Fig. 2 Energy loss per distance traveled for electrons in three different solids as function electron energy. (Crystallographic structure neglected); based on NIST data

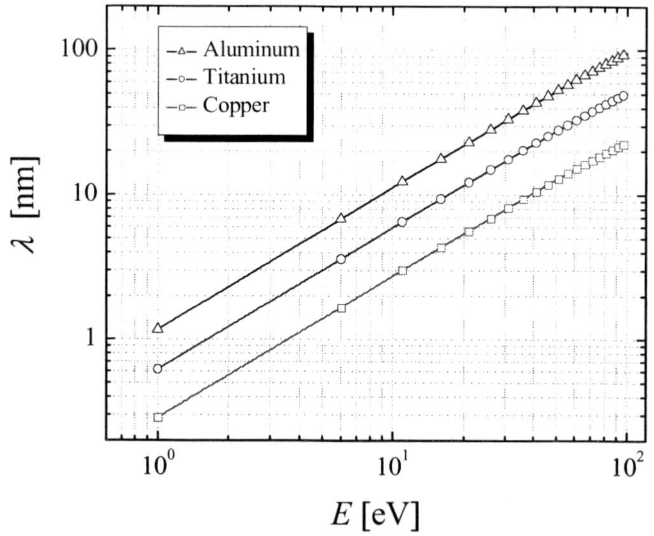

Fig. 1 Mean free path between eleastic scattering events for electrons in three different solids as a function energy. (Crystallographic structure neglected); based on ICRU data

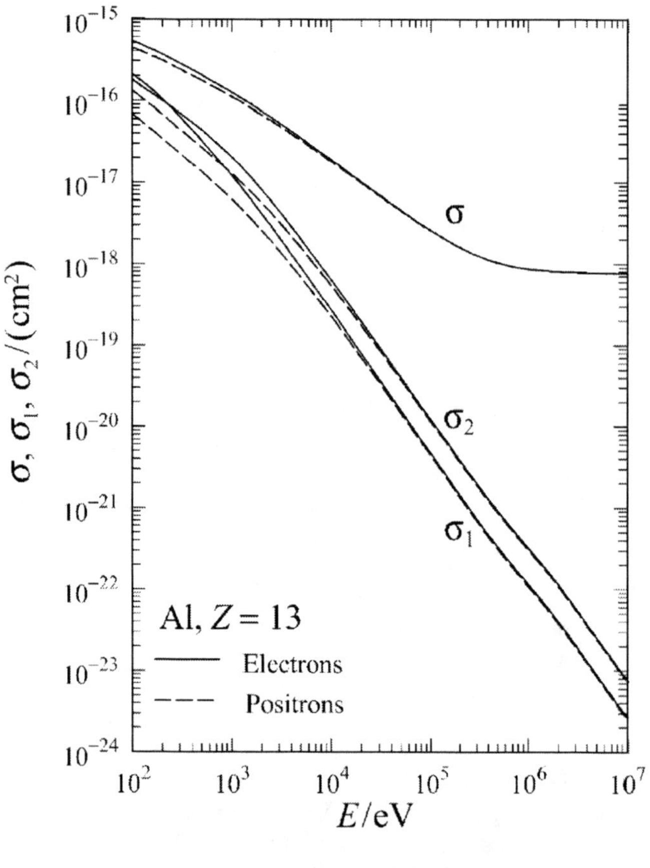

Fig. 3 Total cross-section σ, 1st transport cross-section σ_1, and second transport cross-section σ_2 for elastic scattering of electrons by aluminum. (from [10])

Typically, an electron gun creates an image of its cathode at the surface of a solid foil which acts as transmission window to the atmosphere. The spatial electron current density profile $j(x,y,E)$is modulated during the electron transmission through the solid both spatially and energetically.

The electron trajectory integration regimes are different for vacuum and medium. However, electrons are continuously tracked and scattering events are considered individually, i.e. no averaging or "condensed history" or other simplifying approaches are employed in the simulation.

Fig. 5 and Fig. 6 show examples of trajectories of electrons in Titanium with initial energies of 20keV and 100keV. Notice the distinctly different appearance.

When exiting the solid, and dependent on its thickness, the transmitted electrons have a current density profile $j'(x,y,E)$ with a lower, but typically much wider energy distribution than the original distribution which is almost exclusively determined by the thermal emission process. These distributions constitute then the input for the electron tracing in air. Current density profiles (translating into dose profiles) have been obtained at different planes of models.

Such simulations permit the optimization of electron beam shapes, foil thickness and structure, as well as spatial dose profiles.

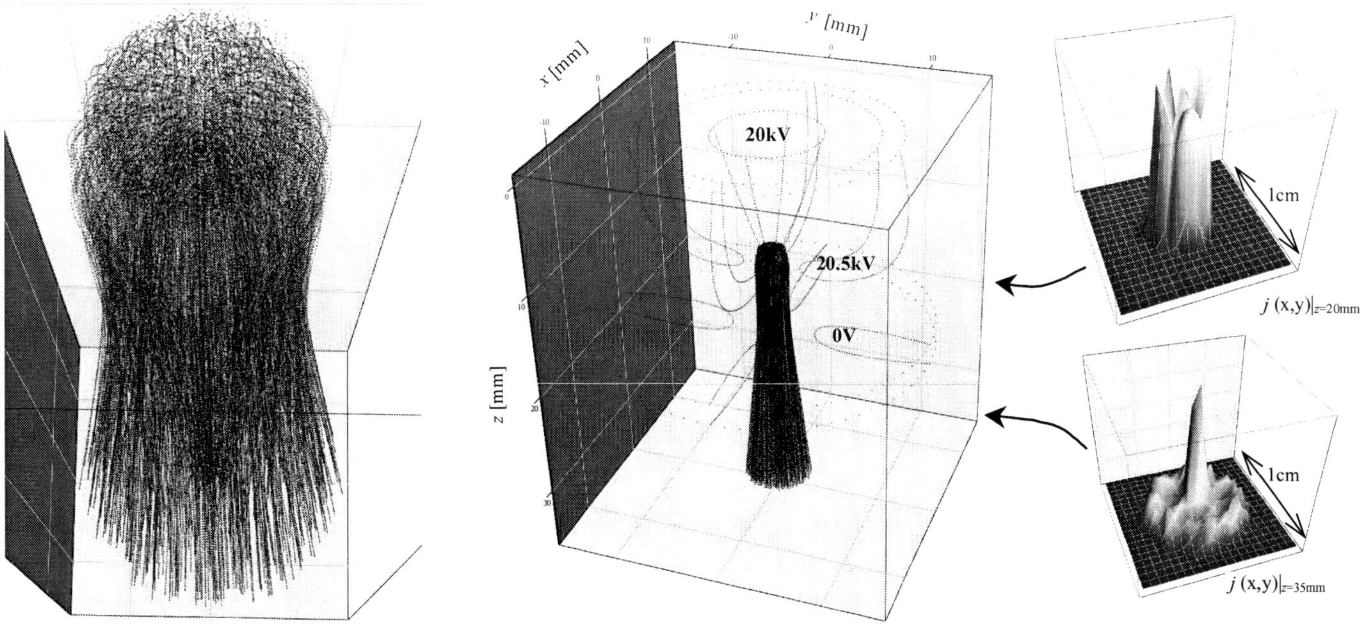

Fig. 4 Monte Carlo simulation of an electron gun; electron energy E_{kin}=20keV; Left and center: trajectories of 10^4 electrons, dotted shapes indicate electrodes, yellow base represents foil through which electron beam will penetrate; right: current densities $j(x,y)$ in planes at z=20mm and z=35mm, current densities plots based on 10^6 electron trajectories

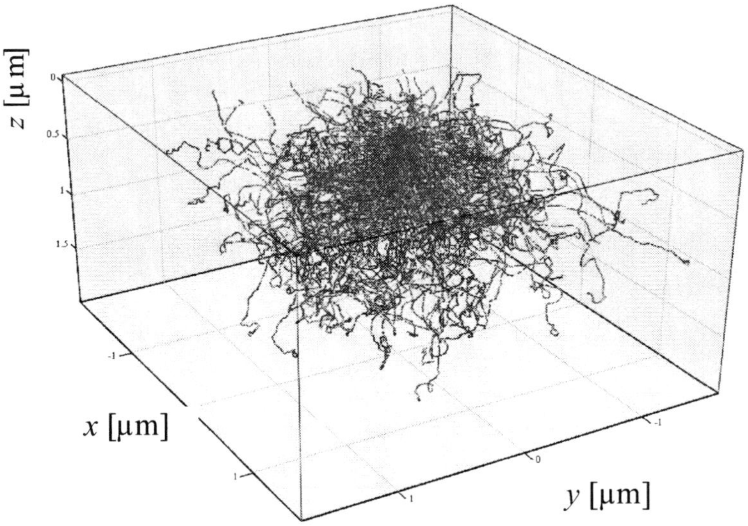

Fig. 5 Monte Carlo simulation of 500 electrons with an initial energy of $_iE_{kin}$=20keV in a 5µm thick Titanium foil. 0% transmission, 25% backscattered, average energy of backscattered electrons: $_bE_{kin}$=13.3keV

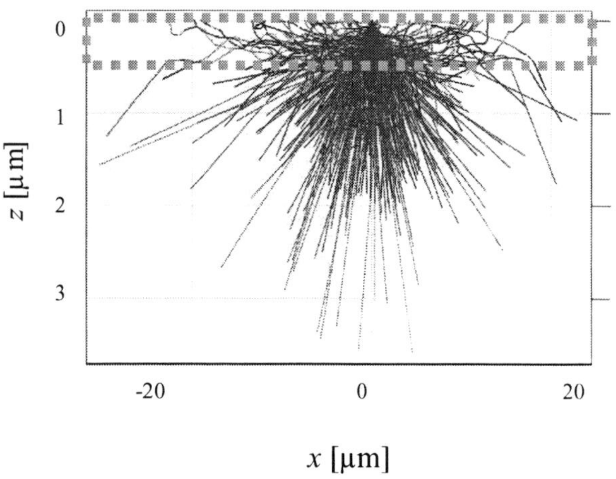

Fig. 6 MC simulation of 500 electrons with an initial energy of $_iE_{kin}$=100keV through a 5µm thick Titanium foil. 90% transmission, average energy of transmitted electrons: $_tE_{kin}$=90keV; 10% backscattered, average energy of backscattered electrons: $_bE_{kin}$=77keV

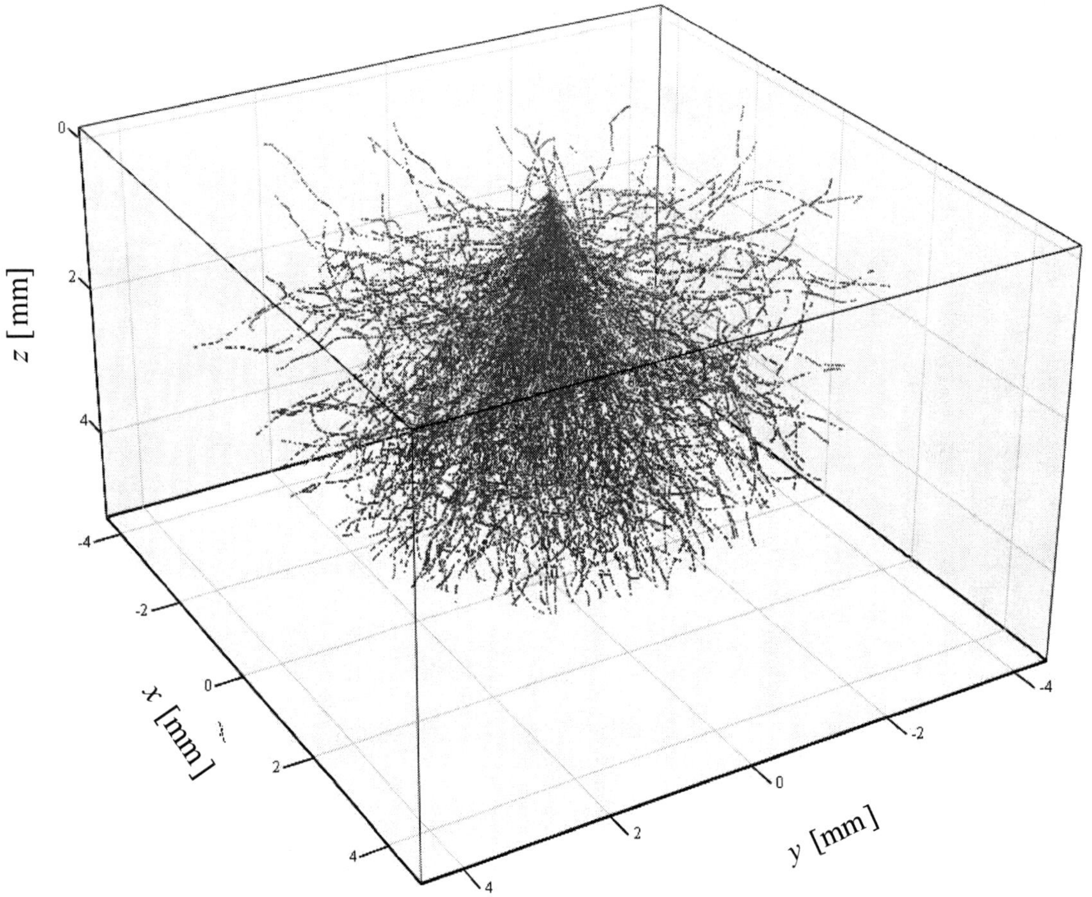

Fig. 7 Monte Carlo simulation of 500 electrons with an initial energy of ${}_iE_{kin}$=20keV shot into air at p=10^5Pa. Simulation terminated at E_{kin}=10keV. Notice the different scale compared to the previous figures.

[1] E. Urgiles et. al. : "Electron beam Irradiation for Microbial Reduction on Spacecraft Components", 2007 IEEE Aerospace Conference, 3-10 March 2007, pp.: 1 – 15

[2] P.R. Chalise et.al.: "Bacterial inactivation using low-energy pulsed-electron beam", IEEE Transactions on Plasma Science, Volume 32, Issue 4, Part: 2 , Aug. 2004, pp.: 1532 - 1539

[3] R. Kowalewicz, T. Redel: "Interaction of a high current polyenergetic electron beam with metal", IEEE Transactions on Plasma Science, Volume 23, Issue 3, Jun 1995, pp.: 270 – 274

[4] D. Emfietzoglou, A. Akkerman, J. Barak: "New Monte Carlo calculations of charged particle track-structure in silicon", IEEE Transactions on Nuclear Science, Volume 51, Issue 5, Part: 3 , Date: Oct. 2004, pp.: 2872 – 2879

[5] David C. Joy: "Monte Carlo Modeling for Electron Microscopy and Microanalysis", Oxford University Press (New York) 1995.

[6] L. G. H. Huxley, R. W. Crompton: "The Drift and Diffusion of Electrons in Gases". Wiley Interscience (New York), 1973

[7] G.R. Govinda Raju, M.S. Dincer: "Monte Carlo simulation of electron swarms in nitrogen in uniform E×B fields", IEEE Transactions on Plasma Science, Volume 18, Issue 5, Date: Oct 1990, pp.: 819 – 825

[8] W.N.G. Hitchon, G.J. Parker, J.E. Lawler: "Accurate models of collisions in glow discharge simulations" IEEE Transactions on Plasma Science, Volume 22, Issue 3, Date: Jun 1994, pp.: 267 – 274

[9] Z. M. Raspopovic, S. Sakadzic, S.A. Bzenic, Z.Lj. Petrovic: "Benchmark calculations for Monte Carlo simulations of electron transport", IEEE Transactions on Plasma Science, Volume 27, Issue 5, Oct. 1999, pp.: 1241 – 1248

[10] F. Salvat, A. Jablonski and C. J. Powell: "ELSEPA — Dirac partial-wave calculation of elastic scattering of electrons and positrons by atoms, positive ions and molecules", Computer Physics Communications, Volume 165, Issue 2, 2005, pp.: 157-190

[11] "Elastic scattering of electrons and positrons", Journal of the ICRU, , Vol. 7 No 1 (2007) Report 77, Oxford University Press

von Neumann Entropies Analysis of Nanostructures: PAMAM Dendrimers of Growing Generation

Rodolfo O. Esquivel*, Nelson Flores-Gallegos and Edmundo Carrera

Departamento de Química, Universidad Autónoma Metropolitana-Iztapalapa
San Rafael Atlixco No. 186, Col. Vicentina, C.P. 09340, México D.F.
*esquivel@xanum.uam.mx

ABSTRACT

Quantum Information Theory (QIT) is a new field with potential implications for the conceptual foundations of Quantum Mechanics through density matrices. In particular, information entropies in Hilbert space representation are highly advantageous in contrast with the ones in real space representation since they can be easily calculated for large systems. Quantum information entropies have shown, through quantum mechanics concepts such as entanglement, several interesting critical points which are not present in the energy profile, such as depletion and accumulation of charge, non-nuclear attractors and bond breaking/formation regions. In this work, von Neumann informational entropies are employed to characterize the initial steps (monomer, dimmer, trimer, tetramer) towards growing a nanostructured molecule of Polyamidoamine (PAMAM) dendrimer of generation zero (G0). Structures were geometrically optimized at the b3lyp/6311+G** level of theory.

Keywords: dendrimers, PAMAM, quantum information theory, an initio calculations, entanglement

1 INTRODUCTION

The most interesting technological implications of quantum mechanics are based on the notion of entanglement, which is the essential ingredient for both quantum cryptography, quantum computing, and quantum teleportation [1]. Generally speaking, if two particles are in an entangled state, then even if the particles are physically separated by a great distance, they behave in some respect as a single entity rather than as two separate entities. Entanglement shows up in cases where a former unit dissociates into simpler sub-systems. Corresponding processes are known quite well in chemistry. The real-space partitioning of a molecule into subsystems is still a challenging problem in theoretical chemistry [2-14], because during this process a certain entanglement of the subsystems emerges, and it is very difficult to get rid of it without destroying elementary correlations between the subsystems. So, apart from its evident importance for the foundations of physics, entanglement plays a role in chemistry too. Although information entropies have been used for a variety of studies in quantum chemistry [15-21], applications of entanglement measures in chemical systems are very scarce. Very recently, marginal and non-marginal information measures in Hilbert space have been proposed [22], and applied to small chemical systems, showing than entanglement can be realized in molecules [23].

It is of great interest in Chemistry to understand molecular systems as combination of atoms and molecular fragments. Thus, the concept of AIM has been the focus of great deal of attention [2-14]. Chemical processes involve small changes between atoms and molecular subsystems and it is crucial to understand the interactions (correlation and entanglement) involved in such chemical changes. The main goal of the present study is to show that marginal entropies of bipartite composite systems in Hilbert space [22,23] are able to reveal the growing behavior of PAMAM towards its generation zero. This is achieved by use of novel measures of conditional, mutual, joint information von Neumann entropies computed by means of the spectral decomposition of the first reduced density matrix in the natural atomic orbital-based representation, assuring rotational invariance, and N and v-representability in the AIM scheme [24]. .

Dendrimers [25] are nanostructured molecules which are highly branched, star-shaped macromolecules which are defined by three components: a central core, an interior dendritic structure (the branches), and an exterior surface with functional surface groups. Monodisperse dendrimers are synthesized by step-wise chemical methods to give distinct generations (G0, G1, G2, ...) of molecules with narrow molecular weight distribution, uniform size and shape, and multiple (multivalent) surface groups. In particular, Polyamidoamine (PAMAM) dendrimers are the most common class of dendrimers suitable for many materials science and biotechnology applications which consist of alkyl-diamine core and tertiary amine branches. PAMAM dendrimers represent an exciting new class of macromolecular architecture called "dense star" polymers systematically changing the normal dendrimer features (e.g. generation or surface group).

2 THEORETICAL FRAMEWORK

The uncertainty in a collection of possible observables A_i with corresponding probability distribution $p_i(A)$ is given by its Shannon entropy H(A) [26]

$$H(A) = -\sum_i p_i(A) \ln p_i(A) \qquad (1)$$

This measure is suitable for systems described by classical physics, and is useful to measure uncertainty of observables but it is not suitable for measuring uncertainty of the general state of a quantum system. It is the von Neumann entropy which is appropriate to measure uncertainty of quantum systems since it depends on the density matrix (see below).

Another important concept derived from relative entropy concerns the gathering of information. When one system learns something about another, their states become correlated. How correlated they are, or how much information they have about each other, can be quantified by the mutual information. The *Shannon mutual information* between two random variables *A* and *B*, having a joint probability distribution $p_{ij}(A,B)$ and marginal probability distributions

$$p_i(A) = \sum_j p_{ij}(A,B) \text{ and } p_j(B) = \sum_i p_{ji}(B,A) \qquad (2)$$

defined as

$$H(A:B) = H(A) + H(B) - H(A,B) = \sum_{ij} p_{ij}(A,B) \ln \frac{p_{ij}(A,B)}{p_i(A) p_j(B)} \qquad (3)$$

Where $H(A,B)$ is the joint entropy defined as

$$H(A,B) = -\sum_{ij} p_{ij}(A,B) \ln p_{ij}(A,B) \qquad (4)$$

which measures the uncertainty about the whole system AB. The mutual information $H(A:B)$ can be written in terms of the Shannon relative entropy. In this sense it represents a distance between the distribution $p(A,B)$ and the product of the marginals $p(A) \times p(B)$. Suppose that we wish to know the probability of observing b_j if a_i has been observed. This is called a conditional probability and is given by

$$p_{ij}(A|B) = \frac{p_{ij}(A,B)}{p_j(B)} \text{ and } p_{ji}(B|A) = \frac{p_{ji}(B,A)}{p_i(A)} \qquad (5)$$

Hence the conditional entropy is,

$$H(A|B) = -\sum_{ij} p_{ij}(A,B) \ln p_{ij}(A|B) \qquad (6)$$

This quantity, being positive, tells us how uncertain we are about the value of *B* once we have learned about the value of *A*. Now the Shannon mutual information can be rewritten as

$$H(A:B) = H(A) - H(A|B) \qquad (7)$$

And the joint entropy as

$$H(A,B) = H(B) + H(A|B) \qquad (8)$$

The difference between classical and quantum entropies can be seen in the fact that quantum states are described by a density matrix ρ (and not just probability vectors). The density matrix is a positive semidefinite Hermitian matrix, whose trace is unity. An important class of density matrices is the idempotent one, i.e., $\rho = \rho^2$. The states these matrices represent are called pure states. When there is no uncertainty in the knowledge of the system its state is then pure. Another important concept is that of a composite quantum system, which is one that consists of a number of quantum subsystems. When those subsystems are entangled it is impossible to ascribe a definite state vector to any one of them, unless we deal with a bipartite composite system. The most often cited entangled system is the Einstein-Podolsky-Rosen state (EPR) [27,28], which describes a pair of two photons.

. Since the uncertainty in the probability distribution is naturally described by the Shannon entropy, this classical measure can also be applied in quantum theory. In an entangled system this entropy is related to a single observable. The general state of a quantum system, is described by its density matrix ρ. Let the observables a_i and b_j, pertaining to the subsystems *A* and *B*, respectively, have a discrete and non degenerate spectrum, with probabilities $p(a_i)$ and $p(b_j)$. For simplicity let us define them as $p_i(A)$ and $p_j(B)$ In addition, let the joint probability be $p_{ij}(A,B)$. Then

$$H(A) = -\sum_i p_i(A) \ln p_i(A) = -\sum_{ij} p_{ij}(A,B) \ln \sum_j p_{ij}(A,B) \qquad (9)$$

and similarly for H(B).

An indication of correlation is that the sum of the uncertainties in the individual subsystems is greater than the uncertainty in the total state. Hence, the Shannon mutual information is a good indicator of how much the two given observables are correlated. However, this quantity, as it is inherently classical, describes the correlations between single observables only. The quantity that is related to the correlations in the overall state as a whole is the von Neumann mutual information which depends on the density matrix. The von Neumann entropy [29], may be considered as the proper quantum analog of the Shannon entropy [30] for a system described by a density matrix ρ, and is defined as

$$S(\rho) = -Tr(\rho \ln \rho) \quad (10)$$

The Shannon entropy is equal to the von Neumann entropy only when it describes the uncertainties in the values of the observables that commute with the density matrix, i.e., if ρ is a mixed state composed of orthogonal quantum states, otherwise

$$S(\rho) \leq H(A) \quad (11)$$

where A is any observable of a system described by ρ. This means that there is more uncertainty in a single observable than in the whole of the state [31].

We have recently shown [24] that there is an information-theoretic justification for performing Lowdin symmetric transformations [32] on the atomic Hilbert space, to produce orthonormal atomic orbitals of maximal occupancy for the given wavefunction, which are derived in turn from atomic angular symmetry subblocks of the density matrix, localized on a particular atom and transforming to the angular symmetry of the atoms. The advantages of these kind of atoms-in-molecules (AIM) approaches [33,34] are that the resulting natural atomic orbitals are N- and *v*-representables [24], positively bounded, and rotationally invariant [35,36].

According to the preliminaries above we have recently proposed new measures of correlation and entanglement through marginal (H-type) and non-marginal (R-type) von Neumann entropies [22,23]. In Hilbert space we may define a measure of quantum correlations between molecular fragments for a bipartite system through natural atomic probabilities and their joint probability. In the natural atomic decomposition scheme we employ there are m states pertaining to molecular fragment A, i.e., $\{p_i(A);\ i= 1 \text{ to } m\}$ with n states corresponding to molecular fragment B: $\{p_j(B);\ j= 1 \text{ to } n\}$, thus, we may define the joint entropy through global operations by correlating m x n states providing that the following constraints are met:

$$\sum_i \sum_j P_{ij}(A,B) = \sum_i P_i(A) = \sum_j P_j(B) = \sum_i P_{ij}(A/B) = 1 \quad (12)$$

We are now in position of using definitions above related to the von Neumann entropies, taking into account that in our natural atomic scheme of probabilities, equality in Eq. (11) holds, and instead of referring to observables we deal with subsystems (molecular fragments), that is why von Neumann entropies are adequate for our study, though we keep the H-terminology to emphasize the orthogonal and commuting properties of the subspaces we are dealing with.

3 RESULTS AND CONCLUSION

In the present study we analyze novel von Neumann marginal entropies of bipartite composite systems in Hilbert space (Sec 2) to assess their utility for revealing the growing behavior of recently, ethylenediamine-core (EDA) PAMAM which have a 4-fold multiplicity, i.e, a four branched growing structure. The typical end-groups are primary amines or carboxylate functions depending on if these are full or half-generation PAMAMs. In this work we have calculated several conformations for the monomer, dimer, trimer towards the G0 structure at the b3lyp/6311+G** level of theory, except for the G0 geometry which was obtained from Maiti et al [39]. The electronic structure calculations performed in this study were carried out with the NWchem suite of programs [37] and the natural atomic probabilities were obtained by use of the NBO 5.G program [38].

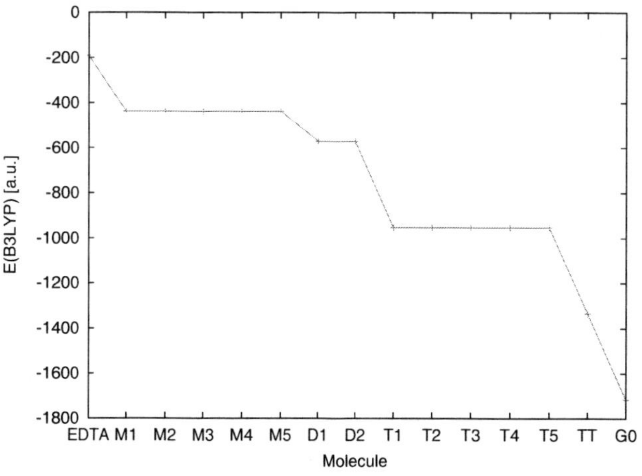

Figure 1: Energies for the monomers M1 through M5, dimers D1 and D2, trimers T1 thorugh T5, tetramer and the G0 structure for PAMAM

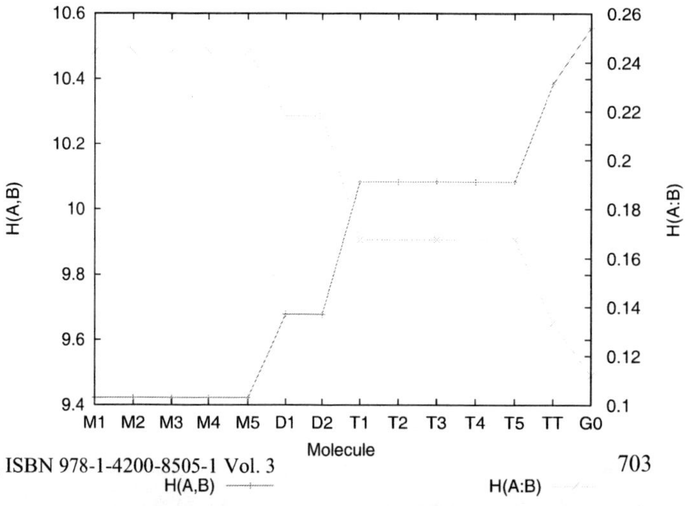

Figure 2: Mutual and joint von Neumann entropies H(A:B) and H(A,B) for the monomers M1 through M5, dimers D1 and D2, trimers T1 thorugh T5, tetramer and the G0 structure for PAMAM

From Figures 1 and 2, wherein the energy and the mutual and joint von Neumann entropies are depicted for all the calculated structures it is clear that QIT entropies describe well the growing behavior of PAMAM dendrimer of G0 type. Ongoing research is currently performes in our laboratory to extend the study to G1-G11 PAMAMs and to incorporate descriptors form statistical complexity theory.

REFERENCES

[1] Bennett CH; Brassard G; Crépeau C; Josza R; Peres A; Wootters WK Phys Rev Lett, 70, 1895, **1993**

[2] Moffitt W. Proc. R. Soc. London, Ser. A, 210, 245, **1951**

[3] Mulliken R. S. J. Chem. Phys., **3**, 573, **1955**

[4] Mulliken R. S. J. Chem. Phys., **23**, 1833, **1955**

[5] Bader R. F. W. *An Introduction to the Structure of Atoms and Molecules* Clarke, Toronto, **1970**.

[6] Bader R. F. W. *Atoms in Molecules* (Oxford, New York, 1994).

[7] Bader R. F. W. and T. T. Nguyen-Dang Adv. Quantum Chem., 14, 63, **1981**

[8] Hirshfeld F. L. Theor. Chim. Acta **44**, 129 (1977).

[9] R. G. Parr; R. A. Donnelly; M. Levy; and W. E. Palke J. Chem. Phys., 68, 3801, **1978**.

[10] Parr R. G. Int. J. Quantum Chem., 26, 687, **1984**

[11] Rychlewski J. and Parr R. G. J. Chem. Phys., 84, 1696, **1986**

[12] Li L. and R. G. Parr J. Chem. Phys., 84, 1704 , **1986**

[13] Cedillo A.; Chattaraj P. K.; and Parr R. G. Int. J. Quantum Chem., 97, 2959, **2000**

[14] Ayers P. J. Chem. Phys., , 113, 10886, **2000**

[15] Esquivel R. O.; Rodriguez A. L.; Sagar R. P.; Ho M.; and Smith, Jr. V. H. Phys. Rev. A, 54, 259, **1996**

[16] Ramírez J. C.; Soriano C.; Esquivel R. O.; Sagar R. P.; Ho M and Smith, Jr. V. H. Phys. Rev. A, 56, 4477, **1997**

[17] Ziesche P.; Gunnarsson O.; John W. Phys. Rev. B, 55,10270, **1997**

[18] Guevara N. L.; Sagar R. P. and Esquivel R.O. Phys. Rev. A, 67,012507, **2003**

[19] Guevara N. L.; Sagar R. P. and Esquivel R. O. J. Chem. Phys., 122, 084101, **2005**

[20] Sen K.D.; Antolin J.; and Angulo J.C. Phys. Rev. A, 76, 032502, **2007**

[21] Sen K.D.; Panos C.P.; Chatsisavvas K. Ch.; and Moustakidis Ch. C. Phys. Lett. A, 364, 286, **2007**

[22]Esquivel R. O. and Flores-Gallegos N. Phys. Rev. A (to be published)

[23] Flores-Gallegos N. and Esquivel R. O. J. Comp. App. Math. (to be published)

[24] Carrera E.; Flores-Gallegos N. and Esquivel R. O. J. Chem. Phys. (to be published)

[25] M. Ballauff, C.N. Lykos, Angew. Chem. Int. Ed. 43, 2998, **2004**

[26] Shannon, C. E.; and W. Weaver **1949**, *The Mathematical Theory of Communication* (University of Illinois, Urbana, IL).

[27]Einstein, A.;. Podolsky B; and Rosen N. Phys. Rev., 47, 777, **1935**

[28] Bell, J., *Speakable and Unspeakable in Quantum Mechanics* **1987** (Cambridge University, Cambridge).

[29] von Neumann, J., *Mathematical Foundations of Quantum Mechanics*, **1955** translated from the German ed. by R. T. Beyer (Princeton University, Princeton).

[30] Wehrl, A., Rev. Mod. Phys., 50, 221, **1978**

[31] Vedral V. Rev Mod. Phys.,74, 197, **2002**

[32]Lowdin P.O. Adv. Quantum Chem., 5, 185, **1970**

[33] Reed A.E. and Weinhold F. J. Chem. Phys., 78, 4006, **1983**

[34] Davidson E.R. J. Chem. Phys., 46, 3320, **1967**

[35] Reed A.E.;. Weinstock R.B; and Weinhold F. J. Chem. Phys., 83, 735, **1985**

[36] Bruhn G.; Davidson E.R.; Mayer I.; and Clark A.E., Int. J. Quantum Chem. 106, 2065, **2006**

[37] E. J. Bylaska, W. A. de Jong, K. Kowalski, T. P. Straatsma, M. Valiev, D. Wang, E. Apra, T. L. Windus, S. Hirata, M. T. Hackler, Y. Zhao, P.-D. Fan, R. J. Harrison, M. Dupuis, D. M. A. Smith, J. Nieplocha, V. Tipparaju, M. Krishnan, A. A. Auer, M. Nooijen, E. Brown, G. Cisneros, G. I. Fann, H. Fruchtl, J. Garza, K. Hirao, R. Kendall, J. A. Nichols, K. Tsemekhman, K. Wolinski, J. Anchell, D. Bernholdt, P. Borowski, T. Clark, D. Clerc, H. Dachsel, M. Deegan, K. Dyall, D. Elwood, E. Glendening, M. Gutowski, A. Hess, J. Jaffe, B. Johnson, J. Ju, R. Kobayashi, R. Kutteh, Z. Lin, R. Littlefield, X. Long, B. Meng, T. Nakajima, S. Niu, L. Pollack, M. Rosing, G. Sandrone, M. Stave, H. Taylor, G. Thomas, J. van Lenthe, A. Wong, and Z. Zhang, "NWChem, A Computational Chemistry Package for Parallel Computers, Version 5.0" (2006), Pacific Northwest National Laboratory, Richland, Washington 99352-0999, USA.

[38]*NBO 5.0.* Glendening E. D.; Badenhoop J, K.; Reed A. E.; Carpenter J. E.; Bohmann J. A.; Morales C. M.; and Weinhold F. Theoretical Chemistry Institute **2001,** University of Wisconsin, Madison

[39]P.K. Maiti, T Cagin, G. Wang, W.A. Goddard III, Macromolecules 37, 6236, **2004**; personal communication.

Probabilistic Models for Damage and Self-Repair in DNA Self-Assembly

U. Majumder*

* Department of Computer Science, Duke University, Durham NC, USA, urmim@cs.duke.edu

ABSTRACT

Since its inception, the focus of DNA self-assembly based nanostructures has mostly been on one-time assembly. However, DNA nanostructures are very fragile and prone to damage. Knowing the extent of damage that can occur under various physical conditions can be useful in making robust designs for self-assembled nanostructures. Thus in this paper, we present simple models for estimating the extent of damage in DNA nanostructures due to various external forces. We note that these models have not been validated against experimental data and are only meant to serve as a basis for designing DNA nanostructures that are robust to external damage. We conclude with a discussion on computing the probability of repair of a damaged nanostructure.

Keywords: dna, self-assembly, damage, self-repair, probability

1 Introduction

DNA self-assembly is an autonomous phenomenon where components (single strands or DNA "tiles"[1]) organize themselves into stable superstructures. In recent years, DNA based self-assembled nanostructures have gone from conceptual design to experimental reality [2]. They can, however, be very fragile. For instance, observe the DNA nanotube in Figure 1 which has been partially opened up by an AFM tip during imaging.

Winfree [1] briefly investigated if a hole in a DNA lattice will repair correctly in a kinetic simulation model. However, to the best of our knowledge, there has been no work on either modeling the extent of damage in DNA nanostructures or any experimental measurement of it. Short of an experimental demonstration, we felt that a realistic model for lattice damage and self-repair would enable us to better evaluate the current designs for self-repairing tile set. We study the extent of damage due to external mechanical forces as well as intense radiation. Furthermore, we compute the probability of self-repair at equilibrium.

[1]DNA tiles are made of several individual DNA strands with single strand overhangs (also called "sticky ends") that allow them to assemble into a lattice.

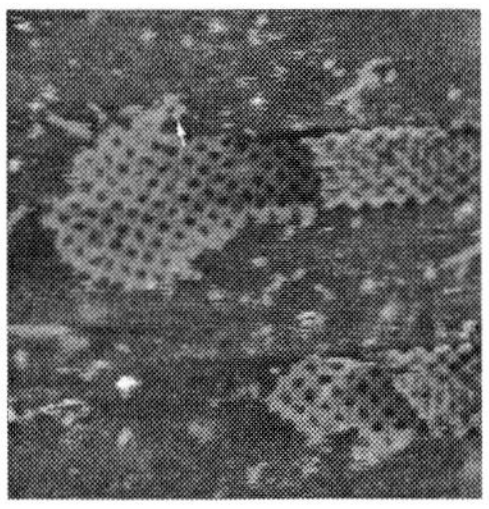

Figure 1: *Damage in DNA nanostructures: broken DNA nanotubes due to repeated AFM scanning.*

2 Mechanical Damage Model

Suppose an AFM tip strikes a tile on a lattice composed of several DNA tiles. If we assume that our lattice is a rigid surface then the effect of the hitting force on a neighboring tile can be modeled as a function of its distance from the source of the impact [3]. In this model, the probability that a tile will fall off is given by the probability that a shock wave from the hitting force propagates to this tile along some path (Probabilistic Damage Model: see Section 2.1). However, if we assume that the lattice is flexible, we can use a mass spring model for the lattice and hence calculate the effect of the hitting force on the lattice [5] (Flexible Lattice Model: see Section 2.2). One important observation that we should make here is that we do not know yet whether these models are appropriate, since they have not been verified experimentally.

2.1 Probabilistic Damage Model

We assume that the force F_1 on a tile located at a distance of r from the tile receiving the impulse is proportional to $\frac{1}{\sqrt{r}}$ [3]. In our model, r is the Manhattan distance of the tile under consideration from the source of the impact. F_2 is the resistive force from the sticky end connections of the tiles. For simplicity, we assume F_2 to be the same for all the tiles. For any tile with $F_1 > F_2$, the probability that a tile gets knocked off the lattice is greater than zero so long as the shock wave has reached it from the origin of damage. To estimate the fraction of the lattice damaged, we first compute the probability of a damage path of length i. A *damage path* is defined as a path that originates in the tile which is

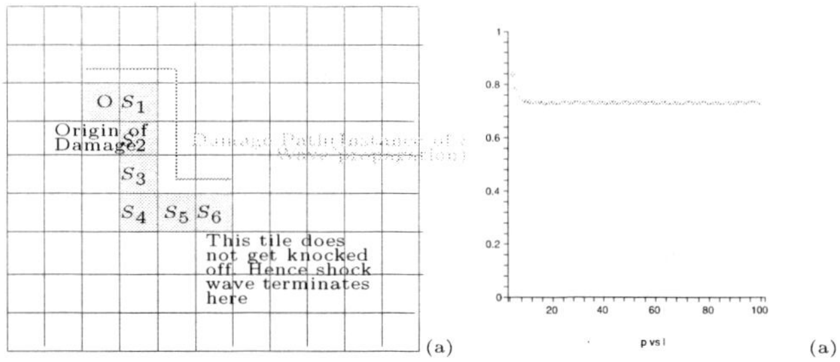

Figure 2: *(a)Instance of shock wave propagation and creation of damage path, (b)Estimation of p*

directly hit by the tip (say O) and meanders outwards through its successors $< S_1, S_2, S_3, \dots S_{i-1} >$ and stops at S_i. In the damage path each tile $O, S_1, S_2, S_3, \dots S_{i-1}$ is knocked off the lattice except for S_i [Figure2(a)]. Let us denote the probability of the damage path that stops at S_i by $P(i)$. Observe that S_i is located at a Manhattan distance of i from O. Since F_1 is proportional to $\frac{1}{\sqrt{i}}$, the probability that a tile will fall off, given that at least one of its neighbors is already knocked off, is given by $\frac{p}{\sqrt{i}}$ where p $(0 < p < 1)$ is the normalization factor and can be evaluated from the probability distribution for a damage path. Then,

$$P(i) = \frac{p^i}{(\sqrt{i-1})!}(1 - \frac{p}{\sqrt{i}}) \quad (1)$$

This is because each tile on this path falls with a probability $\frac{p}{\sqrt{j}}$ where $j = 1, 2, \dots, i-1$ and the shock wave stops at S_i with probability $(1 - \frac{p}{\sqrt{i}})$. Furthermore, $P(i)$ is a probability mass function and hence, if the maximum length of our damage paths is restricted to l we have

$$\sum_{i=1}^{l} P(i) = 1 \quad (2)$$

Now we can estimate the expected fractional damage size $D(n, l)$ for a lattice L with n tiles, by summing the probabilities of a damage path from O to each S_i (maximum damage path length being l) in the lattice. For ease of computation, we achieve this by calculating the expected number of tiles knocked off at a Manhattan distance of i from O, $\forall i, 1 \dots l$.

$$D(n, l) = \frac{(1-p) + \sum_{i=1}^{l} 4i \times P(i)}{n} \quad (3)$$

The first term accounts for the event when the damage path probabilistically stops at O and $4i$ is the number of tiles located at a Manhattan distance i from O.

Simulation Results Based on equation 1 and 2, we can solve for p analytically. However, for large systems,

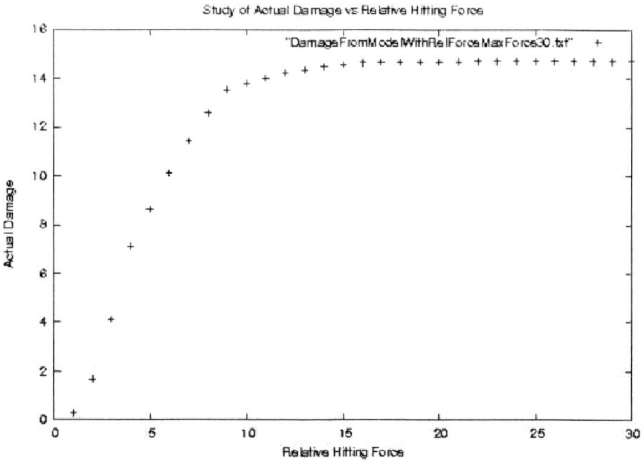

Figure 3: *Plot of Actual Damage vs relative hitting force of the AFM. The plot reveals the pseudo-geometric nature of the probability distribution of damage.*

the problem is too hard and hence we use Monte Carlo simulation to evaluate p [Figure2(b)]. Now our l is determined by the Manhattan distance where $F_1(r) = F_2$. Thus if $F_1(l) = \frac{c}{\sqrt{l}}$, where c is a constant, then

$$l = (\frac{c}{F_2})^2 \quad (4)$$

We call l the *relative hitting force* in our simulation plot [Figure 3] since it approximately measures the ratio between the original impulse $(F_1(0))$ and the resistive binding force between a pair of tiles (F_2). We study the effect of relative hitting force on the lattice in isolation. As the plot in Figure2(b) reveals that the value of p stabilizes to 0.7316 beyond a relative hitting force of 10. Even the drop from an initial value of 0.87 to 0.75 occurs before the relative hitting force even reach a value of 5. Hence for all practical purposes we consider the value of p to be 0.73. Furthermore, the plot in Figure 3verifies the pseudo-geometric probability distribution of a damage path described in equation 3. Evidently, the amount of actual damage reaches nearly a constant value beyond a relative hitting force of 10.

2.2 Flexible Lattice Model

This section considers a different model which views the rectangular DNA lattice of size $m \times n$ as a simple mass spring system similar to cloth dynamics in computer graphics [5]. In this model, each tile is positioned at grid point (i, j), $i = 1, 2, \dots, m$ and $j = 1, 2, \dots n$. For simplicity, assume that the external mechanical impulse F hits the lattice at a single tile location. The internal tension of the spring linking tile $T_{i,j}$ with each of its neighbor $T_{k,l}$ is given by

$$F\_int_{i,j} = - \sum_{(k,l) \in R} K_{i,j,k,l} \left[l_{i,j,k,l} - l^0_{i,j,k,l} \frac{l_{i,j,k,l}}{||l_{i,j,k,l}||} \right] \quad (5)$$

where R is the set of neighbors of $T_{i,j}$. $l_{i,j,k,l} = \overrightarrow{T_{i,j}T_{k,l}}$, $l^0_{i,j,k,l}$ is the natural length of the spring linking tiles and

$K_{i,j,k,l}$ is the stiffness of that spring. With DNA the actual value of $K_{i,j,k,l}$ is dependent on the type of bases that are involved in the sticky ends for the pair of tiles. We make some simplifying assumptions: the stiffness K and the unextended spring length l^0 are the same for every pair of adjacent tiles and known to us *a priori* for the purposes of simulation.

2.3 Simulation Algorithm

Since it is difficult to obtain a closed form solution for the expected number of tiles that are knocked off the lattice because of the impulse, we outline a simple simulation algorithm for estimating the number of tiles removed when an impulse hits the lattice. We assume that the lattice is a connected graph where the state of the tiles is updated every Δt time. To that extent, we start a breadth first search from the tile receiving the original impulse.

The tile under consideration moves for Δt time in the direction of the hitting force, thus extending the springs connecting it to its neighbors. If the extension is beyond a threshold value, the tile snaps off the lattice. In the next level of the breadth first search, a component of the spring force pulls each of the neighbor away from the lattice while a reaction force pushes the tile under consideration to restore its original configuration, if it has not fallen out of the lattice. In the following level, we examine spring extensions for the neighbors of the neighbors of the tile that received the initial impulse. The process is repeated recursively until we reach the boundary. We repeat the breadth first search on the tiles until the velocity of all the tiles are either zero or the whole lattice has fallen apart. The algorithm returns the number of tiles that are knocked off. A cartoon of the lattice before and after an impact is shown in Figure 4.

Some of the important assumptions we make are (1) the velocity of each tile remains constant during the Δt interval, (2) a damping force models collisions between tiles and individual water molecules.

The model approximately captures the overall behavior of a free floating lattice when acted upon by an external impulse. It, however, may be very sensitive to the parameter values such as the spring stiffness constant and the fracture threshold. We also note that this model does not consider collisions between tiles, which may change their relative velocity and, hence, affect the total number of tiles that actually are knocked off the lattice.

3 Thermal Damage Model

Suppose a lattice is being irradiated by a powerful electron beam that raises the local temperature of a part of the lattice beyond the melting temperature of the tiles. The increased temperature will alter the dissociation rate of tiles in the region, since it is temperature dependent. This section develops a model for computing number of tiles removed due to this local temperature increase.

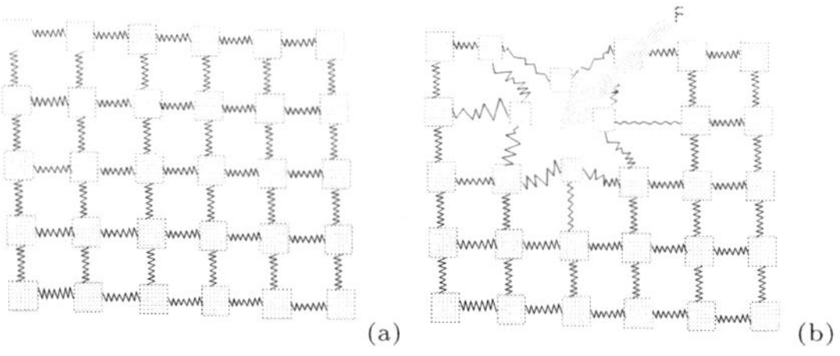

Figure 4: *Cartoon for Mechanical Damage in a Flexible Lattice: (a)Original Lattice in solution, (b)Damaged Lattice due to an external force. The tile that received the immediate impact F (not shown in the figure) is knocked off since one of the springs that connect it to its four neighbors extended beyond the threshold value while some of the neighbors are displaced (The affected tiles have smaller size to indicate that they are below the rest of the tiles in the lattice). However, the magnitude of the force is relatively low and the shock wave from the original impact dies long before it reaches the boundary as is evident from the relatively unchanged configuration of the tiles on the boundary.*

In Winfree's kATM model [4], if there are m empty sites adjacent to the aggregate, then the net "on rate" is given as:

$$k_{on} = m\hat{k}e^{-G_{mc}} \tag{6}$$

where G_{mc} is the entropic cost of fixing the location of a monomer unit while $\hat{k} = 20k_f$, k_f being the forward rate constant. For all occupied sites (i,j) within the aggregate, the net "off rate" is

$$k_{off} = \sum_b k_{off,b}\ ,\ k_{off,b} = \sum_{ij \text{ s.t } b_{ij}=b} \hat{k}_f e^{-b_{ij}G_{se}} \tag{7}$$

where b_{ij} is the total strength for matching labels and $G_{se} = (\frac{4000K}{T} - 11s)$ is the free energy cost of breaking a single sticky end bond, with s being the sticky end length of the oligonucleotide and T being the temperature. In general, for a tile with b matches at a site with the forward rate of association as r_f and the rate of dissociation of a tile with b matches as $r_{r,b}$, we have

$$\frac{r_f}{r_{r,b}} = e^{bG_{se}-G_{mc}} \tag{8}$$

Since G_{se} is inversely proportional to T, if T increases such that $\frac{r_f}{r_{r,b}} < 1$, a hole is created. As an example of how temperature changes lead to reduced lattice, Figure 5 shows the effects of globally increasing temperature by $1.9°C$ after the lattice is fully formed. Here starting with

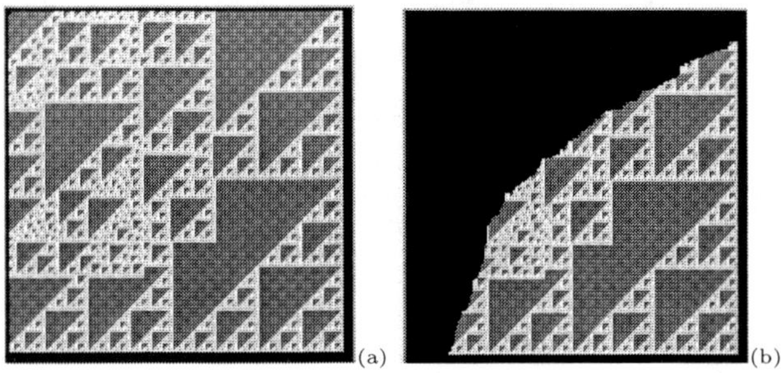

Figure 5: a) Original Lattice, (b) Damaged Lattice due to increased temperature

a Sierpinski Triangle patterned lattice with 62992 tiles assembled at a temperature of $35.969°C$ with $G_{mc} = 19$ and $G_{se} = 9.7$, when heated to $37.892°C$ ($G_{mc} = 19$, $G_{se} = 9.3$) for 2×10^5 sec reduces to 34585 tiles.

Although $\frac{r_f}{r_{r,b}} < 1$ gives us the condition for damage, we are more interested in designing a concrete model to compute the extent of damage. For this purpose we first compute the net rate of both binding and dissociation events k_{any} as

$$k_{any} = k_{on} + k_{off}. \tag{9}$$

In equilibrium, all the rates stabilize and the probability of any event E occurring is given by a Poisson distribution. Furthermore, the probability of an off event O happening (given that some event has happened), is given by $O \sim Binomial(N, k_{off})$ where N is the total number of events. If n is the size of the damage, we can compute the probability $P(O = n|E = i)$ of a hole of size n, given that a total of i events has occurred. Then the probability of dissociating n tiles is $P(O = n)$ where

$$P(O = n) = \sum_{i=0}^{\infty} P(O = n|E = i)P(E = i) \tag{10}$$

and the expected size of the damage in that case is

$$E(O) = \sum_{j=0}^{M} jP(O = j) \tag{11}$$

where M is the size of the initial aggregate. This equation is not easy to solve analytically but we can obtain an estimate for $E(O)$ with Monte Carlo integration. A more accurate estimate can be obtained if we do not assume that the rates k_{on} or k_{off} to be constant. According to kTAM, these rates vary with number of empty sites, concentration and total number of bonds which change as events occur.

4 Self-repair Model

Given a damage model, it is also useful to estimate the likelihood of the lattice reconstructing itself, a phenomenon we call *self-healing*. This can also be modeled probabilistically. We estimate the probability of self-repair using techniques from Ref. [4]. We use the same notation as in Section 3.

Suppose an error-free lattice has a hole of size n to repair. We assume that self-healing is error-free too. Then using the principles of detailed balance, at equilibrium, it has been shown in Ref. [4] that an aggregate A formed by the addition of any sequence of n tiles $T_1, T_2, \ldots, T_n$ with a total strength $b_A = \sum_{i=1}^{n} b_i$. will obey the following equation:

$$\frac{[A]}{[T]} = e^{-((n-1)G_{mc} - b_A G_{se})} \tag{12}$$

where T denotes one of $T_1, T_2, \ldots, T_n$. Furthermore, we will use A_m to denote a lattice with m errors (and A_0 is the error-free self-healed lattice). For small m, there can be $\binom{2n}{m}$ suboptimal aggregates for each perfect aggregate of size n and, hence, the probability of an errorless aggregate A_0 of size n, $P(A_0 = n)$, can be derived as :

$$P(A_0 = n) \sim \frac{[A_0]}{\sum_{m=0}^{\infty} \binom{2n}{m}[A_m]} \sim 1 - 2ne^{-G_{se}} \tag{13}$$

Hence, the probability that a damage of size n will get self-healed is $1 - 2ne^{-G_{se}}$.

5 Summary

This paper presents damage and self-repair models for self-assembled DNA nanostructures that would allow us to design error-resilient nanostructures in the future. As a part of future work, we intend to validate our models with experimental data.

REFERENCES

[1] Erik Winfree and Renat Bekbolatov Proofreading tile sets: Error correction for algorithmic self-assembly *DNA 9*, LNCS 2943, pp:126-144.

[2] J Xu and T. H LaBean and S.L Craig DNA Structures are their Applications in Nanotechnology *ed:A Ciferri*, CRC Press -Taylor & Francis Group, Boca Raton.

[3] Hertzberg R Deformation and Fracture Mechanics of Engineering Materials *John Wiley and Sons*, NY 1996.

[4] E Winfree Simulations of Computing by Self-Assembly *Caltech CS Report* 1998.22.

[5] X Provot Deformation Constraints in a Mass Spring Model to Describe Rigid Cloth Behavior *In Proc. Graphics Interface*, 95, pgs:147-154, 1995.

A Theoretical Investigation of the Electronic Structure and Spin Polarization in CrO_2

Jon D. Swaim, Giuseppe Mallia and Nicholas M. Harrison

Department of Chemistry, Imperial College London, London SW7 2AZ United Kingdom

ABSTRACT

The prospects for devices based on the manipulation of electronic spin rather than charge (spintronics) are based on the identification of materials which support strongly spin-polarized currents. The ferromagnetic material, CrO_2, has found wide spread use in magnetic recording applications and, due to its theoretical 100% spin-polarization, has potential for use in spintronics. Previous band structure calculations predict CrO_2 to be a half-metallic ferromagnet, but the underlying electronic structure of CrO_2 is still controversial. In the current work strong electron-electron interactions are introduced using the hybrid-exchange approximation to DFT using the B3LYP and PBE0 functionals [1-2]. The computed electronic structure is found to be significantly different to that computed within the LDA and to be consistent with recent spectroscopy measurements. In particular there is an improved prediction of the energy gap in minority spin states due to the effects of strong *d*-*d* interactions.

Keywords: CrO2, spintronics, half-metal, ferromagnetism, hybrid, DFT

1 Introduction

CrO_2 has received much attention as a potential spintronic material due to theoretical predictions that it is a half-metallic ferromagnet [3-7]. Half-metallic ferromagnets are completely spin-polarized at the Fermi level (E_F), with a metallic density of majority spin (spin-up) states and an insulating band gap in the minority (spin-down) states. Point-contact Andreev reflection spectroscopy and spin-polarized tunneling experiments have observed a spin-polarization of the conduction electrons that is larger than 95% [6]. CrO_2 therefore has potential for use as a source of spin-polarized electrons in magnetic tunneling applications such as magnetic random access memory [7]. CrO_2 is a room temperature ferromagnet with a Curie temperature, T_C, of 390 K [8] with the strong magnetic coupling thought to be due to double exchange interactions [9]. Although the transport properties are not yet well established, bulk CrO_2 is generally considered to be a poor metal, exhibiting a T^2 dependence on resistivity above a characteristic temperature between 80-100 K and, at low very T, a residual resistivity of 5 $\mu\Omega$cm [10]. The T^2 dependence is indicative of strong electron-electron interactions and its suppression at very low T has been assigned to the absence of minority-spin states (at E_F) and thus the exclusion of spin flip scattering by magnon excitations [11].

CrO_2 crystallizes in the rutile structure (space group D_{4h}^{14}: $P4_2/mnm$) with two formula units per unit cell. The Bravais lattice is tetragonal with lattice constants $a = b = 4.421$ Å and $c = 2.916$ Å[9]. Each Cr atom is surrounded by six O atoms, and the CrO_6 octahedron gives rise to a crystal field potential that removes the degeneracy of the Cr-3d orbitals, splitting them into a t_{2g} triplet (d_{xy}, d_{yz} and d_{zx}) and an e_g doublet ($d_{x^2-y^2}$ and $d_{z^2-r^2}$) [13].

The electronic structure of CrO_2 has been studied previously within DFT using several different approximations to electronic exchange and correlation. Schwarz first predicted CrO_2 to be a half-metallic ferromagnetic using the LDA in 1986 [3]. The LDA was seen to predict a finite density of states (DOS) at E_F for spin-up electrons and a gap of 1.5 eV in the spin-down spectrum. The inadequcies of the LDA and generalized gradient approximations (GGA) for studying strongly correlated systems are, however, well known [4]. One of the reasons is the unphysical self-interaction of electrons implicit in non-orbitally dependent effective potentials. In strongly correlated systems, this self-interaction can delocalize the 3d electrons, underestimate band gaps and incorrectly describe the ground state (e.g., metallic rather than insulating) [5]. We point out that a number of measurements, including x-ray absorption spectroscopy [7], resistivity [11] and optical [12], have suggested that strong correlation effects are present in CrO_2. The LDA+U approach provides an *ad hoc* approach for approximating the electron-electron interactions by introducing an on-site Coloumb potential U. LDA+U band structure calculations by Korotin *et al.* yielded improved semi-quantitative agreement with the DOS measured by x-ray and ultraviolet photoemission experiments [13]. Recently, hybrid density functional methods such as B3LYP and PBE0 have been shown to be quite successful in the description of strongly correlated systems [14]. In these functionals, the non-local Hartree-Fock exchange is mixed with the GGA exchange-correlation energy, reducing the self-interaction error and leading to improved estimates of band gaps [15], magnetic moments [16] and magnetic coupling constants [14]. In this paper we apply the B3LYP and PBE0 hybrid-exchange functionals to the half-metallic ferromagnet CrO_2. In agreement with previous work [17], our results show that correlation effects beyond LDA are essential in understanding the electronic and mangetic properties of CrO_2. We demonstrate that hybrid-exchange functionals can lead to improved prediction of band gaps and d-d correlation effects in CrO_2.

2 Computational Details

All unrestricted Hartree-Fock (UHF) and density functional theory (DFT) calculations were performed using the CRYSTAL06 software [18]. For DFT calculations, the exchange-correlation energy was evaluated in the LDA, GGA as defined by Perdew, Burke and Ernzerhof (PBE), and B3LYP and PBE0 hybrid-exchange approximations. In the B3LYP method, the amount of exact Fock exchange mixed into the GGA exchange-correlation functional (20 %) was determined by fitting theoretical results to thermochemical data for certain atoms and molecules [16]. Interestingly, the functional has turned out to be quite successful in descrbing periodic systems. The PBE0 functional, developed by Perdew, Burke and Erzenrhof [19], is based on fourth-order pertubation theory and includes 25% Fock exchange.

All atoms are described as Bloch functions expanded with atom-centered Gaussians [18]. The Cr atom is represented with a 86-411G* basis set (one s, four sp, and two d shells) adopted from a previous Cr_2O_3 study [20]. The basis set for the O atom is a 8-411G* contraction (one s, three sp, and one d shells) taken from an Al_2O_3 study [21]. In the calculations, a shrinking factor of 8 was adopted to form a reciprocal mesh of k-points in the first Brillouin zone according to the Pack-Monkhorst method [18]. Thresholds of 10^{-7}, 10^{-7}, 10^{-7}, 10^{-7} and 10^{-14} were used as tolerances for the monoelectronic and bielectronic integrals, and the overall self-consistent field procedure was set to converge when ΔE=1x10^{-7} Hartree. The cell parametes and internal coordinates were optimized using analytical energy gradients with respect to the cell vectors and nuclear coordinates, respectively.

3 Results & Discussion

In agreement with experimental work, the ground state energy of the ferromagnetic phase, E_{FM}, was always found to be lower than that of the antiferromagnetic phase, E_{AFM}. The relative stability of the phases is shown in Table 1 as the energy difference in the two phases, $\Delta E = E_{AFM} - E_{FM}$. In periodic UHF and DFT studies based on the Kohn-Sham formalism, the Slater determinants are not eigenfunctions of total square spin operator, but of $\hat{S}_z$, the z component of the spin operator. Hence, the magnetic coupling of two Cr^{4+} cations each with total spin S=1 is described by the Ising Hamiltonian

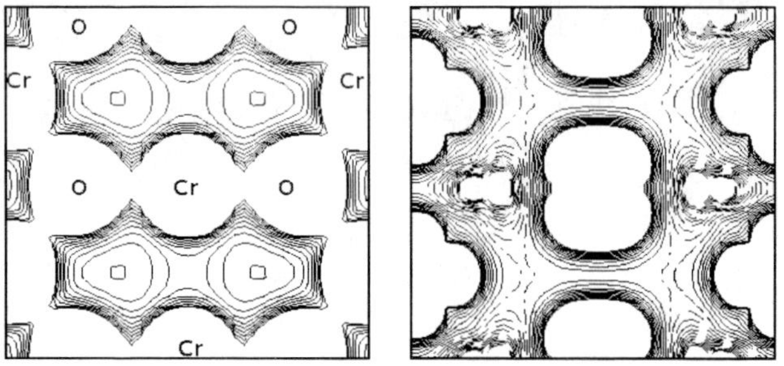

Figure 1: Total (left) and spin (right) density maps in 110 plane. Positive continuous and negative dashes lines represent spin-up and spin-down densities, respectively.

$$\hat{H}_{\text{Ising}} = -\sum_{i,j} J_{ij}\hat{S}_{iz} \cdot \hat{S}_{jz} \quad (1)$$

where $\hat{S}_{iz}$ and $\hat{S}_{jz}$ are the spin operators for sites i and j, respectively, and $J_{ij} < 0$ corresponds to antiferromagnetic order. The extraction of J from the Ising Hamiltonian has been justified on the basis that particular eigenfunctions have equivalent expectation values for both the Heisenberg and Ising Hamiltonians, and thus a mapping exists between the model Hamiltonians [22]. However, in CrO_2, only Z nearest neighbor magnetic ions are considered ($Z = 8$ for CrO_2), so the magnetic coupling constant J can be extracted from the relation $\Delta E = JS^2Z$ [14]. An estimate of the Curie temperature T_C can then be obtained within the mean-field approximation:

$$T_C = \frac{S(S+1)}{3k_B} J \quad (2)$$

where k_B is Boltzman's constant [23]. Values of J and T_C are listed in Table 1, as well as the net atomic charges q and spin moments μ obtained from the Mulliken population analysis. Previous work has shown that hybrid-exchange functionals with 35% Fock exchange can lead to a proper description of magnetic coupling, but other functionals often overestimate J [14]. Indeed, our B3LYP magnetic coupling constant was overestimated by $\sim$ 30%. In the ferromagnetic case, all of the functionals produced Cr spin moments on the order of 2 μ_B. Fig. 1 shows the charge and spin density maps of the ferromagnetic ground state from the B3LYP calculation.

The calculated DOS from LDA and PBE are nearly identical, and indeed bear strong resemblance to those

Table 1: The energy difference ΔE (eV) between FM and AFM phases relative to FM phase, magnetic coupling constant J (eV), mean-field Curie temperature (K), net atomic charges $q = Z - (n_\alpha + n_\beta)$ (e) and spin moments $\mu = n_\alpha - n_\beta$ (μ_B).

				FM				AFM			
				q		μ		q		μ^*	
	ΔE	J	T_C	Cr	O	Cr	O	Cr	O	Cr	O
LDA	0.348	0.044	340.412	1.926	-0.963	2.090	-0.045	1.910	-0.955	1.631	0.006
PBE	0.750	0.094	727.245	1.961	-0.981	2.175	-0.087	1.948	-0.974	1.699	0.010
B3LYP	0.523	0.065	502.882	2.058	-1.029	2.295	-0.148	2.065	-1.033	2.048	0.011
PBE0	0.331	0.041	317.202	2.075	-1.037	2.428	-0.214	2.101	-1.050	2.113	0.019
UHF	0.000			2.575	-1.287	2.151	-0.076	2.224	-1.112	2.957	0.528
Exp.			390-400								

* Magnitudes of spin moments

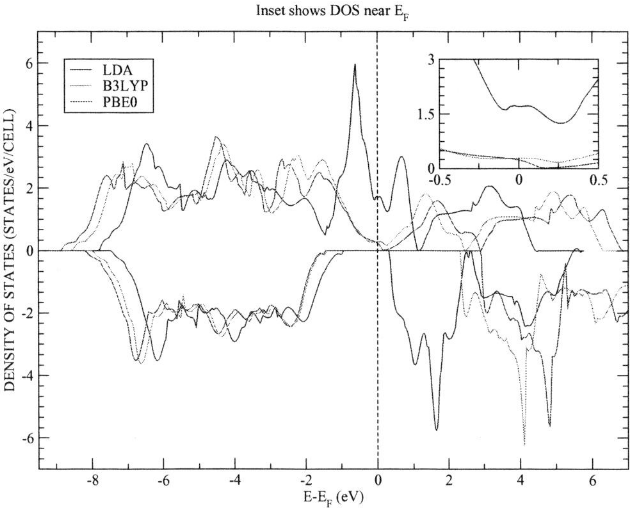

Figure 2: Spin-resolved DOS from LDA, B3LYP and PBE0 functionals. Inset shows DOS near E_F.

produced by previous authors. The notable exception is the fact that PBE predicts a slightly larger spin-down band gap (1.9 eV) than LDA (1.5 eV). To our knowledge, the only experimental study that has reported the size of this band gap is infrared spectroscopy work by Singley *et al.*, in which they ascribe a resonance of $\omega = 270000\text{cm}^{-1} = 3.35$ eV to this gap [12]. As expected, both LDA and PBE calculations underestimate this band gap.

Our results show several important differences in the DOS between the hybrid-exchange and LDA-based functionals. We note that, unlike the previous LDA+U studies, B3LYP and PBE0 show the formation of a psuedogap near E_F in the spin-up electrons (Fig. 2). The small but finite DOS at E_F is in good qualitative agreement with recent photoelectron spectroscopy studies [8]. Moreover, the spin-down band gaps calculated within B3LYP (3.5 eV) and PBE0 (4.3 eV) are in much better agreement with the experimental value than LDA or PBE. Our results agree quite well with the previous work [15] suggesting B3LYP to be an appropriate functional for estimating band gaps. However, all reported theoretical values to date are on the order of eV, and it remains unclear how the magnitude of this band gap relates to the temperature (80-100 K) at which spin-flip scattering is observed in the form of T^2 resistivity dependence. Our band calculations do not fully include many-body correlation effects, so this difference in spin-flip energy will remain to be elucidated for now. However, in the spirit of realizing the potenial of CrO_2 as a spintronic material, the available states at E_F and their dependence on T must be investigated. Recently, Chioncel *et al.* have employed LDA with dynamical mean field theory (LDA+DMFT) as a means to address dynamical correlation effects. They have reported spin-down low-energy excitations at E_F in semi-quantitative agreement with spin-polarization measurments at similar temperatures [17].

Transition metal (TM) oxides can be classified in terms of the on-site d-d Coulomb energy U, p-d charge transfer energy Δ and the ligand p-metal d hybridization energy t, based on a scheme proposed by Zaanen, Sawatzky, and Allen (ZSA) [24]. The Coulomb energy U is defined as the energy required to remove an electron from a d orbtial and place it in another d orbital site, $U = E(d^{n+1}) + E(d^{n-1}) - 2E(d^n)$, where hybridization is neglected, and E always represents the lowest multiplet configuration. The charge transfer energy Δ is the energy required to excite a ligand p electron to a transition metal d orbital, $\Delta = E(d^{n+1}\underline{L}) - E(d^n)$, where L denotes a ligand hole. TM oxides are considered to be Mott-Hubbard insulators when $U < \Delta$ and charge-transfer insulators when $U > \Delta$. In this formalism, we look to the partial Cr and O DOS for evidence of U and Δ in the bulk electronic structure. In the calculated DOS from LDA and PBE, the Cr-3d and

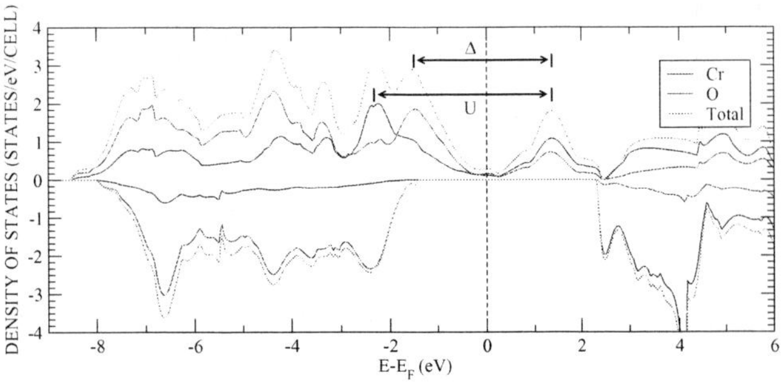

Figure 3: Partial Cr and O DOS from B3LYP.

O-2p peaks below E_F occur at the same energy, i.e., $U \simeq \Delta$, which places CrO_2 on the border on the Mott-Hubbard and charge-transfer regimes, similar to antiferromagnetic Cr_2O_3. However, the hybrid-exchange functionals produce results in which $U > \Delta$, supporing the work by Korotin *et al.* with the classification of CrO_2 as a charg-transfer insulator. Qualitative estimates of U (and Δ) are defined as the energy gaps between the Cr-3d (O-2p) peak below E_F and the Cr-3d peak above E_F (Fig. 3). We also wish to point out that B3LYP places the Cr-3d peak at 2.3 eV below E_F, in excellent agreement with recent photoelectron spectroscopy data by Ventrice *et al.* [8].

The classification of CrO_2 as a charge-transfer insulator is supported by the B3LYP and PBE0 band structures, in which one of the bands crossing E_F is of pure O-2p character. As expected, the band structure from B3LYP (shown in Fig. 4) and PBE0 are typical of a half-metallic ferromagnetic in that they appear 100% spin-polarized with a metallic density of spin-up states at E_F. The hybrid functionals predict more Cr-3d-O-2p hybridization than LDA or PBE, consistent with LDA+U results that led authors to conclude that the ground state of CrO_2 is ferromagnetic through the double exchange mechanism [13].

4 Conclusions

The electronic structure of bulk CrO_2 was investigated using hybrid-exchange density functional theory. We showed that the hybrid-exchange functionals give better theoretical predictions of the energy gap in the minority-spin states and an improved description of strong electron-electron interactions. We emphasize that future theoretical methods should include an accurate treatment of electron-correlation effects in order to properly understand the nature of CrO_2.

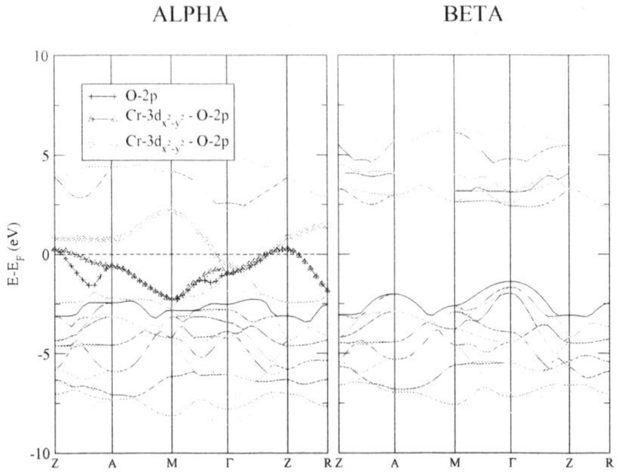

Figure 4: Calculated band structure from B3LYP.

REFERENCES

[1] A. D. Becke, *J. Chem. Phys.*, **98**, 5648 (1993)

[2] C. Adamo and V. Barone, *J. Chem. Phys.* **110**, 6158 (1999)

[3] K. Schwarz, *J. Phys. F: Met. Phys.*, **16** L211-L215 (1986)

[4] Perdew *et al.*, *Rev. Rev. B.*, **23**, 5048-5079 (1981)

[5] W. E. Pickett, *Rev. Mod. Phys.*, **61**, 433 (1989)

[6] Soulen *et al.*, *Science*, **282**, 85-88 (1998)

[7] C. B. Stagarescu *et al.*, *Phys. Rev. B*, **61** 14, R9233-R9236 (2000)

[8] C. A. Ventrice *et al.*, *J. Phys.: Condens. Matter* **19** 315207 (2007)

[9] P. Schottmann, *Phys. Rev. B.* **67** 174419 (2003)

[10] S. M. Watts *et al.*, *Phys. Rev. B* **61** 14 9621-9626 (2000)

[11] K. Suzuki and P. M. Tedrow, *Phys. Rev. B*, **58**, 17 11597-11602 (1998)

[12] E. J. Singley *et al.*, *Phys. Rev. B*, **60**, 6 4126-4130 (1999)

[13] Korotin *et al.*, *Phys. Rev. Lett.* **80**, 19 4305-4308 (1998)

[14] X. B. Feng and N. M. Harrison, *Phys. Rev. B*, **70**, 092402 (2004)

[15] Muscat *et al. Chem. Phys. Lett.* **342**, 3-4 397-401 (2001)

[16] X. B. Feng, *Phys. Rev. B*, **69**, 155107 (2004)

[17] Chioncel *et al. Phys. Rev. B.* **75**, 140406(R) (2007)

[18] V. R. Saunders *et al.*, CRYSTAL06 Users Manual, Universit di Torino, Torino, 2006.

[19] J. P. Perdew *et al. Journal Chem. Phys.* **105**, 12 9982-9985 (1996)

[20] M. Catti *et al. J. Phys. Chem. Solids* **57**, 1735-1741 (1996)

[21] M. Catti *et al. Phys. Rev. B* **49**, 14179 (1994)

[22] I. de P.R. Moreira and F. Illas, *Phys. Rev. B*, **55** 7, 4129-4136 (1997)

[23] N.W. Ashcroft and N.D. Mermin, *Solid State Physics*, Holt, Rinehart, and Winston, New York (1976)

[24] Bocquet *et al. Phys. Rev. B.* **53**, 3 1161-1169 (1996)

A Theoretical Study on Chemical Reaction of Water Molecules under Laser Irradiation: Ultra-Accelerated Quantum Chemical Molecular Dynamics Approach

A. Endou*, A. Nomura*, Y. Sasaki*, K. Chiba*, H. Hata*, K. Okushi*, A. Suzuki**, M. Koyama*, H. Tsuboi*, N. Hatakeyama*, H. Takaba*, M. Kubo*, C.A. Del Carpio*, M. Kitada***, H. Kabashima*** and A. Miyamoto**,*

*Department of Applied Chemistry, Graduate School of Engineering, Tohoku University, 6-6-11-1302 Aoba, Aramaki, Aoba-ku, Sendai 980-8579, Japan

**New Industry Creation Hatchery Center, Tohoku University, 6-6-10 Aoba, Aramaki, Aoba-ku, Sendai 980-8579, Japan

***Automobile R&D Center, Honda R&D Co., Ltd., 4630 Shimotakanezawa, Haga-machi, Haga-gun, Tochigi 321-3393, Japan

ABSTRACT

Recently, we succeeded in realizing ultra acceleration of our tight-binding quantum chemical molecular dynamics (UA-QCMD) simulator, "New-Colors", which is more than 10,000,000 times faster the traditional first-principles molecular dynamics method. It means that a simulation scale of water molecules using UA-QCMD method is easily extended from angstrom scale to nanoscale. It was shown that our New-Colors simulator was effective to perform nanoscale model of water under laser irradiation. This suggests the possibility of the extension of the findings of bond dissociation of water molecules induced by laser irradiation in the electronic level to the macroscopic behavior. Our group also develops a novel multi-level computational chemistry approach for the phase change of water under laser irradiation, by using the kinetic Monte Carlo method.

Keywords: ultra accelerated quantum chemical molecular dynamics, water, infrared laser, bond dissociation

1 INTRODUCTION

Dynamic behavior of water molecules is of paramount interest in various fields not only physicochemical researches but also chemical industry, biochemistry, and nanotechnology. With recent development of laser apparatus, a number of studies have been performed to clarify the dynamics of water molecules under laser irradiation condition. However, there still exist extreme difficulties in the experimental studies because no direct observation method in the electronic and atomistic levels is available experimentally under laser irradiation. Theoretical approaches such as molecular dynamics (MD) and quantum chemistry (QC) may shed light on clarifying and/or understanding the dynamic behaviors of water molecules under laser irradiation. In this sense, the first-principles molecular dynamics (FPMD) method is powerful tool in order to understand a dynamics of matter from the electronic and atomistic points of view. However, FPMD method requires extremely huge computation cost.

On the other hand, our group has developed original tight-binding quantum chemical molecular dynamics (TB-QCMD) simulator, "Colors" [1-5], based on the TB approximation. "Colors" has an advantage in computation time compared to the first-principles molecular dynamics (FPMD). Recently, we succeeded to develop a novel ultra-accelerated QCMD (UA-QCMD) simulator, "New-Colors", which realizes 10,000,000 times faster calculation than the conventional FPMD. This easily enables us the extension of simulation scale from angstrom to nanometer, by using UA-QCMD method. In addition, it is expected that theoretical findings obtained by large-scale electronic/atomistic levels simulation are worthwhile in applying to macroscopic simulations such as kinetic Monte Carlo (KMC) simulation and computational fluid dynamics (CFD) simulation. Our group has attempted to establish the multi-level simulation method based on the large-scale electronic/atomistic simulations, KMC, and CFD simulation. In this paper, we present the development of novel large-scale electronic/atomistic simulation method, UA-QCMD method and its application to the chemical reaction dynamics of water molecules under laser irradiation condition.

2 METHODS

The main part of New-Colors simulator consists of two parts: our TB-QCMD simulator, Colors [1-5], and our MD simulator, New-RYUDO [6]. The former simulator is used for the determination of Morse-type 2-body interatomic potential functions (eq. (1)) between atoms, by performing electronic structure calculations:

$$E_{AB} = D_{AB} \cdot \left\{ \exp\left[-2\beta_{AB}\left(R_{AB} - R_{AB}^{*} \right) \right] - 2\exp\left[-\beta_{AB}\left(R_{AB} - R_{AB}^{*} \right) \right] \right\} \quad (1)$$

where E_{AB}, D_{AB}, β_{AB}, R_{AB}, and R^*_{AB} refers to the interatomic potential energy between atoms A and B, binding energy between atoms A and B, factor for potential curve, interatomic distance between atoms A and B, and equilibrium interatomic distance between atoms A and B, respectively. The latter simulator, New-RYUDO, is used for classical MD simulation.

2.1 Electronic Structure Calculations

In TB-QCMD simulator, the electronic structure calculation is performed by solving the Schrödinger equation (**HC** = ε**SC**; **H**, **C**, ε, and **S** refers to the Hamiltonian matrix, eigenvectors, eigenvalues, and overlap integral matrix, respectively) with the diagonalization condition ($\mathbf{C}^T\mathbf{SC} = \mathbf{I}$; **I** refers to the unit matrix). The level of theory corresponds to the extended Hückel method. To determine the off-diagonal elements of **H**, H_{rs}, the corrected distance-dependent Wolfsberg-Helmholz formula [7] (eq. (2)) was used.

$$H_{rs} = \frac{K}{2} S_{rs} \left(H_{rr} + H_{ss} \right) \tag{2}$$

To solve the Schrödinger equation in this simulator, parameters for Hamiltonian matrix **H** are used, which are derived on the basis of first-principles calculation results. The details of the parameterization based on the first-principles method are published elsewhere [5, 8]. A total energy of a system is obtained by using the following equation:

$$E = \sum_{i=1}^{N} \frac{1}{2} m_i v_i^2 + \sum_{k=1}^{\mathrm{occ}} n_k \varepsilon_k + \sum_{i=1}^{N} \sum_{j=i+1}^{N} \frac{Z_i Z_j e^2}{r_{ij}} + \sum_{i=1}^{N} \sum_{j=i+1}^{N} E_{ij}^{\mathrm{repul}} \left(r_{ij} \right) + \sum_{i=1}^{N} E_{corr} \left(Z_i \right) \tag{3}$$

where the first, second, third, fourth, and fifth term on the right-hand side refers to the summation of the kinetic energy of nuclei, molecular orbital (MO) energy, Coulombic energy, exchange-repulsion energy, and energy correction term, respectively. The second term on the right-hand side of eq. (3) is rewritten as follows:

$$\sum_{k=1}^{\mathrm{occ}} n_k \varepsilon_k = \sum_{k=1}^{\mathrm{occ}} \sum_{r} n_k \left(C_{kr} \right)^2 H_{rr} + \sum_{k=1}^{\mathrm{occ}} \sum_{r} \sum_{s} n_k C_{kr} C_{ks} H_{rs} \tag{4}$$

where the first and second term on the right-hand side refers to the monoatomic contribution to the binding energy and the diatomic contribution to the binding energy, respectively (n_k is the number of electrons occupied in k-th molecular orbital). The binding energy calculated from the second term of eq. (4) is used for the determination of D_{AB} parameter in eq. (1).

Table 1: Phsycochemical property of H_2O molecule.

	Our TB-QCMD	First-principles
Atomic charge		
H	0.16	0.15
O	-0.32	-0.30
Electronic configuration		
H	$(1s)^{0.84}$	$(1s)^{0.85}$
O	$(2s)^{1.66}(2p)^{4.66}$	$(2s)^{1.66}(2p)^{4.65}$
Bond population		
H-O	0.68	0.51
H-H	-0.14	-0.13
Binding energy / kcal mol^{-1}	-244.4	-244.0

2.2 Molecular Dynamics Calculations

Our MD simulator, New-RYUDO [6] can perform classical MD simulations by solving equation of motion for atoms. This simulator was developed based on MXDORTO code [9]. In this MD simulator, Verlet algorithm [10] is employed to integrate equation of motion. Temperature scaling method equivalent to the Woodcock algorithm [11] is implemented. In the present study, we also implemented the scaling method for total energy of a system, in order to realize the addition of external energy to a simulation model.

3 RESULTS AND DISCUSSION

3.1 Parameterization and Justification

First, we decided the parameter for TB-QCMD simulator so as to reproduce the binding energy and electronic structure of water molecule. In the first-principles calculations, we used DMol^3 code [12]. Initial structure of water molecule was prepared by geometry optimization calculation using DMol^3 code. The calculated interatomic distance of O-H bond and bond angle among H-O-H was 0.969 Å and 103.7 °, respectively. Table 1 shows the atomic charges, electronic configuration (atomic orbital population), bond population, and binding energy of H_2O molecule obtained by our TB-QCMD method and DMol^3 code. From this table, it is found that the electronic states and binding energy on water molecules obtained by our TB-QCMD simulator are successfully reproduced those obtained by the first-principles method. We also calculated the normal vibration frequency of water molecule based on the GF matrix method. Calculated frequency for the H-O stretching vibration mode and for the H-O-H bending vibration mode was respectively 3696 and 1703 cm^{-1}, which qualitatively agrees with the experimental values

(3756 and 1595 cm^{-1}). Thus, the parameterization in the present study was justified.

3.2 Dynamic Behavior of Water Molecules at High Temperature

Thermal equilibrium state of water molecules and hydrogen and oxygen ($H_2O = H2 + (1/2)O_2$) can be realized at 1750 K, by rough estimation. We constructed simulation model with 60 molecules of water. The density was adjusted to 1.0 g cm^{-3}. First, MD simulation using New-RYUDO simulator was performed at 1750 K with NVT ensemble. The total number of MD steps and integration time step was 100,000 and 0.1×10^{-15} s, respectively. Next, by using both New-Colors and New-RYUDO, we carried out the UA-QCMD simulation of water molecules without velocity scaling. Our preliminary TB-QCMD study showed the possibility of the formation of dissociated species such as H_3O and OH (detailed data was not shown here). In the present simulation, we also found the similar situation. The collision of these species resulted in the formation of two H_2O molecules. To study the other situation regarding the collision of H_3O and OH species surrounded by water molecules, the dynamic behavior of water molecules at 1750 K, starting with the simulation model that contains H_3O and OH species as shown in Figure 1, was also investigated. By changing the initial configuration of the dissociated species, we obtained the formation dynamics of H_2 molecule from two H_3O species. In such a case, the change in the H-H distance and binding energy with time was analyzed (Table 2). In Table 2, H-H distance and binding energy of H-H at 5×10^{-12} s and 10×10^{-12} s after H_2 formation are summarized. This suggests the existence of H_2 molecule formed by two H_3O species in 10^{-12} s order of time period.

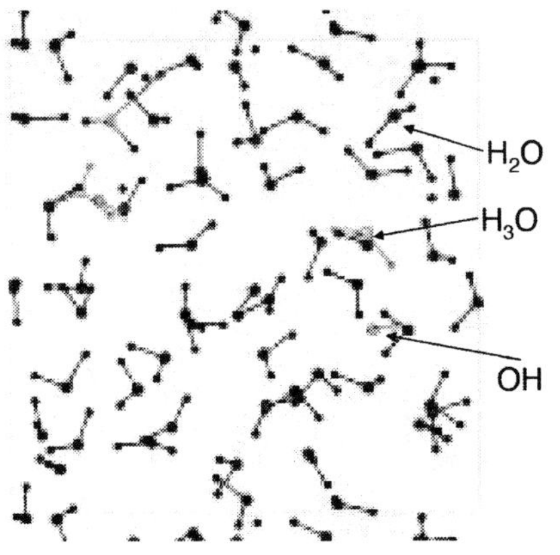

Figure 1: Initial structure for UA-QCMD simulation at 1750 K, including H_3O and OH species.

Table 2: Property of formed H2 molecule.

	5 °— 10^{-12} s after	10 °— 10^{-12} s after
H-H Distance / Å	0.841	0.680
Binding energy / kcal mol-1	-98.1	-78.7

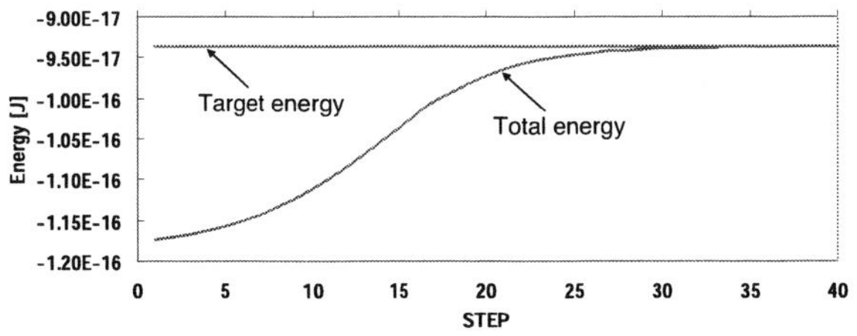

Figure 2: Result of total energy scaling by using New-RYUDO simulator.

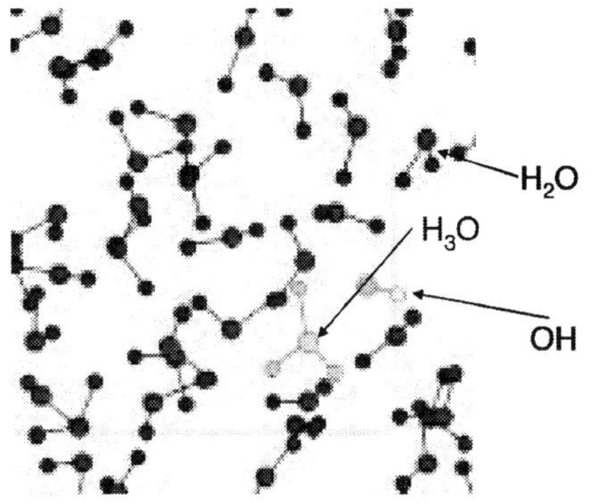

Figure 3: Snapshot of UA-QCMD simulation at 300 K, adding external energy.

3.3 Dynamic Behavior of Water Molecules added External Energy

Next, we studied the dynamic behavior of water molecules added the external energy. Thermal equilibrium state of water molecules and hydrogen and oxygen can be also realized at 300 K, by adding external energy, *ca.* 56.4 kcal mol^{-1}. To construct the simulation model, MD simulation using New-RYUDO simulator was performed at 300 K with NVT ensemble. The total number of MD steps and integration time step was 100,000 and 0.1×10^{-15} s, respectively. After NVT simulation, we added the external energy by using the energy scaling function implemented in New-RYUDO simulator. Figure 2 shows the change in total energy of the present system (water molecules) with the MD steps. From this figure, we can see that the scaling of the total energy of a system works well.

We further performed UA-QCMD simulation for the obtained models of water molecules without velocity scaling. In the present condition, similar to the former section, we observed the formation of H_3O and OH species as shown in Figure 3. In this condition, the exchange of hydrogen between H_3O species and water molecule, just like as proton exchange, was observed. The other UAQCMD simulation in which the initial model contained H_3O and OH species, similar to the former section, was also performed. In this case, as expected, it was also observed the formation of H_2 molecule by collision of two H_3O species.

3.4 Dynamic Behavior of Water Molecules under Laser Irradiation Condition

Finally, the dynamic behavior of water molecules at 300 K under laser irradiation was studied by using our UAQCMD simulator. O-H bond of water molecule was excited vibrationally, by adding the additional velocities on H atom of water molecule. The change in O-H interatomic distances with time was depicted in Figure 4. After collision of two H_2O molecules, the formation of two species (H_3O and OH) was immediately occurred. Interestingly, it was found that the vibrational excitation for the smaller system consisting of 10 water molecules in the present UA-QCMD simulation increased slightly the pressure of the system. It is expected, theoretically, to change a local environment of water molecules in the atomistic level.

In our preliminary TB-QCMD study using the smaller simulation model, it was suggested that the combination of the vibrational excitation and the addition of external energy to water molecules was effective to form the dissociated species such as H_3O and atomic hydrogen. In the present study, we also obtained the same tendency qualitatively. The fate of formed species, H_3O and OH, are in calculating and analyzing, which will be presented in the conference.

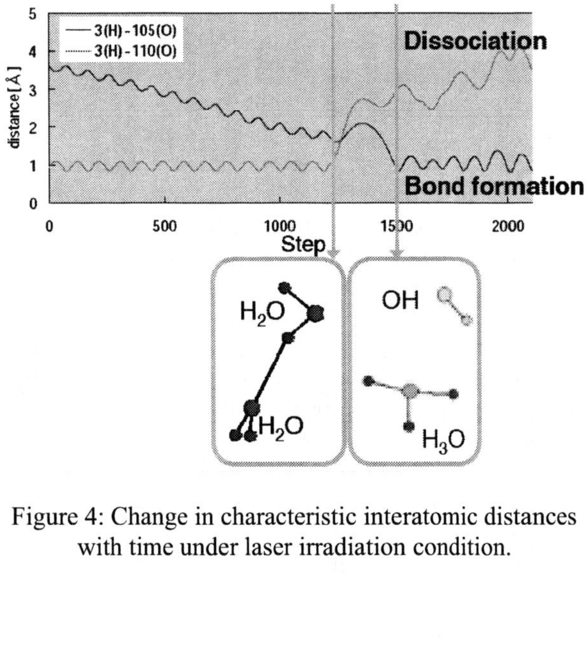

Figure 4: Change in characteristic interatomic distances with time under laser irradiation condition.

4 SUMMARY

In the present paper, our original ultra-accelerated quantum chemical molecular dynamics (UA-QCMD) simulator, "New-Colors", which realizes 10,000,000 times faster calculation than the conventional first-principles molecular dynamics method, and its application to the chemical reaction dynamics of water molecules was introduced. The justification of our UA-QCMD simulator was shown by the excellent agreement of UA-QCMD result with the first-princpiles results for physicochemical properties of water molecule. Dynamic behaviors of water molecules under high temeperature condition, under the addition of external energy, and under the laser irradiation condition were simulated by our UA-QCMD code. These results suggest the formation of dissociated species such as H_3O and OH from water molecules. The details of the chemical reaction dynamics of water under laser irradiation will be presented at the conference.

REFERENCES

[1] M. Elanany, P. Selvam, T. Yokosuka, S. Takami, M. Kubo, A. Imamura, and A. Miyamoto, J. Phys. Chem. B, 107, 1518, 2003.
[2] Y. Luo, Y. Ito, H.-F. Zhong, A. Endou, M. Kubo, S. Manogaran, A. Imamura, and A. Miyamoto, Chem. Phys. Lett., 384, 30, 2004.
[3] R. Ishimoto, C. Jung, H. Tsuboi, M. Koyama, A. Endou, M. Kubo, C.A. Del Carpio, and A. Miyamoto, Appl. Catal. A:Gen., 305, 64, 2006.
[4] K. Kasahara, H. Tsuboi, M. Koyama, A. Endou, M. Kubo, C. A. Del Carpio, and A. Miyamoto, Electrochem. Solid State Lett., 9, A490, 2006.
[5] A. Endou, H. Onuma, S. Jung, R. Ishimoto, H. Tsuboi, M. Koyama, H. Takaba, M. Kubo, C. A. Del Carpio, and A. Miyamoto, Jpn. J. Appl. Phys., 46, 2505, 2007.
[6] P. Selvam, H. Tsuboi, M. Koyama, A. Endou, H. Takaba, M. Kubo, C. A. Del Carpio, and A. Miyamoto, Rev. Eng. Chem., 22, 377, 2006.
[7] G. Carzaferri, L. Forss, I. Kamber, J. Phys. Chem. 93, 5366, 1989.
[8] A. Suzuki, P. Selvam, T. Kusagaya, S. Takami, M. Kubo, A. Imamura, and A. Miyamoto, Int. J. Quant. Chem., 102, 318, 2005.
[9] K. Kawamura, MXDORTO, Japan Chemistry Program Exhange, #029.
[10] L. Verlet, Phys. Rev., 159, 98, 1967.
[11] L. V. Woodcock, Chem. Phys. Lett., 10, 257, 1971.
[12] B. Delley, J. Chem. Phys., 92, 508, 1990; B. Delley,J. Chem. Phys., 113, 7756, 2000.

Density Gradient Quantum Surface Potential

Hedley Morris*, Ellis Cumberbatch*, Henok Abebe**, Vance Tyree** and Darryl Yong***

* Claremont Graduate University, CA, USA,hedley.morris@cgu.edu, ellis.cumberbatch@cgu.edu
** University of Southern California, MOSIS service, CA, USA , abebeh@ISI.edu, tyree@mosis.com
*** Harvey Mudd College, Claremont, CA, USA, dyong@hmc.edu

ABSTRACT

The PSP model [1], a generalized Surface-Potential (SP) model, has been chosen to be an industry standard for the next generation, 60nm, technology. In order to include quantum effects within surface potential models we use asymptotic analysis techniques [2], [3] applied to the the Density-Gradient equations [4]. Our quantum modified surface potential (SP) results from the ability to obtain a first integral of the Poisson equation.

Keywords: Surface Potential, Density Gradient, Quantum corrections, Asymptotic methods

1 The Quantum Problem

For MOSFETs, the ratio of the maximum channel dopant concentration to the intrinsic level is normally very large, therefore following [6], a large parameter λ is introduced as

$$\lambda = \max\left|\frac{N(x_1)}{n_i}\right| = \frac{N_A}{n_i},$$

where N_A is the substrate doping and n_i is the intrinsic carrier density in silicon. To normalize the electron concentration in the device, the following rescaling is performed

$$(\psi, \Phi_n) = (w, \phi)V_{th}\ln\lambda,\ x_1 = xL_D\left(\frac{2\ln\lambda}{\lambda}\right)^{1/2}$$

where V_{th} is the thermal voltage, x_1 is the coordinate perpendicular to the motion of carriers. The quantity $L_D = \left(\frac{kT\varepsilon_{si}}{n_i q^2}\right)^{1/2}$ is the Debye length.

In the DD model, the electron and hole drift potentials are $\phi_n = \phi_p = \phi$, but according to the DG model, ϕ_n and ϕ_p are given by

$$\begin{aligned}\phi_n &= \phi + \phi_{qn},\ \phi_{qn} = 2b_n\frac{\nabla^2\sqrt{n}}{\sqrt{n}}\\ \phi_p &= \phi + \phi_{qp},\ \phi_{qp} = 2b_p\frac{\nabla^2\sqrt{p}}{\sqrt{p}}\\ b_n &= \frac{\hbar^2}{4r_n m_n q},\ b_p = \frac{\hbar^2}{4r_p m_n q}\end{aligned}$$

where n and p are the electron and hole densities respectively, $\hbar$ is Planck's constant, r is a fitting parameter, m is the mass of the electron and q is its charge. The quantum corrections ϕ_{qn} and ϕ_{qp} are derived from the Schrödinger equation, based on the finite curvature (energy) and strict continuity of wave functions.

In the DG model, a microscopic quantum description is used in regions with dominant quantum effects, and a macroscopic (fluid-type) model is employed in subregions where collisional effects are expected to be dominant [5]. In the expression for ϕ_n, the correction ϕ_{qn} to the quasi-Fermi potential is derived using Boltzmann statistics. With the quantum corrections to the DD model, the governing equations become

$$\frac{d^2w}{dx^2} = \frac{n-p}{N_A} + 1, \tag{1a}$$

$$w - \phi = \frac{1}{\ln\lambda}\ln\left(\frac{n}{n_i}\right) - \frac{\lambda\beta^2}{(\ln\lambda)^2}\frac{1}{\sqrt{n}}\frac{d^2\sqrt{n}}{dx^2}, \tag{1b}$$

where $\beta^2 = \frac{2b_n}{V_{th}L_D^2}$. The quantum correction does not affect the boundary conditions for the potential at the Si/SiO_2 interface, but using the well-accepted boundary condition for the electron density, the boundary conditions for (1) are $n(0) = 0$, $w(0) = w_s$, and $w(\infty) = w_\infty \sim -1$, as well as the Robin boundary condition

$$\left.\frac{dw}{dx}\right|_{x=0} = c(w_s - V_{gs}), \tag{2}$$

2 The Subthreshold Case

The subthreshold case corresponds to the weak inversion regime, meaning $0 \le w_s(V_{gs}) \le 1$. In this case, the dominant contribution to the space charge density near the interface arises from the immobile acceptor ions, N_A [6]. Thus, the system of equations for the subthreshold case becomes

$$\frac{d^2w}{dx^2} = 1, \tag{3a}$$

$$w - \phi = \frac{1}{\ln\lambda}\ln\left(\frac{n}{n_i}\right) - \frac{\lambda\beta^2}{(\ln\lambda)^2}\frac{1}{\sqrt{n}}\frac{d^2\sqrt{n}}{dx^2}. \tag{3b}$$

We will determine the potential and electron density using matched asymptotic expansions (MAE's). This is done by introducing a quantum inner layer near the interface governed by (3) and an outer depletion layer governed by the classical model, and properly matching them.

3 Asymptotics

In the region near the interface, the quantum inner layer, the device behavior is drastically different than in the classical case. Solutions in this layer are obtained by formal expansions in a small parameter, ε. Following [7], the quantum inner layer independent variable scaling is

$$X = \frac{x}{\varepsilon} \qquad \text{where} \qquad \varepsilon^2 = \frac{\lambda\beta^2}{2\ln\lambda}.$$

The dependent variables of (3) are also expanded formally in powers of ε as

$$(w, T, Y) = (W_0, T_0, Y_0) + \varepsilon(W_1, T_1, Y_1) + \mathcal{O}(\varepsilon^2),$$

where it is convenient, as in [7], to use the following rescaled variables

$$T = \sqrt{\frac{n}{n_i}} \qquad \text{or} \qquad Y = \frac{1}{\ln\lambda}\ln\left(\frac{n}{n_i}\right). \tag{4}$$

When the scaled independent variable X is substituted into (3a), the equation reads

$$\frac{d^2w}{dX^2} = \varepsilon^2 \qquad \text{thus} \qquad \frac{d^2}{dX^2}(W_0 + \varepsilon W_1 + \mathcal{O}(\varepsilon^2)) = \varepsilon^2.$$

Equation (3b) can be rewritten in terms of either T or Y as

$$\frac{1}{T}\frac{d^2T}{dX^2} - \ln T + \frac{\ln\lambda}{2}(w-\phi) = 0, \tag{5a}$$

$$\frac{d^2Y}{dX^2} + \frac{1}{2}\ln\lambda\left(\frac{dY}{dX}\right)^2 + w - \phi - Y = 0. \tag{5b}$$

The boundary condition is also written in terms of the scaled variables as

$$\frac{d}{dX}(W_0 + \varepsilon W_1 + \mathcal{O}(\varepsilon^2))\bigg|_{X=0} = \varepsilon c(W_{0s} - V_{gs}). \tag{6}$$

The value of the potential at $x = 0$ in this paper is defined as $w_s \equiv W_{0s}$. Separating in orders of ε and using the boundary condition (6) to solve the resulting systems, the inner solution $W(X) = W_0(X) + \varepsilon W_1(X) + \mathcal{O}(\varepsilon^2)$ is

$$W(X) = W_{0s} + \varepsilon c(W_{0s} - V_{gs})X + \mathcal{O}(\varepsilon^2). \tag{7}$$

This quantum inner layer solution represents the device behavior close to the interface, on the order of the reference length, $L_D\sqrt{\ln\lambda/\lambda}$, which has a value of 123 nm for $\lambda = 10^6$ [7]. Equation (5b) will describe the quantum potential's coupling to the electron density in this layer.

Proceeding the quantum inner layer is the depletion outer layer, because in the subthreshold case, the device has not yet gone into inversion, thus no inversion layer is present. The equations for the outer depletion layer are

$$\frac{d^2w}{dx^2} = 1, \tag{8a}$$

$$n(x) = n_i e^{(w-\phi)\ln\lambda}. \tag{8b}$$

Expanding the depletion potential and boundary condition using $w(x) = w_0(x) + \varepsilon w_1(x) + \mathcal{O}(\varepsilon^2)$ and grouping into orders of ε, we arrive at the following depletion layer solution:

$$w(x) = \frac{1}{2}x^2 + (a_0 + \varepsilon a_1)x + (b_0 + \varepsilon b_1) + \mathcal{O}(\varepsilon^2). \tag{9}$$

The coefficients $(a, b) = (a_0, b_0) + \varepsilon(a_1, b_1) + \mathcal{O}(\varepsilon^2)$ are determined by matching the depletion potential on the left with the quantum layer, and on the right with a transition layer that is used to blend the solution with the bulk. To match the depletion layer with the quantum layer, an intermediate variable is introduced:

$$x_\eta = \frac{x}{\eta(\varepsilon)}, \qquad \text{where} \qquad \varepsilon \ll \eta(\varepsilon) \ll 1.$$

This gives $a_0 = c(W_{0s} - v_{gs})$, $b_0 = W_{0s}$, $b_1 = 0$. No information on the coefficient a_1 can be extracted by the matching with the quantum layer, so this information comes by matching the depletion layer with the bulk solution, $w_b(x) = -1 + \mathcal{O}(1/\lambda^2\ln\lambda)$. The matching of the depletion layer with the bulk is done in [6] by introducing a transition layer about some unknown depth x_d (referred to as the depletion width) where the proper scalings are

$$x_t = (x - x_d)(\ln\lambda)^{1/2}, \; w = w_t(x_t) = -1 + \frac{h_0(x_t)}{\ln(\lambda)}.$$

In terms of these new scaled variables, the transition layer equation for h_0 is

$$\frac{d^2h_0}{dx_t^2} = 1 - e^{-h_0},$$

where $h_0(\infty) = 0$ is needed to match to the bulk. It is not possible to explicitly integrate this equation, however a first integral provides the implicit expression

$$-\sqrt{2}x_t = \int_1^{h_0}(e^{-y} + y - 1)^{-1/2}\,dy.$$

Analogous to the matching of the depletion layer with the quantum layer, we define an intermediate variable

$$x_\eta = \frac{x - x_d}{\eta(\lambda)} \qquad \text{where} \qquad \frac{1}{(\ln\lambda)^{1/2}} \ll \eta(\lambda) \ll 1.$$

Expanding the implicit transition layer solution as $x_t \to \infty$ or alternatively as $h_0 \to \infty$ provides

$$w_t(x_\eta) \sim \frac{1}{2}\eta^2 x_\eta^2 - \frac{k\eta x_\eta}{\sqrt{2\ln\lambda}} + \frac{1}{\ln\lambda}\left(\frac{k^2}{4}+1\right) - 1,$$

where

$$k = \int_1^\infty [(y-1)^{-1/2} - (e^{-y}+y-1)^{-1/2}]\,dy \approx 0.81785.$$

See [6] for more details describing this numerical result. Expanding the depletion solution (9) using $w(x_\eta) = w_0(x_\eta) + \varepsilon w_1(x_\eta) + \mathcal{O}(\varepsilon^2)$ in terms of the intermediate variable gives

$$\begin{aligned} w(x_\eta) &= \frac{1}{2}(\eta x_\eta + x_d)^2 + c(W_{0s} - V_{gs})(\eta x_\eta + x_d) \\ &\quad + \varepsilon a_1(\eta x_\eta + x_d) + W_{0s} + \mathcal{O}(\varepsilon^2). \end{aligned}$$

Comparing the depletion and transition solutions to $\mathcal{O}(1)$ and $\mathcal{O}(\eta)$ gives the following equations

$$\begin{aligned} \mathcal{O}(1): &\quad \tfrac{1}{2}x_d^2 + (c(W_{0s} - V_{gs}) + \varepsilon a_1)\,x_d \\ &\quad -\tfrac{1}{\ln\lambda}\left(\tfrac{k^2}{4}+1\right) + W_{0s} + 1 = 0 \\ \mathcal{O}(\eta): &\quad x_d + c(W_{0s} - V_{gs}) + \varepsilon a_1 + \tfrac{k}{\sqrt{2\ln\lambda}} = 0 \end{aligned}$$

Solving this system of equations for a_1 and x_d gives

$$a_1 = -\frac{1}{\varepsilon}\left[\sqrt{2}\left(1 + W_{0s} - \frac{1}{\ln\lambda}\right)^{1/2} + c(W_{0s} - V_{gs})\right]$$

$$x_d = \sqrt{2}\left(1 + W_{0s} - \frac{1}{\ln\lambda}\right)^{1/2} - \frac{k}{\sqrt{2\ln\lambda}}.$$

Thus, we are now able to write the full depletion layer solution $w(x) = w_0(x) + \varepsilon w_1(x) + \mathcal{O}(\varepsilon^2)$ as

$$w(x) = \frac{1}{2}x^2 - \sqrt{2}\left(1 + W_{0s} - \frac{1}{\ln\lambda}\right)^{1/2} x + W_{0s} + \mathcal{O}(\varepsilon^2). \tag{11}$$

This expression for the depletion layer potential agrees with the results obtained in the weak inversion-depletion analysis done in [6]. From this expression we notice that the expansion breaks down near flatband where $w_s = -1 + \mathcal{O}(1/\ln\lambda)$ [6]. At this point, it is possible to use the boundary condition (2) for the depletion potential (11) to derive an expression for surface potential W_{0s} in terms of applied gate voltage V_{gs}. The boundary condition states that

$$\left.\frac{dw}{dx}\right|_{x=0} = c(W_{0s} - V_{gs}),$$

thus,

$$-\sqrt{2}\left(1 + W_{0s} - \frac{1}{\ln\lambda}\right)^{1/2} = c(W_{0s} - V_{gs}).$$

Solving the above expression for $W_{0s}(V_{gs})$ gives

$$W_{0s}(V_{gs}) = V_{gs} + \frac{1}{c^2} - \frac{\sqrt{2}}{c}\left(\frac{1}{2c^2} + 1 + V_{gs} - \frac{1}{\ln\lambda}\right)^{1/2}, \tag{12}$$

which gives an explicit value for the surface potential given any gate voltage. Analogously, solving the classical equation in the depletion region gives

$$w_s(V_{gs}) = \frac{1}{c^2} + V_{gs} - \frac{\sqrt{2}}{c}\left(\frac{1}{2c^2} + 1 + V_{gs}\right)^{1/2}. \tag{13}$$

The equations (5) relating the quantum layer potential and electron density is independent of the solution regime being examined, and consequently the solution is the same as in [7] where the strong inversion case is examined. The solution is obtained by substituting the expansion $T = T_0 + \varepsilon T_1 + \mathcal{O}(\varepsilon^2)$ into (5a) and taking $\mathcal{O}(1)$ terms, giving

$$\frac{d^2T_0}{dX^2} - T_0\ln(T_0) + \frac{\ln\lambda}{2}(W_0 - \phi)T_0 = 0,$$

where from (4) we can write

$$\begin{aligned} T_0 &= \exp(Y_0\ln\lambda/2), \\ \tau_{0s} &\equiv T_0(\infty) = \exp((W_{0s} - \phi)\ln\lambda/2). \end{aligned}$$

Using these expressions with the quantum inner solution $W_0(X) = W_{0s}$ and boundary condition, this $\mathcal{O}(1)$ expression can be written as

$$\frac{d^2S_0}{dX^2} = S_0\ln(S_0) \qquad \text{where} \qquad S_0 = \frac{T_0}{\tau_{0s}}.$$

The solution to this equation is available only in implicit form, but an approximation that yields similar asymptotic results is

$$S_0(X) = \tanh(X/2),$$

which has the required behavior at $X \ll 1$ and asymptotic decay for $X \gg 1$ [7]. From this approximation, the expression for the quantum electron density can be expressed in terms of Y as

$$\begin{aligned} Y_0(X) &= \frac{2}{\ln\lambda}\ln(\tau_{0s}\tanh(X/2)) \\ &= W_{0s} - \phi + \frac{2}{\ln\lambda}\ln(\tanh(X/2)). \end{aligned} \tag{14}$$

In the depletion layer, the electron density behaves classically, with a straightforward expression given by (8b).

To correctly represent the composite solution throughout the entire region, the two separately calculated solutions are added together and the common terms are subtracted out. Following [7],

$$Y = \frac{1}{\ln\lambda}\ln\left(\frac{n}{n_i}\right) = w_0 + \varepsilon w_1 + Y_0 - W_{0s} + \cdots.$$

Thus, using this expression and solving for the electron density $n(x)$, the result is given by

$$n(x) = n_i \exp\left[(w(x) + Y_0(x) - W_{0s}) \ln \lambda\right], \qquad (15)$$

where $w(x)$ and $Y_0(x)$ are given respectively by (11) and (14), and W_{0s} is the common term subtracted out.

4 First Integral

An interesting calculation can be done in which the governing equations in the quantum case (1) can be integrated once to give an exact expression for $w_s(V_{gs})$. This is done by first making the substitution $a = \sqrt{n}$ to simplify the equations, which gives

$$w'' = \frac{a^2 - p}{N_A} + 1, \qquad (16a)$$

$$w - \phi = \frac{1}{\ln \lambda} \ln\left(\frac{a^2}{n_i}\right) - \frac{\lambda \beta^2}{(\ln \lambda)^2} \frac{a''}{a}, \qquad (16b)$$

where $p = n_i \exp(-w \ln \lambda)$. Multiplying (16a) by w' gives

$$a^2 w' = N_A(w'w'' - w') + n_i e^{-w \ln \lambda} w', \qquad (17)$$

and multiplying (16b) by aa' gives

$$aa'w - \left[\frac{a^2\phi}{2}\right]' = \frac{1}{2\ln\lambda}\left[a^2\left(2\ln\left(\frac{a}{\sqrt{n_i}}\right) - 1\right)\right]' - \frac{\lambda\beta^2}{(\ln\lambda)^2}\left[\frac{(a')^2}{2}\right]'. \qquad (18)$$

Finally, we use the identity $(a^2 w)' = 2aa'w + a^2 w'$ to combine (17) and (18) into a single equation in which every term can be integrated once. After integrating, we obtain

$$\frac{1}{2}\left[a^2 w - N_A\left(\frac{1}{2}(w')^2 - w\right) + \frac{n_i}{\ln\lambda} e^{-w\ln\lambda}\right] - \frac{a^2\phi}{2} = \frac{a^2}{2\ln\lambda}\left(2\ln\left(\frac{a}{\sqrt{n_i}}\right) - 1\right) - \frac{\lambda\beta^2}{(\ln\lambda)^2}\frac{(a')^2}{2} + \frac{K}{2}, \qquad (19)$$

where K is an integration constant that is determined using the limiting values of w and a as $x \to \infty$. The constant works out to be

$$\frac{K}{N_A} = w_\infty + \frac{1}{\ln\lambda}\left[\frac{2e^{-w_\infty \ln\lambda}}{\lambda} - 1\right].$$

Now we let $x = 0$ and use the boundary conditions $a(0) = 0$, $w(0) = w_s$, $w'(0) = c(w_s - V_{gs})$ to get

$$c(w_s - V_{gs}) = -\sqrt{2}\left(w_s + \frac{e^{-w_s \ln\lambda}}{\lambda \ln\lambda} + \frac{\lambda\alpha^2\beta^2}{N_A(\ln\lambda)^2} - \frac{K}{N_A}\right)^{1/2},$$

where $\alpha \equiv a'(0)$. One method of computing α without requiring the complete analytic solution to the ODE's (1a) and (1b) is to use an asymptotic expansion in x to determine w and a in a "quantum boundary layer". In [7] matched asymptotics were used to find a uniform asymptotic solution for w and a in the case of inversion, $w_s \geq 1$. Since we only need the value of α, only the solution in this quantum boundary layer is necessary. Even though their results were for the strong inversion case, by using the first term of the asymptotic solution in this quantum layer we found that

$$\alpha = a'(0) \approx \frac{1}{\varepsilon}\sqrt{\frac{n_i \lambda \ln\lambda}{2}}.$$

REFERENCES

[1] G. Gildenblat, X. Li, H. Wang, W. Wu, R. van Langevelde, A.J. Scholten, G.D.J. Smit and D.B.M. Klaassen, "Introduction to PSP MOSFET Model", Nanotech2005.

[2] E. Cumberbatch, H. Abebe, H. C. Morris, Current-Voltage Characteristics from an Asymptotic Analysis of the MOSFET Equations, Journal of Engineering Mathematics 19, 1, 25-46, (2001)

[3] H. Abebe, Modeling the current-voltage (I-V) characteristics of the MOSFET device with quantum mechanical effects due to thin oxide near the Si/SiO2 interface using asymptotic methods, Ph.D. Thesis, Claremont Graduate School, January 2002.

[4] M.G. Ancona, Zhiping Yu, W.-C Lee, R.W. Dutton, P.V. Voorde, Density-gradient simulations of quantum effects in ultra-thin-oxide MOS structures, p. 97-100 SISPAD '97., 1997 International Conference, 1997.

[5] A. El Ayyadi and A. Jüngel, Semiconductor Simulations Using a Coupled Quantum Drift-Diffusion Schrödinger-Poisson Model, SIAM J. Appl. Math 66, 2, 554–572, 2005.

[6] M. J. Ward and F. M. Odeh and D. S. Cohen, Asymptotic Methods for Metal Oxide Semiconductor Field Effect Transistor Modeling, SIAM J. Appl. Math, 50, 4, 1099–1125, 1990.

[7] E. Cumberbatch and S. Uno and H. Abebe, Nano-Scale MOSFET Device Modeling with Quantum Effects, European Journal of Applied Mathematics, to appear, 2006.

Fast Adaptive Computation of Neighboring Atoms

Stephane Redon

Nano-D, INRIA Grenoble - Rhône-Alpes, France, stephane.redon@inria.fr

ABSTRACT

The main cost of a molecular dynamics or Monte Carlo simulation is the computation of the current potential energy or forces resulting from the interaction of atoms composing the molecular system. When a distance cut-off is used to speed up this computation, a fast method is needed to determine pairs of neighboring atoms.

We have recently introduced an adaptive torsion-angle quasi-statics simulation algorithm, which enables users to finely trade between precision and computational cost, while providing some precision guarantees. In that algorithm, proximity queries are adaptively performed using hierarchies of *oriented* bounding boxes.

In this paper, we show that using *axis-aligned* bounding boxes results in faster proximity queries. We thus introduce a semi-adaptive method to determine pairs of neighboring atoms, where all bounding boxes in the hierarchy are updated, but where interaction lists are adaptively updated during the simulation. The new method allows us to perform proximity queries about two orders of magnitude faster than the previous approach.

Keywords: adaptive, simulation, neighbors, lists

1 INTRODUCTION

As is well known, the main cost of a molecular dynamics or Monte Carlo simulation is the computation of the current potential energy or forces resulting from the interaction of atoms composing the molecular system. Since each atom may interact with each other atom, this computation has a quadratic complexity in the number of atoms, which may be too slow for large systems.

One traditional way to speed up this process is to restrict the computation to pairs of *neighboring* atoms, and accept the resulting error. Precisely, a user-defined *distance cut-off* defines how far two atoms can "see" each other. This reduces the number of interacting atoms, and thus the number of interatomic forces that have to be computed. However, a fast method is needed to determine pairs of neighboring atoms.

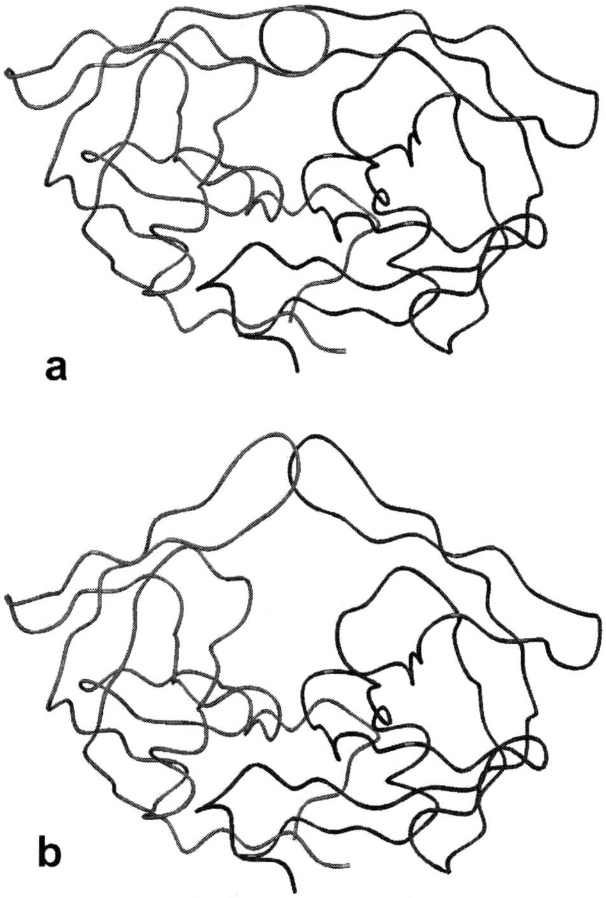

Figure 1: **Creating an open structure for an HIV protease**. In this example, the user opens an HIV protease (**a**, pdb code 2AZ8) with a few mouse clicks (**b**). The model contains two dimers, 1834 atoms, and 816 degrees of freedom.

We have recently introduced an adaptive torsion-angle quasi-statics simulation algorithm [2]. Our adaptive algorithm enables a user to select the *number* of degrees of freedom that should be simulated, or the *precision* at which the computation of the torsion angle accelerations should be performed, and the adaptive algorithm automatically determines the set of most important degrees of freedom under these constraints. The automatic, rigorous trade-off between precision and computational cost allows a user to perform analysis and design of potentially complex molecular systems on low-end computers such as laptops: Figure 1 presents an HIV protease

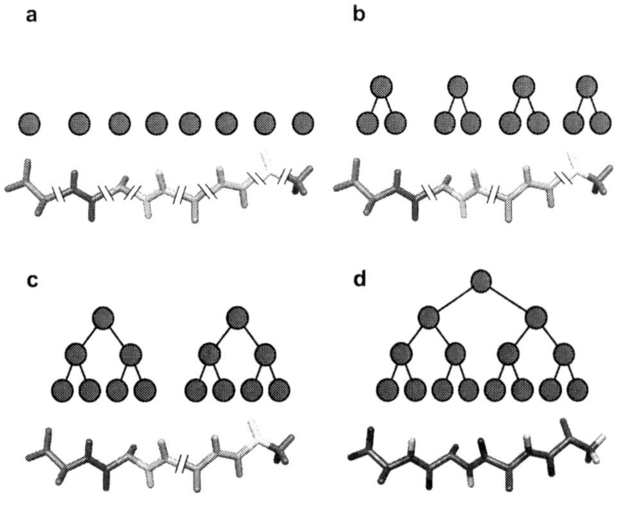

Figure 2: **The assembly tree of a tetra-alanine**. The assembly tree describes a sequence of assembly operations to build the molecular system (see Section 2).

being "opened" with a few mouse clicks by a user using our adaptive tools.

In this paper, we report on how we have managed to speed up the determination of neighboring atoms using an hybrid approach, which combines a linear (non-adaptive) update step of the bounding volumes used to determine neighboring sub-systems, with an adaptive update of the lists of pairs of interacting rigid bodies.

2 ADAPTIVE SIMULATION

Here, we briefly describe our recent work on adaptive molecular simulation for completeness. We refer the reader to Rossi *et al.* [2] for details.

2.1 Adaptive quasi-statics

Our adaptive torsion-angle quasi-statics algorithm is based on a recursive representation of a molecular system: each molecular system is formed by assembling *two* sub-molecular systems, which are in turn each formed of two sub-systems, etc., until we reach single atoms or user-defined groups of atoms that will remain rigid throughout the simulation (the "rigid bodies"). The sequence of assembly operations can thus be described in an *assembly tree*, where leaf nodes represent rigid bodies, internal nodes represent sub-assemblies, and the root node represents the complete molecular system (see Figure 2 for an example). Note that this representation is suitable for both single molecules and groups of molecules.

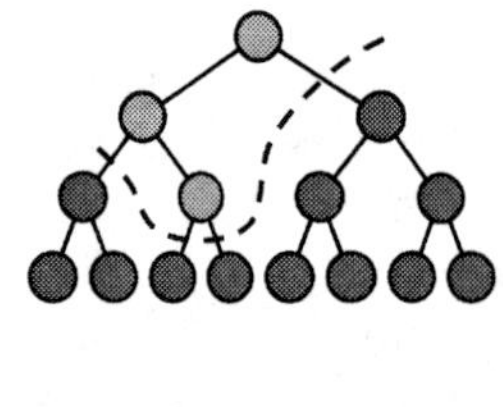

Figure 3: **Active Region.** The active region is the sub-tree which contains the simulated joints.

Each internal node in the assembly tree (including the root node), also represents the relative transformation between the two child molecular systems, so that the goal of the simulation is to compute the accelerations of these relative transformations (in torsion-angle dynamics, the accelerations of the torsion angles).

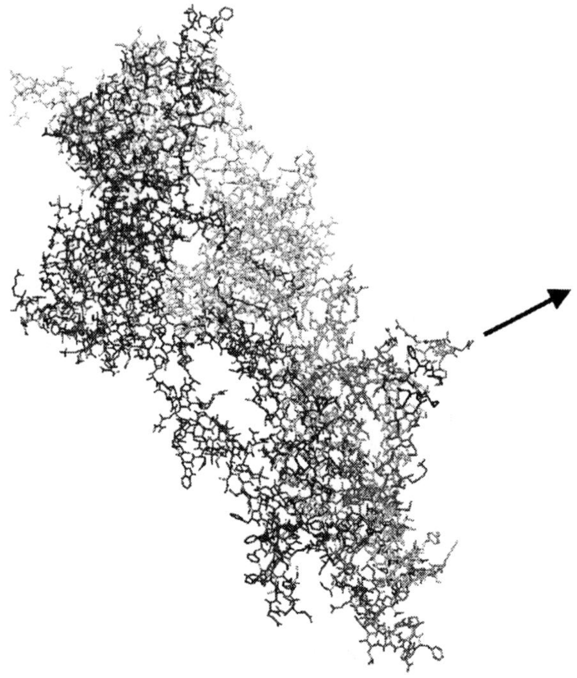

Figure 4: **Adaptive quasi-statics of an ATPase**. The user deforms an ATPase (PDB code 1IW0), while allowing only 10 degrees of freedom out of 4103 present in the system. The adaptive algorithm automatically selects the 10 most relevant degrees of freedom and refines the motion of the ATPase around the point of application of the user force. The resulting rigid bodies are displayed (one color per rigid body).

Our approach to adaptive simulation consists in restricting the computations to a *sub-tree* of the assembly tree. In other words, we simulate a limited subset of degrees of freedom in the molecular system (which form a sub-tree of the assembly tree, see Figure 3). The active degrees of freedom (*i.e.* the most mobile) are determined automatically and rigorously at each time step, which ensures that the user obtains a good approximation of what a complete, non-adaptive simulation would produce, despite potentially limited computational resources. Figure 4 shows an example of an ATPase being interactively deformed by a user allowing 10 degrees of freedom (out of 4103). The adaptive algorithm automatically refines the motion of the ATPase close to the point of application of the user force.

2.2 Adaptive proximity queries

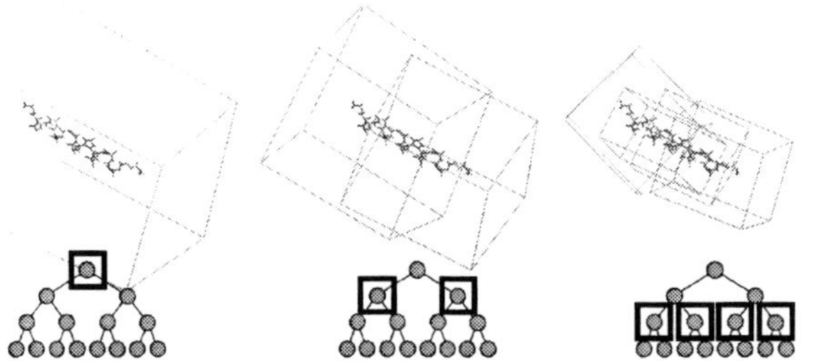

Figure 5: **OBB hierarchy of a poly-alanin.** The figure shows levels 1, 2 and 3 of the OBB hierarchy associated to a poly-alanin.

Freezing and unfreezing relative motions within groups of atoms amounts to cluster the atoms in rigid bodies, whose size depends on the relative amount of local motion. The clusters automatically vary in number and size at each time step, depending on the current configuration of the system, interatomic forces (*e.g.* electrostatic and van der Waals forces), user forces, and user-defined time or precision constraints. One important observation is that all physical quantities which only involve relative positions of atoms in the cluster *do not have to be updated as long as the clusters remain rigid.* This includes the inertia tensor of the cluster, the OBB attached to the cluster, and all interatomic forces which involve atoms in the cluster.

Rossi *et al.* [2] introduces data structures and algorithms which take advantage of this observation to propose a fully adaptive quasi-statics simulation algorithm. In particular, the proximity query algorithm in [2] uses oriented bounding boxes (OBBs) to determine pairs of interacting atoms (or, equivalently, interacting rigid bodies). Specifically, each node of the assembly tree is associated to an OBB which bounds all the atoms in the corresponding sub-assembly (see Figure 5). The OBBs associated to the leaves of the assembly tree are constant, since the leaves correspond to rigid bodies, and they can thus be pre-computed. Each OBB associated to an internal node can be recursively computed from the OBBs of the two children of the node, in constant time, by bounding the two child OBBs. Most importantly, the coordinates of each OBB can be expressed in a reference frame rigidly attached to its corresponding sub-assembly, so that OBBs have to be updated in the active region only. When the atoms move during the simulation, either independently or as part of a rigid group, OBB/OBB overlap tests allow to rapidly detect pairs of interacting atoms, and update *interaction lists* [1], [2]. Interaction lists, stored in each node of the assembly tree, contain the pairs of interactions involving

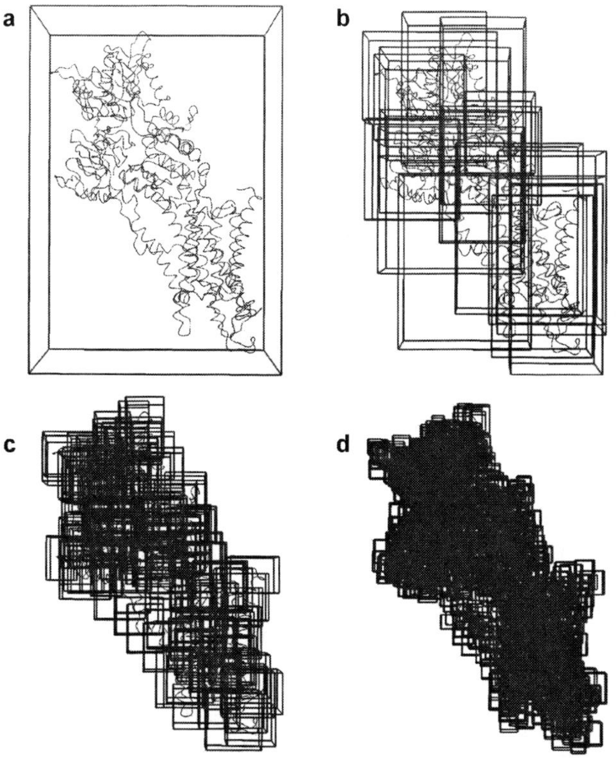

Figure 6: **AABB hierarchy of an ATPase (PDB code 1IW0).** The figure shows levels 1, 5, 8 and 10 of the AABB hierarchy associated to an ATPase.

one descendent of the left child of the node with one descendent of the right child of the node. An interaction involving two rigid bodies A and B is thus stored in the interaction list of the *deepest common parent* of A and B. Because interaction lists only depend on the relative positions of the atoms in sub-assemblies, they have to be updated in the active region only. With the adaptive OBB hierarchy update, this results in a fully adaptive algorithm to detect pairs of interacting atoms.

3 AXIS-ALIGNED BOXES

In order to make the update of the OBB hierarchy adaptive, each OBB has to be built based on the OBBs of the two child nodes, and not from the set of atoms corresponding to the sub-assembly (else, the cost of the update would be proportional to the number of atoms in that sub-assembly). However, to ensure that each OBB encloses all atoms *without* using the atom positions, the OBB has to *bound* the child OBBs themselves. Unfortunately, this tends to produce OBBs which are larger than necessary. As a result, the recursive proximity query tends to perform too many OBB/OBB overlap tests, which may slow down the determination of interacting atoms significantly. Thus, while the update of the OBB hierarchy is adaptive, the complete algorithm isn't as efficient as it could be.

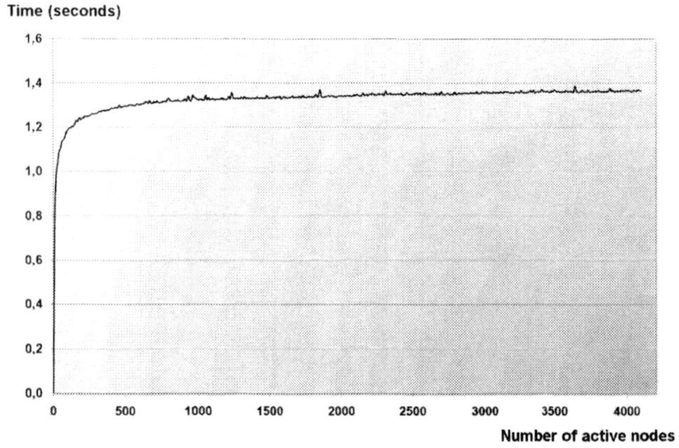

Figure 7: **Cost of proximity queries with OBBs.**

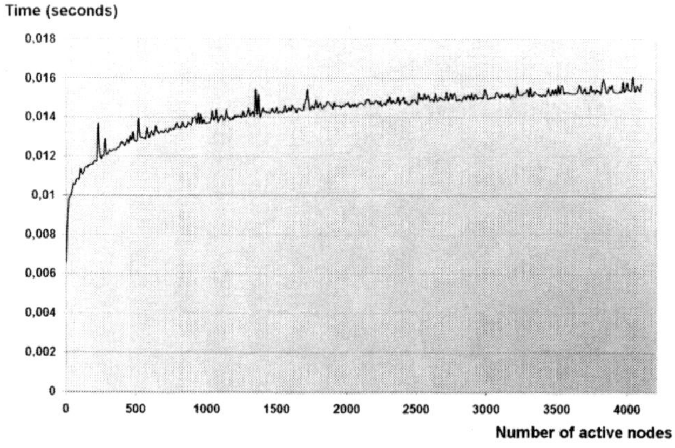

Figure 8: **Cost of proximity queries with AABBs.**

In this work, we thus replace the oriented bounding boxes by axis-aligned bounding boxes (AABBs). As before, one AABB is associated to each node of the assembly tree. However, since by definition all boxes are aligned on the axes of the principal reference frame of the simulation, they all have to be recomputed at each time step. The AABBs of the leaves of the assembly tree (which correspond to the rigid bodies) are computed first, by enclosing all atoms in the rigid body, and enlarging the box by the user-defined distance cutoff. Then, the AABBs of the internal nodes are recursively computed from the bottom up, by bounding the child AABBs. Note that, unlike internal OBBs, internal AABBs are as tight as possible, even though they are directly computed from their child AABBs in constant time. Figure 6 shows levels 1, 5, 8 and 10 of the AABB hierarchy of an ATPase (PDB code 1IW0).

Once all AABBs have been updated, the proximity query algorithm can proceed as before: for each active node, the interaction list of the node is updated by recursively traversing the hierarchy of bounding volumes [2].

4 RESULTS

The method has been implemented in C++ and tested on a 2.8GHz dual core laptop computer with 4GB or RAM (only one core is used, though). The benchmark procedure consists in simulating the motion of an ATPase (PDB code 1IW0, see Figures 4 and 6) and determine the time taken to both (a) update the bounding-volume hierarchy and (b) use the bounding-volume hierarchy to update the interaction lists. The model consists in 4103 internal degrees of freedom (torsion angles) and 9305 atoms. Figure 7 shows the cost of proximity queries depending on the number of active degrees of freedom. As can be seen, this time is zero when no degrees of freedom are allowed (the ATPase behaves as a rigid body), since the method is fully adaptive and nothing has to be updated in this case.

Figure 8 shows that using AABBs results in an about two orders of magnitude speedup over OBBs, despite requiring a complete update of the AABB hierarchy (as can be seen from Figure 8, updating the AABB hierarchy takes about 6.8 milliseconds even when the protease is fully rigid).

Note that in both cases the cost of the proximity queries tends towards its maximum rather quickly when the number of degrees of freedom increases. This is due to the fact that the protein is tightly packed, so that even when few torsion angles are allowed, many interactions may have to be updated. The situation may be quite different when the protein is unfolded (not shown here).

5 CONCLUSION

We have presented a new scheme to compute lists of pairs of neighboring atoms. Our approach uses a hierarchy of axis-aligned bounding boxes — one per assembly node —, which are updated from the bottom up, starting from the atoms. Once the boxes have been updated, we adaptively update the interaction lists for the active assembly nodes only. We have shown that, despite a non-adaptive step required to update all boxes, the overall approach results in faster proximity queries than when using oriented bounding boxes, which were too conservative.

REFERENCES

[1] S. Gottschalk, M. C. Lin, and D. Manocha. Obbtree: a hierarchical structure for rapid interference detection. *In ACM Transactions on Graphics (SIGGRAPH 1996)*, 1996.

[2] R. Rossi, M. Isorce, S. Morin, J. Flocard, K. Arumugam, S. Crouzy, M. Vivaudou, and S. Redon. Adaptive torsion-angle quasi-statics: a general simulation method with applications to protein structure analysis and design. *Bioinformatics*, 23(13):i408–17, 2007.

Control of NEMS Based on Carbon Nanotube

O.V. Ershova*, A.A. Knizhnik**, I.V. Lebedeva*, Yu.E. Lozovik***, A.M. Popov*** and B.V. Potapkin****

* Moscow Institute of Physics and Technology, Dolgoprudny, Russia, lebedeva@kintech.ru
** Kintech Lab Ltd, Moscow, Russia
*** Institute of Spectroscopy, Troitsk, Russia
**** RRC "Kurchatov Institute", Moscow, Russia

ABSTRACT

A new method is proposed for controlling the motion of nanoelectromechanical systems (NEMS) based on carbon nanotubes. In this method chemical adsorption of atoms and molecules at open ends of a single-walled carbon nanotube leads to the appearance of an electric dipole moment. In this case the nanotube can be actuated by a non-uniform electric field. Possibility of the proposed method is shown on the example of the gigahertz oscillator based on a carbon nanotube. Molecular dynamics simulations of the oscillator operation are performed. The simulations reveal for this NEMS considerable thermodynamic fluctuations. The influence of thermodynamic fluctuations on the possibility of controlling the motion of NEMS is investigated.

Keywords: gigahertz oscillator, DWNT, Q-factor, nanotube

1 INTRODUCTION

Ability of free relative sliding and rotation of walls in carbon nanotubes [1], [2] allows using nanotube walls as moveable elements in nanoelectromechanical systems (NEMS). Among these devices there are rotational [3] and plain [2] nanobearings, a nanogear [4], a mechanical nanoswitch [5], a nanoactuator [6], a Brownian motor [7], a nanorelay [8], a nanobolt-nanonut pair [9]–[11], and a gigahertz oscillator [12], [13]. The crucial issue in nanotechnology is actuation of NEMS components in a controllable way. Several methods to control relative sliding of nanotube walls have been proposed recently: by magnetic field in the case of metallic moveable wall [14]; by electric field in the case of moveable wall with metallic ions inside [15]; by pressure of heated gas enclosed between moveable and fixed walls [16].

We propose new method for controlling the motion of NEMS based on carbon nanotubes. More specifically, if electron donors or/and acceptors are adsorbed at the ends of a nanotube wall, the wall has an electric dipole moment. The motion of such a functionalized wall can be controlled by a nonuniform electric field. The possibility of the proposed method is confirmed by molecular dynamics (MD) simulations of the controlled operation of the (5,5)@(10,10)nanotube-based gigahertz oscillator.

As compared to microelectromechanical systems, the principal feature of NEMS connected with the small number of atoms in these systems is that thermodynamic fluctuations in NEMS are significant. These fluctuations can essentially influence the operation of NEMS. Thermodynamic fluctuations were considered in the case of directional motion in NEMS with ratchet (Brownian motors) [7]. Here we present the study of influence of thermodynamic fluctuations on the processes of energy dissipation in NEMS. The controlled motion of the oscillator is examined by MD simulations and using the mechanical model. These investigations indicate a critical influence of thermodynamic fluctuations on the possibility of controlling the NEMS operation.

2 CONTROL OF OSCILLATOR OPERATION

The scheme, operational principles and theory of the gigahertz oscillator based on relative sliding of carbon nanotube walls were considered recently [12], [13]. Upon the telescopic extension of the inner wall outside the outer wall, the Van der Waals force F_W draws the inner wall back into the outer wall. The dependence of the Van der Waals force $F_W(x)$ on the distance x between the centers of the nanotube walls can be approximated by the following relationship [12], [13]:

$$F_W(x) = \begin{cases} F_W, |L-l|/2 < |x| < |L+l|/2 \\ 0, |x| < |L-l|/2 \\ 0, |x| > |L+l|/2, \end{cases} \tag{1}$$

where L and l are the lengths of the outer and inner nanotube walls, respectively. The frequency of the gigahertz oscillator strongly depends on the oscillation amplitude [12]:

$$\omega = \sqrt{\frac{2F_W s}{m}} \frac{\pi}{(L-l)+4s}, \tag{2}$$

where m is the mass of the movable wall and s is the maximum telescopic extension. The molecular dynamics simulations of the gigahertz oscillator show that the oscillation energy dissipates [17]. Thus, the oscillation amplitude decreases with time and the frequency increases with time. To provide the stationary operation of the gigahertz oscillator, it is necessary to compensate the energy dissipation by the work of some external force.

We considered the possibility to compensate the energy dissipation by applying to the moveable wall an external harmonic force $F(t) = F_0 \cos \omega t$ directed along the axis of the oscillator, where ω is the desirable oscillation frequency. In particular, this force can be applied to the wall which has an electric dipole moment using a nonuniform electric field. Let us now assume that, at the instant of time $t = 0$ the inner wall is at the center of the outer wall and moves at its maximum velocity $V_{max} = \sqrt{2F_W s/m}$. The work done by the force F over the oscillation period is given by the integral

$$A_c = \int_0^T V(t)F(t)dt = \frac{4F_W F_0}{m\omega^2} \cos\left(\frac{\omega t_{in}}{2}\right), \quad (3)$$

where $V(t)$ is the time dependence of the velocity of the inner wall, $t_{in} = (L - l)\sqrt{m/2F_W s}$ is the time of the motion of the inner wall inside the outer wall from one end to the other end without the telescopic extension. The work done by the frictional force over the same time is given by $A_f = -F_W s/Q$, where Q is the Q-factor of the oscillator, which is the ratio of the oscillation energy E to the oscillation energy loss ΔE over one oscillation period. The critical amplitude F_{0c} of the external control force, that is the minimum amplitude for which the motion controlling is possible, can be found from the condition $A_c + A_f = 0$. As a result, we obtain the critical amplitude of the control force

$$F_{0c} = \frac{ms\omega^2}{4Q\cos\left(\frac{\omega t_{in}}{2}\right)}. \quad (4)$$

The detailed analysis shows that the critical amplitude F_{0c} of the control force is minimum if the lengths of inner and outer walls are equal $L = l$. In this case the critical amplitude of the control force is given by

$$F_{0c} = \frac{\pi^2 F_W}{32Q}. \quad (5)$$

The estimations of characteristics of the electric field required for controlling the operation of NEMS were performed on the example of the gigahertz oscillator based on the (5,5)@(10,10) double-walled carbon nanotube with the equal lengths of inner and outer walls: $L = l = 3.1$ nm. One end of the inner wall is capped and hydrogen atoms are adsorbed on all dangling bonds at the opposite open end of the inner wall. The electric dipole moment of the inner wall in this case equals $4.5 \cdot 10^{-29}$ C·m [18].

The value of Van der Waals force for the (5,5)@(10,10) nanotube calculated using the Lennard-Jones 12-6 potential [19] is about $F_W = 1100$ pN. For this value of Van der Waals force and the Q-factor Q lying in the range $50 - 250$ (see Table 2) the critical amplitude of the control force is $F_{0c} = 1.4 - 6.8$ pN. This corresponds to voltages from 10 to 80 V at the plates of a spherical capacitor with the radii of 100 and 110 nm.

3 Q-FACTOR CALCULATIONS

As it is shown in section 1, the critical amplitude of the control force depends on the Q-factor of the gigahertz oscillator. We performed microcanonical MD simulations of free oscillations to study Q-factors of gigahertz oscillators. The in-house MD-kMC code was used. The interaction between the atoms of the inner and outer walls was described by the Lennard-Jones 12-6 potential with the parameters obtained from the AMBER database [19]. The parameters for the Lennard-Jones 12-6 potential are $\varepsilon_{CC} = 3.73$ meV and $\sigma_{CC} = 3.40$ Å for carbon-carbon interaction and $\varepsilon_{CH} = 0.65$ meV and $\sigma_{CH} = 2.59$ Å for carbon-hydrogen interaction. The cut-off distance of the Lennard-Jones potential is 12 Å. The empirical Brenner potential was used to describe the covalent carbon-carbon and carbon-hydrogen interactions [20]. The time step was 0.2 fs. At the beginning of the MD simulations the inner wall was pulled out at about 30% of its length and then released with zero initial velocity. The outer wall was kept fixed during the simulations. Nanotubes with walls of equal length were considered.

If no control force is applied, the oscillations are damped because of the interwall friction (see Fig. 1). The frequencies of damped oscillations of the (5,5)@(10,10) nanotube-based oscillator $2.4 - 4.6$ nm in length are $55 - 105$ GHz (see Table 1). This result is in agreement with other MD simulations [14], [17], [21]–[23].

We estimated the instantaneous Q-factor for every half-period $Q_{T/2} = 0.5E/\Delta E_{T/2}$, where E is the oscillation energy and $\Delta E_{T/2}$ is the oscillation energy loss over a half period. Significant fluctuations of Q-factor $Q_{T/2}$ are observed. Thus, $Q_{T/2}$ should be regarded as a statistically distributed quantity. $Q_{T/2}$ is singular for $\Delta E_{T/2} = 0$. Therefore, to obtain the average value of the Q-factor over the full simulation time we averaged the inverse Q-factor $Q^{-1}_{T/2}$. The root-mean-square deviation σ of the inverse Q-factor was also calculated.

The calculated values of the Q-factor and the relative deviation $\delta = \sigma/Q^{-1}_{T/2}$ of the inverse Q-factor at different temperatures and for the oscillators of different length are listed in Tables 1, 2. The considerable decrease of the Q-factor is found for the oscillator that is less than 3 nm in length. For the oscillators of greater length the Q-factor only slightly depends on length (see Table 1). According to Table 2, the Q-factor of the os-

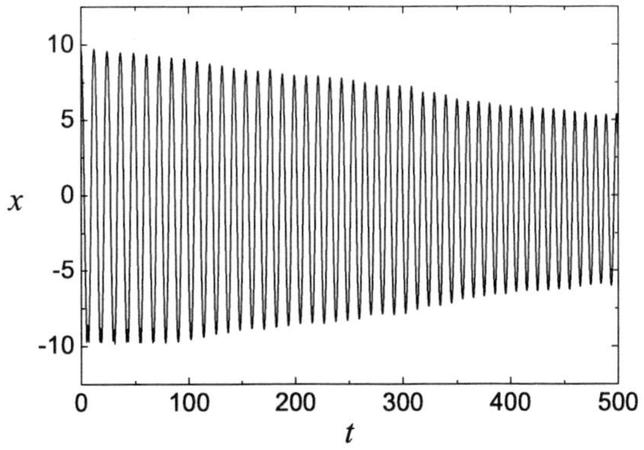

Figure 1: Distance x (in Å) between the mass centers of the walls as a function of time (in ps). The free oscillations of the (5,5)@(10,10) nanotube-based oscillator 3.1 nm in length at temperature 300 K.

cillator strongly increases with decreasing the temperature, which is in agreement with the result in [23].

Table 1: Calculated frequencies, Q-factors and relative deviations δ of the inverse Q-factor of the (5,5)@(10,10) nanotube-based oscillators of different length at temperature 150 K. The both ends of the inner wall are open and not functionalized.

length (nm)	frequency (GHz)	Q	δ
2.4	105	66 ± 9	1.20
3.1	80	120 ± 21	1.32
3.8	65	105 ± 19	1.19
4.6	55	112 ± 17	0.94

4 SIMULATIONS OF CONTROLLED OSCILLATIONS

We also performed the MD simulations of the controlled operation of a nanotube-based oscillator. In these simulations the temperature of the outer wall was kept constant by means of the Berendsen thermostat. The harmonic electric field of the spherical capacitor described in section 1 acted on the inner moveable wall with one end capped and the other end terminated by hydrogen atoms. The frequency of the field was equal to the oscillation frequency at the initial moment of the simulation. The result which confirms the possibility of the operation mode with steady frequency is shown in Fig. 2. The MD simulation predicts that the critical amplitude of the control voltage is less than the value given by Eq. (5). This is because Eq. (5) is derived for a long oscillator with a high oscillation amplitude. In this case

Table 2: Calculated Q-factors and relative deviations δ of the inverse Q-factor of the (5,5)@(10,10) nanotube-based oscillator at different temperatures. The oscillator is 3.1 nm in length. The inner wall has one end capped and the other end terminated by hydrogen atoms.

T (K)	Q	δ
50	253 ± 33	1.19
100	162 ± 25	1.41
150	135 ± 19	1.45
300	55 ± 8	1.45

the expression (1) for the Van der Waals force is adequate. For the oscillator of 3.1 nm length considered here the Van der Waals force can not be taken constant and has less average value.

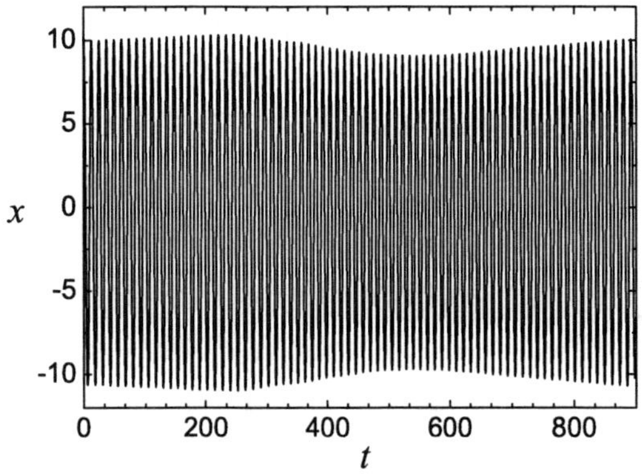

Figure 2: Distance x (in Å) between the mass centers of the walls as a function of time (in ps). The MD simulation of the controlled oscillations of the (5,5)@(10,10) nanotube-based oscillator 3.1 nm in length at temperature 50 K. The voltage 10.9 V is applied.

The MD simulations allow one to study the system behavior only for times of few nanoseconds. To reach longer simulation times and to investigate the influence of the Q-factor fluctuations on the possibility of controlling the oscillator operation the mechanical model was used. The motion equation for the moveable inner wall of the gigahertz oscillator with the fixed outer wall was solved semi-analytically. The frictional force was considered to be proportional to the relative velocity $F_f = -\gamma V$ of the walls. In this case the motion equation for the moveable wall is given by

$$\ddot{x} + \gamma\dot{x} + a\,\mathrm{sign}(x) = b\cos(\omega t), \tag{6}$$

where $a = F_W/m$, $b = F_0/m$ and the frictional coefficient

$$\gamma = \frac{3}{8Q}\sqrt{\frac{F_W}{2sm}}. \tag{7}$$

The value of inverse Q-factor Q^{-1} was changed randomly each half period. The Gaussian distribution was used for the inverse Q-factor. This distribution was cut for the values less than $-9\langle Q^{-1}_{T/2}\rangle$ and greater than $11\langle Q^{-1}_{T/2}\rangle$. Smaller values of the relative deviation δ of inverse Q-factor Q^{-1} should correspond to greater sizes of NEMS (Table 2).

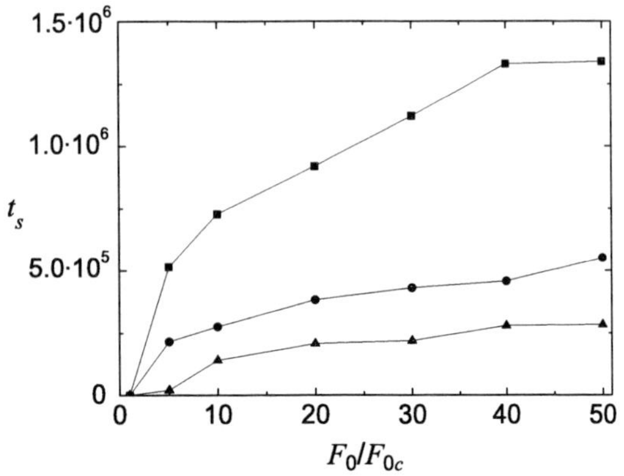

Figure 3: The dependence of the lifetime t_s (in oscillation periods) of the operation mode with steady frequency of the oscillator with Q-factor $Q = 253$ at temperature $T = 50$ K on the ratio of the control force amplitude to critical control force amplitude. The squares, circles and triangles correspond to relative deviations $\delta = 0.4$, $\delta = 0.5$ and $\delta = 1.6$, respectively.

Due to the Q-factor fluctuations the lifetime of the operation mode with steady frequency is not infinite. However, the performed simulations show that for the amplitude of the control force $F_0/F_{0c} \geq 2.0$ and the relative deviation $\delta \leq 0.2$ the lifetime t_s is greater than 10^8 oscillations periods. As can be seen in Fig. 3, the lifetime t_s decreases with increasing δ and only slightly depends on the amplitude F_0 for $F_0/F_{0c} > 20$.

5 CONCLUSIONS

We proposed a new method of controlling the operation of NEMS based on carbon nanotubes by a nonuniform electric field. The molecular dynamics simulations showed the possibility of the method on the example of the nanotube-based gigahertz oscillator. The simulations revealed considerable thermodynamic fluctuations of the oscillator Q-factor. The influence of the Q-factor fluctuations on the possibility of controlling the oscillator operation was investigated with the help of semi-analytical solution of the oscillator motion equation. It was demonstrated that an increase of the Q-factor fluctuations causes a decrease of the average lifetime of the operation mode with steady oscillation frequency. Thermodynamic fluctuations can impose restrictions on sizes and operational temperatures for which the controllability of NEMS is possible.

REFERENCES

[1] M.F. Yu, O. Lourie, M.J. Dyer, K. Moloni, T.F. Kelly, R.S. Ruoff, Science, 287, 637, 2000.

[2] J. Cumings, A. Zettl, Science, 289, 602, 2000.

[3] R.E. Tuzun, D.W. Noid, B.G. Sumpter, Nanotechnology, 6, 52, 1995.

[4] D. Srivastava, Nanotechnology, 8, 186, 1997.

[5] L. Forro, Science, 289, 560, 2000.

[6] Yu.E. Lozovik, A.G. Nikolaev, A.M. Popov, JETP, 103, 449, 2006.

[7] Z.C. Tu, X. Hu, Phys. Rev. B, 72, 033404, 2005.

[8] L. Maslov, Nanotechnology, 17, 2475, 2006.

[9] R. Saito, R. Matsuo, T. Kimura, G. Dresselhaus, M.S. Dresselhaus, Chem. Phys. Lett., bf 348, 187, 2001.

[10] Yu.E. Lozovik, A.V. Minogin, A.M. Popov, Phys. Lett. A, 313, 112, 2003.

[11] Yu.E. Lozovik, A.M. Popov, Fullerenes, Nanotubes and Carbon Nanostructures, 12, 485, 2004.

[12] Q. Zheng, Q. Jiang, Phys. Rev. Lett., 88, 045503, 2002.

[13] Q. Zheng, J.Z. Liu, Q. Jiang, Phys. Rev. B, 65, 245409, 2002.

[14] S. B. Legoas, V. R. Coluci, S. F. Braga, P. Z. Coura, S. O. Dantas, and D. S. Galvao, Nanotechnology, 15, S184, 2004.

[15] J.W. Kang, H.J. Hwang, J. Appl. Phys., 96, 3900, 2004.

[16] J.W. Kang, K.O. Song, O.K. Kwon, H.J. Hwang, Nanotechnology, 16, 2670, 2005.

[17] Y. Zhao, C.-C. Ma, G. Chen, Q. Jiang, Phys. Rev. Lett., 91, 175504, 2003.

[18] O.V. Ershova, Yu.E. Lozovik, A.M. Popov, O.N. Bubel, N.A. Poklonskii, E.F. Kislyakov, Physics of the Solid State, 49, 2010, 2007.

[19] http://amber.scripps.edu//#ff.

[20] D.W. Brenner, Phys. Rev. B, 42, 9458, 1990.

[21] W. Guo, Y. Guo, H. Gao, Q. Zheng, and W. Zhong, Phys. Rev. Lett., 91, 125501, 2003.

[22] P. Tangney, S. G. Louie, and M. L. Cohen, Phys. Rev. Lett., 93, 065503, 2004.

[23] C.-C. Ma, Y. Zhao, Y.-C. Yam, G. Chen, Q. Jiang, Nanotechnology, 16, 1253, 2005.

Field Emission Properties of Carbon Nanotube Arrays with Defects and Impurities

D. Roy Mahapatra*, N. Sinha**, R.V.N. Melnik*** and J.T.W. Yeow**

* Department of Aerospace Engineering, Indian Institute of Science, Bangalore, India
** Department of Systems Design Engineering, University of Waterloo, Waterloo, ON, Canada
*** M^{2}NeT Lab, Wilfrid Laurier University, Waterloo, ON, Canada
rmelnik@wlu.ca

ABSTRACT

It has been found experimentally that the results related to the collective field emission performance of carbon nanotube (CNT) arrays show variability. The emission performance depends on the electronic structure of CNTs (especially their tips). Due to limitations in the synthesis process, production of highly pure and defect free CNTs is very difficult. The presence of defects and impurities affects the electronic structure of CNTs. Therefore, it is essential to analyze the effect of defects on the electronic structure, and hence, the field emission current. In this paper, we develop a modeling approach for evaluating the effect of defects and impurities on the overall field emission performance of a CNT array. We employ a concept of effective stiffness degradation for segments of CNTs, which is due to structural defects. Then, we incorporate the vacancy defects and charge impurity effects in our Green's function based approach. Simulation results indicate decrease in average current due to the presence of such defects and impurities.

Keywords: carbon nanotube, field emission, electron-phonon, defect, impurity, transport.

1 INTRODUCTION

From the time field emission from carbon nanotubes (CNTs) was reported in 1995 [1],[2], their applications in devices, such as field emission displays, gas discharge tubes, electron microscopes, cathode-ray lamps and x-ray tube sources have been demonstrated successfully [3], [4]. In recent years, studies on field emission from CNTs have been growing. CNTs in the form of arrays or thin films give rise to several strongly correlated processes of electromechanical interaction and degradation. Such processes are mainly due to (1) electron-phonon interaction (2) electromechanical force field leading to stretching of CNTs (3) ballistic transport induced thermal spikes, coupled with high dynamic stress, leading to degradation of emission performance at the device scale [5]. Fairly detailed physics based models of CNTs accounting for aspects (1) and (2) above have already been developed, and numerical results indicate good agreement with experimental results [6]-[9]. These studies are based on the electronic structure of an ideal CNT where a CNT is metallic or semiconducting depending on whether $n - m$ is a multiple of three or not. Here n and m are two integral components of the chiral vector. Although studies have reported defects in CNTs in general [10], [11], not much is known as to how these defects affect the field emission property of CNTs. For a better understanding of the field emission phenomenon, a system level modeling approach incorporating structural defects, vacancies or charge impurities is currently missing. This is a practical and important problem due to the fact that degradation of field emission performance is indeed observed in experimental I-V curves. What is not clear from these experiments is whether such degradation in the I-V response is due to dynamic reorientation of CNTs or due to the defects or due to both of these effects combined. Non-equilibrium Green's function based simulations using a tight-binding Hamiltonian for a single CNT segment demonstrate the localization of carrier density at various locations of the CNTs. About 11% decrease in the drive current with steady difference in the drain current in the range of 0.2-0.4V of the gate voltage was reported in [12] when negative charge impurity was introduced at various locations of the CNT over a length of $\approx$ 20nm. In the context of field emission from CNT tips, a simple estimate of defects have been proposed by introducing a correction factor in the Fowler-Nordheim formulae [13]. However, it is clear that a more detailed physics based treatment is required for a better understanding of processes at the device-scale level. The goal of this paper is to develop a model to analyze the effects of defects and impurities on the field emission performance of CNT arrays. This paper is structured as follows: in Section 2, a physics based model is proposed that incorporates structural, vacancy and charge impurity defects. Numerical simulations comparing longitudinal strains and field emission current histories for CNT arrays with and without defects and impurities are presented in Section 3. Section 4 contains concluding remarks.

2 MODEL FORMULATION

The physics of field emission from metallic surfaces is fairly well understood. The current density (J) due to field emission from a metallic surface is usually obtained by applying the Fowler-Nordheim (FN) approximation [14]

$$J = \frac{BE^2}{\Phi} \exp\left[-\frac{C\Phi^{3/2}}{E} \right] , \quad (1)$$

where E is the electric field, Φ is the work function of the cathode material, and B and C are constants.

Based on our previously developed model [6], which describes the degradation of CNTs and the CNT geometry and orientation, the decreased surface area can be expressed as

$$\pi d_t \Delta h = V_{\text{cell}} n_1(t) \left[s(s-a_1)(s-a_2)(s-a_3) \right]^{1/2} , \quad (2)$$

where d_t is the diameter of the CNT, Δh is the decrease in the length of the CNT (aligned vertically or oriented as a segment) over a time interval Δt due to degradation and fragmentation, V_{cell} is the representative volume element, n_1 is the concentration of carbon cluster in the cell, a_1, a_2, a_3 are the lattice constants, and $s = \frac{1}{2}(a_1 + a_2 + a_3)$ (see Fig. 1). The chiral vector for the CNT is expressed as

$$\vec{C}_h = n\vec{a}_1 + m\vec{a}_2 , \quad (3)$$

where n and m are integers ($n \geq |m| \geq 0$) and the pair (n, m) defines the chirality of the CNT. The following properties hold: $\vec{a}_1.\vec{a}_1 = a_1^2$, $\vec{a}_2.\vec{a}_2 = a_2^2$, and $2\vec{a}_1.\vec{a}_2 = a_1^2 + a_2^2 - a_3^2$. With the help of these properties the circumference and the diameter of the CNT can be expressed as, respectively [15],

$$|\vec{C}_h| = \sqrt{n^2 a_1^2 + m^2 a_2^2 + nm(a_1^2 + a_2^2 - a_3^2)} , \quad (4)$$

$$d_t = \frac{|\vec{C}_h|}{\pi} . \quad (5)$$

Let us now introduce the rate of degradation of the CNT or simply the burning rate as $v_{\text{burn}} = \lim_{\Delta t \to 0} \Delta h / \Delta t$. By dividing both side of Eq. (2) by Δt and passing to the limit, we have

$$\pi d_t v_{\text{burn}} = V_{\text{cell}} \frac{dn_1(t)}{dt} \left[s(s-a_1)(s-a_2)(s-a_3) \right]^{1/2} , \quad (6)$$

By combining Eqs. (4)-(6), the rate of degradation of CNTs is finally obtained as

$$v_{\text{burn}} = V_{\text{cell}} \frac{dn_1(t)}{dt} \left[\frac{s(s-a_1)(s-a_2)(s-a_3)}{n^2 a_1^2 + m^2 a_2^2 + nm(a_1^2 + a_2^2 - a_3^2)} \right]^{1/2} . \quad (7)$$

Therefore, at a given time, the length of a CNT can be expressed as $h(t) = h_0 - v_{\text{burn}} t$, where h_0 is the initial average height of the CNTs and d is the distance between the cathode substrate and the anode.

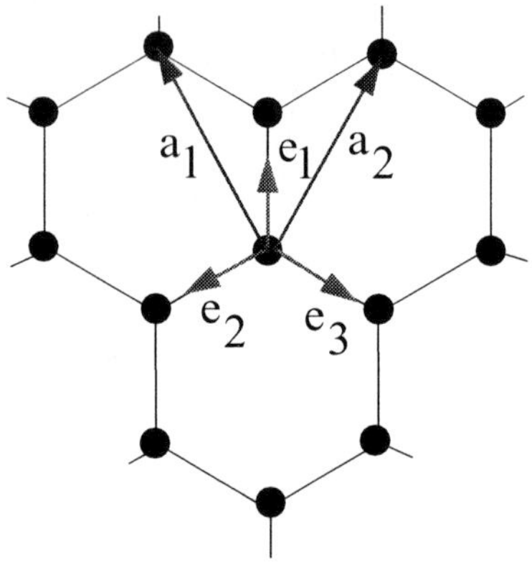

Figure 1: Schematic drawing showing hexagonal arrangement of carbon atoms in a CNT.

We follow the same procedure as in [5] to obtain the effective electric field component for field emission calculation with Eq. (1). The effective electric field component is expressed as

$$E_z = -e^{-1} \frac{d\mathcal{V}(z)}{dz} , \quad (8)$$

where e is the positive electronic charge and $\mathcal{V}$ is the electrostatic potential energy. The total electrostatic potential energy can be expressed as

$$\mathcal{V}(x, z) = -eV_s - e(V_d - V_s)\frac{z}{d} + \sum_j G(i, j)(\hat{n}_j - n) , \quad (9)$$

where V_s is the constant source potential (on the substrate side), V_d is the drain potential (on the anode side), $G(i, j)$ is the Green's function [16] with i being the ring position, and $\hat{n}_j$ denotes the electron density at node position j on the ring. Here, ring is assumed as per unit length of a CNT. The field emission current (I_{cell}) from the anode surface associated with the elemental volume V_{cell} of the CNT array is obtained as

$$I_{\text{cell}} = A_{\text{cell}} \sum_{j=1}^{N} J_j , \quad (10)$$

where A_{cell} is the anode surface area and N is the number of CNTs in the volume element. The total current is obtained by summing the cell-wise current (I_{cell}). This formulation takes into account the effect of CNT tip orientations and one can perform statistical analysis of the device current for randomly distributed and randomly oriented CNTs.

The novelty of our present approach is twofold. Firstly, we employ a concept of effective stiffness degradation for segments of CNTs, which is due to structural defects, by modifying the chiral vector in Eq. (3). The effective stiffness degradation is modeled by introducing an effective chiral vector $\alpha\vec{C}_h$, where $0 < \alpha < 1$ describes the extent of structural defects over the tube circumference at a particular cross-section of a CNT in the array. Secondly, we incorporate the vacancy defects and charge impurity effects in our Green's function based approach. This is done by computing the effective electric potential energy as

$$\mathcal{V}(x,z) = -eV_s - e(V_d - V_s)\frac{z}{d} + \sum_j G(i,j)(\hat{n}_j - n \pm \overline{m}_l + \delta_{kj}) \, , \qquad (11)$$

where $\overline{m}_l$ denotes the charge due to impurity at coordinate l, and δ_{kj} denotes the vacancy at atom position k. The effective electric field is subsequently calculated by Eq. (8) and current density is obtained by using Eq. (1).

3 RESULTS AND DISCUSSIONS

The CNT film considered in this study consists of randomly oriented multiwalled CNTs (MWNTs). The film was grown on a stainless steel substrate. The film surface area (projected on anode) is 49.93 mm^2 and the average height of the film (based on randomly distributed CNTs) is 10-14 μm. As in [5], in the simulation and analysis, the constants B and C in Eq. (1) were taken as $B = (1.4 \times 10^{-6}) \times \exp((9.8929) \times \Phi^{-1/2})$ and $C = 6.5 \times 10^7$. It has been reported in the literature (e.g., [17]) that the work function Φ for CNTs is smaller than the work functions for metal, silicon, and graphite. However, there are significant variations in the experimental values of Φ. The value of Φ depends on the structure, defect, types of CNTs (i.e., SWNT/MWNT), and surface state of CNTs (specifically, cap nature, the edge structure of graphene sheet in an open end). The type of substrate materials has also significant influence on the electronic band-edge potential. All these factors should be taken into consideration together. A comprehensive understanding about the work function of CNTs is still missing. The results reported in this paper are based on computation with $\Phi = 2.2eV$.

An array of 10 vertically aligned and each 12 μm long CNTs is considered for the device scale analysis. Defect regions are introduced randomly over the CNT length with $\alpha = 0.2$ and positive charge density of $\overline{m}_l = 10$. Figure 2 shows the decrease in the longitudinal strain due to defects. Contrary to the expected influence of purely mechanical degradation, this result indicates that the charge impurity, and hence weaker transport, can lead to a different electromechanical force field, which ultimately can reduce the strain. However, there could be significant fluctuations in such strain field due to electron-phonon coupling. The effect of such fluctuations (with defects) can be seen in Fig. 3 where we provide the plot of the field emission current history. The average current also decreases significantly due to such defects.

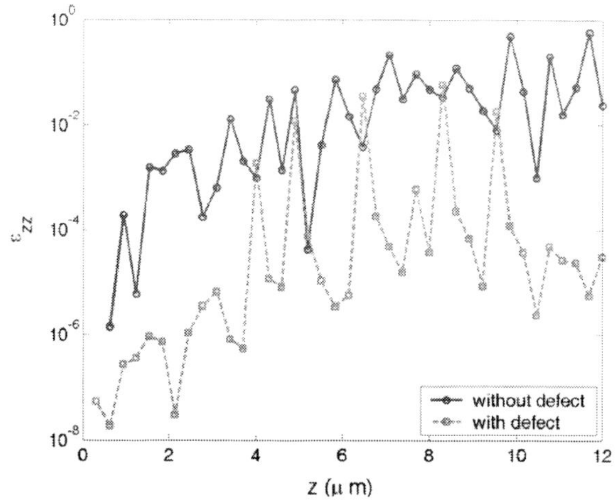

Figure 2: Cross-sectional average longitudinal strain distribution along the CNT at t=0.1s.

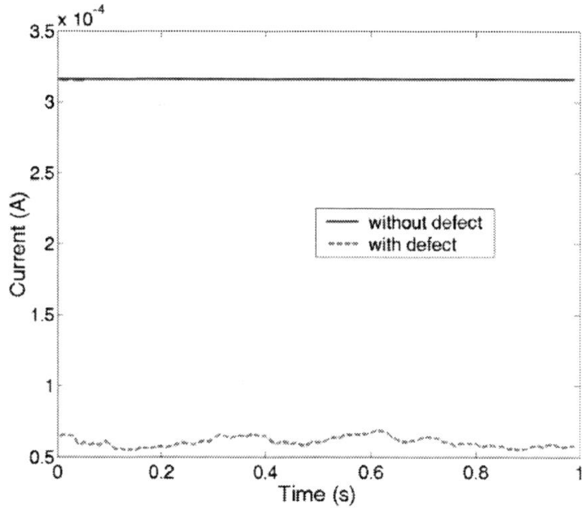

Figure 3: Field emission current history for gate voltage of 600V from an array of 10 CNTs with and without defects.

4 CONCLUSION

In this paper, a model incorporating defects and impurities in CNT based field emission cathodes has been developed from the device design point of view. The model has been incorporated by using a two-step procedure. Firstly, structural defects are considered by modifying the chiral vector. Secondly, vacancy defects and charge impurity effects are introduced in our Green's function based approach. Based on this procedure, the impact of the defects and impurities on the field emission current has been computed. It has been found that the average current decreases significantly due to such defects.

REFERENCES

[1] A.G. Rinzler, J.H. Hafner, P. Nikolaev, L. Lou, S.G. Kim, D. Tomanek, D. Colbert and R.E. Smalley, Science 269, 1550, 1995.

[2] W.A. de Heer, A. Chatelain, and D. Ugrate, Science 270, 1179, 1995.

[3] J.M. Bonard, J.P. Salvetat, T. Stockli, L. Forro and A. Chatelain, Appl. Phys. A 69, 245, 1999.

[4] Y. Saito and S. Uemura, Carbon 38, 169, 2000.

[5] D. Roy Mahapatra, N. Sinha, S.V. Anand, R. Krishnan, Vikram N.V., R.V.N. Melnik and J.T.W. Yeow, Proc. 11th Annual NSTI Nanotech. Conf., 2008. (Accepted)

[6] N. Sinha, D. Roy Mahapatra, J.T.W. Yeow, R.V.N. Melnik and D. A. Jaffray, Proc. IEEE Int. Conf. Nanotech., 673, 2006.

[7] N. Sinha, D. Roy Mahapatra, J.T.W. Yeow, R.V.N. Melnik and D.A. Jaffray, J. comp. Theor. Nanosci. 4, 535, 2007.

[8] N. Sinha, D. Roy Mahapatra, Y. Sun, J.T.W. Yeow, R.V.N. Melnik and D.A. Jaffray, Nanotechnology 19, 25710, 2008.

[9] N. Sinha, D. Roy Mahapatra, J.T.W. Yeow and R.V.N. Melnik, Proc. IEEE Int. Conf. Nanotech., 961, 2007.

[10] J.-C. Charlier, Acc. Chem. Res. 35, 1063, 2002.

[11] Ph. Lambin, A.A. Lucas and J.-C. Charlier, J. Phys. Chem. Solids 58, 1833, 1997.

[12] N. Neophytou, D. Kienle, E. Polizzi and M.P. Anantram, Appl. Phys. Lett. 88, 242106, 2006.

[13] G. Wei, Appl. Phys. Lett. 89, 143111, 2006.

[14] R.H. Fowler and L. Nordheim, Proc. Royal Soc. London A 119, 173, 1928.

[15] H. Jiang, P. Zhang, B. Liu, Y. Huang, P.H. Geubelle, H. Gao and K.C. Hwang, Comp. Mater. Sci. 28, 429, 2003.

[16] A. Svizhenko, M.P. Anantram and T.R. Govindan, IEEE Trans. Nanotech. 4, 557, 2005.

[17] Z.P. Huang, Y. Tu, D.L. Carnahan and Z.F. Ren, "Field emission of carbon nanotubes," Encyclopedia of Nanoscience and Nanotechnology (Ed. H.S. Nalwa) 3, 401-416, 2004.

The Concatenation of the Concurrent Self-replication and Self-organization Processes

S. Wegrzyn [1), 2)] and L. Znamirowski [1), 2)]

[1)] Institute of Informatics, Silesian University of Technology
ul. Akademicka 16, 44-100 Gliwice, Poland, stefan.wegrzyn@polsl.pl, lech.znamirowski@polsl.pl
[2)] Institute of Theoretical and Applied Informatics, Polish Academy of Sciences
ul. Baltycka, 44-100, Gliwice, Poland, wegrzyn@iitis.gliwice.pl, lz@iitis.gliwice.pl

ABSTRACT

In the paper the concatenation of the concurrently running self-replication processes of objects, and concurrently running self-replication and self-organization processes which besides of the multi-stage product synthesis (*gluey matrix* technology) are basis of the *Molecular Nanotechnology of Direct Product Fabrication*, has been analyzed and discussed. Analysis of the interdependent self-replication processes and also, the mutual self-replication and self-organization processes basing on the growth functions, provides facilities for determination of the control of these processes in the sense of stabilization, process extinguishing, and avalanche proliferation. In the paper, the numerical example has been presented.

Keywords: concurrent processes, concatenation, molecular simulation, self-organization, self-replication

1 INTRODUCTION

It is possible to create technical nanosystems of informatics which will be realizing like in the case of biological systems the sets of technological operations which material products are coming into existence and so will be taken to manufacturing objects and products on the basis of the molecular nanotechnology. It leads to the three problems connected with the realization of creating the technical nanosystems of informatics of the direct product fabrication and it concerns: the nanosystems of informatics, the self-replication and the self-organization processes.

In *nanosystems of informatics* the process of manufacturing is based on the extorted connecting of elements of the structure to the form of intermediate product for desirable product and modifications of the intermediate product created till the obtainment of the appropriate final product.

The *self-replication* is being realized in technical systems of informatics through the replication of molecular objects together with molecular programs stored in them. The strategy of construction of products and objects in technical systems of informatics can use the general strategy appearing in biological systems in a form of the embryonic development or can implement the processes of *self-organization*.

In the paper the concatenation of the concurrent self-replication and self-organization processes is discussed.

2 SELF-REPLICATION PROCESSES

Let's assume that there is a set of determined elements or *environment*. Let's assume that the energy provided from outside ensures mobility of the elements and as a result, free movement conditions are fulfilled in the environment area. An *object* means a part of the environment selected in such a way that the exchange of elements between the environment and the object is possible provided that the determined conditions are fulfilled. Due to the above, the object can gather elements necessary for its development and can develop in accordance with its internal program. If in a consequence of the object development, the partition of other objects, which are able to continue successive partition, and descendant objects, which are able to continue their independent development processes, takes place; such a phenomenon is called a *self-replication process*.

2.1 Growth Function in Replication Process

Let us define the measure of a time k to be the integer while the time unit we denote by the symbol 1. Considering the spontaneous self-replication process, we describe it in such a manner, that the number of elements created after the time k is described by the growth function $S(k)$ receiving integer value for dimensionless argument $k = 0, 1, 2, 3,$ ..... which describes the consecutive number of time interval. The two basic possible types of self-replication may arise: linear and avalanche.

2.2 Linear Self-replication

Let $S(k)$ be the growth function of self-replication process described by

$$S(k+1) = S(k) + A, \tag{1}$$

with initial condition $S(0) = B$.

Symbol A denotes a number of elements about the same properties, participating in the process, for which a total number of elements is rising in the k time interval, while B is determining the number of elements in the initial moment of the self-replication process. We will make an assumption, that if $S(k)$ in (1) is zero, then the process stops (and similarly in different kinds of the self-replication).

The growth defined by the function $S(k) = 1 + k$ in Eq. (1), we will call as a *linear growth*.

Example 1

Let be $A = 1$ and $B = 1$; the total number of elements $S(k)$ after time k is

$$S(k) = k + 1, \text{ or more precisely } \int S(k) = 1 + \int k, \tag{2}$$

where a symbol $\int$ denotes a step function [5], which value is changing only for integer values of argument and between them the value of function is staying constant.

2.3 Avalanche Self-replication

Let the growth function $\int S(k)$ of the self-replication process be determined by equation:

$$\int S(k+1) = \int A(n) \cdot S(k)\,, \tag{3}$$

with initial condition $S(0) = B$. Symbol $A(n)$ denotes a function of argument n, determining a number of elements created in the time (*avalanche growth*) unit about the same properties as properties of the element generating them.

Example 2

Let be $A(n) = n+1$ and $B = 1$; the total number of elements $\int S(k)$ after time k is:

$$\int S(k) = (n+1)\int S(k-1) = (n+1)(n+1)\int S(k-2) = \underbrace{(n+1)(n+1) \quad \ldots \quad (n+1)}_{k} \cdot \int S(0)$$

that is to say

$$\int S(k) = \int (n+1)^k\,. \tag{4}$$

Resolving (4) for e.g. $n = 2$ we obtain $\int S(k) = \int 3^k$.

3 SELF-ORGANIZATION PROCESSES

Let us assume that there is a set of N elements. The elements of a set N stay in a chaotic movement and have a such a property, that if will be found in a suitable mutual position, they are connecting together and are already permanently staying in such a position forming the strand as a result of the *self-organization* process. If the set contains N of elements, than the length of longest possible strand amounts d_N. The state, when the all N elements of the set connects together forming the strand about the d_N length, is one of possible steady states finishing this elements' connecting process (we will assume that the closed strands does not appear [3]).

Let the step function $\int f(x)$ determines the number of unconnected elements of a set N in the x moment. Let the measure of the x time will always be an integer positive numbers. We will mark the elementary unit of the time with symbol 1. We will assume that only one connection is created between the elements of a set N in the elementary time 1, so for the initial condition $\int f(0) = N$ we may write

$$\int f(x) = N - \int x\,. \tag{5}$$

Let us go from the x time appointed with moments of connections of elements, to the real time k (for the sake of notation simplification in here the symbol k will be also the positive integer number, determining the number of time interval). Depending on process, in general we may write

$$k_{i+1} = P(k_i), \qquad \text{where } i = 0,\ 1,\ 2,\ 3,\ \ldots\ ,$$

where $P(.)$ depends on process, e.g. for

$$k_{i+1} = k_i + (1+i), \tag{6}$$

(growing up of intervals with arithmetic progression [3, 8]), basing on (5) and (6) we obtain (for proper initial condition) for the rise of the strand length $\int L(k)$

$$\int L(k) = \int \tfrac{1}{2}\left(\sqrt{1+8k} - 1\right). \tag{7}$$

4 CONCATENATION OF THE SELF-REPLICATION PROCESSES

In general, through the term *concatenation* we understand the interdependencies occurring between processes running concurrently.

We consider e.g. two self-replication processes determined by the growth function $\int S(.)$ and $\int T(.)$, which can be written using the following notation:

$$\int S(k_{i+1}) = \int F_1[\int S(k_i), k_i]\,, \tag{8}$$

with initial condition $S(k_0) = W_1$

and

$$\int T(k_{i+1}) = \int F_2[\int T(k_i), k_i]\,, \tag{9}$$

with initial condition $T(k_0) = W_2$,

with that:

$$k_{i+1} = k_i + 1\,, \qquad i = 1,\ 2,\ 3,\ \ldots \tag{10}$$

$$k_0 = 0\,. \tag{11}$$

The functions $\int F_1(.)$ and $\int F_2(.)$ describing processes (8) and (9), fulfill the relations:

$$\int \Delta S = \int S(k_{i+1}) - \int S(k_i) \geq 0 \tag{12a}$$

$$\int \Delta T = \int T(k_{i+1}) - \int T(k_i) \geq 0\,, \tag{12b}$$

for all i, that is to say the growths functions of these processes are non-decreasing, i.e. the number of created elements in the process increases, may be stopped, but in any case cannot decrease. It can be noticed, we analyze the processes determined by the non-decreasing growth functions.

Let's take note, that in general case the time interval number defined in (10), can be incremented with value non-equal 1, similarly, the number of initial moment can be different than 0 as in Eq. (11). The introduction of this manner of denotation of the arguments in (8) and (9), allows to consider much wider class of processes in comparing with simple formalism introduced earlier in Eqs. (1) and (3).

In further, we will find the relation between growth functions of the processes $\int S(.)$ and $\int T(.)$, and a growth function of the process $\int R(.)$, establishing concatenation of processes $\int S(.)$ and $\int T(.)$ defined as follow.

A division of a set of elements $\int R(k_i)$ in the time interval k_i for further divisions inside the running self-replication process, is defined by the parameter $0 < \alpha < 1$ in the following manner:

for continuation of a process $\int S(.)$

$$\int S(k_i) = Ent\ \{\alpha \cdot \int R(k_i)\}\,, \tag{13}$$

for continuation of a process $\int T(.)$, with respect to (13), one may write:

$$\int T(k_i) = \int R(k_i) - Ent\ \{\alpha \cdot \int R(k_i)\}\,. \tag{14}$$

Symbol $Ent\{.\}$ denotes the *entier* function, sometimes expressed as $E(.)$ or [.].

Parameter α may receive all values in the interval (0; 1) with excluding the interval endpoints i.e. $\alpha = 0$ and $\alpha = 1$, because these are the particular cases (independent, alternative process realization of $\int S(.)$ or $\int T(.)$).

Definition 1

The growth function $\int R(k_i)$, caused by concatenation of the growth processes $\int S(k_i)$ and $\int T(k_i)$, is defined on the set of discrete argument k_i in the following manner:

$$\int R(k_{i+1}) = \int F_1[\int S(k_i), k_i] + \int F_2[\int T(k_i), k_i],$$
$$\int S(k_{i+1}) = Ent\{\alpha \cdot \int R(k_{i+1})\}, \quad (15)$$
$$\int T(k_{i+1}) = \int R(k_{i+1}) - \int S(k_{i+1}),$$

for initial condition:

$R(k_0) = A$ (A integer),

$S(k_0) = Ent\{\alpha \cdot R(k_0)\}$, $T(k_0) = R(k_0) - S(k_0)$,

$k_{i+1} = k_i + 1$ for $i = 0, 1, 2, 3, \ldots$ and $k_0 = 0$. ■

Example 3

Let's consider the concatenation of two self-replication processes: linear self-replication (1) and an avalanche (3). The goal of the analysis is to find the growth function $\int R(.)$ of the process establishing the concatenation of the two self-replication processes characterized by the process parameters the same as in the Examples 1 and 2.

So, we consider two processes, for which in case, when they are running independently, their description has the following form:

Process I:

$$\int S(k_{i+1}) = \int S(k_i) + A, \quad (16)$$

Process II:

$$\int T(k_{i+1}) = A(n) \cdot \int T(k_i), \quad (17)$$

with sufficient initial conditions.

Let's assume, that the concatenation of the processes (16) and (17) is performed with the coefficient $\alpha = 0.3$ (Equation (13)). Accepting the particular parameters in Eqs. (16) and (17), the same as in Examples 1 and 2, i.e. $A = 1$, $n = 2$ and $A(n) = n + 1$, we get the process $\int R(.)$ with initial condition $R(k_0) = B = 1$, generated by the concurrently running processes I and II. The process $\int R(.)$ defined by the dependencies (8), (9) and (15) can be developed using the set of following equations:

$$\int R(k_{i+1}) = \int S(k_i) + 1 + 3 \cdot \int T(k_i),$$
$$\int S(k_{i+1}) = Ent\{0.3 \cdot \int R(k_{i+1})\}, \quad (18)$$
$$\int T(k_{i+1}) = \int R(k_{i+1}) - \int S(k_{i+1}),$$

with initial condition:

$$R(k_0) = 1$$

and $S(k_0) = Ent\{0.3 \cdot 1\} = 0$,

$$T(k_0) = 1 - 0 = 1,$$

for the time intervals fixed with the numbers:

$k_{i+1} = k_i + 1$ for $i = 0, 1, 2, 3, \ldots$, and $k_0 = 0$.

Numerical values of the solution of the equations (18) are gathered in Table 1 while the growth function of the process $\int R(.)$ representing the concatenation of two self-replication processes is presented in Fig. 1.

For comparison, in Table 1, in the columns $\int S(k)$ and $\int T(k)$ the values of growth functions of the self-replication processes (16) and (17) has been presented for the independent realization of these processes.

i	$k_i = k$	$\int R(k_i)$	$\int S(k_i)$	$\int T(k_i)$	$\int S(k)$	$\int T(k)$
0	$k_0 = 0$	1	$S(0) + T(0) = 1$		1	1
			0	1		
1	$k_1 = 1$	0+3=3	0	3	2	3
2	$k_2 = 2$	0+9=9	2	7	3	9
3	$k_3 = 3$	3+21=24	7	17	4	27
4	$k_4 = 4$	8+51=59	17	42	5	81
5	$k_5 = 5$	18+126=144	43	101	6	243
6	$k_6 = 6$	44+303=347	104	243	7	729
7	$k_7 = 7$	105+729=834	250	584	8	2187
8	$k_8 = 8$	251+1752=2003	600	1403	9	6561
9	$k_9 = 9$	601+4209=4810	1443	3367	10	19683
10	$k_{10} = 10$	1444+10101= 11545	3463	8082	11	59049
. . .	. . .					

Table 1: Growth function $\int R(k_i)$ of the concatenation of self-replication (16) and (17)

5 CONCATENATION OF THE REPLICATION AND SELF-ORGANIZATION

Let us consider two processes: self-replication with the growth function $\int S(.)$ and self-organization with the growth function $\int U(.)$, described accordingly with the relations:

a) self-replication process:

$$\int S(k_{i+1}) = \int F[\int S(k_i), k_i] \quad (19a)$$

with initial condition $S(k_0) = W_1$, with that:

$$k_{i+1} = k_i + A \quad \text{for } i = 0, 1, 2, 3, \ldots, k_0 = 0. \quad (19b)$$

b) self-organization process:

$$\int U(k_{i+1}) = \int G[\int U(k_i), k_i] \quad (20a)$$

with initial condition $U(k_0) = W_1$, with that:

$$k_{i+1} = k_i + \int \varphi[\int U(k_i), i], \quad i = 0, 1, 2, 3, \ldots, \quad (20b)$$

$k_0 = 0$.

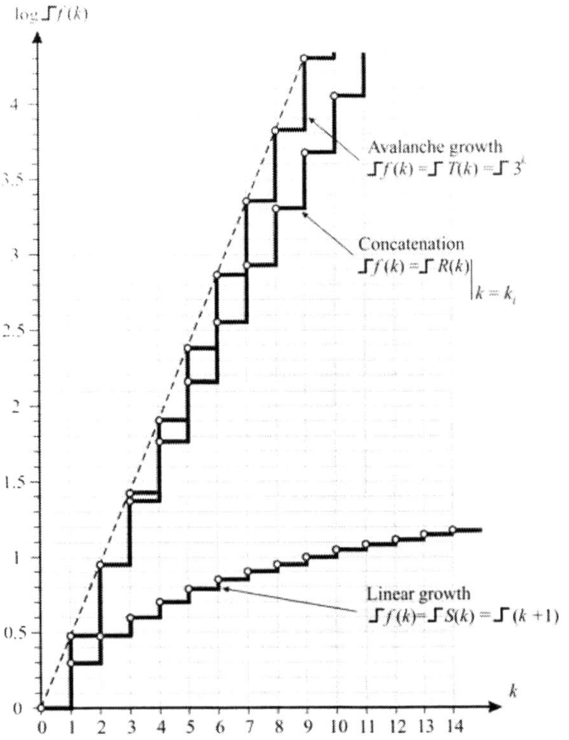

Figure 1: Concatenation of the two processes (α=0.3) of self-replication (16) and (17)

Function $\int\varphi(.)$ enables to modulate the length of the step of self-organization, dependent on the number of free elements existing in the process. Functions $\int F(.)$ and $\int G(.)$, which describe the self-replication and self-organization processes, fulfill the following relations:

$$\int\Delta S = \int S(k_{i+1}) - \int S(k_i) \geq 0, \tag{21a}$$

$$\int\Delta U = \int U(k_{i+1}) - \int U(k_i) \leq 0, \tag{21b}$$

for all i.

Therefore, the growth function of the self-replication process is non-decreasing function, on the contrary, the growth function of the self-organization process is non-increasing function (comp. (12a and b)).

6. Concluding Remarks

Considering the concatenation of the replication process (19a) and self-organization process (22a), on basis of Definition 1, we see, that the growth function of the resulting process has different properties than the growth function of the concatenation of two self-replication processes.

In case of the concatenation of *two self-replication processes*, the resulting process is characterized by the growth function which is determining faster growth of elements number than the growth function of the slower self-replication process but slower growth of elements number than the growth function of the faster self-replication process.

In case of the concatenation of *self-replication and self-organization processes*, the resulting process is characterized by the growth function which determines three possible states: the number of elements is growing up, the number of elements is decreasing till to termination of the self-replication process and in a third case the process reaches a stable state i.e. the number of elements arising in the self-replication process is equal to the number of elements snatched in the self-organization process.

The results of the concatenation of self-replication and self-organization processes can be used in the investigation of a connective self-replication of object with the internal molecular program [4], linking of such a processes, and also in analysis of processes where the very complex nanostructures appear e.g. in the processes of molecular biology.

The growing, extinction, and stabilization in the concatenation of self-replication and self-organization processes can be obtained by the mutual choice of the process parameters (19a) and (20a) including also the cycle A and modulated cycle $\int\varphi(.)$.

REFERENCES

[1] S. Węgrzyn and L. Znamirowski, "Nanotechnological, Two-stage Production Processes", *2003 Third IEEE Conf. on Nanotechnology, IEEE-NANO 2003, Proc. Vol. Two,* San Francisco, 12-14 August 2003, IEEE, pp. 490-493, 2003.

[2] S. Węgrzyn and L. Znamirowski, "Foundations of the Informatic, Molecular Nanotechnologies", *ICCE-15, 15 Ann. Int. Conf. on Comp./Nano Engr., Proc.*, D. Hui (Ed.), Int. Com. Comp./Nano Engr., Univ. of New Orleans, Haikou, China, July 15-21, pp. 1073-1074, 2007.

[3] S. Węgrzyn and L. Znamirowski, *Outline of Nanoscience and Informatic Molecular Nanotechnology*, SUT Press, Gliwice (in Polish), 2007.

[4] L. Znamirowski, "Molecular Synergy in the Nano-networks", *NSTI Nanotech 2007, The Nanotechnology Conf. and Trade Show, Techn. Proc., Vol. 1,* Santa Clara, California, May 20-24 2007, CRC Press, Boca Raton, New York, pp. 638-641, 2007.

[5] S. Węgrzyn, *Fundamentals of Automatic Control*, PWN (Polish Scien. Publ. PWN), Warsaw 1978.

[6] "BioCarta, Charting Pathways of Life" (2006), http://www.biocarta.com/.

[7] AfCS, "Nature the Signaling Gateway" (2006), http://www. signaling-gateway.org/.

[8] S. Nowak, S. Węgrzyn, and L. Znamirowski, "Self-organization Process Inside the Set of Elements Inspired by their Properties, *Theoretical and Applied Informatics (Archiwum Informatyki Teoretycznej i Stosowanej),* T. 17, Z. 4, pp. 231-240, 2005.

[9] L. Znamirowski and E. Zukowska, "Simulation of Environment-Forced Conformations in the Polypeptide Chains", *Proc. of the ICSEE'03*, 2003 Western Multi-Conf., Soc. for M&S, Orlando, FL, January 19-23, pp. 87-91, 2003.

[10] S. Węgrzyn, S. Nowak, J. Klamka and L. Znamirowski, "Systems of Informatics of the Direct Materials Nanofabrication", *3rd Int.. Congr. of Nanotechn. 2006,* 10/30th–11/2nd San Francisco, Calif., Intern. Assoc. of Nanotechn. *Proc.ICNT2006-P-MAT-217,* IANANO, Sacramento, pp. 1-15, 2006.

New Finite Element Method Modeling for contractile Forces of Cardiomyocytes on Hybrid Biopolymer Microcantilevers

Kyounghwan Na, Jinseok Kim, Sungwook Yang, Young Mee Yoon, and Eui-Sung Yoon

Nano-Bio Research Center, Korea Institute of Science and Technology
PO Box 131, Cheongryang, Seoul, 130-650, Korea, naguryong@kist.re.kr

ABSTRACT

New and novel Finite Element Method (FEM) model for contraction of cardiomyocytes on PDMS microcantilevers was studied. In new FEM model, contractile force of cardiomyocytes was considered as internal stress generated by themselves, not as interfacial force exerted externally. Also, random distribution of cardiomyocytes in FEM model was implemented through quantification of real distribution of cardiomyocytes on microcantilever by means of analysis of fluorescent image. From the result of FEM analysis using new model, the vertical deflection of PDMS microcantilever was almost in proportion to the portion of cardiomyocyte on the surface of microcantilever and then improved as compared with odd result of FEM analysis using previous model. The simulated vertical deflection of microcantilever for 2~5nN/um^2 of contractile force using new modeling for FEM coincided well with those of analytical solution and experimental data in tendency and shape of graph.

Keywords: PDMS microcantilever, cardiomyocyte, contractile force sensor, finite element method modeling

1 INTRODUCTION

The measurement of magnitude and beating frequency of cardiomyocyte contraction is significant not only for real-time monitoring of the toxicity of chemicals on cardiomyocyte, but also for understanding of the mechanisms of heart failure. Measuring contractile force of cardiomyocyte have been studied by many research groups and through various methods[1-4]. Previously, we reported on the hybrid biopolymer microcantilever for measurement of contractile force of cardiomyocyte[1]. Through comparison between vertical deflection of fabricated PDMS microcantilever and that from finite element method analysis, contractile force of cardiomyocytes of neonatal rat was figured out. However, FEM analysis was used not as verification tool, but as estimation tool, and therefore, the suitability of FEM model to real behavior of cardiomyocyte is critical to reliability of measured contractile force. In this paper, we proposed new and novel FEM model for contractile force of cardiomyocyte on hybrid biopolymer microcantilever. Based on measured deflection of microcantilever, the validity of new model was estimated.

2 CONTRACTILE FORCE SENSOR

Previously, we proposed the measurement method of contractile force of cardiomyocyte by using the hybrid biopolymer microcantilever[1]. Polydimethylsiloxane (PDMS) is proper for application to contractile force sensor of cardiomyocyte due to its flexibility and biocompatibility and it is relatively easy to fabricate microcantilever structure with PDMS by micromolding process[5]. As culturing cardiomyocyte on the surface of PDMS microcantilever, contractile force of cardiomyocyte exert shear force on the surface of PDMS microcantilever, and then it is deflected vertically due to bending moment generated by shear force as seen in Figure 1 below.

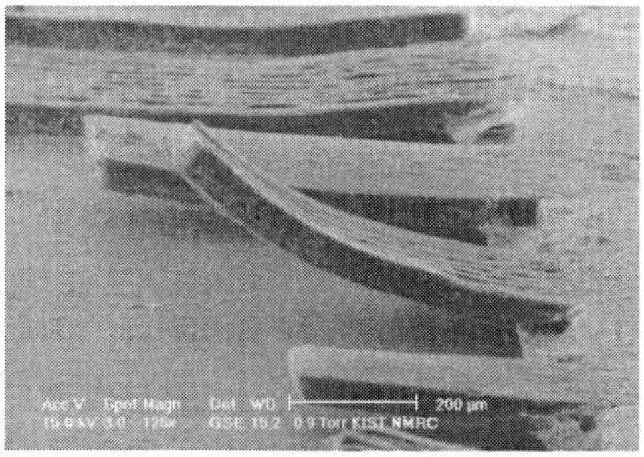

Figure 1: Vertical deflection of microcantilever by contractile force

These values of vertical deflection of microcantilever can be measured by scanning electron microscope(SEM) image analysis. On the other hand, the relation between contractile force and microcantilever deflection can be predicted by FEM simulation. Contractile force of cardiomyocyte is extracted through comparison between measured and FEM-predicted vertical deflection of microcantilever. Therefore, reliability of measured contractile force is seriously dependent on matching accuracy of measured and FEM-predicted vertical deflection of microcantilever. However, previous FEM model[1] did not resemble the contractile behaviors of real cardiomyocytes. So, we tried to build new FEM model which is more coincident with the contractile behaviors of real cardiomyocytes and can bring more reliable results.

3 MODEL DISCRIPTION

As mentioned above, we present new and novel FEM model for contractile forces of cardiomyocytes on hybrid biopolymer microcantilevers to complement problems of previous model and enhance the reliability of contractile force sensor. There are mainly two differences between previous and new model. One is the type of forces exerted by cardiomyocytes and the other is random distribution of cardiomyocytes on the surface of microcantilever. Basic formation of FEM model was considered as a double-layered microcantilever of which substrate is 20um thick PDMS and thin film is 10um thick cardiomyocytes, just like in previous model. The area of cardiomyocyte was assumed to be 20*40 um^2. We considered only longitudinal direction of forces associated with vertical deflection of microcantilever and identified that forces of other directions have little effect on vertical deflection of microcantilever.

3.1 Contractile Force of Cardiomyocytes

When cardiomyocytes are cultured on the plane, they adhere to the plane at several focal points and their contractile forces are not identical to each other. In other words, the contraction of cardiomyocyte is applied to PDMS structure not as continuously distributed stress, but discontinuously distributed shear force. For that reason, contractile force of cardiomyocytes was considered as gradient and discontinuous shear force exerted at the interface between cardiomyocytes and microcantilever structure, as described in Figure 2.

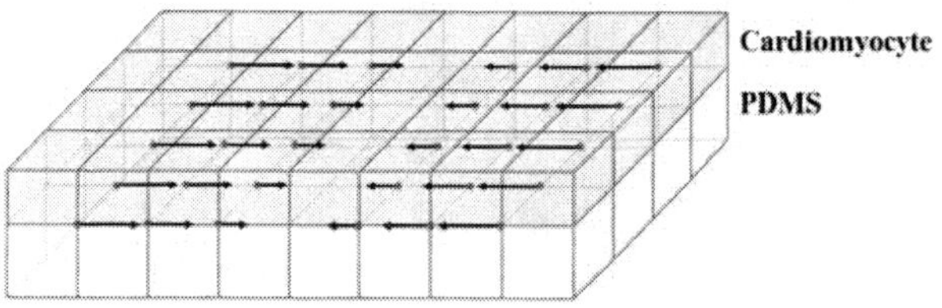

Figure 2: The schematic of previous FEM model

In this case, bending moment generated by shear stress at the interface which means contractile force of cardiomyocytes induces vertical deflection of the whole structure that consist of PDMS microcantilever and cardiomyocytes. But cardiomyocytes contract spontaneously, and vertical deflection of PDMS microcantilever is induced by shear stress imparted to it via the interface, in turn. Thus as described in Figure 3, we considered contractile force of cardiomyocytes as internal stress generated by themselves, not as interfacial force generated externally, and then contraction of cardiomyocytes affected PDMS microcantilever adhered by them. We introduced the concept of thermal stress in order to implement the internal stress yielding volumetric shrinkage in simulation tool, just like as the deflection of bimetal.

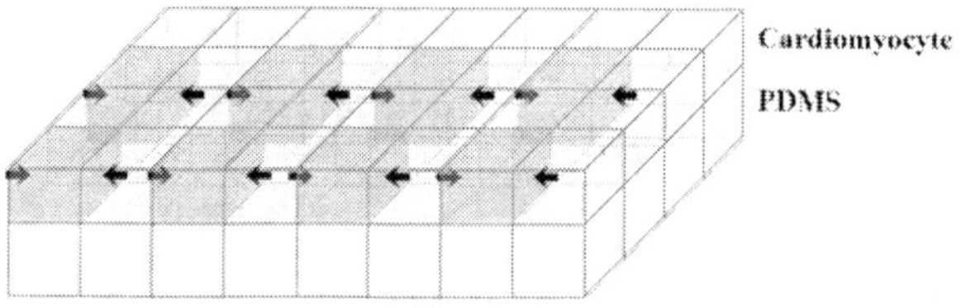

Figure 3: The schematic of new FEM model

3.2 Random Distribution of Cardiomyocytes

In new FEM model, random distribution of real cardiomyocytes on the microcantilever was implemented so that new FEM model could be more similar to the behavior of real cardiomyocytes. Based on the observation of fluorescent image in Figure 4, we tried to quantitate the distribution of cardiomyocytes on microcantilever. We divided the image of microcantilever surface into 20*40um^2 area, that is, 4 times of assumed cell area of unit regions, and then classified each regions of whole surface as non-nuclear, mono-nuclear and poly-nuclear region according to the number of nucleus in the region(0,1,and more than 2, respectively). As a result of quantitative analysis for 400*1200um^2 area of microcantilever, non-nuclear, mono-nuclear or poly-nuclear region occupied 23.67%, 65.67% or 10.67% of total regions, respectively.

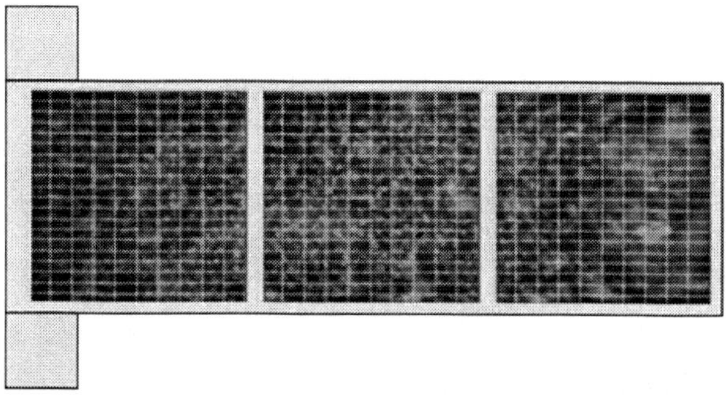

Figure 4: Fluorescent image of nucleus on microcantilever

In FEM model, cardiomyocyte layer consisted of 40*60 blocks of 10*20um^2 area and 4 adjacent blocks were defined as one region. 10.67% of all regions were selected as poly-nuclear region, and then 65.67% of all regions were selected as mono-nuclear region among the remains by means of simple computer program coded for random selection. And 2 of 4 blocks of poly-nuclear region and 1of 4 blocks of mono-nuclear region were selected again as cardiomyocytes. The rest blocks except selected ones were designated as buffer, of which mechanical properties such as Young's modulus and Poisson's ratio were identical to cardiomyocyte but which has no internal stress. It was assumed that Young's modulus of PDMS and cardiomyocyte were 750kPa and 188kPa, respectively, and Poisson's ratios of both materials were 0.49.

4 RESULT AND DISCUSSION

FEM analysis using new FEM model for cardiomyocyte was performed and result of that compared with analytical solution and experimental data in our previous report[1]. Also, the results of FEM analysis using previous and new FEM model was compared and analyzed. FEM analysis was performed with ANSYS program from ANSYS Incorporation. Figure 5~7 show the deflection results for 2~5nN/um^2 of contractile force from analytical solution, FEM analysis using previous and new model, respectively, compared with experimental data.

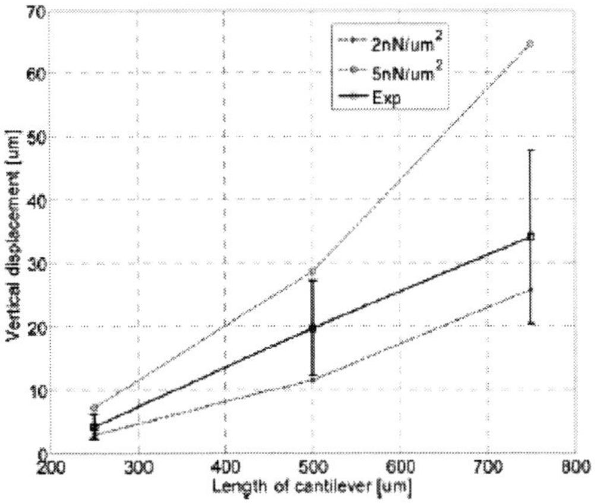

Figure 5: Deflection of micro cantilever from analytical solution and experimental data

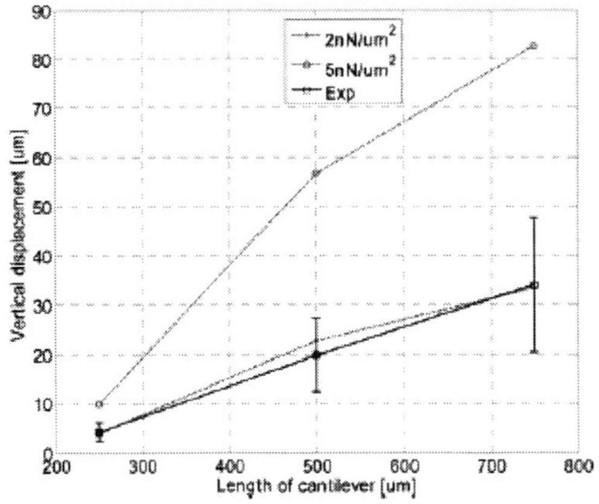

Figure 6: Deflection of microcantilever from FEM simulation with previous model and experimental data

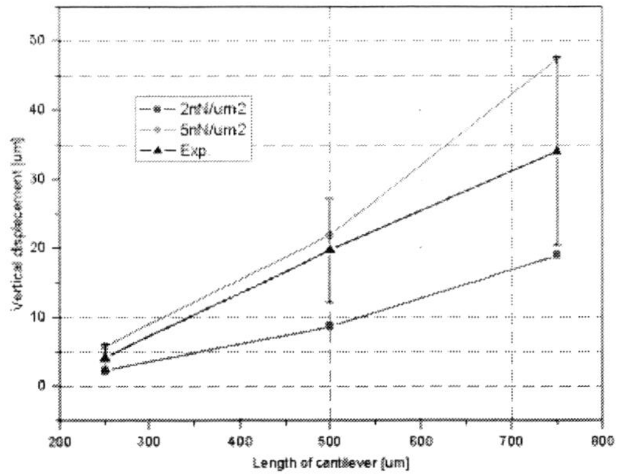

Figure 7 Deflection of microcantilever from FEM simulation with new model and experimental data

According to the simulation results using previous FEM model shown in Figure 5, vertical deflection of microcantilever increased with the lengths of microcantilevers. But the graph of result was considerably different from that of experimental data and analytical solution in Figure 6 in shape. However, the simulation results using new FEM model in Figure 7 agreed well with analytical solution and experimental data. Through Figure 7, we could see that FEM results for contractile force using new model coincided very well with 2~5nN/um^2 of the preceding result for contractile force of cardiomyocyte, also[3]. Moreover, under random cell distribution uniform, the higher the portion of non-nuclear region was, the more the microcantilever was deflected, and the higher the portion of poly-nuclear region was, the less the microcantilever was deflected, in previous model. In other words, the vertical deflection of microcantilever was decreased as increasing the number of cardiomyocytes adhered on the microcantilever, oddly. However, the vertical deflection of microcantilever by new FEM model was nearly proportional to the portion of cardiomyocyte as shown in Figure 8 and the problem of previous model described above was solved.

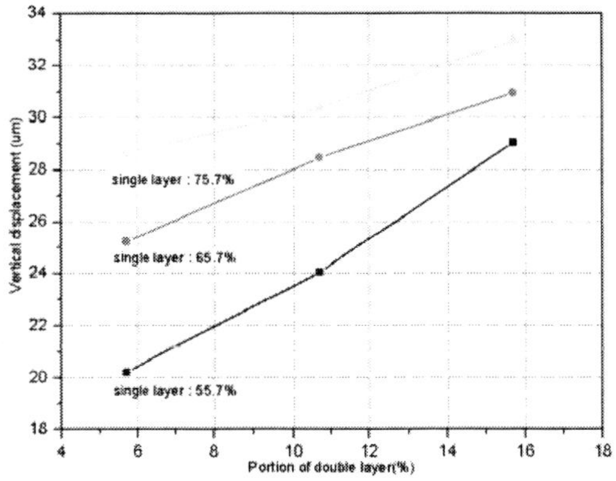

Figure 8: Deflection of microcantilever according to portions of each region

5 CONCLUSION

We presented new Finite Element Method (FEM) model for cardiomyocytes in order to enhance the reliability of contractile force sensor for cardiomyocyte using PDMS. The shear force for contractile force was replaced with internal stress and random distribution of cardiomyocytes on the surface of microcantilever was modeled, also. As a result of FEM analysis using new model, the odd result from previous FEM model was solved, in which the vertical deflection of PDMS microcantilever was decreased as increasing the number of cardiomyocytes on the surface of microcantilever. And result from FEM analysis using new model was agreed well with the analytical solution, the experimental data and the preceding result for contractile force of cardiomyocyte.

ACKNOWLEGMENTS

This research was supported by the Intelligent Microsystems Center (http://www.microsystem.re.kr), which carries out one of the 21st Century's Frontier R&D Projects sponsored by the Korea Ministry of Commerce, Industry and Energy also by the Nano Bioelectronics and Systems Research Center of Seoul National University, which is an ERC supported by the Korean Science and Engineering Foundation (KOSEF).

REFERENCES

[1] J. Park et al., "Real-Time Measurement of the Contractile Forces of Self-Organized Cardiomyocytes on HybridBiopolymer Microcantilevers", Analytical Chemistry, 77, 6571-6580, 2005.

[2] Jan Domke et al., "Mapping the mechanical pulse of single cardiomyocytes with the atomic force microscope", Eur. Biophys. J, 28, 179-186, 1999

[3] Balaban, N. Q. et al., "Force and focal adhesion assembly: a close relationship studied using elastic micropatterned substrates", Nature Cell Biology, 3, 466-472, 2001.

[4] N. Tymchenko, "A Novel Cell Force Sensor for Quantification of Traction during Cell Spreading and Contact Guidance", Biophysical Journal 93, 335–345, 2007

[5] J. Park et al., "Fabrication of complex 3D polymerstructures for cell–polymer hybridsystems", Journal of Micromechanics and Microengineering, 16, 1614-1619, 2006

Multiscale approach to nanocapsule design

Z. Shreif and P. Ortoleva

Center for Cell and Virus Theory, Department of Chemistry,

Indiana University, Bloomington, Indiana, USA, ortoleva@indiana.edu

ABSTRACT

The design of nanocapsules for targeted delivery of therapeutics presents many, often seemingly contradictory, constraints. Considering the variation in the nature of the payload and surrounding medium, software for predicting the rate of drug release for given nanocapsule structure under specific conditions in the microenvironment would be a valuable asset. An algorithm for such software using a novel all-atom, multiscale technique is presented. The method takes into consideration the atomistic effects; a necessary condition to obtain a realistic model. Other advantages of this method include the ability to (1) develop a model that doesn't require recalibration with each new application and (2) predict the supra-nanometer scale behavior such as timed payload release. Multiscale techniques are used to derive equations for the stochastic dynamics of therapeutic delivery. Application to liposomal doxorubin release is presented.

Keywords: nanomedicine; nanocapsule; therapeutic delivery; computer-aided nanocapsule design

1 INTRODUCTION

The delivery of drugs, siRNA, or genes via a functionalized nanocapsule (e.g. viral capsides and liposomes) is of great interest. Requirements for therapeutic delivery nanocapsules include (1) the ability to deliver payload at the target site while minimizing release at non-target tissues in order to reduce toxicity and increase efficacy, (2) the ability to control release of the payload over a long period of time with constant concentration, and (3) have a sufficient circulation time.

Considering the variation in the nature of the payload and the thermal and chemical environments that nanocapsules must address, it would be a great advantage to have a general physico-chemical simulator that can be used in computer-aided nanocapsule therapeutic delivery. For example, the prediction of the rate of drug release for given nanocapsule structure and conditions in the microenvironment based on a parameter-free model of supra-molecular structures that would optimize payload targeting would be a valuable asset. Some theoretical work has been presented previously [1-6] which provided some insights to certain aspects of the problem. These models are either empirical or mechanistic. Empirical models [3-4] only take into consideration the overall order of the payload release rate law, while mechanistic ones [5-6] take into account the specific processes involved such as diffusion, swelling, and erosion. However, these models are macroscopic and, therefore, do not take into consideration atomistic effects and, furthermore, require recalibration with each new application.

Release of payloads is a phenomenon that occurs over long and wide timescales; while release takes seconds to hours, atomic collisions/vibrations take place on the 10^{-12} second scale. Molecular dynamics codes, while powerful at the small scale, are impractical when it comes to dealing with nanometer scale problems spanning a timescale of a millisecond and more. An all-atom multiscale approach has been developed recently for simulating the migration and structural transitions of nanoparticles and other nanoscale phenomena [7-10]. This formulation allows for the use of an interatomic force field, making the approach universal, avoiding recalibration with each new application. In this work, this multiscale approach is applied to the nanocapsule delivery problem. The presented computational method preserves key atomic-scale behaviors needed to make predictions of interactions of functionalized nanocapsules with the cell surface receptors, drug, siRNA, gene, or other payload.

In this study, we introduce novel technical advances that capture key aspects of the nanoscale structures needed for therapeutic delivery analysis. In section 2, a variety of order parameters (characterizing nanoscale features of the capsule and its surroundings) are introduced to enable a multiscale analysis of a complex system. The final result is a Fokker-Planck (FP) equation governing the rate of stochastic payload release and structural changes and migration accompanying it. In section 3, key parameters which minimize the need for calibration are identified and predicted drug release scenarios are presented. Conclusions are drawn in section 4.

2 DERIVING THE STOCHASTIC MODEL

The specific aim of this study is to provide a starting point for a computer-aided nanocapsule design strategy. Consider a system consisting of the nanocapsule, payload, and host medium. We introduce four order parameters and their conjugate momenta which we prove to be slowly varying via Newton's equations. These order parameters are the center of mass position of the nanocapsule, $\vec{R}$, that of the drug, $\vec{R}_d$, a measure of the capsule dilatation, Φ, and the dispersal (i.e. spatial extent of the cloud of payload molecules), Λ:

$$\vec{R} = \sum_{i=1}^{N} \frac{m_i \vec{r}_i}{m^*} \Theta_i \tag{1}$$

$$\vec{R}_d = \sum_{i=1}^{N} \frac{m_i \vec{r}_i}{m_d^*} \Theta_i^d \tag{2}$$

$$\Phi = \sum_{i=1}^{N} m_i \vec{s}_i \cdot \bar{\bar{X}} \hat{s}_i^0 \Theta_i / m^* \tag{3}$$

$$\Lambda = \sum_{i=1}^{N} m_i s_i^d \Theta_i^d / m_d^* \tag{4}$$

where m_i and $\vec{r}_i$ are the mass and position of the ith atom; $m^* = \sum_{i=1}^{N} m_i \Theta_i$, and $m_d^* = \sum_{i=1}^{N} m_i \Theta_i^d$ are the total mass of the nanocapsule and payload; $\Theta_i = 1$ when i is in the nanocapsule and zero otherwise, and similarly with Θ_i^d for the payload; $\vec{s}_i$ is the position of atom i relative to $\vec{R}$; $\bar{\bar{X}}$ is a length-preserving rotation matrix that depends on a set of three Euler angles specifying nanocapsule orientation; $\hat{s}_i^0 = \vec{s}_i^0 / s_i^0$ where s_i^0 is the length of $\vec{s}_i^0$ and the superscript 0 indicates a reference nanocapsule structure; $\vec{s}_i^d$ is the position of atom i relative to $\vec{R}_d$, and s_i^d is its length. Newton's equations imply $d\vec{R}/dt = -\mathcal{L}\vec{R}$, $d\vec{R}_d/dt = -\mathcal{L}\vec{R}_d$, $d\Phi/dt = -\mathcal{L}\Phi$, and $d\Lambda/dt = -\mathcal{L}\Lambda$, where $\mathcal{L}$ is the Liouville operator

$$\mathcal{L} = -\sum_{i=1}^{N} \left[\frac{\vec{p}_i}{m_i} \cdot \frac{\partial}{\partial \vec{r}_i} + \vec{F}_i \cdot \frac{\partial}{\partial \vec{p}_i} \right] \tag{5}$$

With this, and introducing a smallness parameter ε, such that $\varepsilon^2 = m/m^* = m_d/m_d^*$ (m is the typical mass of a capsule atom, and $m_d \equiv m m_d^*/m^*$ is on the order of the mass of a typical payload atom), we get

$$d\vec{R}/dt = \varepsilon \vec{P}/m \tag{6}$$

$$d\vec{R}_d/dt = \varepsilon \vec{P}_d/m_d \tag{7}$$

$$d\Phi/dt = \varepsilon \Pi/m \tag{8}$$

$$d\Lambda/dt = \varepsilon \Pi_d/m_d \tag{9}$$

where $\vec{P}$, $\vec{P}_d$, Π, Π_d are the conjugate momenta of $\vec{R}$, $\vec{R}_d$, Φ, and Λ, and are defined as $\vec{P} = \varepsilon \sum_{i=1}^{N} \vec{p}_i \Theta_i$, $\vec{P}_d = \varepsilon \sum_{i=1}^{N} \vec{p}_i \Theta_i^d$, $\Pi = \varepsilon \sum_{i=1}^{N} \vec{\pi}_i \cdot \bar{\bar{X}} \hat{s}_i^0 \Theta_i$, $\Pi_d = \varepsilon \sum_{i=1}^{N} \vec{\pi}_i^d \cdot \hat{s}_i^d \Theta_i^d$ where $\vec{\pi}_i$ and $\vec{\pi}_i^d$ are the relative velocities of the capsule and payload atoms. The conjugate momenta are also found to be slowly varying. Applying Newton's equation, we get

$$\frac{d\vec{P}}{dt} = \varepsilon \vec{f}, \quad \vec{f} = \sum_{i=1}^{N} \vec{F}_i \Theta_i \tag{10}$$

$$\frac{d\vec{P}_d}{dt} = \varepsilon \vec{f}_d, \quad \vec{f}_d = \sum_{i=1}^{N} \vec{F}_i \Theta_i^d \tag{11}$$

$$\frac{d\Pi}{dt} = \varepsilon g, \quad g = \sum_{i=1}^{N} \vec{F}_i \cdot \bar{\bar{X}} \hat{s}_i^0 \Theta_i \tag{12}$$

$$\frac{d\Pi_d}{dt} = \varepsilon h, \quad h = \sum_{i=1}^{N} \vec{F}_i \cdot \hat{s}_i^d \Theta_i^d \tag{13}$$

where $\vec{f}$ and $\vec{f}_d$ are the net force on the nanocapsule and that on the payload, g is the "dilatation force", and h is the "dispersal force".

We suggest that this set of order parameters constitutes a minimal description capturing many nanocapsule delivery phenomena. With this, we follow the multiscale approach of Refs [7-11] to derive an FP equation of stochastic dynamics for the order parameters. Starting from the Liouville equation describing the evolution of the *N*-atom probability density, we arrive at an FP equation describing the evolution of the reduced probability density, W

$$\frac{\partial W}{\partial t} = \varepsilon \mathcal{D}' W \tag{14}$$

where

$$\mathcal{D}' = \mathcal{D} - \left[\frac{\bar{P}}{m} \bullet \frac{\partial}{\partial \bar{R}} + \bar{f}^{th} \bullet \frac{\partial}{\partial \bar{P}} + \frac{\Pi}{m} \frac{\partial}{\partial \Phi} + g^{th} \frac{\partial}{\partial \Pi} + \frac{\bar{P}_d}{m_d} \bullet \frac{\partial}{\partial \bar{R}_d} + \bar{f}_d^{th} \bullet \frac{\partial}{\partial \bar{P}_d} + \frac{\Pi_d}{m_d} \frac{\partial}{\partial \Lambda} + h^{th} \frac{\partial}{\partial \Pi_d} \right]$$

$\bar{f}^{th}$, g^{th}, $\bar{f}_d^{th}$, h^{th} are the thermal average of the forces, which is also equivalent to the long-time average though Gibbs hypothesis, and

$$\begin{aligned}
\mathcal{D} = \bar{\bar{\gamma}}_{ff} \frac{\partial}{\partial \bar{P}}\left(\beta \frac{\bar{P}}{m} + \frac{\partial}{\partial \bar{P}}\right) + \bar{\gamma}_{fg} \bullet \frac{\partial}{\partial \bar{P}}\left(\beta \frac{\Pi}{m} + \frac{\partial}{\partial \Pi}\right) + \bar{\bar{\gamma}}_{ff_d} \frac{\partial}{\partial \bar{P}}\left(\beta \frac{\bar{P}_d}{m_d} + \frac{\partial}{\partial \bar{P}_d}\right) \\
+ \bar{\gamma}_{fh} \bullet \frac{\partial}{\partial \bar{P}}\left(\beta \frac{\Pi_d}{m_d} + \frac{\partial}{\partial \Pi_d}\right) + \bar{\gamma}_{gf} \bullet \frac{\partial}{\partial \Pi}\left(\beta \frac{\bar{P}}{m} + \frac{\partial}{\partial \bar{P}}\right) + \gamma_{gg} \frac{\partial}{\partial \Pi}\left(\beta \frac{\Pi}{m} + \frac{\partial}{\partial \Pi}\right) \\
+ \bar{\gamma}_{gf_d} \bullet \frac{\partial}{\partial \Pi}\left(\beta \frac{\bar{P}_d}{m_d} + \frac{\partial}{\partial \bar{P}_d}\right) + \gamma_{gh} \frac{\partial}{\partial \Pi}\left(\beta \frac{\Pi_d}{m_d} + \frac{\partial}{\partial \Pi_d}\right) + \bar{\bar{\gamma}}_{f_d f} \frac{\partial}{\partial \bar{P}_d}\left(\beta \frac{\bar{P}}{m} + \frac{\partial}{\partial \bar{P}}\right) \\
+ \bar{\gamma}_{f_d g} \bullet \frac{\partial}{\partial \bar{P}_d}\left(\beta \frac{\Pi}{m} + \frac{\partial}{\partial \Pi}\right) + \bar{\bar{\gamma}}_{f_d f_d} \frac{\partial}{\partial \bar{P}_d}\left(\beta \frac{\bar{P}_d}{m_d} + \frac{\partial}{\partial \bar{P}_d}\right) + \bar{\gamma}_{f_d h} \bullet \frac{\partial}{\partial \bar{P}_d}\left(\beta \frac{\Pi_d}{m_d} + \frac{\partial}{\partial \Pi_d}\right) \\
+ \bar{\gamma}_{hf} \bullet \frac{\partial}{\partial \Pi_d}\left(\beta \frac{\bar{P}}{m} + \frac{\partial}{\partial \bar{P}}\right) + \gamma_{hg} \frac{\partial}{\partial \Pi_d}\left(\beta \frac{\Pi}{m} + \frac{\partial}{\partial \Pi}\right) + \bar{\gamma}_{hf_d} \bullet \frac{\partial}{\partial \Pi_d}\left(\beta \frac{\bar{P}_d}{m_d} + \frac{\partial}{\partial \bar{P}_d}\right) \\
+ \gamma_{hh} \frac{\partial}{\partial \Pi_d}\left(\beta \frac{\Pi_d}{m_d} + \frac{\partial}{\partial \Pi_d}\right).
\end{aligned}$$

The γ factors account for the cross frictional effects, expressions for which are to be presented in a following paper [12].

3 SIMULATING PAYLAOD RELEASE

The FP equation (14) is equivalent to a set of Langevin equations wherein the forces and the friction coefficients can be calculated using MD code. Here, we try to illustrate our method by adopting a simplified model wherein the capsule is at the target site and ready for release. With this, only two Langevin equations remain:

$$\frac{d\Lambda}{dt} = \frac{\Pi_d}{m_d} \tag{15}$$

$$\frac{d\Pi_d}{dt} = h^{th} - \gamma_{hh}\Pi_d + A(t) \tag{16}$$

where $A(t)$ is a random force.

The dispersal is related to the concentration profile, $C(r)$, by

$$C(r,\tau) = C_0(\tau) e^{-(r/a\Lambda)^2} \tag{17}$$

where r is the distance from the center of mass of the nanocapsule, α is a constant, and C_0 is the payload concentration at the center of mass of the nanocapsule. In the above equation, payload release is assumed to be spherically symmetric, and the concentration to be a maximum at the center of mass. h^{th} is minus the derivative of the potential energy, U, with respect to the dispersal. The latter is related to the concentration profile by the phenomenological expression

$$U = 4\pi \int_0^\infty u C r^2 \mathrm{d}r, \quad u = \begin{cases} \bar{u} & R_c \le r \le R_o \\ 0 & \text{otherwise} \end{cases} \tag{18}$$

where R_c and R_o are the inner and outer radius of the nanocapsule, respectively.

Table 1 Values used for the simulation

atoms	R_0 (nm)	R_c (nm)	m^* (g)	m_d^* (g)
6.7×10^5	100	92	7.208×10^{-17}	9.01×10^{-18}

In what follows, we show release profiles for different values of friction coefficient (Fig. 1). Values for the radii, masses, and number of atoms (Table 1) are chosen to be consistent with experimental observations on a typical liposome loaded with doxorubicin [13].

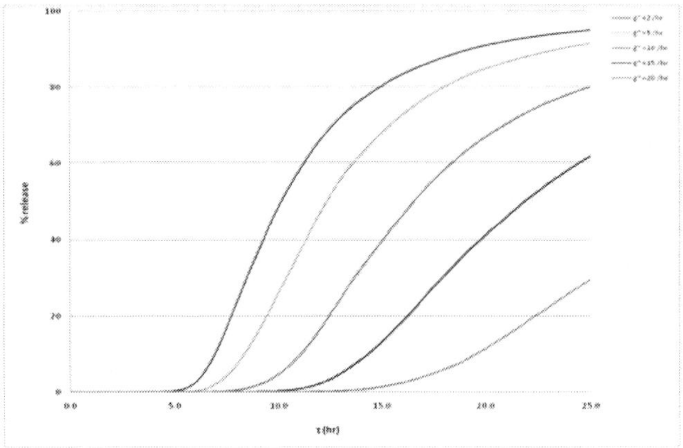

Fig. 1 Release profile simulated using equations (15) and (16) for parameter values as in Table 1

The g^* value indicated in the graphs is a calibrated coefficient that is related to the maximum friction value in the inner shell of the nanocapsule, γ_{max}, via $\gamma_{max} = g^* e^{\bar{u}/RT}$. Thus, changes in γ_{max} can be induced either through changes in the barrier height or friction coefficients, both of which can be modified via temperature changes or interaction with the cell membrane. From Fig. 1, we see that as the barrier height or friction inside the shell increases, the rate of release of drug from the nanocapsule decreases. This is consistent with the fact that increasing the length and/or saturation of the fatty acyl chains comprising a liposome leads to slower release rates. Increasing γ_{max} also leads to longer residence time in the nanocapsule, as shown from the simulation results summarized in Fig. 2.

As can be seen in Figs. 1 and 2, the nature of payload/nanocapsule/medium dictates how long a nanocapsule's membrane can sequester the paylaod. Before the nanocapsule reaches the target site, the barrier height should be larger than the fluctuating forces. Once at the target site, release can be enhanced by lowering the barrier height and friction, or increasing the amplitude of fluctuations. This can be done by either applying an external field such as heat, light, or ultrasound, or by changes in the medium, i.e. the trigger is intrinsic to the system.

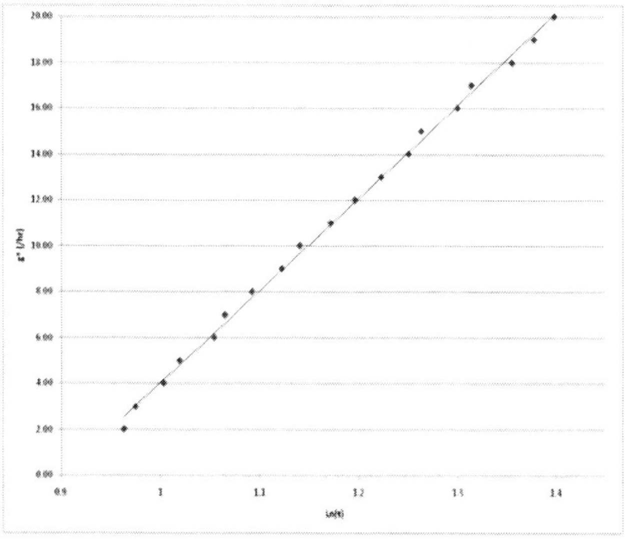

Fig. 2 Residence time in the nanocapsule simulated with different g^* values

4 CONCLUSIONS

An all-atom, multiscale approach for modeling nanocapsule therapeutic delivery systems has been presented. This method allows for the use of an interatomic force field, thereby avoiding the need for recalibration with each new application. Order parameters were introduced to characterize special features of these systems. These order parameters are the center of mass of the payload and nanocapsule, the state of the capsule (a measure of dilatation), and the dispersal (extent of release of the payload). We believe that these order parameters constitute the minimal set needed to describe such systems. Additional order parameters can readily be introduced to account for the presence of a cell surface and other nanoobjects; include other system-specific effects such as externally applied heat, magnetic forces, and ligand properties; or provide a more detailed description of the nanocapsule (i.e. shape, orientation, or distribution of small-scale structure across the nanocapsule). A reduced equation for the stochastic dynamics of these parameters was derived. To illustrate the approach, the time-course of liposomal doxorubicin delivery was simulated for a simplified case wherein the fully loaded capsule starts at the target zone and ready for release. In ongoing work, we are developing modules for estimating friction coefficients and thermal average forces based on statistical mechanical formulas and a universal interatomic force field; the goal being to avoid recalibration with each new application.

REFERENCES

[1] H. Harashima and H. Kiwada, *Advanced Drug Delivery Reviews*, **40**, 39-61, 1999

[2] D.G. Kanjickal and ST. Lopina, *Critical Reviews in Therapeutic Drug Carrier Systems*, **21**, 345-386, 2004

[3] D.O. Cooney, *AIChE Journal*, **18**, 446-449, 1972

[4] N.A. Peppas, *Pharmaceutica Acta Helvetiae*, **60**, 110-111, 1985

[5] P.I. Lee, *Journal of Membrane Science*, **7**, 255-275, 1980

[6] A. Göpferich, *Journal of Controlled Release*, **44**, 271-281, 1997

[7] P. Ortoleva, *Journal of Physical Chemistry B*, **109**, 21258-21266, 2005

[8] Y. Miao and P. Ortoleva, *Journal of Chemical Physics*, **125**, 044901, 2006

[9] Y. Miao and P. Ortoleva, *Journal of Chemical Physics*, **125**, 214901, 2006

[10] Z. Shreif and P. Ortoleva, *Journal of Statistical Physics*, **130**: 669-685, 2008

[11] S. Pankavich, Z. Shreif, and P. Ortoleva, *Physica A*, 2008 (accepted)

[12] Z. Shreif and P. Ortoleva, *Computational and Mathematical Methods in Medicine*, submitted

[13] D.C. Drummond, O. Meyer, K.L. Hong, D.B. Kirpotin, and D. Papahadjopoulos, *Pharmacological Reviews*, **51**, 691-743, 1999

Capacitance Modeling of Short-Channel DG and GAA MOSFETs

H. Børli, S. Kolberg, and T.A. Fjeldly
*UniK – University Graduate Center, Norwegian University of Science and Technology,
N-2021 Kjeller, Norway, {hborli, kolberg, torfj}@unik.no

ABSTRACT

Modeling of the intrinsic capacitances of short-channel, nanoscale DG and GAA MOSFETs is presented, covering a wide range of operation from subthreshold to strong inversion. In subthreshold, the electrostatics is dominated by the inter-electrode capacitive coupling, from which analytical expressions for the charge conserving trans- and self-capacitances of the DG device can readily be derived. Near and above threshold, the influence of the electronic charge is taken into account in a precise, self-consistent manner by combining suitable model expressions with Poisson's equation in the device body. The models are verified by comparison with numerical device simulations.

Keywords: short-channel MOSFET, double-gate, gate-all-around, nanoscale, capacitances, conformal mapping.

1 INTRODUCTION

We have previously presented precise models for the electrostatics and the drain current of short-channel, nanoscale double-gate (DG) and gate-all-around (GAA) MOSFETs [1-7]. These models are based on a procedure where the inter-electrode capacitive coupling between the source, drain, and gates is considered separately as the solution of the Laplace equation for the device body potential. Using conformal mapping techniques, this leads to a precise analytical solution in terms of elliptic integrals for the 2D case of the DG MOSFET [2-4]. We have also shown that the DG results can be successfully applied to the GAA MOSFET by performing an appropriate device scaling to compensate for the difference in gate control between the two devices [6]. These solutions are dominant in subthreshold for low-doped, nanoscale devices.

Near and above threshold, the influence of the electronic charge on the electrostatics is taken into account in a precise, self-consistent manner by combining suitable model expressions with Poisson's equation [7].

Here, we discuss how to derive the intrinsic device capacitances by considering the total, vertical electric displacement field on the gate, source, drain and electrodes from the device electrostatics.

The DG and GAA devices considered have gate length $L = 25$ nm, silicon substrate thickness/diameter $t_{si} = 12$ nm, insulator thickness $t_{ox} = 1.6$ nm, and insulator relative dielectric constant $\varepsilon_{ox} = 7$. The doping density of the *p*-type silicon body is $1\text{x}10^{15}\text{cm}^{-3}$. As gate material, we selected a near-midgap metal with the work function 4.53 eV. Idealized Schottky contacts with a work function of 4.17 eV (corresponding to that of *n*+ silicon) are assumed for the source and drain. This ensures equipotential surfaces on all the device contacts. To simplify the modeling, we replace the insulator by an electrostatically equivalent silicon layer with thickness of $t'_{ox} = t_{ox}\varepsilon_{si}/\varepsilon_{ox}$, where ε_{si} is the relative permittivity of silicon. The device dimensions considered are such that a classical treatment of the electron distribution is warranted.

The modeled capacitances are verified against numerical simulations. Since no fitting parameters are used, the capacitance model together with the corresponding drain current model is scalable over a wide range of geometric and material combinations.

2 ELECTROSTATICS

2.1 Subthreshold

The capacitive coupling between the electrodes of the DG MOSFET is given by Laplace's equation, which can be solved by the technique of conformal mapping [3,4]. The extended four-corner device body of the (x,y)-plane is mapped into the upper half of a complex (u,iv)-plane, as indicated in Fig. 1.

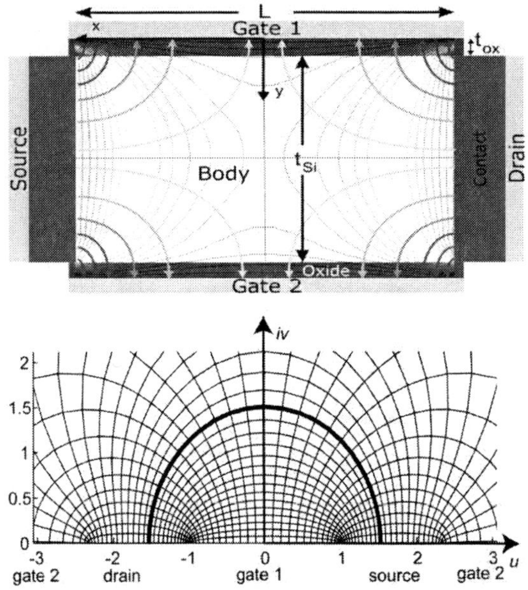

Figure 1: Schematic view of the mapping of a DG MOSFET body (top) into the upper half-plane of the (u,iv)-plane (bottom). A rectangular grid in the (x,y)-plane transforms into the grid shown in the (u,iv)-plane.

The mapping between the two planes is given by the following Schwartz-Christoffel coordinate transformation [3,8]

$$z = x + iy = \frac{L}{2}\frac{F(k,w)}{K(k)} \tag{1}$$

where $w = u+iv$, $F(k,w)$ is the elliptic integral of the first kind, $K(k) = F(k,1)$ is the corresponding complete elliptic integral, and the modulus k is a constant between 0 and 1 determined by the geometric ratio $L/(t_{si} + 2t'_{ox})$. We note that the boundary of the extended, rectangular body in Fig. 1a maps into the real u-axis of the (u,iv)-plane and the four corners map into the position $u = \pm 1$ and $u = \pm 1/k$. Moreover, the vertical imaginary v-axis correspond to the gate-to-gate symmetry axis while the bold semicircle (with the radius $1/\sqrt{k}$) corresponds to the source-to-drain symmetry axis [3].

In the (u,iv)-plane, the inter-electrode contribution to the potential distribution is determined from the Laplace equation as follows

$$\varphi_{DG}^{LP}(u,v) = \frac{v}{\pi}\int_{-\infty}^{\infty}\frac{\varphi_{DG}^{b}(u')}{(u-u')^2+v^2}du' \tag{2}$$

where $\varphi_{DG}^{b}(u')$ is the electrostatic potential along the entire boundary, i.e., along the equipotential surfaces of the source the drain, and the gates, and with minor contributions from the insulator gaps in the four corners. The major terms, corresponding to the case of a very thin insulator, can be expressed analytically in terms of the coordinates u and v [2-5]. The minor terms, which are especially important for determining the capacitances, can be obtained by carefully modeling the potential distribution across the insulator gaps. For this, we apply another conformal mapping procedure suitable for a single corner structure [8], as indicated in Fig. 2. The equipotential lines and the field lines shown for the vicinity of the gap are solutions obtained from this procedure [9]. The bold, dashed line shows the section of the device boundary associated with a single insulator gap. The potential distribution obtained along this line shown in the inset can be approximated by a polynomial of the form

$$\varphi_{ox}(0,y_{1c}) \approx ay_{1c}^6 + by_{1c} + V_{gs} - V_{FB} \tag{3}$$

where the parameters a and b are determined from the potential at source or drain, V_{gs} is the gate-source voltage, V_{FB} is the flat-band voltage of the gate, and y_{1c} is the local coordinate along the boundary in Fig. 2.

The GAA MOSFETs are 3D structures that cannot be analyzed the same way. However, because of the cylindrical symmetry, we observe that many structural similarities exist between the 2D potential distribution obtained for the DG MOSFET and that of a longitudinal cross-section through the axis of the GAA device.

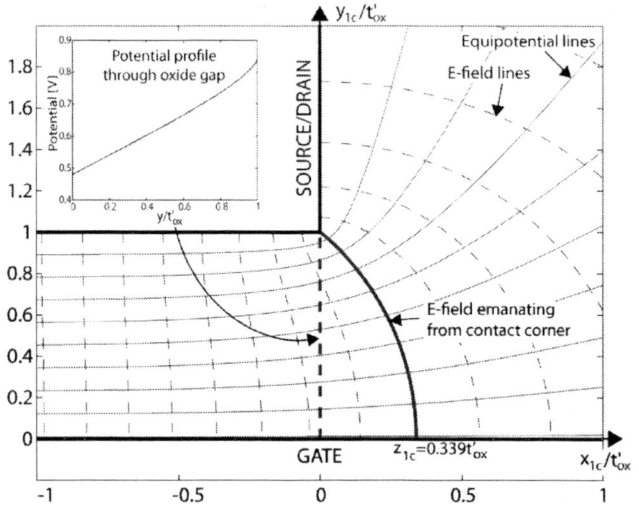

Figure 2: Equipotential lines and field lines from the one-corner analysis with inset showing the potential profile along the boundary through the insulator gap (bold dashed line).

In fact, the major difference between the two is the strength of the gate control. This difference can be expressed in terms of the so-called characteristic lengths, which are a measure of the penetration depth of the contact electrostatic influence along the source-to-drain symmetry axis. The characteristic lengths λ_{DG} and λ_{GAA} for the DG and GAA MOSFETs, respectively, only depend on the silicon and insulator thicknesses and the ratio of their dielectric constants [10,11].

Based on this observation, we propose to approximate the inter-electrode potential distribution of the GAA MOSFET as follows [6]: First we calculate the potential distribution φ_{DG}^{LP} of a DG device with an expanded length $L_{DG} = L\lambda_{DG}/\lambda_{GAA}$, where L is the true length of the GAA device. Next, this potential distribution is compressed uniformly in the longitudinal direction using the scaling factor $\lambda_{GAA}/\lambda_{DG}$ as indicated in Fig. 3. Finally, the resulting distribution is mapped into the central, longitudinal cross-section of the GAA MOSFET.

We emphasize that this procedure for deriving the DG MOSFET inter-electrode potential does not give an exact solution of the 3D Laplace equation for the GAA MOSFET. However, comparisons with numerical calculations show that the error is at most a few millivolts, mostly localized to regions near source and drain.

2.2 Self-consistency

Near and above threshold, the contribution to the body potential from the inversion charge must be included. In this case, Poisson's equation is divided into two superimposed parts, the first of which is the Laplace equation, which describes the inter-electrode capacitive

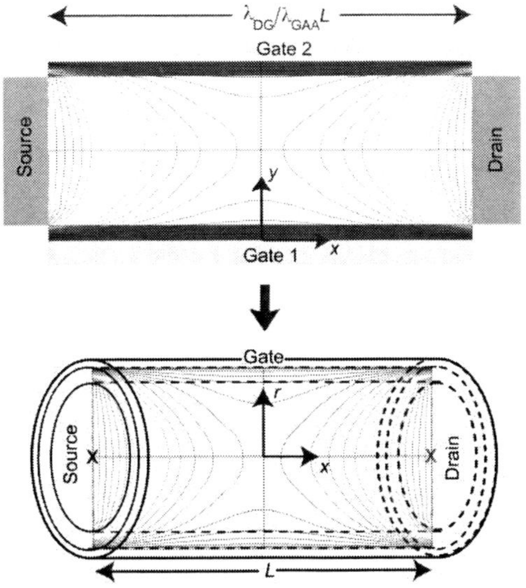

Figure 3: Schematic illustration of the mapping of a DG MOSFET inter-electrode potential distribution for an extended device of length L_{DG} (top) into the longitudinal cross-section of a GAA device of length L (bottom).

coupling discussed in Section 2.1. The second part accounts for the electrostatic effects associated with the charge carriers, which must be derived in a self-consistent manner. With a finite drain bias, the self-consistency also encompasses the quasi-Fermi potential distribution and the drain current.

The self-consistent procedure for modeling the electrostatics and the drain current of both the DG and GAA MOSFETs is described elsewhere [6,7].

3 CAPACITANCE MODELING

From the device electrostatics, it is possible to find the vertical displacement field distributions at the source, drain, and gate electrode surfaces. According to Gauss' law, these fields determine the charges on the electrodes, from which the intrinsic capacitances can be derived.

Taking proper account of all charges, the four-terminal DG MOSFET can be described in terms of 16 trans- and self-capacitances C_{XY}, of which 9 are independent owing to the principle of charge conservation [12]. Here, C_{XY} reflects the change of charge assigned to electrode X for a small variation in voltage applied to terminal Y according to the definition:

$$C_{XY} = \pm \frac{\partial Q_X}{\partial V_Y} \tag{4}$$

where the + sign is used for $X = Y$ (self-capacitances), and the – sign is used for $X \neq Y$ (trans-capacitances). The GAA MOSFET is a tree-terminal device and so is the DG when applying symmetric gate biasing. For this case, the number of capacitances reduces to 9 of which 4 are independent. The equivalent circuit of the three-terminal device is shown in Fig. 4 [13].

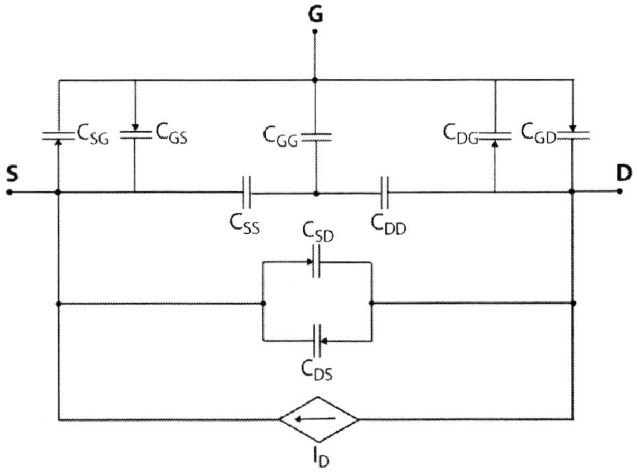

Figure 4: Intrinsic equivalent circuit with charge conserving capacitances for three-terminal MOSFETs.

3.1 Subthreshold capacitances

In subthreshold, the intrinsic capacitances are dominated by the inter-electrode capacitive coupling discussed in Section 2.1. From the major term in (2) (limit of small insulator gaps), we obtain the following expression for the charge on the various electrodes of the DG MOSFET [9]:

$$Q_X = \varepsilon_{si} \int_{z_{\min}}^{z_{\max}} E_\perp dz = i\varepsilon_{si} \int_{u_{\min}}^{u_{\max}} \left. \frac{\partial \varphi_{DG}^{LP}}{\partial v} \right|_{v\to 0} du$$
$$= \frac{i\varepsilon_{si}}{\pi} \left[V_G \ln\left(\frac{(u-1)(ku+1)}{(u+1)(ku-1)} \right) + V_S \ln\left(\frac{u+1}{ku+1} \right) + V_D \ln\left(\frac{u-1}{ku-1} \right) \right] \Bigg|_{u_{\min}}^{u_{\max}} \tag{5}$$

Here, $V_G = V_{gs} - V_{FB}$, $V_S = V_{bi}$, $V_D = V_{ds} + V_{bi}$, where V_{bi} is the built-in voltage of the source and drain contacts and V_{ds} is the drain-source voltage. The limits of integration are the appropriate dimensions of the electrodes over which the intrinsic charges are distributed. For the drain and source electrodes, the integration runs from $y = t'_{ox}$ to $t_{si} + t'_{ox}$ for $x = L/2$ (source) or $x = -L/2$ (drain), or between the corresponding coordinates on the u-axis in the (u, v)-plane. The latter are obtained from the transformation in (1). For the gate electrodes, we notice from Fig. 2 that the charges close to the corners, between the bold solid and dashed lines, correspond to field lines that terminate on the sides of the source/drain electrodes. Therefore, in order to preserve intrinsic total charge neutrality, these charges should be excluded and assigned to the extrinsic capacitances. From the one-corner analysis, we find that the integration for the intrinsic gate charge should run between $x = -L/2 + x_0$ and $L/2 - x_0$ for the two gates, where $x_0 = 0.339t'_{ox}$, or between the corresponding coordinates along the u-axis in the (u,iv)-plane.

Analytical expressions for the subthreshold capacitances are obtained by introducing (5) in (4). We note that since (5) is linear in the applied voltages, these capacitances are voltage independent. However, in order to improve the precision, we should also include the corrections in the electrostatics associated with the finite insulator thickness (see Section 2.1). This is done by applying (3) in (2) to correct Q_X in (5). The adjusted capacitances are used when comparing the model with numerical simulations below.

The modeling of the subthreshold capacitances in the GAA MOSFET follows the same procedure as outlined above for the DG MOSFET, but with a proper account for the cylindrical symmetry of the GAA device.

3.2 Self-consistent capacitances

From near threshold to strong inversion, the effect of the inversion charges on intrinsic device capacitances will steadily increase in importance. From the self-consistent device electrostatics in this regime, we again find the perpendicular electric field on the electrodes, from which the total electrode charges Q_S, Q_D, and Q_G and the intrinsic capacitances are determined. However, especially the mirror charges Q_{Sc} and Q_{Dc} owing to the body inversion charge Q_{Bc} may be difficult to determine precisely this way because of strong corner effects.

An alternative procedure is therefore to calculate Q_{Bc} in addition to Q_G and determine the mirror charge on the gate as $Q_{Gc} = Q_G - Q_{Gi}$, where Q_{Gi} is the contribution from the inter-electrode coupling (see Section 3.1). Hence, from charge conservation we have $Q_{Sc} + Q_{Dc} = -(Q_{Bc} + Q_{Gc})$. At zero drain-source bias, this simplifies to $Q_{Sc0} = Q_{Dc0} = -(Q_{Bc0} + Q_{Gc0})/2$.

With applied drain-source bias, we can approximate Q_{Sc} and Q_{Dc} quite well by requiring overall charge neutrality between the body charge and its mirror charges in the source-side half and in the drain-side half of the device separately. To find the total charges, we have to include the contributions from the inter-electrode coupling.

Based on the above analysis, we again obtain the self-consistent, intrinsic capacitances from (4). However, we note that well above threshold, the inversion charge contribution to the body potential will be dominant and the electrons tend to screen out the effects of the inter-electrode capacitive coupling except for the regions close to source and drain. Hence, the device electrostatics and the capacitances can be modeled according to a simplified strong-inversion, long-channel analysis [14-16].

In Figs. 5 and 6 are shown a comparison of modeled and numerically simulated capacitances versus V_{gs} for the DG MOSFET and the GAA MOSFET, respectively. Two values of the drain bias are used in each case. In all cases, we observe a very satisfactory correspondence between the model and the simulation, considering that no adjustable parameters are used in the modeling.

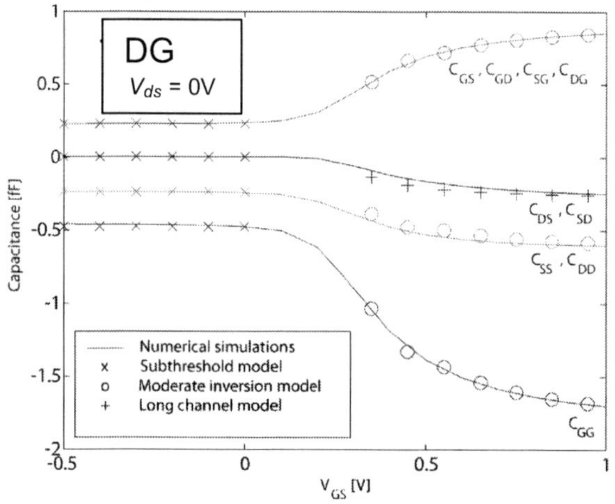

a)

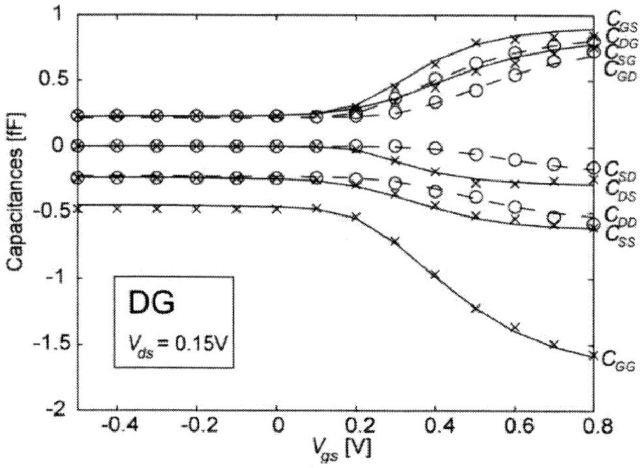

b)

Figure 5: Modeled DG MOSFET capacitances (symbols) for V_{ds}= 0 V (a) and 0.15 V (b). The curves are obtained from numerical simulations using the Silvaco Atlas device simulator.

4 CONCLUSION

We have developed a precise 2D modeling framework for calculating the intrinsic, charge conserving capacitances of nanoscale, short-channel DG and GAA MOSFETs. The 2D modeling is based on conformal mapping techniques and a self-consistent analysis of the device electrostatics that include the effects of both the inter-electrode capacitive coupling between the contacts and the presence of inversion electrons. The modeling framework covers a wide range of bias voltages from subthreshold to strong inversion. The capacitances calculated from the present model show very good agreement with those from numerical simulations (Silvaco Atlas).

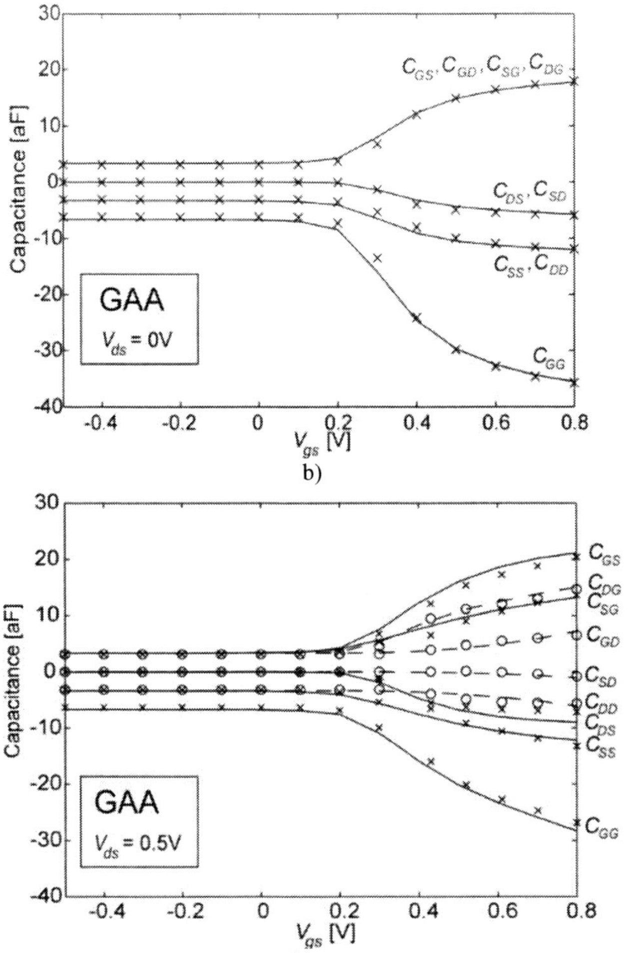

Figure 6: Modeled GAA MOSFET capacitances (symbols) for V_{ds}= 0 V and 0.1 V. The curves are obtained from numerical simulations (Silvaco Atlas).

ACKOWLEDGEMENTS

This work was supported by the European Commission under contract no. 506844 (SINANO) and the Norwegian Research Council under contract No. 159559/130 (SMIDA). We acknowledge the donation of TCAD tools from Silvaco and would like to thank Prof. Benjamin Iniguez from Universitat Rovira i Virgili, Tarragona, Spain, for helpful discussions.

REFERENCES

[1] S. Kolberg and T. A. Fjeldly, J. Comput. Electronics, vol. 5, pp. 217–222, 2006.

[2] S. Kolberg and T. A. Fjeldly, Physica Scripta, vol. T126, pp. 57– 60, 2006.

[3] S. Kolberg, T. A. Fjeldly, and B. Iñiguez, Lecture Notes in Computer Science, vol. 3994, pp. 607 – 614, 2006.

[4] T. A. Fjeldly, S. Kolberg and B. Iñiguez, Proc. NSTI-Nanotech 2007, vol. 3, Boston MA, pp. 668–673, 2006.

[5] B. Iñiguez, T. A. Fjeldly, A. Lazaro, F. Danneville and M. J. Deen, IEEE Trans. Electron Devices, vol. 53, no. 9, pp. 2128-2142, 2006.

[6] H. Børli, S. Kolberg, and T. A. Fjeldly, Proc. NSTI-Nanotech, vol. 3, Santa Clara, CA, pp. 505-509, 2007.

[7] H. Børli, S. Kolberg, T. A. Fjeldly, and B. Iñiguez, submitted to IEEE Trans. Electron Devices.

[8] E. Weber, "Electromagnetic Fields," v. 1, "Mapping of Fields," Wiley, New York, 1950..

[9] H. Børli, S. Kolberg, and T. A. Fjeldly, Proc. IEEE Int. Nanoelectronics Conf. (INEC 2008), Shanghai, China, 2008.

[10] K. Suzuki, Y. Tosaka, T. Tanaka, H. Horie, and Y. Arimoto, IEEE Trans. Electron Devices, vol. 40, pp. 2326-2329, 1993.

[11] C. P. Auth and J. D. Plummer, "Scaling theory for cylindrical fully depleted, surrounding gate MOSFET," *IEEE Electron Device Lett.*, vol. 18, pp. 74-76, 1997.

[12] D. E. Ward and R. W. Dutton, IEEE J. Solid-State Circ., vol. 13, 703, (1978).

[13] M. Nawaz and T. A. Fjeldly, IEEE Trans. Electron Dev., vol. 44, no. 10, pp. 1813-1821, 1997.

[14] Y. Taur, X. Liang, W. Wang, and H. Lu, IEEE Electron Device Lett., vol. 25, no. 2, pp. 107-109, 2004.

[15] B. Iñiguez, D. Jiménez, J. Roig, H. A. Hamid, L. F. Marsal, and J. Pallarès, IEEE Trans. Electron Devices, vol. 52, pp. 1868-1873, 2005.

[16] O. Moldovan, D. Jiménez, J. R. Guitart, F. A. Chaves and B. Iñiguez, IEEE Trans. Electron Devices, vol. 54, 1718, (2007).

New Properties and New Challenges in MOS Compact Modeling

Xing Zhou, Guan Huei See, Guojun Zhu, Zhaomin Zhu, Shihuan Lin, Chengqing Wei, Ashwin Srinivas, and Junbin Zhang

School of Electrical & Electronic Engineering, Nanyang Technological University
Nanyang Avenue, Singapore 639798, exzhou@ntu.edu.sg

ABSTRACT

Conventional (*four*-terminal) bulk-MOS models are based on *unipolar* conduction in a *doped* body with *body contact* and ideal *symmetric* PN-junction source/drain (S/D) contacts. As bulk-MOS technology is approaching its fundamental limit, non-classical devices such as ultra-thin body (UTB) SOI as well as multiple-gate (MG) and gate-all-around/Si-nanowire (GAA/SiNW) MOSFETs emerge as promising candidates for future-generation device building blocks. This trend poses new challenges to developing a compact model suitable for these new device structures and requires a paradigm shift in the core model structure. In MG/NW (including non-body-contacted UTB-SOI) MOSFETs, however, the (*three*-terminal) device has a nearly *undoped* body and without a body contact, and S/D contacts (PN-junction or Schottky) also become an integral part of intrinsic channel. Carrier transport may become *bipolar* and may change from drift-diffusion dominant to tunneling dominant, depending on the S/D contacts. Source–drain asymmetry, either intentional or unintentional, in a theoretically symmetric MOSFET also becomes important to be captured in a compact model, which is nontrivial in a model that depends on terminal S/D swapping at the circuit level. This paper discusses these new challenges and demonstrates solution methods based on the unified regional modeling (URM) approach to the ultimate goal of unification of MOS compact models.

Keywords: bipolar, body contact, compact model, MOSFET, multiple-gate, nanowire, Schottky-barrier, symmetry, ultra-thin body SOI, undoped body, unified regional modeling, unipolar.

1 INTRODUCTION

MOSFET compact model (CM) has been at the heart of VLSI circuit design and chip fabrication over the past several decades. Starting from the most complete and rigorous Pao–Sah formulation [1, 2] of the iterative surface-potential (ϕ_s) voltage-equation solution and double-integral current-equation solution, and its simplest version of fixed bulk-charge threshold-voltage (V_t) and drain-current (I_{ds}) equations [1], history has witnessed generations of MOS compact-model development from first-generation V_t-based to second-generation ϕ_s-based formulations. We have seen the efforts required for the radical changes from one generation to the next, which prompts the need for an extendable core model infrastructure to meet the challenges in developing future-generation models.

The essence of compact modeling is to come up with an analytical equation that faithfully describes the terminal current/charge characteristics of a transistor, including all its higher order derivatives, and scalable over terminal bias, geometry, temperature, frequency variations, which can match fabricated device characteristics and can be used efficiently and accurately in large circuit design. This requires integrating (or "compacting") physically-formulated differential equations for the specific transistor structure and boundary conditions, and adding various higher order effects on top of the intrinsic core model. The ideal long-channel transistor is only in the "linear" channel in the gradual-channel approximation (GCA) as illustrated by the shaded region (with voltages V_s and $V_{d,sat}$) in Fig. 1, whereas the real transistor has to include the velocity-saturation region as well as the built-in voltages of the S/D contacts, which depends on the contact type.

In this paper, we outline various properties of an MOSFET compact model and compare conventional with non-classical MOSFETs in terms of symmetry, body doping and contact, carrier type, and S/D contacts. We demonstrate solution methods to meet the new challenges in unifying MOS compact models with the unified regional modeling (URM) approach.

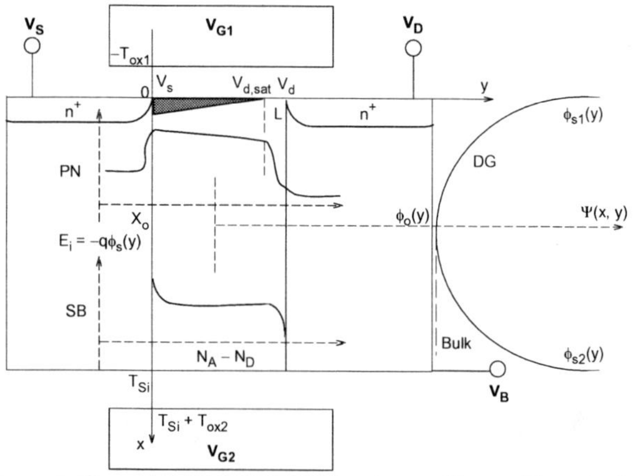

Figure 1: Generic MG-MOSFET structure with or without body contact (V_B) and with PN-junction or SB S/D contacts, showing band diagrams across the channel and potential across the body.

2 CONVENTIONAL AND NEW PROPERTIES OF MOSFETS

2.1 Symmetric vs. asymmetric MOSFET

An ideal MOSFET's physical source and drain are indistinguishable or interchangeable and, in fact, are defined by the convention of current-flow direction. It is due to this property that in all existing simulators, intrinsic (nMOS) model evaluation is all performed with positive S/D bias (V_{ds}) since whenever V_{ds} across the two (physical) terminals of an nMOS becomes negative, its polarity is swapped by the simulator such that the drain current will always flow from "drain" to "source" by *convention*. As a result, in the intrinsic (nMOS) model equations, V_{ds} will never be negative.

To comply with this symmetry property, a compact model has to pass the Gummel symmetry test (GST), which requires that the drain current I_{ds} be strictly an odd function of V_{ds} and no singularities in any higher order derivatives as $V_d = -V_s \to 0$. This is trivial to achieve in a numerical model, but a nontrivial task for a CM with mathematical functions, since in a CM based on convention, saturation always happens at the "drain" end with respect to the source and, together with complications in the use of a smoothing function for the effective drain–source voltage ($V_{ds,eff}$), those models often cannot satisfy the GST with any order of derivatives. One such example is the use of the smoothing function

$$V_{ds,eff} = \frac{V_{ds}}{\left[1+\left(\left|V_{ds}\right|/V_{ds,sat}\right)^{a_x}\right]^{1/a_x}} \quad (1)$$

which, although satisfying the odd-function requirement, will encounter singularities at higher order derivatives when a_x has to be small for short-channel drain-conductance modeling.

We have proposed a paradigm shift in modeling symmetric and asymmetric MOSFETs [3] by describing all physical quantities associated with the respective source and drain relative to the bulk and taking the difference for the effective S/D voltage ($V_{ds,eff} = V_{d,eff} - V_{s,eff}$), where source and drain are the device terminals by *label* (rather than by convention). In this way, I_{ds} will be an exact odd function of V_{ds} and negative V_{ds} can be evaluated in the model. The GST will always be satisfied even with different smoothing functions for $V_{ds,eff}$. Essentially, the intrinsic channel drain–source current ($I_{ds0} = I_d - I_s$)

$$\begin{aligned} I_{ds0} &= \overline{\beta}\left(\overline{q_i} + \overline{A_b} v_{th}\right) V_{ds,eff} \\ &= \overline{\beta}\left(\overline{q_i} + \overline{A_b} v_{th}\right) V_{d,eff} - \overline{\beta}\left(\overline{q_i} + \overline{A_b} v_{th}\right) V_{s,eff} \end{aligned} \quad (2)$$

is being modeled by the difference of the drain (I_d) and source (I_s) (or "forward" and "reverse") currents in a totally symmetric way for a perfectly symmetric-S/D MOSFET (i.e., source and drain are interchangeable). The complete short-channel drain-current model is built strictly as an odd function of V_{ds} [3]:

$$I_{ds} = \frac{g_{vo} I_{ds0}}{1 + R_{sd} I_{ds0}/V_{ds,eff}} . \quad (3)$$

On the other hand, as technology scales, S/D asymmetry in real devices (due to unintentional process and layout variations) becomes an important effect to be captured by the model. "Intentional" asymmetry, such as "true-LDD" [4, 5], dual-material-gate (DMG) [6], asymmetric-halo [7], and self-aligned asymmetric MOSFETs [8], have also been proposed, which always show improved performance over their symmetric counterparts although still not being widely used. For these (unintentional or intentional) asymmetric MOSFETs, the characteristics will be different for positive and negative V_{ds} sweeping on the same device. The only way to capture this difference in a conventional model with terminal V_{ds} swapping is to extract two complete sets of model parameters for the respective forward and reverse mode of operation. Obviously, this is not the physical way of modeling asymmetric MOSFETs, which is restricted by the model framework designed only for symmetric devices.

Our proposed paradigm shift also provides a simple modeling of asymmetric devices since source and drain can now be distinguished from label (or layout). One example is shown in Fig. 2 [3], in which the source side has a thicker junction and a lower doping than those of the drain extension, and only the saturation-velocity parameter for the respective source and drain sides is refitted for the forward and reverse mode. The model captures the different forward/reverse characteristics, including it output resistance (inset).

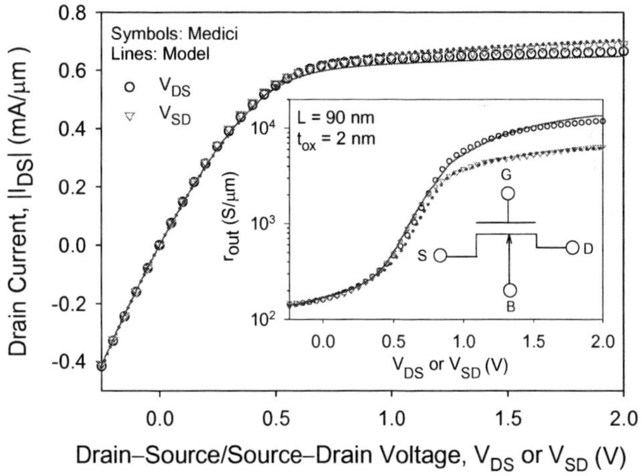

Figure 2: Modeled drain current (lines) of an asymmetric MOSFET with different S/D junction depth and doping with $\pm V_{DS}$ sweeping, and compared with the same numerical device (symbols). The inset shows the corresponding output resistance and the schematic of the asymmetric nMOS. (After [3])

2.2 Body-contacted vs. non-body-contacted MOSFET

Conventional bulk-MOS always has a body (substrate) contact and a bias (V_b) applied to it. This allows the three-terminal (gate, source, drain) voltages to be referenced to bulk, including the channel-to-bulk voltage, or the imref split, defined as $V_{cb} = \phi_{Fn} - \phi_{Fp}$. In conventional MOS modeling, "unipolar" transport is assumed in which holes (for nMOS) are always assumed at equilibrium so that $\phi_{Fp} = \phi_F$ (body Fermi potential), and the electron imref

$$\phi_{Fn}(y) = V_{cb}(y) + \phi_F \tag{4}$$

will change from $V_{sb} + \phi_F$ to $V_{db} + \phi_F$ along the channel from source ($y = 0$) to drain ($y = L$). This still allows I_{ds} to be formulated with bulk reference (B-ref) in a symmetric way.

However, in non-classical double-gate (DG) FinFETs or GAA/SiNWs, there is essentially no place to make a body contact. In some SOI technologies in which there is no body contact, it also belongs to this category. For body-contacted SOI, it would belong to bulk-MOS in terms of hole imref since it will be set by the body bias. In non-body-contacted SOI and MG/SiNW MOSFETs, with unipolar assumption, the hole imref will be determined by the lower of the S/D bias, $\phi_{Fp} \approx \phi_F = V_m \equiv \min(V_s, V_d)$ referenced to ground. It should be noted that this concept applies to all existing SOI modeling without a body contact, and the so-called "floating-body" concept in SOI is flawed since the body silicon film on which the conducting surface/volume channel forms will never be "floating." In this situation, the common and existing practice is to define the channel voltage as the *imref split*, $V_{cs} = \phi_{Fn} - \phi_F$, with source reference (S-ref), which changes from 0 at source to V_{ds} at drain end. This is consistent with the model by "S/D convention," but it would be very difficult to pass the GST due to source referencing.

We propose a fundamental change in core model formulation with "ground reference" (G-ref) [9], in which both hole and electron imrefs (as well as all other terminal voltages) are referenced to ground. (4) is then modified to

$$\phi_{Fn}(y) = V_{cm}(y) + \phi_F = (V_c(y) - V_m) + V_m = V_c(y) \tag{5}$$

which changes from V_s to V_d (G-ref) along the channel from source to drain. This formalism is also consistent with "S/D by label" (V_m can be V_s or V_d, where S/D by label) and forward/reverse currents can be similarly formulated with complete S/D symmetry, and extendable to asymmetric devices as well. A GST for the (three-terminal) undoped symmetric-DG MOSFET is shown in Fig. 3, with up to sixth-order derivatives.

2.3 Doped vs. undoped-body MOSFET

Conventional bulk-MOS devices always have a doped (thick) body. Current transport is through drift and diffusion of unipolar (minority-carrier) conduction in the surface channel of the depleted (majority-carrier) body. Conventional bulk-MOS compact modeling is based on charge-sheet approximation (CSA) [10] and GCA by solving the y-dependent current equation together with the x-dependent voltage equation. This core model formalism is valid as long as the body is doped such that CSA applies in the GCA and current conduction is mainly confined in a thin "surface channel." The voltage equation can be solved exactly by iteration or approximately from regional solutions.

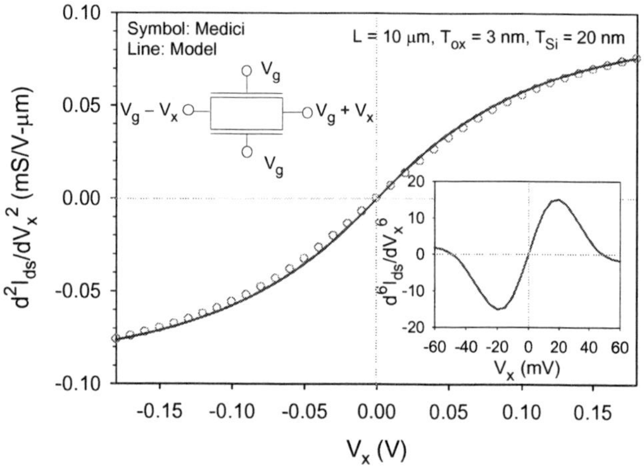

Figure 3: Comparison of the "ground-referenced" model GST second-order derivative (lines) with Medici numerical data (symbols) with constant mobility. Lower inset: model's sixth-order derivative. V_x step size: 10 mV.

In non-classical MG/UTB-SOI MOSFETs, when the body thickness is thin and doping is low, full-depletion (FD) occurs before strong inversion and, hence, the voltage equation governing the surface-potential solutions will be different from bulk-MOS due to different boundary conditions. In the generic MG-MOS devices, there are two unknowns but the Poisson equation cannot be integrated twice if body doping cannot be ignored. CSA also becomes invalid in the "volume-inversion" regime. To circumvent this difficulty, most existing compact models for DG/GAA transistors assume undoped (pure) body, together with unipolar-transport assumption, such that the voltage equation can be solved with the second integral of Poisson equation for two unknowns. However, ideal pure-body devices do not practically exist and even unintentional body doping would cause errors in surface-potential solutions if body doping is ignored.

We have been adopting the URM approach to surface-potential solutions for generic MG-MOS transistors that encompass all different types of device structures (bulk/UTB-SOI/DG/GAA) and operations [11]. Figure 4 shows one example of various regional and unified surface-potential solutions in a doped symmetric-DG MOSFET in comparison with the numerical data. The key is the solution at full-depletion voltage (V_{FD}), which separates depletion and volume-inversion regional solutions. The URM approach makes it possible to obtain a unified solution that scales with doping and body/oxide thicknesses, as shown in Fig. 5 for the derivatives of the surface potential from high doping to zero doping (pure body), with seamless transition in low-doping ranges that are otherwise missed in undoped-body models.

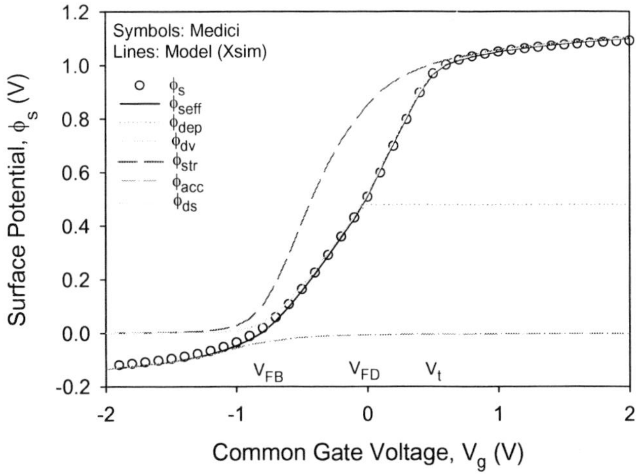

Figure 4: Modeled surface potential of a doped symmetric-DG MOSFET and its regional components (lines) compared with Medici solution (symbol). (After [11])

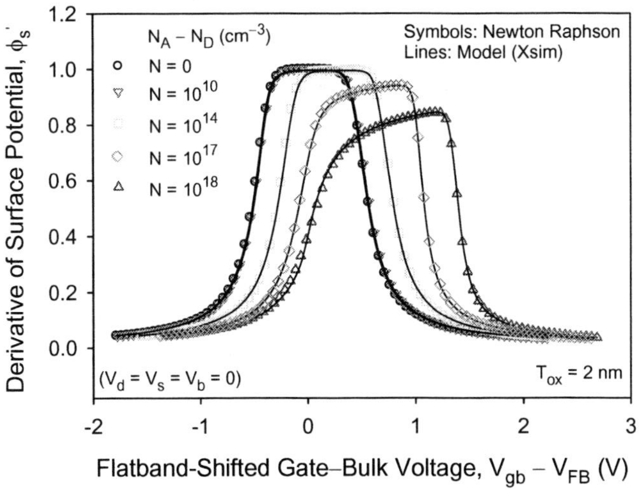

Figure 5: Derivatives of the unified surface potential in a bulk-MOS at various doping as indicated, including below intrinsic and pure body (lines), compared with the exact iterative solutions.

It should be noted that when the net doping term is not included in the Poisson equation, it implies theoretically either pure body ($N_A = N_D = 0$) or completely-compensated body ($N_A - N_D = 0$) that are both practically non-exist. As a matter of fact, "pure" body is different from "intrinsic" body (there has been confusion in some literature) in which net dopant concentration, $N_A - N_D = n_i$, is not zero (and strictly speaking, it needs to be included in the Poisson equation), and can be high at high temperatures. Also, the exact symmetric (bell-shaped) derivative of ϕ_s vs. V_g only happens in pure body, not even with "intrinsic doping," which can be a nontrivial modeling task [12]. This detailed physics has been captured in our URM model, as shown in Fig. 5 [13], with identically symmetric behavior for zero doping but not strictly symmetric for "intrinsic" doping (10^{10} cm^{-3}).

2.4 Unipolar vs. bipolar MOSFET

In almost all existing literature on MOS models since Shockley's unipolar MOS theory [14] until recently [15], unipolar carrier conduction is assumed, which means majority carrier in the body will always be at equilibrium. However, it has been shown [16, 17] that the electron and hole imrefs become bias-voltage dependent for body net doping below intrinsic concentration (n_i), as given by

$$\phi_{Fp} = v_{th} \ln\left[\frac{N_A - N_D}{2n_i} + \sqrt{\left(\frac{N_A - N_D}{2n_i}\right)^2 + e^{-V/v_{th}}}\right] \tag{6}$$

where $V \equiv \phi_{Fn} - \phi_{Fp}$ is the imref split or channel voltage due to non-zero V_{ds}, and for undoped MOSFET, $\phi_{Fp} = -V/2$. Obviously, this new feature would not be observable if the body net doping is higher than the intrinsic doping and, hence, hole imref can still be modeled by the equilibrium Fermi potential [18]:

$$\phi_F = v_{th} \sinh^{-1}\left(\frac{N_A - N_D}{2n_i}\right). \tag{7}$$

However, for (theoretically) pure-body MOSFETs, whether there will be bipolar conduction depends on the S/D contact (see more discussions in the next sub-section).

In all existing literature on undoped DG/GAA surface-potential solutions, holes are ignored in the Poisson solution which, as shown by us recently [19], gives a ~1.5-nV error in the surface potential for symmetric-DG as compared with the rigorous iterative solution of the elliptic-integral voltage equation. Although being able to be ignored at gate voltages well above the flatband (V_{FB}), this would cause a singularity in the derivative of the C_{gg} exactly at $V_{gs} = V_{FB}$. Otherwise, two-carrier solutions still can be obtained from the sum of two single-carrier solutions.

2.5 PN-junction vs. Schottky-barrier MOSFET

The channel-voltage-dependent electron/hole imrefs derived in the above sub-section are still subject to the boundary conditions of the real device when current-transport equation is solved. In undoped thin-body DG/NW MOSFETs, if the S/D extensions are PN junctions, there will be only one type of carriers as the source and sink for current flow, so unipolar transport would still be valid. If, on the other hand, the S/D extensions are Schottky-barrier (SB) contacts that favor both electron and hole injection by tunneling and/or thermionic emission, there would be ambipolar current depending on the polarity of the terminal voltages. However, transport would be essentially tunneling and/or thermionic limited rather than drift-diffusion limited due to the much larger voltage drops across the SB-induced potential barriers. We have recently developed such a compact model for SB-MOSFET based on a compact quasi-2D potential solution [20] and combined tunneling-thermionic current calculation using the Miller-Good [21] approximation, and a sample current plot is shown in Fig. 6. The Medici device was simulated with electron (n) only, hole (p) only, and two-carrier ($n+p$) transport. The tunneling/thermionic compact model without any drift-diffusion transport matched so well to the numerical device, showing the dominant mechanism is due to tunneling and thermionic current (without drift-diffusion) in SB-MOSFETs.

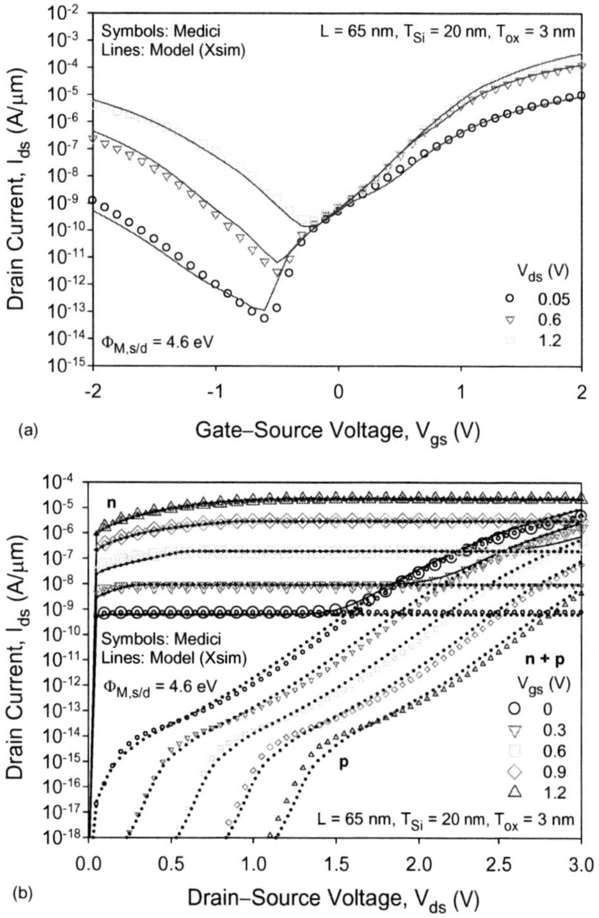

Figure 6: (a) Modeled (line) and Medici (symbols) transfer characteristics of SB-MOSFET showing ambipolar current at three V_{ds}; (b) modeled (lines) and Medici (symbols) output characteristics for one-carrier only (dotted lines/small symbols) and two-carrier current (solid lines/large symbols) at five V_{gs}.

3 KEY CHALLENGES AND CORE MODEL INFRASTRUCTURE

3.1 Key Challenges and Solutions

From the ensuring discussions, it is clear that model symmetry and asymmetry as well as with or without body contact are two different but related challenges. Conventional approaches to symmetry in body-contacted MOSFETs based on convention (terminal swapping) will not solve the new problems in asymmetric and non-body-contacted non-classical MOSFETs. The proposed ground-referenced modeling (with S/D by label) will be the key to solving both problems; of course, actual formulations will not be trivial. Models based on (1) for the effective drain–source voltage, or in general, with V_{ds} appearing in the model equation, will eventually face fundamental limitation in passing the GST.

Body doping, carrier transport, and contact effect are also different but related challenges. In conventional devices in which doping is usually high, unipolar transport is sufficient. In DG/NW devices in which body is usually undoped, Debye screening length is ~26 μm and practical transistors are all "short" length, and conducting carriers depend on the S/D contacts. So, contacts will be an essential part of the intrinsic core model for undoped devices, and transport will be either drift-diffusion or tunneling limited with PN-junction or SB contacts, respectively. In addition, to model practical devices with unintentional doping, and extending to partially/fully-depleted SOI models, the core model with body-doping scaling from high to zero is necessary.

3.2 Unified Regional Modeling Approach

Since it is known that Poisson equation cannot be integrated twice if doping term is considered and the generic MG MOSFET has two unkowns, a "short-cut" to tackle MG devices with undoped body cannot be extended when doping has to be considered. Our demonstrated URM approach can solve for the generic surface potentials with doping scaling, that can also be extended to include poly-doping and quantum-mechanical effects. Although the regional solution is not as accurate as that from iterative or explicit solutions (with nV error), it captures the essential physics, which has otherwise no exact solution (i.e., no physical equation to solve iteratively) when doping has to be considered.

3.3 Unification of MOS Models

The ultimate goal is to unify MOS models of various types and operations in one core infrastructure. As MG device operations vary from one to another seamlessly (e.g., symmetric to asymmetric DG, partially-depleted to fully-depleted SOI), so does the model describing their behaviors. The key methodology is to start with the symmetric-DG with URM approach to surface-potential solutions, which can be extended to common-gate asymmetric-DG as a special case and, eventually, to independent-gate asymmetric-DC in general that also includes fully-depleted UTB-SOI operation. Once the most difficult task is overcome, it will cover all different structures and operations, and it pays for the efforts.

4 SUMMARY AND CONCLUSIONS

In summary, we have discussed new properties and challenges in developing compact models for future-generation non-classical CMOS in comparison with the conventional bulk-MOS device modeling. Looking back the history of compact-model development, it is important to construct a core model infrastructure that is extendable to various device structures and operations. This requires a paradigm shift in intrinsic model description that accounts for physical as well as contact effects with seamless scaling over device geometry, layer thickness, doping and bias. Such a modeling framework should also revert back to the simplest bulk-MOS device, as would a DG FinFET when it is "enlarged" to resemble that of a conventional bulk/SOI device. Then, the model will be able to share similar set of parameters and extraction for progressing technologies, as well as providing accuracy and speed trade-off within a single design, and even for design and verification purposes which require different levels of accuracy.

Acknowledgement: This work was supported in part by Nanyang Technological University under Grant RGM30/03 and in part by Semiconductor Research Corporation under Grant 2004-VJ-1166G. G. H. See, S. H. Lin, and J. B. Zhang are partially supported by the NTU–AAP joint research collaboration project. C. Q. Wei is supported by scholarship from Chartered Semiconductor Ltd.

REFERENCES

[1] C. T. Sah and H. C. Pao, "The effects of fixed bulk charge on the characteristics of metal-oxide-semiconductor transistors," *IEEE Trans. Electron Devices*, vol. ED-13, no. 4, pp. 393–409, Apr. 1966.

[2] H. C. Pao and C. T. Sah, "Effects of diffusion current on characteristics of metal-oxide (insulator)-semiconductor transistors," *Solid-State Electron.*, vol. 9, no. 10, pp. 927–937, 1966.

[3] G. H. See, X. Zhou, K. Chandrasekaran, S. B. Chiah, Z. M. Zhu, C. Q. Wei, S. H. Lin, G. J. Zhu, and G. H. Lim, "A compact model satisfying Gummel symmetry in higher order derivatives and applicable to asymmetric MOSFETs," *IEEE Trans. Electron Devices*, vol. 55, no. 2, pp. 624–631, Feb. 2008.

[4] T. Horiuchi, T. Homma, Y. Murao, and K. Okumura, "An asymmetric sidewall process for high performance LDD MOSFET's," *IEEE Trans. Electron Devices*, vol. 41, no. 2, pp. 186–190, Feb. 1994.

[5] J. F. Chen, J. Tao, P. Fang, and C. Hu "0.35-μm asymmetric and symmetric LDD device comparison using a reliability/speed/power methodology," *IEEE Electron Device Lett.*, vol. 19, no. 7, pp. 216–218, July 1998.

[6] X. Zhou, "Exploring the novel characteristics of hetero-material gate field-effect transistors (HMGFET's) with gate-material engineering," *IEEE Trans. Electron Devices*, vol. 47, no. 1, pp. 113–120, Jan. 2000.

[7] T. N. Buti, S. Ogura, N. Rovedo, K. Tobimatsu, and C. F. Codella, "Asymmetrical halo source GOLD drain (HS-GOLD) deep sub-half micron n-MOSFET design for reliability and performance," in *IEDM Tech. Dig.*, 1989, pp. 617–620.

[8] J. P. Kim, W. Y. Choi, J. Y. Song, S. W. Kim, J. D. Lee, and B.-G. Park, "Design and fabrication of asymmetric MOSFETs using a novel self-aligned structure," *IEEE Trans. Electron Devices*, vol. 54, no. 11, pp. 2969–2974, Nov. 2007.

[9] G. J. Zhu, G. H. See, X. Zhou, Z. M. Zhu, S. H. Lin, C. Q. Wei, J. B. Zhang, and A. Srinivas, "'Ground-referenced' model for three-terminal symmetric double-gate MOSFETs with source–drain symmetry," submitted for publication.

[10] J. R. Brews, "A charge-sheet model of the MOSFET," *Solid-State Electron.*, vol. 21, no. 2, pp. 345–355, 1978.

[11] X. Zhou, G. H. See, G. J. Zhu, K. Chandrasekaran, Z. M. Zhu, S. C. Rustagi, S. H. Lin, C. Q, Wei, and G. H. Lim, "Unified compact model for generic double-gate MOSFETs," in *Proc. NSTI Nanotech*, Santa Clara, CA, May 2007, vol. 3, pp. 538–543.

[12] G. Gildenblat, D. B. M. Klaassen, and C. C. McAndrew, "Comments on 'a physics-based analytic solution to the MOSFET surface-potential from accumulation to strong-inversion region'," and J. He, "Author's reply," *IEEE Trans. Electron Devices*, vol. 54, no. 8, pp. 2061–2062, Aug. 2007.

[13] G. H. See, G. J. Zhu, X. Zhou, and S. H. Lin, "Unified regional surface potential for compact modeling of bulk/double-gate MOSFETs with doping from highly-doped to undoped body," submitted for publication.

[14] W. Shockley, "A unipolar 'field effect' transistor," *Proc. IRE*, vol. 40, no. 11, pp. 1365–1376, Nov. 1952.

[15] C.-T. Sah and B. Jie, "The theory of field-effect transistors: XI. The bipolar electrochemical currents (1-2-MOS-gates on thin-thick pure-impure base)," *Chinese J. Semicond.*, vol. 29, no. 3, pp. 397–409, Mar. 2008.

[16] C. T. Sah, "A history of MOS transistor compact modeling," in *Proc. NSTI Nanotech 2005*, Anaheim, May 2005, vol. WCM, pp. 347–390.

[17] X. Zhou, G. H. See, G. J. Zhu, K. Chandrasekaran, Z. M. Zhu, S. C. Rustagi, S. H. Lin, C. Q, Wei, and G. H. Lim, "Unified compact model for generic double-gate MOSFETs," in *Proc. NSTI Nanotech*, Santa Clara, CA, May 2007, vol. 3, pp. 538–543.

[18] R. H. Kingston and S. F. Neustadter, "Calculation of the space charge, electric field, and free carrier concentration at the surface of a semiconductor," *J. Appl. Phys.*, vol. 26, no. 6, pp. 718–720, Jun. 1955.

[19] X. Zhou, Z. M. Zhu, S. C. Rustagi, G. H. See, G. J. Zhu, S. H. Lin, C. Q. Wei, and G. H. Lim, "Rigorous surface-potential solution for updoped symmetric double-gate MOSFETs considering both electrons and holes at quasi nonequilibrium," *IEEE Trans. Electron Devices*, vol. 55, no. 2, pp. 616–623, Feb. 2008.

[20] G. J. Zhu, G. H. See, X. Zhou, Z. M. Zhu, S. H. Lin, C. Q. Wei, J. B. Zhang, and A. Srinivas, "Quasi-2D surface-potential solution to three-terminal undoped symmetric double-gate Schottky-barrier MOSFETs," to appear in *Proc. NSTI Nanotech 2008*, Boston, Jun. 2008.

[21] S. C. Miller, Jr. and R. H. Good, Jr., "A WKB-type approximation to the Schrödinger equation," *Phys. Rev.*, vol. 91, no. 1, pp. 174–179, Jul. 1953.

Unified Regional Surface Potential for Modeling Common-Gate Symmetric/Asymmetric Double-Gate MOSFETs with Quantum-Mechanical Effects

Guan Huei See, Xing Zhou, Guojun Zhu, Zhaomin Zhu, Shihuan Lin, Chengqing Wei, Junbin Zhang, and Ashwin Srinivas

[1]School of Electrical & Electronic Engineering, Nanyang Technological University
Nanyang Avenue, Singapore 639798, exzhou@ntu.edu.sg

ABSTRACT

Technology scaling has caused the MOSFET device to enter into a regime in which the quantum mechanical effect (QME) cannot be ignored. Besides, double-gate (DG) MOSFET is considered as one of the potential devices that may be replacing the bulk-MOSFET as the next generation transistor. However, there is no simple solution even without considering the quantum-mechanical correction for DG MOSFET compact model. In this work, the explicit surface potential for common-gate asymmetric double-gate (ca-DG) MOSFET model is extended to include QME. The model is physically derived and verified with Medici simulation data for all the essential physical conditions. It matches well with the numerical data. The potential models are ready to be applied to the drain current model as it contains the essential physics that scales with the DG MOSFET structures.

***Index Terms*—Compact model, quantum mechanical effect, double-gate, unified regional surface potential.**

1 INTRODUCTION

As the advanced MOSFET technology continues to scale the oxide thickness in order to gain in the performance of the transistor, the carrier quantization effect can no longer be ignored [1]. It was suggested by van Dort, *et. al.* that the quantization effect of the space charge density can be accounted for by simple modification of intrinsic carrier density [2] in bulk-MOSFET. This has been implemented in most of the compact models [1, 3].

A lot of attention have been given to DG MOSFET lately due to its superior performance as compared to its counterpart bulk MOSFETs [4-6]. DG MOSFET is believed to be the next generation device that is suitable for continued scaling. However, there is no simple solution even without considering the quantum mechanical correction for DG-MOSFET compact modeling. It is impossible to have a second integration when holes, acceptor/donor and electron are considered. Compromise has to be made in order to obtain a solvable solution either by dropping the acceptor/donor terms [7-10] or by the perturbation method [11].

In this work, the explicit surface potential in common-gate asymmetric DG-MOSFET model is extended to include the quantum mechanical correction. The model is physically derived and verified with the Medici simulation data for all the essential physical conditions based on [2].

This paper is organized as follows. Section 2 will discusses the quantum mechanical model implemented in explicit regional form. We will then look into the model behavior and model comparison with the numerical simulation data in Section 3 for a wide range of physical conditions. Finally, we will draw conclusions in Section 4.

2 MODEL FORMULATION

The intrinsic carrier concentration with carrier quantization is modified using

$$n_i^{qm} = n_i f^{qm} \tag{1}$$

with

$$f^{qm} = \exp\left(-\Delta E_g^{qm}/v_{th}\right) \tag{2}$$

where $\Delta E_g^{qm} = \Delta E_g/2q$ (in eV) is the bandgap widening term. The increase of energy bandgap of silicon due to the effect of carrier energy quantization to surface potential well is given as [12]

$$\Delta E_g = \kappa \frac{3\hbar^2}{8qm^*}\left[\frac{12m^* q^2}{\varepsilon_{si}\hbar^2} F_s\right]^{2/3} \tag{3}$$

and

$$F_s = \frac{2C_{ox}\left(V_{gs} - V_{FB} - \phi_s\right)/q}{3} \tag{4}$$

where $\hbar$ is the reduced Plank's constant, m^* is the effective electron/hole mass and the factor κ, was introduced in [13], C_{ox} is oxide capacitance per unit area, q is electronic charge and V_{FB} is the flat-band voltage.

Following the unified regional modeling (URM) approach [14], the unified regional surface potential solutions are first computed to calculate the vertical electric field, F_s from (4). If the spatial dependence of (4) is ignored [1], then the Poisson solution for common-gate DG can be derived as:

$$V_{gs} - V_{FB} - \phi_s^{qm} = -Q_{sc}^{qm}/C_{ox} \tag{5}$$

with quantized channel induced charge, Q_{sc}^{qm} and surface potential, ϕ_s^{qm} in the channel. The induced charge in the channel with the quantization effect is given by

$$Q_{sc}^{qm} = -C_{ox}\Upsilon \operatorname{sgn}\left(\phi_s^{qm}\right)\sqrt{f_\phi^{qm}} \tag{6}$$

where

$$f_{\phi}^{qm}=v_{th}\left[f^{qm}\exp\left(-\frac{\phi_s^{qm}}{v_{th}}\right)-\exp\left(-\frac{\phi_o}{v_{th}}\right)\right]+\left(\phi_s^{qm}-\phi_o\right)\left[1-\exp\left(-\frac{V+2\phi_F}{v_{th}}\right)\right]$$
$$+v_{th}\exp\left(-\frac{V+2\phi_F}{v_{th}}\right)\left[f^{qm}\exp\left(\frac{\phi_s^{qm}}{v_{th}}\right)-\exp\left(\frac{\phi_o}{v_{th}}\right)\right] \quad (7)$$

is quantization factor that modifies the voltage-balance equation (5). The value of this factor for classical theory (i.e., no QME) is 1. Note that the quantization factor only modifies the free carrier near the surface. Therefore, the modifications due to QME are to be done in strong accumulation and strong inversion.

In either strong accumulation or strong inversion, the contribution of the zero-field potential can be ignored. Only a single carrier (holes or electrons) needs to be considered. Therefore, (6) is rearranged for accumulation with $f_{\phi}{}^{qm} = v_{th}\exp(-\phi_s{}^{qm}/v_{th})$,

$$\ln\left(\frac{V_{gsf}^2}{f^{qm}\gamma^2 v_{th}}\right)+\ln\left(1-\frac{2\phi_s^{qm}}{V_{gsf}}+\frac{\left(\phi_s^{qm}\right)^2}{V_{gsf}^2}\right)=-\phi_s^{qm}/v_{th} \quad (8)$$

and strong inversion with $f_{\phi}{}^{qm} = v_{th}\exp[(\phi_s{}^{qm} - 2\phi_F - V)/v_{th})]$

$$\ln\left(\frac{V_{gsf}^2}{f^{qm}\gamma^2 v_{th}}\right)+\ln\left(1-\frac{2\phi_s^{qm}}{V_{gsf}}+\frac{\left(\phi_s^{qm}\right)^2}{V_{gsf}^2}\right)=\left(\phi_s^{qm}-2\phi_F-V\right)/v_{th} \quad (9)$$

Taking a Taylor expansion of the 2nd logarithmic term up to the third-order polynomial, both equations can be expressed as a general cubic equation

$$\left(\phi_s^{qm}\right)^3+p\left(\phi_s^{qm}\right)^2+q\left(\phi_s^{qm}\right)+r=0 \quad (10)$$

which can be solved analytically[14, 15]. All the rest of the regional pieces, namely weak accumulation, depletion, and volume inversion can be derived similarly without the QME, as shown in [14]. The strong accumulation or strong inversion solution is a cubic solution that is given as

$$\phi_s^{qm}=\left(\frac{\sqrt{D}-B}{2}\right)^{1/3}-\frac{A}{3}\left(\frac{2}{\sqrt{D}-B}\right)^{1/3}-\frac{p}{3} \quad (11)$$

where

$$\phi_s^{qm}=\begin{cases}\phi_{cc}^{qm}, & V_{gb}<V_{FB}\\ \phi_{ss}^{qm}, & V_{gb}>V_t\end{cases}. \quad (12)$$

With the help of interpolation/smoothing functions

$$\vartheta_f\{x;\sigma\}\equiv 0.5\left(x+\sqrt{x^2+4\sigma}\right)$$
$$\vartheta_r\{x;\sigma\}\equiv 0.5\left(x-\sqrt{x^2+4\sigma}\right) \quad (13)$$
$$\vartheta_{eff^\pm}\{x_{sat};x,\pm\delta\}\equiv$$
$$x_{sat}-0.5\left[x_{sat}-x\pm\delta+\sqrt{\left(x_{sat}-x\pm\delta\right)^2\mp 4\delta x_{sat}}\right]$$

the regional solutions can be unified to obtain single-piece solutions. The unified regional surface potential solution with QME in strong accumulation is solved by applying the unified reverse flat-band shifted gate-source voltage, V_{GSR} [14],

$$\phi_{str,cc}^{qm}=\phi_{cc}^{qm}\Big|_{V_{gs}-V_{FB}=V_{GSR}} \quad (14)$$

with (2) redefined as

$$f^{qm}=f_{acc}^{qm}=\exp\left(-\frac{\Delta E_g^{qm}}{qv_{th}}\right)-\varepsilon^{qm} \quad (15)$$

where ε^{qm} is introduced as a fitting parameter to adjust the minor flat-band shift due to QME in the van Dort model. The strong inversion regional solution can be solved similarly by applying unified forward flat-band shifted gate-source voltage (13),

$$\phi_{str}^{qm}=\phi_{ss}^{qm}\Big|_{V_{gs}-V_{FB}=V_{GSF}}. \quad (16)$$

Other regions have no QME. Following the similar procedure as in [14], a single-piece solution can be obtained. The model is compared with Medici simulation data for validation.

3 RESULTS AND DISCUSSIONS

The comparison of the surface potential with classical model and with QME for different body doping concentrations is shown in Figure 1 (at the default value $\kappa = 1$ unless specified explicitly). The model follows well for the whole range of practical body doping concentration beginning from ideal pure semiconductor (i.e., $N_A = N_D = 0$) to highly doped body, $N_A = 10^{18}$ cm^{-3}.

Figure 2a shows the comparison of unified regional surface and zero-field potentials with quantum mechanical effect for different κ values, which is taken directly from the Medici simulation for undoped body. Figure 2b shows the corresponding derivatives of the surface and zero-field potentials. Figure 3 shows the similar plot but at high doping concentration, $N_A = 10^{18}$ cm^{-3} and the derivative of the surface potential is included in the inset. The surface potential is following the data physically and smoothly (from the derivatives). Although the zero-field potential over-predicted the data at highly doped body, this has negligible effect in the current model [16], as most current only conduct in the depletion layer at the surface.

Figure 4 and Figure 5 show the prediction for surface potential (and the inset shows the zero-field potential) for different body thickness and asymmetric oxide thickness, respectively. The surface potential is able to predict the variations accordingly.

4 CONCLUSION

In conclusion, DG MOSFETs potential models including the QME have been derived physically using the unified regional modeling approach. The transition from partially-depleted to fully-depleted operation with QME is seamlessly built into the model. The potential solutions can be applied to the drain current model as the essential physics that scales with the DG MOSFET structures are included.

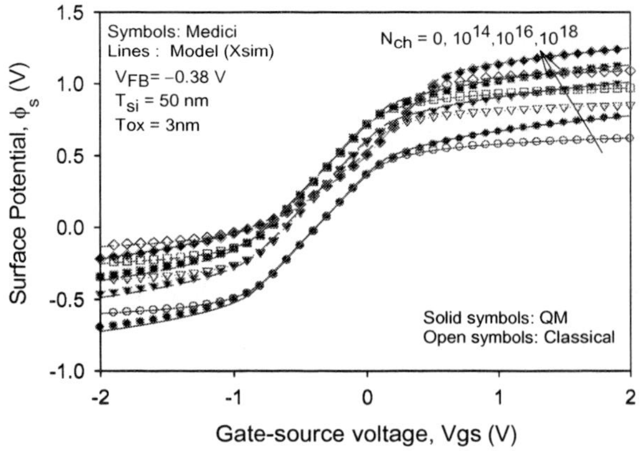

Figure 1: Model comparison of unified regional surface potential for both classical model and with QME for different body doping.

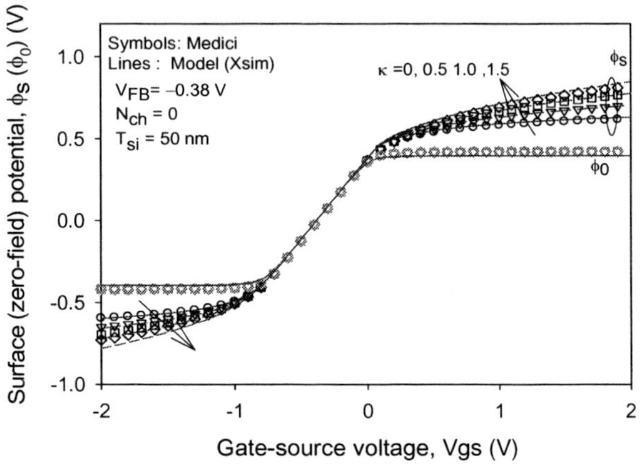

(a)

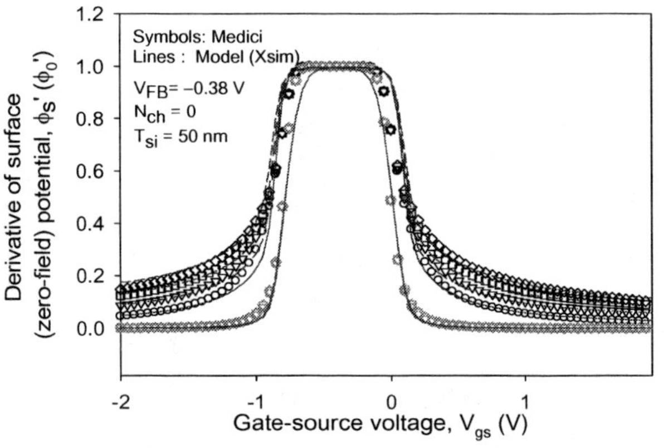

(b)

Figure 2: Model comparison of (a) unified regional surface potential and zero-field potential with QME for undoped body. The parameter κ is taken directly from Medici and used in the model play-back; and (b) the corresponding derivative of the unified surface potential with QME.

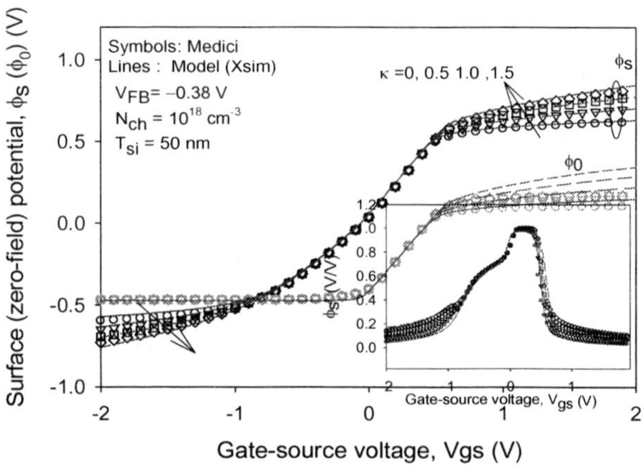

Figure 3: Model comparison of unified regional surface potential and zero-field potential with QME at highly doped body $N_A = 10^{18}$ cm^{-3}. The over-predicting of zero-field potential has negligible effect on the drain current.

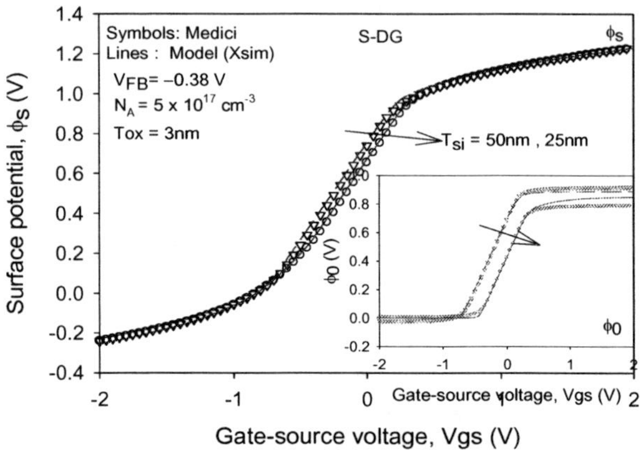

Figure 4: Model comparison of unified regional surface potential (inset: zero-field potential) with QME for different body thickness.

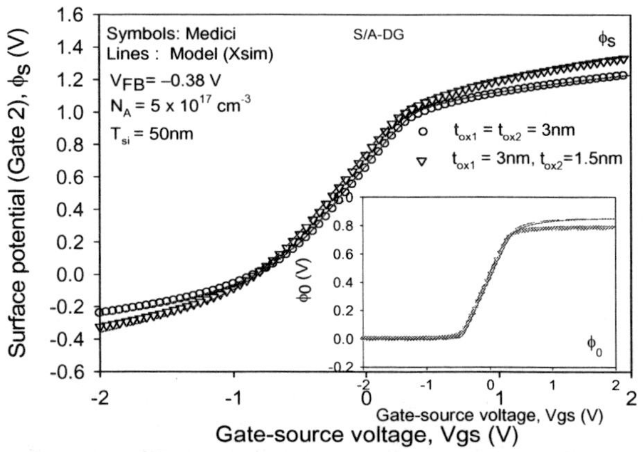

Figure 5: Model comparison of unified regional surface potential (inset: zero-field potential) with QME for asymmetry in gate 2 oxide thickness.

Acknowledgment: This work was supported in part by the Semiconductor Research Corporation under Contract 2004-VJ-1166G and in part by Nanyang Technological University under Grant RGM30/03.

REFERENCES

1. R. Rios, et al. *A physical compact MOSFET model, including quantum mechanical effects, for statistical circuit design applications*. IEDM, 1995.
2. M. J. van Dort, P. H. Woerlee, and A. J. Walker, *A simple model for quantisation effects in heavily-doped silicon MOSFETs at inversion conditions.* Solid-State Electron., **37**(3), pp. 411-414, 1994.
3. G. Gildenblat, T. L. Chen, and P. Bendix, *Closed-form approximation for the perturbation of MOSFET surface potential by quantum-mechanical effects.* Electron. Lett., **36**(12), pp. 1072-1073, 2000.
4. S. Inaba, et al. *FinFET: the prospective multi-gate device for future SoC applications*. ESSDERC 2006.
5. J. P. Colinge, *Multi-gate SOI MOSFETs.* Microelectron. Eng., **84**(9-10), pp. 2071-2076, 2007.
6. D. Hisamoto, et al., *FinFET-a self-aligned double-gate MOSFET scalable to 20 nm.* IEEE Trans. Electron Devices, **47**(12), pp. 2320-2325, 2000.
7. X. Zhou, et al., *Rigorous Surface-Potential Solution for Undoped Symmetric Double-Gate MOSFETs Considering Both Electrons and Holes at Quasi Nonequilibrium.* IEEE Trans. Electron Devices, 55(2), pp. 616-623, 2008.
8. T. Yuan, et al., *A continuous, analytic drain-current model for DG MOSFETs.* IEEE Electron Device Lett., **25**(2), pp. 107-109, 2004.
9. A. Ortiz-Conde, et al., *A Review of Core Compact Models for Undoped Double-Gate SOI MOSFETs.* IEEE Trans. Electron Devices, **54**(1), pp. 131-140, 2007.
10. W. Z. Shangguan, et al., *Surface-Potential Solution for Generic Undoped MOSFETs With Two Gates.* IEEE Trans. Electron Devices, **54**(1), pp. 169-172, 2007.
11. M. V. Dunga, et al., *Modeling Advanced FET Technology in a Compact Model.* IEEE Trans. Electron Devices, **53**(9), pp. 1971-1978, 2006.
12. F. Pregaldiny, C. Lallement, and D. Mathiot, *Accounting for quantum mechanical effects from accumulation to inversion, in a fully analytical surface-potential-based MOSFET model.* Solid-State Electron., **48**(5), pp. 781-787, 2004.
13. P. V. Voorde, et al. *Accurate doping profile determination using TED/QM models extensible to sub-quarter micron nMOSFETs*. IEDM 1996.
14. G. H. See, et al. *Unified Regional Surface Potential for Modeling Common-Gate Symmetric/Asymmetric Double-Gate MOSFETs with Any Body Doping.* Workshop on Compact Modeling, Nanotech 2008.
15. D. Herbison-Evans, *Solving quartics and cubics for graphics*, in *Technical Report TR94-487*. Updated, 22 July 2005, Basser Dept. of Computer Science, University of Sydney Australia.
16. X. Zhou, et al. *Unified Compact Model for Generic Double-Gate MOSFETs.* Workshop of Compact Modeling, Nanotech 2007, pp. 538-543.

Quasi-2D Surface-Potential Solution to Three-Terminal Undoped Symmetric Double-Gate Schottky-Barrier MOSFETs

Guojun Zhu, Guan Huei See, Xing Zhou, Zhaomin Zhu, Shihuan Lin, Chengqing Wei, Junbin Zhang, and Ashwin Srinivas

School of Electrical & Electronic Engineering, Nanyang Technological University
Nanyang Avenue, Singapore 639798, exzhou@ntu.edu.sg

ABSTRACT

This paper presents an analytical quasi-2D surface potential solution to Schottky-Barrier (SB) MOSFETs, with one unified terminal-bias and device-parameter dependent characteristic length, which takes the screening of the gate field by free carriers into account. The proposed model is also valid for *arbitrary* terminal bias with respective to any reference voltage and applicable to *asymmetric* SB-DG devices. Terminal current can be calculated based on the obtained energy band model, which can provide insights into the modeling and design of future advanced MOSFETs with Schottky-barrier source/drain contacts.

Keywords: Quasi-2D, Schottky barrier, undoped body, symmetric double-gate MOSFETs, asymmetry.

1 INTRODUCTION

The first paper for Schottky-barrier (SB) MOSFETs dated back to 1968, which employed PtSi for the source/drain (S/D) regions to pMOS bulk device [1]. As conventional MOSFET is reaching its scaling limit, SB devices attract a renewed interest due to its low parasitic S/D resistance and superior scalability [2]. Recently, novel SB devices, such as SB carbon nanotube transistors (CNT) [3], SB silicon nanowire (SiNW) and double-gate (DG) MOSFETs [4-5], emerge, which show excellent scalability into the sub-45nm regime and novel applications to three-valued static memory [6].

Computational studies by solving two-dimensional (2D) Poisson equation self-consistently with quantum transport equation were done by several authors to simulate devices like SB-DG, SB-CNT, etc., most of them assumed ballistic quantum transport in the channel [5, 7-10].

The quasi-2D method has been used to account for short-channel effects in PN-junction MOSFETs. Recently some authors applied the quasi-2D solution to SB MOSFETs to derive potential profiles [11-12]. In this paper, the quasi-2D solution is revisited and extended to tunneling-dominant regime, which is valid for *arbitrary* terminal bias (with respective to any reference voltage) and applicable to *asymmetric* SB-DG devices as well. A terminal-bias dependent characteristic length is physically derived, which has been missed in all previous literature.

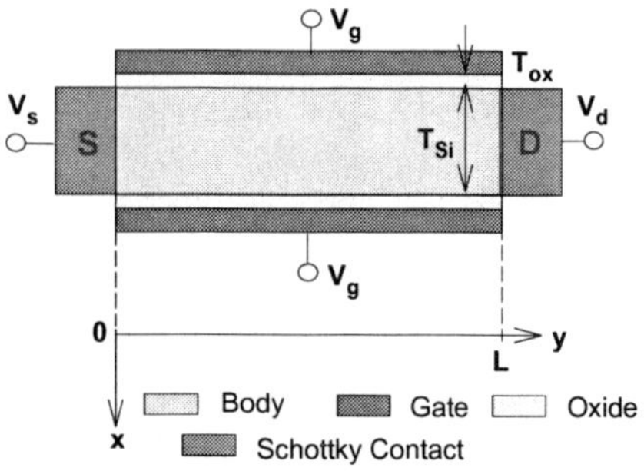

Figure 1. Schematic of the ideal SB-DG MOSFET.

2 MODEL EQUATIONS

The quasi-2D solution is based on the 2D Poisson's equation for the surface potential [13] applied to the body of updoped s-DG device with a Gaussian box of depth $X_o = T_{Si}/2$:

$$C_{ox}\left(V_g - V_{FB} - \phi_s(y)\right) + \frac{\varepsilon_{Si} X_o}{\eta}\frac{d^2\phi_s(y)}{dy^2} = -Q_i \tag{1}$$

where η is a fitting parameter and Q_i is the total charge (per unit area) in the body. When gate voltage is low and SB device is in thermionic-dominant (subthreshold) regime, Q_i is roughly constant and the well-known solution in bulk PN-junction MOSFET [14] applies with the respective S/D characteristic lengths, $\lambda_{s(d)}$, and SB built-in voltages, $V_{bi,s(d)}$, (i.e., workfunction differences between S/D and body):

$$V_s(y) = \phi_s(y) + \delta\phi_{s,s}(y) + \delta\phi_{s,d}(y) \tag{2}$$

$$\delta\phi_{s,s}(y) = \left(V_{bi,s} + V_s - \phi_s\right)\frac{\sinh\left((L-y)/\lambda_s\right)}{\sinh\left(L/\lambda_s\right)} \tag{2a}$$

$$\delta\phi_{s,d}(y) = \left(V_{bi,d} + V_d - \phi_s\right)\frac{\sinh\left(y/\lambda_d\right)}{\sinh\left(L/\lambda_d\right)}. \tag{2b}$$

The built-in voltage is due to the source (drain) to body workfunction difference:

$$V_{bi,s(d)} = -\phi_{MS,s} = -\left(\Phi_{M,s(d)} - \Phi_S\right) = \left(\chi + E_g/2q + \phi_F\right) - \Phi_{M,s(d)} \tag{3}$$

where $\Phi_{M,s(d)}$ is the source (drain) Schottky contact workfunction and ϕ_F is the body Fermi level (zero for pure Si). The long-channel surface potential (ϕ_s) excluding the effect of SB-induced charges near S/D can be taken from any implicit/explicit voltage-equation ϕ_s solutions, e.g., the Lambert W function approximation for undoped s-DG [15].

In the thermionic-dominant (subthreshold) regime in which (2) is obtained, the characteristic length is a constant, given as

$$\lambda_{sub} = \sqrt{\frac{\varepsilon_{Si} T_{ox} X_o}{\eta \varepsilon_{ox}}} \,. \tag{4}$$

However, near-constant-Q_i assumption is not valid in the tunneling-dominant regime in which (4) cannot be used since it has no terminal-bias dependency, thus is not able to account for screening of the gate field by SB-injected charges.

To extend the solution (2) in all regions, a terminal-bias-dependent characteristic length (due to Q_i variations) has to be developed. We model the SB-induced charge density at the source (drain) end as

$$N_{sb,s(d)} = \frac{Q_{i,s(d)}}{qX_o} = \frac{\sigma_{s(d)} C_{ox} \left(V_g - V_{sbfb,s(d)} - \phi_{s,s(d)} \{ V_{sbfb,s(d)} \} \right)}{qX_o} \tag{5}$$

where $V_{sbfb,s(d)}$ is the source (drain) side "SB flat-band voltage," and $\phi_{s,s(d)}\{V_{sbfb,s(d)}\}$ is the same long-channel (Lambert W) surface potential at the source (drain) end except the (gate) flat-band voltage V_{FB} is replaced by $V_{sbfb,s(d)}$ such that the SB-induced charge $Q_{i,s(d)} = 0$ when $V_g = V_{sbfb,s(d)}$, which is given by

$$V_{sbfb,s(d)} = V_{gFB,s(d)} + V_{bi,s(d)} \tag{6}$$

where $V_{gFB,s(d)}$ is the gate voltage required to make the energy band near source (drain) along y flat, which is a fixed value for a given technology and can be easily extracted from the model.

To avoid zero or negative $N_{sb,s(d)}$, the following effective SB-induced charge density is proposed:

$$N_{eff,s(d)} = n_i \exp\left[\sinh^{-1}\left(\frac{N_{sb,s(d)}}{2n_i} \right) \right]. \tag{7}$$

The tunneling-dominant characteristic length is then modeled similar to depletion approximation based on $N_{eff,s(d)}$

$$\lambda_{tun,s(d)} = \sqrt{\frac{2\varepsilon_{Si} \Phi_{B,s(d)}}{qN_{eff,s(d)}}} \tag{8}$$

which includes terminal-bias dependencies.

$$\Phi_{B,s(d)} = \Phi_{M,s(d)} - \chi \tag{9}$$

is the SB height. The final unified characteristic length is obtained by joining the subthreshold and linear pieces using the smoothing function:

$$\lambda_{s(d)} = \frac{1}{2}\Big[\lambda_{tun,s(d)} + \lambda_{sub} + \delta_{s(d)} - \sqrt{\left(\lambda_{tun,s(d)} - \lambda_{sub} - \delta_{s(d)} \right)^2 + 4\delta_{s(d)} \lambda_{tun,s(d)}} \Big] \tag{10}$$

which is used in the quasi-2D solution (2). This novel approach circumvents the difficult problem for variable Q_i in (1) by a single-piece solution for all terminal-bias as well as gate-length, body-thickness, and S/D workfunction variations. The analytical solution also allows easy evaluation of SB-MOS tunneling and thermionic currents in a compact form.

3 RESULTS AND DISCUSSION

Figures 2-3 show the modeled energy bands (lines) in comparison with the solutions extracted from Medici (symbols) at different gate and drain biases. The model matches Medici solution quite well in both positive and negative gate voltages (with electron and hole-dominant current, respectively) as terminal biases are incorporated into the characteristic length model.

Figures 4-7 show that the modeled energy band profiles are able to follow Medici numerical solutions reasonably well at different S/D workfunctions, oxide thicknesses, body thicknesses, and gate lengths.

Due to the separation of S/D terminals, the model is applicable to arbitrary terminal biases as shown in Figure 8. Figure 9 shows another advantage of the proposed model that it can be easily extended to handle asymmetric structure (e.g., different S/D workfunctions), which is becoming more and more popular.

Once an analytical 2D surface potential is obtained and validated with numerical data, it can be easily used in evaluation of drain current based on tunneling and/or thermionic emission mechanisms [11] using, e.g., Miller-Good (MG) [16] model. Due to page limit, drain current formulations will be presented elsewhere, and two sample results are shown in this paper.

Figure 10 shows the terminal electron current characteristics for various drain voltages. For a device with workfunction of 4.3 eV, due to its low SB height for electrons, its behavior is quite similar to conventional PN-junction MOSFETs.

Figure 11 shows the terminal electron current characteristics for different technologies (workfunctions). The modeled current matches reasonable well to Medici numerical solution. A typical multiple-slope behavior, which is unique to SB MOSFETs, is well captured by our drain current model.

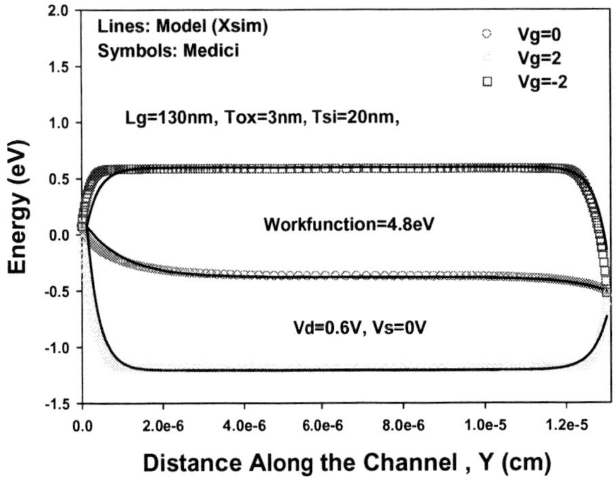

Figure 2. Energy band profiles at different gate biases.

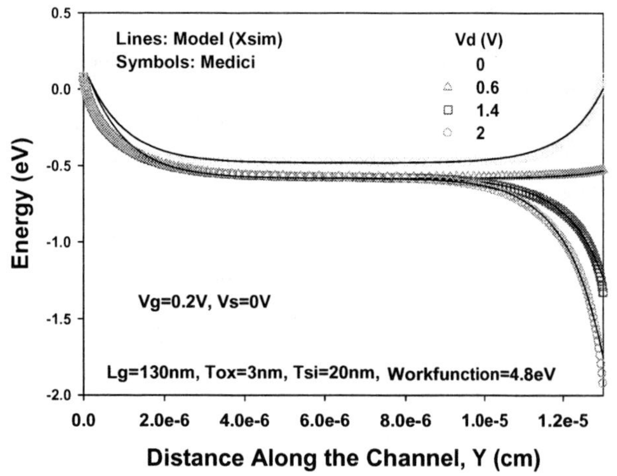

Figure 3. Energy band profiles at different drain biases.

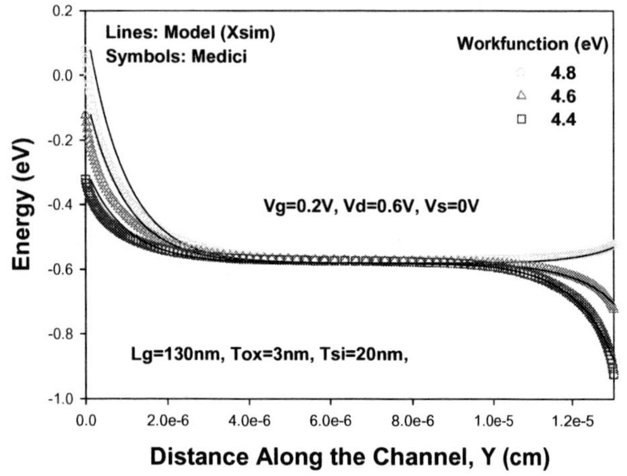

Figure 4. Energy band profiles at different S/D workfunctions.

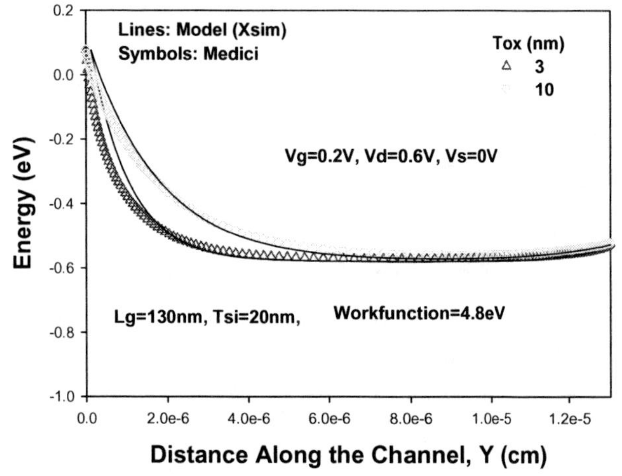

Figure 5. Energy band profiles at different oxide thicknesses.

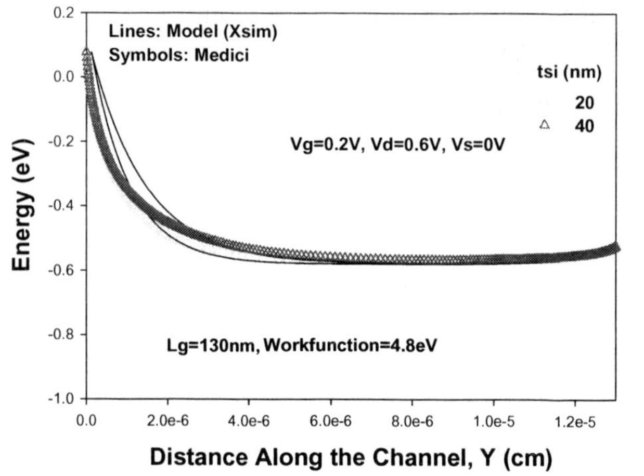

Figure 6. Energy band profiles at different body thicknesses.

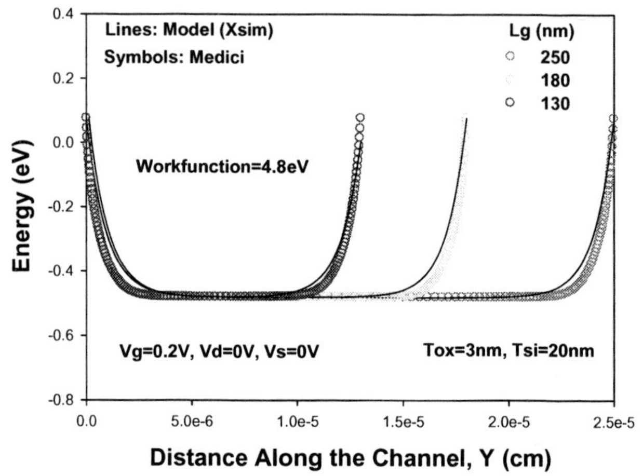

Figure 7. Energy band profiles at different gate lengths.

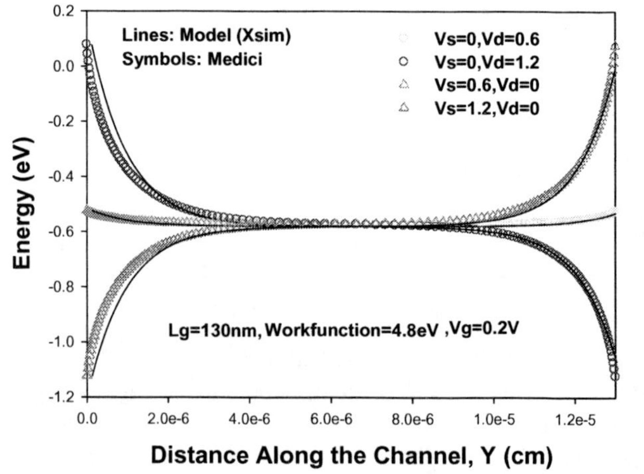

Figure 8. Energy band profiles at various source/drain terminal biases.

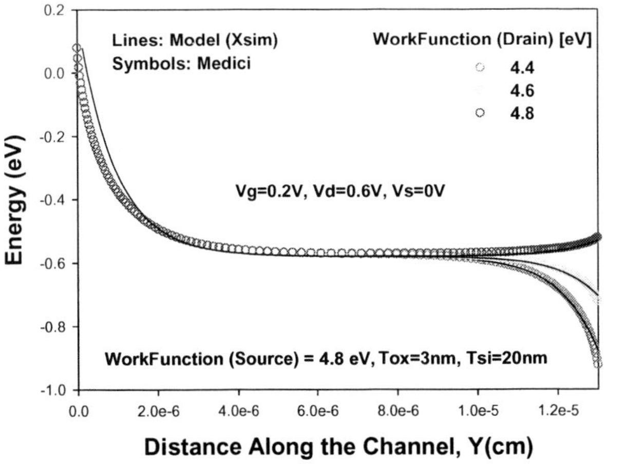

Figure 9. Energy band profiles at different drain workfunctions with fixed source workfunction.

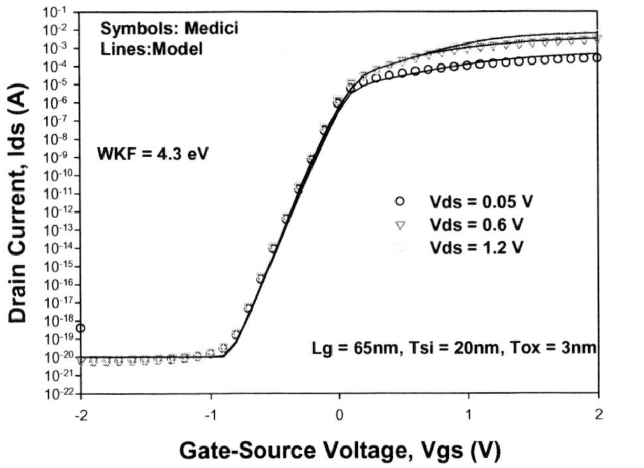

Figure 10. Drain current (electron only) calculated from energy band profile solution at different drain biases for workfunction = 4.3 eV.

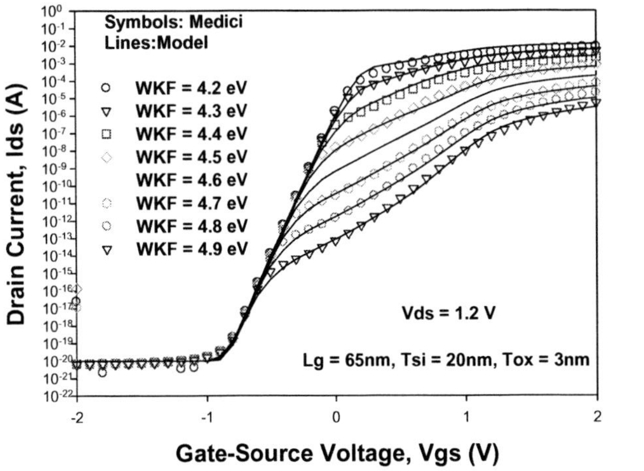

Figure 11. Drain current (electron only) calculated from the energy band profile solution at different workfunctions for fixed drain bias $V_{ds} = 1.2$ V.

4 CONCLUSIONS

In conclusion, the proposed quasi-2D surface-potential solution with a single-piece characteristic length valid for all bias conditions incorporates the SB boundary conditions for undoped DG-MOSFET, scalable with contact and intrinsic device physical parameters, and also independent of source and drain terminals as well as reference voltages.

Due to explicit formulation of the potential profile, position-dependent energy band and electrical field can be easily evaluated. Terminal current can be obtained from the quasi-2D energy-band model, which provides the basis for formulating a compact model for SB MOSFETs.

Acknowledgment: This work was supported by Nanyang Technology University under Grant No. RGM30/03. The first author would like to thank Teck-Seng Lee for helpful discussions on carrier transport in SB MOSFETs.

REFERENCES

[1] M. P. Lepselter, *et al.*, "SB-IGFET: An Insulated-Gate Field-Effect Transistor Using Schottky Barrier Contacts for Source and Drain," *Proc. IEEE*, pp. 1400-1402, Aug. 1968.

[2] J. M. Larson, *et al.*, "Overview and Status of Metal S/D Schottky-Barrier MOSFET Technology," *IEEE Trans. Electron Devices*, Vol. 53, No. 5, pp. 1048-1058, May 2006.

[3] S. Heinze, *et al.*, "Field-Modulated Carrier Transport in Carbon Nanotube Transistors," *Phys. Rev. Lett.* Vol. 89, No. 12, Sep. 2002.

[4] S. M. Koo, *et al.*, "Silicon Nanowires as Enhancement-mode Schottky barrier field-effect transistors," *Nanotechnology*. 16, pp. 1482-1485, 2005.

[5] J. Guo, *et al.*, "A Computational Study of Thin-Body, Double-Gate, Schottky Barrier MOSFETs," *IEEE Trans. Electron Devices*, Vol. 49, No. 11, pp. 1897-1902, Nov. 2002.

[6] A. Raychowdhury, *et al.*, "Design of a Novel Three-Valued Static Memory Using Schottky Barrier Carbon Nanotube FETs," *Proc. 5th IEEE Conference on Nanotechnology*, Jul. 2005.

[7] J. Guo, *et al.*, "A Numerical Study of Scaling Issues for Schottky-Barrier Carbon Nanotube Transistors," *IEEE Trans. Electron Devices*, Vol. 51, No. 2, pp. 172-177, Feb. 2004.

[8] J. Clifford, *et al.*, "Bipolar Conduction and Drain-Induced Barrier Thinning in Carbon Nanotube FETs," *IEEE Trans. Electron Devices*, Vol. 2, No. 3, pp. 181-185, Sep. 2003.

[9] C. Y. Ahn, *et al.*, "Ballistic Quantum Transport in Nano-scale Schottky Barrier Tunnel Transistors," *Proc. 5th IEEE Conference on Nanotechnology*, Jul. 2005.

[10] C. K. Huang, *et al.*, "Two-Dimensional Numerical Simulation of Schottky Barrier MOSFET with Channel Length to 10nm," *IEEE Trans. Electron Devices*, Vol. 45, No. 4, pp. 842-848, Apr. 1998.

[11] R. A. Vega, "On the Modeling and Design of Schottky Field-Effect Transistors," *IEEE Trans. Electron Devices*, Vol. 53, No. 4, pp. 866-874, Apr. 2006.

[12] B. Xu, *et al.*, "An Analytical Potential Model of Double-Gate MOSFETs with Schottky Source/Drain," *8th International Conference on Solid-State and Integrated Circuit Technology*, 2006.

[13] K. W. Terrill, *et al.*, "An Analytical Model for the Channel Electric Field in MOSFET's with Graded-drain Structures," *IEEE Electron Device Lett.*, Vol. EDL-5, No. 11, pp. 440-442, Nov. 1984.

[14] Z. H. Liu, *et al.*, "Threshold Voltage Model for Deep-submicrometer MOSFETs," *IEEE Trans. Electron Devices*, Vol. 40, No. 1, pp. 86-95, Jan. 1993.

[15] G. J. Zhu, *et al.*, "'Ground-Referenced' Model for Three-Terminal Symmetric Double-Gate MOSFETs with Source–Drain Symmetry," submitted for publication.

[16] S. C. Miller, Jr. and R. H. Good, Jr., "A WKB-Type Approximation to the Schrödinger Equation," *Phys. Rev.*, Vol. 91, No. 1, pp. 174-179, Jul. 1953.

Construction of a Compact Modeling Platform and Its Application to the Development of Multi-Gate MOSFET Models for Circuit Simulation

M. Miura-Mattausch, M. Chan*, J. He**, H. Koike***, H. J. Mattausch
T. Nakagawa***, Y. J. Park+, T. Tsutsumi++ and Z. Yu+++

Graduate School of Advanced Sciences of Matter, Hiroshima University
Higashi-Hiroshima, Hiroshima, 739-8530, Japan, mmm@hiroshima-u.ac.jp, hjm@hiroshima-u.ac.jp
*Department of Electronic and Computer Engineering, Hong Kong University of Science & Technology, Hong Kong, mchan@ee.ust.hk
**School of computer & Information Engineering, Shenzhen Graduate School, Peking University
Shenzhen, 518055, P. R. China, frankhe@pku.edu.cn
***Nanoelectronics Research Institute, National Institute of Advanced Industrial Science and Technology, Tsukuba, Ibaraki, 305-8568 Japan, h.koike@aist.go.jp, nakagawa.tadashi@aist.go.jp
+School of Electrical Engineering and Computer Science and NSI-NCRC, Seoul National University
Seoul, Korea, ypark@snu.ac.kr
++Department of Computer Science, School of Science and Technology, Meiji University
Kawasaki, 214-8571, Japan, tsutsumi@cs.meiji.ac.jp
+++Institute of Microelectronics, Tsinghua University, Beijing 100084, China, yuzhip@tsinghua.edu.cn

ABSTRACT

We aim at constructing a common platform for compact model development based on the Verilog-A language for collaboration among different research groups. The project aims in particular at a framework for efficient development of multi-gate MOSFET models for circuit simulation. We have developed several prototypes of multi-gate MOSFET models based on different concepts till now. Phenomena expected to become important for the multi-gate MOSFET generation are modeled on the basis of their physical origins. These phenomena are implemented into each of the specific multi-gate MOSFET models by plugging in modules from the common platform. Parasitic resistive and capacitive contributions are also modularized to represent the complete circuit performance of the multi-gate MOSFET device for efficient circuit development.

Keywords: common platform, quantum effect, multi-gate MOSFET, circuit simulation, parasitic effect

1 INTODUCTION

Many different kinds of basic devices have been developed with different kinds of fabrication technologies for the application in integrated circuits. To efficiently utilize these devices for a specific integrated circuit design, accurate compact models of these devices have to be developed. It is furthermore desired to predict resulting circuit performances at an early stage ideally in parallel to the device-technology development. To realize the necessary rapid compact model development, the availability of a modular compact-modeling platform will be very useful.

We have started the project called Nano Device Modeling Initiative (NDMI) in 2005 October as a collaborative effort among different research groups from different countries.

Subjects being investigated are summarized in Table 1. One of our important aims is to construct a platform for rapid compact model development by providing a prototype consisting of modules for different model functions, which are individual elements calculating different device features, as schematically shown in Fig. 1. This platform will enable the plugging in of developed modules for compact model parts by substituting modules of the prototype or by adding new modules We take the advanced MOSFET model HiSIM [1] as the prototype model for modularization.

The multi-gate MOSFET has been considered as a possible candidate for the next generation device below the 45nm technology node. Phenomena expected to become obvious in this device are investigated based on device physics. Modeling of the multi-gate MOSFETs has been performed in the NDMI project based on different concepts. The SOI-MOSFET modeling has been also investigated as a foregoing study.

Subject	Purpose
1	Common Platform Development
2	Model Development for Microscopic Phenomena
3	Model Development for SOI-MOSFET
4	Model Development for MG-MOSFET

Table1. Subjects aimed at in the NDMI project.

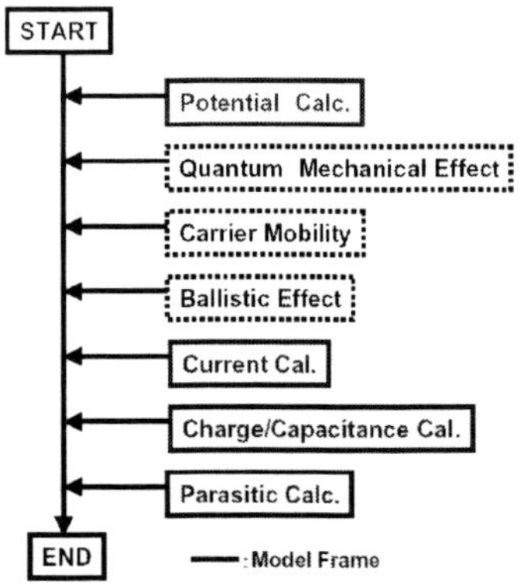

Fig. 1. Schematic representation of the concept for the compact-modeling framework, providing modules of individual functions of the complete compact model.

2 OVERVIEW OF RESULTS OBTAINED

2.1 Modeling of Quantum Phenomena

It is expected that the quantum mechanical effects cannot be ignored for sub-50nm technologies and beyond. The most important quantum effect is observed in the quantized carrier distribution at different energy states as schematically shown in Fig. 2.

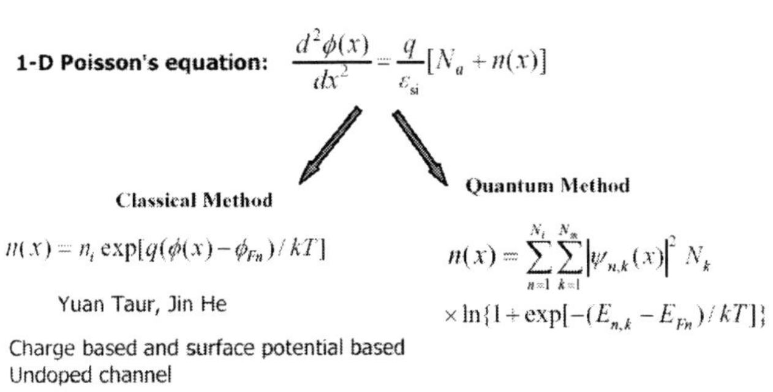

Fig. 2 Illustration of the conceptual differences between the classical and quantum mechanical solution of the Poisson equation.

Figure 3 compares results with developed quantum model for the inversion charge, the Baccarani results, and the Schred results, by solving the Poisson equation and the Schrödinger equation simultaneously [2].

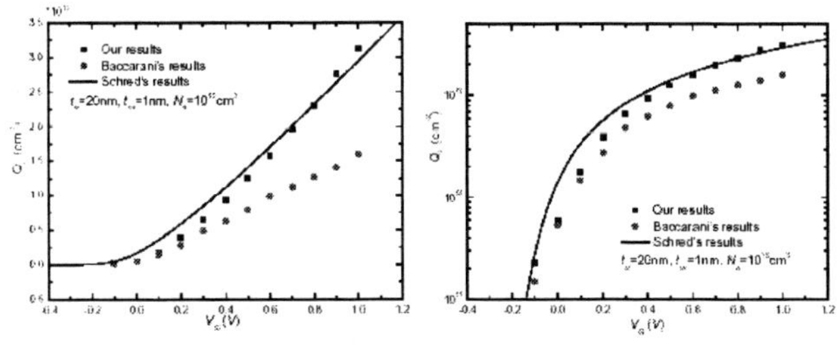

Fig. 3. Calculated inversion charge with the quantum method in comparison to other existing results.

Surface potentials are calculated in our approach by solving the basic device equations with the quantized charge, which, in the conventional planar case, leads to [3]

$$C_{ox}\left(V_{gs}-V_{FB}-\phi_s\right)=Q_n+Q_b \tag{1}$$

$$Q_n=q\sum_i\sum_{j=1}^{3}\frac{m_{xy}^i}{\pi\hbar^2}k_BT\log\left(\exp\left(\frac{E_f-E_j^i}{k_BT}\right)+1\right) \tag{2}$$

where i denotes the valley's index, $m_{xy}=\sqrt{m_xm_y}$ is the two-dimensional DOS effective mass, while the z-axis is directed to the oxide interface, j is the sub-band index, E_f and E_j are the Fermi level and quantization energy levels in the channel, respectively.

The energy spectrum in the conventional planar structure can be calculated with the triangular potential approximation, where the effective field is determined as

$$E_{eff}=\frac{\alpha Q_n+Q_b}{\varepsilon_{Si}}=\alpha C_{ox}\frac{V_{gs}-V_{FB}-\phi_s}{\varepsilon_{Si}}+\frac{1-\alpha}{\varepsilon_{Si}}Q_b(\phi_s) \tag{3}$$

where α is the fitting parameter, which is equal to 0.5 as a nominal value. Then the sub-bands in conventional MOSFET are determined as

$$E_{SG,j}^i=-q\phi_s+\left(\frac{\hbar^2}{2m_z^i}\right)^{\frac{1}{3}}\left[\frac{3}{2}\pi qE_{eff}\left(j-\frac{1}{4}\right)\right]^{\frac{2}{3}} \tag{4}$$

where the intrinsic energy in the substrate is the reference point. Since the effective field is a function of ϕ_s and V_{gs}, we can solve the equations iteratively.

The effective field in the double-gate structure can be defined in the similar manner, and we use the energy spectrum for two lowest energy levels as

$$E_{DG,j}^i\left(\phi_s,V_g\right)=-q\phi_s+\frac{\hbar^2\pi^2}{8m_z^id^2}j^2+\left(\frac{\hbar^2}{2m_z^i}\right)^{\frac{1}{3}}\left[\frac{9}{8}\pi qE_{eff}\right]^{\frac{2}{3}} \tag{5}$$

Once the surface potential is calculated with the quantized condition, the gate current density can be calculated analytically using the effective field of Eq. (3) and a semi-classical approach [3]. We derive the drain current using the drift-diffusion model, where the lateral field is determined by the gradient of the lowest sub-band along the channel, namely

$$I_{ds}=\mu Q_n\frac{1}{q}\frac{E_0}{dy}+\mu\frac{kT}{q}\frac{dQ_n}{dy} \tag{6}$$

where E_0 is the lowest subband.

To show the validity of the model, we compare the simulation results for MOS structures with the Schrödinger-Poisson scheme, which includes the wave function penetration into oxide region [4]. The *C-V* characteristics for a planar structure are shown in Fig. 4, where the substrate doping is $Na = 5e10^{18}cm^{-3}$. The corresponding gate tunneling current densities are shown on Fig. 5.

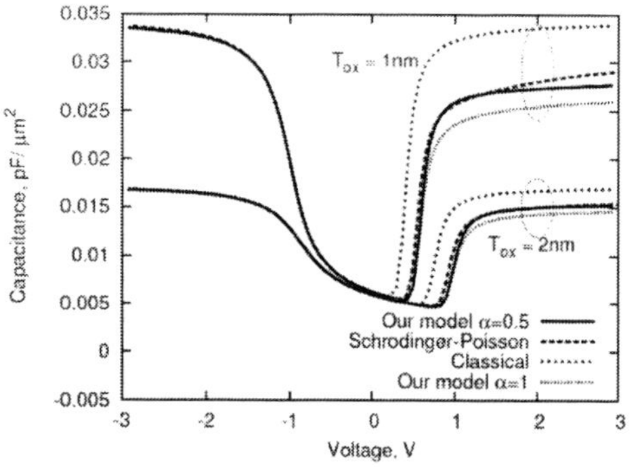

Fig. 4. The dependence of capacitances on the gate voltage.

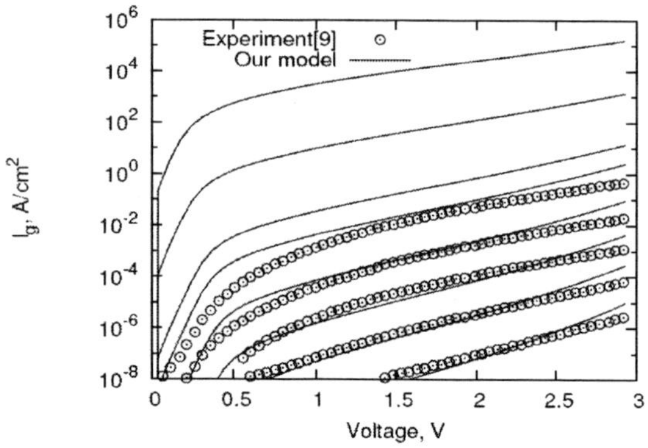

Fig. 5. The gate tunneling current density in a comparison between our model and experimental data.

2.2 Modeling of the SOI-MOSFET

Modeling of the SOI-MOSFET has been undertaken, and is focused on accurately capturing the floating-body effect, which should be also observed in the multi-gate MOSFET. The hole charge stored at the source side is explicitly considered in the Poisson equation, solved iteratively [6,7]. Result is shown in Fig. 6.

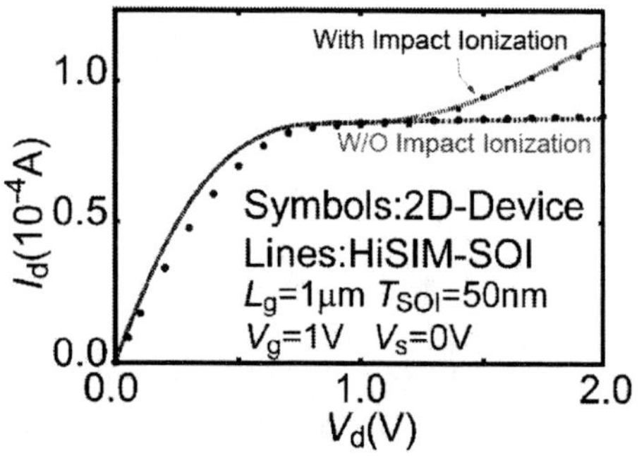

Fig.6. Calculated drain current with the developed SOI model HiSIM-SOI in comparison to 2D-device simulation.

2.3 Modeling of Multi-Gate MOSFETs

Modeling of MG-MOSFET has been investigated intensively at different affiliations (see for example [8,9]). Since it is till a beginning phase of the development, and there are still many unknown elements to be understood. Our approach is split on several different aspects. Here results are summarized.

[Modeling of Short-Channel Effect]

In device design of MG-MOSFET or DG-MOSFETs, the silicon substrate thickness together with channel doping concentration can be determined, if the channel charge is resulted from the surface or volume inversion. Because of the volume inversion, the concept of a surface potential, being a determining factor of the channel inversion charge density (per unit gate area), is no longer valid. It is thus imperative to know the potential distribution across the channel to the depth direction in order to predict the device characteristics.

First the short-channel effect is modeled as shown in Fig, 7 [10] by considering the volume inversion effect. For the modeling the effective oxide thickness is considered for the quantum-mechanical effect.

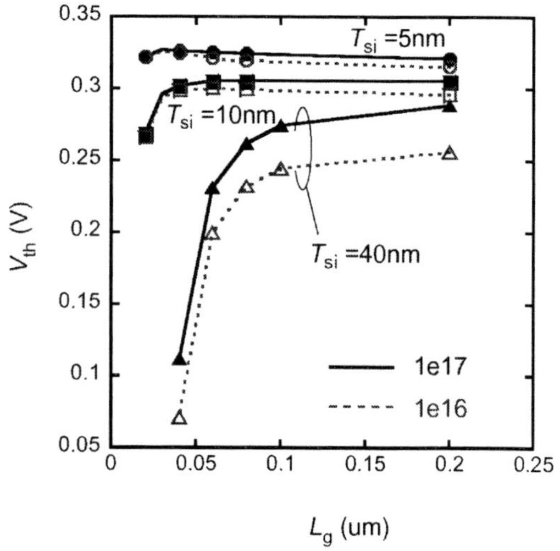

Fig. 7. Calculated threshold voltage V_{th} as a function of the gate length L_g for two impurity concentrations. Symbols are 2D-device simulation results and lines are model results.

[Independent Gate Control]

The DG-MOSFET has been invented 1980 by Sekigawa et al. [11]. The two gates have been controlled independently to obtain an additional feature of freedom, applicable for circuit design as shown in Fig. 8. The developed model focuses on the basic features of the gate control [12]. Results are shown in Fig. 9.

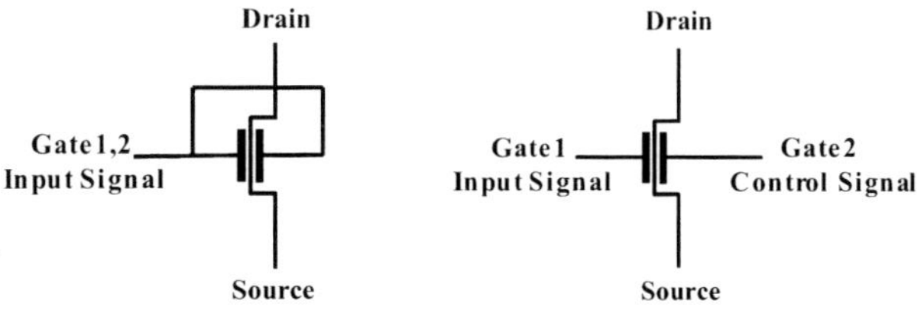

Fig. 8. DG MOSFET schematics with common and independent control of the 2 gates.

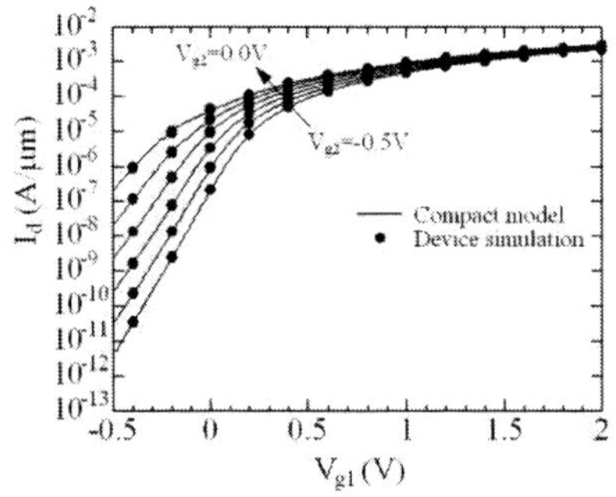

Fig. 9. Calculated drain current Id as a function of the gate voltage V_{g1} at one side of the double gate MOSFET for different gate voltage of the opposite side V_{g2}, based on the charge-sheet approximation.

[Versatile Generic DG Core Model]

A generic core model for the undoped four-terminal double-gate (DG) MOSFET valid for symmetric, asymmetric, SOI, and independent gate operation modes is also developed [13]. Based on the exact solution of the 1-D Poisson's equation of a general DG-MOSFET configuration, a generic drain current model is derived from Pao-Sah's double integral. The core model is verified by extensive comparisons with 2-D numerical simulations under different bias conditions to all four terminals. The concise mathematical formulation allows the unification of various double-gate models into a carrier-based core model for compact DG-MOSFET model development. Some significant results are shown in Figs. 11 and 12.

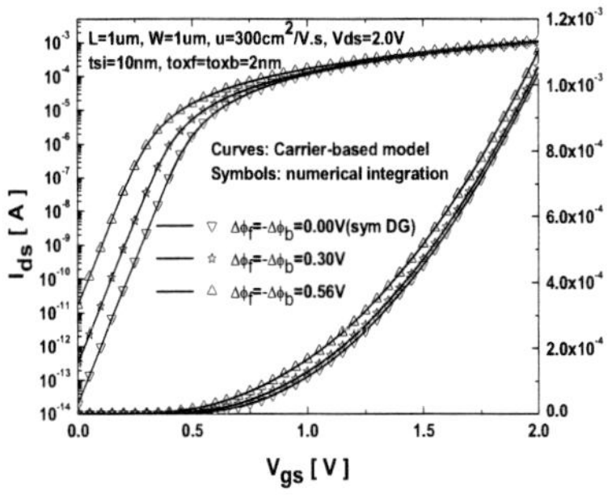

Fig.10. Drain current versus gate voltage for various work-function asymmetry differences.

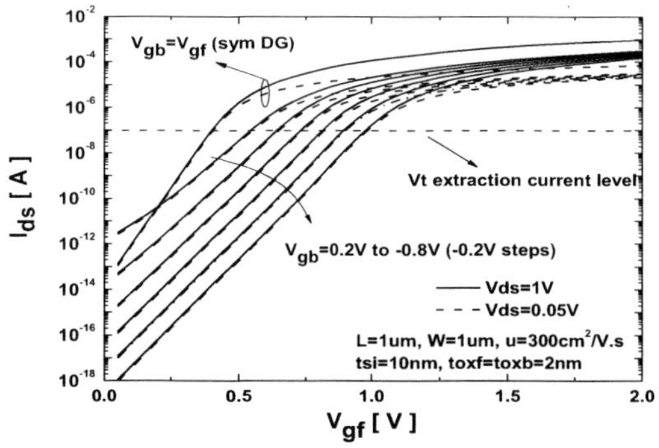

Fig.11. Drain current versus front gate voltage modulated by different V_{gb} in the independent gate DG MOSFET, compared with the symmetric DG-MOSFET.

Furthermore, a unified model for the undoped symmetric Double-Gate (DG) and Surrounding-Gate (SRG) MOS transistors is derived and verified [14]. It is shown that the solutions of the Poisson equation and the Pao-Sah current equation for DG and SRG MOSFETs can be represented by an equivalent mathematical equation.

[Iterative Solution of Potential Distribution]

Another compact DG-MOSFET model prototype called HiSIM-DG considers the volume inversion effect and solves the Poisson equation iteratively including any form of substrate doping [15]. The developed model calculates the bias dependence for not only the surface potential but also for the center potential of the silicon layer accurately. A predictive capability of silicon-layer-thickness dependence is verified by comparison to 2-D device simulation results. It is observed that the volume inversion effect prevents the DG-MOSFET from performance degradation for a reduction of device sizes. Calculated potential values are shown in Fig. 13.

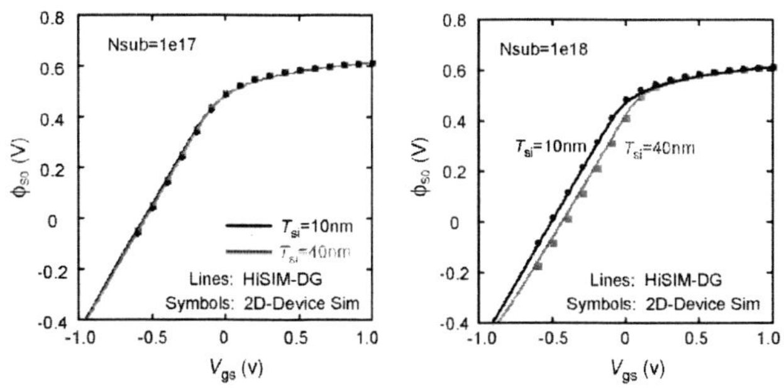

Fig. 13. Calculated surface potential values at the source side ϕ_{S0} and at the drain side ϕ_{SL} for two channel impurity concentrations.

[Derivation of a Closed-Form Equation]

We additionally deal with the closed-form solution of complete 1D Poisson's equation for the DG MOSFET without making any simplification in the constituents in space charge [16]. By integrating the Poisson equation under the gradual channel approximation and applying the symmetry condition at $x = 0$, we obtain:

$$\frac{d\varphi}{dx} = \sqrt{2A\cdot V_t\cdot\alpha + 2B\cdot V_t + 2C\cdot V_t\cdot\beta}\sqrt{e^{\varphi/V_t} - e^{\varphi_0/V_t}} \tag{7}$$

where

$$A = \frac{qn_i}{\varepsilon_{si}}\cdot e^{\frac{q\phi_B}{kT}}\cdot e^{\frac{qV_{ch}}{kT}}, \quad B = \frac{qn_i}{\varepsilon_{si}}\cdot e^{\frac{-q\phi_B}{kT}}\cdot e^{\frac{-qV_{ch}}{kT}}$$

$$C = \frac{qN_A}{\varepsilon_{si}} \quad \text{and} \quad V_t = \frac{kT}{q}$$

$$\alpha(\varphi;\varphi_0) = \left(e^{-\varphi/V_t} - e^{-\varphi_0/V_t}\right)/\left(e^{\varphi/V_t} - e^{\varphi_0/V_t}\right)$$

$$\beta(\varphi;\varphi_0) = \left(\varphi/V_t - \varphi_0/V_t\right)/\left(e^{\varphi/V_t} - e^{\varphi_0/V_t}\right)$$

Notice that the intrinsic Fermi level in the flat band condition, E_i, is defined as the zero reference for the potential, while V_{ch}, which is the quasi-Fermi level for electrons in the given 1D cut with fixed coordinate along the channel, and V_g are all relative to the Fermi-level in the heavily doped source region.

Integrating Eq. (7), with α and β treated as constants, and meanwhile introducing a fitting parameter λ, Eq. (7) yields the following solution:

$$\varphi = \varphi_0 - 2V_t \ln\left[\cos\left(\sqrt{\frac{A}{2V_t}\alpha + \lambda\frac{B}{2V_t} + \frac{C}{2V_t}\beta}\cdot xe^{\frac{\varphi_0}{2V_t}}\right)\right] \tag{8}$$

Fig. 14 shows the comparison of the three model predicted results (our model, Taur's model and a perturbation model) and the numerical simulation data for the potential distribution profile in a DG-MOSFET, with different doping concentrations.

The fitting parameter λ reflects the relative contribution of inversion carriers compared with the ionized dopants and the minority carriers. Fig. 15 shows that λ increases with t_{si} but decreases with increasing V_g.

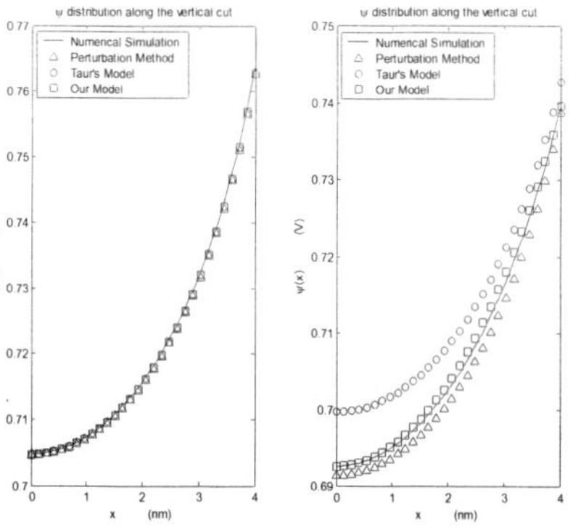

(a) (b)

Fig. 14 Comparison of the three model-predicted potential distributions with numerical simulation for different doping density. t_{ox}=2nm, t_{si}=8nm, V_g- V_{FB}=1.4V, V_{ch}=0.2V, (a) N_A=10^{16}cm^{-3},(b) N_A=10^{18}cm^{-3}.

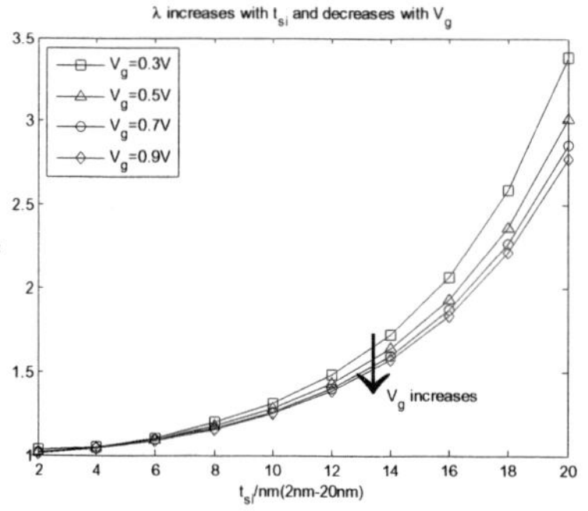

Fig. 15 The change of λ with t_{si} and V_g acting as parameter. Other parameters are t_{ox}=2nm, V_{ch}=0V, N_A=10^{18}cm^{-3}.

[Parasitics of MG MOSFETs]

Geometry-dependent parasitic components including parasitic gate capacitance and gate resistance in multi-fin-MOSFETs are modeled using a distributed RC coupling approach [17]. The accuracy of the model is verified by 2-D and 3-D device simulation. Fig. 16 shows the comparison of calculated capacitance and conductance as a function of fin spacing with 2-D device simulation results.

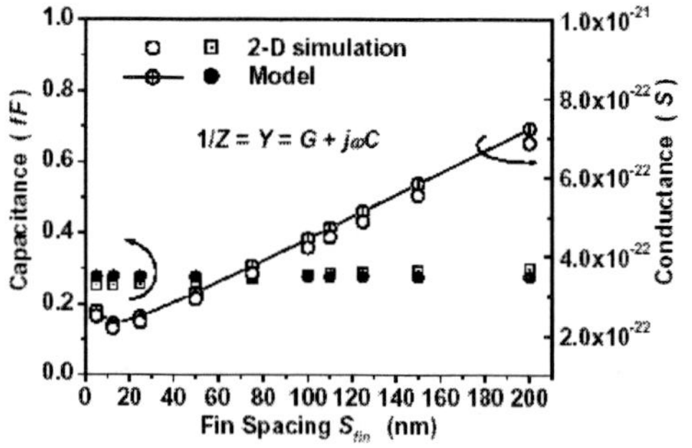

Fig. 9. Modeling of parasitic capacitances of the multi-gate MOSFET. Calculated results are also shown as a function of the finger spacing.

2.4 XMOS Fabrication

To verify the constructed prototype models for the DG-MOSFET, real devices are also developed based on the XMOS technology [18,19], which has been under continuous refinement since more than 20 years at AIST(Japanese national institute of Advanced Industrial Science and Technology). The XMOS structure is schematically shown in Fig. 10.

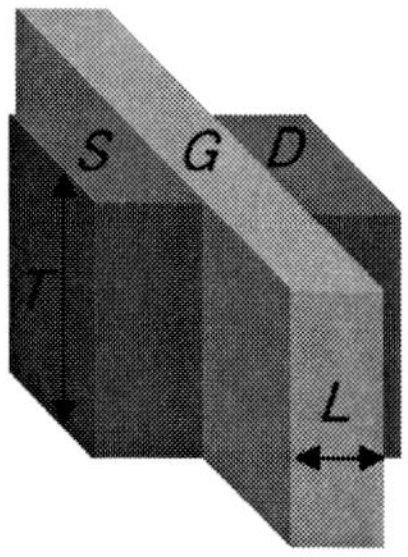

Fig. 10. XMOS structure now under fabrication at AIST(National institute of Advanced Industrial Science and Technology), Japan.

3 CONCLUSION

The development activities of NDMI (Nano Device Modeling Initiative), directed at a common modeling platform and a high quality MG-MOSFET model have been described. Our final goal is to provide a multi-gate MOSFET model by merging all our prototype results. This is performed by plugging in of the developed modular components into the platform. The subjects being undertaken are summarized in Table 1. Development of the SOI-MOSFET model has mainly the purpose to review phenomena observed in thin silicon layer devices.

ACKNOWLEDGEMENTS

This work is supported by the NEDO (New Energy Industrial Technology Development Organization) foundation of the Japanese Government. Authors want to express their sincere thanks for the chance of carrying out such a broad international cooperation.

REFERENCES

[1] M. Miura-Mattausch, et al., *IEEE Transactions on Electron Devices*, Vol. 53, No. 9, pp. 1994-2007, 2006.9.(invited)

[2] Z. Yu, et al., Proc. of Int. Conference on Solid-State and Integrated-Circuit Technology, pp. 1361-1363, 2006.

[3] Y. J. Park, et al., Proc. Int. Simulation of Semiconductor Processes and Devices, pp. 297–300, 2007.

[4] L. F. Register, *et.al.,* Applied Physics Letters, vol. 74, no. 3, pp. 457-459, 1999.

[5] S. Jin, *et.al.,* JSTS, vol. 52, no. 11, pp.2422-2429, 2006.

[6] N. Sadachika, et al., *IEEE Transactions on Electron Devices*, Vol. 53, No. 9, p2017-2024, 2006.

[7] T. Murakami, et al., Jap. J. Appl. Phys., Vol. 47, No. 48, 2008.

[8] Y. Taur, et al., IEEE Electron Device Lett.,vol. 25, no. 2, pp. 107–109, Feb. 2004.

[9] M. V. Dunga, et al., IEEE Trans. Electron Devices, vol. 53, no. 9, pp. 1971–1978, Sep. 2006.

[10] H. Oka, et al., Jap. J. Appl. Phys., Vol. 46, No. 4B, pp.2096-2100, 2007.

[11] Sekigawa, et al., Solid State Electron., vol. 27, pp. 827-828, 1984.

[12] T. Nakagawa, et al., *Tech. Proc. Workshop on Compact Modeling*, pp. 655-658, 2007.

[13] Feng Liu, et al., IEEE Trans. on Electron Devices, TED-55, No. 3, March., pp.816-826, 2008.

[14] Jin He, et al., Semiconductor Science and Technology, Vol.22 （12） , pp.1312-1316, 2007.

[15] N. Sadachika et al., Proc. SISPAD, pp. 289-292, 2007.

[16] J. Tang, et al., Proc. IWCM, Seoul, 2008. Pp. 31-34.

[17] M. Chan, et al., Technical Proceedings of the 2007 Nanotechnology Conference and Trade Show, Vol. 3, pp. 591-594, 2007.

[18] K. Suzuki, et al., IEEE Trans. Electron Devices, vol. 40, pp.2326-2329, 1993.

[19] Y. Liu, et al. IEEE Trans. Nanotechnology, vol., 2, pp. 198-204, 2003.

Unified Regional Surface Potential for Modeling Common-Gate Symmetric/Asymmetric Double-Gate MOSFETs with Any Body Doping

Guan Huei See, Xing Zhou, Guojun Zhu, Zhaomin Zhu, Shihuan Lin, Chengqing Wei, Junbin Zhang, and Ashwin Srinivas

School of Electrical & Electronic Engineering, Nanyang Technological University
Nanyang Avenue, Singapore 639798, exzhou@ntu.edu.sg

ABSTRACT

Double-gate MOSFET is one of the key potential devices to allow further extension of CMOS technology scaling. The compact modeling community faces great challenges to model the physical effects due to the coupling of the two gates. Most assumed that only one-carrier type (i.e., electrons for nMOS) contributes to transport. However, the assumption of ignoring dopant in the Poisson solution is strictly valid only for ideal (pure) semiconductors that do not practically exist. Even for undoped body, unintentional dopants always exist, which would make the second integral of Poisson equation theoretically impossible. In this work, regional surface potentials and mid-gap potentials are solved explicitly and unified for a single-piece solution. The transition from partially-depleted to fully-depleted operation is seamlessly built into the model without any fitting parameters. The potential solutions are then applied in bulk-like drain current equation, in which the models are able to scale with the DG MOSFET structure physically.

***Index Terms*—Compact model, doped double-gate, regional surface potential.**

1 INTRODUCTION

Semiconductor industry is aggressively scaling down the CMOS transistors and the bulk MOSFET will soon reach its scaling limit. Other novel devices, such as FinFETs, multiple-gate (MG), nanowire surrounding-gate MOSFETs, have been explored [1] as alternatives to continue the CMOS scaling according to Moore's law. Double-gate (DG) MOSFET is one of such devices that has the potential due to its stronger immunity to short-channel effects, improved subthreshold slope and higher drive current. These attractive benefits have prompted many researchers for theoretical and experimental studies on DG MOSFETs [2-7], including compact models.

The compact modeling community faces great challenges to model the physical effects due to the coupling of the two gates. The input-voltage equation can be derived from the first integral of the Poisson equation, but it is not sufficient to solve for the two-variable implicit equation. Different groups considered different conditions to compromise [8-11]. Most assumed that only one-carrier type (i.e., electrons for nMOS) contributes to transport [8], although its validity has been studied only recently [12] with a rigorous iterative solution of elliptic-integrals that considered both types of carriers (electrons and holes). However, the assumption of ignoring dopant in the Poisson solution is strictly valid only for ideal (pure) semiconductors that do not practically exist. Even for undoped body in a wafer process, unintentional dopants always exist, which would make the physical derivation with the second integral of Poisson equation theoretically impossible. BSIM-MG [5] has included the effect of body doping in the surface-potential computation through the perturbation approach. The surface potential behavior follows the physical structures but it involves multiple iterations in different regions, which can be computational expensive for compact modeling. In this work, regional surface potentials are solved explicitly and unified for a single-piece solution.

2 MODEL FORMULATION

In DG MOSFET modeling, it is preferable to divide the regions of operation according to prevailing contributions from each term in the Poisson equation: namely, holes (accumulation), acceptor (depletion/weak inversion), and electrons (strong inversion). The first integral of Poisson equation, together with Gauss' law applied to the common-gate symmetric DG (s-DG) configuration at the surface and zero-field locations, yield the input-voltage equation

$$\begin{aligned} &V_{gs}-V_{FB}-\phi_s=\pm\Upsilon\sqrt{f_\phi} \\ &f_\phi=v_{th}\left(e^{-\phi_s/v_{th}}-e^{-\phi_o/v_{th}}\right)+\left(\phi_s-\phi_o\right) \\ &\quad+e^{-(2\phi_{Fp}+V)/v_{th}}\left[v_{th}\left(e^{\phi_s/v_{th}}-e^{\phi_o/v_{th}}\right)-\left(\phi_s-\phi_o\right)\right] \end{aligned} \tag{1}$$

where $\Upsilon=(2q\varepsilon_{si}N_{ch})^{1/2}/C_{ox}$ is the body factor, and $N_{ch}=n_i\exp(\phi_F/v_{th})$ is effective channel doping concentration. ϕ_s and ϕ_o are the surface and zero-field potentials, respectively.

In our unified regional modeling (URM) approach for common asymmetric double-gate (a-DG) MOSFETs, the full-depletion voltage has to be determined first using the full-depletion approximation. The applied (common) gate voltage that causes the sum of two depletion widths at both gates to simultaneously cover the entire body thickness is the full-depletion voltage, $V_{gs,FD}$

$$X_{d1,FD}\left(V_{gs,FD}\right)+X_{d2,FD}\left(V_{gs,FD}\right)=T_{Si} \tag{2}$$

where $X_{d1,FD}$ and $X_{d2,FD}$ are the depletion widths from top and bottom surface, respectively [9].

In the region for each prevailing term, for example, electrons or holes are dominant in strong accumulation or strong inversion, respectively. Under either condition, the contribution of the zero-field potential can be ignored. Rearranging (1) in strong accumulation with $f_\phi = v_{th}\exp(-\phi_s/v_{th})$, we obtain

$$\ln\left(\frac{V_{gsf}^2}{\Upsilon^2 v_{th}}\right)+\ln\left(1-\frac{2\phi_s}{V_{gsf}}+\frac{\phi_s^2}{V_{gsf}^2}\right)=-\phi_s/v_{th}, \tag{3}$$

and strong inversion with $f_\phi = v_{th}\exp[(\phi_s - 2\phi_{Fp} - V)/v_{th})]/v_{th}$

$$\ln\left(\frac{V_{gsf}^2}{\Upsilon^2 v_{th}}\right)+\ln\left(1-\frac{2\phi_s}{V_{gsf}}+\frac{\phi_s^2}{V_{gsf}^2}\right)=\left(\phi_s-2\phi_{Fp}-V\right)/v_{th} \tag{4}$$

where $V_{gsf} = V_{gs} - V_{FB}$. Taking a Taylor expansion of the 2nd logarithmic term up to the third-order polynomial, both equations can be expressed as a general cubic equation

$$\phi_s^3+p\phi_s^2+q\phi_s+r=0 \tag{5}$$

which can be solved analytically [13]. The analytical solution of the cubic surface-potential equation has a unique solution:

$$\phi_s=\left(\frac{\sqrt{D}-B}{2}\right)^{1/3}-\frac{A}{3}\left(\frac{2}{\sqrt{D}-B}\right)^{1/3}-\frac{p}{3} \tag{6}$$

where

$$A=\frac{1}{3}\left(3q-p^2\right) \tag{7}$$

$$B=\frac{1}{27}\left(2p^3-9pq+27r\right) \tag{8}$$

$$D=4\frac{A^3}{27}+B^2 \tag{9}$$

Therefore, the final combined solution is

$$\phi_s=\begin{cases}\phi_{cc}, & V_{gs}<V_{FB}\\ \phi_{ss}, & V_{gs}>V_t\end{cases}, \tag{10}$$

where ϕ_{cc} and ϕ_{ss} are the regional solutions for strong accumulation and strong inversion, respectively. The prevailing term due to acceptor in depletion (before volume inversion) can be solved with

$$\phi_{dd}=\left(-\frac{\Upsilon}{2}+\sqrt{\frac{\Upsilon^2}{4}+V_{gsf}}\right)^2. \tag{11}$$

Similarly, for "weak accumulation" the prevailing term due to acceptor can be extended for $V_{gs} < V_{FB}$ by obtaining the odd function ("conjugate") of (11) and, thus, yielding

$$\phi_{add}=-\left(-\frac{\Upsilon}{2}+\sqrt{\frac{\Upsilon^2}{4}-V_{gsf}}\right)^2. \tag{12}$$

The regional volume inversion solution is given as

$$\phi_{vi}=V_{gsf}-\Upsilon\sqrt{qn_ie^{\phi_{Fp}/v_{th}}X_d^2/2\varepsilon_{Si}}. \tag{13}$$

With the help of interpolation/smoothing functions

$$\begin{aligned}&\vartheta_f\{x;\sigma\}\equiv 0.5\left(x+\sqrt{x^2+4\sigma}\right)\\&\vartheta_r\{x;\sigma\}\equiv 0.5\left(x-\sqrt{x^2+4\sigma}\right)\\&\vartheta_{eff^\pm}\{x_{sat},x;\pm\delta\}\equiv\\&x_{sat}-0.5\left[x_{sat}-x\pm\delta+\sqrt{\left(x_{sat}-x\pm\delta\right)^2\mp 4\delta x_{sat}}\right]\end{aligned} \tag{14}$$

the regional solutions [(5) to (13)] can be unified to obtain a single-piece solution. V_{gsf} in these equations are to be replaced with

$$V_{gsf}=\begin{cases}V_{GSR}=0.5\left(V_{gsf}-\sqrt{V_{gsf}^{\ 2}+4\sigma_a}\right), & V_{gs}<V_{FB}\\ V_{GSF}=0.5\left(V_{gsf}+\sqrt{V_{gsf}^{\ 2}+4\sigma_f}\right), & V_{gs}>V_{FB}\end{cases}. \tag{15}$$

Thus the unified strong accumulation and strong inversion solutions are

$$\phi_s=\begin{cases}\phi_{cc}\rightarrow\phi_{astr}\big|_{V_{gsf}=V_{GSR}}, & V_{gs}<V_{FB}\\ \phi_{ss}\rightarrow\phi_{str}\big|_{V_{gsf}=V_{GSF}}, & V_{gs}>V_t\end{cases} \tag{16}$$

and unified weak accumulation, depletion and volume inversion solutions [9] are

$$\phi_s=\begin{cases}\phi_{add}\rightarrow\phi_{asub}\big|_{V_{gsf}=V_{GSR}}, & V_{gs}<V_{FB}\\ \phi_{dd}\rightarrow\phi_{sub}\big|_{V_{gsf}=V_{GSF}}, & V_{gs}>V_{FB}\\ \phi_{dv}\rightarrow\phi_{vi}\big|_{V_{gsf}=V_{GSF}}, & V_{gs}>V_{FB}\end{cases}. \tag{17}$$

The unified solutions are then combined as

$$\phi_s=\begin{cases}\phi_{acc}=\vartheta_{eff^-}\{\phi_{astr},\phi_{asub};\delta\}, & V_{gs}<V_{FB}\\ \phi_{ds}=\vartheta_{eff^+}\{\phi_{str},\phi_{dv};\delta\}, & V_{gs}>V_{FB}\end{cases}. \tag{18}$$

The final single-piece solution is given as

$$\phi_{seff}=\phi_{acc}+\phi_{ds}. \tag{19}$$

Once ϕ_s is obtained, the zero-field potential ϕ_o can be determined regionally in accumulation/inversion regions

$$\phi_o^{\ \pm}=\mp 2v_{th}\ln\left\{\cos\left[\arccos\left(\sqrt{\frac{B^\pm}{A^\pm}}e^{\phi_s/2v_{th}}\right)+\frac{\sqrt{B^\pm}X_d}{2v_{th}}\right]\right\}\pm v_{th}\ln\frac{B^\pm}{A^\pm} \tag{20}$$

where

$$B^\pm=A^\pm e^{\frac{\mp\phi_{s,str}}{v_{th}}}-\phi'(0)^2 \tag{21}$$

$$A^\pm=\begin{cases}A^-=\dfrac{2qn_ie^{\phi_{Fp}}v_{th}}{\varepsilon_{Si}}, & V_{gs}<V_{FB}\\ A^+=\dfrac{2qn_ie^{-\phi_{Fp}-V}v_{th}}{\varepsilon_{Si}}, & V_{gs}>V_{FB}\end{cases} \tag{22}$$

The drain current is derived similarly to bulk MOSFET, with the normalized inversion charge [14] given as

$$V_{gt,s}=\Upsilon\sqrt{v_{th}e^{\frac{\phi_s\ 2\phi_{Fp}}{v_{th}}}}\sin\left(\frac{\Upsilon C_{ox}}{\varepsilon_{Si}}\frac{X_d}{2v_{th}}\sqrt{v_{th}e^{\frac{\phi_o-2\phi_{Fp}}{v_{th}}}}\right). \tag{23}$$

Finally, the unified surface potential and drain current are obtained and compared with the Medici simulation data.

3 RESULTS AND DISCUSSIONS

Figure 1 shows the individual unified regional surface potential and zero-field potential solutions with undoped (pure) body (i.e., $N_A = N_D = 0$; not "intrinsically doped" with $N_A - N_D = n_i$). In most literature models with undoped body assumed "intrinsic" doping, and it is nontrivial to model "pure" body without any doping. It can be seen that the surface potential can actually be divided to four regions of operation, namely, strong accumulation, weak accumulation, weak inversion and strong inversion, in which our model can capture the behavior in each individual region correctly as well as the combined solution using interpolation and smoothing functions. The first derivative gives an identically symmetric bell-shaped curve that may not be trivial to model with other approaches [15].

Figure 2 shows the similar plot but at high doping concentration, $N_A = 10^{18}$ cm^{-3}. The weak accumulation is not observable but the depletion starts to occur. This is readily taken care of with the smoothing function that will ensure only the right solutions are used in the effective surface potential solutions. The subthreshold region has both the depletion and volume inversion, as usually happens in partially-depleted SOI MOSFETs.

To demonstrate the scalability of the surface potential model, Figure 3 and Figure 4 show the doping scaling for the surface potential and zero-field potential, respectively. The model seamlessly and smoothly (as shown by the corresponding derivatives) changes from fully-depleted to partially-depleted without any fitting parameters using the same set of equations. The surface potential models are now to be used for drain current/charge model evaluation.

The unified regional surface potentials are applied to the drain-current model. It can be seen that both body doping concentration and body/oxide thickness variations are captured physically in the model, as shown in Figure 5.

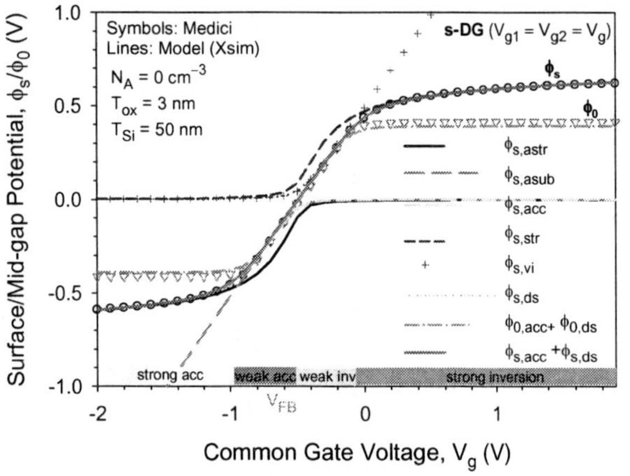

Figure 1: Unified regional surface potential and zero-field potential solutions for s-DG with undoped body (zero doping). It can be seen that the subthreshold is dominated by volume inversion (ideal slope 1). The curve is an exact odd function about flatband voltage.

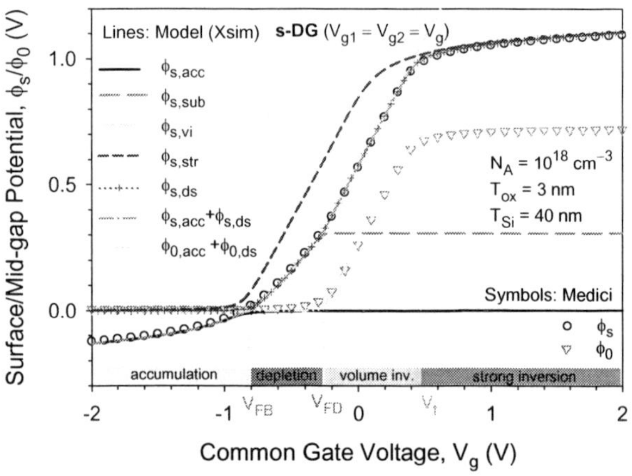

Figure 2: Unified regional surface potential and zero-field potential solutions for s-DG with high body doping. It can be seen that subthreshold region is now dominated by both depletion and volume inversion.

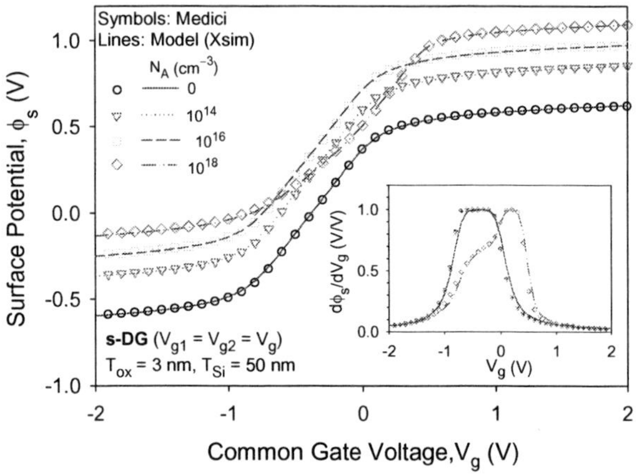

Figure 3: Surface potential scaling from undoped to highly-doped body, compared with Medici numerical data. Inset: The 1st-order derivatives of the corresponding surface potentials.

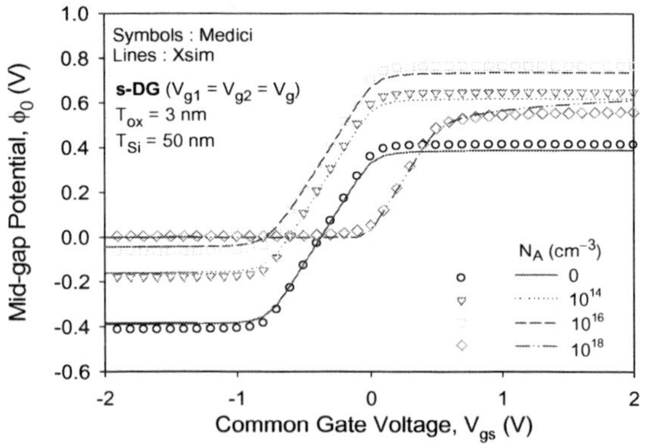

Figure 4: Zero-field potential scaling from undoped to highly-doped body, compared with Medici data.

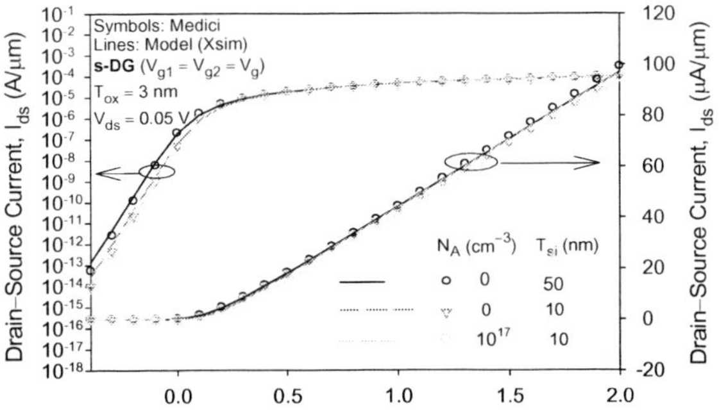

(a)

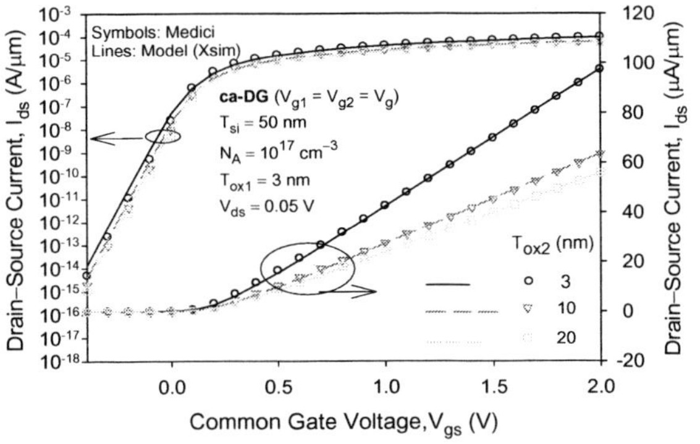

(b)

Figure 5: Drain-current model verification for (a) different body thicknesses and doping concentrations in s-DG and (b) oxide thicknesses in ca-DG MOSFETs.

4 CONCLUSION

In conclusion, regional surface potential and zero-field potential for DG MOSFETs have been derived physically. The model is fully explicit. The transition from partially-depleted to fully-depleted operation is seamlessly built into the model without any fitting parameters. The unified regional surface potential is applied to drain-current model. It shows that it contains the essential physics that scales with the DG MOSFET structures.

Acknowledgment: This work was supported in part by the Semiconductor Research Corporation under Contract 2004-VJ-1166G and in part by Nanyang Technological University under Grant RGM30/03.

REFERENCES

1. J. P. Colinge, *Multi-gate SOI MOSFETs.* Microelectronic Engineering, 84(9-10), pp. 2071-2076, 2007.
2. Y. Bo, et al., *Analytic Charge Model for Surrounding-Gate MOSFETs.* IEEE Trans. Electron Devices, 54(3), pp. 492-496, 2007.
3. D. Hisamoto, et al., *FinFET-a self-aligned double-gate MOSFET scalable to 20 nm.* IEEE Trans. Electron Devices. 47(12), pp. 2320-2325, 2000.
4. B. Dufrene, et al., *Investigation of the four-gate action in G/sup 4/-FETs.* IEEE Trans. Electron Devices, 51(11), pp. 1931-1935, 2004.
5. M. V. Dunga, et al., *Modeling Advanced FET Technology in a Compact Model.* IEEE Trans. Electron Devices, 53(9), pp. 1971-1978, 2006.
6. J. G. Fossum, *Physical insights on nanoscale multi-gate CMOS design.* Solid-State Electron., 51(2), pp. 188-194, 2007.
7. Y. Kyoung Hwan, et al. *Gate-All-Around (GAA) Twin Silicon Nanowire MOSFET (TSNWFET) with 15 nm Length Gate and 4 nm Radius Nanowires. IEDM* 2006.
8. A. Ortiz-Conde, et al., *A Review of Core Compact Models for Undoped Double-Gate SOI MOSFETs.* IEEE Trans. Electron Devices, 54(1), pp. 131-140, 2007.
9. X. Zhou, et al. *Unified Compact Model for Generic Double-Gate MOSFETs. Workshop of Compact Modeling, Nanotech* 2007.
10. T. Yuan, *An analytical solution to a double-gate MOSFET with undoped body.* IEEE Electron Device Lett., 21(5), pp. 245-247, 2000.
11. J. He, et al., *A Carrier-Based Approach for Compact Modeling of the Long-Channel Undoped Symmetric Double-Gate MOSFETs.* IEEE Trans. Electron Devices, 54(5), pp. 1203-1209, 2007.
12. X. Zhou, et al., *Rigorous Surface-Potential Solution for Undoped Symmetric Double-Gate MOSFETs Considering Both Electrons and Holes at Quasi Nonequilibrium.* IEEE Trans. Electron Devices, 55(2), pp. 616-623, 2008.
13. D. Herbison-Evans, *Solving quartics and cubics for graphics*, in *Technical Report TR94-487.* Updated, 22 July 2005, Basser Dept. of Computer Science, University of Sydney Australia.
14. G. H. See, et al., *A Compact Model Satisfying Gummel Symmetry in Higher-order Derivatives and Applicable to Asymmetric MOSFETs.* IEEE Trans. Electron Devices, 55(2), pp. 624-631, 2008.
15. G. Gildenblat, et al., *Comments on "a physics-based analytic solution to the MOSFET surface potential from accumulation to strong-inversion region".* IEEE Trans. Electron Devices, 54(8), pp. 2061-2062, 2007.

Surface Potential versus Voltage Equation from Accumulation to Strong Inversion Region for Undoped Symmetric Double-Gate MOSFETs

Jin He[1,2], Yu Chen[1], Chi Liu[1], Yiqun Wei[1], and Mansun Chan[2]

[1] School of Computer& Information Engineering, Peking University Shenzhen Graduate School, Shenzhen 518055, P.R.China
[2] Department of ECE, Hong Kong University of Science & Technology, Clearwater Bay, Kowloon, Hong Kong
Tel: 86-0755-26032104 Fax: 86-0755-26035377 Email: frankhe@pku.edu.cn

Abstract- **Surface potential is one key variable in DG MOSFET compact modeling. A complete surface potential versus voltage equation and its continuous solution from the accumulation to strong inversion region are presented in this paper for undoped (lightly doped) symmetric double-gate (DG) MOSFETs. The results are based on the exact solution of Poisson's equation with an amending mathematical condition of the continuity, allowing the surface potential to be accurately described from the accumulation region, through sub-threshold, finally to the strong inversion region. From presented surface potential solution, the dependence of the surface potential and corresponding gate capacitance on the quasi-Fermi-potential, silicon film thickness, gate oxide layer, and temperature are tested running different parameters, showing the solution continuity, smoothing, and high accuracy, compared with the 2-D numerical simulation. The presented equation and its solution will be useful in developing a complete surface potential-based DG-MOSFET compact model.**

Index terms- **Surface potential, DG-MOSFETs, device physics, compact modeling, ULSI circuit simulation**

I INTRODUCTION

Since the bulk MOSFET surface potential-based model has been chosen as the next generation industry standard, the analytic surface potential-based model studies on the non-classical multiple-gate devices have also been explored [1-3]. However, the complete surface potential equation for the DG MOSFETs has not been established well so far although various compact models, such as potential-based model, charge-based model and carrier-based model have been established [4-9] on the electrostatics analysis of the DG MOS transistor [10].

In this work, a complete surface potential versus voltage equation is presented and its continuous solution is also discussed for the undoepd symmetric DG MOSFETs from the accumulation to strong inversion region. Using the Newton's iterative method, the surface potential characteristics of DG MOSFETs have been analyzed numerically and the dependences of the surface potential solution on the quasi-Fermi-potential, temperature, the scale of the silicon film and the gate oxide layer have been also demonstrated.

II SOLUTION DEVELOPMENT

We consider an ideal undoped symmetric n-channel MOSFET in which the electrostatics is control by the back and surface gates good enough to neglect the short channel effects associated to two-dimensional effects (gradual channel approximation). Therefore, the electrostatic behaviour of the DG MOSFETs can be described by the 1-D Poisson's equation in the vertical direction.

$$\frac{d^2\phi}{dx^2} = -\frac{\rho}{\varepsilon_{si}} \tag{1}$$

Since in an undoped symmetric n-type DG-MOSFET,

$$N_a = N_d = 0 \tag{2}$$

The space and hole concentrations can be simply given by

$$\rho = q(p-n) \tag{3}$$

Electron and hole concentrations can be expressed by the potential as

$$n = n_i e\left(\frac{q(\phi - V)}{kT}\right) \tag{4}$$

$$p = n_i e\left(-\frac{q(\phi)}{kT}\right) \tag{5}$$

Poisson's equation thus takes the following form:

$$\frac{d^2\phi}{dx^2} = \frac{qn_i}{\varepsilon_{si}}\left[e^{\frac{q(\phi-V)}{kT}} - e^{\frac{-q\phi}{kT}}\right] \tag{6}$$

with the symmetric boundary conditions:

$$\frac{d\phi}{dx}(x=0)=0, \frac{d\phi}{dx}(x=\pm\frac{T_{si}}{2}) = C_{ox}\left(V_{gs}-\phi_s-\Delta\phi_i\right) \tag{7}$$

$$\phi(x=\pm\frac{T_{si}}{2}) = \phi_s, \phi(x=0)=\phi_0 \tag{8}$$

Where ϕ_s and ϕ_0 are the surface potential at the interface between the silicon surface and the gate oxide layer and the centric potential at the silicon film centre as the coordinate origin, respectively. All other symbols have the common meanings.

For the DG-MOSFET operation above flat-band point, the electron dominates the space electron layer. In this case, only electron term is considered while the hole term is negligible is reasonable. Thus, (6) is simplified as

$$\frac{d^2\phi}{dx^2} = \frac{qn_i}{\varepsilon_{si}} e^{\frac{q(\phi-V)}{kT}} \tag{9}$$

With the boundary conditions, (9) is analytically preformed to yield. We obtain the formulation solution of the centric potential and the surface potential.

$$\phi_0 = V + \frac{kT}{q}\ln\left[e^{\frac{q(\phi_s - V)}{kT}} - \frac{C_{ox}^2(V_{gs}-\Delta\phi_i-\phi_s)^2}{2n_i\varepsilon_{si}kT}\right] \quad (10)$$

$$\frac{C_{ox}(V_{gs}-\Delta\phi_i-\phi_s)}{\sqrt{2n_i\varepsilon_{si}kT}}e^{\frac{q(\phi_s-V)}{2kT}} = \sin\left[\frac{T_{si}}{2}\sqrt{\frac{q^2n_i}{2\varepsilon_{si}kT}\left[e^{\frac{q(\phi_s-V)}{kT}} - \frac{C_{ox}^2(V_{gs}-\Delta\phi_i-\phi_s)^2}{2n_i\varepsilon_{si}kT}\right]}\right] \quad (11)$$

We note that (11) does not stand for the electrostatic potential close the flat-band point due to the simplification of Poisson's equation, which causes a discontinuity problem. To keep the continuity of surface potential, Eq. (11) is fixed as

$$\frac{C_{ox}(V_{gs}-\Delta\phi_i-\phi_s)}{\sqrt{2n_i\varepsilon_{si}kT}}\left[e^{\frac{q(\phi_s-V)}{kT}} - e^{\frac{qV}{kT}}\right] = \sin\left[\frac{T_{si}}{2}\sqrt{\frac{q^2n_i}{2\varepsilon_{si}kT}\left[e^{\frac{q(\phi_s-V)}{kT}} - e^{\frac{qV}{kT}} - \frac{C_{ox}^2(V_{gs}-\Delta\phi_i-\phi_s)^2}{2n_i\varepsilon_{si}kT}\right]}\right] \quad (12)$$

Meanwhile, to keep the surface potential derivative continuity, Eq (12) can be amended further as

$$\frac{C_{ox}(V_{gs}-\Delta\phi_i-\phi_s)}{\sqrt{2n_i\varepsilon_{si}kT}}\left[e^{\frac{q(\phi_s-V)}{kT}} - \left(1-\frac{q\phi_s}{kT}\right)e^{\frac{qV}{kT}}\right]^{\frac{1}{2}} = \sin\left[\frac{T_{si}}{2}\sqrt{\frac{q^2n_i}{2\varepsilon_{si}kT}\left[e^{\frac{q(\phi_s-V)}{kT}} - \left(1+\frac{q\phi_s}{kT}\right)e^{\frac{qV}{kT}} - \frac{C_{ox}^2(V_{gs}-\Delta\phi_i-\phi_s)^2}{2n_i\varepsilon_{si}kT}\right]}\right] \quad (13)$$

Below the flat-band point, only the hole is considered. So the expressions of the surface potential can be decided as the follows similarly.

$$\frac{C_{ox}(V_{gs}-\Delta\phi_i-\phi_s)}{\sqrt{2n_i\varepsilon_{si}kT}}e^{\frac{q\phi_s}{2kT}} = -\sin\left[\frac{T_{si}}{2}\sqrt{\frac{q^2n_i}{2\varepsilon_{si}kT}\left[e^{\frac{-q\phi_s}{kT}} - \frac{C_{ox}^2(V_{gs}-\Delta\phi_i-\phi_s)^2}{2n_i\varepsilon_{si}kT}\right]}\right] \quad (14)$$

$$\frac{C_{ox}(V_{gs}-\Delta\phi_i-\phi_s)}{\sqrt{2n_i\varepsilon_{si}kT}}\left[e^{\frac{q\phi_s}{kT}} - 1\right]^{1/2} = -\sin\left[\frac{T_{si}}{2}\sqrt{\frac{q^2n_i}{2\varepsilon_{si}kT}\left[e^{\frac{-q\phi_s}{kT}} - 1 - \frac{C_{ox}^2(V_{gs}-\Delta\phi_i-\phi_s)^2}{2n_i\varepsilon_{si}kT}\right]}\right] \quad (15)$$

$$\frac{C_{ox}(V_{gs}-\Delta\phi_i-\phi_s)}{\sqrt{2n_i\varepsilon_{si}kT}}\left[e^{\frac{q\phi_s}{kT}} - \left(1+\frac{q\phi_s}{kT}\right)\right]^{\frac{1}{2}} = -\sin\left[\frac{T_{si}}{2}\sqrt{\frac{q^2n_i}{2\varepsilon_{si}kT}\left[e^{\frac{-q\phi_s}{kT}} - 1 + \frac{q\phi_s}{kT} - \frac{C_{ox}^2(V_{gs}-\Delta\phi_i-\phi_s)^2}{2n_i\varepsilon_{si}kT}\right]}\right] \quad (16)$$

According to the discussion above, a single surface potential versus voltage equation is also obtained for the symmetric DG MOSFETs from the accumulation to strong inversion region assuming some signum functions are used

$$\frac{C_{ox}(V_{gs}-\Delta\phi-\phi)}{\sqrt{2n_i\varepsilon_{si}kT\theta}}\left[e^{\frac{-\mathrm{sgn}(\phi)q\phi}{kT}} - \left(1-\mathrm{sgn}(\phi)\frac{q\phi}{kT}\right)\right]^{\frac{1}{2}} = \mathrm{sgn}(\phi)\sin\left[\frac{T_{si}}{2}\sqrt{\frac{qn_i}{2\varepsilon_{si}kT}\left[\theta e^{\frac{\mathrm{sgn}(\phi)q\phi}{kT}} - \left(1+\mathrm{sgn}(\phi)\frac{q\phi}{kT}\right)\theta - \frac{C_{ox}^2(V_{gs}-\Delta\phi-\phi)^2}{2n_i\varepsilon_{si}kT}\right]}\right] \quad (17)$$

From a combination of (14) and (23) and

$$\frac{C_{ox}(V_{gs}-\Delta\phi-\phi)}{\sqrt{2n_i\varepsilon_{si}kT\theta}}\left[e^{\frac{-\mathrm{sgn}(\phi)q\phi}{kT}} - \left(1-\mathrm{sgn}(\phi)\frac{q\phi}{kT}\right)\right]^{\frac{1}{2}} = \mathrm{sgn}(\phi)\sin\left[\frac{T_{si}}{2}\sqrt{\frac{qn_i}{2\varepsilon_{si}kT}\left[\theta e^{\frac{\mathrm{sgn}(\phi)q\phi}{kT}} - \left(1+\mathrm{sgn}(\phi)\frac{q\phi}{kT}\right)\theta - \frac{C_{ox}^2(V_{gs}-\Delta\phi-\phi)^2}{2n_i\varepsilon_{si}kT}\right]}\right] \quad (18)$$

where (17) is continuous for surface potential and charge but discontinuous for the derivative of surface potential. In contrast, (18) is continuous for all related quantities, and the sgn(ϕ) is the signum function showed as the follow

$$\theta = \begin{cases} e^{-qV/kT} & \text{above flat-band point} \\ 1 & \text{flat-band or accumulation} \end{cases} \quad (19)$$

III RESULTS AND DISCUSSION

We use a Newton-Raphson iterative method to calculate the surface potential value of undoped DG MOSFET from the accumulation to the strong inversion region. Firstly, a discussion about the results of Eq. (11) & Eq. (14), Eq. (12) & Eq. (15) or Eq. (17) and Eq(13) & Eq(16) or Eq. (18) are performed as follows.

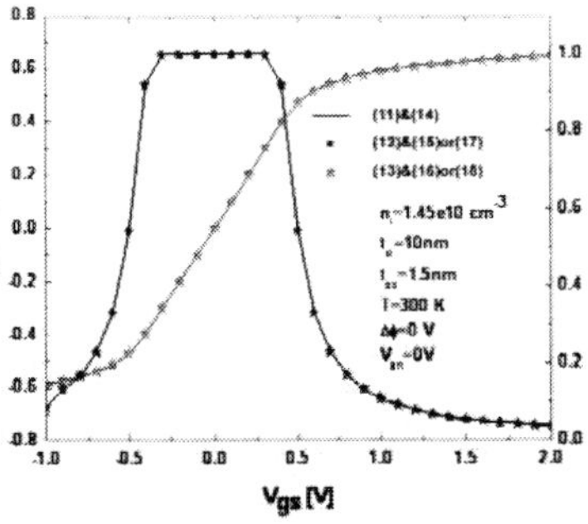

Fig.1 Plot of surface potential and surface potential derivative of the three equation pairs

Fig.1 plots the calculated ϕ_s (left) and $d\phi_s/V_{gs}$ (right) versus V_{gs} for wide gate voltage range from three equation pairs. The results from three equation pairs match well in a large gate voltage range indicating that the modification to Eqs. (11) & (14) is reasonable.

Fig.2 demonstrates the calculated result of ϕ_s versus V_{gs} curve near the flat-band point in a refined gate voltage region. Eq. (11) & (14) results in a discontinuity problem of, thus the calculated surface potential cannot pass through the flat-band point smoothly. ϕ_s calculated by using (17) can pass through the flat-band point smoothly even considering non-zero V_{ch}. (18) also demonstrates continuous characteristic and it fits the result of (17) so well that no difference can be observed.

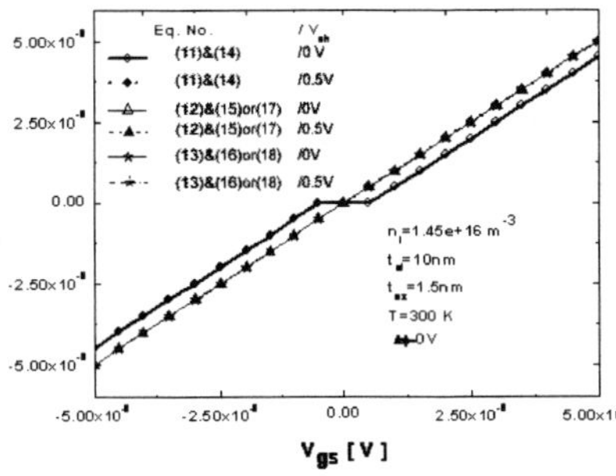

Fig. 2. Plot of ϕ_s versus V_{gs} near flat-band point among three equation pairs for different V_{ch}.

Fig.3 plots the calculation of $d\phi_s / dV_{gs}$ fromEq. (17) and Eq.(18). As expected, Eq.(17) results in a discontinuous derivative close to the flat-band point when $V_{ch} \neq 0$, indicating that Eq.(17) cannot be used to predict the capacitance-voltage characteristics of the DG-MOSFETs. Eq.(18) has to be used and the resultant $d\phi_s / dV_{gs}$ pass the flat-band point smoothly either for V_{ch}=0.0V or V_{ch}=0.5V.

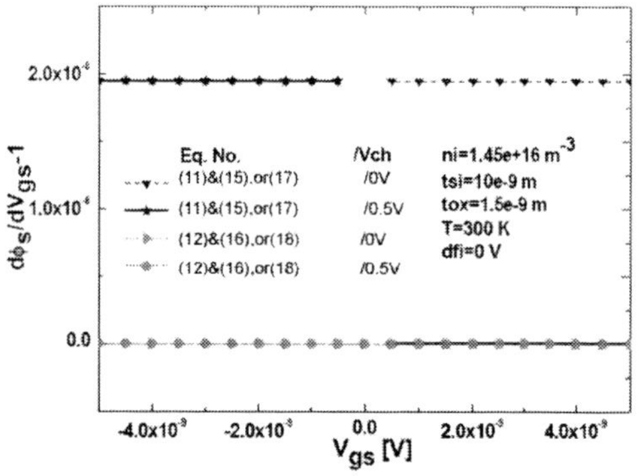

Fig. 3 The comparison of $d\phi_s / dV_{gs} - 1$ versus V_{gs} near the flat-band point between Eq. (17) and (18)

All surface potential equation predictions in the following discussion are based on (18), and the corresponding central potentials (ϕ_0) are from (4) and (11) using Newton-Raphson iterative method. The 3-D numerical simulation is performed from a device simulator, ULTRAS-SRG, from Institute of Microelectronics, Peking University[11].

Fig.4 is the result comparison of ϕ_s and ϕ_0 versus V_{gs} curve between the 2-D simulation and the surface potential equation prediction for different quasi-Fermi-potential. The equation prediction agrees with the 2-D simulation very well. The error is less than 3%, indicating the accuracy of (18). The quasi-Fermi-potential comes into effect only when operation goes into the strong inversion region. The curve of 0φmerges with ϕ_s in the sub-threshold region, but diverges from ϕ_s when $|V_{gs}|$ is large enough, saturating into a constant value.

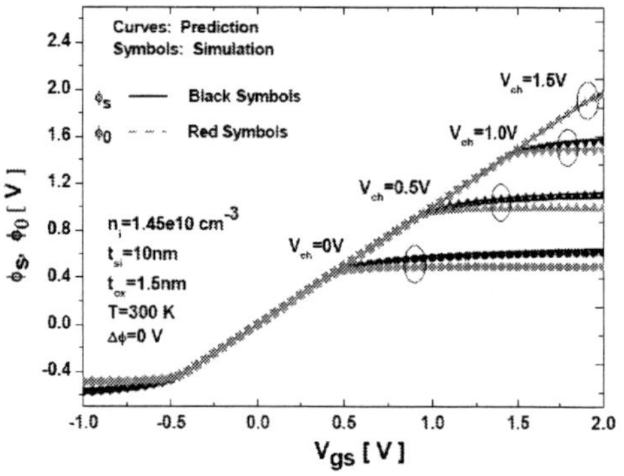

Fig.4 Comparison of ϕ_s and ϕ_0 versus V_{gs} between the 2-D simulation and surface potential equation prediction for different V_{ch} in a DG MOSFET.

Fig.5 plots the comparison of the gate capacitance curves between the surface potential equation and the 2-D simulation for different quasi-Fermi-potential. The equation predicted gate capacitance keeps continuous and smooth in the whole operation region, and matches the simulation very well.

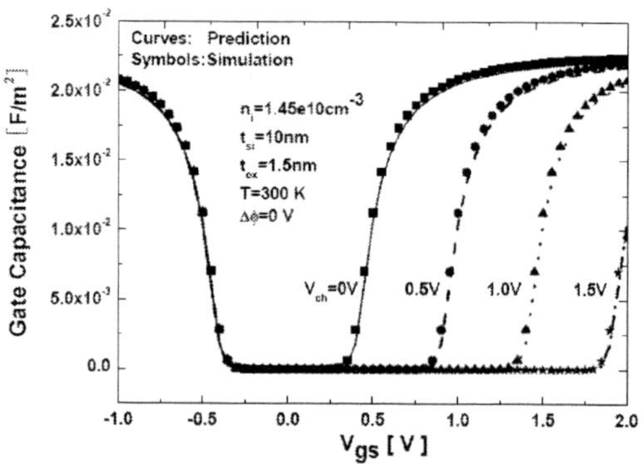

Fig.5 Comparison of the gate capacitance versus V_{gs} between the 2-D simulation and surface potential equation prediction for different V_{ch} in a DG MOSFET.

Fig.6, Fig.7 and Fig. 8 compare ϕ_s versus V_{gs} curves between the 2-D simulation and the surface potential equation prediction for different gate oxide thickness (t_{ox}), silicon film thickness(t_{si}) and temperature(T). The error is less than 4.5%, thus, the presented equation correctly predicts the surface potential dependence on device geometry and the operation temperature. The variation of t_{ox} affects ϕ_s, while ϕ_0 almost remains constant. It is found from Fig.7 that the variation of the silicon thickness makes the surface potential less conspicuous but the central potential decrease with increase of silicon thickness. In Fig. 8, it is also noted a higher temperature results in a higher surface and central potential absolute value either in accumulation region or in the strong inversion region

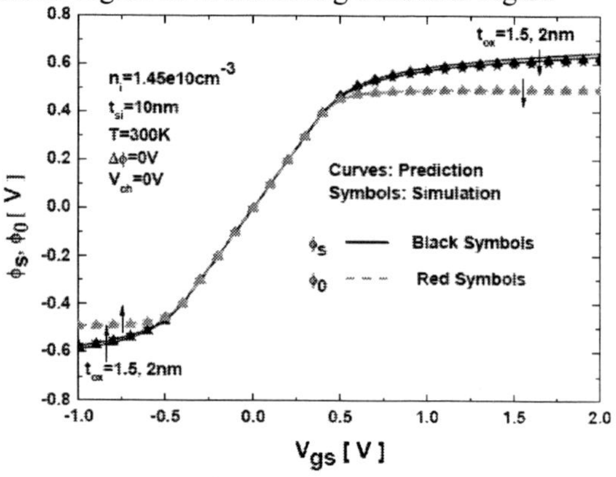

Fig.6 Comparison of ϕ_s and ϕ_0 versus V_{gs} between the 2-D simulation and surface potential equation prediction for different thickness of gate oxide in a DG MOSFET.

Fig. 9 compares the gate capacitance between the 2-D simulation and the surface potential equation prediction for different operation temperature. The results match well, almost overlapping. At a higher temperature, the curve ascends and descends more smoothly, and represents a lower capacitance value. Still, the curves converge when the gate voltage is high or low enough.

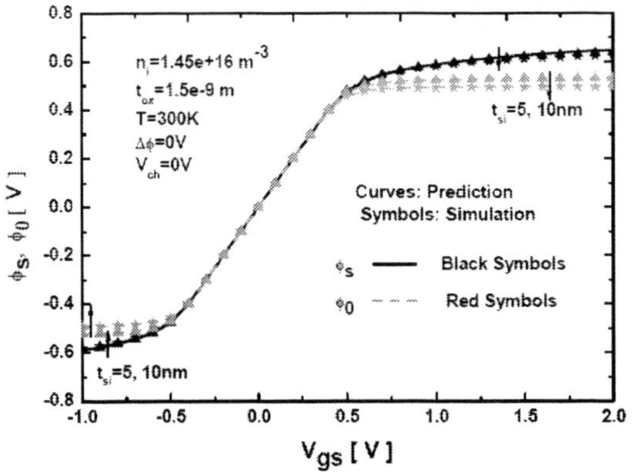

Fig.7 Comparison of ϕ_s and ϕ_0 versus V_{gs} between the 2-D simulation and surface potential equation prediction for different silicon film thickness in a DG MOSFET

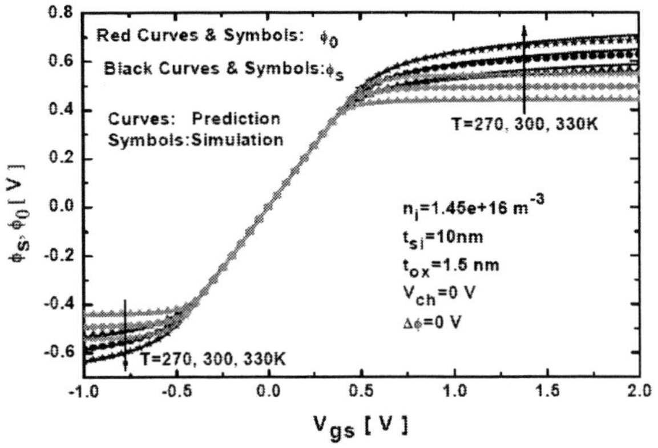

Fig.8 Comparison of ϕ_s and ϕ_0 versus V_{gs} between the 2-D simulation and surface potential equation and the resultant error of ϕ_s for different operation temperature in a DG MOSFET.

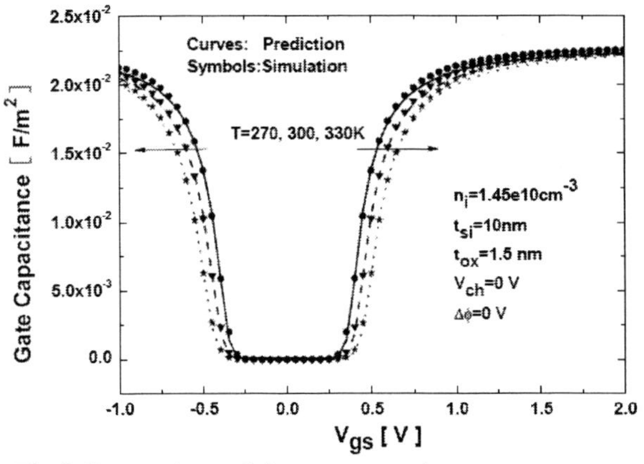

Fig.9 Comparison of the gate capacitance versus gate voltage curves between the 2-D simulation and surface potential equation prediction for different operation temperature in a DG MOSFET.

IV CONCLUSIONS

The complete surface potential versus voltage equation and its continuous solution have been proposed in this paper for the undoped symmetric DG MOSFETs from the accumulation region to the strong inversion region. The equation accuracy and its solution continuity have been widely verified from the surface potential, charge and trans-capacitance calculation. The presented results may be useful for the understating of the DG-MOSFET device physics and developing next generation advanced surface potential model.

Acknowledgements

This work is supported by the Scientific Research Foundation for the Returned Oversea Chinese Scholars, State Education Ministry (SRF for ROCS, SEM), and the National Natural Science Foundation of China (90607017). This work is also partially support by the International Joint Research Program (NEDO Grant) from Japan under the Project Code NEDOO5/06.EG01.

References:

[1] CMC website of Next Generation MOSFET Model Standard Phase-III Evaluation Results. http://www.eigroup.org/CMC.

[2] M.V.Dunga, C.H.Lin, Xuemei Xi, S.Chen, D.Lu, A.M.Niknejad, and Chenming Hu, "BSIM4 and BSIM multi-gate progress." Workshop on Compact Modeling, NSTI-Nanotech, USA, pp.658-661, May 2006, Boston, USA.

[3] Huaxin Lu, Xiaoping Liang, Wei Wang, and Yuan Taur, "Compact Modelling of Short-channel Double-gate MOSFETs." Workshop on Compact Modeling, NSTI-Nanotech, pp.741-744, May 2006, Boston, USA.

[4] Y.Taur, Xiaoping Liang, Wei Wang, and H.Lu, IEEE Electron Device Lett. EDL-25: pp. 107–109, 2004.

[5] A.Oritiz-Conde, F.J.Garcia Sonchez and Juan Muci, Solid State Electronics, Vol.49, pp. 640-647, 2005.

[6] H.Lu and Y.Taur, IEEE Trans. Electron Devices, TED-53, No. 5, pp. 1161-1168, May 2006.

[7] Jin He, Jane Xi, M.Chan, A.M.Niknejad, C.Hu, IEEE International Symposium on Quality of electronic design, Proc. International Symposium on Quality of Electronic Design, pp.45-50, April 2004.

[8] A.S.Roy, J.M.Sallese, and C.C.Enz, Solid State Electron. Vol.50, No. 4, pp. 687-693, Apr. 2006.

[9] Jin He, Feng Liu, Jian Zhang, Jie Feng, Yan Song, and Mansun Chan, IEEE Trans on Electron Devices, TED, Vol.54, No.5, May, 2007.

[10] S.H.Oh, D.Monore, and J.M.Hergenrother, "Analytical description of short-channel effects in fully-depleted double-gate and cylindrical, surrounding-gate MOSFETs." IEEE EDL-21, No.9, pp.445-447, 2000.

[11] Ultras-DG is available by the requirement or on the website page:http://ime.pku.edu.Cn/nano.

Modeling of Floating-Body Devices Based on Complete Potential Description

N. Sadachika, T. Murakami, M. Ando, K. Ishimura, K. Ohyama, M. Miyake, H. J. Mattausch, S. Baba*, H. Oka** and M. Miura-Mattausch

Advanced Sciences of Matter, Hiroshima University
1-3-1 Kagamiyama, Higashi-Hiroshima, Hiroshima, Japan, 739-8530, sadatika@hiroshima-u.ac.jp
*Oki Electric Industry, Hachioji 193-8550, Tokyo, **Fujitsu Laboratories, Akiruno, 197-0833, Tokyo

ABSTRACT

Advanced MOSFETs exploit the carrier confinement to suppress the short-channel effect, which is realized by reducing the bulk layer thickness. The ongoing developments of the multi-gate MOSFET as well as the fully-depleted SOI-MOSFET with ultra thin silicon layer are proved to be applicable beyond the 50nm technology node. However, these advanced devices suffer from the floating-body effect caused by an unfixed body node, which plays an important role for the device performances. Here we present a modeling approach, based on a consistent potential description, for simplifying and solving the Poisson equation with a floating-body node.

Keywords: floating body, HiSIM, compact model, surface potential, circuit simulation

1 INTRODUCTION

The MOSFET is the most widely applied device for integrated circuits, intensively scaled down to improve the circuit performances. However, the conventional bulk MOSFET structure hits the limitation of scaling due to the inevitable short channel effect. To overcome this limitation, possible future variations such as silicon on insulator (SOI) MOSFET and double-gate (DG) MOSFET technologies have been intensively investigated and researched [1,2]. One specific feature of these devices is that they have a thin silicon layer region sandwiched by insulating oxides and that there is no electrode to tie the potential of this silicon body directly to a certain potential in order to maintain the carrier confinement.

Accurate modeling of the thin-layer MOSFETs for circuit simulation is still under development with different approaches to enable accurate circuit design with these devices [3,4,5,6]. An important aspect for the modeling is to fulfill the charge conservation, otherwise suffer from convergence problems in circuit simulations. Due to the unfixed potential value at the back side of the silicon layer, potential-based modeling, solving the potential distribution vertical to the device surface, is the best solution for the floating body devices, securing the charge conservation in a consistent way.

Consequently, we have developed compact models for the SOI-MOSFET and the DG-MOSFET by focusing on the floating-body potential distribution, based on the HiSIM (Hiroshima university STARC IGFET Model) frame work [7,8]. These models are called HiSIM-SOI and HiSIM-DG, respectively.

2 FEATURES OF THIN LAYER MOSFETS

Figure 1a-c compares the structures and band diagrams of bulk MOSFET, SOI-MOSFET and DG-MOSFET, respectively. The potential values on both sides of the oxide are determined explicitly by the applied voltages in the bulk MOSFET. Therefore the surface potential can be calculated explicitly from the boundary conditions. The SOI-MOSFET structure includes two nodes at the front side and the back-side of the thin silicon layer. The backside node can be considered as a "floating node", which is affected by the applied biases V_g and $V_{substrate}$. In DG-MOSFETs, 2 gates control the potential distribution of the silicon layer. The silicon layer thickness is usually less than the sum of two depletion layer thicknesses. Therefore the central potential cannot be determined explicitly and can be considered as a floating node.

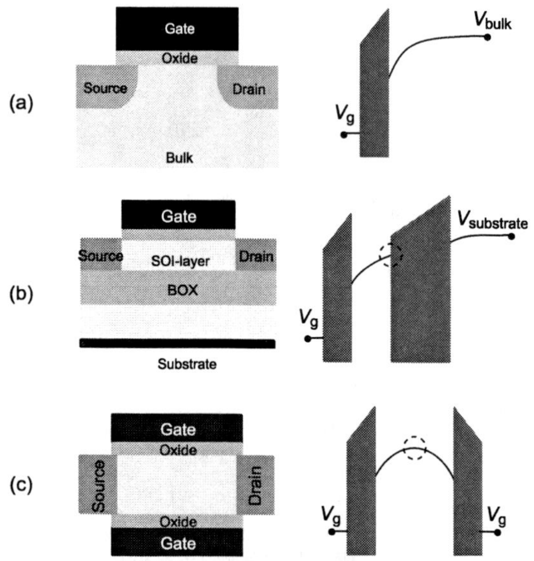

Figure 1: Schematics of structure and band diagram of (a) bulk MOSFET, (b) SOI-MOSFET and (c) DG-MOSFET. Potential values shown with • are tied by electrodes and dashed circles show floating regions in the SOI-MOSFET and DG-MOSFET.

3 BASIC CONCEPT OF THE SURFACE POTENTIAL BASED MODEL

HiSIM is a compact circuit simulation model based on the complete surface-potential description, which is obtained from the Poisson equation solved iteratively at the source and drain end, as shown in figure 2. The calculated potential values determine all device characteristics consistently.

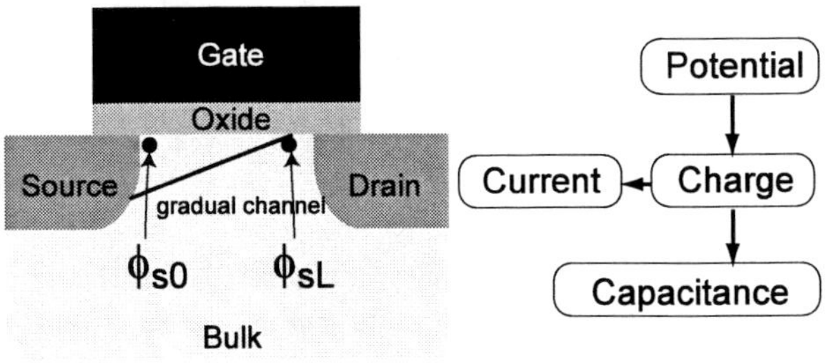

Figure 2: Basic concept of HiSIM

We demonstrate the extension of HiSIM to devices possessing a floating node within the device.

4 MODELING OF THE SOI MOSFET

To extend the frame work of HiSIM to the SOI MOSFET, and to include all device features accurately, HiSIM-SOI determines not only the surface-potential at the channel surface, but also at the 2 surfaces of the BOX (Buried Oxide) self-consistently [9,10]. The total iterative potential calculation for the 3 surfaces requires only about twice as much calculation time as in the bulk-MOSFET case solving just at the channel-surface. Figure 3 demonstrates the accuracy of the three calculated surface potentials in comparison to the results with a 2D-device simulation. Good agreement has been obtained and the floating potential at the bottom of the silicon layer is modeled accurately.

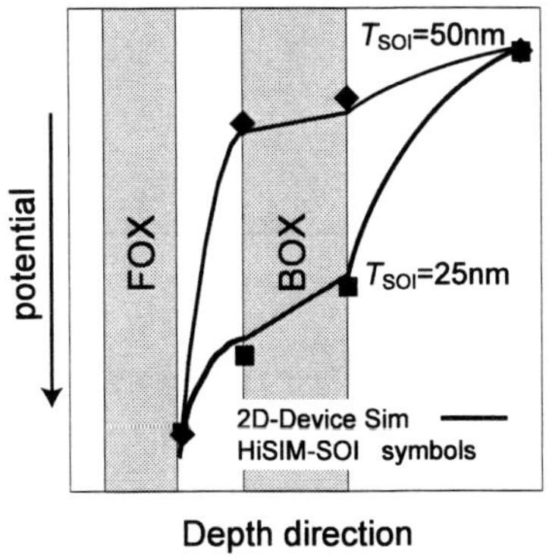

Figure 3: Calculated potential values along the depth (vertical to the channel) direction for the SOI region thickness (T_{SOI}) of 25nm and 50nm.

In the SOI-MOSFET, electron and hole pairs are generated by the impact ionization and holes accumulate in the body region, causing unexpected device behavior, which is known as the floating body effect (see figure 4). This effect greatly affects device characteristics and is modeled in HiSIM-SOI by adding the accumulated holes to the Poisson equation consistently within the iterative calculation. The accumulated holes are proportional to the impact ionization current I_{sub} [11]. Calculation results of the hole concentration are plotted in figure 5 as a function of the drain voltage in comparison with 2D-device simulation results. The acquired potential distribution along the depth direction at the source end is shown in figure 6. Due to the accumulated holes, the potential in the silicon layer is enhanced especially for high drain bias conditions where the impact ionization becomes strong.

Calculated drain currents, using the potential distributions with the accumulated holes in the SOI region (see figure 6), are shown in figure 7. Accurate determination of accumulated holes in the silicon body and consistent solution of the Poisson equation are very important to capture the floating body effect in SOI MOSFETs.

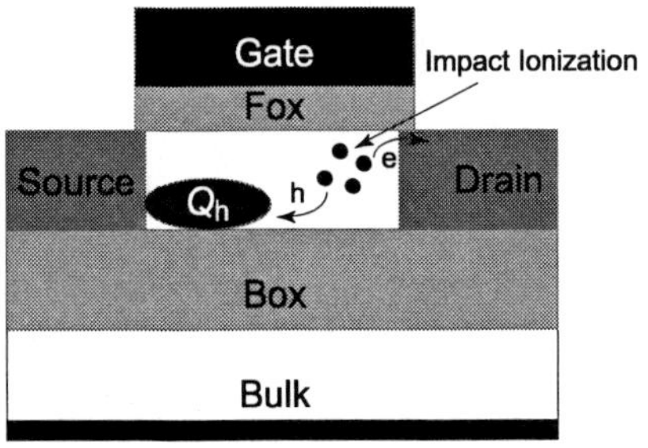

Figure 4: Schematic of the SOI MOSFET with accumulated holes due to impact ionization.

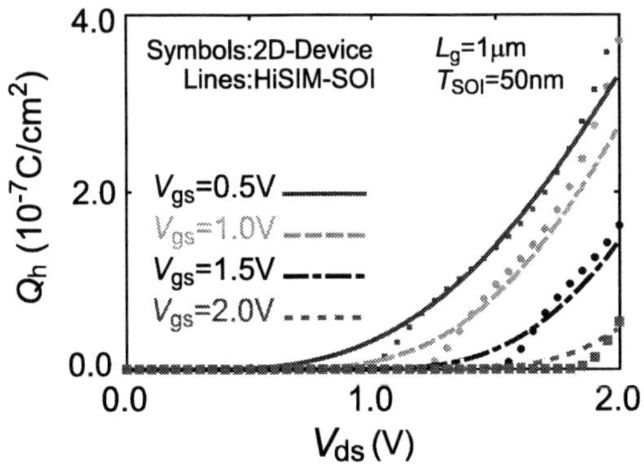

Figure 5: Comparison of the accumulated hole concentration as a function of the drain voltage as calculated by HiSIM-SOI and 2D-Device simulation for different gate voltages.

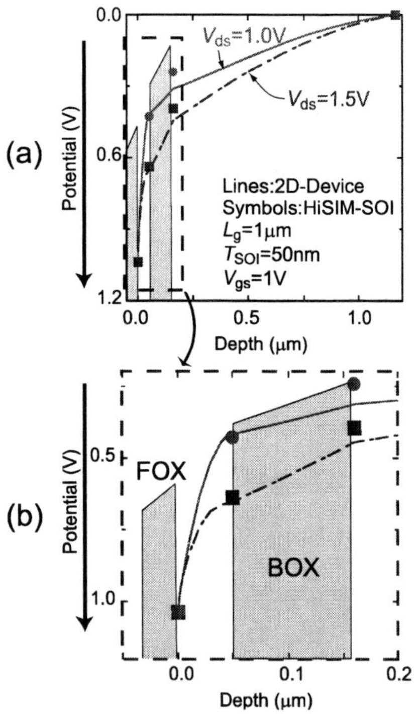

Figure 6: (a) Calculation result of the potential distribution along the depth direction considering the floating body effect for different drain biases and (b) its close up of the SOI region.

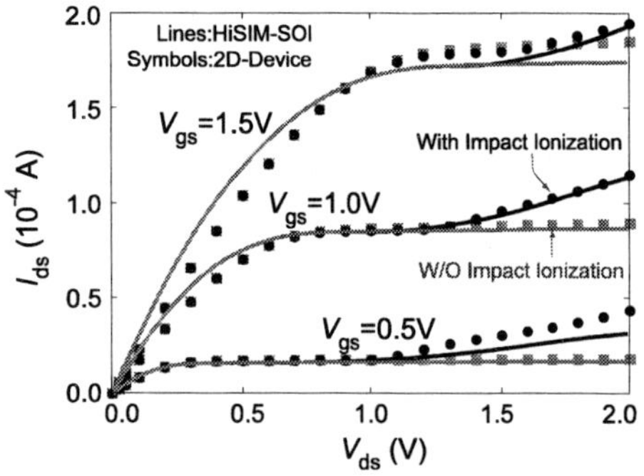

Figure 7: Calculated drain current characteristics of HiSIM SOI including the accumulated holes in the SOI region as compared with 2D-Device simulation results for L_g=1um. Calculation result without considering the impact ionization is also plotted for comparison

5 MODELING OF THE DG MOSFETS

The DG-MOSFET has a thin silicon body with another gate electrode, which enables improved channel control and results in ideal short channel effect immunity. Similar to SOI-MOSFETs, the potential value in the middle of the silicon layer is not fixed by any electrodes, as shown in figure 8 by 2D-device simulation, which makes the modeling of the device difficult. In our approach, to overcome this difficulty, the potential calculation process is divided into 2 parts. First, we solve the potential value at the center of the channel. After that, with the obtained fixed value for a "quasi" body potential, a surface potential calculation can be performed by solving the Poisson equation iteratively. In this solution example the symmetric DG MOSFET is considered, with the same thickness of the 2 gate oxides and the same voltage applied to the 2 gates.

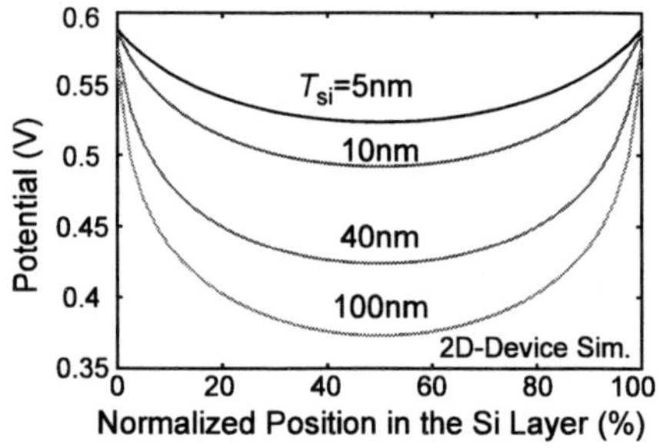

Figure 8: Potential distribution along the normalized vertical direction to the channel calculated by a 2D-device simulator. Increase of the floating node potential at the center of the channel is observed as T_{si} reduces.

Figure 9 shows the calculation result of the floating-body potential at the center, as compared with 2D-device simulation results, for different silicon layer thicknesses from 100nm to 5nm. Good agreements are achieved for this wide range of silicon layer thicknesses. Once the floating potential value is fixed, then the surface potential can be evaluated at both gates. The results are shown in figure 10 a-b as a function of gate voltage. The resulting drain currents are plotted in figure 11. Good accuracy of the potentials is of course the requirement for accurate circuit simulation.

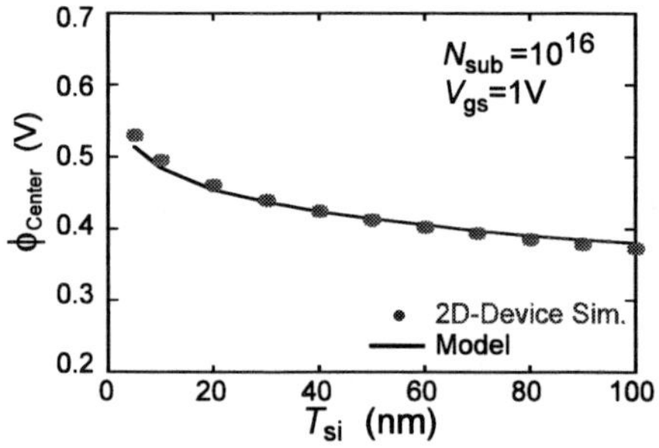

Figure 9: Comparison of the calculated floating potential value at the center of the silicon layer for different silicon layer thicknesses by the developed model and 2D-device simulation.

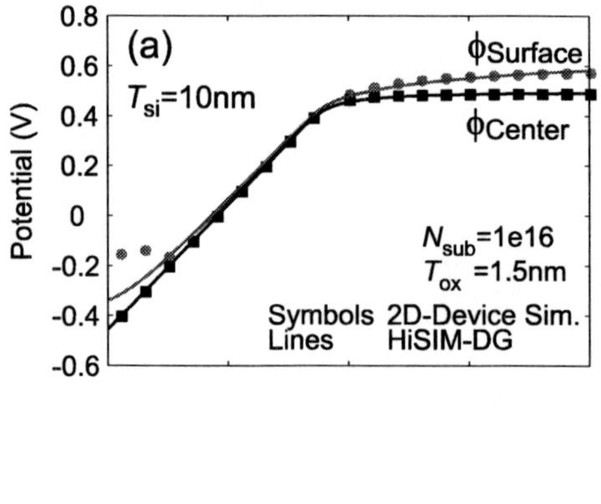

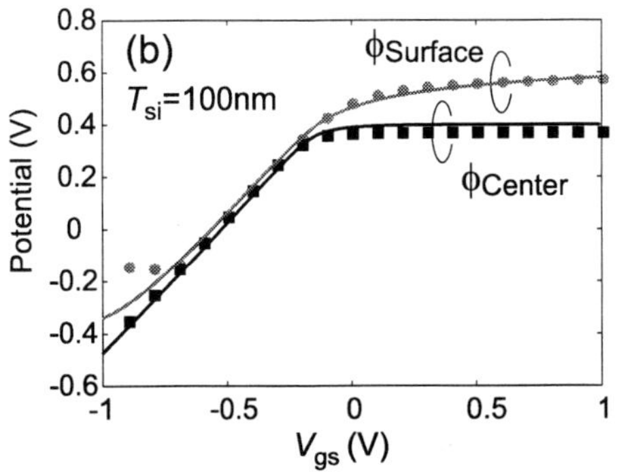

Figure 10: Comparison of potential value at the center of the silicon layer and at source-end surface in the channel as a function of the applied gate voltages by the developed model and 2D-device simulation for (a) T_{si}=10nm and (b) 100nm.

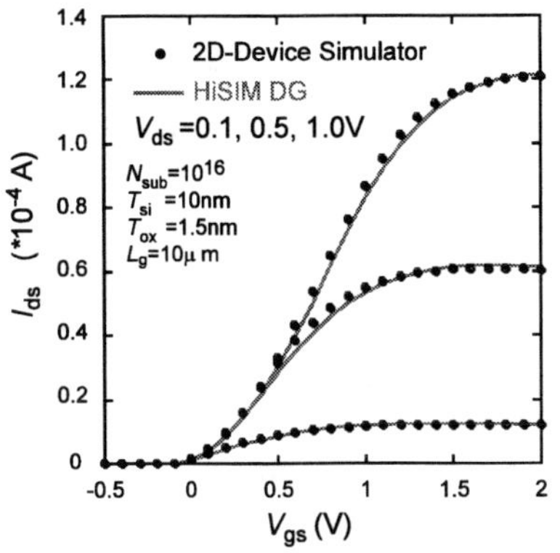

Figure 11: Comparison of calculated current voltage characteristics by HiSIM-DG and 2D-device simulation for channel impurity concentration of 10^{16} cm^{-3} and T_{si}=10nm, T_{ox}=1.5nm and L_g=10μm.

CONCLUSION

Modeling approaches for floating body devices are presented based on complete and consistent potential descriptions, and are applied to the SOI MOSFET and the double gate MOSFET. An accurate modeling of the floating potential is the key and the basis for the whole device model, because it critically affects all device characteristics.

ACKNOWLEDGEMENT

The author would like to thank Semiconductor Technology Academic Research Center (STARC) and New Energy and Industrial Technology Development Organization (NEDO) for their funding.

REFERENCES

[1] Y. Bin, C. Leland, S. Ahmed, W. Haihong, S. Bell, Y. Chih-Yuh, C. Tabery, H. Chau, X. Qi, K. Tsu-Jae, J. Bokor, H. Chenming, L. Ming-Ren and D. Kyser, "FinFET scaling to 10 nm gate length," IEDM Tech. Dig. Dec. 2002, pp. 251 – 254.

[2] Yongxun Liu, S. Kijima, E. Sugimata, M. Masahara, K. Endo, T. Matsukawa, K. Ishii, K. Sakamoto, T. Sekigawa, H. Yamauchi, Y. Takanashi, E. Suzuki, "Investigation of the TiN Gate Electrode With Tunable Work Function and Its Application for FinFET Fabrication," IEEE Trans. Nanotechnology, vol. 5, Issue 6, Nov. 2006 pp.723-730

[3] Lu, Darsen D. Dunga, Mohan V. Lin, Chung-Hsun Niknejad, Ali M, Hu, Chenming, "A Multi-Gate MOSFET Compact Model Featuring Independent-Gate Operation," IEDM Tech. Dig. Dec. 2007, pp. 565 – 568.

[4] T. Nakagawa, T. Sekigawa, M. Hioki, S. Ouchi, H. Koide, T. Tsutsumi, "Charge-Based Capacitance Model for Four-terminal DG MOSFETs," IWCM Tech. Dig. Jan. 2007, pp. 45 – 49.

[5] Huaxin Lu and Yuan Taur, "An Analytical Potential Model for Symmetric and Asymmetric DG MOSFETs," IEEE Trans. Electron Devices, vol. 53, Issue. 5, May. 2006, pp. 1161-1168.

[6] Lixin Ge and Jerry G. Fossum, "Analytical Modeling of Quantization and Volume Inversion in Thin Si-Film DG MOSFETs," IEEE Trans. Electron Devices, vol. 49, Issue. 2, Feb. 2002, pp. 287-294.

[7] M. Miura-Mattausch, H. Ueno, M. Tanaka, H.J. Mattausch, S. Kumashiro, T. Yamaguchi, K. Yamashita, N. Nakayama, "HiSIM: a MOSFET model for circuit simulation connecting circuit performance with technology," IEDM Tech. Dig. Dec. 2002, pp.109 – 112.

[8] M. Miura-Mattausch, N. Sadachika, D. Navarro, G. Suzuki, Y. Takeda, M. Miyake, T. Warabino, Y. Mizukane, R. Inagaki, T. Ezaki, H.J. Mattausch, T. Ohguro T. Iizuka, M. Taguchi, S. Kumashiro, S. Miyamoto, "HiSIM2: Advanced MOSFET Model Valid for RF Circuit Simulation," IEEE Trans. Electron Devices, vol. 53, Issue 9, Sept. 2006 pp.1994-2007

[9] D. Kitamaru, Y. Uetsuji, N. Sadachika and M. Miura-Mattausch, "Complete Surface-Potential-Based Fully-Depleted Silicon-on-Insulator Metal-Oxide-Semiconductor Field-Effect-Transistor Model for Circuit Simulation," J. Appl. Phys., Vol. 43, No. 4B, 2004, pp. 2166-2169.

[10] N. Sadachika, D. Kitamaru, Y. Uetsuji, D. Navarro, M. Mohd Yusoff, T. Ezaki, H. J. Mattausch and M. Miura-Mattausch, "Completely Surface-Potential-Based Compact Model of the Fully Depleted SOI-MOSFET Including Short-Channel Effects," IEEE Trans. Electron Devices, vol. 53, no.9, Sep. 2006, pp. 2017-2024.

[11] T. Murakami, M. Ando, N. Sadachika, T. Yoshida, and M. Miura-Mattausch, "Modeling of Floating-Body Effect in SOI-MOSFET with Complete Surface-Potential Description," JJAP, Vol. 47, No. 48, 2008.

The Driftless Electromigration Theory*,**
(Diffusion-Generation-Recombination-Trapping)

Chih-Tang Sah[1,2,3,+] and Bin B. Jie[2,+]

Abstract

Electromigration (EM) is the transport of atoms and ions in metals at high electrical current-density (>100kA/cm^2) leaving behind voids. It was delineated in 1961 by Huntington [1] in gold wire, and empirically modeled by the 1969-Black formula [2] to fit the Time-To-Failure (TTF) experimental data of metal interconnect lines in integrated circuits with power-law dependences of electron-current density and sample temperature, and a thermal activation energy, $TTF = AJ^{-\beta}T^{\gamma}\exp(-E_{\alpha}/k_BT)$. Tan and Roy recently reviewed the 40-year applications [3]. Since the first Landauer theoretical analysis in 1957 [4], theorists have attempted for 50 years to derive the Black formula by trying to justify the force of the electrons to move an atom, known as electron-wind. Landauer concluded in 1989 [5] that electron wind is untenable even at the most fundamental and complete many-body quantum transport theory. Sah showed in his 1996 homework solution manual for undergraduate device core-course [6] that the Black formula can be derived for a generic void model using the simple classical macroscopic transport theory, including diffusion and the often if not always neglected generation-recombination-trapping (DGRT) of the ions, without the empirical electron-wind force. We review this driftless model in this presentation.

---

* Project supported by CTSAH Associates (CTSA) which was founded by the late Linda Su-Nan Chang Sah.

** This theory was first described as the solution of a home work problem in 1996 by Sah [6].

(1) University of Florida, Gainesville, Florida, USA. Permanent address.

(2) Peking University, China.

(3) Chinese Academy of Sciences, Foreign Member, Beijing, China.

+ Email addresses, Tom_Sah@msn.com and BB_Jie@msn.com.

References

[1] R. B. Huntington and A. R. Grone, "Current-Induced Marker Motion in Gold Wires," J. Phys. Chem. Solids, 20(1/2), 76-87, January 1961, and references cited.

[2] James R. Black, "Electromigration – A Brief Survey and Some Recent Results," IEEE Trans. ED-16(4), 338-347, April 1969, and references cited.

[3] Cher Ming Tan and Arijit Roy, "Electromigration in ULSI Interconnects," Materials Science and Engineering: R: Reports, 58(1-2), 1-76, October 2007. 327 references cited and analyzed. *Electromigration in ULSI Interconnects*, World Scientific Publishing Company, Winter 2008.

[4] Rolf Landauer, "Spatial Variation of Currents and Fields Due to Localized Scatterers in Metallic Conduction," IBM Journal of Research and Development, 1, 223-231, July 1957, and references cited.

[5] Rolf Landauer, "Comment on Lodder's 'Exact' Electromigration Theory," Solid State Communication, 72(9), 867-868, 2 October 1989, and cited references of earlier Landauer papers and others.

[6] Chih-Tang Sah, *Fundamentals of Solid-State Electronics, Solution Manual*, problem P.941.1 and analytical solutions on pages 174-176, World Scientific Publishing Company, Singapore, 1996.

Adaptable Simulator-independent HiSIM2.4 Extractor

Thomas Gneiting[*], Takashi Eguchi[**] and Wladek Grabinski[***]

[*]AdMOS GmbH Advanced Modeling Solutions, Germany
thomas.gneiting@admos.de
[**]Agilent Technologies, Japan, takashi_eguchi@agilent.com
[***]GMC Suisse, wladek@grabinski.ch

ABSTRACT

This paper presents a method and its software implementation to extract Spice parameters of the HiSIM2.4 (Hiroshima-university STARC IGFET Model) surface-potential-based model [1, 3]. The completed flow of dedicated parameter extraction procedures is currently designed for the HiSIM2.4 model and can be a potential base to cover the upcoming HiSIM-LDMOS model which was recently selected by the CMC as a new standard for High Voltage MOSFET devices. A unique feature of this approach is that the underlying measured data base is collected independently and can be accessed to generate any standard MOSFET model. Other models like PSP, BSIM3/4 or EKV3 [10] can be generated within the same framework from the common measurement database [4-9]. The parameter extraction routines are based on highly efficient direct extractions, taking into account all HiSIM2.4 modeled effects important for advanced CMOS devices with extremely reduced device feature sizes in the process nodes 65/45nm and beyond. Moreover, the set of local optimizations and interactive tuners is available for all standard IV, CV curves and S-par characteristics as well as special PCM like diagrams to guaranty a high level of the simulation model scalability. Besides the basic DC and CV parameters, high frequency effects are taken into account to enable the usage of the simulation model in analog/RF designs as well. The presented software package is competed with a reporting module for effective results analysis using graphs and data visualization. Furthermore, the procedures are highly automated to assure the accuracy and quality of released Spice-level models.

Keywords: parameter extraction, compact modeling, PSP, HiSIM, EKV, LDMOS

1 INTRODUCTION

Simulations of advanced CMOS circuits, in particular, for analog/RF applications require technology specific libraries with Spice model parameters. Existing typical procedures determine parameters in sequence and neglect the interactions between target parameters and, as a result, the fit of the model to measured data may be less than optimum. Moreover, the parameters are extracted in relation to a specific device and, consequently, they correspond to different device sizes. The extraction procedures are also generally specific to a particular model, and considerable work is required to change or improve these models.

This paper describes the application of a general-purpose method and its software implementation allowing to obtain a complete set of parameters for any arbitrary model. This extraction method is implemented in the IC-CAP, measurement and parameter extraction software. The HiSIM 2.4 model has been selected to illustrate the method and its flow. After a brief review of the HiSIM 2.4 model, section 2 discusses the measurement data base architecture, section 3 outlines the HiSIM specific extraction flow and presents the results obtained by the proposed extraction method. Section 4 summarizes the conclusions derived in this investigation.

1.1 HiSIM 2.4 Model

HiSIM (Hiroshima-university STARC IGFET Model) is a surface-potential-based MOSFET model developed by Hiroshima University and the STARC organization [1, 3]. Using the charge sheet approximation and the gradual-channel approximation, all device characteristics are described analytically by channel-surface potentials at the source side and at the drain side. The surface potentials are obtained by solving the Poisson equation with the Gauss law using an iterative algorithm. The HiSIM model solves the Poisson equation iteratively without introducing any assumptions, similar to the 2-D numerical simulator and accurately yields the resulting characteristics. All phenomena such as advanced mobility model, short-channel and reverse-short-channel effects are included in the surface potential calculations. The advanced technologies accompanied by aggressive downscaling of device sizes cause various phenomena such as short-channel effects, namely: short-channel and reverse short-channel effect are consisting in the HiSIM 2.4 model. All the observed short channel effects are caused by lateral electric field contributions in the MOSFET channel. The reverse short-channel effect, mainly associated with the pocket implant technology, causes the nonuniform impurity concentration along the channel. The model parameters describing the extension of the pocket are included in HiSIM. Therefore, an important advantage of the HiSIM 2.4 modeling approach is that detailed information about the fabrication technology allows to precisely characterize device variations. Moreover, the short-channel and reverse short-channel effects are integrated as the threshold voltage shift together with polydepletion and narrow width effects. The HiSIM model also preserves technology independent mobility universality conditions with the low field mobility described by three independent mechanisms of coulomb, phonon, and surface roughness scattering.

1.2 HiSIM LDMOS Extension

The HiSIM-LDMOS model preserves all the features of the bulk-MOSFET HiSIM model with extensions specific to the modeling of the drift region [2, 11]. The potential distribution along the LDMOS extension region is described by solving the Poisson equation iteratively, including the resistance effect of the drift region. LDMOS self-heating effect simulations are possible by applying an internal thermal network.

2 DATABASE ARCHITECTURE

A unified environment was developed to enable the generation of several MOS model types based on a common data base of measured data. The basic idea of this concept is described in [9].

2.1 Advanced data representation

To enable an effective parameter extraction it is critical to prepare the measured device data in such a way that a certain effect, e.g. an effect related to channel length or a temperature effect, can be clearly identified in the data set. However, typically the classical MOS related curves (e.g. id vs. vg or id vs. vd) are measured from a huge amount of devices. Therefore, it is necessary to generate the above mentioned data sets with special data base operations [4-9]. An example of such curves can be seen in the Fig. 1 and 2.

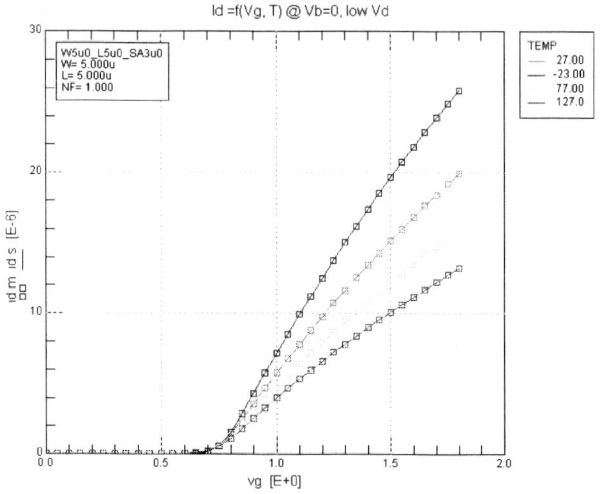

Figure 1: The id vs. vg plots at different temperatures

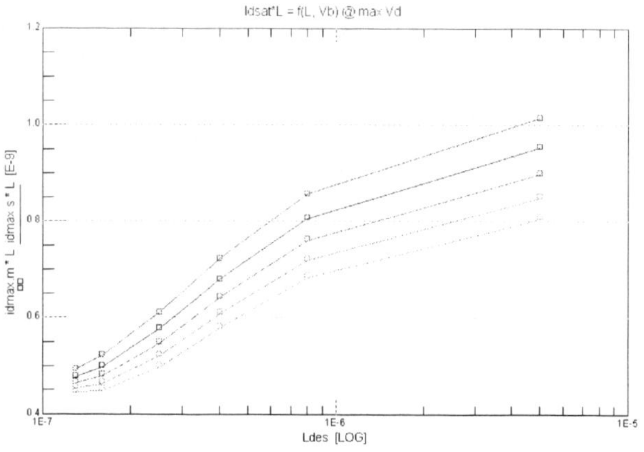

Figure 2: On current I_{on} at different gate lengths

2.2 Data Flow

The following sketch in Figure 3 shows the principal data base operations which are necessary to generate the id vs. vg diagram at different temperatures, given in Figure 1. First, we have a collection of id-vg and id-vd curves for different devices at different temperatures. This is the set of raw data typically measured in the lab. We put all this data into a database and can now derive subsets of the data with a reduced amount of operating conditions (vg, vd, etc.) over a range of temperatures and devices geometries.

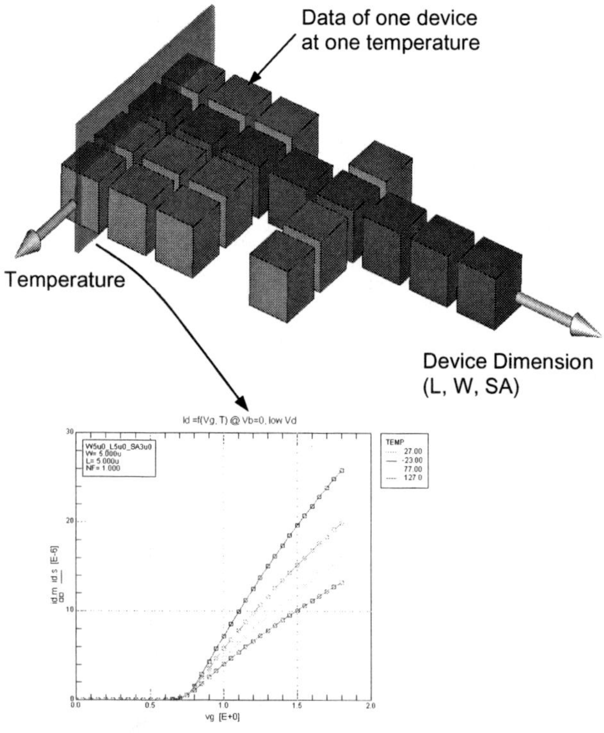

Figure 3: Data base operation to derive an id vs. vg diagram at different temperatures

3 HISIM 2.4 SPECIFIC EXTRACTIONS

3.1 Requirements for Parameter Extractions

The complexity of state of the art CMOS processes requires a very dedicated strategy in extracting model parameters. Due to the large diversity of process flavors e.g. having low threshold voltage (Vth), high Vth and native Vth devices, the amount of simulation models to cover those flavors is steadily increasing. Therefore, the requirements for an effective model parameter extraction program are the following:

- Extractions should be adoptable to different process generations from a minimum feature size of 0.18um down to 32nm
- Extraction functions for certain parameters should be intelligent in such a way that they can identify reasonable regions from given curves (like id vs. vg at low vd) to extract the desired parameters
- The complete extraction procedure should be repeatable and should not depend on the operator.
- Finally, it should be possible to automate the extraction process to enhance productivity in the modeling groups.

We will discuss these requirements in the following sub chapters and show how we solved those issues.

3.2 Special Emphasis on Scalability and Analog Behavior

It is desirable for selected applications (digital, analog/mixed-mode, RF and high voltage) to apply only one set of model parameters for each of the currently used transistors (i.e. the n- and p-channel MOS devices) because circuit simulations can then be performed by only specifying the type and dimensions of the transistors.

The MOS model extraction procedures include steps of the DC and AC parameterization to extract the final model. Model extraction flow is divided into geometrical (in the entire range of L's and W's) and bias regions (sub-threshold, linear and saturation) for different device types applying recommended DC/AC sweeps. The sweeps could be application-specific for CMOS processes targeting digital applications only (i.e. drain current, intrinsic capacitances, leakage); analog/mixed-mode technologies (trans-conductances and output conductances Gm/Gds, higher order derivatives, etc.); RF (maximum operating frequencies, noise performance, etc.); and high voltage (Ids/Vdsmax, Ids/Vgsmax at operation bias conditions).

Finally, both DC and AC characteristics of the extracted model for the available W/L geometry range is used to extract and verify device temperature scaling at least at three different temperatures (e.g., -40°C, 27°C and 125°C for automotive applications) and subsequently repeated for the linear and saturation operation regions.

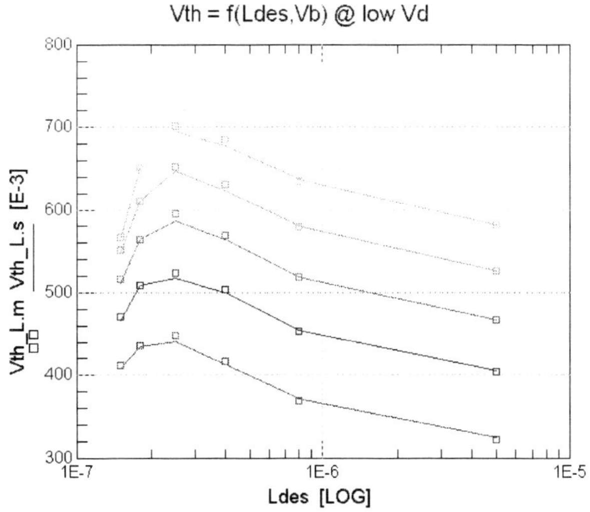

Figure 4: Example of Vth(L) scaling

Extraction of the extrinsic model parameters (i.e. junction and overlap capacitances) or specific RF model extensions are an integral part of the extractor and are available as dedicated modules.

3.3 Direct Extractions, Optimization and Tuner Routines

As the HiSIM device characteristics are strongly dependent on basic device parameter values (i.e. gate dielectric parameters, coefficient of gate length and width modification) it is recommended to start with initial parameter values (not changing during extraction procedure) according to the reference documentation [1, 3].

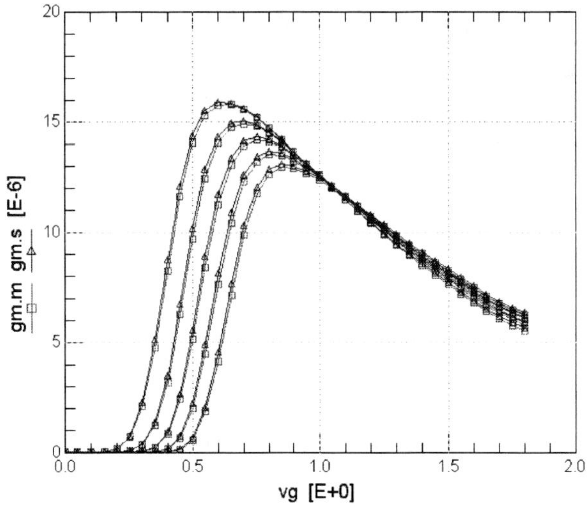

Figure 5: The gm curves used for mobility modeling

Basically, the most important extraction of the HiSIM surface potential-based model are its intrinsic device parameters such as gate oxide thickness (TOX) as well as channel and substrate doping concentrations (NSUBC,

NSUBP), since they mostly determine the device features [3]. Within the extraction package, these parameters can be extracted from measured data or can be set as initial values, if information about device parameters is available before starting extraction. The first step of IV extractions is VFBC and NSUBC parameter determination using back bias dependent Id-Vgs data of the large (long-wide channel) device. VFBC is directly extracted using the relation

Table 1: Main HiSIM Extraction Sequence

Step	Target Data	Extractions
(a) rough extraction around Vth or lower		
1	Large IdVg	VFBC, NSUBC
2	L_Scale IdVg	LP, NSUBP
3		SCP1, (SCP2), SCP3
4		SC1, (SC2), SC3
		PARL2
(b) rough extraction around Vth or higher		
5	Large IdVg	MUECB0, MUECB1
		MUEPH1, MUESR1
6	L_Scale IdVg	XLD
7	W_Scale IdVg	XWD
(c) rough extraction with IdVd in the lin & sat regions		
8	Short IdVd, IdVg	XLD, RS, (RD)
9	Short IdVd	VMAX
		VOVER, VOVERP
10		CLM1, CLM2, CLM3
		CLM5, CLM6
11	Large IdVd	MUEPH1, MUESR1
(d) rough extraction around Vth or higher		
12	Short IdVg	MUEPHL, MUEPLP
		MUESRL, MUESLP
13	W_Scale IdVg	WFC, WVTH0
14	Narrow IdVg	MUEPHW, MUEPWP
		MUESRW, MUESWP
(e) fine extraction with IdVg in the subthreshold regions		
(f) fine extraction with IdVd in the lin & sat regions		

between threshold and flat-band voltage taking the surface potential effect into account. These two parameters basically model the Vbs-dependent drain current of the large device, especially around threshold voltage or at lower Vgs. The mobility model has an impact on the current behavior in the region of Vgs above threshold voltage, where current flows through the channel. Within the extraction package, mobility parameters are initially extracted using the calculated effective mobility curve based on measured data (Fig.5). After initial extraction, these mobility parameters are optimized to fit Id and its derivatives versus Vgs characteristics. For short-channel effect (SCE), LP plays an important role to model the Vth-Lgate characteristic. LP is extracted from measured Lgate dependency of Vth data. LP and NSUBP together with SCPn (n=1 to 3) are modeling the Vth increase with Lgate decrease caused by pocket implant technology. SCn (n=1 to 3) models the Vth decrease with Lgate decrease. These parameters are tuned to fit the measured Vth-Lgate characteristic. Note that SCP2 and SC2 are modeling Vds dependency of the SCE, so other Vds related parameters, such as high-field mobility and channel-length modulation (CLM), are to be tuned together with those two parameters. The main extraction sequence is briefly summarized in Table 1 for the core IV measurements.

4 CONCLUSION

In this paper, a new method and its software implementation has been presented to extract parameters of a compact MOSFET model. In our case, dedicated direct extraction routines as well as optimization and parameter tuning steps have been applied. The approach can be easily generalized to extract multiple features at all available characteristics based on measurement data. The interactive extraction flow can be automated, allowing user independent processes to enhance the productivity.

Although the HiSIM 2.4 model was used to demonstrate the results, any other compact model, i.e. the HiSIM-LDMOS model can be easily implemented in a similar fashion.

REFERENCES

[1] Hiroshima University & STARC, "HiSIM 2.4.0 User's Manual", 2007

[2] Hiroshima University & STARC, "HiSIM-LDMOS User's Manual", 2007

[3] M. Miura-Mattausch, N. Sadachika, D. Navarro, G. Suzuki, Y. Takeda, M. Miyake, T. Warabino, Y. Mizukane, R. Inagaki, T. Ezaki, H.J. Mattausch, T. Ohguro, T. Iizuka, M. Taguchi, S. Kumashiro and S. Miyamoto, "HiSIM2: Advanced MOSFET Model Valid for RF Circuit Simulation", IEEE Transactions On Electron Devices, Vol. 53, No. 9, September 2006

[4] T. Gneiting, "BSIM4, BSIM3v3 and BSIMSOI RF MOS Modeling", RF Modeling and Measurement Workshop, European Microwave Week, Paris 2000

[5] T. Gneiting, "BSIM4 Modeling", Arbeitskreis MOS Modelle, München, 2001

[6] R. Friedrich, T. Gneiting, "BSIM4 Model Parameter Extraction", IC-CAP Nonlinear Device Model Manual, Agilent Technologies., 2001

[7] "Characterization System for Submicron CMOS Technologies", JESSI Reports AC41 94-1 through 94-6, 1994

[8] J. Deen, T. Fjeldly, T. Gneiting, F. Sischka, "CMOS RF Modeling, Characterization and Applications – RF MOS Measurements", World Scientific, ISBN 981-02-4905-5

[9] T. Gneiting, "A Unified Environment for the Modeling of Ultra Deep Submicron MOS Transistors", Nanotech 2003

[10] W.Grabinski, B.Nauwelaers, D.Schreurs (Eds.) "Transistor Level Modeling for Analog/RF IC Design", Springer, 2006, ISBN: 1-4020-4555-7;

[11] W.Grabinski and T.Gneiting (Eds) "Power/HVMOS Devices Compact Modeling", Springer, 2008, (in print)

ADMS – A Fully Customizable Compact Model Compiler

Ben Gu and Laurent Lemaitre
{benjamin.gu,laurent.lemaitre}@freescale.com

Freescale Semiconductor

Abstract

This paper presents an overview of ADMS, a fully customizable compact model compiler. The architecture of ADMS and syntax of ADMST language are introduced. A few optimization techniques which can be employed to build an efficient customized model compiler using ADMS are briefly reviewed, and effects of these techniques were demonstrated using the PSP model.

Keywords: model compiler, compact model, model optiimzation

Introduction

Compact device models are an important element of circuit simulators like SPICE. The accuracy and performance of models to a large extent determines the quality and reliability of simulations. Conventionally, compact models are plugged into a simulator through an application programming interface (API). Using this approach, model developers must code the compact model using the data structures defined in the API and write several interface routines to communicate with the simulator [1]. This approach has proved to be very inefficient. Coding the model equations and the partial derivatives of charges and currents in a programming language can be very tedious and error prone, which makes implementing a model into a simulator often take months [2]. In addition, the approach is quite inflexible: The code developed for simulator A can't be used in simulator B without extensive changes, and any modification to device nodes, model parameters, or equations will result in substantial changes to the code.

Over the last few years, the Verilog-A language has arisen as a new standard for compact model development [3-6]. Developing and delivering models as Verilog-A modules liberates model developers from the tedium of coding equations in a programming language. Further, the error-prone writing of analytical partial derivatives, needed for solving equations using Newton iterations, is handled automatically, which accelerates the development process and improves the quality of compact models. More importantly, models developed in Verilog-A format are ready to run on any circuit simulator which supports the Verilog-A language. Because of these advantages, the Verilog-A language has been adopted by several leading compact model developers, such as the Berkeley BSIM group [7], Arizona State's PSP group [8], and so on.

On the other hand, as more and more models are developed and delivered in Verilog-A format, there comes a new challenge for simulator developers: How should they implement these models into simulators automatically and efficiently? They could just leave the model in Verilog-A, relying on the simulator's Verilog-A implementation. Often, this is an interpretive evaluation, in which the Verilog-A code describing the model equations is evaluated on the fly. This approach is not optimal for large simulations, as interpretation is slow and not competitive with compiled built-in models. Even if the simulator handles Verilog-A through compilation, there will be overhead with the more general Verilog-A language which will not exist for built-in models.

Another approach, which we take here, is a custom-compiled approach. The Verilog-A code is compiled into a common programming language, i.e. C, by some specialized software (compact model compiler), respecting the simulator's compact model API and effectively making the Verilog-A module into a built-in model. Using this approach, the computational efficiency of the models can be brought up to a level which is comparable to built-in models implemented by experienced engineers. Recently there has been significant interest in developing compilers which convert compact models defined in Verilog-A into a programming language (typically C) for implementation into simulators [9-11]. Among those compilers, the Advanced Device Model Synthesizer [12] (ADMS), an open source program developed at Freescale Semiconductor, is becoming increasingly popular. Because of its open source availability and agile architecture, it has been used by several EDA vendors and semiconductor companies for different applications. In this paper, we present an overview of the architecture of ADMS, its data structure, and the syntax of the ADMST language. We also review a few optimization techniques which can be employed to build an efficient customized model compiler using ADMS.

Architecture of ADMS

ADMS is a compact model compiler which users can customize according to their own specifications. The software strategy employed by ADMS is somewhat similar to the compiler-compiler techniques used in the GNU bison program. As illustrated in Fig.1, The ADMS software has two major components: a Verilog-A parser and an ADMST interpreter. The Verilog-A parser reads in a Verilog-A source file describing the compact model and stores information of the model in a data structure which is similar to a XML tree. The ADMST interpreter parses and executes users' definition of compilers given in a ADMS-specific format, ADMST. The ADMST language is very similar to the XSLT and XPATH languages used in XML technologies, and this language can be used to perform operations on the XML data of the model so as to build a fully customized compiler. The ADMST interpreter reads the instructions written in ADMS scripting languages and transforms the XML tree into any other form of presentation (e.g. C, Matlab, html). This software architecture allows users to build their own applications flexibly. For instance, a user can develop different sets of ADMST scripts to implement Verilog-A models into different simulators, i.e. SPICE or any commercial simulator having an exposed model API. Using ADMST scripts, ADMS can also be extended to other applications such as automatically creating test-bench circuits or documentation for models.

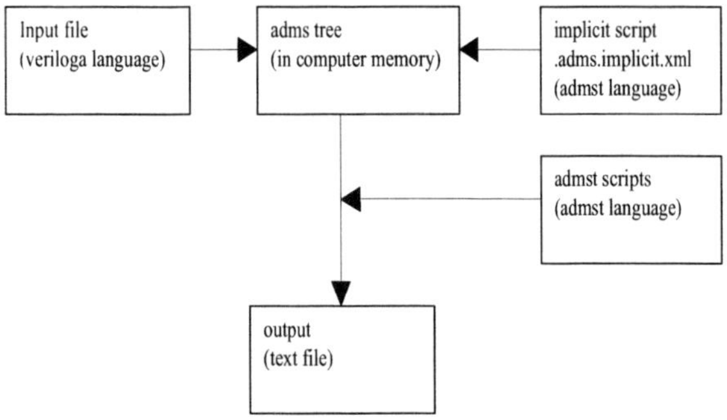

Fig. 1. Architecture of the ADMS system

ADMS Data Tree

After a Verilog-A source file is parsed, a data tree is built in memory. This XML tree is a representation of the model data stored in the Verilog-A file. Figure 2 shows a visual representation of the ADMS data tree that has been created for a simple module. The picture has been simplified; the actual XML tree has many more branches and nodes.

```
//Verilog-AMS code
module myR (p,n);
  parameter real G=1;
  analog begin
    I(p,n)<+G*V(p,n);
  end
endmodule
```

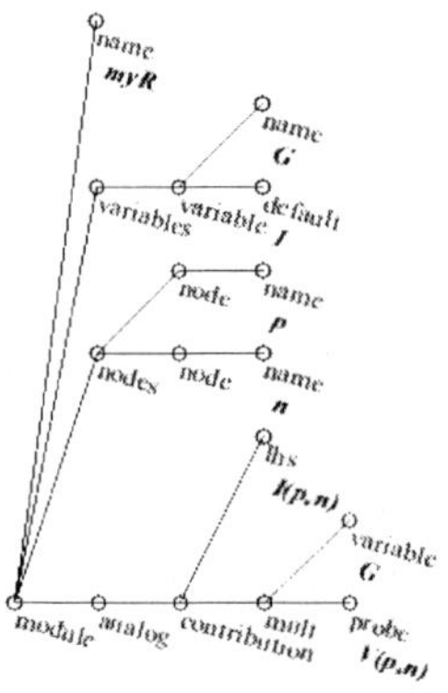

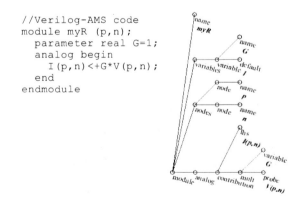

Fig. 2. ADMS Data Tree (Simplified View)

ADMST and ADMSTPATH

The language used to describe the behavior of the compiler is very similar to the XSLT and XPATH languages used in XML technologies. Missing features in the XSLT language for this particular application and the lack of available high-speed open-source XPATH interpreters required us to abandon a fully-XML based approach. We therefore defined an ADMS-specific system. However, a significant level of backward compatibility has been maintained. At some time it will be still possible to move back to the XML system with minimal effort when it is extended to handle capabilities necessary for ADMS.

The "Hello World" routine (German version) is a starting point to illustrate the syntax of the ADMST – the language tells ADMS how to transform the ADMS data tree. Figure 3 gives an implementation of a generic "Hello World" routine (lines #2 to #4).

```
1.  $ cat hello.xml
2.  <admst version="2.2.0">
3.    <admst:text format="Guten Morgen Welt\n"/>
4.  </admst>
5.  $ admsXml -e hello.xml
6.  Guten Morgen Welt
```

Fig. 3. "Hello World" in the ADMST language.

This example illustrates how close the syntax of the ADMST language is to the HTML language – a subset of the XML system. The ADMST language can be defined using following recursive definition:

```
1.  <admst:instruction attr="val">
2.    ...other admst:instructions (children like)
3.  </admst:instruction>
```

ADMST scripts are built by using the above definition recursively. The structure of ADMST scripts is a tree. Each node of the tree is an **admst:instruction**. The XML interpreter of ADMS traverses all instructions. When it enters (or leaves) an **admst:instruction** it executes some procedures. The keyword **instruction** is a placeholder. It can take one of the following self-explanatory values: *text, for-each, if, choose, template, apply-templates, open, when, join, value-of, value-to, push, reverse, message, fatal*, and so on. When the XML interpreter traverses node **adms:text** as shown in Fig. 3 it calls a routine that prints the text of line #6.

The meaning of pairs **attr="val"** is trickier. To keep it simple, each **attr="val"** pair acts as an argument to every **admst: instruction**. At line #3 attribute **format** passes to the ADMST instruction **admst:text** the actual value of the string that will be printed. An almost ubiquitous attribute is the **select** attribute. This attribute specifies the position of the XML interpreter inside the ADMS data tree. The language used to specify the location inside each **admst:instruction** is the ADMSTPATH language.

The script given in Fig. 4 illustrates the concepts introduced so far. Here is the sequence of events that will occur in the course of compilation of the script: file **myfile** is opened for writing. Then all nodes **module** of the ADMS data tree are traversed. Starting from each **module** all nodes

node are in turn traversed. Finally the attribute name of each **node** is printed in file **myfile**.

```
<admst:open file="myfile">
  <admst:for-each select="module">
    <admst:for-each select="node">
      <admst:value-of select="name">
      <admst:text format="node is: %s\n">
    </admst:for-each>
  </admst:for-each>
</admst:open>
```

Fig. 4. Simple script illustrating the ADMST syntax.

Details and a pragmatic introduction to the ADMS system can be found at the home page of ADMS [12].

Optimizing Performance of Generated Code

The ADMST language provides a framework on which users can build their own model compilers. One of the most essential requirements for a model compiler is the capability of generating computationally efficient computer code to evaluate the model described in Verilog-A format. In this section, we review a few techniques users of ADMS can employ to enhance the computational efficiency of the generated code. These techniques has been implemented in the ADMST language in the compact model compiler developed for the Freescale in-house circuit simulator, Mica, and have been used to significantly improve the computational efficiency of the the PSP model in Mica. The PSP model is a surface potential based compact model, which was chosen by the compact model council (CMC) as a next-generation standard model for MOSFET devices [5,8]. The model has always been developed and distributed in the Verilog-A format, and there has been an active discussion in CMC about the model's computational efficiency since it was selected as a standard, which makes it particularly apropos for our discussions in this paper.

Code Partitioning

A compact model typically consists of a large number of equations evaluating currents and charges of a physical device described by the model. These equations are partitioned into three sections: a process-dependent section, a geometry-dependent section and a bias-dependent section. The process-dependent section deals with the computations based only on model parameters that reflect characteristics of the fabrication process. This section is often alternatively referred to as the model initialization section and it is evaluated only once for each individual model in a given circuit during a simulation. The second section does computations involving geometry-dependent quantities, such as width, length, area, multiplicity, and so on. This section is often referred to as instance/geometry initializations and it is evaluated once for every device instance that uses a particular model in the circuit. The last section groups the computation of all quantities that depend on the electrical bias conditions. Evaluation of this block is repeated at every iteration during the simulation process. In the PSP model, the Verilog-A code has been partitioned by the model developers, however an intelligent compiler should efficiently determine an optimized partitioning of the physics-based equations that avoids overloading of the bias-dependent section. After Verilog-A code of the model is carefully partitioned, the primary candidate for code optimization is the bias-dependent partition as this part of the code will be evaluated at every iteration of the simulation.

Bias Dependency Tracking

In order to make the model work with simulators, a compact model compiler should not only generate code to evaluate all the equations presented in the Verilog-A code, but also generate code to analytically compute partial derivatives of those equations with respect to (w.r.t) the solution variables (typically the device bias voltages). Even so, many expressions in the model do not require differentiation, only those which are directly dependent on solution variables. Thus ADMST scripts should track these dependencies and generate the appropriate partial derivatives only when needed, via application of the chain rule.

Reuse of Expensive Math Computations

The quality of the partial derivative code generation is essential to the efficiency of the compiler. Below is an equation in the section evaluating mobility reduction in the Verilog-A code of the PSP model,

```
Mutmp = pow(Eeffm * MUE_i, THEMU_i) + CS_i *
        (Pm / (Pm + Dm + 1.0e-14));
```

For this equation, an unoptimized compiler will generate C code as follows:

```
Mutmp_Vs_bp
  = pow(Eeffm * MUE_i, THEMU_i) *
    MUE_i * Eeffm_Vs_bp / (Eeffm * MUE_i) +
    CS_i * (Pm_Vs_bp * (Pm + Dm + 1.0e-14) -
    Pm * (Pm_Vs_bp + Dm_Vs_bp)) /
    ((Pm+Dm + 1.0e-14) * (Pm + Dm + 1.0e-14));
Mutmp_Vd_s
  = pow(Eeffm * MUE_i, THEMU_i) *
    MUE_i * Eeffm_Vd_s / (Eeffm * MUE_i) +
    CS_i * (Pm_Vd_s * (Pm + Dm + 1.0e-14) -
    Pm * (Pm_Vd_s + Dm_Vd_s)) /
    ((Pm + Dm + 1.0e-14) * (Pm + Dm + 1.0e-14));
Mutmp_Vgp_s
  = pow(Eeffm * MUE_i, THEMU_i) *
    MUE_i * Eeffm_Vgp_s / (Eeffm * MUE_i) +
    CS_i * (Pm_Vgp_s * (Pm + Dm + 1.0e-14) -
    Pm * (Pm_Vgp_s + Dm_Vgp_s)) /
    ((Pm + Dm + 1.0e-14) * (Pm + Dm + 1.0e-14));
Mutmp = pow(Eeffm * MUE_i, THEMU_i)
  + CS_i * Pm / (Pm + Dm + 1.0e-14);
```

The first three lines of code compute the partial derivatives of Mutmp w.r.t to three branch voltages, V(S,BP), V(D,S), and V(GP,S) respectively, and the last line computes Mutmp itself. Obviously the code is far from being efficient. The first thing we notice is that evaluation of the same pow() function is repeated four times in the generated code. This will make the code unnecessarily slow given that evaluating the pow() function is expensive. So one optimization technique could be to evaluate the expensive functions first, such as log, exp, pow, sqrt, and so on, then reuse the computations in the generated code. In addition to these functions, the divide operation is also expensive. On many CPU architectures, a divide operation is as expensive as evaluating a sqrt() function, so opportunities to reuse divide operations can be exploited as well. With these techniques implemented, an optimized compiler could generate code as follows:

```
__pow_0 = pow(Eeffm * MUE_i, THEMU_i - 1.0);
__dF1_pow_0 = (THEMU_i) * __pow_0;
__pow_0 = (Eeffm * MUE_i) * __pow_0;
__dF1_div_1 = 1.0 / (Pm + Dm + 1.0e-14);
__div_1 = (Pm) * __dF1_div_1;
__dF2_div_1 = -__div_1 * __dF1_div_1;

Mutmp_Vs_bp
  = (__dF1_pow_0 * MUE_i * Eeffm_Vs_bp) +
    CS_i * (__dF1_div_1 * Pm_Vs_bp
     + __dF2_div_1 * (Pm_Vs_bp + Dm_Vs_bp));
Mutmp_Vd_s
  = (__dF1_pow_0 * MUE_i * Eeffm_Vd_s) +
    CS_i * (__dF1_div_1 * Pm_Vd_s
    + __dF2_div_1 * (Pm_Vd_s + Dm_Vd_s));
Mutmp_Vgp_s
  = (__dF1_pow_0 * MUE_i * Eeffm_Vgp_s) +
    CS_i * (__dF1_div_1 * Pm_Vgp_s
    + __dF2_div_1 * (Pm_Vgp_s + Dm_Vgp_s));
Mutmp = __pow_0 + CS_i * __div_1;
```

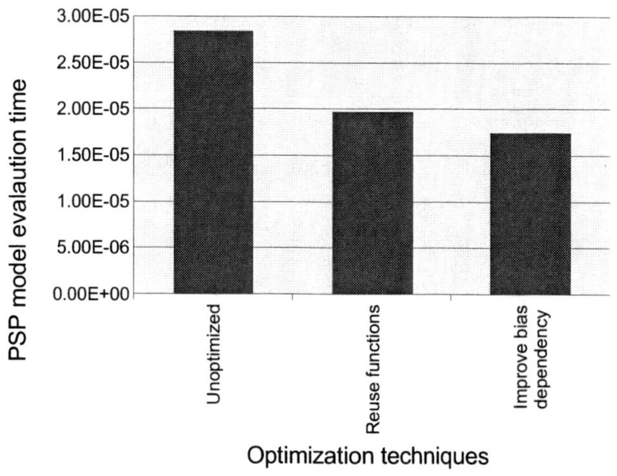

Fig. 5. Performance improvements of the PSP model

Figure 5 summarizes performance improvements of the PSP model achieved by cumulatively using the techniques we have reviewed in this section. The result presented is an average of one million evaluations of the code generated by the customized model compiler for Mica, using ADMS. The .model statement used in the experiment is from one of bulk technologies developed by Freescale, and node voltages at each evaluation are randomly generated from 0 to 3V. As shown in Fig. 5, using these techniques collectively can lead to a respectable result, speeding up the computational performance of the PSP model by nearly 1.6x.

Conclusions

This paper reviews architecture of ADMS, a fully customizable a compact compiler and syntax of ADMST language with which users define their specifications of device compilers. Several common optimization techniques to improve the computational efficiency of compilers built by ADMS are introduced by using the PSP model as an example.

Acknowledgments

Authors are sincerely grateful to Steve Hamm in Freescale for his extremely useful suggestions on improving the manuscript. Also they would like to thank Colin McAndrew and every member of Mica group in Freescale for being wonderful to work with.

References

[1] T. Quarles, "Adding Devices to SPICE3" (UCB/ERL M89/45, April 1989)
[2] K. Kundert, "Automatic Model Compilation, An idea whose time has come", http://www.designers-guide.org/Perspective/modcomp.pdf
[3] L. Lemaitre, et. al. , "Extensions to Verilog-AMS to Support Compact Device Modeling", IEEE BMAS 2003
[4] G. Coram, "How to (and how not to) write a compact model in Verilog-A", IEEE BMAS 2004
[5] Compact Modeling Council meeting minutes, http://www.geia.org/index.asp?bid=597
[6] Verilog-AMS language reference manual http://www.eda-stds.org/verilog-ams/.
[7] http://www-device.eecs.berkeley.edu/
[8] http://pspmodel.asu.edu/
[9] B. Wan, B. Hu, L. Zhou, C. Shi, "MCAST: an abstract-syntax-tree based model compiler for circuit simulation," Proc. IEEE CICC, pp. 249-252, Sept. 2003.
[10] B. Troyanovsky et al., "Analog RF Model Development With Verilog-A," MTT-S 2005
[11] L. Lemaitre et. al. IEEE CICC Proc. pp. 27-30, 2002
[12] http://sourceforge.net/projects/mot-adms/
[13] http://www.gnu.org/software/bison/manual/

Source/Drain Junction Partition in MOS Snapback Modeling for ESD Simulation

Yuanzhong (Paul) Zhou and Jean-Jacques Hajjar

Analog Devices Inc., Wilmington, MA 01887, USA
Paul.Zhou@analog.com

ABSTRACT

An enhancement to the modeling of the 'snapback' in MOS transistors for ESD simulation is presented. The new model uses industry standard models and includes all major physical effects characteristics of snapback. The MOS snapback model is enhanced by partitioning the Drain and Source junctions so that only a portion of them is included in the parasitic BJT. The comparison of simulation and measured data of a Grounded-Gate NMOS shows good agreement for both positive and negative drain voltage stresses. Simulation of the base current of the parasitic bipolar shows significant improvement as well.

Keywords: ESD, SPICE, macro model, snapback, junction partition

1 INTRODUCTION

Time-to-market and cost-to-market are increasingly important in semiconductor integrated circuit manufacturing. Electrostatic Discharge (ESD) failure is one of the major factors resulting in significant costs and time delay in new product development. This is mainly due to the fact that trial-and-error approaches currently still dominates on-chip ESD protection designs.

Predicting the ESD performance of a design prior to manufacturing, just like in regular circuit design, is very appealing. SPICE-type circuit simulation of ESD events with accurate compact models can help reduce design iterations and cost. The simulation provides useful insight on the interaction between ESD and core function circuits and helps ensure that both ESD protection and core circuitry work properly.

Describing the device operation in the ESD voltage/current space through compact models is one of the most critical challenges for ESD circuit simulations. Fig. 1 shows a typical ESD protection scheme used in integrated circuits. ESD structures are connected between I/O pins (IN or OUT) and power supply pins (VDD or VSS) for I/O protection and between VDD and VSS pins as power clamps. ESD protection circuitry/devices are designed to be transparent to the core circuit. They are typically in the off-state under normal operating conditions. They turn-on and provide a low impedance path to Ground in order to dissipate the spurious static charge resulting from an ESD event. MOS devices operating in snapback mode are widely used for ESD protection. A positive ESD stress applied to the drain of the MOS device raises its voltage to a level, *Vt1,* which causes the device to 'snapback' to a lower voltage, *VH*, subsequently creating a low impedance shunting path. Under a negative ESD stress, the forward biased MOS backgate/Drain diode provides such a low impedance path.

Devices under ESD stress operate at currents and voltages that are well beyond the intended operation of the typical devices in a submicron CMOS process technology. Standard SPICE models do not describe high current characteristics and do not address many of the physical phenomena, such as 'snapback' in MOS transistors, particular to the ESD events. Consequently Standard SPICE models cannot predict device behavior under ESD stress conditions.

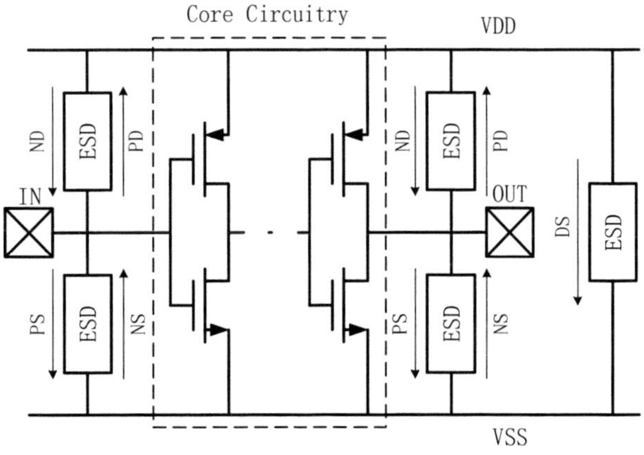

Fig. 1: Typical ESD protection scheme.

ESD simulation using compact models that describe snapback phenomenon in MOS devices has been reported by several groups [1-6]. The models have achieved success in describing the snapback behavior of MOS devices. Among them, the device model originally developed in [5] provides a practical approach in which all major physical effects for snapback phenomenon are intrinsically included and can be used to accurately simulate the high current and high voltage characteristics under an ESD event. However, device characteristic under a negative ESD stress has not been addressed at all or not enough in those models, which may lead to incorrect results in circuit simulation.

In this paper, the device model that has been developed to accurately simulate snapback behavior of MOS devices is to be reviewed. Subsequently Source/Drain junction partition in MOS snapback modeling will be investigated and a new enhancement on snapback modeling based on the partition will be presented.

2 SNAPBACK MODELING

2.1 Snapback in MOS

Snapback in a MOS device is due to the turn on of the parasitic lateral bipolar transistor (BJT) composed of the Drain/Body/Source of the MOS transistor. The BJT is triggered by the substrate current *Isub*. When the drain voltage *Vds* reaches the snapback trigger voltage *Vt1*, *Isub* becomes high enough to cause a voltage drop across the substrate resistance (*Vbe* for the BJT) that turns on the parasitic BJT. Consequently, *Vds* drops and remains at a lower holding voltage. The device behavior is determined by the MOSFET before snapback and the parasitic BJT dominates after snapback.

The substrate current results mainly from impact ionization in the Drain/Body junction of the MOS (or the avalanche in the Base/Collector junction). It is a function of *Vds* and the gate voltage *Vgs* before snapback and is independent of *Vgs* after snapback.

Besides the avalanche current, gate induced drain leakage (GIDL) also contributes to the MOS substrate current. GIDL is caused by band to band tunneling in the drain region underneath the gate and can be the dominant component when *Vgs*=0 in advanced technologies.

The displacement current, defined as the rate at which the charge is dissipated in the drain (i.e. *dV/dt),* also plays a significant role. Another important transient phenomenon is the finite time required for the BJT to turn on. It is often the limiting factor for the turn on speed of ESD protection circuit which is very important for protection performance.

2.2 Basic Snapback Models

A MOS compact model for ESD simulation is an extension of the standard MOS model with three additional components: a BJT, a current source and a resistor (Fig. 2). The BJT is for the parasitic bipolar transistor and is typically modeled by Ebers-Moll (EM) or Gummel-Poon (GP) equations.

The current source is for the substrate current. Its main contributor is the avalanche current which is given by

$$Igen=(M-1)\cdot(Ids+Ic) \qquad (1)$$

or

$$Igen=(M_{MOS}-1)\cdot Ids+(M_{BJT}-1)\cdot I_C \qquad (2)$$

to account for the independence of the current from the gate voltage after snapback [3]. *Ids* is the MOS surface drain current, and *Ic* is the BJT collector current. M, M_{MOS} and M_{BJT} are for the multiplication factor, which is described by the "Miller formula" [7]. In compact models, *M* is often implemented in exponential form:

$$\begin{aligned} M &= \exp\left[k1(Vd - Vdsat - d1)\right] \\ &\quad + \exp\left[k2(Vd - Vdsat - d2)\right] \end{aligned} \qquad (3)$$

where *k1, k2, d1* and *d2* are fitting parameters.

The model implementation includes several different approaches, namely dedicated circuit simulator models, behavior languages such as Verilog-A, and subcircuits containing complicated voltage controlled current sources, either as a behavioral language module or as a SPICE component. Each implementation method has its limitations attributed mostly to the complexity of the explicit current source.

2.3 Macromodel using Standard Devices

A macro model, shown in Fig. 3, that uses advanced industry standard models, eliminates the need to derive a custom model as in [1-4]. In this new approach the explicit current source is removed since all key effects in MOS snapback modeling are intrinsically included in the MOS and BJT compact models. The avalanche and GIDL currents are built in the MOS and BJT models. The displacement current or *dV/dt* effect is modeled by the Collector/Base junction capacitance in the BJT. The transit time of the BJT and separate *M* for the MOS and the BJT are also included.

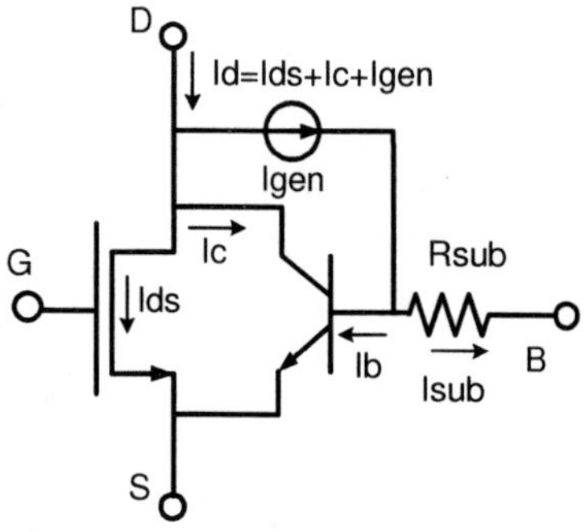

Fig. 2: A typical snapback MOS model

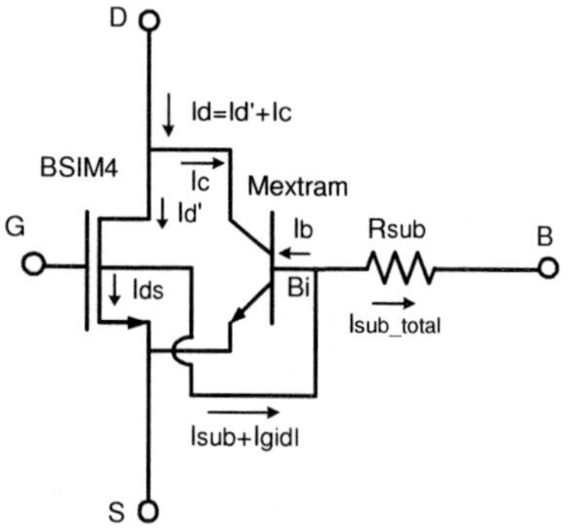

Fig. 3: A new snapback macro model

3 MODEL WITH S/D JUNCTION PARTITION

3.1 Model Discrepancies

The macro model shown in Fig. 3 describes the snapback behavior of MOS devices very well. However, it fails to model accurately the impedance for negative *Vds*. To model snapback correctly for a typical ESD NMOS protection device, the resistance value of the substrate resistor in the equivalent circuit shown in Fig. 2 and Fig. 3 must be in the range of several hundred Ohms. But the typical resistance of the Drain/Body diode is only a few Ohms when it is measured with the junction forward biased.

On the other hand, we have also found discrepancy in the value of the Base/Collector junction capacitance. The capacitance has significant impact on trigger voltage *Vt1* due to the effect of *dV/dt* [8]. The capacitance value obtained from the dependence of *Vt1* on pulse rise time is smaller than the value directly measured from the whole Drain/Body junction.

Snapback holding voltage *VH* depends on various factors. The current gain (beta) of the BJT is the most significant. It has been found that the beta extracted from a Gummel plot needs to be adjusted higher to fit the measured *VH*. This means that the base current of the BJT for correct *VH* modeling needs to be smaller than the measured current from the substrate terminal. Moreover, the base current also indicates much higher resistance in simulation than in measurement.

It is believed that only a portion of the Drain/Body and Source/Body junctions contribute to the parasitic BJT [1]. This could be the direct cause of the above discrepancies.

3.2 Enhanced Model with Junction Partition

To account for their partial contribution to the parasitic BJT, the Drain/Body and the Source/Body junctions need to be partitioned in the model [6]. Two diodes are added to the equivalent circuit in Fig. 3 to implement the partition (Fig.4). The new macro model for NMOS has five basic components: a MOS transistor for the main device, a three terminal parasitic BJT transistor, a resistor for the substrate resistance, and two diodes. The MOS is modeled by BSIM4 and the BJT by a proprietary bipolar model that is very similar to Mextram.

In a typical ESD protection configuration the Body and Source terminals are shorted together. Therefore, when *Vds* is greater than zero, the impact of the two diodes is negligible. When *Vds* is negative, the diode between the drain and the substrate terminals becomes dominant.

The total substrate current for snapback modeling, which is equivalent to *Igen* in Fig. 2, in the new equivalent circuit is

$$Igen = Isub + Igidl + Iavl + Idio1 \qquad (4)$$

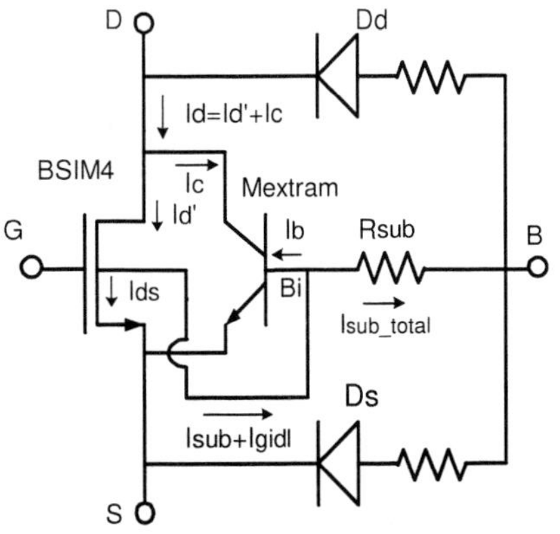

Fig. 4: Macro model with Drain\Source junction partition.

Isub is the impact ionization part of the substrate current in the MOS component, which is an exponential function of (*Vds-Vdseff*) and depends on *Vgs* and *Leff*. The GIDL current *Igidl* depends on *Vds*, *Vgs* and *Vbs*. BSIM4 expresses it as an exponential function of (*Vds-Vgs*) [9].

The avalanche current of the Collector/Base junction in advanced BJT models (*Iave*) is also an exponential function that depends on the voltage drop over the junction [10]. *Iave* is independent of the MOS gate voltage *Vgs*.

Idio1 is the current through the diode Dd. When *Vds*>0 it is negligible and *Igen* is basically the same as the one without the diodes. When *Vds*<0, *Idio1* dominates total *Ids* current.

When the device is measured in a forward Gummel configuration, the total body terminal current or the base current is

$$Ib = Ib' + Idio2 \qquad (4)$$

where *Ib'* is the base current from the BJT and *Idio2* is from the forward biased Source/Body diode Ds.

4 SIMULATION AND DISCUSSION

The simulation data for the model has been correlated with measurement data. The MOS model and the BJT model were first extracted from IV curves in normal operating region using standard model extraction methodology. Then the substrate resistance R_{sub}, the collector capacitance CJC and the current gain BF in the BJT model were determined from the snapback characteristics. In the final step, the parameters in the diodes for junction partition were obtained from the IV curve for Vds<0 and the total drain junction capacitance.

The snapback were measured using transmission line pulse (TLP) technique with a Barth Model 4002 TLP tester. The TLP pulse width used was 100ns and the rise time was varied between 200ps, 2ns and 10ns. Transient simulations were carried out using voltage pulse sequences as the input. The *Va* and *Ia* values in snapback curves were extracted from the region of I, V pulse waveforms where the current and voltage were stable.

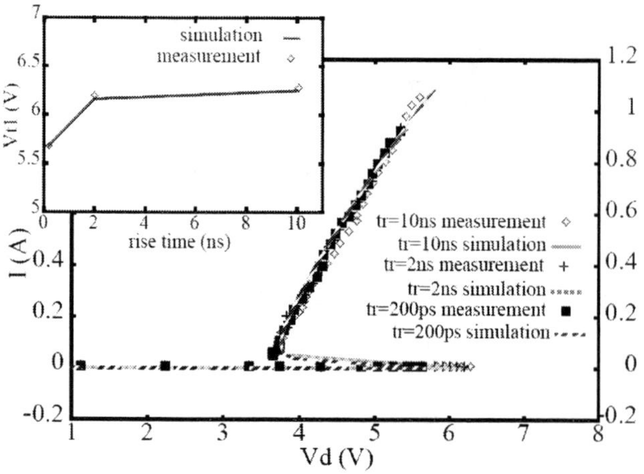

Fig. 5: Snapback curves of a ggNMOS device for different rise times (tr=200ps, 2ns, 10ns)

Fig. 5 depicts a comparison of simulation and TLP measured data of a ggNMOS device for different rise times. *Vt1* values are plotted as a function of the pulse rise time in the inserted plot. When t_{rise} decreases to 200ps *Vt1* reduces by approximately 0.6V from the t_{rise}=10ns value due to the increase of displacement current. The capacitance value from the BJT model that fits the data is lower than the measured value. The ratio for the device in Fig.5 is about 0.9.

Fig. 6 shows a comparison of simulation and quasi-static TLP measured data of a ggNMOS device. The simulation results show very good agreement with the data for both positive and negative Vds. The resistance in the range of Vds<0 is about 5 Ohm, compared with the Rsub value of about 800 Ohm.

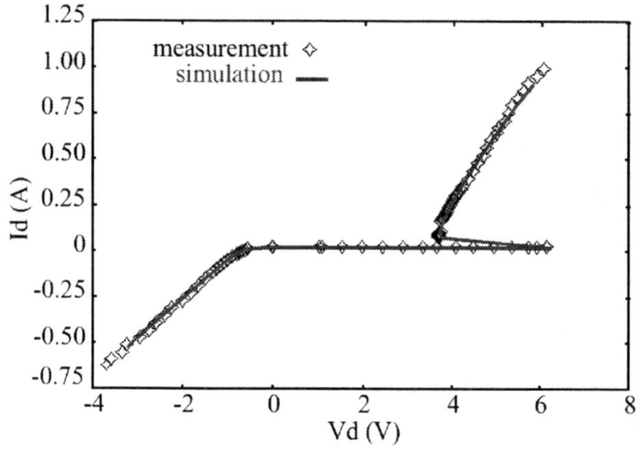

Fig. 6: Comparison of simulation and measurement data of a ggNMOS device.

Fig. 5 depicts simulation and measurement data corresponding to equation (4). It shows that after junction partition the model fits the base current curve of the parasitic bipolar much better.

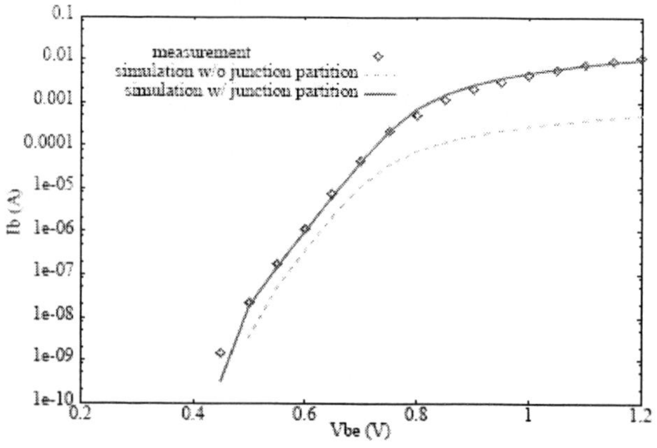

Fig. 7: Base current comparison with and without Drain and Source junction partition.

5 CONCLUSION

Partition of the Drain and Source junctions in MOS snapback model has been discussed using a new macro model approach. The model is based on industry standard models only and intrinsically includes all major physical effects particular to snapback. The partition enhances the MOS snapback model and makes the model valid for both positive and negative drain voltage stresses. The comparison of simulation and measured data of a Grounded-Gate NMOS shows good agreement. The total drain capacitance and the base current curve of the parasitic bipolar shows significant improvement as well.

REFERENCES

[1] Amorasekera, S. Ramaswamy, M. C. Chang M and C. Duvvury, *Proc. IRPS* (1996), p.318

[2] M. Mergens, W. Wilkening, S. Mettler, H. Wolf and W. Fichtner, *Proc. IRPS*, p.167, 1999.

[3] J. Li, S. Joshi, R. Barnes and E. Rosenbaum, *IEEE Trans- CAD,* v. 25, p.1047, 2006

[4] X. F. Gao, J. J. Liou, J. Bernier, G. Croft and A. Ortiz-Conde, *IEEE Trans-CAD,* v. 21, p.1497, 2002

[5] Y. Zhou, D. Connerney, R. Carroll, T. Luk, *Proc. ISQED* (2005), p.476-481

[6] Y. Zhou, T. Weyl, J.-J. Hajjar and K. Lisiak, *Proc. EDSSC* (2007), p.51-56

[7] F. Hsu, P. Ko, S. Tam, and R. Miller, *IEEE Trans. Elec. Dev.*, ED-29, p.1835-1740, 1982.

[8] Y. Zhou, D. Connerney, R. Carroll, T. Luk, *Proc. WCM* (2005), p.476-481

[9] W. Liu, "MOSFET Models for SPICE Simulation including BSIM3v3 and BSIM4", John Wiley, New York, 2001

[10] W. J. Kloosterman and H. C. de Graaff, *Proc. BCTM*, p.5.4, 1988

Improved layout dependent modeling of the base resistance in advanced HBTs

S. Lehmann[1], M. Schroter[1,2]

[1]Chair for Electron Devices and Integrated Circuits, University of Technology Dresden, Germany

[2]ECE Dept., University of California San Diego, USA

Abstract - A physics-based analytical set of equations is applied to calculate the internal and external base resistance of SiGe HBTs for a practically relevant variety of single base contact (SBC) layouts. Through comparison with results from quasi-3D device simulations the accuracy of the equations is verified and validity limits are determined. Additionally, the impact of a broken silicide layer on the base resistance is investigated.

1. INTRODUCTION

High-speed wireline and wireless communication circuits and systems are increasingly realized using Silicon-Germanium (SiGe) HBTs since their performance has benefited tremendously from integration into CMOS processes. One of the most important parameters of HBTs having significant impact on speed and noise behavior is the base resistance r_B. Therefore, accurate analytical equations are required for describing r_B as a function of bias, temperature, and layout. The goal of this paper is to provide insight into layout (i.e. geometry) dependence of the base resistance r_B and the associated current flow in structures with base contact schemes that have become increasingly common in advanced HBTs.

In [1,2] a set of equations was presented which has been widely used for calculating the layout dependence of the base resistance for circuit optimization and statistical design (e.g. [3,4,5,6]). However, bipolar process technology has changed significant since [1,2] were published. Modern SiGe HBT structures have a silicided external base region, which makes single-base contact schemes feasible. In some cases [7], the base is even contacted at the emitter foreside in order to reduce the BC capacitance and the collector resistance. Also, the ratio of the base link sheet resistance r_{Sl} to the internal base sheet resistance r_{SBi} in HBTs, $r_{Sl}/r_{SBi} \approx 1$, is significantly larger than in the BJT processes ($r_{Sl}/r_{SBi} \approx 0.2$) considered in [1,2]. Furthermore, shrinking vertical and lateral dimensions occasionally cause the silicide to break. Although this is a reliability issue that needs to be eliminated in a production process, it is useful to at least obtain a feeling for its impact on r_B.

The applicability of the equations in [1,2] has been evaluated for the above mentioned structures. The observed errors for some cases led to a more comprehensive improved (for double- and single-base contact structures) and extended (for foreside-base structures) set of compact equations [8]. In this paper, well-proven quasi-3D device simulation is employed for investigating a variety of base contact arrangements and layout dimensions. An npn transistor structure in forward d.c. operation is assumed. The term "low-bias" is synonymous with "negligible emitter current crowding". Hence, the internal base sheet resistance already assumes its operating point dependent value at low-bias which can be different from its zero-bias value.

2. METHOD

As shown in [1] the flow lines associated with the base current can be transformed into a 2D plane as shown in Fig. 1. The various base current contributions supporting the current flowing into the internal base are indicated by I_{front}, I_{fore}, I_{back}. The sum of these currents is injected across the BE junction. The treatment of general emitter geometries is based on the hole transport and continuity equation.

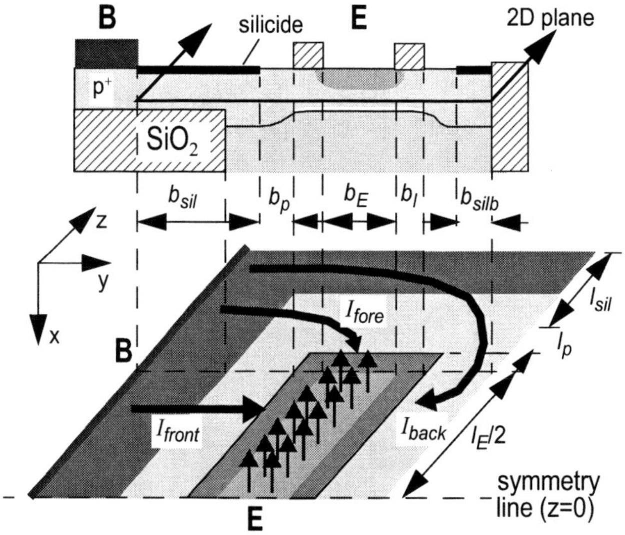

Fig. 1 Schematic cross-section of a single-base structure (with slot contact) and projection of the base current flow path into a 2D plane. The long arrows in the yz-plane indicate the relevant current contributions to the front, fore and back side, while the short arrows pointing out of the internal base region (in -x-direction) represent the base current injected into the emitter.

The quasi-3D simulation method is based on the reasonable assumption that the hole quasi-fermi potential φ_p does not depend on the vertical dimension x over the interval where the hole density significantly contributes to the sheet resistance of a particular region. Then the sheet resistance in the resulting 2D simulation structure can be written as

$$r_S = (q\bar{\mu}_p\bar{N}L_x)^{-1} \tag{1}$$

with L_x as unit dimension in x-direction. In the simulation, the desired sheet resistance value is adjusted by the average mobility value $\bar{\mu}_p$ rather than through the average doping concentration $\bar{N}$ in order to avoid non-physical abrupt energy barriers between the regions. The bias dependence of the internal base sheet resistance r_{SBi} is not included in the simulation (although possible) since the focus is on the geometry dependence of r_B and since its bias dependence is weak for SiGe HBTs.

The base resistance is obtained using eq. (1) in [1]. Compared to the Fourier series solution in, e.g., [9] this approach permits more general and realistic structures as well as an investigation of the high-frequency small-signal behavior.

3. INVESTIGATED STRUCTURES

The layout sketched in Fig. 1 with a single parallel base contact (SBC) were investigated with different dimensions. The corresponding values for the structures presented in this paper are listed in Table 1. In all cases the emitter width is $b_E = 0.25\mu m$. Each structure contains the following regions:

- A silicided region with sheet resistance r_{Ssil} and the dimension b_{sil} at the frontside, l_{sil} at the foreside, and b_{silb} at the backside (SBC structure only); both r_{Ssil} (= 8, 16)Ω/sq as well as l_{sil} and b_{silb} were varied.
- A poly-silicon on mono-silicon region with sheet resistance $r_{Sp} = 25\Omega$/sq and dimension b_p at the frontside as well as l_p at the foreside. This region is missing in those processes where the silicide is defined by the BE spacer.
- A link region surrounding the emitter with sheet resistance $r_{Sl} = 2200\Omega$/sq and width b_l.
- An internal base region (under the emitter) with the sheet resistance $r_{SBi} = 2200\Omega$/sq and dimensions b_E and l_E.

value/μm↓	b_l	$b_p = l_p$	b_{sil}	l_{sil}	b_{silb}	r_{Ssil}
case 1	0.15	0.15	0.2	0.2	0.2	8
case 2	0.15	0.15	0.1	0.1	0.1	8
case 3	0.15	0.15	0.1	0.1	0.1	16

Table 1: Dimensions (in μm) used for the simulated structure. Note that the ratio of the dimensions is important rather than their absolute values; r_{Ssil} values in Ω/sq.

4. SINGLE-BASE CONTACT STRUCTURES

Single-base contact structures are often favored in applications due to the reduced base-collector capacitance and collector resistance. The schematic layout of the simulated structures is shown in Fig. 2 along with the partitioning of the various regions for resistance calculations and with the resulting resistance network. Every node in the equivalent circuit corresponds to an equipotential line in the structure. In order to be able to separate the description of the bias dependent internal base resistance r_{Bi} from the bias independent external base resistance r_{Bx}, the boundary between base link and internal transistor (underneath the emitter) is assumed to be an equipotential line. As can be seen in Fig. 3 this is (still) a fairly good assumption for a structure with an aspect ratio $b_E/l_E = 1/5$, but begins to fail on the backside for longer stripes [8] or higher silicide resistance. The deviation at the corners due to the rounding of the lines will even disappear in fabricated structures due to lithography reasons.

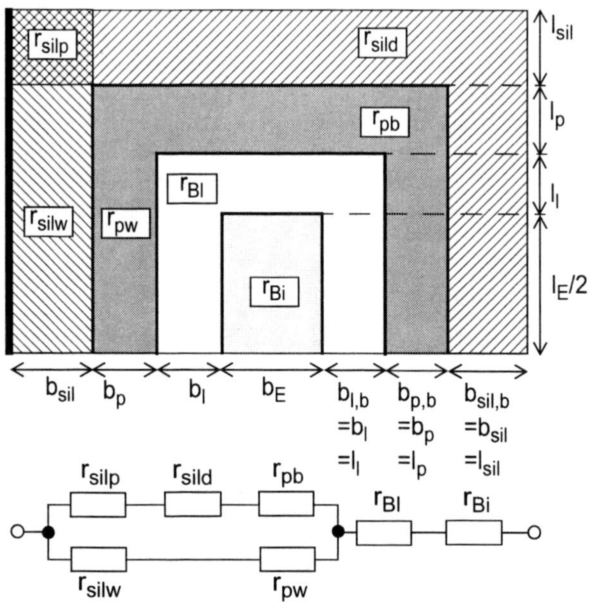

Fig. 2 The various resistance components within the SBC structure and the associated equivalent circuit. The thick lines indicate the assumed equipotential lines.

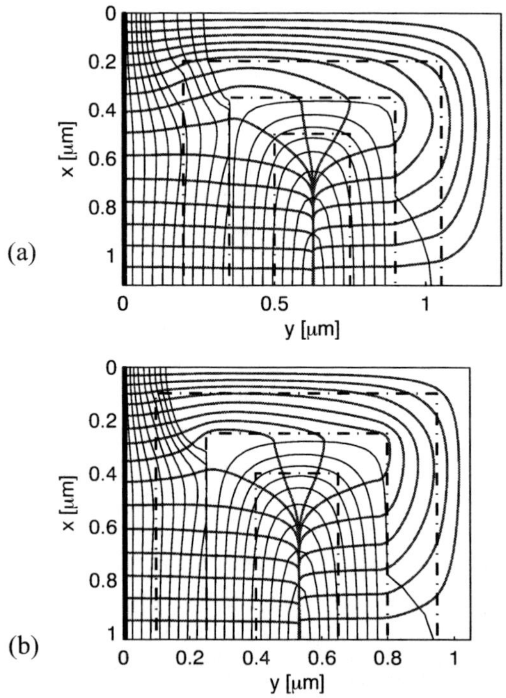

Fig. 3 Current flow and equipotential lines in an SBC structure for the emitter aspect ratio $b_E/l_E = 0.25/1.25$: (a) standard silicide widths (case 1), (b) reduced silicide widths (case 2).

In contrast to BJT processes the base layer epitaxy leads to a link sheet resistance r_{Sl} that is very close to the internal base sheet resistance r_{SBi}. As a consequence, the link resistance r_{Bl} has become the dominant contribution to r_{Bx} and is, besides the silicidation, the main reason why single-base contact structures have become so popular in applications. This has increased the importance for an accurate description of the layout dependence of r_{Bl}. According to Fig. 3, the equipotential lines at the emitter corners look rather like quarter circles, which will be true even more in realistic structures with corner rounding. Assuming an emitter window rounding radius b_{Er} the corner link resistance r_{lc} is given by the expression obtained for a cylindrical resistance structure (e.g. [11])

$$r_{lc,\ln} = \frac{r_{Sl}}{2\pi}\ln\left(1+\frac{b_l}{b_{Er}}\right) \quad \text{for} \quad b_{Er} \le \frac{b_E}{2}. \tag{2}$$

This resistance is the dominant component of r_{Bx} in a square or short structure. A sufficiently accurate approximation without the ln() term can be obtained by assuming an average radius of $(b_{Er}+b_l/2)$,

$$r_{lc} \cong \frac{r_{Sl}}{2\pi}\frac{b_l}{(b_{Er}+b_l/2)} \quad \text{for} \quad b_{Er} \le \frac{b_E}{2}, \tag{3}$$

with $b_{Er} = b_E/2$ for $b_{Er} > b_E/2$. The latter approximation deviates less than 5% from eq. (2) for the chosen b_{Er}=0.125µm (cf. Fig 4).

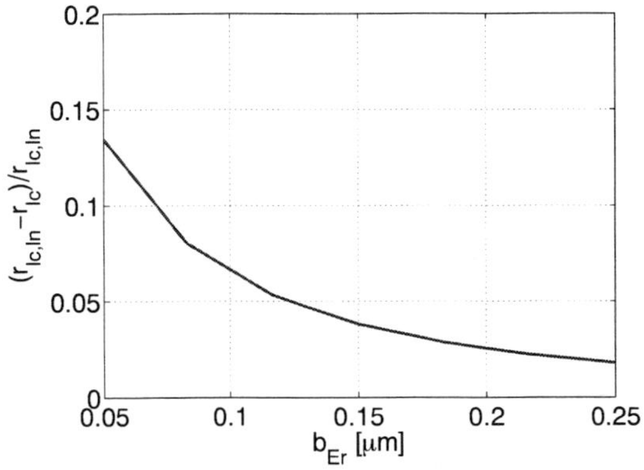

Fig. 4 Relative deviation of the corner resistance value using (3) with the layout dimensions of Table 1.

In the remaining link region at the front, foreside and backside a fairly parallel current flow is observed, allowing the corresponding resistance to be modeled as

$$r_{lw} = \frac{r_{Sl}}{2}\frac{b_l}{l_E+b_E-4b_{Er}} \quad \text{for} \quad l_E \ge b_E \ge 2b_{Er}. \tag{4}$$

The total link resistance is then given by the parallel circuit of the above two components:

$$r_{Bl} = (1/r_{lc}+1/r_{lw})^{-1}. \tag{5}$$

If the dimensions of the external base are reduced, the distributed resistance r_{sild} around the emitter length becomes more important since this region supports the backside of the internal transistor.

The distributed current flow in this region can be considered to consist of two components as sketched in Fig. 5. According to [10] the equivalent resistance for the region y=[1.5a,1.5a+l] can be written

$$R_{eq}(\delta) = r_S\sqrt{\frac{\delta}{1-\delta}}\coth\left[\frac{l}{a\sqrt{\delta(1-\delta)}}\right]. \tag{6}$$

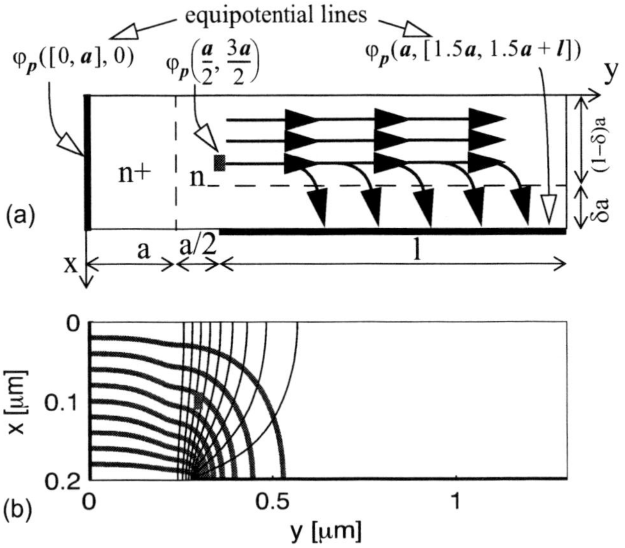

Fig. 5 (a) Schematic simulated test structure for determining the partitioning factor in the distributed resistance calculation, and (b) resulting current flow from the left contact to the lower contact and equipotential lines for l=5a and r_S=8Ω/sq.

To determine δ two different methods for calculating R_{eq} from device simulation have been used. Using a reference quasi-fermipotential and the simulated current I follows

$$R_{eq1} = \frac{\varphi_p\left(\frac{a}{2},\frac{3a}{2}\right)}{I}. \tag{7}$$

According to Fig 5b, assuming an equipotential line in x-direction at y=3a/2 is only an approximation. Thus, an average voltage drop between y=0 and 3a/2 is used to correct the total resistance

$$R_{eq2} = \frac{\varphi_p(0,0)}{I} - \frac{1}{a}\int_{y=0}^{1.5a} r_S(y)dy \tag{8}$$

using the sheet resistance $r_S(y)$ of this area.

The obtained resistance range enclosed by the values of the two applied methods as shown in Fig. 6 confirms the value $\delta = 0.15$ used in [1,2]. This also holds for different sheet resistances.

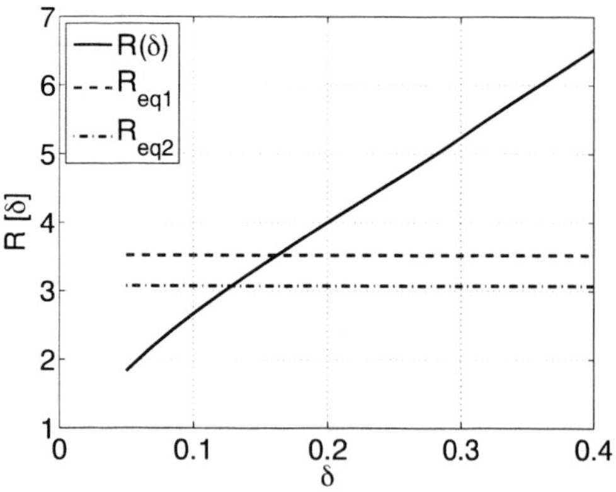

Fig. 6 Distributed resistance respective to the chosen δ (solid) and the results of the two methods for the resistance determination (dash, dash-dot) for $r_S=8\Omega$/sq.

As observed in Fig. 3, a complete separation between silicide and poly-silicon region is not possible. Following the approach in [2] and sketched in Fig. 2, the front region and the remaining fore and back region are modeled separately as parallel current paths. An improved version of the corresponding analytic equations is presented in [8].

For long emitters the current flow into the link and internal base region decreases on the backside. In how far this impacts the internal base resistance is shown in Fig. 7. Here the improved transition function f_i that was originally introduced in [2] is shown versus an also improved transition variable given by the parallel connection of the silicide and poly-Si resistance,

$$u = \frac{\left(\frac{b_{sil,b}}{r_{Ssil}(d_p + 2l_p + b_{sil,b})} + \frac{b_p}{r_{Sp}d_p}\right)^{-1}}{r_{SBi}(b_E/l_E)}, \quad (9)$$

with $d_p = 3b_l + b_E + l_p + l_E/2$ as average fore and back side current path in the poly-silicon layer. The denominator is just a normalization variable. For short emitters or sufficiently low-ohmic silicide regions (i.e. small u) the current flow is similar to that of a double-base contact structure, and so is the resistance. However, for long emitters or relatively high-ohmic (e.g. narrow) silicide regions (large u) r_{Bi} tends to approach the value of a single-base *walled* structure, which is a factor 4 larger than for a double-base structure.

Table 2 contains a comparison of the simulated and modeled resistance values for selected structures using physics-based equations of [1,2,8]. Good agreement is obtained for almost all layout variations. Even for the most critical layouts with narrow silicide width, high silicide resistance and very low b_E/l_E ratios the deviation is below 10%.

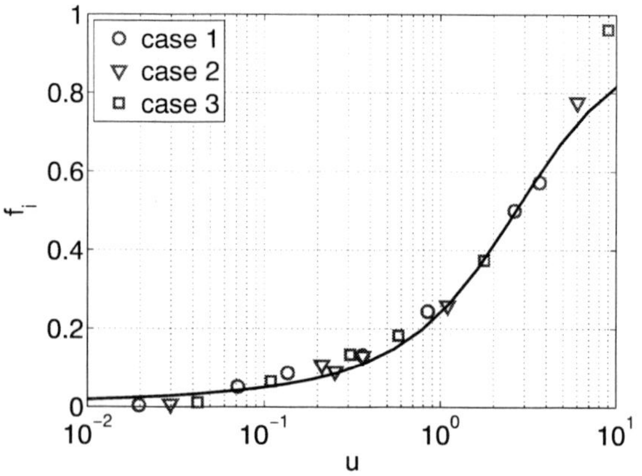

Fig. 7 Transition function f_i vs. transition variable u for all investigated SBC structures. Simulation is indicated by symbols. The solid line represents the analytical equation given in [8].

r_B/Ω \ b_E/l_E	1/1	1/5	1/10	1/20	1/40
case 1: sim	346	146	83.4	47.6	28.8
model	358	142	81.6	47.1	28.6
case 2: sim	348	148	86.1	50.7	31.6
model	360	143	83.9	50.1	30.1
case 3: sim	351	152	89.8	54.3	34.2
model	364	146	87.1	53.3	31.2

Table 2: Total base resistance values for the SBC structure: comparison between device simulation and compact model.

5. FORESIDE-BASE CONTACT STRUCTURES

Placing the base contact at the emitter foreside as shown in Fig. 8 can further reduce the base-collector capacitance and collector resistance. This may come though at the expense of limitation in emitter length due to a significantly increased base resistance if external dimensions continue to shrink and the silicide sheet resistance keeps increasing. The used dimensions and silicide sheet resistance values are listed in Table 1. Fig. 8 also exhibits the intended equivalent circuit which is based again on equipotential line considerations. Since the existing compact equations in [1,2] do not cover this case at all, a new set of equations was developed in [8] with f_i for modeling r_{Bi}.

As shown in Fig. 9(a), for a wider and low-ohmic silicide region the current flow lines in the poly-silicon and link region are almost perpendicular to those in the silicide region, indicating that these regions are more or less entirely supported by the silicide region. If the cross-section of the latter is reduced and the sheet resistance increased, the poly-silicon region starts to take over more of this current and a simple compact description based on perpendicular current flow in the poly region becomes less accurate for ratios $b_E/l_E < 0.2$. As long as $r_{Sp} << r_{Sl}$ though, (6) can then be applied to the poly region as in [1,2]. Generally, an accurate compact solution is more complicated than for SBC structures.

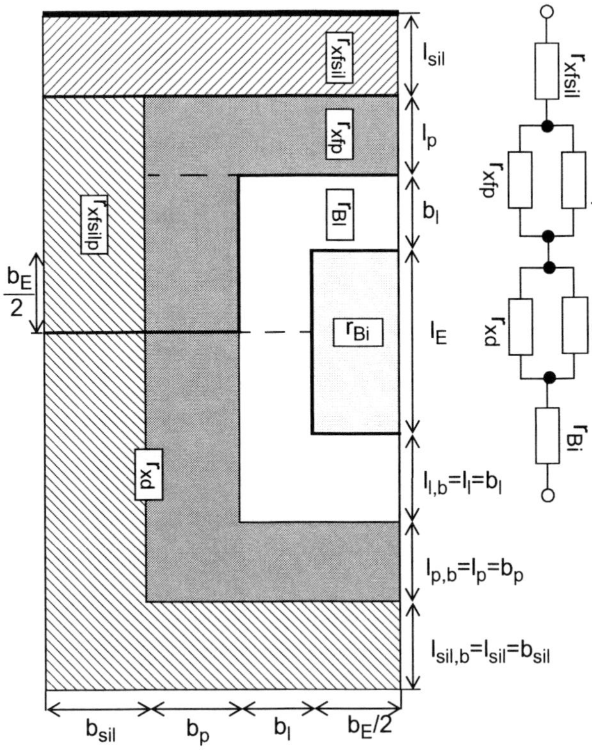

Fig. 8 The various resistance components within the FBC structure and the associated equivalent circuit. The thick lines indicate the assumed equipotential lines. For fabrication reasons, generally $l_p = b_p$.

As the results in Table 3 show the model equations give good results down to a b_E/l_E ratio of 0.2. For smaller b_E/l_E ratios the simulated values drop faster than the modeled ones. However, both show the same trend for even larger b_E/l_E ratios at which the base resistance starts to increase again. This caused by the simultaneous increase of the silicide resistance and decrease of the internal and link resistance with l_E; i.e. their ratio is proportional l_E^2.

b_E/l_E r_B/Ω	1/1	1/5	1/10	1/20	1/40
case 1: sim	346	146	79.6	73.5	75
model	358	142	119	126	127
case 2: sim	348	148	86.5	72.4	106
model	360	143	127	125	146
case 3: sim	351	167	83	91	144
model	364	153	122	140	165

Table 3: Total base resistance values for the FBC structure: comparison between the results of device simulation and compact model.

6. BROKEN SILICIDE

Shrinking dimensions of the silicide region can lead to a "broken" layer disconnecting the remaining portion of the structure from the contact. In order to provide a feeling for the difference in base resistance, the SBC and FBC structures were simulated for case 2 with two different locations of the interruption. The latter forces the flow lines through the higher-ohmic poly-silicon layer, which is shown for the SBC structure in Fig. 10 and for the FBC structure in Fig 11.

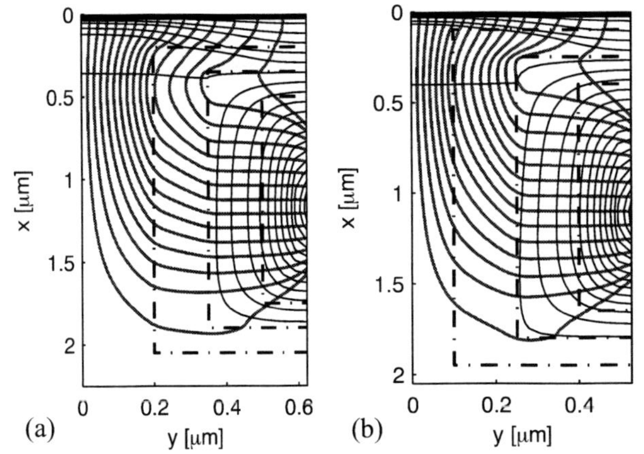

Fig. 9 Current flow and equipotential lines in an FBC structure with n_B=1 for the emitter aspect ratio $b_E/l_E = 0.25/1.25$: (a) standard silicide widths $l_{sil}=b_{sil}=0.2\mu m$, (b) reduced silicide widths $l_{sil}=b_{sil}=0.1\mu m$ and $r_{Ssil}=16\Omega/sq$.

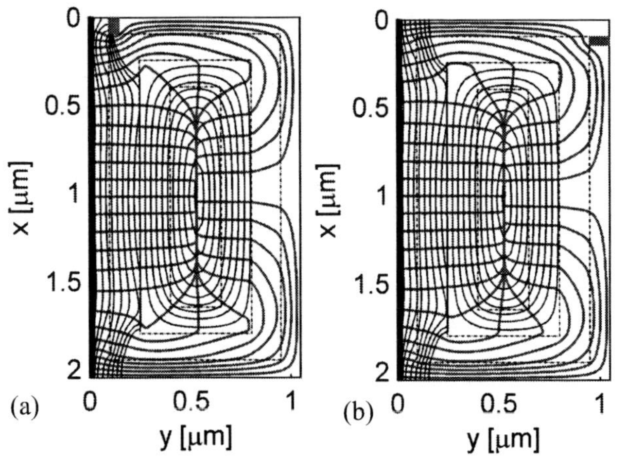

Fig. 10 Current flow and equipotential lines in an SBC structure with different locations of broken silicide layer: (a) "open" at foreside (r_B=148.5Ω), (b) "open" at backside (r_B=148.07Ω). Emitter aspect ratio $b_E/l_E = 0.25/1.25$, $l_{sil}=b_{sil}=0.1\mu m$, $r_{Ssil} = 8\Omega/sq$.

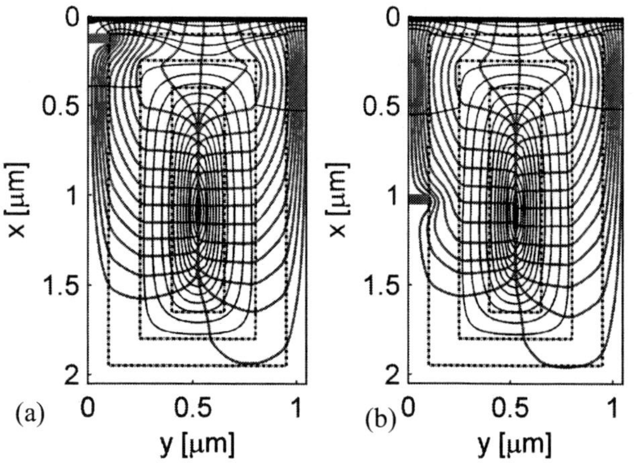

Fig. 11 Current flow and equipotential lines in an FBC structure with different locations of broken silicide layer: (a) one-sided "open" at foreside y = 0 (r_B=159.73Ω), (b) one-sided "open" at y = $l_E/2$ (r_B=158.88Ω). Emitter aspect ratio b_E/l_E = 0.25/1.25, l_{sil}=b_{sil}=0.1µm, r_{Ssil} = 8Ω/sq.

Although the impact of the interruption of the silicide on the current flow lines is clearly visible for the respective two different cases, the results show negligible differences in the total base resistance. This is mostly due to its much lower value compared to that of r_{Sl} and r_{SBi} as well as due to the relatively high aspect ratio (b_E/l_E = 0.2).

7. CONCLUSIONS

Improved physics-based compact analytical formulations presented in [1,2,8] for modeling of the base resistance of SiGe HBTs have been applied to:

- single-base contact in parallel to the emitter;
- base contact *perpendicular* to the emitter finger;
- variations in the external base layout dimensions;
- a large emitter aspect ratio range.

The equation set has been compared to the solution of quasi-3D device simulation. An accurate description of the investigated parallel single base contact structures for a large variety of geometries has been demonstrated. For the foreside contact arrangement the equations are accurate down to an emitter aspect ration b_E/l_E of 0.2. For smaller values they still show the same trend though indicating the fundamental effect is included in the equations.

Simplified formulations for the corner component of the base link resistance and the distributed silicide resistance have been validated.

Finally, the impact of a broken silicide layer on the total base resistance was shown to be negligible at least for emitter aspect ratios down to 0.2. This alleviates process reliability issues.

8. ACKNOWLEDGMENTS

The authors are grateful to STMicroelectronics and the German Research Foundation (DFG grant SCH695/1-2) for financial support.

9. REFERENCES

[1] M. Schröter, "Simulation and modeling of the low-frequency base resistance of bipolar transistors in dependence on current and geometry", IEEE Trans. Electron Dev., Vol. 38, pp. 538-544, 1991.

[2] M. Schröter, „Modeling of the low-frequency base resistance of single base contact bipolar transistors", IEEE Trans. Electron Dev., Vol. 39, pp. 1966-1968, 1992.

[3] K. Walter et al., "A scalable, statistical SPICE Gummel-Poon model for SiGe-HBTs", Proc. BCTM, pp. 32-35, 1997.

[4] S. Yoshitomi, K. Kawakyu, T. Teraguchi, H. Kimijina, K. Yonemura, "Modification of compact bipolar transistor model for DFM (Design for manufacturing) applications", Proc. MIXDES, pp. 125-130, 2006.

[5] M. Rickelt, H.-M. Rein, "A novel transistor model for simulating avalanche-breakdown effects in Si bipolar circuits", IEEE J. Solid-State Circuits, Vol. 37, pp. 1184-1197, 2002.

[6] M. Schroter, H.-M. Rein, W. Rabe, R. Reimann, H.-J. Wassener and A. Koldehoff, „Physics- and process-based bipolar transistor modeling for integrated circuit design", IEEE Journal of Solid-State Circuits, Vol. 34, pp. 1136-1149, 1999. See also TRADICA Manual v5.2.

[7] B. Heinemann, R. Barth, D. Bolze, J. Drews, P. Formanek, T. Grabolla, U. Haak, W. Hoppner, D.K. Kopke, B. Kuck, R. Kurps, S. Marschmeyer, H.H. Richter, H. Rucker, P. Schley, D. Schmidt, W. Winkler, D. Wolansky, H.E. Wulf and Y. Yamamoto, "A Low-Parasitic Collector Construction for High-speed SiGe:C HBTs", IEDM Tech. Dig. pp. 251–254, 2004.

[8] M. Schroter, J. Krause, S. Lehmann, D. Celi, "Compact layout and bias dependent base resistance modeling for advanced SiGe HBTs", IEEE Trans. Electron Dev., submitted for publication.

[9] J. C. J. Paaschens, "Compact Modeling of the Noise of a Bipolar Transistor Under DC and AC Current Crowding Conditions", IEEE Trans. on Electron Dev., Vol. 51, No. 9, 1483-1495, 2004.

[10] H. Murrmann and D. Widmann, "Current crowding on metal contacts to planar devices", IEEE Trans. Electron. Devices, vol. ED-16, pp. 1022-1024, 1969.

The Bipolar Field-Effect Transistor Theory*,**

(A. Summary of Recent Progresses)

Bin B. Jie[2,+] and Chih-Tang Sah[1,2,3,+]

Abstract

Field-effect transistor (FET) was conceived 80 years ago in Lilienfeld 's 1926-1932 patents [1]. Shockley-1952 [2] invented the volume-channel FET 55 years ago using two opposing p/n junctions as gates on the two surfaces of a thin semiconductor film to control the conductance of the thin-film's volume-channel. Atalla-Kahng-1960 [3] demonstrated the surface-channel FET using a single conductor (metal)-on-insulator (oxide) as the gate (MOS or MIS gate) on thick silicon to control the surface-channel conductivity. The initial circuit design simulator SPICE [3] used the 1964-Sah constant gate threshold-voltage MOSFET model [4], improved for voltage-dependent gate threshold-voltage in 1966-Sah-Pao [5]. These are the 25% FET theory since they included only the drift current for one carrier species, missing the diffusion current and the second carrier species, electrons or holes. Diffusion current was included in 1966-Pao-Sah [6] to give the 50% or the Unipolar FET (UniFET) theory. We noticed recently (March 2007) the simultaneous appearance of both electron and hole currents in the experimental data of the latest nanometer dual-MOS gates on thin-base silicon FETs [7], which led us to the development of the 100% Bipolar Field-Effect Transistor Theory (BiFET). This paper summarizes the history and progresses presented in our monthly reports [8-14].

---

* Project supported by CTSAH Associates (CTSA) which was founded by the late Linda Su-Nan Chang Sah.

** This theory was first presented as Late News by us at the Workshop on Compact Modeling (WCM), May 23, 2007, Santa Clara Convention Center, California [7].

(1) Chinese Academy of Sciences, Foreign Member, Beijing, China.

(2) Peking University.

(3) University of Florida, Gainesville, Florida, USA.

+ Email addresses: BB_Jie@msn.com and Tom_Sah@msn.com.

References

[1] Chih-Tang Sah, "Evolution of the MOS Transistor – From Conception to VLSI," Invited Paper, IEEE Proc. 76(10), 1280-1326, October 1988, and references cited.

[2] William Shockley, "A Unipolar 'Field-Effect' Transistor," Proc IRE 40(11), 1365-1376, November 1952.

[3] Chih-Tang Sah and Bin B. Jie, "A History of the MOS Transistor," Keynote at Workshop of Compact Modeling, Proceedings, pp 1-2 and 349-390, May 2005.

[4] Chih-Tang Sah, "Characteristics of the Metal-Oxide-Semiconductor Transistors," IEEE Tran Elec Dev. 11(7), 324-345, July 1964.

[5] Chih-Tang Sah and Henry C. Pao, "The Effects of Fixed Bulk Charge on the Characteristics of Metal-Oxide-Semiconductor Transistor," IEEE Tran Elec Dev. 13(4), 410-415, April 1966.

[6] Henry C. Pao and Chih-Tang Sah, "The Effect of Diffusion Current on the Characteristics of MOS Transistors," Solid-state Electronics 9(10), 927-938, October 1, 1966

[7] Chih-Tang Sah and Bin B. Jie, "Bipolar Theory of MOS Field-Effect Transistors and Experiments," Chinese Journal of Semiconductors 28(10), 1534-1540, October 2007.

[8] Chih-Tang Sah and Bin B. Jie, "The Bipolar Field-Effect Transistor: I. Electrochemical Current Theory (Two-MOS-Gates on Pure-Base),"Chinese Journal of Semiconductors 28(11), 1661-1673, November 2007.

[9] Chih-Tang Sah and Bin B. Jie, "The Bipolar Field-Effect Transistor: II. Drift-Diffusion Current Theory (Two-MOS-Gates on Pure-Base),"Chinese Journal of Semiconductors 28(12), 1849-1859, December 2007.

[10] Binbin Jie and Chih-Tang Sah, "The Bipolar Field-Effect Transistor: III. Short Channel Electrochemical Current Theory (Two-MOS-Gates on Pure-Base)," (Chinese) Journal of Semiconductors 29(1), 1-11, January 2008.

[11] Binbin Jie and Chih-Tang Sah, "The Bipolar Field-Effect Transistor: IV. Short Channel Drift-Diffusion Current Theory (Two-MOS-Gates on Pure-Base)," (Chinese) Journal of Semiconductors 29(2), 193-200, February 2008.
[12] Chih-Tang Sah and Binbin Jie, "The Theory of Field-Effect Transistors: XI. The Bipolar Electrochemical Currents (1-2-MOS-Gates on Thin-Thick Pure-Impure Base)," (Chinese) Journal of Semiconductors 29(3), 397-409, March 2008.
[13] Chih-Tang Sah and Binbin Jie, "The Bipolar Theory of the Field-Effect Transistors: X. The Fundamental Physics and Theory (All Devices Structures)," (Chinese) Journal of Semiconductors 29(4), 613-619, April 2008.
[14] Binbin Jie and Chih-Tang Sah, "The Bipolar Field-Effect Transistor: V. Bipolar Electrochemical Current Theory (Two-MOS-Gates on Pure-Base)," (Chinese) Journal of Semiconductors 29(4), 620-627, April 2008.

The Bipolar Field-Effect Transistor Theory*,**

(B. Latest Advances)

Chih-Tang Sah[1,2,3,+] and Bin B. Jie[2,+]

Abstract

Latest advances are presented on theoretical device and circuit characterizations of the Bipolar Field-effect transistor (BiFET) [1]. The 2-Dimensional (2-D) rectangular geometry of the transistor (uniform in the width direction) is employed to separate the 2-D equations into two surface-electric-potential-coupled 1-D equations, enabling generic baseline solutions, without 2-D features which are then treated as modifications of the 1-D solutions. The 1952-Shockley 2-section volume-channel geometry model of Junction-Gate (JG) FET is applied to the surface-and-volume-channels of the MOS BiFET, designated as an emitter and a collector sections, each can simultaneously have electron and hole, surface or volume channels. The exactly identical (near thermal equilibrium, no hot carriers) electrochemical potential (ECP or quasi-Fermi potential) and drift-diffusion (DD) approaches are employed. Numerical results are readily obtained for the analytical ECP, but tedious for the DD theory requiring analytical approximations. Asymptotic approach to 1-Gate from 2-Gate, and from impure-Base to pure-Base are illustrated. Deviations of DD theory from ECP theory are demonstrated. Two-Dimensional geometric effects near intersections of the four electrodes (Gate 1, Gate 2, Source and Drain) are described. DC characteristics are computed and presented for both the traditional transistor device and also the basic-building-block (BBB [2]) circuit function configurations.

---

* Project supported by CTSAH Associates (CTSA) which was founded by the late Linda Su-Nan Chang Sah.

** This theory was first described as a Late News at the Workshop on Compact Modeling, May 23, 2007, Santa Clara Convention Center, California.

(1) Chinese Academy of Sciences, Foreign Member, Beijing, China.

(2) Peking University.

(3) University of Florida, Gainesville, Florida, USA. Permanent address.

+ Email addresses: Tom_Sah@msn.com and BB_Jie@msn.com.

References

[1] See references given in Bin B. Jie and Chih-Tang Sah, "The Bipolar Field-Effect Transistor Theory (A. Summary of Recent Progresses.)." Preceding paper, this conference.

[2] Chih-Tang Sah, *Fundamentals of Solid-State Electronics*, World Scientific Publishing Company, 1991. See sections 670 to 674 on pages 596 to 643 on Circuit Applications of MOSFET for the Basic-Building-Block (BBB) circuits (DRAM, NMOS-CMOS inverters, SRAM, Nonvolatile-RAM).

An Accurate and Versatile ED- and LD-MOS Model for High-Voltage CMOS IC SPICE Simulation

Bogdan Tudor, Joddy W. Wang, Bo P. Hu, Weidong Liu, and Frank Lee

Synopsys, Inc., 700 E. Middlefield Road, Mountain View CA, U.S.A.
Bogdan.Tudor@synopys.com

ABSTRACT

This paper presents a high-voltage compact MOSFET model that has been proven physically accurate and numerically robust for various generations of high-voltage ED (extended drain) and LD (laterally double diffused) production CMOS process technologies. The model takes into account elegantly in formulation almost all of those physical effects identified for high voltage MOSFET operation. They include, but are not limited to:

- Quasi saturation.
- Bias-dependent, symmetric or asymmetric non-linear source and drain resistances.
- Self heating.
- Impact ionization in the drift region.
- LDMOS-specific charging effects.
- Asymmetrical charge response and transport for the forward- and reverse-mode operations of asymmetrical LDMOS.

Keywords: high-voltage MOSFET, HVMOS, LDMOS, EDMOS, compact model.

1 INTRODUCTION

The denomination "High Voltage MOSFET" encompasses a large class of devices used in many different application areas. These can vary from integrated, medium power, to discrete, very high power devices. Typical biases range from less than 10V to over 1000V. Based on the process, device types are classified as LD, ED or V (vertical) MOSFETs. The significant differences between various structures make the effort of developing a generic compact model particularly challenging. Moreover, the high voltage of operation and the presence of self-heating are susceptible to result in more circuit simulation convergence problems than in the case of low-power CMOS.

The physics and operation of such devices has been well studied (see for example [1, 2]). Several attempts to develop compact models [3, 4, 5] or macro-models [6, 7] have been made in recent years. However, while most of these attempts are oriented towards LD-MOSFETs, symmetric and asymmetric ED devices have received less attention. Also, at higher voltages, the phenomena occurring in the low doped drain (LDD) region become increasingly difficult to model accurately. Such cases often require adding extra circuit elements (macro-modeling), overcomplicating the netlist. In addition, some of the existing models account for the characteristic HV effects mostly in an empirical manner. This in turn leads to model accuracy and scalability problems.

2 MODEL DEVELOPMENT

Our proposed model is constructed to be able to reproduce the particular nature and characteristics of HV MOSFET devices by a set of mathematical and physical formulations obtained from solving the Poisson and drift-diffusion equations. In addition, the process parameters and the topological device structure are maximally incorporated in the model. For instance, the drain drift resistance, a key component in the device design and optimization for the best possible current driving capability and electrical breakdown, is connected to the intrinsic channel through an internal topological node that accurately tracks the dynamic partitions of charges, highly non-linear channel and drain resistances, and impact ionizations, all between the intrinsic and extrinsic device actions. This treatment is demonstrated to be pivotal in satisfying the strict requirements of the modeling accuracy, scalability, and numerical efficiency for production use.

The main goal we set for our model was the ability to handle both ED- and LD-MOSFET devices in all modes of operation, including reverse-mode and self-heating, with comparable accuracy, and in an entirely compact formulation. Thus, the starting point in developing the model was to combine the features of the two main device types and to come up with a unified, conceptual device structure. This was further decomposed in basic physical elements from which the model topology was derived. Figure 1 illustrates a simplified topology that includes an internal MOSFET model and independent models for the drain and source resistances of the device, respectively.

2.1 Model basis

We selected the BSIM4 model [8] as a foundation for the internal MOSFET in the topology. We have decided to do so because BSIM4 is a robust and mature model, having been the standard in the semiconductor industry for many years. This helped us to comprehensively cover the behavior of the internal MOSFET device, including temperature and geometry scaling. Advanced RF-related effects, such as non-quasi-static (NQS) and RF-related bias-dependent RG models are also built in.

2.2 HV-MOS specific model features

A number of extra elements and model equations have been added, to accurately account for the particular physics of the device:

- Self-heating network, with scalable and temperature-dependent elements.
- Asymmetric diode models, including carrier recombination and tunneling components.
- Substrate current, including impact ionization at the drain of the internal MOSFET and impact ionization in the drift region (see Fig. 5a-b).
- Accurate temperature equations.
- Handling of reverse-mode of operation, for asymmetric devices.
- LDMOS-specific charge and capacitance model.

2.3 TCAD simulations

Extensive TCAD simulations were designed, in order to provide one with deeper insights into the microscopical physical processes of electric potential and charge distributions in the high field regions inside the device. By taking advantage of the unique insight of TCAD device simulation, several key quantities of the HV MOSFET were separately tested and validated, and the corresponding model equations have been refined:

- The drain and source resistances and their dependence on bias and temperature (Fig. 2a-b).
- Confirmation of the internal drain and source voltage values (Fig. 3a-c).
- The characteristic capacitance behavior of LDMOS devices (Fig. 4).

3 PARAMETER EXTRACTION

A global parameter extraction approach, customized to our model's features, has been developed. The procedure, based on optimization, simultaneously uses data measured on multiple device geometries, at different temperatures. The flow starts with the extraction of current equation parameters in the linear region, followed by the saturation region. At this point, reverse-mode data can be also used, if available, for a better characterization of asymmetric devices. Substrate current measurements are included here as well. To allow the accurate extraction of self-heating-related parameters, some of the temperature coefficients are optimized together with the room temperature parameters by using data available at all temperatures.

For the best possible accuracy in the extraction of the specific LDMOS capacitance parameters, optimization based on both C-V and I-V data is recommended, some of the key I-V model parameters being refined this way.

In order to preserve the physical character of model equations this global extraction approach is recommended, but binning is also available, as an alternative.

4 MODEL VALIDATION

The model has been validated based on a wide range of devices and technologies, with gate oxide thicknesses ranging from below 10nm to over 100nm and with the drain bias from a few volts up to several hundreds of volts. Symmetrical and asymmetrical ED-MOS and LD-MOS devices have been characterized and extracted with excellent accuracy without having to use binned models. The model proved to be accurate for high voltage devices, with no need for external circuit elements.

Figure 6a-c illustrates the accuracy of the model in handling the typical behavior of HV MOSFETs, including impact ionization, quasi-saturation and self-heating.

5 CONCLUSION

We have developed a compact model that is applicable to both LD- and ED-MOSFET devices. This model has consistently demonstrated very good accuracy and scalability over bias, geometry and temperature, without the need for binning or macro-modeling. The model has already been successfully used in industrial production.

REFERENCES

[1] Y. Kim et al, "New physical insights and models for high-voltage LDMOST IC CAD," IEEE Trans. ED, vol. 38, no. 7, pp. 1641–1649, Jul. 1991.

[2] C.M. Liu, J.B. Kuo, "Quasisaturation Capacitance Behavior of a DMOS Device", IEEE Trans. ED, vol. 44, no. 7, pp. 1117-1123, July 2007.

[3] A. Aarts et al, "A Surface-Potential-Based High-Voltage Compact LDMOS Transistor Model", IEEE Trans. ED, vol.52 no.5, pp. 999-1007, May 2005.

[4] Z. Liu, "High-Voltage Applications Demand a Robust HVMOS Device Model", Avant! Electronics Journal, December 1998, pp. 16-19.

[5] N Hefyene et al, "EKV compact model extension for HV lateral DMOS transistors", Proceedings of ASDAM 2002, pp. 345 – 348.

[6] J. Jang et al, "Circuit model for power LDMOS, including Quasi-Saturation", Proceedings of SISPAD 1999, pp. 15-18.

[7] E.C. Griffith et al, "Capacitance Modelling of LDMOS Transistors", Proceedings of ESSDERC 2000, pp. 624 – 627.

[8] "BSIM4.6.0 MOSFET Model User's Manual", University of California Berkeley, 2006.

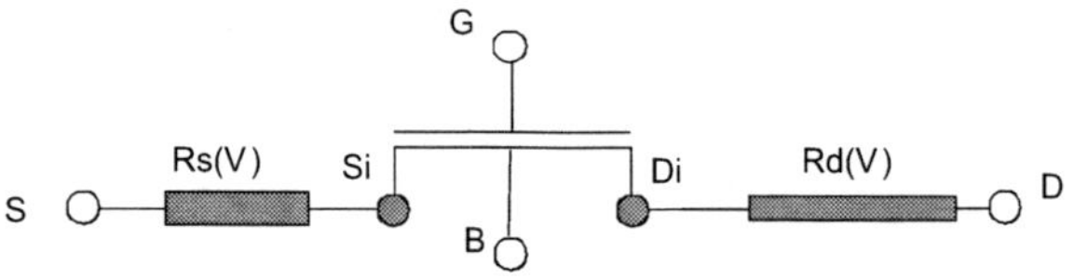

Figure 1. Simplified model topology.

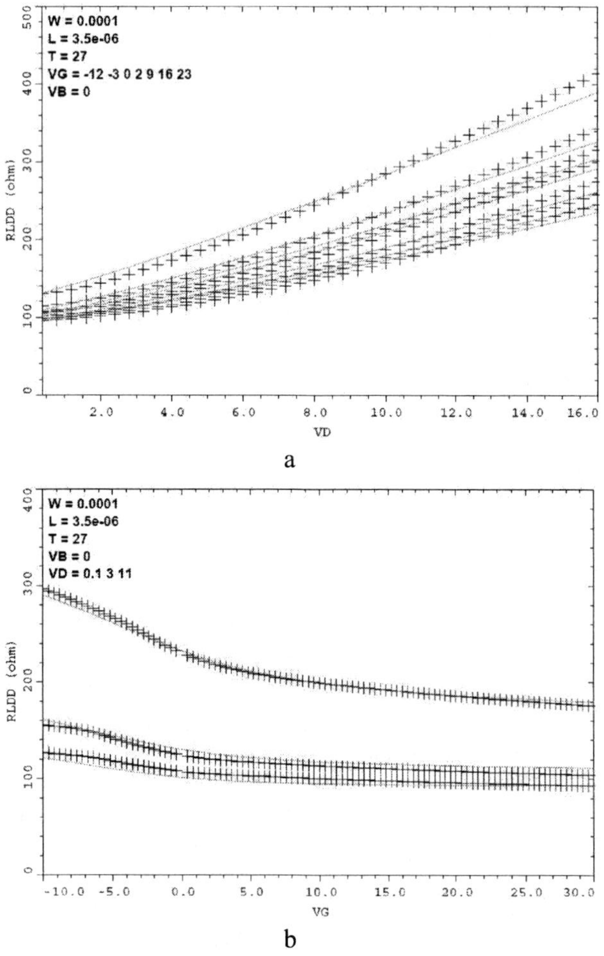

Figure 2. R_{LDD} model versus Medici LDD structure simulation: V_D dependence (a) and V_G dependence (b). Symbols: Medici. Lines: model.

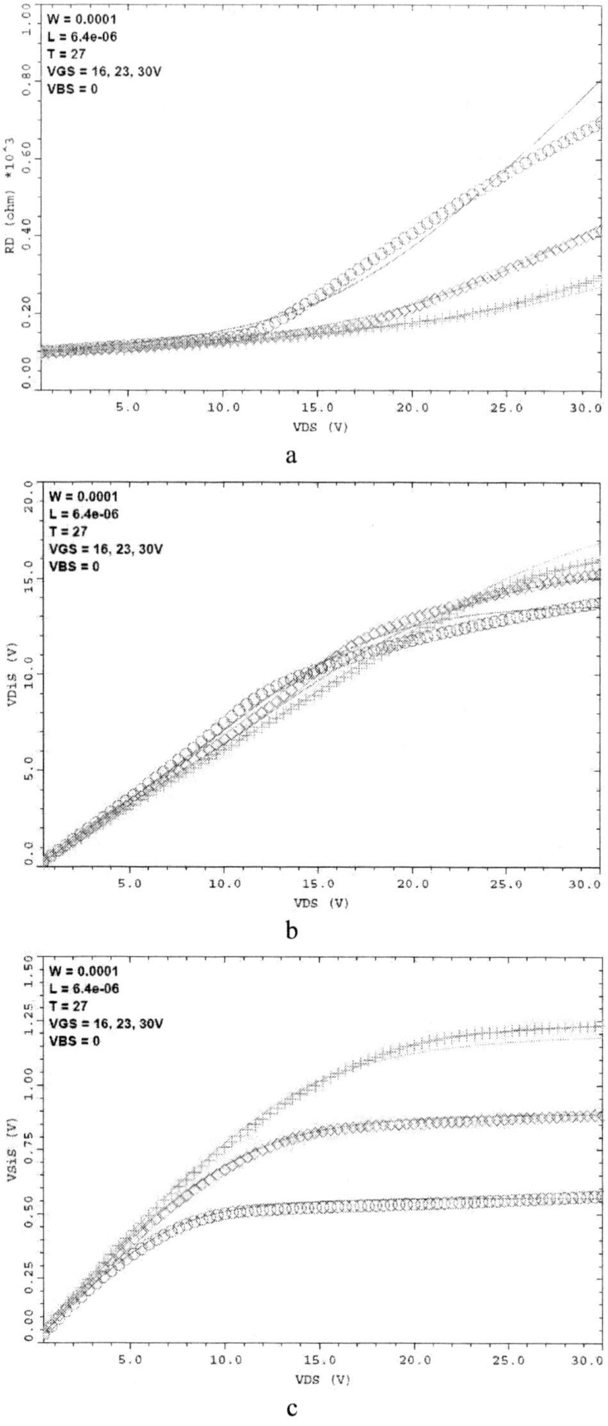

Figure 3. Model verification against Medici HVMOS structure simulation: $R_D(V_{DS})$ (a), $V_{DiD}(V_{DS})$ (b), and $V_{SiS}(V_{DS})$ (c). Lines: model. Symbols: Medici (+: V_{GS}=16V, ◊: V_{GS}=23V, O: V_{GS}=30V).

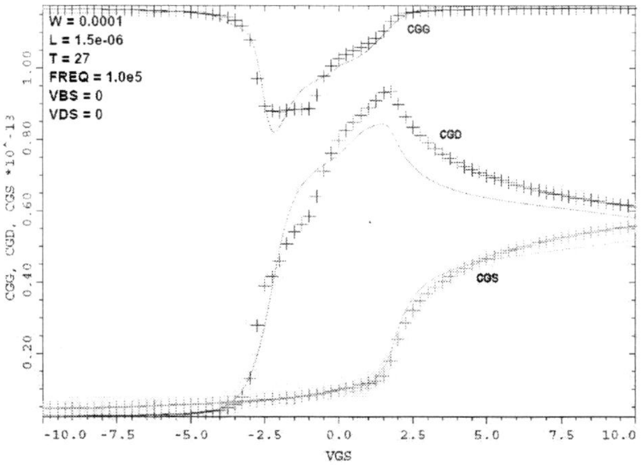

Figure 4. LDMOS C(V) model against Medici simulation. Symbols: Medici. Lines: model.

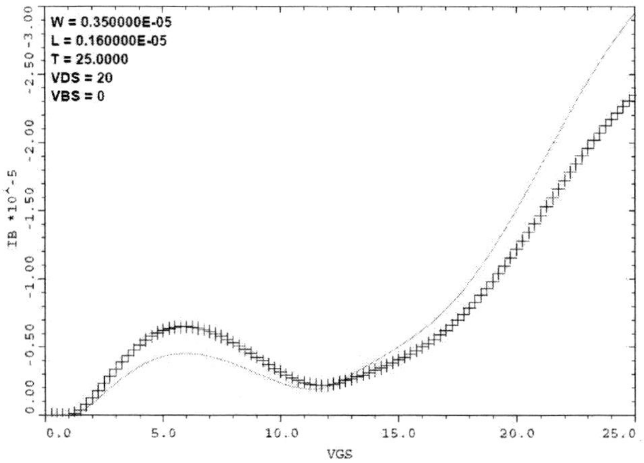

a

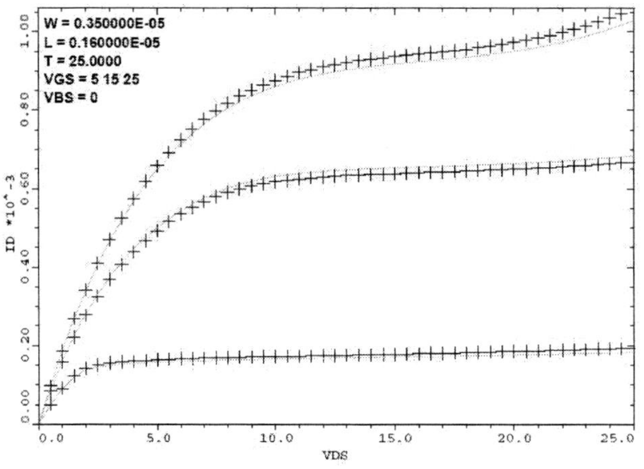

b

Figure 5. Impact ionization in the drift region: effect on I_{SUB} (a) and on I_D (b). Symbols: measured data. Lines: model.

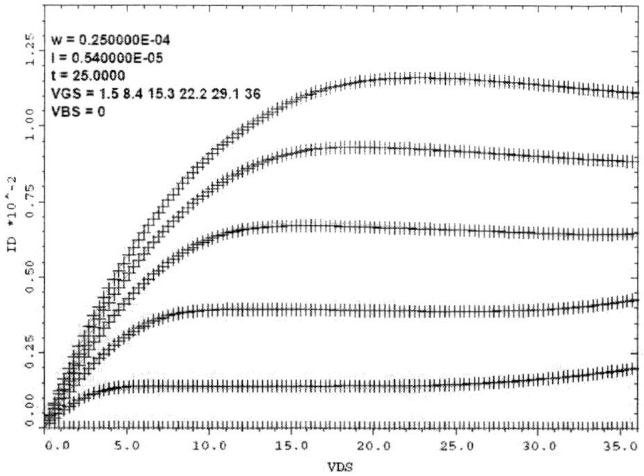

a

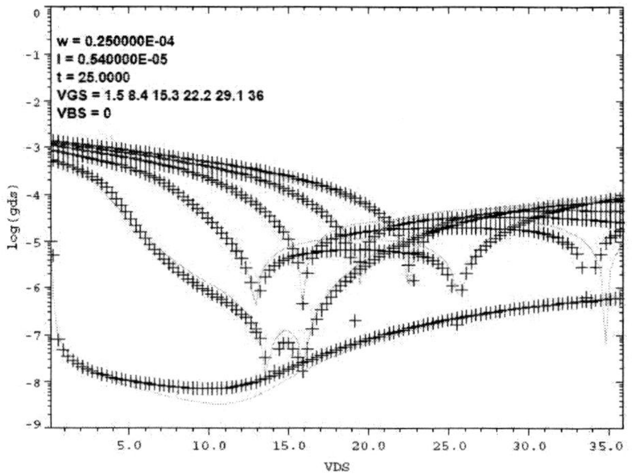

b

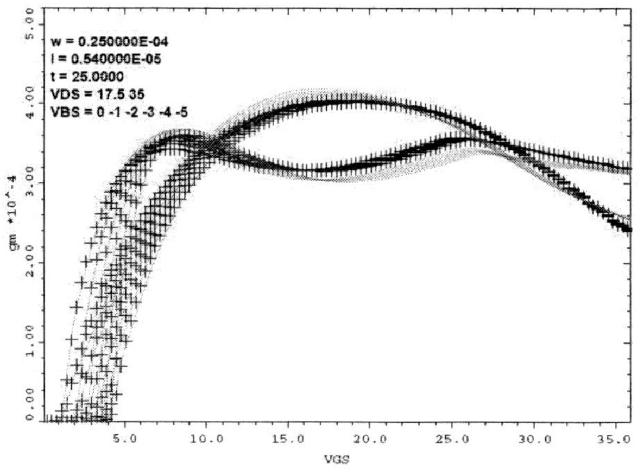

c

Figure 6. Simulated versus measured data: $I_D(V_{DS})$ (a), $g_{ds}(V_{DS})$ (b), and $g_m(V_{GS})$ (c). Symbols: measured data. Lines: model.

Compact Modeling of Noise in Nonuniform Channel MOSFET

A. S. Roy[1], C. C. Enz[1,2], T. C. Lim[3] and F. Danneville[3]

1 Ecole Polytechnique Fédérale de Lausanne (EPFL), anandasankar.roy@epfl.ch
2 Swiss Center for Electronics and Microtechnology (CSEM), christian.enz@csem.ch
3 IEMN, UMR CNRS 8520, France, Francois.Danneville@IEMN.univ-lille1.fr

ABSTRACT

Compact MOSFET noise models are mostly based on the Klaassen-Prins (KP) approach. However, the noise properties of lateral nonuniform MOSFETs are considerably different from the prediction obtained with the conventional KP-based methods which, at low gate voltages, can overestimate the thermal noise by 2-3 orders of magnitude. The presence of lateral nonuniformity makes the vector impedance field (IF) (the quantity responsible for noise propagation) position and bias dependent. This insight clearly explains the observed discrepancy and shows that the bias dependence of the important noise parameters cannot be predicted by conventional KP-based methods. Interestingly, this bias dependence of the noise parameters in the presence of lateral nonuniformity can be effectively used in the channel engineering of MOSFET to optimize the RF noise performance.

Keywords: MOSFET, compact modeling, nonuniform doping, thermal noise

1 INTRODUCTION

MOSFET devices often have nonuniform doping profile along the channel. Lateral double-diffused MOS (LDMOS) devices are well-known examples of MOS devices with a lateral nonuniform channel doping profile. Another important example of this kind of devices are single halo devices where the source end is highly doped to improve the short channel effect. Even the standard MOSFETs are also laterally nonuniform because of the halo implants. Uniformly doped MOSFET noise models are based on Klaassen-Prins (KP) [1] or some equivalent methods [2], which are all denoted here as KP method. Although, there is a trend in the community to directly apply KP method even in the presence of lateral nonuniformity, depending on the noise mechanism and doping profile, this kind of approach can give totally erroneous prediction [3], [4].

In this paper we first present a noise modeling methodology taking lateral nonuniformity into account [3], [4]. The very general nature of the treatment will alow the methodology to be used with any arbitrary doping profile and field-dependent mobility. The main result is that the lateral nonuniform doping makes the vector impedance field for the drain terminal bias dependent. This fact explains the failure of KP-based methods which assumes a constant impedance field

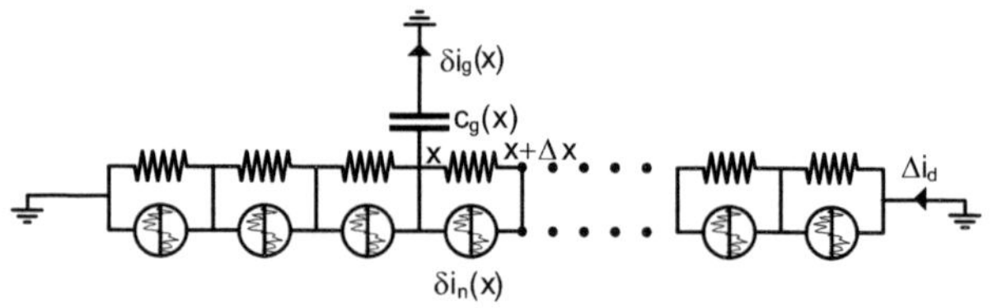

Figure 1: Illustration of noise calculation by using a Langevin approach. The current at any position has two components. One component is set up by the voltage perturbation caused by the noise source (which can be thought of following through the resistances) and other one is the noise current itself (which is represented by the noisy current sources $\delta i_n(x)$). Δi_d is the total noise current flowing through the channel and is constant along the channel. The induced gate current noise $\Delta i_g(x)$ originates due to the fluctuation of the channel potential across the gate capacitance $C_g(x)$ and lies in quadrature with the drain current noise

and predicts unique bias dependence of the correlation coefficient between drain and induced gate noise. Interestingly, this bias dependence can be effectively used in the channel engineering of MOSFET to optimize the RF noise performance. In this paper we will show how to model noise in lateral nonuniform device and how the position and bias dependence of impedance fields can be exploited to improve the RF noise performance of a nonuniformly doped MOSFET.

2 MODEL DESCRIPTION

In a lateral nonuniform MOSFET, the doping varies with position. This kind of doping makes the threshold voltage of the channel position dependent, which causes the inversion charge and mobility to become an *explicit* function of the position. Under drift-diffusion approximation, the current at any position x along the channel can be written as

$$I(x) = W\mu(x, \frac{dV}{dx})(-Q_i(x, V))\frac{dV}{dx} = g(x, V, \frac{dV}{dx})\frac{dV}{dx} \quad (1)$$

where V is the channel potential, W is the width of the device, Q_i is the inversion charge and μ is the mobility. In our analysis we always assume the source to be located at $x = 0$ and the drain at L. Now, presence of a noise current in the channel generates a perturbation v of the channel potential, which then causes a change in transport current. Therefore the total current flowing at position x can be expressed as a sum of transport current (including the effect of perturbation in channel potential) and the noise current itself. Fig. 1 illustrates this situation. So the effect of adding the Langevin noise source $\delta i_n(x)$ in (1) can be written as

$$I_0(x)+i_d(x) = g\left(x, V_0+v, \frac{d(V_0+v)}{dx}\right)\frac{d(V_0+v)}{dx}+\delta i_n(x), \tag{2}$$

where $I_0(x)$ and V_0 are the unperturbed current and voltage in the channel and $i_d(x)$ is the total noise current at the position x. In the following derivation, subscript '0' will be used to denote unperturbed quantities and $\frac{dV}{dx}$ and E will be used interchangeably. Using a perturbation analysis [3], [4], $i_d(x)$ can be expressed as

$$i_d(x) = \frac{g_0 + \frac{\partial g_0}{\partial E_0}E_0}{g_0}\cdot\frac{d}{dx}(g_0 v) - \frac{\partial g_0}{\partial x}v + \delta i_n(x). \tag{3}$$

As $i_d(x) = \Delta i_d$ is constant along the channel, we can express equation (3) as

$$\frac{d}{dx}(g_0 v) - \frac{1}{g_0}\left(\frac{g_0}{g_0+\frac{\partial g_0}{\partial E_0}E_0}\right)\frac{\partial g_0}{\partial x}(g_0 v) = \frac{g_0}{g_0+\frac{\partial g_0}{\partial E_0}E_0}\left(\delta i_n(x) - \Delta i_d\right). \tag{4}$$

This equation can be thought of a first order ODE with respect to $g_0 v$ with an integration factor $R(x)$ given by

$$R(x) = \exp\left(-\int_0^x \frac{1}{g_0}\left(\frac{g_0}{g_0+\frac{\partial g_0}{\partial E_0}E_0}\right)\frac{\partial g_0}{\partial x}dx\right). \tag{5}$$

Multiplying both sides by $R(x)$ and defining $f(x)$ as

$$f(x) = \frac{g_0 R(x)}{g_0+\frac{\partial g_0}{\partial E_0}E_0}, \tag{6}$$

we obtain

$$\frac{d(R(x)g_0 v)}{dx} = f(x)\left(\delta i_n(x) - \Delta i_d\right). \tag{7}$$

Now we integrate both sides from 0 to L. Noticing that Δi_d is constant along the channel and v vanishes at the end points, we obtain the total drain current Δi_d as

$$\Delta i_d = \frac{\int_0^L f(x)\delta i_n(x)dx}{\int_0^L f(x)dx} = \int_0^L \Delta A_d(x)\delta i_n(x)dx, \tag{8}$$

where, IF for drain, $\Delta A_d(x)$, is clearly

$$\Delta A_d(x) = \frac{f(x)}{\int_0^L f(x)dx}. \tag{9}$$

It is easy to check that when there is neither nonuniformity nor mobility degradation, $f(x)$=1 and (8) reduces to

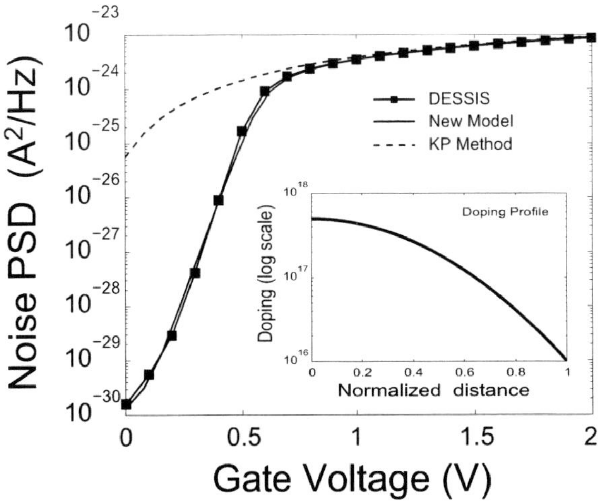

Figure 2: Plot of drain thermal noise PSD versus gate voltage for a lateral nonuniform MOSFET at $V_{DS} = 0$. The structure simulated has a source doping of 5×10^{17} cm^{-3} and a drain doping of 1×10^{16} cm^{-3} (with a gaussian profile [1]), channel length of 2 μm, width of 1 μm, and a oxide thickness of 8 nm

$\Delta i_d = \int_0^L \delta i_n(x)dx/L$ (final result of KP method) and in the presence of mobility degradation only, it reduces to our previous result [2].

To calculate the induced gate current we need to find out the voltage perturbation induced by this elementary noise source as a function of position. A potential fluctuation $v(x)$ between x and $x+\Delta x$ causes a fluctuation of the gate current by capacitive coupling. Therefore, the gate noise current Δi_g is given by

$$\Delta i_g = -j\omega W\int_0^L C_g(x)v(x)dx, \tag{10}$$

where $C_g = \frac{dQ_g}{dV}$, Q_g is the charge stored per unit area in the gate. Integrating (4) we obtain $v(x)$ as

$$v(x) = \frac{1}{R(x)g_0}\int_0^x f(x_1)\left(\delta i_n(x_1) - \Delta i_d\right)dx_1. \tag{11}$$

Therefore Δi_g becomes

$$\Delta i_g = -j\omega W\int_0^L \frac{C_g(x)}{R(x)g_0}\int_0^x f(x_1)\left(\delta i_n(x_1) - \Delta i_d\right)dx_1 dx. \tag{12}$$

We introduce the notation

$$\lambda(x) = \int^x \frac{C_g(x)}{R(x)g_0}dx, \tag{13}$$

and integrating by parts we obtain

$$\Delta i_g = -j\omega W\left(\lambda(L)\int_0^L f(x)\delta i_n(x)dx - \int_0^L \lambda(x)f(x)\delta i_n(x)dx - \lambda(L)\Delta i_d\int_0^L f(x) + \Delta i_d\int_0^L \lambda(x)f(x)dx\right). \tag{14}$$

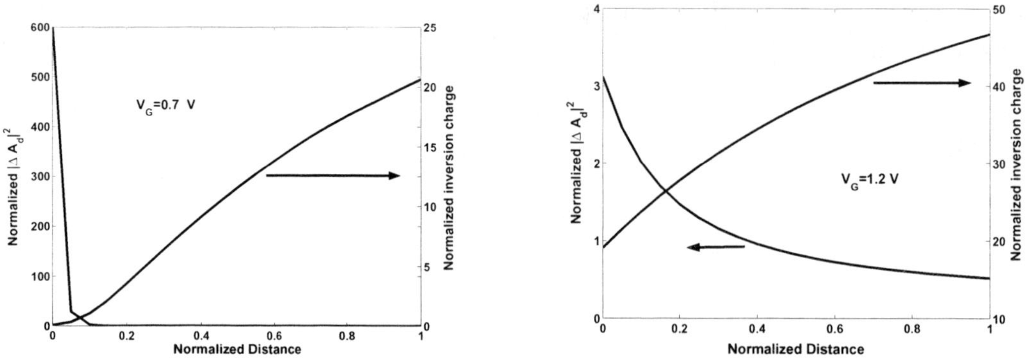

(a) (b)

Figure 3: Profile of squared drain IF (normalized with respect to the inverse of channel length) and inversion charge density (normalized by $C_{ox}U_T$) at $V_{DS} = 0$ (a) for low gate voltage ($V_G = 0.7$ V) and (b) for high gate voltage ($V_G = 1.2$ V). The PSD of the local noise source is proportional to the inversion charge. For normal MOSFET the normalized IF is equal to 1 but it is peaked near the source for lateral nonuniform MOSFET. This considerably changes the noise behavior of lateral nonuniform MOSFET

To clearly understand why conventional KP-based methods fail for a nonuniform MOSFET, let us ignore the effect of field-dependent mobility, which implies $\partial g/\partial E = 0$. Therefore, $f(x)$ now becomes

$$f(x) = \exp\left(-\int_0^x \frac{1}{g}\frac{\partial g}{\partial x}dx\right). \tag{15}$$

Once the ΔA_ks are calculated, PSDs and cross PSDs are easily obtained. We will assume that the noise sources are spatially uncorrelated. So the PSD of the local noise source $S_{\delta i_n^2}(x, x')$ can be written as

$$S_{\delta i_n^2}(x_1, x_2) = S_{\delta i_n^2}(x_1)\delta(x_1 - x_2). \tag{16}$$

Therefore the drain current PSD $S_{i_d^2}$, gate PSD $S_{i_g^2}$, and cross PSD $S_{i_d i_g}$ become

$$S_{i_d^2} = \int_0^L |\Delta A_d|^2 S_{\delta i_n^2}\,dx, \tag{17}$$

$$S_{i_g^2} = \int_0^L |\Delta A_g|^2 S_{\delta i_n^2}\,dx, \tag{18}$$

$$S_{i_g i_d} = \int_0^L \Delta A_g \Delta A_d^* S_{\delta i_n^2}\,dx. \tag{19}$$

The analytical method we just developed is applicable to any kind of noise mechanism. However, the PSD of the local noise source $S_{\delta i_n^2}$ depends on the noise mechanism and needs to be chosen carefully. In the rest of the paper we will consider mainly thermal noise. For thermal noise $S_{\delta i_n^2} = 4 \cdot q \cdot W \cdot Q_{inv} \cdot D_n$, where D_n is the noise diffusivity [5], [6]. Details about the definition of $S_{\delta i_n^2}$ and D_n can be found in [5]–[7]. For simple electron mobility model this definition of $S_{\delta i_n^2}$ reduces to the one used in [8] i.e. $S_{\delta i_n^2} = 4kT_L g$, where T_L is the lattice temperature and k is the Boltzmann's constant.

3 EFFECT OF LATERAL NONUNIFORM DOPING ON DRAIN NOISE

In a lateral nonuniformly doped MOSFET, the doping varies with position. This kind of doping makes the threshold voltage of the channel position dependent and application of KP-based method in this situation can give totally erroneous prediction. Fig. 2 illustrates this by a plot of the drain current PSD for thermal noise in equilibrium versus the gate voltage obtained from a 2-D device noise simulation (DESSIS) and the noise PSD predicted by the KP method. The figure clearly shows that even in equilibrium, at low value of gate voltages the KP method grossly overestimates (by 2-3 orders of magnitude) the noise.

In a uniformly doped MOSFET $\partial g/\partial x = 0$, which results in $f(x) = 1$ and hence $\Delta A_d(x) = 1/L$ (see (9)). The point is, IF in a uniformly doped MOSFET is independent of position and bias, whereas in nonuniform MOSFET it depends on both position and bias (because of the non vanishing $\partial g/\partial x$). Now the question is why is this effect so much pronounced at low gate voltages? The answer lies in the doping profile of these devices. The source end has a much higher doping than the drain end which results in a higher threshold voltage at the source end compared to the drain end. At low gate voltages the source will be in weak inversion and the drain will be in strong inversion. In this case, since g is very small near the source, the function $f(x)$ levels off very rapidly near the source (because of the $1/g$ term present in the expression of $f(x)$). This causes ΔA_d to highly peak near the source end. Eq. (17) reveals that the contribution to the drain terminal noise from any point gets determined by the product of two terms. The first one is the $|\Delta A_d|^2$, which represents the noise propagation and the second term is the local noise source PSD, which is proportional to the inversion charge. Fig. 3(a) shows the plot of $|\Delta A_d|^2$ and inversion charge density versus nor-

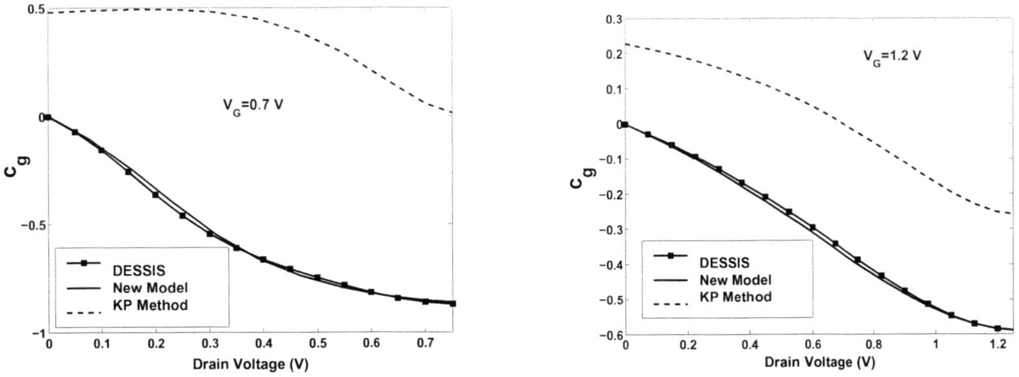

(a) (b)

Figure 4: Plot of the imaginary part of c_g versus drain voltage at (a) low gate voltage ($V_G = 0.7$ V) and (b) at high gate voltage ($V_G = 1.2$). KP method is totally incapable of predicting the behavior and even gives a wrong sign

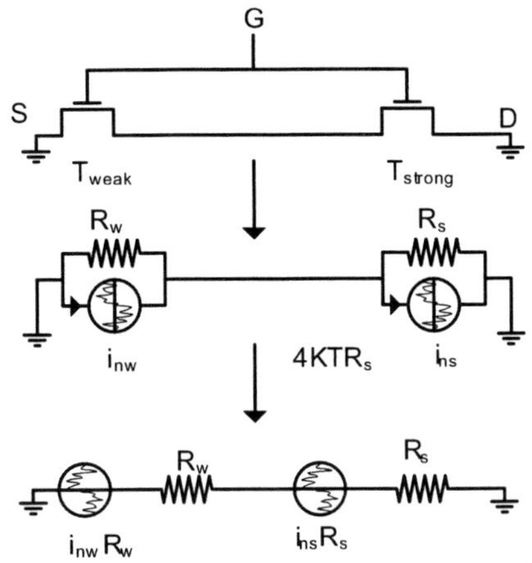

Figure 5: Intuitive explanation of the noise behavior in lateral nonuniform MOSFET. If the effect of distributed doping is lumped in to two transistor, the transistor near the source will have a higher threshold voltage(because of higher doping). At low gate voltages, the source end will be weakly inverted and transistor near drain end will be strongly inverted. As the weakly inverted transistor has a much higher resistance, it follows that the noise current of the total combination will be determined by the weakly inverted transistor

malized position at low gate voltage. Since the drain end is more strongly inverted compared to the source end, the inversion charge towards the drain end is much higher. But these charges do not contribute to the drain PSD because $|\Delta A_d|^2$ near the drain end is negligible. On the other hand, KP method predicts a ΔA_d independent of position (equal to the inverse of the channel length) which assigns the same weight to both strongly inverted region near the drain and weak/moderately inverted region near the source. As the strongly inverted drain region incorrectly gets a much higher weight, KP grossly overestimates the drain noise. Note that when the bias dependent IF is included, our model satisfactorily matches the device simulation as shown in Fig. 2.

4 EFFECT OF LATERAL NONUNIFORM DOPING ON CORRELATION COEFFICIENT AND RF NOISE PERFORMANCE

In order to validate our induced gate noise modeling approach, we plot the correlation between drain and induced gate noise, c_g, defined as $c_g = S_{i_d i_g}/\sqrt{S_{i_d^2} S_{i_g^2}}$, as a function of the drain voltage. Fig. 4(a) and Fig. 4(b) show the plots of the imaginary part of c_g for low and high gate voltages respectively and our model again gives a very good match. Note that this term also behaves considerably different from conventional MOSFET where c_g saturates to 0.6 in weak inversion [9] and to 0.4 in strong inversion. From these plots, we make two very interesting observations. First one is that KP-based methods [2], [8] for induced gate noise produces even a sign error in the correlation coefficient. This happens because the induced gate current changes sign as one moves from source to drain [2]. Here also a conventional method incorrectly puts a lower weight to the source end, and, since the charge is much higher near the drain end, the total contribution gets dominated by the drain end. But in reality, the gate noise gets dominated by the source end (note that ΔA_d is highly peaked near the source, it causes the product of ΔA_d and ΔA_g determine the sign of c_g, see (19)). As ΔA_g changes sign from source to drain, it is evident that KP method will cause a sign error. The second important observation is that the value of c_g, especially in weak inversion and high drain voltage, is higher than that of a uniformly doped MOSFET. This fact has tremendous impact on the RF noise performance of MOSFET.

The minimum noise figure F_{min} of a MOSFET is given by [10], [11]

$$F_{min} = 1 + 2\frac{\omega}{\omega_t}\sqrt{\gamma\beta}\sqrt{1-|c_g|^2}, \quad (20)$$

where γ, β are the excess noise factor for drain and gate respectively [10], [11]. From (20), it is clear that as $|c_g|$ gets

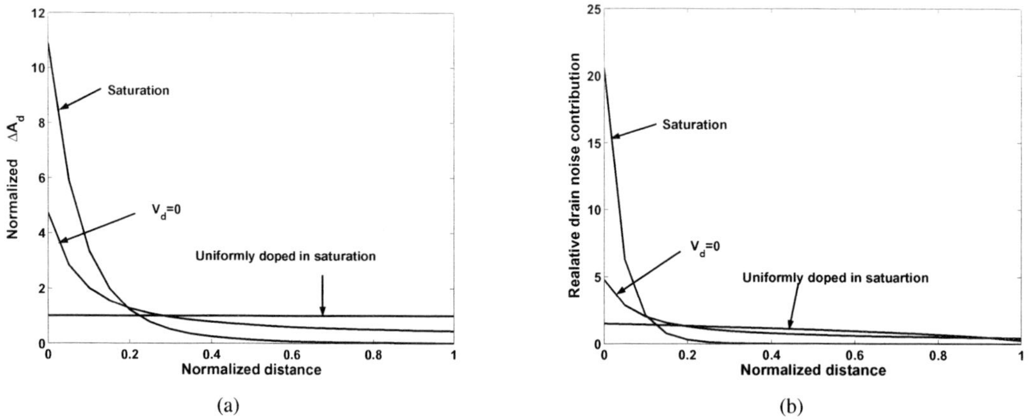

Figure 6: Profile of a) drain impedance field (normalized to the inverse of channel length) b) relative contribution to drain noise over the position ($|\Delta A_d|^2 \quad S_{\delta i_n^2}/S_{i_d^2}$) for lateral nonuniformly and uniformly doped MOSFET at $V_G = 0.7$ V. The plots shows that compared to uniformly doped MOSFET, impedance field is highly localized near the source end and the localization increases as the drain voltage is applied. As a result, in a lateral nonuniform device, only a portion near the source actually generates noise and as the drain voltage increases the fraction of the channel that actually contributes to the noise shrinks

closer to unity in a lateral nonuniform device, this kind of device will have a much smaller minimum noise figure as demonstrated in [12].

In the rest of this Section, we will explain the physical mechanism which causes a very high correlation between drain and gate noise in lateral nonuniform devices. If the effect of distributed doping is lumped into two transistors as shown in Fig. 5, then the transistor near the source will be weakly inverted and the transistor near the drain will be strongly inverted. If the noise current of the weakly inverted transistor is i_{nw} and the noise current of the strongly inverted transistor is i_{ns}, the total terminal noise current i_{nt} would be,

$$i_{nt} = \frac{R_w}{R_w + R_s} i_{nw} + \frac{R_s}{R_w + R_s} i_{ns}, \qquad (21)$$

where R_w and R_s are the channel resistances of the weakly and strongly inverted transistor, respectively.

Now, as $R_w >> R_s$, the terminal current mainly gets determined by the transistor near the source end which would imply that the impedance field ΔA_d is localized near the source. Now what happens if the drain voltage is increased? As $R_w >> R_s$, the channel potential drops mainly across the weakly inverted transistor. The resistance of the weakly inverted channel depends exponentially on the channel potential whereas the strongly inverted transistor has a linear dependence. As a result, the change in the resistance of the weakly inverted transistor ΔR_w is much greater than the change in the strongly inverted transistor $\Delta R_w >> \Delta R_s$. It makes the impedance field even more strongly localized near the source (see Fig. 6). Fig. 6 also shows the relative contribution to the drain noise from position x (i.e. $|\Delta A_d|^2 S_{\delta i_n^2}/S_{i_d^2}$), which clearly indicates that in a lateral nonuniform device only a portion near the source end actually generates noise and as the drain voltage increases, the fraction of the channel that actually contributes to the noise, shrinks.

Induced gate current originates because the drain noise current generated from a local noise source sets up a fluctuation in the surface potential. Because the nonuniform doping drastically reduces the impedance field ΔA_d near the drain, the noise sources located at the drain end are not very efficient in creating a drain noise current which is the cause of induced gate current. As a result ΔA_g also becomes localized near the source and this localization increases as the drain voltage increases. Fig. 7 shows the profile of ΔA_g and the relative contribution to the gate noise (i.e. $|\Delta A_g|^2 S_{\delta i_n^2}/S_{i_g^2}$) from position x. This figure clearly indicates that the gate noise contribution is also localized near the source. Notice that the relative gate and drain noise contributions plotted along the channel (Fig. 6(b) and Fig. 7(b)) are almost of the same magnitude and shape, which is the signature of a high correlation between the induced gate and drain noise currents.

Up to this, we have understood that in a lateral nonuniform MOSFET both ΔA_d and ΔA_g are strongly localized near the source and this localization increases as the drain bias increases. Now consider what happens to c_g if ΔA_d and ΔA_g are localized at some point x_0. Then the contribution to the PSDs and cross PSDs comes mainly from a region Δx_0 around x_0. Therefore, $S_{i_d^2} = |\Delta A_d(x_0)|^2 S_{\delta i_n^2}(x_0)\Delta x_0$, $S_{i_g^2} = |\Delta A_g(x_0)|^2 S_{\delta i_n^2}(x_0)\Delta x_0$ and $S_{i_g i_d} = \Delta A_g(x_0)\Delta A_d^*(x_0) S_{\delta i_n^2}(x_0)\Delta x_0$. From the definition of c_g, it follows that $c_g \approx -j$, meaning that induced gate and drain noise currents are fully correlated (c_g is purely imaginary because of the capacitive coupling occurring between channel noise and the gate). This result is also very intuitive. Had there been a single noise source in the channel, drain and gate noise would have been fully correlated (because they originate from the same noise source) and localized impedance fields actually turns the situation in the channel closer to this ideal one.

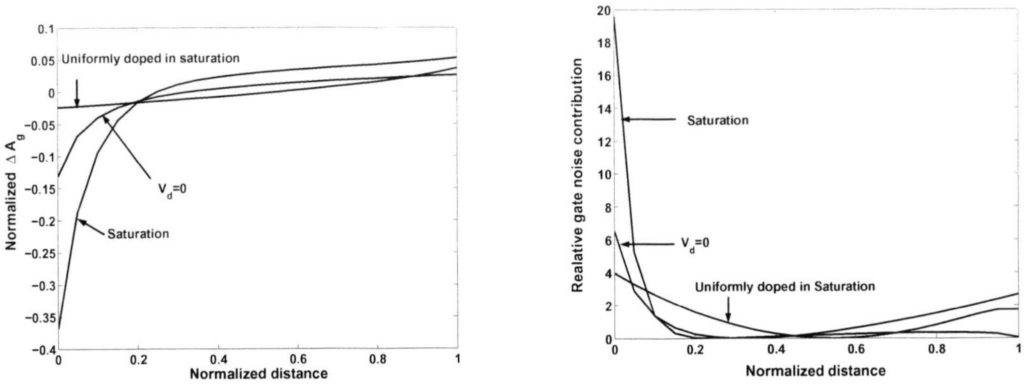

(a) (b)

Figure 7: Profile of a) gate impedance field (normalized with respect to $-j\omega L/\mu U_T$) b)relative contribution to gate noise over the position ($|\Delta A_g|^2 \quad S_{\delta i_n^2}/S_{i_g^2}$) for a lateral nonuniformly and uniformly doped MOSFET at $V_G = 0.7$ V. The plots shows that compared to uniformly doped MOSFET, impedance field is highly localized near the source end and the localization increases as the drain voltage is applied. As a result, in a lateral nonuniform device, only a portion near the source end actually generates noise and as the drain voltage increases, the fraction of the channel that actually contributes to the noise shrinks

5 CONCLUSION

In this work we have presented a general analytical noise modeling methodology accounting for both lateral nonuniformity and field-dependent mobility. This methodology is applicable to any kind of noise mechanism, doping profile and field-dependent mobility. It is shown that conventional KP-based methods, at low gate voltages, can overestimate the thermal noise by 2-3 orders of magnitude. In addition, a very high correlation between drain and induced gate noise improves the RF noise properties of the MOSFET. The high correlation between drain and gate noise essentially arises from the strong localization of drain and gate impedance field, the quantity responsible for noise propagation. This high localization of impedance fields occurs because of the fundamental fact that the presence of nonuniform doping makes the drain impedance field, contrary to the uniformly doped MOSFET, position and bias dependent.

REFERENCES

[1] F. M. Klaassen and J. Prins, "Thermal Noise of MOS Transistors," *Philips Res. Repts*, vol. 22, pp. 505–514, Oct. 1967.

[2] A. S. Roy, C. C. Enz, and J.-M. Sallese, "Noise Modeling Methodologies in the Presence of Mobility Degradation and Their Equivalence," *IEEE Trans. Electron Devices*, vol. 53, no. 2, pp. 348–355, Feb. 2006.

[3] A. S. Roy, Y. S. Chauhan, C. C. Enz, and J.-M. Sallese, "Noise Modeling in Lateral Asymmetric MOSFET," in *IEEE International Electron Devices Meeting*, Dec. 2006, pp. 751 – 754.

[4] A. S. Roy, C. C. Enz, and J.-M. Sallese, "Noise Modeling in Lateral Nonuniform MOSFET," *IEEE Trans. Electron Devices*, vol. 54, no. 8, pp. 1994–2001, Aug. 2007.

[5] J. P. Nougier, "Noise and diffusion of hot carriers ," in *Physics of nonlinear transport in semiconductors*, D. K. Ferry, J. R. Barker, and C. Jacoboni, Eds. Plenum Press, 1980, pp. 415–477.

[6] ——, "Fluctuations and Noise of Hot Carriers in Semiconductor Materials and Devices," *IEEE Trans. Electron Devices*, vol. 41, no. 11, pp. 2034–2048, Nov. 1994.

[7] A. S. Roy and C. Enz, "Compact Modeling of Thermal Noise in the MOS Transistor," *IEEE Trans. Electron Devices*, vol. 52, no. 4, pp. 611–614, Apr. 2005.

[8] J. C. J. Paasschens, A. J. Scholten, and R. van Langevelde, "Generalisations of the Klaassen-Prins Equation for Calculating the Noise of Semiconductor Devices," *IEEE Trans. Electron Devices*, vol. 52, no. 11, pp. 2463– 2472, Nov. 2005.

[9] A.-S. Porret and C. C. Enz, "Non-Quasi-Static (NQS) Thermal Noise Modelling of the MOS Transistor," *IEE Proceedings Circuits, Devices and Systems*, vol. 151, no. 2, pp. 155–166, April 2004.

[10] A. Cappy, "Noise Modeling and Measurement Techniques," *IEEE Trans. Microwave Theory Tech.*, vol. 36, no. 1, pp. 1–10, Jan. 1988.

[11] C. Enz and Y. Cheng, "MOS Transistor Modeling for RF IC Design," *IEEE Journal of Solid-State Circuits*, vol. 35, no. 2, pp. 186–201, Feb. 2000.

[12] T. C. Lim, R. Valentin, G. Dambrine, and F. Danneville, "MOSFETs RF Noise Optimization via Channel Engineering," *IEEE Electron Device Letters*, vol. 29, no. 1, pp. 118–121, Jan. 2008.

An Iterative Approach to Characterize Various Advanced Non-Uniformly Doped Channel Profiles

R. Kaur*, R. Chaujar*, M. Saxena** and R. S. Gupta*

*Semiconductor Devices Research Laboratory, Department of Electronic Science, University of Delhi, South Campus, New Delhi-110021, India, ravneetsawhney13@rediffmail.com,rishuchaujar@rediffmail.com,rsgu@bol.net.in
**Department of Electronics, Deen Dayal Upadhyaya College, University of Delhi, New Delhi, India, saxena_manoj77@yahoo.co.in

ABSTRACT

In this paper, an efficient drain current model for sub-100nm channel engineered LDD, halo and their combination, has been presented. Drain Induced Barrier Lowering (DIBL) effect has been incorporated in the analytical model through Voltage Doping Transformation (VDT) method, which replaces the influence of the lateral drain-source field by an equivalent reduction in the channel doping concentration. The analytical assessment require an iterative approach for evaluation of the short channel depletion width for regions of different doping, from which the surface potential, surface electric field and hence drain current model have been evolved. Comparisons have been drawn among the studied devices based on their improved subthreshold and on-state performances.

Keywords: ATLAS-2D, DIBL, hot carrier reliability, NUDC, and VDT

1 INTRODUCTION

Since past three decades, in the pursuit of superior performances relative to high-speed circuits and packing density, miniaturization of device dimensions has been adopted as a powerful tool. Gradually, as device feature sizes move into sub-100nm regime, the device characteristics degrade due to the emergence of typical Short Channel Effects (SCEs) such as steep threshold voltage roll-off, increased off-state leakage current, DIBL, hot carrier effect and the degradation of carrier mobility. To combat these SCEs, the design of ultra-small devices necessitates the use of various channel engineered [1-2] architectures using Non-Uniform Doping Channel (NUDC) profiles.

In this paper, an efficient drain current model for sub-100nm channel engineered LDD, halo and their combination, has been presented. Using Poisson's equation, a simple 2D potential distribution model in the channel for NUDC MOSFETs as shown in Fig.1, has been developed using which the drain current model is obtained. The model incorporates DIBL effect using Voltage Doping Transformation (VDT) [3] method, which replaces the influence of the lateral drain-source field by an equivalent reduction in the channel doping concentration. The analytical evaluation of d –the short channel depletion width of differently doped region, requires iterative solution and the process flow to develop the drain current model is described in the flowchart given in Fig. 2. Comparisons have been drawn among the studied devices based on their improved subthreshold and on-state performances.

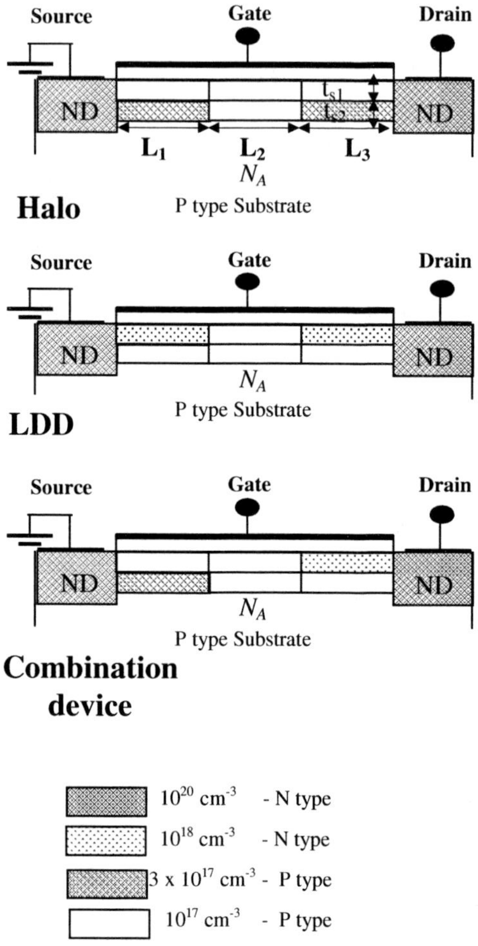

Fig.1: Schematic Design of various devices under consideration with their doping profiles. Channel Length = L (L_1:L_2:L_3=1:1:1), W=1000nm, ε_{si} = 3.9, t_{ox}= 3nm, t_{s1} = 10nm, t_{s2} = 10nm, X_j=30nm and $q\Phi_M$ = 4.77 eV.

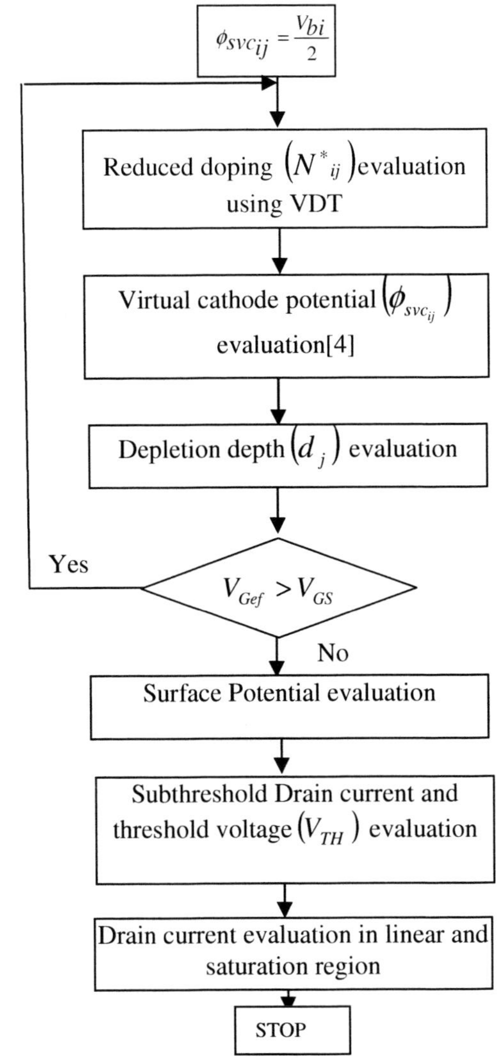

Fig.2 Flow chart for analytical model evaluation. (i=1,2 corresponds to region I, II, III and IV, V, VI respectively and j=1,2,3 corresponds to region I and IV; II and V; III and VI respectively.)

2 MODEL FORMULATION

The lateral influence of source and drain field onto the vertical growth of substrate depletion region is provided by the concept of VDT [2]. The lateral field influence initiated by the *S/D* junctions causes an equivalent reduction in effective substrate doping and the effective doping in the channel region is calculated as

$$N'_{Aij} = N_{Aij} - \frac{2\varepsilon_{si}}{qLc_{1j}} V^*_{Dij} \qquad (1)$$

$$V^*_{Dij} = \begin{cases} V_{i1} = V_{bi} - \phi_{sL1} + 2\left(V_{bi} - \phi_{svc_{i1}}\right) + \\ 2\sqrt{\left(V_{bi} - \phi_{svc_{i1}}\right)\left(\phi_{sL1} - \phi_{svc_{i1}}\right)} & 0 < x \le L_1 \\ V_{i2} = \phi_{sL1} - \phi_{sL2} + 2\left(\phi_{sL1} - \phi_{svc_{i2}}\right) + \\ 2\sqrt{\left(\phi_{sL1} - \phi_{svc_{i2}}\right)\left(\phi_{sL2} - \phi_{svc_{i2}}\right)} & L_1 < x \le L_1 + L_2 \\ V_{i3} = \phi_{sL2} - \left(V_{bi} + V_{DS}\right) + 2\left(\phi_{sL2} - \phi_{svc_{i3}}\right) + \\ 2\sqrt{\left(\phi_{sL2} - \phi_{svc_{i3}}\right)\left(V_{bi} + V_{DS} - \phi_{svc_{i3}}\right)} & L_1 + L_2 < x \le L_1 + L_2 + L_3 \\ \hline t_1 < y \le t_2 \left(= t_{s1} + t_{s2}\right) & \\ 2V_{bi} + V_{DS} - 2\sqrt{V_{bi}\left(V_{bi} + V_{DS}\right)} & 0 < x \le L \ \& \ y > t_2 \end{cases} \qquad (2)$$

such that

$$\phi_{sL1} = -\left(\frac{R_1}{R_1 + R_2 + R_3}\right) V_{DS} + \left[V'_{GS_1} + \frac{qN_{11}\varepsilon_{si}}{\varepsilon_{ox}^2} t_{ox}^2 \left(1 + \sqrt{1 + \frac{2}{qN_{11}\varepsilon_{si}\varepsilon_{ox}^2} t_{ox}^2 \left(V'_{GS_1} + V_{bi}\right)}\right)\right] + V_1 \qquad (3)$$

$$\phi_{sL2} = -\left(\frac{R_1 + R_2}{R_1 + R_2 + R_3}\right) V_{DS} + \left[V'_{GS_3} + \frac{qN_{13}\varepsilon_{si}}{\varepsilon_{ox}^2} t_{ox}^2 \left(1 + \sqrt{1 + \frac{2}{qN_{13}\varepsilon_{si}\varepsilon_{ox}^2} t_{ox}^2 \left(V'_{GS_3} + V_{bi}\right)}\right)\right] + V_2 \qquad (4)$$

$$R_j = \frac{L_j}{q\mu_{1j} N_{1j} t_{s1} W(width)} \qquad \text{for } j=1,2,3 \qquad (5)$$

ϕ_{svc} is the potential distribution along *y-axis* and is obtained using $\phi_{vc}(y) = \psi(x=0, y)$.[4]

$$Lc_{ij} = \begin{cases} Lc_{1j} = L_j & 0 < y \le t_1 (= t_{s1}) \\ Lc_{2j} = L_j + \alpha_j & t_1 < y \le t_2 (= t_{s1} + t_{s2}) \\ Lc_3 = \sum_{j=1}^{3} L_j + \beta_j & y > t_2 \quad \text{for } i=1,2 \ \ j=1,2,3 \end{cases} \qquad (6)$$

here α_j and β_j are the extra lengths of current lines in the second layer and bulk.

So, the expression for depletion depth d_j is given as

$$d_j = t_{s1} + t_{s2} + U_j \qquad \text{for } j=1,2,3 \qquad (7)$$

where $U_j = \sqrt{\left(t_{s1} + t_{s2} + t_{ox}\frac{\varepsilon_{si}}{\varepsilon_{ox}}\right)^2 + \frac{2\varepsilon_{si}}{qN^*_3}\left(V'_{GS_j} - V_{gj}\right)} - \left(t_{s1} + t_{s2} + t_{ox}\frac{\varepsilon_{si}}{\varepsilon_{ox}}\right)$ (8)

$$V_{gj} = \frac{q}{\varepsilon_{si}} N^{*}{}_{2j} t_{s2} \left(t_{s1} + \frac{t_{s2}}{2} + \varepsilon_{si} \frac{t_{ox}}{\varepsilon_{ox}} \right) + \frac{q}{\varepsilon_{si}} N^{*}{}_{1j} t_{s1} \left(\frac{t_{s1}}{2} + \varepsilon_{si} \frac{t_{ox}}{\varepsilon_{ox}} \right) \tag{9}$$

$$W_d = \begin{cases} d_1 & 0 \geq x \geq L_1 \\ d_2 & L_1 > x \geq L_1 + L_2 \\ d_3 & L_1 + L_2 > x \geq L_1 + L_2 + L_3 \end{cases} \tag{10}$$

Using the superposition principle, the 2D analytical solutions for potential is obtained as

$$\xi(x,y) = \xi_l(y) + \xi_s(x,y)\,;\ 0 \leq x \leq L \ \&\ 0 \leq y \leq W_d \tag{11}$$

where $\xi_l(y) = \dfrac{qN'_A(y - W_d)^2}{2\varepsilon_{si}} + V_{sub}$;

$$\xi_s(x,y) = \xi_{s1}(x,y) + \xi_{s2}(x,y)$$

$$\xi_{s1}(x,y) = \sum_{r=1}^{\infty} \sin\left(\frac{r\pi x}{L}\right) \times \left(\frac{\chi 1_r \sinh\left(\frac{r\pi y}{L}\right) + \chi 2_r \sinh\left(\frac{r\pi (W_t - y)}{L}\right)}{\sinh\left(\frac{r\pi W_t}{L}\right)} \right) \tag{12}$$

$$\xi_{s2}(x,y) = \sum_{r=1}^{\infty} \sin\left(\frac{r\pi y}{W_t}\right) \times \left(\frac{\chi 3_r \sinh\left(\frac{r\pi x}{W_t}\right) + \chi 4_r \sinh\left(\frac{r\pi (L_1 - x)}{W_t}\right)}{\sinh\left(\frac{r\pi L}{W_t}\right)} \right) \tag{13}$$

N'_A is the reduced doping (in the respective region) and W_d is the depletion depth obtained using VDT. 2D-channel potential is used to model subthreshold drain current from which threshold voltage (V_{TH}) is evaluated. Using V_{TH} drain current in the linear and saturation region can be calculated analytically as a function of V_{GS} and V_{DS} as

$$I_{DLinear} = \frac{\mu W C_{ox} \left((V_{GS} - V_{TH}) V_{DS} - \frac{1}{2} a_o V_{DS}{}^2 \right)}{\left(1 + \frac{V_{DS}}{L E_c} \right) L} \tag{14}$$

$$I_{DSAT} = \frac{\mu W C_{ox} \left((V_{GS} - V_{TH}) V_{DSAT} - \frac{1}{2} a_o V_{DSAT}{}^2 \right)}{\left(1 + \frac{V_{DSAT}}{(L - \Delta L) E_c} \right) (L - \Delta L)} \tag{15}$$

3 RESULTS AND DISCUSSION

Fig. 3 presents the comparison between the modeled and simulated drain current (I_{DS}) variation (on linear and log scale) with V_{GS} and with V_{DS} for different structures. The obtained analytical results are in fair agreement with the simulated results obtained using ATLAS 2D: device simulation software.

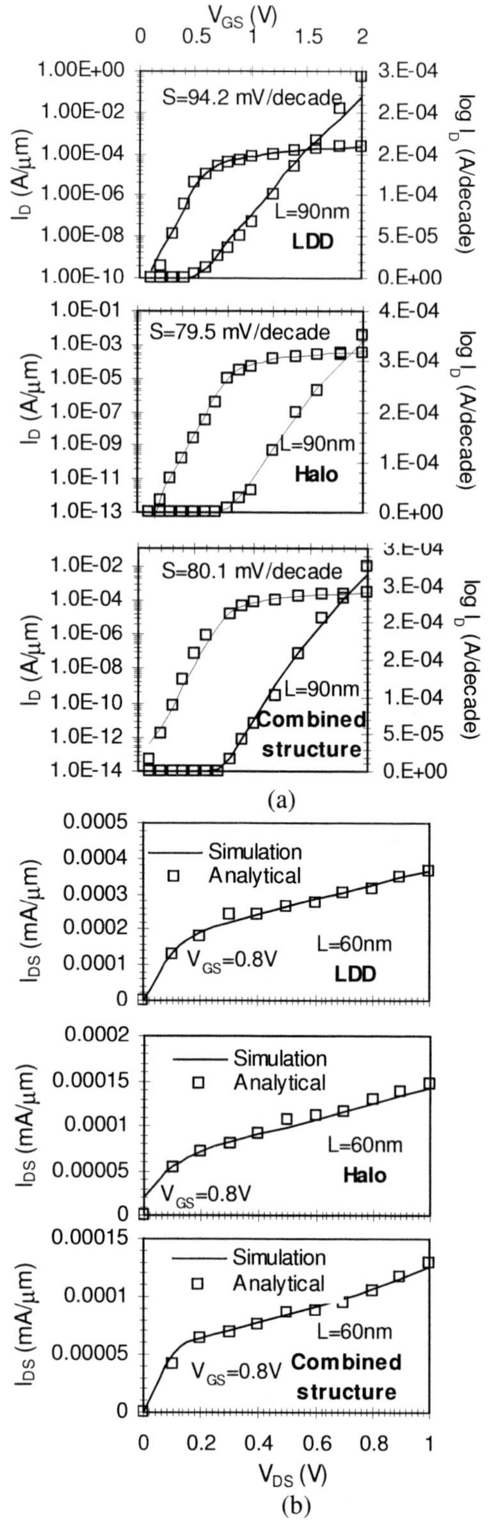

Fig. 3 Drain current variation along the channel for Halo, LDD and combined structure. $q\Phi_M = 4.77$ eV, V_{GS}=0.8V and V_{DS}=0.05V.

Fig. 4 presents the variation of the I_{on}, I_{off} and their ratio, I_{on}/I_{off} with the channel length down to 60nm. It has been observed that off-state current (I_{off}) is lowest in halo due to the presence of highly doped regions near the S/D junctions that prevents dopant out diffusion in subthreshold regime of device operation. LDD, on the other hand demonstrates higher on-state current (I_{on}) on account of having lower threshold voltage as compared to other structures. Hence, the combination bearing the benefits of both halo and LDD, exhibits I_{off} and I_{on} values in between the range, set by the two devices and hence, presents higher I_{on}/I_{off} ratio – an essential figure of merit for digital performance.

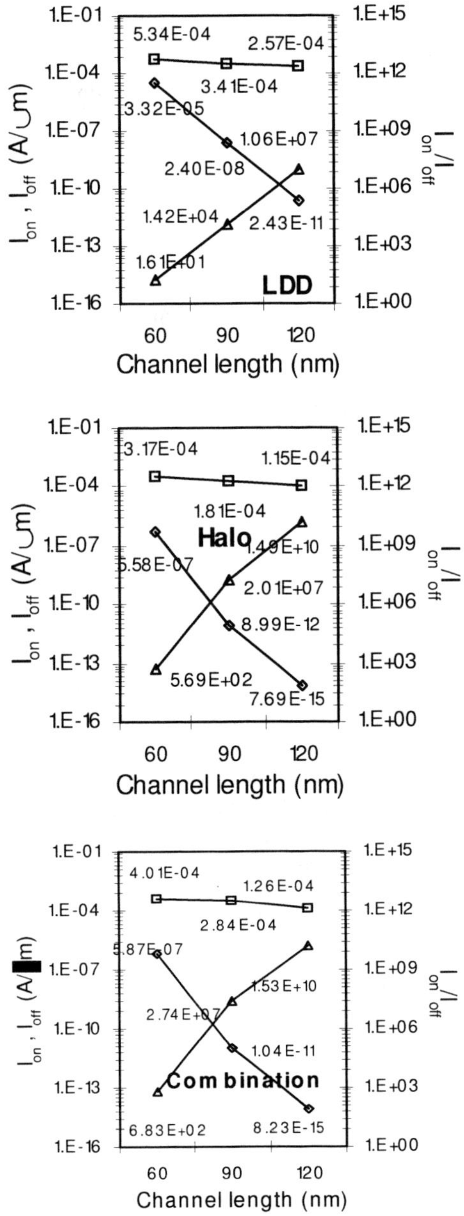

Fig.4 Variation of I_{on} and I_{off} with channel length for Halo, LDD and combination architecture. V_{DS} = 1.0Vand $q\Phi_M$ = 4.77 eV. (□□- I_{on}, ◇◇- I_{off} and ΔΔ - I_{on}/I_{off}).

4 CONCLUSION

A new simple and computationally efficient two-dimensional analytical model that can accurately model various advanced MOSFET structures incorporating the effect of DIBL via VDT has been presented and verified. The combined structure, incorporating the beneficial effect of both asymmetric halo and LDD, proves out to be the best candidate for switching applications.

ACKNOWLEDGMENTS

Authors are grateful to Defence Research and Development Organisation (DRDO) Ministry of Defence, Government of India and Rishu Chaujar is grateful to University Grants Commission (UGC) for providing the necessary financial assistance to carry out this research work.

REFERENCES

[1] B. Yu, C. H. J. Wann, E.D. Nowak, K. Noda and C. Hu, " Short-Channel Effect Improved by Lateral Channel-Engineering in Deep-Submicronmeter MOSFET's," IEEE Trans. Electron Devices, Vol. 44, No. 4, pp. 627-634, April 1997.

[2] S. C. Williams, R. B. Hulfachor, K. W. Kim, M. A. Littlejohn and W. C. Holton, "Scaling Trends for Device Performance and Reliability in Channel-Engineered n-MOSFET's," IEEE Trans. Electron Devices, Vol. 45, No. 1, pp. 254–260, January 1998.

[3] T. Skotnicki, G. Merckel, and T. Pedron, "The Voltage-Doping Transformation: A New Approach to the Modeling of MOSFET Short-Channel Effects," IEEE Trans. Electron Devices, Vol. 9, No. 3, pp. 109–112, March 1988.

[4] R.R. Troutman, "Ion-implanted threshold tailoring for insulated gate fieldeffect transistors," IEEE Trans. Electron Devices, Vol. 24, No.3, pp. 182-192, March 1977.

Modeling of Spatial Correlations in Process, Device, and Circuit Variations

Ning Lu
IBM Semiconductor Research and Development Center, Systems and Technology Group,
Essex Junction, VT 05452 USA lun@us.ibm.com

ABSTRACT

We present an innovative method to model the spatial correlations in semiconductor process and device variations or in VLSI circuit variations. Without using the commonly adopted PCA technique, we give a very compact expression to represent a given spatial correlations among a set of similar statistical variables/instances located at different places on a chip/die. Our compact expression is easy for implementation in a SPICE model and is efficient in circuit simulations.

Keywords: Spice modeling, statistical modeling, spatial correlations, statistical static timing analysis

1 INTRODUCTION

In semiconductor processes and devices as well as in VLSI circuits, the degree of correlation between any two intra-die instances of a process or device parameter or of a circuit decreases with increasing separation between them. Examples include CMOS field effect transistor (FET) channel length, FET channel width, diode current, diffused or poly resistor's resistance value, MIMCAP's capacitance value, interconnect resistance or capacitance values, ring oscillator's period/speed, the speed of other logic circuits (e.g., NAND, NOR, etc.), and many other device parameters. Various measured hardware data has revealed such a gradual de-correlation of spatial correlation over distance (see Fig. 1; also [1–2]). In statisitcal static timing analysis for VLSI circuits, spatial correlation modelling has been studied [1, 3–5]. However, all of these papers use the principal component analysis (PCA) technique, resulting a large correlation matrix for a set of spatially correlated instances of a process/device/ciruit parameter located across a part of a chip or the whole chip. Subsequent matrix diagonalization yields many eigenvalues and eigenvectors.

In this paper, we present an innovative method to model the spatial correlations in semiconductor process and device variations or in VLSI circuit variations. Without using the commonly adopted PCA technique, we give a very compact expression to represent a given spatial correlation among a set of similar statistical variables/instances located at different places on a chip/die. Our compact expression is easy for implementation in a SPICE model and is efficient in circuit simulations. Describing intra-die variations using spatial (i.e., distance-dependent) correlations unifies various descriptions of intra-die variations, such as mismatch, across chip variations, random uncorrelated variations, and random correlated variations, etc.

2 MODELLING ONE-DIMENSIONAL SPATIAL CORRELATIONS

Consider spatial correlation within a strip of a chip region. All instances of a device parameter characteristics have the same mean value x_0 and the same standard deviation σ, but the correlation between any two instances vary (e.g., decrease) with the separation/distance between them. We devide the strip into I sub-regions (Fig. 2), and I can be a very large integer. All instances within the same sub-region are treated as perfectly correlated, but any two instances from different sub-regions are treated as either partially correlated or have no correlation. Let c_{ij} be the correlation coefficient between a device instance in sub-region i and another device instance in sub-region j.

I. We start from a special nearest-neighbor-only correlation problem,

$$c_{ii} = 1, i = 1, 2, \cdots, I; \;\; c_{i,i+1} = c_{i+1,i} = \pm\frac{1}{2}, i = 1, 2, \cdots, I-1;$$
$$c_{ij} = 0, \;\; i, j = 1, 2, \cdots, I, \;\; |i-j| \geq 2. \qquad (1)$$

The size (I x I) of the correlation matrix can be very larg. A set of compact solutions is simply

$$x_i = x_0 + \sigma(g_i \pm g_{i+1})/\sqrt{2}, \quad i = 1, 2, \cdots, I, \qquad (2)$$

where (and throughout the paper) each of g_1, …, g_{I+1} is an indepenent stochastic/random variable of mean zero and standard deviation one. Solution (2) uses (I + 1) indepenent stochastic variables, but each x_i has only two stochastic (i.e., g_j) terms and thus is very compact and very suitable for fast SPICE simulations. If the PCA technique were used to find a solution of (1), then I indepenent stochastic variables would be used and each x_i would depend on many stochastic terms.

II. When the degree of correlation in (1) is more general,

$$c_{i,i+1} = c_{i+1,i} = r, \quad i = 1, 2, \cdots, I-1, \quad |r| \leq 1/2, \qquad (3)$$

then the solution (2) is replaced by

$$x_i = x_0 + \sigma[a_1 g_i + a_2(r) g_{i+1}], \quad i = 1, 2, \cdots, I, \qquad (4a)$$

$$a_2(r) = \mathrm{sgn}(r)\sqrt{(1+\sqrt{1-4r^2})/2}, \quad a_1 = \sqrt{1-a_2^2}. \qquad (4b)$$

III. We next consider a long-distance partial correlation in which the degree of correlation decreases linearly with the distance between sub-regions,

$$c_{i,i+m} = c_{i+m,i} = 1-(m/M), \; i = 1, \cdots, I, \; j = 0, 1, \cdots, M,$$

$$c_{ij} = 0\,,\quad i, j = 1, 2, \cdots, I, \quad |i - j| \geq M\,. \tag{5}$$

A set of compact solutions is found to be

$$x_i = x_0 + \frac{\sigma}{\sqrt{M}} \sum_{k=1}^{M} g_{i+k-1}\,,\quad i = 1, 2, \cdots, I\,, \tag{6}$$

which uses $(I + M - 1)$ indepenent stochastic variables, but each instance x_i uses only M stochastic terms.

IV. We now consider a general spatial correlation of correlation length $(M - 1)$ units,

$c_{i,i\pm m} = f(M,m)$, $m = 0, 1, \ldots, M$,

$f(M, 0) = 1$, $f(M, M) = 0$, and all other $c_{ij} = 0$. (7)

We use $(I + M - 1)$ indepenent stochastic variables $g_1, \ldots, g_{I+M-1}$ to represent I correlated instances of a process, device, or circuit parameter characteristics,

$$x_i = x_0 + \sigma \sum_{k=1}^{M} a_k g_{i+k-1}\,,\quad i = 1, 2, \cdots, I\,. \tag{8}$$

Here M weighting coefficients a_k satisfy M equations,

$$f(M,m) = \sum_{k=1}^{M-m} a_k a_{k+m}\,,\quad m = 0, 1, \cdots, M - 1\,. \tag{9}$$

$m = 0$ case is the normalization condition, $\sum_{k=1}^{M} a_k^2 = 1$.

(i). Often, it is easier to first try an expression for a_k and then find corresponding spatial correlation. An example: Let a_k vary linearly with k,

$$a_k = k\sqrt{6/[M(M+1)(2M+1)]}\,,\quad k = 1, 2, \cdots, M\,. \tag{10}$$

Then, the corresponding spatial correlation is found analytically,

$$f(M,m) = (1 - m/M)[1 - m/(M+1)][1 + m/(2M+1)]. \tag{11}$$

Analytic solutions can also be obtained when a_k is proportional to k^n with n being a positive integer.

(ii) In practice, it is straightforward to first select a family of solution a_k curves and then calculate a family of corresponding spatial correlations numerically. Figure 4(a) shows a family of solution a_k curves in the power form,

$$a_k = \beta k^p\,,\quad k = 1, 2, \cdots, M\,, \tag{12}$$

and Fig. 4(b) plots a family of corresponding spatial correlations. In Eq. (12) and Eqs. (13)–(15) in the following, p is a real value, and β is a normalization constant. All spatial correlations in this family have a sharp peak at the center $m = 0$. Changing a_k in Eq. (12) to a symmetric form about the middle k value, one has

$$a_k = \beta(k_0 - |k - k_0|)^p\,,\quad k_0 = \tfrac{1}{2}(M+1)\,. \tag{13}$$

Figures 5(a) and 5(b) show solution a_k curves in Eq. (13) and corresponding spatial correlations for several power p values, respectively. Some of spatial correlations in this family have a smoother peak at the center $m = 0$. Various other forms of a_k curves are easily constructed. Here are two additional families of a_k curves. Figures 6(a) and 6(b) show the following family of asymmetric solution a_k curves,

$$a_k = \beta[3(k/M)^2 - 2(k/M)^3]^p\,,\quad k = 1, 2, \cdots, M\,, \tag{14}$$

and corresponding spatial correlations for several power p values, respectively. All spatial correlations in this family have a sharp peak at the center $m = 0$. Figures 7(a) and 7(b) plot the following family of symmetric solution a_k curves,

$$a_k = \beta \sin^{2p}[\pi k/(M+1)]\,,\quad k = 1, 2, \cdots, M\,, \tag{15}$$

and corresponding spatial correlations for several power p values, respectively. Most of spatial correlations in this family have a smoother peak at the center $m = 0$. Last, we select one of spatial correlations that is closest to a given spatial correlation, and use the corresponding set of a_k in a SPICE model or in a statistical static timing analysis.

3 MODELLING TWO-DIMENSIONAL SPATIAL CORRELATIONS

We now consider spatial correlation in a region of a chip or in the whole chip. A chip/die is divided into $(I \times J)$ sub-regions. Let $C(i, j; k, l)$ denote the correlation coefficient between a device in sub-region (i, j) and another device in sub-region (k, l) (Fig. 8).

I. We first treat a nearest-neighbor-only correlation problem: Correlation degree is r, and its range is 1 unit in either x or y direction but not in diagonal directions: $C(i, j; i \pm 1, j) = C(i, j; i, j \pm 1) = r$, but $C(i, j; i \pm 1, j \pm 1) = 0$. We use $(IJ + I + J)$ independent stochastic variables to model IJ correlated instances of a device parameter characteristics,

$$x_{i,j} = x_0 + \sigma\sqrt{1 - a_2^2}\, g_{i,j} + \sigma a_2(\sqrt{2}r)(g_{i+1,j} + g_{i,j+1})/\sqrt{2},$$

with $|r| \leq \sqrt{2}/4$, where a_2 function is given in Eq. (4b).

II. We next study a general two-dimensional correlation problem of correlation length $(M - 1)$ $(N - 1)$ units in the x (y) direction, $C(i, j; i \pm m, j \pm n) = F(M, m; N, n)$, $m = 0, 1, 2, \ldots, M$, $n = 0, 1, 2, \ldots, N$, $F(M, M; N, n) = 0$, $F(M, m; N, N) = 0$, and all other $C(i, j; k, l) = 0$. We use $(I + M - 1)(J + N - 1)$ independent stochastic variables to represent IJ correlated process/device/circuit instances,

$$x_{ij} = x_0 + \sigma \sum_{k=1}^{M} \sum_{l=1}^{N} A_{kl}\, g_{i+k-1,\,j+l-1}\,,\ i = 1, \ldots, I,\ j = 1, \ldots, J.$$

Here, MN weighting coefficients satisfy MN relations

$$F(M, m; N, n) = \sum_{k=1}^{M-m} \sum_{l=1}^{N-n} A_{k,l} A_{k+m,l+n} ,$$

$$m = 0, 1, \ldots, M-1, n = 0, 1, \ldots, N-1. \qquad (16)$$

The relation at $m = n = 0$ is the normalization condition, $\sum_{k=1}^{M} \sum_{l=1}^{N} A_{kl}^2 = 1$. For any one-dimensional spatial correlation $f(M, m)$ and corresponding solution a_k, we can generate corresponding two-dimensional spatial correlation and solution by a decomposition method in Eq. (16): $A_{kl} = a_k a_l$, $F(M, m; N, n) = f(M, m) f(N, n)$. For example, the one-dimensional linear-decay correlation relation in Eq. (5) (Fig. 3) becomes a bilinear decay spatial correlation in two dimensions (Fig. 9).

4 SUMMARY

We have presented a compact solution to the problem of modeling spatial correlations in semiconductor process, device, and circuit variations. By seeking solutions in a higher dimensional space, we have obtained a much compact solution than the commonly used PCA technique would give. We have also presented a method to obtain a closest solution for a desired spatial correlation. Our compact expression is easy for implementation in a SPICE model and is efficient in circuit simulations. Also, our compact solution is good for statistical static timing analysis, where the spatial correlation is an important consideration.

ACKNOWLEDGMENT

The author would like to thank IBM management for suppport.

REFERENCES

[1] H. J.-S. Doh, et al., *Proceedings of International Conference on Simulation of Semiconductor Processes and Devices*, 2005, pp. 131–134.

[2] A. Gattiker, M. Bhushan, and M. B. Ketchen, *Proc. IEEE International Test Conference,* 2006, pp. 1–10.

[3] H. Chang and S. S. Sapatnekar, *IEEE Transactions on Computer-Aided Design of Integrated Circuits and Systems*, Vol. **24**, pp. 1467–1482 (2005).

[4] A. Agarwal, et al., *Proceedings of the Asia and South Pacific Design Automation Conference (ASP-DAC)* 2003, pp. 271–276.

[5] H. Mangassarian, M. Anis, *Proceedings of Design, Automation and Test in Europe*, 2005, pp. 132–137.

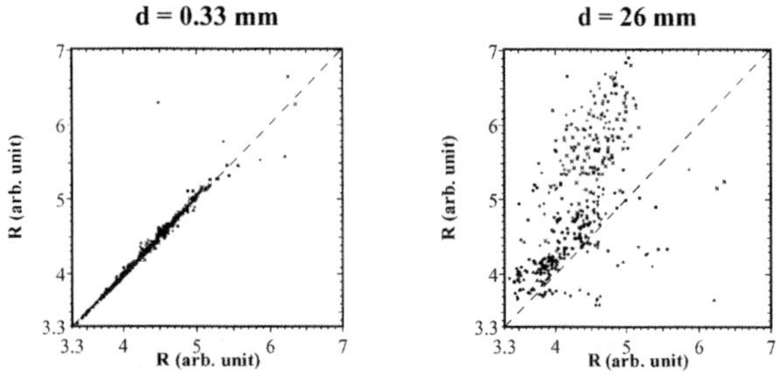

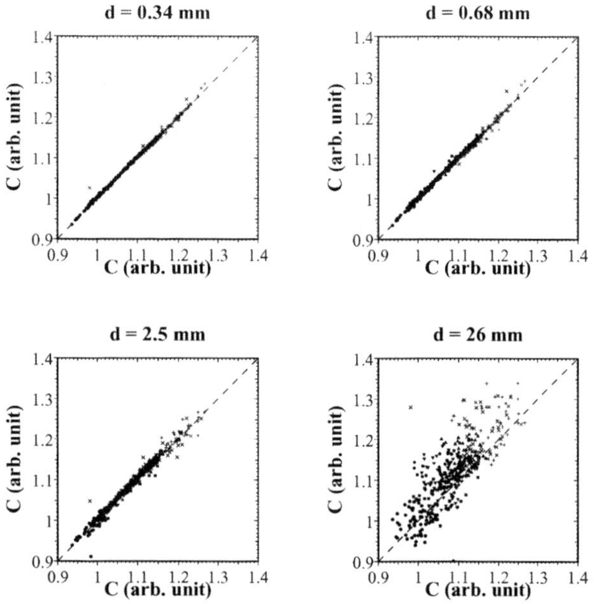

Fig. 1. Scatter plots show the correlation between measured interconnect resistance (R) and capacitance (C) of same wire width/space (but distance d apart) from an IBM 45 nm technology.

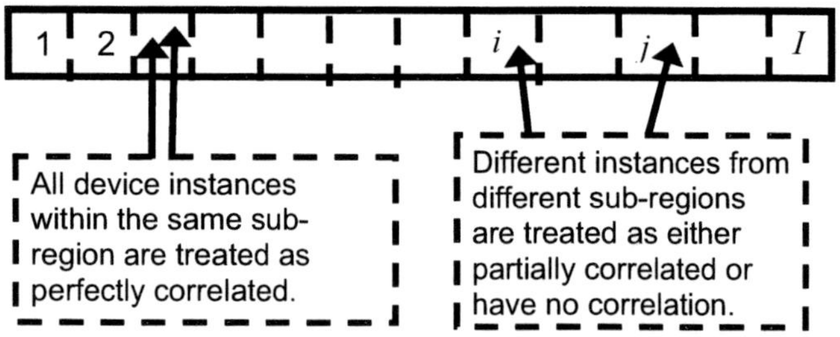

Fig. 2. One-dimensional case: A strip of a chip region is divided into I sub-regions. c_{ij} is the correlation coefficient between a device instance in sub-region i and another device instance in sub-region j.

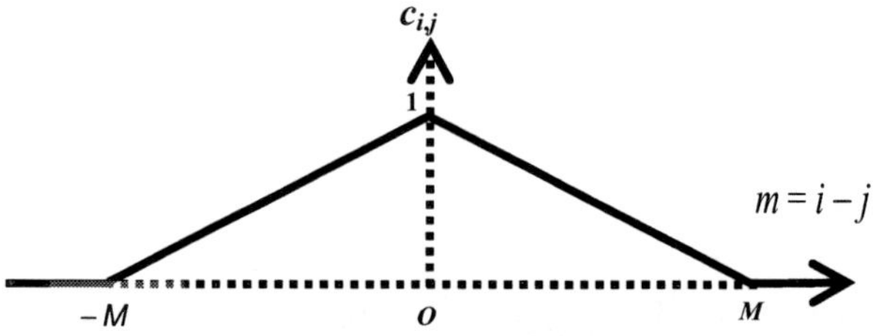

Fig. 3. One-dimensional linear-decay spatial correlation.

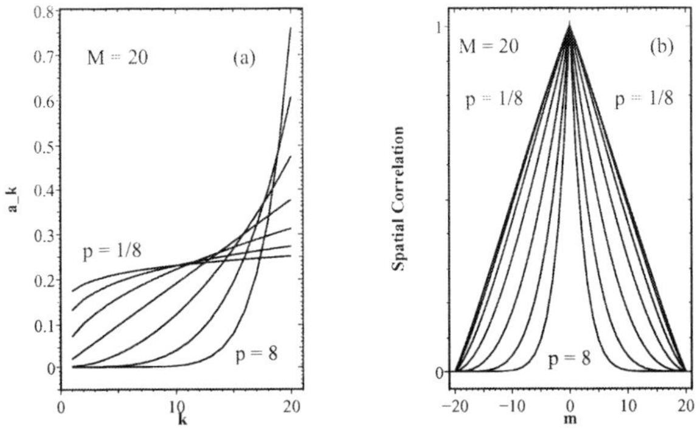

Fig. 4. (a) shows a family of asymmetric solution curves a_k, Eq. (12), and (b) shows a corresponding family of spatial correlation curves for several real power p values: p = 1/8, ¼, ½, 1, 2, 4, and 8.

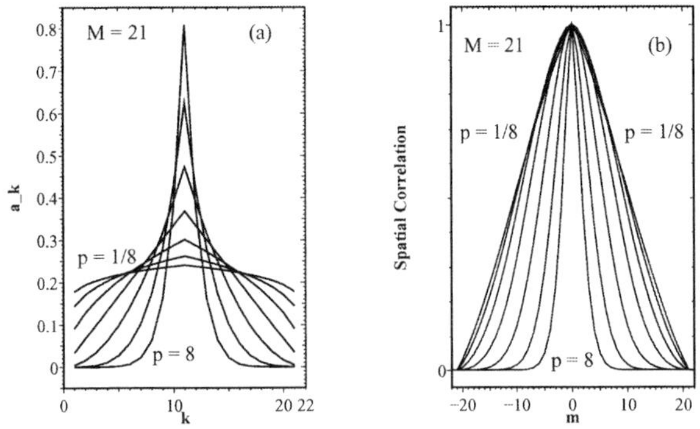

Fig. 5. (a) shows a family of symmetric solution curves a_k, Eq. (13), and (b) shows a corresponding family of spatial correlation curves for several real power p values: p = 1/8, ¼, ½, 1, 2, 4, and 8.

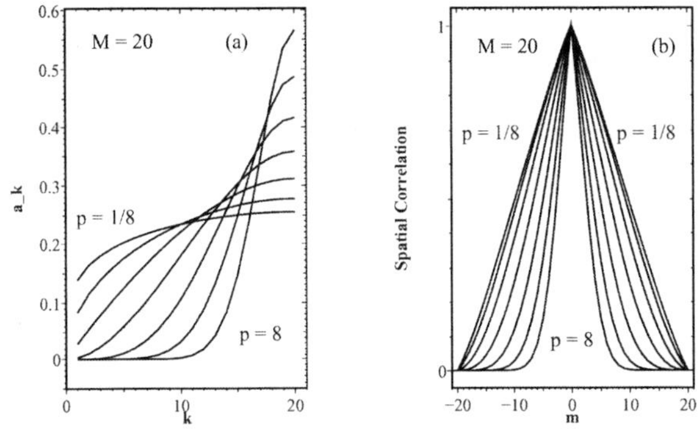

Fig. 6. (a) shows a family of asymmetric solution curves a_k, Eq. (14), and (b) shows a corresponding family of spatial correlation curves for several real power p values: p = 1/8, ¼, ½, 1, 2, 4, and 8.

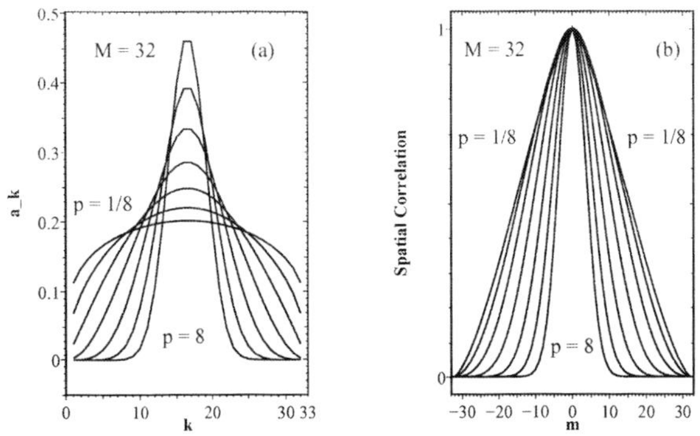

Fig. 7. (a) shows a family of symmetric solution curves a_k, Eq. (15), and (b) shows a corresponding family of spatial correlation curves for several real power p values: p = 1/8, ¼, ½, 1, 2, 4, and 8.

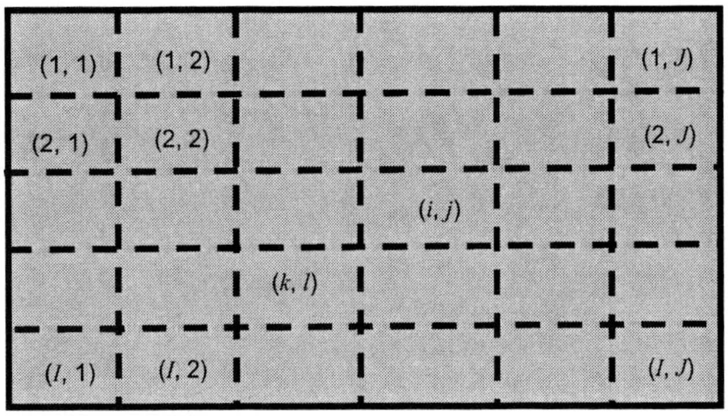

Fig. 8. Two-dimensional case: A chip/die is divided into (I x J) sub-regions. $C(i, j; k, l)$ denotes the correlation coefficient between a device in sub-region (i, j) and another device in sub-region (k, l).

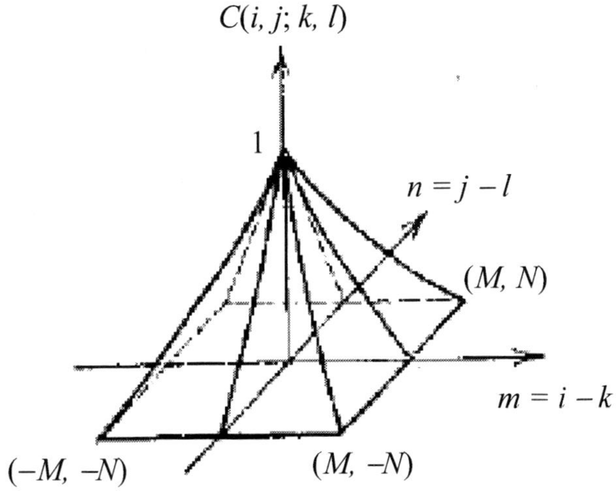

Fig. 9. Two-dimensional bilinear-decay spatial correlation.

Model Implementation for Accurate Variation Estimation of Analog Parameters in Advanced SOI Technologies

Sushant Suryagandh, Niraj Subba, Vineet Wason, Priyanka Chiney, Zhi-Yuan Wu, Brian Q. Chen, Srinath Krishnan, Manuj Rathor, Ali Icel

Advanced Micro Devices
One AMD Place, MS 79, Sunnyvale, CA 94086-3453, USA
Sushant.suryagandh@amd.com

ABSTRACT

Analog design uses transistors with longer channel length for high performance. Transconductance (g_m), Output Resistance (R_{out}) and Intrinsic Gain (g_m x R_{out}) form the metric to gauge this performance. It is critical to design these circuits for manufacturing variability. This work presents a systematic compact modeling approach to capture analog variation in compact models. We show that the variation in R_{out} is very well correlated to the variation in DIBL, especially if the transistors are biased at around 100-200mV gate-overdrive. We observe that this correlation becomes stronger as the channel length increases. We propose a sub-circuit based approach to model DIBL variation in the corner models. This method provides accurate variability estimates of g_m and R_{out} in the compact models for 65nm and 45nm technologies.

1 INTRODUCTION

Device design in advanced microprocessors is driven by the need of high I_{on}/I_{off} ratio since the majority of the chip deals with digital operation involving logic levels 0 and 1. Channel length scaling along-with appropriate T_{ox}, doping density and junction depth scaling is done to achieve higher drive current at a given leakage level for these logic circuits. The device requirements are often conflicting between digital and analog circuits [1]. For analog circuits, high g_m, R_{out} and intrinsic gain (g_mxR_{out}) are important DC parameters. g_m increases and R_{out} decreases with increasing channel length [2]. However, the degradation in Rout is severe than the increase in g_m so intrinsic gain decreases with reduction in the channel length (add figure here from that data). Also, R_{out} for the floating body (FB) PDSOI transistors is lower than the body-tied (BT) transistor due to kink and capacitive coupling [1]. Therefore, long channel BT transistors are used in the analog circuitry in a microprocessor.

Normally, the circuits are designed to tolerate manufacturing variability for high yield. Usually, different process corners are provided in the compact models for the designers to design for robustness. In general, these corners are very useful for estimating parameters like drive current, threshold voltage etc. There hasn't been much work done to understand the sources of variation for analog parameters such as g_m, R_{out} and their model-to-silicon correlation. This paper is divided into two sections. The first section focuses on the observations based on the data. The assessment of current variation models, their deficiency and new methodology for better accuracy are discussed in the second section.

2 DEVICE CHARACTERIZATION

DC device parametric has been characterized for analog specific BT transistors in AMD technology, described by Horstmann et. al. [3]. The sample space is big enough for accurate statistical conclusions. g_m and R_{out} are measured at gate-overdrive of 100mV, 200mV and drain bias of $0.5V_{dd}$.

3 RESULTS & DISCUSSION

3.1 Data Analysis

Throughout this paper, the variation in the data is defined as the ratio of 1 sigma deviation of the distribution to its median. Fig. 1 shows the variation of g_m, R_{out} for 65nm and 45nm technologies. Suffix 1 and 2 represent parameters measured at a fixed drain bias and 100mV, 200mV gate-overdrives respectively. R_{out} is a strong function of channel length [2]. Therefore, it is intuitive to expect R_{out} variation to reduce from shorter to longer channel lengths. However, as shown in Fig. 1, the R_{out} variation behaves non-monotonically with the channel length. This is due to halo doping in a MOSFET. For halo-based processes, R_{out} for long channel devices is determined by DITS [5], [6]. This is evident from Fig.2 which shows normalized R_{out} as a function of normalized DIBL. Channel length increases from L1 to L5. It can be seen that the impact of DIBL variation on R_{out} is the strongest for the longest channel length.

From Fig. 1, it can also be inferred that the effect of DITS variation is reduced as the device moves into stronger inversion. This is reflected by lower variability for

$R_{out}2$ compared to that of $R_{out}1$. DIBL variability has minimum impact on g_m variation. The variation in g_m is small and independent of gate bias.

3.2 New Modeling Approach

Fig. 3 shows model-silicon comparison for DIBL. The model variation is generated with the MC simulations using length, width and threshold voltage as random variables. It can be seen that the model doesn't produce any variations in DIBL especially at long channel lengths; whereas the data shows a significant amount of variation. This model deficiency results in underestimating R_{out} variations in the model compared to the data as shown in Fig. 4.

In order to provide more accurate R_{out} estimates, we propose a new and practical approach to incorporate large long channel DIBL in variation models. We provide a sub-circuit model with a VCVS in series with the gate. This voltage source imitates DIBL by providing additional gate-overdrive linearly proportional to the drain bias. Fig. 5 explains this concept through I_d-V_d curve. The VCVS impacts high V_{ds} region without affecting the linear region. The constant of proportionality for the voltage source is obtained from the data. With this new method, we are able to match DIBL and R_{out} variability in the data through our models as shown in Fig. 6 and 7. This method has an advantage over other methods as the same sub-circuit can be used on top of the core model.

4 CONCLUSIONS

Out data analysis for 65nm and 45nm technologies showed that the variation in Rout at long channel depends very strongly on the variation in DIBL. We propose a new method to implement this DIBL variation in the corner models for accurate Rout estimation for analog design. The advantage of this method is that it is independent of the underlying BSIM parameters for a particular device. It is also very straightforward to adjust the spread in DIBL if any of the process/integration elements change it.

5 REFERENCES

[1] Suryagandh et. al. *IEEE TED*, 1122 – 1128, July 2004

[2] Suryagandh et. al., *Proc.ESSDERC*, pp. 119 – 122, Sept. 2007

[3] M. Horstmann et. al., *Proc. IEDM Tech Dig.*, 233-236, Dec. 2005

[4] Suryagandh et. al., *Proc ESSDERC*, pp. 423 – 426, Sept. 2003

[5] K. M. Cao et. al. *Proc IEDM*, pp. 171 – 174, Dec. 1999

[6] S. Mudanai et. al.., *IEEE TED*, pp. 2091 – 2097, Sept. 2006

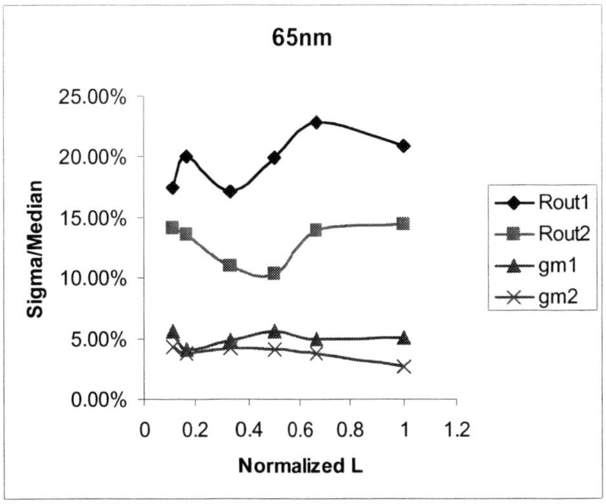

(a)

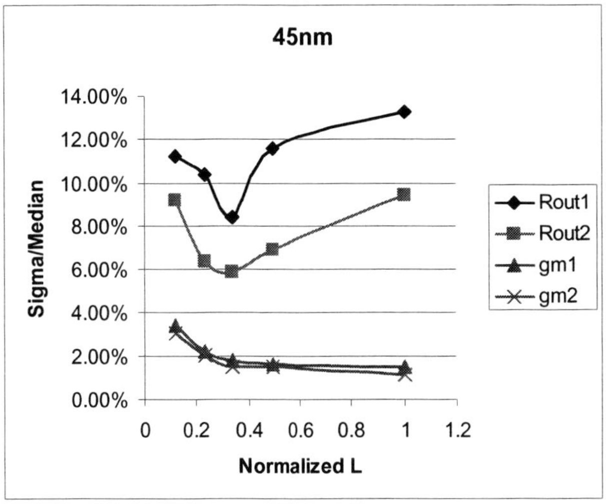

(b)

Fig. 1: Variability of g_m and R_{out} in 45nm and 65nm technologies.

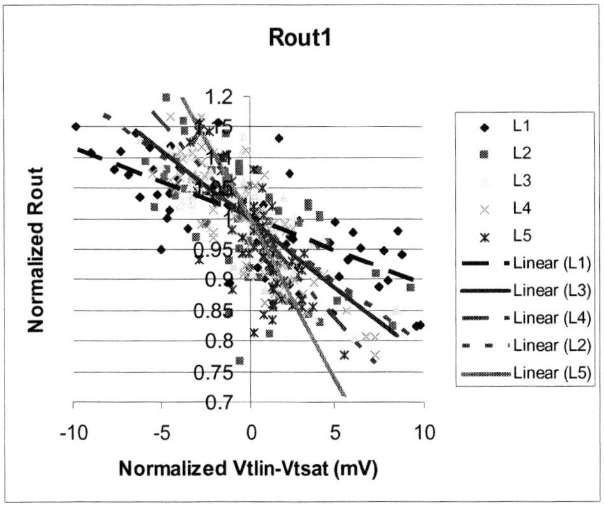

Fig. 2: Variation in R_{out} data plotted as a function of DIBL variation.

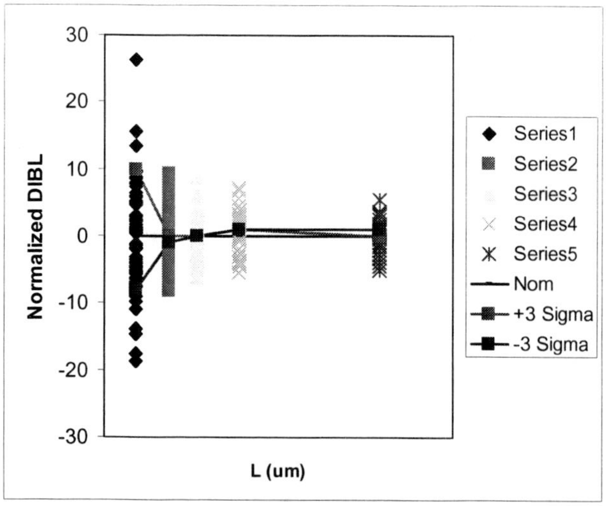

Fig. 3: Model-Si correlation of DIBL with existing variation modeling methodology.

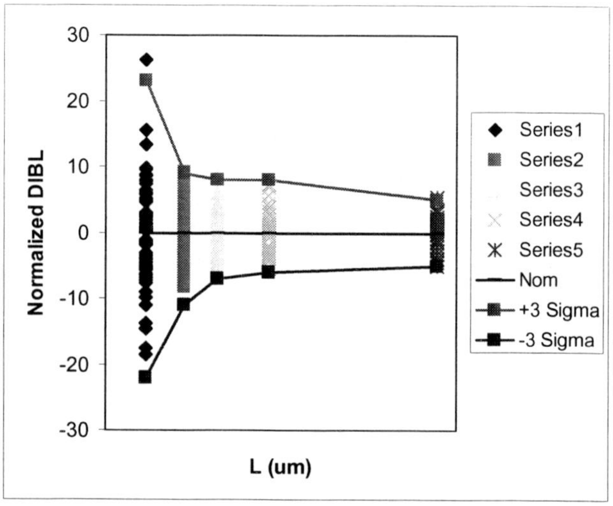

Fig. 6: Model-Si correlation of DIBL with new variation modeling methodology.

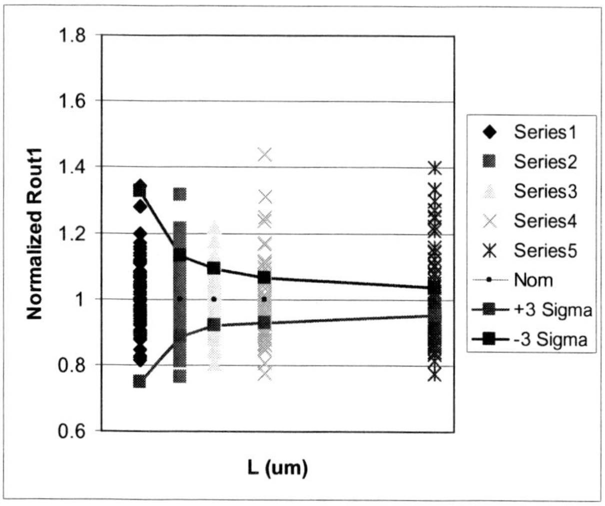

Fig. 4: Model-Si correlation of $R_{out}1$ with existing variation modeling methodology.

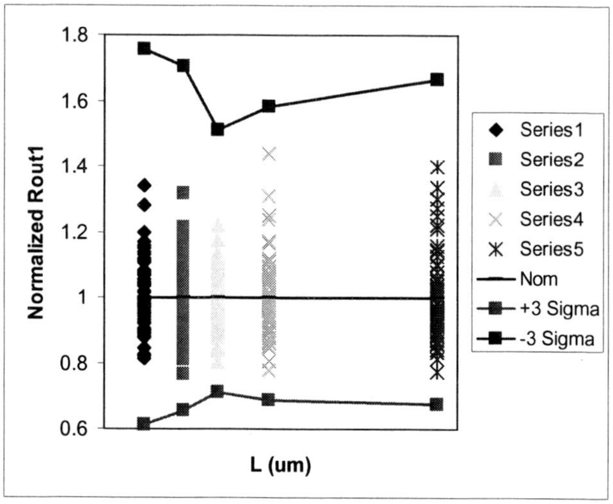

Fig. 7: Model-Si correlation of $R_{out}1$ with new variation modeling methodology.

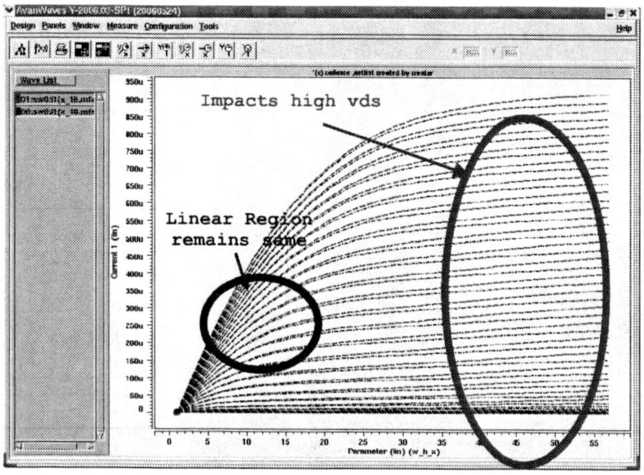

Fig.5: VCVS for the implementation of new variation modeling methodology. It imitates DIBL by providing additional gate voltage for high V_{ds} region.

NSTI-Nanotech 2008, www.nsti.org, ISBN 978-1-4200-8505-1 Vol. 3

Modeling of gain in advanced CMOS technologies

A. Spessot[*,+], F. Gattel[*], P. Fantini[*] and A. Marmiroli[*]

[*]STMicroelectronics, Advanced R&D, NVMTD-FTM, via Olivetti 2, 20041 Agrate Brianza, Italy,
[+]alessio.spessot@st.com

ABSTRACT

The impressive downscaling of CMOS technology and its more and more massive introduction in System-on-Chip (SoC) oriented applications require a modeling approach able to describe new physical effects paying attention to such different environments (digital and analog) coexisting in a single technological platform for SoC. In particular, in this paper we deep insight the modeling of gain, a key parameter ruling the analog performances in relation with their layout dependence affected by the shallow trench isolation (STI) induced mechanical stress. Both n- and p-channel have been investigated as a function of temperature. The W scaling modeling has been dealt with for the first time. A 90 nm CMOS advanced technology for Embedded Flash applications has been characterized.

Keywords: gain, compact modeling, mobility, temperature, *W* scaling

1 MODELING OF GAIN

The gain parameter of MOSFET plays a crucial role in analog design. It is so defined in linear region:

$$g_m = \mu_0 \cdot \frac{\varepsilon_{OX}}{t_{OX}} \cdot \frac{W - \Delta W}{L - \Delta L} \cdot V_{DS} \qquad (1)$$

where μ_0 is the low-field mobility, ε_{ox} the relative oxide dielectric constant, t_{ox} the electrical oxide thickness, W and L the MOSFET channel width and length, ΔW and ΔL the width and length variations, respectively. V_{DS} is the drain-source bias. The accurate definition of transconductance becomes related with the capability to well describe μ_0, that is modulated by temperature and geometrical features (modulating the mechanical stress) and doping, being other terms, in principle, constant. However, in ultra-scaled CMOS technologies, it is not a trivial problem to discriminate the correlations between ΔW, ΔL terms and μ_0. A 90 nm CMOS technology for Embedded Flash applications has been characterized.

This work starts dealing with the correct characterization and modeling of the channel geometrical definition (ΔW and ΔL), then the low-field mobility dependence as a function of temperature, layout, type of channel doping. A novel modeling trend on the gain versus W is tackled. Conclusions follow.

1.1 ΔL and ΔW extraction

Let's start to consider the extraction of ΔL and ΔW parameters. The well known low-field mobility modification as a function of SA layout parameter has been recently introduced in BSIM 4 compact model [1].

Fig. 1 shows the ΔL values obtained from the typical extraction plot reporting $1/g_m$ vs. L as a function of different SA parameters (see inset of Fig. 1)..

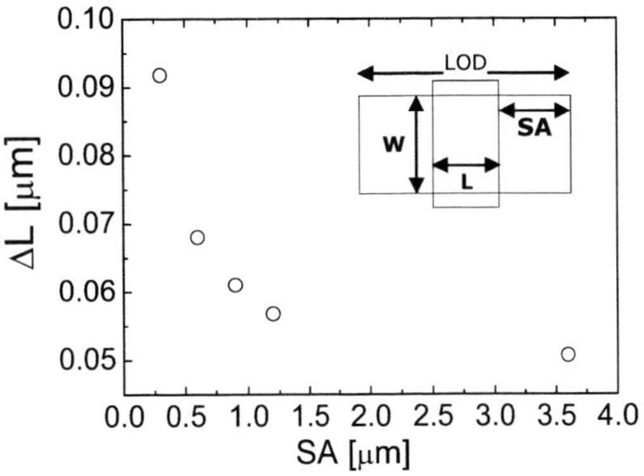

Fig. 1: Extracted ΔL parameters vs different SA parameters with the 'classical' method.

Different ΔL values are obtained for various SA bringing into question the real value of a so obtained ΔL. Recently a new methodology to extract the layout value based on the capacitance measure was proposed [2]. To avoid the complication related with that approach (i.e. the parasitic capacitance determination and the bias dependence), we propose the simpler method of the linear regression of the inverse of gain vs L scaling, but using STI-stress free test structure of Fig. 2 annihilating the effect of strain on the low-field mobility. Fig. 3 shows the good comparison of results obtained by using the proposed strategy and the C-V method. This result has been also validated by means of the comparison with device TCAD simulations through which we have estimated ΔL when the I-V experimental curve of the short channel MOSFET was reproduced via numerical simulation after the low-field mobility calibration in the long-channel device.

experimental finding is to be taken into account when delay chain containing both *n*- and *p*-channel MOSFETs must be calibrated as a function of temperature. A more linear trend of holes versus temperature in comparison with electrons can also be observed considering *p*-well against *n*-well resistors [4].

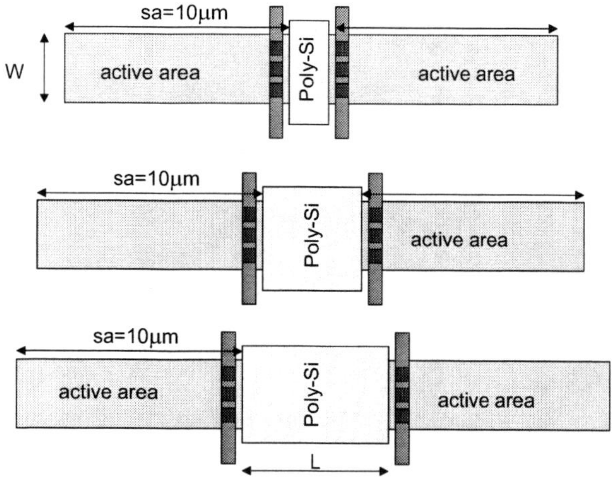

Fig. 2: Layout of the proposed STI-stress free.

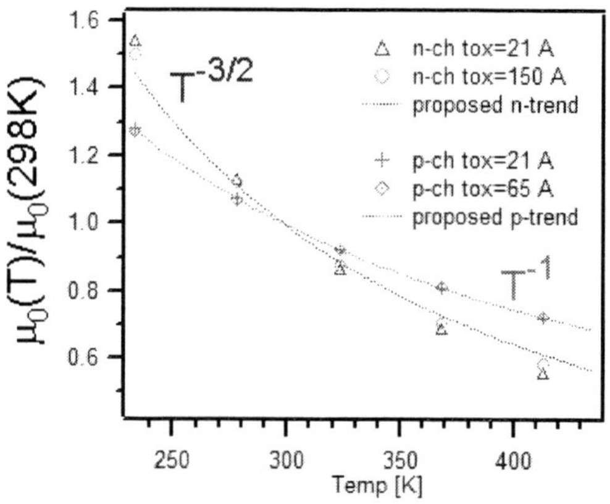

Fig. 4: Normalized temperature trend of low-field mobility for both *n*- and *p*-channel showing unchanged trend considering different technologies

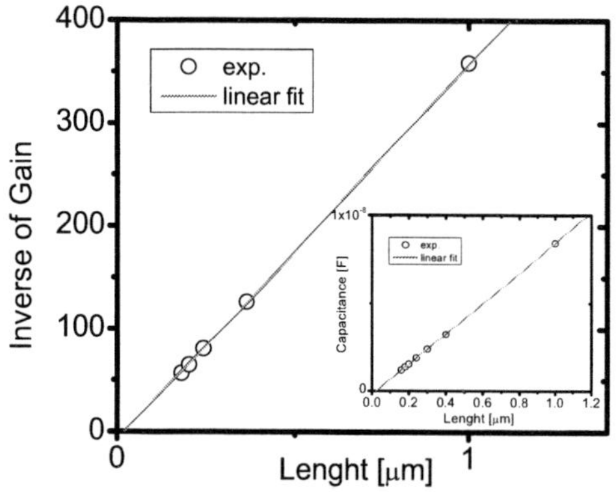

Fig. 3: Extracted ΔL parameters from both STI-stress free structures and C-V method.

1.2 Low field mobility: temperature

Let's move to consider the low-field mobility (μ_0). It is well-known that it strongly depends on temperature (Fig. 4) reports the mobility trend in the usual temperature operative range of different families of devices showing that, as rule of thumb: $\mu_0(T) \propto T^{-3/2}$ for the *n*-channel and $\mu_0(T) \propto T^{-1}$ for *p*-channel. These trends are coherent with the more pronounced phonon scattering dependence for electrons and coulombic scattering for holes. In fact, holes, being slower than electrons, are more affected by the Rutherford scattering with ion impurities [3]. This

1.3 Low field mobility: Mechanical stress effects

Then, STI-induced mechanical stress affects the low-field mobility of both electrons and holes as a consequence of the strain-altered band structure [5]. As it concerns the conduction band modification we take for granted the deformation-potential theory defining a rigid shift of its bottom edge produced by a homogeneous strain [6]. In particular, the energy shift of the six conduction band valleys induced by the presence of stress can be written as a linear combination of the strain tensor components and the deformation potential constants:

$$\delta E_{nk} = \sum_{j=1}^{6} \Xi_j \cdot u_j \qquad (2)$$

where δE_{nk} is the shift of the energy at the band edge point *k* due to a strain with components u_j referred to the crystallographic axes. In summary, the effect of strain can be reviewed as a rigid energetic shift without altering, in the framework of this theory, the dispersion shape of the conduction band (CB). This shift causes the conduction band repopulation and so, in terms of MOSFET performance, a change in low-field mobility. In fact, this will generate a size modulation of ellopsoidal isoenergy surface corresponding to the CB minimum closed to X position of the Brillouin zone and, consequently, an effective mass change that will vary the device transport properties. Using a classical approach (Boltzmann distribution: $n_{0nk}=N_c \cdot e^{-(E_{0nk}-E_F)/kT}$) the fractional change in electron density for each valley with δE_{nk} energetic shift is given by:

$$\frac{\delta n_{nk}}{n_{0nk}} = -\frac{1}{kT} \cdot \delta E_{nk} \qquad (3)$$

where δE_{nk} is calculated from (2).

Indeed, there are two factors leading to the increased mobility: reduced effective mass and reduced intervalley scattering. Scattering is reduced because low energy electrons only access to iso-energy valleys.

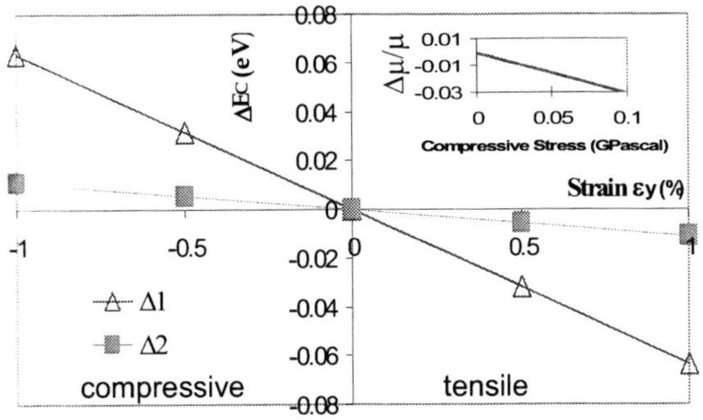

Fig. 5: Shift of the conduction band edge under a uniaxial stress applied along the W direction for the 4 valleys (Δ1) of the (001) plane and for the Δ2 valleys. Inset: the related mobility reduction.

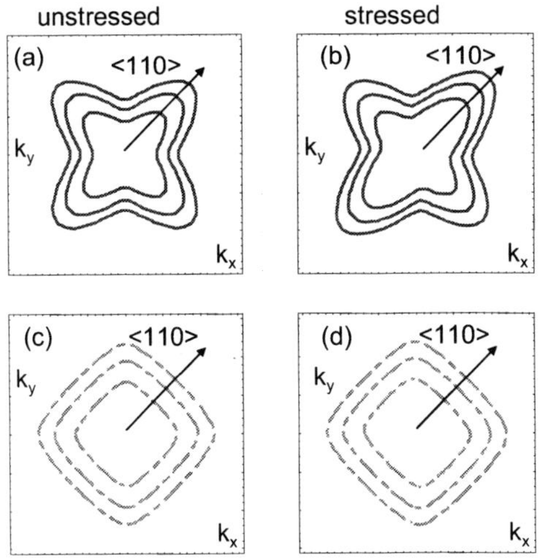

Fig. 6: a) Relaxed Silicon: isoenergy surface plot of the Heavy Holes (HH) at 10,20,30 meV with respect to the top of the valence band b) Compression of 1 Gpascal along the W direction; the band structure of HH are shifted and modified, reducing the mobility. c) Light Holes band, relaxed silicon; d) Same compression of b) for LH

Therefore, we can study the impact of stress constrain in *n*-MOS device as a consequence of the minimum conduction band energy shift induced by the mechanical stress presence. Fig 5 reports the so determined energetic shift for the conduction band. Values of Ξ_j can be deduced from ref. 6 and, in the inset, the related mobility variation.

For the valence band, we choose to describe the Bir-Pikus theory considering both the degeneracy lift between light and heavy holes and the bands dispersion deformation [7]. In Fig. 6a) and 6c) the constant energy surface of both heavy holes (HH) and light holes (LH) for relaxed silicon is pictured. The "matrix element" parameters for the valence band computation have been captured from literature [2] together with the deformation-potential values [3]. In fig. 6b) and 6d) the light and heavy holes band energy dispersion in the k_x, k_y conduction plane has been plotted for the strained silicon under a compression in the *W* direction of 1 GPascal. As previously noted [9], in the case of the conduction band the stress-induced enhancement of the subbands shift cannot explain the mobility modification. We note that for uni-axial compression the increased occupation along the [-110] direction (where the density of states is higher) leads to decreased the transport effective mass for conduction in the [110] direction because of the steeper gradient of the band structure in this direction, and consequently a decreased mobility and drive current.

Fig 7 reports the so determined energetic shift the strain-altered valence band in the [110] channel direction.

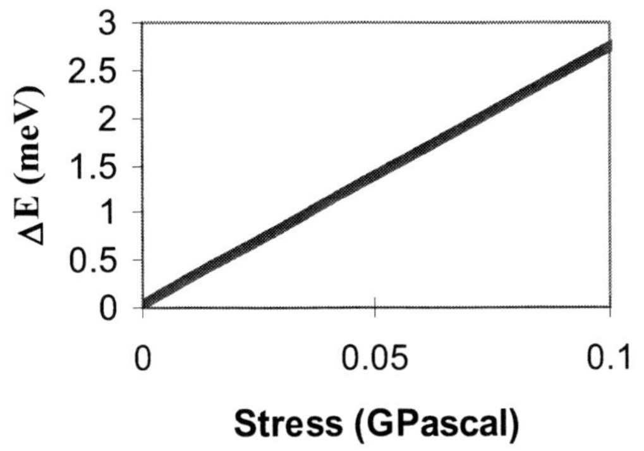

Fig. 7: Energetical separation between the top of the HH and the LH bands under a uniaxial stress applied along the *W*. The hole mobility reduction cannot be explained only by this phenomena

1.4 Transconductance trend vs *W*

Figs 8 and 9 show a non monotonic trend of the *W* normalized gain as a function of *W* for both *n*-channel and *p*-channel, respectively.

The initial gain lowering with W scaling up to around 1 μm can be ascribed to the low-field mobility decreasing by the lateral STI-induced stress coherently with the altered band structure shown in Figs. 5, 6. This behavior can be taken into account with the introduction of a monotonic μ_0 dependence vs *W* for both electrons and holes (dashed lines in Figs 8 and 9).

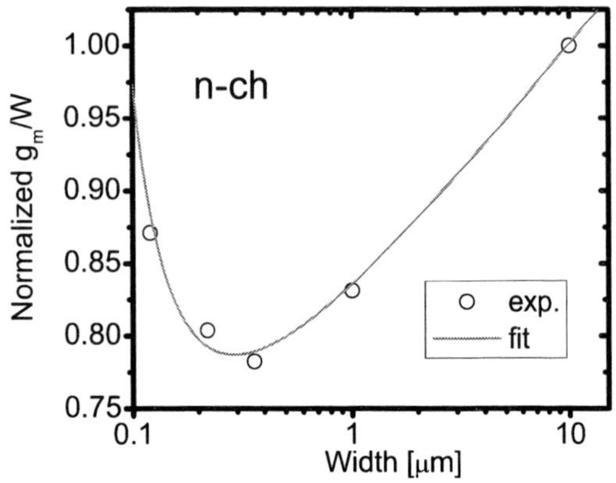

Fig. 8: *n*-channel normalized ratio of g_m/W as a function of *W* showing two regimes: decreasing for STI stress and increasing for ΔW. A good match with the proposed dependence.

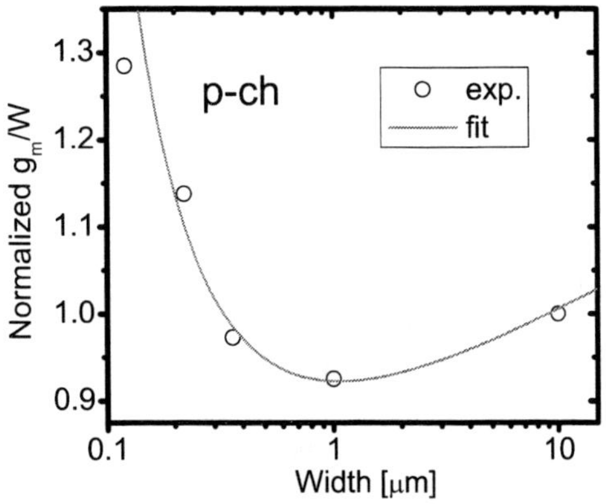

Fig. 9: *p*-channel normalized ratio of g_m/W as a function of *W* showing two regimes: decreasing for STI stress and increasing for ΔW. A good match with the proposed dependence.

The non-monotonic experimentally observed can reproduced, with the ΔW introduction that plays an opposite role (enhancing the gain) for the narrower structures. This also correlates with ID_{SAT} trend just observed in [10]. Then, we suggest to improve the low-field mobility expression with the introduction of *W* dependence in standard compact models. Finally, in a standard parameters flow extraction we suggest to fit this gain behavior to compute the value of ΔW. On the contrary, it is important to avoid obtaining the ΔW value from the gain scaling versus *W*, since STI mechanical stress gets dirty the extraction.

2 CONCLUSIONS

Summarizing, we investigated the modeling of gain in advanced CMOS technologies. We addressed a new methodology to pay an accurate attention to the layout dependence, due to the role of STI mechanical stress, in the 'classical' ΔL determination. With the support of both strain-altered band structure modeling and experimental data, a new expression for the low-field mobility dependence on *W* has been proposed. The method should push a strong improvement in the next generation mixed-signal design.

REFERENCES

[1] http://www-device.eecs.berkeley.edu/~bsim3/bsim4.html

[2] D. Fleury, A. Cros, K Romanjek, D. Roy, F. Perrier, B. Dumont and H. Brut, ICMTS 2007 Proc, 89.

[3] P.Y. Yu and M. Cardona, Fundamental of Semiconductors, Ed. Springer 1996.

[4] I. Aureli, D. Ventrice, C. Coregoni and P. Fantini, ICMTS 2007 Proc., 268

[5] M. V. Fischetti and S.E. Laux, JAP **80** (1996), 2247

[6] P. Fantini, A. Ghetti, G.P. Carnevale, E. Bonera and D. Rideau, IEDM Tech. Dig. (2005), 992.

[7] C. Herring and E. Vogt, *Phys. Rev.* **101** (1955) 944-961

[8] G.E. Pikus and G.L. Bir, *Sov. Phys. Tech. Phys.*, vol. **3**, p.2194 (1958).

[9] S. Kobayashi, M. Saitoh, Y. Nakabayashi and K. Uchida, APL **91** (2007), 203506

[10] P.B.Y. Tan, A.V. Kordesch and O. Sidek, APM 2005, 43

Effective Drive Current in CMOS Inverters for Sub-45nm Technologies

Jenny Hu*, Jae-Eun Park[†], Greg Freeman[†], Richard Wachnik[†], H.-S. Philip Wong*
* Department of Electrical Engineering, Stanford University, CA, USA jennyhu@stanford.edu
[†] IBM, East Fishkill, NY, USA

ABSTRACT

We propose a new model for the effective drive current (I_{eff}) of CMOS inverters, where the maximum FET current obtained during inverter switching (I_{PEAK}) is a key parameter. I_{eff} is commonly defined as the average between I_H and I_L, where $I_H=I_{ds}(V_{gs}=V_{DD}, V_{ds}=0.5V_{DD})$ and $I_L = I_{ds}(V_{gs}=0.5V_{DD}, V_{ds}=V_{DD})$. In the past, this I_{eff} definition has been accurate in modeling the inverter delay. However, we find that as devices are scaled further into the nanoscale regime, the maximum transient current can deviate severely from I_H, in which case, another metric should be used. The deviation of I_{PEAK} from I_H is found to increase as delay decreases or as device overdrive voltage increases. We define $I_{eff} = (I_{PEAK}+I_M+I_L)/3$, where $I_M = I_{ds}(V_{gs}=0.75V_{DD}, V_{ds}=0.75V_{DD})$. We evaluate our model against others by comparing the analytical and HSPICE extracted I_{eff} ratios across devices of varying threshold voltages, V_{TH}. Our model is shown to better capture changes in V_{TH}/V_{DD}, which are important since V_{DD} and V_{TH} will be key parameters for optimizing device performances for target applications (low power or high performance) in sub-45nm technologies.

Keywords: CMOS, inverter, delay, performance

1. INTRODUCTION

As we continue to scale CMOS devices further into the nanoscale regime, performance metrics valid for older technologies need to be re-evaluated for their suitable application with the current technology. FET performance is typically measured through the CMOS inverter delay $\tau = CV/I$, where C is the load capacitance, V is the power supply V_{DD}, and I is the saturation on-current $I_{ON}= I_{ds}(V_{gs}= V_{ds}= V_{DD})$. In recent years, the concept of an effective drive current has been proposed as a better representation for calculating the inverter delay, due to the fact that I_{ON} is never reached during switching [1 – 4]. However, as devices are scaled, current-voltage characteristics continue to deviate from the predicted behavior of ideally scaled MOSFET due to minimal scaling of V_{TH} and V_{DD}, and non-idealities such as parasitic series resistances, velocity saturation, and DIBL. To more accurately assess the performance of sub-45nm CMOS, a more representative effective drive current is needed.

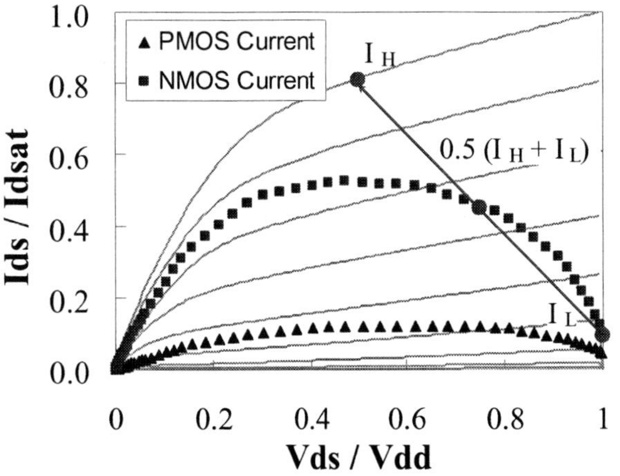

Figure 1. Pull-down switching trajectory of low V_{TH} device, with bias points plotted in equal time intervals. Arrow indicates assumed trajectory in the two-point I_{eff}.

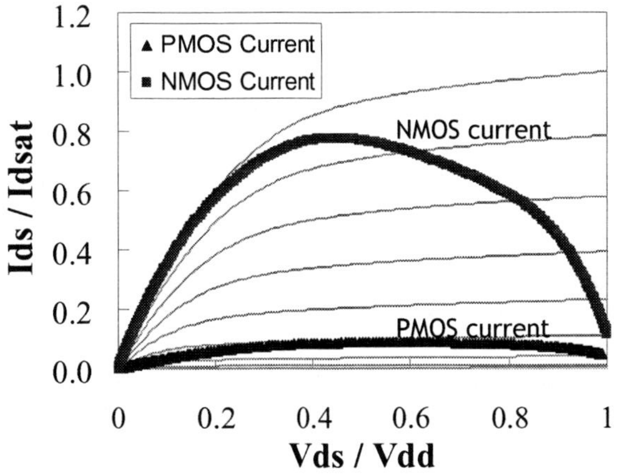

Figure 2. Pull-down switching trajectory of 250 nm technology, where I_{PEAK} occurs closer to I_H.

2. TRANSIENT CURRENT TRAJECTORY

In determining a suitable method to evaluate the inverter delay, it is important to first examine the transient current trajectory to better understand approximations used in I_{eff}. Fig. 1 illustrates the transient inverter current trajectory of a pull-down transition plotted over the DC current characteristics of an NMOS, with current normalized to I_{ON}. It is noted that the peak current (I_{PEAK}) obtained during switching falls far below I_H, and even

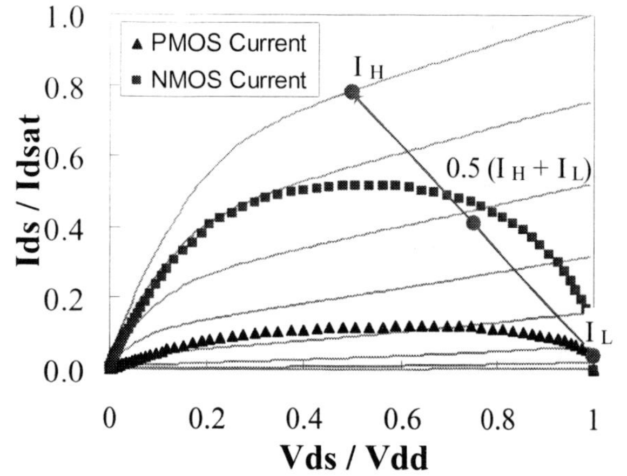

Figure 3. Pull-down switching trajectory of a higher V_{TH} device, showing a larger V_{PEAK} than in Figure 1, but I_{PEAK} is still far from I_H.

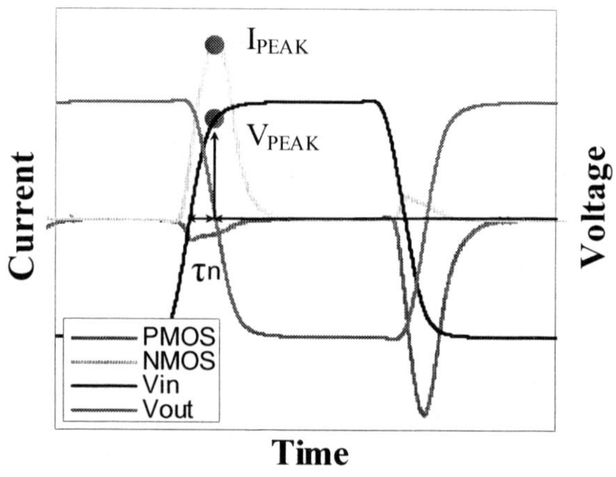

Figure 4. Transient response of both a pull-up and pull-down transition, showing $V_{PEAK} < V_{dd}$ and the dependence upon delay.

further from I_{ON}. This deviates greatly from previous technologies, where I_{PEAK} approached I_H as shown in Fig.2 for a 250nm technology. With the inverter delay defined as the time between $V_{IN}=V_{DD}/2$ to $V_{OUT}=V_{DD}/2$ during switching, the pull-down delay can be viewed as the integration of CV/I , where I ideally is the trajectory current. From Na et al. [2]:

$$\tau_{PD} = \int_{V_{GS}=V_{DD}/2}^{V_{DS}=V_{DD}/2} \frac{C_{LOAD}}{I_{DS}\left(V_{DS}, V_{GS}\right)} dV_{DS} \qquad (1)$$

From the equation it is apparent that the more closely the I_{eff} equation follows the current trajectory path, the more accurately it will reflect the device performance. The current trajectory has a parabolic shape, so a minimum of three points would be needed to more precisely capture the trends in current.

In addition to examining the current trajectory across technology nodes, we examine the changes in the trajectory as we scale the threshold voltage V_{TH} of devices in a given technology. Designing devices for a variety of applications, whether low power or high performance, results in devices with a wide range of V_{TH}, thus making it important to study the effect of scaling V_{TH}/V_{DD} on I_{eff}. The current trajectory of a higher V_{TH} device is plotted in Fig. 3. When comparing Fig. 1 and 3, it is significant to note the differences in the current trajectory paths taken by the devices with two separate V_{TH}. As the V_{TH} of a device decreases, the peak current in the trajectory occurs at a lower V_{gs}, further away from I_H. This behavior can be explained by Fig. 4, which illustrates the transient response of a ring oscillator. For a given V_{DD}, as V_{TH} decreases, the current overdrive is greater, thus decreasing the delay. With a shorter delay, the output falls from V_{DD} to $V_{DD}/2$ faster than the input can rise to V_{DD}, causing I_{PEAK} to occur at a V_{gs} much lower than V_{DD}. Therefore, a decrease in delay also decreases V_{PEAK}, defined as the gate voltage at which I_{PEAK} occurs. This can be equivalently interpreted as the current trajectory following a lower V_{gs} curve.

This change in current trajectories suggests that evaluating I_H at the same bias regardless of the device overdrive is not sufficient. Also, the simplification of the parabolic current trajectory as a linear path from I_L to I_H [1] is no longer a good approximation if the peak current is far below I_H. In previous technologies, larger τ allowed $I_{PEAK} \approx I_H$ and $V_{PEAK} \approx V_{DD}$, but this is changing as devices and power supplies are scaled. We propose an I_{eff} model that includes more than two points of current and takes into account of I_{PEAK}.

3. EFFECTIVE DRIVE CURRENT MODEL

Na et al. [1] defined the effective drive current by a two-point model, $I_{eff} = (I_H + I_L)/2$, where $I_L = I_{ds}(V_{gs}=0.5\ V_{DD}, V_{ds}= V_{DD})$ and $I_H = I_{ds}(V_{gs}= V_{DD}, V_{ds}=0.5\ V_{DD})$. This effectively approximates the current trajectory as a linear path between I_H and I_L. Deng et al. [2] introduced a three-point model of $I_{eff} = (I_H + I_M + I_L)/3$, where $I_M = I_{ds}(V_{gs}=0.75\ V_{DD}, V_{ds}=0.75\ V_{DD})$. A four-point model was also proposed, where current through the device that should be off is accounted for. The four-point model was tested, however, we did not find significant benefit over the three-point model in a complexity and accuracy tradeoff.

We propose another three-point model of $I_{eff} = (I_{PEAK} + I_M + I_L)/3$, where I_{PEAK} is used to replaced I_H. Including I_{PEAK} will allow Ieff to include the effects of changes in switching trajectories between devices.

To analyze these I_{eff} models, the analytical I_{eff} is compared to the simulated I_{eff} extracted from HSPICE simulations of a nine-stage ring oscillator using IBM's 45nm bulk and SOI models for several devices designed for a range of V_{TH}. These models take into account critical doping changes between the various V_{TH} devices.

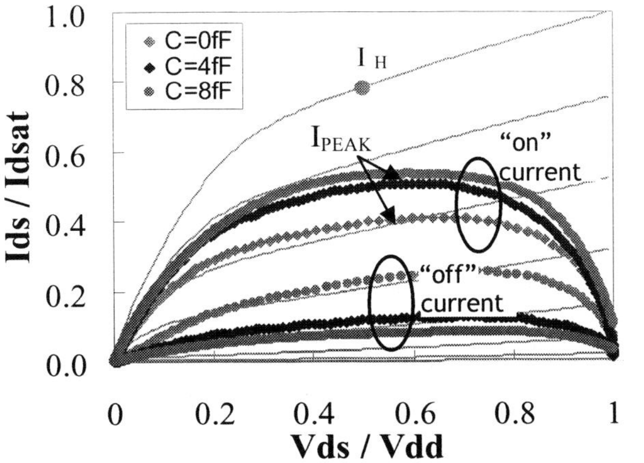

Figure 5. Dependence of I_{PEAK} on the load capacitance. A smaller load capacitance has a smaller I_{PEAK} and greater short circuit current in the device that is turning off.

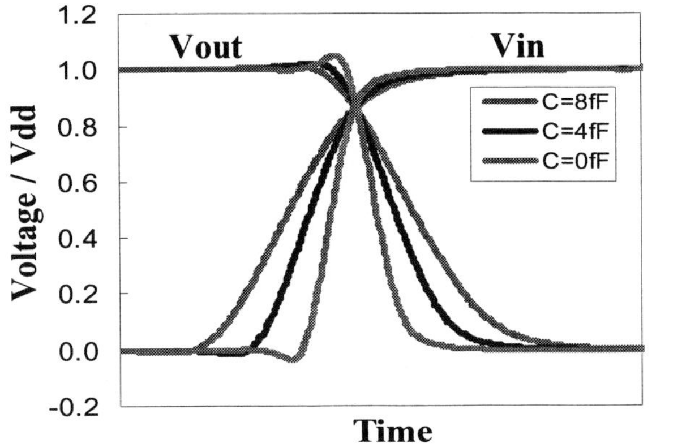

Figure 6. A smaller load capacitance has a larger short circuit current in the device that is shutting off is due to greater overlap of voltage range that both Vgs and Vds are significant. A large load capacitance is used to accentuate the effect for illustration purposes.

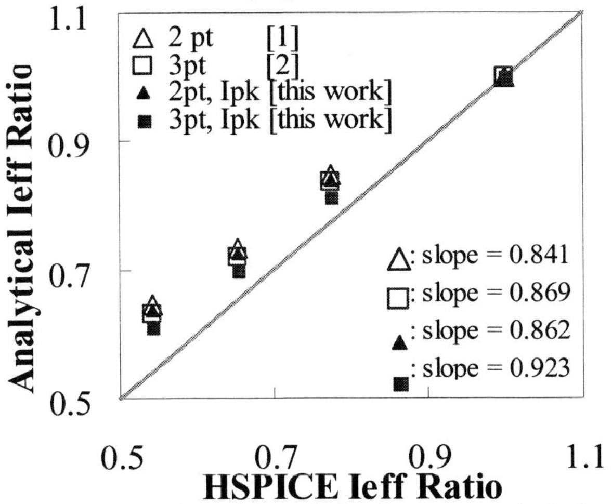

Figure 7. Evaluation of the relation between analytical and HSPICE I_{eff} ratios for several definitions of I_{eff}. Ratios are taken relative to the I_{eff} of the device with the lowest V_{TH}. For each definition of I_{eff}, the slope of the best fit line is compared, with a slope of 1 being ideal. The maximum slope occurs in the three-point definition using I_{PEAK}.

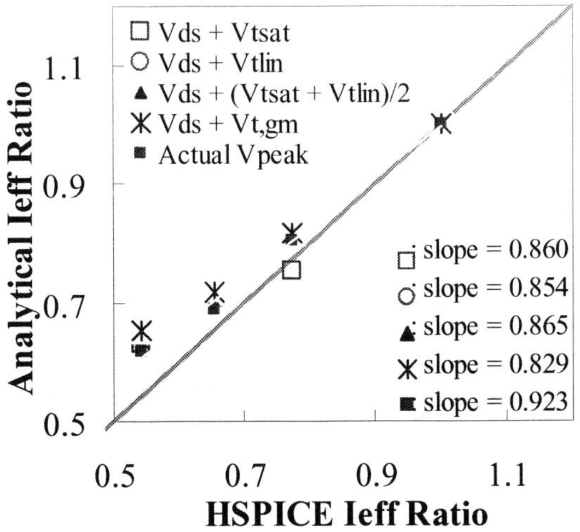

Figure 8. Evaluation of the correlation between analytical and HSPICE I_{eff} ratios for several definitions of V_{PEAK}. Maximum slope is for $V_{PEAK} = 0.5V_{dd} + (V_{TSAT}+V_{TLIN})/2$

4. PEAK CURRENT

From the ring oscillator simulations, we found the dependence I_{PEAK} has on the load capacitance. As illustrated in Fig. 5, in the pull-down case, as the load capacitance decreases, I_{PEAK} of the NMOS also decreases. This can be explained by Fig. 4, due to a decrease in the delay with reduced load capacitance. More notable is the increase in the short circuit current of the device that is turning off, or off-current, which is the PMOS during a pull-down transition. The origin of this off-current is further elaborated in Fig. 6. Consider the waveform of Vin and Vout. We note that for smaller load capacitances, the slope of the voltages is sharper, resulting in greater overlap of voltages in the range where both the NMOS and PMOS are on, thus a larger off current. Although we illustrated the dependence on the load capacitance, the underlying reason is a shorter delay, so this same dependence would appear in any situation where the delay is sufficiently short. In cases where this current approaches I_{PEAK} through smaller load capacitance or more advanced technologies, it will be critical to employ a four-point model to account for this off current. However, in the 45nm technology evaluated in this work, for the reasonable load capacitances considered, a three-point model is found to be adequate.

5. RESULTS

Our model is shown to better capture changes in V_{TH}/V_{DD}. Fig. 7 summarizes the evaluation of several I_{eff} definitions across four devices of varying V_{TH}. The ratios of both the analytical and simulated I_{eff} are taken with respect to the device with the corresponding largest I_{eff}, or lowest V_{TH}. For each definition of I_{eff}, the slope of the best fit line for the four V_{TH} devices is compared, with a slope of 1 being ideal. It is noted that regardless of I_H or I_{PEAK}, when the two-point definition is replaced with the three-point, there is an improvement in the correlation. $I_{eff} = (I_{PEAK} + I_M + I_L)/3$ results in the best correlation between the analytical and HSPICE extracted I_{eff}, with a slope of 0.923 compared to the 0.841 slope of the two-point I_{eff} definition.

In each case where I_H is replaced with I_{PEAK}, there is a noted improvement in correlation, with an overall 7.2% improvement from the two-point I_{eff}.

However, evaluation of I_{PEAK} becomes useful only when the value can be predicted from DC current characteristics. We propose using a simple estimate of I_{PEAK}, disregarding the effects of the load capacitance mentioned in Fig. 6. From Hamoui et al. [4], the maximum transient current occurs roughly when the NFET leaves saturation and enters triode operation, at $V_{gs} = V_{PEAK} = m*V_{ds} + V_{TH}$, where m is the body factor. Since inverter delay is defined as the time between $V_{gs}= V_{DD}/2$ to $V_{ds}= V_{DD}/2$, it is logical to evaluate I_{PEAK} using $V_{ds}= V_{DD}/2$. Threshold voltage can be defined in many ways, so we evaluate several V_{TH} definitions to determine the most suitable for V_{PEAK}: 1.V_{TSAT}, 2.V_{TLIN}, 3.$(V_{TSAT}+V_{TLIN})/2$, and 4.$V_{TH,PEAK\ gm}$. Fig. 8 illustrates $(V_{TSAT}+V_{TLIN})/2$ has the most promise, although there is a lower correlation than directly extracting V_{PEAK}, there is still a higher correlation compared to the traditional two-point I_{eff}.

6. CONCLUSION

In summary, we propose a new metric for I_{eff}, using $(I_{PEAK} + I_M + I_L) / 3$. We propose using the peak current attained during an inverter switching trajectory, as an alternative to I_H, to better correlate with ring oscillator simulation extracted I_{eff}. Use of I_{PEAK} becomes imperative in cases where the peak current trajectory deviates significantly from I_H, which becomes increasingly important as delay decreases or overdrive voltage increases, often as we scale beyond the 45nm technology node.

ACKNOWLEDGEMENTS

This work is supported in part by the Focus Center Research Program (FCRP) (C2S2) and NSF (ECS-0501096). J. Hu is additionally supported by the Stanford Graduate Fellowship and the National Defense Science and Engineering Graduate (NDSEG) Fellowship. This work was also performed in part at IBM.

REFERENCES

[1] K. K. Ng, C. S. Rafferty, and H. Cong, "Effective On-current of MOSFETS for Large-signal Speed Consideration," IEDM Tech. Dig., pp. 693 – 696, 2001.

[2] M. H. Na, E. J. Nowak, W. Haensch, and J. Cai, "The Effective Drive Current in CMOS Inverters," IEDM Tech. Dig., pp.121–124, 2002.

[3] J. Deng and H.-S. P. Wong, "Metrics for Performance Benchmarking of Nanoscale Si and Carbon Nanotube FETs Including Device Nonidealities," IEEE Trans. Elec. Dev., Vol.53, No.6, pp.1317–1322, June 2006

[4] K. Arnim, C. Pacha, K. Hofmann, T. Schulz, K. Schrufer, and J. Berthold, "An Effective Switching Current Methodology to Predict the Performance of Complex Digital Circuits," IEDM Tech. Dig., pp.483 – 486, 2007.

[5] A.Hamoui and N. Rumin, "An Analytical Model for current, Delay, and Power Analysis of Submicron CMOS Logic Circuits," IEEE Trans. on Circuits and Systems II, Vol.47, No.10, pp.999–1006, October 2000.

Process Aware Compact Model Parameter Extraction for 45 nm Process Flow

Aditya P. Karmarkar[*], V. K. Dasarapu[*], A. R. Saha[*], G. Braun[**], S. Krishnamurthy[***] and X.-W. Lin[***]

[*]Synopsys (India) Private Limited, My Home Tycoon, Block-A,
Begumpet, Hyderabad 500016, Andhra Pradesh, India, AdityaP.Karmarkar@synopsys.com
[**]Synopsys Switzerland, LLC, Zurich, Switzerland, braun@synopsys.com
[***]Synopsys. Inc., Mountain View, CA, USA, Sathya.Krishnamurthy@synopsys.com

ABSTRACT

CMOS devices are simulated using a 45-nm process flow that uses advanced techniques to achieve the requisite performance. The process parameters with maximum impact on the device characteristics are identified and analyzed. Global and process-aware model parameters are extracted for the 45-nm process. A five-stage ring oscillator is examined to demonstrate the effects of process variability on circuit performance. Good agreement between the model and the numeric simulations is observed demonstrating the robustness of the extraction methodology and the process-aware model parameters.

Keywords: parameter extraction, process-aware compact models, design for manufacturability

1 INTRODUCTION

In the semiconductors industry, the device geometries decrease and the integration densities increase with each technology node. Designing and manufacturing circuits with smaller device geometries is a major challenge for the microelectronics industry because of the process variability impact on device and circuit performance, leading to lower reliability and yield [1], [2].

To design robust circuits using deep sub-micron devices, the effects of process variability on the circuit model parameters must be examined in detail. A thorough assessment of the process variability impact on the circuit model parameters leads to better designs, improved manufacturability and higher yield. In this paper, methodologies to extract circuit model parameters that account for process variability are demonstrated. Strategies to account for the process variability induced circuit performance variation are also developed.

2 NUMERIC SIMULATIONS

The process simulations for the CMOS devices are performed using a 45-nm process flow that uses high-k gate dielectric with an effective oxide thickness (EOT) of 0.812 nm, halo and source/drain implants, stress engineering and spike and laser anneals to attain the requisite performance. Table 1 shows key parameters for the devices simulated using the 45-nm process flow.

	NMOS	**PMOS**
L_{gate} [nm]	45	45
I_{on} [mA/μm]	1.207	-0.3482
I_{off} [μA/μm])	1.407×10^{-3}	-8.595×10^{-4}
G_m [mS/μm]	0.5610	0.1292
V_{th} [V]	0.3453	-0.4381
V_{dd} [V]	1.0	-1.0

Table 1: 45 nm technology parameters

The process parameters for the NMOS and PMOS devices are adjusted to match the electrical performance. Electrical characteristics are simulated using simple drift diffusion models [3]. It can be observed from the device characteristics presented in Table 1 that these devices are appropriate for low power applications, since the PMOS characteristics exhibit a low on current (I_{on}) accompanied by a significantly lower off current (I_{off}).

The compact model parameters are extracted using the BSIM4 MOSFET model [4]. Two sets of process simulations are performed to estimate the global and process-aware model parameters. Global parameters are extracted for the 45-nm process with drawn gate lengths from 32 nm to 1 μm. The process-aware model parameters are extracted from simulations where process parameters like gate length, gate taper angle, halo dose and halo energy are varied. These parameters are selected to model the process variability because of their greater impact on the electrical characteristics. The gate oxide thickness is not varied as the gate oxide is a combination of SiO_2 and HfO_2. Parameter extraction is performed using a specialized parameter extraction tool that can directly generate process-aware compact models [5].

3 PARAMETER EXTRACTION

Process and device simulations are performed using a 45-nm process flow to extract the global model parameters. For these simulations, the drawn gate length is varied from 32 nm to 1 μm and the other process parameters are kept constant. Figure 1 shows the variation of the threshold voltage (V_{th}), calculated in the linear region, and the transconductance (G_m) with respect to gate length; and Figure 2 shows the variation of the on and off currents (I_{on} and I_{off}) with gate length for PMOS and NMOS devices.

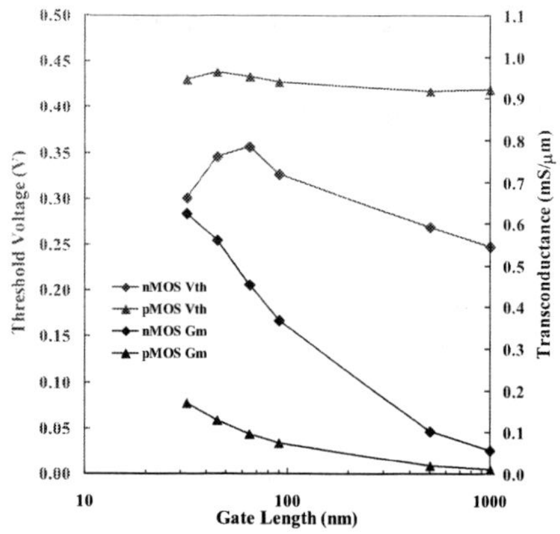

Figure 1: V_{th} and G_m with respect to L_{gate}.

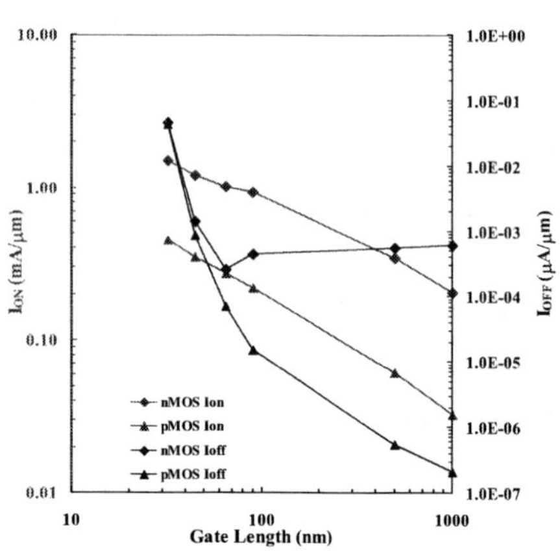

Figure 2: I_{on} and I_{off} as functions of L_{gate}.

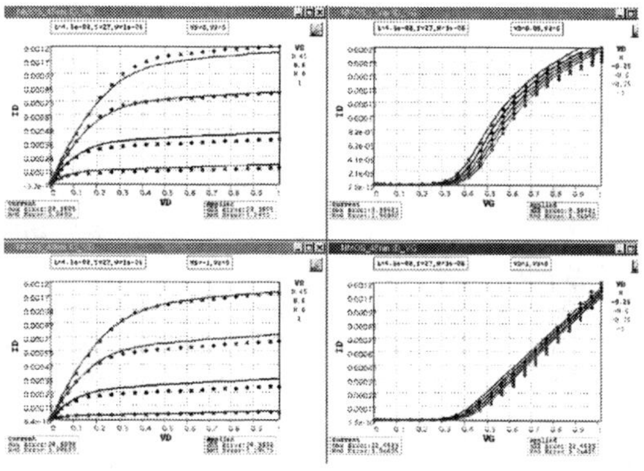

Figure 3: I-V characteristics for NMOS with L_{gate} = 45 nm.

The global model parameters represent the nominal process conditions and various drawn gate lengths. Figure 3 shows the current-voltage characteristics for a 45-nm NMOS device. The points show the numeric simulation data and the solid lines show the electrical characteristics generated by the global SPICE model. Similarly, Figure 4 shows the comparison between the simulation results and the global model for a 45-nm PMOS device. The global SPICE model extracted here shows an RMS error of ~4%.

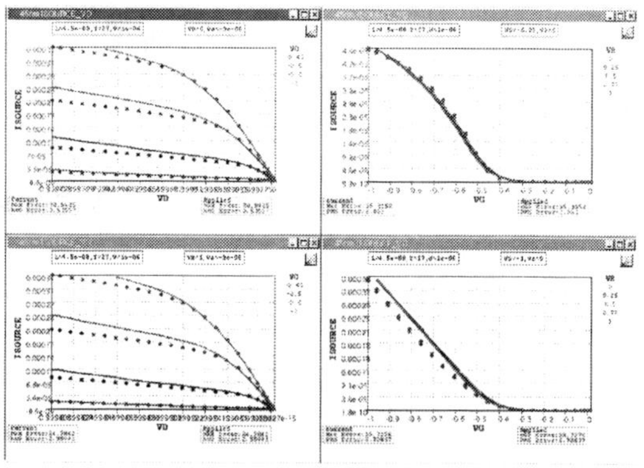

Figure 4: I–V characteristics for PMOS with L_{gate} = 45 nm.

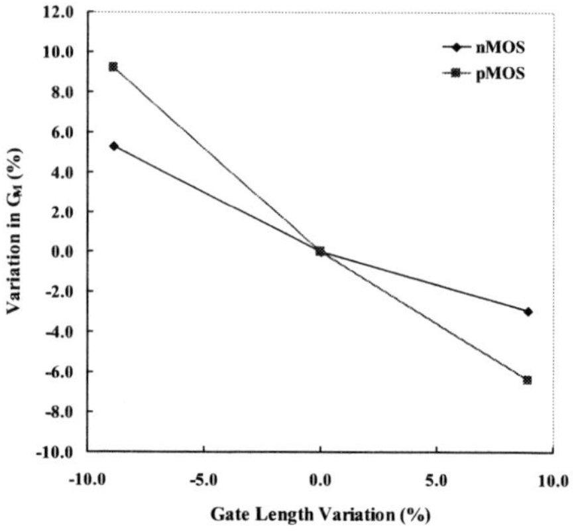

Figure 5: G_m variation with respect to L_{gate} variation.

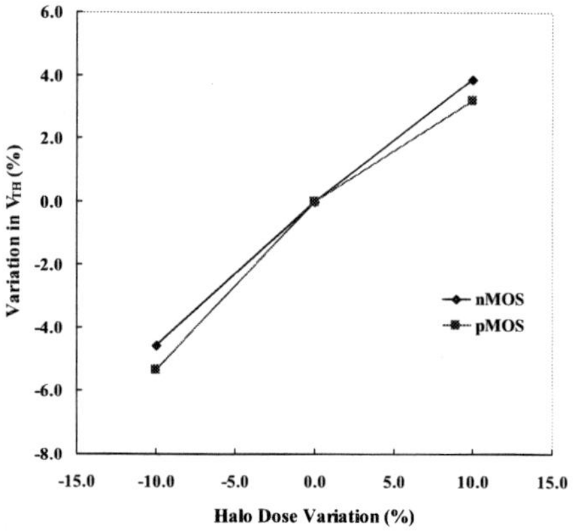

Figure 6: V_{th} variation with respect to halo dose variation.

To extract the process-aware model parameters, gate length, gate taper angle, halo dose and halo energy are varied around their nominal values. Figure 5 shows the variation in the NMOS and PMOS transconductance with respect to the gate length variation. Similarly, Figure 6 shows the threshold voltage variation with respect to the halo dose variation. These figures clearly show the impact of process variability on the device characteristics. The process-aware model is based on the global model and the process variability induced performance variation.

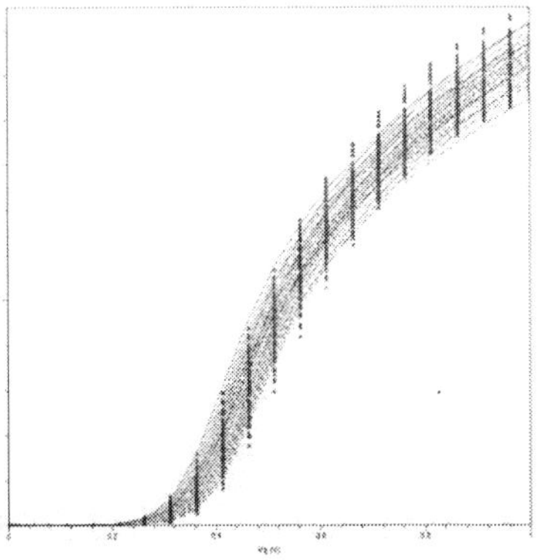

Figure 7: NMOS I_d–V_g at $V_d = 0.05$ V and $V_b = 0.0$ V.

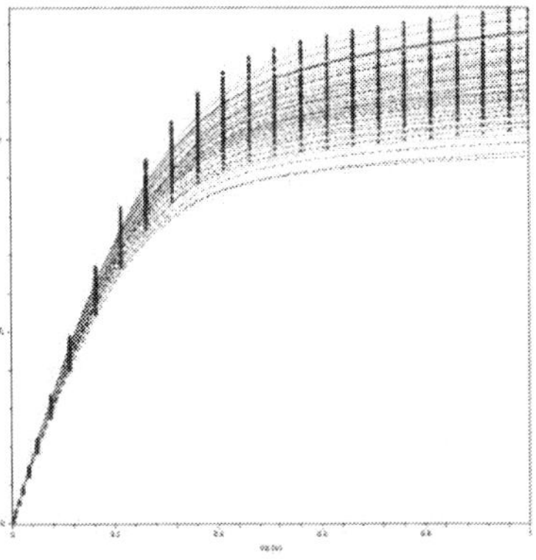

Figure 8: NMOS I_d–V_d at $V_g = 1.0$ V and $V_b = 0.0$ V.

Figure 7 shows the I_d-V_g characteristics of the NMOS devices with a drain voltage of 0.05 V and a bulk voltage of 0.0 V. The lines show the electrical characteristics obtained from the TCAD simulations and the dots show the behavior predicted by the process-aware model. Similarly, Figure 8 shows the I_d-V_d characteristics at a gate voltage of 1.0 V and bulk voltage of 0.0 V. These figures clearly indicate that the process-aware model developed here can account for process variability induced performance variation.

4 CIRCUIT MODELING

Simple digital circuits, like the five-stage ring oscillator shown in Figure 9, are simulated to assess the accuracy of the extracted circuit model parameters. Mixed mode TCAD simulations are compared with circuit simulations performed using the process-aware model parameters [6]. In addition, the absolute model error is calculated using:

$$Error = \frac{Q_{SPICE} - Q_{TCAD}}{\sqrt{\sum (Q_{TCAD})^2 / N}}, \qquad (1)$$

which indicates the model accuracy. These studies demonstrate the accuracy and the robustness of the process-aware circuit model parameters.

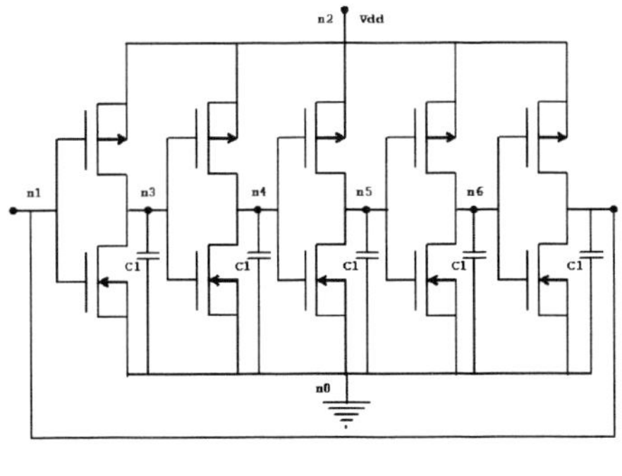

Figure 9: Five-stage ring oscillator.

For the five-stage ring oscillator studied here, mixed mode TCAD simulations are compared with HSPICE® simulations performed using the process-aware model parameters. For the HSPICE® simulations, a load capacitor of 5.0×10^{-15} F is added at each inverter output to account for the device capacitances. Figure 10 shows the variation of the ring oscillator power dissipation with respect to the gate length. The error bars represent an error of 15% in the model.

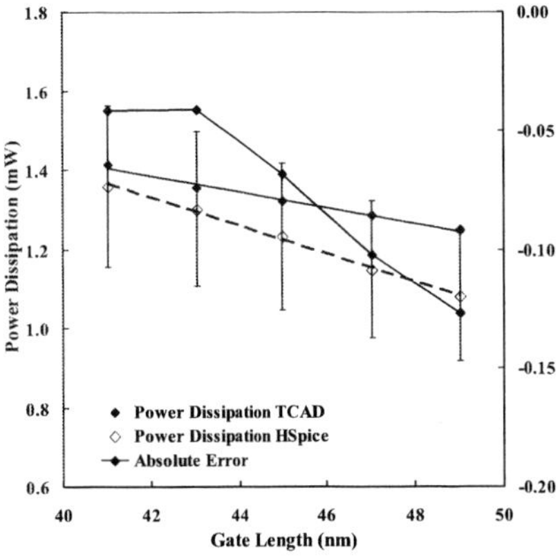

Figure 10: Power dissipation as a function of gate length.

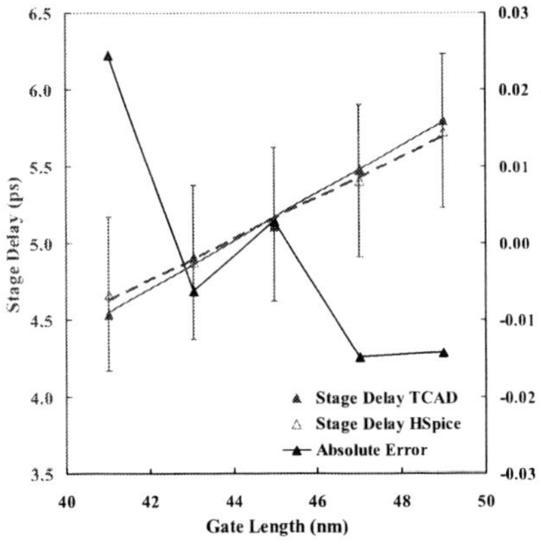

Figure 11: Stage delay as a function of gate length.

Figure 11 shows the variation of the stage delay with respect to the gate length. Figure 12 shows the variation of ring oscillator frequency with respect to the gate length. In case of Figures 11 – 12, the error bars represent an error of 5% in the model.

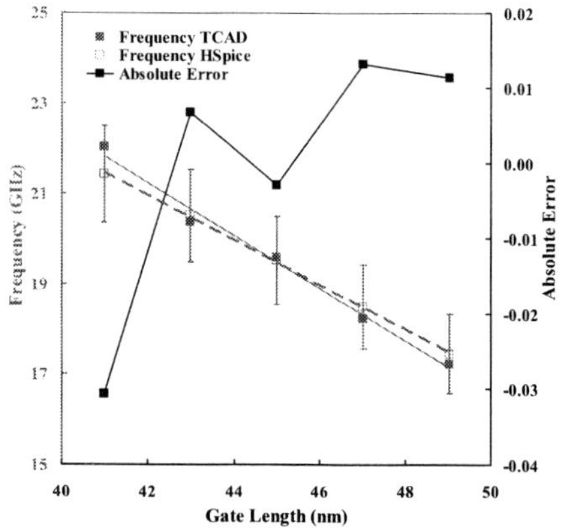

Figure 12: Frequency as a function of gate length.

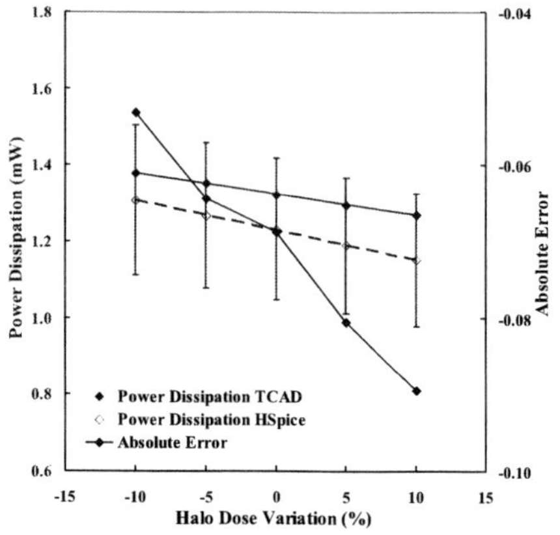

Figure 13: Power dissipation versus halo-dose variation.

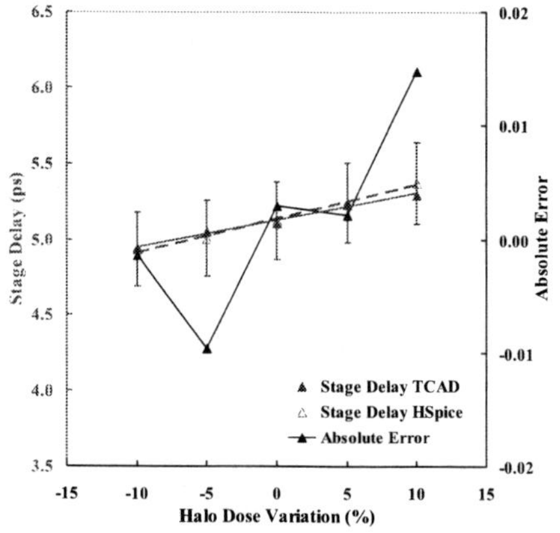

Figure 14: Stage delay as a function of halo dose variation.

Simulations are also carried out for variations in halo dose. Figure 13 shows the variation of ring oscillator power dissipation as a function of halo dose variation. The error bars show an error of 15% in the model. Figure 14 and Figure 15 show the variation of stage delay and frequency as functions of halo dose variation, respectively. Here, the error bars show an error of 5% in the model.

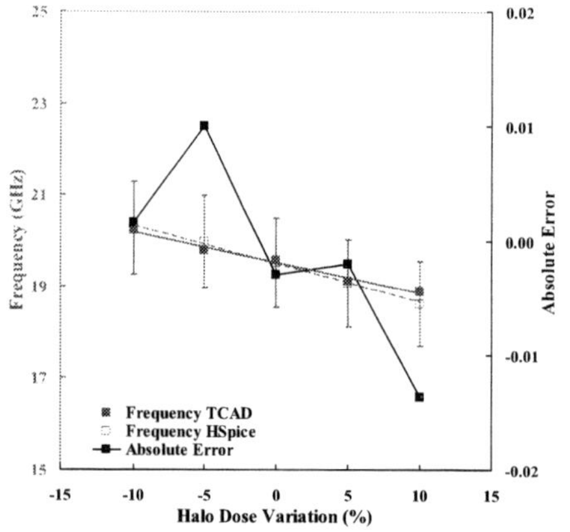

Figure 15: Frequency with respect to halo dose variation.

For Figure 10 – 15, the solid markers represent the mixed mode TCAD results and the hollow markers show the HSPICE® results. The absolute error is also plotted for the five-stage ring oscillator studied here. The results discussed so far show that the process-aware model can accurately predict the behavior of standard cells as well as more complex circuit elements.

5 CONCLUSIONS

Manufacturing process variability is a major cause of parametric yield loss. The process variability needs to be taken into account in order to improve the overall yield. Here, a methodology to extract process-aware model parameters and to create robust circuit designs that account for process variability is presented. The ring oscillator results discussed here show that the process-aware models can accurately predict the process variability impact on the performance of complex circuit elements.

REFERENCES

[1] Borges, R. et al., in 8th *Intl. Conf. on Solid-State and Integrated Circuit Tech. (ICSICT)*, Shanghai, China, pp. 1853–1856, Oct. 2006.

[2] Papanikolaou, A. et al., in *IEEE Custom Integrated Circuits Conf.*, San Jose, CA, USA pp. 773–778, Sep. 2007.

[3] TCAD Sentaurus user guides, Version Z-2007.03, Mountain View, CA: Synopsys, Inc., 2007.

[4] M. V. Dunga et al., *BSIM4.6.0 MOSFET Model – User's Manual*, Berkeley: *University of California*, 2006.

[5] Paramos User Guide, Version A-2007.09, Mountain View, CA: Synopsys, Inc., 2007.

[6] HSPICE® user guides and reference manuals, Version A-2007.12, Mountain View, CA: Synopsys, Inc., 2007

Analytical modelling and performance analysis of Double-Gate MOSFET-based circuit including ballistic/quasi-ballistic effects

S. Martinie[*,**], D. Munteanu[**], G. Le Carval[*], M.-A. Jaud[*], J.L. Autran[**,***]

[*]CEA-LETI MINATEC, 17 rue des Martyrs, 38054 Grenoble, Cedex 9, France.
[**] L2MP-CNRS, UMR CNRS 6137, Bât. IRPHE, 49 rue Joliot Curie, BP 146, 13384 Marseille Cedex 13, France.
[***] University Institute of France (IUF), 75000 Paris, France.
e-mail: sebastien.martinie@cea.fr, Phone: +33 4 38 78 18 40, Fax:+33 4 38 78 51 40

ABSTRACT

In this paper we present a compact model of Double-Gate MOSFET architecture including ballistic and quasi-ballistic transport down to 20 nm channel length. In addition, this original model takes into account short channel effects (SCE/DIBL) by a simple analytical approach. The quasi-ballistic transport description is based on Lundstrom's backscattering coefficient given by the so-called flux method. We also include an original description of scattering of processes by introducing the "dynamical mean free path" formalism. Finally, we implemented our model in a Verilog-A environment, and applied it to the simulation of circuit elements such as CMOS inverters and Ring Oscillators to analyze the impact of ballistic and quasi-ballistic transport on circuit performances.

Keywords: Double-Gate MOSFET, ballistic/quasi-ballistic transport, compact model, Ring Oscillator.

1 INTRODUCTION

As the MOSFET continues to shrink rapidly, emerging physical phenomena, such as ballistic transport, have to be considered in the modelling and simulation of ultra-scaled devices. Future Double-Gate MOSFETs (DGMOS Fig. 1), designed with channel lengths in the decananometer scale, are expected to be more ballistic or quasi-ballistic than diffusive. At this level of miniaturisation is essential to directly evaluate the impact of ballistic and quasi-ballistic transport at circuit level through simulation of several circuit demonstrators. The implementation of compact models in Verilog-A environment offers the opportunity to describe as accurately as possible the physics of transport and to analyze its impact on various circuit elements.

Several analytical models based on the Drift-Diffusion formalism demonstrate that is possible to introduce the diffusive transport in compact modelling. However, when the channel length approaches the value of mean free path, the mobility definition can no more strictly explain the electronic transport in the device. In this case we use the flux theory and the main parameter of this approach is the backscattering coefficient, which expresses the ballistic and the quasi-ballistic transport. Some well-known works performed by Lundstrom *et al* [1-2] demonstrate the usefulness of the flux theory in qualitatively describing quasi-ballistic transport in compact modelling.

In this work we demonstrate the feasibility of a simulation study of ballistic/quasi-ballistic transport at circuit level and we show the impact of this advanced transport on the commutation of CMOS inverter and the oscillation frequency of ring oscillator. This paper is organized as follows: the section I explains the model physics and the corresponding assumptions. In part II, we explain our device simulation in ballistic and quasi-ballistic case. Finally, we highlight the qualitative connection between physics of quasi-ballistic transport and its impact on circuit performance is thoroughly analyzed.

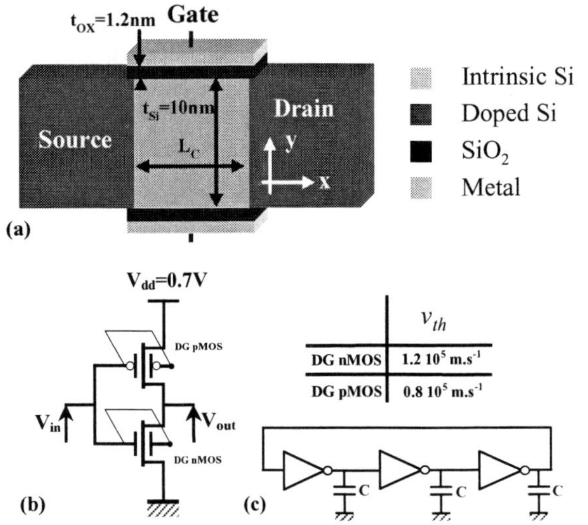

Figure 1: Double-Gate MOSFET (a), CMOS inverter (b), ring oscillator (c) and definition of the main geometrical and electrical parameters.

2 DG MOSFET MODEL

The proposed analytical model (implemented in Verilog-A environment) is based on the well-known work of Natori [3] and Lundstrom [1]. This formula describes the ballistic and quasi-ballistic current:

$$I_D = W.C_{ox}.(V_{GS} - V_T).v_{th}.\left(\frac{1-R}{1-R}\right).\left(\frac{1-e^{-qV_{DS}/k.T}}{1+\left(\frac{1-R}{1-R}\right)e^{-qV_{DS}/k.T}}\right) \qquad (1)$$

where W is the gate width, C_{ox} is the gate oxide capacitance, V_{GS} is the gate to source voltage, V_{DS} is the drain to source voltage, V_T is the threshold voltage, k is the Boltzmann constant, q is the electron charge and T is the lattice temperature.

In addition, Lundstrom *et al.* developed physical compact models describing the device operation in quasi-ballistic regimes using the backscattering coefficient R [3]:

$$R = \frac{L_{kT}}{\lambda + L_{kT}}\ ;\ L_{kT} = L_c.\frac{k.T}{q.V_{DS}} \qquad (2)$$

where λ is the mean free path and L_{kT} is the distance over which the channel potential drops by kT/q compared to the peak value of the source to channel barrier. Physically, L_{kT} represents the critical distance over which scattering events modify the current; L_{kT} depends on both source-to-drain drop voltage and gate length [2].

We consider here the Dynamical mean Free Path which is a "local free path of ballistic carriers" [4]. This characteristic length represents the average distance to be crossed before the next scattering event. This approach considers that each carrier is ballistic as long as any event does not disturb its trajectory. The dynamic mean free path connects the ballistic velocity and all interactions (coulomb and acoustic/optical phonon interaction) experienced by carriers crossing the channel. The scattering process with impurities (τ_{imp}) and phonon interactions (τ_{ph}) are calculated as in [4] and the value of dfp used here is 27 nm in the intrinsic silicon channel. In practice, *dfp* replaces λ to describe quasi-ballistic transport:

$$dfp = v_{bal}.\tau_{tot}\ ;\ \begin{cases} \tau_{tot}^{-1} = \tau_{imp}^{-1} + \tau_{ph}^{-1} \\ v_{bal} = \sqrt{\dfrac{2.\varepsilon_{bal}}{m^*}} \\ \varepsilon_{bal} = \dfrac{3}{2}.k.T + q.V_{DS} \end{cases} \qquad (3)$$

where m^* is the mass in direction of transport, v_{bal} the ballistic velocity, τ_{tot} the total scattering rate and ε_{bal} the carrier energy.

To obtain an accurate model and describe all electrostatic effects, we have also introduced Short Channel Effect and Drain Induced Barrier Lowering (SCE/DIBL) using the Suzuki's model [5] for V_T. Thus, V_T in equation (11) is modified by ΔV_T:

$$V_T = V_{th} - \Delta V_T \qquad (4)$$

where V_{th} and ΔV_T are the long channel threshold voltage and its variation due to SCE/DIBL.

The above-threshold regime is linked to the subthreshold regime using an interpolation function based on the subthreshold swing S parameter also defined by Suzuki in [5]. This assures the perfect continuity of our model between on-state current (I_{on}) and off-state current. The definition of V_{th}, ΔV_T and S will be not detail in this paper, but can be found in [5].

3 DEVICE SIMULATIONS

After implementation in Verilog-A environment, the model has been used to simulate the DGMOS structure schematically presented in Figure 1. The source and drain regions are heavily doped (1×10^{20} cm^{-3}) and an intrinsic thin silicon channel is considered. The channel length varies from 10 nm to 200 nm; a gate oxide of 1.2 nm thick and a midgap metal gate are also considered.

It is well-known that the ballistic current is independent of channel length [3] except when SCE or DIBL appears. In order to clearly confirm this point, simulations have been performed for several length (20, 25, 30, 40, 50, 100 and 200 nm) and considering two types of transport (quasi-ballistic and ballistic; Fig. 2). Note that for the ballistic case, the mean free path value has been chosen to be extremely large compared to the channel length. In contrast to the ballistic case, the quasi-ballistic transport has the same behaviour as that of diffusive transport and the form of the output characteristics depends on L_c.

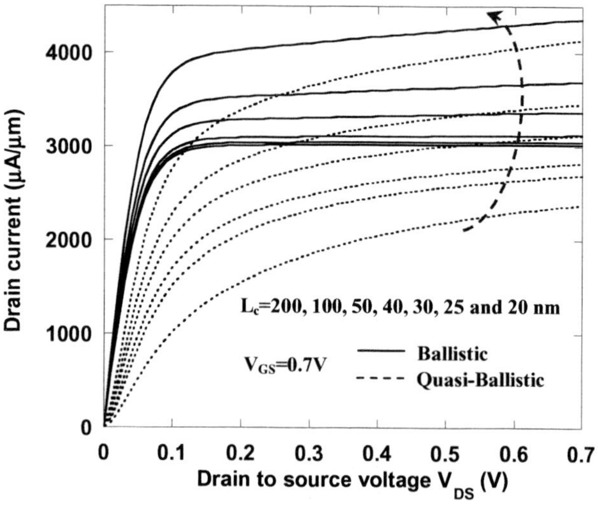

Figure 2: Drain current versus V_{DS} for L_c=200, 100, 50, 40, 30, 25 and 20 nm and V_{GS}.

Figure 3 shows drain current versus the gate voltage characteristics for the DGnMOS and DGpMOS simulated devices at V_{DS}=0.7V. In this approach, we suppose that the transport description (for ballistic and quasi-ballistic case) for holes is identical to that of electrons, with uniquely changing the thermal velocity value in non-degenerate conditions [6]. As expected, the ballistic and quasi-ballistic current shows a perfect continuity between the above and the subthreshold regime (illustrate on Fig. 3a). Finally, figure 3b shows that DGnMOS and DGpMOS have the same behaviour in terms of transport, with different current levels due to the different values of thermal velocity.

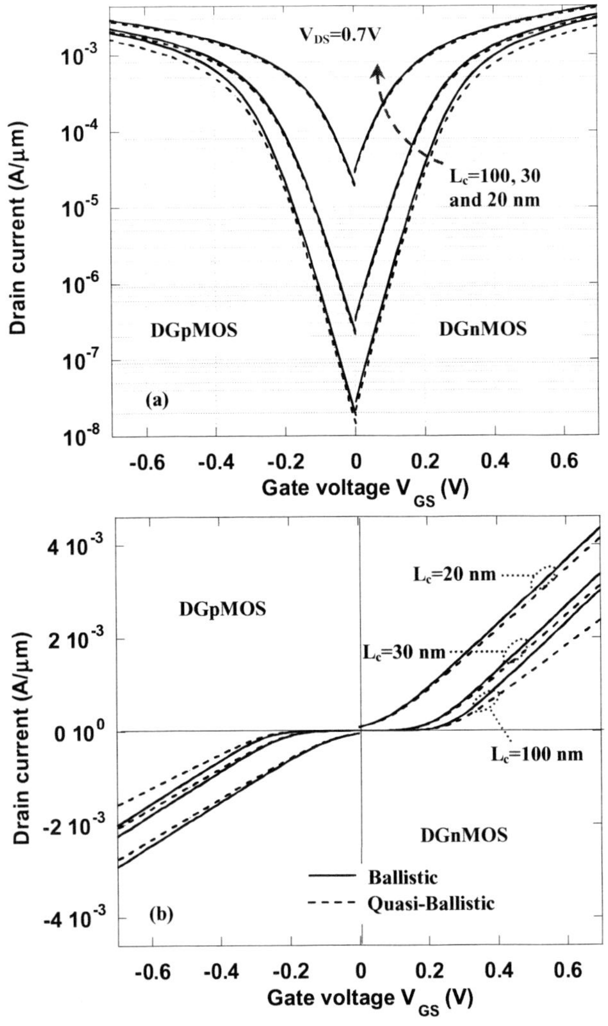

Figure 3: Drain current ((a) log and (b) lin scale) versus V_{GS} for L_c=100, 30 and 20 nm. Solid line for ballistic transport and dashed line for quasi-ballistic transport.

4 CIRCUIT SIMULATIONS

In addition to the simulation of single device operation, we have simulated different circuit elements such as CMOS inverters and ring oscillators (Fig. 1b and 1c) to show the impact of ballistic/quasi-ballistic transport at circuit level.

The output voltage (V_{out}) of the CMOS inverter switches more sharply from the "1" state to the "0" state in the ballistic case than in quasi-ballistic transport (Fig. 4). The commutation of the CMOS inverter depends on the limit between linear and saturation region, which controls the switch between transistors. When SCE/DIBL occur, the transition between linear and saturate regime is modified, and the switch from the "1" state to the "0" state is less sharp. In the quasi-ballistic case, the abruptness of the CMOS characteristic is strongly deteriorated. In conclusion, these results prove that the ballistic transport improves the commutation and the static performances of the CMOS inverter.

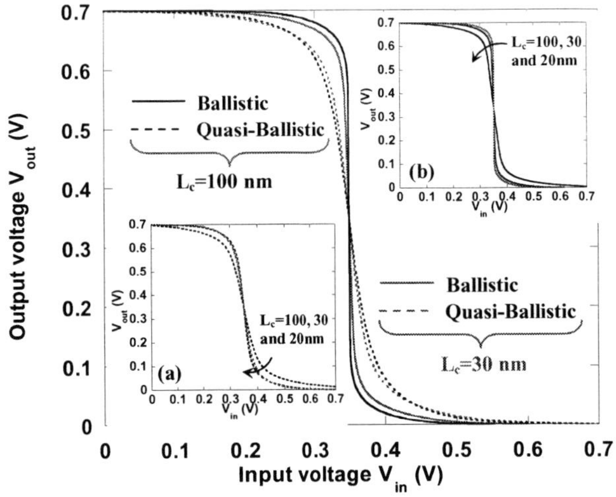

Figure 4: V_{out} versus V_{in} in a CMOS inverter. [Inset (a) and (b): transfer curve in ballistic and quasi-ballistic case for L_c=100, 30 and 20nm].

Figure 5 shows the oscillation frequency as a function of the charge capacitance for two channel lengths: 100 and 30 nm. As expected the oscillation frequency is reduced when the charge capacitance increases, due to variation of the propagation time through the inverters. We can also note the strong influence of short channel effects that increase the current value and reduce the difference between the oscillation frequencies in quasi-ballistic transport compared with ballistic transport.

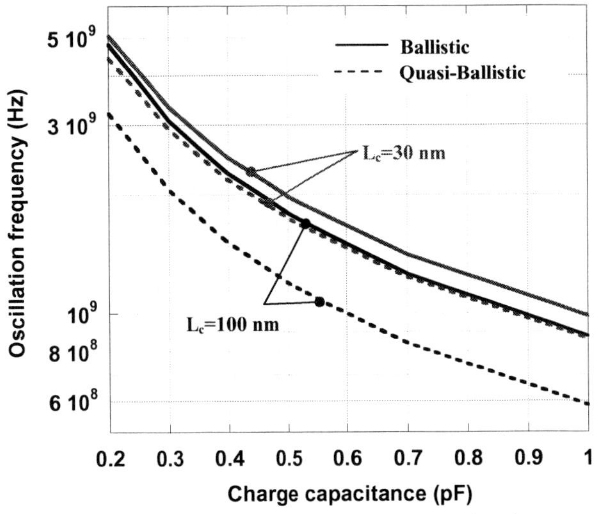

Figure 5: Oscillation frequency versus the charge capacitance for Lc=100 and 30nm

Moreover, we thoroughly studied the influence of parasitic elements and phenomena on circuit performances. Figure 6.a illustrates the impact of two charge capacitances on the oscillation frequency versus channel length. Figure 6.b shows the I_{on} current and V_T versus the channel length.

This last figure demonstrates the strong influence of SCE/DIBL effect on the transient performance. The explanation is that when SCE/DIBL effects occur:

- Firstly, the benefit of ballistic transport (versus quasi-ballistic) on the commutation of the CMOS inverters is hidden (inset (a) and (b) of figure 4).
- Secondly, the I_{on} current is strongly increased (figure 6.b).

Consequently the oscillation frequency in ballistic and quasi-ballistic case is less influenced by the value of the dynamic mean free path.

These results show that the oscillation frequency is directly influenced by the type of transport (ballistic versus quasi-ballistic) which changes strongly the static and the transient performances. But parasitic phenomena such as interconnect capacitances or SCE/DIBL are essential parameters in the analysis of circuit performances even when the intrinsic behaviour of transistors is dominated by ballistic transport.

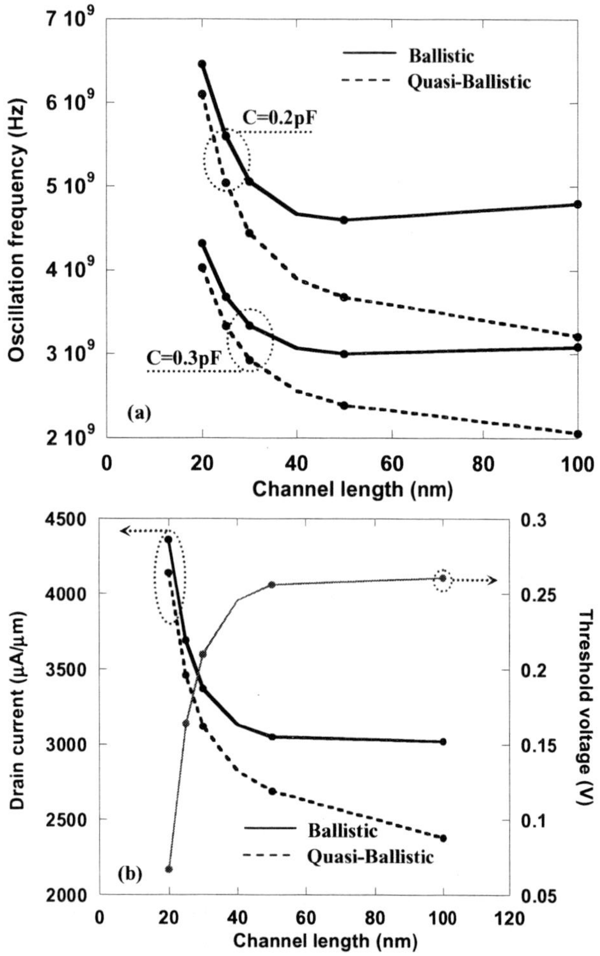

Figure 6: (a) Oscillation frequency versus the channel length for C=0.2 and 0.3 pF. (b) Drain current at $V_{DS}=V_{GS}=0.7V$ and V_T versus the channel length.

5 CONCLUSION

In this work, a compact model for DGMOS taking into account ballistic and quasi-ballistic transport has been proposed and implemented in Verilog-A environment. Short channel effects and an interpolation function to link the above and the subthreshold voltage have been included to obtain a complete description of current characteristics. The dynamical mean free path definition was used to describe scattering processes with impurities and phonons. Finally, the model has been used to simulate two different small-circuits (CMOS inverter and ring oscillators) and to show the significant impact of ballistic/quasi-ballistic transport on the commutation of CMOS inverter and the oscillation frequency of ring oscillator. Our simulation results prove that the ballistic transport improves the commutation and the static performances of the CMOS inverter, and increases the oscillation frequency of ring oscillators.

This work also demonstrates the feasibility of a simulation study of ballistic/quasi-ballistic transport at circuit level and highlights the direct relation between the type of transport and static or transient performances of small-circuits.

ACKNOWLEDGEMENTS

This work was supported by the French Ministry of Research (ANR PNANO project "MODERN", ANR-05-NANO-002).

REFERENCES

[1] M. Lundstrom and Z. Ren, "Essential physics of carrier transport in nanoscale MOSFETs", *IEEE Trans. Electron Devices*,vol. 49, no. 1, pp. 131-141, Jan. 2002.

[2] M. Lundstrom, "Elementary scattering theory of the Si MOSFET", *IEEE Trans. Electron Device Lett.*, vol. 18, no 7,pp. 361-363, Jul. 1997.

[3] K. Natori, "Ballistic metal-oxide-semiconductor field effect transistor ", *J. Appl. Phys.*, vol. 76, no. 8, pp. 4879-4890, Oct. 1994.

[4] E. Fuch *et al*, "A new bascattering model giving a description of the quasi-ballistic transport in Nano-MOSFET", *IEEE Trans. Electron Devices*, vol. 52, no. 10, pp. 2280-2289 Oct. 2005.

[5] K. Suzuki *et al*, "Analytical threshold voltage for short channel n+-p+ double-gate SOI MOSFETs", *IEEE Trans. Electron Devices*,vol 43, no 5, pp732-738, May. 1996.

[6] F.Assad *et al*, "Performance limits of silicon MOSFET's", *IEEE IEDM Tech. Dig*, 1999.

An Improved Impact Ionization Model for SOI Circuit Simulation

Xuemei (Jane) Xi, Fei Li, Bogdan Tudor, Wenyuan (Joddy) Wang, Weidong Liu, Frank Lee, Ping Wang*, Niraj Subba**, Jung-Suk Goo**

Synopsys Inc., * Innovative Silicon, **AMD
Email: Jane.Xi@Synopsys.Com

ABSTRACT

The impact ionization (II) model accuracy issue in industry standard SOI MOSFET is discussed in the paper. Based on Medici 2D simulation study, an improved impact ionization model is proposed which can capture the voltage and geometry dependence for impact ionization current contributed by the parasitic BJT collector current. The model is implemented into the BSIMSOI infrastructure in HSPICE. A reasonably good fit of the model to the 2D simulation results is obtained.

Keywords: impact ionization, SOI MOSFET model, parasitic BJT effect.

1 INTRODUCTION

Generally the device substrate current arising from the impact ionization process near the drain edge can provide important guidance in estimating device reliability [1]. In the floating body SOI (Silicon-On-Insulator) device, impact ionization current plays an important role in determining the floating body potential [2-4]. Therefore, an accurate impact ionization model is very crucial, especially for advanced technology where thin gate oxide, short channel length, etc, which increase the electric field near drain region, hence substrate current significantly.

State-of-the-art impact ionization compact model[2] has been validated over various biases, and geometry range, and has been successfully applied to digital/analog CMOS/SOI IC design. However, in the recent interest on single transistor memory, such as Z-RAM [5-6] that utilizes the often detrimental floating body effect for charge storage, impact ionization effect contributed by the parasitic BJT collector current becomes important in determining memory state where device operates in subthreshold region. Our research found that impact ionization model in the standard BSIMSOI model shows a deviation from 2D device simulation in this operation region.

In this paper, Medici 2D simulation is used to study the impact ionization behavior in all the operation ranges. Model accuracy issue in industry standard SOI MOSFET is discussed and an improved model is proposed to capture the voltage/geometry dependence for impact ionization current contributed by the parasitic BJT collector current. The model is then verified using 2D simulation results and a good agreement is achieved.

2 EXISTING MODEL STUDY

In the industrial standard BSIMSOI model [2], impact ionization current components due to MOSFET and parasitic BJT are both considered, where a same bias dependence for impact ionization rate was used for these two components. This approximation generally won't cause accuracy problem since the majority of impact ionization current in the interested operation regions is contributed by MOSFET drain current. Fig.1 gives the total impact ionization current 2D simulation result for device channel length of 90nm, 65nm, 45nm SOI MOSFETs. As shorter length can significantly increase BJT current and therefore the II current due to BJT, data show that for worst case of 45nm, II current contributed by parasitic BJT is more than one order lower than that of MOSFET contribution when device operates in strong inversion region, so it is negligible.

When SOI MOSFET device operates in subthreshold to accumulation regions, parasitic BJT effect starts to dominant nodal drain current at high drain bias. Medici 2D simulation shows that the impact ionization current in these regions has a near linear dependence on SOI body thickness (Fig.2). From modern BJT model studies [7-8], when BJT characteristics is dominated by emitter area, its collector current is proportional to this effective emitter area. In the case of parasitic BJT in MOSFET structure when using unit width in the simulation, the parasitic BJT collect current should then be proportional to silicon thickness for SOI MOSFET, hence the same dependence for its impact ionization current. The 2D simulation result here indicates that the impact ionization current so generated in subthreshold and/or accumulation regions comes from parasitic BJT current contribution.

Regarding the impact ionization current bias dependence, from 2D simulation results, this II current has very weak dependence on Vgs in subthreshold to accumulation regions (Fig.2), where existing BSIMSOI II model tends to give a strong Vgs dependence for parasitic BJT contributed impact ionization current. As to the model drain bias dependence, parasitic BJT contributed impact ionization current should be explicit function of Vdb rather than the voltage difference between source and drain (Vds), as was the case in BSIMSOI where it is included through

Vdiff. A new model is proposed in the following section to overcome these issues.

3 IMPROVED MODEL DEVELOPMENT

In the new model derivation, Vgs dependence of II current is ignored for simplicity and reasonable accuracy. Impact ionization is well studied in BJT device which generates electron-hole pairs in the depletion layer of the reverse biased drain-substrate (collector/base) junction due to the high electrical field and weak avalanche model is developed in BJT [7-8] to address this effect. We modified the weak avalanche model used in [7] with special consideration of the parasitic BJT effect in MOSFET. Due to the lateral nature of this parasitic BJT, channel length dependency of the impact ionization rate is incorporated in the model. The temperature dependence is also improved. The equations are:

$$I_{ii\_BJT} = \frac{CBJTII + EBJTII \cdot L_{eff}}{L_{eff}} \cdot I_C \cdot (Vbci - V_{bd}) \cdot \exp\left(-ABJTII \cdot (Vbci - V_{bd})^{(MBJTII-1)}\right) \quad (1)$$

$$Vbci = VBCI \cdot \left(1 + TVBCI \cdot \left(\frac{T}{TNOM} - 1\right)\right) \quad (2)$$

This model has been implemented into the BSIMSOI infrastructure in HSPICE. The model parameters and their default values are listed in table I.

4 MODEL VERIFICATION AND DISCUSSION

The model is verified over several geometries and bias range with 2D Medici simulation and proved to be able to present the correct physics of parasitic BJT impact ionization effect. To obtain the pure impact ionization current, 2D simulation was applied twice where one model turns off the impact ionization model and the other turns it on. The body current difference from these two simulations will then be calculated which turns out to be the pure impact ionization current. Impact ionization rate is then calculated as the ratio of impact ionization current vs. total drain current for simplicity. Fig. 3 shows the II rate against drain voltage at Vgs=-1V for L=45nm. Both linear and logarithm curve are given. A good agreement is obtained. Fig. 4 shows the results for L=65nm. And Fig. 5 gives the results for L=32nm. By using the same set of parameters, but keeping the term $\left(\frac{CBJTII + EBJTII \cdot L_{eff}}{L_{eff}}\right)$ as another variable for parameter extraction, the length dependence of this variable can be captured by the proposed expression in equation (1) with result shown in Fig. 6.

5 CONCLUSION

An improved impact ionization model for SOI circuit simulation has been developed and implemented into standard BSIMSOI in HSPICE, which facilitates the BSIMSOI application into deep subthreshold to accumulation regions where accurate II model is a prerequisite for novel memory circuit simulation.

REFERENCES

[1] P. K. Ko, R. S. Muller, and C. Hu, IEDM Tech. Dig., 1981, pp. 600–603.

[2] http://www-device.eecs.berkeley.edu/~bsimsoi

[3] P. Su, S. Fung, H. Wan, A. Niknejad, M. Chan and C. Hu, 2002 IEEE International SOI Conference Proceedings, Williamsburg, VA, Oct. 2002, pp. 201-202.

[4] W. Wu, X. Li, H. Wang, G. Gildenblat, G. Workman, S. Veeraraghavan and C. McAndrew, IEEE 2005 Custom Integrated Circuits Conference, pp. 819-822, 2005

[5] S. Okhonin, M. Nagoga, E. Carman, R. Beffa, E. Faraoni, "New Generation of Z-RAM," IEDM, Dec. 2007, pp. 925-928.

[6] C. Kuo, T. King, and C. Hu, IEEE Trans. Electron Devices, vol. 50, no. 12, pp. 2408-2416, Dec. 2003

[7] McAndrew, C. C., Seitchik, J., Bowers, D., Dunn, M., Foisy, M., Getreu, I., McSwain, M., Moinian, S., Parker, J., Roulston, D. J., Schroter, M., van Wijnen, P., and Wagner, L., IEEE Journal of Solid-State Circuits, vol. 31, no. 10, pp. 1476-1483, Oct. 1996

[8] HICUM http://www.iee.et.tu-dresden.de/iee/eb/hic_new

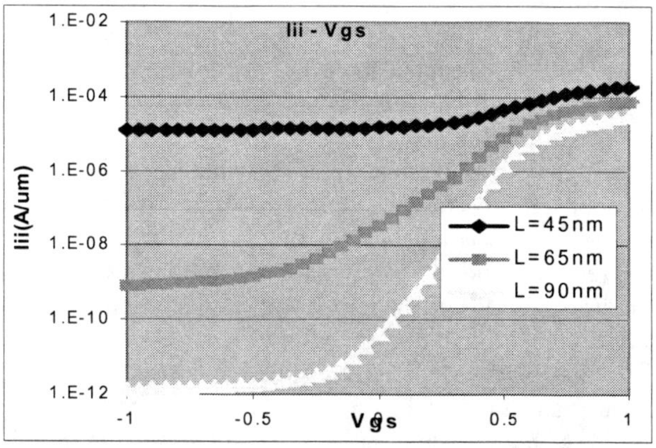

Fig.1 Iii vs. Vgs at various channel length

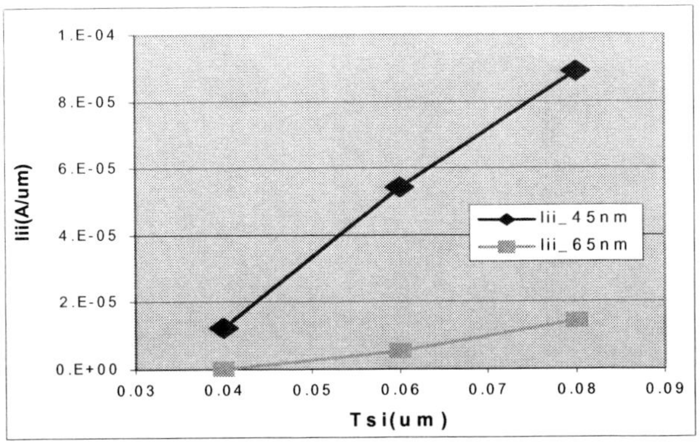

Fig.2 Iii vs. Tsi at various channel length

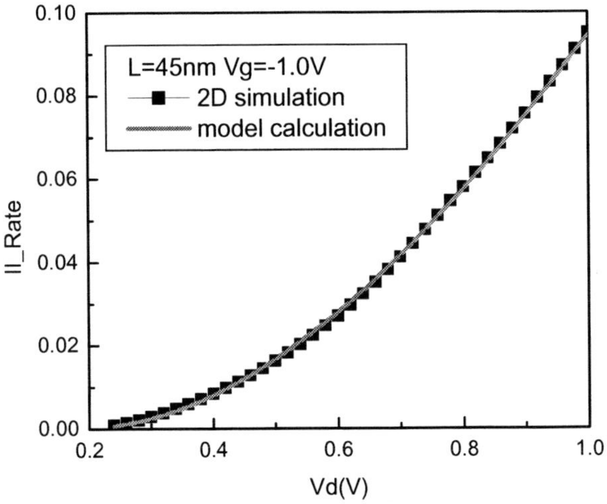

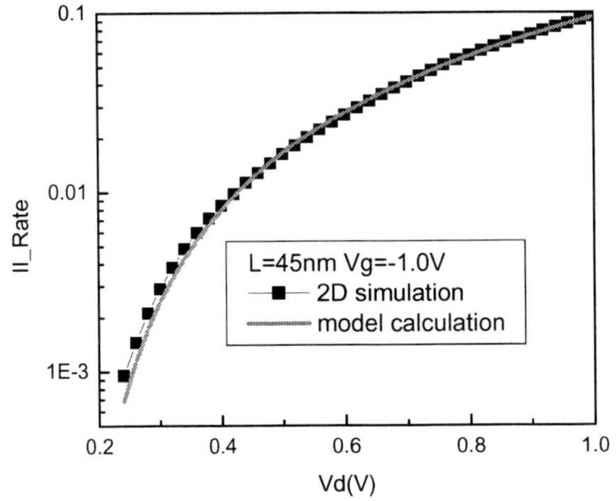

Fig.3 II rate vs. Vd at Vb=0.4V for L=45nm

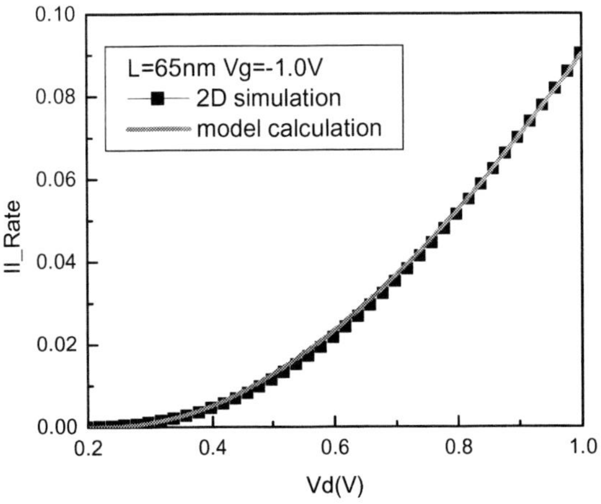

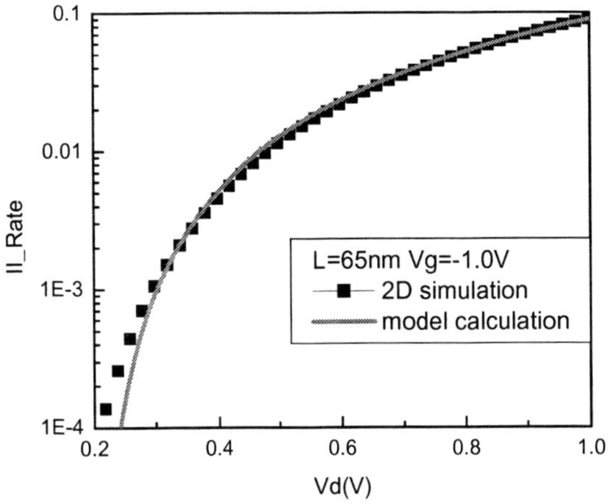

Fig.4 II rate vs. Vd at Vb=0.4V for L=65nm

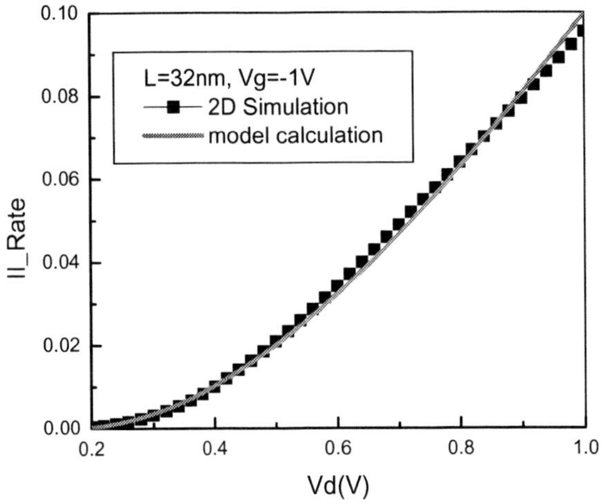

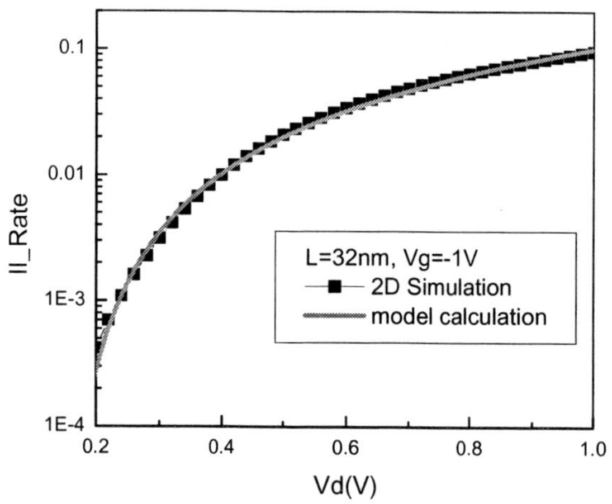

Fig. 5 II rate vs. Vd at Vb=0.4V for L=32nm

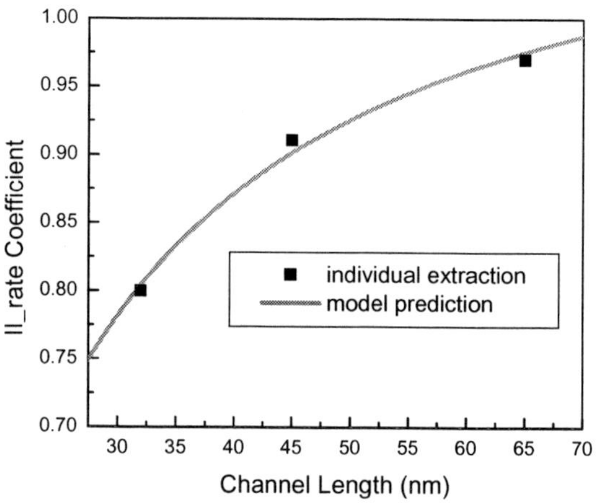

Fig. 6 II Rate channel length dependence

Parameter	Description	Unit	Default
IIMOD	impact ionization model selector	-	0 (original II model)
EBJTII	impact ionization parameter for BJT part	1/V	0.0
CBJTII	Length scaling parameter for II BJT part	m/V	0.0
VBCI	Internal B-C built-in potential	V	0.7
ABJTII	Exponent factor for avalanche current	-	0.0
MBJTII	Internal B-C grading coefficient		0.4
TVBCI	Temperature coefficient for VBCI		0.0

Table I: New parameters enhanced in the new II model

Parameter Extraction for Advanced MOSFET Model using Particle Swarm Optimization

R. A. Thakker, M. B. Patil, and K. G. Anil

Department of Electrical Engineering, IIT Bombay, India, rajesht@ee.iitb.ac.in

ABSTRACT

In this paper, parameter extraction for PSP MOSFET model is demonstrated using Particle Swarm Optimization (PSO) algorithm. *I-V* measurements are taken on 65 nm technology NMOS devices. For the purpose of comparison, parameter extraction is also carried out using Genetic Algorithm (GA). It is shown that PSO algorithm gives better agreement between measurements and model in comparison to GA and with less computational effort. A novel "memory loss (ML)" operation is introduced in the PSO algorithm for the first time, which further improves algorithm efficiency. The PSO algorithm with ML operation has taken 81 minutes on average to extract parameters of two devices of channel lengths, 70 nm and 1 μm.

Keywords: MOSFET parameter extraction, particle swarm optimization, genetic algorithm

1 INTRODUCTION

In the sub 100-nm regime, MOSFET parameter extraction has become a challenging task. Commonly used gradient based methods have many difficulties such as good initial guess requirement, singularities in objective functions, redundancy of parameters, etc. Genetic Algorithm (GA), which does not suffer from these problems, is recently reported for parameter extraction of various state-of-the-art MOSFET models [1]-[2]. Compared to GA, particle swarm optimization (PSO) algorithm [3] is reported to be more efficient for several applications and is also shown to perform better for MOSFET parameter extraction [4]. In this paper, PSO algorithm with a novel "memory loss (ML)" feature is explored for MOSFET parameter extraction. The state-of-the-art PSP MOSFET model is considered in this work. The parameter extraction is performed using PSO algorithm, PSO algorithm with ML operation, and genetic algorithm, and results are compared.

This paper is organized as follows. PSO algorithm and the memory loss operation are described in Sec. 2. Parameter extraction strategy is discussed in Sec. 3, and the results are presented in Sec. 4.

2 PSO ALGORITHM

Many variants of the PSO algorithm have appeared in the literature. In this work, we have used basic PSO algorithm with ML operation for MOSFET parameter extraction.

2.1 Basic PSO Algorithm

PSO algorithm, which mimics behavior of birds flocking in search of food, uses a cooperative approach among randomly generated "particles" to find the globally optimum solution [3]. For a problem with n variables $(x_1,x_2,...,x_n)$, a population of particles is initially generated by randomly assigning positions and velocities to each particle for each variable. In MOSFET parameter extraction, variables correspond to the model parameters to be extracted. If we denote the position and velocity of the ith particle by $\mathbf{x}_i$ and $\mathbf{v}_i$ respectively, then

$$\mathbf{x}_i \equiv (x_i^1, x_i^2, ..., x_i^n), \text{and } \mathbf{v}_i \equiv (v_i^1, v_i^2, ..., v_i^n) \quad (1)$$

Each particle in the population is a candidate for the solution, and the particles are moved towards the fittest particle (i.e., closer to the solution) in the PSO algorithm. In this process, the algorithm finds a better solution, and it is expected to reach the desired solution over time. Each particle keeps in memory the best position (denoted by $\bar{\mathbf{x}}_i$) it has attained during its trajectory. The velocity of a particle is updated on the basis of three vectors: (i) particle's own velocity, (ii) the displacement of the particle from its past best position (i.e., $\bar{\mathbf{x}}_i - \mathbf{x}_i$), (iii) the displacement of the particle from the globally best particle (i.e., $\bar{\mathbf{x}}_g - \mathbf{x}_i$). The particle moves in a direction which is a weighted addition of these three vectors. The velocity and position update of a particle at time $t + \Delta t$ is mathematically represented as follows.

$$\mathbf{v}_i(t+\Delta t) = w\,\mathbf{v}_i(t) + p_1 r_1(\bar{\mathbf{x}}_i - \mathbf{x}_i) + p_2 r_2(\bar{\mathbf{x}}_g - \mathbf{x}_i),$$

$$\mathbf{x}_i(t+\Delta t) = \mathbf{x}_i(t) + \mathbf{v}_i(t+\Delta t)\,\Delta t, \qquad \Delta t = 1 \quad (2)$$

Here, i is the particle index and t is the time. In actual implementation, t is generally taken to be the same as the iteration number, and therefore Δt is numerically equal to 1. The multiplicative constants w, p_1, and p_2 are parameters of the PSO algorithm, and r_1 and r_2 are random numbers uniformly distributed in the range [0, 1]. The constant w is called "inertia" of the particle, since it represents the influence of the previous velocity of the particle on its new velocity. p_1 and p_2 are acceleration coefficients. The positions $\bar{x}_i$ and $\bar{x}_g$ represent the best position attained by i^{th} particle and the globally fittest particle during their trajectory, respectively, up to time t.

The inertia parameter (w, Eq. 2), which is used to update the velocity of particles, can be either kept constant, or varied linearly or adaptively. The most commonly used linear approach to update inertia is implemented in this work and given below,

$$w(t) = (w_i - w_f)(t_{max} - t)/t_{max} + w_f \tag{3}$$

where w_i and w_f represent the initial and final values of w, respectively. t is the current iteration number and t_{max} is the maximum number of iterations. The commonly used range for w parameter is 0.9 to 0.4. The parameters p_1 and p_2 are assigned with 1.494.

2.2 Memory Loss (ML) Operation

The strength of the PSO algorithm is that each particle has a memory in the sense that it remembers the best position attained by it during its trajectory. However, in some cases, this feature makes the population vulnerable to the possibility of getting stuck in a local minimum. In the genetic algorithm, the mutation operation is generally used to bring the population out of a local minimum. In past, the mutation operator has also been applied on the particle's velocity vector in the PSO algorithm and was found to be useful. In this work, the mutation operator is used to perform a "memory loss" operation, as follows.

In each iteration, the particles are made to undergo the memory loss operation with a small probability. A random number is generated for each particle, and if it is less than certain specified probability (typically, 1%), a randomly selected component of its past best position (i.e., its memory) is replaced by a new random value. The following comments may be made about the memory loss operation.

1. It will be effective in changing the orientation of the search space in the PSO algorithm.
2. The memory loss operation is expected to continuously add diversity to the population at a small rate, thus possibly preventing the algorithm from getting stuck in a local minimum.
3. The ML operation is very easy to incorporate in the basic PSO algorithm.

In addition, it was studied by authors that the ML operation allows the inertia range to be reduced from "0.9 to 0.4" to "0.729 to 0.4" , which further makes PSO algorithm more efficient.

3 PSP MOSFET MODEL

Several advanced MOSFET models are currently in use for circuit simulation. They tend to be very complex in order to account for various short-channel effects in modern devices. These models involve generally a large number of parameters to enable accurate representation of experimental characteristics, thus making parameter extraction a challenging task. We have considered PSP MOSFET model [5], which has approximately 70 parameters to confine the behavior of one device of deep sub-micron MOSFET technology.

In the following, the procedure used for parameter extraction is described. In this context, it should be noted that it is generally not possible to extract all of the desired parameters in one optimization step because the *I-V* characteristics in different regions of device operation are affected by different set of parameters. For this reason, the parameters have to be extracted in several extraction steps. This step-by-step procedure will be referred to as the "parameter extraction strategy" in the following.

There are three types of parameters in PSP model.

(a) Geometry-independent (global) parameters (e.g., the oxide thickness, *tox*).

(b) Geometry-dependent (local) parameters whose values would depend on the device width W and length L (e.g. velocity saturation, *vsat*).

(c) Interpolation parameters, which can be used to compute the local parameters (in (b)) using an interpolating formula for given W and L values. For example, drain-induced barrier lowering (DIBL) local parameter (CF) in the PSP model is give by Eq. 4.

$$CF = CFL \left[\frac{L_{EN}}{L_E} \right]^{CFLEXP} \left[1 + CFW \frac{W_{EN}}{W_E} \right] \tag{4}$$

where CFL, CFLEXP, and CFW are interpolation parameters. In this work, we have restricted to parameters in (a) and (b). The extraction of the interpolation parameters (in (c)) is currently under investigation.

We have extracted 34 parameters which affect the DC current-voltage characteristics. Of these, 16 parameters are geometry-independent (i.e. type (a) above) and others are of type (b). The extraction steps used are as follows. *I-V* characteristics of devices with different widths (W1, W2, W3), and different lengths (L1,

L2, ..., L8) are used. W1 and L1 being the smallest device, and W8 and L8 being the largest device. For a given W and L, the following steps describe the extraction procedure.

1. Doping and body bias dependent parameters are extracted from subthreshold region of $I_d - V_g$ with V_{ds} = 0.05, and V_{bs} = 0.0, -0.45, -0.9 V.

2. Mobility and series resistance parameters are extracted from $I_d - V_g$ (V_{ds} = 0.05 V).

3. DIBL and below-threshold channel-length modulation (CLM) parameter are extracted from subthreshold region of $I_d - V_g$ for V_{ds} = 0.9.

4. CLM and velocity saturation parameters are extracted from $I_d - V_d$ with V_{gs} = 0.9, 0.65, 0.4 and V_{bs} = 0.0, -0.45, -0.9 V.

5. Parameters extracted in step 3 are refined, using a narrower range for the relevant parameters.

6. Gate leakage parameters are extracted from $I_g - V_g$ (V_{bs} = 0.0, V_{ds} = 0.0, 0.45, 0.9 V).

We start with the device with W = W3 and L = L8, and extract parameters using steps 1-6 described above. Next, the parameters for the device with W = W3 and L = L1 (shortest channel) are extracted, keeping some of the parameters (those which do not have physical scaling, type (a)) fixed to values obtained from previous extraction step. Now, the parameters of devices with intermediate channel lengths are extracted.

4 RESULTS AND DISCUSSION

The parameter extraction for PSP MOSFET model is carried out as per the parameter extraction strategy discussed in Sec. 3. 34 parameters of PSP model are extracted from measured $I_d - V_g$, $I_d - V_d$, and $I_g - V_g$ data of NMOS devices with channel length (L) ranging from 70 nm to 1 μm and width (W) equal to 10 μm. PSP parameters are extracted using basic PSO algorithm, PSO with ML operation (PSO-ML), and GA. Population and maximum number of iterations (t_{max}) are set to 100 and 1000, respectively. Inertia (w) parameter is assigned range 0.9-0.4 in case of basic PSO and 0.729-0.4 for PSO-ML. p_1 and p_2 are equal to 1.494. For GA, crossover and mutation probabilities are set to 0.84 and 0.15, respectively.

Twenty independent runs are carried out to examine the consistency of the algorithms. The RMS error between measured and model-generated data is computed. The $I_g - V_g$ and $I_d - V_d$ characteristics of the smallest device (70 nm) are taken as benchmark to compare algorithms. The time taken by the algorithms to extract parameters of two extreme devices (70 nm and 1 m) is also noted. All runs are performed on a machine with AMD 2.2 GHz Opteron processor and 16 GB RAM. The results averaged over 20 runs are shown in Table 1.

Both PSO algorithms give comparable results in terms of RMS error (see Table I). But, the PSO with ML operation (PSO-ML) has taken 10 minutes less. The results of parameter extraction using GA are also reported in the table. GA has taken more time and also gave larger RMS error. The comparison between experimental data and model results for the PSO-ML algorithm are shown in Figs. 1-8 and in all cases, the agreement with experimental data is found to be very good.

In conclusion, the PSO algorithm with memory loss operation is found to be very effective for MOSFET pa-

Table 1: RMS error and CPU time averaged over 20 runs.

Characteristics	PSO	PSO-ML	GA
	RMS Error %		
$I_d - V_g$ (V_{ds} = 50 mV)	7.88	8.18	10.53
$I_d - V_g$ (V_{ds} = 0.9 V)	8.83	8.29	8.52
$I_d - V_d$ (V_{gs} = 0.4, 0.65, 0.9 V)	3.89	3.65	4.07
	CPU Time Min.		
	91	81	112

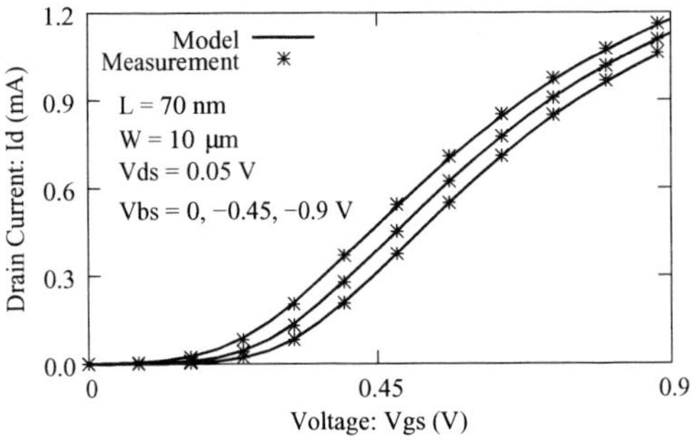

Figure 1: I_d-V_g characteristics for W/L = 10 μm/70 nm.

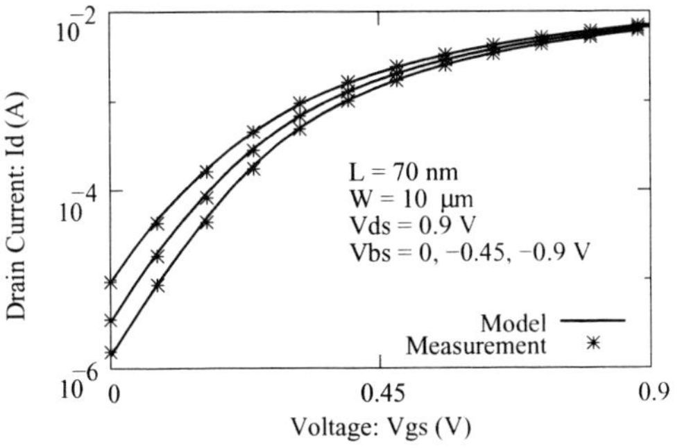

Figure 2: I_d-V_g characteristics (subthreshold) for W/L = 10 μm/70 nm.

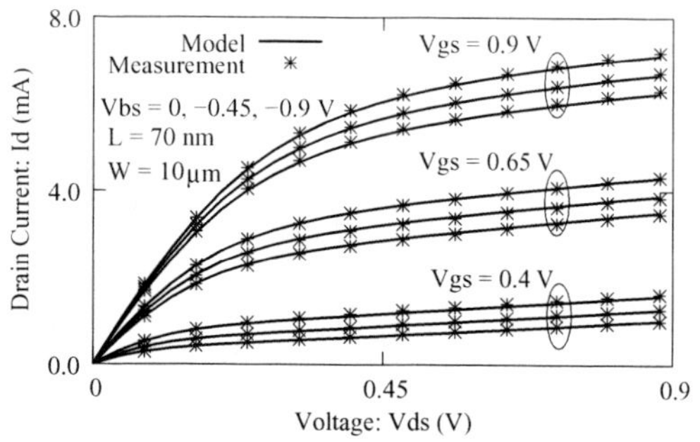

Figure 3: I_d-V_d characteristics for W/L = 10 μm/70 nm.

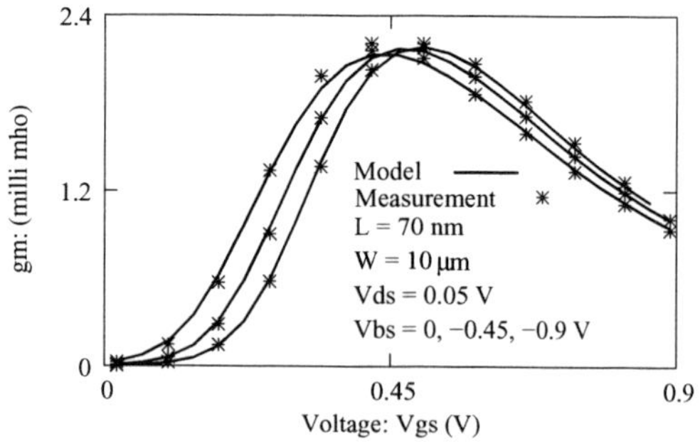

Figure 4: g_m-V_g characteristics for W/L = 10 μm/70 nm.

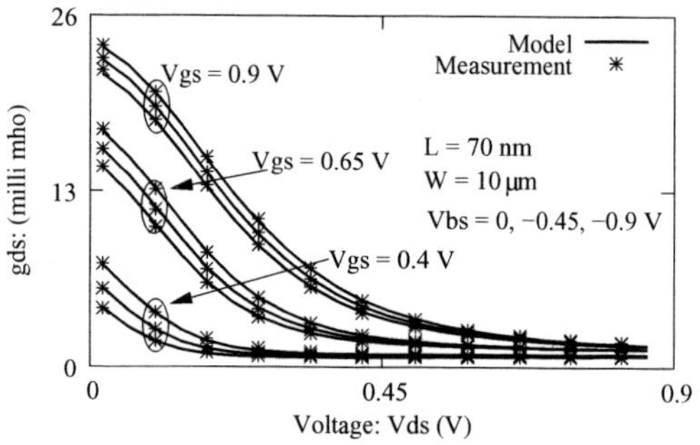

Figure 5: g_{ds}-V_d characteristics for W/L = 10 μm/70 nm.

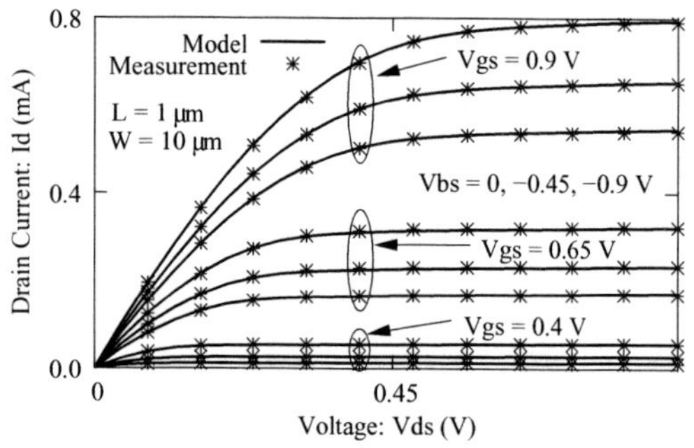

Figure 6: I_d-V_d characteristics for W/L = 10 μm/1 μm.

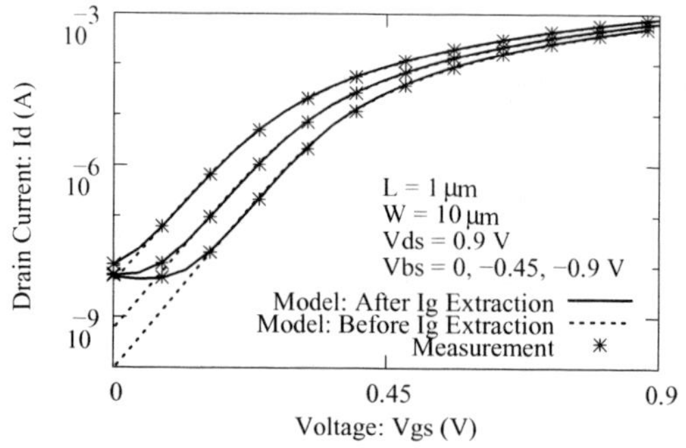

Figure 7: I_d-V_g characteristics for W/L = 10 μm/1 μm showing the effect of gate current parameter extraction.

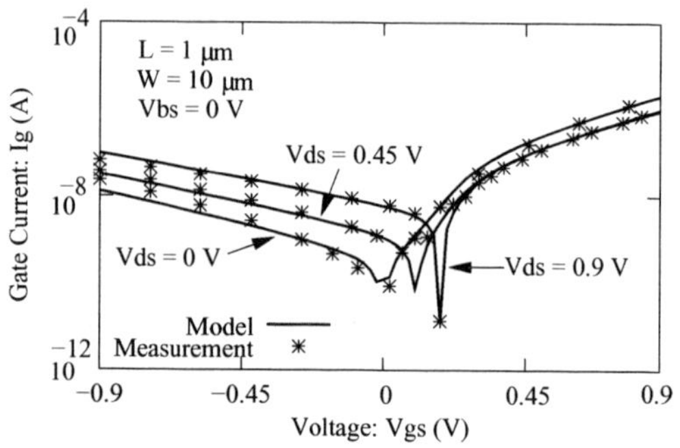

Figure 8: I_g-V_g characteristics for W/L = 10 μm/1 μm.

rameter extraction and substantially better than the genetic algorithm.

This work was performed in part at the "Centre of Excellence for Nanoelectronics (CEN), IIT Bombay" which is supported by the Ministry of Communications and Information Technology, Govt. of India.

The authors would also like to thank IMEC, Belgium for providing the devices used in this study.

REFERENCES

[1] Y. Li, *Micro. Engg.*, vol. 84, Issue 2, pp. 260-272, Feb. 2007.

[2] J. Watts, C. Bittner, D. Heaberlin, and J. Hoffman, *Proc. of Int. Conf. Mod. and Sim. of Micro.*, pp. 176-179, Apr. 1999.

[3] J. Kennedy and R. Eberhart., NJ, pp. 1942-1948, 1995

[4] R. Thakker, N. Gandhi, M. Patil, and K. Anil, *Proc. Int. Workshop Phy. of Semi. Cond. Dev.*, pp. 130-133, Dec. - 2007.

[5] G. Gildenblat et al., PSP 102.0 *User's manual.*

Compact Models for Double Gate MOSFET with Quantum Mechanical Effects Using Lambert Function

H. Abebe*, E. Cumberbatch**, H. Morris** and V. Tyree*

*USC Viterbi School of Engineering, Information Sciences Institute, MOSIS service, Marina del Rey, CA 90292, USA. Tel: (310) 448-8740, Fax: (310) 823-5624, e-mail: abebeh@mosis.com and tyree@mosis.com

**Claremont Graduate University, School of Mathematical Sciences, 710 N College Ave, Claremont, CA 91711, USA. Tel: (909) 607-3369, Fax: (909) 621-8390, e-mail: ellis.cumberbatch@cgu.edu and hedley.morris@cgu.edu

ABSTRACT

This paper is a continuation of the work presented in [5]. Iterative compact device models with quantum mechanical effects for a Double Gate (DG) MOSFET are presented using the Lambert function approach [4, 5]. The quantum model is based on the triangular potential and band gap widening approximations on the intrinsic electron density [1, 2]. The channel current model simulation results are shown in Figure 2-3 and in Figure 4 the charge and capacitance simulations are compared with the Schrödinger-Poisson one dimensional numerical results that are generated from SCHRED [6].

Keywords: device modeling, DG MOSFET, quantum effect, SPICE

1 INTRODUCTION

Taur and Lu, [3], have derived expressions for the I-V characteristics for DG MOSFET device which has the geometry shown in figure 1. The formula obtained in [3] requires the solution of a transcendental equation for an intermediate function β. In this paper this transcendental equation, (3), is solved iteratively with very fast convergence using the rational functions listed in the last section (see the Appendix). The work also includes quantum mechanical effects for nano-scale DG MOSFET.

An accurate model for quantum confinement effects in a nano-scale DG MOSFET device can be achieved by solving the coupled Schrödinger and Poisson equations using self-consistent numerical methods [1, 6]. However, some approximations must be made on the electrostatic potential near the silicon/oxide interface to get a compact analytical model. The approximations used for an analytical solution usually come in the form of triangular potential well profile, with effective surface field [1, 2, 9]. The triangular potential approximation is used here for modeling the channel current and total gate capacitance.

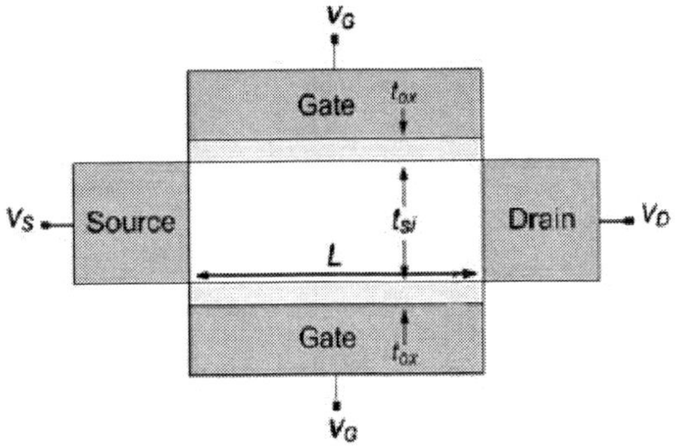

Figure 1. The structure of a double gate (DG) MOSFET

2 COMPACT DEVICE MODELS WITH QUANTUM MECHANICAL EFFECT

The channel current for undoped (or lightly doped) long channel DG MOSFET with very thin oxide is derived in [3] as

$$I_{ds} = I_{ds0}[\beta \tan\beta - \frac{1}{2}\beta^2 + r\beta^2 \tan^2 \beta]_{\beta_D}^{\beta_S} \tag{1}$$

where $I_{ds0} = \mu \frac{W}{L}\frac{4\varepsilon_{si}}{t_{si}}(\frac{2kT}{q})^2, r = \frac{\varepsilon_{si}t_{ox}}{\varepsilon_{ox}t_{si}}$ and β is a solution of the equation

$$\ln(\beta \sec\beta) + 2r\beta \tan\beta = v$$

$$v = \frac{q(V_g - \Delta\phi - V)}{2kT} - \ln(\frac{2}{t_{si}}\sqrt{\frac{2\varepsilon_{si}kT}{q^2 n_i}}) \tag{2}$$

The electron density is $n = n_i e^{(\psi - V)/V_{th}}$. The total mobile charge per unit gate area is given by $Q = 2\varepsilon_{si}(d\psi / dx)_{x=t_{si}/2} = 2\varepsilon_{si}(2kT/q)(2\beta / t_{si})\tan\beta$.

The parameters:

q represents electron charge.

ψ electrostatic potential.

ε_{si} semiconductor permittivity.

ε_{ox} silicon-oxide permittivity.

t_{si} silicon thickness.

t_{ox} oxide thickness.

n_i intrinsic density.

V_g the gate voltage.

$V_{th} = kT / q$ thermal voltage.

k Boltzmann constant.

T temperature.

V quasi-Fermi potential, where *V=0* at the source and $V=V_{ds}$ at the drain.

Equation (2) can be recast into the form

$$(2rz)e^{2rz} = x \tag{3}$$

where $z = \beta \tan\beta$ and $x = (2r\sin\beta)e^{v}$

The solution of the above equation (3) is

$$z = \frac{1}{2r} LambertW(x) \tag{4}$$

From (3) and (4) we can write

$$\beta = \Phi[\frac{1}{2r} LambertW(x)] \tag{5}$$

where $\Phi(z)$ is the solution of the equation $z = \Phi \tan\Phi$ and *LambertW(x)* function is the solution of $x = We^{W}$ (see [7]). As $0 \le \beta \le \pi / 2$ the argument of *LambertW(x)* remains positive. Accurate approximations for *LambertW(x)* and the function $\Phi(z)$ are given in the Appendix. An iterative solution can easily be found to equation (5) by using the low voltage approximation as an initial solution to β. To determine the low voltage solution it is necessary to rewrite (3) as

$$\beta e^{-v} = e^{-2r\beta\tan\beta}\cos\beta \tag{6}$$

The right hand side of (6) has the Taylor expansion

$$e^{-2r\beta\tan\beta}\cos\beta = 1 - (\frac{4r+1}{2})\beta^2 + O(\beta^4) \tag{7}$$

Equations (6) and (7) give a quadratic in β and by means of the quadratic formula we can get the recursion

$$\beta_{(n+1)} = \Phi[\frac{1}{2r} LambertW(2r\sin\beta_n e^{v})] \tag{8}$$

where *n=0, 1, 2 ...* with the initial estimate

$\beta_0 = \frac{e^{-v}}{(4r+1)}(-1 + \sqrt{1 + 2(4r+1)e^{2v}})$. Using only four iterations one obtains excellent results compared to the numerical (see figures 2 and 3).

Quantum effect: One of the reasons that most of the modern silicon MOSFETs are fabricated on <100> oriented substrates is due to the smallest interface-trap density compared to <111> and <110> crystal plane orientations. In this section <100> silicon orientation is considered. The quantum effect near the silicon/oxide interface can be described by solving the 1D Schrödinger equations (for the longitudinal effective mass and again for the transverse effective mass). However, more than 90% of the electrons are concentrated in the ground state and the remaining less than 10% are in the first excited state. The computation that determines the charge density is simplified by considering the Schrödinger equation only along the longitudinal direction

$$\frac{d^2\psi_j}{dx^2} + \frac{2m^*}{\hbar^2}[E_j - U]\psi_j = 0 \tag{9}$$

where ψ_j is the electron wave function with the corresponding energy eigenvalue E_j, $\hbar$ is Planck's constant divided by 2π, $m^*=0.916m_e$ is electron effective mass in the direction perpendicular to the transistor channel surface and m_e is the free electron mass. The boundary conditions for the wave function used in this work are: $\psi_j(x=0) = \psi_j(x \to \infty) = 0$.

The band bending near the silicon/oxide interface at strong inversion confines the carriers to a narrow surface channel and an electron in the semiconductor conduction band is bounded and its energy is quantized. A constant density of states assumption and the Fermi-Dirac statistics give the electron concentration in j^{th} sub-band as

$$N_j = \frac{0.38m_e}{\pi\hbar^2}\int_{E_j}^{\infty}\frac{dE}{1+e^{(E-E_f)/kT}} = 0.38m_e(\frac{kT}{\pi\hbar^2})\ln(1+e^{(E_f-E_j)/kT}) \tag{10}$$

where *j=0, 1,2 ...*, E_f is the Fermi energy,

Substituting the triangular well approximation for *U* in (9) and solving the eigenvalue problem gives the ***Airy functions*** as solutions for the Schrödinger equation with energy eigenvalue

$$E_j = \left(\frac{\hbar^2}{2m^*}\right)^{1/3}\left((3/2)\pi q F_s (j + \frac{3}{4})\right)^{2/3} \tag{11}$$

where F_s is the surface electric field.

It is also possible to get a similar analytical expression to (11) using the Wentzel-Kramer-Brillouin (WKB) method of asymptotic approximation (see [8]).

Stern, [1], pointed out that (11) is a good approximation when the MOS device is at depletion but overestimates the ground state eigenvalues at strong inversion by 6% compared to the numerical result. In [9] it is shown that replacing F_s by the effective surface field F_{seff} using the weighting coefficient can significantly improve the accuracy of (11) at strong inversion.

The classical description of the semiconductor intrinsic density is

$$n_i^c = 2\left(\frac{kT}{2\pi\hbar^2}\right)^{\frac{3}{2}} (m_h m_e)^{\frac{3}{4}} e^{-E_g/2kT)} \qquad (12)$$

where E_g is the semiconductor energy gap and m_h is mass of the hole.

As a result of the quantum confinement effect near the interface, the semiconductor band gap will increase by ΔE_g and this increase can be estimated from the energy difference of the electron ground state and the edge of conduction band, [2, 10]:

$$\Delta E_g = \frac{13}{9}\beta^*(\varepsilon_{si}/4kT)^{1/3}(F_{seff})^{2/3} \qquad (13)$$

where $\beta^* = 6.6X10^{-29} J\cdot m, F_{seff} = \eta(V_g + V_t)/t_{ox}$, the threshold voltage $V_t = V_{t0} + 2V_{th}\ln[q(V_g - V_{t0})/4rkT]$, $V_{t0} = \Delta\phi + 2V_{th}\ln[\frac{2}{t_{si}}\sqrt{2\varepsilon_{si}kT/q^2 n_i^c}]$ and the weighting coefficient η=0.5.

The quantum correction for the intrinsic density becomes

$$n_i^q = e^{-\Delta E_g/2kT} n_i^c \qquad (14)$$

3 RESULTS AND DISCUSSIONS

In this section we present comparisons of our simulation results with the numerical and SCHRED. The simulation results in figures 2-4 are generated using (1), (8) and (14). The compact iteration approach gives excellent results compared to the exact numerical for $L=W$ square device (see figures 2 and 3). In figure 4 our quantum confinement simulation results give good comparison with the self-consistent Schrödinger-Poisson numerical results, SCHRED. The compact quantum simulation results, dash lines, show a significant reduction of the channel current and gate capacitance from the classical simulations, which are the solid lines.

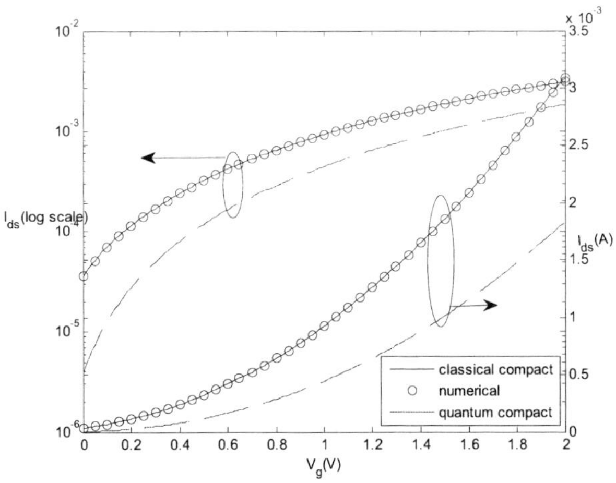

Figure 2: Channel current versus gate voltage for V_{ds}=2V, $\Delta\phi = -0.75V$, $\mu = 300\,\text{cm}^2/V\cdot s$, DG MOSFET with 5nm silicon and 1.5nm oxide thicknesses.

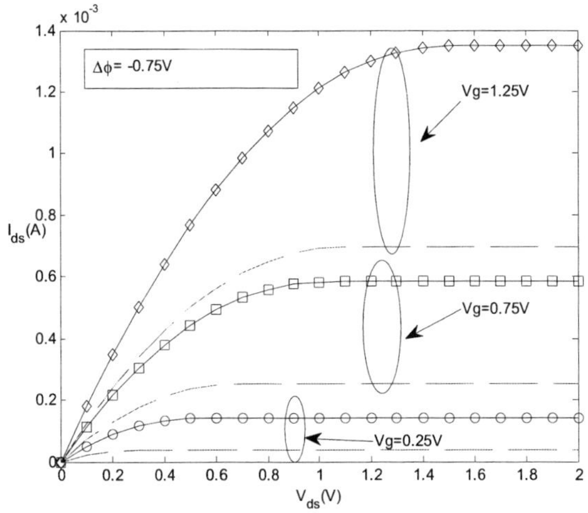

Figure 3: Channel current versus source-drain voltage for $\mu = 300\,\text{cm}^2/V\cdot s$, DG MOSFET with 5nm silicon and 1.5nm oxide thicknesses.

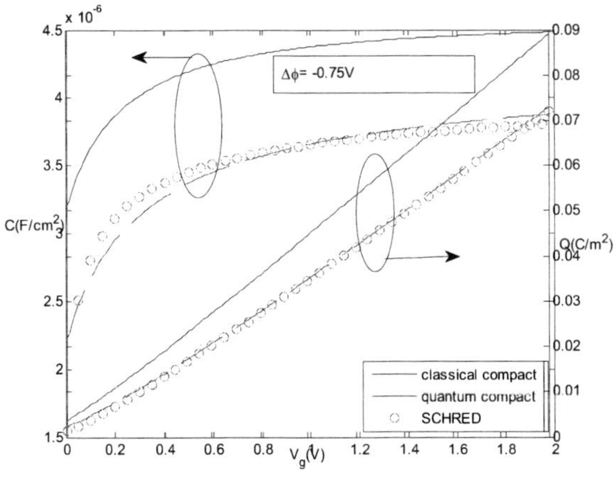

Figure 4: Total mobile charge and gate capacitance versus gate voltage plots for $\mu = 300\,\text{cm}^2/V\cdot s$, DG MOSFET with 5nm silicon and 1.5nm oxide thicknesses.

REFERENCES

[1] M. Stern, "Self-consistent result for n-type Si inversion layers." *Physical Review B,* vol. 5, No. 12, pp. 4891-4899, 15 June (1972).

[2] M. J. Van Dort, P. H. Woerlee and A. J. Walker, "A simple model for quantization effects in heavily-doped silicon MOSFETs at inversion conditions," *Solid State Electronics,* Vol. 37, No. 3, pp. 411-414, (1994).

[3] H. Lu and Y. Taur, "Physics-Based, Non-Charge-Sheet Compact Modeling of Double Gate MOSFETs," *Nanotech Proceedings, WCM,* pp. 58-62, May 8-12, (2005), Anaheim, CA.

[4] A. Ortiz-Conde, F. J. Garcia Sanchez, M.Guzman, "Exact Analytical Solution of Channel Surface Potential as an Explicit Function of Gate Voltage in Undoped-body MOSFETs Using the Lambert *W* function and a Threshold Voltage Definition," *Solid-State Electronics,* Vol. 47, No. 11, pp. 2067-2074, (2003).

[5] H. Morris, E. Cumberbatch, H. Abebe and V. Tyree, "Compact modeling for the I-V characteristics of double gate and surround gate MOSFETs," *IEEE UGIM Proceedings,* pp. 117-121, June 25-28, (2006), San Jose, CA.

[6] SCHRED, http://www.nanohub.org

[7] R. Corless, G. Gonnet, D. Hare, D. Jeffrey, and D. Knuth, "On the Lambert W function", Advances in Computational Mathematics 5(4): 329-359 (1996).

[8] H. Abebe, E. Cumberbatch, H. Morris and V. Tyree, "Analytical models for quantized sub-band energy levels and inversion charge centroid for MOS structures derived from asymptotic and WKB approximations," *Proceedings 2006 Nanotechnology Conference*, Vol. 3, pp. 519-522, May 7-11, (2006), Boston, Massachusetts, USA.

[9] Yutao Ma, Litian Liu, Zhiping Yu, and Zhijian Li, "Validity and applicability of triangular potential well approximation in modeling of MOS structure inversion and accumulation layers," *IEEE Trans. on Electron Devices*, Vol. 47, No. 9, September (2000).

[10] Zhiping Yu, Robert W. Dutton, and Richard A. Kiehl, "Circuit/Device modeling at the quantum level," *IEEE, Computational Electronics, Six international workshop on* July (1998), pp. 222-229.

APPENDIX

I. Approximation of the function $\Phi(z)$:

The function $\Phi(z)$ is the solution of the equation $\Phi \tan\Phi = z$ and $\tan\Phi$ can be approximated by the rational expression

$$\tan\Phi \cong \frac{\Phi_0 + (\frac{1}{3} - \gamma)\Phi_0^3}{1 - \gamma\Phi_0^2 + \delta\Phi_0^4}$$

where $\gamma = \dfrac{40}{9\pi^2}$ and $\delta = \dfrac{16}{9\pi^4}$

Using the above rational expression, the first approximation of $\Phi(z)$ becomes

$$\Phi_0 = \sqrt{\frac{(1+z\gamma) - \sqrt{(z\gamma-1)^2 - 4z(z\delta - 1/3)}}{2(z\delta + \gamma - 1/3)}}$$

The first order Taylor expansion near $z_0 = \Phi_0 \tan\Phi_0$ gives

$$\Phi(z) = \Phi_0 - \frac{(\Phi_0 \tan\Phi_0 - z)}{\tan\Phi_0 + \Phi_0 \sec^2\Phi_0}$$

II. Approximation of the *Lambert W(x)* function

The following rational functions are given in [7] for *Lambert W* function approximation:

$$y = z - \frac{z(z-1)}{1+z} + \frac{z(z-1)^2}{2(1+z)^3} - \frac{(z-1)^3(z-2z^2)}{6(1+z)^5} + \frac{z(6z^2-8z+1)(z-1)^4}{24(1+z)^7} - \frac{z(24z^3-58z^2+22z-1)(z-1)^5}{120(1+z)^9} + \frac{z(120z^4-444z^3+328z^2-52z+1)}{720(1+z)^{11}} - \frac{z(720z^5-3708z^4+4400z^3-1452z^2+114z-1)}{5040(1+z)^{13}}$$

with $z = x/e$, provides a good approximation to $\mathrm{LambertW}(x)$ for $x < 8$ and the function

$$y = L_1 - L_2 + \frac{L_2}{L_1} + \frac{L_2(-2+L_2)}{2L_1^2} + \frac{L_2(6-9L_2+2L_2^2)}{6L_1^3} + \frac{L_2(-12+36L_2-22L_2^2+3L_2^3)}{12L_1^4} + \frac{L_2(60-300L_2+350L_2^2-125L_2^3+12L_2^4)}{60L_1^5}$$

with $L_1 = \ln(x)$ and $L_2 = \ln\ln(x)$, is a good approximation to $\mathrm{LambertW}(x)$ for $x \geq 8$. These two formulae together provide a good compact formula for $\mathrm{LambertW}(x)$ over the entire positive domain.

Neural Computational Approach for FinFET Modeling and Nano-Circuit Simulation

*M. S. Alam, **A. Kranti and **G.A. Armstrong
*Department of Electronics Engineering, A.M.U., Aligarh-202002, INDIA, m.alam@ee.qub.ac.uk
**Queen's University of Belfast, Northern Ireland, BT9 5AH, U. K., a.kranti@ ee.qub.ac.uk

ABSTRACT

The present paper demonstrates the suitability of artificial neural network (ANN) for non-linear modeling of a FinFET for RF applications. FinFET used in this work is designed using careful engineering of source drain extension (SDE) regions, which simultaneously improves transit frequency (f_T) and dc gain $A_{V0} = (g_m / g_{ds})$. The framework for the ANN based FinFET model is a common source large-signal equivalent circuit, where the dependence of intrinsic capacitances, resistances and dc drain current (I_d) on drain-source (V_{ds}) and gate-source (V_{gs}) bias is derived by two-layered neural network architecture. All extrinsic components of the FinFET model are treated as bias independent. The model is implemented in a circuit simulator and verified by the model ability to generate acceptable I_d and circuit parameters to the excitations not used during training. FinFET based on low noise amplifier (LNA) at ~15μA/μm gives ~18.5dB improvement in A_{V0} and nearly identical third-order-intercept (IIP_3) when compared to an identical size bulk MOSFET.

Keywords: Artificial Neural Network, FinFET, Non-linear Modeling, Nano-Circuit, Low Noise amplifier.

1 INTRODUCTION

FinFETs are very promising for low power, low voltage portable applications [1]. For nano-circuit simulation, a suitable FinFET model is required. As the MOSFET dimensions are scaled below 50 nm, it is very difficult to develop physics based compact models [2]. In recent years artificial neural network (ANN) has been used for modeling of variety of transistors [3] [4], as it avoids repeated solving of the complex transcendental equations of a traditional compact physical model, but still offers sufficient accuracy. An ANN model is either based on direct [5] or indirect equivalent circuit approach [3][4]. In the equivalent circuit shown in Fig. 1, both non-linear dc and dynamic (ac) behavior can be modeled simultaneously. While a similar indirect approach has been used for other devices [3] [4], this is the first application for FinFET modeling. 3D ATLAS [6] has been used to simulate a FinFET, which comprises of gate length (L_g) = 60 nm, fin height (H_{fin}) = 60 nm, fin width (T_{fin}) = 42 nm and T_{OX} = 2.2 nm as shown in Fig. 2. Details of measured results have been outlined [7]. This FinFET has since been redesigned for optimal low power performance, with careful engineering of gate–source/drain underlap region, to simultaneously maximize both f_T and dc gain A_{V0} [8]. Source/drain profile was modeled using the expression $N_{SD}(x) = (N_{SD}(x))_{peak} \exp(-x^2/\sigma^2)$, where $(N_{SD}(x))_{peak}$ is the peak source/drain doping, σ (lateral straggle) defines the roll-off [9] of source/drain profile as $\sigma = \sqrt{2sd/\ln(10)}$, where s is the spacer width and d the source/drain doping gradient [9] [10] evaluated at the gate edge $d = \left(\frac{1}{|dN_{SD}(x)/dx|}\right)$

Modeling of nanoscale SDE engineered devices for circuit applications involves formulation of short-channel effects (SCEs), velocity saturation, gate voltage dependent effective gate length (L_{eff}), source/drain (S/D) series resistance and voltage dependent mobility effects [2]. These physical effects are difficult to formulate specially for nano-scale devices. Therefore, in this work artificial neural networks (ANNs) are applied for FinFET modeling

The developed ANN model is implemented in circuit simulator. At both device and circuit level ANN model accuracy has been successfully demonstrated. At device level f_T of FinFET is comparable to the bulk MOSFET, whereas maximum frequency of oscillation (f_{max}) is nearly two times higher due to lower drain-to-source conductance (g_{ds}) and lower capacitance compared to bulk MOSFETs. However, FinFET LNA give better circuit performance in terms of A_{V0} and comparable linearity compared to bulk. Bulk MOSFET model is based on BSIM4 [11] model parameters, which is derived form nano-predictive technology model (PTM) [12] and used in this work for comparison.

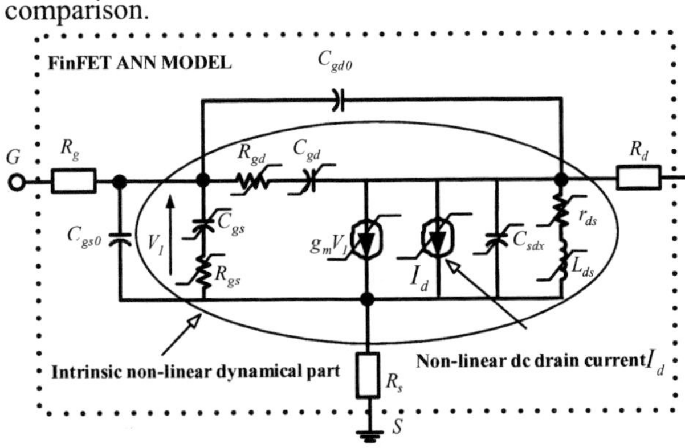

Fig. 1 Non-linear model for FinFET

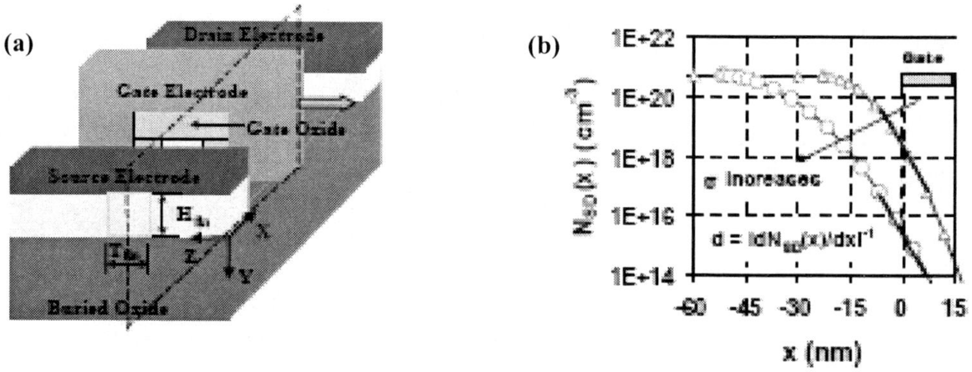

Fig. 2. (a) Schematic diagram of a FinFET analyzed in the present work, and **(b)** Variation of source doping profile for various σ values along the cut-plane along the channel as indicated by dashed lines.

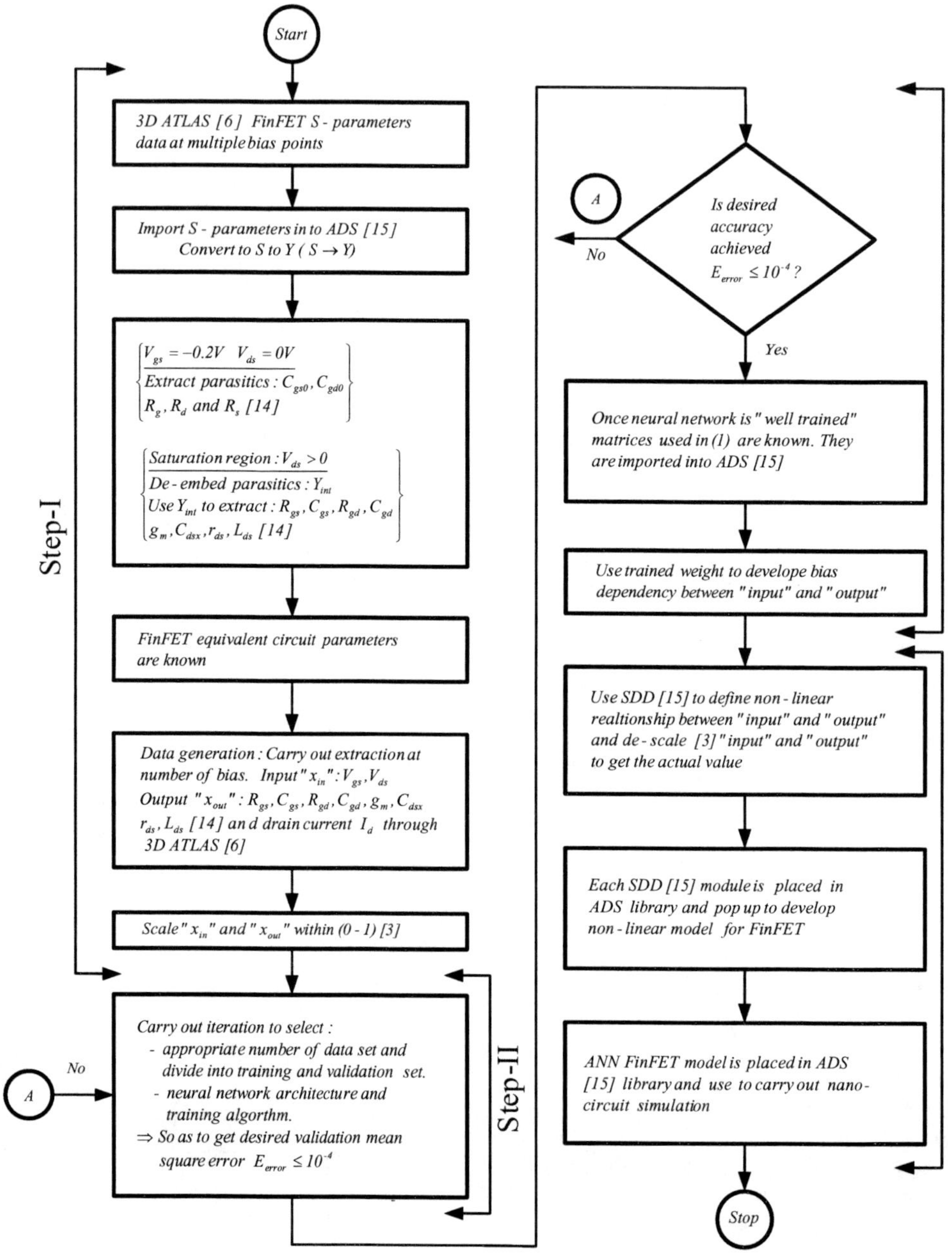

Fig. 3. Basic steps involved in FinFET model development and nano-circuit simulation

2 NEURAL COMPUTATIONAL APPROACH

ANNs are mathematical tools used to represent the non-linear relationship between set of input data and a set of output data. In general, due to its architecture and choice of activation functions, an ANN model suffers less from convergence problems than the alternative polynomial functions [13]. Non-linear relation between bias voltages input x (V_{gs} and V_{ds}) and circuit parameters and I_d outputs y (R_{gs}, R_{gd}, C_{gd}, g_m, C_{dsx}, g_{ds}, L_{ds} and I_d) shown in Fig.1 is modeled using 2-layer ANN [3][4]. A total of 200 "normalized" samples of x and y (drain current I_d), scaled to lie within range 0 to 1, are split into training (125 samples) and testing (= validation) (75 samples) categories. Training and validation steps were run to find the optimal weight values for each parameter to minimize mean square error ($E_{rror} \leq 10^{-4}$) between ANN prediction and original data in the bias range: V_g : -0.2 to 0.4 V and V_d : 0 V to 1.2V generated through 3D ATLAS. Using the same procedure as was done for drain current I_d, circuit parameters bias dependency were derived using 2 layered architecture. Test data samples for circuit parameters were created by carrying out y – parameter extraction [14] generated from 3D ATLAS simulation of a 60 nm FinFET with appropriate source-drain engineering [8]. Trained weights were imported into Advanced Design System (ADS) [15] where its Symbolically Defined Devices (SDD) feature was used to develop the non-linear relationship between x and y. The SDD modules for each circuit parameter were saved in ADS [15] library, and appropriately accessed from the top level to develop the non-linear model. The basic steps involved in ANN based model development is given in flow chart given in Fig. 3.

3 RESULTS AND DISCUSSION

To validate model, we first compared 3D ATLAS and modeled I_d for p- and n-channels FinFET. Fig. 4 compares predicted results (from ANN model) and 3D ATLAS [6] for current voltage characteristics $I_d - V_{gs}$ and $I_d - V_{ds}$. As can be seen from Fig. 4 that very good agreement between them is obtained over full bias range.

Table 1 gives an example of the excellent match between extracted [14] and ANN predicted circuit parameters. Fig. 5 compares FinFET f_T and f_{max} versus drain current density J_{ds} (μA/μm). Extrinsic resistances for source/drain and gate of 20Ω and 5Ω [7], [16], external to the ANN model based on detailed simulation and measurement, represent the best estimate of parasitic resistance for a 300 finger device. A similar design based on the "Predictive Technology Model"(PTM) for bulk MOSFET [12] with same parasitics gives higher f_T and f_{max} in FinFET technology. For an optimal FinFET design, a predicted two-fold enhancement in f_{max} over traditional bulk technology is due to lower g_{ds} and lower C_{gd}. Discrete measurement points in Fig. 5 represent reported values of f_T and f_{max} for a non-underlap SDE FinFET design [7]. Further improvement in f_{max} has been achieved from a more optimal underlap FinFET design [8].

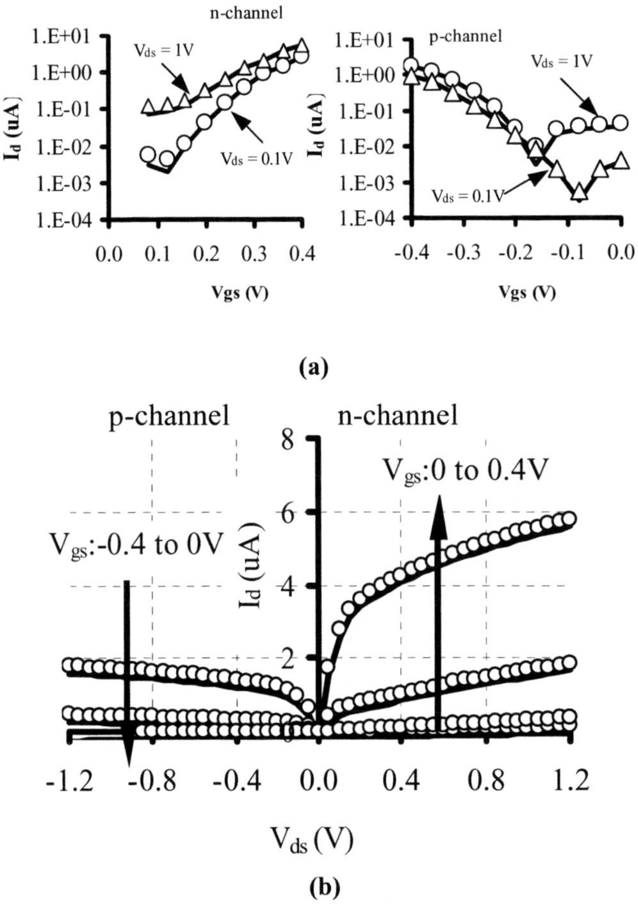

Fig. 4 DC current-voltage characteristics **(a)** $I_d - V_{gs}$ **(b)**. $I_d - V_{ds}$; Solid lines are results from ANN model and symbol from 3D ATLAS.

The ANN model has been utilized to design a 5GHz low noise amplifier (LNA) [18] matched to 50Ω source and load [19]. A FinFET design offers improvement at ~15μA/μm in both third-order-intercept IIP_3 (~1.5 dB) and A_{V0} (~18.5dB), compared to a bulk technology as illustrated in Fig. 6. This is due to higher $P_{IP3} = \frac{4}{50}\left(\frac{dI_d/dV_{gs}}{dI_d^3/dV_{gs}^3}\right)$ (i.e. better g_m linearity) [19] and lower g_{ds} in FinFET compared to bulk as given in Table 2. However, A_{V0} enhancement in FinFET degrades at high drain current due to high series resistance.

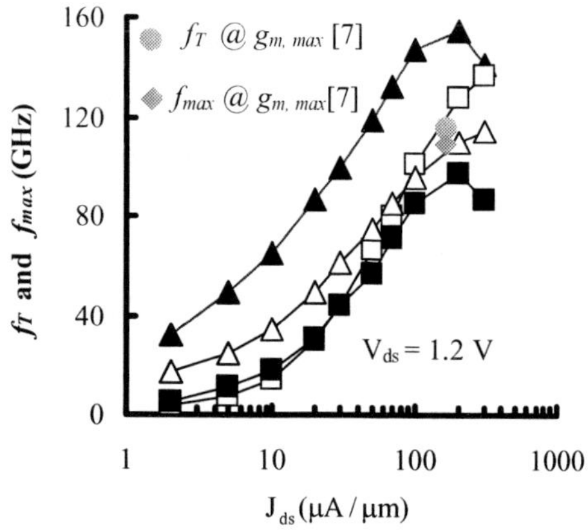

Fig. 5. Comparison of measured and simulated f_T and f_{max} from ANN for 300 fingers. Filled Symbol = FinFET; Open Symbol = Bulk; Line = 3D ATLAS

Table 1 : Extracted circuit -parameters for single finger FinFET

Overlap capacitances (V_{gs} = -0.2 V V_{ds} = 0.0 V)		
$C_{gd0} = C_{gs0}$	0.015 fF	
Circuit parameters (V_{gs} = 0.2 V V_{ds} = 0.2V)		
Parameters	Extracted [14]	ANN Model
R_{gd} (kΩ)	6.37	6.34
R_{gs} (kΩ)	5.47	5.43
100×C_{gd} (fF)	1.06	1.02
100×C_{gs} (fF)	2.27	2.25
100×g_{ds} (mS)	0.0134	0.013
100×g_m (mS)	0.178	0.176
L_{ds} (nH)	4339	4330
100×C_{dsx} (fF)	0.834	0.830

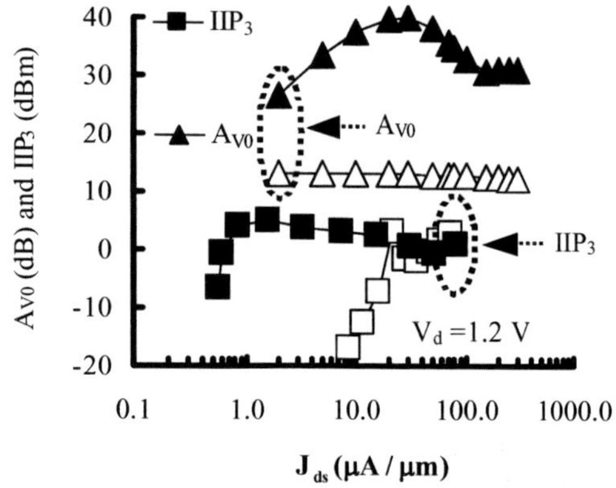

Fig. 4. Comparison of A_{V0} and IIP_3 for FinFET and bulk for 300 fingers. Filled Symbol = FinFET; Open Symbol =Bulk MOSFET

Table 2: Comparison of FinFET and bulk circuit parameters for 300 fingers (I_d =15μA/μm and V_{ds} = 1.2V)

Parameters	FinFET	Bulk
$g_{m1} = dI_{ds} / dV_{gs}$(A/V)	0.01928	0.0071
$g_{m3} = d^3I_{ds}/d^3V_{gs}$(A/V^3)	0.586	0.235
$P_{IP3} = (g_{m1} / g_{m3})\times(4/50)$	2.63×10^{-3}	2.41×10^{-3}
$g_{ds} = dI_{ds} / dV_{ds}$ (A/V)	0.00265	0.0465

4 CONCLUSIONS

A behavioural model based on a neural network has been incorporated in a circuit simulator for simulation of FinFET based nano-circuits. The ANN model has been trained to predict y -parameters, which agrees closed with those extracted from precise 3D device simulations. A high frequency demonstrator LNA circuit based on the ANN model has been designed and simulated. Improved A_{V0} has been observed when design is based on FinFET, as opposed to bulk MOSFET. Performance enhancement can be directly attributed to lower g_{ds} and lower capacitance in FinFET.

ACKNOWLEDGEMENT

M. S. Alam is grateful to the fund received under project entitled "Design and Modeling of Nano-Scale SOI MOSFETs" from All India Council for Technical Education (AICTE), Govt. of India. This work was partially supported by Engineering and Physical Science Research Council, UK.

REFERENCES

[1] A. Kranti et. al., *IEEE Electron Device Letter*, **28**, pp. 139-141, 2007.

[2] G. Pei et. al., *IEEE Trans. Electron Device*, **50**, pp. 2135–43, 2003.

[3] M. S. Alam et., al., *Journal of Active and Passive Electronic Devices*, **2, pp. 1-19,** 2007.

[4] M. S. Alam et., al., *Microwave Optical Technology Letters*, **37**, pp. 53-56, 2003.

[5] F. Djeffal et. al., *Solid-State Electronics*, **51**, pp. 48-56, 2007.

[6] ATLAS Users Maual, 2007, *SILVACO* (USA).

[7] D. Lederer et. al., *IEEE Conference on RF Systems*, pp. 8-11, 2006.

[8] A. Kranti et. al., *IEEE Trans. Electron Devices*, 54, pp. 3308-3316, 2007.

[9] A. Kranti et. al., *Solid-State Electronics*, **50**, pp. 437-447, 2006.

[10] A. Kranti et. al., *Semiconductor Science and Technology*, **21**, pp. 409-421, 2006.

[11] BSIM4 Manual, University of California, Berkeley, 2003

[12] W. Zhao et. al. , *IEEE Trans Electron Devices*; **53**: pp. 2816–2823, 2006.

[13] H. Taher et. al., *In Proc of 12th GAAS Symposium*, Amsrterdam, p.427, 2004.

[14 I.M. Kang et. al., *IEEE Transaction on Nanotechnology*, **5**, pp. 205-210, 2006.

[15] Advanced Design System (ADS)-2005A

[16] Wen Wu, et. al., *IEEE Electron Device Letter*, **27**, pp. 68-70, 2006.

[17] W. Zhao et., al., *IEEE Trans. Electron Devices*, **53**, pp. 2816-2863, 2006.

[18] V. Subramanian et. al., *IEDM Technical Digest*, pp. 919-922, 2005.

[19] Wei Ma et. al., *Solid-State Electronics*, **48**, pp. 1741-46, 2004.

Closed Form Current and Conductance Model for Symmetric Double-Gate MOSFETs using Field-dependent Mobility and Body Doping

V. Hariharan, R. Thakker, M. B. Patil, J. Vasi and V. Ramgopal Rao

Center for Nanoelectronics, Dept. of Electrical Engg., Indian Institute of Technology Bombay
Powai, Mumbai 400076, India. (Email: vharihar@ee.iitb.ac.in / rrao@ee.iitb.ac.in)

ABSTRACT

A closed-form inversion charge-based long-channel drain current model is developed for a symmetrically driven, lightly doped Symmetric Double-Gate MOSFET (SDGFET). It is based on the drift-diffusion transport mechanism and considers velocity saturation using the Caughey-Thomas model with exponent $n=2$, vertical field mobility degradation and body doping. It is valid in sub-threshold as well as above-threshold. Its main feature is that the physical model for velocity saturation has been retained as an integral part of the model derivation, instead of adding its effect at the end by considering an averaged electric field. The model is also extended to model the Channel Length Modulation effect in the post-velocity saturation regime. Comparisons of currents and conductances are made with 2D device simulation results and a reasonable match is shown all the way from sub-threshold to strong inversion.

Keywords: double-gate, mosfet, field, mobility, doping

1 INTRODUCTION

Technology scaling of the conventional MOSFET is reaching a point where there are numerous problems with it going forward, and any suggested work-around has some other problem linked to it. As a result, alternate structures have been studied for quite a while. One such structure is the Double Gate MOSFET (DGFET), a practical realization of which is via the Double-Gate FinFET. DGFETs are more amenable to scaling compared to the conventional MOSFETs, by virtue of their better electrostatics [1, 2]. Also, as devices shrink, adjusting their threshold voltage by heavy doping in the channel is not an acceptable option because of problems like random dopant fluctuations and degraded channel mobility. Hence, it is of special interest to model lightly doped DGFETs. A DGFET with identical material and thickness for the front and back gate electrodes and dielectric, is called a symmetric DGFET (SDGFET).

There have been many efforts to model the drain current for DGFETs. References [3, 4] were based on charge-sheet-models. References [4-13] assumed a constant mobility. References [3, 14] considered velocity saturation effects using the Caughey-Thomas model or its variants with exponent $n=1$ (the variants (eg. [15]) differing in the way the critical electric field E_c relates to v_{sat}, but all of them nevertheless using an exponent $n=1$). References [16, 17] considered velocity saturation effects using an exponent $n=2$ but they considered a *spatially averaged* lateral electric field as the driving field for velocity saturation. The key differentiator in the present work is that the velocity saturation effects are included as an integral part of the model derivation where the *spatial* variation of the lateral electric field driving the velocity saturation effect is represented accurately. Hence, the model is expected to be physically more accurate. Using an exponent $n=2$ has been found to yield a better match with experimental data for N-channel devices [18]. Further, it has been suggested [19] that using an exponent $n=1$ or any odd number for that matter, would yield a model that would fail the Gummel symmetry test at $V_{DS}=0$.

Our model also considers low-field mobility degradation and body doping. We describe next the high-level approach used in deriving the model, and we finally show a sampling of the final results in the form of I_D-V_G, g_m-V_G, I_D-V_D and g_{ds}-V_D plots showing analytical versus 2D device simulation results.

2 MODEL DERIVATION

The schematic of an ideal SDGFET is shown in Fig. 1.

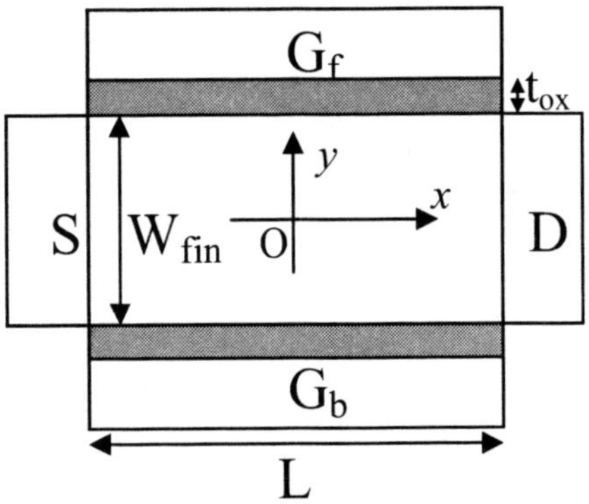

Fig. 1: Schematic of a SDGFET

Under the GCA and neglecting the body doping term initially, the 1D Poisson equation can be written as:

$$\frac{\partial^2 \psi}{\partial y^2} = \frac{q n_i}{\varepsilon} e^{(\psi - \phi_{fn})/\phi_t} \tag{1}$$

where ψ is the electrostatic potential and φ_{fn} is the electron quasi-fermi potential with respect to the φ_{fn} value deep in the source end.

Proceeding as in [5], this can be solved to yield:

$$\frac{4\varepsilon\phi_t\beta_1\tan(\beta_1)}{W_{fin}C_{ox}}+\phi_{fn}+2\phi_t\ln\left(\frac{2\beta_1\sec(\beta_1)}{\beta W_{fin}}\right)-(V_{GS}-\Delta\phi)=f(\beta_1)=0 \quad (2)$$

where β is given by:

$$\beta=\sqrt{\frac{qn_i}{2\varepsilon\phi_t}} \quad (3)$$

and β_1 is a state variable and is related to the inversion charge areal density:

$$Q_i=\frac{-8\varepsilon\phi_t\beta_1\tan(\beta_1)}{W_{fin}} \quad (4)$$

Now, in the drift-diffusion model, the drain current *per unit fin height* is:

$$I_{DS}=-\mu_{eff}(x)Q_i(x)\frac{d\phi_{fn}}{dx} \quad (5)$$

We describe next the approach adopted to incorporate the various physical effects.

2.1 Velocity Saturation

Velocity saturation is modeled using the Caughey-Thomas model [20] with exponent $n=2$:

$$\mu_{eff}(x)=\frac{\mu_0}{\sqrt{1+\frac{\mu_0^2E_{xs}^2}{v_{sat}^2}}} \quad (6)$$

In (6), we choose to model the driving field E_x as being the lateral field at the *oxide-silicon interface* E_{xs}. This is not unreasonable, since even though charge sheet models are invalid in DGFETs [5] and there is non-negligible current flowing even far from the oxide-silicon interface, the current at the interface is still dominant (except in sub-threshold regime [21] where the leakiest path is along the fin center).

Proceeding then on the same lines as in [5], we finally get (7), where β_2 is given by (8). Equations (2) and (7) are the key equations that need to be solved. At this point, we make some very valid approximations, and derive a drain current model in terms of β_{2s} and β_{2d}, where the suffixes s and d denote its values at the source and drain ends respectively.

$$\int_{\beta_{1s}}^{\beta_1}\left\{\begin{bmatrix}\frac{16\phi_t^2\beta_2^2}{I_{DS}^2}\left[\frac{2\varepsilon}{W_{fin}C_{ox}}\left(\beta_1\sec^2\beta_1+\tan\beta_1\right)+\frac{1+\beta_2}{\beta_1}\right]^2\\-\frac{\left(\beta_1\sec^2\beta_1+\tan\beta_1\right)^2}{C_{ox}^2v_{sat}^2}\end{bmatrix}^{1/2}\right\}d\beta_1=\frac{-W_{fin}}{4\mu_0\varepsilon\phi_t}\left(x+\frac{L}{2}\right) \quad (7)$$

$$\beta_2=\beta_1\tan\beta_1 \quad (8)$$

As can be seen from (4), β_2 is proportional to the inversion charge areal density.

Again, making some approximations in (2) and (8), we bypass solving for β_1 and instead derive an approximate closed form solution for β_2 in terms of ($V_{GS}-\varphi_{fn}-\Delta\varphi$). Setting $\varphi_{fn}=0$ and V_{DS} therein yields β_{2s} and β_{2d} respectively.

To calculate the drain saturation voltage V_{DSat}, we set $\partial I_{DS}/\partial V_{DS}=0$, and make suitable approximations to get a closed form expression for β_{2dsat}. This results in a minimum V_{GS} below which a V_{DSat} does not exist (corresponding to the constraint that $\beta_{2d}>0$, to be physically meaningful). We call this V_{GS} as **VGSC**. This result is a direct consequence of considering drift as well as diffusion components in our approach. It is not possible to derive a closed-form exact expression for **VGSC**. So instead **VGSC** is deemed as a model parameter.

Having found β_{2dsat} for a given V_{GS}, V_{DSat} can be calculated using the approximated expressions discussed earlier.

To model channel-length-modulation (CLM) in the post-velocity saturation regime, we have used an approach similar to [22, 18] and modified it for a DGFET.

2.2 Body doping

Body doping is expected to be low in FinFETs and is therefore expected to make a difference only in the sub-threshold regime. We therefore solve the 1D Poisson equation in the sub-threshold regime, this time considering the body doping term. By comparing the expressions obtained with and without considering body doping, we then approximate a *merged* model for the electrostatic potential and electric field, such that they collapse to the correct respective expressions in the extreme cases of sub-threshold and above-threshold, and that they also collapse to the corresponding expressions in the extreme cases of zero and non-zero body doping. The final result is the addition of a N_a dependant spatially constant term in (2).

2.3 Low-field mobility degradation

To add support for vertical field mobility degradation in our core model, we use the same engineering model as that used in the PSP model [17], suitably modified for a DGFET for the depletion charge term, viz.

$$\mu_{eff} = \frac{\mu_0}{1 + \left(\frac{|E_{eff}|}{E_0}\right)^{\theta} + \frac{CS}{\left(1 + \frac{|\overline{Q_i}|}{qN_a W_{fin}}\right)^2}} \qquad (9)$$

where:

$$|E_{eff}| = \frac{\eta|\overline{Q_i}| + qN_a W_{fin}}{2\varepsilon} \qquad (10)$$

In doing so, we have introduced the following model parameters: $\mathbf{\mu_0}$, $\mathbf{\theta}$, $\mathbf{\eta}$, $\mathbf{E_0}$ and **CS**. The term $|avg(Q_i)|$ is a spatial average of Q_i taken along the source-drain direction.

3 DEVICE SIMULATIONS

2D device simulations were done using Synopsis Sentaurus Device assuming abrupt source-body and drain-body junctions. In order to enable Coulomb scattering in the mobility calculation, the University of Bologna mobility model was used instead of the default Lombardi model. Simulations were done for 2 geometries, viz. *(i)* L_g=0.8um, W_{fin}=20nm, T_{ox}=1.4nm; and *(ii)* L_g=0.4um, W_{fin}=10nm, T_{ox}=1nm. For the 0.8um device, simulations were done for 4 different body dopings, viz. 1e16, 1e17, 1e18 and 2e18 cm^{-3}.

The device simulation results were used as virtual experimental data, and model parameters were extracted from it using a parameter extraction program developed at IIT Bombay [23] that is based on a stochastic method, viz. the Particle Swarm Optimization (PSO) algorithm.

4 PARAMETER EXTRACTION

The model has 16 parameters, of which 4 were set to known values (**WFIN**, **L**, **TOX** and **NA**) and the others were extracted in a 3-step fashion from I_D-V_G (including g_m-V_G) and I_D-V_D TCAD data.

In the first step, the work function difference **Δφ** was extracted from I_D-V_G and g_m-V_G data, limiting focus to V_{DS}=50mV and the sub-threshold regime (V_{GS} < 0.4V). Reasonable default values were used for the other model parameters in this step.

In the second step, the low-field mobility degradation parameters were extracted from I_D-V_G and g_m-V_G data, considering the **Δφ** value extracted in the first step and giving it 5% freedom to vary in this step (called parameter refinement). This step continued to limit focus to V_{DS}=50mV but looked at the entire range of V_{GS}.

In the third step, the velocity saturation and CLM parameters were extracted from I_D-V_D data considering the parameters extracted up until the prior steps. However the low-field mobility parameters were refined by up to 3% in this step. This step focused on I_D-V_D data for V_{GS} >= 0.5V.

The **VGSC** parameter was extracted in all 3 steps, though it makes its presence felt only in the third step. It is extracted in the first two steps in order to avoid complex number evaluations in those steps (which could happen in the V_{DSat} expressions, if an appropriate default value for it is not chosen).

5 RESULTS

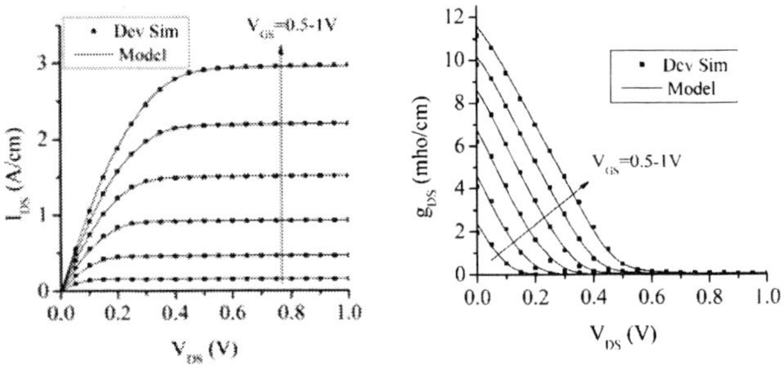

Fig. 2: I_d-V_d (left) and g_{ds}-V_d (right) for the 0.8um device with 1e15 body doping

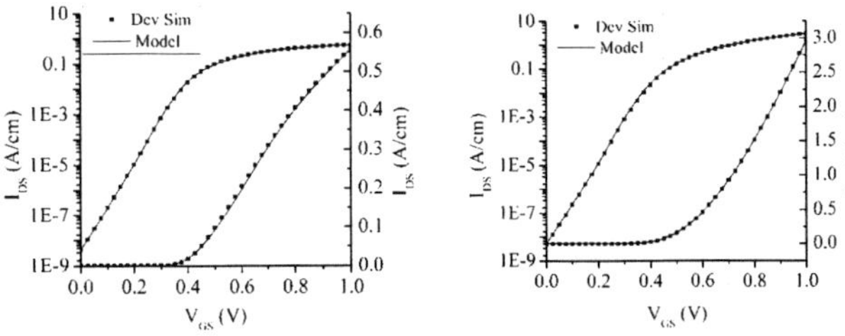

Fig. 3: I_d-V_g @ V_d=50mV (left) and 1V (right) for the 0.8um device with 1e15 body doping

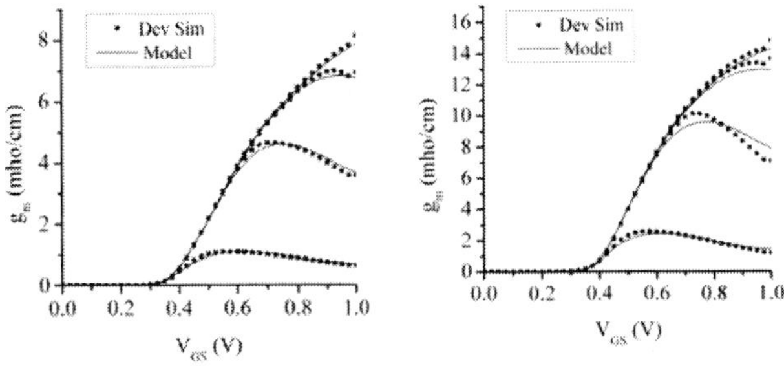

Fig. 4: g_m-V_g for the 0.8um (left) and 0.4um (right) device with 1e15 body doping, for V_d=50mV, 0.24V, 0.43V and 1V

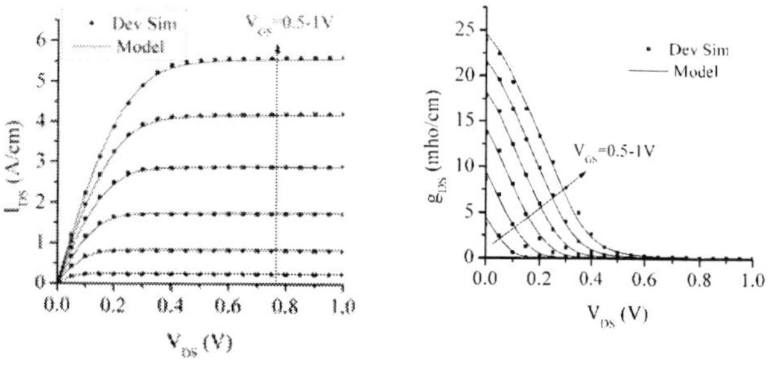

Fig. 5: I_d-V_d (left) and g_{ds}-V_d (right) for the 0.4um device with 1e15 body doping

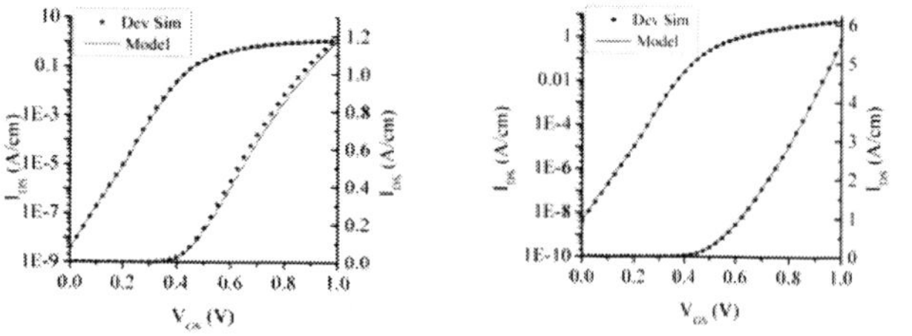

Fig. 6: I_d-V_g @ V_d=50mV (left) and 1V (right) for the 0.4um device with 1e15 body doping

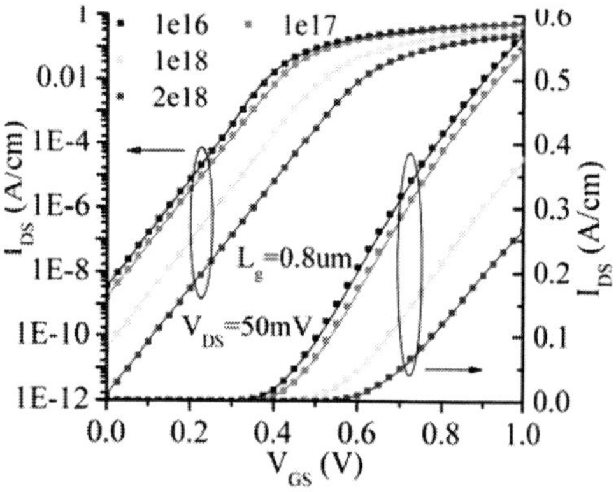

Fig. 7: I_d-V_g for various body dopings. The dots are device simulation data, the lines of the corresponding color are model data

Fig. 2-6 show a sampling of various I_D-V_D, g_{ds}-V_D, I_D-V_G and g_m-V_G curves showing analytical versus device simulation results. Fig. 7 shows the analytical versus device-simulation I_D-V_G curves for various body dopings. All quantities are shown per unit fin height. As can be seen, the match is very good.

6 CONCLUSIONS

A closed-form inversion charge-based long-channel drain current model has been developed for a symmetrically driven, lightly doped Symmetric Double-Gate MOSFET (SDGFET), considering field-dependent mobility and body doping.

ACKNOWLEDGMENT

This work was performed at the Centre of Excellence for Nanoelectronics (CEN), IIT Bombay, which is supported by the Ministry of Communications & Information Technology, Govt. of India. Synopsys Inc. is also gratefully acknowledged for their TCAD tool support.

REFERENCES

[1] P. M. Solomon et al., *IEEE CDM*, pp. 48-62, Jan 2003.
[2] E. J. Nowak et al., *IEEE CDM*, pp. 20-31, Jan/Feb. 2004.
[3] G. Pei, W. Ni, A. V. Kammula, B. A. Minch, and E. C-C. Kan, *IEEE TED*, vol. 50, pp. 2135-2143, Oct. 2003.
[4] M. V. Dunga et al., *IEEE TED*, vol. 53, pp. 1971-1978, Sep. 2006.
[5] Y. Taur, X. Liang, W. Wang, and H. Lu, *IEEE EDL*, vol. 25, pp. 107-109, Feb. 2004.
[6] Jin He et al., *Proc. ISQED*, p. 45, 2004.
[7] J. M. Sallese et al., *SSE*, vol. 49, pp. 485-489, Mar. 2005.
[8] A. S. Roy, J. M. Sallese, C.C. Enz, *SSE*, vol. 50, pp. 687-693, Apr. 2006.
[9] H. Lu and Y. Taur, *IEEE TED*, vol. 53, pp. 1161-1168, May. 2006.
[10] A. Ortiz-Conde, F. J. G. Sanchez, J. Muci, *SSE*, vol. 49, pp. 640-647, Apr. 2005.
[11] Jin He et al., *IEEE TED*, vol. 54, pp. 1203-1209, May 2007.
[12] Z. Zhu et al., *Elec. Lett.*, vol. 43, pp. 1464-1466, Dec. 2007.
[13] Z. Zhu, X. Zhou, K. Chandrasekaran, S. C. Rustagi, and G. H. See, *JJAP*, vol. 46, No. 4B, 2007, pp. 2067-2072.
[14] M. Wong and X. Shi, *IEEE TED*, vol. 53, pp. 1389-1397, Jun. 2006.
[15] C. G. Sodini, Ping-Keung Ko and J. L. Moll, *IEEE TED*, vol. 31, pp. 1386 – 1393, Oct. 1984.
[16] G. D. J. Smit et al., *NSTI-Nanotech*, vol. 3, pp. 520-525, 2007.
[17] G. Gildenblat, et al., *IEEE TED*, vol. 53, pp. 1979-1993, Sep. 2006.
[18] Y. Taur and T. Ning, *Fundamentals of Modern VLSI Devices*, Cambridge Univ. Press, 2003.
[19] K. Joardar, K. K. Gullapalli, C. C. McAndrew, M. E. Burnham, and A. Wild, *IEEE TED*, vol. 45, pp. 134-148, Jan. 1998.
[20] D. M. Caughey and R. E. Thomas, *Proc. IEEE*, vol. 55, pp. 2192 - 2193, Dec. 1967.
[21] G. Pei, J. Kedzierski, P. Oldiges, M. Ieong, and E. C-C. Kan, *IEEE TED*, vol. 49, pp. 1411-1419, Aug. 2002.
[22] P. K. Ko, et al, *IEDM*, pp. 600-603, 1981.
[23] R. Thakker, N. Gandhi, M. Patil, and K. Anil, *IWPSD*, pp. 130-133, 2007.

Comparison of Four-Terminal DG MOSFET Compact Model with Thin Si Channel FinFET Devices

T. Nakagawa*, S. O'uchi**, Y. Liu**, T. Sekigawa*, T. Tsutsumi*,***, M. Hioki*, and H. Koike*

*Electroinformatics Group, **Silicon Nanoscale Devices Group
Nanoelectronics Research Institute
National Institute of Advanced Industrial Science and Technology (AIST)
1-1-1 Umezono, Tsukuba, Ibaraki, 305-8568 Japan, nakagawa.tadashi@aist.go.jp
***Meiji University

ABSTRACT

We discuss a compact model for four-terminal double-gate (DG) MOSFETs based on double charge-sheet drift-diffusion transport, with carrier-velocity saturation. In this model, large V_D at the saturation region is attributed to the infiltrated drain electric field. The latter overrides the drift-diffusion model at the transition point where the quasi-Fermi level is smoothly connected. The model can handle asymmetric gate structure, as well as independent gate voltage for two gates. This approach provides physics-based carrier profile. Comparison with the result of 2D device simulator shows that accurate intrinsic capacitance model has been obtained by the analytical derivative of channel carrier with respect to the terminal voltages. The model was compared with the FinFET samples fabricated by using lightly doped p-type (110) SOI wafers. By fitting five geometrical factors, two flat-band voltages, four transport-related parameters, two extrinsic resistances, the model gives a quite acceptable device model.

Keywords: Compact model, MOSFET, Double-gate, FinFET

1 INTRODUCTION

Ever since the gate length of MOSFET approached 1µm, short channel effect (SCE) has been an annoying problem for device developers. A lot of methods have been introduced to avoid SCE and to keep the pace of design-rule reduction unhindered. Today, exhaustion of such counter-measures is again discussed. To significantly suppress SCE, double-gate (DG) MOSFET structure was proposed [1]. The structure has long been overlooked mainly because of its difficulty for manufacturing. But as the new approach to suppress SCE becomes an imminent issue, the FinFET, which is one of double-gate structures, has gained much attention.

Today, development of new device structures should go with development of its compact model. We proposed a compact model of the DG MOSFET [2]. The model can handle asymmetric gate-structure and four-terminal operation. A new drift-diffusion transport equation was adopted to cope with the carrier-velocity saturation effect directly [3]. The model is now written by Verilog-A, a circuit description language which has recently extended to facilitate subcircuit-style description of the device compact model [4].

In this paper, we first describe present status of the compact model by comparing with the device-simulator results. We then discuss on the comparison with fabricated devices.

2 MODEL

2.1 Transport Equation

As a start point, we assume undoped Si-channel DG MOSFET. Gradual channel approximation and charge-sheet approximation are adopted. Since there are two Si/oxide interfaces, we assume double charge-sheet when perpendicular electric field has a zero-point inside the channel. The absence of space-charge ions means there is linear relationship between i-th charge-sheet carrier densities n_i and the surface potential ψ_j for j-th interface as

$$\psi_j(x) = \sum_{i=1,2} A_{ij} n_i(x) + \text{const}_j \,. \tag{1}$$

By introducing several approximations to decouple two charge-sheets [5], we get the following equation:

$$\psi_i(x) = (q/C_i)n_i(x) + \text{const} \,, \tag{2}$$

where C_i is an effective capacitance.

At high longitudinal electric field $E_{//}$, carrier velocity saturates. This is often expressed as the decrease of carrier mobility μ. In the case of electrons, the replacement of mobility with the following form gives good result.

$$\mu = \mu_S \Bigg/ \sqrt{1 + \left(\frac{E_{//}(y)}{v_{sat}/\mu_S}\right)^2} \tag{3}$$

where μ_S is the surface mobility, v_{sat} is the saturation velocity. Although μ_S is a function of position y along the channel, its variation is not large, and here we neglect it. On the other hand, $E_{//}$ dependence of the mobility should be included to treat carrier-velocity saturation rigorously. The resultant transport equation is as follows:

$$-q\mu_S\left(\frac{q}{C}\frac{dn}{dy}n+\frac{kT}{q}\frac{dn}{dy}\right)=I\sqrt{1+\left(\frac{E_{//}}{v_{sat}/\mu_S}\right)^2}\,. \quad (4)$$

The solution gives the relation between channel length L and carrier density as:

$$L=F(n_a(L))-F(n_a(0))\,, \quad (5)$$

$$F(n_a(y))=\frac{n_b^2\log(n_a+\sqrt{n_a^2-n_b^2})-n_a\sqrt{n_a^2-n_b^2}}{2\beta(v_{sat}/\mu_S)n_b}\,, \quad (6)$$

where $n_a=n/n_{th}+1$, $n_a=(I/qv_{sat})/n_{th}$, and $n_{th}=C/q\beta$.

2.2 Transition Point

In real devices, carrier density at the source and the drain is equal to the doping density N_D. Consequently, carrier density of the charge-sheet at the source/drain end should be n_D, which is the 2D-equivalent of N_D. Therefore, the next step that should be done is to embed the above transport-controlled region within the channel. The widely accepted approach is to set three regions inside the channel [6].

Region I: where carrier density decreases from n_D to $n(0)$, forming the built-in potential barrier. Change in quasi-Fermi level is negligible. Thickness is ignored.
Region II: where the transport equation dominates. The carrier density changes from $n(0)$ to $n(L)$.
Region III: where carrier density increases from $n(L)$ to n_D. Change in quasi-Fermi level is negligible. Thickness is often changed to vary the *effective* channel length.

The drain voltage V_D is then expressed as

$$V_D=\psi(L)-\psi(0)+\beta^{-1}\log(n(0)/n(L))\,. \quad (7)$$

Although this approach works rather well, it requires arbitrarily small $n(L)$ to produce large V_D. This feature is unacceptable for the velocity-saturation model, because limited carrier velocity implies lower limit of carrier density for a given drain current.

In real devices, such a situation is avoided by a so-called pinch-off condition. When larger drain voltage is applied, the drain electric field, in stead of the lateral electric field formed by the carrier density gradient, start to drag the carriers.

In our model, another region IIb is placed between region II and III, where the drain electric field drags the carrier, and where carrier density is assumed to be constant. The boundary between region II and IIb is defined as a point where the first and the second derivative of the quasi-Fermi level match for both sides. Since the thickness of region IIb expresses channel length modulation effect, thickness of region III is now neglected.

The drain electric field can be assumed to decay exponentially in the channel with the characteristic length λ [7]. If we ignore k-difference between silicon and the oxide, $\lambda=(T_{OX1}+T_{OX2}+T_{Si})/\pi$.

2.3 DIBL

DIBL in DG MOSFETs has similar property to that in SOI MOSFETs, where DIBL is caused by exponentially decaying source/drain electric field [8]. We modeled that the surface potential changes as

$$\Delta\psi(x)=\Delta\psi_S(x)\exp(-x/\lambda)+\Delta\psi_D(x)\exp((x-L)/\lambda) \quad (8)$$

where $\Delta\psi_S$ ($\Delta\psi_D$) is the potential difference between the surface potential without channel carrier and the source/drain potential. The minimum barrier lowering $\Delta\psi_{MIN}$ gives the DIBL. Although the position for $\Delta\psi_{MIN}$ is away from the source, it is assumed to be the source end.

3 COMPARISON WITH SIMULATOR

3.1 Drain Currents

The comparison with the ATLAS device simulator was made to assure the accuracy of the model. The model parameters were fit to get good agreements both for long and ultra-short channel. Figure 1 shows the comparison of the simulator and the model at L_G=1μm device. Although the matching is good, the model gives lower drain current in some gate1 voltages. Discrepancy may come from different mobility modeling implementation.

The adopted mobility model is remotely related to the CVT model. It includes phonon and surface-roughness mobility as functions of average perpendicular field $E_{\perp av}$.

$$\mu_S^{-1}=\mu_0^{-1}+\left(A_{PH}E_{\perp eff}^{-1/3}\right)^{-1}+\left(A_{SR}E_{\perp eff}^{-2}\right)^{-1} \quad (9)$$
$$E_{\perp eff}=\sqrt{E_{\perp av}^2+(kT/qT_{Si})^2}$$

While this mobility is a function of y in the charge sheet, it is not the function of x (which is the distance from the interface), which is the same as the experimentally extracted mobility. The CVT mobility for the device simulator, on the other hand, is often the function of both x and y. These two implementations result in different models [9]. The difference of these two approaches is exacerbated that the strength of perpendicular electric field in the DG MOSFET channel ranges from zero to a finite value. The necessity of a modified mobility model will be clarified only with the comparison of the compact model to a long-channel, real device. Since the sample to be used in this paper is ultra-short channel device, little will be known about the adequacy of the mobility model.

Figure 2 shows the comparison with the simulator for ultra-short (30nm) channel structure. In the device, carrier-velocity saturation plays significant role, and discrepancy in the mobility model becomes not prominent.

3.2 Capacitances

Since our compact model incorporates carrier-velocity saturation effect in the transport equation, obtained carrier profile is expected much closer to the real device. Consequently, a good intrinsic capacitance model is expected simply by dedifferentiating gate charges and source/drain charges.

Figure 3 shows the analytically calculated capacitances for a short channel (200nm) device. Since the model does not yet includes fringe capacitances, deviation from the device simulator results are observed in gate-OFF condition. Beside this feature, agreements are excellent in spite of no additional fitting parameter for capacitance modeling.

4 COMPARISON WITH REAL DEVICES

FinFET-style DG MOSFETs were fabricated by using lightly doped p-type (110) SOI wafers with 300-nm thick buried oxide layer. To obtain flat interfaces, the Si-channel was formed by anisotropic wet etching process using a tetramethylammonium hydroxide (TMAH) solution. As a result, (111)-oriented, atomically flat surfaces are formed on both sides of a silicon fin structure. The physical gate length L_G was 105nm and the channel thickness T_{Si} was 13nm.

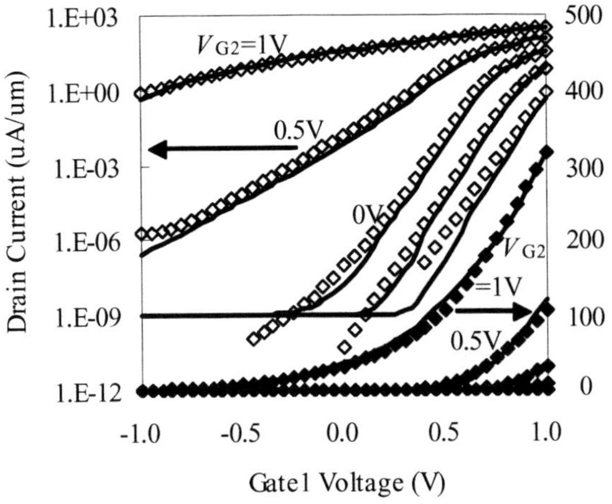

Figure 1: Simulator (marks) and model (lines) results of long (1μm) channel DG MOSFET. Leak current in the model is limited by Gmin.

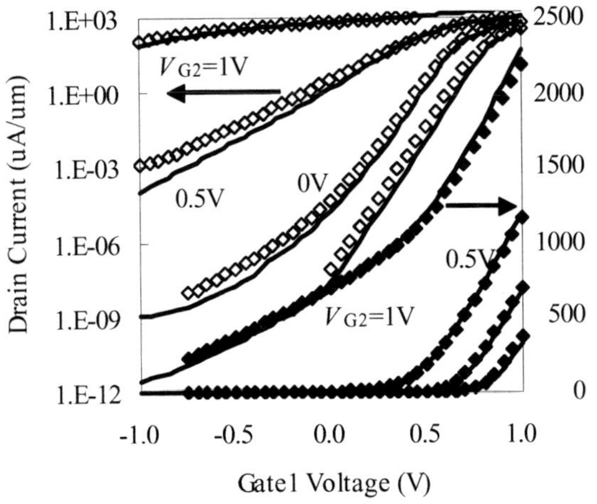

Figure 2: Simulator (marks) and model (lines) results of short (30nm) channel DG MOSFET.

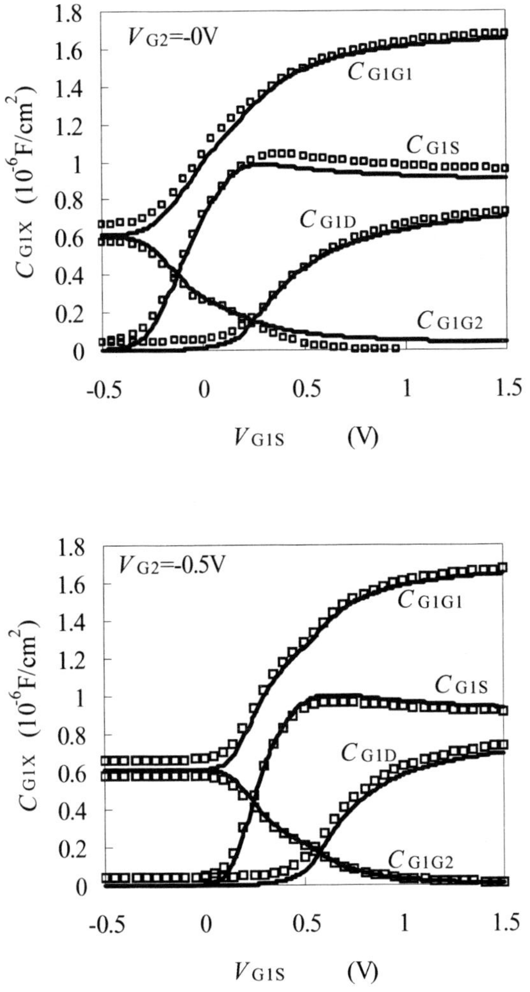

Figure 3: Intrinsic capacitances of four-terminal DG MOSFET. Solid curves are for the model, marks are for the simulator. For both figures, gate length is 200nm, oxide thickness is 2nm, silicon channel thickness is 5nm.

The best fit value for the electrical channel length was 65 nm for the simulator with abrupt source/drain doping and 55nm for the compact model. The difference indicates the definition of channel length in the presence of inevitable doping profile changes slightly between these two methods. The source/drain resistance was 230 Ωμm; a rather reasonable value considering the fabrication process.

No fitting effort was necessary for the channel thickness T_{Si} and oxide k. Oxide thickness was slightly (0.4nm) thinned from the measured value (3.2nm). Other parameters were the gate flat-band voltage, the bulk (111)-orientation surface mobility (715.5cm^2/Vs), saturation velocity (1.18x10^{7}cm/s), and three factors to tune the strength of phonon scattering, surface-roughness scattering and DIBL. These are the exhaustive list of fitting parameters.

Figure 3 shows the comparison of I_D-V_G curve. Solid lines are for model results, and the marks are for experimental data. The model is in good agreement except the absence of GIDL current in the model. To get the best fitting for the sample V_{th} roll-off caused by the DIBL is 20% increased compared to the theoretical value.

Figure 4 shows the comparison of I_D-V_D curve. Again, solid lines are for model results, and the marks are for experimental data. Agreement is still rather good considering the small number of fitting parameters.

5 SUMMARY

A compact model for four-terminal double-gate MOSFETs based on double charge-sheet model has been discussed. The basic equations have been explained. The transport equation accommodates carrier velocity saturation explicitly. Comparison with 2D device simulator shows that accurate intrinsic capacitance model has been obtained simply by the analytical derivative of channel carrier with respect to the terminal voltage. Comparison between the model and the data from the fabricated FinFET devices has been in good agreement. Best fit channel lengths for the simulator and the compact model have been similar.

REFERENCES

[1] T. Sekigawa and Y. Hayashi, *Solid-State Electron.*, Vol. 27, pp. 827-828, 1984.

[2] T. Nakagawa, T. Sekigawa, T. Tsutsumi, E. Suzuki, and H. Koike, *Tech. Digest Nanotech 2003*, Vol. 2, pp. 330-333, 2003.

[3] T. Nakagawa, T. Sekigawa, T. Tsutsumi, M. Hioki, E. Suzuki and H. Koike, *Tech. Proc., WCM 2005*, pp. 183-186, 2005.

[4] T. Nakagawa, T. Sekigawa, M. Hioki, S. O'uchi, T. Tsutsumi, and H. Koike, *Tech. Digest IWCM'08*, pp. 17-23, 2008.

[5] T. Nakagawa, T. Sekigawa, T. Tsutsumi, M. Hioki, E. Suzuki, and H. Koike, *Tech. Digest Nanotech 2004*, Vol. 2, pp. 159-162, 2004.

[6] J. R. Brews, *Solid-State Electron.*, Vol. 21, pp. 345-355, 1978.

[7] D. J. Frank, Y. Taur, and H-S. P. Wong, *IEEE Electron Dev. Letts.*, Vol. 19, 1998.

[8] R-H. Yan, A. Ourmazd, and K. F. Lee, *IEEE Trans. Electron Dev.*, Vol. 39, pp. 1704-1710.

[9] H. Shin, A. F. Tasch, C. M. Maziar, and S. K. Banerjee, *IEEE Trans. Electron Dev.*, Vol. 36, pp. 1117-1124, 1989.

[10] Y.X. Liu, K. Ishii, T. Tsutsumi, M. Masahara, and E. Suzuki Y.X. Liu et al., *IEEE Electron Dev. Letts.*, Vol. 24, pp. 484-486, 2003.

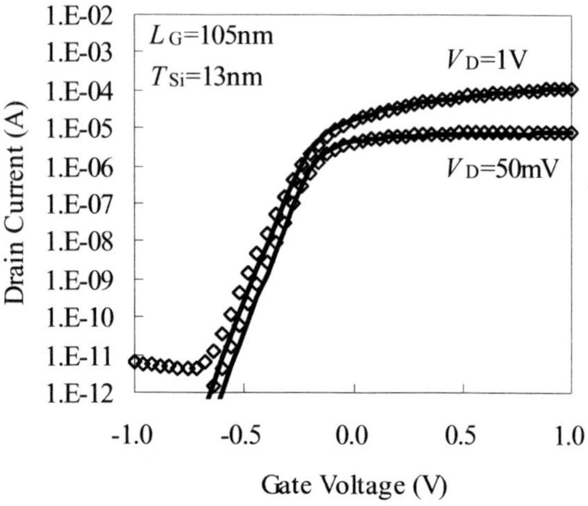

Figure 4: I_D-V_G characteristics of the SOI FinFET (marks) and the model (solid lines).

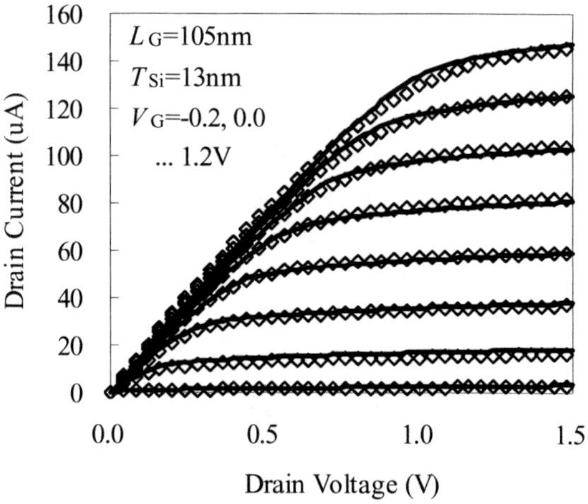

Figure 5: I_D-V_D characteristics of the SOI FinFET (marks) and the model (solid lines).

MOSFET Compact Modeling Issues for Low Temperature (77 K - 200 K) Operation

P. Martin*, M. Cavelier*, R. Fascio* and G. Ghibaudo**

*CEA-LETI, Minatec, 17 rue des Martyrs, 38054 Grenoble Cedex 9, France, patrick.martin@cea.fr
**IMEP, Minatec, 3 parvis Louis Néel, 38016 Grenoble Cedex 1, France, ghibaudo@minatec.inpg.fr

ABSTRACT

Advanced compact models are evaluated for simulation of mixed analog-digital circuits working at low temperature (77 to 200 K). This evaluation is performed on a dual gate oxide CMOS technology with 0.18 µm / 1.8 V and 0.35 µm / 3.3 V MOSFET transistors. A detailed temperature analysis of some physical effects is performed. Specific effects, such as anomalous narrow width effect or quantization of the inversion charge, are observed at low or intermediate temperature. Some improvements of compact models will allow a very accurate description of MOS transistors at low temperature.

Keywords: MOSFET, compact model, low temperature

1 INTRODUCTION

This paper deals with the simulation of hybrid CMOS readout circuits working at low or at intermediate temperature, typically 77 K, 130 K or 200 K, such as those used in high performance infrared image sensors [1]. SPICE parameters of transistors, including DC, AC, 1/f noise and matching parameters, are mandatory for design and simulation of such circuits. However CMOS foundries do not provide full sets of MOSFET parameters for simulation at temperature lower than -55°C (≈ 220 K). So these MOSFET parameters must be extracted prior to circuit simulation. Furthermore, the used MOSFET compact model must address the important weak and moderate inversion regimes where most of the analog transistors of the CMOS readout circuits are operating. A low temperature version of the EKV2.6 model was previously described and was successfully applied to different CMOS bulk processes such as 0.7 µm / 5 V, 0.5 - 0.35 µm / 3.3 V and 0.21 µm / 1.8 V from different foundries at cryogenic temperature [2]. In particular this model incorporates a charge-based mobility model including Coulomb, phonon and surface roughness scattering mechanisms as well as velocity saturation.

In this work, we access different compact models, such as the standard version of the EKV3 charge model [3], for a commercial process, namely a dual gate oxide CMOS technology, with 0.18 µm / 1.8 V and 0.35 µm / 3.3 V MOSFET transistors.

2 CMOS PROCESS

We characterize a commercial mixed mode / RFCMOS process optimized for room temperature operation. In this dual gate oxide process, two kinds of MOSFET transistors are available: (1) transistors with a physical gate thickness of 3.3 nm, a minimum channel length of 0.18 µm and operating at a maximum recommended voltage of 1.8 V and (2) transistors with a physical gate thickness of 6.5 nm, a minimum channel length of 0.35 µm and operating at 3.3 V. p-MOSFET are made in a NWELL. Depending on the channel doping, transistors with different threshold voltages are provided. p-MOSFET transistors may be done either with a standard threshold voltage (STD) or with a low threshold voltage (LVT). n-MOSFET transistors are made either in a P-substrate (also referred as the PWELL) or in a triple well (TWELL). n-MOSFET transistors have either a standard threshold voltage (STD), a low threshold voltage (LVT) or a zero-volt threshold voltage (ZVT). As a whole, twelve MOSFET transistors are allowed for mixed analog-digital circuit design. This process is a dual gate process and uses two kinds of polysilicon gates, N^+ doped polysilicon gate for n-MOSFET and P^+ doped polysilicon gate for p-MOSFET. As a consequence both NMOS and PMOS are surface channel transistors and the threshold voltage temperature coefficients are nearly equal (0.8 ~ 1.2 mV/K). This deep submicron process has pocket implants to combat short channel effects. Finally we have to mention that this process has LDD zones to diminish hot-carrier degradation and use Shallow-Trench Isolation (STI).

3 CHARACTERIZATION

Prior to parameter extraction using a compact model, a fine characterization of the different transistors at different temperatures is needed. This step is mandatory to examine different effects, such as short channel and narrow width effects.

NMOS and PMOS present a reverse short channel (RSCE) effect and the Vth roll-up is observed (Fig. 1). The Vth roll-off due to both charge sharing and Drain Induced Barrier Lowering (DIBL) is not observed. The threshold voltage shift [DVth = Vth - Vth (L = 20 µm)] is rather insensitive to the bulk voltage Vbs.

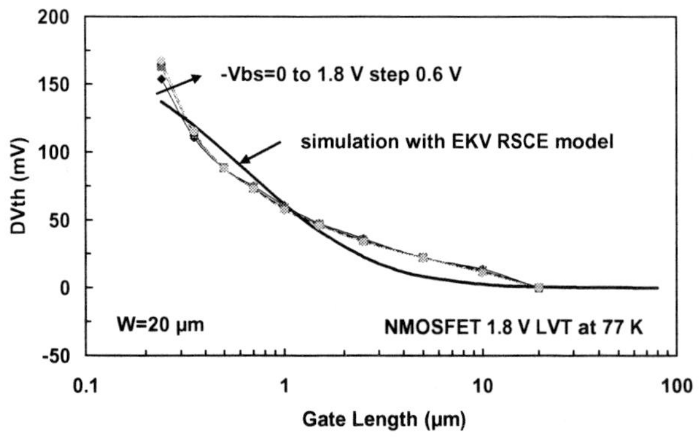

Figure 1: Measured RSCE threshold voltage shifts (symbols) and simulated curve (solid line)

A significant reduction of this RSCE effect with temperature is observed (Fig. 2). This reduction has been previously observed and interpreted as the variation of the threshold voltage with the Fermi potential [4]. The RSCE effect is well modeled in the EKV2.6 and EKV3 compact models at a given temperature as shown in Fig. 1 by using two parameters: LR and QLR, where LR is a characteristic length and QLR a charge density. In order to obtain a very precise low temperature modeling, a quadratic temperature dependency of the QLR parameter has to be introduced as shown in Fig. 3.

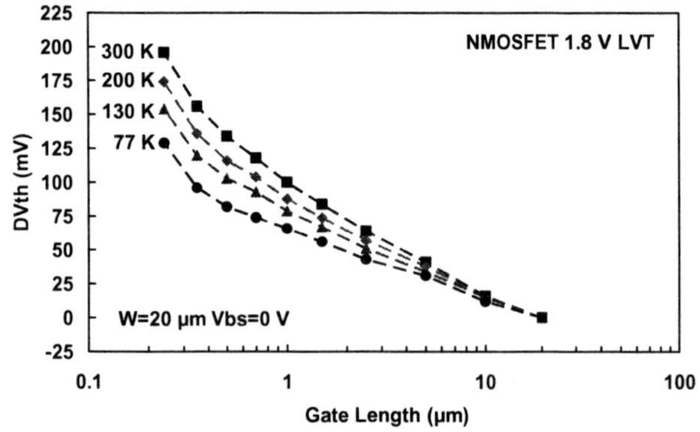

Figure 2: Measured RSCE threshold voltage shifts

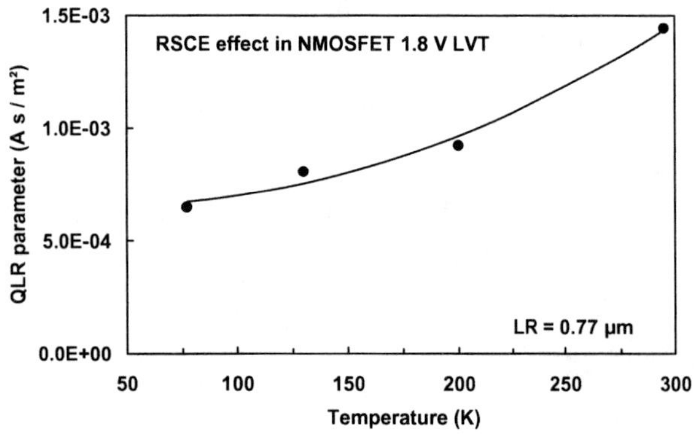

Figure 3: Variation of the QLR parameter with temperature

Concerning the narrow width effect (NWE), PMOS transistors of this process present an inverse narrow width (INWE) effect due to shallow trench isolation (STI). A significant change of this INWE effect with temperature is observed (Fig. 4). As for the RSCE effect, and in order to obtain a precise low temperature modeling, a temperature dependency of this effect has to be introduced.

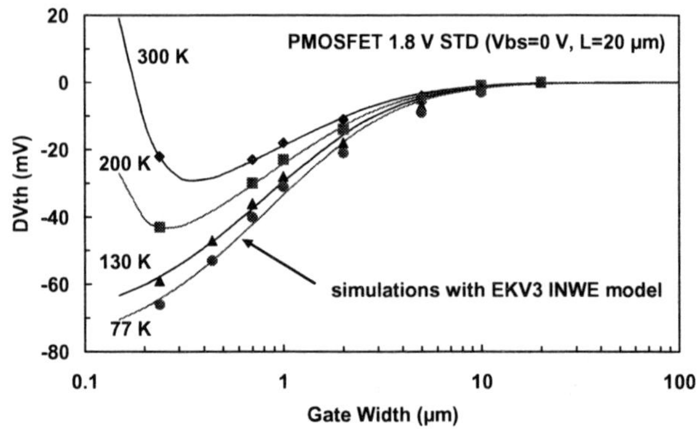

Figure 4: INWE effect at different temperatures: experimental (symbols) and simulated curves (solid lines)

In NMOS transistors, we observed an anomalous narrow width effect (Fig. 5). These transistors do not present a classical NWE effect, neither an INWE effect. The threshold voltage shift DVth is sensitive to the bulk voltage. This anomalous NWE effect is not modeled in the EKV3 compact model.

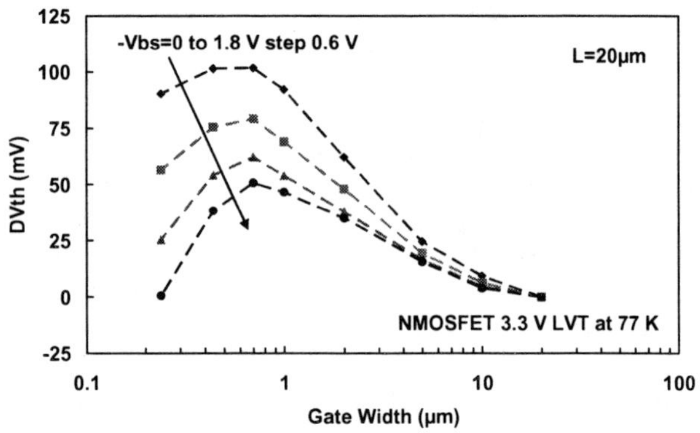

Figure 5: Anomalous NWE effect at 77 K

In lowly doped 1.8 V LVT NMOS transistors, we observed a peak in the gate transconductance characteristic, Gm vs. Vgs, at temperature lower than 200 K (Fig. 6). This peak is observed at the beginning of strong inversion. It is observed only on long channels and not on transistors shorter than 2 µm (Fig. 7). It is not observed on 1.8 V STD transistors, even with long channels.

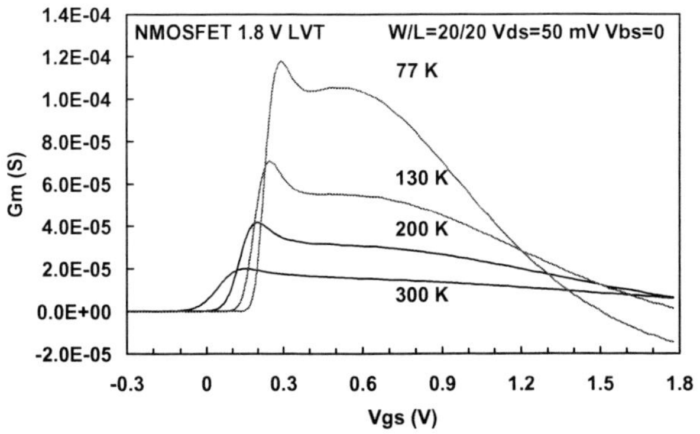

Figure 6: Gm behavior at different temperatures for a long channel 1.8 V LVT NMOSFET

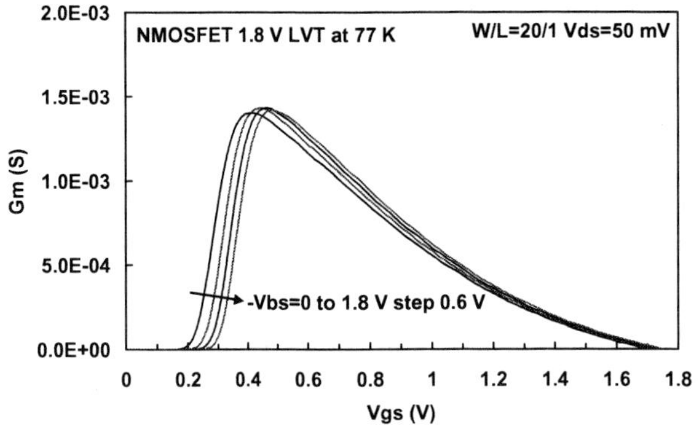

Figure 7: Gm behavior at 77 K for a short channel transistor

To understand this phenomenon, complementary measurements were done by varying the bulk voltage (Fig. 8). As shown in Fig. 9, the peak-to-valley ratio in Gm is attenuated and tends to unity as the bulk voltage increases.

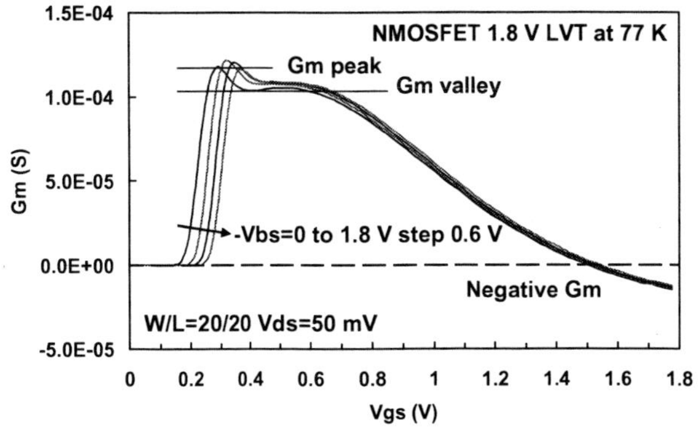

Figure 8: Gm behavior at different Vbs

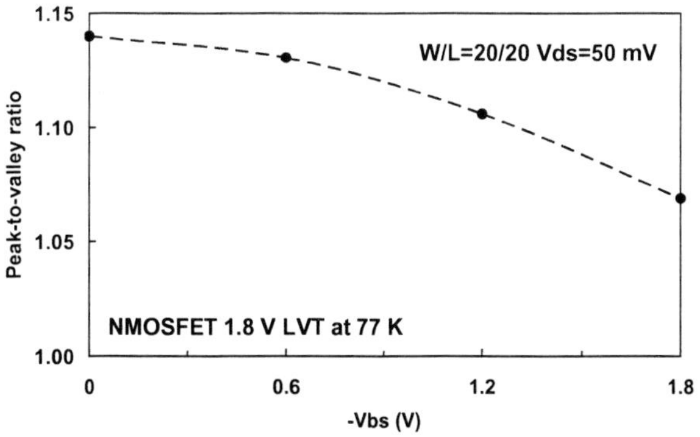

Figure 9: Gm peak-to-valley ratio vs. Vbs

Such a peak has been already observed at intermediate temperature (77 – 120 K) on some CMOS processes [5]. It has been interpreted as the result of quantization effects in the inversion layer. Energy subband separation effects are enhanced at very low temperature or at very high effective transverse field E_{eff} at the Si/SiO_2 interface given by:

$$E_{eff} = \frac{\eta\, Qi + Qb}{\varepsilon_{Si}} \qquad (1)$$

Where Qi is the inversion charge and Qb the bulk charge, η = 1/2 for NMOS and 1/3 for PMOS.

At not too low temperature, different subbands with different mobility are filled. The occupancy of these subbands is modified by E_{eff}. As E_{eff} is proportional not only to the inversion charge Qi but also to the depletion charge Qb, subbands occupancy can also be varied by the channel doping Nch or by the bulk voltage as Qb is given by eq. 2:

$$Qb = \sqrt{2\, q\, \varepsilon_{Si}\, Nch\, (\psi_s + Vsb)} \qquad (2)$$

In order to observe the peak in the Gm vs. Vgs characteristic, the first subband must be filled. So the temperature must not be too low, otherwise only the fundamental subband will be filled. The effective transverse electrical field must also be low. This last condition is achieved for low Vgs (low Qi), and/or low Nch or low Vbs (low Qb). This interpretation is consistent with the fact that we observe this phenomenon only on low doped transistors, i.e. on LVT transistors, and not on STD transistors where the channel doping is higher. It is also coherent with the fact that we observe it only on long channel. As a matter of fact, due to the RSCE effect existing in these transistors (see Fig. 1), the effective channel doping increases as the channel length decreases and the Gm peak is no more observed. This quantum mechanical effect is not taken into account in any advanced compact models, such as EKV3, PSP or HiSIM.

Another specific effect was observed. We found that as temperature is lowered, the measured subthreshold swing of long transistors is higher than that predicted by simulation. This is due to interface traps whose density Nit increases when temperature decreases. In the EKV3, an empirical parameter N0 is introduced, having an ideal value of N0 = 1, corresponding to the absence of interface states. This parameter, which does not affect the strong inversion regime, allows minimizing the error in weak inversion (Fig. 10).

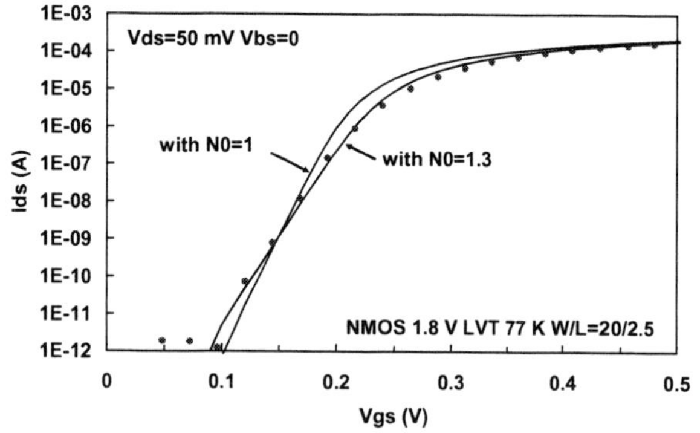

Figure 10: Influence of interface traps on the subthreshold slope, experimental (symbols) and EKV3 simulations (solid lines) at 77 K

As shown in Fig. 8, the gate transconductance measured in the ohmic mode (Vds = 50 mV) could be negative on cooled NMOS transistors at high vertical field. The explanation is that, when increasing Vgs, the increase of the inversion charge could not compensate the high mobility attenuation. The drain current will decrease, so Gm becomes negative. The improved charge-based mobility model introduced in the EKV3 model allows to reproduce accurately drain-source current and gate transconductance in the linear regime as shown in Fig. 11.

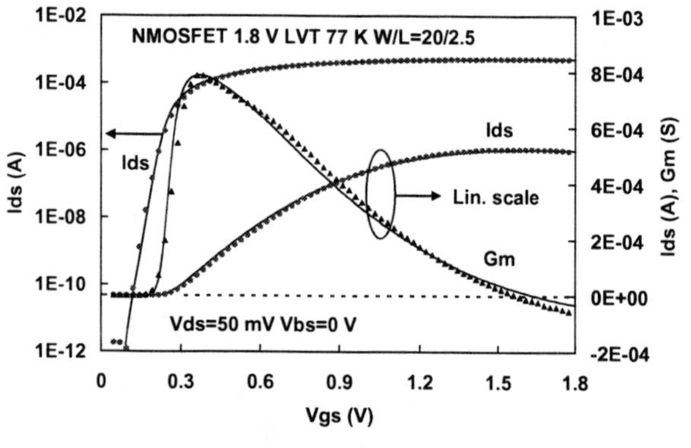

Figure 11: Experimental (symbols) and simulated (solid lines) drain-source current (Ids) and gate transconductance (Gm) with EKV3 at 77 K

4 CONCLUSION

In conclusion, better MOSFET compact models with new mobility law and accounting for weak inversion slope degradation due to interface traps and for different effects such as RSCE, INWE, DIBL and DITS (Drain Induced Threshold Voltage Shift) are necessary for precise analog modeling at low temperature. Specific effects, such as anomalous narrow width or quantization effects, observed only in some transistors, are not yet well understood. It is expected that further improvements will allow a very accurate MOSFET modeling at low temperature.

5 ACKNOWLEDGEMENTS

The authors would like to thank Patrick Maillart from Sofradir for design of test structures and Professor Matthias Bucher from Technical University of Crete for the EKV3 Verilog-A code.

REFERENCES

[1] F. Guellec, P. Villard, F. Rothan, L. Alacoque, C. Chancel, P. Martin, P. Castelein, P. Maillart, F. Pistone and P. Costa, "Sigma-delta Column-wise A/D Conversion for Cooled ROIC", SPIE Proceedings Vol. 6542, Infrared Technology and Applications XXXIII, 2007.

[2] P. Martin and M. Bucher, "Comparaison of 0.35 and 0.21 µm CMOS Technologies for Low Temperature Operation (77 K - 200 K) and Analog Circuit Design", 6th Workshop on Low Temperature Electronics (WOLTE-6), ESTEC, Noordwijk, The Netherlands, 2004.

[3] A. Bazigos, M. Bucher, F. Krummenacher, J.-M. Sallese, A.-S. Roy, C. Enz, "EKV3 Compact MOSFET Model Documentation, Model Version 301.01", Technical Report, Technical University of Crete, November 23, 2007.

[4] B. Szelag, F. Balestra and G. Ghibaudo, "Comprehensive Analysis of Reverse Short-Channel Effect in Silicon MOSFET's from Low-Temperature Operation", IEEE Electron Devices Letters, vol. 19, n° 12, pp. 511-513, 1998.

[5] A. Emrani, "Propriétés électriques et modèles physiques des composants MOS à basse température pour la cryo-microélectronique", PhD Thesis, INPG, Grenoble, 1992.

Interface-trap Charges on Recombination DC Current-Voltage Characteristics in MOS Transistors

Zuhui Chen*, Bin B Jie** and Chih-Tang Sah***

*School of EEE, Nanyang Technological University, Singapore, zhchen@ntu.edu.sg
**IME, Peking University, China, bb_jie@msn.com
***University of Florida, USA, tom_sah@msn.com

Abstract

Steady-state Shockley-Read-Hall kinetics is employed to study the interface-trap charges at the SiO_2/Si interface on the electron-hole recombination direct-current current-voltage (R-DCIV) properties in MOS field-effect transistors. The analysis includes device parameter variations of neutral interface-trap density, dopant impurity concentration, oxide thickness, and forward source/drain junction bias. It shows that the R-DCIV curve is increasingly distorted as the increasing of interface-trap charges. The result suggests that the lineshape distortion observed in the past experiments, previously attributed to spatial variation of surface dopant impurity concentration, can also arise from interface-trap charges along the surface channel region.

Keywords: interface traps, interface-trap charges, MOS transistors, recombination DCIV

1. Introduction

Electron and hole generation, recombination and trapping at the SiO_2/Si interfacial electronic traps create additional currents and limit device performance characteristics, such as low stand-by dissipation-power, long operating-lifetime, low-noise amplification, and long-endurance memory. Due to the technical importance, extensive researches have been undertaken to study interfacial electronic traps at the SiO_2/Si interface in order to delineate their microscopic origins [1-3]. The generation of interface traps degrades the usable applied voltage and therefore limits device performance, and shortens the device lifetime at a given condition. Device reliability degradation may be exacerbated by strong electric field, high temperature and age since interface traps increase with time. The increasing of interface traps and interface-trap charges would eventually cause the transistor to cease functioning within the designed parameters. Thus, interface-trap charges are directly related to a transistor's reliability and critical to the proper performance of transistors.

The recombination DCIV electrical methodology was proposed as a simple and powerful tool to diagnose the operation reliability and extract submicron MOS transistor parameters with nanometer spatial resolutions [4-5]. The principle of the R-DCIV method is the use of a surface-potential-controlling gate terminal voltage to modulate the base-terminal DC current from electron-hole recombination at the SiO_2/Si interface traps to give the device and material properties. In this paper, we will explore the effect of interface-trap charges on the MOS transistor characteristics using the R-DCIV methodology. Since interface-trap density increases with the increasing of transistor write-erase cycles, investigations will focus on the effect of interface-trap charges at high interface-trap density on device characteristics, especially important for the floating-gate MOS memory transistors, which are basic units in the chips of the flash memory stick, cell phones and other portable electronics equipments. As proposed and explained by Sah [6-8], the random variations of Si-O and Si-Si bond angles and lengths along the SiO_2/Si interface generate localized perturbations which would then simultaneously create the neural interface traps with 0 or -1 charge state for neutral electron traps and 0 or +1 charge state for neutral hole traps. Thus, a paired-linear distribution of neutral interface traps with a linear energy distribution of neutral electron traps and a linear energy distribution of neutral hole traps will be theoretically studied in order to show interface-trap charge effect on transistor characteristics.

2. Theory of the R-DCIV Method

The steady-state areal rate of electron-hole recombination, R_{SS}, at a discrete energy level of interface traps with an areal density N_{IT} (trap/cm^2), is given by the Shockley-Read-Hall (SRH) formula [4-5]:

$$R_{SS} = \frac{c_{ns}c_{ps}N_S P_S - e_{ns}e_{ps}}{c_{ns}N_S + e_{ns} + c_{ps}P_S + e_{ps}} N_{IT} \qquad (1)$$

Here, c_{ns}, c_{ps}, e_{ns} and e_{ps} are the electron-hole capture-emission rate coefficients at the interface traps. N_S and P_S are respectively the surface electron and hole concentrations. The basewell-terminal recombination current I_B is obtained by integrating the SRH steady-state electron-hole recombination rate at interface traps, R_{SS}, over the channel area dydz [4-5]

$$I_B(V_{GB}) = q\iint R_{SS}(V_{GB}, y)dydz$$

$$= \frac{q(c_{ns}c_{ps})^{1/2} n_i W}{2} \int \frac{\{\exp[U_{PN}(y)] - 1\} N_{IT}(y)dy}{\exp[U_{PN}(y)/2]\cosh[U_S^*(y)] + \cosh(U_{TI}^*)} \qquad (2)$$

U_S^* is effective surface potential and U_{TI}^* is effective interface-trap energy level. $U_{PN}=V_{PN}/(kT/q)$ is the normalized forward bias applied the p/n junctions.

For a top-emitter bias configuration with $V_{DS}=0$, the gate voltage equation is given by:

$$V_{GB} = V_S + V_{FB} - Q_{IT} / C_{OX} + \varepsilon_S \times E_S / C_{OX} \quad (3)$$

Here, C_{OX} is the oxide capacitance per unit area and E_S is the electric field on the semiconductor side of the SiO_2/Si interface. The charge neutrality condition with the interface-trap charges is expressed by:

$$\rho = q \times [P - N - P_{IM} + Q_{IT} / q] = 0 \quad (4)$$

The Mass Action Law with voltage bias applied at the source and drain junctions is given by $P \times N = n_i^2 \exp(U_{PN})$. Thus, the surface electron and hole concentrations can be solved with:

$$N_S = n_i \exp(U_S - U_N)$$
$$= n_i \{[(\frac{P_{IM} - Q_{IT}/q}{2n_i})^2 + \exp(U_{PN})]^{1/2} - (\frac{P_{IM} - Q_{IT}/q}{2n_i})\}\exp(U_S) \quad (5a)$$

$$P_S = n_i \exp(U_P - U_S)$$
$$= n_i \{[(\frac{P_{IM} - Q_{IT}/q}{2n_i})^2 + \exp(U_{PN})]^{1/2} + (\frac{P_{IM} - Q_{IT}/q}{2n_i})\}\exp(-U_S) \quad (5b)$$

The electron and hole quasi-Fermi potentials, U_N and U_P, are a function of the dopant impurity ion concentration P_{IM} and the voltage U_{PN} applied to the drain, source or basewell p/n junctions in MOS transistors.

The neutral interface traps are defined as an electrically neutral trapping potential that can bind only one electron or hole. The charge formulae of the neutral electron trap Q_{ET} and neutral hole traps Q_{PT} are respectively given by [9]

$$Q_{ET} = \frac{-qN_{ET} \times (c_{ns} N_S + e_{ps})}{c_{ns} N_S + e_{ns} + c_{ps} P_S + e_{ps}} \quad (6a)$$

$$Q_{PT} = \frac{qN_{PT} \times (c_{ps} P_S + e_{ns})}{c_{ns} N_S + e_{ns} + c_{ps} P_S + e_{ps}} \quad (6b)$$

N_{ET} and N_{PT} are respectively neutral electron-trap and hole-trap concentrations. The total charge Q_{IT} is the sum of Q_{ET} and Q_{PT}, i.e., $Q_{IT}=Q_{ET}+Q_{PT}$. Poisson equation is changed to include the interface-trap charges Q_{IT} given by:

$$\varepsilon_S \nabla E = q \times [P - N - P_{IM} + Q_{IT} / q] \quad (7)$$

Then, we obtain the semiconductor surface electric field E_S:

$$E_S = \sqrt{\frac{2kTn_i}{\varepsilon_S}} \{[e^{-U_S} + U_S - 1]e^{U_P} + [e^{U_S} - U_S - 1]e^{-U_N} - \frac{Q_{IT}}{qn_i}\} \quad (8)$$

The E_S is extended to include the minority carriers from the forward bias applied p/n junctions.

The contribution of interface-trap charges is very small by comparing with P_{IM} and can be ignored even if there is one-species trap and all the traps are occupied by either electrons or holes for doped devices, which can be anticipated by in (4) and (7). However, for un-doped devices such as pure double-gate transistors, the interface-trap charges would play a very important role in charge neutrality condition and Poisson equation since interface-trap density is comparable to that of intrinsic carriers. A detailed report on interface traps in un-doped devices will be presented in a future article.

3. R-DCIV Lineshape Analysis

In the following discussion, we assume a paired-linear interface traps which includes a linear energy distribution of neutral electron traps N_{ET} with a decreasing of trap density from the conduction band edge to the valence band edge, and a linear energy distribution of neutral hole traps N_{PT} with a increasing of trap density from the conduction band edge to the valence band edge. 55 energy levels with an energy step of 20emV are used to simulate a linear energy distribution of interface traps in the silicon energy gap. For three dimension bulk model, the ratio of neutral electron traps and neutral hole traps is given by [10]:

$$R = \frac{D_{ET0}}{D_{PT0}} = \frac{1.09412}{0.5228205} = 2.095526 \quad (9)$$

D_{ET0} and D_{PT0} are the density of state respectively at conduction band edge and valence band edge. The densities of N_{ET} and N_{PT} can be respectively obtained with:

$$N_{ET} = \int D_{ET0} \times E \times dE = \frac{D_{ET0}}{2} \times E_{ET}^{2} \quad (10a)$$

$$N_{PT} = \int D_{PT0} \times E \times dE = \frac{D_{PT0}}{2} \times (1 - E_{PT})^2 \quad (10b)$$

E_{ET} and E_{PT} are respectively for the energy levels of the electron and hole traps. The total interface trap density N_{IT} is the sum of N_{ET} and N_{EP}, i.e., $N_{IT}=N_{ET}+N_{PT}$. D_{ET0} is determined by the given total electron-trap density N_{ET} in (10a), while D_{PT0} is solved by (9) and the hole-trap density N_{PT} can be obtained from (10b). We assume that electron and hole capture rates are equal with $c_{ns}=c_{ps}=10^{-8}cm^{-3}/s$ and transistor temperature T is 296.57K at which the intrinsic carrier concentration is $n_i = 10^{10}cm^{-3}$. A metal gate is also assumed, and the other parameters are listed on the figures.

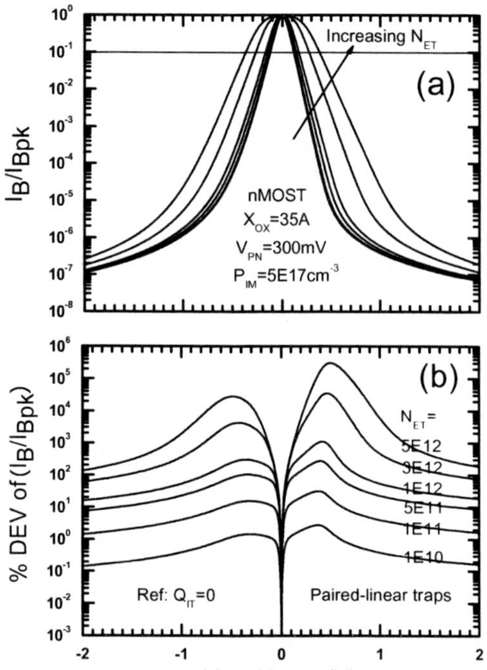

Fig. 1 Interface-trap charges on R-DCIV lineshape as a function of interface-trap density: (a) Normalized I_B vs. V_{GB} and (b) Percentage deviation. $N_{ET}=1.0\times10^{10}$, 1.0×10^{11}, 5.0×10^{11}, 1.0×10^{12}, 3.0×10^{12} and $5.0\times10^{12}cm^{-2}$.

The R-DCIV lineshape is primarily determined by the dopant impurity concentration P_{IM} and oxide thickness X_{OX}, and their spatial profiles at the SiO_2/Si interface [7]. However, these results were based on assumption that the energy level of interface trap is at the midgap, i.e., E_{TI}=0eV, and the interface-trap charge is zero, i.e. Q_{IT}=0. The quantitative proofs have not been reported on the effect of interface-trap charges on the R-DCIV curves. Fig. 1 shows interface-trap charges on R-DCIV lineshape as a function of interface-trap density. The neutral electron-trap density N_{ET} varies from 1.0×10^{10} to $5.0\times10^{12}cm^{-2}$ along surface channel region and the neutral hole-trap density varies from 0.477×10^{8} to $2.386\times10^{12}cm^{-2}$ since R= N_{ET}/N_{PT}=2.095526 for 3-D bulk model in (9). The normalized R-DCIV lineshape broadens as the increasing of N_{ET} as shown Fig. 1(a). Fig. 1(b) shows the percentage deviation of I_B/I_{Bpk} using the reference curve with Q_{IT}=0. The 90% base current range, defined by $I_B/I_{B\text{-peak}}$ = 0.1 to 1.0 shown by the horizontal line in Fig. 1(a), is covered by a gate voltage range from −0.1V to +0.1V at low N_{ET}, such as $N_{ET}=1.0\times10^{10}cm^{-2}$, and from −0.3V to +0.3V at high N_{ET}, such as $N_{ET}=5.0\times10^{12}cm^{-2}$. The percentage deviation is over 200 for $N_{ET}=1.0\times10^{12}cm^{-2}$ when comparing the I_B-V_{GB} curves using the range from I_B/I_{Bpk}=0.1 to 1.0. The normalized R-DCIV curves and percentage deviations show that R-DCIV lineshape is significantly affected by the trap-density dependence of the trap charge Q_{IT}.

The recombination DCIV dependences of interface-trap charges on two most important MOS transistor parameters, dopant impurity concentration P_{IM} and gate oxide thickness X_{OX}, are shown in Fig. 2 and Fig. 3. P_{IM} varies from 1×10^{15} to 1×10^{18} cm^{-3} and X_{OX} varies from 12A to 100A with $N_{ET}=1.0\times10^{12}cm^{-2}$. Such a high interface-trap concentration is generated during repeated program-erase cycling of non-volatile floating-gate and SNONS memory transistors, recently reported by Victor Kuo of Taiwan Power Semiconductor Corporation [11]. The solid-line curves are for the device with interface-trap charge Q_{IT} while the dash-line curves correspond to the transistor without Q_{IT}, i.e., Q_{IT}=0. The percentage deviation from curves with Q_{IT}=0 is used to evaluate the interface-trap charges on device characteristics. Both figures show large deviations from curves without Q_{IT} when comparing the I_B-V_{GB} curves using the range from I_B/I_{Bpk}=0.1 to 1.0. For dopant impurity concentration $P_{IM}=5.0\times10^{17}cm^{-3}$, which is in practical range, the % deviation is over 200%. These percentage deviations indicate that the effect of interface-trap charges is not negligible when using R-DCIV method to extract device and materials parameters.

Figure 4 illustrates the interface-trap charges on the R-DCIV lineshape as a function of the forward bias (1mV to 600mV) or the injected minority carrier concentration. For small injected minority carrier concentration, the 90% base current range of R-DCIV curves covers a gate voltage range from −0.2V to +0.2V for injection V_{PN}=600mV as shown in Fig. 4(a). Figure 4(b) shows that the % deviation of interface-trap charges from zero interface-trap charge is

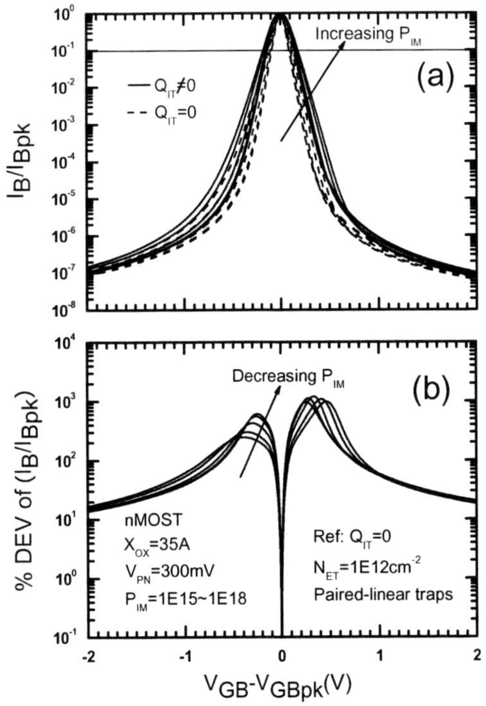

Fig. 2 Interface-trap charges on R-DCIV lineshape as a function of dopant concentration P_{IM}: (a) Normalized I_B vs. V_{GB} and (b) Percentage deviation. $P_{IM}=1.0\times10^{15}$, 1.0×10^{16}, 1.0×10^{17}, 5.0×10^{17} and 1.0×10^{18} cm^{-3}.

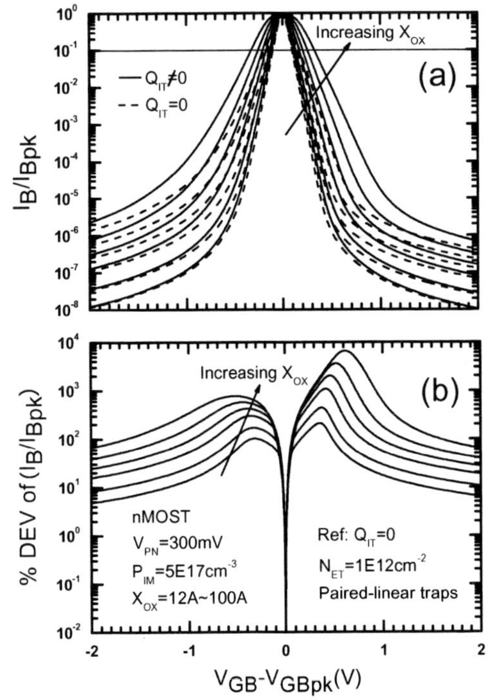

Fig. 3 Interface-trap charges on R-DCIV lineshape as a function of gate oxide thickness X_{OX}: (a) Normalized I_B vs. V_{GB} and (b) Percentage deviation. X_{OX}=12, 20, 35, 50, 70 and 100A.

over 50% for very small forward bias such as V_{PN}=1mV while % deviation is over 500% for V_{PN}=600mV when matching 90% base current range of the R-DCIV curves. The value of % deviation suggests that interface-trap charge has significant effect on the R-DCIV lineshape when forward bias V_{PN} is in practical range ($V_{PN} \leq \sim$600mV).

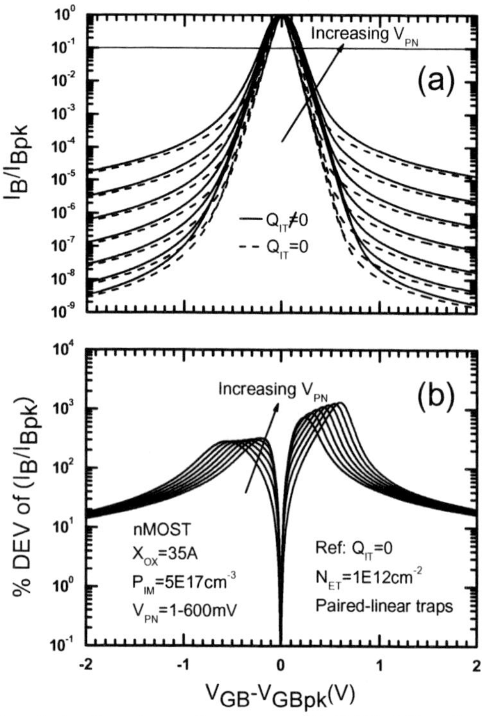

Fig. 4 Interface-trap charges on R-DCIV lineshape as a function of forward bias V_{PN}: (a) Normalized I_B vs. V_{GB} and (b) Percentage deviation. V_{PN}=1, 100, 200, 300, 400, 500 and 600mV.

The preceding analysis shows that interface-trap charges has significantly impact the R-DCIV lineshape at room temperature. Thus, the use of gate voltage formula without including interface-trap charges would cause a large loss of accuracy when extracting the parameters, especially for the mid-life devices with high interface-trap density from the R-DCIV experimental data. The fundamental reason is that the interface charge Q_{IT}, which is proportional to the interface-trap density N_{IT}, plays important role in the gate voltage equation as indicated in (3). The larger Q_{IT} would lead to greater lineshape distortion. Only for the transistors with low Q_{IT}, the R-DCIV lineshape distortion is negligible to characterize the fundamental device properties at the SiO_2/Si interface.

4. Conclusion

This paper presents a study of the effect of interface-trap charges in MOS transistors on the R-DCIV lineshape (Recombination DC- Current-Voltage) curves using the Schockley-Read-Hall thermal recombination kinetics at room temperature. Four device and material parameters over the practical range are computed to illustrate the effects on R-DCIV lineshape. Based on the theoretical analysis, we conclude that interface-trap charges significantly impact MOS transistor characteristics. The broadened R-DCIV lineshape observed in experiments can be accounted for partially by the density and profile of interface-trap charges along surface channel region, not just by the spatial variation of surface dopant impurity concentration. Thus, the gate voltage formula should include the term of interface-trap charges when using recombination DCIV methodology to extract the parameters of stressed transistors.

References

[1] O. Leistiko, A. S. Grove and C.-T. Sah, "Redistribution of Acceptor and Donor Impurities during Thermal Oxidation of Silicon," J. Applied Physics, vol. 35, pp.2695-2701, Sep. 1964.

[2] B. E. Deal, A. S. Grove, E. H. Snow, and C.-T. Sah, "Observation of Impurity Redistribution During Thermal Oxidation of Silicon Using the MOS Structure," J. Electrochem. Soc., vol. 112, pp.308-314, March 1965.

[3] Chih-Tang Sah, "Evolution of the MOS Transistor-From Conception to VLSI," Proc. IEEE, vol. 76, no. 10, pp 1280-1326, Oct. 1988.

[4] Chih-Tang Sah, "DCIV Diagnosis for Submicron MOS Transistor Design, Process, Reliability and Manufacturing," in Proceedings of the 6th International Conference on Solid-state and Integrated Circuit Technology (ICSICT), 1 (2001).

[5] Yih Wang and Chih-Tang Sah, "Lateral profiling of impurity surface concentration in submicron metal-oxide-silicon transistors," J. Appl. Phys. Vol. 90, No. 7, Oct. 2001.

[6] Chih-Tang Sah, "Insulating layers on silicon substrate," in Properties of Silicon, p. 497 (INSPEC, IEE, 1988).

[7] Chih-Tang Sah, Fundamentals of Solid-State Electronics-Solution Manual, Section 912, p. 107-108, World Scientific-Singapore, 1996.

[8] Zuhui Chen, Bin B. Jie and Chih-Tang Sah, "Effects of Energy Distribution of Interface Traps on Recombination DC Current-Voltage Lineshape," J. Appl. Phys. 100, 114511(2006).

[9] Chih-Tang Sah, "Equivalent Circuit Models in Semiconductor Transport for Thermal, Optical, Auger-Impact, and Tunneling Recombination-Generation – Trapping Processes," Phys. Stat. Sol. (a) 7, 541-549 (1971).

[10] Chih-Tang Sah, private communication by email to Zuhui Chen on April 25, 2007. This Sah model was based on the perturbation from random change of bond angle and bond length at the SiO_2/Si interface, to be published.

[11] Victor Chao-Wei Kuo, et al, "Detailed comparison of program, erase and data retention characteristics between p+poly and n+poly SONOS NAND flash memory," Proc. MTDT Conference, Aug. 2006, Taipei.

Compact Analytical Threshold Voltage Model for Nanoscale Multi-Layered-Gate Electrode Workfunction Engineered Recessed Channel (MLGEWE-RC) MOSFET

R.Chaujar*, R.Kaur*, M.Saxena**, M.Gupta* and R.S.Gupta*

* Semiconductor Devices Research Laboratory, Department of Electronic Science, University of Delhi, South Campus, New Delhi, India, rishuchaujar@rediffmail.com and ravneetsawhney13@rediffmail.com, rsgu@bol.net.in

**Department of Electronics, Deen Dayal Upadhyaya College, University of Delhi, Karampura, New Delhi, India, saxena_manoj77@yahoo.co.in

ABSTRACT

In this paper, a compact analytical threshold voltage model for multi-layered-gate electrode workfunction engineered recessed channel (MLGEWE-RC) MOSFET is presented and investigated using ATLAS device simulator. The novel device integrates the merits of recessed channel, gate electrode workfunction engineered (GEWE) architecture and multi-layered gate dielectric design. Our model includes the evaluation of surface potential, electric field distribution along the channel and threshold voltage. We demonstrate that MLGEWE-RC MOSFET design exhibits significant enhancement in terms of improved hot carrier effect immunity, carrier transport efficiency and reduced short channel effects (SCEs) proving its efficacy for high-speed integration circuits and analog design. The accuracy of the results obtained using our analytical model is verified using 2-D device simulations.

Keywords: ATLAS, high-K, MLGEWE-RC, surface potential and threshold voltage.

1 INTRODUCTION

For the last few decades, Si CMOS technology has been driven by device scaling to increase performance, as well as reduce cost and maintain low power consumption. With scaling of the device dimensions into the sub-100-nm regime, the gate oxide thickness of CMOS transistors has also been scaled down constantly. As continual CMOS scaling requirements and many digital and mixed-signal applications are driven by standby power consumption, the gate leakage current is becoming one of the critical elements of sub-65 nm CMOS technology development. A paradigm shift has been occurring in the industry, where materials innovation, rather than scaling, is becoming the primary enabler for performance enhancement in CMOS technology. For gate materials, traditional SiO_2 is being replaced by high-K dielectrics to reduce the gate leakage current.

A high-K gate dielectric becomes a key in providing the increased physical gate dielectric thickness, $t_{ox}(=t_{ox1}+t_{ox2})$, keeping effective oxide thickness (t_{oxeff}) same; without compromising the direct tunneling gate leakage current. Increasing physical gate dielectric thickness, however, results in a higher gate-fringing field, thereby reducing the gate control and hence, aggravating short channel effects (SCEs) [1-2]. An ultra thin SiO_2 interlayer between the high-K layer and silicon substrate is, thus, introduced, in order to improve the interface quality and stability. The CMOS transistors designed with the multi-layered high-k gate dielectrics achieve the expected high drive current performance and lower leakage current, thereby proving its efficacy for high performance CMOS logic applications.

Recessed Channel (RC) MOSFETs [3-4] are known to alleviate the short channel effects and the hot carrier effects appreciably due to the presence of a groove separating the source and the drain regions; and consequentially their respective depletion regions. Further integration of RC MOSFETs with dual material gate architecture [5-6] and multi-layered gate dielectric architecture enhances the driving current capability, gate control over the channel and suppresses the short channel effects. The resultant integrated device structure is, thus, referred to as MLGEWE-RC MOSFET.

The primary objective here is to apply the dual gate material architecture and multi-layered gate dielectric concept to RC MOSFET and explain the unique features offered by this structure. An analytical model using Poisson's equation has also been presented for the surface potential, electric field and threshold model for the MLGEWE-RC MOSFET and its validation is done with ATLAS device simulation results [7]. A very close match has been found between the model and simulation results. The results are then compared with the conventional RC and GEWE-RC MOSFET designs; reflecting enhancement in device characteristics of the proposed MLGEWE-RC MOSFET design.

2 MODEL FORMULATION

A schematic structure of MLGEWE-RC MOSFET is shown in Fig.1 with M_1 and M_2 of lengths L_{g1} and L_{g2} respectively. In MLGEWE-RC MOSFET, the gate consists of multi-layered-gate dielectrics having a thickness t_{ox1} and t_{ox2} of the lower and the upper gate dielectrics with the corresponding permittivites, ε_{ox1} and ε_{ox2}, respectively. The source/drain (S/D) regions are rectangular and uniformly doped at 10^{20} cm^{-3}. The channel doping concentration (or substrate doping density), N_A, is also uniform. Assuming the impurity density in the channel region to be uniform, the potential distribution $\zeta(x,y)$ in silicon film in the weak inversion region can be given as

$$\frac{\partial^2\zeta(x,y)}{\partial x^2}+\frac{\partial^2\zeta(x,y)}{\partial y^2}=\frac{qN_A}{\varepsilon_{si}};$$

for $0 \le x \le L_{eff}$ and $(d + t_{oxeff}) \le y \le (d + t_{oxeff} + Y_D)$ (1)

where ε_{si} is the dielectric constant of silicon, q is the electronic charge, Y_D is the depletion layer thickness, t_{oxeff} is the effective gate oxide thickness ($=t_{ox1}+(\varepsilon_{ox1}/\varepsilon_{ox2})t_{ox2}$), d is the groove depth and L_{eff} is the effective channel length which is given by,

$$L_{eff}=L_g+2(t_{oxeff}+NJD) \qquad (2)$$

where L_g is the gate length ($L_g=L_{g1}+L_{g2}$); NJD is the negative junction depth and Y_D is the depletion layer thickness.

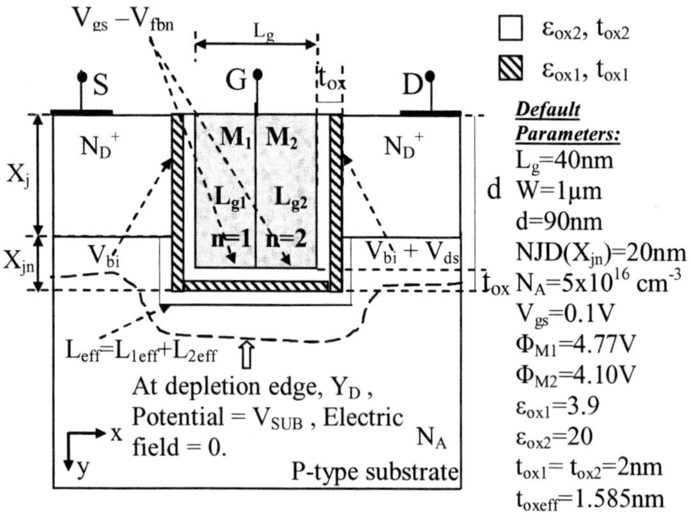

Figure 1: Schematic cross-section of MLGEWE-RC MOSFET design.

In the present analysis, the channel region has been divided into two parts, in which the potential under M_1 and M_2 can be represented as

$$\zeta_1(x,y)=\zeta_{S1}(x,d+t_{oxeff})+\sum_{r=1}^{3}P_{r1}(x)y^r$$

for $0 < x \le L_{1eff}$; $(d + t_{oxeff}) \le y \le (d + t_{oxeff} + Y_D)$ (3)

$$\zeta_2(x,y)=\zeta_{S2}(x,d+t_{oxeff})+\sum_{r=1}^{3}P_{r2}(x)y^r$$

for $L_{1eff} < x \le (L_{1eff}+L_{2eff})$;$(d + t_{oxeff}) \le y \le (d + t_{oxeff} + Y_D)$ (4)

where

$$L_{1eff}=L_{g1}+(NJD+t_{oxeff});\qquad L_{2eff}=L_{g2}+(NJD+t_{oxeff})$$

$\zeta_1(x,y)=\zeta_{S1}(x,d+t_{oxeff})$ and $\zeta_2(x,y)=\zeta_{S2}(x,d+t_{oxeff})$ are surface potentials under regions M_1 and M_2, and $P_{r1}(x)$ and $P_{r2}(x)$ are the arbitrary coefficients. The Poisson equation is solved separately under the two regions (M_1 and M_2) using the boundary conditions shown in Fig.1. Further, substituting $\zeta_{S1}(x_{min},d+t_{oxeff})=2\zeta_f$ *and* $V_{gs}=V_{th}$ in the expression for minimum surface potential, an expression for threshold voltage is obtained as

$$V_{TH}=\frac{2\zeta_F+\left[\frac{qN_A}{\varepsilon_{si}}+\left(\frac{2Y_D^2+8dY_D+8t_{oxeff}Y_D-8(\alpha)}{(\alpha)^2(\beta)}\right)\cdot V_{SUB}\right]\cdot\lambda^2-\delta_1(x_{min})}{1+\left(\frac{2Y_D^2+8dY_D+8t_{oxeff}Y_D-8(\alpha)}{(\alpha)^2(\beta)}\right)\cdot\lambda^2}+V_{FB1} \qquad (5)$$

3 RESULTS AND DICUSSION

Fig. 2 reflects a higher step near the drain end for the proposed MLGEWE-RC MOSFET design in comparison to the other two designs, thus resulting in better screening of the channel region from drain bias variations.

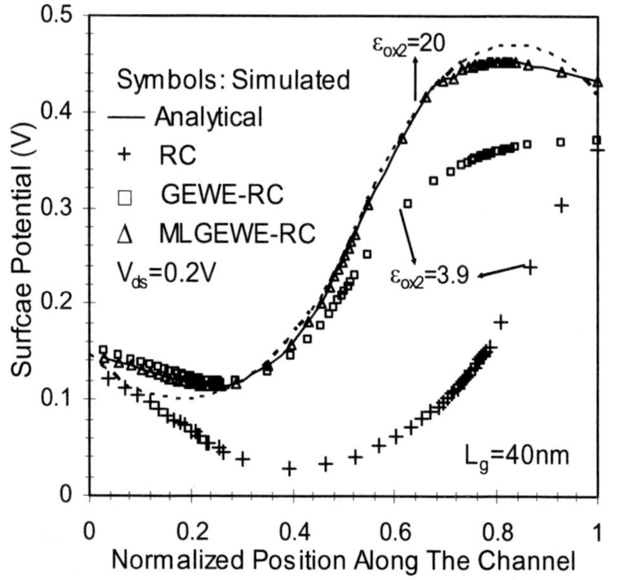

Figure 2: Surface potential variation with the normalized channel position comparing 3 different designs.

NSTI-Nanotech 2008, www.nsti.org, ISBN 978-1-4200-8505-1 Vol. 3

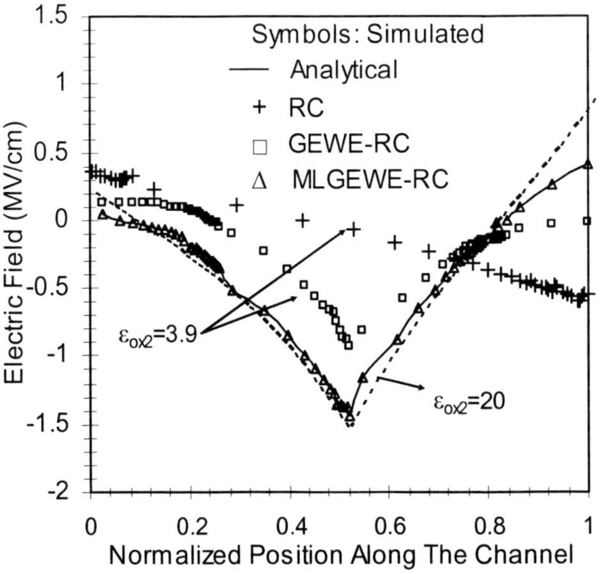

Figure 3: Electric Field variation with the normalized channel position comparing 3 different designs.

Further, the lateral field penetration from Source/Drain region is reduced significantly due to the recessed gate nature of the device, minimizing the DIBL and punchthrough effects. This lateral penetration of the field is further reduced with the use of gate stack, as is evident from Fig.3. Results reflect a higher electric field peak, for the proposed design, at the interface of two metal gates; thus causing more uniformity of electric field in the channel. This, hence, results in higher carrier transport efficiency. Further, Fig.4 reflects that V_{TH} reduces as upper gate oxide permittivity increases, thereby, enhancing the drive current.

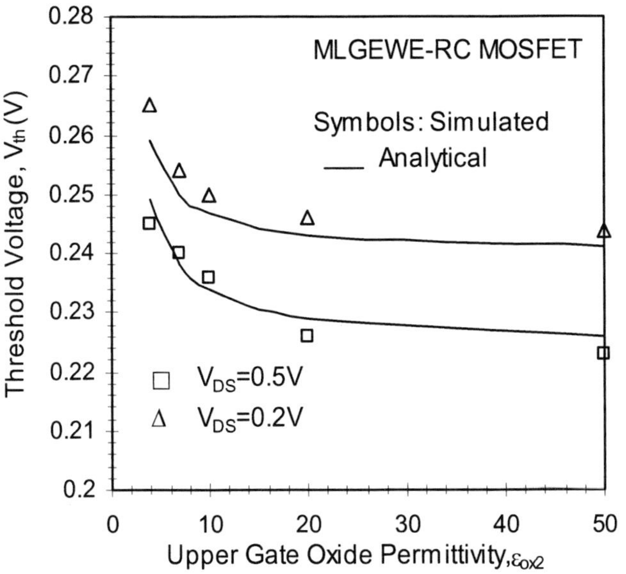

Figure 4: Threshold voltage variation as a function of upper gate oxide permittivity for 2 different drain bias.

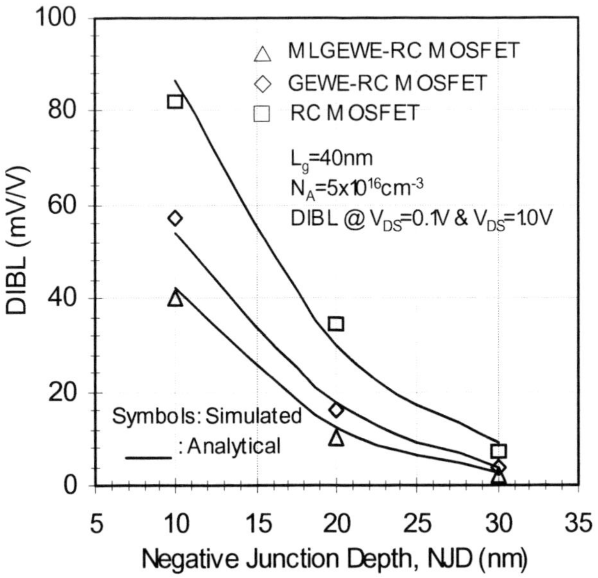

Figure 5: DIBL variation with negative junction depth, NJD, for RC, GEWE and MLGEWE-RC MOSFETs.

Fig.5 depicts that MLGEWE-RC exhibits the minimum DIBL due to the step potential profile, attributed to GEWE architecture and improved gate control over the channel, attributed to the multi-layered gate oxide architecture. Further, higher the negative junction depth (NJD), better is the screening of the channel region from the drain bias variations. It is seen that MLGEWE-RC MOSFET exhibits significant enhancement in terms improved gate control over the channel, carrier transport efficiency and hence, the driving current and hot carrier effect immunity.

4 CONCLUSION

The effectiveness of the GEWE and multi-layered gate dielectric concept to the RC structure has been examined for the first time by developing a 2-D analytical model. The results obtained have been compared with ATLAS simulations. The model results agree well with the simulated results. It also emphasizes that the proposed structure improves gate controllability over the channel and further leads to a reduced short channel effects as the surface potential profile shows a step at the interface of the two metals. Thus the short channel behavior of the RC MOSFETs is further enhanced with the introduction of the GEWE and gate stack structure over their single gate counterparts.

5 ACKNOWLEDGMENTS

Authors are grateful to Defense Research and Development Organization (DRDO), Ministry of Defense, Government of India, and one of the authors (Rishu Chaujar) is grateful to University Grant Commission (UGC) for providing the necessary financial assistance to carry out this research work.

REFERENCES

[1] B.Cheng, M.Cao, R.Rao, A.Inani, P.V.Voorde, W.M.Greene, et.al, "The Impact of High-k Gate Dielectrics and Metal Gate Electrodes on Sub-100 nm MOSFET's", IEEE Trans. Electron Devices, 46, 1537-1544, 1999.

[2] R.Chau, S.Datta, M.Doczy, B.Doyle, J.Kavalieros and M.Metz, "High-k/Metal-Gate Stack and Its MOSFET Characteristics", IEEE Elect. Device Lett., 25, 408-410, 2004.

[3] P.H.Bricout and E.Dubois, "Short-Channel Effect Immunity and Current Capability of Sub-0.1-Micron MOSFET's Using a Recessed Channel", IEEE Trans. Electron Devices, 43, 1251-1255,1996.

[4] H.Ren and Y.Hao, "The influence of geometric structure on the hot-carrier-effect immunity for deep-sub-micron grooved gate PMOSFET", Solid-State Electronics, 46, 665-673, 2002.

[5] W.Long, H.Ou, J.M.Kuo and K.K.Chin, "Dual-Material Gate (DMG) Field Effect Transistor", IEEE Trans. Electron Devices, 46, 865-70, 1999.

[6] X.Zhou, "Exploring the novel characteristics of hetero-material gate field-effect transistors (HMGFET's) with gate-material engineering," IEEE Trans. Electron Devices, 47, 113–120, 2000.

[7] ATLAS: Device simulation software, 2002.

Compact Model of the Ballistic Subthreshold Current in Independent Double-Gate MOSFETs

D. Munteanu*, M. Moreau*, J.L. Autran*,**

* L2MP-CNRS, UMR CNRS 6137, 49 rue Joliot-Curie, 13384 Marseille, France
** University Institute of France (IUF), 75000 Paris, France
E-mail: daniela.munteanu@l2mp.fr, Phone: +33 4 96 13 98 19, Fax: +33 4 96 13 97 09

ABSTRACT

We present an analytical model for the subthreshold characteristic of ultra-thin Independent Double-Gate transistors working in the ballistic regime. This model takes into account short-channel effects, quantization effects and source-to-drain tunneling (WKB approximation) in the expression of the subthreshold drain current. Important device parameters, such as off-state current or subthreshold swing, can be easily evaluated through this full analytical approach. The model can be successfuly implemented in a TCAD circuit simulator for the simulation of IDG MOSFET based-circuits.

Keywords: Independently Driven Double-Gate MOSFET, ballistic transport, quantum effects, subthreshold current model

1 INTRODUCTION

Double-Gate (DG) MOSFETs are extensively investigated because of their promising performances with respect to the ITRS specifications for deca-nanometer channel lengths. In spite of excellent electrical performances due to its multiple conduction surfaces, conventional DG MOSFET allows only three-terminal operation because the two gate electrodes, i.e. the front gate and the back gate, are generally tied together. DG structures with independent gates have been proposed [1]-[2], allowing a four terminal operation. Independent Double-Gate (IDG) MOSFETs offer additional potentialities, such as a dynamic threshold voltage control by one of the two gates, transconductance modulation, signal mixer, in addition to the conventional switching operation. Thus, IDG MOSFETs are promising for future high performance and low power consumption very large scale integrated circuits. However, one of the identified challenges for IDG MOSFET optimization remains the development of compact models [3]-[6] taking into account the main physical phenomena governing the devices at this scale of integration. In this work, an analytical subthreshold model of ultra-thin IDG MOSFETs working in the ballistic regime is presented. The present approach captures the essential physics of such ultimate devices: short-channel effects, quantum confinement, thermionic current and tunneling of carriers through the source-to-drain barrier. Important device parameters, such as the off-state current (I_{off}) or the subthreshold swing, can be easily evaluated through this full analytical approach which also provides a complete set of equations for developing equivalent-circuit model used in ICs simulation.

2 DRAIN CURRENT MODELING

Figure 1 shows the schematic n-channel IDG MOSFET considered in this work. Carrier transport in the ultra-thin silicon film (thickness t_{Si}) is considered 1D in the x-direction and the resulting current is controlled by both the front and back gate-to-source (V_{G1} and V_{G2}) and drain-to-source (V_D) voltages which impact the shape as well as the amplitude of the source-to-drain energy barrier. In the subthreshold regime, minority carriers can be neglected and Poisson's equation is analytically solved in the x-direction with explicit boundary conditions at the two oxide/silicon interfaces taking into account the electrostatic influence of V_{G1} and V_{G2}. The expression of $\Psi(x)$ is obtained by applying the Gauss's law to the particular closed dashed surface shown in Fig. 1 [7]:

$$-E(x)\frac{t_{Si}}{2}+E(x+dx)\frac{t_{Si}}{2}-E_{S1}(x)=-\frac{qN_A t_{Si}dx}{2\varepsilon_{Si}} \quad (1)$$

where E_{S1} is the electric field at the front interface given by:

$$E_{S1}=-\frac{(V_{FB1}-V_{FB2})-(V_{G1}-V_{G2})}{t_{Si}+2\gamma t_{ox}} \quad (2)$$

where the electric field $E(x)\approx-\frac{d\psi(x)}{dx}$. After some algebraic manipulations, the following differential equation is obtained for the electrostatic potential in the silicon film:

$$\frac{d^2\Psi}{dx^2}-\frac{2C_{ox}}{\varepsilon_{Si}t_{Si}}\Psi=\frac{1}{\varepsilon_{Si}t_{Si}}[qN_A t_{Si}+2C_{ox}(V_{G1}-V_{FB1}-\phi_F)]+\frac{1}{\varepsilon_{Si}t_{Si}}[2C_{ox}(V_{G2}-V_{FB2}-\phi_F)] \quad (3)$$

The solution of Eq. (3) is:

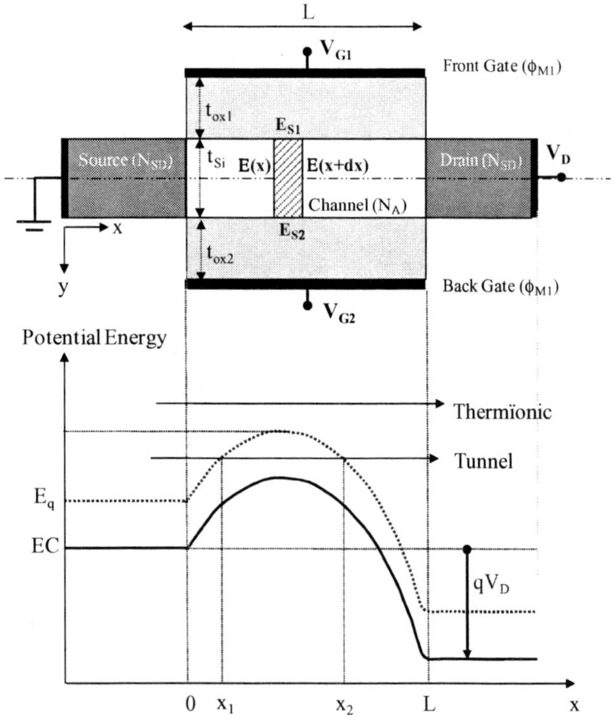

Figure 1: Schematic ultra-thin IDG MOSFET and its technological and electrical parameters considered in this work. The first energy subband profile $E_1(x)$ obtained from Eq. (9) is also represented (dotted line).

$$\psi(x) = C_1 \exp(\alpha x) + C_2 \exp(-\alpha x) - \frac{R}{\alpha^2} \tag{4}$$

where

$$C_{1,2} = \pm \frac{\phi_S [1 - \exp(\mp \alpha L)] + V_D + R \dfrac{1 - \exp(\mp \alpha L)}{\alpha^2}}{2 \sinh(\alpha L)} \tag{5}$$

$$\alpha = \sqrt{\frac{2C_{ox}}{\varepsilon_{Si} t_{Si}}} \tag{6}$$

$$R = \frac{1}{\varepsilon_{Si} t_{Si}} \left(qN_A t_{Si} + 2C_{ox}(V_{G1} - V_{FB1}) \frac{\gamma t_{ox} + t_{Si}}{2\gamma t_{ox} + t_{Si}} \right) + \frac{1}{\varepsilon_{Si} t_{Si}} \left(2C_{ox}(V_{G2} - V_{FB2}) \frac{\gamma t_{ox}}{2\gamma t_{ox} + t_{Si}} \right) \tag{7}$$

$$\phi_S = \frac{kT}{q} \ln\left(\frac{N_A N_{SD}}{n_i^2}\right) \quad \phi_F = \frac{kT}{q} \ln\left(\frac{N_A}{n_i}\right) \quad \gamma = \frac{\varepsilon_{Si}}{\varepsilon_{ox}} \tag{8}$$

Considering the limit case of an ultra-thin Silicon film, we can assume only one energy subband for the vertical confinement of carriers. The first energy subband profile $E_1(x)$ can be easily derived as:

$$E_1(x) = q(\phi_S - \psi(x)) + E_q \tag{9}$$

where E_q is the first energy level calculated using the model developed in [8]:

$$E_q \cong \left(\frac{\hbar^2}{2m_l}\right) \times \left[\left(\frac{\pi}{t_{Si}}\right)^2 + A_l^2 \times \left(3 - \frac{4}{3} \frac{1}{1 + (A_l t_{Si} / \pi)^2}\right)\right] \tag{10}$$

$$A_l \cong \left(\frac{3}{4} \times \frac{2m_l q E_{S1}}{\hbar^2}\right)^{1/3} \tag{11}$$

where m_l is the electron longitudinal effective mass (vertical confinement). Once $E_1(x)$ is known as a function of V_{G1}, V_{G2} and V_D, the total ballistic current (per device width unit) can be evaluated as follows:

$$I_{DS} = I_{Therm} + I_{Tun} \tag{12}$$

where I_{Term} and I_{Tun} are the thermionic and the tunneling components of the ballistic current, respectively. For a two-dimensional gas of electrons [9], I_{therm} and I_{tun} are given by:

$$I_{Therm} = \frac{2q}{\pi^2 \hbar} \int_{-\infty}^{+\infty} dk_z \times \int_{E_{1,max}}^{+\infty} [f(E, E_{FS}) - f(E, E_{FD})] dE_x \tag{13}$$

$$I_{Tun} = \frac{2q}{\pi^2 \hbar} \int_{-\infty}^{+\infty} dk_z \times \int_{0}^{E_{1,max}} [f(E, E_{FS}) - f(E, E_{FD})] T(E_x) dE_x \tag{14}$$

where $f(E_F,E_{FS})$ is the Fermi–Dirac distribution function, E_{FS} and E_{FD} are the Fermi-level in the source and drain reservoirs respectively ($E_{FD} = E_{FS} - qV_{DS}$), k_z is the electron wave vector component in the z direction, the factor 2 accounts for the two Silicon valleys characterized by m_l in the confinement direction (y-direction) and $T(E_x)$ is the barrier transparency for electrons and E is the total energy of carriers in source and drain reservoirs given by:

$$E = E_1 + E_x + \frac{\hbar^2 k_z^2}{2m_t} \tag{15}$$

where E_1 is the energy level of the first subband (given by Eq. (9)), E_x is the carrier energy in the direction of the current, m_t is the electron transverse effective mass. In Eq. (13) and Eq. (14) $E_{1,max}$ is the maximum of the source-to-drain energy barrier given by:

$$E_{1,max} = E_1(x_{max}) \tag{16}$$

where x_{max} is obtained from Eq. 3 with the condition:

$$\frac{d\Psi(x)}{dx} = 0 \quad for \quad x = x_{max} \tag{17}$$

$$x_{max} = \frac{1}{2\alpha} \ln\left(\frac{C_2}{C_1}\right) \tag{18}$$

In Eq. (14), the barrier transparency is calculated using the WKB approximation:

$$T(E_x) = exp\left(-2\int_{x_1}^{x_2}\sqrt{\frac{2m_t(E_1(x)-E_x)}{\hbar^2}}dx\right) \quad (19)$$

where x_1 and x_2 are the coordinates of the turning points (Fig. 1). x_1 and x_2 have literal expressions due to the analytical character of the barrier:

$$x_{1,2}(E_x) = \frac{1}{\alpha} ln\left[\frac{A \pm \sqrt{\Delta}}{2C_1}\right] \quad (20)$$

where the quantities A and Δ are defined as follows:

$$A = \phi_S + \frac{R}{\alpha^2} + \frac{E_q}{q} - \frac{E_x}{q} \quad (21)$$

$$\Delta = A^2 - 4C_1C_2 \quad (22)$$

The WKB approximation has the main advantage to be CPU inexpensive and reasonably accurate for channel lengths down to a few nanometers. Moreover, it has been shown in [10] that differences between results obtained considering the WKB approximation and full quantum treatment (tight-binding scheme) are surprisingly small (typically a few percents), which confers to the WKB approach a reasonable accuracy in the frame of the present analysis.

3 RESULTS AND MODEL VALIDATION

Figure 2 shows the first energy subband profile in the Si film calculated with the model in a L=10 nm intrinsic channel IDG MOSFET (t_{Si} = 2 nm, t_{ox} = 0.6 nm). In order to test the validity of the model, we compare these profiles with those obtained with a self-consistent Poisson-Schrödinger solver based on a real-space Non-Equilibrium Green's Function (NEGF) approach (DGGREEN2D [11]).

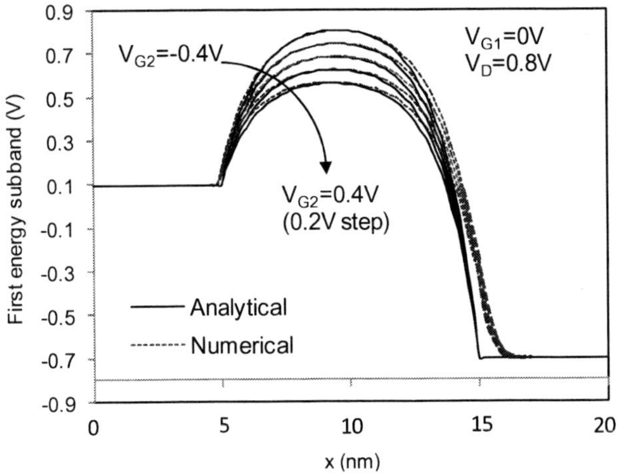

Figure 2: First energy subband profiles in the silicon film calculated with the analytical model (continuous lines) and with the 2D numerical code DGGREEN2D [11] (dotted lines). L=10 nm, V_D=0.8 V, V_{G1}=0 V.

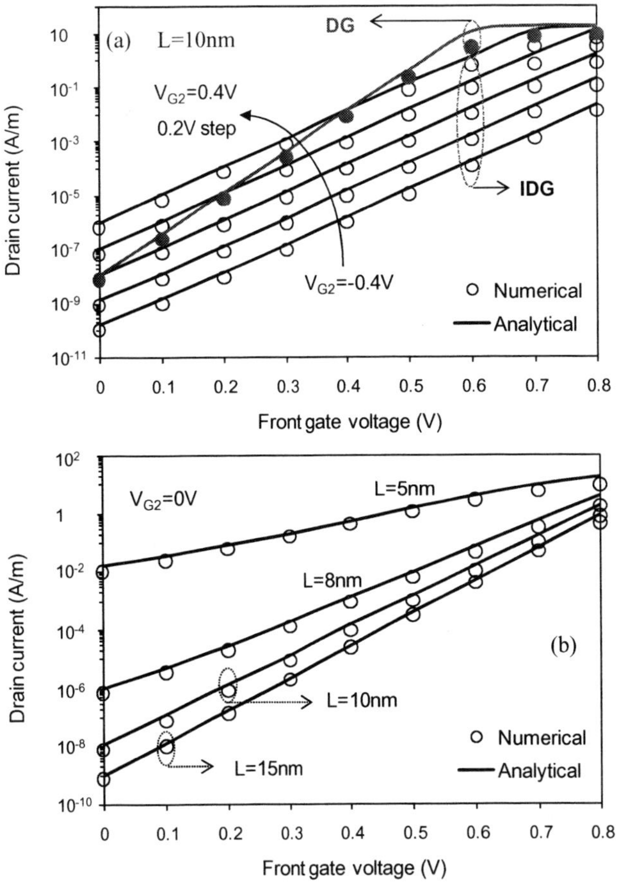

Figure 3: Subthreshold $I_{DS}(V_{G1})$ characteristics calculated with the analytical model (lines). Values obtained with the 2D numerical code DGGREEN2D [11] are also reported for comparison (symbols). DG=double-gate with connected gates. Other device parameters are the same as in Fig. 2.

As shown in Fig. 2 a good agreement is obtained between the two barriers in the subthreshold regime. In particular, we note an excellent agreement for the positions of the maximum as well as the amplitude of the barrier between the analytical and numerical curves. The slight difference in the barrier width is due to the electric field penetration in the source and drain regions, only taken into account in the numerical approach. Subthreshold $I_D(V_{G1})$ curves calculated with the analytical model are shown in Fig. 3a (for L = 10 nm and different V_{G2}) and Fig. 3b (for different channel lengths and V_{G2} = 0 V). The curves very well fit numerical data obtained with DGGREEN2D for devices in the deca-nanometer range.

4 DISCUSSION

In the following we use the proposed model to analyze the impact of source-to-drain quantum tunnelling on the IDG MOSFET operation. Figure 4 shows the source to drain energy barrier for different channel lengths. We note that below 8 nm the width of the channel barrier decreases

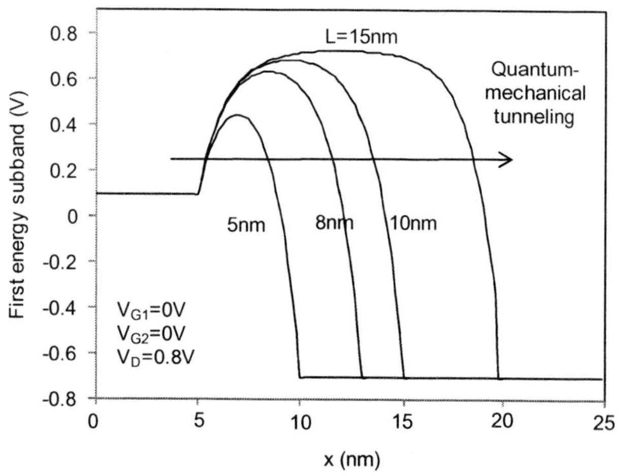

Figure 4: Variation of the source to drain energy barrier in the channel when decreasing the channel length ($V_{G1}=V_{G2}=0$ V, $V_D=0.8$ V).

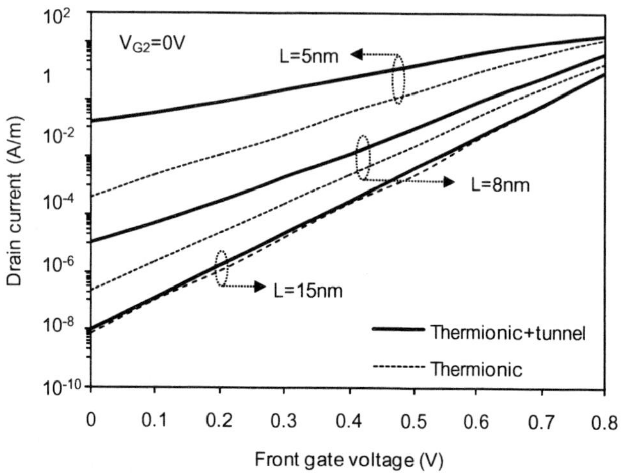

Figure 5: Subthreshold $I_D(V_{G1})$ characteristics calculated with the analytical model when considering or not the WKB tunneling component in the ballistic current. Device parameters are the same as in Fig. 2.

markedly, increasing the impact of the quantum tunneling on the device characteristics. We have analyzed the impact of carrier tunneling on device performances through a detailed comparison between simulations with and without quantum mechanical tunneling. Two cases are considered: (1) thermionic emission for $E_x > E_{max}$ and $T(E_x) = 0$ for $E_x < E_{max}$; (2) thermionic emission for $E_x > E_{max}$ and quantum tunneling with $T(E_x)$ given by the WKB approximation. Figure 5 shows subthreshold $I_D(V_{G1})$ characteristics calculated with the analytical model with and without WKB tunnelling component for channel lengths from 5 nm to 15 nm. These results highlight the dramatic impact of the source-to-drain tunneling on the subthreshold slope and also on the I_{OFF} current. In this subthreshold regime, the carrier transmission by thermionic emission is reduced or even suppressed due to the high channel barrier; as a consequence, when the channel length decreases the tunneling becomes dominant and constitutes the main physical phenomenon limiting the devices scaling, typically below channel lengths of ~ 8 nm. Quantum mechanical tunneling significantly degrades the off-state current especially in short channels, where the off-state current increases by more than two decades (L = 5 nm). However, the leakage current should be slightly reduced for the shorter geometries since it is calculated assuming perfectly ballistic transport and thus ignoring the partial reflection of electron wave functions on the source barrier. The subthreshold swing also increases (with about 30% for L = 8 nm with respect to L = 15 nm) due to quantum mechanical tunneling. As previously indicated for the leakage current, these results can be considered as an upper limit, since we assume perfectly ballistic transport without wave function reflection at the source barrier. Finally, the threshold voltage roll-off, is also notably affected by the carrier tunneling for channels shorter than 8 nm.

5 CONCLUSION

An analytical model for the subthreshold drain current in ultra-thin independent Double-Gate MOSFETs working in the ballistic regime is presented. The model is particularly well-adapted for ultra-short IDG transistors in the decananometer scale since it accounts for the main physical phenomena related to these ultimate devices: 2D short channel effects, quantum vertical confinement as well as carrier transmission by both thermionic emission and quantum tunneling through the source-to-drain barrier. The model is used to predict essential subthreshold parameters and can successfully be included in circuit models for the simulation of IDG MOSFET-based ICs.

Acknowledgements

This work was supported by the French Ministry of Research (ANR PNANO project "MULTIGRILLES").

References

[1]. M. Masahara et al., IEEE Trans. El. Dev. **52** (2005) 2046.
[2]. L. Mathew et al., in Proc. IEEE Int. SOI Conf. 2004, 187.
[3]. W. Zhang et al., IEEE Trans. El. Dev., **52** (2005) 2198.
[4]. G. Pei and E.C. Kan, IEEE Trans. El. Dev. **51** (2004) 2086.
[5]. G. Pei et al., IEEE Trans. El. Dev., **50**, (2003) 2135.
[6]. X. Loussier et al., Proc. Nanotech-WCM 2006, p. 808.
[7]. D. Munteanu et al., Proc. Nanotech-WCM 2007, p. 574.
[8] V. Trivedi and J. Fossum, IEEE El. Dev. Lett. **26** (2005) 579.
[9]. D.K. Ferry and S.M. Goodnick. Cambridge University Press, 1997.
[10] M. Städele, in Proc.ESSDERC 2002, p. 135.
[11]. J.L. Autran and D. Munteanu, J. Comput. Theo. Nanosci. (2007), in press.

Improved Carrier Mobility in Compact Model of Independent Double Gate MOSFET

M. Reyboz, P. Martin, O. Rozeau and T. Poiroux.

CEA/LETI-Minatec, 17 rue des Martyrs,
38054 Grenoble Cedex 9, France, marina.reyboz@cea.fr

ABSTRACT

We have already developed an explicit threshold voltage based compact model of independent double gate (IDG) MOSFET which works well for gate length between 30 nm and 1µm, or more [1]. However, the mobility was assumed constant. In this paper, a model with adapted carrier mobility degradation due to the transverse electrical field and velocity saturation is presented.

Keywords: compact model, mobility degradation, velocity saturation, independent double gate MOSFET.

1 INTRODUCTION

IDG MOSFET is a particularly promising device, which is expected for sub-32nm node. To take advantage of the second gate, which can be driven independently and to create new circuits, a compact model including mobility degradation and saturation velocity is essential.

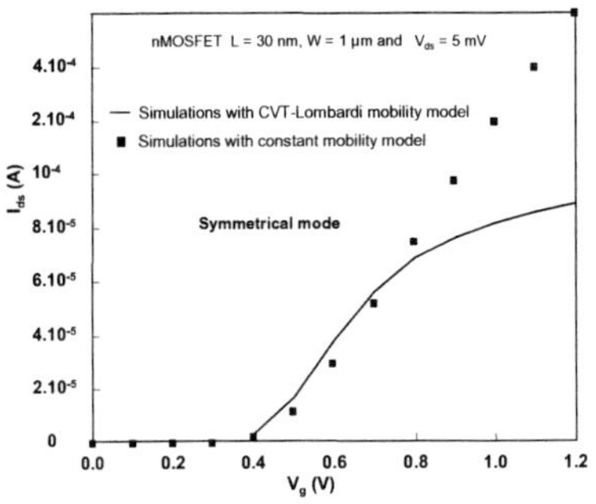

Figure 1: comparison of the drain current with and without a mobility degradation model.

Indeed, Fig. 1 shows a comparison of the drain current versus the gate voltage in symmetrical mode with constant and degradation mobility models. The difference at high transverse electrical field is very important. That is why a mobility model which takes into account the impact of the transverse fields is necessary. Same conclusion is drawn at high lateral field. Identical phenomenon is observed when gates are independently driven.

Consequently, an explicit 2D compact model of IDG MOSFET with an adapted mobility model is developed. The model is a threshold-voltage (V_{th}) based compact model. Firstly, the threshold-voltage based compact model is briefly quoted. Then, we show how mobility degradation and saturation velocity are added in the core of the model. Finally, the model is confronted to Atlas numerical simulations [2] to show and prove its accuracy.

2 EXPLICIT V_{TH} MODEL

Fig. 2 shows the IDG MOSFET structure. L is the gate length, T_{si} is the silicon film (or body) thickness, T_{ox1} and T_{ox2} are the front and the back gate oxide thicknesses. V_{g1} and V_{g2} are the front and the back gate voltages, respectively. $\Delta\Phi_{m1}$ and $\Delta\Phi_{m2}$, which are the work function differences between the front (respectively back) gate and the intrinsic silicon are supposed zero. The silicon film is supposed undoped.

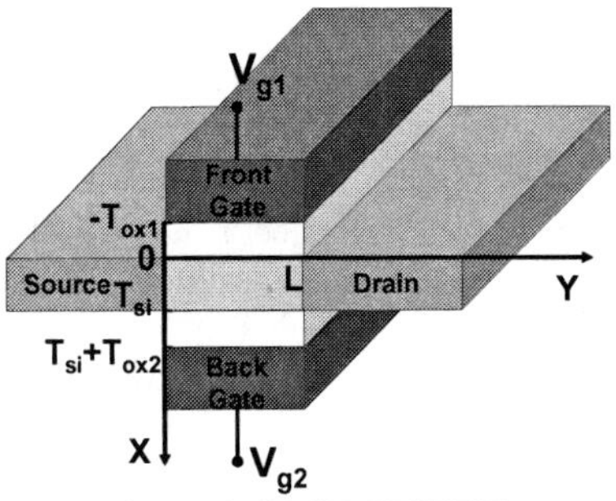

Figure 2: IDG MOSFET.

We begin with the core of the IDG compact model developed in recent years [1,3]. It is a totally explicit threshold voltage based model written in Verilog-A language. Moreover, this model is continuous from weak to strong inversion and from linear to saturation regime. It also takes into account physical short channel effects in the case of independently driven gates. Finally, this model also addressed the case of symmetrical devices when $V_{g1}=V_{g2}$, $T_{ox1}=T_{ox2}$ and $\Delta\Phi_{m1}=\Delta\Phi_{m2}$ and asymmetrical ones when $T_{ox1}\neq T_{ox2}$ and/or $\Delta\Phi_{m1}\neq\Delta\Phi_{m2}$.

3 ADAPTED CARRIER MOBILITY MODEL INCLUDING VELOCITY SATURATION

3.1 Empirical model

We start with the Lombardi mobility model [4], the same used in the ATLAS numerical simulations [2]. This model includes optical intervalley and acoustic phonons, surface roughness and velocity saturation. Optical and acoustic phonons decrease the mobility for a middle transverse electrical field $E_\perp$ whereas the surface roughness acts at high field.
The model starts from the Matthiessen rule:

$$\frac{1}{\mu}=\frac{1}{\mu_{AC}}+\frac{1}{\mu_b}+\frac{1}{\mu_{sr}} \tag{1}$$

where μ_{AC} accounts for the acoustic phonons, μ_b for the bulk mobility and μ_{sr} for the surface roughness.
The mobility due to the transverse electrical is empirically described as:

$$\frac{1}{\mu}=\frac{1}{\dfrac{4.75\times10^7}{E_\perp}+\dfrac{1.5\times10^4}{E_\perp^{1/3}}}+\frac{1}{1417}+\frac{1}{\dfrac{5.82\times10^{14}}{E_\perp^{2}}} \tag{2}$$

These parameters are used in the numerical simulations for electrons, at a temperature of 300 K and intrinsic doping. In these simulations, the transverse electrical field is the local one. It means it is derived for each point of the mesh, which is impossible to do in a compact model. Note that for the value of transverse field in a DG MOSFET, $E_\perp^{1/3}$ is negligible.

Moreover, carriers are accelerated by a lateral electrical field. Their velocity begins saturating when the field reaches a limit value. It is modeled by an equation which provides a smooth transition between low field and high field behavior [2].

3.2 Mobility model due to transverse electrical field

The mobility model should take into account same effects (phonons and surface roughness) but making them compact. To adapt this model to the independent double gate MOSFET, two mobility equations are necessary, one for the front and the other one for the back interface, μ_1 and μ_2. Besides, the mobility degradation due to the transverse electrical field is modeled thanks to front and back effective transverse electrical fields E_{eff1} and E_{eff2}. They are defined by:

$$E_{eff1}=E_{s1}+E_{cpl} \tag{3a}$$

$$E_{eff2}=E_{s2}-E_{cpl} \tag{3b}$$

where E_{si} is the front or the back surface field. i is an index: 1 for the front interface and 2 for the back one.

$$E_{si}=-\frac{Q_{invi}}{2\varepsilon_{ox}}\Rightarrow E_{si}=-\frac{V_{gsi}}{2T_{oxi}} \tag{4a}$$

$$V_{gsi}=2u_t n_i \ln\left[1+\frac{\exp\left(\dfrac{V_{gi}-V_{thi}-V_{offi}}{2u_t n_i}\right)}{1+2\exp\left(-\dfrac{V_{gi}-V_{thi}}{2u_t n_i}\right)}\right] \tag{4b}$$

V_{gsi} allows a continuous description from weak to strong inversion and V_{offi} is a correction factor describing the non total screening of the inversion layer in strong inversion. All parameters are explained in [3].

E_{cpl} is the coupling field characterizing the interaction between interfaces. This is the particularity of the presented model: thanks to E_{cpl}, which allows an accurate description of the transverse field when one interface is in weak inversion and the other one in strong inversion, the mobility model will be precise and physical.

To include (2) in our compact model, we define front and back mobility as:

$$\frac{1}{\mu_1}=\frac{1}{\mu_{01}}+\frac{1}{\dfrac{a_1}{E_{s1}+b_1E_{cpl}}}+\frac{1}{\dfrac{c_1}{E_{eff1}^{2}}} \tag{5a}$$

$$\frac{1}{\mu_2}=\frac{1}{\mu_{02}}+\frac{1}{\dfrac{a_2}{E_{s2}-b_2E_{cpl}}}+\frac{1}{\dfrac{c_2}{E_{eff2}^{2}}} \tag{5b}$$

μ_{0i}, a_i, b_i, c_i are fitting parameters. These new equations are introduced in our compact model instead of the constant mobility parameters.

3.3 Velocity saturation model

The carrier saturation velocity is an important effect that affects short channel transistors. When the lateral electrical field is high, the carrier velocity saturates at v_{sat}. The chosen model is adapted from [5] defining two saturation fields E_{sat1} and E_{sat2}:

$$E_{sati} = 2\frac{v_{sati}}{\mu_i} \quad (6)$$

where v_{sati} is the saturation velocity, which is a fitting parameter. E_{sati} is included in the saturation drain voltage V_{dsati} given by:

$$V_{dsati} = n_i \frac{E_{sati} L\left(V_{gi} - V_{thi} + 2u_t\right)}{E_{sati} L + \left(V_{gi} - V_{thi} + 2u_t\right)} \quad (7)$$

n_i is the coupling factor, also defined in [3]. Finally, V_{dasti} is included in the drain current equation of our compact model [1].

$$I_{ds} = I_{ds1} + I_{ds2} \quad (8)$$

4 RESULTS

This model was confronted to ATLAS numerical simulations with the CVT-Lombardi model [2]. As reported in the previous section, model and simulation assumptions are the same. The silicon thickness is 10 nm and the front and the back gate oxide thicknesses are 1 nm.

Results are shown in Figs 3 to 10. Figs 3 to 5 are for a long channel MOSFET (130 nm). Drain current for different voltages are drawn showing an excellent matching between the developed model and the numerical simulations.

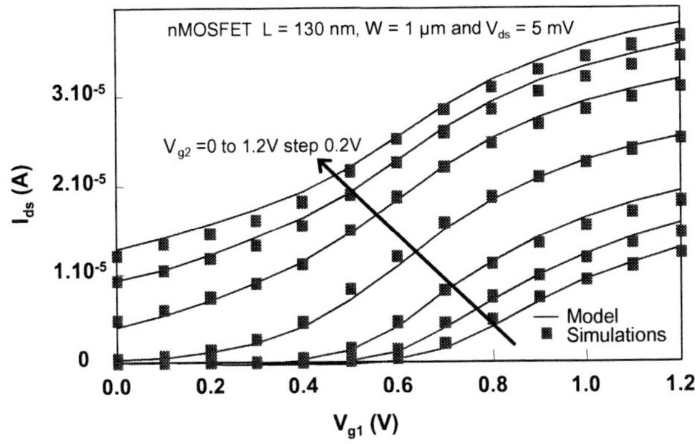

Figure 3: Drain current of a 130 nm gate length transistor versus front gate voltage for different back gate voltages (from 0 V to 1.2 V) in linear scale, in linear regime for a drain voltage of 5 mV.

Fig. 3 presents the drain current versus the front gate voltage for different back gate voltages in linear regime and in linear scale whereas Fig. 4 illustrates the same characteristics in saturation regime and in logarithmic scale in order to prove the accuracy of the developed model whatever the operating mode.

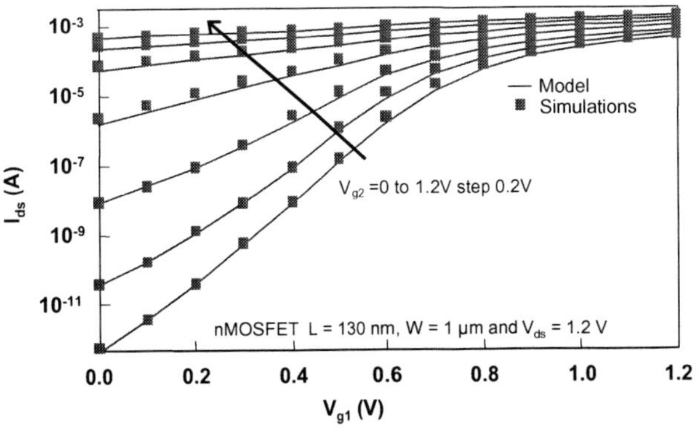

Figure 4: Drain current of a 130 nm gate length transistor versus front gate voltage for different back gate voltages (from 0 V to 1.2 V) in logarithmic scale for a drain voltage of 1.2 V.

Fig. 5 shows the drain current versus the drain voltage. It is also well modeled whatever the front drain voltage. It means that the coupling field is well taken into account in the carrier mobility model.

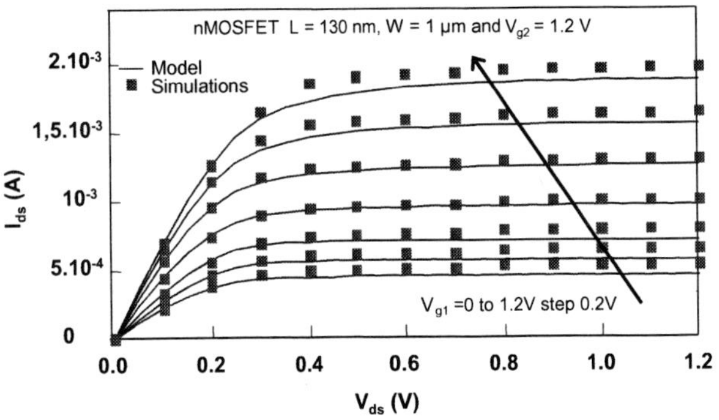

Figure 5: Drain current of a 130 nm gate length transistor versus drain voltage for different front gate voltages (from 0 to 1.2 V) for a back gate voltage of 1.2 V.

Figs. 6 to 10 are for a short channel MOSFET (30 nm). Same figures as for a long channel are shown.

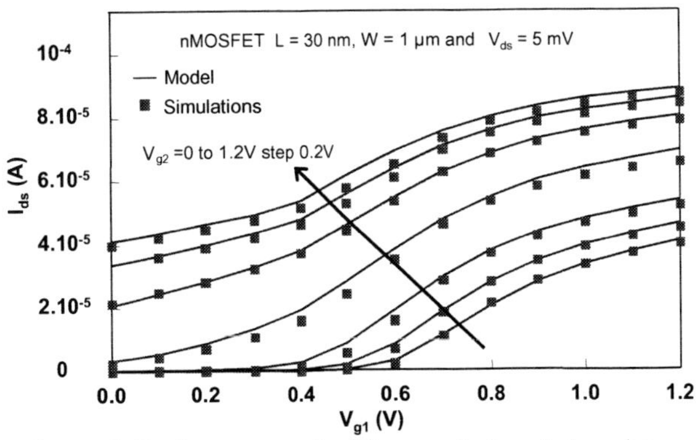

Figure 6: Drain current of a 30 nm gate length transistor versus front gate voltage for different back gate voltages (from 0 to 1.2 V) in linear scale for V_{ds}=5 mV.

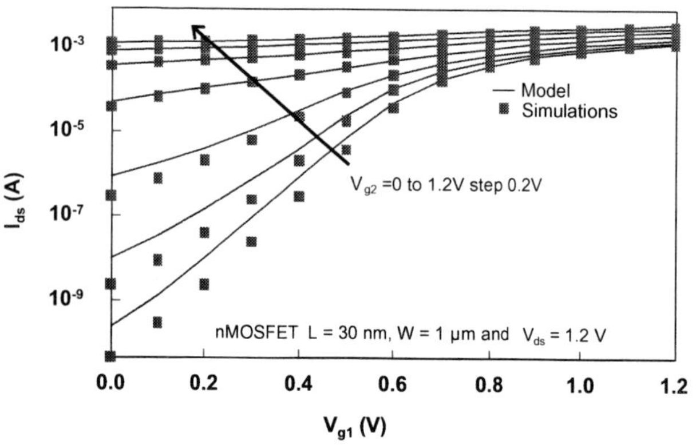

Figure 7: Drain current of a 30 nm gate length transistor versus front gate voltage for different back gate voltages (from 0 to 1.2 V) in logarithmic scale for V_{ds}=1.2 V.

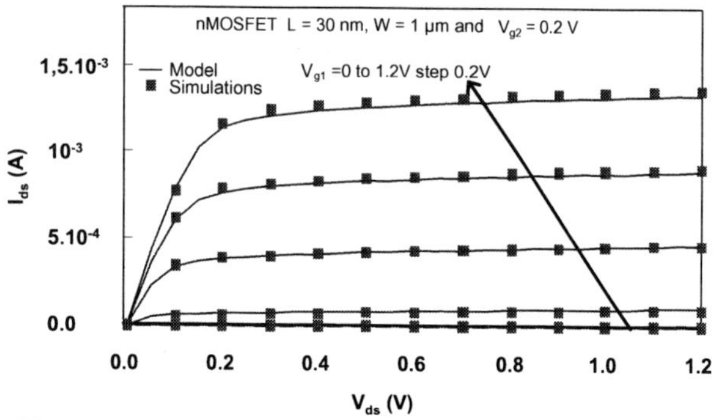

Figure 8: Drain current of a 30 nm gate length transistor versus drain voltage for different front gate voltages (from 0 to 1.2 V) for a back gate voltage of 0.2 V.

Very good results are also obtained. Only in weak inversion at both interfaces and in saturation regime the model is less exact. The front gate transconductance and the drain conductance are also displayed on Figs. 9 and 10. They prove that not only the drain current matches well the numerical simulations, but also the derivatives.

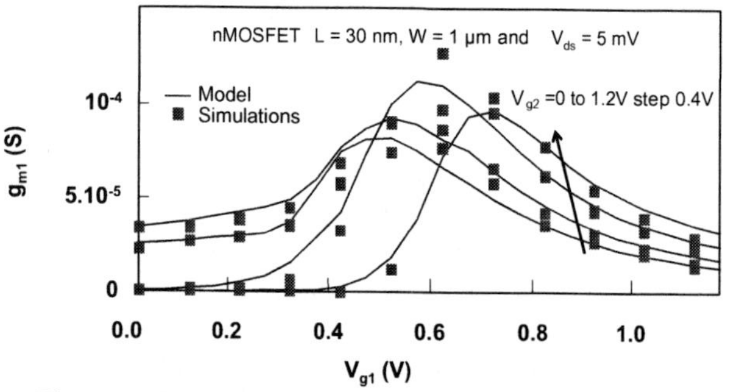

Figure 9: Front gate transconductance of a 30 nm gate length transistor versus front gate voltage for different back gate voltages (from 0 to 1.2 V) for V_{ds}=5 mV.

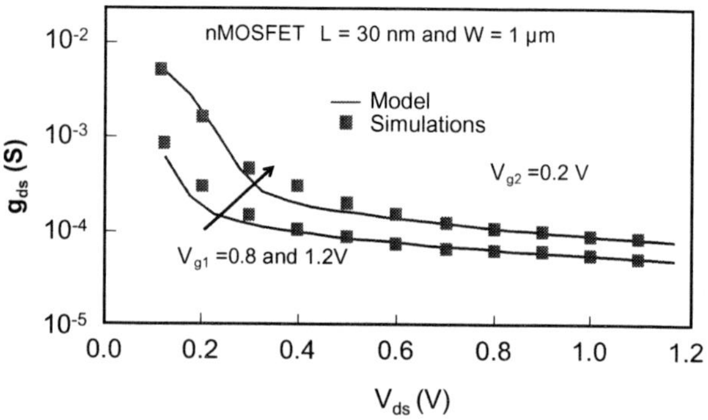

Figure 10: Drain conductance of a 30 nm gate length transistor versus drain voltage for different front gate voltages for a back gate voltage of 0.2 V.

5 CONCLUSION

An explicit threshold voltage based compact model of IDG MOSFET with a physically-based carrier mobility model is presented. This model takes into account the transverse electrical field as well as the velocity saturation. The mobility degradation due to the transverse electrical field model is adapted to the independent double gate MOSFET thanks to the inclusion of a coupling field between both interfaces. That is why whatever the operating mode (weak /weak, weak/strong or strong/strong inversion), the mobility model is adapted to the IDG MOSFET.

Confrontations with Atlas simulations prove its excellent accuracy for a long and a short channel device whatever the polarization.

Another interest of this model is that different mobility models can be implemented at each interface.

REFERENCES

[1] M. Reyboz, O. Rozeau, T. Poiroux, P. Martin, M. Cavelier and J. Jomaah, “Explicit Short Channel Compact Model of Independent Double Gate MOSFET”, NSTI Nanotech, WCM 2007.

[2] ATLAS User’s Manual – Device Simulation Software, SILVACO International Inc.

[3] M. Reyboz, “Modélisation analytique de transistors double grille à effet de champ en technologie sub-45 nm”, INP Grenoble, PhD thesis 2007.

[4] C. Lombardi, S. Manzini, A. Saporito and M. Vanzi, “A Physically Based Mobility Model for Numerical Simulation of Nonplanar Devices”, IEEE trans. on CAD, vol.7, n°11, Nov. 1988.

[5] BSIM4.5.0 MOSFET Model User’s Manual – University of California, Berkeley.

A Technique for Constructing RTS Noise Model Based on Statistical Analysis

Cheng-Qing Wei[1,2,3], Yong-Zhong Xiong[2], Xing Zhou[1], Lap Chan[3]

[1]School of Electrical & Electronic Engineering, Nanyang Technological University
Nanyang Avenue, Singapore 639798, exzhou@ntu.edu.sg

[2]Institute of Microelectronics, Agency for Science, Technology and Research
11 Science Park Road, Science Park II, Singapore 117685, yongzhong@ime.a-star.edu.sg

[3]Chartered Semiconductor Manufacturing
60 Woodlands Industrial Park D, Street Two, Singapore 738406, chanlap@charteredsemi.com

ABSTRACT

Low frequency noise such as random telegraph signal (RTS) noise is more and more important as device shrinks down. A significant part of the high frequency phase noise is generated due to up-converted low frequency noise. RF CMOS circuit performance and the device reliability will be limited and negatively impacted. Therefore, a noise model which accurately predicts the noise characteristics of deep-submicron devices is crucial for the low noise RFIC design. In this paper, the RTS noise parameters, i.e., pulse amplitude, pulse width and pulse delay, are obtained from the Schottky diode under different biases, and a technique for constructing the RTS noise model based on the statistical analysis of these noise data is introduced. From the comparisons of the noise data distributions, it is shown that the difference between the modeled RTS and measured RTS is quite small, and this model is useful in generating noise data because it is closely matched to the measured noise.

Keywords: random telegraph signal, low frequency noise, statistical noise model

1 INTRODUCTION

Throughout the last two decades, simple two-level random telegraph signals (RTS) have been observed in different types of devices, including reverse-biased diodes, bipolar junction transistors, junction field-effect transistors, and metal-oxide-semiconductor field-effect transistors (MOSFETs), etc. [1-3]. With device shrinking in dimensions, low-frequency noise is dominated by RTS noise. In the high-frequency regime, Schottky diodes with high speed advantage used in mixers and microwave oscillators have significant phase noise due to up-converted low frequency noise. This will negatively affect the circuit performance of analog and RF CMOS.

The Schottky diode used here is an n-doped Ti-Si Schottky diode with silicon doping density of $10^{18}cm^{-3}$. The total diode area is $0.67\mu m^2$ with $0.22\mu m^2$ contact area. The fabrication process is almost identical to the one proposed in [4, 5]. The standard noise measurement procedure was used at room temperature in a shielded room. The diode was connected to a battery-sourced voltage supply unit and to SR570 LN current preamplifier. The output was connected to a HP 35670A dynamic signal analyzer to collect the RTS noise signal in the time domain. The observed RTS noise pulse is from 0.48V to 0.68V in this study, and based on the statistical analysis, a RTS noise model is built with Matlab.

2 MEASUREMENT RESULTS AND DISCUSSIONS

Three parameters are needed to fully characterize one RTS pulse. They are pulse amplitude, pulse width and pulse delay. Hence, the RTS noise model in the time domain can be reconstructed if the statistical distributions of all three parameters are known for a particular bias. The details of the distribution of these three parameters at each bias are as follows:

Pulse amplitude distribution:
- 0.49V-0.57V: normal distribution
- 0.58V-0.68V: normal distribution with two peaks

Pulse width distribution:
- 0.49V-0.57V: lognormal distribution
- 0.58V-0.63V: lognormal distribution with two peaks
- 0.64V-0.68V: Gamma distribution

Pulse delay distribution:
- Exponential distribution for all biases

The normal distribution with single peak of pulse amplitude at lower bias is easy to be understood, because two-level RTS pulse is normally generated by a single active trap and each trap can capture only one electron, hence the magnitude of the current fluctuation due to a single trap should be the same, as indicated by the single peak in the amplitude distribution, while at higher biases, some high energy trap may be activated, leading to more peaks in the pulse amplitude distribution.

As shown in Fig.1, T_A is the trap in silicon along the current flow path and T_B is the trap at the interface near the Schottky contact. Since the electrons trapped at interface take much shorter path and less time to be recombined at Schottky contact compared to those trapped in silicon along the conduction path, the recombination rate of electrons trapped by interface is higher than those trapped by silicon. Hence, it is easier for interface trapped electrons to be trapped or de-trapped, causing current fluctuation. Based on this understanding, at low biases, the lognormal distribution of the pulse width signifies the presence and dominance of interface traps that have smaller pulse width due to larger recombination rate. At larger biases, the current will increase and the current conduction path may widen and more silicon traps along the conduction path will be activated and contribute to the current fluctuation, which is reflected by the two-peak appearance in the lognormal distribution of pulse width from 0.58V to 0.63V. However, at very large bias, the electron trapped at the interface recombines too fast to be detected, so that current fluctuation due to interface trap can hardly be observed. In this situation, the current fluctuation is dominated by those traps in the silicon along the current flow path, hence, the observed Gamma distribution of pulse width from 0.64V to 0.68V. The exponential distribution of pulse delay within the whole bias range is also reasonable, since pulse delay time is closely related to the pulse width, given the fixed total record time length. The pulse delay time is basically very small in both lower bias and higher bias regions. In the lower bias region, this is due to the fast trapping-detrapping of interface traps dominated in this bias region, while in the higher bias region, it is due to the large pulse width of the silicon traps dominated in this bias region.

3 STATISTICAL MODEL DEVELOPMENT AND VERIFICATION

The RTS generator model built by Matlab Simulink is shown in Fig. 2, and this Simulink generator can generate random data for pulse amplitude, pulse width and pulse delay based on the respective statistical distribution assigned. The RTS noise reconstruction algorithm is as follows (see Fig. 3):

Step 1: RTS pulse amplitude (a randomly generated variable of a particular distribution) is generated at t=0, having a value of c_1 with the dimension of current.

Step 2: RTS pulse width (a randomly generated variable of a particular distribution) is generated at t=0, having a value of a_1 with the dimension of time.

Step 3: As time is increased, if the time elapsed is less than a_1, the signal keeps output as c_1.

Step 4: When the time elapsed is greater than a_1, a RTS pulse delay (a randomly generated variable of a particular distribution) having a value of b_1 with dimension of time is generated instantaneously.

Step 5: When the time elapsed is greater than a_1 but less than a_1+b_1, the signal keeps output as 0.

Step 6: When the time elapsed is greater than a_1+b_1, new random data for RTS pulse amplitude and pulse width having values of c_2 and a_2 respectively are generated again, based on their respective statistical distributions.

The simulation time period was set to 156.097ms, since for each set of data at each bias, the time record length is 7.8045ms, and total of 20 sets of data were taken for each bias. This loop continues until the end of the simulated period set. A comparison between the distributions of three parameters of the model generated RTS and measured RTS data needs to be done to check the model accuracy. As shown in Figs. 4-5, the distribution of the modeled RTS and measured RTS are nearly the same, the distribution type can be recognized as the same type and the difference of the mean and variance values are quite small between modeled RTS and measured RTS, respectively, thus this model can be used to simulate the RTS noise of this device. The frequency spectrum of measured RTS and modeled RTS can be found using Matlab code, which further proves the accuracy and usefulness of this model in generating noise spectrum which is closely matched to the FFT analysis of measured noise.

4 CONCLUSION

In this paper, a technique for constructing the RTS noise model based on the statistical analysis of the noise data obtained from the Schottky diode under different biases is introduced. The three RTS parameters: pulse amplitude, pulse width and pulse delay are very important, since the RTS noise model is built based on the statistical distributions of these three parameters. From the comparisons of the distributions, it is shown that the difference between the modeled RTS and measured RTS is quite small, and this model is useful in generating noise data because it is closely matched to the measured noise.

REFERENCES

[1] Kirton MJ, *et al.*, "Noise in solid-state microstructures: a new perspective on individual defects, interface states, and low frequency noise," *Adv Phys*, Vol. 38, pp. 367, 1989.

[2] Hung KK, *et al.*, "Random telegraph noise of deep-submicrometer MOSFETs," *IEEE Electron Dev Lett*, Vol 11, pp. 90, 1990.

[3] Simoen E, *et al.*, "Explaining the amplitude of RTS noise in submicrometer MOSFETs," *IEEE Trans Electron Dev*, Vol 39, pp. 422, 1992.

[4] A. M. Cowley, *et al.*, "Titanium-silicon Schottky barrier diodes," *Solid State Electronics*, Vol 12, pp. 403-414, 1970.

[5] Y.Z. Xiong, G.Q. Lo, J.L. Shi, M.B. Yu, W.Y. Loh and D.L. Kwong, "Low-Frequency and RF Performance of Schottky-Diode for RFIC Applications and the Observation of RTS Noise Characteristics," accepted by 2006 NSTI Nanotechnology Conference and Trade Show, May 7-11, 2006, Boston, Massachusetts, U.S.A.

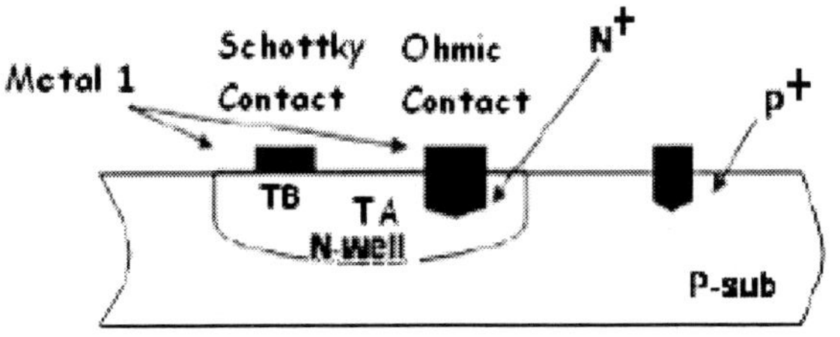

Figure 1: Side View of the Schottky Diode [4].

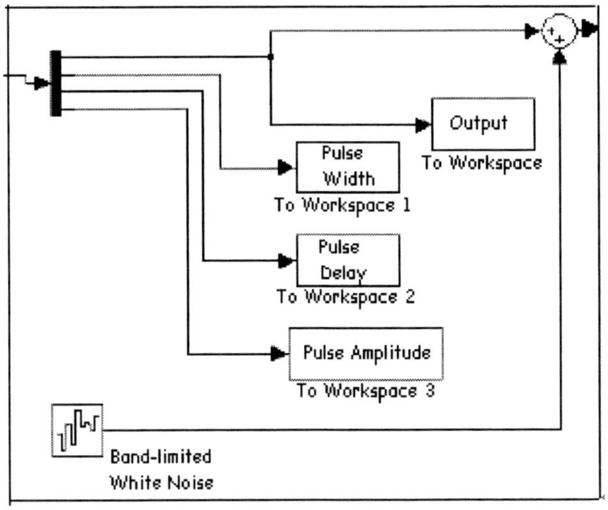

Figure 2: RTS Generator Model by Matlab Simulink.

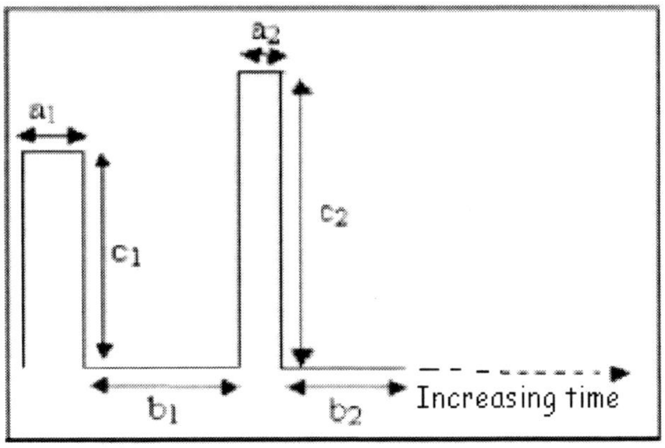

Figure 3: RTS Noise Reconstruction Algorithm.

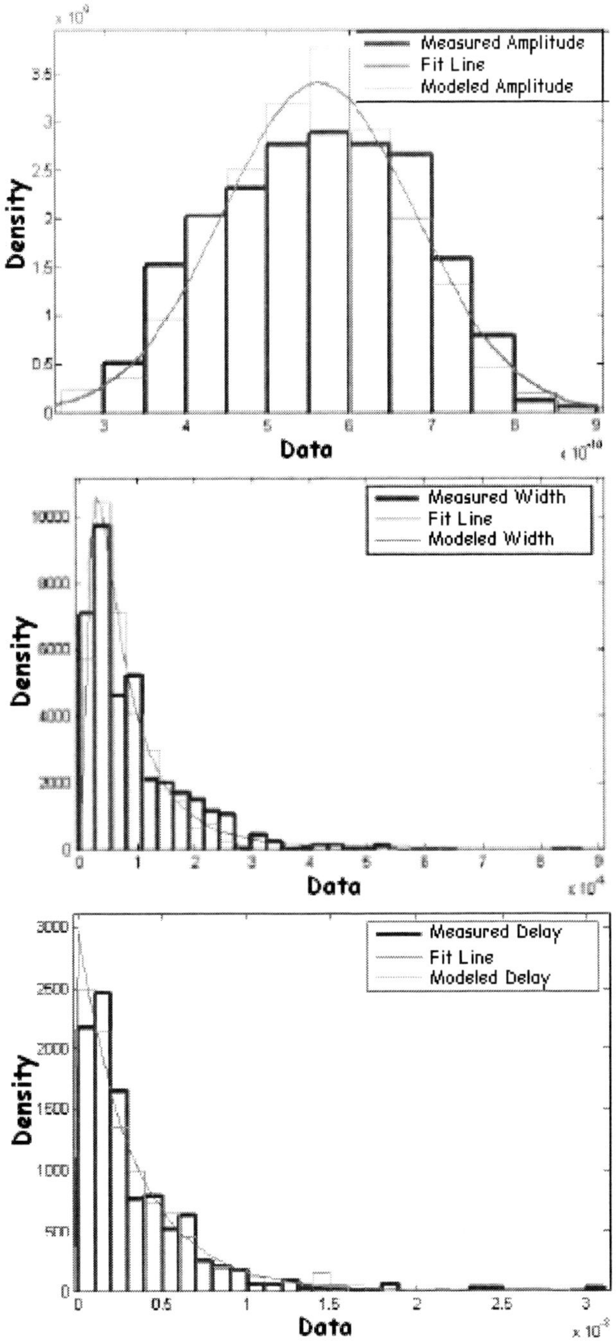

Figure 4: Comparison of the plot of three parameters between modeled RTS and Measured RTS.

Amplitude

Measured RTS

Distribution:	Normal
Log likelihood:	7592.85
Domain:	-Inf < y < Inf
Mean:	5.62524e-010
Variance:	1.37604e-020

Generated RTS

Distribution:	Normal
Log likelihood:	21506
Domain:	-Inf < y < Inf
Mean:	5.5747e-010
Variance:	1.22492e-020

Pulse Width

Measured RTS

Distribution:	Lognormal
Log likelihood:	2975.15
Domain:	0 < y < Inf
Mean:	9.12436e-005
Variance:	9.18801e-009

Generated RTS

Distribution:	Lognormal
Log likelihood:	8447.68
Domain:	0 < y < Inf
Mean:	8.82324e-005
Variance:	7.73223e-009

Pulse Delay

Measured RTS

Distribution:	Exponential
Log likelihood:	2554.06
Domain:	0 <= y < Inf
Mean:	0.000336326
Variance:	1.13115e-007

Generated RTS

Distribution:	Exponential
Log likelihood:	6983.76
Domain:	0 <= y < Inf
Mean:	0.000340956
Variance:	1.16251e-007

Figure 5: Statistical comparison of three parameters between modeled RTS and Measured RTS.

Impact of Non-Uniformly Doped and Multilayered Asymmetric Gate Stack Design on Device Characteristics of Surrounding Gate MOSFETs

H. Kaur*, S. Kabra*, S. Haldar** and R. S. Gupta*

* Semiconductor Devices Research Laboratory, Department of Electronic Science, University of Delhi, South Campus, New Delhi, India, harsupreetkaur@gmail.com, rsgu@bol.net.in
**[1]Department of Physics, Motilal Nehru College, University of Delhi, South Campus, New Delhi, India

ABSTRACT

In the present work, a new structural concept, non-uniformly doped multilayered asymmetric gate stack (ND-MAG) surrounding gate MOSFET has been proposed and it has been demonstrated using analytical modeling and simulation that ND-MAG SGT leads to suppression of short channel and hot carrier effects besides also improving the transport efficiency and gate controllability as compared to UD devices.

Keywords: non-uniformly doped, multilayered asymmetric gate stack, surrounding gate MOSFET, short channel effects, gate controllability

1 INTRODUCTION

The evolution of MOSFET technology has been governed largely by device scaling over the past twenty years. One of the key issues concerning present CMOS design is whether MOSFET devices can be scaled to 0.1μm channel length and beyond for continuing density and performance improvement as continued miniaturization has led to short channel effects (SCEs), hot electron effects and low carrier transport efficiency. In order to overcome the scaling limitations and to enhance the device performance various nonclassical structures such as Pi gate MOSFETs, Omega MOSFET, Cylindrical/Surrounding gate MOSFETs have been proposed. Among these, the surrounding gate MOSFET [1-4], in particular, has drawn a great deal of attention as it offers high packing density, steep subthreshold characteristics and higher current drive. Another remarkable feature of this structure is that the gate surrounds the silicon pillar completely and therefore controls the channel potential in a more effective manner resulting in increased short channel immunity. All these features make the SGT a potential candidate to succeed the classical planar MOSFET.

However, in the nanoscale regime incorporation of alternative device designs is also necessary to improve device performance. In order to address the issues related to short channel degradation and improvement in device performance, the use of non-uniformly doped channel design was suggested as a possible solution for reducing the short channel effects present in deep sub micrometer devices [5-6]. Furthermore, in order to increase the transistor performance, every new technology node requires the reduction of the gate oxide thickness. However, the extent to which gate oxide thickness can be scaled down is limited by direct tunneling. An alternative could be the use of insulator material with a higher permittivity than SiO_2, which would allow a thicker gate insulator. The major advantage of using high-*k* dielectrics for the gate insulator comes from the fact that while scaling, the significant parameter for constant electric field scaling is not the physical thickness of the gate insulator but rather the capacitance per unit area. However, studies reported that threshold voltage (V_{th}) roll-off, DIBL and subthreshold slope (S) increase with an increase in dielectric permittivity (or thickness) implying that short channel performance degrades [7-8]. This can be attributed to the loss of gate control owing to the increased fringing fields. In order to overcome these detrimental issues, the multilayered asymmetric gate stack oxide design along with the non-uniformly doped design has been incorporated in surrounding gate MOSFET (ND-MAG) SGT. Using modeling and simulation [9], it is demonstrated that ND-MAG provides an effective solution to these drawbacks. It has been demonstrated that incorporation of ND and MAG design leads to an improvement in short channel immunity and hot carrier reliability while also enhancing the gate controllability and carrier transport efficiency and thus ensures better performance as compared to conventional devices. Thus, the critical issues of short channel effects, hot carrier effects and gate leakage can be addressed by incorporating the non-uniformly doped and asymmetric gate stack architectures.

2 MODEL FORMULATION

Fig.1. shows the cross-section view of ND-MAG SGT. As can be seen the channel has two regions, the one near the source is heavily doped and the one near drain is low-doped. Also there is an asymmetric gate stack, i.e., multilayered gate stack is present near the drain and single gate oxide is present near the drain. Thus there is a single gate oxide (SGO) region near the source and a gate stack oxide (GSO) region near the drain.

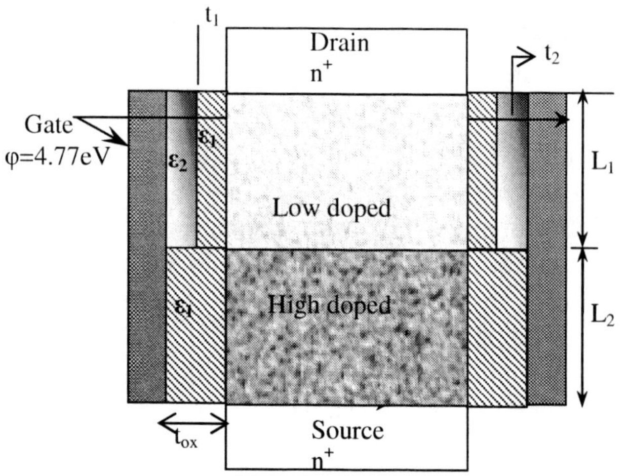

Fig.1. Cross-sectional view of ND-MAG SGT

Assuming the impurity density in the channel to be uniform, the Poisson equation in cylindrical coordinates for the two regions can be written as:

$$\frac{1}{r}\frac{\partial}{\partial r}\left(r\frac{\partial \psi_i(r,z)}{\partial r}\right)+\frac{\partial^2 \psi_i(r,z)}{\partial z^2}=\frac{qN_{ai}}{\varepsilon_{si}} \qquad (1)$$

where, $\psi_i(r,z)$ is the potential distribution in the silicon film, N_{ai} is the doping concentration with $i=1$ for region 1 near the source and $i=2$ for region 2 near the drain end.

The potential distribution in the silicon film in the two regions is assumed to be a parabolic profile in the radial direction [10] and on solving the Poisson equation in the two regions separately using the boundary conditions [11-12], the surface potential in the two regions is obtained as:

$$\psi_{s1}(z)=A\exp\left(\frac{-z}{\lambda_1}\right)+B\exp\left(\frac{z}{\lambda_1}\right)-v_1 \qquad (2)$$

$$\psi_{s2}(z)=C\exp\left(\frac{-(z-L_1)}{\lambda_2}\right)+D\exp\left(\frac{(z-L_1)}{\lambda_2}\right)-v_2 \qquad (3)$$

Where,

$$v_1=\frac{qN_{aH}\lambda_1^{\,2}}{\varepsilon_{si}}-\psi_{gs1} \qquad (4)$$

$$v_2=\frac{qN_{aL}\lambda_2^{\,2}}{\varepsilon_{si}}-\psi_{gs2} \qquad (5)$$

where $\psi_{gsi}=V_{gs}-V_{fbi}$ is the gate potential, V_{fbi} is the flatband voltage and ψ_{si} is the potential at the surface of silicon film for regions 1 and 2 respectively and λ_1 and λ_2 are the characteristic lengths given by:

$$\lambda_1=\sqrt{\frac{\eta t_{si}^{\,2}\ln\left(1+\frac{2t_{ox}}{t_{si}}\right)}{8}} \qquad (6)$$

$$\lambda_2=\sqrt{\frac{\eta t_{si}^{\,2}\ln\left(1+\frac{2t_{oxeff}}{t_{si}}\right)}{8}} \qquad (7)$$

where, $\eta=\varepsilon_{Si}/\varepsilon_{ox}$, t_{si} is the thickness of silicon film, t_{ox} is the oxide layer thickness of SGO region and t_{oxeff} is the effective oxide layer thickness of the GSO region in terms of the corresponding thickness of the SiO_2 layer and is defined as:

$$t_{oxeff}=t_1+\frac{\varepsilon_1}{\varepsilon_2}t_2 \qquad (8)$$

where, t_1 is the thickness of the SiO_2 layer and t_2 is the thickness of the high-*k* layer. The coefficients *A*, *B*, *C* and *D* have been obtained using the conditions of continuity for potential and electric field at the interface of high and low doped regions.

The position of minimum surface potential, z_{min} is calculated by differentiating (2) w.r.t. z and equating the resulting expression to zero.

$$z_{\min}=\frac{\lambda_1}{2}\ln\left(\frac{A}{B}\right) \qquad (9)$$

The minimum surface potential is then obtained from (4) as:

$$\psi_{s1\min}=2\sqrt{AB}-v_1 \qquad (10)$$

The electric field distribution in the high and low doped region can be obtained by differentiating $\psi_{s1(z)}$ and $\psi_{s2(z)}$ respectively.

The threshold voltage, V_{th} can be obtained by equating the minimum surface potential ψ_{s1min} to $2\psi_{f1}$ and the expression for threshold voltage is obtained as:

$$V_{th}=\frac{-b+\sqrt{b^2-4ac}}{2a} \qquad (11)$$

where, a, b and c are the various coefficients obtained in the analysis.

Subthreshold slope, *S*, is defined as:

$$S=\frac{kT}{q}\ln(10)\frac{1}{\frac{\partial \phi_s(z_{\min})}{\partial V_{gs}}} \qquad (12)$$

Thus, *S* can be obtained by using (10) in (12).

of DIBL with channel length and it is seen that DIBL is lowest for ND-MAG as compared to UD devices.

3 RESULTS AND DISCUSSION

Fig.2 shows the variation of surface potential and electric field with channel length. It is seen that ND-MAG exhibits a step function in the surface potential profile as compared to UD device which screens the region near the source from variations in drain voltage and hence ensures more reduction in DIBL. Moreover, average electric field under the gate is higher for ND-MAG which improves the carrier transport efficiency. The peak electric field near drain is lower implying reduction in impact ionization.

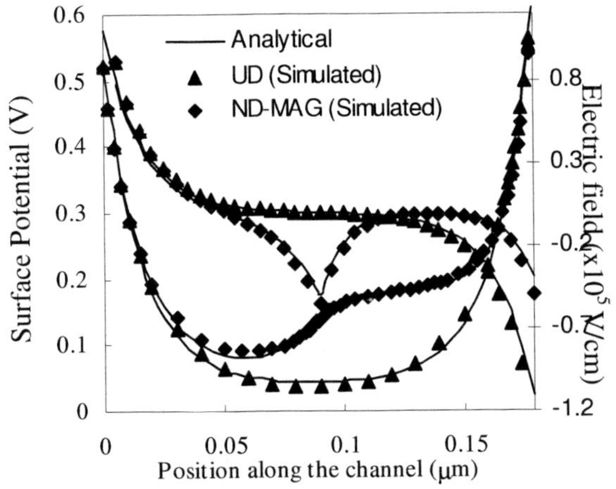

Fig.2. Variation of surface potential and electric field along the channel for L=180nm, L_1=90nm, t_{ox}=5nm, t_{Si}=75nm, $N_H=3x10^{23}m^{-3}$, $N_L=6x10^{22}m^{-3}$, t_1=1nm, t_2=4nm, ε_2=20, V_{gs}=0.2V, V_{ds}=0.1V.

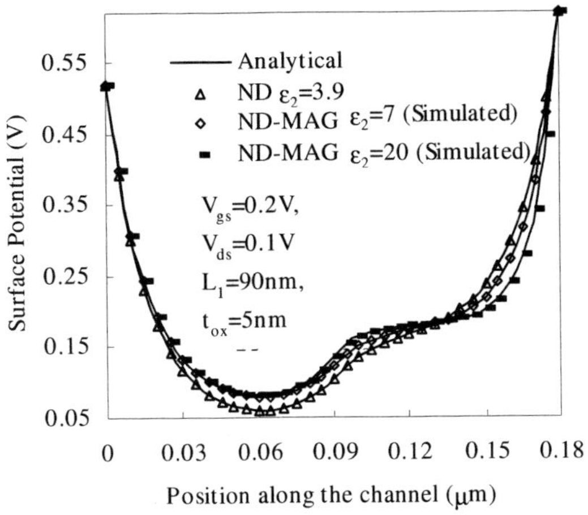

Fig.3 Variation of surface potential along channel for different values of the dielectric constant of the upper dielectric layer, ε_2

Fig.3 shows that as ε_2 increases, the minimum surface potential decreases implying a better gate controllability and therefore increased short channel immunity. Fig.4 shows that V_{th} roll-off is considerably reduced by incorporating ND-MAG design. Fig.5 shows the variation of DIBL with channel length and it is seen that DIBL is lowest for ND-MAG as compared to UD devices.

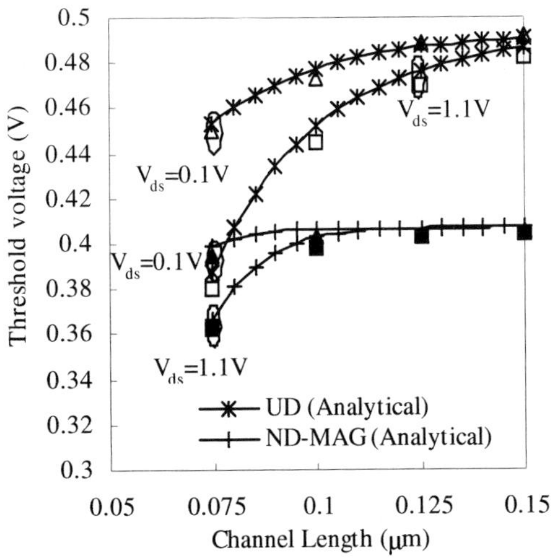

Fig.4 Variation of threshold voltage with channel length.

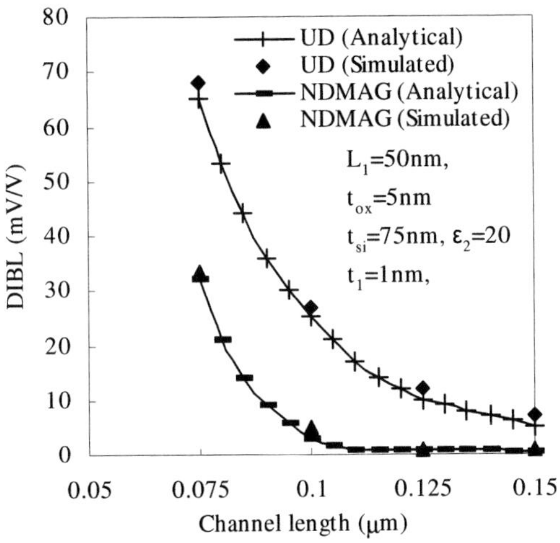

Fig.5 DIBL variation with channel length for UDand ND-MAG devices.

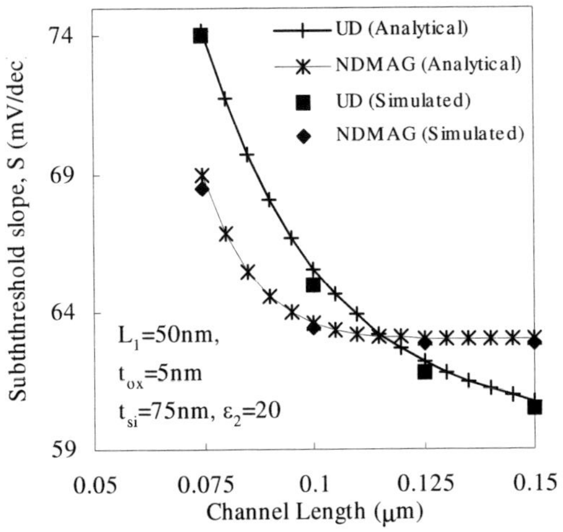

Fig.6 Variation of subthreshold slope with channel length.

In fig.6, the analytical and simulated results showing the variation of subthreshold slope with channel length has been plotted for the devices. It is found that on increasing the dielectric constant of the upper dielectric layer, ε_2 of the gate stack region, that is, on introducing MAG design, the subthreshold slope reduces considerably which again confirms that short channel immunity can be improved by incorporating MAG design.

4 CONCLUSION

A two-dimensional analytical model for a new structural concept ND-MAG (non-uniformly doped multilayered asymmetric gate stack) SGT has been developed and its impact on the device characteristics has been analyzed. The analytical results so obtained have been compared with the simulated results obtained from the device simulator ATLAS and have been found to be in good agreement. It has been demonstrated that the magnitude of the positive offset voltage is higher for ND-MAG as compared to UD devices which ensures better screening from drain bias variations leading to a reduction in DIBL. The effectiveness of ND-MAG design can also be seen as a reduction in V_{th} roll-off and subthreshold slope. Moreover, the peak in the electric field distribution under the gate is higher for ND-MAG as compared to UD devices, which ensures uniformity in the average drift velocity of the electrons in the channel which results in an improvement in the carrier transport efficiency. In addition, ND-MAG design also leads to a reduction in the peak electric field near the drain end as compared to UD devices which implies a reduction in hot carrier effects. The study thus affirms the fact that incorporation of ND and MAG designs ensures better performance as compared to UD devices.

5 ACKNOWLEDGMENT

The authors are grateful to the Defense Research and Development Organization (DRDO) for providing the necessary financial assistance to carry out the present research work.

REFERENCES

[1] H. Takato, K. Sunouchi, N. Okabe, A. Nitayama, K. Hieda, F. Horiguchi, F. Masuoka, "Impact of Surrounding Gate Transistor (SGT) for Ultra-High-density LSI's", IEEE Trans. Electron Devices, 38(3), 573-578, 1991

[2] A. Nitayami, H. Takato, N. Okabe, K. Sunouchi, K. Hieda, F. Horiguchi, F. Masuoka, "Multi-Pillar Surrounding Gate Transistor (M-SGT) for Compact and High-Speed Circuits", IEEE Trans. Electron Devices , 38(3), 579-583, 1991.

[3] S. Watanabe, K. Tsuchida, D. Takashima, Y. Oowaki, A. Nitayama, K. Hieda, H. Takato, K. Sunouchi, F. Horiguchi, K. Ohuchi, F. Masuoka H. Hara, "A Novel Circuit Technology with Surrounding Gate Transistors (SGT's) for Ultra High Density DRAM's", IEEE Journal of Solid-State Circuits, 30(9), 960-970, 1995.

[4] S. Maeda, S. Maegawa, T. Ipposhi, H. Nishimura, H. Kuriyama, O. Tanina, Y. Inoue, T. Nishimura, N. Tsubouchi, "Impact of a Vertical Φ-Shape Transistor (VΦT) Cell for 1 Gbit DRAM and Beyond", IEEE Trans. Electron Devices, 42(12), 2117-2124, 1995.

[5] H. Kaur, S. Kabra, S. Bindra, S. Haldar, R. S. Gupta, "Impact of graded channel (GC) design in fully depleted cylindrical/surrounding gate MOSFET (FD CGT/SGT) for improved short channel immunity and hot carrier reliability", Solid State Electronics, 51, 398-404, 2007.

[6] H. Kaur, S. Kabra, S. Haldar, R. S. Gupta, "An Analytical Drain Current Model for Graded Channel Cylindrical/Surrounding Gate MOSFET", Microelectronics Journal, 38, 352-359, 2007.

[7] B. Cheng, M. Cao, R. Rao, A. Inani, P. V. Voorde, W. M. Greene, J. M. C. Stork, Z. Yu Z, Zeitzoff P M, Woo J C S. The impact of high-k dielectrics and metal gate electrodes on sub-100nm MOSFET's. IEEE Trans. Electron. Devices, 46(7), 1537-1543, 1999.

[8] A. Inani, R. V. Rao, B. Cheng, J. Woo, "Gate stack architecture analysis and channel engineering in deep sub-micron MOSFETs", Jpn. J. Appl. Phys , 38(4B), 2266- 2271, 1999.

[9] SILVACO International, "ATLAS Users Manual 2000.

[10] K. Young, "Short Channel Effect in Fully Depleted SOI MOSFET's," IEEE Trans. Electron Devices, Vol.36, pp.399-402, 1989.

[11] C.P Auth and J. D.Plummer, "Scaling Theory for Cylindrical, Fully-Depleted, Surrounding-Gate MOSFET's," IEEE Electron Device Letters, Vol.18, pp.74-76, 1997.

[12] A.Kranti, S. Haldar and R. S. Gupta, "Analytical model for threshold voltage and I-V characteristics of fully depleted short channel cylindrical/surrounding gate MOSFET," Microelectronic Engineering, Vol.56, pp.241-259, 2001.

HiSIM-LDMOS/HV: A Complete Surface-Potential-Based MOSFET Model for High Voltage Applications

Y. Oritsuki, M. Yokomiti, T. Sakuda, N. Sadachika, M. Miyake, T. Kajiwara, H. Kikuchihara
T. Yoshida*, U. Feldmann, H. J. Mattausch, M. Miura-Mattausch
Hiroshima University, 1-3-1 Kagamiyama, Higashi-Hiroshima, Hisishima, 739-8530, Japan
*NEC Information Systems, 1753 Shimonumabe, Nakahara, Kawasaki, Kanagawa, 211-866, Japan

ABSTRACT

We present here the high-voltage MOSFET model HiSIM-LDMOS/HV based on the complete surface-potential description. The model is valid both for symmetrical and asymmetrical device structures, has accurate scaling properties for structural variations such as gate wide, gate length or drift-region length and is valid for a wide range of bias conditions. The predictability of the developed model is discussed with examples of various device parameter variations.

Keywords: surface potential, resistance, LDMOS, HVMOS, circuit simulation

1 INTRODUCTION

High-voltage MOSFETs are becoming important due to expanding utilization of integrated circuits for high-voltage requirements such as in many automotive of mobile applications. Necessary voltage-handling capabilities extend from a few volts to several hundred volts. This wide range of the operation conditions is realized with a low impurity concentration region, called drift region. To achieve the capabilities for high voltage applications, two structures have been developed. One is the laterally diffused asymmetric MOSFET (LDMOS) as shown in Fig. 1a [1]. The other is the symmetrical high-voltage MOSFET (HVMOS) as shown in Fig. 1b.

It has been observed that the capacitances of LDMOS devices show specific complicated features as a function of applied voltages. Namely, strong peaks are observed in C_{gg} as a function of the gate voltage V_{gs} (see Fig. 2). Previously, modeling has been done either by introducing an internal node at the channel/drift junction [2] or with a resistance in a macro model [3]. The former approach solves the node potential iteratively until the channel current and the current in the drift region at the internal node become equal. Recently, the LDMOS features have been modeled by considering the potential drop in the drift region in a consistent way by directly solving the device equations with an iterative approach [4]. Our investigation here mainly aims at extending the LDMOS model to including the symmetrical case, which is also important for high-voltage MOSFET optimization. The extended model is called HiSIM-LDMOS/HV.

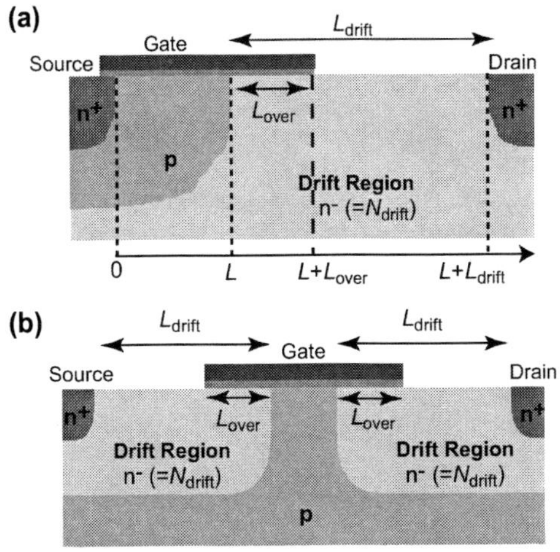

Fig. 1. Cross-sections of the studied LDMOS (a) and HVMOS (b) devices.

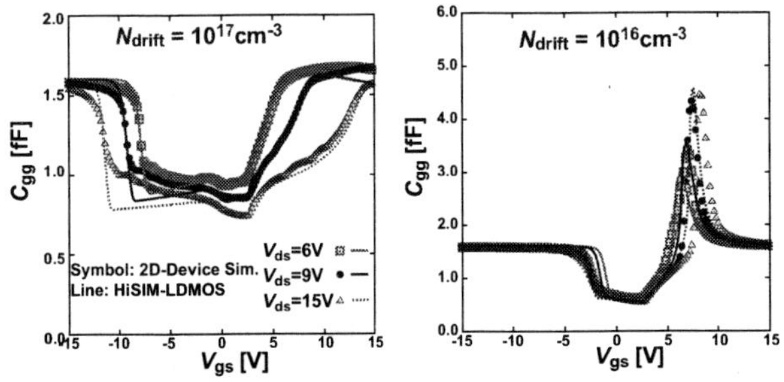

Fig. 2. Capacitance comparison between 2D-device simulator (symbols) and developed compact model (lines) results for the LDMOS structure with different drift-region doping of 10^{17}cm^{-3} (left) and 10^{16}cm^{-3} (right).

2 LDMOS CHARACTERISTICS

The channel part of the LDMOS device is formed by out-diffusion of the impurity concentration from the source contact region. The drift region is usually formed by the substrate. On the other hand, HVMOS has the conventional MOSFET structure, but with long drift regions at both source and drain. The resistive drift region for providing the high-voltage capability, namely its length L_{drift} and its doping concentration N_{drift}, determine important operational properties of the LDMOS/HVMOS device.

Figs. 3a-c show comparisons of *I-V* and g_m characteristics for two impurity concentrations in the drift region N_{drift}, while keeping the length L_{drift} and other model parameters the same [4]. Corresponding C_{gg} characteristics are compared for the two different N_{drift} values in Fig. 2. The anomalies observed for reduced N_{drift} in C_{gg} are caused by the increased resistance effect in the drift region due to the lower impurity concentration, which coincide with drastic reduction of g_m, as obvious from Fig. 3b.

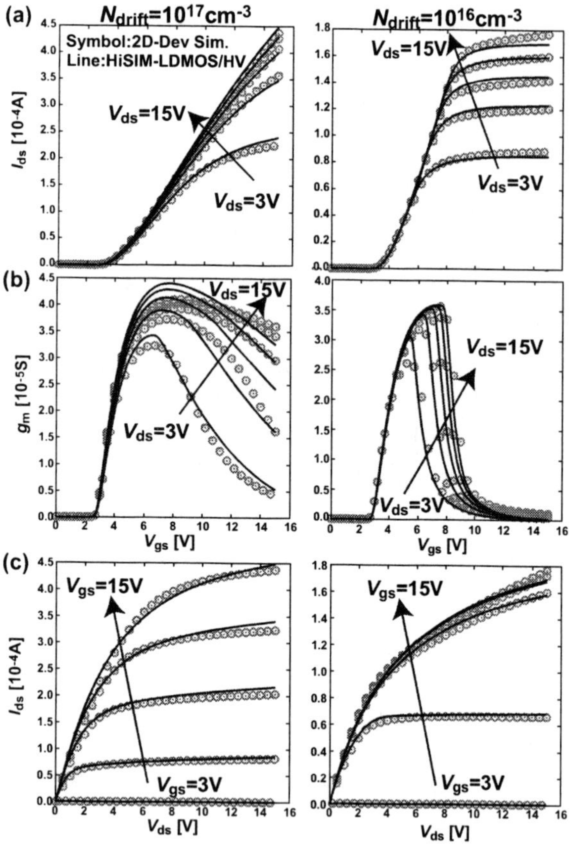

Fig. 3. *I-V* and g_m comparison between 2D-device simulator (symbols) and developed compact model (lines) results for the LDMOS structure with different drift-region doping of $10^{17}cm^{-3}$ (left) and $10^{16}cm^{-3}$ (right).

To model the resistance effect of the drift region in a consistent way, the iterative Poisson-equation solution for the surface potentials in the MOSFET channel is extended to include the resistance effect of the drift region. The respective potential drop is written as

$$\Delta V = I_{ds} * R_{drift} \tag{1}$$

where the resistance in the drift region is denoted as R_{drift} [5]. This potential drop is treated as the reduction of the applied voltages. Fig. 4 shows a calculated potential distribution denoting also the potential drop in the drift region. The influence of the resistance on the potential node at the channel/drift junction is summarized in Fig. 5. The specific features of the LDMOS structure can be exactly seen in this comparison. With negligible resistance in the drain contact, the potential $\phi_{s(\Delta L)}$ at the junction between MOSFET channel and drift region increases nearly linearly up to $\phi_{s0} + V_{ds}$, where ϕ_{s0} is the potential at source and is about equal to unity under the strong inversion condition [6]. The resistive drift region causes a potential drop as I_{ds} increases with larger V_{gs}. Under this condition, the potential value $\phi_{s(\Delta L)}$ even decreases while V_{gs} is increasing as can be seen in Fig. 5a. The described effect for the potential causes the anomalous features of the LDMOS device behavior.

Fig. 5b compares calculated $\phi_{s(\Delta L)}$ with HiSIM-LDMOS/HV and with a 2D-device simulator as a function of V_{ds}. Under the saturation condition the potential is expected to saturate. However, $\phi_{s(\Delta L)}$ continues to increases gradually due to the resistance effect. This continuous increase of $\phi_{s(\Delta L)}$ under the saturation condition results in the quasi-saturation behavior of LDMOS devices, with gradually increasing drain current, even beyond the saturation condition for the MOSFET.

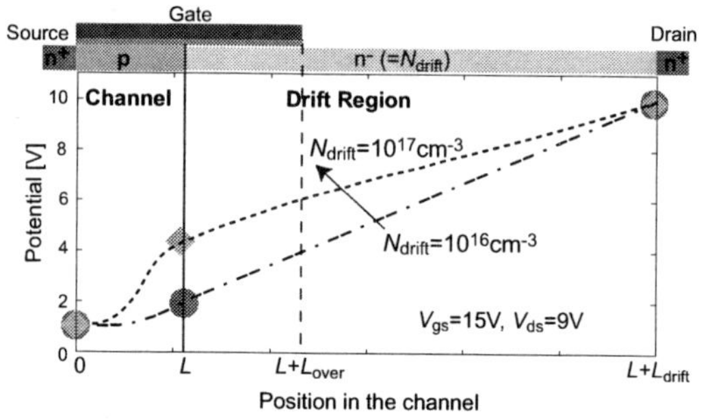

Fig. 4. Calculated potential distribution along the channel with the developed HiSIM- LDMOS/HV model (shown with symbols) for two N_{drift} concentrations. Lines are 2D-device simulation results.

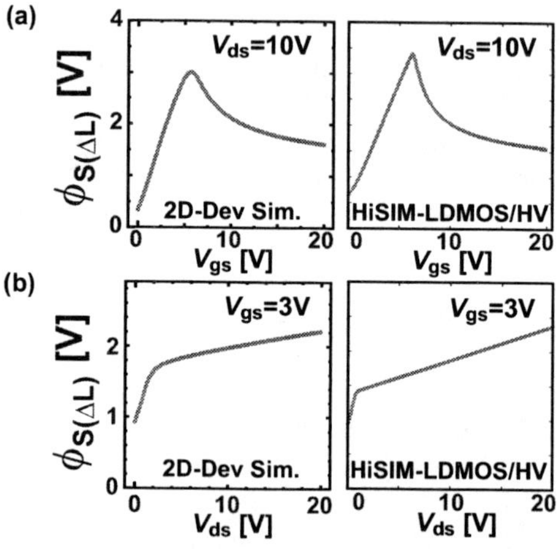

Fig. 5. Comparison of calculated surface potential at the channel / drift region junction with a 2D-device simulator and HiSIM-LDMOS/HV, (a) as a function of V_{gs}, and (b) as a function of V_{ds}.

3 EXTENSION TO THE SYMMETRICAL HVMOS DIVECE

For symmetrical HVMOS modeling, the resistance model is applied to the source side as well. Fig. 6 shows a calculated potential drop within the source causing the reduction of V_{gs} to V_{gseff}. This additional potential drop results in the reduction of V_{ds} and V_{bs} as well. Therefore, the influence of the source resistance is expected to be drastic.

Accurate modeling of the overlap charge, Q_{over}, becomes more important for the symmetrical HVMOS device due to its increased contribution to the operational characteristics. For this accuracy purpose the bias dependent surface potentials within the overlap region have to be considered in describing the formation of the accumulation, the depletion or the inversion condition underneath the gate overlap region, which now depend in a complicated way dynamically on bias conditions. There modeling tasks are achieved by solving the Poisson equation in the same way as in the channel. The overlap charges are determined with the calculated surface potential distribution under the approximation that the potential variation along the overlap region is negligible. The surface-potential values are of course a function of N_{drift}, which determines also the flat-band voltage within the overlap region. Calculated overlap capacitances are shown in Fig. 7 as a function of V_{gs}.

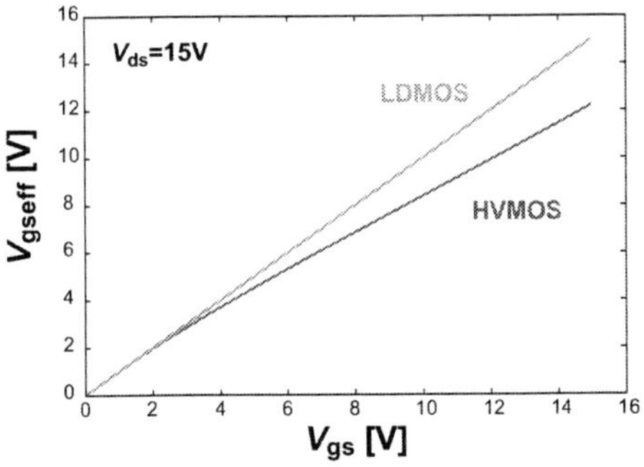

Fig. 6. Comparison of the effective gate-source voltage (V_{gseff}) as a function of the applied gate-source voltage (V_{gs}) for the symmetrical HVMOS and the LDMOS device structures.

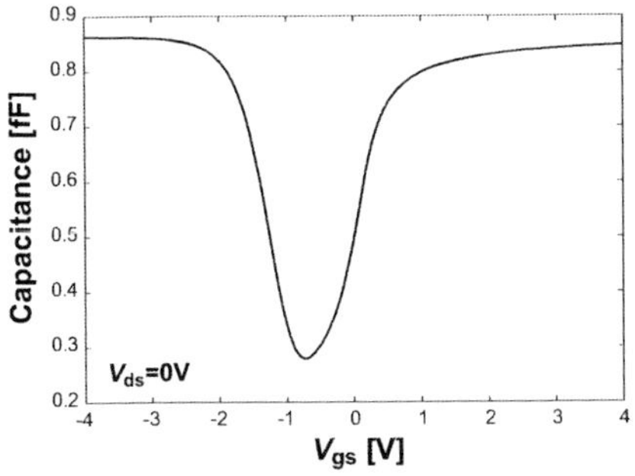

Fig. 7. Calculated overlap capacitance at the drain side with HiSIM-LDMOS/HV at V_{ds}=0V.

4 RESULTS AND DISCUSSIONS

Figs. 8a,b compare calculated *I-V* characteristics of HiSIM-LDMOS/HV for LDMOS and symmetrical HVMOS with 2D-device simulation results, respectively. The high resistance effect of the drift region causes a reduction of the potential increase in the channel, which results also in a drastical reduction of the drain current. The reduction is much more enhanced for the symmetrical HVMOS case due to the potential drop in the L_{drift} region at the source side. On the contrary the LDMOS device shows gradually increasing current behavior due to the reduction of R_{drift} for increased carrier concentration in the drift region. Thus, it is verified that all specific features of LDMOS/HVMOS devices can be well reproduced with the single model HiSIM-LDMOS/HV. This is an advantage of the modeling based on the surface potential, which additionally secures the consistency of the overall model description.

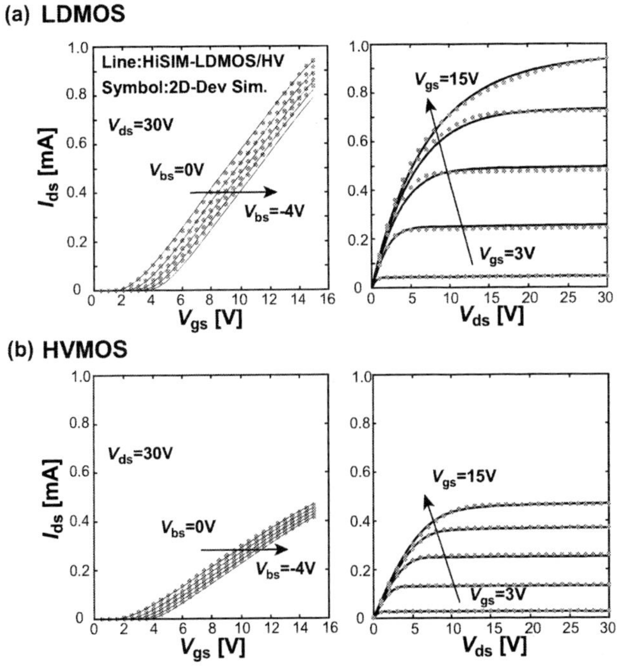

Fig. 8. Current comparison, (left-hand side) as a function of V_{gs} and (right-hand side) as a function of V_{ds}, between LDMOS and HVMOS structures, respectively. Results from HiSIM-LDMOS/HV (lines) and a 2D-device simulator (symbols) are in very good agreement.

Figs. 9a,b compare calculated capacitances for LDMOS and symmetrical HVMOS devices with HiSIM-LDMOS/HV and 2D-device simulation results, respectively. Results agree well for both device structures. The origin for the shoulders in the overall capacitances comes from the overlap capacitances, which are non-negligible for high-voltage MOSFETs.

The scalability of the developed compact high-voltage LDMOS model with the dimensions of the MOSFET part is assured by the application of the surface-potential MOSFET model HiSIM2, which has been verified to accurately reproduce MOSFET properties for all channel length and channel width with a single parameter set [5, 6]. The methodology of HiSIM-LDMOS/HV for consistently determining the potential distribution in MOS channel and drift region furthermore leads to the scaling property with the drift-region length L_{drift}, again with a single model-parameter set for the complete HiSIM-LDMOS/HV model. Fig. 10 verifies these scaling properties with the I_{ds}-V_{gs} characteristics for 3 different L_{drift} values at small and large drain-source bias voltages V_{ds}. It can be seen that the compact model results agree very well with the results of 2D-device simulation at different L_{drift} values.

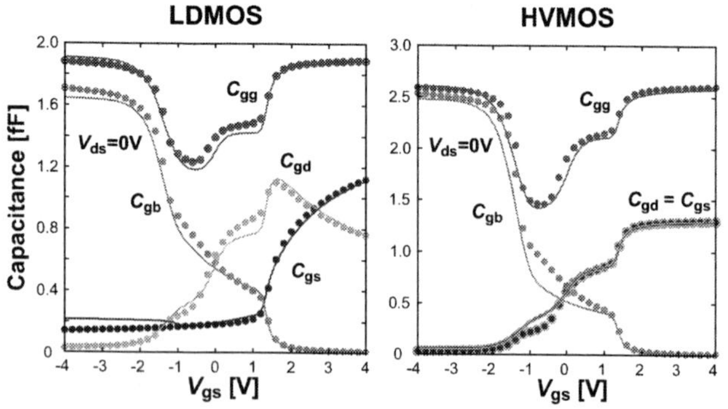

Fig. 9. Comparison of capacitances calculated for the LDMOS and symmetrical HVMOS structures with HiSIM-LDMOS/HV (lines) and a 2D-device simulator (symbols). Again HiSIM-LDMOS/HV is verified to be in good agreement with 2D-device simulation.

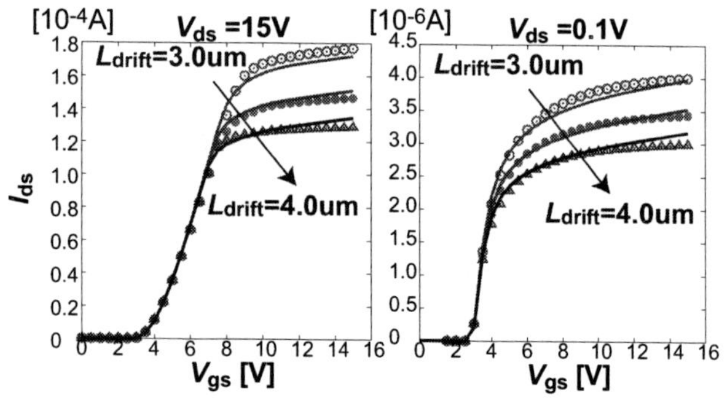

Fig. 10. Scalability of the developed compact high-voltage HiSIM-LDMOS/HV model with drift-region length L_{drift}. The plots show the I_{ds}-V_{gs} characteristics at high and low drain bias in a comparison of 2D-device simulation (symbols) and compact model (lines) results.

5 CONCLUSIONS

We have developed a compact model for high-voltage MOSFETs based on the surface potential distribution in the MOSFET core and its consistent extension to the drift region. The model solves the entire LDMOS/HV structure without relying on any form of macro or subcircuit formulation, accurately reproducing all structure-dependent LDMOS/HV features.

ACKNOWLEDGEMENT

STARC (Semiconductor Technology Academic Research Center) and NEDO (New Energy Development Organization), Japan, have provided support and cooperation for this research.

REFERENCES

[1] C.Y. Tsai, T. Efland, S. Pendharkar, J. Mitros, A. Tessmer, J. Smith, J. Erdeljac, and L. Hutter, "16-60 V rated LDMOS show advanced performance in a 0.72 μm evolution BiCMOS power technology," Tech. Digest IEDM, p.367-370, 1997.

[2] A. Aarts, N. D'Halleweyn, "A Surface-Potential-Based High-Voltage Compact LDMOS Transistor Model," IEEE Trans Electron Devices, Vol. 52, No. 5, p. 999-1007, 2005.

[3] Y.S. Chauhan, F. Krummenacher, C. Anghel, R. Gillon, B. Bakeroot, M. Declercq, and A. M. Ionescu, "Analysis and Modeling of Lateral Non-Uniform Doping in High-Voltage MOSFETs," Tech. Digest IEDM, p.213-216, 2006.

[4] M. Yokomiti, N. Sadachika, M. Miyake, T. Kajiwara H. J. Matttausch, M. Miura-Mattausch, "LDMOS Model for Device and Circuit Optimization," Jap. J. Appl. Phys., vol. 47, published in April 2008.

[5] HiISM-LDMOS/HV User's Manual, Hiroshima University/STRAC, March, 2008

[6] M. Miura-Mattausch, N. Sadachika, D. Navarro, G. Suzuki, Y. Takeda, M. Miyake, T. Warabino, Y. Mizukane, R. Inagaki, T. Ezaki, H. J. Matttausch, T. Ohguro, T. Iizuka, M. Taguchi, S. Kumashiro, and S. Miyamoto, "HiSIM2: Advanced MOSFET Model Valid for RF Circuit Simulation," IEEE Trans. Electron Devices, vol. 53, No. 9, p. 1994-2007, 2006.

Si-Based Process Aware SPICE Models for Statistical Circuit Analysis

Sathya Krishnamurthy*, Vinay K. Dasarapu*, Yuri Mahotin*, Ross Ryles**, Frederic Roger**, Sumit Uppal**, Puspita Mukherjee**, Alan Cuthbertson**, Xi-wei Lin*

*Synopsys Inc. 700 E, Middlefield Road, Mountain View, CA 94041, USA,
Sathya.Krishnamurthy@synopsys.com
**Atmel North Tyneside Ltd, UK.

ABSTRACT

Process independent design methodologies are no longer going to be productive with the scaling of the device technology nodes into nanometer regime. Design and process are more tightly integrated for the better manufacturability. Critical link between the process and design is the compact model of the devices being used in a particular technology design, which will in-turn serves as a virtual fabrication house for the designer. This is possible, only if the compact model has process dependency information. In this paper, we will describe the methodology to create the process-aware compact models and validate the model using the 130 nm technology Silicon data at a) device level and b) circuit level. We have achieved good results in matching the silicon-measured data of a 3481 stage Ring-oscillator circuit with that of the extracted process-aware compact model for various process conditions.

Keywords: MOSFET Modeling, Process-Aware Compact Models, Circuit Simulation, TCAD.

1 INTRODUCTION

As CMOS technology is scaled, design dependent yield loss becomes increasingly important due to increasing interactions between design and manufacturing. The manufacturing variability can be too large to achieve performance goals by designing a chip only to SPICE corner models. This leads to a requirement of SPICE models that include process variations during manufacturing. Currently, process variations are imposed on models via statistical distribution of SPICE parameters. This approach suffers from several fundamental flaws: (a) the actual SPICE model parameters may deviate far from their underlying physics, as they often end up as fitting parameters to silicon data, (b) it is erroneous to treat SPICE parameters as statistically independent of each other and (c) SPICE parameters cannot be directly linked to any one specific process variation. The situation is worse, if principle component analysis (PCA) is used, as it represents a further abstraction of process variability and it may not be possible to give feedback to the process engineers on how a particular process variable is affecting some of the critical electrical parameters of the device.

A methodology of extracting a process aware model using TCAD simulations was presented in [2] and [3]. Pertinent compact model parameters were obtained as a polynomial function of process parameters. However, TCAD models are approximate and do not completely capture the underlying real process variations.

In this paper, we present a methodology that allows for process splits and actual silicon data to be used to extract SPICE model parameters as a polynomial function of process parameter variations i.e. silicon based "process aware" SPICE models were extracted. Validation of the constructed process-aware compact model is done at the device level and as well as at the circuit level by using a 3481 stage Ring Oscillator silicon results.

2 METHODOLOGY

In order to construct a process-aware compact model for circuit simulations, the first thing we need to have is a global compact model which will match all the I-V and C-V silicon data at all the bias conditions and device geometries of interest at nominal process conditions. Once we have a global compact model with acceptable accuracy, one can convert the global compact model into a process-aware compact model by following the below steps:

1. Select key model parameters which are directly related to the high sensitive process parameters for a given technology.
2. Add an extra term $f_{mpar}(P_1,P_2,P_3,\ldots,P_N)$ to the nominal value of the model parameter *mpar*; for example, in BSIM3 compact model, threshold voltage parameter VTH0 can be written as:

$$Vth0 = Vth0\_nominal + \Sigma\Sigma a_i^{(n)} P_i^n \quad (1)$$

Where a_i (i=1,2,3...) are the coefficients whose values to be extracted by fitting the given I-V data for different process conditions, and the value of "n" can be selected depending on the variation of the electrical quantities with respect to the process conditions. For example, n=1 gives the linear dependence of process parameters on vth0 as given in Eq. 1, and n=2 gives the quadratic dependency.

3. Once we have extracted the process dependent key model parameters, validate the model with the silicon data for arbitrary values of the process parameters within the valid range of process parameters.
4. Next step is to check the predictability of the extracted compact model. This test should be done at device level and as well as at the circuit level.
   a) At device level, compare the Silicon I-V data with I-V obtained from circuit simulations using the extracted process-aware library file at process conditions different from the process conditions used for extraction of the model.
   b) Ring oscillator circuit delay is a perfect metric to be used for validating any compact model. Compare the delay obtained from the measurements on a fabricated ring oscillator with that of the simulated ones.
5. Once above tests are successfully completed then only the compact model can be used for designing circuits while considering the process variations. This way, one can design the circuits robustly with respect to process variations.

3 RESULTS

In this work, we have used Silicon I-V data of 130 nm technology node, and the compact model selected is BSIM3v3. For the global parameter extraction, we have used the devices with dimensions listed in Table 1 for both NMOS and PMOS devices.

W/L (um)	10	0.8	0.35	0.13
10.0	X	X	X	X
0.6	X	X	X	X
0.15	X	X	X	X

Table 1: Lists the value of Width and Length of the transistors being used for extracting the global model. "X" defines that we have included that device I-V data in the extraction process.

The extraction accuracy of the global model parameter extraction has been checked using RMS error on all the curves of all the devices being used. We have achieved total RMS error value less than 4 %. After extracting the model parameters by fitting all the given I-V data at various bias conditions for the given device dimensions, next step is to construct the process-aware compact model. For this purpose, we have used the I-V Silicon data for the process conditions mentioned in Table 2. Here, we have selected PolyCD and TOX as the process parameters for both NMOS and PMOS devices. Selection of the process parameters are done based on the sensitivity analysis of the process variables on the critical electrical parameters. In this particular case, we have found the variation of PolyCD and TOX values have maximum impact on the device characteristics compared to the variations on the other process parameters.

Device	NMOS		PMOS	
S. No.	PolyCD (nm)	TOX (A)	PolyCD (nm)	TOX (A)
P1	125	23.5	125	22
P2	108	23.5	125	24.5
P3	125	26	125	30
P4	125	29	125	27.5
P5	125	31.5	108	22
P6	142	23.5	142	22

Table 2 lists the process conditions used for the splits for a 0.13μm CMOS process.

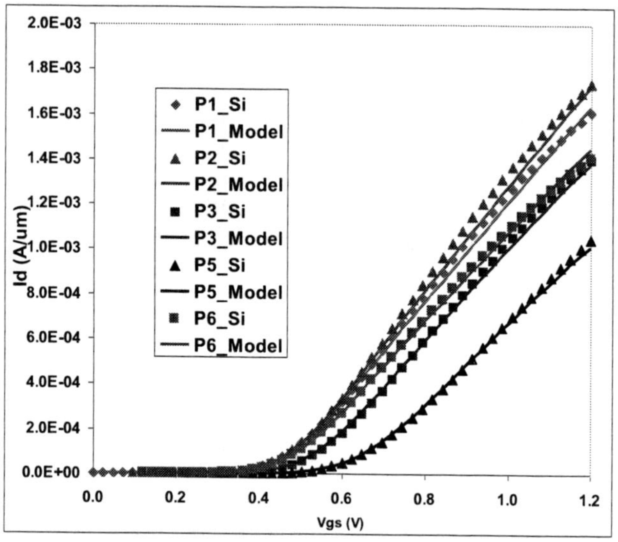

Fig. 1: IdVg (Vds=0.1 V) Model vs. Measured for process conditions shown in Table 1; NMOS 10x0.13.device.

We have chosen the quadratic dependence of process parameters on the model parameters (i.e., n=2 in Eq. 1). We have extracted the coefficients for the chosen key model parameters by fitting the I-V data for the process conditions given in Table 2.

The quality of the extraction process is shown in Figs. 1 and 2. Fig.1 shows the IdVg curves in linear region of operation for the process conditions given in Table 2 for NMOS device of W/L 10x0.13 um. Similarly, Fig.2 shows the IdVg curves in linear region of PMOS device. As seen from the figures, the extraction quality has been reasonable good, the error between the simulated and Silicon data is less than 3 %.

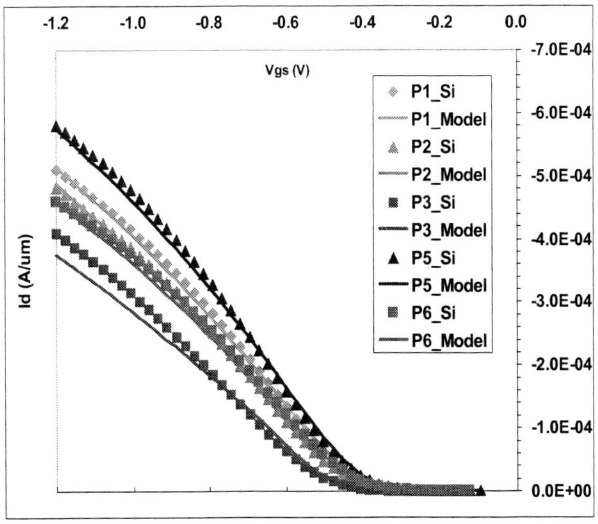

Fig. 2: IdVg (Vds=-0.1 V) Model vs. Measured for process conditions shown in Table 1; PMOS 10x0.13.

The robustness of this methodology was tested by the quality of fits to the silicon devices and data for process conditions not used in the extraction as shown in Figures 3-6 for both PMOS and NMOS devices. Figs. 3 and 4 shows the IdVg and IdVd characteristics, respectively, of NMOS device with PolyCD values 117 nm and 134 nm. Note that the devices with these PolyCD values were not used in the extraction process. However, we see a good match between the Silicon data and the simulated I-V curves. This shows the predictability of the process-aware model at the device level. Similarly, Figs. 5 and 6 show the I-V curves of PMOS device at two different PolyCD values. Good accuracy in matching with the Silicon I-V data can be observed here.

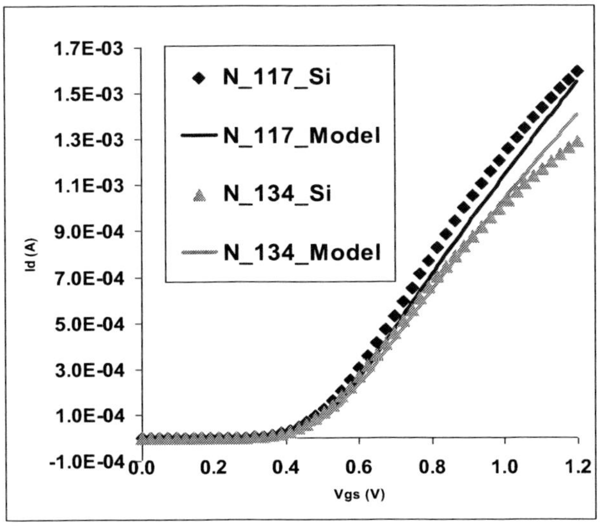

Fig. 3: IdVg (Vds=0.1 V) Model vs. Measured; NMOS for different PolyCD values.

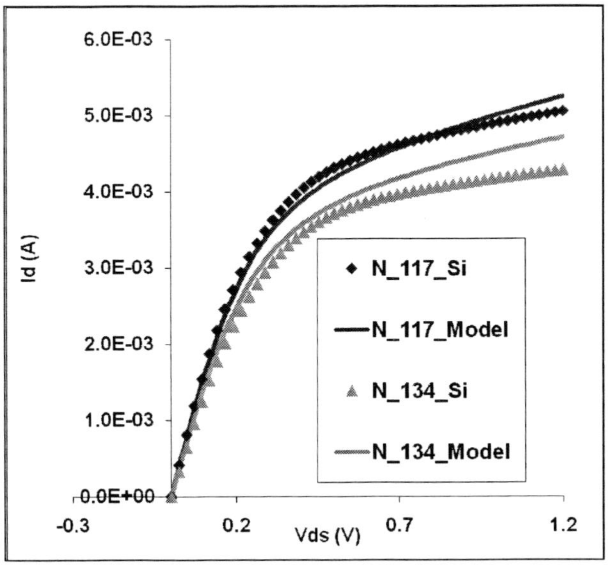

Fig. 4: IdVd (Vgs=1.2 V) Model vs. Measured; NMOS for different PolyCD values.

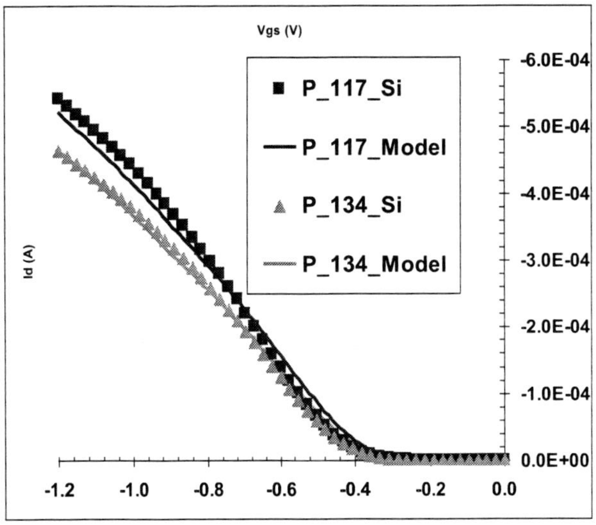

Fig. 5: IdVg (Vds=-0.1 V) Model vs. Measured; PMOS for different PolyCDvalues.

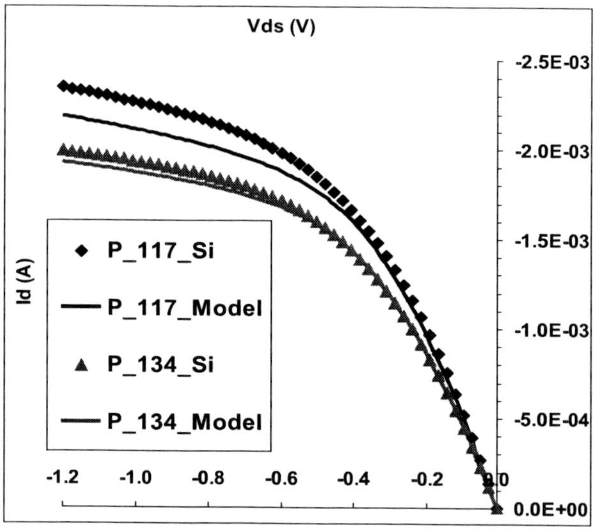

Fig. 6: IdVd (Vgs=-1.2 V) Model vs. Measured; PMOS for different PolyCD values.

A further test of quality of this approach was based on the predictive capability of the model on the performance of a simple circuit. Figure 7 shows the excellent fit between the measured period and the prediction from the process aware SPICE model for a 3481-stage Ring Oscillator circuit.

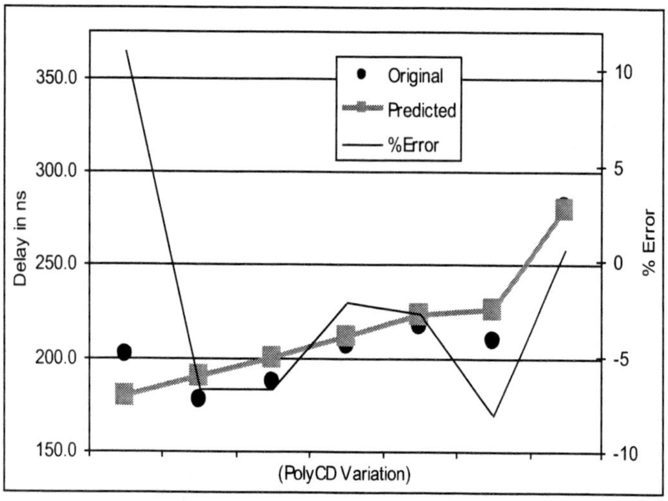

Fig. 7: 3481 Stage Ring Oscillator Circuit time period as obtained from measurements and HSPICE simulations using the process-aware compact model for different PolyCD; Right Y-axis shows the % error between the simulations and measurements.

4 SUMMARY

In this paper, we have demonstrated the methodology to construct the process-aware compact models. Also, we have presented the validation of the extracted model using the Silicon data at device and circuit. Process-aware SPICE models enable engineers to optimize process and design and minimize split lot experiments to identify maximum performance point for a given design. It also provides more control over the existing area, power and performance related constraints, to help optimize for a robust design.

REFERENCES

[1] K.A. Bowman et al., "Impact of die-to-die and within-die parameter fluctuations on the maximum clock frequency distribution for giga-scale integration," JSSC, vol. 37, no. 2, pp. 183-190, 2002.

[2] S. Tirumala et al., "Bringing manufacturing into design via process-dependent SPICE models", Proc. of ISQED, 2006.

[3] Y. Mahotin et al., "Process Aware Hybrid SPICE Models using TCAD and Silicon Data", Technical Proceedings of the 2007 NSTI Nanotechnology Conference.

Index of Authors

Index of Keywords